Geometry

(A = area, B = area of base, C = circumference, S = lateral area or surface area, V = volume)

1. Triangle

$$A = \frac{1}{2}bh$$

2. Similar Triangles

$$\frac{a'}{a} = \frac{b'}{b} = \frac{c'}{c}$$

3. Pythagorean Theorem

$$a^2 + b^2 = c^2$$

4. Parallelogram

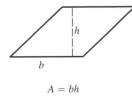

$$A = bh$$

5. Trapezoid

$$A = \frac{1}{2}(a + b)h$$

6. Circle

$$A = \pi r^2,$$
$$C = 2\pi r$$

7. Any Cylinder or Prism with Parallel Bases

$$V = Bh$$

8. Right Circular Cylinder

$$V = \pi r^2 h, \; S = 2\pi rh$$

9. Any Cone or Pyramid

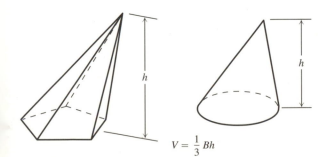

$$V = \frac{1}{3}Bh$$

10. Right Circular Cone

$$V = \frac{1}{3}\pi r^2 h, \; S = \pi rs$$

11. Sphere

$$V = \frac{4}{3}\pi r^3, \; S = 4\pi r^2$$

9TH EDITION

Calculus and Analytic Geometry

George B. Thomas, Jr.
Massachusetts Institute of Technology

Ross L. Finney

With the collaboration of
Maurice D. Weir
Naval Postgraduate School

 Addison-Wesley Publishing Company

Reading, Massachusetts • Menlo Park, California • New York
Don Mills, Ontario • Wokingham, England • Amsterdam
Bonn • Sydney • Singapore • Tokyo • Madrid
San Juan • Milan • Paris

Acquisitions Editor	Laurie Rosatone	*Production Editorial Services*	Barbara Pendergast
Development Editor	Marianne Lepp	*Art Editors*	Susan London-Payne, Connie Hulse
Managing Editor	Karen Guardino	*Copy Editor*	Barbara Flanagan
Senior Production Supervisor	Jennifer Bagdigian	*Proofreader*	Joyce Grandy
Senior Marketing Manager	Andrew Fisher	*Text Design*	Martha Podren, Podren Design;
Marketing Coordinator	Benjamin Rivera		Geri Davis, Quadrata, Inc.
Prepress Buying Manager	Sarah McCracken	*Cover Design*	Marshall Henrichs
Art Buyer	Joseph Vetere	*Cover Photo*	John Lund/Tony Stone Worldwide
Senior Manufacturing Manager	Roy Logan	*Composition*	TSI Graphics, Inc.
Manufacturing Coordinator	Evelyn Beaton	*Technical Illustration*	Tech Graphics

Photo Credits: **142, 238, 408, 633, 722, 875, 899,** From *PSSC Physics 2/e,* 1965; D.C. Heath & Co. with Education Development Center, Inc., Newton, MA. Reprinted with permission **186,** AP/Wide World Photos **266,** Scott A. Burns, Urbana, IL **287,** Joshua E. Barnes, University of Hawaii **354,** Marshall Henrichs **398,** © Richard F. Voss/IBM Research **442,** © Susan Van Etten **872,** AP/Wide World Photos **889,** © 1994 Nelson L. Max, University of California/Biological Photo Service; Graphic by Alfred Gray **938,** ND Roger-Viollet **1068,** NASA/Jet Propulsion Laboratory

Reprinted with corrections, June, 1998.
Library of Congress Cataloging-in-Publication Data
Thomas, George Brinton, 1914–
    Calculus and analytic geometry / George B. Thomas, Ross L. Finney.
    —9th ed.
    p. cm.
    Includes index.
    ISBN 0-201-53174-7
    1. Calculus. 2. Geometry, Analytic. I. Finney, Ross L.
II. Title.
QA303.T42 1996
515'. 15—dc20

94-30543
CIP

11 12 13 14 15 – VH – 99

Contents

4   Integration

5   Applications of Integrals

6   Transcendental Functions

7   Techniques of Integration

CAS Explorations and Projects
(Listed by chapter and section)

To the Instructor

This Is a Major Revision

Throughout the 40 years that it has been in print, Thomas/Finney has been used to support a variety of teaching methods from traditional to experimental. In response to the many exciting currents in teaching calculus in the 1990s, the new edition is the most extensive revision of Thomas/Finney ever. We have built on the traditional strengths of the book—excellent exercises, sound mathematics, variety in applications—to produce a flexible text that contains all the elements needed to teach the many different kinds of courses that exist today.

A book does not make a course: The instructor and the students do. With this in mind we have added features to Thomas/Finney 9th edition to make it the most flexible calculus teaching resource yet.

- The exercises have been reorganized to facilitate assigning a subset of the material in a section.
- The grapher explorations, all accessible with any graphing calculator, many suitable for in-class and group work, have been expanded.
- New Computer Algebra System (CAS) explorations and projects that require a CAS have been included. Some of these can be done quickly while others require several hours. All are suitable for either individual or group work. You will find a list of CAS exercise topics following the Table of Contents.
- Technology Connection notes appear throughout the text suggesting experiments students might do with a grapher to supplement their understanding of a given topic. These notes are meant to encourage students to think of their grapher as a casually available tool, like a pencil.
- We revised the entire first semester and large parts of the second and third semesters to provide what we believe is a cleaner, more visual, and more accessible book.

With all these changes, we have not compromised our belief that the fundamental goal of a calculus book is to prepare students to enter the scientific community.

Students Will Find Even More Support for Creative Problem Solving

Throughout this book, we have included examples and discussions that encourage students to think visually and numerically. Almost every exercise set has easy to

67

[0, 3] by [0, 3]

Keeping *x* between 1.75 and 2.28 will keep *y* between 1.8 and 2.2.

Technology   *Target Values*   You can experiment with target values on a graphing utility. Graph the function together with a target interval defined by horizontal lines above and below the proposed limit. Adjust the range or use zoom until the function's behavior inside the target interval is clear. Then observe what happens when you try to find an interval of *x*-values that will keep the function values within the target interval. (See also Exercises 7–14 and CAS Exercises 61–64.)

For example, try this for $f(x) = \sqrt{3x - 2}$ and the target interval (1.8, 2.2) on the *y*-axis. That is, graph $y_1 = f(x)$ and the lines $y_2 = 1.8$, $y_3 = 2.2$. Then try the target intervals (1.98, 2.02) and (1.9998, 2.0002).

mid-level exercises that require students to generate and interpret graphs as a tool for understanding mathematical or real-world relationships. Many sections also contain a few more challenging problems to extend the range of the mathematically curious.

This edition has more than 2300 figures to appeal to the students' geometric intuition. Drawing lessons aid students with difficult 3-dimensional sketches, enhancing their ability to think in 3-space. In this edition we have increased the use of visualization internal to the discussion. The burden of exposition is shared by art in the body of the text when we feel that pictures and text together will convey ideas better than words alone.

Throughout the text, students are asked to experiment, investigate, and explain. Writing exercises are placed throughout the text. In addition, each chapter end contains a list of questions that ask students to review and summarize what they have learned. Many of these exercises make good writing assignments.

32. *Recovering a function from its derivative*

a)   Use the following information to graph the function *f* over the closed interval $[-2, 5]$.

  i)   The graph of *f* is made of closed line segments joined end to end.
  ii) The graph starts at the point $(-2, 3)$.
  iii) The derivative of *f* is the step function in Fig. 2.13.

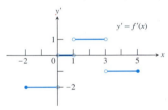

2.13 The derivative graph for Exercise 32.

b)   Repeat part (a) assuming that the graph starts at $(-2, 0)$ instead of $(-2, 3)$.

118

L meets *C* only at *P* but is not tangent to *C*.

L is tangent to *C* at *P* but meets *C* at several points.

L is tangent to *C* at *P* but lies on two sides of *C*, crossing *C* at *P*.

1.49 Exploding myths about tangent lines.

98

Students Will Master Techniques

Problem Solving Strategies   We believe that the students learn best when procedural techniques are laid out as clearly as possible. To this end we have revisited the summaries of the steps used to solve problems, adding some where necessary, deleting some where a thought process rather than a technique was at issue, and making each one clear and useful. As always, we are especially careful that examples in the text follow the steps outlined by the discussion.

Exercises   Every exercise set has been reviewed and revised. Exercises are now *grouped by topic,* with special sections for grapher explorations. Many sections also have a set of Computer Algebra System (CAS) Explorations and Projects, a new feature for this edition. Within each group, the exercises are graded and paired. Within this framework, the exercises generally follow the order of presentation of the text.

Exercises that require a calculator or computer are identified by icons: ▦ calculator exercise, ⧏ graphing utility (such as graphing calculator) exercise, and ✲Computer Algebra System exercise.

Hidden Behavior

Sometimes graphing f' or f'' will suggest where to zoom in on a computer generated graph of f to reveal behavior hidden in the grapher's original picture.

Checklist for Graphing a Function $y = f(x)$

1. Look for symmetry.
   Is the function even? odd?
2. Is the function a shift of a known function?
3. Analyze dominant terms.
   Divide rational functions into polynomial + remainder.
4. Check for asymptotes and removable discontinuities.
   Is there a zero denominator at any point?
   What happens as $x \to \pm\infty$?
5. Compute f' and solve $f' = 0$. Identify critical points and determine intervals of rise and fall.
6. Compute f'' to determine concavity and inflection points.
7. Sketch the graph's general shape.
8. Evaluate f at special values (endpoints, critical points, intercepts).
9. Graph f, using dominant terms, general shape, and special points for guidance.

230

Within the exercise sets, we have practice exercises, exercises that encourage critical thinking, more challenging exercises (in subsections marked "Applications and Theory"), and exercises that require writing in English about concepts. Writing exercises are placed both throughout the exercise sets, and in an end-of-chapter feature called "Questions to Guide Your Review."

Chapter End   At the end of each chapter are three features with questions that summarize the chapter in different ways.

Questions to Guide Your Review ask students to think about concepts and verbalize their understanding without trying to calculate numeric answers. These are, as always, suitable for writing exercises.

Practice Exercises provide a review of the techniques, ideas, and key applications.

Additional Exercises—Theory, Examples, Applications supply challenging applications and theoretic problems that deepen the understanding of mathematical ideas.

Applications, Technology, History—Features That Bring Calculus to Life

Applications and Examples   It has been a hallmark of this book through the years that we illustrate applications of calculus with real data based on already familiar situations or situations students are likely to encounter soon. Throughout the text, we cite sources for the data and/or articles from which the applications are drawn, helping students understand that calculus is a current, dynamic field. Most of these appli-

17. *A sailboat's displacement.* To find the volume of water displaced by a sailboat, the common practice is to partition the waterline into 10 subintervals of equal length, measure the cross section area $A(x)$ of the submerged portion of the hull at each partition point, and then use Simpson's rule to estimate the integral of $A(x)$ from one end of the waterline to the other. The table here lists the area measurements at "Stations" 0 through 10, as the partition points are called, for the cruising sloop *Pipedream,* shown here. The common subinterval length (distance between consecutive stations) is $h = 2.54$ ft (about 2′ 6 1/2″, chosen for the convenience of the builder).

$$-1 \quad 0 \quad 1 \quad 2 \quad 3 \quad 4 \quad 5 \quad 6 \quad 7 \quad 8 \quad 9 \quad 10 \quad 11 \quad 12$$

a) Estimate *Pipedream*'s displacement volume to the nearest cubic foot.

Station	Submerged area (ft²)
0	0
1	1.07
2	3.84
3	7.82
4	12.20
5	15.18
6	16.14
7	14.00
8	9.21
9	3.24
10	0

b) The figures in the table are for seawater, which weighs 64 lb/ft³. How many pounds of water does *Pipedream* displace? (Displacement is given in pounds for small craft, and long tons [1 long ton = 2240 lb] for larger vessels.)

443

cations are directed toward science and engineering, but there are many from biology and the social sciences as well.

Technology: Graphing Calculator and Computer Algebra Systems Explorations *Virtually every section* of the text contains calculator exercises that explore numerical patterns and/or graphing calculator exercises that ask students to generate and interpret graphs as a tool to understanding mathematical and real-world relationships. Many of the calculator and graphing calculator exercises are suitable for classroom demonstration or for group work by students in or out of class.

Computer Algebra System (CAS) exercises have been added to every chapter. These exercises, 160 in all, have been tested on both Mathematica and Maple. A full list of CAS exercise topics follows the Table of Contents.

As in previous editions, $\sec^{-1}x$ has been <u>defined</u> so that its range,$[0, \pi/2) \cup (\pi/2, \pi]$, and derivative, $1/(|x| \sqrt{x^2 - 1})$, agree with the results returned by Computer Algebra Systems and scientific calculators.

Notes appear throughout the text encouraging students to explore with graphers.

History Any student is enriched by seeing the human side of mathematics. As in earlier editions, we feature history boxes that describe the origins of ideas, conflicts concerning ownership of ideas, and interesting sidelights into popular topics such as fractals and chaos.

The Many Faces of This Book

Mathematics Is a Formal and Beautiful Language A good part of the beauty of the calculus lies in the fact that it is a stunning creation of the human mind. As in previous editions we have been careful to say only what is true and mathematically sound. In this edition we reviewed every definition, theorem, corollary, and proof for clarity and mathematical correctness.

Even Better Suited to Be the Reference Text in a Reform Course Whether calculus is taught by a traditional lecture or entirely in labs with individual and group learning which focuses on numeric and graphical experimentation, ideas and techniques need to be articulated clearly. This book provides the exercises for computer and grapher experiments and group learning and, in a traditional format, the summation of the lesson—the formal statement of the mathematics and the clear presentation of the technique.

Students Will Learn from This Book for Many Years to Come We provide far more material than any one instructor would want to teach. We do this intentionally. Students can continue to learn calculus from this book long after the class has ended. It provides an accessible review of the calculus a student has already studied. It is a resource for the working engineer or scientist.

Content Features of the Ninth Edition

Preliminary Material

- Lines, functions, and graphs are reviewed briefly.
- Trigonometric functions (formerly treated in an Appendix) are included.

Limits

- The limit is introduced through rates of change (Section 1.1) but defined before the derivative.
- The initial discussion of the limit is intuitive, using numeric and graphic examples of rates of change.
- The basic rules for working with limits are presented in Section 1.2.
- The limit is presented formally in Section 1.3, using input/output control systems to motivate the ε - δ definition. Covering the formal definition of the limit is optional.
- Chapter 1 concludes with the definition of the tangent line and instantaneous rate of change at a point, bringing to a close the investigation begun in Section 1.1.

Derivatives

- Chapter 2 opens with the concept and definition of the derivative as a function.
- The treatment of implicit differentiation has been revised (Section 2.6).
- The treatment of related rates has been moved earlier in the text (Section 2.7).

Applications of the Derivative

- Extrema (Section 3.1) and the Mean Value Theorem (Section 3.2) are now treated in separate sections. The first section presents the motivating problems of maximization and antidifferentiation. The second section provides motivating questions about antidifferentiation, and the Mean Value Theorem provides the answers. Testing critical points is the subject of Section 3.3.
- The graphing sections (Sections 3.4, 3.5) have been revised to emphasize the qualitative reading of graphs.
- Section 3.5 on asymptotes and dominant terms has been rewritten to present a unified approach to graphing rational functions.
- Presentation of L'Hôpital's Rule is postponed to Chapter 6 where it is applied to comparisons with exponential and logarithmic functions.
- Quadratic approximation, formerly in Chapter 3, has been included with Taylor polynomials in Chapter 8.

Integration

- As in previous editions, the indefinite integral is covered before the definite integral (Section 4.1).
- Differential equations and initial value problems are presented immediately after the indefinite integral (Section 4.2).
- Substitution is introduced for indefinite integrals in Section 4.3 and discussed again for definite integrals in Section 4.8.
- The definite integral is motivated by estimating with finite sums (Section 4.4).
- Some techniques of integration have been moved into Chapter 6 (Transcendental Functions), making Chapter 7 (Techniques of Integration) a shorter, more focused chapter.
- Integration using a Computer Algebra System (CAS) is covered in Section 7.5 along with integral tables.

Sequences and Series

- The introduction to sequences has been spread over two sections (Sections 8.1, 8.2), providing more time for this idea.
- Chapter 8 has been reorganized to allow one section per lecture. (See the Table of Contents.)
- Power series are applied to solve differential equations and initial value problems (Section 8.11).

Conic Sections

- The geometry of conic sections is treated in Section 9.1.
- Eccentricity is covered separately in Section 9.2, where it is used to classify the conics.

Differential Equations

- Differential equations and initial value problems are previewed in Section 3.2 by introducing students to the idea of determining functions from derivatives and initial values. Section 3.4 continues the preparation for differential equations by showing how to sketch the graph of a function given the formula for its first derivative and a point through which the graph must pass.
- Differential equations, initial value problems, and their applications then become the central topics of the following sections:

4.2	Differential Equations, Initial Value Problems, and Mathematical Modeling
6.5	Growth and Decay (the initial value problem $y' = ky$, $y(0) = y_0$ and its applications)
6.11	First Order Differential Equations
6.12	Euler's Numerical Method; Slope Fields

  Solutions of differential equations and initial value problems appear as appropriate in exercise sets throughout the remainder of the text.
- Section 4.7 solves initial value problems using the Fundamental Theorem of Calculus, and Section 8.11 solves differential equations and initial value problems with power series.

Multivariate Calculus

- The treatments of curvature and torsion and the **TNB**-frame in Chapter 11 (Vector-Valued Functions and Motion in Space) has been simplified and unified.
- Chapter 14 (Integration in Vector Fields) has been reorganized to place all the material on line integrals before the material on surface integrals.
- There is now a section (Section 14.6) on integration over parametrized surfaces.

Supplements for the Instructor

OmniTest[3] in DOS-Based Format: This easy-to-use software is developed exclusively for Addison-Wesley by ips Publishing, a leader in computerized testing and assessment. Among its features are the following.

- **DOS interface is easy to learn and operate.** The windows look-alike interface makes it easy to choose and control the items as well as the format for each test.
- **You can easily create make-up exams, customized homework assignments, and multiple test forms to prevent plagiarism.** OmniTest[3] is

algorithm driven—meaning the program can automatically insert new numbers into the same equation—creating hundreds of variations of that equation. The numbers are constrained to keep answers reasonable. This allows you to create a virtually endless supply of parallel versions of the same test. This new version of OmniTest also allows you to "lock in" the values shown in the model problem, if you wish.

- **Test items are keyed by section** to the text. Within the section, you can select questions that test individual objectives from that section.
- **You can enter your own questions** by way of OmniTest[3]'s sophisticated editor—complete with mathematical notation.

Instructor's Solutions Manual   by Maurice D. Weir (Naval Postgraduate School). This two-volume supplement contains the worked-out solutions for *all* the exercises in the text.

Answer Book   contains short answers to most exercises in the text.

Supplements for the Instructor and Student

Student Study Guide   by Maurice D. Weir (Naval Postgraduate School). Organized to correspond with the text, this workbook in a semiprogrammed format increases student proficiency with study tips and additional practice.

Student Solutions Manual   by Maurice D. Weir (Naval Postgraduate School). This manual is designed for the student and contains carefully worked-out solutions to all of the odd-numbered exercises in the text.

Differential Equations Primer   A short, supplementary manual containing approximately a chapter's-worth of material. Available should the instructor choose to cover this material within the calculus sequence.

Technology-Related Supplements

Analyzer[*]   This program is a tool for exploring functions in calculus and many other disciplines. It can graph a function of a single variable and overlay graphs of other functions. It can differentiate, integrate, or iterate a function. It can find roots, maxima and minima, and inflection points, as well as vertical asymptotes. In addition, Analyzer[*] can compose functions, graph polar and parametric equations, make families of curves, and make animated sequences with changing parameters. It exploits the unique flexibility of the Macintosh wherever possible, allowing input to be either numeric (from the keyboard) or graphic (with a mouse). Analyzer[*] runs on Macintosh II, Plus, or better.

The Calculus Explorer   Consisting of 27 programs ranging from functions to vector fields, this software enables the instructor and student to use the computer as an "electronic chalkboard." The Explorer is highly interactive and allows for manipulation of variables and equations to provide graphical visualization of mathematical relationships that are not intuitively obvious. The Explorer provides user-friendly operation through an easy-to-use menu-driven system, extensive on-line documentation, superior graphics capability, and fast operation. An accompanying manual includes sections covering each program, with appropriate examples and exercises. Available for IBM PC/compatibles.

InSight   A calculus demonstration software program that enhances understanding of calculus concepts graphically. The program consists of ten simulations. Each presents an application and takes the user through the solution visually. The format is interactive. Available for IBM PC/compatibles.

Laboratories for Calculus I Using Mathematica   By Margaret Höft, The University of Michigan–Dearborn. An inexpensive collection of *Mathematica* lab experiments consisting of material usually covered in the first term of the calculus sequence.

Math Explorations Series   Each manual provides problems and explorations in calculus. Intended for self-paced and laboratory settings, these books are an excellent complement to the text.

> **Exploring Calculus with a Graphing Calculator, Second Edition,** by Charlene E. Beckmann and Ted Sundstrom of Grand Valley State University.
>
> **Exploring Calculus with Mathematica,** by James K. Finch and Millianne Lehmann of the University of San Francisco.
>
> **Exploring Calculus with Derive,** by David C. Arney of the United States Military Academy at West Point.
>
> **Exploring Calculus with Maple,** by Mark H. Holmes, Joseph G. Ecker, William E. Boyce, and William L. Seigmann of Rensselaer Polytechnic Institute.
>
> **Exploring Calculus with Analyzer*,** by Richard E. Sours of Wilkes University.
>
> **Exploring Calculus with the IBM PC Version 2.0,** by John B. Fraleigh and Lewis I. Pakula of the University of Rhode Island.

Acknowledgments

We would like to express our thanks for the many valuable contributions of the people who reviewed this book as it developed through its various stages:

Manuscript Reviewers

Erol Barbut, *University of Idaho*
Neil E. Berger, *University of Illinois at Chicago*
George Bradley, *Duquesne University*
Thomas R. Caplinger, *Memphis State University*
Curtis L. Card, *Black Hills State University*
James C. Chesla, *Grand Rapids Community College*
P.M. Dearing, *Clemson University*
Maureen H. Fenrick, *Mankato State University*
Stuart Goldenberg, *CA Polytechnic State University*
Johnny L. Henderson, *Auburn University*
James V. Herod, *Georgia Institute of Technology*
Paul Hess, *Grand Rapids Community College*
Alice J. Kelly, Santa Clara University
Jeuel G. LaTorre, *Clemson University*
Pamela Lowry, *Lawrence Technological University*
John E. Martin, III, *Santa Rosa Junior College*
James Martino, *Johns Hopkins University*
James R. McKinney, *California State Polytechnic University*
Jeff Morgan, *Texas A & M University*
F. J. Papp, *University of Michigan—Dearborn*
Peter Ross, *Santa Clara University*
Rouben Rostamian, *University of Maryland—Baltimore County*
William L. Siegmann, *Rensselaer Polytechnic Institute*
John R. Smart, *University of Wisconsin—Madison*
Dennis C. Smolarski, S. J., *Santa Clara University*
Bobby N. Winters, *Pittsburgh State University*

Technology Notes Reviewers

Lynn Kamstra Ipina, *University of Wyoming*
Robert Flagg, *University of Southern Maine*
Jeffrey Stephen Fox, *University of Colorado at Boulder*
James Martino, *Johns Hopkins University*
Carl W. Morris, *University of Missouri—Columbia*
Robert G. Stein, *California State University—San Bernardino*

Accuracy Checkers

Steven R. Finch, *Massachusetts Bay Community College*
Paul R. Lorczak, *MathSoft, Inc.*
John R. Martin, *Tarrant County Junior College*
Jeffrey D. Oldham, *Stanford University*

Exercises

In addition, we thank the following people who reviewed the exercise sets for content and balance and contributed many of the interesting new exercises:

Jennifer Earles Szydlik, *University of Wisconsin—Madison*
Aparna W. Higgins, *University of Dayton*
William Higgins, *Wittenberg University*
Leonard F. Klosinski, *Santa Clara University*
David Mann, *Naval Postgraduate School*
Kirby C. Smith, *Texas A & M University*

Kirby Smith was also a pre-revision reviewer and we wish to thank him for his many helpful suggestions.

We would like to express our appreciation to David Canright, Naval Postgraduate School, for his advice and his contributions to the CAS exercise sets, and Gladwin Bartel, at Otero Junior College, for his many helpful suggestions.

Answers

We would like to thank Cynthia Hutcherson for providing answers for exercises in some of the chapters in this edition. We also appreciate the work of an outstanding team of graduate students at Stanford University, who checked every answer in the text for accuracy: Miguel Abreu, David Cardon, Tanya Kalich, Jeffrey D. Oldham, and Julie Roskies. Jeffrey D. Oldham also tested all the CAS exercises, and we thank him for his many helpful suggestions.

Other Contributors

We are particularly grateful to Maurice D. Weir, Naval Postgraduate School, who shared his teaching ideas throughout the preparation of this book. He produced the final exercise sets and wrote most of the CAS exercises for this edition. We appreciate his constant encouragement and thoughtful advice.

We thank Richard A. Askey, University of Wisconsin—Madison, David McKay, Oregon State University, and Richard G. Montgomery, Southern Oregon State College, for sharing their teaching ideas for this edition.

We are also grateful to Erich Laurence Hauenstein, College of DuPage, for generously providing an improved treatment of chaos in Newton's method, and to Robert Carlson, University of Colorado, Colorado Springs, for improving the exposition in the section on relative rates of growth of functions.

To the Student

What Is Calculus?

Calculus is the mathematics of motion and change. Where there is motion or growth, where variable forces are at work producing acceleration, calculus is the right mathematics to apply. This was true in the beginnings of the subject, and it is true today.

Calculus was first invented to meet the mathematical needs of the scientists of the sixteenth and seventeenth centuries, needs that were mainly mechanical in nature. Differential calculus dealt with the problem of calculating rates of change. It enabled people to define slopes of curves, to calculate velocities and accelerations of moving bodies, to find firing angles that would give cannons their greatest range, and to predict the times when planets would be closest together or farthest apart. Integral calculus dealt with the problem of determining a function from information about its rate of change. It enabled people to calculate the future location of a body from its present position and a knowledge of the forces acting on it, to find the areas of irregular regions in the plane, to measure the lengths of curves, and to find the volumes and masses of arbitrary solids.

Today, calculus and its extensions in mathematical analysis are far reaching indeed, and the physicists, mathematicians, and astronomers who first invented the subject would surely be amazed and delighted, as we hope you will be, to see what a profusion of problems it solves and what a range of fields now use it in the mathematical models that bring understanding about the universe and the world around us. The goal of this edition is to present a modern view of calculus enhanced by the use of technology.

How to Learn Calculus

Learning calculus is not the same as learning arithmetic, algebra, and geometry. In those subjects, you learn primarily how to calculate with numbers, how to simplify algebraic expressions and calculate with variables, and how to reason about points, lines, and figures in the plane. Calculus involves those techniques and skills but develops others as well, with greater precision and at a deeper level. Calculus introduces so many new concepts and computational operations, in fact, that you will no longer be able to learn everything you need in class. You will have to learn a fair amount on your own or by working with other students. What should you do to learn?

1. Read the text. You will not be able to learn all the meanings and connections you need just by attempting the exercises. You will need to read relevant

passages in the book and work through examples step by step. Speed reading will not work here. You are reading and searching for detail in a step-by-step logical fashion. This kind of reading, required by any deep and technical content, takes attention, patience, and practice.

2. Do the homework, keeping the following principles in mind.
   a) Sketch diagrams whenever possible.
   b) Write your solutions in a connected step-by-step logical fashion, as if you were explaining to someone else.
   c) Think about why each exercise is there. Why was it assigned? How is it related to the other assigned exercises?

3. Use your calculator and computer whenever possible. Complete as many grapher and CAS (Computer Algebra System) exercises as you can, *even if they are not assigned.* Graphs provide insight and visual representations of important concepts and relationships. Numbers can reveal important patterns. A CAS gives you the freedom to explore realistic problems and examples that involve calculations that are too difficult or lengthy to do by hand.

4. Try on your own to write short descriptions of the key points each time you complete a section of the text. If you succeed, you probably understand the material. If you do not, you will know where there is a gap in your understanding.

Learning calculus is a process—it does not come all at once. Be patient, persevere, ask questions, discuss ideas and work with classmates, and seek help when you need it, right away. The rewards of learning calculus will be very satisfying, both intellectually and professionally.

G.B.T., Jr., *State College, PA*
R.L.F., *Monterey, CA*

Preliminaries

Overview   This chapter reviews the main things you need to know to start calculus. The topics include the real number system, Cartesian coordinates in the plane, straight lines, parabolas, circles, functions, and trigonometry.

1   Real Numbers and the Real Line

This section reviews real numbers, inequalities, intervals, and absolute values.

Real Numbers and the Real Line

Much of calculus is based on properties of the real number system. **Real numbers** are numbers that can be expressed as decimals, such as

$$-\frac{3}{4} = -0.75000\ldots$$

$$\frac{1}{3} = 0.33333\ldots$$

$$\sqrt{2} = 1.4142\ldots$$

The dots ... in each case indicate that the sequence of decimal digits goes on forever.

The real numbers can be represented geometrically as points on a number line called the **real line**.

The symbol $\mathbb{R}$ donotes either the real number system or, equivalently, the real line.

Properties of Real Numbers

The properties of the real number system fall into three categories: algebraic properties, order properties, and completeness. The algebraic properties say that the real numbers can be added, subtracted, multiplied, and divided (except by 0) to produce more real numbers under the usual rules of arithmetic. *You can never divide by* 0.

The order properties of real numbers are summarized in the following list.

The symbol $\Rightarrow$ means "implies."

Rules for Inequalities

If a, b, and c are real numbers, then:

1. $a < b \Rightarrow a + c < b + c$
2. $a < b \Rightarrow a - c < b - c$
3. $a < b$ and $c > 0 \Rightarrow ac < bc$
4. $a < b$ and $c < 0 \Rightarrow bc < ac$
   Special case: $a < b \Rightarrow -b < -a$
5. $a > 0 \Rightarrow \dfrac{1}{a} > 0$
6. If a and b are both positive or both negative, then $a < b \Rightarrow \dfrac{1}{b} < \dfrac{1}{a}$

Notice the rules for multiplying an inequality by a number. Multiplying by a positive number preserves the inequality; multiplying by a negative number reverses the inequality. Also, reciprocation reverses the inequality for numbers of the same sign.

The completeness property of the real number system is deeper and harder to define precisely. Roughly speaking, it says that there are enough real numbers to "complete" the real number line, in the sense that there are no "holes" or "gaps" in it. Many of the theorems of calculus would fail if the real number system were not complete, and the nature of the connection is important. The topic is best saved for a more advanced course, however, and we will not pursue it.

Subsets of $\mathbb{R}$

We distinguish three special subsets of real numbers.

1. The **natural numbers**, namely $1, 2, 3, 4, \ldots$
2. The **integers**, namely $0, \pm 1, \pm 2, \pm 3, \ldots$
3. The **rational numbers**, namely the numbers that can be expressed in the form of a fraction m/n, where m and n are integers and $n \neq 0$. Examples are

$$\frac{1}{3}, \quad -\frac{4}{9}, \quad \frac{200}{13}, \quad \text{and} \quad 57 = \frac{57}{1}.$$

The rational numbers are precisely the real numbers with decimal expansions that are either

a) terminating (ending in an infinite string of zeros), for example,

$$\frac{3}{4} = 0.75000\ldots = 0.75 \qquad \text{or}$$

b) repeating (ending with a block of digits that repeats over and over), for example

$$\frac{23}{11} = 2.090909\ldots = 2.\overline{09}.$$

The bar indicates the block of repeating digits.

The set of rational numbers has all the algebraic and order properties of the real numbers but lacks the completeness property. For example, there is no rational number whose square is 2; there is a "hole" in the rational line where $\sqrt{2}$ should be.

Real numbers that are not rational are called **irrational numbers**. They are characterized by having nonterminating and nonrepeating decimal expansions. Examples are π, $\sqrt{2}$, $\sqrt[3]{5}$, and $\log_{10} 3$.

Intervals

A subset of the real line is called an **interval** if it contains at least two numbers and contains all the real numbers lying between any two of its elements. For example, the set of all real numbers x such that $x > 6$ is an interval, as is the set of all x such that $-2 \le x \le 5$. The set of all nonzero real numbers is not an interval; since 0 is absent, the set fails to contain every real number between -1 and 1 (for example).

Geometrically, intervals correspond to rays and line segments on the real line, along with the real line itself. Intervals of numbers corresponding to line segments are **finite intervals**; intervals corresponding to rays and the real line are **infinite intervals**.

A finite interval is said to be **closed** if it contains both of its endpoints, **half-open** if it contains one endpoint but not the other, and **open** if it contains neither endpoint. The endpoints are also called **boundary points**; they make up the interval's **boundary**. The remaining points of the interval are **interior points** and together make up what is called the interval's **interior**.

Table 1 Types of intervals

	Notation	Set	Graph
Finite:	(a, b)	$\{x \mid a < x < b\}$	
	$[a, b]$	$\{x \mid a \le x \le b\}$	
	$[a, b)$	$\{x \mid a \le x < b\}$	
	$(a, b]$	$\{x \mid a < x \le b\}$	
Infinite:	(a, ∞)	$\{x \mid x > a\}$	
	$[a, \infty)$	$\{x \mid x \ge a\}$	
	$(-\infty, b)$	$\{x \mid x < b\}$	
	$(-\infty, b]$	$\{x \mid x \le b\}$	
	$(-\infty, \infty)$	$\mathbb{R}$ (set of all real numbers)	

Solving Inequalities

The process of finding the interval or intervals of numbers that satisfy an inequality in x is called **solving** the inequality.

EXAMPLE 1   Solve the following inequalities and graph their solution sets on the real line.

a) $2x - 1 < x + 3$ **b)** $-\dfrac{x}{3} < 2x + 1$ **c)** $\dfrac{6}{x - 1} \geq 5$

Solution

a)

$$2x - 1 < x + 3$$

$$2x < x + 4 \qquad \text{Add 1 to both sides.}$$

$$x < 4 \qquad \text{Subtract } x \text{ from both sides.}$$

The solution set is the interval $(-\infty, 4)$ (Fig. 1a).

b)

$$-\frac{x}{3} < 2x + 1$$

$$-x < 6x + 3 \qquad \text{Multiply both sides by 3.}$$

$$0 < 7x + 3 \qquad \text{Add } x \text{ to both sides.}$$

$$-3 < 7x \qquad \text{Subtract 3 from both sides.}$$

$$-\frac{3}{7} < x \qquad \text{Divide by 7.}$$

The solution set is the interval $(-3/7, \infty)$ (Fig. 1b).

c)   The inequality $6/(x - 1) \geq 5$ can hold only if $x > 1$, because otherwise $6/(x - 1)$ is undefined or negative. Therefore, the inequality will be preserved if we multiply both sides by $(x - 1)$, and we have

$$\frac{6}{x - 1} \geq 5$$

$$6 \geq 5x - 5 \qquad \text{Multiply both sides by } (x - 1).$$

$$11 \geq 5x \qquad \text{Add 5 to both sides.}$$

$$\frac{11}{5} \geq x. \qquad \text{Or } x \leq \frac{11}{5}.$$

The solution set is the half-open interval $(1, 11/5]$ (Fig. 1c).   ❑

Figures in left margin:

1 Solutions for Example 1.

Absolute Value

The **absolute value** of a number x, denoted by $|x|$, is defined by the formula

$$|x| = \begin{cases} x, & x \geq 0 \\ -x, & x < 0. \end{cases}$$

EXAMPLE 2 $|3| = 3, \quad |0| = 0, \quad |-5| = -(-5) = 5, \quad \big|-|a|\big| = |a|$ ❑

Notice that $|x| \geq 0$ for every real number x, and $|x| = 0$ if and only if $x = 0$.

Since the symbol $\sqrt{a}$ always denotes the *nonnegative* square root of a, an alternate definition of $|x|$ is

$$|x| = \sqrt{x^2}.$$

Geometrically, $|x|$ represents the distance from x to the origin 0 on the real line. More generally (Fig. 2)

$$|x - y| = \text{the distance between } x \text{ and } y.$$

The absolute value has the following properties.

It is important to remember that $\sqrt{a^2} = |a|$. Do not write $\sqrt{a^2} = a$ unless you already know that $a \geq 0$.

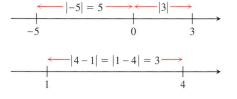

2 Absolute values give distances between points on the number line.

Absolute Value Properties

1. $|-a| = |a|$ A number and its negative have the same absolute value.

2. $|ab| = |a||b|$ The absolute value of a product is the product of the absolute values.

3. $\left|\dfrac{a}{b}\right| = \dfrac{|a|}{|b|}$ The absolute value of a quotient is the quotient of the absolute values.

4. $|a + b| \leq |a| + |b|$ **The triangle inequality** The absolute value of the sum of two numbers is less than or equal to the sum of their absolute values.

If a and b differ in sign, then $|a + b|$ is less than $|a| + |b|$. In all other cases, $|a + b|$ equals $|a| + |b|$.

Notice that absolute value bars in expressions like $|-3 + 5|$ also work like parentheses: We do the arithmetic inside *before* taking the absolute value.

EXAMPLE 3

$$|-3 + 5| = |2| = 2 < |-3| + |5| = 8$$

$$|3 + 5| = |8| = |3| + |5|$$

$$|-3 - 5| = |-8| = 8 = |-3| + |-5|$$ ❑

EXAMPLE 4   Solve the equation $|2x - 3| = 7$.

Solution   The equation says that $2x - 3 = \pm 7$, so there are two possibilities:

$$2x - 3 = 7 \qquad\quad 2x - 3 = -7 \qquad \text{Equivalent equations without absolute values}$$

$$2x = 10 \qquad\qquad 2x = -4 \qquad \text{Solve as usual.}$$

$$x = 5 \qquad\qquad\quad x = -2$$

The solutions of $|2x - 3| = 7$ are $x = 5$ and $x = -2$. ❑

Inequalities Involving Absolute Values

The inequality $|a| < D$ says that the distance from a to 0 is less than D. Therefore, a must lie between D and $-D$.

The symbol ⇔ means "if and only if," or "implies and is implied by."

Intervals and Absolute Values

If D is any positive number, then

$$|a| < D \quad \Leftrightarrow \quad -D < a < D, \tag{1}$$

$$|a| \leq D \quad \Leftrightarrow \quad -D \leq a \leq D. \tag{2}$$

EXAMPLE 5   Solve the inequality $|x - 5| < 9$ and graph the solution set on the real line.

Solution
$$|x - 5| < 9$$
$$-9 < x - 5 < 9 \qquad \text{Eq. (1)}$$
$$-9 + 5 < x < 9 + 5 \qquad \text{Add 5 to each part to isolate } x.$$
$$-4 < x < 14$$

3 The solution set of the inequality $|x - 5| < 9$ is the interval $(-4, 14)$ graphed here (Example 5).

The solution set is the open interval $(-4, 14)$ (Fig. 3).   ❑

EXAMPLE 6   Solve the inequality $\left|5 - \dfrac{2}{x}\right| < 1$.

Solution   We have

$$\left|5 - \frac{2}{x}\right| < 1 \quad \Leftrightarrow \quad -1 < 5 - \frac{2}{x} < 1 \qquad \text{Eq. (1)}$$

$$\Leftrightarrow \quad -6 < -\frac{2}{x} < -4 \qquad \text{Subtract 5.}$$

$$\Leftrightarrow \quad 3 > \frac{1}{x} > 2 \qquad \text{Multiply by } -\frac{1}{2}.$$

$$\Leftrightarrow \quad \frac{1}{3} < x < \frac{1}{2}. \qquad \text{Take reciprocals.}$$

Notice how the various rules for inequalities were used here. Multiplying by a negative number reverses the inequality. So does taking reciprocals in an inequality in which both sides are positive. The original inequality holds if and only if $(1/3) < x < (1/2)$. The solution set is the open interval $(1/3, 1/2)$.   ❑

EXAMPLE 7   Solve the inequality and graph the solution set:

a) $|2x - 3| \leq 1$ 　　　　　　　　　　　　**b)** $|2x - 3| \geq 1$

Solution

a)
$$|2x - 3| \leq 1$$
$$-1 \leq 2x - 3 \leq 1 \qquad \text{Eq. (2)}$$
$$2 \leq 2x \leq 4 \qquad \text{Add 3.}$$
$$1 \leq x \leq 2 \qquad \text{Divide by 2.}$$

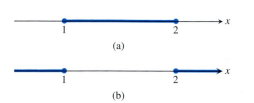

4 Graphs of the solution sets (a) $[1, 2]$ and (b) $(-\infty, 1] \cup [2, \infty)$ in Example 7.

The solution set is the closed interval $[1, 2]$ (Fig. 4a).

b)
$$|2x - 3| \geq 1$$

$2x - 3 \geq 1$	or	$-(2x - 3) \geq 1$
$2x - 3 \geq 1$	or	$2x - 3 \leq -1$
$x - \dfrac{3}{2} \geq \dfrac{1}{2}$	or	$x - \dfrac{3}{2} \leq -\dfrac{1}{2}$
$x \geq 2$	or	$x \leq 1$

Multiply second inequality by -1.

Divide by 2.

Add $\dfrac{3}{2}$.

The solution set is $(-\infty, 1] \cup [2, \infty)$ (Fig. 4b).　　❑

Union and intersection

Notice the use of the symbol $\cup$ to denote the union of intervals. A number lies in the **union** of two sets if it lies in either set. Similarly we use the symbol $\cap$ to denote intersection. A number lies in the **intersection** $I \cap J$ of two sets if it lies in *both* sets I and J. For example, $[1, 3) \cap [2, 4] = [2, 3)$.

Exercises 1

Decimal Representations

1. Express $1/9$ as a repeating decimal, using a bar to indicate the repeating digits. What are the decimal representations of $2/9$? $3/9$? $8/9$?

2. Express $1/11$ as a repeating decimal, using a bar to indicate the repeating digits. What are the decimal representations of $2/11$? $3/11$? $9/11$?

Inequalities

3. If $2 < x < 6$, which of the following statements about x are necessarily true, and which are not necessarily true?

a) $0 < x < 4$

b) $0 < x - 2 < 4$

c) $1 < \dfrac{x}{2} < 3$

d) $\dfrac{1}{6} < \dfrac{1}{x} < \dfrac{1}{2}$

e) $1 < \dfrac{6}{x} < 3$

f) $|x - 4| < 2$

g) $-6 < -x < 2$

h) $-6 < -x < -2$

4. If $-1 < y - 5 < 1$, which of the following statements about y are necessarily true, and which are not necessarily true?

a) $4 < y < 6$

b) $-6 < y < -4$

c) $y > 4$

d) $y < 6$

e) $0 < y - 4 < 2$

f) $2 < \dfrac{y}{2} < 3$

g) $\dfrac{1}{6} < \dfrac{1}{y} < \dfrac{1}{4}$

h) $|y - 5| < 1$

In Exercises 5–12, solve the inequalities and graph the solution sets.

5. $-2x > 4$

6. $8 - 3x \geq 5$

7. $5x - 3 \leq 7 - 3x$

8. $3(2 - x) > 2(3 + x)$

9. $2x - \dfrac{1}{2} \geq 7x + \dfrac{7}{6}$

10. $\dfrac{6 - x}{4} < \dfrac{3x - 4}{2}$

11. $\dfrac{4}{5}(x - 2) < \dfrac{1}{3}(x - 6)$

12. $-\dfrac{x + 5}{2} \leq \dfrac{12 + 3x}{4}$

Absolute Value

Solve the equations in Exercises 13–18.

13. $|y| = 3$

14. $|y - 3| = 7$

15. $|2t + 5| = 4$

16. $|1 - t| = 1$

17. $|8 - 3s| = \dfrac{9}{2}$

18. $\left| \dfrac{s}{2} - 1 \right| = 1$

Solve the inequalities in Exercises 19–34, expressing the solution sets as intervals or unions of intervals. Also, graph each solution set on the real line.

19. $|x| < 2$

20. $|x| \leq 2$

21. $|t - 1| \leq 3$

22. $|t + 2| < 1$

23. $|3y - 7| < 4$

24. $|2y + 5| < 1$

25. $\left| \dfrac{z}{5} - 1 \right| \leq 1$

26. $\left| \dfrac{3}{2}z - 1 \right| \leq 2$

27. $\left| 3 - \dfrac{1}{x} \right| < \dfrac{1}{2}$

28. $\left| \dfrac{2}{x} - 4 \right| < 3$

29. $|2s| \geq 4$

30. $|s + 3| \geq \dfrac{1}{2}$

31. $|1 - x| > 1$

32. $|2 - 3x| > 5$

33. $\left| \dfrac{r + 1}{2} \right| \geq 1$

34. $\left| \dfrac{3r}{5} - 1 \right| > \dfrac{2}{5}$

Quadratic Inequalities

Solve the inequalities in Exercises 35–42. Express the solution sets as intervals or unions of intervals and graph them. Use the result $\sqrt{a^2} = |a|$ as appropriate.

35. $x^2 < 2$

36. $4 \leq x^2$

37. $4 < x^2 < 9$

38. $\dfrac{1}{9} < x^2 < \dfrac{1}{4}$

39. $(x - 1)^2 < 4$

40. $(x + 3)^2 < 2$

41. $x^2 - x < 0$

42. $x^2 - x - 2 \geq 0$

Theory and Examples

43. Do not fall into the trap $|-a| = a$. For what real numbers a is this equation true? For what real numbers is it false?

44. Solve the equation $|x - 1| = 1 - x$.

45. *A proof of the triangle inequality.* Give the reason justifying each of the numbered steps in the following proof of the triangle inequality.

$$|a + b|^2 = (a + b)^2 \qquad (1)$$
$$= a^2 + 2ab + b^2$$
$$\leq a^2 + 2|a||b| + b^2 \qquad (2)$$
$$\leq |a|^2 + 2|a||b| + |b|^2 \qquad (3)$$
$$= (|a| + |b|)^2$$
$$|a + b| \leq |a| + |b| \qquad (4)$$

46. Prove that $|ab| = |a||b|$ for any numbers a and b.

47. If $|x| \leq 3$ and $x > -1/2$, what can you say about x?

48. Graph the inequality $|x| + |y| \leq 1$.

49. GRAPHER

   a) Graph the functions $f(x) = x/2$ and $g(x) = 1 + (4/x)$ together to identify the values of x for which
$$\frac{x}{2} > 1 + \frac{4}{x}.$$

   b) Confirm your findings in (a) algebraically.

50. GRAPHER

   a) Graph the functions $f(x) = 3/(x - 1)$ and $g(x) = 2/(x + 1)$ together to identify the values of x for which
$$\frac{3}{x - 1} < \frac{2}{x + 1}.$$

   b) Confirm your findings in (a) algebraically.

---

2 Coordinates, Lines, and Increments

This section reviews coordinates and lines and discusses the notion of increment.

Cartesian Coordinates in the Plane

The positions of all points in the plane can be measured with respect to two perpendicular real lines in the plane intersecting in the 0-point of each (Fig. 5). These lines are called **coordinate axes** in the plane. On the horizontal x-axis, numbers are denoted by x and increase to the right. On the vertical y-axis, numbers are denoted by y and increase upward. The point where x and y are both 0 is the **origin** of the coordinate system, often denoted by the letter O.

If P is any point in the plane, we can draw lines through P perpendicular to the two coordinate axes. If the lines meet the x-axis at a and the y-axis at b, then a is the **x-coordinate** of P, and b is the **y-coordinate**. The ordered pair (a, b) is the point's **coordinate pair**. The x-coordinate of every point on the y-axis is 0. The y-coordinate of every point on the x-axis is 0. The origin is the point $(0, 0)$.

The origin divides the x-axis into the **positive x-axis** to the right and the **negative x-axis** to the left. It divides the y-axis into the **positive** and **negative y-axis** above and below. The axes divide the plane into four regions called **quadrants**, numbered counterclockwise as in Fig. 6.

5 Cartesian coordinates.

A Word About Scales

When we plot data in the coordinate plane or graph formulas whose variables have different units of measure, we do not need to use the same scale on the two axes. If we plot time vs. thrust for a rocket motor, for example, there is no reason to place the mark that shows 1 sec on the time axis the same distance from the origin as the mark that shows 1 lb on the thrust axis.

When we graph functions whose variables do not represent physical measurements and when we draw figures in the coordinate plane to study their geometry and trigonometry, we try to make the scales on the axes identical. A vertical unit

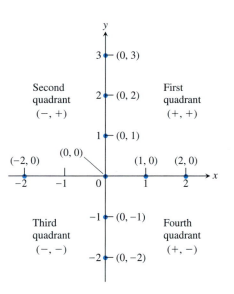

6 The points on the axes all have coordinate pairs, but we usually label them with single numbers. Notice the coordinate sign patterns in the quadrants.

of distance then looks the same as a horizontal unit. As on a surveyor's map or a scale drawing, line segments that are supposed to have the same length will look as if they do and angles that are supposed to be congruent will look congruent.

Computer displays and calculator displays are another matter. The vertical and horizontal scales on machine-generated graphs usually differ, and there are corresponding distortions in distances, slopes, and angles. Circles may look like ellipses, rectangles may look like squares, right angles may appear to be acute or obtuse, and so on. Circumstances like these require us to take extra care in interpreting what we see. High-quality computer software usually allows you to compensate for such scale problems by adjusting the *aspect ratio* (ratio of vertical to horizontal scale). Some computer screens also allow adjustment within a narrow range. When you use a grapher, try to make the aspect ratio 1, or close to it.

Increments and Distance

When a particle moves from one point in the plane to another, the net changes in its coordinates are called *increments*. They are calculated by subtracting the coordinates of the starting point from the coordinates of the ending point.

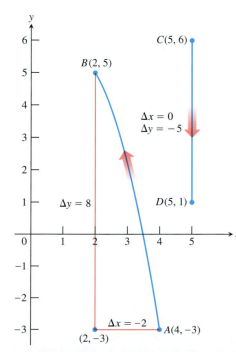

7 Coordinate increments may be positive, negative, or zero.

EXAMPLE 1 In going from the point $A(4, -3)$ to the point $B(2, 5)$ (Fig. 7), the increments in the x- and y-coordinates are

$$\Delta x = 2 - 4 = -2, \qquad \Delta y = 5 - (-3) = 8. \qquad \square$$

Definition

An **increment** in a variable is a net change in that variable. If x changes from x_1 to x_2, the increment in x is

$$\Delta x = x_2 - x_1.$$

EXAMPLE 2 From $C(5, 6)$ to $D(5, 1)$ (Fig. 7) the coordinate increments are

$$\Delta x = 5 - 5 = 0, \qquad \Delta y = 1 - 6 = -5. \qquad \square$$

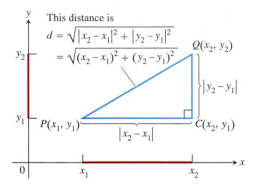

8 To calculate the distance between $P(x_1, y_1)$ and $Q(x_2, y_2)$, apply the Pythagorean theorem to triangle PCQ.

The distance between points in the plane is calculated with a formula that comes from the Pythagorean theorem (Fig. 8).

Distance Formula for Points in the Plane

The distance between $P(x_1, y_1)$ and $Q(x_2, y_2)$ is

$$d = \sqrt{(\Delta x)^2 + (\Delta y)^2} = \sqrt{(x_2 - x_1)^2 + (y_2 - y_1)^2}.$$

EXAMPLE 3

a)   The distance between $P(-1, 2)$ and $Q(3, 4)$ is

$$\sqrt{(3 - (-1))^2 + (4 - 2)^2} = \sqrt{(4)^2 + (2)^2} = \sqrt{20} = \sqrt{4 \cdot 5} = 2\sqrt{5}.$$

b)   The distance from the origin to $P(x, y)$ is

$$\sqrt{(x - 0)^2 + (y - 0)^2} = \sqrt{x^2 + y^2}. \qquad \square$$

Graphs

The graph of an equation or inequality involving the variables x and y is the set of all points $P(x, y)$ whose coordinates satisfy the equation or inequality.

EXAMPLE 4   *Circles centered at the origin*

a)   If $a > 0$, the equation $x^2 + y^2 = a^2$ represents all points $P(x, y)$ whose distance from the origin is $\sqrt{x^2 + y^2} = \sqrt{a^2} = a$. These points lie on the circle of radius a centered at the origin. This circle is the graph of the equation $x^2 + y^2 = a^2$ (Fig. 9a).

b)   Points (x, y) whose coordinates satisfy the inequality $x^2 + y^2 \leq a^2$ all have distance $\leq a$ from the origin. The graph of the inequality is therefore the circle of radius a centered at the origin together with its interior (Fig. 9b).

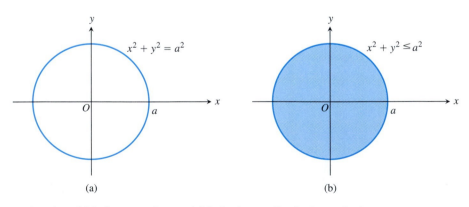

(a)   (b)

9 Graphs of (a) the equation and (b) the inequality in Example 4.

$\square$

The circle of radius 1 unit centered at the origin is called the **unit circle**.

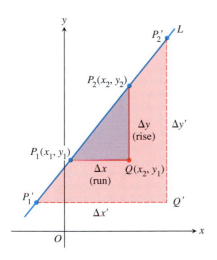

10 The parabola $y = x^2$.

11 Triangles P_1QP_2 and $P_1'Q'P_2'$ are similar, so

$$\frac{\Delta y'}{\Delta x'} = \frac{\Delta y}{\Delta x} = m.$$

12 The slope of L_1 is

$$m = \frac{\Delta y}{\Delta x} = \frac{6 - (-2)}{3 - 0} = \frac{8}{3}.$$

That is, y increases 8 units every time x increases 3 units. The slope of L_2 is

$$m = \frac{\Delta y}{\Delta x} = \frac{2 - 5}{4 - 0} = \frac{-3}{4}.$$

That is, y decreases 3 units every time x increases 4 units.

EXAMPLE 5   Consider the equation $y = x^2$. Some points whose coordinates satisfy this equation are $(0, 0)$, $(1, 1)$, $(-1, 1)$, $(2, 4)$, and $(-2, 4)$. These points (and all others satisfying the equation) make up a smooth curve called a parabola (Fig. 10). □

Straight Lines

Given two points $P_1(x_1, y_1)$ and $P_2(x_2, y_2)$ in the plane, we call the increments $\Delta x = x_2 - x_1$ and $\Delta y = y_2 - y_1$ the **run** and the **rise**, respectively, between P_1 and P_2. Two such points always determine a unique straight line (usually called simply a line) passing through them both. We call the line P_1P_2.

Any nonvertical line in the plane has the property that the ratio

$$m = \frac{\text{rise}}{\text{run}} = \frac{\Delta y}{\Delta x} = \frac{y_2 - y_1}{x_2 - x_1}$$

has the same value for every choice of the two points $P_1(x_1, y_1)$ and $P_2(x_2, y_2)$ on the line (Fig. 11).

Definition

The constant

$$m = \frac{\text{rise}}{\text{run}} = \frac{\Delta y}{\Delta x} = \frac{y_2 - y_1}{x_2 - x_1}$$

is the **slope** of the nonvertical line P_1P_2.

The slope tells us the direction (uphill, downhill) and steepness of a line. A line with positive slope rises uphill to the right; one with negative slope falls downhill to the right (Fig. 12). The greater the absolute value of the slope, the more rapid the rise or fall. The slope of a vertical line is *undefined*. Since the run Δx is zero for a vertical line, we cannot form the ratio m.

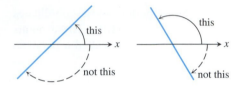

13 Angles of inclination are measured counterclockwise from the x-axis.

The direction and steepness of a line can also be measured with an angle. The **angle of inclination (inclination)** of a line that crosses the x-axis is the smallest counterclockwise angle from the x-axis to the line (Fig. 13). The inclination of a horizontal line is $0°$. The inclination of a vertical line is $90°$. If ϕ (the Greek letter phi) is the inclination of a line, then $0 \leq \phi < 180°$.

The relationship between the slope m of a nonvertical line and the line's inclination ϕ is shown in Fig. 14:

$$m = \tan \phi.$$

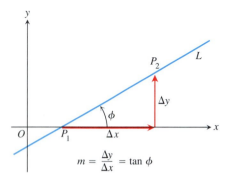

14 The slope of a nonvertical line is the tangent of its angle of inclination.

Parallel and Perpendicular Lines

Lines that are parallel have equal angles of inclination. Hence, they have the same slope (if they are not vertical). Conversely, lines with equal slopes have equal angles of inclination and so are parallel.

If two nonvertical lines L_1 and L_2 are perpendicular, their slopes m_1 and m_2 satisfy $m_1 m_2 = -1$, so each slope is the *negative reciprocal* of the other:

$$m_1 = -\frac{1}{m_2}, \qquad m_2 = -\frac{1}{m_1}.$$

The argument goes like this: In the notation of Fig. 15, $m_1 = \tan \phi_1 = a/h$, while $m_2 = \tan \phi_2 = -h/a$. Hence, $m_1 m_2 = (a/h)(-h/a) = -1$.

Equations of Lines

Straight lines have relatively simple equations. All points on the *vertical line* through the point a on the x-axis have x-coordinates equal to a. Thus, $x = a$ is an equation for the vertical line. Similarly, $y = b$ is an equation for the *horizontal line* meeting the y-axis at b.

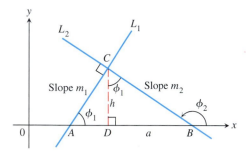

15 $\triangle ADC$ is similar to $\triangle CDB$. Hence ϕ_1 is also the upper angle in $\triangle CDB$. From the sides of $\triangle CDB$, we read $\tan \phi_1 = a/h$.

EXAMPLE 6   The vertical and horizontal lines through the point $(2, 3)$ have equations $x = 2$ and $y = 3$, respectively (Fig. 16).

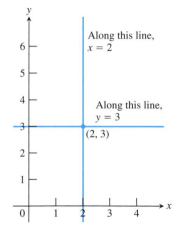

16 The standard equations for the vertical and horizontal lines through (2, 3) are $x = 2$ and $y = 3$.

We can write an equation for a nonvertical straight line L if we know its slope m and the coordinates of one point $P_1(x_1, y_1)$ on it. If $P(x, y)$ is *any* other point on L, then

$$\frac{y - y_1}{x - x_1} = m,$$

so that

$$y - y_1 = m(x - x_1) \qquad \text{or} \qquad y = y_1 + m(x - x_1).$$

> **Definition**
>
> The equation
>
> $$y = y_1 + m(x - x_1)$$
>
> is the **point–slope equation** of the line that passes through the point (x_1, y_1) and has slope m.

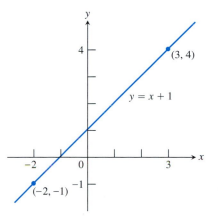

16 The line in Example 8.

EXAMPLE 7 Write an equation for the line through the point $(2, 3)$ with slope $-3/2$.

Solution We substitute $x_1 = 2$, $y_1 = 3$, and $m = -3/2$ into the point–slope equation and obtain

$$y = 3 - \frac{3}{2}(x - 2), \qquad \text{or} \qquad y = -\frac{3}{2}x + 6.$$

EXAMPLE 8 Write an equation for the line through $(-2, -1)$ and $(3, 4)$.

Solution The line's slope is

$$m = \frac{-1 - 4}{-2 - 3} = \frac{-5}{-5} = 1.$$

We can use this slope with either of the two given points in the point–slope equation:

With $(x_1, y_1) = (-2, -1)$	**With $(x_1, y_1) = (3, 4)$**
$y = -1 + 1 \cdot (x - (-2))$	$y = 4 + 1 \cdot (x - 3)$
$y = -1 + x + 2$	$y = 4 + x - 3$
$y = x + 1$	$y = x + 1$

Same result

Either way, $y = x + 1$ is an equation for the line (Fig. 17).

The y-coordinate of the point where a nonvertical line intersects the y-axis is called the **y-intercept** of the line. Similarly, the **x-intercept** of a nonhorizontal line is the x-coordinate of the point where it crosses the x-axis (Fig. 18). A line with slope m and y-intercept b passes through the point $(0, b)$, so it has equation

17 Line L has x-intercept a and y-intercept b.

$$y = b + m(x - 0), \qquad \text{or, more simply,} \qquad y = mx + b.$$

Definition

The equation

$$y = mx + b$$

is called the **slope–intercept equation** of the line with slope m and y-intercept b.

EXAMPLE 9   The line $y = 2x - 5$ has slope 2 and y-intercept -5.   ❑

The equation

$$Ax + By = C \qquad (A \text{ and } B \text{ not both } 0)$$

is called the **general linear equation** in x and y because its graph always represents a line and every line has an equation in this form (including lines with undefined slope).

EXAMPLE 10   Find the slope and y-intercept of the line $8x + 5y = 20$.

Solution   Solve the equation for y to put it in slope–intercept form. Then read the slope and y-intercept from the equation:

$$8x + 5y = 20$$
$$5y = -8x + 20$$
$$y = -\frac{8}{5}x + 4.$$

The slope is $m = -8/5$. The y-intercept is $b = 4$.   ❑

EXAMPLE 11   *Lines through the origin*

Lines with equations of the form $y = mx$ have y-intercept 0 and so pass through the origin. Several examples are shown in Fig. 19.   ❑

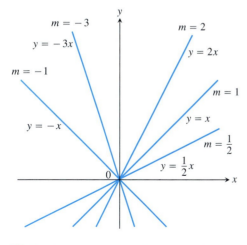

19 The line $y = mx$ has slope m and passes through the origin.

Applications—The Importance of Lines and Slopes

Light travels along lines, as do bodies falling from rest in a planet's gravitational field or coasting under their own momentum (like a hockey puck gliding across the ice). We often use the equations of lines (called **linear equations**) to study such motions.

Many important quantities are related by linear equations. Once we know that a relationship between two variables is linear, we can find it from any two pairs of corresponding values just as we find the equation of a line from the coordinates of two points.

Slope is important because it gives us a way to say how steep something is (roadbeds, roofs, stairs). The notion of slope also enables us to describe how rapidly things are changing. For this reason it will play an important role in calculus.

EXAMPLE 12   *Celsius vs. Fahrenheit*

Fahrenheit temperature (F) and Celsius temperature (C) are related by a linear equation of the form $F = mC + b$. The freezing point of water is $F = 32°$ or $C = 0°$, while the boiling point is $F = 212°$ or $C = 100°$. Thus

$$32 = 0m + b, \qquad \text{and} \qquad 212 = 100m + b,$$

so $b = 32$ and $m = (212 - 32)/100 = 9/5$. Therefore,

$$F = \frac{9}{5}C + 32, \qquad \text{or} \qquad C = \frac{5}{9}(F - 32).$$

Exercises 2

Increments and Distance

In Exercises 1–4, a particle moves from A to B in the coordinate plane. Find the increments Δx and Δy in the particle's coordinates. Also find the distance from A to B.

1. $A(-3, 2)$, $B(-1, -2)$
2. $A(-1, -2)$, $B(-3, 2)$
3. $A(-3.2, -2)$, $B(-8.1, -2)$
4. $A(\sqrt{2}, 4)$, $B(0, 1.5)$

Describe the graphs of the equations in Exercises 5–8.

5. $x^2 + y^2 = 1$
6. $x^2 + y^2 = 2$
7. $x^2 + y^2 \leq 3$
8. $x^2 + y^2 = 0$

Slopes, Lines, and Intercepts

Plot the points in Exercises 9–12 and find the slope (if any) of the line they determine. Also find the common slope (if any) of the lines perpendicular to line AB.

9. $A(-1, 2)$, $B(-2, -1)$
10. $A(-2, 1)$, $B(2, -2)$
11. $A(2, 3)$, $B(-1, 3)$
12. $A(-2, 0)$, $B(-2, -2)$

In Exercises 13–16, find an equation for (a) the vertical line and (b) the horizontal line through the given point.

13. $(-1, 4/3)$
14. $(\sqrt{2}, -1.3)$
15. $(0, -\sqrt{2})$
16. $(-\pi, 0)$

In Exercises 17–30, write an equation for each line described.

17. Passes through $(-1, 1)$ with slope -1
18. Passes through $(2, -3)$ with slope $1/2$
19. Passes through $(3, 4)$ and $(-2, 5)$
20. Passes through $(-8, 0)$ and $(-1, 3)$
21. Has slope $-5/4$ and y-intercept 6
22. Has slope $1/2$ and y-intercept -3
23. Passes through $(-12, -9)$ and has slope 0

24. Passes through $(1/3, 4)$ and has no slope
25. Has y-intercept 4 and x-intercept -1
26. Has y-intercept -6 and x-intercept 2
27. Passes through $(5, -1)$ and is parallel to the line $2x + 5y = 15$
28. Passes through $(-\sqrt{2}, 2)$ parallel to the line $\sqrt{2}x + 5y = \sqrt{3}$
29. Passes through $(4, 10)$ and is perpendicular to the line $6x - 3y = 5$
30. Passes through $(0, 1)$ and is perpendicular to the line $8x - 13y = 13$

In Exercises 31–34, find the line's x- and y-intercepts and use this information to graph the line.

31. $3x + 4y = 12$
32. $x + 2y = -4$
33. $\sqrt{2}x - \sqrt{3}y = \sqrt{6}$
34. $1.5x - y = -3$

35. Is there anything special about the relationship between the lines $Ax + By = C_1$ and $Bx - Ay = C_2$ ($A \neq 0, B \neq 0$)? Give reasons for your answer.

36. Is there anything special about the relationship between the lines $Ax + By = C_1$ and $Ax + By = C_2$ ($A \neq 0, B \neq 0$)? Give reasons for your answer.

Increments and Motion

37. A particle starts at $A(-2, 3)$ and its coordinates change by increments $\Delta x = 5, \Delta y = -6$. Find its new position.

38. A particle starts at $A(6, 0)$ and its coordinates change by increments $\Delta x = -6, \Delta y = 0$. Find its new position.

39. The coordinates of a particle change by $\Delta x = 5$ and $\Delta y = 6$ as it moves from $A(x, y)$ to $B(3, -3)$. Find x and y.

40. A particle started at $A(1, 0)$, circled the origin once counterclockwise, and returned to $A(1, 0)$. What were the net changes in its coordinates?

Applications

41. *Insulation.* By measuring slopes in Fig. 20, estimate the temperature change in degrees per inch for (a) the gypsum wallboard; (b) the fiberglass insulation; (c) the wood sheathing. (Graphs can shift in printing, so your answers may differ slightly from those in the back of the book.)

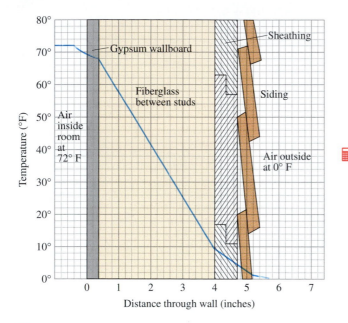

20 The temperature changes in the wall in Exercises 41 and 42. (Source: *Differentiation,* by W. U. Walton et al., Project CALC, Education Development Center, Inc., Newton, Mass. [1975], p. 25.)

42. *Insulation.* According to Fig. 20, which of the materials in Exercise 41 is the best insulator? the poorest? Explain.

43. *Pressure under water.* The pressure p experienced by a diver under water is related to the diver's depth d by an equation of the form $p = kd + 1$ (k a constant). At the surface, the pressure is 1 atmosphere. The pressure at 100 meters is about 10.94 atmospheres. Find the pressure at 50 meters.

44. *Reflected light.* A ray of light comes in along the line $x + y = 1$ from the second quadrant and reflects off the x-axis (Fig. 21). The angle of incidence is equal to the angle of reflection. Write an equation for the line along which the departing light travels.

45. *Fahrenheit vs. Celsius.* In the FC-plane, sketch the graph of the equation

$$C = \frac{5}{9}(F - 32)$$

linking Fahrenheit and Celsius temperatures (Example 12). On the same graph sketch the line $C = F$. Is there a temperature at which a Celsius thermometer gives the same numerical reading as a Fahrenheit thermometer? If so, find it.

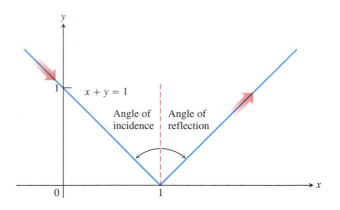

21 The path of the light ray in Exercise 44. Angles of incidence and reflection are measured from the perpendicular.

46. *The Mt. Washington Cog Railway.* Civil engineers calculate the slope of roadbed as the ratio of the distance it rises or falls to the distance it runs horizontally. They call this ratio the **grade** of the roadbed, usually written as a percentage. Along the coast, commercial railroad grades are usually less than 2%. In the mountains, they may go as high as 4%. Highway grades are usually less than 5%.

The steepest part of the Mt. Washington Cog Railway in New Hampshire has an exceptional 37.1% grade. Along this part of the track, the seats in the front of the car are 14 ft above those in the rear. About how far apart are the front and rear rows of seats?

Theory and Examples

47. By calculating the lengths of its sides, show that the triangle with vertices at the points $A(1, 2)$, $B(5, 5)$, and $C(4, -2)$ is isosceles but not equilateral.

48. Show that the triangle with vertices $A(0, 0)$, $B(1, \sqrt{3})$, and $C(2, 0)$ is equilateral.

49. Show that the points $A(2, -1)$, $B(1, 3)$, and $C(-3, 2)$ are vertices of a square, and find the fourth vertex.

50. The rectangle shown here has sides parallel to the axes. It is three times as long as it is wide, and its perimeter is 56 units. Find the coordinates of the vertices A, B, and C.

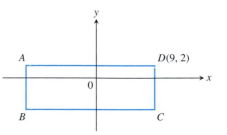

51. Three different parallelograms have vertices at $(-1, 1)$, $(2, 0)$, and $(2, 3)$. Sketch them and find the coordinates of the fourth vertex of each.

52. A 90° rotation counterclockwise about the origin takes $(2, 0)$ to $(0, 2)$, and $(0, 3)$ to $(-3, 0)$, as shown in Fig. 22. Where does it take each of the following points?

a) $(4, 1)$ b) $(-2, -3)$ c) $(2, -5)$
d) $(x, 0)$ e) $(0, y)$ f) (x, y)
g) What point is taken to $(10, 3)$?

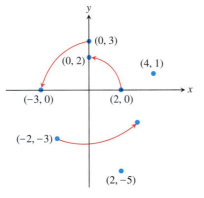

22 The points moved by the 90° rotation in Exercise 52.

53. For what value of k is the line $2x + ky = 3$ perpendicular to the line $4x + y = 1$? For what value of k are the lines parallel?

54. Find the line that passes through the point $(1, 2)$ and through the point of intersection of the two lines $x + 2y = 3$ and $2x - 3y = -1$.

55. Show that the point with coordinates

$$\left(\frac{x_1 + x_2}{2}, \frac{y_1 + y_2}{2} \right)$$

is the midpoint of the line segment joining $P(x_1, y_1)$ to $Q(x_2, y_2)$.

56. *The distance from a point to a line.* We can find the distance from a point $P(x_0, y_0)$ to a line $L: Ax + By = C$ by taking the following steps (there is a somewhat faster method in Section 10.5):

1. Find an equation for the line M through P perpendicular to L.
2. Find the coordinates of the point Q in which M and L intersect.
3. Find the distance from P to Q.

Use these steps to find the distance from P to L in each of the following cases.

a) $P(2, 1)$, $L: y = x + 2$
b) $P(4, 6)$, $L: 4x + 3y = 12$
c) $P(a, b)$, $L: x = -1$
d) $P(x_0, y_0)$, $L: Ax + By = C$

---

3   Functions

Functions are the major tools for describing the real world in mathematical terms. This section reviews the notion of function and discusses some of the functions that arise in calculus.

Functions

The temperature at which water boils depends on the elevation above sea level (the boiling point drops as you ascend). The interest paid on a cash investment depends on the length of time the investment is held. In each case, the value of one variable quantity, which we might call y, depends on the value of another variable quantity, which we might call x. Since the value of y is completely determined by the value of x, we say that y is a function of x.

The letters used for variable quantities may come from what is being described. When we study circles, we usually call the area A and the radius r. Since $A = \pi r^2$, we say that A is a function of r. The equation $A = \pi r^2$ is a *rule* that tells how to calculate a *unique* (single) output value of A for each possible input value of the radius r.

The set of all possible input values for the radius is called the **domain** of the function. The set of all output values of the area is the **range** of the function. Since circles cannot have negative radii or areas, the domain and range of the circle area function are both the interval $[0, \infty)$, consisting of all nonnegative real numbers.

The domain and range of a mathematical function can be any sets of objects; they do not have to consist of numbers. Most of the domains and ranges we will encounter in this book, however, will be sets of real numbers.

Leonhard Euler (1707–1783)

Leonhard Euler, the dominant mathematical figure of his century and the most prolific mathematician who ever lived, was also an astronomer, physicist, botanist, chemist, and expert in Oriental languages. He was the first scientist to give the function concept the prominence in his work that it has in mathematics today. Euler's collected books and papers fill 70 volumes. His introductory algebra text, written originally in German (Euler was Swiss), is still read in English translation.

In calculus we often want to refer to a generic function without having any particular formula in mind. Euler invented a symbolic way to say "y is a function of x" by writing

$$y = f(x) \qquad (\text{"}y \text{ equals } f \text{ of } x\text{"})$$

In this notation, the symbol f represents the function. The letter x, called the **independent variable,** represents an input value from the domain of f, and y, the **dependent variable,** represents the corresponding output value $f(x)$ in the range of f. Here is the formal definition of *function*.

Definition

A **function** from a set D to a set R is a rule that assigns a *unique* element $f(x)$ in R to each element x in D.

In this definition, $D = D(f)$ (read "D of f") is the domain of the function f and R is a set *containing* the range of f. See Fig. 23.

Think of a function f as a kind of machine that produces an output value $f(x)$ in its range whenever we feed it an input value x from its domain (Fig. 24).

In this book we will usually define functions in one of two ways:

1. by giving a formula such as $y = x^2$ that uses a dependent variable y to denote the value of the function, or
2. by giving a formula such as $f(x) = x^2$ that defines a function symbol f to name the function.

Strictly speaking, we should call the function f and not $f(x)$, as the latter denotes the value of the function at the point x. However, as is common usage, we will often refer to the function as $f(x)$ in order to name the variable on which f depends.

It is sometimes convenient to use a single letter to denote both a function and its dependent variable. For instance, we might say that the area A of a circle of radius r is given by the function $A(r) = \pi r^2$.

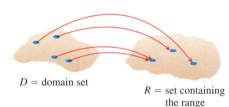

$D = $ domain set

$R = $ set containing the range

23 A function from a set D to a set R assigns a unique element of R to each element in D.

x — Input (Domain) — f — $f(x)$ — Output (Range)

24 A "machine" diagram for a function.

Evaluation

As we said earlier, most of the functions in this book will be **real-valued functions** of a **real variable,** functions whose domains and ranges are sets of real numbers. We evaluate such functions by substituting particular values from the domain into the function's defining rule to calculate the corresponding values in the range.

EXAMPLE 1 The volume V of a ball (solid sphere) of radius r is given by the function

$$V(r) = \frac{4}{3}\pi r^3.$$

The volume of a ball of radius 3 m is

$$V(3) = \frac{4}{3}\pi (3)^3 = 36\pi \ \text{m}^3.$$

EXAMPLE 2   Suppose that the function F is defined for all real numbers t by the formula

$$F(t) = 2(t - 1) + 3.$$

Evaluate F at the input values 0, 2, $x + 2$, and $F(2)$.

Solution   In each case we substitute the given input value for t into the formula for F:

$$F(0) = 2(0 - 1) + 3 = -2 + 3 = 1$$

$$F(2) = 2(2 - 1) + 3 = 2 + 3 = 5$$

$$F(x + 2) = 2(x + 2 - 1) + 3 = 2x + 5$$

$$F(F(2)) = F(5) = 2(5 - 1) + 3 = 11.$$

The Domain Convention

When we define a function $y = f(x)$ with a formula and the domain is not stated explicitly, the domain is assumed to be the largest set of x-values for which the formula gives real y-values. This is the function's so-called **natural domain.** If we want the domain to be restricted in some way, we must say so.

The domain of the function $y = x^2$ is understood to be the entire set of real numbers. The formula gives a real y-value for every real number x. If we want to restrict the domain to values of x greater than or equal to 2, we must write "$y = x^2, x \geq 2$."

Changing the domain to which we apply a formula usually changes the range as well. The range of $y = x^2$ is $[0, \infty)$. The range of $y = x^2, x \geq 2$, is the set of all numbers obtained by squaring numbers greater than or equal to 2. In symbols, the range is $\{x^2 | x \geq 2\}$ or $\{y | y \geq 4\}$ or $[4, \infty)$.

Most of the functions we encounter will have domains that are either intervals or unions of intervals.

EXAMPLE 3

Function	Domain (x)	Range (y)
$y = \sqrt{1 - x^2}$	$[-1, 1]$	$[0, 1]$
$y = \dfrac{1}{x}$	$(-\infty, 0) \cup (0, \infty)$	$(-\infty, 0) \cup (0, \infty)$
$y = \sqrt{x}$	$[0, \infty)$	$[0, \infty)$
$y = \sqrt{4 - x}$	$(-\infty, 4]$	$[0, \infty)$

The formula $y = \sqrt{1 - x^2}$ gives a real y-value for every x in the closed interval from -1 to 1. Beyond this domain, $1 - x^2$ is negative and its square root is not a real number. The values of $1 - x^2$ vary from 0 to 1 on the given domain, and the square roots of these values do the same. The range of $\sqrt{1 - x^2}$ is $[0, 1]$.

The formula $y = 1/x$ gives a real y-value for every x except $x = 0$. *We cannot divide any number by zero.* The range of $y = 1/x$, the set of reciprocals of all nonzero real numbers, is precisely the set of all nonzero real numbers.

The formula $y = \sqrt{x}$ gives a real y-value only if $x \geq 0$. The range of $y = \sqrt{x}$ is $[0, \infty)$ because every nonnegative number is some number's square root (namely, it is the square root of its own square).

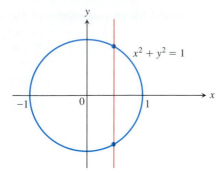

25 This circle is not the graph of a function $y = f(x)$; it fails the vertical line test.

In $y = \sqrt{4-x}$, the quantity $4-x$ cannot be negative. That is, $4 - x \geq 0$, or $x \leq 4$. The formula gives real y-values for all $x \leq 4$. The range of $\sqrt{4-x}$ is $[0, \infty)$, the set of all square roots of nonnegative numbers. ☐

Graphs of Functions

The **graph** of a function f is the graph of the equation $y = f(x)$. It consists of the points in the Cartesian plane whose coordinates (x, y) are input–output pairs for f.

Not every curve you draw is the graph of a function. A function f can have only one value $f(x)$ for each x in its domain, so no *vertical line* can intersect the graph of a function more than once. Thus, a circle cannot be the graph of a function since some vertical lines intersect the circle twice (Fig. 25). If a is in the domain of a function f, then the vertical line $x = a$ will intersect the graph of f in the single point $(a, f(a))$.

EXAMPLE 4   Graph the function $y = x^2$ over the interval $[-2, 2]$.

Solution

Step 1: Make a table of xy-pairs that satisfy the function rule, in this case the equation $y = x^2$.

x	$y = x^2$
-2	4
-1	1
0	0
1	1
2	4

Step 2: Plot the points (x, y) whose coordinates appear in the table.

Step 3: Draw a smooth curve through the plotted points. Label the curve with its equation.

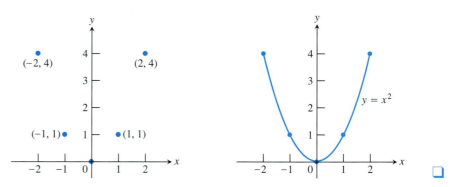

How do we know that the graph of $y = x^2$ doesn't look like one of these curves?

Computers and graphing calculators graph functions in much this way—by stringing together plotted points—and the same question arises.

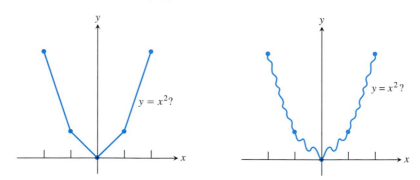

To find out, we could plot more points. But how would we then connect *them*? The basic question still remains: How do we know for sure what the graph looks like between the points we plot? The answer lies in calculus, as we will see in Chapter 3. There we will use a marvelous mathematical tool called the *derivative* to find a curve's shape between plotted points. Meanwhile we will have to settle for plotting points and connecting them as best we can.

Figure 26 shows the graphs of several functions frequently encountered in calculus. It is a good idea to learn the shapes of these graphs so that you can recognize them or sketch them when the need arises.

26 Useful graphs.

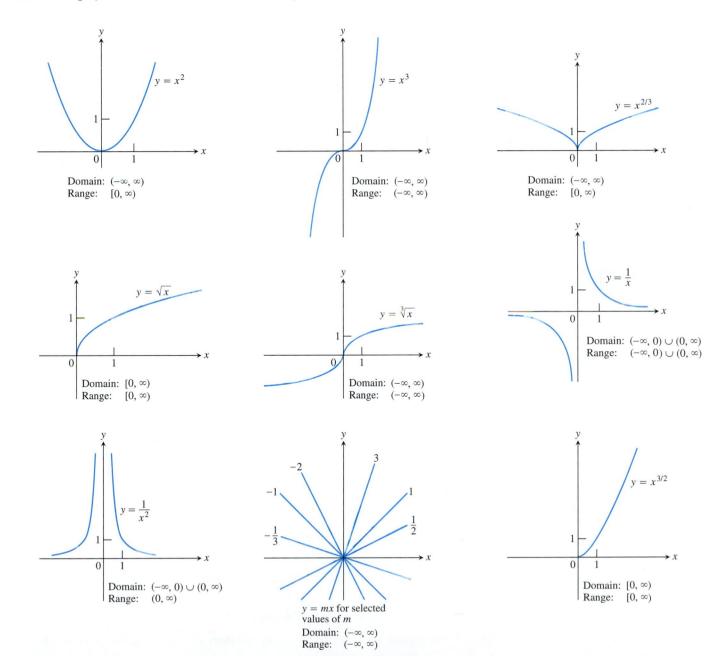

Sums, Differences, Products, and Quotients

Like numbers, functions can be added, subtracted, multiplied, and divided (except where the denominator is zero) to produce new functions. If f and g are functions, then for every x that belongs to the domains of both f and g, we define functions $f + g$, $f - g$, and fg by the formulas

$$(f + g)(x) = f(x) + g(x)$$

$$(f - g)(x) = f(x) - g(x)$$

$$(fg)(x) = f(x)g(x).$$

At any point of $D(f) \cap D(g)$ at which $g(x) \neq 0$, we can also define the function f/g by the formula

$$\left(\frac{f}{g}\right)(x) = \frac{f(x)}{g(x)} \qquad \text{(where } g(x) \neq 0\text{)}.$$

Functions can also be multiplied by constants: If c is a real number, then the function cf is defined for all x in the domain of f by

$$(cf)(x) = cf(x).$$

EXAMPLE 5

Function	Formula	Domain
f	$f(x) = \sqrt{x}$	$[0, \infty)$
g	$g(x) = \sqrt{1 - x}$	$(-\infty, 1]$
$3g$	$3g(x) = 3\sqrt{1 - x}$	$(-\infty, 1]$
$f + g$	$(f + g)(x) = \sqrt{x} + \sqrt{1 - x}$	$[0, 1] = D(f) \cap D(g)$
$f - g$	$(f - g)(x) = \sqrt{x} - \sqrt{1 - x}$	$[0, 1]$
$g - f$	$(g - f)(x) = \sqrt{1 - x} - \sqrt{x}$	$[0, 1]$
$f \cdot g$	$(f \cdot g)(x) = f(x)g(x) = \sqrt{x(1 - x)}$	$[0, 1]$
f/g	$\dfrac{f}{g}(x) = \dfrac{f(x)}{g(x)} = \sqrt{\dfrac{x}{1 - x}}$	$[0, 1)$ $(x = 1$ excluded$)$
g/f	$\dfrac{g}{f}(x) = \dfrac{g(x)}{f(x)} = \sqrt{\dfrac{1 - x}{x}}$	$(0, 1]$ $(x = 0$ excluded$)$

Composite Functions

Composition is another method for combining functions.

Definition

If f and g are functions, the **composite** function $f \circ g$ ("f circle g") is defined by

$$(f \circ g)(x) = f(g(x)).$$

The domain of $f \circ g$ consists of the numbers x in the domain of g for which $g(x)$ lies in the domain of f.

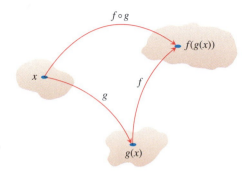

27 The relation of $f \circ g$ to g and f.

The definition says that two functions can be composed when the range of the first lies in the domain of the second (Fig. 27). To find $(f \circ g)(x)$, we *first find* $g(x)$ and *second find* $f(g(x))$.

To evaluate the composite function $g \circ f$ (when defined), we reverse the order, finding $f(x)$ first and then $g(f(x))$. The domain of $g \circ f$ is the set of numbers x in the domain of f such that $f(x)$ lies in the domain of g.

The functions $f \circ g$ and $g \circ f$ are usually quite different.

EXAMPLE 6   If $f(x) = \sqrt{x}$ and $g(x) = x + 1$, find

a) $(f \circ g)(x)$ **b)** $(g \circ f)(x)$ **c)** $(f \circ f)(x)$ **d)** $(g \circ g)(x)$.

Solution

Composite	Domain
a) $(f \circ g)(x) = f(g(x)) = \sqrt{g(x)} = \sqrt{x+1}$	$[-1, \infty)$
b) $(g \circ f)(x) = g(f(x)) = f(x) + 1 = \sqrt{x} + 1$	$[0, \infty)$
c) $(f \circ f)(x) = f(f(x)) = \sqrt{f(x)} = \sqrt{\sqrt{x}} = x^{1/4}$	$[0, \infty)$
d) $(g \circ g)(x) = g(g(x)) = g(x) + 1 = (x+1) + 1 = x + 2$	$\mathbb{R}$ or $(-\infty, \infty)$

To see why the domain of $f \circ g$ is $[-1, \infty)$, notice that $g(x) = x + 1$ is defined for all real x but belongs to the domain of f only if $x + 1 \geq 0$, that is to say, if $x \geq -1$. ∎

Even Functions and Odd Functions—Symmetry

A function $y = f(x)$ is **even** if $f(-x) = f(x)$ for every number x in the domain of f. Notice that this implies that both x and $-x$ must be in the domain of f. The function $f(x) = x^2$ is even because $f(-x) = (-x)^2 = x^2 = f(x)$.

The graph of an even function $y = f(x)$ is symmetric about the y-axis. Since $f(-x) = f(x)$, the point (x, y) lies on the graph if and only if the point $(-x, y)$ lies on the graph (Fig. 28a). Once we know the graph on one side of the y-axis, we automatically know it on the other side.

A function $y = f(x)$ is **odd** if $f(-x) = -f(x)$ for every number x in the domain of f. Again, both x and $-x$ must lie in the domain of f. The function $f(x) = x^3$ is odd because $f(-x) = (-x)^3 = -x^3 = -f(x)$.

The graph of an odd function $y = f(x)$ is symmetric about the origin. Since $f(-x) = -f(x)$, the point (x, y) lies on the graph if and only if the point $(-x, -y)$ lies on the graph (Fig. 28b). Here again, once we know the graph of f on one side of the y-axis, we know it on both sides.

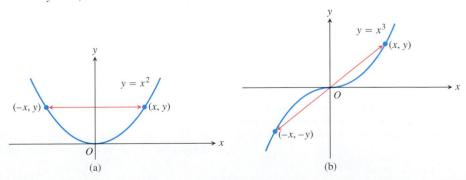

28 (a) Symmetry about the y-axis. If (x, y) is on the graph, so is $(-x, y)$. (b) Symmetry about the origin. If (x, y) is on the graph, so is $(-x, -y)$.

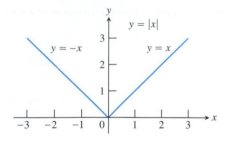

29 The absolute value function.

Piecewise Defined Functions

Sometimes a function uses different formulas on different parts of its domain. One example is the absolute value function

$$|x| = \begin{cases} x, & x \geq 0 \\ -x, & x < 0, \end{cases}$$

whose graph is given in Fig. 29. Here are some examples.

EXAMPLE 7   The function

$$f(x) = \begin{cases} -x, & x < 0 \\ x^2, & 0 \leq x \leq 1 \\ 1, & x > 1 \end{cases}$$

is defined on the entire real line but has values given by different formulas depending on the position of x (Fig. 30).   ❏

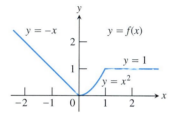

30 To graph the function $y = f(x)$ shown here, we apply different formulas to different parts of its domain (Example 7).

EXAMPLE 8   *The greatest integer function*

The function whose value at any number x is the *greatest integer less than or equal to x* is called the **greatest integer function** or the **integer floor function**. It is denoted $\lfloor x \rfloor$, or, in some books, $[x]$ or $[[x]]$. Figure 31 shows the graph. Observe that

$$\lfloor 2.4 \rfloor = 2, \qquad \lfloor 1.9 \rfloor = 1, \qquad \lfloor 0 \rfloor = 0, \qquad \lfloor -1.2 \rfloor = -2,$$
$$\lfloor 2 \rfloor = 2, \qquad \lfloor 0.2 \rfloor = 0, \qquad \lfloor -0.3 \rfloor = -1 \qquad \lfloor -2 \rfloor = -2.$$ ❏

EXAMPLE 9   *The least integer function*

The function whose value at any number x is the *smallest integer greater than or equal to x* is called the **least integer function** or the **integer ceiling function**. It is denoted $\lceil x \rceil$. Figure 32 shows the graph. For positive values of x, this function might represent, for example, the cost of parking x hours in a parking lot which charges \$1 for each hour or part of an hour.

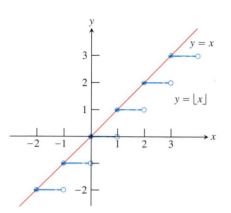

31 The graph of the greatest integer function $y = \lfloor x \rfloor$ lies on or below the line $y = x$, so it provides an integer floor for x.

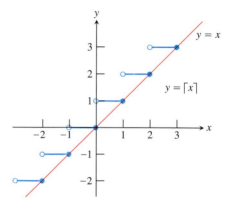

32 The graph of the least integer function $y = \lceil x \rceil$ lies on or above the line $y = x$, so it provides an integer ceiling for x.   ❏

Exercises 3

Functions

In Exercises 1–6, find the domain and range of each function.

1. $f(x) = 1 + x^2$

2. $f(x) = 1 - \sqrt{x}$

3. $F(t) = \dfrac{1}{\sqrt{t}}$

4. $F(t) = \dfrac{1}{1 + \sqrt{t}}$

5. $g(z) = \sqrt{4 - z^2}$

6. $g(z) = \dfrac{1}{\sqrt{4 - z^2}}$

In Exercises 7 and 8, which of the graphs are graphs of functions of x, and which are not? Give reasons for your answers.

7. a)   **b)**

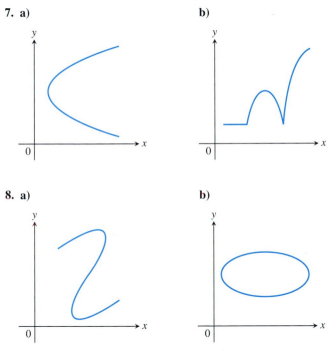

8. a)   **b)**

Finding Formulas for Functions

9. Express the area and perimeter of an equilateral triangle as a function of the triangle's side length x.

10. Express the side length of a square as a function of the length d of the square's diagonal. Then express the area as a function of the diagonal length.

11. Express the edge length of a cube as a function of the cube's diagonal length d. Then express the surface area and volume of the cube as a function of the diagonal length.

12. A point P in the first quadrant lies on the graph of the function $f(x) = \sqrt{x}$. Express the coordinates of P as functions of the slope of the line joining P to the origin.

Functions and Graphs

Graph the functions in Exercises 13–24. What symmetries, if any, do the graphs have? Use the graphs in Fig. 26 for guidance, as needed.

13. $y = -x^3$

14. $y = -\dfrac{1}{x^2}$

15. $y = -\dfrac{1}{x}$

16. $y = \dfrac{1}{|x|}$

17. $y = \sqrt{|x|}$

18. $y = \sqrt{-x}$

19. $y = x^3/8$

20. $y = -4\sqrt{x}$

21. $y = -x^{3/2}$

22. $y = (-x)^{3/2}$

23. $y = (-x)^{2/3}$

24. $y = -x^{2/3}$

25. Graph the following equations and explain why they are not graphs of functions of x.

   a) $|y| = x$ **b)** $y^2 = x^2$

26. Graph the following equations and explain why they are not graphs of functions of x.

   a) $|x| + |y| = 1$ **b)** $|x + y| = 1$

Even and Odd Functions

In Exercises 27–38, say whether the function is even, odd, or neither.

27. $f(x) = 3$

28. $f(x) = x^{-5}$

29. $f(x) = x^2 + 1$

30. $f(x) = x^2 + x$

31. $g(x) = x^3 + x$

32. $g(x) = x^4 + 3x^2 - 1$

33. $g(x) = \dfrac{1}{x^2 - 1}$

34. $g(x) = \dfrac{x}{x^2 - 1}$

35. $h(t) = \dfrac{1}{t - 1}$

36. $h(t) = |t^3|$

37. $h(t) = 2t + 1$

38. $h(t) = 2|t| + 1$

Sums, Differences, Products, and Quotients

In Exercises 39 and 40, find the domains and ranges of $f, g, f + g$, and $f \cdot g$.

39. $f(x) = x,$ $g(x) = \sqrt{x - 1}$

40. $f(x) = \sqrt{x + 1},$ $g(x) = \sqrt{x - 1}$

In Exercises 41 and 42, find the domains and ranges of $f, g, f/g$, and g/f.

41. $f(x) = 2,$ $g(x) = x^2 + 1$

42. $f(x) = 1,$ $g(x) = 1 + \sqrt{x}$

Composites of Functions

43. If $f(x) = x + 5$ and $g(x) = x^2 - 3$, find the following.

a) $f(g(0))$ b) $g(f(0))$
c) $f(g(x))$ d) $g(f(x))$
e) $f(f(-5))$ f) $g(g(2))$
g) $f(f(x))$ h) $g(g(x))$

44. If $f(x) = x - 1$ and $g(x) = 1/(x + 1)$, find the following.

a) $f(g(1/2))$ b) $g(f(1/2))$
c) $f(g(x))$ d) $g(f(x))$
e) $f(f(2))$ f) $g(g(2))$
g) $f(f(x))$ h) $g(g(x))$

45. If $u(x) = 4x - 5$, $v(x) = x^2$, and $f(x) = 1/x$, find formulas for the following.

a) $u(v(f(x)))$ b) $u(f(v(x)))$
c) $v(u(f(x)))$ d) $v(f(u(x)))$
e) $f(u(v(x)))$ f) $f(v(u(x)))$

46. If $f(x) = \sqrt{x}$, $g(x) = x/4$, and $h(x) = 4x - 8$, find formulas for the following.

a) $h(g(f(x)))$ b) $h(f(g(x)))$
c) $g(h(f(x)))$ d) $g(f(h(x)))$
e) $f(g(h(x)))$ f) $f(h(g(x)))$

Let $f(x) = x - 3$, $g(x) = \sqrt{x}$, $h(x) = x^3$, and $j(x) = 2x$. Express each of the functions in Exercises 47 and 48 as a composite involving one or more of f, g, h, and j.

47. a) $y = \sqrt{x} - 3$ b) $y = 2\sqrt{x}$
c) $y = x^{1/4}$ d) $y = 4x$
e) $y = \sqrt{(x-3)^3}$ f) $y = (2x - 6)^3$

48. a) $y = 2x - 3$ b) $y = x^{3/2}$
c) $y = x^9$ d) $y = x - 6$
e) $y = 2\sqrt{x - 3}$ f) $y = \sqrt{x^3 - 3}$

49. Copy and complete the following table.

	$g(x)$	$f(x)$	$(f \circ g)(x)$
a)	$x - 7$	$\sqrt{x}$	
b)	$x + 2$	$3x$	
c)		$\sqrt{x - 5}$	$\sqrt{x^2 - 5}$
d)	$\dfrac{x}{x - 1}$	$\dfrac{x}{x - 1}$	
e)		$1 + \dfrac{1}{x}$	x
f)	$\dfrac{1}{x}$		x

50. *A magic trick.* You may have heard of a magic trick that goes like this: Take any number. Add 5. Double the result. Subtract 6. Divide by 2. Subtract 2. Now tell me your answer, and I'll tell you what you started with.

    Pick a number and try it.

You can see what is going on if you let x be your original number and follow the steps to make a formula $f(x)$ for the number you end up with.

Piecewise Defined Functions

Graph the functions in Exercises 51–54.

51. $f(x) = \begin{cases} x, & 0 \le x \le 1 \\ 2 - x, & 1 < x \le 2 \end{cases}$

52. $g(x) = \begin{cases} 1 - x, & 0 \le x \le 1 \\ 2 - x, & 1 < x \le 2 \end{cases}$

53. $F(x) = \begin{cases} 3 - x, & x \le 1 \\ 2x, & x > 1 \end{cases}$

54. $G(x) = \begin{cases} 1/x, & x < 0 \\ x, & 0 \le x \end{cases}$

55. Find a formula for each function graphed.

a)   b)

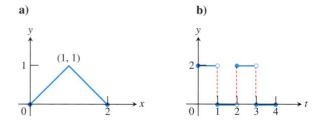

56. Find a formula for each function graphed.

a)   b)

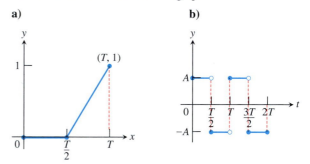

The Greatest and Least Integer Functions

57. For what values of x is (a) $\lfloor x \rfloor = 0$? (b) $\lceil x \rceil = 0$?

58. What real numbers x satisfy the equation $\lfloor x \rfloor = \lceil x \rceil$?

59. Does $\lceil -x \rceil = -\lfloor x \rfloor$ for all real x? Give reasons for your answer.

60. Graph the function
$$f(x) = \begin{cases} \lfloor x \rfloor, & x \ge 0 \\ \lceil x \rceil, & x < 0 \end{cases}$$

Why is $f(x)$ called the *integer part* of x?

Even and Odd Functions

61. Assume that f is an even function, g is an odd function, and both f and g are defined on the entire real line $\mathbb{R}$. Which of the following (where defined) are even? odd?

a) fg b) f/g c) g/f
d) $f^2 = ff$ e) $g^2 = gg$ f) $f \circ g$
g) $g \circ f$ h) $f \circ f$ i) $g \circ g$

62. Can a function be both even and odd? Give reasons for your answer.

▦ Grapher

63. (*Continuation of Example 5.*) Graph the functions $f(x) = \sqrt{x}$ and $g(x) = \sqrt{1-x}$ together with their (a) sum, (b) product, (c) two differences, (d) two quotients.

64. Let $f(x) = x - 7$ and $g(x) = x^2$. Graph f and g together with $f \circ g$ and $g \circ f$.

4

Shifting Graphs

This section shows how to change an equation to shift its graph up or down or to the right or left. Knowing about this can help us spot familiar graphs in new locations. It can also help us graph unfamiliar equations more quickly. We practice mostly with circles and parabolas (because they make useful examples in calculus), but the methods apply to other curves as well. We will revisit parabolas and circles in Chapter 9.

How to Shift a Graph

To shift the graph of a function $y = f(x)$ straight up, we add a positive constant to the right-hand side of the formula $y = f(x)$.

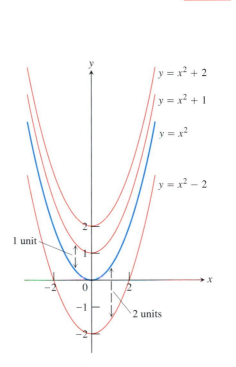

EXAMPLE 1   Adding 1 to the right-hand side of the formula $y = x^2$ to get $y = x^2 + 1$ shifts the graph up 1 unit (Fig. 33).   ❑

To shift the graph of a function $y = f(x)$ straight down, we add a negative constant to the right-hand side of the formula $y = f(x)$.

EXAMPLE 2   Adding -2 to the right-hand side of the formula $y = x^2$ to get $y = x^2 - 2$ shifts the graph down 2 units (Fig. 33).   ❑

To shift the graph of $y = f(x)$ to the left, we add a positive constant to x.

33 To shift the graph of $f(x) = x^2$ up (or down), we add positive (or negative) constants to the formula for f.

EXAMPLE 3   Adding 3 to x in $y = x^2$ to get $y = (x + 3)^2$ shifts the graph 3 units to the left (Fig. 34).   ❑

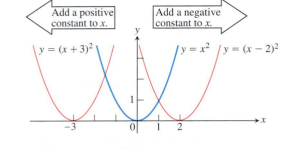

34 To shift the graph of $y = x^2$ to the left, we add a positive constant to x. To shift the graph to the right, we add a negative constant to x.

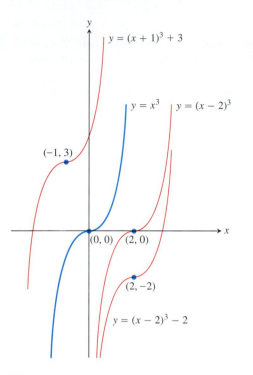

35 The graph of $y = x^3$ shifted to three new positions in the xy-plane.

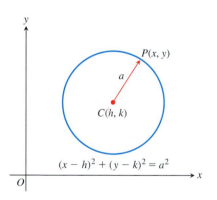

36 A circle of radius a in the xy-plane, with center at (h, k).

To shift the graph of $y = f(x)$ to the right, we add a negative constant to x.

EXAMPLE 4   Adding -2 to x in $y = x^2$ to get $y = (x - 2)^2$ shifts the graph 2 units to the right (Fig. 34).   ❏

Shift Formulas

VERTICAL SHIFTS

$y - k = f(x)$ or   Shifts the graph *up* k units if $k > 0$

$\qquad y = f(x) + k$ Shifts it *down* $|k|$ units if $k < 0$

HORIZONTAL SHIFTS

$y = f(x - h)$ Shifts the graph *right* h units *if* $h > 0$

Shifts it *left* $|h|$ units *if* $h < 0$

EXAMPLE 5   The graph of $y = (x - 2)^3 - 2$ is the graph of $y = x^3$ shifted 2 units to the right and 2 units down. The graph of $y = (x + 1)^3 + 3$ is the graph of $y = x^3$ shifted 1 unit to the left and 3 units up (Fig. 35).   ❏

Equations for Circles

A **circle** is the set of points in a plane whose distance from a given fixed point in the plane is constant (Fig. 36). The fixed point is the **center** of the circle; the constant distance is the **radius**. We saw in Section 2, Example 4, that the circle of radius a centered at the origin has equation $x^2 + y^2 = a^2$. If we shift the circle to place its center at the point (h, k), its equation becomes $(x - h)^2 + (y - k)^2 = a^2$.

The Standard Equation for the Circle of Radius a Centered at the Point (h, k)

$$(x - h)^2 + (y - k)^2 = a^2 \tag{1}$$

EXAMPLE 6   If the circle $x^2 + y^2 = 25$ is shifted 2 units to the left and 3 units up, its new equation is $(x + 2)^2 + (y - 3)^2 = 25$. As Eq. (1) says it should be, this is the equation of the circle of radius 5 centered at $(h, k) = (-2, 3)$.   ❏

EXAMPLE 7   The standard equation for the circle of radius 2 centered at $(3, 4)$ is

$$(x - 3)^2 + (y - 4)^2 = (2)^2$$

or

$$(x - 3)^2 + (y - 4)^2 = 4.$$

There is no need to square out the x- and y-terms in this equation. In fact, it is better not to do so. The present form reveals the circle's center and radius.   ❏

EXAMPLE 8   Find the center and radius of the circle

$$(x - 1)^2 + (y + 5)^2 = 3.$$

Solution   Comparing

$$(x - h)^2 + (y - k)^2 = a^2$$

with

$$(x - 1)^2 + (y + 5)^2 = 3$$

shows that $h = 1, k = -5$, and $a = \sqrt{3}$. The center is the point $(h, k) = (1, -5)$; the radius is $a = \sqrt{3}$.   ❏

Technology   *Square Windows*   We use the term "square window" when the units or scalings on both axes are the same. In a square window graphs are true in shape. They are distorted in a nonsquare window.

The term square window does not refer to the shape of the graphic display. Graphing calculators usually have rectangular displays. The displays of Computer Algebra Systems are usually square. When a graph is displayed, the x-unit may differ from the y-unit in order to fit the graph in the display, resulting in a distorted picture. The graphing window can be made square by shrinking or stretching the units on one axis to match the scale on the other, giving the true graph. Many systems have built-in functions to make the window "square." If yours does not, you will have to do some calculations and set the window size manually to get a square window, or bring to your viewing some foreknowledge of the true picture.

On your graphing utility, compare the perpendicular lines $y_1 = x$ and $y_2 = -x + 4$ in a square window and a nonsquare one such as $[-10, 10]$ by $[10, 10]$. Graph the semicircle $y = \sqrt{8 - x^2}$ in the same windows.

Two perpendicular lines and a semicircle graphed distorted by a rectangular window.

If an equation for a circle is not in standard form, we can find the circle's center and radius by first converting the equation to standard form. The algebraic technique for doing so is *completing the square* (see inside front cover).

EXAMPLE 9   Find the center and radius of the circle

$$x^2 + y^2 + 4x - 6y - 3 = 0.$$

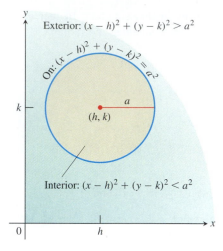

37 The interior and exterior of the circle $(x - h)^2 + (y - k)^2 = a^2$.

Solution We convert the equation to standard form by completing the squares in x and y:

$$x^2 + y^2 + 4x - 6y - 3 = 0 \qquad \text{Start with the given equation.}$$

$$(x^2 + 4x \quad) + (y^2 - 6y \quad) = 3 \qquad \text{Gather terms. Move the constant to the right-hand side.}$$

$$\left(x^2 + 4x + \left(\frac{4}{2}\right)^2\right) + \left(y^2 - 6y + \left(\frac{-6}{2}\right)^2\right) = \qquad \text{Add the square of half the coefficient of } x \text{ to each side of the equation. Do the same for } y. \text{ The parenthetical expressions on the left-hand side are now perfect squares.}$$

$$3 + \left(\frac{4}{2}\right)^2 + \left(\frac{-6}{2}\right)^2$$

$$(x^2 + 4x + 4) + (y^2 - 6y + 9) = 3 + 4 + 9$$

$$(x + 2)^2 + (y - 3)^2 = 16 \qquad \text{Write each quadratic as a squared linear expression.}$$

With the equation now in standard form, we read off the center's coordinates and the radius: $(h, k) = (-2, 3)$ and $a = 4$. ☐

Interior and Exterior

The points that lie inside the circle $(x - h)^2 + (y - k)^2 = a^2$ are the points less than a units from (h, k). They satisfy the inequality

$$(x - h)^2 + (y - k)^2 < a^2.$$

They make up the region we call the **interior** of the circle (Fig. 37).

The circle's **exterior** consists of the points that lie more than a units from (h, k). These points satisfy the inequality

$$(x - h)^2 + (y - k)^2 > a^2.$$

EXAMPLE 10

Inequality	Region
$x^2 + y^2 < 1$	Interior of the unit circle
$x^2 + y^2 \leq 1$	Unit circle plus its interior
$x^2 + y^2 > 1$	Exterior of the unit circle
$x^2 + y^2 \geq 1$	Unit circle plus its exterior

☐

Parabolic Graphs

The graph of an equation like $y = 3x^2$ or $y = -5x^2$ that has the form

$$y = ax^2$$

is a **parabola** whose **axis** (axis of symmetry) is the y-axis. The parabola's **vertex** (point where the parabola and axis cross) lies at the origin. The parabola opens upward if $a > 0$ and downward if $a < 0$. The larger the value of $|a|$, the narrower the parabola (Fig. 38).

If we interchange x and y in the formula $y = ax^2$, we obtain the equation

$$x = ay^2.$$

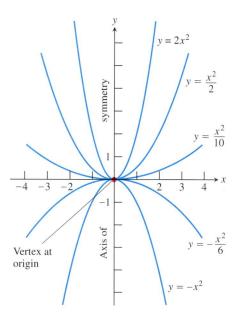

38 Besides determining the direction in which the parabola $y = ax^2$ opens, the number a is a scaling factor. The parabola widens as a approaches zero and narrows as $|a|$ becomes large.

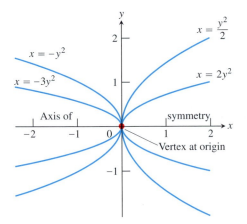

39 The parabola $x = ay^2$ is symmetric about the x-axis. It opens to the right if $a > 0$ and to the left if $a < 0$.

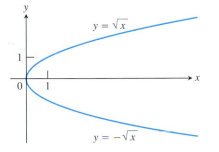

40 The graphs of the functions $y = \sqrt{x}$ and $y = -\sqrt{x}$ join at the origin to make the graph of the equation $x = y^2$ (Example 11).

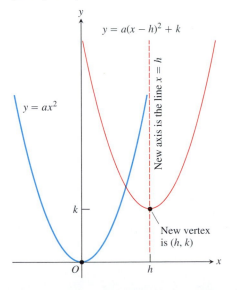

41 The parabola $y = ax^2$, $a > 0$, shifted h units to the right and k units up.

With x and y now reversed, the graph is a parabola whose axis is the x-axis and whose vertex lies at the origin (Fig. 39).

EXAMPLE 11   The formula $x = y^2$ gives x as a function of y but does *not* give y as a function of x. If we solve for y, we find that $y = \pm\sqrt{x}$. For each positive value of x we get *two* values of y instead of the required single value.

When taken separately, the formulas $y = \sqrt{x}$ and $y = -\sqrt{x}$ do define functions of x. Each formula gives exactly one value of y for each possible value of x. The graph of $y = \sqrt{x}$ is the upper half of the parabola $x = y^2$. The graph of $y = -\sqrt{x}$ is the lower half (Fig. 40). ⬜

The Quadratic Equation $y = ax^2 + bx + c$, $a \neq 0$

To shift the parabola $y = ax^2$ horizontally, we rewrite the equation as

$$y = a(x - h)^2.$$

To shift it vertically as well, we change the equation to

$$y - k = a(x - h)^2. \tag{2}$$

The combined shifts place the vertex at the point (h, k) and the axis along the line $x = h$ (Fig. 41).

Normally there would be no point in multiplying out the right-hand side of Eq. (2). In this case, however, we can learn something from doing so because the resulting equation, when rearranged, takes the form

$$y = ax^2 + bx + c. \tag{3}$$

This tells us that the graph of every equation of the form $y = ax^2 + bx + c$, $a \neq 0$, is the graph of $y = ax^2$ shifted somewhere else. Why? Because the steps that take us from Eq. (2) to Eq. (3) can be reversed to take us from (3) back to (2). The curve $y = ax^2 + bx + c$ has the same shape and orientation as the curve $y = ax^2$.

The axis of the parabola $y = ax^2 + bx + c$ turns out to be the line $x = -b/(2a)$. The y-intercept, $y = c$, is obtained by setting $x = 0$.

The Graph of $y = ax^2 + bx + c$, $a \neq 0$

The graph of the equation $y = ax^2 + bx + c$, $a \neq 0$, is a parabola. The parabola opens upward if $a > 0$ and downward if $a < 0$. The axis is the line

$$x = -\frac{b}{2a}. \tag{4}$$

The vertex of the parabola is the point where the axis and parabola intersect. Its x-coordinate is $x = -b/2a$; its y-coordinate is found by substituting $x = -b/2a$ in the parabola's equation.

EXAMPLE 12   *Graphing a parabola*

Graph the equation $y = -\frac{1}{2}x^2 - x + 4$.

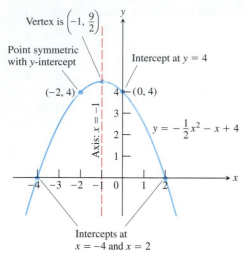

Vertex is $\left(-1, \frac{9}{2}\right)$

Point symmetric with y-intercept

$(-2, 4)$

Intercept at $y = 4$

$(0, 4)$

Axis: $x = -1$

$y = -\frac{1}{2}x^2 - x + 4$

Intercepts at $x = -4$ and $x = 2$

42 The parabola in Example 12.

Solution We take the following steps.

Step 1: *Compare the equation with* $y = ax^2 + bx + c$ *to identify* a, b, *and* c.

$$a = -\frac{1}{2}, \qquad b = -1, \qquad c = 4$$

Step 2: *Find the direction of opening.* Down, because $a < 0$.

Step 3: *Find the axis and vertex.* The axis is the line

$$x = -\frac{b}{2a} = -\frac{(-1)}{2(-1/2)} = -1, \qquad \text{Eq. (4)}$$

so the x-coordinate of the vertex is -1. The y-coordinate is

$$y = -\frac{1}{2}(-1)^2 - (-1) + 4 = \frac{9}{2}.$$

The vertex is $(-1, 9/2)$.

Step 4: *Find the x-intercepts (if any).*

$$-\frac{1}{2}x^2 - x + 4 = 0 \qquad \text{Set } y = 0 \text{ in the parabola's equation.}$$

$$x^2 + 2x - 8 = 0 \qquad \text{Solve as usual.}$$

$$(x - 2)(x + 4) = 0$$

$$x = 2, \qquad x = -4$$

Step 5: *Sketch the graph.* We plot points, sketch the axis (lightly), and use what we know about symmetry and the direction of opening to complete the graph (Fig. 42). ❑

Exercises 4

Shifting Graphs

1. Figure 43 shows the graph of $y = -x^2$ shifted to two new positions. Write equations for the new graphs.

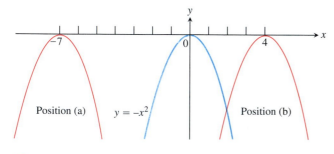

-7

0

4

Position (a)

$y = -x^2$

Position (b)

43 The parabolas in Exercise 1.

2. Figure 44 shows the graph of $y = x^2$ shifted to two new positions. Write equations for the new graphs.

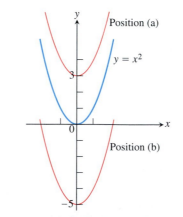

Position (a)

$y = x^2$

3

0

Position (b)

-5

44 The parabolas in Exercise 2.

3. Match the equations listed in (a)–(d) to the graphs in Fig. 45.

a) $y = (x - 1)^2 - 4$ **b)** $y = (x - 2)^2 + 2$

c) $y = (x + 2)^2 + 2$ **d)** $y = (x + 3)^2 - 2$

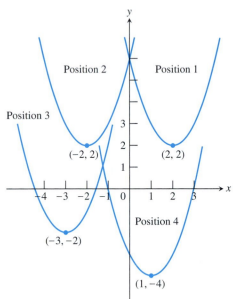

45 The parabolas in Exercise 3.

4. Figure 46 shows the graph of $y = -x^2$ shifted to four new positions. Write an equation for each new graph.

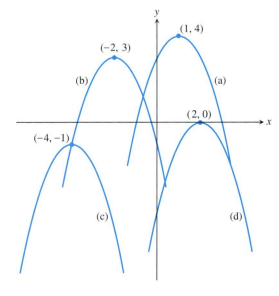

46 The parabolas in Exercise 4.

Exercises 5–16 tell how many units and in what directions the graphs of the given equations are to be shifted. Give an equation for the shifted graph. Then sketch the original and shifted graphs together, labeling each graph with its equation. Use the graphs in Fig. 26 for reference as needed.

5. $x^2 + y^2 = 49$ Down 3, left 2

6. $x^2 + y^2 = 25$ Up 3, left 4

7. $y = x^3$ Left 1, down 1

8. $y = x^{2/3}$ Right 1, down 1

9. $y = \sqrt{x}$ Left 0.81

10. $y = -\sqrt{x}$ Right 3

11. $y = 2x - 7$ Up 7

12. $y = \dfrac{1}{2}(x + 1) + 5$ Down 5, right 1

13. $x = y^2$ Left 1   **14.** $x = -3y^2$ Up 2, right 3

15. $y = 1/x$ Up 1, right 1   **16.** $y = 1/x^2$ Left 2, down 1

Graph the functions in Exercises 17–36. Use the graphs in Fig. 26 for reference as needed.

17. $y = \sqrt{x + 4}$ **18.** $y = \sqrt{9 - x}$

19. $y = |x - 2|$ **20.** $y = |1 - x| - 1$

21. $y = 1 + \sqrt{x - 1}$ **22.** $y = 1 - \sqrt{x}$

23. $y = (x + 1)^{2/3}$ **24.** $y = (x - 8)^{2/3}$

25. $y = 1 - x^{2/3}$ **26.** $y + 4 = x^{2/3}$

27. $y = \sqrt[3]{x - 1} - 1$ **28.** $y = (x + 2)^{3/2} + 1$

29. $y = \dfrac{1}{x - 2}$ **30.** $y = \dfrac{1}{x} - 2$

31. $y = \dfrac{1}{x} + 2$ **32.** $y = \dfrac{1}{x + 2}$

33. $y = \dfrac{1}{(x - 1)^2}$ **34.** $y = \dfrac{1}{x^2} - 1$

35. $y = \dfrac{1}{x^2} + 1$ **36.** $y = \dfrac{1}{(x + 1)^2}$

37. The accompanying figure shows the graph of a function $f(x)$ with domain $[0, 2]$ and range $[0, 1]$. Find the domains and ranges of the following functions, and sketch their graphs.

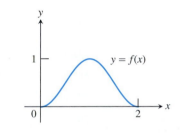

a) $f(x) + 2$ **b)** $f(x) - 1$

c) $2f(x)$ **d)** $-f(x)$

e) $f(x + 2)$ **f)** $f(x - 1)$

g) $f(-x)$ **h)** $-f(x + 1) + 1$

38. The accompanying figure shows the graph of a function $g(t)$ with domain $[-4, 0]$ and range $[-3, 0]$. Find the domains and ranges of the following functions, and sketch their graphs.

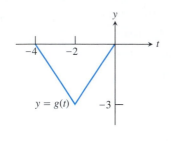

a) $g(-t)$
b) $-g(t)$
c) $g(t) + 3$
d) $1 - g(t)$
e) $g(-t + 2)$
f) $g(t - 2)$
g) $g(1 - t)$
h) $-g(t - 4)$

Circles

In Exercises 39–44, find an equation for the circle with the given center $C(h, k)$ and radius a. Then sketch the circle in the xy-plane. Include the circle's center in your sketch. Also, label the circle's x- and y-intercepts, if any, with their coordinate pairs.

39. $C(0, 2)$, $a = 2$
40. $C(-3, 0)$, $a = 3$
41. $C(-1, 5)$, $a = \sqrt{10}$
42. $C(1, 1)$, $a = \sqrt{2}$
43. $C(-\sqrt{3}, -2)$, $a = 2$
44. $C(3, 1/2)$, $a = 5$

Graph the circles whose equations are given in Exercises 45–50. Label each circle's center and intercepts (if any) with their coordinate pairs.

45. $x^2 + y^2 + 4x - 4y + 4 = 0$
46. $x^2 + y^2 - 8x + 4y + 16 = 0$
47. $x^2 + y^2 - 3y - 4 = 0$
48. $x^2 + y^2 - 4x - (9/4) = 0$
49. $x^2 + y^2 - 4x + 4y = 0$
50. $x^2 + y^2 + 2x = 3$

Parabolas

Graph the parabolas in Exercises 51–58. Label the vertex, axis, and intercepts in each case.

51. $y = x^2 - 2x - 3$
52. $y = x^2 + 4x + 3$
53. $y = -x^2 + 4x$
54. $y = -x^2 + 4x - 5$
55. $y = -x^2 - 6x - 5$
56. $y = 2x^2 - x + 3$
57. $y = \frac{1}{2}x^2 + x + 4$
58. $y = -\frac{1}{4}x^2 + 2x + 4$

59. Graph the parabola $y = x - x^2$. Then find the domain and range of $f(x) = \sqrt{x - x^2}$.

60. Graph the parabola $y = 3 - 2x - x^2$. Then find the domain and range of $g(x) = \sqrt{3 - 2x - x^2}$.

Inequalities

Describe the regions defined by the inequalities and pairs of inequalities in Exercises 61–68.

61. $x^2 + y^2 > 7$
62. $x^2 + y^2 < 5$
63. $(x - 1)^2 + y^2 \le 4$
64. $x^2 + (y - 2)^2 \ge 4$
65. $x^2 + y^2 > 1$, $x^2 + y^2 < 4$
66. $x^2 + y^2 \le 4$, $(x + 2)^2 + y^2 \le 4$
67. $x^2 + y^2 + 6y < 0$, $y > -3$
68. $x^2 + y^2 - 4x + 2y > 4$, $x > 2$

69. Write an inequality that describes the points that lie inside the circle with center $(-2, 1)$ and radius $\sqrt{6}$.

70. Write an inequality that describes the points that lie outside the circle with center $(-4, 2)$ and radius 4.

71. Write a pair of inequalities that describe the points that lie inside or on the circle with center $(0, 0)$ and radius $\sqrt{2}$, and on or to the right of the vertical line through $(1, 0)$.

72. Write a pair of inequalities that describe the points that lie outside the circle with center $(0, 0)$ and radius 2, and inside the circle that has center $(1, 3)$ and passes through the origin.

Shifting Lines

73. The line $y = mx$, which passes through the origin, is shifted vertically and horizontally to pass through the point (x_0, y_0). Find an equation for the new line. (This equation is called the line's *point–slope equation*.)

74. The line $y = mx$ is shifted vertically to pass through the point $(0, b)$. What is the new line's equation?

Intersecting Lines, Circles, and Parabolas

In Exercises 75–82, graph the two equations and find the points in which the graphs intersect.

75. $y = 2x$, $x^2 + y^2 = 1$
76. $x + y = 1$, $(x - 1)^2 + y^2 = 1$
77. $y - x = 1$, $y = x^2$
78. $x + y = 0$, $y = -(x - 1)^2$
79. $y = -x^2$, $y = 2x^2 - 1$
80. $y = \frac{1}{4}x^2$, $y = (x - 1)^2$
81. $x^2 + y^2 = 1$, $(x - 1)^2 + y^2 = 1$
82. $x^2 + y^2 = 1$, $x^2 + y = 1$

✦ CAS Explorations and Projects

In Exercises 83–86, you will explore graphically what happens to the graph of $y = f(ax)$ as you change the value of the constant a. Use a CAS or computer grapher to perform the following steps.

a) Plot the function $y = f(x)$ together with the function $y = f(ax)$ for $a = 2, 3$, and 10 over the specified interval. Describe what happens to the graph as a increases through positive values.

b) Plot the function $y = f(x)$ and $y = f(ax)$ for the negative values $a = -2, -3$. What happens to the graph in this situation?

c) Plot the function $y = f(x)$ and $y = f(ax)$ for the fractional values $a = 1/2, 1/3, 1/4$. Describe what happens to the graph when $|a| < 1$.

83. $f(x) = \dfrac{5x}{x^2 + 4}$, $[-10, 10]$

84. $f(x) = \dfrac{2x(x - 1)}{x^2 + 1}$, $[-3, 2]$

85. $f(x) = \dfrac{x + 1}{2x^2 + 1}$, $[-2, 2]$

86. $f(x) = \dfrac{x^4 - 4x^3 + 10}{x^2 + 4}$, $[-1, 4]$

---

5 Trigonometric Functions

This section reviews radian measure, trigonometric functions, periodicity, and basic trigonometric identities.

Radian Measure

In navigation and astronomy, angles are measured in degrees, but in calculus it is best to use units called radians because of the way they simplify later calculations (Section 2.4).

Let ACB be a central angle in a **unit circle** (circle of radius 1), as in Fig. 47.

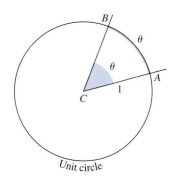

Unit circle

47 The radian measure of angle ACB is the length of the arc AB.

The **radian measure** θ of angle ACB is defined to be the length of the circular arc AB. Since the circumference of the circle is 2π and one complete revolution of a circle is $360°$, the relation between radians and degrees is given by the following equation.

$$\pi \text{ radians} = 180°$$

Degrees	Radians

48 The angles of two common triangles, in degrees and radians.

EXAMPLE 1 *Conversions (Fig. 48)*

Convert 45° to radians: $\qquad 45 \cdot \dfrac{\pi}{180} = \dfrac{\pi}{4}$ rad

Convert $\dfrac{\pi}{6}$ rad to degrees: $\qquad \dfrac{\pi}{6} \cdot \dfrac{180}{\pi} = 30°$

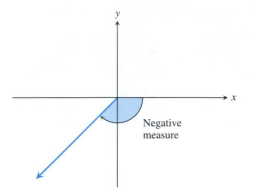

49 Angles in standard position in the *xy*-plane.

Conversion formulas

1 degree $= \dfrac{\pi}{180}$ (≈ 0.02) radians

Degrees to radians: multiply by $\dfrac{\pi}{180}$

1 radian $= \dfrac{180}{\pi}$ (≈ 57) degrees

Radians to degrees: multiply by $\dfrac{180}{\pi}$

An angle in the *xy*-plane is said to be in **standard position** if its vertex lies at the origin and its initial ray lies along the positive *x*-axis (Fig. 49). Angles measured counterclockwise from the positive *x*-axis are assigned positive measures; angles measured clockwise are assigned negative measures.

When angles are used to describe counterclockwise rotations, our measurements can go arbitrarily far beyond 2π radians or 360°. Similarly, angles describing clockwise rotations can have negative measures of all sizes (Fig. 50).

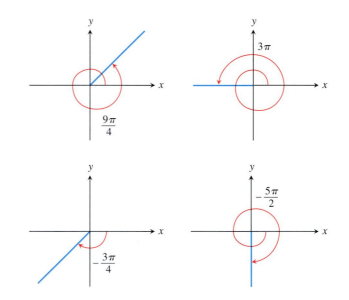

50 Nonzero radian measures can be positive or negative.

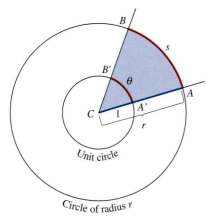

51 The radian measure of angle *ACB* is the length θ of arc $A'B'$ on the unit circle centered at *C*. The value of θ can be found from any other circle as *s/r*.

There is a useful relationship between the length *s* of an arc *AB* on a circle of radius *r* and the radian measure θ of the angle the arc subtends at the circle's center *C* (Fig. 51). If we draw a unit circle with the same center *C*, the arc $A'B'$ cut by the angle will have length θ, by the definition of radian measure. From the similarity of the circular sectors *ACB* and *A'CB'*, we then have $s/r = \theta/1$.

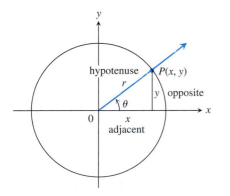

$$\sin \theta = \frac{\text{opp}}{\text{hyp}} \qquad \csc \theta = \frac{\text{hyp}}{\text{opp}}$$

$$\cos \theta = \frac{\text{adj}}{\text{hyp}} \qquad \sec \theta = \frac{\text{hyp}}{\text{adj}}$$

$$\tan \theta = \frac{\text{opp}}{\text{adj}} \qquad \cot \theta = \frac{\text{adj}}{\text{opp}}$$

52 Trigonometric ratios of an acute angle.

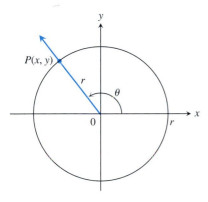

53 The trigonometric functions of a general angle θ are defined in terms of x, y, and r.

54 The new and old definitions agree for acute angles.

> **Radian Measure and Arc Length**
>
> $$\frac{s}{r} = \theta, \qquad \text{or} \qquad s = r\theta$$

Notice that these equalities hold precisely because we are measuring the angle in radians.

> **Angle Convention: Use Radians**
>
> From now on in this book it is assumed that all angles are measured in radians unless degrees or some other unit is stated explicitly. When we talk about the angle $\pi/3$, we mean $\pi/3$ radians (which is $60°$), not $\pi/3$ degrees. When you do calculus, keep your calculator in radian mode.

EXAMPLE 2   Consider a circle of radius 8. (a) Find the central angle subtended by an arc of length 2π on the circle. (b) Find the length of an arc subtending a central angle of $3\pi/4$.

Solution

a) $\theta = \dfrac{s}{r} = \dfrac{2\pi}{8} = \dfrac{\pi}{4}$ b) $s = r\theta = 8\left(\dfrac{3\pi}{4}\right) = 6\pi$

The Six Basic Trigonometric Functions

You are probably familiar with defining the trigonometric functions of an acute angle in terms of the sides of a right triangle (Fig. 52). We extend this definition to obtuse and negative angles by first placing the angle in standard position in a circle of radius r. We then define the trigonometric functions in terms of the coordinates of the point $P(x, y)$ where the angle's terminal ray intersects the circle (Fig. 53).

Sine:	$\sin \theta = \dfrac{y}{r}$	Cosecant:	$\csc \theta = \dfrac{r}{y}$
Cosine:	$\cos \theta = \dfrac{x}{r}$	Secant:	$\sec \theta = \dfrac{r}{x}$
Tangent:	$\tan \theta = \dfrac{y}{x}$	Cotangent:	$\cot \theta = \dfrac{x}{y}$

These extended definitions agree with the right-triangle definitions when the angle is acute (Fig. 54).

As you can see, $\tan \theta$ and $\sec \theta$ are not defined if $x = 0$. This means they are

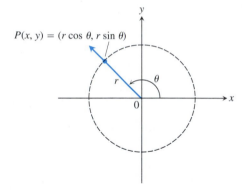

55 The Cartesian coordinates of a point in the plane expressed in terms of r and θ.

not defined if θ is $\pm\pi/2$, $\pm 3\pi/2$, ... Similarly, $\cot\theta$ and $\csc\theta$ are not defined for values of θ for which $y = 0$, namely $\theta = 0$, $\pm\pi$, $\pm 2\pi$, ...

Notice also the following definitions, whenever the quotients are defined.

$$\tan\theta = \frac{\sin\theta}{\cos\theta} \qquad \cot\theta = \frac{1}{\tan\theta}$$

$$\sec\theta = \frac{1}{\cos\theta} \qquad \csc\theta = \frac{1}{\sin\theta}$$

The coordinates of any point $P(x, y)$ in the plane can now be expressed in terms of the point's distance from the origin and the angle that ray OP makes with the positive x-axis (Fig. 55). Since $x/r = \cos\theta$ and $y/r = \sin\theta$, we have

$$x = r\cos\theta, \qquad y = r\sin\theta. \tag{1}$$

Values of Trigonometric Functions

If the circle in Fig. 53 has radius $r = 1$, the equations defining $\sin\theta$ and $\cos\theta$ become

$$\cos\theta = x, \qquad \sin\theta = y.$$

We can then calculate the values of the cosine and sine directly from the coordinates of P, if we happen to know them, or indirectly from the acute reference triangle made by dropping a perpendicular from P to the x-axis (Fig. 56). We read the magnitudes of x and y from the triangle's sides. The signs of x and y are determined by the quadrant in which the triangle lies.

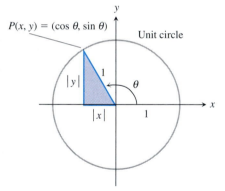

56 The acute reference triangle for an angle θ.

EXAMPLE 3   Find the sine and cosine of $2\pi/3$ radians.

Solution

Step 1: Draw the angle in standard position in the unit circle and write in the lengths of the sides of the reference triangle (Fig. 57).

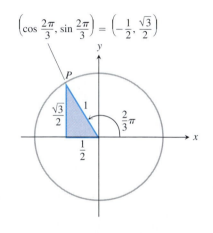

57 The triangle for calculating the sine and cosine of $2\pi/3$ radians (Example 3).

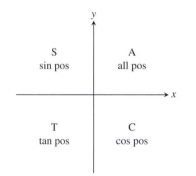

58 The CAST rule.

S
sin pos

A
all pos

T
tan pos

C
cos pos

Step 2: Find the coordinates of the point P where the angle's terminal ray cuts the circle:

$$\cos \frac{2\pi}{3} = x\text{-coordinate of } P = -\frac{1}{2}$$

$$\sin \frac{2\pi}{3} = y\text{-coordinate of } P = \frac{\sqrt{3}}{2}.$$ ❏

A useful rule for remembering when the basic trigonometric functions are positive and negative is the CAST rule (Fig. 58).

EXAMPLE 4   Find the sine and cosine of $-\pi/4$ radians.

Solution

Step 1: Draw the angle in standard position in the unit circle and write in the lengths of the sides of the reference triangle (Fig. 59).

$$\left(\frac{\sqrt{2}}{2}, -\frac{\sqrt{2}}{2}\right) = \left(\cos\left(-\frac{\pi}{4}\right), \sin\left(-\frac{\pi}{4}\right)\right)$$

59 The triangle for calculating the sine and cosine of $-\pi/4$ radians (Example 4).

Step 2: Find the coordinates of the point P where the angle's terminal ray cuts the circle:

$$\cos\left(-\frac{\pi}{4}\right) = x\text{-coordinate of } P = \frac{\sqrt{2}}{2},$$

$$\sin\left(-\frac{\pi}{4}\right) = y\text{-coordinate of } P = -\frac{\sqrt{2}}{2}.$$ ❏

Calculations similar to those in Examples 3 and 4 allow us to fill in Table 2.

Table 2 Values of sin θ, cos θ, and tan θ for selected values of θ

Degrees	−180	−135	−90	−45	0	30	45	60	90	135	180
θ (radians)	$-\pi$	$-3\pi/4$	$-\pi/2$	$-\pi/4$	0	$\pi/6$	$\pi/4$	$\pi/3$	$\pi/2$	$3\pi/4$	π
sin θ	0	$-\sqrt{2}/2$	−1	$-\sqrt{2}/2$	0	1/2	$\sqrt{2}/2$	$\sqrt{3}/2$	1	$\sqrt{2}/2$	0
cos θ	−1	$-\sqrt{2}/2$	0	$\sqrt{2}/2$	1	$\sqrt{3}/2$	$\sqrt{2}/2$	1/2	0	$-\sqrt{2}/2$	−1
tan θ	0	1		−1	0	$\sqrt{3}/3$	1	$\sqrt{3}$		−1	0

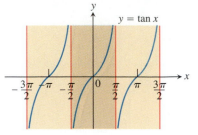

Domain: $(-\infty, \infty)$
Range: $[-1, 1]$

Domain: $(-\infty, \infty)$
Range: $[-1, 1]$

Domain: All real numbers except odd integer multiples of $\pi/2$
Range: $(-\infty, \infty)$

Domain: $x \neq \pm\dfrac{\pi}{2}, \pm\dfrac{3\pi}{2}, \ldots$
Range: $(-\infty, -1] \cup [1, \infty)$

Domain: $x \neq 0, \pm\pi, \pm2\pi, \ldots$
Range: $(-\infty, -1] \cup [1, \infty)$

Domain: $x \neq 0, \pm\pi, \pm2\pi, \ldots$
Range: $(-\infty, \infty)$

60 The graphs of the six basic trigonometric functions as functions of radian measure. Each function's periodicity shows clearly in its graph.

Graphs

When we graph trigonometric functions in the coordinate plane, we usually denote the independent variable by x instead of θ. See Fig. 60.

Periodicity

When an angle of measure x and an angle of measure $x + 2\pi$ are in standard position, their terminal rays coincide. The two angles therefore have the same trigonometric values. For example, $\cos(x + 2\pi) = \cos x$. Functions like the trigonometric functions whose values repeat at regular intervals are called periodic.

> ### Definition
>
> A function $f(x)$ is **periodic** if there is a positive number p such that $f(x + p) = f(x)$ for all x. The smallest such value of p is the **period** of f.

As we can see in Fig. 60, the tangent and cotangent functions have period $p = \pi$. The other four functions have period 2π.

Figure 61 shows graphs of $y = \cos 2x$ and $y = \cos(x/2)$ plotted against the graph of $y = \cos x$. Multiplying x by a number greater than 1 speeds up a trigonometric function (increases the frequency) and shortens its period. Multiplying x by a positive number less than 1 slows a trigonometric function down and lengthens its period.

Periods of trigonometric functions

Period π:
$$\tan(x + \pi) = \tan x$$
$$\cot(x + \pi) = \cot x$$

Period 2π:
$$\sin(x + 2\pi) = \sin x$$
$$\cos(x + 2\pi) = \cos x$$
$$\sec(x + 2\pi) = \sec x$$
$$\csc(x + 2\pi) = \csc x$$

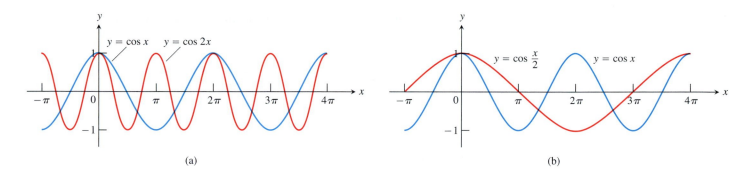

(a)   (b)

61 (a) Shorter period: cos 2x. (b) Longer period: cos (x/2)

The importance of periodic functions stems from the fact that much of the behavior we study in science is periodic. Brain waves and heartbeats are periodic, as are household voltage and electric current. The electromagnetic field that heats food in a microwave oven is periodic, as are cash flows in seasonal businesses and the behavior of rotational machinery. The seasons are periodic—so is the weather. The phases of the moon are periodic, as are the motions of the planets. There is strong evidence that the ice ages are periodic, with a period of 90,000–100,000 years.

If so many things are periodic, why limit our discussion to trigonometric functions? The answer lies in a surprising and beautiful theorem from advanced calculus that says that every periodic function we want to use in mathematical modeling can be written as an algebraic combination of sines and cosines. Thus, once we learn the calculus of sines and cosines, we will know everything we need to know to model the mathematical behavior of periodic phenomena.

Even vs. Odd

The symmetries in the graphs in Fig. 60 reveal that the cosine and secant functions are even and the other four functions are odd:

Even	Odd
$\cos(-x) = \cos x$	$\sin(-x) = -\sin x$
$\sec(-x) = \sec x$	$\tan(-x) = -\tan x$
	$\csc(-x) = -\csc x$
	$\cot(-x) = -\cot x$

Identities

Applying the Pythagorean theorem to the reference right triangle we obtain by dropping a perpendicular from the point $P(\cos\theta, \sin\theta)$ on the unit circle to the x-axis (Fig. 62) gives

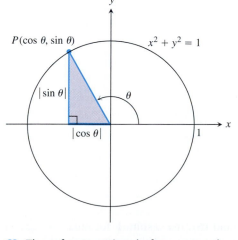

62 The reference triangle for a general angle θ.

$$\cos^2\theta + \sin^2\theta = 1. \qquad (2)$$

This equation, true for all values of θ, is probably the most frequently used identity in trigonometry.

Dividing Eq. (2) in turn by $\cos^2 \theta$ and $\sin^2 \theta$ gives the identities

$$1 + \tan^2 \theta = \sec^2 \theta,$$

$$1 + \cot^2 \theta = \csc^2 \theta.$$

You may recall the following identities from an earlier course.

All the trigonometric identities you will need in this book derive from Eqs. (2) and (3).

Angle Sum Formulas

$$\cos (A + B) = \cos A \cos B - \sin A \sin B$$

$$\sin (A + B) = \sin A \cos B + \cos A \sin B$$

(3)

These formulas hold for all angles A and B. There are similar formulas for $\cos (A - B)$ and $\sin (A - B)$ (Exercises 35 and 36).

Substituting θ for both A and B in the angle sum formulas gives two more useful identities:

Instead of memorizing Eqs. (3) you might find it helpful to remember Eqs. (4), and then recall where they came from.

Double-angle Formulas

$$\cos 2\theta = \cos^2 \theta - \sin^2 \theta$$

$$\sin 2\theta = 2 \sin \theta \cos \theta$$

(4)

Additional formulas come from combining the equations

$$\cos^2 \theta + \sin^2 \theta = 1, \qquad \cos^2 \theta - \sin^2 \theta = \cos 2\theta.$$

We add the two equations to get $2 \cos^2 \theta = 1 + \cos 2\theta$ and subtract the second from the first to get $2 \sin^2 \theta = 1 - \cos 2\theta$.

Additional Double-angle Formulas

$$\cos^2 \theta = \frac{1 + \cos 2\theta}{2}$$

(5)

$$\sin^2 \theta = \frac{1 - \cos 2\theta}{2}$$

(6)

When θ is replaced by $\theta/2$ in Eqs. (5) and (6), the resulting formulas are called **half-angle** formulas. Some books refer to Eqs. (5) and (6) by this name as well.

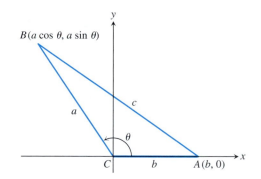

63 The square of the distance between A and B gives the law of cosines.

The Law of Cosines

If a, b, and c are sides of a triangle ABC and if θ is the angle opposite c, then

$$c^2 = a^2 + b^2 - 2ab\cos\theta. \tag{7}$$

This equation is called the **law of cosines.**

We can see why the law holds if we introduce coordinate axes with the origin at C and the positive x-axis along one side of the triangle, as in Fig. 63. The coordinates of A are $(b, 0)$; the coordinates of B are $(a\cos\theta, a\sin\theta)$. The square of the distance between A and B is therefore

$$c^2 = (a\cos\theta - b)^2 + (a\sin\theta)^2$$
$$= a^2\underbrace{(\cos^2\theta + \sin^2\theta)}_{1} + b^2 - 2ab\cos\theta$$
$$= a^2 + b^2 - 2ab\cos\theta.$$

Combining these equalities gives the law of cosines.

The law of cosines generalizes the Pythagorean theorem. If $\theta = \pi/2$, then $\cos\theta = 0$ and $c^2 = a^2 + b^2$.

Exercises 5

Radians, Degrees, and Circular Arcs

1. On a circle of radius 10 m, how long is an arc that subtends a central angle of (a) $4\pi/5$ radians? (b) $110°$?

2. A central angle in a circle of radius 8 is subtended by an arc of length 10π. Find the angle's radian and degree measures.

3. CALCULATOR You want to make an $80°$ angle by marking an arc on the perimeter of a 12-in.-diameter disk and drawing lines from the ends of the arc to the disk's center. To the nearest tenth of an inch, how long should the arc be?

4. CALCULATOR If you roll a 1-m-diameter wheel forward 30 cm over level ground, through what angle will the wheel turn? Answer in radians (to the nearest tenth) and degrees (to the nearest degree).

Evaluating Trigonometric Functions

5. Copy and complete the table of function values shown on the following page. If the function is undefined at a given angle, enter "UND." Do not use a calculator or tables.

θ	$-\pi$	$-2\pi/3$	0	$\pi/2$	$3\pi/4$
$\sin \theta$					
$\cos \theta$					
$\tan \theta$					
$\cot \theta$					
$\sec \theta$					
$\csc \theta$					

6. Copy and complete the following table of function values. If the function is undefined at a given angle, enter "UND." Do not use a calculator or tables.

θ	$-3\pi/2$	$-\pi/3$	$-\pi/6$	$\pi/4$	$5\pi/6$
$\sin \theta$					
$\cos \theta$					
$\tan \theta$					
$\cot \theta$					
$\sec \theta$					
$\csc \theta$					

In Exercises 7–12, one of $\sin x$, $\cos x$, and $\tan x$ is given. Find the other two if x lies in the specified interval.

7. $\sin x = \dfrac{3}{5}, \quad x \text{ in } \left[\dfrac{\pi}{2}, \pi\right]$

8. $\tan x = 2, \quad x \text{ in } \left[0, \dfrac{\pi}{2}\right]$

9. $\cos x = \dfrac{1}{3}, \quad x \text{ in } \left[-\dfrac{\pi}{2}, 0\right]$

10. $\cos x = -\dfrac{5}{13}, \quad x \text{ in } \left[\dfrac{\pi}{2}, \pi\right]$

11. $\tan x = \dfrac{1}{2}, \quad x \text{ in } \left[\pi, \dfrac{3\pi}{2}\right]$

12. $\sin x = -\dfrac{1}{2}, \quad x \text{ in } \left[\pi, \dfrac{3\pi}{2}\right]$

Graphing Trigonometric Functions

Graph the functions in Exercises 13–22. What is the period of each function?

13. $\sin 2x$

14. $\sin (x/2)$

15. $\cos \pi x$

16. $\cos \dfrac{\pi x}{2}$

17. $-\sin \dfrac{\pi x}{3}$

18. $-\cos 2\pi x$

19. $\cos \left(x - \dfrac{\pi}{2}\right)$

20. $\sin \left(x + \dfrac{\pi}{2}\right)$

21. $\sin \left(x - \dfrac{\pi}{4}\right) + 1$

22. $\cos \left(x + \dfrac{\pi}{4}\right) - 1$

Graph the functions in Exercises 23–26 in the ts-plane (t-axis horizontal, s-axis vertical). What is the period of each function? What symmetries do the graphs have?

23. $s = \cot \ 2t$

24. $s = -\tan \ \pi t$

25. $s = \sec \left(\dfrac{\pi t}{2}\right)$

26. $s = \csc \left(\dfrac{t}{2}\right)$

27. GRAPHER

a) Graph $y = \cos x$ and $y = \sec x$ together for $-3\pi/2 \le x \le 3\pi/2$. Comment on the behavior of $\sec x$ in relation to the signs and values of $\cos x$.

b) Graph $y = \sin x$ and $y = \csc x$ together for $-\pi \le x \le 2\pi$. Comment on the behavior of $\csc x$ in relation to the signs and values of $\sin x$.

28. GRAPHER Graph $y = \tan x$ and $y = \cot x$ together for $-7 \le x \le 7$. Comment on the behavior of $\cot x$ in relation to the signs and values of $\tan x$.

29. Graph $y = \sin x$ and $y = \lfloor \sin x \rfloor$ together. What are the domain and range of $\lfloor \sin x \rfloor$?

30. Graph $y = \sin x$ and $y = \lceil \sin x \rceil$ together. What are the domain and range of $\lceil \sin x \rceil$?

Additional Trigonometric Identities

Use the angle sum formulas to derive the identities in Exercises 31–36.

31. $\cos \left(x - \dfrac{\pi}{2}\right) = \sin x$

32. $\cos \left(x + \dfrac{\pi}{2}\right) = -\sin x$

33. $\sin \left(x + \dfrac{\pi}{2}\right) = \cos x$

34. $\sin \left(x - \dfrac{\pi}{2}\right) = -\cos x$

35. $\cos (A - B) = \cos A \cos B + \sin A \sin B$

36. $\sin (A - B) = \sin A \cos B - \cos A \sin B$

37. What happens if you take $B = A$ in the identity $\cos (A - B) = \cos A \cos B + \sin A \sin B$? Does the result agree with something you already know?

38. What happens if you take $B = 2\pi$ in the angle sum formulas? Do the results agree with something you already know?

Using the Angle Sum Formulas

In Exercises 39–42, express the given quantity in terms of $\sin x$ and $\cos x$.

39. $\cos (\pi + x)$

40. $\sin (2\pi - x)$

41. $\sin \left(\dfrac{3\pi}{2} - x\right)$

42. $\cos \left(\dfrac{3\pi}{2} + x\right)$

43. Evaluate $\sin \dfrac{7\pi}{12}$ as $\sin \left(\dfrac{\pi}{4} + \dfrac{\pi}{3}\right)$.

44. Evaluate $\cos \dfrac{11\pi}{12}$ as $\cos \left(\dfrac{\pi}{4} + \dfrac{2\pi}{3} \right)$.

45. Evaluate $\cos \dfrac{\pi}{12}$.

46. Evaluate $\sin \dfrac{5\pi}{12}$.

Using the Double-angle Formulas

Find the function values in Exercises 47–50.

47. $\cos^2 \dfrac{\pi}{8}$

48. $\cos^2 \dfrac{\pi}{12}$

49. $\sin^2 \dfrac{\pi}{12}$

50. $\sin^2 \dfrac{\pi}{8}$

Theory and Examples

51. *The tangent sum formula.* The standard formula for the tangent of the sum of two angles is

$$\tan (A + B) = \frac{\tan A + \tan B}{1 - \tan A \tan B}.$$

Derive the formula.

52. *(Continuation of Exercise 51.)* Derive a formula for $\tan (A - B)$.

53. Apply the law of cosines to the triangle in the accompanying figure to derive the formula for $\cos (A - B)$.

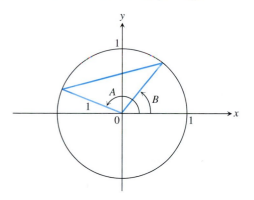

54. When applied to a figure similar to the one in Exercise 53, the law of cosines leads directly to the formula for $\cos (A + B)$. What is that figure and how does the derivation go?

55. CALCULATOR A triangle has sides $a = 2$ and $b = 3$ and angle $C = 60°$. Find the length of side c.

56. CALCULATOR A triangle has sides $a = 2$ and $b = 3$ and angle $C = 40°$. Find the length of side c.

57. *The law of sines.* The **law of sines** says that if a, b, and c are the sides opposite the angles A, B, and C in a triangle, then

$$\frac{\sin A}{a} = \frac{\sin B}{b} = \frac{\sin C}{c}.$$

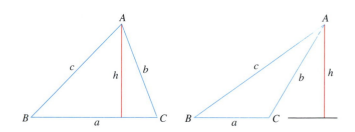

Use the accompanying figures and the identity $\sin (\pi - \theta) = \sin \theta$, if required, to derive the law.

58. CALCULATOR A triangle has sides $a = 2$ and $b = 3$ and angle $C = 60°$ (as in Exercise 55). Find the sine of angle B using the law of sines.

59. CALCULATOR A triangle has side $c = 2$ and angles $A = \pi/4$ and $B = \pi/3$. Find the length a of the side opposite A.

60. *The approximation sin x ≈ x.* It is often useful to know that, when x is measured in radians, $\sin x \approx x$ for numerically small values of x. In Section 3.7, we will see why the approximation holds. The approximation error is less than 1 in 5000 if $|x| < 0.1$.

   a) With your grapher in radian mode, graph $y = \sin x$ and $y = x$ together in a viewing window about the origin. What do you see happening as x nears the origin?

   b) With your grapher in degree mode, graph $y = \sin x$ and $y = x$ together about the origin again. How is the picture different from the one obtained with radian mode?

   c) *A quick radian mode check.* Is your calculator in radian mode? Evaluate $\sin x$ at a value of x near the origin, say $x = 0.1$. If $\sin x \approx x$, the calculator is in radian mode; if not, it isn't. Try it.

General Sine Curves

Figure 64 on the following page shows the graph of a **general sine function** of the form

$$f(x) = A \sin \left(\frac{2\pi}{B}(x - C) \right) + D,$$

where $|A|$ is the *amplitude*, $|B|$ is the *period*, C is the *horizontal shift*, and D is the *vertical shift*. Identify A, B, C, and D for the sine functions in Exercises 61–64 and sketch their graphs.

61. $y = 2 \sin (x + \pi) - 1$

62. $y = \dfrac{1}{2} \sin (\pi x - \pi) + \dfrac{1}{2}$

63. $y = -\dfrac{2}{\pi} \sin \left(\dfrac{\pi}{-2}t \right) + \dfrac{1}{\pi}$

64. $y = \dfrac{L}{2\pi} \sin \dfrac{2\pi t}{L}, \quad L > 0$

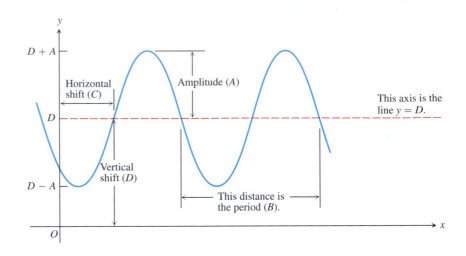

64 The general sine curve

$$y = A \sin\left[(2\pi/B)(x - C)\right] + D,$$

shown for A, B, C, and D positive.

The Trans-Alaska Pipeline

The builders of the Trans-Alaska Pipeline used insulated pads to keep the heat from the hot oil in the pipeline from melting the permanently frozen soil beneath. To design the pads, it was necessary to take into account the variation in air temperature throughout the year. Figure 65 shows how we can use a general sine function, defined in the introduction to Exercises 61–64, to represent temperature data. The data points in the figure are plots of the mean air temperature for Fairbanks, Alaska, based on records of the National Weather Service from 1941 to 1970. The sine function used to fit the data is

$$f(x) = 37 \sin\left(\frac{2\pi}{365}(x - 101)\right) + 25,$$

where f is temperature in degrees Fahrenheit and x is the number of the day counting from the beginning of the year. The fit is remarkably good.

65. *Temperature in Fairbanks, Alaska.* Find the (a) amplitude, (b) period, (c) horizontal shift, and (d) vertical shift of the general sine function

$$f(x) = 37 \sin\left(\frac{2\pi}{365}(x - 101)\right) + 25.$$

66. *Temperature in Fairbanks, Alaska.* Use the equation in Exercise 65 to approximate the answers to the following questions about the temperature in Fairbanks, Alaska, shown in Fig. 65. Assume that the year has 365 days.

a) What are the highest and lowest mean daily temperatures shown?

b) What is the average of the highest and lowest mean daily temperatures shown? Why is this average the vertical shift of the function?

⚙ CAS Explorations and Projects

In Exercises 67–70, you will explore graphically the general sine function

$$f(x) = A \sin\left(\frac{2\pi}{B}(x - C)\right) + D$$

as you change the values of the constants A, B, C, and D. Use a CAS or computer grapher to perform the steps in the exercises.

67. *The period B.* Set the constants $A = 3, C = D = 0$.

a) Plot $f(x)$ for the values $B = 1, 3, 2\pi, 5\pi$ over the interval

65 Normal mean air temperature at Fairbanks, Alaska, plotted as data points. The approximating sine function is

$$f(x) = 37 \sin\left(\frac{2\pi}{365}(x - 101)\right) + 25.$$

(Source: "Is the Curve of Temperature Variation a Sine Curve?" by B. M. Lando and C. A. Lando, *The Mathematics Teacher*, 7:6, Fig. 2, p. 535 [September 1977].)

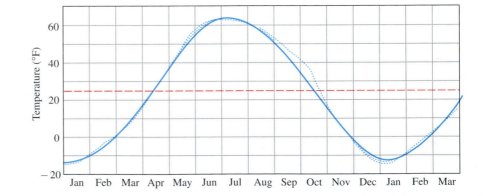

$-4\pi \leq x \leq 4\pi$. Describe what happens to the graph of the general sine function as the period increases.

b) What happens to the graph for negative values of B? Try it with $B = -3$ and $B = -2\pi$.

68. *The horizontal shift C.* Set the constants $A = 3$, $B = 6$, $D = 0$.

a) Plot $f(x)$ for the values $C = 0$, 1, and 2 over the interval $-4\pi \leq x \leq 4\pi$. Describe what happens to the graph of the general sine function as C increases through positive values.

b) What happens to the graph for negative values of C?

c) What smallest positive value should be assigned to C so the graph exhibits no horizontal shift? Confirm your answer with a plot.

69. *The vertical shift D.* Set the constants $A = 3$, $B = 6$, $C = 0$.

a) Plot $f(x)$ for the values $D = 0$, 1, and 3 over the interval $-4\pi \leq x \leq 4\pi$. Describe what happens to the graph of the general sine function as D increases through positive values.

b) What happens to the graph for negative values of D?

70. *The amplitude A.* Set the constants $B = 6$, $C = D = 0$.

a) Describe what happens to the graph of the general sine function as A increases through positive values. Confirm your answer by plotting $f(x)$ for the values $A = 1$, 5, and 9.

b) What happens to the graph for negative values of A?

PRELIMINARIES

QUESTIONS TO GUIDE YOUR REVIEW

---

1. What are the order properties of the real numbers? How are they used in solving inequalities?

2. What is a number's absolute value? Give examples. How are $|-a|$, $|ab|$, $|a/b|$, and $|a + b|$ related to $|a|$ and $|b|$?

3. How are absolute values used to describe intervals or unions of intervals? Give examples.

4. How do you find the distance between two points in the coordinate plane?

5. How can you write an equation for a line if you know the coordinates of two points on the line? the line's slope and the coordinates of one point on the line? the line's slope and y-intercept? Give examples.

6. What are the standard equations for lines perpendicular to the coordinate axes?

7. How are the slopes of mutually perpendicular lines related? What about parallel lines? Give examples.

8. When a line is not vertical, what is the relation between its slope and its angle of inclination?

9. What is a function? Give examples. How do you graph a real-valued function of a real variable?

10. Name some typical algebraic and trigonometric functions and draw their graphs.

11. What is an even function? an odd function? What geometric properties do the graphs of such functions have? What advantage can we take of this? Give an example of a function that is neither even nor odd. What, if anything, can you say about sums, products, quotients, and composites involving even and odd functions?

12. If f and g are real-valued functions, how are the domains of $f + g$, $f - g$, fg, and f/g related to the domains of f and g? Give examples.

13. When is it possible to compose one function with another? Give examples of composites and their values at various points. Does the order in which functions are composed ever matter?

14. How do you change the equation $y = f(x)$ to shift its graph up or down? to the left or right? Give examples.

15. Describe the steps you would take to graph the circle $x^2 + y^2 + 4x - 6y + 12 = 0$.

16. If a, b, and c are constants and $a \neq 0$, what can you say about the graph of the equation $y = ax^2 + bx + c$? In particular, how would you go about sketching the curve $y = 2x^2 + 4x$?

17. What inequality describes the points in the coordinate plane that lie inside the circle of radius a centered at the point (h, k)? that lie inside or on the circle? that lie outside the circle? that lie outside or on the circle?

18. What is radian measure? How do you convert from radians to degrees? degrees to radians?

19. Graph the six basic trigonometric functions. What symmetries do the graphs have?

20. How can you sometimes find the values of trigonometric functions from triangles? Give examples.

21. What is a periodic function? Give examples. What are the periods of the six basic trigonometric functions?

22. Starting with the identity $\cos^2 \theta + \sin^2 \theta = 1$ and the formulas for $\cos(A + B)$ and $\sin(A + B)$, show how a variety of other trigonometric identities may be derived.

PRELIMINARIES   PRACTICE EXERCISES

Geometry

1. A particle in the plane moved from $A(-2, 5)$ to the y-axis in such a way that Δy equaled $3\,\Delta x$. What were the particle's new coordinates?

2. **a)** Plot the points $A(8, 1)$, $B(2, 10)$, $C(-4, 6)$, $D(2, -3)$, and $E(14/3, 6)$.
   b) Find the slopes of the lines AB, BC, CD, DA, CE, and BD.
   c) Do any four of the five points A, B, C, D, and E form a parallelogram?
   d) Are any three of the five points collinear? How do you know?
   e) Which of the lines determined by the five points pass through the origin?

3. Do the points $A(6, 4)$, $B(4, -3)$, and $C(-2, 3)$ form an isosceles triangle? a right triangle? How do you know?

4. Find the coordinates of the point on the line $y = 3x + 1$ that is equidistant from $(0, 0)$ and $(-3, 4)$.

Functions and Graphs

5. Express the area and circumference of a circle as functions of the circle's radius. Then express the area as a function of the circumference.

6. Express the radius of a sphere as a function of the sphere's surface area. Then express the surface area as a function of the volume.

7. A point P in the first quadrant lies on the parabola $y = x^2$. Express the coordinates of P as functions of the angle of inclination of the line joining P to the origin.

8. A hot-air balloon rising straight up from a level field is tracked by a range finder located 500 ft from the point of lift-off. Express the balloon's height as a function of the angle the line from the range finder to the balloon makes with the ground.

Composition with absolute values. In Exercises 9–14, graph f_1 and f_2 together. Then describe how applying the absolute value function before applying f_1 affects the graph.

$f_1(x)$	$f_2(x) = f_1(\lvert x\rvert)$
9. x	$\lvert x\rvert$
10. x^3	$\lvert x\rvert^3$
11. x^2	$\lvert x\rvert^2$
12. $\dfrac{1}{x}$	$\dfrac{1}{\lvert x\rvert}$
13. $\sqrt{x}$	$\sqrt{\lvert x\rvert}$
14. $\sin x$	$\sin \lvert x\rvert$

Composition with absolute values. In Exercises 15–20, graph g_1 and g_2 together. Then describe how taking absolute values after applying g_1 affects the graph.

$g_1(x)$	$g_2(x) = \lvert g_1(x)\rvert$
15. x^3	$\lvert x^3\rvert$
16. $\sqrt{x}$	$\lvert\sqrt{x}\rvert$
17. $\dfrac{1}{x}$	$\left\lvert \dfrac{1}{x}\right\rvert$
18. $4 - x^2$	$\lvert 4 - x^2\rvert$
19. $x^2 + x$	$\lvert x^2 + x\rvert$
20. $\sin x$	$\lvert \sin x\rvert$

Trigonometry

In Exercises 21–24, sketch the graph of the given function. What is the period of the function?

21. $y = \cos 2x$

22. $y = \sin \dfrac{x}{2}$

23. $y = \sin \pi x$

24. $y = \cos \dfrac{\pi x}{2}$

25. Sketch the graph $y = 2\cos\left(x - \dfrac{\pi}{3}\right)$.

26. Sketch the graph $y = 1 + \sin\left(x + \dfrac{\pi}{4}\right)$.

In Exercises 27–30, ABC is a right triangle with the right angle at C. The sides opposite angles A, B, and C are a, b, and c, respectively.

27. **a)** Find a and b if $c = 2$, $B = \pi/3$.
    b) Find a and c if $b = 2$, $B = \pi/3$.

28. **a)** Express a in terms of A and c.
    b) Express a in terms of A and b.

29. **a)** Express a in terms of B and b.
    b) Express c in terms of A and a.

30. **a)** Express $\sin A$ in terms of a and c.
    b) Express $\sin A$ in terms of b and c.

31. **CALCULATOR** Two guy wires stretch from the top T of a vertical pole to points B and C on the ground, where C is 10 m closer to the base of the pole than is B. If wire BT makes an angle of $35°$ with the horizontal, and wire CT makes an angle of $50°$ with the horizontal, how high is the pole?

32. **CALCULATOR** Observers at positions A and B 2 km apart simultaneously measure the angle of elevation of a weather balloon to

be 40° and 70°, respectively. If the balloon is directly above a point on the line segment between A and B, find the height of the balloon.

33. Express $\sin 3x$ in terms of $\sin x$ and $\cos x$.

34. Express $\cos 3x$ in terms of $\sin x$ and $\cos x$.

35. a) GRAPHER Graph the function $f(x) = \sin x + \cos(x/2)$.
    b) What appears to be the period of this function?
    c) Confirm your finding in (b) algebraically.

36. a) GRAPHER Graph $f(x) = \sin(1/x)$.
    b) What are the domain and range of f?
    c) Is f periodic? Give reasons for your answer.

PRELIMINARIES   ADDITIONAL EXERCISES–THEORY, EXAMPLES, APPLICATIONS

Geometry

1. An object's center of mass moves at a constant velocity v along a straight line past the origin. The accompanying figure shows the coordinate system and the line of motion. The dots show positions that are 1 sec apart. Why are the areas $A_1, A_2, \ldots, A_5$ in the figure all equal? As in Kepler's equal area law (see Section 11.5), the line that joins the object's center of mass to the origin sweeps out equal areas in equal times.

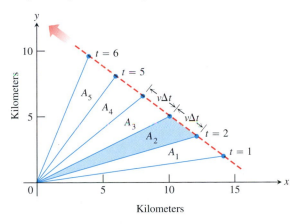

2. a) Find the slope of the line from the origin to the midpoint P of side AB in the triangle in the accompanying figure $(a, b > 0)$.

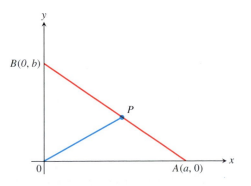

    b) When is OP perpendicular to AB?

Functions and Graphs

3. Are there two functions f and g such that $f \circ g = g \circ f$? Give reasons for your answer.

4. Are there two functions f and g with the following property? The graphs of f and g are not straight lines but the graph of $f \circ g$ *is* a straight line. Give reasons for your answer.

5. If $f(x)$ is odd, can anything be said of $g(x) = f(x) - 2$? What if f is even instead? Give reasons for your answer.

6. If $g(x)$ is an odd function defined for all values of x, can anything be said about $g(0)$? Give reasons for your answer.

7. Graph the equation $|x| + |y| = 1 + x$.

8. Graph the equation $y + |y| = x + |x|$.

Trigonometry

In Exercises 9–14, ABC is an arbitrary triangle with sides a, b, and c opposite angles A, B, and C, respectively.

9. Find b if $a = \sqrt{3}, A = \pi/3, B = \pi/4$.

10. Find $\sin B$ if $a = 4, b = 3, A = \pi/4$.

11. Find $\cos A$ if $a = 2, b = 2, c = 3$.

12. Find c if $a = 2, b = 3, C = \pi/4$.

13. Find $\sin B$ if $a = 2, b = 3, c = 4$.

14. Find $\sin C$ if $a = 2, b = 4, c = 5$.

Derivations and Proofs

15. Prove the following identities.

    a) $\dfrac{1 - \cos x}{\sin x} = \dfrac{\sin x}{1 + \cos x}$

    b) $\dfrac{1 - \cos x}{1 + \cos x} = \tan^2 \dfrac{x}{2}$

16. Explain the following "proof without words" of the law of cosines. (Source: "Proof without Words: The Law of Cosines," Sidney H. Kung, *Mathematics Magazine,* Vol. 63, No. 5, Dec. 1990, p. 342.)

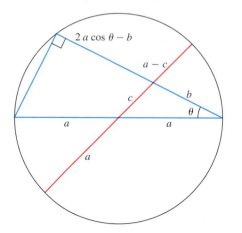

17. Show that the area of triangle ABC is given by $(1/2)ab \sin C = (1/2)bc \sin A = (1/2)ca \sin B$.

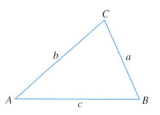

* 18. Show that the area of triangle ABC is given by $\sqrt{s(s-a)(s-b)(s-c)}$ where $s = (a+b+c)/2$ is the semiperimeter of the triangle.*

19. *Properties of inequalities.* If a and b are real numbers, we say that a **is less than** b and write $a < b$ if (and only if) $b - a$ is positive. Use this definition to prove the following properties of inequalities.

If a, b, and c are real numbers, then:

1. $a < b \implies a + c < b + c$
2. $a < b \implies a - c < b - c$
3. $a < b$ and $c > 0 \implies ac < bc$
4. $a < b$ and $c < 0 \implies bc < ac$
   (Special case: $a < b \implies -b < -a$)
5. $a > 0 \implies \dfrac{1}{a} > 0$
6. $0 < a < b \implies \dfrac{1}{b} < \dfrac{1}{a}$
7. $a < b < 0 \implies \dfrac{1}{b} < \dfrac{1}{a}$

20. *Properties of absolute values.* Prove the following properties of absolute values of real numbers.

a) $|-a| = |a|$
b) $\left|\dfrac{a}{b}\right| = \dfrac{|a|}{|b|}$

21. Prove that the following inequalities hold for any real numbers a and b.

a) $|a| < |b|$ if and only if $a^2 < b^2$
b) $|a - b| \geq ||a| - |b||$

22. *Generalizing the triangle inequality.* Prove by mathematical induction that the following inequalities hold for any n real numbers $a_1, a_2, \ldots, a_n$. (Mathematical induction is reviewed in Appendix 1.)

a) $|a_1 + a_2 + \cdots + a_n| \leq |a_1| + |a_2| + \cdots + |a_n|$
b) $|a_1 + a_2 + \cdots + a_n| \geq |a_1| - |a_2| - \cdots - |a_n|$

23. Show that if f is both even and odd, then $f(x) = 0$ for every x in the domain of f.

24. a) *Even-odd decompositions.* Let f be a function whose domain is symmetric about the origin, that is, $-x$ belongs to the domain whenever x does. Show that f is the sum of an even function and an odd function:
$$f(x) = E(x) + O(x),$$
where E is an even function and O is an odd function. (*Hint:* Let $E(x) = (f(x) + f(-x))/2$. Show that $E(-x) = E(x)$, so that E is even. Then show that $O(x) = f(x) - E(x)$ is odd.)

b) *Uniqueness.* Show that there is only one way to write f as the sum of an even and an odd function. (*Hint:* One way is given in part (a). If also $f(x) = E_1(x) + O_1(x)$ where E_1 is even and O_1 is odd, show that $E - E_1 = O_1 - O$. Then use Exercise 23 to show that $E = E_1$ and $O = O_1$.)

Grapher Explorations—Effects of Parameters

25. What happens to the graph of $y = ax^2 + bx + c$ as

a) a changes while b and c remain fixed?
b) b changes (a and c fixed, $a \neq 0$)?
c) c changes (a and b fixed, $a \neq 0$)?

26. What happens to the graph of $y = a(x + b)^3 + c$ as

a) a changes while b and c remain fixed?
b) b changes (a and c fixed, $a \neq 0$)?
c) c changes (a and b fixed, $a \neq 0$)?

27. Find all values of the slope of the line $y = mx + 2$ for which the x-intercept exceeds $1/2$.

---

*Asterisk denotes more challenging problem.

CHAPTER 1

Limits and Continuity

OVERVIEW The concept of limit of a function is one of the fundamental ideas that distinguishes calculus from algebra and trigonometry.

In this chapter we develop the limit, first intuitively and then formally. We use limits to describe the way a function f varies. Some functions vary continuously; small changes in x produce only small changes in $f(x)$. Other functions can have values that jump or vary erratically. We also use limits to define tangent lines to graphs of functions. This geometric application leads at once to the important concept of derivative of a function. The derivative, which we investigate thoroughly in Chapter 2, quantifies the way a function's values change.

1.1 Rates of Change and Limits

In this section we introduce two rates of change, speed and population growth. This leads to the main idea of the section, the idea of limit.

Speed

A moving body's **average speed** over any particular time interval is the amount of distance covered during the interval divided by the length of the interval.

EXAMPLE 1 A rock falls from the top of a 150-ft cliff. What is its average speed (a) during the first 2 sec of fall? (b) during the 1-sec interval between second 1 and second 2?

Solution Physical experiments show that a solid object dropped from rest to fall freely near the surface of the earth will fall

$$y = 16t^2 \text{ ft}$$

during the first t sec. The average speed of the rock during a given time interval is the change in distance, Δy, divided by the length of the time interval, Δt.

a) For the first 2 sec: $\quad \dfrac{\Delta y}{\Delta t} = \dfrac{16(2)^2 - 16(0)^2}{2 - 0} = 32 \dfrac{\text{ft}}{\text{sec}}$

b) From second 1 to second 2: $\quad \dfrac{\Delta y}{\Delta t} = \dfrac{16(2)^2 - 16(1)^2}{2 - 1} = 48 \dfrac{\text{ft}}{\text{sec}}$

Free fall

Near the surface of the earth, all bodies fall with the same constant acceleration. The distance a body falls after it is released from rest is a constant multiple of the square of the time elapsed. At least, that is what happens when the body falls in a vacuum, where there is no air to slow it down. The square-of-time rule also holds for dense, heavy objects like rocks, ball bearings, and steel tools during the first few seconds of their fall through air, before their velocities build up to where air resistance begins to matter. When air resistance is absent or insignificant and the only force acting on a falling body is the force of gravity, we call the way the body falls *free fall*.

Table 1.1 Average speeds over short time intervals

Average speed: $\dfrac{\Delta y}{\Delta t} = \dfrac{16(t_0 + h)^2 - 16t_0{}^2}{h}$		
Length of time interval h	**Average speed over interval of length h starting at $t_0 = 1$**	**Average speed over interval of length h starting at $t_0 = 2$**
1	48	80
0.1	33.6	65.6
0.01	32.16	64.16
0.001	32.016	64.016
0.0001	32.0016	64.0016

EXAMPLE 2   Find the speed of the rock at $t = 1$ and $t = 2$ sec.

Solution   We can calculate the average speed of the rock over a time interval $[t_0, t_0 + h]$, having length $\Delta t = h$, as

$$\frac{\Delta y}{\Delta t} = \frac{16(t_0 + h)^2 - 16t_0{}^2}{h}.$$

We cannot use this formula to calculate the "instantaneous" speed at t_0 by substituting $h = 0$, because we cannot divide by zero. But we *can* use it to calculate average speeds over increasingly short time intervals starting at $t_0 = 1$ and $t_0 = 2$. When we do so, we see a pattern (Table 1.1).

The average speed on intervals starting at $t_0 = 1$ seems to approach a limiting value of 32 as the length of the interval decreases. This suggests that the rock is falling at a speed of 32 ft/sec at $t_0 = 1$ sec. Similarly, the rock's speed at $t_0 = 2$ sec would appear to be 64 ft/sec.   ❑

Average Rates of Change and Secant Lines

Given an arbitrary function $y = f(x)$, we calculate the average rate of change of y with respect to x over the interval $[x_1, x_2]$ by dividing the change in value of y, $\Delta y = f(x_2) - f(x_1)$, by the length of the interval $\Delta x = x_2 - x_1 = h$ over which the change occurred.

Definition

The **average rate of change** of $y = f(x)$ with respect to x over the interval $[x_1, x_2]$ is

$$\frac{\Delta y}{\Delta x} = \frac{f(x_2) - f(x_1)}{x_2 - x_1} = \frac{f(x_1 + h) - f(x_1)}{h}.$$

Geometrically, an average rate of change is a secant slope.

Notice that the average rate of change of f over $[x_1, x_2]$ is the slope of the line through the points $P(x_1, f(x_1))$ and $Q(x_2, f(x_2))$ (Fig. 1.1). In geometry, a line joining two points of a curve is called a **secant** to the curve. Thus, the average rate of change of f from x_1 to x_2 is identical with the slope of secant PQ.

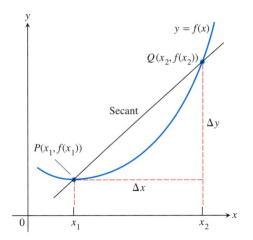

1.1 A secant to the graph $y = f(x)$. Its slope is $\Delta y / \Delta x$, the average rate of change of f over the interval $[x_1, x_2]$.

Experimental biologists often want to know the rates at which populations grow under controlled laboratory conditions.

EXAMPLE 3 *The average growth rate of a laboratory population*

Figure 1.2 shows how a population of fruit flies *(Drosophila)* grew in a 50-day experiment. The number of flies was counted at regular intervals, the counted values plotted with respect to time, and the points joined by a smooth curve. Find the average growth rate from day 23 to day 45.

Solution There were 150 flies on day 23 and 340 flies on day 45. Thus the number of flies increased by $340 - 150 = 190$ in $45 - 23 = 22$ days. The average rate of change of the population from day 23 to day 45 was

$$\text{Average rate of change:} \quad \frac{\Delta p}{\Delta t} = \frac{340 - 150}{45 - 23} = \frac{190}{22} \approx 8.6 \text{ flies/day.}$$

This average is the slope of the secant through the points P and Q on the graph in Fig. 1.2. ❏

The average rate of change from day 23 to day 45 calculated in Example 3 does not tell us how fast the population was changing on day 23 itself. For that we need to examine time intervals closer to the day in question.

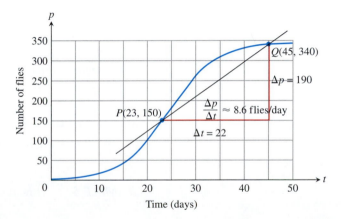

1.2 Growth of a fruit fly population in a controlled experiment. (Source: *Elements of Mathematical Biology* by A. J. Lotka, 1956, Dover, New York, p. 69.)

Q	Slope of $PQ = \Delta p/\Delta t$ (flies/day)
(45, 340)	$\dfrac{340 - 150}{45 - 23} \approx 8.6$
(40, 330)	$\dfrac{330 - 150}{40 - 23} \approx 10.6$
(35, 310)	$\dfrac{310 - 150}{35 - 23} \approx 13.3$
(30, 265)	$\dfrac{265 - 150}{30 - 23} \approx 16.4$

1.3 The positions and slopes of four secants through the point P on the fruit fly graph.

EXAMPLE 4   How fast was the number of flies in the population of Example 3 growing on day 23 itself?

Solution   To answer this question, we examine the average rates of change over increasingly short time intervals starting at day 23. In geometric terms, we find these rates by calculating the slopes of secants from P to Q, for a sequence of points Q approaching P along the curve (Fig. 1.3).

The values in the table show that the secant slopes rise from 8.6 to 16.4 as the t-coordinate of Q decreases from 45 to 30, and we would expect the slopes to rise slightly higher as t continued on toward 23. Geometrically, the secants rotate about P and seem to approach the red line in the figure, a line that goes through P in the same direction that the curve goes through P. We will see that this line is called the *tangent* to the curve at P. Since the line appears to pass through the points (14, 0) and (35, 350), it has slope

$$\frac{350 - 0}{35 - 14} = 16.7 \text{ flies/day (approximately).}$$

On day 23 the population was increasing at a rate of about 16.7 flies/day. ❑

The rates at which the rock in Example 2 was falling at the instants $t = 1$ and $t = 2$ and the rate at which the population in Example 4 was changing on day $t = 23$ are called *instantaneous rates of change*. As the examples suggest, we find instantaneous rates as limiting values of average rates. In Example 4, we also pictured the tangent line to the population curve on day 23 as a limiting position of secant lines. Instantaneous rates and tangent lines, intimately connected, appear in many other contexts. To talk about the two constructively, and to understand the connection further, we need to investigate the process by which we determine limiting values, or *limits*, as we will soon call them.

Limits of Function Values

Before we give a definition of limit, let us look at another example.

EXAMPLE 5   How does the function $f(x) = \dfrac{x^2 - 1}{x - 1}$ behave near $x = 1$?

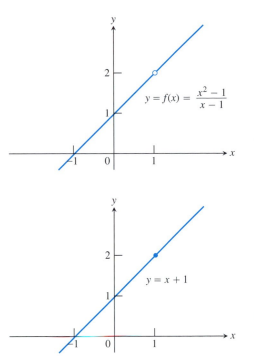

1.4 The graph of *f* is identical with the line $y = x + 1$ except at $x = 1$, where *f* is not defined.

Solution The given formula defines *f* for all real numbers *x* except $x = 1$ (we cannot divide by zero). For any $x \neq 1$ we can simplify the formula by factoring the numerator and canceling common factors:

$$f(x) = \frac{(x - 1)(x + 1)}{x - 1} = x + 1 \quad \text{for} \quad x \neq 1.$$

The graph of *f* is thus the line $y = x + 1$ with one point removed, namely the point $(1, 2)$. This removed point is shown as a "hole" in Fig. 1.4. Even though $f(1)$ is not defined, it is clear that we can make the value of $f(x)$ *as close as we want* to 2 by choosing *x* *close enough* to 1 (Table 1.2).

We say that $f(x)$ approaches arbitrarily close to 2 as *x* approaches 1, or, more simply, $f(x)$ approaches the *limit* 2 as *x* approaches 1. We write this as

$$\lim_{x \to 1} f(x) = 2, \quad \text{or} \quad \lim_{x \to 1} \frac{x^2 - 1}{x - 1} = 2.$$

Table 1.2 The closer *x* gets to 1, the closer $f(x) = (x^2 - 1)/(x - 1)$ seems to get to 2.

Values of *x* below and above 1	$f(x) = \dfrac{x^2 - 1}{x - 1} = x + 1, \quad x \neq 1$
0.9	1.9
1.1	2.1
0.99	1.99
1.01	2.01
0.999	1.999
1.001	2.001
0.999999	1.999999
1.000001	2.000001

Definition

Informal Definition of Limit

Let $f(x)$ be defined on an open interval about x_0, *except possibly at x_0 itself.* If $f(x)$ gets arbitrarily close to *L* for all *x* sufficiently close to x_0, we say that *f* approaches the **limit** *L* as *x* approaches x_0, and we write

$$\lim_{x \to x_0} f(x) = L.$$

This definition is "informal" because phrases like *arbitrarily close* and *sufficiently close* are imprecise; their meaning depends on the context. To a machinist manufacturing a piston, *close* may mean *within a few thousandths of an inch*. To an astronomer studying distant galaxies, *close* may mean *within a few thousand light-years*. The definition is clear enough, however, to enable us to recognize and evaluate limits of specific functions. We will need the more precise definition of Section 1.3, however, when we set out to prove theorems about limits.

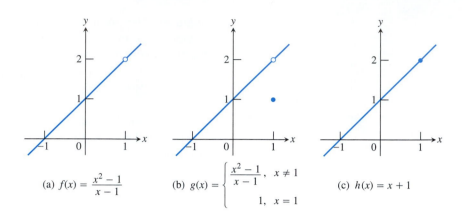

(a) $f(x) = \dfrac{x^2 - 1}{x - 1}$

(b) $g(x) = \begin{cases} \dfrac{x^2 - 1}{x - 1}, & x \ne 1 \\ 1, & x = 1 \end{cases}$

(c) $h(x) = x + 1$

1.5 $\lim\limits_{x \to 1} f(x) = \lim\limits_{x \to 1} g(x) = \lim\limits_{x \to 1} h(x) = 2.$

EXAMPLE 6   The existence of a limit as $x \to x_0$ does not depend on how the function may be defined at x_0. The function f in Fig 1.5 has limit 2 as $x \to 1$ even though f is not defined at $x = 1$. The function g has limit 2 as $x \to 1$ even though $2 \ne g(1)$. The function h is the only one whose limit as $x \to 1$ equals its value at $x = 1$. For h we have $\lim_{x \to 1} h(x) = h(1)$. This kind of equality of limit and function value is special, and we will return to it in Section 1.5.   ❏

Sometimes $\lim_{x \to x_0} f(x)$ can be evaluated by calculating $f(x_0)$. This holds, for example, whenever $f(x)$ is an algebraic combination of polynomials and trigonometric functions for which $f(x_0)$ is defined. (We will say more about this in Sections 1.2 and 1.5.)

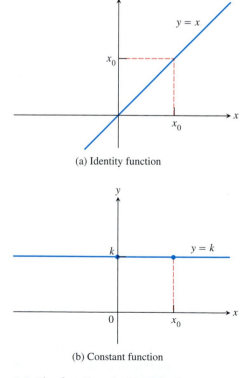

(a) Identity function

EXAMPLE 7

a) $\lim\limits_{x \to 2} (4) = 4$

b) $\lim\limits_{x \to -13} (4) = 4$

c) $\lim\limits_{x \to 3} x = 3$

d) $\lim\limits_{x \to 2} (5x - 3) = 10 - 3 = 7$

e) $\lim\limits_{x \to -2} \dfrac{3x + 4}{x + 5} = \dfrac{-6 + 4}{-2 + 5} = -\dfrac{2}{3}$ ❏

EXAMPLE 8

a) If f is the **identity function** $f(x) = x$, then for any value of x_0 (Fig. 1.6a),

$$\lim_{x \to x_0} f(x) = \lim_{x \to x_0} x = x_0.$$

b) If f is the **constant function** $f(x) = k$ (function with the constant value k), then for any value of x_0 (Fig. 1.6b),

$$\lim_{x \to x_0} f(x) = \lim_{x \to x_0} k = k$$ ❏

(b) Constant function

1.6 The functions in Example 8.

Some ways that limits can fail to exist are illustrated in Fig. 1.7 and described in the next example.

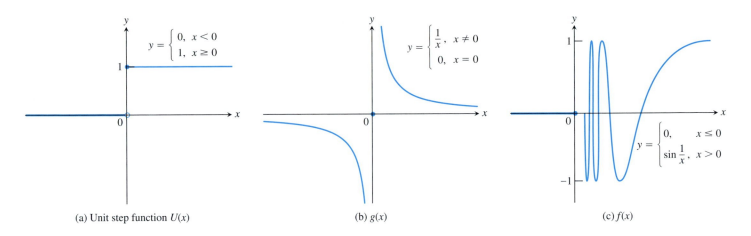

(a) Unit step function $U(x)$ (b) $g(x)$ (c) $f(x)$

1.7 The functions in Example 9.

EXAMPLE 9   *A function may fail to have a limit at a point in its domain.*

Discuss the behavior of the following functions as $x \to 0$.

a) $U(x) = \begin{cases} 0, & x < 0 \\ 1, & x \geq 0 \end{cases}$

b) $g(x) = \begin{cases} 1/x, & x \neq 0 \\ 0, & x = 0 \end{cases}$

c) $f(x) = \begin{cases} 0, & x \leq 0 \\ \sin \dfrac{1}{x}, & x > 0 \end{cases}$

Solution

a) It jumps: The **unit step function** $U(x)$ has no limit as $x \to 0$ because its values jump at $x = 0$. For negative values of x arbitrarily close to zero, $U(x) = 0$. For positive values of x arbitrarily close to zero, $U(x) = 1$. There is no *single* value L approached by $U(x)$ as $x \to 0$ (Fig. 1.7a).

b) It grows too large: $g(x)$ has no limits as $x \to 0$ because the values of g grow arbitrarily large in absolute value as $x \to 0$ and do not stay close to *any* real number (Fig. 1.7b).

c) It oscillates too much: $f(x)$ has no limit as $x \to 0$ because the function's values oscillate between $+1$ and -1 in every open interval containing 0. The values do not stay close to any one number as $x \to 0$ (Fig. 1.7c).   ❑

Exercises 1.1

Limits from Graphs

1. For the function $g(x)$ graphed here, find the following limits or explain why they do not exist.

a) $\lim\limits_{x \to 1} g(x)$ b) $\lim\limits_{x \to 2} g(x)$ c) $\lim\limits_{x \to 3} g(x)$

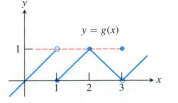

2. For the function $f(t)$ graphed here, find the following limits or explain why they do not exist.

a) $\lim\limits_{t \to -2} f(t)$ **b)** $\lim\limits_{t \to -1} f(t)$ **c)** $\lim\limits_{t \to 0} f(t)$

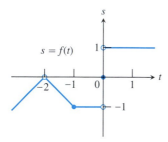

3. Which of the following statements about the function $y = f(x)$ graphed here are true, and which are false?

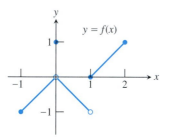

a) $\lim\limits_{x \to 0} f(x)$ exists   **b)** $\lim\limits_{x \to 0} f(x) = 0$

c) $\lim\limits_{x \to 0} f(x) = 1$ **d)** $\lim\limits_{x \to 1} f(x) = 1$

e) $\lim\limits_{x \to 1} f(x) = 0$

f) $\lim\limits_{x \to x_0} f(x)$ exists at every point x_0 in $(-1, 1)$

4. Which of the following statements about the function $y = f(x)$ graphed here are true, and which are false?

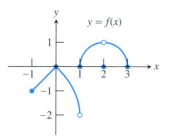

a) $\lim\limits_{x \to 2} f(x)$ does not exist

b) $\lim\limits_{x \to 2} f(x) = 2$

c) $\lim\limits_{x \to 1} f(x)$ does not exist

d) $\lim\limits_{x \to x_0} f(x)$ exists at every point x_0 in $(-1, 1)$

e) $\lim\limits_{x \to x_0} f(x)$ exists at every point x_0 in $(1, 3)$

Existence of Limits

In Exercises 5 and 6, explain why the limits do not exist.

5. $\lim\limits_{x \to 0} \dfrac{x}{|x|}$ **6.** $\lim\limits_{x \to 1} \dfrac{1}{x - 1}$

7. Suppose that a function $f(x)$ is defined for all real values of x except $x = x_0$. Can anything be said about the existence of $\lim_{x \to x_0} f(x)$? Give reasons for your answer.

8. Suppose that a function $f(x)$ is defined for all x in $[-1, 1]$. Can anything be said about the existence of $\lim_{x \to 0} f(x)$? Give reasons for your answer.

9. If $\lim_{x \to 1} f(x) = 5$, must f be defined at $x = 1$? If it is, must $f(1) = 5$? Can we conclude *anything* about the values of f at $x = 1$? Explain.

10. If $f(1) = 5$, must $\lim_{x \to 1} f(x)$ exist? If it does, then must $\lim_{x \to 1} f(x) = 5$? Can we conclude *anything* about $\lim_{x \to 1} f(x)$? Explain.

Calculator/Grapher Exercises—Estimating Limits

11. Let $f(x) = (x^2 - 9)/(x + 3)$.

a)   CALCULATOR Make a table of the values of f at the points $x = -3.1, -3.01, -3.001$, and so on as far as your calculator can go. Then estimate $\lim_{x \to -3} f(x)$. What estimate do you arrive at if you evaluate f at $x = -2.9, -2.99, -2.999, \ldots$ instead?

b)   GRAPHER Support your conclusions in (a) by graphing f near $x_0 = -3$ and using ZOOM and TRACE to estimate y-values on the graph as $x \to -3$.

c)   Find $\lim_{x \to -3} f(x)$ algebraically.

12. Let $g(x) = (x^2 - 2)/(x - \sqrt{2})$.

a)   CALCULATOR Make a table of the values of g at the points $x = 1.4, 1.41, 1.414$, and so on through successive decimal approximations of $\sqrt{2}$. Estimate $\lim_{x \to \sqrt{2}} g(x)$.

b)   GRAPHER Support your conclusion in (a) by graphing g near $x_0 = \sqrt{2}$ and using ZOOM and TRACE to estimate y-values on the graph as $x \to \sqrt{2}$.

c)   Find $\lim_{x \to \sqrt{2}} g(x)$ algebraically.

13. Let $G(x) = (x + 6)/(x^2 + 4x - 12)$.

a)   CALCULATOR Make a table of the values of G at $x = -5.9, -5.99, -5.999 \ldots$. Then estimate $\lim_{x \to -6} G(x)$. What estimate do you arrive at if you evaluate G at $x = -6.1, -6.01, -6.001, \ldots$ instead?

b)   GRAPHER Support your conclusions in (a) by graphing G and using ZOOM and TRACE to estimate y-values on the graph as $x \to -6$.

c)   Find $\lim_{x \to -6} G(x)$ algebraically.

14. Let $h(x) = (x^2 - 2x - 3)/(x^2 - 4x + 3)$.

a)   CALCULATOR Make a table of the values of h at $x = 2.9, 2.99, 2.999$, and so on. Then estimate $\lim_{x \to 3} h(x)$. What estimate do you arrive at if you evaluate h at $x = 3.1, 3.01, 3.001, \ldots$ instead?

b)   GRAPHER Support your conclusions in (a) by graphing h near $x_0 = 3$ and using ZOOM and TRACE to estimate y-values on the graph as $x \to 3$.

c)   Find $\lim_{x \to 3} h(x)$ algebraically.

15. Let $f(x) = (x^2 - 1)/(|x| - 1)$.

a)   CALCULATOR Make tables of the values of f at values of x that approach $x_0 = -1$ from above and below. Then estimate $\lim_{x \to -1} f(x)$.

b)   GRAPHER Support your conclusion in (a) by graphing f near $x_0 = -1$ and using ZOOM and TRACE to estimate y-values on the graph as $x \to -1$.

c)   Find $\lim_{x \to -1} f(x)$ algebraically.

16. Let $F(x) = (x^2 + 3x + 2)/(2 - |x|)$.

a)   CALCULATOR Make tables of values of F at values of x that approach $x_0 = -2$ from above and below. Then estimate $\lim_{x \to -2} F(x)$.

b)   GRAPHER Support your conclusion in (a) by graphing F near $x_0 = -2$ and using ZOOM and TRACE to estimate y-values on the graph as $x \to -2$.

c)   Find $\lim_{x \to -2} F(x)$ algebraically.

17. Let $g(\theta) = (\sin \theta)/\theta$.

a)   CALCULATOR Make tables of values of g at values of θ that approach $\theta_0 = 0$ from above and below. Then estimate $\lim_{\theta \to 0} g(\theta)$.

b)   GRAPHER Support your conclusion in (a) by graphing g near $\theta_0 = 0$.

18. Let $G(t) = (1 - \cos t)/t^2$.

a)   CALCULATOR Make tables of values of G at values of t that approach $t_0 = 0$ from above and below. Then estimate $\lim_{t \to 0} G(t)$.

b)   GRAPHER Support your conclusion in (a) by graphing G near $t_0 = 0$.

19. Let $f(x) = x^{1/(1-x)}$.

a)   CALCULATOR Make tables of values of f at values of x that approach $x_0 = 1$ from above and below. Does f appear to have a limit as $x \to 1$? If so, what is it? If not, why not?

b)   GRAPHER Support your conclusions in (a) by graphing f near $x_0 = 1$.

20. Let $f(x) = (3^x - 1)/x$.

a)   CALCULATOR Make tables of values of f at values of x that approach $x_0 = 0$ from above and below. Does f appear to have a limit as $x \to 0$? If so, what is it? If not, why not?

b)   GRAPHER Support your conclusions in (a) by graphing f near $x_0 = 0$.

Limits by Substitution

In Exercises 21–28, find the limits by substitution. *Support your answers with a grapher or calculator if available.*

21. $\lim_{x \to 2} 2x$

22. $\lim_{x \to 0} 2x$

23. $\lim_{x \to 1/3} (3x - 1)$

24. $\lim_{x \to 1} \dfrac{-1}{(3x - 1)}$

25. $\lim_{x \to -1} 3x(2x - 1)$

26. $\lim_{x \to -1} \dfrac{3x^2}{2x - 1}$

27. $\lim_{x \to \pi/2} x \sin x$

28. $\lim_{x \to \pi} \dfrac{\cos x}{1 - \pi}$

Average Rates of Change

In Exercises 29–34, find the average rate of change of the function over the given interval or intervals.

29. $f(x) = x^3 + 1$;
(a) $[2, 3]$,   (b) $[-1, 1]$

30. $g(x) = x^2$;
(a) $[-1, 1]$,   (b) $[-2, 0]$

31. $h(t) = \cot t$;
(a) $[\pi/4, 3\pi/4]$,   (b) $[\pi/6, \pi/2]$

32. $g(t) = 2 + \cos t$;
(a) $[0, \pi]$,   (b) $[-\pi, \pi]$

33. $R(\theta) = \sqrt{4\theta + 1}$; $[0, 2]$

34. $P(\theta) = \theta^3 - 4\theta^2 + 5\theta$; $[1, 2]$

35. Figure 1.8 shows the time-to-distance graph for a 1994 Ford Mustang Cobra accelerating from a standstill.

a)   Estimate the slopes of secants PQ_1, PQ_2, PQ_3, and PQ_4, arranging them in order in a table. What are the appropriate units for these slopes?

b)   Then estimate the Cobra's speed at time $t = 20$ sec.

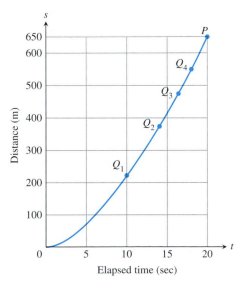

1.8 The time-to-distance graph for Exercise 35.

36. Figure 1.9 shows the plot of distance fallen (m) vs. time for a wrench that fell from the top platform of a communications mast on the moon to the station roof 80 m below.

a) Estimate the slopes of the secants $PQ_1, PQ_2, PQ_3,$ and PQ_4, arranging them in a table like the one in Fig. 1.3.

b) About how fast was the wrench going when it hit the roof?

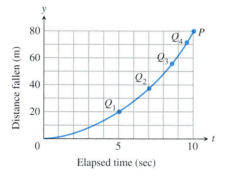

1.9 The time-to-distance graph for Exercise 36.

37. CALCULATOR The profits of a small company for each of the first five years of its operation are given in the following table:

Year	Profit in $1000s
1990	6
1991	27
1992	62
1993	111
1994	174

a) Plot points representing the profit as a function of year, and join them by as smooth a curve as you can.

b) What is the average rate of increase of the profits between 1992 and 1994?

c) Use your graph to estimate the rate at which the profits were changing in 1992.

38. CALCULATOR Make a table of values for the function $F(x) = (x + 2)/(x - 2)$ at the points $x = 2, x = 11/10, x = 101/100,$ $x = 1001/1000, x = 10001/10000,$ and $x = 1$.

a) Find the average rate of change of $F(x)$ over the intervals $[1, x]$ for each $x \neq 1$ in your table.

b) Extending the table if necessary, try to determine the rate of change of $F(x)$ at $x = 1$.

39. CALCULATOR Let $g(x) = \sqrt{x}$ for $x \geq 0$.

a) Find the average rate of change of $g(x)$ with respect to x over the intervals $[1, 2], [1, 1.5],$ and $[1, 1 + h]$.

b) Make a table of values of the average rate of change of g with respect to x over the interval $[1, 1 + h]$ for some values of h approaching zero, say $h = 0.1, 0.01, 0.001, 0.0001, 0.00001,$ and 0.000001.

c) What does your table indicate is the rate of change of $g(x)$ with respect to x at $x = 1$?

d) Calculate the limit as h approaches zero of the average rate of change of $g(x)$ with respect to x over the interval $[1, 1 + h]$.

40. CALCULATOR Let $f(t) = 1/t$ for $t \neq 0$.

a) Find the average rate of change of f with respect to t over the intervals (i) from $t = 2$ to $t = 3$, and (ii) from $t = 2$ to $t = T$.

b) Make a table of values of the average rate of change of f with respect to t over the interval $[2, T]$, for some values of T approaching 2, say $T = 2.1, 2.01, 2.001, 2.0001, 2.00001,$ and 2.000001.

c) What does your table indicate is the rate of change of f with respect to t at $t = 2$?

d) Calculate the limit as T approaches 2 of the average rate of change of f with respect to t over the interval from 2 to T. You will have to do some algebra before you can substitute $T = 2$.

✿ CAS Explorations and Projects

In Exercises 41–46, use a CAS to perform the following steps:

a) Plot the function near the point x_0 being approached.

b) From your plot guess the value of the limit.

c) Evaluate the limit symbolically. How close was your guess?

41. $\lim_{x \to 2} \dfrac{x^4 - 16}{x - 2}$

42. $\lim_{x \to -1} \dfrac{x^3 - x^2 - 5x - 3}{(x + 1)^2}$

43. $\lim_{x \to 0} \dfrac{\sqrt[3]{1 + x} - 1}{x}$

44. $\lim_{x \to 3} \dfrac{x^2 - 9}{\sqrt{x^2 + 7} - 4}$

45. $\lim_{x \to 0} \dfrac{1 - \cos x}{x \sin x}$

46. $\lim_{x \to 0} \dfrac{2x^2}{3 - 3 \cos x}$

1.2

Rules for Finding Limits

This section presents theorems for calculating limits. The first three let us build on the results of Example 8 in the preceding section to find limits of polynomials, rational functions, and powers. The fourth prepares for calculations later in the text.

Limits of Powers and Algebraic Combinations

Theorem 1
Properties of Limits

The following rules hold if $\lim_{x \to c} f(x) = L$ and $\lim_{x \to c} g(x) = M$ (L and M real numbers).

1. *Sum Rule:* $\qquad\qquad\qquad\quad \lim_{x \to c} [f(x) + g(x)] = L + M$

2. *Difference Rule:* $\qquad\qquad\quad \lim_{x \to c} [f(x) - g(x)] = L - M$

3. *Product Rule:* $\qquad\qquad\quad\; \lim_{x \to c} f(x) \cdot g(x) = L \cdot M$

4. *Constant Multiple Rule:* $\qquad \lim_{x \to c} k f(x) = kL \quad$ (any number k)

5. *Quotient Rule:* $\qquad\qquad\quad\; \lim_{x \to c} \dfrac{f(x)}{g(x)} = \dfrac{L}{M}, \quad M \neq 0$

6. *Power Rule:* $\qquad\qquad\qquad$ If m and n are integers, then

$$\lim_{x \to c} [f(x)]^{m/n} = L^{m/n},$$

provided $L^{m/n}$ is a real number.

In words, the formulas in Theorem 1 say:

1. The limit of the sum of two functions is the sum of their limits.
2. The limit of the difference of two functions is the difference of their limits.
3. The limit of the product of two functions is the product of their limits.
4. The limit of a constant times a function is that constant times the limit of the function.
5. The limit of the quotient of two functions is the quotient of their limits, provided the limit of the denominator is not zero.
6. The limit of any rational power of a function is that power of the limit of the function, provided the latter is a real number.

We will prove the Sum Rule in Section 1.3. Rules 2–5 are proved in Appendix 2. Rule 6 is proved in more advanced texts.

EXAMPLE 1   Find $\lim_{x \to c} \dfrac{x^3 + 4x^2 - 3}{x^2 + 5}$.

Solution   Starting with the limits $\lim_{x \to c} x = c$ and $\lim_{x \to c} k = k$ from Section 1.1, Example 8, and combining them using various parts of Theorem 1, we obtain:

a) $\lim\limits_{x \to c} x^2 = \left(\lim\limits_{x \to c} x\right)\left(\lim\limits_{x \to c} x\right) = c \cdot c = c^2$ Product or Power

b) $\lim\limits_{x \to c} (x^2 + 5) = \lim\limits_{x \to c} x^2 + \lim\limits_{x \to c} 5 = c^2 + 5$ Sum and (a)

c) $\lim\limits_{x \to c} 4x^2 = 4 \lim\limits_{x \to c} x^2 = 4c^2$ Constant Multiple and (a)

d) $\lim\limits_{x \to c} (4x^2 - 3) = \lim\limits_{x \to c} 4x^2 - \lim\limits_{x \to c} 3 = 4c^2 - 3$ Difference and (c)

e) $\lim\limits_{x \to c} x^3 = \left(\lim\limits_{x \to c} x^2\right)\left(\lim\limits_{x \to c} x\right) = c^2 \cdot c = c^3$ Product and (a), or Power

f) $\lim\limits_{x \to c} (x^3 + 4x - 3) = \lim\limits_{x \to c} x^3 + \lim\limits_{x \to c} (4x^2 - 3)$ Sum

$\qquad\qquad\qquad\quad = c^3 + 4c^2 - 3$ (d) and (e)

g) $\lim\limits_{x \to c} \dfrac{x^3 + 4x^2 - 3}{x^2 + 5} = \dfrac{\lim\limits_{x \to c} (x^3 + 4x^2 - 3)}{\lim\limits_{x \to c} (x^2 + 5)}$ Quotient

$\qquad\qquad\qquad\quad = \dfrac{c^3 + 4c^2 - 3}{c^2 + 5}$ (f) and (b)   ❑

EXAMPLE 2   Find $\lim\limits_{x \to -2} \sqrt{4x^2 - 3}$.

Solution

$$\lim\limits_{x \to -2} \sqrt{4x^2 - 3} = \sqrt{4(-2)^2 - 3}$$ Example 1(d) and Power Rule with $n = 1/2$

$$= \sqrt{16 - 3}$$

$$= \sqrt{13}$$ ❑

Two consequences of Theorem 1 further simplify the task of calculating limits of polynomials and rational functions. To evaluate the limit of a polynomial function as x approaches c, merely substitute c for x in the formula for the function. To evaluate the limit of a rational function as x approaches a point c *at which the denominator is not zero,* substitute c for x in the formula for the function.

Theorem 2
Limits of Polynomials Can Be Found by Substitution
If $P(x) = a_n x^n + a_{n-1} x^{n-1} + \cdots + a_0$, then

$$\lim\limits_{x \to c} P(x) = P(c) = a_n c^n + a_{n-1} c^{n-1} + \cdots + a_0.$$

Theorem 3
Limits of Rational Functions Can Be Found by Substitution If the Limit of the Denominator Is Not Zero
If $P(x)$ and $Q(x)$ are polynomials and $Q(c) \neq 0$, then

$$\lim\limits_{x \to c} \frac{P(x)}{Q(x)} = \frac{P(c)}{Q(c)}.$$

EXAMPLE 3

$$\lim_{x \to -1} \frac{x^3 + 4x^2 - 3}{x^2 + 5} = \frac{(-1)^3 + 4(-1)^2 - 3}{(-1)^2 + 5} = \frac{0}{6} = 0.$$

This is the limit in Example 1 with $c = -1$, now done in one step.   ❏

Identifying common factors

It can be shown that if $Q(x)$ is a polynomial and $Q(c) = 0$, then $(x - c)$ is a factor of $Q(x)$. Thus, if the numerator and denominator of a rational function of x are both zero at $x = c$, then $(x - c)$ is a common factor.

Eliminating Zero Denominators Algebraically

Theorem 3 applies only when the denominator of the rational function is not zero at the limit point c. If the denominator is zero, canceling common factors in the numerator and denominator will sometimes reduce the fraction to one whose denominator is no longer zero at c. When this happens, we can find the limit by substitution in the simplified fraction.

EXAMPLE 4   *Canceling a common factor*

Evaluate $\lim_{x \to 1} \dfrac{x^2 + x - 2}{x^2 - x}$.

Solution We cannot just substitute $x = 1$, because it makes the denominator zero. However, we can factor the numerator and denominator and cancel the common factor to obtain

$$\frac{x^2 + x - 2}{x^2 - x} = \frac{(x - 1)(x + 2)}{x(x - 1)} = \frac{x + 2}{x}, \quad \text{if } x \neq 1.$$

Thus

$$\lim_{x \to 1} \frac{x^2 + x - 2}{x^2 - x} = \lim_{x \to 1} \frac{x + 2}{x} = \frac{1 + 2}{1} = 3.$$

See Fig. 1.10.   ❏

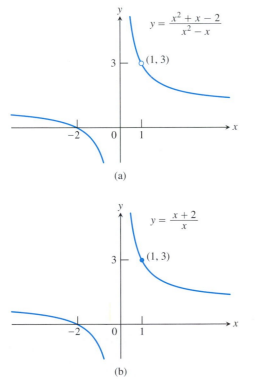

EXAMPLE 5   *Creating and canceling a common factor*

Find $\lim_{h \to 0} \dfrac{\sqrt{2 + h} - \sqrt{2}}{h}$.

Solution We cannot find the limit by substituting $h = 0$, and the numerator and denominator do not have obvious factors. However, we can create a common factor in the numerator by multiplying it (and the denominator) by the so-called *conjugate expression* $\sqrt{2 + h} + \sqrt{2}$, obtained by changing the sign between the square roots:

$$\frac{\sqrt{2 + h} - \sqrt{2}}{h} = \frac{\sqrt{2 + h} - \sqrt{2}}{h} \cdot \frac{\sqrt{2 + h} + \sqrt{2}}{\sqrt{2 + h} + \sqrt{2}}$$

$$= \frac{2 + h - 2}{h(\sqrt{2 + h} + \sqrt{2})}$$

$$= \frac{h}{h(\sqrt{2 + h} + \sqrt{2})} \qquad \text{We have created a common factor of } h \ldots$$

$$= \frac{1}{\sqrt{2 + h} + \sqrt{2}} \qquad \ldots \text{which we cancel.}$$

1.10 The graph of $f(x) = (x^2 + x - 2)/(x^2 - x)$ in (a) is the same as the graph of $g(x) = (x + 2)/x$ in (b) except at $x = 1$, where f is undefined. The functions have the same limit as $x \to 1$.

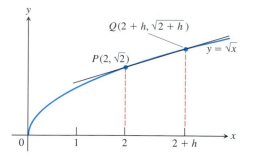

1.11 The limit of the slope of secant PQ as $Q \to P$ along the curve is $1/(2\sqrt{2})$ (Example 5).

Therefore,

$$\lim_{h \to 0} \frac{\sqrt{2+h} - \sqrt{2}}{h} = \lim_{h \to 0} \frac{1}{\sqrt{2+h} + \sqrt{2}}$$

$$= \frac{1}{\sqrt{2+0} + \sqrt{2}} \qquad \text{The denominator is no longer 0 at } h = 0, \text{ so we can substitute.}$$

$$= \frac{1}{2\sqrt{2}}.$$

Notice that the fraction $(\sqrt{2+h} - \sqrt{2})/h$ is the slope of the secant through the point $P(2, \sqrt{2})$ and the point $Q(2+h, \sqrt{2+h})$ nearby on the curve $y = \sqrt{x}$. Figure 1.11 shows the secant for $h > 0$. Our calculation shows that the limiting value of this slope as $Q \to P$ along the curve from either side is $1/(2\sqrt{2})$. ❏

The Sandwich Theorem

The following theorem will enable us to calculate a variety of limits in subsequent chapters. It is called the Sandwich Theorem because it refers to a function f whose values are sandwiched between the values of two other functions g and h that have the same limit L at a point c. Being trapped between the values of two functions that approach L, the values of f must also approach L (Fig. 1.12). You will find a proof in Appendix 2.

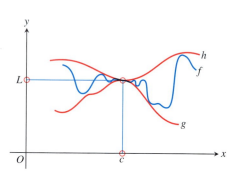

1.12 The graph of f is sandwiched between the graphs of g and h.

> **Theorem 4**
> **The Sandwich Theorem**
>
> Suppose that $g(x) \le f(x) \le h(x)$ for all x in some open interval containing c, except possibly at $x = c$ itself. Suppose also that
>
> $$\lim_{x \to c} g(x) = \lim_{x \to c} h(x) = L.$$
>
> Then $\lim_{x \to c} f(x) = L$.

EXAMPLE 6 Given that

$$1 - \frac{x^2}{4} \le u(x) \le 1 + \frac{x^2}{2} \quad \text{for all } x \ne 0,$$

find $\lim_{x \to 0} u(x)$.

Solution Since

$$\lim_{x \to 0} (1 - (x^2/4)) = 1 \quad \text{and} \quad \lim_{x \to 0} (1 + (x^2/2)) = 1,$$

the Sandwich Theorem implies that $\lim_{x \to 0} u(x) = 1$ (Fig. 1.13). ❏

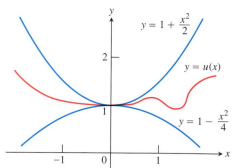

1.13 Any function $u(x)$ whose graph lies in the region between $y = 1 + (x^2/2)$ and $y = 1 - (x^2/4)$ has limit 1 as $x \to 0$.

EXAMPLE 7 Show that if $\lim_{x \to c} |f(x)| = 0$, then $\lim_{x \to c} f(x) = 0$.

Solution Since $-|f(x)| \le f(x) \le |f(x)|$, and $-|f(x)|$ and $|f(x)|$ both have limit 0 as x approaches c, $\lim_{x \to c} f(x) = 0$ by the Sandwich Theorem. ❏

Exercises 1.2

Limit Calculations

Find the limits in Exercises 1–16.

1. $\lim\limits_{x \to -7} (2x + 5)$

2. $\lim\limits_{x \to 12} (10 - 3x)$

3. $\lim\limits_{x \to 2} (-x^2 + 5x - 2)$

4. $\lim\limits_{x \to -2} (x^3 - 2x^2 + 4x + 8)$

5. $\lim\limits_{t \to 6} 8(t - 5)(t - 7)$

6. $\lim\limits_{s \to 2/3} 3s(2s - 1)$

7. $\lim\limits_{x \to 2} \dfrac{x + 3}{x + 6}$

8. $\lim\limits_{x \to 5} \dfrac{4}{x - 7}$

9. $\lim\limits_{y \to -5} \dfrac{y^2}{5 - y}$

10. $\lim\limits_{y \to 2} \dfrac{y + 2}{y^2 + 5y + 6}$

11. $\lim\limits_{x \to -1} 3(2x - 1)^2$

12. $\lim\limits_{x \to -4} (x + 3)^{1984}$

13. $\lim\limits_{y \to -3} (5 - y)^{4/3}$

14. $\lim\limits_{z \to 0} (2z - 8)^{1/3}$

15. $\lim\limits_{h \to 0} \dfrac{3}{\sqrt{3h + 1} + 1}$

16. $\lim\limits_{h \to 0} \dfrac{5}{\sqrt{5h + 4} + 2}$

Find the limits in Exercises 17–30.

17. $\lim\limits_{x \to 5} \dfrac{x - 5}{x^2 - 25}$

18. $\lim\limits_{x \to -3} \dfrac{x + 3}{x^2 + 4x + 3}$

19. $\lim\limits_{x \to -5} \dfrac{x^2 + 3x - 10}{x + 5}$

20. $\lim\limits_{x \to 2} \dfrac{x^2 - 7x + 10}{x - 2}$

21. $\lim\limits_{t \to 1} \dfrac{t^2 + t - 2}{t^2 - 1}$

22. $\lim\limits_{t \to -1} \dfrac{t^2 + 3t + 2}{t^2 - t - 2}$

23. $\lim\limits_{x \to -2} \dfrac{-2x - 4}{x^3 + 2x^2}$

24. $\lim\limits_{y \to 0} \dfrac{5y^3 + 8y^2}{3y^4 - 16y^2}$

25. $\lim\limits_{u \to 1} \dfrac{u^4 - 1}{u^3 - 1}$

26. $\lim\limits_{v \to 2} \dfrac{v^3 - 8}{v^4 - 16}$

27. $\lim\limits_{x \to 9} \dfrac{\sqrt{x} - 3}{x - 9}$

28. $\lim\limits_{x \to 4} \dfrac{4x - x^2}{2 - \sqrt{x}}$

29. $\lim\limits_{x \to 1} \dfrac{x - 1}{\sqrt{x + 3} - 2}$

30. $\lim\limits_{x \to -1} \dfrac{\sqrt{x^2 + 8} - 3}{x + 1}$

Using Limit Rules

31. Suppose $\lim\limits_{x \to 0} f(x) = 1$ and $\lim\limits_{x \to 0} g(x) = -5$. Name the rules in Theorem 1 that are used to accomplish steps (a), (b), and (c) of the following calculation.

$$\lim\limits_{x \to 0} \frac{2f(x) - g(x)}{(f(x) + 7)^{2/3}} = \frac{\lim\limits_{x \to 0} (2f(x) - g(x))}{\lim\limits_{x \to 0} (f(x) + 7)^{2/3}} \qquad \text{(a)}$$

$$= \frac{\lim\limits_{x \to 0} 2f(x) - \lim\limits_{x \to 0} g(x)}{\left(\lim\limits_{x \to 0} (f(x) + 7) \right)^{2/3}} \qquad \text{(b)}$$

$$= \frac{2 \lim\limits_{x \to 0} f(x) - \lim\limits_{x \to 0} g(x)}{\left(\lim\limits_{x \to 0} f(x) + \lim\limits_{x \to 0} 7 \right)^{2/3}} \qquad \text{(c)}$$

$$= \frac{(2)(1) - (-5)}{(1 + 7)^{2/3}} = \frac{7}{4}$$

32. Let $\lim\limits_{x \to 1} h(x) = 5, \lim\limits_{x \to 1} p(x) = 1$, and $\lim\limits_{x \to 1} r(x) = 2$. Name the rules in Theorem 1 that are used to accomplish steps (a), (b), and (c) of the following calculation.

$$\lim\limits_{x \to 1} \frac{\sqrt{5h(x)}}{p(x)(4 - r(x))} = \frac{\lim\limits_{x \to 1} \sqrt{5h(x)}}{\lim\limits_{x \to 1} (p(x)(4 - r(x)))} \qquad \text{(a)}$$

$$= \frac{\sqrt{\lim\limits_{x \to 1} 5h(x)}}{\left(\lim\limits_{x \to 1} p(x) \right) \left(\lim\limits_{x \to 1} (4 - r(x)) \right)} \qquad \text{(b)}$$

$$= \frac{\sqrt{5 \lim\limits_{x \to 1} h(x)}}{\left(\lim\limits_{x \to 1} p(x) \right) \left(\lim\limits_{x \to 1} 4 - \lim\limits_{x \to 1} r(x) \right)} \qquad \text{(c)}$$

$$= \frac{\sqrt{(5)(5)}}{(1)(4 - 2)} = \frac{5}{2}$$

33. Suppose $\lim\limits_{x \to c} f(x) = 5$ and $\lim\limits_{x \to c} g(x) = -2$. Find

a) $\lim\limits_{x \to c} f(x)g(x)$

b) $\lim\limits_{x \to c} 2f(x)g(x)$

c) $\lim\limits_{x \to c} (f(x) + 3g(x))$

d) $\lim\limits_{x \to c} \dfrac{f(x)}{f(x) - g(x)}$

34. Suppose $\lim\limits_{x \to 4} f(x) = 0$ and $\lim\limits_{x \to 4} g(x) = -3$. Find

a) $\lim\limits_{x \to 4} (g(x) + 3)$

b) $\lim\limits_{x \to 4} xf(x)$

c) $\lim\limits_{x \to 4} (g(x))^2$

d) $\lim\limits_{x \to 4} \dfrac{g(x)}{f(x) - 1}$

35. Suppose $\lim\limits_{x \to b} f(x) = 7$ and $\lim\limits_{x \to b} g(x) = -3$. Find

a) $\lim\limits_{x \to b} (f(x) + g(x))$

b) $\lim\limits_{x \to b} f(x) \cdot g(x)$

c) $\lim\limits_{x \to b} 4g(x)$

d) $\lim\limits_{x \to b} f(x)/g(x)$

36. Suppose that $\lim\limits_{x \to -2} p(x) = 4, \lim\limits_{x \to -2} r(x) = 0$, and $\lim\limits_{x \to -2} s(x) = -3$. Find

a) $\lim\limits_{x \to -2} (p(x) + r(x) + s(x))$

b) $\lim\limits_{x \to -2} p(x) \cdot r(x) \cdot s(x)$

c) $\lim\limits_{x \to -2} (-4p(x) + 5r(x))/s(x)$

Limits of Average Rates of Change

Because of their connection with secant lines, tangents, and instantaneous rates, limits of the form

$$\lim_{h \to 0} \frac{f(x+h) - f(x)}{h}$$

occur frequently in calculus. In Exercises 37–42, evaluate this limit for the given value of x and function f.

37. $f(x) = x^2, \quad x = 1$

38. $f(x) = x^2, \quad x = -2$

39. $f(x) = 3x - 4, \quad x = 2$

40. $f(x) = 1/x, \quad x = -2$

41. $f(x) = \sqrt{x}, \quad x = 7$

42. $f(x) = \sqrt{3x + 1}, \quad x = 0$

Using the Sandwich Theorem

43. If $\sqrt{5 - 2x^2} \le f(x) \le \sqrt{5 - x^2}$ for $-1 \le x \le 1$, find $\lim_{x \to 0} f(x)$.

44. If $2 - x^2 \le g(x) \le 2 \cos x$ for all x, find $\lim_{x \to 0} g(x)$.

45. a) It can be shown that the inequalities

$$1 - \frac{x^2}{6} < \frac{x \sin x}{2 - 2 \cos x} < 1$$

hold for all values of x close to zero. What, if anything, does this tell you about

$$\lim_{x \to 0} \frac{x \sin x}{2 - 2 \cos x}?$$

Give reasons for your answer.

b) GRAPHER Graph

$$y = 1 - (x^2/6), \quad y = (x \sin x)/(2 - 2 \cos x), \text{ and } y = 1$$

together for $-2 \le x \le 2$. Comment on the behavior of the graphs as $x \to 0$.

46. a) Suppose that the inequalities

$$\frac{1}{2} - \frac{x^2}{24} < \frac{1 - \cos x}{x^2} < \frac{1}{2}$$

hold for values of x close to zero. (They do, as you will see in Section 8.10.) What, if anything, does this tell you about

$$\lim_{x \to 0} \frac{1 - \cos x}{x^2}?$$

Give reasons for your answer.

b) GRAPHER Graph the equations $y = (1/2) - (x^2/24)$, $y = (1 - \cos x)/x^2$, and $y = 1/2$ together for $-2 \le x \le 2$. Comment on the behavior of the graphs as $x \to 0$.

Theory and Examples

47. If $x^4 \le f(x) \le x^2$ for x in $[-1, 1]$ and $x^2 \le f(x) \le x^4$ for $x < -1$ and $x > 1$, at what points c do you automatically know $\lim_{x \to c} f(x)$? What can you say about the value of the limit at these points?

48. Suppose that $g(x) \le f(x) \le h(x)$ for all $x \ne 2$ and suppose that

$$\lim_{x \to 2} g(x) = \lim_{x \to 2} h(x) = -5.$$

Can we conclude anything about the values of f, g, and h at $x = 2$? Could $f(2) = 0$? Could $\lim_{x \to 2} f(x) = 0$? Give reasons for your answers.

49. If $\lim_{x \to 4} \dfrac{f(x) - 5}{x - 2} = 1$, find $\lim_{x \to 4} f(x)$.

50. If $\lim_{x \to -2} \dfrac{f(x)}{x^2} = 1$, find (a) $\lim_{x \to -2} f(x)$ and (b) $\lim_{x \to -2} \dfrac{f(x)}{x}$.

51. a) If $\lim_{x \to 2} \dfrac{f(x) - 5}{x - 2} = 3$, find $\lim_{x \to 2} f(x)$.

b) If $\lim_{x \to 2} \dfrac{f(x) - 5}{x - 2} = 4$, find $\lim_{x \to 2} f(x)$.

52. If $\lim_{x \to 0} \dfrac{f(x)}{x^2} = 1$, find (a) $\lim_{x \to 0} f(x)$ and (b) $\lim_{x \to 0} \dfrac{f(x)}{x}$.

53. a) GRAPHER Graph $g(x) = x \sin (1/x)$ to estimate $\lim_{x \to 0} g(x)$, zooming in on the origin as necessary.

b) Confirm your estimate in (a) with a proof.

54. a) GRAPHER Graph $h(x) = x^2 \cos (1/x^3)$ to estimate $\lim_{x \to 0} h(x)$, zooming in on the origin as necessary.

b) Confirm your estimate in (a) with a proof.

1.3

Target Values and Formal Definitions of Limits

In this section we give a formal definition of the limit introduced in the previous two sections. We replace vague phrases like "gets arbitrarily close" in the informal definition with specific conditions that can be applied to any particular example. To do this we first examine how to control the input of a function to ensure that the output is kept within preset bounds.

Keeping Outputs near Target Values

We sometimes need to know what input values x will result in output values of the function $y = f(x)$ near a particular target value. How near depends on the context.

A gas station attendant, asked for $5.00 worth of gas, will try to pump a volume of gas worth $5.00 to the nearest cent. An automobile mechanic grinding a 3.385-in. cylinder will not let the bore exceed this value by more than 0.002 in. A pharmacist making ointments will measure ingredients to the nearest milligram.

EXAMPLE 1   *Controlling a linear function*

How close to $x_0 = 4$ must we hold the input x to be sure that the output $y = 2x - 1$ lies within 2 units of $y_0 = 7$?

Solution   We are asked: For what values of x is $|y - 7| < 2$? To find the answer we first express $|y - 7|$ in terms of x:

$$|y - 7| = |(2x - 1) - 7| = |2x - 8|.$$

The question then becomes: What values of x satisfy the inequality $|2x - 8| < 2$? To find out, we solve the inequality:

$$|2x - 8| < 2$$
$$-2 < 2x - 8 < 2$$
$$6 < 2x < 10$$
$$3 < x < 5$$
$$-1 < x - 4 < 1.$$

Keeping x within 1 unit of $x_0 = 4$ will keep y within 2 units of $y_0 = 7$ (Fig. 1.14).

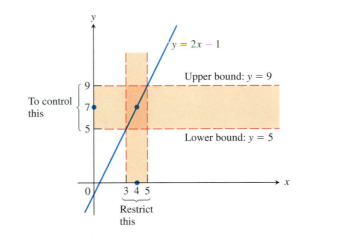

1.14 Keeping x within 1 unit of $x_0 = 4$ will keep y within 2 units of $y_0 = 7$.

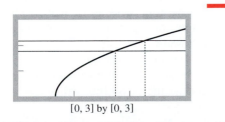

[0, 3] by [0, 3]

Keeping x between 1.75 and 2.28 will keep y between 1.8 and 2.2.

Technology   *Target Values*   You can experiment with target values on a graphing utility. Graph the function together with a target interval defined by horizontal lines above and below the proposed limit. Adjust the range or use zoom until the function's behavior inside the target interval is clear. Then observe what happens when you try to find an interval of x-values that will keep the function values within the target interval. (See also Exercises 7–14 and CAS Exercises 61–64.)

For example, try this for $f(x) = \sqrt{3x - 2}$ and the target interval (1.8, 2.2) on the y-axis. That is, graph $y_1 = f(x)$ and the lines $y_2 = 1.8$, $y_3 = 2.2$. Then try the target intervals (1.98, 2.02) and (1.9998, 2.0002).

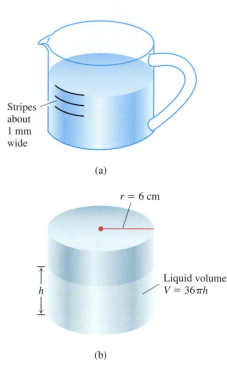

Stripes
about
1 mm
wide

(a)

$r = 6$ cm

h

Liquid volume
$V = 36\pi h$

(b)

1.15 A 1-L measuring cup (a), modeled as a right circular cylinder (b) of radius $r = 6$ cm (Example 2).

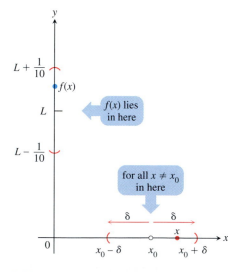

y

$L + \dfrac{1}{10}$

$f(x)$

L

$f(x)$ lies
in here

$L - \dfrac{1}{10}$

for all $x \neq x_0$
in here

δ δ

x

0 $x_0 - \delta$ x_0 $x_0 + \delta$ x

1.16 A preliminary stage in the development of the definition of limit.

EXAMPLE 2   *Why the stripes on a 1-liter kitchen measuring cup are about a millimeter wide*

The interior of a typical 1-L measuring cup is a right circular cylinder of radius 6 cm (Fig. 1.15). The volume of water we put in the cup is therefore a function of the level h to which the cup is filled, the formula being

$$V = \pi 6^2 h = 36\pi h.$$

How closely must we measure h to measure out 1 L of water (1000 cm³) with an error of no more than 1% (10 cm³)?

Solution   We want to know in what interval to hold values of h to make V satisfy the inequality

$$|V - 1000| = |36\pi h - 1000| \leq 10.$$

To find out, we solve the inequality:

$$|36\pi h - 1000| \leq 10$$

$$-10 \leq 36\pi h - 1000 \leq 10$$

$$990 \leq 36\pi h \leq 1010$$

$$\frac{990}{36\pi} \leq h \leq \frac{1010}{36\pi}$$

$$8.8 \leq h \leq 8.9$$

rounded up,   rounded down,
to be safe   to be safe

The interval in which we should hold h is about $8.9 - 8.8 = 0.1$ cm wide (1 mm). With stripes 1 mm wide, we can expect to measure a liter of water with an accuracy of 1%, which is more than enough accuracy for cooking.   ☐

The Precise Definition of Limit

In a target-value problem, we determine how close to hold a variable x to a particular value x_0 to ensure that the outputs $f(x)$ of some function lie within a prescribed interval about a target value L. To show that the limit of $f(x)$ as $x \to x_0$ actually equals L, we must be able to show that the gap between $f(x)$ and L can be made less than *any prescribed error*, no matter how small, by holding x close enough to x_0.

Suppose we are watching the values of a function $f(x)$ as x approaches x_0 (without taking on the value of x_0 itself). Certainly we want to be able to say that $f(x)$ stays within one-tenth of a unit of L as soon as x stays within some distance δ of x_0 (Fig. 1.16). But that in itself is not enough, because as x continues on its course toward x_0, what is to prevent $f(x)$ from jittering about within the interval from $L - 1/10$ to $L + 1/10$ without tending toward L?

We can be told that the error can be no more than 1/100 or 1/1000 or 1/100,000. Each time, we find a new δ-interval about x_0 so that keeping x within that interval satisfies the new error tolerance. And each time the possibility exists that $f(x)$ jitters away from L at the last minute.

The following figures illustrate the problem. You can think of this as a quarrel between a skeptic and a scholar. The skeptic presents ϵ-challenges to prove that the limit does not exist or, more precisely, that there is room for doubt, and the scholar answers every challenge with a δ-interval around x_0.

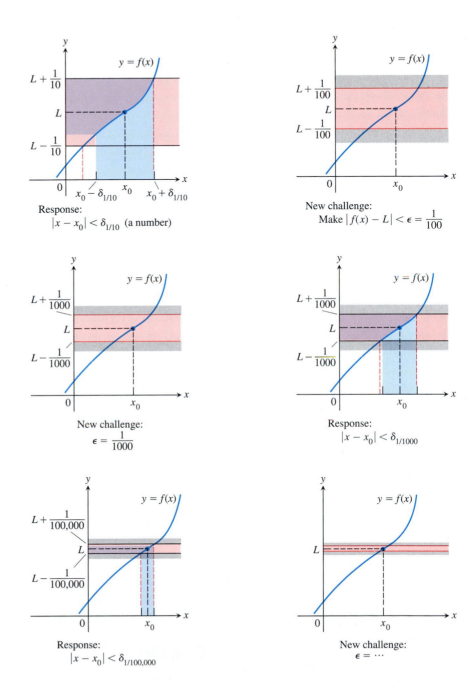

How do we stop this seemingly endless series of challenges and responses? By proving that for every error tolerance ϵ that the challenger can produce, we can find, calculate, or conjure a matching distance δ that keeps x "close enough" to x_0 to keep $f(x)$ within that tolerance of L (Fig. 1.17 on the following page).

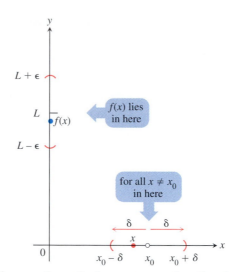

1.17 The relation of δ and ϵ in the definition of limit.

Here, at last, is a mathematical way to say that the closer x gets to x_0, the closer $y = f(x)$ gets to L.

Definition
A Formal Definition of Limit

Let $f(x)$ be defined on an open interval about x_0, except possibly at x_0 itself. We say that $f(x)$ approaches the **limit** L **as** x **approaches** x_0, and write

$$\lim_{x \to x_0} f(x) = L,$$

if, for every number $\epsilon > 0$, there exists a corresponding number $\delta > 0$ such that for all x

$$0 < |x - x_0| < \delta \implies |f(x) - L| < \epsilon.$$

The Weierstrass definition

The concepts of limit and continuity (and, indeed, real number and function) did not enter mathematics overnight with the great discoveries of Sir Isaac Newton (1642–1727) and Baron Gottfried Wilhelm Leibniz (1646–1716). Mathematicians had an imperfect understanding of these fundamental ideas even as late as the last century. Definitions of the limit given by French mathematician Augustin-Louis Cauchy (1789–1857) and others referred to variables "approaching indefinitely" a fixed value and frequently made use of "infinitesimals," quantities that become infinitely small but not zero. The now accepted ϵ-δ definition of limit was formulated by German mathematician Karl Weierstrass (1815–1897) in the middle of the nineteenth century as part of his attempt to put mathematical analysis on a sound logical foundation.

To return to the idea of target values, suppose you are machining a generator shaft to a close tolerance. You may try for diameter L, but since nothing is perfect, you must be satisfied with a diameter $f(x)$ somewhere between $L - \epsilon$ and $L + \epsilon$. The δ is the measure of how accurate your control setting for x must be to guarantee this degree of accuracy in the diameter of the shaft. Notice that as the tolerance for error becomes stricter, you may have to adjust δ. That is, the value of δ, how tight your control setting must be, depends on the value of ϵ, the error tolerance.

Examples: Testing the Definition

The formal definition of limit does not tell how to find the limit of a function, but it enables us to verify that a suspected limit is correct. The following examples show how the definition can be used to verify limit statements for specific functions. (The first two examples correspond to parts of Examples 7 and 8 in Section 1.1.) However, the real purpose of the definition is not to do calculations like this, but rather to prove general theorems so that the calculation of specific limits can be simplified.

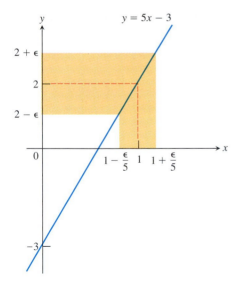

1.18 If $f(x) = 5x - 3$, then $0 < |x - 1| < \epsilon/5$ guarantees that $|f(x) - 2| < \epsilon$ (Example 3).

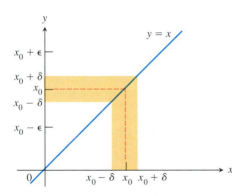

1.19 For the function $f(x) = x$, we find that $0 < |x - x_0| < \delta$ will guarantee $|f(x) - x_0| < \epsilon$ whenever $\delta \leq \epsilon$ (Example 4a).

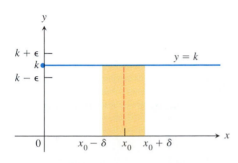

1.20 For the function $f(x) = k$, we find that $|f(x) - k| < \epsilon$ for any positive δ (Example 4b).

EXAMPLE 3   Show that $\lim_{x \to 1} (5x - 3) = 2$.

Solution   Set $x_0 = 1$, $f(x) = 5x - 3$, and $L = 2$ in the definition of limit. For any given $\epsilon > 0$ we have to find a suitable $\delta > 0$ so that if $x \neq 1$ and x is within distance δ of $x_0 = 1$, that is, if

$$0 < |x - 1| < \delta,$$

then $f(x)$ is within distance ϵ of $L = 2$, that is

$$|f(x) - 2| < \epsilon.$$

We find δ by working backwards from the ϵ-inequality:

$$|(5x - 3) - 2| = |5x - 5| < \epsilon$$

$$5|x - 1| < \epsilon$$

$$|x - 1| < \epsilon/5$$

Thus we can take $\delta = \epsilon/5$ (Fig. 1.18). If $0 < |x - 1| < \delta = \epsilon/5$, then

$$|(5x - 3) - 2| = |5x - 5| = 5|x - 1| < 5(\epsilon/5) = \epsilon.$$

This proves that $\lim_{x \to 1} (5x - 3) = 2$.

The value of $\delta = \epsilon/5$ is not the only value that will make $0 < |x - 1| < \delta$ imply $|5x - 5| < \epsilon$. Any smaller positive δ will do as well. The definition does not ask for a "best" positive δ, just one that will work. ❏

EXAMPLE 4   *Two important limits*

Verify: (a) $\lim_{x \to x_0} x = x_0$ (b) $\lim_{x \to x_0} k = k$ (k constant).

Solution

a) Let $\epsilon > 0$ be given. We must find $\delta > 0$ such that for all x

$$0 < |x - x_0| < \delta \qquad \text{implies} \qquad |x - x_0| < \epsilon.$$

The implication will hold if δ equals ϵ or any smaller positive number (Fig. 1.19). This proves that $\lim_{x \to x_0} x = x_0$.

b) Let $\epsilon > 0$ be given. We must find $\delta > 0$ such that for all x

$$0 < |x - x_0| < \delta \qquad \text{implies} \qquad |k - k| < \epsilon.$$

Since $k - k = 0$, we can use any positive number for δ and the implication will hold (Fig. 1.20). This proves that $\lim_{x \to x_0} k = k$. ❏

Finding Deltas Algebraically for Given Epsilons

In Examples 3 and 4, the interval of values about x_0 for which $|f(x) - L|$ was less than ϵ was symmetric about x_0 and we could take δ to be half the length of the interval. When such symmetry is absent, as it usually is, we can take δ to be the distance from x_0 to the interval's nearer endpoint.

EXAMPLE 5   For the limit $\lim_{x \to 5} \sqrt{x - 1} = 2$, find a $\delta > 0$ that works for $\epsilon = 1$. That is, find a $\delta > 0$ such that for all x

$$0 < |x - 5| < \delta \implies |\sqrt{x - 1} - 2| < 1.$$

Solution We organize the search into two steps. First we solve the inequality $|\sqrt{x-1}-2| < 1$ to find an interval (a, b) about $x_0 = 5$ on which the inequality holds for all $x \neq x_0$. Then we find a value of $\delta > 0$ that places the interval $5 - \delta < x < 5 + \delta$ (centered at $x_0 = 5$) inside the interval (a, b).

Step 1: *Solve the inequality* $|\sqrt{x-1}-2| < 1$ *to find an interval about* $x_0 = 5$ *on which the inequality holds for all* $x \neq x_0$.

$$|\sqrt{x-1}-2| < 1$$

$$-1 < \sqrt{x-1} - 2 < 1$$

$$1 < \sqrt{x-1} < 3$$

$$1 < x - 1 < 9$$

$$2 < x < 10$$

The inequality holds for all x in the open interval $(2, 10)$, so it holds for all $x \neq 5$ in this interval as well.

Step 2: *Find a value of* $\delta > 0$ *that places the centered interval* $5 - \delta < x < 5 + \delta$ *inside the interval* $(2, 10)$. The distance from 5 to the nearer endpoint of $(2, 10)$ is 3 (Fig. 1.21). If we take $\delta = 3$ or any smaller positive number, then the inequality $0 < |x - 5| < \delta$ will automatically place x between 2 and 10 to make $|\sqrt{x-1}-2| < 1$ (Fig. 1.22):

$$0 < |x-5| < 3 \quad \Longrightarrow \quad |\sqrt{x-1}-2| < 1. \qquad \square$$

1.21 An open interval of radius 3 about $x_0 = 5$ will lie inside the open interval $(2, 10)$.

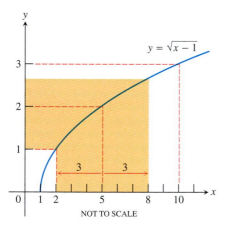

1.22 The function and intervals in Example 5.

How to Find a δ for a Given f, L, x_0, and $\epsilon > 0$ Algebraically

The process of finding a $\delta > 0$ such that for all x

$$0 < |x - x_0| < \delta \quad \Longrightarrow \quad |f(x) - L| < \epsilon$$

can be accomplished in two steps.

Step 1 *Solve the inequality* $|f(x) - L| < \epsilon$ *to find an open interval* (a, b) *about* x_0 *on which the inequality holds for all* $x \neq x_0$.

Step 2 *Find a value of* $\delta > 0$ *that places the open interval* $(x_0 - \delta, x_0 + \delta)$ *centered at* x_0 *inside the interval* (a, b). *The inequality* $|f(x) - L| < \epsilon$ *will hold for all* $x \neq x_0$ *in this* δ-*interval.*

EXAMPLE 6 Prove that $\lim_{x \to 2} f(x) = 4$ if

$$f(x) = \begin{cases} x^2, & x \neq 2 \\ 1, & x = 2. \end{cases}$$

Solution Our task is to show that given $\epsilon > 0$ there exists a $\delta > 0$ such that for all x

$$0 < |x - 2| < \delta \quad \Longrightarrow \quad |f(x) - 4| < \epsilon.$$

Step 1: *Solve the inequality* $|f(x) - 4| < \epsilon$ *to find an open interval about* $x_0 = 2$ *on which the inequality holds for all* $x \neq x_0$.

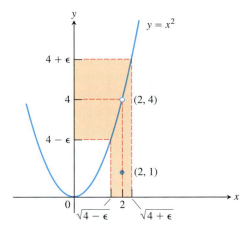

1.23 The function in Example 6.

For $x \neq x_0 = 2$, we have $f(x) = x^2$, and the inequality to solve is $|x^2 - 4| < \epsilon$:

$$|x^2 - 4| < \epsilon$$

$$-\epsilon < x^2 - 4 < \epsilon$$

$$4 - \epsilon < x^2 < 4 + \epsilon$$

$$\sqrt{4 - \epsilon} < |x| < \sqrt{4 + \epsilon} \qquad \text{Assumes } \epsilon < 4; \text{ see below.}$$

$$\sqrt{4 - \epsilon} < x < \sqrt{4 + \epsilon}. \qquad \begin{array}{l}\text{An open interval about } x_0 = 2 \\ \text{that solves the inequality}\end{array}$$

The inequality $|f(x) - 4| < \epsilon$ holds for all $x \neq 2$ in the open interval $(\sqrt{4 - \epsilon}, \sqrt{4 + \epsilon})$ (Fig. 1.23).

Step 2: *Find a value of $\delta > 0$ that places the centered interval $(2 - \delta, 2 + \delta)$ inside the interval $(\sqrt{4 - \epsilon}, \sqrt{4 + \epsilon})$.*

Take δ to be the distance from $x_0 = 2$ to the nearer endpoint of $(\sqrt{4 - \epsilon}, \sqrt{4 + \epsilon})$. In other words, take $\delta = \min\ \{2 - \sqrt{4 - \epsilon}, \sqrt{4 + \epsilon} - 2\}$, the *minimum* (the smaller) of the two numbers $2 - \sqrt{4 - \epsilon}$ and $\sqrt{4 + \epsilon} - 2$. If δ has this or any smaller positive value, the inequality $0 < |x - 2| < \delta$ will automatically place x between $\sqrt{4 - \epsilon}$ and $\sqrt{4 + \epsilon}$ to make $|f(x) - 4| < \epsilon$. For all x,

$$0 < |x - 2| < \delta \qquad \Longrightarrow \qquad |f(x) - 4| < \epsilon.$$

This completes the proof.

Why was it all right to assume $\epsilon < 4$? Because, in finding a δ such that for all x, $0 < |x - 2| < \delta$ implied $|f(x) - 4| < \epsilon < 4$, we found a δ that would work for any larger ϵ as well.

Finally, notice the freedom we gained in letting $\delta = \min\ \{2 - \sqrt{4 - \epsilon}, \sqrt{4 + \epsilon} - 2\}$. We did not have to spend time deciding which, if either, number was the smaller of the two. We just let δ represent the smaller and went on to finish the argument. ❑

Using the Definition to Prove Theorems

We do not usually rely on the formal definition of limit to verify specific limits such as those in the preceding examples. Rather we appeal to general theorems about limits, in particular the theorems of Section 1.2. The definition is used to prove these theorems. As an example, we prove part 1 of Theorem 1, the Sum Rule.

EXAMPLE 7   *Proving the rule for the limit of a sum*

Given that $\lim_{x \to c} f(x) = L$ and $\lim_{x \to c} g(x) = M$, prove that

$$\lim_{x \to c} (f(x) + g(x)) = L + M.$$

Solution   Let $\epsilon > 0$ be given. We want to find a positive number δ such that for all x

$$0 < |x - c| < \delta \qquad \Longrightarrow \qquad |f(x) + g(x) - (L + M)| < \epsilon.$$

Regrouping terms, we get

$$|f(x) + g(x) - (L + M)| = |(f(x) - L) + (g(x) - M)|$$

$$\leq |f(x) - L| + |g(x) - M|. \qquad \begin{array}{l}\text{Triangle Inequality:} \\ |a + b| \leq |a| + |b|\end{array}$$

Since $\lim_{x \to c} f(x) = L$, there exists a number $\delta_1 > 0$ such that for all x

$$0 < |x - c| < \delta_1 \quad \Longrightarrow \quad |f(x) - L| < \epsilon/2.$$

Similarly, since $\lim_{x \to c} g(x) = M$, there exists a number $\delta_2 > 0$ such that for all x

$$0 < |x - c| < \delta_2 \quad \Longrightarrow \quad |g(x) - M| < \epsilon/2.$$

Let $\delta = \min\{\delta_1, \delta_2\}$, the smaller of δ_1 and δ_2. If $0 < |x - c| < \delta$ then $|x - c| < \delta_1$, so $|f(x) - L| < \epsilon/2$, and $|x - c| < \delta_2$, so $|g(x) - M| < \epsilon/2$. Therefore

$$|f(x) + g(x) - (L + M)| < \frac{\epsilon}{2} + \frac{\epsilon}{2} = \epsilon.$$

This shows that $\lim_{x \to c} (f(x) + g(x)) = L + M.$ ◻

Exercises 1.3

Centering Intervals About a Point

In Exercises 1–6, sketch the interval (a, b) on the x-axis with the point x_0 inside. Then find a value of $\delta > 0$ such that for all x, $0 < |x - x_0| < \delta \Longrightarrow a < x < b$.

1. $a = 1, \quad b = 7, \quad x_0 = 5$

2. $a = 1, \quad b = 7, \quad x_0 = 2$

3. $a = -7/2, \quad b = -1/2, \quad x_0 = -3$

4. $a = -7/2, \quad b = -1/2, \quad x_0 = -3/2$

5. $a = 4/9, \quad b = 4/7, \quad x_0 = 1/2$

6. $a = 2.7591, \quad b = 3.2391, \quad x_0 = 3$

Finding Deltas Graphically

In Exercises 7–14, use the graphs to find a $\delta > 0$ such that for all x

$$0 < |x - x_0| < \delta \quad \Longrightarrow \quad |f(x) - L| < \epsilon.$$

9.

10.

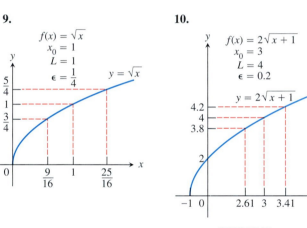

NOT TO SCALE

7.

8.

11.

12.

NOT TO SCALE

NOT TO SCALE

13.

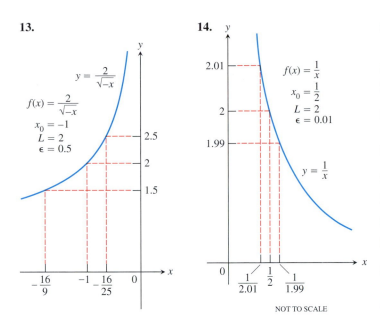

$$y = \frac{2}{\sqrt{-x}}$$

$$f(x) = \frac{2}{\sqrt{-x}}$$
$$x_0 = -1$$
$$L = 2$$
$$\epsilon = 0.5$$

14.

$$f(x) = \frac{1}{x}$$
$$x_0 = \frac{1}{2}$$
$$L = 2$$
$$\epsilon = 0.01$$

$$y = \frac{1}{x}$$

NOT TO SCALE

Finding Deltas Algebraically

Each of Exercises 15–30 gives a function $f(x)$ and numbers L, x_0, and $\epsilon > 0$. In each case, find an open interval about x_0 on which the inequality $|f(x) - L| < \epsilon$ holds. Then give a value for $\delta > 0$ such that for all x satisfying $0 < |x - x_0| < \delta$ the inequality $|f(x) - L| < \epsilon$ holds.

15. $f(x) = x + 1, \quad L = 5, \quad x_0 = 4, \quad \epsilon = 0.01$

16. $f(x) = 2x - 2, \quad L = -6, \quad x_0 = -2, \quad \epsilon = 0.02$

17. $f(x) = \sqrt{x + 1}, \quad L = 1, \quad x_0 = 0, \quad \epsilon = 0.1$

18. $f(x) = \sqrt{x}, \quad L = 1/2, \quad x_0 = 1/4, \quad \epsilon = 0.1$

19. $f(x) = \sqrt{19 - x}, \quad L = 3, \quad x_0 = 10, \quad \epsilon = 1$

20. $f(x) = \sqrt{x - 7}, \quad L = 4, \quad x_0 = 23, \quad \epsilon = 1$

21. $f(x) = 1/x, \quad L = 1/4, \quad x_0 = 4, \quad \epsilon = 0.05$

22. $f(x) = x^2, \quad L = 3, \quad x_0 = \sqrt{3}, \quad \epsilon = 0.1$

23. $f(x) = x^2, \quad L = 4, \quad x_0 = -2, \quad \epsilon = 0.5$

24. $f(x) = 1/x, \quad L = -1, \quad x_0 = -1, \quad \epsilon = 0.1$

25. $f(x) = x^2 - 5, \quad L = 11, \quad x_0 = 4, \quad \epsilon = 1$

26. $f(x) = 120/x, \quad L = 5, \quad x_0 = 24, \quad \epsilon = 1$

27. $f(x) = mx, \quad m > 0, \quad L = 2m, \quad x_0 = 2, \quad \epsilon = 0.03$

28. $f(x) = mx, \quad m > 0, \quad L = 3m, \quad x_0 = 3, \quad \epsilon = c > 0$

29. $f(x) = mx + b, \quad m > 0, \quad L = (m/2) + b, \quad x_0 = 1/2,$
  $\epsilon = c > 0$

30. $f(x) = mx + b, \quad m > 0, \quad L = m + b, \quad x_0 = 1, \quad \epsilon = 0.05$

More on Formal Limits

Each of Exercises 31–36 gives a function $f(x)$, a point x_0, and a positive number ϵ. Find $L = \lim\limits_{x \to x_0} f(x)$. Then find a number $\delta > 0$

such that for all x

$$0 < |x - x_0| < \delta \quad \Rightarrow \quad |f(x) - L| < \epsilon.$$

31. $f(x) = 3 - 2x, \quad x_0 = 3, \quad \epsilon = 0.02$

32. $f(x) = -3x - 2, \quad x_0 = -1, \quad \epsilon = 0.03$

33. $f(x) = \dfrac{x^2 - 4}{x - 2}, \quad x_0 = 2, \quad \epsilon = 0.05$

34. $f(x) = \dfrac{x^2 + 6x + 5}{x + 5}, \quad x_0 = -5, \quad \epsilon = 0.05$

35. $f(x) = \sqrt{1 - 5x}, \quad x_0 = -3, \quad \epsilon = 0.5$

36. $f(x) = 4/x, \quad x_0 = 2, \quad \epsilon = 0.4$

Prove the limit statements in Exercises 37–50.

37. $\lim\limits_{x \to 4} (9 - x) = 5$

38. $\lim\limits_{x \to 3} (3x - 7) = 2$

39. $\lim\limits_{x \to 9} \sqrt{x - 5} = 2$

40. $\lim\limits_{x \to 0} \sqrt{4 - x} = 2$

41. $\lim\limits_{x \to 1} f(x) = 1 \quad \text{if} \quad f(x) = \begin{cases} x^2, & x \neq 1 \\ 2, & x = 1 \end{cases}$

42. $\lim\limits_{x \to -2} f(x) = 4 \quad \text{if} \quad f(x) = \begin{cases} x^2, & x \neq -2 \\ 1, & x = -2 \end{cases}$

43. $\lim\limits_{x \to 1} \dfrac{1}{x} = 1$

44. $\lim\limits_{x \to \sqrt{3}} \dfrac{1}{x^2} = \dfrac{1}{3}$

45. $\lim\limits_{x \to -3} \dfrac{x^2 - 9}{x + 3} = -6$

46. $\lim\limits_{x \to 1} \dfrac{x^2 - 1}{x - 1} = 2$

47. $\lim\limits_{x \to 1} f(x) = 2 \quad \text{if} \quad f(x) = \begin{cases} 4 - 2x, & x < 1 \\ 6x - 4, & x \geq 1 \end{cases}$

48. $\lim\limits_{x \to 0} f(x) = 0 \quad \text{if} \quad f(x) = \begin{cases} 2x, & x < 0 \\ x/2, & x \geq 0 \end{cases}$

49. $\lim\limits_{x \to 0} x \sin \dfrac{1}{x} = 0$

$$y = x \sin \frac{1}{x}$$

(Generated by Mathematica)

50. $\lim_{x \to 0} x^2 \sin \dfrac{1}{x} = 0$

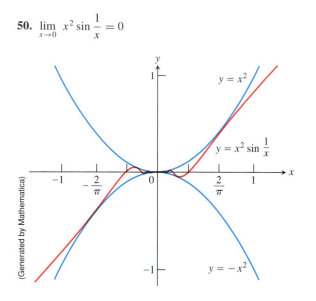

(Generated by Mathematica)

Theory and Examples

51. Define what it means to say that $\lim_{x \to 2} f(x) = 5$.

52. Define what it means to say that $\lim_{x \to 0} g(x) = k$.

53. *A wrong statement about limits.* Show by example that the following statement is wrong.

> The number L is the limit of $f(x)$ as x approaches x_0 if $f(x)$ gets closer to L as x approaches x_0.

Explain why the function in your example does not have the given value of L as a limit as $x \to x_0$.

54. *Another wrong statement about limits.* Show by example that the following statement is wrong.

> The number L is the limit of $f(x)$ as x approaches x_0 if, given any $\epsilon > 0$, there exists a value of x for which $|f(x) - L| < \epsilon$.

Explain why the function in your example does not have the given value of L as a limit as $x \to x_0$.

■ 55. *Grinding engine cylinders.* Before contracting to grind engine cylinders to a cross-section area of 9 in^2, you need to know how much deviation from the ideal cylinder diameter of $x_0 = 3.385$ in. you can allow and still have the area come within 0.01 in^2 of the required 9 in^2. To find out, you let $A = \pi(x/2)^2$ and look for the interval in which you must hold x to make $|A - 9| \le 0.01$. What interval do you find?

56. *Manufacturing electrical resistors.* Ohm's law for electrical circuits like the one shown in Fig. 1.24 states that $V = RI$. In this equation, V is a constant voltage, I is the current in amperes, and R is the resistance in ohms. Your firm has been asked to supply the resistors for a circuit in which V will be 120 volts and

I is to be 5 ± 0.1 amp. In what interval does R have to lie for I to be within 0.1 amp of the target value $I_0 = 5$?

1.24 The circuit in Exercise 56.

When Is a Number *L Not* the Limit of *f(x)* as $x \to x_0$?

We can prove that $\lim_{x \to x_0} f(x) \ne L$ by providing an $\epsilon > 0$ such that no possible $\delta > 0$ satisfies the condition

For all x, $0 < |x - x_0| < \delta \implies |f(x) - L| < \epsilon$.

We accomplish this for our candidate ϵ by showing that for each $\delta > 0$ there exists a value of x such that

$$0 < |x - x_0| < \delta \quad \text{and} \quad |f(x) - L| \ge \epsilon.$$

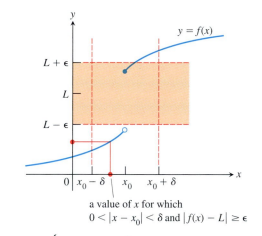

a value of x for which
$0 < |x - x_0| < \delta$ and $|f(x) - L| \ge \epsilon$

57. Let $f(x) = \begin{cases} x, & x < 1 \\ x + 1, & x > 1. \end{cases}$

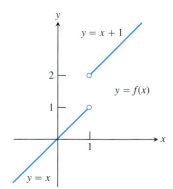

a) Let $\epsilon = 1/2$. Show that no possible $\delta > 0$ satisfies the following condition:

For all x, $0 < |x - 1| < \delta$ $\implies$ $|f(x) - 2| < 1/2$.

That is, for each $\delta > 0$ show that there is a value of x such that

$$0 < |x - 1| < \delta \quad \text{and} \quad |f(x) - 2| \geq 1/2.$$

This will show that $\lim_{x \to 1} f(x) \neq 2$.

b) Show that $\lim_{x \to 1} f(x) \neq 1$.

c) Show that $\lim_{x \to 1} f(x) \neq 1.5$.

58. Let $h(x) = \begin{cases} x^2, & x < 2 \\ 3, & x = 2 \\ 2, & x > 2. \end{cases}$

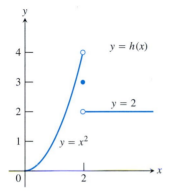

Show that

a) $\lim_{x \to 2} h(x) \neq 4$

b) $\lim_{x \to 2} h(x) \neq 3$

c) $\lim_{x \to 2} h(x) \neq 2$

59. For the function graphed here, show that

a) $\lim_{x \to 3} f(x) \neq 4$

b) $\lim_{x \to 3} f(x) \neq 4.8$

c) $\lim_{x \to 3} f(x) \neq 3$

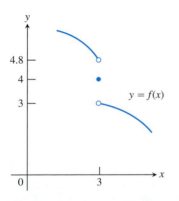

60. a) For the function graphed here, show that $\lim_{x \to -1} g(x) \neq 2$.

b) Does $\lim_{x \to -1} g(x)$ appear to exist? If so, what is the value of the limit? If not, why not?

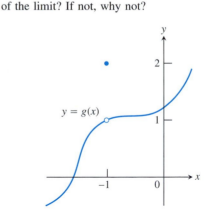

✪ CAS Explorations and Projects

In Exercises 61–66, you will further explore finding deltas graphically. Use a CAS to perform the following steps:

a) Plot the function $y = f(x)$ near the point x_0 being approached.

b) Guess the value of the limit L and then evaluate the limit symbolically to see if you guessed correctly.

c) Using the value $\epsilon = 0.2$, graph the banding lines $y_1 = L - \epsilon$ and $y_2 = L + \epsilon$ together with the function f near x_0.

d) From your graph in part (c), estimate a $\delta > 0$ such that for all x

$$0 < |x - x_0| < \delta \quad \implies \quad |f(x) - L| < \epsilon.$$

Test your estimate by plotting f, y_1, and y_2 over the interval $0 < |x - x_0| < \delta$. For your viewing window use $x_0 - 2\delta \leq x \leq x_0 + 2\delta$ and $L - 2\epsilon \leq y \leq L + 2\epsilon$. If any function values lie outside the interval $[L - \epsilon, L + \epsilon]$, your choice of δ was too large. Try again with a smaller estimate.

e) Repeat parts (c) and (d) successively for $\epsilon = 0.1, 0.05,$ and 0.001.

61. $f(x) = \dfrac{x^4 - 81}{x - 3}, \quad x_0 = 3$

62. $f(x) = \dfrac{5x^3 + 9x^2}{2x^5 + 3x^2}, \quad x_0 = 0$

63. $f(x) = \dfrac{\sin 2x}{3x}, \quad x_0 = 0$

64. $f(x) = \dfrac{x(1 - \cos x)}{x - \sin x}, \quad x_0 = 0$

65. $f(x) = \dfrac{\sqrt[3]{x} - 1}{x - 1}, \quad x_0 = 1$

66. $f(x) = \dfrac{3x^2 - (7x + 1)\sqrt{x} + 5}{x - 1}, \quad x_0 = 1$

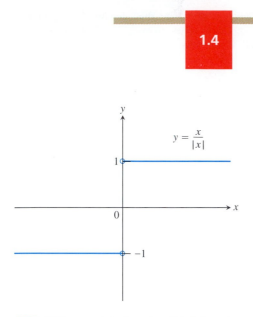

1.25 Different right-hand and left-hand limits at the origin.

1.4

Extensions of the Limit Concept

In this section we extend the concept of limit to

1. *one-sided limits,* which are limits as x approaches a from the left-hand side or the right-hand side only,
2. *infinite limits,* which are not really limits at all, but provide useful symbols and language for describing the behavior of functions whose values become arbitrarily large, positive or negative.

One-Sided Limits

To have a limit L as x approaches a, a function f must be defined on *both sides* of a, and its values $f(x)$ must approach L as x approaches a from either side. Because of this, ordinary limits are sometimes called **two-sided** limits.

It is possible for a function to approach a limiting value as x approaches a from only one side, either from the right or from the left. In this case we say that f has a **one-sided** (either right-hand or left-hand) limit at a. The function $f(x) = x/|x|$ graphed in Fig. 1.25 has limit 1 as x approaches zero from the right, and limit -1 as x approaches zero from the left.

The "+" and "−"

The significance of the signs in the notation for one-sided limits is this:

$x \to a^-$ means x approaches a from the negative side of a, through values less than a.

$x \to a^+$ means x approaches a from the positive side of a, through values greater than a.

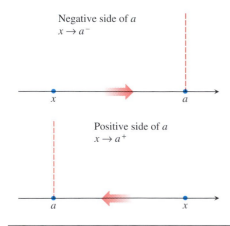

> ### Definition
> ### Informal Definition of Right-hand and Left-hand Limits
> Let $f(x)$ be defined on an interval (a, b) where $a < b$. If $f(x)$ approaches arbitrarily close to L as x approaches a from within that interval, then we say that f has **right-hand limit** L at a, and we write
>
> $$\lim_{x \to a^+} f(x) = L.$$
>
> Let $f(x)$ be defined on an interval (c, a) where $c < a$. If $f(x)$ approaches arbitrarily close to M as x approaches a from within the interval (c, a), then we say that f has **left-hand limit** M at a, and we write
>
> $$\lim_{x \to a^-} f(x) = M.$$
>
> For the function $f(x) = x/|x|$ in Fig. 1.25, we have
>
> $$\lim_{x \to 0^+} f(x) = 1 \qquad \text{and} \qquad \lim_{x \to 0^-} f(x) = -1.$$

A function cannot have an ordinary limit at an endpoint of its domain, but it can have a one-sided limit.

EXAMPLE 1 The domain of $f(x) = \sqrt{4 - x^2}$ is $[-2, 2]$; its graph is the semicircle in Fig. 1.26. We have

$$\lim_{x \to -2^+} \sqrt{4 - x^2} = 0 \qquad \text{and} \qquad \lim_{x \to 2^-} \sqrt{4 - x^2} = 0.$$

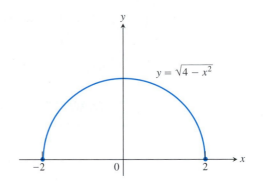

1.26 $\lim\limits_{x \to 2^-} \sqrt{4 - x^2} = 0, \quad \lim\limits_{x \to -2^+} \sqrt{4 - x^2} = 0.$

The function does not have a left-hand limit at $x = -2$ or a right-hand limit at $x = 2$. It does not have ordinary two-sided limits at either -2 or 2. ❏

One-sided limits have all the limit properties listed in Theorem 1, Section 1.2. The right-hand limit of the sum of two functions is the sum of their right-hand limits, and so on. The theorems for limits of polynomials and rational functions hold with one-sided limits, as does the Sandwich Theorem.

The connection between one-sided and two-sided limits is stated in the following theorem (proved at the end of this section).

Theorem 5
One-sided vs. Two-sided Limits

A function $f(x)$ has a limit as x approaches c if and only if it has left-hand and right-hand limits there, and these one-sided limits are equal:

$$\lim_{x \to c} f(x) = L \quad \Leftrightarrow \quad \lim_{x \to c^-} f(x) = L \quad \text{and} \quad \lim_{x \to c^+} f(x) = L.$$

EXAMPLE 2   All of the following statements about the function graphed in Figure 1.27 are true.

At $x = 0$: $\lim_{x \to 0^+} f(x) = 1$,
$\lim_{x \to 0^-} f(x)$ and $\lim_{x \to 0} f(x)$ do not exist. (The function is not defined to the left of $x = 0$.)

At $x = 1$: $\lim_{x \to 1^-} f(x) = 0$ even though $f(1) = 1$,
$\lim_{x \to 1^+} f(x) = 1$,
$\lim_{x \to 1} f(x)$ does not exist. (The right- and left-hand limits are not equal.)

At $x = 2$: $\lim_{x \to 2^-} f(x) = 1$,
$\lim_{x \to 2^+} f(x) = 1$,
$\lim_{x \to 2} f(x) = 1$ even though $f(2) = 2$.

At $x = 3$: $\lim_{x \to 3^-} f(x) = \lim_{x \to 3^+} f(x) = \lim_{x \to 3} f(x) = f(3) = 2$

At $x = 4$: $\lim_{x \to 4^-} f(x) = 1$ even though $f(4) \neq 1$,
$\lim_{x \to 4^+} f(x)$ and $\lim_{x \to 4} f(x)$ do not exist. (The function is not defined to the right of $x = 4$.)

At every other point a in $[0, 4]$, $f(x)$ has limit $f(a)$. ❏

1.27 Graph of the function in Example 2.

In the examples so far in this section, the functions that failed to have a limit at some point at least had one existing one-sided limit there. The function in the following example has neither a left-hand limit nor a right-hand limit at $x = 0$ even though it is defined everywhere except at $x = 0$.

EXAMPLE 3   Show that $y = \sin(1/x)$ has no limit as x approaches zero from either side (Fig. 1.28).

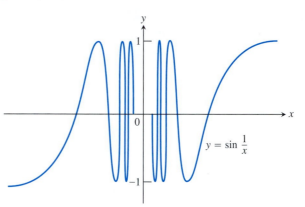

1.28 The function $y = \sin(1/x)$ has neither a right-hand nor a left-hand limit as x approaches zero (Example 3).

Solution   As x approaches zero, its reciprocal, $1/x$, grows without bound and the values of $\sin(1/x)$ cycle repeatedly from -1 to 1. There is no single number L that the function's values stay increasingly close to as x approaches zero. This is true even if we restrict x to positive values or to negative values. The function has neither a right-hand limit nor a left-hand limit at $x = 0$.   ❑

Infinite Limits

Let us look closely at the function $f(x) = 1/x$ that drives the sine in Example 3. As $x \to 0^+$, the values of f grow without bound, eventually reaching and surpassing every positive real number. That is, given any positive real number B, however large, the values of f become larger still (Fig. 1.29). Thus, f has no limit as $x \to 0^+$. It is nevertheless convenient to describe the behavior of f by saying that $f(x)$ approaches ∞ as $x \to 0^+$. We write

$$\lim_{x \to 0^+} f(x) = \lim_{x \to 0^+} \frac{1}{x} = \infty.$$

In writing this, we are *not* saying that the limit exists. Nor are we saying that there is a real number ∞, for there is no such number. Rather, we are saying that $\lim_{x \to 0^+}(1/x)$ *does not exist because* $1/x$ *becomes arbitrarily large and positive as* $x \to 0^+$.

As $x \to 0^-$, the values of $f(x) = 1/x$ become arbitrarily large and negative. Given any negative real number $-B$, the values of f eventually lie below $-B$. (See Fig. 1.29.) We write

$$\lim_{x \to 0^-} f(x) = \lim_{x \to 0^-} \frac{1}{x} = -\infty.$$

Again, we are not saying that the limit exists and equals the number $-\infty$. There *is* no real number $-\infty$. We are describing the behavior of a function whose limit as $x \to 0^-$ *does not exist because its values become arbitrarily large and negative.*

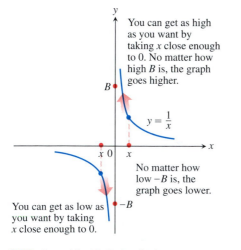

You can get as high as you want by taking x close enough to 0. No matter how high B is, the graph goes higher.

$y = \dfrac{1}{x}$

No matter how low $-B$ is, the graph goes lower.

You can get as low as you want by taking x close enough to 0.

1.29 One-sided infinite limits:

$$\lim_{x \to 0^+} \frac{1}{x} = \infty \quad \text{and} \quad \lim_{x \to 0^-} \frac{1}{x} = -\infty.$$

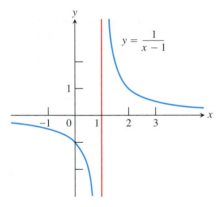

$y = \dfrac{1}{x-1}$

1.30 Near $x = 1$, the function $y = 1/(x - 1)$ behaves the way the function $y = 1/x$ behaves near $x = 0$. Its graph is the graph of $y = 1/x$ shifted 1 unit to the right.

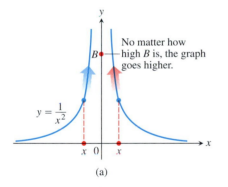

No matter how high B is, the graph goes higher.

$y = \dfrac{1}{x^2}$

(a)

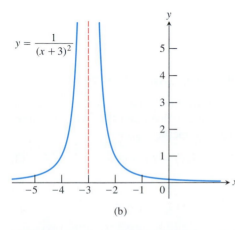

$y = \dfrac{1}{(x+3)^2}$

(b)

1.31 The graphs of the functions in Example 5.

EXAMPLE 4   *One-sided infinite limits*

Find $\displaystyle\lim_{x \to 1^+} \frac{1}{x - 1}$ and $\displaystyle\lim_{x \to 1^-} \frac{1}{x - 1}$.

Geometric Solution   The graph of $y = 1/(x - 1)$ is the graph of $y = 1/x$ shifted 1 unit to the right (Fig. 1.30). Therefore, $y = 1/(x - 1)$ behaves near 1 exactly the way $y = 1/x$ behaves near 0:

$$\lim_{x \to 1^+} \frac{1}{x - 1} = \infty \quad \text{and} \quad \lim_{x \to 1^-} \frac{1}{x - 1} = -\infty.$$

Analytic Solution   Think about the number $x - 1$ and its reciprocal. As $x \to 1^+$, we have $(x - 1) \to 0^+$ and $1/(x - 1) \to \infty$. As $x \to 1^-$, we have $(x - 1) \to 0^-$ and $1/(x - 1) \to -\infty$.   ❏

EXAMPLE 5   *Two-sided infinite limits*

Discuss the behavior of

a) $f(x) = \dfrac{1}{x^2}$ near $x = 0$,

b) $g(x) = \dfrac{1}{(x + 3)^2}$ near $x = -3$.

Solution

a)   As x approaches zero from either side, the values of $1/x^2$ are positive and become arbitrarily large (Fig. 1.31a):

$$\lim_{x \to 0} f(x) = \lim_{x \to 0} \frac{1}{x^2} = \infty.$$

b)   The graph of $g(x) = 1/(x + 3)^2$ is the graph of $f(x) = 1/x^2$ shifted 3 units to the left (Fig. 1.31b). Therefore, g behaves near -3 exactly the way f behaves near 0.

$$\lim_{x \to -3} g(x) = \lim_{x \to -3} \frac{1}{(x + 3)^2} = \infty.$$ ❏

The function $y = 1/x$ shows no consistent behavior as $x \to 0$. We have $1/x \to \infty$ if $x \to 0^+$, but $1/x \to -\infty$ if $x \to 0^-$. All we can say about $\lim_{x \to 0} (1/x)$ is that it does not exist. The function $y = 1/x^2$ is different. Its values approach infinity as x approaches zero from either side, so we can say that $\lim_{x \to 0} (1/x^2) = \infty$.

EXAMPLE 6   *Rational functions can behave in various ways near zeros of their denominators.*

a) $\displaystyle\lim_{x \to 2} \frac{(x - 2)^2}{x^2 - 4} = \lim_{x \to 2} \frac{(x - 2)^2}{(x - 2)(x + 2)} = \lim_{x \to 2} \frac{x - 2}{x + 2} = 0$

b) $\displaystyle\lim_{x \to 2} \frac{x - 2}{x^2 - 4} = \lim_{x \to 2} \frac{x - 2}{(x - 2)(x + 2)} = \lim_{x \to 2} \frac{1}{x + 2} = \frac{1}{4}$

c) $\displaystyle\lim_{x \to 2^+} \frac{x - 3}{x^2 - 4} = \lim_{x \to 2^+} \frac{x - 3}{(x - 2)(x + 2)} = -\infty$ The values are negative for $x > 2$, x near 2.

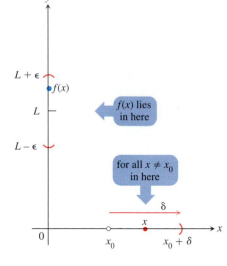

1.32 Diagram for the definition of right-hand limit.

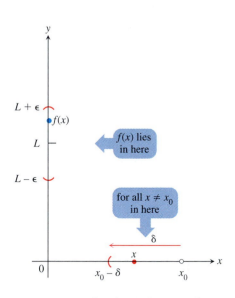

1.33 Diagram for the definition of left-hand limit.

d) $\lim\limits_{x \to 2^-} \dfrac{x-3}{x^2-4} = \lim\limits_{x \to 2^-} \dfrac{x-3}{(x-2)(x+2)} = \infty$ The values are positive for $x < 2$, x near 2.

e) $\lim\limits_{x \to 2} \dfrac{x-3}{x^2-4} = \lim\limits_{x \to 2} \dfrac{x-3}{(x-2)(x+2)}$ does not exist.   See (c) and (d).

f) $\lim\limits_{x \to 2} \dfrac{2-x}{(x-2)^3} = \lim\limits_{x \to 2} \dfrac{-(x-2)}{(x-2)^3} = \lim\limits_{x \to 2} \dfrac{-1}{(x-2)^2} = -\infty$

In parts (a) and (b) the effect of the zero in the denominator at $x = 2$ is canceled because the numerator is zero there also. Thus a finite limit exists. This is not true in part (f), where cancellation still leaves a zero in the denominator.   ❑

Precise Definitions of One-sided Limits

The formal definition of two-sided limit in Section 1.3 is readily modified for one-sided limits.

> **Definitions**
> **Right-hand Limit**
> We say that $f(x)$ has right-hand limit L at x_0, and write
> $$\lim_{x \to x_0^+} f(x) = L \qquad \text{(See Fig. 1.32)}$$
> if for every number $\epsilon > 0$ there exists a corresponding number $\delta > 0$ such that for all x
> $$x_0 < x < x_0 + \delta \quad \Rightarrow \quad |f(x) - L| < \epsilon. \tag{1}$$
>
> **Left-hand Limit**
> We say that f has left-hand limit L at x_0, and write
> $$\lim_{x \to x_0^-} f(x) = L \qquad \text{(See Fig. 1.33)}$$
> if for every number $\epsilon > 0$ there exists a corresponding number $\delta > 0$ such that for all x
> $$x_0 - \delta < x < x_0 \quad \Rightarrow \quad |f(x) - L| < \epsilon. \tag{2}$$

The Relation Between One- and Two-sided Limits

If we subtract x_0 from the δ-inequalities in implications (1) and (2), we can see the logical relation between the one-sided limits just defined and the two-sided limit defined in Section 1.3. For right-hand limits, subtracting x_0 gives

$$0 < x - x_0 < \delta \quad \Rightarrow \quad |f(x) - L| < \epsilon; \tag{3}$$

for left-hand limits we get

$$-\delta < x - x_0 < 0 \quad \Rightarrow \quad |f(x) - L| < \epsilon. \tag{4}$$

Together, (3) and (4) say the same thing as

$$0 < |x - x_0| < \delta \quad \Rightarrow \quad |f(x) - L| < \epsilon, \tag{5}$$

the implication required for two-sided limit. Thus, f has limit L at x_0 if and only if f has right-hand limit L and left-hand limit L at x_0.

Precise Definitions of Infinite Limits

Instead of requiring $f(x)$ to lie arbitrarily close to a finite number L for all x sufficiently close to x_0, the definitions of infinite limits require $f(x)$ to lie arbitrarily far from the origin. Except for this change, the language is identical with what we have seen before. Figures 1.34 and 1.35 accompany these definitions.

1.34 $\lim\limits_{x \to x_0} f(x) = \infty.$

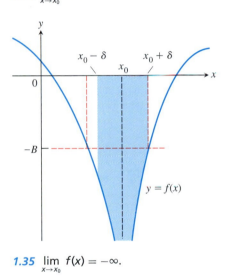

1.35 $\lim\limits_{x \to x_0} f(x) = -\infty.$

> **Definitions**
> **Infinite Limits**
>
> 1. We say that $f(x)$ approaches infinity as x approaches x_0, and write
>
> $$\lim_{x \to x_0} f(x) = \infty,$$
>
> if for every positive real number B there exists a corresponding $\delta > 0$ such that for all x
>
> $$0 < |x - x_0| < \delta \quad \Rightarrow \quad f(x) > B.$$
>
> 2. We say that $f(x)$ approaches minus infinity as x approaches x_0, and write
>
> $$\lim_{x \to x_0} f(x) = -\infty,$$
>
> if for every negative real number $-B$ there exists a corresponding $\delta > 0$ such that for all x
>
> $$0 < |x - x_0| < \delta \quad \Rightarrow \quad f(x) < -B.$$

The precise definitions of one-sided infinite limits at x_0 are similar and are stated in the exercises.

Exercises 1.4

Finding Limits Graphically

1. Which of the following statements about the function $y = f(x)$ graphed here are true, and which are false?

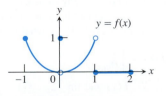

a) $\lim\limits_{x \to -1^+} f(x) = 1$

b) $\lim\limits_{x \to 0^-} f(x) = 0$

c) $\lim\limits_{x \to 0^-} f(x) = 1$

d) $\lim\limits_{x \to 0^-} f(x) = \lim\limits_{x \to 0^+} f(x)$

e) $\lim\limits_{x \to 0} f(x)$ exists

f) $\lim\limits_{x \to 0} f(x) = 0$

g) $\lim\limits_{x \to 0} f(x) = 1$

h) $\lim\limits_{x \to 1} f(x) = 1$

i) $\lim\limits_{x \to 1} f(x) = 0$

j) $\lim\limits_{x \to 2^-} f(x) = 2$

k) $\lim\limits_{x \to -1^-} f(x)$ does not exist.

l) $\lim\limits_{x \to 2^+} f(x) = 0$

2. Which of the following statements about the function $y = f(x)$ graphed here are true, and which are false?

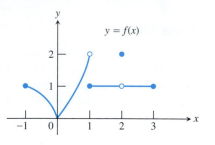

a) $\lim_{x \to -1^+} f(x) = 1$

b) $\lim_{x \to 2} f(x)$ does not exist.

c) $\lim_{x \to 2} f(x) = 2$

d) $\lim_{x \to 1^-} f(x) = 2$

e) $\lim_{x \to 1^+} f(x) = 1$

f) $\lim_{x \to 1} f(x)$ does not exist.

g) $\lim_{x \to 0^+} f(x) = \lim_{x \to 0^-} f(x)$

h) $\lim_{x \to c} f(x)$ exists at every c in the open interval $(-1, 1)$.

i) $\lim_{x \to c} f(x)$ exists at every c in the open interval $(1, 3)$.

j) $\lim_{x \to -1^-} f(x) = 0$

k) $\lim_{x \to 3^+} f(x)$ does not exist.

3. Let $f(x) = \begin{cases} 3 - x, & x < 2 \\ \dfrac{x}{2} + 1, & x > 2 \end{cases}$

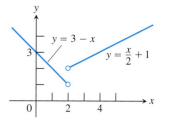

a) Find $\lim_{x \to 2^+} f(x)$ and $\lim_{x \to 2^-} f(x)$.

b) Does $\lim_{x \to 2} f(x)$ exist? If so, what is it? If not, why not?

c) Find $\lim_{x \to 4^-} f(x)$ and $\lim_{x \to 4^+} f(x)$.

d) Does $\lim_{x \to 4} f(x)$ exist? If so, what is it? If not, why not?

4. Let $f(x) = \begin{cases} 3 - x, & x < 2 \\ 2, & x = 2 \\ \dfrac{x}{2}, & x > 2. \end{cases}$

a) Find $\lim_{x \to 2^+} f(x)$, $\lim_{x \to 2^-} f(x)$, and $f(2)$.

b) Does $\lim_{x \to 2} f(x)$ exist? If so, what is it? If not, why not?

c) Find $\lim_{x \to -1^-} f(x)$ and $\lim_{x \to -1^+} f(x)$.

d) Does $\lim_{x \to -1} f(x)$ exist? If so, what is it? If not, why not?

5. Let $f(x) = \begin{cases} 0, & x \le 0 \\ \sin \dfrac{1}{x}, & x > 0. \end{cases}$

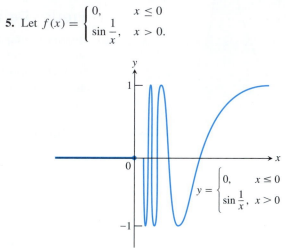

a) Does $\lim_{x \to 0^+} f(x)$ exist? If so, what is it? If not, why not?

b) Does $\lim_{x \to 0^-} f(x)$ exist? If so, what is it? If not, why not?

c) Does $\lim_{x \to 0} f(x)$ exist? If so, what is it? If not, why not?

6. Let $g(x) = \sqrt{x} \sin(1/x)$.

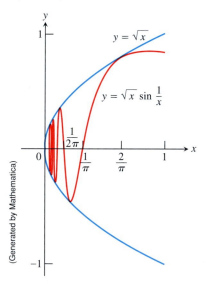

(Generated by Mathematica)

a) Does $\lim_{x \to 0^+} g(x)$ exist? If so, what is it? If not, why not?

b) Does $\lim_{x \to 0^-} g(x)$ exist? If so, what is it? If not, why not?

c) Does $\lim_{x \to 0} g(x)$ exist? If so, what is it? If not, why not?

7. a) Graph $f(x) = \begin{cases} x^3, & x \ne 1 \\ 0, & x = 1. \end{cases}$

b) Find $\lim_{x \to 1^-} f(x)$ and $\lim_{x \to 1^+} f(x)$.

c) Does $\lim_{x \to 1} f(x)$ exist? If so, what is it? If not, why not?

8. a) Graph $f(x) = \begin{cases} 1 - x^2, & x \neq 1 \\ 2, & x = 1. \end{cases}$

b) Find $\lim_{x \to 1^+} f(x)$ and $\lim_{x \to 1^-} f(x)$.
c) Does $\lim_{x \to 1} f(x)$ exist? If so, what is it? If not, why not?

Graph the functions in Exercises 9 and 10. Then answer these questions.

a) What are the domain and range of f?
b) At what points c, if any, does $\lim_{x \to c} f(x)$ exist?
c) At what points does only the left-hand limit exist?
d) At what points does only the right-hand limit exist?

9. $f(x) = \begin{cases} \sqrt{1 - x^2} & \text{if} \quad 0 \leq x < 1 \\ 1 & \text{if} \quad 1 \leq x < 2 \\ 2 & \text{if} \quad x = 2 \end{cases}$

10. $f(x) = \begin{cases} x & \text{if} \quad -1 \leq x < 0, \quad \text{or} \quad 0 < x \leq 1 \\ 1 & \text{if} \quad x = 0 \\ 0 & \text{if} \quad x < -1, \quad \text{or} \quad x > 1 \end{cases}$

Finding Limits Algebraically

Find the limits in Exercises 11–20.

11. $\lim_{x \to -0.5^-} \sqrt{\dfrac{x + 2}{x + 1}}$
12. $\lim_{x \to 1^+} \sqrt{\dfrac{x - 1}{x + 2}}$

13. $\lim_{x \to -2^+} \left(\dfrac{x}{x + 1} \right) \left(\dfrac{2x + 5}{x^2 + x} \right)$

14. $\lim_{x \to 1^-} \left(\dfrac{1}{x + 1} \right) \left(\dfrac{x + 6}{x} \right) \left(\dfrac{3 - x}{7} \right)$

15. $\lim_{h \to 0^+} \dfrac{\sqrt{h^2 + 4h + 5} - \sqrt{5}}{h}$

16. $\lim_{h \to 0^-} \dfrac{\sqrt{6} - \sqrt{5h^2 + 11h + 6}}{h}$

17. a) $\lim_{x \to -2^+} (x + 3) \dfrac{|x + 2|}{x + 2}$
  b) $\lim_{x \to -2^-} (x + 3) \dfrac{|x + 2|}{x + 2}$

18. a) $\lim_{x \to 1^+} \dfrac{\sqrt{2x}(x - 1)}{|x - 1|}$
  b) $\lim_{x \to 1^-} \dfrac{\sqrt{2x}(x - 1)}{|x - 1|}$

19. a) $\lim_{\theta \to 3^+} \dfrac{\lfloor \theta \rfloor}{\theta}$
  b) $\lim_{\theta \to 3^-} \dfrac{\lfloor \theta \rfloor}{\theta}$

20. a) $\lim_{t \to 4^+} (t - \lfloor t \rfloor)$
  b) $\lim_{t \to 4^-} (t - \lfloor t \rfloor)$

Infinite Limits

Find the limits in Exercises 21–32.

21. $\lim_{x \to 0^+} \dfrac{1}{3x}$
22. $\lim_{x \to 0^-} \dfrac{5}{2x}$

23. $\lim_{x \to 2^-} \dfrac{3}{x - 2}$
24. $\lim_{x \to 3^+} \dfrac{1}{x - 3}$

25. $\lim_{x \to -8^+} \dfrac{2x}{x + 8}$
26. $\lim_{x \to -5^-} \dfrac{3x}{2x + 10}$

27. $\lim_{x \to 7} \dfrac{4}{(x - 7)^2}$
28. $\lim_{x \to 0} \dfrac{-1}{x^2(x + 1)}$

29. a) $\lim_{x \to 0^+} \dfrac{2}{3x^{1/3}}$
  b) $\lim_{x \to 0^-} \dfrac{2}{3x^{1/3}}$

30. a) $\lim_{x \to 0^+} \dfrac{2}{x^{1/5}}$
  b) $\lim_{x \to 0^-} \dfrac{2}{x^{1/5}}$

31. $\lim_{x \to 0} \dfrac{4}{x^{2/5}}$
32. $\lim_{x \to 0} \dfrac{1}{x^{2/3}}$

Find the limits in Exercises 33–36.

33. $\lim_{x \to (\pi/2)^-} \tan x$
34. $\lim_{x \to (-\pi/2)^+} \sec x$

35. $\lim_{\theta \to 0^-} (1 + \csc \theta)$
36. $\lim_{\theta \to 0} (2 - \cot \theta)$

Additional Calculations

Find the limits in Exercises 37–42.

37. $\lim \dfrac{1}{x^2 - 4}$ as

a) $x \to 2^+$ **b)** $x \to 2^-$
c) $x \to -2^+$ **d)** $x \to -2^-$

38. $\lim \dfrac{x}{x^2 - 1}$ as

a) $x \to 1^+$ **b)** $x \to 1^-$
c) $x \to -1^+$ **d)** $x \to -1^-$

39. $\lim \left(\dfrac{x^2}{2} - \dfrac{1}{x} \right)$ as

a) $x \to 0^+$ **b)** $x \to 0^-$
c) $x \to \sqrt[3]{2}$ **d)** $x \to -1$

40. $\lim \dfrac{x^2 - 1}{2x + 4}$ as

a) $x \to -2^+$ **b)** $x \to -2^-$
c) $x \to 1^+$ **d)** $x \to 0^-$

41. $\lim \dfrac{x^2 - 3x + 2}{x^3 - 2x^2}$ as

a) $x \to 0^+$ **b)** $x \to 2^+$
c) $x \to 2^-$ **d)** $x \to 2$
e) What, if anything, can be said about the limit as $x \to 0$?

42. $\lim \dfrac{x^2 - 3x + 2}{x^3 - 4x}$ as

a) $x \to 2^+$ **b)** $x \to -2^+$
c) $x \to 0^-$ **d)** $x \to 1^+$
e) What, if anything, can be said about the limit as $x \to 0$?

Find the limits in Exercises 43–46.

43. $\lim \left(2 - \dfrac{3}{t^{1/3}} \right)$ as

a) $t \to 0^+$ **b)** $t \to 0^-$

44. $\lim \left(\dfrac{1}{t^{3/5}} + 7 \right)$ as

a) $t \to 0^+$
b) $t \to 0^-$

45. $\lim \left(\dfrac{1}{x^{2/3}} + \dfrac{2}{(x-1)^{2/3}} \right)$ as

a) $x \to 0^+$
b) $x \to 0^-$
c) $x \to 1^+$
d) $x \to 1^-$

46. $\lim \left(\dfrac{1}{x^{1/3}} - \dfrac{1}{(x-1)^{4/3}} \right)$ as

a) $x \to 0^+$
b) $x \to 0^-$
c) $x \to 1^+$
d) $x \to 1^-$

Theory and Examples

47. Once you know $\lim_{x \to a^+} f(x)$ and $\lim_{x \to a^-} f(x)$ at an interior point of the domain of f, do you then know $\lim_{x \to a} f(x)$? Give reasons for your answer.

48. If you know that $\lim_{x \to c} f(x)$ exists, can you find its value by calculating $\lim_{x \to c^+} f(x)$? Give reasons for your answer.

49. Suppose that f is an odd function of x. Does knowing that $\lim_{x \to 0^+} f(x) = 3$ tell you anything about $\lim_{x \to 0^-} f(x)$? Give reasons for your answer.

50. Suppose that f is an even function of x. Does knowing that $\lim_{x \to 2^-} f(x) = 7$ tell you anything about either $\lim_{x \to -2^-} f(x)$ or $\lim_{x \to -2^+} f(x)$? Give reasons for your answer.

Formal Definitions of One-sided Limits

51. Given $\epsilon > 0$, find an interval $I = (5, 5 + \delta)$, $\delta > 0$, such that if x lies in I, then $\sqrt{x - 5} < \epsilon$. What limit is being verified and what is its value?

52. Given $\epsilon > 0$, find an interval $I = (4 - \delta, 4)$, $\delta > 0$, such that if x lies in I, then $\sqrt{4 - x} < \epsilon$. What limit is being verified and what is its value?

Use the definitions of right-hand and left-hand limits to prove the limit statements in Exercises 53 and 54.

53. $\lim_{x \to 0^-} \dfrac{x}{|x|} = -1$

54. $\lim_{x \to 2^+} \dfrac{x - 2}{|x - 2|} = 1$

55. Find (a) $\lim_{x \to 400^+} \lfloor x \rfloor$ and (b) $\lim_{x \to 400^-} \lfloor x \rfloor$; then use limit definitions to verify your findings. (c) Based on your conclusions in (a) and (b), can anything be said about $\lim_{x \to 400} \lfloor x \rfloor$? Give reasons for your answers.

56. Let $f(x) = \begin{cases} x^2 \sin(1/x), & x < 0 \\ \sqrt{x}, & x > 0. \end{cases}$

Find (a) $\lim_{x \to 0^+} f(x)$ and (b) $\lim_{x \to 0^-} f(x)$; then use limit definitions to verify your findings. (c) Based on your conclusions in (a) and (b), can anything be said about $\lim_{x \to 0} f(x)$? Give reasons for your answer.

The Formal Definition of Infinite Limit

Use formal definitions to prove the limit statements in Exercises 57–60.

57. $\lim_{x \to 0} \dfrac{1}{x^2} = \infty$

58. $\lim_{x \to 0} \dfrac{-1}{x^2} = -\infty$

59. $\lim_{x \to 3} \dfrac{-2}{(x-3)^2} = -\infty$

60. $\lim_{x \to -5} \dfrac{1}{(x+5)^2} = \infty$

Formal Definitions of Infinite One-sided Limits

61. Here is the definition of **infinite right-hand limit.**

We say that $f(x)$ approaches infinity as x approaches x_0 from the right, and write

$$\lim_{x \to x_0^+} f(x) = \infty,$$

if, for every positive real number B, there exists a corresponding number $\delta > 0$ such that for all x

$$x_0 < x < x_0 + \delta \qquad \Rightarrow \qquad f(x) > B.$$

Modify the definition to cover the following cases.

a) $\lim_{x \to x_0^-} f(x) = \infty$
b) $\lim_{x \to x_0^+} f(x) = -\infty$
c) $\lim_{x \to x_0^-} f(x) = -\infty$

Use the formal definitions from Exercise 61 to prove the limit statements in Exercises 62–67.

62. $\lim_{x \to 0^+} \dfrac{1}{x} = \infty$

63. $\lim_{x \to 0^-} \dfrac{1}{x} = -\infty$

64. $\lim_{x \to 2^-} \dfrac{1}{x - 2} = -\infty$

65. $\lim_{x \to 2^+} \dfrac{1}{x - 2} = \infty$

66. $\lim_{x \to 1^+} \dfrac{1}{1 - x^2} = -\infty$

67. $\lim_{x \to 1^-} \dfrac{1}{1 - x^2} = \infty$

1.5

Continuity

When we plot function values generated in the laboratory or collected in the field, we often connect the plotted points with an unbroken curve to show what the function's values are likely to have been at the times we did not measure. In doing so, we are assuming that we are working with a continuous function, a function whose outputs vary continuously with the inputs and do not jump from one value to another without taking on the values in between.

So many physical processes proceed continuously that throughout the eighteenth and nineteenth centuries it rarely occurred to anyone to look for any other kind of behavior. It came as quite a surprise when the physicists of the 1920s discovered that the vibrating atoms in a hydrogen molecule can oscillate only at discrete energy levels, that light comes in particles, and that, when heated, atoms emit light at discrete frequencies and not in continuous spectra. As a result of these and other discoveries, and because of the heavy use of discrete functions in computer science and statistics, the issue of continuity has become one of practical as well as theoretical importance.

In this section, we define continuity, show how to tell whether a function is continuous at a given point, and examine the intermediate value property of continuous functions.

Continuity at a Point

In practice, most functions of a real variable have domains that are intervals or unions of separate intervals, and it is natural to restrict our study of continuity to functions with these domains. This leaves us with only three kinds of points to consider: **interior points** (points that lie in an open interval in the domain), **left endpoints,** and **right endpoints.**

Definition

A function f is **continuous at an interior point** $x = c$ of its domain if

$$\lim_{x \to c} f(x) = f(c).$$

In Fig. 1.36 on the following page, the first function is continuous at $x = 0$. The function in (b) would be continuous if it had $f(0) = 1$. The function in (c) would be continuous if $f(0)$ were 1 instead of 2. The discontinuities in (b) and (c) are **removable.** Each function has a limit as $x \to 0$, and we can remove the discontinuity by setting $f(0)$ equal to this limit.

The discontinuities in parts (d)–(f) of Fig. 1.36 are more serious: $\lim_{x \to 0} f(x)$ does not exist and there is no way to improve the situation by changing f at 0. The step function in (d) has a **jump discontinuity:** the one-sided limits exist but have different values. The function $f(x) = 1/x^2$ in (e) has an **infinite discontinuity.** Jumps and infinite discontinuities are the ones most frequently encountered, but there are others. The function in (f) is discontinuous at the origin because it oscillates too much to have a limit as $x \to 0$.

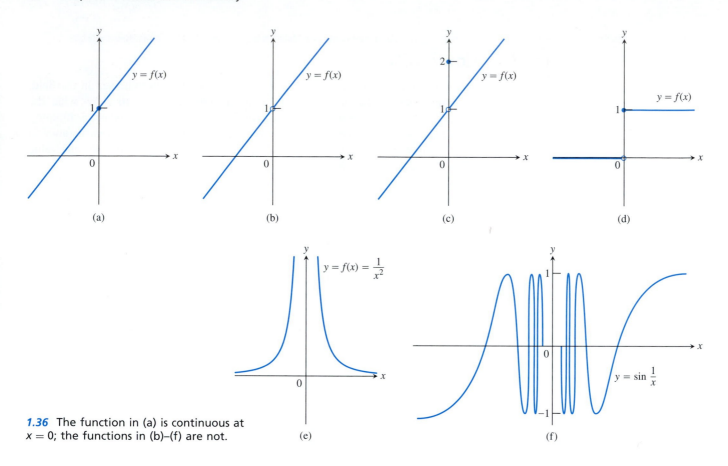

1.36 The function in (a) is continuous at $x = 0$; the functions in (b)–(f) are not.

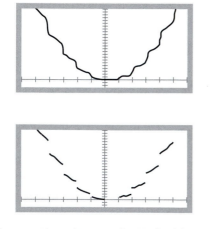

a) $y_1 = x^*\text{int } x$ incorrectly graphed in connected mode.

b) $y_1 = x^*\text{int } x$ correctly graphed in dot mode.

Technology *Deceptive Pictures* A graphing utility (calculator or Computer Algebra System—CAS*) plots a graph much as you do when plotting by hand: by plotting points, or *pixels*, and then connecting them in succession. The resulting picture may be misleading when points on opposite sides of a point of discontinuity in the graph are incorrectly connected. To avoid incorrect connections some systems allow you to use a "dot mode," which plots only the points. Dot mode, however, may not reveal enough information to portray the true behavior of the graph. Try the following four functions on your graphing device. If you can, plot them in both "connected" and "dot" modes.

$y_1 = x^*\text{int } x$ at $x = 2$ jump discontinuity

$y_2 = \sin \dfrac{1}{x}$ at $x = 0$ oscillating discontinuity

$y_3 = \dfrac{1}{x - 2}$ at $x = 2$ infinite discontinuity

$y_4 = \dfrac{x^2 - 2}{x - \sqrt{2}}$ at $x = 2$ removable discontinuity

*Rhymes with class.

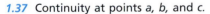

1.37 Continuity at points *a*, *b*, and *c*.

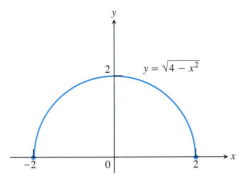

1.38 Continuous at every domain point.

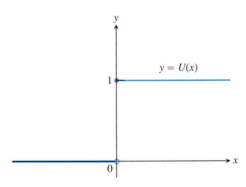

1.39 Right-continuous at the origin.

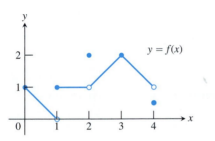

1.40 This function, defined on the closed interval [0, 4], is discontinuous at *x* = 1, 2, and 4. It is continuous at all other points of its domain.

Continuity at endpoints is defined by taking one-sided limits.

> **Definition**
>
> A function *f* is **continuous at a left endpoint** $x = a$ of its domain if
>
> $$\lim_{x \to a^+} f(x) = f(a)$$
>
> and **continuous at a right endpoint** $x = b$ of its domain if
>
> $$\lim_{x \to b^-} f(x) = f(b).$$

In general, a function *f* is **right-continuous (continuous from the right)** at a point $x = c$ in its domain if $\lim_{x \to c^+} f(x) = f(c)$. It is **left-continuous (continuous from the left)** at *c* if $\lim_{x \to c^-} f(x) = f(c)$. Thus, a function is continuous at a left endpoint *a* of its domain if it is right-continuous at *a* and continuous at a right endpoint *b* of its domain if it is left-continuous at *b*. A function is continuous at an interior point *c* of its domain if and only if it is both right-continuous and left-continuous at *c* (Fig. 1.37).

EXAMPLE 1   The function $f(x) = \sqrt{4 - x^2}$ is continuous at every point of its domain, [−2, 2] (Fig. 1.38). This includes $x = -2$, where *f* is right-continuous, and $x = 2$, where *f* is left-continuous.   ❏

EXAMPLE 2   The unit step function $U(x)$, graphed in Fig. 1.39, is right-continuous at $x = 0$, but is neither left-continuous nor continuous there.   ❏

We summarize continuity at a point in the form of a test.

> **Continuity Test**
>
> A function $f(x)$ is continuous at $x = c$ if and only if it meets the following three conditions.
>
> **1.** $f(c)$ exists   (*c* lies in the domain of *f*)
> **2.** $\lim_{x \to c} f(x)$ exists   (*f* has a limit as $x \to c$)
> **3.** $\lim_{x \to c} f(x) = f(c)$ (the limit equals the function value)

For one-sided continuity and continuity at an endpoint, the limits in parts 2 and 3 of the test should be replaced by the appropriate one-sided limits.

EXAMPLE 3   Consider the function $y = f(x)$ in Fig. 1.40, whose domain is the closed interval [0, 4]. Discuss the continuity of *f* at $x = 0, 1, 2, 3,$ and 4.

Solution The continuity test gives the following results:

a) f is continuous at $x = 0$ because

   i) $f(0)$ exists ($f(0) = 1$),

   ii) $\lim_{x \to 0^+} f(x) = 1$ (the right-hand limit exists at this left endpoint),

   iii) $\lim_{x \to 0^+} f(x) = f(0)$ (the limit equals the function value).

b) f is discontinuous at $x = 1$ because $\lim_{x \to 1} f(x)$ does not exist. Part 2 of the test fails: f has different right- and left-hand limits at the interior point $x = 1$. However, f *is* right-continuous at $x = 1$ because

   i) $f(1)$ exists ($f(1) = 1$),

   ii) $\lim_{x \to 1^+} f(x) = 1$ (the right-hand limit exists at $x = 1$),

   iii) $\lim_{x \to 1^+} f(x) = f(1)$ (the right-hand limit equals the function value).

c) f is discontinuous at $x = 2$ because $\lim_{x \to 2} f(x) \neq f(2)$. Part 3 of the test fails.

d) f is continuous at $x = 3$ because

   i) $f(3)$ exists ($f(3) = 2$),

   ii) $\lim_{x \to 3} f(x) = 2$ (the limit exists at $x = 2$),

   iii) $\lim_{x \to 3} f(x) = f(3)$ (the limit equals the function value).

e) f is discontinuous at the right endpoint $x = 4$ because $\lim_{x \to 4^-} f(x) \neq f(4)$. The right-endpoint version of Part 3 of the test fails.   ❑

Rules of Continuity

It follows from Theorem 1 in Section 1.2 that if two functions are continuous at a point, then various algebraic combinations of those functions are continuous at that point.

Theorem 6

Continuity of Algebraic Combinations

If functions f and g are continuous at $x = c$, then the following functions are continuous at $x = c$:

1. $f + g$ and $f - g$
2. fg
3. kf, where k is any number
4. f/g (provided $g(c) \neq 0$)
5. $(f(x))^{m/n}$ (provided $f(x)^{m/n}$ is defined on an interval containing c, and m and n are integers)

As a consequence, polynomials and rational functions are continuous at every point where they are defined.

Theorem 7
Continuity of Polynomials and Rational Functions

Every polynomial is continuous at every point of the real line. Every rational function is continuous at every point where its denominator is different from zero.

EXAMPLE 4   The functions $f(x) = x^4 + 20$ and $g(x) = 5x(x - 2)$ are continuous at every value of x. The function

$$r(x) = \frac{f(x)}{g(x)} = \frac{x^4 + 20}{5x(x - 2)}$$

is continuous at every value of x except $x = 0$ and $x = 2$, where the denominator is 0. ❏

EXAMPLE 5   *Continuity of f(x) = |x|*

The function $f(x) = |x|$ is continuous at every value of x (Fig. 1.41). If $x > 0$, we have $f(x) = x$, a polynomial. If $x < 0$, we have $f(x) = -x$, another polynomial. Finally, at the origin, $\lim_{x \to 0} |x| = 0 = |0|$. ❏

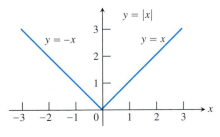

1.41 The sharp corner does not prevent the function from being continuous at the origin (Example 5).

EXAMPLE 6   *Continuity of trigonometric functions*

We will show in Chapter 2 that the functions $\sin x$ and $\cos x$ are continuous at every value of x. Accordingly, the quotients

$$\tan x = \frac{\sin x}{\cos x} \qquad \cot x = \frac{\cos x}{\sin x}$$

$$\sec x = \frac{1}{\cos x} \qquad \csc x = \frac{1}{\sin x}$$

are continuous at every point where they are defined. ❏

Theorem 8 tells us that continuity is preserved under the operation of composition.

Theorem 8
Continuity of Composites

If f is continuous at c, and g is continuous at $f(c)$, then $g \circ f$ is continuous at c (see Fig. 1.42).

1.42 The continuity of composites.

The continuity of composites holds for any finite number of functions. The only requirement is that each function be continuous where it is applied. For an outline of the proof of Theorem 8, see Exercise 6 in Appendix 2.

EXAMPLE 7   The following functions are continuous everywhere on their respective domains.

a) $y = \sqrt{x}$ Theorems 6 and 7 (rational power of a polynomial)

b) $y = \sqrt{x^2 - 2x - 5}$ Theorems 6 and 7, or (a) plus Theorems 7 and 8 (power of a polynomial or composition with the square root)

c) $y = \dfrac{x \cos(x^{2/3})}{1 + x^4}$ Theorems 6, 7, and 8 (power, composite, product, polynomial, and quotient)

d) $y = \left| \dfrac{x - 2}{x^2 - 2} \right|$ Theorems 7 and 8 (composite of absolute value and a rational function)

Continuous Extension to a Point

As we saw in Section 1.2, a rational function may have a limit even at a point where its denominator is zero. If $f(c)$ is not defined, but $\lim_{x \to c} f(x) = L$ exists, we can define a new function $F(x)$ by the rule

$$F(x) = \begin{cases} f(x) & \text{if } x \text{ is in the domain of } f \\ L & \text{if } x = c. \end{cases}$$

The function F is continuous at $x = c$. It is called the **continuous extension** of f to $x = c$. For rational functions f, continuous extensions are usually found by canceling common factors.

EXAMPLE 8   Show that

$$f(x) = \frac{x^2 + x - 6}{x^2 - 4}$$

has a continuous extension to $x = 2$, and find that extension.

Solution   Although $f(2)$ is not defined, if $x \neq 2$ we have

$$f(x) = \frac{x^2 + x - 6}{x^2 - 4} = \frac{(x - 2)(x + 3)}{(x - 2)(x + 2)} = \frac{x + 3}{x + 2}.$$

The function

$$F(x) = \frac{x + 3}{x + 2}$$

is equal to $f(x)$ for $x \neq 2$, but is also continuous at $x = 2$, having there the value of 5/4. Thus F is the continuous extension of f to $x = 2$, and

$$\lim_{x \to 2} \frac{x^2 + x - 6}{x^2 - 4} = \lim_{x \to 2} f(x) = \frac{5}{4}.$$

The graph of f is shown in Fig. 1.43. The continuous extension F has the same graph except with no hole at $(2, 5/4)$.

Continuity on Intervals

A function is called **continuous** if it is continuous everywhere in its domain. A function that is not continuous throughout its entire domain may still be continuous when restricted to particular intervals within the domain.

1.43 (a) The graph of

$$f(x) = \frac{x^2 + x - 6}{x^2 - 4}$$

and (b) the graph of its continuous extension

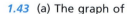

$$F(x) = \frac{x + 3}{x + 2} = \begin{cases} \dfrac{x^2 + x - 6}{x^2 - 4}, & x \neq 2 \\ \dfrac{5}{4}, & x = 2 \end{cases}$$

(Example 8).

A function f is said to be **continuous on an interval** I in its domain if $\lim_{x \to c} f(x) = f(c)$ at every interior point c and if the appropriate one-sided limits equal the function values at any endpoints I may contain. A function continuous on an interval I is automatically continuous on any interval contained in I. Polynomials are continuous on every interval, and rational functions are continuous on every interval on which they are defined.

EXAMPLE 9 *Functions continuous on intervals*

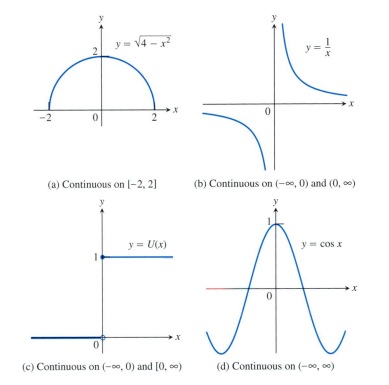

(a) Continuous on [−2, 2]   (b) Continuous on (−∞, 0) and (0, ∞)

(c) Continuous on (−∞, 0) and [0, ∞)   (d) Continuous on (−∞, ∞)

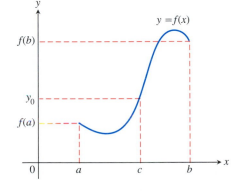

1.44 The function f, being continuous on [a, b], takes on every value between $f(a)$ and $f(b)$.

Functions that are continuous on intervals have properties that make them particularly useful in mathematics and its applications. One of these is the intermediate value property. A function is said to have the **intermediate value property** if it never takes on two values without taking on all the values in between.

Theorem 9
The Intermediate Value Theorem
Suppose $f(x)$ is continuous on an interval I, and a and b are any two points of I. Then if y_0 is a number between $f(a)$ and $f(b)$, there exists a number c between a and b such that $f(c) = y_0$ (Fig. 1.44).

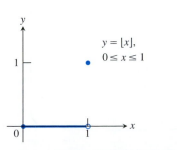

1.45 The function $f(x) = \lfloor x \rfloor$, $0 \le x \le 1$, does not take on any value between $f(0) = 0$ and $f(1) = 1$.

The proof of the Intermediate Value Theorem depends on the completeness property of the real number system and can be found in more advanced texts.

The continuity of f on I is essential to the theorem. If f is discontinuous at even one point of I, the theorem's conclusion may fail, as it does for the function graphed in Fig. 1.45.

A Consequence for Graphing: Connectivity   Theorem 9 is the reason the graph of a function continuous on an interval *I* cannot have any breaks. It will be **connected,** a single, unbroken curve, like the graph of sin *x*. It will not have jumps like the graph of the greatest integer function $\lfloor x \rfloor$ or separate branches like the graph of $1/x$.

The Consequence for Root Finding   We call a solution of the equation $f(x) = 0$ a **root** or **zero** of the function f. The Intermediate Value Theorem tells us that if f is continuous, then any interval on which f changes sign must contain a zero of the function.

This observation is the basis of the way we solve equations of the form $f(x) = 0$ with a graphing calculator or computer grapher (when f is continuous). The solutions are the x-intercepts of the graph of f. We graph the function $y = f(x)$ over a large interval to see roughly where its zeros are. Then we zoom in on the intersection points one at a time to estimate their coordinates. Figure 1.46 shows a typical sequence of steps in a graphical solution of the equation $x^3 - x - 1 = 0$.

Graphical procedures for solving equations and finding zeros of functions, while instructive, are relatively slow. We usually get faster results from numerical methods, as you will see in Section 3.8.

1.46 A graphical solution of the equation $x^3 - x - 1 = 0$. We graph the function $f(x) = x^3 - x - 1$ and, with successive screen enlargements, estimate the coordinates of the point where the graph crosses the x-axis.

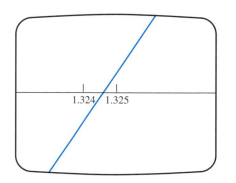

First we make a graph with a relatively large scale. It reveals a root (zero) between $x = 1$ and $x = 2$.

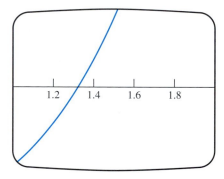

We change the viewing window to $1 \leq x \leq 2, -1 \leq y \leq 1$. We now see that the root lies between 1.3 and 1.4.

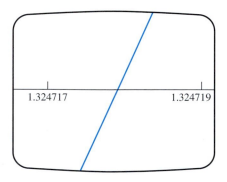

We change the window to $1.3 \leq x \leq 1.35, -0.1 \leq y \leq 0.1$. The root lies between 1.32 and 1.33.

We change the window to $1.32 \leq x \leq 1.33, -0.01 \leq y \leq 0.01$. The root lies between 1.324 and 1.325.

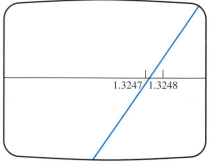

We change the window to $1.324 \leq x \leq 1.325, -0.001 \leq y \leq 0.001$. The root lies between 1.3247 and 1.3248.

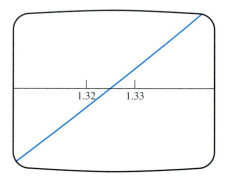

After two more enlargements, we arrive at a screen that shows the root to be approximately 1.324718.

EXAMPLE 10   Is any real number exactly 1 less than its cube?

Solution   This is the question that gave rise to the equation we just solved. Any such number must satisfy the equation $x = x^3 - 1$ or $x^3 - x - 1 = 0$. Hence, we are looking for a zero of $f(x) = x^3 - x - 1$. By trial, we find that $f(1) = -1$ and $f(2) = 5$ and conclude from Theorem 9 that there is at least one number in $[1, 2]$ where f is zero. So, yes, there is a number that is 1 less than its cube, and we just estimated its value graphically to be about 1.3247 18.   ❑

Exercises 1.5

Continuity from Graphs

In Exercises 1–4, say whether the function graphed is continuous on $[-1, 3]$. If not, where does it fail to be continuous and why?

1.   **2.**

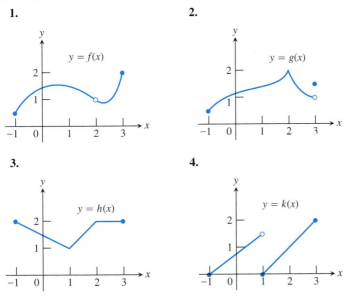

3.   **4.**

Exercises 5–10 are about the function

$$f(x) = \begin{cases} x^2 - 1, & -1 \le x < 0 \\ 2x, & 0 < x < 1 \\ 1, & x = 1 \\ -2x + 4, & 1 < x < 2 \\ 0, & 2 < x < 3 \end{cases}$$

graphed in Fig. 1.47.

5. a)   Does $f(-1)$ exist?
  b)   Does $\lim_{x \to -1^+} f(x)$ exist?
  c)   Does $\lim_{x \to -1^+} f(x) = f(-1)$?
  d)   Is f continuous at $x = -1$?

6. a)   Does $f(1)$ exist?
  b)   Does $\lim_{x \to 1} f(x)$ exist?
  c)   Does $\lim_{x \to 1} f(x) = f(1)$?
  d)   Is f continuous at $x = 1$?

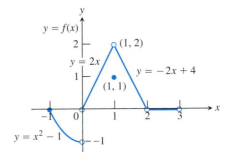

1.47 The graph for Exercises 5–10.

7. a)   Is f defined at $x = 2$? (Look at the definition of f.)
  b)   Is f continuous at $x = 2$?

8. At what values of x is f continuous?

9. What value should be assigned to $f(2)$ to make the extended function continuous at $x = 2$?

10. To what new value should $f(1)$ be changed to remove the discontinuity?

Applying the Continuity Test

At which points do the functions in the following exercises fail to be continuous? At which points, if any, are the discontinuities removable? not removable? Give reasons for your answers.

11. Exercise 1, Section 1.4   **12.** Exercise 2, Section 1.4

At what points are the functions in Exercises 13–28 continuous?

13. $y = \dfrac{1}{x - 2} - 3x$

14. $y = \dfrac{1}{(x + 2)^2} + 4$

15. $y = \dfrac{x + 1}{x^2 - 4x + 3}$

16. $y = \dfrac{x + 3}{x^2 - 3x - 10}$

17. $y = |x - 1| + \sin x$

18. $y = \dfrac{1}{|x| + 1} - \dfrac{x^2}{2}$

19. $y = \dfrac{\cos x}{x}$

20. $y = \dfrac{x + 2}{\cos x}$

21. $y = \csc 2x$

22. $y = \tan \dfrac{\pi x}{2}$

23. $y = \dfrac{x \tan x}{x^2 + 1}$

24. $y = \dfrac{\sqrt{x^4 + 1}}{1 + \sin^2 x}$

25. $y = \sqrt{2x + 3}$

26. $y = \sqrt[4]{3x - 1}$

27. $y = (2x - 1)^{1/3}$

28. $y = (2 - x)^{1/5}$

Limits of Composite Functions

Find the limits in Exercises 29–34.

29. $\lim\limits_{x \to \pi} \sin (x - \sin x)$

30. $\lim\limits_{t \to 0} \sin \left(\dfrac{\pi}{2} \cos(\tan t) \right)$

31. $\lim\limits_{y \to 1} \sec (y \sec^2 y - \tan^2 y - 1)$

32. $\lim\limits_{x \to 0} \tan \left(\dfrac{\pi}{4} \cos (\sin x^{1/3}) \right)$

33. $\lim\limits_{t \to 0} \cos \left(\dfrac{\pi}{\sqrt{19 - 3 \sec 2t}} \right)$

34. $\lim\limits_{x \to \pi/6} \sqrt{\csc^2 x + 5 \sqrt{3} \tan x}$

Continuous Extensions

35. Define $g(3)$ in a way that extends $g(x) = (x^2 - 9)/(x - 3)$ to be continuous at $x = 3$.

36. Define $h(2)$ in a way that extends $h(t) = (t^2 + 3t - 10)/(t - 2)$ to be continuous at $t = 2$.

37. Define $f(1)$ in a way that extends $f(s) = (s^3 - 1)/(s^2 - 1)$ to be continuous at $s = 1$.

38. Define $g(4)$ in a way that extends $g(x) = (x^2 - 16)/(x^2 - 3x - 4)$ to be continuous at $x = 4$.

39. For what value of a is

$$f(x) = \begin{cases} x^2 - 1, & x < 3 \\ 2ax, & x \ge 3 \end{cases}$$

continuous at every x?

40. For what value of b is

$$g(x) = \begin{cases} x, & x < -2 \\ bx^2, & x \ge -2 \end{cases}$$

continuous at every x?

Grapher Explorations—Continuous Extension

In Exercises 41–44, graph the function f to see whether it appears to have a continuous extension to the origin. If it does, use TRACE and ZOOM to find a good candidate for the extended function's value at $x = 0$. If the function does not appear to have a continuous extension, can it be extended to be continuous at the origin from the right or from the left? If so, what do you think the extended function's value(s) should be?

41. $f(x) = \dfrac{10^x - 1}{x}$

42. $f(x) = \dfrac{10^{|x|} - 1}{x}$

43. $f(x) = \dfrac{\sin x}{|x|}$

44. $f(x) = (1 + 2x)^{1/x}$

Theory and Examples

45. A continuous function $y = f(x)$ is known to be negative at $x = 0$ and positive at $x = 1$. Why does the equation $f(x) = 0$ have at least one solution between $x = 0$ and $x = 1$? Illustrate with a sketch.

46. Explain why the equation $\cos x = x$ has at least one solution.

47. Show that the equation $x^3 - 15x + 1 = 0$ has three solutions in the interval $[-4, 4]$.

48. Show that the function $F(x) = (x - a)^2(x - b)^2 + x$ takes on the value $(a + b)/2$ for some value of x.

49. If $f(x) = x^3 - 8x + 10$, show that there are values c for which $f(c)$ equals (a) π; (b) $-\sqrt{3}$; (c) 5,000,000.

50. Explain why the following five statements ask for the same information.

  a) Find the roots of $f(x) = x^3 - 3x - 1$.

  b) Find the x-coordinates of the points where the curve $y = x^3$ crosses the line $y = 3x + 1$.

  c) Find all the values of x for which $x^3 - 3x = 1$.

  d) Find the x-coordinates of the points where the cubic curve $y = x^3 - 3x$ crosses the line $y = 1$.

  e) Solve the equation $x^3 - 3x - 1 = 0$.

51. Give an example of a function $f(x)$ that is continuous for all values of x except $x = 2$, where it has a removable discontinuity. Explain how you know that f is discontinuous at $x = 2$, and how you know the discontinuity is removable.

52. Give an example of a function $g(x)$ that is continuous for all values of x except $x = -1$, where it has a nonremovable discontinuity. Explain how you know that g is discontinuous there and why the discontinuity is not removable.

* 53. * *A function discontinuous at every point.*

  a) Use the fact that every nonempty interval of real numbers contains both rational and irrational numbers to show that the function

$$f(x) = \begin{cases} 1 & \text{if } x \text{ is rational} \\ 0 & \text{if } x \text{ is irrational} \end{cases}$$

  is discontinuous at every point.

  b) Is f right-continuous or left-continuous at any point?

54. If functions $f(x)$ and $g(x)$ are continuous for $0 \le x \le 1$, could $f(x)/g(x)$ possibly be discontinuous at a point of $[0, 1]$? Give reasons for your answer.

55. If the product function $h(x) = f(x) \cdot g(x)$ is continuous at $x = 0$, must $f(x)$ and $g(x)$ be continuous at $x = 0$? Give reasons for your answer.

---

*Asterisk denotes a challenging problem.

56. Give an example of functions f and g, both continuous at $x = 0$, for which the composite $f \circ g$ is discontinuous at $x = 0$. Does this contradict Theorem 8? Give reasons for your answer.

57. Is it true that a continuous function that is never zero on an interval never changes sign on that interval? Give reasons for your answer.

58. Is it true that if you stretch a rubber band by moving one end to the right and the other to the left, some point of the band will end up in its original position? Give reasons for your answer.

59. *A fixed point theorem.* Suppose that a function f is continuous on the closed interval $[0, 1]$ and that $0 \le f(x) \le 1$ for every x in $[0, 1]$. Show that there must exist a number c in $[0, 1]$ such that $f(c) = c$ (c is called a **fixed point** of f).

60. *The sign-preserving property of continuous functions.* Let f be defined on an interval (a, b) and suppose that $f(c) \neq 0$ at some c where f is continuous. Show that there is an interval $(c - \delta, \ c + \delta)$ about c where f has the same sign as $f(c)$. Notice how remarkable this conclusion is. Although f is defined throughout (a, b), it is not required to be continuous at any point except c. That and the condition $f(c) \neq 0$ are enough to make f different from zero (positive or negative) throughout an entire interval.

61. Explain how Theorem 6 follows from Theorem 1 in Section 1.2.

62. Explain how Theorem 7 follows from Theorems 2 and 3 in Section 1.2.

Solving Equations Graphically

Use a graphing calculator or computer grapher to solve the equations in Exercises 63–70.

63. $x^3 - 3x - 1 = 0$

64. $2x^3 - 2x^2 - 2x + 1 = 0$

65. $x(x - 1)^2 = 1$ (one root)

66. $x^x = 2$

67. $\sqrt{x} + \sqrt{1 + x} = 4$

68. $x^3 - 15x + 1 = 0$ (three roots)

69. $\cos x = x$ (one root). Make sure you are using radian mode.

70. $2 \sin x = x$ (three roots). Make sure you are using radian mode.

1.6

Tangent Lines

This section continues the discussion of secants and tangents begun in Section 1.1. We calculate limits of secant slopes to find tangents to curves.

What *Is* a Tangent to a Curve?

For circles, tangency is straightforward. A line L is tangent to a circle at a point P if L passes through P perpendicular to the radius at P (Fig. 1.48). Such a line just *touches* the circle. But what does it mean to say that a line L is tangent to some other curve C at a point P? Generalizing from the geometry of the circle, we might say that it means one of the following.

1. L passes through P perpendicular to the line from P to the center of C.
2. L passes through only one point of C, namely P.
3. L passes through P and lies on one side of C only.

While these statements are valid if C is a circle, none of them work consistently for more general curves. Most curves do not have centers, and a line we may want to call tangent may intersect C at other points or cross C at the point of tangency (Fig. 1.49 on the following page).

To define tangency for general curves, we need a dynamic approach that takes into account the behavior of the secants through P and nearby points Q as Q moves toward P along the curve (Fig. 1.50 on the following page). It goes like this:

1. We start with what we *can* calculate, namely the slope of the secant PQ.
2. Investigate the limit of the secant slope as Q approaches P along the curve.
3. If the limit exists, take it to be the slope of the curve at P and define the tangent to the curve at P to be the line through P with this slope.

This is what we were doing in the fruit fly example in Section 1.1.

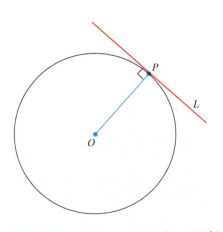

1.48 L is tangent to the circle at P if it passes through P perpendicular to radius OP.

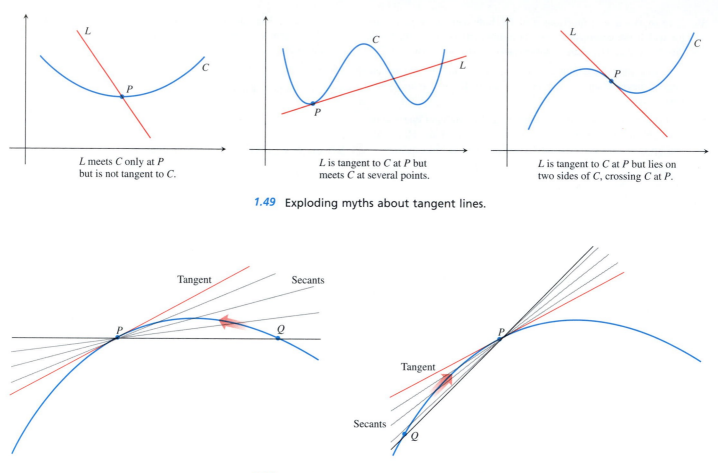

L meets C only at P
but is not tangent to C.

L is tangent to C at P but
meets C at several points.

L is tangent to C at P but lies on
two sides of C, crossing C at P.

1.49 Exploding myths about tangent lines.

Tangent   Secants

Tangent

Secants

1.50 The dynamic approach to tangency. The tangent to the curve at P is the line through P whose slope is the limit of the secant slopes as $Q \to P$ from either side.

How do you find a tangent to a curve?

This was the dominant mathematical question of the early seventeenth century and it is hard to overestimate how badly the scientists of the day wanted to know the answer. In optics, the tangent determined the angle at which a ray of light entered a curved lens. In mechanics, the tangent determined the direction of a body's motion at every point along its path. In geometry, the tangents to two curves at a point of intersection determined the angle at which the curves intersected. Descartes went so far as to say that the problem of finding a tangent to a curve was "the most useful and most general problem not only that I know but even that I have any desire to know."

EXAMPLE 1   Find the slope of the parabola $y = x^2$ at the point $P(2, 4)$. Write an equation for the tangent to the parabola at this point.

Solution   We begin with a secant line through $P(2, 4)$ and $Q(2 + h, (2 + h)^2)$ nearby. We then write an expression for the slope of the secant PQ and investigate what happens to the slope as Q approaches P along the curve:

$$\text{Secant slope} = \frac{\Delta y}{\Delta x} = \frac{(2 + h)^2 - 2^2}{h} = \frac{h^2 + 4h + 4 - 4}{h}$$

$$= \frac{h^2 + 4h}{h} = h + 4.$$

If $h > 0$, Q lies above and to the right of P, as in Fig. 1.51. If $h < 0$, Q lies to the left of P (not shown). In either case, as Q approaches P along the curve, h approaches zero and the secant slope approaches 4:

$$\lim_{h \to 0} (h + 4) = 4.$$

We take 4 to be the parabola's slope at P.

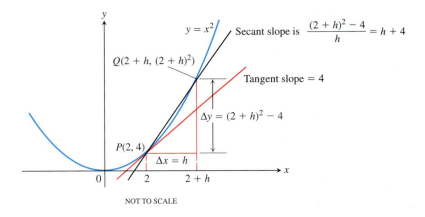

NOT TO SCALE

1.51 Diagram for finding the slope of the parabola $y = x^2$ at the point $P(2, 4)$ (Example 1).

The tangent to the parabola at P is the line through P with slope 4:

$$y = 4 + 4(x - 2) \qquad \text{Point–slope equation}$$

$$y = 4x - 4.$$

Finding a Tangent to the Graph of a Function

To find a tangent to an arbitrary curve $y = f(x)$ at a point $P(x_0, f(x_0))$ we use the same dynamic procedure. We calculate the slope of the secant through P and a point $Q(x_0 + h, f(x_0 + h))$. We then investigate the limit of the slope as $h \to 0$ (Fig. 1.52). If the limit exists, we call it the slope of the curve at P and define the tangent at P to be the line through P having this slope.

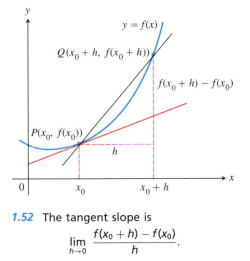

1.52 The tangent slope is

$$\lim_{h \to 0} \frac{f(x_0 + h) - f(x_0)}{h}.$$

> **Definitions**
>
> The **slope of the curve** $y = f(x)$ at the point $P(x_0, f(x_0))$ is the number
>
> $$m = \lim_{h \to 0} \frac{f(x_0 + h) - f(x_0)}{h} \qquad \text{(provided the limit exists).}$$
>
> The **tangent line** to the curve at P is the line through P with this slope.

Whenever we make a new definition it is a good idea to try it on familiar objects to be sure it gives the results we want in familiar cases. The next example shows that the new definition of slope agrees with the old definition when we apply it to nonvertical lines.

> **How to Find the Tangent to the Curve** $y = f(x)$ **at** (x_0, y_0)
>
> 1. Calculate $f(x_0)$ and $f(x_0 + h)$.
> 2. Calculate the slope
> $$m = \lim_{h \to 0} \frac{f(x_0 + h) - f(x_0)}{h}.$$
> 3. If the limit exists, find the tangent line as $y = y_0 + m(x - x_0)$.

EXAMPLE 2 *Testing the definition*

Show that the line $y = mx + b$ is its own tangent at any point $(x_0, mx_0 + b)$.

Solution We let $f(x) = mx + b$ and organize the work into three steps.

Step 1: *Find* $f(x_0)$ *and* $f(x_0 + h)$.

$$f(x_0) = mx_0 + b$$

$$f(x_0 + h) = m(x_0 + h) + b = mx_0 + mh + b$$

Pierre de Fermat (1601–1665)

The dynamic approach to tangency, invented by Fermat in 1629, proved to be one of the seventeenth century's major contributions to calculus.

Fermat, a skilled linguist and one of his century's greatest mathematicians, tended to confine his writing to professional correspondence and to papers written for personal friends. He rarely wrote completed descriptions of his work, even for his personal use. His famous "last theorem" (that $a^n + b^n = c^n$ has no positive integer solutions for a, b, and c if n is an integer greater than 2) is known only from a note he jotted in the margin of a book. His name slipped into relative obscurity until the late 1800s, and it was only from a four-volume edition of his works published at the beginning of this century that the true importance of his many achievements became clear.

Besides the work in physics and number theory for which he is best known, Fermat found the areas under curves as limits of sums of rectangle areas (as we do today) and developed a method for finding the centroids of shapes bounded by curves in the plane. The standard formula for the first derivative of a polynomial function, the formulas for calculating arc length and for finding the area of a surface of revolution, and the second derivative test for extreme values of functions can all be found in his papers. We will see what these are as the text continues.

Step 2: *Find the slope* $\lim\limits_{h\to 0} \ (f(x_0 + h) - f(x_0))/h$.

$$\lim_{h\to 0} \frac{f(x_0 + h) - f(x_0)}{h} = \lim_{h\to 0} \frac{(mx_0 + mh + b) - (mx_0 + b)}{h}$$

$$= \lim_{h\to 0} \frac{mh}{h} = m$$

Step 3: *Find the tangent line using the point–slope equation.* The tangent line at the point $(x, mx_0 + b)$ is

$$y = (mx_0 + b) + m(x - x_0)$$

$$y = mx_0 + b + mx - mx_0$$

$$y = mx + b. \qquad \square$$

EXAMPLE 3

a)   Find the slope of the curve $y = 1/x$ at $x = a$.
b)   Where does the slope equal $-1/4$?
c)   What happens to the tangent to the curve at the point $(a, 1/a)$ as a changes?

Solution

a)   Here $f(x) = 1/x$. The slope at $(a, 1/a)$ is

$$\lim_{h\to 0} \frac{f(a + h) - f(a)}{h} = \lim_{h\to 0} \frac{\dfrac{1}{a+h} - \dfrac{1}{a}}{h}$$

$$= \lim_{h\to 0} \frac{1}{h} \frac{a - (a + h)}{a(a + h)}$$

$$= \lim_{h\to 0} \frac{-h}{ha(a + h)}$$

$$= \lim_{h\to 0} \frac{-1}{a(a + h)} = -\frac{1}{a^2}.$$

Notice how we had to keep writing "$\lim_{h\to 0}$" at the beginning of each line until the stage where we could evaluate the limit by substituting $h = 0$.

b)   The slope of $y = 1/x$ at the point where $x = a$ is $-1/a^2$. It will be $-1/4$ provided

$$-\frac{1}{a^2} = -\frac{1}{4}.$$

This equation is equivalent to $a^2 = 4$, so $a = 2$ or $a = -2$. The curve has slope $-1/4$ at the two points $(2, 1/2)$ and $(-2, -1/2)$ (Fig. 1.53).

c)   Notice that the slope $-1/a^2$ is always negative. As $a \to 0^+$, the slope approaches $-\infty$ and the tangent becomes increasingly steep (Fig. 1.54). We see this again as $a \to 0^-$. As a moves away from the origin, the slope approaches 0^- and the tangent levels off. $\qquad \square$

Rates of Change

The expression

$$\frac{f(x_0 + h) - f(x_0)}{h}$$

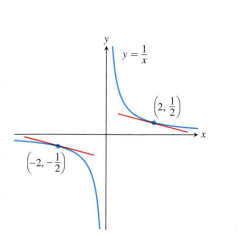

1.53 The two tangent lines to $y = 1/x$ having slope $-1/4$.

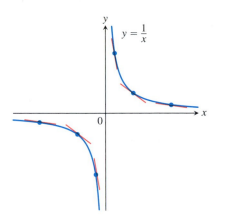

1.54 The tangent slopes, steep near the origin, become more gradual as the point of tangency moves away.

All of these refer to the same thing.

1. The slope of $y = f(x)$ at $x = x_0$
2. The slope of the tangent to $y = f(x)$ at $x = x_0$
3. The rate of change of $f(x)$ with respect to x at $x = x_0$
4. The derivative of f at $x = x_0$
5. $\displaystyle \lim_{h \to 0} \frac{f(x_0 + h) - f(x_0)}{h}$

---

is called the **difference quotient of f at x_0**. If the difference quotient has a limit as h approaches zero, that limit is called the **derivative of f at x_0**. If we interpret the difference quotient as a secant slope, the derivative gives the slope of the curve and tangent at the point where $x = x_0$. If we interpret the difference quotient as an average rate of change, as we did in Section 1.1, the derivative gives the function's rate of change with respect to x at the point $x = x_0$. The derivative is one of the two most important mathematical objects considered in calculus. We will begin a thorough study of it in Chapter 2.

EXAMPLE 4 *Instantaneous speed (Continuation of Section 1.1, Examples 1 and 2)*

In Examples 1 and 2 in Section 1.1, we studied the speed of a rock falling freely from rest near the surface of the earth. We knew that the rock fell $y = 16t^2$ feet during the first t seconds, and we used a sequence of average rates over increasingly short intervals to estimate the rock's speed at the instant $t = 1$. Exactly what *was* the rock's speed at this time?

Solution We let $f(t) = 16t^2$. The average speed of the rock over the interval between $t = 1$ and $t = 1 + h$ seconds was

$$\frac{f(1+h) - f(1)}{h} = \frac{16(1+h)^2 - 16(1)^2}{h} = \frac{16(h^2 + 2h)}{h} = 16(h + 2).$$

The rock's speed at the instant $t = 1$ was

$$\lim_{h \to 0} 16(h + 2) = 16(0 + 2) = 32 \text{ ft/sec}.$$

Our original estimate of 32 ft/sec was right.   ☐

Exercises 1.6

Slopes and Tangent Lines

In Exercises 1–4, use the grid and a straight edge to make a rough estimate of the slope of the curve (in y-units per x-unit) at the points P_1 and P_2. Graphs can shift during a press run, so your estimates may be somewhat different from those in the back of the book.

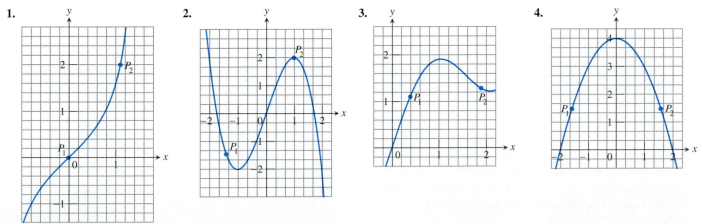

In Exercises 5–10, find an equation for the tangent to the curve at the given point. Then sketch the curve and tangent together.

5. $y = 4 - x^2$, $(-1, 3)$

6. $y = (x - 1)^2 + 1$, $(1, 1)$

7. $y = 2\sqrt{x}$, $(1, 2)$

8. $y = \dfrac{1}{x^2}$, $(-1, 1)$

9. $y = x^3$, $(-2, -8)$

10. $y = \dfrac{1}{x^3}$, $\left(-2, -\dfrac{1}{8}\right)$

In Exercises 11–18, find the slope of the function's graph at the given point. Then find an equation for the line tangent to the graph there.

11. $f(x) = x^2 + 1$, $(2, 5)$

12. $f(x) = x - 2x^2$, $(1, -1)$

13. $g(x) = \dfrac{x}{x - 2}$, $(3, 3)$

14. $g(x) = \dfrac{8}{x^2}$, $(2, 2)$

15. $h(t) = t^3$, $(2, 8)$

16. $h(t) = t^3 + 3t$, $(1, 4)$

17. $f(x) = \sqrt{x}$, $(4, 2)$

18. $f(x) = \sqrt{x + 1}$, $(8, 3)$

In Exercises 19–22, find the slope of the curve at the point indicated.

19. $y = 5x^2$, $x = -1$

20. $y = 1 - x^2$, $x = 2$

21. $y = \dfrac{1}{x - 1}$, $x = 3$

22. $y = \dfrac{x - 1}{x + 1}$, $x = 0$

Tangent Lines with Specified Slopes

At what points do the graphs of the functions in Exercises 23 and 24 have horizontal tangents?

23. $f(x) = x^2 + 4x - 1$

24. $g(x) = x^3 - 3x$

25. Find equations of all lines having slope -1 that are tangent to the curve $y = 1/(x - 1)$.

26. Find an equation of the straight line having slope $1/4$ that is tangent to the curve $y = \sqrt{x}$.

Rates of Change

27. An object is dropped from the top of a 100-m-high tower. Its height aboveground after t seconds is $100 - 4.9t^2$ m. How fast is it falling 2 sec after it is dropped?

28. At t seconds after lift-off, the height of a rocket is $3t^2$ ft. How fast is the rocket climbing after 10 sec?

29. What is the rate of change of the area of a circle ($A = \pi r^2$) with respect to its radius when the radius is $r = 3$?

30. What is the rate of change of the volume of a ball ($V = (4/3)\pi r^3$) with respect to the radius when the radius is $r = 2$?

Testing for Tangents

31. Does the graph of

$$f(x) = \begin{cases} x^2 \sin\ (1/x), & x \neq 0 \\ 0, & x = 0 \end{cases}$$

have a tangent at the origin? Give reasons for your answer.

32. Does the graph of

$$g(x) = \begin{cases} x \sin\ (1/x), & x \neq 0 \\ 0, & x = 0 \end{cases}$$

have a tangent at the origin? Give reasons for your answer.

Vertical Tangents

We say that the curve $y = f(x)$ has a **vertical tangent** at the point where $x = x_0$ if $\lim_{h \to 0}\ (f(x_0 + h) - f(x_0))/h = \infty$ or $-\infty$.

Vertical tangent at $x = 0$:

$$\lim_{h \to 0} \frac{f(0 + h) - f(0)}{h} = \lim_{h \to 0} \frac{h^{1/3} - 0}{h}$$

$$= \lim_{h \to 0} \frac{1}{h^{2/3}} = \infty$$

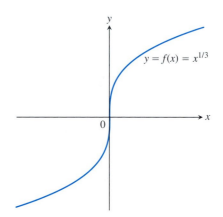

No vertical tangent at $x = 0$:

$$\lim_{h \to 0} \frac{g(0 + h) - g(0)}{h} = \lim_{h \to 0} \frac{h^{2/3} - 0}{h}$$

$$= \lim_{h \to 0} \frac{1}{h^{1/3}}$$

does not exist, because the limit is ∞ from the right and $-\infty$ from the left.

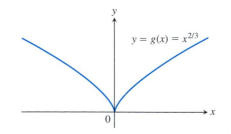

33. Does the graph of

$$f(x) = \begin{cases} -1, & x < 0 \\ 0, & x = 0 \\ 1, & x > 0 \end{cases}$$

have a vertical tangent at the origin? Give reasons for your answer.

34. Does the graph of

$$U(x) = \begin{cases} 0, & x < 0 \\ 1, & x \geq 0 \end{cases}$$

have a vertical tangent at the point $(0, 1)$? Give reasons for your answer.

▦ Grapher Explorations—Vertical Tangents

a) Graph the curves in Exercises 35–44. Where do the graphs appear to have vertical tangents?

b) Confirm your findings in (a) with limit calculations.

35. $y = x^{2/5}$

36. $y = x^{4/5}$

37. $y = x^{1/5}$

38. $y = x^{3/5}$

39. $y = 4x^{2/5} - 2x$

40. $y = x^{5/3} - 5x^{2/3}$

41. $y = x^{2/3} - (x - 1)^{1/3}$

42. $y = x^{1/3} + (x - 1)^{1/3}$

43. $y = \begin{cases} -\sqrt{|x|}, & x \leq 0 \\ \sqrt{x}, & x > 0 \end{cases}$

44. $y = \sqrt{|4 - x|}$

✪ CAS Explorations and Projects

Use a CAS to perform the following steps for the functions in Exercises 45–48.

a) Plot $y = f(x)$ over the interval $x_0 - \dfrac{1}{2} \leq x \leq x_0 + 3$.

b) Define the difference quotient q at x_0 as a function of the general step size h.

c) Find the limit of q as $h \to 0$.

d) Define the secant lines $y = f(x_0) + q^*(x - x_0)$ for $h = 3, 2$, and 1. Graph them together with f and the tangent line over the interval in part (a).

45. $f(x) = x^3 + 2x, \quad x_0 = 0$

46. $f(x) = x + \dfrac{5}{x}, \quad x_0 = 1$

47. $f(x) = x + \sin\ (2x), \quad x_0 = \pi/2$

48. $f(x) = \cos\ x + 4\sin\ (2x), \quad x_0 = \pi$

CHAPTER 1 QUESTIONS TO GUIDE YOUR REVIEW

1. What is the average rate of change of the function $g(t)$ over the interval from $t = a$ to $t = b$? How is it related to a secant line?

2. What limit must be calculated to find the rate of change of a function $g(t)$ at $t = t_0$?

3. Does the existence and value of the limit of a function $f(x)$ as x approaches c ever depend on what happens at $x = c$? Explain, and give examples.

4. What theorems are available for calculating limits? Give examples of how the theorems are used.

5. How are one-sided limits related to limits? How can this relationship sometimes be used to calculate a limit or prove it does not exist? Give examples.

6. How is the problem of controlling the input x of a function f so that the output $y = f(x)$ will be within a certain specified tolerance ϵ of a target value $y_0 = f(x_0)$ related to the problem of proving that f has limit y_0 as $x \to x_0$?

7. What exactly does $\lim_{x \to x_0} f(x) = L$ mean? Give an example in which you find a $\delta > 0$ for a given f, L, x_0, and $\epsilon > 0$ in the formal definition of limit.

8. Give formal definitions of the following statements.

a) $\lim_{x \to 2^-}\ f(x) = 5$

b) $\lim_{x \to 2^+}\ f(x) = 5$

c) $\lim_{x \to 2}\ f(x) = \infty$

d) $\lim_{x \to 2}\ f(x) = -\infty$

9. What conditions must be satisfied by a function if it is to be continuous at an interior point of its domain? at an endpoint?

10. How can looking at the graph of a function help you tell where the function is continuous?

11. What does it mean for a function to be right-continuous at a point? left-continuous? How are continuity and one-sided continuity related?

12. What can be said about the continuity of polynomials? of rational functions? of trigonometric functions? of rational powers and algebraic combinations of functions? of composites of functions? of absolute values of functions?

13. Under what circumstances can you extend a function $f(x)$ to be continuous at a point $x = c$? Give an example.

14. What does it mean for a function to be continuous on an interval?

15. What does it mean for a function to be continuous? Give examples to illustrate the fact that a function that is not continuous on its entire domain may still be continuous on selected intervals within the domain.

16. What property must a function f that is continuous on an interval $[a, b]$ have? Show by examples that f need not have this property if it is discontinuous at some point of the interval.

17. It is often said that a function is continuous if you can draw its graph without having to lift your pen from the paper. Why is that?

18. What does continuity have to do with solving equations?

19. When is a line tangent to a curve C at a point P?

20. What is the significance of the formula
$$\lim_{h \to 0} \frac{f(x+h) - f(x)}{h}?$$

CHAPTER 1 PRACTICE EXERCISES

Limit Calculations and Continuity

1. Graph the function
$$f(x) = \begin{cases} 1, & x \le -1 \\ -x, & -1 < x < 0 \\ 1, & x = 0 \\ -x, & 0 < x < 1 \\ 1, & x \ge 1. \end{cases}$$

Then discuss, in complete detail, limits, one-sided limits, continuity, and one-sided continuity of f at each of the points $x = -1, 0$, and 1. Are any of the discontinuities removable? Explain.

2. Repeat the instructions of Exercise 1 for
$$f(x) = \begin{cases} 0, & x \le -1 \\ 1/x, & 0 < |x| < 1 \\ 0, & x = 1 \\ 1, & x > 1. \end{cases}$$

3. Suppose that $f(x)$ and $g(x)$ are defined for all x and that $\lim_{x \to c} f(x) = -7$ and $\lim_{x \to c} g(x) = 0$. Find the limit as $x \to c$ of the following functions.

a) $3f(x)$

b) $(f(x))^2$

c) $f(x) \cdot g(x)$

d) $\dfrac{f(x)}{g(x) - 7}$

e) $\cos(g(x))$

f) $|f(x)|$

4. Suppose that $f(x)$ and $g(x)$ are defined for all x and that $\lim_{x \to 0} f(x) = 1/2$ and $\lim_{x \to 0} g(x) = \sqrt{2}$. Find the limits as $x \to 0$ of the following functions.

a) $-g(x)$

b) $g(x) \cdot f(x)$

c) $f(x) + g(x)$

d) $1/f(x)$

e) $x + f(x)$

f) $\dfrac{f(x) \cdot \cos x}{x - 1}$

In Exercises 5 and 6, find the value that $\lim_{x \to 0} g(x)$ must have if the given limit statements hold.

5. $\lim_{x \to 0} \left(\dfrac{4 - g(x)}{x} \right) = 1$

6. $\lim_{x \to -4} \left(x \lim_{x \to 0} g(x) \right) = 2$

In Exercises 7–10, find the limit of $g(x)$ as x approaches the indicated value.

7. $\lim_{x \to 0^+} (4g(x))^{1/3} = 2$

8. $\lim_{x \to \sqrt{5}} \dfrac{1}{x + g(x)} = 2$

9. $\lim_{x \to 1} \dfrac{3x^2 + 1}{g(x)} = \infty$

10. $\lim_{x \to -2} \dfrac{5 - x^2}{\sqrt{g(x)}} = 0$

In Exercises 11–18, find the limit or explain why it does not exist.

11. $\lim \dfrac{x^2 - 4x + 4}{x^3 + 5x^2 - 14x}$ (a) as $x \to 0$, (b) as $x \to 2$

12. $\lim \dfrac{x^2 + x}{x^5 + 2x^4 + x^3}$ (a) as $x \to 0$, (b) as $x \to -1$

13. $\lim_{x \to 1} \dfrac{1 - \sqrt{x}}{1 - x}$

14. $\lim_{x \to a} \dfrac{x^2 - a^2}{x^4 - a^4}$

15. $\lim_{h \to 0} \dfrac{(x+h)^2 - x^2}{h}$

16. $\lim_{x \to 0} \dfrac{(x+h)^2 - x^2}{h}$

17. $\lim_{x \to 0} \dfrac{\dfrac{1}{2+x} - \dfrac{1}{2}}{x}$

18. $\lim_{x \to 0} \dfrac{(2+x)^3 - 8}{x}$

19. On what intervals are the following functions continuous?

a) $f(x) = x^{1/3}$

b) $g(x) = x^{3/4}$

c) $h(x) = x^{-2/3}$

d) $k(x) = x^{-1/6}$

20. Can $f(x) = x(x^2 - 1)/|x^2 - 1|$ be extended to be continuous at $x = 1$ or -1? Give reasons for your answers. (Graph the function—you will find the graph interesting.)

▦ Grapher Explorations—Continuous Extensions

In Exercises 21–24, graph the function to see whether it appears to have a continuous extension to the given point a. If it does, use TRACE and ZOOM to find a good candidate for the extended function's value at a. If the function does not appear to have a continuous extension, can it be extended to be continuous from the right or left? If so, what do you think the extended function's value should be?

21. $f(x) = \dfrac{x-1}{x - \sqrt[4]{x}}, \quad a = 1$

22. $g(\theta) = \dfrac{5\cos\theta}{4\theta - 2\pi}, \quad a = \pi/2$

23. $h(t) = (1 + |t|)^{1/t}, \quad a = 0$

24. $k(x) = \dfrac{x}{1 - 2^{|x|}}, \quad a = 0$

Grapher Explorations—Roots

25. Let $f(x) = x^3 - x - 1$.

    a)   Show that f must have a zero between -1 and 2.

    b)   Solve the equation $f(x) = 0$ graphically with an error of at most 10^{-8}.

    c)   It can be shown that the exact value of the solution in (b)

is

$$\left(\frac{1}{2} + \frac{\sqrt{69}}{18}\right)^{1/3} + \left(\frac{1}{2} - \frac{\sqrt{69}}{18}\right)^{1/3}.$$

Evaluate this exact answer and compare it with the value determined in (b).

26. Let $f(x) = x^3 - 2x + 2$.

    a)   Show that f must have a zero between -2 and 0.

    b)   Solve the equation $f(x) = 0$ graphically with an error of at most 10^{-4}.

    c)   It can be shown that the exact value of the solution in (b) is

$$\left(\sqrt{\frac{19}{27}} - 1\right)^{1/3} - \left(\sqrt{\frac{19}{27}} + 1\right)^{1/3}.$$

Evaluate this exact answer and compare it with the value determined in (b).

CHAPTER   **1**   ADDITIONAL EXERCISES–THEORY, EXAMPLES, APPLICATIONS

---

1. a) If $\lim_{x \to c} f(x) = 5$, must $f(c) = 5$?

   b) If $f(c) = 5$, must $\lim_{x \to c} f(x) = 5$?

Give reasons for your answers.

2. Can $\lim_{x \to c} (f(x)/g(x))$ exist even if $\lim_{x \to c} f(c) = 0$ and $\lim_{x \to c} g(x) = 0$? Give reasons for your answer.

3. *Assigning a value to 0^0.* The rules of exponents tell us that $a^0 = 1$ if a is any number different from zero. They also tell us that $0^n = 0$ if n is any positive number.

    If we tried to extend these rules to include the case 0^0, we would get conflicting results. The first rule would say $0^0 = 1$ while the second would say $0^0 = 0$.

    We are not dealing with a question of right or wrong here. Neither rule applies as it stands, so there is no contradiction. We could, in fact, define 0^0 to have any value we wanted as long as we could persuade others to agree.

    What value would you like 0^0 to have? Here are two examples that might help you to decide. (See Exercise 4 for another example.)

   a) CALCULATOR Calculate x^x for $x = 0.1, 0.01, 0.001$, and so on as far as your calculator can go. Write down the value you get each time. What pattern do you see?

   b) GRAPHER Graph the function $y = x^x$ (as $y = x\hat{\ }x$) for $0 \le x \le 1$. Even though the function is not defined for $x \le 0$, the graph will approach the y-axis from the right. Toward what y-value does it seem to be headed? Zoom in to estimate the value more closely. What do you think it is?

4. *A reason you might want 0^0 to be something other than 0 or 1.* As the number x increases through positive values, the numbers $1/x$ and $1/(\ln\, x)$ both approach zero. What happens to the number

$$f(x) = \left(\frac{1}{x}\right)^{1/(\ln\, x)}$$

as x increases? Here are two ways to find out.

   a) CALCULATOR Evaluate f for $x = 10, 100, 1000$, and so on, as far as your calculator can reasonably go. What pattern do you see?

   b) GRAPHER Graph f in a variety of graphing windows, including windows that contain the origin. What do you see? Use TRACE to read y-values along the graph. What do you find? Chapter 6 will explain what is going on.

5. *Lorentz contraction.* In relativity theory the length of an object, say a rocket, appears, to an observer, to depend on the speed at which the object is traveling with respect to the observer. If the observer measures the rocket's length as L_0 at rest, then at speed v the rocket's length will appear to be

$$L = L_0 \sqrt{1 - \frac{v^2}{c^2}}.$$ The Lorentz contraction formula.

Here, $c \approx 3 \times 10^8$ m/sec is the speed of light in a vacuum. What happens to L as v increases? Find $\lim_{v \to c^-} L$. Why was the left-hand limit needed?

6. *Roots of a quadratic equation that is almost linear.* The equation $ax^2 + 2x - 1 = 0$, where a is a constant, has two roots if $a > -1$ and $a \neq 0$, one positive and one negative:

$$r_+(a) = \frac{-1 + \sqrt{1+a}}{a}, \quad r_-(a) = \frac{-1 - \sqrt{1+a}}{a}.$$

a) What happens to $r_+(a)$ as $a \to 0$? as $a \to -1^+$?

b) What happens to $r_-(a)$ as $a \to 0$? as $a \to -1^+$?

c) **GRAPHER** Support your conclusions by graphing $r_+(a)$ and $r_-(a)$ as functions of a. Describe what you see.

d) **GRAPHER** For added support, graph $f(x) = ax^2 + 2x - 1$ simultaneously for $a = 1, 0.5, 0.2, 0.1,$ and 0.05.

7. If $\lim_{x \to 0^+} f(x) = A$ and $\lim_{x \to 0^-} f(x) = B$, find

a) $\lim_{x \to 0^+} f(x^3 - x)$

b) $\lim_{x \to 0^-} f(x^3 - x)$

c) $\lim_{x \to 0^+} f(x^2 - x^4)$

d) $\lim_{x \to 0^-} f(x^2 - x^4)$

8. Which of the following statements are true, and which are false? If true, say why; if false, give a counterexample (that is, an example confirming the falsehood).

a) If $\lim_{x \to a} f(x)$ exists but $\lim_{x \to a} g(x)$ does not exist, then $\lim_{x \to a} (f(x) + g(x))$ does not exist.

b) If neither $\lim_{x \to a} f(x)$ nor $\lim_{x \to a} g(x)$ exists, then $\lim_{x \to a} (f(x) + g(x))$ does not exist.

c) If f is continuous at a, then so is $|f|$.

d) If $|f|$ is continuous at a, then so is f.

9. Show that the equation $x + 2\cos x = 0$ has at least one solution.

10. Explain why the function $f(x) = \sin(1/x)$ has no continuous extension to $x = 0$.

11. *Controlling the flow from a draining tank.* Torricelli's law says that if you drain a tank like the one in the figure below, the rate y at which water runs out is a constant times the square root of the water's depth x. The constant depends on the size of the exit valve. Suppose that $y = \sqrt{x}/2$ for a certain tank. You are trying to maintain a fairly constant exit rate by pouring more water into the tank with a hose from time to time. How deep must you keep the water if you want to maintain the exit rate (a) within 0.2 ft³/min of the rate $y_0 = 1$ ft³/min? (b) within 0.1 ft³/min of the rate $y_0 = 1$ ft³/min?

Exit rate y ft³/min

12. *Thermal expansion in precise equipment.* As you may know, most metals expand when heated and contract when cooled. The dimensions of a piece of laboratory equipment are sometimes so critical that the temperature in the shop where it is made and the laboratory where it is used must not be allowed to vary. A typical aluminum bar that is 10 cm wide at 70°F will be

$$y = 10 + (t - 70) \times 10^{-4}$$

centimeters wide at a nearby temperature t. Suppose you are using a bar like this in a gravity wave detector, where its width must stay within 0.0005 cm of the ideal 10 cm. How close to $t_0 = 70°F$ must you maintain the temperature to ensure that this tolerance is not exceeded?

13. *Antipodal points.* Is there any reason to believe that there is always a pair of antipodal (diametrically opposite) points on the earth's equator where the temperatures are the same? Explain.

14. *Uniqueness of limits.* Show that a function cannot have two different limits at the same point. That is, if $\lim_{x \to x_0} f(x) = L_1$ and $\lim_{x \to x_0} f(x) = L_2$, then $L_1 = L_2$.

In Exercises 15–18, use the formal definition of limit to prove that the function is continuous at x_0.

15. $f(x) = x^2 - 7, \quad x_0 = 1$

16. $g(x) = 1/(2x), \quad x_0 = 1/4$

17. $h(x) = \sqrt{2x - 3}, \quad x_0 = 2$

18. $F(x) = \sqrt{9 - x}, \quad x_0 = 5$

In Exercises 19 and 20, use the formal definition of limit to prove that the function has a continuous extension to the given value of x.

19. $f(x) = \dfrac{x^2 - 1}{x + 1}, \quad x = -1$

20. $g(x) = \dfrac{x^2 - 2x - 3}{2x - 6}, \quad x = 3$

21. *Max {a, b} and min {a, b}.*

a) Show that the expression

$$\max \; \{a, b\} = \frac{a + b}{2} + \frac{|a - b|}{2}$$

equals a if $a \geq b$ and equals b if $b \geq a$. In other words, max (a, b) gives the larger of the two numbers a and b.

b) Find a similar expression for min {a, b}, the smaller of a and b.

*22. *A function continuous at only one point.* Let

$$f(x) = \begin{cases} x & \text{if } x \text{ is rational} \\ 0 & \text{if } x \text{ is irrational.} \end{cases}$$

a) Show that f is continuous at $x = 0$.

b) Use the fact that every nonempty open interval of real numbers contains both rational and irrational numbers to show that f is not continuous at any nonzero value of x.

*23. *Bounded functions.* A real-valued function f is **bounded from above** on a set D if there exists a number N such that $f(x) \le N$ for all x in D. We call N, when it exists, an **upper bound** for f on D and say that f is bounded from above by N. In a similar manner, we say that f is **bounded from below** on D if there exists a number M such that $f(x) \ge M$ for all x in D. We call M, when it exists, a **lower bound** for f on D and say that f is bounded from below by M. We say that f is **bounded** on D if it is bounded from both above and below.

a) Show that f is bounded on D if and only if there exists a number B such that $|f(x)| \le B$ for all x in D.

b) Suppose that f is bounded from above by N. Show that if $\lim_{x \to x_0} f(x) = L$, then $L \le N$.

c) Suppose that f is bounded from below by M. Show that if $\lim_{x \to x_0} f(x) = L$, then $L \ge M$.

*24. *The Dirichlet ruler function.* If x is a rational number, then x can be written in a unique way as a quotient of integers m/n where $n > 0$ and m and n have no common factors greater than 1. (We say that such a fraction is in *lowest terms*. For example, $6/4$ written in lowest terms is $3/2$.) Let $f(x)$ be defined for all x in the interval $[0, 1]$ by

$$f(x) = \begin{cases} 1/n & \text{if } x = m/n \text{ is a rational number in lowest terms} \\ 0 & \text{if } x \text{ is irrational.} \end{cases}$$

For instance, $f(0) = f(1) = 1$, $f(1/2) = 1/2$, $f(1/3) = f(2/3) = 1/3$, $f(1/4) = f(3/4) = 1/4$, and so on.

a) Show that f is discontinuous at every rational number in $[0, 1]$.

b) Show that f is continuous at every irrational number in $[0, 1]$. (*Hint:* If ϵ is a given positive number, show that there are only finitely many rational numbers r in $[0, 1]$ such that $f(r) \ge \epsilon$.)

c) Sketch the graph of f. Why do you think f is called the "ruler function"?

CHAPTER 2

Derivatives

OVERVIEW   In Chapter 1 we defined the slope of a curve at a point as the limit of secant slopes. This limit, called a derivative, measures the rate at which a function changes and is one of the most important ideas in calculus. Derivatives are used widely in science, economics, medicine, and computer science to calculate velocity and acceleration, to explain the behavior of machinery, to estimate the drop in water levels as water is pumped out of a tank, and to predict the consequences of making errors in measurements. Finding derivatives by evaluating limits can be lengthy and difficult. In this chapter we develop techniques to make calculating derivatives easier.

2.1 The Derivative of a Function

At the end of Chapter 1, we defined the slope of a curve $y = f(x)$ at the point where $x = x_0$ to be

$$m = \lim_{h \to 0} \frac{f(x_0 + h) - f(x_0)}{h}.$$

We called this limit, when it existed, the derivative of f at x_0. In this section, we investigate the derivative as a *function* derived from f by considering the limit at each point of f's domain.

> **Definition**
>
> The **derivative** of the function f with respect to the variable x is the function f' whose value at x is
>
> $$f'(x) = \lim_{h \to 0} \frac{f(x + h) - f(x)}{h},$$
>
> provided the limit exists.

The domain of f', the set of points in the domain of f for which the limit exists, may be smaller than the domain of f. If $f'(x)$ exists, we say that f **has a derivative (is differentiable)** at x.

109

Notation

There are many ways to denote the derivative of a function $y = f(x)$. Besides $f'(x)$, the most common notations are these:

y'	"y prime"	Nice and brief but does not name the independent variable
$\dfrac{dy}{dx}$	"dy dx"	Names the variables and uses d for derivative
$\dfrac{df}{dx}$	"df dx"	Emphasizes the function's name
$\dfrac{d}{dx} f(x)$	"ddx of f(x)"	Emphasizes the idea that differentiation is an operation performed on f (Fig. 2.1)
$D_x f$	"dx of f"	A common operator notation
$\dot{y}$	"y dot"	One of Newton's notations, now common for time derivatives

We also read dy/dx as "the derivative of y with respect to x," and df/dx and $(d/dx) f(x)$ as "the derivative of f with respect to x."

Why all these notations?

The "prime" notations y' and f' come from notations that Newton used for derivatives. The d/dx notations are similar to those used by Leibniz. Each has its own strengths and weaknesses.

Input: function $y = f(x)$ → Operation $\dfrac{d}{dx}$ → Output: derivative $y' = \dfrac{df}{dx}$

2.1 Flow diagram for the operation of taking a derivative with respect to x.

Calculating Derivatives from the Definition

The process of calculating a derivative is called **differentiation**. Examples 2 and 3 of Section 1.6 illustrate the process for the functions $y = mx + b$ and $y = 1/x$. Example 2 shows that

$$\frac{d}{dx}(mx + b) = m.$$

In Example 3, we see that

$$\frac{d}{dx}\left(\frac{1}{x}\right) = -\frac{1}{x^2}.$$

Here are two more examples.

Steps for Calculating $f'(x)$ from the Definition of Derivative

1. Write expressions for $f(x)$ and $f(x + h)$.
2. Expand and simplify the difference quotient
$$\frac{f(x + h) - f(x)}{h}.$$
3. Using the simplified quotient, find $f'(x)$ by evaluating the limit
$$f'(x) = \lim_{h \to 0} \frac{f(x + h) - f(x)}{h}.$$

EXAMPLE 1

a) Differentiate $f(x) = \dfrac{x}{x - 1}$.

b) Where does the curve $y = f(x)$ have slope -1?

Solution

a) We take the three steps listed in the margin.

Step 1: Here we have $f(x) = \dfrac{x}{x - 1}$

and

$$f(x + h) = \frac{(x + h)}{(x + h) - 1}, \text{ so}$$

2.2 $y' = -1$ at $x = 0$ and $x = 2$.

Step 2: $\dfrac{f(x+h) - f(x)}{h} = \dfrac{\dfrac{x+h}{x+h-1} - \dfrac{x}{x-1}}{h}$

$\qquad = \dfrac{1}{h} \cdot \dfrac{(x+h)(x-1) - x(x+h-1)}{(x+h-1)(x-1)}$

$\qquad\qquad\qquad\qquad\qquad\qquad\qquad \dfrac{a}{b} - \dfrac{c}{d} = \dfrac{ad-cb}{bd}$

$\qquad = \dfrac{1}{h} \cdot \dfrac{-h}{(x+h-1)(x-1)}$, and

Step 3: $f'(x) = \lim\limits_{h\to 0} \dfrac{-1}{(x+h-1)(x-1)} = \dfrac{-1}{(x-1)^2}$.

b) The slope of $y = f(x)$ will be -1 provided

$$-\frac{1}{(x-1)^2} = -1.$$

This equation is equivalent to $(x-1)^2 = 1$, so $x = 2$ or $x = 0$ (Fig. 2.2). ❑

EXAMPLE 2

a) Find the derivative of $y = \sqrt{x}$ for $x > 0$.

b) Find the tangent line to the curve $y = \sqrt{x}$ at $x = 4$.

Solution

You will often need to know the derivative of $\sqrt{x}$ for $x > 0$:

$$\frac{d}{dx}\sqrt{x} = \frac{1}{2\sqrt{x}}.$$

Try to remember it.

a) **Step 1:** $f(x) = \sqrt{x}$ and $f(x+h) = \sqrt{x+h}$

Step 2: $\dfrac{f(x+h) - f(x)}{h} = \dfrac{\sqrt{x+h} - \sqrt{x}}{h}$

$\qquad\qquad\qquad\qquad$ Multiply by $\dfrac{\sqrt{x+h} + \sqrt{x}}{\sqrt{x+h} + \sqrt{x}}$

$\qquad = \dfrac{(x+h) - x}{h(\sqrt{x+h} + \sqrt{x})}$

$\qquad = \dfrac{1}{\sqrt{x+h} + \sqrt{x}}$

Step 3: $f'(x) = \lim\limits_{h\to 0} \dfrac{1}{\sqrt{x+h} + \sqrt{x}} = \dfrac{1}{2\sqrt{x}}$

See Fig. 2.3.

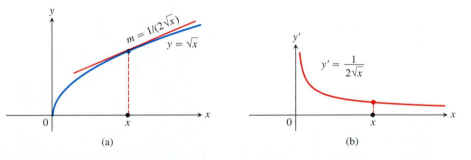

(a)   (b)

2.3 The graphs of (a) $y = \sqrt{x}$ and (b) $y' = 1/(2\sqrt{x})$, $x > 0$ (Example 2). The function is defined at $x = 0$, but its derivative is not.

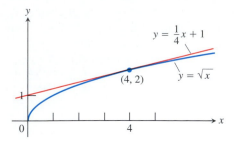

2.4 The curve $y = \sqrt{x}$ and its tangent at (4, 2). The tangent's slope is found by evaluating dy/dx at $x = 4$ (Example 2).

The symbol for evaluation

In addition to

$$f'(a) = \lim_{h \to 0} \frac{f(a+h) - f(a)}{h},$$

the value of the derivative of $y = f(x)$ with respect to x at $x = a$ can be denoted in the following ways:

$$y' \big|_{x=a} = \frac{dy}{dx}\bigg|_{x=a} = \frac{d}{dx} f(x)\bigg|_{x=a}.$$

Here the symbol $|_{x=a}$, called an **evaluation symbol,** tells us to evaluate the expression to its left at $x = a$.

---

b) The slope of the curve at $x = 4$ is

$$\frac{dy}{dx}\bigg|_{x=4} = \frac{1}{2\sqrt{x}}\bigg|_{x=4} = \frac{1}{4}.$$

The tangent is the line through the point (4, 2) with slope 1/4 (Fig. 2.4).

$$y = 2 + \frac{1}{4}(x - 4)$$

$$y = \frac{1}{4}x + 1$$

Graphing f' from Estimated Values

When we measure the values of a function $y = f(x)$ in the laboratory or in the field (pressure vs. temperature, say, or population vs. time) we usually connect the data points with lines or curves to picture the graph of f. We can often make a reasonable plot of f' by estimating slopes on this graph. The following examples show how this is done and what can be learned from the process.

EXAMPLE 3 *Medicine*

On April 23, 1988, the human-powered airplane *Daedalus* flew a record-breaking 119 km from Crete to the island of Santorini in the Aegean Sea, southeast of mainland Greece. During the 6-h endurance tests before the flight, researchers monitored the prospective pilots' blood-sugar concentrations. The concentration graph for one of the athlete-pilots is shown in Fig. 2.5(a), where the concentration in milligrams/deciliter is plotted against time in hours.

The graph is made of line segments connecting data points. The constant slope of each segment gives an estimate of the derivative of the concentration between measurements. We calculated the slope of each segment from the coordinate grid and plotted the derivative as a step function in Fig. 2.5(b). To make the plot for the first hour, for instance, we observed that the concentration increased from about 79 mg/dL to 93 mg/dL. The net increase was $\Delta y = 93 - 79 = 14$ mg/dL. Dividing this by $\Delta t = 1$ h gave the rate of change as

$$\frac{\Delta y}{\Delta t} = \frac{14}{1} = 14 \text{ mg/dL per h.}$$

Daedalus's flight path on April 23, 1988.

2.1   The Derivative of a Function   **113**

(a)

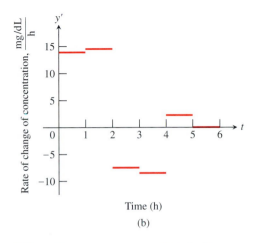

Time (h)

(b)

2.5 (a) The sugar concentration in the blood of a *Daedalus* pilot during a 6-h preflight endurance test. (b) The derivative of the pilot's blood-sugar concentration shows how rapidly the concentration rose and fell during various portions of the test. (Source: *The Daedalus Project: Physiological Problems and Solutions* by Ethan R. Nadel and Steven R. Bussolari, *American Scientist*, Vol. 76, No. 4, July–August 1988, p. 358.)

2.6 We made the graph of $y' = f'(x)$ in (b) by plotting slopes from the graph of $y = f(x)$ in (a). The vertical coordinate of B' is the slope at B, and so on. The graph of $y' = f'(x)$ is a visual record of how the slope of f changes with x.

Notice that we can make no estimate of the concentration's rate of change at times $t = 1, 2, \ldots, 5$, where the graph we have drawn for the concentration has a corner and no slope. The derivative step function is not defined at these times.

When we have so many data that the graph we get by connecting the data points resembles a smooth curve, we may wish to plot the derivative as a smooth curve. The next example shows how this is done.

EXAMPLE 4   Graph the derivative of the function $y = f(x)$ in Fig. 2.6(a).

Solution   We draw a pair of axes, marking the horizontal axis in x-units and the vertical axis in y'-units (Fig. 2.6b). Next we sketch tangents to the graph of f at frequent intervals and use their slopes to estimate the values of $y' = f'(x)$ at these points. We plot the corresponding (x, y') pairs and connect them with a smooth curve.

From the graph of $y' = f'(x)$ we see at a glance

1. where f's rate of change is positive, negative, or zero;
2. the rough size of the growth rate at any x and its size in relation to the size of $f(x)$;
3. where the rate of change itself is increasing or decreasing.   □

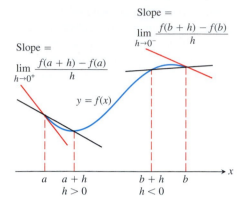

Slope $=$
$$\lim_{h \to 0^-} \frac{f(b + h) - f(b)}{h}$$

Slope $=$
$$\lim_{h \to 0^+} \frac{f(a + h) - f(a)}{h}$$

$y = f(x)$

$a \quad a + h$
$h > 0$

$b + h \quad b$
$h < 0$

2.7 Derivatives at endpoints are one-sided limits.

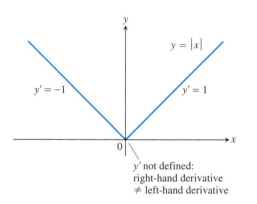

y

$y = |x|$

$y' = -1$

$y' = 1$

0

y' not defined:
right-hand derivative
$\neq$ left-hand derivative

2.8 Not differentiable at the origin.

Differentiable on an Interval; One-sided Derivatives

A function $y = f(x)$ is **differentiable** on an open interval (finite or infinite) if it has a derivative at each point of the interval. It is differentiable on a closed interval $[a, b]$ if it is differentiable on the interior (a, b) and if the limits

$$\lim_{h \to 0^+} \frac{f(a + h) - f(a)}{h} \qquad \textbf{Right-hand derivative at } a$$

$$\lim_{h \to 0^-} \frac{f(b + h) - f(b)}{h} \qquad \textbf{Left-hand derivative at } b$$

exist at the endpoints (Fig. 2.7).

Right-hand and left-hand derivatives may be defined at any point of a function's domain. The usual relation between one-sided and two-sided limits holds for these derivatives. Because of Theorem 5, Section 1.4, a function has a derivative at a point if and only if it has left-hand and right-hand derivatives there, and these one-sided derivatives are equal.

EXAMPLE 5   The function $y = |x|$ is differentiable on $(-\infty, 0)$ and $(0, \infty)$ but has no derivative at $x = 0$. To the right of the origin,

$$\frac{d}{dx}(|x|) = \frac{d}{dx}(x) = \frac{d}{dx}(1 \cdot x) = 1. \qquad \frac{d}{dx}(mx + b) = m$$

To the left,

$$\frac{d}{dx}(|x|) = \frac{d}{dx}(-x) = \frac{d}{dx}(-1 \cdot x) = -1$$

(Fig. 2.8). There can be no derivative at the origin because the one-sided derivatives differ there:

$$\text{Right-hand derivative of } |x| \text{ at zero} = \lim_{h \to 0^+} \frac{|0 + h| - |0|}{h} = \lim_{h \to 0^+} \frac{|h|}{h}$$

$$= \lim_{h \to 0^+} \frac{h}{h} \qquad |h| = h \text{ when } h > 0$$

$$= \lim_{h \to 0^+} 1 = 1$$

$$\text{Left-hand derivative of } |x| \text{ at zero} = \lim_{h \to 0^-} \frac{|0 + h| - |0|}{h} = \lim_{h \to 0^-} \frac{|h|}{h}$$

$$= \lim_{h \to 0^-} \frac{-h}{h} \qquad |h| = -h \text{ when } h < 0$$

$$= \lim_{h \to 0^-} -1 = -1. \qquad \square$$

When Does a Function *Not* Have a Derivative at a Point?

A function has a derivative at a point x_0 if the slopes of the secant lines through $P(x_0, f(x_0))$ and a nearby point Q on the graph approach a limit as Q approaches P. Whenever the secants fail to take up a limiting position or become vertical as Q approaches P, the derivative does not exist. A function whose graph is otherwise smooth will fail to have a derivative at a point where the graph has

1. a *corner,* where the one-sided derivatives differ

2. a *cusp,* where the slope of PQ approaches ∞ from one side and $-\infty$ from the other

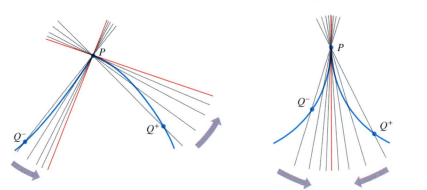

3. a *vertical tangent,* where the slope of PQ approaches ∞ from both sides or approaches $-\infty$ from both sides (here, $-\infty$)

4. a *discontinuity.*

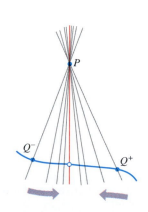

How rough can the graph of a continuous function be?

The absolute value function fails to be differentiable at a single point. Using a similar idea, we can use a sawtooth graph to define a continuous function that fails to have a derivative at infinitely many points.

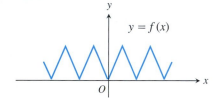

But can a continuous function fail to have a derivative at *every* point?

The answer, surprisingly enough, is yes, as Karl Weierstrass (1815–1897) found in 1872. One of his formulas (there are many like it) was

$$f(x) = \sum_{n=0}^{\infty} \left(\frac{2}{3}\right)^n \cos(9^n \pi x),$$

a formula that expresses f as an infinite sum of cosines with increasingly higher frequencies. By adding wiggles to wiggles infinitely many times, so to speak, the formula produces a graph that is too bumpy in the limit to have a tangent anywhere.

Continuous curves that fail to have a tangent anywhere play a useful role in chaos theory, in part because there is no way to assign a finite length to such a curve. We will see what length has to do with derivatives when we get to Section 5.5.

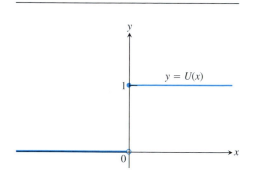

2.9 The unit step function does not have the intermediate value property and cannot be the derivative of a function on the real line.

Differentiable Functions Are Continuous

A function is continuous at every point where it has a derivative.

Theorem 1

If f has a derivative at $x = c$, then f is continuous at $x = c$.

Proof Given that $f'(c)$ exists, we must show that $\lim_{x \to c} f(x) = f(c)$, or, equivalently, that $\lim_{h \to 0} f(c + h) = f(c)$. If $h \neq 0$, then

$$f(c + h) = f(c) + (f(c + h) - f(c))$$

$$= f(c) + \frac{f(c + h) - f(c)}{h} \cdot h.$$

Now take limits as $h \to 0$. By Theorem 1 of Section 1.2,

$$\lim_{h \to 0} f(c + h) = \lim_{h \to 0} f(c) + \lim_{h \to 0} \frac{f(c + h) - f(c)}{h} \cdot \lim_{h \to 0} h$$

$$= f(c) + f'(c) \cdot 0$$

$$= f(c) + 0$$

$$= f(c). \qquad \square$$

Similar arguments with one-sided limits show that if f has a derivative from one side (right or left) at $x = c$, then f is continuous from that side at $x = c$.

Caution The converse of Theorem 1 is false. A function need not have a derivative at a point where it is continuous, as we saw in Example 5.

The Intermediate Value Property of Derivatives

Not every function can be some function's derivative, as we see from the following theorem.

Theorem 2

If a and b are any two points in an interval on which f is differentiable, then f' takes on every value between $f'(a)$ and $f'(b)$.

Theorem 2 (which we will not prove) says that a function cannot *be* a derivative on an interval unless it has the intermediate value property there (Fig. 2.9). The question of when a function is a derivative is one of the central questions in all calculus, and Newton's and Leibniz's answer to this question revolutionized the world of mathematics. We will see what their answer was when we reach Chapter 4.

Exercises 2.1

Finding Derivative Functions and Values

Using the definition, calculate the derivatives of the functions in Exercises 1–6. Then find the values of the derivatives as specified.

1. $f(x) = 4 - x^2$; $f'(-3), f'(0), f'(1)$

2. $F(x) = (x - 1)^2 + 1$; $F'(-1), F'(0), F'(2)$

3. $g(t) = \dfrac{1}{t^2}$; $g'(-1), g'(2), g'(\sqrt{3})$

4. $k(z) = \dfrac{1 - z}{2z}$; $k'(-1), k'(1), k'(\sqrt{2})$

5. $p(\theta) = \sqrt{3\theta}$; $p'(1), p'(3), p'(2/3)$

6. $r(s) = \sqrt{2s + 1}$; $r'(0), r'(1), r'(1/2)$

In Exercises 7–12, find the indicated derivatives.

7. $\dfrac{dy}{dx}$ if $y = 2x^3$

8. $\dfrac{dr}{ds}$ if $r = \dfrac{s^3}{2} + 1$

9. $\dfrac{ds}{dt}$ if $s = \dfrac{t}{2t + 1}$

10. $\dfrac{dv}{dt}$ if $v = t - \dfrac{1}{t}$

11. $\dfrac{dp}{dq}$ if $p = \dfrac{1}{\sqrt{q + 1}}$

12. $\dfrac{dz}{dw}$ if $z = \dfrac{1}{\sqrt{3w - 2}}$

Slopes and Tangent Lines

In Exercises 13–16, differentiate the functions and find the slope of the tangent line at the given value of the independent variable.

13. $f(x) = x + \dfrac{9}{x}$, $x = -3$

14. $k(x) = \dfrac{1}{2 + x}$, $x = 2$

15. $s = t^3 - t^2$, $t = -1$

16. $y = (x + 1)^3$, $x = -2$

In Exercises 17–18, differentiate the functions. Then find an equation of the tangent line at the indicated point on the graph of the function.

17. $y = f(x) = \dfrac{8}{\sqrt{x - 2}}$, $(x, y) = (6, 4)$

18. $w = g(z) = 1 + \sqrt{4 - z}$, $(z, w) = (3, 2)$

In Exercises 19–22, find the values of the derivatives.

19. $\left.\dfrac{ds}{dt}\right|_{t=-1}$ if $s = 1 - 3t^2$

20. $\left.\dfrac{dy}{dx}\right|_{x=\sqrt{3}}$ if $y = 1 - \dfrac{1}{x}$

21. $\left.\dfrac{dr}{d\theta}\right|_{\theta=0}$ if $r = \dfrac{2}{\sqrt{4 - \theta}}$

22. $\left.\dfrac{dw}{dz}\right|_{z=4}$ if $w = z + \sqrt{z}$

An Alternative Formula for Calculating Derivatives

The formula for the secant slope whose limit leads to the derivative depends on how the points involved are labeled. In the notation of Fig. 2.10, the secant slope is $(f(x) - f(c))/(x - c)$ and the slope of the curve at P is

$$f'(c) = \lim_{x \to c} \frac{f(x) - f(c)}{x - c}.$$

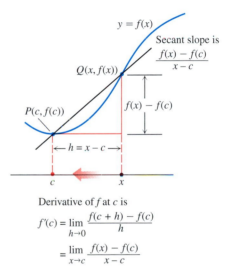

Derivative of f at c is

$$f'(c) = \lim_{h \to 0} \frac{f(c + h) - f(c)}{h}$$

$$= \lim_{x \to c} \frac{f(x) - f(c)}{x - c}$$

2.10 The way we write the difference quotient for the derivative of a function f depends on how we label the points involved.

The use of this formula simplifies some derivative calculations. Use it in Exercises 23–26 to find the derivative of the function at the given value of c.

23. $f(x) = \dfrac{1}{x + 2}$, $c = -1$

24. $f(x) = \dfrac{1}{(x - 1)^2}$, $c = 2$

25. $g(t) = \dfrac{t}{t - 1}$, $c = 3$

26. $k(s) = 1 + \sqrt{s}$, $c = 9$

Graphs

Match the functions graphed in Exercises 27–30 with the derivatives graphed in Fig. 2.11.

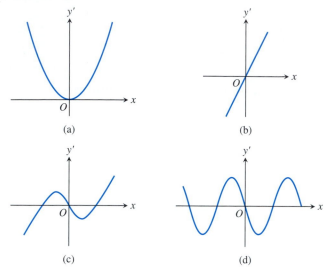

(a)   (b)

(c)   (d)

2.11 The derivative graphs for Exercises 27–30.

27.   **28.**

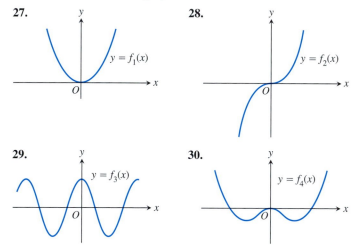

$y = f_1(x)$ $y = f_2(x)$

29.   **30.**

$y = f_3(x)$ $y = f_4(x)$

31. a) The graph in Fig. 2.12 is made of line segments joined end to end. At which points of the interval $[-4, 6]$ is f' not defined? Give reasons for your answer.

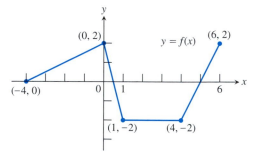

$(0, 2)$ $y = f(x)$ $(6, 2)$

$(-4, 0)$ 0 1 6

$(1, -2)$ $(4, -2)$

2.12 The graph for Exercise 31.

b) Graph the derivative of f. Call the vertical axis the y'-axis. The graph should show a step function.

32. *Recovering a function from its derivative*

a) Use the following information to graph the function f over the closed interval $[-2, 5]$.

   i) The graph of f is made of closed line segments joined end to end.

   ii) The graph starts at the point $(-2, 3)$.

   iii) The derivative of f is the step function in Fig. 2.13.

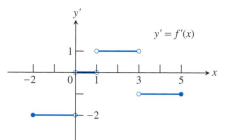

$y' = f'(x)$

2.13 The derivative graph for Exercise 32.

b) Repeat part (a) assuming that the graph starts at $(-2, 0)$ instead of $(-2, 3)$.

33. *Growth in the economy.* The graph in Fig. 2.14 shows the average annual percentage change $y = f(t)$ in the U.S. gross national product (GNP) for the years 1983–1988. Graph dy/dt (where defined). (Source: *Statistical Abstracts of the United States,* 110th Edition, U.S. Department of Commerce, p. 427.)

2.14 The graph for Exercise 33.

34. *Fruit flies.* (*Continuation of Example 3, Section 1.1.*) Populations starting out in closed environments grow slowly at first, when there are relatively few members, then more rapidly as the number of reproducing individuals increases and resources are still abundant, then slowly again as the population reaches the carrying capacity of the environment.

a) Use the graphical technique of Example 4 to graph the derivative of the fruit fly population introduced in Section 1.1. The graph of the population is reproduced here as Fig. 2.15. What units should be used on the horizontal and vertical axes for the derivative's graph?

2.15 The graph for Exercise 34.

b) During what days does the population seem to be increasing fastest? slowest?

Compare the right-hand and left-hand derivatives to show that the functions in Exercises 35–38 are not differentiable at the point *P*.

35.

36.

37.

38.

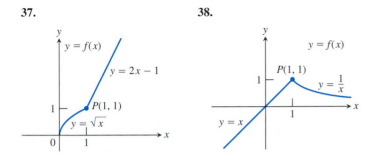

Each figure in Exercises 39–44 shows the graph of a function over a closed interval *D*. At what domain points does the function appear to be

a) differentiable?
b) continuous but not differentiable?
c) neither continuous nor differentiable?

Give reasons for your answers.

39.

40.

41.

42.

43.

44.

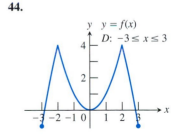

Theory and Examples

In Exercises 45–48,

a) Find the derivative $y' = f'(x)$ of the given function $y = f(x)$.
b) Graph $y = f(x)$ and $y' = f'(x)$ side by side using separate sets of coordinate axes, and answer the following questions.
c) For what values of *x*, if any, is y' positive? zero? negative?
d) Over what intervals of *x*-values, if any, does the function $y = f(x)$ increase as *x* increases? decrease as *x* increases? How is this related to what you found in (c)? (We will say more about this relationship in Chapter 3.)

45. $y = -x^2$ **46.** $y = -1/x$
47. $y = x^3/3$ **48.** $y = x^4/4$

49. Does the curve $y = x^3$ ever have a negative slope? If so, where? Give reasons for your answer.

50. Does the curve $y = 2\sqrt{x}$ have any horizontal tangents? If so, where? Give reasons for your answer.

51. Does the parabola $y = 2x^2 - 13x + 5$ have a tangent whose slope is -1? If so, find an equation for the line and the point of tangency. If not, why not?

52. Does any tangent to the curve $y = \sqrt{x}$ cross the x-axis at $x = -1$? If so, find an equation for the line and the point of tangency. If not, why not?

53. Does any function differentiable on $(-\infty, \infty)$ have $y = \lfloor x \rfloor$ as its derivative? Give reasons for your answer.

54. Graph the derivative of $f(x) = |x|$. Then graph $y = (|x| - 0)/(x - 0) = |x|/x$. What can you conclude?

55. Does knowing that a function $f(x)$ is differentiable at $x = x_0$ tell you anything about the differentiability of the function $-f$ at $x = x_0$? Give reasons for your answer.

56. Does knowing that a function $g(t)$ is differentiable at $t = 7$ tell you anything about the differentiability of the function $3g$ at $t = 7$? Give reasons for your answer.

57. Suppose that functions $g(t)$ and $h(t)$ are defined for all values of t and that $g(0) = h(0) = 0$. Can $\lim_{t \to 0}(g(t))/(h(t))$ exist? If it does exist, must it equal zero? Give reasons for your answers.

58. a) Let $f(x)$ be a function satisfying $|f(x)| \le x^2$ for $-1 \le x \le 1$. Show that f is differentiable at $x = 0$ and find $f'(0)$.

b) Show that

$$f(x) = \begin{cases} x^2 \sin \dfrac{1}{x}, & x \neq 0 \\ 0, & x = 0 \end{cases}$$

is differentiable at $x = 0$ and find $f'(0)$.

▦ Grapher Explorations

59. Graph $y = 1/(2\sqrt{x})$ in a window that has $0 \le x \le 2$. Then, on the same screen, graph

$$y = \frac{\sqrt{x + h} - \sqrt{x}}{h}$$

for $h = 1, 0.5, 0.1$. Then try $h = -1, -0.5, -0.1$. Explain what is going on.

60. Graph $y = 3x^2$ in a window that has $-2 \le x \le 2, 0 \le y \le 3$. Then, on the same screen, graph

$$y = \frac{(x + h)^3 - x^3}{h}$$

for $h = 2, 1, 0.2$. Then try $h = -2, -1, -0.2$. Explain what is going on.

61. *Weierstrass's nowhere differentiable continuous function.* The sum of the first eight terms of the Weierstrass function $f(x) = \sum_{n=0}^{\infty}(2/3)^n \cos(9^n \pi x)$ is

$$g(x) = \cos(\pi x) + \left(\frac{2}{3}\right)^1 \cos(9\pi x) + \left(\frac{2}{3}\right)^2 \cos(9^2 \pi x)$$

$$+ \left(\frac{2}{3}\right)^3 \cos(9^3 \pi x) + \cdots + \left(\frac{2}{3}\right)^7 \cos(9^7 \pi x).$$

Graph this sum. Zoom in several times. How wiggly and bumpy is this graph? Specify a viewing window in which the displayed portion of the graph is smooth.

✺ CAS Explorations and Projects

Use a CAS to perform the following steps for the functions in Exercises 62–67.

a) Plot $y = f(x)$ to see that function's global behavior.

b) Define the difference quotient q at a general point x, with general stepsize h.

c) Take the limit as $h \to 0$. What formula does this give?

d) Substitute the value $x = x_0$ and plot the function together with its tangent line at that point.

e) Substitute various values for x larger and smaller than x_0 into the formula obtained in part (c). Do the numbers make sense with your picture?

f) Graph the formula obtained in part (c). What does it mean when its values are negative? zero? positive? Does this make sense with your plot from part (a)? Give reasons for your answer.

62. $f(x) = x^3 + x^2 - x, \quad x_0 = 1$

63. $f(x) = x^{1/3} + x^{2/3}, \quad x_0 = 1$

64. $f(x) = \dfrac{4x}{x^2 + 1}, \quad x_0 = 2$ **65.** $f(x) = \dfrac{x - 1}{3x^2 + 1}, \quad x_0 = -1$

66. $f(x) = \sin 2x, \quad x_0 = \pi/2$ **67.** $f(x) = x^2 \cos x, \quad x_0 = \pi/4$

2.2 Differentiation Rules

This section shows how to differentiate functions without having to apply the definition each time.

Powers, Multiples, Sums, and Differences

The first rule of differentiation is that the derivative of every constant function is zero.

Rule 1 Derivative of a Constant

If c is constant, then $\dfrac{d}{dx} c = 0$.

EXAMPLE 1 $\dfrac{d}{dx}(8) = 0, \qquad \dfrac{d}{dx}\left(-\dfrac{1}{2}\right) = 0, \qquad \dfrac{d}{dx}\left(\sqrt{3}\right) = 0$ ❑

Proof of Rule 1 We apply the definition of derivative to $f(x) = c$, the function whose outputs have the constant value c (Fig. 2.16). At every value of x, we find that

$$f'(x) = \lim_{h \to 0} \frac{f(x+h) - f(x)}{h} = \lim_{h \to 0} \frac{c - c}{h} = \lim_{h \to 0} 0 = 0.$$ ❑

The next rule tells how to differentiate x^n if n is a positive integer.

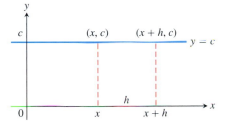

2.16 The rule $(d/dx)(c) = 0$ is another way to say that the values of constant functions never change and that the slope of a horizontal line is zero at every point.

Rule 2 Power Rule for Positive Integers

If n is a positive integer, then

$$\frac{d}{dx} x^n = n x^{n-1}.$$

To apply the Power Rule, we subtract 1 from the original exponent (n) and multiply the result by n.

EXAMPLE 2

f	x	x^2	x^3	x^4	$\cdots$
f'	1	$2x$	$3x^2$	$4x^3$	$\cdots$

❑

Proof of Rule 2 If $f(x) = x^n$, then $f(x+h) = (x+h)^n$. Since n is a positive integer, we can use the fact that

$$a^n - b^n = (a - b)(a^{n-1} + a^{n-2}b + \cdots + ab^{n-2} + b^{n-1})$$

to simplify the difference quotient for f. Taking $x + h = a$ and $x = b$, we have $a - b = h$. Thus

$$\frac{f(x + h) - f(x)}{h} = \frac{(x + h)^n - x^n}{h}$$

$$= \frac{(h)[(x + h)^{n-1} + (x + h)^{n-2}x + \cdots + (x + h)x^{n-2} + x^{n-1}]}{h}$$

$$= \underbrace{(x + h)^{n-1} + (x + h)^{n-2}x + \cdots + (x + h)x^{n-2} + x^{n-1}}_{n \text{ terms, each with limit } x^{n-1} \text{ as } h \to 0}.$$

Hence

$$\frac{d}{dx}x^n = \lim_{h \to 0} \frac{f(x + h) - f(x)}{h} = nx^{n-1}. \qquad \square$$

The next rule says that when a differentiable function is multiplied by a constant, its derivative is multiplied by the same constant.

> **Rule 3   The Constant Multiple Rule**
>
> If u is a differentiable function of x, and c is a constant, then
>
> $$\frac{d}{dx}(cu) = c\frac{du}{dx}.$$

In particular, if n is a positive integer, then

$$\frac{d}{dx}(cx^n) = cn\, x^{n-1}.$$

EXAMPLE 3   The derivative formula

$$\frac{d}{dx}(3x^2) = 3 \cdot 2x = 6x$$

says that if we rescale the graph of $y = x^2$ by multiplying each y-coordinate by 3, then we multiply the slope at each point by 3 (Fig. 2.17). $\qquad \square$

EXAMPLE 4   *A useful special case*

The derivative of the negative of a differentiable function is the negative of the function's derivative. Rule 3 with $c = -1$ gives

$$\frac{d}{dx}(-u) = \frac{d}{dx}(-1 \cdot u) = -1 \cdot \frac{d}{dx}(u) = -\frac{du}{dx}. \qquad \square$$

Proof of Rule 3

$$\frac{d}{dx}cu = \lim_{h \to 0} \frac{cu(x + h) - cu(x)}{h} \qquad \text{Derivative definition with } f(x) = cu(x)$$

$$= c \lim_{h \to 0} \frac{u(x + h) - u(x)}{h} \qquad \text{Limit property}$$

$$= c\frac{du}{dx} \qquad u \text{ is differentiable.} \qquad \square$$

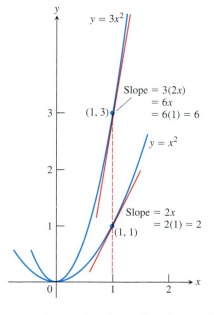

2.17 The graphs of $y = x^2$ and $y = 3x^2$. Tripling the y-coordinates triples the slope (Example 3).

The next rule says that the derivative of the sum of two differentiable functions is the sum of their derivatives.

Denoting functions by *u* and *v*

The functions we are working with when we need a differentiation formula are likely to be denoted by letters like *f* and *g*. When we apply the formula, we do not want to find it using these same letters in some other way. To guard against this, we denote the functions in differentiation rules by letters like *u* and *v* that are not likely to be already in use.

Rule 4   The Sum Rule

If u and v are differentiable functions of x, then their sum $u + v$ is differentiable at every point where u and v are both differentiable. At such points,

$$\frac{d}{dx}(u + v) = \frac{du}{dx} + \frac{dv}{dx}.$$

Combining the Sum Rule with the Constant Multiple Rule gives the equivalent **Difference Rule,** which says that the derivative of a *difference* of differentiable functions is the difference of their derivatives.

$$\frac{d}{dx}(u - v) = \frac{d}{dx}[u + (-1)v] = \frac{du}{dx} + (-1)\frac{dv}{dx} = \frac{du}{dx} - \frac{dv}{dx}$$

The Sum Rule also extends to sums of more than two functions, as long as there are only finitely many functions in the sum. If $u_1, u_2, \ldots, u_n$ are differentiable at x, then so is $u_1 + u_2 + \cdots + u_n$, and

$$\frac{d}{dx}(u_1 + u_2 + \cdots + u_n) = \frac{du_1}{dx} + \frac{du_2}{dx} + \cdots + \frac{du_n}{dx}.$$

EXAMPLE 5

a) $y = x^4 + 12x$

$$\frac{dy}{dx} = \frac{d}{dx}(x^4) + \frac{d}{dx}(12x)$$

$$= 4x^3 + 12$$

b) $y = x^3 + \frac{4}{3}x^2 - 5x + 1$

$$\frac{dy}{dx} = \frac{d}{dx}x^3 + \frac{d}{dx}\left(\frac{4}{3}x^2\right) - \frac{d}{dx}(5x) + \frac{d}{dx}(1)$$

$$= 3x^2 + \frac{4}{3} \cdot 2x - 5 + 0$$

$$= 3x^2 + \frac{8}{3}x - 5 \qquad \square$$

Notice that we can differentiate any polynomial term by term, the way we differentiated the polynomials in Example 5.

Proof of Rule 4   We apply the definition of derivative to $f(x) = u(x) + v(x)$:

$$\frac{d}{dx}[u(x) + v(x)] = \lim_{h \to 0} \frac{[u(x + h) + v(x + h)] - [u(x) + v(x)]}{h}$$

$$= \lim_{h \to 0} \left[\frac{u(x + h) - u(x)}{h} + \frac{v(x + h) - v(x)}{h}\right]$$

$$= \lim_{h \to 0} \frac{u(x + h) - u(x)}{h} + \lim_{h \to 0} \frac{v(x + h) - v(x)}{h} = \frac{du}{dx} + \frac{dv}{dx}. \qquad \square$$

Proof by mathematical induction

Many formulas can be shown to hold for every positive integer n greater than or equal to some lowest integer n_0 by applying an axiom called the *mathematical induction principle*. A proof using this axiom is called a *proof by mathematical induction* or a *proof by induction*. The steps in proving a formula by induction are

1. Check that it holds for $n = n_0$.
2. Prove that if it holds for any positive integer $n = k \geq n_0$, then it holds for $n = k + 1$.

Once these steps are completed, the axiom says, we know that the formula holds for all $n \geq n_0$. For more mathematical induction, see Appendix 1.

Proof of the Sum Rule for Sums of More Than Two Functions We prove the statement

$$\frac{d}{dx}(u_1 + u_2 + \cdots + u_n) = \frac{du_1}{dx} + \frac{du_2}{dx} + \cdots + \frac{du_n}{dx}$$

by mathematical induction. The statement is true for $n = 2$, as was just proved. This is step 1 of the induction proof.

Step 2 is to show that if the statement is true for any positive integer $n = k$, where $k \geq n_0 = 2$, then it is also true for $n = k + 1$. So suppose that

$$\frac{d}{dx}(u_1 + u_2 + \cdots + u_k) = \frac{du_1}{dx} + \frac{du_2}{dx} + \cdots + \frac{du_k}{dx}. \qquad (1)$$

Then

$$\frac{d}{dx} \underbrace{(u_1 + u_2 + \cdots + u_k}_{\substack{\text{Call the function} \\ \text{defined by this sum } u.}} + \underbrace{u_{k+1})}_{\substack{\text{Call this} \\ \text{function } v.}}$$

$$= \frac{d}{dx}(u_1 + u_2 + \cdots + u_k) + \frac{du_{k+1}}{dx} \qquad \text{Rule 4 for } \frac{d}{dx}(u + v)$$

$$= \frac{du_1}{dx} + \frac{du_2}{dx} + \cdots + \frac{du_k}{dx} + \frac{du_{k+1}}{dx}. \qquad \text{Eq. (1)}$$

With these steps verified, the mathematical induction principle now guarantees the Sum Rule for every integer $n \geq 2$. ❏

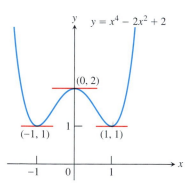

2.18 The curve $y = x^4 - 2x^2 + 2$ and its horizontal tangents (Example 6).

EXAMPLE 6   Does the curve $y = x^4 - 2x^2 + 2$ have any horizontal tangents? If so, where?

Solution   The horizontal tangents, if any, occur where the slope dy/dx is zero. To find these points, we

1. Calculate dy/dx: $\dfrac{dy}{dx} = \dfrac{d}{dx}(x^4 - 2x^2 + 2) = 4x^3 - 4x$

2. Solve the equation $\dfrac{dy}{dx} = 0$ for x: $4x^3 - 4x = 0$
$$4x(x^2 - 1) = 0$$
$$x = 0, 1, -1$$

The curve $y = x^4 - 2x^2 + 2$ has horizontal tangents at $x = 0, 1$, and -1. The corresponding points on the curve are $(0, 2)$, $(1, 1)$ and $(-1, 1)$. See Fig. 2.18. ❏

Products and Quotients

While the derivative of the sum of two functions is the sum of their derivatives, the derivative of the product of two functions is *not* the product of their derivatives. For instance,

$$\frac{d}{dx}(x \cdot x) = \frac{d}{dx}(x^2) = 2x, \qquad \text{while} \qquad \frac{d}{dx}(x) \cdot \frac{d}{dx}(x) = 1 \cdot 1 = 1.$$

The derivative of a product of two functions is the sum of *two* products, as we now explain.

> **Rule 5   The Product Rule**
>
> If u and v are differentiable at x, then so is their product uv, and
>
> $$\frac{d}{dx}(uv) = u\frac{dv}{dx} + v\frac{du}{dx}.$$

The derivative of the product uv is u times the derivative of v plus v times the derivative of u. In *prime notation*, $(uv)' = uv' + vu'$.

Proof of Rule 5

$$\frac{d}{dx}(uv) = \lim_{h\to 0} \frac{u(x+h)v(x+h) - u(x)v(x)}{h}$$

To change this fraction into an equivalent one that contains difference quotients for the derivatives of u and v, we subtract and add $u(x+h)v(x)$ in the numerator:

$$\frac{d}{dx}(uv) = \lim_{h\to 0} \frac{u(x+h)v(x+h) - u(x+h)v(x) + u(x+h)v(x) - u(x)v(x)}{h}$$

$$= \lim_{h\to 0} \left[u(x+h)\frac{v(x+h) - v(x)}{h} + v(x)\frac{u(x+h) - u(x)}{h} \right]$$

$$= \lim_{h\to 0} u(x+h) \cdot \lim_{h\to 0} \frac{v(x+h) - v(x)}{h} + v(x) \cdot \lim_{h\to 0} \frac{u(x+h) - u(x)}{h}.$$

As h approaches zero, $u(x+h)$ approaches $u(x)$ because u, being differentiable at x, is continuous at x. The two fractions approach the values of dv/dx at x and du/dx at x. In short,

$$\frac{d}{dx}(uv) = u\frac{dv}{dx} + v\frac{du}{dx}. \qquad \square$$

EXAMPLE 7   Find the derivative of $y = (x^2 + 1)(x^3 + 3)$.

Solution   From the Product Rule with $u = x^2 + 1$ and $v = x^3 + 3$, we find

$$\frac{d}{dx}[(x^2 + 1)(x^3 + 3)] = (x^2 + 1)(3x^2) + (x^3 + 3)(2x)$$

$$= 3x^4 + 3x^2 + 2x^4 + 6x$$

$$= 5x^4 + 3x^2 + 6x. \qquad \square$$

Example 7 can be done as well (perhaps better) by multiplying out the original expression for y and differentiating the resulting polynomial. We now check:

$$y = (x^2 + 1)(x^3 + 3) = x^5 + x^3 + 3x^2 + 3$$

$$\frac{dy}{dx} = 5x^4 + 3x^2 + 6x.$$

This is in agreement with our first calculation.

Picturing the product rule

If $u(x)$ and $v(x)$ are positive and increase when x increases, and if $h > 0$,

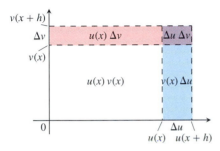

the total shaded area in the picture is

$$u(x+h)v(x+h) - u(x)v(x) =$$

$$u(x+h)\Delta v + v(x+h)\Delta u - \Delta u\Delta v.$$

Dividing both sides of this equation by h gives

$$\frac{u(x+h)v(x+h) - u(x)v(x)}{h}$$

$$= u(x+h)\frac{\Delta v}{h} + v(x+h)\frac{\Delta u}{h} - \Delta u\frac{\Delta v}{h}.$$

As $h \to 0^+$, $\Delta u \cdot \dfrac{\Delta v}{h} \to 0 \cdot \dfrac{dv}{dx} = 0$, leaving

$$\frac{d}{dx}(uv) = u\frac{dv}{dx} + v\frac{du}{dx}.$$

There are times, however, when the Product Rule *must* be used. In the following example, we have only numerical values to work with.

EXAMPLE 8 Let $y = uv$ be the product of the functions u and v. Find $y'(2)$ if

$$u(2) = 3, \qquad u'(2) = -4, \qquad v(2) = 1, \qquad \text{and} \qquad v'(2) = 2.$$

Solution From the Product Rule, in the form

$$y' = (uv)' = uv' + vu',$$

we have

$$y'(2) = u(2)v'(2) + v(2)u'(2)$$

$$= (3)(2) + (1)(-4) = 6 - 4 = 2. \qquad \square$$

Quotients

Just as the derivative of the product of two differentiable functions is not the product of their derivatives, the derivative of the quotient of two functions is not the quotient of their derivatives. What happens instead is this:

Rule 6   The Quotient Rule

If u and v are differentiable at x, and $v(x) \neq 0$, then the quotient u/v is differentiable at x, and

$$\frac{d}{dx}\left(\frac{u}{v}\right) = \frac{v\dfrac{du}{dx} - u\dfrac{dv}{dx}}{v^2}.$$

Proof of Rule 6

$$\frac{d}{dx}\left(\frac{u}{v}\right) = \lim_{h \to 0} \frac{\dfrac{u(x+h)}{v(x+h)} - \dfrac{u(x)}{v(x)}}{h}$$

$$= \lim_{h \to 0} \frac{v(x)u(x+h) - u(x)v(x+h)}{hv(x+h)v(x)}$$

To change the last fraction into an equivalent one that contains the difference quotients for the derivatives of u and v, we subtract and add $v(x)u(x)$ in the numerator. We then get

$$\frac{d}{dx}\left(\frac{u}{v}\right) = \lim_{h \to 0} \frac{v(x)u(x+h) - v(x)u(x) + v(x)u(x) - u(x)v(x+h)}{hv(x+h)v(x)}$$

$$= \lim_{h \to 0} \frac{v(x)\dfrac{u(x+h) - u(x)}{h} - u(x)\dfrac{v(x+h) - v(x)}{h}}{v(x+h)v(x)}.$$

Taking the limit in the numerator and denominator now gives the Quotient Rule.

$\square$

EXAMPLE 9 Find the derivative of $y = \dfrac{t^2 - 1}{t^2 + 1}$.

Solution We apply the Quotient Rule with $u = t^2 - 1$ and $v = t^2 + 1$:

$$\frac{dy}{dt} = \frac{(t^2 + 1) \cdot 2t - (t^2 - 1) \cdot 2t}{(t^2 + 1)^2} \qquad \frac{d}{dt}\left(\frac{u}{v}\right) = \frac{v(du/dt) - u(dv/dt)}{v^2}$$

$$= \frac{2t^3 + 2t - 2t^3 + 2t}{(t^2 + 1)^2}$$

$$= \frac{4t}{(t^2 + 1)^2}. \qquad \square$$

The Power Rule for Negative Integers

The Power Rule for negative integers is the same as the rule for positive integers.

Rule 7 Power Rule for Negative Integers

If n is a negative integer and $x \neq 0$, then

$$\frac{d}{dx}(x^n) = nx^{n-1}.$$

Proof of Rule 7 The proof uses the Quotient Rule in a clever way. If n is a negative integer, then $n = -m$ where m is a positive integer. Hence, $x^n = x^{-m} = 1/x^m$ and

$$\frac{d}{dx}(x^n) = \frac{d}{dx}\left(\frac{1}{x^m}\right)$$

$$= \frac{x^m \cdot \dfrac{d}{dx}(1) - 1 \cdot \dfrac{d}{dx}(x^m)}{(x^m)^2} \qquad \begin{array}{l}\text{Quotient Rule with}\\ u = 1 \text{ and } v = x^m\end{array}$$

$$= \frac{0 - mx^{m-1}}{x^{2m}} \qquad \begin{array}{l}\text{Since } m > 0,\\ \dfrac{d}{dx}(x^m) = mx^{m-1}\end{array}$$

$$= -mx^{-m-1}$$

$$= nx^{n-1}. \qquad \text{Since } -m = n \quad \square$$

EXAMPLE 10

$$\frac{d}{dx}\left(\frac{1}{x}\right) = \frac{d}{dx}(x^{-1}) = (-1)x^{-2} = -\frac{1}{x^2}$$

$$\frac{d}{dx}\left(\frac{4}{x^3}\right) = 4\frac{d}{dx}(x^{-3}) = 4(-3)x^{-4} = -\frac{12}{x^4} \qquad \square$$

EXAMPLE 11 Find an equation for the tangent to the curve

$$y = x + \frac{2}{x}$$

at the point $(1, 3)$ (Fig. 2.19).

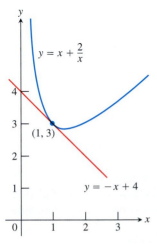

2.19 The tangent to the curve $y = x + (2/x)$ at $(1, 3)$. The curve has a third-quadrant portion not shown here. We will see how to graph functions like this in Chapter 3.

Solution   The slope of the curve is

$$\frac{dy}{dx} = \frac{d}{dx}(x) + 2\frac{d}{dx}\left(\frac{1}{x}\right) = 1 + 2\left(-\frac{1}{x^2}\right) = 1 - \frac{2}{x^2}.$$

The slope at $x = 1$ is

$$\frac{dy}{dx}\bigg|_{x=1} = \left[1 - \frac{2}{x^2}\right]_{x=1} = 1 - 2 = -1.$$

The line through $(1, 3)$ with slope $m = -1$ is

$$y - 3 = (-1)(x - 1) \qquad \text{Point–slope equation}$$

$$y = -x + 1 + 3$$

$$y = -x + 4.$$

Choosing Which Rules to Use

The choice of which rules to use in solving a differentiation problem can make a difference in how much work you have to do. Here is an example.

EXAMPLE 12   Rather than using the Quotient Rule to find the derivative of

$$y = \frac{(x - 1)(x^2 - 2x)}{x^4},$$

expand the numerator and divide by x^4:

$$y = \frac{(x - 1)(x^2 - 2x)}{x^4} = \frac{x^3 - 3x^2 + 2x}{x^4} = x^{-1} - 3x^{-2} + 2x^{-3}.$$

Then use the Sum and Power Rules:

$$\frac{dy}{dx} = -x^{-2} - 3(-2)x^{-3} + 2(-3)x^{-4}$$

$$= -\frac{1}{x^2} + \frac{6}{x^3} - \frac{6}{x^4}.$$

Second and Higher Order Derivatives

The derivative $y' = dy/dx$ is the **first (first order) derivative** of y with respect to x. This derivative may itself be a differentiable function of x; if so, its derivative

$$y'' = \frac{dy'}{dx} = \frac{d}{dx}\left(\frac{dy}{dx}\right) = \frac{d^2y}{dx^2}$$

is called the **second (second order) derivative** of y with respect to x.

If y'' is differentiable, its derivative, $y''' = dy''/dx = d^3y/dx^3$ is the **third (third order) derivative** of y with respect to x. The names continue as you imagine, with

$$y^{(n)} = \frac{d}{dx}y^{(n-1)}$$

denoting the **nth (nth order) derivative** of y with respect to x, for any positive integer n.

Notice that

$$\frac{d}{dx}\left(\frac{dy}{dx}\right)$$

does not mean multiplication. It means "the derivative of the derivative."

How to read the symbols for derivatives

y' "y prime" y'' "y double prime"

$\dfrac{d^2 y}{dx^2}$ "d squared y dx squared"

y''' "y triple prime"

$y^{(n)}$ "y super n"

$\dfrac{d^n y}{dx^n}$ "d to the n of y by dx to the n"

EXAMPLE 13   The first four derivatives of $y = x^3 - 3x^2 + 2$ are

First derivative: $y' = 3x^2 - 6x$

Second derivative: $y'' = 6x - 6$

Third derivative: $y''' = 6$

Fourth derivative: $y^{(4)} = 0.$

The function has derivatives of all orders, the fifth and later derivatives all being zero.   ☐

Exercises 2.2

Derivative Calculations

In Exercises 1–12, find the first and second derivatives.

1. $y = -x^2 + 3$

2. $y = x^2 + x + 8$

3. $s = 5t^3 - 3t^5$

4. $w = 3z^7 - 7z^3 + 21z^2$

5. $y = \dfrac{4x^3}{3} - x$

6. $y = \dfrac{x^3}{3} + \dfrac{x^2}{2} + \dfrac{x}{4}$

7. $w = 3z^{-2} - \dfrac{1}{z}$

8. $s = -2t^{-1} + \dfrac{4}{t^2}$

9. $y = 6x^2 - 10x - 5x^{-2}$

10. $y = 4 - 2x - x^{-3}$

11. $r = \dfrac{1}{3s^2} - \dfrac{5}{2s}$

12. $r = \dfrac{12}{\theta} - \dfrac{4}{\theta^3} + \dfrac{1}{\theta^4}$

In Exercises 13–16, find y' (a) by applying the Product Rule and (b) by multiplying the factors to produce a sum of simpler terms to differentiate.

13. $y = (3 - x^2)(x^3 - x + 1)$

14. $y = (x - 1)(x^2 + x + 1)$

15. $y = (x^2 + 1)\left(x + 5 + \dfrac{1}{x}\right)$

16. $y = \left(x + \dfrac{1}{x}\right)\left(x - \dfrac{1}{x} + 1\right)$

Find the derivatives of the functions in Exercises 17–28.

17. $y = \dfrac{2x + 5}{3x - 2}$

18. $z = \dfrac{2x + 1}{x^2 - 1}$

19. $g(x) = \dfrac{x^2 - 4}{x + 0.5}$

20. $f(t) = \dfrac{t^2 - 1}{t^2 + t - 2}$

21. $v = (1 - t)(1 + t^2)^{-1}$

22. $w = (2x - 7)^{-1}(x + 5)$

23. $f(s) = \dfrac{\sqrt{s} - 1}{\sqrt{s} + 1}$

24. $u = \dfrac{5x + 1}{2\sqrt{x}}$

25. $v = \dfrac{1 + x - 4\sqrt{x}}{x}$

26. $r = 2\left(\dfrac{1}{\sqrt{\theta}} + \sqrt{\theta}\right)$

27. $y = \dfrac{1}{(x^2 - 1)(x^2 + x + 1)}$

28. $y = \dfrac{(x + 1)(x + 2)}{(x - 1)(x - 2)}$

Find the derivatives of all orders of the functions in Exercises 29 and 30.

29. $y = \dfrac{x^4}{2} - \dfrac{3}{2}x^2 - x$

30. $y = \dfrac{x^5}{120}$

Find the first and second derivatives of the functions in Exercises 31–38.

31. $y = \dfrac{x^3 + 7}{x}$

32. $s = \dfrac{t^2 + 5t - 1}{t^2}$

33. $r = \dfrac{(\theta - 1)(\theta^2 + \theta + 1)}{\theta^3}$

34. $u = \dfrac{(x^2 + x)(x^2 - x + 1)}{x^4}$

35. $w = \left(\dfrac{1 + 3z}{3z}\right)(3 - z)$

36. $w = (z + 1)(z - 1)(z^2 + 1)$

37. $p = \left(\dfrac{q^2 + 3}{12q}\right)\left(\dfrac{q^4 - 1}{q^3}\right)$

38. $p = \dfrac{q^2 + 3}{(q - 1)^3 + (q + 1)^3}$

Using Numerical Values

39. Suppose u and v are functions of x that are differentiable at $x = 0$ and that

$$u(0) = 5, \quad u'(0) = -3, \quad v(0) = -1, \quad v'(0) = 2.$$

Find the values of the following derivatives at $x = 0$.

a) $\dfrac{d}{dx}(uv)$ **b)** $\dfrac{d}{dx}\left(\dfrac{u}{v}\right)$ **c)** $\dfrac{d}{dx}\left(\dfrac{v}{u}\right)$ **d)** $\dfrac{d}{dx}(7v - 2u)$

40. Suppose u and v are differentiable functions of x and that

$$u(1) = 2, \quad u'(1) = 0, \quad v(1) = 5, \quad v'(1) = -1.$$

Find the values of the following derivatives at $x = 1$.

a) $\dfrac{d}{dx}(uv)$ **b)** $\dfrac{d}{dx}\left(\dfrac{u}{v}\right)$ **c)** $\dfrac{d}{dx}\left(\dfrac{v}{u}\right)$ **d)** $\dfrac{d}{dx}(7v - 2u)$

Slopes and Tangents

41. a) Find an equation for the line perpendicular to the tangent to the curve $y = x^3 - 4x + 1$ at the point $(2, 1)$.

b) What is the smallest slope on the curve? At what point on the curve does the curve have this slope?

c) Find equations for the tangents to the curve at the points where the slope of the curve is 8.

42. a) Find equations for the horizontal tangents to the curve $y = x^3 - 3x - 2$. Also find equations for the lines that are perpendicular to these tangents at the points of tangency.

b) What is the smallest slope on the curve? At what point on the curve does the curve have this slope? Find an equation for the line that is perpendicular to the curve's tangent at this point.

43. Find the tangents to *Newton's Serpentine* (graphed here) at the origin and the point $(1, 2)$.

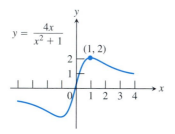

$y = \dfrac{4x}{x^2 + 1}$

$(1, 2)$

44. Find the tangent to the *Witch of Agnesi* (graphed here) at the point $(2, 1)$. There is a nice story about the name of this curve in the marginal note on Agnesi in Section 9.4.

$y = \dfrac{8}{x^2 + 4}$

$(2, 1)$

45. The curve $y = ax^2 + bx + c$ passes through the point $(1, 2)$ and is tangent to the line $y = x$ at the origin. Find a, b, and c.

46. The curves $y = x^2 + ax + b$ and $y = cx - x^2$ have a common tangent line at the point $(1, 0)$. Find a, b, and c.

47. a) Find an equation for the line that is tangent to the curve $y = x^3 - x$ at the point $(-1, 0)$.

b) GRAPHER Graph the curve and tangent line together. The tangent intersects the curve at another point. Use ZOOM and TRACE to estimate the point's coordinates.

c) GRAPHER Confirm your estimates of the coordinates of the second intersection point by solving the equations for the curve and tangent simultaneously (SOLVER key).

48. a) Find an equation for the line that is tangent to the curve $y = x^3 - 6x^2 + 5x$ at the origin.

b) GRAPHER Graph the curve and tangent together. The tangent intersects the curve at another point. Use ZOOM and TRACE to estimate the point's coordinates.

c) GRAPHER Confirm your estimates of the coordinates of the second intersection point by solving the equations for the curve and tangent simultaneously (SOLVER key).

Physical Applications

49. *Pressure and volume.* If the gas in a closed container is maintained at a constant temperature T, the pressure P is related to the volume V by a formula of the form

$$P = \frac{nRT}{V - nb} - \frac{an^2}{V^2},$$

in which a, b, n, and R are constants. Find dP/dV.

50. *The body's reaction to medicine.* The reaction of the body to a dose of medicine can sometimes be represented by an equation of the form

$$R = M^2 \left(\frac{C}{2} - \frac{M}{3} \right),$$

where C is a positive constant and M is the amount of medicine absorbed in the blood. If the reaction is a change in blood pressure, R is measured in millimeters of mercury. If the reaction is a change in temperature, R is measured in degrees, and so on.

Find dR/dM. This derivative, as a function of M, is called the sensitivity of the body to the medicine. In Section 3.6, we will see how to find the amount of medicine to which the body is most sensitive. (Source: *Some Mathematical Models in Biology*, Revised Edition, R. M Thrall, J. A. Mortimer, K. R. Rebman, R. F. Baum, eds., December 1967, PB-202 364, p. 221; distributed by NTIS, U.S. Department of Commerce.)

Theory and Examples

51. Suppose that the function v in the Product Rule has a constant value c. What does the Product Rule then say? What does this say about the Constant Multiple Rule?

52. *The Reciprocal Rule*

a) The **Reciprocal Rule** says that at any point where the function $v(x)$ is differentiable and different from zero,

$$\frac{d}{dx}\left(\frac{1}{v} \right) = -\frac{1}{v^2}\frac{dv}{dx}.$$

Show that the Reciprocal Rule is a special case of the Quotient Rule.

b) Show that the Reciprocal Rule and the Product Rule together imply the Quotient Rule.

53. *Another proof of the Power Rule for positive integers.* Use the algebra formula

$$x^n - c^n = (x - c)(x^{n-1} + x^{n-2}c + \cdots + xc^{n-2} + c^{n-1})$$

together with the derivative formula

$$f'(c) = \lim_{x \to c} \frac{f(x) - f(c)}{x - c}$$

from Exercises 2.1 to show that $(d/dx)(x^n) = nx^{n-1}$.

54. *Generalizing the Product Rule.* The Product Rule gives the formula

$$\frac{d}{dx}(uv) = u\frac{dv}{dx} + v\frac{du}{dx}$$

for the derivative of the product uv of two differentiable functions of x.

a) What is the analogous formula for the derivative of the product uvw of *three* differentiable functions of x?

b) What is the formula for the derivative of the product $u_1u_2u_3u_4$ of *four* differentiable functions of x?

c) What is the formula for the derivative of a product $u_1u_2u_3 \cdots u_n$ of a finite number n of differentiable functions of x?

55. *Rational Powers*

a) Find $\frac{d}{dx}(x^{3/2})$ by writing $x^{3/2}$ as $x \cdot x^{1/2}$ and using the Product Rule. Express your answer as a rational number times a rational power of x. Work parts (b) and (c) by a similar method.

b) Find $\frac{d}{dx}(x^{5/2})$.

c) Find $\frac{d}{dx}(x^{7/2})$.

d) What patterns do you see in your answers to (a), (b), and (c)? Rational powers are one of the topics in Section 2.6.

2.3

Rates of Change

In this section we examine some applications in which derivatives are used to represent and interpret the rates at which things change in the world around us. It is natural to think of change in terms of dependence on time, such as the position, velocity, and acceleration of a moving object, but there is no need to be so restrictive. Change with respect to variables other than time can be treated in the same way. For example, a physician may want to know how small changes in dosage can affect the body's response to a drug. An economist may want to study how investment changes with respect to variations in interest rates. These questions can all be expressed in terms of the rate of change of a function with respect to a variable.

Average and Instantaneous Rates of Change

We start by recalling the concept of average rate of change of a function over an interval, introduced in Section 1.1. The derivative of the function is the limit of this average rate as the length of the interval goes to zero.

Definitions

The **average rate of change** of a function $f(x)$ with respect to x over the interval from x_0 to $x_0 + h$ is

$$\text{Average rate of change} = \frac{f(x_0 + h) - f(x_0)}{h}.$$

The **(instantaneous) rate of change** of f with respect to x at x_0 is the derivative

$$f'(x_0) = \lim_{h \to 0} \frac{f(x_0 + h) - f(x_0)}{h},$$

provided the limit exists.

It is conventional to use the word *instantaneous* even when x does not represent time. The word is, however, frequently omitted. When we say *rate of change*, we mean *instantaneous rate of change*.

EXAMPLE 1   The area A of a circle is related to its diameter by the equation

$$A = \frac{\pi}{4}D^2.$$

How fast is the area changing with respect to the diameter when the diameter is 10 m?

Solution   The (instantaneous) rate of change of the area with respect to the diameter is

$$\frac{dA}{dD} = \frac{\pi}{4}2D = \frac{\pi D}{2}.$$

When $D = 10$ m, the area is changing at rate $(\pi/2)10 = 5\pi$ m^2/m. This means that a small change ΔD m in the diameter would result in a change of about $5\pi\,\Delta D$ m^2 in the area of the circle.   ❏

Position at time t …   and at time $t + \Delta t$

$s = f(t)$ $s + \Delta s = f(t + \Delta t)$

2.20 The positions of a body moving along a coordinate line at time t and shortly later at time $t + \Delta t$.

Motion Along a Line—Displacement, Velocity, Speed, and Acceleration

Suppose that an object is moving along a coordinate line (say an s-axis) so that we know its position s on that line as a function of time t:

$$s = f(t).$$

The **displacement** of the object over the time interval from t to $t + \Delta t$ (Fig. 2.20) is

$$\Delta s = f(t + \Delta t) - f(t),$$

and the **average velocity** of the object over that time interval is

$$v_{av} = \frac{\text{displacement}}{\text{travel time}} = \frac{\Delta s}{\Delta t} = \frac{f(t + \Delta t) - f(t)}{\Delta t}.$$

To find the body's velocity at the exact instant t, we take the limit of the average velocity over the interval from t to $t + \Delta t$ as Δt shrinks to zero. This limit is the derivative of f with respect to t.

> **Definition**
>
> The **(instantaneous) velocity** is the derivative of the position function $s = f(t)$ with respect to time. At time t the velocity is
>
> $$v(t) = \frac{ds}{dt} = \lim_{\Delta t \to 0} \frac{f(t + \Delta t) - f(t)}{\Delta t}.$$

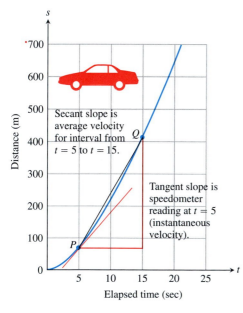

2.21 The time-to-distance data for Example 2.

EXAMPLE 2   Figure 2.21 shows a distance–time graph of a 1994 Ford Mustang Cobra. The slope of the secant PQ is the average velocity for the 10-sec interval from $t = 5$ to $t = 15$ sec, in this case 35.5 m/sec or 128 km/h. The slope of the tangent at P is the speedometer reading at $t = 5$ sec, about 20 m/sec or 72 km/h. The car's top speed is 220 km/h (about 137 mph). (Source: *Car & Driver*, April 1994.)   ❏

Technology *Parametric Functions* To graph curves $y = f(x)$, where y is a function of x, your graphing utility should be set in *function mode*. Not all curves can be represented in that mode, so most graphing utilities have a *parametric mode* as well. In this mode you plot the points $(x(t), y(t))$ whose coordinates are functions of the varying "time" parameter t. Thus you can think of the curve as the path of a moving particle as it changes its (x, y) position over time (see Section 9.4). A curve $y = f(x)$ can be graphed in parametric mode using the equations $x = t$, $y = f(t)$. Set your graphing utility to parametric mode and try the following equations.

Relation	Parametrization
$y = x^2$ (y a function of x)	$x(t) = t, \quad y(t) = t^2, \quad -\infty < t < \infty$
$x^2 + y^2 = 4$ (y not a function of x)	$x(t) = 2\cos t, \quad y(t) = 2\sin t,$
	$0 \le t \le 2\pi$

The parabola $x(t) = t$, $y(t) = t^2$, for $t \ge -2$

Besides telling us how fast the object is moving, the velocity also tells us in what direction it is moving. When the object is moving forward (s increasing) the velocity is positive; when the body is moving backward (s decreasing) the velocity is negative (Fig. 2.22).

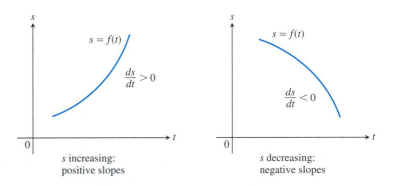

s increasing: positive slopes

s decreasing: negative slopes

2.22 $v = ds/dt$ is positive when s increases and negative when s decreases.

If we drive to a friend's house and back at 30 mph, say, the speedometer will show 30 on the way over but it will not show -30 on the way back, even though our distance from home is decreasing. The speedometer always shows speed, which is the absolute value of velocity. Speed measures the rate of forward progress regardless of direction.

Definition

Speed is the absolute value of velocity.

$$\text{Speed} \; = |v(t)| = \left| \frac{ds}{dt} \right|$$

EXAMPLE 3   Figure 2.23 shows the velocity $v = f'(t)$ of a particle moving on a coordinate line. The particle moves forward for the first 3 seconds, moves backward for the next 2 seconds, stands still for a second, and moves forward again. Notice that the particle achieves its greatest speed at time $t = 4$, while moving backward.

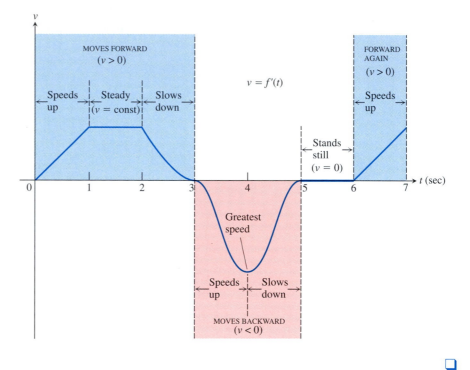

2.23 The velocity graph for Example 3.

The rate at which a body's velocity changes is called the body's acceleration. The acceleration measures how quickly the body picks up or loses speed.

Definition

Acceleration is the derivative of velocity with respect to time. If a body's position at time t is $s = f(t)$, then the body's acceleration at time t is

$$a(t) = \frac{dv}{dt} = \frac{d^2s}{dt^2}.$$

We can illustrate all this with free fall. As we mentioned at the beginning of Chapter 1, near the surface of the earth all bodies fall with the same constant acceleration. When air resistance is absent or insignificant and the only force acting on a falling body is the force of gravity, we call the way the body falls **free fall.** The mathematical description of this type of motion captured the imagination of many great scientists, including Aristotle, Galileo, and Newton. Experimental and theoretical investigations revealed that the distance a body released from rest falls in time t is proportional to the square of the amount of time it has fallen. We express this by saying that

$$s = \frac{1}{2}gt^2,$$

where s is distance and g is the acceleration due to Earth's gravity. This equation holds in a vacuum, where there is no air resistance, but it closely models the fall of dense, heavy objects, such as rocks or steel tools, for the first few seconds of their fall, before air resistance starts to slow them down.

The value of g in the equation $s = (1/2)gt^2$ depends on the units used to measure t and s. With t in seconds (the usual unit), we have the following values:

Free-Fall Equations (Earth)

English units: $g = 32\dfrac{\text{ft}}{\text{sec}^2}$, $s = \dfrac{1}{2}(32)t^2 = 16t^2$ (s in feet)

Metric units: $g = 9.8\dfrac{\text{m}}{\text{sec}^2}$, $s = \dfrac{1}{2}(9.8)t^2 = 4.9t^2$ (s in meters)

The abbreviation ft/sec^2 is read "feet per second squared" or "feet per second per second," and m/sec^2 is read "meters per second squared."

This description allows us to answer many questions concerning the position and velocity of a falling object.

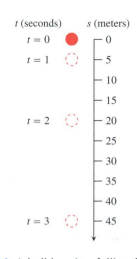

t (seconds)   s (meters)

$t = 0$ ● ┌ 0

$t = 1$ ○ ├ 5

├ 10

├ 15

$t = 2$ ○ ├ 20

├ 25

├ 30

├ 35

├ 40

$t = 3$ ○ ├ 45

2.24 A ball bearing falling from rest (Example 4).

EXAMPLE 4   Figure 2.24 shows the free fall of a heavy ball bearing released from rest at time $t = 0$ sec.

a) How many meters does the ball fall in the first 2 sec?
b) What is its velocity, speed, and acceleration then?

Solution

a) The metric free-fall equation is $s = 4.9t^2$. During the first 2 sec, the ball falls

$$s(2) = 4.9(2)^2 = 19.6 \text{ m}.$$

b) At any time t, *velocity* is the derivative of displacement:

$$v(t) = s'(t) = \frac{d}{dt}(4.9t^2) = 9.8t.$$

At $t = 2$, the velocity is

$$v(2) = 19.6 \text{ m/sec}$$

in the downward (increasing s) direction. The *speed* at $t = 2$ is

$$\text{speed} = |v(2)| = 19.6 \text{ m/sec}.$$

2.25 (a) The rock in Example 5. (b) The graphs of *s* and *v* as functions of time; *s* is largest when *v* = *ds*/*dt* = 0. The graph of *s* is *not* the path of the rock: it is a plot of height vs. time. The slope is the rock's velocity.

The *acceleration* at any time *t* is

$$a(t) = v'(t) = s''(t) = 9.8 \text{ m/sec}^2.$$

At *t* = 2, the acceleration is 9.8 m/sec². ❑

EXAMPLE 5   A dynamite blast blows a heavy rock straight up with a launch velocity of 160 ft/sec (about 109 mph) (Fig. 2.25a). It reaches a height of *s* = $160t - 16t^2$ ft after *t* sec.

a)   How high does the rock go?

b)   What is the velocity and speed of the rock when it is 256 ft above the ground on the way up? on the way down?

c)   What is the acceleration of the rock at any time *t* during its flight (after the blast)?

d)   When does the rock hit the ground again?

Solution

a)   In the coordinate system we have chosen, *s* measures height from the ground up, so the velocity is positive on the way up and negative on the way down. The instant the rock is at its highest point is the one instant during the flight when the velocity is 0. Therefore, to find the maximum height, all we need to do is to find when *v* = 0 and evaluate *s* at this time.

   At any time *t*, the velocity is

$$v = \frac{ds}{dt} = \frac{d}{dt}(160t - 16t^2) = 160 - 32t \text{ ft/sec}.$$

The velocity is zero when

$$160 - 32t = 0, \quad \text{or} \quad t = 5 \text{ sec}.$$

The rock's height at *t* = 5 sec is

$$s_{\text{max}} = s(5) = 160(5) - 16(5)^2 = 800 - 400 = 400 \text{ ft}.$$

See Fig. 2.25(b).

b)   To find the rock's velocity at 256 ft on the way up and again on the way down, we find the two values of *t* for which

$$s(t) = 160t - 16t^2 = 256.$$

To solve this equation we write

$$16t^2 - 160t + 256 = 0$$

$$16(t^2 - 10t + 16) = 0$$

$$(t - 2)(t - 8) = 0$$

$$t = 2 \text{ sec}, \quad t = 8 \text{ sec}.$$

The rock is 256 ft above the ground 2 sec after the explosion and again 8 sec after the explosion. The rock's velocities at these times are

$$v(2) = 160 - 32(2) = 160 - 64 = 96 \text{ ft/sec},$$

$$v(8) = 160 - 32(8) = 160 - 256 = -96 \text{ ft/sec}.$$

At both instants, the rock's speed is 96 ft/sec.

c)   At any time during its flight following the explosion, the rock's acceleration is

$$a = \frac{dv}{dt} = \frac{d}{dt}(160 - 32t) = -32 \text{ ft/sec}^2.$$

The acceleration is always downward. When the rock is rising, it is slowing down; when it is falling, it is speeding up.

d)   The rock hits the ground at the positive time t for which $s = 0$. The equation $160t - 16t^2 = 0$ factors to give $16t(10 - t) = 0$, so it has solutions $t = 0$ and $t = 10$. At $t = 0$ the blast occurred and the rock was thrown upward. It returned to the ground 10 seconds later.   ❑

Technology   *Simulation of Motion on a Vertical Line*   The parametric equations

$$x(t) = c, \qquad y(t) = f(t)$$

will illuminate pixels along the vertical line $x = c$. If $f(t)$ denotes the height of a moving body at time t, graphing $(x(t), y(t)) = (c, f(t))$ will simulate the actual motion. Try it for the rock in Example 5 with $x(t) = 2$, say, and $y(t) = 160t - 16t^2$, in dot mode with tStep $= 0.1$. Why does the spacing of the dots vary? Why does the grapher seem to stop after it reaches the top? (Try the plots for $0 \le t \le 5$ and $5 \le t \le 10$ separately.)

For a second experiment, plot the parametric equations

$$x(t) = t, \qquad y(t) = 160t - 16t^2$$

together with the vertical line simulation of the motion, again in dot mode. Use what you know about the behavior of the rock from the calculations of Example 5 to select a window size that will display all the interesting behavior.

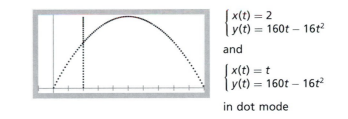

$$\begin{cases} x(t) = 2 \\ y(t) = 160t - 16t^2 \end{cases}$$

and

$$\begin{cases} x(t) = t \\ y(t) = 160t - 16t^2 \end{cases}$$

in dot mode

Sensitivity to Change

When a small change in x produces a large change in the value of a function $f(x)$, we say that the function is relatively **sensitive** to changes in x. The derivative $f'(x)$ is a measure of the sensitivity to change at x.

Why peas wrinkle

British geneticists have recently discovered that the wrinkling trait comes from an extra piece of DNA that prevents the gene that directs starch synthesis from functioning properly. With the plant's starch conversion impaired, sucrose and water build up in the young seeds. As the seeds mature, they lose much of this water, and the shrinkage leaves them wrinkled.

EXAMPLE 6   *Sensitivity to change*

The Austrian monk Gregor Johann Mendel (1822–1884), working with garden peas and other plants, provided the first scientific explanation of hybridization. His careful records showed that if p (a number between 0 and 1) is the frequency of the gene for smooth skin in peas (dominant) and $(1 - p)$ is the frequency of the gene for wrinkled skin in peas, then the proportion of smooth-skinned peas in the population at large is

$$y = 2p(1 - p) + p^2 = 2p - p^2.$$

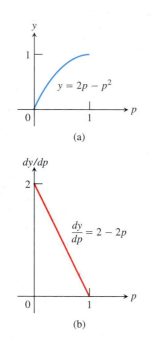

2.26 (a) The graph of $y = 2p - p^2$, describing the proportion of smooth-skinned peas. (b) The graph of dy/dp.

The graph of y versus p in Fig. 2.26(a) suggests that the value of y is more sensitive to a change in p when p is small than when p is large. Indeed, this is borne out by the derivative graph in Fig. 2.26(b), which shows that dy/dp is close to 2 when p is near 0 and close to 0 when p is near 1.

We will say more about sensitivity in Section 3.7.   ❑

Derivatives in Economics

Engineers use the terms *velocity* and *acceleration* to refer to the derivatives of functions describing motion. Economists, too, have a specialized vocabulary for rates of change and derivatives. They call them *marginals*.

In a manufacturing operation, the *cost of production* $c(x)$ is a function of x, the number of units produced. The *marginal cost of production* is the rate of change of cost (c) with respect to level of production (x), so it is dc/dx.

For example, let $c(x)$ represent the dollars needed to produce x tons of steel in one week. It costs more to produce $x + h$ units, and the cost difference, divided by h, is the average increase in cost per ton per week:

$$\frac{c(x+h) - c(x)}{h} = \begin{array}{l} \text{average increase in cost/ton/wk} \\ \text{to produce the next } h \text{ tons of steel} \end{array}$$

The limit of this ratio as $h \to 0$ is the *marginal cost* of producing more steel when the current production level is x tons (Fig. 2.27):

$$\frac{dc}{dx} = \lim_{h \to 0} \frac{c(x+h) - c(x)}{h} = \text{marginal cost of production.}$$

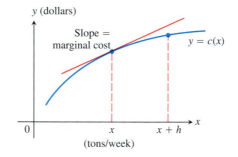

2.27 Weekly steel production: $c(x)$ is the cost of producing x tons per week. The cost of producing an additional h tons is $c(x + h) - c(x)$.

2.28 The marginal cost dc/dx is approximately the extra cost Δc of producing $\Delta x = 1$ more unit.

Sometimes the marginal cost of production is loosely defined to be the extra cost of producing one unit:

$$\frac{\Delta c}{\Delta x} = \frac{c(x+1) - c(x)}{1},$$

which is approximately the value of dc/dx at x. To see why this is an acceptable approximation, observe that if the slope of c does not change quickly near x, then the difference quotient will be close to its limit, the derivative dc/dx, even if $\Delta x = 1$ (Fig. 2.28). In practice, the approximation works best for large values of x.

EXAMPLE 7   *Marginal cost*

Suppose it costs

$$c(x) = x^3 - 6x^2 + 15x$$

Choosing functions to illustrate economics

In case you are wondering why economists use polynomials of low degree to illustrate complicated phenomena like cost and revenue, here is the rationale: While formulas for real phenomena are rarely available in any given instance, the theory of economics can still provide valuable guidance. The functions about which theory speaks can often be illustrated with low degree polynomials on relevant intervals. Cubic polynomials provide a good balance between being easy to work with and being complicated enough to illustrate important points.

dollars to produce x radiators when 8 to 30 radiators are produced. Your shop currently produces 10 radiators a day. About how much extra will it cost to produce one more radiator a day?

Solution   The cost of producing one more radiator a day when 10 are produced is about $c'(10)$:

$$c'(x) = \frac{d}{dx}(x^3 - 6x^2 + 15x) = 3x^2 - 12x + 15$$

$$c'(10) = 3(100) - 12(10) + 15 = 195.$$

The additional cost will be about $195.   ❑

EXAMPLE 8   *Marginal tax rate*

To get some feel for the language of marginal rates, consider marginal tax rates. If your marginal income tax rate is 28% and your income increases by $1,000, you can expect to have to pay an extra $280 in income taxes. This does not mean that you pay 28% of your entire income in taxes. It just means that at your current income level I, the rate of increase of taxes T with respect to income is $dT/dI = 0.28$. You will pay $0.28 out of every extra dollar you earn in taxes. Of course, if you earn a lot more, you may land in a higher tax bracket and your marginal rate will increase.   ❑

EXAMPLE 9   *Marginal revenue*

If

$$r(x) = x^3 - 3x^2 + 12x$$

gives the dollar revenue from selling x thousand candy bars, $5 \le x \le 20$, the marginal revenue when x thousand are sold is

$$r'(x) = \frac{d}{dx}(x^3 - 3x^2 + 12x) = 3x^2 - 6x + 12.$$

As with marginal cost, the marginal revenue function estimates the increase in revenue that will result from selling one additional unit. If you currently sell 10 thousand candy bars a week, you can expect your revenue to increase by about

$$r'(10) = 3(100) - 6(10) + 12 = \$252$$

if you increase sales to 11 thousand bars a week.   ❑

Exercises 2.3

Motion Along a Coordinate Line

Exercises 1–6 give the position $s = f(t)$ of a body moving on a coordinate line for $a \le t \le b$, with s in meters and t in seconds.

a) Find the body's displacement and average velocity for the given time interval.

b) Find the body's speed and acceleration at the endpoints of the interval.

c) When during the interval does the body change direction (if ever)?

1. $s = 0.8t^2$, $0 \le t \le 10$ (free fall on the moon)

2. $s = 1.86t^2$, $0 \le t \le 0.5$ (free fall on Mars)

3. $s = -t^3 + 3t^2 - 3t$, $0 \le t \le 3$

4. $s = (t^4/4) - t^3 + t^2$, $0 \le t \le 2$

5. $s = \dfrac{25}{t^2} - \dfrac{5}{t}$, $1 \leq t \leq 5$

6. $s = \dfrac{25}{t+5}$, $-4 \leq t \leq 0$

7. At time t, the position of a body moving along the s-axis is $s = t^3 - 6t^2 + 9t$ m. (a) Find the body's acceleration each time the velocity is zero. (b) Find the body's speed each time the acceleration is zero. (c) Find the total distance traveled by the body from $t = 0$ to $t = 2$.

8. At time $t \geq 0$, the velocity of a body moving along the s-axis is $v = t^2 - 4t + 3$. (a) Find the body's acceleration each time the velocity is zero. (b) When is the body moving forward? moving backward? (c) When is the body's velocity increasing? decreasing?

Free-Fall Applications

9. The equations for free fall at the surfaces of Mars and Jupiter (s in meters, t in seconds) are $s = 1.86t^2$ on Mars, $s = 11.44t^2$ on Jupiter. How long would it take a rock falling from rest to reach a velocity of 27.8 m/sec (about 100 km/h) on each planet?

10. A rock thrown vertically upward from the surface of the moon at a velocity of 24 m/sec (about 86 km/h) reaches a height of $s = 24t - 0.8t^2$ meters in t seconds.

    a) Find the rock's velocity and acceleration at time t. (The acceleration in this case is the acceleration of gravity on the moon.)
    b) How long does it take the rock to reach its highest point?
    c) How high does the rock go?
    d) How long does it take the rock to reach half its maximum height?
    e) How long is the rock aloft?

11. On Earth, in the absence of air, the rock in Exercise 10 would reach a height of $s = 24t - 4.9t^2$ meters in t seconds.

    a) Find the rock's velocity and acceleration at time t.
    b) How long would it take the rock to reach its highest point?
    c) How high would the rock go?
    d) How long would it take the rock to reach half its maximum height?
    e) How long would the rock be aloft?

12. Explorers on a small airless planet used a spring gun to launch a ball bearing vertically upward from the surface at a launch velocity of 15 m/sec. Because the acceleration of gravity at the planet's surface was g_s m/sec^2, the explorers expected the ball bearing to reach a height of $s = 15t - (1/2)g_st^2$ meters t seconds later. The ball bearing reached its maximum height 20 sec after being launched. What was the value of g_s?

13. A 45-caliber bullet fired straight up from the surface of the moon would reach a height of $s = 832t - 2.6t^2$ feet after t seconds. On Earth, in the absence of air, its height would be $s = 832t - 16t^2$ feet after t seconds. How long will the bullet be aloft in each case? How high would the bullet go?

14. (Continuation of Exercise 13.) On Jupiter, in the absence of air,

the bullet's height would be $s = 832t - 37.53t^2$ feet after t seconds. On Mars it would be $s = 832t - 6.1t^2$ feet after t seconds. How high would the bullet go in each case?

15. *Galileo's free-fall formula.* Galileo developed a formula for a body's velocity during free fall by rolling balls from rest down increasingly steep inclined planks and looking for a limiting formula that would predict a ball's behavior when the plank was vertical and the ball fell freely (part a of the accompanying figure). He found that, for any given angle of the plank, the ball's velocity t seconds into the motion was a constant multiple of t. That is, the velocity was given by a formula of the form $v = kt$. The value of the constant k depended on the inclination of the plank.

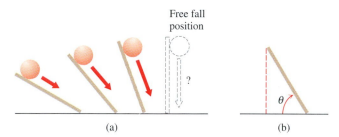

In modern notation (part b of the figure), with distance in meters and time in seconds, what Galileo determined by experiment was that, for any given angle θ, the ball's velocity t seconds into the roll was

$$v = 9.8(\sin \theta)t \text{ m/sec.}$$

    a) What is the equation for the ball's velocity during free fall?
    b) Building on your work in (a), what constant acceleration does a freely falling body experience near the surface of the earth?

16. *Free fall from the tower of Pisa.* Had Galileo dropped a cannonball from the tower of Pisa, 179 ft above the ground, the ball's height aboveground t seconds into the fall would have been $s = 179 - 16t^2$.

    a) What would have been the ball's velocity, speed, and acceleration at time t?
    b) About how long would it have taken the ball to hit the ground?
    c) What would have been the ball's velocity at the moment of impact?

Conclusions About Motion from Graphs

17. The accompanying figure shows the velocity $v = ds/dt = f(t)$ (m/sec) of a body moving along a coordinate line.

a) When does the body reverse direction?

b) When (approximately) is the body moving at a constant speed?

c) Graph the body's speed for $0 \le t \le 10$.

d) Graph the acceleration, where defined.

18. A particle P moves on the number line shown in part (a) of the accompanying figure. Part (b) shows the position of P as a function of time t.

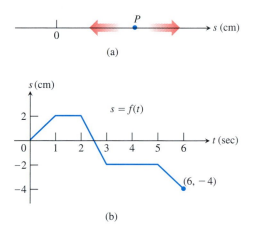

(a)

(b)

a) When is P moving to the left? moving to the right? standing still?

b) Graph the particle's velocity and speed (where defined).

19. When a model rocket is launched, the propellant burns for a few seconds, accelerating the rocket upward. After burnout, the rocket coasts upward for a while and then begins to fall. A small explosive charge pops out a parachute shortly after the rocket starts down. The parachute slows the rocket to keep it from breaking when it lands.

The figure here shows velocity data from the flight of the model rocket. Use the data to answer the following.

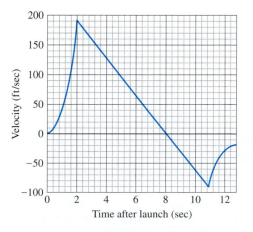

a) How fast was the rocket climbing when the engine stopped?

b) For how many seconds did the engine burn?

c) When did the rocket reach its highest point? What was its velocity then?

d) When did the parachute pop out? How fast was the rocket falling then?

e) How long did the rocket fall before the parachute opened?

f) When was the rocket's acceleration greatest?

g) When was the acceleration constant? What was its value then (to the nearest integer)?

20. The accompanying figure shows the velocity $v = f(t)$ of a particle moving on a coordinate line.

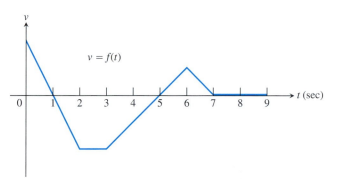

a) When does the particle move forward? move backward? speed up? slow down?

b) When is the particle's acceleration positive? negative? zero?

c) When does the particle move at its greatest speed?

d) When does the particle stand still for more than an instant?

21. The graph here shows the position s of a truck traveling on a highway. The truck starts at $t = 0$ and returns 15 hours later at $t = 15$.

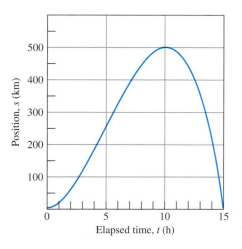

a) Use the technique described in Section 2.1, Example 4, to graph the truck's velocity $v = ds/dt$ for $0 \le t \le 15$. Then repeat the process, with the velocity curve, to graph the truck's acceleration dv/dt.

b) Suppose $s = 15t^2 - t^3$. Graph ds/dt and d^2s/dt^2 and compare your graphs with those in (a).

22. The multiflash photograph in Fig. 2.29 on the following page shows two balls falling from rest. The vertical rulers are marked in centimeters. Use the equation $s = 490t^2$ (the free-fall equation for s in centimeters and t in seconds) to answer the following questions.

a) How long did it take the balls to fall the first 160 cm? What was their average velocity for the period?

b) How fast were the balls falling when they reached the 160-cm mark? What was their acceleration then?

c) About how fast was the light flashing (flashes per second)?

2.29 Two balls falling from rest (Exercise 22).

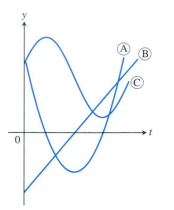

2.30 The graphs for Exercise 23.

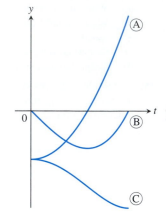

2.31 The graphs for Exercise 24.

23. The graphs in Fig. 2.30 show the position s, velocity $v = ds/dt$, and acceleration $a = d^2s/dt^2$ of a body moving along a coordinate line as functions of time t. Which graph is which? Give reasons for your answers.

24. The graphs in Fig. 2.31 show the position s, the velocity $v = ds/dt$, and the acceleration $a = d^2s/dt^2$ of a body moving along the coordinate line as functions of time t. Which graph is which? Give reasons for your answers.

Economics

25. *Marginal cost.* Suppose that the dollar cost of producing x washing machines is $c(x) = 2000 + 100x - 0.1x^2$.

a) Find the average cost per machine of producing the first 100 washing machines.

b) Find the marginal cost when 100 washing machines are produced.

c) Show that the marginal cost when 100 washing machines are produced is approximately the cost of producing one more washing machine after the first 100 have been made, by calculating the latter cost directly.

26. *Marginal revenue.* Suppose the revenue from selling x custom-made office desks is

$$r(x) = 2000 \left(1 - \frac{1}{x+1} \right)$$

dollars.

a) Find the marginal revenue when x desks are produced.

b) Use the function $r'(x)$ to estimate the increase in revenue that will result from increasing production from 5 desks a week to 6 desks a week.

c) Find the limit of $r'(x)$ as $x \to \infty$. How would you interpret this number?

Additional Applications

27. When a bactericide was added to a nutrient broth in which bacteria were growing, the bacterium population continued to grow for a while, but then stopped growing and began to decline. The size of the population at time t (hours) was $b = 10^6 + 10^4 t - 10^3 t^2$. Find the growth rates at (a) $t = 0$; (b) $t = 5$; and (c) $t = 10$ hours.

28. The number of gallons of water in a tank t minutes after the tank has started to drain is $Q(t) = 200(30 - t)^2$. How fast is the water running out at the end of 10 min? What is the average rate at which the water flows out during the first 10 min?

29. It takes 12 hours to drain a storage tank by opening the valve at the bottom. The depth y of fluid in the tank t hours after the valve is opened is given by the formula

$$y = 6 \left(1 - \frac{t}{12} \right)^2 \text{ m.}$$

a) Find the rate dy/dt (m/h) at which the tank is draining at time t.

b) When is the fluid level in the tank falling fastest? slowest? What are the values of dy/dt at these times?

c) GRAPHER Graph y and dy/dt together and discuss the behavior of y in relation to the signs and values of dy/dt.

30. The volume $V = (4/3)\pi r^3$ of a spherical balloon changes with the radius.

a) At what rate does the volume change with respect to the radius when $r = 2$ ft?

b) By approximately how much does the volume increase when the radius changes from 2 to 2.2 ft?

31. Suppose that the distance an aircraft travels along a runway before takeoff is given by $D = (10/9)t^2$, where D is measured in meters from the starting point and t is measured in seconds from the time the brakes are released. If the aircraft will become airborne when its speed reaches 200 km/hr, how long will it take to become airborne, and what distance will it travel in that time?

32. *Volcanic lava fountains.* Although the November 1959 Kilauea Iki eruption on the island of Hawaii began with a line of fountains along the wall of the crater, activity was later confined to a single vent in the crater's floor, which at one point shot lava 1900 ft straight into the air (a world record). What was the lava's exit velocity in feet per second? in miles per hour?

(Hint: If v_0 is the exit velocity of a particle of lava, its height t seconds later will be $s = v_0 t - 16t^2$ feet. Begin by finding the time at which $ds/dt = 0$. Neglect air resistance.)

Grapher Explorations

Exercises 33–36 give the position function $s = f(t)$ of a body moving along the s-axis as a function of time t. Graph f together with the velocity function $v(t) = ds/dt = f'(t)$ and the acceleration function $a(t) = d^2s/dt^2 = f''(t)$. Comment on the body's behavior in relation to the signs and values v and a. Include in your commentary such topics as the following.

a) When is the body momentarily at rest?

b) When does it move to the left (down) or to the right (up)?

c) When does it change direction?

d) When does it speed up and slow down?

e) When is it moving fastest (highest speed)? slowest?

f) When is it farthest from the axis origin?

33. $s = 200t - 16t^2, \quad 0 \le t \le 12.5$ (A heavy object fired straight up from the earth's surface at 200 ft/sec)

34. $s = t^2 - 3t + 2, \quad 0 \le t \le 5$

35. $s = t^3 - 6t^2 + 7t, \quad 0 \le t \le 4$

36. $s = 4 - 7t + 6t^2 - t^3, \quad 0 \le t \le 4$

2.4 Derivatives of Trigonometric Functions

Trigonometric functions are important because so many of the phenomena we want information about are periodic (electromagnetic fields, heart rhythms, tides, weather). A surprising and beautiful theorem from advanced calculus says that every periodic function we are likely to use in mathematical modeling can be written as an algebraic combination of sines and cosines, so the derivatives of sines and cosines play a key role in describing important changes. This section shows how to differentiate the six basic trigonometric functions.

Some Special Limits

Our first step is to establish some inequalities and limits. It is assumed throughout that angles are measured in radians.

Theorem 3

If θ is measured in radians, then

$$-|\theta| < \sin\theta < |\theta| \quad \text{and} \quad -|\theta| < 1 - \cos\theta < |\theta|.$$

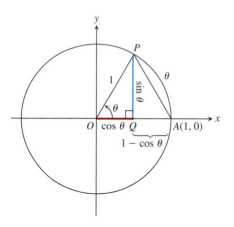

2.32 From the geometry of this figure, drawn for $\theta > 0$, we get the inequality $\sin^2\theta + (1 - \cos\theta)^2 < \theta^2$.

Proof To establish these inequalities, we picture θ as an angle in standard position (Fig. 2.32). The circle in the figure is a unit circle, so $|\theta|$ equals the length of the circular arc AP. The length of line segment AP is therefore less than $|\theta|$.

Triangle APQ is a right triangle with sides of length

$$QP = |\sin\theta|, \qquad AQ = 1 - \cos\theta.$$

From the Pythagorean theorem and the fact that $AP < |\theta|$, we get

$$\sin^2\theta + (1 - \cos\theta)^2 = (AP)^2 < \theta^2. \qquad (1)$$

The terms on the left side of Eq. (1) are both positive, so each is smaller than their sum and hence is less than θ^2:

$$\sin^2\theta < \theta^2 \qquad \text{and} \qquad (1 - \cos\theta)^2 < \theta^2.$$

By taking square roots, we can see that this is equivalent to saying that

$$|\sin\theta| < |\theta| \qquad \text{and} \qquad |1 - \cos\theta| < |\theta|$$

or

$$-|\theta| < \sin\theta < |\theta| \qquad \text{and} \qquad -|\theta| < 1 - \cos\theta < |\theta|. \qquad \square$$

EXAMPLE 1 Show that $\sin\theta$ and $\cos\theta$ are continuous at $\theta = 0$. That is,

$$\lim_{\theta \to 0} \sin\theta = 0 \qquad \text{and} \qquad \lim_{\theta \to 0} \cos\theta = 1.$$

Solution As $\theta \to 0$, both $|\theta|$ and $-|\theta|$ approach 0. The values of the limits therefore follow immediately from Theorem 3 and the Sandwich Theorem. $\square$

The function $f(\theta) = (\sin\theta)/\theta$ graphed in Fig. 2.33 appears to have a removable discontinuity at $\theta = 0$. As the figure suggests, $\lim_{\theta \to 0} f(\theta) = 1$.

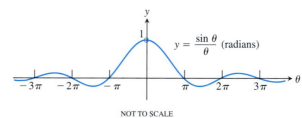

NOT TO SCALE

2.33 The graph of $f(\theta) = (\sin\theta)/\theta$.

Theorem 4

$$\lim_{\theta \to 0} \frac{\sin\theta}{\theta} = 1 \qquad (\theta \text{ in radians}) \qquad (2)$$

Proof The plan is to show that the right-hand and left-hand limits are both 1. Then we will know that the two-sided limit is 1 as well.

To show that the right-hand limit is 1, we begin with values of θ that are positive and less than $\pi/2$ (Fig. 2.34). Notice that

$$\text{Area } \triangle OAP < \text{area sector } OAP < \text{area } \triangle OAT.$$

We can express these areas in terms of θ as follows:

$$\text{Area } \triangle OAP = \frac{1}{2}\text{base} \times \text{height} = \frac{1}{2}(1)(\sin\theta) = \frac{1}{2}\sin\theta$$

$$\text{Area sector } OAP = \frac{1}{2}r^2\theta = \frac{1}{2}(1)^2\theta = \frac{\theta}{2} \qquad (3)$$

$$\text{Area } \triangle OAT = \frac{1}{2}\text{base} \times \text{height} = \frac{1}{2}(1)(\tan\theta) = \frac{1}{2}\tan\theta,$$

so

$$\frac{1}{2}\sin\theta < \frac{1}{2}\theta < \frac{1}{2}\tan\theta.$$

This last inequality will go the same way if we divide all three terms by the positive number $(1/2)\sin\theta$:

$$1 < \frac{\theta}{\sin\theta} < \frac{1}{\cos\theta}.$$

We next take reciprocals, which reverses the inequalities:

$$1 > \frac{\sin\theta}{\theta} > \cos\theta.$$

Since $\lim_{\theta\to 0^+}\cos\theta = 1$, the Sandwich Theorem gives

$$\lim_{\theta\to 0^+}\frac{\sin\theta}{\theta} = 1.$$

Finally, observe that $\sin\theta$ and θ are both *odd functions*. Therefore, $f(\theta) = (\sin\theta)/\theta$ is an *even function*, with a graph symmetric about the y-axis (see Fig. 2.33). This symmetry implies that the left-hand limit at 0 exists and has the same value as the right-hand limit:

$$\lim_{\theta\to 0^-}\frac{\sin\theta}{\theta} = 1 = \lim_{\theta\to 0^+}\frac{\sin\theta}{\theta},$$

so $\lim_{\theta\to 0}(\sin\theta)/\theta = 1$ by Theorem 5 of Section 1.4.   ❑

Theorem 4 can be combined with limit rules and known trigonometric identities to yield other trigonometric limits.

EXAMPLE 2   Show that $\lim_{h\to 0}\dfrac{\cos h - 1}{h} = 0$.

Solution   Using the half-angle formula $\cos h = 1 - 2\sin^2(h/2)$, we calculate

$$\lim_{h\to 0}\frac{\cos h - 1}{h} = \lim_{h\to 0} -\frac{2\sin^2(h/2)}{h}$$

$$= -\lim_{\theta\to 0}\frac{\sin\theta}{\theta}\sin\theta \qquad \text{Let } \theta = h/2.$$

$$= -(1)(0) = 0.$$ ❑

Equation (3) is where radian measure comes in: The area of sector OAP is $\theta/2$ only if θ is measured in radians.

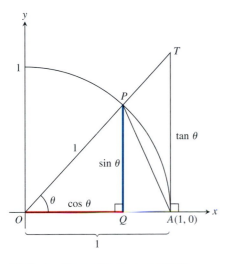

2.34 The figure for the proof of Theorem 4. $TA/OA = \tan\theta$, but $OA = 1$, so $TA = \tan\theta$.

$y_1 = \sin x,\ -2\pi \le x \le 2\pi$
$y_2 = d(y_1)/dx,\ -2\pi \le x \le 2\pi$

Technology   *Conjectures Based on Grapher Images*   What you see in the window of a graphing utility can suggest conjectures, sometimes rather strongly. Graph the functions

$$y_1 = \sin x$$

$$y_2 = d(y_1)/dx \qquad \text{(This is computed by a built-in differentiation utility.)}$$

Does the graph of y_2 look familiar? What function do you think it is? Test your conjecture by adding the function's graph to the screen.

The Derivative of the Sine

To calculate the derivative of $y = \sin x$, we combine the limits in Example 2 and Theorem 4 with the addition formula

$$\sin(x + h) = \sin x \cos h + \cos x \sin h. \tag{4}$$

We have

$$\frac{dy}{dx} = \lim_{h \to 0} \frac{\sin(x + h) - \sin x}{h} \qquad \text{Derivative definition}$$

$$= \lim_{h \to 0} \frac{(\sin x \cos h + \cos x \sin h) - \sin x}{h} \qquad \text{Eq. (4)}$$

$$= \lim_{h \to 0} \frac{\sin x (\cos h - 1) + \cos x \sin h}{h}$$

$$= \lim_{h \to 0} \left(\sin x \cdot \frac{\cos h - 1}{h} \right) + \lim_{h \to 0} \left(\cos x \cdot \frac{\sin h}{h} \right)$$

$$= \sin x \cdot \lim_{h \to 0} \frac{\cos h - 1}{h} + \cos x \cdot \lim_{h \to 0} \frac{\sin h}{h} \qquad \text{Example 2 and Theorem 4}$$

$$= \sin x \cdot 0 + \cos x \cdot 1$$

$$= \cos x.$$

In short, the derivative of the sine is the cosine.

$$\frac{d}{dx}(\sin x) = \cos x$$

EXAMPLE 3

a) $y = x^2 - \sin x$: $\dfrac{dy}{dx} = 2x - \dfrac{d}{dx}(\sin x)$ Difference Rule

$$= 2x - \cos x$$

b) $y = x^2 \sin x$: $\dfrac{dy}{dx} = x^2 \dfrac{d}{dx}(\sin x) + 2x \sin x$ Product Rule

$$= x^2 \cos x + 2x \sin x$$

Radian measure in calculus

In case you are wondering why calculus uses radian measure when the rest of the world seems to use degrees, the answer lies in the argument that the derivative of the sine is the cosine. The derivative of $\sin x$ is $\cos x$ *only* if x is measured in radians. The argument requires that when h is a small increment in x,

$$\lim_{h \to 0} (\sin h)/h = 1.$$

This is true only for radian measure, as we saw during the proof of Theorem 4. You will see what the degree-mode derivatives of the sine and cosine are if you do Exercise 76.

c) $\quad y = \dfrac{\sin x}{x}:$

$$\frac{dy}{dx} = \frac{x \cdot \dfrac{d}{dx}(\sin x) - \sin x \cdot 1}{x^2} \qquad \text{Quotient Rule}$$

$$= \frac{x \cos x - \sin x}{x^2} \qquad\qquad\quad \square$$

The Derivative of the Cosine

With the help of the addition formula,

$$\cos(x + h) = \cos x \cos h - \sin x \sin h, \tag{5}$$

we have

$$\frac{d}{dx}(\cos x) = \lim_{h \to 0} \frac{\cos(x + h) - \cos x}{h} \qquad \begin{array}{l}\text{Derivative}\\\text{definition}\end{array}$$

$$= \lim_{h \to 0} \frac{(\cos x \cos h - \sin x \sin h) - \cos x}{h} \qquad \text{Eq. (5)}$$

$$= \lim_{h \to 0} \frac{\cos x (\cos h - 1) - \sin x \sin h}{h}$$

$$= \lim_{h \to 0} \cos x \cdot \frac{\cos h - 1}{h} - \lim_{h \to 0} \sin x \cdot \frac{\sin h}{h}$$

$$= \cos x \cdot \lim_{h \to 0} \frac{\cos h - 1}{h} - \sin x \cdot \lim_{h \to 0} \frac{\sin h}{h}$$

$$= \cos x \cdot 0 - \sin x \cdot 1 \qquad\qquad \begin{array}{l}\text{Example 2 and}\\\text{Theorem 4}\end{array}$$

$$= -\sin x.$$

In short, the derivative of the cosine is the negative of the sine.

$$\boxed{\frac{d}{dx}(\cos x) = -\sin x}$$

Figure 2.35 shows another way to visualize this result.

2.35 The curve $y' = -\sin x$ as the graph of the slopes of the tangents to the curve $y = \cos x$.

EXAMPLE 4

a) $\quad y = 5x + \cos x$

$$\frac{dy}{dx} = \frac{d}{dx}(5x) + \frac{d}{dx}(\cos x) \qquad \text{Sum Rule}$$

$$= 5 - \sin x$$

b) $\quad y = \sin x \cos x$

$$\frac{dy}{dx} = \sin x \frac{d}{dx}(\cos x) + \cos x \frac{d}{dx}(\sin x) \qquad \text{Product Rule}$$

$$= \sin x (-\sin x) + \cos x (\cos x)$$

$$= \cos^2 x - \sin^2 x$$

c) $\quad y = \dfrac{\cos x}{1 - \sin x}$

$$\frac{dy}{dx} = \frac{(1 - \sin x)\dfrac{d}{dx}(\cos x) - \cos x \dfrac{d}{dx}(1 - \sin x)}{(1 - \sin x)^2} \qquad \text{Quotient Rule}$$

$$= \frac{(1 - \sin x)(- \sin x) - \cos x(0 - \cos x)}{(1 - \sin x)^2}$$

$$= \frac{1 - \sin x}{(1 - \sin x)^2} \qquad \sin^2 x + \cos^2 x = 1$$

$$= \frac{1}{1 - \sin x} \qquad\qquad\qquad \square$$

Simple Harmonic Motion

The motion of a body bobbing up and down on the end of a spring is an example of *simple harmonic motion*. The next example describes a case in which there are no opposing forces like friction or buoyancy to slow the motion down.

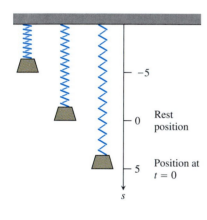

2.36 The body in Example 5.

EXAMPLE 5   A body hanging from a spring (Fig. 2.36) is stretched 5 units beyond its rest position and released at time $t = 0$ to bob up and down. Its position at any later time t is

$$s = 5 \cos t.$$

What are its velocity and acceleration at time t?

Solution   We have

Position: $s = 5 \cos t$

Velocity: $v = \dfrac{ds}{dt} = \dfrac{d}{dt}(5 \cos t) = 5\dfrac{d}{dt}(\cos t) = -5 \sin t$

Acceleration: $a = \dfrac{dv}{dt} = \dfrac{d}{dt}(-5 \sin t) = -5\dfrac{d}{dt}(\sin t) = -5 \cos t.$

Here is what we can learn from these equations:

1. As time passes, the body moves up and down between $s = 5$ and $s = -5$ on the s-axis. The amplitude of the motion is 5. The period of the motion is 2π, the period of $\cos t$.

2. The function $\sin t$ attains its greatest magnitude (1) when $\cos t = 0$, as the graphs of the sine and cosine show (Fig. 2.37). Hence, the body's speed, $|v| = 5|\sin t|$, is greatest every time $\cos t = 0$, i.e., every time the body passes its rest position.

   The body's speed is zero when $\sin t = 0$. This occurs at the endpoints of the interval of motion, when $\cos t = \pm 1$.

3. The acceleration, $a = -5 \cos t$, is zero only at the rest position, where the cosine is zero. When the body is anywhere else, the spring is either pulling on it or pushing on it. The acceleration is greatest in magnitude at the points farthest from the origin, where $\cos t = \pm 1$. $\square$

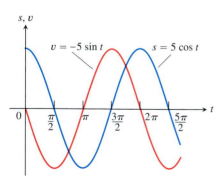

2.37 The graphs of the position and velocity of the body in Example 5.

Jerk

A sudden change in acceleration is called a "jerk." When a ride in a car or a bus is jerky, it is not that the accelerations involved are necessarily large but that the changes in acceleration are abrupt. Jerk is what spills your soft drink. The derivative responsible for jerk is d^3s/dt^3.

Definition

Jerk is the derivative of acceleration. If a body's position at time t is $s = f(t)$, the body's jerk at time t is

$$j = \frac{da}{dt} = \frac{d^3s}{dt^3}.$$

Recent tests have shown that motion sickness comes from accelerations whose changes in magnitude or direction take us by surprise. Keeping an eye on the road helps us to see the changes coming. A driver is less likely to become sick than a passenger reading in the backseat.

EXAMPLE 6

a) The jerk of the constant acceleration of gravity ($g = 32$ ft/sec^2) is zero:

$$j = \frac{d}{dt}(g) = 0.$$

We don't experience motion sickness if we are just sitting around.

b) The jerk of the simple harmonic motion in Example 5 is

$$j = \frac{da}{dt} = \frac{d}{dt}(-5\cos t)$$

$$= 5\sin t.$$

It has its greatest magnitude when $\sin t = \pm 1$, not at the extremes of the displacement but at the origin, where the acceleration changes direction and sign. ☐

The Derivatives of the Other Basic Functions

Because $\sin x$ and $\cos x$ are differentiable functions of x, the related functions

$$\tan x = \frac{\sin x}{\cos x} \qquad \sec x = \frac{1}{\cos x}$$

$$\cot x = \frac{\cos x}{\sin x} \qquad \csc x = \frac{1}{\sin x}$$

are differentiable at every value of x at which they are defined. Their derivatives, calculated from the Quotient Rule, are given by the following formulas.

Notice the minus signs in the derivative formulas for the cofunctions.

$$\frac{d}{dx}(\tan x) = \sec^2 x \qquad (6) \qquad\qquad \frac{d}{dx}(\sec x) = \sec x \tan x \qquad (7)$$

$$\frac{d}{dx}(\cot x) = -\csc^2 x \qquad (8) \qquad\qquad \frac{d}{dx}(\csc x) = -\csc x \cot x \qquad (9)$$

To show how a typical calculation goes, we derive Eq. (6). The other derivations are left to Exercises 67 and 68.

EXAMPLE 7   Find dy/dx if $y = \tan x$.

Solution

$$\frac{d}{dx}(\tan x) = \frac{d}{dx}\left(\frac{\sin x}{\cos x}\right) = \frac{\cos x \dfrac{d}{dx}(\sin x) - \sin x \dfrac{d}{dx}(\cos x)}{\cos^2 x} \qquad \text{Quotient Rule}$$

$$= \frac{\cos x \cos x - \sin x\,(-\sin x)}{\cos^2 x}$$

$$= \frac{\cos^2 x + \sin^2 x}{\cos^2 x}$$

$$= \frac{1}{\cos^2 x} = \sec^2 x \qquad\qquad\qquad □$$

EXAMPLE 8   Find y'' if $y = \sec x$.

Solution

$$y = \sec x$$

$$y' = \sec x \tan x \qquad\qquad\qquad \text{Eq. (7)}$$

$$y'' = \frac{d}{dx}(\sec x \tan x)$$

$$= \sec x \frac{d}{dx}(\tan x) + \tan x \frac{d}{dx}(\sec x) \qquad \text{Product Rule}$$

$$= \sec x\,(\sec^2 x) + \tan x\,(\sec x \tan x)$$

$$= \sec^3 x + \sec x \tan^2 x \qquad\qquad\qquad □$$

EXAMPLE 9

a) $\dfrac{d}{dx}(3x + \cot x) = 3 + \dfrac{d}{dx}(\cot x) = 3 - \csc^2 x$

b) $\dfrac{d}{dx}\left(\dfrac{2}{\sin x}\right) = \dfrac{d}{dx}(2\csc x) = 2\dfrac{d}{dx}(\csc x)$

$$= 2(-\csc x \cot x) = -2\csc x \cot x \qquad\qquad\qquad □$$

Continuity of Trigonometric Functions

Since the six basic trigonometric functions are differentiable throughout their domains they are also continuous throughout their domains by Theorem 1, Section 2.1. This means that $\sin x$ and $\cos x$ are continuous for all x, that $\sec x$ and $\tan x$ are continuous except when x is a nonzero integer multiple of $\pi/2$, and that $\csc x$ and $\cot x$ are continuous except when x is an integer multiple of π. For each function, $\lim_{x \to c} f(x) = f(c)$ whenever $f(c)$ is defined. As a result, we can calculate the limits of many algebraic combinations and composites of trigonometric functions by direct substitution.

EXAMPLE 10

$$\lim_{x \to 0} \frac{\sqrt{2 + \sec x}}{\cos(\pi - \tan x)} = \frac{\sqrt{2 + \sec 0}}{\cos(\pi - \tan 0)} = \frac{\sqrt{2 + 1}}{\cos(\pi - 0)} = \frac{\sqrt{3}}{-1} = -\sqrt{3}$$ ❏

Other Limits Calculated with Theorem 4

The equation $\lim_{\theta \to 0} (\sin \theta)/\theta = 1$ holds no matter how θ may be expressed:

$$\lim_{x \to 0} \frac{\sin x}{x} = 1, \quad \theta = x; \qquad \lim_{x \to 0} \frac{\sin 7x}{7x} = 1, \quad \theta = 7x;$$

As $x \to 0$, $\theta \to 0$ As $x \to 0$, $\theta \to 0$

$$\lim_{x \to 0} \frac{\sin (2/3)x}{(2/3)x} = 1, \quad \theta = (2/3)x$$

As $x \to 0$, $\theta \to 0$

Knowing this helps us calculate related limits involving angles in radian measure.

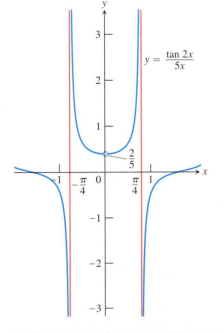

$y = \dfrac{\tan 2x}{5x}$

2.38 The graph of $y = (\tan 2x)/5x$ steps across the y-axis at $y = 2/5$ (Example 11).

EXAMPLE 11

a) $\displaystyle \lim_{x \to 0} \frac{\sin 2x}{5x} = \lim_{x \to 0} \frac{(2/5) \cdot \sin 2x}{(2/5) \cdot 5x}$

$\qquad\qquad\qquad$ Eq. (2) does not apply to the original fraction. We need a $2x$ in the denominator, not a $5x$. We produce it by multiplying numerator and denominator by 2/5.

$\displaystyle = \frac{2}{5} \lim_{x \to 0} \frac{\sin 2x}{2x}$

$\qquad\qquad\qquad$ Now Eq. (2) applies

$\displaystyle = \frac{2}{5}(1) = \frac{2}{5}$

b) $\displaystyle \lim_{x \to 0} \frac{\tan 2x}{5x} = \lim_{x \to 0} \left(\frac{\sin 2x}{5x} \cdot \frac{1}{\cos 2x} \right)$

$\qquad\qquad\qquad$ $\tan 2x = \dfrac{\sin 2x}{\cos 2x}$

$\displaystyle = \left(\lim_{x \to 0} \frac{\sin 2x}{5x} \right) \left(\lim_{x \to 0} \frac{1}{\cos 2x} \right)$

$\displaystyle = \left(\frac{2}{5} \right) \left(\frac{1}{\cos 0} \right) = \frac{2}{5}$

$\qquad\qquad\qquad$ Part (a) and continuity of $\cos x$

See Fig. 2.38.   ❏

Applications

The occurrence of the function $(\sin x)/x$ in calculus is not an isolated event. The function arises in such diverse fields as quantum physics (where it appears in solutions of the wave equation) and electrical engineering (in signal analysis and signal filter design) as well as in the mathematical fields of differential equations and probability theory.

EXAMPLE 12

$$\lim_{t \to (\pi/2)} \frac{\sin\left(t - \dfrac{\pi}{2}\right)}{t - \dfrac{\pi}{2}} \qquad \begin{array}{l} \text{Set } \theta = t - (\pi/2). \\ \text{Then } \theta \to 0 \text{ as} \\ t \to (\pi/2). \end{array}$$

$$= \lim_{\theta \to 0} \frac{\sin \theta}{\theta} = 1 \qquad \blacksquare$$

Exercises 2.4

Derivatives

In Exercises 1–12, find dy/dx.

1. $y = -10x + 3\cos x$

2. $y = \dfrac{3}{x} + 5\sin x$

3. $y = \csc x - 4\sqrt{x} + 7$

4. $y = x^2 \cot x - \dfrac{1}{x^2}$

5. $y = (\sec x + \tan x)(\sec x - \tan x)$

6. $y = (\sin x + \cos x)\sec x$

7. $y = \dfrac{\cot x}{1 + \cot x}$

8. $y = \dfrac{\cos x}{1 + \sin x}$

9. $y = \dfrac{4}{\cos x} + \dfrac{1}{\tan x}$

10. $y = \dfrac{\cos x}{x} + \dfrac{x}{\cos x}$

11. $y = x^2 \sin x + 2x \cos x - 2\sin x$

12. $y = x^2 \cos x - 2x \sin x - 2\cos x$

In Exercises 13–16, find ds/dt.

13. $s = \tan t - t$

14. $s = t^2 - \sec t + 1$

15. $s = \dfrac{1 + \csc t}{1 - \csc t}$

16. $s = \dfrac{\sin t}{1 - \cos t}$

In Exercises 17–20, find $dr/d\theta$.

17. $r = 4 - \theta^2 \sin \theta$

18. $r = \theta \sin \theta + \cos \theta$

19. $r = \sec \theta \csc \theta$

20. $r = (1 + \sec \theta)\sin \theta$

In Exercises 21–24, find dp/dq.

21. $p = 5 + \dfrac{1}{\cot q}$

22. $p = (1 + \csc q)\cos q$

23. $p = \dfrac{\sin q + \cos q}{\cos q}$

24. $p = \dfrac{\tan q}{1 + \tan q}$

25. Find y'' if (a) $y = \csc x$, (b) $y = \sec x$.

26. Find $y^{(4)} = d^4y/dx^4$ if (a) $y = -2\sin x$, (b) $y = 9\cos x$.

Limits

Find the limits in Exercises 27–32.

27. $\displaystyle\lim_{x \to 2} \sin\left(\dfrac{1}{x} - \dfrac{1}{2}\right)$

28. $\displaystyle\lim_{x \to -\pi/6} \sqrt{1 + \cos(\pi \csc x)}$

29. $\displaystyle\lim_{x \to 0} \sec\left[\cos x + \pi \tan\left(\dfrac{\pi}{4\sec x}\right) - 1\right]$

30. $\displaystyle\lim_{x \to 0} \sin\left(\dfrac{\pi + \tan x}{\tan x - 2\sec x}\right)$

31. $\displaystyle\lim_{t \to 0} \tan\left(1 - \dfrac{\sin t}{t}\right)$

32. $\displaystyle\lim_{\theta \to 0} \cos\left(\dfrac{\pi \theta}{\sin \theta}\right)$

Find the limits in Exercises 33–48.

33. $\displaystyle\lim_{\theta \to 0} \dfrac{\sin \sqrt{2\theta}}{\sqrt{2\theta}}$

34. $\displaystyle\lim_{t \to 0} \dfrac{\sin kt}{t} \quad (k \text{ constant})$

35. $\displaystyle\lim_{y \to 0} \dfrac{\sin 3y}{4y}$

36. $\displaystyle\lim_{h \to 0^-} \dfrac{h}{\sin 3h}$

37. $\displaystyle\lim_{x \to 0} \dfrac{\tan 2x}{x}$

38. $\displaystyle\lim_{t \to 0} \dfrac{2t}{\tan t}$

39. $\displaystyle\lim_{x \to 0} \dfrac{x \csc 2x}{\cos 5x}$

40. $\displaystyle\lim_{x \to 0} 6x^2(\cot x)(\csc 2x)$

41. $\displaystyle\lim_{x \to 0} \dfrac{x + x \cos x}{\sin x \cos x}$

42. $\displaystyle\lim_{x \to 0} \dfrac{x^2 - x + \sin x}{2x}$

43. $\displaystyle\lim_{t \to 0} \dfrac{\sin(1 - \cos t)}{1 - \cos t}$

44. $\displaystyle\lim_{h \to 0} \dfrac{\sin(\sin h)}{\sin h}$

45. $\displaystyle\lim_{\theta \to 0} \dfrac{\sin \theta}{\sin 2\theta}$

46. $\displaystyle\lim_{x \to 0} \dfrac{\sin 5x}{\sin 4x}$

47. $\lim\limits_{x \to 0} \dfrac{\tan 3x}{\sin 8x}$

48. $\lim\limits_{y \to 0} \dfrac{\sin 3y \cot 5y}{y \cot 4y}$

Tangent Lines

In Exercises 49–52, graph the curves over the given intervals, together with their tangents at the given values of x. Label each curve and tangent with its equation.

49. $y = \sin x$, $\;-3\pi/2 \le x \le 2\pi$
$x = -\pi,\, 0,\, 3\pi/2$

50. $y = \tan x$, $\;-\pi/2 < x < \pi/2$
$x = -\pi/3,\, 0,\, \pi/3$

51. $y = \sec x$, $\;-\pi/2 < x < \pi/2$
$x = -\pi/3,\, \pi/4$

52. $y = 1 + \cos x$, $\;-3\pi/2 \le x \le 2\pi$
$x = -\pi/3,\, 3\pi/2$

Do the graphs of the functions in Exercises 53–56 have any horizontal tangents in the interval $0 \le x \le 2\pi$? If so, where? If not, why not? You may want to visualize your findings by graphing the functions with a grapher.

53. $y = x + \sin x$ **54.** $y = 2x + \sin x$

55. $y = x - \cot x$ **56.** $y = x + 2\cos x$

57. Find all points on the curve $y = \tan x$, $-\pi/2 < x < \pi/2$, where the tangent line is parallel to the line $y = 2x$. Sketch the curve and tangent(s) together, labeling each with its equation.

58. Find all points on the curve $y = \cot x$, $0 < x < \pi$, where the tangent line is parallel to the line $y = -x$. Sketch the curve and tangent(s) together, labeling each with its equation.

In Exercises 59 and 60, find an equation for (a) the tangent to the curve at P and (b) the horizontal tangent to the curve at Q.

59.   **60.**

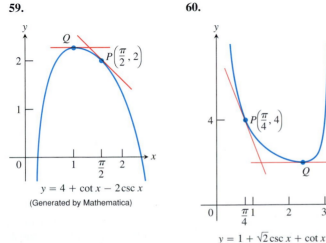

$y = 4 + \cot x - 2\csc x$

(Generated by Mathematica)

$y = 1 + \sqrt{2}\,\csc x + \cot x$

(Generated by Mathematica)

Simple Harmonic Motion

The equations in Exercises 61 and 62 give the position $s = f(t)$ of a body moving on a coordinate line (s in meters, t in seconds). Find the body's velocity, speed, acceleration, and jerk at time $t = \pi/4$ sec.

61. $s = 2 - 2\sin t$ **62.** $s = \sin t + \cos t$

Theory and Examples

63. Is there a value of c that will make

$$f(x) = \begin{cases} \dfrac{\sin^2 3x}{x^2}, & x \ne 0 \\[2mm] c, & x = 0 \end{cases}$$

continuous at $x = 0$? Give reasons for your answer.

64. Is there a value of b that will make

$$g(x) = \begin{cases} x + b, & x < 0 \\ \cos x, & x \ge 0 \end{cases}$$

continuous at $x = 0$? differentiable at $x = 0$? Give reasons for your answers.

65. Find $\dfrac{d^{999}}{dx^{999}}(\cos x)$ **66.** Find $\dfrac{d^{725}}{dx^{725}}(\sin x)$

67. Derive the formula for the derivative with respect to x of

    a) $\sec x$ **b)** $\csc x$.

68. Derive the formula for the derivative with respect to x of $\cot x$.

Grapher Explorations

69. Graph $y = \cos x$ for $-\pi \le x \le 2\pi$. On the same screen, graph

$$y = \frac{\sin(x + h) - \sin x}{h}$$

for $h = 1, 0.5, 0.3$, and 0.1. Then, in a new window, try $h = -1, -0.5$, and -0.3. What happens as $h \to 0^+$? as $h \to 0^-$? What phenomenon is being illustrated here?

70. Graph $y = -\sin x$ for $-\pi \le x \le 2\pi$. On the same screen, graph

$$y = \frac{\cos(x + h) - \cos x}{h}$$

for $h = 1, 0.5, 0.3$, and 0.1. Then, in a new window, try $h = -1, -0.5$, and -0.3. What happens as $h \to 0^+$? as $h \to 0^-$? What phenomenon is being illustrated here?

71. *Centered difference quotients.* The **centered difference quotient**

$$\frac{f(x + h) - f(x - h)}{2h}$$

is used to approximate $f'(x)$ in numerical work because (1) its limit as $h \to 0$ equals $f'(x)$ when $f'(x)$ exists, and (2) it usually gives a better approximation of $f'(x)$ for a given value of h than Fermat's difference quotient

$$\frac{f(x + h) - f(x)}{h}.$$

See the figure below.

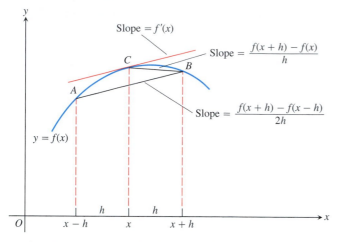

a) To see how rapidly the centered difference quotient for $f(x) = \sin x$ converges to $f'(x) = \cos x$, graph $y = \cos x$ together with

$$y = \frac{\sin(x+h) - \sin(x-h)}{2h}$$

over the interval $[-\pi, 2\pi]$ for $h = 1, 0.5$, and 0.3. Compare the results with those obtained in Exercise 69 for the same values of h.

b) To see how rapidly the centered difference quotient for $f(x) = \cos x$ converges to $f'(x) = -\sin x$, graph $y = -\sin x$ together with

$$y = \frac{\cos(x+h) - \cos(x-h)}{2h}$$

over the interval $[-\pi, 2\pi]$ for $h = 1, 0.5$, and 0.3. Compare the results with those obtained in Exercise 70 for the same values of h.

72. *A caution about centered difference quotients. (Continuation of Exercise 71.)* The quotient

$$\frac{f(x+h) - f(x-h)}{2h}$$

may have a limit as $h \to 0$ when f has no derivative at x. As a case in point, take $f(x) = |x|$ and calculate

$$\lim_{h \to 0} \frac{|0+h| - |0-h|}{2h}.$$

As you will see, the limit exists even though $f(x) = |x|$ has no derivative at $x = 0$.

73. Graph $y = \tan x$ and its derivative together on $(-\pi/2, \pi/2)$. Does the graph of the tangent function appear to have a smallest slope? a largest slope? Is the slope ever negative? Give reasons for your answers.

74. Graph $y = \cot x$ and its derivative together for $0 < x < \pi$. Does the graph of the cotangent function appear to have a smallest slope? a largest slope? Is the slope ever positive? Give reasons for your answers.

75. Graph $y = (\sin x)/x$, $y = (\sin 2x)/x$, and $y = (\sin 4x)/x$ together over the interval $-2 \le x \le 2$. Where does each graph appear to cross the y-axis? Do the graphs really intersect the axis? What would you expect the graphs of $y = (\sin 5x)/x$ and $y = (\sin(-3x))/x$ to do as $x \to 0$? Why? What about the graph of $y = (\sin kx)/x$ for other values of k? Give reasons for your answers.

76. *Radians vs. degrees.* What happens to the derivatives of $\sin x$ and $\cos x$ if x is measured in degrees instead of radians? To find out, take the following steps.

a) With your graphing calculator or computer grapher *in degree mode,* graph

$$f(h) = \frac{\sin h}{h}$$

and estimate $\lim_{h \to 0} f(h)$. Compare your estimate with $\pi/180$. Is there any reason to believe the limit *should* be $\pi/180$?

b) With your grapher still in degree mode, estimate

$$\lim_{h \to 0} \frac{\cos h - 1}{h}.$$

c) Now go back to the derivation of the formula for the derivative of $\sin x$ in the text and carry out the steps of the derivation using degree-mode limits. What formula do you obtain for the derivative?

d) Work through the derivation of the formula for the derivative of $\cos x$ using degree-mode limits. What formula do you obtain for the derivative?

e) The disadvantages of the degree-mode formulas become apparent as you start taking derivatives of higher order. Try it. What are the second and third degree-mode derivatives of $\sin x$ and $\cos x$?

2.5

The Chain Rule

We now know how to differentiate $\sin x$ and $x^2 - 4$, but how do we differentiate a composite like $\sin(x^2 - 4)$? The answer is, with the Chain Rule, which says that the derivative of the composite of two differentiable functions is the product of their derivatives evaluated at appropriate points. The Chain Rule is probably the most widely used differentiation rule in mathematics. This section describes the rule and how to use it. We begin with examples.

EXAMPLE 1   The function $y = 6x - 10 = 2(3x - 5)$ is the composite of the functions $y = 2u$ and $u = 3x - 5$. How are the derivatives of these three functions related?

Solution   We have

$$\frac{dy}{dx} = 6, \qquad \frac{dy}{du} = 2, \qquad \frac{du}{dx} = 3.$$

Since $6 = 2 \cdot 3$,

$$\frac{dy}{dx} = \frac{dy}{du} \cdot \frac{du}{dx}.$$ ❑

Is it an accident that

$$\frac{dy}{dx} = \frac{dy}{du} \cdot \frac{du}{dx}?$$

If we think of the derivative as a rate of change, our intuition allows us to see that this relationship is reasonable. For $y = f(u)$ and $u = g(x)$, if y changes twice as fast as u and u changes three times as fast as x, then we expect y to change six times as fast as x. This is much like the effect of a multiple gear train (Fig. 2.39).

Let us try this again on another function.

EXAMPLE 2

$$y = 9x^4 + 6x^2 + 1 = (3x^2 + 1)^2$$

is the composite of $y = u^2$ and $u = 3x^2 + 1$. Calculating derivatives, we see that

$$\frac{dy}{du} \cdot \frac{du}{dx} = 2u \cdot 6x$$

$$= 2(3x^2 + 1) \cdot 6x$$

$$= 36x^3 + 12x$$

and

$$\frac{dy}{dx} = \frac{d}{dx}(9x^4 + 6x^2 + 1)$$

$$= 36x^3 + 12x.$$

Once again,

$$\frac{dy}{du} \cdot \frac{du}{dx} = \frac{dy}{dx}.$$ ❑

The derivative of the composite function $f(g(x))$ at x is the derivative of f at $g(x)$ times the derivative of g at x. This is known as the Chain Rule (Fig. 2.40).

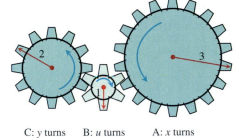

C: y turns   B: u turns   A: x turns

2.39 When gear A makes x turns, gear B makes u turns and gear C makes y turns. By comparing circumferences or counting teeth, we see that $y = u/2$ and $u = 3x$, so $y = 3x/2$. Thus $dy/du = 1/2$, $du/dx = 3$, and $dy/dx = 3/2 = (dy/du)(du/dx)$.

2.40 Rates of change multiply: the derivative of $f \circ g$ at x is the derivative of f at the point $g(x)$ times the derivative of g at x.

Composite $f \circ g$

Rate of change at x is $f'(g(x)) \cdot g'(x)$

g

f

Rate of change at x is $g'(x)$

Rate of change at $g(x)$ is $f'(g(x))$

x $u = g(x)$ $y = f(u) = f(g(x))$

Theorem 5

The Chain Rule

If $f(u)$ is differentiable at the point $u = g(x)$, and $g(x)$ is differentiable at x, then the composite function $(f \circ g)(x) = f(g(x))$ is differentiable at x, and

$$(f \circ g)'(x) = f'(g(x)) \cdot g'(x). \tag{1}$$

In Leibniz notation, if $y = f(u)$ and $u = g(x)$, then

$$\frac{dy}{dx} = \frac{dy}{du} \cdot \frac{du}{dx}, \tag{2}$$

where dy/du is evaluated at $u = g(x)$.

It would be tempting to try to prove the Chain Rule by writing

$$\frac{\Delta y}{\Delta x} = \frac{\Delta y}{\Delta u} \cdot \frac{\Delta u}{\Delta x}$$

and taking the limit as $\Delta x \to 0$. This would work if we knew that Δu, the change in u, was nonzero, but we do not know this. A small change in x could conceivably produce no change in u. The proof requires a different approach, using ideas in Section 3.7. We will return to it when the time comes.

EXAMPLE 3   Find the derivative of $y = \sqrt{x^2 + 1}$.

Solution   Here $y = f(g(x))$, where $f(u) = \sqrt{u}$ and $g(x) = x^2 + 1$. Since the derivatives of f and g are

$$f'(u) = \frac{1}{2\sqrt{u}} \qquad \text{and} \qquad g'(x) = 2x,$$

the Chain Rule gives

$$\frac{dy}{dx} = \frac{d}{dx} f(g(x)) = f'(g(x)) \cdot g'(x)$$

$$= \frac{1}{2\sqrt{g(x)}} \cdot g'(x) = \frac{1}{2\sqrt{x^2 + 1}} \cdot (2x)$$

$$= \frac{x}{\sqrt{x^2 + 1}}.$$

The "Outside-Inside" Rule

It sometimes helps to think about the Chain Rule the following way. If $y = f(g(x))$, Eq. (2) tells us that

$$\frac{dy}{dx} = f'[g(x)] \cdot g'(x). \tag{3}$$

In words, Eq. (3) says: To find dy/dx, differentiate the "outside" function f and leave the "inside" $g(x)$ alone; then multiply by the derivative of the inside.

EXAMPLE 4

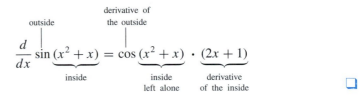

Repeated Use of the Chain Rule

We sometimes have to use the Chain Rule two or more times to find a derivative. Here is an example.

EXAMPLE 5   Find the derivative of $g(t) = \tan{(5 - \sin{2t})}$.

Solution

$$g'(t) = \frac{d}{dt}(\tan{(5 - \sin{2t})})$$

$$= \sec^2(5 - \sin{2t}) \cdot \frac{d}{dt}(5 - \sin{2t})$$ Derivative of $\tan{u}$ with $u = 5 - \sin{2t}$

$$= \sec^2(5 - \sin{2t}) \cdot \left(0 - (\cos{2t}) \cdot \frac{d}{dt}(2t)\right)$$ Derivative of $5 - \sin{u}$ with $u = 2t$

$$= \sec^2(5 - \sin{2t}) \cdot (-\cos{2t}) \cdot 2$$

$$= -2(\cos{2t})\sec^2(5 - \sin{2t})$$

Differentiation Formulas That Include the Chain Rule

Many of the differentiation formulas you will encounter in your scientific work already include the Chain Rule.

If f is a differentiable function of u, and u is a differentiable function of x, then substituting $y = f(u)$ in the Chain Rule formula

$$\frac{dy}{dx} = \frac{dy}{du} \cdot \frac{du}{dx}$$

leads to the formula

$$\frac{d}{dx}f(u) = f'(u)\frac{du}{dx}. \tag{4}$$

For example, if u is a differentiable function of x, n is an integer, and $y = u^n$, then the Chain Rule gives

$$\frac{dy}{dx} = \frac{d}{du}(u^n) \cdot \frac{du}{dx}$$

$$= nu^{n-1}\frac{du}{dx}.$$ Differentiating u^n with respect to u itself gives nu^{n-1}.

> **Power Chain Rule**
>
> If $u(x)$ is a differentiable function and n is an integer, then u^n is differentiable and
>
> $$\frac{d}{dx}u^n = nu^{n-1}\frac{du}{dx}. \qquad (5)$$

$\sin^n x$ is short for $(\sin x)^n, \quad n \neq -1$.

EXAMPLE 6

a) $\dfrac{d}{dx}\sin^5 x = 5\sin^4 x\,\dfrac{d}{dx}(\sin x)$ Eq. (5) with $u = \sin x, n = 5$

$\qquad\qquad = 5\sin^4 x \cos x$

b) $\dfrac{d}{dx}(2x + 1)^{-3} = -3(2x + 1)^{-4}\dfrac{d}{dx}(2x + 1)$ Eq. (5) with $u = 2x + 1, n = -3$

$\qquad\qquad\qquad = -3(2x + 1)^{-4}\,(2)$

$\qquad\qquad\qquad = -6(2x + 1)^{-4}$

c) $\dfrac{d}{dx}(5x^3 - x^4)^7 = 7(5x^3 - x^4)^6\dfrac{d}{dx}(5x^3 - x^4)$ Eq. (5) with $u = 5x^3 - x^4, n = 7$

$\qquad\qquad\qquad = 7(5x^3 - x^4)^6\,(5 \cdot 3x^2 - 4x^3)$

$\qquad\qquad\qquad = 7(5x^3 - x^4)^6\,(15x^2 - 4x^3)$

d) $\dfrac{d}{dx}\left(\dfrac{1}{3x - 2}\right) = \dfrac{d}{dx}(3x - 2)^{-1}$ Eq. (5) with $u = 3x - 2, n = -1$

$\qquad\qquad\qquad = -1(3x - 2)^{-2}\dfrac{d}{dx}(3x - 2)$

$\qquad\qquad\qquad = -1(3x - 2)^{-2}\,(3)$

$\qquad\qquad\qquad = -\dfrac{3}{(3x - 2)^2}$

In part (d) we could also have found the derivative with the Quotient Rule. ❏

EXAMPLE 7   *Radians vs. degrees*

It is important to remember that the formulas for the derivatives of $\sin x$ and $\cos x$ were obtained under the assumption that x is measured in radians, *not* degrees. The Chain Rule brings new understanding to the difference between the two. Since $180° = \pi$ radians, $x° = \pi x/180$ radians. By the Chain Rule,

$$\frac{d}{dx}\sin (x°) = \frac{d}{dx}\sin\left(\frac{\pi x}{180}\right) = \frac{\pi}{180}\cos\left(\frac{\pi x}{180}\right) = \frac{\pi}{180}\cos (x°).$$

See Fig. 2.41. Similarly, the derivative of $\cos (x°)$ is $-(\pi/180)\sin (x°)$.

The factor $\pi/180$, annoying in the first derivative, would compound with repeated differentiation. We see at a glance the compelling reason for the use of radian measure. ❏

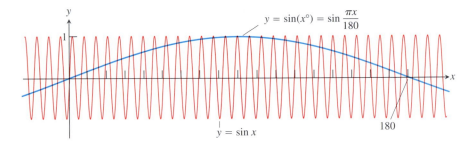

$$y = \sin(x°) = \sin \frac{\pi x}{180}$$

$$y = \sin x$$

180

2.41 sin (x°) oscillates only π/180 times as often as sin x oscillates. Its maximum slope is π/180.

❋ Melting Ice Cubes

In mathematics, we tend to use letters like f, g, x, y, and u for functions and variables. However, other fields use letters like V, for volume, and s, for side, that come from the names of the things being modeled. The letters in the Chain Rule then change too, as in the next example.

EXAMPLE 8   *The melting ice cube*

How long will it take an ice cube to melt?

Solution   As with all applications to science, we start with a mathematical model. We assume that the cube retains its cubical shape as it melts. We call its side length s, so its volume is $V = s^3$. We assume that V and s are differentiable functions of time t. We assume also that the cube's volume decreases at a rate that is proportional to its surface area. This latter assumption seems reasonable enough when we think that the melting takes place at the surface: Changing the amount of surface changes the amount of ice exposed to melt. In mathematical terms,

$$\frac{dV}{dt} = -k(6s^2), \qquad k > 0.$$

The minus sign indicates that the volume is decreasing. We assume that the proportionality factor k is constant. (It probably depends on many things, however, such as the relative humidity of the surrounding air, the air temperature, and the incidence or absence of sunlight, to name only a few.)

Finally, we need at least one more piece of information: How long will it take a specific percentage of the ice cube to melt? We have nothing to guide us unless we make one or more observations, but now let us assume a particular set of conditions in which the cube lost 1/4 of its volume during the first hour. (You could use letters instead of particular numbers: say $n\%$ in r hours. Then your answer would be in terms of n and r.)

Mathematically, we now have the following problem.

Given: $V = s^3$ and $\dfrac{dV}{dt} = -k(6s^2)$

$\qquad\qquad V = V_0$ when $t = 0$

$\qquad\qquad V = (3/4)V_0$ when $t = 1$ h

Find: The value of t when $V = 0$

We apply the Chain Rule to differentiate $V = s^3$ with respect to t:

$$\frac{dV}{dt} = 3s^2 \frac{ds}{dt}.$$

We set this equal to the given rate, $-k(6s^2)$, to get

$$3s^2 \frac{ds}{dt} = -6ks^2$$

$$\frac{ds}{dt} = -2k.$$

The side length is *decreasing* at the constant rate of $2k$ units per hour. Thus, if the initial length of the cube's side is s_0, the length of its side one hour later is $s_1 = s_0 - 2k$. This equation tells us that

$$2k = s_0 - s_1.$$

The melting time is the value of t that makes $2kt = s_0$. Hence,

$$t_{\text{melt}} = \frac{s_0}{2k} = \frac{s_0}{s_0 - s_1} = \frac{1}{1 - (s_1/s_0)}.$$

But

$$\frac{s_1}{s_0} = \frac{\left(\frac{3}{4}V_0\right)^{1/3}}{(V_0)^{1/3}} = \left(\frac{3}{4}\right)^{1/3} \approx 0.91.$$

Therefore,

$$t_{\text{melt}} = \frac{1}{1 - 0.91} \approx 11 \text{ h}.$$

If 1/4 of the cube melts in 1 h, it will take about 10 h more for the rest of it to melt.

❏

If we were natural scientists interested in testing the assumptions on which our mathematical model is based, our next step would be to run a number of experiments and compare their outcomes with the model's predictions. One practical application might lie in analyzing the proposal to tow large icebergs from polar waters to offshore locations near southern California, where the melting ice could provide fresh water. As a first approximation, we might imagine the iceberg to be a large cube or rectangular solid, or perhaps a pyramid. We will say more about mathematical modeling in Section 4.2.

Exercises 2.5

Derivative Calculations

In Exercises 1–8, given $y = f(u)$ and $u = g(x)$, find $dy/dx = f'(g(x))g'(x)$.

1. $y = 6u - 9, \quad u = (1/2)x^4$

2. $y = 2u^3, \quad u = 8x - 1$

3. $y = \sin u, \quad u = 3x + 1$

4. $y = \cos u, \quad u = -x/3$

5. $y = \cos u, \quad u = \sin x$

6. $y = \sin u, \quad u = x - \cos x$

7. $y = \tan u, \quad u = 10x - 5$

8. $y = -\sec u, \quad u = x^2 + 7x$

In Exercises 9–18, write the function in the form $y = f(u)$ and $u = g(x)$. Then find dy/dx as a function of x.

9. $y = (2x + 1)^5$

10. $y = (4 - 3x)^9$

11. $y = \left(1 - \dfrac{x}{7}\right)^{-7}$

12. $y = \left(\dfrac{x}{2} - 1\right)^{-10}$

13. $y = \left(\dfrac{x^2}{8} + x - \dfrac{1}{x}\right)^4$

14. $y = \left(\dfrac{x}{5} + \dfrac{1}{5x}\right)^5$

15. $y = \sec (\tan x)$

16. $y = \cot\left(\pi - \dfrac{1}{x}\right)$

17. $y = \sin^3 x$

18. $y = 5 \cos^{-4} x$

Find the derivatives of the functions in Exercises 19–38.

19. $p = \sqrt{3 - t}$

20. $q = \sqrt{2r - r^2}$

21. $s = \dfrac{4}{3\pi} \sin 3t + \dfrac{4}{5\pi} \cos 5t$

22. $s = \sin\left(\dfrac{3\pi t}{2}\right) + \cos\left(\dfrac{3\pi t}{2}\right)$

23. $r = (\csc \theta + \cot \theta)^{-1}$

24. $r = -(\sec \theta + \tan \theta)^{-1}$

25. $y = x^2 \sin^4 x + x \cos^{-2} x$

26. $y = \dfrac{1}{x} \sin^{-5} x - \dfrac{x}{3} \cos^3 x$

27. $y = \dfrac{1}{21}(3x - 2)^7 + \left(4 - \dfrac{1}{2x^2}\right)^{-1}$

28. $y = (5 - 2x)^{-3} + \dfrac{1}{8}\left(\dfrac{2}{x} + 1\right)^4$

29. $y = (4x + 3)^4 (x + 1)^{-3}$

30. $y = (2x - 5)^{-1}(x^2 - 5x)^6$

31. $h(x) = x \tan (2\sqrt{x}) + 7$

32. $k(x) = x^2 \sec\left(\dfrac{1}{x}\right)$

33. $f(\theta) = \left(\dfrac{\sin \theta}{1 + \cos \theta}\right)^2$

34. $g(t) = \left(\dfrac{1 + \cos t}{\sin t}\right)^{-1}$

35. $r = \sin (\theta^2) \cos (2\theta)$

36. $r = \sec \sqrt{\theta} \tan\left(\dfrac{1}{\theta}\right)$

37. $q = \sin\left(\dfrac{t}{\sqrt{t + 1}}\right)$

38. $q = \cot\left(\dfrac{\sin t}{t}\right)$

In Exercises 39–48, find dy/dt.

39. $y = \sin^2 (\pi t - 2)$

40. $y = \sec^2 \pi t$

41. $y = (1 + \cos 2t)^{-4}$

42. $y = (1 + \cot (t/2))^{-2}$

43. $y = \sin (\cos (2t - 5))$

44. $y = \cos\left(5 \sin\left(\dfrac{t}{3}\right)\right)$

45. $y = \left(1 + \tan^4\left(\dfrac{t}{12}\right)\right)^3$

46. $y = \dfrac{1}{6}\left(1 + \cos^2 (7t)\right)^3$

47. $y = \sqrt{1 + \cos (t^2)}$

48. $y = 4 \sin\left(\sqrt{1 + \sqrt{t}}\right)$

Find y'' in Exercises 49–52.

49. $y = \left(1 + \dfrac{1}{x}\right)^3$

50. $y = (1 - \sqrt{x})^{-1}$

51. $y = \dfrac{1}{9} \cot (3x - 1)$

52. $y = 9 \tan\left(\dfrac{x}{3}\right)$

Finding Numerical Values of Derivatives

In Exercises 53–58, find the value of $(f \circ g)'$ at the given value of x.

53. $f(u) = u^5 + 1, \quad u = g(x) = \sqrt{x}, \quad x = 1$

54. $f(u) = 1 - \dfrac{1}{u}, \quad u = g(x) = \dfrac{1}{1 - x}, \quad x = -1$

55. $f(u) = \cot \dfrac{\pi u}{10}, \quad u = g(x) = 5\sqrt{x}, \quad x = 1$

56. $f(u) = u + \dfrac{1}{\cos^2 u}, \quad u = g(x) = \pi x, \quad x = 1/4$

57. $f(u) = \dfrac{2u}{u^2 + 1}, \quad u = g(x) = 10x^2 + x + 1, \quad x = 0$

58. $f(u) = \left(\dfrac{u - 1}{u + 1}\right)^2, \quad u = g(x) = \dfrac{1}{x^2} - 1, \quad x = -1$

59. Suppose that functions f and g and their derivatives with respect to x have the following values at $x = 2$ and $x = 3$.

x	$f(x)$	$g(x)$	$f'(x)$	$g'(x)$
2	8	2	1/3	−3
3	3	−4	2π	5

Find the derivatives with respect to x of the following combinations at the given value of x.

a) $2f(x), \quad x = 2$

b) $f(x) + g(x), \quad x = 3$

c) $f(x) \cdot g(x), \quad x = 3$

d) $f(x)/g(x), \quad x = 2$

e) $f(g(x)), \quad x = 2$

f) $\sqrt{f(x)}, \quad x = 2$

g) $1/g^2(x), \quad x = 3$

h) $\sqrt{f^2(x) + g^2(x)}, \quad x = 2$

60. Suppose that the functions f and g and their derivatives with respect to x have the following values at $x = 0$ and $x = 1$.

x	$f(x)$	$g(x)$	$f'(x)$	$g'(x)$
0	1	1	5	1/3
1	3	−4	−1/3	−8/3

Find the derivatives with respect to x of the following combinations at the given value of x,

a) $5f(x) - g(x), \quad x = 1$

b) $f(x)g^3(x), \quad x = 0$

c) $\dfrac{f(x)}{g(x) + 1}, \quad x = 1$

d) $f(g(x)), \quad x = 0$

e) $g(f(x)), \quad x = 0$

f) $(x^{11} + f(x))^{-2}, \quad x = 1$

g) $f(x + g(x)), \quad x = 0$

61. Find ds/dt when $\theta = 3\pi/2$ if $s = \cos\theta$ and $d\theta/dt = 5$.

62. Find dy/dt when $x = 1$ if $y = x^2 + 7x - 5$ and $dx/dt = 1/3$.

Choices in Composition

What happens if you can write a function as a composite in different ways? Do you get the same derivative each time? The Chain Rule says you should. Try it with the functions in Exercises 63 and 64.

63. Find dy/dx if $y = x$ by using the Chain Rule with y as a composite of

   a) $y = (u/5) + 7$ and $u = 5x - 35$
   b) $y = 1 + (1/u)$ and $u = 1/(x - 1)$.

64. Find dy/dx if $y = x^{3/2}$ by using the Chain Rule with y as a composite of

   a) $y = u^3$ and $u = \sqrt{x}$
   b) $y = \sqrt{u}$ and $u = x^3$.

Tangents and Slopes

65. a) Find the tangent to the curve $y = 2\tan(\pi x/4)$ at $x = 1$.
   b) What is the smallest value the slope of the curve can ever have on the interval $-2 < x < 2$? Give reasons for your answer.

66. a) Find equations for the tangents to the curves $y = \sin 2x$ and $y = -\sin(x/2)$ at the origin. Is there anything special about how the tangents are related? Give reasons for your answer.
   b) Can anything be said about the tangents to the curves $y = \sin mx$ and $y = -\sin(x/m)$ at the origin (m a constant $\neq 0$)? Give reasons for your answer.
   c) For a given m, what are the largest values the slopes of the curves $y = \sin mx$ and $y = -\sin(x/m)$ can ever have? Give reasons for your answer.
   d) The function $y = \sin x$ completes one period on the interval $[0, 2\pi]$, the function $y = \sin 2x$ completes two periods, the function $y = \sin(x/2)$ completes half a period, and so on. Is there any relation between the number of periods $y = \sin mx$ completes on $[0, 2\pi]$ and the slope of the curve $y = \sin mx$ at the origin? Give reasons for your answer.

Theory, Examples, and Applications

67. *Running machinery too fast.* Suppose that a piston is moving straight up and down and that its position at time t seconds is

$$s = A\cos(2\pi bt),$$

with A and b positive. The value of A is the amplitude of the motion, and b is the frequency (number of times the piston moves up and down each second). What effect does doubling the frequency have on the piston's velocity, acceleration, and jerk? (Once you find out, you will know why machinery breaks when you run it too fast.)

68. *Temperatures in Fairbanks, Alaska.* The graph in Fig. 2.42 shows the average Fahrenheit temperature in Fairbanks, Alaska, during a typical 365-day year. The equation that approximates the temperature on day x is

$$y = 37\sin\left[\frac{2\pi}{365}(x - 101)\right] + 25.$$

   a) On what day is the temperature increasing the fastest?
   b) About how many degrees per day is the temperature increasing when it is increasing at its fastest?

69. The position of a particle moving along a coordinate line is $s = \sqrt{1 + 4t}$, with s in meters and t in seconds. Find the particle's velocity and acceleration at $t = 6$ sec.

70. Suppose the velocity of a falling body is $v = k\sqrt{s}$ m/sec (k a constant) at the instant the body has fallen s meters from its starting point. Show that the body's acceleration is constant.

71. The velocity of a heavy meteorite entering the earth's atmosphere is inversely proportional to $\sqrt{s}$ when it is s kilometers from the earth's center. Show that the meteorite's acceleration is inversely proportional to s^2.

72. A particle moves along the x-axis with velocity $dx/dt = f(x)$. Show that the particle's acceleration is $f(x)f'(x)$.

73. *Temperature and the period of a pendulum.* For oscillations of small amplitude (short swings), we may safely model the relationship between the period T and the length L of a simple

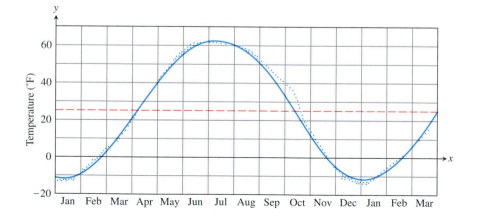

2.42 Normal mean air temperatures at Fairbanks, Alaska, plotted as data points. The approximating sine function is

$$f(x) = 37\sin\left[\frac{2\pi}{365}(x - 101)\right] + 25$$

(Exercise 68).

pendulum with the equation

$$T = 2\pi\sqrt{\frac{L}{g}},$$

where g is the constant acceleration of gravity at the pendulum's location. If we measure g in centimeters per second squared, we measure L in centimeters and T in seconds. If the pendulum is made of metal, its length will vary with temperature, either increasing or decreasing at a rate that is roughly proportional to L. In symbols, with u being temperature and k the proportionality constant

$$\frac{dL}{du} = kL.$$

Assuming this to be the case, show that the rate at which the period changes with respect to temperature is $kT/2$.

74. Suppose that $f(x) = x^2$ and $g(x) = |x|$. Then the composites

$$(f \circ g)(x) = |x|^2 = x^2 \quad \text{and} \quad (g \circ f)(x) = |x^2| = x^2$$

are both differentiable at $x = 0$ even though g itself is not differentiable at $x = 0$. Does this contradict the Chain Rule? Explain.

75. Suppose that $u = g(x)$ is differentiable at $x = 1$ and that $y = f(u)$ is differentiable at $u = g(1)$. If the graph of $y = f(g(x))$ has a horizontal tangent at $x = 1$, can we conclude anything about the tangent to the graph of g at $x = 1$ or the tangent to the graph of f at $u = g(1)$? Give reasons for your answer.

76. Suppose $u = g(x)$ is differentiable at $x = -5$, $y = f(u)$ is differentiable at $u = g(-5)$, and $(f \circ g)'(-5)$ is negative. What, if anything, can be said about the values of $g'(-5)$ and $f'(g(-5))$?

Using the Chain Rule, show that the power rule $(d/dx)x^n = nx^{n-1}$ holds for the functions x^n in Exercises 77 and 78.

77. $x^{1/4} = \sqrt{\sqrt{x}}$

78. $x^{3/4} = \sqrt{x\sqrt{x}}$

⊞ Grapher Explorations

79. *The derivative of* $\sin 2x$. Graph the function $y = 2\cos 2x$ for $-2 \le x \le 3.5$. Then, on the same screen, graph

$$y = \frac{\sin 2(x+h) - \sin 2x}{h}$$

for $h = 1.0, 0.5,$ and 0.2. Experiment with other values of h, including negative values. What do you see happening as $h \to 0$? Explain this behavior.

80. *The derivative of* $\cos(x^2)$. Graph $y = -2x\sin(x^2)$ for $-2 \le x \le 3$. Then, on the same screen, graph

$$y = \frac{\cos[(x+h)^2] - \cos(x^2)}{h}$$

for $h = 1.0, 0.7,$ and 0.3. Experiment with other values of h. What do you see happening as $h \to 0$? Explain this behavior.

✿ CAS Explorations and Projects

81. As Fig. 2.43 shows, the trigonometric "polynomial"

$$s = f(t) = 0.78540 - 0.63662\cos 2t - 0.07074\cos 6t -$$
$$0.02546\cos 10t - 0.01299\cos 14t$$

gives a good approximation of the sawtooth function $s = g(t)$ on the interval $[-\pi, \pi]$. How well does the derivative of f approximate the derivative of g at the points where dg/dt is defined? To find out, carry out the following steps.

a) Graph dg/dt (where defined) over $[-\pi, \pi]$.

b) Find df/dt.

c) Graph df/dt. Where does the approximation of dg/dt by df/dt seem to be best? least good? Approximations by trigonometric polynomials are important in the theories of heat and oscillation, but we must not expect too much of them, as we see in the next exercise.

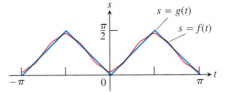

2.43 The approximation of a sawtooth function by a trigonometric "polynomial" (Exercise 81).

82. (*Continuation of Exercise 81.*)   In Exercise 81, the trigonometric polynomial $f(t)$ that approximated the sawtooth function $g(t)$ on $[-\pi, \pi]$ had a derivative that approximated the derivative of the sawtooth function. It is possible, however, for a trigonometric polynomial to approximate a function in a reasonable way without its derivative approximating the function's derivative at all well. As a case in point, the "polynomial"

$$s = h(t) = 1.2732\sin 2t + 0.4244\sin 6t + 0.25465\sin 10t$$
$$+ 0.18186\sin 14t + 0.14147\sin 18t$$

graphed in Fig. 2.44 approximates the step function $s = k(t)$ shown there. Yet the derivative of h is nothing like the derivative of k.

a) Graph dk/dt (where defined) over $[-\pi, \pi]$.

b) Find dh/dt.

c) Graph dh/dt to see how badly the graph fits the graph of dk/dt. Comment on what you see.

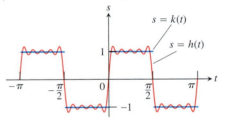

2.44 The approximation of a step function by a trigonometric "polynomial" (Exercise 82).

Implicit Differentiation and Rational Exponents

When we cannot put an equation $F(x, y) = 0$ in the form $y = f(x)$ to differentiate in the usual way, we may still be able to find dy/dx by *implicit differentiation*. This section describes the technique and uses it to extend the Power Rule for differentiation to include all rational exponents.

When are the functions defined by $F(x, y) = 0$ differentiable?

When may we expect the functions of x defined by an equation of the form $F(x, y) = 0$, where $F(x, y)$ denotes an expression in x and y, to be differentiable? A theorem in advanced calculus guarantees this to be the case if F is continuous (in a sense to be described in Chapter 12) and the first derivatives of F with respect to each variable, with the other held constant, are continuous, and the derivative with respect to y is nonzero. The functions you will encounter in this section all meet these criteria.

Implicit Differentiation

The graph of the equation $x^3 + y^3 - 9xy = 0$ (Fig. 2.45) has a well-defined slope at nearly every point because it is the union of the graphs of the functions $y = f_1(x)$, $y = f_2(x)$, and $y = f_3(x)$, which are differentiable except at O and A. But how do we find the slope when we cannot conveniently solve the equation to find the functions? The answer is to treat y as a differentiable function of x and differentiate both sides of the equation with respect to x, using the differentiation rules for powers, sums, products, and quotients and the Chain Rule. Then solve for dy/dx in terms of x and y *together* to obtain a formula that calculates the slope at any point (x, y) on the graph from the values of x and y.

The process by which we find dy/dx is called **implicit differentiation.** The phrase derives from the fact that the equation $x^3 + y^3 - 9xy = 0$ defines the functions f_1, f_2, and f_3 that give the graph's slope *implicitly* (i.e., hidden inside the equation), without giving us *explicit* formulas to work with.

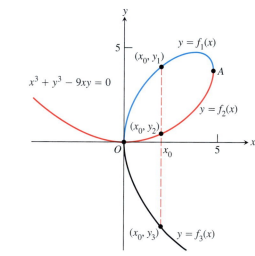

2.45 The curve $x^3 + y^3 - 9xy = 0$ is not the graph of any one function of x. However, the curve can be divided into separate arcs that *are* the graphs of functions of x. This particular curve, called a *folium*, dates to Descartes in 1638.

EXAMPLE 1   Find dy/dx if $y^2 = x$.

Solution The equation $y^2 = x$ defines two differentiable functions of x that we can actually find, namely $y_1 = \sqrt{x}$ and $y_2 = -\sqrt{x}$ (Fig. 2.46). We know how to calculate the derivative of each of these for $x > 0$:

$$\frac{dy_1}{dx} = \frac{1}{2\sqrt{x}}, \qquad \text{and} \qquad \frac{dy_2}{dx} = -\frac{1}{2\sqrt{x}}.$$

But suppose we knew only that the equation $y^2 = x$ defined y as one or more differentiable functions of x for $x > 0$ without knowing exactly what these functions were. Could we still find dy/dx?

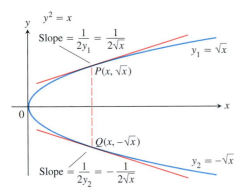

2.46 The equation $y^2 - x = 0$, or $y^2 = x$ as it is usually written, defines two differentiable functions of x on the interval $x \geq 0$. Example 1 shows how to find the derivatives of these functions without solving the equation $y^2 = x$ for y.

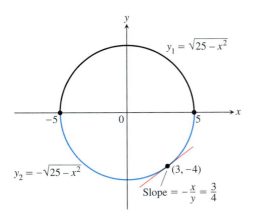

2.47 The circle combines the graphs of two functions. The graph of y_2 is the lower semicircle and passes through $(3, -4)$.

Solving polynomial equations in x and y

The quadratic formula enables us to solve a second degree equation like $y^2 - 2xy + 3x^2 = 0$ for y in terms of x. There are somewhat more complicated formulas for solving equations of degree three and four. But there are no general formulas for solving equations of degree five or higher. Finding slopes on curves defined by such equations usually requires implicit differentiation.

The answer is yes. To find dy/dx we simply differentiate both sides of the equation $y^2 = x$ with respect to x, treating $y = f(x)$ as a differentiable function of x:

$$y^2 = x$$

$$2y \frac{dy}{dx} = 1$$

$$\frac{dy}{dx} = \frac{1}{2y}.$$

The Chain Rule gives $\frac{d}{dx} y^2 = \frac{d}{dx}[f(x)]^2 = 2f(x)f'(x) = 2y \frac{dy}{dx}.$

This one formula gives the derivatives we calculated for *both* of the explicit solutions $y_1 = \sqrt{x}$ and $y_2 = -\sqrt{x}$:

$$\frac{dy_1}{dx} = \frac{1}{2y_1} = \frac{1}{2\sqrt{x}}, \qquad \frac{dy_2}{dx} = \frac{1}{2y_2} = \frac{1}{2(-\sqrt{x})} = -\frac{1}{2\sqrt{x}}.$$

EXAMPLE 2   Find the slope of circle $x^2 + y^2 = 25$ at the point $(3, -4)$.

Solution   The circle is not the graph of a single function of x. Rather it is the combined graphs of two differentiable functions, $y_1 = \sqrt{25 - x^2}$ and $y_2 = -\sqrt{25 - x^2}$ (Fig. 2.47). The point $(3, -4)$ lies on the graph of y_2, so we can find the slope by calculating explicitly:

$$\left.\frac{dy_2}{dx}\right|_{x=3} = -\left.\frac{-2x}{2\sqrt{25 - x^2}}\right|_{x=3} = -\frac{-6}{2\sqrt{25 - 9}} = \frac{3}{4}. \tag{1}$$

But we can also solve the problem more easily by differentiating the given equation of the circle implicitly with respect to x:

$$\frac{d}{dx}(x^2) + \frac{d}{dx}(y^2) = \frac{d}{dx}(25)$$

$$2x + 2y \frac{dy}{dx} = 0$$

$$\frac{dy}{dx} = -\frac{x}{y}.$$

The slope at $(3, -4)$ is $\left.-\frac{x}{y}\right|_{(3, -4)} = -\frac{3}{-4} = \frac{3}{4}.$

Notice that unlike the slope formula in Eq. (1), which applies only to points below the x-axis, the formula $dy/dx = -x/y$ applies everywhere the circle has a slope. Notice also that the derivative involves *both* variables x and y, not just the independent variable x.

To calculate the derivatives of other implicitly defined functions, we proceed as in Examples 1 and 2: We treat y as a differentiable implicit function of x and apply the usual rules to differentiate both sides of the defining equation.

EXAMPLE 3   Find dy/dx if $2y = x^2 + \sin y$.

Solution

$$2y = x^2 + \sin y$$

$$\frac{d}{dx}(2y) = \frac{d}{dx}(x^2 + \sin y) \qquad \text{Differentiate both sides with respect to } x \ldots$$

$$= \frac{d}{dx}(x^2) + \frac{d}{dy}(\sin y) \qquad \ldots \text{treating } y \text{ as a function of } x \text{ and using the Chain Rule.}$$

$$2\frac{dy}{dx} = 2x + \cos y \frac{dy}{dx}$$

$$2\frac{dy}{dx} - \cos y \frac{dy}{dx} = 2x \qquad \text{Collect terms with } dy/dx \ldots$$

$$(2 - \cos y)\frac{dy}{dx} = 2x \qquad \ldots \text{and factor out } dy/dx.$$

$$\frac{dy}{dx} = \frac{2x}{2 - \cos y} \qquad \text{Solve for } dy/dx \text{ by dividing.}$$

❑

Lenses, Tangents, and Normal Lines

In the law that describes how light changes direction as it enters a lens, the important angles are the angles the light makes with the line perpendicular to the surface of the lens at the point of entry (angles A and B in Fig. 2.48). This line is called the *normal* to the surface at the point of entry. In a profile view of a lens like the one in Fig. 2.48, the normal is the line perpendicular to the tangent to the profile curve at the point of entry.

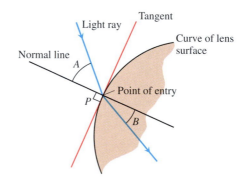

2.48 The profile of a lens, showing the bending (refraction) of a ray of light as it passes through the lens surface.

Definition

A line is **normal** to a curve at a point if it is perpendicular to the curve's tangent there. The line is called the **normal** to the curve at that point.

The profiles of lenses are often described by quadratic curves. When they are, we can use implicit differentiation to find the tangents and normals.

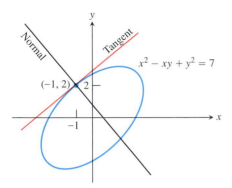

2.49 The graph of $x^2 - xy + y^2 = 7$ is an ellipse. Example 4 shows how to find equations for the tangent and normal lines at the point $(-1, 2)$.

Helga von Koch's snowflake curve (1904)

Start with an equilateral triangle, calling it curve 1. On the middle third of each side, build an equilateral triangle pointing outward. Then erase the interiors of the old middle thirds. Call the expanded curve curve 2. Now put equilateral triangles, again pointing outward, on the middle thirds of the sides of curve 2. Erase the interiors of the old middle thirds to make curve 3. Repeat the process, as shown, to define an infinite sequence of plane curves. The limit curve of the sequence is Koch's snowflake curve.

The snowflake curve is too rough to have a tangent at any point. In other words, the equation $F(x, y) = 0$ defining the curve does not define y as a differentiable function of x or x as a differentiable function of y at any point. We will encounter the snowflake again when we study length in Section 5.5.

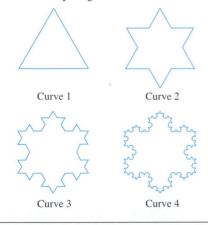

Curve 1   Curve 2

Curve 3   Curve 4

EXAMPLE 4   Find the tangent and normal to the curve $x^2 - xy + y^2 = 7$ at the point $(-1, 2)$ (Fig. 2.49).

Solution   We first use implicit differentiation to find dy/dx:

$$x^2 - xy + y^2 = 7$$

$$\frac{d}{dx}(x^2) - \frac{d}{dx}(xy) + \frac{d}{dx}(y^2) = \frac{d}{dx}(7) \qquad \text{Differentiate both sides with respect to } x, \dots$$

$$2x - \left(x\frac{dy}{dx} + y\frac{dx}{dx}\right) + 2y\frac{dy}{dx} = 0 \qquad \dots \text{treating } xy \text{ as a product and } y \text{ as a function of } x.$$

$$(2y - x)\frac{dy}{dx} = y - 2x \qquad \text{Collect terms.}$$

$$\frac{dy}{dx} = \frac{y - 2x}{2y - x}. \qquad \text{Solve for } dy/dx.$$

We then evaluate the derivative at $(x, y) = (-1, 2)$ to obtain

$$\frac{dy}{dx}\bigg|_{(-1,2)} = \frac{y - 2x}{2y - x}\bigg|_{(-1,2)} = \frac{2 - 2(-1)}{2(2) - (-1)} = \frac{4}{5}.$$

The tangent to the curve at $(-1, 2)$ is the line

$$y = 2 + \frac{4}{5}(x - (-1))$$

$$y = \frac{4}{5}x + \frac{14}{5}.$$

The normal to the curve at $(-1, 2)$ is

$$y = 2 - \frac{5}{4}(x - (-1))$$

$$y = -\frac{5}{4}x + \frac{3}{4}.$$

Using Implicit Differentiation to Find Derivatives of Higher Order

Implicit differentiation can also produce derivatives of higher order.

EXAMPLE 5   Find d^2y/dx^2 if $2x^3 - 3y^2 = 7$.

Solution   To start, we differentiate both sides of the equation with respect to x to find $y' = dy/dx$:

$$2x^3 - 3y^2 = 7$$

$$\frac{d}{dx}(2x^3) - \frac{d}{dx}(3y^2) = \frac{d}{dx}(7)$$

$$6x^2 - 6yy' = 0$$

$$x^2 - yy' = 0$$

$$y' = \frac{x^2}{y} \qquad \text{(if } y \neq 0).$$

We differentiate the equation $x^2 - yy' = 0$ again to find y'':

$$\frac{d}{dx}(x^2 - yy') = \frac{d}{dx}(0)$$

$$2x - y'y' - yy'' = 0 \qquad \text{Product Rule with } u = y, v = y'$$

$$yy'' = 2x - (y')^2$$

$$y'' = \frac{2x}{y} - \frac{(y')^2}{y} \qquad (y \neq 0).$$

Finally, we substitute $y' = x^2/y$ to express y'' in terms of x and y:

$$y'' = \frac{2x}{y} - \frac{(x^2/y)^2}{y} = \frac{2x}{y} - \frac{x^4}{y^3} \qquad (y \neq 0).$$

Rational Powers of Differentiable Functions

We know that the Power Rule

$$\frac{d}{dx}x^n = nx^{n-1} \tag{2}$$

holds when n is an integer. We can now show that it holds when n is any rational number.

Theorem 6

Power Rule for Rational Powers

If n is a rational number, then x^n is differentiable at every interior point x of the domain of x^{n-1}, and

$$\frac{d}{dx}x^n = nx^{n-1}. \tag{3}$$

Proof Let p and q be integers with $q > 0$ and suppose that $y = \sqrt[q]{x^p} = x^{p/q}$. Then

$$y^q = x^p.$$

This equation is an algebraic combination of powers of x and y, so the advanced theorem we mentioned at the beginning of the section assures us that y is a differentiable function of x. Since p and q are integers (for which we already have the Power Rule), we can differentiate both sides of the equation implicitly with respect to x and obtain

$$qy^{q-1}\frac{dy}{dx} = px^{p-1}. \tag{4}$$

If $y \neq 0$, we can then divide both sides of Eq. (4) by qy^{q-1} to solve for dy/dx, obtaining

$$\frac{dy}{dx} = \frac{px^{p-1}}{qy^{q-1}} \qquad \text{Eq. (4) divided by } qy^{q-1}$$

$$= \frac{p}{q} \cdot \frac{x^{p-1}}{(x^{(p/q)})^{q-1}} \qquad y = x^{p/q}$$

$$= \frac{p}{q} \cdot \frac{x^{p-1}}{x^{p-p/q}} \qquad \frac{p}{q}(q-1) = p - \frac{p}{q}$$

$$= \frac{p}{q} \cdot x^{(p-1)-(p-p/q)} \qquad \text{A law of exponents}$$

$$= \frac{p}{q} \cdot x^{(p/q)-1}.$$

This proves the rule. ❑

EXAMPLE 6

a) $\dfrac{d}{dx}(x^{1/2}) = \dfrac{1}{2}x^{-1/2} = \dfrac{1}{2\sqrt{x}}$ Eq. (3) with $n = \dfrac{1}{2}$

function derivative defined
defined for $x \geq 0$ only for $x > 0$

b) $\dfrac{d}{dx}(x^{1/5}) = \dfrac{1}{5}x^{-4/5}$ Eq. (3) with $n = \dfrac{1}{5}$

function derivative not
defined for all x defined at $x = 0$ ❑

A version of the Power Rule with a built-in application of the Chain Rule states that if n is a rational number, u is differentiable at x, and $(u(x))^{n-1}$ is defined, then u^n is differentiable at x, and

$$\frac{d}{dx}u^n = nu^{n-1}\frac{du}{dx}. \qquad (5)$$

EXAMPLE 7

a) $\dfrac{d}{dx}(1-x^2)^{1/4} = \dfrac{1}{4}(1-x^2)^{-3/4}(-2x)$ Eq. (5) with $u = 1 - x^2$ and $n = 1/4$

function defined
on $[-1, 1]$

$$= \frac{-x}{2(1-x^2)^{3/4}}$$

derivative defined
only on $(-1, 1)$

b) $\dfrac{d}{dx}(\cos x)^{-1/5} = -\dfrac{1}{5}(\cos x)^{-6/5}\dfrac{d}{dx}(\cos x)$

$$= -\frac{1}{5}(\cos x)^{-6/5}(-\sin x)$$

$$= \frac{1}{5}\sin x(\cos x)^{-6/5}$$

❑

Exercises 2.6

Derivatives of Rational Powers

Find dy/dx in Exercises 1–10.

1. $y = x^{9/4}$ **2.** $y = x^{-3/5}$

3. $y = \sqrt[3]{2x}$ **4.** $y = \sqrt[4]{5x}$

5. $y = 7\sqrt{x+6}$ **6.** $y = -2\sqrt{x-1}$

7. $y = (2x+5)^{-1/2}$ **8.** $y = (1-6x)^{2/3}$

9. $y = x(x^2+1)^{1/2}$ **10.** $y = x(x^2+1)^{-1/2}$

Find the first derivatives of the functions in Exercises 11–18.

11. $s = \sqrt[7]{t^2}$ **12.** $r = \sqrt[4]{\theta^{-3}}$

13. $y = \sin\left[(2t+5)^{-2/3}\right]$ **14.** $z = \cos\left[(1-6t)^{2/3}\right]$

15. $f(x) = \sqrt{1 - \sqrt{x}}$ **16.** $g(x) = 2(2x^{-1/2}+1)^{-1/3}$

17. $h(\theta) = \sqrt[3]{1 + \cos(2\theta)}$ **18.** $k(\theta) = (\sin(\theta+5))^{5/4}$

Differentiating Implicitly

Use implicit differentiation to find dy/dx in Exercises 19–32.

19. $x^2y + xy^2 = 6$ **20.** $x^3 + y^3 = 18xy$

21. $2xy + y^2 = x + y$ **22.** $x^3 - xy + y^3 = 1$

23. $x^2(x-y)^2 = x^2 - y^2$ **24.** $(3xy+7)^2 = 6y$

25. $y^2 = \dfrac{x-1}{x+1}$ **26.** $x^2 = \dfrac{x-y}{x+y}$

27. $x = \tan y$ **28.** $x = \sin y$

29. $x + \tan(xy) = 0$ **30.** $x + \sin y = xy$

31. $y \sin\left(\dfrac{1}{y}\right) = 1 - xy$ **32.** $y^2 \cos\left(\dfrac{1}{y}\right) = 2x + 2y$

Find $dr/d\theta$ in Exercises 33–36.

33. $\theta^{1/2} + r^{1/2} = 1$ **34.** $r - 2\sqrt{\theta} = \dfrac{3}{2}\theta^{2/3} + \dfrac{4}{3}\theta^{3/4}$

35. $\sin(r\theta) = \dfrac{1}{2}$ **36.** $\cos r + \cos \theta = r\theta$

Higher Derivatives

In Exercises 37–42, use implicit differentiation to find dy/dx and then d^2y/dx^2.

37. $x^2 + y^2 = 1$ **38.** $x^{2/3} + y^{2/3} = 1$

39. $y^2 = x^2 + 2x$ **40.** $y^2 - 2x = 1 - 2y$

41. $2\sqrt{y} = x - y$ **42.** $xy + y^2 = 1$

43. If $x^3 + y^3 = 16$, find the value of d^2y/dx^2 at the point $(2, 2)$.

44. If $xy + y^2 = 1$, find the value of d^2y/dx^2 at the point $(0, -1)$.

Slopes, Tangents, and Normals

In Exercises 45 and 46, find the slope of the curve at the given points.

45. $y^2 + x^2 = y^4 - 2x$ at $(-2, 1)$ and $(-2, -1)$

46. $(x^2 + y^2)^2 = (x - y)^2$ at $(1, 0)$ and $(1, -1)$

In Exercises 47–56, verify that the given point is on the curve and find the lines that are (a) tangent and (b) normal to the curve at the given point.

47. $x^2 + xy - y^2 = 1$, $(2, 3)$

48. $x^2 + y^2 = 25$, $(3, -4)$

49. $x^2y^2 = 9$, $(-1, 3)$

50. $y^2 - 2x - 4y - 1 = 0$, $(-2, 1)$

51. $6x^2 + 3xy + 2y^2 + 17y - 6 = 0$, $(-1, 0)$

52. $x^2 - \sqrt{3}xy + 2y^2 = 5$, $(\sqrt{3}, 2)$

53. $2xy + \pi \sin y = 2\pi$, $(1, \pi/2)$

54. $x \sin 2y = y \cos 2x$, $(\pi/4, \pi/2)$

55. $y = 2 \sin(\pi x - y)$, $(1, 0)$

56. $x^2 \cos^2 y - \sin y = 0$, $(0, \pi)$

57. Find the two points where the curve $x^2 + xy + y^2 = 7$ crosses the x-axis, and show that the tangents to the curve at these points are parallel. What is the common slope of these tangents?

58. Find points on the curve $x^2 + xy + y^2 = 7$ (a) where the tangent is parallel to the x-axis and (b) where the tangent is parallel to the y-axis. In the latter case, dy/dx is not defined, but dx/dy is. What value does dx/dy have at these points?

59. *The eight curve.* Find the slopes of the curve $y^4 = y^2 - x^2$ at the two points shown here.

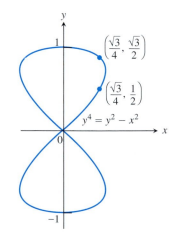

60. *The cissoid of Diocles (from about* 200 B.C.*).* Find equations for the tangent and normal to the cissoid of Diocles $y^2(2-x)=x^3$ at $(1, 1)$.

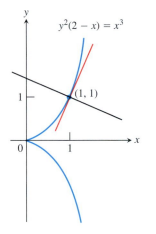

61. *The devil's curve (Gabriel Cramer [the Cramer of Cramer's rule], 1750).* Find the slopes of the devil's curve $y^4 - 4y^2 = x^4 - 9x^2$ at the four indicated points.

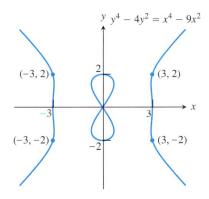

62. *The folium of Descartes.* (See Fig. 2.45.)

a) Find the slope of the folium of Descartes, $x^3 + y^3 - 9xy = 0$ at the points $(4, 2)$ and $(2, 4)$.

b) At what point other than the origin does the folium have a horizontal tangent?

c) Find the coordinates of the point A in Fig. 2.45, where the folium has a vertical tangent.

Theory and Examples

63. Which of the following could be true if $f''(x) = x^{-1/3}$?

a) $f(x) = \dfrac{3}{2}x^{2/3} - 3$ **b)** $f(x) = \dfrac{9}{10}x^{5/3} - 7$

c) $f'''(x) = -\dfrac{1}{3}x^{-4/3}$ **d)** $f'(x) = \dfrac{3}{2}x^{2/3} + 6$

64. Is there anything special about the tangents to the curves $2x^2 + 3y^2 = 5$ and $y^2 = x^3$ at the points $(1, \pm 1)$? Give reasons for your answer.

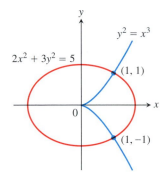

65. The line that is normal to the curve $x^2 + 2xy - 3y^2 = 0$ at $(1, 1)$ intersects the curve at what other point?

66. Find the normals to the curve $xy + 2x - y = 0$ that are parallel to the line $2x + y = 0$.

67. Show that if it is possible to draw these three normals from the point $(a, 0)$ to the parabola $x = y^2$ shown here, then a must be greater than $1/2$. One of the normals is the x-axis. For what value of a are the other two normals perpendicular?

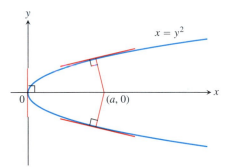

68. What is the geometry behind the restrictions on the domains of the derivatives in Example 6 and Example 7(a)?

In Exercises 69 and 70 find both dy/dx (treating y as a function of x) and dx/dy (treating x as a function of y). How do dy/dx and dx/dy seem to be related? Can you explain the relationship geometrically in terms of the graphs?

69. $xy^3 + x^2y = 6$ **70.** $x^3 + y^2 = \sin^2 y$

⊞ Grapher Explorations

71. a) Given that $x^4 + 4y^2 = 1$, find dy/dx two ways: (1) by solving for y and differentiating the resulting functions in the usual way and (2) by implicit differentiation. Do you get the same result each way?

b) Solve the equation $x^4 + 4y^2 = 1$ for y and graph the resulting functions together to produce a complete graph of the equation $x^4 + 4y^2 = 1$. Then add the graphs of the first derivatives of these functions to your display. Could you have predicted the general behavior of the derivative graphs from looking at the graph of $x^4 + 4y^2 = 1$? Could you have predicted the general behavior of the graph of $x^4 + 4y^2 = 1$ by looking at the derivative graphs? Give reasons for your answers.

72. a) Given that $(x - 2)^2 + y^2 = 4$, find dy/dx two ways: (1) by solving for y and differentiating the resulting functions with respect to x and (2) by implicit differentiation. Do you get the same result each way?

b) Solve the equation $(x - 2)^2 + y^2 = 4$ for y and graph the resulting functions together to produce a complete graph of the equation $(x - 2)^2 + y^2 = 4$. Then add the graphs of the functions' first derivatives to your picture. Could you have predicted the general behavior of the derivative graphs from looking at the graph of $(x - 2)^2 + y^2 = 4$? Could you have predicted the general behavior of the graph of $(x - 2)^2 + y^2 = 4$ by looking at the derivative graphs? Give reasons for your answers.

✲ CAS Explorations and Projects

Use a CAS to perform the following steps in Exercises 73–80.

a) Plot the equation with the implicit plotter of CAS. Check to see that the given point P satisfies the equation.

b) Using implicit differentiation find a formula for the derivative dy/dx and evaluate it at the given point P.

c) Use the slope found in part (b) to define the equation of the tangent line to the curve at P. Then plot the implicit curve and tangent line together on a single graph.

73. $x^3 - xy + y^3 = 7, \quad P(2, 1)$

74. $x^5 + y^3x + yx^2 + y^4 = 4, \quad P(1, 1)$

75. $y^2 + y = \dfrac{2 + x}{1 - x}, \quad P(0, 1)$

76. $y^3 + \cos xy = x^2, \quad P(1, 0)$

77. $x + \tan\left(\dfrac{y}{x}\right) = 2, \quad P\left(1, \dfrac{\pi}{4}\right)$

78. $xy^3 + \tan(x + y) = 1, \quad P\left(\dfrac{\pi}{4}, 0\right)$

79. $2y^2 + (xy)^{1/3} = x^2 + 2, \quad P(1, 1)$

80. $x\sqrt{1 + 2y} + y = x^2, \quad P(1, 0)$

2.7

Related Rates of Change

How rapidly will the fluid level inside a vertical cylindrical storage tank drop if we pump the fluid out at the rate of 3000 L/min?

A question like this asks us to calculate a rate that we cannot measure directly from a rate that we can. To do so, we write an equation that relates the variables involved and differentiate it to get an equation that relates the rate we seek to the rate we know.

EXAMPLE 1   *Pumping out a tank*

How rapidly will the fluid level inside a vertical cylindrical tank drop if we pump the fluid out at the rate of 3000 L/min?

Solution We draw a picture of a partially filled vertical cylindrical tank, calling its radius r and the height of the fluid h (Fig. 2.50). Call the volume of the fluid V.

As time passes, the radius remains constant, but V and h change. We think of V and h as differentiable functions of time and use t to represent time. We are told that

$$\frac{dV}{dt} = -3000.$$

We pump out at the rate of 3000 L/min. The rate is negative because the volume is decreasing.

We are asked to find

$$\frac{dh}{dt}.$$

How fast will the fluid level drop?

$\dfrac{dh}{dt} = ?$

h

$\dfrac{dV}{dt} = -3000 \text{ L/min}$

2.50 The cylindrical tank in Example 1.

Reminder

Rates of change are represented by derivatives. If a quantity is increasing, its derivative with respect to time is positive; if a quantity is decreasing, its derivative is negative.

To find dh/dt, we first write an equation that relates h to V. The equation depends on the units chosen for V, r, and h. With V in liters and r and h in meters, the appropriate equation for the cylinder's volume is

$$V = 1000\pi r^2 h$$

because a cubic meter contains 1000 liters.

Since V and h are differentiable functions of t, we can differentiate both sides of the equation $V = 1000\pi r^2 h$ with respect to t to get an equation that relates dh/dt to dV/dt:

$$\frac{dV}{dt} = 1000\pi r^2 \frac{dh}{dt}. \qquad \text{\footnotesize\textit{r is a constant.}}$$

We substitute the known value $dV/dt = -3000$ and solve for dh/dt:

$$\frac{dh}{dt} = \frac{-3000}{1000\pi r^2} = -\frac{3}{\pi r^2}. \qquad (1)$$

The fluid level will drop at the rate of $3/(\pi r^2)$ m/min. ❏

Equation (1) shows how the rate at which the fluid level drops depends on the tank's radius. If r is small, dh/dt will be large; if r is large, dh/dt will be small.

If $r = 1$ m: $\qquad \dfrac{dh}{dt} = -\dfrac{3}{\pi} \approx -0.95$ m/min $= -95$ cm/min

If $r = 10$ m: $\qquad \dfrac{dh}{dt} = -\dfrac{3}{100\pi} \approx -0.0095$ m/min $= -0.95$ cm/min

EXAMPLE 2   *A rising balloon*

A hot-air balloon rising straight up from a level field is tracked by a range finder 500 ft from the lift-off point. At the moment the range finder's elevation angle is $\pi/4$, the angle is increasing at the rate of 0.14 rad/min. How fast is the balloon rising at that moment?

Solution   We answer the question in six steps.

Step 1: *Draw a picture and name the variables and constants* (Fig. 2.51). The variables in the picture are

$\theta =$ the angle the range finder makes with the ground (radians)

$y =$ the height of the balloon (feet).

We let t represent time and assume θ and y to be differentiable functions of t.

The one constant in the picture is the distance from the range finder to the lift-off point (500 ft). There is no need to give it a special symbol.

Step 2: *Write down the additional numerical information.*

$$\frac{d\theta}{dt} = 0.14 \text{ rad/min} \qquad \text{when} \qquad \theta = \frac{\pi}{4}$$

Step 3: *Write down what we are asked to find.* We want dy/dt when $\theta = \pi/4$.

Step 4: *Write an equation that relates the variables y and θ.*

$$\frac{y}{500} = \tan\theta, \qquad \text{or} \qquad y = 500\tan\theta$$

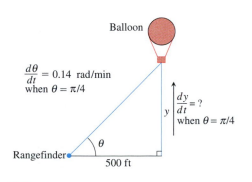

$\dfrac{d\theta}{dt} = 0.14$ rad/min when $\theta = \pi/4$

$\dfrac{dy}{dt} = ?$ when $\theta = \pi/4$

Balloon

Rangefinder

500 ft

θ

y

2.51 The balloon in Example 2.

Step 5: *Differentiate with respect to t using the Chain Rule.* The result tells how dy/dt (which we want) is related to $d\theta/dt$ (which we know).

$$\frac{dy}{dt} = 500 \sec^2 \theta \, \frac{d\theta}{dt}$$

Step 6: *Evaluate with $\theta = \pi/4$ and $d\theta/dt = 0.14$ to find dy/dt.*

$$\frac{dy}{dt} = 500(\sqrt{2})^2(0.14) = (1000)(0.14) = 140 \qquad \sec\frac{\pi}{4} = \sqrt{2}$$

At the moment in question, the balloon is rising at the rate of 140 ft/min. ❏

Strategy for Solving Related Rate Problems

1. *Draw a picture and name the variables and constants.* Use t for time. Assume all variables are differentiable functions of t.
2. *Write down the numerical information* (in terms of the symbols you have chosen).
3. *Write down what you are asked to find* (usually a rate, expressed as a derivative).
4. *Write an equation that relates the variables.* You may have to combine two or more equations to get a single equation that relates the variable whose rate you want to the variable whose rate you know.
5. *Differentiate with respect to t.* Then express the rate you want in terms of the rate and variables whose values you know.
6. *Evaluate.* Use known values to find the unknown rate.

EXAMPLE 3   *A highway chase*

A police cruiser, approaching a right-angled intersection from the north, is chasing a speeding car that has turned the corner and is now moving straight east. When the cruiser is 0.6 mi north of the intersection and the car is 0.8 mi to the east, the police determine with radar that the distance between them and the car is increasing at 20 mph. If the cruiser is moving at 60 mph at the instant of measurement, what is the speed of the car?

Solution   We carry out the steps of the basic strategy.

Step 1: *Picture and variables.* We picture the car and cruiser in the coordinate plane, using the positive x-axis as the eastbound highway and the positive y-axis as the southbound highway (Fig. 2.52). We let t represent time and set

$x =$ position of car at time t,

$y =$ position of cruiser at time t,

$s =$ distance between car and cruiser at time t.

We assume x, y, and s to be differentiable functions of t.

Step 2: *Numerical information.* At the instant in question,

$$x = 0.8 \text{ mi}, \qquad y = 0.6 \text{ mi}, \qquad \frac{dy}{dt} = -60 \text{ mph}, \qquad \frac{ds}{dt} = 20 \text{ mph}.$$

(dy/dt is negative because y is decreasing.)

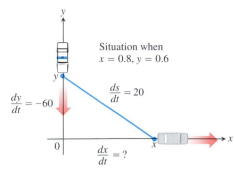

2.52 Figure for Example 3.

Step 3: *To find:* $\dfrac{dx}{dt}$

Step 4: *How the variables are related:* $s^2 = x^2 + y^2$ Pythagorean theorem
(The equation $s = \sqrt{x^2 + y^2}$ would also work.)

Step 5: *Differentiate with respect to t.*

$$2s\,\frac{ds}{dt} = 2x\,\frac{dx}{dt} + 2y\,\frac{dy}{dt} \qquad\text{Chain Rule}$$

$$\frac{ds}{dt} = \frac{1}{s}\left(x\,\frac{dx}{dt} + y\,\frac{dy}{dt} \right)$$

$$= \frac{1}{\sqrt{x^2 + y^2}}\left(x\,\frac{dx}{dt} + y\,\frac{dy}{dt} \right)$$

Step 6: *Evaluate, with $x = 0.8$, $y = 0.6$, $dy/dt = -60$, $ds/dt = 20$, and solve for* $dx/dt.$

$$20 = \underbrace{\frac{1}{\sqrt{(0.8)^2 + (0.6)^2}}}_{1}\left(0.8\,\frac{dx}{dt} + (0.6)(-60) \right)$$

$$20 = 0.8\,\frac{dx}{dt} - 36$$

$$\frac{dx}{dt} = \frac{20 + 36}{0.8} = 70$$

At the moment in question, the car's speed is 70 mph.   ❏

EXAMPLE 4   Water runs into a conical tank at the rate of 9 ft³/min. The tank stands point down and has a height of 10 ft and a base radius of 5 ft. How fast is the water level rising when the water is 6 ft deep?

Solution   We carry out the steps of the basic strategy.

Step 1: *Picture and variables.* We draw a picture of a partially filled conical tank (Fig. 2.53). The variables in the problem are

> V = volume (ft³) of water in the tank at time t (min),
>
> x = radius (ft) of the surface of the water at time t,
>
> y = depth (ft) of water in the tank at time t.

We assume V, x, and y to be differentiable functions of t. The constants are the dimensions of the tank.

Step 2: *Numerical information.* At the time in question,

$$y = 6 \text{ ft}, \qquad \frac{dV}{dt} = 9 \text{ ft}^3/\text{min}.$$

Step 3: *To find:* $\dfrac{dy}{dt}.$

Step 4: *How the variables are related.*

$$V = \frac{1}{3}\pi x^2 y \qquad\text{Cone volume formula} \qquad (2)$$

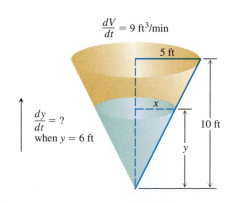

$\dfrac{dV}{dt} = 9$ ft³/min

5 ft

x

$\dfrac{dy}{dt} = ?$
when $y = 6$ ft

10 ft

y

2.53 The conical tank in Example 4.

This equation involves x as well as V and y. Because no information is given about x and dx/dt at the time in question, we need to eliminate x. Using similar triangles (Fig. 2.53) gives us a way to express x in terms of y:

$$\frac{x}{y} = \frac{5}{10}, \quad \text{or} \quad x = \frac{y}{2}.$$

Therefore,

$$V = \frac{1}{3}\pi \left(\frac{y}{2}\right)^2 y = \frac{\pi}{12}y^3. \tag{3}$$

Step 5: *Differentiate with respect to t.* We differentiate Eq. (3), getting

$$\frac{dV}{dt} = \frac{\pi}{12} \cdot 3y^2 \frac{dy}{dt} = \frac{\pi}{4}y^2 \frac{dy}{dt}. \tag{4}$$

We then solve for dy/dt to express the rate we want (dy/dt) in terms of the rate we know (dV/dt):

$$\frac{dy}{dt} = \frac{4}{\pi y^2}\frac{dV}{dt}.$$

Step 6: *Evaluate, with $y = 6$ and $dV/dt = 9$.*

$$\frac{dy}{dt} = \frac{4}{\pi(6)^2} \cdot 9 = \frac{1}{\pi} \approx 0.32 \text{ ft/min}$$

At the moment in question, the water level is rising at about 0.32 ft/min.   ❑

Exercises 2.7

1. Suppose that the radius r and area $A = \pi r^2$ of a circle are differentiable functions of t. Write an equation that relates dA/dt to dr/dt.

2. Suppose that the radius r and surface area $S = 4\pi r^2$ of a sphere are differentiable functions of t. Write an equation that relates dS/dt to dr/dt.

3. The radius r and height h of a right circular cylinder are related to the cylinder's volume V by the formula $V = \pi r^2 h$.

   a) How is dV/dt related to dh/dt if r is constant?
   b) How is dV/dt related to dr/dt if h is constant?
   c) How is dV/dt related to dr/dt and dh/dt if neither r nor h is constant?

4. The radius r and height h of a right circular cone are related to the cone's volume V by the equation $V = (1/3)\pi r^2 h$.

   a) How is dV/dt related to dh/dt if r is constant?
   b) How is dV/dt related to dr/dt if h is constant?
   c) How is dV/dt related to dr/dt and dh/dt if neither r nor h is constant?

5. *Changing voltage.* The voltage V (volts), current I (amperes), and resistance R (ohms) of an electric circuit like the one shown here are related by the equation $V = IR$. Suppose that V is increasing at the rate of 1 volt/sec while I is decreasing at the rate of 1/3 amp/sec. Let t denote time in seconds.

   a) What is the value of dV/dt?
   b) What is the value of dI/dt?
   c) What equation relates dR/dt to dV/dt and dI/dt?
   d) Find the rate at which R is changing when $V = 12$ volts and $I = 2$ amp. Is R increasing, or decreasing?

6. The power P (watts) of an electric circuit is related to the circuit's resistance R (ohms) and current i (amperes) by the equation $P = Ri^2$.

   a) How are dP/dt, dR/dt, and di/dt related if none of P, R, and i are constant?
   b) How is dR/dt related to di/dt if P is constant?

7. Let x and y be differentiable functions of t and let $s = \sqrt{x^2 + y^2}$

be the distance between the points $(x, 0)$ and $(0, y)$ in the xy-plane.

a) How is ds/dt related to dx/dt if y is constant?

b) How is ds/dt related to dx/dt and dy/dt if neither x nor y is constant?

c) How is dx/dt related to dy/dt if s is constant?

8. If x, y, and z are lengths of the edges of a rectangular box, the common length of the box's diagonals is $s = \sqrt{x^2 + y^2 + z^2}$.

a) Assuming that x, y, and z are differentiable functions of t, how is ds/dt related to $dx/dt, dy/dt$, and dz/dt?

b) How is ds/dt related to dy/dt and dz/dt if x is constant?

c) How are $dx/dt, dy/dt$, and dz/dt related if s is constant?

9. The area A of a triangle with sides of lengths a and b enclosing an angle of measure θ is

$$A = \frac{1}{2}ab \sin \theta.$$

a) How is dA/dt related to $d\theta/dt$ if a and b are constant?

b) How is dA/dt related to $d\theta/dt$ and da/dt if only b is constant?

c) How is dA/dt related to $d\theta/dt, da/dt$, and db/dt if none of a, b, and θ are constant?

10. *Heating a plate.* When a circular plate of metal is heated in an oven, its radius increases at the rate of 0.01 cm/min. At what rate is the plate's area increasing when the radius is 50 cm?

11. *Changing dimensions in a rectangle.* The length l of a rectangle is decreasing at the rate of 2 cm/sec while the width w is increasing at the rate of 2 cm/sec. When $l = 12$ cm and $w = 5$ cm, find the rates of change of (a) the area, (b) the perimeter, and (c) the lengths of the diagonals of the rectangle. Which of these quantities are decreasing, and which are increasing?

12. *Changing dimensions in a rectangular box.* Suppose that the edge lengths x, y, and z of a closed rectangular box are changing at the following rates:

$$\frac{dx}{dt} = 1 \text{ m/sec}, \qquad \frac{dy}{dt} = -2 \text{ m/sec}, \qquad \frac{dz}{dt} = 1 \text{ m/sec}.$$

Find the rates at which the box's (a) volume, (b) surface area, and (c) diagonal length $s = \sqrt{x^2 + y^2 + z^2}$ are changing at the instant when $x = 4$, $y = 3$, and $z = 2$.

13. *A sliding ladder.* A 13-ft ladder is leaning against a house when its base starts to slide away. By the time the base is 12 ft from the house, the base is moving at the rate of 5 ft/sec.

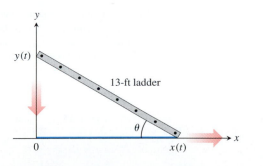

a) How fast is the top of the ladder sliding down the wall then?

b) At what rate is the area of the triangle formed by the ladder, wall, and ground changing then?

c) At what rate is the angle θ between the ladder and the ground changing then?

14. *Commercial air traffic.* Two commercial airplanes are flying at 40,000 ft along straight-line courses that intersect at right angles. Plane A is approaching the intersection point at a speed of 442 knots (nautical miles per hour; a nautical mile is 2000 yd). Plane B is approaching the intersection at 481 knots. At what rate is the distance between the planes changing when A is 5 nautical miles from the intersection point and B is 12 nautical miles from the intersection point?

15. *Flying a kite.* A girl flies a kite at a height of 300 ft, the wind carrying the kite horizontally away from her at a rate of 25 ft/sec. How fast must she let out the string when the kite is 500 ft away from her?

16. *Boring a cylinder.* The mechanics at Lincoln Automotive are reboring a 6-in.-deep cylinder to fit a new piston. The machine they are using increases the cylinder's radius one-thousandth of an inch every 3 min. How rapidly is the cylinder volume increasing when the bore (diameter) is 3.800 in.?

17. *A growing sand pile.* Sand falls from a conveyor belt at the rate of 10 m³/min onto the top of a conical pile. The height of the pile is always three-eighths of the base diameter. How fast are the (a) height and (b) radius changing when the pile is 4 m high? Answer in cm/min.

18. *A draining conical reservoir.* Water is flowing at the rate of 50 m³/min from a shallow concrete conical reservoir (vertex down) of base radius 45 m and height 6 m. (a) How fast is the water level falling when the water is 5 m deep? (b) How fast is the radius of the water's surface changing then? Answer in cm/min.

19. *A draining hemispherical reservoir.* Water is flowing at the rate of 6 m³/min from a reservoir shaped like a hemispherical bowl of radius 13 m, shown here in profile. Answer the following questions, given that the volume of water in a hemispherical bowl of radius R is $V = (\pi/3)y^2(3R - y)$ when the water is y units deep.

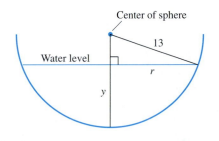

a) At what rate is the water level changing when the water is 8 m deep?

b) What is the radius r of the water's surface when the water is y m deep?

c) At what rate is the radius r changing when the water is 8 m deep?

20. *A growing raindrop.* Suppose that a drop of mist is a perfect sphere and that, through condensation, the drop picks up moisture at a rate proportional to its surface area. Show that under these circumstances the drop's radius increases at a constant rate.

21. *The radius of an inflating balloon.* A spherical balloon is inflated with helium at the rate of 100π ft³/min. How fast is the balloon's radius increasing at the instant the radius is 5 ft? How fast is the surface area increasing?

22. *Hauling in a dinghy.* A dinghy is pulled toward a dock by a rope from the bow through a ring on the dock 6 ft above the bow. The rope is hauled in at the rate of 2 ft/sec. (a) How fast is the boat approaching the dock when 10 ft of rope are out? (b) At what rate is angle θ changing then (see the figure)?

Ring at edge of dock

23. *A balloon and a bicycle.* A balloon is rising vertically above a level, straight road at a constant rate of 1 ft/sec. Just when the balloon is 65 ft above the ground, a bicycle moving at a constant rate of 17 ft/sec passes under it. How fast is the distance between the bicycle and balloon increasing 3 sec later?

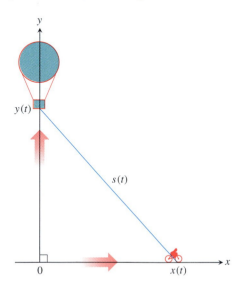

24. *Making coffee.* Coffee is draining from a conical filter into a cylindrical coffeepot at the rate of 10 in³/min. (a) How fast is the level in the pot rising when the coffee in the cone is 5 in. deep? (b) How fast is the level in the cone falling then?

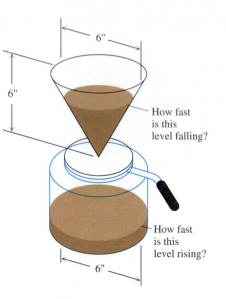

How fast is this level falling?

How fast is this level rising?

25. *Cardiac output.* In the late 1860s, Adolf Fick, a professor of physiology in the Faculty of Medicine in Würtzberg, Germany, developed one of the methods we use today for measuring how much blood your heart pumps in a minute. Your cardiac output as you read this sentence is probably about 7 liters a minute. At rest it is likely to be a bit under 6 L/min. If you are a trained marathon runner running a marathon, your cardiac output can be as high as 30 L/min.

Your cardiac output can be calculated with the formula

$$y = \frac{Q}{D},$$

where Q is the number of milliliters of CO_2 you exhale in a minute and D is the difference between the CO_2 concentration (ml/L) in the blood pumped to the lungs and the CO_2 concentration in the blood returning from the lungs. With $Q = 233$ ml/min and $D = 97 - 56 = 41$ ml/L,

$$y = \frac{233 \text{ ml/min}}{41 \text{ ml/L}} \approx 5.68 \text{ L/min},$$

fairly close to the 6 L/min that most people have at basal (resting) conditions. (Data courtesy of J. Kenneth Herd, M.D., Quillan College of Medicine, East Tennessee State University.)

Suppose that when $Q = 233$ and $D = 41$, we also know that D is decreasing at the rate of 2 units a minute but that Q remains unchanged. What is happening to the cardiac output?

26. *Cost, revenue, and profit.* A company can manufacture x items at a cost of $c(x)$ dollars, a sales revenue of $r(x)$ dollars, and a profit of $p(x) = r(x) - c(x)$ dollars (everything in thousands). Find dc/dt, dr/dt, and dp/dt for the following values of x and dx/dt.

a) $r(x) = 9x$, $c(x) = x^3 - 6x^2 + 15x$, and $dx/dt = 0.1$ when $x = 2$

b) $r(x) = 70x$, $c(x) = x^3 - 6x^2 + 45/x$, and $dx/dt = 0.05$ when $x = 1.5$

27. *Moving along a parabola.* A particle moves along the parabola $y = x^2$ in the first quadrant in such a way that its x-coordinate (measured in meters) increases at a steady 10 m/sec. How fast is the angle of inclination θ of the line joining the particle to the origin changing when $x = 3$ m?

28. *Moving along another parabola.* A particle moves from right to left along the parabola $y = \sqrt{-x}$ in such a way that its x-coordinate (measured in meters) decreases at the rate of 8 m/sec. How fast is the angle of inclination θ of the line joining the particle to the origin changing when $x = -4$?

29. *Motion in the plane.* The coordinates of a particle in the metric xy-plane are differentiable functions of time t with $dx/dt = -1$ m/sec and $dy/dt = -5$ m/sec. How fast is the particle's distance from the origin changing as it passes through the point (5, 12)?

30. *A moving shadow.* A man 6 ft tall walks at the rate of 5 ft/sec toward a streetlight that is 16 ft above the ground. At what rate is the tip of his shadow moving? At what rate is the length of his shadow changing when he is 10 ft from the base of the light?

31. *Another moving shadow.* A light shines from the top of a pole 50 ft high. A ball is dropped from the same height from a point 30 ft away from the light. How fast is the shadow of the ball moving along the ground 1/2 sec later? (Assume the ball falls a distance $s = 16t^2$ ft in t sec.)

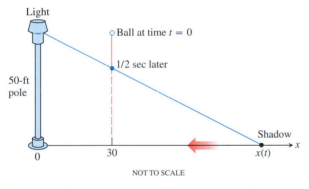

NOT TO SCALE

32. You are videotaping a race from a stand 132 ft from the track, following a car that is moving at 180 mph (264 ft/sec). How fast will your camera angle θ be changing when the car is right in front of you? A half second later?

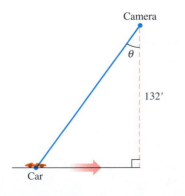

33. *A melting ice layer.* A spherical iron ball 8 in. in diameter is coated with a layer of ice of uniform thickness. If the ice melts at the rate of 10 in³/min, how fast is the thickness of the ice decreasing when it is 2 in. thick? How fast is the outer surface area of ice decreasing?

34. *Highway patrol.* A highway patrol plane flies 3 mi above a level, straight road at a steady 120 mi/h. The pilot sees an oncoming car and with radar determines that at the instant the line-of-sight distance from plane to car is 5 mi the line-of-sight distance is decreasing at the rate of 160 mi/h. Find the car's speed along the highway.

35. *A building's shadow.* On a morning of a day when the sun will pass directly overhead, the shadow of an 80-ft building on level ground is 60 ft long. At the moment in question, the angle θ the sun makes with the ground is increasing at the rate of 0.27°/min. At what rate is the shadow decreasing? (Remember to use radians. Express your answer in inches per minute, to the nearest tenth.)

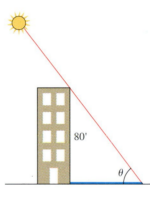

36. *Walkers.* A and B are walking on straight streets that meet at right angles. A approaches the intersection at 2 m/sec; B moves away from the intersection 1 m/sec. At what rate is the angle θ changing when A is 10 m from the intersection and B is 20 m from the intersection? Express your answer in degrees per second to the nearest degree.

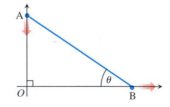

37. A baseball diamond is a square 90 ft on a side. A player runs from first base to second at a rate of 16 ft/sec.

   a) At what rate is the player's distance from third base changing when the player is 30 ft from first base?

   b) At what rates are angles θ_1 and θ_2 (see the figure) changing at that time?

c) The player slides into second base at the rate of 15 ft/sec. At what rates are angles θ_1 and θ_2 changing as the player touches base?

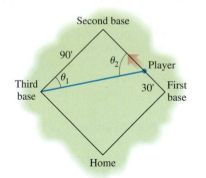

Second base

90'

θ_2

Player

θ_1

Third base

30'

First base

Home

38. *A second hand.* At what rate is the distance between the tip of the second hand and the 12 o'clock mark changing when the second hand points to 4 o'clock?

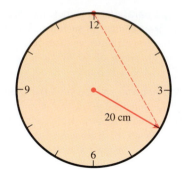

12

9

3

20 cm

6

39. *Ships.* Two ships are steaming straight away from a point O along routes that make a 120° angle. Ship A moves at 14 knots (nautical miles per hour; a nautical mile is 2000 yd). Ship B moves at 21 knots. How fast are the ships moving apart when $OA = 5$ and $OB = 3$ nautical miles?

CHAPTER 2 QUESTIONS TO GUIDE YOUR REVIEW

1. What is the derivative of a function f? How is its domain related to the domain of f? Give examples.

2. What role does the derivative play in defining slopes, tangents, and rates of change?

3. How can you sometimes graph the derivative of a function when all you have is a table of the function's values?

4. What does it mean for a function to be differentiable on an open interval? on a closed interval?

5. How are derivatives and one-sided derivatives related?

6. Describe geometrically when a function typically does *not* have a derivative at a point.

7. How is a function's differentiability at a point related to its continuity there, if at all?

8. Could the unit step function

$$U(x) = \begin{cases} 0, & x < 0 \\ 1, & x \geq 0 \end{cases}$$

possibly be the derivative of some other function on $[-1, 1]$? Explain.

9. What rules do you know for calculating derivatives? Give some examples.

10. Explain how the three formulas

a) $\dfrac{d}{dx}(x^n) = nx^{n-1}$,

b) $\dfrac{d}{dx}(cu) = c\dfrac{du}{dx}$,

c) $\dfrac{d}{dx}(u_1 + u_2 + \cdots + u_n) = \dfrac{du_1}{dx} + \dfrac{du_2}{dx} + \cdots + \dfrac{du_n}{dx}$

enable us to differentiate any polynomial.

11. What formula do we need, in addition to the three listed in question 10, to differentiate rational functions?

12. What is a second derivative? a third derivative? How many derivatives do the functions you know have? Give examples.

13. What is the relationship between a function's average and instantaneous rates of change? Give an example.

14. How do derivatives arise in the study of motion? What can you learn about a body's motion along a line by examining the derivatives of the body's position function? Give examples.

15. How can derivatives arise in economics?

16. Give examples of still other applications of derivatives.

17. What is the value of $\lim_{\theta \to 0}(\sin \theta)/\theta$? Does it matter whether θ is measured in degrees or radians? Explain.

18. What do the limits $\lim_{h \to 0}(\sin h)/h$ and $\lim_{h \to 0}(\cos h - 1)/h$ have to do with the derivatives of the sine and cosine functions? What *are* the derivatives of these functions?

19. Once you know the derivatives of $\sin x$ and $\cos x$, how can you find the derivatives of tan x, cot x, sec x, and csc x? What *are* the derivatives of these functions?

20. At what points are the six basic trigonometric functions continuous? How do you know?

21. What is the rule for calculating the derivative of a composite of two differentiable functions? How is such a derivative evaluated? Give examples.

22. If u is a differentiable function of x, how do you find $(d/dx)(u^n)$ if n is an integer? if n is a rational number? Give examples.

23. What is implicit differentiation? When do you need it? Give examples.

24. How do related rate problems arise? Give examples.

25. Outline a strategy for solving related rate problems. Illustrate with an example.

CHAPTER 2 PRACTICE EXERCISES

Derivatives of Functions

Find the derivatives of the functions in Exercises 1–36.

1. $y = x^5 - 0.125x^2 + 0.25x$

2. $y = 3 - 0.7x^3 + 0.3x^7$

3. $y = x^3 - 3(x^2 + \pi^2)$

4. $y = x^7 + \sqrt{7}x - \dfrac{1}{\pi + 1}$

5. $y = (x + 1)^2(x^2 + 2x)$

6. $y = (2x - 5)(4 - x)^{-1}$

7. $y = (\theta^2 + \sec\theta + 1)^3$

8. $y = \left(-1 - \dfrac{\csc\theta}{2} - \dfrac{\theta^2}{4}\right)^2$

9. $s = \dfrac{\sqrt{t}}{1 + \sqrt{t}}$

10. $s = \dfrac{1}{\sqrt{t} - 1}$

11. $y = 2\tan^2 x - \sec^2 x$

12. $y = \dfrac{1}{\sin^2 x} - \dfrac{2}{\sin x}$

13. $s = \cos^4(1 - 2t)$

14. $s = \cot^3\left(\dfrac{2}{t}\right)$

15. $s = (\sec t + \tan t)^5$

16. $s = \csc^5(1 - t + 3t^2)$

17. $r = \sqrt{2\theta \sin\theta}$

18. $r = 2\theta\sqrt{\cos\theta}$

19. $r = \sin\sqrt{2\theta}$

20. $r = \sin(\theta + \sqrt{\theta + 1})$

21. $y = \dfrac{1}{2}x^2 \csc\dfrac{2}{x}$

22. $y = 2\sqrt{x}\sin\sqrt{x}$

23. $y = x^{-1/2}\sec(2x)^2$

24. $y = \sqrt{x}\csc(x + 1)^3$

25. $y = 5\cot x^2$

26. $y = x^2\cot 5x$

27. $y = x^2\sin^2(2x^2)$

28. $y = x^{-2}\sin^2(x^3)$

29. $s = \left(\dfrac{4t}{t + 1}\right)^{-2}$

30. $s = \dfrac{-1}{15(15t - 1)^3}$

31. $y = \left(\dfrac{\sqrt{x}}{1 + x}\right)^2$

32. $y = \left(\dfrac{2\sqrt{x}}{2\sqrt{x} + 1}\right)^2$

33. $y = \sqrt{\dfrac{x^2 + x}{x^2}}$

34. $y = 4x\sqrt{x + \sqrt{x}}$

35. $r = \left(\dfrac{\sin\theta}{\cos\theta - 1}\right)^2$

36. $r = \left(\dfrac{1 + \sin\theta}{1 - \cos\theta}\right)^2$

In Exercises 37–48, find dy/dx.

37. $y = (2x + 1)\sqrt{2x + 1}$

38. $y = 20(3x - 4)^{1/4}(3x - 4)^{-1/5}$

39. $y = \dfrac{3}{(5x^2 + \sin 2x)^{3/2}}$

40. $y = (3 + \cos^3 3x)^{-1/3}$

41. $xy + 2x + 3y = 1$

42. $x^2 + xy + y^2 - 5x = 2$

43. $x^3 + 4xy - 3y^{4/3} = 2x$

44. $5x^{4/5} + 10y^{6/5} = 15$

45. $\sqrt{xy} = 1$

46. $x^2y^2 = 1$

47. $y^2 = \dfrac{x}{x + 1}$

48. $y^2 = \sqrt{\dfrac{1 + x}{1 - x}}$

In Exercises 49 and 50, find dp/dq.

49. $p^3 + 4pq - 3q^2 = 2$

50. $q = (5p^2 + 2p)^{-3/2}$

In Exercises 51 and 52, find dr/ds.

51. $r\cos 2s + \sin^2 s = \pi$

52. $2rs - r - s + s^2 = -3$

53. Find d^2y/dx^2 by implicit differentiation:

    a) $x^3 + y^3 = 1$ **b)** $y^2 = 1 - \dfrac{2}{x}$

54. a) By differentiating $x^2 - y^2 = 1$ implicitly, show that $dy/dx = x/y$.

    b) Then show that $d^2y/dx^2 = -1/y^3$.

Numerical Values of Derivatives

55. Suppose that functions $f(x)$ and $g(x)$ and their first derivatives have the following values at $x = 0$ and $x = 1$.

x	$f(x)$	$g(x)$	$f'(x)$	$g'(x)$
0	1	1	5	1/3
1	3	-4	$-1/3$	$-8/3$

Find the first derivatives of the following combinations at the given value of x.

a) $5f(x) - g(x)$, $x = 1$

b) $f(x)g^3(x)$, $x = 0$

c) $\dfrac{f(x)}{g(x) + 1}$, $x = 1$

d) $f(g(x))$, $x = 0$

e) $g(f(x))$, $x = 0$

f) $(x + f(x))^{3/2}$, $x = 1$

g) $f(x + g(x))$, $x = 0$

56. Suppose that the function $f(x)$ and its first derivative have the following values at $x = 0$ and $x = 1$.

x	$f(x)$	$f'(x)$
0	9	-2
1	-3	1/5

Find the first derivatives of the following combinations at the given value of x.

a) $\sqrt{x}\, f(x)$, $x = 1$

b) $\sqrt{f(x)}$, $x = 0$

c) $f\left(\sqrt{x}\right)$, $x = 1$

d) $f(1 - 5\tan x)$, $x = 0$

e) $\dfrac{f(x)}{2 + \cos x}$, $x = 0$

f) $10 \sin\left(\dfrac{\pi x}{2}\right) f^2(x)$, $x = 1$

57. Find the value of dy/dt at $t = 0$ if $y = 3 \sin 2x$ and $x = t^2 + \pi$.

58. Find the value of ds/du at $u = 2$ if $s = t^2 + 5t$ and $t = (u^2 + 2u)^{1/3}$.

59. Find the value of dw/ds at $s = 0$ if $w = \sin\left(\sqrt{r} - 2\right)$ and $r = 8 \sin\left(s + \pi/6\right)$.

60. Find the value of dr/dt at $t = 0$ if $r = (\theta^2 + 7)^{1/3}$ and $\theta^2 t + \theta = 1$.

61. If $y^3 + y = 2 \cos x$, find the value of d^2y/dx^2 at the point $(0, 1)$.

62. If $x^{1/3} + y^{1/3} = 4$, find d^2y/dx^2 at the point $(8, 8)$.

Derivative Definition

In Exercises 63 and 64, find the derivative using the definition.

63. $f(t) = \dfrac{1}{2t + 1}$

64. $g(x) = 2x^2 + 1$

65. a) Graph the function
$$f(x) = \begin{cases} x^2, & -1 \le x < 0 \\ -x^2, & 0 \le x \le 1. \end{cases}$$

b) Is f continuous at $x = 0$?

c) Is f differentiable at $x = 0$?

Give reasons for your answers.

66. a) Graph the function
$$f(x) = \begin{cases} x, & -1 \le x < 0 \\ \tan x, & 0 \le x \le \pi/4. \end{cases}$$

b) Is f continuous at $x = 0$?

c) Is f differentiable at $x = 0$?

Give reasons for your answers.

67. a) Graph the function
$$f(x) = \begin{cases} x, & 0 \le x \le 1 \\ 2 - x, & 1 < x \le 2. \end{cases}$$

b) Is f continuous at $x = 1$?

c) Is f differentiable at $x = 1$?

Give reasons for your answers.

68. For what value or values of the constant m, if any, is
$$f(x) = \begin{cases} \sin 2x, & x \le 0 \\ mx, & x > 0 \end{cases}$$

a) continuous at $x = 0$?

b) differentiable at $x = 0$?

Give reasons for your answers.

Slopes, Tangents, and Normals

69. Are there any points on the curve $y = (x/2) + 1/(2x - 4)$ where the slope is $-3/2$? If so, find them.

70. Are there any points on the curve $y = x - 1/(2x)$ where the slope is 3? If so, find them.

71. Find the points on the curve $y = 2x^3 - 3x^2 - 12x + 20$ where the tangent is parallel to the x-axis.

72. Find the x- and y-intercepts of the line that is tangent to the curve $y = x^3$ at the point $(-2, -8)$.

73. Find the points on the curve $y = 2x^3 - 3x^2 - 12x + 20$ where the tangent is

a) perpendicular to the line $y = 1 - (x/24)$;

b) parallel to the line $y = \sqrt{2} - 12x$.

74. Show that the tangents to the curve $y = (\pi \sin x)/x$ at $x = \pi$ and $x = -\pi$ intersect at right angles.

75. Find the points on the curve $y = \tan x$, $-\pi/2 < x < \pi/2$, where the normal is parallel to the line $y = -x/2$. Sketch the curve and normals together, labeling each with its equation.

76. Find equations for the tangent and normal to the curve $y = 1 + \cos x$ at the point $(\pi/2, 1)$. Sketch the curve, tangent, and normal together, labeling each with its equation.

77. The parabola $y = x^2 + C$ is to be tangent to the line $y = x$. Find C.

78. Show that the tangent to the curve $y = x^3$ at any point (a, a^3) meets the curve again at a point where the slope is four times the slope at (a, a^3).

79. For what value of c is the curve $y = c/(x + 1)$ tangent to the line through the points $(0, 3)$ and $(5, -2)$?

80. Show that the normal line at any point of the circle $x^2 + y^2 = a^2$ passes through the origin.

In Exercises 81–86, find equations for the lines that are tangent and normal to the curve at the given point.

81. $x^2 + 2y^2 = 9$, $(1, 2)$ **82.** $x^3 + y^2 = 2$, $(1, 1)$

83. $xy + 2x - 5y = 2$, $(3, 2)$ **84.** $(y - x)^2 = 2x + 4$, $(6, 2)$

85. $x + \sqrt{xy} = 6$, $(4, 1)$ **86.** $x^{3/2} + 2y^{3/2} = 17$, $(1, 4)$

87. Find the slope of the curve $x^3 y^3 + y^2 = x + y$ at the points $(1, 1)$ and $(1, -1)$.

88. The graph below suggests that the curve $y = \sin(x - \sin x)$ might have horizontal tangents at the x-axis. Does it? Give reasons for your answer.

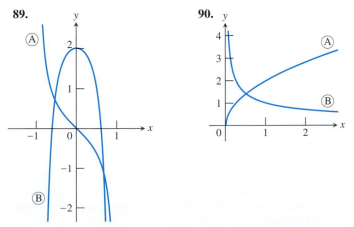

(Generated by Mathematica)

Analyzing Graphs

Each of the figures in Exercises 89 and 90 shows two graphs, the graph of a function $y = f(x)$ together with the graph of its derivative $f'(x)$. Which graph is which? How do you know?

89.   **90.**

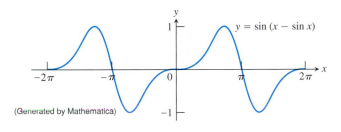

91. Use the following information to graph the function $y = f(x)$ for $-1 \le x \le 6$.

   i) The graph of f is made of line segments joined end to end.
   ii) The graph starts at the point $(-1, 2)$.
   iii) The derivative of f, where defined, agrees with the step function shown here.

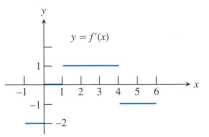

92. Repeat Exercise 91, supposing that the graph starts at $(-1, 0)$ instead of $(-1, 2)$.

Exercises 93 and 94 are about the graphs in Fig. 2.54. The graphs in part (a) show the numbers of rabbits and foxes in a small arctic

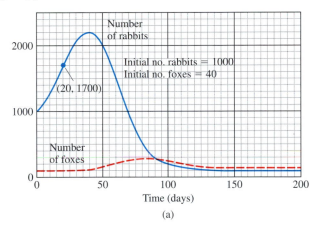

Initial no. rabbits = 1000
Initial no. foxes = 40

(a)

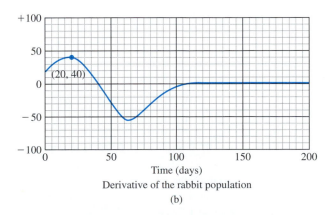

Derivative of the rabbit population

(b)

2.54 Rabbits and foxes in an arctic predator-prey food chain. (Source: *Differentiation* by W. U. Walton et al., Project CALC, Education Development Center, Inc., Newton, Mass, 1975, p. 86.)

population. They are plotted as functions of time for 200 days. The number of rabbits increases at first, as the rabbits reproduce. But the foxes prey on the rabbits and, as the number of foxes increases, the rabbit population levels off and then drops. Figure 2.54(b) shows the graph of the derivative of the rabbit population. We made it by plotting slopes, as in Example 4 in Section 2.1.

93. a) What is the value of the derivative of the rabbit population in Fig. 2.54 when the number of rabbits is largest? smallest?

b) What is the size of the rabbit population in Fig. 2.54 when its derivative is largest? smallest?

94. In what units should the slopes of the rabbit and fox population curves be measured?

Limits

Find the limits in Exercises 95–104.

95. $\lim\limits_{s \to 0} \dfrac{\sin (s/2)}{s/3}$

96. $\lim\limits_{\theta \to -\pi} \dfrac{\sin^2 (\theta + \pi)}{\theta + \pi}$

97. $\lim\limits_{x \to 0} \dfrac{\sin x}{2x^2 - x}$

98. $\lim\limits_{x \to 0} \dfrac{3x - \tan 7x}{2x}$

99. $\lim\limits_{r \to 0} \dfrac{\sin r}{\tan 2r}$

100. $\lim\limits_{\theta \to 0} \dfrac{\sin (\sin \theta)}{\theta}$

101. $\lim\limits_{\theta \to (\pi/2)^-} \dfrac{4 \tan^2 \theta + \tan \theta + 1}{\tan^2 \theta + 5}$

102. $\lim\limits_{\theta \to 0^+} \dfrac{1 - 2 \cot^2 \theta}{5 \cot^2 \theta - 7 \cot \theta - 8}$

103. $\lim\limits_{x \to 0} \dfrac{x \sin x}{2 - 2 \cos x}$

104. $\lim\limits_{\theta \to 0} \dfrac{1 - \cos \theta}{\theta^2}$

Show how to extend the functions in Exercises 105 and 106 to be continuous at the origin.

105. $g(x) = \dfrac{\tan (\tan x)}{\tan x}$

106. $f(x) = \dfrac{\tan (\tan x)}{\sin (\sin x)}$

107. Is there any value of k that will make

$$f(x) = \begin{cases} \dfrac{\sin x}{2x}, & x \neq 0 \\ k, & x = 0 \end{cases}$$

continuous at $x = 0$? If so, what is it? Give reasons for your answer.

108. a) GRAPHER Graph the function

$$f(x) = \begin{cases} \dfrac{x^2}{\sin^2 2x}, & x \neq 0 \\ c, & x = 0. \end{cases}$$

b) Find a value of c that makes f continuous at $x = 0$. Justify your answer.

Related Rates

109. The total surface area S of a right circular cylinder is related to the base radius r and height h by the equation $S = 2\pi r^2 + 2\pi rh$.

a) How is dS/dt related to dr/dt if h is constant?

b) How is dS/dt related to dh/dt if r is constant?

c) How is dS/dt related to dr/dt and dh/dt if neither r nor h is constant?

d) How is dr/dt related to dh/dt if S is constant?

110. The lateral surface area S of a right circular cone is related to the base radius r and height h by the equation $S = \pi r \sqrt{r^2 + h^2}$.

a) How is dS/dt related to dr/dt if h is constant?

b) How is dS/dt related to dh/dt if r is constant?

c) How is dS/dt related to dr/dt and dh/dt if neither r nor h is constant?

111. The radius of a circle is changing at the rate of $-2/\pi$ m/sec. At what rate is the circle's area changing when $r = 10$ m?

112. The volume of a cube is increasing at the rate of 1200 cm³/min at the instant its edges are 20 cm long. At what rate are the edges changing at that instant?

113. If two resistors of R_1 and R_2 ohms are connected in parallel in an electric circuit to make an R-ohm resistor, the value of R can be found from the equation

$$\frac{1}{R} = \frac{1}{R_1} + \frac{1}{R_2}.$$

If R_1 is decreasing at the rate of 1 ohm/sec and R_2 is increasing at the rate of 0.5 ohm/sec, at what rate is R changing when $R_1 = 75$ ohms and $R_2 = 50$ ohms?

114. The impedance Z (ohms) in a series circuit is related to the resistance R (ohms) and reactance X (ohms) by the equation $Z = \sqrt{R^2 + X^2}$. If R is increasing at 3 ohms/sec and X is decreasing at 2 ohms/sec, at what rate is Z changing when $R = 10$ ohms and $X = 20$ ohms?

115. The coordinates of a particle moving in the metric xy-plane are differentiable functions of time t with $dx/dt = -1$ m/sec and $dy/dt = -5$ m/sec. How fast is the particle approaching the origin as it passes through the point $(5, 12)$?

116. A particle moves along the curve $y = x^{3/2}$ in the first quadrant in such a way that its distance from the origin increases at the rate of 11 units per second. Find dx/dt when $x = 3$.

117. Water drains from the conical tank shown in Fig. 2.55 at the rate of 5 ft³/min. (a) What is the relation between the variables h and r in the figure? (b) How fast is the water level dropping when $h = 6$ ft?

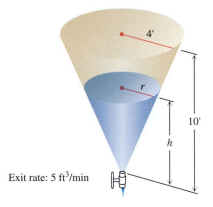

2.55 The conical tank in Exercise 117.

Exit rate: 5 ft³/min

2.56 The television cable in Exercise 118.

118. As television cable is pulled from a large spool to be strung from the telephone poles along a street, it unwinds from the spool in layers of constant radius (see Fig. 2.56). If the truck pulling the cable moves at a steady 6 ft/sec (a touch over 4 mph), use the equation $s = r\theta$ to find how fast (rad/sec) the spool is turning when the layer of radius 1.2 ft is being unwound.

119. The figure below shows a boat 1 km offshore, sweeping the shore with a searchlight. The light turns at a constant rate, $d\theta/dt = -0.6$ rad/sec.

a) How fast is the light moving along the shore when it reaches point A?

b) How many revolutions per minute is 0.6 rad/sec?

120. Points A and B move along the x- and y-axes, respectively, in such a way that the distance r (meters) along the perpendicular from the origin to line AB remains constant. How fast is OA changing, and is it increasing, or decreasing, when $OB = 2r$ and B is moving toward O at the rate of $0.3r$ m/sec?

CHAPTER	**2**	ADDITIONAL EXERCISES–THEORY, EXAMPLES, APPLICATIONS

1. An equation like $\sin^2 \theta + \cos^2 \theta = 1$ is called an **identity** because it holds for all values of θ. An equation like $\sin \theta = 0.5$ is not an identity because it holds only for selected values of θ, not all. If you differentiate both sides of a trigonometric identity in θ with respect to θ, the resulting new equation will also be an identity.

Differentiate the following to show that the resulting equations hold for all θ.

a) $\sin 2\theta = 2 \sin \theta \cos \theta$

b) $\cos 2\theta = \cos^2 \theta - \sin^2 \theta$

2. If the identity $\sin(x + a) = \sin x \cos a + \cos x \sin a$ is differentiated with respect to x, is the resulting equation also an identity? Does this principle apply to the equation $x^2 - 2x - 8 = 0$? Explain.

3. a) Find values for the constants a, b, and c that will make

$$f(x) = \cos x \quad \text{and} \quad g(x) = a + bx + cx^2$$

satisfy the conditions

$$f(0) = g(0), \quad f'(0) = g'(0), \quad \text{and} \quad f''(0) = g''(0).$$

a) Find values for b and c that will make

$$f(x) = \sin(x + a) \quad \text{and} \quad g(x) = b \sin x + c \cos x$$

satisfy the conditions

$$f(0) = g(0) \quad \text{and} \quad f'(0) = g'(0).$$

b) For the determined values of a, b, and c, what happens for the third and fourth derivatives of f and g in each of parts (a) and (b)?

4. a) Show that $y = \sin x$, $y = \cos x$, and $y = a \cos x + b \sin x$ (a and b constants) all satisfy the equation

$$y'' + y = 0.$$

b) How would you modify the functions in (a) to satisfy the equation

$$y'' + 4y = 0?$$

Generalize this result.

5. *An osculating circle.* Find the values of $h, k,$ and a that make the circle $(x - h)^2 + (y - k)^2 = a^2$ tangent to the parabola $y = x^2 + 1$ at the point $(1, 2)$ and that also make the second derivatives d^2y/dx^2 have the same value on both curves there. Circles like this one that are tangent to a curve and have the same second derivative as the curve at the point of tangency are called *osculating circles* (from the Latin *osculari* meaning "to kiss"). We will encounter them again in Chapter 11.

6. *Marginal revenue.* A bus will hold 60 people. The number x of people per trip who use the bus is related to the fare charged (p dollars) by the law $p = [3 - (x/40)]^2$. Write an expression for the total revenue $r(x)$ per trip received by the bus company. What number of people per trip will make the marginal revenue dr/dx equal to zero? What is the corresponding fare? (This is the fare that maximizes the revenue, so the bus company should probably rethink its fare policy.)

7. *Industrial production*

a) Economists often use the expression "rate of growth" in relative rather than absolute terms. For example, let $u = f(t)$ be the number of people in the labor force at time t in a given industry. (We treat this function as though it were differentiable even though it is an integer-valued step function.)

Let $v = g(t)$ be the average production per person in the labor force at time t. The total production is then $y = uv$. If the labor force is growing at the rate of 4% per year $(du/dt = 0.04u)$ and the production per worker is growing at the rate of 5% per year $(dv/dt = 0.05v)$, find the rate of growth of the total production, y.

b) Suppose that the labor force in (a) is decreasing at the rate of 2% per year while the production per person is increasing at the rate of 3% per year. Is the total production increasing, or is it decreasing, and at what rate?

8. The designer of a 30-ft-diameter spherical hot-air balloon wants to suspend the gondola 8 ft below the bottom of the balloon with cables tangent to the surface of the balloon (Fig. 2.57). Two of the cables are shown running from the top edges of the gondola to their points of tangency, $(-12, -9)$ and $(12, -9)$. How wide should the gondola be?

9. *Pisa by parachute.* The accompanying photograph shows Mike McCarthy parachuting from the top of the Tower of Pisa on August 5, 1988. Make a rough sketch to show the shape of the graph of his speed during the jump.

10. The position at time $t \geq 0$ of a particle moving along a coordinate line is

$$s = 10 \cos (t + \pi/4).$$

a) What is the particle's starting position $(t = 0)$?

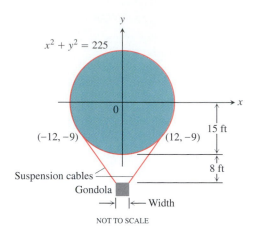

2.57 The balloon and gondola in Exercise 8.

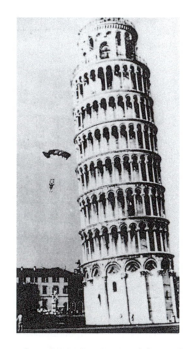

Mike McCarthy of London jumped from the Tower of Pisa and then opened his parachute in what he said was a world record low-level parachute jump of 179 feet. Source: *Boston Globe,* Aug. 6, 1988.

b) What are the points farthest to the left and right of the origin reached by the particle?

c) Find the particle's velocity and acceleration at the points in question (b).

d) When does the particle first reach the origin? What are its velocity, speed, and acceleration then?

11. On Earth, you can easily shoot a paper clip 64 ft straight up into the air with a rubber band. In t seconds after firing, the paper clip is $s = 64t - 16t^2$ ft above your hand.

   a) How long does it take the paper clip to reach its maximum height? With what velocity does it leave your hand?

   b) On the moon, the same acceleration will send the paper clip to a height of $s = 64t - 2.6t^2$ ft in t seconds. About how long will it take the paper clip to reach its maximum height and how high will it go?

12. At time t sec, the positions of two particles on a coordinate line are $s_1 = 3t^3 - 12t^2 + 18t + 5$ m and $s_2 = -t^3 + 9t^2 - 12t$ m. When do the particles have the same velocities?

13. A particle of constant mass m moves along the x-axis. Its velocity v and position x satisfy the equation

$$\frac{1}{2}m\,(v^2 - v_0{}^2) = \frac{1}{2}k(x_0{}^2 - x^2),$$

where k, v_0, and x_0 are constants. Show that whenever $v \neq 0$,

$$m\frac{dv}{dt} = -kx.$$

14. a) Show that if the position x of a moving point is given by a quadratic function of t, $x = At^2 + Bt + C$, then the average velocity over any time interval $[t_1, t_2]$ is equal to the instantaneous velocity at the midpoint of the time interval.

   b) What is the geometric significance of the result in (a)?

15. Find all values of the constants m and b for which the function

$$y = \begin{cases} \sin x & \text{for } x < \pi \\ mx + b & \text{for } x \geq \pi, \end{cases}$$

is (a) continuous at $x = \pi$; (b) differentiable at $x = \pi$.

16. Does the function

$$f(x) = \begin{cases} \dfrac{1 - \cos x}{x} & \text{for } x \neq 0, \\ 0 & \text{for } x = 0, \end{cases}$$

have a derivative at $x = 0$? Explain.

17. a) For what values of a and b will

$$f(x) = \begin{cases} ax, & x < 2 \\ ax^2 - bx + 3, & x \geq 2 \end{cases}$$

be differentiable for all values of x?

   b) Discuss the geometry of the resulting graph of f.

18. a) For what values of a and b will

$$g(x) = \begin{cases} ax + b, & x \leq -1 \\ ax^3 + x + 2b, & x > -1 \end{cases}$$

be differentiable for all values of x?

   b) Discuss the geometry of the resulting graph of g.

19. Is there anything special about the derivative of an odd differentiable function of x? Give reasons for your answer.

20. Is there anything special about the derivative of an even differentiable function of x? Give reasons for your answer.

21. *A surprising result.* Suppose that the functions f and g are defined throughout an open interval containing the point x_0, that f is differentiable at x_0, that $f(x_0) = 0$, and that g is continuous at x_0. Show that the product fg is differentiable at x_0. This shows, for example, that while $|x|$ is not differentiable at $x = 0$, the product $x|x|$ *is* differentiable at $x = 0$.

22. *(Continuation of Exercise 21.)* Use the result of Exercise 21 to show that the following functions are differentiable at $x = 0$.

   a) $|x| \sin x$ **b)** $x^{2/3} \sin x$ **c)** $\sqrt[3]{x}(1 - \cos x)$

   d) $h(x) = \begin{cases} x^2 \sin(1/x), & x \neq 0 \\ 0, & x = 0 \end{cases}$

23. Is the derivative of

$$h(x) = \begin{cases} x^2 \sin(1/x), & x \neq 0 \\ 0, & x = 0 \end{cases}$$

derived at $x = 0$? continuous at $x = 0$? How about the derivative of $k(x) = xh(x)$? Give reasons for your answers.

24. Suppose that a function f satisfies the following conditions for all real values of x and y:

   i) $f(x + y) = f(x) \cdot f(y)$;

   ii) $f(x) = 1 + xg(x)$, where $\lim\limits_{x \to 0} g(x) = 1$.

Show that the derivative $f'(x)$ exists at every value of x and that $f'(x) = f(x)$.

25. *The generalized product rule.* Use mathematical induction (Appendix 1) to prove that if $y = u_1 u_2 \cdots u_n$ is a finite product of differentiable functions, then y is differentiable on their common domain and

$$\frac{dy}{dx} = \frac{du_1}{dx}u_2 \cdots u_n + u_1\frac{du_2}{dx}\cdots u_n + \cdots + u_1 u_2 \cdots u_{n-1}\frac{du_n}{dx}.$$

26. *Leibniz's rule for higher order derivatives of products.* Leibniz's rule for higher order derivatives of products of differentiable functions says that

   a) $\dfrac{d^2(uv)}{dx^2} = \dfrac{d^2u}{dx^2}v + 2\dfrac{du}{dx}\dfrac{dv}{dx} + u\dfrac{d^2v}{dx^2},$

   b) $\dfrac{d^3(uv)}{dx^3} = \dfrac{d^3u}{dx^3}v + 3\dfrac{d^2u}{dx^2}\dfrac{dv}{dx} + 3\dfrac{du}{dx}\dfrac{d^2v}{dx^2} + u\dfrac{d^3v}{dx^3},$

   c) $\dfrac{d^n(uv)}{dx^n} = \dfrac{d^nu}{dx^n}v + n\dfrac{d^{n-1}u}{dx^{n-1}}\dfrac{dv}{dx} + \cdots$

   $\qquad + \dfrac{n(n-1)\cdots(n-k+1)}{k!}\dfrac{d^{n-k}u}{dx^{n-k}}\dfrac{d^kv}{dx^k} + \cdots + u\dfrac{d^nv}{dx^n}.$

The equations in (a) and (b) are special cases of the equation in (c). Derive the equation in (c) by mathematical induction, using the fact that

$$\binom{m}{k} + \binom{m}{k+1} = \frac{m!}{k!(m-k)!} + \frac{m!}{(k+1)!(m-k-1)!}.$$

Applications of Derivatives

OVERVIEW This chapter shows how to draw conclusions from derivatives. We use derivatives to find extreme values of functions, to predict and analyze the shapes of graphs, to find replacements for complicated formulas, to determine how sensitive formulas are to errors in measurement, and to find the zeros of functions numerically. The key to many of these accomplishments is the Mean Value Theorem, a theorem whose corollaries provide the gateway to integral calculus in Chapter 4.

3.1 Extreme Values of Functions

This section shows how to locate and identify extreme values of continuous functions.

The Max-Min Theorem

A function that is continuous at every point of a closed interval has an absolute maximum and an absolute minimum value on the interval. We always look for these values when we graph a function, and we will see the role they play in problem solving (this chapter) and in the development of the integral calculus (Chapters 4 and 5).

> **Theorem 1**
> **The Max-Min Theorem for Continuous Functions**
> If f is continuous at every point of a closed interval I, then f assumes both an absolute maximum value M and an absolute minimum value m somewhere in I. That is, there are numbers x_1 and x_2 in I with $f(x_1) = m$, $f(x_2) = M$, and $m \leq f(x) \leq M$ for every other x in I (Fig. 3.1 on the following page).

The proof of Theorem 1 requires a detailed knowledge of the real number system and we will not give it here.

3.1 Typical arrangements of a continuous function's absolute maxima and minima on a closed interval $[a, b]$.

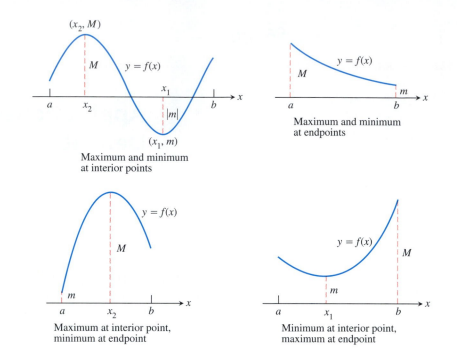

Maximum and minimum at interior points

Maximum and minimum at endpoints

Maximum at interior point, minimum at endpoint

Minimum at interior point, maximum at endpoint

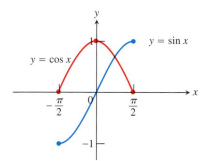

3.2 Figure for Example 1.

EXAMPLE 1   On $[-\pi/2, \pi/2]$, $f(x) = \cos x$ takes on a maximum value of 1 (once) and a minimum value of 0 (twice). The function $g(x) = \sin x$ takes on a maximum value of 1 and a minimum value of -1 (Fig. 3.2).   ❏

As Figs. 3.3 and 3.4 show, the requirements that the interval be closed and the function continuous are key ingredients of Theorem 1. Without them, the conclusion of the theorem need not hold.

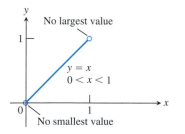

3.3 On an open interval, a continuous function need not have either a maximum or a minimum value. The function $f(x) = x$ has neither a largest nor a smallest value on $(0, 1)$.

3.4 Even a single point of discontinuity can keep a function from having either a maximum or a minimum value on a closed interval. The function

$$y = \begin{cases} x + 1, & -1 \le x < 0 \\ 0, & x = 0 \\ x - 1, & 0 < x \le 1 \end{cases}$$

is continuous at every point of $[-1, 1]$ except $x = 0$, yet its graph over $[-1, 1]$ has neither a highest nor a lowest point.

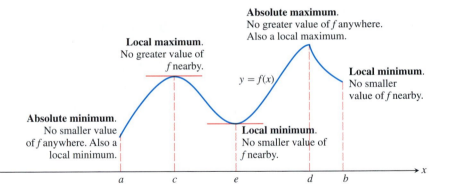

3.5 How to classify maxima and minima.

Local vs. Absolute (Global) Extrema

Figure 3.5 shows a graph with five extreme points. The function's absolute minimum occurs at *a* even though at *e* the function's value is smaller than at any other point *nearby*. The curve rises to the left and falls to the right around *c*, making $f(c)$ a maximum locally. The function attains its absolute maximum at *d*.

Definition

Absolute Extreme Values

Let *f* be a function with domain *D*. Then *f* has an **absolute maximum** value on *D* at a point *c* if

$$f(x) \le f(c) \qquad \text{for all } x \text{ in } D$$

and an **absolute minimum** value on *D* at *c* if

$$f(x) \ge f(c) \qquad \text{for all } x \text{ in } D.$$

Absolute maximum and minimum values are called absolute **extrema** (plural of the Latin *extremum*). Absolute extrema are also called **global** extrema.

Functions with the same defining rule can have different extrema, depending on the domain.

EXAMPLE 2   (See Fig. 3.6.)

	Function rule	Domain *D*	Absolute extrema on *D* (if any)
a)	$y = x^2$	$(-\infty, \infty)$	No absolute maximum. Absolute minimum of 0 at $x = 0$.
b)	$y = x^2$	$[0, 2]$	Absolute maximum of $(2)^2 = 4$ at $x = 2$. Absolute minimum of 0 at $x = 0$.
c)	$y = x^2$	$(0, 2]$	Absolute maximum of 4 at $x = 2$. No absolute minimum.
d)	$y = x^2$	$(0, 2)$	No absolute extrema. ❑

(a) abs min only

(b) abs max and min

(c) abs max only

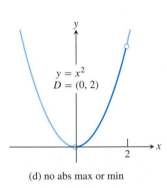

(d) no abs max or min

3.6 Graphs for Example 2.

Definition

Local Extreme Values

A function f has a **local maximum** value at an interior point c of its domain if

$$f(x) \le f(c) \qquad \text{for all } x \text{ in some open interval containing } c.$$

A function f has a **local minimum** value at an interior point c of its domain if

$$f(x) \ge f(c) \qquad \text{for all } x \text{ in some open interval containing } c.$$

We can extend the definitions of local extrema to the endpoints of intervals by defining f to have a **local maximum** or **local minimum** value *at an endpoint c* if the appropriate inequality holds for all x in some half-open interval in its domain containing c. In Fig. 3.5, the function f has local maxima at c and d and local minima at a, e, and b.

An absolute maximum is also a local maximum. Being the largest value overall, it is also the largest value in its immediate neighborhood. Hence, *a list of all local maxima will automatically include the absolute maximum if there is one. Similarly, a list of all local minima will include the absolute minimum if there is one.*

Finding Extrema

The next theorem explains why we usually need to investigate only a few values to find a function's extrema.

Theorem 2

The First Derivative Theorem for Local Extreme Values

If f has a local maximum or minimum value at an interior point c of its domain, and if f' is defined at c, then

$$f'(c) = 0.$$

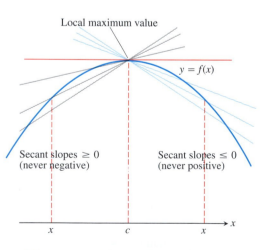

3.7 A curve with a local maximum value. The slope at c, simultaneously the limit of nonpositive numbers and nonnegative numbers, is zero.

Local maximum value

$y = f(x)$

Secant slopes ≥ 0 (never negative)

Secant slopes ≤ 0 (never positive)

Proof To show that $f'(c)$ is zero at a local extremum, we show first that $f'(c)$ cannot be positive and second that $f'(c)$ cannot be negative. The only number that is neither positive nor negative is zero, so that is what $f'(c)$ must be.

To begin, suppose that f has a local maximum value at $x = c$ (Fig. 3.7) so that $f(x) - f(c) \le 0$ for all values of x near enough to c. Since c is an interior point of f's domain, $f'(c)$ is defined by the two-sided limit

$$\lim_{x \to c} \frac{f(x) - f(c)}{x - c}.$$

This means that the right-hand and left-hand limits both exist at $x = c$ and equal $f'(c)$. When we examine these limits separately, we find that

$$f'(c) = \lim_{x \to c^+} \frac{f(x) - f(c)}{x - c} \le 0. \qquad \text{Because } (x - c) > 0 \text{ and } f(x) \le f(c) \qquad (1)$$

Similarly,

$$f'(c) = \lim_{x \to c^-} \frac{f(x) - f(c)}{x - c} \geq 0. \qquad \text{Because } (x - c) < 0 \\ \text{and } f(x) \leq f(c) \qquad (2)$$

Together, (1) and (2) imply $f'(c) = 0$.

This proves the theorem for local maximum values. To prove it for local minimum values, we simply use $f(x) \geq f(c)$, which reverses the inequalities in (1) and (2). ❑

Theorem 2 says that a function's first derivative is always zero at an interior point where the function has a local extreme value and the derivative is defined. Hence the only places where a function f can possibly have an extreme value (local or global) are

1. interior points where $f' = 0$,
2. interior points where f' is undefined,
3. endpoints of the domain of f.

The following definition helps us to summarize.

Definition

An interior point of the domain of a function f where f' is zero or undefined is a **critical point** of f.

Summary

The only domain points where a function can assume extreme values are critical points and endpoints.

Most quests for extreme values call for finding the absolute extrema of a continuous function on a closed interval. Theorem 1 assures us that such values exist; Theorem 2 tells us that they are taken on only at critical points and endpoints. These points are often so few in number that we can simply list them and calculate the corresponding function values to see what the largest and smallest are.

How to Find the Absolute Extrema of a Continuous Function f on a Closed Interval

1. Evaluate f at all critical points and endpoints.
2. Take the largest and smallest of these values.

EXAMPLE 3   Find the absolute maximum and minimum values of $f(x) = x^2$ on $[-2, 1]$.

Solution   The function is differentiable over its entire domain, so the only critical point is where $f'(x) = 2x = 0$, namely $x = 0$. We need to check the function's values at $x = 0$ and at the endpoints $x = -2$ and $x = 1$:

Critical point value:	$f(0) = 0$
Endpoint values:	$f(-2) = 4$
	$f(1) = 1$

The function has an absolute maximum value of 4 at $x = -2$ and an absolute minimum value of 0 at $x = 0$. ❑

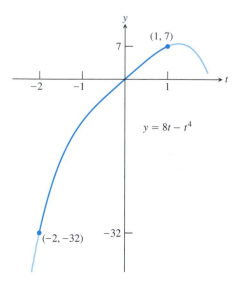

3.8 The extreme values of $g(t) = 8t - t^4$ on $[-2, 1]$ (Example 4).

EXAMPLE 4   Find the absolute extrema values of $g(t) = 8t - t^4$ on $[-2, 1]$.

Solution   The function is differentiable on its entire domain, so the only critical points occur where $g'(t) = 0$. Solving this equation gives

$$8 - 4t^3 = 0$$

$$t^3 = 2$$

$$t = 2^{1/3},$$

a point not in the given domain. The function's local extrema therefore occur at the endpoints, where we find

$$g(-2) = -32 \qquad \text{(Absolute minimum)}$$

$$g(1) = 7. \qquad \text{(Absolute maximum)}$$

See Fig. 3.8.   ❑

EXAMPLE 5   Find the absolute extrema of $h(x) = x^{2/3}$ on $[-2, 3]$.

Solution   The first derivative

$$h'(x) = \frac{2}{3}x^{-1/3} = \frac{2}{3x^{1/3}}$$

has no zeros but is undefined at $x = 0$. The values of h at this one critical point and at the endpoints $x = -2$ and $x = 3$ are

$$h(0) = 0$$

$$h(-2) = (-2)^{2/3} = 4^{1/3}$$

$$h(3) = (3)^{2/3} = 9^{1/3}.$$

The absolute maximum value is $9^{1/3}$, assumed at $x = 3$; the absolute minimum is 0, assumed at $x = 0$ (Fig. 3.9).   ❑

While a function's extrema can occur only at critical points and endpoints, not every critical point or endpoint signals the presence of an extreme value. Figures 3.10 and 3.11 illustrate this for interior points, and Exercise 34 asks you for a function that fails to assume an extreme value at an endpoint of its domain.

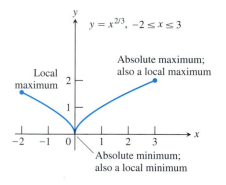

3.9 The extreme values of $h(x) = x^{2/3}$ on $[-2, 3]$ occur at $x = 0$ and $x = 3$ (Example 5).

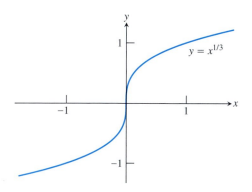

3.10 $f(x) = x^{1/3}$ has no extremum at $x = 0$, even though $f'(x) = (1/3)x^{-2/3}$ is undefined at $x = 0$.

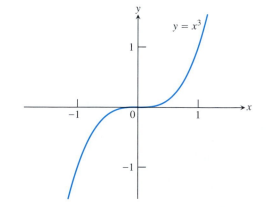

3.11 $g(x) = x^3$ has no extremum at $x = 0$ even though $g'(x) = 3x^2$ is zero at $x = 0$.

As we will see in Section 3.3, we can determine the behavior of a function f at a critical point c by further examining f', but we must look beyond what f' does at c itself.

Exercises 3.1

Finding Extrema from Graphs

In Exercises 1–6, determine from the graph whether the function has any absolute extreme values on $[a, b]$. Then explain how your answer is consistent with Theorem 1.

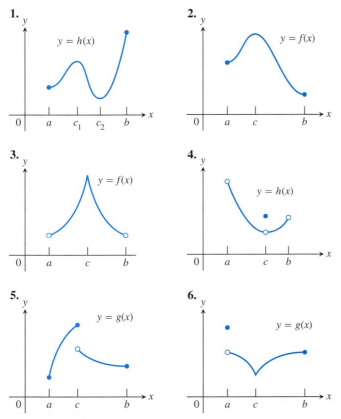

1.

2.

3.

4.

5.

6.

Absolute Extrema on Closed Intervals

In Exercises 7–22, find the absolute maximum and minimum values of each function on the given interval. Then graph the function. Identify the points on the graph where the absolute extrema occur, and include their coordinates.

7. $f(x) = \dfrac{2}{3}x - 5, \quad -2 \le x \le 3$

8. $f(x) = -x - 4, \quad -4 \le x \le 1$

9. $f(x) = x^2 - 1, \quad -1 \le x \le 2$

10. $f(x) = 4 - x^2, \quad -3 \le x \le 1$

11. $F(x) = -\dfrac{1}{x^2}, \quad 0.5 \le x \le 2$

12. $F(x) = -\dfrac{1}{x}, \quad -2 \le x \le -1$

13. $h(x) = \sqrt[3]{x}, \quad -1 \le x \le 8$

14. $h(x) = -3x^{2/3}, \quad -1 \le x \le 1$

15. $g(x) = \sqrt{4 - x^2}, \quad -2 \le x \le 1$

16. $g(x) = -\sqrt{5 - x^2}, \quad -\sqrt{5} \le x \le 0$

17. $f(\theta) = \sin\theta, \quad -\dfrac{\pi}{2} \le \theta \le \dfrac{5\pi}{6}$

18. $f(\theta) = \tan\theta, \quad -\dfrac{\pi}{3} \le \theta \le \dfrac{\pi}{4}$

19. $g(x) = \csc x, \quad \dfrac{\pi}{3} \le x \le \dfrac{2\pi}{3}$

20. $g(x) = \sec x, \quad -\dfrac{\pi}{3} \le x \le \dfrac{\pi}{6}$

21. $f(t) = 2 - |t|, \quad -1 \le t \le 3$

22. $f(t) = |t - 5|, \quad 4 \le t \le 7$

In Exercises 23–26, find the function's absolute maximum and minimum values and say where they are assumed.

23. $f(x) = x^{4/3}, \quad -1 \le x \le 8$

24. $f(x) = x^{5/3}, \quad -1 \le x \le 8$

25. $g(\theta) = \theta^{3/5}, \quad -32 \le \theta \le 1$

26. $h(\theta) = 3\theta^{2/3}, \quad -27 \le \theta \le 8$

Local Extrema in the Domain

In Exercises 27 and 28, find the values of any local maxima and minima the functions may have on the given domains, and say where they are assumed. Which extrema, if any, are absolute for the given domain?

27. **a)** $f(x) = x^2 - 4, \quad -2 \le x \le 2$
    b) $g(x) = x^2 - 4, \quad -2 \le x < 2$
    c) $h(x) = x^2 - 4, \quad -2 < x < 2$
    d) $k(x) = x^2 - 4, \quad -2 \le x < \infty$
    e) $l(x) = x^2 - 4, \quad 0 < x < \infty$

28. **a)** $f(x) = 2 - 2x^2$, $-1 \le x \le 1$
    b) $g(x) = 2 - 2x^2$, $-1 < x \le 1$
    c) $h(x) = 2 - 2x^2$, $-1 < x < 1$
    d) $k(x) = 2 - 2x^2$, $-\infty < x \le 1$
    e) $l(x) = 2 - 2x^2$, $-\infty < x < 0$

Theory and Examples

29. The function $f(x) = |x|$ has an absolute minimum value at $x = 0$ even though f is not differentiable at $x = 0$. Is this consistent with Theorem 2? Give reasons for your answer.

30. Why can't the conclusion of Theorem 2 be expected to hold if c is an endpoint of the function's domain?

31. If an even function $f(x)$ has a local maximum value at $x = c$, can anything be said about the value of f at $x = -c$? Give reasons for your answer.

32. If an odd function $g(x)$ has a local minimum value at $x = c$, can anything be said about the value of g at $x = -c$? Give reasons for your answer.

33. We know how to find the extreme values of a continuous function $f(x)$ by investigating its values at critical points and endpoints. But what if there *are* no critical points or endpoints? What happens then? Do such functions really exist? Give reasons for your answers.

34. Give an example of a function defined on [0, 1] that has neither a local maximum nor a local minimum value at 0.

✲ CAS Explorations and Projects

In Exercises 35–40, you will use a CAS to help find the absolute extrema of the given function over the specified closed interval. Perform the following steps:

a) Plot the function over the interval to see general behavior there.

b) Find the interior points where $f' = 0$. (In some exercises you may have to use the numerical equation solver to approximate a solution.) You may want to plot f' as well.

c) Find the interior points where f' does not exist.

d) Evaluate the function at all points found in parts (b) and (c) and at the endpoints of the interval.

e) Find the function's absolute extreme values on the interval and identify where they occur.

35. $f(x) = x^4 - 8x^2 + 4x + 2$, $\left[-\dfrac{20}{25}, \dfrac{64}{25}\right]$

36. $f(x) = -x^4 + 4x^3 - 4x + 1$, $\left[-\dfrac{3}{4}, 3\right]$

37. $f(x) = x^{2/3}(3 - x)$, $[-2, 2]$

38. $f(x) = 2 + 2x - 3x^{2/3}$, $\left[-1, \dfrac{10}{3}\right]$

39. $f(x) = \sqrt{x} + \cos x$, $[0, 2\pi]$

40. $f(x) = x^{3/4} - \sin x + \dfrac{1}{2}$, $[0, 2\pi]$

3.2

The Mean Value Theorem

If a body falls freely from rest near the surface of the earth, its position t seconds into the fall is $s = 4.9t^2$ m. From this we deduce that the body's velocity and acceleration are $v = ds/dt = 9.8t$ m/sec and $a = d^2s/dt^2 = 9.8$ m/sec^2. But suppose we started with the body's acceleration. Could we work backward to find its velocity and displacement functions?

What we are really asking here is what functions can have a given derivative. More generally, we might ask what kind of function can have a particular *kind* of derivative. What kind of function has a positive derivative, for instance, or a negative derivative, or a derivative that is always zero? We answer these questions by applying corollaries of the Mean Value Theorem.

When the French mathematician Michel Rolle published his theorem in 1691, his goal was to show that between every two zeros of a polynomial function there always lies a zero of the polynomial we now know to be the function's derivative. (The modern version of the theorem is not restricted to polynomials.)

Rolle distrusted the new methods of calculus, however, and spent a great deal of time and energy denouncing their use and attacking l'Hôpital's all too popular (he felt) calculus book. It is ironic that Rolle is known today only for his inadvertent contribution to a field he tried to suppress.

Rolle's Theorem

There is strong geometric evidence that between any two points where a differentiable curve crosses the x-axis there is a point on the curve where the tangent is horizontal. A 300-year-old theorem of Michel Rolle (1652–1719) assures us that this is indeed the case.

(a)

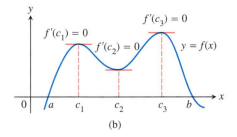

(b)

3.12 Rolle's theorem says that a differentiable curve has at least one horizontal tangent between any two points where it crosses the *x*-axis. It may have just one (a), or it may have more (b).

Theorem 3
Rolle's Theorem

Suppose that $y = f(x)$ is continuous at every point of the closed interval $[a, b]$ and differentiable at every point of its interior (a, b). If

$$f(a) = f(b) = 0,$$

then there is at least one number c in (a, b) at which

$$f'(c) = 0.$$

See Fig. 3.12.

Proof   Being continuous, f assumes absolute maximum and minimum values on $[a, b]$. These can occur only

1. at interior points where f' is zero,
2. at interior points where f' does not exist,
3. at the endpoints of the function's domain, in this case a and b.

By hypothesis, f has a derivative at every interior point. That rules out (2), leaving us with interior points where $f' = 0$ and with the two endpoints a and b.

If either the maximum or the minimum occurs at a point c inside the interval, then $f'(c) = 0$ by Theorem 2 in Section 3.1, and we have found a point for Rolle's theorem.

If both maximum and minimum are at a or b, then f is constant, $f' = 0$, and c can be taken anywhere in the interval. This completes the proof.   ❏

The hypotheses of Theorem 3 are essential. If they fail at even one point, the graph may not have a horizontal tangent (Fig. 3.13).

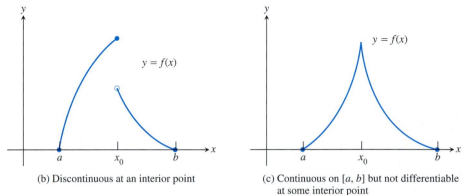

(a) Discontinuous at an endpoint

(b) Discontinuous at an interior point

(c) Continuous on $[a, b]$ but not differentiable at some interior point

3.13 No horizontal tangent.

EXAMPLE 1   The polynomial function

$$f(x) = \frac{x^3}{3} - 3x$$

graphed in Fig. 3.14 (on the following page) is continuous at every point of $[-3, 3]$ and is differentiable at every point of $(-3, 3)$. Since $f(-3) = f(3) = 0$, Rolle's

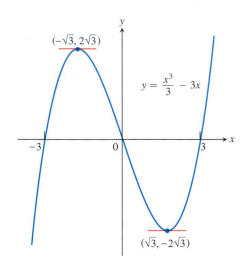

3.14 As predicted by Rolle's theorem, this curve has horizontal tangents between the points where it crosses the x-axis (Example 1).

theorem says that f' must be zero at least once in the open interval between $a = -3$ and $b = 3$. In fact, $f'(x) = x^2 - 3$ is zero twice in this interval, once at $x = -\sqrt{3}$ and again at $x = \sqrt{3}$. ∎

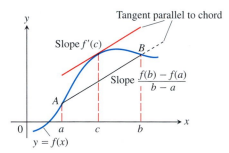

3.15 Geometrically, the Mean Value Theorem says that somewhere between A and B the curve has at least one tangent parallel to chord AB.

The Mean Value Theorem

The Mean Value Theorem is a slanted version of Rolle's theorem (Fig. 3.15). There is a point where the tangent is parallel to chord AB.

> **Theorem 4**
> **The Mean Value Theorem**
> Suppose $y = f(x)$ is continuous on a closed interval $[a, b]$ and differentiable on the interval's interior (a, b). Then there is at least one point c in (a, b) at which
> $$\frac{f(b) - f(a)}{b - a} = f'(c). \tag{1}$$

Proof We picture the graph of f as a curve in the plane and draw a line through the points $A(a, f(a))$ and $B(b, f(b))$ (see Fig. 3.16). The line is the graph of the

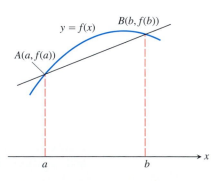

3.16 The graph of f and the chord AB over the interval $[a, b]$.

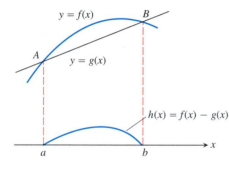

3.17 The chord *AB* in Fig. 3.16 is the graph of the function *g*(*x*). The function *h*(*x*) = *f*(*x*) − *g*(*x*) gives the vertical distance between the graphs of *f* and *g* at *x*.

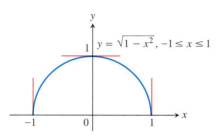

3.18 The function $f(x) = \sqrt{1 - x^2}$ satisfies the hypotheses (and conclusion) of the Mean Value Theorem on [−1, 1] even though *f* is not differentiable at −1 and 1.

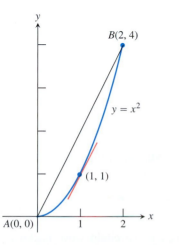

3.19 As we find in Example 2, *c* = 1 is where the tangent is parallel to the chord.

function

$$g(x) = f(a) + \frac{f(b) - f(a)}{b - a}(x - a) \tag{2}$$

(point–slope equation). The vertical difference between the graphs of *f* and *g* at *x* is

$$h(x) = f(x) - g(x)$$
$$= f(x) - f(a) - \frac{f(b) - f(a)}{b - a}(x - a). \tag{3}$$

Figure 3.17 shows the graphs of *f*, *g*, and *h* together.

The function *h* satisfies the hypotheses of Rolle's theorem on [*a*, *b*]. It is continuous on [*a*, *b*] and differentiable on (*a*, *b*) because both *f* and *g* are. Also, *h*(*a*) = *h*(*b*) = 0 because the graphs of *f* and *g* both pass through *A* and *B*. Therefore, *h*′ = 0 at some point *c* in (*a*, *b*). This is the point we want for Eq. (1).

To verify Eq. (1), we differentiate both sides of Eq. (3) with respect to *x* and then set *x* = *c*:

$$h'(x) = f'(x) - \frac{f(b) - f(a)}{b - a} \qquad \text{Derivative of Eq. (3) ...}$$

$$h'(c) = f'(c) - \frac{f(b) - f(a)}{b - a} \qquad \text{... with } x = c$$

$$0 = f'(c) - \frac{f(b) - f(a)}{b - a} \qquad h'(c) = 0$$

$$f'(c) = \frac{f(b) - f(a)}{b - a}, \qquad \text{Rearranged}$$

which is what we set out to prove. ❏

Notice that the hypotheses of the Mean Value Theorem do not require *f* to be differentiable at either *a* or *b*. Continuity at *a* and *b* is enough (Fig. 3.18).

We usually do not know any more about the number *c* than the theorem tells, which is that *c* exists. In a few cases we can satisfy our curiosity about the identity of *c*, as in the next example. However, our ability to identify *c* is the exception rather than the rule, and the importance of the theorem lies elsewhere.

EXAMPLE 2   The function $f(x) = x^2$ (Fig. 3.19) is continuous for $0 \le x \le 2$ and differentiable for $0 < x < 2$. Since $f(0) = 0$ and $f(2) = 4$, the Mean Value Theorem says that at some point *c* in the interval, the derivative $f'(x) = 2x$ must have the value $(4 - 0)/(2 - 0) = 2$. In this (exceptional) case we can identify *c* by solving the equation $2c = 2$ to get $c = 1$. ❏

Physical Interpretations

If we think of the number $(f(b) - f(a))/(b - a)$ as the average change in *f* over [*a*, *b*] and $f'(c)$ as an instantaneous change, then the Mean Value Theorem says that at some interior point the instantaneous change must equal the average change over the entire interval.

EXAMPLE 3   If a car accelerating from zero takes 8 sec to go 352 ft, its average velocity for the 8-sec interval is $352/8 = 44$ ft/sec. At some point during the acceleration, the Mean Value Theorem says, the speedometer must read exactly 30 mph (44 ft/sec) (Fig. 3.20). □

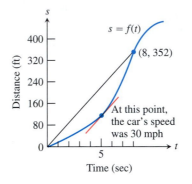

3.20 Distance vs. elapsed time for the car in Example 3.

Corollaries and Some Answers

At the beginning of the section, we asked what kind of function has a zero derivative. The first corollary of the Mean Value Theorem provides the answer.

> ### Corollary 1
> **Functions with Zero Derivatives Are Constant**
> If $f'(x) = 0$ at each point of an interval I, then $f(x) = C$ for all x in I, where C is a constant.

We know that if a function f has a constant value on an interval I, then f is differentiable on I and $f'(x) = 0$ for all x in I. Corollary 1 provides the converse.

Proof of Corollary 1   We want to show that f has a constant value on I. We do so by showing that if x_1 and x_2 are any two points in I, then $f(x_1) = f(x_2)$.

Suppose that x_1 and x_2 are two points in I, numbered from left to right so that $x_1 < x_2$. Then f satisfies the hypotheses of the Mean Value Theorem on $[x_1, x_2]$: It is differentiable at every point of $[x_1, x_2]$, and hence continuous at every point as well. Therefore,

$$\frac{f(x_2) - f(x_1)}{x_2 - x_1} = f'(c)$$

at some point c between x_1 and x_2. Since $f' = 0$ throughout I, this equation translates successively into

$$\frac{f(x_2) - f(x_1)}{x_2 - x_1} = 0, \quad f(x_2) - f(x_1) = 0, \quad \text{and} \quad f(x_1) = f(x_2).$$
□

At the beginning of the section, we also asked if we could work backward from the acceleration of a body falling freely from rest to find the body's velocity and displacement functions. The answer is yes, and it is a consequence of the next corollary.

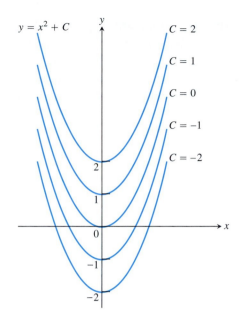

$y = x^2 + C$

$C = 2$

$C = 1$

$C = 0$

$C = -1$

$C = -2$

3.21 From a geometric point of view, Corollary 2 of the Mean Value Theorem says that the graphs of functions with identical derivatives can differ only by a vertical shift. The graphs of the functions with derivative 2x are the parabolas $y = x^2 + C$, shown here for selected values of C.

Corollary 2
Functions with the Same Derivative Differ by a Constant
If $f'(x) = g'(x)$ at each point of an interval I, then there exists a constant C such that $f(x) = g(x) + C$ for all x in I.

Proof At each point x in I the derivative of the difference function $h = f - g$ is

$$h'(x) = f'(x) - g'(x) = 0.$$

Thus, $h(x) = C$ on I (Corollary 1). That is, $f(x) - g(x) = C$ on I, so $f(x) = g(x) + C$. ❑

Corollary 2 says that functions can have identical derivatives on an interval only if their values on the interval have a constant difference. We know, for instance, that the derivative of $f(x) = x^2$ on $(-\infty, \infty)$ is $2x$. Any other function with derivative $2x$ on $(-\infty, \infty)$ must have the formula $x^2 + C$ for some value of C (Fig. 3.21).

EXAMPLE 4   Find the function $f(x)$ whose derivative is $\sin x$ and whose graph passes through the point $(0, 2)$.

Solution Since $f(x)$ has the same derivative as $g(x) = -\cos x$, we know that $f(x) = -\cos x + C$ for some constant C. The value of C can be determined from the condition that $f(0) = 2$ (the graph of f passes through $(0, 2)$):

$$f(0) = -\cos(0) + C = 2, \qquad \text{so} \qquad C = 3.$$

The formula for f is $f(x) = -\cos x + 3$. ❑

Finding Velocity and Position from Acceleration

Here is how to find the velocity and displacement functions of a body falling freely from rest with acceleration 9.8 m/sec^2.

We know that $v(t)$ is some function whose derivative is 9.8. We also know that the derivative of $g(t) = 9.8t$ is 9.8. By Corollary 2,

$$v(t) = 9.8t + C \tag{4}$$

for some constant C. Since the body falls from rest, $v(0) = 0$. Thus

$$9.8(0) + C = 0, \qquad \text{and} \qquad C = 0.$$

The velocity function must be $v(t) = 9.8t$. How about the position function $s(t)$?

We know that $s(t)$ is some function whose derivative is $9.8t$. We also know that the derivative of $h(t) = 4.9t^2$ is $9.8t$. By Corollary 2,

$$s(t) = 4.9t^2 + C \tag{5}$$

for some constant C. Since $s(0) = 0$,

$$4.9(0)^2 + C = 0, \qquad \text{and} \qquad C = 0.$$

The position function must be $s(t) = 4.9t^2$.

The ability to find functions from their rates of change is one of the great powers we gain from calculus. As we will see, it lies at the heart of the mathematical developments in Chapter 4. We will continue the story there.

Increasing Functions and Decreasing Functions

At the beginning of the section we asked what kinds of functions have positive derivatives or negative derivatives. The answer, provided by the Mean Value Theorem's third corollary, is this: The only functions with positive derivatives are increasing functions; the only functions with negative derivatives are decreasing functions.

Definitions

Let f be a function defined on an interval I and let x_1 and x_2 be any two points in I.

1. f **increases** on I if $x_1 < x_2 \Rightarrow f(x_1) < f(x_2)$.
2. f **decreases** on I if $x_1 < x_2 \Rightarrow f(x_2) < f(x_1)$.

Corollary 3

The First Derivative Test for Increasing and Decreasing

Suppose that f is continuous on $[a, b]$ and differentiable on (a, b).

If $f' > 0$ at each point of (a, b), then f increases on $[a, b]$.

If $f' < 0$ at each point of (a, b), then f decreases on $[a, b]$.

Proof Let x_1 and x_2 be two points in $[a, b]$ with $x_1 < x_2$. The Mean Value Theorem applied to f on $[x_1, x_2]$ says that

$$f(x_2) - f(x_1) = f'(c)(x_2 - x_1) \tag{6}$$

for some c between x_1 and x_2. The sign of the right-hand side of Eq. (6) is the same as the sign of $f'(c)$ because $x_2 - x_1$ is positive. Therefore, $f(x_2) > f(x_1)$ if f' is positive on (a, b), and $f(x_2) < f(x_1)$ if f' is negative on (a, b). ❑

EXAMPLE 5 The function $f(x) = x^2$ decreases on $(-\infty, 0)$, where $f'(x) = 2x < 0$. It increases on $(0, \infty)$, where $f'(x) = 2x > 0$ (Fig. 3.22). ❑

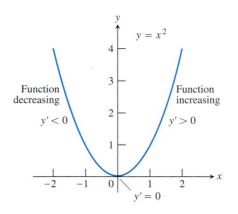

3.22 The graph for Example 5.

Exercises 3.2

Finding c in the Mean Value Theorem

Find the value or values of c that satisfy the equation

$$\frac{f(b) - f(a)}{b - a} = f'(c)$$

in the conclusion of the Mean Value Theorem for the functions and intervals in Exercises 1–4.

1. $f(x) = x^2 + 2x - 1, \quad [0, 1]$

2. $f(x) = x^{2/3}, \quad [0, 1]$

3. $f(x) = x + \dfrac{1}{x}, \quad \left[\dfrac{1}{2}, 2\right]$

4. $f(x) = \sqrt{x - 1}, \quad [1, 3]$

Checking and Using Hypotheses

Which of the functions in Exercises 5–8 satisfy the hypotheses of the Mean Value Theorem on the given interval, and which do not? Give reasons for your answers.

5. $f(x) = x^{2/3}, \quad [-1, 8]$ 6. $f(x) = x^{4/5}, \quad [0, 1]$

7. $f(x) = \sqrt{x(1 - x)}, \quad [0, 1]$

8. $f(x) = \begin{cases} \dfrac{\sin x}{x}, & -\pi \le x < 0 \\ 0, & x = 0 \end{cases}$

9. The function

$$f(x) = \begin{cases} x, & 0 \le x < 1 \\ 0, & x = 1 \end{cases}$$

is zero at $x = 0$ and $x = 1$ and differentiable on $(0, 1)$, but its derivative on $(0, 1)$ is never zero. How can this be? Doesn't Rolle's theorem say the derivative has to be zero somewhere in $(0, 1)$? Give reasons for your answer.

10. For what values of a, m, and b does the function

$$f(x) = \begin{cases} 3, & x = 0 \\ -x^2 + 3x + a, & 0 < x < 1 \\ mx + b, & 1 \le x \le 2 \end{cases}$$

satisfy the hypotheses of the Mean Value Theorem on the interval $[0, 2]$?

Roots (Zeros)

11. a) Plot the zeros of each polynomial on a line together with the zeros of its first derivative.

    i) $y = x^2 - 4$

    ii) $y = x^2 + 8x + 15$

    iii) $y = x^3 - 3x^2 + 4 = (x + 1)(x - 2)^2$

    iv) $y = x^3 - 33x^2 + 216x = x(x - 9)(x - 24)$

b) Use Rolle's theorem to prove that between every two zeros of $x^n + a_{n-1}x^{n-1} + \cdots + a_1x + a_0$ there lies a zero of

$$nx^{n-1} + (n - 1)a_{n-1}x^{n-2} + \cdots + a_1.$$

12. Suppose that f'' is continuous on $[a, b]$ and that f has three zeros in the interval. Show that f'' has at least one zero in (a, b). Generalize this result.

13. Show that if $f'' > 0$ throughout an interval $[a, b]$, then f' has at most one zero in $[a, b]$. What if $f'' < 0$ throughout $[a, b]$ instead?

14. Show that a cubic polynomial can have at most three real zeros.

Theory and Examples

15. Show that at some instant during a 2-h automobile trip the car's speedometer reading will equal the average speed for the trip.

16. *Temperature change.* It took 14 sec for a thermometer to rise from $-19°C$ to $100°C$ when it was taken from a freezer and placed in boiling water. Show that somewhere along the way the mercury was rising at exactly $8.5°C/\text{sec}$.

17. Suppose that f is differentiable on $[0, 1]$ and that its derivative is never zero. Show that $f(0) \ne f(1)$.

18. Show that $|\sin b - \sin a| \le |b - a|$ for any numbers a and b.

19. Suppose that f is differentiable on $[a, b]$ and that $f(b) < f(a)$. Can you then say anything about the values of f' on $[a, b]$?

20. Suppose that f and g are differentiable on $[a, b]$ and that $f(a) = g(a)$ and $f(b) = g(b)$. Show that there is at least one point between a and b where the tangents to the graphs of f and g are parallel.

21. Let f be differentiable at every value of x and suppose that $f(1) = 1$, that $f' < 0$ on $(-\infty, 1)$, and that $f' > 0$ on $(1, \infty)$.

    a) Show that $f(x) \ge 1$ for all x.

    b) Must $f'(1) = 0$? Explain.

22. Let $f(x) = px^2 + qx + r$ be a quadratic function defined on a closed interval $[a, b]$. Show that there is exactly one point c in (a, b) at which f satisfies the conclusion of the Mean Value Theorem.

23. *A surprising graph.* Graph the function

$$f(x) = \sin x \sin(x + 2) - \sin^2(x + 1).$$

What does the graph do? Why does the function behave this way? Give reasons for your answers.

24. If the graphs of two functions $f(x)$ and $g(x)$ start at the same point in the plane and the functions have the same rate of change at every point, do the graphs have to be identical? Give reasons for your answer.

25. a) Show that $g(x) = 1/x$ decreases on every interval in its domain.

b) If the conclusion in (a) is really true, how do you explain the fact that $g(1) = 1$ is actually greater than $g(-1) = -1$?

26. Let f be a function defined on an interval $[a, b]$. What conditions could you place on f to guarantee that

$$\min f' \le \frac{f(b) - f(a)}{b - a} \le \max f',$$

where $\min f'$ and $\max f'$ refer to the minimum and maximum values of f' on $[a, b]$? Give reasons for your answer.

27. CALCULATOR Use the inequalities in Exercise 26 to estimate $f(0.1)$ if $f'(x) = 1/(1 + x^4 \cos x)$ for $0 \le x \le 0.1$ and $f(0) = 1$.

28. CALCULATOR Use the inequalities in Exercise 26 to estimate $f(0.1)$ if $f'(x) = 1/(1 - x^4)$ for $0 \le x \le 0.1$ and $f(0) = 2$.

29. *The geometric mean of a and b.* The **geometric mean** of two positive numbers a and b is the number $\sqrt{ab}$. Show that the value of c in the conclusion of the Mean Value Theorem for $f(x) = 1/x$ on an interval $[a, b]$ of positive numbers is $c = \sqrt{ab}$.

30. *The arithmetic mean of a and b.* The **arithmetic mean** of two numbers a and b is the number $(a + b)/2$. Show that the value of c in the conclusion of the Mean Value Theorem for $f(x) = x^2$ on any interval $[a, b]$ is $c = (a + b)/2$.

Finding Functions from Derivatives

31. Suppose that $f(-1) = 3$ and that $f'(x) = 0$ for all x. Must $f(x) = 3$ for all x? Give reasons for your answer.

32. Suppose that $f(0) = 5$ and that $f'(x) = 2$ for all x. Must $f(x) = 2x + 5$ for all x? Give reasons for your answer.

33. Suppose that $f'(x) = 2x$ for all x. Find $f(2)$ if

a) $f(0) = 0$ **b)** $f(1) = 0$ **c)** $f(-2) = 3$.

34. What can be said about functions whose derivatives are constant? Give reasons for your answer.

In Exercises 35–40, find all possible functions with the given derivative.

35. a) $y' = x$ **b)** $y' = x^2$ **c)** $y' = x^3$

36. a) $y' = 2x$
    b) $y' = 2x - 1$
    c) $y' = 3x^2 + 2x - 1$

37. a) $y' = -\dfrac{1}{x^2}$

    b) $y' = 1 - \dfrac{1}{x^2}$

    c) $y' = 5 + \dfrac{1}{x^2}$

38. a) $y' = \dfrac{1}{2\sqrt{x}}$

    b) $y' = \dfrac{1}{\sqrt{x}}$

    c) $y' = 4x - \dfrac{1}{\sqrt{x}}$

39. a) $y' = \sin 2t$ **b)** $y' = \cos \dfrac{t}{2}$

    c) $y' = \sin 2t + \cos \dfrac{t}{2}$

40. a) $y' = \sec^2 \theta$ **b)** $y' = \sqrt{\theta}$

    c) $y' = \sqrt{\theta} - \sec^2 \theta$

In Exercises 41–44, find the function with the given derivative whose graph passes through the point P.

41. $f'(x) = 2x - 1$, $P(0, 0)$

42. $g'(x) = \dfrac{1}{x^2} + 2x$, $P(-1, 1)$

43. $r'(\theta) = 8 - \csc^2 \theta$, $P\left(\dfrac{\pi}{4}, 0\right)$

44. $r'(t) = \sec t \tan t - 1$, $P(0, 0)$

Counting Zeros

When we solve an equation $f(x) = 0$ numerically, we usually want to know beforehand how many solutions to look for in a given interval. With the help of Corollary 3 we can sometimes find out.

Suppose that

1. f is continuous on $[a, b]$ and differentiable on (a, b),
2. $f(a)$ and $f(b)$ have opposite signs,
3. $f' > 0$ on (a, b) or $f' < 0$ on (a, b).

Then f has exactly one zero between a and b: It cannot have more than one because it is either increasing on $[a, b]$ or decreasing on $[a, b]$. Yet it has at least one, by the Intermediate Value Theorem (Section 1.5). For example, $f(x) = x^3 + 3x + 1$ has exactly one zero on $[-1, 1]$ because f is differentiable on $[-1, 1]$, $f(-1) = -3$ and $f(1) = 5$ have opposite signs, and $f'(x) = 3x^2 + 3 > 0$ for all x (Fig. 3.23).

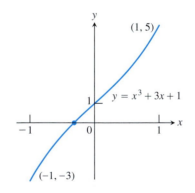

3.23 The only real zero of the polynomial $y = x^3 + 3x + 1$ is the one shown here between -1 and 0.

Show that the functions in Exercises 45–52 have exactly one zero in the given interval.

45. $f(x) = x^4 + 3x + 1$, $[-2, -1]$

46. $f(x) = x^3 + \dfrac{4}{x^2} + 7$, $(-\infty, 0)$

47. $g(t) = \sqrt{t} + \sqrt{1+t} - 4, \quad (0, \infty)$

48. $g(t) = \dfrac{1}{1-t} + \sqrt{1+t} - 3.1, \quad (-1, 1)$

49. $r(\theta) = \theta + \sin^2\left(\dfrac{\theta}{3}\right) - 8, \quad (-\infty, \infty)$

50. $r(\theta) = 2\theta - \cos^2\theta + \sqrt{2}, \quad (-\infty, \infty)$

51. $r(\theta) = \sec\theta - \dfrac{1}{\theta^3} + 5, \quad (0, \pi/2)$

52. $r(\theta) = \tan\theta - \cot\theta - \theta, \quad (0, \pi/2)$

☸ CAS Exploration

53. *Rolle's original theorem*

a) Construct a polynomial $f(x)$ that has zeros at $x = -2, -1,$ 0, 1, and 2.

b) Graph f and its derivative f' together. How is what you see related to Rolle's original theorem? (See the marginal note on Rolle.)

c) Do $g(x) = \sin x$ and its derivative g' illustrate the same phenomenon?

d) How would you state and prove Rolle's original theorem in light of what we know today?

| 3.3 | # The First Derivative Test for Local Extreme Values |

This section shows how to test a function's critical points for the presence of local extreme values.

The Test

As we see once again in Fig. 3.24, a function f may have local extrema at some critical points while failing to have local extrema at others. The key is the sign of f' in the point's immediate vicinity. As x moves from left to right, the values of f increase where $f' > 0$ and decrease where $f' < 0$.

At the points where f has a minimum value, we see that $f' < 0$ on the interval immediately to the left and $f' > 0$ on the interval immediately to the right. (If the point is an endpoint, there is only the interval on the appropriate side to consider.) This means that the curve is falling (values decreasing) on the left of the minimum value and rising (values increasing) on its right. Similarly, at the points where f has a maximum value, $f' > 0$ on the interval immediately to the left and $f' < 0$ on the interval immediately to the right. This means that the curve is rising (values increasing) on the left of the maximum value and falling (values decreasing) on its right.

These observations lead to a test for the presence of local extreme values.

3.24 A function's first derivative tells how the graph rises and falls.

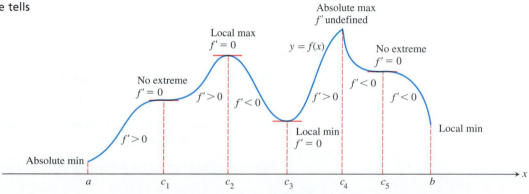

Theorem 5
The First Derivative Test for Local Extreme Values
The following test applies to a continuous function $f(x)$.

At a critical point c:

1. If f' changes from positive to negative at c ($f' > 0$ for $x < c$ and $f' < 0$ for $x > c$), then f has a local maximum value at c.

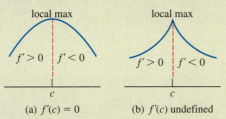

2. If f' changes from negative to positive at c ($f' < 0$ for $x < c$ and $f' > 0$ for $x > c$), then f has a local minimum value at c.

3. If f' does not change sign at c (f' has the same sign on both sides of c), then f has no local extreme value at c.

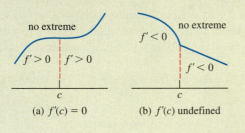

At a left endpoint a:

If $f' < 0$ ($f' > 0$) for $x > a$, then f has a local maximum (minimum) value at a.

At a right endpoint b:

If $f' < 0$ ($f' > 0$) for $x < b$, then f has a local minimum (maximum) value at b.

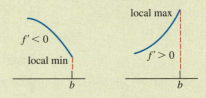

EXAMPLE 1   Find the critical points of

$$f(x) = x^{1/3}(x - 4) = x^{4/3} - 4x^{1/3}.$$

Identify the intervals on which f is increasing and decreasing. Find the function's local and absolute extreme values.

Solution   The function f is defined for all real numbers and is continuous (Fig. 3.25).

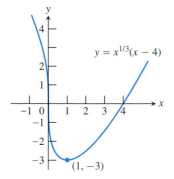

3.25 The graph of $y = x^{1/3}(x - 4)$ (Example 1).

The first derivative

$$f'(x) = \frac{d}{dx}\left(x^{4/3} - 4x^{1/3}\right) = \frac{4}{3}x^{1/3} - \frac{4}{3}x^{-2/3}$$

$$= \frac{4}{3}x^{-2/3}(x - 1) = \frac{4(x - 1)}{3x^{2/3}}$$

is zero at $x = 1$ and undefined at $x = 0$. There are no endpoints in f's domain, so the critical points, $x = 0$ and $x = 1$, are the only places where f might have an extreme value of any kind.

These critical points divide the x-axis into intervals on which f' is either positive or negative. The sign pattern of f' reveals the behavior of f both between and at the critical points. We can display the information in a picture like the following.

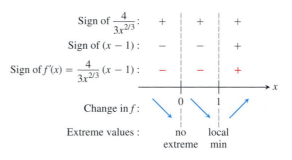

To make the picture, we marked the critical points on the x-axis, noted the sign of each factor of f' on the intervals between the points, and "multiplied" the signs of the factors to find the sign of f'. We then applied Corollary 3 of the Mean Value Theorem to determine that f decreases ($\searrow$) on $(-\infty, 0)$, decreases on $(0, 1)$, and increases ($\nearrow$) on $(1, \infty)$. Theorem 5 tells us that f has no extreme at $x = 0$ (f' does not change sign) and that f has a local minimum at $x = 1$ (f' changes from negative to positive).

The value of the local minimum is $f(1) = 1^{1/3}(1 - 4) = -3$. This is also an absolute minimum because the function's values fall toward it from the left and rise away from it on the right. Figure 3.25 shows this value in relation to the function's graph.   ❑

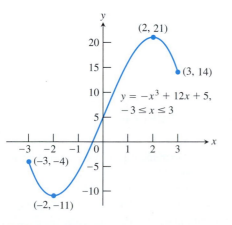

3.26 The graph of $g(x) = -x^3 + 12x + 5$, $-3 \le x \le 3$ (Example 2).

EXAMPLE 2   Find the intervals on which

$$g(x) = -x^3 + 12x + 5, \qquad -3 \le x \le 3$$

is increasing and decreasing. Where does the function assume extreme values and what are these values?

Solution   The function f is continuous on its domain, $[-3, 3]$ (Fig. 3.26). The first derivative

$$g'(x) = -3x^2 + 12 = -3(x^2 - 4)$$

$$= -3(x + 2)(x - 2),$$

defined at all points of $[-3, 3]$, is zero at $x = -2$ and $x = 2$. These critical points divide the domain of g into intervals on which g' is either positive or negative. We analyze the behavior of g by picturing the sign pattern of g':

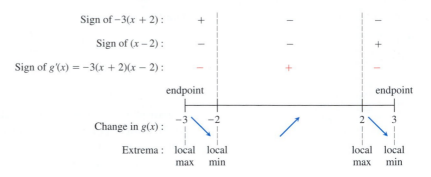

We conclude that g has local maxima at $x = -3$ and $x = 2$ and local minima at $x = -2$ and $x = 3$. The corresponding values of $g(x) = -x^3 + 12x + 5$ are

Local maxima: $g(-3) = -4$, $g(2) = 21$

Local minima: $g(-2) = -11$, $g(3) = 14$.

Since g is defined on a closed interval, we also know that $g(-2)$ is the absolute minimum and $g(2)$ is the absolute maximum. Figure 3.26 shows these values in relation to the function's graph. ❏

Exercises 3.3

Analyzing f Given f'

Answer the following questions about the functions whose derivatives are given in Exercises 1–8:

a) What are the critical points of f?
b) On what intervals is f increasing or decreasing?
c) At what points, if any, does f assume local maximum and minimum values?

1. $f'(x) = x(x - 1)$

2. $f'(x) = (x - 1)(x + 2)$

3. $f'(x) = (x - 1)^2(x + 2)$

4. $f'(x) = (x - 1)^2(x + 2)^2$

5. $f'(x) = (x - 1)(x + 2)(x - 3)$

6. $f'(x) = (x - 7)(x + 1)(x + 5)$

7. $f'(x) = x^{-1/3}(x + 2)$

8. $f'(x) = x^{-1/2}(x - 3)$

Extremes of Given Functions

In Exercises 9–28:

a) Find the intervals on which the function is increasing and decreasing.
b) Then identify the function's local extreme values, if any, saying where they are taken on.
c) Which, if any, of the extreme values are absolute?
d) GRAPHER You may wish to support your findings with a graphing calculator or computer grapher.

9. $g(t) = -t^2 - 3t + 3$

10. $g(t) = -3t^2 + 9t + 5$

11. $h(x) = -x^3 + 2x^2$

12. $h(x) = 2x^3 - 18x$

13. $f(\theta) = 3\theta^2 - 4\theta^3$

14. $f(\theta) = 6\theta - \theta^3$

15. $f(r) = 3r^3 + 16r$

16. $h(r) = (r + 7)^3$

17. $f(x) = x^4 - 8x^2 + 16$

18. $g(x) = x^4 - 4x^3 + 4x^2$

19. $H(t) = \dfrac{3}{2}t^4 - t^6$

20. $K(t) = 15t^3 - t^5$

21. $g(x) = x\sqrt{8 - x^2}$

22. $g(x) = x^2\sqrt{5 - x}$

23. $f(x) = \dfrac{x^2 - 3}{x - 2}, \quad x \neq 2$

24. $f(x) = \dfrac{x^3}{3x^2 + 1}$

25. $f(x) = x^{1/3}(x + 8)$

26. $g(x) = x^{2/3}(x + 5)$

27. $h(x) = x^{1/3}(x^2 - 4)$

28. $k(x) = x^{2/3}(x^2 - 4)$

Extremes on Half-Open Intervals

In Exercises 29–36:

a) Identify the function's local extreme values in the given domain, and say where they are assumed.
b) Which of the extreme values, if any, are absolute?
c) GRAPHER You may wish to support your findings with a graphing calculator or computer grapher.

29. $f(x) = 2x - x^2, \quad -\infty < x \leq 2$

30. $f(x) = (x + 1)^2, \quad -\infty < x \leq 0$

31. $g(x) = x^2 - 4x + 4, \quad 1 \le x < \infty$

32. $g(x) = -x^2 - 6x - 9, \quad -4 \le x < \infty$

33. $f(t) = 12t - t^3, \quad -3 \le t < \infty$

34. $f(t) = t^3 - 3t^2, \quad -\infty < t \le 3$

35. $h(x) = \dfrac{x^3}{3} - 2x^2 + 4x, \quad 0 \le x < \infty$

36. $k(x) = x^3 + 3x^2 + 3x + 1, \quad -\infty < x \le 0$

Graphing Calculator or Computer Grapher

In Exercises 37–40:

a) Find the local extrema of each function on the given interval, and say where they are assumed.

b) **GRAPHER** Graph the function and its derivative together. Comment on the behavior of f in relation to the signs and values of f'.

37. $f(x) = \dfrac{x}{2} - 2\sin\dfrac{x}{2}, \quad 0 \le x \le 2\pi$

38. $f(x) = -2\cos x - \cos^2 x, \quad -\pi \le x \le \pi$

39. $f(x) = \csc^2 x - 2\cot x, \quad 0 < x < \pi$

40. $f(x) = \sec^2 x - 2\tan x, \quad \dfrac{-\pi}{2} < x < \dfrac{\pi}{2}$

Theory and Examples

Show that the functions in Exercises 41 and 42 have local extreme values at the given values of θ, and say which kind of local extreme the function has.

41. $h(\theta) = 3\cos\dfrac{\theta}{2}, \quad 0 \le \theta \le 2\pi, \quad$ at $\theta = 0$ and $\theta = 2\pi$

42. $h(\theta) = 5\sin\dfrac{\theta}{2}, \quad 0 \le \theta \le \pi, \quad$ at $\theta = 0$ and $\theta = \pi$

43. Sketch the graph of a differentiable function $y = f(x)$ through the point $(1, 1)$ if $f'(1) = 0$ and

a) $f'(x) > 0$ for $x < 1$ and $f'(x) < 0$ for $x > 1$;

b) $f'(x) < 0$ for $x < 1$ and $f'(x) > 0$ for $x > 1$;

c) $f'(x) > 0$ for $x \ne 1$;

d) $f'(x) < 0$ for $x \ne 1$.

44. Sketch the graph of a differentiable function $y = f(x)$ that has

a) a local minimum at $(1, 1)$ and a local maximum at $(3, 3)$;

b) a local maximum at $(1, 1)$ and a local minimum at $(3, 3)$;

c) local maxima at $(1, 1)$ and $(3, 3)$;

d) local minima at $(1, 1)$ and $(3, 3)$.

45. Sketch the graph of a continuous function $y = g(x)$ such that

a) $g(2) = 2, \quad 0 < g' < 1$ for $x < 2$, $g'(x) \to 1^-$ as $x \to 2^-$, $-1 < g' < 0$ for $x > 2$, and $g'(x) \to -1^+$ as $x \to 2^+$;

b) $g(2) = 2, \quad g' < 0$ for $x < 2$, $\quad g'(x) \to -\infty$ as $x \to 2^-$, $g' > 0$ for $x > 2$, and $g'(x) \to \infty$ as $x \to 2^+$.

46. Sketch the graph of a continuous function $y = h(x)$ such that

a) $h(0) = 0, -2 \le h(x) \le 2$ for all x, $h'(x) \to \infty$ as $x \to 0^-$, and $h'(x) \to -\infty$ as $x \to 0^+$;

b) $h(0) = 0, -2 \le h(x) \le 0$ for all x, $h'(x) \to \infty$ as $x \to 0^-$, and $h'(x) \to -\infty$ as $x \to 0^+$.

47. As x moves from left to right through the point $c = 2$, is the graph of $f(x) = x^3 - 3x + 2$ rising, or is it falling? Give reasons for your answer.

48. Find the intervals on which the function $f(x) = ax^2 + bx + c$, $a \ne 0$, is increasing and decreasing. Describe the reasoning behind your answer.

| 3.4 | # Graphing with y' and y'' |

In Section 3.1, we saw the role played by the first derivative in locating a function's extreme values. A function can have extreme values only at the endpoints of its domain and at its critical points. We also saw that critical points do not necessarily yield extreme values. In Section 3.2, we saw that almost all the information about a differentiable function is contained in its derivative. To recover the function completely, the only additional information we need is the value of the function at any one single point. If a function's derivative is $2x$ and the graph passes through the origin, the function must be x^2. If a function's derivative is $2x$ and the graph passes through the point $(0, 4)$, the function must be $x^2 + 4$.

In Section 3.3, we extended our ability to recover information from a function's first derivative by showing how to use it to tell exactly what happens at a critical point. We can tell whether there really is an extreme value there or whether the graph just continues to rise or fall.

In the present section, we show how to determine the way the graph of a function $y = f(x)$ bends or turns. We know that the information must be contained in y', but how do we find it? The answer, for functions that are twice differentiable except perhaps at isolated points, is to differentiate y'. Together y' and y'' tell us the shape of the function's graph. We will see in Chapter 4 how this enables us to sketch solutions of differential equations and initial value problems.

Concavity

As you can see in Fig. 3.27, the curve $y = x^3$ rises as x increases, but the portions defined on the intervals $(-\infty, 0)$ and $(0, \infty)$ turn in different ways. As we come in from the left toward the origin along the curve, the curve turns to our right and falls below its tangents. As we leave the origin, the curve turns to our left and rises above its tangents.

To put it another way, the slopes of the tangents decrease as the curve approaches the origin from the left and increase as the curve moves from the origin into the first quadrant.

Definition

The graph of a differentiable function $y = f(x)$ is **concave up** on an interval where y' is increasing and **concave down** on an interval where y' is decreasing.

If $y = f(x)$ has a second derivative, we can apply Corollary 3 of the Mean Value Theorem to conclude that y' increases if $y'' > 0$ and decreases if $y'' < 0$.

The Second Derivative Test for Concavity

Let $y = f(x)$ be twice differentiable on an interval I.

1. If $y'' > 0$ on I, the graph of f over I is concave up.
2. If $y'' < 0$ on I, the graph of f over I is concave down.

EXAMPLE 1

a) The curve $y = x^3$ (Fig. 3.27) is concave down on $(-\infty, 0)$ where $y'' = 6x < 0$ and concave up on $(0, \infty)$ where $y'' = 6x > 0$.

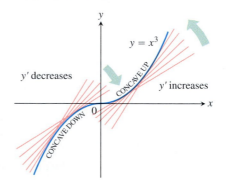

3.27 The graph of $f(x) = x^3$ is concave down on $(-\infty, 0)$ and concave up on $(0, \infty)$.

b) The parabola $y = x^2$ (Fig. 3.28) is concave up on every interval because $y'' = 2 > 0$.

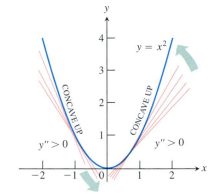

3.28 The graph of $f(x) = x^2$ on any interval is concave up.

Points of Inflection

To study the motion of a body moving along a line, we often graph the body's position as a function of time. One reason for doing so is to reveal where the body's acceleration, given by the second derivative, changes sign. On the graph, these are the points where the concavity changes.

> **Definition**
>
> A point where the graph of a function has a tangent line and where the concavity changes is called a **point of inflection.**

Thus a point of inflection on a curve is a point where y'' is positive on one side and negative on the other. At such a point, y'' is either zero (because derivatives have the intermediate value property) or undefined.

> On the graph of a twice-differentiable function, $y'' = 0$ at a point of inflection.

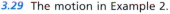

3.29 The motion in Example 2.

EXAMPLE 2   *Simple harmonic motion*

The graph of $s = 2 + \cos t$, $t \geq 0$ (Fig. 3.29), changes concavity at $t = \pi/2, 3\pi/2,$ ..., where the acceleration $s'' = -\cos t$ is zero.

EXAMPLE 3   *Marginal cost*

Inflection points have applications in some areas of economics. Suppose that $y = c(x)$ is the total cost of producing x units of something (Fig. 3.30). The point of inflection at P is then the point at which the marginal cost (the approximate cost of producing one more unit) changes from decreasing to increasing.

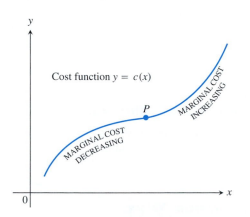

3.30 The point of inflection on a typical cost curve separates the interval of decreasing marginal cost from the interval of increasing marginal cost. This is the point where the marginal cost is smallest (Example 3).

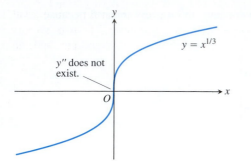

3.31 A point where y'' fails to exist can be a point of inflection.

EXAMPLE 4   *An inflection point where y'' does not exist*

The curve $y = x^{1/3}$ has a point of inflection at $x = 0$ (Fig. 3.31), but y'' does not exist there.

$$y'' = \frac{d^2}{dx^2}\left(x^{1/3}\right) = \frac{d}{dx}\left(\frac{1}{3}x^{-2/3}\right) = -\frac{2}{9}x^{-5/3}$$

EXAMPLE 5   *No inflection where $y''= 0$*

The curve $y = x^4$ has no inflection point at $x = 0$ (Fig. 3.32). Even though $y'' = 12x^2$ is zero there, it does not change sign.

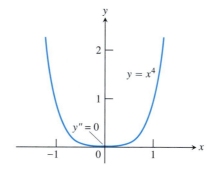

3.32 The graph of $y = x^4$ has no inflection point at the origin, even though $y'' = 0$ there.

Technology   *Graphing a Function with Its Derivatives*   When we graph a function $y = f(x)$, it may be difficult to identify the inflection points exactly by zooming in. Try it on the curve $y = 2\cos x - \sqrt{2}x$, $-\pi \le x \le 3\pi/2$. Adding the graph of f' to the display can help to identify inflection points more closely, but the strongest visual evidence comes from graphing f and f'' together. It is interesting to watch all three functions, f, f', and f'', being graphed simultaneously.

The graph of $y = 2\cos x - \sqrt{2}\,x$ and its first derivative.

The Second Derivative Test for Local Extreme Values

Instead of examining y' for sign changes at a critical point, we can sometimes use the following test to determine the presence of a local extremum.

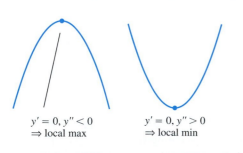

$y' = 0, y'' < 0$
$\Rightarrow$ local max

$y' = 0, y'' > 0$
$\Rightarrow$ local min

The Second Derivative Test for Local Extreme Values

If $f'(c) = 0$ and $f''(c) < 0$, then f has a local maximum at $x = c$.

If $f'(c) = 0$ and $f''(c) > 0$, then f has a local minimum at $x = c$.

Notice that the test requires us to know y'' only at c itself, and not in an interval about c. This makes the test easy to apply. That's the good news. The bad news is that the test is inconclusive if $y'' = 0$ or if y'' does not exist. When this happens, use the first derivative test for local extreme values.

Graphing with y' and y''

We now apply what we have learned to sketch the graphs of functions.

Testing the critical points in Example 6

As a quick test to see if any of the critical points are local extreme values, we could try the second derivative test.

At $x = 3$, $y'' > 0$:
We now know that this point is definitely a local minimum.

At $x = 0$, $y'' = 0$:
Test fails, and so we will need to check the signs of y' to know whether this point gives a local extreme value.

EXAMPLE 6   Graph the function

$$y = x^4 - 4x^3 + 10.$$

Solution

Step 1: *Find y' and y''.*

$$y = x^4 - 4x^3 + 10$$

$$y' = 4x^3 - 12x^2 = 4x^2(x - 3)$$

$$y'' = 12x^2 - 24x = 12x(x - 2)$$

Critical points: $x = 0$, $x = 3$

Possible inflection points: $x = 0$, $x = 2$

Step 2: *Rise and fall.* Sketch the sign pattern for y' and use it to describe the behavior of y.

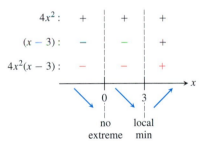

Step 3: *Concavity.* Sketch the sign pattern for y'' and use it to describe the way the graph bends.

Step 4: *Summary and general shape.* Summarize the information from steps 2 and 3. Show the shape over each interval. Then combine the shapes to show the curve's general form.

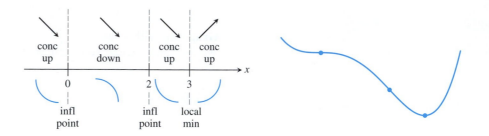

Step 5: *Specific points and curve.* Plot the curve's intercepts (if convenient) and the points where y' and y'' are zero. Indicate any local extreme values and inflection points. Use the general shape in step 4 as a guide to sketch the curve. (Plot additional points as needed.)

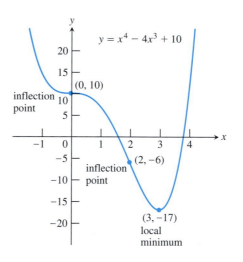

The steps in Example 6 give a general procedure for graphing by hand.

Strategy for Graphing $y = f(x)$

1. Find y' and y''.
2. Find the rise and fall of the curve.
3. Determine the concavity of the curve.
4. Make a summary and show the curve's general shape.
5. Plot specific points and sketch the curve.

EXAMPLE 7   Graph $y = x^{5/3} - 5x^{2/3}$.

Solution

Step 1: *Find y' and y''.*

$$y = x^{5/3} - 5x^{2/3} = x^{2/3}(x - 5)$$

$$y' = \frac{5}{3}x^{2/3} - \frac{10}{3}x^{-1/3} = \frac{5}{3}x^{-1/3}(x - 2)$$

$$y'' = \frac{10}{9}x^{-1/3} + \frac{10}{9}x^{-4/3} = \frac{10}{9}x^{-4/3}(x + 1)$$

The x-intercepts are at $x = 0$ and $x = 5$.

Critical points: $x = 0$, $x = 2$

Possible inflection points: $x = 0$, $x = -1$

Step 2: *Rise and fall.*

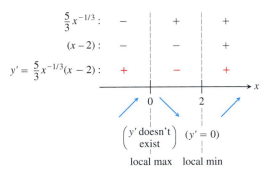

Cusps

The graph of a continuous function $y = f(x)$ has a *cusp* at a point $x = c$ if the concavity is the same on both sides of c and either

1. $\lim\limits_{x \to c^-} f'(x) = \infty$ and $\lim\limits_{x \to c^+} f'(x) = -\infty$

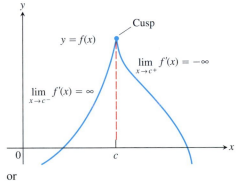

or

2. $\lim\limits_{x \to c^-} f'(x) = -\infty$ and $\lim\limits_{x \to c^+} f'(x) = \infty$.

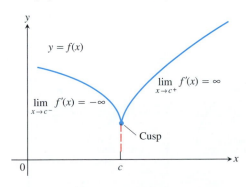

A cusp can be either a local maximum (1) or a local minimum (2).

Step 3: *Concavity.*

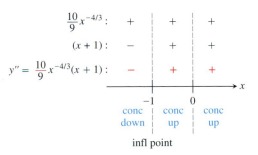

From the sign pattern for y'', we see that there is an inflection point at $x = -1$, but not at $x = 0$. However, knowing that

1. the function $y = x^{5/3} - 5x^{2/3}$ is continuous,
2. $y' \to \infty$ as $x \to 0^-$ and $y' \to -\infty$ as $x \to 0^+$ (see the formula for y' in step 2), and
3. the concavity does not change at $x = 0$ (step 3) tells us that the graph has a *cusp* at $x = 0$.

Step 4: *Summary.*   *General shape.*

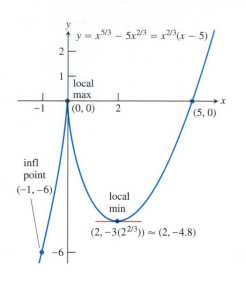

Step 5: *Specific points and curve.* See the figure to the left.

Learning About Functions from Derivatives

Pause for a moment to see how remarkable the conclusions in Examples 6 and 7 really are. In each case, we have been able to recover almost everything we need to know about a differentiable function $y = f(x)$ by examining y'. We can find where the graph rises and falls and where the local extremes are assumed. We can differentiate y' to learn how the graph bends as it passes over the intervals of rise and fall. We can determine the shape of the function's graph. The only information we cannot get from the derivative is how to place the graph in the xy-plane. That requires evaluating the formula for f at various points. Or so it seems. But as we saw in Section 3.2, even *that* is nearly superfluous. All we really need, in addition to y', is the value of f at a single point.

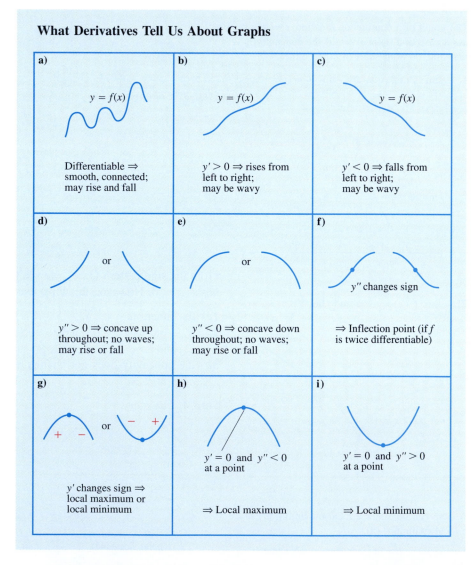

Exercises 3.4

Analyzing Graphed Functions

Identify the inflection points and local maxima and minima of the functions graphed in Exercises 1–8. Identify the intervals on which the functions are concave up and concave down.

1.

$$y = \frac{x^3}{3} - \frac{x^2}{2} - 2x + \frac{1}{3}$$

2.

$$y = \frac{x^4}{4} - 2x^2 + 4$$

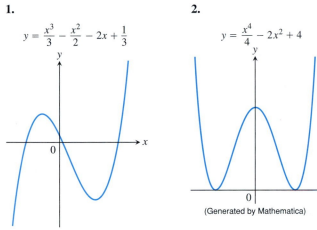

(Generated by Mathematica)

(Generated by Mathematica)

3.

$$y = \frac{3}{4}(x^2 - 1)^{2/3}$$

4.

$$y = \frac{9}{14}x^{1/3}(x^2 - 7)$$

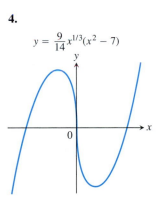

5.

$$y = x + \sin 2x, \quad -\frac{2\pi}{3} \le x \le \frac{2\pi}{3}$$

6.

$$y = \tan x - 4x, \quad -\frac{\pi}{2} < x < \frac{\pi}{2}$$

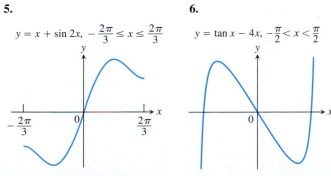

(Generated by Mathematica)

7.

$$y = \sin|x|, -2\pi \le x \le 2\pi$$

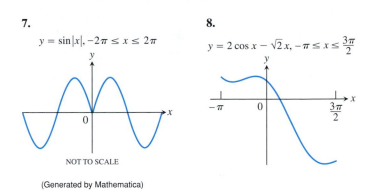

NOT TO SCALE

(Generated by Mathematica)

8.

$$y = 2\cos x - \sqrt{2}x, \quad -\pi \le x \le \frac{3\pi}{2}$$

Graphing Equations

Use the steps of the graphing procedure on page 214 to graph the equations in Exercises 9–40. Include the coordinates of any local extreme points and inflection points.

9. $y = x^2 - 4x + 3$

10. $y = 6 - 2x - x^2$

11. $y = x^3 - 3x + 3$

12. $y = x(6 - 2x)^2$

13. $y = -2x^3 + 6x^2 - 3$

14. $y = 1 - 9x - 6x^2 - x^3$

15. $y = (x - 2)^3 + 1$

16. $y = 1 - (x + 1)^3$

17. $y = x^4 - 2x^2 = x^2(x^2 - 2)$

18. $y = -x^4 + 6x^2 - 4 = x^2(6 - x^2) - 4$

19. $y = 4x^3 - x^4 = x^3(4 - x)$

20. $y = x^4 + 2x^3 = x^3(x + 2)$

21. $y = x^5 - 5x^4 = x^4(x - 5)$

22. $y = x\left(\frac{x}{2} - 5\right)^4$

23. $y = x + \sin x, \quad 0 \le x \le 2\pi$

24. $y = x - \sin x, \quad 0 \le x \le 2\pi$

25. $y = x^{1/5}$

26. $y = x^{3/5}$

27. $y = x^{2/5}$

28. $y = x^{4/5}$

29. $y = 2x - 3x^{2/3}$

30. $y = 5x^{2/5} - 2x$

31. $y = x^{2/3}\left(\frac{5}{2} - x\right)$

32. $y = x^{2/3}(x - 5)$

33. $y = x\sqrt{8 - x^2}$

34. $y = (2 - x^2)^{3/2}$

35. $y = \frac{x^2 - 3}{x - 2}, \quad x \ne 2$

36. $y = \frac{x^3}{3x^2 + 1}$

37. $y = |x^2 - 1|$

38. $y = |x^2 - 2x|$

39. $y = \sqrt{|x|} = \begin{cases} \sqrt{-x}, & x \le 0 \\ \sqrt{x}, & x > 0 \end{cases}$

40. $y = \sqrt{|x - 4|}$

Sketching the General Shape Knowing y'

Each of Exercises 41–62 gives the first derivative of a continuous function $y = f(x)$. Find y'' and then use steps 2–4 of the graphing procedure on page 214 to sketch the general shape of the graph of f.

41. $y' = 2 + x - x^2$

42. $y' = x^2 - x - 6$

43. $y' = x(x - 3)^2$

44. $y' = x^2(2 - x)$

45. $y' = x(x^2 - 12)$

46. $y' = (x - 1)^2(2x + 3)$

47. $y' = (8x - 5x^2)(4 - x)^2$

48. $y' = (x^2 - 2x)(x - 5)^2$

49. $y' = \sec^2 x, \quad -\dfrac{\pi}{2} < x < \dfrac{\pi}{2}$

50. $y' = \tan x, \quad -\dfrac{\pi}{2} < x < \dfrac{\pi}{2}$

51. $y' = \cot \dfrac{\theta}{2}, \quad 0 < \theta < 2\pi$

52. $y' = \csc^2 \dfrac{\theta}{2}, \quad 0 < \theta < 2\pi$

53. $y' = \tan^2 \theta - 1, \quad -\dfrac{\pi}{2} < \theta < \dfrac{\pi}{2}$

54. $y' = 1 - \cot^2 \theta, \quad 0 < \theta < \pi$

55. $y' = \cos t, \quad 0 \le t \le 2\pi$

56. $y' = \sin t, \quad 0 \le t \le 2\pi$

57. $y' = (x + 1)^{-2/3}$

58. $y' = (x - 2)^{-1/3}$

59. $y' = x^{-2/3}(x - 1)$

60. $y' = x^{-4/5}(x + 1)$

61. $y' = 2|x| = \begin{cases} -2x, & x \le 0 \\ 2x, & x > 0 \end{cases}$

62. $y' = \begin{cases} -x^2, & x \le 0 \\ x^2, & x > 0 \end{cases}$

Sketching y from Graphs of y' and y''

Each of Exercises 63–66 shows the graphs of the first and second derivatives of a function $y = f(x)$. Copy the picture and add to it a sketch of the approximate graph of f, given that the graph passes through the point P.

63.

64.

65.

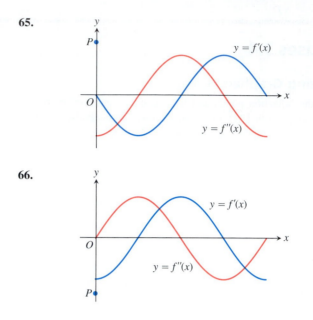

66.

Theory and Examples

67. The accompanying figure shows a portion of the graph of a twice-differentiable function $y = f(x)$. At each of the five labeled points, classify y' and y'' as positive, negative, or zero.

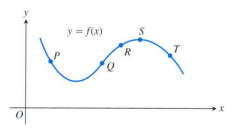

68. Sketch a smooth connected curve $y = f(x)$ with

$f(-2) = 8,$	$f'(2) = f'(-2) = 0,$
$f(0) = 4,$	$f'(x) < 0 \quad \text{for} \quad |x| < 2,$
$f(2) = 0,$	$f''(x) < 0 \quad \text{for} \quad x < 0,$
$f'(x) > 0 \quad \text{for} \quad |x| > 2,$	$f''(x) > 0 \quad \text{for} \quad x > 0.$

69. Sketch the graph of a twice-differentiable function $y = f(x)$ with the following properties. Label coordinates where possible.

x	y	Derivatives
$x < 2$		$y' < 0, \quad y'' > 0$
2	1	$y' = 0, \quad y'' > 0$
$2 < x < 4$		$y' > 0, \quad y'' > 0$
4	4	$y' > 0, \quad y'' = 0$
$4 < x < 6$		$y' > 0, \quad y'' < 0$
6	7	$y' = 0, \quad y'' < 0$
$x > 6$		$y' < 0, \quad y'' < 0$

70. Sketch the graph of a twice-differentiable function $y = f(x)$ that passes through the points $(-2, 2)$, $(-1, 1)$, $(0, 0)$, $(1, 1)$ and $(2, 2)$ and whose first two derivatives have the following sign patterns:

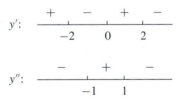

Velocity and acceleration. The graphs in Exercises 71 and 72 show the position $s = f(t)$ of a body moving back and forth on a coordinate line. (a) When is the body moving away from the origin? toward the origin? At approximately what times is the (b) velocity equal to zero? (c) acceleration equal to zero? (d) When is the acceleration positive? negative?

71.

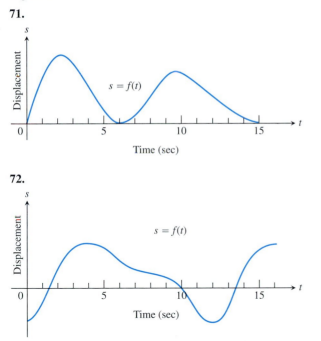

72.

73. *Marginal cost.* The accompanying graph shows the hypothetical cost $c = f(x)$ of manufacturing x items. At approximately what production level does the marginal cost change from decreasing to increasing?

74. The accompanying graph shows the monthly revenue of the Widget Corporation for the last twelve years. During approximately what time intervals was the marginal revenue increasing? decreasing?

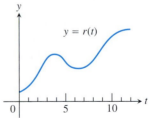

75. Suppose the derivative of the function $y = f(x)$ is

$$y' = (x - 1)^2(x - 2).$$

At what points, if any, does the graph of f have a local minimum, local maximum, or point of inflection? (*Hint:* Draw the sign pattern for y'.)

76. Suppose the derivative of the function $y = f(x)$ is

$$y' = (x - 1)^2(x - 2)(x - 4).$$

At what points, if any, does the graph of f have a local minimum, local maximum, or point of inflection?

77. For $x > 0$, sketch a curve $y = f(x)$ that has $f(1) = 0$ and $f'(x) = 1/x$. Can anything be said about the concavity of such a curve? Give reasons for your answer.

78. Can anything be said about the graph of a function $y = f(x)$ that has a continuous second derivative that is never zero? Give reasons for your answer.

79. If b, c, and d are constants, for what value of b will the curve $y = x^3 + bx^2 + cx + d$ have a point of inflection at $x = 1$? Give reasons for your answer.

80. *Horizontal tangents.* True, or false? Explain.

a) The graph of every polynomial of even degree (largest exponent even) has at least one horizontal tangent.

b) The graph of every polynomial of odd degree (largest exponent odd) has at least one horizontal tangent.

81. *Parabolas*

a) Find the coordinates of the vertex of the parabola $y = ax^2 + bx + c$, $a \neq 0$.

b) When is the parabola concave up? concave down? Give reasons for your answers.

82. Is it true that the concavity of the graph of a twice-differentiable function $y = f(x)$ changes every time $f''(x) = 0$? Give reasons for your answer.

83. *Quadratic curves.* What can you say about the inflection points of a quadratic curve $y = ax^2 + bx + c$, $a \neq 0$? Give reasons for your answer.

84. *Cubic curves.* What can you say about the inflection points of a cubic curve $y = ax^3 + bx^2 + cx + d$, $a \neq 0$? Give reasons for your answer.

▦ Grapher Explorations

In Exercises 85–88, find the inflection points (if any) on the graph of the function and the coordinates of the points on the graph where the function has a local maximum or local minimum value. Then graph the function in a region large enough to show all these points simultaneously. Add to your picture the graphs of the function's first and second derivatives. How are the values at which these graphs intersect the x-axis related to the graph of the function? In what other ways are the graphs of the derivatives related to the graph of the function?

85. $y = x^5 - 5x^4 - 240$

86. $y = x^3 - 12x^2$

87. $y = \frac{4}{5}x^5 + 16x^2 - 25$

88. $y = \frac{x^4}{4} - \frac{x^3}{3} - 4x^2 + 12x + 20$

89. Graph $f(x) = 2x^4 - 4x^2 + 1$ and its first two derivatives together. Comment on the behavior of f in relation to the signs and values of f' and f''.

90. Graph $f(x) = x \cos x$ and its second derivative together for $0 \le x \le 2\pi$. Comment on the behavior of the graph of f in relation to the signs and values of f''.

91. a) On a common screen, graph $f(x) = x^3 + kx$ for $k = 0$ and nearby positive and negative values of k. How does the value of k seem to affect the shape of the graph?

b) Find $f'(x)$. As you will see, $f'(x)$ is a quadratic function of x. Find the discriminant of the quadratic (the discriminant of $ax^2 + bx + c$ is $b^2 - 4ac$). For what values of k is the discriminant positive? zero? negative? For what values of

k does f' have two zeros? one or no zeros? Now explain what the value of k has to do with the shape of the graph of f.

c) Experiment with other values of k. What appears to happen as $k \to -\infty$? as $k \to \infty$?

92. a) On a common screen, graph $f(x) = x^4 + kx^3 + 6x^2, -1 \le x \le 4$ for $k = -4$, and some nearby values of k. How does the value of k seem to affect the shape of the graph?

b) Find $f''(x)$. As you will see, $f''(x)$ is a quadratic function of x. What is the discriminant of this quadratic (see Exercise 91b)? For what values of k is the discriminant positive? zero? negative? For what values of k does $f''(x)$ have two zeros? one or no zeros? Now explain what the value of k has to do with the shape of the graph of f.

93. a) Graph $y = x^{2/3}(x^2 - 2)$ for $-3 \le x \le 3$. Then use calculus to confirm what the screen shows about concavity, rise, and fall. (Depending on your grapher, you may have to enter $x^{2/3}$ as $(x^2)^{1/3}$ to obtain a plot for negative values of x.)

b) Does the curve have a cusp at $x = 0$, or does it just have a corner with different right-hand and left-hand derivatives?

94. a) Graph $y = 9x^{2/3}(x - 1)$ for $-0.5 \le x \le 1.5$. Then use calculus to confirm what the screen shows about concavity, rise, and fall. What concavity does the curve have to the left of the origin? (Depending on your grapher, you may have to enter $x^{2/3}$ as $(x^2)^{1/3}$ to obtain a plot for negative values of x.)

b) Does the curve have a cusp at $x = 0$, or does it just have a corner with different right-hand and left-hand derivatives?

95. Does the curve $y = x^2 + 3 \sin 2x$ have a horizontal tangent near $x = -3$? Give reasons for your answer.

| 3.5 | # Limits as $x \to \pm\infty$, Asymptotes, and Dominant Terms |

In this section, we analyze the graphs of rational functions (quotients of polynomial functions), as well as other functions with interesting limit behavior as $x \to \pm\infty$. Among the tools we use are asymptotes and dominant terms.

Limits as $x \to \pm\infty$

The function $f(x) = 1/x$ is defined for all $x \ne 0$ (Fig. 3.33). When x is positive and becomes increasingly large, $1/x$ becomes increasingly small. When x is negative and its magnitude becomes increasingly large, $1/x$ again becomes small. We summarize these observations by saying that $f(x) = 1/x$ has limit 0 as $x \to \pm\infty$.

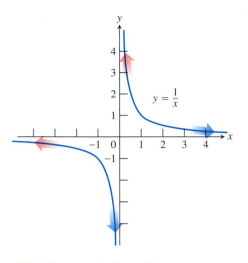

3.33 The graph of $y = 1/x$.

Definitions

1. We say that $f(x)$ has the **limit L as x approaches infinity** and write

$$\lim_{x\to\infty} f(x) = L$$

if, for every number $\epsilon > 0$, there exists a corresponding number M such that for all x

$$x > M \quad \Rightarrow \quad |f(x) - L| < \epsilon.$$

2. We say that $f(x)$ has the **limit L as x approaches minus infinity** and write

$$\lim_{x\to-\infty} f(x) = L$$

if, for every number $\epsilon > 0$, there exists a corresponding number N such that for all x

$$x < N \quad \Rightarrow \quad |f(x) - L| < \epsilon.$$

The symbol infinity (∞)

As always, the symbol ∞ does not represent a real number and we cannot use it in arithmetic in the usual way.

The strategy for calculating limits of functions as $x \to \pm\infty$ is similar to the one for finite limits in Section 1.2. There, we first found the limits of the constant and identity functions $y = k$ and $y = x$. We then extended these results to other functions by applying a theorem about limits of algebraic combinations. Here we do the same thing, except that the starting functions are $y = k$ and $y = 1/x$ instead of $y = k$ and $y = x$.

The basic facts to be verified by applying the formal definition are

$$\lim_{x\to\pm\infty} k = k \quad \text{and} \quad \lim_{x\to\pm\infty} \frac{1}{x} = 0. \tag{1}$$

We prove the latter and leave the former to Exercises 87 and 88.

EXAMPLE 1 Show that

a) $\displaystyle\lim_{x\to\infty} \frac{1}{x} = 0$ b) $\displaystyle\lim_{x\to-\infty} \frac{1}{x} = 0.$

Solution

a) Let $\epsilon > 0$ be given. We must find a number M such that for all x

$$x > M \quad \Rightarrow \quad \left|\frac{1}{x} - 0\right| = \left|\frac{1}{x}\right| < \epsilon.$$

The implication will hold if $M = 1/\epsilon$ or any larger positive number (Fig. 3.34). This proves $\lim_{x\to\infty} (1/x) = 0$.

b) Let $\epsilon > 0$ be given. We must find a number N such that for all x

$$x < N \quad \Rightarrow \quad \left|\frac{1}{x} - 0\right| = \left|\frac{1}{x}\right| < \epsilon.$$

The implication will hold if $N = -1/\epsilon$ or any number less than $-1/\epsilon$ (Fig. 3.34). This proves $\lim_{x\to-\infty} (1/x) = 0$. ◻

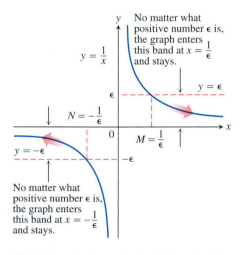

3.34 The geometry behind the argument in Example 1.

The following theorem enables us to build on Eqs. (1) to calculate other limits.

> **Theorem 6**
> **Properties of Limits as $x \to \pm\infty$**
> The following rules hold if $\lim_{x\to\pm\infty} f(x) = L$ and $\lim_{x\to\pm\infty} g(x) = M$ (L and M real numbers).
>
> 1. *Sum Rule:* $\qquad\qquad\qquad$ $\lim_{x\to\pm\infty} [f(x) + g(x)] = L + M$
>
> 2. *Difference Rule:* $\qquad\qquad$ $\lim_{x\to\pm\infty} [f(x) - g(x)] = L - M$
>
> 3. *Product Rule:* $\qquad\qquad\;\;$ $\lim_{x\to\pm\infty} f(x) \cdot g(x) = L \cdot M$
>
> 4. *Constant Multiple Rule:* $\quad\;$ $\lim_{x\to\pm\infty} kf(x) = kL \qquad$ (any number k)
>
> 5. *Quotient Rule:* $\qquad\qquad\;\;$ $\lim_{x\to\pm\infty} \dfrac{f(x)}{g(x)} = \dfrac{L}{M}, \qquad$ if $M \neq 0$
>
> 6. *Power Rule:* $\qquad\qquad\quad\;$ If m and n are integers, then $\lim_{x\to\pm\infty} [f(x)]^{m/n}$
> $\qquad\qquad\qquad\qquad\qquad\qquad = L^{m/n}$ provided $L^{m/n}$ is a real number.

These properties are just like the properties in Theorem 1, Section 1.2, and we use them the same way.

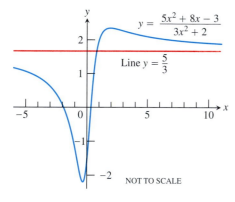

3.35 The function in Example 3.

EXAMPLE 2

a) $\lim_{x\to\infty} \left(5 + \dfrac{1}{x}\right) = \lim_{x\to\infty} 5 + \lim_{x\to\infty} \dfrac{1}{x}$ $\qquad$ Sum Rule

$\qquad\qquad\qquad\qquad = 5 + 0 = 5$ $\qquad$ Known values

b) $\lim_{x\to-\infty} \dfrac{\pi\sqrt{3}}{x^2} = \lim_{x\to-\infty} \pi\sqrt{3} \cdot \dfrac{1}{x} \cdot \dfrac{1}{x}$

$\qquad\qquad = \lim_{x\to-\infty} \pi\sqrt{3} \cdot \lim_{x\to-\infty} \dfrac{1}{x} \cdot \lim_{x\to-\infty} \dfrac{1}{x}$ $\qquad$ Product Rule

$\qquad\qquad = \pi\sqrt{3} \cdot 0 \cdot 0 = 0$ $\qquad$ Known values $\qquad$ ❏

Limits of Rational Functions as $x \to \pm\infty$

To determine the limit of a rational function as $x \to \pm\infty$, we can divide the numerator and denominator by the highest power of x in the denominator. What happens then depends on the degrees of the polynomials involved.

The **degree** of the polynomial

$$a_n x^n + a_{n-1} x^{n-1} + \cdots + a_1 x + a_0,$$

$a_n \neq 0$, is n, the largest exponent.

EXAMPLE 3 *Numerator and denominator of same degree*

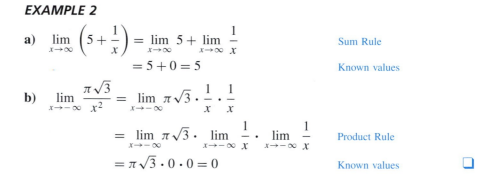

$$\lim_{x\to\infty} \frac{5x^2 + 8x - 3}{3x^2 + 2} = \lim_{x\to\infty} \frac{5 + (8/x) - (3/x^2)}{3 + (2/x^2)} \qquad \text{Divide numerator and denominator by } x^2.$$

$$= \frac{5 + 0 - 0}{3 + 0} = \frac{5}{3} \qquad \text{See Fig. 3.35.}$$

❏

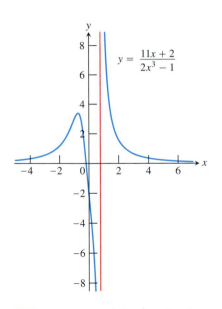

3.36 The graph of the function in Example 4. The graph approaches the x-axis as $|x|$ increases.

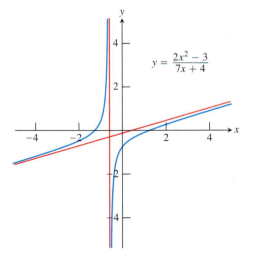

3.37 The function in Example 5(a).

The **leading coefficient** of the polynomial $a_n x^n + a_{n-1} x^{n-1} + \cdots + a_1 x + a_0$, $a_n \neq 0$, is a_n, the coefficient of the highest-powered term.

EXAMPLE 4   *Degree of numerator less than degree of denominator*

$$\lim_{x \to -\infty} \frac{11x + 2}{2x^3 - 1} = \lim_{x \to -\infty} \frac{(11/x^2) + (2/x^3)}{2 - (1/x^3)}$$

Divide numerator and denominator by x^3.

$$= \frac{0 + 0}{2 - 0} = 0$$

See Fig. 3.36. ❏

EXAMPLE 5   *Degree of numerator greater than degree of denominator*

a) $\displaystyle\lim_{x \to -\infty} \frac{2x^2 - 3}{7x + 4} = \lim_{x \to -\infty} \frac{2x - (3/x)}{7 + (4/x)}$

Divide numerator and denominator by x.

$$= -\infty$$

The numerator now approaches $-\infty$ while the denominator approaches 7, so the ratio $\to -\infty$. See Fig. 3.37.

b) $\displaystyle\lim_{x \to -\infty} \frac{-4x^3 + 7x}{2x^2 - 3x - 10} = \lim_{x \to -\infty} \frac{-4x + (7/x)}{2 - (3/x) - (10/x^2)}$

Divide numerator and denominator by x^2.

$$= \infty$$

Numerator $\to \infty$. Denominator $\to 2$. Ratio $\to \infty$.

❏

Examples 3–5 reveal a pattern for finding limits of rational functions as $x \to \pm\infty$.

1.   If the numerator and the denominator have the same degree, the limit is the ratio of the polynomials' leading coefficients (Example 3).
2.   If the degree of the numerator is less than the degree of the denominator, the limit is zero (Example 4).
3.   If the degree of the numerator is greater than the degree of the denominator, the limit is $+\infty$ or $-\infty$, depending on the signs assumed by the numerator and denominator as $|x|$ becomes large (Example 5).

Summary for Rational Functions

1.   If deg $(f) = $ deg (g), $\displaystyle\lim_{x \to \pm\infty} \frac{f(x)}{g(x)} = \frac{a_n}{b_n}$, the ratio of the leading coefficients of f and g.

2.   If deg $(f) < $ deg (g), $\displaystyle\lim_{x \to \pm\infty} \frac{f(x)}{g(x)} = 0.$

3.   If deg $(f) > $ deg (g), $\displaystyle\lim_{x \to \pm\infty} \frac{f(x)}{g(x)} = \pm\infty$, depending on the signs of numerator and denominator.

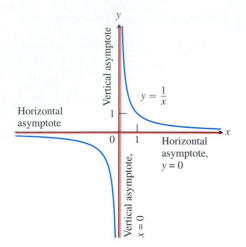

3.38 The coordinate axes are asymptotes of both branches of the hyperbola $y = 1/x$.

Horizontal and Vertical Asymptotes

If the distance between the graph of a function and some fixed line approaches zero as the graph moves increasingly far from the origin, we say that the graph approaches the line asymptotically and that the line is an *asymptote* of the graph.

EXAMPLE 6   The coordinate axes are asymptotes of the curve $y = 1/x$ (Fig. 3.38). The x-axis is an asymptote of the curve on the right because

$$\lim_{x \to \infty} \frac{1}{x} = 0$$

and on the left because

$$\lim_{x \to -\infty} \frac{1}{x} = 0.$$

The y-axis is an asymptote of the curve both above and below because

$$\lim_{x \to 0^+} \frac{1}{x} = \infty \quad \text{and} \quad \lim_{x \to 0^-} \frac{1}{x} = -\infty.$$

Notice that the denominator is zero at $x = 0$ and the function is undefined. ❏

Definitions

A line $y = b$ is a **horizontal asymptote** of the graph of a function $y = f(x)$ if either

$$\lim_{x \to \infty} f(x) = b \quad \text{or} \quad \lim_{x \to -\infty} f(x) = b.$$

A line $x = a$ is a **vertical asymptote** of the graph if either

$$\lim_{x \to a^+} f(x) = \pm\infty \quad \text{or} \quad \lim_{x \to a^-} f(x) = \pm\infty.$$

EXAMPLE 7   The curves

$$y = \sec x = \frac{1}{\cos x} \quad \text{and} \quad y = \tan x = \frac{\sin x}{\cos x}$$

both have vertical asymptotes at odd-integer multiples of $\pi/2$, where $\cos x = 0$ (Fig. 3.39).

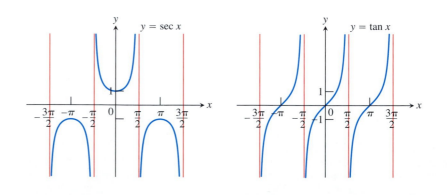

3.39 The graphs of sec x and tan x (Example 7).

The graphs of

$$y = \csc x = \frac{1}{\sin x} \qquad \text{and} \qquad y = \cot x = \frac{\cos x}{\sin x}$$

have vertical asymptotes at integer multiples of π, where $\sin x = 0$ (Fig. 3.40).

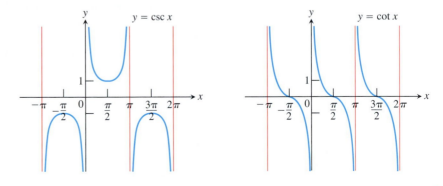

3.40 The graphs of csc x and cot x (Example 7).

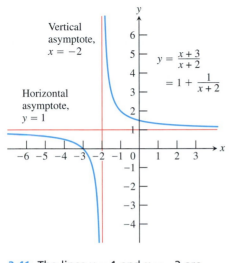

Vertical asymptote, $x = -2$

$y = \frac{x + 3}{x + 2}$

$= 1 + \frac{1}{x + 2}$

Horizontal asymptote, $y = 1$

3.41 The lines $y = 1$ and $x = -2$ are asymptotes of the curve $y = (x + 3)/(x + 2)$ (Example 8).

EXAMPLE 8   Find the asymptotes of the curve

$$y = \frac{x + 3}{x + 2}.$$

Solution   We are interested in the behavior as $x \to \pm\infty$ and as $x \to -2$, where the denominator is zero.

The asymptotes are quickly revealed if we recast the rational function as a polynomial with a remainder, by dividing $(x + 2)$ into $(x + 3)$.

$$\begin{array}{r} 1 \\ x + 2 \overline{\smash{\big)}\, x + 3} \\ \underline{x + 2} \\ 1 \end{array}$$

This enables us to rewrite y:

$$y = 1 + \frac{1}{x + 2}$$

From this we see that the curve in question is the graph of $y = 1/x$ shifted 1 unit up and 2 units left (Fig. 3.41). The asymptotes, instead of being the coordinate axes, are now the lines $y = 1$ and $x = -2$.

EXAMPLE 9   Find the asymptotes of the graph of

$$f(x) = -\frac{8}{x^2 - 4}.$$

Solution   We are interested in the behavior as $x \to \pm\infty$ and as $x \to \pm 2$, where the denominator is zero. Notice that f is an even function of x, so its graph is symmetric with respect to the y-axis.

The behavior as $x \to \pm\infty$.   Since $\lim_{x \to \infty} f(x) = 0$, the line $y = 0$ is an asymptote of the graph to the right. By symmetry it is an asymptote to the left as well (Fig. 3.42).

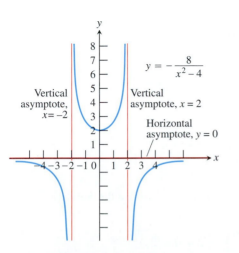

Vertical asymptote, $x = -2$

$y = -\frac{8}{x^2 - 4}$

Vertical asymptote, $x = 2$

Horizontal asymptote, $y = 0$

3.42 The graph of $y = -8/(x^2 - 4)$ (Example 9). Notice that the curve approaches the x-axis from only one side. Asymptotes do not have to be two-sided.

The behavior as $x \to \pm 2$. Since

$$\lim_{x \to 2^+} f(x) = -\infty \qquad \text{and} \qquad \lim_{x \to 2^-} f(x) = \infty,$$

the line $x = 2$ is an asymptote both from the right and from the left. By symmetry, the same holds for the line $x = -2$.

There are no other asymptotes because f has a finite limit at every other point.

❑

We might be tempted at this point to say that rational functions have vertical asymptotes where their denominators are zero. That is nearly true, but not quite. What is true is that rational functions *reduced to lowest terms* have vertical asymptotes where their denominators are zero.

EXAMPLE 10   *A removable discontinuity at a zero of the denominator*

The graph of

$$f(x) = \frac{x^3 - 1}{x^2 - 1}$$

has a vertical asymptote at $x = -1$ but not at $x = 1$. Since

$$\frac{x^3 - 1}{x^2 - 1} = \frac{(x - 1)(x^2 + x + 1)}{(x - 1)(x + 1)} = \frac{x^2 + x + 1}{x + 1},$$

the function has a finite limit ($3/2$) as $x \to 1$ and the discontinuity is removable (Fig. 3.43).   ❑

The Sandwich Theorem (Section 1.2, Theorem 4) also holds for limits as $x \to \pm\infty$. Here is a typical application.

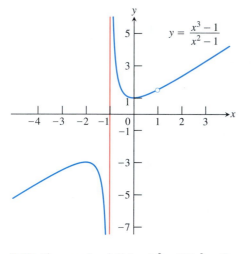

3.43 The graph of $f(x) = (x^3 - 1)/(x^2 - 1)$ has one vertical asymptote, not two. The discontinuity at $x = 1$ is removable.

EXAMPLE 11   Using the Sandwich Theorem, find the asymptotes of the curve

$$y = 2 + \frac{\sin x}{x}.$$

Solution   We are interested in the behavior as $x \to \pm\infty$ and as $x \to 0$, where the denominator is zero.

The behavior as $x \to 0$.   We know that $\lim_{x \to 0} (\sin x)/x = 1$, so there is no asymptote at the origin.

The behavior as $x \to \pm\infty$.   Since

$$0 \le \left| \frac{\sin x}{x} \right| \le \left| \frac{1}{x} \right|,$$

and $\lim_{x \to \pm\infty} |1/x| = 0$, we have $\lim_{x \to \pm\infty} (\sin x)/x = 0$ by the Sandwich Theorem. Hence,

$$\lim_{x \to \pm\infty} \left(2 + \frac{\sin x}{x} \right) = 2 + 0 = 2,$$

and the line $y = 2$ is an asymptote of the curve on both left and right (Fig. 3.44).   ❑

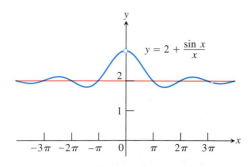

3.44 A curve may cross one of its asymptotes infinitely often (Example 11).

Oblique Asymptotes

If the degree of the numerator of a rational function is one greater than the degree of the denominator, the graph has an **oblique asymptote,** that is, a linear asymptote that is neither vertical nor horizontal.

EXAMPLE 12   Find the asymptotes of the graph of

$$f(x) = \frac{x^2 - 3}{2x - 4}.$$

Solution   We are interested in the behavior as $x \to \pm\infty$ and also as $x \to 2$, where the denominator is zero. We divide $(2x - 4)$ into $(x^2 - 3)$:

$$
\begin{array}{r}
\frac{x}{2} + 1 \\
2x - 4 \overline{\smash{\big)}\, x^2 - 3 } \\
\underline{x^2 - 2x } \\
2x - 3 \\
\underline{2x - 4} \\
1
\end{array}
$$

This tells us that

$$f(x) = \frac{x^2 - 3}{2x - 4} = \underbrace{\frac{x}{2} + 1}_{\text{linear}} + \underbrace{\frac{1}{2x - 4}}_{\text{remainder}}. \qquad (2)$$

Since $\lim_{x \to 2^+} f(x) = \infty$ and $\lim_{x \to 2^-} f(x) = -\infty$, the line $x = 2$ is a two-sided asymptote. As $x \to \pm\infty$, the remainder approaches 0 and $f(x) \to (x/2) + 1$. The line $y = (x/2) + 1$ is an asymptote both to the right and to the left (Fig. 3.45). ☐

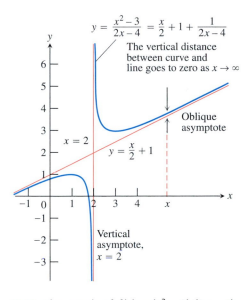

$$y = \frac{x^2 - 3}{2x - 4} = \frac{x}{2} + 1 + \frac{1}{2x - 4}$$

The vertical distance between curve and line goes to zero as $x \to \infty$

Oblique asymptote

$x = 2$

$y = \frac{x}{2} + 1$

Vertical asymptote, $x = 2$

3.45 The graph of $f(x) = (x^2 - 3)/(2x - 4)$ (Example 12).

Graphing with Asymptotes and Dominant Terms

Of all the observations we can make quickly about the function

$$f(x) = \frac{x^2 - 3}{2x - 4}$$

in Example 12, probably the most useful is that

$$f(x) = \frac{x}{2} + 1 + \frac{1}{2x - 4}.$$

This tells us immediately that

$$f(x) \approx \frac{x}{2} + 1 \qquad \text{for } x \text{ numerically large}$$

$$f(x) \approx \frac{1}{2x - 4} \qquad \text{for } x \text{ near 2}$$

If we want to know how f behaves, this is the way to find out. It behaves like $y = (x/2) + 1$ when x is numerically large and the contribution of $1/(2x - 4)$ to the total value of f is insignificant. It behaves like $1/(2x - 4)$ when x is so close to 2 that $1/(2x - 4)$ makes the dominant contribution.

We say that $(x/2) + 1$ **dominates** when x is numerically large, and we say that $1/(2x - 4)$ dominates when x is near 2. **Dominant terms** like these are the key to predicting a function's behavior.

EXAMPLE 13   Graph the function

$$y = \frac{x^3 + 1}{x}.$$

Solution   We investigate symmetry, dominant terms, asymptotes, rise, fall, extreme values, and concavity.

Step 1: *Symmetry.*   There is none.

Step 2: *Find any dominant terms and asymptotes.*   We write the rational function as a polynomial plus remainder:

$$y = x^2 + \frac{1}{x}. \tag{3}$$

For $|x|$ large, $y \approx x^2$. For x near zero, $y \approx 1/x$.

Equation (3) reveals a vertical asymptote at $x = 0$, where the denominator of the remainder is zero.

Step 3: *Find y' and analyze the function's critical points. Where does the curve rise and fall?*

The first derivative

$$y' = 2x - \frac{1}{x^2} = \frac{2x^3 - 1}{x^2} \qquad \text{From Eq. (3)}$$

is undefined at $x = 0$ and zero when

$$2x - \frac{1}{x^2} = 0$$

$$2x^3 - 1 = 0$$

$$x^3 = \frac{1}{2}$$

$$x = \frac{1}{\sqrt[3]{2}} \approx 0.8.$$

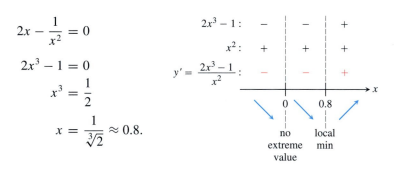

Step 4: *Find y'' and determine the curve's concavity.*   The second derivative

$$y'' = 2 + \frac{2}{x^3} = \frac{2x^3 + 2}{x^3}$$

is undefined at $x = 0$ and zero when

$$2 + \frac{2}{x^3} = 0$$

$$2x^3 + 2 = 0$$

$$x^3 = -1$$

$$x = -1.$$

Step 5: *Summarize the information from the preceding steps and sketch the curve's general shape.*

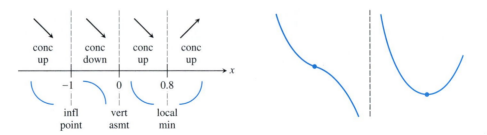

Step 6: *Plot the curve's intercepts, mark any horizontal tangents, and graph the dominant terms.* See Fig. 3.46. This provides a framework for graphing the curve.

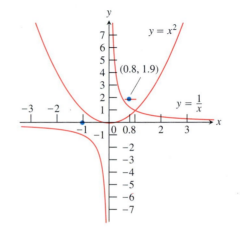

3.46 The dominant terms and horizontal tangent provide a framework for graphing the function.

Step 7: *Now add the final curve to your figure, using the framework and the curve's general shape as guides.* See Fig. 3.47.   ❏

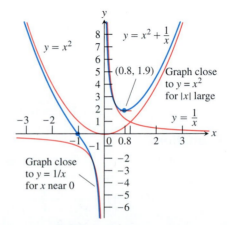

3.47 The function, graphed with the aid of the framework in Fig. 3.46.

Hidden Behavior

Sometimes graphing f' or f'' will suggest where to zoom in on a computer generated graph of f to reveal behavior hidden in the grapher's original picture.

Checklist for Graphing a Function $y = f(x)$

1. Look for symmetry.
    Is the function even? odd?
2. Is the function a shift of a known function?
3. Analyze dominant terms.
    Divide rational functions into polynomial + remainder.
4. Check for asymptotes and removable discontinuities.
    Is there a zero denominator at any point?
    What happens as $x \to \pm\infty$?
5. Compute f' and solve $f' = 0$. Identify critical points and determine intervals of rise and fall.
6. Compute f'' to determine concavity and inflection points.
7. Sketch the graph's general shape.
8. Evaluate f at special values (endpoints, critical points, intercepts).
9. Graph f, using dominant terms, general shape, and special points for guidance.

Exercises 3.5

Calculating Limits as $x \to \pm\infty$

In Exercises 1–6, find the limit of each function (a) as $x \to \infty$ and (b) as $x \to -\infty$. (You may wish to visualize your answer with a grapher.)

1. $f(x) = \dfrac{2}{x} - 3$

2. $f(x) = \pi - \dfrac{2}{x^2}$

3. $g(x) = \dfrac{1}{2 + (1/x)}$

4. $g(x) = \dfrac{1}{8 - (5/x^2)}$

5. $h(x) = \dfrac{-5 + (7/x)}{3 - (1/x^2)}$

6. $h(x) = \dfrac{3 - (2/x)}{4 + \left(\sqrt{2}/x^2\right)}$

Find the limits in Exercises 7–10.

7. $\displaystyle\lim_{x \to \infty} \dfrac{\sin 2x}{x}$

8. $\displaystyle\lim_{\theta \to -\infty} \dfrac{\cos\theta}{3\theta}$

9. $\displaystyle\lim_{t \to -\infty} \dfrac{2 - t + \sin t}{t + \cos t}$

10. $\displaystyle\lim_{r \to \infty} \dfrac{r + \sin r}{2r + 7 - 5\sin r}$

Limits of Rational Functions

In Exercises 11–24, find the limit of each rational function (a) as $x \to \infty$ and (b) as $x \to -\infty$.

11. $f(x) = \dfrac{2x + 3}{5x + 7}$

12. $f(x) = \dfrac{2x^3 + 7}{x^3 - x^2 + x + 7}$

13. $f(x) = \dfrac{x + 1}{x^2 + 3}$

14. $f(x) = \dfrac{3x + 7}{x^2 - 2}$

15. $f(x) = \dfrac{1 - 12x^3}{4x^2 + 12}$

16. $g(x) = \dfrac{1}{x^3 - 4x + 1}$

17. $h(x) = \dfrac{7x^3}{x^3 - 3x^2 + 6x}$

18. $g(x) = \dfrac{3x^2 - 6x}{4x - 8}$

19. $f(x) = \dfrac{2x^5 + 3}{-x^2 + x}$

20. $g(x) = \dfrac{10x^5 + x^4 + 31}{x^6}$

21. $g(x) = \dfrac{x^4}{x^3 + 1}$

22. $h(x) = \dfrac{9x^4 + x}{2x^4 + 5x^2 - x + 6}$

23. $h(x) = \dfrac{-2x^3 - 2x + 3}{3x^3 + 3x^2 - 5x}$

24. $h(x) = \dfrac{-x^4}{x^4 - 7x^3 + 7x^2 + 9}$

Limits with Noninteger or Negative Powers

The process by which we determine limits of rational functions applies equally well to ratios containing noninteger or negative powers of x: divide numerator and denominator by the highest power of x in the denominator and proceed from there. Find the limits in Exercises 25–30.

25. $\displaystyle\lim_{x \to \infty} \dfrac{2\sqrt{x} + x^{-1}}{3x - 7}$

26. $\displaystyle\lim_{x \to \infty} \dfrac{2 + \sqrt{x}}{2 - \sqrt{x}}$

27. $\displaystyle\lim_{x \to -\infty} \dfrac{\sqrt[3]{x} - \sqrt[5]{x}}{\sqrt[3]{x} + \sqrt[5]{x}}$

28. $\displaystyle\lim_{x \to \infty} \dfrac{x^{-1} + x^{-4}}{x^{-2} - x^{-3}}$

29. $\lim\limits_{x\to\infty} \dfrac{2x^{5/3} - x^{1/3} + 7}{x^{8/5} + 3x + \sqrt{x}}$

30. $\lim\limits_{x\to-\infty} \dfrac{\sqrt[3]{x} - 5x + 3}{2x + x^{2/3} - 4}$

Inventing Graphs from Values and Limits

In Exercises 31–34, sketch the graph of a function $y = f(x)$ that satisfies the given conditions. No formulas are required—just label the coordinate axes and sketch an appropriate graph. (The answers are not unique, so your graphs may not be exactly like those in the answer section.)

31. $f(0) = 0$, $f(1) = 2$, $f(-1) = -2$, $\lim\limits_{x\to-\infty} f(x) = -1$, and $\lim\limits_{x\to\infty} f(x) = 1$

32. $f(0) = 0$, $\lim\limits_{x\to\pm\infty} f(x) = 0$, $\lim\limits_{x\to 0^+} f(x) = 2$, and $\lim\limits_{x\to 0^-} f(x) = -2$

33. $f(0) = 0$, $\lim\limits_{x\to\pm\infty} f(x) = 0$, $\lim\limits_{x\to 1^-} f(x) = \lim\limits_{x\to -1^+} f(x) = \infty$, $\lim\limits_{x\to 1^+} f(x) = -\infty$, and $\lim\limits_{x\to -1^-} f(x) = -\infty$

34. $f(2) = 1$, $f(-1) = 0$, $\lim\limits_{x\to\infty} f(x) = 0$, $\lim\limits_{x\to 0^+} f(x) = \infty$, $\lim\limits_{x\to 0^-} f(x) = -\infty$, and $\lim\limits_{x\to -\infty} f(x) = 1$

Inventing Functions

In Exercises 35–38, find a function that satisfies the given conditions and sketch its graph. (The answers here are not unique. Any function that satisfies the conditions is acceptable. Feel free to use formulas defined in pieces if that will help.)

35. $\lim\limits_{x\to\pm\infty} f(x) = 0$, $\lim\limits_{x\to 2^-} f(x) = \infty$, and $\lim\limits_{x\to 2^+} f(x) = \infty$

36. $\lim\limits_{x\to\pm\infty} g(x) = 0$, $\lim\limits_{x\to 3^-} g(x) = -\infty$, and $\lim\limits_{x\to 3^+} g(x) = \infty$

37. $\lim\limits_{x\to-\infty} h(x) = -1$, $\lim\limits_{x\to\infty} h(x) = 1$, $\lim\limits_{x\to 0^-} h(x) = -1$, and $\lim\limits_{x\to 0^+} h(x) = 1$

38. $\lim\limits_{x\to\pm\infty} k(x) = 1$, $\lim\limits_{x\to 1^-} k(x) = \infty$, and $\lim\limits_{x\to 1^+} k(x) = -\infty$

Graphing Rational Functions

Graph the rational functions in Exercises 39–66. Include the graphs and equations of the asymptotes and dominant terms.

39. $y = \dfrac{1}{x - 1}$

40. $y = \dfrac{1}{x + 1}$

41. $y = \dfrac{1}{2x + 4}$

42. $y = \dfrac{-3}{x - 3}$

43. $y = \dfrac{x + 3}{x + 2}$

44. $y = \dfrac{2x}{x + 1}$

45. $y = \dfrac{2x^2 + x - 1}{x^2 - 1}$

46. $y = \dfrac{x^2 - 49}{x^2 + 5x - 14}$

47. $y = \dfrac{x^2 - 1}{x}$

48. $y = \dfrac{x^2 + 4}{2x}$

49. $y = \dfrac{x^4 + 1}{x^2}$

50. $y = \dfrac{x^3 + 1}{x^2}$

51. $y = \dfrac{1}{x^2 - 1}$

52. $y = \dfrac{x^2}{x^2 - 1}$

53. $y = -\dfrac{x^2 - 2}{x^2 - 1}$

54. $y = \dfrac{x^2 - 4}{x^2 - 2}$

55. $y = \dfrac{x^2}{x - 1}$

56. $y = -\dfrac{x^2}{x + 1}$

57. $y = \dfrac{x^2 - 4}{x - 1}$

58. $y = -\dfrac{x^2 - 4}{x + 1}$

59. $y = \dfrac{x^2 - x + 1}{x - 1}$

60. $y = -\dfrac{x^2 - x + 1}{x - 1}$

61. $y = \dfrac{x^3 - 3x^2 + 3x - 1}{x^2 + x - 2}$

62. $y = \dfrac{x^3 + x - 2}{x - x^2}$

63. $y = \dfrac{x}{x^2 - 1}$

64. $y = \dfrac{x - 1}{x^2(x - 2)}$

65. $y = \dfrac{8}{x^2 + 4}$ (Agnesi's witch)

66. $y = \dfrac{4x}{x^2 + 4}$ (Newton's serpentine)

Grapher Explorations

Graph the curves in Exercises 67–72 and explain the relation between the curve's formula and what you see.

67. $y = \dfrac{x}{\sqrt{4 - x^2}}$

68. $y = \dfrac{-1}{\sqrt{4 - x^2}}$

69. $y = x^{2/3} + \dfrac{1}{x^{1/3}}$

70. $y = 2\sqrt{x} + \dfrac{2}{\sqrt{x}} - 3$

71. $y = \sin\left(\dfrac{\pi}{x^2 + 1}\right)$

72. $y = -\cos\left(\dfrac{\pi}{x^2 + 1}\right)$

Graphing Terms

Each of the functions in Exercises 73–76 is given as the sum or difference of two terms. First graph the terms (with the same set of axes). Then, using these graphs as guides, sketch in the graph of the function.

73. $y = \sec x + \dfrac{1}{x}$, $-\dfrac{\pi}{2} < x < \dfrac{\pi}{2}$

74. $y = \sec x - \dfrac{1}{x^2}$, $-\dfrac{\pi}{2} < x < \dfrac{\pi}{2}$

75. $y = \tan x + \dfrac{1}{x^2}$, $-\dfrac{\pi}{2} < x < \dfrac{\pi}{2}$

76. $y = \dfrac{1}{x} - \tan x$, $-\dfrac{\pi}{2} < x < \dfrac{\pi}{2}$

Theory and Examples

77. Let $f(x) = (x^3 + x^2)/(x^2 + 1)$. Show that there is a value of c for which $f(c)$ equals

a) -2 b) $\cos 3$ c) $5{,}000{,}000.$

78. Find $\lim_{x \to \infty} \left(\sqrt{x^2 + x} - \sqrt{x^2 - x} \right)$.

79. *Symmetry.* Suppose an odd function is known to be increasing on the interval $x > 0$. What can be said of its behavior on the interval $x < 0$?

80. *Symmetry.* Suppose an even function is known to be increasing on the interval $x < 0$. What can be said of its behavior on the interval $x > 0$?

81. Suppose that $f(x)$ and $g(x)$ are polynomials in x and that $\lim_{x \to \infty} (f(x)/g(x)) = 2$. Can you conclude anything about $\lim_{x \to -\infty} (f(x)/g(x))$? Give reasons for your answer.

82. Suppose that $f(x)$ and $g(x)$ are polynomials in x. Can the graph of $f(x)/g(x)$ have an asymptote if $g(x)$ is never zero? Give reasons for your answer.

83. How many horizontal asymptotes can the graph of a given rational function have? Give reasons for your answer.

84. How many vertical asymptotes can the graph of a given rational function have? Give reasons for your answer.

85. a) The word *asymptote* derives from an old Greek word for "never touching." In practice, however, a curve may cross one of its asymptotes a finite number of times or even infinitely often, as does the curve $y = 2 + (\sin x)/x$ in Example 11. Show that the slope of the curve nevertheless approaches the slope of the asymptote as $x \to \infty$.

   b) Give an example of a function $f(x)$ with the following properties:

   i) f is differentiable for $x > 0$.

   ii) $\lim_{x \to \infty} f(x) = 2$.

   iii) $\lim_{x \to \infty} f'(x)$ does not exist.

86. *A puzzle.* Suppose we want to identify the oblique asymptote of the curve

$$y = \frac{x^2 + 3x + 7}{x + 2}.$$

If we recast the rational function as a polynomial plus remainder, we find

$$\frac{x^2 + 3x + 7}{x + 2} = x + 1 + \frac{5}{x + 2}.$$

The oblique asymptote should be the line $y = x + 1$.

But suppose, instead, that we divide both numerator and denominator by x, to obtain

$$\frac{x^2 + 3x + 7}{x + 2} = \frac{x + 3 + (7/x)}{1 + (2/x)}.$$

For x large, the value is approximately $x + 3$, so the asymptote should be the line $y = x + 3$.

Which line is the real asymptote, and why?

Use the formal definitions of limits as $x \to \pm\infty$ to establish the limits in Exercises 87 and 88.

87. If f has the constant value $f(x) = k$, then $\lim_{x \to \infty} f(x) = k$.

88. If f has the constant value $f(x) = k$, then $\lim_{x \to -\infty} f(x) = k$.

More Grapher Explorations

Graph the functions in Exercises 89–92. What asymptotes do the graphs have? Why are the asymptotes located where they are?

89. $y = -\dfrac{x^2 - 4}{x + 1}$

90. $y = \dfrac{x^2 + x - 6}{2x - 2}$

91. $y = \dfrac{x^3 - x^2 - 1}{x^2 - 1}$

92. $y = \dfrac{x^3 - 2x^2 + x + 1}{x - x^2}$

Graph the functions in Exercises 93–98 together with their dominant terms. Comment on the relation of the graphs of the dominant terms to the graphs of the functions.

93. $y = x^3 + \dfrac{3}{x}$

94. $y = x^3 - \dfrac{3}{x}$

95. $y = 2 \sin x + \dfrac{1}{x}$

96. $y = 2 \cos x - \dfrac{1}{x}$

97. $y = \dfrac{x^2}{2} + 3 \sin 2x$

98. $y = (x - 1)^{11} + 2 \sin 2\pi x$

Graph the functions in Exercises 99 and 100. Then answer the following questions.

   a) How does the graph behave as $x \to 0^+$? as $x \to 0^-$?

   b) How does the graph behave as $x \to \pm\infty$?

   c) How does the graph behave as $x = 1$ and $x = -1$?

Give reasons for your answers.

99. $y = \dfrac{3}{2} \left(x - \dfrac{1}{x} \right)^{2/3}$

100. $y = \dfrac{3}{2} \left(\dfrac{x}{x - 1} \right)^{2/3}$

101. Graph the function

$$y = -\frac{x^3 - 2}{x^2 + 1}$$

over the following intervals.

   a) $-9 \le x \le 9$

   b) $-90 \le x \le 90$

   c) $-900 \le x \le 900$

The graph in (a) should be good. The graph in (b) may indicate some activity near the origin but will not show what. The graph in (c) will look just like the graph of the line $y = -x$. Why?

102. Graph the function $y = x^{2/3}/(x^2 - 1)$ over the interval $-2 \le x \le 2$. The curve will appear to be concave down between $x = -1$ and $x = 1$, with no sign of a cusp at the origin. Zoom in on the origin and watch the true shape of the graph emerge. Why do you think the cusp fails to appear in the first view of the graph?

▨ Grapher Explorations—"Seeing" Limits at Infinity

Sometimes a change of variable can change an unfamiliar expression into one whose limit we know how to find. For example,

$$\lim_{x \to \infty} \sin \frac{1}{x} = \lim_{\theta \to 0^+} \sin \theta \qquad \text{Substitute } \theta = 1/x$$
$$= 0.$$

This suggests a creative way to "see" limits at infinity. Describe the procedure and use it to picture and determine limits in Exercises 103–108.

103. $\lim\limits_{x \to \pm\infty} x \sin \dfrac{1}{x}$

104. $\lim\limits_{x \to -\infty} \dfrac{\cos (1/x)}{1 + (1/x)}$

105. $\lim\limits_{x \to \pm\infty} \dfrac{3x + 4}{2x - 5}$

106. $\lim\limits_{x \to \infty} \left(\dfrac{1}{x}\right)^{1/x}$

107. $\lim\limits_{x \to \pm\infty} \left(3 + \dfrac{2}{x}\right)\left(\cos \dfrac{1}{x}\right)$

108. $\lim\limits_{x \to \infty} \left(\dfrac{3}{x^2} - \cos \dfrac{1}{x}\right)\left(1 + \sin \dfrac{1}{x}\right)$

3.6 Optimization

To optimize something means to maximize or minimize some aspect of it. What is the size of the most profitable production run? What is the least expensive shape for an oil can? What is the stiffest beam we can cut from a 12-inch log? In the mathematical models in which we use functions to describe the things that interest us, we usually answer such questions by finding the greatest or smallest value of a differentiable function.

Examples from Business and Industry

EXAMPLE 1 *Metal fabrication*

An open-top box is to be made by cutting small congruent squares from the corners of a 12-by-12-in. sheet of tin and bending up the sides. How large should the squares cut from the corners be to make the box hold as much as possible?

Solution We start with a picture (Fig. 3.48). In the figure, the corner squares are x inches on a side. The volume of the box is a function of this variable:

$$V(x) = x(12 - 2x)^2 = 144x - 48x^2 + 4x^3. \qquad V = hlw$$

3.48 An open box made by cutting the corners from a square sheet of tin.

(a)

(b)

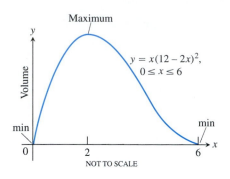

3.49 The volume of the box in Fig. 3.48 graphed as a function of x.

Since the sides of the sheet of tin are only 12 in. long, $x \leq 6$ and the domain of V is the interval $0 \leq x \leq 6$.

A graph of V (Fig. 3.49) suggests a minimum value of 0 at $x = 0$ and $x = 6$ and a maximum near $x = 2$. To learn more, we examine the first derivative of V with respect to x:

$$\frac{dV}{dx} = 144 - 96x + 12x^2 = 12(12 - 8x + x^2) = 12(2 - x)(6 - x).$$

Of the two zeros, $x = 2$ and $x = 6$, only $x = 2$ lies in the interior of the function's domain and makes the critical-point list. The values of V at this one critical point and two endpoints are

Critical-point value: $V(2) = 128$

Endpoint values: $V(0) = 0, \quad V(6) = 0.$

The maximum volume is 128 in³. The cut-out squares should be 2 in. on a side.

❑

EXAMPLE 2   *Product design*

You have been asked to design a 1-L oil can shaped like a right circular cylinder. What dimensions will use the least material?

Solution   We picture the can as a right circular cylinder with height h and diameter $2r$ (Fig. 3.50). If r and h are measured in centimeters and the volume is expressed as 1000 cm³, then r and h are related by the equation

$$\pi r^2 h = 1000. \qquad \text{1 L = 1000 cm}^3 \qquad (1)$$

How shall we interpret the phrase "least material"? One possibility is to ignore the thickness of the material and the waste in manufacturing. Then we ask for dimensions r and h that make the total surface area

$$A = \underbrace{2\pi r^2}_{\substack{\text{cylinder} \\ \text{ends}}} + \underbrace{2\pi rh}_{\substack{\text{cylinder} \\ \text{wall}}} \qquad (2)$$

3.50 This 1-L can uses the least material when $h = 2r$ (Example 2).

as small as possible while satisfying the constraint $\pi r^2 h = 1000$. (Exercise 18 describes one way we might take waste into account.)

We are not quite ready to find critical points because Eq. (2) gives A as a function of two variables and our procedure calls for A to be a function of a single variable. However, Eq. (1) can be solved to express either r or h in terms of the other.

Solving for h is easier, so we take

$$h = \frac{1000}{\pi r^2}.$$

This changes the formula for A to

$$A = 2\pi r^2 + 2\pi rh = 2\pi r^2 + 2\pi r \frac{1000}{\pi r^2} = 2\pi r^2 + \frac{2000}{r}.$$

For small r (a tall thin container, like a pipe), the term $2000/r$ dominates and A is large. For larger r (a short wide container, like a pizza pan), the term $2\pi r^2$

dominates and A is again large. If A has a minimum, it must be at a value of r that is neither too large nor too small.

Since A is differentiable throughout its domain $(0, \infty)$ and the domain has no endpoints, A can have a minimum only where $dA/dr = 0$.

$$A = 2\pi r^2 + \frac{2000}{r}$$

$$\frac{dA}{dr} = 4\pi r - \frac{2000}{r^2} \qquad \text{\color{blue}Find } dA/dr.$$

$$4\pi r - \frac{2000}{r^2} = 0 \qquad \text{\color{blue}Set it equal to 0.}$$

$$4\pi r^3 = 2000 \qquad \text{\color{blue}Solve for } r.$$

$$r = \sqrt[3]{\frac{500}{\pi}} \qquad \text{\color{blue}Critical point}$$

So something happens at $r = \sqrt[3]{500/\pi}$, but what?

If the domain of A were a closed interval, we could find out by evaluating A at this critical point and the endpoints and comparing the results. But the domain is not a closed interval, so we must learn what is happening at $r = \sqrt[3]{500/\pi}$ by determining the shape of A's graph. We can do this by investigating the second derivative, d^2A/dr^2:

$$\frac{dA}{dr} = 4\pi r - \frac{2000}{r^2}$$

$$\frac{d^2A}{dr^2} = 4\pi + \frac{4000}{r^3}.$$

The second derivative is positive throughout the domain of A. The value of A at $r = \sqrt[3]{500/\pi}$ is therefore an absolute minimum because the graph of A is concave up (Fig. 3.51).

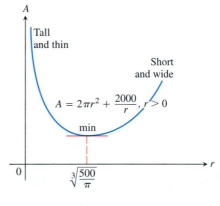

3.51 The graph of $A = 2\pi r^2 + 2000/r$ is concave up.

When

$$r = \sqrt[3]{500/\pi},$$

$$h = \frac{1000}{\pi r^2} = 2\sqrt[3]{500/\pi} = 2r. \qquad \text{\color{blue}After some arithmetic} \quad \textbf{(3)}$$

Equation (3) tells us that the most efficient can has its height equal to its diameter. With a calculator we find

$$r \approx 5.42 \text{ cm}, \qquad h \approx 10.84 \text{ cm}.$$

Examples from Mathematics

EXAMPLE 3   *Products of numbers*

Find two positive numbers whose sum is 20 and whose product is as large as possible.

Solution   If one number is x, the other is $(20 - x)$. Their product is

$$f(x) = x(20 - x) = 20x - x^2.$$

We want the value or values of x that make $f(x)$ as large as possible. The domain of f is the closed interval $0 \leq x \leq 20$.

     We evaluate f at the critical points and endpoints. The first derivative,

$$f'(x) = 20 - 2x,$$

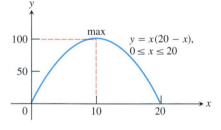

is defined at every point of the interval $0 \leq x \leq 20$ and is zero only at $x = 10$. Listing the values of f at this one critical point and the endpoints gives

     Critical-point value: $f(10) = 20(10) - (10)^2 = 100$

     Endpoint values: $f(0) = 0,$ $f(20) = 0.$

We conclude that the maximum value is $f(10) = 100$. The corresponding numbers are $x = 10$ and $(20 - 10) = 10$ (Fig. 3.52).   ❑

3.52 The product of x and $(20 - x)$ reaches a maximum value of 100 when $x = 10$ (Example 3).

EXAMPLE 4   *Geometry*

A rectangle is to be inscribed in a semicircle of radius 2. What is the largest area the rectangle can have, and what are its dimensions?

Solution   To describe the dimensions of the rectangle, we place the circle and rectangle in the coordinate plane (Fig. 3.53). The length, height, and area of the rectangle can then be expressed in terms of the position x of the lower right-hand corner:

     Length: $2x$ Height: $\sqrt{4 - x^2}$ Area: $2x \cdot \sqrt{4 - x^2}.$

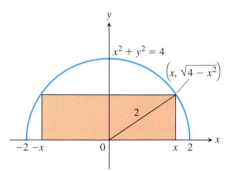

3.53 The rectangle and semicircle in Example 4.

Notice that the values of x are to be found in the interval $0 \leq x \leq 2$, where the selected corner of the rectangle lies.

Our mathematical goal is now to find the absolute maximum value of the continuous function

$$A(x) = 2x\sqrt{4 - x^2}$$

on the domain $[0, 2]$. We do this by examining the values of A at the critical points and endpoints. The derivative

$$\frac{dA}{dx} = \frac{-2x^2}{\sqrt{4 - x^2}} + 2\sqrt{4 - x^2}$$

is not defined when $x = 2$ and is equal to zero when

$$\frac{-2x^2}{\sqrt{4 - x^2}} + 2\sqrt{4 - x^2} = 0$$

$$-2x^2 + 2(4 - x^2) = 0 \qquad \text{Multiply both sides by } \sqrt{4 - x^2}.$$

$$8 - 4x^2 = 0$$

$$x^2 = 2$$

$$x = \pm\sqrt{2}.$$

Of the two zeros, $x = \sqrt{2}$ and $x = -\sqrt{2}$, only $x = \sqrt{2}$ lies in the interior of A's domain and makes the critical-point list. The values of A at the endpoints and at this one critical point are

Critical-point value: $\quad A(\sqrt{2}) = 2\sqrt{2}\sqrt{4 - 2} = 4$
Endpoint values: $\qquad\quad A(0) = 0, \qquad A(2) = 0.$

The area has a maximum value of 4 when the rectangle is $\sqrt{4 - x^2} = \sqrt{2}$ units high and $2x = 2\sqrt{2}$ units long. ∎

✳ Fermat's Principle and Snell's Law

The speed of light depends on the medium through which it travels and tends to be slower in denser media. In a vacuum, it travels at the famous speed $c = 3 \times 10^8$ m/sec, but in the earth's atmosphere it travels slightly slower than that, and in glass slower still (about two-thirds as fast).

Fermat's principle in optics states that light always travels from one point to another along the quickest route. This observation enables us to predict the path light will take when it travels from a point in one medium (air, say) to a point in another medium (say, glass or water).

EXAMPLE 5   Find the path that a ray of light will follow in going from a point A in a medium where the speed of light is c_1 across a straight boundary to a point B in a medium where the speed of light is c_2.

Solution   Since light traveling from A to B will do so by the quickest route, we look for a path that will minimize the travel time.

We assume that A and B lie in the xy-plane and that the line separating the two media is the x-axis (Fig. 3.54).

In a uniform medium, where the speed of light remains constant, "shortest time" means "shortest path," and the ray of light will follow a straight line. Hence

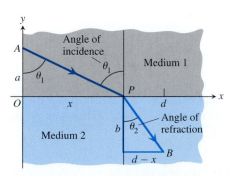

3.54 A light ray refracted (deflected from its path) as it passes from one medium to another. θ_1 is the angle of incidence and θ_2 is the angle of refraction.

the path from A to B will consist of a line segment from A to a boundary point P, followed by another line segment from P to B. From the formula distance equals rate times time, we have

$$\text{time} = \frac{\text{distance}}{\text{rate}}.$$

The time required for light to travel from A to P is therefore

$$t_1 = \frac{AP}{c_1} = \frac{\sqrt{a^2 + x^2}}{c_1}.$$

From P to B the time is

$$t_2 = \frac{PB}{c_2} = \frac{\sqrt{b^2 + (d - x)^2}}{c_2}.$$

The time from A to B is the sum of these:

$$t = t_1 + t_2 = \frac{\sqrt{a^2 + x^2}}{c_1} + \frac{\sqrt{b^2 + (d - x)^2}}{c_2}. \tag{4}$$

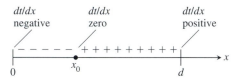

dt/dx negative dt/dx zero dt/dx positive

3.55 The sign pattern of dt/dx in Example 5.

Equation (4) expresses t as a differentiable function of x whose domain is $[0, d]$, and we want to find the absolute minimum value of t on this closed interval. We find

$$\frac{dt}{dx} = \frac{x}{c_1 \sqrt{a^2 + x^2}} - \frac{(d - x)}{c_2 \sqrt{b^2 + (d - x)^2}}. \tag{5}$$

In terms of the angles θ_1 and θ_2 in Fig. 3.54,

$$\frac{dt}{dx} = \frac{\sin \theta_1}{c_1} - \frac{\sin \theta_2}{c_2}. \tag{6}$$

We can see from Eq. (5) that $dt/dx < 0$ at $x = 0$ and $dt/dx > 0$ at $x = d$. Hence, $dt/dx = 0$ at some point x_0 in between (Fig. 3.55). There is only one such point because dt/dx is an increasing function of x (Exercise 52). At this point,

$$\frac{\sin \theta_1}{c_1} = \frac{\sin \theta_2}{c_2}.$$

This equation is **Snell's law** or the **law of refraction.**

We conclude that the path the ray of light follows is the one described by Snell's law. Figure 3.56 shows how this works for air and water.   ◻

3.56 For air and water at room temperature, the light velocity ratio is 1.33 and Snell's law becomes $\sin \theta_1 = 1.33 \sin \theta_2$. In this laboratory photograph, $\theta_1 = 35.5°, \theta_2 = 26°$, and $(\sin 35.5°/\sin 26°) \approx 0.581/0.438 \approx 1.33$, as predicted.

This photograph also illustrates that angle of reflection = angle of incidence (Exercise 39).

Cost and Revenue in Economics

Here we want to point out two of the many places where calculus makes a contribution to economic theory. The first has to do with the relationship between profit, revenue (money received), and cost.

Suppose that

$r(x) = $ the revenue from selling x items

$c(x) = $ the cost of producing the x items

$p(x) = r(x) - c(x) = $ the profit from selling x items.

Developing a physical law

In developing a physical law, we typically observe an effect, measure values and list them in a table, and then try to find a rule by which one thing can be connected with another. The Alexandrian Greek Claudius Ptolemy (c. 100–c. 170 A.D.) tried to do this for the refraction of light by water. He made a table of angles of incidence and corresponding angles of refraction, with values very close to the ones we find for air and water today.

Angle in air (degrees)	Ptolemy's angle in water (degrees)	Modern angle in water (degrees)
10	8	7.5
20	15.5	15
30	22.5	22
40	28	29
50	35	35
60	40.5	40.5
70	45	45
80	50	47.6

The rule that connected these angles, however, eluded him, as it did everyone else for the next 1400 years. The Dutch mathematician Willebrord Snell (1580–1626) found it in 1621.

Finding a rule is nice, but the real glory of science is finding a way of thinking that makes the rule evident. Fermat discovered it around 1650. His idea was this: Of all the paths light might take to get from one point to another, it follows the path that takes the shortest time. In Example 5, you see how this principle leads to Snell's law. The derivation we give is Fermat's own.

For more on marginal revenue and cost, see the end of Section 2.3.

The marginal revenue and cost at this production level (x items) are

$$\frac{dr}{dx} = \text{marginal revenue}$$

$$\frac{dc}{dx} = \text{marginal cost.}$$

The first theorem is about the relationship of the profit p to these derivatives.

Theorem 7

Maximum profit (if any) occurs at a production level at which marginal revenue equals marginal cost.

Proof   We assume that $r(x)$ and $c(x)$ are differentiable for all $x > 0$, so if $p(x) = r(x) - c(x)$ has a maximum value, it occurs at a production level at which $p'(x) = 0$. Since $p'(x) = r'(x) - c'(x)$, $p'(x) = 0$ implies

$$r'(x) - c'(x) = 0 \qquad \text{or} \qquad r'(x) = c'(x).$$

This concludes the proof (Fig. 3.57).

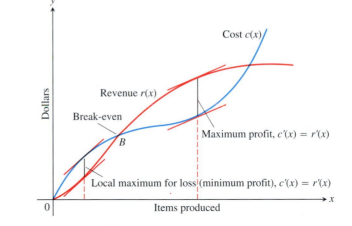

3.57 The graph of a typical cost function starts concave down and later turns concave up. It crosses the revenue curve at the break-even point *B*. To the left of *B*, the company operates at a loss. To the right, the company operates at a profit, with the maximum profit occurring where $c'(x) = r'(x)$. Farther to the right, cost exceeds revenue (perhaps because of a combination of market saturation and rising labor and material costs) and production levels become unprofitable again.

❑

What guidance do we get from Theorem 7? We know that a production level at which $p'(x) = 0$ need not be a level of maximum profit. It might be a level of minimum profit, for example. But if we are making financial projections for our company, we should look for production levels at which marginal cost seems to equal marginal revenue. If there is a most profitable production level, it will be one of these.

EXAMPLE 6   The cost and revenue functions at American Gadget are

$$r(x) = 9x \qquad \text{and} \qquad c(x) = x^3 - 6x^2 + 15x,$$

where x represents thousands of gadgets. Is there a production level that will maximize American Gadget's profit? If so, what is it?

Solution

$$r(x) = 9x, \qquad c(x) = x^3 - 6x^2 + 15x \qquad \text{\color{blue}{Find }} r'(x) \text{ \color{blue}{and} } c'(x).$$

$$r'(x) = 9, \qquad c'(x) = 3x^2 - 12x + 15$$

$$3x^2 - 12x + 15 = 9 \qquad\qquad\qquad\qquad\quad \text{\color{blue}{Set them equal.}}$$

$$3x^2 - 12x + 6 = 0 \qquad\qquad\qquad\qquad\quad \text{\color{blue}{Rearrange.}}$$

$$x^2 - 4x + 2 = 0$$

$$x = \frac{4 \pm \sqrt{16 - 4 \cdot 2}}{2} \qquad\qquad \text{\color{blue}{Solve for } } x \text{ \color{blue}{with the} } \\ \text{\color{blue}{quadratic formula.}}$$

$$= \frac{4 \pm 2\sqrt{2}}{2}$$

$$= 2 \pm \sqrt{2}$$

The possible production levels for maximum profit are $x = 2 + \sqrt{2}$ thousand units and $x = 2 - \sqrt{2}$ thousand units. A quick glance at the graphs in Fig. 3.58 or at the corresponding values of r and c shows $x = 2 + \sqrt{2}$ to be a point of maximum profit and $x = 2 - \sqrt{2}$ to be a local maximum for loss.

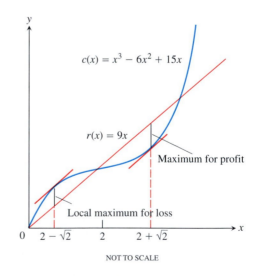

NOT TO SCALE

3.58 The cost and revenue curves for Example 6.

Another way to look for optimal production levels is to look for levels that minimize the average cost of the units produced. The next theorem helps us to find them.

Theorem 8

The production level (if any) at which average cost is smallest is a level at which the average cost equals the marginal cost.

Proof   We start with

$$c(x) = \text{cost of producing } x \text{ items}, \; x > 0$$

$$\frac{c(x)}{x} = \text{average cost of producing } x \text{ items},$$

assumed differentiable.

If the average cost can be minimized, it will be at a production level at which

$$\frac{d}{dx}\left(\frac{c(x)}{x}\right) = 0$$

$$\frac{xc'(x) - c(x)}{x^2} = 0 \qquad \text{Quotient Rule}$$

$$xc'(x) - c(x) = 0 \qquad \text{Multiplied by } x^2$$

$$\underbrace{c'(x)}_{\substack{\text{marginal} \\ \text{cost}}} = \underbrace{\frac{c(x)}{x}}_{\substack{\text{average} \\ \text{cost}}}.$$

This completes the proof.   ❑

Again we have to be careful about what Theorem 8 does and does not say. It does not say that there is a production level of minimum average cost—it says where to look to see if there is one. Look for production levels at which average cost and marginal cost are equal. Then check to see if any of them gives a minimum average cost.

EXAMPLE 7   The cost function at American Gadget is $c(x) = x^3 - 6x^2 + 15x$ (x in thousands of units). Is there a production level that minimizes average cost? If so, what is it?

Solution   We look for levels at which average cost equals marginal cost.

Cost: $c(x) = x^3 - 6x^2 + 15x$

Marginal cost: $c'(x) = 3x^2 - 12x + 15$

Average cost: $\dfrac{c(x)}{x} = x^2 - 6x + 15$

$$3x^2 - 12x + 15 = x^2 - 6x + 15 \qquad \text{MC = AC}$$

$$2x^2 - 6x = 0$$

$$2x(x - 3) = 0$$

$$x = 0 \quad \text{or} \quad x = 3$$

Since $x > 0$, the only production level that might minimize average cost is $x = 3$ thousand units.

We check the derivatives:

$$\frac{c(x)}{x} = x^2 - 6x + 15 \qquad \text{Average cost}$$

$$\frac{d}{dx}\left(\frac{c(x)}{x}\right) = 2x - 6$$

$$\frac{d^2}{dx^2}\left(\frac{c(x)}{x}\right) = 2 > 0.$$

The second derivative is positive, so $x = 3$ gives an absolute minimum. ❏

Modeling Discrete Phenomena with Differentiable Functions

In case you are wondering how we can use differentiable functions $c(x)$ and $r(x)$ to describe the cost and revenue that come from producing a number of items x, which can only be an integer, here is the rationale.

When x is large, we can reasonably fit the cost and revenue data with smooth curves $c(x)$ and $r(x)$ that are defined not only at integer values of x but at the values in between. Once we have these differentiable functions, which are supposed to behave like the real cost and revenue when x is an integer, we can apply calculus to draw conclusions about their values. We then translate these mathematical conclusions into inferences about the real world that we hope will have predictive value. When they do, as is the case with the economic theory here, we say that the functions give a good model of reality.

What do we do when our calculus tells us that the best production level is a value of x that isn't an integer, as it did in Example 6 when it said that $x = 2 + \sqrt{2}$ thousand units would be the production level for maximum profit? The practical answer is to use the nearest convenient integer. For $x = 2 + \sqrt{2}$ thousand, we might use 3414, or perhaps 3410 or 3420 if we ship in boxes of 10.

Exercises 3.6

If you have a grapher, this is a good place to use it. We have included some specific grapher exercises but there is something to be learned from graphing in most of the other exercises as well.

Whenever you are maximizing or minimizing a function of a single variable, we urge you to graph it over the domain that is appropriate to the problem you are solving. The graph will provide insight before you calculate and will furnish a visual context for understanding your answer.

Applications in Geometry

1. A sector shaped like a slice of pie is cut from a circle of radius r. The outer circular arc of the sector has length s. If the sector's total perimeter $(2r + s)$ is to be 100 m, what values of r and s will maximize the sector's area?

2. What is the largest possible area for a right triangle whose hypotenuse is 5 cm long?

3. What is the smallest perimeter possible for a rectangle whose area is 16 in^2?

4. Show that among all rectangles with a given perimeter, the one with the largest area is a square.

5. The figure shown here shows a rectangle inscribed in an isosceles right triangle whose hypotenuse is 2 units long.

   a) Express the y-coordinate of P in terms of x. (You might start by writing an equation for the line AB.)

b) Express the area of the rectangle in terms of x.

c) What is the largest area the rectangle can have?

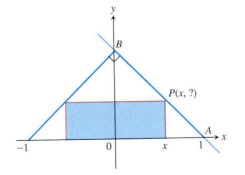

6. A rectangle has its base on the x-axis and its upper two vertices on the parabola $y = 12 - x^2$. What is the largest area the rectangle can have?

7. You are planning to make an open rectangular box from an 8-by-15-in. piece of cardboard by cutting squares from the corners and folding up the sides. What are the dimensions of the box of largest volume you can make this way?

8. You are planning to close off a corner of the first quadrant with a line segment 20 units long running from $(a, 0)$ to $(0, b)$. Show that the area of the triangle enclosed by the segment is largest when $a = b$.

9. A rectangular plot of farmland will be bounded on one side by a river and on the other three sides by a single-strand electric fence. With 800 m of wire at your disposal, what is the largest area you can enclose?

10. A 216-m² rectangular pea patch is to be enclosed by a fence and divided into two equal parts by another fence parallel to one of the sides. What dimensions for the outer rectangle will require the smallest total length of fence? How much fence will be needed?

11. *The lightest steel holding tank.* Your iron works has contracted to design and build a 500-ft³, square-based, open-top, rectangular steel holding tank for a paper company. The tank is to be made by welding $\frac{1}{2}$-in.-thick stainless steel plates together along their edges. As the production engineer, your job is to find dimensions for the base and height that will make the tank weigh as little as possible. What dimensions do you tell the shop to use?

12. *Catching rainwater.* An 1125-ft³ open-top rectangular tank with a square base x ft on a side and y ft deep is to be built with its top flush with the ground to catch runoff water. The costs associated with the tank involve not only the material from which the tank is made but also an excavation charge proportional to the product xy. If the cost is

$$c = 5(x^2 + 4xy) + 10xy,$$

what values of x and y will minimize it?

13. You are designing a poster to contain 50 in² of printing with margins of 4 in. each at top and bottom and 2 in. at each side. What overall dimensions will minimize the amount of paper used?

14. Find the volume of the largest right circular cone that can be inscribed in a sphere of radius 3.

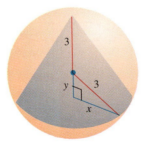

15. Two sides of a triangle have lengths a and b, and the angle between them is θ. What value of θ will maximize the triangle's area? (*Hint:* $A = (1/2)ab \sin \theta$.)

16. Find the largest possible value of $s = 2x + y$ if x and y are side lengths in a right triangle whose hypotenuse is $\sqrt{5}$ units long.

17. What are the dimensions of the lightest (least material) open-top right circular cylindrical can that will hold a volume of 1000 cm³? Compare the result here with the result in Example 2.

18. You are designing 1000-cm³ right circular cylindrical cans whose manufacture will take waste into account. There is no waste in cutting the aluminum for the sides, but the tops and bottoms of radius r will be cut from squares that measure $2r$ units on a side. The total amount of aluminum used by each can will therefore be

$$A = 8r^2 + 2\pi rh$$

rather than the $A = 2\pi r^2 + 2\pi rh$ in Example 2. In Example 2 the ratio of h to r for the most economical cans was 2 to 1. What is the ratio now?

19. a) The U.S. Postal Service will accept a box for domestic shipment only if the sum of its length and girth (distance around) does not exceed 108 in. What dimensions will give a box with a square end the largest possible volume?

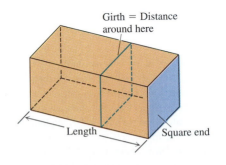

Girth = Distance around here

Length   Square end

b) GRAPHER Graph the volume of a 108-in. box (length plus girth equals 108 in.) as a function of its length, and compare what you see with your answer in (a).

20. (*Continuation of Exercise 19.*) Suppose that instead of having a box with square ends you have a box with square sides so that its dimensions are h by h by w and the girth is $2h + 2w$. What dimensions will give the box its largest volume now?

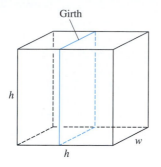

Girth

21. Compare the answers to the following two construction problems.

   a) A rectangular sheet of perimeter 36 cm and dimensions x cm by y cm is to be rolled into the cylinder shown here (a). What values of x and y give the largest volume?

   b) The rectangular sheet of perimeter 36 cm and dimensions x by y is to be revolved about one of the sides of length y to sweep out the cyclinder shown here (b). What values of x and y give the largest volume?

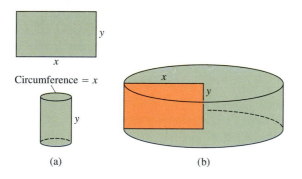

Circumference = x

(a)   (b)

22. A right triangle whose hypotenuse is $\sqrt{3}$ m long is revolved about one of its legs to generate a right circular cone. Find the radius, height, and volume of the cone of greatest volume that can be made this way.

23. *Circle vs. square*

   a) A 4-m length of wire is available for making a circle and a square. How should the wire be distributed between the two shapes to maximize the sum of the enclosed areas?

   b) GRAPHER Graph the total area enclosed by the wire as a function of the circle's radius. Reconcile what you see with your answer in (a).

   c) GRAPHER Now graph the total area enclosed by the wire as a function of the square's side length. Again, reconcile what you see with your answer in (a).

24. If the sum of the surface areas of a cube and a sphere is held constant, what ratio of an edge of the cube to the radius of the sphere will make the sum of the volumes (a) as small as possible, (b) as large as possible?

25. A window is in the form of a rectangle surmounted by a semicircle. The rectangle is of clear glass while the semicircle is of tinted glass that transmits only half as much light per unit area as clear glass does. The total perimeter is fixed. Find the proportions of the window that will admit the most light. Neglect the thickness of the frame.

26. A silo (base not included) is to be constructed in the form of a cylinder surmounted by a hemisphere. The cost of construction per square unit of surface area is twice as great for the hemisphere as it is for the cylindrical sidewall. Determine the dimensions to be used if the volume is fixed and the cost of construction is to be kept to a minimum. Neglect the thickness of the silo and waste in construction.

27. The trough here is to be made to the dimensions shown. Only the angle θ can be varied. What value of θ will maximize the trough's volume?

28. A rectangular sheet of $8\frac{1}{2}$-by-11-in. paper shown here is placed on a flat surface, and one of the corners is placed on the opposite longer edge. The other corners are held in their original positions. With all four corners now held fixed, the paper is smoothed flat. The problem is to make the length of the crease as small as possible. Call the length L.

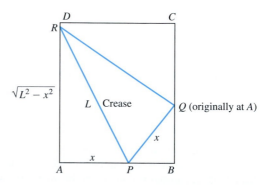

a)   Try it with paper.

b)   Show that $L^2 = 2x^3/(2x - 8.5)$.

c)   What value of x minimizes L^2?

d)   CALCULATOR Find the minimum value of L to the nearest tenth of an inch.

e)   GRAPHER Graph L as a function of x and compare what you see with your answer in (d).

Physical Applications

29. The height of a body moving vertically is given by

$$s = -\frac{1}{2}gt^2 + v_0t + s_0, \quad g > 0,$$

with s in meters and t in seconds. Find the body's maximum height.

30. CALCULATOR The 8-ft wall shown here stands 27 ft from the building. Find the length of the shortest straight beam that will reach to the side of the building from the ground outside the wall.

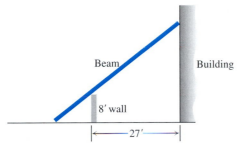

31. *The strength of a beam.* The strength S of a rectangular wooden beam is proportional to its width w times the square of its depth d.

a)   Find the dimensions of the strongest beam that can be cut from a 12-in.-diameter cylindrical log.

b)   GRAPHER Graph S as a function of the beam's width w, assuming the proportionality constant to be $k = 1$. Reconcile what you see with your answer in (a).

c)   GRAPHER On the same screen, or on a separate screen, graph S as a function of the beam's depth d, again taking $k = 1$. Compare the graphs with one another and with your answer in (a). What would be the effect of changing to some other value of k? Try it.

32. *The stiffness of a beam.* The stiffness S of a rectangular beam is proportional to its width times the cube of its depth.

a)   Find the dimensions of the stiffest beam that can be cut from a 12-in.-diameter log.

b)   GRAPHER Graph S as a function of the beam's width w, assuming the proportionality constant to be $k = 1$. Reconcile what you see with your answer in (a).

c)   GRAPHER On the screen, or on a separate screen, graph S as a function of the beam's depth d, again taking $k = 1$. Compare the graphs with one another and with your answer in (a). What would be the effect of changing to some other value of k? Try it.

33. Suppose that at any given time t (sec) the current i (amp) in an alternating current circuit is $i = 2\cos t + 2\sin t$. What is the peak current for this circuit (largest magnitude)?

34. A small frictionless cart, attached to the wall by a spring, is pulled 10 cm from its rest position and released at time $t = 0$ to roll back and forth for 4 sec. Its position at time t is $s = 10\cos \pi t$.

a)   What is the cart's maximum speed? When is the cart moving that fast? Where is it then? What is the magnitude of the acceleration then?

b)   Where is the cart when the magnitude of the acceleration is greatest? What is the cart's speed then?

35. Two masses hanging side by side from springs have positions $s_1 = 2\sin t$ and $s_2 = \sin 2t$, respectively.

a)   At what times in the interval $0 < t$ do the masses pass each other? (*Hint:* $\sin 2t = 2\sin t \cos t$.)

b)   When in the interval $0 \le t \le 2\pi$ is the vertical distance between the masses the greatest? What is this distance? (*Hint:* $\cos 2t = 2\cos^2 t - 1$.)

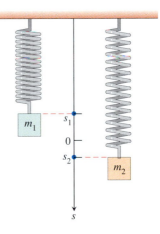

36. The positions of two particles on the s-axis are $s_1 = \sin t$ and $s_2 = \sin (t + \pi/3)$.

a)   At what time(s) in the interval $0 \le t \le 2\pi$ do the particles meet?

b)   What is the farthest apart the particles ever get?

c)   When in the interval $0 \le t \le 2\pi$ is the distance between the particles changing the fastest?

37. Suppose that at time $t \ge 0$ the position of a particle moving on the x-axis is $x = (t - 1)(t - 4)^4$.

a)   When is the particle at rest?

b)   During what time interval does the particle move to the left?

c) What is the fastest the particle goes while moving to the left?

d) **GRAPHER** Graph x as a function of t for $0 \leq t \leq 6$. Graph dx/dt over the same interval, in another color if possible. Compare the graphs with one another and with your answers in (a)–(c).

38. At noon, ship A was 12 nautical miles due north of ship B. Ship A was sailing south at 12 knots (nautical miles per hour—a nautical mile is 2000 yd) and continued to do so all day. Ship B was sailing east at 8 knots and continued to do so all day.

a) Start counting time with $t = 0$ at noon and express the distance s between the ships as a function of t.

b) How rapidly was the distance between the ships changing at noon? One hour later?

c) **CALCULATOR** The visibility that day was 5 nautical miles. Did the ships ever sight each other?

d) **GRAPHER** Graph s and ds/dt together as functions of t for $-1 \leq t \leq 3$, using different colors if possible. Compare the graphs and reconcile what you see with your answers in (b) and (c).

e) The graph of ds/dt looks as if it might have a horizontal asymptote in the first quadrant. This in turn suggests that ds/dt approaches a limiting value at $t \to \infty$. What is this value? What is its relation to the ships' individual speeds?

39. Fermat's principle in optics states that light always travels from one point to another along a path that minimizes the travel time. Figure 3.59 shows light from a source A reflected by a plane mirror to a receiver at point B. Show that for the light to obey Fermat's principle, the angle of incidence must equal the angle of reflection. (This result can also be derived without calculus. There is a purely geometric argument, which you may prefer.)

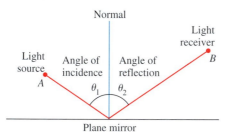

3.59 In studies of light reflection, the angles of incidence and reflection are measured from the line normal to the reflecting surface. Exercise 39 asks you to show that if light obeys Fermat's "least-time" principle, then $\theta_1 = \theta_2$.

40. *Tin pest.* Metallic tin, when kept below 13°C for a while, becomes brittle and crumbles to a gray powder. Tin objects eventually crumble to this gray powder spontaneously if kept in a cold climate for years. The Europeans who saw the tin organ pipes in their churches crumble away years ago called the change *tin pest* because it seemed to be contagious. And indeed it was, for the gray powder is a catalyst for its own formation.

A *catalyst* for a chemical reaction is a substance that controls the rate of the reaction without undergoing any permanent change

in itself. An *autocatalytic reaction* is one whose product is a catalyst for its own formation. Such a reaction may proceed slowly at first if the amount of catalyst present is small and slowly again at the end, when most of the original substance is used up. But in between, when both the substance and its catalyst product are abundant, the reaction proceeds at a faster pace.

In some cases it is reasonable to assume that the rate $v = dx/dt$ of the reaction is proportional both to the amount of the original substance present and to the amount of product. That is, v may be considered to be a function of x alone, and

$$v = kx(a - x) = kax - kx^2,$$

where

$x =$ the amount of product

$a =$ the amount of substance at the beginning

$k =$ a positive constant.

At what value of x does the rate v have a maximum? What is the maximum value of v?

Mathematical Applications

41. Is the function $f(x) = x^2 - x + 1$ ever negative? Explain.

42. You have been asked to determine whether the function $f(x) = 3 + 4 \cos x + \cos 2x$ is ever negative.

a) Explain why you need consider values of x only in the interval $[0, 2\pi]$.

b) Is f ever negative? Explain.

43. Find the points on the curve $y = \sqrt{x}$ nearest the point $(c, 0)$

a) if $c \geq 1/2$

b) if $c < 1/2$.

44. What value of a makes $f(x) = x^2 + (a/x)$ have (a) a local minimum of $x = 2$; (b) a point of inflection at $x = 1$?

45. What values of a and b make

$$f(x) = x^3 + ax^2 + bx$$

have (a) a local maximum at $x = -1$ and a local minimum at $x = 3$; (b) a local minimum at $x = 4$ and a point of inflection at $x = 1$?

46. Show that $f(x) = x^2 + (a/x)$ cannot have a local maximum for any value of a.

47. a) The function $y = \cot x - \sqrt{2} \csc x$ has an absolute maximum value on the interval $0 < x < \pi$. Find it.

b) **GRAPHER** Graph the function and compare what you see with your answer in (a).

48. a) The function $y = \tan x + 3 \cot x$ has an absolute minimum value on the interval $0 < x < \pi/2$. Find it.

b) **GRAPHER** Graph the function and compare what you see with your answer in (a).

49. How close does the curve $y = \sqrt{x}$ come to the point $(1/2, 16)$?

50. Let $f(x)$ and $g(x)$ be the differentiable functions graphed here. Point c is the point where the vertical distance between the curves is the greatest. Is there anything special about the tangents to the two curves at *c?* Give reasons for your answer.

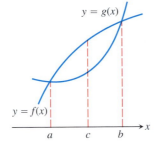

51. Show that if *a, b, c,* and *d* are positive integers, then

$$\frac{(a^2 + 1)(b^2 + 1)(c^2 + 1)(d^2 + 1)}{abcd} \geq 16.$$

52. *The derivative dt/dx in Example 5*

a) Show that

$$f(x) = \frac{x}{\sqrt{a^2 + x^2}}$$

is an increasing function of *x.*

b) Show that

$$g(x) = \frac{d - x}{\sqrt{b^2 + (d - x)^2}}$$

is a decreasing function of *x.*

c) Show that

$$\frac{dt}{dx} = \frac{x}{c_1 \sqrt{a^2 + x^2}} - \frac{d - x}{c_2 \sqrt{b^2 + (d - x)^2}}$$

is an increasing function of *x.*

Medicine

53. *Sensitivity to medicine (Continuation of Exercise 50, Section 2.2).* Find the amount of medicine to which the body is most sensitive by finding the value of *M* that maximizes the derivative *dR/dM,* where

$$R = M^2 \left(\frac{C}{2} - \frac{M}{3}\right)$$

and *C* is a constant.

54. *How we cough*

a) When we cough, the trachea (windpipe) contracts to increase the velocity of the air going out. This raises the questions of how much it should contract to maximize the velocity and whether it really contracts that much when we cough.

   Under reasonable assumptions about the elasticity of the tracheal wall and about how the air near the wall is slowed by friction, the average flow velocity *v* can be modeled by the equation

$$v = c(r_0 - r)r^2 \text{ cm/sec}, \quad \frac{r_0}{2} \leq r \leq r_0,$$

where r_0 is the rest radius of the trachea in centimeters and *c* is a positive constant whose value depends in part on the length of the trachea.

   Show that *v* is greatest when $r = (2/3)r_0$, that is, when the trachea is about 33% contracted. The remarkable fact is that x-ray photographs confirm that the trachea contracts about this much during a cough.

b) **GRAPHER** Take r_0 to be 0.5 and *c* to be 1, and graph *v* over the interval $0 \leq r \leq 0.5$. Compare what you see to the claim that *v* is at a maximum when $r = (2/3)r_0$.

Business and Economics

55. It costs you *c* dollars each to manufacture and distribute backpacks. If the backpacks sell at *x* dollars each, the number sold is given by $n = a/(x - c) + b(100 - x)$, where *a* and *b* are certain positive constants. What selling price will bring a maximum profit?

56. You operate a tour service that offers the following rates:

a) $200 per person if 50 people (the minimum number to book the tour) go on the tour.

b) For each additional person, up to a maximum of 80 people total, everyone's charge is reduced by $2.

   It costs $6000 (a fixed cost) plus $32 per person to conduct the tour. How many people does it take to maximize your profit?

57. *The best quantity to order.* One of the formulas for inventory management says that the average weekly cost of ordering, paying for, and holding merchandise is

$$A(q) = \frac{km}{q} + cm + \frac{hq}{2},$$

where *q* is the quantity you order when things run low (shoes, radios, brooms, or whatever the item might be), *k* is the cost of placing an order (the same, no matter how often you order), *c* is the cost of one item (a constant), *m* is the number of items sold each week (a constant), and *h* is the weekly holding cost per item (a constant that takes into account things such as space, utilities, insurance, and security). Your job, as the inventory manager for your store, is to find the quantity that will minimize *A(q)*. What is it? (The formula you get for the answer is called the *Wilson lot size formula.*)

58. (*Continuation of Exercise 57.*) Shipping costs sometimes depend on order size. When they do, it is more realistic to replace *k* by $k + bq$, the sum of *k* and a constant multiple of *q*. What is the most economical quantity to order now?

59. Show that if $r(x) = 6x$ and $c(x) = x^3 - 6x^2 + 15x$ are your revenue and cost functions, then the best you can do is break even (have revenue equal cost).

60. Suppose $c(x) = x^3 - 20x^2 + 20{,}000\,x$ is the cost of manufacturing *x* items. Find a production level that will minimize the average cost of making *x* items.

3.7

Linearization and Differentials

Sometimes we can approximate complicated functions with simpler ones that give the accuracy we want for specific applications and are easier to work with. The approximating functions discussed in this section are called *linearizations*. They are based on tangent lines.

We introduce new variables dx and dy and define them in a way that gives new meaning to the Leibniz notation dy/dx. We will use dy to estimate error in measurement and sensitivity to change.

Linear Approximations

As you can see in Fig. 3.60, the tangent to a curve $y = f(x)$ lies close to the curve near the point of tangency. For a brief interval to either side, the y-values along the tangent line give a good approximation to the y-values on the curve.

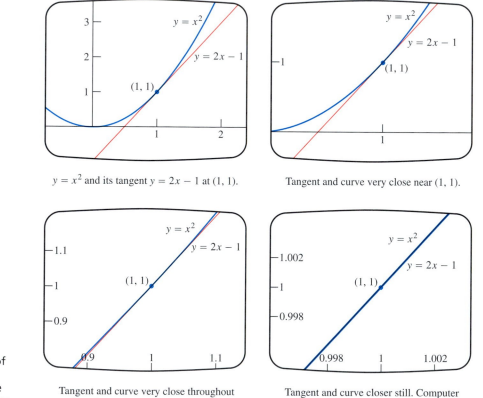

$y = x^2$ and its tangent $y = 2x - 1$ at $(1, 1)$.

Tangent and curve very close near $(1, 1)$.

Tangent and curve very close throughout entire x-interval shown.

Tangent and curve closer still. Computer screen cannot distinguish tangent from curve on this x-interval.

3.60 The more we magnify the graph of a function near a point where the function is differentiable, the flatter the graph becomes and the more it resembles its tangent.

In the notation of Fig. 3.61, the tangent passes through the point $(a, f(a))$, so its point–slope equation is

$$y = f(a) + f'(a)(x - a).$$

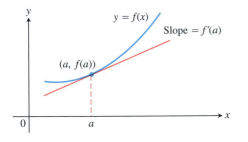

3.61 The equation of the tangent line is $y = f(a) + f'(a)(x - a)$.

Thus, the tangent is the graph of the function

$$L(x) = f(a) + f'(a)(x - a).$$

For as long as the line remains close to the graph of f, $L(x)$ gives a good approximation to $f(x)$.

> **Definitions**
>
> If f is differentiable at $x = a$, then the approximating function
>
> $$L(x) = f(a) + f'(a)(x - a) \qquad (1)$$
>
> is the **linearization** of f at a. The approximation
>
> $$f(x) \approx L(x)$$
>
> of f by L is the **standard linear approximation** of f at a. The point $x = a$ is the **center** of the approximation.

EXAMPLE 1   Find the linearization of $f(x) = \sqrt{1 + x}$ at $x = 0$.

Solution   We evaluate Eq. (1) for f at $a = 0$. With

$$f'(x) = \frac{1}{2}(1 + x)^{-1/2},$$

we have $f(0) = 1$, $f'(0) = 1/2$, and

$$L(x) = f(a) + f'(a)(x - a) = 1 + \frac{1}{2}(x - 0) = 1 + \frac{x}{2}.$$

See Fig. 3.62.

3.62 The graph of $y = \sqrt{1+x}$ and its linearizations at $x = 0$ and $x = 3$. Figure 3.63 shows a magnified view of the small window about 1 on the y-axis.

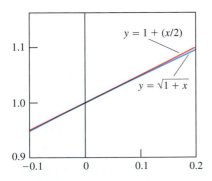

3.63 Magnified view of the window in Fig. 3.62.

The approximation $\sqrt{1 + x} \approx 1 + (x/2)$ (Fig. 3.63) gives

$$\sqrt{1.2} \approx 1 + \frac{0.2}{2} = 1.10, \qquad \text{Accurate to 2 decimals}$$

$$\sqrt{1.05} \approx 1 + \frac{0.05}{2} = 1.025, \qquad \text{Accurate to 3 decimals}$$

$$\sqrt{1.005} \approx 1 + \frac{0.005}{2} = 1.00250. \qquad \text{Accurate to 5 decimals}$$

Do not be misled by these calculations into thinking that whatever we do with a linearization is better done with a calculator. In practice, we would never use a linearization to find a particular square root. The utility of a linearization is its ability to replace a complicated formula by a simpler one over an entire interval of values. If we have to work with $\sqrt{1+x}$ for x close to 0 and can tolerate the small amount of error involved, we can work with $1 + (x/2)$ instead. Of course, we then need to know how much error there is. We will touch on this toward the end of the section but will not have the full story until Chapter 8.

A linear approximation normally loses accuracy away from its center. As Fig. 3.62 suggests, the approximation $\sqrt{1+x} \approx 1 + (x/2)$ will probably be too crude to be useful near $x = 3$. There, we need the linearization at $x = 3$.

EXAMPLE 2   Find the linearization of $f(x) = \sqrt{1+x}$ at $x = 3$.

Solution   We evaluate Eq. (1) for f at $a = 3$. With

$$f(3) = 2, \qquad f'(3) = \frac{1}{2}(1+x)^{-1/2}\Big|_{x=3} = \frac{1}{4},$$

we have

$$L(x) = 2 + \frac{1}{4}(x - 3) = \frac{5}{4} + \frac{x}{4}.$$ ❑

At $x = 3.2$, the linearization in Example 2 gives

$$\sqrt{1+x} = \sqrt{1 + 3.2} \approx \frac{5}{4} + \frac{3.2}{4} = 1.250 + 0.800 = 2.050,$$

which differs from the true value $\sqrt{4.2} \approx 2.04939$ by less than one one-thousandth. The linearization in Example 1 gives

$$\sqrt{1+x} = \sqrt{1 + 3.2} \approx 1 + \frac{3.2}{2} = 1 + 1.6 = 2.6,$$

a result that is off by more than 25%.

EXAMPLE 3   The most important linear approximation for roots and powers is

$$(1+x)^k \approx 1 + kx \qquad (x \approx 0; \text{ any number } k) \qquad (2)$$

(Exercise 20). This approximation, good for values of x sufficiently close to zero, has broad application.

Common linear approximations, $x \approx 0$	Approximation ($x \approx 0$)	Source: Eq. (2) with ...
$\sin x \approx x$	$\sqrt{1+x} \approx 1 + \dfrac{x}{2}$	$k = 1/2$
$\cos x \approx 1$		
$\tan x \approx x$	$\dfrac{1}{1-x} = (1-x)^{-1} \approx 1 + (-1)(-x) = 1 + x$	$k = -1; -x$ in place of x
$(1+x)^k \approx 1 + kx$		
(See the Exercises.)	$\sqrt[3]{1 + 5x^4} = (1 + 5x^4)^{1/3} \approx 1 + \dfrac{1}{3}(5x^4) = 1 + \dfrac{5}{3}x^4$	$k = 1/3; 5x^4$ in place of x
	$\dfrac{1}{\sqrt{1-x^2}} = (1-x^2)^{-1/2} \approx 1 + \left(-\dfrac{1}{2}\right)\left(-x^2\right) = 1 + \dfrac{x^2}{2}$	$k = -1/2; -x^2$ in place of x

❑

3.64 The graph of $f(x) = \cos x$ and its linearization at $x = \pi/2$. Near $x = \pi/2$, $\cos x \approx -x + (\pi/2)$.

EXAMPLE 4   Find the linearization of $f(x) = \cos x$ at $x = \pi/2$ (Fig. 3.64).

Solution   With

$$f(\pi/2) = \cos(\pi/2) = 0 \quad \text{and} \quad f'(\pi/2) = -\sin(\pi/2) = -1,$$

we have

$$L(x) = f(a) + f'(a)(x - a)$$

$$= 0 + (-1)\left(x - \frac{\pi}{2}\right)$$

$$= -x + \frac{\pi}{2}.$$

Differentials

The meaning of *dx* and *dy*

In most contexts, the differential dx of the independent variable is its change Δx, but we do not impose this restriction on the definition.

Unlike the independent variable dx, the variable dy is always a dependent variable. It depends on both x and dx.

Definitions

Let $y = f(x)$ be a differentiable function. The **differential** dx is an independent variable. The **differential** dy is

$$dy = f'(x)\, dx.$$

EXAMPLE 5   Find dy if

a) $y = x^5 + 37x$ **b)** $y = \sin 3x$.

Solution

a) $dy = (5x^4 + 37)\, dx$ **b)** $dy = (3\cos 3x)\, dx$

If $dx \neq 0$ and we divide both sides of the equation $dy = f'(x)\, dx$ by dx, we obtain the familiar equation

$$\frac{dy}{dx} = f'(x).$$

This equation says that when $dx \neq 0$, we can regard the derivative dy/dx as a quotient of differentials.

We sometimes write

$$df = f'(x)\, dx$$

in place of $dy = f'(x)\, dx$, and call df the **differential** of f. For instance, if $f(x) = 3x^2 - 6$, then

$$df = d(3x^2 - 6) = 6x\, dx.$$

Every differentiation formula like

$$\frac{d(u+v)}{dx} = \frac{du}{dx} + \frac{dv}{dx}$$

has a corresponding differential form like

$$d(u+v) = du + dv,$$

obtained by multiplying both sides by dx (Table 3.1).

Table 3.1 Formulas for differentials

$$dc = 0$$
$$d(cu) = c\, du$$
$$d(u + v) = du + dv$$
$$d(uv) = u\, dv + v\, du$$
$$d\left(\frac{u}{v}\right) = \frac{v\, du - u\, dv}{v^2}$$
$$d(u^n) = nu^{n-1}\, du$$
$$d(\sin u) = \cos u\, du$$
$$d(\cos u) = -\sin u\, du$$
$$d(\tan u) = \sec^2 u\, du$$
$$d(\cot u) = -\csc^2 u\, du$$
$$d(\sec u) = \sec u \tan u\, du$$
$$d(\csc u) = -\csc u \cot u\, du$$

EXAMPLE 6

a) $d(\tan 2x) = \sec^2(2x)\, d(2x) = 2\sec^2 2x\, dx$

b) $d\left(\dfrac{x}{x+1}\right) = \dfrac{(x+1)\, dx - x\, d\,(x+1)}{(x+1)^2} = \dfrac{x\, dx + dx - x\, dx}{(x+1)^2} = \dfrac{dx}{(x+1)^2}$ $\square$

Estimating Change with Differentials

Suppose we know the value of a differentiable function $f(x)$ at a point x_0 and we want to predict how much this value will change if we move to a nearby point $x_0 + dx$. If dx is small, f and its linearization L at x_0 will change by nearly the same amount. Since the values of L are simple to calculate, calculating the change in L offers a practical way to estimate the change in f.

In the notation of Fig. 3.65, the change in f is

$$\Delta f = f(x_0 + dx) - f(x_0).$$

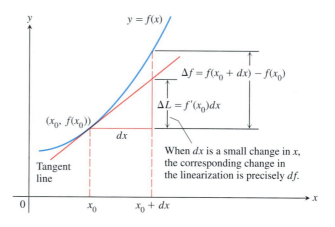

3.65 If dx is small, the change in the linearization of f is nearly the same as the change in f.

The corresponding change in L is

$$\Delta L = L(x_0 + dx) - L(x_0)$$

$$= \underbrace{f(x_0) + f'(x_0)\Big[(x_0 + dx) - x_0\Big]}_{L(x_0+dx)} - \underbrace{f(x_0)}_{L(x_0)=f(x_0)}$$

$$= f'(x_0)\, dx.$$

Thus, the differential $df = f'(x)\, dx$ has a geometric interpretation: When df is evaluated at $x = x_0$, $df = \Delta L$, the change in the linearization of f corresponding to the change dx.

The Differential Estimate of Change

Let $f(x)$ be differentiable at $x = x_0$. The approximate change in the value of f when x changes from x_0 to $x_0 + dx$ is

$$df = f'(x_0)\, dx.$$

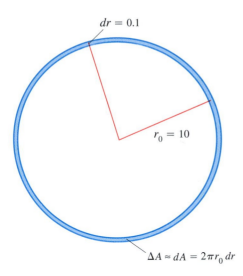

$dr = 0.1$

$r_0 = 10$

$\Delta A \approx dA = 2\pi r_0 \, dr$

3.66 When dr is small compared with r_0, as it is when $dr = 0.1$ and $r_0 = 10$, the differential $dA = 2\pi r_0 dr$ gives a good estimate of ΔA (Example 7).

EXAMPLE 7 The radius r of a circle increases from $r_0 = 10$ m to 10.1 m (Fig. 3.66). Estimate the increase in the circle's area A by calculating dA. Compare this with the true change ΔA.

Solution Since $A = \pi r^2$, the estimated increase is

$$dA = A'(r_0)\, dr = 2\pi r_0\, dr = 2\pi(10)(0.1) = 2\pi \text{ m}^2.$$

The true change is

$$\Delta A = \pi(10.1)^2 - \pi(10)^2 = (102.01 - 100)\pi = \underbrace{2\pi}_{dA} + \underbrace{0.01\pi}_{\text{error}}.$$

Absolute, Relative, and Percentage Change

As we move from x_0 to a nearby point $x_0 + dx$, we can describe the change in f in three ways:

	True	**Estimated**
Absolute change	$\Delta f = f(x_0 + dx) - f(x_0)$	$df = f'(x_0)\, dx$
Relative change	$\dfrac{\Delta f}{f(x_0)}$	$\dfrac{df}{f(x_0)}$
Percentage change	$\dfrac{\Delta f}{f(x_0)} \times 100$	$\dfrac{df}{f(x_0)} \times 100$

EXAMPLE 8 The estimated percentage change in the area of the circle in Exercise 7 is

$$\frac{dA}{A(r_0)} \times 100 = \frac{2\pi}{100\pi} \times 100 = 2\%.$$

EXAMPLE 9 *The earth's surface area*

Suppose the earth were a perfect sphere and we determined its radius to be 3959 ± 0.1 miles. What effect would the tolerance of ± 0.1 have on our estimate of the earth's surface area?

Solution The surface area of a sphere of radius r is $S = 4\pi r^2$. The uncertainty in the calculation of S that arises from measuring r with a tolerance of dr miles is about

$$dS = \left(\frac{dS}{dr}\right) dr = 8\pi r\, dr.$$

With $r = 3959$ and $dr = 0.1$, our estimate of S could be off by as much as

$$dS = 8\pi(3959)(0.1) \approx 9950 \text{ mi}^2,$$

to the nearest square mile, which is about the area of the state of Maryland.

If we underestimated the radius of the earth by 528 ft during a calculation of the earth's surface area, we would leave out an area the size of the state of Maryland.

EXAMPLE 10 About how accurately should we measure the radius r of a sphere to calculate the surface area $S = 4\pi r^2$ within 1% of its true value?

Solution We want any inaccuracy in our measurement to be small enough to make the corresponding increment ΔS in the surface area satisfy the inequality

$$|\Delta S| \le \frac{1}{100} S = \frac{4\pi r^2}{100}.$$

We replace ΔS in this inequality with

$$dS = \left(\frac{dS}{dr}\right) dr = 8\pi r \, dr.$$

This gives

$$|8\pi r \, dr| \le \frac{4\pi r^2}{100}, \qquad \text{or} \qquad |dr| \le \frac{1}{8\pi r} \cdot \frac{4\pi r^2}{100} = \frac{1}{2}\frac{r}{100}.$$

We should measure r with an error dr that is no more than 0.5% of the true value. ❑

Angiography: An opaque dye is injected into a partially blocked artery to make the inside visible under x-rays. This reveals the location and severity of the blockage.

Angioplasty: A balloon-tipped catheter is inflated inside the artery to widen it at the blockage site.

EXAMPLE 11 *Unclogging arteries*

In the late 1830s, the French physiologist Jean Poiseuille ("pwa-*zoy*") discovered the formula we use today to predict how much the radius of a partially clogged artery has to be expanded to restore normal flow. His formula,

$$V = kr^4,$$

says that the volume V of fluid flowing through a small pipe or tube in a unit of time at a fixed pressure is a constant times the fourth power of the tube's radius r. How will a 10% increase in r affect V?

Solution The differentials of r and V are related by the equation

$$dV = \frac{dV}{dr} \, dr = 4kr^3 dr.$$

Hence,

$$\frac{dV}{V} = \frac{4kr^3 dr}{kr^4} = 4\frac{dr}{r}. \qquad \text{\color{blue}Dividing by } V = kr^4$$

The relative change in V is 4 times the relative change in r, so a 10% increase in r will produce a 40% increase in the flow. ❑

Sensitivity

The equation $df = f'(x) \, dx$ tells how sensitive the output of f is to a change in input at different values of x. The larger the value of f' at x, the greater is the effect of a given change dx.

EXAMPLE 12 You want to calculate the height of a bridge from the equation $s = 16t^2$ by timing how long it takes a heavy stone you drop to splash into the water below. How sensitive will your calculation be to a 0.1-sec error in measuring the time?

Solution The size of ds in the equation

$$ds = 32t \, dt$$

depends on how big t is. If $t = 2$ sec, the error caused by $dt = 0.1$ is only

$$ds = 32(2)(0.1) = 6.4 \text{ ft.}$$

Three seconds later, at $t = 5$ sec, the error caused by the same dt is

$$ds = 32(5)(0.1) = 16 \text{ ft.} \qquad \square$$

The Error in the Approximation $\Delta f \approx df$

Let $f(x)$ be differentiable at $x = x_0$ and suppose that Δx is an increment of x. We have two ways to describe the change in f as x changes from x_0 to $x_0 + \Delta x$:

The true change: $\Delta f = f(x_0 + \Delta x) - f(x_0)$

The differential estimate: $df = f'(x_0)\Delta x.$

How well does df approximate Δf?

We measure the approximation error by subtracting df from Δf:

$$\text{Approximation error} = \Delta f - df$$
$$= \Delta f - f'(x_0)\Delta x$$
$$= \underbrace{f(x_0 + \Delta x) - f(x_0)}_{\Delta f} - f'(x_0)\Delta x$$
$$= \underbrace{\left(\frac{f(x_0 + \Delta x) - f(x_0)}{\Delta x} - f'(x_0) \right)}_{\text{Call this part } \epsilon} \Delta x$$

$$= \epsilon \cdot \Delta x.$$

As $\Delta x \to 0$, the difference quotient

$$\frac{f(x_0 + \Delta x) - f(x_0)}{\Delta x}$$

approaches $f'(x_0)$ (remember the definition of $f'(x_0)$), so the quantity in parentheses becomes a very small number (which is why we called it ϵ). In fact, $\epsilon \to 0$ as $\Delta x \to 0$. When Δx is small, the approximation error $\epsilon \Delta x$ is smaller still.

$$\underbrace{\Delta f}_{\substack{\text{true} \\ \text{change}}} = \underbrace{f'(x_0)\Delta x}_{\substack{\text{estimated} \\ \text{change}}} + \underbrace{\epsilon \Delta x}_{\text{error}}$$

While we do not know exactly how small the error is and will not be able to make much progress on this front until Chapter 8, there is something worth noting here, namely the *form* taken by the equation.

If $y = f(x)$ is differentiable at $x = x_0$, and x changes from x_0 to $x_0 + \Delta x$, the change Δy in f is given by an equation of the form

$$\Delta y = f'(x_0)\Delta x + \epsilon \Delta x \qquad (3)$$

in which $\epsilon \to 0$ as $\Delta x \to 0$.

Surprising as it may seem, just knowing the form of Eq. (3) enables us to bring the proof of the Chain Rule to a successful conclusion.

Proof of the Chain Rule

You may recall our saying in Section 2.5 that the proof we wanted to give for the Chain Rule depended on ideas in Section 3.7, the present section. We were referring to Eq. (3), and here is the proof:

Our goal is to show that if $f(u)$ is a differentiable function of u and $u = g(x)$ is a differentiable function of x, then the composite $y = f(g(x))$ is a differentiable function of x. More precisely, if g is differentiable at x_0 and f is differentiable at $g(x_0)$, then the composite is differentiable at x_0 and

$$\left.\frac{dy}{dx}\right|_{x=x_0} = f'(g(x_0)) \cdot g'(x_0).$$

Let Δx be an increment in x and let Δu and Δy be the corresponding increments in u and y. As you can see in Fig. 3.67,

$$\left.\frac{dy}{dx}\right|_{x=x_0} = \lim_{\Delta x \to 0} \frac{\Delta y}{\Delta x},$$

so our goal is to show that this limit is $f'(g(x_0)) \cdot g'(x_0)$.

By Eq. (3),

$$\Delta u = g'(x_0)\Delta x + \epsilon_1 \Delta x = (g'(x_0) + \epsilon_1)\Delta x,$$

where $\epsilon_1 \to 0$ as $\Delta x \to 0$. Similarly,

$$\Delta y = f'(u_0)\Delta u + \epsilon_2 \Delta u = (f'(u_0) + \epsilon_2)\Delta u,$$

where $\epsilon_2 \to 0$ as $\Delta u \to 0$. Notice also that $\Delta u \to 0$ as $\Delta x \to 0$. Combining the equations for Δu and Δy gives

$$\Delta y = (f'(u_0) + \epsilon_2)(g'(x_0) + \epsilon_1)\Delta x,$$

so

$$\frac{\Delta y}{\Delta x} = f'(u_0)g'(x_0) + \epsilon_2 \, g'(x_0) + f'(u_0)\epsilon_1 + \epsilon_2\epsilon_1.$$

Since ϵ_1 and ϵ_2 go to zero as Δx goes to zero, three of the four terms on the right vanish in the limit, leaving

$$\lim_{\Delta x \to 0} \frac{\Delta y}{\Delta x} = f'(u_0)g'(x_0) = f'(g(x_0)) \cdot g'(x_0).$$

This concludes the proof.   ❏

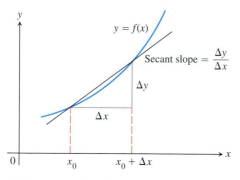

3.67 The graph of y as a function of x. The derivative of y with respect to x at $x = x_0$ is $\lim\limits_{\Delta x \to 0} \Delta y / \Delta x$.

✳ The Conversion of Mass to Energy

Here is an example of how the approximation

$$\frac{1}{\sqrt{1 - x^2}} \approx 1 + \frac{1}{2}x^2 \tag{4}$$

from Example 3 is used in an applied problem.

Newton's second law,

$$F = \frac{d}{dt}(mv) = m\frac{dv}{dt} = ma,$$

is stated with the assumption that mass is constant, but we know this is not strictly true because the mass of a body increases with velocity. In Einstein's corrected formula, mass has the value

$$m = \frac{m_0}{\sqrt{1 - v^2/c^2}},$$ (5)

where the "rest mass" m_0 represents the mass of a body that is not moving and c is the speed of light, which is about 300,000 km/sec. When v is very small compared with c, v^2/c^2 is close to zero and it is safe to use the approximation

$$\frac{1}{\sqrt{1 - v^2/c^2}} \approx 1 + \frac{1}{2}\left(\frac{v^2}{c^2}\right)$$

(Eq. 4 with $x = v/c$) to write

$$m = \frac{m_0}{\sqrt{1 - v^2/c^2}} \approx m_0\left[1 + \frac{1}{2}\left(\frac{v^2}{c^2}\right)\right] = m_0 + \frac{1}{2}m_0v^2\left(\frac{1}{c^2}\right),$$

or

$$m \approx m_0 + \frac{1}{2}m_0v^2\left(\frac{1}{c^2}\right).$$ (6)

Equation (6) expresses the increase in mass that results from the added velocity v.

In Newtonian physics, $(1/2)m_0v^2$ is the kinetic energy (KE) of the body, and if we rewrite Eq. (6) in the form

$$(m - m_0)c^2 \approx \frac{1}{2}m_0v^2,$$

we see that

$$(m - m_0)c^2 \approx \frac{1}{2}m_0v^2 = \frac{1}{2}m_0v^2 - \frac{1}{2}m_0(0)^2 = \Delta(\text{KE}),$$

or

$$(\Delta m)c^2 \approx \Delta(\text{KE}).$$ (7)

In other words, the change in kinetic energy $\Delta(\text{KE})$ in going from velocity 0 to velocity v is approximately equal to $(\Delta m)c^2$.

With c equal to 3×10^8 m/sec, Eq. (7) becomes

$$\Delta(\text{KE}) \approx 90,000,000,000,000,000 \, \Delta m \text{ joules} \qquad \text{mass in kilograms}$$

and we see that a small change in mass can create a large change in energy. The energy released by exploding a 20-kiloton atomic bomb, for instance, is the result of converting only 1 gram of mass to energy. The products of the explosion weigh only 1 gram less than the material exploded. A U.S. penny weighs about 3 grams.

Exercises 3.7

Finding Linearizations

In Exercises 1–6, find the linearization $L(x)$ of $f(x)$ at $x = a$.

1. $f(x) = x^4$ at $x = 1$

2. $f(x) = x^{-1}$ at $x = 2$

3. $f(x) = x^3 - x$ at $x = 1$

4. $f(x) = x^3 - 2x + 3$ at $x = 2$

5. $f(x) = \sqrt{x}$ at $x = 4$

6. $f(x) = \sqrt{x^2 + 9}$ at $x = -4$

You want linearizations that will replace the functions in Exercises 7–12 over intervals that include the given points x_0. To make your subsequent work as simple as possible, you want to center each linearization not at x_0 but at a nearby integer $x = a$ at which the given function and its derivative are easy to evaluate. What linearization do you use in each case?

7. $f(x) = x^2 + 2x$, $x_0 = 0.1$

8. $f(x) = x^{-1}$, $x_0 = 0.6$

9. $f(x) = 2x^2 + 4x - 3$, $x_0 = -0.9$

10. $f(x) = 1 + x$, $x_0 = 8.1$

11. $f(x) = \sqrt[3]{x}$, $x_0 = 8.5$

12. $f(x) = \dfrac{x}{x+1}$, $x_0 = 1.3$

Linearizing Trigonometric Functions

In Exercises 13–16, find the linearization of f at $x = a$. Then graph the linearization and f together.

13. $f(x) = \sin x$ at (a) $x = 0$, (b) $x = \pi$

14. $f(x) = \cos x$ at (a) $x = 0$, (b) $x = -\pi/2$

15. $f(x) = \sec x$ at (a) $x = 0$, (b) $x = -\pi/3$

16. $f(x) = \tan x$ at (a) $x = 0$, (b) $x = \pi/4$

The Approximation $(1 + x)^k \approx 1 + kx$

17. Use the formula $(1 + x)^k \approx 1 + kx$ to find linear approximations of the following functions for values of x near zero.

a) $f(x) = (1 + x)^2$

b) $f(x) = \dfrac{1}{(1 + x)^5}$

c) $g(x) = \dfrac{2}{1 - x}$

d) $g(x) = (1 - x)^6$

e) $h(x) = 3(1 + x)^{1/3}$

f) $h(x) = \dfrac{1}{\sqrt{1 + x}}$

18. *Faster than a calculator.* Use the approximation $(1 + x)^k \approx 1 + kx$ to estimate

a) $(1.0002)^{50}$

b) $\sqrt[3]{1.009}$.

19. Find the linearization of $f(x) = \sqrt{x + 1} + \sin x$ at $x = 0$. How is it related to the individual linearizations for $\sqrt{x + 1}$ and $\sin x$?

20. We know from the Power Rule that the equation

$$\frac{d}{dx}(1 + x)^k = k(1 + x)^{k-1}$$

holds for every rational number k. In Chapter 6, we will show

that it holds for every irrational number as well. Assuming this result for now, show that the linearization of $f(x) = (1 + x)^k$ at $x = 0$ is $L(x) = 1 + kx$ for any number k.

Derivatives in Differential Form

In Exercises 21–32, find dy.

21. $y = x^3 - 3\sqrt{x}$

22. $y = x\sqrt{1 - x^2}$

23. $y = \dfrac{2x}{1 + x^2}$

24. $y = \dfrac{2\sqrt{x}}{3(1 + \sqrt{x})}$

25. $2y^{3/2} + xy - x = 0$

26. $xy^2 - 4x^{3/2} - y = 0$

27. $y = \sin(5\sqrt{x})$

28. $y = \cos(x^2)$

29. $y = 4\tan(x^3/3)$

30. $y = \sec(x^2 - 1)$

31. $y = 3\csc(1 - 2\sqrt{x})$

32. $y = 2\cot\left(\dfrac{1}{\sqrt{x}}\right)$

Approximation Error

In Exercises 33–38, each function $f(x)$ changes value when x changes from x_0 to $x_0 + dx$. Find

a) the change $\Delta f = f(x_0 + dx) - f(x_0)$;

b) the value of the estimate $df = f'(x_0)\,dx$; and

c) the approximation error $|\Delta f - df|$.

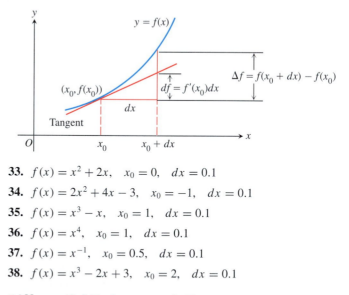

33. $f(x) = x^2 + 2x$, $x_0 = 0$, $dx = 0.1$

34. $f(x) = 2x^2 + 4x - 3$, $x_0 = -1$, $dx = 0.1$

35. $f(x) = x^3 - x$, $x_0 = 1$, $dx = 0.1$

36. $f(x) = x^4$, $x_0 = 1$, $dx = 0.1$

37. $f(x) = x^{-1}$, $x_0 = 0.5$, $dx = 0.1$

38. $f(x) = x^3 - 2x + 3$, $x_0 = 2$, $dx = 0.1$

Differential Estimates of Change

In Exercises 39–44, write a differential formula that estimates the given change in volume or surface area.

39. The change in the volume $V = (4/3)\pi r^3$ of a sphere when the radius changes from r_0 to $r_0 + dr$

40. The change in the volume $V = x^3$ of a cube when the edge lengths change from x_0 to $x_0 + dx$

41. The change in the surface area $S = 6x^2$ of a cube when the edge lengths change from x_0 to $x_0 + dx$

42. The change in the lateral surface area $S = \pi r \sqrt{r^2 + h^2}$ of a right circular cone when the radius changes from r_0 to $r_0 + dr$ and the height does not change

43. The change in the volume $V = \pi r^2 h$ of a right circular cylinder when the radius changes from r_0 to $r_0 + dr$ and the height does not change

44. The change in the lateral surface area $S = 2\pi r h$ of a right circular cylinder when the height changes from h_0 to $h_0 + dh$ and the radius does not change

Applications

45. The radius of a circle is increased from 2.00 to 2.02 m.

   a) Estimate the resulting change in area.
   b) Express the estimate in (a) as a percentage of the circle's original area.

46. The diameter of a tree was 10 in. During the following year, the circumference grew 2 in. About how much did the tree's diameter grow? the tree's cross-section area?

47. The edge of a cube is measured as 10 cm with an error of 1%. The cube's volume is to be calculated from this measurement. Estimate the percentage error in the volume calculation.

48. About how accurately should you measure the side of a square to be sure of calculating the area within 2% of its true value?

49. The diameter of a sphere is measured as 100 ± 1 cm and the volume is calculated from this measurement. Estimate the percentage error in the volume calculation.

50. Estimate the allowable percentage error in measuring the diameter D of a sphere if the volume is to be calculated correctly to within 3%.

51. The height and radius of a right circular cylinder are equal, so the cylinder's volume is $V = \pi h^3$. The volume is to be calculated from a measurement of h and must be calculated with an error of no more than 1% of the true value. Find approximately the greatest error that can be tolerated in the measurement of h, expressed as a percentage of h.

52. a) About how accurately must the interior diameter of a 10-m-high cylindrical storage tank be measured to calculate the tank's volume to within 1% of its true value?
   b) About how accurately must the tank's exterior diameter be measured to calculate the amount of paint it will take to paint the side of the tank within 5% of the true amount?

53. A manufacturer contracts to mint coins for the federal government. How much variation dr in the radius of the coins can be tolerated if the coins are to weigh within 1/1000 of their ideal weight? Assume that the thickness does not vary.

54. (*Continuation of Example 11.*) By what percentage should r be increased to increase V by 50%?

55. (*Continuation of Example 12.*) Show that a 5% error in measuring t will cause about a 10% error in calculating s from the equation $s = 16t^2$.

56. *The effect of flight maneuvers on the heart.* The amount of work done in a unit of time by the heart's main pumping chamber, the left ventricle, is given by the equation

$$W = PV + \frac{V\delta v^2}{2g},$$

where W is the work, P is the average blood pressure, V is the volume of blood pumped out during the unit of time, δ is the density of the blood, v is the average velocity of the exiting blood, and g is the acceleration of gravity.

When P, V, δ, and v remain constant, W becomes a function of g and the equation takes the simplified form

$$W = a + \frac{b}{g} \qquad (a, b \text{ constant}). \tag{8}$$

As a member of NASA's medical team, you want to know how sensitive W is to apparent changes in g caused by flight maneuvers, and this depends on the initial value of g. As part of your investigation, you decide to compare the effect on W of a given change dg on the moon, where $g = 5.2$ ft/sec^2, with the effect the same change dg would have on Earth, where $g = 32$ ft/sec^2. You use Eq. (8) to find the ratio of dW_{moon} to dW_{Earth}. What do you conclude?

57. *Sketching the change in a cube's volume.* The volume $V = x^3$ of a cube with edges of length x increases by an amount ΔV when x increases by an amount Δx. Show with a sketch how to represent ΔV geometrically as the sum of the volumes of

   a) three slabs of dimensions x by x by Δx;
   b) three bars of dimensions x by Δx by Δx;
   c) one cube of dimensions Δx by Δx by Δx.

The differential formula $dV = 3x^2 dx$ estimates the change in V with the three slabs.

58. *Measuring the acceleration of gravity.* When the length L of a clock pendulum is held constant by controlling its temperature, the pendulum's period T depends on the acceleration of gravity g. The period will therefore vary slightly as the clock is moved from place to place on the earth's surface, depending on the change in g. By keeping track of ΔT, we can estimate the variation in g from the equation $T = 2\pi(L/g)^{1/2}$ that relates T, g, and L.

   a) With L held constant and g as the independent variable, calculate dT and use it to answer (b) and (c).
   b) If g increases, will T increase, or decrease? Will a pendulum clock speed up, or slow down? Explain.
   c) A clock with a 100-cm pendulum is moved from a location where $g = 980$ cm/sec^2 to a new location. This increases the period by $dT = 0.001$ sec. Find dg and estimate the value of g at the new location.

Theory and Examples

59. Show that the approximation of $\sqrt{1 + x}$ by its linearization at the origin must improve as $x \to 0$ by showing that

$$\lim_{x \to 0} \frac{\sqrt{1 + x}}{1 + (x/2)} = 1.$$

60. Show that the approximation of $\tan x$ by its linearization at the origin must improve as $x \to 0$ by showing that

$$\lim_{x \to 0} \frac{\tan x}{x} = 1.$$

61. Suppose that the graph of a differentiable function $f(x)$ has a horizontal tangent at $x = a$. Can anything be said about the linearization of f at $x = a$? Give reasons for your answer.

62. *Reading derivatives from graphs.* The idea that differentiable curves flatten out when magnified can be used to estimate the values of the derivatives of functions at particular points. We magnify the curve until the portion we see looks like a straight line through the point in question, and then we use the screen's coordinate grid to read the slope of the curve as the slope of the line it resembles.

a) To see how the process works, try it first with the function $y = x^2$ at $x = 1$. The slope you read should be 2.

b) Then try it with the curve $y = e^x$ at $x = 1$, $x = 0$, and $x = -1$. In each case, compare your estimate of the derivative with the value of e^x at the point. What pattern do you see? Test it with other values of x. Chapter 6 will explain what is going on.

63. *Linearizations at inflection points.* As Fig. 3.64 suggests, linearizations fit particularly well at inflection points. You will understand why if you do Exercise 40 in Section 8.10 later in the book. As another example, graph *Newton's serpentine,* $f(x) = 4x/(x^2 + 1)$, together with its linearizations at $x = 0$ and $x = \sqrt{3}$.

64. *The linearization is the best linear approximation.* (This is why we use the linearization.) Suppose that $y = f(x)$ is differentiable at $x = a$ and that $g(x) = m(x - a) + c$ is a linear function in which m and c are constants. If the error $E(x) = f(x) - g(x)$ were small enough near $x = a$, we might think of using g as a linear approximation of f instead of the linearization $L(x) = f(a) + f'(a)(x - a)$. Show that if we impose on g the conditions

1. $E(a) = 0$ The approximation error is zero at $x = a$.

2. $\displaystyle\lim_{x \to a} \frac{E(x)}{x - a} = 0$ The error is negligible when compared with $x - a$.

then $g(x) = f(a) + f'(a)(x - a)$. Thus, the linearization $L(x)$ gives the only linear approximation whose error is both zero at $x = a$ and negligible in comparison with $x - a$.

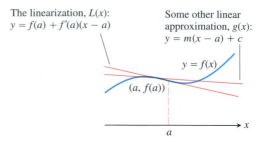

The linearization, $L(x)$:
$y = f(a) + f'(a)(x - a)$

Some other linear approximation, $g(x)$:
$y = m(x - a) + c$

$y = f(x)$

$(a, f(a))$

65. **CALCULATOR** Enter 2 in your calculator and take successive square roots by pressing the square root key repeatedly (or raising the displayed number repeatedly to the 0.5 power). What pattern do you see emerging? Explain what is going on. What happens if you take successive tenth roots instead?

66. **CALCULATOR** Repeat Exercise 65 with 0.5 in place of 2 as the original entry. What happens now? Can you use any positive number x in place of 2? Explain what is going on.

CAS Explorations and Projects

In Exercises 67–70, you will use a CAS to estimate the magnitude of the error in using the linearization in place of the function over a specified interval I. Perform the following steps:

a) Plot the function f over I.

b) Find the linearization L of the function at the point a.

c) Plot f and L together on a single graph.

d) Plot the absolute error $|f(x) - L(x)|$ over I and find its maximum value.

e) From your graph in part (d), estimate as large a $\delta > 0$ as you can, satisfying

$$|x - a| < \delta \Rightarrow |f(x) - L(x)| < \epsilon$$

for $\epsilon = 0.5, 0.1$, and 0.01. Then check graphically to see if your δ-estimate holds true.

67. $f(x) = x^3 + x^2 - 2x$, $[-1, 2]$, $a = 1$

68. $f(x) = \dfrac{x - 1}{4x^2 + 1}$, $\left[-\dfrac{3}{4}, 1\right]$, $a = \dfrac{1}{2}$

69. $f(x) = x^{2/3}(x - 2)$, $[-2, 3]$, $a = 2$

70. $f(x) = \sqrt{x} - \sin x$, $[0, 2\pi]$, $a = 2$

3.8

Newton's Method

We know simple formulas for solving linear and quadratic equations, and there are somewhat more complicated formulas for cubic and quartic equations (equations of degree three and four). At one time it was hoped that similar formulas might be found for quintic and higher degree equations, but the Norwegian mathematician

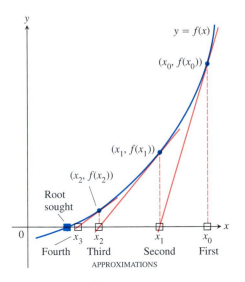

3.68 Newton's method starts with an initial guess x_0 and (under favorable circumstances) improves the guess one step at a time.

Neils Henrik Abel (1802–1829) showed that no formulas like these are possible for polynomial equations of degree greater than four.

When exact formulas for solving an equation $f(x) = 0$ are not available, we can turn to numerical techniques from calculus to approximate the solutions we seek. One of these techniques is *Newton's method* or, as it is more accurately called, the *Newton-Raphson method*. It is based on the idea of using tangent lines to replace the graph of $y = f(x)$ near the points where f is zero. Once again, linearization is the key to solving a practical problem.

The Theory

The goal of Newton's method for estimating a solution of an equation $f(x) = 0$ is to produce a sequence of approximations that approach the solution. We pick the first number x_0 of the sequence. Then, under favorable circumstances, the method does the rest by moving step by step toward a point where the graph of f crosses the x-axis (Fig. 3.68).

The initial estimate, x_0, may be found by graphing or just plain guessing. The method then uses the tangent to the curve $y = f(x)$ at $(x_0, f(x_0))$ to approximate the curve, calling the point where the tangent meets the x-axis x_1. The number x_1 is usually a better approximation to the solution than is x_0. The point x_2 where the tangent to the curve at $(x_1, f(x_1))$ crosses the x-axis is the next approximation in the sequence. We continue on, using each approximation to generate the next, until we are close enough to the root to stop.

We can derive a formula for generating the successive approximations in the following way. Given the approximation x_n, the point–slope equation for the tangent to the curve at $(x_n, f(x_n))$ is

$$y - f(x_n) = f'(x_n)(x - x_n) \tag{1}$$

(Fig. 3.69). We find where the tangent crosses the x-axis by setting y equal to 0 in this equation and solving for x, giving, in turn,

$$0 - f(x_n) = f'(x_n)(x - x_n) \qquad \text{Eq. (1) with } y = 0$$

$$-f(x_n) = f'(x_n)x - f'(x_n)x_n$$

$$f'(x_n)x = f'(x_n)x_n - f(x_n)$$

$$x = x_n - \frac{f(x_n)}{f'(x_n)}. \qquad \text{Assuming } f'(x_n) \neq 0$$

This value of x is the next approximation, x_{n+1}.

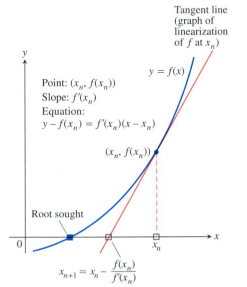

3.69 The geometry of the successive steps of Newton's method. From x_n we go up to the curve and follow the tangent line down to find x_{n+1}.

The Strategy for Newton's Method

1. Guess a first approximation to a root of the equation $f(x) = 0$. A graph of $y = f(x)$ will help.
2. Use the first approximation to get a second, the second to get a third, and so on, using the formula

$$x_{n+1} = x_n - \frac{f(x_n)}{f'(x_n)}, \qquad (f'(x_n) \neq 0) \tag{2}$$

where $f'(x_n)$ is the derivative of f at x_n.

The Practice

In our first example we find decimal approximations to $\sqrt{2}$ by estimating the positive root of the equation $f(x) = x^2 - 2 = 0$.

EXAMPLE 1   Find the positive root of the equation

$$f(x) = x^2 - 2 = 0.$$

Solution   With $f(x) = x^2 - 2$ and $f'(x) = 2x$, Eq. (2) becomes

$$x_{n+1} = x_n - \frac{x_n{}^2 - 2}{2x_n}.$$

To use our calculator efficiently, we rewrite this equation in a form that uses fewer arithmetic operations:

$$x_{n+1} = x_n - \frac{x_n}{2} + \frac{1}{x_n}$$

$$= \frac{x_n}{2} + \frac{1}{x_n}.$$

The equation

$$x_{n+1} = \frac{x_n}{2} + \frac{1}{x_n}$$

enables us to go from each approximation to the next with just a few keystrokes. With the starting value $x_0 = 1$, we get the results in the first column of the following table. (To 5 decimal places, $\sqrt{2} = 1.41421$.)

	Error	Number of correct figures
$x_0 = 1$	-0.41421	1
$x_1 = 1.5$	0.08579	1
$x_2 = 1.41667$	0.00246	3
$x_3 = 1.41422$	0.00001	5

Newton's method is the method used by most calculators to calculate roots because it converges so fast (more about this later). If the arithmetic in the table in Example 1 had been carried to 13 decimal places instead of 5, then going one step further would have given $\sqrt{2}$ correctly to more than 10 decimal places.

EXAMPLE 2   Find the x-coordinate of the point where the curve $y = x^3 - x$ crosses the horizontal line $y = 1$.

Solution   The curve crosses the line when $x^3 - x = 1$ or $x^3 - x - 1 = 0$. When does $f(x) = x^3 - x - 1$ equal zero? The graph of f (Fig. 3.70) shows a single root, located between $x = 1$ and $x = 2$. We apply Newton's method to f with the starting value $x_0 = 1$. The results are displayed in Table 3.2 and Fig. 3.71.

At $n = 5$ we come to the result $x_6 = x_5 = 1.3247\ 17957$. When $x_{n+1} = x_n$, Eq. (2) shows that $f(x_n) = 0$. We have found a solution of $f(x) = 0$ to 9 decimals.

Algorithm and iteration

It is customary to call a specified sequence of computational steps like the one in Newton's method an *algorithm*. When an algorithm proceeds by repeating a given set of steps over and over, using the answer from the previous step as the input for the next, the algorithm is called *iterative* and each repetition is called an *iteration*. Newton's method is one of the really fast iterative techniques for finding roots.

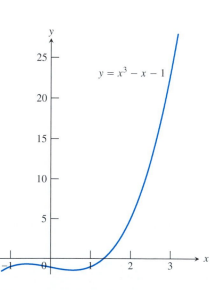

$y = x^3 - x - 1$

3.70 The graph of $f(x) = x^3 - x - 1$ crosses the x-axis between $x = 1$ and $x = 2$.

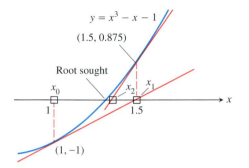

3.71 The first three x-values in Table 3.2.

Table 3.2 The result of applying Newton's method to $f(x) = x^3 - x - 1$ with $x_0 = 1$

n	x_n		$f(x_n)$		$f'(x_n)$		$x_{n+1} = x_n - \dfrac{f(x_n)}{f'(x_n)}$	
0	1		-1		2		1.5	
1	1.5		0.875		5.75		1.3478	26087
2	1.3478	26087	0.1006	82173	4.4499	05482	1.3252	00399
3	1.3252	00399	0.0020	58362	4.2684	68293	1.3247	18174
4	1.3247	18174	0.0000	00924	4.2646	34722	1.3247	17957
5	1.3247	17957	-1.0437E-9		4.2646	32997	1.3247	17957

The equation $x^3 - x - 1 = 0$ is the equation we solved graphically in Section 1.5. Notice how much more rapidly and accurately we find the solution here. ❑

In Fig. 3.72, we have indicated that the process in Example 2 might have started at the point $B_0(3, 23)$ on the curve, with $x_0 = 3$. Point B_0 is quite far from the x-axis, but the tangent at B_0 crosses the x-axis at about $(2.11, 0)$, so x_1 is still an improvement over x_0. If we use Eq. (2) repeatedly as before, with $f(x) = x^3 - x - 1$ and $f'(x) = 3x^2 - 1$, we confirm the 9-place solution $x_6 = x_5 = 1.3247\ 17957$ in six steps.

The curve in Fig. 3.72 has a local maximum at $x = -1/\sqrt{3}$ and a local minimum at $x = +1/\sqrt{3}$. We would not expect good results from Newton's method if we were to start with x_0 between these points, but we can start any place to the right of $x = 1/\sqrt{3}$ and get the answer. It would not be very clever to do so, but we could even begin far to the right of B_0, for example with $x_0 = 10$. It takes a bit longer, but the process still converges to the same answer as before.

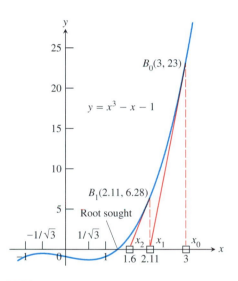

3.72 Any starting value x_0 to the right of $x = 1/\sqrt{3}$ will lead to the root.

Convergence Is Usually Assured

In practice, Newton's method usually converges with impressive speed, but since this is not guaranteed you must test that convergence is actually taking place. One way to do this would be to begin by graphing the function to find a good starting value for x_0. It is important to test that you are getting closer to a zero of the function, by evaluating $|f(x_n)|$, and to check that the method is converging, by evaluating $|x_n - x_{n+1}|$.

Theory does provide some help, however. A theorem from advanced calculus says that if

$$\left| \frac{f(x)f''(x)}{[f'(x)]^2} \right| < 1 \tag{3}$$

for all x in an interval about a root r, then the method will converge to r for any starting value x_0 in that interval. In practice, the theorem is somewhat hard to apply and convergence is evaluated by calculating $f(x_n)$ and $|x_n - x_{n+1}|$.

Inequality (3) is a *sufficient* but not a *necessary* condition. The method can and does converge in some cases where there is no interval about r on which the inequality holds. Newton's method always converges if the curve $y = f(x)$ is convex ("bulges") toward the x-axis in the interval between x_0 and the root sought. See Fig. 3.73.

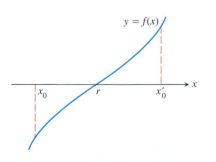

3.73 Newton's method will converge to r from either starting point.

Under favorable circumstances, the speed with which Newton's method converges to r is expressed by the advanced calculus formula

$$\underbrace{\left|x_{n+1} - r\right|}_{\text{error } e_{n+1}} \leq \frac{\max |f''|}{2 \min |f'|} \left|x_n - r\right|^2 = \text{constant} \cdot \underbrace{\left|x_n - r\right|^2}_{\text{error } e_n}, \qquad (4)$$

where max and min refer to the maximum and minimum values in an interval surrounding r. The formula says that the error in step $n + 1$ is no greater than a constant times the square of the error in step n. This may not seem like much, but think of what it says. If the constant is less than or equal to 1, and $|x_n - r| < 10^{-3}$, then $|x_{n+1} - r| < 10^{-6}$. *In a single step* the method moves from three decimal places of accuracy to six!

The results in (3) and (4) both assume that f is "nice." In the case of (4), this means that f has only a single root at r, so that $f'(r) \neq 0$. If f has a multiple root at r, the convergence may be slower.

But Things Can Go Wrong

(x_n, f(x_n))

x_n

3.74 If $f'(x_n) = 0$, there is no intersection point to define x_{n+1}.

Newton's method stops if $f'(x_n) = 0$ (Fig. 3.74). In that case, try a new starting point. Of course, f and f' may have a common root. To detect whether this is so, you could first find the solutions of $f'(x) = 0$ and check f at those values. Or you could graph f and f' together.

Newton's method does not always converge. For instance, if

$$f(x) = \begin{cases} -\sqrt{r - x}, & x < r \\ \sqrt{x - r}, & x \geq r, \end{cases} \qquad (5)$$

the graph will be like the one in Fig. 3.75. If we begin with $x_0 = r - h$, we get $x_1 = r + h$, and successive approximations go back and forth between these two values. No amount of iteration brings us closer to the root than our first guess.

If Newton's method does converge, it converges to a root. In theory, that is. In practice, there are situations in which the method appears to converge but there is no root there. Fortunately, such situations are rare.

When Newton's method converges to a root, it may not be the root you have in mind. Figure 3.76 shows two ways this can happen.

The solution then is to use everything you know about the curve—from graphs drawn by computer or from calculus-based analysis—to get a feeling for the shape of the curve near r and to choose an x_0 close to r. Use Newton's method and test its convergence as you go along. The chances are you will have no problems.

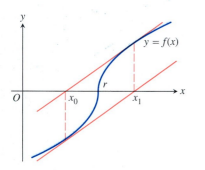

3.75 Newton's method fails to converge.

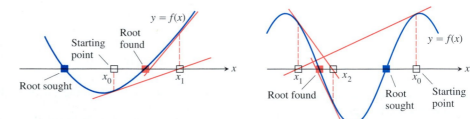

3.76 Newton's method may miss the root you want if you start too far away.

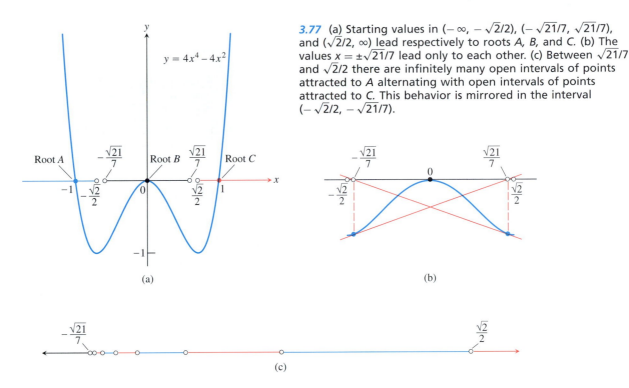

3.77 (a) Starting values in $(-\infty, -\sqrt{2}/2)$, $(-\sqrt{21}/7, \sqrt{21}/7)$, and $(\sqrt{2}/2, \infty)$ lead respectively to roots A, B, and C. (b) The values $x = \pm\sqrt{21}/7$ lead only to each other. (c) Between $\sqrt{21}/7$ and $\sqrt{2}/2$ there are infinitely many open intervals of points attracted to A alternating with open intervals of points attracted to C. This behavior is mirrored in the interval $(-\sqrt{2}/2, -\sqrt{21}/7)$.

✳ Chaos in Newton's Method

The process of finding roots by Newton's method can be chaotic, meaning that for some equations the final outcome can be extremely sensitive to the starting value's location.

The equation $4x^4 - 4x^2 = 0$ is a case in point (Fig. 3.77a). Starting values in the blue zone on the x-axis lead to root A. Starting values in the black lead to root B, and starting values in the red zone lead to root C. The points $\pm\sqrt{2}/2$ give horizontal tangents. The points $\pm\sqrt{21}/7$ "cycle," each leading to the other, and back (Fig. 3.77b).

The interval between $\sqrt{21}/7$ and $\sqrt{2}/2$ contains infinitely many open intervals of points leading to root A, alternating with intervals of points leading to root C (Fig. 3.77c). The boundary points separating consecutive intervals (there are infinitely many) do not lead to roots, but cycle back and forth from one to another.

Here is where the "chaos" is truly manifested. As we select points that approach $\sqrt{21}/7$ from the right it becomes increasingly difficult to distinguish which lead to root A and which to root C. On the same side of $\sqrt{21}/7$, we find arbitrarily close together points whose ultimate destinations are far apart.

If we think of the roots as "attractors" of other points, the coloring in Fig. 3.77 shows the intervals of the points they attract (the "intervals of attraction"). You might think that points between roots A and B would be attracted to either A or B, but, as we see, that is not the case. Between A and B there are infinitely many intervals of points attracted to C. Similarly, between B and C lie infinitely many intervals of points attracted to A.

We encounter an even more dramatic example of chaotic behavior when we apply Newton's method to solve the complex-number equation $z^6 - 1 = 0$. It has six solutions: 1, −1, and the four numbers $\pm(1/2) \pm (\sqrt{3}/2)i$. As Fig. 3.78 (on the following page) suggests, each of the six roots has infinitely many "basins"

3.78 This computer-generated initial value portrait uses color to show where different points in the complex plane end up when they are used as starting values in applying Newton's method to solve the equation $z^6 - 1 = 0$. Red points go to 1, green points to $(1/2) + (\sqrt{3}/2)i$, dark blue points to $(-1/2) + (\sqrt{3}/2)i$, and so on. Starting values that generate sequences that do not arrive within 0.1 units of a root after 32 steps are colored black.

of attraction in the complex plane (Appendix 3). Starting points in red basins are attracted to the root 1, those in the green basin to the root $(1/2) + (\sqrt{3}/2)i$, and so on. Each basin has a boundary whose complicated pattern repeats without end under successive magnifications.

Exercises 3.8

Root Finding

1. Use Newton's method to estimate the solutions of the equation $x^2 + x - 1 = 0$. Start with $x_0 = -1$ for the left-hand solution and with $x_0 = 1$ for the solution on the right. Then, in each case, find x_2.

2. Use Newton's method to estimate the one real solution of $x^3 + 3x + 1 = 0$. Start with $x_0 = 0$ and then find x_2.

3. Use Newton's method to estimate the two zeros of the function $f(x) = x^4 + x - 3$. Start with $x_0 = -1$ for the left-hand zero and with $x_0 = 1$ for the zero on the right. Then, in each case, find x_2.

4. Use Newton's method to estimate the two zeros of the function $f(x) = 2x - x^2 + 1$. Start with $x_0 = 0$ for the left-hand zero and with $x_0 = 2$ for the zero on the right. Then, in each case, find x_2.

5. Use Newton's method to find the positive fourth root of 2 by solving the equation $x^4 - 2 = 0$. Start with $x_0 = 1$ and find x_2.

6. Use Newton's method to find the negative fourth root of 2 by solving the equation $x^4 - 2 = 0$. Start with $x_0 = -1$ and find x_2.

7. CALCULATOR At what value(s) of x does $\cos x = 2x$?

8. CALCULATOR At what value(s) of x does $\cos x = -x$?

9. CALCULATOR Use the Intermediate Value Theorem from Section 1.5 to show that $f(x) = x^3 + 2x - 4$ has a root between $x = 1$ and $x = 2$. Then find the root to 5 decimal places.

10. CALCULATOR Estimate π to as many decimal places as your calculator will display by using Newton's method to solve the equation $\tan x = 0$ with $x_0 = 3$.

Theory, Examples, and Applications

11. Suppose your first guess is lucky, in the sense that x_0 is a root of $f(x) = 0$. Assuming that $f'(x_0)$ is defined and not 0, what happens to x_1 and later approximations?

12. You plan to estimate $\pi/2$ to 5 decimal places by using Newton's method to solve the equation $\cos x = 0$. Does it matter what your starting value is? Give reasons for your answer.

13. *Oscillation.* Show that if $h > 0$, applying Newton's method to

$$f(x) = \begin{cases} \sqrt{x}, & x \geq 0 \\ \sqrt{-x}, & x < 0 \end{cases}$$

leads to $x_1 = -h$ if $x_0 = h$ and to $x_1 = h$ if $x_0 = -h$. Draw a picture that shows what is going on.

14. *Approximations that get worse and worse.* Apply Newton's method to $f(x) = x^{1/3}$ with $x_0 = 1$, and calculate $x_1, x_2, x_3,$ and x_4. Find a formula for $|x_n|$. What happens to $|x_n|$ as $n \to \infty$? Draw a picture that shows what is going on.

15. a) Explain why the following four statements ask for the same information:

i) Find the roots of $f(x) = x^3 - 3x - 1$.

ii) Find the x-coordinates of the intersections of the curve $y = x^3$ with the line $y = 3x + 1$.

iii) Find the x-coordinates of the points where the curve $y = x^3 - 3x$ crosses the horizontal line $y = 1$.

iv) Find the values of x where the derivative of $g(x) = (1/4)x^4 - (3/2)x^2 - x + 5$ equals zero.

a) CALCULATOR Use Newton's method to find the two negative zeros of $f(x) = x^3 - 3x - 1$ to 5 decimal places.

b) GRAPHER Graph $f(x) = x^3 - 3x - 1$ for $-2 \leq x \leq 2.5$. Use ZOOM and TRACE to estimate the zeros of f to 5 decimal places.

c) GRAPHER Graph $g(x) = 0.25x^4 - 1.5x^2 - x + 5$. Use ZOOM and TRACE with appropriate rescaling to find, to 5 decimal places, the values of x where the graph has horizontal tangents.

16. *Locating a planet.* To calculate a planet's space coordinates, we have to solve equations like $x = 1 + 0.5 \sin x$. Graphing the function $f(x) = x - 1 - 0.5 \sin x$ suggests that the function has a root near $x = 1.5$. Use one application of Newton's method to improve this estimate. That is, start with $x_0 = 1.5$ and find x_1. (The value of the root is 1.49870 to 5 decimal places.) Remember to use radians.

17. *Finding an ion concentration.* While trying to find the acidity of a saturated solution of magnesium hydroxide in hydrochloric acid, you derive the equation

$$\frac{3.64 \times 10^{-11}}{[H_3O^+]^2} = [H_3O^+] + 3.6 \times 10^{-4}$$

for the hydronium ion concentration $[H_3O^+]$. To find the value of $[H_3O^+]$, you set $x = 10^4 [H_3O^+]$ and convert the equation to

$$x^3 + 3.6x^2 - 36.4 = 0.$$

You then solve this by Newton's method. What do you get for x? (Make it good to 2 decimal places.) For $[H_3O^+]$?

18. Show that Newton's method cannot converge to a point $x = c$ where the function's graph has an upward pointing cusp above the x-axis like the one in the margin on p. 215.

⊞ ✹ Computer or Programmable Calculator

Exercises 19–28 require a computer or programmable calculator.

19. The curve $y = \tan x$ crosses the line $y = 2x$ between $x = 0$ and $x = \pi/2$. Use Newton's method to find where.

20. Use Newton's method to find the two real solutions of the equation $x^4 - 2x^3 - x^2 - 2x + 2 = 0$.

21. a) How many solutions does the equation $\sin 3x = 0.99 - x^2$ have?

b) Use Newton's method to find them.

22. a) Does $\cos 3x$ ever equal x?

b) Use Newton's method to find where.

23. Find the four real zeros of the function $f(x) = 2x^4 - 4x^2 + 1$.

24. *The sonobuoy problem.* In submarine location problems it is often necessary to find a submarine's closest point of approach (CPA) to a sonobuoy (sound detector) in the water. Suppose that the submarine travels on a parabolic path $y = x^2$ and that the buoy is located at the point $(2, -1/2)$.

a) Show that the value of x that minimizes the distance between the submarine and the buoy is a solution of the equation $x = 1/(x^2 + 1)$.

b) Solve the equation $x = 1/(x^2 + 1)$ with Newton's method.

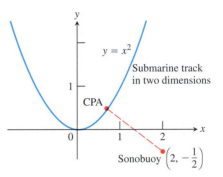

(Source: *The Contraction Mapping Principle,* by C. O. Wilde, UMAP Unit 326, Arlington, MA, COMAP, Inc.)

25. *Curves that are nearly flat at the root.* Some curves are so flat that, in practice, Newton's method stops too far from the root to give a useful estimate. Try Newton's method on $f(x) = (x - 1)^{40}$ with a starting value of $x_0 = 2$ to see how close your machine comes to the root $x = 1$.

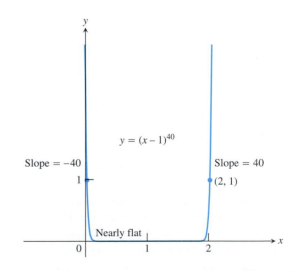

26. *Finding a root different from the one sought.* All three roots of $f(x) = 4x^4 - 4x^2$ can be found by starting Newton's method near $x = \sqrt{21}/7$. Try it. See Fig. 3.77.

27. Find the approximate values of r_1 through r_4 in the factorization

$$8x^4 - 14x^3 - 9x^2 + 11x - 1$$
$$= 8(x - r_1)(x - r_2)(x - r_3)(x - r_4).$$

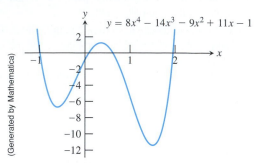

28. *Chaos in Newton's method.* If you have a computer or a calculator that can be programmed to do complex-number arithmetic, experiment with Newton's method to solve the equation $z^6 - 1 = 0$. The recursion relation to use is

$$z_{n+1} = z_n - \frac{z_n^6 - 1}{6z_n^5} \qquad \text{or} \qquad z_{n+1} = \frac{5}{6}z_n + \frac{1}{6z_n^5}.$$

Try these starting values (among others): $2, i, \sqrt{3} + i$.

CHAPTER   3   QUESTIONS TO GUIDE YOUR REVIEW

1. What can be said about the values of a function that is continuous on a closed interval?

2. What does it mean for a function to have a local extreme value on its domain? An absolute extreme value? How are local and absolute extreme values related, if at all? Give examples.

3. What is the First Derivative Theorem for Local Extreme Values? How does it lead to a procedure for finding a function's local extreme values?

4. How do you find the absolute extrema of a continuous function on a closed interval? Give examples.

5. What are the hypotheses and conclusion of Rolle's theorem? Are the hypotheses really necessary? Explain.

6. What are the hypotheses and conclusion of the Mean Value Theorem? What physical interpretations might the theorem have?

7. State the Mean Value Theorem's three corollaries.

8. How can you sometimes identify a function $f(x)$ by knowing f' and knowing the value of f at a point $x = x_0$? Give an example.

9. What is the First Derivative Test for Local Extreme Values? Give examples of how it is applied.

10. How do you test a twice-differentiable function to determine where its graph is concave up or concave down? Give examples.

11. What is an inflection point? Give an example. What physical significance do inflection points sometimes have?

12. What is the Second Derivative Test for Local Extreme Values? Give examples of how it is applied.

13. What do the derivatives of a function tell you about the shape of its graph?

14. List the steps you would take to graph a polynomial function. Illustrate with an example.

15. What is a cusp? Give examples.

16. What exactly do $\lim_{x \to \infty} f(x) = L$ and $\lim_{x \to -\infty} f(x) = L$ mean? Give examples.

17. What are $\lim_{x \to \pm\infty} k$ (k a constant) and $\lim_{x \to \pm\infty} (1/x)$? How do you extend these results to other functions? Give examples.

18. How do you find the limit of a rational function as $x \to \pm\infty$? What are the three basic possibilities? Give examples.

19. List the steps you would take to graph a rational function. Illustrate with an example.

20. Outline a general strategy for solving max-min problems. Give examples.

21. What is the linearization $L(x)$ of a function $f(x)$ at a point $x = a$? What is required of f at a for the linearization to exist? How are linearizations used? Give examples.

22. If x moves from x_0 to a nearby value $x_0 + dx$, how do you estimate the corresponding change in the value of a differentiable function $f(x)$? How do you estimate the relative change? The percentage change? Give an example.

23. How do you estimate the error in a linear approximation? Give an example.

24. Describe Newton's method for solving equations. Give an example. What is the theory behind the method? What are some of the things to watch out for when you use the method?

CHAPTER 3 PRACTICE EXERCISES

Existence of Extreme Values

1. Does $f(x) = x^3 + 2x + \tan x$ have any local maximum or minimum values? Give reasons for your answer.

2. Does $g(x) = \csc x + 2 \cot x$ have any local maximum values? Give reasons for your answer.

3. Does $f(x) = (7 + x)(11 - 3x)^{1/3}$ have an absolute minimum value? An absolute maximum? If so, find them or give reasons why they fail to exist. List all critical points of f.

4. Find values of a and b such that the function

$$f(x) = \frac{ax + b}{x^2 - 1}$$

has a local extreme value of 1 at $x = 3$. Is this extreme value a local maximum, or a local minimum? Give reasons for your answer.

5. The greatest integer function $f(x) = \lfloor x \rfloor$, defined for all values of x, assumes a local maximum value of 0 at each point of $[0, 1)$. Could any of these local maximum values also be local minimum values of f? Give reasons for your answer.

6. a) Give an example of a differentiable function f whose first derivative is zero at some point c even though f has neither a local maximum nor a local minimum at c.

   b) How is this consistent with Theorem 2 in Section 3.1? Give reasons for your answer.

7. The function $y = 1/x$ does not take on either a maximum or a minimum on the interval $0 < x < 1$ even though the function is continuous on this interval. Does this contradict the Max-Min Theorem for continuous functions? Why?

8. What are the maximum and minimum values of the function $y = |x|$ on the interval $-1 \le x < 1$? Notice that the interval is not closed. Is this consistent with the Max-Min Theorem for continuous functions? Why?

9. **Grapher** A graph that is large enough to show a function's global behavior may fail to reveal important local features. The graph of $f(x) = (x^8/8) - (x^6/2) - x^5 + 5x^3$ is a case in point.

   a) Graph f over the interval $-2.5 \le x \le 2.5$. Where does the graph appear to have local extreme values or points of inflection?

   b) Now factor $f'(x)$ and show that f has a local maximum at $x = \sqrt[5]{5} \approx 1.70998$ and local minima at $x = \pm\sqrt{3} \approx \pm1.73205$.

   c) Zoom in on the graph to find a viewing window that shows the presence of the extreme values at $x = \sqrt[5]{5}$ and $x = \sqrt{3}$.

   The moral here is that without calculus the existence of two of the three extreme values would probably have gone unnoticed.

On any normal graph of the function, the values would lie close enough together to fall within the dimensions of a single pixel on the screen.

(Source: *Uses of Technology in the Mathematics Curriculum*, by Benny Evans and Jerry Johnson, Oklahoma State University, published in 1990 under National Science Foundation Grant USE-8950044.)

10. (*Continuation of Exercise 9.*)

   a) Graph $f(x) = (x^8/8) - (2/5)x^5 - 5x - (5/x^2) + 11$ over the interval $-2 \le x \le 2$. Where does the graph appear to have local extreme values or points of inflection?

   b) Show that f has a local maximum value at $x = \sqrt[7]{5} \approx 1.2585$ and a local minimum value at $x = \sqrt[3]{2} \approx 1.2599$.

   c) Zoom in to find a viewing window that shows the presence of the extreme values at $x = \sqrt[7]{5}$ and $x = \sqrt[3]{2}$.

The Mean Value Theorem

11. a) Show that $g(t) = \sin^2 t - 3t$ decreases on every interval in its domain.

    b) How many solutions does the equation $\sin^2 t - 3t = 5$ have? Give reasons for your answer.

12. a) Show that $y = \tan\theta$ increases on every interval in its domain.

    b) If the conclusion in (a) is really correct, how do you explain the fact that $\tan \pi = 0$ is less than $\tan(\pi/4) = 1$?

13. a) Show that the equation $x^4 + 2x^2 - 2 = 0$ has exactly one solution on $[0, 1]$.

    b) **CALCULATOR** Find the solution to as many decimal places as you can.

14. a) Show that $f(x) = x/(x + 1)$ increases on every interval in its domain.

    b) Show that $f(x) = x^3 + 2x$ has no local maximum or minimum values.

15. **CALCULATOR** As a result of a heavy rain, the volume of water in a reservoir increased by 1400 acre-ft in 24 h. Show that at some instant during that period the reservoir's volume was increasing at a rate in excess of 225,000 gal/min. (An acre-foot is 43,560 ft^3, the volume that would cover one acre to the depth of one foot. A cubic foot holds 7.48 gal.)

16. The formula $F(x) = 3x + C$ gives a different function for each value of C. All of these functions, however, have the same derivative with respect to x, namely $F'(x) = 3$. Are these the only differentiable functions whose derivative is 3? Could there be any others? Give reasons for your answers.

17. Show that

$$\frac{d}{dx}\left(\frac{x}{x+1}\right) = \frac{d}{dx}\left(-\frac{1}{x+1}\right)$$

even though

$$\frac{x}{x+1} \neq -\frac{1}{x+1}.$$

Doesn't this contradict Corollary 2 of the Mean Value Theorem? Give reasons for your answer.

18. Calculate the first derivatives of $f(x) = x^2/(x^2 + 1)$ and $g(x) = -1/(x^2 + 1)$. What can you conclude about the graphs of these functions?

Graphs and Graphing

Graph the curves in Exercises 19–28.

19. $y = x^2 - (x^3/6)$

20. $y = x^3 - 3x^2 + 3$

21. $y = -x^3 + 6x^2 - 9x + 3$

22. $y = (1/8)(x^3 + 3x^2 - 9x - 27)$

23. $y = x^3(8 - x)$

24. $y = x^2(2x^2 - 9)$

25. $y = x - 3x^{2/3}$

26. $y = x^{1/3}(x - 4)$

27. $y = x\sqrt{3 - x}$

28. $y = x\sqrt{4 - x^2}$

Each of Exercises 29–34 gives the first derivative of a function $y = f(x)$. (a) At what points, if any, does the graph of f have a local maximum, local minimum, or inflection point? (b) Sketch the general shape of the graph.

29. $y' = 16 - x^2$

30. $y' = x^2 - x - 6$

31. $y' = 6x(x + 1)(x - 2)$

32. $y' = x^2(6 - 4x)$

33. $y' = x^4 - 2x^2$

34. $y' = 4x^2 - x^4$

 GRAPHER In Exercises 35–38, graph each function. Then use the function's first derivative to explain what you see.

35. $y = x^{2/3} + (x - 1)^{1/3}$

36. $y = x^{2/3} + (x - 1)^{2/3}$

37. $y = x^{1/3} + (x - 1)^{1/3}$

38. $y = x^{2/3} - (x - 1)^{1/3}$

Sketch the graphs of the functions in Exercises 39–46.

39. $y = \dfrac{x + 1}{x - 3}$

40. $y = \dfrac{2x}{x + 5}$

41. $y = \dfrac{x^2 + 1}{x}$

42. $y = \dfrac{x^2 - x + 1}{x}$

43. $y = \dfrac{x^3 + 2}{2x}$

44. $y = \dfrac{x^4 - 1}{x^2}$

45. $y = \dfrac{x^2 - 4}{x^2 - 3}$

46. $y = \dfrac{x^2}{x^2 - 4}$

Using the graphs of the dominant terms as a guide, sketch the graphs of the equations in Exercises 47 and 48.

47. $y = \csc x - \dfrac{1}{x^2}$, $0 < x < \pi$

48. $y = \tan x - \dfrac{2}{x}$, $-\dfrac{\pi}{2} < x < 0$

Drawing Conclusions About Motion from Graphs

Each of the graphs in Exercises 49 and 50 is the graph of the position function $s = f(t)$ of a body moving on a coordinate line (t represents time). At approximately what times (if any) is each body's (a) velocity equal to zero? (b) acceleration equal to zero? During approximately what time intervals does the body move (c) forward? (d) backward?

49.

50.

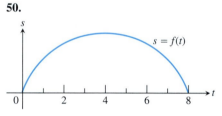

Limits

Find the limits in Exercises 51–60.

51. $\displaystyle\lim_{x\to\infty} \frac{2x + 3}{5x + 7}$

52. $\displaystyle\lim_{x\to-\infty} \frac{2x^2 + 3}{5x^2 + 7}$

53. $\displaystyle\lim_{x\to-\infty} \frac{x^2 - 4x + 8}{3x^3}$

54. $\displaystyle\lim_{x\to\infty} \frac{1}{x^2 - 7x + 1}$

55. $\displaystyle\lim_{x\to-\infty} \frac{x^2 - 7x}{x + 1}$

56. $\displaystyle\lim_{x\to\infty} \frac{x^4 + x^3}{12x^3 + 128}$

57. $\displaystyle\lim_{x\to 0} \frac{\sin x}{\lfloor x \rfloor}$ $\left(\begin{array}{l}\text{If you have a grapher, try graphing}\\ \text{the function for } -5 \leq x \leq 5.\end{array}\right)$

58. $\displaystyle\lim_{\theta\to\infty} \frac{\cos\theta - 1}{\theta}$ $\left(\begin{array}{l}\text{If you have a grapher, try graphing}\\ f(x) = x(\cos(1/x) - 1) \text{ near the}\\ \text{origin to "see" the limit at infinity.}\end{array}\right)$

59. $\displaystyle\lim_{x\to\infty} \frac{x + \sin x + 2\sqrt{x}}{x + \sin x}$

60. $\displaystyle\lim_{x\to\infty} \frac{x^{2/3} + x^{-1}}{x^{2/3} + \cos^2 x}$

Optimization

61. The sum of two nonnegative numbers is 36. Find the numbers if (a) the difference of their square roots is to be as large as possible, (b) the sum of their square roots is to be as large as possible.

62. The sum of two nonnegative numbers is 20. Find the numbers

   a) if the product of one number and the square root of the other is to be as large as possible;

b) if one number plus the square root of the other is to be as large as possible.

63. An isosceles triangle has its vertex at the origin and its base parallel to the x-axis with the vertices above the axis on the curve $y = 27 - x^2$. Find the largest area the triangle can have.

64. A customer has asked you to design an open-top rectangular stainless steel vat. It is to have a square base and a volume of 32 ft^3, to be welded from quarter-inch plate, and to weigh no more than necessary. What dimensions do you recommend?

65. Find the height and radius of the largest right circular cylinder that can be put in a sphere of radius $\sqrt{3}$.

66. The figure here shows two right circular cones, one upside down inside the other. The two bases are parallel, and the vertex of the smaller cone lies at the center of the larger cone's base. What values of r and h will give the smaller cone the largest possible volume?

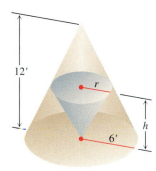

67. A drilling rig 12 mi offshore is to be connected by a pipe to a refinery onshore, 20 mi down the coast from the rig. If underwater pipe costs \$50,000 per mile and land-based pipe costs \$30,000 per mile, what values of x and y give the least expensive connection?

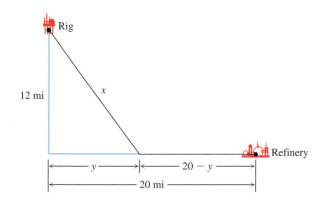

68. An athletic field is to be built in the shape of a rectangle x units long capped by semicircular regions of radius r at the two ends. The field is to be bounded by a 400-m racetrack. What values of x and r give the rectangle the largest possible area?

69. Your company can manufacture x hundred grade A tires and y hundred grade B tires a day, where $0 \le x \le 4$ and
$$y = \frac{40 - 10x}{5 - x}.$$
Your profit on grade A tires is twice your profit on grade B tires. Find the most profitable number of each kind of tire to make.

70. Suppose a manufacturer can sell x items a week for a revenue of $r = 200x - 0.01x^2$ cents, and it costs $c = 50x + 20{,}000$ cents to make x items. Is there a most profitable number of items to make each week? If so, what is it? Explain.

Linearization

71. Find the linearizations of

a) $\tan x$ at $x = -\pi/4$ **b)** $\sec x$ at $x = -\pi/4$.

Graph the curves and linearizations together.

72. We can obtain a useful linear approximation of the function $f(x) = 1/(1 + \tan x)$ at $x = 0$ by combining the approximations
$$\frac{1}{1+x} \approx 1 - x \quad \text{and} \quad \tan x \approx x$$
to get
$$\frac{1}{1 + \tan x} \approx 1 - x.$$
Show that this result is the standard linear approximation of $1/(1 + \tan x)$ at $x = 0$.

73. Find the linearization of $f(x) = \sqrt{1+x} + \sin x - 0.5$ at $x = 0$.

74. Find the linearization of $f(x) = 2/(1-x) + \sqrt{1+x} - 3.1$ at $x = 0$.

Differential Estimates of Change

75. Write a formula that estimates the change that occurs in the volume of a right circular cone when the radius changes from r_0 to $r_0 + dr$ and the height does not change.

$V = \frac{1}{3}\pi r^2 h$
$S = \pi r \sqrt{r^2 + h^2}$
(Lateral surface area)

76. Write a formula that estimates the change that occurs in the lateral surface area of a cone when the height changes from h_0 to $h_0 + dh$ and the radius does not change.

Applications of Differentials

77. a) How accurately should you measure the edge of a cube to be reasonably sure of calculating the cube's surface area with an error of no more than 2%?

b) Suppose the edge is measured with the accuracy required in (a). About how accurately can the cube's volume be calculated from the edge measurement? To find out, estimate the percentage error in the volume calculation that would result from using the edge measurement.

78. The circumference of a great circle of a sphere is measured as 10 cm with a possible error of 0.4 cm. The measurement is then used to calculate the radius. The radius is then used to calculate the surface area and volume of the sphere. Estimate the percentage errors in the calculated values of (a) the radius, (b) the surface area, and (c) the volume.

79. To find the height of a tree, you measure the angle from the ground to the treetop from a point 100 ft away from the base. The best figure you can get with the equipment at hand is $30° \pm 1°$. About how much error could the tolerance of $\pm 1°$ create in the calculated height? Remember to work in radians.

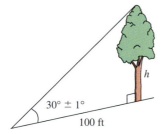

80. To find the height of a lamppost, you stand a 6-ft pole 20 ft from the lamp and measure the length a of its shadow. The figure you get for a is 15 ft, give or take an inch. Calculate the height of the lamppost from the value $a = 15$ and estimate the possible error in the result.

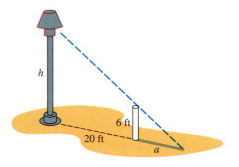

Newton's Method

81. CALCULATOR Let $f(x) = 3x - x^3$. Show that the equation $f(x) = -4$ has a solution in the interval [2, 3] and use Newton's method to find it.

82. Let $f(x) = x^4 - x^3$. Show that the equation $f(x) = 75$ has a solution in the interval [3, 4] and use Newton's method to find it.

CHAPTER	3	ADDITIONAL EXERCISES–THEORY, EXAMPLES, APPLICATIONS

1. What can you say about a function whose maximum and minimum values on an interval are equal? Give reasons for your answer.

2. Is it true that a discontinuous function cannot have both an absolute maximum and an absolute minimum value on a closed interval? Give reasons for your answer.

3. Can you conclude anything about the extreme values of a continuous function on an open interval? on a half-open interval? Give reasons for your answer.

4. Use the sign pattern for the derivative

$$\frac{df}{dx} = 6(x - 1)(x - 2)^2(x - 3)^3(x - 4)^4$$

to identify the points where f has local maximum and minimum values.

5. a) Suppose that the first derivative of $y = f(x)$ is

$$y' = 6(x + 1)(x - 2)^2.$$

At what points, if any, does the graph of f have a local maximum, local minimum, or point of inflection?

b) Suppose that the first derivative of $y = f(x)$ is

$$y' = 6x(x + 1)(x - 2).$$

At what points, if any, does the graph of f have a local maximum, local minimum, or point of inflection?

6. If $f'(x) \leq 2$ for all x, what is the most the values of f can increase on [0, 6]? Give reasons for your answer.

7. Suppose that f is continuous on [a, b] and that c is an interior point of the interval. Show that if $f'(x) \leq 0$ on [a, c) and $f'(x) \geq 0$ on (c, b], then $f(x)$ is never less than $f(c)$ on [a, b].

8. a) Show that $-1/2 \le x/(1 + x^2) \le 1/2$ for every value of x.
  b) Suppose that f is a function whose derivative is $f'(x) = x/(1 + x^2)$. Use the result in (a) to show that

$$|f(b) - f(a)| \le \frac{1}{2}|b - a|$$

  for any a and b.

9. The derivative of $f(x) = x^2$ is zero at $x = 0$, but f is not a constant function. Doesn't this contradict the corollary of the Mean Value Theorem that says that functions with zero derivatives are constant? Give reasons for your answer.

10. Let $h = fg$ be the product of two differentiable functions of x.

  a) If f and g are positive, with local maxima at $x = a$, and if f' and g' change sign at a, does h have local maximum at a?
  b) If the graphs of f and g have inflection points at $x = a$, does the graph of h have an inflection point at a?

  In either case, if the answer is yes, give a proof. If the answer is no, give a counterexample.

11. Use the following information to find the values of a, b, and c in the formula $f(x) = (x + a)/(bx^2 + cx + 2)$.

  i) The values of a, b, and c are either 0 or 1.
  ii) The graph of f passes through the point $(-1, 0)$.
  iii) The line $y = 1$ is an asymptote of the graph of f.

12. For what value or values of the constant k will the curve $y = x^3 + kx^2 + 3x - 4$ have exactly one horizontal tangent?

13. Points A and B lie at the ends of a diameter of a unit circle and point C lies on the circumference. Is it true that the perimeter of triangle ABC is largest when the triangle is isosceles? How do you know?

14. *The ladder problem.* What is the approximate length (ft) of the longest ladder you can carry horizontally around the corner of the corridor shown here? Round your answer down to the nearest foot.

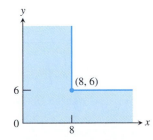

15. You want to bore a hole in the side of the tank shown here at a height that will make the stream of water coming out hit the ground as far from the tank as possible. If you drill the hole near the top, where the pressure is low, the water will exit slowly but spend a relatively long time in the air. If you drill the hole near the bottom, the water will exit at a higher velocity but have only a short time to fall. Where is the best place, if any, for the hole? (*Hint:* How long will it take an exiting particle of water to fall from height y to the ground?)

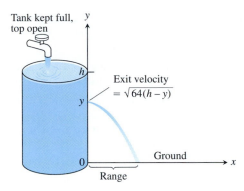

16. An American football player wants to kick a field goal with the ball being on a right hash mark. Assume that the goal posts are b feet apart and that the hash mark line is a distance $a > 0$ feet from the right goal post. (See the accompanying figure.) Find the distance h from the goal post line that gives the kicker his largest angle β. Assume the football field is flat.

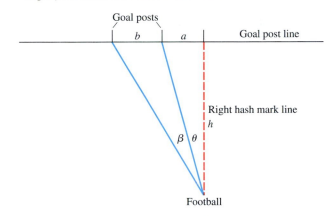

17. *A max-min problem with a variable answer.* Sometimes the solution of a max-min problem depends on the proportions of the shapes involved. As a case in point, suppose that a right circular cylinder of radius r and height h is inscribed in a right circular cone of radius R and height H, as shown here. Find the value of r (in terms of R and H) that maximizes the total surface area of the cylinder (including top and bottom). As you will see, the solution depends on whether $H \le 2R$ or $H > 2R$.

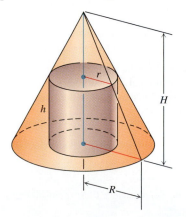

18. Find the smallest value of the positive constant m that will make $mx - 1 + (1/x)$ greater than or equal to zero for all positive values of x.

19. *The second derivative test.* The second derivative test for local maxima and minima (Section 3.4) says:

  a) f has a local maximum value at $x = c$ if $f'(c) = 0$ and $f''(c) < 0$;

  b) f has a local minimum value at $x = c$ if $f'(c) = 0$ and $f''(c) > 0$.

To prove statement (a), let $\epsilon = (1/2)|f''(c)|$. Then use the fact that

$$f''(c) = \lim_{h \to 0} \frac{f'(c+h) - f'(c)}{h} = \lim_{h \to 0} \frac{f'(c+h)}{h}$$

to conclude that for some $\delta > 0$,

$$0 < |h| < \delta \Rightarrow \frac{f'(c+h)}{h} < f''(c) + \epsilon < 0.$$

Thus $f'(c+h)$ is positive for $-\delta < h < 0$ and negative for $0 < h < \delta$. Prove statement (b) in a similar way.

20. *Schwarz's inequality*

  a)   Show that if $a > 0$, then $f(x) = ax^2 + 2bx + c \geq 0$ for all (real) x if and only if $b^2 \leq ac$.

  b)   Derive **Schwarz's inequality,**

$$(a_1 b_1 + a_2 b_2 + \cdots + a_n b_n)^2 \leq$$
$$(a_1{}^2 + a_2{}^2 + \cdots + a_n{}^2)(b_1{}^2 + b_2{}^2 + \cdots + b_n{}^2),$$

by applying what you learned in (a) to the sum

$$(a_1 x + b_1)^2 + (a_2 x + b_2)^2 + \cdots + (a_n x + b_n)^2.$$

  c)   Show that equality holds in Schwarz's inequality only if there exists a real number x that makes $a_i x$ equal $-b_i$ for every value of i from 1 to n.

21. *The period of a clock pendulum.* The period T of a clock pendulum (time for one full swing and back) is given by the formula $T^2 = 4\pi^2 L/g$, where T is measured in seconds, $g = 32.2$ ft/sec^2, and L, the length of the pendulum, is measured in feet. Find approximately

  a)   the length of a clock pendulum whose period is $T = 1$ sec;

  b)   the change dT in T if the pendulum in (a) is lengthened 0.01 ft; and

  c)   the amount the clock gains or loses in a day as a result of the period's changing by the amount dT found in (b).

22. *Estimating reciprocals without division.* You can estimate the value of the reciprocal of a number a without ever dividing by a if you apply Newton's method to the function $f(x) = (1/x) - a$.

For example, if $a = 3$, the function involved is $f(x) = (1/x) - 3$.

  a)   Graph $y = (1/x) - 3$. Where does the graph cross the x-axis?

  b)   Show that the recursion formula in this case is

$$x_{n+1} = x_n(2 - 3x_n),$$

so there is no need for division.

End Behavior Models

We call the function $y = 0$ an end behavior model for $f(x) = 1/x$ in the sense that $y = 0$ is a simpler function that behaves virtually the same way for $|x|$ large.

Definition

The function g is an **end behavior model** for f if

1. $\lim_{x \to \pm\infty} f/g = 1$ when $g(x) \neq 0$ for $|x|$ large,

   or

2. $\lim_{x \to \pm\infty} f(x) = 0$ when $g(x) = 0$.

For instance, $g(x) = 2$ is an end behavior model for $f(x) = 2 + (\sin x)/x$.

23. Show that $y = 3x^4$ is an end behavior model for $f(x) = 3x^4 - 2x^3 + 5x + 1$.

24. *Polynomial end behavior*

  a)   Show that $y = a_n x^n$ is an end behavior model for

$$f(x) = a_n x^n + a_{n-1} x^{n-1} + \cdots + a_1 x + a_0. \quad (a_n \neq 0)$$

  b)   Then show that for $n \geq 1$ there are only four types of polynomial end behavior models.

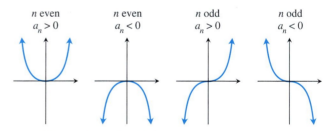

| n even
$a_n > 0$ | n even
$a_n < 0$ | n odd
$a_n > 0$ | n odd
$a_n < 0$ |

  c)   *Polynomials of odd degree.* Show that the polynomial function

$$f(x) = a_n x^n + a_{n-1} x^{n-1} + \cdots + a_1 x + a_0 \quad (a_n \neq 0)$$

has at least one zero if n is odd and $n > 1$.

CHAPTER 4

Integration

OVERVIEW   This chapter examines two processes and their relation to one another. One is the process by which we determine functions from their derivatives. The other is the process by which we arrive at exact formulas for such things as volume and area through successive approximations. Both processes are called integration.

Integration and differentiation are intimately connected. The nature of the connection is one of the most important ideas in all mathematics, and its independent discovery by Leibniz and Newton still constitutes one of the greatest technical advances of modern times.

4.1 Indefinite Integrals

One of the early accomplishments of calculus was predicting the future position of a moving body from one of its known locations and a formula for its velocity function. Today we view this as one of a number of occasions on which we determine a function from one of its known values and a formula for its rate of change. It is a routine process today, thanks to calculus, to calculate how fast a space vehicle needs to be going at a certain point to escape the earth's gravitational field or to predict the useful life of a sample of radioactive polonium-210 from its present level of activity and its rate of decay.

The process of determining a function from one of its known values and its derivative $f(x)$ has two steps. The first is to find a formula that gives us all the functions that could possibly have f as a derivative. These functions are the so-called antiderivatives of f, and the formula that gives them all is called the indefinite integral of f. The second step is to use the known function value to select the particular antiderivative we want from the indefinite integral. The first step is the subject of the present section; the second is the subject of the next.

Finding a formula that gives all of a function's antiderivatives might seem like an impossible task, or at least to require a little magic. But this is not the case at all. If we can find even one of a function's antiderivatives we can find them all, because of the first two corollaries of the Mean Value Theorem of Section 3.2.

Finding Antiderivatives—Indefinite Integrals

> **Definitions**
>
> A function $F(x)$ is an **antiderivative** of a function $f(x)$ if
>
> $$F'(x) = f(x)$$
>
> for all x in the domain of f. The set of all antiderivatives of f is the **indefinite integral** of f with respect to x, denoted by
>
> $$\int f(x)\,dx.$$
>
> The symbol $\int$ is an **integral sign.** The function f is the **integrand** of the integral and x is the **variable of integration.**

According to Corollary 2 of the Mean Value Theorem (Section 3.2), once we have found one antiderivative F of a function f, the other antiderivatives of f differ from F by a constant. We indicate this in integral notation in the following way:

$$\int f(x)\,dx = F(x) + C. \tag{1}$$

The constant C is the **constant of integration** or **arbitrary constant.** Equation (1) is read, "The indefinite integral of f with respect to x is $F(x) + C$." When we find $F(x) + C$, we say that we have **integrated** f and **evaluated** the integral.

EXAMPLE 1   Evaluate $\int 2x\,dx$.

Solution

$$\int 2x\,dx = x^2 + C$$

an antiderivative of $2x$

the arbitrary constant

The formula $x^2 + C$ generates all the antiderivatives of the function $2x$. The functions $x^2 + 1$, $x^2 - \pi$, and $x^2 + \sqrt{2}$ are all antiderivatives of the function $2x$, as you can check by differentiation. $\quad\blacksquare$

Many of the indefinite integrals needed in scientific work are found by reversing derivative formulas. You will see what we mean if you look at Table 4.1, which lists a number of standard integral forms side by side with their derivative-formula sources.

In case you are wondering why the integrals of the tangent, cotangent, secant, and cosecant do not appear in the table, the answer is that the usual formulas for them require logarithms. In Section 4.7, we will see that these functions do have antiderivatives, but we will have to wait until Chapters 6 and 7 to see what they are.

Table 4.1 Integral formulas

Indefinite integral	Reversed derivative formula
1. $\displaystyle\int x^n\,dx = \frac{x^{n+1}}{n+1} + C, \quad n \neq -1, \; n \text{ rational}$	$\displaystyle\frac{d}{dx}\left(\frac{x^{n+1}}{n+1}\right) = x^n$
$\displaystyle\int dx = \int 1\,dx = x + C \quad \text{(special case)}$	$\displaystyle\frac{d}{dx}(x) = 1$
2. $\displaystyle\int \sin kx\,dx = -\frac{\cos kx}{k} + C$	$\displaystyle\frac{d}{dx}\left(-\frac{\cos kx}{k}\right) = \sin kx$
3. $\displaystyle\int \cos kx\,dx = \frac{\sin kx}{k} + C$	$\displaystyle\frac{d}{dx}\left(\frac{\sin kx}{k}\right) = \cos kx$
4. $\displaystyle\int \sec^2 x\,dx = \tan x + C$	$\displaystyle\frac{d}{dx}\tan x = \sec^2 x$
5. $\displaystyle\int \csc^2 x\,dx = -\cot x + C$	$\displaystyle\frac{d}{dx}(-\cot x) = \csc^2 x$
6. $\displaystyle\int \sec x \tan x\,dx = \sec x + C$	$\displaystyle\frac{d}{dx}\sec x = \sec x \tan x$
7. $\displaystyle\int \csc x \cot x\,dx = -\csc x + C$	$\displaystyle\frac{d}{dx}(-\csc x) = \csc x \cot x$

EXAMPLE 2 *Selected integrals from Table 4.1*

a) $\displaystyle\int x^5\,dx = \frac{x^6}{6} + C$ Formula 1 with $n = 5$

b) $\displaystyle\int \frac{1}{\sqrt{x}}\,dx = \int x^{-1/2}\,dx = 2x^{1/2} + C = 2\sqrt{x} + C$ Formula 1 with $n = -1/2$

c) $\displaystyle\int \sin 2x\,dx = -\frac{\cos 2x}{2} + C$ Formula 2 with $k = 2$

d) $\displaystyle\int \cos \frac{x}{2}\,dx = \int \cos \frac{1}{2}x\,dx = \frac{\sin(1/2)\,x}{1/2} + C = 2\sin\frac{x}{2} + C$ Formula 3 with $k = 1/2$

       Finding an integral formula can sometimes be difficult, but checking it, once found, is relatively easy: differentiate the right-hand side. The derivative should be the integrand.

EXAMPLE 3

   Right: $\displaystyle\int x \cos x\,dx = x \sin x + \cos x + C$

Reason: The derivative of the right-hand side is the integrand:

$$\frac{d}{dx}(x \sin x + \cos x + C) = x \cos x + \sin x - \sin x + 0 = x \cos x.$$

Wrong: $\int x \cos x \, dx = x \sin x + C$

Reason: The derivative of the right-hand side is not the integrand:

$$\frac{d}{dx}(x \sin x + C) = x \cos x + \sin x + 0 \neq x \cos x.$$ ❑

Do not worry about how to derive the correct integral formula in Example 3. We will present a technique for doing so in Chapter 7.

Rules of Algebra for Antiderivatives

Among the things we know about antiderivatives are these:

1. A function is an antiderivative of a constant multiple kf of a function f if and only if it is k times an antiderivative of f.
2. In particular, a function is an antiderivative of $-f$ if and only if it is the negative of an antiderivative of f.
3. A function is an antiderivative of a sum or difference $f \pm g$ if and only if it is the sum or difference of an antiderivative of f and an antiderivative of g.

When we express these observations in integral notation, we get the standard arithmetic rules for indefinite integration (Table 4.2).

Table 4.2 Rules for indefinite integration

1. *Constant Multiple Rule:* $\int kf(x) \, dx = k \int f(x) \, dx$

   (Does not work if k varies with x.)

2. *Rule for Negatives:* $\int -f(x) \, dx = -\int f(x) \, dx$

   (Rule 1 with $k = -1$)

3. *Sum and Difference Rule:* $\int [f(x) \pm g(x)] \, dx = \int f(x) \, dx \pm \int g(x) \, dx$

EXAMPLE 4   *Rewriting the constant of integration*

$$\int 5 \sec x \tan x \, dx = 5 \int \sec x \tan x \, dx \qquad \text{Table 4.2, Rule 1}$$

$$= 5(\sec x + C) \qquad \text{Table 4.1, Formula 6}$$

$$= 5 \sec x + 5C \qquad \text{First form}$$

$$= 5 \sec x + C' \qquad \text{Shorter form, where } C' \text{ is } 5C$$

$$= 5 \sec x + C \qquad \text{Usual form—no prime. Since 5 times an arbitrary constant is an arbitrary constant, we rename } C'.$$ ❑

What about all the different forms in Example 4? Each one gives all the antiderivatives of $f(x) = 5 \sec x \tan x$, so each answer is correct. But the least complicated of the three, and the usual choice, is

$$\int 5 \sec x \tan x \, dx = 5 \sec x + C.$$

Just as the Sum and Difference Rule for differentiation enables us to differentiate expressions term by term, the Sum and Difference Rule for integration enables us to integrate expressions term by term. When we do so, we combine the individual constants of integration into a single arbitrary constant at the end.

EXAMPLE 5 *Term-by-term integration*

Evaluate

$$\int (x^2 - 2x + 5) \, dx.$$

Solution If we recognize that $(x^3/3) - x^2 + 5x$ is an antiderivative of $x^2 - 2x + 5$, we can evaluate the integral as

$$\int (x^2 - 2x + 5) \, dx = \underbrace{\frac{x^3}{3} - x^2 + 5x}_{\text{antiderivative}} + \overset{\text{arbitrary constant}}{C}.$$

If we do not recognize the antiderivative right away, we can generate it term by term with the Sum and Difference Rule:

$$\int (x^2 - 2x + 5) \, dx = \int x^2 \, dx - \int 2x \, dx + \int 5 \, dx$$

$$= \frac{x^3}{3} + C_1 - x^2 + C_2 + 5x + C_3.$$

This formula is more complicated than it needs to be. If we combine C_1, C_2, and C_3 into a single constant $C = C_1 + C_2 + C_3$, the formula simplifies to

$$\frac{x^3}{3} - x^2 + 5x + C$$

and *still* gives all the antiderivatives there are. For this reason we recommend that you go right to the final form even if you elect to integrate term by term. Write

$$\int (x^2 - 2x + 5) \, dx = \int x^2 \, dx - \int 2x \, dx + \int 5 \, dx$$

$$= \frac{x^3}{3} - x^2 + 5x + C.$$

Find the simplest antiderivative you can for each part and add the constant at the end. ☐

The Integrals of sin² x and cos² x

We can sometimes use trigonometric identities to transform integrals we do not know how to evaluate into integrals we do know how to evaluate. The integral formulas for $\sin^2 x$ and $\cos^2 x$ arise frequently in applications.

EXAMPLE 6

a) $\displaystyle\int \sin^2 x \, dx = \int \frac{1 - \cos 2x}{2} \, dx$ $\sin^2 x = \dfrac{1 - \cos 2x}{2}$

$$= \frac{1}{2} \int (1 - \cos 2x) \, dx = \frac{1}{2} \int dx - \frac{1}{2} \int \cos 2x \, dx$$

$$= \frac{1}{2}x - \frac{1}{2}\frac{\sin 2x}{2} + C = \frac{x}{2} - \frac{\sin 2x}{4} + C$$

b) $\displaystyle\int \cos^2 x \, dx = \int \frac{1 + \cos 2x}{2} \, dx$ $\cos^2 x = \dfrac{1 + \cos 2x}{2}$

$$= \frac{x}{2} + \frac{\sin 2x}{4} + C$$ As in part (a), but with a sign change   ❑

Exercises 4.1

Finding Antiderivatives

In Exercises 1–18, find an antiderivative for each function. Do as many as you can mentally. Check your answers by differentiation.

1. a) $2x$ **b)** x^2 **c)** $x^2 - 2x + 1$

2. a) $6x$ **b)** x^7 **c)** $x^7 - 6x + 8$

3. a) $-3x^{-4}$ **b)** x^{-4} **c)** $x^{-4} + 2x + 3$

4. a) $2x^{-3}$ **b)** $\dfrac{x^{-3}}{2} + x^2$ **c)** $-x^{-3} + x - 1$

5. a) $\dfrac{1}{x^2}$ **b)** $\dfrac{5}{x^2}$ **c)** $2 - \dfrac{5}{x^2}$

6. a) $-\dfrac{2}{x^3}$ **b)** $\dfrac{1}{2x^3}$ **c)** $x^3 - \dfrac{1}{x^3}$

7. a) $\dfrac{3}{2}\sqrt{x}$ **b)** $\dfrac{1}{2\sqrt{x}}$ **c)** $\sqrt{x} + \dfrac{1}{\sqrt{x}}$

8. a) $\dfrac{4}{3}\sqrt[3]{x}$ **b)** $\dfrac{1}{3\sqrt[3]{x}}$ **c)** $\sqrt[3]{x} + \dfrac{1}{\sqrt[3]{x}}$

9. a) $\dfrac{2}{3}x^{-1/3}$ **b)** $\dfrac{1}{3}x^{-2/3}$ **c)** $-\dfrac{1}{3}x^{-4/3}$

10. a) $\dfrac{1}{2}x^{-1/2}$ **b)** $-\dfrac{1}{2}x^{-3/2}$ **c)** $-\dfrac{3}{2}x^{-5/2}$

11. a) $-\pi \sin \pi x$ **b)** $3 \sin x$ **c)** $\sin \pi x - 3 \sin 3x$

12. a) $\pi \cos \pi x$ **b)** $\dfrac{\pi}{2}\cos \dfrac{\pi x}{2}$ **c)** $\cos \dfrac{\pi x}{2} + \pi \cos x$

13. a) $\sec^2 x$ **b)** $\dfrac{2}{3}\sec^2 \dfrac{x}{3}$ **c)** $-\sec^2 \dfrac{3x}{2}$

14. a) $\csc^2 x$ **b)** $-\dfrac{3}{2}\csc^2 \dfrac{3x}{2}$ **c)** $1 - 8 \csc^2 2x$

15. a) $\csc x \cot x$ **b)** $-\csc 5x \cot 5x$

  c) $-\pi \csc \dfrac{\pi x}{2} \cot \dfrac{\pi x}{2}$

16. a) $\sec x \tan x$ **b)** $4 \sec 3x \tan 3x$

  c) $\sec \dfrac{\pi x}{2} \tan \dfrac{\pi x}{2}$

17. $(\sin x - \cos x)^2$ **18.** $(1 + 2\cos x)^2$

Evaluating Integrals

Evaluate the integrals in Exercises 19–58. Check your answers by differentiation.

19. $\displaystyle\int (x + 1) \, dx$ **20.** $\displaystyle\int (5 - 6x) \, dx$

21. $\displaystyle\int \left(3t^2 + \frac{t}{2}\right) dt$ **22.** $\displaystyle\int \left(\frac{t^2}{2} + 4t^3\right) dt$

23. $\displaystyle\int (2x^3 - 5x + 7) \, dx$ **24.** $\displaystyle\int (1 - x^2 - 3x^5) \, dx$

25. $\displaystyle\int \left(\frac{1}{x^2} - x^2 - \frac{1}{3}\right) dx$ **26.** $\displaystyle\int \left(\frac{1}{5} - \frac{2}{x^3} + 2x\right) dx$

27. $\displaystyle\int x^{-1/3} \, dx$ **28.** $\displaystyle\int x^{-5/4} \, dx$

29. $\displaystyle\int \left(\sqrt{x} + \sqrt[3]{x}\right) dx$ **30.** $\displaystyle\int \left(\frac{\sqrt{x}}{2} + \frac{2}{\sqrt{x}}\right) dx$

31. $\displaystyle\int \left(8y - \frac{2}{y^{1/4}}\right) dy$ **32.** $\displaystyle\int \left(\frac{1}{7} - \frac{1}{y^{5/4}}\right) dy$

33. $\int 2x \left(1 - x^{-3}\right) dx$

34. $\int x^{-3}(x + 1) \, dx$

35. $\int \frac{t \sqrt{t} + \sqrt{t}}{t^2} \, dt$

36. $\int \frac{4 + \sqrt{t}}{t^3} \, dt$

37. $\int (-2 \cos t) \, dt$

38. $\int (-5 \sin t) \, dt$

39. $\int 7 \sin \frac{\theta}{3} \, d\theta$

40. $\int 3 \cos 5\theta \, d\theta$

41. $\int (-3 \csc^2 x) \, dx$

42. $\int \left(-\frac{\sec^2 x}{3}\right) dx$

43. $\int \frac{\csc \theta \cot \theta}{2} \, d\theta$

44. $\int \frac{2}{5} \sec \theta \tan \theta \, d\theta$

45. $\int (4 \sec x \tan x - 2 \sec^2 x) \, dx$

46. $\int \frac{1}{2} (\csc^2 x - \csc x \cot x) \, dx$

47. $\int (\sin 2x - \csc^2 x) \, dx$

48. $\int (2 \cos 2x - 3 \sin 3x) \, dx$

49. $\int 4 \sin^2 y \, dy$

50. $\int \frac{\cos^2 y}{7} \, dy$

51. $\int \frac{1 + \cos 4t}{2} \, dt$

52. $\int \frac{1 - \cos 6t}{2} \, dt$

53. $\int (1 + \tan^2 \theta) \, d\theta$

54. $\int (2 + \tan^2 \theta) \, d\theta$

(*Hint:* $1 + \tan^2 \theta = \sec^2 \theta$)

55. $\int \cot^2 x \, dx$

56. $\int (1 - \cot^2 x) \, dx$

(*Hint:* $1 + \cot^2 x = \csc^2 x$)

57. $\int \cos \theta (\tan \theta + \sec \theta) \, d\theta$

58. $\int \frac{\csc \theta}{\csc \theta - \sin \theta} \, d\theta$

Checking Integration Formulas

Verify the integral formulas in Exercises 59–64 by differentiation. In Section 4.3, we will see where formulas like these come from.

59. $\int (7x - 2)^3 \, dx = \frac{(7x - 2)^4}{28} + C$

60. $\int (3x + 5)^{-2} \, dx = -\frac{(3x + 5)^{-1}}{3} + C$

61. $\int \sec^2 (5x - 1) \, dx = \frac{1}{5} \tan (5x - 1) + C$

62. $\int \csc^2 \left(\frac{x - 1}{3}\right) dx = -3 \cot \left(\frac{x - 1}{3}\right) + C$

63. $\int \frac{1}{(x + 1)^2} \, dx = -\frac{1}{x + 1} + C$

64. $\int \frac{1}{(x + 1)^2} \, dx = \frac{x}{x + 1} + C$

65. Right, or wrong? Say which for each formula and give a brief reason for each answer.

a) $\int x \sin x \, dx = \frac{x^2}{2} \sin x + C$

b) $\int x \sin x \, dx = -x \cos x + C$

c) $\int x \sin x \, dx = -x \cos x + \sin x + C$

66. Right, or wrong? Say which for each formula and give a brief reason for each answer.

a) $\int \tan \theta \sec^2 \theta \, d\theta = \frac{\sec^3 \theta}{3} + C$

b) $\int \tan \theta \sec^2 \theta \, d\theta = \frac{1}{2} \tan^2 \theta + C$

c) $\int \tan \theta \sec^2 \theta \, d\theta = \frac{1}{2} \sec^2 \theta + C$

67. Right, or wrong? Say which for each formula and give a brief reason for each answer.

a) $\int (2x + 1)^2 \, dx = \frac{(2x + 1)^3}{3} + C$

b) $\int 3(2x + 1)^2 \, dx = (2x + 1)^3 + C$

c) $\int 6(2x + 1)^2 \, dx = (2x + 1)^3 + C$

68. Right, or wrong? Say which for each formula and give a brief reason for each answer.

a) $\int \sqrt{2x + 1} \, dx = \sqrt{x^2 + x} + C$

b) $\int \sqrt{2x + 1} \, dx = \sqrt{x^2 + x} + C$

c) $\int \sqrt{2x + 1} \, dx = \frac{1}{3} \left(\sqrt{2x + 1}\right)^3 + C$

Theory and Examples

69. Suppose that

$$f(x) = \frac{d}{dx}(1 - \sqrt{x}) \quad \text{and} \quad g(x) = \frac{d}{dx}(x + 2).$$

Find:

a) $\int f(x) \, dx$

b) $\int g(x) \, dx$

c) $\int [-f(x)] \, dx$

d) $\int [-g(x)] \, dx$

e) $\int [f(x) + g(x)] \, dx$

f) $\int [f(x) - g(x)] \, dx$

g) $\displaystyle\int [x + f(x)]\, dx$

h) $\displaystyle\int [g(x) - 4]\, dx$

70. Repeat Exercise 69, assuming that

$$f(x) = \frac{d}{dx}\, e^x \quad \text{and} \quad g(x) = \frac{d}{dx}\,(x \sin x).$$

4.2 Differential Equations, Initial Value Problems, and Mathematical Modeling

This section shows how to use a known value of a function to select a particular antiderivative from the functions in an indefinite integral. The ability to do this is important in mathematical modeling, the process by which we, as scientists, use mathematics to learn about reality.

Initial Value Problems

An equation like

$$\frac{dy}{dx} = f(x)$$

that has a derivative in it is called a **differential equation.** The problem of finding a function y of x when we know its derivative and its value y_0 at a particular point x_0 is called an **initial value problem.** We solve such a problem in two steps, as demonstrated in Example 1.

EXAMPLE 1 *Finding a body's velocity from its acceleration and initial velocity*

The acceleration of gravity near the surface of the earth is 9.8 m/sec². This means that the velocity v of a body falling freely in a vacuum changes at the rate of

$$\frac{dv}{dt} = 9.8 \text{ m/sec}^2.$$

If the body is dropped from rest, what will its velocity be t seconds after it is released?

Solution In mathematical terms, we want to solve the initial value problem that consists of

The differential equation: $\quad \dfrac{dv}{dt} = 9.8$

The initial condition: $\quad v = 0$ when $t = 0$ (abbreviated as $v(0) = 0$)

We first solve the differential equation by integrating both sides with respect to t:

$$\frac{dv}{dt} = 9.8 \qquad\qquad \text{\color{blue}The differential equation}$$

$$\int \frac{dv}{dt}\, dt = \int 9.8\, dt \qquad \text{\color{blue}Integrate with respect to } t.$$

$$v + C_1 = 9.8t + C_2 \qquad \text{\color{blue}Integrals evaluated}$$

$$v = 9.8t + C. \qquad\qquad \text{\color{blue}Constants combined as one}$$

This last equation tells us that the body's velocity t seconds into the fall is $9.8t + C$ m/sec for some value of C. What value? We find out from the initial condition:

$$v = 9.8t + C$$

$$0 = 9.8(0) + C \qquad v(0) = 0$$

$$C = 0.$$

Conclusion: The body's velocity t seconds into the fall is

$$v = 9.8t + 0 = 9.8t \text{ m/sec.} \qquad \square$$

The indefinite integral $F(x) + C$ of the function $f(x)$ gives the **general solution** $y = F(x) + C$ of the differential equation $dy/dx = f(x)$. The general solution gives all the solutions of the equation (there are infinitely many, one for each value of C). We **solve** the differential equation by finding its general solution. We then solve the initial value problem by finding the **particular solution** that satisfies the initial condition $y(x_0) = y_0$ (y has the value y_0 when $x = x_0$).

EXAMPLE 2 *Finding a curve from its slope function and a point*

Find the curve whose slope at the point (x, y) is $3x^2$ if the curve is required to pass through the point $(1, -1)$.

Solution In mathematical language, we are asked to solve the initial value problem that consists of

The differential equation: $\qquad \dfrac{dy}{dx} = 3x^2 \qquad$ The curve's slope is $3x^2$.

The initial condition: $\qquad y(1) = -1.$

To solve it we first solve the differential equation:

$$\frac{dy}{dx} = 3x^2$$

$$\int \frac{dy}{dx}\, dx = \int 3x^2 \, dx$$

$$y = x^3 + C. \qquad \begin{array}{l}\text{Constants of integration}\\ \text{combined, giving the general}\\ \text{solution}\end{array}$$

This tells us that y equals $x^3 + C$ for some value of C. We find that value from the condition $y(1) = -1$:

$$y = x^3 + C$$

$$-1 = (1)^3 + C$$

$$C = -2. \qquad \square$$

The curve we want is $y = x^3 - 2$ (Fig. 4.1).

In the next example, we have to integrate a second derivative twice to find the function we are looking for. The first integration,

$$\int \frac{d^2 s}{dt^2}\, dt = \frac{ds}{dt} + C,$$

gives the function's first derivative. The second integration gives the function.

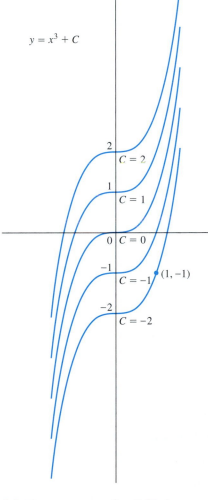

4.1 The curves $y = x^3 + C$ fill the coordinate plane without overlapping. In Example 2 we identify the curve $y = x^3 - 2$ as the one that passes through the given point $(1, -1)$.

4.2 The sketch for modeling the projectile motion in Example 3.

EXAMPLE 3 *Finding a projectile's height from its acceleration, initial velocity, and initial position*

A heavy projectile is fired straight up from a platform 3 m above the ground, with an initial velocity of 160 m/sec. Assume that the only force affecting the projectile during its flight is from gravity, which produces a downward acceleration of 9.8 m/sec². Find an equation for the projectile's height above the ground as a function of time t if $t = 0$ when the projectile is fired. How high above the ground is the projectile 3 sec after firing?

Solution To model the problem, we draw a figure (Fig. 4.2) and let s denote the projectile's height above the ground at time t. We assume s to be a twice-differentiable function of t and represent the projectile's velocity and acceleration with the derivatives

$$v = \frac{ds}{dt} \quad \text{and} \quad a = \frac{dv}{dt} = \frac{d^2s}{dt^2}.$$

Since gravity acts in the direction of *decreasing s* in our model, the initial value problem to solve is the following:

The differential equation: $\quad \dfrac{d^2s}{dt^2} = -9.8$

The initial conditions: $\quad \dfrac{ds}{dt}(0) = 160 \quad \text{and} \quad s(0) = 3.$

We integrate the differential equation with respect to t to find ds/dt:

$$\int \frac{d^2s}{dt^2}\,dt = \int (-9.8)\,dt$$

$$\frac{ds}{dt} = -9.8t + C_1.$$

We apply the first initial condition to find C_1:

$$160 = -9.8(0) + C_1 \qquad \frac{ds}{dt}(0) = 160$$

$$C_1 = 160.$$

This completes the formula for ds/dt:

$$\frac{ds}{dt} = -9.8t + 160.$$

We integrate ds/dt with respect to t to find s:

$$\int \frac{ds}{dt}\,dt = \int (-9.8t + 160)\,dt$$

$$s = -4.9t^2 + 160t + C_2.$$

We apply the second initial condition to find C_2:

$$3 = -4.9(0)^2 + 160(0) + C_2 \qquad s(0) = 3$$

$$C_2 = 3.$$

This completes the formula for s as a function of t:

$$s = -4.9t^2 + 160t + 3.$$

To find the projectile's height 3 sec into the flight, we set $t = 3$ in the formula for s. The height is

$$s = -4.9(3)^2 + 160(3) + 3 = 438.9 \text{ m.}$$ ❑

When we find a function from its first derivative, we have one arbitrary constant, as in Examples 1 and 2. When we find a function from its second derivative, we have to deal with two constants, one from each antidifferentiation, as in Example 3. To find a function from its third derivative would require us to find the values of three constants, and so on. In each case, the values of the constants are determined by the problem's initial conditions. Each time we find an antiderivative, we need an initial condition to tell us the value of C.

Sketching Solution Curves

The graph of a solution of a differential equation is called a **solution curve (integral curve)**. The curves $y = x^3 + C$ in Fig. 4.1 are solution curves of the differential equation $dy/dx = 3x^2$. When we cannot find explicit formulas for the solution curves of an equation $dy/dx = f(x)$ (that is, we cannot find an antiderivative of f), we may still be able to find their general shape by examining derivatives.

EXAMPLE 4   Sketch the solutions of the differential equation

$$y' = \frac{1}{x^2 + 1}.$$

Solution

Step 1: y' and y''. As in Section 3.4, the curve's general shape is determined by y' and y''. We already know y':

$$y' = \frac{1}{x^2 + 1}.$$

We find y'' by differentiation, in the usual way:

$$y'' = \frac{d}{dx}(y') = \frac{d}{dx}\left(\frac{1}{x^2 + 1}\right)$$

$$= \frac{-2x}{(x^2 + 1)^2}.$$

Step 3: *Concavity.* The second derivative changes from $(+)$ to $(-)$ at $x = 0$, so the curves all have an inflection point at $x = 0$.

Step 2: *Rise and fall.* The domain of y' is $(-\infty, \infty)$. There are no critical points, so the solution curves have no cusps or extrema. The curves rise from left to right because $y' > 0$. At $x = 0$, the curves have slope 1.

Step 4: *Summary:*

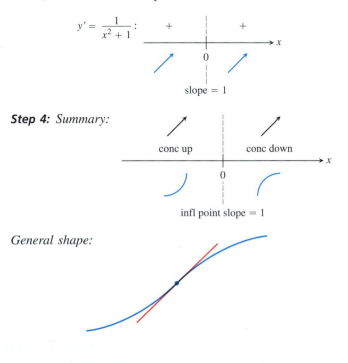

General shape:

The first derivative tells us still more:

$$\lim_{x \to \pm\infty} y' = \lim_{x \to \pm\infty} \frac{1}{x^2 + 1} = 0,$$

so the curves level off as $x \to \pm\infty$.

Step 5: *Specific points and solution curves.* We plot an assortment of points on the y-axis where we know the curves' slope (it is 1 at $x = 0$), mark tangents with that slope for guidance, and sketch "parallel" curves of the right general shape.

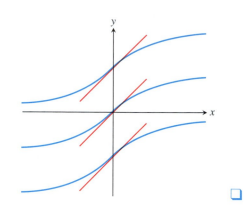

EXAMPLE 5   Sketch the solution of the initial value problem

Differential equation: $y' = \dfrac{1}{x^2 + 1}$

Initial condition: $y = 0$ when $x = 0$.

Solution   We find the solution's general shape (Example 4) and sketch the solution curve that passes through the point $(0, 0)$ (Fig. 4.3).

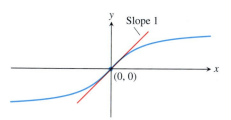

4.3 The solution curve in Example 5.

The technique we have learned for sketching solutions is particularly helpful when we are faced with an equation $dy/dx = f(x)$ that involves a function whose antiderivatives have no elementary formula. The antiderivatives of the function $f(x) = 1/(x^2 + 1)$ in Example 4 do have an elementary formula, as we will see in Chapter 6, but the antiderivatives of $g(x) = \sqrt{1 + x^4}$ do not. To solve the equation $dy/dx = \sqrt{1 + x^4}$, we must proceed either graphically or numerically.

Mathematical Modeling

The development of a mathematical model usually takes four steps: First we observe something in the real world (a ball bearing falling from rest or the trachea contracting during a cough, for example) and construct a system of mathematical variables and relationships that imitate some of its important features. We build a mathematical metaphor for what we see. Next we apply (usually) existing mathematics to the variables and relationships in the model to draw conclusions about them. After that we translate the mathematical conclusions into information about the system under study. Finally we check the information against observation to see if the model has predictive value. We also investigate the possibility that the model applies to other systems. The really good models are the ones that lead to conclusions that are consistent with observation, that have predictive value and broad application, and that are not too hard to use.

The natural cycle of mathematical imitation, deduction, interpretation, and confirmation is shown in the diagrams on the following page.

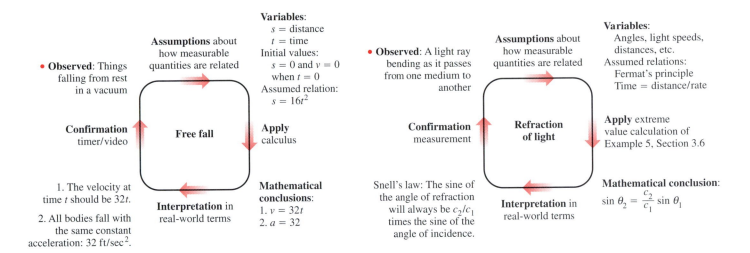

Variables:
s = distance
t = time
Initial values:
$s = 0$ and $v = 0$
when $t = 0$
Assumed relation:
$s = 16t^2$

• **Observed:** Things falling from rest in a vacuum

Assumptions about how measurable quantities are related

Confirmation timer/video

Free fall

Apply calculus

1. The velocity at time t should be $32t$.

2. All bodies fall with the same constant acceleration: 32 ft/sec².

Interpretation in real-world terms

Mathematical conclusions:
1. $v = 32t$
2. $a = 32$

• **Observed:** A light ray bending as it passes from one medium to another

Assumptions about how measurable quantities are related

Confirmation measurement

Refraction of light

Snell's law: The sine of the angle of refraction will always be c_2/c_1 times the sine of the angle of incidence.

Interpretation in real-world terms

Variables:
Angles, light speeds, distances, etc.
Assumed relations:
Fermat's principle
Time = distance/rate

Apply extreme value calculation of Example 5, Section 3.6

Mathematical conclusion:
$$\sin \theta_2 = \frac{c_2}{c_1} \sin \theta_1$$

Computer Simulation

When a system we want to study is complicated, we can sometimes experiment first to see how the system behaves under different circumstances. But if this is not possible (the experiments might be expensive, time-consuming, or dangerous), we might run a series of simulated experiments on a computer—experiments that behave like the real thing, without the disadvantages. Thus we might model the effects of atomic war, the effect of waiting a year longer to harvest trees, the effect of crossing particular breeds of cattle, or the effect of reducing atmospheric ozone by 1%, all without having to pay the consequences or wait to see how things work out naturally.

We also bring computers in when the model we want to use has too many calculations to be practical any other way. NASA's space flight models are run on computers—they have to be to generate course corrections on time. If you want to model the behavior of galaxies that contain billions and billions of stars, a computer offers the only possible way. One of the most spectacular computer simulations in recent years, carried out by Alar Toomre at MIT, explained a peculiar galactic shape that was not consistent with our previous ideas about how galaxies are formed. The galaxies had acquired their odd shapes, Toomre concluded, by passing through one another (Fig. 4.4).

4.4 (a) The modeling cycle for the shapes of colliding galaxies. (b) The computer's image of how galaxies are reshaped by the collision.

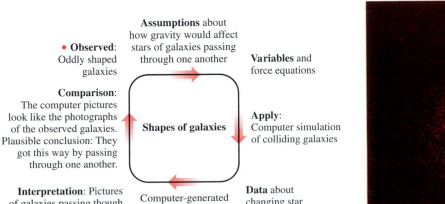

• **Observed:** Oddly shaped galaxies

Assumptions about how gravity would affect stars of galaxies passing through one another

Variables and force equations

Comparison: The computer pictures look like the photographs of the observed galaxies. Plausible conclusion: They got this way by passing through one another.

Shapes of galaxies

Apply: Computer simulation of colliding galaxies

Interpretation: Pictures of galaxies passing though one another

Computer-generated visual images

Data about changing star positions

(a)

(b)

Exercises 4.2

Initial Value Problems

1. Which of the following graphs shows the solution of the initial value problem

$$\frac{dy}{dx} = 2x, \quad y = 4 \text{ when } x = 1?$$

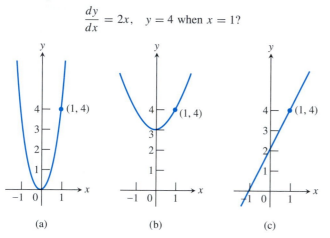

   (a)   (b)   (c)

   Give reasons for your answer.

2. Which of the following graphs shows the solution of the initial value problem

$$\frac{dy}{dx} = -x, \quad y = 1 \text{ when } x = -1?$$

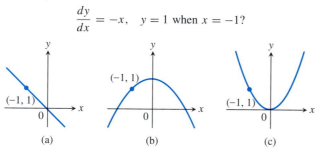

   (a)   (b)   (c)

   Give reasons for your answer.

Solve the initial value problems in Exercises 3–22.

3. $\dfrac{dy}{dx} = 2x - 7, \quad y(2) = 0$

4. $\dfrac{dy}{dx} = 10 - x, \quad y(0) = -1$

5. $\dfrac{dy}{dx} = \dfrac{1}{x^2} + x, \quad x > 0; \quad y(2) = 1$

6. $\dfrac{dy}{dx} = 9x^2 - 4x + 5, \quad y(-1) = 0$

7. $\dfrac{dy}{dx} = 3x^{-2/3}, \quad y(-1) = -5$

8. $\dfrac{dy}{dx} = \dfrac{1}{2\sqrt{x}}, \quad y(4) = 0$

9. $\dfrac{ds}{dt} = 1 + \cos t, \quad s(0) = 4$

10. $\dfrac{ds}{dt} = \cos t + \sin t, \quad s(\pi) = 1$

11. $\dfrac{dr}{d\theta} = -\pi \sin \pi\theta, \quad r(0) = 0$

12. $\dfrac{dr}{d\theta} = \cos \pi\theta, \quad r(0) = 1$

13. $\dfrac{dv}{dt} = \dfrac{1}{2} \sec t \tan t, \quad v(0) = 1$

14. $\dfrac{dv}{dt} = 8t + \csc^2 t, \quad v\left(\dfrac{\pi}{2}\right) = -7$

15. $\dfrac{d^2y}{dx^2} = 2 - 6x; \quad y'(0) = 4, \quad y(0) = 1$

16. $\dfrac{d^2y}{dx^2} = 0; \quad y'(0) = 2, \quad y(0) = 0$

17. $\dfrac{d^2r}{dt^2} = \dfrac{2}{t^3}; \quad \dfrac{dr}{dt}\bigg|_{t=1} = 1, \quad r(1) = 1$

18. $\dfrac{d^2s}{dt^2} = \dfrac{3t}{8}; \quad \dfrac{ds}{dt}\bigg|_{t=4} = 3, \quad s(4) = 4$

19. $\dfrac{d^3y}{dx^3} = 6; \quad y''(0) = -8, \quad y'(0) = 0, \quad y(0) = 5$

20. $\dfrac{d^3\theta}{dt^3} = 0; \quad \theta''(0) = -2, \quad \theta'(0) = -\dfrac{1}{2}, \quad \theta(0) = \sqrt{2}$

21. $y^{(4)} = -\sin t + \cos t;$
    $y'''(0) = 7, \quad y''(0) = y'(0) = -1, \quad y(0) = 0$

22. $y^{(4)} = -\cos x + 8 \sin 2x;$
    $y'''(0) = 0, \quad y''(0) = y'(0) = 1, \quad y(0) = 3$

Finding Position from Velocity

Exercises 23–26 give the velocity $v = ds/dt$ and initial position of a body moving along a coordinate line. Find the body's position at time t.

23. $v = 9.8t + 5, \quad s(0) = 10$

24. $v = 32t - 2, \quad s(1/2) = 4$

25. $v = \sin \pi t, \quad s(0) = 0$

26. $v = \dfrac{2}{\pi} \cos \dfrac{2t}{\pi}, \quad s(\pi^2) = 1$

Finding Position from Acceleration

Exercises 27–30 give the acceleration $a = d^2s/dt^2$, initial velocity, and initial position of a body moving on a coordinate line. Find the body's position at time t.

27. $a = 32; \quad v(0) = 20, \quad s(0) = 5$

28. $a = 9.8;$ $v(0) = -3,$ $s(0) = 0$

29. $a = -4 \sin 2t;$ $v(0) = 2,$ $s(0) = -3$

30. $a = \dfrac{9}{\pi^2} \cos \dfrac{3t}{\pi};$ $v(0) = 0,$ $s(0) = -1$

Finding Curves

31. Find the curve $y = f(x)$ in the xy-plane that passes through the point $(9, 4)$ and whose slope at each point is $3\sqrt{x}$.

32. **a)** Find a curve $y = f(x)$ with the following properties:

   i) $\dfrac{d^2y}{dx^2} = 6x$

   ii) Its graph passes through the point $(0, 1)$ and has a horizontal tangent there.

   b) How many curves like this are there? How do you know?

Solution (Integral) Curves

Exercises 33–36 show solution curves of differential equations. In each exercise, find an equation for the curve through the labeled point.

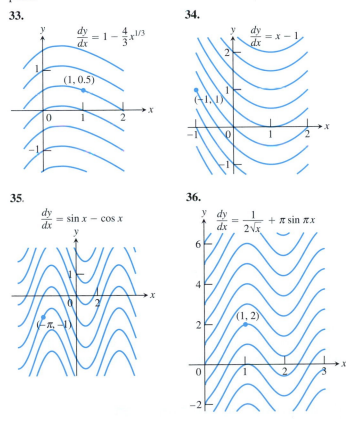

33.

$\dfrac{dy}{dx} = 1 - \dfrac{4}{3}x^{1/3}$

$(1, 0.5)$

34.

$\dfrac{dy}{dx} = x - 1$

$(-1, 1)$

35.

$\dfrac{dy}{dx} = \sin x - \cos x$

$(-\pi, -1)$

36.

$\dfrac{dy}{dx} = \dfrac{1}{2\sqrt{x}} + \pi \sin \pi x$

$(1, 2)$

Use the technique described in Example 4 to sketch some of the solutions of the differential equations in Exercises 37–40. Then solve the equations to check on how well you did.

37. $\dfrac{dy}{dx} = 2x$

38. $\dfrac{dy}{dx} = -2x + 2$

39. $\dfrac{dy}{dx} = 1 - 3x^2$

40. $\dfrac{dy}{dx} = x^2$

Use the technique described in Examples 4 and 5 to sketch the solutions of the initial value problems in Exercises 41–44.

41. $\dfrac{dy}{dx} = \dfrac{1}{\sqrt{1 - x^2}},$ $-1 < x < 1;$ $y(0) = 0$

42. $\dfrac{dy}{dx} = \sqrt{1 + x^4},$ $y(0) = 1$

43. $\dfrac{dy}{dx} = \dfrac{1}{x^2 + 1} - 1,$ $y(0) = 1$

44. $\dfrac{dy}{dx} = \dfrac{x}{x^2 + 1},$ $y(0) = 0$

Applications

45. On the moon the acceleration of gravity is 1.6 m/sec². If a rock is dropped into a crevasse, how fast will it be going just before it hits bottom 30 sec later?

46. A rocket lifts off the surface of Earth with a constant acceleration of 20 m/sec². How fast will the rocket be going 1 min later?

47. With approximately what velocity do you enter the water if you dive from a 10-m platform? (Use $g = 9.8$ m/sec².)

48. **CALCULATOR** The acceleration of gravity near the surface of Mars is 3.72 m/sec². If a rock is blasted straight up from the surface with an initial velocity of 93 m/sec (about 208 mph), how high does it go? (*Hint:* When is the velocity zero?)

49. *Stopping a car in time.* You are driving along a highway at a steady 60 mph (88 ft/sec) when you see an accident ahead and slam on the brakes. What constant deceleration is required to stop your car in 242 ft? To find out, carry out the following steps.

Step 1: Solve the initial value problem

   Differential equation: $\dfrac{d^2s}{dt^2} = -k$ (k constant)

   Initial conditions: $\dfrac{ds}{dt} = 88$ and $s = 0$ when $t = 0$.

   Measuring time and distance from when the brakes are applied

Step 2: Find the value of t that makes $ds/dt = 0$. (The answer will involve k.)

Step 3: Find the value of k that makes $s = 242$ for the value of t you found in step 2.

50. *Stopping a motorcycle.* The State of Illinois Cycle Rider Safety Program requires riders to be able to brake from 30 mph (44 ft/sec) to 0 in 45 ft. What constant deceleration does it take to do that?

51. *Motion along a coordinate line.* A particle moves on a co-ordinate line with acceleration $a = d^2s/dt^2 = 15\sqrt{t} - (3/\sqrt{t})$, subject to the conditions that $ds/dt = 4$ and $s = 0$ when $t = 1$. Find

   a) the velocity $v = ds/dt$ in terms of t,

   b) the position s in terms of t.

52. *The hammer and the feather.* When *Apollo 15* astronaut David Scott dropped a hammer and a feather on the moon to demonstrate that in a vacuum all bodies fall with the same (constant) acceleration, he dropped them from about 4 ft above the ground. The television footage of the event shows the hammer and feather falling more slowly than on Earth, where, in a vacuum, they would have taken only half a second to fall the 4 ft. How long did it take the hammer and feather to fall 4 ft on the moon? To find out, solve the following initial value problem for *s* as a function of *t*. Then find the value of *t* that makes *s* equal to 0.

$$\text{Differential equation:} \quad \frac{d^2s}{dt^2} = -5.2 \text{ ft/sec}^2$$

$$\text{Initial conditions:} \quad \frac{ds}{dt} = 0 \text{ and } s = 4 \text{ when } t = 0$$

53. *Motion with constant acceleration.* The standard equation for the position *s* of a body moving with a constant acceleration *a* along a coordinate line is

$$s = \frac{a}{2}t^2 + v_0 t + s_0, \tag{1}$$

where v_0 and s_0 are the body's velocity and position at time $t = 0$. Derive this equation by solving the initial value problem

$$\text{Differential equation:} \quad \frac{d^2s}{dt^2} = a$$

$$\text{Initial conditions:} \quad \frac{ds}{dt} = v_0 \text{ and } s = s_0 \text{ when } t = 0$$

54. (*Continuation of Exercise 53.*) *Free fall near the surface of a planet.* For free fall near the surface of a planet where the acceleration of gravity has a constant magnitude of *g* length-units/sec², Eq. (1) takes the form

$$s = -\frac{1}{2}gt^2 + v_0 t + s_0, \tag{2}$$

where *s* is the body's height above the surface. The equation has a minus sign because the acceleration acts downward, in the direction of decreasing *s*. The velocity v_0 is positive if the object is rising at time $t = 0$, and negative if the object is falling.

Instead of using the result of Exercise 53, you can derive Eq. (2) directly by solving an appropriate initial value problem. What initial value problem? Solve it to be sure you have the right one, explaining the solution steps as you go along.

Theory and Examples

55. *Finding displacement from an antiderivative of velocity*

a) Suppose that the velocity of a body moving along the *s*-axis is

$$\frac{ds}{dt} = v = 9.8t - 3.$$

1) Find the body's displacement over the time interval from $t = 1$ to $t = 3$ given that $s = 5$ when $t = 0$.
2) Find the body's displacement from $t = 1$ to $t = 3$ given that $s = -2$ when $t = 0$.
3) Now find the body's displacement from $t = 1$ to $t = 3$ given that $s = s_0$ when $t = 0$.

b) Suppose the position *s* of a body moving along a coordinate line is a differentiable function of time *t*. Is it true that once you know an antiderivative of the velocity function ds/dt you can find the body's displacement from $t = a$ to $t = b$ even if you do not know the body's exact position at either of those times? Give reasons for your answer.

56. *Uniqueness of solutions.* If differentiable functions $y = F(x)$ and $y = G(x)$ both solve the initial value problem

$$\frac{dy}{dx} = f(x), \quad y(x_0) = y_0,$$

on an interval *I*, must $F(x) = G(x)$ for every *x* in *I*? Give reasons for your answer.

4.3

Integration by Substitution—Running the Chain Rule Backward

A change of variable can often turn an unfamiliar integral into one we can evaluate. The method for doing this is called the substitution method of integration. It is one of the principal methods for evaluating integrals. This section shows how and why the method works.

The Generalized Power Rule in Integral Form

When *u* is a differentiable function of *x* and *n* is a rational number different from −1, the Chain Rule tells us

$$\frac{d}{dx}\left(\frac{u^{n+1}}{n+1}\right) = u^n \frac{du}{dx}.$$

This same equation, from another point of view, says that $u^{n+1}/(n+1)$ is one of the antiderivatives of the function $u^n(du/dx)$. Therefore,

$$\int \left(u^n \frac{du}{dx} \right) dx = \frac{u^{n+1}}{n+1} + C.$$

The integral on the left-hand side of this equation is usually written in the simpler "differential" form,

$$\int u^n \, du,$$

obtained by treating the dx's as differentials that cancel. Combining the last two equations gives the following rule.

Equation (1) actually holds for any real exponent $n \neq -1$, as we will see in Chapter 6.

> If u is any differentiable function,
>
> $$\int u^n \, du = \frac{u^{n+1}}{n+1} + C. \qquad (n \neq -1, \ n \text{ rational}) \qquad (1)$$

In deriving Eq. (1) we assumed u to be a differentiable function of the variable x, but the name of the variable does not matter and does not appear in the final formula. We could have represented the variable with θ, t, y, or any other letter. Equation (1) says that whenever we can cast an integral in the form

$$\int u^n \, du, \qquad (n \neq -1)$$

with u a differentiable function and du its differential, we can evaluate the integral as $\left[u^{n+1}/(n+1) \right] + C$.

EXAMPLE 1   Evaluate $\int (x+2)^5 \, dx$.

Solution   We can put the integral in the form

$$\int u^n \, du$$

by substituting

$$u = x + 2, \qquad du = d(x+2) = \frac{d}{dx}(x+2) \cdot dx$$

$$= 1 \cdot dx = dx.$$

Then

$$\int (x+2)^5 dx = \int u^5 \, du \qquad \qquad u = x+2, \quad du = dx$$

$$= \frac{u^6}{6} + C \qquad \qquad \text{Integrate, using Eq. (1)}$$
$$\qquad \qquad \qquad \text{with } n = 5.$$

$$= \frac{(x+2)^6}{6} + C. \qquad \text{Replace } u \text{ by } x+2.$$

EXAMPLE 2

$$\int \sqrt{1 + y^2} \cdot 2y \, dy = \int u^{1/2} \, du$$

Let $u = 1 + y^2$,
$du = 2y \, dy$.

$$= \frac{u^{(1/2)+1}}{(1/2) + 1} + C$$

Integrate, using
Eq. (1) with
$n = 1/2$.

$$= \frac{2}{3} u^{3/2} + C$$

Simpler form

$$= \frac{2}{3} (1 + y^2)^{3/2} + C$$

Replace u by
$1 + y^2$.

EXAMPLE 3   *Adjusting the integrand by a constant*

$$\int \sqrt{4t - 1} \, dt = \int u^{1/2} \cdot \frac{1}{4} du$$

Let $u = 4t - 1$,
$du = 4 \, dt$,
$(1/4) \, du = dt$.

$$= \frac{1}{4} \int u^{1/2} du$$

With the 1/4 out front,
the integral is now in
standard form.

$$= \frac{1}{4} \cdot \frac{u^{3/2}}{3/2} + C$$

Integrate, using Eq. (1)
with $n = 1/2$.

$$= \frac{1}{6} u^{3/2} + C$$

Simpler form

$$= \frac{1}{6} (4t - 1)^{3/2} + C$$

Replace u by $4t - 1$.

Trigonometric Functions

If u is a differentiable function of x, then $\sin u$ is a differentiable function of x. The Chain Rule gives the derivative of $\sin u$ as

$$\frac{d}{dx} \sin u = \cos u \frac{du}{dx}.$$

From another point of view, however, this same equation says that $\sin u$ is one of the antiderivatives of the product $\cos u \cdot (du/dx)$. Therefore,

$$\int \left(\cos u \frac{du}{dx} \right) dx = \sin u + C.$$

A formal cancellation of the dx's in the integral on the left leads to the following rule.

If u is a differentiable function, then

$$\int \cos u \, du = \sin u + C. \tag{2}$$

Equation (2) says that whenever we can cast an integral in the form

$$\int \cos u \, du,$$

we can integrate with respect to u to evaluate the integral as $\sin u + C$.

EXAMPLE 4

$$\int \cos (7\theta + 5) \, d\theta = \int \cos u \cdot \frac{1}{7} du \qquad \begin{array}{l} \text{Let } u = 7\theta + 5, \\ du = 7 \, d\theta, \\ (1/7) \, du = d\theta. \end{array}$$

$$= \frac{1}{7} \int \cos u \, du \qquad \begin{array}{l} \text{With } (1/7) \text{ out front,} \\ \text{the integral is now} \\ \text{in standard form.} \end{array}$$

$$= \frac{1}{7} \sin u + C \qquad \begin{array}{l} \text{Integrate with} \\ \text{respect to } u. \end{array}$$

$$= \frac{1}{7} \sin(7\theta + 5) + C \qquad \begin{array}{l} \text{Replace } u \text{ by} \\ 7\theta + 5. \end{array} \qquad \square$$

The companion formula for the integral of $\sin u$ when u is a differentiable function is

$$\int \sin u \, du = -\cos u + C. \qquad (3)$$

EXAMPLE 5

$$\int x^2 \sin (x^3) \, dx = \int \sin(x^3) \cdot x^2 dx$$

$$= \int \sin u \cdot \frac{1}{3} du \qquad \begin{array}{l} \text{Let } u = x^3 \\ du = 3x^2 \, dx \\ (1/3) du = x^2 \, dx. \end{array}$$

$$= \frac{1}{3} \int \sin u \, du$$

$$= \frac{1}{3}(-\cos u) + C \qquad \begin{array}{l} \text{Integrate with respect} \\ \text{to } u. \end{array}$$

$$= -\frac{1}{3} \cos (x^3) + C \qquad \text{Replace } u \text{ by } x^3. \qquad \square$$

The Chain Rule formulas for the derivatives of the tangent, cotangent, secant, and cosecant of a differentiable function u lead to the following integrals.

$$\int \sec^2 u \, du = \tan u + C \quad (4) \qquad \int \sec u \tan u \, du = \sec u + C \quad (6)$$

$$\int \csc^2 u \, du = -\cot u + C \quad (5) \qquad \int \csc u \cot u \, du = -\csc u + C \quad (7)$$

In each formula, u is a differentiable function of a real variable. Each formula can be checked by differentiating the right-hand side with respect to that variable. In each case, the Chain Rule applies to produce the integrand on the left.

EXAMPLE 6

$$\int \frac{1}{\cos^2 2\theta} \, d\theta = \int \sec^2 2\theta \, d\theta \qquad \sec 2\theta = \frac{1}{\cos 2\theta}$$

$$= \int \sec^2 u \cdot \frac{1}{2} \, du \qquad \begin{array}{l} \text{Let } u = 2\theta, \\ du = 2 \, d\theta, \\ d\theta = (1/2) \, du. \end{array}$$

$$= \frac{1}{2} \int \sec^2 u \, du$$

$$= \frac{1}{2} \tan u + C \qquad \text{Integrate, using Eq. (4).}$$

$$= \frac{1}{2} \tan 2\theta + C \qquad \text{Replace } u \text{ by } 2\theta.$$

Check:

$$\frac{d}{d\theta} \left(\frac{1}{2} \tan 2\theta + C \right) = \frac{1}{2} \cdot \frac{d}{d\theta} (\tan 2\theta) + 0$$

$$= \frac{1}{2} \cdot \left(\sec^2 2\theta \cdot \frac{d}{d\theta} (2\theta) \right) \qquad \text{Chain Rule}$$

$$= \frac{1}{2} \cdot \sec^2 2\theta \cdot 2 = \frac{1}{\cos^2 2\theta}. \qquad \blacksquare$$

The Substitution Method of Integration

The substitutions in the preceding examples are all instances of the following general rule.

$$\int f(g(x)) \cdot g'(x) \, dx = \int f(u) \, du \qquad \begin{array}{l} \textbf{1. Substitute } u = g(x), \\ du = g'(x) \, dx. \end{array}$$

$$= F(u) + C \qquad \begin{array}{l} \textbf{2. Evaluate by finding an} \\ \text{antiderivative } F(u) \text{ of} \\ f(u). \text{ (Any one will do.)} \end{array}$$

$$= F(g(x)) + C \qquad \textbf{3. Replace } u \text{ by } g(x).$$

> **The Substitution Method of Integration**
>
> Take these steps to evaluate the integral
>
> $$\int f(g(x))g'(x) \, dx,$$
>
> when f and g' are continuous functions:
>
> **Step 1:** Substitute $u = g(x)$ and $du = g'(x) \, dx$ to obtain the integral
>
> $$\int f(u) \, du.$$
>
> **Step 2:** Integrate with respect to u.
> **Step 3:** Replace u by $g(x)$ in the result.

These three steps are the steps of the substitution method of integration. The method works because $F(g(x))$ is an antiderivative of $f(g(x)) \cdot g'(x)$ whenever F is an antiderivative of f:

$$\frac{d}{dx} F(g(x)) = F'(g(x)) \cdot g'(x) \qquad \text{Chain Rule}$$

$$= f(g(x)) \cdot g'(x) \qquad \text{Because } F' = f$$

Implicit in the substitution method is the assumption that we are replacing x by a function of u. Thus, the substitution $u = g(x)$ must be solvable for x to give x as a function $x = g^{-1}(u)$ ("g inverse of u"). The domains of u and x may need to be restricted on occasion to make this possible. You need not be concerned with this issue at the moment. We will discuss inverses in Section 6.1 and treat the theory of substitutions in greater detail in Sections 7.4 and 13.7.

EXAMPLE 7

$$\int (x^2 + 2x - 3)^2 (x + 1)\, dx = \int u^2 \cdot \frac{1}{2}\, du$$

Let $u = x^2 + 2x - 3$,
$du = 2x\, dx + 2\, dx$
$= 2(x + 1)\, dx$,
$(1/2)\, du = (x + 1)\, dx$.

$$= \frac{1}{2}\int u^2\, du$$

$$= \frac{1}{2} \cdot \frac{u^3}{3} + C = \frac{1}{6} u^3 + C$$

Integrate with respect to u.

$$= \frac{1}{6}(x^2 + 2x - 3)^3 + C$$

Replace u.   ❑

EXAMPLE 8

$$\int \sin^4 t \cos t\, dt = \int u^4\, du$$

Let $u = \sin t$,
$du = \cos t\, dt$.

$$= \frac{u^5}{5} + C$$

Integrate with respect to u.

$$= \frac{\sin^5 t}{5} + C$$

Replace u.   ❑

The success of the substitution method depends on finding a substitution that will change an integral we cannot evaluate directly into one that we can. If the first substitution fails, we can try to simplify the integrand further with an additional substitution or two. (You will see what we mean if you do Exercises 47 and 48.) Alternatively, we can start afresh. There can be more than one good way to start, as in the next example.

EXAMPLE 9   Evaluate

$$\int \frac{2z\, dz}{\sqrt[3]{z^2 + 1}}.$$

Solution   We can use the substitution method of integration as an exploratory tool: substitute for the most troublesome part of the integrand and see how things work out. For the integral here, we might try $u = z^2 + 1$ or we might even press our luck and take u to be the entire cube root. Here is what happens in each case.

Solution 1   Substitute $u = z^2 + 1$.

$$\int \frac{2z\, dz}{\sqrt[3]{z^2 + 1}} = \int \frac{du}{u^{1/3}}$$

Let $u = z^2 + 1$,
$du = 2z\, dz$.

$$= \int u^{-1/3}\, du$$

In the form $\int u^n\, du$

$$= \frac{u^{2/3}}{2/3} + C$$

Integrate with respect to u.

$$= \frac{3}{2} u^{2/3} + C$$

$$= \frac{3}{2}(z^2 + 1)^{2/3} + C$$

Replace u by $z^2 + 1$.

Solution 2 Substitute $u = \sqrt[3]{z^2 + 1}$ instead.

$$\int \frac{2z\,dz}{\sqrt[3]{z^2+1}} = \int \frac{3u^2\,du}{u} \qquad \begin{array}{l}\text{Let } u = \sqrt[3]{z^2+1},\\ u^3 = z^2 + 1,\\ 3u^2\,du = 2z\,dz.\end{array}$$

$$= 3\int u\,du$$

$$= 3 \cdot \frac{u^2}{2} + C \qquad \begin{array}{l}\text{Integrate with respect}\\ \text{to } u.\end{array}$$

$$= \frac{3}{2}(z^2 + 1)^{2/3} + C \qquad \begin{array}{l}\text{Replace } u \text{ by}\\ (z^2 + 1)^{1/3}.\end{array}$$

Exercises 4.3

Evaluating Integrals

Evaluate the indefinite integrals in Exercises 1–12 by using the given substitutions to reduce the integrals to standard form.

1. $\int \sin 3x\,dx, \quad u = 3x$

2. $\int x \sin (2x^2)\,dx, \quad u = 2x^2$

3. $\int \sec 2t \tan 2t\,dt, \quad u = 2t$

4. $\int \left(1 - \cos \frac{t}{2}\right)^2 \sin \frac{t}{2}\,dt, \quad u = 1 - \cos \frac{t}{2}$

5. $\int 28(7x - 2)^{-5}\,dx, \quad u = 7x - 2$

6. $\int x^3(x^4 - 1)^2\,dx, \quad u = x^4 - 1$

7. $\int \frac{9r^2\,dr}{\sqrt{1 - r^3}}, \quad u = 1 - r^3$

8. $\int 12(y^4 + 4y^2 + 1)^2(y^3 + 2y)\,dy, \quad u = y^4 + 4y^2 + 1$

9. $\int \sqrt{x} \sin^2(x^{3/2} - 1)\,dx, \quad u = x^{3/2} - 1$

10. $\int \frac{1}{x^2} \cos^2 \left(\frac{1}{x}\right)\,dx, \quad u = -\frac{1}{x}$

11. $\int \csc^2 2\theta \cot 2\theta\,d\theta$

    a) Using $u = \cot 2\theta$ **b)** Using $u = \csc 2\theta$

12. $\int \frac{dx}{\sqrt{5x + 8}}$

    a) Using $u = 5x + 8$ **b)** Using $u = \sqrt{5x + 8}$

Evaluate the integrals in Exercises 13–46.

13. $\int \sqrt{3 - 2s}\,ds$

14. $\int (2x + 1)^3\,dx$

15. $\int \frac{1}{\sqrt{5s + 4}}\,ds$

16. $\int \frac{3\,dx}{(2 - x)^2}$

17. $\int \theta \sqrt[4]{1 - \theta^2}\,d\theta$

18. $\int 8\theta \sqrt[3]{\theta^2 - 1}\,d\theta$

19. $\int 3y\sqrt{7 - 3y^2}\,dy$

20. $\int \frac{4y\,dy}{\sqrt{2y^2 + 1}}$

21. $\int \frac{1}{\sqrt{x}(1 + \sqrt{x})^2}\,dx$

22. $\int \frac{(1 + \sqrt{x})^3}{\sqrt{x}}\,dx$

23. $\int \cos (3z + 4)\,dz$

24. $\int \sin (8z - 5)\,dz$

25. $\int \sec^2(3x + 2)\,dx$

26. $\int \tan^2 x \sec^2 x\,dx$

27. $\int \sin^5 \frac{x}{3} \cos \frac{x}{3}\,dx$

28. $\int \tan^7 \frac{x}{2} \sec^2 \frac{x}{2}\,dx$

29. $\int r^2 \left(\frac{r^3}{18} - 1\right)^5\,dr$

30. $\int r^4 \left(7 - \frac{r^5}{10}\right)^3\,dr$

31. $\int x^{1/2} \sin (x^{3/2} + 1)\,dx$

32. $\int x^{1/3} \sin (x^{4/3} - 8)\,dx$

33. $\int \sec\left(v + \frac{\pi}{2}\right)\tan\left(v + \frac{\pi}{2}\right)dv$

34. $\int \csc\left(\frac{v - \pi}{2}\right)\cot\left(\frac{v - \pi}{2}\right)dv$

35. $\int \dfrac{\sin(2t + 1)}{\cos^2(2t + 1)}\,dt$

36. $\int \dfrac{6\cos t}{(2 + \sin t)^3}\,dt$

37. $\int \sqrt{\cot y}\,\csc^2 y\,dy$

38. $\int \dfrac{\sec z \tan z}{\sqrt{\sec z}}\,dz$

39. $\int \dfrac{1}{t^2}\cos\left(\dfrac{1}{t} - 1\right)dt$

40. $\int \dfrac{1}{\sqrt{t}}\cos(\sqrt{t} + 3)\,dt$

41. $\int \dfrac{1}{\theta^2}\sin\dfrac{1}{\theta}\cos\dfrac{1}{\theta}\,d\theta$

42. $\int \dfrac{\cos\sqrt{\theta}}{\sqrt{\theta}\,\sin^2\sqrt{\theta}}\,d\theta$

43. $\int (s^3 + 2s^2 - 5s + 5)(3s^2 + 4s - 5)\,ds$

44. $\int (\theta^4 - 2\theta^2 + 8\theta - 2)(\theta^3 - \theta + 2)\,d\theta$

45. $\int t^3(1 + t^4)^3\,dt$

46. $\int \sqrt{\dfrac{x - 1}{x^5}}\,dx$

Simplifying Integrals Step by Step

If you do not know what substitution to make, try reducing the integral step by step, using a trial substitution to simplify the integral a bit and then another to simplify it some more. You will see what we mean if you try the sequences of substitutions in Exercises 47 and 48.

47. $\int \dfrac{18\tan^2 x \sec^2 x}{(2 + \tan^3 x)^2}\,dx$

   a) $u = \tan x$, followed by $v = u^3$, then by $w = 2 + v$
   b) $u = \tan^3 x$, followed by $v = 2 + u$
   c) $u = 2 + \tan^3 x$

48. $\int \sqrt{1 + \sin^2(x - 1)}\,\sin(x - 1)\cos(x - 1)\,dx$

   a) $u = x - 1$, followed by $v = \sin u$, then by $w = 1 + v^2$
   b) $u = \sin(x - 1)$, followed by $v = 1 + u^2$
   c) $u = 1 + \sin^2(x - 1)$

Evaluate the integrals in Exercises 49 and 50.

49. $\int \dfrac{(2r - 1)\cos\sqrt{3(2r - 1)^2 + 6}}{\sqrt{3(2r - 1)^2 + 6}}\,dr$

50. $\int \dfrac{\sin\sqrt{\theta}}{\sqrt{\theta}\cos^3\sqrt{\theta}}\,d\theta$

Initial Value Problems

Solve the initial value problems in Exercises 51–56.

51. $\dfrac{ds}{dt} = 12t\,(3t^2 - 1)^3, \quad s(1) = 3$

52. $\dfrac{dy}{dx} = 4x\,(x^2 + 8)^{-1/3}, \quad y(0) = 0$

53. $\dfrac{ds}{dt} = 8\sin^2\left(t + \dfrac{\pi}{12}\right), \quad s(0) = 8$

54. $\dfrac{dr}{d\theta} = 3\cos^2\left(\dfrac{\pi}{4} - \theta\right), \quad r(0) = \dfrac{\pi}{8}$

55. $\dfrac{d^2s}{dt^2} = -4\sin\left(2t - \dfrac{\pi}{2}\right), \quad s'(0) = 100,\ s(0) = 0$

56. $\dfrac{d^2y}{dx^2} = 4\sec^2 2x\,\tan 2x, \quad y'(0) = 4,\ y(0) = -1$

57. The velocity of a particle moving back and forth on a line is $v = ds/dt = 6\sin 2t$ m/sec for all t. If $s = 0$ when $t = 0$, find the value of s when $t = \pi/2$ sec.

58. The acceleration of a particle moving back and forth on a line is $a = d^2s/dt^2 = \pi^2\cos \pi t$ m/sec^2 for all t. If $s = 0$ and $v = 8$ m/sec when $t = 0$, find s when $t = 1$ sec.

Theory and Examples

59. It looks as if we can integrate $2\sin x \cos x$ with respect to x in three different ways:

   a) $\displaystyle\int 2\sin x \cos x\,dx = \int 2u\,du \qquad u = \sin x,$

               $= u^2 + C_1 = \sin^2 x + C_1$

   b) $\displaystyle\int 2\sin x \cos x\,dx = \int -2u\,du \qquad u = \cos x,$

               $= -u^2 + C_2 = -\cos^2 x + C_2$

   c) $\displaystyle\int 2\sin x \cos x\,dx = \int \sin 2x\,dx \qquad 2\sin x \cos x = \sin 2x$

               $= -\dfrac{\cos 2x}{2} + C_3.$

Can all three integrations be correct? Give reasons for your answer.

60. The substitution $u = \tan x$ gives

$$\int \sec^2 x \tan x\,dx = \int u\,du = \frac{u^2}{2} + C = \frac{\tan^2 x}{2} + C.$$

The substitution $u = \sec x$ gives

$$\int \sec^2 x \tan x\,dx = \int u\,du = \frac{u^2}{2} + C = \frac{\sec^2 x}{2} + C.$$

Can both integrations be correct? Give reasons for your answer.

Estimating with Finite Sums

This section shows how practical questions can lead in natural ways to approximations by finite sums.

Area and Cardiac Output

The number of liters of blood your heart pumps in a minute is called your *cardiac output*. For a person at rest, the rate might be 5 or 6 liters per minute. During strenuous exercise the rate might be as high as 30 liters per minute. It might also be altered significantly by disease.

Instead of measuring a patient's cardiac output with exhaled carbon dioxide, as in Exercise 25 in Section 2.7, a doctor may prefer to use the dye-dilution technique described here. You inject 5 to 10 mg of dye in a main vein near the heart. The dye is drawn into the right side of the heart and pumped through the lungs and out the left side of the heart into the aorta, where its concentration can be measured every few seconds as the blood flows past. The data in Table 4.3 and the plot in Fig. 4.5 show the response of a healthy, resting patient to an injection of 5.6 mg of dye.

To calculate the patient's cardiac output, we divide the amount of dye by the area under the dye concentration curve and multiply the result by 60:

$$\text{Cardiac output} = \frac{\text{amount of dye}}{\text{area under curve}} \times 60. \qquad (1)$$

You can see why the formula works if you check the units in which the various quantities are measured. The amount of dye is in milligrams and the area is in (milligrams/liter) × seconds, which gives cardiac output in liters/minute:

$$\frac{\text{mg}}{\dfrac{\text{mg}}{\text{L}} \cdot \text{sec}} \cdot \frac{\text{sec}}{\text{min}} = \text{mg} \cdot \frac{\text{L}}{\text{mg} \cdot \text{sec}} \cdot \frac{\text{sec}}{\text{min}} = \frac{\text{L}}{\text{min}}.$$

In the example that follows, we estimate the area under the concentration curve in Fig. 4.5 and find the patient's cardiac output.

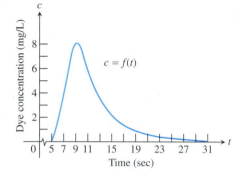

4.5 The dye concentrations from Table 4.3, plotted and fitted with a smooth curve. Time is measured with $t = 0$ at the time of injection. The dye concentrations are zero at the beginning, while the dye passes through the lungs. They then rise to a maximum at about $t = 9$ sec and taper to zero by $t = 31$ sec.

Table 4.3 Dye-dilution data

Seconds after injection t	Dye concentration (adjusted for recirculation) c	Seconds after injection t	Dye concentration (adjusted for recirculation) c
5	0	19	0.91
7	3.8	21	0.57
9	8.0	23	0.36
11	6.1	25	0.23
13	3.6	27	0.14
15	2.3	29	0.09
17	1.45	31	0

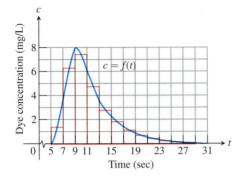

4.6 The region under the concentration curve of Fig. 4.5 is approximated with rectangles. We ignore the portion from $t = 29$ to $t = 31$; its concentration is negligible.

EXAMPLE 1   Find the cardiac output of the patient whose data appear in Table 4.3 and Fig. 4.5.

Solution   We know the amount of dye to use in Eq. (1) (it is 5.6 mg), so all we need is the area under the concentration curve. None of the area formulas we know can be used for this irregularly shaped region. But we can get a good estimate of this area by approximating the region between the curve and the t-axis with rectangles and adding the areas of the rectangles (Fig. 4.6). Each rectangle omits some of the area under the curve but includes area from outside the curve, which compensates. In Fig. 4.6 each rectangle has a base 2 units long and a height that is equal to the height of the curve above the midpoint of the base. The rectangle's height acts as a sort of average value of the function over the time interval on which the rectangle stands. After reading rectangle heights from the curve, we multiply each rectangle's height and base to find its area, and then get the following estimate:

$$\text{Area under curve} \approx \text{sum of rectangle areas}$$

$$\approx f(6) \cdot 2 + f(8) \cdot 2 + f(10) \cdot 2 + \cdots + f(28) \cdot 2$$

$$\approx (1.4)(2) + (6.3)(2) + (7.5)(2) + \cdots + (0.1)(2)$$

$$\approx (28.8)(2) = 57.6 \text{ mg} \cdot \text{sec/L}. \tag{2}$$

Dividing this figure into the amount of dye and multiplying by 60 gives a corresponding estimate of the cardiac output:

$$\text{Cardiac output} \approx \frac{\text{amount of dye}}{\text{area estimate}} \times 60 = \frac{5.6}{57.6} \times 60 \approx 5.8 \text{ L/min}.$$

The patient's cardiac output is about 5.8 L/min.   ❑

Technology   *Using a Grapher to Calculate Finite Sums*   If your graphing utility has a method for evaluating sums, you might want to use it in this section. Later in the chapter, you will find it useful for approximating "definite" integrals. There will be other uses still later in your study of calculus.

Distance Traveled

Suppose we know the velocity function $v = ds/dt = f(t)$ m/sec of a car moving down a highway and want to know how far the car will travel in the time interval $a \leq t \leq b$. If we know an antiderivative F of f, we can find the car's position function $s = F(t) + C$ and calculate the distance traveled as the difference between the car's positions at times $t = a$ and $t = b$ (as in Section 4.2, Exercise 55).

If we do not know an antiderivative of $v = f(t)$, we can approximate the answer with a sum in the following way. We partition $[a, b]$ into short time intervals *on each of which v is fairly constant*. Since velocity is the rate at which the car is traveling, we approximate the distance traveled on each time interval with the formula

$$\text{Distance} = \text{rate} \times \text{time} = f(t) \cdot \Delta t$$

and add the results across $[a, b]$. To be specific, suppose the partitioned interval looks like this with the subintervals all of length Δt. Let t_1 be a point in the first subinterval. If the interval is short enough so the rate is almost constant, the car will move about $f(t_1)\Delta t$ m during that interval. If t_2 is a point in the second interval, the car will move an additional $f(t_2)\Delta t$ m during that interval, and so on. The sum of these products approximates the total distance D traveled from $t = a$ to $t = b$. If we use n subintervals, then

$$D \approx f(t_1)\,\Delta t + f(t_2)\,\Delta t + \cdots + f(t_n)\,\Delta t. \qquad (3)$$

Let's try this on the projectile in Example 3, Section 4.2. The projectile was fired straight into the air. Its velocity t sec into the flight was $v = f(t) = 160 - 9.8t$ and it rose 435.9 m from a height of 3 m to a height of 438.9 m during the first 3 sec of flight.

EXAMPLE 2   The velocity function of a projectile fired straight into the air is $f(t) = 160 - 9.8t$. Use the summation technique just described to estimate how far the projectile rises during the first 3 sec. How close do the sums come to the exact figure of 435.9 m?

Solution   We explore the results for different numbers of intervals and different choices of evaluation points.

3 *subintervals of length* 1, *with f evaluated at left-hand endpoints:*

With f evaluated at $t = 0$, 1, and 2, we have

$$D \approx f(t_1)\,\Delta t + f(t_2)\,\Delta t + f(t_3)\,\Delta t \qquad \text{Eq. (1)}$$

$$\approx [160 - 9.8(0)](1) + [160 - 9.8(1)](1) + [160 - 9.8(2)](1)$$

$$\approx 450.6.$$

3 *subintervals of length* 1, *with f evaluated at right-hand endpoints:*

With f evaluated at $t = 1$, 2, and 3, we have

$$D \approx f(t_1)\,\Delta t + f(t_2)\,\Delta t + f(t_3)\,\Delta t \qquad \text{Eq. (1)}$$

$$\approx [160 - 9.8(1)](1) + [160 - 9.8(2)](1) + [160 - 9.8(3)](1)$$

$$\approx 421.2.$$

With 6 *subintervals of length* 1/2, *we get*

Using left-hand endpoints: $D \approx 443.25$.

Using right-hand endpoints: $D \approx 428.55$.

These six-interval estimates are somewhat closer than the three-interval estimates. The results improve as the subintervals get shorter.

Table 4.4 Travel-distance estimates

Number of subintervals	Length of each subinterval	Left-endpoint sum	Right-endpoint sum
3	1	450.6	421.2
6	0.5	443.25	428.55
12	0.25	439.58	432.23
24	0.125	437.74	434.06
48	0.0625	436.82	434.98
96	0.03125	436.36	435.44
192	0.015625	436.13	435.67

Error magnitude = |true value − calculated value|

As we can see in Table 4.4, the left-endpoint sums approach the true value 435.9 from above while the right-endpoint sums approach it from below. The true value lies between these upper and lower sums. The magnitude of the error in the closest entries is 0.23, a small percentage of the true value.

$$\text{Error percentage} = \frac{0.23}{435.9} \approx 0.05\%.$$

It would be safe to conclude from the table's last entries that the projectile rose about 436 m during its first 3 sec of flight.   ❑

Notice the mathematical similarity between Examples 1 and 2. In each case, we have a function f defined on a closed interval and estimate what we want to know with a sum of function values multiplied by interval lengths. We can use similar sums to estimate volumes.

Volume

Here are two examples using finite sums to estimate volumes.

EXAMPLE 3   A solid lies between planes perpendicular to the x-axis at $x = -2$ and $x = 2$. The cross sections of the solid perpendicular to the axis between these planes are vertical squares whose base edges run from the semicircle $y = -\sqrt{9 - x^2}$ to the semicircle $y = \sqrt{9 - x^2}$ (Fig. 4.7a, on the following page). The height of the square at x is $2\sqrt{9 - x^2}$. Estimate the volume of the solid.

Solution   We partition the interval $[-2, \ 2]$ on the x-axis into four subintervals of length $\Delta x = 1$. The solid's cross section at the left-hand endpoint of each subinterval is a square (Fig. 4.7b). On each of these squares we construct a right cylinder (square slab) of height 1 extending to the right (Fig. 4.7c). We add the cylinders' volumes to estimate the volume of the solid.

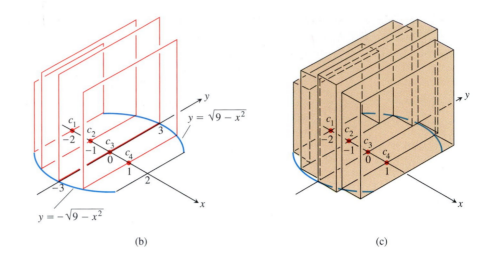

(a) (b) (c)

4.7 (a) The solid in Example 3. (b) Square cross sections of the solid at $x = -2, -1, 0,$ and 1. (c) Rectangular cylinders (slabs) based on the cross sections to approximate the solid.

We calculate the volume of each cylinder with the formula $V = Ah$ (base area × height). The area of the solid's cross section at x is $A(x) = (\text{side})^2 = (2\sqrt{9 - x^2})^2 = 4(9 - x^2)$, so the sum of the volumes of the cylinders is

$$S_4 = A(c_1)\,\Delta x + A(c_2)\,\Delta x + A(c_3)\,\Delta x + A(c_4)\,\Delta x$$

$$= 4(9 - c_1{}^2)(1) + 4(9 - c_2{}^2)(1) + 4(9 - c_3{}^2)(1) + 4(9 - c_4{}^2)(1)$$

$$= 4\big[(9 - (-2)^2) + (9 - (-1)^2) + (9 - (0)^2) + (9 - (1)^2)\big]$$

$$= 4\big[(9 - 4) + (9 - 1) + (9 - 0) + (9 - 1)\big]$$

$$= 4(36 - 6) = 120.$$

This compares favorably with the solid's true volume $V = 368/3 \approx 122.67$ (we will see how to calculate V in Section 4.7). The difference between S and V is a small percentage of V:

$$\text{Error percentage} = \frac{|V - S_4|}{V} = \frac{(368/3) - 120}{(368/3)}$$

$$= \frac{8}{368} \approx 2.2\%.$$

With a finer partition (more subintervals) the approximation would be even better.

❏

EXAMPLE 4   Estimate the volume of a solid sphere of radius 4.

Solution   We picture the sphere as if its surface were generated by revolving the graph of the function $f(x) = \sqrt{16 - x^2}$ about the x-axis (Fig. 4.8a). We partition the interval $-4 \leq x \leq 4$ into 8 subintervals of length $\Delta x = 1$. We then approximate the solid with right circular cylinders based on cross sections of the solid by planes perpendicular to the x-axis at the subintervals' left-hand endpoints (Fig. 4.8b). (The cylinder at $x = -4$ is degenerate because the cross section there is just a point.) We add the cylinders' volumes to estimate the volume of a sphere.

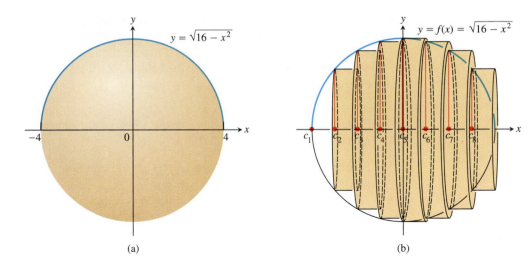

4.8 (a) The semicircle $y = \sqrt{16 - x^2}$ revolved about the x-axis to outline a sphere. (b) The solid sphere approximated with cross-section-based cylinders.

We calculate the volume of each cylinder with the formula $V = \pi r^2 h$. The sum of the eight cylinders' volumes is

$$S_8 = \pi \, [f(c_1)]^2 \, \Delta x + \pi \, [f(c_2)]^2 \, \Delta x + \pi \, [f(c_3)]^2 \, \Delta x + \cdots + \pi \, [f(c_8)]^2 \, \Delta x$$

$$= \pi \left[\sqrt{16 - c_1{}^2} \right]^2 \Delta x + \pi \left[\sqrt{16 - c_2{}^2} \right]^2 \Delta x + \pi \left[\sqrt{16 - c_3{}^2} \right]^2 \Delta x$$

$$+ \cdots + \pi \left[\sqrt{16 - c_8{}^2} \right]^2 \Delta x$$

$$= \pi \left[(16 - (-4)^2) + (16 - (-3)^2) + (16 - (-2)^2) + \cdots + (16 - (3)^2) \right]$$

$$= \pi \, [0 + 7 + 12 + 15 + 16 + 15 + 12 + 7]$$

$$= 84 \, \pi.$$

This compares favorably with the sphere's true volume,

$$V = \frac{4}{3} \, \pi \, r^3 = \frac{4}{3} \, \pi \, (4)^3 = \frac{256\pi}{3}.$$

The difference between S_8 and V is a small percentage of V:

$$\text{Error percentage} = \frac{|V - S_8|}{V} = \frac{(256/3)\pi - 84\pi}{(256/3)\pi}$$

$$= \frac{256 - 252}{256} = \frac{1}{64} \approx 1.6\%. \qquad \square$$

The Average Value of a Nonnegative Function

To find the average of a finite set of values, we add them and divide by the number of values added. But what happens if we want to find the average of an infinite number of values? For example, what is the average value of the function $f(x) = x^2$ on the interval $[-1, 1]$? To see what this kind of "continuous" average might mean, imagine that we are pollsters sampling the function. We pick random x's between -1 and 1, square them, and average the squares. As we take larger samples, we expect this average to approach some number, which seems reasonable to call the *average of f over* $[-1, 1]$.

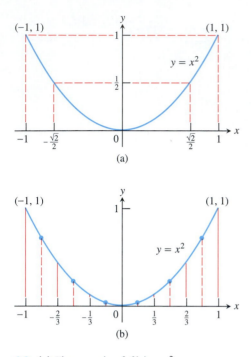

(a)

(b)

4.9 (a) The graph of $f(x) = x^2$, $-1 \le x \le 1$. (b) Values of f sampled at regular intervals.

The graph in Fig. 4.9(a) suggests that the average square should be less than 1/2, because numbers with squares less than 1/2 make up more than 70% of the interval $[-1, 1]$. If we had a computer to generate random numbers, we could carry out the sampling experiment described above, but it is much easier to estimate the average value with a finite sum.

EXAMPLE 5   Estimate the average value of the function $f(x) = x^2$ on the interval $[-1, 1]$.

Solution   We look at the graph of $y = x^2$ and partition the interval $[0, 1]$ into 6 subintervals of length $\Delta x = 1/3$ (Fig. 4.9b).

It appears that a good estimate for the average square on each subinterval is the square of the midpoint of the subinterval. Since the subintervals have the same length, we can average these six estimates to get a final estimate for the average value over $[-1, 1]$.

$$
\text{Average value} \approx \frac{\left(-\dfrac{5}{6}\right)^2 + \left(-\dfrac{3}{6}\right)^2 + \left(-\dfrac{1}{6}\right)^2 + \left(\dfrac{1}{6}\right)^2 + \left(\dfrac{3}{6}\right)^2 + \left(\dfrac{5}{6}\right)^2}{6}
$$

$$
\approx \frac{1}{6} \cdot \frac{25 + 9 + 1 + 1 + 9 + 25}{36} = \frac{70}{216} \approx 0.324
$$

We will be able to show later that the average value is 1/3.
Notice that

$$
\frac{\left(-\dfrac{5}{6}\right)^2 + \left(-\dfrac{3}{6}\right)^2 + \left(-\dfrac{1}{6}\right)^2 + \left(\dfrac{1}{6}\right)^2 + \left(\dfrac{3}{6}\right)^2 + \left(\dfrac{5}{6}\right)^2}{6}
$$

$$
= \frac{1}{2}\left[\left(-\frac{5}{6}\right)^2 \cdot \frac{1}{3} + \left(-\frac{3}{6}\right)^2 \cdot \frac{1}{3} + \cdots + \left(\frac{5}{6}\right)^2 \cdot \frac{1}{3}\right]
$$

$$
= \frac{1}{\text{length of } [-1, 1]} \cdot \left[f\left(-\frac{5}{6}\right) \cdot \frac{1}{3} + f\left(-\frac{3}{6}\right) \cdot \frac{1}{3} + \cdots + f\left(\frac{5}{6}\right) \cdot \frac{1}{3}\right]
$$

$$
= \frac{1}{\text{length of } [-1, 1]} \cdot \left[\begin{array}{c}\text{a sum of function values}\\ \text{multiplied by interval lengths}\end{array}\right].
$$

Once again our estimate has been achieved by multiplying function values by interval lengths and summing the results for all the intervals.   ▢

Conclusion

The examples in this section describe instances in which sums of function values multiplied by interval lengths provide approximations that are good enough to answer practical questions. You will find additional examples in the exercises.

The distance approximations in Example 2 improved as the intervals involved became shorter and more numerous. We knew this because we had already found the exact answer with antiderivatives in Section 4.2. If we had made our partitions of the time interval still finer, would the sums have approached the exact answer as a limit? Is the connection between the sums and the antiderivative in this case just a coincidence? Could we have calculated the area in Example 1, the volumes in

Examples 3 and 4, and the average value in Example 5 with antiderivatives as well? As we will see, the answers are "Yes, they would have," "No, it is not a coincidence," and "Yes, we could have."

Exercises 4.4

Cardiac Output

1. The table below gives dye concentrations for a dye-dilution cardiac-output determination like the one in Example 1. The amount of dye injected in this case was 5 mg instead of 5.6 mg. Use rectangles to estimate the area under the dye concentration curve and then go on to estimate the patient's cardiac output.

Seconds after injection t	Dye concentration (adjusted for recirculation) c
2	0
4	0.6
6	1.4
8	2.7
10	3.7
12	4.1
14	3.8
16	2.9
18	1.7
20	1.0
22	0.5
24	0

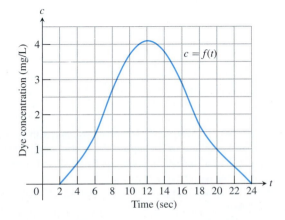

2. The accompanying table gives dye concentrations for a cardiac-output determination like the one in Example 1. The amount of dye injected in this case was 10 mg. Plot the data and connect the data points with a smooth curve. Estimate the area under the curve and calculate the cardiac output from this estimate.

Seconds after injection t	Dye concentration (adjusted for recirculation) c	Seconds after injection t	Dye concentration (adjusted for recirculation) c
0	0	16	7.9
2	0	18	7.8
4	0.1	20	6.1
6	0.6	22	4.7
8	2.0	24	3.5
10	4.2	26	2.1
12	6.3	28	0.7
14	7.5	30	0

Distance

3. The table below shows the velocity of a model train engine moving along a track for 10 sec. Estimate the distance traveled by the engine using 10 subintervals of length 1 with (a) left-endpoint values and (b) right-endpoint values.

Time (sec)	Velocity (in./sec)	Time (sec)	Velocity (in./sec)
0	0	6	11
1	12	7	6
2	22	8	2
3	10	9	6
4	5	10	0
5	13		

4. You are sitting on the bank of a tidal river watching the incoming tide carry a bottle upstream. You record the velocity of the flow every five minutes for an hour, with the results shown in the table on the following page. About how far upstream did the bottle travel during that hour? Find an estimate using 12 subintervals of length 5 with (a) left-endpoint values and (b) right-endpoint values.

Time (min)	Velocity (m/sec)	Time (min)	Velocity (m/sec)
0	1	35	1.2
5	1.2	40	1.0
10	1.7	45	1.8
15	2.0	50	1.5
20	1.8	55	1.2
25	1.6	60	0
30	1.4		

5. You and a companion are about to drive a twisty stretch of dirt road in a car whose speedometer works but whose odometer (mileage counter) is broken. To find out how long this particular stretch of road is, you record the car's velocity at 10-sec intervals, with the results shown in the table below. Estimate the length of the road (a) using left-endpoint values and (b) using right-endpoint values.

Time (sec)	Velocity (converted to ft/sec) (30 mi/h = 44 ft/sec)	Time (sec)	Velocity (converted to ft/sec) (30 mi/h = 44 ft/sec)
0	0	70	15
10	44	80	22
20	15	90	35
30	35	100	44
40	30	110	30
50	44	120	35
60	35		

6. The table below gives data for the velocity of a vintage sports car accelerating from 0 to 142 mi/h in 36 sec (10 thousandths of an hour).

Time (h)	Velocity (mi/h)	Time (h)	Velocity (mi/h)
0.0	0	0.006	116
0.001	40	0.007	125
0.002	62	0.008	132
0.003	82	0.009	137
0.004	96	0.010	142
0.005	108		

a) Use rectangles to estimate how far the car traveled during the 36 sec it took to reach 142 mi/h.

b) Roughly how many seconds did it take the car to reach the halfway point? About how fast was the car going then?

Volume

7. *(Continuation of Example 3.)* Suppose we use only two square cylinders to estimate the volume V of the solid in Example 3, as shown in profile in the figure here.

a) Find the sum S_2 of the volumes of the cylinders.

b) Express $|V - S_2|$ as a percentage of V to the nearest percent.

8. *(Continuation of Example 3.)* Suppose we use six square cylinders to estimate the volume V of the solid in Example 3, as shown in the accompanying profile view.

a) Find the sum S_6 of the volumes of the cylinders.

b) Express $|V - S_6|$ as a percentage of V to the nearest percent.

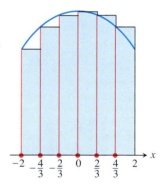

9. *(Continuation of Example 4.)* Suppose we approximate the volume V of the sphere in Example 4 by partitioning the interval $-4 \le x \le 4$ into four subintervals of length 2 and using cylinders based on the cross sections at the subintervals' left-hand endpoints. (As in Example 4, the leftmost cylinder will have a zero radius.)

a) Find the sum S_4 of the volumes of the cylinders.

b) Express $|V - S_4|$ as a percentage of V to the nearest percent.

10. To estimate the volume V of a solid sphere of radius 5 you partition its diameter into five subintervals of length 2. You then slice the sphere with planes perpendicular to the diameter at the subintervals' left-hand endpoints and add the volumes of cylinders of height 2 based on the cross sections of the sphere determined by these planes.

a)   Find the sum S_5 of the volumes of the cylinders.

b)   Express $|V - S_5|$ as a percentage of V to the nearest percent.

11. To estimate the volume V of a solid hemisphere of radius 4, imagine its axis of symmetry to be the interval $[0, 4]$ on the x-axis. Partition $[0, 4]$ into eight subintervals of equal length and approximate the solid with cylinders based on the circular cross sections of the hemisphere perpendicular to the x-axis at the subintervals' left-hand endpoints. (See the accompanying profile view.)

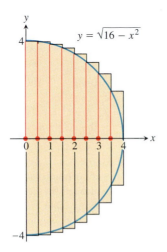

$$y = \sqrt{16 - x^2}$$

a)   Find the sum S_8 of the volumes of the cylinders. Do you expect S_8 to overestimate V, or to underestimate V? Give reasons for your answer.

b)   Express $|V - S_8|$ as a percentage of V to the nearest percent.

12. Repeat Exercise 11 using cylinders based on cross sections at the *right-hand* endpoints of the subintervals.

13. *Estimates with large error.* A solid lies between planes perpendicular to the x-axis at $x = 0$ and $x = 4$. The cross sections of the solid perpendicular to the axis between these planes are vertical squares whose base edges run from the parabolic curve $y = -\sqrt{x}$ to the parabolic curve $y = \sqrt{x}$.

$$x = y^2$$

a)   Find the sum S_4 of the volumes of the cylinders obtained by partitioning $0 \le x \le 4$ into four subintervals of length 1

based on the cross sections at the subinterval's right-hand endpoints.

b)   The true volume is $V = 32$. Express $|V - S_4|$ as a percentage of V to the nearest percent.

c)   Repeat parts (a) and (b) for the sum S_8.

14. *Estimates with large error.* A solid lies between planes perpendicular to the x-axis at $x = 0$ and $x = 4$. The cross sections of the solid perpendicular to the axis between these planes are vertical equilateral triangles whose base edges run from the parabolic curve $y = -\sqrt{x}$ to the parabolic curve $y = \sqrt{x}$.

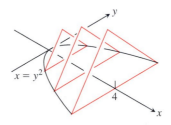

$$x = y^2$$

a)   Find the sum S_4 of the volumes of the cylinders obtained by partitioning $0 \le x \le 4$ into four subintervals of length 1 based on the cross sections at the subinterval's left-hand endpoints.

b)   The true volume is $V = 8\sqrt{3}$. Express $|V - S_4|$ as a percentage of V to the nearest percent.

c)   CALCULATOR Repeat parts (a) and (b) for the sum S_8.

15. A reservoir shaped like a hemispherical bowl of radius 8 m is filled with water to a depth of 4 m. (a) Find an estimate S of the water's volume by approximating the water with eight circumscribed solid cylinders. (b) As you will see in Section 4.7, Exercise 71, the water's volume is $V = 320\pi/3$ m^3. Find the error $|V - S|$ as a percentage of V to the nearest percent.

16. A rectangular swimming pool is 30 ft wide and 50 ft long. The table below shows the depth $h(x)$ of the water at 5-ft intervals from one end of the pool to the other. Estimate the volume of water in the pool using (a) left-endpoint values of h; (b) right-endpoint values of h.

Position x ft	Depth $h(x)$ ft	Position x ft	Depth $h(x)$ ft
0	6.0	30	11.5
5	8.2	35	11.9
10	9.1	40	12.3
15	9.9	45	12.7
20	10.5	50	13.0
25	11.0		

17. The nose "cone" of a rocket is a paraboloid obtained by revolving the curve $y = \sqrt{x}$, $0 \le x \le 5$, about the x-axis, where x is measured in feet. To estimate the volume V of the nose cone, we partition [0, 5] into five subintervals of equal length, slice the cone with planes perpendicular to the x-axis at the subintervals' left-hand endpoints, and construct cylinders of height 1 based on cross sections at these points. (See the accompanying figure.)

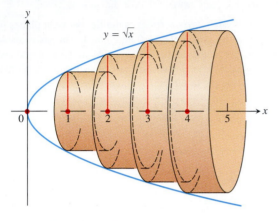

a) Find the sum S_5 of the volumes of the cylinders. Do you expect S_5 to overestimate V, or to underestimate V? Give reasons for your answer.

b) As you will see in Section 4.7, Exercise 72, the volume of the nose cone is $V = 25\pi/2$ ft³. Express $|V - S_5|$ as a percentage of V to the nearest percent.

18. Repeat Exercise 17 using cylinders based on cross sections at the *right-hand* endpoints of the subintervals.

Average Value of a Function

In Exercises 19–22, use a finite sum to estimate the average value of f on the given interval by partitioning the interval into four subintervals of equal length and evaluating f at the subinterval midpoints.

19. $f(x) = x^3$ on $[0, 2]$ 20. $f(x) = 1/x$ on $[1, 9]$

21. $f(t) = (1/2) + \sin^2 \pi t$ on $[0, 2]$

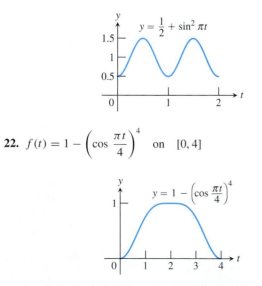

22. $f(t) = 1 - \left(\cos \dfrac{\pi t}{4}\right)^4$ on $[0, 4]$

Velocity and Distance

23. An object is dropped straight down from an airplane. The object falls faster and faster but the acceleration is decreasing over time because of air resistance. The acceleration is measured in ft/sec² and recorded every second after the drop for 5 sec, as shown in the following table.

t	0	1	2	3	4	5
a	32.00	19.41	11.77	7.14	4.33	2.63

a) Find an upper estimate for the speed when $t = 5$.
b) Find a lower estimate for the speed when $t = 5$.
c) Find an upper estimate for the distance fallen when $t = 3$.

24. An object is shot straight upward from sea level with an initial velocity of 400 ft/sec. Assuming gravity is the only force acting on the object, give an upper estimate for its speed after 5 sec have elapsed. Use $g = 32$ ft/sec² for the gravitational constant. Find a lower estimate for the height attained after 5 sec.

Pollution Control

25. Oil is leaking out of a tanker damaged at sea. The damage to the tanker is worsening as evidenced by the increased leakage each hour, recorded in the following table.

Time (hours)	0	1	2	3	4
Leakage (gal/hr)	50	70	97	136	190
Time (hours)	5	6	7	8	
Leakage (gal/hr)	265	369	516	720	

a) Give an upper and a lower estimate of the total quantity of oil that has escaped after 5 hours.
b) Repeat part (a) for the quantity of oil that has escaped after 8 hours.
c) The tanker continues to leak 720 gal/h after the first 8 hours. If the tanker originally contained 25,000 gal of oil, approximately how many more hours will elapse in the worst case before all of the oil has spilled? in the best case?

26. A power plant generates electricity by burning oil. Pollutants produced as a result of the burning process are removed by scrubbers in the smoke stacks. Over time the scrubbers become less efficient and eventually they must be replaced when the amount of pollution released exceeds government standards. Measurements are taken at the end of each month determining the rate at which pollutants are released into the atmosphere, recorded as follows.

Month	Jan	Feb	Mar	Apr	May	Jun	Jul	Aug	Sep	Oct	Nov	Dec
Pollutant release rate (tons/day)	0.2	0.25	0.27	0.34	0.45	0.52	0.63	0.70	0.81	0.85	0.89	0.95

a) Assuming a 30-day month and that new scrubbers allow only 0.05 tons/day released, give an upper estimate of the total tonnage of pollutant released by the end of June. What is a lower estimate?

b) In the best case, approximately when will a total of 125 tons of pollutant have been released into the atmosphere?

☼ CAS Explorations and Projects

In Exercises 27–30, use a CAS to perform the following steps:

a) Plot the functions over the given interval.

b) Partition the interval into $n = 100$, 200, and 1000 subintervals of equal length, and evaluate the function at the midpoint of each subinterval.

c) Compute the average value of the function values generated in part (b).

d) Solve the equation $f(x) = $ (average value) for x using the average value calculated in (c) for the $n = 1000$ partitioning.

27. $f(x) = \sin x$ on $[0, \pi]$ **28.** $f(x) = \sin^2 x$ on $[0, \pi]$

29. $f(x) = x \sin \dfrac{1}{x}$ on $\left[\dfrac{\pi}{4}, \pi\right]$

30. $f(x) = x \sin^2 \dfrac{1}{x}$ on $\left[\dfrac{\pi}{4}, \pi\right]$

4.5

Riemann Sums and Definite Integrals

In the preceding section, we estimated distances, areas, volumes, and average values with finite sums. The terms in the sums were obtained by multiplying selected function values by the lengths of intervals. In this section, we say what it means for sums like these to approach a limit as the intervals involved become more numerous and shorter. We begin by introducing a compact notation for sums that contain large numbers of terms.

Sigma Notation for Finite Sums

We use the capital Greek letter Σ ("sigma") to write an abbreviation for the sum

$$f(t_1)\,\Delta t + f(t_2)\,\Delta t + \cdots + f(t_n)\,\Delta t$$

as $\sum_{k=1}^{n} f(t_k)\,\Delta t$, "the sum from k equals 1 to n of f of t_k times delta t." When we write a sum this way, we say that we have written it in sigma notation.

Definitions
Sigma Notation for Finite Sums

The symbol $\sum_{k=1}^{n} a_k$ denotes the sum $a_1 + a_2 + \cdots + a_n$. The a's are the **terms** of the sum: a_1 is the first term, a_2 is the second term, a_k is the **kth term,** and a_n is the nth and last term. The variable k is the **index of summation.** The values of k run through the integers from 1 to n. The number 1 is the **lower limit of summation;** the number n is the **upper limit of summation.**

EXAMPLE 1

The sum in sigma notation	The sum written out—one term for each value of k	The value of the sum
$\displaystyle\sum_{k=1}^{5} k$	$1 + 2 + 3 + 4 + 5$	15
$\displaystyle\sum_{k=1}^{3} (-1)^k k$	$(-1)^1(1) + (-1)^2(2) + (-1)^3(3)$	$-1 + 2 - 3 = -2$
$\displaystyle\sum_{k=1}^{2} \frac{k}{k+1}$	$\dfrac{1}{1+1} + \dfrac{2}{2+1}$	$\dfrac{1}{2} + \dfrac{2}{3} = \dfrac{7}{6}$

The lower limit of summation does not have to be 1; it can be any integer.

EXAMPLE 2   Express the sum $1 + 3 + 5 + 7 + 9$ in sigma notation.

Solution

Starting with $k = 2$: $1 + 3 + 5 + 7 + 9 = \displaystyle\sum_{k=2}^{6} (2k - 3)$

Starting with $k = -3$: $1 + 3 + 5 + 7 + 9 = \displaystyle\sum_{k=-3}^{1} (2k + 7)$

The formula generating the terms changes with the lower limit of summation, but the terms generated remain the same. It is often simplest to start with $k = 0$ or $k = 1$.

Starting with $k = 0$: $1 + 3 + 5 + 7 + 9 = \displaystyle\sum_{k=0}^{4} (2k + 1)$

Starting with $k = 1$: $1 + 3 + 5 + 7 + 9 = \displaystyle\sum_{k=1}^{5} (2k - 1)$

Algebra with Finite Sums

We can use the following rules whenever we work with finite sums.

Algebra Rules for Finite Sums

1.   *Sum Rule:* $\displaystyle\sum_{k=1}^{n} (a_k + b_k) = \sum_{k=1}^{n} a_k + \sum_{k=1}^{n} b_k$

2.   *Difference Rule:* $\displaystyle\sum_{k=1}^{n} (a_k - b_k) = \sum_{k=1}^{n} a_k - \sum_{k=1}^{n} b_k$

3.   *Constant Multiple Rule:* $\displaystyle\sum_{k=1}^{n} ca_k = c \cdot \sum_{k=1}^{n} a_k$ (Any number c)

4.   *Constant Value Rule:* $\displaystyle\sum_{k=1}^{n} c = n \cdot c$ (c is any constant value.)

There are no surprises in this list. The formal proofs can be done by mathematical induction (Appendix 1).

EXAMPLE 3

a) $\sum_{k=1}^{n}(3k - k^2) = 3\sum_{k=1}^{n}k - \sum_{k=1}^{n}k^2$ — Difference Rule and Constant Multiple Rule

b) $\sum_{k=1}^{n}(-a_k) = \sum_{k=1}^{n}(-1)\cdot a_k = -1\cdot\sum_{k=1}^{n}a_k = -\sum_{k=1}^{n}a_k$ — Constant Multiple Rule

c) $\sum_{k=1}^{3}(k+4) = \sum_{k=1}^{3}k + \sum_{k=1}^{3}4$ — Sum Rule

$= (1+2+3) + (3\cdot 4)$ — Constant Value Rule

$= 6 + 12 = 18$

Sum Formulas for Positive Integers

Over the years people have discovered a variety of formulas for the values of finite sums. The most famous of these are the formula for the sum of the first n integers (Gauss discovered it at age 5) and the formulas for the sums of the squares and cubes of the first n integers.

The first n integers: $\sum_{k=1}^{n}k = \dfrac{n(n+1)}{2}$ (1)

The first n squares: $\sum_{k=1}^{n}k^2 = \dfrac{n(n+1)(2n+1)}{6}$ (2)

The first n cubes: $\sum_{k=1}^{n}k^3 = \left(\dfrac{n(n+1)}{2}\right)^2$ (3)

EXAMPLE 4 Evaluate $\sum_{k=1}^{4}(k^2 - 3k)$.

Solution We can use the algebra rules and known formulas to evaluate the sum without writing out the terms.

$\sum_{k=1}^{4}(k^2 - 3k) = \sum_{k=1}^{4}k^2 - 3\sum_{k=1}^{4}k$ — Difference Rule and Constant Multiple Rule

$= \dfrac{4(4+1)(8+1)}{6} - 3\left(\dfrac{4(4+1)}{2}\right)$ — Eqs. (2) and (1) with $n=4$

$= 30 - 30 = 0$

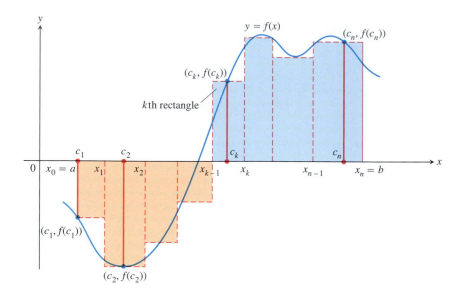

4.10 The graph of a typical function $y = f(x)$ over a closed interval $[a, b]$. The rectangles approximate the region between the graph of the function and the x-axis.

Riemann Sums

The approximating sums in Section 4.4 are examples of a more general kind of sum called a *Riemann* ("*ree*-mahn") *sum*. The functions in the examples had nonnegative values, but the more general notion has no such restriction. Given an arbitrary continuous function $y = f(x)$ on an interval $[a, b]$ (Fig. 4.10), we partition the interval into n subintervals by choosing $n - 1$ points, say $x_1, x_2, \ldots, x_{n-1}$, between a and b subject only to the condition that

$$a < x_1 < x_2 < \cdots < x_{n-1} < b.$$

To make the notation consistent, we usually denote a by x_0 and b by x_n. The set

$$P = \{x_0, x_1, \ldots, x_n\}$$

is called a **partition** of $[a, b]$.

The partition P defines n closed **subintervals**

$$[x_0, x_1], [x_1, x_2], \ldots, [x_{n-1}, x_n].$$

The typical closed subinterval $[x_{k-1}, x_k]$ is called the **kth subinterval** of P.

The length of the kth subinterval is $\Delta x_k = x_k - x_{k-1}$.

In each subinterval $[x_{k-1}, x_k]$, we select a point c_k and construct a vertical rectangle from the subinterval to the point $(c_k, f(c_k))$ on the curve $y = f(x)$. The choice of c_k does not matter as long as it lies in $[x_{k-1}, x_k]$. See Fig. 4.10 again.

If $f(c_k)$ is positive, the number $f(c_k) \, \Delta x_k = $ height $\times$ base is the area of the

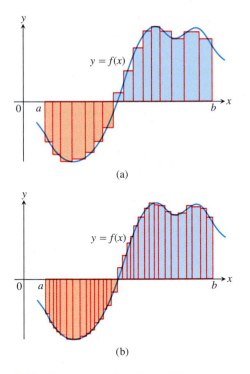

4.11 The curve of Fig. 4.10 with rectangles from finer partitions of $[a, b]$. Finer partitions create more rectangles with shorter bases.

rectangle. If $f(c_k)$ is negative, then $f(c_k)\,\Delta x_k$ is the negative of the area. In any case, we add the n products $f(c_k)\,\Delta x_k$ to form the sum

$$S_P = \sum_{k=1}^{n} f(c_k)\,\Delta x_k.$$

This sum, which depends on P and the choice of the numbers c_k, is called a **Riemann sum for f on the interval $[a, b]$,** after German mathematician Georg Friedrich Bernhard Riemann (1826–1866), who studied the limits of such sums.

As the partitions of $[a, b]$ become finer, the rectangles defined by the partition approximate the region between the x-axis and the graph of f with increasing accuracy (Fig. 4.11). So we expect the associated Riemann sums to have a limiting value. To test this expectation, we need to develop a numerical way to say that partitions become finer and to determine whether the corresponding sums have a limit. We accomplish this with the following definitions.

The **norm** of a partition P is the partition's longest subinterval length. It is denoted by

$$\|P\| \qquad \text{(read "the norm of } P\text{").}$$

The way to say that successive partitions of an interval become finer is to say that the norms of these partitions approach zero. As the norms go to zero, the subintervals become shorter and their number approaches infinity.

EXAMPLE 5   The set $P = \{0, 0.2, 0.6, 1, 1.5, 2\}$ is a partition of $[0, 2]$. There are five subintervals of P: $[0, 0.2]$, $[0.2, 0.6]$, $[0.6, 1]$, $[1, 1.5]$, and $[1.5, 2]$.

The lengths of the subintervals are $\Delta x_1 = 0.2$, $\Delta x_2 = 0.4$, $\Delta x_3 = 0.4$, $\Delta x_4 = 0.5$, and $\Delta x_5 = 0.5$. The longest subinterval length is 0.5, so the norm of the partition is $\|P\| = 0.5$. In this example, there are two subintervals of this length.   ❑

Definition

The Definite Integral as a Limit of Riemann Sums

Let $f(x)$ be a function defined on a closed interval $[a, b]$. We say that the **limit** of the Riemann sums $\sum_{k=1}^{n} f(c_k)\,\Delta x_k$ on $[a, b]$ as $\|P\| \to 0$ is the number I if the following condition is satisfied:

Given any number $\epsilon > 0$, there exists a corresponding number $\delta > 0$ such that for every partition P of $[a, b]$

$$\|P\| < \delta \qquad \Rightarrow \qquad \left| \sum_{k=1}^{n} f(c_k)\,\Delta x_k - I \right| < \epsilon$$

for any choice of the numbers c_k in the subintervals $[x_{k-1}, x_k]$.

If the limit exists, we write

$$\lim_{\|P\| \to 0} \sum_{k=1}^{n} f(c_k)\,\Delta x_k = I.$$

We call I the **definite integral** of f over $[a, b]$, we say that f is **integrable** over $[a, b]$, and we say that the Riemann sums of f on $[a, b]$ **converge** to the number I.

We usually write I as $\int_a^b f(x)\, dx$, which is read "integral of f from a to b." Thus, if the limit exists,

$$\lim_{\|P\| \to 0} \sum_{k=1}^{n} f(c_k)\, \Delta x_k = \int_a^b f(x)\, dx.$$

The amazing fact is that despite the variety in the Riemann sums $\Sigma f(c_k)\, \Delta x_k$ as the partitions change and the arbitrary choice of c_k's in the intervals of each new partition, the sums always have the same limit as $\|P\| \to 0$ as long as f is continuous. The need to establish the existence of this limit became clear as the nineteenth century progressed, and it was finally established when Riemann proved the following theorem in 1854. You can find a current version of Riemann's proof in most advanced calculus books.

Theorem 1

The Existence of Definite Integrals

All continuous functions are integrable. That is, if a function f is continuous on an interval $[a, b]$, then its definite integral over $[a, b]$ exists.

Why should we expect such a theorem to hold? Imagine a typical partition P of the interval $[a, b]$. The function f, being continuous, has a minimum value $\min_k$ ("min kay") and a maximum value $\max_k$ ("max kay") on each subinterval. The products $\min_k \Delta x_k$ associated with the minimum values (Fig. 4.12a) add up to what we call the **lower sum** for f on P:

$$L = \min_1 \Delta x_1 + \min_2 \Delta x_2 + \cdots + \min_n \Delta x_n.$$

The products $\max_k \Delta x_k$ obtained from the maximum values (Fig. 4.12b) add up to the **upper sum** for f on P:

$$U = \max_1 \Delta x_1 + \max_2 \Delta x_2 + \cdots + \max_n \Delta x_n.$$

The difference $U - L$ between the upper and lower sums is the sum of the areas of the shaded blocks in Fig. 4.12(c). As $\|P\| \to 0$, the blocks in Fig. 4.12(c) become more numerous, narrower, and shorter. As Fig. 4.12(d) suggests, we can make the nonnegative number $U - L$ less than any prescribed positive ϵ by taking $\|P\|$ close enough to zero. In other words,

$$\lim_{\|P\| \to 0} (U - L) = 0, \tag{4}$$

and, as shown in more advanced texts,

$$\lim_{\|P\| \to 0} L = \lim_{\|P\| \to 0} U. \tag{5}$$

The fact that Eqs. (4) and (5) hold for any continuous function is a consequence of a special property, called *uniform continuity,* that continuous functions have on closed intervals. This property guarantees that as $\|P\| \to 0$ the blocks that make up the difference between U and L in Fig. 4.12(c) become less tall as they become less wide and that we can make them all as short as we please by making them narrow enough. Passing over the ϵ - δ arguments associated with uniform continuity

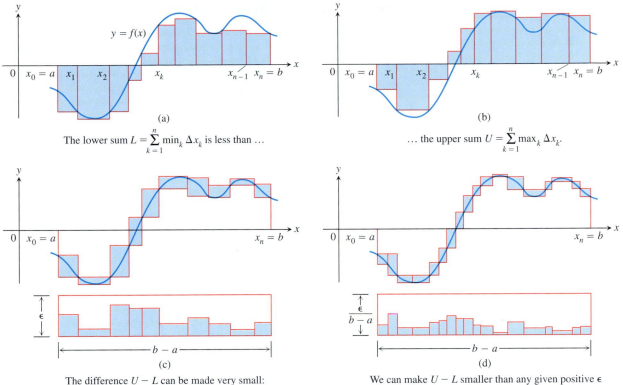

(a)

The lower sum $L = \sum_{k=1}^{n} \min_k \Delta x_k$ is less than ...

(b)

... the upper sum $U = \sum_{k=1}^{n} \max_k \Delta x_k$.

(c)

The difference $U - L$ can be made very small: less than $\epsilon \cdot (b - a)$.

(d)

We can make $U - L$ smaller than any given positive ϵ by making $\|P\|$ small enough.

4.12 The difference between upper and lower sums.

keeps our derivation of Eq. (5) from being a proof. But the argument is right in spirit and gives a faithful portrait of the proof.

Assuming that Eq. (5) holds for any continuous function f on $[a, b]$, suppose we choose a point c_k from each subinterval $[x_{k-1}, x_k]$ of P and form the Riemann sum $\sum_{k=1}^{n} f(c_k) \Delta x_k$. Then $\min_k \leq f(c_k) \leq \max_k$ for each k, so

$$L \leq \sum_{k=1}^{n} f(c_k) \Delta x_k \leq U.$$

The Riemann sum for f is sandwiched between L and U. By a modified version of the Sandwich Theorem of Section 1.2, the limit of the Riemann sums as $\|P\| \to 0$ exists and equals the common limit of U and L:

$$\lim_{\|P\| \to 0} L = \lim_{\|P\| \to 0} \sum_{k=1}^{n} f(c_k) \Delta x_k = \lim_{\|P\| \to 0} U.$$

Pause for a moment to see how remarkable this conclusion really is. It says that no matter how we choose the points c_k to form the Riemann sums as $\|P\| \to 0$, the limit is always the same. We can take every $f(c_k)$ to be the minimum value of f on $[x_{k-1}, x_k]$. The limit is the same. We can take every $f(c_k)$ to be the maximum value of f on $[x_{k-1}, x_k]$. The limit is the same. We can choose every c_k at random. The limit is the same.

Although we stated the integral existence theorem specifically for continuous functions, many discontinuous functions are integrable as well. We treat the integration of bounded piecewise continuous functions in Additional Exercises 11–18 at the end of this chapter. We explore the integration of unbounded functions in Section 7.6.

Functions with No Riemann Integral

While some discontinuous functions are integrable, others are not. The function

$$f(x) = \begin{cases} 1 & \text{when } x \text{ is rational} \\ 0 & \text{when } x \text{ is irrational,} \end{cases}$$

for example, has no Riemann integral over [0, 1]. For any partition P of [0, 1], the upper and lower sums are

$$U = \sum \max_k \Delta x_k = \sum 1 \cdot \Delta x_k = \sum \Delta x_k = 1, \qquad \text{Every subinterval contains a rational number}$$

$$L = \sum \min_k \Delta x_k = \sum 0 \cdot \Delta x_k = 0. \qquad \text{Every subinterval contains an irrational number.}$$

For the integral of f to exist over [0, 1], U and L would have to have the same limit as $\|P\| \to 0$. But they do not:

$$\lim_{\|P\| \to 0} L = 0 \qquad \text{while} \qquad \lim_{\|P\| \to 0} U = 1.$$

Therefore, f has no integral on [0, 1]. No constant multiple kf has an integral either, unless k is zero.

Terminology

There is a fair amount of terminology associated with the symbol $\int_a^b f(x)\,dx$.

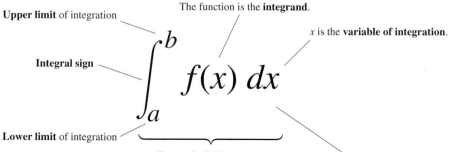

Upper limit of integration

The function is the **integrand**.

x is the **variable of integration**.

Integral sign

$$\int_a^b f(x)\,dx$$

Lower limit of integration

Integral of f from a to b

When you find the value of the integral, you have **evaluated** the integral.

The value of the definite integral of a function over any particular interval depends on the function and not on the letter we choose to represent its independent variable. If we decide to use t or u instead of x, we simply write the integral as

$$\int_a^b f(t)\,dt \qquad \text{or} \qquad \int_a^b f(u)\,du \qquad \text{instead of} \qquad \int_a^b f(x)\,dx.$$

No matter how we write the integral, it is still the same number, defined as a limit of Riemann sums. Since it does not matter what letter we use, the variable of integration is called a **dummy variable.**

EXAMPLE 6   Express the limit of Riemann sums

$$\lim_{\|P\| \to 0} \sum_{k=1}^{n} \left(3c_k^2 - 2c_k + 5 \right) \Delta x_k$$

as an integral if P denotes a partition of the interval $[-1, 3]$.

Solution   The function being evaluated at c_k in each term of the sum is $f(x) = 3x^2 - 2x + 5$. The interval being partitioned is $[-1, 3]$. The limit is therefore the integral of f from -1 to 3:

$$\lim_{\|P\| \to 0} \sum_{k=1}^{n} (3c_k{}^2 - 2c_k + 5)\, \Delta x_k = \int_{-1}^{3} (3x^2 - 2x + 5)\, dx.$$ ☐

Constant Functions

Theorem 1 says nothing about how to *calculate* definite integrals. Except for a few special cases, that takes another theorem (Section 4.7). Among the exceptions are constant functions. Suppose that f has the constant value $f(x) = c$ over $[a, b]$. Then, no matter how the c_k's are chosen,

$$\sum_{k=1}^{n} f(c_k)\, \Delta x_k = \sum_{k=1}^{n} c \cdot \Delta x_k \qquad \text{$f(c_k)$ always equals c.}$$

$$= c \cdot \sum_{k=1}^{n} \Delta x_k \qquad \text{Constant Multiple Rule for Sums}$$

$$= c(b - a). \qquad \sum_{k=1}^{n} \Delta x_k = \text{length of interval } [a, b] = b - a$$

Since the sums all have the value $c(b - a)$, their limit, the integral, does too.

If $f(x)$ has the constant value c on $[a, b]$, then

$$\int_{a}^{b} f(x)\, dx = \int_{a}^{b} c\, dx = c(b - a).$$

EXAMPLE 7

a) $\displaystyle \int_{-1}^{4} 3\, dx = 3(4 - (-1)) = (3)(5) = 15$

b) $\displaystyle \int_{-1}^{4} (-3)\, dx = -3(4 - (-1)) = (-3)(5) = -15$ ☐

The Area Under the Graph of a Nonnegative Function

The sums we used to estimate the height of the projectile in Section 4.4, Example 2, were Riemann sums for the projectile's velocity function

$$v = f(t) = 160 - 9.8t$$

on the interval $[0, 3]$. We can see from Fig. 4.13 how the associated rectangles approximate the trapezoid between the t-axis and the curve $v = 160 - 9.8t$. As the norm of the partition goes to zero, the rectangles fit the trapezoid with increasing accuracy and the sum of the areas they enclose approaches the trapezoid's area, which is

$$\text{Trapezoid area} = h \cdot \frac{b_1 + b_2}{2} = 3 \cdot \frac{160 + 130.6}{2} = 435.9.$$

Region is a trapezoid with height = 3
base (top) = 130.6
base (bottom) = 160.

4.13 Rectangles for a Riemann sum of the velocity function $f(t) = 160 - 9.8t$ over the interval $[0, 3]$.

This confirms our suspicion that the sums we were constructing in Section 4.4, Example 2, approached a limit of 435.9. Since the limit of these sums is also the integral of f from 0 to 3, we now know the value of the integral as well:

$$\int_0^3 (160 - 9.8t)\, dt = \text{trapezoid area} = 435.9.$$

We can exploit the connection between integrals and area in two ways. When we know a formula for the area of the region between the x-axis and the graph of a continuous nonnegative function $y = f(x)$, we can use it to evaluate the function's integral. When we do not know the region's area, we can use the function's integral to define and calculate the area.

Definition

Let $f(x) \geq 0$ be continuous on $[a, b]$. The **area** of the region between the graph of f and the x-axis is

$$A = \int_a^b f(x)\, dx.$$

Whenever we make a new definition, as we have here, consistency becomes an issue. Does the definition that we have just developed for nonstandard shapes give correct results for standard shapes? The answer is yes, but the proof is complicated and we will not go into it.

EXAMPLE 8   *Using an area to evaluate a definite integral*

Evaluate

$$\int_a^b x\, dx, \qquad 0 < a < b.$$

Solution   We sketch the region under the curve $y = x$, $a \leq x \leq b$ (Fig. 4.14), and see that it is a trapezoid with height $(b - a)$ and bases a and b. The value of the integral is the area of this trapezoid:

$$\int_a^b x\, dx = (b - a) \cdot \frac{a + b}{2} = \frac{b^2}{2} - \frac{a^2}{2}.$$

Thus,

$$\int_1^{\sqrt{5}} x\, dx = \frac{(\sqrt{5})^2}{2} - \frac{(1)^2}{2} = 2$$

and so on.

Notice that $x^2/2$ is an antiderivative of x, further evidence of a connection between antiderivatives and summation.   ❑

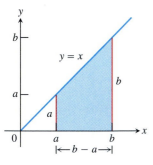

4.14 The region in Example 8.

EXAMPLE 9   *Using a definite integral to find an area*

Find the area of the region between the parabola $y = x^2$ and the x-axis on the interval $[0, b]$.

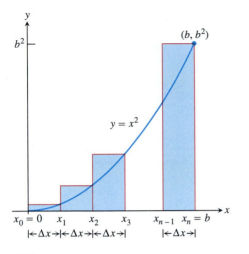

4.15 The rectangles of the Riemann sums in Example 9.

Notice that $x^3/3$ is an antiderivative of x^2.

Solution We evaluate the integral for the area as a limit of Riemann sums.

We sketch the region (a nonstandard shape) (Fig. 4.15) and partition $[0, b]$ into n subintervals of length $\Delta x = (b - 0)/n = b/n$. The points of the partition are

$$x_0 = 0, \quad x_1 = \Delta x, \quad x_2 = 2\Delta x, \quad \cdots, \quad x_{n-1} = (n - 1)\,\Delta x, \quad x_n = n\Delta x = b.$$

We are free to choose the c_k's any way we please. We choose each c_k to be the right-hand endpoint of its subinterval, a choice that leads to manageable arithmetic. Thus, $c_1 = x_1, c_2 = x_2$, and so on. The rectangles defined by these choices have areas

$$f(c_1)\,\Delta x = f(\Delta x)\,\Delta x = (\Delta x)^2\,\Delta x = (1^2)(\Delta x)^3$$

$$f(c_2)\,\Delta x = f(2\Delta x)\,\Delta x = (2\Delta x)^2\,\Delta x = (2^2)(\Delta x)^3$$

$$\vdots$$

$$f(c_n)\,\Delta x = f(n\Delta x)\,\Delta x = (n\Delta x)^2\,\Delta x = (n^2)(\Delta x)^3.$$

The sum of these areas is

$$S_n = \sum_{k=1}^{n} f(c_k)\,\Delta x$$

$$= \sum_{k=1}^{n} k^2(\Delta x)^3$$

$$= (\Delta x)^3 \sum_{k=1}^{n} k^2 \qquad (\Delta x)^3 \text{ is a constant.}$$

$$= \frac{b^3}{n^3} \cdot \frac{n(n + 1)(2n + 1)}{6} \qquad \Delta x = b/n, \text{ and Eq. (2)}$$

$$= \frac{b^3}{6} \cdot \frac{(n + 1)(2n + 1)}{n^2}$$

$$= \frac{b^3}{6} \cdot \frac{2n^2 + 3n + 1}{n^2}$$

$$= \frac{b^3}{6} \cdot \left(2 + \frac{3}{n} + \frac{1}{n^2}\right). \qquad (6)$$

We can now use the definition of definite integral

$$\int_a^b f(x)\,dx = \lim_{\|P\| \to 0} \sum_{k=1}^{n} f(c_k)\,\Delta x$$

to find the area under the parabola from $x = 0$ to $x = b$ as

$$\int_0^b x^2\,dx = \lim_{n \to \infty} S_n \qquad \text{In this example,} \\ \|P\| \to 0 \text{ is equivalent} \\ \text{to } n \to \infty.$$

$$= \lim_{n \to \infty} \frac{b^3}{6} \cdot \left(2 + \frac{3}{n} + \frac{1}{n^2}\right) \qquad \text{Eq. (6)}$$

$$= \frac{b^3}{6} \cdot (2 + 0 + 0) = \frac{b^3}{3}.$$

With different values of b, we get

$$\int_0^1 x^2 \, dx = \frac{1^3}{3} = \frac{1}{3}, \qquad \int_0^{1.5} x^2 \, dx = \frac{(1.5)^3}{3} = \frac{3.375}{3} = 1.125,$$

and so on.

$\square$

Exercises 4.5

Sigma Notation

Write the sums in Exercises 1–6 without sigma notation. Then evaluate them.

1. $\displaystyle\sum_{k=1}^{2} \frac{6k}{k+1}$

2. $\displaystyle\sum_{k=1}^{3} \frac{k-1}{k}$

3. $\displaystyle\sum_{k=1}^{4} \cos k\pi$

4. $\displaystyle\sum_{k=1}^{5} \sin k\pi$

5. $\displaystyle\sum_{k=1}^{3} (-1)^{k+1} \sin \frac{\pi}{k}$

6. $\displaystyle\sum_{k=1}^{4} (-1)^k \cos k\pi$

7. Which of the following express $1 + 2 + 4 + 8 + 16 + 32$ in sigma notation?

a) $\displaystyle\sum_{k=1}^{6} 2^{k-1}$ **b)** $\displaystyle\sum_{k=0}^{5} 2^k$ **c)** $\displaystyle\sum_{k=-1}^{4} 2^{k+1}$

8. Which of the following express $1 - 2 + 4 - 8 + 16 - 32$ in sigma notation?

a) $\displaystyle\sum_{k=1}^{6} (-2)^{k-1}$ **b)** $\displaystyle\sum_{k=0}^{5} (-1)^k 2^k$

c) $\displaystyle\sum_{k=-2}^{3} (-1)^{k+1} 2^{k+2}$

9. Which formula is not equivalent to the other two?

a) $\displaystyle\sum_{k=2}^{4} \frac{(-1)^{k-1}}{k-1}$ **b)** $\displaystyle\sum_{k=0}^{2} \frac{(-1)^k}{k+1}$ **c)** $\displaystyle\sum_{k=-1}^{1} \frac{(-1)^k}{k+2}$

10. Which formula is not equivalent to the other two?

a) $\displaystyle\sum_{k=1}^{4} (k-1)^2$ **b)** $\displaystyle\sum_{k=-1}^{3} (k+1)^2$ **c)** $\displaystyle\sum_{k=-3}^{-1} k^2$

Express the sums in Exercises 11–16 in sigma notation. The form of your answer will depend on your choice of the lower limit of summation.

11. $1 + 2 + 3 + 4 + 5 + 6$

12. $1 + 4 + 9 + 16$

13. $\dfrac{1}{2} + \dfrac{1}{4} + \dfrac{1}{8} + \dfrac{1}{16}$

14. $2 + 4 + 6 + 8 + 10$

15. $1 - \dfrac{1}{2} + \dfrac{1}{3} - \dfrac{1}{4} + \dfrac{1}{5}$

16. $-\dfrac{1}{5} + \dfrac{2}{5} - \dfrac{3}{5} + \dfrac{4}{5} - \dfrac{5}{5}$

Values of Finite Sums

17. Suppose that $\displaystyle\sum_{k=1}^{n} a_k = -5$ and $\displaystyle\sum_{k=1}^{n} b_k = 6$. Find the values of

a) $\displaystyle\sum_{k=1}^{n} 3a_k$ **b)** $\displaystyle\sum_{k=1}^{n} \frac{b_k}{6}$ **c)** $\displaystyle\sum_{k=1}^{n} (a_k + b_k)$

d) $\displaystyle\sum_{k=1}^{n} (a_k - b_k)$ **e)** $\displaystyle\sum_{k=1}^{n} (b_k - 2a_k)$

18. Suppose that $\displaystyle\sum_{k=1}^{n} a_k = 0$ and $\displaystyle\sum_{k=1}^{n} b_k = 1$. Find the values of

a) $\displaystyle\sum_{k=1}^{n} 8a_k$ **b)** $\displaystyle\sum_{k=1}^{n} 250b_k$

c) $\displaystyle\sum_{k=1}^{n} (a_k + 1)$ **d)** $\displaystyle\sum_{k=1}^{n} (b_k - 1)$

Use the algebra rules on p. 310 and the formulas in Eqs. (1)–(3) to evaluate the sums in Exercises 19–28.

19. a) $\displaystyle\sum_{k=1}^{10} k$ **b)** $\displaystyle\sum_{k=1}^{10} k^2$ **c)** $\displaystyle\sum_{k=1}^{10} k^3$

20. a) $\displaystyle\sum_{k=1}^{13} k$ **b)** $\displaystyle\sum_{k=1}^{13} k^2$ **c)** $\displaystyle\sum_{k=1}^{13} k^3$

21. $\displaystyle\sum_{k=1}^{7} (-2k)$

22. $\displaystyle\sum_{k=1}^{5} \frac{\pi k}{15}$

23. $\displaystyle\sum_{k=1}^{6} (3 - k^2)$

24. $\displaystyle\sum_{k=1}^{6} (k^2 - 5)$

25. $\displaystyle\sum_{k=1}^{5} k(3k + 5)$

26. $\displaystyle\sum_{k=1}^{7} k(2k + 1)$

27. $\displaystyle\sum_{k=1}^{5} \frac{k^3}{225} + \left(\sum_{k=1}^{5} k\right)^3$

28. $\displaystyle\left(\sum_{k=1}^{7} k\right)^2 - \sum_{k=1}^{7} \frac{k^3}{4}$

Rectangles for Riemann Sums

In Exercises 29–32, graph each function $f(x)$ over the given interval. Partition the interval into four subintervals of equal length. Then add to your sketch the rectangles associated with the Riemann sum $\sum_{k=1}^{4} f(c_k) \, \Delta x_k$, given that c_k is the (a) left-hand endpoint, (b) right-hand endpoint, (c) midpoint of the kth subinterval. (Make a separate sketch for each set of rectangles.)

29. $f(x) = x^2 - 1$, $[0, 2]$

30. $f(x) = -x^2$, $[0, 1]$

31. $f(x) = \sin x$, $[-\pi, \pi]$

32. $f(x) = \sin x + 1$, $[-\pi, \pi]$

33. Find the norm of the partition $P = \{0, 1.2, 1.5, 2.3, 2.6, 3\}$.

34. Find the norm of the partition $P = \{-2, -1.6, -0.5, 0, 0.8, 1\}$.

Expressing Limits as Integrals

Express the limits in Exercises 35–42 as definite integrals.

35. $\displaystyle\lim_{\|P\|\to 0} \sum_{k=1}^{n} c_k^2 \Delta x_k$, where P is a partition of $[0, 2]$

36. $\displaystyle\lim_{\|P\|\to 0} \sum_{k=1}^{n} 2c_k^3 \Delta x_k$, where P is a partition of $[-1, 0]$

37. $\displaystyle\lim_{\|P\|\to 0} \sum_{k=1}^{n} (c_k^2 - 3c_k) \Delta x_k$, where P is a partition of $[-7, 5]$

38. $\displaystyle\lim_{\|P\|\to 0} \sum_{k=1}^{n} \left(\frac{1}{c_k}\right) \Delta x_k$, where P is a partition of $[1, 4]$

39. $\displaystyle\lim_{\|P\|\to 0} \sum_{k=1}^{n} \frac{1}{1 - c_k} \Delta x_k$, where P is a partition of $[2, 3]$

40. $\displaystyle\lim_{\|P\|\to 0} \sum_{k=1}^{n} \sqrt{4 - c_k^2}\, \Delta x_k$, where P is a partition of $[0, 1]$

41. $\displaystyle\lim_{\|P\|\to 0} \sum_{k=1}^{n} (\sec c_k)\, \Delta x_k$, where P is a partition of $[-\pi/4, 0]$

42. $\displaystyle\lim_{\|P\|\to 0} \sum_{k=1}^{n} (\tan c_k)\, \Delta x_k$, where P is a partition of $[0, \pi/4]$

Constant Functions

Evaluate the integrals in Exercises 43–48.

43. $\displaystyle\int_{-2}^{1} 5\,dx$

44. $\displaystyle\int_{3}^{7} (-20)\,dx$

45. $\displaystyle\int_{0}^{3} (-160)\,dt$

46. $\displaystyle\int_{-4}^{-1} \frac{\pi}{2}\,d\theta$

47. $\displaystyle\int_{-2.1}^{3.4} 0.5\,ds$

48. $\displaystyle\int_{\sqrt{2}}^{\sqrt{18}} \sqrt{2}\,dr$

Using Area to Evaluate Integrals

In Exercises 49–56, graph the integrands and use areas to evaluate the integrals.

49. $\displaystyle\int_{-2}^{4} \left(\frac{x}{2} + 3\right) dx$

50. $\displaystyle\int_{1/2}^{3/2} (-2x + 4)\,dx$

51. $\displaystyle\int_{-3}^{3} \sqrt{9 - x^2}\,dx$

52. $\displaystyle\int_{-4}^{0} \sqrt{16 - x^2}\,dx$

53. $\displaystyle\int_{-2}^{1} |x|\,dx$

54. $\displaystyle\int_{-1}^{1} (1 - |x|)\,dx$

55. $\displaystyle\int_{-1}^{1} (2 - |x|)\,dx$

56. $\displaystyle\int_{-1}^{1} \left(1 + \sqrt{1 - x^2}\right) dx$

Use areas to evaluate the integrals in Exercises 57–60.

57. $\displaystyle\int_{0}^{b} x\,dx$, $b > 0$

58. $\displaystyle\int_{0}^{b} 4x\,dx$, $b > 0$

59. $\displaystyle\int_{a}^{b} 2s\,ds$, $0 < a < b$

60. $\displaystyle\int_{a}^{b} 3t\,dt$, $0 < a < b$

Evaluations

Use the results of Examples 8 and 9 to evaluate the integrals in Exercises 61–72.

61. $\displaystyle\int_{1}^{\sqrt{2}} x\,dx$

62. $\displaystyle\int_{0.5}^{2.5} x\,dx$

63. $\displaystyle\int_{\pi}^{2\pi} \theta\,d\theta$

64. $\displaystyle\int_{\sqrt{2}}^{5\sqrt{2}} r\,dr$

65. $\displaystyle\int_{0}^{\sqrt[3]{7}} x^2\,dx$

66. $\displaystyle\int_{0}^{0.3} s^2\,ds$

67. $\displaystyle\int_{0}^{1/2} t^2\,dt$

68. $\displaystyle\int_{0}^{\pi/2} \theta^2\,d\theta$

69. $\displaystyle\int_{a}^{2a} x\,dx$

70. $\displaystyle\int_{a}^{\sqrt{3}a} x\,dx$

71. $\displaystyle\int_{0}^{\sqrt[3]{b}} x^2\,dx$

72. $\displaystyle\int_{0}^{3b} x^2\,dx$

Finding Area

In Exercises 73–76, use a definite integral to find the area of the region between the given curve and the x-axis on the interval $[0, b]$, as in Example 9.

73. $y = 3x^2$

74. $y = \pi x^2$

75. $y = 2x$

76. $y = \dfrac{x}{2} + 1$

Theory and Examples

77. What values of a and b maximize the value of

$$\int_{a}^{b} (x - x^2)\,dx?$$

(*Hint:* Where is the integrand positive?)

78. What values of a and b minimize the value of

$$\int_{a}^{b} (x^4 - 2x^2)\,dx?$$

79. *Upper and lower sums for increasing functions*

   a) Suppose the graph of a continuous function $f(x)$ rises steadily as x moves from left to right across an interval $[a, b]$. Let P be a partition of $[a, b]$ into n subintervals of length $\Delta x = (b - a)/n$. Show by referring to the accompanying figure that the difference between the upper and lower sums for f on this partition can be represented graphically as the area of a rectangle R whose dimensions are $[f(b) - f(a)]$ by Δx. (*Hint:* The difference $U - L$ is the sum of areas of rectangles whose diagonals $Q_0Q_1, Q_1Q_2, \ldots, Q_{n-1}Q_n$ lie along the curve. There is no overlapping when these rectangles are shifted horizontally onto R.)

b) Suppose that instead of being equal, the lengths Δx_k of the subintervals of the partition of $[a, b]$ vary in size. Show that

$$U - L \leq |f(b) - f(a)|\Delta x_{max},$$

where Δx_{max} is the norm of P, and hence that $\lim_{\|P\| \to 0} (U - L) = 0$.

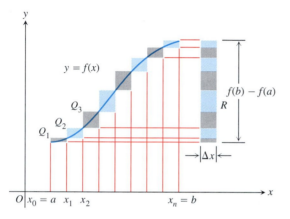

80. *Upper and lower sums for decreasing functions* (Continuation of Exercise 79)

a) Draw a figure like the one in Exercise 79 for a continuous function $f(x)$ whose values decrease steadily as x moves from left to right across the interval $[a, b]$. Let P be a partition of $[a, b]$ into subintervals of equal length. Find an expression for $U - L$ that is analogous to the one you found for $U - L$ in Exercise 79(a).

b) Suppose that instead of being equal, the lengths Δx_k of the subintervals of P vary in size. Show that the inequality

$$U - L \leq |f(b) - f(a)|\Delta x_{max}$$

of Exercise 79(b) still holds and hence that $\lim_{\|P\| \to 0} (U - L) = 0$.

81. Evaluate $\int_0^b x^2\,dx, b > 0$, by carrying out the calculations of Example 9 with inscribed rectangles, as shown here, instead of circumscribed rectangles.

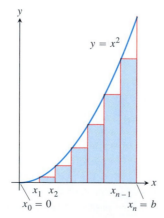

82. Let

$$S_n = \frac{1}{n}\left[\frac{1}{n} + \frac{2}{n} + \frac{3}{n} + \cdots + \frac{n-1}{n}\right].$$

Calculate $\lim_{n \to \infty} S_n$ by showing that S_n is an approximating sum of the integral

$$\int_0^1 x\,dx,$$

whose value we know from Example 8. (*Hint:* Partition $[0, 1]$ into n intervals of equal length and write out the approximating sum for inscribed rectangles.)

83. Let

$$S_n = \frac{1^2}{n^3} + \frac{2^2}{n^3} + \cdots + \frac{(n-1)^2}{n^3}.$$

To calculate $\lim_{n \to \infty} S_n$, show that

$$S_n = \frac{1}{n}\left[\left(\frac{1}{n}\right)^2 + \left(\frac{2}{n}\right)^2 + \cdots + \left(\frac{n-1}{n}\right)^2\right]$$

and interpret S_n as an approximating sum of the integral

$$\int_0^1 x^2\,dx,$$

whose value we know from Example 9. (*Hint:* Partition $[0, 1]$ into n intervals of equal length and write out the approximating sum for inscribed rectangles.)

84. Use the formula

$$\sin h + \sin 2h + \sin 3h + \cdots + \sin mh$$
$$= \frac{\cos(h/2) - \cos((m + (1/2))h)}{2\sin(h/2)}$$

to find the area under the curve $y = \sin x$ from $x = 0$ to $x = \pi/2$, in two steps:

a) Partition the interval $[0, \pi/2]$ into n subintervals of equal length and calculate the corresponding upper sum U; then

b) Find the limit of U as $n \to \infty$ and $\Delta x = (b - a)/n \to 0$.

✪ CAS Explorations and Projects

If your CAS can draw rectangles associated with Riemann sums, use it to draw rectangles associated with Riemann sums that converge to the integrals in Exercises 85–90. Use $n = 4, 10, 20$, and 50 subintervals of equal length in each case.

85. $\displaystyle\int_0^1 (1 - x)\,dx = \frac{1}{2}$

86. $\displaystyle\int_0^1 (x^2 + 1)\,dx = \frac{4}{3}$

87. $\displaystyle\int_{-\pi}^{\pi} \cos x\,dx = 0$

88. $\displaystyle\int_0^{\pi/4} \sec^2 x\,dx = 1$

89. $\int_{-1}^{1} |x|\, dx = 1$

90. $\int_{1}^{2} \frac{1}{x}\, dx$ (The integral's value is ln 2.)

91. a) Write the sum S_n in Exercise 82 in sigma notation and use your CAS to find $\lim_{n\to\infty} S_n$.
   b) Do the same for the sum S_n in Exercise 83.

92. Write the sum $\sin h + \sin 2h + \cdots + \sin mh$ in Exercise 84 in sigma notation and use your CAS to find $\lim_{n\to\infty} S_n$.

93. (*Continuation of Section 4.4, Example 3.*) In sigma notation, the left-endpoint sum in Example 3, Section 4.4, is

$$S_4 = \sum_{k=1}^{4} 4\left[9 - (-2 + (k-1))^2\right].$$

a) Use sigma notation to write the analogous left-endpoint sums S_8 for eight subintervals of length 4/8 and S_{25} for 25 subintervals of length 4/25.

b) Use sigma notation to write the left-endpoint sum S_n for n subintervals of length 4/n.

c) Find $\lim_{n\to\infty} S_n$. How does this limit appear to be related to the volume of the solid?

94. (*Continuation of Section 4.4 Example 4.*) In sigma notation, the left-endpoint sum in Example 4, Section 4.4, is

$$S_8 = \sum_{k=1}^{8} \pi\left[16 - (-4 + (k-1))^2\right].$$

a) Use sigma notation to write the analogous left-endpoint sums S_{16} for 16 subintervals of length 1/2 and S_{80} for 80 subintervals of length 1/10.

b) Use sigma notation to write the left-endpoint sum S_n for n subintervals of length 8/n.

c) Find $\lim_{n\to\infty} S_n$. How does this limit appear to be related to the volume of the sphere?

4.6

Properties, Area, and the Mean Value Theorem

This section describes working rules for integrals, examines the relationship between the integral of an arbitrary continuous function and area, and takes a fresh look at average value.

Properties of Definite Integrals

We often want to add and subtract definite integrals, multiply their integrands by constants, and compare them with other definite integrals. We do this with the rules in Table 4.5 (on the following page). All the rules except the first two follow from the way integrals are defined with Riemann sums. You might think that this would make them relatively easy to prove. After all, we might argue, sums have these properties so their limits should have them, too. But when we get down to the details we find that most of the proofs require complicated ϵ-δ arguments with norms of subdivisions and are not easy at all. We omit all but two of the proofs. The remaining proofs can be found in more advanced texts.

Notice that Rule 1 is a definition. We want every integral over an interval of zero length to be zero. Rule 1 extends the definition of definite integral to allow for the case $a = b$. Rule 2, also a definition, extends the definition of definite integral to allow for the case $b < a$. Rules 3 and 4 are like the analogous rules for limits and indefinite integrals. Once we know the integrals of two functions, we automatically know the integrals of all constant multiples of these functions and their sums and differences. We can also use Rules 3 and 4 repeatedly to evaluate integrals of arbitrary finite linear combinations of integrable functions term by term. For any

Table 4.5 Rules for definite integrals

1. *Zero:* $$\int_a^a f(x)\,dx = 0 \qquad \text{(A definition)}$$

2. *Order of Integration:* $$\int_b^a f(x)\,dx = -\int_a^b f(x)\,dx \qquad \text{(Also a definition)}$$

3. *Constant Multiples:* $$\int_a^b kf(x)\,dx = k\int_a^b f(x)\,dx \qquad \text{(Any number } k)$$

$$\int_a^b -f(x)\,dx = -\int_a^b f(x)\,dx \qquad (k = -1)$$

4. *Sums and Differences:* $$\int_a^b (f(x) \pm g(x))\,dx = \int_a^b f(x)\,dx \pm \int_a^b g(x)\,dx$$

5. *Additivity:* $$\int_a^b f(x)\,dx + \int_b^c f(x)\,dx = \int_a^c f(x)\,dx$$

6. *Max-Min Inequality:* If max f and min f are the maximum and minimum values of f on $[a, b]$, then

$$\min f \cdot (b - a) \le \int_a^b f(x)\,dx \le \max f \cdot (b - a).$$

7. *Domination:* $$f(x) \ge g(x) \quad \text{on} \quad [a, b] \quad \Rightarrow \quad \int_a^b f(x)\,dx \ge \int_a^b g(x)\,dx$$

$$f(x) \ge 0 \quad \text{on} \quad [a, b] \quad \Rightarrow \quad \int_a^b f(x)\,dx \ge 0$$

$$\text{(Special case)}$$

constants $c_1, \ldots, c_n$, regardless of sign, and functions $f_1(x), \ldots, f_n(x)$, integrable on $[a, b]$,

$$\int_a^b (c_1 f_1(x) + \cdots + c_n f_n(x))\,dx = c_1 \int_a^b f_1(x)\,dx + \cdots + c_n \int_a^b f_n(x)\,dx.$$

The proof, omitted, comes from mathematical induction.

Figure 4.16 illustrates Rule 5 with a positive function, but the rule applies to any integrable function.

Proof of Rule 3 Rule 3 says that the integral of k times a function is k times the integral of the function. This is true because

$$\int_a^b kf(x)\,dx = \lim_{\|P\|\to 0} \sum_{i=1}^n kf(c_i)\Delta x_i$$

$$= \lim_{\|P\|\to 0} k \sum_{i=1}^n f(c_i)\Delta x_i$$

$$= k \lim_{\|P\|\to 0} \sum_{i=1}^n f(c_i)\Delta x_i = k \int_a^b f(x)\,dx. \qquad \square$$

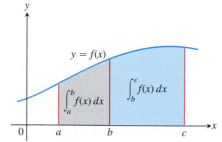

4.16 Additivity for definite integrals:

$$\int_a^b f(x)\,dx + \int_b^c f(x)\,dx = \int_a^c f(x)\,dx$$

$$\int_b^c f(x)\,dx = \int_a^c f(x)\,dx - \int_a^b f(x)\,dx.$$

Proof of Rule 6 Rule 6 says that the integral of f over $[a, b]$ is never smaller than the minimum value of f times the length of the interval and never larger than the maximum value of f times the length of the interval. The reason is that for every partition of $[a, b]$ and for every choice of the points c_k,

$$\min f \cdot (b - a) = \min f \cdot \sum_{k=1}^{n} \Delta x_k \qquad \sum_{k=1}^{n} \Delta x_k = b - a$$

$$= \sum_{k=1}^{n} \min f \cdot \Delta x_k$$

$$\leq \sum_{k=1}^{n} f(c_k) \, \Delta x_k \qquad \min f \leq f(c_k)$$

$$\leq \sum_{k=1}^{n} \max f \cdot \Delta x_k \qquad f(c_k) \leq \max f$$

$$= \max f \cdot \sum_{k=1}^{n} \Delta x_k$$

$$= \max f \cdot (b - a).$$

In short, all Riemann sums for f on $[a, b]$ satisfy the inequality

$$\min f \cdot (b - a) \leq \sum_{k=1}^{n} f(c_k) \, \Delta x_k \leq \max f \cdot (b - a).$$

Hence their limit, the integral, does too.   ❏

EXAMPLE 1   Suppose that

$$\int_{-1}^{1} f(x) \, dx = 5, \qquad \int_{1}^{4} f(x) \, dx = -2, \qquad \int_{-1}^{1} h(x) \, dx = 7.$$

Then

1. $\displaystyle \int_{4}^{1} f(x) \, dx = -\int_{1}^{4} f(x) \, dx = -(-2) = 2$ Rule 2

2. $\displaystyle \int_{-1}^{1} [2f(x) + 3h(x)] \, dx = 2 \int_{-1}^{1} f(x) \, dx + 3 \int_{-1}^{1} h(x) \, dx$

   $$= 2\,(5) + 3\,(7) = 31$$ Rules 3 and 4

3. $\displaystyle \int_{-1}^{4} f(x) \, dx = \int_{-1}^{1} f(x) \, dx + \int_{1}^{4} f(x) \, dx = 5 + (-2) = 3$ Rule 5   ❏

In Section 4.5 we learned to evaluate three general integrals:

$$\int_{a}^{b} c \, dx = c(b - a) \qquad \text{(Any constant } c\text{)} \tag{1}$$

$$\int_{a}^{b} x \, dx = \frac{b^2}{2} - \frac{a^2}{2} \qquad (0 < a < b) \tag{2}$$

$$\int_{0}^{b} x^2 \, dx = \frac{b^3}{3} \qquad (b > 0). \tag{3}$$

The rules in Table 4.5 enable us to build on these results.

EXAMPLE 2   Evaluate $\int_0^2 \left(\frac{t^2}{4} - 7t + 5 \right) dt.$

Solution

$$\int_0^2 \left(\frac{t^2}{4} - 7t + 5 \right) dt = \frac{1}{4} \int_0^2 t^2 \, dt - 7 \int_0^2 t \, dt + \int_0^2 5 \, dt \qquad \text{Rules 3 and 4}$$

$$= \frac{1}{4} \left(\frac{(2)^3}{3} \right) - 7 \left(\frac{(2)^2}{2} - \frac{(0)^2}{2} \right) + 5(2 - 0) \qquad \text{Eqs. (1)–(3)}$$

$$= \frac{2}{3} - 14 + 10 = -\frac{10}{3} \qquad \qquad ❑$$

EXAMPLE 3   Evaluate $\int_2^3 x^2 \, dx.$

Solution   We cannot apply Eq. (3) directly because the lower limit of integration is different from 0. We can, however, use the Additivity Rule to express $\int_2^3 x^2 \, dx$ as a difference of two integrals that *can* be evaluated with Eq. (3):

$$\int_0^2 x^2 \, dx + \int_2^3 x^2 \, dx = \int_0^3 x^2 \, dx \qquad \text{Rule 5}$$

$$\int_2^3 x^2 \, dx = \int_0^3 x^2 \, dx - \int_0^2 x^2 \, dx \qquad \begin{array}{l}\text{Solve for} \\ \int_2^3 x^2 dx.\end{array}$$

$$= \frac{(3)^3}{3} - \frac{(2)^3}{3} \qquad \begin{array}{l}\text{Eq. (3) now} \\ \text{applies.}\end{array}$$

$$= \frac{27}{3} - \frac{8}{3} = \frac{19}{3}.$$

In Section 4.7, we will see how to evaluate $\int_2^3 x^2 \, dx$ in a more direct way.   ❑

The Max-Min Inequality for definite integrals (Rule 6) says that min $f \cdot (b - a)$ is a **lower bound** for the value of $\int_a^b f(x) \, dx$ and that max $f \cdot (b - a)$ is an **upper bound**.

EXAMPLE 4   Show that the value of

$$\int_0^1 \sqrt{1 + \cos x} \, dx$$

cannot possibly be 2.

Solution   The maximum value of $\sqrt{1 + \cos x}$ on [0, 1] is $\sqrt{1 + 1} = \sqrt{2}$, so

$$\int_0^1 \sqrt{1 + \cos x} \, dx \leq \max \sqrt{1 + \cos x} \cdot (1 - 0) \qquad \begin{array}{l}\text{Table 4.5,} \\ \text{Rule 6}\end{array}$$

$$\leq \sqrt{2} \cdot 1 = \sqrt{2}.$$

The integral cannot exceed $\sqrt{2}$, so it cannot possibly equal 2.   ❑

EXAMPLE 5   Use the inequality $\cos x \geq (1 - x^2/2)$, which holds for all x, to find a lower bound for the value of $\int_0^1 \cos x \, dx$.

Solution

$$\int_0^1 \cos x \, dx \geq \int_0^1 \left(1 - \frac{x^2}{2} \right) dx \qquad \text{Rule 7}$$

$$\geq \int_0^1 1 \, dx - \frac{1}{2} \int_0^1 x^2 \, dx \qquad \text{Rules 3 and 4}$$

$$\geq 1 \cdot (1 - 0) - \frac{1}{2} \cdot \frac{(1)^3}{3} = \frac{5}{6} \approx 0.83.$$

The value of the integral is at least 5/6.

Integrals and Total Area

If an integrable function $y = f(x)$ has both positive and negative values on an interval $[a, b]$, then the Riemann sums for f on $[a, b]$ add the areas of the rectangles that lie above the x-axis to the negatives of the areas of the rectangles that lie below it (Fig. 4.17). The resulting cancellation reduces the sums, so their limiting value is a number whose magnitude is less than the total area between the curve and the x-axis. The value of the integral is the area above the axis minus the area below the axis.

This means that we must take special care in finding areas by integration.

4.17 (a) The Riemann sums are algebraic sums of areas and so is the integral to which they converge. (b) The value of the integral of f from a to b is

$$\int_a^b f(x) \, dx = \int_a^{x_1} f(x) \, dx + \int_{x_1}^{x_2} f(x) \, dx$$
$$+ \int_{x_2}^b f(x) \, dx = A_1 - A_2 + A_3.$$

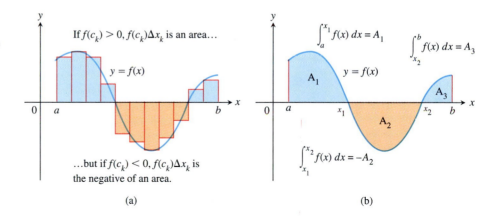

(a)

(b)

EXAMPLE 6   Find the area of the region between the curve $y = 4 - x^2$, $0 \leq x \leq 3$, and the x-axis.

Solution   The x-intercept of the curve partitions $[0, 3]$ into subintervals on which $f(x) = 4 - x^2$ has the same sign (Fig. 4.18). To find the area of the region between the graph of f and the x-axis, we integrate f over each subinterval and add the absolute values of the results.

Integral over $[0, 2]$:

$$\int_0^2 (4 - x^2) \, dx = \int_0^2 4 \, dx - \int_0^2 x^2 \, dx$$

$$= 4(2 - 0) - \frac{(2)^3}{3} \qquad \text{Eqs. (1) and (3)}$$

$$= 8 - \frac{8}{3} = \frac{16}{3}$$

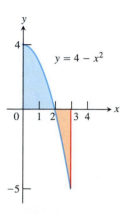

4.18 Part of the region in Example 6 lies below the x-axis.

<table>
<tr><td>

How to Find the Area of the Region Between a Curve $y = f(x)$, $a \leq x \leq b$, and the x-axis

1. Partition $[a, b]$ with the zeros of f.
2. Integrate f over each subinterval.
3. Add the absolute values of the integrals.

</td></tr>
</table>

Integral over [2, 3]:

$$\int_2^3 (4 - x^2)\, dx = \int_2^3 4\, dx - \int_2^3 x^2\, dx$$

$$= 4(3 - 2) - \left(\frac{(3)^3}{3} - \frac{(2)^3}{3} \right) \qquad \text{Eq. (1) and Example 3}$$

$$= 4 - \frac{19}{3} = -\frac{7}{3}$$

The region's area: $\qquad \text{Area} = \frac{16}{3} + \left| -\frac{7}{3} \right| = \frac{23}{3}.$ ❑

The Average Value of an Arbitrary Continuous Function

In Section 4.4, Example 5, we discussed the average value of a nonnegative continuous function. We are now ready to define average value without requiring f to be nonnegative, and to show that every continuous function assumes its average value at least once.

We start once again with the idea from arithmetic that the average of n numbers is the sum of the numbers divided by n. For a continuous function f on a closed interval $[a, b]$ there may be infinitely many values to consider, but we can sample them in an orderly way. We partition $[a, b]$ into n subintervals of equal length (the length is $\Delta x = (b - a)/n$) and evaluate f at a point c_k in each subinterval (Fig. 4.19). The average of the n sampled values is

$$\frac{f(c_1) + f(c_2) + \cdots + f(c_n)}{n} = \frac{1}{n} \cdot \sum_{k=1}^{n} f(c_k) \qquad \text{The sum in sigma notation}$$

$$= \frac{\Delta x}{b - a} \cdot \sum_{k=1}^{n} f(c_k) \qquad \Delta x = \frac{b - a}{n}$$

$$= \frac{1}{b - a} \cdot \underbrace{\sum_{k=1}^{n} f(c_k)\, \Delta x}_{\text{a Riemann sum for } f \text{ on } [a, b]}$$

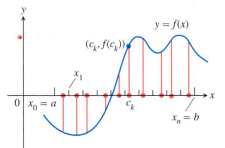

4.19 A sample of values of a function on an interval $[a, b]$.

Thus, the average of the sampled values is always $1/(b - a)$ times a Riemann sum for f on $[a, b]$. As we increase the size of the sample and let the norm of the partition approach zero, the average must approach $(1/(b - a)) \int_a^b f(x)\, dx$. We are led by this remarkable fact to the following definition.

<table>
<tr><td>

Definition

If f is integrable on $[a, b]$, its **average (mean) value** on $[a, b]$ is

$$\text{av}(f) = \frac{1}{b - a} \int_a^b f(x)\, dx.$$

</td></tr>
</table>

EXAMPLE 7   Find the average value of $f(x) = 4 - x^2$ on $[0, 3]$. Does f actually take on this value at some point in the given domain?

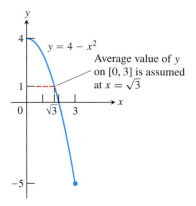

4.20 The average value of $f(x) = 4 - x^2$ on $[0, 3]$ occurs at $x = \sqrt{3}$ (Example 7).

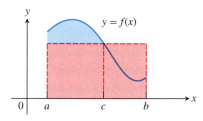

4.21 Theorem 2 for a positive function: At some point c in $[a, b]$,

$$f(c) \cdot (b - a) = \int_a^b f(x)\, dx.$$

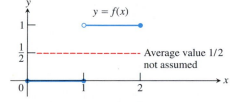

4.22 A discontinuous function need not assume its average value.

Solution

$$\text{av}\,(f) = \frac{1}{b - a} \int_a^b f(x)\, dx$$

$$= \frac{1}{3 - 0} \int_0^3 (4 - x^2)\, dx = \frac{1}{3} \left(\int_0^3 4\, dx - \int_0^3 x^2\, dx \right)$$

$$= \frac{1}{3} \left(4(3 - 0) - \frac{(3)^3}{3} \right) = \frac{1}{3}(12 - 9) = 1$$

The average value of $f(x) = 4 - x^2$ over the interval $[0, 3]$ is 1. The function assumes this value when $4 - x^2 = 1$ or $x = \pm\sqrt{3}$. Since one of these points, $x = \sqrt{3}$, lies in $[0, 3]$, the function does assume its average value in the given domain (Fig. 4.20). ❑

The Mean Value Theorem for Definite Integrals

The statement that a continuous function on a closed interval assumes its average value at least once in the interval is known as the Mean Value Theorem for Definite Integrals.

> **Theorem 2**
> **The Mean Value Theorem for Definite Integrals**
> If f is continuous on $[a, b]$, then at some point c in $[a, b]$,
>
> $$f(c) = \frac{1}{b - a} \int_a^b f(x)\, dx.$$
>
> (Fig. 4.21).

In Example 7, we found a point where f assumed its average value by setting $f(x)$ equal to the calculated average value and solving for x. But this does not prove that such a point will always exist. It proves only that it existed in Example 7. To prove Theorem 2, we need a more general argument.

Proof of Theorem 2   If we divide both sides of the Max-Min Inequality (Rule 6) by $(b - a)$, we obtain

$$\min f \leq \frac{1}{b - a} \int_a^b f(x)\, dx \leq \max f.$$

Since f is continuous, the Intermediate Value Theorem for Continuous Functions (Section 1.5) says that f must assume every value between $\min f$ and $\max f$. It must therefore assume the value $(1/(b - a)) \int_a^b f(x)\, dx$ at some point c in $[a, b]$. ❑

The continuity of f is important here. A discontinuous function can step over its average value (Fig. 4.22).

What else can we learn from Theorem 2? Here is an example.

EXAMPLE 8   Show that if f is continuous on $[a, b]$, $a \neq b$, and if

$$\int_a^b f(x)\, dx = 0,$$

then $f(x) = 0$ at least once in $[a, b]$.

Solution   The average value of f on $[a, b]$ is

$$\operatorname{av}(f) = \frac{1}{b-a} \int_a^b f(x)\, dx = \frac{1}{b-a} \cdot 0 = 0.$$

By Theorem 2, f assumes this value at some point c in $[a, b]$.   ▢

Exercises 4.6

Using Properties and Known Values to Find Other Integrals

1. Suppose that f and g are continuous and that

$$\int_1^2 f(x)\, dx = -4, \quad \int_1^5 f(x)\, dx = 6, \quad \int_1^5 g(x)\, dx = 8.$$

Use the rules in Table 4.5 to find

a) $\displaystyle \int_2^2 g(x)\, dx$ **b)** $\displaystyle \int_5^1 g(x)\, dx$

c) $\displaystyle \int_1^2 3f(x)\, dx$ **d)** $\displaystyle \int_2^5 f(x)\, dx$

e) $\displaystyle \int_1^5 [f(x) - g(x)]\, dx$ **f)** $\displaystyle \int_1^5 [4f(x) - g(x)]\, dx$

2. Suppose that f and h are continuous and that

$$\int_1^9 f(x)\, dx = -1, \quad \int_7^9 f(x)\, dx = 5, \quad \int_7^9 h(x)\, dx = 4.$$

Use the rules in Table 4.5 to find

a) $\displaystyle \int_1^9 -2f(x)\, dx$ **b)** $\displaystyle \int_7^9 [f(x) + h(x)]\, dx$

c) $\displaystyle \int_7^9 [2f(x) - 3h(x)]\, dx$ **d)** $\displaystyle \int_9^1 f(x)\, dx$

e) $\displaystyle \int_1^7 f(x)\, dx$ **f)** $\displaystyle \int_9^7 [h(x) - f(x)]\, dx$

3. Suppose that $\int_1^2 f(x)\, dx = 5$. Find

a) $\displaystyle \int_1^2 f(u)\, du$ **b)** $\displaystyle \int_1^2 \sqrt{3} f(z)\, dz$

c) $\displaystyle \int_2^1 f(t)\, dt$ **d)** $\displaystyle \int_1^2 [-f(x)]\, dx$

4. Suppose that $\int_{-3}^0 g(t)\, dt = \sqrt{2}$. Find

a) $\displaystyle \int_0^{-3} g(t)\, dt$ **b)** $\displaystyle \int_{-3}^0 g(u)\, du$

c) $\displaystyle \int_{-3}^0 [-g(x)]\, dx$ **d)** $\displaystyle \int_{-3}^0 \frac{g(r)}{\sqrt{2}}\, dr$

5. Suppose that f is continuous and that $\int_0^3 f(z)\, dz = 3$ and $\int_0^4 f(z)\, dz = 7$. Find

a) $\displaystyle \int_3^4 f(z)\, dz$ **b)** $\displaystyle \int_4^3 f(t)\, dt$

6. Suppose that h is continuous and that $\int_{-1}^1 h(r)\, dr = 0$ and $\int_{-1}^3 h(r)\, dr = 6$. Find

a) $\displaystyle \int_1^3 h(r)\, dr$ **b)** $\displaystyle -\int_3^1 h(u)\, du$

Evaluate the integrals in Exercises 7–18.

7. $\displaystyle \int_3^1 7\, dx$ **8.** $\displaystyle \int_0^{-2} \sqrt{2}\, dx$

9. $\displaystyle \int_0^2 5x\, dx$ **10.** $\displaystyle \int_3^5 \frac{x}{8}\, dx$

11. $\displaystyle \int_0^2 (2t - 3)\, dt$ **12.** $\displaystyle \int_0^{\sqrt{2}} \left(t - \sqrt{2}\right)\, dt$

13. $\displaystyle \int_2^1 \left(1 + \frac{z}{2}\right)\, dz$ **14.** $\displaystyle \int_3^0 (2z - 3)\, dz$

15. $\displaystyle \int_1^2 3u^2\, du$ **16.** $\displaystyle \int_{1/2}^1 24u^2\, du$

17. $\displaystyle \int_0^2 (3x^2 + x - 5)\, dx$ **18.** $\displaystyle \int_1^0 (3x^2 + x - 5)\, dx$

Area

In Exercises 19–22, find the total shaded area.

19.

20.

21.

22.

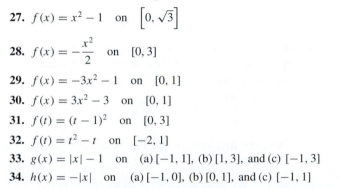

In Exercises 23–26, graph the function over the given interval. Then
(a) integrate the function over the interval and (b) find the area of the
region between the graph and the x-axis.

23. $y = x^2 - 6x + 8$, $[0, 3]$

24. $y = -x^2 + 5x - 4$, $[0, 2]$

25. $y = 2x - x^2$, $[0, 3]$

26. $y = x^2 - 4x$, $[0, 5]$

Average Value

In Exercises 27–34, graph the function and find its average value over
the given interval. At what point or points in the given interval does
the function assume its average value?

27. $f(x) = x^2 - 1$ on $\left[0, \sqrt{3}\right]$

28. $f(x) = -\dfrac{x^2}{2}$ on $[0, 3]$

29. $f(x) = -3x^2 - 1$ on $[0, 1]$

30. $f(x) = 3x^2 - 3$ on $[0, 1]$

31. $f(t) = (t - 1)^2$ on $[0, 3]$

32. $f(t) = t^2 - t$ on $[-2, 1]$

33. $g(x) = |x| - 1$ on   (a) $[-1, 1]$, (b) $[1, 3]$, and (c) $[-1, 3]$

34. $h(x) = -|x|$ on   (a) $[-1, 0]$, (b) $[0, 1]$, and (c) $[-1, 1]$

In Exercises 35–38, find the average value of the function over the
given interval from the graph of f (without integrating).

35. $f(x) = \begin{cases} x + 4, & -4 \le x \le -1 \\ -x + 2, & -1 < x \le 2 \end{cases}$ on $[-4, 2]$

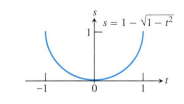

36. $f(t) = 1 - \sqrt{1 - t^2}$ on $[-1, 1]$

37. $f(t) = \sin t$ on $[0, 2\pi]$

38. $f(\theta) = \tan \theta$ on $\left[\dfrac{-\pi}{4}, \dfrac{\pi}{4}\right]$

Theory and Examples

39. Use the Max-Min Inequality to find upper and lower bounds for
the value of

$$\int_0^1 \frac{1}{1 + x^2} \, dx.$$

40. (*Continuation of Exercise 39.*) Use the Max-Min Inequality to
find upper and lower bounds for

$$\int_0^{0.5} \frac{1}{1 + x^2} \, dx \quad \text{and} \quad \int_{0.5}^1 \frac{1}{1 + x^2} \, dx.$$

Add these to arrive at an improved estimate of

$$\int_0^1 \frac{1}{1 + x^2} \, dx.$$

41. Show that the value of $\int_0^1 \sin(x^2) \, dx$ cannot possibly be 2.

42. Show that the value of $\int_0^1 \sqrt{x + 8} \, dx$ lies between $2\sqrt{2} \approx 2.8$
and 3.

43. Suppose that f is continuous and that $\int_1^2 f(x) \, dx = 4$. Show
that $f(x) = 4$ at least once on $[1, 2]$.

44. Suppose that f and g are continuous on $[a, b]$, $a \neq b$, and that
$\int_a^b (f(x) - g(x)) \, dx = 0$. Show that $f(x) = g(x)$ at least once
in $[a, b]$.

45. *Integrals of nonnegative functions.* Use the Max-Min Inequality to show that if f is integrable then

$$f(x) \ge 0 \quad \text{on} \quad [a, b] \quad \Rightarrow \quad \int_a^b f(x) \, dx \ge 0.$$

46. *Integrals of nonpositive functions.* Show that if f is integrable then

$$f(x) \le 0 \quad \text{on} \quad [a, b] \quad \Rightarrow \quad \int_a^b f(x)\,dx \le 0.$$

47. Use the inequality $\sin x \le x$, which holds for $x \ge 0$, to find an upper bound for the value of $\int_0^1 \sin x\,dx$.

48. The inequality $\sec x \ge 1 + (x^2/2)$ holds on $(-\pi/2, \pi/2)$. Use it to find a lower bound for the value of $\int_0^1 \sec x\,dx$.

49. If av (f) really is a typical value of the integrable function $f(x)$ on $[a, b]$, then the number av(f) should have the same integral over $[a, b]$ that f does. Does it? That is, does

$$\int_a^b \text{av}(f)\,dx = \int_a^b f(x)\,dx?$$

Give reasons for your answer.

50. It would be nice if average values of integrable functions obeyed the following rules on an interval $[a, b]$:

a) av $(f + g) = $ av $(f) + $ av (g)
b) av $(kf) = k$ av (f) (any number k)
c) av $(f) \le$ av (g) if $f(x) \le g(x)$ on $[a, b]$.

Do these rules ever hold? Give reasons for your answers.

51. If you average 30 mi/h on a 150-mi trip and then return over the same 150 mi at the rate of 50 mi/h, what is your average speed for the trip? Give reasons for your answer. (Source: David H. Pleacher, *The Mathematics Teacher,* Vol. 85, No. 6, pp. 445–446, September 1992.)

52. A dam released 1000 m³ of water at 10 m³/min and then released another 1000 m³ at 20 m³/min. What was the average rate at which the water was released? Give reasons for your answer.

4.7

The Fundamental Theorem

This section presents the Fundamental Theorem of Integral Calculus. The independent discovery by Leibniz and Newton of this astonishing connection between integration and differentiation started the mathematical developments that fueled the scientific revolution for the next two hundred years and constitutes what is still regarded as the most important computational discovery in the history of the world.

The Fundamental Theorem, Part 1

If $f(t)$ is an integrable function, the integral from any fixed number a to another number x defines a function F whose value at x is

$$F(x) = \int_a^x f(t)\,dt. \tag{1}$$

For example, if f is nonnegative and x lies to the right of a, $F(x)$ is the area under the graph from a to x. The variable x is the upper limit of integration of an integral, but F is just like any other real-valued function of a real variable. For each value of the input x there is a well-defined numerical output, in this case the integral of f from a to x.

Equation (1) gives an important way to define new functions and to describe solutions of differential equations (more about this later). The reason for mentioning Eq. (1) now, however, is the connection it makes between integrals and derivatives. For if f is any continuous function whatever, then F is a differentiable function of x whose derivative is f itself. At every value of x,

$$\frac{d}{dx} F(x) = \frac{d}{dx} \int_a^x f(t)\,dt = f(x).$$

This idea is so important that it is the first part of the Fundamental Theorem of Calculus.

> **Theorem 3**
>
> **The Fundamental Theorem of Calculus, Part 1**
>
> If f is continuous on $[a, b]$, then $F(x) = \int_a^x f(t)\, dt$ has a derivative at every point of $[a, b]$ and
>
> $$\frac{dF}{dx} = \frac{d}{dx} \int_a^x f(t)\, dt = f(x), \qquad a \le x \le b. \tag{2}$$

This conclusion is beautiful, powerful, deep, and surprising, and Eq. (2) may well be the most important equation in mathematics. It says that the differential equation $dF/dx = f$ has a solution for every continuous function f. It says that every continuous function f is the derivative of some other function, namely $\int_a^x f(t)\, dt$. It says that every continuous function has an antiderivative. And it says that the processes of integration and differentiation are inverses of one another.

Proof of Theorem 3 We prove Theorem 3 by applying the definition of derivative directly to the function $F(x)$. This means writing out the difference quotient

$$\frac{F(x + h) - F(x)}{h} \tag{3}$$

and showing that its limit as $h \to 0$ is the number $f(x)$.

When we replace $F(x + h)$ and $F(x)$ by their defining integrals, the numerator in Eq. (3) becomes

$$F(x + h) - F(x) = \int_a^{x+h} f(t)\, dt - \int_a^x f(t)\, dt.$$

The Additivity Rule for integrals (Table 4.5 in Section 4.6) simplifies the right-hand side to

$$\int_x^{x+h} f(t)\, dt,$$

so that Eq. (3) becomes

$$\frac{F(x + h) - F(x)}{h} = \frac{1}{h} [F(x + h) - F(x)]$$

$$= \frac{1}{h} \int_x^{x+h} f(t)\, dt. \tag{4}$$

According to the Mean Value Theorem for Definite Integrals (Theorem 2 in the preceding section), the value of the last expression in Eq. (4) is one of the values taken on by f in the interval joining x and $x + h$. That is, for some number c in this interval,

$$\frac{1}{h} \int_x^{x+h} f(t)\, dt = f(c). \tag{5}$$

We can therefore find out what happens to $(1/h)$ times the integral as $h \to 0$ by watching what happens to $f(c)$ as $h \to 0$.

What does happen to $f(c)$ as $h \to 0$? As $h \to 0$, the endpoint $x + h$ approaches x, pushing c ahead of it like a bead on a wire:

So c approaches x, and, since f is continuous at x, $f(c)$ approaches $f(x)$:

$$\lim_{h \to 0} f(c) = f(x). \tag{6}$$

Going back to the beginning, then, we have

$$\frac{dF}{dx} = \lim_{h \to 0} \frac{F(x + h) - F(x)}{h} \qquad \text{Definition of derivative}$$

$$= \lim_{h \to 0} \frac{1}{h} \int_x^{x+h} f(t)\, dt \qquad \text{Eq. (4)}$$

$$= \lim_{h \to 0} f(c) \qquad \text{Eq. (5)}$$

$$= f(x). \qquad \text{Eq. (6)}$$

This concludes the proof. ❏

If the values of f are positive, the equation

$$\frac{d}{dx} \int_a^x f(t)\, dt = f(x)$$

has a nice geometric interpretation. For then the integral of f from a to x is the area $A(x)$ of the region between the graph of f and the x-axis from a to x. Imagine covering this region from left to right by unrolling a carpet of variable width $f(t)$ (Fig. 4.23). As the carpet rolls past x, the rate at which the floor is being covered is $f(x)$.

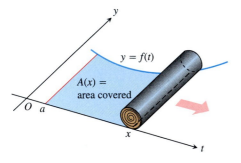

4.23 The rate at which the carpet covers the floor at the point x is the width of the carpet's leading edge as it rolls past x. In symbols, $dA/dx = f(x)$.

EXAMPLE 1

$$\frac{d}{dx} \int_{-\pi}^x \cos t\, dt = \cos x \qquad \text{Eq. (2) with } f(t) = \cos t$$

$$\frac{d}{dx} \int_0^x \frac{1}{1 + t^2}\, dt = \frac{1}{1 + x^2} \qquad \text{Eq. (2) with } f(t) = \frac{1}{1 + t^2}$$

❏

EXAMPLE 2 Find dy/dx if

$$y = \int_1^{x^2} \cos t\, dt.$$

Solution Notice that the upper limit of integration is not x but x^2. To find dy/dx we must therefore treat y as the composite of

$$y = \int_1^u \cos t\, dt \qquad \text{and} \qquad u = x^2$$

and apply the Chain Rule:

$$\frac{dy}{dx} = \frac{dy}{du}\frac{du}{dx}$$ Chain Rule

$$= \frac{d}{du}\int_{1}^{u} \cos t\, dt \cdot \frac{du}{dx}$$ Substitute the formula for y.

$$= \cos u \cdot \frac{du}{dx}$$ Eq. (2) with $f(t) = \cos t$

$$= \cos x^2 \cdot 2x$$ $u = x^2$

$$= 2x \cos x^2.$$ Usual form   ❏

EXAMPLE 3   Express the solution of the following initial value problem as an integral.

Differential equation: $\dfrac{dy}{dx} = \tan x$

Initial condition: $y(1) = 5$

Solution   The function

$$F(x) = \int_{1}^{x} \tan t\, dt$$

is an antiderivative of $\tan x$. Hence the general solution of the equation is

$$y = \int_{1}^{x} \tan t\, dt + C.$$

As always, the initial condition determines the value of C:

$$5 = \int_{1}^{1} \tan t\, dt + C   \qquad y(1) = 5$$

$$5 = 0 + C   \tag{7}$$

$$C = 5.$$

The solution of the initial value problem is

$$y = \int_{1}^{x} \tan t\, dt + 5.$$

How did we know where to start integrating when we constructed $F(x)$? We could have started anywhere, but the best value to start with is the initial value of x (in this case $x = 1$). Then the integral will be zero when we apply the initial condition (as it was in Eq. 7) and C will automatically be the initial value of y.

❏

The Evaluation of Definite Integrals

We now come to the second part of the Fundamental Theorem of Calculus, the part that describes how to evaluate definite integrals.

> **Theorem 4**
> **The Fundamental Theorem of Calculus, Part 2**
> If f is continuous at every point of $[a, b]$ and F is any antiderivative of f on $[a, b]$, then
> $$\int_a^b f(x)\,dx = F(b) - F(a). \tag{8}$$

How to Evaluate $\displaystyle\int_a^b f(x)\,dx$

1. Find an antiderivative F of f. Any antiderivative will do, so pick the simplest one you can.
2. Calculate the number $F(b) - F(a)$.

This number will be $\displaystyle\int_a^b f(x)\,dx$.

Theorem 4 says that to evaluate the definite integral of a continuous function f from a to b, all we need do is find an antiderivative F of f and calculate the number $F(b) - F(a)$. The existence of the antiderivative is assured by the first part of the Fundamental Theorem.

Proof of Theorem 4 To prove Theorem 4, we use the fact that functions with identical derivatives differ only by a constant. We already know one function whose derivative equals f, namely,

$$G(x) = \int_a^x f(t)\,dt.$$

Therefore, if F is any other such function, then

$$F(x) = G(x) + C \tag{9}$$

throughout $[a, b]$ for some constant C. When we use Eq. (9) to calculate $F(b) - F(a)$, we find that

$$
\begin{aligned}
F(b) - F(a) &= [G(b) + C] - [G(a) + C] \\
&= G(b) - G(a) \\
&= \int_a^b f(t)\,dt - \int_a^a f(t)\,dt \\
&= \int_a^b f(t)\,dt - 0 = \int_a^b f(t)\,dt.
\end{aligned}
$$

This establishes Eq. (8) and concludes the proof. ❑

Notation

The usual notation for the number $F(b) - F(a)$ is $F(x)\big]_a^b$ when $F(x)$ has a single term, or $[F(x)]_a^b$ for $F(b) - F(a)$ when $F(x)$ has more than one term.

EXAMPLE 4

a) $\displaystyle\int_0^\pi \cos x\,dx = \sin x\,\bigg]_0^\pi = \sin \pi - \sin 0 = 0 - 0 = 0$

b) $\displaystyle\int_{-\pi/4}^0 \sec x \tan x\,dx = \sec x\,\bigg]_{-\pi/4}^0 = \sec 0 - \sec\left(-\frac{\pi}{4}\right) = 1 - \sqrt{2}$

c) $\displaystyle\int_1^4 \left(\frac{3}{2}\sqrt{x} - \frac{4}{x^2}\right) dx = \left[x^{3/2} + \frac{4}{x}\right]_1^4$

$$= \left[(4)^{3/2} + \frac{4}{4}\right] - \left[(1)^{3/2} + \frac{4}{1}\right]$$

$$= [8 + 1] - [5] = 4. \qquad\square$$

Theorem 4 explains the formulas we derived for the integrals of x and x^2 in Section 4.5. We can now see that without any restriction on the signs of a and b,

$$\int_a^b x \, dx = \frac{x^2}{2}\bigg]_a^b = \frac{b^2}{2} - \frac{a^2}{2} \qquad \text{Because } x^2/2 \text{ is an antiderivative of } x$$

$$\int_a^b x^2 \, dx = \frac{x^3}{3}\bigg]_a^b = \frac{b^3}{3} - \frac{a^3}{3} \qquad \text{Because } x^3/3 \text{ is an antiderivative of } x^2$$

EXAMPLE 5 Find the area of the region between the x-axis and the graph of $f(x) = x^3 - x^2 - 2x$, $-1 \le x \le 2$.

Solution First find the zeros of f. Since

$$f(x) = x^3 - x^2 - 2x = x(x^2 - x - 2) = x(x + 1)(x - 2),$$

the zeros are $x = 0$, -1, and 2 (Fig. 4.24). The zeros partition $[-1, 2]$ into two subintervals: $[-1, 0]$, on which $f \ge 0$ and $[0, 2]$, on which $f \le 0$. We integrate f over each subinterval and add the absolute values of the calculated values.

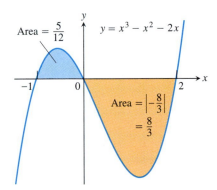

4.24 The region between the curve $y = x^3 - x^2 - 2x$ and the x-axis (Example 5).

Integral over $[-1, 0]$:
$$\int_{-1}^0 (x^3 - x^2 - 2x) \, dx = \left[\frac{x^4}{4} - \frac{x^3}{3} - x^2\right]_{-1}^0$$
$$= 0 - \left[\frac{1}{4} + \frac{1}{3} - 1\right] = \frac{5}{12}$$

Integral over $[0, 2]$:
$$\int_0^2 (x^3 - x^2 - 2x) \, dx = \left[\frac{x^4}{4} - \frac{x^3}{3} - x^2\right]_0^2$$
$$= \left[4 - \frac{8}{3} - 4\right] - 0 = -\frac{8}{3}$$

Enclosed area: Total enclosed area $= \dfrac{5}{12} + \left|-\dfrac{8}{3}\right| = \dfrac{37}{12}$ ❑

EXAMPLE 6 *Household electricity*

We model the voltage in our home wiring with the sine function

$$V = V_{\max} \sin 120\pi t,$$

which expresses the voltage V in volts as a function of time t in seconds. The function runs through 60 cycles each second (its frequency is 60 hertz, or 60 Hz). The positive constant $V_{\max}$ ("vee max") is the **peak voltage.**

The average value of V over a half-cycle (duration 1/120 sec; see Fig. 4.25) is

$$V_{\text{av}} = \frac{1}{(1/120) - 0} \int_0^{1/120} V_{\max} \sin 120\pi t \, dt$$
$$= 120 V_{\max} \left[-\frac{1}{120\pi} \cos 120\pi t\right]_0^{1/120}$$
$$= \frac{V_{\max}}{\pi} [-\cos \pi + \cos 0]$$
$$= \frac{2V_{\max}}{\pi}.$$

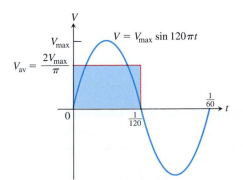

4.25 The graph of the household voltage $V = V_{\max} \sin 120\pi t$ over a full cycle. Its average value over a half-cycle is $2V_{\max}/\pi$. Its average value over a full cycle is zero.

The average value of the voltage over a full cycle, as we can see from Fig. 4.25, is zero. (Also see Exercise 64.) If we measured the voltage with a standard moving-coil galvanometer, the meter would read zero.

To measure the voltage effectively, we use an instrument that measures the square root of the average value of the square of the voltage, namely

$$V_{\text{rms}} = \sqrt{(V^2)_{\text{av}}}.$$

The subscript "rms" (read the letters separately) stands for "root mean square." Since the average value of $V^2 = (V_{\text{max}})^2 \sin^2 120\pi t$ over a cycle is

$$(V^2)_{\text{av}} = \frac{1}{(1/60) - 0} \int_0^{1/60} (V_{\text{max}})^2 \sin^2 120\pi t \, dt = \frac{(V_{\text{max}})^2}{2} \qquad (10)$$

(Exercise 64c), the rms voltage is

$$V_{\text{rms}} = \sqrt{\frac{(V_{\text{max}})^2}{2}} = \frac{V_{\text{max}}}{\sqrt{2}}. \qquad (11)$$

The values given for household currents and voltages are always rms values. Thus, "115 volts ac" means that the rms voltage is 115. The peak voltage,

$$V_{\text{max}} = \sqrt{2} \, V_{\text{rms}} = \sqrt{2} \cdot 115 \approx 163 \text{ volts,}$$

obtained from Eq. (11), is considerably higher. ❑

Exercises 4.7

Evaluating Integrals

Evaluate the integrals in Exercises 1–26.

1. $\int_{-2}^{0} (2x + 5) \, dx$

2. $\int_{-3}^{4} \left(5 - \frac{x}{2}\right) dx$

3. $\int_{0}^{4} \left(3x - \frac{x^3}{4}\right) dx$

4. $\int_{-2}^{2} (x^3 - 2x + 3) \, dx$

5. $\int_{0}^{1} (x^2 + \sqrt{x}) \, dx$

6. $\int_{0}^{5} x^{3/2} \, dx$

7. $\int_{1}^{32} x^{-6/5} \, dx$

8. $\int_{-2}^{-1} \frac{2}{x^2} \, dx$

9. $\int_{0}^{\pi} \sin x \, dx$

10. $\int_{0}^{\pi} (1 + \cos x) \, dx$

11. $\int_{0}^{\pi/3} 2 \sec^2 x \, dx$

12. $\int_{\pi/6}^{5\pi/6} \csc^2 x \, dx$

13. $\int_{\pi/4}^{3\pi/4} \csc \theta \cot \theta \, d\theta$

14. $\int_{0}^{\pi/3} 4 \sec u \tan u \, du$

15. $\int_{\pi/2}^{0} \frac{1 + \cos 2t}{2} \, dt$

16. $\int_{-\pi/3}^{\pi/3} \frac{1 - \cos 2t}{2} \, dt$

17. $\int_{-\pi/2}^{\pi/2} (8y^2 + \sin y) \, dy$

18. $\int_{-\pi/3}^{-\pi/4} \left(4 \sec^2 t + \frac{\pi}{t^2}\right) dt$

19. $\int_{1}^{-1} (r + 1)^2 \, dr$

20. $\int_{-\sqrt{3}}^{\sqrt{3}} (t + 1)(t^2 + 4) \, dt$

21. $\int_{\sqrt{2}}^{1} \left(\frac{u^7}{2} - \frac{1}{u^5}\right) du$

22. $\int_{1/2}^{1} \left(\frac{1}{v^3} - \frac{1}{v^4}\right) dv$

23. $\int_{1}^{\sqrt{2}} \frac{s^2 + \sqrt{s}}{s^2} \, ds$

24. $\int_{9}^{4} \frac{1 - \sqrt{u}}{\sqrt{u}} \, du$

25. $\int_{-4}^{4} |x| \, dx$

26. $\int_{0}^{\pi} \frac{1}{2} (\cos x + |\cos x|) \, dx$

Evaluating Integrals Using Substitutions

In Exercises 27–34, use a substitution to find an antiderivative and then apply the Fundamental Theorem to evaluate the integral.

27. $\int_{0}^{1} (1 - 2x)^3 \, dx$

28. $\int_{1}^{2} \sqrt{3x + 1} \, dx$

29. $\int_{0}^{1} t\sqrt{t^2 + 1} \, dt$

30. $\int_{-1}^{2} \frac{t \, dt}{\sqrt{2t^2 + 8}}$

31. $\displaystyle\int_0^\pi \sin^2\left(1+\frac{\theta}{2}\right)d\theta$

32. $\displaystyle\int_{3\pi/8}^{\pi/2} \sec^2\left(\pi-2\theta\right)d\theta$

33. $\displaystyle\int_0^\pi \sin^2\frac{x}{4}\cos\frac{x}{4}\,dx$

34. $\displaystyle\int_{2\pi/3}^\pi \tan^3\frac{x}{4}\sec^2\frac{x}{4}\,dx$

Area

In Exercises 35–40, find the total area between the region and the x-axis.

35. $y = -x^2 - 2x, \quad -3 \le x \le 2$

36. $y = 3x^2 - 3, \quad -2 \le x \le 2$

37. $y = x^3 - 3x^2 + 2x, \quad 0 \le x \le 2$

38. $y = x^3 - 4x, \quad -2 \le x \le 2$

39. $y = x^{1/3}, \quad -1 \le x \le 8$

40. $y = x^{1/3} - x, \quad -1 \le x \le 8$

Find the areas of the shaded regions in Exercises 41–44.

41.

42.

43.

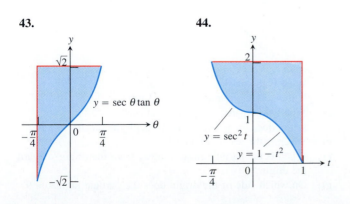

44.

Derivatives of Integrals

Find the derivatives in Exercises 45–48 (a) by evaluating the integral and differentiating the result and (b) by differentiating the integral directly.

45. $\displaystyle\frac{d}{dx}\int_0^{\sqrt{x}} \cos t\,dt$

46. $\displaystyle\frac{d}{dx}\int_1^{\sin x} 3t^2\,dt$

47. $\displaystyle\frac{d}{dt}\int_0^{t^4} \sqrt{u}\,du$

48. $\displaystyle\frac{d}{d\theta}\int_0^{\tan\theta} \sec^2 y\,dy$

Find dy/dx in Exercises 49–54.

49. $\displaystyle y = \int_0^x \sqrt{1+t^2}\,dt$

50. $\displaystyle y = \int_1^x \frac{1}{t}\,dt, \quad x > 0$

51. $\displaystyle y = \int_0^{\sqrt{x}} \sin(t^2)\,dt$

52. $\displaystyle y = \int_0^{x^2} \cos\sqrt{t}\,dt$

53. $\displaystyle y = \int_0^{\sin x} \frac{dt}{\sqrt{1-t^2}}, \quad |x| < \frac{\pi}{2}$

54. $\displaystyle y = \int_0^{\tan x} \frac{dt}{1+t^2}$

Initial Value Problems

Each of the following functions solves one of the initial value problems in Exercises 55–58. Which function solves which problem? Give brief reasons for your answers.

a) $\displaystyle y = \int_1^x \frac{1}{t}\,dt - 3$

b) $\displaystyle y = \int_0^x \sec t\,dt + 4$

c) $\displaystyle y = \int_{-1}^x \sec t\,dt + 4$

d) $\displaystyle y = \int_\pi^x \frac{1}{t}\,dt - 3$

55. $\displaystyle\frac{dy}{dx} = \frac{1}{x}, \quad y(\pi) = -3$

56. $y' = \sec x, \quad y(-1) = 4$

57. $y' = \sec x, \quad y(0) = 4$

58. $\displaystyle y' = \frac{1}{x}, \quad y(1) = -3$

Express the solutions of the initial value problems in Exercises 59–62 in terms of integrals.

59. $\displaystyle\frac{dy}{dx} = \sec x, \quad y(2) = 3$

60. $\displaystyle\frac{dy}{dx} = \sqrt{1+x^2}, \quad y(1) = -2$

61. $\displaystyle\frac{ds}{dt} = f(t), \quad s(t_0) = s_0$

62. $\displaystyle\frac{dv}{dt} = g(t), \quad v(t_0) = v_0$

Applications

63. *Archimedes' area formula for parabolas.* Archimedes (287–212 B.C.), inventor, military engineer, physicist, and the greatest mathematician of classical times in the western world, discovered that the area under a parabolic arch is two-thirds the base times the height.

a) Use an integral to find the area under the arch

$$y = 6 - x - x^2, \quad -3 \le x \le 2.$$

b) Find the height of the arch.

c) Show that the area is two-thirds the base b times the height h.

d) Sketch the parabolic arch $y = h - (4h/b^2)x^2$, $-b/2 \le x \le b/2$, assuming that h and b are positive. Then use calculus to find the area of the region enclosed between the arch and the x-axis.

64. (*Continuation of Example 6.*)

a) Show by evaluating the integral in the expression

$$\frac{1}{(1/60) - 0} \int_0^{1/60} V_{\max} \sin 120\pi t \, dt$$

that the average value of $V = V_{\max} \sin 120\pi t$ over a full cycle is zero.

b) The circuit that runs your electric stove is rated 240 volts rms. What is the peak value of the allowable voltage?

c) Show that

$$\int_0^{1/60} (V_{\max})^2 \sin^2 120\pi t \, dt = \frac{(V_{\max})^2}{120}.$$

65. *Cost from marginal cost.* The marginal cost of printing a poster when x posters have been printed is

$$\frac{dc}{dx} = \frac{1}{2\sqrt{x}}$$

dollars. Find (a) $c(100) - c(1)$, the cost of printing posters 2–100; (b) $c(400) - c(100)$, the cost of printing posters 101–400.

66. *Revenue from marginal revenue.* Suppose that a company's marginal revenue from the manufacture and sale of egg beaters is

$$\frac{dr}{dx} = 2 - 2/(x + 1)^2,$$

where r is measured in thousands of dollars and x in thousands of units. How much money should the company expect from a production run of $x = 3$ thousand egg beaters? To find out, integrate the marginal revenue from $x = 0$ to $x = 3$.

Drawing Conclusions about Motion from Graphs

67. Suppose that f is the differentiable function shown in the accompanying graph and that the position at time t (sec) of a particle moving along a coordinate axis is

$$s = \int_0^t f(x) \, dx$$

meters. Use the graph to answer the following questions. Give reasons for your answers.

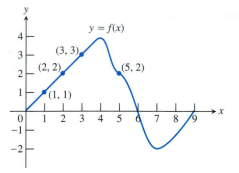

a) What is the particle's velocity at time $t = 5$?

b) Is the acceleration of the particle at time $t = 5$ positive, or negative?

c) What is the particle's position at time $t = 3$?

d) At what time during the first 9 sec does s have its largest value?

e) Approximately when is the acceleration zero?

f) When is the particle moving toward the origin? away from the origin?

g) On which side of the origin does the particle lie at time $t = 9$?

68. Suppose that g is the differentiable function graphed here and that the position at time t (sec) of a particle moving along a coordinate axis is

$$s = \int_0^t g(x) \, dx$$

meters. Use the graph to answer the following questions. Give reasons for your answers.

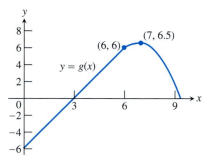

a) What is the particle's velocity at $t = 3$?

b) Is the acceleration at time $t = 3$ positive, or negative?

c) What is the particle's position at time $t = 3$?

d) When does the particle pass through the origin?

e) When is the acceleration zero?

f) When is the particle moving away from the origin? toward the origin?

g) On which side of the origin does the particle lie at $t = 9$?

Volumes from Section 4.4

69. (*Continuation of Section 4.4, Example 3.*) The approximating sum for the volume of the solid in Example 3, Section 4.4, was a Riemann sum for an integral. What integral? Evaluate it to find the volume.

70. (*Continuation of Section 4.4, Example 4.*) The approximating sum for the volume of the sphere in Example 4, Section 4.4, was a Riemann sum for an integral. What integral? Evaluate it to find the volume.

71. (*Continuation of Section 4.4, Exercise 15.*) The approximating sums for the volume of water in Exercise 15, Section 4.4, are Riemann sums for an integral. What integral? Evaluate it to find the volume.

72. (*Continuation of Section 4.4, Exercise 17.*) The approximating sums for the volume of the rocket nose cone in Exercise 17, Section 4.4, is a Riemann sum for an integral. What integral? Evaluate it to find the volume.

Theory and Examples

73. Show that if k is a positive constant, then the area between the x-axis and one arch of the curve $y = \sin kx$ is $2/k$.

74. Find

$$\lim_{x \to 0} \frac{1}{x^3} \int_0^x \frac{t^2}{t^4 + 1} \, dt.$$

75. Suppose $\int_1^x f(t) \, dt = x^2 - 2x + 1$. Find $f(x)$.

76. Find $f(4)$ if $\int_0^x f(t) \, dt = x \cos \pi x$.

77. Find the linearization of

$$f(x) = 2 - \int_2^{x+1} \frac{9}{1+t} \, dt$$

at $x = 1$.

78. Find the linearization of

$$g(x) = 3 + \int_1^{x^2} \sec (t - 1) \, dt$$

at $x = -1$.

79. Suppose that f has a positive derivative for all values of x and that $f(1) = 0$. Which of the following statements must be true of the function

$$g(x) = \int_0^x f(t) \, dt?$$

Give reasons for your answers.

a) g is a differentiable function of x.
b) g is a continuous function of x.
c) The graph of g has a horizontal tangent at $x = 1$.
d) g has a local maximum at $x = 1$.
e) g has a local minimum at $x = 1$.
f) The graph of g has an inflection point at $x = 1$.
g) The graph of dg/dx crosses the x-axis at $x = 1$.

80. Suppose that f has a negative derivative for all values of x and that $f(1) = 0$. Which of the following statements must be true of the function

$$h(x) = \int_0^x f(t) \, dt?$$

Give reasons for your answers.

a) h is a twice-differentiable function of x.
b) h and dh/dx are both continuous.
c) The graph of h has a horizontal tangent at $x = 1$.
d) h has a local maximum at $x = 1$.
e) h has a local minimum at $x = 1$.
f) The graph of h has an inflection point at $x = 1$.
g) The graph of dh/dx crosses the x-axis at $x = 1$.

▦ Grapher Explorations

81. *The Fundamental Theorem.* If f is continuous, we expect

$$\lim_{h \to 0} \frac{1}{h} \int_x^{x+h} f(t) \, dt$$

to equal $f(x)$, as in the proof of Part 1 of the Fundamental Theorem. For instance, if $f(t) = \cos t$, then

$$\frac{1}{h} \int_x^{x+h} \cos t \, dt = \frac{\sin (x + h) - \sin x}{h}. \qquad (12)$$

The right-hand side of Eq. (12) is the difference quotient for the derivative of the sine, and we expect its limit as $h \to 0$ to be $\cos x$.

Graph $\cos x$ for $-\pi \le x \le 2\pi$. Then, in a different color if possible, graph the right-hand side of Eq. (12) as a function of x for $h = 2, 1, 0.5$, and 0.1. Watch how the latter curves converge to the graph of the cosine as $h \to 0$.

82. Repeat Exercise 81 for $f(t) = 3t^2$. What is

$$\lim_{h \to 0} \frac{1}{h} \int_x^{x+h} 3t^2 \, dt = \lim_{h \to 0} \frac{(x + h)^3 - x^3}{h}?$$

Graph $f(x) = 3x^2$ for $-1 \le x \le 1$. Then graph the quotient $((x + h)^3 - x^3)/h$ as a function of x for $h = 1, 0.5, 0.2$, and 0.1. Watch how the latter curves converge to the graph of $3x^2$ as $h \to 0$.

⚙ CAS Explorations and Projects

In Exercises 83–86, let $F(x) = \int_a^x f(t) \, dt$ for the specified function f and interval $[a, b]$. Use a CAS to perform the following steps and answer the questions posed.

a) Plot the functions f and F together over $[a, b]$.
b) Solve the equation $F'(x) = 0$. What can you see to be true about the graphs of f and F at points where $F'(x) = 0$? Is your observation borne out by Part 1 of the Fundamental Theorem coupled with information provided by the first derivative? Explain your answer.
c) Over what intervals (approximately) is the function F increasing and decreasing? What is true about f over those intervals?

d) Calculate the derivative f' and plot it together with F. What can you see to be true about the graph of F at points where $f'(x) = 0$? Is your observation borne out by Part 1 of the Fundamental Theorem? Explain your answer.

83. $f(x) = x^3 - 4x^2 + 3x$, $[0, 4]$

84. $f(x) = 2x^4 - 17x^3 + 46x^2 - 43x + 12$, $\left[0, \dfrac{9}{2}\right]$

85. $f(x) = \sin 2x \cos \dfrac{x}{3}$, $[0, 2\pi]$

86. $f(x) = x \cos \pi x$, $[0, 2\pi]$

In Exercises 87–90, let $F(x) = \int_a^{u(x)} f(t)\,dt$ for the specified a, u, and f. Use a CAS to perform the following steps and answer the questions posed.

a) Find the domain of F.

b) Calculate $F'(x)$ and determine its zeros. For what points in its domain is F increasing? decreasing?

c) Calculate $F''(x)$ and determine its zero. Identify the local extrema and the points of inflection of F.

d) Using the information from parts (a)–(c), draw a rough hand-sketch of $y = F(x)$ over its domain. Then graph $F(x)$ on your CAS to support your sketch.

87. $a = 1$, $u(x) = x^2$, $f(x) = \sqrt{1 - x^2}$

88. $a = 0$, $u(x) = x^2$, $f(x) = \sqrt{1 - x^2}$

89. $a = 0$, $u(x) = 1 - x$, $f(x) = x^2 - 2x - 3$

90. $a = 0$, $u(x) = 1 - x^2$, $f(x) = x^2 - 2x - 3$

91. Calculate $\dfrac{d}{dx}\displaystyle\int_a^{u(x)} f(t)\,dt$ and check your answer using a CAS.

92. Calculate $\dfrac{d^2}{dx^2}\displaystyle\int_a^{u(x)} f(t)\,dt$ and check your answer using a CAS.

4.8

Substitution in Definite Integrals

There are two methods for evaluating a definite integral by substitution, and they both work well. One is to find the corresponding indefinite integral by substitution and use one of the resulting antiderivatives to evaluate the definite integral by the Fundamental Theorem. The other is to use the following formula.

This formula first appeared in a book written by Isaac Barrow (1630–1677), Newton's teacher and predecessor at Cambridge University.

Substitution in Definite Integrals

THE FORMULA

$$\int_a^b f(g(x)) \cdot g'(x)\,dx = \int_{g(a)}^{g(b)} f(u)\,du \qquad (1)$$

HOW TO USE IT

Substitute $u = g(x)$, $du = g'(x)\,dx$, and integrate from $g(a)$ to $g(b)$.

To use the formula, make the same u-substitution you would use to evaluate the corresponding indefinite integral. Then integrate with respect to u from the value u has at $x = a$ to the value u has at $x = b$.

EXAMPLE 1   Evaluate $\displaystyle\int_{-1}^{1} 3x^2\sqrt{x^3 + 1}\,dx$.

Solution   We have two choices.

Method 1: Transform the integral as an indefinite integral, integrate, change back to x, and use the original x-limits.

$$\int 3x^2\sqrt{x^3+1}\,dx = \int \sqrt{u}\,du \qquad \text{Let } u = x^3 + 1, du = 3x^2\,dx.$$

$$= \frac{2}{3}u^{3/2} + C \qquad \text{Integrate with respect to } u.$$

$$= \frac{2}{3}(x^3+1)^{3/2} + C \qquad \text{Replace } u \text{ by } x^3 + 1.$$

$$\int_{-1}^{1} 3x^2\sqrt{x^3+1}\,dx = \frac{2}{3}(x^3+1)^{3/2}\Big]_{-1}^{1} \qquad \begin{array}{l}\text{Use the integral just found,}\\ \text{with limits of integration for } x.\end{array}$$

$$= \frac{2}{3}\left[((1)^3+1)^{3/2} - ((-1)^3+1)^{3/2}\right]$$

$$= \frac{2}{3}\left[2^{3/2} - 0^{3/2}\right] = \frac{2}{3}\left[2\sqrt{2}\right] = \frac{4\sqrt{2}}{3}$$

Method 2: Transform the integral and evaluate the transformed integral with the transformed limits given by Eq. (1).

$$\int_{-1}^{1} 3x^2\sqrt{x^3+1}\,dx$$

$$= \int_{0}^{2} \sqrt{u}\,du \qquad \begin{array}{l}\text{Let } u = x^3 + 1, du = 3x^2\,dx.\\ \text{When } x = -1, u = (-1)^3 + 1 = 0.\\ \text{When } x = 1, u = (1)^3 + 1 = 2.\end{array}$$

$$= \frac{2}{3}u^{3/2}\Big]_{0}^{2} \qquad \text{Evaluate the new definite integral.}$$

$$= \frac{2}{3}\left[2^{3/2} - 0^{3/2}\right] = \frac{2}{3}\left[2\sqrt{2}\right] = \frac{4\sqrt{2}}{3} \qquad \square$$

Which method is better—transforming the integral, integrating, and transforming back to use the original limits of integration, or evaluating the transformed integral with transformed limits? In Example 1, the second method seems easier, but that is not always the case. As a rule, it is best to know both methods and to use whichever one seems better at the time.

Here is another example of evaluating a transformed integral with transformed limits.

EXAMPLE 2

$$\int_{\pi/4}^{\pi/2} \cot\theta \, \csc^2\theta\,d\theta = \int_{1}^{0} u \cdot (-du) \qquad \begin{array}{l}\text{Let } u = \cot\theta, du = -\csc^2\theta\,d\theta\\ \quad -du = \csc^2\theta\,d\theta.\\ \text{When } \theta = \pi/4, u = \cot(\pi/4) = 1.\\ \text{When } \theta = \pi/2, u = \cot(\pi/2) = 0.\end{array}$$

$$= -\int_{1}^{0} u\,du$$

$$= -\left[\frac{u^2}{2}\right]_{1}^{0}$$

$$= -\left[\frac{(0)^2}{2} - \frac{(1)^2}{2}\right] = \frac{1}{2} \qquad \square$$

Technology *Visualizing Integrals with Elusive Antiderivatives* Many integrable functions, such as the important

$$f(x) = e^{-x^2}$$

from probability theory, *do not* have antiderivatives that can be expressed in terms of elementary functions. Nevertheless, we know the antiderivative of f exists by Part 1 of the Fundamental Theorem of Calculus. Use your graphing utility to visualize the integral function

$$F(x) = \int_0^x e^{-t^2}\,dt.$$

What can you say about $F(x)$? Where is it increasing and decreasing? Where are its extreme values, if any? What can you say about the concavity of its graph?

Exercises 4.8

Evaluating Definite Integrals

Evaluate the integrals in Exercises 1–24.

1. a) $\displaystyle\int_0^3 \sqrt{y+1}\,dy$

b) $\displaystyle\int_{-1}^0 \sqrt{y+1}\,dy$

2. a) $\displaystyle\int_0^1 r\sqrt{1-r^2}\,dr$

b) $\displaystyle\int_{-1}^1 r\sqrt{1-r^2}\,dr$

3. a) $\displaystyle\int_0^{\pi/4} \tan x \sec^2 x\,dx$

b) $\displaystyle\int_{-\pi/4}^0 \tan x \sec^2 x\,dx$

4. a) $\displaystyle\int_0^{\pi} 3\cos^2 x \sin x\,dx$

b) $\displaystyle\int_{2\pi}^{3\pi} 3\cos^2 x \sin x\,dx$

5. a) $\displaystyle\int_0^1 t^3(1+t^4)^3\,dt$

b) $\displaystyle\int_{-1}^1 t^3(1+t^4)^3\,dt$

6. a) $\displaystyle\int_0^{\sqrt{7}} t(t^2+1)^{1/3}\,dt$

b) $\displaystyle\int_{-\sqrt{7}}^0 t(t^2+1)^{1/3}\,dt$

7. a) $\displaystyle\int_{-1}^1 \frac{5r}{(4+r^2)^2}\,dr$

b) $\displaystyle\int_0^1 \frac{5r}{(4+r^2)^2}\,dr$

8. a) $\displaystyle\int_0^1 \frac{10\sqrt{v}}{(1+v^{3/2})^2}\,dv$

b) $\displaystyle\int_1^4 \frac{10\sqrt{v}}{(1+v^{3/2})^2}\,dv$

9. a) $\displaystyle\int_0^{\sqrt{3}} \frac{4x}{\sqrt{x^2+1}}\,dx$

b) $\displaystyle\int_{-\sqrt{3}}^{\sqrt{3}} \frac{4x}{\sqrt{x^2+1}}\,dx$

10. a) $\displaystyle\int_0^1 \frac{x^3}{\sqrt{x^4+9}}\,dx$

b) $\displaystyle\int_{-1}^0 \frac{x^3}{\sqrt{x^4+9}}\,dx$

11. a) $\displaystyle\int_0^{\pi/6} (1-\cos 3t)\sin 3t\,dt$

b) $\displaystyle\int_{\pi/6}^{\pi/3} (1-\cos 3t)\sin 3t\,dt$

12. a) $\displaystyle\int_{-\pi/2}^0 \left(2+\tan\frac{t}{2}\right)\sec^2\frac{t}{2}\,dt$

b) $\displaystyle\int_{-\pi/2}^{\pi/2} \left(2+\tan\frac{t}{2}\right)\sec^2\frac{t}{2}\,dt$

13. a) $\displaystyle\int_0^{2\pi} \frac{\cos z}{\sqrt{4+3\sin z}}\,dz$

b) $\displaystyle\int_{-\pi}^{\pi} \frac{\cos z}{\sqrt{4+3\sin z}}\,dz$

14. a) $\displaystyle\int_{-\pi/2}^0 \frac{\sin w}{(3+2\cos w)^2}\,dw$

b) $\displaystyle\int_0^{\pi/2} \frac{\sin w}{(3+2\cos w)^2}\,dw$

15. $\displaystyle\int_0^1 \sqrt{t^5+2t}\,(5t^4+2)\,dt$

16. $\displaystyle\int_1^4 \frac{dy}{2\sqrt{y}(1+\sqrt{y})^2}$

17. $\displaystyle\int_0^{\pi/6} \cos^{-3} 2\theta \sin 2\theta\,d\theta$

18. $\displaystyle\int_{\pi}^{3\pi/2} \cot^5\left(\frac{\theta}{6}\right)\sec^2\left(\frac{\theta}{6}\right)\,d\theta$

19. $\displaystyle\int_0^{\pi} 5(5-4\cos t)^{1/4}\sin t\,dt$

20. $\displaystyle\int_0^{\pi/4} (1-\sin 2t)^{3/2}\cos 2t\,dt$

21. $\displaystyle\int_0^1 (4y-y^2+4y^3+1)^{-2/3}(12y^2-2y+4)\,dy$

22. $\displaystyle\int_0^1 (y^3 + 6y^2 - 12y + 9)^{-1/2} (y^2 + 4y - 4)\, dy$

23. $\displaystyle\int_0^{\sqrt[3]{\pi^2}} \sqrt{\theta}\, \cos^2 (\theta^{3/2})\, d\theta$ **24.** $\displaystyle\int_{-1}^{-1/2} t^{-2} \sin^2 \left(1 + \frac{1}{t} \right) dt$

Area

Find the total areas of the shaded regions in Exercises 25–28.

25.

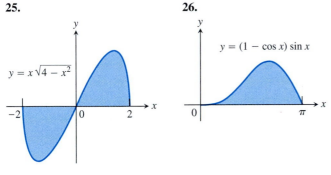

$y = x\sqrt{4 - x^2}$

26.

$y = (1 - \cos x) \sin x$

27.

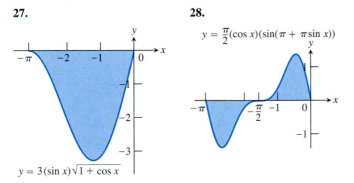

$y = 3(\sin x)\sqrt{1 + \cos x}$

28.

$y = \frac{\pi}{2}(\cos x)(\sin(\pi + \pi \sin x))$

Theory and Examples

29. Suppose that $F(x)$ is an antiderivative of $f(x) = (\sin x)/x$, $x > 0$. Express

$$\int_1^3 \frac{\sin 2x}{x}\, dx$$

in terms of F.

30. Show that if f is continuous, then

$$\int_0^1 f(x)\, dx = \int_0^1 f(1 - x)\, dx.$$

31. Suppose that

$$\int_0^1 f(x)\, dx = 3.$$

Find

$$\int_{-1}^0 f(x)\, dx$$

if (a) f is odd, (b) f is even.

32. a)   Show that

$$\int_{-a}^a h(x)\, dx = \begin{cases} 0 & \text{if } h \text{ is odd} \\ 2\displaystyle\int_0^a h(x)\, dx & \text{if } h \text{ is even.} \end{cases}$$

b)   Test the result in part (a) with $h(x) = \sin x$ and with $h(x) = \cos x$, taking $a = \pi/2$ in each case.

33. If f is a continuous function, find the value of the integral

$$I = \int_0^a \frac{f(x)\, dx}{f(x) + f(a - x)}$$

by making the substitution $u = a - x$ and adding the resulting integral to I.

34. By using a substitution, prove that for all positive numbers x and y,

$$\int_x^{xy} \frac{1}{t}\, dt = \int_1^y \frac{1}{t}\, dt.$$

The Shift Property for Definite Integrals

A basic property of definite integrals is their invariance under translation, as expressed by the equation.

$$\int_a^b f(x)\, dx = \int_{a-c}^{b-c} f(x + c)\, dx. \tag{2}$$

The equation holds whenever f is integrable and defined for the necessary values of x. For example (Fig. 4.26),

$$\int_{-2}^{-1} (x + 2)^3\, dx = \int_0^1 x^3\, dx. \tag{3}$$

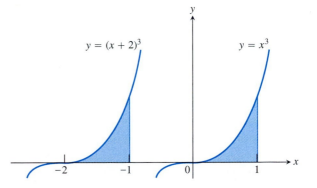

$y = (x + 2)^3$ $y = x^3$

4.26 The integrations in Eq. (3). The shaded regions, being congruent, have equal areas.

35. Use a substitution to verify Eq. (2).

36. For each of the following functions, graph $f(x)$ over $[a, b]$ and $f(x + c)$ over $[a - c, b - c]$ to convince yourself that Eq. (2) is reasonable.

a) $f(x) = x^2$, $a = 0$, $b = 1$, $c = 1$
b) $f(x) = \sin x$, $a = 0$, $b = \pi$, $c = \pi/2$
c) $f(x) = \sqrt{x - 4}$, $a = 4$, $b = 8$, $c = 5$

Numerical Integration

As we have seen, the ideal way to evaluate a definite integral $\int_a^b f(x)\,dx$ is to find a formula $F(x)$ for one of the antiderivatives of $f(x)$ and calculate the number $F(b) - F(a)$. But some antiderivatives are hard to find, and still others, like the antiderivatives of $(\sin x)/x$ and $\sqrt{1 + x^4}$, have no elementary formulas. We do not mean merely that no one has yet succeeded in finding elementary formulas for the antiderivatives of $(\sin x)/x$ and $\sqrt{1 + x^4}$. We mean it has been proved that no such formulas exist.

Whatever the reason, when we cannot evaluate a definite integral with an antiderivative, we turn to numerical methods such as the trapezoidal rule and Simpson's rule, described in this section.

The Trapezoidal Rule

When we cannot find a workable antiderivative for a function f that we have to integrate, we partition the interval of integration, replace f by a closely fitting polynomial on each subinterval, integrate the polynomials, and add the results to approximate the integral of f. The higher the degrees of the polynomials for a given partition, the better the results. For a given degree, the finer the partition, the better the results, until we reach limits imposed by round-off and truncation errors.

The polynomials do not need to be of high degree to be effective. Even line segments (graphs of polynomials of degree 1) give good approximations if we use enough of them. To see why, suppose we partition the domain $[a, b]$ of f into n subintervals of length $\Delta x = h = (b - a)/n$ and join the corresponding points on the curve with line segments (Fig. 4.27). The vertical lines from the ends of the segments to the partition points create a collection of trapezoids that approximate the region between the curve and the x-axis. We add the areas of the trapezoids, counting area above the x-axis as positive and area below the axis as negative:

$$T = \frac{1}{2}(y_0 + y_1)h + \frac{1}{2}(y_1 + y_2)h + \cdots + \frac{1}{2}(y_{n-2} + y_{n-1})h + \frac{1}{2}(y_{n-1} + y_n)h$$

$$= h\left(\frac{1}{2}y_0 + y_1 + y_2 + \cdots + y_{n-1} + \frac{1}{2}y_n\right)$$

$$= \frac{h}{2}(y_0 + 2y_1 + 2y_2 + \cdots + 2y_{n-1} + y_n),$$

where

$$y_0 = f(a), \quad y_1 = f(x_1), \quad \ldots, \quad y_{n-1} = f(x_{n-1}), \quad y_n = f(b).$$

The trapezoidal rule says: Use T to estimate the integral of f from a to b.

The length $h = (b - a)/n$ is called the **step size.** It is conventional to use h in this context instead of Δx.

The Trapezoidal Rule

To approximate $\int_a^b f(x)\,dx$, use

$$T = \frac{h}{2}(y_0 + 2y_1 + 2y_2 + \cdots + 2y_{n-1} + y_n) \tag{1}$$

(for n subintervals of length $h = (b - a)/n$ and $y_k = f(x_k)$).

4.27 The trapezoidal rule approximates short stretches of the curve $y = f(x)$ with line segments. To estimate the integral of f from a to b, we add the "signed" areas of the trapezoids made by joining the ends of the segments to the x-axis.

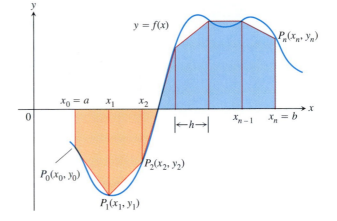

Table 4.6

x	$y = x^2$
1	1
$\dfrac{5}{4}$	$\dfrac{25}{16}$
$\dfrac{6}{4}$	$\dfrac{36}{16}$
$\dfrac{7}{4}$	$\dfrac{49}{16}$
2	4

4.28 The trapezoidal approximation of the area under the graph of $y = x^2$ from $x = 1$ to $x = 2$ is a slight overestimate.

EXAMPLE 1   Use the trapezoidal rule with $n = 4$ to estimate

$$\int_1^2 x^2 \, dx.$$

Compare the estimate with the exact value of the integral.

Solution   To find the trapezoidal approximation, we divide the interval of integration into four subintervals of equal length and list the values of $y = x^2$ at the endpoints and partition points (see Table 4.6). We then evaluate Eq. (1) with $n = 4$ and $h = 1/4$:

$$T = \frac{h}{2}(y_0 + 2y_1 + 2y_2 + 2y_3 + y_4)$$

$$= \frac{1}{8}\left(1 + 2\left(\frac{25}{16}\right) + 2\left(\frac{36}{16}\right) + 2\left(\frac{49}{16}\right) + 4\right) = \frac{75}{32}$$

$$= 2.34375.$$

The exact value of the integral is

$$\int_1^2 x^2 \, dx = \frac{x^3}{3}\Bigg]_1^2 = \frac{8}{3} - \frac{1}{3} = \frac{7}{3} = 2.\overline{3}.$$

The approximation is a slight overestimate. Each trapezoid contains slightly more than the corresponding strip under the curve (Fig. 4.28).   ❑

Controlling the Error in the Trapezoidal Approximation

Pictures suggest that the magnitude of the error

$$E_T = \int_a^b f(x) \, dx - T \tag{2}$$

in the trapezoidal approximation will decrease as the **step size** h decreases, because the trapezoids fit the curve better as their number increases. A theorem from advanced calculus assures us that this will be the case if f has a continuous second derivative.

The Error Estimate for the Trapezoidal Rule

If f'' is continuous and M is any upper bound for the values of $|f''|$ on $[a, b]$, then

$$|E_T| \leq \frac{b-a}{12} h^2 M. \tag{3}$$

Although theory tells us there will always be a smallest safe value of M, in practice we can hardly ever find it. Instead, we find the best value we can and go on from there to estimate $|E_T|$. This may seem sloppy, but it works. To make $|E_T|$ small for a given M, we make h small.

EXAMPLE 2   Find an upper bound for error in the approximation found in Example 1 for the value of

$$\int_1^2 x^2 \, dx.$$

Solution   We first find an upper bound M for the magnitude of the second derivative of $f(x) = x^2$ on the interval $1 \leq x \leq 2$. Since $f''(x) = 2$ for all x, we may safely take $M = 2$. With $b - a = 1$ and $h = 1/4$, Eq. (3) gives

$$|E_T| \leq \frac{b-a}{12} h^2 M = \frac{1}{12} \left(\frac{1}{4}\right)^2 (2) = \frac{1}{96}.$$

This is precisely what we find when we subtract $T = 75/32$ from $\int_1^2 x^2 \, dx = 7/3$, since $|7/3 - 75/32| = |-1/96|$. Here our estimate gave the error's magnitude *exactly*, but this is exceptional.   ☐

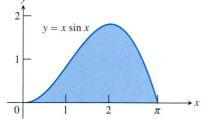

4.29 Graph of the integrand in Example 3.

EXAMPLE 3   Find an upper bound for the error incurred in estimating

$$\int_0^\pi x \sin x \, dx$$

with the trapezoidal rule with $n = 10$ steps (Fig. 4.29).

Solution   With $a = 0$, $b = \pi$, and $h = (b - a)/n = \pi/10$, Eq. (3) gives

$$|E_T| \leq \frac{b-a}{12} h^2 M = \frac{\pi}{12} \left(\frac{\pi}{10}\right)^2 M = \frac{\pi^3}{1200} M.$$

The number M can be any upper bound for the magnitude of the second derivative of $f(x) = x \sin x$ on $[0, \pi]$. A routine calculation gives

$$f''(x) = 2 \cos x - x \sin x,$$

so

$$|f''(x)| = |2 \cos x - x \sin x|$$

$$\leq 2|\cos x| + |x||\sin x| \qquad \text{Triangle inequality:}$$
$$|a + b| \leq |a| + |b|$$

$$\leq 2 \cdot 1 + \pi \cdot 1 = 2 + \pi. \qquad |\cos x| \text{ and } |\sin x| \text{ never}$$
$$\text{exceed 1, and } 0 \leq x \leq \pi.$$

We can safely take $M = 2 + \pi$. Therefore,

$$|E_T| \leq \frac{\pi^3}{1200} M = \frac{\pi^3(2 + \pi)}{1200} < 0.133. \qquad \text{Rounded up to be safe}$$

The absolute error is no greater than 0.133.

For greater accuracy, we would not try to improve M but would take more steps. With $n = 100$ steps, for example, $h = \pi/100$ and

$$|E_T| \leq \frac{\pi}{12} \left(\frac{\pi}{100}\right)^2 M = \frac{\pi^3(2 + \pi)}{120{,}000} < 0.00133 = 1.33 \times 10^{-3}. \qquad \square$$

EXAMPLE 4   As we will see in Chapter 6, the value of $\ln 2$ can be calculated from the integral

$$\ln 2 = \int_1^2 \frac{1}{x}\, dx.$$

How many subintervals (steps) should be used in the trapezoidal rule to approximate the integral with an error of magnitude less than 10^{-4}?

Solution   To determine n, the number of subintervals, we use Eq. (3) with

$$b - a = 2 - 1 = 1, \qquad h = \frac{b - a}{n} = \frac{1}{n},$$

$$f''(x) = \frac{d^2}{dx^2}(x^{-1}) = 2x^{-3} = \frac{2}{x^3}.$$

Then

$$\left|E_T\right| \leq \frac{b - a}{12} h^2 \max\left|f''(x)\right| = \frac{1}{12}\left(\frac{1}{n}\right)^2 \max\left|\frac{2}{x^3}\right|,$$

where max refers to the interval $[1, 2]$.

This is one of the rare cases where we can find the exact value of $\max|f''|$. On $[1, 2]$, $y = 2/x^3$ decreases steadily from a maximum of $y = 2$ to a minimum of $y = 1/4$. Therefore,

$$|E_T| \leq \frac{1}{12}\left(\frac{1}{n}\right)^2 \cdot 2 = \frac{1}{6n^2}.$$

The error's absolute value will therefore be less than 10^{-4} if

$$\frac{1}{6n^2} < 10^{-4},$$

$$\frac{10^4}{6} < n^2, \qquad \text{Multiply both sides by } 10^4 n^2.$$

$$\frac{100}{\sqrt{6}} < |n|, \qquad \text{Square roots of both sides}$$

$$\frac{100}{\sqrt{6}} < n, \qquad n \text{ is positive.}$$

$$40.83 < n. \qquad \text{Rounded up, to be safe}$$

Simpson's one-third rule

The idea of using the formula

$$A = \frac{h}{3}(y_0 + 4y_1 + y_2)$$

to estimate the area under a curve is known as Simpson's one-third rule. But the rule was in use long before Thomas Simpson (1720–1761) was born. It is another of history's beautiful quirks that one of the ablest mathematicians of eighteenth-century England is remembered not for his successful texts and his contributions to mathematical analysis but for a rule that was never his, that he never laid claim to, and that bears his name only because he happened to mention it in a book he wrote.

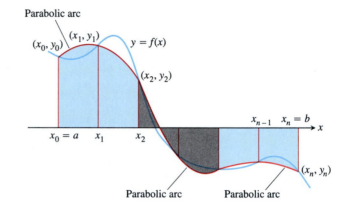

4.30 Simpson's rule approximates short stretches of curve with parabolic arcs.

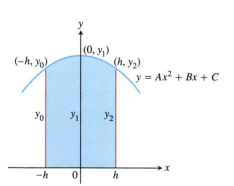

4.31 By integrating from $-h$ to h, we find the shaded area to be

$$\frac{h}{3}(y_0 + 4y_1 + y_2).$$

The first integer beyond 40.83 is $n = 41$. With $n = 41$ subintervals we can guarantee calculating ln 2 with an error of magnitude less than 10^{-4}. Any larger n will work, too. ☐

Simpson's Rule

Simpson's rule for approximating $\int_a^b f(x)\,dx$ is based on approximating f with quadratic polynomials instead of linear polynomials. We approximate the graph with parabolic arcs instead of line segments (Fig. 4.30).

The integral of the quadratic polynomial $y = Ax^2 + Bx + C$ in Fig. 4.31 from $x = -h$ to $x = h$ is

$$\int_{-h}^{h} (Ax^2 + Bx + C)\,dx = \frac{h}{3}(y_0 + 4y_1 + y_2) \tag{4}$$

(Appendix 4). Simpson's rule follows from partitioning $[a, b]$ into an even number of subintervals of equal length h, applying Eq. (4) to successive interval pairs, and adding the results.

Simpson's Rule

To approximate $\int_a^b f(x)\,dx$, use

$$S = \frac{h}{3}(y_0 + 4y_1 + 2y_2 + 4y_3 + \cdots + 2y_{n-2} + 4y_{n-1} + y_n). \tag{5}$$

The y's are the values of f at the partition points

$$x_0 = a, \; x_1 = a + h, \; x_2 = a + 2h, \; \ldots, \; x_{n-1} = a + (n-1)h, \; x_n = b.$$

The number n is even, and $h = (b - a)/n$.

Error Control for Simpson's Rule

The magnitude of the Simpson's rule error,

$$E_S = \int_a^b f(x)\,dx - S, \tag{6}$$

decreases with the step size, as we would expect from our experience with the trapezoidal rule. The inequality for controlling the Simpson's rule error, however, assumes f to have a continuous fourth derivative instead of merely a continuous second derivative. The formula, once again from advanced calculus, is this:

The Error Estimate for Simpson's Rule

If $f^{(4)}$ is continuous and M is any upper bound for the values of $|f^{(4)}|$ on $[a, b]$, then

$$|E_S| \le \frac{b-a}{180} h^4 M. \tag{7}$$

As with the trapezoidal rule, we can almost never find the smallest possible value of M. We just find the best value we can and go on from there to estimate $|E_S|$.

EXAMPLE 5 Use Simpson's rule with $n = 4$ to approximate

$$\int_0^1 5x^4 \, dx.$$

What estimate does Eq. (7) give for the error in the approximation?

Solution Again we have chosen an integral whose exact value we can calculate directly:

$$\int_0^1 5x^4 \, dx = x^5 \Big]_0^1 = 1.$$

To find the Simpson approximation, we partition the interval of integration into four subintervals and evaluate $f(x) = 5x^4$ at the partition points (Table 4.7). We then evaluate Eq. (5) with $n = 4$ and $h = 1/4$:

$$S = \frac{h}{3} (y_0 + 4y_1 + 2y_2 + 4y_3 + y_4)$$

$$= \frac{1}{12} \left(0 + 4 \left(\frac{5}{256} \right) + 2 \left(\frac{80}{256} \right) + 4 \left(\frac{405}{256} \right) + 5 \right) \approx 1.00260.$$

To estimate the error, we first find an upper bound M for the magnitude of the fourth derivative of $f(x) = 5x^4$ on the interval $0 \le x \le 1$. Since the fourth derivative has the constant value $f^{(4)}(x) = 120$, we may safely take $M = 120$. With $b - a = 1$ and $h = 1/4$, Eq. (7) gives

$$|E_S| \le \frac{b-a}{180} h^4 M = \frac{1}{180} \left(\frac{1}{4} \right)^4 (120) = \frac{1}{384} < 0.00261.$$

Table 4.7

x	$y = 5x^4$
0	0
$\frac{1}{4}$	$\frac{5}{256}$
$\frac{2}{4}$	$\frac{80}{256}$
$\frac{3}{4}$	$\frac{405}{256}$
1	5

Which Rule Gives Better Results?

The answer lies in the error-control formulas

$$|E_T| \le \frac{b-a}{12} h^2 M, \qquad |E_S| \le \frac{b-a}{180} h^4 M.$$

Trapezoidal vs. Simpson

If Simpson's rule is more accurate, why bother with the trapezoidal rule? There are two reasons. First, the trapezoidal rule is useful in a number of specific applications because it leads to much simpler expressions. Second, the trapezoidal rule is the basis for *Rhomberg integration,* one of the most satisfactory machine methods when high precision is required.

The M's of course mean different things, the first being an upper bound on $|f''|$ and the second an upper bound on $|f^{(4)}|$. But there is more. The factor $(b-a)/180$ in the Simpson formula is one-fifteenth of the factor $(b-a)/12$ in the trapezoidal formula. More important still, the Simpson formula has an h^4 while the trapezoidal formula has only an h^2. If h is one-tenth, then h^2 is one-hundredth but h^4 is only one ten-thousandth. If both M's are 1, for example, and $b-a=1$, then, with $h=1/10$,

$$|E_T| \le \frac{1}{12}\left(\frac{1}{10}\right)^2 \cdot 1 = \frac{1}{1200},$$

while

$$|E_S| \le \frac{1}{180}\left(\frac{1}{10}\right)^4 \cdot 1 = \frac{1}{1,800,000} = \frac{1}{1500} \cdot \frac{1}{1200}.$$

For roughly the same amount of computational effort, we get better accuracy with Simpson's rule—at least in this case.

The h^2 versus h^4 is the key. If h is less than 1, then h^4 can be significantly smaller than h^2. On the other hand, if h equals 1, there is no difference between h^2 and h^4. If h is greater than 1, the value of h^4 may be significantly larger than the value of h^2. In the latter two cases, the error-control formulas offer little help. We have to go back to the geometry of the curve $y = f(x)$ to see whether trapezoids or parabolas, if either, are going to give the results we want.

Working with Numerical Data

The next example shows how we can use Simpson's rule to estimate the integral of a function from values measured in the laboratory or in the field even when we have no formula for the function. We can use the trapezoidal rule the same way.

EXAMPLE 6   A town wants to drain and fill a small polluted swamp (Fig. 4.32). The swamp averages 5 ft deep. About how many cubic yards of dirt will it take to fill the area after the swamp is drained?

Solution   To calculate the volume of the swamp, we estimate the surface area and multiply by 5. To estimate the area, we use Simpson's rule with $h = 20$ ft and the y's equal to the distances measured across the swamp, as shown in Fig. 4.32.

$$S = \frac{h}{3}(y_0 + 4y_1 + 2y_2 + 4y_3 + 2y_4 + 4y_5 + y_6)$$

$$= \frac{20}{3}(146 + 488 + 152 + 216 + 80 + 120 + 13) = 8100.$$

The volume is about $(8100)(5) = 40{,}500$ ft^3 or 1500 yd^3. ☐

Horizontal spacing = 20 ft

4.32 The swamp in Example 6.

Round-off Errors

Although decreasing the step size h reduces the error in the Simpson and trapezoidal approximations in theory, it may fail to do so in practice. When h is very small, say $h = 10^{-5}$, the round-off errors in the arithmetic required to evaluate S and T may accumulate to such an extent that the error formulas no longer describe what is going on. Shrinking h below a certain size can actually make things worse. While this will not be an issue in the present book, you should consult a text on numerical analysis for alternative methods if you are having problems with round-off.

Exercises 4.9

Estimating Integrals

The instructions for the integrals in Exercises 1–10 have two parts, one for the trapezoidal rule and one for Simpson's rule.

I. *Using the trapezoidal rule*

    a) Estimate the integral with $n = 4$ steps and use Eq. (3) to find an upper bound for $|E_T|$.

    b) Evaluate the integral directly, and use Eq. (2) to find $|E_T|$.

    c) **CALCULATOR** Use the formula $(|E_T|/\text{true value}) \times 100$ to express $|E_T|$ as a percentage of the integral's true value.

II. *Using Simpson's rule*

    a) Estimate the integral with $n = 4$ steps and use Eq. (7) to find an upper bound for $|E_S|$.

    b) Evaluate the integral directly, and use Eq. (6) to find $|E_S|$.

    c) **CALCULATOR** Use the formula $(|E_S|/\text{true value}) \times 100$ to express $|E_S|$ as a percentage of the integral's true value.

1. $\displaystyle\int_{1}^{2} x \, dx$

2. $\displaystyle\int_{1}^{3} (2x - 1) \, dx$

3. $\displaystyle\int_{-1}^{1} (x^2 + 1) \, dx$

4. $\displaystyle\int_{-2}^{0} (x^2 - 1) \, dx$

5. $\displaystyle\int_{0}^{2} (t^3 + t) \, dt$

6. $\displaystyle\int_{-1}^{1} (t^3 + 1) \, dt$

7. $\displaystyle\int_{1}^{2} \frac{1}{s^2} \, ds$

8. $\displaystyle\int_{2}^{4} \frac{1}{(s-1)^2} \, ds$

9. $\displaystyle\int_{0}^{\pi} \sin t \, dt$

10. $\displaystyle\int_{0}^{1} \sin \pi t \, dt$

In Exercises 11–14, use the tabulated values of the integrand to estimate the integral with (a) the trapezoidal rule and (b) Simpson's rule with $n = 8$ steps. Round your answers to 5 decimal places. Then (c) find the integral's exact value and the approximation error E_T or E_S, as appropriate, from Eqs. (2) and (6).

11. $\displaystyle\int_{0}^{1} x\sqrt{1 - x^2} \, dx$

x	$x\sqrt{1 - x^2}$
0	0.0
0.125	0.12402
0.25	0.24206
0.375	0.34763
0.5	0.43301
0.625	0.48789
0.75	0.49608
0.875	0.42361
1.0	0

12. $\displaystyle\int_{0}^{3} \frac{\theta}{\sqrt{16 + \theta^2}} \, d\theta$

θ	$\theta/\sqrt{16 + \theta^2}$
0	0.0
0.375	0.09334
0.75	0.18429
1.125	0.27075
1.5	0.35112
1.875	0.42443
2.25	0.49026
2.625	0.58466
3.0	0.6

13. $\displaystyle\int_{-\pi/2}^{\pi/2} \frac{3 \cos t}{(2 + \sin t)^2} \, dt$

t	$(3 \cos t)/(2 + \sin t)^2$
−1.57080	0.0
−1.17810	0.99138
−0.78540	1.26906
−0.39270	1.05961
0	0.75
0.39270	0.48821
0.78540	0.28946
1.17810	0.13429
1.57080	0

14. $\displaystyle\int_{\pi/4}^{\pi/2} (\csc^2 y)\sqrt{\cot y} \, dy$

y	$(\csc^2 y)\sqrt{\cot y}$
0.78540	2.0
0.88357	1.51606
0.98175	1.18237
1.07992	0.93998
1.17810	0.75402
1.27627	0.60145
1.37445	0.46364
1.47262	0.31688
1.57080	0

The Minimum Number of Subintervals

In Exercises 15–26, use Eqs. (3) and (7), as appropriate, to estimate the minimum number of subintervals needed to approximate the integrals with an error of magnitude less than 10^{-4} by (a) the trapezoidal rule and (b) Simpson's rule. (The integrals in Exercises 15–22 are the integrals from Exercises 1–8.)

15. $\displaystyle\int_{1}^{2} x \, dx$

16. $\displaystyle\int_{1}^{3} (2x - 1) \, dx$

17. $\displaystyle\int_{-1}^{1} (x^2 + 1) \, dx$

18. $\displaystyle\int_{-2}^{0} (x^2 - 1) \, dx$

19. $\int_0^2 (t^3 + t)\, dt$

20. $\int_{-1}^1 (t^3 + 1)\, dt$

21. $\int_1^2 \frac{1}{s^2}\, ds$

22. $\int_2^4 \frac{1}{(s-1)^2}\, ds$

23. $\int_0^3 \sqrt{x+1}\, dx$

24. $\int_0^3 \frac{1}{\sqrt{x+1}}\, dx$

25. $\int_0^2 \sin(x+1)\, dx$

26. $\int_{-1}^1 \cos(x+\pi)\, dx$

Applications

27. As the fish-and-game warden of your township, you are responsible for stocking the town pond with fish before fishing season. The average depth of the pond is 20 ft. You plan to start the season with one fish per 1000 ft³. You intend to have at least 25% of the opening day's fish population left at the end of the season. What is the maximum number of licenses the town can sell if the average seasonal catch is 20 fish per license?

0 ft
520 ft
800 ft
1000 ft
1140 ft
1160 ft
1110 ft
860 ft
0 ft

Vertical spacing = 200 ft

28. CALCULATOR The design of a new airplane requires a gasoline tank of constant cross-section area in each wing. A scale drawing of a cross section is shown here. The tank must hold 5000 lb of gasoline, which has a density of 42 lb/ft³. Estimate the length of the tank.

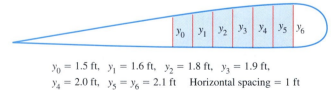

$y_0 = 1.5$ ft, $y_1 = 1.6$ ft, $y_2 = 1.8$ ft, $y_3 = 1.9$ ft,
$y_4 = 2.0$ ft, $y_5 = y_6 = 2.1$ ft   Horizontal spacing = 1 ft

29. CALCULATOR A vehicle's aerodynamic drag is determined in part by its cross-section area and, all other things being equal, engineers try to make this area as small as possible. Use Simpson's rule to estimate the cross-section area of James Worden's solar-powered Solectria car at MIT (Fig. 4.33).

4.33 Solectria cars are produced by Selectron Corp., Arlington, MA (Exercise 29).

30. The accompanying table shows time-to-speed data for a 1994 Ford Mustang Cobra accelerating from rest to 130 mph. How far had the Mustang traveled by the time it reached this speed?

Speed change	Seconds
Zero to 30 mph	2.2
40 mph	3.2
50 mph	4.5
60 mph	5.9
70 mph	7.8
80 mph	10.2
90 mph	12.7
100 mph	16.0
110 mph	20.6
120 mph	26.2
130 mph	37.1

Source: *Car and Driver,* April 1994.

Theory and Examples

31. *Polynomials of low degree.* The magnitude of the error in the trapezoidal approximation of $\int_a^b f(x)\,dx$ is

$$|E_T| = \frac{b-a}{12}\,h^2\,|f''(c)|,$$

where c is some point (usually unidentified) in (a, b). If f is a linear function of x, then $f''(c) = 0$, so $E_T = 0$ and T gives the exact value of the integral for any value of h. This is no surprise, really, for if f is linear, the line segments approximating the graph of f fit the graph exactly. The surprise comes with Simpson's rule. The magnitude of the error in Simpson's rule is

$$|E_S| = \frac{b-a}{180}\,h^4\,|f^{(4)}(c)|,$$

where once again c lies in (a, b). If f is a polynomial of degree less than 4, then $f^{(4)} = 0$ no matter what c is, so $E_S = 0$ and S gives the integral's exact value—even if we use only two steps. As a case in point, use Simpson's rule with $n = 2$ to estimate

$$\int_0^2 x^3\,dx.$$

Compare your answer with the integral's exact value.

32. *Usable values of the sine-integral function.* The sine-integral function,

$$\text{Si}(x) = \int_0^x \frac{\sin t}{t}\,dt,\qquad \text{"Sine integral of } x\text{"}$$

is one of the many functions in engineering whose formulas cannot be simplified. There is no elementary formula for the antiderivative of $(\sin t)/t$. The values of $\text{Si}(x)$, however, are readily estimated by numerical integration.

Although the notation does not show it explicitly, the function being integrated is

$$f(t) = \begin{cases} \dfrac{\sin t}{t}, & t \neq 0 \\ 1, & t = 0, \end{cases}$$

the continuous extension of $(\sin t)/t$ to the interval $[0, x]$. The function has derivatives of all orders at every point of its domain. Its graph is smooth (Fig. 4.34) and you can expect good results from Simpson's rule.

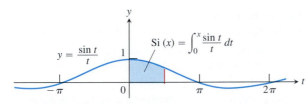

4.34 The continuous extension of $y = (\sin t)/t$. The sine-integral function Si(x) is the subject of Exercise 32.

a) Use the fact that $|f^{(4)}| \leq 1$ on $[0, \pi/2]$ to give an upper

bound for the error that will occur if

$$\text{Si}\left(\frac{\pi}{2}\right) = \int_0^{\pi/2} \frac{\sin t}{t}\,dt$$

is estimated by Simpson's rule with $n = 4$.

b) Estimate $\text{Si}(\pi/2)$ by Simpson's rule with $n = 4$.

c) Express the error bound you found in (a) as a percentage of the value you found in (b).

33. (*Continuation of Example 3.*) The error bounds in Eqs. (3) and (7) are "worst case" estimates, and the trapezoidal and Simpson rules are often more accurate than the bounds suggest. The trapezoidal rule estimate of

$$\int_0^\pi x \sin x\,dx$$

in Example 3 is a case in point.

x	$x \sin x$
0	0
$(0.1)\pi$	0.09708
$(0.2)\pi$	0.36932
$(0.3)\pi$	0.76248
$(0.4)\pi$	1.19513
$(0.5)\pi$	1.57080
$(0.6)\pi$	1.79270
$(0.7)\pi$	1.77912
$(0.8)\pi$	1.47727
$(0.9)\pi$	0.87372
π	0

a) Use the trapezoidal rule with $n = 10$ to approximate the value of the integral. The table to the right gives the necessary y-values.

b) Find the magnitude of the difference between π, the integral's value, and your approximation in (a). You will find the difference to be considerably less than the upper bound of 0.133 calculated with $n = 10$ in Example 3.

c) GRAPHER The upper bound of 0.133 for $|E_T|$ in Example 3 could have been improved somewhat by having a better bound for

$$|f''(x)| = |2\cos x - x\sin x|$$

on $[0, \pi]$. The upper bound we used was $2 + \pi$. Graph f'' over $[0, \pi]$ and use TRACE or ZOOM to improve this upper bound.

Use the improved upper bound as M in Eq. (3) to make an improved estimate of $|E_T|$. Notice that the trapezoidal rule approximation in (a) is also better than this improved estimate would suggest.

34. CALCULATOR (*Continuation of Exercise 33*)

a) GRAPHER Show that the fourth derivative of $f(x) = x \sin x$ is

$$f^{(4)}(x) = -4\cos x + x\sin x.$$

Use TRACE or ZOOM to find an upper bound M for the values of $|f^{(4)}|$ on $[0, \pi]$.

b) Use the value of M from (a) together with Eq. (7) to obtain an upper bound for the magnitude of the error in estimating the value of

$$\int_0^\pi x \sin x\,dx$$

with Simpson's rule with $n = 10$ steps.

c) Use the data in the table in Exercise 33 to estimate $\int_0^\pi x \sin x \, dx$ with Simpson's rule with $n = 10$ steps.

d) To 6 decimal places, find the magnitude of the difference between your estimate in (c) and the integral's true value, π. You will find the error estimate obtained in (b) to be quite good.

You are planning to use Simpson's rule to estimate the values of the integrals in Exercises 35 and 36. Before proceeding, you turn to Eq. (7) to determine the step size h needed to assure the accuracy you want. What happens? Can this be avoided by using the trapezoidal rule and Eq. (3) instead? Give reasons for your answers.

35. $\int_0^4 x^{3/2} \, dx$

36. $\int_0^1 x^{5/2} \, dx$

Numerical Integrator

As we mentioned at the beginning of the section, the definite integrals of many continuous functions cannot be evaluated with the Fundamental Theorem of Calculus because their antiderivatives lack elementary formulas. Numerical integration offers a practical way to estimate the values of these so-called *nonelementary integrals*. If your calculator or computer has a numerical integration routine, try it on the integrals in Exercises 37–40.

37. $\int_0^1 \sqrt{1 + x^4} \, dx$ — A nonelementary integral that came up in Newton's research

38. $\int_0^{\pi/2} \frac{\sin x}{x} \, dx$ — The integral from Exercise 32. To avoid division by zero, you may have to start the integration at a small positive number like 10^{-6} instead of 0.

39. $\int_0^{\pi/2} \sin(x^2) \, dx$ — An integral associated with the diffraction of light

40. $\int_0^{\pi/2} 40\sqrt{1 - 0.64 \cos^2 t} \, dt$ — The length of the ellipse $(x^2/25) + (y^2/9) = 1$

CHAPTER 4 QUESTIONS TO GUIDE YOUR REVIEW

1. Can a function have more than one antiderivative? If so, how are the antiderivatives related? Explain.

2. What is an indefinite integral? How do you evaluate one? What general formulas do you know for evaluating indefinite integrals?

3. How can you sometimes use a trigonometric identity to transform an unfamiliar intregal into one you know how to evaluate?

4. How can you sometimes solve a differential equation of the form $dy/dx = f(x)$?

5. What is an initial value problem? How do you solve one? Give an example.

6. If you know the acceleration of a body moving along a coordinate line as a function of time, what more do you need to know to find the body's position function? Give an example.

7. How do you sketch the solutions of a differential equation $dy/dx = f(x)$ when you do not know an antiderivative of f? How would you sketch the solution of an initial value problem $dy/dx = f(x), y(x_0) = y_0$ under these circumstances?

8. How can you sometimes evaluate indefinite integrals by substitution? Give examples.

9. How can you sometimes estimate quantities like distance traveled, area, volume, and average value with finite sums? Why might you want to do so?

10. What is sigma notation? What advantage does it offer? Give examples.

11. What rules are available for calculating with sigma notation?

12. What is a Riemann sum? Why might you want to consider such a sum?

13. What is the norm of a partition of a closed interval?

14. What is the definite integral of a function f over a closed interval $[a, b]$? When can you be sure it exists?

15. What is the relation between definite integrals and area? Describe some other interpretations of definite integrals.

16. Describe the rules for working with definite integrals (Table 4.5). Give examples.

17. What is the average value of an integrable function over a closed interval? Must the function assume its average value? Explain.

18. What does a function's average value have to do with sampling a function's values?

19. What is the Fundamental Theorem of Calculus? Why is it so important? Illustrate each part of the theorem with an example.

20. How does the Fundamental Theorem provide a solution to the initial value problem $dy/dx = f(x), y(x_0) = y_0$, when f is continuous?

21. How does the method of substitution work for definite integrals? Give examples.

22. How is integration by substitution related to the Chain Rule?

23. You are collaborating to produce a short "how-to" manual for

numerical integration, and you are writing about the trapezoidal rule. (a) What would you say about the rule itself and how to use it? how to achieve accuracy? (b) What would you say if you were writing about Simpson's rule instead?

24. How would you compare the relative merits of Simpson's rule and the trapezoidal rule?

CHAPTER 4 PRACTICE EXERCISES

Finite Sums and Estimates

1. The accompanying figure shows the graph of the velocity (ft/sec) of a model rocket for the first 8 sec after launch. The rocket accelerated straight up for the first 2 sec and then coasted to reach its maximum height at $t = 8$ sec.

Time after launch (sec)

a) Assuming that the rocket was launched from ground level, about how high did it go? (This is the rocket in Section 2.3, Exercise 19, but you do not need to do Exercise 19 to do the exercise here.)

b) Sketch a graph of the rocket's height aboveground as a function of time for $0 \le t \le 8$.

2. a) The accompanying figure shows the velocity (m/sec) of a body moving along the s-axis during the time interval from $t = 0$ to $t = 10$ sec. About how far did the body travel during those 10 sec?

b) Sketch a graph of s as a function of t for $0 \le t \le 10$ assuming $s(0) = 0$.

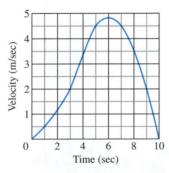

Time (sec)

3. Suppose that $\sum_{k=1}^{10} a_k = -2$ and $\sum_{k=1}^{10} b_k = 25$. Find the value of

a) $\sum_{k=1}^{10} \dfrac{a_k}{4}$

b) $\sum_{k=1}^{10} (b_k - 3a_k)$

c) $\sum_{k=1}^{10} (a_k + b_k - 1)$

d) $\sum_{k=1}^{10} \left(\dfrac{5}{2} - b_k \right)$

4. Suppose that $\sum_{k=1}^{20} a_k = 0$ and $\sum_{k=1}^{20} b_k = 7$. Find the values of

a) $\sum_{k=1}^{20} 3a_k$

b) $\sum_{k=1}^{20} (a_k + b_k)$

c) $\sum_{k=1}^{20} \left(\dfrac{1}{2} - \dfrac{2b_k}{7} \right)$

d) $\sum_{k=1}^{20} (a_k - 2)$

Definite Integrals

In Exercises 5–8, express each limit as a definite integral. Then evaluate the integral to find the value of the limit. In each case, P is a partition of the given interval and the numbers c_k are chosen from the subintervals of P.

5. $\lim\limits_{\|P\| \to 0} \sum\limits_{k=1}^{n} (2c_k - 1)^{-1/2} \, \Delta x_k$, where P is a partition of $[1, 5]$

6. $\lim\limits_{\|P\| \to 0} \sum\limits_{k=1}^{n} c_k (c_k^2 - 1)^{1/3} \, \Delta x_k$, where P is a partition of $[1, 3]$

7. $\lim\limits_{\|P\| \to 0} \sum\limits_{k=1}^{n} \left(\cos \left(\dfrac{c_k}{2} \right) \right) \Delta x_k$, where P is a partition of $[-\pi, 0]$

8. $\lim\limits_{\|P\| \to 0} \sum\limits_{k=1}^{n} (\sin c_k)(\cos c_k) \, \Delta x_k$, where P is a partition of $[0, \pi/2]$

9. If $\int_{-2}^{2} 3 f(x) \, dx = 12$, $\int_{-2}^{5} f(x) \, dx = 6$, and $\int_{-2}^{5} g(x) \, dx = 2$, find the values of the following.

a) $\displaystyle\int_{-2}^{2} f(x) \, dx$

b) $\displaystyle\int_{2}^{5} f(x) \, dx$

c) $\displaystyle\int_{5}^{-2} g(x) \, dx$

d) $\displaystyle\int_{-2}^{5} (-\pi g(x)) \, dx$

e) $\displaystyle\int_{-2}^{5} \left(\dfrac{f(x) + g(x)}{5} \right) dx$

10. If $\int_0^2 f(x)\,dx = \pi$, $\int_0^2 7g(x)\,dx = 7$, and $\int_0^1 g(x)\,dx = 2$, find the values of the following.

a) $\displaystyle\int_0^2 g(x)\,dx$

b) $\displaystyle\int_1^2 g(x)\,dx$

c) $\displaystyle\int_2^0 f(x)\,dx$

d) $\displaystyle\int_0^2 \sqrt{2}\,f(x)\,dx$

e) $\displaystyle\int_0^2 (g(x) - 3f(x))\,dx$

Area

In Exercises 11–14, find the total area of the region between the graph of f and the x-axis.

11. $f(x) = x^2 - 4x + 3, \quad 0 \le x \le 3$

12. $f(x) = 1 - (x^2/4), \quad -2 \le x \le 3$

13. $f(x) = 5 - 5x^{2/3}, \quad -1 \le x \le 8$

14. $f(x) = 1 - \sqrt{x}, \quad 0 \le x \le 4$

Initial Value Problems

Solve the initial value problems in Exercises 15–18.

15. $\dfrac{dy}{dx} = \dfrac{x^2 + 1}{x^2}, \quad y(1) = -1$

16. $\dfrac{dy}{dx} = \left(x + \dfrac{1}{x}\right)^2, \quad y(1) = 1$

17. $\dfrac{d^2 r}{dt^2} = 15\sqrt{t} + \dfrac{3}{\sqrt{t}}; \quad r'(1) = 8, \quad r(1) = 0$

18. $\dfrac{d^3 r}{dt^3} = -\cos t; \quad r''(0) = r'(0) = 0, \quad r(0) = -1$

19. Show that $y = x^2 + \displaystyle\int_1^x \dfrac{1}{t}\,dt$ solves the initial value problem

$$\dfrac{d^2 y}{dx^2} = 2 - \dfrac{1}{x^2}; \quad y'(1) = 3, \quad y(1) = 1.$$

20. Show that $y = \int_0^x (1 + 2\sqrt{\sec t}\,)\,dt$ solves the initial value problem

$$\dfrac{d^2 y}{dx^2} = \sqrt{\sec x}\,\tan x; \quad y'(0) = 3, \quad y(0) = 0.$$

Express the solutions of the initial value problems in Exercises 21 and 22 in terms of integrals.

21. $\dfrac{dy}{dx} = \dfrac{\sin x}{x}, \quad y(5) = -3$

22. $\dfrac{dy}{dx} = \sqrt{2 - \sin^2 x}, \quad y(-1) = 2$

Evaluating Indefinite Integrals

Evaluate the integrals in Exercises 23–44.

23. $\displaystyle\int (x^3 + 5x - 7)\,dx$

24. $\displaystyle\int \left(8t^3 - \dfrac{t^2}{2} + t\right)\,dt$

25. $\displaystyle\int \left(3\sqrt{t} + \dfrac{4}{t^2}\right)\,dt$

26. $\displaystyle\int \left(\dfrac{1}{2\sqrt{t}} - \dfrac{3}{t^4}\right)\,dt$

27. $\displaystyle\int \dfrac{r\,dr}{(r^2 + 5)^2}$

28. $\displaystyle\int \dfrac{6r^2\,dr}{(r^3 - \sqrt{2})^3}$

29. $\displaystyle\int 3\theta\sqrt{2 - \theta^2}\,d\theta$

30. $\displaystyle\int \dfrac{\theta^2}{9\sqrt{73 + \theta^3}}\,d\theta$

31. $\displaystyle\int x^3 (1 + x^4)^{-1/4}\,dx$

32. $\displaystyle\int (2 - x)^{3/5}\,dx$

33. $\displaystyle\int \sec^2 \dfrac{s}{10}\,ds$

34. $\displaystyle\int \csc^2 \pi s\,ds$

35. $\displaystyle\int \csc \sqrt{2}\theta \cot \sqrt{2}\theta\,d\theta$

36. $\displaystyle\int \sec \dfrac{\theta}{3} \tan \dfrac{\theta}{3}\,d\theta$

37. $\displaystyle\int \sin^2 \dfrac{x}{4}\,dx$

38. $\displaystyle\int \cos^2 \dfrac{x}{2}\,dx$

39. $\displaystyle\int 2(\cos x)^{-1/2} \sin x\,dx$

40. $\displaystyle\int (\tan x)^{-3/2} \sec^2 x\,dx$

41. $\displaystyle\int (2\theta + 1 + 2\cos(2\theta + 1))\,d\theta$

42. $\displaystyle\int \left(\dfrac{1}{\sqrt{2\theta - \pi}} + 2\sec^2(2\theta - \pi)\right)\,d\theta$

43. $\displaystyle\int \left(t - \dfrac{2}{t}\right)\left(t + \dfrac{2}{t}\right)\,dt$

44. $\displaystyle\int \dfrac{(t + 1)^2 - 1}{t^4}\,dt$

Evaluating Definite Integrals

Evaluate the integrals in Exercises 45–70.

45. $\displaystyle\int_{-1}^1 (3x^2 - 4x + 7)\,dx$

46. $\displaystyle\int_0^1 (8s^3 - 12s^2 + 5)\,ds$

47. $\displaystyle\int_1^2 \dfrac{4}{v^2}\,dv$

48. $\displaystyle\int_1^{27} x^{-4/3}\,dx$

49. $\displaystyle\int_1^4 \dfrac{dt}{t\sqrt{t}}$

50. $\displaystyle\int_1^4 \dfrac{(1 + \sqrt{u})^{1/2}}{\sqrt{u}}\,du$

51. $\displaystyle\int_0^1 \dfrac{36\,dx}{(2x + 1)^3}$

52. $\displaystyle\int_0^1 \dfrac{dr}{\sqrt[3]{(7 - 5r)^2}}$

53. $\displaystyle\int_{1/8}^1 x^{-1/3}(1 - x^{2/3})^{3/2}\,dx$

54. $\displaystyle\int_0^{1/2} x^3 (1 + 9x^4)^{-3/2}\,dx$

55. $\displaystyle\int_0^{\pi} \sin^2 5r\,dr$

56. $\displaystyle\int_0^{\pi/4} \cos^2\left(4t - \dfrac{\pi}{4}\right)\,dt$

57. $\displaystyle\int_0^{\pi/3} \sec^2 \theta\,d\theta$

58. $\displaystyle\int_{\pi/4}^{3\pi/4} \csc^2 x\,dx$

59. $\int_{\pi}^{3\pi} \cot^2 \frac{x}{6} \, dx$

60. $\int_{0}^{\pi} \tan^2 \frac{\theta}{3} \, d\theta$

61. $\int_{-\pi/3}^{0} \sec x \tan x \, dx$

62. $\int_{\pi/4}^{3\pi/4} \csc z \cot z \, dz$

63. $\int_{0}^{\pi/2} 5(\sin x)^{3/2} \cos x \, dx$

64. $\int_{-1}^{1} 2x \sin (1 - x^2) \, dx$

65. $\int_{-\pi/2}^{\pi/2} 15 \sin^4 3x \cos 3x \, dx$

66. $\int_{0}^{2\pi/3} \cos^{-4} \left(\frac{x}{2} \right) \sin \left(\frac{x}{2} \right) dx$

67. $\int_{0}^{\pi/2} \frac{3 \sin x \cos x}{\sqrt{1 + 3 \sin^2 x}} \, dx$

68. $\int_{0}^{\pi/4} \frac{\sec^2 x}{(1 + 7 \tan x)^{2/3}} \, dx$

69. $\int_{0}^{\pi/3} \frac{\tan \theta}{\sqrt{2} \sec \theta} \, d\theta$

70. $\int_{\pi^2/36}^{\pi^2/4} \frac{\cos \sqrt{t}}{\sqrt{t} \sin \sqrt{t}} \, dt$

Average Values

71. Find the average value of $f(x) = mx + b$

  a) over $[-1, 1]$
  b) over $[-k, k]$

72. Find the average value of

  a) $y = \sqrt{3x}$ over $[0, 3]$
  b) $y = \sqrt{ax}$ over $[0, a]$

73. Let f be a function that is differentiable on $[a, b]$. In Chapter 1 we defined the average rate of change of f over $[a, b]$ to be

$$\frac{f(b) - f(a)}{b - a}$$

and the instantaneous rate of change of f at x to be $f'(x)$. In this chapter we defined the average value of a function. For the new definition of average to be consistent with the old one, we should have

$$\frac{f(b) - f(a)}{b - a} = \text{average value of } f' \text{ on } [a, b].$$

Is this the case? Give reasons for your answer.

74. Is it true that the average value of an integrable function over an interval of length 2 is half the function's integral over the interval? Give reasons for your answer.

Numerical Integration

75. CALCULATOR According to the error-bound formula for Simpson's rule, how many subintervals should you use to be sure of estimating the value of

$$\ln 3 = \int_{1}^{3} \frac{1}{x} \, dx$$

by Simpson's rule with an error of no more than 10^{-4} in absolute value? (Remember that for Simpson's rule, the number of subintervals has to be even.)

76. A brief calculation shows that if $0 \le x \le 1$, then the second derivative of $f(x) = \sqrt{1 + x^4}$ lies between 0 and 8. Based on this, about how many subdivisions would you need to estimate the integral of f from 0 to 1 with an error no greater than 10^{-3} in absolute value using the trapezoidal rule?

77. A direct calculation shows that

$$\int_{0}^{\pi} 2 \sin^2 x \, dx = \pi.$$

How close do you come to this value by using the trapezoidal rule with $n = 6$? Simpson's rule with $n = 6$? Try them and find out.

78. You are planning to use Simpson's rule to estimate the value of the integral

$$\int_{1}^{2} f(x) \, dx$$

with an error magnitude less than 10^{-5}. You have determined that $|f^{(4)}(x)| \le 3$ throughout the interval of integration. How many subintervals should you use to assure the required accuracy? (Remember that for Simpson's rule the number has to be even.)

79. CALCULATOR Compute the average value of the temperature function

$$f(x) = 37 \sin \left(\frac{2\pi}{365} (x - 101) \right) + 25$$

for a 365-day year. This is one way to estimate the annual mean air temperature in Fairbanks, Alaska. The National Weather Service's official figure, a numerical average of the daily normal mean air temperatures for the year, is 25.7°F, which is slightly higher than the average value of $f(x)$. Figure 2.42 shows why.

80. *Specific heat of a gas.* Specific heat C_v is the amount of heat required to raise the temperature of a given mass of gas with constant volume by 1°C, measured in units of cal/deg-mole (calories per degree gram molecule). The specific heat of oxygen depends on its temperature T and satisfies the formula

$$C_v = 8.27 + 10^{-5} (26T - 1.87T^2).$$

Find the average value of C_v for $20° \le T \le 675°$C and the temperature at which it is attained.

Theory and Examples

81. Is it true that every function $y = f(x)$ that is differentiable on $[a, b]$ is itself the derivative of some function on $[a, b]$? Give reasons for your answer.

82. Suppose that $F(x)$ is an antiderivative of $f(x) = \sqrt{1 + x^4}$. Express $\int_{0}^{1} \sqrt{1 + x^4} \, dx$ in terms of F and give a reason for your answer.

83. Find dy/dx if $y = \int_{x}^{1} \sqrt{1 + t^2} \, dt$. Explain the main steps in your calculation.

84. Find dy/dx if $y = \int_{\cos x}^{0} (1/(1 - t^2)) \, dt$. Explain the main steps in your calculation.

85. *A new parking lot.* To meet the demand for parking, your town has allocated the area shown here. As the town engineer, you have been asked by the town council to find out if the lot can be built for $11,000. The cost to clear the land will be $0.10 a square foot, and the lot will cost $2.00 a square foot to pave. Can the job be done for $11,000?

0 ft

36 ft

54 ft

51 ft

49.5 ft

54 ft

64.4 ft

67.5 ft

42 ft

Ignored

Vertical spacing = 15 ft

86. Skydivers A and B are in a helicopter hovering at 6400 ft. Skydiver A jumps and descends for 4 sec before opening her parachute. The helicopter then climbs to 7000 ft and hovers there. Forty-five seconds after A leaves the aircraft, B jumps and descends for 13 sec before opening her parachute. Both skydivers descend at 16 ft/sec with parachute open. Assume that the skydivers fall freely (no effective air resistance) before their parachutes open.

a) At what altitude does A's parachute open?
b) At what altitude does B's parachute open?
c) Which skydiver lands first?

Average Daily Inventory

Average value is used in economics to study such things as average daily inventory. If $I(t)$ is the number of radios, tires, shoes, or whatever product a firm has on hand on day t (we call I an **inventory function**), the average value of I over a time period $[0, T]$ is called the firm's average daily inventory for the period.

$$\textbf{Average daily inventory} = \text{av}(I) = \frac{1}{T} \int_0^T I(t)\, dt.$$

If h is the dollar cost of holding one item per day, the product $\text{av}(I) \cdot h$ is the **average daily holding cost** for the period.

87. As a wholesaler, Tracey Burr Distributors receives a shipment of 1200 cases of chocolate bars every 30 days. TBD sells the chocolate to retailers at a steady rate, and t days after a shipment arrives, its inventory of cases on hand is $I(t) = 1200 - 40t, 0 \le t \le 30$. What is TBD's average daily inventory for the 30-day period? What is its average daily holding cost if the cost of holding one case is 3¢ a day?

88. Rich Wholesale Foods, a manufacturer of cookies, stores its cases of cookies in an air-conditioned warehouse for shipment every 14 days. Rich tries to keep 600 cases on reserve to meet occasional peaks in demand, so a typical 14-day inventory function is $I(t) = 600 + 600t, 0 \le t \le 14$. The daily holding cost for each case is 4¢ per day. Find Rich's average daily inventory and average daily holding cost.

89. Solon Container receives 450 drums of plastic pellets every 30 days. The inventory function (drums on hand as a function of days) is $I(t) = 450 - t^2/2$. Find the average daily inventory. If the holding cost for one drum is 2¢ per day, find the average daily holding cost.

90. Mitchell Mailorder receives a shipment of 600 cases of athletic socks every 60 days. The number of cases on hand t days after the shipment arrives is $I(t) = 600 - 20\sqrt{15t}$. Find the average daily inventory. If the holding cost for one case is 1/2¢ per day, find the average daily holding cost.

CHAPTER   **4**   ADDITIONAL EXERCISES–THEORY, EXAMPLES, APPLICATIONS

Theory and Examples

1. a) If $\int_0^1 7 f(x)\, dx = 7$, does $\int_0^1 f(x)\, dx = 1$?

b) If $\int_0^1 f(x)\, dx = 4$ and $f(x) \ge 0$, does $\int_0^1 \sqrt{f(x)}\, dx = \sqrt{4} = 2$?

Give reasons for your answers.

2. Suppose $\int_{-2}^2 f(x)\, dx = 4, \int_2^5 f(x)\, dx = 3, \int_{-2}^5 g(x)\, dx = 2$.

Which, if any, of the following statements are true?

a) $\int_5^2 f(x)\, dx = -3$ **b)** $\int_{-2}^5 (f(x) + g(x)) = 9$

c) $f(x) \le g(x)$ on the interval $-2 \le x \le 5$

3. Show that

$$y = \frac{1}{a} \int_0^x f(t) \sin a(x - t) \, dt$$

solves the initial value problem

$$\frac{d^2 y}{dx^2} + a^2 y = f(x), \quad \frac{dy}{dx} = 0 \quad \text{and} \quad y = 0 \text{ when } x = 0.$$

(*Hint:* $\sin(ax - at) = \sin ax \cos at - \cos ax \sin at$.)

4. Suppose x and y are related by the equation

$$x = \int_0^y \frac{1}{\sqrt{1 + 4t^2}} \, dt.$$

Show that $d^2 y / dx^2$ is proportional to y and find the constant of proportionality.

5. Find $f(4)$ if

a) $\displaystyle \int_0^{x^2} f(t) \, dt = x \cos \pi x,$

b) $\displaystyle \int_0^{f(x)} t^2 \, dt = x \cos \pi x.$

6. Find $f(\pi/2)$ from the following information.

i) f is positive and continuous.

ii) The area under the curve $y = f(x)$ from $x = 0$ to $x = a$ is

$$\frac{a^2}{2} + \frac{a}{2} \sin a + \frac{\pi}{2} \cos a.$$

7. The area of the region in the xy-plane enclosed by the x-axis, the curve $y = f(x)$, $f(x) \geq 0$, and the lines $x = 1$ and $x = b$ is equal to $\sqrt{b^2 + 1} - \sqrt{2}$ for all $b > 1$. Find $f(x)$.

8. Prove that

$$\int_0^x \left(\int_0^u f(t) \, dt \right) du = \int_0^x f(u)(x - u) \, du.$$

(*Hint:* Express the integral on the right-hand side as the difference of two integrals. Then show that both sides of the equation have the same derivative with respect to x.)

9. Find the equation for the curve in the xy-plane that passes through the point $(1, -1)$ if its slope at x is always $3x^2 + 2$.

10. You sling a shovelful of dirt up from the bottom of a hole with an initial velocity of 32 ft/sec. The dirt must rise 17 ft above the release point to clear the edge of the hole. Is that enough speed to get the dirt out, or had you better duck?

Bounded Piecewise Continuous Functions

Although we are mainly interested in continous functions, many functions in applications are piecewise continuous. All bounded piecewise continuous functions are integrable (as are many unbounded functions, as we will see in Chapter 7). **Bounded** on an interval I means that for some finite constant M, $|f(x)| \leq M$ for all x in I. **Piecewise continuous** on I means that I can be partitioned into open or half open subintervals on which f is continuous. To integrate a bounded piecewise continuous function that has a continuous extension to each

closed subinterval of the partition, we integrate the individual extensions and add the results. The integral of the function

$$f(x) = \begin{cases} 1 - x, & -1 \leq x < 0 \\ x^2, & 0 \leq x < 2 \\ -1, & 2 \leq x \leq 3, \end{cases}$$

(Fig. 4.35) over $[-1, 3]$ is

$$\int_{-1}^3 f(x) \, dx = \int_{-1}^0 (1 - x) \, dx + \int_0^2 x^2 \, dx + \int_2^3 (-1) \, dx$$

$$= \left[x - \frac{x^2}{2} \right]_{-1}^0 + \left[\frac{x^3}{3} \right]_0^2 + \left[-x \right]_2^3$$

$$= \frac{3}{2} + \frac{8}{3} - 1 = \frac{19}{6}.$$

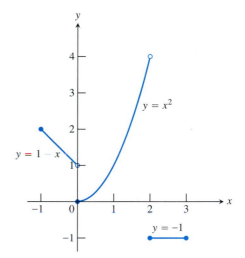

4.35 Piecewise continuous functions like this are integrated piece by piece.

The Fundamental Theorem applies to bounded piecewise continuous functions with the restriction that $(d/dx) \int_a^x f(t) \, dt$ is expected to equal $f(x)$ only at values of x at which f is continuous. There is a similar restriction on Leibniz's rule below.

Graph the functions in Exercises 11–16 and integrate them over their domains.

11. $f(x) = \begin{cases} x^{2/3}, & -8 \leq x < 0 \\ -4, & 0 \leq x \leq 3, \end{cases}$

12. $f(x) = \begin{cases} \sqrt{-x}, & -4 \leq x < 0 \\ x^2 - 4, & 0 \leq x \leq 3 \end{cases}$

13. $g(t) = \begin{cases} t, & 0 \leq t < 1 \\ \sin \pi t, & 1 \leq t \leq 2 \end{cases}$

14. $h(z) = \begin{cases} \sqrt{1 - z}, & 0 \leq z < 1 \\ (7z - 6)^{-1/3}, & 1 \leq z \leq 2 \end{cases}$

15. $f(x) = \begin{cases} 1, & -2 \le x < -1 \\ 1 - x^2, & -1 \le x < 1 \\ 2, & 1 \le x \le 2 \end{cases}$

16. $h(r) = \begin{cases} r, & -1 \le r < 0 \\ 1 - r^2, & 0 \le r < 1 \\ 1, & 1 \le r \le 2 \end{cases}$

17. Find the average value of the function graphed in Fig. 4.36(a).

18. Find the average value of the function graphed in Fig. 4.36(b).

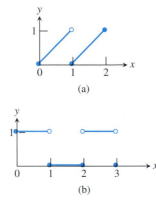

4.36 The graphs for Exercises 17 and 18.

Leibniz's Rule

In applications, we sometimes encounter functions like

$$f(x) = \int_{\sin x}^{x^2} (1 + t)\, dt \quad \text{and} \quad g(x) = \int_{\sqrt{x}}^{2\sqrt{x}} \sin t^2\, dt,$$

defined by integrals that have variable upper limits of integration and variable lower limits of integration at the same time. The first integral can be evaluated directly but the second cannot. We may find the derivative of either integral, however, by a formula called **Leibniz's rule:**

Leibniz's Rule

If f is continuous on $[a, b]$, and $u(x)$ and $v(x)$ are differentiable functions of x whose values lie in $[a, b]$, then

$$\frac{d}{dx} \int_{u(x)}^{v(x)} f(t)\, dt = f(v(x)) \frac{dv}{dx} - f(u(x)) \frac{du}{dx}.$$

Figure 4.37 gives a geometric interpretation of Leibniz's rule. It shows a carpet of variable width $f(t)$ that is being rolled up at the left at the same time x as it is being unrolled at the right. (In this interpretation time is x, not t.) At time x, the floor is covered from $u(x)$ to $v(x)$. The rate du/dx at which the carpet is being rolled up need not be the same as the rate dv/dx at which the carpet is being

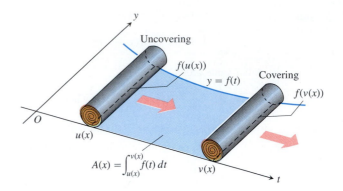

4.37 Rolling and unrolling a carpet: a geometric interpretation of Leibniz's rule:

$$\frac{dA}{dx} = f(v(x)) \frac{dv}{dx} - f(u(x)) \frac{du}{dx}.$$

laid down. At any given time x, the area covered by carpet is

$$A(x) = \int_{u(x)}^{v(x)} f(t)\, dt.$$

At what rate is the covered area changing? At the instant x, $A(x)$ is increasing by the width $f(v(x))$ of the unrolling carpet times the rate dv/dx at which the carpet is being unrolled. That is, $A(x)$ is being increased at the rate

$$f(v(x)) \frac{dv}{dx}.$$

At the same time, A is being decreased at the rate

$$f(u(x)) \frac{du}{dx},$$

the width at the end that is being rolled up times the rate du/dx. The net rate of change in A is

$$\frac{dA}{dx} = f(v(x)) \frac{dv}{dx} - f(u(x)) \frac{du}{dx},$$

which is precisely Leibniz's rule.

To prove the rule, let F be an antiderivative of f on $[a, b]$. Then

$$\int_{u(x)}^{v(x)} f(t)\, dt = F(v(x)) - F(u(x)). \qquad (1)$$

Differentiating both sides of this equation with respect to x gives the equation we want:

$$\frac{d}{dx} \int_{u(x)}^{v(x)} f(t)\, dt = \frac{d}{dx} \left[F(v(x)) - F(u(x)) \right]$$

$$= F'(v(x)) \frac{dv}{dx} - F'(u(x)) \frac{du}{dx} \qquad \text{Chain Rule}$$

$$= f(v(x)) \frac{dv}{dx} - f(u(x)) \frac{du}{dx}.$$

You will see another way to derive the rule in Chapter 12, Additional Exercise 3.

Use Leibniz's rule to find the derivatives of the functions in Exercises 19–21.

19. $f(x) = \int_{1/x}^{x} \frac{1}{t} \, dt$

20. $f(x) = \int_{\cos x}^{\sin x} \frac{1}{1 - t^2} \, dt$

21. $g(y) = \int_{\sqrt{y}}^{2\sqrt{y}} \sin t^2 \, dt$

22. Use Leibniz's rule to find the value of x that maximizes the value of the integral

$$\int_{x}^{x+3} t(5 - t) \, dt.$$

Problems like this arise in the mathematical theory of political elections. See "The Entry Problem in a Political Race," by Steven J. Brams and Philip D. Straffin, Jr., in *Political Equilibrium,* Peter Ordeshook and Kenneth Shepfle, Editors, Kluwer-Nijhoff, Boston, 1982, pp. 181–195.

Approximating Finite Sums with Integrals

In many applications of calculus, integrals are used to approximate finite sums—the reverse of the usual procedure of using finite sums to approximate integrals. Here is an example.

EXAMPLE 7   Estimate the sum of the square roots of the first n positive integers, $\sqrt{1} + \sqrt{2} + \cdots + \sqrt{n}$.

Solution   See Fig. 4.38. The integral

$$\int_{0}^{1} \sqrt{x} \, dx = \frac{2}{3} x^{3/2} \Big]_{0}^{1} = \frac{2}{3}$$

is the limit of the sums

$$S_n = \sqrt{\frac{1}{n}} \cdot \frac{1}{n} + \sqrt{\frac{2}{n}} \cdot \frac{1}{n} + \cdots + \sqrt{\frac{n}{n}} \cdot \frac{1}{n}$$

$$= \frac{\sqrt{1} + \sqrt{2} + \cdots + \sqrt{n}}{n^{3/2}}.$$

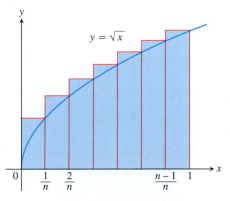

4.38 The relation of the circumscribed rectangles to the integral $\int_0^1 \sqrt{x} \, dx$ leads to an estimate of the sum $\sqrt{1} + \sqrt{2} + \sqrt{3} + \cdots + \sqrt{n}$.

Therefore, when n is large, S_n will be close to 2/3 and we will have

$$\text{Root sum} = \sqrt{1} + \sqrt{2} + \cdots + \sqrt{n} = S_n \cdot n^{3/2} \approx \frac{2}{3} n^{3/2}.$$

The following table shows how good the approximation can be.

n	Root sum	$(2/3)n^{3/2}$	Relative error
10	22.468	21.082	1.386/22.468 ≈ 6%
50	239.04	235.70	1.4%
100	671.46	666.67	0.7%
1000	21,097	21,082	0.07%

23. Evaluate

$$\lim_{n \to \infty} \frac{1^5 + 2^5 + 3^5 + \cdots + n^5}{n^6}$$

by showing that the limit is

$$\int_{0}^{1} x^5 \, dx$$

and evaluating the integral.

24. See Exercise 23. Evaluate

$$\lim_{n \to \infty} \frac{1}{n^4} (1^3 + 2^3 + 3^3 + \cdots + n^3).$$

25. Let $f(x)$ be a continuous function. Express

$$\lim_{n \to \infty} \frac{1}{n} \left[f\left(\frac{1}{n}\right) + f\left(\frac{2}{n}\right) + \cdots + f\left(\frac{n}{n}\right) \right]$$

as a definite integral.

26. Use the result of Exercise 25 to evaluate

a) $\lim_{n \to \infty} \frac{1}{n^2} (2 + 4 + 6 + \cdots + 2n),$

b) $\lim_{n \to \infty} \frac{1}{n^{16}} (1^{15} + 2^{15} + 3^{15} + \cdots + n^{15}),$

c) $\lim_{n \to \infty} \frac{1}{n} \left(\sin \frac{\pi}{n} + \sin \frac{2\pi}{n} + \sin \frac{3\pi}{n} + \cdots + \sin \frac{n\pi}{n} \right).$

What can be said about the following limits?

d) $\lim_{n \to \infty} \frac{1}{n^{17}} (1^{15} + 2^{15} + 3^{15} + \cdots + n^{15})$

e) $\lim_{n \to \infty} \frac{1}{n^{15}} (1^{15} + 2^{15} + 3^{15} + \cdots + n^{15})$

27. a)   Show that the area A_n of an n-sided regular polygon in a circle of radius r is

$$A_n = \frac{nr^2}{2} \sin \frac{2\pi}{n}.$$

b)   Find the limit of A_n as $n \to \infty$. Is this answer consistent with what you know about the area of a circle?

28. *The error function.* The error function,

$$\text{erf}\,(x) = \frac{2}{\sqrt{\pi}} \int_0^x e^{-t^2}\,dt,$$

important in probability and in the theories of heat flow and signal transmission, must be evaluated numerically because there is no elementary expression for the antiderivative of e^{-t^2}.

a) Use Simpson's rule with $n = 10$ to estimate erf(1).

b) In [0, 1],

$$\left| \frac{d^4}{dt^4}\left(e^{-t^2}\right) \right| \le 12.$$

Give an upper bound for the magnitude of the error of the estimate in (a).

CHAPTER 5

Applications of Integrals

OVERVIEW Many things we want to know can be calculated with integrals: the areas between curves, the volumes and surface areas of solids, the lengths of curves, the amount of work it takes to pump liquids from belowground, the forces against floodgates, the coordinates of the points where solid objects will balance. We define all of these as limits of Riemann sums of continuous functions on closed intervals, that is, as integrals, and evaluate these limits with calculus.

There is a pattern to how we define the integrals in applications, a pattern that, once learned, enables us to define new integrals when we need them. We look at specific applications first, then examine the pattern and show how it leads to integrals in new situations.

5.1 Areas Between Curves

This section shows how to find the areas of regions in the coordinate plane by integrating the functions that define the regions' boundaries.

The Basic Formula as a Limit of Riemann Sums

Suppose we want to find the area of a region that is bounded above by the curve $y = f(x)$, below by the curve $y = g(x)$, and on the left and right by the lines $x = a$ and $x = b$ (Fig. 5.1). The region might accidentally have a shape whose area we could find with geometry, but if f and g are arbitrary continuous functions we usually have to find the area with an integral.

To see what the integral should be, we first approximate the region with n vertical rectangles based on a partition $P = \{x_0, x_1, \ldots, x_n\}$ of $[a, b]$ (Fig. 5.2, on the following page). The area of the kth rectangle (Fig. 5.3, on the following page) is

$$\Delta A_k = \text{height} \times \text{width} = [f(c_k) - g(c_k)] \Delta x_k.$$

We then approximate the area of the region by adding the areas of the n rectangles:

$$A \approx \sum_{k=1}^{n} \Delta A_k = \sum_{k=1}^{n} [f(c_k) - g(c_k)] \Delta x_k. \qquad \text{Riemann sum}$$

As $\|P\| \to 0$ the sums on the right approach the limit $\int_a^b [f(x) - g(x)] \, dx$ because

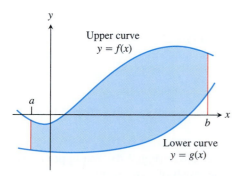

5.1 The region between $y = f(x)$ and $y = g(x)$ and the lines $x = a$ and $x = b$.

Upper curve
$y = f(x)$

Lower curve
$y = g(x)$

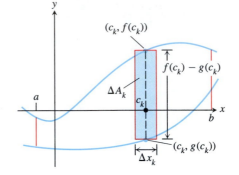

5.2 We approximate the region with rectangles perpendicular to the x-axis.

5.3 ΔA_k = area of kth rectangle, $f(c_k) - g(c_k)$ = height, Δx_k = width

f and g are continuous. We take the area of the region to be the value of this integral. That is,

$$A = \lim_{\|P\| \to 0} \sum_{k=1}^{n} [f(c_k) - g(c_k)] \Delta x_k = \int_a^b [f(x) - g(x)] \, dx.$$

Definition

If f and g are continuous with $f(x) \geq g(x)$ throughout $[a, b]$, then the **area** of the region between the curves $y = f(x)$ and $y = g(x)$ from a to b is the integral of $[f - g]$ from a to b:

$$A = \int_a^b [f(x) - g(x)] \, dx. \tag{1}$$

To apply Eq. (1) we take the following steps.

How to Find the Area Between Two Curves

1. *Graph the curves and draw a representative rectangle.* This reveals which curve is f (upper curve) and which is g (lower curve). It also helps find the limits of integration if you do not already know them.
2. *Find the limits of integration.*
3. *Write a formula for $f(x) - g(x)$.* Simplify it if you can.
4. *Integrate $[f(x) - g(x)]$ from a to b.* The number you get is the area.

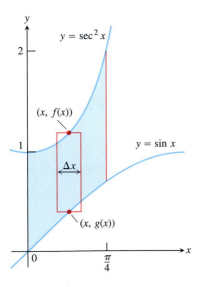

5.4 The region in Example 1 with a typical approximating rectangle.

EXAMPLE 1   Find the area between $y = \sec^2 x$ and $y = \sin x$ from 0 to $\pi/4$.

Solution

Step 1: We sketch the curves and a vertical rectangle (Fig. 5.4). The upper curve is the graph of $f(x) = \sec^2 x$; the lower is the graph of $g(x) = \sin x$.

Step 2: The limits of integration are already given: $a = 0, b = \pi/4$.

Step 3: $f(x) - g(x) = \sec^2 x - \sin x$

Step 4:

$$A = \int_0^{\pi/4} (\sec^2 x - \sin x)\, dx = \Big[\tan x + \cos x\Big]_0^{\pi/4}$$

$$= \left[1 + \frac{\sqrt{2}}{2}\right] - [0 + 1] = \frac{\sqrt{2}}{2}$$

Curves That Intersect

When a region is determined by curves that intersect, the intersection points give the limits of integration.

EXAMPLE 2   Find the area of the region enclosed by the parabola $y = 2 - x^2$ and the line $y = -x$.

Solution

Step 1: Sketch the curves and a vertical rectangle (Fig. 5.5). Identifying the upper and the lower curves, we take $f(x) = 2 - x^2$ and $g(x) = -x$. The x-coordinates of the intersection points are the limits of integration.

Step 2: We find the limits of integration by solving $y = 2 - x^2$ and $y = -x$ simultaneously for x:

$$2 - x^2 = -x \qquad \text{Equate } f(x) \text{ and } g(x).$$

$$x^2 - x - 2 = 0 \qquad \text{Rewrite.}$$

$$(x + 1)(x - 2) = 0 \qquad \text{Factor.}$$

$$x = -1, \qquad x = 2. \qquad \text{Solve.}$$

The region runs from $x = -1$ to $x = 2$. The limits of integration are $a = -1, b = 2$.

Step 3:

$$f(x) - g(x) = (2 - x^2) - (-x) = 2 - x^2 + x \qquad \text{Rearrangement}$$

$$= 2 + x - x^2 \qquad \text{a matter of taste}$$

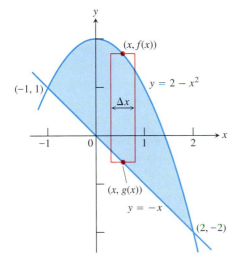

5.5 The region in Example 2 with a typical approximating rectangle.

Step 4:

$$A = \int_a^b [f(x) - g(x)]\, dx = \int_{-1}^2 (2 + x - x^2)\, dx = \left[2x + \frac{x^2}{2} - \frac{x^3}{3}\right]_{-1}^2$$

$$= \left(4 + \frac{4}{2} - \frac{8}{3}\right) - \left(-2 + \frac{1}{2} + \frac{1}{3}\right)$$

$$= 6 + \frac{3}{2} - \frac{9}{3} = \frac{9}{2}$$

Technology   *The Intersection of Two Graphs*   One of the difficult and sometimes frustrating parts of integration applications is finding the limits of integration. To do this you often have to find the zeroes of a function or the intersection points of two curves.

To solve the equation $f(x) = g(x)$ using a graphing utility, you enter

$$y_1 = f(x) \quad \text{and} \quad y_2 = g(x)$$

and use the grapher routine to find the points of intersection. Alternatively, you can solve the equation $f(x) - g(x) = 0$ with a root finder. Try both procedures with

$$f(x) = \ln x \quad \text{and} \quad g(x) = 3 - x.$$

When points of intersection are not clearly revealed or you suspect hidden behavior, additional work with the graphing utility or further use of calculus may be necessary.

a) The intersecting curves $y_1 = \ln x$ and $y_2 = 3 - x$, using a built-in function to find the intersection
b) Using a built-in root finder to find the zero of $f(x) = \ln x - 3 + x$

Boundaries with Changing Formulas

If the formula for a bounding curve changes at one or more points, we partition the region into subregions that correspond to the formula changes and apply Eq. (1) to each subregion.

EXAMPLE 3 Find the area of the region in the first quadrant that is bounded above by $y = \sqrt{x}$ and below by the x-axis and the line $y = x - 2$.

Solution

Step 1: The sketch (Fig. 5.6) shows that the region's upper boundary is the graph of $f(x) = \sqrt{x}$. The lower boundary changes from $g(x) = 0$ for $0 \le x \le 2$ to $g(x) = x - 2$ for $2 \le x \le 4$ (there is agreement at $x = 2$). We partition the region at $x = 2$ into subregions A and B and sketch a representative rectangle for each subregion.

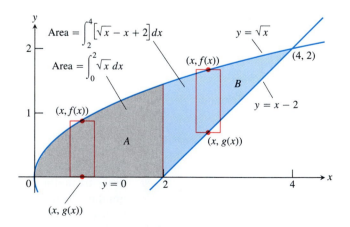

5.6 When the formula for a bounding curve changes, the area integral changes to match (Example 3).

Step 2: The limits of integration for region A are $a = 0$ and $b = 2$. The left-hand limit for region B is $a = 2$. To find the right-hand limit, we solve the equations

$y = \sqrt{x}$ and $y = x - 2$ simultaneously for x:

$$\sqrt{x} = x - 2 \qquad \text{Equate } f(x) \text{ and } g(x).$$

$$x = (x - 2)^2 = x^2 - 4x + 4 \qquad \text{Square both sides.}$$

$$x^2 - 5x + 4 = 0 \qquad \text{Rewrite.}$$

$$(x - 1)(x - 4) = 0 \qquad \text{Factor.}$$

$$x = 1, \qquad x = 4. \qquad \text{Solve.}$$

Only the value $x = 4$ satisfies the equation $\sqrt{x} = x - 2$. The value $x = 1$ is an extraneous root introduced by squaring. The right-hand limit is $b = 4$.

Step 3: For $0 \leq x \leq 2$: $\qquad f(x) - g(x) = \sqrt{x} - 0 = \sqrt{x}$

For $2 \leq x \leq 4$: $\qquad f(x) - g(x) = \sqrt{x} - (x - 2) = \sqrt{x} - x + 2$

Step 4: We add the area of subregions A and B to find the total area:

$$\text{Total area} = \underbrace{\int_0^2 \sqrt{x}\, dx}_{\text{area of } A} + \underbrace{\int_2^4 (\sqrt{x} - x + 2)\, dx}_{\text{area of } B}$$

$$= \left[\frac{2}{3}x^{3/2}\right]_0^2 + \left[\frac{2}{3}x^{3/2} - \frac{x^2}{2} + 2x\right]_2^4$$

$$= \frac{2}{3}(2)^{3/2} - 0 + \left(\frac{2}{3}(4)^{3/2} - 8 + 8\right) - \left(\frac{2}{3}(2)^{3/2} - 2 + 4\right)$$

$$= \frac{2}{3}(8) - 2 = \frac{10}{3}. \qquad \square$$

Integration with Respect to y

If a region's bounding curves are described by functions of y, the approximating rectangles are horizontal instead of vertical and the basic formula has y in place of x.

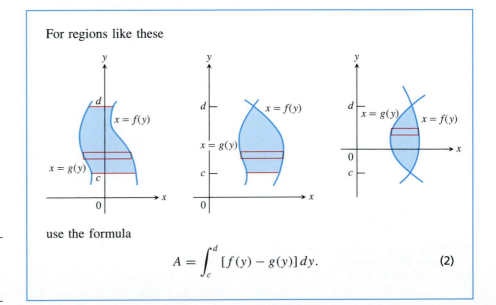

For regions like these

use the formula

$$A = \int_c^d [f(y) - g(y)]\, dy. \tag{2}$$

In Eq. (2), f always denotes the right-hand curve and g the left-hand curve, so $f(y) - g(y)$ is nonnegative.

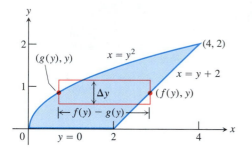

5.7 It takes two integrations to find the area of this region if we integrate with respect to x. It takes only one if we integrate with respect to y (Example 4).

EXAMPLE 4   Find the area of the region in Example 3 by integrating with respect to y.

Solution

Step 1: We sketch the region and a typical *horizontal* rectangle based on a partition of an interval of y-values (Fig. 5.7). The region's right-hand boundary is the line $x = y + 2$, so $f(y) = y + 2$. The left-hand boundary is the curve $x = y^2$, so $g(y) = y^2$.

Step 2: The lower limit of integration is $y = 0$. We find the upper limit by solving $x = y + 2$ and $x = y^2$ simultaneously for y:

$$y + 2 = y^2 \qquad \text{Equate } f(y) = y + 2 \text{ and } g(y) = y^2.$$

$$y^2 - y - 2 = 0 \qquad \text{Rewrite.}$$

$$(y + 1)(y - 2) = 0 \qquad \text{Factor.}$$

$$y = -1, \qquad y = 2 \qquad \text{Solve.}$$

The upper limit of integration is $b = 2$. (The value $y = -1$ gives a point of intersection *below* the x-axis.)

Step 3:

$$f(y) - g(y) = y + 2 - y^2 = 2 + y - y^2 \qquad \begin{array}{l}\text{Rearrangement a}\\\text{matter of taste}\end{array}$$

Step 4:

$$A = \int_a^b [f(y) - g(y)]\,dy = \int_0^2 [2 + y - y^2]\,dy$$

$$= \left[2y + \frac{y^2}{2} - \frac{y^3}{3}\right]_0^2$$

$$= 4 + \frac{4}{2} - \frac{8}{3} = \frac{10}{3}$$

This is the result of Example 3, found with less work.   ❑

Combining Integrals with Formulas from Geometry

The fastest way to find an area may be to combine calculus and geometry.

EXAMPLE 5   *The Area of the Region in Example 3 Found the Fastest Way*

Find the area of the region in Example 3.

Solution   The area we want is the area between the curve $y = \sqrt{x}, 0 \leq x \leq 4$, and the x-axis, *minus* the area of a triangle with base 2 and height 2 (Fig. 5.8):

$$\text{Area} = \int_0^4 \sqrt{x}\,dx - \frac{1}{2}(2)(2)$$

$$= \frac{2}{3}x^{3/2}\Big]_0^4 - 2$$

$$= \frac{2}{3}(8) - 0 - 2 = \frac{10}{3}.$$ ❑

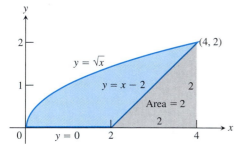

5.8 The area of the blue region is the area under the parabola $y = \sqrt{x}$ minus the area of the triangle.

Moral of Examples 3–5   It is sometimes easier to find the area between two curves by integrating with respect to y instead of x. Also, it may help to combine geometry and calculus. After sketching the region, take a moment to determine the best way to proceed.

Exercises 5.1

Find the areas of the shaded regions in Exercises 1–8.

1.

2.

3.

4.

5.

NOT TO SCALE

6.

7.

8.

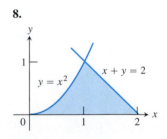

In Exercises 9–12, find the total shaded area.

9.

10.

11.

12.

Find the areas of the regions enclosed by the lines and curves in Exercises 13–22.

13. $y = x^2 - 2$ and $y = 2$

14. $y = 2x - x^2$ and $y = -3$

15. $y = x^4$ and $y = 8x$

16. $y = x^2 - 2x$ and $y = x$

17. $y = x^2$ and $y = -x^2 + 4x$

18. $y = 7 - 2x^2$ and $y = x^2 + 4$

19. $y = x^4 - 4x^2 + 4$ and $y = x^2$

20. $y = x\sqrt{a^2 - x^2}$, $a > 0$, and $y = 0$

21. $y = \sqrt{|x|}$ and $5y = x + 6$ (How many intersection points are there?)

22. $y = |x^2 - 4|$ and $y = (x^2/2) + 4$

Find the areas of the regions enclosed by the lines and curves in Exercises 23–30.

23. $x = 2y^2$, $x = 0$, and $y = 3$

24. $x = y^2$ and $x = y + 2$

25. $y^2 - 4x = 4$ and $4x - y = 16$

26. $x - y^2 = 0$ and $x + 2y^2 = 3$

27. $x + y^2 = 0$ and $x + 3y^2 = 2$

28. $x - y^{2/3} = 0$ and $x + y^4 = 2$

29. $x = y^2 - 1$ and $x = |y|\sqrt{1 - y^2}$

30. $x = y^3 - y^2$ and $x = 2y$

Find the areas of the regions enclosed by the curves in Exercises 31–34.

31. $4x^2 + y = 4$ and $x^4 - y = 1$

32. $x^3 - y = 0$ and $3x^2 - y = 4$

33. $x + 4y^2 = 4$ and $x + y^4 = 1$, for $x \geq 0$

34. $x + y^2 = 3$ and $4x + y^2 = 0$

Find the areas of the regions enclosed by the lines and curves in Exercises 35–42.

35. $y = 2 \sin x$ and $y = \sin 2x$, $0 \leq x \leq \pi$

36. $y = 8 \cos x$ and $y = \sec^2 x$, $-\pi/3 \leq x \leq \pi/3$

37. $y = \cos(\pi x/2)$ and $y = 1 - x^2$

38. $y = \sin(\pi x/2)$ and $y = x$

39. $y = \sec^2 x$, $y = \tan^2 x$, $x = -\pi/4$, and $x = \pi/4$

40. $x = \tan^2 y$ and $x = -\tan^2 y$, $-\pi/4 \leq y \leq \pi/4$

41. $x = 3 \sin y \sqrt{\cos y}$ and $x = 0$, $0 \leq y \leq \pi/2$

42. $y = \sec^2(\pi x/3)$ and $y = x^{1/3}$, $-1 \leq x \leq 1$

43. Find the area of the propeller-shaped region enclosed by the curve $x - y^3 = 0$ and the line $x - y = 0$.

44. Find the area of the propeller-shaped region enclosed by the curves $x - y^{1/3} = 0$ and $x - y^{1/5} = 0$.

45. Find the area of the region in the first quadrant bounded by the line $y = x$, the line $x = 2$, the curve $y = 1/x^2$, and the x-axis.

46. Find the area of the "triangular" region in the first quadrant bounded on the left by the y-axis and on the right by the curves $y = \sin x$ and $y = \cos x$.

47. The region bounded below by the parabola $y = x^2$ and above by the line $y = 4$ is to be partitioned into two subsections of equal area by cutting across it with the horizontal line $y = c$.

a) Sketch the region and draw a line $y = c$ across it that looks about right. In terms of c, what are the coordinates of the points where the line and parabola intersect? Add them to your figure.

b) Find c by integrating with respect to y. (This puts c in the limits of integration.)

c) Find c by integrating with respect to x. (This puts c into the integrand as well.)

48. Find the area of the region between the curve $y = 3 - x^2$ and the line $y = -1$ by integrating with respect to (a) x, (b) y.

49. Find the area of the region in the first quadrant bounded on the left by the y-axis, below by the line $y = x/4$, above left by the curve $y = 1 + \sqrt{x}$, and above right by the curve $y = 2/\sqrt{x}$.

50. Find the area of the region in the first quadrant bounded on the left by the y-axis, below by the curve $x = 2\sqrt{y}$, above left by the curve $x = (y - 1)^2$, and above right by the line $x = 3 - y$.

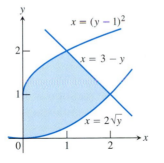

51. The figure here shows triangle AOC inscribed in the region cut from the parabola $y = x^2$ by the line $y = a^2$. Find the limit of the ratio of the area of the triangle to the area of the parabolic region as a approaches zero.

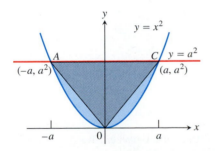

52. Suppose the area of the region between the graph of a positive continuous function f and the x-axis from $x = a$ to $x = b$ is 4 square units. Find the area between the curves $y = f(x)$ and $y = 2f(x)$ from $x = a$ to $x = b$.

53. Which of the following integrals, if either, calculates the area of the shaded region shown here? Give reasons for your answer.

a) $\displaystyle\int_{-1}^{1} (x - (-x)) \, dx = \int_{-1}^{1} 2x \, dx$

b) $\displaystyle\int_{-1}^{1} (-x - (x)) \, dx = \int_{-1}^{1} -2x \, dx$

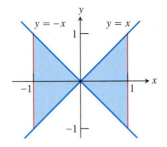

54. True, sometimes true, or never true? The area of the region between the graphs of the continuous functions $y = f(x)$ and $y = g(x)$ and the vertical lines $x = a$ and $x = b$ $(a < b)$ is

$$\int_a^b [f(x) - g(x)] \, dx.$$

Give reasons for your answer.

❂ CAS Explorations and Projects

In Exercises 55–58, you will find the area between curves in the plane when you cannot find their points of intersection using simple algebra. Use a CAS to perform the following steps:

a) Plot the curves together to see what they look like and how many points of intersection they have.

b) Use the numerical equation solver in your CAS to find all the points of intersection.

c) Integrate $|f(x) - g(x)|$ over consecutive pairs of intersection values.

d) Sum together the integrals found in part (c).

55. $f(x) = \dfrac{x^3}{3} - \dfrac{x^2}{2} - 2x + \dfrac{1}{3}, \quad g(x) = x - 1$

56. $f(x) = \dfrac{x^4}{2} - 3x^3 + 10, \quad g(x) = 8 - 12x$

57. $f(x) = x + \sin(2x), \quad g(x) = x^3$

58. $f(x) = x^2 \cos x, \quad g(x) = x^3 - x$

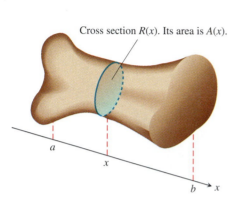

Cross section $R(x)$. Its area is $A(x)$.

5.9 If the area $A(x)$ of the cross section $R(x)$ is a continuous function of x, we can find the volume of the solid by integrating $A(x)$ from a to b.

Finding Volumes by Slicing

From the areas of regions with curved boundaries, we can calculate the volumes of cylinders with curved bases by multiplying base area by height. From the volumes of such cylinders, we can calculate the volumes of other solids.

Slicing

Suppose we want to find the volume of a solid like the one shown in Fig. 5.9. At each point x in the closed interval $[a, b]$ the cross section of the solid is a region $R(x)$ whose area is $A(x)$. This makes A a real-valued function of x. If it is also a continuous function of x, we can use it to define and calculate the volume of the solid as an integral in the following way.

We partition the interval $[a, b]$ along the x-axis in the usual manner and slice the solid, as we would a loaf of bread, by planes perpendicular to the x-axis at the partition points. The kth slice, the one between the planes at x_{k-1} and x_k, has approximately the same volume as the cylinder between these two planes based on the region $R(x_k)$ (Fig. 5.10). The volume of this cylinder is

$$V_k = \text{base area} \times \text{height}$$

$$= A(x_k) \times (\text{distance between the planes at } x_{k-1} \text{ and } x_k)$$

$$= A(x_k)\Delta x_k.$$

The volume of the solid is therefore approximated by the cylinder volume sum

$$\sum_{k=1}^{n} A(x_k)\Delta x_k.$$

This is a Riemann sum for the function $A(x)$ on $[a, b]$. We expect the approximations from these sums to improve as the norm of the partition of $[a, b]$ goes to zero, so we define their limiting integral to be the volume of the solid.

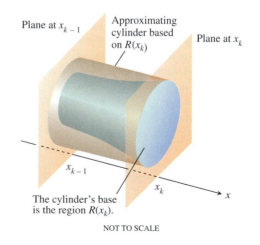

Plane at x_{k-1}

Approximating cylinder based on $R(x_k)$

Plane at x_k

x_{k-1}

x_k

x

The cylinder's base is the region $R(x_k)$.

NOT TO SCALE

5.10 Enlarged view of the slice of the solid between the planes at x_{k-1} and x_k and its approximating cylinder.

Definition

The **volume** of a solid of known integrable cross-section area $A(x)$ from $x = a$ to $x = b$ is the integral of A from a to b:

$$V = \int_a^b A(x)\, dx. \tag{1}$$

To apply Eq. (1), we take the following steps.

How to Find Volumes by the Method of Slicing

1. Sketch the solid and a typical cross section.
2. Find a formula for $A(x)$.
3. Find the limits of integration.
4. Integrate $A(x)$ to find the volume.

EXAMPLE 1   A pyramid 3 m high has a square base that is 3 m on a side. The cross section of the pyramid perpendicular to the altitude x m down from the vertex is a square x m on a side. Find the volume of the pyramid.

Solution

Step 1: *A sketch.* We draw the pyramid with its altitude along the x-axis and its vertex at the origin and include a typical cross section (Fig. 5.11).

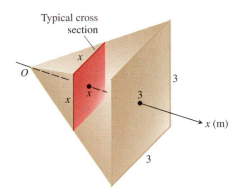

5.11 The cross sections of the pyramid in Example 1 are squares.

Step 2: *A formula for $A(x)$.* The cross section at x is a square x meters on a side, so its area is

$$A(x) = x^2.$$

Step 3: *The limits of integration.* The squares go from $x = 0$ to $x = 3$.

Step 4: *The volume.*

$$V = \int_a^b A(x)\, dx = \int_0^3 x^2\, dx = \frac{x^3}{3}\Big]_0^3 = 9.$$

The volume is 9 m^3.

Bonaventura Cavalieri (1598–1647)

Cavalieri, a student of Galileo's, discovered that if two plane regions can be arranged to lie over the same interval of the x-axis in such a way that they have identical vertical cross sections at every point, then the regions have the same area. The theorem (and a letter of recommendation from Galileo) were enough to win Cavalieri a chair at the University of Bologna in 1629. The solid geometry version in Example 3, which Cavalieri never proved, was given his name by later geometers.

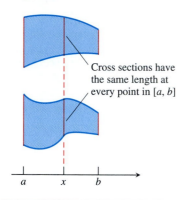

Cross sections have the same length at every point in [a, b]

EXAMPLE 2   A curved wedge is cut from a cylinder of radius 3 by two planes. One plane is perpendicular to the axis of the cylinder. The second plane crosses the first plane at a 45° angle at the center of the cylinder. Find the volume of the wedge.

Solution

Step 1: *A sketch.* We draw the wedge and sketch a typical cross section perpendicular to the x-axis (Fig. 5.12).

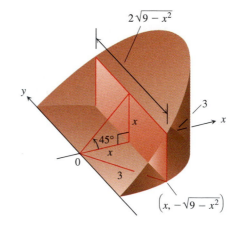

5.12 The wedge of Example 2, sliced perpendicular to the x-axis. The cross sections are rectangles.

Step 2: *The formula for* $A(x)$*.* The cross section at x is a rectangle of area

$$A(x) = (\text{height})(\text{width}) = (x)\left(2\sqrt{9 - x^2}\right)$$

$$= 2x\sqrt{9 - x^2}.$$

Step 3: *The limits of integration.* The rectangles run from $x = 0$ to $x = 3$.

Step 4: *The volume.*

$$V = \int_a^b A(x)\,dx = \int_0^3 2x\sqrt{9 - x^2}\,dx$$

$$= -\frac{2}{3}(9 - x^2)^{3/2}\Big]_0^3$$

$$= 0 + \frac{2}{3}(9)^{3/2} \quad \text{Let } u = 9 - x^2,$$
$$\phantom{= 0 + \frac{2}{3}(9)^{3/2}} \quad du = -2x\,dx, \text{ integrate,}$$
$$\phantom{= 0 + \frac{2}{3}(9)^{3/2}} \quad \text{and substitute back.}$$

$$= 18. \qquad \qquad \qquad \Box$$

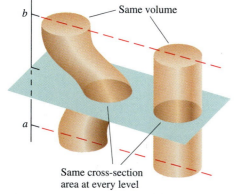

5.13 *Cavalieri's theorem:* These solids have the same volume. You can illustrate this yourself with stacks of coins.

EXAMPLE 3   *Cavalieri's Theorem*

Cavalieri's theorem says that solids with equal altitudes and identical parallel cross-section areas have the same volume (Fig. 5.13). We can see this immediately from Eq. (1) because the cross-section area function $A(x)$ is the same in each case. $\Box$

Exercises 5.2

Cross-Section Areas

In Exercises 1 and 2, find a formula for the area $A(x)$ of the cross sections of the solid perpendicular to the *x*-axis.

1. The solid lies between planes perpendicular to the *x*-axis at $x = -1$ and $x = 1$. In each case, the cross sections perpendicular to the *x*-axis between these planes run from the semicircle $y = -\sqrt{1 - x^2}$ to the semicircle $y = \sqrt{1 - x^2}$.

a) The cross sections are circular disks with diameters in the *xy*-plane.

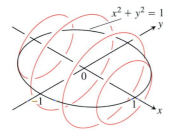

b) The cross sections are squares with bases in the *xy*-plane.

c) The cross sections are squares with diagonals in the *xy*-plane. (The length of a square's diagonal is $\sqrt{2}$ times the length of its sides.)

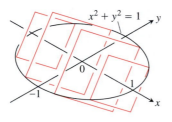

d) The cross sections are equilateral triangles with bases in the *xy*-plane.

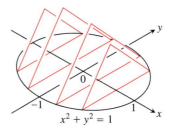

2. The solid lies between planes perpendicular to the *x*-axis at $x = 0$ and $x = 4$. The cross sections perpendicular to the *x*-axis between these planes run from the parabola $y = -\sqrt{x}$ to the parabola $y = \sqrt{x}$.

a) The cross sections are circular disks with diameters in the *xy*-plane.

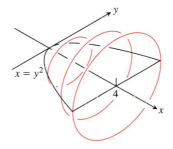

b) The cross sections are squares with bases in the *xy*-plane.

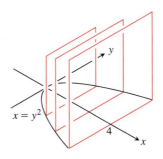

c) The cross sections are squares with diagonals in the *xy*-plane.

d) The cross sections are equilateral triangles with bases in the *xy*-plane.

Volumes by Slicing

Find the volumes of the solids in Exercises 3–12.

3. The solid lies between planes perpendicular to the *x*-axis at $x = 0$ and $x = 4$. The cross sections perpendicular to the axis on the interval $0 \le x \le 4$ are squares whose diagonals run from the parabola $y = -\sqrt{x}$ to the parabola $y = \sqrt{x}$.

4. The solid lies between planes perpendicular to the x-axis at $x = -1$ and $x = 1$. The cross sections perpendicular to the x-axis are circular disks whose diameters run from the parabola $y = x^2$ to the parabola $y = 2 - x^2$.

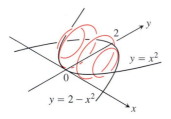

5. The solid lies between planes perpendicular to the x-axis at $x = -1$ and $x = 1$. The cross sections perpendicular to the axis between these planes are vertical squares whose base edges run from the semicircle $y = -\sqrt{1 - x^2}$ to the semicircle $y = \sqrt{1 - x^2}$.

6. The solid lies between planes perpendicular to the x-axis at $x = -1$ and $x = 1$. The cross sections perpendicular to the x-axis between these planes are squares whose diagonals run from the semicircle $y = -\sqrt{1 - x^2}$ to the semicircle $y = \sqrt{1 - x^2}$. (The length of a square's diagonal is $\sqrt{2}$ times the length of its sides.)

7. The base of the solid is the region between the curve $y = 2\sqrt{\sin x}$ and the interval $[0, \pi]$ on the x-axis. The cross sections perpendicular to the x-axis are

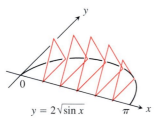

a) vertical equilateral triangles with bases running from the x-axis to the curve;

b) vertical squares with bases running from the x-axis to the curve.

8. The solid lies between planes perpendicular to the x-axis at $x = -\pi/3$ and $x = \pi/3$. The cross sections perpendicular to the x-axis are

a) circular disks with diameters running from the curve $y = \tan x$ to the curve $y = \sec x$;

b) vertical squares whose base edges run from the curve $y = \tan x$ to the curve $y = \sec x$.

9. The solid lies between planes perpendicular to the y-axis at $y = 0$ and $y = 2$. The cross sections perpendicular to the y-axis are circular disks with diameters running from the y-axis to the parabola $x = \sqrt{5}y^2$.

10. The base of the solid is the disk $x^2 + y^2 \le 1$. The cross sections by planes perpendicular to the y-axis between $y = -1$ and $y = 1$ are isosceles right triangles with one leg in the disk.

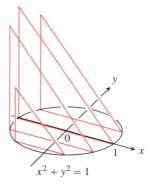

$x^2 + y^2 = 1$

Cavalieri's Theorem

11. *A twisted solid.* A square of side length s lies in a plane perpendicular to a line L. One vertex of the square lies on L. As this square moves a distance h along L, the square turns one revolution about L to generate a corkscrew-like column with square cross sections.

a) Find the volume of the column.

b) What will the volume be if the square turns twice instead of once? Give reasons for your answer.

12. A solid lies between planes perpendicular to the x-axis at $x = 0$ and $x = 12$. The cross sections by planes perpendicular to the x-axis are circular disks whose diameters run from the line $y = x/2$ to the line $y = x$. Explain why the solid has the same volume as a right circular cone with base radius 3 and height 12.

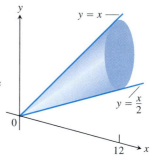

13. *Cavalieri's original theorem.* Prove Cavalieri's original theorem (marginal note, page 376), assuming that each region is bounded above and below by the graphs of continuous functions.

14. *The volume of a hemisphere (a classical application of Cavalieri's theorem).* Derive the formula $V = (2/3)\pi R^3$ for the volume of a hemisphere of radius R by comparing its cross sections with the cross sections of a solid right circular cylinder of radius R and height R from which a solid right circular cone of base radius R and height R has been removed.

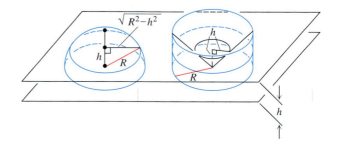

5.3 Volumes of Solids of Revolution—Disks and Washers

The most common application of the method of slicing is to solids of revolution. **Solids of revolution** are solids whose shapes can be generated by revolving plane regions about axes. Thread spools are solids of revolution; so are hand weights and billiard balls. Solids of revolution sometimes have volumes we can find with formulas from geometry, as in the case of a billiard ball. But when we want to find the volume of a blimp or to predict the weight of a part we are going to have turned on a lathe, formulas from geometry are of little help and we turn to calculus for the answers.

If we can arrange for the region to be the region between the graph of a continuous function $y = R(x), a \leq x \leq b$, and the x-axis, and for the axis of revolution to be the x-axis (Fig. 5.14), we can find the solid's volume in the following way.

The typical cross section of the solid perpendicular to the axis of revolution is a disk of radius $R(x)$ and area

$$A(x) = \pi(\text{radius})^2 = \pi[R(x)]^2.$$

The solid's volume, being the integral of A from $x = a$ to $x = b$, is the integral of $\pi[R(x)]^2$ from a to b.

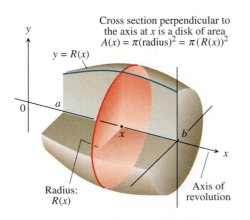

5.14 The solid generated by revolving the region between the curve $y = R(x)$ and the x-axis from a to b about the x-axis.

Volume of a Solid of Revolution (Rotation About the x-axis)

The volume of the solid generated by revolving about the x-axis the region between the x-axis and the graph of the continuous function $y = R(x), a \leq x \leq b$, is

$$V = \int_a^b \pi[\text{radius}]^2 \, dx = \int_a^b \pi[R(x)]^2 \, dx. \qquad (1)$$

EXAMPLE 1   The region between the curve $y = \sqrt{x}, 0 \leq x \leq 4$, and the x-axis is revolved about the x-axis to generate a solid. Find its volume.

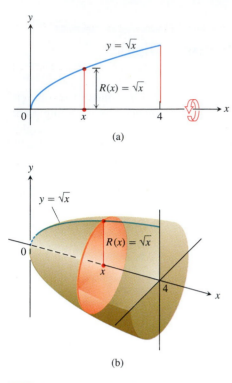

5.15 The region (a) and solid (b) in Example 1.

Solution We draw figures showing the region, a typical radius, and the generated solid (Fig. 5.15). The volume is

$$V = \int_a^b \pi [R(x)]^2 \, dx \qquad \text{Eq. (1)}$$

$$= \int_0^4 \pi \left[\sqrt{x} \right]^2 dx \qquad R(x) = \sqrt{x}$$

$$= \pi \int_0^4 x \, dx = \pi \frac{x^2}{2} \Bigg]_0^4 = \pi \frac{(4)^2}{2} = 8\pi.$$

How to Find Volumes Using Eq. (1)

1. Draw the region and identify the radius function $R(x)$.
2. Square $R(x)$ and multiply by π.
3. Integrate to find the volume.

The axis of revolution in the next example is not the x-axis, but the rule for calculating the volume is the same: Integrate $\pi(\text{radius})^2$ between appropriate limits.

EXAMPLE 2   Find the volume of the solid generated by revolving the region bounded by $y = \sqrt{x}$ and the lines $y = 1, x = 4$ about the line $y = 1$.

Solution We draw figures showing the region, a typical radius, and the generated solid (Fig. 5.16). The volume is

$$V = \int_1^4 \pi [R(x)]^2 \, dx \qquad \text{Eq. (1)}$$

$$= \int_1^4 \pi \left[\sqrt{x} - 1 \right]^2 dx \qquad R(x) = \sqrt{x} - 1$$

$$= \pi \int_1^4 \left[x - 2\sqrt{x} + 1 \right] dx$$

$$= \pi \left[\frac{x^2}{2} - 2 \cdot \frac{2}{3} x^{3/2} + x \right]_1^4 = \frac{7\pi}{6}.$$

To find the volume of a solid generated by revolving a region between the y-axis and a curve $x = R(y), c \le y \le d$, about the y-axis, we use Eq. (1) with x replaced by y.

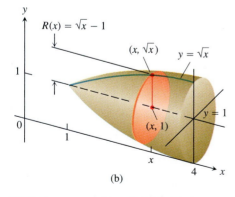

5.16 The region (a) and solid (b) in Example 2.

Volume of a Solid of Revolution (Rotation About the y-axis)

$$V = \int_c^d \pi (\text{radius})^2 \, dy = \int_c^d \pi [R(y)]^2 \, dy \qquad (2)$$

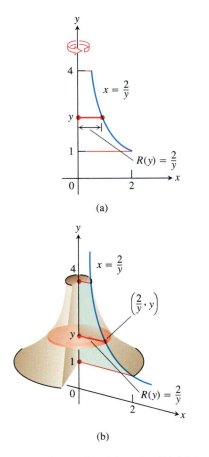

5.17 The region (a) and solid (b) in Example 3.

EXAMPLE 3 Find the volume of the solid generated by revolving the region between the y-axis and the curve $x = 2/y$, $1 \le y \le 4$, about the y-axis.

Solution We draw figures showing the region, a typical radius, and the generated solid (Fig. 5.17). The volume is

$$V = \int_1^4 \pi [R(y)]^2 \, dy \qquad \text{Eq. (2)}$$

$$= \int_1^4 \pi \left(\frac{2}{y}\right)^2 \, dy \qquad R(y) = \frac{2}{y}$$

$$= \pi \int_1^4 \frac{4}{y^2} \, dy = 4\pi \left[-\frac{1}{y}\right]_1^4 = 4\pi \left[\frac{3}{4}\right]$$

$$= 3\pi. \qquad \qquad \Box$$

EXAMPLE 4 Find the volume of the solid generated by revolving the region between the parabola $x = y^2 + 1$ and the line $x = 3$ about the line $x = 3$.

Solution We draw figures showing the region, a typical radius, and the generated solid (Fig. 5.18). The volume is

$$V = \int_{-\sqrt{2}}^{\sqrt{2}} \pi [R(y)]^2 \, dy \qquad \text{Eq. (2)}$$

$$= \int_{-\sqrt{2}}^{\sqrt{2}} \pi [2 - y^2]^2 \, dy \qquad \begin{array}{l} R(y) = 3 - (y^2 + 1) \\ = 2 - y^2 \end{array}$$

$$= \pi \int_{-\sqrt{2}}^{\sqrt{2}} [4 - 4y^2 + y^4] \, dy$$

$$= \pi \left[4y - \frac{4}{3}y^3 + \frac{y^5}{5}\right]_{-\sqrt{2}}^{\sqrt{2}}$$

$$= \frac{64\pi\sqrt{2}}{15}. \qquad \qquad \Box$$

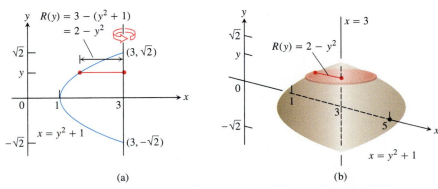

5.18 The region (a) and solid (b) in Example 4.

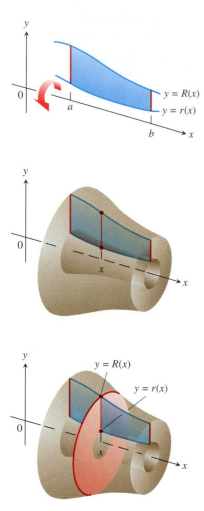

5.19 The cross sections of the solid of revolution generated here are washers, not disks, so the integral $\int_a^b A(x)\,dx$ leads to a slightly different formula.

The Washer Method

If the region we revolve to generate a solid does not border on or cross the axis of revolution, the solid has a hole in it (Fig. 5.19). The cross sections perpendicular to the axis of revolution are washers instead of disks. The dimensions of a typical washer are

$$\text{Outer radius:} \quad R(x)$$

$$\text{Inner radius:} \quad r(x)$$

The washer's area is

$$A(x) = \pi [R(x)]^2 - \pi [r(x)]^2 = \pi \left([R(x)]^2 - [r(x)]^2 \right).$$

The Washer Formula for Finding Volumes

$$V = \int_a^b \pi \left([R(x)]^2 - [r(x)]^2 \right) dx \qquad (3)$$

$$\underset{\substack{\text{outer} \\ \text{radius} \\ \text{squared}}}{|} \qquad \underset{\substack{\text{inner} \\ \text{radius} \\ \text{squared}}}{|}$$

Notice that the function integrated in Eq. (3) is $\pi(R^2 - r^2)$, not $\pi(R - r)^2$. Also notice that Eq. (3) gives the disk method formula if $r(x)$ is zero throughout $[a, b]$. Thus, the disk method is a special case of the washer method.

EXAMPLE 5   The region bounded by the curve $y = x^2 + 1$ and the line $y = -x + 3$ is revolved about the x-axis to generate a solid. Find the volume of the solid.

Solution

Step 1: Draw the region and sketch a line segment across it perpendicular to the axis of revolution (the red segment in Fig. 5.20).

Step 2: Find the limits of integration by finding the x-coordinates of the intersection points.

$$x^2 + 1 = -x + 3$$
$$x^2 + x - 2 = 0$$
$$(x + 2)(x - 1) = 0$$
$$x = -2, \qquad x = 1$$

Step 3: Find the outer and inner radii of the washer that would be swept out by the line segment if it were revolved about the x-axis along with the region. (We drew the washer in Fig. 5.21, but in your own work you need not do that.) These radii are the distances of the ends of the line segment from the axis of revolution.

$$\text{Outer radius:} \quad R(x) = -x + 3$$

$$\text{Inner radius:} \quad r(x) = x^2 + 1$$

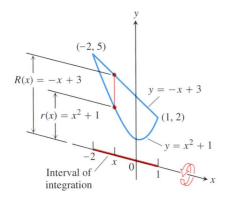

5.20 The region in Example 5 spanned by a line segment perpendicular to the axis of revolution. When the region is revolved about the *x*-axis, the line segment will generate a washer.

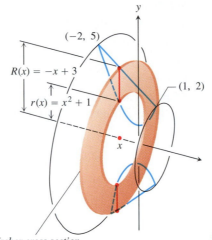

Washer cross section
Outer radius: $R(x) = -x + 3$
Inner radius: $r(x) = x^2 + 1$

5.21 The inner and outer radii of the washer swept out by the line segment in Fig. 5.20.

Step 4: Evaluate the volume integral.

$$V = \int_a^b \pi \left([R(x)]^2 - [r(x)]^2 \right) dx \qquad \text{Eq. (3)}$$

$$= \int_{-2}^1 \pi ((-x + 3)^2 - (x^2 + 1)^2) \, dx \qquad \text{Values from steps 2 and 3}$$

$$= \int_{-2}^1 \pi (8 - 6x - x^2 - x^4) \, dx \qquad \text{Expressions squared and combined}$$

$$= \pi \left[8x - 3x^2 - \frac{x^3}{3} - \frac{x^5}{5} \right]_{-2}^1 = \frac{117\pi}{5}$$

How to Find Volumes by the Washer Method

1. *Draw the region and sketch a line segment across it perpendicular to the axis of revolution.* When the region is revolved, this segment will generate a typical washer cross section of the generated solid.
2. *Find the limits of integration.*
3. *Find the outer and inner radii* of the washer swept out by the line segment.
4. *Integrate* to find the volume.

To find the volume of a solid generated by revolving a region about the *y*-axis, we use the steps listed above but integrate with respect to *y* instead of *x*.

EXAMPLE 6   The region bounded by the parabola $y = x^2$ and the line $y = 2x$ in the first quadrant is revolved about the *y*-axis to generate a solid. Find the volume of the solid.

Solution

Step 1: Draw the region and sketch a line segment across it perpendicular to the axis of revolution, in this case the y-axis (Fig. 5.22).

Step 2: The line and parabola intersect at $y = 0$ and $y = 4$, so the limits of integration are $c = 0$ and $d = 4$.

Step 3: The radii of the washer swept out by the line segment are $R(y) = \sqrt{y}$, $r(y) = y/2$ (Figs. 5.22 and 5.23).

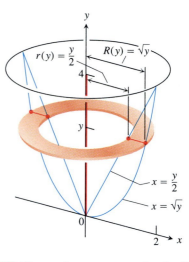

5.22 The region, limits of integration, and radii in Example 6.

5.23 The washer swept out by the line segment in Fig. 5.22.

Step 4:

$$V = \int_c^d \pi \left([R(y)]^2 - [r(y)]^2\right) dy \qquad \text{Eq. (3) with } y \text{ in place of } x$$

$$= \int_0^4 \pi \left(\left[\sqrt{y}\right]^2 - \left[\frac{y}{2}\right]^2\right) dy \qquad \text{Values from steps 2 and 3}$$

$$= \pi \int_0^4 \left(y - \frac{y^2}{4}\right) dy = \pi \left[\frac{y^2}{2} - \frac{y^3}{12}\right]_0^4 = \frac{8}{3}\pi \qquad \square$$

EXAMPLE 7 The region in the first quadrant enclosed by the parabola $y = x^2$, the y-axis, and the line $y = 1$ is revolved about the line $x = 3/2$ to generate a solid. Find the volume of the solid.

Solution

Step 1: Draw the region and sketch a line segment across it perpendicular to the axis of revolution, in this case the line $x = 3/2$ (Fig. 5.24).

Step 2: The limits of integration are $y = 0$ to $y = 1$.

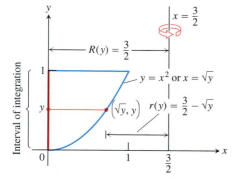

5.24 The region, limits of integration, and radii in Example 7.

Step 3: The radii of the washer swept out by the line segment are $R(y) = 3/2$, $r(y) = (3/2) - \sqrt{y}$ (Figs. 5.24 and 5.25).

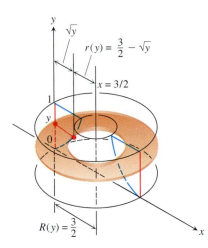

5.25 The washer swept out by the line segment in Fig. 5.24.

Step 4:

$$V = \int_c^d \pi \left([R(y)]^2 - [r(y)]^2\right) dy$$

Eq. (3) with y in place of x

$$= \int_0^1 \pi \left(\left[\frac{3}{2}\right]^2 - \left[\frac{3}{2} - \sqrt{y}\right]^2\right) dy$$

$$= \pi \int_0^1 (3\sqrt{y} - y) \, dy = \pi \left[2y^{3/2} - \frac{y^2}{2}\right]_0^1 = \frac{3\pi}{2}$$ ❑

Exercises 5.3

Volumes by the Disk Method

In Exercises 1–4, find the volume of the solid generated by revolving the shaded region about the given axis.

1. About the x-axis

2. About the y-axis

3. About the y-axis

4. About the x-axis

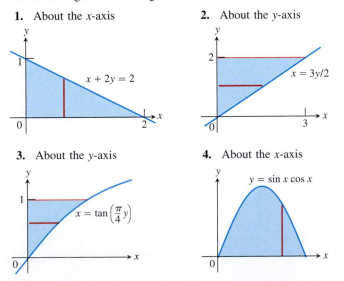

Find the volumes of the solids generated by revolving the regions bounded by the lines and curves in Exercises 5–10 about the x-axis.

5. $y = x^2$, $y = 0$, $x = 2$

6. $y = x^3$, $y = 0$, $x = 2$

7. $y = \sqrt{9 - x^2}$, $y = 0$

8. $y = x - x^2$, $y = 0$

9. $y = \sqrt{\cos x}$, $0 \le x \le \pi/2$, $y = 0$, $x = 0$

10. $y = \sec x$, $y = 0$, $x = -\pi/4$, $x = \pi/4$

In Exercises 11 and 12, find the volume of the solid generated by revolving the region about the given line.

11. The region in the first quadrant bounded above by the line $y = \sqrt{2}$, below by the curve $y = \sec x \tan x$, and on the left by the y-axis, about the line $y = \sqrt{2}$

12. The region in the first quadrant bounded above by the line $y = 2$, below by the curve $y = 2 \sin x$, $0 \le x \le \pi/2$, and on the left by the y-axis, about the line $y = 2$

Find the volumes of the solids generated by revolving the regions bounded by the lines and curves in Exercises 13–18 about the y-axis.

13. $x = \sqrt{5}\, y^2$, $x = 0$, $y = -1$, $y = 1$

14. $x = y^{3/2}$, $x = 0$, $y = 2$

15. $x = \sqrt{2 \sin 2y}, \quad 0 \le y \le \pi/2, \quad x = 0$

16. $x = \sqrt{\cos (\pi y/4)}, \quad -2 \le y \le 0, \quad x = 0$

17. $x = 2/(y + 1), \quad x = 0, \quad y = 0, \quad y = 3$

18. $x = \sqrt{2y}/(y^2 + 1), \quad x = 0, \quad y = 1$

Volumes by the Washer Method

Find the volumes of the solids generated by revolving the shaded regions in Exercises 19 and 20 about the indicated axes.

19. The x-axis

20. The y-axis

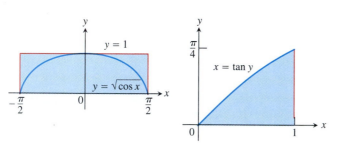

Find the volumes of the solids generated by revolving the regions bounded by the lines and curves in Exercises 21–28 about the x-axis.

21. $y = x, \quad y = 1, \quad x = 0$ **22.** $y = 2x, \quad y = x, \quad x = 1$

23. $y = 2\sqrt{x}, \quad y = 2, \quad x = 0$

24. $y = -\sqrt{x}, \quad y = -2, \quad x = 0$

25. $y = x^2 + 1, \quad y = x + 3$

26. $y = 4 - x^2, \quad y = 2 - x$

27. $y = \sec x, \quad y = \sqrt{2}, \quad -\pi/4 \le x \le \pi/4$

28. $y = \sec x, \quad y = \tan x, \quad x = 0, \quad x = 1$

In Exercises 29–34, find the volume of the solid generated by revolving each region about the y-axis.

29. The region enclosed by the triangle with vertices $(1, 0)$, $(2, 1)$, and $(1, 1)$

30. The region enclosed by the triangle with vertices $(0, 1)$, $(1, 0)$, and $(1, 1)$

31. The region in the first quadrant bounded above by the parabola $y = x^2$, below by the x-axis, and on the right by the line $x = 2$

32. The region bounded above by the curve $y = \sqrt{x}$ and below by the line $y = x$

33. The region in the first quadrant bounded on the left by the circle $x^2 + y^2 = 3$, on the right by the line $x = \sqrt{3}$, and above by the line $y = \sqrt{3}$

34. The region bounded on the left by the line $x = 4$ and on the right by the circle $x^2 + y^2 = 25$

In Exercises 35 and 36, find the volume of the solid generated by revolving each region about the given axis.

35. The region in the first quadrant bounded above by the curve

$y = x^2$, below by the x-axis, and on the right by the line $x = 1$, about the line $x = -1$

36. The region in the second quadrant bounded above by the curve $y = -x^3$, below by the x-axis, and on the left by the line $x = -1$, about the line $x = -2$

Volumes of Solids of Revolution

37. Find the volume of the solid generated by revolving the region bounded by $y = \sqrt{x}$ and the lines $y = 2$ and $x = 0$ about

  a)   the x-axis;

  b)   the y-axis;

  c)   the line $y = 2$;

  d)   the line $x = 4$.

38. Find the volume of the solid generated by revolving the triangular region bounded by the lines $y = 2x$, $y = 0$, and $x = 1$ about

  a)   the line $x = 1$;

  b)   the line $x = 2$.

39. Find the volume of the solid generated by revolving the region bounded by the parabola $y = x^2$ and the line $y = 1$ about

  a)   the line $y = 1$;

  b)   the line $y = 2$;

  c)   the line $y = -1$.

40. By integration, find the volume of the solid generated by revolving the triangular region with vertices $(0, 0)$, $(b, 0)$, $(0, h)$ about

  a)   the x-axis;

  b)   the y-axis.

41. *Designing a wok.* You are designing a wok frying pan that will be shaped like a spherical bowl with handles. A bit of experimentation at home persuades you that you can get one that holds about 3 L if you make it 9 cm deep and give the sphere a radius of 16 cm. To be sure, you picture the wok as a solid of revolution, as shown here, and calculate its volume with an integral. To the nearest cubic centimeter, what volume do you really get? $(1 \text{ L} = 1000 \text{ cm}^3)$

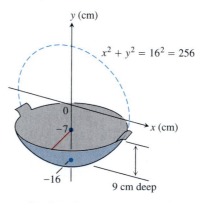

42. *Designing a plumb bob.* Having been asked to design a brass plumb bob that will weigh in the neighborhood of 190 g, you decide to shape it like the solid of revolution shown here. Find the plumb bob's volume. If you specify a brass that weighs 8.5 g/cm³, how much will the plumb bob weigh (to the nearest gram)?

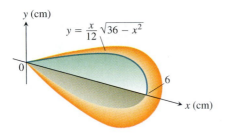

$$y = \frac{x}{12}\sqrt{36 - x^2}$$

43. The arch $y = \sin x$, $0 \le x \le \pi$, is revolved about the line $y = c$, $0 \le c \le 1$, to generate the solid in Fig. 5.26.

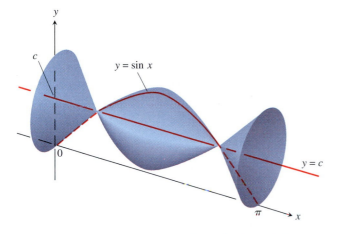

$y = \sin x$

$y = c$

5.26 Exercise 43 asks for the value of c that minimizes the volume of this solid.

a) Find the value of c that minimizes the volume of the solid. What is the minimum value?

b) What value of c in [0, 1] maximizes the volume of the solid?

c) **GRAPHER** Graph the solid's volume as a function of c, first for $0 \le c \le 1$ and then on a larger domain. What happens to the volume of the solid as c moves away from [0, 1]? Does this make sense physically? Give reasons for your answer.

44. *An auxiliary fuel tank.* You are designing an auxiliary fuel tank that will fit under a helicopter's fuselage to extend its range. After some experimentation at your drawing board, you decide to shape the tank like the surface generated by revolving the curve $y = 1 - (x^2/16)$, $-4 \le x \le 4$, about the x-axis (dimensions in feet).

a) How many cubic feet of fuel will the tank hold (to the nearest cubic foot)?

b) A cubic foot holds 7.481 gal. If the helicopter gets 2 mi to the gallon, how many additional miles will the helicopter be able to fly once the tank is installed (to the nearest mile)?

45. *The volume of a torus.* The disk $x^2 + y^2 \le a^2$ is revolved about the line $x = b$ $(b > a)$ to generate a solid shaped like a doughnut and called a *torus*. Find its volume. (*Hint:* $\int_{-a}^{a} \sqrt{a^2 - y^2}\, dy = \pi a^2/2$, since it is the area of a semicircle of radius a.)

46. a) A hemispherical bowl of radius a contains water to a depth h. Find the volume of water in the bowl.

b) (Related rates) Water runs into a sunken concrete hemispherical bowl of radius 5 m at the rate of 0.2 m³/sec. How fast is the water level in the bowl rising when the water is 4 m deep?

47. *Testing the consistency of the calculus definition of volume.* The volume formulas in this section are all consistent with the standard formulas from geometry.

a) As a case in point, show that if you revolve the region enclosed by the semicircle $y = \sqrt{a^2 - x^2}$ and the x-axis about the x-axis to generate a solid sphere, the disk formula for volume (Eq. 1) will give $(4/3)\pi a^3$ just as it should.

b) Use calculus to find the volume of a right circular cone of height h and base radius r.

5.4

Cylindrical Shells

When we need to find the volume of a solid of revolution, cylindrical shells sometimes work better than washers (Fig. 5.27, on the following page). In part, the reason is that the formula they lead to does not require squaring.

The Shell Formula

Suppose we revolve the tinted region in Fig. 5.28 (on the following page) about the y-axis to generate a solid. To estimate the volume of the solid, we can approximate the region with rectangles based on a partition P of the interval $[a, b]$ over which the region stands. The typical approximating rectangle is Δx_k units wide by $f(c_k)$ units high, where c_k is the midpoint of the rectangle's base. A formula from geometry tells

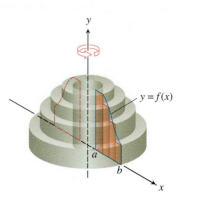

5.27 A solid of revolution approximated by cylindrical shells.

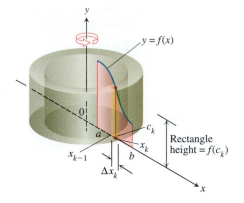

5.28 The shell swept out by the kth rectangle.

us that the volume of the shell swept out by the rectangle is

$$\Delta V_k = 2\pi \times \text{average shell radius} \times \text{shell height} \times \text{thickness},$$

which in our case is

$$\Delta V_k = 2\pi \times c_k \times f(c_k) \times \Delta x_k.$$

We approximate the volume of the solid by adding the volumes of the shells swept out by the n rectangles based on P:

$$V \approx \sum_{k=1}^{n} \Delta V_k = \sum_{k=1}^{n} 2\pi c_k f(c_k) \Delta x_k. \qquad \text{A Riemann sum}$$

The limit of this sum as $\|P\| \to 0$ gives the volume of the solid:

$$V = \lim_{\|P\| \to 0} \sum_{k=1}^{n} 2\pi c_k f(c_k) \Delta x_k = \int_{a}^{b} 2\pi x f(x) \, dx.$$

The Shell Formula for Revolution About the y-axis

The volume of the solid generated by revolving the region between the x-axis and the graph of a continuous function $y = f(x) \geq 0, 0 \leq a \leq x \leq b$, about the y-axis is

$$V = \int_{a}^{b} 2\pi \left(\begin{array}{c} \text{shell} \\ \text{radius} \end{array} \right) \left(\begin{array}{c} \text{shell} \\ \text{height} \end{array} \right) dx = \int_{a}^{b} 2\pi x f(x) \, dx. \qquad (1)$$

5.29 The region, shell dimensions, and interval of integration in Example 1.

EXAMPLE 1   The region bounded by the curve $y = \sqrt{x}$, the x-axis, and the line $x = 4$ is revolved about the y-axis to generate a solid. Find the volume of the solid.

Solution

Step 1: Sketch the region and draw a line segment across it *parallel* to the axis of revolution (Fig. 5.29). Label the segment's height (shell height) and distance from

One way to remember Eq. (1) is to imagine cutting and unrolling a cylindrical shell to get a (nearly) flat rectangular solid.

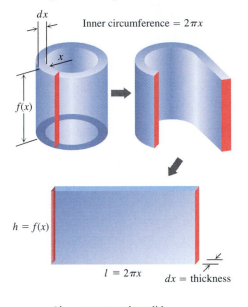

Almost a rectangular solid
$V \approx$ length $\times$ height $\times$ thickness
$\approx 2\pi x \cdot f(x) \cdot dx$

the axis of revolution (shell radius). The width of the segment is the shell thickness dx. (We drew the shell in Fig. 5.30, but you need not do that.)

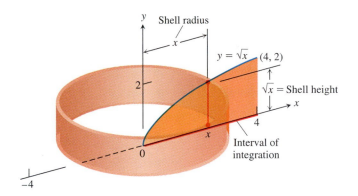

5.30 The shell swept out by the line segment in Fig. 5.29.

Step 2: Find the limits of integration: x runs from $a = 0$ to $b = 4$.

$$V = \int_a^b 2\pi \left(\begin{array}{c} \text{shell} \\ \text{radius} \end{array} \right) \left(\begin{array}{c} \text{shell} \\ \text{height} \end{array} \right) dx \qquad \text{Eq. (1)}$$

$$= \int_0^4 2\pi (x)(\sqrt{x})\, dx \qquad \begin{array}{c} \text{Values from steps} \\ \text{1 and 2} \end{array}$$

$$= 2\pi \int_0^4 x^{3/2}\, dx = 2\pi \left[\frac{2}{5} x^{5/2} \right]_0^4 = \frac{128\pi}{5}$$

Equation (1) is for vertical axes of revolution. For horizontal axes, we replace the x's with y's.

The Shell Formula for Revolution About the x-axis

$$V = \int_c^d 2\pi \left(\begin{array}{c} \text{shell} \\ \text{radius} \end{array} \right) \left(\begin{array}{c} \text{shell} \\ \text{height} \end{array} \right) dy = \int_c^d 2\pi y f(y)\, dy \qquad (2)$$

(for $f(y) \geq 0$ and $0 \leq c \leq y \leq d$)

EXAMPLE 2   The region bounded by the curve $y = \sqrt{x}$, the x-axis, and the line $x = 4$ is revolved about the x-axis to generate a solid. Find the volume of the solid.

Solution

Step 1: Sketch the region and draw a line segment across it parallel to the axis of revolution (Fig. 5.31). Label the segment's length (shell height) and distance from the axis of revolution (shell radius). The width of the segment is the shell thickness dy. (We drew the shell in Fig. 5.32, shown on the following page, but you need not do that.)

5.31 The region, shell dimensions, and interval of integration in Example 2.

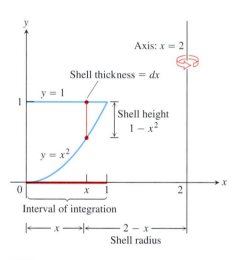

Step 2: Identify the limits of integration: y runs from $c = 0$ to $d = 2$.

Step 3: Integrate to find the volume.

$$V = \int_c^d 2\pi \left(\begin{array}{c} \text{shell} \\ \text{radius} \end{array} \right) \left(\begin{array}{c} \text{shell} \\ \text{height} \end{array} \right) dy \qquad \text{Eq. (2)}$$

$$= \int_0^2 2\pi (y)(4 - y^2) \, dy \qquad \text{Values from steps 1 and 2}$$

$$= 2\pi \left[2y^2 - \frac{y^4}{4} \right]_0^2 = 8\pi$$

This agrees with the disk method of calculation in Section 5.3, Example 1. ❏

5.32 The shell swept out by the line segment in Fig. 5.31.

How to Use the Shell Method

Regardless of the position of the axis of revolution (horizontal or vertical), the steps for implementing the shell method are these:

1. *Draw the region and sketch a line segment* across it *parallel* to the axis of revolution. *Label* the segment's height or length (shell height), distance from the axis of revolution (shell radius), and width (shell thickness).
2. *Find* the limits of integration.
3. *Integrate* the product 2π (shell radius) (shell height) with respect to the appropriate variable (x or y) to find the volume.

5.33 The region, shell dimensions, and interval of integration in Example 3.

In the next example, the axis of revolution is the vertical line $x = 2$.

EXAMPLE 3   The region in the first quadrant bounded by the parabola $y = x^2$, the y-axis, and the line $y = 1$ is revolved about the line $x = 2$ to generate a solid. Find the volume of the solid.

Solution

Step 1: Draw a line segment across the region parallel to the axis of revolution (the line $x = 2$) (Fig. 5.33). Label the segment's height (shell height), distance from the axis of revolution (shell radius), and width (in this case, dx). (We drew the shell in Fig. 5.34, but you need not do that.)

Step 2: The limits of integration: x runs from $a = 0$ to $b = 1$.

Step 3:

$$V = \int_a^b 2\pi \left(\begin{array}{c} \text{shell} \\ \text{radius} \end{array} \right) \left(\begin{array}{c} \text{shell} \\ \text{height} \end{array} \right) dx \qquad \text{Eq. (1)}$$

$$= \int_0^1 2\pi (2 - x)(1 - x^2) \, dx \qquad \text{Values from steps 1 and 2}$$

$$= 2\pi \int_0^1 (2 - x - 2x^2 + x^3) \, dx$$

$$= \frac{13\pi}{6}$$

❏

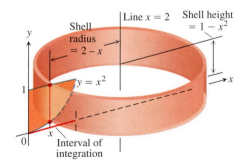

5.34 The shell swept out by the line segment in Fig. 5.33.

Table 5.1 summarizes the washer and shell methods for the solid generated by revolving the region bounded by $y = x$ and $y = x^2$ about the coordinate axes. For this particular region, both methods work well for both axes of revolution. But this is not always the case. When a region is revolved about the y-axis, for example, and washers are used, we must integrate with respect to y. However, it may not be possible to express the integrand in terms of y. In such a case, the shell method allows us to integrate with respect to x instead.

The washer and shell methods for calculating volumes of solids of revolution always agree. In Section 6.1 (Exercise 52), we will be able to prove the equivalence for a broad class of solids.

Table 5.1 Washers vs. shells

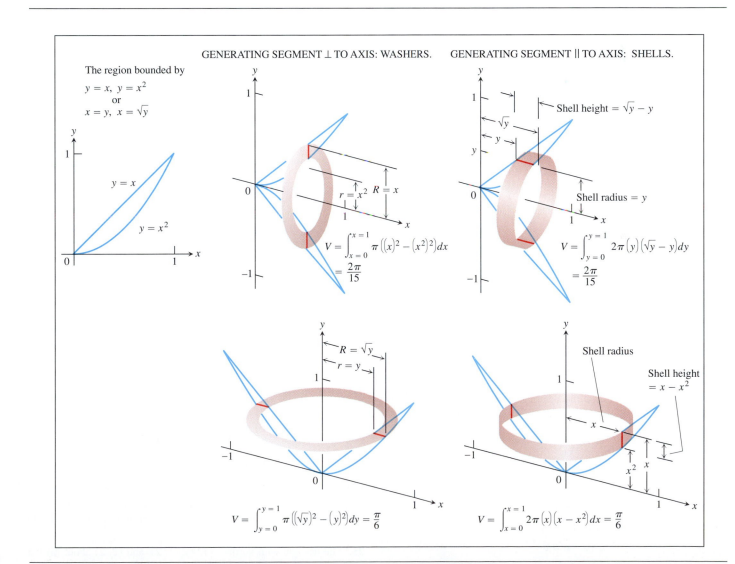

GENERATING SEGMENT ⊥ TO AXIS: WASHERS.   GENERATING SEGMENT ∥ TO AXIS: SHELLS.

The region bounded by
$y = x,\ y = x^2$
or
$x = y,\ x = \sqrt{y}$

$y = x$

$y = x^2$

$r = x^2\quad R = x$

$$V = \int_{x=0}^{x=1} \pi\left((x)^2 - (x^2)^2\right)dx$$
$$= \frac{2\pi}{15}$$

Shell height $= \sqrt{y} - y$

Shell radius $= y$

$$V = \int_{y=0}^{y=1} 2\pi\,(y)\left(\sqrt{y} - y\right)dy$$
$$= \frac{2\pi}{15}$$

$R = \sqrt{y}$
$r = y$

$$V = \int_{y=0}^{y=1} \pi\left((\sqrt{y})^2 - (y)^2\right)dy = \frac{\pi}{6}$$

Shell radius

Shell height
$= x - x^2$

$$V = \int_{x=0}^{x=1} 2\pi\,(x)\left(x - x^2\right)dx = \frac{\pi}{6}$$

Exercises 5.4

In Exercises 1–6, use the shell method to find the volumes of the solids generated by revolving the shaded region about the indicated axis.

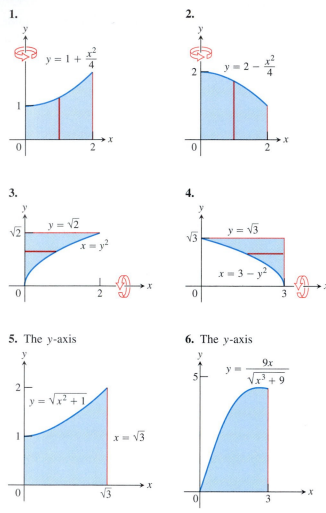

1.

$y = 1 + \dfrac{x^2}{4}$

2.

$y = 2 - \dfrac{x^2}{4}$

3.

$y = \sqrt{2}$

$x = y^2$

4.

$y = \sqrt{3}$

$x = 3 - y^2$

5. The y-axis

$y = \sqrt{x^2 + 1}$

$x = \sqrt{3}$

6. The y-axis

$y = \dfrac{9x}{\sqrt{x^3 + 9}}$

Use the shell method to find the volumes of the solids generated by revolving the regions bounded by the curves and lines in Exercises 7–14 about the y-axis.

7. $y = x$, $y = -x/2$, $x = 2$

8. $y = 2x$, $y = x/2$, $x = 1$

9. $y = x^2$, $y = 2 - x$, $x = 0$, for $x \geq 0$

10. $y = 2 - x^2$, $y = x^2$, $x = 0$

11. $y = \sqrt{x}$, $y = 0$, $x = 4$

12. $y = 2x - 1$, $y = \sqrt{x}$, $x = 0$

13. $y = 1/x$, $y = 0$, $x = 1/2$, $x = 2$

14. $y = 3/(2\sqrt{x})$, $y = 0$, $x = 1$, $x = 4$

Use the shell method to find the volumes of the solids generated by revolving the regions bounded by the curves and lines in Exercises 15–22 about the x-axis.

15. $x = \sqrt{y}$, $x = -y$, $y = 2$

16. $x = y^2$, $x = -y$, $y = 2$

17. $x = 2y - y^2$, $x = 0$ **18.** $x = 2y - y^2$, $x = y$

19. $y = |x|$, $y = 1$ **20.** $y = x$, $y = 2x$, $y = 2$

21. $y = \sqrt{x}$, $y = 0$, $y = x - 2$

22. $y = \sqrt{x}$, $y = 0$, $y = 2 - x$

In Exercises 23 and 24, use the shell method to find the volumes of the solids generated by revolving the shaded regions about the indicated axes.

23. a) The x-axis
   b) The line $y = 1$
   c) The line $y = 8/5$
   d) The line $y = -2/5$

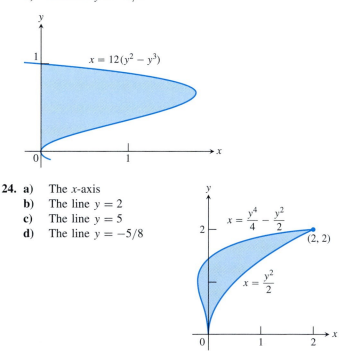

$x = 12(y^2 - y^3)$

24. a) The x-axis
   b) The line $y = 2$
   c) The line $y = 5$
   d) The line $y = -5/8$

$x = \dfrac{y^4}{4} - \dfrac{y^2}{2}$

$x = \dfrac{y^2}{2}$

$(2, 2)$

In Exercises 25–32, find the volumes of the solids generated by revolving the regions about the given axes. If you think it would be better to use disks or washers in any given instance, feel free to do so.

25. The triangle with vertices $(1, 1)$, $(1, 2)$, and $(2, 2)$ about (a) the x-axis; (b) the y-axis; (c) the line $x = 10/3$; (d) the line $y = 1$

26. The region in the first quadrant bounded by the curve $x = y - y^3$ and the y-axis about (a) the x-axis; (b) the line $y = 1$

27. The region in the first quadrant bounded by $x = y - y^3$, $x = 1$, and $y = 1$ about (a) the x-axis; (b) the y-axis; (c) the line $x = 1$; (d) the line $y = 1$

28. The triangular region bounded by the lines $2y = x + 4$, $y = x$, and $x = 0$ about (a) the x-axis; (b) the y-axis; (c) the line $x = 4$; (d) the line $y = 8$

29. The region in the first quadrant bounded by $y = x^3$ and $y = 4x$ about (a) the x-axis; (b) the line $y = 8$

30. The region bounded by $y = \sqrt{x}$ and $y = x^2/8$ about (a) the x-axis; (b) the y-axis

31. The region bounded by $y = 2x - x^2$ and $y = x$ about (a) the y-axis; (b) the line $x = 1$

32. The region bounded by $y = \sqrt{x}$, $y = 2$, $x = 0$ about (a) the x-axis; (b) the y-axis; (c) the line $x = 4$; (d) the line $y = 2$

33. The region in the first quadrant that is bounded above by the curve $y = 1/x^{1/4}$, on the left by the line $x = 1/16$, and below by the line $y = 1$, is revolved about the x-axis to generate a solid. Find the volume of the solid by (a) the washer method; (b) the shell method.

34. The region in the first quadrant that is bounded above by the curve $y = 1/\sqrt{x}$, on the left by the line $y = 1/4$, and below by the line $y = 1$ is revolved about the y-axis to generate a solid. Find the volume of the solid by (a) the washer method; (b) the shell method.

35. Let $f(x) = \begin{cases} (\sin x)/x, & 0 < x \le \pi \\ 1, & x = 0. \end{cases}$

   a) Show that $xf(x) = \sin x$, $0 \le x \le \pi$.
   b) Find the volume of the solid generated by revolving the shaded region about the y-axis.

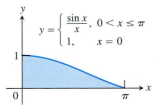

$$y = \begin{cases} \dfrac{\sin x}{x}, & 0 < x \le \pi \\ 1, & x = 0 \end{cases}$$

36. Let $g(x) = \begin{cases} (\tan x)^2/x, & 0 < x \le \pi/4 \\ 0, & x = 0. \end{cases}$

   a) Show that $xg(x) = (\tan x)^2$, $0 \le x \le \pi/4$.
   b) Find the volume of the solid generated by revolving the shaded region about the y-axis.

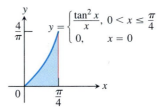

$$y = \begin{cases} \dfrac{\tan^2 x}{x}, & 0 < x \le \dfrac{\pi}{4} \\ 0, & x = 0 \end{cases}$$

37. The region shown here is to be revolved about the x-axis to generate a solid. Which of the methods (disk, washer, shell) could you use to find the volume of the solid? How many integrals would be required in each case? Explain.

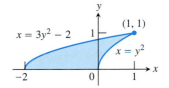

38. The region shown here is to be revolved about the y-axis to generate a solid. Which of the methods (disk, washer, shell) could you use to find the volume of the solid? How many integrals would be required in each case? Give reasons for your answers.

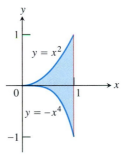

39. Suppose that the function $f(x)$ is nonnegative and continuous for $x \ge 0$. Suppose also that, for every positive number b, revolving the region enclosed by the graph of f, the coordinate axes, and the line $x = b$ about the y-axis generates a solid of volume $2\pi b^3$. Find $f(x)$.

5.5

Lengths of Plane Curves

We approximate the length of a curved path in the plane the way we use a ruler to estimate the length of a curved road on a map, by measuring from point to point with straight-line segments and adding the results. There is a limit to the accuracy of such an estimate, however, imposed in part by how accurately we measure and in part by how many line segments we use.

With calculus we can usually do a better job because we can imagine using straight-line segments as short as we please, each set of segments making a polygonal path that fits the curve more tightly than before. When we proceed this way, with a smooth curve, the lengths of the polygonal paths approach a limit we can calculate with an integral.

The Basic Formula

Suppose we want to find the length of the curve $y = f(x)$ from $x = a$ to $x = b$. We partition $[a, b]$ in the usual way and connect the corresponding points on the curve with line segments to form a polygonal path that approximates the curve (Fig. 5.35). If we can find a formula for the length of the path, we will have a formula for approximating the length of the curve.

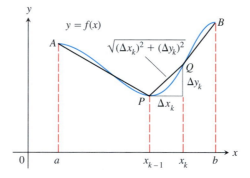

5.35 A typical segment PQ of a polygonal path approximating the curve AB.

The length of a typical line segment PQ (see the figure) is $\sqrt{(\Delta x_k)^2 + (\Delta y_k)^2}$. The length of the curve is therefore approximated by the sum

$$\sum_{k=1}^{n} \sqrt{(\Delta x_k)^2 + (\Delta y_k)^2}. \tag{1}$$

We expect the approximation to improve as the partition of $[a, b]$ becomes finer, and we would like to show that the sums in (1) approach a calculable limit as the norm of the partition goes to zero. To show this, we rewrite the sum in (1) in a form to which we can apply the Integral Existence Theorem from Chapter 4. Our starting point is the Mean Value Theorem for derivatives.

Definition

A function with a continuous first derivative is said to be **smooth** and its graph is called a **smooth curve.**

If f is smooth, by the Mean Value Theorem there is a point $(c_k, f(c_k))$ on the curve between P and Q where the tangent is parallel to the segment PQ (Fig. 5.36). At this point

$$f'(c_k) = \frac{\Delta y_k}{\Delta x_k}, \qquad \text{or} \qquad \Delta y_k = f'(c_k)\Delta x_k.$$

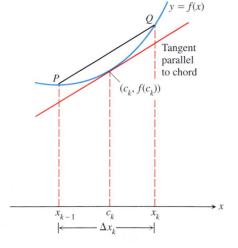

5.36 Enlargement of the arc PQ in Fig. 5.35.

With this substitution for Δy_k, the sums in (1) take the form

$$\sum_{k=1}^{n} \sqrt{(\Delta x_k)^2 + (f'(c_k)\Delta x_k)^2} = \sum_{k=1}^{n} \sqrt{1 + (f'(c_k))^2}\,\Delta x_k. \qquad \text{A Riemann sum}$$

Because $\sqrt{1 + (f'(x))^2}$ is continuous on $[a, b]$, the limit of the sums on the right as the norm of the partition goes to zero is $\int_a^b \sqrt{1 + (f'(x))^2}\,dx$. We define the length of the curve to be the value of this integral.

Definition

If f is smooth on $[a, b]$, the **length** of the curve $y = f(x)$ from a to b is

$$L = \int_a^b \sqrt{1 + \left(\frac{dy}{dx}\right)^2}\,dx = \int_a^b \sqrt{1 + (f'(x))^2}\,dy. \qquad (2)$$

EXAMPLE 1   Find the length of the curve

$$y = \frac{4\sqrt{2}}{3}x^{3/2} - 1, \qquad 0 \le x \le 1.$$

Solution   We use Eq. (2) with $a = 0, b = 1$, and

$$y = \frac{4\sqrt{2}}{3}x^{3/2} - 1$$

$$\frac{dy}{dx} = \frac{4\sqrt{2}}{3} \cdot \frac{3}{2}x^{1/2} = 2\sqrt{2}x^{1/2}$$

$$\left(\frac{dy}{dx}\right)^2 = \left(2\sqrt{2}x^{1/2}\right)^2 = 8x.$$

The length of the curve from $x = 0$ to $x = 1$ is

$$L = \int_0^1 \sqrt{1 + \left(\frac{dy}{dx}\right)^2}\,dx = \int_0^1 \sqrt{1 + 8x}\,dx \qquad \begin{array}{l}\text{Eq. (2) with} \\ a = 0, b = 1\end{array}$$

$$= \frac{2}{3} \cdot \frac{1}{8}(1 + 8x)^{3/2}\Big]_0^1 = \frac{13}{6}. \qquad \begin{array}{l}\text{Let } u = 1 + 8x, \\ \text{integrate, and} \\ \text{replace } u \text{ by} \\ 1 + 8x.\end{array}$$

Dealing with Discontinuities in *dy/dx*

At a point on a curve where dy/dx fails to exist, dx/dy may exist and we may be able to find the curve's length by expressing x as a function of y and applying the following analogue of Eq. (2):

Formula for the Length of a Smooth Curve $x = g(y)$, $c \le y \le d$

$$L = \int_c^d \sqrt{1 + \left(\frac{dx}{dy}\right)^2}\,dy = \int_c^d \sqrt{1 + (g'(y))^2}\,dy. \qquad (3)$$

5.37 The graph of $y = (x/2)^{2/3}$ from $x = 0$ to $x = 2$ is also the graph of $x = 2y^{3/2}$ from $y = 0$ to $y = 1$.

EXAMPLE 2   Find the length of the curve $y = (x/2)^{2/3}$ from $x = 0$ to $x = 2$.

Solution   The derivative

$$\frac{dy}{dx} = \frac{2}{3}\left(\frac{x}{2}\right)^{-1/3}\left(\frac{1}{2}\right) = \frac{1}{3}\left(\frac{2}{x}\right)^{1/3}$$

is not defined at $x = 0$, so we cannot find the curve's length with Eq. (2). We therefore rewrite the equation to express x in terms of y:

$$y = \left(\frac{x}{2}\right)^{2/3}$$

$$y^{3/2} = \frac{x}{2} \qquad \begin{array}{l}\text{Raise both sides}\\\text{to the power 3/2.}\end{array}$$

$$x = 2y^{3/2}. \qquad \text{Solve for } x.$$

From this we see that the curve whose length we want is also the graph of $x = 2y^{3/2}$ from $y = 0$ to $y = 1$ (Fig. 5.37).

The derivative

$$\frac{dx}{dy} = 2\left(\frac{3}{2}\right)y^{1/2} = 3y^{1/2}$$

is continuous on $[0, 1]$. We may therefore use Eq. (3) to find the curve's length:

$$L = \int_c^d \sqrt{1 + \left(\frac{dx}{dy}\right)^2}\,dy = \int_0^1 \sqrt{1 + 9y}\,dy \qquad \begin{array}{l}\text{Eq. (3) with}\\ c = 0, d = 1\end{array}$$

$$= \frac{1}{9} \cdot \frac{2}{3}(1 + 9y)^{3/2}\Big]_0^1 \qquad \begin{array}{l}\text{Let } u = 1 + 9y,\\ du/9 = dy,\\ \text{integrate, and}\\ \text{substitute back.}\end{array}$$

$$= \frac{2}{27}(10\sqrt{10} - 1) \approx 2.27. \qquad \square$$

The Short Differential Formula

The equations

$$L = \int_a^b \sqrt{1 + \left(\frac{dy}{dx}\right)^2}\,dx \qquad \text{and} \qquad L = \int_c^d \sqrt{1 + \left(\frac{dx}{dy}\right)^2}\,dy \qquad (4)$$

are often written with differentials instead of derivatives. This is done formally by thinking of the derivatives as quotients of differentials and bringing the dx and dy inside the radicals to cancel the denominators. In the first integral we have

$$\sqrt{1 + \left(\frac{dy}{dx}\right)^2}\,dx = \sqrt{1 + \frac{dy^2}{dx^2}}\,dx = \sqrt{dx^2 + \frac{dy^2}{dx^2}dx^2} = \sqrt{dx^2 + dy^2}.$$

In the second integral we have

$$\sqrt{1 + \left(\frac{dx}{dy}\right)^2}\,dy = \sqrt{1 + \frac{dx^2}{dy^2}}\,dy = \sqrt{dy^2 + \frac{dx^2}{dy^2}dy^2} = \sqrt{dx^2 + dy^2}.$$

Thus the integrals in (4) reduce to the same differential formula:

$$L = \int_a^b \sqrt{dx^2 + dy^2}. \qquad (5)$$

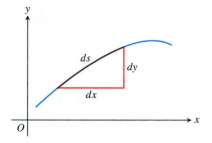

5.38 Diagram for remembering the equation $ds = \sqrt{dx^2 + dy^2}$.

Of course, dx and dy must be expressed in terms of a common variable, and appropriate limits of integration must be found before the integration in Eq. (5) is performed.

We can shorten Eq. (5) still further. Think of dx and dy as two sides of a small triangle whose "hypotenuse" is $ds = \sqrt{dx^2 + dy^2}$ (Fig. 5.38). The differential ds is then regarded as a differential of arc length that can be integrated between appropriate limits to give the length of the curve. With $\sqrt{dx^2 + dy^2}$ set equal to ds, the integral in Eq. (5) simply becomes the integral of ds.

Definition

The Arc Length Differential and the Differential Formula for Arc Length

$$ds = \sqrt{dx^2 + dy^2} \qquad\qquad L = \int ds$$

arc length differential   differential formula for arc length

---

✳ Curves with Infinite Length

Curve 1   Curve 2   Curve 3   Curve 4

5.39 The first four polygonal approximations in the construction of Helga von Koch's snowflake.

As you may recall from Section 2.6, Helga von Koch's snowflake curve K is the limit curve of an infinite sequence $C_1, C_2, \ldots, C_n, \ldots$ of "triangular" polygonal curves. Figure 5.39 shows the first four curves in the sequence. Each time we introduce a new vertex in the construction process, it remains as a vertex in all subsequent curves and becomes a point on the limit curve K. This means that each of the C's is itself a polygonal approximation of K—the endpoints of its sides all belonging to K. The length of K should therefore be the limit of the lengths of the curves C_n. At least, that is what it should be if we apply the definition of length we developed for smooth curves.

What, then, is the limit of the lengths of the curves C_n? If the original equilateral triangle C_1 has sides of length 1, the total length of C_1 is 3. To make C_2 from C_1, we replace each side of C_1 by four segments, each of which is one-third as long as the original side. The total length of C_2 is therefore 3(4/3). To get the length of C_3, we multiply by 4/3 again. We do so again to get the length of C_4. By the time we get out to C_n, we have a curve of length $3(4/3)^{n-1}$.

Curve Number	1 2	3	$\cdots\; n$	$\cdots$
Length	$3\;\; 3\left(\dfrac{4}{3}\right)$	$3\left(\dfrac{4}{3}\right)^2$	$\cdots\; 3\left(\dfrac{4}{3}\right)^{n-1}$	$\cdots$

The length of C_{10} is nearly 40 and the length of C_{100} is greater than 7,000,000,000,000. The lengths grow too rapidly to have a finite limit. Therefore the snowflake curve has no length, or, if you prefer, infinite length.

What went wrong? Nothing. The formulas we derived for length are for the graphs of smooth functions, curves that are smooth enough to have a continuously turning tangent at every point. Helga von Koch's snowflake curve is too rough for that, and our derivative-based formulas do not apply.

Benoit Mandelbrot's theory of fractals has proved to be a rich source of curves with infinite length, curves that when magnified prove to be as rough and varied as they looked before magnification. Like coastlines on an ocean, such curves cannot be smoothed out by magnification (Fig. 5.40, on the following page).

5.40 Repeated magnifications of a fractal coastline. Like Helga Von Koch's snowflake curve, coasts like these are too rough to have a measurable length.

Exercises 5.5

Finding Integrals for Lengths of Curves

In Exercises 1–8:

a) Set up an integral for the length of the curve.

b) Graph the curve to see what it looks like.

c) Use your grapher's or computer's integral evaluator to find the curve's length numerically.

1. $y = x^2$, $-1 \le x \le 2$

2. $y = \tan x$, $-\pi/3 \le x \le 0$

3. $x = \sin y$, $0 \le y \le \pi$

4. $x = \sqrt{1 - y^2}$, $-1/2 \le y \le 1/2$

5. $y^2 + 2y = 2x + 1$ from $(-1, -1)$ to $(7, 3)$

6. $y = \sin x - x \cos x$, $0 \le x \le \pi$

7. $y = \int_0^x \tan t \, dt$, $0 \le x \le \pi/6$

8. $x = \int_0^y \sqrt{\sec^2 t - 1} \, dt$, $-\pi/3 \le y \le \pi/4$

Finding Lengths of Curves

Find the lengths of the curves in Exercises 9–18. If you have a grapher, you may want to graph these curves to see what they look like.

9. $y = (1/3)(x^2 + 2)^{3/2}$ from $x = 0$ to $x = 3$

10. $y = x^{3/2}$ from $x = 0$ to $x = 4$

11. $x = (y^3/3) + 1/(4y)$ from $y = 1$ to $y = 3$
   (*Hint:* $1 + (dx/dy)^2$ is a perfect square.)

12. $x = (y^{3/2}/3) - y^{1/2}$ from $y = 1$ to $y = 9$
   (*Hint:* $1 + (dx/dy)^2$ is a perfect square.)

13. $x = (y^4/4) + 1/(8y^2)$ from $y = 1$ to $y = 2$
   (*Hint:* $1 + (dx/dy)^2$ is a perfect square.)

14. $x = (y^3/6) + 1/(2y)$ from $y = 2$ to $y = 3$
   (*Hint:* $1 + (dx/dy)^2$ is a perfect square.)

15. $y = (3/4)x^{4/3} - (3/8)x^{2/3} + 5$, $1 \le x \le 8$

16. $y = (x^3/3) + x^2 + x + 1/(4x + 4)$, $0 \le x \le 2$

17. $x = \int_0^y \sqrt{\sec^4 t - 1} \, dt$, $-\pi/4 \le y \le \pi/4$

18. $y = \int_{-2}^x \sqrt{3t^4 - 1} \, dt$, $-2 \le x \le -1$

19. a)   Find a curve through the point $(1, 1)$ whose length integral (Eq. 2) is

$$L = \int_1^4 \sqrt{1 + \frac{1}{4x}} \, dx.$$

b)   How many such curves are there? Give reasons for your answer.

20. a) Find a curve through the point $(0, 1)$ whose length integral (Eq. 3) is

$$L = \int_1^2 \sqrt{1 + \frac{1}{y^4}} \, dy.$$

b) How many such curves are there? Give reasons for your answer.

21. Find the length of the curve

$$y = \int_0^x \sqrt{\cos 2t} \, dt$$

from $x = 0$ to $x = \pi/4$.

22. *The length of an astroid.* The graph of the equation $x^{2/3} + y^{2/3} = 1$ is one of a family of curves called *astroids* (not "asteroids") because of their starlike appearance (see the accompanying figure). Find the length of this particular astroid by finding the length of half the first-quadrant portion, $y = (1 - x^{2/3})^{3/2}$, $\sqrt{2}/4 \le x \le 1$, and multiplying by 8.

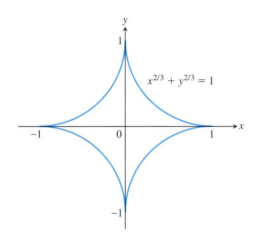

$x^{2/3} + y^{2/3} = 1$

Numerical Integration

You may have wondered why so many of the curves we have been working with have unusual formulas. The reason is that the square root $\sqrt{1 + (dy/dx)^2}$ that appears in the integrals for length and surface area almost never leads to a function whose antiderivative we can find. In fact, the square root itself is a well-known source of nonelementary integrals. Most integrals for length and surface area have to be evaluated numerically, as in Exercises 23 and 24.

23. Your metal fabrication company is bidding for a contract to make sheets of corrugated iron roofing like the one shown here. The cross sections of the corrugated sheets are to conform to the curve

$$y = \sin \frac{3\pi}{20} x, \quad 0 \le x \le 20 \text{ in.}$$

If the roofing is to be stamped from flat sheets by a process that does not stretch the material, how wide should the original material be? To find out, use numerical integration to approximate the length of the sine curve to 2 decimal places.

Original sheet   Corrugated sheet

20 in.

$y = \sin \dfrac{3\pi}{20} x$ x (in.)

24. Your engineering firm is bidding for the contract to construct the tunnel shown here. The tunnel is 300 ft long and 50 ft wide at the base. The cross section is shaped like one arch of the curve $y = 25 \cos (\pi x/50)$. Upon completion, the tunnel's inside surface (excluding the roadway) will be treated with a waterproof sealer that costs \$1.75 per square foot to apply. How much will it cost to apply the sealer? (*Hint:* Use numerical integration to find the length of the cosine curve.)

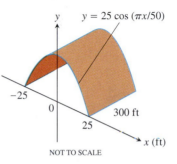

$y = 25 \cos (\pi x/50)$

300 ft

x (ft)

NOT TO SCALE

Theory and Examples

25. Is there a smooth curve $y = f(x)$ whose length over the interval $0 \le x \le a$ is always $\sqrt{2}a$? Give reasons for your answer.

26. *Using tangent fins to derive the length formula for curves.* Assume f is smooth on $[a, b]$ and partition the interval $[a, b]$ in the usual way. In each subinterval $[x_{k-1}, x_k]$ construct the *tangent fin* at the point $(x_{k-1}, f(x_{k-1}))$, shown in the figure.

a) Show that the length of the kth tangent fin over the interval $[x_{k-1}, x_k]$ equals $\sqrt{(\Delta x_k)^2 + (f'(x_{k-1})\Delta x_k)^2}$.

b) Show that

$$\lim_{n \to \infty} \sum_{k=1}^n (\text{length of } k\text{th tangent fin}) = \int_a^b \sqrt{1 + (f'(x))^2} \, dx,$$

which is the length L of the curve $y = f(x)$ from a to b.

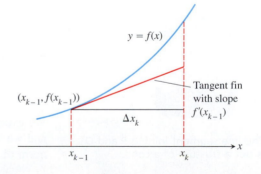

$y = f(x)$

Tangent fin with slope $f'(x_{k-1})$

$(x_{k-1}, f(x_{k-1}))$

Δx_k

x_{k-1} x_k

✿ CAS Explorations and Projects

In Exercises 27–32, use a CAS to perform the following steps for the given curve over the closed interval.

a) Plot the curve together with the polygonal path approximations for $n = 2, 4, 8$ partition points over the interval. (See Fig. 5.35.)

b) Find the corresponding approximation to the length of the curve by summing the lengths of the line segments.

c) Evaluate the length of the curve using an integral. Compare your approximations for $n = 2, 4, 8$ to the actual length given by the integral. How does the actual length compare with the approximations as n increases? Explain your answer.

27. $f(x) = \sqrt{1 - x^2}, \quad -1 \le x \le 1$

28. $f(x) = x^{1/3} + x^{2/3}, \quad 0 \le x \le 2$

29. $f(x) = \sin(\pi x^2), \quad 0 \le x \le \sqrt{2}$

30. $f(x) = x^2 \cos x, \quad 0 \le x \le \pi$

31. $f(x) = \dfrac{x - 1}{4x^2 + 1}, \quad -\dfrac{1}{2} \le x \le 1$

32. $f(x) = x^3 - x^2, \quad -1 \le x \le 1$

5.6	# Areas of Surfaces of Revolution

When you jump rope, the rope sweeps out a surface in the space around you, a surface called a surface of revolution. As you can imagine, the area of this surface depends on the rope's length and on how far away each segment of the rope swings. This section explores the relation between the area of a surface of revolution and the length and reach of the curve that generates it. The areas of more complicated surfaces will be treated in Chapter 14.

The Basic Formula

Suppose we want to find the area of the surface swept out by revolving the graph of a nonnegative function $y = f(x), a \le x \le b$, about the x-axis. We partition $[a, b]$ in the usual way and use the points in the partition to partition the graph into short arcs. Figure 5.41 shows a typical arc PQ and the band it sweeps out as part of the graph of f.

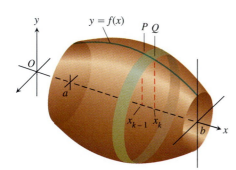

5.41 The surface generated by revolving the graph of a nonnegative function $y = f(x), a \le x \le b$, about the x-axis. The surface is a union of bands like the one swept out by the arc PQ.

As the arc PQ revolves about the x-axis, the line segment joining P and Q sweeps out part of a cone whose axis lies along the x-axis (magnified view in Fig. 5.42). A piece of a cone like this is called a *frustum* of the cone, *frustum* being Latin for "piece." The surface area of the frustum approximates the surface area of the band swept out by the arc PQ.

The surface area of the frustum of a cone (see Fig. 5.43) is 2π times the average of the base radii times the slant height:

$$\text{Frustum surface area} = 2\pi \cdot \frac{r_1 + r_2}{2} \cdot L = \pi(r_1 + r_2)L.$$

For the frustum swept out by the segment PQ (Fig. 5.44), this works out to be

$$\text{Frustum surface area} = \pi(f(x_{k-1}) + f(x_k))\sqrt{(\Delta x_k)^2 + (\Delta y_k)^2}.$$

The area of the original surface, being the sum of the areas of the bands swept out by arcs like arc PQ, is approximated by the frustum area sum

$$\sum_{k=1}^{n} \pi(f(x_{k-1}) + f(x_k))\sqrt{(\Delta x_k)^2 + (\Delta y_k)^2}. \tag{1}$$

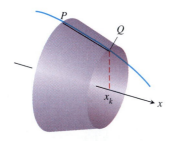

5.42 The line segment joining P and Q sweeps out a frustum of a cone.

We expect the approximation to improve as the partition of $[a, b]$ becomes finer, and we would like to show that the sums in (1) approach a calculable limit as the norm of the partition goes to zero.

5.43 The important dimensions of the frustum in Fig. 5.42.

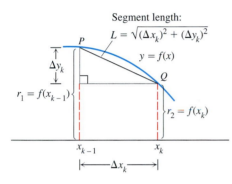

5.44 Dimensions associated with the arc and segment PQ.

To show this, we try to rewrite the sum in (1) as the Riemann sum of some function over the interval from a to b. As in the calculation of arc length, we begin by appealing to the Mean Value Theorem for derivatives.

If f is smooth, then by the Mean Value Theorem, there is a point $(c_k, f(c_k))$ on the curve between P and Q where the tangent is parallel to the segment PQ (Fig. 5.45). At this point,

$$f'(c_k) = \frac{\Delta y_k}{\Delta x_k},$$

$$\Delta y_k = f'(c_k)\Delta x_k.$$

With this substitution for Δy_k, the sums in (1) take the form

$$\sum_{k=1}^{n} \pi(f(x_{k-1}) + f(x_k))\sqrt{(\Delta x_k)^2 + (f'(c_k)\Delta x_k)^2}$$

$$= \sum_{k=1}^{n} \pi(f(x_{k-1}) + f(x_k))\sqrt{1 + (f'(c_k))^2}\,\Delta x_k. \quad (2)$$

At this point there is both good news and bad news.

The bad news is that the sums in (2) are not the Riemann sums of any function because the points x_{k-1}, x_k, and c_k are not the same and there is no way to make them the same. The good news is that this does not matter. A theorem called Bliss's theorem, from advanced calculus, assures us that as the norm of the partition of [a, b] goes to zero, the sums in Eq. (2) converge to

$$\int_a^b 2\pi f(x)\sqrt{1 + (f'(x))^2}\,dx$$

just the way we want them to. We therefore define this integral to be the area of the surface swept out by the graph of f from a to b.

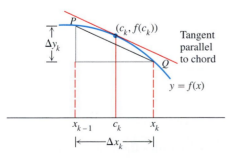

5.45 If f is smooth, the Mean Value Theorem guarantees the existence of a point on arc PQ where the tangent is parallel to segment PQ.

Definition
The Surface Area Formula for the Revolution About the x-axis

If the function $f(x) \geq 0$ is smooth on [a, b], the **area** of the surface generated by revolving the curve $y = f(x)$ about the x-axis is

$$S = \int_a^b 2\pi y\sqrt{1 + \left(\frac{dy}{dx}\right)^2}\,dx = \int_a^b 2\pi f(x)\sqrt{1 + (f'(x))^2}\,dx. \quad (3)$$

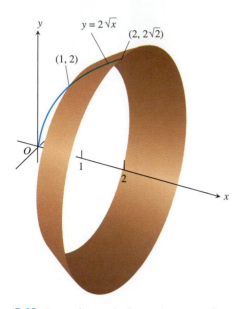

5.46 Example 1 calculates the area of this surface.

The square root in Eq. (3) is the same one that appears in the formula for the length of the generating curve.

EXAMPLE 1   Find the area of the surface generated by revolving the curve $y = 2\sqrt{x}$, $1 \le x \le 2$, about the x-axis (Fig. 5.46).

Solution   We evaluate the formula

$$S = \int_a^b 2\pi y \sqrt{1 + \left(\frac{dy}{dx}\right)^2}\, dx \qquad \text{Eq. (3)}$$

with

$$a = 1, \qquad b = 2, \qquad y = 2\sqrt{x}, \qquad \frac{dy}{dx} = \frac{1}{\sqrt{x}},$$

$$\sqrt{1 + \left(\frac{dy}{dx}\right)^2} = \sqrt{1 + \left(\frac{1}{\sqrt{x}}\right)^2}$$

$$= \sqrt{1 + \frac{1}{x}} = \sqrt{\frac{x+1}{x}} = \frac{\sqrt{x+1}}{\sqrt{x}}.$$

With these substitutions,

$$S = \int_1^2 2\pi \cdot 2\sqrt{x}\, \frac{\sqrt{x+1}}{\sqrt{x}}\, dx = 4\pi \int_1^2 \sqrt{x+1}\, dx$$

$$= 4\pi \cdot \frac{2}{3}(x+1)^{3/2} \Big]_1^2 = \frac{8\pi}{3}(3\sqrt{3} - 2\sqrt{2}).$$

Revolution About the y-axis

For revolution about the y-axis, we interchange x and y in Eq. (3).

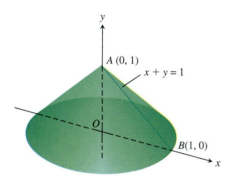

5.47 Revolving line segment *AB* about the *y*-axis generates a cone whose lateral surface area we can now calculate in two different ways (Example 2).

Surface Area Formula for Revolution About the y-axis

If $x = g(y) \ge 0$ is smooth on $[c, d]$, the area of the surface generated by revolving the curve $x = g(y)$ about the y-axis is

$$S = \int_c^d 2\pi x \sqrt{1 + \left(\frac{dx}{dy}\right)^2}\, dy = \int_c^d 2\pi g(y)\sqrt{1 + (g'(y))^2}\, dy. \qquad \textbf{(4)}$$

EXAMPLE 2   The line segment $x = 1 - y$, $0 \le y \le 1$, is revolved about the y-axis to generate the cone in Fig. 5.47. Find its lateral surface area.

Solution   Here we have a calculation we can check with a formula from geometry:

$$\text{Lateral surface area} = \frac{\text{base circumference}}{2} \times \text{slant height} = \pi\sqrt{2}.$$

To see how Eq. (4) gives the same result, we take

$$c = 0, \qquad d = 1, \qquad x = 1 - y, \qquad \frac{dx}{dy} = -1,$$

$$\sqrt{1 + \left(\frac{dx}{dy}\right)^2} = \sqrt{1 + (-1)^2} = \sqrt{2}$$

and calculate

$$S = \int_c^d 2\pi x \sqrt{1 + \left(\frac{dx}{dy}\right)^2}\, dy = \int_0^1 2\pi (1 - y)\sqrt{2}\, dy$$

$$= 2\pi \sqrt{2}\left[y - \frac{y^2}{2}\right]_0^1 = 2\pi \sqrt{2}\left(1 - \frac{1}{2}\right)$$

$$= \pi \sqrt{2}.$$

The results agree, as they should.

The Short Differential Form

The equations

$$S = \int_a^b 2\pi y \sqrt{1 + \left(\frac{dy}{dx}\right)^2}\, dx \qquad \text{and} \qquad S = \int_c^d 2\pi x \sqrt{1 + \left(\frac{dx}{dy}\right)^2}\, dy$$

are often written in terms of the arc length differential $ds = \sqrt{dx^2 + dy^2}$ as

$$S = \int_a^b 2\pi y\, ds \qquad \text{and} \qquad S = \int_c^d 2\pi x\, ds.$$

In the first of these, y is the distance from the x-axis to an element of arc length ds. In the second, x is the distance from the y-axis to an element of arc length ds. Both integrals have the form

$$S = \int 2\pi (\text{radius})(\text{band width}) = \int 2\pi \rho\, ds,$$

where ρ is the radius from the axis of revolution to an element of arc length ds (Fig. 5.48).

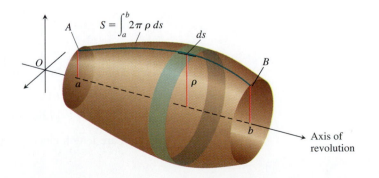

5.48 The area of the surface swept out by revolving arc *AB* about the axis shown here is $\int_a^b 2\pi \rho\, ds$. The exact expression depends on the formulas for ρ and *ds*.

If you wish to remember only one formula for surface area, you might make it the short differential form.

Short Differential Form

$$S = \int 2\pi \rho \, ds$$

In any particular problem, you would then express the radius function ρ and the arc length differential ds in terms of a common variable and supply limits of integration for that variable.

EXAMPLE 3 Find the area of the surface generated by revolving the curve $y = x^3, 0 \le x \le 1/2$, about the x-axis (Fig. 5.49).

Solution We start with the short differential form:

$$S = \int 2\pi \rho \, ds$$

$$= \int 2\pi y \, ds \qquad \text{For revolution about the } x\text{-axis, the radius function is } \rho = y.$$

$$= \int 2\pi y \sqrt{dx^2 + dy^2}. \qquad ds = \sqrt{dx^2 + dy^2}$$

We then decide whether to express dy in terms of dx or dx in terms of dy. The original form of the equation, $y = x^3$, makes it easier to express dy in terms of dx, so we continue the calculation with

$$y = x^3, \qquad dy = 3x^2 \, dx, \qquad \text{and} \qquad \sqrt{dx^2 + dy^2} = \sqrt{dx^2 + (3x^2 \, dx)^2}$$

$$= \sqrt{1 + 9x^4} \, dx.$$

With these substitutions, x becomes the variable of integration and

$$S = \int_{x=0}^{x=1/2} 2\pi y \sqrt{dx^2 + dy^2}$$

$$= \int_0^{1/2} 2\pi x^3 \sqrt{1 + 9x^4} \, dx$$

$$= 2\pi \left(\frac{1}{36}\right)\left(\frac{2}{3}\right)(1 + 9x^4)^{3/2} \bigg]_0^{1/2} \qquad \text{Substitute } u = 1 + 9x^4, du/36 = x^3 dx, \text{ integrate, and substitute back.}$$

$$= \frac{\pi}{27}\left[\left(1 + \frac{9}{16}\right)^{3/2} - 1\right]$$

$$= \frac{\pi}{27}\left[\left(\frac{25}{16}\right)^{3/2} - 1\right] = \frac{\pi}{27}\left(\frac{125}{64} - 1\right)$$

$$= \frac{61\pi}{1728}.$$

As with arc length calculations, even the simplest curves can provide a workout. ❑

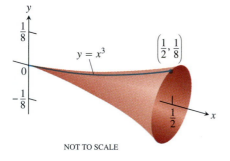

5.49 The surface generated by revolving the curve $y = x^3, 0 \le x \le 1/2$, about the x-axis could be the design for a champagne glass (Example 3).

NOT TO SCALE

$y = x^3$

$\left(\frac{1}{2}, \frac{1}{8}\right)$

Exercises 5.6

Finding Integrals for Surface Area

In Exercises 1–8:

a) Set up an integral for the area of the surface generated by revolving the given curve about the indicated axis.

b) Graph the curve to see what it looks like. If you can, graph the surface, too.

c) Use your grapher's or computer's integral evaluator to find the surface's area numerically.

1. $y = \tan x, \quad 0 \le x \le \pi/4; \quad$ x-axis

2. $y = x^2, \quad 0 \le x \le 2; \quad$ x-axis

3. $xy = 1, \quad 1 \le y \le 2; \quad$ y-axis

4. $x = \sin y, \quad 0 \le y \le \pi; \quad$ y-axis

5. $x^{1/2} + y^{1/2} = 3$ from $(4, 1)$ to $(1, 4); \quad$ x-axis

6. $y + 2\sqrt{y} = x, \quad 1 \le y \le 2; \quad$ y-axis

7. $x = \displaystyle\int_0^y \tan t\, dt, \quad 0 \le y \le \pi/3; \quad$ y-axis

8. $y = \displaystyle\int_1^x \sqrt{t^2 - 1}\, dt, \quad 1 \le x \le \sqrt{5}; \quad$ x-axis

Finding Surface Areas

9. Find the lateral (side) surface area of the cone generated by revolving the line segment $y = x/2, 0 \le x \le 4$, about the x-axis. Check your answer with the geometry formula

$$\text{Lateral surface area} = \frac{1}{2} \times \text{base circumference} \times \text{slant height}.$$

10. Find the lateral surface area of the cone generated by revolving the line segment $y = x/2, 0 \le x \le 4$ about the y-axis. Check your answer with the geometry formula

$$\text{Lateral surface area} = \frac{1}{2} \times \text{base circumference} \times \text{slant height}.$$

11. Find the surface area of the cone frustum generated by revolving the line segment $y = (x/2) + (1/2), 1 \le x \le 3$, about the x-axis. Check your result with the geometry formula

$$\text{Frustum surface area} = \pi(r_1 + r_2) \times \text{slant height}.$$

12. Find the surface area of the cone frustum generated by revolving the line segment $y = (x/2) + (1/2), 1 \le x \le 3$, about the y-axis. Check your result with the geometry formula

$$\text{Frustum surface area} = \pi(r_1 + r_2) \times \text{slant height}.$$

Find the areas of the surfaces generated by revolving the curves in Exercises 13–22 about the indicated axes. If you have a grapher, you may want to graph these curves to see what they look like.

13. $y = x^3/9, \quad 0 \le x \le 2; \quad$ x-axis

14. $y = \sqrt{x}, \quad 3/4 \le x \le 15/4; \quad$ x-axis

15. $y = \sqrt{2x - x^2}, \quad 0.5 \le x \le 1.5; \quad$ x-axis

16. $y = \sqrt{x + 1}, \quad 1 \le x \le 5; \quad$ x-axis

17. $x = y^3/3, \quad 0 \le y \le 1; \quad$ y-axis

18. $x = (1/3)y^{3/2} - y^{1/2}, \quad 1 \le y \le 3; \quad$ y-axis

19. $x = 2\sqrt{4 - y}, \quad 0 \le y \le 15/4; \quad$ y-axis

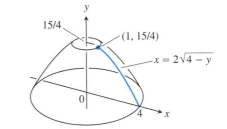

20. $x = \sqrt{2y - 1}, \quad 5/8 \le y \le 1; \quad$ y-axis

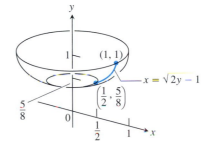

21. $x = (y^4/4) + 1/(8y^2), \quad 1 \le y \le 2; \quad$ x-axis (*Hint:* Express $ds = \sqrt{dx^2 + dy^2}$ in terms of dy, and evaluate the integral $S = \int 2\pi y\, ds$ with appropriate limits.)

22. $y = (1/3)(x^2 + 2)^{3/2}, \quad 0 \le x \le \sqrt{2}; \quad$ y-axis (*Hint:* Express $ds = \sqrt{dx^2 + dy^2}$ in terms of dx, and evaluate the integral $S = \int 2\pi x\, ds$ with appropriate limits.)

23. *Testing the new definition.* Show that the surface area of a sphere of radius a is still $4\pi a^2$ by using Eq. (3) to find the area of the surface generated by revolving the curve $y = \sqrt{a^2 - x^2}$, $-a \le x \le a$, about the x-axis.

24. *Testing the new definition.* The lateral (side) surface area of a cone of height h and base radius r should be $\pi r \sqrt{r^2 + h^2}$, the semiperimeter of the base times the slant height. Show that this is still the case by finding the area of the surface generated by revolving the line segment $y = (r/h)x, 0 \le x \le h$, about the x-axis.

25. a) Write an integral for the area of the surface generated by revolving the curve $y = \cos x, -\pi/2 \le x \le \pi/2$, about the x-axis. In Section 7.4 we will see how to evaluate such integrals.

   b) CALCULATOR Find the surface area numerically.

26. *The surface of an astroid.* Find the area of the surface generated by revolving about the x-axis the portion of the astroid $x^{2/3} + y^{2/3} = 1$ shown here. (*Hint:* Revolve the first-quadrant portion $y = (1 - x^{2/3})^{3/2}, 0 \le x \le 1$, about the x-axis and double your result.)

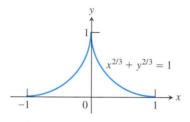

27. *Enameling woks.* Your company decided to put out a deluxe version of the successful wok you designed in Section 5.3, Exercise 41. The plan is to coat it inside with white enamel and outside with blue enamel. Each enamel will be sprayed on 0.5 mm thick before baking. (See diagram here.) Your manufacturing department wants to know how much enamel to have on hand for a production run of 5000 woks. What do you tell them? (Neglect waste and unused material and give your answer in liters. Remember that $1 \text{ cm}^3 = 1 \text{ mL}$, so $1 \text{ L} = 1000 \text{ cm}^3$.)

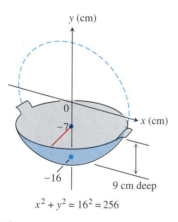

$$x^2 + y^2 = 16^2 = 256$$

28. *Slicing bread.* Did you know that if you cut a spherical loaf of bread into slices of equal width, each slice will have the same amount of crust? To see why, suppose the semicircle $y = \sqrt{r^2 - x^2}$ shown here is revolved about the x-axis to generate a sphere. Let AB be an arc of the semicircle that lies above an interval of length h on the x-axis. Show that the area swept out by AB does not depend on the location of the interval. (It does depend on the length of the interval.)

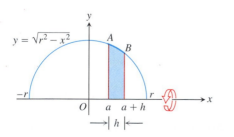

29. The shaded band shown here is cut from a sphere of radius R by parallel planes h units apart. Show that the surface area of the band is $2\pi Rh$.

30. Here is a schematic drawing of the 90-ft dome used by the U.S. National Weather Service to house radar in Bozeman, Mont.

a) How much outside surface is there to paint (not counting the bottom)?

b) CALCULATOR Express the answer to the nearest square foot.

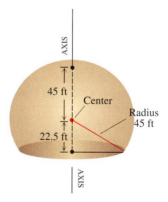

31. *Surfaces generated by curves that cross the axis of revolution.* The surface area formula in Eq. (3) was developed under the assumption that the function f whose graph generated the surface was nonnegative over the interval $[a, b]$. For curves that cross the axis of revolution, we replace Eq. (3) with the absolute value formula

$$S = \int 2\pi\rho \, ds = \int 2\pi |f(x)| \, ds. \tag{5}$$

Use Eq. (5) to find the surface area of the double cone generated by revolving the line segment $y = x, -1 \le x \le 2$, about the x-axis.

32. *(Exercise 31, continued.)* Find the area of the surface generated by revolving the curve $y = x^3/9, -\sqrt{3} \le x \le \sqrt{3}$, about the x-axis. What do you think will happen if you drop the absolute value bars from Eq. (5) and attempt to find the surface area with the formula $S = \int 2\pi f(x) \, ds$ instead? Try it.

Numerical Integration

Find, to 2 decimal places, the areas of the surfaces generated by revolving the curves in Exercises 33–36 about the *x*-axis.

33. $y = \sin x, \quad 0 \le x \le \pi$

34. $y = x^2/4, \quad 0 \le x \le 2$

35. $y = x + \sin 2x, \quad -2\pi/3 \le x \le 2\pi/3$ (the curve in Section 3.4, Exercise 5)

36. $y = \dfrac{x}{12}\sqrt{36 - x^2}, 0 \le x \le 6$ (the surface of the plumb bob in Section 5.3, Exercise 44)

37. *An alternative derivation of the surface area formula.* Assume *f* is smooth on [*a, b*] and partition [*a, b*] in the usual way. In the *k*th subinterval [x_{k-1}, x_k] construct the tangent line to the curve at the midpoint $m_k = (x_{k-1} + x_k)/2$, as in the figure here.

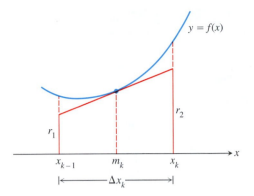

a)   Show that $r_1 = f(m_k) - f'(m_k)\dfrac{\Delta x_k}{2}$ and $r_2 = f(m_k) + f'(m_k)\dfrac{\Delta x_k}{2}$.

b)   Show that the length L_k of the tangent line segment in the *k*th subinterval is $L_k = \sqrt{(\Delta x_k)^2 + (f'(m_k)\Delta x_k)^2}$.

c)   Show that the lateral surface area of the frustum of the cone swept out by the tangent line segment as it revolves about the *x*-axis is $2\pi f(m_k)\sqrt{1 + (f'(m_k))^2}\,\Delta x_k$.

d)   Show that the area of the surface generated by revolving $y = f(x)$ about the *x*-axis over [*a, b*] is

$$\lim_{n \to \infty} \sum_{k=1}^{n} \left(\begin{array}{c} \text{lateral surface area} \\ \text{of } k\text{th frustum} \end{array} \right) = \int_{a}^{b} 2\pi f(x)\sqrt{1 + (f'(x))^2}\, dx.$$

5.7   Moments and Centers of Mass

Many structures and mechanical systems behave as if their masses were concentrated at a single point, called the center of mass (Fig. 5.50, on the following page). It is important to know how to locate this point, and doing so is basically a mathematical enterprise. For the moment we deal with one- and two-dimensional objects. Three-dimensional objects are best done with the multiple integrals of Chapter 13.

Masses Along a Line

We develop our mathematical model in stages. The first stage is to imagine masses m_1, m_2, and m_3 on a rigid *x*-axis supported by a fulcrum at the origin.

The resulting system might balance, or it might not. It depends on how large the masses are and how they are arranged.

Each mass m_k exerts a downward force $m_k g$ equal to the magnitude of the mass times the acceleration of gravity. Each of these forces has a tendency to turn the axis about the origin, the way you turn a seesaw. This turning effect, called a **torque,** is measured by multiplying the force $m_k g$ by the signed distance x_k from the point of application to the origin. Masses to the left of the origin exert negative (counterclockwise) torque. Masses to the right of the origin exert positive (clockwise) torque.

Mass vs. weight

Weight is the force that results from gravity pulling on a mass. If an object of mass *m* is placed in a location where the acceleration of gravity is *g*, the object's weight there is

$$F = mg$$

(as in Newton's second law).

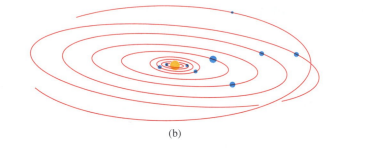

(b)

5.50 (a) The motion of this wrench gliding on ice seems haphazard until we notice that the wrench is simply turning about its center of mass as the center glides in a straight line. (b) The planets, asteroids, and comets of our solar system revolve about their collective center of mass. (It lies inside the sun.)

The sum of the torques measures the tendency of a system to rotate about the origin. This sum is called the **system torque.**

$$\text{System torque} = m_1 g x_1 + m_2 g x_2 + m_3 g x_3 \qquad (1)$$

The system will balance if and only if its torque is zero.

If we factor out the g in Eq. (1), we see that the system torque is

$$g(m_1 x_1 + m_2 x_2 + m_3 x_3).$$

a feature of the environment   a feature of the system

Thus the torque is the product of the gravitional acceleration g, which is a feature of the environment in which the system happens to reside, and the number $(m_1 x_1 + m_2 x_2 + m_3 x_3)$, which is a feature of the system itself, a constant that stays the same no matter where the system is placed.

The number $(m_1 x_1 + m_2 x_2 + m_3 x_3)$ is called the **moment of the system about the origin.** It is the sum of the **moments** $m_1 x_1, m_2 x_2, m_3 x_3$ of the individual masses.

$$M_O = \text{Moment of system about origin} = \sum m_k x_k$$

(We shift to sigma notation here to allow for sums with more terms. For $\sum m_k x_k$, read "summation m_k times x_k.")

We usually want to know where to place the fulcrum to make the system balance, that is, at what point $\overline{x}$ to place it to make the torque zero.

Special location for balance

The torque of each mass about the fulcrum in this special location is

$$\text{Torque of } m_k \text{ about } \overline{x} = \left(\begin{array}{c} \text{signed distance} \\ \text{of } m_k \text{ from } \overline{x} \end{array} \right) \left(\begin{array}{c} \text{downward} \\ \text{force} \end{array} \right)$$

$$= (x_k - \overline{x}) m_k g.$$

When we write the equation that says that the sum of these torques is zero, we get an equation we can solve for $\overline{x}$:

$$\sum (x_k - \overline{x})m_k g = 0$$ Sum of the torques equals zero

$$g \sum (x_k - \overline{x})m_k = 0$$ Constant Multiple Rule for Sums

$$\sum (m_k x_k - \overline{x}m_k) = 0$$ g divided out, m_k distributed

$$\sum m_k x_k - \sum \overline{x}m_k = 0$$ Difference Rule for Sums

$$\sum m_k x_k = \overline{x} \sum m_k$$ Rearranged, Constant Multiple Rule again

$$\overline{x} = \frac{\sum m_k x_k}{\sum m_k}.$$ Solved for $\overline{x}$

This last equation tells us to find $\overline{x}$ by dividing the system's moment about the origin by the system's total mass:

$$\overline{x} = \frac{\sum x_k m_k}{\sum m_k} = \frac{\text{system moment about origin}}{\text{system mass}}.$$

The point $\overline{x}$ is called the system's **center of mass.**

Wires and Thin Rods

In many applications, we want to know the center of mass of a rod or a thin strip of metal. In cases like these where we can model the distribution of mass with a continuous function, the summation signs in our formulas become integrals in a manner we now describe.

Imagine a long, thin strip lying along the x-axis from $x = a$ to $x = b$ and cut into small pieces of mass Δm_k by a partition of the interval $[a, b]$.

The kth piece is Δx_k units long and lies approximately x_k units from the origin. Now observe three things.

First, the strip's center of mass $\overline{x}$ is nearly the same as that of the system of point masses we would get by attaching each mass Δm_k to the point x_k:

$$\overline{x} \approx \frac{\text{system moment}}{\text{system mass}}.$$

Second, the moment of each piece of the strip about the origin is approximately $x_k \Delta m_k$, so the system moment is approximately the sum of the $x_k \Delta m_k$:

$$\text{System moment} \approx \sum x_k \Delta m_k.$$

Third, if the density of the strip at x_k is $\delta(x_k)$, expressed in terms of mass per unit length, and δ is continuous, then Δm_k is approximately equal to $\delta(x_k)\Delta x_k$ (mass per unit length times length):

$$\Delta m_k \approx \delta(x_k)\Delta x_k.$$

Combining these three observations gives

$$\overline{x} \approx \frac{\text{system moment}}{\text{system mass}} \approx \frac{\sum x_k \Delta m_k}{\sum \Delta m_k} \approx \frac{\sum x_k \delta(x_k)\Delta x_k}{\sum \delta(x_k)\Delta x_k}. \tag{2}$$

Density

A material's density is its mass per unit volume. In practice, however, we tend to use units we can conveniently measure. For wires, rods, and narrow strips we use mass per unit length. For flat sheets and plates we use mass per unit area.

The sum in the last numerator in Eq. (2) is a Riemann sum for the continuous function $x\delta(x)$ over the closed interval $[a, b]$. The sum in the denominator is a Riemann sum for the function $\delta(x)$ over this interval. We expect the approximations in (2) to improve as the strip is partitioned more finely, and we are led to the equation

$$\overline{x} = \frac{\int_a^b x\delta(x)\,dx}{\int_a^b \delta(x)\,dx}.$$

This is the formula we use to find $\overline{x}$.

Moment, Mass, and Center of Mass of a Thin Rod or Strip Along the x-axis with Density Function $\delta(x)$

Moment about the origin: $\quad M_O = \displaystyle\int_a^b x\delta(x)\,dx \qquad$ (3a)

Mass: $\quad M = \displaystyle\int_a^b \delta(x)\,dx \qquad$ (3b)

Center of mass: $\quad \overline{x} = \dfrac{M_O}{M} \qquad$ (3c)

To find a center of mass, divide moment by mass.

EXAMPLE 1 *Strips and rods of constant density*

Show that the center of mass of a straight, thin strip or rod of constant density lies halfway between its two ends.

Solution We model the strip as a portion of the x-axis from $x = a$ to $x = b$ (Fig. 5.51). Our goal is to show that $\overline{x} = (a + b)/2$, the point halfway between a and b.

The key is the density's having a constant value. This enables us to regard the function $\delta(x)$ in the integrals in Eqs. (3) as a constant (call it δ), with the result that

$$M_O = \int_a^b \delta x\,dx = \delta\int_a^b x\,dx = \delta\left[\frac{1}{2}x^2\right]_a^b = \frac{\delta}{2}(b^2 - a^2)$$

$$M = \int_a^b \delta\,dx = \delta\int_a^b dx = \delta\left[x\right]_a^b = \delta(b - a)$$

$$\overline{x} = \frac{M_O}{M} = \frac{\frac{\delta}{2}(b^2 - a^2)}{\delta(b - a)}$$

$$= \frac{a + b}{2}. \qquad \text{The } \delta\text{'s cancel in the formula for } \overline{x}.$$

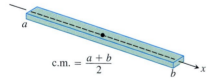

5.51 The center of mass of a straight, thin rod or strip of constant density lies halfway between its ends.

c.m. $= \dfrac{a+b}{2}$

5.52 We can treat a rod of variable thickness as a rod of variable density. See Example 2.

EXAMPLE 2 *A variable density*

The 10-m-long rod in Fig. 5.52 thickens from left to right so that its density, instead of being constant, is $\delta(x) = 1 + (x/10)$ kg/m. Find the rod's center of mass.

Solution The rod's moment about the origin (Eq. 3a) is

$$M_O = \int_0^{10} x\delta(x)\,dx = \int_0^{10} x\left(1 + \frac{x}{10}\right)dx = \int_0^{10}\left(x + \frac{x^2}{10}\right)dx$$

$$= \left[\frac{x^2}{2} + \frac{x^3}{30}\right]_0^{10} = 50 + \frac{100}{3} = \frac{250}{3} \text{ kg} \cdot \text{m}. \qquad \text{\color{blue}The units of a moment are mass} \times \text{length.}$$

The rod's mass (Eq. 3b) is

$$M = \int_0^{10} \delta(x)\,dx = \int_0^{10}\left(1 + \frac{x}{10}\right)dx = \left[x + \frac{x^2}{20}\right]_0^{10} = 10 + 5 = 15 \text{ kg}.$$

The center of mass (Eq. 3c) is located at the point

$$\bar{x} = \frac{M_O}{M} = \frac{250}{3} \cdot \frac{1}{15} = \frac{50}{9} \approx 5.56 \text{ m}. \qquad \square$$

Masses Distributed over a Plane Region

Suppose we have a finite collection of masses located in the plane, with mass m_k at the point (x_k, y_k) (see Fig. 5.53). The mass of the system is

$$\text{System mass:} \qquad M = \sum m_k.$$

Each mass m_k has a moment about each axis. Its moment about the x-axis is $m_k y_k$, and its moment about the y-axis is $m_k x_k$. The moments of the entire system about the two axes are

$$\text{Moment about } x\text{-axis:} \qquad M_x = \sum m_k y_k,$$

$$\text{Moment about } y\text{-axis:} \qquad M_y = \sum m_k x_k.$$

The x-coordinate of the system's center of mass is defined to be

$$\bar{x} = \frac{M_y}{M} = \frac{\sum m_k x_k}{\sum m_k}. \qquad (4)$$

With this choice of $\bar{x}$, as in the one-dimensional case, the system balances about the line $x = \bar{x}$ (Fig. 5.54).

The y-coordinate of the system's center of mass is defined to be

$$\bar{y} = \frac{M_x}{M} = \frac{\sum m_k y_k}{\sum m_k}. \qquad (5)$$

With this choice of $\bar{y}$, the system balances about the line $y = \bar{y}$ as well. The torques exerted by the masses about the line $y = \bar{y}$ cancel out. Thus, as far as balance is concerned, the system behaves as if all its mass were at the single point $(\bar{x}, \bar{y})$. We call this point the system's *center of mass*.

Thin, Flat Plates

In many applications, we need to find the center of mass of a thin, flat plate: a disk of aluminum, say, or a triangular sheet of steel. In such cases we assume the distribution of mass to be continuous, and the formulas we use to calculate $\bar{x}$ and $\bar{y}$ contain integrals instead of finite sums. The integrals arise in the following way.

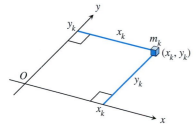

5.53 Each mass m_k has a moment about each axis.

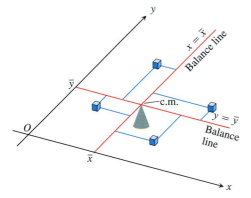

5.54 A two-dimensional array of masses balances on its center of mass.

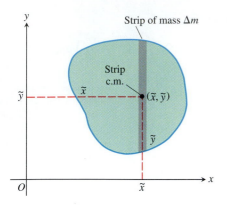

5.55 A plate cut into thin strips parallel to the *y*-axis. The moment exerted by a typical strip about each axis is the moment its mass Δm would exert if concentrated at the strip's center of mass $(\tilde{x}, \tilde{y})$.

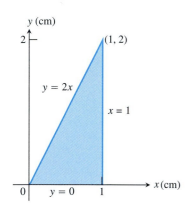

5.56 The plate in Example 3.

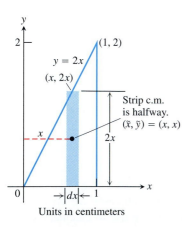

Units in centimeters

5.57 Modeling the plate in Example 3 with vertical strips.

Imagine the plate occupying a region in the *xy*-plane, cut into thin strips parallel to one of the axes (in Fig. 5.55, the *y*-axis). The center of mass of a typical strip is $(\tilde{x}, \tilde{y})$. We treat the strip's mass Δm as if it were concentrated at $(\tilde{x}, \tilde{y})$. The moment of the strip about the *y*-axis is then $\tilde{x}\Delta m$. The moment of the strip about the *x*-axis is $\tilde{y}\Delta m$. Equations (4) and (5) then become

$$\overline{x} = \frac{M_y}{M} = \frac{\sum \tilde{x}\,\Delta m}{\sum \Delta m}, \qquad \overline{y} = \frac{M_x}{M} = \frac{\sum \tilde{y}\,\Delta m}{\sum \Delta m}.$$

As in the one-dimensional case, the sums are Riemann sums for integrals and approach these integrals as limiting values as the strips into which the plate is cut become narrower and narrower. We write these integrals symbolically as

$$\overline{x} = \frac{\int \tilde{x}\,dm}{\int dm} \qquad \text{and} \qquad \overline{y} = \frac{\int \tilde{y}\,dm}{\int dm}.$$

Moments, Mass, and Center of Mass of a Thin Plate Covering a Region in the *xy*-plane

Moment about the *x*-axis: $\quad M_x = \displaystyle\int \tilde{y}\,dm$

Moment about the *y*-axis: $\quad M_y = \displaystyle\int \tilde{x}\,dm \qquad$ (6)

Mass: $\quad M = \displaystyle\int dm$

Center of mass: $\quad \overline{x} = \dfrac{M_y}{M}, \qquad \overline{y} = \dfrac{M_x}{M}$

To evaluate these integrals, we picture the plate in the coordinate plane and sketch a strip of mass parallel to one of the coordinates axes. We then express the strip's mass dm and the coordinates $(\tilde{x}, \tilde{y})$ of the strip's center of mass in terms of *x* or *y*. Finally, we integrate $\tilde{y}\,dm$, $\tilde{x}\,dm$, and dm between limits of integration determined by the plate's location in the plane.

EXAMPLE 3   The triangular plate shown in Fig. 5.56 has a constant density of $\delta = 3$ g/cm². Find (a) the plate's moment M_y about the *y*-axis, (b) the plate's mass M, and (c) the *x*-coordinate of the plate's center of mass (c.m.).

Solution

Method 1: *Vertical strips* (Fig. 5.57).

a)   The moment M_y: The typical vertical strip has

center of mass (c.m.): $\quad (\tilde{x}, \tilde{y}) = (x, x),$

length: $2x$,   area: $dA = 2x\,dx,$
width: dx,   mass: $dm = \delta\,dA = 3 \cdot 2x\,dx = 6x\,dx,$
distance of c.m. from *y*-axis: $\tilde{x} = x.$

The moment of the strip about the y-axis is

$$\tilde{x}\, dm = x \cdot 6x\, dx = 6x^2\, dx.$$

The moment of the plate about the y-axis is therefore

$$M_y = \int \tilde{x}\, dm = \int_0^1 6x^2\, dx = 2x^3 \Big]_0^1 = 2 \text{ g} \cdot \text{cm}.$$

b) The plate's mass:

$$M = \int dm = \int_0^1 6x\, dx = 3x^2 \Big]_0^1 = 3 \text{ g}.$$

c) The x-coordinate of the plate's center of mass:

$$\overline{x} = \frac{M_y}{M} = \frac{2 \text{ g} \cdot \text{cm}}{3 \text{ g}} = \frac{2}{3} \text{ cm}.$$

By a similar computation we could find M_x and $\overline{y} = M_x/M$.

Method 2: *Horizontal strips* (Fig. 5.58).

a) The moment M_y: The y-coordinate of the center of mass of a typical horizontal strip is y (see the figure), so

$$\tilde{y} = y.$$

The x-coordinate is the x-coordinate of the point halfway across the triangle. This makes it the average of $y/2$ (the strip's left-hand x-value) and 1 (the strip's right-hand x-value):

$$\tilde{x} = \frac{(y/2) + 1}{2} = \frac{y}{4} + \frac{1}{2} = \frac{y + 2}{4}.$$

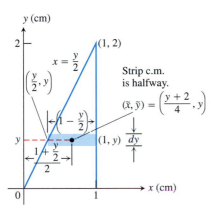

y (cm)

2 ─ (1, 2)

$x = \dfrac{y}{2}$

$\left(\dfrac{y}{2}, y\right)$

Strip c.m. is halfway.

$(\tilde{x}, \tilde{y}) = \left(\dfrac{y + 2}{4}, y\right)$

$\leftarrow\!\left(1 - \dfrac{y}{2}\right)\!\rightarrow$

y ─ ─ ─ ─ (1, y) dy

$\leftarrow\dfrac{1 + \dfrac{y}{2}}{2}\rightarrow$

0 ── 1 ──→ x (cm)

5.58 Modeling the plate in Example 3 with horizontal strips.

We also have

length: $1 - \dfrac{y}{2} = \dfrac{2 - y}{2},$

width: $dy,$

area: $dA = \dfrac{2 - y}{2}\, dy,$

mass: $dm = \delta\, dA = 3 \cdot \dfrac{2 - y}{2}\, dy,$

distance of c.m. to y-axis: $\tilde{x} = \dfrac{y + 2}{4}.$

The moment of the strip about the y-axis is

$$\tilde{x}\, dm = \frac{y + 2}{4} \cdot 3 \cdot \frac{2 - y}{2}\, dy = \frac{3}{8}(4 - y^2)\, dy.$$

The moment of the plate about the y-axis is

$$M_y = \int \tilde{x}\, dm = \int_0^2 \frac{3}{8}(4 - y^2)\, dy = \frac{3}{8}\left[4y - \frac{y^3}{3}\right]_0^2 = \frac{3}{8}\left(\frac{16}{3}\right) = 2 \text{ g} \cdot \text{cm}.$$

b) The plate's mass:

$$M = \int dm = \int_0^2 \frac{3}{2}(2 - y)\, dy = \frac{3}{2}\left[2y - \frac{y^2}{2}\right]_0^2 = \frac{3}{2}(4 - 2) = 3 \text{ g}.$$

How to Find a Plate's Center of Mass

1. Picture the plate in the xy-plane.
2. Sketch a strip of mass parallel to one of the coordinate axes and find its dimensions.
3. Find the strip's mass dm and center of mass $(\tilde{x}, \tilde{y})$.
4. Integrate $\tilde{y}\, dm$, $\tilde{x}\, dm$, and dm to find M_x, M_y, and M.
5. Divide the moments by the mass to calculate $\overline{x}$ and $\overline{y}$.

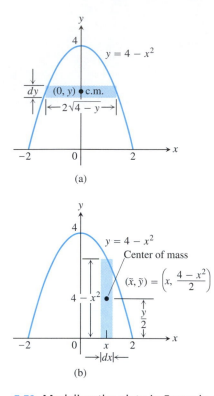

(a)

(b)

5.59 Modeling the plate in Example 4 with (a) horizontal strips leads to an inconvenient integration, so we model with (b) vertical strips instead.

c) The x-coordinate of the plate's center of mass:

$$\bar{x} = \frac{M_y}{M} = \frac{2 \text{ g} \cdot \text{cm}}{3 \text{ g}} = \frac{2}{3} \text{ cm}.$$

By a similar computation, we could find M_x and $\bar{y}$. □

If the distribution of mass in a thin, flat plate has an axis of symmetry, the center of mass will lie on this axis. If there are two axes of symmetry, the center of mass will lie at their intersection. These facts often help to simplify our work.

EXAMPLE 4   Find the center of mass of a thin plate of constant density δ covering the region bounded above by the parabola $y = 4 - x^2$ and below by the x-axis (Fig. 5.59).

Solution   Since the plate is symmetric about the y-axis and its density is constant, the distribution of mass is symmetric about the y-axis and the center of mass lies on the y-axis. This means that $\bar{x} = 0$. It remains to find $\bar{y} = M_x/M$.

A trial calculation with horizontal strips (Fig. 5.59a) leads to an inconvenient integration

$$M_x = \int_0^4 2\delta \, y\sqrt{4 - y} \, dy.$$

We therefore model the distribution of mass with vertical strips instead (Fig. 5.59b). The typical vertical strip has

center of mass (c.m): $(\tilde{x}, \tilde{y}) = \left(x, \dfrac{4 - x^2}{2} \right)$,

length: $4 - x^2$,

width: dx,

area: $dA = (4 - x^2) \, dx$,

mass: $dm = \delta \, dA = \delta(4 - x^2) \, dx$,

distance from c.m to x-axis: $\tilde{y} = \dfrac{4 - x^2}{2}$.

The moment of the strip about the x-axis is

$$\tilde{y} \, dm = \frac{4 - x^2}{2} \cdot \delta(4 - x^2) \, dx = \frac{\delta}{2}(4 - x^2)^2 \, dx.$$

The moment of the plate about the x-axis is

$$M_x = \int \tilde{y} \, dm = \int_{-2}^{2} \frac{\delta}{2}(4 - x^2)^2 \, dx$$

$$= \frac{\delta}{2} \int_{-2}^{2} (16 - 8x^2 + x^4) \, dx = \frac{256}{15}\delta. \tag{7}$$

The mass of the plate is

$$M = \int dm = \int_{-2}^{2} \delta(4 - x^2) \, dx = \frac{32}{3}\delta. \tag{8}$$

Therefore,

$$\bar{y} = \frac{M_x}{M} = \frac{(256/15)\,\delta}{(32/3)\,\delta} = \frac{8}{5}.$$

The plate's center of mass is the point

$$(\bar{x}, \bar{y}) = \left(0, \frac{8}{5}\right).$$

EXAMPLE 5   *Variable density*

Find the center of mass of the plate in Example 4 if the density at the point (x, y) is $\delta = 2x^2$, twice the square of the distance from the point to the y-axis.

Solution   The mass distribution is still symmetric about the y-axis, so $\bar{x} = 0$. With $\delta = 2x^2$, Eqs. (7) and (8) become

$$M_x = \int \tilde{y}\,dm = \int_{-2}^{2} \frac{\delta}{2}(4 - x^2)^2\,dx = \int_{-2}^{2} x^2(4 - x^2)^2\,dx$$

$$= \int_{-2}^{2} (16x^2 - 8x^4 + x^6)\,dx = \frac{2048}{105}, \tag{7'}$$

$$M = \int dm = \int_{-2}^{2} \delta(4 - x^2)\,dx = \int_{-2}^{2} 2x^2(4 - x^2)\,dx$$

$$= \int_{-2}^{2} (8x^2 - 2x^4)\,dx = \frac{256}{15}. \tag{8'}$$

Therefore,

$$\bar{y} = \frac{M_x}{M} = \frac{2048}{105} \cdot \frac{15}{256} = \frac{8}{7}.$$

The plate's new center of mass is

$$(\bar{x}, \bar{y}) = \left(0, \frac{8}{7}\right).$$

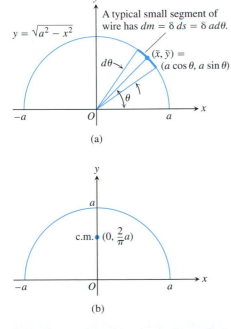

(a)

(b)

5.60 The semicircular wire in Example 6. (a) The dimensions and variables used in finding the center of mass. (b) The center of mass does not lie on the wire.

EXAMPLE 6   Find the center of mass of a wire of constant density δ shaped like a semicircle of radius a.

Solution   We model the wire with the semicircle $y = \sqrt{a^2 - x^2}$ (Fig. 5.60). The distribution of mass is symmetric about the y-axis, so $\bar{x} = 0$. To find $\bar{y}$, we imagine the wire divided into short segments. The typical segment (Fig. 5.60a) has

length: $ds = a\,d\theta$,

mass: $dm = \delta\,ds = \delta\,a\,d\theta$,   Mass per unit length times length

distance of c.m. to x-axis: $\tilde{y} = a \sin\theta$.

Hence,

$$\bar{y} = \frac{\int \tilde{y}\, dm}{\int dm} = \frac{\int_0^\pi a \sin\theta \cdot \delta a\, d\theta}{\int_0^\pi \delta a\, d\theta} = \frac{\delta a^2 [-\cos\theta]_0^\pi}{\delta a \pi} = \frac{2}{\pi} a.$$

The center of mass lies on the axis of symmetry at the point $(0,\ 2a/\pi)$, about two-thirds of the way up from the origin (Fig. 5.60b).   ❑

Centroids

When the density function is constant, it cancels out of the numerator and denominator of the formulas for $\bar{x}$ and $\bar{y}$. This happened in nearly every example in this section. As far as $\bar{x}$ and $\bar{y}$ were concerned, δ might as well have been 1. Thus, when the density is constant, the location of the center of mass is a feature of the geometry of the object and not of the material from which it is made. In such cases engineers may call the center of mass the **centroid** of the shape, as in "Find the centroid of a triangle or a solid cone." To do so, just set δ equal to 1 and proceed to find $\bar{x}$ and $\bar{y}$ as before, by dividing moments by masses.

Exercises 5.7

Thin Rods

1. An 80-lb child and a 100-lb child are balancing on a seesaw. The 80-lb child is 5 ft from the fulcrum. How far from the fulcrum is the 100-lb child?

2. The ends of a log are placed on two scales. One scale reads 100 kg and the other 200 kg. Where is the log's center of mass?

3. The ends of two thin steel rods of equal length are welded together to make a right-angled frame. Locate the frame's center of mass. (*Hint:* Where is the center of mass of each rod?)

4. You weld the ends of two steel rods into a right-angled frame. One rod is twice the length of the other. Where is the frame's center of mass? (*Hint:* Where is the center of mass of each rod?)

Exercises 5–12 give density functions of thin rods lying along various intervals of the x-axis. Use Eqs. (3a–c) to find each rod's moment about the origin, mass, and center of mass.

5. $\delta(x) = 4, \quad 0 \le x \le 2$

6. $\delta(x) = 4, \quad 1 \le x \le 3$

7. $\delta(x) = 1 + (x/3), \quad 0 \le x \le 3$

8. $\delta(x) = 2 - (x/4), \quad 0 \le x \le 4$

9. $\delta(x) = 1 + (1/\sqrt{x}), \quad 1 \le x \le 4$

10. $\delta(x) = 3(x^{-3/2} + x^{-5/2}), \quad 0.25 \le x \le 1$

11. $\delta(x) = \begin{cases} 2 - x, & 0 \le x < 1 \\ x, & 1 \le x \le 2 \end{cases}$

12. $\delta(x) = \begin{cases} x + 1, & 0 \le x < 1 \\ 2, & 1 \le x \le 2 \end{cases}$

Thin Plates with Constant Density

In Exercises 13–24, find the center of mass of a thin plate of constant density δ covering the given region.

13. The region bounded by the parabola $y = x^2$ and the line $y = 4$

14. The region bounded by the parabola $y = 25 - x^2$ and the x-axis

15. The region bounded by the parabola $y = x - x^2$ and the line $y = -x$

16. The region enclosed by the parabolas $y = x^2 - 3$ and $y = -2x^2$

17. The region bounded by the y-axis and the curve $x = y - y^3$, $0 \le y \le 1$

18. The region bounded by the parabola $x = y^2 - y$ and the line $y = x$

19. The region bounded by the x-axis and the curve $y = \cos x$, $-\pi/2 \le x \le \pi/2$

20. The region between the x-axis and the curve $y = \sec^2 x$, $-\pi/4 \le x \le \pi/4$

21. The region bounded by the parabolas $y = 2x^2 - 4x$ and $y = 2x - x^2$

22. a) The region cut from the first quadrant by the circle $x^2 + y^2 = 9$

    b) The region bounded by the x-axis and the semicircle $y = \sqrt{9 - x^2}$

    Compare your answer with the answer in (a).

23. The "triangular" region in the first quadrant between the circle $x^2 + y^2 = 9$ and the lines $x = 3$ and $y = 3$. (*Hint:* Use geometry to find the area.)

24. The region bounded above by the curve $y = 1/x^3$, below by the curve $y = -1/x^3$, and on the left and right by the lines $x = 1$ and $x = a > 1$. Also, find $\lim_{a \to \infty} \overline{x}$.

Thin Plates with Varying Density

25. Find the center of mass of a thin plate covering the region between the x-axis and the curve $y = 2/x^2$, $1 \le x \le 2$, if the plate's density at the point (x, y) is $\delta(x) = x^2$.

26. Find the center of mass of a thin plate covering the region bounded below by the parabola $y = x^2$ and above by the line $y = x$ if the plate's density at the point (x, y) is $\delta(x) = 12x$.

27. The region bounded by the curves $y = \pm 4/\sqrt{x}$ and the lines $x = 1$ and $x = 4$ is revolved about the y-axis to generate a solid.

    a) Find the volume of the solid.

    b) Find the center of mass of a thin plate covering the region if the plate's density at the point (x, y) is $\delta(x) = 1/x$.

    c) Sketch the plate and show the center of mass in your sketch.

28. The region between the curve $y = 2/x$ and the x-axis from $x = 1$ to $x = 4$ is revolved about the x-axis to generate a solid.

    a) Find the volume of the solid.

    b) Find the center of mass of a thin plate covering the region if the plate's density at the point (x, y) is $\delta(x) = \sqrt{x}$.

    c) Sketch the plate and show the center of mass in your sketch.

Centroids of Triangles

29. *The centroid of a triangle lies at the intersection of the triangle's medians (Fig. 5.61a).* You may recall that the point inside a triangle that lies one-third of the way from each side toward the opposite vertex is the point where the triangle's three medians intersect. Show that the centroid lies at the intersection of the medians by showing that it too lies one-third of the way from each side toward the opposite vertex. To do so, take the following steps.

    1. Stand one side of the triangle on the x-axis as in Fig. 5.61(b). Express dm in terms of L and dy.

    2. Use similar triangles to show that $L = (b/h)(h - y)$. Substitute this expression for L in your formula for dm.

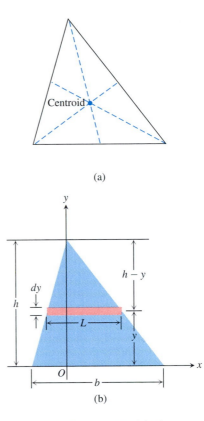

5.61 The triangle in Exercise 29. (a) The centroid. (b) The dimensions and variables to use in locating the center of mass.

    3. Show that $\overline{y} = h/3$.

    4. Extend the argument to the other sides.

Use the result in Exercise 29 to find the centroids of the triangles whose vertices appear in Exercises 30–34. (*Hint:* Draw each triangle first.)

30. $(-1, 0)$, $(1, 0)$, $(0, 3)$ **31.** $(0, 0)$, $(1, 0)$, $(0, 1)$

32. $(0, 0)$, $(a, 0)$, $(0, a)$ **33.** $(0, 0)$, $(a, 0)$, $(0, b)$

34. $(0, 0)$, $(a, 0)$, $(a/2, b)$

Thin Wires

35. Find the moment about the x-axis of a wire of constant density that lies along the curve $y = \sqrt{x}$ from $x = 0$ to $x = 2$.

36. Find the moment about the x-axis of a wire of constant density that lies along the curve $y = x^3$ from $x = 0$ to $x = 1$.

37. Suppose the density of the wire in Example 6 is $\delta = k \sin \theta$ (k constant). Find the center of mass.

38. Suppose the density of the wire in Example 6 is $\delta = 1 + k |\cos \theta|$ (k constant). Find the center of mass.

Engineering Formulas

Verify the statements and formulas in Exercises 39–42.

39. The coordinates of the centroid of a differentiable plane curve are

$$\bar{x} = \frac{\int x \, ds}{\text{length}}, \qquad \bar{y} = \frac{\int y \, ds}{\text{length}}.$$

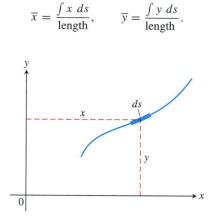

40. Whatever the value of $p > 0$ in the equation $y = x^2/(4p)$, the y-coordinate of the centroid of the parabolic segment shown here is $\bar{y} = (3/5)a$.

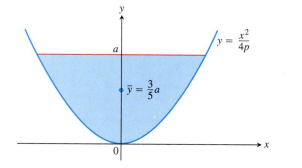

41. For wires and thin rods of constant density shaped like circular arcs centered at the origin and symmetric about the y-axis, the y-coordinate of the center of mass is

$$\bar{y} = \frac{a \sin \alpha}{\alpha} = \frac{ac}{s}.$$

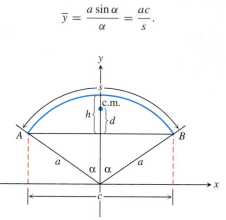

42. (*Continuation of Exercise 41*)

a) Show that when α is small, the distance d from the centroid to chord AB is about $2h/3$ (in the notation of the figure here) by taking the following steps.

1. Show that

$$\frac{d}{h} = \frac{\sin \alpha - \alpha \cos \alpha}{\alpha - \alpha \cos \alpha}. \tag{9}$$

2. GRAPHER Graph

$$f(\alpha) = \frac{\sin \alpha - \alpha \cos \alpha}{\alpha - \alpha \cos \alpha}$$

and use TRACE to show that $\lim_{\alpha \to 0^+} f(\alpha) \approx 2/3$. (You will be able to confirm the suggested equality in Section 6.6, Exercise 74.)

b) CALCULATOR The error (difference between d and $2h/3$) is small even for angles greater than $45°$. See for yourself by evaluating the right-hand side of Eq. (9) for $\alpha = 0.2$, 0.4, 0.6, 0.8, and 1.0 rad.

5.8

Work

In everyday life, *work* means an activity that requires muscular or mental effort. In science, the term refers specifically to a force acting on a body and the body's subsequent displacement. This section shows how to calculate work. The applications run from compressing railroad car springs and emptying subterranean tanks to forcing electrons together and lifting satellites into orbit.

Work Done by a Constant Force

When a body moves a distance d along a straight line as a result of being acted on by a force of constant magnitude F in the direction of motion, we calculate the

Joules

The joule, abbreviated J and pronounced "jewel," is named after the English physicist James Prescott Joule (1818–1889). The defining equation is

$$1 \text{ joule} = (1 \text{ newton})(1 \text{ meter}).$$

In symbols, $1 \text{ J} = 1 \text{ N} \cdot \text{m}$.

It takes a force of about 1 N to lift an apple from a table. If you lift it 1 m you have done about 1 J of work on the apple. If you then eat the apple you will have consumed about 80 food calories, the heat equivalent of nearly 335,000 joules. If this energy were directly useful for mechanical work, it would enable you to lift 335,000 more apples up 1 m.

work W done by the force on the body with the formula

$$W = Fd \qquad \text{(Constant-force formula for work).} \qquad (1)$$

Right away we can see a considerable difference between what we are used to calling work and what this formula says work is. If you push a car down the street, you will be doing work on the car, both by your own reckoning and by Eq. (1). But if you push against the car and the car does not move, Eq. (1) says you will do no work on the car, even if you push for an hour.

From Eq. (1) we see that the unit of work in any system is the unit of force multiplied by the unit of distance. In SI units (SI stands for *Système International*, or International System), the unit of force is a newton, the unit of distance is a meter, and the unit of work is a newton-meter (N · m). This combination appears so often it has a special name, the **joule.** In the British system, the unit of work is the foot-pound, a unit frequently used by engineers.

EXAMPLE 1   If you jack up the side of a 2000-lb car 1.25 ft to change a tire (you have to apply a constant vertical force of about 1000 lb) you will perform $1000 \times 1.25 = 1250$ ft-lb of work on the car. In SI units, you have applied a force of 4448 N through a distance of 0.381 m to do $4448 \times 0.381 \approx 1695$ J of work.

Work Done by a Variable Force

If the force you apply varies along the way, as it will if you are lifting a leaking bucket or compressing a spring, the formula $W = Fd$ has to be replaced by an integral formula that takes the variation in F into account.

Suppose that the force performing the work acts along a line that we can model with the x-axis and that its magnitude F is a continuous function of the position. We want to find the work done over the interval from $x = a$ to $x = b$. We partition $[a, b]$ in the usual way and choose an arbitrary point c_k in each subinterval $[x_{k-1}, x_k]$. If the subinterval is short enough, F, being continuous, will not vary much from x_{k-1} to x_k. The amount of work done across the interval will be about $F(c_k)$ times the distance Δx_k, the same as it would be if F were constant and we could apply Eq. (1). The total work done from a to b is therefore approximated by the Riemann sum

$$\sum_{k=1}^{n} F(c_k) \Delta x_k. \qquad (2)$$

We expect the approximation to improve as the norm of the partition goes to zero, so we define the work done by the force from a to b to be the integral of F from a to b.

Definition

The **work** done by a variable force $F(x)$ directed along the x-axis from $x = a$ to $x = b$ is

$$W = \int_a^b F(x)\, dx. \qquad (3)$$

x

20

0

5.62 The leaky bucket in Example 3.

The units of the integral are joules if F is in newtons and x is in meters, and foot-pounds if F is in pounds and x in feet.

EXAMPLE 2   The work done by a force of $F(x) = 1/x^2$ N along the x-axis from $x = 1$ m to $x = 10$ m is

$$W = \int_1^{10} \frac{1}{x^2}\, dx = -\frac{1}{x}\Big]_1^{10} = -\frac{1}{10} + 1 = 0.9 \text{ J.}$$

EXAMPLE 3   A leaky 5-lb bucket is lifted from the ground into the air by pulling in 20 ft of rope at a constant speed (Fig. 5.62). The rope weighs 0.08 lb/ft. The bucket starts with 2 gal of water (16 lb) and leaks at a constant rate. It finishes draining just as it reaches the top. How much work was spent

a)   lifting the water alone;
b)   lifting the water and bucket together;
c)   lifting the water, bucket, and rope?

Solution

a)   *The water alone.* The force required to lift the water is equal to the water's weight, which varies steadily from 16 to 0 lb over the 20-ft lift. When the bucket is x ft off the ground, the water weighs

$$F(x) = 16\left(\frac{20-x}{20}\right) = 16\left(1 - \frac{x}{20}\right) = 16 - \frac{4x}{5} \text{ lb.}$$

original weight   proportion left
of water   at elevation x

The work done is

$$W = \int_a^b F(x)\, dx \qquad \text{Use Eq. (3) for variable forces.}$$

$$= \int_0^{20} \left(16 - \frac{4x}{5}\right) dx = \left[16x - \frac{2x^2}{5}\right]_0^{20} = 320 - 160 = 160 \text{ ft} \cdot \text{lb.}$$

b)   *The water and bucket together.* According to Eq. (1), it takes $5 \times 20 = 100$ ft · lb to lift a 5-lb weight 20 ft. Therefore

$$160 + 100 = 260 \text{ ft} \cdot \text{lb}$$

of work were spent lifting the water and bucket together.

c)   *The water, bucket, and rope.* Now the total weight at level x is

lb/ft   ft

$$F(x) = \left(16 - \frac{4x}{5}\right) + 5 + (0.08)(20 - x).$$

variable   constant   weight of rope
weight   weight   paid out at
of water   of bucket   elevation x

The work lifting the rope is

$$\text{Work on rope} = \int_0^{20} (0.08)(20 - x)\, dx = \int_0^{20} (1.6 - 0.08x)\, dx$$

$$= \left[1.6x - 0.04x^2\right]_0^{20} = 32 - 16 = 16 \text{ ft} \cdot \text{lb}.$$

The total work for the water, bucket, and rope combined is

$$160 + 100 + 16 = 276 \text{ ft} \cdot \text{lb}. \qquad \square$$

Hooke's Law for Springs: *F = kx*

Hooke's law says that the force it takes to stretch or compress a spring x length units from its natural (unstressed) length is proportional to x. In symbols,

$$F = kx. \qquad (4)$$

The constant k, measured in force units per unit length, is a characteristic of the spring, called the **force constant** (or spring constant) of the spring. Hooke's law (Eq. 4) gives good results as long as the force doesn't distort the metal in the spring. We assume that the forces in this section are too small to do that.

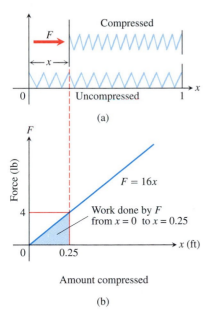

(a)

(b)

5.63 The force *F* needed to hold a spring under compression increases linearly as the spring is compressed.

EXAMPLE 4 Find the work required to compress a spring from its natural length of 1 ft to a length of 0.75 ft if the force constant is $k = 16$ lb/ft.

Solution We picture the uncompressed spring laid out along the x-axis with its movable end at the origin and its fixed end at $x = 1$ ft (Fig. 5.63). This enables us to describe the force required to compress the spring from 0 to x with the formula $F = 16x$. To compress the spring from 0 to 0.25 ft, the force must increase from

$$F(0) = 16 \cdot 0 = 0 \text{ lb} \qquad \text{to} \qquad F(0.25) = 16 \cdot 0.25 = 4 \text{ lb}.$$

The work done by F over this interval is

$$W = \int_0^{0.25} 16x\, dx = 8x^2 \Big]_0^{0.25} = 0.5 \text{ ft} \cdot \text{lb}. \qquad \text{Eq. (3) with } a = 0, \ b = 0.25, F(x) = 16x$$

$\square$

EXAMPLE 5 A spring has a natural length of 1 m. A force of 24 N stretches the spring to a length of 1.8 m.

a) Find the force constant k.
b) How much work will it take to stretch the spring 2 m beyond its natural length?
c) How far will a 45-N force stretch the spring?

Solution

a) *The force constant.* We find the force constant from Eq. (4). A force of 24 N stretches the spring 0.8 m, so

$$24 = k(0.8) \qquad \text{Eq. (4) with } F = 24, x = 0.8$$

$$k = 24/0.8 = 30 \text{ N/m}.$$

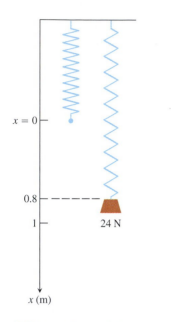

5.64 A 24-N weight stretches this spring 0.8 m beyond its unstressed length.

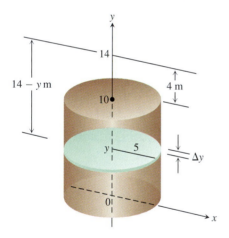

5.65 To find the work it takes to pump the water from a tank, think of lifting the water one thin slab at a time.

How to Find Work Done During Pumping

1. Draw a figure with a coordinate system.
2. Find the weight F of a thin horizontal slab of liquid.
3. Find the work ΔW it takes to lift the slab to its destination.
4. Integrate the work expression from the base to the surface of the liquid.

b) *The work to stretch the spring* 2 m. We imagine the unstressed spring hanging along the *x*-axis with its free end at $x = 0$ (Fig. 5.64). The force required to stretch the spring x m beyond its natural length is the force required to pull the free end of the spring x units from the origin. Hooke's law with $k = 30$ says that this force is

$$F(x) = 30x.$$

The work done by F on the spring from $x = 0$ m to $x = 2$ m is

$$W = \int_0^2 30x \, dx = 15x^2 \Big]_0^2 = 60 \text{ J.}$$

c) *How far will a 45-N force stretch the spring?* We substitute $F = 45$ in the equation $F = 30x$ to find

$$45 = 30x, \quad \text{or} \quad x = 1.5 \text{ m.}$$

A 45-N force will stretch the spring 1.5 m. No calculus is required to find this.   ❑

Pumping Liquids from Containers

How much work does it take to pump all or part of the liquid from a container? To find out, we imagine lifting the liquid out one thin horizontal slab at a time and applying the equation $W = Fd$ to each slab. We then evaluate the integral this leads to as the slabs become thinner and more numerous. The integral we get each time depends on the weight of the liquid and the dimensions of the container, but the way we find the integral is always the same. The next examples show what to do.

EXAMPLE 6   How much work does it take to pump the water from a full upright circular cylindrical tank of radius 5 m and height 10 m to a level of 4 m above the top of the tank?

Solution   We draw the tank (Fig. 5.65), add coordinate axes, and imagine the water divided into thin horizontal slabs by planes perpendicular to the *y*-axis at the points of a partition P of the interval [0, 10].

The typical slab between the planes at y and $y + \Delta y$ has a volume of

$$\Delta V = \pi (\text{radius})^2 (\text{thickness}) = \pi (5)^2 \Delta y = 25\pi \, \Delta y \text{ m}^3.$$

The force F required to lift the slab is equal to its weight,

$$F = 9800 \Delta V$$

Water weighs 9800 N/m³.

$$= 9800(25\pi \, \Delta y) = 245{,}000\pi \, \Delta y \text{ N.}$$

The distance through which F must act is about $(14 - y)$ m, so the work done lifting the slab is about

$$\Delta W = \text{force} \times \text{distance} = 245{,}000\pi (14 - y) \Delta y \text{ J.}$$

The work it takes to lift all the water is approximately

$$W \approx \sum_0^{10} \Delta W = \sum_0^{10} 245{,}000\pi (14 - y) \Delta y \text{ J.}$$

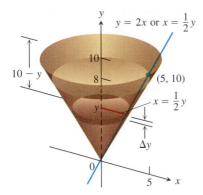

5.66 The olive oil in Example 7.

This is a Riemann sum for the function $245{,}000\pi(14 - y)$ over the interval $0 \le y \le 10$. The work of pumping the tank dry is the limit of these sums as $\|P\| \to 0$:

$$W = \int_0^{10} 245{,}000\pi(14 - y)\,dy = 245{,}000\pi \int_0^{10} (14 - y)\,dy$$

$$= 245{,}000\pi \left[14y - \frac{y^2}{2} \right]_0^{10} = 245{,}000\pi[90]$$

$$\approx 69{,}272{,}118 \approx 69.3 \times 10^6 \text{ J}.$$

A 1-horsepower output motor rated at 746 J/sec could empty the tank in a little less than 26 h. $\square$

EXAMPLE 7   The conical tank in Fig. 5.66 is filled to within 2 ft of the top with olive oil weighing 57 lb/ft^3. How much work does it take to pump the oil to the rim of the tank?

Solution   We imagine the oil divided into thin slabs by planes perpendicular to the y-axis at the points of a partition of the interval [0, 8].

The typical slab between the planes at y and $y + \Delta y$ has a volume of about

$$\Delta V = \pi (\text{radius})^2 (\text{thickness}) = \pi \left(\frac{1}{2}y \right)^2 \Delta y = \frac{\pi}{4}y^2 \Delta y \text{ ft}^3.$$

The force $F(y)$ required to lift this slab is equal to its weight,

$$F(y) = 57\,\Delta V = \frac{57\pi}{4}y^2 \Delta y \text{ lb.} \qquad \begin{array}{l} \text{Weight = weight per unit} \\ \text{volume} \times \text{volume} \end{array}$$

The distance through which $F(y)$ must act to lift this slab to the level of the rim of the cone is about $(10 - y)$ ft, so the work done lifting the slab is about

$$\Delta W = \frac{57\pi}{4}(10 - y)y^2 \Delta y \text{ ft} \cdot \text{lb.}$$

The work done lifting all the slabs from $y = 0$ to $y = 8$ to the rim is approximately

$$W \approx \sum_0^8 \frac{57\pi}{4}(10 - y)y^2 \Delta y \text{ ft} \cdot \text{lb.}$$

This is a Riemann sum for the function $(57\pi/4)(10 - y)y^2$ on the interval from $y = 0$ to $y = 8$. The work of pumping the oil to the rim is the limit of these sums as the norm of the partition goes to zero.

$$W = \int_0^8 \frac{57\pi}{4}(10 - y)y^2\,dy$$

$$= \frac{57\pi}{4} \int_0^8 (10y^2 - y^3)\,dy$$

$$= \frac{57\pi}{4} \left[\frac{10y^3}{3} - \frac{y^4}{4} \right]_0^8 \approx 30{,}561 \text{ ft} \cdot \text{lb.} \qquad \square$$

Exercises 5.8

Work Done by a Variable Force

1. The workers in Example 3 changed to a larger bucket that held 5 gal (40 lb) of water, but the new bucket had an even larger leak so that it, too, was empty by the time it reached the top. Assuming that the water leaked out at a steady rate, how much work was done lifting the water? (Do not include the rope and bucket.)

2. The bucket in Example 3 is hauled up twice as fast so that there is still 1 gal (8 lb) of water left when the bucket reaches the top. How much work is done lifting the water this time? (Do not include the rope and bucket.)

3. A mountain climber is about to haul up a 50-m length of hanging rope. How much work will it take if the rope weighs 0.624 N/m?

4. A bag of sand originally weighing 144 lb was lifted at a constant rate. As it rose, sand also leaked out at a constant rate. The sand was half gone by the time the bag had been lifted 18 ft. How much work was done lifting the sand this far? (Neglect the weight of the bag and lifting equipment.)

5. An electric elevator with a motor at the top has a multistrand cable weighing 4.5 lb/ft. When the car is at the first floor, 180 ft of cable are paid out, and effectively 0 ft are out when the car is at the top floor. How much work does the motor do just lifting the cable when it takes the car from the first floor to the top?

6. When a particle of mass m is at $(x, 0)$, it is attracted toward the origin with a force whose magnitude is k/x^2. If the particle starts from rest at $x = b$ and is acted on by no other forces, find the work done on it by the time it reaches $x = a, 0 < a < b$.

7. Suppose that the gas in a circular cylinder of cross-section area A is being compressed by a piston. If p is the pressure of the gas in pounds per square inch and V is the volume in cubic inches, show that the work done in compressing the gas from state (p_1, V_1) to state (p_2, V_2) is given by the equation

$$\text{Work} = \int_{(p_1, V_1)}^{(p_2, V_2)} p \, dV.$$

(*Hint:* In the coordinates suggested in the figure here, $dV = A \, dx$. The force against the piston is pA.)

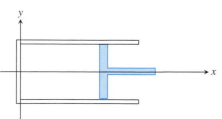

8. (*Continuation of Exercise 7.*) Use the integral in Exercise 7 to find the work done in compressing the gas from $V_1 = 243 \text{ in}^3$

to $V_2 = 32 \text{ in}^3$ if $p_1 = 50 \text{ lb/in}^3$ and p and V obey the gas law $pV^{1.4} = \text{constant}$ (for adiabatic processes).

Springs

9. It took 1800 J of work to stretch a spring from its natural length of 2 m to a length of 5 m. Find the spring's force constant.

10. A spring has a natural length of 10 in. An 800-lb force stretches the spring to 14 in. (a) Find the force constant. (b) How much work is done in stretching the spring from 10 in. to 12 in.? (c) How far beyond its natural length will a 1600-lb force stretch the spring?

11. A force of 2 N will stretch a rubber band 2 cm (0.02 m). Assuming Hooke's law applies, how far will a 4-N force stretch the rubber band? How much work does it take to stretch the rubber band this far?

12. If a force of 90 N stretches a spring 1 m beyond its natural length, how much work does it take to stretch the spring 5 m beyond its natural length?

13. *Subway car springs.* It takes a force of 21,714 lb to compress a coil spring assembly on a New York City Transit Authority subway car from its free height of 8 in. to its fully compressed height of 5 in.

    a)   What is the assembly's force constant?
    b)   How much work does it take to compress the assembly the first half inch? the second half inch? Answer to the nearest in · lb.

    (Data courtesy of Bombardier, Inc., Mass Transit Division, for spring assemblies in subway cars delivered to the New York City Transit Authority from 1985 to 1987.)

14. A bathroom scale is compressed 1/16 in. when a 150-lb person stands on it. Assuming the scale behaves like a spring that obeys Hooke's law, how much does someone who compresses the scale 1/8 in. weigh? How much work is done compressing the scale 1/8 in.?

Pumping Liquids from Containers

The Weight of Water
Because of variations in the earth's gravitational field, the weight of a cubic foot of water at sea level can vary from about 62.26 lb at the equator to as much as 62.59 lb near the poles, a variation of about 0.5%. A cubic foot that weighs about 62.4 lb in Melbourne and New York City will weigh 62.5 lb in Juneau and Stockholm. While 62.4 is a typical figure and a common textbook value, there is considerable variation.

15. The rectangular tank shown here, with its top at ground level, is used to catch runoff water. Assume that the water weighs 62.4 lb/ft³.

   a) How much work does it take to empty the tank by pumping the water back to ground level once the tank is full?

   b) If the water is pumped to ground level with a (5/11)-hp motor (work output 250 ft · lb/sec), how long will it take to empty the full tank (to the nearest minute)?

   c) Show that the pump in part (b) will lower the water level 10 ft (halfway) during the first 25 min of pumping.

   d) What are the answers to parts (a) and (b) in a location where water weighs 62.26 lb/ft³? 62.59 lb/ft³?

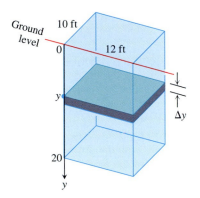

16. The rectangular cistern (storage tank for rainwater) shown here has its top 10 ft below ground level. The cistern, currently full, is to be emptied for inspection by pumping its contents to ground level.

   a) How much work will it take to empty the cistern?

   b) How long will it take a (1/2)-hp pump, rated at 275 ft · lb/sec, to pump the tank dry?

   c) How long will it take the pump in part (b) to empty the tank halfway? (It will be less than half the time required to empty the tank completely.)

   d) What are the answers to parts (a)–(c) in a location where water weighs 62.26 lb/ft³? 62.59 lb/ft³?

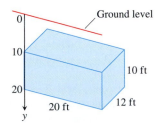

17. How much work would it take to pump the water from the tank in Example 6 to the level of the top of the tank (instead of 4 m higher)?

18. Suppose that, instead of being full, the tank in Example 6 is only half full. How much work does it take to pump the remaining water to a level 4 m above the top of the tank?

19. A vertical right circular cylindrical tank measures 30 ft high and 20 ft in diameter. It is full of kerosene weighing 51.2 lb/ft³. How much work does it take to pump the kerosene to the level of the top of the tank?

20. The cylindrical tank shown here can be filled by pumping water from a lake 15 ft below the bottom of the tank. There are two ways to go about it. One is to pump the water through a hose attached to a valve in the bottom of the tank. The other is to attach the hose to the rim of the tank and let the water pour in. Which way will be faster? Give reasons for your answer.

21. CALCULATOR The truncated conical container shown here is full of strawberry milkshake that weighs 4/9 oz/in³. As you can see, the container is 7 in. deep, 2.5 in. across at the base, and 3.5 in. across at the top (a standard size at Brigham's in Boston). The straw sticks up an inch above the top. About how much work does it take to suck up the milkshake through the straw (neglecting friction)? Answer in inch-ounces.

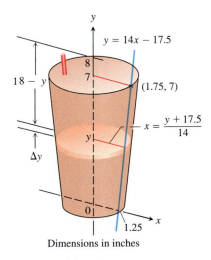

Dimensions in inches

22. a) Suppose the conical container in Example 7 contains milk (weighing 64.5 lb/ft³) instead of olive oil. How much work will it take to pump the contents to the rim?

   b) How much work will it take to pump the oil in Example 7 to a level 3 ft above the cone's rim?

23. To design the interior surface of a huge stainless steel tank, you revolve the curve $y = x^2$, $0 \le x \le 4$, about the y-axis. The container, with dimensions in meters, is to be filled with seawater, which weighs 10,000 N/m³. How much work will it take to empty the tank by pumping the water to the tank's top?

24. We model pumping from spherical containers the way we do from other containers, with the axis of integration along the vertical axis of the sphere. Use the figure here to find how much work it takes to empty a full hemispherical water reservoir of radius 5 m by pumping the water to a height of 4 m above the top of the reservoir. Water weighs 9800 N/m³.

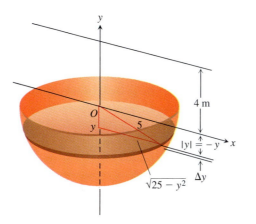

25. You are in charge of the evacuation and repair of the storage tank shown here. The tank is a hemisphere of radius 10 ft and is full of benzene weighing 56 lb/ft³. A firm you contacted says it can empty the tank for 1/2¢ per foot-pound of work. Find the work required to empty the tank by pumping the benzene to an outlet 2 ft above the top of the tank. If you have $5000 budgeted for the job, can you afford to hire the firm?

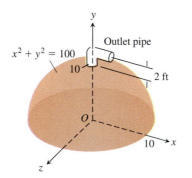

26. Your town has decided to drill a well to increase its water supply. As the town engineer, you have determined that a water tower will be necessary to provide the pressure needed for distribution, and you have designed the system shown here. The water is to be pumped from a 300-ft well through a vertical 4-in. pipe into the base of a cylindrical tank 20 ft in diameter and 25 ft high. The base of the tank will be 60 ft aboveground. The pump is a 3-hp pump, rated at 1650 ft · lb/sec. To the nearest hour, how

long will it take to fill the tank the first time? (Include the time it takes to fill the pipe.) Assume water weighs 62.4 lb/ft³.

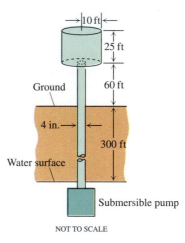

NOT TO SCALE

Other Applications

27. *Putting a satellite in orbit.* The strength of the earth's gravitational field varies with the distance r from the earth's center, and the magnitude of the gravitational force experienced by a satellite of mass m during and after launch is

$$F(r) = \frac{mMG}{r^2}.$$

Here, $M = 5.975 \times 10^{24}$ kg is the earth's mass, $G = 6.6720 \times 10^{-11}$ N · m²kg⁻² is the universal gravitational constant, and r is measured in meters. The work it takes to lift a 1000-kg satellite from the earth's surface to a circular orbit 35,780 km above the earth's center is therefore given by the integral

$$\text{Work} = \int_{6,370,000}^{35,780,000} \frac{1000MG}{r^2} \, dr \text{ joules.}$$

Evaluate the integral. The lower limit of integration is the earth's radius in meters at the launch site. (This calculation does not take into account energy spent lifting the launch vehicle or energy spent bringing the satellite to orbit velocity.)

28. *Forcing electrons together.* Two electrons r meters apart repel each other with a force of

$$F = \frac{23 \times 10^{-29}}{r^2} \text{ newtons.}$$

a) Suppose one electron is held fixed at the point $(1, 0)$ on the x-axis (units in meters). How much work does it take to move a second electron along the x-axis from the point $(-1, 0)$ to the origin?

b) Suppose an electron is held fixed at each of the points $(-1, 0)$ and $(1, 0)$. How much work does it take to move a third electron along the x-axis from $(5, 0)$ to $(3, 0)$?

Work and Kinetic Energy

29. If a variable force of magnitude $F(x)$ moves a body of mass

m along the x-axis from x_1 to x_2, the body's velocity v can be written as dx/dt (where t represents time). Use Newton's Second Law of Motion $F = m(dv/dt)$ and the Chain Rule

$$\frac{dv}{dt} = \frac{dv}{dx}\frac{dx}{dt} = v\frac{dv}{dx}$$

to show that the net work done by the force in moving the body from x_1 to x_2 is

$$W = \int_{x_1}^{x_2} F(x)\,dx = \frac{1}{2}mv_2^2 - \frac{1}{2}mv_1^2,$$

where v_1 and v_2 are the body's velocities at x_1 and x_2. In physics the expression $(1/2)mv^2$ is called the *kinetic energy* of the body moving with velocity v. Therefore, *the work done by the force equals the change in the body's kinetic energy,* and we can find the work by calculating this change.

In Exercises 30–36, use the result of Exercise 29.

30. *Tennis.* A 2-oz tennis ball was served at 160 ft/sec (about 109 mph). How much work was done on the ball to make it go this fast? (To find the ball's mass from its weight, express the weight in pounds and divide by 32 ft/sec², the acceleration of gravity.)

31. *Baseball.* How many foot-pounds of work does it take to throw a baseball 90 mph? A baseball weighs 5 oz $= 0.3125$ lb.

32. *Golf.* A 1.6-oz golf ball is driven off the tee at a speed of 280 ft/sec (about 191 mph). How many foot-pounds of work are done getting the ball into the air?

33. *Tennis.* During the match in which Pete Sampras won the 1990 U.S. Open men's tennis championship, Sampras hit a serve that

was clocked at a phenomenal 124 mph. How much work did Sampras have to do on the 2-oz ball to get it to that speed?

34. *Football.* A quarterback threw a 14.5-oz football 88 ft/sec (60 mph). How many foot-pounds of work were done on the ball to get it to this speed?

35. *Softball.* How much work has to be performed on a 6.5-oz softball to pitch it 132 ft/sec (90 mph)?

36. *A ball bearing.* A 2-oz steel ball bearing is placed on a vertical spring whose force constant is $k = 18$ lb/ft. The spring is compressed 3 inches and released. About how high does the ball bearing go?

Weight vs. Mass

Weight is the force that results from gravity pulling on a mass. The two are related by the equation in Newton's second law,

$$\text{Weight} = \text{mass} \times \text{acceleration.}$$

Thus,

$$\text{Newtons} = \text{kilograms} \times \text{m/sec}^2,$$

$$\text{Pounds} = \text{slugs} \times \text{ft/sec}^2.$$

To convert mass to weight, multiply by the acceleration of gravity. To convert weight to mass, divide by the acceleration of gravity.

5.9 | **Fluid Pressures and Forces**

We make dams thicker at the bottom than at the top (Fig. 5.67) because the pressure against them increases with depth. It is a remarkable fact that the pressure at any point on a dam depends only on how far below the surface the point is and not on how much the surface of the dam happens to be tilted at that point. The pressure, in pounds per square foot at a point h feet below the surface, is always $62.4h$. The number 62.4 is the weight-density of water in pounds per cubic foot.

The formula, pressure $= 62.4h$, makes sense when you think of the units involved:

$$\frac{\text{lb}}{\text{ft}^2} = \frac{\text{lb}}{\text{ft}^3} \times \text{ft.}$$

5.67 To withstand the increasing pressure, dams are built thicker as they go down.

As you can see, this equation depends only on units and not on the fluid involved. The pressure h feet below the surface of any fluid is the fluid's weight-density times h.

Weight-density

A fluid's weight-density is its weight per unit volume. Typical values (lb/ft^3) are

Gasoline	42
Mercury	849
Milk	64.5
Molasses	100
Olive oil	57
Seawater	64
Water	62.4

5.68 These containers are filled with water to the same depth and have the same base area. The total force is therefore the same on the bottom of each container. The containers' shapes do not matter here.

5.69 Schematic drawing of the molasses tank in Example 1. How much force did the lowest foot of the vertical wall have to withstand when the tank was full? It takes an integral to find out. Notice that the proportions of the tank were ideal.

The Pressure-Depth Equation

In a fluid that is standing still, the pressure p at depth h is the fluid's weight-density w times h:

$$p = wh. \qquad (1)$$

In this section we use the equation $p = wh$ to derive a formula for the total force exerted by a fluid against all or part of a vertical or horizontal containing wall.

The Constant-Depth Formula for Fluid Force

In a container of fluid with a flat horizontal base, the total force exerted by the fluid against the base can be calculated by multiplying the area of the base by the pressure at the base. We can do this because total force equals force per unit area (pressure) times area. (See Fig. 5.68.) If F, p, and A are the total force, pressure, and area, then

$$F = \text{total force} = \text{force per unit area} \times \text{area}$$

$$= \text{pressure} \times \text{area} = pA$$

$$= whA. \qquad \text{\small $p = wh$ from Eq. (1)}$$

Fluid Force on a Constant-Depth Surface

$$F = pA = whA \qquad (2)$$

EXAMPLE 1   *The Great Molasses Flood*

At 1:00 P.M. on January 15, 1919, an unusually warm day, a 90-ft-high, 90-ft-diameter cylindrical metal tank in which the Puritan Distilling Company was storing molasses at the corner of Foster and Commercial streets in Boston's North End exploded. The molasses flooded into the streets, 30 ft deep, trapping pedestrians and horses, knocking down buildings, and oozing into homes. It was eventually tracked all over town and even made its way into the suburbs (on trolley cars and people's shoes). It took weeks to clean up.

Given that the molasses weighed 100 lb/ft^3, what was the total force exerted by the molasses against the bottom of the tank at the time it blew? Assuming the tank was full, we can find out from Eq. (2):

$$\text{Total force} = whA = (100)(90)(\pi(45)^2) \approx 57{,}255{,}526 \text{ lb.} \qquad \square$$

How about the force against the walls of the tank? For example, what was the total force against the bottom foot-wide band of tank wall (Fig. 5.69)? The area of the band was

$$A = 2\pi rh = 2\pi(45)(1) = 90\pi \ \text{ft}^2.$$

y

Surface of fluid

Submerged vertical
plate

Strip
depth

Δy

b

y

a

$L(y)$

Strip length at level y

5.70 The force exerted by a fluid against one side of a thin horizontal strip is about ΔF = pressure × area = w × (strip depth) × $L(y)\,\Delta y$. The plate here is flat, but it might have been curved instead, like the vertical wall of a cylindrical tank. Whatever the case, the strip length is measured along the surface of the plate.

The tank was 90 ft deep, so the pressure near the bottom was approximately

$$p = wh = (100)(90) = 9000 \text{ lb/ft}^2.$$

Therefore the total force against the band was approximately

$$F = whA = (9000)(90\pi) \approx 2,544,690 \text{ lb}.$$

But this is not exactly right. The top of the band was 89 ft below the surface, not 90, and the pressure there was less. To find out exactly what the force on the band was, we need to take into account the variation of the pressure across the band.

The Variable-Depth Formula

Suppose we want to know the force exerted by a fluid against one side of a vertical plate submerged in a fluid of weight-density w. To find it, we model the plate as a region extending from $y = a$ to $y = b$ in the xy-plane (Fig. 5.70). We partition $[a, b]$ in the usual way and imagine the region to be cut into thin horizontal strips by planes perpendicular to the y-axis at the partition points. The typical strip from y to $y + \Delta y$ is Δy units wide by $L(y)$ units long. We assume $L(y)$ to be a continuous function of y.

The pressure varies across the strip from top to bottom, just as it did in the molasses tank. But if the strip is narrow enough, the pressure will remain close to its bottom-edge value of w × (strip depth). The force exerted by the fluid against one side of the strip will be about

$$\Delta F = (\text{pressure along bottom edge}) \times (\text{area})$$

$$= w \times (\text{strip depth}) \times L(y)\Delta y.$$

The force against the entire plate will be about

$$\sum_{a}^{b} \Delta F = \sum_{a}^{b} (w \times (\text{strip depth}) \times L(y)\Delta y). \qquad (3)$$

The sum in (3) is a Riemann sum for a continuous function on $[a, b]$, and we expect the approximations to improve as the norm of the partition goes to zero. We define the force against the plate to be the limit of these sums.

Definition
The Integral for Fluid Force
Suppose that a plate submerged vertically in fluid of weight-density w runs from $y = a$ to $y = b$ on the y-axis. Let $L(y)$ be the length of the horizontal strip measured from left to right along the surface of the plate at level y. Then the force exerted by the fluid against one side of the plate is

$$F = \int_{a}^{b} w \cdot (\text{strip depth}) \cdot L(y)\,dy. \qquad (4)$$

EXAMPLE 2   A flat isosceles right triangular plate with base 6 ft and height 3 ft is submerged vertically, base up, 2 ft below the surface of a swimming pool. Find the force exerted by the water against one side of the plate.

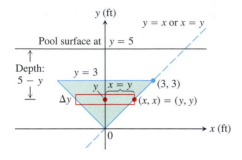

5.71 To find the force on one side of the submerged plate in Example 2, we can use a coordinate system like the one here.

Solution We establish a coordinate system to work in by placing the origin at the plate's bottom vertex and running the y-axis upward along the plate's axis of symmetry (Fig. 5.71). (We will look at other coordinate systems in Exercises 3 and 4.) The surface of the pool lies along the line $y = 5$ and the plate's top edge along the line $y = 3$. The plate's right-hand edge lies along the line $y = x$, with the upper right vertex at $(3, 3)$. The length of a thin strip at level y is

$$L(y) = 2x = 2y.$$

The depth of the strip beneath the surface is $(5 - y)$. The force exerted by the water against one side of the plate is therefore

$$F = \int_a^b w \times \left(\begin{array}{c} \text{strip} \\ \text{depth} \end{array} \right) \times L(y)\, dy \qquad \text{Eq. (4)}$$

$$= \int_0^3 62.4(5 - y)\, 2y\, dy$$

$$= 124.8 \int_0^3 (5y - y^2)\, dy$$

$$= 124.8 \left[\frac{5}{2} y^2 - \frac{y^3}{3} \right]_0^3 = 1684.8 \text{ lb.}$$

How to Find Fluid Force

Whatever coordinate system you use, you can find the fluid force against one side of a submerged vertical plate or wall by taking these steps.

1. *Find expressions* for the length and depth of a typical thin horizontal strip.
2. *Multiply their product by the fluid's weight-density w and integrate* over the interval of depths occupied by the plate or wall.

5.72 The molasses tank with the coordinate origin at the bottom (Example 3).

EXAMPLE 3 We can now calculate exactly the force exerted by the molasses against the bottom 1-ft band of the Puritan Distilling Company's storage tank when the tank was full.

The tank was a right circular cylindrical tank 90 ft high and 90 ft in diameter. Using a coordinate system with the origin at the bottom of the tank and the y-axis pointing up (Fig. 5.72), we find that the typical horizontal strip at level y has

Strip depth: $90 - y,$

Strip length: $\pi \times \text{tank diameter} = 90\pi.$

The force against the band is therefore

$$\text{Force} = \int_0^1 w(\text{depth})(\text{length})\, dy = \int_0^1 100(90 - y)(90\pi)\, dy \qquad \begin{array}{l} \text{For molasses,} \\ w = 100 \end{array}$$

$$= 9000\pi \int_0^1 (90 - y)\, dy \approx 2{,}530{,}553 \text{ lb.}$$

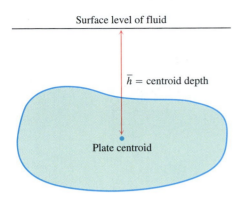

Surface level of fluid

$\bar{h}$ = centroid depth

Plate centroid

5.73 The force against one side of the plate is $w \cdot \bar{h} \cdot$ plate area.

As expected, the force is slightly less than the constant-depth estimate following Example 1.

Fluid Forces and Centroids

If we know the location of the centroid of a submerged flat vertical plate (Fig. 5.73), we can take a shortcut to find the force against one side of the plate. From Eq. (4),

$$F = \int_a^b w \times (\text{strip depth}) \times L(y)\, dy$$

$$= w \int_a^b (\text{strip depth}) \times L(y)\, dy$$

$$= w \times (\text{moment about surface level line of region occupied by plate})$$

$$= w \times (\text{depth of plate's centroid}) \times (\text{area of plate}).$$

Fluid Forces and Centroids

The force of a fluid of weight-density w against one side of a submerged flat vertical plate is the product of w, the distance $\bar{h}$ from the plate's centroid to the fluid surface, and the plate's area:

$$F = w\bar{h}\, A. \tag{5}$$

EXAMPLE 4   Use Eq. (5) to find the force in Example 2.

Solution   The centroid of the triangle (Fig. 5.71) lies on the y-axis, one-third of the way from the base to the vertex, so $\bar{h} = 3$. The triangle's area is

$$A = \frac{1}{2}(\text{base})(\text{height})$$

$$= \frac{1}{2}(6)(3) = 9.$$

Hence,

$$F = w\bar{h}\, A = (62.4)(3)(9)$$

$$= 1684.8 \text{ lb.} \qquad \square$$

Equation (5) says that the fluid force on one side of a submerged flat vertical plate is the same as it would be if the plate's entire area lay $\bar{h}$ units beneath the surface. For many shapes, the location of the centroid can be found in a table, and Eq. (5) gives a practical way to find F. Of course, the centroid's location was found by someone who performed an integration equivalent to evaluating the integral in Eq. (4). We recommend for now that you practice your mathematical modeling by drawing pictures and thinking things through the way we did when we developed Eq. (4). Then check your results, when you conveniently can, with Eq. (5).

Exercises 5.9

The weight-densities of the fluids in the following exercises can be found in the table on page 428.

1. What was the total fluid force against the cylindrical inside wall of the molasses tank in Example 1 when the tank was full? half full?

2. What was the total fluid force against the bottom 1-ft band of the inside wall of the molasses tank in Example 1 when the tank was half full?

3. Calculate the fluid force on one side of the plate in Example 2 using the coordinate system shown here.

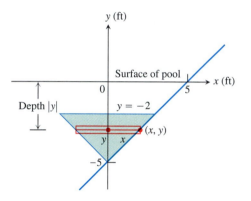

4. Calculate the fluid force on one side of the plate in Example 2 using the coordinate system shown here.

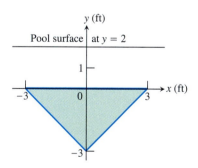

5. The plate in Example 2 is lowered another 2 ft into the water. What is the fluid force on one side of the plate now?

6. The plate in Example 2 is raised to put its top edge at the surface of the pool. What is the fluid force on one side of the plate now?

7. The isosceles triangular plate shown here is submerged vertically 1 ft below the surface of a freshwater lake.

   a) Find the fluid force against one face of the plate.
   b) What would be the fluid force on one side of the plate if the water were seawater instead of freshwater?

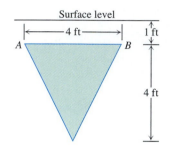

8. The plate in Exercise 7 is revolved 180° about line *AB* so that part of the plate sticks out of the lake, as shown here. What force does the water exert on one face of the plate now?

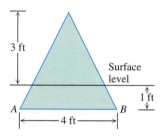

9. The vertical ends of a watering trough are isosceles triangles like the one shown here (dimensions in feet).

   a) Find the fluid force against the ends when the trough is full.
   b) CALCULATOR How many inches do you have to lower the water level in the trough to cut the fluid force on the ends in half? (Answer to the nearest half inch.)
   c) Does it matter how long the trough is? Give reasons for your answer.

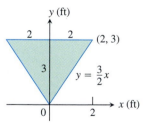

10. The vertical ends of a watering trough are squares 3 ft on a side.

   a) Find the fluid force against the ends when the trough is full.
   b) CALCULATOR How many inches do you have to lower the water level in the trough to reduce the fluid force by 25%?
   c) Does it matter how long the trough is? Give reasons for your answer.

11. The viewing portion of the rectangular glass window in a typical fish tank at the New England Aquarium in Boston is 63 in. wide and runs from 0.5 in. below the water's surface to 33.5 in. below the surface. Find the fluid force against this portion of the window. The weight-density of seawater is 64 lb/ft³. (In case you were wondering, the glass is 3/4 in. thick and the tank walls extend 4 in. above the water to keep the fish from jumping out.)

12. A horizontal rectangular freshwater fish tank with base 2 × 4 ft and height 2 ft (interior dimensions) is filled to within 2 in. of the top.

   a) Find the fluid force against each side and end of the tank.
   b) If the tank is sealed and stood on end (without spilling), so that one of the square ends is the base, what does that do to the fluid forces on the rectangular sides?

13. **CALCULATOR** A rectangular milk carton measures 3.75 × 3.75 in. at the base and is 7.75 in. tall. Find the force of the milk on one side when the carton is full.

14. **CALCULATOR** A standard olive oil can measures 5.75 by 3.5 in. at the base and is 10 in. tall. Find the fluid force against the base and each side when the can is full.

15. A semicircular plate 2 ft in diameter sticks straight down into fresh water with the diameter along the surface. Find the force exerted by the water on one side of the plate.

16. A tank truck hauls milk in a 6-ft-diameter horizontal right circular cylindrical tank. How much force does the milk exert on each end of the tank when the tank is half full?

17. The cubical metal tank shown here has a parabolic gate, held in place by bolts and designed to withstand a fluid force of 160 lb without rupturing. The liquid you plan to store has a weight-density of 50 lb/ft³.

   a) What is the fluid force on the gate when the liquid is 2 ft deep?
   b) What is the maximum height to which the container can be filled without exceeding its design limitation?

Parabolic gate

Enlarged view of parabolic gate

18. The rectangular tank shown here has a 1 ft × 1 ft square window 1 ft above the base. The window is designed to withstand a fluid force of 312 lb without cracking.

   a) What fluid force will the window have to withstand if the tank is filled with water to a depth of 3 ft?
   b) To what level can the tank be filled with water without exceeding the window's design limitation?

19. **CALCULATOR** The end plates of the trough shown here were designed to withstand a fluid force of 6667 lb. How many cubic feet of water can the tank hold without exceeding this limitation? Round down to the nearest cubic foot.

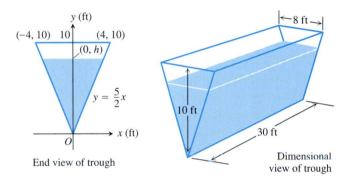

End view of trough

Dimensional view of trough

20. Water is running into the rectangular swimming pool shown here at the rate of 1000 ft³/h.

   a) Find the fluid force against the triangular drain plate after 9 h of filling.
   b) The drain plate is designed to withstand a fluid force of 520 lb. How high can you fill the pool without exceeding this limitation?

Triangular drain plate

Enlarged view of drain plate

21. A vertical rectangular plate a units long by b units wide is submerged in a fluid of weight density w with its long edges parallel to the fluid's surface. Find the average value of the pressure along the vertical dimension of the plate. Explain your answer.

22. *(Continuation of Exercise 21.)* Show that the force exerted by the fluid on one side of the plate is the average value of the pressure (found in Exercise 21) times the area of the plate.

23. Water pours into the tank here at the rate of 4 ft^3/min. The tank's cross sections are 4-ft-diameter semicircles. One end of the tank is movable, but moving it to increase the volume compresses a spring. The spring constant is $k = 100$ lb/ft. If the end of the tank moves 5 ft against the spring, the water will drain out of a safety hole in the bottom at the rate of 5 ft^3/min. Will the movable end reach the hole before the tank overflows?

5.10

The Basic Pattern and Other Modeling Applications

There is a pattern to what we did in the preceding sections. In each section we wanted to measure something that was modeled or described by one or more continuous functions. In Section 5.1 it was the area between the graphs of two continuous functions. In Section 5.2 it was the volume of a solid. In Section 5.8 it was the work done by a force whose magnitude was a continuous function, and so on. In each case we responded by partitioning the interval on which the function or functions were defined and approximating what we wanted to measure with Riemann sums over the interval. We used the integral defined by the limit of the Riemann sums to define and calculate what we wanted to measure. Table 5.2 shows the pattern.

Literally thousands of things in biology, chemistry, economics, engineering, finance, geology, medicine, and other fields (the list would fill pages) are modeled and calculated by exactly this process.

This section reviews the process and looks at a few more of the integrals it leads to.

Displacement vs. Distance Traveled

If a body with position function $s(t)$ moves along a coordinate line without changing direction, we can calculate the total distance it travels from $t = a$ to $t = b$ by integrating its velocity function $v(t)$ from $t = a$ to $t = b$, as we did in Chapter 4. If the body changes direction one or more times during the trip, we need to integrate the body's *speed* $|v(t)|$ to find the total distance traveled. Integrating the velocity will give only the body's **displacement,** $s(b) - s(a)$, the difference between its initial and final positions.

To see why, partition the time interval $a \le t \le b$ into subintervals in the usual way and let Δt_k denote the length of the kth interval. If Δt_k is small enough, the body's velocity $v(t)$ will not change much from t_{k-1} to t_k and the right-hand

Table 5.2 The phases of developing an integral to calculate something

Phase 1	Phase 2	Phase 3
We describe or model something we want to measure in terms of one or more continuous functions defined on a closed interval $[a, b]$.	We partition $[a, b]$ into subintervals of length Δx_k and choose a point c_k in each subinterval. We approximate what we want to measure with a finite sum. We identify the sum as a Riemann sum of a continuous function over $[a, b]$.	The approximations improve as the norm of the partition goes to zero. The Riemann sums approach a limiting integral. We use the integral to define and calculate what we originally wanted to measure.
The area between the curves $y = f(x)$, $y = g(x)$ on $[a, b]$ when $f(x) \geq g(x)$ 	$\sum [f(c_k) - g(c_k)] \Delta x_k$	$A = \lim\limits_{\|P\| \to 0} \sum [f(c_k) - g(c_k)] \Delta x_k$ $= \displaystyle\int_a^b [f(x) - g(x)]\, dx$
The volume of the solid defined by revolving the curve $y = R(x)$, $a \leq x \leq b$, about the x-axis. 	$\sum \pi [R(c_k)]^2 \Delta x_k$	$V = \lim\limits_{\|P\| \to 0} \sum \pi [R(c_k)]^2 \Delta x_k$ $= \displaystyle\int_a^b \pi [R(x)]^2\, dx$
The work done by a continuous variable force of magnitude $F(x)$ directed along the x-axis from a to b 	$\sum F(c_k) \Delta x_k$	$W = \lim\limits_{\|P\| \to 0} \sum F(c_k) \Delta x_k$ $= \displaystyle\int_a^b F(x)\, dx$

endpoint value $v(t_k)$ will give a good approximation of the velocity throughout the interval. Accordingly, the change in the body's position coordinate during the kth time interval will be about

$$v(t_k)\Delta t_k.$$

The change will be positive if $v(t_k)$ is positive and negative if $v(t_k)$ is negative.

In either case, the distance traveled during the kth interval will be about

$$|v(t_k)|\Delta t_k.$$

The total trip distance will be approximately

$$\sum_{k=1}^{n} |v(t_k)|\Delta t_k. \tag{1}$$

The sum in Eq. (1) is a Riemann sum for the speed $|v(t)|$ on the interval $[a, b]$. We expect the approximations to improve as the norm of the partition of $[a, b]$ goes to zero. It therefore looks as if we should be able to calculate the total distance traveled by the body by integrating the body's speed from a to b. In practice, this turns out to be the right thing to do. The mathematical model predicts the distance correctly every time.

$$\text{Distance traveled } = \int_a^b |v(t)|\, dt$$

If we wish to predict how far up or down the line from its initial position a body will end up when a trip is over, we integrate v instead of its absolute value.

To see why, let $s(t)$ be the body's position at time t and let F be an antiderivative of v. Then

$$s(t) = F(t) + C$$

for some constant C. The displacement caused by the trip from $t = a$ to $t = b$ is

$$s(b) - s(a) = (F(b) + C) - (F(a) + C)$$

$$= F(b) - F(a) = \int_a^b v(t)\, dt.$$

$$\text{Displacement } = \int_a^b v(t)\, dt$$

EXAMPLE 1   The velocity of a body moving along a line from $t = 0$ to $t = 3\pi/2$ sec was

$$v(t) = 5\cos t \text{ m/sec.}$$

Find the total distance traveled and the body's displacement.

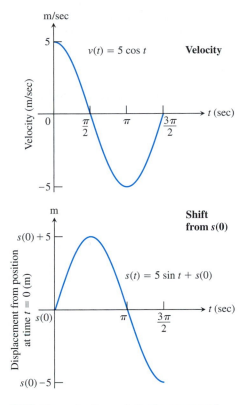

5.74 The velocity and displacement of the body in Example 1.

5.75 The steps leading to Delesse's rule: (a) a slice through a sample cube; (b) the granular material in the slice; (c) the slab between consecutive slices determined by a partition of $[0, L]$.

Solution

$$\text{Distance traveled} = \int_0^{3\pi/2} |5 \cos t| \, dt \qquad \textcolor{blue}{\text{Distance is the integral of speed.}}$$

$$= \int_0^{\pi/2} 5 \cos t \, dt + \int_{\pi/2}^{3\pi/2} (-5 \cos t) \, dt$$

$$= 5 \sin t \Big]_0^{\pi/2} - 5 \sin t \Big]_{\pi/2}^{3\pi/2}$$

$$= 5(1 - 0) - 5(-1 - 1) = 5 + 10 = 15 \text{ m}$$

$$\text{Displacement} = \int_0^{3\pi/2} 5 \cos t \, dt \qquad \textcolor{blue}{\text{Displacement is the integral of velocity.}}$$

$$= 5 \sin t \Big]_0^{3\pi/2} = 5(-1) - 5(0) = -5 \text{ m}$$

During the trip, the body traveled 5 m forward and 10 m backward for a total distance of 15 m. This displaced the body 5 m to the left (Fig. 5.74). ∎

Delesse's Rule

As you may know, the sugar in an apple starts turning into starch as soon as the apple is picked, and the longer the apple sits around, the starchier it becomes. You can tell fresh apples from stale by both flavor and consistency.

To find out how much starch is in a given apple, we can look at a thin slice under a microscope. The cross sections of the starch granules will show up clearly, and it is easy to estimate the proportion of the viewing area they occupy. This two-dimensional proportion will be the same as the three-dimensional proportion of uncut starch granules in the apple itself. The apparently magical equality of these proportions was first discovered by a French geologist, Achille Ernest Delesse, in the 1840s. Its explanation lies in the notion of average value.

Suppose we want to find the proportion of some granular material in a solid and that the sample we have chosen to analyze is a cube whose edges have length L. We picture the cube with an x-axis along one edge and imagine slicing the cube with planes perpendicular to points of the interval $[0, L]$ (Fig. 5.75). Call the proportion of the area of the slice at x occupied by the granular material of interest (starch, in our apple example) $r(x)$ and assume r is a continuous function of x.

(a)

(b)

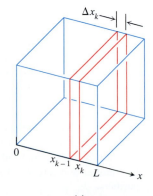

(c)

Now partition the interval $[0, L]$ into subintervals in the usual way. Imagine the cube sliced into thin slices by planes at the subdivision points. The length Δx_k of the kth subinterval is the distance between the planes at x_{k-1} and x_k. If the planes are close enough together, the sections cut from the grains by the planes will resemble cylinders with bases in the plane at x_k. The proportion of granular material between the planes will be about the same as the proportion of cylinder base area in the plane at x_k, which in turn will be about $r(x_k)$. Thus the amount of granular material in the slab between the two planes will be about

$$(\text{Proportion}) \times (\text{slab volume}) = r(x_k)L^2 \, \Delta x_k.$$

The amount of granular material in the entire sample cube will be about

$$\sum_{k=1}^{n} r(x_k)L^2 \, \Delta x_k.$$

This sum is a Riemann sum for the function $r(x)L^2$ over the interval $[0, L]$. We expect the approximations by sums like these to improve as the norm of the subdivision of $[0, L]$ goes to zero and therefore expect the integral

$$\int_0^L r(x)L^2 \, dx$$

to give the amount of granular material in the sample cube.

We can obtain the proportion of granular material in the sample by dividing this amount by the cube's volume, L^3. If we have chosen our sample well, this will also be the proportion of granular material in the solid from which the sample was taken. Putting it all together, we get

$$\begin{array}{l}\text{Proportion of granular}\\ \text{material in solid}\end{array} = \begin{array}{l}\text{Proportion of granular}\\ \text{material in the sample cube}\end{array}$$

$$= \frac{\int_0^L r(x)L^2 \, dx}{L^3} = \frac{L^2 \int_0^L r(x) \, dx}{L^3} = \frac{1}{L} \int_0^L r(x) \, dx$$

$$= \text{average value of } r(x) \text{ over } [0, L]$$

$$= \begin{array}{l}\text{proportion of area occupied by granular}\\ \text{material in a typical cross section.}\end{array}$$

This is Delesse's rule. Once we have found $\bar{r}$, the average of $r(x)$ over $[0, L]$, we have found the proportions of granular material in the solid.

In practice, $\bar{r}$ is found by averaging over a number of cross sections. There are several things to watch out for in the process. In addition to the possibility that the granules cluster in ways that make representative samples difficult to find, there is the possibility that we might not recognize a granule's trace for what it is. Some cross sections of normal red blood cells look like disks and ovals, while others look surprisingly like dumbbells. We do not want to dismiss the dumbbells as experimental error the way one research group did a few years ago.

Useless Integrals — Bad Models

Some of the integrals we get from forming Riemann sums do what we want, but others do not. It all depends on how we choose to model the problems we want to solve. Some choices are good; others are not. Here is an example.

Delesse's rule

Achille Ernest Delesse was a mid-nineteenth-century mining engineer interested in determining the composition of rocks. To find out how much of a particular mineral a rock contained, he cut it through, polished an exposed face, and covered the face with transparent waxed paper, trimmed to size. He then traced on the paper the exposed portions of the mineral that interested him. After weighing the paper, he cut out the mineral traces and weighed them. The ratio of the weights gave not only the proportion of the surface occupied by the mineral but, more important, the proportion of the entire rock occupied by the mineral. This rule is still used by petroleum geologists today. A two-dimensional analogue of it is used to determine the porosities of the ceramic filters that extract organic molecules in chemistry laboratories and screen out microbes in water purifiers.

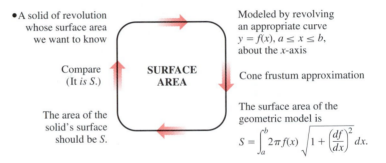

- A solid of revolution whose surface area we want to know

Modeled by revolving an appropriate curve $y = f(x)$, $a \le x \le b$, about the x-axis

Compare (It *is* S.)

SURFACE AREA

Cone frustum approximation

The area of the solid's surface should be S.

The surface area of the geometric model is

$$S = \int_a^b 2\pi f(x) \sqrt{1 + \left(\frac{df}{dx}\right)^2} \, dx.$$

5.76 The modeling cycle for surface area.

We use the surface area formula

$$S = \int_a^b 2\pi f(x) \sqrt{1 + \left(\frac{df}{dx}\right)^2} \, dx \tag{2}$$

because it has predictive value and always gives results consistent with information from other sources. In other words, the model we used to derive the formula (Fig. 5.76) was a good one.

Why not find the surface area by approximating with cylindrical bands instead of conical bands, as suggested in Fig. 5.77? The Riemann sums we get this way converge just as nicely as the ones based on conical bands, and the resulting integral is simpler. Instead of Eq. (2), we get

$$S = \int_a^b 2\pi f(x) \, dx. \tag{3}$$

After all, we might argue, we used cylinders to derive good volume formulas, so why not use them again to derive surface area formulas?

The answer is that the formula in Eq. (3) has no predictive value and almost never gives results consistent with other calculations. The comparison step in the modeling process fails for this formula.

There is a moral here: Just because we end up with a nice-looking integral does not mean it will do what we want. Constructing an integral is not enough—we have to test it too (Exercises 15 and 16).

The Theorems of Pappus

In the third century, an Alexandrian Greek named Pappus discovered two formulas that relate centroids to surfaces and solids of revolution. The formulas provide shortcuts to a number of otherwise lengthy calculations.

(a)

(b)

5.77 Why not use (a) cylindrical bands instead of (b) conical bands to approximate surface area?

Theorem 1

Pappus's Theorem for Volumes

If a plane region is revolved once about a line in the plane that does not cut through the region's interior, then the volume of the solid it generates is equal to the region's area times the distance traveled by the region's centroid during the revolution. If ρ is the distance from the axis of revolution to the centroid, then

$$V = 2\pi\rho A. \tag{4}$$

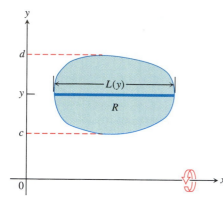

5.78 The region R is to be revolved (once) about the x-axis to generate a solid. A 1700-year-old theorem says that the solid's volume can be calculated by multiplying the region's area by the distance traveled by its centroid during the revolution.

Proof We draw the axis of revolution as the x-axis with the region R in the first quadrant (Fig. 5.78). We let $L(y)$ denote the length of the cross section of R perpendicular to the y-axis at y. We assume $L(y)$ to be continuous.

By the method of cylindrical shells, the volume of the solid generated by revolving the region about the x-axis is

$$V = \int_c^d 2\pi (\text{shell radius})(\text{shell height})\,dy = 2\pi \int_c^d y\,L(y)\,dy. \tag{5}$$

The y-coordinate of R's centroid is

$$\overline{y} = \frac{\int_c^d \tilde{y}\,dA}{A} = \frac{\int_c^d y\,L(y)\,dy}{A},$$

so that

$$\int_c^d y\,L(y)\,dy = A\overline{y}.$$

Substituting $A\overline{y}$ for the last integral in Eq. (5) gives $V = 2\pi\overline{y}A$. With ρ equal to $\overline{y}$, we have $V = 2\pi\rho A$. ❑

EXAMPLE 2 The volume of the torus (doughnut) generated by revolving a circular disk of radius a about an axis in its plane at a distance $b \geq a$ from its center (Fig. 5.79) is

$$V = 2\pi(b)(\pi a^2) = 2\pi^2 b a^2. \qquad ❑$$

5.79 With Pappus's first theorem, we can find the volume of a torus without having to integrate (Example 2).

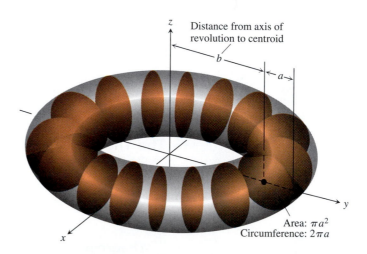

Distance from axis of revolution to centroid

Area: πa^2
Circumference: $2\pi a$

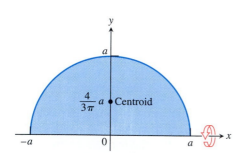

5.80 With Pappus's first theorem, we can locate the centroid of a semicircular region without having to integrate (Example 3).

EXAMPLE 3 Locate the centroid of a semicircular region.

Solution We model the region as the region between the semicircle $y = \sqrt{a^2 - x^2}$ (Fig. 5.80) and the x-axis and imagine revolving the region about the x-axis to generate a solid sphere. By symmetry, the x-coordinate of the centroid is $\overline{x} = 0$. With $\overline{y} = \rho$ in Eq. (4), we have

$$\overline{y} = \frac{V}{2\pi A} = \frac{(4/3)\pi a^3}{2\pi (1/2)\pi a^2} = \frac{4}{3\pi}a. \qquad ❑$$

> **Theorem 2**
> **Pappus's Theorem for Surface Areas**
> If an arc of a smooth plane curve is revolved once about a line in the plane that does not cut through the arc's interior, then the area of the surface generated by the arc equals the length of the arc times the distance traveled by the arc's centroid during the revolution. If ρ is the distance from the axis of revolution to the centroid, then
>
> $$S = 2\pi\rho L. \tag{6}$$

The proof we give assumes that we can model the axis of revolution as the *x*-axis and the arc as the graph of a smooth function of *x*.

Proof We draw the axis of revolution as the *x*-axis with the arc extending from $x = a$ to $x = b$ in the first quadrant (Fig. 5.81). The area of the surface generated by the arc is

$$S = \int_{x=a}^{x=b} 2\pi y \, ds = 2\pi \int_{x=a}^{x=b} y \, ds. \tag{7}$$

The *y*-coordinate of the arc's centroid is

$$\bar{y} = \frac{\displaystyle\int_{x=a}^{x=b} \tilde{y} \, ds}{\displaystyle\int_{x=a}^{x=b} ds} = \frac{\displaystyle\int_{x=a}^{x=b} y \, ds}{L}. \qquad \begin{array}{l} L = \int ds \text{ is the arc's} \\ \text{length and } \tilde{y} = y. \end{array}$$

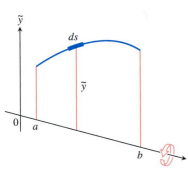

5.81 Figure for Pappus's area theorem.

Hence

$$\int_{x=a}^{x=b} y \, ds = \bar{y} L.$$

Substituting $\bar{y}L$ for the last integral in Eq. (7) gives $S = 2\pi\bar{y}L$. With ρ equal to $\bar{y}$, we have $S = 2\pi\rho L$.

EXAMPLE 4   The surface area of the torus in Example 2 is

$$S = 2\pi(b)(2\pi a) = 4\pi^2 ba. \qquad \square$$

Exercises 5.10

Distance and Displacement

In Exercises 1–8, the function $v(t)$ is the velocity in meters per second of a body moving along a coordinate line. (a) Graph v to see where it is positive and negative. Then find (b) the total distance traveled by the body during the given time interval and (c) the body's displacement.

1. $v(t) = 5\cos t, \quad 0 \le t \le 2\pi$

2. $v(t) = \sin \pi t, \quad 0 \le t \le 2$

3. $v(t) = 6\sin 3t, \quad 0 \le t \le \pi/2$

4. $v(t) = 4\cos 2t, \quad 0 \le t \le \pi$

5. $v(t) = 49 - 9.8t$, $0 \le t \le 10$

6. $v(t) = 8 - 1.6t$, $0 \le t \le 10$

7. $v(t) = 6t^2 - 18t + 12 = 6(t-1)(t-2)$, $0 \le t \le 2$

8. $v(t) = 6t^2 - 18t + 12 = 6(t-1)(t-2)$, $0 \le t \le 3$

9. The function $s = (1/3)t^3 - 3t^2 + 8t$ gives the position of a body moving on the horizontal s-axis at time $t \ge 0$ (s in meters, t in seconds).

   a) Show that the body is moving to the right at time $t = 0$.

   b) When does the body move to the left?

   c) What is the body's position at time $t = 3$?

   d) When $t = 3$, what is the total distance the body has traveled?

   e) GRAPHER Graph s as a function of t and comment on the relationship of the graph to the body's motion.

10. The function $s = -t^3 + 6t^2 - 9t$ gives the position of a body moving on the horizontal s-axis at time $t \ge 0$ (s in meters, t in seconds).

   a) Show that the body is moving to the left at $t = 0$.

   b) When does the body move to the right?

   c) Does the body ever move to the right of the origin? Give reasons for your answer.

   d) What is the body's position at time $t = 3$?

   e) What is the total distance the particle has traveled by the time $t = 3$?

   f) GRAPHER Graph s as a function of t and comment on the relationship of the graph to the body's motion.

11. Here are the velocity graphs of two bodies moving on a coordinate line. Find the total distance traveled and the body's displacement for the given time interval.

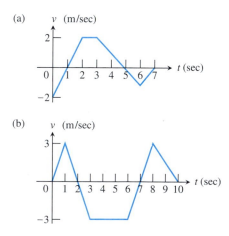

12. CALCULATOR The table at the top of the next column shows the velocity of a model train engine moving back and forth on a track for 10 sec. Use Simpson's rule to find the resulting displacement and total distance traveled.

Time (sec)	Velocity (in./sec)	Time (sec)	Velocity (in./sec)
0	0	6	−11
1	12	7	− 6
2	22	8	2
3	10	9	6
4	− 5	10	0
5	−13		

Delesse's Rule

13. The photograph here shows a grid superimposed on the polished face of a piece of granite. Use the grid and Delesse's rule to estimate the proportion of shrimp-colored granular material in the rock.

14. The photograph here shows a grid superimposed on a microscopic view of a stained section of human lung tissue. The clear spaces between the cells are cross sections of the lung's air sacks (called *alveoli*, accent on the second syllable). Use the grid and Delesse's rule to estimate the proportion of air space in the lung.

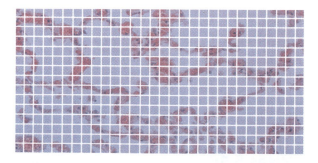

Modeling Surface Area

15. *Modeling surface area.* The lateral surface area of the cone swept out by revolving the line segment $y = x/\sqrt{3}$, $0 \le x \le \sqrt{3}$, about the x-axis should be (1/2)(base circumference)(slant height) $= (1/2)(2\pi)(2) = 2\pi$. What do you get if you use Eq. (3) with $f(x) = x/\sqrt{3}$?

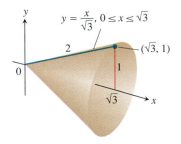

$$y = \frac{x}{\sqrt{3}}, \ 0 \le x \le \sqrt{3}$$

$(\sqrt{3}, 1)$

16. *Modeling surface area.* The only surface for which Eq. (3) gives the area we want is a cylinder. Show that Eq. (3) gives $S = 2\pi r h$ for the cylinder swept out by revolving the line segment $y = r, 0 \le x \le h$, about the x-axis.

17. *A sailboat's displacement.* To find the volume of water displaced by a sailboat, the common practice is to partition the waterline into 10 subintervals of equal length, measure the cross section area $A(x)$ of the submerged portion of the hull at each partition point, and then use Simpson's rule to estimate the integral of $A(x)$ from one end of the waterline to the other. The table here lists the area measurements at "Stations" 0 through 10, as the partition points are called, for the cruising sloop *Pipedream,* shown here. The common subinterval length (distance between consecutive stations) is $h = 2.54$ ft (about $2'$ 6 $1/2''$, chosen for the convenience of the builder).

a) Estimate *Pipedream*'s displacement volume to the nearest cubic foot.

Station	Submerged area (ft²)
0	0
1	1.07
2	3.84
3	7.82
4	12.20
5	15.18
6	16.14
7	14.00
8	9.21
9	3.24
10	0

b) The figures in the table are for seawater, which weighs

64 lb/ft³. How many pounds of water does *Pipedream* displace? (Displacement is given in pounds for small craft, and long tons [1 long ton = 2240 lb] for larger vessels.)

(Data from *Skene's Elements of Yacht Design*, Francis S. Kinney, Dodd, Mead & Company, Inc., 1962)

18. *Prismatic coefficients* (*Continuation of Exercise 17*). A boat's prismatic coefficient is the ratio of the displacement volume to the volume of a prism whose height equals the boat's waterline length and whose base equals the area of the boat's largest submerged cross section. The best sailboats have prismatic coefficients between 0.51 and 0.54. Find *Pipedream*'s prismatic coefficient, given a waterline length of 25.4 ft and a largest submerged cross section area of 16.14 ft² (at Station 6).

The Theorems of Pappus

19. The square region with vertices $(0, 2)$, $(2, 0)$, $(4, 2)$, and $(2, 4)$ is revolved about the x-axis to generate a solid. Find the volume and surface area of the solid.

20. Use a theorem of Pappus to find the volume generated by revolving about the line $x = 5$ the triangular region bounded by the coordinate axes and the line $2x + y = 6$. (As you saw in Exercise 31 of Section 5.7, the centroid of a triangle lies at the intersection of the medians, one-third of the way from the midpoint of each side toward the opposite vertex.)

21. Find the volume of the torus generated by revolving the circle $(x - 2)^2 + y^2 = 1$ about the y-axis.

22. Use the theorems of Pappus to find the lateral surface area and the volume of a right circular cone.

23. Use the second theorem of Pappus and the fact that the surface area of a sphere of radius a is $4\pi a^2$ to find the centroid of the semicircle $y = \sqrt{a^2 - x^2}$.

24. As found in Exercise 23, the centroid of the semicircle $y = \sqrt{a^2 - x^2}$ lies at the point $(0, 2a/\pi)$. Find the area of the surface swept out by revolving the semicircle about the line $y = a$.

25. The area of the region R enclosed by the semiellipse $y = (b/a)\sqrt{a^2 - x^2}$ and the x-axis is $(1/2)\pi \, ab$ and the volume of the ellipsoid generated by revolving R about the x-axis is $(4/3)\pi \, ab^2$. Find the centroid of R. Notice the remarkable fact that the location is independent of a.

26. As found in Example 3, the centroid of the region enclosed by the x-axis and the semicircle $y = \sqrt{a^2 - x^2}$ lies at the point $(0, 4a/3\pi)$. Find the volume of the solid generated by revolving this region about the line $y = -a$.

27. The region of Exercise 26 is revolved about the line $y = x - a$ to generate a solid. Find the volume of the solid.

28. As found in Exercise 23, the centroid of the semicircle $y = \sqrt{a^2 - x^2}$ lies at the point $(0, 2a/\pi)$. Find the area of the surface generated by revolving the semicircle about the line $y = x - a$.

29. Find the moment about the x-axis of the semicircular region in Example 3. If you use results already known, you will not need to integrate.

CHAPTER   **5**   QUESTIONS TO GUIDE YOUR REVIEW

1. How do you define and calculate the area of the region between the graphs of two continuous functions? Give an example.

2. How do you define and calculate the volumes of solids by the method of slicing? Give an example.

3. How are the disk and washer methods for calculating volumes derived from the method of slicing? Give examples of volume calculations by these methods.

4. Describe the method of cylindrical shells. Give an example.

5. How do you define and calculate the length of the graph of a smooth function over a closed interval? Give an example. What about functions that do not have continuous first derivatives?

6. How do you define and calculate the area of the surface swept out by revolving the graph of a smooth function $y = f(x), a \le x \le b$, about the x-axis? Give an example.

7. What is a center of mass?

8. How do you locate the center of mass of a straight, narrow rod or strip of material? Give an example. If the density of the material is constant, you can tell right away where the center of mass is. Where is it?

9. How do you locate the center of mass of a thin flat plate of material? Give an example.

10. How do you define and calculate the work done by a variable force directed along a portion of the x-axis? How do you calculate the work it takes to pump a liquid from a tank? Give examples.

11. How do you calculate the force exerted by a liquid against a portion of a vertical wall? Give an example.

12. Suppose you know the velocity function $v(t)$ of a body that will be moving back and forth along a coordinate line from time $t = a$ to time $t = b$. How can you predict how much the motion will shift the body's position? How can you predict the total distance the body will travel?

13. What does Delesse's rule say? Give an example.

14. What do Pappus's two theorems say? Give examples of how they are used to calculate surface areas and volumes and to locate centroids.

15. There is a basic pattern to the way we constructed integrals in this chapter. What is it? Give examples.

CHAPTER   **5**   PRACTICE EXERCISES

Areas

Find the areas of the regions enclosed by the curves and lines in Exercises 1–12.

1. $y = x, \quad y = 1/x^2, \quad x = 2$

2. $y = x, \quad y = 1/\sqrt{x}, \quad x = 2$

3. $\sqrt{x} + \sqrt{y} = 1, \quad x = 0, \quad y = 0$

4. $x^3 + \sqrt{y} = 1, \quad x = 0, \quad y = 0, \quad \text{for} \quad 0 \le x \le 1$

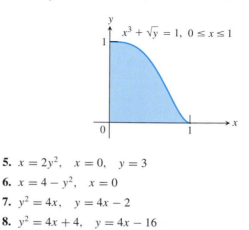

5. $x = 2y^2, \quad x = 0, \quad y = 3$

6. $x = 4 - y^2, \quad x = 0$

7. $y^2 = 4x, \quad y = 4x - 2$

8. $y^2 = 4x + 4, \quad y = 4x - 16$

9. $y = \sin x, \quad y = x, \quad 0 \le x \le \pi/4$

10. $y = |\sin x|, \quad y = 1, \quad -\pi/2 \le x \le \pi/2$

11. $y = 2\sin x, \quad y = \sin 2x, \quad 0 \le x \le \pi$

12. $y = 8\cos x, \quad y = \sec^2 x, \quad -\pi/3 \le x \le \pi/3$

13. Find the area of the "triangular" region bounded on the left by $x + y = 2$, on the right by $y = x^2$, and above by $y = 2$.

14. Find the area of the "triangular" region bounded on the left by $y = \sqrt{x}$, on the right by $y = 6 - x$, and below by $y = 1$.

15. Find the extreme values of $f(x) = x^3 - 3x^2$ and find the area of the region enclosed by the graph of f and the x-axis.

16. Find the area of the region cut from the first quadrant by the curve $x^{1/2} + y^{1/2} = a^{1/2}$.

17. Find the total area of the region enclosed by the curve $x = y^{2/3}$ and the lines $x = y$ and $y = -1$.

18. Find the total area of the region between the curves $y = \sin x$ and $y = \cos x$ for $0 \le x \le 3\pi/2$.

Volumes

Find the volumes of the solids in Exercises 19–24.

19. The solid lies between planes perpendicular to the x-axis at $x = 0$ and $x = 1$. The cross sections perpendicular to the x-axis between these planes are circular disks whose diameters run from the parabola $y = x^2$ to the parabola $y = \sqrt{x}$.

20. The base of the solid is the region in the first quadrant between the line $y = x$ and the parabola $y = 2\sqrt{x}$. The cross sections of the solid perpendicular to the x-axis are equilateral triangles whose bases stretch from the line to the curve.

21. The solid lies between planes perpendicular to the x-axis at $x = \pi/4$ and $x = 5\pi/4$. The cross sections between these planes are circular disks whose diameters run from the curve $y = 2\cos x$ to the curve $y = 2\sin x$.

22. The solid lies between planes perpendicular to the x-axis at $x = 0$ and $x = 6$. The cross sections between these planes are squares whose bases run from the x-axis up to the curve $x^{1/2} + y^{1/2} = \sqrt{6}$.

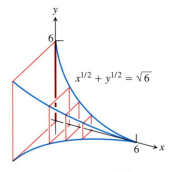

$x^{1/2} + y^{1/2} = \sqrt{6}$

23. The solid lies between planes perpendicular to the x-axis at $x = 0$ and $x = 4$. The cross sections of the solid perpendicular to the x-axis between these planes are circular disks whose diameters run from the curve $x^2 = 4y$ to the curve $y^2 = 4x$.

24. The base of the solid is the region bounded by the parabola $y^2 = 4x$ and the line $x = 1$ in the xy-plane. Each cross section perpendicular to the x-axis is an equilateral triangle with one edge in the plane. (The triangles all lie on the same side of the plane.)

25. Find the volume of the solid generated by revolving the region bounded by the x-axis, the curve $y = 3x^4$, and the lines $x = 1$ and $x = -1$ about (a) the x-axis; (b) the y-axis; (c) the line $x = 1$; (d) the line $y = 3$.

26. Find the volume of the solid generated by revolving the "triangular" region bounded by the curve $y = 4/x^3$ and the lines $x = 1$ and $y = 1/2$ about (a) the x-axis; (b) the y-axis; (c) the line $x = 2$; (d) the line $y = 4$.

27. Find the volume of the solid generated by revolving the region bounded on the left by the parabola $x = y^2 + 1$ and on the right by the line $x = 5$ about (a) the x-axis; (b) the y-axis; (c) the line $x = 5$.

28. Find the volume of the solid generated by revolving the region bounded by the parabola $y^2 = 4x$ and the line $y = x$ about (a) the x-axis; (b) the y-axis; (c) the line $x = 4$; (d) the line $y = 4$.

29. Find the volume of the solid generated by revolving the "triangular" region bounded by the x-axis, the line $x = \pi/3$, and the curve $y = \tan x$ in the first quadrant about the x-axis.

30. Find the volume of the solid generated by revolving the region bounded by the curve $y = \sin x$ and the lines $x = 0$, $x = \pi$, and $y = 2$ about the line $y = 2$.

31. Find the volume of the solid generated by revolving the region between the x-axis and the curve $y = x^2 - 2x$ about (a) the x-axis; (b) the line $y = -1$; (c) the line $x = 2$; (d) the line $y = 2$.

32. Find the volume of the solid generated by revolving about the x-axis the region bounded by $y = 2\tan x$, $y = 0$, $x = -\pi/4$, and $x = \pi/4$. (The region lies in the first and third quadrants and resembles a skewed bow tie.)

33. A round hole of radius $\sqrt{3}$ ft is bored through the center of a solid sphere of radius 2 ft. Find the volume of material removed from the sphere.

34. **CALCULATOR** The profile of a football resembles the ellipse shown here. Find the football's volume to the nearest cubic inch.

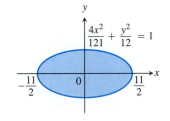

$$\frac{4x^2}{121} + \frac{y^2}{12} = 1$$

Lengths of Curves

Find the lengths of the curves in Exercises 35–38.

35. $y = x^{1/2} - (1/3)x^{3/2}, \quad 1 \le x \le 4$

36. $x = y^{2/3}, \quad 1 \le y \le 8$

37. $y = (5/12)x^{6/5} - (5/8)x^{4/5}, \quad 1 \le x \le 32$

38. $x = (y^3/12) + (1/y), \quad 1 \le y \le 2$

Areas of Surfaces of Revolution

In Exercises 39–42, find the areas of the surfaces generated by revolving the curves about the given axes.

39. $y = \sqrt{2x + 1}, \quad 0 \le x \le 3, \quad x\text{-axis}$

40. $y = x^3/3, \quad 0 \le x \le 1, \quad x\text{-axis}$

41. $x = \sqrt{4y - y^2}, \quad 1 \le y \le 2, \quad y\text{-axis}$

42. $x = \sqrt{y}, \quad 2 \le y \le 6, \quad y\text{-axis}$

Centroids and Centers of Mass

43. Find the centroid of a thin, flat plate covering the region enclosed by the parabolas $y = 2x^2$ and $y = 3 - x^2$.

44. Find the centroid of a thin, flat plate covering the region enclosed by the x-axis, the lines $x = 2$ and $x = -2$, and the parabola $y = x^2$.

45. Find the centroid of a thin, flat plate covering the "triangular" region in the first quadrant bounded by the y-axis, the parabola $y = x^2/4$, and the line $y = 4$.

46. Find the centroid of a thin, flat plate covering the region enclosed by the parabola $y^2 = x$ and the line $x = 2y$.

47. Find the center of mass of a thin, flat plate covering the region enclosed by the parabola $y^2 = x$ and the line $x = 2y$ if the density function is $\delta(y) = 1 + y$. (Use horizontal strips.)

48. a) Find the center of mass of a thin plate of constant density covering the region between the curve $y = 3/x^{3/2}$ and the x-axis from $x = 1$ to $x = 9$.

 b) Find the plate's center of mass if, instead of being constant, the density is $\delta(x) = x$. (Use vertical strips.)

Work

49. A rock climber is about to haul up 100 N (about 22.5 lb) of equipment that has been hanging beneath her on 40 m of rope that weighs 0.8 newton per meter. How much work will it take? (*Hint:* Solve for the rope and equipment separately; then add.)

50. You drove an 800-gal tank truck from the base of Mt. Washington to the summit and discovered on arrival that the tank was only half full. You started with a full tank, climbed at a steady rate, and accomplished the 4750-ft elevation change in 50 min. Assuming that the water leaked out at a steady rate, how much work was spent in carrying water to the top? Do not count the work done in getting yourself and the truck there. Water weighs 8 lb/U.S. gal.

51. If a force of 20 lb is required to hold a spring 1 ft beyond its unstressed length, how much work does it take to stretch the spring this far? an additional foot?

52. A force of 200 N will stretch a garage door spring 0.8 m beyond its unstressed length. How far will a 300-N force stretch the spring? How much work does it take to stretch the spring this far?

53. A reservoir shaped like a right circular cone, point down, 20 ft across the top and 8 ft deep, is full of water. How much work does it take to pump the water to a level 6 ft above the top?

54. (*Continuation of Exercise 53.*) The reservoir is filled to a depth of 5 ft, and the water is to be pumped to the same level as the top. How much work does it take?

55. A right circular conical tank, point down, with top radius 5 ft and height 10 ft is filled with a liquid whose weight-density is 60 lb/ft³. How much work does it take to pump the liquid to a point 2 ft above the tank? If the pump is driven by a motor rated at 275 ft · lb/sec (1/2-hp), how long will it take to empty the tank?

56. A storage tank is a right circular cylinder 20 ft long and 8 ft in diameter with its axis horizontal. If the tank is half full of olive oil weighing 57 lb/ft³, find the work done in emptying it through a pipe that runs from the bottom of the tank to an outlet that is 6 ft above the top of the tank.

Fluid Force

57. The vertical triangular plate shown here is the end plate of a trough full of water ($w = 62.4$). What is the fluid force against the plate?

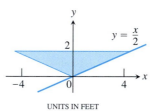

UNITS IN FEET

58. The vertical trapezoidal plate shown here is the end plate of a trough full of maple syrup weighing 75 lb/ft³. What is the force exerted by the syrup against the end plate of the trough when the syrup is 10 in. deep?

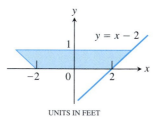

UNITS IN FEET

59. A flat vertical gate in the face of a dam is shaped like the parabolic region between the curve $y = 4x^2$ and the line $y = 4$, with measurements in feet. The top of the gate lies 5 ft below the surface of the water. Find the force exerted by the water against the gate ($w = 62.4$).

60. CALCULATOR You plan to store mercury ($w = 849$ lb/ft³) in a vertical right circular cylindrical tank of radius 1 ft whose interior side wall can withstand a total fluid force of 40,000 lb. About how many cubic feet of mercury can you store in the tank at any one time?

61. The container profiled in Fig. 5.82 is filled with two nonmixing liquids of weight density w_1 and w_2. Find the fluid force on one side of the vertical square plate $ABCD$. The points B and D lie in the boundary layer and the square is $6\sqrt{2}$ ft on a side.

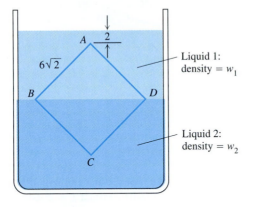

5.82 Profile of the container in Exercise 61.

62. The isosceles trapezoidal plate shown here is submerged vertically in water ($w = 62.4$) with its upper edge 4 ft below the surface. Find the fluid force on one side of the plate in two different ways:

a) By evaluating an integral.

b) By dividing the plate into a parallelogram and an isosceles triangle, locating their centroids, and using the equation $F = w\overline{h}A$ from Section 5.9.

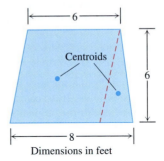

Dimensions in feet

Distance and Displacement

In Exercises 63–66, the function $v = f(t)$ is the velocity (m/sec) of a body moving along a coordinate line. Find (a) the total distance the body travels during the given time interval and (b) the body's displacement.

63. $v = t^2 - 8t + 12, \quad 0 \le t \le 6$

64. $v = t^3 - 3t^2 + 2t, \quad 0 \le t \le 2$

65. $v = 5\cos t, \quad 0 \le t \le 3\pi/2$

66. $v = -\pi \sin \pi t, \quad 0 \le t \le 3/2$

CHAPTER 5 — ADDITIONAL EXERCISES–THEORY, EXAMPLES, APPLICATIONS

Volume and Length

1. A solid is generated by revolving about the x-axis the region bounded by the graph of the continuous function $y = f(x)$, the x-axis, and the fixed line $x = a$ and the variable line $x = b$, $b > a$. Its volume, for all b, is $b^2 - ab$. Find $f(x)$.

2. A solid is generated by revolving about the x-axis the region bounded by the graph of the continuous function $y = f(x)$, the x-axis, and the lines $x = 0$ and $x = a$. Its volume, for all $a > 0$, is $a^2 + a$. Find $f(x)$.

3. Suppose that the increasing function $f(x)$ is smooth for $x \ge 0$ and that $f(0) = a$. Let $s(x)$ denote the length of the graph of f from $(0, a)$ to $(x, f(x))$, $x > 0$. Find $f(x)$ if $s(x) = Cx$ for some constant C. What are the allowable values for C?

4. a) Show that for $0 < \alpha \le \pi/2$,

$$\int_0^\alpha \sqrt{1 + \cos^2 \theta} \, d\theta > \sqrt{\alpha^2 + \sin^2 \alpha}.$$

b) Generalize the result in (a).

Moments and Centers of Mass

5. Find the centroid of the region bounded below by the x-axis and above by the curve $y = 1 - x^n$, n an even positive integer. What is the limiting position of the centroid as $n \to \infty$?

6. CALCULATOR If you haul a telephone pole on a two-wheeled carriage behind a truck, you want the wheels to be three feet or so behind the pole's center of mass to provide an adequate "tongue" weight. NYNEX's class 1 40-ft wooden poles have a 27-in. circumference at the top and a 43.5-in. circumference at the base. About how far from the top is the center of mass?

7. Suppose that a thin metal plate of area A and constant density δ occupies a region R in the xy-plane, and let M_y be the plate's moment about the y-axis. Show that the plate's moment about the line $x = b$ is

a) $M_y - b\,\delta\,A$ if the plate lies to the right of the line, and

b) $b\,\delta\,A - M_y$ if the plate lies to the left of the line.

8. Find the center of mass of a thin plate covering the region bounded by the curve $y^2 = 4ax$ and the line $x = a$, a = positive constant, if the density at (x, y) is directly proportional to (a) x, (b) $|y|$.

9. **a)** Find the centroid of the region in the first quadrant bounded by two concentric circles and the coordinate axes, if the circles have radii a and b, $0 < a < b$, and their centers are at the origin.

   b) Find the limits of the coordinates of the centroid as a approaches b and discuss the meaning of the result.

10. A triangular corner is cut from a square 1 ft on a side. The area of the triangle removed is 36 in². If the centroid of the remaining region is 7 in. from one side of the original square, how far is it from the remaining sides?

Surface Area

11. At points on the curve $y = 2\sqrt{x}$, line segments of length $h = y$ are drawn perpendicular to the xy-plane (Fig. 5.83). Find the area of the surface formed by these perpendiculars from $(0, 0)$ to $(3, 2\sqrt{3})$.

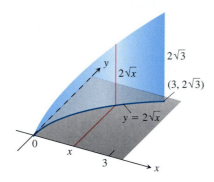

5.83 The surface in Exercise 11.

12. At points on a circle of radius a, line segments are drawn perpendicular to the plane of the circle, the perpendicular at each point P being of length ks, where s is the length of the arc of the circle measured counterclockwise from $(a, 0)$ to P and k is a positive constant, as shown here. Find the area of the surface formed by the perpendiculars along the arc beginning at $(a, 0)$ and extending once around the circle.

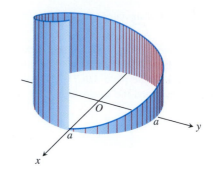

Work

13. A particle of mass m starts from rest at time $t = 0$ and is moved along the x-axis with constant acceleration a from $x = 0$ to $x = h$ against a variable force of magnitude $F(t) = t^2$. Find the work done.

14. *Work and kinetic energy.* Suppose a 1.6-oz golf ball is placed on a vertical spring with force constant $k = 2$ lb/in. The spring is compressed 6 in. and released. About how high does the ball go (measured from the spring's rest position)?

Fluid Force

15. A triangular plate ABC is submerged in water with its plane vertical. The side AB, 4 ft long, is 6 ft below the surface of the water, while the vertex C is 2 ft below the surface. Find the force exerted by the water on one side of the plate.

16. A vertical rectangular plate is submerged in a fluid with its top edge parallel to the fluid's surface. Show that the force exerted by the fluid on one side of the plate equals the average value of the pressure up and down the plate times the area of the plate.

17. The *center of pressure* on one side of a plane region submerged in a fluid is defined to be the point at which the total force exerted by the fluid can be applied without changing its total moment about any axis in the plane. Find the depth to the center of pressure (a) on a vertical rectangle of height h and width b if its upper edge is in the surface of the fluid; (b) on a vertical triangle of height h and base b if the vertex opposite b is a ft and the base b is $(a + h)$ ft below the surface of the fluid.

CHAPTER 6

Transcendental Functions

OVERVIEW Many of the functions in mathematics and science are inverses of one another. The functions $\ln x$ and e^x are probably the best-known function–inverse pair, but others are nearly as important. The trigonometric functions, when suitably restricted, have important inverses, and there are other useful pairs of logarithmic and exponential functions. Less widely known are the hyperbolic functions and their inverses, functions that arise in the study of hanging cables, heat flow, and the friction encountered by objects falling through the air. We describe all of these functions in this chapter and look at the kinds of problems they solve.

6.1 Inverse Functions and Their Derivatives

In this section, we define what it means for functions to be inverses of one another and look at what this says about the formulas, graphs, and derivatives of function–inverse pairs.

One-to-One Functions

A function is a rule that assigns a value from its range to each point in its domain. Some functions assign the same value to more than one point. The squares of -1 and 1 are both 1; the sines of $\pi/3$ and $2\pi/3$ are both $\sqrt{3}/2$. Other functions never assume a given value more than once. The square roots and cubes of different numbers are always different. A function that has distinct values at distinct points is called one-to-one.

> **Definition**
> A function $f(x)$ is **one-to-one** on a domain D if $f(x_1) \neq f(x_2)$ whenever $x_1 \neq x_2$.

EXAMPLE 1 $f(x) = \sqrt{x}$ is one-to-one on any domain of nonnegative numbers because $\sqrt{x_1} \neq \sqrt{x_2}$ whenever $x_1 \neq x_2$. ❑

449

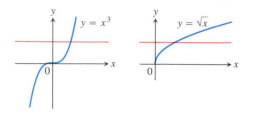

One-to-one: Graph meets each
horizontal line at most once.

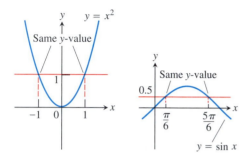

Not one-to-one: Graph meets one or
more horizontal lines more than once.

6.1 Using the horizontal line test, we see
that $y = x^3$ and $y = \sqrt{x}$ are one-to-one,
but $y = x^2$ and $y = \sin x$ are not.

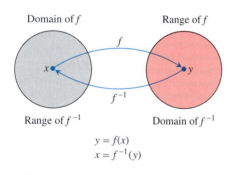

$$y = f(x)$$
$$x = f^{-1}(y)$$

6.2 The inverse of a function f sends
each output back to the input from
which it came.

EXAMPLE 2 $g(x) = \sin x$ *is not* one-to-one on the interval $[0, \pi]$ because
$\sin(\pi/6) = \sin(5\pi/6)$. The sine *is* one-to-one on $[0, \pi/2]$, however, because sines
of angles in the first quadrant are distinct.   ∎

The graph of a one-to-one function $y = f(x)$ can intersect a given horizontal
line at most once. If it intersects the line more than once it assumes the same
y-value more than once, and is therefore not one-to-one (Fig. 6.1).

The Horizontal Line Test

A function $y = f(x)$ is one-to-one if and only if its graph intersects each
horizontal line at most once.

Inverses

Since each output of a one-to-one function comes from just one input, a one-to-
one function can be reversed to send the outputs back to the inputs from which
they came. The function defined by reversing a one-to-one function f is called the
inverse of f. The symbol for the inverse of f is $\boldsymbol{f^{-1}}$, read "f inverse" (Fig. 6.2).
The -1 in f^{-1} is *not* an exponent: $f^{-1}(x)$ does not mean $1/f(x)$.

As Fig. 6.2 suggests, the result of composing f and f^{-1} in either order is the
identity function, the function that assigns each number to itself. This gives a way
to test whether two functions f and g are inverses of one another. Compute $f \circ g$
and $g \circ f$. If $(f \circ g)(x) = (g \circ f)(x) = x$, then f and g are inverse of one another;
otherwise they are not. If f cubes every number in its domain, g had better take
cube roots or it isn't the inverse of f.

Functions f and g are an inverse pair if and only if

$$f(g(x)) = x \qquad \text{and} \qquad g(f(x)) = x.$$

In this case, $g = f^{-1}$ and $f = g^{-1}$.

A function has an inverse if and only if it is one-to-one. This means, for
example, that increasing functions have inverses and decreasing functions have
inverses (Exercise 39). Functions with positive derivatives have inverses because
they increase throughout their domains (Corollary 3 of the Mean Value Theorem,
Section 3.2). Similarly, because they decrease throughout their domains, functions
with negative derivatives have inverses.

Finding Inverses

How is the graph of the inverse of a function related to the graph of the function?
If the function is increasing, say, its graph rises from left to right, like the graph in
Fig. 6.3(a). To read the graph, we start at the point x on the x-axis, go up to the
graph, and then move over to the y-axis to read the value of y. If we start with y
and want to find the x from which it came, we reverse the process (Fig. 6.3b).

The graph of f is the graph of f^{-1} with the input–output pairs reversed. To
display the graph in the usual way, we have to reverse the pairs by reflecting the

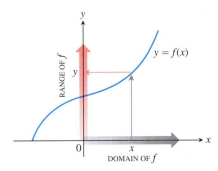

(a) To find the value of f at x, we start at x and go up to the curve and over to the y-axis.

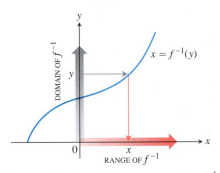

(b) The graph of f can also serve as a graph of f^{-1}. To find the x that gave y, we start at y and go over to the curve and down to the x-axis. The domain of f^{-1} is the range of f. The range of f^{-1} is the domain of f.

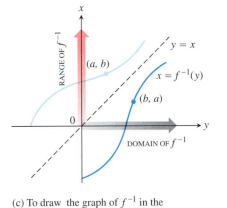

(c) To draw the graph of f^{-1} in the usual way, we reflect it in the line $y = x$.

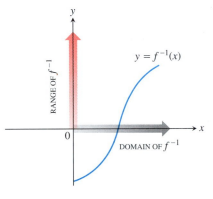

(d) Then we interchange the letters x and y. We now have a graph of f^{-1} as a function of x.

6.3 The graph of $f^{-1}(x)$.

graph in the 45° line $y = x$ (Fig. 6.3c) and interchanging the letters x and y (Fig. 6.3d). This puts the independent variable, now called x, on the horizontal axis and the dependent variable, now called y, on the vertical axis. The graphs of $f(x)$ and $f^{-1}(x)$ are symmetric about the line $y = x$.

The pictures in Fig. 6.3 tell us how to express f^{-1} as a function of x, which is stated at the left.

How to Express f^{-1} as a Function of x

Step 1: Solve the equation $y = f(x)$ for x in terms of y.

Step 2: Interchange x and y. The resulting formula will be $y = f^{-1}(x)$.

EXAMPLE 3   Find the inverse of $y = \dfrac{1}{2}x + 1$, expressed as a function of x.

Solution

Step 1: Solve for x in terms of y: $y = \dfrac{1}{2}x + 1$

$$2y = x + 2$$

$$x = 2y - 2.$$

Step 2: Interchange x and y: $y = 2x - 2.$

The inverse of the function $f(x) = (1/2)x + 1$ is the function $f^{-1}(x) = 2x - 2.$

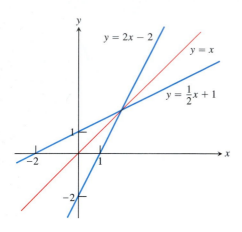

6.4 Graphing $f(x) = (1/2)x + 1$ and $f^{-1}(x) = 2x - 2$ together shows the graphs' symmetry with respect to the line $y = x$.

To check, we verify that both composites give the identity function:

$$f^{-1}(f(x)) = 2\left(\frac{1}{2}x + 1\right) - 2 = x + 2 - 2 = x$$

$$f(f^{-1}(x)) = \frac{1}{2}(2x - 2) + 1 = x - 1 + 1 = x.$$

See Fig. 6.4. ❏

EXAMPLE 4 Find the inverse of the function $y = x^2, x \geq 0$, expressed as a function of x.

Solution

Step 1: *Solve for x in terms of y:*

$$y = x^2$$

$$\sqrt{y} = \sqrt{x^2} = |x| = x \qquad |x| = x \text{ because } x \geq 0$$

Step 2: *Interchange x and y:* $y = \sqrt{x}$.

The inverse of the function $y = x^2$, $x \geq 0$, is the function $y = \sqrt{x}$. See Fig. 6.5.

Notice that, unlike the restricted function $y = x^2$, $x \geq 0$, the unrestricted function $y = x^2$ is not one-to-one and therefore has no inverse. ❏

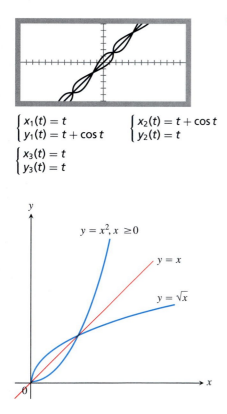

$$\begin{cases} x_1(t) = t \\ y_1(t) = t + \cos t \end{cases} \qquad \begin{cases} x_2(t) = t + \cos t \\ y_2(t) = t \end{cases}$$

$$\begin{cases} x_3(t) = t \\ y_3(t) = t \end{cases}$$

6.5 The functions $y = \sqrt{x}$ and $y = x^2$, $x \geq 0$, are inverses of one another.

> **Technology** *Using Parametric Equations to Graph Inverses* (See the Technology Notes in Section 2.3 for a discussion of parametric mode.) It is easy to graph the inverse of the function $y = f(x)$, using the parametric form
>
> $$x(t) = f(t), \qquad y(t) = t.$$
>
> You can graph the function and its inverse together, using
>
> $$x_1(t) = t, \qquad y_1(t) = f(t) \qquad \text{(the function)}$$
> $$x_2(t) = f(t), \qquad y_2(t) = t \qquad \text{(its inverse)}$$
>
> Even better, graph the function, its inverse, *and* the identity function $y = x$, expressed parametrically as
>
> $$x_3(t) = t, \qquad y_3(t) = t \qquad \text{(the identity function)}$$
>
> The graphing is particularly effective if done simultaneously.
> Try it on the functions $y = x^5/(x^2 + 1)$ and $y = x + \cos x$. You will see the symmetry best if you use a square window (one in which the x- and y-axes are identically scaled).

Derivatives of Inverses of Differentiable Functions

If we calculate the derivatives of $f(x) = (1/2)x + 1$ and its inverse $f^{-1}(x) = 2x - 2$ from Example 3, we see that

$$\frac{d}{dx} f(x) = \frac{d}{dx} \left(\frac{1}{2}x + 1\right) = \frac{1}{2}$$

$$\frac{d}{dx} f^{-1}(x) = \frac{d}{dx}(2x - 2) = 2.$$

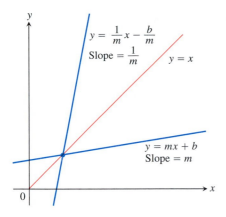

6.6 The slopes of nonvertical lines reflected across the line $y = x$ are reciprocals of one another.

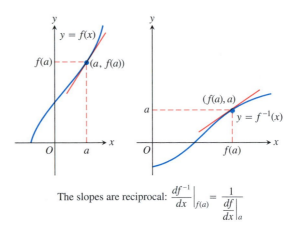

The slopes are reciprocal: $\dfrac{df^{-1}}{dx}\Big|_{f(a)} = \dfrac{1}{\dfrac{df}{dx}\Big|_a}$

6.7 The graphs of inverse functions have reciprocal slopes at corresponding points.

The derivatives are reciprocals of one another. The graph of f is the line $y = (1/2)x + 1$, and the graph of f^{-1} is the line $y = 2x - 2$ (Fig. 6.4). Their slopes are reciprocals of one another.

This is not a special case. Reflecting any nonhorizontal or nonvertical line across the line $y = x$ always inverts the line's slope. If the original line has slope $m \neq 0$ (Fig. 6.6), the reflected line has slope $1/m$ (Exercise 36).

The reciprocal relation between the slopes of graphs of inverses holds for other functions as well. If the slope of $y = f(x)$ at the point $(a, f(a))$ is $f'(a) \neq 0$, then the slope of $y = f^{-1}(x)$ at the corresponding point $(f(a), a)$ is $1/f'(a)$ (Fig. 6.7). Thus, the derivative of f^{-1} at $f(a)$ equals the reciprocal of the derivative of f at a. As you might imagine, we have to impose some mathematical conditions on f to be sure this conclusion holds. The usual conditions, from advanced calculus, are stated in Theorem 1.

Theorem 1
The Derivative Rule for Inverses

If f is differentiable at every point of an interval I and df/dx is never zero on I, then f^{-1} is differentiable at every point of the interval $f(I)$. The value of df^{-1}/dx at any particular point $f(a)$ is the reciprocal of the value of df/dx at a:

$$\left(\frac{df^{-1}}{dx}\right)_{x=f(a)} = \frac{1}{\left(\dfrac{df}{dx}\right)_{x=a}}. \tag{1}$$

In short,

$$\left(f^{-1}\right)' = \frac{1}{f'}. \tag{2}$$

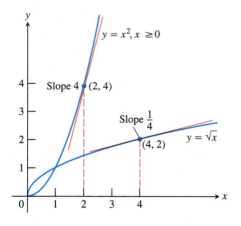

6.8 The derivative of $f^{-1}(x) = \sqrt{x}$ at the point (4, 2) is the reciprocal of the derivative of $f(x) = x^2$ at (2, 4).

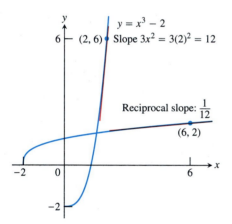

6.9 The derivative of $f(x) = x^3 - 2$ at $x = 2$ tells us the derivative of f^{-1} at $x = 6$.

EXAMPLE 5   For $f(x) = x^2$, $x \geq 0$, and its inverse $f^{-1}(x) = \sqrt{x}$ (Fig. 6.8), we have

$$\frac{df}{dx} = \frac{d}{dx}(x^2) = 2x \quad \text{and} \quad \frac{df^{-1}}{dx} = \frac{d}{dx}\sqrt{x} = \frac{1}{2\sqrt{x}}, \quad x > 0.$$

The point (4, 2) is the mirror image of the point (2, 4) across the line $y = x$.

At the point (2, 4): $\qquad \dfrac{df}{dx} = 2x = 2(2) = 4.$

At the point (4, 2): $\qquad \dfrac{df^{-1}}{dx} = \dfrac{1}{2\sqrt{x}} = \dfrac{1}{2\sqrt{4}} = \dfrac{1}{4} = \dfrac{1}{df/dx}.$ □

Equation (1) sometimes enables us to find specific values of df^{-1}/dx without knowing a formula for f^{-1}.

EXAMPLE 6   Let $f(x) = x^3 - 2$. Find the value of df^{-1}/dx at $x = 6 = f(2)$ without finding a formula for $f^{-1}(x)$.

Solution

$$\left.\frac{df}{dx}\right|_{x=2} = \left. 3x^2 \right|_{x=2} = 12$$

$$\left.\frac{df^{-1}}{dx}\right|_{x=f(2)} = \frac{1}{12} \qquad \text{Eq. (1)}$$

See Fig. 6.9. □

Another Way to Look at Theorem 1

If $y = f(x)$ is differentiable at $x = a$ and we change x by a small amount dx, the corresponding change in y is approximately

$$dy = f'(a)\, dx.$$

This means that y changes about $f'(a)$ times as fast as x and that x changes about $1/f'(a)$ times as fast as y.

Exercises 6.1

Identifying One-to-One Functions Graphically

Which of the functions graphed in Exercises 1–6 are one-to-one, and which are not?

5.

6.

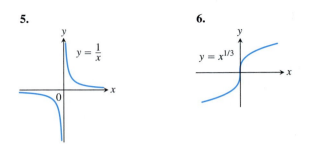

Graphing Inverse Functions

Each of Exercises 7–10 shows the graph of a function $y = f(x)$. Copy the graph and draw in the line $y = x$. Then use symmetry with respect to the line $y = x$ to add the graph of f^{-1} to your sketch. (It is not necessary to find a formula for f^{-1}.) Identify the domain and range of f^{-1}.

7.

8.

9.

10.

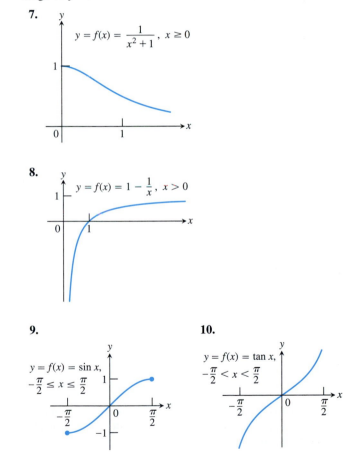

11. a) Graph the function $f(x) = \sqrt{1 - x^2}, 0 \le x \le 1$. What symmetry does the graph have?

   b) Show that f is its own inverse. (Remember that $\sqrt{x^2} = x$ if $x \ge 0$.)

12. a) Graph the function $f(x) = 1/x$. What symmetry does the graph have?

   b) Show that f is its own inverse.

Formulas for Inverse Functions

Each of Exercises 13–18 gives a formula for a function $y = f(x)$ and shows the graphs of f and f^{-1}. Find a formula for f^{-1} in each case.

13. $f(x) = x^2 + 1, \quad x \ge 0$

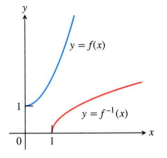

14. $f(x) = x^2, \quad x \le 0$

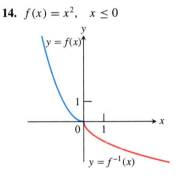

15. $f(x) = x^3 - 1$

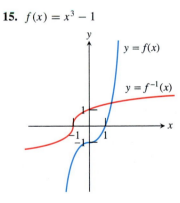

16. $f(x) = x^2 - 2x + 1, \quad x \ge 1$

17. $f(x) = (x+1)^2$, $x \geq -1$ **18.** $f(x) = x^{2/3}$, $x \geq 0$

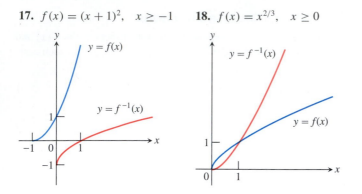

Each of Exercises 19–24 gives a formula for a function $y = f(x)$. In each case, find $f^{-1}(x)$ and identify the domain and range of f^{-1}. As a check, show that $f(f^{-1}(x)) = f^{-1}(f(x)) = x$.

19. $f(x) = x^5$ **20.** $f(x) = x^4$, $x \geq 0$

21. $f(x) = x^3 + 1$ **22.** $f(x) = (1/2)x - 7/2$

23. $f(x) = 1/x^2$, $x > 0$ **24.** $f(x) = 1/x^3$, $x \neq 0$

Derivatives of Inverse Functions

In Exercises 25–28:

a) Find $f^{-1}(x)$.
b) Graph f and f^{-1} together.
c) Evaluate df/dx at $x = a$ and df^{-1}/dx at $x = f(a)$ to show that at these points $df^{-1}/dx = 1/(df/dx)$.

25. $f(x) = 2x + 3$, $a = -1$

26. $f(x) = (1/5)x + 7$, $a = -1$

27. $f(x) = 5 - 4x$, $a = 1/2$

28. $f(x) = 2x^2$, $x \geq 0$, $a = 5$

29. a) Show that $f(x) = x^3$ and $g(x) = \sqrt[3]{x}$ are inverses of one another.
b) Graph f and g over an x-interval large enough to show the graphs intersecting at $(1, 1)$ and $(-1, -1)$. Be sure the picture shows the required symmetry in the line $y = x$.
c) Find the slopes of the tangents to the graphs of f and g at $(1, 1)$ and $(-1, -1)$ (four tangents in all).
d) What lines are tangent to the curves at the origin?

30. a) Show that $h(x) = x^3/4$ and $k(x) = (4x)^{1/3}$ are inverses of one another.
b) Graph h and k over an x-interval large enough to show the graphs intersecting at $(2, 2)$ and $(-2, -2)$. Be sure the picture shows the required symmetry about the line $y = x$.
c) Find the slopes of the tangents to the graphs at h and k at $(2, 2)$ and $(-2, -2)$.
d) What lines are tangent to the curves at the origin?

31. Let $f(x) = x^3 - 3x^2 - 1$, $x \geq 2$. Find the value of df^{-1}/dx at the point $x = -1 = f(3)$.

32. Let $f(x) = x^2 - 4x - 5$, $x > 2$. Find the value of df^{-1}/dx at the point $x = 0 = f(5)$.

33. Suppose that the differentiable function $y = f(x)$ has an inverse and that the graph of f passes through the point $(2, 4)$ and has a slope of $1/3$ there. Find the value of df^{-1}/dx at $x = 4$.

34. Suppose that the differentiable function $y = g(x)$ has an inverse and that the graph of g passes through the origin with slope 2. Find the slope of the graph of g^{-1} at the origin.

35. a) Find the inverse of the function $f(x) = mx$, where m is a constant different from zero.
b) What can you conclude about the inverse of a function $y = f(x)$ whose graph is a line through the origin with a nonzero slope m?

36. Show that the graph of the inverse of $f(x) = mx + b$, where m and b are constants and $m \neq 0$, is a line with slope $1/m$ and y-intercept $-b/m$.

37. a) Find the inverse of $f(x) = x + 1$. Graph f and its inverse together. Add the line $y = x$ to your sketch, drawing it with dashes or dots for contrast.
b) Find the inverse of $f(x) = x + b$ (b constant). How is the graph of f^{-1} related to the graph of f?
c) What can you conclude about the inverses of functions whose graphs are lines parallel to the line $y = x$?

38. a) Find the inverse of $f(x) = -x + 1$. Graph the line $y = -x + 1$ together with the line $y = x$. At what angle do the lines intersect?
b) Find the inverse of $f(x) = -x + b$ (b constant). What angle does the line $y = -x + b$ make with the line $y = x$?
c) What can you conclude about the inverses of functions whose graphs are lines perpendicular to the line $y = x$?

Increasing and Decreasing Functions

39. *Increasing functions and decreasing functions.* As in Section 3.2, a function $f(x)$ increases on an interval I if for any two points x_1 and x_2 in I,

$$x_2 > x_1 \quad \Rightarrow \quad f(x_2) > f(x_1).$$

Similarly, a function decreases on I if for any two points x_1 and x_2 in I,

$$x_2 > x_1 \quad \Rightarrow \quad f(x_2) < f(x_1).$$

Show that increasing functions and decreasing functions are one-to-one. That is, show that for any x_1 and x_2 in I, $x_2 \neq x_1$ implies $f(x_2) \neq f(x_1)$.

Use the results of Exercise 39 to show that the functions in Exercises 40–44 have inverses over their domains. Find a formula for df^{-1}/dx using Theorem 1.

40. $f(x) = (1/3)x + (5/6)$

41. $f(x) = 27x^3$

42. $f(x) = 1 - 8x^3$

43. $f(x) = (1 - x)^3$

44. $f(x) = x^{5/3}$

Theory and Applications

45. If $f(x)$ is one-to-one, can anything be said about $g(x) = -f(x)$? Give reasons for your answer.

46. If $f(x)$ is one-to-one and $f(x)$ is never 0, can anything be said about $h(x) = 1/f(x)$? Give reasons for your answer.

47. Suppose that the range of g lies in the domain of f so that the composite $f \circ g$ is defined. If f and g are one-to-one, can anything be said about $f \circ g$? Give reasons for your answer.

48. If a composite $f \circ g$ is one-to-one, must g be one-to-one? Give reasons for your answer.

49. Suppose $f(x)$ is positive, continuous, and increasing over the interval $[a, b]$. By interpreting the graph of f show that

$$\int_a^b f(x)\, dx + \int_{f(a)}^{f(b)} f^{-1}(y)\, dy = bf(b) - af(a).$$

50. Determine conditions on the constants a, b, c, and d so that the rational function

$$f(x) = \frac{ax + b}{cx + d}$$

has an inverse.

51. *Still another way to view Theorem 1.* If we write $g(x)$ for $f^{-1}(x)$, Eq. (1) can be written as

$$g'(f(a)) = \frac{1}{f'(a)}, \quad \text{or} \quad g'(f(a)) \cdot f'(a) = 1.$$

If we then write x for a, we get

$$g'(f(x)) \cdot f'(x) = 1.$$

The latter equation may remind you of the Chain Rule, and indeed there is a connection.

Assume that f and g are differentiable functions that are inverses of one another, so that $(g \circ f)(x) = x$. Differentiate both sides of this equation with respect to x, using the Chain Rule to express $(g \circ f)'(x)$ as a product of derivatives of g and f. What do you find? (This is not a proof of Theorem 1 because we assume here the theorem's conclusion that $g = f^{-1}$ is differentiable.)

52. *Equivalence of the washer and shell methods for finding volume.* Let f be differentiable on the interval $a \leq x \leq b$, with $a > 0$, and suppose that f has a differentiable inverse, f^{-1}. Revolve about the y-axis the region bounded by the graph of f and the lines $x = a$ and $y = f(b)$ to generate a solid. Then the values of the integrals given by the washer and shell methods for the volume have identical values:

$$\int_{f(a)}^{f(b)} \pi\left((f^{-1}(y))^2 - a^2\right) dy = \int_a^b 2\pi x(f(b) - f(x))\, dx.$$

To prove this equality, define

$$W(t) = \int_{f(a)}^{f(t)} \pi\left((f^{-1}(y))^2 - a^2\right) dy$$

$$S(t) = \int_a^t 2\pi x(f(t) - f(x))\, dx.$$

Then show that the functions W and S agree at a point of $[a, b]$ and have identical derivatives on $[a, b]$. As you saw in Section 4.2, Exercise 56, this will guarantee $W(t) = S(t)$ for all t in $[a, b]$. In particular, $W(b) = S(b)$. (Source: "Disks and Shells Revisited," by Walter Carlip, *American Mathematical Monthly*, Vol. 98, No. 2, February 1991, pp. 154–156.)

⊗ CAS Explorations and Projects

In Exercises 53–60, you will explore some functions and their inverses together with their derivatives and linear approximating functions at specified points. Perform the following steps using your CAS:

a) Plot the function $y = f(x)$ together with its derivative over the given interval. Explain why you know that f is one-to-one over the interval.

b) Solve the equation $y = f(x)$ for x as a function of y, and name the resulting inverse function g.

c) Find the equation for the tangent line to f at the specified point $(x_0, f(x_0))$.

d) Find the equation for the tangent line to g at the point $(f(x_0), x_0)$ located symmetrically across the 45° line $y = x$ (which is the graph of the identity function). Use Theorem 1 to find the slope of this tangent line.

e) Plot the functions f and g, the identity, the two tangent lines, and the line segment joining the points $(x_0, f(x_0))$ and $(f(x_0), x_0)$. Discuss the symmetries you see across the main diagonal.

53. $y = \sqrt{3x - 2}, \quad \frac{2}{3} \leq x \leq 4, \quad x_0 = 3$

54. $y = \dfrac{3x + 2}{2x - 11}, \quad -2 \leq x \leq 2, \quad x_0 = 1/2$

55. $y = \dfrac{4x}{x^2 + 1}, \quad -1 \leq x \leq 1, \quad x_0 = 1/2$

56. $y = \dfrac{x^3}{x^2 + 1}, \quad -1 \leq x \leq 1, \quad x_0 = 1/2$

57. $y = x^3 - 3x^2 - 1, \quad 2 \leq x \leq 5, \quad x_0 = \dfrac{27}{10}$

58. $y = 2 - x - x^3, \quad -2 \leq x \leq 2, \quad x_0 = \dfrac{3}{2}$

59. $y = e^x, \quad -3 \leq x \leq 5, \quad x_0 = 1$

60. $y = \sin x, \quad -\dfrac{\pi}{2} \leq x \leq \dfrac{\pi}{2}, \quad x_0 = 1$

In Exercises 61 and 62, repeat the steps above to solve for the functions $y = f(x)$ and $x = f^{-1}(y)$ defined implicitly by the given equations over the interval.

61. $y^{1/3} - 1 = (x + 2)^3, \quad -5 \leq x \leq 5, \quad x_0 = -3/2$

62. $\cos y = x^{1/5}, \quad 0 \leq x \leq 1, \quad x_0 = 1/2$

Natural Logarithms

The most important function–inverse pair in mathematics and science is the pair consisting of the natural logarithm function ln x and the exponential function e^x. The key to understanding e^x is ln x, so we introduce ln x first. The importance of logarithms came at first from the improvement they brought to arithmetic. The revolutionary properties of logarithms made possible the calculations of the great seventeenth-century advances in offshore navigation and celestial mechanics. Nowadays we do complicated arithmetic with calculators, but the properties of logarithms remain as important as ever.

The Natural Logarithm Function

The natural logarithm of a positive number x, written as ln x, is the value of an integral.

> **Definition**
> **The Natural Logarithm Function**
> $$\ln x = \int_1^x \frac{1}{t}\,dt, \qquad x > 0$$

If $x > 1$, then ln x is the area under the curve $y = 1/t$ from $t = 1$ to $t = x$ (Fig. 6.10). For $0 < x < 1$, ln x gives the negative of the area under the curve from

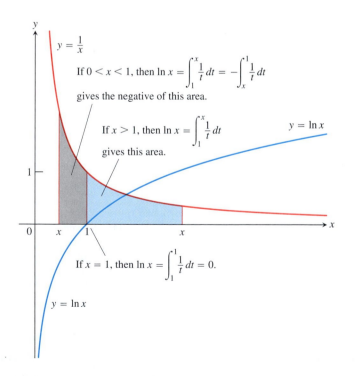

6.10 The graph of $y = \ln x$ and its relation to the function $y = 1/x$, $x > 0$. The graph of the logarithm rises above the x-axis as x moves from 1 to the right, and it falls below the axis as x moves from 1 to the left.

Typical 2-place values of ln x

x	ln x
0	undefined
0.05	−3.00
0.5	−0.69
1	0
2	0.69
3	1.10
4	1.39
10	2.30

x to 1. The function is not defined for $x \leq 0$. We also have

$$\ln 1 = \int_1^1 \frac{1}{t}\, dt = 0. \qquad \text{Upper and lower limits equal}$$

Notice that we show the graph of $y = 1/x$ in Fig. 6.10 but use $y = 1/t$ in the integral. Using x for everything would have us writing

$$\ln x = \int_1^x \frac{1}{x}\, dx,$$

with x meaning two different things. So we change the variable of integration to t.

The Derivative of $y = \ln x$

By the first part of the Fundamental Theorem of Calculus (in Section 4.6),

$$\frac{d}{dx} \ln x = \frac{d}{dx} \int_1^x \frac{1}{t}\, dt = \frac{1}{x}.$$

For every positive value of x, therefore,

$$\frac{d}{dx} \ln x = \frac{1}{x}.$$

If u is a differentiable function of x whose values are positive, so that $\ln u$ is defined, then applying the Chain Rule

$$\frac{dy}{dx} = \frac{dy}{du}\frac{du}{dx}$$

to the function $y = \ln u$ gives

$$\frac{d}{dx} \ln u = \frac{d}{du} \ln u \cdot \frac{du}{dx} = \frac{1}{u}\frac{du}{dx}.$$

$$\frac{d}{dx} \ln u = \frac{1}{u}\frac{du}{dx}, \qquad u > 0 \qquad (1)$$

EXAMPLE 1

$$\frac{d}{dx} \ln 2x = \frac{1}{2x}\frac{d}{dx}(2x) = \frac{1}{2x}(2) = \frac{1}{x} \qquad \square$$

Notice the remarkable occurrence in Example 1. The function $y = \ln 2x$ has the same derivative as the function $y = \ln x$. This is true of $y = \ln ax$ for any number a:

$$\frac{d}{dx} \ln ax = \frac{1}{ax} \cdot \frac{d}{dx}(ax) = \frac{1}{ax}(a) = \frac{1}{x}. \qquad (2)$$

EXAMPLE 2 Equation (1) with $u = x^2 + 3$ gives

$$\frac{d}{dx} \ln (x^2 + 3) = \frac{1}{x^2 + 3} \cdot \frac{d}{dx}(x^2 + 3) = \frac{1}{x^2 + 3} \cdot 2x = \frac{2x}{x^2 + 3}. \qquad \square$$

In the late 1500s, a Scottish baron, John Napier, invented a device called the *logarithm* that simplified arithmetic by replacing multiplication by addition. The equation that accomplished this was

$$\ln ax = \ln a + \ln x.$$

To multiply two positive numbers a and x, you looked up their logarithms in a table, added the logarithms, found the sum in the body of the table, and read the table backward to find the product ax.

Having the table was the key, of course, and Napier spent the last 20 years of his life working on a table he never finished (while the astronomer Tycho Brahe waited in vain for the information he needed to speed his calculations). The table was completed after Napier's death (and Brahe's) by Napier's friend Henry Briggs in London. Base 10 logarithms subsequently became known as Briggs's logarithms (what else?) and some books on navigation still refer to them this way.

Napier also invented an artillery piece that could hit a cow a mile away. Horrified by the weapon's accuracy, he stopped production and suppressed the cannon's design.

Properties of Logarithms

The properties that made logarithms the single most important improvement in arithmetic before the advent of modern computers are listed in Table 6.1. The properties made it possible to replace multiplication of positive numbers by addition, and division of positive numbers by subtraction. They also made it possible to replace exponentiation by multiplication. For the moment, we add the restriction that the exponent n in Rule 4 be a rational number. You will see why when we prove the rule.

EXAMPLE 3

a) $\ln 6 = \ln (2 \cdot 3) = \ln 2 + \ln 3$ Product

b) $\ln 4 - \ln 5 = \ln \dfrac{4}{5} = \ln 0.8$ Quotient

c) $\ln \dfrac{1}{8} = -\ln 8$ Reciprocal

   $= -\ln 2^3 = -3 \ln 2$ Power $\square$

EXAMPLE 4

a) $\ln 4 + \ln \sin x = \ln (4 \sin x)$ Product

b) $\ln \dfrac{x + 1}{2x - 3} = \ln (x + 1) - \ln (2x - 3)$ Quotient

c) $\ln \sec x = \ln \dfrac{1}{\cos x} = -\ln \cos x$ Reciprocal

d) $\ln \sqrt[3]{x + 1} = \ln (x + 1)^{1/3} = \dfrac{1}{3} \ln (x + 1)$ Power $\square$

Proof that ln ax = ln a + ln x The argument is unusual—and elegant. It starts by observing that $\ln ax$ and $\ln x$ have the same derivative (Eq. 2). According to Corollary 1 of the Mean Value Theorem, then, the functions must differ by a

Table 6.1 Properties of natural logarithms

For any numbers $a > 0$ and $x > 0$,

1.	Product Rule:	$\ln ax = \ln a + \ln x$	
2.	Quotient Rule:	$\ln \dfrac{a}{x} = \ln a - \ln x$	
3.	Reciprocal Rule:	$\ln \dfrac{1}{x} = -\ln x$	Rule 2 with $a = 1$
4.	Power Rule:	$\ln x^n = n \ln x$	

constant, which means that

$$\ln ax = \ln x + C \tag{3}$$

for some C. With this much accomplished, it remains only to show that C equals $\ln a$.

Equation (3) holds for all positive values of x, so it must hold for $x = 1$. Hence,

$$\ln (a \cdot 1) = \ln 1 + C$$

$$\ln a = 0 + C \qquad \ln 1 = 0$$

$$C = \ln a. \qquad \text{Rearranged}$$

Substituting $C = \ln a$ in Eq. (3) gives the equation we wanted to prove:

$$\ln ax = \ln a + \ln x. \tag{4}$$

❑

Proof that ln (a/x) = ln a − ln x   We get this from Eq. (4) in two stages. Equation (4) with a replaced by $1/x$ gives

$$\ln \frac{1}{x} + \ln x = \ln \left(\frac{1}{x} \cdot x \right)$$

$$= \ln 1 = 0,$$

so that

$$\ln \frac{1}{x} = -\ln x.$$

Equation (4) with x replaced by $1/x$ then gives

$$\ln \frac{a}{x} = \ln \left(a \cdot \frac{1}{x} \right) = \ln a + \ln \frac{1}{x}$$

$$= \ln a - \ln x.$$

❑

Proof that ln x^n = n ln x (assuming n rational)   We use the same-derivative argument again. For all positive values of x,

$$\frac{d}{dx} \ln x^n = \frac{1}{x^n} \frac{d}{dx} (x^n) \qquad \text{Eq. (1) with } u = x^n$$

$$= \frac{1}{x^n} n x^{n-1}$$

Here is where we need n to be rational, at least for now. We have proved the Power Rule only for rational exponents.

$$= n \cdot \frac{1}{x} = \frac{d}{dx} (n \ln x).$$

Since $\ln x^n$ and $n \ln x$ have the same derivative,

$$\ln x^n = n \ln x + C$$

for some constant C. Taking x to be 1 identifies C as zero, and we're done.   ❑

As for using the rule $\ln x^n = n \ln x$ for irrational values of n, go right ahead and do so. It does hold for all n, and there is no need to pretend otherwise. From the point of view of mathematical development, however, we want you to be aware that the rule is far from proved.

The Graph and Range of ln x

The derivative $d(\ln x)/dx = 1/x$ is positive for $x > 0$, so $\ln x$ is an increasing function of x. The second derivative, $-1/x^2$, is negative, so the graph of $\ln x$ is concave down.

We can estimate $\ln 2$ by numerical integration to be about 0.69. We therefore know that

$$\ln 2^n = n \ln 2 > n \left(\frac{1}{2}\right) = \frac{n}{2}$$

and

$$\ln 2^{-n} = -n \ln 2 < -n \left(\frac{1}{2}\right) = -\frac{n}{2}.$$

It follows that

$$\lim_{x \to \infty} \ln x = \infty \qquad \text{and} \qquad \lim_{x \to 0^+} \ln x = -\infty.$$

The domain of $\ln x$ is the set of positive real numbers; the range is the entire real line.

Logarithmic Differentiation

The derivatives of positive functions given by formulas that involve products, quotients, and powers can often be found more quickly if we take the natural logarithm of both sides before differentiating. This enables us to use the rules in Table 6.1 to simplify the formulas before differentiating. The process, called **logarithmic differentiation,** is illustrated in the next example.

EXAMPLE 5 Find dy/dx if $y = \dfrac{(x^2 + 1)(x + 3)^{1/2}}{x - 1}, \quad x > 1.$

Solution We take the natural logarithm of both sides and simplify the result with the rules in Table 6.1:

$$\ln y = \ln \frac{(x^2 + 1)(x + 3)^{1/2}}{x - 1}$$

$$= \ln \left((x^2 + 1)(x + 3)^{1/2}\right) - \ln (x - 1) \qquad \text{Quotient Rule}$$

$$= \ln (x^2 + 1) + \ln (x + 3)^{1/2} - \ln (x - 1) \qquad \text{Product Rule}$$

$$= \ln (x^2 + 1) + \frac{1}{2} \ln (x + 3) - \ln (x - 1). \qquad \text{Power Rule}$$

We then take derivatives of both sides with respect to x, using Eq. (1) on the left:

$$\frac{1}{y}\frac{dy}{dx} = \frac{1}{x^2 + 1} \cdot 2x + \frac{1}{2} \cdot \frac{1}{x + 3} - \frac{1}{x - 1}.$$

Next we solve for dy/dx:

$$\frac{dy}{dx} = y \left(\frac{2x}{x^2 + 1} + \frac{1}{2x + 6} - \frac{1}{x - 1} \right).$$

Finally, we substitute for y:

$$\frac{dy}{dx} = \frac{(x^2+1)(x+3)^{1/2}}{x-1}\left(\frac{2x}{x^2+1} + \frac{1}{2x+6} - \frac{1}{x-1}\right).$$ ❏

How to Differentiate y = f(x) > 0 by Logarithmic Differentiation

1. $\ln y = \ln f(x)$ Take logs of both sides.

2. $\dfrac{d}{dx}\ln y = \dfrac{d}{dx}(\ln f(x))$ Differentiate both sides . . .

3. $\dfrac{1}{y}\dfrac{dy}{dx} = \dfrac{d}{dx}(\ln f(x))$ . . . using Eq. (1) on the left.

4. $\dfrac{dy}{dx} = y\dfrac{d}{dx}(\ln f(x))$ Solve for dy/dx.

5. $\dfrac{dy}{dx} = f(x)\dfrac{d}{dx}(\ln f(x))$ Substitute $y = f(x)$.

The Integral ∫ (1/u) du

Equation (1) leads to the integral formula

$$\int \frac{1}{u}\,du = \ln u + C \tag{5}$$

when u is a positive differentiable function, but what if u is negative? If u is negative, then $-u$ is positive and

$$\int \frac{1}{u}\,du = \int \frac{1}{(-u)}\,d(-u) \tag{6}$$

$$= \ln(-u) + C. \quad \text{Eq. (5) with } u \text{ replaced by } -u$$

We can combine Eqs. (5) and (6) into a single formula by noticing that in each case the expression on the right is $\ln|u| + C$. In Eq. (5), $\ln u = \ln|u|$ because $u > 0$; in Eq. (6), $\ln(-u) = \ln|u|$ because $u < 0$. Whether u is positive or negative, the integral of $(1/u)\,du$ is $\ln|u| + C$.

If u is a nonzero differentiable function,

$$\int \frac{1}{u}\,du = \ln|u| + C. \tag{7}$$

We know that

$$\int u^n\,du = \frac{u^{n+1}}{n+1} + C, \qquad n \neq -1.$$

Equation (7) explains what to do when n equals -1.

Equation (7) says that integrals of a certain *form* lead to logarithms. That is,

$$\int \frac{f'(x)}{f(x)} dx = \ln |f(x)| + C$$

whenever $f(x)$ is a differentiable function that maintains a constant sign on the domain given for it.

EXAMPLE 6

$$\int_0^2 \frac{2x}{x^2 - 5} dx = \int_{-5}^{-1} \frac{du}{u} = \ln |u| \Big]_{-5}^{-1} \qquad \begin{array}{l} u = x^2 - 5, \quad du = 2x \, dx, \\ u(0) = -5, \quad u(2) = -1 \end{array}$$

$$= \ln |-1| - \ln |-5| = \ln 1 - \ln 5 = -\ln 5 \qquad \square$$

EXAMPLE 7

$$\int_{-\pi/2}^{\pi/2} \frac{4 \cos \theta}{3 + 2 \sin \theta} d\theta = \int_1^5 \frac{2}{u} du \qquad \begin{array}{l} u = 3 + 2 \sin \theta, \quad du = 2 \cos \theta \, d\theta, \\ u(-\pi/2) = 1, \quad u(\pi/2) = 5 \end{array}$$

$$= 2 \ln |u| \Big]_1^5$$

$$= 2 \ln |5| - 2 \ln |1| = 2 \ln 5 \qquad \square$$

The Integrals of tan *x* and cot *x*

Equation (7) tells us at last how to integrate the tangent and cotangent functions. For the tangent,

$$\int \tan x \, dx = \int \frac{\sin x}{\cos x} dx = \int \frac{-du}{u} \qquad \begin{array}{l} u = \cos x, \\ du = -\sin x \, dx \end{array}$$

$$= -\int \frac{du}{u} = -\ln |u| + C \qquad \text{Eq. (7)}$$

$$= -\ln |\cos x| + C = \ln \frac{1}{|\cos x|} + C \qquad \text{Reciprocal Rule}$$

$$= \ln |\sec x| + C.$$

For the cotangent,

$$\int \cot x \, dx = \int \frac{\cos x \, dx}{\sin x} = \int \frac{du}{u} \qquad \begin{array}{l} u = \sin x, \\ du = \cos x \, dx \end{array}$$

$$= \ln |u| + C = \ln |\sin x| + C = -\ln |\csc x| + C.$$

$$\int \tan u \, du = -\ln |\cos u| + C = \ln |\sec u| + C$$

$$\int \cot u \, du = \ln |\sin u| + C = -\ln |\csc x| + C$$

EXAMPLE 8

$$\int_0^{\pi/6} \tan 2x \, dx = \int_0^{\pi/3} \tan u \cdot \frac{du}{2} = \frac{1}{2} \int_0^{\pi/3} \tan u \, du \qquad \begin{array}{l} \text{Substitute } u = 2x, \\ dx = du/2, \\ u(0) = 0, \\ u(\pi/6) = \pi/3 \end{array}$$

$$= \frac{1}{2} \ln |\sec u| \Big]_0^{\pi/3} = \frac{1}{2}(\ln 2 - \ln 1) = \frac{1}{2} \ln 2$$

Exercises 6.2

Using the Properties of Logarithms

1. Express the following logarithms in terms of ln 2 and ln 3.

   a) ln 0.75   b) ln (4/9)   c) ln (1/2)

   d) $\ln \sqrt[3]{9}$ e) $\ln 3\sqrt{2}$ f) $\ln \sqrt{13.5}$

2. Express the following logarithms in terms of ln 5 and ln 7.

   a) ln (1/125)   b) ln 9.8   c) $\ln 7\sqrt{7}$

   d) ln 1225   e) ln 0.056

   f)   (ln 35 + ln (1/7))/(ln 25)

Use the properties of logarithms to simplify the expressions in Exercises 3 and 4.

3. a) $\ln \sin \theta - \ln \left(\dfrac{\sin \theta}{5} \right)$

   b) $\ln (3x^2 - 9x) + \ln \left(\dfrac{1}{3x} \right)$

   c) $\dfrac{1}{2}\ln (4t^4) - \ln 2$

4. a) $\ln \sec \theta + \ln \cos \theta$

   b) $\ln (8x + 4) - 2 \ln 2$

   c) $3 \ln \sqrt[3]{t^2 - 1} - \ln (t + 1)$

Derivatives of Logarithms

In Exercises 5–36, find the derivative of y with respect to x, t, or θ, as appropriate.

5. $y = \ln 3x$

6. $y = \ln kx$, k constant

7. $y = \ln (t^2)$

8. $y = \ln (t^{3/2})$

9. $y = \ln \dfrac{3}{x}$

10. $y = \ln \dfrac{10}{x}$

11. $y = \ln (\theta + 1)$

12. $y = \ln (2\theta + 2)$

13. $y = \ln x^3$

14. $y = (\ln x)^3$

15. $y = t(\ln t)^2$

16. $y = t\sqrt{\ln t}$

17. $y = \dfrac{x^4}{4}\ln x - \dfrac{x^4}{16}$

18. $y = \dfrac{x^3}{3}\ln x - \dfrac{x^3}{9}$

19. $y = \dfrac{\ln t}{t}$

20. $y = \dfrac{1 + \ln t}{t}$

21. $y = \dfrac{\ln x}{1 + \ln x}$

22. $y = \dfrac{x \ln x}{1 + \ln x}$

23. $y = \ln (\ln x)$

24. $y = \ln (\ln (\ln x))$

25. $y = \theta(\sin (\ln \theta) + \cos (\ln \theta))$

26. $y = \ln (\sec \theta + \tan \theta)$

27. $y = \ln \dfrac{1}{x\sqrt{x + 1}}$

28. $y = \dfrac{1}{2} \ln \dfrac{1 + x}{1 - x}$

29. $y = \dfrac{1 + \ln t}{1 - \ln t}$

30. $y = \sqrt{\ln \sqrt{t}}$

31. $y = \ln (\sec(\ln \theta))$

32. $y = \ln \left(\dfrac{\sqrt{\sin \theta \cos \theta}}{1 + 2 \ln \theta} \right)$

33. $y = \ln \left(\dfrac{(x^2 + 1)^5}{\sqrt{1 - x}} \right)$

34. $y = \ln \sqrt{\dfrac{(x + 1)^5}{(x + 2)^{20}}}$

35. $y = \displaystyle\int_{x^2/2}^{x^2} \ln \sqrt{t} \, dt$

36. $y = \displaystyle\int_{\sqrt{x}}^{\sqrt[3]{x}} \ln t \, dt$

Logarithmic Differentiation

In Exercises 37–50, use logarithmic differentiation to find the derivative of y with respect to the given independent variable.

37. $y = \sqrt{x(x + 1)}$

38. $y = \sqrt{(x^2 + 1)(x - 1)^2}$

39. $y = \sqrt{\dfrac{t}{t + 1}}$

40. $y = \sqrt{\dfrac{1}{t(t + 1)}}$

41. $y = \sqrt{\theta + 3} \sin \theta$

42. $y = (\tan \theta) \sqrt{2\theta + 1}$

43. $y = t(t + 1)(t + 2)$

44. $y = \dfrac{1}{t(t + 1)(t + 2)}$

45. $y = \dfrac{\theta + 5}{\theta \cos \theta}$

46. $y = \dfrac{\theta \sin \theta}{\sqrt{\sec \theta}}$

47. $y = \dfrac{x\sqrt{x^2 + 1}}{(x + 1)^{2/3}}$

48. $y = \sqrt{\dfrac{(x + 1)^{10}}{(2x + 1)^5}}$

49. $y = \sqrt[3]{\dfrac{x(x - 2)}{x^2 + 1}}$

50. $y = \sqrt[3]{\dfrac{x(x + 1)(x - 2)}{(x^2 + 1)(2x + 3)}}$

Integration

Evaluate the integrals in Exercises 51–68.

51. $\displaystyle\int_{-3}^{-2} \frac{dx}{x}$

52. $\displaystyle\int_{-1}^{0} \frac{3\,dx}{3x-2}$

53. $\displaystyle\int \frac{2y\,dy}{y^2-25}$

54. $\displaystyle\int \frac{8r\,dr}{4r^2-5}$

55. $\displaystyle\int_{0}^{\pi} \frac{\sin t}{2-\cos t}\,dt$

56. $\displaystyle\int_{0}^{\pi/3} \frac{4\sin\theta}{1-4\cos\theta}\,d\theta$

57. $\displaystyle\int_{1}^{2} \frac{2\ln x}{x}\,dx$

58. $\displaystyle\int_{2}^{4} \frac{dx}{x\ln x}$

59. $\displaystyle\int_{2}^{4} \frac{dx}{x(\ln x)^2}$

60. $\displaystyle\int_{2}^{16} \frac{dx}{2x\sqrt{\ln x}}$

61. $\displaystyle\int \frac{3\sec^2 t}{6+3\tan t}\,dt$

62. $\displaystyle\int \frac{\sec y\tan y}{2+\sec y}\,dy$

63. $\displaystyle\int_{0}^{\pi/2} \tan\frac{x}{2}\,dx$

64. $\displaystyle\int_{\pi/4}^{\pi/2} \cot t\,dt$

65. $\displaystyle\int_{\pi/2}^{\pi} 2\cot\frac{\theta}{3}\,d\theta$

66. $\displaystyle\int_{0}^{\pi/12} 6\tan 3x\,dx$

67. $\displaystyle\int \frac{dx}{2\sqrt{x}+2x}$

68. $\displaystyle\int \frac{\sec x\,dx}{\sqrt{\ln(\sec x+\tan x)}}$

Theory and Applications

69. Locate and identify the absolute extreme values of

   a) $\ln(\cos x)$ on $[-\pi/4, \pi/3]$,

   b) $\cos(\ln x)$ on $[1/2, 2]$.

70. a) Prove that $f(x) = x - \ln x$ is increasing for $x > 1$.

   b) Using part (a), show that $\ln x < x$ if $x > 1$.

71. Find the area between the curves $y = \ln x$ and $y = \ln 2x$ from $x = 1$ to $x = 5$.

72. Find the area between the curve $y = \tan x$ and the x-axis from $x = -\pi/4$ to $x = \pi/3$.

73. The region in the first quadrant bounded by the coordinate axes, the line $y = 3$, and the curve $x = 2/\sqrt{y+1}$ is revolved about the y-axis to generate a solid. Find the volume of the solid.

74. The region between the curve $y = \sqrt{\cot x}$ and the x-axis from $x = \pi/6$ to $x = \pi/2$ is revolved about the x-axis to generate a solid. Find the volume of the solid.

75. The region between the curve $y = 1/x^2$ and the x-axis from $x = 1/2$ to $x = 2$ is revolved about the y-axis to generate a solid. Find the volume of the solid.

76. In Section 5.4, Exercise 6, we revolved about the y-axis the region between the curve $y = 9x/\sqrt{x^3+9}$ and the x-axis from $x = 0$ to $x = 3$ to generate a solid of volume 36π. What volume do you get if you revolve the region about the x-axis instead? (See Section 5.4, Exercise 6, for a graph.)

77. Find the lengths of the following curves.

   a) $y = (x^2/8) - \ln x, \quad 4 \le x \le 8$

   b) $x = (y/4)^2 - 2\ln(y/4), \quad 4 \le y \le 12$

78. Find a curve through the point $(1, 0)$ whose length from $x = 1$ to $x = 2$ is

$$L = \int_{1}^{2} \sqrt{1 + \frac{1}{x^2}}\,dx.$$

79. CALCULATOR

   a) Find the centroid of the region between the curve $y = 1/x$ and the x-axis from $x = 1$ to $x = 2$. Give the coordinates to 2 decimal places.

   b) Sketch the region and show the centroid in your sketch.

80. a) Find the center of mass of a thin plate of constant density covering the region between the curve $y = 1/\sqrt{x}$ and the x-axis from $x = 1$ to $x = 16$.

   b) Find the center of mass if, instead of being constant, the density function is $\delta(x) = 4/\sqrt{x}$.

Solve the initial value problems in Exercises 81 and 82.

81. $\dfrac{dy}{dx} = 1 + \dfrac{1}{x}, \quad y(1) = 3$

82. $\dfrac{d^2y}{dx^2} = \sec^2 x, \quad y(0) = 0 \quad \text{and} \quad y'(0) = 1$

83. *The linearization of $\ln(1+x)$ at $x = 0$.* Instead of approximating x near $x = 1$, we approximate $\ln(1+x)$ near $x = 0$. We get a simpler formula this way.

   a) Derive the linearization $\ln(1+x) \approx x$ at $x = 0$.

   b) CALCULATOR Estimate to 5 decimal places the error involved in replacing $\ln(1+x)$ by x on the interval $[0, 0.1]$.

   c) GRAPHER Graph $\ln(1+x)$ and x together for $0 \le x \le 0.5$. Use different colors, if available. At what points does the approximation of $\ln(1+x)$ seem best? least good? By reading coordinates from the graphs, find as good an upper bound for the error as your grapher will allow.

84. *Estimating values of $\ln x$ with Simpson's rule.* Although linearizations are good for replacing the logarithmic function over short intervals, Simpson's rule is better for estimating *particular* values of $\ln x$.

   As a case in point, the values of $\ln(1.2)$ and $\ln(0.8)$ to 5 places are

$$\ln(1.2) = 0.18232, \quad \ln(0.8) = -0.22314.$$

Estimate $\ln(1.2)$ and $\ln(0.8)$ first with the formula $\ln(1+x) \approx x$ and then use Simpson's rule with $n = 2$. (Impressive, isn't it?)

85. Find

$$\lim_{x\to\infty} \frac{\ln(x^2)}{\ln x}.$$

Generalize this result.

86. *The derivative of ln kx.* Could $y = \ln 2x$ and $y = \ln 3x$ possibly have the same derivative at each point? (Differentiate them to find out.) What about $y = \ln kx$, for other positive values of the constant k? Give reasons for your answer.

Grapher Explorations

87. Graph $\ln x$, $\ln 2x$, $\ln 4x$, $\ln 8x$, and $\ln 16x$ (as many as you can) together for $0 < x \le 10$. What is going on? Explain.

88. Graph $y = \ln|\sin x|$ in the window $0 \le x \le 22, -2 \le y \le 0$. Explain what you see. How could you change the formula to turn the arches upside down?

89. a) Graph $y = \sin x$ and the curves $y = \ln(a + \sin x)$ for $a = 2, 4, 8, 20,$ and 50 together for $0 \le x \le 23$.

   b) Why do the curves flatten as a increases? (*Hint:* Find an a-dependent upper bound for $|y'|$.)

90. Does the graph of $y = \sqrt{x} - \ln x$, $x > 0$, have an inflection point? Try to answer the question (a) by graphing, (b) by using calculus.

6.3

The Exponential Function

Whenever we have a quantity y whose rate of change over time is proportional to the amount of y present, we have a function that satisfies the differential equation

$$\frac{dy}{dt} = ky.$$

If, in addition, $y = y_0$ when $t = 0$, the function is the exponential function $y = y_0 e^{kt}$. This section defines the exponential function (it is the inverse of $\ln x$) and explores the properties that account for the amazing frequency with which the function appears in mathematics and its applications. We will look at some of these applications in Section 6.5.

The Inverse of ln x and the Number e

The function $\ln x$, being an increasing function of x with domain $(0, \infty)$ and range $(-\infty, \infty)$, has an inverse $\ln^{-1} x$ with domain $(-\infty, \infty)$ and range $(0, \infty)$. The graph of $\ln^{-1} x$ is the graph of $\ln x$ reflected across the line $y = x$. As you can see,

$$\lim_{x \to \infty} \ln^{-1} x = \infty \qquad \text{and} \qquad \lim_{x \to -\infty} \ln^{-1} x = 0.$$

The number $\ln^{-1} 1$ is denoted by the letter e (Fig. 6.11).

Definition

$$e = \ln^{-1} 1$$

Although e is not a rational number, we will see in Chapter 8 that it is possible to find its value with a computer to as many places as we want with the formula

$$e = \lim_{n \to \infty} \left(1 + 1 + \frac{1}{2} + \frac{1}{6} + \cdots + \frac{1}{n!}\right).$$

To 15 places,

$$e = 2.7\ 1828\ 1828\ 45\ 90\ 45.$$

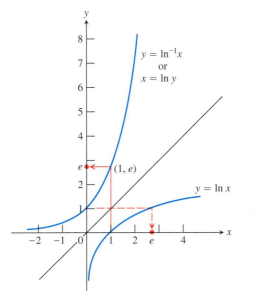

6.11 The graphs of $y = \ln x$ and $y = \ln^{-1} x$. The number e is $\ln^{-1} 1$.

The Function $y = e^x$

We can raise the number e to a rational power x in the usual way:

$$e^2 = e \cdot e, \qquad e^{-2} = \frac{1}{e^2}, \qquad e^{1/2} = \sqrt{e},$$

and so on. Since e is positive, e^x is positive too. This means that e^x has a logarithm. When we take the logarithm we find that

$$\ln e^x = x \ln e = x \cdot 1 = x. \tag{1}$$

Since $\ln x$ is one-to-one and $\ln(\ln^{-1}x) = x$, Eq. (1) tells us that

$$e^x = \ln^{-1}x \qquad \text{for } x \text{ rational.} \tag{2}$$

Equation (2) provides a way to extend the definition of e^x to irrational values of x. The function $\ln^{-1}x$ is defined for all x, so we can use it to assign a value to e^x at every point where e^x had no previous value.

Typical Values of e^x

x	e^x (rounded)
-1	0.37
0	1
1	2.72
2	7.39
10	22026
100	2.6881×10^{43}

Definition

For every real number x, $e^x = \ln^{-1}x$.

Equations Involving $\ln x$ and e^x

Since $\ln x$ and e^x are inverses of one another, we have

Inverse Equations for e^x and $\ln x$

$$e^{\ln x} = x \qquad (\text{all } x > 0) \tag{3}$$

$$\ln(e^x) = x \qquad (\text{all } x) \tag{4}$$

You might want to do parts of the next example on your calculator.

EXAMPLE 1

a) $\ln e^2 = 2$

b) $\ln e^{-1} = -1$

c) $\ln \sqrt{e} = \dfrac{1}{2}$

d) $\ln e^{\sin x} = \sin x$

e) $e^{\ln 2} = 2$

f) $e^{\ln(x^2+1)} = x^2 + 1$

g) $e^{3\ln 2} = e^{\ln 2^3} = e^{\ln 8} = 8$ One way

h) $e^{3\ln 2} = (e^{\ln 2})^3 = 2^3 = 8$ Another way

Useful Operating Rules

1. To remove logarithms from an equation, exponentiate both sides.
2. To remove exponentials, take the logarithm of both sides.

EXAMPLE 2 Find y if $\ln y = 3t + 5$.

Solution Exponentiate both sides:

$$e^{\ln y} = e^{3t+5}$$

$$y = e^{3t+5}. \qquad \text{Eq. (3)} \qquad \square$$

EXAMPLE 3 Find k if $e^{2k} = 10$.

Solution Take the natural logarithm of both sides:

$$e^{2k} = 10$$

$$\ln e^{2k} = \ln 10$$

$$2k = \ln 10 \qquad \text{Eq. (4)}$$

$$k = \frac{1}{2} \ln 10. \qquad \square$$

Laws of Exponents

Table 6.2 Laws of exponents for e^x

For all numbers x, x_1, and x_2,

1. $e^{x_1} \cdot e^{x_2} = e^{x_1 + x_2}$
2. $e^{-x} = \dfrac{1}{e^x}$
3. $\dfrac{e^{x_1}}{e^{x_2}} = e^{x_1 - x_2}$
4. $(e^{x_1})^{x_2} = e^{x_1 x_2} = (e^{x_2})^{x_1}$

Even though e^x is defined in a seemingly roundabout way as $\ln^{-1} x$, it obeys the familiar laws of exponents from algebra (Table 6.2).

Proof of Law 1 Let

$$y_1 = e^{x_1} \qquad \text{and} \qquad y_2 = e^{x_2}. \qquad (5)$$

Then

$$x_1 = \ln y_1 \quad \text{and} \quad x_2 = \ln y_2 \qquad \text{Take logs of both sides of Eqs. (5).}$$

$$x_1 + x_2 = \ln y_1 + \ln y_2$$

$$= \ln y_1 y_2 \qquad \text{Product Rule}$$

$$e^{x_1 + x_2} = e^{\ln y_1 y_2} \qquad \text{Exponentiate.}$$

$$= y_1 y_2 \qquad e^{\ln u} = u$$

$$= e^{x_1} e^{x_2}. \qquad \square$$

The proof of Law 4 is similar. Laws 2 and 3 follow from Law 1 (Exercise 78).

EXAMPLE 4

a) $e^{x + \ln 2} = e^x \cdot e^{\ln 2} = 2e^x$ \qquad Law 1

b) $e^{-\ln x} = \dfrac{1}{e^{\ln x}} = \dfrac{1}{x}$ \qquad Law 2

c) $\dfrac{e^{2x}}{e} = e^{2x-1}$ \qquad Law 3

d) $(e^3)^x = e^{3x} = (e^x)^3$ \qquad Law 4 \qquad $\square$

The Derivative and Integral of e^x

The exponential function is differentiable because it is the inverse of a differentiable function whose derivative is never zero. Starting with $y = e^x$, we have, in order,

$$y = e^x$$

$$\ln y = x \qquad \text{Logarithms of both sides}$$

$$\frac{1}{y}\frac{dy}{dx} = 1 \qquad \text{Derivatives of both sides with respect to } x$$

$$\frac{dy}{dx} = y$$

$$\frac{dy}{dx} = e^x. \qquad y \text{ replaced by } e^x$$

The startling conclusion we draw from this sequence of equations is that e^x is its own derivative.

As we will see in Section 6.5, the only functions that behave this way are constant multiples of e^x.

$$\boxed{\frac{d}{dx}e^x = e^x} \qquad (6)$$

EXAMPLE 5

$$\frac{d}{dx}(5e^x) = 5\frac{d}{dx}e^x$$

$$= 5e^x \qquad \square$$

The Chain Rule extends Eq. (6) in the usual way to a more general form.

If u is any differentiable function of x, then

$$\boxed{\frac{d}{dx}e^u = e^u\frac{du}{dx}.} \qquad (7)$$

EXAMPLE 6

a) $\dfrac{d}{dx}e^{-x} = e^{-x}\dfrac{d}{dx}(-x) = e^{-x}(-1) = -e^{-x} \qquad$ Eq. (7) with $u = -x$

b) $\dfrac{d}{dx}e^{\sin x} = e^{\sin x}\dfrac{d}{dx}(\sin x) = e^{\sin x} \cdot \cos x \qquad$ Eq. (7) with $u = \sin x$ $\qquad \square$

The integral equivalent of Eq. (7) is

Transcendental numbers and transcendental functions

Numbers that are solutions of polynomial equations with rational coefficients are called **algebraic:** -2 is algebraic because it satisfies the equation $x + 2 = 0$, and $\sqrt{3}$ is algebraic because it satisfies the equation $x^2 - 3 = 0$. Numbers that are not algebraic are called **transcendental,** a term coined by Euler to describe numbers, like e and π, that appeared to "transcend the power of algebraic methods." But it was not until a hundred years after Euler's death (1873) that Charles Hermite proved the transcendence of e in the sense that we describe. A few years later (1882), C. L. F. Lindemann proved the transcendence of π.

Today we call a function $y = f(x)$ algebraic if it satisfies an equation of the form

$$P_n y^n + \cdots + P_1 y + P_0 = 0$$

in which the P's are polynomials in x with rational coefficients. The function $y = 1/\sqrt{x+1}$ is algebraic because it satisfies the equation $(x + 1)y^2 - 1 = 0$. Here the polynomials are $P_2 = x + 1$, $P_1 = 0$, and $P_0 = -1$. Polynomials and rational functions with rational coefficients are algebraic, as are all sums, products, quotients, rational powers, and rational roots of algebraic functions.

Functions that are not algebraic are called transcendental. The six basic trigonometric functions are transcendental, as are the inverses of the trigonometric functions and the exponential and logarithmic functions that are the main subject of the present chapter.

$$\int e^u \, du = e^u + C.$$

EXAMPLE 7

$$\int_0^{\ln 2} e^{3x} \, dx = \int_0^{\ln 8} e^u \cdot \frac{1}{3} \, du \qquad u = 3x, \quad \frac{1}{3} \, du = dx, \, u(0) = 0,$$

$$u(\ln 2) = 3 \ln 2 = \ln 2^3 = \ln 8$$

$$= \frac{1}{3} \int_0^{\ln 8} e^u \, du$$

$$= \frac{1}{3} e^u \Big]_0^{\ln 8}$$

$$= \frac{1}{3}[8 - 1] = \frac{7}{3} \qquad \square$$

EXAMPLE 8

$$\int_0^{\pi/2} e^{\sin x} \cos x \, dx = e^{\sin x} \Big]_0^{\pi/2} \qquad \text{Antiderivative from Example 6}$$

$$= e^1 - e^0 = e - 1 \qquad \square$$

EXAMPLE 9 *Solving an initial value problem*

Solve the initial value problem

$$e^y \frac{dy}{dx} = 2x, \qquad x > \sqrt{3}; \qquad y(2) = 0.$$

Solution We integrate both sides of the differential equation with respect to x to obtain

$$e^y = x^2 + C.$$

We use the initial condition to determine C:

$$C = e^0 - (2)^2$$

$$= 1 - 4 = -3.$$

This completes the formula for e^y:

$$e^y = x^2 - 3. \tag{8}$$

To find y, we take logarithms of both sides:

$$\ln e^y = \ln (x^2 - 3)$$

$$y = \ln (x^2 - 3). \tag{9}$$

Notice that the solution is valid for $x > \sqrt{3}$.

It is always a good idea to check a solution in the original equation. From Eqs. (8) and (9), we have

$$e^y \frac{dy}{dx} = e^y \frac{d}{dx} \ln (x^2 - 3) \qquad \text{Eq. (9)}$$

$$= e^y \frac{2x}{x^2 - 3}$$

$$= (x^2 - 3) \frac{2x}{x^2 - 3} \qquad \text{Eq. (8)}$$

$$= 2x.$$

The solution checks.

Exercises 6.3

Algebraic Calculations with the Exponential and Logarithm

Find simpler expressions for the quantities in Exercises 1–4.

1. a) $e^{\ln 7.2}$ b) $e^{-\ln x^2}$ c) $e^{\ln x - \ln y}$

2. a) $e^{\ln (x^2+y^2)}$ b) $e^{-\ln 0.3}$ c) $e^{\ln \pi x - \ln 2}$

3. a) $2 \ln \sqrt{e}$ b) $\ln (\ln e^e)$ c) $\ln (e^{-x^2-y^2})$

4. a) $\ln (e^{\sec \theta})$ b) $\ln (e^{(e^x)})$ c) $\ln (e^{2 \ln x})$

Solving Equations with Logarithmic or Exponential Terms

In Exercises 5–10, solve for y in terms of t or x, as appropriate.

5. $\ln y = 2t + 4$

6. $\ln y = -t + 5$

7. $\ln (y - 40) = 5t$

8. $\ln (1 - 2y) = t$

9. $\ln (y - 1) - \ln 2 = x + \ln x$

10. $\ln (y^2 - 1) - \ln (y + 1) = \ln (\sin x)$

In Exercises 11 and 12, solve for k.

11. a) $e^{2k} = 4$ b) $100e^{10k} = 200$ c) $e^{k/1000} = a$

12. a) $e^{5k} = \frac{1}{4}$ b) $80e^k = 1$ c) $e^{(\ln 0.8)k} = 0.8$

In Exercises 13–16, solve for t.

13. a) $e^{-0.3t} = 27$ b) $e^{kt} = \frac{1}{2}$ c) $e^{(\ln 0.2)t} = 0.4$

14. a) $e^{-0.01t} = 1000$ b) $e^{kt} = \frac{1}{10}$ c) $e^{(\ln 2)t} = \frac{1}{2}$

15. $e^{\sqrt{t}} = x^2$

16. $e^{(x^2)}e^{(2x+1)} = e^t$

Derivatives

In Exercises 17–36, find the derivative of y with respect to x, t, or θ, as appropriate.

17. $y = e^{-5x}$ 18. $y = e^{2x/3}$

19. $y = e^{5-7x}$ 20. $y = e^{(4\sqrt{x}+x^2)}$

21. $y = xe^x - e^x$ 22. $y = (1 + 2x)e^{-2x}$

23. $y = (x^2 - 2x + 2)e^x$ 24. $y = (9x^2 - 6x + 2)e^{3x}$

25. $y = e^\theta (\sin \theta + \cos \theta)$ 26. $y = \ln (3\theta e^{-\theta})$

27. $y = \cos (e^{-\theta^2})$ 28. $y = \theta^3 e^{-2\theta} \cos 5\theta$

29. $y = \ln (3te^{-t})$ 30. $y = \ln (2e^{-t} \sin t)$

31. $y = \ln \left(\dfrac{e^\theta}{1 + e^\theta} \right)$ 32. $y = \ln \left(\dfrac{\sqrt{\theta}}{1 + \sqrt{\theta}} \right)$

33. $y = e^{(\cos t + \ln t)}$ 34. $y = e^{\sin t} (\ln t^2 + 1)$

35. $y = \displaystyle\int_0^{\ln x} \sin e^t \, dt$ 36. $y = \displaystyle\int_{e^{4\sqrt{x}}}^{e^{2x}} \ln t \, dt$

In Exercises 37–40, find dy/dx.

37. $\ln y = e^y \sin x$ 38. $\ln xy = e^{x+y}$

39. $e^{2x} = \sin (x + 3y)$ 40. $\tan y = e^x + \ln x$

Integrals

Evaluate the integrals in Exercises 41–62.

41. $\displaystyle\int (e^{3x} + 5e^{-x}) \, dx$ 42. $\displaystyle\int (2e^x - 3e^{-2x}) \, dx$

43. $\displaystyle\int_{\ln 2}^{\ln 3} e^x \, dx$ 44. $\displaystyle\int_{-\ln 2}^0 e^{-x} \, dx$

45. $\displaystyle\int 8e^{(x+1)}\,dx$

46. $\displaystyle\int 2e^{(2x-1)}\,dx$

47. $\displaystyle\int_{\ln 4}^{\ln 9} e^{x/2}\,dx$

48. $\displaystyle\int_{0}^{\ln 16} e^{x/4}\,dx$

49. $\displaystyle\int \frac{e^{\sqrt{r}}}{\sqrt{r}}\,dr$

50. $\displaystyle\int \frac{e^{-\sqrt{r}}}{\sqrt{r}}\,dr$

51. $\displaystyle\int 2t\,e^{-t^2}\,dt$

52. $\displaystyle\int t^3 e^{(t^4)}\,dt$

53. $\displaystyle\int \frac{e^{1/x}}{x^2}\,dx$

54. $\displaystyle\int \frac{e^{-1/x^2}}{x^3}\,dx$

55. $\displaystyle\int_{0}^{\pi/4} (1+e^{\tan\theta})\sec^2\theta\,d\theta$

56. $\displaystyle\int_{\pi/4}^{\pi/2} (1+e^{\cot\theta})\csc^2\theta\,d\theta$

57. $\displaystyle\int e^{\sec \pi t}\sec \pi t \tan \pi t\,dt$

58. $\displaystyle\int e^{\csc (\pi+t)}\csc (\pi+t)\cot (\pi+t)\,dt$

59. $\displaystyle\int_{\ln (\pi/6)}^{\ln (\pi/2)} 2e^v \cos e^v\,dv$

60. $\displaystyle\int_{0}^{\sqrt{\ln \pi}} 2x\,e^{x^2}\cos (e^{x^2})\,dx$

61. $\displaystyle\int \frac{e^r}{1+e^r}\,dr$

62. $\displaystyle\int \frac{dx}{1+e^x}$

Initial Value Problems

Solve the initial value problems in Exercises 63–66.

63. $\dfrac{dy}{dt} = e^t \sin (e^t - 2), \quad y\,(\ln 2) = 0$

64. $\dfrac{dy}{dt} = e^{-t}\sec^2(\pi e^{-t}), \quad y(\ln 4) = 2/\pi$

65. $\dfrac{d^2 y}{dx^2} = 2e^{-x}, \quad y(0) = 1 \quad \text{and} \quad y'(0) = 0$

66. $\dfrac{d^2 y}{dt^2} = 1 - e^{2t}, \quad y(1) = -1 \quad \text{and} \quad y'(1) = 0$

Theory and Applications

67. Find the absolute maximum and minimum values of $f(x) = e^x - 2x$ on $[0, 1]$.

68. Where does the periodic function $f(x) = 2e^{\sin (x/2)}$ take on its extreme values and what are these values?

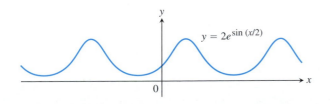

$y = 2e^{\sin (x/2)}$

69. Find the absolute maximum value of $f(x) = x^2 \ln (1/x)$ and say where it is assumed.

70. GRAPHER Graph $f(x) = (x - 3)^2 e^x$ and its first derivative together. Comment on the behavior of f in relation to the signs and values of f'. Identify significant points on the graphs with calculus, as necessary.

71. Find the area of the "triangular" region in the first quadrant that is bounded above by the curve $y = e^{2x}$, below by the curve $y = e^x$, and on the right by the line $x = \ln 3$.

72. Find the area of the "triangular" region in the first quadrant that is bounded above by the curve $y = e^{x/2}$, below by the curve $y = e^{-x/2}$, and on the right by the line $x = 2 \ln 2$.

73. Find a curve through the origin in the xy-plane whose length from $x = 0$ to $x = 1$ is

$$L = \int_{0}^{1} \sqrt{1 + \frac{1}{4}e^x}\,dx.$$

74. Find the area of the surface generated by revolving the curve $x = (e^y + e^{-y})/2, 0 \le y \le \ln 2$, about the y-axis.

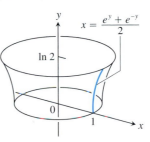

75. a) Show that $\int \ln x\,dx = x \ln x - x + C$.
    b) Find the average value of $\ln x$ over $[1, e]$.

76. Find the average value of $f(x) = 1/x$ on $[1, 2]$.

77. *The linearization of e^x at $x = 0$*
    a) Derive the linear approximation $e^x \approx 1 + x$ at $x = 0$.
    b) CALCULATOR Estimate to 5 decimal places the magnitude of the error involved in replacing e^x by $1 + x$ on the interval $[0, 0.2]$.
    c) GRAPHER Graph e^x and $1 + x$ together for $-2 \le x \le 2$. Use different colors, if available. On what intervals does the approximation appear to overestimate e^x? underestimate e^x?

78. *Laws of Exponents.*
    a) Starting with the equation $e^{x_1}e^{x_2} = e^{x_1+x_2}$, derived in the text, show that $e^{-x} = 1/e^x$ for any real number x. Then show that $e^{x_1}/e^{x_2} = e^{x_1-x_2}$ for any numbers x_1 and x_2.
    b) Show that $(e^{x_1})^{x_2} = e^{x_1 x_2} = (e^{x_2})^{x_1}$ for any numbers x_1 and x_2.

79. *A decimal representation of e.* Find e to as many decimal places as your calculator allows by solving the equation $\ln x = 1$.

80. *The inverse relation between e^x and $\ln x$.* Find out how good your calculator is at evaluating the composites

$$e^{\ln x} \quad \text{and} \quad \ln (e^x).$$

81. Show that for any number $a > 1$

$$\int_{1}^{a} \ln x\, dx + \int_{0}^{\ln a} e^{y}\, dy = a \ln a.$$

(See accompanying figure.)

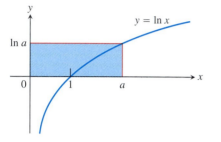

82. *The geometric, logarithmic, and arithmetic mean inequality*

a) Show that the graph of e^{x} is concave up over every interval of x-values.

b) Show, by reference to the accompanying figure, that if $0 < a < b$ then

$$e^{(\ln a + \ln b)/2} \cdot (\ln b - \ln a) < \int_{\ln a}^{\ln b} e^{x}\, dx <$$

$$\frac{e^{\ln a} + e^{\ln b}}{2} \cdot (\ln b - \ln a).$$

c) Use the inequality in (b) to conclude that

$$\sqrt{ab} < \frac{b - a}{\ln b - \ln a} < \frac{a + b}{2}.$$

This inequality says that the geometric mean of two positive numbers is less than their logarithmic mean, which in turn is less than their arithmetic mean.

(For more about this inequality, see "The Geometric, Logarithmic, and Arithmetic Mean Inequality" by Frank Burk, *American Mathematical Monthly,* Vol. 94, No. 6, June–July 1987, pp. 527–528.)

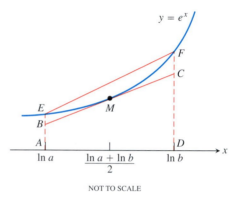

NOT TO SCALE

| 6.4 | a^{x} and $\log_{a} x$ |

While we have not yet devised a way to raise positive numbers to any but rational powers, we have an exception in the number e. The definition $e^{x} = \ln^{-1} x$ defines e^{x} for every real value of x, irrational as well as rational. In this section, we show how this enables us to raise any other positive number to an arbitrary power and thus to define an exponential function $y = a^{x}$ for any positive number a. We also prove the Power Rule for differentiation in its final form (good for all exponents) and define functions like x^{x} and $(\sin x)^{\tan x}$ that involve raising the values of one function to powers given by another.

Just as e^{x} is but one of many exponential functions, $\ln x$ is one of many logarithmic functions, the others being the inverses of the function a^{x}. These logarithmic functions have important applications in science and engineering.

The Function a^{x}

Since $a = e^{\ln a}$ for any positive number a, we can think of a^{x} as $(e^{\ln a})^{x} = e^{x \ln a}$. We therefore make the following definition.

> **Definition**
> For any numbers $a > 0$ and x,
> $$a^{x} = e^{x \ln a}. \tag{1}$$

Table 6.3 Laws of exponents

For $a > 0$, and any x and y:

1. $a^x \cdot a^y = a^{x+y}$

2. $a^{-x} = \dfrac{1}{a^x}$

3. $\dfrac{a^x}{a^y} = a^{x-y}$

4. $(a^x)^y = a^{xy} = (a^y)^x$

EXAMPLE 1

a) $2^{\sqrt{3}} = e^{\sqrt{3}\ln 2}$

b) $2^{\pi} = e^{\pi \ln 2}$ ❑

The function a^x obeys the usual laws of exponents (Table 6.3). We omit the proofs.

The Power Rule (Final Form)

We can now define x^n for any $x > 0$ and any real number n as $x^n = e^{n \ln x}$. Therefore, the n in the equation $\ln x^n = n \ln x$ no longer needs to be rational—it can be any number as long as $x > 0$:

$$\ln x^n = \ln (e^{n \ln x}) = n \ln x \cdot \ln e \qquad \ln e^u = u, \text{ any } u$$

$$= n \ln x.$$

Together, the law $a^x/a^y = a^{x-y}$ and the definition $x^n = e^{n \ln x}$ enable us to establish the Power Rule for differentiation in its final form. Differentiating x^n with respect to x gives

$$\frac{d}{dx} x^n = \frac{d}{dx} e^{n \ln x} \qquad \text{Definition of } x^n,\ x > 0$$

$$= e^{n \ln x} \cdot \frac{d}{dx}(n \ln x) \qquad \text{Chain Rule for } e^u$$

$$= x^n \cdot \frac{n}{x} \qquad \text{The definition again}$$

$$= n x^{n-1}. \qquad \text{Table 6.3, Law 3}$$

In short, as long as $x > 0$,

$$\frac{d}{dx} x^n = n x^{n-1}.$$

The Chain Rule extends this equation to the Power Rule's final form.

Power Rule (Final Form)

If u is a positive differentiable function of x and n is any real number, then u^n is a differentiable function of x and

$$\frac{d}{dx} u^n = n u^{n-1} \frac{du}{dx}.$$

EXAMPLE 2

a) $\dfrac{d}{dx} x^{\sqrt{2}} = \sqrt{2} x^{\sqrt{2}-1} \qquad (x > 0)$

b) $\dfrac{d}{dx} (\sin x)^{\pi} = \pi (\sin x)^{\pi-1} \cos x \qquad (\sin x > 0)$ ❑

The Derivative of a^x

We start with the definition $a^x = e^{x \ln a}$:

$$\frac{d}{dx}a^x = \frac{d}{dx}e^{x \ln a} = e^{x \ln a} \cdot \frac{d}{dx}(x \ln a) \qquad \text{\color{blue}Chain Rule}$$

$$= a^x \ln a.$$

If $a > 0$, then

$$\frac{d}{dx}a^x = a^x \ln a.$$

With the Chain Rule, we get a more general form.

If $a > 0$ and u is a differentiable function of x, then a^u is a differentiable function of x and

$$\frac{d}{dx}a^u = a^u \ln a \frac{du}{dx}. \tag{2}$$

Equation (2) shows why e^x is the exponential function preferred in calculus. If $a = e$, then $\ln a = 1$ and Eq. (2) simplifies to

$$\frac{d}{dx}e^x = e^x \ln e = e^x.$$

EXAMPLE 3

a) $\quad \dfrac{d}{dx}3^x = 3^x \ln 3$

b) $\quad \dfrac{d}{dx}3^{-x} = 3^{-x} \ln 3 \dfrac{d}{dx}(-x) = -3^{-x} \ln 3$

c) $\quad \dfrac{d}{dx}3^{\sin x} = 3^{\sin x} \ln 3 \dfrac{d}{dx}(\sin x) = 3^{\sin x}(\ln 3)\cos x$ ◻

From Eq. (2), we see that the derivative of a^x is positive if $\ln a > 0$, or $a > 1$, and negative if $\ln a < 0$, or $0 < a < 1$. Thus, a^x is an increasing function of x if $a > 1$ and a decreasing function of x if $0 < a < 1$. In each case, a^x is one-to-one. The second derivative

$$\frac{d^2}{dx^2}(a^x) = \frac{d}{dx}(a^x \ln a) = (\ln a)^2 a^x$$

is positive for all x, so the graph of a^x is concave up on every interval of the real line (Fig. 6.12).

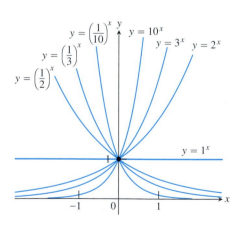

6.12 Exponential functions decrease if $0 < a < 1$ and increase if $a > 1$. As $x \to \infty$, we have $a^x \to 0$ if $0 < a < 1$ and $a^x \to \infty$ if $a > 1$. As $x \to -\infty$, we have $a^x \to \infty$ if $0 < a < 1$ and $a^x \to 0$ if $a > 1$.

Other Power Functions

The ability to raise positive numbers to arbitrary real powers makes it possible to define functions like x^x and $x^{\ln x}$ for $x > 0$. We find the derivatives of such functions by rewriting the functions as powers of e.

EXAMPLE 4   Find dy/dx if $y = x^x$, $x > 0$.

Solution   Write x^x as a power of e:

$$y = x^x = e^{x \ln x}.$$ Eq. (1) with $a = x$

Then differentiate as usual:

$$\frac{dy}{dx} = \frac{d}{dx} e^{x \ln x}$$

$$= e^{x \ln x} \frac{d}{dx} (x \ln x)$$

$$= x^x \left(x \cdot \frac{1}{x} + \ln x \right)$$

$$= x^x (1 + \ln x).$$ ❑

The Integral of a^u

If $a \neq 1$, so that $\ln a \neq 0$, we can divide both sides of Eq. (2) by $\ln a$ to obtain

$$a^u \frac{du}{dx} = \frac{1}{\ln a} \frac{d}{dx} (a^u).$$

Integrating with respect to x then gives

$$\int a^u \frac{du}{dx} dx = \int \frac{1}{\ln a} \frac{d}{dx} (a^u) \, dx = \frac{1}{\ln a} \int \frac{d}{dx} (a^u) \, dx = \frac{1}{\ln a} a^u + C.$$

Writing the first integral in differential form gives

$$\int a^u \, du = \frac{a^u}{\ln a} + C.$$ (3)

EXAMPLE 5

a) $\displaystyle \int 2^x \, dx = \frac{2^x}{\ln 2} + C$ Eq. (3) with $a = 2$, $u = x$

b) $\displaystyle \int 2^{\sin x} \cos x \, dx$

$$= \int 2^u \, du = \frac{2^u}{\ln 2} + C$$

$$= \frac{2^{\sin x}}{\ln 2} + C$$ $u = \sin x$ in Eq. (3)   ❑

Logarithms with Base a

As we saw earlier, if a is any positive number other than 1, the function a^x is one-to-one and has a nonzero derivative at every point. It therefore has a differentiable inverse. We call the inverse the **logarithm of x with base a** and denote it by **$\log_a x$.**

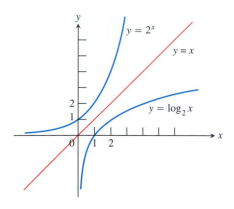

6.13 The graph of 2^x and its inverse, $\log_2 x$.

Definition

For any positive number $a \neq 1$,

$$\log_a x = \text{inverse of } a^x.$$

The graph of $y = \log_a x$ can be obtained by reflecting the graph of $y = a^x$ across the line $y = x$ (Fig. 6.13).

Since $\log_a x$ and a^x are inverses of one another, composing them in either order gives the identity function.

Inverse Equations for a^x and $\log_a x$

$$a^{\log_a x} = x \qquad (x > 0) \tag{4}$$

$$\log_a (a^x) = x \qquad (\text{all } x) \tag{5}$$

EXAMPLE 6

a) $\log_2 (2^5) = 5$

b) $\log_{10} (10^{-7}) = -7$

c) $2^{\log_2 (3)} = 3$

d) $10^{\log_{10}(4)} = 4$ ☐

The Evaluation of $\log_a x$

The evaluation of $\log_a x$ is simplified by the observation that $\log_a x$ is a numerical multiple of $\ln x$.

$$\log_a x = \frac{1}{\ln a} \cdot \ln x = \frac{\ln x}{\ln a} \tag{6}$$

We can derive Eq. (6) from Eq. (4):

$$a^{\log_a (x)} = x \qquad \text{Eq. (4)}$$

$$\ln a^{\log_a (x)} = \ln x \qquad \text{Take the natural logarithm of both sides.}$$

$$\log_a (x) \cdot \ln a = \ln x \qquad \text{The Power Rule in Table 6.1}$$

$$\log_a x = \frac{\ln x}{\ln a} \qquad \text{Solve for } \log_a x.$$

EXAMPLE 7

$$\log_{10} 2 = \frac{\ln 2}{\ln 10} \approx \frac{0.69315}{2.30259} \approx 0.30103$$ ☐

Table 6.4
Properties of base a logarithms

For any numbers $x > 0$ and $y > 0$,

1. *Product Rule:*
$$\log_a xy = \log_a x + \log_a y$$

2. *Quotient Rule:*
$$\log_a \frac{x}{y} = \log_a x - \log_a y$$

3. *Reciprocal Rule:*
$$\log_a \frac{1}{y} = -\log_a y$$

4. *Power Rule:*
$$\log_a x^y = y \log_a x$$

The arithmetic properties of $\log_a x$ are the same as the ones for $\ln x$ (Table 6.4). These rules can be proved by dividing the corresponding rules for the natural logarithm function by $\ln a$. For example,

$$\ln xy = \ln x + \ln y \qquad \text{Rule 1 for natural logarithms ...}$$

$$\frac{\ln xy}{\ln a} = \frac{\ln x}{\ln a} + \frac{\ln y}{\ln a} \qquad \text{... divided by } \ln a \text{ ...}$$

$$\log_a xy = \log_a x + \log_a y. \qquad \text{... gives Rule 1 for base } a \text{ logarithms.}$$

The Derivative of $\log_a u$

To find the derivative of a base a logarithm, we first convert it to a natural logarithm. If u is a positive differentiable function of x, then

$$\frac{d}{dx}(\log_a u) = \frac{d}{dx}\left(\frac{\ln u}{\ln a}\right) = \frac{1}{\ln a}\frac{d}{dx}(\ln u) = \frac{1}{\ln a} \cdot \frac{1}{u}\frac{du}{dx}.$$

$$\frac{d}{dx}(\log_a u) = \frac{1}{\ln a} \cdot \frac{1}{u}\frac{du}{dx} \qquad (7)$$

EXAMPLE 8

$$\frac{d}{dx}\log_{10}(3x+1) = \frac{1}{\ln 10} \cdot \frac{1}{3x+1}\frac{d}{dx}(3x+1) = \frac{3}{(\ln 10)(3x+1)} \qquad \square$$

Integrals Involving $\log_a x$

To evaluate integrals involving base a logarithms, we convert them to natural logarithms.

EXAMPLE 9

$$\int \frac{\log_2 x}{x}dx = \frac{1}{\ln 2}\int \frac{\ln x}{x}dx \qquad \log_2 x = \frac{\ln x}{\ln 2}$$

$$= \frac{1}{\ln 2}\int u\,du \qquad u = \ln x, \quad du = \frac{1}{x}dx$$

$$= \frac{1}{\ln 2}\frac{u^2}{2} + C = \frac{1}{\ln 2}\frac{(\ln x)^2}{2} + C = \frac{(\ln x)^2}{2\ln 2} + C \qquad \square$$

*Base 10 Logarithms

Base 10 logarithms, often called **common logarithms,** appear in many scientific formulas. For example, earthquake intensity is often reported on the logarithmic **Richter scale.** Here the formula is

$$\text{Magnitude } R = \log_{10}\left(\frac{a}{T}\right) + B,$$

where a is the amplitude of the ground motion in microns at the receiving station, T is the period of the seismic wave in seconds, and B is an empirical factor that allows for the weakening of the seismic wave with increasing distance from the epicenter of the earthquake.

EXAMPLE 10   For an earthquake 10,000 km from the receiving station, $B = 6.8$. If the recorded vertical ground motion is $a = 10$ microns and the period is $T = 1$ sec, the earthquake's magnitude is

$$R = \log_{10}\left(\frac{10}{1}\right) + 6.8 = 1 + 6.8 = 7.8.$$

An earthquake of this magnitude does great damage near its epicenter.   ❏

The **pH scale** for measuring the acidity of a solution is a base 10 logarithmic scale. The pH value (hydrogen potential) of the solution is the common logarithm of the reciprocal of the solution's hydronium ion concentration, $[H_3O^+]$:

$$pH = \log_{10}\frac{1}{[H_3O^+]} = -\log_{10}[H_3O^+].$$

The hydronium ion concentration is measured in moles per liter. Vinegar has a pH of 3, distilled water a pH of 7, seawater a pH of 8.15, and household ammonia a pH of 12. The total scale ranges from about 0.1 for normal hydrochloric acid to 14 for a normal (1 N) solution of sodium hydroxide.

Another example of the use of common logarithms is the **decibel** or db ("dee bee") **scale** for measuring loudness. If I is the **intensity** of sound in watts per square meter, the decibel level of the sound is

$$\textbf{Sound level} = 10 \ \log_{10}(I \times 10^{12}) \ \text{db}. \tag{8}$$

If you ever wondered why doubling the power of your audio amplifier increases the sound level by only a few decibels, Eq. (8) provides the answer. As the following example shows, doubling I adds only about 3 db.

EXAMPLE 11   Doubling I in Eq. (8) adds about 3 db. Writing log for $\log_{10}$ (a common practice), we have

$$
\begin{aligned}
\text{Sound level with } I \text{ doubled} &= 10 \log(2I \times 10^{12}) &&\text{Eq. (8) with } 2I \text{ for } I\\
&= 10 \log(2 \cdot I \times 10^{12}) \\
&= 10 \log 2 + 10 \ \log(I \times 10^{12}) \\
&= \text{original sound level} + 10 \log 2 \\
&\approx \text{original sound level} + 3. &&\log_{10} 2 \approx 0.30
\end{aligned}
$$

❏

Most foods are acidic (pH < 7).

Food	pH Value
Bananas	4.5–4.7
Grapefruit	3.0–3.3
Oranges	3.0–4.0
Limes	1.8–2.0
Milk	6.3–6.6
Soft drinks	2.0–4.0
Spinach	5.1–5.7

Typical sound levels

Threshold of hearing	0 db
Rustle of leaves	10 db
Average whisper	20 db
Quiet automobile	50 db
Ordinary conversation	65 db
Pneumatic drill 10 feet away	90 db
Threshold of pain	120 db

Exercises 6.4

Algebraic Calculations

Simplify the expressions in Exercises 1–4.

1. a) $5^{\log_5 7}$ **b)** $8^{\log_8 \sqrt{2}}$ **c)** $1.3^{\log_{1.3} 75}$

d) $\log_4 16$ **e)** $\log_3 \sqrt{3}$ **f)** $\log_4\left(\frac{1}{4}\right)$

2. a) $2^{\log_2 3}$ **b)** $10^{\log_{10}(1/2)}$ **c)** $\pi^{\log_\pi 7}$

d) $\log_{11} 121$ **e)** $\log_{121} 11$ **f)** $\log_3\left(\dfrac{1}{9}\right)$

3. a) $2^{\log_4 x}$ **b)** $9^{\log_3 x}$ **c)** $\log_2(e^{(\ln 2)(\sin x)})$

4. a) $25^{\log_5 (3x^2)}$ **b)** $\log_e(e^x)$ **c)** $\log_4(2^{e^x \sin x})$

Express the ratios in Exercises 5 and 6 as ratios of natural logarithms and simplify.

5. a) $\dfrac{\log_2 x}{\log_3 x}$ **b)** $\dfrac{\log_2 x}{\log_8 x}$ **c)** $\dfrac{\log_x a}{\log_{x^2} a}$

6. a) $\dfrac{\log_9 x}{\log_3 x}$ **b)** $\dfrac{\log_{\sqrt{10}} x}{\log_{\sqrt{2}} x}$ **c)** $\dfrac{\log_a b}{\log_b a}$

Solve the equations in Exercises 7–10 for x.

7. $3^{\log_3 (7)} + 2^{\log_2 (5)} = 5^{\log_5 (x)}$

8. $8^{\log_8 (3)} - e^{\ln 5} = x^2 - 7^{\log_7 (3x)}$

9. $3^{\log_3 (x^2)} = 5e^{\ln x} - 3 \cdot 10^{\log_{10} (2)}$

10. $\ln e + 4^{-2\log_4(x)} = \dfrac{1}{x}\log_{10}(100)$

Derivatives

In Exercises 11–38, find the derivative of y with respect to the given independent variable.

11. $y = 2^x$ **12.** $y = 3^{-x}$

13. $y = 5^{\sqrt{s}}$ **14.** $y = 2^{(s^2)}$

15. $y = x^\pi$ **16.** $y = t^{1-e}$

17. $y = (\cos \theta)^{\sqrt{2}}$ **18.** $y = (\ln \theta)^\pi$

19. $y = 7^{\sec \theta} \ln 7$ **20.** $y = 3^{\tan \theta} \ln 3$

21. $y = 2^{\sin 3t}$ **22.** $y = 5^{-\cos 2t}$

23. $y = \log_2 5\theta$ **24.** $y = \log_3(1 + \theta \ln 3)$

25. $y = \log_4 x + \log_4 x^2$ **26.** $y = \log_{25} e^x - \log_5 \sqrt{x}$

27. $y = \log_2 r \cdot \log_4 r$ **28.** $y = \log_3 r \cdot \log_9 r$

29. $y = \log_3\left(\left(\dfrac{x+1}{x-1}\right)^{\ln 3}\right)$ **30.** $y = \log_5\sqrt{\left(\dfrac{7x}{3x+2}\right)^{\ln 5}}$

31. $y = \theta \sin(\log_7 \theta)$ **32.** $y = \log_7\left(\dfrac{\sin \theta \cos \theta}{e^\theta \, 2^\theta}\right)$

33. $y = \log_5 e^x$ **34.** $y = \log_2\left(\dfrac{x^2 e^2}{2\sqrt{x+1}}\right)$

35. $y = 3^{\log_2 t}$ **36.** $y = 3 \log_8(\log_2 t)$

37. $y = \log_2(8t^{\ln 2})$ **38.** $y = t \log_3\left(e^{(\sin t)(\ln 3)}\right)$

Logarithmic Differentiation

In Exercises 39–46, use logarithmic differentiation to find the derivative of y with respect to the given independent variable.

39. $y = (x + 1)^x$ **40.** $y = x^{(x+1)}$

41. $y = (\sqrt{t})^t$ **42.** $y = t^{\sqrt{t}}$

43. $y = (\sin x)^x$ **44.** $y = x^{\sin x}$

45. $y = x^{\ln x}$ **46.** $y = (\ln x)^{\ln x}$

Integration

Evaluate the integrals in Exercises 47–56

47. $\displaystyle\int 5^x \, dx$ **48.** $\displaystyle\int (1.3)^x \, dx$

49. $\displaystyle\int_0^1 2^{-\theta} \, d\theta$ **50.** $\displaystyle\int_{-2}^0 5^{-\theta} \, d\theta$

51. $\displaystyle\int_1^{\sqrt{2}} x2^{(x^2)} \, dx$ **52.** $\displaystyle\int_1^4 \dfrac{2\sqrt{x}}{\sqrt{x}} \, dx$

53. $\displaystyle\int_0^{\pi/2} 7^{\cos t} \sin t \, dt$ **54.** $\displaystyle\int_0^{\pi/4} \left(\dfrac{1}{3}\right)^{\tan t} \sec^2 t \, dt$

55. $\displaystyle\int_2^4 x^{2x}(1 + \ln x) \, dx$ **56.** $\displaystyle\int_1^2 \dfrac{2^{\ln x}}{x} \, dx$

Evaluate the integrals in Exercises 57–60.

57. $\displaystyle\int 3x^{\sqrt{3}} dx$ **58.** $\displaystyle\int x^{\sqrt{2}-1} dx$

59. $\displaystyle\int_0^3 (\sqrt{2}+1)x^{\sqrt{2}} dx$ **60.** $\displaystyle\int_1^e x^{(\ln 2)-1} dx$

Evaluate the integrals in Exercises 61–70.

61. $\displaystyle\int \dfrac{\log_{10} x}{x} \, dx$ **62.** $\displaystyle\int_1^4 \dfrac{\log_2 x}{x} \, dx$

63. $\displaystyle\int_1^4 \dfrac{\ln 2 \log_2 x}{x} \, dx$ **64.** $\displaystyle\int_1^e \dfrac{2 \ln 10 \log_{10} x}{x} \, dx$

65. $\displaystyle\int_0^2 \dfrac{\log_2(x+2)}{x+2} \, dx$ **66.** $\displaystyle\int_{1/10}^{10} \dfrac{\log_{10}(10x)}{x} \, dx$

67. $\displaystyle\int_0^9 \dfrac{2 \log_{10}(x+1)}{x+1} \, dx$ **68.** $\displaystyle\int_2^3 \dfrac{2 \log_2(x-1)}{x-1} \, dx$

69. $\displaystyle\int \dfrac{dx}{x \log_{10} x}$ **70.** $\displaystyle\int \dfrac{dx}{x(\log_8 x)^2}$

Evaluate the integrals in Exercises 71–74.

71. $\displaystyle\int_1^{\ln x} \dfrac{1}{t} \, dt, \quad x > 1$ **72.** $\displaystyle\int_1^{e^x} \dfrac{1}{t} \, dt$

73. $\displaystyle\int_1^{1/x} \dfrac{1}{t} \, dt, \quad x > 0$ **74.** $\dfrac{1}{\ln a}\displaystyle\int_1^x \dfrac{1}{t} \, dt, \quad x > 0$

Theory and Applications

75. Find the area of the region between the curve $y = 2x/(1 + x^2)$ and the interval $-2 \le x \le 2$ of the x-axis.

76. Find the area of the region between the curve $y = 2^{1-x}$ and the interval $-1 \le x \le 1$ of the x-axis.

77. Blood pH. The pH of human blood normally falls between 7.37 and 7.44. Find the corresponding bounds for $[H_3O^+]$.

78. Brain fluid pH. The cerebrospinal fluid in the brain has a hydronium ion concentration of about $[H_3O^+] = 4.8 \times 10^{-8}$ moles per liter. What is the pH?

79. Audio amplifiers. By what factor k do you have to multiply the intensity of I of the sound from your audio amplifier to add 10 db to the sound level?

80. Audio amplifiers. You multiplied the intensity of the sound of your audio system by a factor of 10. By how many decibels did this increase the sound level?

81. In any solution, the product of the hydronium ion concentration $[H_3O^+]$ (moles/L) and the hydroxyl ion concentration $[OH^-]$ (moles/L) is about 10^{-14}.

   a) What value of $[H_3O^+]$ minimizes the sum of the concentrations, $S = [H_3O^+] + [OH^-]$? (*Hint:* Change notation. Let $x = [H_3O^+]$.)

   b) What is the pH of a solution in which S has this minimum value?

   c) What ratio of $[H_3O^+]$ to $[OH^-]$ minimizes S?

82. Could $\log_a b$ possibly equal $1/\log_b a$? Give reasons for your answer.

Grapher Explorations

83. The equation $x^2 = 2^x$ has three solutions: $x = 2$, $x = 4$, and one other. Estimate the third solution as accurately as you can by graphing.

84. Could $x^{\ln 2}$ possibly be the same as $2^{\ln x}$ for $x > 0$? Graph the two functions and explain what you see.

85. The linearization of 2^x

   a) Find the linearization of $f(x) = 2^x$ at $x = 0$. Then round its coefficients to 2 decimal places.

   b) Graph the linearization and function together for $-3 \le x \le 3$ and $-1 \le x \le 1$.

86. The linearization of $\log_3 x$

   a) Find the linearization of $f(x) = \log_3 x$ at $x = 3$. Then round its coefficients to 2 decimal places.

   b) Graph the linearization and function together in the window $0 \le x \le 8$ and $2 \le x \le 4$.

Calculations with Other Bases

87. CALCULATOR Most scientific calculators have keys for $\log_{10} x$ and $\ln x$. To find logarithms to other bases, we use the equation $\log_a x = (\ln x)/(\ln a)$.

> To find $\log_2 x$, find $\ln x$ and divide by $\ln 2$: $\qquad \log_2 5 = \dfrac{\ln 5}{\ln 2} \approx 2.3219$.
>
> To find $\ln x$ given $\log_2 x$, multiply by $\ln 2$: $\qquad \ln 5 = \log_2 5 \cdot \ln 2 \approx 1.6094$.

Find the following logarithms to 5-decimal places.

   a) $\log_3 8$ 　　　　　　　　　　　**b)** $\log_7 0.5$

   c) $\log_{20} 17$ 　　　　　　　　　　**d)** $\log_{0.5} 7$

   e) $\ln x$, given that $\log_{10} x = 2.3$

   f) $\ln x$, given that $\log_2 x = 1.4$

   g) $\ln x$, given that $\log_2 x = -1.5$

   h) $\ln x$, given that $\log_{10} x = -0.7$

88. Conversion factors

   a) Show that the equation for converting base 10 logarithms to base 2 logarithms is

$$\log_2 x = \frac{\ln 10}{\ln 2} \log_{10} x.$$

   b) Show that the equation for converting base a logarithms to base b logarithms is

$$\log_b x = \frac{\ln a}{\ln b} \log_a x.$$

6.5   Growth and Decay

In this section, we derive the law of exponential change and describe some of the applications that account for the importance of logarithmic and exponential functions.

The Law of Exponential Change

To set the stage once again, suppose we are interested in a quantity y (velocity, temperature, electric current, whatever) that increases or decreases at a rate that at any given time t is proportional to the amount present. If we also know the amount present at time $t = 0$, call it y_0, we can find y as a function of t by solving the following initial value problem:

Differential equation: $\dfrac{dy}{dt} = ky$

Initial condition: $y = y_0$ when $t = 0$.

(1)

If y is positive and increasing, then k is positive, and we use Eq. (1) to say that the rate of growth is proportional to what has already been accumulated. If y is positive and decreasing, then k is negative, and we use Eq. (1) to say that the rate of decay is proportional to the amount still left.

We see right away that the constant function $y = 0$ is a solution of Eq. (1). To find the nonzero solutions, we divide Eq. (1) by y:

$$\frac{1}{y} \cdot \frac{dy}{dt} = k$$

$\ln |y| = kt + C$ Integrate with respect to t; $\int (1/u)\,du = \ln |u| + C$.

$|y| = e^{kt+C}$ Exponentiate.

$|y| = e^C \cdot e^{kt}$ $e^{a+b} = e^a \cdot e^b$

$y = \pm e^C e^{kt}$ If $|y| = r$, then $y = \pm r$.

$y = Ae^{kt}$. A is a more convenient name for $\pm e^C$.

By allowing A to take on the value 0 in addition to all possible values $\pm e^C$, we can include the solution $y = 0$ in the formula.

We find the right value of A for the initial value problem by solving for A when $y = y_0$ and $t = 0$:

$$y_0 = Ae^{k \cdot 0} = A.$$

The solution of the initial value problem is therefore $y = y_0\, e^{kt}$.

The Law of Exponential Change

$$y = y_0\, e^{kt} \tag{2}$$

Growth: $k > 0$ Decay: $k < 0$

The number k is the **rate constant** of the equation.

The derivation of Eq. (2) shows that the only functions that are their own derivatives are constant multiples of the exponential function.

Population Growth

Strictly speaking, the number of individuals in a population (of people, plants, foxes, or bacteria, for example) is a discontinuous function of time because it takes on discrete values. However, as soon as the number of individuals becomes large enough, it can safely be described with a continuous or even differentiable function.

If we assume that the proportion of reproducing individuals remains constant and assume a constant fertility, then at any instant t the birth rate is proportional to the number $y(t)$ of individuals present. If, further, we neglect departures, arrivals, and deaths, the growth rate dy/dt will be the same as the birth rate ky. In other words, $dy/dt = ky$, so that $y = y_0 e^{kt}$. As with all kinds of growth, there may be limitations imposed by the surrounding environment, but we will not go into these here.

EXAMPLE 1   One model for the way diseases spread assumes that the rate dy/dt at which the number of infected people changes is proportional to the number y. The more infected people there are, the faster the disease will spread. The fewer there are, the slower it will spread.

Suppose that in the course of any given year the number of cases of a disease is reduced by 20%. If there are 10,000 cases today, how many years will it take to reduce the number to 1000?

Solution   We use the equation $y = y_0 e^{kt}$. There are three things to find:

1. the value of y_0,
2. the value of k,
3. the value of t that makes $y = 1000$.

Step 1: *The value of y_0.* We are free to count time beginning anywhere we want. If we count from today, then $y = 10,000$ when $t = 0$, so $y_0 = 10,000$. Our equation is now

$$y = 10,000 \, e^{kt}. \tag{3}$$

Step 2: *The value of k.* When $t = 1$ year, the number of cases will be 80% of its present value, or 8000. Hence,

$$8000 = 10,000 \, e^{k(1)} \qquad \text{Eq. (3) with } t = 1 \text{ and } y = 8000$$

$$e^k = 0.8$$

$$\ln(e^k) = \ln 0.8$$

$$k = \ln 0.8.$$

At any given time t,

$$y = 10,000 \, e^{(\ln 0.8)t}. \tag{4}$$

Step 3: *The value of t that makes $y = 1000$.* We set y equal to 1000 in Eq. (4) and solve for t:

$$1000 = 10,000 \, e^{(\ln 0.8)t}$$

$$e^{(\ln 0.8)t} = 0.1$$

$$(\ln 0.8)t = \ln 0.1 \qquad \text{Logs of both sides}$$

$$t = \frac{\ln 0.1}{\ln 0.8} \approx 10.32 \text{ years.}$$

It will take a little more than 10 years to reduce the number of cases to 1000. ❏

Continuously Compounded Interest

If you invest an amount A_0 of money at a fixed annual interest rate r (expressed as a decimal) and if interest is added to your account k times a year, it turns out that the amount of money you will have at the end of t years is

$$A_t = A_0 \left(1 + \frac{r}{k}\right)^{kt}. \tag{5}$$

The interest might be added ("compounded," bankers say) monthly ($k = 12$), weekly ($k = 52$), daily ($k = 365$), or even more frequently, say by the hour or by the minute. But there is still a limit to how much you will earn that way, and the limit is

$$\lim_{k \to \infty} A_t = \lim_{k \to \infty} A_0 \left(1 + \frac{r}{k}\right)^{kt}$$

$$= A_0 e^{rt}.$$

The resulting formula for the amount of money in your account after t years is

$$A(t) = A_0 e^{rt}. \tag{6}$$

Interest paid according to this formula is said to be **compounded continuously.** The number r is called the **continuous interest rate.**

Evaluating

$$\lim_{k \to \infty} A_0 \left(1 + \frac{r}{k}\right)^{kt}$$

involves what is called the indeterminate form 1^∞. We will see how to evaluate limits of this type in Section 6.6.

EXAMPLE 2   Suppose you deposit $621 in a bank account that pays 6% compounded continuously. How much money will you have 8 years later?

Solution   We use Eq. (6) with $A_0 = 621$, $r = 0.06$, and $t = 8$:

$$A(8) = 621 \, e^{(0.06)(8)} = 621 \, e^{0.48} = 1003.58 \qquad \text{Nearest cent}$$

Had the bank paid interest quarterly ($k = 4$ in Eq. (5)), the amount in your account would have been $1000.01. Thus the effect of continuous compounding, as compared with quarterly compounding, has been an addition of $3.57. A bank might decide it would be worth this additional amount to be able to advertise, "We compound interest every second, night and day—better yet, we compound the interest continuously." ❑

Radioactivity

When an atom emits some of its mass as radiation, the remainder of the atom re-forms to make an atom of some new element. This process of radiation and change is called **radioactive decay,** and an element whose atoms go spontaneously through this process is called **radioactive.** Thus, radioactive carbon-14 decays into nitrogen; radium, through a number of intervening radioactive steps, decays into lead.

Experiments have shown that at any given time the rate at which a radioactive element decays (as measured by the number of nuclei that change per unit time) is approximately proportional to the number of radioactive nuclei present. Thus, the decay of a radioactive element is described by the equation $dy/dt = -ky$, $k > 0$. If y_0 is the number of radioactive nuclei present at time zero, the number still present at any later time t will be

$$y = y_0 e^{-kt}, \qquad k > 0.$$

For radon-222 gas, t is measured in days and $k = 0.18$. For radium-226, which used to be painted on watch dials to make them glow at night (a dangerous practice), t is measured in years and $k = 4.3 \times 10^{-4}$. The decay of radium in the earth's crust is the source of the radon we sometimes find in our basements.

It is conventional to use $-k \, (k > 0)$ here instead of $k \, (k < 0)$ to emphasize that y is decreasing.

EXAMPLE 3 *Half-life*

The **half-life** of a radioactive element is the time required for half of the radioactive nuclei present in a sample to decay. It is a remarkable fact that the half-life is a constant that does not depend on the number of radioactive nuclei initially present in the sample, but only on the radioactive substance.

To see why, let y_0 be the number of radioactive nuclei initially present in the sample. Then the number y present at any later time t will be $y = y_0 e^{-kt}$. We seek the value of t at which the number of radioactive nuclei present equals half the original number:

$$y_0 e^{-kt} = \frac{1}{2} y_0$$

$$e^{-kt} = \frac{1}{2}$$

$$-kt = \ln \frac{1}{2} = -\ln 2 \qquad \text{Reciprocal Rule for logarithms}$$

$$t = \frac{\ln 2}{k}$$

This value of t is the half-life of the element. It depends only on the value of k; the number y_0 does not enter in. ❑

$$\text{Half-life} = \frac{\ln 2}{k} \qquad\qquad (7)$$

EXAMPLE 4 *Polonium-210*

The effective radioactive lifetime of polonium-210 is so short we measure it in days rather than years. The number of radioactive atoms remaining after t days in a sample that starts with y_0 radioactive atoms is

$$y = y_0 e^{-5 \times 10^{-3} t}.$$

Find the element's half-life.

Solution

$$\text{Half-life} = \frac{\ln 2}{k} \qquad \text{Eq. (7)}$$

$$= \frac{\ln 2}{5 \times 10^{-3}} \qquad \text{The } k \text{ from polonium's decay equation}$$

$$\approx 139 \text{ days} \qquad\qquad ❑$$

EXAMPLE 5 *Carbon-14*

People who do carbon-14 dating use a figure of 5700 years for its half-life (more about carbon-14 dating in the exercises). Find the age of a sample in which 10% of the radioactive nuclei originally present have decayed.

Carbon-14 dating

The decay of radioactive elements can sometimes be used to date events from the Earth's past. The ages of rocks more than 2 billion years old have been measured by the extent of the radioactive decay of uranium (half-life 4.5 billion years!). In a living organism, the ratio of radioactive carbon, carbon-14, to ordinary carbon stays fairly constant during the lifetime of the organism, being approximately equal to the ratio in the organism's surroundings at the time. After the organism's death, however, no new carbon is ingested, and the proportion of carbon-14 in the organism's remains decreases as the carbon-14 decays. It is possible to estimate the ages of fairly old organic remains by comparing the proportion of carbon-14 they contain with the proportion assumed to have been in the organism's environment at the time it lived. Archaeologists have dated shells (which contain $CaCO_3$), seeds, and wooden artifacts this way. The estimate of 15,500 years for the age of the cave paintings at Lascaux, France, is based on carbon-14 dating. After generations of controversy, the Shroud of Turin, long believed by many to be the burial cloth of Christ, was shown by carbon-14 dating in 1988 to have been made after A.D. 1200.

Solution We use the decay equation $y = y_0 e^{-kt}$. There are two things to find:

1. the value of k,
2. the value of t when $y_0 e^{-kt} = 0.9 \, y_0$, or $e^{-kt} = 0.9$ 90% of the radioactive nuclei still present

Step 1: *The value of k.* We use the half-life equation:

$$k = \frac{\ln 2}{\text{half-life}} = \frac{\ln 2}{5700} \qquad \text{(about } 1.2 \times 10^{-4})$$

Step 2: *The value of t that makes* $e^{-kt} = 0.9$.

$$e^{-kt} = 0.9$$
$$e^{-(\ln 2/5700)t} = 0.9$$
$$-\frac{\ln 2}{5700}t = \ln 0.9 \qquad \text{Logs of both sides}$$
$$t = -\frac{5700 \ln 0.9}{\ln 2} \approx 866 \text{ years.}$$

The sample is about 866 years old.   ☐

Heat Transfer: Newton's Law of Cooling

Soup left in a tin cup cools to the temperature of the surrounding air. A hot silver ingot immersed in water cools to the temperature of the surrounding water. In situations like these, the rate at which an object's temperature is changing at any given time is roughly proportional to the difference between its temperature and the temperature of the surrounding medium. This observation is called *Newton's law of cooling,* although it applies to warming as well, and there is an equation for it.

If T is the temperature of the object at time t, and T_S is the surrounding temperature, then

$$\frac{dT}{dt} = -k(T - T_S). \tag{8}$$

If we substitute y for $(T - T_S)$, then

$$\frac{dy}{dt} = \frac{d}{dt}(T - T_S) = \frac{dT}{dt} - \frac{d}{dt}(T_S)$$
$$= \frac{dT}{dt} - 0 \qquad T_S \text{ is a constant.}$$
$$= \frac{dT}{dt}$$

In terms of y, Eq. (8) therefore reads

$$\frac{dy}{dt} = -ky,$$

and we know that the solution to this differential equation is

$$y = y_0 e^{-kt}.$$

Thus, **Newton's law of cooling** is

$$T - T_S = (T_0 - T_S)e^{-kt}, \tag{9}$$

where T_0 is the value of T at time zero.

EXAMPLE 6   A hard-boiled egg at 98°C is put in a sink of 18°C water. After 5 minutes, the egg's temperature is 38°C. Assuming that the water has not warmed appreciably, how much longer will it take the egg to reach 20°C?

Solution   We find how long it would take the egg to cool from 98°C to 20°C and subtract the 5 minutes that have already elapsed.

According to Eq. (9), the egg's temperature t minutes after it is put in the sink is

$$T = 18 + (98 - 18)e^{-kt} = 18 + 80e^{-kt}.$$

To find k, we use the information that $T = 38$ when $t = 5$:

$$38 = 18 + 80e^{-5k}$$

$$e^{-5k} = \frac{1}{4}$$

$$-5k = \ln \frac{1}{4} = -\ln 4$$

$$k = \frac{1}{5} \ln 4 = 0.2 \ \ln 4 \qquad \text{(about 0.28)}.$$

The egg's temperature at time t is $T = 18 + 80e^{-(0.2 \ln 4)t}$. Now find the time t when $T = 20$:

$$20 = 18 + 80e^{-(0.2 \ln 4)t}$$

$$80e^{-(0.2 \ln 4)t} = 2$$

$$e^{-(0.2 \ln 4)t} = \frac{1}{40}$$

$$-(0.2 \ln 4)t = \ln \frac{1}{40} = -\ln 40$$

$$t = \frac{\ln 40}{0.2 \ \ln 4} \approx 13 \text{ min.}$$

The egg's temperature will reach 20°C about 13 min after it is put in water to cool. Since it took 5 min to reach 38°C, it will take about 8 min more to reach 20°C.

❑

Exercises 6.5

The answers to most of the following exercises are in terms of logarithms and exponentials. A calculator can be helpful, enabling you to express the answers in decimal form.

1. *Human evolution continues.* The analysis of tooth shrinkage by C. Loring Brace and colleagues at the University of Michigan's Museum of Anthropology indicates that human tooth size is continuing to decrease and that the evolutionary process did not come to a halt some 30,000 years ago as many scientists contend. In northern Europeans, for example, tooth size reduction now has a rate of 1% per 1000 years.

a)   If t represents time in years and y represents tooth size, use the condition that $y = 0.99y_0$ when $t = 1000$ to find the value of k in the equation $y = y_0 e^{kt}$. Then use this value of k to answer the following questions.

b)   In about how many years will human teeth be 90% of their present size?

c)   What will be our descendants' tooth size 20,000 years from now (as a percentage of our present tooth size)?

(Source: *LSA Magazine,* Spring 1989, Vol. 12, No. 2, p. 19, Ann Arbor, MI.)

2. *Atmospheric pressure.* The earth's atmospheric pressure p is often modeled by assuming that the rate dp/dh at which p changes with the altitude h above sea level is proportional to p. Suppose that the pressure at sea level is 1013 millibars (about 14.7 pounds per square inch) and that the pressure at an altitude of 20 km is 90 millibars.

   a)   Solve the initial value problem

      Differential equation: $dp/dh = kp$ (k a constant)

      Initial condition: $p = p_0$ when $h = 0$

      to express p in terms of h. Determine the values of p_0 and k from the given altitude-pressure data.

   b)   What is the atmospheric pressure at $h = 50$ km?

   c)   At what altitude does the pressure equal 900 millibars?

3. *First order chemical reactions.* In some chemical reactions, the rate at which the amount of a substance changes with time is proportional to the amount present. For the change of δ-glucono lactone into gluconic acid, for example,

$$\frac{dy}{dt} = -0.6y$$

when t is measured in hours. If there are 100 grams of δ-glucono lactone present when $t = 0$, how many grams will be left after the first hour?

4. *The inversion of sugar.* The processing of raw sugar has a step called "inversion" that changes the sugar's molecular structure. Once the process has begun, the rate of change of the amount of raw sugar is proportional to the amount of raw sugar remaining. If 1000 kg of raw sugar reduces to 800 kg of raw sugar during the first 10 h, how much raw sugar will remain after another 14 h?

5. *Working underwater.* The intensity $L(x)$ of light x feet beneath the surface of the ocean satisfies the differential equation

$$\frac{dL}{dx} = -kL.$$

As a diver, you know from experience that diving to 18 ft in the Caribbean Sea cuts the intensity in half. You cannot work without artificial light when the intensity falls below one-tenth of the surface value. About how deep can you expect to work without artificial light?

6. *Voltage in a discharging capacitor.* Suppose that electricity is draining from a capacitor at a rate that is proportional to the voltage V across its terminals and that, if t is measured in seconds,

$$\frac{dV}{dt} = -\frac{1}{40}V.$$

Solve this equation for V, using V_0 to denote the value of V when $t = 0$. How long will it take the voltage to drop to 10% of its original value?

7. *Cholera bacteria.* Suppose that the bacteria in a colony can grow unchecked, by the law of exponential change. The colony starts with 1 bacterium and doubles every half hour. How many bacteria will the colony contain at the end of 24 h? (Under favorable laboratory conditions, the number of cholera bacteria can double

every 30 min. In an infected person, many bacteria are destroyed, but this example helps explain why a person who feels well in the morning may be dangerously ill by evening.)

8. *Growth of bacteria.* A colony of bacteria is grown under ideal conditions in a laboratory so that the population increases exponentially with time. At the end of 3 h there are 10,000 bacteria. At the end of 5 h there are 40,000. How many bacteria were present initially?

9. *The incidence of a disease* (*Continuation of Example* 1). Suppose that in any given year the number of cases can be reduced by 25% instead of 20%.

   a)   How long will it take to reduce the number of cases to 1000?

   b)   How long will it take to eradicate the disease, that is, reduce the number of cases to less than 1?

10. *The U.S. population.* The Museum of Science in Boston displays a running total of the U.S. population. On May 11, 1993, the total was increasing at the rate of 1 person every 14 sec. The displayed population figure for 3:45 P.M. that day was 257,313,431.

   a)   Assuming exponential growth at a constant rate, find the rate constant for the population's growth (people per 365-day year).

   b)   At this rate, what will the U.S. population be at 3:45 P.M. Boston time on May 11, 2001?

11. *Oil depletion.* Suppose the amount of oil pumped from one of the canyon wells in Whittier, California, decreases at the continuous rate of 10% per year. When will the well's output fall to one-fifth of its present value?

12. *Continuous price discounting.* To encourage buyers to place 100-unit orders, your firm's sales department applies a continuous discount that makes the unit price a function $p(x)$ of the number of units x ordered. The discount decreases the price at the rate of $0.01 per unit ordered. The price per unit for a 100-unit order is $p(100) = \$20.09$.

   a)   Find $p(x)$ by solving the following initial value problem:

      Differential equation: $\dfrac{dp}{dx} = -\dfrac{1}{100}p$

      Initial condition: $p(100) = 20.09.$

   b)   Find the unit price $p(10)$ for a 10-unit order and the unit price $p(90)$ for a 90-unit order.

   c)   The sales department has asked you to find out if it is discounting so much that the firm's revenue, $r(x) = x \cdot p(x)$, will actually be less for a 100-unit order than, say, for a 90-unit order. Reassure them by showing that r has its maximum value at $x = 100$.

   d)   GRAPHER Graph the revenue function $r(x) = xp(x)$ for $0 \le x \le 200$.

13. *Continuously compounded interest.* You have just placed A_0 dollars in a bank account that pays 4% interest, compounded continuously.

   a)   How much money will you have in the account in 5 years?

   b)   How long will it take your money to double? to triple?

14. *John Napier's question.* John Napier (1550–1617), the Scottish laird who invented logarithms, was the first person to answer the question What happens if you invest an amount of money at 100% interest, compounded continuously?

a) What does happen?

b) How long does it take to triple your money?

c) How much can you earn in a year?

Give reasons for your answers.

15. *Benjamin Franklin's will.* The Franklin Technical Institute of Boston owes its existence to a provision in a codicil to Benjamin Franklin's will. In part the codicil reads:

> I wish to be useful even after my Death, if possible, in forming and advancing other young men that may be serviceable to their Country in both Boston and Philadelphia. To this end I devote Two thousand Pounds Sterling, which I give, one thousand thereof to the Inhabitants of the Town of Boston in Massachusetts, and the other thousand to the inhabitants of the City of Philadelphia, in Trust and for the Uses, Interests and Purposes hereinafter mentioned and declared.

Franklin's plan was to lend money to young apprentices at 5% interest with the provision that each borrower should pay each year along

> . . . with the yearly Interest, one tenth part of the Principal, which sums of Principal and Interest shall be again let to fresh Borrowers. . . . If this plan is executed and succeeds as projected without interruption for one hundred Years, the Sum will then be one hundred and thirty-one thousand Pounds of which I would have the Managers of the Donation to the Inhabitants of the Town of Boston, then lay out at their discretion one hundred thousand Pounds in Public Works. . . . The remaining thirty-one thousand Pounds, I would have continued to be let out on Interest in the manner above directed for another hundred Years. . . . At the end of this second term if no unfortunate accident has prevented the operation the sum will be Four Millions and Sixty-one Thousand Pounds.

It was not always possible to find as many borrowers as Franklin had planned, but the managers of the trust did the best they could. At the end of 100 years from the reception of the Franklin gift, in January 1894, the fund had grown from 1000 pounds to almost exactly 90,000 pounds. In 100 years the original capital had multiplied about 90 times instead of the 131 times Franklin had imagined.

What rate of interest, compounded continuously for 100 years, would have multiplied Benjamin Franklin's original capital by 90?

16. (*Continuation of Exercise 15.*) In Benjamin Franklin's estimate that the original 1000 pounds would grow to 131,000 in 100 years, he was using an annual rate of 5% and compounding once each year. What rate of interest per year when compounded continuously for 100 years would multiply the original amount by 131?

17. *Radon-222.* The decay equation for radon-222 gas is known to be $y = y_0 e^{-0.18t}$, with t in days. About how long will it take the radon in a sealed sample of air to fall to 90% of its original value?

18. *Polonium-210.* The half-life of polonium is 139 days, but your sample will not be useful to you after 95% of the radioactive nuclei present on the day the sample arrives has disintegrated. For about how many days after the sample arrives will you be able to use the polonium?

19. *The mean life of a radioactive nucleus.* Physicists using the radioactivity equation $y = y_0 e^{-kt}$ call the number $1/k$ the *mean life* of a radioactive nucleus. The mean life of a radon nucleus is about $1/0.18 = 5.6$ days. The mean life of a carbon-14 nucleus is more than 8000 years. Show that 95% of the radioactive nuclei originally present in a sample will disintegrate within three mean lifetimes, i.e., by time $t = 3/k$. Thus, the mean life of a nucleus gives a quick way to estimate how long the radioactivity of a sample will last.

20. *Californium-252.* What costs $27 million per gram and can be used to treat brain cancer, analyze coal for its sulfur content, and detect explosives in luggage? The answer is californium-252, a radioactive isotope so rare that only 8 g of it have been made in the western world since its discovery by Glenn Seaborg in 1950. The half-life of the isotope is 2.645 years—long enough for a useful service life and short enough to have a high radioactivity per unit mass. One microgram of the isotope releases 170 million neutrons per second.

a) What is the value of k in the decay equation for this isotope?

b) What is the isotope's mean life? (See Exercise 19.)

c) How long will it take 95% of a sample's radioactive nuclei to disintegrate?

21. *Cooling soup.* Suppose that a cup of soup cooled from 90°C to 60°C after 10 minutes in a room whose temperature was 20°C. Use Newton's law of cooling to answer the following questions.

a) How much longer would it take the soup to cool to 35°C?

b) Instead of being left to stand in the room, the cup of 90°C soup is put in a freezer whose temperature is −15°C. How long will it take the soup to cool from 90°C to 35°C?

22. *A beam of unknown temperature.* An aluminum beam was brought from the outside cold into a machine shop where the temperature was held at 65°. After 10 minutes, the beam warmed to 35°F and after another 10 minutes it was 50°F. Use Newton's law of cooling to estimate the beam's initial temperature.

23. *Surrounding medium of unknown temperature.* A pan of warm water (46°C) was put in a refrigerator. Ten minutes later, the water's temperature was 39°C; 10 minutes after that, it was 33°C. Use Newton's law of cooling to estimate how cold the refrigerator was.

24. *Silver cooling in air.* The temperature of an ingot of silver is 60°C above room temperature right now. Twenty minutes ago, it was 70°C above room temperature. How far above room temperature will the silver be

a) 15 minutes from now?

b) two hours from now?

c) When will the silver be 10°C above room temperature?

25. *The age of Crater Lake.* The charcoal from a tree killed in the volcanic eruption that formed Crater Lake in Oregon contained 44.5% of the carbon-14 found in living matter. About how old is Crater Lake?

26. *The sensitivity of carbon-14 dating to measurement.* To see the effect of a relatively small error in the estimate of the amount of carbon-14 in a sample being dated, consider this hypothetical situation:

a) A fossilized bone found in central Illinois in the year A.D. 2000 contains 17% of its original carbon-14 content. Estimate the year the animal died.

b) Repeat (a) assuming 18% instead of 17%.

c) Repeat (a) assuming 16% instead of 17%.

27. *Art forgery.* A painting attributed to Vermeer (1632–1675), which should contain no more than 96.2% of its original carbon-14, contains 99.5% instead. About how old is the forgery?

6.6

L'Hôpital's Rule

In the late seventeenth century, John Bernoulli discovered a rule for calculating limits of fractions whose numerators and denominators both approach zero. The rule is known today as **l'Hôpital's rule,** after Guillaume François Antoine de l'Hôpital (1661–1704), Marquis de St. Mesme, a French nobleman who wrote the first introductory differential calculus text, where the rule first appeared in print.

Indeterminate Quotients

If functions $f(x)$ and $g(x)$ are both zero at $x = a$, then $\lim_{x \to a} f(x)/g(x)$ cannot be found by substituting $x = a$. The substitution produces $0/0$, a meaningless expression known as an **indeterminate form.** Our experience so far has been that limits that lead to indeterminate forms may or may not be hard to find. It took a lot of work to find $\lim_{x \to 0} (\sin x)/x$ in Section 2.4. But we have had remarkable success with the limit

$$f'(a) = \lim_{x \to a} \frac{f(x) - f(a)}{x - a}$$

from which we calculate derivatives and which always produces $0/0$. L'Hôpital's rule enables us to draw on our success with derivatives to evaluate limits that lead to indeterminate forms.

Theorem 2

L'Hôpital's Rule (First Form)

Suppose that $f(a) = g(a) = 0$, that $f'(a)$ and $g'(a)$ exist, and that $g'(a) \neq 0$. Then

$$\lim_{x \to a} \frac{f(x)}{g(x)} = \frac{f'(a)}{g'(a)}. \tag{1}$$

Proof   Working backward from $f'(a)$ and $g'(a)$, which are themselves limits, we have

$$\frac{f'(a)}{g'(a)} = \frac{\displaystyle\lim_{x\to a} \frac{f(x) - f(a)}{x - a}}{\displaystyle\lim_{x\to a} \frac{g(x) - g(a)}{x - a}} = \lim_{x\to a} \frac{\dfrac{f(x) - f(a)}{x - a}}{\dfrac{g(x) - g(a)}{x - a}}$$

$$= \lim_{x\to a} \frac{f(x) - f(a)}{g(x) - g(a)}$$

$$= \lim_{x\to a} \frac{f(x) - 0}{g(x) - 0}$$

$$= \lim_{x\to a} \frac{f(x)}{g(x)}. \qquad \blacksquare$$

Caution

To apply l'Hôpital's rule to f/g, divide the derivative of f by the derivative of g. Do not fall into the trap of taking the derivative of f/g. The quotient to use is f'/g', not $(f/g)'$.

A misnamed rule and the first differential calculus text

In 1694 John Bernoulli agreed to accept a retainer of 300 pounds per year from his former student l'Hôpital to solve problems for him and keep him up to date on calculus. One of the problems was the so-called 0/0 problem, which Bernoulli solved as agreed. When l'Hôpital published his notes on calculus in book form in 1696, the 0/0 rule appeared as a theorem. L'Hôpital acknowledged his debt to Bernoulli and, to avoid claiming authorship of the book's entire contents, had the book published anonymously. Bernoulli nevertheless accused l'Hôpital of plagiarism, an accusation inadvertently supported after l'Hôpital's death in 1704 by the publisher's promotion of the book as l'Hôpital's. By 1721, Bernoulli, a man so jealous he once threw his son Daniel out of the house for accepting a mathematics prize from the French Academy of Sciences, claimed to have been the author of the entire work. As puzzling and fickle as ever, history accepted Bernoulli's claim (until recently), but still named the rule after l'Hôpital.

EXAMPLE 1

a) $\displaystyle\lim_{x\to 0} \frac{3x - \sin x}{x} = \frac{3 - \cos x}{1}\bigg|_{x=0} = 2$

b) $\displaystyle\lim_{x\to 0} \frac{\sqrt{1 + x} - 1}{x} = \frac{\dfrac{1}{2\sqrt{1 + x}}}{1}\bigg|_{x=0} = \frac{1}{2}$

c) $\displaystyle\lim_{x\to 0} \frac{x - \sin x}{x^3} = \frac{1 - \cos x}{3x^2}\bigg|_{x=0} = ?$ Still $\dfrac{0}{0}$ $\blacksquare$

What can we do about the limit in Example 1(c)? A stronger form of l'Hôpital's rule says that whenever the rule gives 0/0 we can apply it again, repeating the process until we get a different result. With this stronger rule we get

$$\lim_{x\to 0} \frac{x - \sin x}{x^3} = \lim_{x\to 0} \frac{1 - \cos x}{3x^2} \qquad \text{Still } \frac{0}{0}\text{; apply the rule again.}$$

$$= \lim_{x\to 0} \frac{\sin x}{6x} \qquad \text{Still } \frac{0}{0}\text{; apply the rule again.}$$

$$= \lim_{x\to 0} \frac{\cos x}{6} = \frac{1}{6}. \qquad \text{A different result. Stop.}$$

Theorem 3
L'Hôpital's Rule (Stronger Form)

Suppose that $f(a) = g(a) = 0$ and that f and g are differentiable on an open interval I containing a. Suppose also that $g'(x) \neq 0$ on I if $x \neq a$. Then

$$\lim_{x\to a} \frac{f(x)}{g(x)} = \lim_{x\to a} \frac{f'(x)}{g'(x)}, \tag{2}$$

if the limit on the right exists (or is ∞ or $-\infty$).

You will find a proof of the finite-limit case of Theorem 3 in Appendix 5.

EXAMPLE 2

$$\lim_{x \to 0} \frac{\sqrt{1+x} - 1 - (x/2)}{x^2} \qquad \frac{0}{0}$$

$$= \lim_{x \to 0} \frac{(1/2)(1+x)^{-1/2} - (1/2)}{2x} \qquad \text{Still } \frac{0}{0}$$

$$= \lim_{x \to 0} \frac{-(1/4)(1+x)^{-3/2}}{2} = -\frac{1}{8} \qquad \text{Not } \frac{0}{0}; \text{ limit is found}$$

When you apply l'Hôpital's rule, look for a change from 0/0 to something else. This is where the limit is revealed.

EXAMPLE 3

$$\lim_{x \to 0} \frac{1 - \cos x}{x + x^2} \qquad \frac{0}{0}$$

$$= \lim_{x \to 0} \frac{\sin x}{1 + 2x} = \frac{0}{1} = 0 \qquad \text{Not } \frac{0}{0}; \text{ limit is found.}$$

If we continue to differentiate in an attempt to apply l'Hôpital's rule once more, we get

$$\lim_{x \to 0} \frac{1 - \cos x}{x + x^2} = \lim_{x \to 0} \frac{\sin x}{1 + 2x} = \lim_{x \to 0} \frac{\cos x}{2} = \frac{1}{2},$$

which is wrong.

EXAMPLE 4

$$\lim_{x \to 0^+} \frac{\sin x}{x^2} \qquad \frac{0}{0}$$

$$= \lim_{x \to 0^+} \frac{\cos x}{2x} = \infty \qquad \text{Not } \frac{0}{0}; \text{ answer is found.}$$

L'Hôpital's rule also applies to quotients that lead to the indeterminate form ∞/∞. If $f(x)$ and $g(x)$ both approach infinity as $x \to a$, then

$$\lim_{x \to a} \frac{f(x)}{g(x)} = \lim_{x \to a} \frac{f'(x)}{g'(x)},$$

provided the latter limit exists. The a here may itself be either finite or infinite.

EXAMPLE 5

a) $$\lim_{x \to (\pi/2)^-} \frac{\sec x}{1 + \tan x} \qquad \frac{\infty}{\infty}$$

$$= \lim_{x \to (\pi/2)^-} \frac{\sec x \tan x}{\sec^2 x} = \lim_{x \to (\pi/2)^-} \sin x = 1$$

b) $$\lim_{x \to \infty} \frac{\ln x}{2\sqrt{x}} = \lim_{x \to \infty} \frac{1/x}{1/\sqrt{x}} = \lim_{x \to \infty} \frac{1}{\sqrt{x}} = 0$$

Indeterminate Products and Differences

We can sometimes handle the indeterminate forms $0 \cdot \infty$ and $\infty - \infty$ by using algebra to get $0/0$ or ∞/∞ instead. Here again, we do not mean to suggest that there is a number $0 \cdot \infty$ or $\infty - \infty$ any more than we mean to suggest that there is a number $0/0$ or ∞/∞. These forms are not numbers but descriptions of function behavior.

EXAMPLE 6

$$\lim_{x \to 0^+} x \cot x \qquad\qquad 0 \cdot \infty; \text{ rewrite } x \cot x.$$

$$= \lim_{x \to 0^+} x \cdot \frac{1}{\tan x} \qquad\qquad \cot x = \frac{1}{\tan x}$$

$$= \lim_{x \to 0^+} \frac{x}{\tan x} \qquad\qquad \text{Now } \frac{0}{0}$$

$$= \lim_{x \to 0^+} \frac{1}{\sec^2 x} = \frac{1}{1} = 1 \qquad\qquad \square$$

EXAMPLE 7 Find $\displaystyle\lim_{x \to 0} \left(\frac{1}{\sin x} - \frac{1}{x} \right)$.

Solution If $x \to 0^+$, then $\sin x \to 0^+$ and

$$\frac{1}{\sin x} - \frac{1}{x} \to \infty - \infty.$$

Similarly, if $x \to 0^-$, then $\sin x \to 0^-$ and

$$\frac{1}{\sin x} - \frac{1}{x} \to -\infty - (-\infty) = -\infty + \infty.$$

Neither form reveals what happens in the limit. To find out, we first combine the fractions.

$$\frac{1}{\sin x} - \frac{1}{x} = \frac{x - \sin x}{x \sin x}, \qquad \text{Common denominator is } x \sin x.$$

and then apply l'Hôpital's rule to the result:

$$\lim_{x \to 0} \left(\frac{1}{\sin x} - \frac{1}{x} \right) = \lim_{x \to 0} \frac{x - \sin x}{x \sin x} \qquad\qquad \frac{0}{0}$$

$$= \lim_{x \to 0} \frac{1 - \cos x}{\sin x + x \cos x} \qquad\qquad \text{Still } \frac{0}{0}$$

$$= \lim_{x \to 0} \frac{\sin x}{2 \cos x - x \sin x} = \frac{0}{2} = 0 \qquad\qquad \square$$

Indeterminate Powers

Limits that lead to the indeterminate forms 1^∞, 0^0, and ∞^0 can sometimes be handled by taking logarithms first. We use l'Hôpital's rule to find the limit of the logarithm and then exponentiate to find the original function behavior.

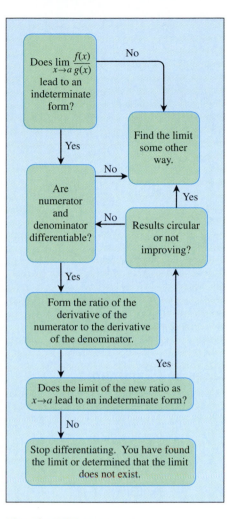

Flowchart 6.1 L'Hôpital's rule

If $\lim_{x \to a} \ln f(x) = L$, then

$$\lim_{x \to a} f(x) = \lim_{x \to a} e^{\ln f(x)} = e^{L}.$$

Here a may be either finite or infinite.

EXAMPLE 8   Show that $\lim_{x \to 0^+} (1 + x)^{1/x} = e$.

Solution   The limit leads to the indeterminate form 1^∞. We let $f(x) = (1 + x)^{1/x}$ and find $\lim_{x \to 0^+} \ln f(x)$. Since

$$\ln f(x) = \ln (1 + x)^{1/x}$$
$$= \frac{1}{x} \ln (1 + x),$$

l'Hôpital's rule now applies to give

$$\lim_{x \to 0^+} \ln f(x) = \lim_{x \to 0^+} \frac{\ln (1 + x)}{x} \qquad \frac{0}{0}$$
$$= \lim_{x \to 0^+} \frac{\frac{1}{1 + x}}{1}$$
$$= \frac{1}{1} = 1.$$

Therefore,

$$\lim_{x \to 0^+} (1 + x)^{1/x} = \lim_{x \to 0^+} f(x) = \lim_{x \to 0^+} e^{\ln f(x)} = e^1 = e. \qquad \square$$

EXAMPLE 9   Find $\lim_{x \to \infty} x^{1/x}$.

Solution   The limit leads to the indeterminate form ∞^0. We let $f(x) = x^{1/x}$ and find $\lim_{x \to \infty} \ln f(x)$. Since

$$\ln f(x) = \ln x^{1/x}$$
$$= \frac{\ln x}{x},$$

l'Hôpital's rule gives

$$\lim_{x \to \infty} \ln f(x) = \lim_{x \to \infty} \frac{\ln x}{x} \qquad \frac{\infty}{\infty}$$
$$= \lim_{x \to \infty} \frac{1/x}{1}$$
$$= \frac{0}{1} = 0.$$

Therefore,

$$\lim_{x \to \infty} x^{1/x} = \lim_{x \to \infty} f(x) = \lim_{x \to \infty} e^{\ln f(x)} = e^0 = 1. \qquad \square$$

Exercises 6.6

Applying l'Hôpital's Rule

Use l'Hôpital's rule to find the limits in Exercises 1–42.

1. $\displaystyle\lim_{x\to 2}\frac{x-2}{x^2-4}$

2. $\displaystyle\lim_{x\to -5}\frac{x^2-25}{x+5}$

3. $\displaystyle\lim_{t\to -3}\frac{t^3-4t+15}{t^2-t-12}$

4. $\displaystyle\lim_{t\to 1}\frac{t^3-1}{4t^3-t-3}$

5. $\displaystyle\lim_{x\to\infty}\frac{5x^2-3x}{7x^2+1}$

6. $\displaystyle\lim_{x\to\infty}\frac{x-8x^2}{12x^2+5x}$

7. $\displaystyle\lim_{t\to 0}\frac{\sin t^2}{t}$

8. $\displaystyle\lim_{t\to 0}\frac{\sin 5t}{t}$

9. $\displaystyle\lim_{x\to 0}\frac{8x^2}{\cos x-1}$

10. $\displaystyle\lim_{x\to 0}\frac{\sin x-x}{x^3}$

11. $\displaystyle\lim_{\theta\to\pi/2}\frac{2\theta-\pi}{\cos(2\pi-\theta)}$

12. $\displaystyle\lim_{\theta\to -\pi/3}\frac{3\theta+\pi}{\sin(\theta+(\pi/3))}$

13. $\displaystyle\lim_{\theta\to\pi/2}\frac{1-\sin\theta}{1+\cos 2\theta}$

14. $\displaystyle\lim_{x\to 1}\frac{x-1}{\ln x-\sin\pi x}$

15. $\displaystyle\lim_{x\to 0}\frac{x^2}{\ln(\sec x)}$

16. $\displaystyle\lim_{x\to\pi/2}\frac{\ln(\csc x)}{(x-(\pi/2))^2}$

17. $\displaystyle\lim_{t\to 0}\frac{t(1-\cos t)}{t-\sin t}$

18. $\displaystyle\lim_{t\to 0}\frac{t\sin t}{1-\cos t}$

19. $\displaystyle\lim_{x\to(\pi/2)^-}\left(x-\frac{\pi}{2}\right)\sec x$

20. $\displaystyle\lim_{x\to(\pi/2)^-}\left(\frac{\pi}{2}-x\right)\tan x$

21. $\displaystyle\lim_{\theta\to 0}\frac{3^{\sin\theta}-1}{\theta}$

22. $\displaystyle\lim_{\theta\to 0}\frac{(1/2)^\theta-1}{\theta}$

23. $\displaystyle\lim_{x\to 0}\frac{x2^x}{2^x-1}$

24. $\displaystyle\lim_{x\to 0}\frac{3^x-1}{2^x-1}$

25. $\displaystyle\lim_{x\to\infty}\frac{\ln(x+1)}{\log_2 x}$

26. $\displaystyle\lim_{x\to\infty}\frac{\log_2 x}{\log_3(x+3)}$

27. $\displaystyle\lim_{x\to 0^+}\frac{\ln(x^2+2x)}{\ln x}$

28. $\displaystyle\lim_{x\to 0^+}\frac{\ln(e^x-1)}{\ln x}$

29. $\displaystyle\lim_{y\to 0}\frac{\sqrt{5y+25}-5}{y}$

30. $\displaystyle\lim_{y\to 0}\frac{\sqrt{ay+a^2}-a}{y},\quad a>0$

31. $\displaystyle\lim_{x\to\infty}(\ln 2x-\ln(x+1))$

32. $\displaystyle\lim_{x\to 0^+}(\ln x-\ln\sin x)$

33. $\displaystyle\lim_{x\to 0^+}\left(\frac{1}{x}-\frac{1}{\sin x}\right)$

34. $\displaystyle\lim_{x\to 0^+}\left(\frac{3x+1}{x}-\frac{1}{\sin x}\right)$

35. $\displaystyle\lim_{x\to 1^+}\left(\frac{1}{x-1}-\frac{1}{\ln x}\right)$

36. $\displaystyle\lim_{x\to 0^+}(\csc x-\cot x+\cos x)$

37. $\displaystyle\lim_{x\to\infty}\int_x^{2x}\frac{1}{t}\,dt$

38. $\displaystyle\lim_{x\to\infty}\frac{1}{x\ln x}\int_1^x\ln t\,dt$

39. $\displaystyle\lim_{\theta\to 0}\frac{\cos\theta-1}{e^\theta-\theta-1}$

40. $\displaystyle\lim_{h\to 0}\frac{e^h-(1+h)}{h^2}$

41. $\displaystyle\lim_{t\to\infty}\frac{e^t+t^2}{e^t-t}$

42. $\displaystyle\lim_{x\to\infty}x^2e^{-x}$

Limits Involving Bases and Exponents

Find the limits in Exercises 43–52.

43. $\displaystyle\lim_{x\to 1^+}x^{1/(1-x)}$

44. $\displaystyle\lim_{x\to 1^+}x^{1/(x-1)}$

45. $\displaystyle\lim_{x\to\infty}(\ln x)^{1/x}$

46. $\displaystyle\lim_{x\to e^+}(\ln x)^{1/(x-e)}$

47. $\displaystyle\lim_{x\to 0^+}x^{-1/\ln x}$

48. $\displaystyle\lim_{x\to\infty}x^{1/\ln x}$

49. $\displaystyle\lim_{x\to\infty}(1+2x)^{1/(2\ln x)}$

50. $\displaystyle\lim_{x\to 0}(e^x+x)^{1/x}$

51. $\displaystyle\lim_{x\to 0^+}x^x$

52. $\displaystyle\lim_{x\to 0^+}\left(1+\frac{1}{x}\right)^x$

Theory and Applications

L'Hôpital's rule does not help with the limits in Exercises 53–56. Try it—you just keep on cycling. Find the limits some other way.

53. $\displaystyle\lim_{x\to\infty}\frac{\sqrt{9x+1}}{\sqrt{x+1}}$

54. $\displaystyle\lim_{x\to 0^+}\frac{\sqrt{x}}{\sqrt{\sin x}}$

55. $\displaystyle\lim_{x\to(\pi/2)^-}\frac{\sec x}{\tan x}$

56. $\displaystyle\lim_{x\to 0^+}\frac{\cot x}{\csc x}$

57. Which one is correct, and which one is wrong? Give reasons for your answers.

    a) $\displaystyle\lim_{x\to 3}\frac{x-3}{x^2-3}=\lim_{x\to 3}\frac{1}{2x}=\frac{1}{6}$

    b) $\displaystyle\lim_{x\to 3}\frac{x-3}{x^2-3}=\frac{0}{6}=0$

58. Which one is correct, and which one is wrong? Give reasons for your answers.

a) $\lim\limits_{x\to 0}\dfrac{x^2-2x}{x^2-\sin x}=\lim\limits_{x\to 0}\dfrac{2x-2}{2x-\cos x}$

$\qquad\qquad =\lim\limits_{x\to 0}\dfrac{2}{2+\sin x}=\dfrac{2}{2+0}=1$

b) $\lim\limits_{x\to 0}\dfrac{x^2-2x}{x^2-\sin x}=\lim\limits_{x\to 0}\dfrac{2x-2}{2x-\cos x}=\dfrac{-2}{0-1}=2$

59. Only one of these calculations is correct. Which one? Why are the others wrong? Give reasons for your answers.

a) $\lim\limits_{x\to 0^+}x\ln x=0\cdot(-\infty)=0$

b) $\lim\limits_{x\to 0^+}x\ln x=0\cdot(-\infty)=-\infty$

c) $\lim\limits_{x\to 0^+}x\ln x=\lim\limits_{x\to 0^+}\dfrac{\ln x}{(1/x)}=\dfrac{-\infty}{\infty}=-1$

d) $\lim\limits_{x\to 0^+}x\ln x=\lim\limits_{x\to 0^+}\dfrac{\ln x}{(1/x)}$

$\qquad\qquad =\lim\limits_{x\to 0^+}\dfrac{(1/x)}{(-1/x^2)}=\lim\limits_{x\to 0^+}(-x)=0$

60. Let

$$f(x)=\begin{cases} x+2, & x\neq 0\\ 0, & x=0\end{cases}$$

$$g(x)=\begin{cases} x+1, & x\neq 0\\ 0, & x=0.\end{cases}$$

Show that

$$\lim_{x\to 0}\frac{f'(x)}{g'(x)}=1\quad\text{but that}\quad\lim_{x\to 0}\frac{f(x)}{g(x)}=2.$$

Doesn't this contradict l'Hôpital's rule? Give reasons for your answers.

61. Find a value of c that makes the function

$$f(x)=\begin{cases} \dfrac{9x-3\sin 3x}{5x^3}, & x\neq 0\\ c, & x=0\end{cases}$$

continuous at $x=0$. Explain why your value of c works.

62. Find a value of c that makes the function

$$g(\theta)=\begin{cases} \dfrac{(\tan\theta)^2}{\sin(4\theta^2/\pi)}, & \theta\neq 0\\ c, & \theta=0\end{cases}$$

continuous from the right at $\theta=0$. Explain why your value of c works.

63. *The continuous compound interest formula.* In deriving the formula $A(t)=A_0\,e^{rt}$ in Section 6.5, we claimed that

$$\lim_{k\to\infty}A_0\left(1+\frac{r}{k}\right)^{kt}=A_0\,e^{rt}.$$

This equation will hold if

$$\lim_{k\to\infty}\left(1+\frac{r}{k}\right)^{kt}=e^{rt},$$

and this, in turn, will hold if

$$\lim_{k\to\infty}\left(1+\frac{r}{k}\right)^{k}=e^{r}.$$

As you can see, the limit leads to the indeterminate form 1^{∞}. Verify the limit using l'Hôpital's rule.

64. Given that $x>0$, find the maximum value, if any, of

a) $x^{1/x}$

b) x^{1/x^2}

c) x^{1/x^n} (n a positive integer)

d) Show that $\lim\limits_{x\to\infty}x^{1/x^n}=1$ for every positive integer n.

▦ Grapher Explorations

65. *Determining the value of e.*

a) Use l'Hôpital's rule to show that

$$\lim_{x\to\infty}\left(1+\frac{1}{x}\right)^{x}=e.$$

b) **CALCULATOR** See how close you can come to

$$e=2.7\ 1828\ 1828\ 45\ 90\ 45$$

by evaluating $f(x)=(1+(1/x))^x$ for $x=10$, 10^2, 10^3, …and so on. You can expect the approximations to approach e at first, but on some calculators they will move away again as round-off errors take their toll.

c) If you have a grapher, you may prefer to do part (b) by graphing $f(x)=(1+(1/x))^x$ for large values of x, using TRACE to display the coordinates along the graph. Again, you may expect to find decreasing accuracy as x increases and, beyond $x=10^{10}$ or so, erratic behavior.

66. This exercise explores the difference between the limit

$$\lim_{x\to\infty}\left(1+\frac{1}{x^2}\right)^{x}$$

and the limit

$$\lim_{x\to\infty}\left(1+\frac{1}{x}\right)^{x}=e,$$

studied in Exercise 65.

a) Graph

$$f(x)=\left(1+\frac{1}{x^2}\right)^{x}\quad\text{and}\quad g(x)=\left(1+\frac{1}{x}\right)^{x}$$

together for $x\geq 0$. How does the behavior of f compare with that of g? Estimate the value of $\lim\limits_{x\to\infty}f(x)$.

b) Confirm your estimate of $\lim\limits_{x\to\infty}f(x)$ by calculating it with l'Hôpital's rule.

67. a) Estimate the value of

$$\lim_{x\to\infty}\left(x-\sqrt{x^2+x}\,\right)$$

by graphing $f(x)=x-\sqrt{x^2+x}$ over a suitably large interval of x-values.

b) Now confirm your estimate by finding the limit with l'Hôpital's rule. As the first step, multiply $f(x)$ by the fraction $(x + \sqrt{x^2 + x})/(x + \sqrt{x^2 + x})$ and simplify the new numerator.

68. Estimate the value of

$$\lim_{x \to 2} \frac{x^2 - 4}{\sqrt{x^2 + 5} - 3}$$

by graphing. Then confirm your estimate with l'Hôpital's rule.

69. Estimate the value of

$$\lim_{x \to 1} \frac{2x^2 - (3x + 1)\sqrt{x} + 2}{x - 1}$$

by graphing. Then confirm your estimate with l'Hôpital's rule.

70. a) Estimate the value of

$$\lim_{x \to 1} \frac{(x - 1)^2}{x \ln x - x - \cos \pi x}$$

by graphing $f(x) = (x - 1)^2/(x \ln x - x - \cos \pi x)$ near $x = 1$. Then confirm your estimate with l'Hôpital's rule.

b) Graph f for $0 < x \le 11$.

71. *The continuous extension of $(\sin x)^x$ to $[0, \pi]$*

a) Graph $f(x) = (\sin x)^x$ on the interval $0 \le x \le \pi$. What value would you assign to f to make it continuous at $x = 0$?

b) Verify your conclusion in (a) by finding $\lim_{x \to 0^+} f(x)$ with l'Hôpital's rule.

c) Returning to the graph, estimate the maximum value of f on $[0, \pi]$. About where is max f taken on?

d) Sharpen your estimate in (c) by graphing f' in the same window to see where its graph crosses the x-axis. To simplify your work, you might want to delete the exponential factor from the expression for f' and graph just the factor that has a zero.

e) Sharpen your estimate of the location of max f further still by solving the equation $f' = 0$ numerically.

f) CALCULATOR Estimate max f by evaluating f at the locations you found in (c), (d), and (e). What is your best value for max f?

72. *The function $(\sin x)^{\tan x}$.* (*Continuation of Exercise 71.*)

a) Graph $f(x) = (\sin x)^{\tan x}$ on the interval $-7 \le x \le 7$. How

do you account for the gaps in the graph? How wide are the gaps?

b) Now graph f on the interval $0 \le x \le \pi$. The function is not defined at $x = \pi/2$, but the graph has no break at this point. What is going on? What value does the graph appear to give for f at $x = \pi/2$? (*Hint:* Use l'Hôpital's rule to find lim f as $x \to (\pi/2)^-$ and $x \to (\pi/2)^+$.)

c) Continuing with the graphs in (b), find max f and min f as accurately as you can and estimate the values of x at which they are taken on.

73. *The place of ln x among the powers of x.* The natural logarithm

$$\ln x = \int_1^x \frac{1}{t} \, dt$$

fills the gap in the set of formulas

$$\int t^{k-1} dt = \frac{t^k}{k} + C, \quad k \ne 0, \tag{3}$$

but the formulas themselves do not reveal how well the logarithm fits in. We can see the nice fit graphically if we select from Eq. (3) the specific antiderivatives

$$\int_1^x t^{k-1} dt = \frac{x^k - 1}{k}, \quad x > 0,$$

and compare their graphs with the graph of ln x.

a) Graph the functions $f(x) = (x^k - 1)/k$ together with ln x on the interval $0 \le x \le 50$ for $k = \pm 1, \pm 0.5, \pm 0.1,$ and ± 0.05.

b) Show that

$$\lim_{k \to 0} \frac{x^k - 1}{k} = \ln x.$$

(Based on "The Place of ln x Among the Powers of x" by Henry C. Finlayson, *American Mathematical Monthly,* Vol. 94, No. 5, May 1987, p. 450.)

74. *Confirmation of the limit in Section 5.7, Exercise 42.* Estimate the value of

$$\lim_{\alpha \to 0^+} \frac{\sin \alpha - \alpha \cos \alpha}{\alpha - \alpha \cos \alpha}$$

as closely as you can by graphing. Then confirm your estimate with l'Hôpital's rule.

6.7

Relative Rates of Growth

This section shows how to compare the rates at which functions of x grow as x becomes large and introduces the so-called little-oh and big-oh notation sometimes used to describe the results of these comparisons. *We restrict our attention to functions whose values eventually become and remain positive as $x \to \infty$.*

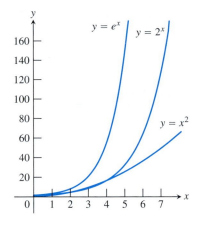

6.14 The graphs of e^x, 2^x, and x^2.

6.15 Scale drawings of the graphs of e^x and ln x.

Relatives Rates of Growth

You may have noticed that exponential functions like 2^x and e^x seem to grow more rapidly as x gets large than the polynomials and rational functions we graphed in Chapter 3. These exponentials certainly grow more rapidly than x itself, and you can see 2^x outgrowing x^2 as x increases in Fig. 6.14. In fact, as $x \to \infty$, the functions 2^x and e^x grow faster than any power of x, even $x^{1,000,000}$ (Exercise 19).

To get a feeling for how rapidly the values of $y = e^x$ grow with increasing x, think of graphing the function on a large blackboard, with the axes scaled in centimeters. At $x = 1$ cm, the graph is $e^1 \approx 3$ cm above the x-axis. At $x = 6$ cm, the graph is $e^6 \approx 403$ cm ≈ 4 m high (it is about to go through the ceiling if it hasn't done so already). At $x = 10$ cm, the graph is $e^{10} \approx 22{,}026$ cm ≈ 220 m high, higher than most buildings. At $x = 24$ cm, the graph is more than halfway to the moon, and at $x = 43$ cm from the origin, the graph is high enough to reach past the sun's closest stellar neighbor, the red dwarf star Proxima Centauri:

$$e^{43} \approx 4.73 \times 10^{18} \text{ cm}$$

$$= 4.73 \times 10^{13} \text{ km}$$

$$\approx 1.58 \times 10^8 \text{ light-seconds} \qquad \text{\color{blue}In a vacuum, light travels at 300,000 km/sec.}$$

$$\approx 5.0 \text{ light-years}$$

The distance to Proxima Centauri is about 4.22 light-years. Yet with $x = 43$ cm from the origin, the graph is still less than 2 feet to the right of the y-axis.

In contrast, logarithmic functions like $y = \log_2 x$ and $y = \ln x$ grow more slowly as $x \to \infty$ than any positive power of x (Excrcise 21). With axes scaled in centimeters, you have to go nearly 5 light-years out on the x-axis to find a point where the graph of $y = \ln x$ is even $y = 43$ cm high. See Fig. 6.15.

These important comparisons of exponential, polynomial, and logarithmic functions can be made precise by defining what it means for a function $f(x)$ to grow faster than a function $g(x)$ as $x \to \infty$.

Definition

Rates of Growth as $x \to \infty$

Let $f(x)$ and $g(x)$ be positive for x sufficiently large.

1. f **grows faster than** g as $x \to \infty$ if

$$\lim_{x \to \infty} \frac{f(x)}{g(x)} = \infty$$

or, equivalently, if

$$\lim_{x \to \infty} \frac{g(x)}{f(x)} = 0.$$

We also say that g **grows slower than** f as $x \to \infty$.

2. f and g **grow at the same rate** as $x \to \infty$ if

$$\lim_{x \to \infty} \frac{f(x)}{g(x)} = L \neq 0. \qquad \text{\color{blue}L finite and not zero}$$

According to these definitions, $y = 2x$ does not grow faster than $y = x$. The two functions grow at the same rate because

$$\lim_{x \to \infty} \frac{2x}{x} = \lim_{x \to \infty} 2 = 2,$$

which is a finite, nonzero limit. The reason for this apparent disregard of common sense is that we want "f grows faster than g" to mean that for large x-values g is negligible when compared with f.

EXAMPLE 1 e^x grows faster than x^2 as $x \to \infty$ because

$$\underbrace{\lim_{x \to \infty} \frac{e^x}{x^2}}_{\infty/\infty} = \underbrace{\lim_{x \to \infty} \frac{e^x}{2x}}_{\infty/\infty} = \lim_{x \to \infty} \frac{e^x}{2} = \infty. \qquad \text{Using l'Hôpital's rule twice}$$

EXAMPLE 2

a) 3^x grows faster than 2^x as $x \to \infty$ because

$$\lim_{x \to \infty} \frac{3^x}{2^x} = \lim_{x \to \infty} \left(\frac{3}{2}\right)^x = \infty.$$

b) As part (a) suggests, exponential functions with different bases never grow at the same rate as $x \to \infty$. If $a > b > 0$, then a^x grows faster than b^x. Since $(a/b) > 1$,

$$\lim_{x \to \infty} \frac{a^x}{b^x} = \lim_{x \to \infty} \left(\frac{a}{b}\right)^x = \infty.$$

EXAMPLE 3 x^2 grows faster than $\ln x$ as $x \to \infty$ because

$$\lim_{x \to \infty} \frac{x^2}{\ln x} = \lim_{x \to \infty} \frac{2x}{1/x} = \lim_{x \to \infty} 2x^2 = \infty. \qquad \text{l'Hôpital's rule}$$

EXAMPLE 4 $\ln x$ grows slower than x as $x \to \infty$ because

$$\lim_{x \to \infty} \frac{\ln x}{x} = \lim_{x \to \infty} \frac{1/x}{1} \qquad \text{l'Hôpital's rule}$$

$$= \lim_{x \to \infty} \frac{1}{x} = 0.$$

EXAMPLE 5 In contrast to exponential functions, logarithmic functions with different bases a and b always grow at the same rate as $x \to \infty$:

$$\lim_{x \to \infty} \frac{\log_a x}{\log_b x} = \lim_{x \to \infty} \frac{\ln x / \ln a}{\ln x / \ln b} = \frac{\ln b}{\ln a}.$$

The limiting ratio is always finite and never zero.

If f grows at the same rate as g as $x \to \infty$, and g grows at the same rate as h as $x \to \infty$, then f grows at the same rate as h as $x \to \infty$. The reason is that

$$\lim_{x \to \infty} \frac{f}{g} = L_1 \qquad \text{and} \qquad \lim_{x \to \infty} \frac{g}{h} = L_2$$

together imply

$$\lim_{x \to \infty} \frac{f}{h} = \lim_{x \to \infty} \frac{f}{g} \cdot \frac{g}{h} = L_1 L_2.$$

If L_1 and L_2 are finite and nonzero, then so is $L_1 L_2$.

EXAMPLE 6   Show that $\sqrt{x^2 + 5}$ and $(2\sqrt{x} - 1)^2$ grow at the same rate as $x \to \infty$.

Solution   We show that the functions grow at the same rate by showing that they both grow at the same rate as the function x:

$$\lim_{x \to \infty} \frac{\sqrt{x^2 + 5}}{x} = \lim_{x \to \infty} \sqrt{1 + \frac{5}{x^2}} = 1,$$

$$\lim_{x \to \infty} \frac{(2\sqrt{x} - 1)^2}{x} = \lim_{x \to \infty} \left(\frac{2\sqrt{x} - 1}{\sqrt{x}} \right)^2 = \lim_{x \to \infty} \left(2 - \frac{1}{\sqrt{x}} \right)^2 = 4.$$ ❑

Order and Oh-Notation

Here we introduce the "little-oh" and "big-oh" notation invented by number theorists a hundred years ago and now commonplace in mathematical analysis and computer science.

> **Definition**
> A function f is **of smaller order than** g as $x \to \infty$ if $\lim\limits_{x \to \infty} \dfrac{f(x)}{g(x)} = 0$. We indicate this by writing $f = o(g)$ ("f is little-oh of g").

Notice that saying $f = o(g)$ as $x \to \infty$ is another way to say that f grows slower than g as $x \to \infty$.

EXAMPLE 7

$$\ln x = o(x) \text{ as } x \to \infty \quad \text{because} \quad \lim_{x \to \infty} \frac{\ln x}{x} = 0$$

$$x^2 = o(x^3 + 1) \text{ as } x \to \infty \quad \text{because} \quad \lim_{x \to \infty} \frac{x^2}{x^3 + 1} = 0$$ ❑

> **Definition**
> Let $f(x)$ and $g(x)$ be positive for x sufficiently large. Then f is **of at most the order of** g as $x \to \infty$ if there is a positive integer M for which
> $$\frac{f(x)}{g(x)} \le M,$$
> for x sufficiently large. We indicate this by writing $f = O(g)$ ("f is big-oh of g").

EXAMPLE 8

$$x + \sin x = O(x) \text{ as } x \to \infty \quad \text{because} \quad \frac{x + \sin x}{x} \leq 2 \text{ for } x \text{ sufficiently large.}$$

❑

EXAMPLE 9

$$e^x + x^2 = O(e^x) \text{ as } x \to \infty \quad \text{because} \quad \frac{e^x + x^2}{e^x} \to 1 \text{ as } x \to \infty,$$

$$x = O(e^x) \text{ as } x \to \infty \quad \text{because} \quad \frac{x}{e^x} \to 0 \text{ as } x \to \infty.$$

❑

If you look at the definitions again, you will see that $f = o(g)$ implies $f = O(g)$ for functions that are positive for x sufficiently large. Also, if f and g grow at the same rate, then $f = O(g)$ and $g = O(f)$ (Exercise 11).

Sequential vs. Binary Search

Computer scientists sometimes measure the efficiency of an algorithm by counting the number of steps a computer must take to make the algorithm do something. There can be significant differences in how efficiently algorithms perform, even if they are designed to accomplish the same task. These differences are often described in big-oh notation. Here is an example.

Webster's Third New International Dictionary lists about 26,000 words that begin with the letter a. One way to look up a word, or to learn if it is not there, is to read through the list one word at a time until you either find the word or determine that it is not there. This method, called sequential search, makes no particular use of the words' alphabetical arrangement. You are sure to get an answer, but it might take 26,000 steps.

Another way to find the word or to learn it is not there is to go straight to the middle of the list (give or take a few words). If you do not find the word, then go to the middle of the half that contains it and forget about the half that does not. (You know which half contains it because you know the list is ordered alphabetically.) This method eliminates roughly 13,000 words in a single step. If you do not find the word on the second try, then jump to the middle of the half that contains it. Continue this way until you have either found the word or divided the list in half so many times there are no words left. How many times do you have to divide the list to find the word or learn that it is not there? At most 15, because

$$(26{,}000/2^{15}) < 1.$$

That certainly beats a possible 26,000 steps.

For a list of length n, a sequential search algorithm takes on the order of n steps to find a word or determine that it is not in the list. A binary search, as the second algorithm is called, takes on the order of $\log_2 n$ steps. The reason is that if $2^{m-1} < n \leq 2^m$, then $m - 1 < \log_2 n \leq m$, and the number of bisections required to narrow the list to one word will be at most $m = \lceil \log_2 n \rceil$, the integer ceiling for $\log_2 n$.

Big-oh notation provides a compact way to say all this. The number of steps in a sequential search of an ordered list is $O(n)$; the number of steps in a binary search is $O(\log_2 n)$. In our example, there is a big difference between the two (26,000 vs. 15), and the difference can only increase with n because n grows faster than $\log_2 n$ as $n \to \infty$.

> To find an item in a list of length n:
>
> A sequential search takes $O(n)$ steps.
>
> A binary search takes $O(\log_2 n)$ steps.

Exercises 6.7

Comparisons with the Exponential e^x

1. Which of the following functions grow faster than e^x as $x \to \infty$? Which grow at the same rate as e^x? Which grow slower?

  a) $x + 3$ **b)** $x^3 + \sin^2 x$

  c) $\sqrt{x}$ **d)** 4^x

  e) $(3/2)^x$ **f)** $e^{x/2}$

  g) $e^x/2$ **h)** $\log_{10} x$

2. Which of the following functions grow faster than e^x as $x \to \infty$? Which grow at the same rate as e^x? Which grow slower?

  a) $10x^4 + 30x + 1$ **b)** $x \ln x - x$

  c) $\sqrt{1 + x^4}$ **d)** $(5/2)^x$

  e) e^{-x} **f)** xe^x

  g) $e^{\cos x}$ **h)** e^{x-1}

Comparisons with the Power x^2

3. Which of the following functions grow faster than x^2 as $x \to \infty$? Which grow at the same rate as x^2? Which grow slower?

  a) $x^2 + 4x$ **b)** $x^5 - x^2$

  c) $\sqrt{x^4 + x^3}$ **d)** $(x + 3)^2$

  e) $x \ln x$ **f)** 2^x

  g) $x^3 e^{-x}$ **h)** $8x^2$

4. Which of the following functions grow faster than x^2 as $x \to \infty$? Which grow at the same rate as x^2? Which grow slower?

  a) $x^2 + \sqrt{x}$ **b)** $10x^2$

  c) $x^2 e^{-x}$ **d)** $\log_{10}(x^2)$

  e) $x^3 - x^2$ **f)** $(1/10)^x$

  g) $(1.1)^x$ **h)** $x^2 + 100x$

Comparisons with the Logarithm $\ln x$

5. Which of the following functions grow faster than $\ln x$ as $x \to \infty$? Which grow at the same rate as $\ln x$? Which grow slower?

  a) $\log_3 x$ **b)** $\ln 2x$

  c) $\ln \sqrt{x}$ **d)** $\sqrt{x}$

  e) x **f)** $5 \ln x$

  g) $1/x$ **h)** e^x

6. Which of the following functions grow faster than $\ln x$ as $x \to \infty$? Which grow at the same rate as $\ln x$? Which grow slower?

  a) $\log_2(x^2)$ **b)** $\log_{10} 10x$

  c) $1/\sqrt{x}$ **d)** $1/x^2$

  e) $x - 2 \ln x$ **f)** e^{-x}

  g) $\ln (\ln x)$ **h)** $\ln (2x + 5)$

Ordering Functions by Growth Rates

7. Order the following functions from slowest growing to fastest growing as $x \to \infty$.

  a) e^x **b)** x^x **c)** $(\ln x)^x$ **d)** $e^{x/2}$

8. Order the following functions from slowest growing to fastest growing as $x \to \infty$.

  a) 2^x **b)** x^2 **c)** $(\ln 2)^x$ **d)** e^x

Big-oh and Little-oh; Order

9. True, or false? As $x \to \infty$,

  a) $x = o(x)$ **b)** $x = o(x + 5)$

  c) $x = O(x + 5)$ **d)** $x = O(2x)$

  e) $e^x = o(e^{2x})$ **f)** $x + \ln x = O(x)$

  g) $\ln x = o(\ln 2x)$ **h)** $\sqrt{x^2 + 5} = O(x)$

10. True, or false? As $x \to \infty$,

  a) $\dfrac{1}{x + 3} = o\left(\dfrac{1}{x}\right)$ **b)** $\dfrac{1}{x} + \dfrac{1}{x^2} = o\left(\dfrac{1}{x}\right)$

  c) $\dfrac{1}{x} - \dfrac{1}{x^2} = o\left(\dfrac{1}{x}\right)$ **d)** $2 + \cos x = O(2)$

  e) $e^x + x = O(e^x)$ **f)** $x \ln x = o(x^2)$

  g) $\ln (\ln x) = O(\ln x)$ **h)** $\ln (x) = o(\ln (x^2 + 1))$

11. Show that if positive functions $f(x)$ and $g(x)$ grow at the same rate as $x \to \infty$, then $f = O(g)$ and $g = O(f)$.

12. When is a polynomial $f(x)$ of smaller order than a polynomial $g(x)$ as $x \to \infty$? Give reasons for your answer.

13. When is a polynomial $f(x)$ of at most the order of a polynomial $g(x)$ as $x \to \infty$? Give reasons for your anwer.

14. *Simpson's rule and the trapedzoidal rule.* The definitions in the present section can be made more general by lifting the restriction that $x \to \infty$ and considering limits as $x \to a$ for any real number a. Show that the error E_S in the Simpson's rule approximation of a definite integral is $O(h^4)$ as $h \to 0$ while the error E_T in the trapezoidal rule approximation is $O(h^2)$. This gives another way to explain the relative accuracies of the two approximation methods.

Other Comparisons

15. What do the conclusions we drew in Section 3.5 about the limits of rational functions tell us about the relative growth of polynomials as $x \to \infty$?

16. GRAPHER

a) Investigate

$$\lim_{x \to \infty} \frac{\ln (x + 1)}{\ln x} \quad \text{and} \quad \lim_{x \to \infty} \frac{\ln (x + 999)}{\ln x}.$$

Then use l'Hôpital's rule to explain what you find.

b) Show that the value of

$$\lim_{x \to \infty} \frac{\ln (x + a)}{\ln x}$$

is the same no matter what value you assign to the constant a. What does this say about the relative rates at which the functions $f(x) = \ln (x + a)$ and $g(x) = \ln x$ grow?

17. Show that $\sqrt{10x + 1}$ and $\sqrt{x + 1}$ grow at the same rate as $x \to \infty$ by showing that they both grow at the same rate as $\sqrt{x}$ as $x \to \infty$.

18. Show that $\sqrt{x^4 + x}$ and $\sqrt{x^4 - x^3}$ grow at the same rate as $x \to \infty$ by showing that they both grow at the same rate as x^2 as $x \to \infty$.

19. Show that e^x grows faster as $x \to \infty$ than x^n for any positive integer n, even $x^{1,000,000}$. (*Hint:* What is the nth derivative of x^n?)

20. *The function e^x outgrows any polynomial.* Show that e^x grows faster as $x \to \infty$ than any polynomial

$$a_n x^n + a_{n-1} x^{n-1} + \cdots + a_1 x + a_0.$$

21. a) Show that $\ln x$ grows slower as $x \to \infty$ than $x^{1/n}$ for any positive integer n, even $x^{1/1,000,000}$.

b) CALCULATOR Although the values of $x^{1/1,000,000}$ eventu-

ally overtake the values of $\ln x$, you have to go way out on the x-axis before this happens. Find a value of x greater than 1 for which $x^{1/1,000,000} > \ln x$. You might start by observing that when $x > 1$ the equation $\ln x = x^{1/1,000,000}$ is equivalent to the equation $\ln (\ln x) = (\ln x)/1,000,000$.

c) CALCULATOR Even $x^{1/10}$ takes a long time to overtake $\ln x$. Experiment with a calculator to find the value of x at which the graphs of $x^{1/10}$ and $\ln x$ cross, or, equivalently, at which $\ln x = 10 \ln (\ln x)$. Bracket the crossing point between powers of 10 and then close in by successive halving.

d) GRAPHER (*Continuation of part c.*) The value of x at which $\ln x = 10 \ln (\ln x)$ is too far out for some graphers and root finders to identify. Try it on the equipment available to you and see what happens.

22. *The function $\ln x$ grows slower than any polynomial.* Show that $\ln x$ grows slower as $x \to \infty$ than any nonconstant polynomial.

Algorithms and Searches

23. a) Suppose you have three different algorithms for solving the same problem and each algorithm takes a number of steps that is of the order of one of the functions listed here:

$$n \log_2 n, \quad n^{3/2}, \quad n(\log_2 n)^2.$$

Which of the algorithms is the most efficient in the long run? Give reasons for your answer.

b) GRAPHER Graph the functions in part (a) together to get a sense of how rapidly each one grows.

24. Repeat Exercise 23 for the functions

$$n, \quad \sqrt{n} \log_2 n, \quad (\log_2 n)^2.$$

25. CALCULATOR Suppose you are looking for an item in an ordered list one million items long. How many steps might it take to find that item with a sequential search? a binary search?

26. CALCULATOR You are looking for an item in an ordered list 450,000 items long (the length of *Webster's Third New International Dictionary*). How many steps might it take to find the item with a sequential search? a binary search?

6.8

Inverse Trigonometric Functions

Inverse trigonometric functions arise when we want to calculate angles from side measurements in triangles. They also provide useful antiderivatives and appear frequently in the solutions of differential equations. This section shows how the functions are defined, graphed, and evaluated.

Defining the Inverses

The six basic trigonometric functions are not one-to-one (their values repeat), but we can restrict their domains to intervals on which they are one-to-one.

Domain Restrictions That Make the Trigonometric Functions One-to-One

Function	Domain	Range
$\sin x$	$[-\pi/2 , \pi/2]$	$[-1, 1]$
$\cos x$	$[0, \pi]$	$[-1, 1]$
$\tan x$	$(-\pi/2, \pi/2)$	$(-\infty, \infty)$
$\cot x$	$(0, \pi)$	$(-\infty, \infty)$
$\sec x$	$[0, \pi/2) \cup (\pi/2, \pi]$	$(-\infty, -1] \cup [1, \infty)$
$\csc x$	$[-\pi/2, 0) \cup (0, \pi/2]$	$(-\infty, -1] \cup [1, \infty)$

Since these restricted functions are now one-to-one, they have inverses, which we denote by

$$y = \sin^{-1} x \qquad \text{or} \qquad y = \text{arc } \sin x$$

$$y = \cos^{-1} x \qquad \text{or} \qquad y = \text{arc } \cos x$$

$$y = \tan^{-1} x \qquad \text{or} \qquad y = \text{arc } \tan x$$

$$y = \cot^{-1} x \qquad \text{or} \qquad y = \text{arc } \cot x$$

$$y = \sec^{-1} x \qquad \text{or} \qquad y = \text{arc } \sec x$$

$$y = \csc^{-1} x \qquad \text{or} \qquad y = \text{arc } \csc x$$

These equations are read "y equals the arc sine of x" or "y equals arc sin x" and so on.

Caution   The -1 in the expressions for the inverse means "inverse." It does *not* mean reciprocal. For example, the *reciprocal* of $\sin x$ is $(\sin x)^{-1} = 1/\sin x = \csc x$.

The domains of the inverses are chosen to satisfy the following relationships.

$$\sec^{-1} x = \cos^{-1}(1/x) \tag{1}$$

$$\csc^{-1} x = \sin^{-1}(1/x) \tag{2}$$

$$\cot^{-1} x = \pi/2 - \tan^{-1} x \tag{3}$$

We can use these relationships to find values of $\sec^{-1} x$, $\csc^{-1} x$, and $\cot^{-1} x$ on calculators that give only $\cos^{-1} x$, $\sin^{-1} x$, and $\tan^{-1} x$. As in some of the examples that follow, we can also find a few of the more common values of $\sec^{-1} x$, $\csc^{-1} x$, and $\cot^{-1} x$ using reference right triangles.

The Arc Sine and Arc Cosine

The arc sine of x is an angle whose sine is x. The arc cosine is an angle whose cosine is x.

6.16 The graphs of (a) $y = \sin x$, $-\pi/2 \le x \le \pi/2$, and (b) its inverse, $y = \sin^{-1} x$. The graph of $\sin^{-1} x$, obtained by reflection across the line $y = x$, is a portion of the curve $x = \sin y$.

6.17 The graphs of (a) $y = \cos x$, $0 \le x \le \pi$, and (b) its inverse, $y = \cos^{-1} x$. The graph of $\cos^{-1} x$, obtained by reflection across the line $y = x$, is a portion of the curve $x = \cos y$.

Definition

$y = \sin^{-1} x$ is the number in $[-\pi/2, \pi/2]$ for which $\sin y = x$.

$y = \cos^{-1} x$ is the number in $[0, \pi]$ for which $\cos y = x$.

The graph of $y = \sin^{-1} x$ (Fig. 6.16) is symmetric about the origin (it lies along the graph of $x = \sin y$). The arc sine is therefore an odd function:

$$\sin^{-1}(-x) = -\sin^{-1} x. \tag{4}$$

The graph of $y = \cos^{-1} x$ (Fig. 6.17) has no such symmetry.

EXAMPLE 1   *Common values of $\sin^{-1} x$*

x	$\sin^{-1} x$
$\sqrt{3}/2$	$\pi/3$
$\sqrt{2}/2$	$\pi/4$
$1/2$	$\pi/6$
$-1/2$	$-\pi/6$
$-\sqrt{2}/2$	$-\pi/4$
$-\sqrt{3}/2$	$-\pi/3$

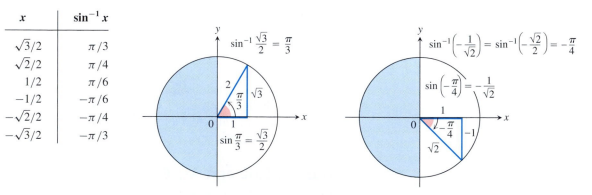

The angles come from the first and fourth quadrants because the range of $\sin^{-1} x$ is $[-\pi/2, \pi/2]$. ☐

EXAMPLE 2   *Common values of $\cos^{-1} x$*

x	$\cos^{-1} x$
$\sqrt{3}/2$	$\pi/6$
$\sqrt{2}/2$	$\pi/4$
$1/2$	$\pi/3$
$-1/2$	$2\pi/3$
$-\sqrt{2}/2$	$3\pi/4$
$-\sqrt{3}/2$	$5\pi/6$

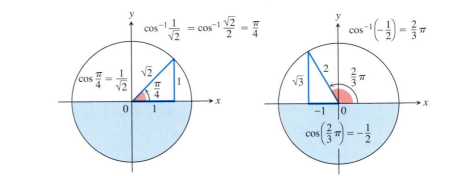

The angles come from the first and second quadrants because the range of $\cos^{-1} x$ is $[0, \pi]$. ❏

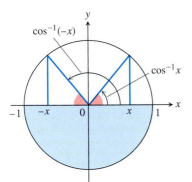

6.18 $\cos^{-1} x + \cos^{-1}(-x) = \pi$

Identities Involving Arc Sine and Arc Cosine

As we can see from Fig. 6.18, the arc cosine of x satisfies the identity

$$\cos^{-1} x + \cos^{-1}(-x) = \pi, \tag{5}$$

or

$$\cos^{-1}(-x) = \pi - \cos^{-1} x. \tag{6}$$

And we can see from the triangle in Fig. 6.19 that for $x > 0$,

$$\sin^{-1} x + \cos^{-1} x = \pi/2. \tag{7}$$

Equation (7) holds for the other values of x in $[-1, 1]$ as well, but we cannot conclude this from the triangle in Fig. 6.19. It is, however, a consequence of Eqs. (4) and (6) (Exercise 55).

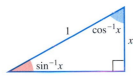

6.19 In this figure,

$$\sin^{-1} x + \cos^{-1} x = \pi/2.$$

Inverses of tan x, cot x, sec x, and csc x

The arc tangent of x is an angle whose tangent is x. The arc cotangent of x is an angle whose cotangent is x.

> **Definition**
>
> $y = \tan^{-1} x$ is the number in $(-\pi/2, \pi/2)$ for which $\tan y = x$.
>
> $y = \cot^{-1} x$ is the number in $(0, \pi)$ for which $\cot y = x$.

6.20 The graph of $y = \tan^{-1} x$.

We use open intervals to avoid values where the tangent and cotangent are undefined.

The graph of $y = \tan^{-1} x$ is symmetric about the origin because it is a branch of the graph $x = \tan y$ that is symmetric about the origin (Fig. 6.20). Algebraically this means that

$$\tan^{-1}(-x) = -\tan^{-1} x; \tag{8}$$

the arc tangent is an odd function. The graph of $y = \cot^{-1} x$ has no such symmetry (Fig. 6.21).

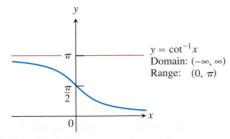

6.21 The graph of $y = \cot^{-1} x$.

6.22 The graph of $y = \sec^{-1} x$.

The inverses of the restricted forms of $\sec x$ and $\csc x$ are chosen to be the functions graphed in Figs. 6.22 and 6.23.

Caution   There is no general agreement about how to define $\sec^{-1} x$ for negative values of x. We chose angles in the second quadrant between $\pi/2$ and π. This choice makes $\sec^{-1} x = \cos^{-1}(1/x)$. It also makes $\sec^{-1} x$ an increasing function on each interval of its domain. Some tables choose $\sec^{-1} x$ to lie in $[-\pi, -\pi/2)$ for $x < 0$ and some texts choose it to lie in $[\pi, 3\pi/2)$ (Fig. 6.24). These choices simplify the formula for the derivative (our formula needs absolute value signs) but fail to satisfy the computational equation $\sec^{-1} x = \cos^{-1}(1/x)$.

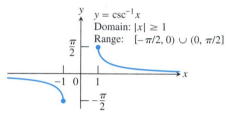

6.23 The graph of $y = \csc^{-1} x$.

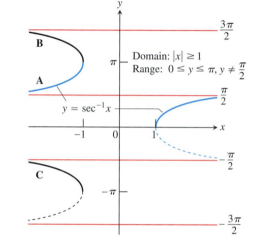

6.24 There are several logical choices for the left-hand branch of $y = \sec^{-1} x$. With choice **A**, Eq. (1) holds, but the formula for the derivative of the arc secant is complicated by absolute value bars. Choices **B** and **C** lead to a simpler derivative formula, but Eq. (1) no longer holds. Most calculators use Eq. (1), so we chose **A**.

EXAMPLE 3   *Common values of* $\tan^{-1} x$

x	$\tan^{-1} x$
$\sqrt{3}$	$\pi/3$
1	$\pi/4$
$\sqrt{3}/3$	$\pi/6$
$-\sqrt{3}/3$	$-\pi/6$
-1	$-\pi/4$
$-\sqrt{3}$	$-\pi/3$

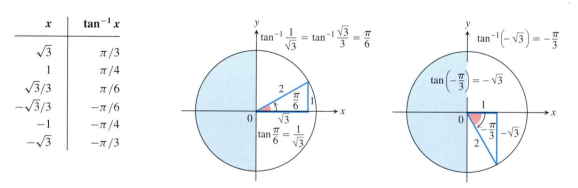

The angles come from the first and fourth quadrants because the range of $\tan^{-1} x$ is $[-\pi/2, \pi/2]$.

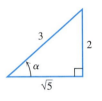

6.25 If $\alpha = \sin^{-1}(2/3)$, then the values of the other basic trigonometric functions of α can be read from this triangle (Example 4).

EXAMPLE 4 Find $\cos\alpha$, $\tan\alpha$, $\sec\alpha$, $\csc\alpha$, and $\cot\alpha$ if

$$\alpha = \sin^{-1}\frac{2}{3}. \tag{12}$$

Solution Equation (12) says that $\sin\alpha = 2/3$. We picture α as an angle in a right triangle with opposite side 2 and hypotenuse 3 (Fig. 6.25). The length of the remaining side is

$$\sqrt{(3)^2 - (2)^2} = \sqrt{9 - 4} = \sqrt{5}. \qquad \text{Pythagorean theorem}$$

We add this information to the figure and then read the values we want from the completed triangle:

$$\cos\alpha = \frac{\sqrt{5}}{3}, \quad \tan\alpha = \frac{2}{\sqrt{5}}, \quad \sec\alpha = \frac{3}{\sqrt{5}}, \quad \csc\alpha = \frac{3}{2}, \quad \cot\alpha = \frac{\sqrt{5}}{2}. \quad \square$$

The "arc" in arc sine and arc cosine

In case you are wondering about the "arc," look at the accompanying figure. It gives a geometric interpretation of $y = \sin^{-1} x$ and $y = \cos^{-1} x$ for angles in the first quadrant. For a unit circle, the equation $s = r\theta$ becomes $s = \theta$, so central angles and the arcs they subtend have the same measure. If $x = \sin y$, then, in addition to being the angle whose sine is x, y is also the length of arc on the unit circle that subtends an angle whose sine is x. So we call y "the arc whose sine is x." When angles were measured by intercepted arc lengths, as they once were, this was a natural way to speak. Today it can sound a bit strange, but the language has stayed with us. The arc cosine has a similar interpretation.

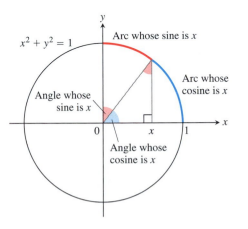

EXAMPLE 5 Find $\cot\left(\sec^{-1}\left(-\frac{2}{\sqrt{3}}\right) + \csc^{-1}(-2)\right)$.

Solution We work from inside out, using reference triangles to exhibit ratios and angles.

Step 1: Negative values of the secant come from second-quadrant angles:

$$\sec^{-1}\left(-\frac{2}{\sqrt{3}}\right) = \sec^{-1}\left(\frac{2}{-\sqrt{3}}\right)$$
$$= \frac{5\pi}{6}.$$

Step 2: Negative values of the cosecant come from fourth-quadrant angles:

$$\csc^{-1}(-2) = \csc^{-1}\left(\frac{2}{-1}\right)$$
$$= -\frac{\pi}{6}.$$

Step 3:

$$\cot\left(\sec^{-1}\left(-\frac{2}{\sqrt{3}}\right) + \csc^{-1}(-2)\right)$$
$$= \cot\left(\frac{5\pi}{6} - \frac{\pi}{6}\right)$$
$$= \cot\left(\frac{2\pi}{3}\right)$$
$$= -\frac{1}{\sqrt{3}} \qquad \square$$

EXAMPLE 6 Find $\sec\left(\tan^{-1}\dfrac{x}{3}\right)$.

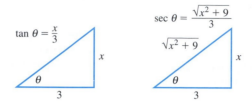

Solution We let $\theta = \tan^{-1}(x/3)$ (to give the angle a name) and picture θ in a right triangle with

$$\tan\theta = \text{opposite/adjacent} = x/3.$$

The length of the triangle's hypotenuse is

$$\sqrt{x^2 + 3^2} = \sqrt{x^2 + 9}.$$

Thus,

$$\sec\left(\tan^{-1}\frac{x}{3}\right) = \sec\theta$$

$$= \frac{\sqrt{x^2 + 9}}{3}. \qquad \sec\theta = \frac{\text{hypotenuse}}{\text{adjacent}} \qquad \square$$

EXAMPLE 7 *Drift correction*

During an airplane flight from Chicago to St. Louis the navigator determines that the plane is 12 mi off course, as shown in Fig. 6.26. Find the angle a for a course parallel to the original, correct course, the angle b, and the correction angle $c = a + b$.

Solution

$$a = \sin^{-1}\frac{12}{180} \approx 0.067 \text{ radian} \approx 3.8°$$

$$b = \sin^{-1}\frac{12}{62} \approx 0.195 \text{ radian} \approx 11.2°$$

$$c = a + b \approx 15°. \qquad\qquad \square$$

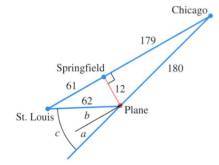

6.26 Diagram for drift correction (Example 7), with distances rounded to the nearest mile (drawing not to scale).

Exercises 6.8

Common Values of Inverse Trignonometric Functions

Use reference triangles like those in Examples 1–3 to find the angles in Exercises 1–12.

1. a) $\tan^{-1} 1$ **b)** $\tan^{-1}(-\sqrt{3})$ **c)** $\tan^{-1}\left(\dfrac{1}{\sqrt{3}}\right)$

2. a) $\tan^{-1}(-1)$ **b)** $\tan^{-1}\sqrt{3}$ **c)** $\tan^{-1}\left(\dfrac{-1}{\sqrt{3}}\right)$

3. a) $\sin^{-1}\left(\dfrac{-1}{2}\right)$ **b)** $\sin^{-1}\left(\dfrac{1}{\sqrt{2}}\right)$ **c)** $\sin^{-1}\left(\dfrac{-\sqrt{3}}{2}\right)$

4. a) $\sin^{-1}\left(\dfrac{1}{2}\right)$ **b)** $\sin^{-1}\left(\dfrac{-1}{\sqrt{2}}\right)$ **c)** $\sin^{-1}\left(\dfrac{\sqrt{3}}{2}\right)$

5. a) $\cos^{-1}\left(\dfrac{1}{2}\right)$ **b)** $\cos^{-1}\left(\dfrac{-1}{\sqrt{2}}\right)$ **c)** $\cos^{-1}\left(\dfrac{\sqrt{3}}{2}\right)$

6. a) $\cos^{-1}\left(\dfrac{-1}{2}\right)$ **b)** $\cos^{-1}\left(\dfrac{1}{\sqrt{2}}\right)$ **c)** $\cos^{-1}\left(\dfrac{-\sqrt{3}}{2}\right)$

7. a) $\sec^{-1}(-\sqrt{2})$ **b)** $\sec^{-1}\left(\dfrac{2}{\sqrt{3}}\right)$ **c)** $\sec^{-1}(-2)$

8. a) $\sec^{-1}\sqrt{2}$ **b)** $\sec^{-1}\left(\dfrac{-2}{\sqrt{3}}\right)$ **c)** $\sec^{-1} 2$

9. a) $\csc^{-1}\sqrt{2}$ **b)** $\csc^{-1}\left(\dfrac{-2}{\sqrt{3}}\right)$ **c)** $\csc^{-1} 2$

10. a) $\csc^{-1}(-\sqrt{2})$ **b)** $\csc^{-1}\left(\dfrac{2}{\sqrt{3}}\right)$ **c)** $\csc^{-1}(-2)$

11. a) $\cot^{-1}(-1)$ **b)** $\cot^{-1}\sqrt{3}$ **c)** $\cot^{-1}\left(\dfrac{-1}{\sqrt{3}}\right)$

12. a) $\cot^{-1} 1$ **b)** $\cot^{-1}(-\sqrt{3})$ **c)** $\cot^{-1}\left(\dfrac{1}{\sqrt{3}}\right)$

Trigonometric Function Values

13. Given that $\alpha = \sin^{-1}(5/13)$, find $\cos\alpha$, $\tan\alpha$, $\sec\alpha$, $\csc\alpha$, and $\cot\alpha$.

14. Given that $\alpha = \tan^{-1}(4/3)$, find $\sin\alpha$, $\cos\alpha$, $\sec\alpha$, $\csc\alpha$, and $\cot\alpha$.

15. Given that $\alpha = \sec^{-1}(-\sqrt{5})$, find $\sin\alpha$, $\cos\alpha$, $\tan\alpha$, $\csc\alpha$, and $\cot\alpha$.

16. Given that $\alpha = \sec^{-1}(-\sqrt{13}/2)$, find $\sin\alpha$, $\cos\alpha$, $\tan\alpha$, $\csc\alpha$, and $\cot\alpha$.

Evaluating Trigonometric and Inverse Trigonometric Terms

Find the values in Exercises 17–28.

17. $\sin\left(\cos^{-1}\dfrac{\sqrt{2}}{2}\right)$ **18.** $\sec\left(\cos^{-1}\dfrac{1}{2}\right)$

19. $\tan\left(\sin^{-1}\left(-\dfrac{1}{2}\right)\right)$ **20.** $\cot\left(\sin^{-1}\left(-\dfrac{\sqrt{3}}{2}\right)\right)$

21. $\csc(\sec^{-1} 2) + \cos(\tan^{-1}(-\sqrt{3}))$

22. $\tan(\sec^{-1} 1) + \sin(\csc^{-1}(-2))$

23. $\sin\left(\sin^{-1}\left(-\dfrac{1}{2}\right) + \cos^{-1}\left(-\dfrac{1}{2}\right)\right)$

24. $\cot\left(\sin^{-1}\left(-\dfrac{1}{2}\right) - \sec^{-1} 2\right)$

25. $\sec(\tan^{-1} 1 + \csc^{-1} 1)$

26. $\sec(\cot^{-1}\sqrt{3} + \csc^{-1}(-1))$

27. $\sec^{-1}\left(\sec\left(-\dfrac{\pi}{6}\right)\right)$ (The answer is *not* $-\pi/6$.)

28. $\cot^{-1}\left(\cot\left(-\dfrac{\pi}{4}\right)\right)$ (The answer is *not* $-\pi/4$.)

Finding Trigonometric Expressions

Evaluate the expressions in Exercises 29–40.

29. $\sec\left(\tan^{-1}\dfrac{x}{2}\right)$ **30.** $\sec(\tan^{-1} 2x)$

31. $\tan(\sec^{-1} 3y)$ **32.** $\tan\left(\sec^{-1}\dfrac{y}{5}\right)$

33. $\cos(\sin^{-1} x)$ **34.** $\tan(\cos^{-1} x)$

35. $\sin(\tan^{-1}\sqrt{x^2 - 2x})$, $x \geq 2$

36. $\sin\left(\tan^{-1}\dfrac{x}{\sqrt{x^2 + 1}}\right)$

37. $\cos\left(\sin^{-1}\dfrac{2y}{3}\right)$ **38.** $\cos\left(\sin^{-1}\dfrac{y}{5}\right)$

39. $\sin\left(\sec^{-1}\dfrac{x}{4}\right)$ **40.** $\sin\left(\sec^{-1}\dfrac{\sqrt{x^2 + 4}}{x}\right)$

Limits

Find the limits in Exercises 41–48. (If in doubt, look at the function's graph.)

41. $\lim\limits_{x \to 1^-} \sin^{-1} x$ **42.** $\lim\limits_{x \to -1^+} \cos^{-1} x$

43. $\lim\limits_{x \to \infty} \tan^{-1} x$ **44.** $\lim\limits_{x \to -\infty} \tan^{-1} x$

45. $\lim\limits_{x \to \infty} \sec^{-1} x$ **46.** $\lim\limits_{x \to -\infty} \sec^{-1} x$

47. $\lim\limits_{x \to \infty} \csc^{-1} x$ **48.** $\lim\limits_{x \to -\infty} \csc^{-1} x$

Applications and Theory

49. You are sitting in a classroom next to the wall looking at the blackboard at the front of the room. The blackboard is 12 ft long and starts 3 ft from the wall you are sitting next to. Show that your viewing angle is

$$\alpha = \cot^{-1}\dfrac{x}{15} - \cot^{-1}\dfrac{x}{3}$$

if you are x ft from the front wall.

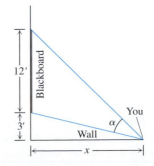

50. The region between the curve $y = \sec^{-1} x$ and the x-axis from $x = 1$ to $x = 2$ (shown here) is revolved about the y-axis to generate a solid. Find the volume of the solid.

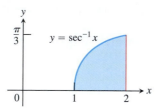

51. The slant height of the cone shown here is 3 m. How large should the indicated angle be to maximize the cone's volume?

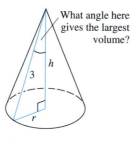

What angle here gives the largest volume?

52. Find the angle α.

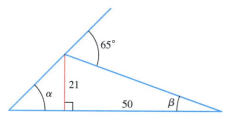

53. Here is an informal proof that $\tan^{-1} 1 + \tan^{-1} 2 + \tan^{-1} 3 = \pi$. Explain what is going on.

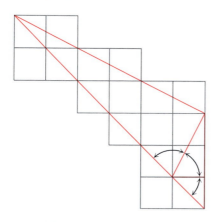

54. *Two derivations of the identity* $\sec^{-1}(-x) = \pi - \sec^{-1} x$.

a) (*Geometric*) Here is a pictorial proof that $\sec^{-1}(-x) = \pi - \sec^{-1} x$. See if you can tell what is going on.

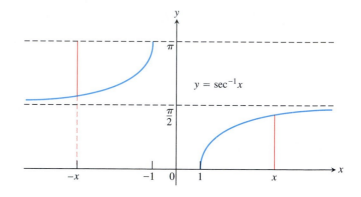

b) (*Algebraic*) Derive the identity $\sec^{-1}(-x) = \pi - \sec^{-1} x$ by combining the following two equations from the text:

$$\cos^{-1}(-x) = \pi - \cos^{-1} x \qquad \text{Eq. (6)}$$
$$\sec^{-1} x = \cos^{-1}(1/x) \qquad \text{Eq. (1)}$$

55. *The identity* $\sin^{-1} x + \cos^{-1} x = \pi/2$. Figure 6.19 establishes the identity for $0 < x < 1$. To establish it for the rest of $[-1, 1]$, verify by direct calculation that it holds for $x = 1, 0$, and -1. Then, for values of x in $(-1, 0)$, let $x = -a$, $a > 0$, and apply Eqs. (4) and (6) to the sum $\sin^{-1}(-a) + \cos^{-1}(-a)$.

56. Show that the sum $\tan^{-1} x + \tan^{-1}(1/x)$ is constant.

Which of the expressions in Exercises 57–60 are defined, and which are not? Give reasons for your answers.

57. a) $\tan^{-1} 2$ **b)** $\cos^{-1} 2$

58. a) $\csc^{-1} \dfrac{1}{2}$ **b)** $\csc^{-1} 2$

59. a) $\sec^{-1} 0$ **b)** $\sin^{-1} \sqrt{2}$

60. a) $\cot^{-1}\left(-\dfrac{1}{2}\right)$ **b)** $\cos^{-1}(-5)$

▦ Calculator Explorations

61. Find the values of

a) $\sec^{-1} 1.5$ **b)** $\csc^{-1}(-1.5)$ **c)** $\cot^{-1} 2$

62. Find the values of

a) $\sec^{-1}(-3)$ **b)** $\csc^{-1} 1.7$ **c)** $\cot^{-1}(-2)$

▦ Grapher Explorations

In Exercises 63–65, find the domain and range of each composite function. Then graph the composites on separate screens. Do the graphs make sense in each case? Give reasons for your answers. Comment on any differences you see.

63. a) $y = \tan^{-1}(\tan x)$ **b)** $y = \tan(\tan^{-1} x)$

64. a) $y = \sin^{-1}(\sin x)$ **b)** $y = \sin(\sin^{-1} x)$

65. a) $y = \cos^{-1}(\cos x)$ **b)** $y = \cos(\cos^{-1} x)$

66. Graph $y = \sec(\sec^{-1} x) = \sec(\cos^{-1}(1/x))$. Explain what you see.

67. *Newton's serpentine.* Graph Newton's serpentine, $y = 4x/(x^2 + 1)$. Then graph $y = 2\sin(2\tan^{-1} x)$ in the same graphing window. What do you see? Explain.

68. Graph the rational function $y = (2 - x^2)/x^2$. Then graph $y = \cos(2\sec^{-1} x)$ in the same graphing window. What do you see? Explain.

| **6.9** | # Derivatives of Inverse Trigonometric Functions; Integrals |

Inverse trigonometric functions provide antiderivatives for a variety of functions that arise in mathematics, engineering, and physics. In this section we find the derivatives of the inverse trigonometric functions (Table 6.5) and discuss related integrals.

EXAMPLE 1

a) $\dfrac{d}{dx}\sin^{-1}(x^2) = \dfrac{1}{\sqrt{1 - (x^2)^2}} \cdot \dfrac{d}{dx}(x^2) = \dfrac{2x}{\sqrt{1 - x^4}}$

b) $\dfrac{d}{dx}\tan^{-1}\sqrt{x + 1} = \dfrac{1}{1 + (\sqrt{x + 1})^2} \cdot \dfrac{d}{dx}(\sqrt{x + 1})$

$\qquad = \dfrac{1}{x + 2} \cdot \dfrac{1}{2\sqrt{x + 1}} = \dfrac{1}{2\sqrt{x + 1}(x + 2)}$

Table 6.5 Derivatives of the inverse trigonometric functions

1.	$\dfrac{d(\sin^{-1} u)}{dx} = \dfrac{du/dx}{\sqrt{1 - u^2}},\quad \|u\| < 1$
2.	$\dfrac{d(\cos^{-1} u)}{dx} = -\dfrac{du/dx}{\sqrt{1 - u^2}},\quad \|u\| < 1$
3.	$\dfrac{d(\tan^{-1} u)}{dx} = \dfrac{du/dx}{1 + u^2}$
4.	$\dfrac{d(\cot^{-1} u)}{dx} = -\dfrac{du/dx}{1 + u^2}$
5.	$\dfrac{d(\sec^{-1} u)}{dx} = \dfrac{du/dx}{\|u\|\sqrt{u^2 - 1}},\quad \|u\| > 1$
6.	$\dfrac{d(\csc^{-1} u)}{dx} = \dfrac{-du/dx}{\|u\|\sqrt{u^2 - 1}},\quad \|u\| > 1$

c) $\dfrac{d}{dx}\sec^{-1}(-3x) = \dfrac{1}{|-3x|\sqrt{(-3x)^2 - 1}} \cdot \dfrac{d}{dx}(-3x)$

$\qquad = \dfrac{-3}{|3x|\sqrt{9x^2 - 1}} = \dfrac{-1}{|x|\sqrt{9x^2 - 1}}. \qquad \square$

EXAMPLE 2

$$\int_0^1 \frac{e^{\tan^{-1} x}}{1 + x^2}\,dx = \int_0^{\pi/4} e^u\,du \qquad \begin{aligned} u &= \tan^{-1} x, \quad du = \dfrac{dx}{1 + x^2}, \\ u(0) &= 0, \quad u(1) = \pi/4 \end{aligned}$$

$$= e^u \Big]_0^{\pi/4} = e^{\pi/4} - 1 \qquad \square$$

We derive Formulas 1 and 5 from Table 6.5. The derivation of Formula 3 is similar. Formulas 2, 4, and 6 can be derived from Formulas 1, 3, and 5 by differentiating appropriate identities (Exercises 81–83).

The Derivative of $y \sin^{-1} u$

We know that the function $x = \sin y$ is differentiable in the interval $-\pi/2 < y < \pi/2$ and that its derivative, the cosine, is positive there. Theorem 1 in Section 6.1 therefore assures us that the inverse function $y = \sin^{-1} x$ is differentiable throughout

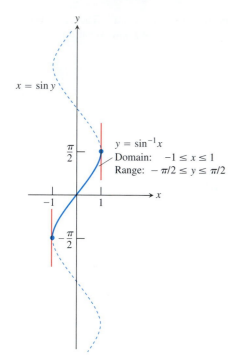

6.27 The graph of $y = \sin^{-1} x$ has vertical tangents at $x = -1$ and $x = 1$.

the interval $-1 < x < 1$. We cannot expect it to be differentiable at $x = 1$ or $x = -1$ because the tangents to the graph are vertical at these points (see Fig. 6.27).

We find the derivative of $y = \sin^{-1} x$ as follows:

$$\sin y = x$$

$y = \sin^{-1} x \quad \Leftrightarrow \quad \sin y = x$

$$\frac{d}{dx}(\sin y) = 1$$

Derivative of both sides with respect to x

$$\cos y \frac{dy}{dx} = 1$$

Chain Rule

$$\frac{dy}{dx} = \frac{1}{\cos y}$$

We can divide because $\cos y > 0$ for $-\pi/2 < y < \pi/2$.

$$= \frac{1}{\sqrt{1 - x^2}}$$

Fig. 6.28

The derivative of $y = \sin^{-1} x$ with respect to x is

$$\frac{d}{dx}(\sin^{-1} x) = \frac{1}{\sqrt{1 - x^2}}.$$

If u is a differentiable function of x with $|u| < 1$, we apply the Chain Rule

$$\frac{dy}{dx} = \frac{dy}{du}\frac{du}{dx}$$

to $y = \sin^{-1} u$ to obtain

$$\frac{d}{dx}(\sin^{-1} u) = \frac{1}{\sqrt{1 - u^2}}\frac{du}{dx}, \qquad |u| < 1.$$

The Derivative of $y = \sec^{-1} u$

We find the derivative of $y = \sec^{-1} x$, $|x| > 1$, in a similar way.

$$\sec y = x$$

$y = \sec^{-1} x \quad \Leftrightarrow \quad \sec y = x$

$$\frac{d}{dx}(\sec y) = 1$$

Derivative of both sides with respect to x

$$\sec y \tan y \frac{dy}{dx} = 1$$

Chain Rule

$$\frac{dy}{dx} = \frac{1}{\sec y \tan y}$$

Since $|x| > 1$, y lies in $(0, \pi/2) \cup (\pi/2, \pi)$ and $\sec y \tan y \neq 0$.

$$= \pm\frac{1}{x\sqrt{x^2 - 1}}$$

Fig. 6.29

6.28 In the reference right triangle above,

$$\sin y = \frac{x}{1} = x,$$

$$\cos y = \frac{\sqrt{1 - x^2}}{1} = \sqrt{1 - x^2}.$$

6.29 In both quadrants, $\sec y = x$. In the first quadrant,

$$\tan y = \sqrt{x^2 - 1}/1 = \sqrt{x^2 - 1}.$$

In the second quadrant,

$$\tan y = \sqrt{x^2 - 1}/(-1) = -\sqrt{x^2 - 1}.$$

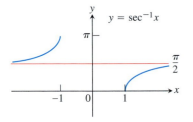

$y = \sec^{-1}x$

6.30 The slope of the curve $y = \sec^{-1}x$ is positive for both $x < -1$ and $x > 1$.

What do we do about the sign? A glance at Fig. 6.30 shows that for $|x| > 1$ the slope of the graph of $y = \sec^{-1}x$ is always positive. Therefore,

$$\frac{d}{dx}(\sec^{-1}x) = \begin{cases} \dfrac{1}{x\sqrt{x^2-1}} & \text{if } x > 1 \\[2ex] -\dfrac{1}{x\sqrt{x^2-1}} & \text{if } x < -1. \end{cases} \tag{1}$$

With absolute values, we can write Eq. (1) as a single formula:

$$\frac{d}{dx}(\sec^{-1}x) = \frac{1}{|x|\sqrt{x^2-1}}, \qquad |x| > 1.$$

If u is a differentiable function of x with $|u| > 1$, we can then apply the Chain Rule to obtain

$$\frac{d}{dx}(\sec^{-1}u) = \frac{1}{|u|\sqrt{u^2-1}}\frac{du}{dx}, \qquad |u| > 1.$$

Integration Formulas

The derivative formulas in Table 6.5 yield three useful integration formulas in Table 6.6.

Table 6.6 Integrals evaluated with inverse trigonometric functions

The following formulas hold for any constant $a \neq 0$.

1. $\displaystyle\int \frac{du}{\sqrt{a^2-u^2}} = \sin^{-1}\left(\frac{u}{a}\right) + C$ (Valid for $u^2 < a^2$)   (2)

2. $\displaystyle\int \frac{du}{a^2+u^2} = \frac{1}{a}\tan^{-1}\left(\frac{u}{a}\right) + C$ (Valid for all u)   (3)

3. $\displaystyle\int \frac{du}{u\sqrt{u^2-a^2}} = \frac{1}{a}\sec^{-1}\left|\frac{u}{a}\right| + C$ (Valid for $u^2 > a^2$)   (4)

The derivative formulas in Table 6.5 have $a = 1$, but in most integrations $a \neq 1$, and the formulas in Table 6.6 are more useful. They are readily verified by differentiating the functions on the right-hand sides.

EXAMPLE 3

a) $\displaystyle\int_{\sqrt{2}/2}^{\sqrt{3}/2} \frac{dx}{\sqrt{1-x^2}} = \sin^{-1}(x)\Big]_{\sqrt{2}/2}^{\sqrt{3}/2}$

$$= \sin^{-1}\left(\frac{\sqrt{3}}{2}\right) - \sin^{-1}\left(\frac{\sqrt{2}}{2}\right) = \frac{\pi}{3} - \frac{\pi}{4} = \frac{\pi}{12}$$

b) $\displaystyle\int_0^1 \frac{dx}{1+x^2} = \tan^{-1}(x)\Big]_0^1 = \tan^{-1}(1) - \tan^{-1}(0) = \frac{\pi}{4} - 0 = \frac{\pi}{4}$

c) $\displaystyle\int_{2/\sqrt{3}}^{\sqrt{2}} \frac{dx}{x\sqrt{x^2-1}} = \sec^{-1}(x)\Big]_{2/\sqrt{3}}^{\sqrt{2}} = \frac{\pi}{4} - \frac{\pi}{6} = \frac{\pi}{12}$

EXAMPLE 4

a) $\displaystyle\int \frac{dx}{\sqrt{9-x^2}} = \int \frac{dx}{\sqrt{(3)^2-x^2}} = \sin^{-1}\left(\frac{x}{3}\right) + C$ $\qquad$ Eq. (2) with $a = 3, \quad u = x$

b) $\displaystyle\int \frac{dx}{\sqrt{3-4x^2}} = \frac{1}{2}\int \frac{du}{\sqrt{a^2-u^2}}$ $\qquad$ $a = \sqrt{3}, \quad u = 2x,$ and $du/2 = dx$

$\qquad\qquad\qquad = \frac{1}{2}\sin^{-1}\left(\frac{u}{a}\right) + C$ $\qquad$ Eq. (2)

$\qquad\qquad\qquad = \frac{1}{2}\sin^{-1}\left(\frac{2x}{\sqrt{3}}\right) + C$

EXAMPLE 5 Evaluate $\displaystyle\int \frac{dx}{\sqrt{4x-x^2}}.$

For more about completing the square, see the end papers of this book.

Solution The expression $\sqrt{4x-x^2}$ does not match any of the formulas in Table 6.6, so we first rewrite $4x-x^2$ by completing the square:

$$4x - x^2 = -(x^2 - 4x) = -(x^2 - 4x + 4) + 4 = 4 - (x-2)^2.$$

Then we substitute $a = 2$, $u = x - 2$, and $du = dx$ to get

$$\int \frac{dx}{\sqrt{4x-x^2}} = \int \frac{dx}{\sqrt{4-(x-2)^2}}$$

$$= \int \frac{du}{\sqrt{a^2-u^2}}$$ $\qquad$ $a = 2, \quad u = x - 2,$ and $du = dx$

$$= \sin^{-1}\left(\frac{u}{a}\right) + C$$ $\qquad$ Eq. (2)

$$= \sin^{-1}\left(\frac{x-2}{2}\right) + C$$

EXAMPLE 6

a) $\displaystyle\int \frac{dx}{10+x^2} = \frac{1}{\sqrt{10}}\tan^{-1}\left(\frac{x}{\sqrt{10}}\right) + C$ $\qquad$ Eq. (3) with $a = \sqrt{10}, \quad u = x$

b) $\displaystyle\int \frac{dx}{7+3x^2} = \frac{1}{\sqrt{3}}\int \frac{du}{a^2+u^2}$ $\qquad$ $a = \sqrt{7}, u = \sqrt{3}x,$ and $du/\sqrt{3} = dx$

$\qquad\qquad\qquad = \frac{1}{\sqrt{3}} \cdot \frac{1}{a}\tan^{-1}\left(\frac{u}{a}\right) + C$ $\qquad$ Eq. (3)

$\qquad\qquad\qquad = \frac{1}{\sqrt{3}} \cdot \frac{1}{\sqrt{7}}\tan^{-1}\left(\frac{\sqrt{3}x}{\sqrt{7}}\right) + C$

$\qquad\qquad\qquad = \frac{1}{\sqrt{21}}\tan^{-1}\left(\frac{\sqrt{3}x}{\sqrt{7}}\right) + C$

EXAMPLE 7 Evaluate $\displaystyle\int \frac{dx}{4x^2+4x+2}.$

Solution We complete the square on the binomial $4x^2 + 4x$:

$$4x^2 + 4x + 2 = 4(x^2 + x) + 2 = 4\left(x^2 + x + \frac{1}{4}\right) + 2 - \frac{4}{4}$$

$$= 4\left(x + \frac{1}{2}\right)^2 + 1 = (2x + 1)^2 + 1.$$

Then we substitute $a = 1$, $u = 2x + 1$, and $du/2 = dx$ to get

$$\int \frac{dx}{4x^2 + 4x + 2} = \int \frac{dx}{(2x + 1)^2 + 1} = \frac{1}{2} \int \frac{du}{u^2 + a^2} \qquad \begin{array}{l} a = 1, \\ u = 2x + 1, \text{ and} \\ du/2 = dx \end{array}$$

$$= \frac{1}{2} \cdot \frac{1}{a} \tan^{-1}\left(\frac{u}{a}\right) \qquad \text{Eq. (3)}$$

$$= \frac{1}{2} \tan^{-1}(2x + 1) + C \qquad \begin{array}{l} a = 1, \\ u = 2x + 1 \end{array}$$

EXAMPLE 8 Evaluate $\displaystyle\int \frac{dx}{x\sqrt{4x^2 - 5}}$.

Solution

$$\int \frac{dx}{x\sqrt{4x^2 - 5}} = \int \frac{\dfrac{du}{2}}{\dfrac{u}{2}\sqrt{u^2 - a^2}} \qquad \begin{array}{l} u = 2x, \ x = u/2, \\ dx = du/2, \\ a = \sqrt{5} \end{array}$$

$$= \int \frac{du}{u\sqrt{u^2 - a^2}} \qquad \text{The 2's cancel.}$$

$$= \frac{1}{a} \sec^{-1}\left|\frac{u}{a}\right| + C \qquad \text{Eq. (4)}$$

$$= \frac{1}{\sqrt{5}} \sec^{-1}\left(\frac{2|x|}{\sqrt{5}}\right) + C \qquad a = \sqrt{5}, \ u = 2x$$

EXAMPLE 9 Evaluate $\displaystyle\int \frac{dx}{\sqrt{e^{2x} - 6}}$.

Solution

$$\int \frac{dx}{\sqrt{e^{2x} - 6}} = \int \frac{du/u}{\sqrt{u^2 - a^2}} \qquad \begin{array}{l} u = e^x, \\ du = e^x dx, \\ dx = du/e^x = du/u, \\ a = \sqrt{6} \end{array}$$

$$= \int \frac{du}{u\sqrt{u^2 - a^2}}$$

$$= \frac{1}{a} \sec^{-1}\left|\frac{u}{a}\right| + C \qquad \text{Eq. (4)}$$

$$= \frac{1}{\sqrt{6}} \sec^{-1}\left(\frac{e^x}{\sqrt{6}}\right) + C$$

Exercises 6.9

Finding Derivatives

In Exercises 1–22, find the derivative of y with respect to the appropriate variable.

1. $y = \cos^{-1}(x^2)$

2. $y = \cos^{-1}(1/x)$

3. $y = \sin^{-1}\sqrt{2}\,t$

4. $y = \sin^{-1}(1-t)$

5. $y = \sec^{-1}(2s+1)$

6. $y = \sec^{-1} 5s$

7. $y = \csc^{-1}(x^2+1), \quad x > 0$

8. $y = \csc^{-1}\dfrac{x}{2}$

9. $y = \sec^{-1}\dfrac{1}{t}, \quad 0 < t < 1$

10. $y = \sin^{-1}\dfrac{3}{t^2}$

11. $y = \cot^{-1}\sqrt{t}$

12. $y = \cot^{-1}\sqrt{t-1}$

13. $y = \ln(\tan^{-1}x)$

14. $y = \tan^{-1}(\ln x)$

15. $y = \csc^{-1}(e^t)$

16. $y = \cos^{-1}(e^{-t})$

17. $y = s\sqrt{1-s^2} + \cos^{-1}s$

18. $y = \sqrt{s^2-1} - \sec^{-1}s$

19. $y = \tan^{-1}\sqrt{x^2-1} + \csc^{-1}x, \quad x > 1$

20. $y = \cot^{-1}\dfrac{1}{x} - \tan^{-1}x$

21. $y = x\sin^{-1}x + \sqrt{1-x^2}$

22. $y = \ln(x^2+4) - x\tan^{-1}\left(\dfrac{x}{2}\right)$

Evaluating Integrals

Evaluate the integrals in Exercises 23–46.

23. $\displaystyle\int \frac{dx}{\sqrt{9-x^2}}$

24. $\displaystyle\int \frac{dx}{\sqrt{1-4x^2}}$

25. $\displaystyle\int \frac{dx}{17+x^2}$

26. $\displaystyle\int \frac{dx}{9+3x^2}$

27. $\displaystyle\int \frac{dx}{x\sqrt{25x^2-2}}$

28. $\displaystyle\int \frac{dx}{x\sqrt{5x^2-4}}$

29. $\displaystyle\int_0^1 \frac{4\,ds}{\sqrt{4-s^2}}$

30. $\displaystyle\int_0^{3\sqrt{2}/4} \frac{ds}{\sqrt{9-4s^2}}$

31. $\displaystyle\int_0^2 \frac{dt}{8+2t^2}$

32. $\displaystyle\int_{-2}^2 \frac{dt}{4+3t^2}$

33. $\displaystyle\int_{-1}^{-\sqrt{2}/2} \frac{dy}{y\sqrt{4y^2-1}}$

34. $\displaystyle\int_{-2/3}^{-\sqrt{2}/3} \frac{dy}{y\sqrt{9y^2-1}}$

35. $\displaystyle\int \frac{3\,dr}{\sqrt{1-4(r-1)^2}}$

36. $\displaystyle\int \frac{6\,dr}{\sqrt{4-(r+1)^2}}$

37. $\displaystyle\int \frac{dx}{2+(x-1)^2}$

38. $\displaystyle\int \frac{dx}{1+(3x+1)^2}$

39. $\displaystyle\int \frac{dx}{(2x-1)\sqrt{(2x-1)^2-4}}$

40. $\displaystyle\int \frac{dx}{(x+3)\sqrt{(x+3)^2-25}}$

41. $\displaystyle\int_{-\pi/2}^{\pi/2} \frac{2\cos\theta\,d\theta}{1+(\sin\theta)^2}$

42. $\displaystyle\int_{\pi/6}^{\pi/4} \frac{\csc^2 x\,dx}{1+(\cot x)^2}$

43. $\displaystyle\int_0^{\ln\sqrt{3}} \frac{e^x\,dx}{1+e^{2x}}$

44. $\displaystyle\int_1^{e^{\pi/4}} \frac{4\,dt}{t(1+\ln^2 t)}$

45. $\displaystyle\int \frac{y\,dy}{\sqrt{1-y^4}}$

46. $\displaystyle\int \frac{\sec^2 y\,dy}{\sqrt{1-\tan^2 y}}$

Evaluate the integrals in Exercises 47–56.

47. $\displaystyle\int \frac{dx}{\sqrt{-x^2+4x-3}}$

48. $\displaystyle\int \frac{dx}{\sqrt{2x-x^2}}$

49. $\displaystyle\int_{-1}^0 \frac{6\,dt}{\sqrt{3-2t-t^2}}$

50. $\displaystyle\int_{1/2}^1 \frac{6\,dt}{\sqrt{3+4t-4t^2}}$

51. $\displaystyle\int \frac{dy}{y^2-2y+5}$

52. $\displaystyle\int \frac{dy}{y^2+6y+10}$

53. $\displaystyle\int_1^2 \frac{8\,dx}{x^2-2x+2}$

54. $\displaystyle\int_2^4 \frac{2\,dx}{x^2-6x+10}$

55. $\displaystyle\int \frac{dx}{(x+1)\sqrt{x^2+2x}}$

56. $\displaystyle\int \frac{dx}{(x-2)\sqrt{x^2-4x+3}}$

Evaluate the integrals in Exercises 57–64.

57. $\displaystyle\int \frac{e^{\sin^{-1}x}\,dx}{\sqrt{1-x^2}}$

58. $\displaystyle\int \frac{e^{\cos^{-1}x}\,dx}{\sqrt{1-x^2}}$

59. $\displaystyle\int \frac{(\sin^{-1}x)^2\,dx}{\sqrt{1-x^2}}$

60. $\displaystyle\int \frac{\sqrt{\tan^{-1}x}\,dx}{1+x^2}$

61. $\displaystyle\int \frac{dy}{(\tan^{-1}y)(1+y^2)}$

62. $\displaystyle\int \frac{dy}{(\sin^{-1}y)\sqrt{1-y^2}}$

63. $\displaystyle\int_{\sqrt{2}}^2 \frac{\sec^2(\sec^{-1}x)\,dx}{x\sqrt{x^2-1}}$

64. $\displaystyle\int_{2/\sqrt{3}}^2 \frac{\cos(\sec^{-1}x)\,dx}{x\sqrt{x^2-1}}$

Limits

Find the limits in Exercises 65–68.

65. $\displaystyle\lim_{x\to 0} \frac{\sin^{-1}5x}{x}$

66. $\displaystyle\lim_{x\to 1^+} \frac{\sqrt{x^2-1}}{\sec^{-1}x}$

67. $\displaystyle\lim_{x\to\infty} x\tan^{-1}\frac{2}{x}$

68. $\displaystyle\lim_{x\to 0} \frac{2\tan^{-1}3x^2}{7x^2}$

Integration Formulas

Verify the integration formulas in Exercises 69–72.

69. $\displaystyle\int \frac{\tan^{-1}x}{x^2}\,dx = \ln x - \frac{1}{2}\ln(1+x^2) - \frac{\tan^{-1}x}{x} + C$

70. $\int x^3 \cos^{-1} 5x \, dx = \dfrac{x^4}{4} \cos^{-1} 5x + \dfrac{5}{4} \int \dfrac{x^4 \, dx}{\sqrt{1 - 25 x^2}}$

71. $\int (\sin^{-1} x)^2 \, dx = x(\sin^{-1} x)^2 - 2x + 2\sqrt{1 - x^2} \sin^{-1} x + C$

72. $\int \ln (a^2 + x^2) \, dx = x \ln (a^2 + x^2) - 2x + 2a \tan^{-1} \dfrac{x}{a} + C$

Initial Value Problems

Solve the initial value problems in Exercises 73–76.

73. $\dfrac{dy}{dx} = \dfrac{1}{\sqrt{1 - x^2}}, \quad y(0) = 0$

74. $\dfrac{dy}{dx} = \dfrac{1}{x^2 + 1} - 1, \quad y(0) = 1$

75. $\dfrac{dy}{dx} = \dfrac{1}{x\sqrt{x^2 - 1}}, \quad x > 1; \quad y(2) = \pi$

76. $\dfrac{dy}{dx} = \dfrac{1}{1 + x^2} - \dfrac{2}{\sqrt{1 - x^2}}, \quad y(0) = 2$

Theory and Examples

77. (*Continuation of Exercise 49, Section 6.8.*) You want to position your chair along the wall to maximize your viewing angle α. How far from the front of the room should you sit?

78. What value of x maximizes the angle θ shown here? How large is θ at that point? Begin by showing that $\theta = \pi - \cot^{-1} x - \cot^{-1}(2 - x)$.

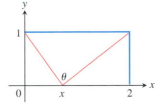

79. Can the integrations in (a) and (b) both be correct? Explain.

a) $\int \dfrac{dx}{\sqrt{1 - x^2}} = \sin^{-1} x + C$

b) $\int \dfrac{dx}{\sqrt{1 - x^2}} = - \int -\dfrac{dx}{\sqrt{1 - x^2}} = -\cos^{-1} x + C$

80. Can the integrations in (a) and (b) both be correct? Explain.

a) $\int \dfrac{dx}{\sqrt{1 - x^2}} = - \int -\dfrac{dx}{\sqrt{1 - x^2}} = -\cos^{-1} x + C$

b) $\int \dfrac{dx}{\sqrt{1 - x^2}} = \int \dfrac{-du}{\sqrt{1 - (-u)^2}} \qquad \begin{aligned} x &= -u, \\ dx &= -du \end{aligned}$

$\qquad = \int \dfrac{-du}{\sqrt{1 - u^2}}$

$\qquad = \cos^{-1} u + C$

$\qquad = \cos^{-1}(-x) + C \qquad u = -x$

81. Use the identity

$$\cos^{-1} u = \dfrac{\pi}{2} - \sin^{-1} u$$

to derive the formula for the derivative of $\cos^{-1} u$ in Table 6.5 from the formula for the derivative of $\sin^{-1} u$.

82. Use the identity

$$\cot^{-1} u = \dfrac{\pi}{2} - \tan^{-1} u$$

to derive the formula for the derivative of $\cot^{-1} u$ in Table 6.5 from the formula for the derivative of $\tan^{-1} u$.

83. Use the identity

$$\csc^{-1} u = \dfrac{\pi}{2} - \sec^{-1} u$$

to derive the formula for the derivative of $\csc^{-1} u$ in Table 6.5 from the formula for the derivative of $\sec^{-1} u$.

84. Derive the formula

$$\dfrac{dy}{dx} = \dfrac{1}{1 + x^2}$$

for the derivative of $y = \tan^{-1} x$ by differentiating both sides of the equivalent equation $\tan y = x$.

85. Use the Derivative Rule in Section 6.1, Theorem 1, to derive

$$\dfrac{d}{dx} \sin^{-1} x = \dfrac{1}{\sqrt{1 - x^2}}, \quad -1 < x < 1.$$

86. Use the Derivative Rule in Section 6.1, Theorem 1, to derive

$$\dfrac{d}{dx} \tan^{-1} x = \dfrac{1}{1 + x^2}.$$

87. What is special about the functions

$$f(x) = \sin^{-1} \dfrac{x - 1}{x + 1}, \quad x \geq 0, \quad \text{and} \quad g(x) = 2 \tan^{-1} \sqrt{x}?$$

Explain.

88. What is special about the functions

$$f(x) = \sin^{-1} \dfrac{1}{\sqrt{x^2 + 1}} \quad \text{and} \quad g(x) = \tan^{-1} \dfrac{1}{x}?$$

Explain.

89. Find the volume of the solid of revolution shown here.

90. Find the length of the curve $y = \sqrt{1 - x^2}$, $-1/2 \le x \le 1/2$.

Volumes by Slicing

Find the volumes of the solids in Exercises 91 and 92.

91. The solid lies between planes perpendicular to the x-axis at $x = -1$ and $x = 1$. The cross sections perpendicular to the x-axis are

    a) circles whose diameters stretch from the curve $y = -1/\sqrt{1 + x^2}$ to the curve $y = 1/\sqrt{1 + x^2}$;

    b) vertical squares whose base edges run from the curve $y = -1/\sqrt{1 + x^2}$ to the curve $y = 1/\sqrt{1 + x^2}$.

92. The solid lies between planes perpendicular to the x-axis at $x = -\sqrt{2}/2$ and $x = \sqrt{2}/2$. The cross sections are

    a) circles whose diameters stretch from the x-axis to the curve $y = 2/\sqrt[4]{1 - x^2}$.

    b) squares whose diagonals stretch from the x-axis to the curve $y = 2/\sqrt[4]{1 - x^2}$.

Calculator and Grapher Explorations

93. CALCULATOR Use numerical integration to estimate the value of

$$\sin^{-1} 0.6 = \int_0^{0.6} \frac{dx}{\sqrt{1 - x^2}}.$$

For reference, $\sin^{-1} 0.6 = 0.64350$ to 5 places.

94. CALCULATOR Use numerical integration to estimate the value of

$$\pi = 4 \int_0^1 \frac{1}{1 + x^2} dx.$$

95. GRAPHER Graph $f(x) = \sin^{-1} x$ together with its first two derivatives. Comment on the behavior of f and the shape of its graph in relation to the signs and values of f' and f''.

96. GRAPHER Graph $f(x) = \tan^{-1} x$ together with its first two derivatives. Comment on the behavior of f and the shape of its graph in relation to the signs and values of f' and f''.

6.10

Hyperbolic Functions

Every function f that is defined on an interval centered at the origin can be written in a unique way as the sum of one even function and one odd function. The decomposition is

$$f(x) = \underbrace{\frac{f(x) + f(-x)}{2}}_{\text{even part}} + \underbrace{\frac{f(x) - f(-x)}{2}}_{\text{odd part}}.$$

If we write e^x this way, we get

$$e^x = \underbrace{\frac{e^x + e^{-x}}{2}}_{\text{even part}} + \underbrace{\frac{e^x - e^{-x}}{2}}_{\text{odd part}}.$$

The notation cosh x is often read "kosh x," rhyming with either "gosh x" or "gauche x," and sinh x is pronounced as if spelled "cinch x" or "shine x."

The even and odd parts of e^x, called the hyperbolic cosine and hyperbolic sine of x, respectively, are useful in their own right. They describe the motions of waves in elastic solids, the shapes of hanging electric power lines, and the temperature distributions in metal cooling fins. The center line of the Gateway Arch to the West in St. Louis is a weighted hyperbolic cosine curve.

Definitions and Identities

The hyperbolic cosine and hyperbolic sine functions are defined by the first two equations in Table 6.7. The table also lists the definitions of the hyperbolic tangent, cotangent, secant, and cosecant. As we will see, the hyperbolic functions bear a number of similarities to the trigonometric functions after which they are named. (See Exercise 86 as well.)

Table 6.7 The six basic hyperbolic functions (See Fig. 6.31 for graphs.)

Hyperbolic cosine of x:	$\cosh x = \dfrac{e^x + e^{-x}}{2}$
Hyperbolic sine of x:	$\sinh x = \dfrac{e^x - e^{-x}}{2}$
Hyperbolic tangent:	$\tanh x = \dfrac{\sinh x}{\cosh x} = \dfrac{e^x - e^{-x}}{e^x + e^{-x}}$
Hyperbolic cotangent:	$\coth x = \dfrac{\cosh x}{\sinh x} = \dfrac{e^x + e^{-x}}{e^x - e^{-x}}$
Hyperbolic secant:	$\operatorname{sech} x = \dfrac{1}{\cosh x} = \dfrac{2}{e^x + e^{-x}}$
Hyperbolic cosecant:	$\operatorname{csch} x = \dfrac{1}{\sinh x} = \dfrac{2}{e^x - e^{-x}}$

Table 6.8 Identities for hyperbolic functions

$$\sinh 2x = 2 \sinh x \cosh x$$
$$\cosh 2x = \cosh^2 x + \sinh^2 x$$
$$\cosh^2 x = \frac{\cosh 2x + 1}{2}$$
$$\sinh^2 x = \frac{\cosh 2x - 1}{2}$$
$$\cosh^2 x - \sinh^2 x = 1$$
$$\tanh^2 x = 1 - \operatorname{sech}^2 x$$
$$\coth^2 x = 1 + \operatorname{csch}^2 x$$

Table 6.9 Derivatives of hyperbolic functions

$$\frac{d}{dx}(\sinh u) = \cosh u \, \frac{du}{dx}$$
$$\frac{d}{dx}(\cosh u) = \sinh u \, \frac{du}{dx}$$
$$\frac{d}{dx}(\tanh u) = \operatorname{sech}^2 u \, \frac{du}{dx}$$
$$\frac{d}{dx}(\coth u) = -\operatorname{csch}^2 u \, \frac{du}{dx}$$
$$\frac{d}{dx}(\operatorname{sech} u) = -\operatorname{sech} u \tanh u \, \frac{du}{dx}$$
$$\frac{d}{dx}(\operatorname{csch} u) = -\operatorname{csch} u \coth u \, \frac{du}{dx}$$

Identities

Hyperbolic functions satisfy the identities in Table 6.8. Except for differences in sign, these are identities we already know for trigonometric functions.

Derivatives and Integrals

The six hyperbolic functions, being rational combinations of the differentiable functions e^x and e^{-x}, have derivatives at every point at which they are defined (Table 6.9). Again, there are similarities with trigonometric functions. The derivative formulas in Table 6.9 lead to the integral formulas in Table 6.10.

EXAMPLE 1

$$\frac{d}{dt}\left(\tanh\sqrt{1+t^2}\right) = \operatorname{sech}^2\sqrt{1+t^2} \cdot \frac{d}{dt}\left(\sqrt{1+t^2}\right)$$
$$= \frac{t}{\sqrt{1+t^2}}\operatorname{sech}^2\sqrt{1+t^2}$$

Table 6.10 Integral formulas for hyperbolic functions

$$\int \sinh u \, du = \cosh u + C$$
$$\int \cosh u \, du = \sinh u + C$$
$$\int \operatorname{sech}^2 u \, du = \tanh u + C$$
$$\int \operatorname{csch}^2 u \, du = -\coth u + C$$
$$\int \operatorname{sech} u \tanh u \, du = -\operatorname{sech} u + C$$
$$\int \operatorname{csch} u \coth u \, du = -\operatorname{csch} u + C$$

EXAMPLE 2

$$\int \coth 5x \, dx = \int \frac{\cosh 5x}{\sinh 5x} dx = \frac{1}{5}\int \frac{du}{u} \qquad \begin{matrix} u = \sinh 5x, \\ du = 5\cosh 5x \, dx \end{matrix}$$
$$= \frac{1}{5}\ln|u| + C = \frac{1}{5}\ln|\sinh 5x| + C$$

EXAMPLE 3

$$\int_0^1 \sinh^2 x \, dx = \int_0^1 \frac{\cosh 2x - 1}{2} dx \qquad \text{Table 6.8}$$
$$= \frac{1}{2}\int_0^1 (\cosh 2x - 1)dx = \frac{1}{2}\left[\frac{\sinh 2x}{2} - x\right]_0^1$$
$$= \frac{\sinh 2}{4} - \frac{1}{2} \approx 0.40672$$

Evaluating hyperbolic functions

Like many standard functions, hyperbolic functions and their inverses are easily evaluated with calculators, which have special keys or keystroke sequences for that purpose.

EXAMPLE 4

$$\int_0^{\ln 2} 4e^x \sinh x \, dx = \int_0^{\ln 2} 4e^x \frac{e^x - e^{-x}}{2} dx = \int_0^{\ln 2} (2e^{2x} - 2) \, dx$$

$$= [e^{2x} - 2x]_0^{\ln 2} = (e^{2 \ln 2} - 2 \ln 2) - (1 - 0)$$

$$= 4 - 2 \ln 2 - 1$$

$$\approx 1.6137 \qquad \square$$

The Inverse Hyperbolic Functions

We use the inverses of the six basic hyperbolic functions in integration. Since $d(\sinh x)/dx = \cosh x > 0$, the hyperbolic sine is an increasing function of x. We denote its inverse by

$$y = \sinh^{-1} x.$$

For every value of x in the interval $-\infty < x < \infty$, the value of $y = \sinh^{-1} x$ is the number whose hyperbolic sine is x. The graphs of $y = \sinh x$ and $y = \sinh^{-1} x$ are shown in Fig. 6.32(a).

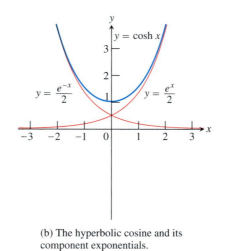

(a) The hyperbolic sine and its component exponentials.

(b) The hyperbolic cosine and its component exponentials.

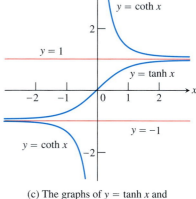

(c) The graphs of $y = \tanh x$ and $y = \coth x = 1/\tanh x$.

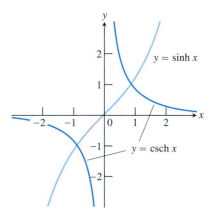

(d) The graphs of $y = \cosh x$ and $y = \text{sech } x = 1/\cosh x$.

(e) The graphs of $y = \sinh x$ and $y = \text{csch } x = 1/\sinh x$.

6.31 The graphs of the six hyperbolic functions.

6.32 The graphs of the inverse hyperbolic sine, cosine, and secant of x. Notice the symmetries about the line $y = x$.

The function $y = \cosh x$ is not one-to-one, as we can see from the graph in Fig. 6.31. But the restricted function $y = \cosh x$, $x \geq 0$, is one-to-one and therefore has an inverse, denoted by

$$y = \cosh^{-1}x.$$

For every value of $x \geq 1$, $y = \cosh^{-1}x$ is the number in the interval $0 \leq y < \infty$ whose hyperbolic cosine is x. The graphs of $y = \cosh x$, $x \geq 0$, and $y = \cosh^{-1} x$ are shown in Fig. 6.32(b).

Like $y = \cosh x$, the function $y = \operatorname{sech} x = 1/\cosh x$ fails to be one-to-one, but its restriction to nonnegative values of x does have an inverse, denoted by

$$y = \operatorname{sech}^{-1}x.$$

For every value of x in the interval $(0, 1]$, $y = \operatorname{sech}^{-1}x$ is the nonnegative number whose hyperbolic secant is x. The graphs of $y = \operatorname{sech} x$, $x \geq 0$, and $y = \operatorname{sech}^{-1}x$ are shown in Fig. 6.32(c).

The hyperbolic tangent, cotangent, and cosecant are one-to-one on their domains and therefore have inverses, denoted by

$$y = \tanh^{-1}x, \qquad y = \coth^{-1}x, \qquad y = \operatorname{csch}^{-1}x.$$

6.33 The graphs of the inverse hyperbolic tangent, cotangent, and cosecant of x.

These functions are graphed in Fig. 6.33.

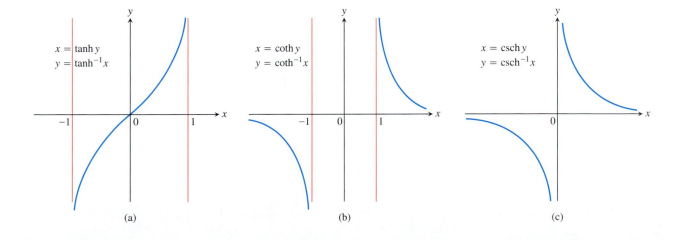

Table 6.11 Identities for inverse hyperbolic functions

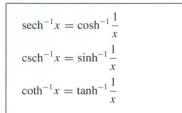

$$\mathrm{sech}^{-1}x = \cosh^{-1}\frac{1}{x}$$

$$\mathrm{csch}^{-1}x = \sinh^{-1}\frac{1}{x}$$

$$\coth^{-1}x = \tanh^{-1}\frac{1}{x}$$

Table 6.12 Derivatives of inverse hyperbolic functions

$$\frac{d(\sinh^{-1}u)}{dx} = \frac{1}{\sqrt{1+u^2}}\frac{du}{dx}$$

$$\frac{d(\cosh^{-1}u)}{dx} = \frac{1}{\sqrt{u^2-1}}\frac{du}{dx}, \quad u > 1$$

$$\frac{d(\tanh^{-1}u)}{dx} = \frac{1}{1-u^2}\frac{du}{dx}, \quad |u| < 1$$

$$\frac{d(\coth^{-1}u)}{dx} = \frac{1}{1-u^2}\frac{du}{dx}, \quad |u| > 1$$

$$\frac{d(\mathrm{sech}^{-1}u)}{dx} = \frac{-du/dx}{u\sqrt{1-u^2}}, \quad 0 < u < 1$$

$$\frac{d(\mathrm{csch}^{-1}u)}{dx} = \frac{-du/dx}{|u|\sqrt{1+u^2}}, \quad u \neq 0$$

Useful Identities

We use the identities in Table 6.11 to calculate the values of $\mathrm{sech}^{-1}x$, $\mathrm{csch}^{-1}x$, and $\coth^{-1}x$ on calculators that give only $\cosh^{-1}x$, $\sinh^{-1}x$, and $\tanh^{-1}x$.

Derivatives and Integrals

The chief use of inverse hyperbolic functions lies in integrations that reverse the derivative formulas in Table 6.12.

The restrictions $|u| < 1$ and $|u| > 1$ on the derivative formulas for $\tanh^{-1}u$ and $\coth^{-1}u$ come from the natural restrictions on the values of these functions. (See Figs. 6.33a and b.) The distinction between $|u| < 1$ and $|u| > 1$ becomes important when we convert the derivative formulas into integral formulas. If $|u| < 1$, the integral of $1/(1 - u^2)$ is $\tanh^{-1}u + C$. If $|u| > 1$, the integral is $\coth^{-1}u + C$.

EXAMPLE 5   Show that if u is a differentiable function of x whose values are greater than 1, then

$$\frac{d}{dx}(\cosh^{-1}u) = \frac{1}{\sqrt{u^2-1}}\frac{du}{dx}.$$

Solution   First we find the derivative of $y = \cosh^{-1}x$ for $x > 1$:

$$y = \cosh^{-1}x$$

$$x = \cosh y \qquad \text{Equivalent equation}$$

$$1 = \sinh y \frac{dy}{dx} \qquad \text{Differentiation with respect to } x$$

$$\frac{dy}{dx} = \frac{1}{\sinh y} = \frac{1}{\sqrt{\cosh^2 y - 1}} \qquad \text{Since } x > 1, y > 0 \text{ and } \sinh y > 0$$

$$= \frac{1}{\sqrt{x^2-1}} \qquad \cosh y = x$$

In short, $\dfrac{d}{dx}(\cosh^{-1}x) = \dfrac{1}{\sqrt{x^2-1}}$. The Chain Rule gives the final result:

$$\frac{d}{dx}(\cosh^{-1}u) = \frac{1}{\sqrt{u^2-1}}\frac{du}{dx}. \qquad \square$$

With appropriate substitutions, the derivative formulas in Table 6.12 lead to the integration formulas in Table 6.13.

EXAMPLE 6   Evaluate $\displaystyle\int_0^1 \frac{2\,dx}{\sqrt{3+4x^2}}$.

Solution   The indefinite integral is

$$\int \frac{2\,dx}{\sqrt{3+4x^2}} = \int \frac{du}{\sqrt{a^2+u^2}} \qquad u = 2x, \quad du = 2\,dx, \quad a = \sqrt{3}$$

$$= \sinh^{-1}\left(\frac{u}{a}\right) + C \qquad \text{Formula from Table 6.13}$$

$$= \sinh^{-1}\left(\frac{2x}{\sqrt{3}}\right) + C.$$

Table 6.13 Integrals leading to inverse hyperbolic functions

1. $\displaystyle\int \frac{du}{\sqrt{a^2+u^2}} = \sinh^{-1}\left(\frac{u}{a}\right) + C, \quad a > 0$

2. $\displaystyle\int \frac{du}{\sqrt{u^2-a^2}} = \cosh^{-1}\left(\frac{u}{a}\right) + C, \quad u > a > 0$

3. $\displaystyle\int \frac{du}{a^2-u^2} = \begin{cases} \dfrac{1}{a}\tanh^{-1}\left(\dfrac{u}{a}\right) + C & \text{if } u^2 < a^2 \\[2mm] \dfrac{1}{a}\coth^{-1}\left(\dfrac{u}{a}\right) + C & \text{if } u^2 > a^2 \end{cases}$

4. $\displaystyle\int \frac{du}{u\sqrt{a^2-u^2}} = -\frac{1}{a}\operatorname{sech}^{-1}\left(\frac{u}{a}\right) + C, \quad 0 < u < a$

5. $\displaystyle\int \frac{du}{u\sqrt{a^2+u^2}} = -\frac{1}{a}\operatorname{csch}^{-1}\left|\frac{u}{a}\right| + C, \quad u \neq 0$

Therefore,

$$\int_0^1 \frac{2\,dx}{\sqrt{3+4x^2}} = \sinh^{-1}\left(\frac{2x}{\sqrt{3}}\right)\Bigg]_0^1 = \sinh^{-1}\left(\frac{2}{\sqrt{3}}\right) - \sinh^{-1}(0)$$

$$= \sinh^{-1}\left(\frac{2}{\sqrt{3}}\right) - 0 \approx 0.98665.$$

Exercises 6.10

Hyperbolic Function Values and Identities

Each of Exercises 1–4 gives a value of sinh x or cosh x. Use the definitions and the identity $\cosh^2 x - \sinh^2 x = 1$ to find the values of the remaining five hyperbolic functions.

1. $\sinh x = -\dfrac{3}{4}$

2. $\sinh x = \dfrac{4}{3}$

3. $\cosh x = \dfrac{17}{15}, \quad x > 0$

4. $\cosh x = \dfrac{13}{5}, \quad x > 0$

Rewrite the expressions in Exercises 5–10 in terms of exponentials and simplify the results as much as you can.

5. $2\cosh(\ln x)$

6. $\sinh(2\ln x)$

7. $\cosh 5x + \sinh 5x$

8. $\cosh 3x - \sinh 3x$

9. $(\sinh x + \cosh x)^4$

10. $\ln(\cosh x + \sinh x) + \ln(\cosh x - \sinh x)$

11. Use the identities

$$\sinh(x+y) = \sinh x \cosh y + \cosh x \sinh y$$
$$\cosh(x+y) = \cosh x \cosh y + \sinh x \sinh y$$

to show that

a) $\sinh 2x = 2\sinh x \cosh x$;

b) $\cosh 2x = \cosh^2 x + \sinh^2 x$.

12. Use the definitions of cosh x and sinh x to show that

$$\cosh^2 x - \sinh^2 x = 1.$$

Derivatives

In Exercises 13–24, find the derivative of y with respect to the appropriate variable.

13. $y = 6\sinh\dfrac{x}{3}$

14. $y = \dfrac{1}{2}\sinh(2x+1)$

15. $y = 2\sqrt{t}\tanh\sqrt{t}$

16. $y = t^2\tanh\dfrac{1}{t}$

17. $y = \ln(\sinh z)$

18. $y = \ln(\cosh z)$

19. $y = \operatorname{sech}\theta(1 - \ln\operatorname{sech}\theta)$

20. $y = \operatorname{csch}\theta(1 - \ln\operatorname{csch}\theta)$

21. $y = \ln\cosh v - \dfrac{1}{2}\tanh^2 v$

22. $y = \ln\sinh v - \dfrac{1}{2}\coth^2 v$

23. $y = (x^2 + 1) \text{ sech } (\ln x)$

(*Hint:* Before differentiating, express in terms of exponentials and simplify.)

24. $y = (4x^2 - 1) \text{ csch } (\ln 2x)$

In Exercises 25–36, find the derivative of y with respect to the appropriate variable.

25. $y = \sinh^{-1} \sqrt{x}$ **26.** $y = \cosh^{-1} 2\sqrt{x+1}$

27. $y = (1 - \theta) \tanh^{-1} \theta$ **28.** $y = (\theta^2 + 2\theta) \tanh^{-1}(\theta + 1)$

29. $y = (1 - t) \coth^{-1} \sqrt{t}$ **30.** $y = (1 - t^2) \coth^{-1} t$

31. $y = \cos^{-1} x - x \text{ sech}^{-1} x$ **32.** $y = \ln x + \sqrt{1 - x^2} \text{ sech}^{-1} x$

33. $y = \text{csch}^{-1} \left(\dfrac{1}{2}\right)^{\theta}$ **34.** $y = \text{csch}^{-1} 2^{\theta}$

35. $y = \sinh^{-1}(\tan x)$

36. $y = \cosh^{-1}(\sec x), \quad 0 < x < \pi/2$

Integration Formulas

Verify the integration formulas in Exercises 37–40.

37. a) $\displaystyle \int \text{sech } x \, dx = \tan^{-1}(\sinh x) + C$

b) $\displaystyle \int \text{sech } x \, dx = \sin^{-1}(\tanh x) + C$

38. $\displaystyle \int x \text{ sech}^{-1} x \, dx = \dfrac{x^2}{2} \text{ sech}^{-1} x - \dfrac{1}{2}\sqrt{1 - x^2} + C$

39. $\displaystyle \int x \coth^{-1} x \, dx = \dfrac{x^2 - 1}{2} \coth^{-1} x + \dfrac{x}{2} + C$

40. $\displaystyle \int \tanh^{-1} x \, dx = x \tanh^{-1} x + \dfrac{1}{2} \ln (1 - x^2) + C$

Indefinite Integrals

Evaluate the integrals in Exercises 41–50.

41. $\displaystyle \int \sinh 2x \, dx$ **42.** $\displaystyle \int \sinh \dfrac{x}{5} \, dx$

43. $\displaystyle \int 6 \cosh \left(\dfrac{x}{2} - \ln 3\right) dx$ **44.** $\displaystyle \int 4 \cosh (3x - \ln 2) \, dx$

45. $\displaystyle \int \tanh \dfrac{x}{7} \, dx$ **46.** $\displaystyle \int \coth \dfrac{\theta}{\sqrt{3}} \, d\theta$

47. $\displaystyle \int \text{sech}^2 \left(x - \dfrac{1}{2}\right) dx$ **48.** $\displaystyle \int \text{csch}^2 (5 - x) \, dx$

49. $\displaystyle \int \dfrac{\text{sech } \sqrt{t} \tanh \sqrt{t} \, dt}{\sqrt{t}}$ **50.** $\displaystyle \int \dfrac{\text{csch } (\ln t) \coth (\ln t) \, dt}{t}$

Definite Integrals

Evaluate the integrals in Exercises 51–60.

51. $\displaystyle \int_{\ln 2}^{\ln 4} \coth x \, dx$ **52.** $\displaystyle \int_{0}^{\ln 2} \tanh 2x \, dx$

53. $\displaystyle \int_{-\ln 4}^{-\ln 2} 2e^{\theta} \cosh \theta \, d\theta$ **54.** $\displaystyle \int_{0}^{\ln 2} 4e^{-\theta} \sinh \theta \, d\theta$

55. $\displaystyle \int_{-\pi/4}^{\pi/4} \cosh (\tan \theta) \sec^2 \theta \, d\theta$

56. $\displaystyle \int_{0}^{\pi/2} 2 \sinh (\sin \theta) \cos \theta \, d\theta$

57. $\displaystyle \int_{1}^{2} \dfrac{\cosh (\ln t)}{t} \, dt$ **58.** $\displaystyle \int_{1}^{4} \dfrac{8 \cosh \sqrt{x}}{\sqrt{x}} \, dx$

59. $\displaystyle \int_{-\ln 2}^{0} \cosh^2 \left(\dfrac{x}{2}\right) dx$ **60.** $\displaystyle \int_{0}^{\ln 10} 4 \sinh^2 \left(\dfrac{x}{2}\right) dx$

Evaluating Inverse Hyperbolic Functions and Related Integrals

When hyperbolic function keys are not available on a calculator, it is still possible to evaluate the inverse hyperbolic functions by expressing them as logarithms as shown in the table below.

$$\sinh^{-1} x = \ln \left(x + \sqrt{x^2 + 1}\right), \quad -\infty < x < \infty$$

$$\cosh^{-1} x = \ln \left(x + \sqrt{x^2 - 1}\right), \quad x \geq 1$$

$$\tanh^{-1} x = \dfrac{1}{2} \ln \dfrac{1 + x}{1 - x}, \quad |x| < 1$$

$$\text{sech}^{-1} x = \ln \left(\dfrac{1 + \sqrt{1 - x^2}}{x}\right), \quad 0 < x \leq 1$$

$$\text{csch}^{-1} x = \ln \left(\dfrac{1}{x} + \dfrac{\sqrt{1 + x^2}}{|x|}\right), \quad x \neq 0$$

$$\coth^{-1} x = \dfrac{1}{2} \ln \dfrac{x + 1}{x - 1}, \quad |x| > 1$$

Use the formulas in the table here to express the numbers in Exercises 61–66 in terms of natural logarithms.

61. $\sinh^{-1}(-5/12)$ **62.** $\cosh^{-1}(5/3)$

63. $\tanh^{-1}(-1/2)$ **64.** $\coth^{-1}(5/4)$

65. $\text{sech}^{-1}(3/5)$ **66.** $\text{csch}^{-1}\left(-1/\sqrt{3}\right)$

Evaluate the integrals in Exercises 67–74 in terms of (a) inverse hyperbolic functions, (b) natural logarithms.

67. $\displaystyle \int_{0}^{2\sqrt{3}} \dfrac{dx}{\sqrt{4 + x^2}}$ **68.** $\displaystyle \int_{0}^{1/3} \dfrac{6 \, dx}{\sqrt{1 + 9x^2}}$

69. $\displaystyle \int_{5/4}^{2} \dfrac{dx}{1 - x^2}$ **70.** $\displaystyle \int_{0}^{1/2} \dfrac{dx}{1 - x^2}$

71. $\displaystyle \int_{1/5}^{3/13} \dfrac{dx}{\sqrt{1 - 16x^2}}$ **72.** $\displaystyle \int_{1}^{2} \dfrac{dx}{x\sqrt{4 + x^2}}$

73. $\displaystyle\int_0^{\pi} \frac{\cos x \, dx}{\sqrt{1 + \sin^2 x}}$

74. $\displaystyle\int_1^{e} \frac{dx}{x\sqrt{1 + (\ln x)^2}}$

Applications and Theory

75. a) Show that if a function f is defined on an interval symmetric about the origin (so that f is defined at $-x$ whenever it is defined at x), then

$$f(x) = \frac{f(x) + f(-x)}{2} + \frac{f(x) - f(-x)}{2}. \qquad (1)$$

Then show that $(f(x) + f(-x))/2$ is even and that $(f(x) - f(-x))/2$ is odd.

   b) Equation (1) simplifies considerably if f itself is (i) even or (ii) odd. What are the new equations? Give reasons for your answers.

76. Derive the formula $\sinh^{-1} x = \ln\left(x + \sqrt{x^2 + 1}\right), -\infty < x < \infty$. Explain in your derivation why the plus sign is used with the square root instead of the minus sign.

77. *Skydiving.* If a body of mass m falling from rest under the action of gravity encounters an air resistance proportional to the square of the velocity, then the body's velocity t seconds into the fall satisfies the differential equation

$$m\frac{dv}{dt} = mg - kv^2,$$

where k is a constant that depends on the body's aerodynamic properties and the density of the air. (We assume that the fall is short enough so that the variation in the air's density will not affect the outcome.)

   a) Show that

$$v = \sqrt{\frac{mg}{k}} \tanh\left(\sqrt{\frac{gk}{m}}\, t\right)$$

satisfies the differential equation and the initial condition that $v = 0$ when $t = 0$.

   b) Find the body's *limiting velocity,* $\lim_{t\to\infty} v$.

   c) **CALCULATOR** For a 160-lb skydiver ($mg = 160$), with time in seconds and distance in feet, a typical value for k is 0.005. What is the diver's limiting velocity?

78. *Accelerations whose magnitudes are proportional to displacement.* Suppose that the position of a body moving along a coordinate line at time t is

   a) $s = a\cos kt + b\sin kt$,

   b) $s = a\cosh kt + b\sinh kt$.

Show in both cases that the acceleration d^2s/dt^2 is proportional to s but that in the first case it is directed toward the origin while in the second case it is directed away from the origin.

79. *Tractor trailers and the tractrix.* When a tractor trailer turns into a cross street or driveway, its rear wheels follow a curve like the one shown here. (This is why the rear wheels sometimes ride up over the curb.) We can find an equation for the curve if we picture the rear wheels as a mass M at the point $(1, 0)$ on the x-axis attached by a rod of unit length to a point P representing the cab at the origin. As the point P moves up the y-axis, it drags M along behind it. The curve traced by M, called a *tractrix* from the Latin word *tractum* for "drag," can be shown to be the graph of the function $y = f(x)$ that solves the initial value problem

Differential equation: $\quad\dfrac{dy}{dx} = -\dfrac{1}{x\sqrt{1 - x^2}} + \dfrac{x}{\sqrt{1 - x^2}},$

Initial condition: $\qquad y = 0 \quad \text{when} \quad x = 1.$

Solve the initial value problem to find an equation for the curve. (You need an inverse hyperbolic function.)

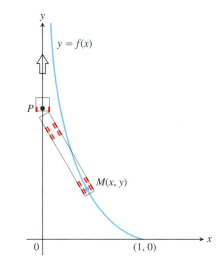

80. Show that the area of the region in the first quadrant enclosed by the curve $y = (1/a)\cosh ax$, the coordinate axes, and the line $x = b$ is the same as the area of a rectangle of height $1/a$ and length s, where s is the length of the curve from $x = 0$ to $x = b$.

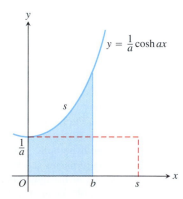

81. A region in the first quadrant is bounded above by the curve $y = \cosh x$, below by the curve $y = \sinh x$, and on the left and right by the y-axis and the line $x = 2$, respectively. Find the volume of the solid generated by revolving the region about the x-axis.

82. The region enclosed by the curve $y = \text{sech}\, x$, the x-axis, and the lines $x = \pm \ln \sqrt{3}$ is revolved about the x-axis to generate a solid. Find the volume of the solid.

83. a) Find the length of the segment of the curve $y = (1/2) \cosh 2x$ from $x = 0$ to $x = \ln \sqrt{5}$.

    b) Find the length of the segment of the curve $y = (1/a) \cosh ax$ from $x = 0$ to $x = b > 0$.

84. *A minimal surface.* Find the area of the surface swept out by revolving about the x-axis the curve $y = 4 \cosh (x/4), -\ln 16 \le x \le \ln 81$.

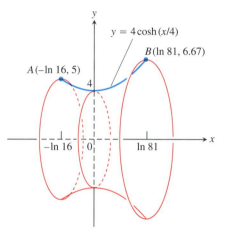

It can be shown that, of all continuously differentiable curves joining points A and B in the figure, the curve $y = 4 \cosh (x/4)$ generates the surface of least area. If you made a rigid wire frame of the end-circles through A and B and dipped them in a soap-film solution, the surface spanning the circles would be the one generated by the curve.

85. a) Find the centroid of the curve $y = \cosh x, -\ln 2 \le x \le \ln 2$.

    b) **CALCULATOR** Evaluate the coordinates to 2 decimal places. Then sketch the curve and plot the centroid to show its relation to the curve.

86. *The hyperbolic in hyperbolic functions.* In case you are wondering where the name *hyperbolic* comes from, here is the answer: Just as $x = \cos u$ and $y = \sin u$ are identified with points (x, y) on the unit circle, the functions $x = \cosh u$ and $y = \sinh u$ are identified with points (x, y) on the right-hand branch of the unit hyperbola, $x^2 - y^2 = 1$ (Fig. 6.34).

Another analogy between hyperbolic and circular functions is that the variable u in the coordinates $(\cosh u, \sinh u)$ for the points of the right-hand branch of the hyperbola $x^2 - y^2 = 1$ is twice the area of the sector AOP pictured in Fig. 6.35. To see why this is so, carry out the following steps.

    a) Show that the area $A(u)$ of sector AOP is given by the formula

$$A(u) = \frac{1}{2} \cosh u \, \sinh u - \int_1^{\cosh u} \sqrt{x^2 - 1} \, dx.$$

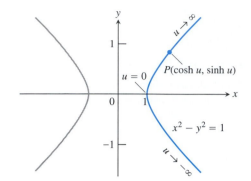

6.34 Since $\cosh^2 u - \sinh^2 u = 1$, the point $(\cosh u, \sinh u)$ lies on the right-hand branch of the hyperbola $x^2 - y^2 = 1$ for every value of u (Exercise 86).

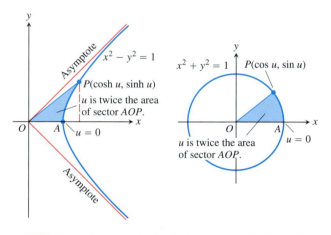

6.35 One of the analogies between hyperbolic and circular functions is revealed by these two diagrams (Exercise 86).

)

    a) Differentiate both sides of the equation in (a) with respect to u to show that

$$A'(u) = \frac{1}{2}.$$

    b) Solve this last equation for $A(u)$. What is the value of $A(0)$? What is the value of the constant of integration C in your solution? With C determined, what does your solution say about the relationship of u to $A(u)$?

Hanging Cables

87. Imagine a cable, like a telephone line or TV cable, strung from one support to another and hanging freely. The cable's weight per unit length is w and the horizontal tension at its lowest point is a vector of length H. If we choose a coordinate system for the plane of the cable in which the x-axis is horizontal, the force of gravity is straight down, the positive y-axis points straight up, and the lowest point of the cable lies at the point $y = H/w$ on the y-axis (Fig. 6.36), then it can be shown that the cable lies

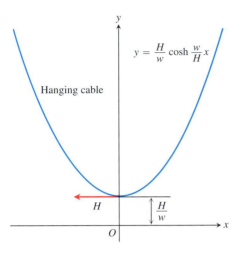

6.36 In a coordinate system chosen to match H and w in the manner shown, a hanging cable lies along the hyperbolic cosine $y = (H/w) \cosh (wx/H)$.

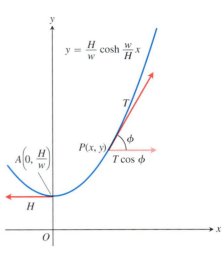

6.37 As discussed in Exercise 87, $T = wy$ in this coordinate system.

along the graph of the hyperbolic cosine

$$y = \frac{H}{w} \cosh \frac{w}{H} x.$$

Such a curve is sometimes called a **chain curve** or a **catenary**, the latter deriving from the Latin *catena,* meaning "chain."

a) Let $P(x, y)$ denote an arbitrary point on the cable. Figure 6.37 displays the tension at P as a vector of length (magnitude) T, as well as the tension H at the lowest point A. Show that the cable's slope at P is

$$\tan \phi = \frac{dy}{dx} = \sinh \frac{w}{H} x.$$

b) Using the result from part (a) and the fact that the tension at P must equal H (the cable is not moving), show that $T = wy$. This means that the magnitude of the tension at $P(x, y)$ is exactly equal to the weight of y units of cable.

88. (*Continuation of Exercise 87.*) The length of arc AP in Fig. 6.37 is $s = (1/a) \sinh ax$, where $a = w/H$. Show that the coordinates of P may be expressed in terms of s as

$$x = \frac{1}{a} \sinh^{-1} as, \quad y = \sqrt{s^2 + \frac{1}{a^2}}.$$

89. *The sag and horizontal tension in a cable.* The ends of a cable 32 ft long and weighing 2 lb/ft are fastened at the same level to posts 30 ft apart.

a) Model the cable with the equation

$$y = \frac{1}{a} \cosh ax, \quad -15 \le x \le 15.$$

Use information from Exercise 88 to show that a satisfies the equation

$$16a = \sinh 15a. \tag{2}$$

b) GRAPHER Solve Eq. (2) graphically by estimating the coordinates of the points where the graphs of the equations $y = 16a$ and $y = \sinh 15a$ intersect in the ay-plane.

c) EQUATION SOLVER or ROOT FINDER Solve Eq. (2) for a numerically. Compare your solution with the value you found in (b).

d) Estimate the horizontal tension in the cable at the cable's lowest point.

e) GRAPHER Graph the catenary

$$y = \frac{1}{a} \cosh ax$$

over the interval $-15 \le x \le 15$. Estimate the sag in the cable at its center.

6.11 First Order Differential Equations

In Section 6.5 we derived the law of exponential change, $y = y_0 e^{kt}$, as the solution of the initial value problem $dy/dt = ky$, $y(0) = y_0$. As we saw, this problem models population growth, radioactive decay, heat transfer, and a great many other phenomena. In the present section, we study initial value problems based on the equation

$dy/dx = f(x, y)$, in which f is a function of both the independent and dependent variables. The applications of this equation, a generalization of $dy/dt = ky$ (think of t as x), are broader still.

First Order Differential Equations

A **first order** differential equation is a relation

$$\frac{dy}{dx} = f(x, y) \tag{1}$$

in which $f(x, y)$ is a function of two variables defined on a region in the xy-plane. A **solution** of Eq. (1) is a differentiable function $y = y(x)$ defined on an interval of x-values (perhaps infinite) such that

$$\frac{d}{dx} y(x) = f(x, y(x))$$

on that interval. The initial condition that $y(x_0) = y_0$ amounts to requiring the solution curve $y = y(x)$ to pass through the point (x_0, y_0).

EXAMPLE 1   The equation

$$\frac{dy}{dx} = 1 - \frac{y}{x}$$

is a first order differential equation in which $f(x, y) = 1 - (y/x)$.   ❏

EXAMPLE 2   Show that the function

$$y = \frac{1}{x} + \frac{x}{2}$$

is a solution of the initial value problem

$$\frac{dy}{dx} = 1 - \frac{y}{x}, \qquad y(2) = \frac{3}{2}.$$

Solution   The given function satisfies the initial condition because

$$y(2) = \left(\frac{1}{x} + \frac{x}{2} \right)_{x=2} = \frac{1}{2} + \frac{2}{2} = \frac{3}{2}.$$

To show that it satisfies the differential equation, we show that the two sides of the equation agree when we substitute $(1/x) + (x/2)$ for y.

On the left: $\dfrac{dy}{dx} = \dfrac{d}{dx} \left(\dfrac{1}{x} + \dfrac{x}{2} \right) = -\dfrac{1}{x^2} + \dfrac{1}{2}$

On the right: $1 - \dfrac{y}{x} = 1 - \dfrac{1}{x} \left(\dfrac{1}{x} + \dfrac{x}{2} \right)$

$$= 1 - \frac{1}{x^2} - \frac{1}{2} = -\frac{1}{x^2} + \frac{1}{2}$$

The function $y = (1/x) + (x/2)$ satisfies both the differential equation and the initial condition, which is what we needed to show.   ❏

We sometimes write $y' = f(x, y)$ for $dy/dx = f(x, y)$.

Separable Equations

The equation $y' = f(x, y)$ is **separable** if f can be expressed as a product of a function of x and a function of y. The differential equation then has the form

$$\frac{dy}{dx} = g(x)h(y).$$

If $h(y) \neq 0$, we can **separate the variables** by dividing both sides by h and multiplying both sides by dx, obtaining

$$\frac{1}{h(y)} \, dy = g(x) \, dx.$$

This groups the y-terms with dy on the left and the x-terms with dx on the right. We then integrate both sides, obtaining

$$\int \frac{1}{h(y)} \, dy = \int g(x) \, dx.$$

The integrated equation provides the solutions we seek by expressing y either explicitly or implicitly as a function of x, up to an arbitrary constant.

EXAMPLE 3 Solve the differential equation

$$\frac{dy}{dx} = (1 + y^2) \, e^x.$$

Solution Since $1 + y^2$ is never zero, we can solve the equation by separating the variables.

$$\frac{dy}{dx} = (1 + y^2) \, e^x$$

$$dy = (1 + y^2) \, e^x \, dx \qquad \text{Treat } dy/dx \text{ as a quotient of differentials and multiply both sides by } dx.$$

$$\frac{dy}{1 + y^2} = e^x \, dx \qquad \text{Divide by } (1 + y^2).$$

$$\int \frac{dy}{1 + y^2} = \int e^x \, dx \qquad \text{Integrate both sides.}$$

$$\tan^{-1} y = e^x + C \qquad C \text{ represents the combined constants of integration.}$$

The equation $\tan^{-1} y = e^x + C$ gives y as an implicit function of x. In this case, we can solve for y as an explicit function of x by taking the tangent of both sides:

$$\tan(\tan^{-1} y) = \tan(e^x + C)$$

$$y = \tan(e^x + C). \qquad \square$$

Linear First Order Equations

A first order differential equation that can be written in the form

$$\frac{dy}{dx} + P(x)y = Q(x), \tag{2}$$

where P and Q are functions of x, is a **linear** first order equation. Equation (2) is the equation's **standard form**.

EXAMPLE 4   Put the following equation in standard form

$$x\frac{dy}{dx} = x^2 + 3y, \qquad x > 0$$

Solution

$$x\frac{dy}{dx} = x^2 + 3y$$

$$\frac{dy}{dx} = x + \frac{3}{x}y \qquad \text{Divide by } x.$$

$$\frac{dy}{dx} - \frac{3}{x}y = x \qquad \begin{array}{l}\text{Standard form with}\\ P(x) = -3/x \text{ and } Q(x) = x\end{array}$$

Notice that $P(x)$ is $-3/x$, not $+3/x$. The standard form is $y' + P(x)y = Q(x)$, so the minus sign is part of the formula for $P(x)$. ❏

EXAMPLE 5   The equation

$$\frac{dy}{dx} = ky$$

with which we modeled bacterial growth, radioactive decay, and temperature change in Section 6.5 is a linear first order equation. Its standard form is

$$\frac{dy}{dx} - ky = 0. \qquad P(x) = -k \text{ and } Q(x) = 0 \qquad ❏$$

We solve the equation

$$\frac{dy}{dx} + P(x)y = Q(x) \tag{3}$$

by multiplying both sides by a positive function $v(x)$ that transforms the left-hand side into the derivative of the product $v(x) \cdot y$. We will show how to find v in a moment, but first we want to show how, once found, it provides the solution we seek.

Here is why multiplying by v works:

$$\frac{dy}{dx} + P(x)y = Q(x) \qquad \begin{array}{l}\text{Original equation}\\ \text{is in standard form.}\end{array}$$

$$v(x)\frac{dy}{dx} + P(x)v(x)y = v(x)Q(x) \qquad \text{Multiply by } v(x).$$

We call $v(x)$ an **integrating factor** for Eq. (3) because its presence makes the equation integrable.

$$\frac{d}{dx}(v(x) \cdot y) = v(x)Q(x) \qquad \begin{array}{l}v(x) \text{ is chosen to make}\\ v\dfrac{dy}{dx} + Pvy = \dfrac{d}{dx}(v \cdot y)\end{array}$$

$$v(x) \cdot y = \int v(x)Q(x)\,dx \qquad \begin{array}{l}\text{Integrate with}\\ \text{respect to } x.\end{array}$$

$$y = \frac{1}{v(x)}\int v(x)Q(x)\,dx \qquad \text{Solve for } y. \tag{4}$$

Equation (4) expresses the solution of Eq. (3) in terms of the functions $v(x)$ and $Q(x)$.

Why doesn't the formula for $P(x)$ appear in the solution as well? It does, but indirectly, in the construction of the positive function $v(x)$. We have

$$\frac{d}{dx}(vy) = v\frac{dy}{dx} + Pvy \qquad \text{Condition imposed on } v$$

$$v\frac{dy}{dx} + y\frac{dv}{dx} = v\frac{dy}{dx} + Pvy \qquad \text{Product Rule for derivatives}$$

$$y\frac{dv}{dx} = Pvy \qquad \text{The terms } v\frac{dy}{dx} \text{ cancel.}$$

This last equation will hold if

$$\frac{dv}{dx} = Pv$$

$$\frac{dv}{v} = P\,dx \qquad \text{Variables separated}$$

$$\int \frac{dv}{v} = \int P\,dx \qquad \text{Integrate both sides.}$$

$$\ln v = \int P\,dx \qquad \begin{array}{l}\text{Since } v > 0, \text{ we do not need}\\ \text{absolute value signs in } \ln v.\end{array}$$

$$e^{\ln v} = e^{\int P\,dx} \qquad \begin{array}{l}\text{Exponentiate both sides to solve}\\ \text{for } v.\end{array}$$

$$v = e^{\int P\,dx} \tag{5}$$

From this, we see that any function v that satisfies Eq. (5) will enable us to solve Eq. (3) with the formula in Eq. (4). We do not need the most general possible v, only one that will work. Therefore, it will do no harm to simplify our lives by choosing the simplest possible antiderivative of P for $\int P\,dx$.

Theorem 4

The solution of the equation

$$\frac{dy}{dx} + P(x)y = Q(x) \tag{6}$$

is

$$y = \frac{1}{v(x)} \int v(x)\,Q(x)\,dx, \tag{7}$$

where

$$v(x) = e^{\int P(x)\,dx}. \tag{8}$$

In the formula for v, we do not need the most general antiderivative of $P(x)$. Any antiderivative will do.

EXAMPLE 6   Solve the equation

$$x\frac{dy}{dx} = x^2 + 3y, \quad x > 0.$$

How to Solve a Linear First Order Equation

1. Put it in standard form.
2. Find an antiderivative of $P(x)$.
3. Find $v(x) = e^{\int P(x)\,dx}$.
4. Use Eq. (7) to find y.

Solution   We solve the equation in four steps.

Step 1: *Put the equation in standard form to identify P and Q.*

$$\frac{dy}{dx} - \frac{3}{x}y = x, \qquad P(x) = -\frac{3}{x}, \qquad Q(x) = x. \qquad \text{Example 4}$$

Step 2: *Find an antiderivative of P(x) (any one will do).*

$$\int P(x)\,dx = \int -\frac{3}{x}\,dx = -3\int \frac{1}{x}\,dx = -3\ln|x| = -3\ln x \qquad (x > 0)$$

Step 3: *Find the integrating factor v(x).*

$$v(x) = e^{\int P(x)\,dx} = e^{-3\ln x} = e^{\ln x^{-3}} = \frac{1}{x^3} \qquad \text{Eq. (8)}$$

Step 4: *Find the solution.*

$$y = \frac{1}{v(x)} \int v(x)Q(x)\,dx \qquad \text{Eq. (7)}$$

$$= \frac{1}{(1/x^3)} \int \left(\frac{1}{x^3}\right)(x)\,dx \qquad \text{Values from steps 1–3}$$

$$= x^3 \cdot \int \frac{1}{x^2}\,dx$$

$$= x^3 \left(-\frac{1}{x} + C\right) \qquad \text{Don't forget the } C \ldots$$

$$= -x^2 + Cx^3 \qquad \begin{array}{l}\ldots\text{it provides part of the}\\ \text{answer.}\end{array}$$

The solution is $y = -x^2 + Cx^3, \quad x > 0$. ❑

EXAMPLE 7   Solve the equation

$$xy' = x^2 + 3y, \qquad x > 0,$$

given the intial condition $y(1) = 2$.

Solution   We first solve the differential equation (Example 6), obtaining

$$y = -x^2 + Cx^3, \qquad x > 0.$$

We then use the initial condition to find the right value for C:

$$y = -x^2 + Cx^3$$

$$2 = -(1)^2 + C(1)^3 \qquad y = 2 \quad \text{when} \quad x = 1$$

$$C = 2 + (1)^2 = 3.$$

The solution of the initial value problem is the function $y = -x^2 + 3x^3$. ❑

Resistance Proportional to Velocity

In some cases it makes sense to assume that, other forces being absent, the resistance encountered by a moving object, like a car coasting to a stop, is proportional to the object's velocity. The slower the object moves, the less its forward progress is resisted by the air through which it passes. We can describe this in mathematical terms if we picture the object as a mass m moving along a coordinate line with position s and velocity v at time t. The resisting force opposing the motion is mass $\times$ acceleration $= m(dv/dt)$, and we can write

$$m\frac{dv}{dt} = -kv \qquad (k > 0) \tag{9}$$

to say that the force decreases in proportion to velocity. If we rewrite (9) as

$$\frac{dv}{dt} + \frac{k}{m}v = 0 \qquad \text{Standard form} \tag{10}$$

and let v_0 denote the object's velocity at time $t = 0$, we can apply Theorem 4 to arrive at the solution

$$v = v_0\, e^{-(k/m)t} \tag{11}$$

(Exercise 42).

What can we learn from Eq. (11)? For one thing, we can see that if m is something large, like the mass of a 20,000-ton ore boat in Lake Erie, it will take a long time for the velocity to approach zero. For another, we can integrate the equation to find s as a function of t.

Suppose a body is coasting to a stop and the only force acting on it is a resistance proportional to its speed. How far will it coast? To find out, we start with Eq. (11) and solve the initial value problem

$$\frac{ds}{dt} = v_0\, e^{-(k/m)t}, \qquad s(0) = 0.$$

Integrating with respect to t gives

$$s = -\frac{v_0 m}{k}e^{-(k/m)t} + C.$$

Substituting $s = 0$ when $t = 0$ gives

$$0 = -\frac{v_0 m}{k} + C \qquad \text{and} \qquad C = \frac{v_0 m}{k}.$$

The body's position at time t is therefore

$$s(t) = -\frac{v_0 m}{k}e^{-(k/m)t} + \frac{v_0 m}{k} = \frac{v_0 m}{k}(1 - e^{-(k/m)t}).$$

To find how far the body will coast, we find the limit of $s(t)$ as $t \to \infty$. Since $-(k/m) < 0$, we know that $e^{-(k/m)t} \to 0$ as $t \to \infty$, so that

$$\lim_{t \to \infty} s(t) = \lim_{t \to \infty} \frac{v_0 m}{k}(1 - e^{-(k/m)t})$$

$$= \frac{v_0 m}{k}(1 - 0) = \frac{v_0 m}{k}.$$

Thus,

$$\text{Distance coasted} = \frac{v_0 m}{k}. \tag{12}$$

This is an ideal figure, of course. Only in mathematics can time stretch to infinity. The number $v_0\, m/k$ is only an upper bound (albeit a useful one). It is true to life in one respect, at least—if m is large, it will take a lot of energy to stop the body. That is why ocean liners have to be docked by tugboats. Any liner of conventional design entering a slip with enough speed to steer would smash into the pier before it could stop.

Weight vs. mass

Weight is the force that results from gravity pulling on a mass. The two are related by the equation in Newton's second law,

$$\text{Weight} = \text{mass} \times \text{acceleration.}$$

To convert mass to weight, multiply by the acceleration of gravity. To convert weight to mass, divide by the acceleration of gravity. In the metric system,

$$\text{Newtons} = \text{kilograms} \times 9.8$$

and

$$\text{Newtons}/9.8 = \text{kilograms.}$$

In the English system, where weight is measured in pounds, mass is measured in **slugs.** Thus,

$$\text{Pounds} = \text{slugs} \times 32$$

and

$$\text{Pounds}/32 = \text{slugs.}$$

A skater weighing 192 lb has a mass of

$$192/32 = 6 \text{ slugs.}$$

EXAMPLE 8   For a 192-lb ice skater, the k in Eq. (11) is about 1/3 slug/sec and $m = 192/32 = 6$ slugs. How long will it take the skater to coast from 11 ft/sec (7.5 mph) to 1 ft/sec? How far will the skater coast before coming to a complete stop?

Solution   We answer the first question by solving Eq. (11) for t:

$$11e^{-t/18} = 1 \qquad \text{Eq. (11) with } k = 1/3, \\ m = 6, \; v_0 = 11, \quad v = 1$$

$$e^{-t/18} = 1/11$$

$$-t/18 = \ln (1/11) = -\ln 11$$

$$t = 18 \ln 11 \approx 43 \text{ sec.}$$

We answer the second question with Eq. (12):

$$\text{Distance coasted} = \frac{v_0 \, m}{k} = \frac{11 \cdot 6}{1/3}$$

$$= 198 \text{ ft.} \qquad \square$$

RL Circuits

The diagram in Fig. 6.38 represents an electrical circuit whose total resistance is a constant R ohms and whose self-inductance, shown as a coil, is L henries, also a constant. There is a switch whose terminals at a and b can be closed to connect a constant electrical source of V volts.

Ohm's law, $V = RI$, has to be modified for such a circuit. The modified form is

$$L\frac{di}{dt} + Ri = V, \tag{13}$$

where i is the intensity of the current in amperes and t is the time in seconds. By solving this equation, we can predict how the current will flow after the switch is closed.

EXAMPLE 9   The switch in the RL circuit in Fig. 6.38 is closed at time $t = 0$. How will the current flow as a function of time?

Solution   Equation (13) is a linear first order differential equation for i as a function of t. Its standard form is

$$\frac{di}{dt} + \frac{R}{L}i = \frac{V}{L}, \tag{14}$$

and the corresponding solution, from Theorem 4, given that $i = 0$ when $t = 0$, is

$$i = \frac{V}{R} - \frac{V}{R}e^{-(R/L)t} \tag{15}$$

6.38 The RL circuit in Example 9.

6.39 The growth of the current in the *RL* circuit in Example 9. *I* is the current's steady state value. The number $t = L/R$ is the time constant of the circuit. The current gets to within 5% of its steady state value in 3 time constants (Exercise 53).

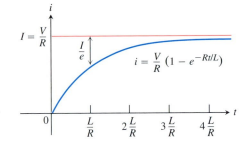

(Exercise 54). Since R and L are positive, $-(R/L)$ is negative and $e^{-(R/L)t} \to 0$ as $t \to \infty$. Thus,

$$\lim_{t\to\infty} i = \lim_{t\to\infty} \left(\frac{V}{R} - \frac{V}{R} e^{-(R/L)t} \right) = \frac{V}{R} - \frac{V}{R} \cdot 0 = \frac{V}{R}.$$

At any given time, the current is theoretically less than V/R, but as time passes the current approaches the **steady state value** V/R. According to the equation

$$L\frac{di}{dt} + Ri = V,$$

$I = V/R$ is the current that will flow in the circuit if either $L = 0$ (no inductance) or $di/dt = 0$ (steady current, $i =$ constant) (Fig. 6.39).

Equation (15) expresses the solution of Eq. (14) as the sum of two terms: a **steady state solution** V/R and a **transient solution** $-(V/R)e^{-(R/L)t}$ that tends to zero as $t \to \infty$. ❑

Exercises 6.11

Verifying Solutions

In Exercises 1 and 2, show that each function $y = f(x)$ is a solution of the accompanying differential equation.

1. $2y' + 3y = e^{-x}$

  a) $y = e^{-x}$
  b) $y = e^{-x} + e^{-(3/2)x}$
  c) $y = e^{-x} + Ce^{-(3/2)x}$

2. $y' = y^2$

  a) $y = -\dfrac{1}{x}$ **b)** $y = -\dfrac{1}{x+3}$ **c)** $y = -\dfrac{1}{x+C}$

In Exercises 3 and 4, show that the function $y = f(x)$ is a solution of the given differential equation.

3. $y = \dfrac{1}{x} \displaystyle\int_1^x \dfrac{e^t}{t} \, dt, \quad x^2 y' + xy = e^x$

4. $y = \dfrac{1}{\sqrt{1+x^4}} \displaystyle\int_1^x \sqrt{1+t^4} \, dt, \quad y' + \dfrac{2x^3}{1+x^4} y = 1$

In Exercises 5–8, show that each function is a solution of the given initial value problem.

Differential equation	Initial condition	Solution candidate
5. $y' + y = \dfrac{2}{1+4e^{2x}}$	$y(-\ln 2) = \dfrac{\pi}{2}$	$y = e^{-x}\tan^{-1}(2e^x)$
6. $y' = e^{-x^2} - 2xy$	$y(2) = 0$	$y = (x-2)e^{-x^2}$
7. $xy' + y = -\sin x$, $x > 0$	$y\left(\dfrac{\pi}{2}\right) = 0$	$y = \dfrac{\cos x}{x}$
8. $x^2 y' = xy - y^2$, $x > 1$	$y(e) = e$	$y = \dfrac{x}{\ln x}$

Separable Equations

Solve the differential equations in Exercises 9–14.

9. $\dfrac{dy}{dx} = 2(x + y^2 x)$ **10.** $(y+1)\dfrac{dy}{dx} = y(x-1)$

11. $2\sqrt{xy}\,\dfrac{dy}{dx} = 1$, $x, y > 0$

12. $\dfrac{dy}{dx} = x^2\sqrt{y}$, $y > 0$

13. $\dfrac{dy}{dx} = e^{x-y}$

14. $\dfrac{dy}{dx} = \dfrac{2x^2 + 1}{xe^y}$, $x > 0$

Linear First Order Equations

Solve the differential equations in Exercises 15–20.

15. $x\,\dfrac{dy}{dx} + y = e^x$, $x > 0$

16. $e^x\,\dfrac{dy}{dx} + 2e^x y = 1$

17. $xy' + 3y = \dfrac{\sin x}{x^2}$, $x > 0$

18. $y' + (\tan x)y = \cos^2 x$, $-\pi/2 < x < \pi/2$

19. $x\,\dfrac{dy}{dx} + 2y = 1 - \dfrac{1}{x}$, $x > 0$

20. $(1+x)y' + y = \sqrt{x}$

First Order Equations

Solve the differential equations in Exercises 21–34.

21. $2y' = e^{x/2} + y$

22. $\sqrt{x}\,\dfrac{dy}{dx} = e^{y+\sqrt{x}}$, $x > 0$

23. $e^{2x}y' + 2e^{2x}y = 2x$

24. $xy' - y = 2x\ln x$

25. $\sec x\,\dfrac{dy}{dx} = e^{y+\sin x}$

26. $x\,\dfrac{dy}{dx} = \dfrac{\cos x}{x} - 2y$, $x > 0$

27. $(t-1)^3\,\dfrac{ds}{dt} + 4(t-1)^2 s = t+1$, $t > 1$

28. $(t+1)\,\dfrac{ds}{dt} + 2s = 3(t+1) + \dfrac{1}{(t+1)^2}$, $t > -1$

29. $(\sec^2 \sqrt{x})\,\dfrac{dx}{dt} = \sqrt{x}$

30. $\sin t - (x\cos^2 t)\,\dfrac{dx}{dt} = 0$, $-\dfrac{\pi}{2} < t < \dfrac{\pi}{2}$

31. $\sin\theta\,\dfrac{dr}{d\theta} + (\cos\theta)r = \tan\theta$, $0 < \theta < \pi/2$

32. $\tan\theta\,\dfrac{dr}{d\theta} + r = \sin^2\theta$, $0 < \theta < \pi/2$

33. $\cosh x\,\dfrac{dy}{dx} + (\sinh x)y = e^{-x}$

34. $\sinh x\,\dfrac{dy}{dx} + 3(\cosh x)y = \cosh x \sinh x$

Solving Initial Value Problems

Solve the initial value problems in Exercises 35–40.

Differential equation	Initial condition
35. $\dfrac{dy}{dt} + 2y = 3$	$y(0) = 1$
36. $t\,\dfrac{dy}{dt} + 2y = t^3$, $t > 0$	$y(2) = 1$
37. $\theta\,\dfrac{dy}{d\theta} + y = \sin\theta$, $\theta > 0$	$y(\pi/2) = 1$
38. $\theta\,\dfrac{dy}{d\theta} - 2y = \theta^3 \sec\theta \tan\theta$, $\theta > 0$	$y(\pi/3) = 2$
39. $(x+1)\,\dfrac{dy}{dx} - 2(x^2 + x)y = \dfrac{e^{x^2}}{x+1}$, $x > -1$	$y(0) = 5$
40. $\dfrac{dy}{dx} + xy = x$	$y(0) = -6$

41. What do you get when you use Theorem 4 to solve the following initial value problem for y as a function of t?

$$\frac{dy}{dt} = ky \quad (k \text{ constant}), \quad y(0) = y_0$$

42. Use Theorem 4 to solve the following initial value problem for v as a function of t.

$$\frac{dv}{dt} + \frac{k}{m}v = 0 \quad (k \text{ and } m \text{ positive constants}), \quad v(0) = v_0$$

Theory and Examples

43. Is either of the following equations correct? Give reasons for your answers.

a) $x\displaystyle\int \frac{1}{x}\,dx = x\ln|x| + C$

b) $x\displaystyle\int \frac{1}{x}\,dx = x\ln|x| + Cx$

44. Is either of the following equations correct? Give reasons for your answers.

a) $\dfrac{1}{\cos x}\displaystyle\int \cos x\,dx = \tan x + C$

b) $\dfrac{1}{\cos x}\displaystyle\int \cos x\,dx = \tan x + \dfrac{C}{\cos x}$

45. *Blood sugar.* If glucose is fed intravenously at a constant rate, the change in the overall concentration $c(t)$ of glucose in the blood with respect to time may be described by the differential equation

$$\frac{dc}{dt} = \frac{G}{100V} - kc.$$

In this equation, G, V, and k are positive constants, G being the rate at which glucose is admitted, in milligrams per minute, and V the volume of blood in the body, in liters (around 5 liters for an adult). The concentration $c(t)$ is measured in milligrams per centiliter. The term $-kc$ is included because the glucose is assumed to be changing continually into other molecules at a rate proportional to its concentration.

a) Solve the equation for $c(t)$, using c_0 to denote $c(0)$.

b) Find the steady state concentration, $\lim_{t\to\infty} c(t)$.

46. *Continuous compounding.* You have $1000 with which to open an account and plan to add $1000 per year. All funds in the account will earn 10% interest per year, compounded continuously. If the added deposits are also credited to your account continuously, the number of dollars x in your account at time t (years) will satisfy the initial value problem

$$\frac{dx}{dt} = 1000 + 0.10x, \quad x(0) = 1000.$$

a) Solve the initial value problem for x as a function of t.

b) **CALCULATOR** About how many years will it take for the amount in your account to reach $100,000?

47. *How long will it take a tank to drain?* If we drain the water from a vertical cylindrical tank by opening a valve at the base of the tank, the water will flow fast when the tank is full but slow down as the tank drains. It turns out that the rate at which the water level drops is proportional to the square root of the water's depth, y. This means that

$$\frac{dy}{dt} = -k\sqrt{y}.$$

The value of k depends on the acceleration of gravity, the shape of the hole, the fluid, and the cross-section areas of the tank and drain hole.

Suppose t is measured in minutes and $k = 1/10$. How long does it take the tank to drain if the water is 9 ft deep to start with?

48. *Escape velocity.* The gravitational attraction F exerted by an airless moon on a body of mass m at a distance s from the moon's center is given by the equation $F = -mg\,R^2 s^{-2}$, where g is the acceleration of gravity at the moon's surface and R is the moon's radius (Fig. 6.40). The force F is negative because it acts in the direction of decreasing s.

a) If the body is projected vertically upward from the moon's surface with an initial velocity v_0 at time $t = 0$, use Newton's second law, $F = ma$, to show that the body's velocity at position s is given by the equation

$$v^2 = \frac{2gR^2}{s} + v_0^2 - 2gR.$$

Thus, the velocity remains positive as long as $v_0 \geq \sqrt{2gR}$. The velocity $v_0 = \sqrt{2gR}$ is the moon's **escape velocity**. A body projected upward with this velocity or a greater one will escape from the moon's gravitational pull.

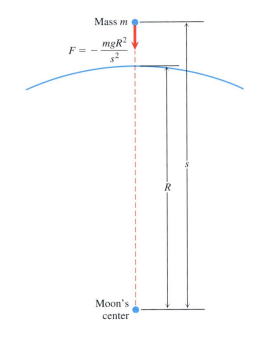

6.40 Diagram for Exercise 48.

b) Show that if $v_0 = \sqrt{2gR}$, then

$$s = R\left(1 + \frac{3v_0}{2R}t\right)^{2/3}.$$

RESISTANCE PROPORTIONAL TO VELOCITY

49. For a 145-lb cyclist on a 15-lb bicycle on level ground, the k in Eq. (11) is about 1/5 slug/sec and $m = 160/32 = 5$ slugs. The cyclist starts coasting at 22 ft/sec (15 mph).

a) About how far will the cyclist coast before reaching a complete stop?

b) To the nearest second, about how long will it take the cyclist's speed to drop to 1 ft/sec?

50. For a 56,000-ton Iowa class battleship, $m = 1,750,000$ slugs and the k in Eq. (11) might be 3000 slugs/sec. Suppose the battleship loses power when it is moving at a speed of 22 ft/sec (13.2 knots).

a) About how far will the ship coast before it stops?

b) About how long will it take the ship's speed to drop to 1 ft/sec?

RL CIRCUITS

51. *Current in a closed RL circuit.* How many seconds after the switch in an RL circuit is closed will it take the current i to reach half of its steady state value? Notice that the time depends on R and L and not on how much voltage is applied.

52. *Current in an open RL circuit.* If the switch is thrown open after the current in an RL circuit has built up to its steady state value, the decaying current (graphed here) obeys the equation

$$L\frac{di}{dt} + Ri = 0, \tag{16}$$

which is Eq. (13) with $V = 0$.

a) Solve Eq. (16) to express i as a function of t.
b) How long after the switch is thrown will it take the current to fall to half its original value?
c) What is the value of the current when $t = L/R$? (The significance of this time is explained in the next exercise.)

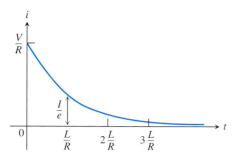

53. *Time constants.* Engineers call the number L/R the *time constant* of the RL circuit in Fig. 6.39. The significance of the time constant is that the current will reach 95% of its final value within 3 time constants of the time the switch is closed (Fig. 6.39). Thus, the time constant gives a built-in measure of how rapidly an individual circuit will reach equilibrium.

a) Find the value of i in Eq. (15) that corresponds to $t = 3L/R$ and show that it is about 95% of the steady state value $I = V/R$.
b) Approximately what percentage of the steady state current will be flowing in the circuit 2 time constants after the switch is closed (i.e., when $t = 2L/R$)?

54. (*Derivation of Eq. (15) in Example 9.*)

a) Use Theorem 4 to show that the solution of the equation

$$\frac{di}{dt} + \frac{R}{L}i = \frac{V}{L}$$

is

$$i = \frac{V}{R} + Ce^{-(R/L)t}.$$

b) Then use the initial condition $i(0) = 0$ to determine the value of C. This will complete the derivation of Eq. (15).
c) Show that $i = V/R$ is a solution of Eq. (14) and that $i = Ce^{-(R/L)t}$ satisfies the equation

$$\frac{di}{dt} + \frac{R}{L}i = 0.$$

MIXTURE PROBLEMS

A chemical in a liquid solution (or dispersed in a gas) runs into a container holding the liquid (or the gas) with, possibly, a specified

amount of the chemical dissolved as well. The mixture is kept uniform by stirring and flows out of the container at a known rate. In this process it is often important to know the concentration of the chemical in the container at any given time. The differential equation describing the process is based on the formula

$$\begin{pmatrix}\text{Rate of change} \\ \text{of amount} \\ \text{in container}\end{pmatrix} = \begin{pmatrix}\text{rate at which} \\ \text{chemical} \\ \text{arrives}\end{pmatrix} - \begin{pmatrix}\text{rate at which} \\ \text{chemical} \\ \text{departs.}\end{pmatrix} \tag{17}$$

If $y(t)$ is the amount of chemical in the container at time t and $V(t)$ is the total volume of liquid in the container at time t, then the departure rate of the chemical at time t is

$$\text{Departure rate} = \frac{y(t)}{V(t)} \cdot (\text{outflow rate})$$

$$= \begin{pmatrix}\text{concentration in} \\ \text{container at time } t\end{pmatrix} \cdot (\text{outflow rate}). \tag{18}$$

Accordingly, Eq. (17) becomes

$$\frac{dy}{dt} = (\text{chemical's arrival rate}) - \frac{y(t)}{V(t)} \cdot (\text{outflow rate}). \tag{19}$$

If, say, y is measured in pounds, V in gallons, and t in minutes, the units in Eq. (19) are

$$\frac{\text{pounds}}{\text{min}} = \frac{\text{pounds}}{\text{min}} - \frac{\text{pounds}}{\text{gal}} \cdot \frac{\text{gal}}{\text{min}}.$$

55. A tank initially contains 100 gal of brine in which 50 lb of salt are dissolved. A brine containing 2 lb/gal of salt runs into the tank at the rate of 5 gal/min. The mixture is kept uniform by stirring and flows out of the tank at the rate of 4 gal/min.

a) At what rate (lb/min) does salt enter the tank at time t?
b) What is the volume of brine in the tank at time t?
c) At what rate (lb/min) does salt leave the tank at time t?
d) Write down and solve the initial value problem describing the mixing process.
e) Find the concentration of salt in the tank 25 min after the process starts.

56. In an oil refinery a storage tank contains 2000 gal of gasoline that initially has 100 lb of an additive dissolved in it. In preparation for winter weather, gasoline containing 2 lb of additive per gallon is pumped into the tank at a rate of 40 gal/min. The well-mixed solution is pumped out at a rate of 45 gal/min. Find the amount of additive in the tank 20 min after the process starts.

57. A tank contains 100 gal of fresh water. A solution containing 1 lb/gal of soluble lawn fertilizer runs into the tank at the rate of 1 gal/min, and the mixture is pumped out of the tank at the rate of 3 gal/min. Find the maximum amount of fertilizer in the tank and the time required to reach the maximum.

58. An executive conference room of a corporation contains 4500 cubic feet of air initially free of carbon monoxide. Starting at time $t = 0$, cigarette smoke containing 4% carbon monoxide is blown into the room at the rate of 0.3 ft^3/min. A ceiling fan keeps the air in the room well circulated and the air leaves the room at the same rate of 0.3 ft^3/min. Find the time when the concentration of carbon monoxide in the room reaches 0.01%.

6.12   Euler's Numerical Method; Slope Fields

If we do not require or cannot immediately find an *exact* solution for an initial value problem $y' = f(x, y)$, $y(x_0) = y_0$, we can probably use a computer to generate a table of approximate numerical values of y for values of x in an appropriate interval. Such a table is called a **numerical solution** of the problem and the method by which we generate the table is called a **numerical method.** Numerical methods are generally fast and accurate and are often the methods of choice when exact formulas are unnecessary, unavailable, or overly complicated. In the present section, we study one such method, called Euler's method, upon which all other numerical methods are based.

Slope Fields

Each time we specify an initial condition $y(x_0) = y_0$ for the solution of a differential equation $y' = f(x, y)$, the solution curve is required to pass through the point (x_0, y_0) and to have slope $f(x_0, y_0)$ there. We can picture these slopes graphically by drawing short line segments of slope $f(x, y)$ at selected points (x, y) in the domain of f. Each segment has the same slope as the solution curve through (x, y) and so is tangent to the curve there. We see how the curves behave by following these tangents (Fig. 6.41).

Constructing a slope field with pencil and paper can be quite tedious. All our examples were generated by a computer. Let us see how a computer might obtain one of the solution curves.

6.41 Slope fields (top row) and selected solution curves (bottom row). In computer renditions, slope segments are sometimes portrayed with vectors, as they are here. This is not to be taken as an indication that slopes have directions, however, for they do not.

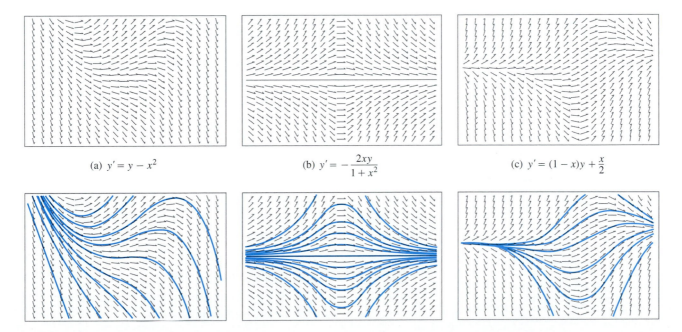

(a) $y' = y - x^2$ (b) $y' = -\dfrac{2xy}{1 + x^2}$ (c) $y' = (1 - x)y + \dfrac{x}{2}$

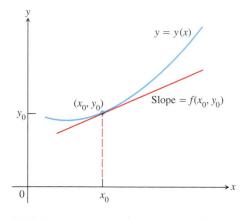

6.42 The equation of the tangent line is $y = L(x) = y_0 + f(x_0, y_0)(x - x_0)$.

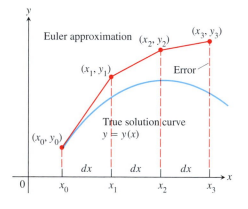

6.43 Three steps in the Euler approximation to the solution of the initial value problem $y' = f(x, y)$, $y = y_0$ when $x = x_0$. The errors involved usually accumulate as we take more steps.

Using Linearizations

If we are given a differential equation $dy/dx = f(x, y)$ and an initial condition $y(x_0) = y_0$, we can approximate the solution curve $y = y(x)$ by its linearization

$$L(x) = y(x_0) + \frac{dy}{dx}\bigg|_{x=x_0} (x - x_0)$$

or

$$L(x) = y_0 + f(x_0, y_0)(x - x_0). \tag{1}$$

The function $L(x)$ will give a good approximation to the solution $y(x)$ in a short interval about x_0 (Fig. 6.42). The basis of Euler's method is to patch together a string of linearizations to approximate the curve over a longer stretch. Here is how the method works.

We know the point (x_0, y_0) lies on the solution curve. Suppose we specify a new value for the independent variable to be $x_1 = x_0 + dx$. If the increment dx is small, then

$$y_1 = L(x_1) = y_0 + f(x_0, y_0)\, dx$$

is a good approximation to the exact solution value $y = y(x_1)$. So from the point (x_0, y_0), which lies *exactly* on the solution curve, we have obtained the point (x_1, y_1), which lies very close to the point $(x_1, y(x_1))$ on the solution curve.

Using the point (x_1, y_1) and the slope $f(x_1, y_1)$, we take a second step. Setting $x_2 = x_1 + dx$, we calculate

$$y_2 = y_1 + f(x_1, y_1)\, dx,$$

to obtain another approximation (x_2, y_2) to values along the solution curve $y = y(x)$ (Fig. 6.43). Continuing in this fashion, we take a third step from the point (x_2, y_2) with slope $f(x_2, y_2)$ to obtain the next approximation

$$y_3 = y_2 + f(x_2, y_2)\, dx,$$

and so on.

EXAMPLE 1   Find the first three approximations y_1, y_2, y_3 using the Euler approximation for the initial value problem

$$y' = 1 + y, \qquad y(0) = 1,$$

starting at $x_0 = 0$ and using $dx = 0.1$.

Solution

First:	$\begin{aligned} y_1 &= y_0 + f(x_0, y_0)\, dx \\ &= y_0 + (1 + y_0)\, dx \\ &= 1 + (1 + 1)(0.1) = 1.2 \end{aligned}$
Second:	$\begin{aligned} y_2 &= y_1 + f(x_1, y_1)\, dx \\ &= y_1 + (1 + y_1)\, dx \\ &= 1.2 + (1 + 1.2)(0.1) = 1.42 \end{aligned}$
Third:	$\begin{aligned} y_3 &= y_2 + f(x_2, y_2)\, dx \\ &= 1.42 + (1 + 1.42)(0.1) = 1.662 \end{aligned}$

The Euler Method

To continue our discussion, the Euler method is a numerical process for generating a table of approximate values of the function that solves the initial value problem

$$y' = f(x, y), \qquad y(x_0) = y_0.$$

If we use equally spaced values for the independent variable in the table and generate n of them, we first set

$$x_1 = x_0 + dx,$$
$$x_2 = x_1 + dx,$$
$$\vdots$$
$$x_n = x_{n-1} + dx. \tag{2}$$

Then we calculate the solution approximations in turn:

$$y_1 = y_0 + f(x_0, y_0)\, dx,$$
$$y_2 = y_1 + f(x_1, y_1)\, dx,$$
$$\vdots$$
$$y_n = y_{n-1} + f(x_{n-1}, y_{n-1})\, dx. \tag{3}$$

The number n of steps can be as large as we like, but errors may accumulate if n is too large.

EXAMPLE 2   Investigate the accuracy of the Euler approximation method for the initial value problem

$$y' = 1 + y, \qquad y(0) = 1$$

in Example 1 over the interval $0 \le x \le 1$, starting at $x_0 = 0$ and taking $dx = 0.1$.

Solution   The exact solution to the initial value problem is $y = 2e^x - 1$ (using either method discussed in Section 6.11). Table 6.14 shows the results of the Euler approximation method using Eqs. (2) and (3) and compares them to the exact results

Table 6.14 Euler solution of $y' = 1 + y$, $y(0) = 1$, increment size $dx = 0.1$

x	y (approx)	y (exact)	Error = y (exact) $-y$ (approx)
0	1	1	0
0.1	1.2	1.2103	0.0103
0.2	1.42	1.4428	0.0228
0.3	1.662	1.6997	0.0377
0.4	1.9282	1.9836	0.0554
0.5	2.2210	2.2974	0.0764
0.6	2.5431	2.6442	0.1011
0.7	2.8974	3.0275	0.1301
0.8	3.2872	3.4511	0.1639
0.9	3.7159	3.9192	0.2033
1.0	4.1875	4.4366	0.2491

rounded to 4 decimal places. By the time we reach $x = 1$ (after 10 steps), the error is about 5.6%. ❏

EXAMPLE 3   Investigate the accuracy of the Euler method for the initial value problem

$$y' = 1 + y, \qquad y(0) = 1$$

over the interval $0 \leq x \leq 1$, starting at $x_0 = 0$ and taking $dx = 0.05$.

Solution   Table 6.15 shows the results and their comparisons with the exact solution. Notice that in doubling the number of steps from 10 to 20 we have reduced the error. This time when we reach $x = 1$ the error is only about 2.9%. ❏

It might be tempting to reduce the increment size even further to obtain greater accuracy. However, each additional calculation not only requires additional computer time but more importantly adds to the buildup of round-off errors due to the approximate representations of numbers in the calculations.

The analysis of error and the investigation of methods to reduce it when making numerical calculations is important, but appropriate for a more advanced course. There are numerical methods that are more accurate than Euler's method, as you will see when you study differential equations. In the exercises you will have the opportunity to explore the trade-offs involved in trying to reduce error by taking more but smaller increment steps.

Table 6.15 Euler solution of $y' = 1 + y$, $y(0) = 1$, increment size $dx = 0.05$

x	y (approx)	y (exact)	Error = y (exact) $- y$ (approx)
0	1	1	0.
0.05	1.1	1.1025	0.0025
0.10	1.205	1.2103	0.0053
0.15	1.3153	1.3237	0.0084
0.20	1.4310	1.4428	0.0118
0.25	1.5526	1.5681	0.0155
0.30	1.6802	1.6997	0.0195
0.35	1.8142	1.8381	0.0239
0.40	1.9549	1.9836	0.0287
0.45	2.1027	2.1366	0.0339
0.50	2.2578	2.2974	0.0396
0.55	2.4207	2.4665	0.0458
0.60	2.5917	2.6442	0.0525
0.65	2.7713	2.8311	0.0598
0.70	2.9599	3.0275	0.0676
0.75	3.1579	3.2340	0.0761
0.80	3.3657	3.4511	0.0854
0.85	3.5840	3.6793	0.0953
0.90	3.8132	3.9192	0.1060
0.95	4.0539	4.1714	0.1175
1.00	4.3066	4.4366	0.1300

Exercises 6.12

Calculating Euler Approximations

In Exercises 1–6, use Euler's method to calculate the first three approximations to the given initial value problem for the specified increment size. Calculate the exact solution and investigate the accuracy of your approximations. Round your results to 4 decimal places.

1. $y' = 1 - \dfrac{y}{x}, \quad y(2) = -1, \quad dx = 0.5$

2. $y' = x(1 - y), \quad y(1) = 0, \quad dx = 0.2$

3. $y' = 2xy + 2y, \quad y(0) = 3, \quad dx = 0.2$

4. $y' = y^2(1 + 2x), \quad y(-1) = 1, \quad dx = 0.5$

5. CALCULATOR $y' = 2xe^{x^2}, \quad y(0) = 2, \quad dx = 0.1$

6. CALCULATOR $y' = y + e^x - 2, \quad y(0) = 2, \quad dx = 0.5$

7. Use the Euler method with $dx = 0.2$ to estimate $y(1)$ if $y' = y$ and $y(0) = 1$. What is the exact value of $y(1)$?

8. Use the Euler method with $dx = 0.2$ to estimate $y(2)$ if $y' = y/x$ and $y(1) = 2$. What is the exact value of $y(2)$?

9. CALCULATOR Use the Euler method with $dx = 0.5$ to estimate $y(5)$ if $y' = y^2/\sqrt{x}$ and $y(1) = -1$. What is the exact value of $y(5)$?

10. CALCULATOR Use the Euler method with $dx = 1/3$ to estimate $y(2)$ if $y' = y - e^{2x}$ and $y(0) = 1$. What is the exact value of $y(2)$?

Slope Fields

In Exercises 11–14, match the differential equations with the solution curves sketched below in the slope fields (a)–(d).

11. $y' = xy$

12. $y' = x + y$

13. $y' = x$

14. $y' = x - y$

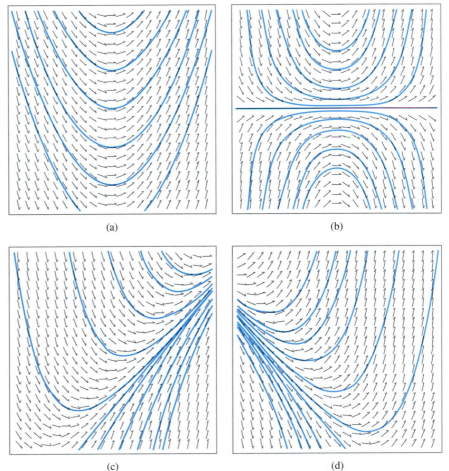

(a)

(b)

(c)

(d)

In Exercises 15 and 16, copy the slope fields and sketch in some of the solution curves.

15. $y' = (y + 2)(y - 2)$

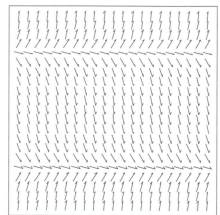

16. $y' = y(y + 1)(y - 1)$

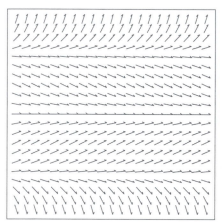

In Exercises 17–20, sketch part of the slope field. Using the slope field, sketch the solution curves that pass through the given points.

17. $y' = y$ with   (a) (0, 1),   (b) (0, 2),   (c) (0, −1)

18. $y' = 2(y - 4)$ with   (a) (0, 1),   (b) (0, 4),   (c) (0, 5)

19. $y' = y(2 - y)$ with   (a) (0, 1/2),   (b) (0, 3/2),   (c) (0, 2),
(d) (0, 3)

20. $y' = y^2$ with   (a) (0, 1),   (b) (0, 2),   (c) (0, −1),   (d) (0, 0)

✹ CAS Explorations and Projects

Use a CAS to explore graphically each of the differential equations in Exercises 21–24. Perform the following steps to help with your explorations.

a) Plot a slope field for the differential equation in the given xy-window.

b) Find the general solution of the differential equation using your CAS DE solver.

c) Graph the solutions for the values of the arbitrary constant $C = -2, -1, 0, 1, 2$ superimposed on your slope field plot.

d) Find and graph the solution that satisfies the specified initial condition over the interval $[0, b]$.

e) Find the Euler numerical approximation to the solution of the initial value problem with 4 subintervals of the x-interval and plot the Euler approximation superimposed on the graph produced in part (d).

f) Repeat part (e) for 8, 16, and 32 subintervals. Plot these three Euler approximations superimposed on the graph from part (e).

g) Find the error y (exact) $-y$ (Euler) at the specified point $x = b$ for each of your four Euler approximations. Discuss the improvement in the percentage error.

21. $y' = x + y$, $y(0) = -7/10$; $-4 \le x \le 4$, $-4 \le y \le 4$; $b = 1$

22. $y' = -x/y$, $y(0) = 2$; $-3 \le x \le 3$, $-3 \le y \le 3$; $b = 2$

23. *A logistic equation.* $y' = y(2 - y)$, $y(0) = 1/2$; $0 \le x \le 4$, $0 \le y \le 3$; $b = 3$

24. $y' = (\sin x)(\sin y)$, $y(0) = 2$; $-6 \le x \le 6$, $-6 \le y \le 6$; $b = 3\pi/2$

Exercises 25 and 26 have no explicit solution in terms of elementary functions. Use a CAS to explore graphically each of the differential equations, performing as many of the steps (a)–(g) above as possible.

25. $y' = \cos(2x - y)$, $y(0) = 2$; $0 \le x \le 5$, $0 \le y \le 5$; $y(2)$

26. *A Gompertz equation.* $y' = y(1/2 - \ln y)$, $y(0) = 1/3$; $0 \le x \le 4$, $0 \le y \le 3$; $y(3)$

27. Use a CAS to find the solutions of $y' + y = f(x)$ subject to the initial condition $y(0) = 0$, if $f(x)$ is

a) $2x$

b) $\sin 2x$

c) $3e^{x/2}$

d) $2e^{-x/2} \cos 2x$.

Graph all four solutions over the interval $-2 \le x \le 6$ to compare the results.

28. a)   Use a CAS to plot the slope field of the differential equation

$$y' = \frac{3x^2 + 4x + 2}{2(y - 1)}$$

over the region $-3 \le x \le 3$ and $-3 \le y \le 3$.

b)   Separate the variables and use a CAS integrator to find the general solution in implicit form.

c)   Using a CAS implicit function grapher, plot solution curves for the arbitrary constant values $C = -6, -4, -2, 0, 2, 4, 6$.

d)   Find and graph the solution that satisfies the initial condition $y(0) = -1$.

CHAPTER 6 QUESTIONS TO GUIDE YOUR REVIEW

1. What functions have inverses? How do you know if two functions f and g are inverses of one another? Give examples of functions that are (are not) inverses of one another.

2. How are the domains, ranges, and graphs of functions and their inverses related? Give an example.

3. How can you sometimes express the inverse of a function of x as a function of x?

4. Under what circumstances can you be sure that the inverse of a function f is differentiable? How are the derivatives of f and f^{-1} related?

5. What is the natural logarithm function? What are its domain, range, and derivative? What arithmetic properties does it have? Comment on its graph.

6. What is logarithmic differentiation? Give an example.

7. What integrals lead to logarithms? Give examples. What are the integrals of $\tan x$ and $\cot x$?

8. How is the exponential function e^x defined? What are its domain, range, and derivative? What laws of exponents does it obey? Comment on its graph.

9. How are the functions a^x and $\log_a x$ defined? Are there any restrictions on a? How is the graph of $\log_a x$ related to the graph of $\ln x$? What truth is there in the statement that there is really only one exponential function and one logarithmic function?

10. Describe some of the applications of base 10 logarithms.

11. What is the law of exponential change? How can it be derived from an initial value problem? What are some of the applications of the law?

12. How do you compare the growth rates of positive functions as $x \to \infty$?

13. What roles do the functions e^x and $\ln x$ play in growth comparisons?

14. Describe big-oh and little-oh notation. Give examples.

15. Which is more efficient—a sequential search or a binary search? Explain.

16. How are the inverse trigonometric functions defined? How can you sometimes use right triangles to find values of these functions? Give examples.

17. How can you find values of $\sec^{-1} x$, $\csc^{-1} x$, and $\cot^{-1} x$ using a calculator's keys for $\cos^{-1} x$, $\sin^{-1} x$, and $\tan^{-1} x$?

18. What are the derivatives of the inverse trigonometric functions? How do the domains of the derivatives compare with the domains of the functions?

19. What integrals lead to inverse trigonometric functions? How do substitution and completing the square broaden the application of these integrals?

20. What are the six basic hyperbolic functions? Comment on their domains, ranges, and graphs. What are some of the identities relating them?

21. What are the derivatives of the six basic hyperbolic functions? What are the corresponding integral formulas? What similarities do you see here with the six basic trigonometric functions?

22. How are the inverse hyperbolic functions defined? Comment on their domains, ranges, and graphs. How can you find values of $\text{sech}^{-1}x$, $\text{csch}^{-1}x$, and $\coth^{-1}x$ using a calculator's keys for $\cosh^{-1}x$, $\sinh^{-1}x$, and $\tanh^{-1}x$?

23. What integrals lead naturally to inverse hyperbolic functions?

24. What is a first order differential equation? When is a function a solution of such an equation?

25. How do you solve separable first order differential equations?

26. How do you solve linear first order differential equations?

27. What is the slope field of a differential equation $y' = f(x, y)$? What can we learn from such fields?

28. Describe Euler's method for solving the initial value problem $y' = f(x, y)$, $y(x_0) = y_0$ numerically. Give an example. Comment on the method's accuracy. Why might you want to solve an initial value problem numerically?

CHAPTER 6   PRACTICE EXERCISES

Differentiation

In Exercises 1–24, find the derivative of y with respect to the appropriate variable.

1. $y = 10e^{-x/5}$

2. $y = \sqrt{2}\,e^{\sqrt{2}x}$

3. $y = \dfrac{1}{4}xe^{4x} - \dfrac{1}{16}e^{4x}$

4. $y = x^2 e^{-2/x}$

5. $y = \ln(\sin^2 \theta)$

6. $y = \ln(\sec^2 \theta)$

7. $y = \log_2(x^2/2)$

8. $y = \log_5(3x - 7)$

9. $y = 8^{-t}$

10. $y = 9^{2t}$

11. $y = 5x^{3.6}$

12. $y = \sqrt{2}\,x^{-\sqrt{2}}$

13. $y = (x + 2)^{x+2}$

14. $y = 2(\ln x)^{x/2}$

15. $y = \sin^{-1}\sqrt{1 - u^2}, \quad 0 < u < 1$

16. $y = \sin^{-1}\left(\dfrac{1}{\sqrt{v}}\right), \quad v > 1$

17. $y = \ln\cos^{-1} x$

18. $y = z\cos^{-1} z - \sqrt{1 - z^2}$

19. $y = t\tan^{-1} t - \dfrac{1}{2}\ln t$

20. $y = (1 + t^2)\cot^{-1} 2t$

21. $y = z\sec^{-1} z - \sqrt{z^2 - 1}, \quad z > 1$

22. $y = 2\sqrt{x - 1}\,\sec^{-1}\sqrt{x}$

23. $y = \csc^{-1}(\sec \theta), \quad 0 < \theta < \pi/2$

24. $y = (1 + x^2)e^{\tan^{-1} x}$

Logarithmic Differentiation

In Exercises 25–30, use logarithmic differentiation to find the derivative of y with respect to the appropriate variable.

25. $y = \dfrac{2(x^2 + 1)}{\sqrt{\cos 2x}}$

26. $y = \sqrt[10]{\dfrac{3x + 4}{2x - 4}}$

27. $y = \left(\dfrac{(t + 1)(t - 1)}{(t - 2)(t + 3)}\right)^5, \quad t > 2$

28. $y = \dfrac{2u2^u}{\sqrt{u^2 + 1}}$

29. $y = (\sin \theta)^{\sqrt{\theta}}$

30. $y = (\ln x)^{1/(\ln x)}$

Integration

Evaluate the integrals in Exercises 31–50.

31. $\displaystyle\int e^x \sin(e^x)\,dx$

32. $\displaystyle\int e^t \cos(3e^t - 2)\,dt$

33. $\displaystyle\int e^x \sec^2(e^x - 7)\,dx$

34. $\displaystyle\int e^y \csc(e^y + 1)\cot(e^y + 1)\,dy$

35. $\displaystyle\int \sec^2(x)e^{\tan x}\,dx$

36. $\displaystyle\int \csc^2 x\, e^{\cot x}\,dx$

37. $\displaystyle\int_{-1}^{1} \dfrac{dx}{3x - 4}$

38. $\displaystyle\int_{1}^{e} \dfrac{\sqrt{\ln x}}{x}\,dx$

39. $\displaystyle\int_{0}^{\pi} \tan\dfrac{x}{3}\,dx$

40. $\displaystyle\int_{1/6}^{1/4} 2\cot \pi x\,dx$

41. $\displaystyle\int_{0}^{4} \dfrac{2t}{t^2 - 25}\,dt$

42. $\displaystyle\int_{-\pi/2}^{\pi/6} \dfrac{\cos t}{1 - \sin t}\,dt$

43. $\displaystyle\int \dfrac{\tan(\ln v)}{v}\,dv$

44. $\displaystyle\int \dfrac{dv}{v\ln v}$

45. $\displaystyle\int \dfrac{(\ln x)^{-3}}{x}\,dx$

46. $\displaystyle\int \dfrac{\ln(x - 5)}{x - 5}\,dx$

47. $\displaystyle\int \dfrac{1}{r}\csc^2(1 + \ln r)\,dr$

48. $\displaystyle\int \dfrac{\cos(1 - \ln v)}{v}\,dv$

49. $\displaystyle\int x3^{x^2}\,dx$

50. $\displaystyle\int 2^{\tan x}\sec^2 x\,dx$

Evaluate the integrals in Exercises 51–64.

51. $\displaystyle\int_{1}^{7} \dfrac{3}{x}\,dx$

52. $\displaystyle\int_{1}^{32} \dfrac{1}{5x}\,dx$

53. $\displaystyle\int_{1}^{4}\left(\dfrac{x}{8} + \dfrac{1}{2x}\right)dx$

54. $\displaystyle\int_{1}^{8}\left(\dfrac{2}{3x} - \dfrac{8}{x^2}\right)dx$

55. $\displaystyle\int_{-2}^{-1} e^{-(x+1)}\,dx$

56. $\displaystyle\int_{-\ln 2}^{0} e^{2w}\,dw$

57. $\displaystyle\int_{0}^{\ln 5} e^r(3e^r + 1)^{-3/2}\,dr$

58. $\displaystyle\int_{0}^{\ln 9} e^\theta(e^\theta - 1)^{1/2}\,d\theta$

59. $\displaystyle\int_{1}^{e} \dfrac{1}{x}(1 + 7\ln x)^{-1/3}\,dx$

60. $\displaystyle\int_{e}^{e^2} \dfrac{1}{x\sqrt{\ln x}}\,dx$

61. $\displaystyle\int_{1}^{3} \dfrac{(\ln(v + 1))^2}{v + 1}\,dv$

62. $\displaystyle\int_{2}^{4} (1 + \ln t)t\ln t\,dt$

63. $\displaystyle\int_{1}^{8} \dfrac{\log_4 \theta}{\theta}\,d\theta$

64. $\displaystyle\int_{1}^{e} \dfrac{8\ln 3\log_3 \theta}{\theta}\,d\theta$

Evalute the integrals in Exercises 65–78.

65. $\displaystyle\int_{-3/4}^{3/4} \dfrac{6\,dx}{\sqrt{9 - 4x^2}}$

66. $\displaystyle\int_{-1/5}^{1/5} \dfrac{6\,dx}{\sqrt{4 - 25x^2}}$

67. $\displaystyle\int_{-2}^{2} \frac{3\,dt}{4+3t^2}$

68. $\displaystyle\int_{\sqrt{3}}^{3} \frac{dt}{3+t^2}$

69. $\displaystyle\int \frac{dy}{y\sqrt{4y^2-1}}$

70. $\displaystyle\int \frac{24\,dy}{y\sqrt{y^2-16}}$

71. $\displaystyle\int_{\sqrt{2}/3}^{2/3} \frac{dy}{|y|\sqrt{9y^2-1}}$

72. $\displaystyle\int_{-2/\sqrt{5}}^{-\sqrt{6}/\sqrt{5}} \frac{dy}{|y|\sqrt{5y^2-3}}$

73. $\displaystyle\int \frac{dx}{\sqrt{-2x-x^2}}$

74. $\displaystyle\int \frac{dx}{\sqrt{-x^2+4x-1}}$

75. $\displaystyle\int_{-2}^{-1} \frac{2\,dv}{v^2+4v+5}$

76. $\displaystyle\int_{-1}^{1} \frac{3\,dv}{4v^2+4v+4}$

77. $\displaystyle\int \frac{dt}{(t+1)\sqrt{t^2+2t-8}}$

78. $\displaystyle\int \frac{dt}{(3t+1)\sqrt{9t^2+6t}}$

Solving Equations with Logarithmic or Exponential Terms

In Exercises 79–84, solve for y.

79. $3^y = 2^{y+1}$

80. $4^{-y} = 3^{y+2}$

81. $9e^{2y} = x^2$

82. $3^y = 3\ln x$

83. $\ln(y-1) = x + \ln y$

84. $\ln(10\ln y) = \ln 5x$

Evaluating Limits

Find the limits in Exercises 85–96.

85. $\displaystyle\lim_{x\to 0} \frac{10^x - 1}{x}$

86. $\displaystyle\lim_{\theta\to 0} \frac{3^\theta - 1}{\theta}$

87. $\displaystyle\lim_{x\to 0} \frac{2^{\sin x} - 1}{e^x - 1}$

88. $\displaystyle\lim_{x\to 0} \frac{2^{-\sin x} - 1}{e^x - 1}$

89. $\displaystyle\lim_{x\to 0} \frac{5 - 5\cos x}{e^x - x - 1}$

90. $\displaystyle\lim_{x\to 0} \frac{4 - 4e^x}{xe^x}$

91. $\displaystyle\lim_{t\to 0^+} \frac{t - \ln(1+2t)}{t^2}$

92. $\displaystyle\lim_{x\to 4} \frac{\sin^2(\pi x)}{e^{x-4} + 3 - x}$

93. $\displaystyle\lim_{t\to 0^+} \left(\frac{e^t}{t} - \frac{1}{t}\right)$

94. $\displaystyle\lim_{y\to 0^+} e^{-1/y} \ln y$

95. $\displaystyle\lim_{x\to\infty} \left(1 + \frac{3}{x}\right)^x$

96. $\displaystyle\lim_{x\to 0^+} \left(1 + \frac{3}{x}\right)^x$

Comparing Growth Rates of Functions

97. Does f grow faster, slower, or at the same rate as g as $x \to \infty$? Give reasons for your answers.

a) $f(x) = \log_2 x, \quad g(x) = \log_3 x$

b) $f(x) = x, \quad g(x) = x + \dfrac{1}{x}$

c) $f(x) = x/100, \quad g(x) = xe^{-x}$

d) $f(x) = x, \quad g(x) = \tan^{-1} x$

e) $f(x) = \csc^{-1} x, \quad g(x) = 1/x$

f) $f(x) = \sinh x, \quad g(x) = e^x$

98. Does f grow faster, slower, or at the same rate as g as $x \to \infty$? Give reasons for your answers.

a) $f(x) = 3^{-x}, \quad g(x) = 2^{-x}$

b) $f(x) = \ln 2x, \quad g(x) = \ln x^2$

c) $f(x) = 10x^3 + 2x^2, \quad g(x) = e^x$

d) $f(x) = \tan^{-1}(1/x), \quad g(x) = 1/x$

e) $f(x) = \sin^{-1}(1/x), \quad g(x) = 1/x^2$

f) $f(x) = \operatorname{sech} x, \quad g(x) = e^{-x}$

99. True, or false? Give reasons for your answers.

a) $\dfrac{1}{x^2} + \dfrac{1}{x^4} = O\left(\dfrac{1}{x^2}\right)$

b) $\dfrac{1}{x^2} + \dfrac{1}{x^4} = O\left(\dfrac{1}{x^4}\right)$

c) $x = o(x + \ln x)$

d) $\ln(\ln x) = o(\ln x)$

e) $\tan^{-1} x = O(1)$

f) $\cosh x = O(e^x)$

100. True, or false? Give reasons for your answers.

a) $\dfrac{1}{x^4} = O\left(\dfrac{1}{x^2} + \dfrac{1}{x^4}\right)$

b) $\dfrac{1}{x^4} = o\left(\dfrac{1}{x^2} + \dfrac{1}{x^4}\right)$

c) $\ln x = o(x + 1)$

d) $\ln 2x = O(\ln x)$

e) $\sec^{-1} x = O(1)$

f) $\sinh x = O(e^x)$

Theory and Applications

101. The function $f(x) = e^x + x$, being differentiable and one-to-one, has a differentiable inverse $f^{-1}(x)$. Find the value of df^{-1}/dx at the point $f(\ln 2)$.

102. Find the inverse of the function $f(x) = 1 + (1/x)$, $x \neq 0$. Then show that $f^{-1}(f(x)) = f(f^{-1}(x)) = x$ and that

$$\left.\frac{df^{-1}}{dx}\right|_{f(x)} = \frac{1}{f'(x)}.$$

In Exercises 103 and 104, find the absolute maximum and minimum values of each function on the given interval.

103. $y = x\ln 2x - x, \quad \left[\dfrac{1}{2e}, \dfrac{e}{2}\right]$

104. $y = 10x(2 - \ln x), \quad (0, e^2]$

105. Find the area between the curve $y = 2(\ln x)/x$ and the x-axis from $x = 1$ to $x = e$.

106. a) Show that the area between the curve $y = 1/x$ and the x-axis from $x = 10$ to $x = 20$ is the same as the area between the curve and the x-axis from $x = 1$ to $x = 2$.

b) Show that the area between the curve $y = 1/x$ and the x-axis from ka to kb is the same as the area between the curve and the x-axis from $x = a$ to $x = b$ $(0 < a < b, k > 0)$.

107. A particle is traveling upward and to the right along the curve $y = \ln x$. Its x-coordinate is increasing at the rate $(dx/dt) = \sqrt{x}$ m/sec. At what rate is the y-coordinate changing at the point $(e^2, 2)$?

108. A girl is sliding down a slide shaped like the curve $y = 9e^{-x/3}$. Her y-coordinate is changing at the rate $dy/dt = (-1/4)\sqrt{9-y}$ ft/sec. At approximately what rate is her x-coordinate changing when she reaches the bottom of the slide at $x = 9$ ft? (Take e^3 to be 20 and round your answer to the nearest ft/sec.)

109. The rectangle shown here has one side on the positive y-axis, one side on the positive x-axis, and its upper right-hand vertex on the curve $y = e^{-x^2}$. What dimensions give the rectangle its largest area, and what is that area?

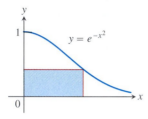

110. The rectangle shown here has one side on the positive y-axis, one side on the positive x-axis, and its upper right-hand vertex on the curve $y = (\ln x)/x^2$. What dimensions give the rectangle its largest area, and what is that area?

111. The functions $f(x) = \ln 5x$ and $g(x) = \ln 3x$ differ by a constant. What constant? Give reasons for your answer.

112. a) If $(\ln x)/x = (\ln 2)/2$, must $x = 2$?
    b) If $(\ln x)/x = -2 \ln 2$, must $x = 1/2$?

Give reasons for your answers.

113. The quotient $(\log_4 x)/(\log_2 x)$ has a constant value. What value? Give reasons for your answer.

114. $\log_x (2)$ vs. $\log_2 (x)$. How does $f(x) = \log_x (2)$ compare with $g(x) = \log_2(x)$? Here is one way to find out:

    a) Use the equation $\log_a b = (\ln b)/(\ln a)$ to express $f(x)$ and $g(x)$ in terms of natural logarithms.

    b) Graph f and g together. Comment on the behavior of f in relation to the signs and values of g.

115. GRAPHER Graph the following functions and use what you see to locate and estimate the extreme values, identify the coordinates of the inflection points, and identify the intervals on which the graphs are concave up and concave down. Then confirm your estimates by working with the functions' derivatives.

    a) $y = (\ln x)/\sqrt{x}$ **b)** $y = e^{-x^2}$
    c) $y = (1 + x)e^{-x}$

116. GRAPHER Graph $f(x) = x \ln x$. Does the function appear to have an absolute minimum value? Confirm your answer with calculus.

117. CALCULATOR What is the age of a sample of charcoal in which 90% of the carbon-14 originally present has decayed?

118. *Cooling a pie.* A deep-dish apple pie, whose internal temperature was 220°F when removed from the oven, was set out on a breezy 40°F porch to cool. Fifteen minutes later, the pie's internal temperature was 180°F. How long did it take the pie to cool from there to 70°F?

119. *Locating a solar station.* You are under contract to build a solar station at ground level on the east–west line between the two buildings shown here. How far from the taller building should you place the station to maximize the number of hours it will be in the sun on a day when the sun passes directly overhead? Begin by observing that

$$\theta = \pi - \cot^{-1} \frac{x}{60} - \cot^{-1} \frac{50 - x}{30}.$$

Then find the value of x that maximizes θ.

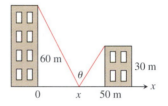

120. A round underwater transmission cable consists of a core of copper wires surrounded by nonconducting insulation. If x denotes the ratio of the radius of the core to the thickness of the insulation, it is known that the speed of the transmission signal is given by the equation $v = x^2 \ln (1/x)$. If the radius of the core is 1 cm, what insulation thickness h will allow the greatest transmission speed?

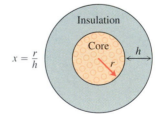

Initial Value Problems

Solve the initial value problems in Exercises 121–124.

Differential equation	Initial condition
121. $\dfrac{dy}{dx} = e^{-x-y-2}$	$y(0) = -2$
122. $\dfrac{dy}{dx} = -\dfrac{y \ln y}{1 + x^2}$	$y(0) = e^2$
123. $(x + 1)\dfrac{dy}{dx} + 2y = x, \quad x > -1$	$y(0) = 1$
124. $x\dfrac{dy}{dx} + 2y = x^2 + 1, \quad x > 0$	$y(1) = 1$

Slope Fields and Euler's Method

In Exercises 125–128, sketch part of the equation's slope field. Then add to your sketch the solution curve that passes through the point $P(1, -1)$. Use Euler's method with $x_0 = 1$ and $dx = 0.2$ to estimate $y(2)$. Round your answers to 4 decimal places. Find the exact value of $y(2)$ for comparison.

125. $y' = x$

126. $y' = 1/x$

127. $y' = xy$

128. $y' = 1/y$

CHAPTER 6 ADDITIONAL EXERCISES–THEORY, EXAMPLES, APPLICATIONS

Limits

Find the limits in Exercises 1–6

1. $\displaystyle \lim_{b \to 1^-} \int_0^b \frac{dx}{\sqrt{1 - x^2}}$

2. $\displaystyle \lim_{x \to \infty} \frac{1}{x} \int_0^x \tan^{-1} t \, dt$

3. $\displaystyle \lim_{x \to 0^+} (\cos \sqrt{x})^{1/x}$

4. $\displaystyle \lim_{x \to \infty} (x + e^x)^{2/x}$

5. $\displaystyle \lim_{n \to \infty} \left(\frac{1}{n+1} + \frac{1}{n+2} + \cdots + \frac{1}{2n} \right)$

6. $\displaystyle \lim_{n \to \infty} \frac{1}{n} \left(e^{1/n} + e^{2/n} + \cdots + e^{(n-1)/n} + e^{n/n} \right)$

7. Let $A(t)$ be the area of the region in the first quadrant enclosed by the coordinate axes, the curve $y = e^{-x}$, and the vertical line $x = t, \ t > 0$. Let $V(t)$ be the volume of the solid generated by revolving the region about the x-axis. Find the following limits.

a) $\displaystyle \lim_{t \to \infty} A(t)$ b) $\displaystyle \lim_{t \to \infty} V(t)/A(t)$ c) $\displaystyle \lim_{t \to 0^+} V(t)/A(t)$

8. *Varying a logarithm's base*

a) Find $\lim \log_a 2$ as $a \to 0^+$, 1^-, 1^+, and ∞.

b) GRAPHER Graph $y = \log_a 2$ as a function of a over the interval $0 < a \le 4$.

Determining Parameter Values

9. Find values of a and b for which

$$\lim_{x \to 0} \frac{\sin ax + bx}{x^3} = -\frac{4}{3}.$$

10. Find values of a and b for which

$$\lim_{x \to 0} \frac{a \cos x - \cos bx}{x^2} = 4.$$

Theory and Examples

11. Find the areas between the curves $y = 2(\log_2 x)/x$ and $y = 2(\log_4 x)/x$ and the x-axis from $x = 1$ to $x = e$. What is the ratio of the larger area to the smaller?

12. GRAPHER Graph $f(x) = \tan^{-1} x + \tan^{-1}(1/x)$ for $-5 \le x \le 5$. Then use calculus to explain what you see. How would you expect f to behave beyond the interval $[-5, 5]$? Give reasons for your answer.

13. For what $x > 0$ does $x^{(x^x)} = (x^x)^x$? Give reasons for your answer.

14. GRAPHER Graph $f(x) = (\sin x)^{\sin x}$ over $[0, 3\pi]$. Explain what you see.

15. Find $f'(2)$ if $f(x) = e^{g(x)}$ and $g(x) = \displaystyle\int_2^x \frac{t}{1 + t^4} \, dt$.

16. a)   Find df/dx if

$$f(x) = \int_1^{e^x} \frac{2 \ln t}{t} \, dt.$$

b)   Find $f(0)$.

c)   What can you conclude about the graph of f? Give reasons for your answer.

17. The figure here shows an informal proof that

$$\tan^{-1} \frac{1}{2} + \tan^{-1} \frac{1}{3} = \frac{\pi}{4}.$$

How does the argument go? (Source: "Behold! Sums of Arctan," by Edward M. Harris, *College Mathematics Journal,* Vol. 18, No. 2, March 1987, p. 141.)

18. $\pi^e < e^\pi$

a)   Why does Fig. 6.44 (on the following page) "prove" that $\pi^e < e^\pi$? (Source: "Proof Without Words," by Fouad Nakhil, *Mathematics Magazine,* Vol. 60, No. 3, June 1987, p. 165.)

b) Figure 6.44 assumes that $f(x) = (\ln x)/x$ has an absolute maximum value at $x = e$. How do you know it does?

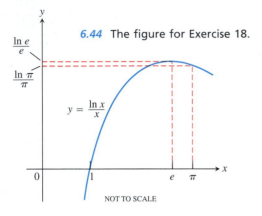

6.44 The figure for Exercise 18.

NOT TO SCALE

19. Use the accompanying figure to show that

$$\int_0^{\pi/2} \sin x \, dx = \frac{\pi}{2} - \int_0^1 \sin^{-1} x \, dx.$$

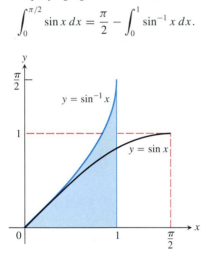

20. *Napier's inequality.* Here are two pictorial proofs that

$$b > a > 0 \quad \Rightarrow \quad \frac{1}{b} < \frac{\ln b - \ln a}{b - a} < \frac{1}{a}.$$

Explain what is going on in each case.

a)

b)

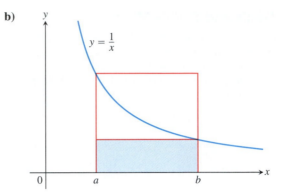

(Source: Roger B. Nelson, *College Mathematics Journal,* Vol. 24, No. 2, March 1993, p. 165.)

21. *Even–odd decompositions*

a) Suppose that g is an even function of x and h is an odd function of x. Show that if $g(x) + h(x) = 0$ for all x then $g(x) = 0$ for all x and $h(x) = 0$ for all x.

b) Use the result in (a) to show that if $f(x) = f_E(x) + f_O(x)$ is the sum of an even function $f_E(x)$ and an odd function $f_O(x)$, then

$$f_E(x) = (f(x) + f(-x))/2 \quad \text{and} \quad f_O(x) = (f(x) - f(-x))/2.$$

c) What is the significance of the result in (b)?

22. Let g be a function that is differentiable throughout an open interval containing the origin. Suppose g has the following properties:

i) $g(x + y) = \dfrac{g(x) + g(y)}{1 - g(x)\,g(y)}$ for all real numbers x, y, and $x + y$ in the domain of g.

ii) $\lim\limits_{h \to 0} g(h) = 0$

iii) $\lim\limits_{h \to 0} \dfrac{g(h)}{h} = 1$

a) Show that $g(0) = 0$.

b) Show that $g'(x) = 1 + [g(x)]^2$.

c) Find $g(x)$ by solving the differential equation in (b).

Applications

23. Find the center of mass of a thin plate of constant density covering the region in the first and fourth quadrants enclosed by the curves $y = 1/(1 + x^2)$ and $y = -1/(1 + x^2)$ and by the lines $x = 0$ and $x = 1$.

24. The region between the curve $y = 1/(2\sqrt{x})$ and the x-axis from $x = 1/4$ to $x = 4$ is revolved about the x-axis to generate a solid.

a) Find the volume of the solid.

b) Find the centroid of the region.

25. *The Rule of 70.* If you use the approximation $\ln 2 \approx 0.70$ (in place of $0.69314\ldots$), you can derive a rule of thumb that says, "To estimate how many years it will take an amount of money to double when invested at r percent compounded continuously, divide r into 70." For instance, an amount of money invested at 5% will double in about $70/5 = 14$ years. If you want it to double in 10 years instead, you have to invest it at $70/10 = 7\%$. Show how the Rule of 70 is derived. (A similar "Rule of 72" uses 72 instead of 70, because 72 has more integer factors.)

26. *Free fall in the fourteenth century.* In the middle of the fourteenth century, Albert of Saxony (1316–1390) proposed a model of free fall that assumed that the velocity of a falling body was proportional to the distance fallen. It seemed reasonable to think that a body that had fallen 20 ft might be moving twice as fast as a body that had fallen 10 ft. And besides, none of the instruments in use at the time were accurate enough to prove otherwise. Today we can see just how far off Albert of Saxony's model was by solving the initial value problem implicit in his model. Solve the problem and compare your solution graphically with the equation $s = 16t^2$. You will see that it describes a motion that starts too slowly at first and then becomes too fast too soon to be realistic.

27. *The best branching angles for blood vessels and pipes.* When a smaller pipe branches off from a larger one in a flow system, we may want it to run off at an angle that is best from some energy-saving point of view. We might require, for instance, that energy loss due to friction be minimized along the section AOB shown in Fig. 6.45. In this diagram, B is a given point to be reached by the smaller pipe, A is a point in the larger pipe upstream from B, and O is the point where the branching occurs. A law due to Poiseuille states that the loss of energy due to friction in nonturbulent flow is proportional to the length of the path and inversely proportional to the fourth power of the radius. Thus, the loss along AO is $(kd_1)/R^4$ and along OB is $(kd_2)/r^4$, where k is a constant, d_1 is the length of AO, d_2 is the length of OB, R is the radius of the larger pipe, and r is the radius of the smaller pipe. The angle θ is to be chosen to minimize the sum of these two losses:

$$L = k\,\frac{d_1}{R^4} + k\,\frac{d_2}{r^4}.$$

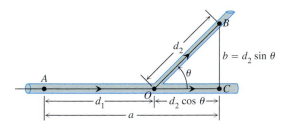

6.45 Diagram for Exercise 27.

In our model, we assume that $AC = a$ and $BC = b$ are fixed. Thus we have the relations

$$d_1 + d_2 \cos\theta = a \quad d_2 \sin\theta = b,$$

so that

$$d_2 = b\,\csc\theta,$$

$$d_1 = a - d_2 \cos\theta = a - b\,\cot\theta.$$

We can express the total loss L as a function of θ:

$$L = k\left(\frac{a - b\,\cot\theta}{R^4} + \frac{b\,\csc\theta}{r^4}\right).$$

a) Show that the critical value of θ for which $dL/d\theta$ equals zero is

$$\theta_c = \cos^{-1}\frac{r^4}{R^4}.$$

b) CALCULATOR If the ratio of the pipe radii is $r/R = 5/6$, estimate to the nearest degree the optimal branching angle given in part (a).

The mathematical analysis described here is also used to explain the angles at which arteries branch in an animal's body. (See *Introduction to Mathematics for Life Scientists,* Second Edition, by E. Batschelet [New York: Springer-Verlag, 1976].)

28. *Group blood testing.* During World War II it was necessary to administer blood tests to large numbers of recruits. There are two standard ways to administer a blood test to N people. In method 1, each person is tested separately. In method 2, the blood samples of x people are pooled and tested as one large sample. If the test is negative, this one test is enough for all x people. If the test is positive, then each of the x people is tested separately, requiring a total of $x + 1$ tests. Using the second method and some probability theory it can be shown that, on the average, the total number of tests y will be

$$y = N\left(1 - q^x + \frac{1}{x}\right).$$

With $q = 0.99$ and $N = 1000$, find the integer value of x that minimizes y. Also find the integer value of x that maximizes y. (This second result is not important to the real-life situation.) The group testing method was used in World War II with a savings of 80% over the individual testing method, but not with the given value of q.

29. *Transport through a cell membrane.* Under some conditions the result of the movement of a dissolved substance across a cell's membrane is described by the equation

$$\frac{dy}{dt} = k\,\frac{A}{V}(c - y).$$

In this equation, y is the concentration of the substance inside the cell and dy/dt is the rate at which y changes over time. The letters k, A, V, and c stand for constants, k being the *permeability coefficient* (a property of the membrane), A the surface area of the membrane, V the cell's volume, and c the concentration of the substance outside the cell. The equation says that the rate at which the concentration changes within the cell is proportional to the difference between it and the outside concentration.

a) Solve the equation for $y(t)$, using y_0 to denote $y(0)$.

b) Find the steady state concentration, $\lim_{t\to\infty} y(t)$.
(Based on *Some Mathematical Models in Biology* by R. M. Thrall, J. A. Mortimer, K. R. Rebman, R. F. Baum, Eds., Revised Edition, December 1967, PB-202 364, pp. 101–103; distributed by N.T.I.S., U.S. Department of Commerce.)

Techniques of Integration

OVERVIEW   We have seen how integrals arise in modeling real phenomena and in measuring objects in the world around us, and we know in theory how integrals are evaluated with antiderivatives. The more sophisticated our models become, however, the more involved our integrals become. We need to know how to change these more involved integrals into forms we can work with. The goal of this chapter is to show how to change unfamiliar integrals into integrals we can recognize, find in a table, or evaluate with a computer.

7.1 Basic Integration Formulas

As we saw in Section 4.1, we evaluate an indefinite integral by finding an antiderivative of the integrand and adding an arbitrary constant. Table 7.1 (on the following page) shows the basic forms of the integrals we have evaluated so far. There is a more extensive table at the back of the book; we will discuss it in Section 7.5.

Algebraic Procedures

We often have to rewrite an integral to match it to a standard formula.

EXAMPLE 1   *A simplifying substitution*

Evaluate $\int \dfrac{2x - 9}{\sqrt{x^2 - 9x + 1}} \, dx$.

Solution

$$\int \frac{2x - 9}{\sqrt{x^2 - 9x + 1}} \, dx = \int \frac{du}{\sqrt{u}} \qquad \begin{aligned} u &= x^2 - 9x + 1 \\ du &= (2x - 9)\, dx \end{aligned}$$

$$= \int u^{-1/2} \, du$$

$$= \frac{u^{(-1/2)+1}}{(-1/2) + 1} + C \qquad \begin{aligned} &\text{Table 7.1,} \\ &\text{Formula 4, with} \\ &n = -1/2 \end{aligned}$$

$$= 2u^{1/2} + C$$

$$= 2\sqrt{x^2 - 9x + 1} + C \qquad \square$$

Table 7.1 Basic integration formulas

1. $\int du = u + C$

2. $\int k\,du = ku + C$ (any number k)

3. $\int (du + dv) = \int du + \int dv$

4. $\int u^n\,du = \dfrac{u^{n+1}}{n+1} + C$ $(n \neq -1)$

5. $\int \dfrac{du}{u} = \ln |u| + C$

6. $\int \sin u\,du = -\cos u + C$

7. $\int \cos u\,du = \sin u + C$

8. $\int \sec^2 u\,du = \tan u + C$

9. $\int \csc^2 u\,du = -\cot u + C$

10. $\int \sec u \tan u\,du = \sec u + C$

11. $\int \csc u \cot u\,du = -\csc u + C$

12. $\int \tan u\,du = -\ln |\cos u| + C$

      $= \ln |\sec u| + C$

13. $\int \cot u\,du = \ln |\sin u| + C$

      $= -\ln |\csc u| + C$

14. $\int e^u\,du = e^u + C$

15. $\int a^u\,du = \dfrac{a^u}{\ln a} + C$ $(a > 0, \quad a \neq 1)$

16. $\int \dfrac{du}{\sqrt{a^2 - u^2}} = \sin^{-1}\left(\dfrac{u}{a}\right) + C$

17. $\int \dfrac{du}{a^2 + u^2} = \dfrac{1}{a} \tan^{-1}\left(\dfrac{u}{a}\right) + C$

18. $\int \dfrac{du}{u\sqrt{u^2 - a^2}} = \dfrac{1}{a} \sec^{-1}\left|\dfrac{u}{a}\right| + C$

EXAMPLE 2   *Completing the square*

Evaluate $\displaystyle\int \dfrac{dx}{\sqrt{8x - x^2}}$.

Solution   We complete the square to write the radicand as

$$8x - x^2 = -(x^2 - 8x) = -(x^2 - 8x + 16 - 16)$$
$$= -(x^2 - 8x + 16) + 16 = 16 - (x - 4)^2.$$

Then

$$\int \frac{dx}{\sqrt{8x - x^2}} = \int \frac{dx}{\sqrt{16 - (x - 4)^2}}$$

$$= \int \frac{du}{\sqrt{a^2 - u^2}} \qquad a = 4, \quad u = (x - 4)$$
$$\qquad\qquad\qquad\qquad\quad du = dx$$

$$= \sin^{-1}\left(\frac{u}{a}\right) + C \qquad \text{Table 7.1, Formula 16}$$

$$= \sin^{-1}\left(\frac{x - 4}{4}\right) + C.$$

EXAMPLE 3   *Expanding a power and using a trigonometric identity*

Evaluate $\int (\sec x + \tan x)^2 \, dx$.

Solution   We expand the integrand and get

$$(\sec x + \tan x)^2 = \sec^2 x + 2 \sec x \tan x + \tan^2 x.$$

The first two terms on the right-hand side of this equation are old friends; we can integrate them at once. How about $\tan^2 x$? There is an identity that connects it with $\sec^2 x$:

$$\tan^2 x + 1 = \sec^2 x, \qquad \tan^2 x = \sec^2 x - 1.$$

We replace $\tan^2 x$ by $\sec^2 x - 1$ and get

$$\int (\sec x + \tan x)^2 \, dx = \int (\sec^2 x + 2 \sec x \tan x + \sec^2 x - 1) \, dx$$

$$= 2 \int \sec^2 x \, dx + 2 \int \sec x \tan x \, dx - \int 1 \, dx$$

$$= 2 \tan x + 2 \sec x - x + C. \qquad \square$$

EXAMPLE 4   *Eliminating a square root*

Evaluate $\int_0^{\pi/4} \sqrt{1 + \cos 4x} \, dx$.

Solution   We use the identity

$$\cos^2 \theta = \frac{1 + \cos 2\theta}{2}, \qquad \text{or} \qquad 1 + \cos 2\theta = 2 \cos^2 \theta.$$

With $\theta = 2x$, this becomes

$$1 + \cos 4x = 2 \cos^2 2x.$$

Hence,

$$\int_0^{\pi/4} \sqrt{1 + \cos 4x} \, dx = \int_0^{\pi/4} \sqrt{2} \sqrt{\cos^2 2x} \, dx$$

$$= \sqrt{2} \int_0^{\pi/4} |\cos 2x| \, dx \qquad \sqrt{u^2} = |u|$$

$$= \sqrt{2} \int_0^{\pi/4} \cos 2x \, dx \qquad \text{On } [0, \pi/4], \cos 2x \geq 0$$
$$\text{so } |\cos 2x| = \cos 2x.$$

$$= \sqrt{2} \left[\frac{\sin 2x}{2} \right]_0^{\pi/4}$$

$$= \sqrt{2} \left[\frac{1}{2} - 0 \right] = \frac{\sqrt{2}}{2}. \qquad \square$$

EXAMPLE 5   *Reducing an improper fraction*

Evaluate $\int \dfrac{3x^2 - 7x}{3x + 2} \, dx$.

$$\begin{array}{r}
x \quad - 3 \\
3x + 2 \overline{)\, 3x^2 - 7x} \\
\underline{3x^2 + 2x} \\
- 9x \\
\underline{- 9x - 6} \\
+ 6
\end{array}$$

Solution   The integrand is an improper fraction (degree of numerator greater than or equal to degree of denominator). To integrate it, we divide first, getting a quotient plus a remainder that is a proper fraction:

$$\frac{3x^2 - 7x}{3x + 2} = x - 3 + \frac{6}{3x + 2}.$$

Therefore,

$$\int \frac{3x^2 - 7x}{3x + 2}\, dx = \int \left(x - 3 + \frac{6}{3x + 2} \right) dx = \frac{x^2}{2} - 3x + 2 \ln |3x + 2| + C.$$

Reducing an improper fraction by long division (Example 5) does not always lead to an expression we can integrate directly. We will see what to do about that in Section 7.3.

EXAMPLE 6   *Separating a fraction*

Evaluate $\displaystyle\int \frac{3x + 2}{\sqrt{1 - x^2}}\, dx.$

Solution   We first separate the integrand to get

$$\int \frac{3x + 2}{\sqrt{1 - x^2}}\, dx = 3 \int \frac{x\, dx}{\sqrt{1 - x^2}} + 2 \int \frac{dx}{\sqrt{1 - x^2}}.$$

In the first of these new integrals we substitute

$$u = 1 - x^2, \qquad du = -2x\, dx, \qquad \text{and} \qquad x\, dx = -\frac{1}{2}\, du.$$

$$3 \int \frac{x\, dx}{\sqrt{1 - x^2}} = 3 \int \frac{(-1/2)\, du}{\sqrt{u}} = -\frac{3}{2} \int u^{-1/2}\, du$$

$$= -\frac{3}{2} \cdot \frac{u^{1/2}}{1/2} + C_1 = -3\sqrt{1 - x^2} + C_1.$$

The second of the new integrals is a standard form,

$$2 \int \frac{dx}{\sqrt{1 - x^2}} = 2 \sin^{-1} x + C_2.$$

Combining these results and renaming $C_1 + C_2$ as C gives

$$\int \frac{3x + 2}{\sqrt{1 - x^2}}\, dx = -3\sqrt{1 - x^2} + 2 \sin^{-1} x + C.$$

EXAMPLE 7   *Multiplying by a form of 1*

Evaluate $\displaystyle\int \sec x\, dx.$

Solution

$$\int \sec x\, dx = \int (\sec x)(1)\, dx = \int \sec x \cdot \frac{\sec x + \tan x}{\sec x + \tan x}\, dx$$

$$= \int \frac{\sec^2 x + \sec x \tan x}{\sec x + \tan x}\, dx$$

$$= \int \frac{du}{u} \qquad \begin{aligned} u &= \tan x + \sec x \\ du &= (\sec^2 x + \sec x \tan x)\, dx \end{aligned}$$

$$= \ln |u| + C = \ln |\sec x + \tan x| + C \qquad \square$$

Table 7.2 The secant and cosecant integrals

1. $\displaystyle \int \sec u \, du = \ln |\sec u + \tan u| + C$

2. $\displaystyle \int \csc u \, du = -\ln |\csc u + \cot u| + C$

With cosecants and cotangents in place of secants and tangents, the method of Example 7 leads to a companion formula for the integral of the cosecant (see Exercise 95).

Procedures for Matching Integrals to Basic Formulas

Procedure	Example
Making a simplifying substitution	$\dfrac{2x-9}{\sqrt{x^2-9x+1}}\, dx = \dfrac{du}{\sqrt{u}}$
Completing the square	$\sqrt{8x-x^2} = \sqrt{16-(x-4)^2}$
Using a trigonometric identity	$(\sec x + \tan x)^2 = \sec^2 x + 2\sec x \tan x + \tan^2 x$ $= \sec^2 x + 2\sec x \tan x + (\sec^2 x - 1)$ $= 2\sec^2 x + 2\sec x \tan x - 1$
Eliminating a square root	$\sqrt{1 + \cos 4x} = \sqrt{2\cos^2 2x} = \sqrt{2}\,|\cos 2x|$
Reducing an improper fraction	$\dfrac{3x^2 - 7x}{3x + 2} = x - 3 + \dfrac{6}{3x + 2}$
Separating a fraction	$\dfrac{3x + 2}{\sqrt{1 - x^2}} = \dfrac{3x}{\sqrt{1 - x^2}} + \dfrac{2}{\sqrt{1 - x^2}}$
Multiplying by a form of 1	$\sec x = \sec x \cdot \dfrac{\sec x + \tan x}{\sec x + \tan x}$ $= \dfrac{\sec^2 x + \sec x \tan x}{\sec x + \tan x}$

Exercises 7.1

Basic Substitutions

Evaluate the integrals in Exercises 1–36 by using substitutions that reduce them to standard forms.

1. $\displaystyle\int \frac{16x\,dx}{\sqrt{8x^2+1}}$

2. $\displaystyle\int \frac{3\cos x\,dx}{\sqrt{1+3\sin x}}$

3. $\displaystyle\int 3\sqrt{\sin v}\cos v\,dv$

4. $\displaystyle\int \cot^3 y\csc^2 y\,dy$

5. $\displaystyle\int_0^1 \frac{16x\,dx}{8x^2+2}$

6. $\displaystyle\int_{\pi/4}^{\pi/3} \frac{\sec^2 z}{\tan z}\,dz$

7. $\displaystyle\int \frac{dx}{\sqrt{x}(\sqrt{x}+1)}$

8. $\displaystyle\int \frac{dx}{x-\sqrt{x}}$

9. $\displaystyle\int \cot(3-7x)\,dx$

10. $\displaystyle\int \csc(\pi x-1)\,dx$

11. $\displaystyle\int e^\theta \csc(e^\theta+1)\,d\theta$

12. $\displaystyle\int \frac{\cot(3+\ln x)}{x}\,dx$

13. $\displaystyle\int \sec\frac{t}{3}\,dt$

14. $\displaystyle\int x\sec(x^2-5)\,dx$

15. $\displaystyle\int \csc(s-\pi)\,ds$

16. $\displaystyle\int \frac{1}{\theta^2}\csc\frac{1}{\theta}\,d\theta$

17. $\displaystyle\int_0^{\sqrt{\ln 2}} 2xe^{x^2}\,dx$

18. $\displaystyle\int_{\pi/2}^{\pi} \sin(y)e^{\cos y}\,dy$

19. $\displaystyle\int e^{\tan v}\sec^2 v\,dv$

20. $\displaystyle\int \frac{e^{\sqrt{t}}\,dt}{\sqrt{t}}$

21. $\displaystyle\int 3^{x+1}\,dx$

22. $\displaystyle\int \frac{2^{\ln x}}{x}\,dx$

23. $\displaystyle\int \frac{2^{\sqrt{w}}\,dw}{2\sqrt{w}}$

24. $\displaystyle\int 10^{2\theta}\,d\theta$

25. $\displaystyle\int \frac{9\,du}{1+9u^2}$

26. $\displaystyle\int \frac{4\,dx}{1+(2x+1)^2}$

27. $\displaystyle\int_0^{1/6} \frac{dx}{\sqrt{1-9x^2}}$

28. $\displaystyle\int_0^1 \frac{dt}{\sqrt{4-t^2}}$

29. $\displaystyle\int \frac{2s\,ds}{\sqrt{1-s^4}}$

30. $\displaystyle\int \frac{2\,dx}{x\sqrt{1-4\ln^2 x}}$

31. $\displaystyle\int \frac{6\,dx}{x\sqrt{25x^2-1}}$

32. $\displaystyle\int \frac{dr}{r\sqrt{r^2-9}}$

33. $\displaystyle\int \frac{dx}{e^x+e^{-x}}$

34. $\displaystyle\int \frac{dy}{\sqrt{e^{2y}-1}}$

35. $\displaystyle\int_1^{e^{\pi/3}} \frac{dx}{x\cos(\ln x)}$

36. $\displaystyle\int \frac{\ln x\,dx}{x+4x\ln^2 x}$

Completing the Square

Evaluate the integrals in Exercises 37–42 by completing the square and using substitutions to reduce them to standard forms.

37. $\displaystyle\int_1^2 \frac{8\,dx}{x^2-2x+2}$

38. $\displaystyle\int_2^4 \frac{2\,dx}{x^2-6x+10}$

39. $\displaystyle\int \frac{dt}{\sqrt{-t^2+4t-3}}$

40. $\displaystyle\int \frac{d\theta}{\sqrt{2\theta-\theta^2}}$

41. $\displaystyle\int \frac{dx}{(x+1)\sqrt{x^2+2x}}$

42. $\displaystyle\int \frac{dx}{(x-2)\sqrt{x^2-4x+3}}$

Trigonometric Identities

Evaluate the integrals in Exercises 43–46 by using trigonometric identities and substitutions to reduce them to standard forms.

43. $\displaystyle\int (\sec x+\cot x)^2\,dx$

44. $\displaystyle\int (\csc x-\tan x)^2\,dx$

45. $\displaystyle\int \csc x\sin 3x\,dx$

46. $\displaystyle\int (\sin 3x\cos 2x-\cos 3x\sin 2x)\,dx$

Improper Fractions

Evaluate each integral in Exercises 47–52 by reducing the improper fraction and using a substitution (if necessary) to reduce it to standard form.

47. $\displaystyle\int \frac{x}{x+1}\,dx$

48. $\displaystyle\int \frac{x^2}{x^2+1}\,dx$

49. $\displaystyle\int_{\sqrt{2}}^3 \frac{2x^3}{x^2-1}\,dx$

50. $\displaystyle\int_{-1}^3 \frac{4x^2-7}{2x+3}\,dx$

51. $\displaystyle\int \frac{4t^3-t^2+16t}{t^2+4}\,dt$

52. $\displaystyle\int \frac{2\theta^3-7\theta^2+7\theta}{2\theta-5}\,d\theta$

Separating Fractions

Evaluate each integral in Exercises 53–56 by separating the fraction and using a substitution (if necessary) to reduce it to standard form.

53. $\displaystyle\int \frac{1-x}{\sqrt{1-x^2}}\,dx$

54. $\displaystyle\int \frac{x+2\sqrt{x-1}}{2x\sqrt{x-1}}\,dx$

55. $\displaystyle\int_0^{\pi/4} \frac{1+\sin x}{\cos^2 x}\,dx$

56. $\displaystyle\int_0^{1/2} \frac{2-8x}{1+4x^2}\,dx$

Multiplying by a Form of 1

Evaluate each integral in Exercises 57–62 by multiplying by a form of 1 and using a substitution (if necessary) to reduce it to standard form.

57. $\int \dfrac{1}{1 + \sin x} dx$

58. $\int \dfrac{1}{1 + \cos x} dx$

59. $\int \dfrac{1}{\sec \theta + \tan \theta} d\theta$

60. $\int \dfrac{1}{\csc \theta + \cot \theta} d\theta$

61. $\int \dfrac{1}{1 - \sec x} dx$

62. $\int \dfrac{1}{1 - \csc x} dx$

Eliminating Square Roots

Evaluate each integral in Exercises 63–70 by eliminating the square root.

63. $\displaystyle\int_0^{2\pi} \sqrt{\dfrac{1 - \cos x}{2}} \, dx$

64. $\displaystyle\int_0^{\pi} \sqrt{1 - \cos 2x} \, dx$

65. $\displaystyle\int_{\pi/2}^{\pi} \sqrt{1 + \cos 2t} \, dt$

66. $\displaystyle\int_{-\pi}^{0} \sqrt{1 + \cos t} \, dt$

67. $\displaystyle\int_{-\pi}^{0} \sqrt{1 - \cos^2 \theta} \, d\theta$

68. $\displaystyle\int_{\pi/2}^{\pi} \sqrt{1 - \sin^2 \theta} \, d\theta$

69. $\displaystyle\int_{-\pi/4}^{\pi/4} \sqrt{1 + \tan^2 y} \, dy$

70. $\displaystyle\int_{-\pi/4}^{0} \sqrt{\sec^2 y - 1} \, dy$

Assorted Integrations

Evaluate the integrals in Exercises 71–82 using any technique you think is appropriate.

71. $\displaystyle\int_{\pi/4}^{3\pi/4} (\csc x - \cot x)^2 \, dx$

72. $\displaystyle\int_0^{\pi/4} (\sec x + 4 \cos x)^2 \, dx$

73. $\int \cos \theta \csc (\sin \theta) \, d\theta$

74. $\int \left(1 + \dfrac{1}{x}\right) \cot (x + \ln x) \, dx$

75. $\int (\csc x - \sec x)(\sin x + \cos x) \, dx$

76. $\int (\csc x + \sec x)(\tan x + \cot x) \, dx$

77. $\int \dfrac{6 \, dy}{\sqrt{y}(1 + y)}$

78. $\int \dfrac{dx}{x\sqrt{4x^2 - 1}}$

79. $\int \dfrac{7 \, dx}{(x - 1)\sqrt{x^2 - 2x - 48}}$

80. $\int \dfrac{dx}{(2x + 1)\sqrt{4x^2 + 4x}}$

81. $\int \sec^2 t \tan (\tan t) \, dt$

82. $\int \dfrac{\tan \theta \, d\theta}{2 \sec \theta + 1}$

Trigonometric Powers

83. a) Evaluate $\int \cos^3 \theta \, d\theta$. (*Hint:* $\cos^2 \theta = 1 - \sin^2 \theta$.)
   b) Evaluate $\int \cos^5 \theta \, d\theta$.
   c) Without actually evaluating the integral, explain how you would evaluate $\int \cos^9 \theta \, d\theta$.

84. a) Evaluate $\int \sin^3 \theta \, d\theta$. (*Hint:* $\sin^2 \theta = 1 - \cos^2 \theta$.)
   b) Evaluate $\int \sin^5 \theta \, d\theta$.
   c) Evaluate $\int \sin^7 \theta \, d\theta$.

d) Without actually evaluating the integral, explain how you would evaluate $\int \sin^{13} \theta \, d\theta$.

85. a) Express $\int \tan^3 \theta \, d\theta$ in terms of $\int \tan \theta \, d\theta$. Then evaluate $\int \tan^3 \theta \, d\theta$. (*Hint:* $\tan^2 \theta = \sec^2 \theta - 1$.)
   b) Express $\int \tan^5 \theta \, d\theta$ in terms of $\int \tan^3 \theta \, d\theta$.
   c) Express $\int \tan^7 \theta \, d\theta$ in terms of $\int \tan^5 \theta \, d\theta$.
   d) Express $\int \tan^{2k+1} \theta \, d\theta$, where k is a positive integer, in terms of $\int \tan^{2k-1} \theta \, d\theta$.

86. a) Express $\int \cot^3 \theta \, d\theta$ in terms of $\int \cot \theta \, d\theta$. Then evaluate $\int \cot^3 \theta \, d\theta$. (*Hint:* $\cot^2 \theta = \csc^2 \theta - 1$.)
   b) Express $\int \cot^5 \theta \, d\theta$ in terms of $\int \cot^3 \theta \, d\theta$.
   c) Express $\int \cot^7 \theta \, d\theta$ in terms of $\int \cot^5 \theta \, d\theta$.
   d) Express $\int \cot^{2k+1} \theta \, d\theta$, where k is a positive integer, in terms of $\int \cot^{2k-1} \theta \, d\theta$.

Theory and Examples

87. Find the area of the region bounded above by $y = 2 \cos x$ and below by $y = \sec x$, $-\pi/4 \le x \le \pi/4$.

88. Find the area of the "triangular" region that is bounded from above and below by the curves $y = \csc x$ and $y = \sin x$, $\pi/6 \le x \le \pi/2$, and on the left by the line $x = \pi/6$.

89. Find the volume of the solid generated by revolving the region in Exercise 87 about the x-axis.

90. Find the volume of the solid generated by revolving the region in Exercise 88 about the x-axis.

91. Find the length of the curve $y = \ln (\cos x)$, $0 \le x \le \pi/3$.

92. Find the length of the curve $y = \ln (\sec x)$, $0 \le x \le \pi/4$.

93. Find the centroid of the region bounded by the x-axis, the curve $y = \sec x$, and the lines $x = -\pi/4$, $x = \pi/4$.

94. Find the centroid of the region that is bounded by the x-axis, the curve $y = \csc x$, and the lines $x = \pi/6$, $x = 5\pi/6$.

95. *The integral of* $\csc x$. Repeat the derivation in Example 7, using cofunctions, to show that

$$\int \csc x \, dx = -\ln |\csc x + \cot x| + C.$$

96. Show that the integral

$$\int \left((x^2 - 1)(x + 1)\right)^{-2/3} dx$$

can be evaluated with any of the following substitutions.

a) $u = 1/(x + 1)$
b) $u = ((x - 1)/(x + 1))^k$
   for $k = 1, 1/2, 1/3, -1/3, -2/3,$ and -1
c) $u = \tan^{-1} x$
d) $u = \tan^{-1} \sqrt{x}$ **e)** $u = \tan^{-1} ((x - 1)/2)$
f) $u = \cos^{-1} x$ **g)** $u = \cosh^{-1} x$

What is the value of the integral? (From "Problems and Solutions," *College Mathematics Journal*, Vol. 21, No. 5, Nov. 1990, pp. 425–426.)

7.2

Integration by Parts

Integration by parts is a technique for simplifying integrals of the form

$$\int f(x)g(x)\, dx \qquad (1)$$

in which f can be differentiated repeatedly and g can be integrated repeatedly without difficulty. The integral

$$\int xe^x\, dx$$

is such an integral because $f(x) = x$ can be differentiated twice to become zero and $g(x) = e^x$ can be integrated repeatedly without difficulty. Integration by parts also applies to integrals like

$$\int e^x \sin x\, dx,$$

in which each part of the integrand appears again after repeated differentiation or integration.

In this section, we describe integration by parts and show how to apply it.

The Formula

The formula for integration by parts comes from the Product Rule,

$$\frac{d}{dx}(uv) = u\frac{dv}{dx} + v\frac{du}{dx}.$$

In its differential form, the rule becomes

$$d(uv) = u\, dv + v\, du,$$

which is then written as

$$u\, dv = d(uv) - v\, du$$

and integrated to give the following formula.

The Integration-by-Parts Formula

$$\int u\, dv = uv - \int v\, du. \qquad (2)$$

The integration-by-parts formula expresses one integral, $\int u\, dv$, in terms of a second integral, $\int v\, du$. With a proper choice of u and v, the second integral may be easier to evaluate than the first. This is the reason for the importance of the formula. When faced with an integral we cannot handle, we can replace it by one with which we might have more success.

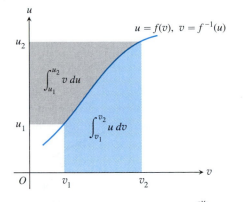

7.1 The area of the blue region, $\int_{v_1}^{v_2} u\,dv$, equals the area of the large rectangle, u_2v_2, minus the areas of the small rectangle, u_1v_1, and the gray region,

$$\int_{u_1}^{u_2} v\,du.$$

In symbols,

$$\int_{v_1}^{v_2} u\,dv = (u_2v_2 - u_1v_1) - \int_{u_1}^{u_2} v\,du.$$

When and How to Use Integration by Parts

When: If substitution doesn't work, try integration by parts.

How: Start with an integral of the form

$$\int f(x)g(x)\,dx.$$

Match this with an integral of the form

$$\int u\,dv$$

by choosing dv to be part of the integrand including dx and possibly $f(x)$ or $g(x)$.

Guideline for choosing u and dv: The formula

$$\int u\,dv = uv - \int v\,du$$

gives a new integral on the right side of the equation. If the new integral is more complex than the original one, try a different choice for u and dv.

The equivalent formula for definite integrals is

$$\int_{v_1}^{v_2} u\,dv = (u_2\,v_2 - u_1\,v_1) - \int_{u_1}^{u_2} v\,du. \tag{3}$$

Figure 7.1 shows how the different parts of the formula may be interpreted as areas.

EXAMPLE 1 Find $\int x\cos x\,dx$.

Solution We use the formula $\int u\,dv = uv - \int v\,du$ with

$$u = x, \qquad dv = \cos x\,dx,$$
$$du = dx, \qquad v = \sin x. \qquad \text{\color{blue}Simplest antiderivative of } \cos x$$

Then

$$\int x\cos x\,dx = x\sin x - \int \sin x\,dx = x\sin x + \cos x + C.$$

Let us examine the choices available for u and dv in Example 1.

EXAMPLE 2 *Example 1 revisited*

To apply integration by parts to

$$\int x\cos x\,dx = \int u\,dv$$

we have four possible choices:

1. Let $u = 1$ and $dv = x\cos x\,dx$.   **2.** Let $u = x$ and $dv = \cos x\,dx$.
3. Let $u = x\cos x$ and $dv = dx$.   **4.** Let $u = \cos x$ and $dv = x\,dx$.

Let's examine these one at a time.

   Choice 1 won't do because we don't know how to integrate $dv = x\cos x\,dx$ to get v.

   Choice 2 works well, as we saw in Example 1.

   Choice 3 leads to

$$u = x\cos x, \qquad\qquad dv = dx,$$
$$du = (\cos x - x\sin x)\,dx, \qquad v = x,$$

and the new integral

$$\int v\,du = \int (x\cos x - x^2\sin x)\,dx.$$

This is worse than the integral we started with.

   Choice 4 leads to

$$u = \cos x, \qquad dv = x\,dx,$$
$$du = -\sin x\,dx, \qquad v = x^2/2,$$

so the new integral is

$$\int v\,du = -\int \frac{x^2}{2}\sin x\,dx.$$

This, too, is worse.

Summary. Keep in mind that the object is to go from $\int u\,dv$ to a new integral that is simpler. Integration by parts does not always work, so we cannot always achieve the goal. ❏

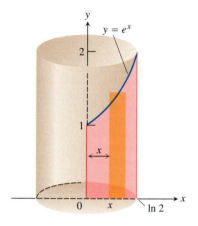

7.2 The solid in Example 3.

EXAMPLE 3   Find the volume of the solid generated by revolving about the y-axis the region in the first quadrant enclosed by the coordinate axes, the curve $y = e^x$ and the line $x = \ln 2$ (Fig. 7.2).

Solution   Using the method of cylindrical shells, we find

$$V = \int_a^b 2\pi x\, f(x)\,dx \qquad \text{The shell volume formula}$$

$$= 2\pi \int_0^{\ln 2} x\,e^x\,dx.$$

To evaluate the integral, we use the formula $\int u\,dv = uv - \int v\,du$ with

$$u = x, \qquad dv = e^x\,dx$$

$$du = dx, \qquad v = e^x. \qquad \text{Simplest antiderivative of } e^x$$

Then

$$\int x\,e^x\,dx = x\,e^x - \int e^x\,dx,$$

so

$$\int_0^{\ln 2} x\,e^x\,dx = x\,e^x\Big]_0^{\ln 2} - \int_0^{\ln 2} e^x\,dx$$

$$= \left[\ln 2\,e^{\ln 2} - 0\right] - \left[e^x\right]_0^{\ln 2}$$

$$= 2\ln 2 - [2 - 1]$$

$$= 2\ln 2 - 1.$$

The solid's volume is therefore

$$V = 2\pi \int_0^{\ln 2} x\,e^x\,dx$$

$$= 2\pi(2\ln 2 - 1). \qquad ❏$$

Integration by parts can be useful even when the integrand has only a single factor. For example, we can use this method to find $\int \ln x\,dx$ (next example) or $\int \cos^{-1} x\,dx$ (Exercise 47).

EXAMPLE 4   Find $\int \ln x\,dx$.

Solution Since $\int \ln x\, dx$ can be written as $\int \ln x \cdot 1\ dx$, we use the formula $\int u\, dv = uv - \int v\, du$ with

$$u = \ln x \qquad \text{Simplifies when differentiated} \qquad dv = dx \qquad \text{Easy to integrate}$$

$$du = \frac{1}{x}dx \qquad\qquad\qquad\qquad\qquad\qquad v = x. \qquad \text{Simplest antiderivative}$$

Then

$$\int \ln x\, dx = x \ln x - \int x \cdot \frac{1}{x}dx = x \ln x - \int dx = x \ln x - x + C. \qquad \square$$

Repeated Use

Sometimes we have to use integration by parts more than once to obtain an answer.

EXAMPLE 5   Find $\int x^2 e^x\, dx$.

Solution We use the formula $\int u\, dv = uv - \int v\, du$ with

$$u = x^2, \qquad dv = e^x\, dx, \qquad v = e^x, \qquad du = 2x\, dx.$$

This gives

$$\int x^2 e^x\, dx = x^2 e^x - 2\int x e^x\, dx.$$

It takes a second integration by parts to find the integral on the right. As in Example 3, its value is $x e^x - e^x + C'$. Hence

$$\int x^2 e^x\, dx = x^2 e^x - 2x e^x + 2e^x + C. \qquad \square$$

Solving for the Unknown Integral

Integrals like the one in the next example occur in electrical engineering. Their evaluation requires two integrations by parts, followed by solving for the unknown integral.

EXAMPLE 6   Find $\int e^x \cos x\, dx$.

Solution We first use the formula $\int u\, dv = uv - \int v\, du$ with

$$u = e^x, \qquad dv = \cos x\, dx, \qquad v = \sin x, \qquad du = e^x dx.$$

Then

$$\int e^x \cos x\, dx = e^x \sin x - \int e^x \sin x\, dx. \qquad (4)$$

The second integral is like the first, except it has $\sin x$ in place of $\cos x$. To evaluate it, we use integration by parts with

$$u = e^x, \qquad dv = \sin x\, dx, \qquad v = -\cos x, \qquad du = e^x\, dx.$$

Then

$$\int e^x \cos x \, dx = e^x \sin x - \left(-e^x \cos x - \int (-\cos x)(e^x \, dx) \right)$$

$$= e^x \sin x + e^x \cos x - \int e^x \cos x \, dx.$$

The unknown integral now appears on both sides of the equation. Combining the two expressions gives

$$2 \int e^x \cos x \, dx = e^x \sin x + e^x \cos x + C.$$

Dividing by 2 and renaming the constant of integration gives

$$\int e^x \cos x \, dx = \frac{e^x \sin x + e^x \cos x}{2} + C'.$$

The choice of $u = e^x$ and $dv = \sin x \, dx$ in the second integration may have seemed arbitrary but it wasn't. In theory, we could have chosen $u = \sin x$ and $dv = e^x \, dx$. Doing so, however, would have turned Eq. (4) into

$$\int e^x \cos x \, dx = e^x \sin x - \left(e^x \sin x - \int e^x \cos x \, dx \right)$$

$$= \int e^x \cos x \, dx.$$

The resulting identity is correct, but useless. *Moral:* Once you have decided on what to differentiate and integrate in circumstances like these, stick with them. Formulas for the integrals of $e^{ax} \cos bx$ and the closely related $e^{ax} \sin bx$ can be found in the integral table at the end of this book. ❑

Tabular Integration

We have seen that integrals of the form $\int f(x)g(x) \, dx$, in which f can be differentiated repeatedly to become zero and g can be integrated repeatedly without difficulty, are natural candidates for integration by parts. However, if many repetitions are required, the calculations can be cumbersome. In situations like this, there is a way to organize the calculations that saves a great deal of work. It is called **tabular integration** and is illustrated in the following examples.

EXAMPLE 7 Find $\int x^2 e^x \, dx$ by tabular integration.

Solution With $f(x) = x^2$ and $g(x) = e^x$, we list

$f(x)$ and its derivatives		$g(x)$ and its integrals
x^2	$(+)$	e^x
$2x$	$(-)$	e^x
2	$(+)$	e^x
0		$e^x.$

We add the products of the functions connected by the arrows, with the middle sign changed, to obtain

$$\int x^2 e^x \, dx = x^2 e^x - 2x e^x + 2e^x + C.$$

EXAMPLE 8   Find $\displaystyle\int x^3 \sin x \, dx$ by tabular integration.

Solution   With $f(x) = x^3$ and $g(x) = \sin x$, we list

$f(x)$ and its derivatives		$g(x)$ and its integrals
x^3	(+)	$\sin x$
$3x^2$	(−)	$-\cos x$
$6x$	(+)	$-\sin x$
6	(−)	$\cos x$
0		$\sin x.$

For more about tabular integration, see the Additional Exercises at the end of this chapter.

Again we add the products of the functions connected by the arrows, with every other sign changed, to obtain

$$\int x^3 \sin x \, dx = -x^3 \cos x + 3x^2 \sin x + 6x \cos x - 6 \sin x + C.$$

Exercises 7.2

Integration by Parts

Evaluate the integrals in Exercises 1–24.

1. $\displaystyle\int x \sin \frac{x}{2} \, dx$

2. $\displaystyle\int \theta \cos \pi\theta \, d\theta$

3. $\displaystyle\int t^2 \cos t \, dt$

4. $\displaystyle\int x^2 \sin x \, dx$

5. $\displaystyle\int_1^2 x \ln x \, dx$

6. $\displaystyle\int_1^e x^3 \ln x \, dx$

7. $\displaystyle\int \tan^{-1} y \, dy$

8. $\displaystyle\int \sin^{-1} y \, dy$

9. $\displaystyle\int x \sec^2 x \, dx$

10. $\displaystyle\int 4x \sec^2 2x \, dx$

11. $\displaystyle\int x^3 e^x \, dx$

12. $\displaystyle\int p^4 e^{-p} \, dp$

13. $\displaystyle\int (x^2 - 5x)e^x \, dx$

14. $\displaystyle\int (r^2 + r + 1)e^r \, dr$

15. $\displaystyle\int x^5 e^x \, dx$

16. $\displaystyle\int t^2 e^{4t} \, dt$

17. $\displaystyle\int_0^{\pi/2} \theta^2 \sin 2\theta \, d\theta$

18. $\displaystyle\int_0^{\pi/2} x^3 \cos 2x \, dx$

19. $\displaystyle\int_{2/\sqrt{3}}^2 t \sec^{-1} t \, dt$

20. $\displaystyle\int_0^{1/\sqrt{2}} 2x \sin^{-1}(x^2) \, dx$

21. $\displaystyle\int e^\theta \sin \theta \, d\theta$

22. $\displaystyle\int e^{-y} \cos y \, dy$

23. $\displaystyle\int e^{2x} \cos 3x \, dx$

24. $\displaystyle\int e^{-2x} \sin 2x \, dx$

Substitution and Integration by Parts

Evaluate the integrals in Exercises 25–30 by using a substitution prior to integration by parts.

25. $\displaystyle\int e^{\sqrt{3s+9}} \, ds$

26. $\displaystyle\int_0^1 x\sqrt{1-x} \, dx$

27. $\displaystyle\int_0^{\pi/3} x \tan^2 x \, dx$

28. $\displaystyle\int \ln(x + x^2) \, dx$

29. $\displaystyle\int \sin(\ln x) \, dx$

30. $\displaystyle\int z (\ln z)^2 \, dz$

Theory and Examples

31. Find the area of the region enclosed by the curve $y = x \sin x$ and the x-axis for (a) $0 \leq x \leq \pi$, (b) $\pi \leq x \leq 2\pi$, (c) $2\pi \leq x \leq 3\pi$. (d) What pattern do you see here? What is the area between the curve and the x-axis for $n\pi \leq x \leq (n+1)\pi$, n an arbitrary nonnegative integer? Give reasons for your answer.

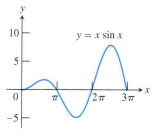

32. Find the area of the region enclosed by the curve $y = x \cos x$ and the x-axis (see the figure below) for

a) $\pi/2 \leq x \leq 3\pi/2$,

b) $3\pi/2 \leq x \leq 5\pi/2$,

c) $5\pi/2 \leq x \leq 7\pi/2$.

d) What pattern do you see? What is the area between the curve and the x-axis for
$$\left(\frac{2n-1}{2}\right)\pi \leq x \leq \left(\frac{2n+1}{2}\right)\pi,$$
n an arbitrary positive integer? Give reasons for your answer.

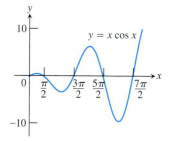

33. Find the volume of the solid generated by revolving the region in the first quadrant bounded by the coordinate axes, the curve $y = e^x$, and the line $x = \ln 2$ about the line $x = \ln 2$.

34. Find the volume of the solid generated by revolving the region in the first quadrant bounded by the coordinate axes, the curve $y = e^{-x}$, and the line $x = 1$ (a) about the y-axis, (b) about the line $x = 1$.

35. Find the volume of the solid generated by revolving the region in the first quadrant bounded by the coordinate axes and the curve $y = \cos x$, $0 \leq x \leq \pi/2$, about (a) the y-axis, (b) the line $x = \pi/2$.

36. Find the volume of the solid generated by revolving the region bounded by the x-axis and the curve $y = x \sin x$, $0 \leq x \leq \pi$, about (a) the y-axis, (b) the line $x = \pi$. (See Exercise 31 for a graph.)

37. a) Find the centroid of a thin plate of constant density covering the region in the first quadrant enclosed by the curve $y = x^2 e^x$, the x-axis, and the line $x = 1$.

b) CALCULATOR Find the coordinates of the centroid to 2 decimal places. Show the center of mass in a rough sketch of the plate.

38. a) Find the centroid of a thin plate of constant density covering the region enclosed by the curve $y = \ln x$, the x-axis, and the line $x = e$.

b) CALCULATOR Find the coordinates of the centroid to 2 decimal places. Show the centroid in a rough sketch of the plate.

39. Find the moment about the y-axis of a thin plate of density $\delta = 1 + x$ covering the region bounded by the x-axis and the curve $y = \sin x$, $0 \leq x \leq \pi$.

40. Although we usually drop the constant of integration in determining v as $\int dv$ in integration by parts, choosing the constant to be different from zero can occasionally be helpful. As a case in point, evaluate
$$\int x \, \tan^{-1} x \, dx,$$
with $u = \tan^{-1} x$ and $v = (x^2/2) + C$, and find a value of C that simplifies the resulting formula.

41. A retarding force, symbolized by the dashpot in the accompanying figure, slows the motion of the weighted spring so that the mass's position at time t is
$$y = 2e^{-t} \cos t, \quad t \geq 0.$$

a) Find the average value of y over the interval $0 \leq t \leq 2\pi$.

b) GRAPHER Graph y over the interval $0 \leq t \leq 2\pi$. Copy the graph and mark the average value of y as a point on the y-axis.

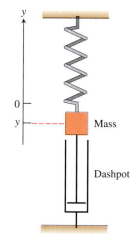

42. In a mass-spring-dashpot system like the one in Exercise 41, the mass's position at time t is
$$y = 4e^{-t}(\sin t - \cos t), \quad t \geq 0.$$

a) Find the average value of y over the interval $0 \le t \le 2\pi$.

b) GRAPHER Graph y over the interval $0 \le t \le 2\pi$. Copy the graph and mark the average value of y as a point on the y-axis.

Integrating Inverses of Functions

Integration by parts leads to a rule for integrating inverses that usually gives good results:

$$\int f^{-1}(x)\,dx = \int yf'(y)\,dy \qquad \begin{array}{l} y = f^{-1}(x), \quad x = f(y) \\ dx = f'(y)\,dy \end{array}$$

$$= yf(y) - \int f(y)\,dy \qquad \begin{array}{l} \text{Integration by parts with} \\ u = y, \; dv = f'(y)\,dy \end{array}$$

$$= xf^{-1}(x) - \int f(y)\,dy$$

The idea is to take the most complicated part of the integral, in this case $f^{-1}(x)$, and simplify it first. For the integral of $\ln x$, we get

$$\int \ln x\,dx = \int ye^y\,dy \qquad \begin{array}{l} y = \ln x, \quad x = e^y \\ dx = e^y\,dy \end{array}$$

$$= ye^y - e^y + C$$

$$= x \ln x - x + C.$$

For the integral of $\cos^{-1} x$ we get

$$\int \cos^{-1} x\,dx = x \cos^{-1} x - \int \cos y\,dy \qquad y = \cos^{-1} x$$

$$= x \cos^{-1} x - \sin y + C$$

$$= x \cos^{-1} x - \sin(\cos^{-1} x) + C.$$

Use the formula

$$\int f^{-1}(x)\,dx = xf^{-1}(x) - \int f(y)\,dy \qquad y = f^{-1}(x) \qquad (5)$$

to evaluate the integrals in Exercises 43–46. Express your answers in terms of x.

43. $\displaystyle\int \sin^{-1} x\,dx$

44. $\displaystyle\int \tan^{-1} x\,dx$

45. $\displaystyle\int \sec^{-1} x\,dx$

46. $\displaystyle\int \log_2 x\,dx$

Another way to integrate $f^{-1}(x)$ (when f^{-1} is integrable, of course) is to use integration by parts with $u = f^{-1}(x)$ and $dv = dx$ to rewrite the integral of f^{-1} as

$$\int f^{-1}(x)\,dx = xf^{-1}(x) - \int x \left(\frac{d}{dx} f^{-1}(x) \right) dx. \qquad (6)$$

Exercises 47 and 48 compare the results of using Eqs. (5) and (6).

47. Equations (5) and (6) give different formulas for the integral of $\cos^{-1} x$:

a) $\displaystyle\int \cos^{-1} x\,dx = x \cos^{-1} x - \sin(\cos^{-1} x) + C$ Eq. (5)

b) $\displaystyle\int \cos^{-1} x\,dx = x \cos^{-1} x - \sqrt{1 - x^2} + C$ Eq. (6)

Can both integrations be correct? Explain.

48. Equations (5) and (6) lead to different formulas for the integral of $\tan^{-1} x$:

a) $\displaystyle\int \tan^{-1} x\,dx = x \tan^{-1} x - \ln \sec(\tan^{-1} x) + C$ Eq. (5)

b) $\displaystyle\int \tan^{-1} x\,dx = x \tan^{-1} x - \ln \sqrt{1 + x^2} + C$ Eq. (6)

Can both integrations be correct? Explain.

Evaluate the integrals in Exercises 49 and 50 with (a) Eq. (5) and (b) Eq. (6). In each case, check your work by differentiating your answer with respect to x.

49. $\displaystyle\int \sinh^{-1} x\,dx$

50. $\displaystyle\int \tanh^{-1} x\,dx$

7.3 Partial Fractions

A theorem from advanced algebra (mentioned later in more detail) says that every rational function, no matter how complicated, can be rewritten as a sum of simpler fractions that we can integrate with techniques we already know. For instance,

$$\frac{5x - 3}{x^2 - 2x - 3} = \frac{2}{x + 1} + \frac{3}{x - 3}, \qquad (1)$$

so we can integrate the rational function on the left by integrating the fractions on the right instead.

The method for rewriting rational functions this way is called the **method of partial fractions.** In this particular case, it consists of finding constants A and B

such that

$$\frac{5x - 3}{x^2 - 2x - 3} = \frac{A}{x + 1} + \frac{B}{x - 3}. \tag{2}$$

(Pretend for a moment that we do not know that $A = 2$ and $B = 3$ will work.) We call the fractions $A/(x + 1)$ and $B/(x - 3)$ **partial fractions** because their denominators are only part of the original denominator $x^2 - 2x - 3$. We call A and B **undetermined coefficients** until proper values for them have been found.

To find A and B, we first clear Eq. (2) of fractions, obtaining

$$5x - 3 = A(x - 3) + B(x + 1) = (A + B)x - 3A + B.$$

This will be an identity in x if and only if the coefficients of like powers of x on the two sides are equal:

$$A + B = 5, \qquad -3A + B = -3.$$

Solving these equations simultaneously gives $A = 2$ and $B = 3$.

EXAMPLE 1   *Two distinct linear factors in the denominator*

Find

$$\int \frac{5x - 3}{(x + 1)(x - 3)} \, dx.$$

Solution   From the preceding discussion,

$$\int \frac{5x - 3}{(x + 1)(x - 3)} \, dx = \int \frac{2}{x + 1} \, dx + \int \frac{3}{x - 3} \, dx$$

$$= 2 \ln |x + 1| + 3 \ln |x - 3| + C. \qquad \square$$

EXAMPLE 2   *A repeated linear factor in the denominator*

Express

$$\frac{6x + 7}{(x + 2)^2}$$

as a sum of partial fractions.

Solution   Since the denominator has a repeated linear factor, $(x + 2)^2$, we must express the fraction in the form

$$\frac{6x + 7}{(x + 2)^2} = \frac{A}{x + 2} + \frac{B}{(x + 2)^2}. \tag{3}$$

Clearing Eq. (3) of fractions gives

$$6x + 7 = A(x + 2) + B = Ax + (2A + B).$$

Matching coefficients of like terms gives $A = 6$ and

$$7 = 2A + B = 12 + B, \qquad \text{or} \qquad B = -5.$$

Hence,

$$\frac{6x + 7}{(x + 2)^2} = \frac{6}{x + 2} - \frac{5}{(x + 2)^2}. \qquad \square$$

How to Evaluate Undetermined Coefficients

1. Clear the given equation of fractions.
2. Equate the coefficients of like terms (powers of x).
3. Solve the resulting equations for the coefficients.

EXAMPLE 3 *An improper fraction*

Express

$$\frac{2x^3 - 4x^2 - x - 3}{x^2 - 2x - 3}$$

as a sum of partial fractions.

Solution First we divide the denominator into the numerator to get a polynomial plus a proper fraction. Then we write the proper fraction as a sum of partial fractions. Long division gives

$$
\begin{array}{r}
2x \\
x^2 - 2x - 3\overline{)2x^3 - 4x^2 - x - 3} \\
2x^3 - 4x^2 - 6x \\
\hline
5x - 3
\end{array}
$$

Hence,

$$\frac{2x^3 - 4x^2 - x - 3}{x^2 - 2x - 3} = 2x + \frac{5x - 3}{x^2 - 2x - 3} \qquad \text{Result of the division}$$

$$= 2x + \frac{2}{x + 1} + \frac{3}{x - 3}. \qquad \begin{array}{l}\text{Proper fraction expanded}\\\text{as in Example 1}\end{array} \qquad ❏$$

EXAMPLE 4 *An irreducible quadratic factor in the denominator*

Express

$$\frac{-2x + 4}{(x^2 + 1)(x - 1)^2}$$

as a sum of partial fractions.

A quadratic polynomial is **irreducible** if it cannot be written as the product of two linear factors with real coefficients.

Solution The denominator has an irreducible quadratic factor as well as a repeated linear factor, so we write

$$\frac{-2x + 4}{(x^2 + 1)(x - 1)^2} = \frac{Ax + B}{x^2 + 1} + \frac{C}{x - 1} + \frac{D}{(x - 1)^2}. \qquad (4)$$

Notice the numerator over $x^2 + 1$: For quadratic factors, we use first degree numerators, not constant numerators. Clearing the equation of fractions gives

$$-2x + 4 = (Ax + B)(x - 1)^2 + C(x - 1)(x^2 + 1) + D(x^2 + 1)$$

$$= (A + C)x^3 + (-2A + B - C + D)x^2$$

$$+ (A - 2B + C)x + (B - C + D).$$

Equating coefficients of like terms gives

Coefficients of x^3: $0 = A + C$

Coefficients of x^2: $0 = -2A + B - C + D$

Coefficients of x^1: $-2 = A - 2B + C$

Coefficients of x^0: $4 = B - C + D$

We solve these equations simultaneously to find the values of A, B, C, and D:

$$-4 = -2A, \quad A = 2 \qquad \text{Subtract fourth equation from second.}$$

$$C = -A = -2 \qquad \text{From the first equation}$$

$$B = 1 \qquad A = 2 \text{ and } C = -2 \text{ in third equation.}$$

$$D = 4 - B + C = 1. \qquad \text{From the fourth equation}$$

We substitute these values into Eq. (4), obtaining

$$\frac{-2x + 4}{(x^2 + 1)(x - 1)^2} = \frac{2x + 1}{x^2 + 1} - \frac{2}{x - 1} + \frac{1}{(x - 1)^2}.$$ ❑

EXAMPLE 5   Evaluate $\displaystyle\int \frac{-2x + 4}{(x^2 + 1)(x - 1)^2}\, dx$.

Solution   We expand the integrand by partial fractions, as in Example 4, and integrate the terms of the expansion:

$$\int \frac{-2x + 4}{(x^2 + 1)(x - 1)^2}\, dx = \int \left(\frac{2x + 1}{x^2 + 1} - \frac{2}{x - 1} + \frac{1}{(x - 1)^2} \right) dx \qquad \text{Example 4}$$

$$= \int \left(\frac{2x}{x^2 + 1} + \frac{1}{x^2 + 1} - \frac{2}{x - 1} + \frac{1}{(x - 1)^2} \right) dx$$

$$= \ln(x^2 + 1) + \tan^{-1} x - 2\ln |x - 1| - \frac{1}{x - 1} + C.$$ ❑

General Description of the Method

Success in writing a rational function $f(x)/g(x)$ as a sum of partial fractions depends on two things:

1. *The degree of $f(x)$ must be less than the degree of $g(x)$.* (If it isn't, divide and work with the remainder term.)
2. *We must know the factors of $g(x)$.* (In theory, any polynomial with real coefficients can be written as a product of real linear factors and real quadratic factors. In practice, the factors may be hard to find.)

A theorem from advanced algebra says that when these two conditions are met, we may write $f(x)/g(x)$ as the sum of partial fractions by taking these steps.

Cases discussed so far

Proper fraction	*Decomposition*

$$\frac{\text{numerator}}{(x + p)(x + q)} = \frac{A}{(x + p)} + \frac{B}{(x + q)}$$

$$\frac{\text{numerator}}{(x + p)^2} = \frac{A}{(x + p)} + \frac{B}{(x + p)^2}$$

$$\frac{\text{numerator}}{(x^2 + p)(x + q)^2} = \frac{Ax + B}{x^2 + p} + \frac{C}{x + q}$$
$$+ \frac{D}{(x + q)^2}$$

The Method of Partial Fractions ($f(x)/g(x)$ Proper)

Step 1   Let $x - r$ be a linear factor of $g(x)$. Suppose $(x - r)^m$ is the highest power of $x - r$ that divides $g(x)$. Then assign the sum of m partial fractions to this factor, as follows:

$$\frac{A_1}{x - r} + \frac{A_2}{(x - r)^2} + \cdots + \frac{A_m}{(x - r)^m}.$$

Do this for each distinct linear factor of $g(x)$.

Step 2 Let $x^2 + px + q$ be an irreducible quadratic factor of $g(x)$. Suppose $(x^2 + px + q)^n$ is the highest power of this factor that divides $g(x)$. Then to this factor assign the sum of the n partial fractions:

$$\frac{B_1x + C_1}{x^2 + px + q} + \frac{B_2x + C_2}{(x^2 + px + q)^2} + \cdots + \frac{B_nx + C_n}{(x^2 + px + q)^n}.$$

Do this for each distinct quadratic factor of $g(x)$ that cannot be factored into linear factors with real coefficients.

Step 3 Set the original fraction $f(x)/g(x)$ equal to the sum of all these partial fractions. Clear the resulting equation of fractions and arrange the terms in decreasing powers of x.

Step 4 Equate the coefficients of corresponding powers of x and solve the resulting equations for the undetermined coefficients.

✳ The Heaviside "Cover-up" Method for Linear Factors

When the degree of the polynomial $f(x)$ is less than the degree of $g(x)$, and

$$g(x) = (x - r_1)(x - r_2) \cdots (x - r_n)$$

is a product of n distinct linear factors, each raised to the first power, there is a quick way to expand $f(x)/g(x)$ by partial fractions.

EXAMPLE 6 Find A, B, and C in the partial-fraction expansion

$$\frac{x^2 + 1}{(x - 1)(x - 2)(x - 3)} = \frac{A}{x - 1} + \frac{B}{x - 2} + \frac{C}{x - 3}. \tag{5}$$

Solution If we multiply both sides of Eq. (5) by $(x - 1)$ to get

$$\frac{x^2 + 1}{(x - 2)(x - 3)} = A + \frac{B(x - 1)}{x - 2} + \frac{C(x - 1)}{x - 3}$$

and set $x = 1$, the resulting equation gives the value of A:

$$\frac{(1)^2 + 1}{(1 - 2)(1 - 3)} = A + 0 + 0,$$
$$A = 1.$$

Thus, the value of A is the number we would have obtained if we had covered the factor $(x - 1)$ in the denominator of the original fraction

$$\frac{x^2 + 1}{(x - 1)(x - 2)(x - 3)} \tag{6}$$

and evaluated the rest at $x = 1$:

$$A = \frac{(1)^2 + 1}{\boxed{(x - 1)} \ (1 - 2)(1 - 3)} = \frac{2}{(-1)(-2)} = 1.$$

⇧
Cover

Similarly, we find the value of B in Eq. (5) by covering the factor $(x - 2)$ in (6) and evaluating the rest at $x = 2$:

$$B = \frac{(2)^2 + 1}{(2 - 1)\ \boxed{(x - 2)}\ (2 - 3)} = \frac{5}{(1)(-1)} = -5.$$

⇧
Cover

Finally, C is found by covering the $(x - 3)$ in (6) and evaluating the rest at $x = 3$:

$$C = \frac{(3)^2 + 1}{(3 - 1)(3 - 2)\ \boxed{(x - 3)}} = \frac{10}{(2)(1)} = 5.$$

⇧
Cover

❑

The steps in the cover-up method are these:

Step 1: Write the quotient with $g(x)$ factored:

$$\frac{f(x)}{g(x)} = \frac{f(x)}{(x - r_1)(x - r_2) \cdots (x - r_n)}. \tag{7}$$

Step 2: Cover the factors $(x - r_i)$ of $g(x)$ in (7) one at a time, each time replacing all the uncovered x's by the number r_i. This gives a number A_i for each root r_i:

$$A_1 = \frac{f(r_1)}{(r_1 - r_2) \cdots (r_1 - r_n)},$$

$$A_2 = \frac{f(r_2)}{(r_2 - r_1)(r_2 - r_3) \cdots (r_2 - r_n)},$$

$$\vdots$$

$$A_n = \frac{f(r_n)}{(r_n - r_1)(r_n - r_2) \cdots (r_n - r_{n-1})}.$$

Step 3: Write the partial-fraction expansion of $f(x)/g(x)$ as

$$\frac{f(x)}{g(x)} = \frac{A_1}{(x - r_1)} + \frac{A_2}{(x - r_2)} + \cdots + \frac{A_n}{(x - r_n)}.$$

EXAMPLE 7 Evaluate

$$\int \frac{x + 4}{x^3 + 3x^2 - 10x}\, dx.$$

Solution The degree of $f(x) = x + 4$ is less than the degree of $g(x) = x^3 + 3x^2 - 10x$, and, with $g(x)$ factored,

$$\frac{x + 4}{x^3 + 3x^2 - 10x} = \frac{x + 4}{x(x - 2)(x + 5)}.$$

The roots of $g(x)$ are $r_1 = 0$, $r_2 = 2$, and $r_3 = -5$. We find

$$A_1 = \frac{0 + 4}{\boxed{x}\ (0 - 2)(0 + 5)} = \frac{4}{(-2)(5)} = -\frac{2}{5},$$

⇧
Cover

$$A_2 = \frac{2+4}{2 \boxed{(x-2)} (2+5)} = \frac{6}{(2)(7)} = \frac{3}{7},$$

⇧
Cover

$$A_3 = \frac{-5+4}{(-5)(-5-2) \boxed{(x+5)}} = \frac{-1}{(-5)(-7)} = -\frac{1}{35}.$$

⇧
Cover

Therefore,

$$\frac{x+4}{x(x-2)(x+5)} = -\frac{2}{5x} + \frac{3}{7(x-2)} - \frac{1}{35(x+5)},$$

and

$$\int \frac{x+4}{x(x-2)(x+5)} \, dx = -\frac{2}{5} \ln|x| + \frac{3}{7} \ln|x-2| - \frac{1}{35} \ln|x+5| + C. \quad \square$$

Other Ways to Determine the Constants

Another way to determine the constants that appear in partial fractions is to differentiate, as in the next example. Still another is to assign selected numerical values to x.

EXAMPLE 8   *Differentiation*

Find A, B, and C in the equation

$$\frac{x-1}{(x+1)^3} = \frac{A}{x+1} + \frac{B}{(x+1)^2} + \frac{C}{(x+1)^3}.$$

Solution   We first clear of fractions:

$$x - 1 = A(x+1)^2 + B(x+1) + C.$$

Substituting $x = -1$ shows $C = -2$. We then differentiate both sides with respect to x, obtaining

$$1 = 2A(x+1) + B.$$

Substituting $x = -1$ shows $B = 1$. We differentiate again to get $0 = 2A$, which shows $A = 0$. Hence

$$\frac{x-1}{(x+1)^3} = \frac{1}{(x+1)^2} - \frac{2}{(x+1)^3}. \quad \square$$

In some problems, assigning small values to x such as $x = 0, \pm 1, \pm 2$, to get equations in A, B, and C provides a fast alternative to other methods.

EXAMPLE 9   *Assigning numerical values to x*

Find A, B, and C in

$$\frac{x^2+1}{(x-1)(x-2)(x-3)} = \frac{A}{x-1} + \frac{B}{x-2} + \frac{C}{x-3}.$$

Solution Clear of fractions to get

$$x^2 + 1 = A(x-2)(x-3) + B(x-1)(x-3) + C(x-1)(x-2).$$

Then let $x = 1, 2, 3$ successively to find A, B, and C:

$$x = 1: \quad (1)^2 + 1 = A(-1)(-2) + B(0) + C(0)$$
$$2 = 2A$$
$$A = 1$$
$$x = 2: \quad (2)^2 + 1 = A(0) + B(1)(-1) + C(0)$$
$$5 = -B$$
$$B = -5$$
$$x = 3: \quad (3)^2 + 1 = A(0) + B(0) + C(2)(1)$$
$$10 = 2C$$
$$C = 5.$$

Conclusion:

$$\frac{x^2 + 1}{(x-1)(x-2)(x-3)} = \frac{1}{x-1} - \frac{5}{x-2} + \frac{5}{x-3}.$$

Exercises 7.3

Expanding Quotients into Partial Fractions

Expand the quotients in Exercises 1–8 by partial fractions.

1. $\dfrac{5x - 13}{(x-3)(x-2)}$

2. $\dfrac{5x - 7}{x^2 - 3x + 2}$

3. $\dfrac{x + 4}{(x+1)^2}$

4. $\dfrac{2x + 2}{x^2 - 2x + 1}$

5. $\dfrac{z + 1}{z^2(z-1)}$

6. $\dfrac{z}{z^3 - z^2 - 6z}$

7. $\dfrac{t^2 + 8}{t^2 - 5t + 6}$

8. $\dfrac{t^4 + 9}{t^4 + 9t^2}$

Nonrepeated Linear Factors

In Exercises 9–16, express the integrands as a sum of partial fractions and evaluate the integrals.

9. $\displaystyle\int \frac{dx}{1 - x^2}$

10. $\displaystyle\int \frac{dx}{x^2 + 2x}$

11. $\displaystyle\int \frac{x + 4}{x^2 + 5x - 6}\, dx$

12. $\displaystyle\int \frac{2x + 1}{x^2 - 7x + 12}\, dx$

13. $\displaystyle\int_4^8 \frac{y\, dy}{y^2 - 2y - 3}$

14. $\displaystyle\int_{1/2}^1 \frac{y + 4}{y^2 + y}\, dy$

15. $\displaystyle\int \frac{dt}{t^3 + t^2 - 2t}$

16. $\displaystyle\int \frac{x + 3}{2x^3 - 8x}\, dx$

Repeated Linear Factors

In Exercises 17–20, express the integrands as a sum of partial fractions and evaluate the integrals.

17. $\displaystyle\int_0^1 \frac{x^3\, dx}{x^2 + 2x + 1}$

18. $\displaystyle\int_{-1}^0 \frac{x^3\, dx}{x^2 - 2x + 1}$

19. $\displaystyle\int \frac{dx}{(x^2 - 1)^2}$

20. $\displaystyle\int \frac{x^2\, dx}{(x-1)(x^2 + 2x + 1)}$

Irreducible Quadratic Factors

In Exercises 21–28, express the integrands as a sum of partial fractions and evaluate the integrals.

21. $\displaystyle\int_0^1 \frac{dx}{(x+1)(x^2+1)}$

22. $\displaystyle\int_1^{\sqrt{3}} \frac{3t^2 + t + 4}{t^3 + t}\, dt$

23. $\displaystyle\int \frac{y^2 + 2y + 1}{(y^2 + 1)^2}\, dy$

24. $\displaystyle\int \frac{8x^2 + 8x + 2}{(4x^2 + 1)^2}\, dx$

25. $\displaystyle\int \frac{2s + 2}{(s^2 + 1)(s-1)^3}\, ds$

26. $\displaystyle\int \frac{s^4 + 81}{s(s^2 + 9)^2}\, ds$

27. $\displaystyle\int \frac{2\theta^3 + 5\theta^2 + 8\theta + 4}{(\theta^2 + 2\theta + 2)^2}\, d\theta$

28. $\displaystyle\int \frac{\theta^4 - 4\theta^3 + 2\theta^2 - 3\theta + 1}{(\theta^2 + 1)^3}\, d\theta$

Improper Fractions

In Exercises 29–34, perform long division on the integrand, write the proper fraction as a sum of partial fractions, and then evaluate the integral.

29. $\displaystyle\int \frac{2x^3 - 2x^2 + 1}{x^2 - x}\, dx$ **30.** $\displaystyle\int \frac{x^4}{x^2 - 1}\, dx$

31. $\displaystyle\int \frac{9x^2 - 3x + 1}{x^3 - x^2}\, dx$ **32.** $\displaystyle\int \frac{16x^3}{4x^2 - 4x + 1}\, dx$

33. $\displaystyle\int \frac{y^4 + y^2 - 1}{y^3 + y}\, dy$ **34.** $\displaystyle\int \frac{2y^4}{y^3 - y^2 + y - 1}\, dy$

Evaluating Integrals

Evaluate the integrals in Exercises 35–40.

35. $\displaystyle\int \frac{e^t\, dt}{e^{2t} + 3e^t + 2}$ **36.** $\displaystyle\int \frac{e^{4t} + 2e^{2t} - e^t}{e^{2t} + 1}\, dt$

37. $\displaystyle\int \frac{\cos y\, dy}{\sin^2 y + \sin y - 6}$ **38.** $\displaystyle\int \frac{\sin \theta\, d\theta}{\cos^2 \theta + \cos \theta - 2}$

39. $\displaystyle\int \frac{(x - 2)^2 \tan^{-1}(2x) - 12x^3 - 3x}{(4x^2 + 1)(x - 2)^2}\, dx$

40. $\displaystyle\int \frac{(x + 1)^2 \tan^{-1}(3x) + 9x^3 + x}{(9x^2 + 1)(x + 1)^2}\, dx$

Initial Value Problems

Solve the initial value problems in Exercises 41–44 for x as a function of t.

41. $(t^2 - 3t + 2)\dfrac{dx}{dt} = 1 \quad (t > 2), \quad x(3) = 0$

42. $(3t^4 + 4t^2 + 1)\dfrac{dx}{dt} = 2\sqrt{3}, \quad x(1) = -\pi\sqrt{3}/4$

43. $(t^2 + 2t)\dfrac{dx}{dt} = 2x + 2 \quad (t, x > 0), \quad x(1) = 1$

44. $(t + 1)\dfrac{dx}{dt} = x^2 + 1 \quad (t > -1), \quad x(0) = \pi/4$

Applications and Examples

In Exercises 45 and 46, find the volume of the solid generated by revolving the shaded region about the indicated axis.

45. The x-axis

46. The y-axis

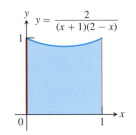

47. CALCULATOR Find, to 2 decimal places, the x-coordinate of the centroid of the region in the first quadrant bounded by the x-axis, the curve $y = \tan^{-1} x$, and the line $x = \sqrt{3}$.

48. CALCULATOR Find the x-coordinate of the centroid of this region to 2 decimal places.

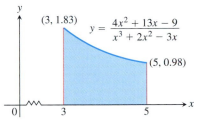

49. *Social diffusion.* Sociologists sometimes use the phrase "social diffusion" to describe the way information spreads through a population. The information might be a rumor, a cultural fad, or news about a technical innovation. In a sufficiently large population, the number of people x who have the information is treated as a differentiable function of time t, and the rate of diffusion, dx/dt, is assumed to be proportional to the number of people who have the information times the number of people who do not. This leads to the equation

$$\frac{dx}{dt} = kx(N - x),$$

where N is the number of people in the population.

Suppose t is in days, $k = 1/250$, and two people start a rumor at time $t = 0$ in a population of $N = 1000$ people.

a) Find x as a function of t.

b) When will half the population have heard the rumor? (This is when the rumor will be spreading the fastest.)

50. *Second order chemical reactions.* Many chemical reactions are the result of the interaction of two molecules that undergo a change to produce a new product. The rate of the reaction typically depends on the concentrations of the two kinds of molecules. If a is the amount of substance A and b is the amount of substance B at time $t = 0$, and if x is the amount of product at time t, then the rate of formation of x may be given by the differential equation

$$\frac{dx}{dt} = k(a - x)(b - x),$$

or

$$\frac{1}{(a-x)(b-x)}\frac{dx}{dt} = k,$$

where k is a constant for the reaction. Integrate both sides of this equation to obtain a relation between x and t (a) if $a = b$, and (b) if $a \neq b$. Assume in each case that $x = 0$ when $t = 0$.

51. *An integral connecting π to the approximation 22/7*

a) Evaluate $\displaystyle\int_0^1 \frac{x^4(x-1)^4}{x^2+1}dx$.

b) **CALCULATOR** How good is the approximation $\pi \approx 22/7$? Find out by expressing $(\pi - 22/7)$ as a percentage of π.

c) **GRAPHER** Graph the function $y = \dfrac{x^4(x-1)^4}{x^2+1}$ for $0 \leq x \leq 1$. Experiment with the range on the y-axis set between 0 and 1, then between 0 and 0.5, and then decreasing the range until the graph can be seen. What do you conclude about the area under the curve?

52. Find the second degree polynomial $P(x)$ such that $P(0) = 1$, $P'(0) = 0$, and

$$\int \frac{P(x)}{x^3(x-1)^2}dx$$

is a rational function.

7.4 Trigonometric Substitutions

Trigonometric substitutions enable us to replace the binomials $a^2 + x^2, a^2 - x^2$, and $x^2 - a^2$ by single squared terms and thereby transform a number of integrals containing square roots into integrals we can evaluate directly.

Three Basic Substitutions

The most common substitutions are $x = a \tan \theta$, $x = a \sin \theta$, and $x = a \sec \theta$. They come from the reference right triangles in Fig. 7.3.

With $x = a \tan \theta$,

$$a^2 + x^2 = a^2 + a^2 \tan^2 \theta = a^2(1 + \tan^2 \theta) = a^2 \sec^2 \theta. \tag{1}$$

With $x = a \sin \theta$,

$$a^2 - x^2 = a^2 - a^2 \sin^2 \theta = a^2(1 - \sin^2 \theta) = a^2 \cos^2 \theta. \tag{2}$$

With $x = a \sec \theta$,

$$x^2 - a^2 = a^2 \sec^2 \theta - a^2 = a^2(\sec^2 \theta - 1) = a^2 \tan^2 \theta. \tag{3}$$

Trigonometric Substitutions

1. $x = a \tan \theta$ replaces $a^2 + x^2$ by $a^2 \sec^2 \theta$.
2. $x = a \sin \theta$ replaces $a^2 - x^2$ by $a^2 \cos^2 \theta$.
3. $x = a \sec \theta$ replaces $x^2 - a^2$ by $a^2 \tan^2 \theta$.

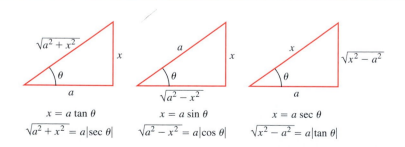

7.3 Reference triangles for trigonometric substitutions that change binomials into single squared terms.

$x = a \tan \theta$
$\sqrt{a^2 + x^2} = a|\sec \theta|$

$x = a \sin \theta$
$\sqrt{a^2 - x^2} = a|\cos \theta|$

$x = a \sec \theta$
$\sqrt{x^2 - a^2} = a|\tan \theta|$

We want any substitution we use in an integration to be reversible so that we can change back to the original variable afterward. For example, if $x = a \tan \theta$, we want to be able to set $\theta = \tan^{-1}(x/a)$ after the integration takes place. If $x = a \sin \theta$, we want to be able to set $\theta = \sin^{-1}(x/a)$ when we're done, and similarly for $x = a \sec \theta$.

As we know from Section 6.8, the functions in these substitutions have inverses only for selected values of θ (Fig. 7.4). For reversibility,

$$x = a \tan \theta \quad \text{requires} \quad \theta = \tan^{-1}\left(\frac{x}{a}\right) \quad \text{with} \quad -\frac{\pi}{2} < \theta < \frac{\pi}{2},$$

$$x = a \sin \theta \quad \text{requires} \quad \theta = \sin^{-1}\left(\frac{x}{a}\right) \quad \text{with} \quad -\frac{\pi}{2} \leq \theta \leq \frac{\pi}{2},$$

$$x = a \sec \theta \quad \text{requires} \quad \theta = \sec^{-1}\left(\frac{x}{a}\right) \quad \text{with} \quad \begin{cases} 0 \leq \theta < \dfrac{\pi}{2} & \text{if} \quad \dfrac{x}{a} \geq 1, \\[2mm] \dfrac{\pi}{2} < \theta \leq \pi & \text{if} \quad \dfrac{x}{a} \leq -1. \end{cases}$$

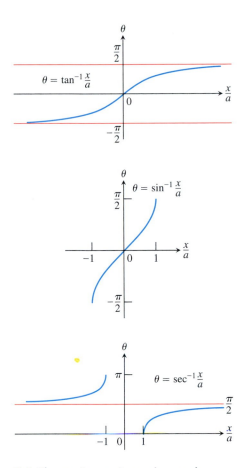

7.4 The arc tangent, arc sine, and arc secant of x/a, graphed as functions of x/a.

To simplify calculations with the substitution $x = a \sec \theta$, we will restrict its use to integrals in which $x/a \geq 1$. This will place θ in $[0, \pi/2)$ and make $\tan \theta \geq 0$. We will then have $\sqrt{x^2 - a^2} = \sqrt{a^2 \tan^2 \theta} = |a \tan \theta| = a \tan \theta$, free of absolute values, provided $a > 0$.

EXAMPLE 1   Evaluate $\displaystyle\int \frac{dx}{\sqrt{4 + x^2}}$.

Solution   We set

$$x = 2 \tan \theta, \qquad dx = 2 \sec^2 \theta \, d\theta, \qquad -\frac{\pi}{2} < \theta < \frac{\pi}{2},$$

$$4 + x^2 = 4 + 4 \tan^2 \theta = 4(1 + \tan^2 \theta) = 4 \sec^2 \theta.$$

Then

$$\int \frac{dx}{\sqrt{4 + x^2}} = \int \frac{2 \sec^2 \theta \, d\theta}{\sqrt{4 \sec^2 \theta}} = \int \frac{\sec^2 \theta \, d\theta}{|\sec \theta|} \qquad \sqrt{\sec^2 \theta} = |\sec \theta|$$

$$= \int \sec \theta \, d\theta \qquad\qquad \begin{matrix} \sec \theta > 0 \text{ for} \\ -\dfrac{\pi}{2} < \theta < \dfrac{\pi}{2} \end{matrix}$$

$$= \ln |\sec \theta + \tan \theta| + C$$

$$= \ln \left| \frac{\sqrt{4 + x^2}}{2} + \frac{x}{2} \right| + C \qquad \text{From Fig. 7.5}$$

$$= \ln \left| \sqrt{4 + x^2} + x \right| + C'. \qquad \text{Taking } C' = C - \ln 2$$

Notice how we expressed $\ln |\sec \theta + \tan \theta|$ in terms of x: We drew a reference triangle for the original substitution $x = 2 \tan \theta$ (Fig. 7.5) and read the ratios from the triangle. ❑

7.5 Reference triangle for $x = 2 \tan \theta$ (Example 1):

$$\tan \theta = \frac{x}{2}$$

and

$$\sec \theta = \frac{\sqrt{4 + x^2}}{2}.$$

EXAMPLE 2   Evaluate $\displaystyle\int \frac{x^2 \, dx}{\sqrt{9 - x^2}}$.

Solution To replace $9 - x^2$ by a single squared term, we set

$$x = 3 \sin \theta, \qquad dx = 3 \cos \theta \, d\theta, \qquad -\frac{\pi}{2} < \theta < \frac{\pi}{2},$$

$$9 - x^2 = 9(1 - \sin^2 \theta) = 9 \cos^2 \theta.$$

Then

$$\int \frac{x^2 \, dx}{\sqrt{9 - x^2}} = \int \frac{9 \sin^2 \theta \cdot 3 \cos \theta \, d\theta}{|3 \cos \theta|}$$

$$= 9 \int \sin^2 \theta \, d\theta \qquad\qquad \cos \theta > 0 \text{ for } -\frac{\pi}{2} < \theta < \frac{\pi}{2}$$

$$= 9 \int \frac{1 - \cos 2\theta}{2} \, d\theta$$

$$= \frac{9}{2} \left(\theta - \frac{\sin 2\theta}{2} \right) + C$$

$$= \frac{9}{2} (\theta - \sin \theta \cos \theta) + C \qquad\qquad \sin 2\theta = 2 \sin \theta \cos \theta$$

$$= \frac{9}{2} \left(\sin^{-1} \frac{x}{3} - \frac{x}{3} \cdot \frac{\sqrt{9 - x^2}}{3} \right) + C \qquad \text{Fig. 7.6}$$

$$= \frac{9}{2} \sin^{-1} \frac{x}{3} - \frac{x}{2} \sqrt{9 - x^2} + C.$$

7.6 Reference triangle for $x = 3 \sin \theta$ (Example 2):

$$\sin \theta = \frac{x}{3}$$

and

$$\cos \theta = \frac{\sqrt{9 - x^2}}{3}.$$

EXAMPLE 3 Evaluate $\displaystyle \int \frac{dx}{\sqrt{25x^2 - 4}}, \quad x > \frac{2}{5}.$

Solution We first rewrite the radical as

$$\sqrt{25x^2 - 4} = \sqrt{25 \left(x^2 - \frac{4}{25} \right)}$$

$$= 5 \sqrt{x^2 - \left(\frac{2}{5} \right)^2}$$

to put the radicand in the form $x^2 - a^2$. We then substitute

$$x = \frac{2}{5} \sec \theta, \qquad dx = \frac{2}{5} \sec \theta \tan \theta \, d\theta, \qquad 0 < \theta < \frac{\pi}{2}$$

$$x^2 - \left(\frac{2}{5} \right)^2 = \frac{4}{25} \sec^2 \theta - \frac{4}{25}$$

$$= \frac{4}{25} (\sec^2 \theta - 1) = \frac{4}{25} \tan^2 \theta,$$

$$\sqrt{x^2 - \left(\frac{2}{5} \right)^2} = \frac{2}{5} |\tan \theta| = \frac{2}{5} \tan \theta. \qquad\qquad \tan \theta > 0 \text{ for } 0 < \theta < \pi/2$$

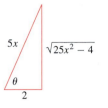

7.7 If $x = (2/5) \sec \theta$, $0 \le \theta < \pi/2$, then $\theta = \sec^{-1}(5x/2)$ and we can read the values of the other trigonometric functions of θ from this right triangle.

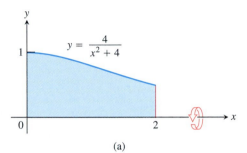

(a)

With these substitutions, we have

$$\int \frac{dx}{\sqrt{25x^2 - 4}} = \int \frac{dx}{5\sqrt{x^2 - (4/25)}} = \int \frac{(2/5) \sec \theta \tan \theta \, d\theta}{5 \cdot (2/5) \tan \theta}$$

$$= \frac{1}{5} \int \sec \theta \, d\theta = \frac{1}{5} \ln |\sec \theta + \tan \theta| + C$$

$$= \frac{1}{5} \ln \left| \frac{5x}{2} + \frac{\sqrt{25x^2 - 4}}{2} \right| + C \qquad \text{Fig. 7.7}$$

A trigonometric substitution can sometimes help us to evaluate an integral containing an integer power of a quadratic binomial, as in the next example.

EXAMPLE 4   Find the volume of the solid generated by revolving about the x-axis the region bounded by the curve $y = 4/(x^2 + 4)$, the x-axis, and the lines $x = 0$ and $x = 2$.

Solution   We sketch the region (Fig. 7.8) and use the disk method (Section 5.3):

$$V = \int_0^2 \pi [R(x)]^2 \, dx = 16\pi \int_0^2 \frac{dx}{(x^2 + 4)^2}. \qquad R(x) = \frac{4}{x^2 + 4}$$

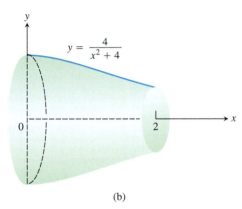

(b)

7.8 The region (a) and solid (b) in Example 4.

To evaluate the integral, we set

$$x = 2 \tan \theta, \qquad dx = 2 \sec^2 \theta \, d\theta, \qquad \theta = \tan^{-1} \frac{x}{2},$$

$$x^2 + 4 = 4 \tan^2 \theta + 4 = 4(\tan^2 \theta + 1) = 4 \sec^2 \theta$$

(Fig. 7.9). With these substitutions,

$$V = 16\pi \int_0^2 \frac{dx}{(x^2 + 4)^2}$$

$$= 16\pi \int_0^{\pi/4} \frac{2 \sec^2 \theta \, d\theta}{(4 \sec^2 \theta)^2} \qquad \begin{matrix} \theta = 0 \\ \text{when } x = 0; \\ \theta = \pi/4 \\ \text{when } x = 2 \end{matrix}$$

$$= 16\pi \int_0^{\pi/4} \frac{2 \sec^2 \theta \, d\theta}{16 \sec^4 \theta} = \pi \int_0^{\pi/4} 2 \cos^2 \theta \, d\theta$$

$$= \pi \int_0^{\pi/4} (1 + \cos 2\theta) d\theta = \pi \left[\theta + \frac{\sin 2\theta}{2} \right]_0^{\pi/4} \qquad \begin{matrix} 2 \cos^2 \theta = \\ 1 + \cos 2\theta \end{matrix}$$

$$= \pi \left[\frac{\pi}{4} + \frac{1}{2} \right] \approx 4.04.$$

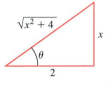

7.9 Reference triangle for $x = 2 \tan \theta$ (Example 4).

Exercises 7.4

Basic Trigonometric Substitutions

Evaluate the integrals in Exercises 1–28.

1. $\displaystyle\int \frac{dy}{\sqrt{9 + y^2}}$

2. $\displaystyle\int \frac{3\,dy}{\sqrt{1 + 9y^2}}$

3. $\displaystyle\int_{-2}^{2} \frac{dx}{4 + x^2}$

4. $\displaystyle\int_{0}^{2} \frac{dx}{8 + 2x^2}$

5. $\displaystyle\int_{0}^{3/2} \frac{dx}{\sqrt{9 - x^2}}$

6. $\displaystyle\int_{0}^{1/2\sqrt{2}} \frac{2\,dx}{\sqrt{1 - 4x^2}}$

7. $\displaystyle\int \sqrt{25 - t^2}\,dt$

8. $\displaystyle\int \sqrt{1 - 9t^2}\,dt$

9. $\displaystyle\int \frac{dx}{\sqrt{4x^2 - 49}}, \quad x > \frac{7}{2}$

10. $\displaystyle\int \frac{5\,dx}{\sqrt{25x^2 - 9}}, \quad x > \frac{3}{5}$

11. $\displaystyle\int \frac{\sqrt{y^2 - 49}}{y}\,dy, \quad y > 7$

12. $\displaystyle\int \frac{\sqrt{y^2 - 25}}{y^3}\,dy, \quad y > 5$

13. $\displaystyle\int \frac{dx}{x^2\sqrt{x^2 - 1}}, \quad x > 1$

14. $\displaystyle\int \frac{2\,dx}{x^3\sqrt{x^2 - 1}}, \quad x > 1$

15. $\displaystyle\int \frac{x^3\,dx}{\sqrt{x^2 + 4}}$

16. $\displaystyle\int \frac{dx}{x^2\sqrt{x^2 + 1}}$

17. $\displaystyle\int \frac{8\,dw}{w^2\sqrt{4 - w^2}}$

18. $\displaystyle\int \frac{\sqrt{9 - w^2}}{w^2}\,dw$

19. $\displaystyle\int_{0}^{\sqrt{3}/2} \frac{4x^2\,dx}{(1 - x^2)^{3/2}}$

20. $\displaystyle\int_{0}^{1} \frac{dx}{(4 - x^2)^{3/2}}$

21. $\displaystyle\int \frac{dx}{(x^2 - 1)^{3/2}}, \quad x > 1$

22. $\displaystyle\int \frac{x^2\,dx}{(x^2 - 1)^{5/2}}, \quad x > 1$

23. $\displaystyle\int \frac{(1 - x^2)^{3/2}}{x^6}\,dx$

24. $\displaystyle\int \frac{(1 - x^2)^{1/2}}{x^4}\,dx$

25. $\displaystyle\int \frac{8\,dx}{(4x^2 + 1)^2}$

26. $\displaystyle\int \frac{6\,dt}{(9t^2 + 1)^2}$

27. $\displaystyle\int \frac{v^2\,dv}{(1 - v^2)^{5/2}}$

28. $\displaystyle\int \frac{(1 - r^2)^{5/2}}{r^8}\,dr$

In Exercises 29–36, use an appropriate substitution and then a trigonometric substitution to evaluate the integrals.

29. $\displaystyle\int_{0}^{\ln 4} \frac{e^t\,dt}{\sqrt{e^{2t} + 9}}$

30. $\displaystyle\int_{\ln(3/4)}^{\ln(4/3)} \frac{e^t\,dt}{(1 + e^{2t})^{3/2}}$

31. $\displaystyle\int_{1/12}^{1/4} \frac{2\,dt}{\sqrt{t} + 4t\sqrt{t}}$

32. $\displaystyle\int_{1}^{e} \frac{dy}{y\sqrt{1 + (\ln y)^2}}$

33. $\displaystyle\int \frac{dx}{x\sqrt{x^2 - 1}}$

34. $\displaystyle\int \frac{dx}{1 + x^2}$

35. $\displaystyle\int \frac{x\,dx}{\sqrt{x^2 - 1}}$

36. $\displaystyle\int \frac{dx}{\sqrt{1 - x^2}}$

Initial Value Problems

Solve the initial value problems in Exercises 37–40 for y as a function of x.

37. $x\dfrac{dy}{dx} = \sqrt{x^2 - 4}, \quad x \ge 2, \quad y(2) = 0$

38. $\sqrt{x^2 - 9}\dfrac{dy}{dx} = 1, \quad x > 3, \quad y(5) = \ln 3$

39. $(x^2 + 4)\dfrac{dy}{dx} = 3, \quad y(2) = 0$

40. $(x^2 + 1)^2\dfrac{dy}{dx} = \sqrt{x^2 + 1}, \quad y(0) = 1$

Applications

41. Find the area of the region in the first quadrant that is enclosed by the coordinate axes and the curve $y = \sqrt{9 - x^2}/3$.

42. Find the volume of the solid generated by revolving about the x-axis the region in the first quadrant enclosed by the coordinate axes, the curve $y = 2/(1 + x^2)$, and the line $x = 1$.

The Substitution $z = \tan\,(x/2)$

The substitution

$$z = \tan \frac{x}{2} \tag{4}$$

reduces the problem of integrating a rational expression in $\sin x$ and $\cos x$ to a problem of integrating a rational function of z. This in turn can be integrated by partial fractions. Thus the substitution (4) is a powerful tool. It is cumbersome, however, and is used only when simpler methods fail.

Figure 7.10 shows how $\tan\,(x/2)$ expresses a rational function of $\sin x$ and $\cos x$. To see the effect of the substitution, we calculate

$$\cos x = 2\cos^2\left(\frac{x}{2}\right) - 1 = \frac{2}{\sec^2(x/2)} - 1$$

$$= \frac{2}{1 + \tan^2(x/2)} - 1 = \frac{2}{1 + z^2} - 1$$

$$\cos x = \frac{1 - z^2}{1 + z^2}, \tag{5}$$

and

$$\sin x = 2\sin\frac{x}{2}\cos\frac{x}{2} = 2\frac{\sin(x/2)}{\cos(x/2)} \cdot \cos^2\left(\frac{x}{2}\right)$$

$$= 2\tan\frac{x}{2} \cdot \frac{1}{\sec^2(x/2)} = \frac{2\tan(x/2)}{1 + \tan^2(x/2)}$$

$$\sin x = \frac{2z}{1 + z^2}. \tag{6}$$

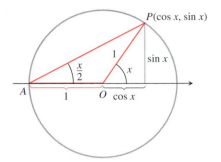

7.10 From this figure, we can read the relation

$$\tan \frac{x}{2} = \frac{\sin x}{1 + \cos x}.$$

Finally, $x = 2 \tan^{-1} z$, so

$$dx = \frac{2\, dz}{1 + z^2}. \tag{7}$$

EXAMPLE

a) $\displaystyle \int \frac{1}{1 + \cos x}\, dx = \int \frac{1 + z^2}{2} \frac{2\, dz}{1 + z^2}$

$$= \int dz = z + C$$

$$= \tan\left(\frac{x}{2}\right) + C$$

b) $\displaystyle \int \frac{1}{2 + \sin x}\, dx = \int \frac{1 + z^2}{2 + 2z + 2z^2} \frac{2\, dz}{1 + z^2}$

$$= \int \frac{dz}{z^2 + z + 1} = \int \frac{dz}{(z + (1/2))^2 + 3/4}$$

$$= \int \frac{du}{u^2 + a^2}$$

$$= \frac{1}{a} \tan^{-1}\left(\frac{u}{a}\right) + C$$

$$= \frac{2}{\sqrt{3}} \tan^{-1} \frac{2z + 1}{\sqrt{3}} + C$$

$$= \frac{2}{\sqrt{3}} \tan^{-1} \frac{1 + 2 \tan(x/2)}{\sqrt{3}} + C \qquad \square$$

Use the substitutions in Eqs. (4)–(7) to evaluate the integrals in Exercises 43–50. Integrals like these arise in calculating the average angular velocity of the output shaft of a universal joint when the input and output shafts are not aligned.

43. $\displaystyle \int \frac{dx}{1 - \sin x}$ **44.** $\displaystyle \int \frac{dx}{1 + \sin x + \cos x}$

45. $\displaystyle \int_0^{\pi/2} \frac{dx}{1 + \sin x}$ **46.** $\displaystyle \int_{\pi/3}^{\pi/2} \frac{dx}{1 - \cos x}$

47. $\displaystyle \int_0^{\pi/2} \frac{d\theta}{2 + \cos \theta}$ **48.** $\displaystyle \int_{\pi/2}^{2\pi/3} \frac{\cos \theta\, d\theta}{\sin \theta \cos \theta + \sin \theta}$

49. $\displaystyle \int \frac{dt}{\sin t - \cos t}$ **50.** $\displaystyle \int \frac{\cos t\, dt}{1 - \cos t}$

Use the substitution $z = \tan(\theta/2)$ to evaluate the integrals in Exercises 51 and 52.

51. $\displaystyle \int \sec \theta\, d\theta$ **52.** $\displaystyle \int \csc \theta\, d\theta$

7.5

Integral Tables and CAS

As you know, the basic techniques of integration are substitution and integration by parts. We apply these techniques to transform unfamiliar integrals into integrals whose forms we recognize or can find in a table. But where do the integrals in the tables come from? They come from applying substitutions and integration by parts. We could derive them all from scratch if we had to, but having the table saves us the trouble of repeating laborious calculations. When an integral matches an integral in the table or can be changed into one of the tabulated integrals with some appropriate combination of algebra, trigonometry, substitution, and calculus, we have a ready-made solution for the problem at hand. The examples and exercises of this section show how the formulas in integral tables are derived and used. The emphasis is on use. The integration formulas at the back of this book are stated in terms of constants a, b, c, m, n, and so on. These constants can usually assume any real value and need not be integers. Occasional limitations on their values are stated with the formulas. Formula 5 requires $n \neq -1$, for example, and Formula 11 requires $n \neq -2$.

The formulas also assume that the constants do not take on values that require dividing by zero or taking even roots of negative numbers. For example, Formula 8 assumes $a \neq 0$, and Formula 13(a) cannot be used unless b is negative.

Many indefinite integrals can also be evaluated with a Computer Algebra System (CAS). These systems are generally faster than tables and usually do not require you to rewrite integrals in special recognizable forms first. We discuss computer algebra systems in the last third of the section.

Integration with Tables

EXAMPLE 1   Find $\displaystyle\int x(2x + 5)^{-1}\,dx$.

Solution   We use Formula 8 (not 7, which requires $n \neq -1$):

$$\int x(ax + b)^{-1}\,dx = \frac{x}{a} - \frac{b}{a^2} \ln |ax + b| + C.$$

With $a = 2$ and $b = 5$, we have

$$\int x(2x + 5)^{-1}\,dx = \frac{x}{2} - \frac{5}{4} \ln |2x + 5| + C. \qquad \square$$

EXAMPLE 2   Find $\displaystyle\int \frac{dx}{x\sqrt{2x + 4}}$.

Solution   We use Formula 13(b):

$$\int \frac{dx}{x\sqrt{ax + b}} = \frac{1}{\sqrt{b}} \ln \left| \frac{\sqrt{ax + b} - \sqrt{b}}{\sqrt{ax + b} + \sqrt{b}} \right| + C, \qquad \text{if } b > 0.$$

With $a = 2$ and $b = 4$, we have

$$\int \frac{dx}{x\sqrt{2x + 4}} = \frac{1}{\sqrt{4}} \ln \left| \frac{\sqrt{2x + 4} - \sqrt{4}}{\sqrt{2x + 4} + \sqrt{4}} \right| + C$$

$$= \frac{1}{2} \ln \left| \frac{\sqrt{2x + 4} - 2}{\sqrt{2x + 4} + 2} \right| + C.$$

Formula 13(a), which requires $b < 0$, would not have been appropriate here. It *is* appropriate, however, in the next example. $\qquad \square$

EXAMPLE 3   Find $\displaystyle\int \frac{dx}{x\sqrt{2x - 4}}$.

Solution   We use Formula 13(a):

$$\int \frac{dx}{x\sqrt{ax - b}} = \frac{2}{\sqrt{b}} \tan^{-1} \sqrt{\frac{ax - b}{b}} + C.$$

With $a = 2$ and $b = 4$, we have

$$\int \frac{dx}{x\sqrt{2x-4}} = \frac{2}{\sqrt{4}} \tan^{-1} \sqrt{\frac{2x-4}{4}} + C = \tan^{-1} \sqrt{\frac{x-2}{2}} + C.$$

EXAMPLE 4   Find $\displaystyle\int \frac{dx}{x^2\sqrt{2x-4}}$.

Solution   We begin with Formula 15:

$$\int \frac{dx}{x^2\sqrt{ax+b}} = -\frac{\sqrt{ax+b}}{bx} - \frac{a}{2b} \int \frac{dx}{x\sqrt{ax+b}} + C.$$

With $a = 2$ and $b = -4$, we have

$$\int \frac{dx}{x^2\sqrt{2x-4}} = -\frac{\sqrt{2x-4}}{-4x} + \frac{2}{2\cdot 4} \int \frac{dx}{x\sqrt{2x-4}} + C.$$

We then use Formula 13(a) to evaluate the integral on the right (Example 3) to obtain

$$\int \frac{dx}{x^2\sqrt{2x-4}} = \frac{\sqrt{2x-4}}{4x} + \frac{1}{4} \tan^{-1} \sqrt{\frac{x-2}{2}} + C.$$

EXAMPLE 5   Find $\displaystyle\int x \sin^{-1} x \, dx$.

Solution   We use Formula 99:

$$\int x^n \sin^{-1} ax \, dx = \frac{x^{n+1}}{n+1} \sin^{-1} ax - \frac{a}{n+1} \int \frac{x^{n+1} \, dx}{\sqrt{1-a^2x^2}}, \qquad n \neq -1.$$

With $n = 1$ and $a = 1$, we have

$$\int x \sin^{-1} x \, dx = \frac{x^2}{2} \sin^{-1} x - \frac{1}{2} \int \frac{x^2 \, dx}{\sqrt{1-x^2}}.$$

The integral on the right is found in the table as Formula 33:

$$\int \frac{x^2}{\sqrt{a^2-x^2}} \, dx = \frac{a^2}{2} \sin^{-1} \left(\frac{x}{a}\right) - \frac{1}{2} x \sqrt{a^2-x^2} + C.$$

With $a = 1$,

$$\int \frac{x^2 \, dx}{\sqrt{1-x^2}} = \frac{1}{2} \sin^{-1} x - \frac{1}{2} x \sqrt{1-x^2} + C.$$

The combined result is

$$\int x \sin^{-1} x \, dx = \frac{x^2}{2} \sin^{-1} x - \frac{1}{2} \left(\frac{1}{2} \sin^{-1} x - \frac{1}{2} x \sqrt{1-x^2}\right) + C'$$

$$= \left(\frac{x^2}{2} - \frac{1}{4}\right) \sin^{-1} x + \frac{1}{4} x \sqrt{1-x^2} + C'.$$

Reduction Formulas

The time required for repeated integrations by parts can sometimes be shortened by applying formulas like

$$\int \tan^n x \, dx = \frac{1}{n-1} \tan^{n-1} x - \int \tan^{n-2} x \, dx \tag{1}$$

$$\int (\ln x)^n \, dx = x(\ln x)^n - n \int (\ln x)^{n-1} \, dx \tag{2}$$

$$\int \sin^n x \cos^m x \, dx = -\frac{\sin^{n-1} x \cos^{m+1} x}{m+n} +$$

$$\frac{n-1}{m+n} \int \sin^{n-2} x \cos^m x \, dx \qquad (n \neq -m). \tag{3}$$

Formulas like these are called **reduction formulas** because they replace an integral containing some power of a function with an integral of the same form with the power reduced. By applying such a formula repeatedly, we can eventually express the original integral in terms of a power low enough to be evaluated directly.

EXAMPLE 6   Find $\int \tan^5 x \, dx.$

Solution   We apply Eq. (1) with $n = 5$ to get

$$\int \tan^5 x \, dx = \frac{1}{4} \tan^4 x - \int \tan^3 x \, dx.$$

We then apply Eq. (1) again, with $n = 3$, to evaluate the remaining integral:

$$\int \tan^3 x \, dx = \frac{1}{2} \tan^2 x - \int \tan x \, dx = \frac{1}{2} \tan^2 x + \ln |\cos x| + C.$$

The combined result is

$$\int \tan^5 x \, dx = \frac{1}{4} \tan^4 x - \frac{1}{2} \tan^2 x - \ln |\cos x| + C'. \qquad \square$$

As their form suggests, reduction formulas are derived by integration by parts.

EXAMPLE 7   *Deriving a reduction formula*

Show that for any positive integer n,

$$\int (\ln x)^n \, dx = x(\ln x)^n - n \int (\ln x)^{n-1} \, dx.$$

Solution   We use the integration by parts formula

$$\int u \, dv = uv - \int v \, du$$

with

$$u = (\ln x)^n, \qquad du = n(\ln x)^{n-1} \frac{dx}{x}, \qquad dv = dx, \qquad v = x,$$

to obtain

$$\int (\ln x)^n dx = x(\ln x)^n - n \int (\ln x)^{n-1} dx.$$

☐

Sometimes two reduction formulas come into play.

EXAMPLE 8   Find $\displaystyle\int \sin^2 x \cos^3 x \, dx.$

Solution 1   We apply Eq. (3) with $n = 2$ and $m = 3$ to get

$$\int \sin^2 x \cos^3 x \, dx = -\frac{\sin x \cos^4 x}{2+3} + \frac{1}{2+3} \int \sin^0 x \cos^3 x \, dx$$

$$= -\frac{\sin x \cos^4 x}{5} + \frac{1}{5} \int \cos^3 x \, dx.$$

We can evaluate the remaining integral with Formula 61 (another reduction formula):

$$\int \cos^n ax \, dx = \frac{\cos^{n-1} ax \sin ax}{na} + \frac{n-1}{n} \int \cos^{n-2} ax \, dx.$$

With $n = 3$ and $a = 1$, we have

$$\int \cos^3 x \, dx = \frac{\cos^2 x \sin x}{3} + \frac{2}{3} \int \cos x \, dx$$

$$= \frac{\cos^2 x \sin x}{3} + \frac{2}{3} \sin x + C.$$

The combined result is

$$\int \sin^2 x \cos^3 x \, dx = -\frac{\sin x \cos^4 x}{5} + \frac{1}{5} \left(\frac{\cos^2 x \sin x}{3} + \frac{2}{3} \sin x + C \right)$$

$$= -\frac{\sin x \cos^4 x}{5} + \frac{\cos^2 x \sin x}{15} + \frac{2}{15} \sin x + C'.$$

Solution 2   Equation (3) corresponds to Formula 68 in the table, but there is another formula we might use, namely Formula 69. With $a = 1$, Formula 69 gives

$$\int \sin^n x \cos^m x \, dx = \frac{\sin^{n+1} x \cos^{m-1} x}{m+n} + \frac{m-1}{m+n} \int \sin^n x \cos^{m-2} x \, dx.$$

In our case, $n = 2$ and $m = 3$, so that

$$\int \sin^2 x \cos^3 x \, dx = \frac{\sin^3 x \cos^2 x}{5} + \frac{2}{5} \int \sin^2 x \cos x \, dx$$

$$= \frac{\sin^3 x \cos^2 x}{5} + \frac{2}{5} \left(\frac{\sin^3 x}{3} \right) + C$$

$$= \frac{\sin^3 x \cos^2 x}{5} + \frac{2}{15} \sin^3 x + C.$$

As you can see, it is faster to use Formula 69, but we often cannot tell beforehand how things will work out. Do not spend a lot of time looking for the "best" formula. Just find one that will work and forge ahead.

Notice also that Formulas 68 (Solution 1) and 69 (Solution 2) lead to different-looking answers. That is often the case with trigonometric integrals and is no cause for concern. The results are equivalent, and we may use whichever one we please.

Nonelementary Integrals

The development of computers and calculators that find antiderivatives by symbolic manipulation has led to a renewed interest in determining which antiderivatives can be expressed as finite combinations of elementary functions (the functions we have been studying) and which cannot. Integrals of functions that do not have elementary antiderivatives are called **nonelementary** integrals. They require infinite series (Chapter 8) or numerical methods for their evaluation. Examples of the latter include the error function

$$\text{erf}(x) = \frac{2}{\sqrt{\pi}} \int_0^x e^{-t^2} dt$$

and integrals such as

$$\int \sin x^2 \, dx \qquad \text{and} \qquad \int \sqrt{1 + x^4} \, dx$$

that arise in engineering and physics. These and a number of others, such as

$$\int \frac{e^x}{x} \, dx, \qquad \int e^{(e^x)} \, dx, \qquad \int \frac{1}{\ln x} \, dx, \qquad \int \ln(\ln x) \, dx, \qquad \int \frac{\sin x}{x} \, dx,$$

$$\int \sqrt{1 - k^2 \sin^2 x} \, dx, \qquad 0 < k < 1,$$

look so easy they tempt us to try them just to see how they turn out. It can be proved, however, that there is no way to express these integrals as finite combinations of elementary functions. The same applies to integrals that can be changed into these by substitution. The integrands all have antiderivatives—they are, after all, continuous—but none of the antiderivatives is elementary.

None of the integrals you are asked to evaluate in the present chapter falls into this category, but you may encounter nonelementary integrals from time to time in your other work.

A General Procedure for Indefinite Integration

While there is no surefire way to evaluate all indefinite integrals, the procedure in Flowchart 7.1 may help.

Integration with a Computer Algebra System (CAS)

A powerful capability of Computer Algebra Systems is their facility to integrate symbolically. This is performed with the **integrate command** specified by the particular system (e.g., **int** in Maple, **Integrate** in Mathematica).

EXAMPLE 9   Suppose you want to evaluate the indefinite integral of the function

$$f(x) = x^2 \sqrt{a^2 + x^2}.$$

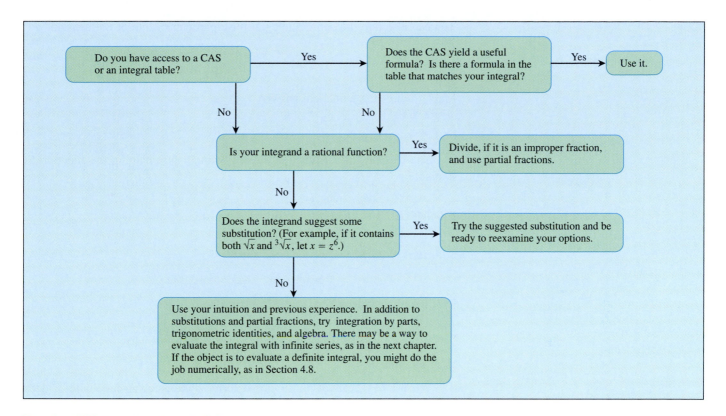

Flowchart 7.1 Procedure for indefinite integration

Using Maple you first define the function:

$$> f := x^2 * \text{sqrt}(a^2 + x^2);$$

Then you use the integrate command on f, identifying the variable of integration:

$$> \text{int}(f, x);$$

Maple returns the answer

$$\frac{1}{4} x(a^2 + x^2)^{3/2} - \frac{1}{8} a^2 x\sqrt{a^2 + x^2} - \frac{1}{8} a^4 \ln\left(x + \sqrt{a^2 + x^2}\right).$$

If you want to see if the answer can be simplified, enter

$$> \text{simplify}(");$$

Maple returns

$$\frac{1}{8} a^2 x\sqrt{a^2 + x^2} + \frac{1}{4} x^3\sqrt{a^2 + x^2} - \frac{1}{8} a^4 \ln\left(x + \sqrt{a^2 + x^2}\right).$$

If you want the definite integral for $0 \le x \le \pi/2$, you can use the format

$$> \text{int}(f, x = 0..\text{Pi}/2);$$

Maple (Version 3.0) will return the expression

$$\frac{1}{64}(4a^2 + \pi^2)^{3/2}\pi - \frac{1}{8} a^4 \ln\left(\frac{1}{2}\pi + \frac{1}{2}\sqrt{4a^2 + \pi^2}\right) - \frac{1}{32} a^2\sqrt{4a^2 + \pi^2}\,\pi$$

$$+ \frac{1}{8} a^4 \ln\left(\sqrt{a^2}\right).$$

You can also find the definite integral for a particular value of the constant a:

$$> a := 1;$$

$$> int(f, x = 0..1);$$

Maple returns the numerical answer

$$\frac{3}{8}\sqrt{2} - \frac{1}{8} \ln (1 + \sqrt{2})$$ ❏

You can integrate a function directly without first naming the function as in Example 9.

EXAMPLE 10   Use a CAS to find $\int \sin^2 x \cos^3 x \, dx$.

Solution   With Maple we have the entry

$$> int ((\sin \hat{\ } 2)(x) * (\cos \hat{\ } 3)(x),x);$$

with the immediate return

$$-\frac{1}{5} \sin (x) \cos (x)^4 + \frac{1}{15} \cos (x)^2 \sin (x) + \frac{2}{15} \sin (x)$$

as in Example 8.   ❏

When a CAS cannot find a closed form solution for an indefinite or definite integral it just returns the integral expression you asked for.

EXAMPLE 11   Use a CAS to find $\int (\cos^{-1} ax)^2 \, dx$.

Solution   Using Maple we enter

$$> int ((arccos(a*x))\hat{\ } 2, x);$$

and Maple returns the expression

$$\int \arccos (ax)^2 \, dx$$

indicating it does not have a closed form solution. In the next chapter you will see how series expansion may help to evaluate such an integral.   ❏

Computer Algebra Systems vary in how they process integrations. We used Maple in Examples 9–11. Mathematica would have returned somewhat different results:

1. In Example 9, given

$$In [1]: = \text{Integrate} [x \hat{\ } 2 * \text{Sqrt} [a \hat{\ } 2 + x \hat{\ } 2], x]$$

Mathematica returns

$$\text{Out} [1] = \text{Sqrt} [a^2 + x^2] \left(\frac{a^2 x}{8} + \frac{x^3}{4} \right) - \frac{a^4 \text{Log} [x + \text{Sqrt} [a^2 + x^2]]}{8}$$

without having to simplify an intermediate result. The answer is close to Formula 22 in the integral tables.

2. The Mathematica answer to the integral

$$\text{In}\,[2]: = \text{Integrate}\,[\text{Sin}\,[x]\,\hat{}\,2 * \text{Cos}\,[x]\,\hat{}\,3, x]$$

in Example 10 is

$$\text{Out}\,[2] = \frac{30\,\text{Sin}\,[x] - 5\,\text{Sin}\,[3x] - 3\,\text{Sin}\,[5x]}{240}$$

differing from both the Maple answer and the answers in Example 8.

3. Mathematica does give a result for the integration

$$\text{In}\,[3]: = \text{Integrate}\,[\text{ArcCos}\,[a * x]\,\hat{}\,2, x]$$

in Example 11:

$$\text{Out}\,[3] = -2x - \frac{2\,\text{Sqrt}\,[1 - a^2 x^2]\,\text{ArcCos}\,[ax]}{a} + x\,\text{ArcCos}\,[ax]^2$$

Although a CAS is very powerful and can aid us in solving difficult problems, each CAS has its own limitations. There are even situations where a CAS may further complicate a problem (in the sense of producing an answer that is extremely difficult to use or interpret). On the other hand, a little mathematical thinking on your part may reduce the problem to one that is quite easy to handle. We provide an example in Exercise 111.

Exercises 7.5

Using Integral Tables

Use the table of integrals at the back of the book to evaluate the integrals in Exercises 1–38.

1. $\displaystyle\int \frac{dx}{x\sqrt{x-3}}$

2. $\displaystyle\int \frac{dx}{x\sqrt{x+4}}$

3. $\displaystyle\int \frac{x\,dx}{\sqrt{x-2}}$

4. $\displaystyle\int \frac{x\,dx}{(2x+3)^{3/2}}$

5. $\displaystyle\int x\sqrt{2x-3}\,dx$

6. $\displaystyle\int x(7x+5)^{3/2}\,dx$

7. $\displaystyle\int \frac{\sqrt{9-4x}}{x^2}\,dx$

8. $\displaystyle\int \frac{dx}{x^2\sqrt{4x-9}}$

9. $\displaystyle\int x\sqrt{4x-x^2}\,dx$

10. $\displaystyle\int \frac{\sqrt{x-x^2}}{x}\,dx$

11. $\displaystyle\int \frac{dx}{x\sqrt{7+x^2}}$

12. $\displaystyle\int \frac{dx}{x\sqrt{7-x^2}}$

13. $\displaystyle\int \frac{\sqrt{4-x^2}}{x}\,dx$

14. $\displaystyle\int \frac{\sqrt{x^2-4}}{x}\,dx$

15. $\displaystyle\int \sqrt{25-p^2}\,dp$

16. $\displaystyle\int q^2\sqrt{25-q^2}\,dq$

17. $\displaystyle\int \frac{r^2}{\sqrt{4-r^2}}\,dr$

18. $\displaystyle\int \frac{ds}{\sqrt{s^2-2}}$

19. $\displaystyle\int \frac{d\theta}{5+4\sin 2\theta}$

20. $\displaystyle\int \frac{d\theta}{4+5\sin 2\theta}$

21. $\displaystyle\int e^{2t}\cos 3t\,dt$

22. $\displaystyle\int e^{-3t}\sin 4t\,dt$

23. $\displaystyle\int x\cos^{-1}x\,dx$

24. $\displaystyle\int x\sin^{-1}x\,dx$

25. $\displaystyle\int \frac{ds}{(9-s^2)^2}$

26. $\displaystyle\int \frac{d\theta}{(2-\theta^2)^2}$

27. $\displaystyle\int \frac{\sqrt{4x+9}}{x^2}\,dx$

28. $\displaystyle\int \frac{\sqrt{9x-4}}{x^2}\,dx$

29. $\displaystyle\int \frac{\sqrt{3t-4}}{t}\,dt$

30. $\displaystyle\int \frac{\sqrt{3t+9}}{t}\,dt$

31. $\displaystyle\int x^2\tan^{-1}x\,dx$

32. $\displaystyle\int \frac{\tan^{-1}x}{x^2}\,dx$

33. $\displaystyle\int \sin 3x\cos 2x\,dx$

34. $\displaystyle\int \sin 2x\cos 3x\,dx$

35. $\int 8 \sin 4t \sin \dfrac{t}{2} \, dt$

36. $\int \sin \dfrac{t}{3} \sin \dfrac{t}{6} \, dt$

37. $\int \cos \dfrac{\theta}{3} \cos \dfrac{\theta}{4} \, d\theta$

38. $\int \cos \dfrac{\theta}{2} \cos 7\theta \, d\theta$

Substitution and Integral Tables

In Exercises 39–52, use a substitution to change the integral into one you can find in the table. Then evaluate the integral.

39. $\int \dfrac{x^3 + x + 1}{(x^2 + 1)^2} \, dx$

40. $\int \dfrac{x^2 + 6x}{(x^2 + 3)^2} \, dx$

41. $\int \sin^{-1} \sqrt{x} \, dx$

42. $\int \dfrac{\cos^{-1} \sqrt{x}}{\sqrt{x}} \, dx$

43. $\int \dfrac{\sqrt{x}}{\sqrt{1-x}} \, dx$

44. $\int \dfrac{\sqrt{2-x}}{\sqrt{x}} \, dx$

45. $\int \cot t \sqrt{1 - \sin^2 t} \, dt, \ 0 < t < \pi/2$

46. $\int \dfrac{dt}{\tan t \sqrt{4 - \sin^2 t}}$

47. $\int \dfrac{dy}{y \sqrt{3 + (\ln y)^2}}$

48. $\int \dfrac{\cos \theta \, d\theta}{\sqrt{5 + \sin^2 \theta}}$

49. $\int \dfrac{3 \, dr}{\sqrt{9r^2 - 1}}$

50. $\int \dfrac{3 \, dy}{\sqrt{1 + 9y^2}}$

51. $\int \cos^{-1} \sqrt{x} \, dx$

52. $\int \tan^{-1} \sqrt{y} \, dy$

Using Reduction Formulas

Use reduction formulas to evaluate the integrals in Exercises 53–72.

53. $\int \sin^5 2x \, dx$

54. $\int \sin^5 \dfrac{\theta}{2} \, d\theta$

55. $\int 8 \cos^4 2\pi t \, dt$

56. $\int 3 \cos^5 3y \, dy$

57. $\int \sin^2 2\theta \cos^3 2\theta \, d\theta$

58. $\int 9 \sin^3 \theta \cos^{3/2} \theta \, d\theta$

59. $\int 2 \sin^2 t \sec^4 t \, dt$

60. $\int \csc^2 y \cos^5 y \, dy$

61. $\int 4 \tan^3 2x \, dx$

62. $\int \tan^4 \left(\dfrac{x}{2} \right) dx$

63. $\int 8 \cot^4 t \, dt$

64. $\int 4 \cot^3 2t \, dt$

65. $\int 2 \sec^3 \pi x \, dx$

66. $\int \dfrac{1}{2} \csc^3 \dfrac{x}{2} \, dx$

67. $\int 3 \sec^4 3x \, dx$

68. $\int \csc^4 \dfrac{\theta}{3} \, d\theta$

69. $\int \csc^5 x \, dx$

70. $\int \sec^5 x \, dx$

71. $\int 16x^3 (\ln x)^2 \, dx$

72. $\int (\ln x)^3 \, dx$

Powers of x Times Exponentials

Evaluate the integrals in Exercises 73–80 using table Formulas 103–106. These integrals can also be evaluated using tabular integration (Section 7.2).

73. $\int x e^{3x} \, dx$

74. $\int x e^{-2x} \, dx$

75. $\int x^3 e^{x/2} \, dx$

76. $\int x^2 e^{\pi x} \, dx$

77. $\int x^2 2^x \, dx$

78. $\int x^2 2^{-x} \, dx$

79. $\int x \pi^x \, dx$

80. $\int x 2^{\sqrt{2} x} \, dx$

Substitutions with Reduction Formulas

Evaluate the integrals in Exercises 81–86 by making a substitution (possibly trigonometric) and then applying a reduction formula.

81. $\int e^t \sec^3 (e^t - 1) \, dt$

82. $\int \dfrac{\csc^3 \sqrt{\theta}}{\sqrt{\theta}} \, d\theta$

83. $\int_0^1 2\sqrt{x^2 + 1} \, dx$

84. $\int_0^{\sqrt{3}/2} \dfrac{dy}{(1 - y^2)^{5/2}}$

85. $\int_1^2 \dfrac{(r^2 - 1)^{3/2}}{r} \, dr$

86. $\int_0^{1/\sqrt{3}} \dfrac{dt}{(t^2 + 1)^{7/2}}$

Hyperbolic Functions

Use the integral tables to evaluate the integrals in Exercises 87–92.

87. $\int \dfrac{1}{8} \sinh^5 3x \, dx$

88. $\int \dfrac{\cosh^4 \sqrt{x}}{\sqrt{x}} \, dx$

89. $\int x^2 \cosh 3x \, dx$

90. $\int x \sinh 5x \, dx$

91. $\int \operatorname{sech}^7 x \tanh x \, dx$

92. $\int \operatorname{csch}^3 2x \coth 2x \, dx$

Theory and Examples

Exercises 93–100 refer to formulas in the table of integrals at the back of the book.

93. Derive Formula 9 by using the substitution $u = ax + b$ to evaluate

$$\int \dfrac{x}{(ax + b)^2} \, dx.$$

94. Derive Formula 17 by using a trigonometric substitution to evaluate

$$\int \dfrac{dx}{(a^2 + x^2)^2}.$$

95. Derive Formula 29 by using a trigonometric substitution to evaluate

$$\int \sqrt{a^2 - x^2}\, dx.$$

96. Derive Formula 46 by using a trigonometric substitution to evaluate

$$\int \frac{dx}{x^2\sqrt{x^2 - a^2}}.$$

97. Derive Formula 80 by evaluating

$$\int x^n \sin ax\, dx$$

by integration by parts.

98. Derive Formula 110 by evaluating

$$\int x^n (\ln ax)^m\, dx$$

by integration by parts.

99. Derive Formula 99 by evaluating

$$\int x^n \sin^{-1} ax\, dx$$

by integration by parts.

100. Derive Formula 101 by evaluating

$$\int x^n \tan^{-1} ax\, dx$$

by integration by parts.

101. Find the area of the surface generated by revolving the curve $y = \sqrt{x^2 + 2}$, $0 \le x \le \sqrt{2}$, about the x-axis.

102. Find the length of the curve $y = x^2$, $0 \le x \le \sqrt{3}/2$.

103. Find the centroid of the region cut from the first quadrant by the curve $y = 1/\sqrt{x + 1}$ and the line $x = 3$.

104. A thin plate of constant density $\delta = 1$ occupies the region enclosed by the curve $y = 36/(2x + 3)$ and the line $x = 3$ in the first quadrant. Find the moment of the plate about the y-axis.

105. CALCULATOR Use the integral table and a calculator to find to 2 decimal places the area of the surface generated by revolving the curve $y = x^2$, $-1 \le x \le 1$, about the x-axis.

106. The head of your firm's accounting department has asked you to find a formula she can use in a computer program to calculate the year-end inventory of gasoline in the company's tanks. A typical tank is shaped like a right circular cylinder of radius r and length L, mounted horizontally, as shown here. The data come to the accounting office as depth measurements taken with a vertical measuring stick marked in centimeters.

a) Show, in the notation of the figure here, that the volume of gasoline that fills the tank to a depth d is

$$V = 2L \int_{-r}^{-r + d} \sqrt{r^2 - y^2}\, dy.$$

b) Evaluate the integral.

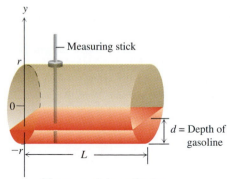

Measurements in centimeters

107. What is the largest value

$$\int_a^b \sqrt{x - x^2}\, dx$$

can have for any a and b? Give reasons for your answer.

108. What is the largest value

$$\int_a^b x\sqrt{2x - x^2}\, dx$$

can have for any a and b? Give reasons for your answer.

✸ CAS Explorations and Projects

In Exercises 109 and 110, use a CAS to perform the integrations.

109. Evaluate the integrals

a) $\displaystyle\int x \ln x\, dx$ **b)** $\displaystyle\int x^2 \ln x\, dx$

c) $\displaystyle\int x^3 \ln x\, dx.$

d) What pattern do you see? Predict the formula for $\int x^4 \ln x\, dx$ and then see if you are correct by evaluating it with a CAS.

e) What is the formula for $\int x^n \ln x\, dx$, $n \ge 1$? Check your answer using a CAS.

110. Evaluate the integrals

a) $\displaystyle\int \frac{\ln x}{x^2}\, dx$ **b)** $\displaystyle\int \frac{\ln x}{x^3}\, dx$

c) $\displaystyle\int \frac{\ln x}{x^4}\, dx.$

d) What pattern do you see? Predict the formula for

$$\int \frac{\ln x}{x^5}\, dx$$

and then see if you are correct by evaluating it with a CAS.

e) What is the formula for

$$\int \frac{\ln x}{x^n}\, dx, n \ge 2?$$

Check your answer using a CAS.

111. a) Use a CAS to evaluate

$$\int_0^{\pi/2} \frac{\sin^n x}{\sin^n x + \cos^n x}\, dx$$

where n is an arbitrary positive integer. Does your CAS find the result?

b) In succession, find the integral when $n = 1, 2, 3, 5, 7.$ Comment on the complexity of the results.

c) Now substitute $x = (\pi/2) - u$ and add the new and old integrals. What is the value of

$$\int_0^{\pi/2} \frac{\sin^n x}{\sin^n x + \cos^n x}\, dx?$$

This exercise illustrates how a little mathematical ingenuity solves a problem not immediately amenable to solution by a CAS.

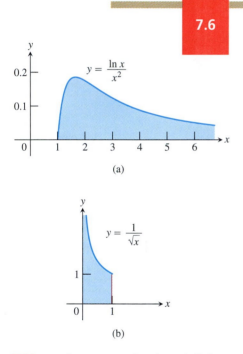

7.11 Are the areas under these infinite curves finite? See Examples 1 and 2.

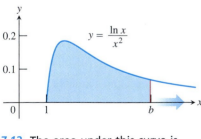

7.12 The area under this curve is

$$\lim_{b \to \infty} \int_1^b ((\ln x)/x^2)\, dx$$

(Example 1).

7.6 Improper Integrals

Up to now, we have required our definite integrals to have two properties. First, that the domain of integration, from a to b, be finite. Second, that the range of the integrand be finite on this domain. In practice, however, we frequently encounter problems that fail to meet one or both of these conditions. As an example of an infinite domain, we might want to consider the area under the curve $y = (\ln x)/x^2$ from $x = 1$ to $x = \infty$ (Fig. 7.11a). As an example of an infinite range, we might want to consider the area under the curve $y = 1/\sqrt{x}$ between $x = 0$ and $x = 1$ (Fig. 7.11b). We treat both examples in the same reasonable way. We ask, "What is the integral when the domain is slightly less?" and examine the answer as the domain increases to the limit. We do the finite case and then see what happens as we approach infinity.

EXAMPLE 1 Is the area under the curve $y = (\ln x)/x^2$ from $x = 1$ to $x = \infty$ finite? If so, what is it?

Solution We find the area under the curve from $x = 1$ to $x = b$ and examine the limit as $b \to \infty$. If the limit is finite, we take it to be the area under the infinite curve (Fig. 7.12). The area from 1 to b is

$$\int_1^b \frac{\ln x}{x^2}\, dx = \left[(\ln x)\left(-\frac{1}{x}\right) \right]_1^b - \int_1^b \left(-\frac{1}{x}\right)\left(\frac{1}{x}\right) dx \qquad \begin{array}{l}\text{Integration by parts with}\\ u = \ln x, \quad dv = dx/x^2,\\ du = dx/x, \quad v = -1/x\end{array}$$

$$= -\frac{\ln b}{b} - \left[\frac{1}{x} \right]_1^b$$

$$= -\frac{\ln b}{b} - \frac{1}{b} + 1.$$

The limit of the area as $b \to \infty$ is

$$\lim_{b \to \infty} \left[-\frac{\ln b}{b} - \frac{1}{b} + 1 \right] = -\left[\lim_{b \to \infty} \frac{\ln b}{b} \right] - 0 + 1$$

$$= -\left[\lim_{b \to \infty} \frac{1/b}{1} \right] + 1 = 0 + 1 = 1. \qquad \begin{array}{l}\text{l'Hôpital's}\\ \text{rule}\end{array}$$

In integral notation, the area under the infinite curve from 1 to ∞ is

$$\int_1^\infty \frac{\ln x}{x^2}\, dx = \lim_{b \to \infty} \int_1^b \frac{\ln x}{x^2}\, dx = 1.$$

❑

7.13 The area under this curve is

$$\lim_{a \to 0^+} \int_a^1 (1/\sqrt{x})\, dx$$

(Example 2).

EXAMPLE 2   Is the area under the curve $y = 1/\sqrt{x}$ from $x = 0$ to $x = 1$ finite? If so, what is it?

Solution   We find the area under the curve from a to 1 and examine the limit as $a \to 0^+$. If the limit is finite, we take it to be the area under the infinite curve (Fig. 7.13). The area from a to 1 is

$$\int_a^1 \frac{1}{\sqrt{x}}\, dx = 2\sqrt{x}\,\Big]_a^1 = 2 - 2\sqrt{a}.$$

The limit as $a \to 0^+$ is

$$\lim_{a \to 0^+} (2 - 2\sqrt{a}) = 2 - 0 = 2.$$

In integral notation, the area under the infinite curve from 0 to 1 is

$$\int_0^1 \frac{1}{\sqrt{x}}\, dx = \lim_{a \to 0^+} \int_a^1 \frac{1}{\sqrt{x}}\, dx = 2.$$

Improper Integrals

The integrals for the areas in Examples 1 and 2 are improper integrals.

Definition

Integrals with infinite limits of integration and integrals of functions that become infinite at a point within the interval of integration are **improper integrals.** When the limits involved exist, we evaluate such integrals with the following definitions:

1. If f is continuous on $[a, \infty)$, then

$$\int_a^\infty f(x)\, dx = \lim_{b \to \infty} \int_a^b f(x)\, dx. \tag{1}$$

2. If f is continuous on $(-\infty, b]$, then

$$\int_{-\infty}^b f(x)\, dx = \lim_{a \to -\infty} \int_a^b f(x)\, dx. \tag{2}$$

3. If f is continuous on $(a, b]$, then

$$\int_a^b f(x)\, dx = \lim_{c \to a^+} \int_c^b f(x)\, dx. \tag{3}$$

4. If f is continuous on $[a, b)$, then

$$\int_a^b f(x)\, dx = \lim_{c \to b^-} \int_a^c f(x)\, dx. \tag{4}$$

In each case, if the limit is finite we say that the improper integral **converges** and that the limit is the **value** of the improper integral. If the limit fails to exist the improper integral **diverges.**

Example 1 illustrates Part 1 of the definition:

$$\int_1^\infty \frac{\ln x}{x^2}\, dx = \lim_{b \to \infty} \int_1^b \frac{\ln x}{x^2}\, dx = 1 \qquad \text{Infinite upper limit of integration}$$

7.14 If the limit exists

$\int_0^1 (1/(1-x))\, dx = \lim_{b \to 1^-} \int_0^b (1/(1-x))\, dx$

(Example 3).

Example 2 illustrates Part 3 of the definition:

$$\int_0^1 \frac{1}{\sqrt{x}}\, dx = \lim_{a \to 0^+} \int_a^1 \frac{1}{\sqrt{x}}\, dx = 2$$

Integrand becomes infinite at lower limit of integration

In each case, the integral converges. The integral in the next example diverges.

EXAMPLE 3   *A divergent improper integral*

Investigate the convergence of

$$\int_0^1 \frac{1}{1-x}\, dx.$$

Solution   The integrand $f(x) = 1/(1-x)$ is continuous on $[0, 1)$ but becomes infinite as $x \to 1^-$ (Fig. 7.14). We evaluate the integral as

$$\lim_{b \to 1^-} \int_0^b \frac{1}{1-x}\, dx = \lim_{b \to 1^-} \left[-\ln|1-x| \right]_0^b$$

$$= \lim_{b \to 1^-} \left[-\ln(1-b) + 0 \right] = \infty.$$

The limit is infinite, so the integral diverges.   ❑

The list in the preceding definition extends in a natural way to integrals with two infinite limits of integration. We will treat these later in the section. The list also extends to integrals of functions that become infinite at an interior point d of the interval of integration. In this case, we define the integral from a to b to be the sum of the integrals from a to d and d to b.

> **Definition**
>
> If f becomes infinite at an interior point d of $[a,b]$, then
>
> $$\int_a^b f(x)\, dx = \int_a^d f(x)\, dx + \int_d^b f(x)\, dx. \tag{5}$$
>
> The integral from a to b **converges** if the integrals from a to d and d to b both converge. Otherwise, the integral from a to b **diverges.**

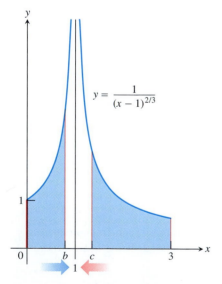

7.15 Example 4 investigates the convergence of

$\int_0^3 (1/(x-1)^{2/3})\, dx.$

EXAMPLE 4   *Infinite at an interior point*

Investigate the convergence of

$$\int_0^3 \frac{dx}{(x-1)^{2/3}}.$$

Solution   The integrand $f(x) = 1/(x-1)^{2/3}$ becomes infinite at $x = 1$ but is continuous on $[0, 1)$ and $(1, 3]$ (Fig. 7.15). The convergence of the integral over $[0, 3]$ depends on the integrals from 0 to 1 and 1 to 3. On $[0, 1]$ we have

$$\int_0^1 \frac{dx}{(x-1)^{2/3}} = \lim_{b \to 1^-} \int_0^b \frac{dx}{(x-1)^{2/3}}$$

$$= \lim_{b \to 1^-} [3(b-1)^{1/3} - 3(0-1)^{1/3}] = 3.$$

On [1, 3] we have

$$\int_1^3 \frac{dx}{(x-1)^{2/3}} = \lim_{c \to 1^+} \int_c^3 \frac{dx}{(x-1)^{2/3}}$$

$$= \lim_{c \to 1^+} [3(3-1)^{1/3} - 3(c-1)^{1/3}] = 3\sqrt[3]{2}.$$

Both limits are finite, so the integral of f from 0 to 3 converges and its value is $3 + 3\sqrt[3]{2}$.   ❑

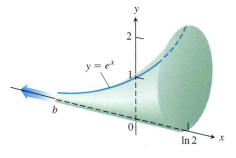

7.16 The calculation in Example 5 shows that this infinite horn has a finite volume.

EXAMPLE 5   The cross sections of the solid horn in Fig. 7.16 perpendicular to the x-axis are circular disks with diameters reaching from the x-axis to the curve $y = e^x$, $-\infty < x \le \ln 2$. Find the volume of the horn.

Solution   The area of a typical cross section is

$$A(x) = \pi (\text{radius})^2 = \pi \left(\frac{1}{2}y\right)^2 = \frac{\pi}{4}e^{2x}.$$

We define the volume of the horn to be the limit as $b \to -\infty$ of the volume of the portion from b to $\ln 2$. As in Section 5.2 (the method of slicing), the volume of this portion is

$$V = \int_b^{\ln 2} A(x)\, dx = \int_b^{\ln 2} \frac{\pi}{4}e^{2x}\, dx = \frac{\pi}{8}e^{2x}\Big]_b^{\ln 2}$$

$$= \frac{\pi}{8}(e^{\ln 4} - e^{2b}) = \frac{\pi}{8}(4 - e^{2b}).$$

As $b \to -\infty$, $e^{2b} \to 0$ and $V \to (\pi/8)(4-0) = \pi/2$. The volume of the horn is $\pi/2$.   ❑

EXAMPLE 6   Evaluate $\displaystyle\int_2^\infty \frac{x+3}{(x-1)(x^2+1)}\, dx$.

Solution

$$\int_2^\infty \frac{x+3}{(x-1)(x^2+1)}\, dx = \lim_{b \to \infty} \int_2^b \frac{x+3}{(x-1)(x^2+1)}\, dx$$

$$= \lim_{b \to \infty} \int_2^b \left(\frac{2}{x-1} - \frac{2x+1}{x^2+1}\right) dx \qquad \text{Partial fractions}$$

$$= \lim_{b \to \infty} \left[2\ln(x-1) - \ln(x^2+1) - \tan^{-1} x\right]_2^b$$

$$= \lim_{b \to \infty} \left[\ln \frac{(x-1)^2}{x^2+1} - \tan^{-1} x\right]_2^b \qquad \begin{array}{l}\text{Combine the}\\ \text{logarithms.}\end{array}$$

$$= \lim_{b \to \infty} \left[\ln \left(\frac{(b-1)^2}{b^2+1}\right) - \tan^{-1} b\right] - \ln \left(\frac{1}{5}\right) + \tan^{-1} 2$$

$$= 0 - \frac{\pi}{2} + \ln 5 + \tan^{-1} 2 \approx 1.1458$$

Notice that we combined the logarithms in the antiderivative *before* we calculated the limit as $b \to \infty$. Had we not done so, we would have encountered the indeterminate form

$$\lim_{b \to \infty} (2 \ln (b - 1) - \ln (b^2 + 1)) = \infty - \infty.$$

The way to evaluate the indeterminate form, of course, is to combine the logarithms, so we would have arrived at the same answer in the end. But our original route was shorter. ❑

Integrals from $-\infty$ to ∞

In the mathematics underlying studies of light, electricity, and sound we encounter integrals with two infinite limits of integration. The next definition addresses the convergence of such integrals.

Definition

If f is continuous on $(-\infty, \infty)$ and if $\int_{-\infty}^{a} f(x) \, dx$ and $\int_{a}^{\infty} f(x) \, dx$ both converge, we say that $\int_{-\infty}^{\infty} f(x) \, dx$ **converges** and define its value to be

$$\int_{-\infty}^{\infty} f(x) \, dx = \int_{-\infty}^{a} f(x) \, dx + \int_{a}^{\infty} f(x) \, dx. \qquad (6)$$

If either or both of the integrals on the right-hand side of this equation diverge, the integral of f from $-\infty$ to ∞ **diverges.**

It can be shown that the choice of a in Eq. (6) is unimportant. We can evaluate or determine the convergence of $\int_{-\infty}^{\infty} f(x) \, dx$ with any convenient choice.

The integral of f from $-\infty$ to ∞ need not equal $\lim_{b \to \infty} \int_{-b}^{b} f(x) \, dx$, which may exist even if $\int_{-\infty}^{\infty} f(x) \, dx$ does not converge (Exercise 75).

EXAMPLE 7

$$\int_{-\infty}^{\infty} \frac{dx}{1 + x^2} = \int_{-\infty}^{0} \frac{dx}{1 + x^2} + \int_{0}^{\infty} \frac{dx}{1 + x^2} \qquad \text{Eq. (6) with } a = 0$$

$$= \lim_{b \to -\infty} \left[\tan^{-1} x \right]_{b}^{0} + \lim_{c \to \infty} \left[\tan^{-1} x \right]_{0}^{c}$$

$$= \lim_{b \to -\infty} \left[\tan^{-1} 0 - \tan^{-1} b \right] + \lim_{c \to \infty} \left[\tan^{-1} c - \tan^{-1} 0 \right]$$

$$= 0 - \left(-\frac{\pi}{2} \right) + \frac{\pi}{2} - 0 = \pi.$$

We interpret the integral as the area of the infinite region between the curve $y = 1/(1 + x^2)$ and the x-axis (Fig. 7.17). ❑

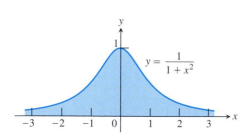

7.17 The area under this "doubly" infinite curve is finite (Example 7).

$$y = \frac{1}{1 + x^2}$$

The Integral $\int_{1}^{\infty} dx / x^p$

The convergence of the integral $\int_{1}^{\infty} dx/x^p$ depends on p. The next example illustrates this with $p = 1$ and $p = 2$.

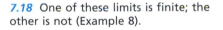

(a)

(b)

7.18 One of these limits is finite; the other is not (Example 8).

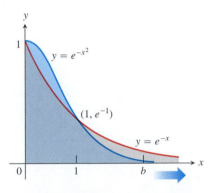

7.19 The graph of e^{-x^2} lies below the graph of e^{-x} for $x > 1$ (Example 9).

EXAMPLE 8   Investigate the convergence of

$$\int_1^\infty \frac{dx}{x} \qquad \text{and} \qquad \int_1^\infty \frac{dx}{x^2}.$$

Solution   The functions involved are continuous on $[1, \infty)$ and their graphs both approach the x-axis as $x \to \infty$ (Fig. 7.18), so it is reasonable to think that the areas under these infinite curves might be finite. In the first case,

$$\int_1^\infty \frac{dx}{x} = \lim_{b \to \infty} \int_1^b \frac{dx}{x} = \lim_{b \to \infty} (\ln b - \ln 1) = \infty,$$

so the integral diverges. In the second case,

$$\int_1^\infty \frac{dx}{x^2} = \lim_{b \to \infty} \int_1^b \frac{dx}{x^2} = \lim_{b \to \infty} \left(-\frac{1}{b} + 1 \right) = 1,$$

so the integral converges and its value is 1.   ❑

Generally, $\int_1^\infty dx/x^p$ converges if $p > 1$ and diverges if $p \leq 1$ (Exercise 67).

Tests for Convergence and Divergence

When an improper integral cannot be evaluated directly (often the case in practice) we turn to the two-step procedure of first establishing the fact of convergence and then approximating the integral numerically. The principal tests for convergence are the direct comparison and limit comparison tests.

EXAMPLE 9   Investigate the convergence of $\int_1^\infty e^{-x^2} dx$.

Solution   By definition,

$$\int_1^\infty e^{-x^2} dx = \lim_{b \to \infty} \int_1^b e^{-x^2} dx.$$

We cannot evaluate the latter integral directly because it is nonelementary. But we *can* show that its limit as $b \to \infty$ is finite. We know that $\int_1^b e^{-x^2} dx$ is an increasing function of b. Therefore either it becomes infinite as $b \to \infty$ or it has a finite limit as $b \to \infty$. It does not become infinite: For every value of $x \geq 1$ we have $e^{-x^2} \leq e^{-x}$ (Fig. 7.19), so that

$$\int_1^b e^{-x^2} dx \leq \int_1^b e^{-x} dx = -e^{-b} + e^{-1} < e^{-1} \approx 0.36788.$$

Hence

$$\int_1^\infty e^{-x^2} dx = \lim_{b \to \infty} \int_1^b e^{-x^2} dx$$

converges to some definite finite value. We do not know exactly what the value is except that it is something less than 0.37.   ❑

The comparison of e^{-x^2} and e^{-x} in Example 9 is a special case of the following test.

> **Theorem 1**
> **Direct Comparison Test**
> Let f and g be continuous on $[a, \infty)$ and suppose that $0 \le f(x) \le g(x)$ for all $x \ge a$. Then
>
> 1. $\displaystyle\int_a^\infty f(x)\,dx$ converges if $\displaystyle\int_a^\infty g(x)\,dx$ converges.
>
> 2. $\displaystyle\int_a^\infty g(x)\,dx$ diverges if $\displaystyle\int_a^\infty f(x)\,dx$ diverges.

EXAMPLE 10

a) $\displaystyle\int_1^\infty \frac{\sin^2 x}{x^2}\,dx$ converges because $0 \le \dfrac{\sin^2 x}{x^2} \le \dfrac{1}{x^2}$ on $[1, \infty)$ and $\displaystyle\int_1^\infty \frac{1}{x^2}\,dx$ converges.

b) $\displaystyle\int_1^\infty \frac{1}{\sqrt{x^2 - 0.1}}\,dx$ diverges because $\dfrac{1}{\sqrt{x^2 - 0.1}} \ge \dfrac{1}{x}$ on $[1, \infty)$

and $\displaystyle\int_1^\infty \frac{1}{x}\,dx$ diverges.

> **Theorem 2**
> **Limit Comparison Test**
> If the positive functions f and g are continuous on $[a, \infty)$ and if
>
> $$\lim_{x\to\infty} \frac{f(x)}{g(x)} = L \qquad (0 < L < \infty),$$
>
> then $\int_a^\infty f(x)\,dx$ and $\int_a^\infty g(x)\,dx$ both converge or both diverge.

In the language of Section 6.7, Theorem 2 says that if two positive functions grow at the same rate as $x \to \infty$, then their integrals from a to ∞ behave alike: They both converge or both diverge. This does not mean that their integrals have the same value, however, as the next example shows.

EXAMPLE 11   Compare

$$\int_1^\infty \frac{dx}{x^2} \qquad \text{and} \qquad \int_1^\infty \frac{dx}{1 + x^2}$$

with the Limit Comparison Test.

Solution   With $f(x) = 1/x^2$ and $g(x) = 1/(1 + x^2)$, we have

$$\lim_{x\to\infty} \frac{f(x)}{g(x)} = \lim_{x\to\infty} \frac{1/x^2}{1/(1 + x^2)}$$

$$= \lim_{x\to\infty} \frac{1 + x^2}{x^2} = \lim_{x\to\infty}\left(\frac{1}{x^2} + 1\right) = 0 + 1 = 1,$$

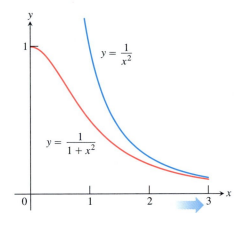

7.20 The functions in Example 11.

a positive finite limit (Fig. 7.20). Therefore, $\int_1^\infty \dfrac{dx}{1+x^2}$ converges because $\int_1^\infty \dfrac{dx}{x^2}$ converges.

The integrals converge to different values, however.

$$\int_1^\infty \frac{dx}{x^2} = 1, \qquad \text{Example 8}$$

and

$$\int_1^\infty \frac{dx}{1+x^2} = \lim_{b\to\infty} \int_1^b \frac{dx}{1+x^2}$$

$$= \lim_{b\to\infty} [\tan^{-1} b - \tan^{-1} 1] = \frac{\pi}{2} - \frac{\pi}{4} = \frac{\pi}{4}. \qquad \square$$

EXAMPLE 12

$$\int_1^\infty \frac{3}{e^x + 5}\, dx \quad \text{converges because} \quad \int_1^\infty \frac{1}{e^x}\, dx \text{ converges}$$

and

$$\lim_{x\to\infty} \frac{1/e^x}{3/(e^x + 5)} = \lim_{x\to\infty} \frac{e^x + 5}{3e^x}$$

$$= \lim_{x\to\infty} \left(\frac{1}{3} + \frac{5}{3e^x} \right) = \frac{1}{3} + 0 = \frac{1}{3},$$

a positive finite limit. As far as the convergence of the improper integral is concerned, $3/(e^x + 5)$ behaves like $1/e^x$. $\square$

Computer Algebra Systems

Computer Algebra Systems can evaluate many convergent improper integrals.

EXAMPLE 13 Evaluate the integral $\displaystyle\int_2^\infty \frac{x+3}{(x-1)(x^2+1)}\, dx$ from Example 6.

Solution Using Maple, enter

$$> \text{f: } = (x+3)/((x-1)^*(x^2+1));$$

Then use the integration command

$$> \text{int}\,(f,\ x{=}2..\text{infinity});$$

Maple returns the answer

$$-\frac{1}{2}\pi + \ln(5) + \arctan(2).$$

To obtain a numerical result use the evaluation command **evalf** and specify the number of digits, as follows:

$$> \text{evalf}(",\ 6);$$

The ditto symbol (") instructs the computer to evaluate the last expression on the screen, in this case $-\dfrac{1}{2}\pi + \ln(5) + \text{arc tan}(2)$. Maple returns 1.14579.

Using Mathematica, entering

In [1]: = Integrate [$(x + 3)/((x - 1)(x^2 + 1))$, {$x, 2, $ Infinity}]

returns

$$\text{Out } [1] = \frac{-\text{Pi}}{2} + \text{ArcTan } [2] + \text{Log } [5].$$

To obtain a numerical result with six digits, use the command "N[%, 6]" which also yields 1.14579.

Types of Improper Integrals Discussed in This Section

INFINITE LIMITS OF INTEGRATION

1. Upper limit

$$\int_1^\infty \frac{\ln x}{x^2}\, dx = \lim_{b \to \infty} \int_1^b \frac{\ln x}{x^2}\, dx$$

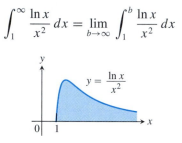

2. Lower limit

$$\int_{-\infty}^0 \frac{dx}{1 + x^2} = \lim_{a \to -\infty} \int_a^0 \frac{dx}{1 + x^2}$$

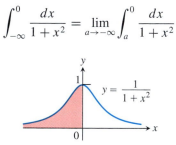

3. Both limits

$$\int_{-\infty}^\infty \frac{dx}{1 + x^2} = \lim_{b \to -\infty} \int_b^0 \frac{dx}{1 + x^2} + \lim_{c \to \infty} \int_0^c \frac{dx}{1 + x^2}$$

INTEGRAND BECOMES INFINITE

4. Upper endpoint

$$\int_0^1 \frac{dx}{(x - 1)^{2/3}} = \lim_{b \to 1^-} \int_0^b \frac{dx}{(x - 1)^{2/3}}$$

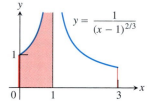

5. Lower endpoint

$$\int_1^3 \frac{dx}{(x - 1)^{2/3}} = \lim_{d \to 1^+} \int_d^3 \frac{dx}{(x - 1)^{2/3}}$$

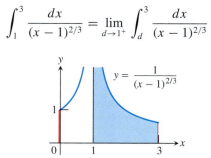

6. Interior point

$$\int_0^3 \frac{dx}{(x - 1)^{2/3}} = \int_0^1 \frac{dx}{(x - 1)^{2/3}} + \int_1^3 \frac{dx}{(x - 1)^{2/3}}$$

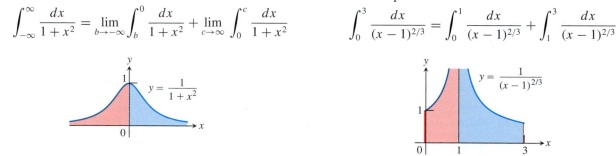

Exercises 7.6

Evaluating Improper Integrals

Evaluate the integrals in Exercises 1–34 without using tables.

1. $\displaystyle\int_0^\infty \frac{dx}{x^2+1}$

2. $\displaystyle\int_1^\infty \frac{dx}{x^{1.001}}$

3. $\displaystyle\int_0^1 \frac{dx}{\sqrt{x}}$

4. $\displaystyle\int_0^4 \frac{dx}{\sqrt{4-x}}$

5. $\displaystyle\int_{-1}^1 \frac{dx}{x^{2/3}}$

6. $\displaystyle\int_{-8}^1 \frac{dx}{x^{1/3}}$

7. $\displaystyle\int_0^1 \frac{dx}{\sqrt{1-x^2}}$

8. $\displaystyle\int_0^1 \frac{dr}{r^{0.999}}$

9. $\displaystyle\int_{-\infty}^{-2} \frac{2\,dx}{x^2-1}$

10. $\displaystyle\int_{-\infty}^2 \frac{2\,dx}{x^2+4}$

11. $\displaystyle\int_2^\infty \frac{2}{v^2-v}\,dv$

12. $\displaystyle\int_2^\infty \frac{2\,dt}{t^2-1}$

13. $\displaystyle\int_{-\infty}^\infty \frac{2x\,dx}{(x^2+1)^2}$

14. $\displaystyle\int_{-\infty}^\infty \frac{x\,dx}{(x^2+4)^{3/2}}$

15. $\displaystyle\int_0^1 \frac{\theta+1}{\sqrt{\theta^2+2\theta}}\,d\theta$

16. $\displaystyle\int_0^2 \frac{s+1}{\sqrt{4-s^2}}\,ds$

17. $\displaystyle\int_0^\infty \frac{dx}{(1+x)\sqrt{x}}$

18. $\displaystyle\int_1^\infty \frac{1}{x\sqrt{x^2-1}}\,dx$

19. $\displaystyle\int_0^\infty \frac{dv}{(1+v^2)(1+\tan^{-1} v)}$

20. $\displaystyle\int_0^\infty \frac{16\tan^{-1} x}{1+x^2}\,dx$

21. $\displaystyle\int_{-\infty}^0 \theta e^\theta\,d\theta$

22. $\displaystyle\int_0^\infty 2e^{-\theta}\sin\theta\,d\theta$

23. $\displaystyle\int_{-\infty}^\infty e^{-|x|}\,dx$

24. $\displaystyle\int_{-\infty}^\infty 2xe^{-x^2}\,dx$

25. $\displaystyle\int_0^1 x\ln x\,dx$

26. $\displaystyle\int_0^1 (-\ln x)\,dx$

27. $\displaystyle\int_0^2 \frac{ds}{\sqrt{4-s^2}}$

28. $\displaystyle\int_0^1 \frac{4r\,dr}{\sqrt{1-r^4}}$

29. $\displaystyle\int_1^2 \frac{ds}{s\sqrt{s^2-1}}$

30. $\displaystyle\int_2^4 \frac{dt}{t\sqrt{t^2-4}}$

31. $\displaystyle\int_{-1}^4 \frac{dx}{\sqrt{|x|}}$

32. $\displaystyle\int_0^2 \frac{dx}{\sqrt{|x-1|}}$

33. $\displaystyle\int_{-1}^\infty \frac{d\theta}{\theta^2+5\theta+6}$

34. $\displaystyle\int_0^\infty \frac{dx}{(x+1)(x^2+1)}$

35. $\displaystyle\int_0^{\pi/2} \tan\theta\,d\theta$

36. $\displaystyle\int_0^{\pi/2} \cot\theta\,d\theta$

37. $\displaystyle\int_0^\pi \frac{\sin\theta\,d\theta}{\sqrt{\pi-\theta}}$

38. $\displaystyle\int_{-\pi/2}^{\pi/2} \frac{\cos\theta\,d\theta}{(\pi-2\theta)^{1/3}}$

39. $\displaystyle\int_0^{\ln 2} x^{-2}e^{-1/x}\,dx$

40. $\displaystyle\int_0^1 \frac{e^{-\sqrt{x}}}{\sqrt{x}}\,dx$

41. $\displaystyle\int_0^\pi \frac{dt}{\sqrt{t}+\sin t}$

42. $\displaystyle\int_0^1 \frac{dt}{t-\sin t}$ $\left(\begin{array}{l}\text{Hint: } t \geq \sin t \\ \text{for } t \geq 0\end{array}\right)$

43. $\displaystyle\int_0^2 \frac{dx}{1-x^2}$

44. $\displaystyle\int_0^2 \frac{dx}{1-x}$

45. $\displaystyle\int_{-1}^1 \ln|x|\,dx$

46. $\displaystyle\int_{-1}^1 -x\ln|x|\,dx$

47. $\displaystyle\int_1^\infty \frac{dx}{x^3+1}$

48. $\displaystyle\int_4^\infty \frac{dx}{\sqrt{x}-1}$

49. $\displaystyle\int_2^\infty \frac{dv}{\sqrt{v}-1}$

50. $\displaystyle\int_0^\infty \frac{d\theta}{1+e^\theta}$

51. $\displaystyle\int_0^\infty \frac{dx}{\sqrt{x^6+1}}$

52. $\displaystyle\int_2^\infty \frac{dx}{\sqrt{x^2-1}}$

53. $\displaystyle\int_1^\infty \frac{\sqrt{x+1}}{x^2}\,dx$

54. $\displaystyle\int_2^\infty \frac{x\,dx}{\sqrt{x^4-1}}$

55. $\displaystyle\int_\pi^\infty \frac{2+\cos x}{x}\,dx$

56. $\displaystyle\int_\pi^\infty \frac{1+\sin x}{x^2}\,dx$

Testing for Convergence

In Exercises 35–64, use integration, the Direct Comparison Test, or the Limit Comparison Test to test the integrals for convergence. If more than one method applies, use whatever method you prefer.

57. $\displaystyle\int_{4}^{\infty} \frac{2\,dt}{t^{3/2} - 1}$

58. $\displaystyle\int_{2}^{\infty} \frac{1}{\ln x}\,dx$

59. $\displaystyle\int_{1}^{\infty} \frac{e^x}{x}\,dx$

60. $\displaystyle\int_{e^e}^{\infty} \ln(\ln x)\,dx$

61. $\displaystyle\int_{1}^{\infty} \frac{1}{\sqrt{e^x - x}}\,dx$

62. $\displaystyle\int_{1}^{\infty} \frac{1}{e^x - 2^x}\,dx$

63. $\displaystyle\int_{-\infty}^{\infty} \frac{dx}{\sqrt{x^4 + 1}}$

64. $\displaystyle\int_{-\infty}^{\infty} \frac{dx}{e^x + e^{-x}}$

Theory and Examples

65. *Estimating the value of a convergent improper integral whose domain is infinite*

a) Show that

$$\int_{3}^{\infty} e^{-3x}\,dx = \frac{1}{3}e^{-9} < 0.000042,$$

and hence that $\int_{3}^{\infty} e^{-x^2}\,dx < 0.000042$. Explain why this means that $\int_{0}^{\infty} e^{-x^2}\,dx$ can be replaced by $\int_{0}^{3} e^{-x^2}\,dx$ without introducing an error of magnitude greater than 0.000042.

b) NUMERICAL INTEGRATOR Evaluate $\int_{0}^{3} e^{-x^2}\,dx$ numerically.

66. *The infinite paint can or Gabriel's horn.* As Example 8 shows, the integral $\int_{1}^{\infty} (dx/x)$ diverges. This means that the integral

$$\int_{1}^{\infty} 2\pi \frac{1}{x}\sqrt{1 + \frac{1}{x^4}}\,dx,$$

which measures the *surface area* of the solid of revolution traced out by revolving the curve $y = 1/x$, $1 \le x$, about the x-axis, diverges also. By comparing the two integrals, we see that, for every finite value $b > 1$,

$$\int_{1}^{b} 2\pi \frac{1}{x}\sqrt{1 + \frac{1}{x^4}}\,dx > 2\pi \int_{1}^{b} \frac{1}{x}\,dx.$$

$y = \dfrac{1}{x}$

However, the integral

$$\int_{1}^{\infty} \pi \left(\frac{1}{x}\right)^2 dx$$

for the *volume* of the solid converges. (a) Calculate it. (b) This solid of revolution is sometimes described as a can that does not

hold enough paint to cover its own interior. Think about that for a moment. It is common sense that a finite amount of paint cannot cover an infinite surface. But if we fill the horn with paint (a finite amount), then we *will* have covered an infinite surface. Explain the apparent contradiction.

67. a) Show that

$$\int_{1}^{\infty} \frac{dx}{x^p} = \frac{1}{p - 1}$$

if $p > 1$ but that the integral is infinite if $p < 1$. Example 8 shows what happens if $p = 1$.

b) Show that

$$\int_{0}^{1} \frac{dx}{x^p} = \frac{1}{1 - p}$$

if $p < 1$ but that the integral diverges if $p \ge 1$.

68. Find the values of p for which each integral converges:

a) $\displaystyle\int_{1}^{2} \frac{dx}{x\,(\ln x)^p}$,

b) $\displaystyle\int_{2}^{\infty} \frac{dx}{x\,(\ln x)^p}$.

Exercises 69–72 are about the infinite region in the first quadrant between the curve $y = e^{-x}$ and the x-axis.

69. Find the area of the region.

70. Find the centroid of the region.

71. Find the volume of the solid generated by revolving the region about the y-axis.

72. Find the volume of the solid generated by revolving the region about the x-axis.

73. Find the area of the region that lies between the curves $y = \sec x$ and $y = \tan x$ from $x = 0$ to $x = \pi/2$.

74. The region in Exercise 73 is revolved about the x-axis to generate a solid.

a) Find the volume of the solid.

b) Show that the inner and outer surfaces of the solid have infinite area.

75. $\displaystyle\int_{-\infty}^{\infty} f(x)\,dx$ *may not equal* $\displaystyle\lim_{b\to\infty} \int_{-b}^{b} f(x)\,dx$. Show that

$$\int_{0}^{\infty} \frac{2x\,dx}{x^2 + 1}$$

diverges and hence that

$$\int_{-\infty}^{\infty} \frac{2x\,dx}{x^2 + 1}$$

diverges. Then show that

$$\lim_{b\to\infty} \int_{-b}^{b} \frac{2x\,dx}{x^2 + 1} = 0.$$

76. Here is an argument that $\ln 3$ equals $\infty - \infty$. Where does the argument go wrong? Give reasons for your answer.

$$\ln 3 = \ln 1 + \ln 3 = \ln 1 - \ln \frac{1}{3}$$

$$= \lim_{b \to \infty} \ln \left(\frac{b - 2}{b} \right) - \ln \frac{1}{3}$$

$$= \lim_{b \to \infty} \left[\ln \frac{x - 2}{x} \right]_3^b$$

$$= \lim_{b \to \infty} \left[\ln (x - 2) - \ln x \right]_3^b$$

$$= \lim_{b \to \infty} \int_3^b \left(\frac{1}{x - 2} - \frac{1}{x} \right) dx$$

$$= \int_3^\infty \left(\frac{1}{x - 2} - \frac{1}{x} \right) dx$$

$$= \int_3^\infty \frac{1}{x - 2} dx - \int_3^\infty \frac{1}{x} dx$$

$$= \lim_{b \to \infty} \left[\ln (x - 2) \right]_3^b - \lim_{b \to \infty} \left[\ln x \right]_3^b$$

$$= \infty - \infty.$$

77. Show that if $f(x)$ is integrable on every interval of real numbers and a and b are real numbers with $a < b$, then

a) $\int_{-\infty}^a f(x)\, dx$ and $\int_a^\infty f(x)\, dx$ both converge if and only if $\int_{-\infty}^b f(x)\, dx$ and $\int_b^\infty f(x)\, dx$ both converge.

b) $\int_{-\infty}^a f(x)\, dx + \int_a^\infty f(x)\, dx = \int_{-\infty}^b f(x)\, dx + \int_b^\infty f(x)\, dx$

when the integrals involved converge.

78. a) Show that if f is even and the necessary integrals exist, then

$$\int_{-\infty}^\infty f(x)\, dx = 2 \int_0^\infty f(x)\, dx.$$

b) Show that if f is odd and the necessary integrals exist, then

$$\int_{-\infty}^\infty f(x)\, dx = 0.$$

Use direct evaluation, the comparison tests, and the results in Exercise 78, as appropriate, to determine the convergence or divergence of the integrals in Exercises 79–86. If more than one method applies, use whatever method you prefer.

79. $\displaystyle\int_{-\infty}^\infty \frac{dx}{\sqrt{x^2 + 1}}$

80. $\displaystyle\int_{-\infty}^\infty \frac{dx}{\sqrt{x^6 + 1}}$

81. $\displaystyle\int_{-\infty}^\infty \frac{dx}{e^x + e^{-x}}$

82. $\displaystyle\int_{-\infty}^\infty \frac{e^{-x}\, dx}{x^2 + 1}$

83. $\displaystyle\int_{-\infty}^\infty e^{-|x|}\, dx$

84. $\displaystyle\int_{-\infty}^\infty \frac{dx}{(x + 1)^2}$

85. $\displaystyle\int_{-\infty}^\infty \frac{|\sin x| + |\cos x|}{|x| + 1}\, dx$

(*Hint:* $|\sin \theta| + |\cos \theta| \geq \sin^2 \theta + \cos^2 \theta$.)

86. $\displaystyle\int_{-\infty}^\infty \frac{x\, dx}{(x^2 + 1)(x^2 + 2)}$

✿ **CAS Explorations and Projects**

In Exercises 87–90, use a CAS to explore the integrals for various values of p (include noninteger values). For what values of p does the integral converge? What is the value of the integral when it does converge? Plot the integrand for various values of p.

87. $\displaystyle\int_0^e x^p \ln x\, dx$

88. $\displaystyle\int_e^\infty x^p \ln x\, dx$

89. $\displaystyle\int_0^\infty x^p \ln x\, dx$

90. $\displaystyle\int_{-\infty}^\infty x^p \ln |x|\, dx$

91. The integral

$$\mathrm{Si}(x) = \int_0^x \frac{\sin t}{t}\, dt,$$

called the **sine-integral function**, has important applications in optics.

a) Plot the integrand $(\sin t)/t$ for $t > 0$. Is the Si function everywhere increasing or decreasing? Do you think $\mathrm{Si}(x) = 0$ for $x > 0$? Check your answers by graphing the function $\mathrm{Si}(x)$ for $0 \leq x \leq 25$.

b) Explore the convergence of

$$\int_0^\infty \frac{\sin t}{t}\, dt.$$

If it converges, what is its value?

92. The function

$$\mathrm{erf}(x) = \int_0^x \frac{2e^{-t^2}}{\sqrt{\pi}}\, dt,$$

called the **error function**, has important applications in probability and statistics.

a) Plot the error function for $0 \leq x \leq 25$.

b) Explore the convergence of

$$\int_0^\infty \frac{2e^{-t^2}}{\sqrt{\pi}}\, dt.$$

If it converges, what appears to be its value? You will see how to confirm your estimate in Section 13.3, Exercise 37.

CHAPTER 7 QUESTIONS TO GUIDE YOUR REVIEW

1. What basic integration formulas do you know?

2. What procedures do you know for matching integrals to basic formulas?

3. What is the formula for integration by parts? Where does it come from? Why might you want to use it?

4. When applying the formula for integration by parts, how do you choose the u and dv? How can you apply integration by parts to an integral of the form $\int f(x)\, dx$?

5. What is tabular integration? Give an example.

6. What is the goal of the method of partial fractions?

7. When the degree of a polynomial $f(x)$ is less than the degree of a polynomial $g(x)$, how do you write $f(x)/g(x)$ as a sum of partial fractions if $g(x)$

   a) is a product of distinct linear factors?

   b) consists of a repeated linear factor?

   c) contains an irreducible quadratic factor?

What do you do if the degree of f is *not* less than the degree of g?

8. What substitutions are sometimes used to change quadratic binomials into single squared terms? Why might you want to make such a change?

9. What restrictions can you place on the variables involved in the three basic trigonometric substitutions to make sure the substitutions are reversible (have inverses)?

10. What is a reduction formula? How are reduction formulas typically derived? How are reduction formulas used? Give an example.

11. How are integral tables typically used? What do you do if a particular integral you want to evaluate is not listed in the table?

12. What is an improper integral? How are the values of various types of improper integrals defined? Give examples.

13. What tests are available for determining the convergence and divergence of improper integrals that cannot be evaluated directly? Give examples of their use.

CHAPTER 7 PRACTICE EXERCISES

Integration Using Substitutions

Evaluate the integrals in Exercises 1–82. To transform each integral into a recognizable basic form, it may be necessary to use one or more of the techniques of algebraic substitution, completing the square, separating fractions, long division, or trigonometric substitution.

1. $\displaystyle\int x\sqrt{4x^2 - 9}\, dx$

2. $\displaystyle\int 6x\sqrt{3x^2 + 5}\, dx$

3. $\displaystyle\int x(2x + 1)^{1/2}\, dx$

4. $\displaystyle\int x(1 - x)^{-1/2}\, dx$

5. $\displaystyle\int \frac{x\, dx}{\sqrt{8x^2 + 1}}$

6. $\displaystyle\int \frac{x\, dx}{\sqrt{9 - 4x^2}}$

7. $\displaystyle\int \frac{y\, dy}{25 + y^2}$

8. $\displaystyle\int \frac{y^3\, dy}{4 + y^4}$

9. $\displaystyle\int \frac{t^3\, dt}{\sqrt{9 - 4t^4}}$

10. $\displaystyle\int \frac{2t\, dt}{t^4 + 1}$

11. $\displaystyle\int z^{2/3}(z^{5/3} + 1)^{2/3}\, dz$

12. $\displaystyle\int z^{-1/5}(1 + z^{4/5})^{-1/2}\, dz$

13. $\displaystyle\int \frac{\sin 2\theta\, d\theta}{(1 - \cos 2\theta)^2}$

14. $\displaystyle\int \frac{\cos\theta\, d\theta}{(1 + \sin\theta)^{1/2}}$

15. $\displaystyle\int \frac{\sin t}{3 + 4\cos t}\, dt$

16. $\displaystyle\int \frac{\cos 2t}{1 + \sin 2t}\, dt$

17. $\displaystyle\int \sin 2x\, e^{\cos 2x}\, dx$

18. $\displaystyle\int \sec x \tan x\, e^{\sec x}\, dx$

19. $\displaystyle\int e^{\theta} \sin(e^{\theta}) \cos^2(e^{\theta})\, d\theta$

20. $\displaystyle\int e^{\theta} \sec^2(e^{\theta})\, d\theta$

21. $\displaystyle\int 2^{x-1}\, dx$

22. $\displaystyle\int 5^{x\sqrt{2}}\, dx$

23. $\displaystyle\int \frac{dv}{v \ln v}$

24. $\displaystyle\int \frac{dv}{v(2 + \ln v)}$

25. $\displaystyle\int \frac{dx}{(x^2 + 1)(2 + \tan^{-1} x)}$

26. $\displaystyle\int \frac{\sin^{-1} x}{\sqrt{1 - x^2}}\, dx$

27. $\displaystyle\int \frac{2\, dx}{\sqrt{1 - 4x^2}}$

28. $\displaystyle\int \frac{dx}{\sqrt{49 - x^2}}$

29. $\displaystyle\int \frac{dt}{\sqrt{16 - 9t^2}}$

30. $\displaystyle\int \frac{dt}{\sqrt{9 - 4t^2}}$

31. $\displaystyle\int \frac{dt}{9 + t^2}$

32. $\displaystyle\int \frac{dt}{1 + 25t^2}$

33. $\displaystyle\int \frac{4\,dx}{5x\sqrt{25x^2 - 16}}$

34. $\displaystyle\int \frac{6\,dx}{x\sqrt{4x^2 - 9}}$

35. $\displaystyle\int \frac{dx}{\sqrt{4x - x^2}}$

36. $\displaystyle\int \frac{dx}{\sqrt{4x - x^2 - 3}}$

37. $\displaystyle\int \frac{dy}{y^2 - 4y + 8}$

38. $\displaystyle\int \frac{dt}{t^2 + 4t + 5}$

39. $\displaystyle\int \frac{dx}{(x - 1)\sqrt{x^2 - 2x}}$

40. $\displaystyle\int \frac{dv}{(v + 1)\sqrt{v^2 + 2v}}$

41. $\displaystyle\int \sin^2 x\,dx$

42. $\displaystyle\int \cos^2 3x\,dx$

43. $\displaystyle\int \sin^3 \frac{\theta}{2}\,d\theta$

44. $\displaystyle\int \sin^3 \theta \cos^2 \theta\,d\theta$

45. $\displaystyle\int \tan^3 2t\,dt$

46. $\displaystyle\int 6\sec^4 t\,dt$

47. $\displaystyle\int \frac{dx}{2\sin x \cos x}$

48. $\displaystyle\int \frac{2\,dx}{\cos^2 x - \sin^2 x}$

49. $\displaystyle\int_{\pi/4}^{\pi/2} \sqrt{\csc^2 y - 1}\,dy$

50. $\displaystyle\int_{\pi/4}^{3\pi/4} \sqrt{\cot^2 t + 1}\,dt$

51. $\displaystyle\int_0^{\pi} \sqrt{1 - \cos^2 2x}\,dx$

52. $\displaystyle\int_0^{2\pi} \sqrt{1 - \sin^2 \frac{x}{2}}\,dx$

53. $\displaystyle\int_{-\pi/2}^{\pi/2} \sqrt{1 - \cos 2t}\,dt$

54. $\displaystyle\int_{\pi}^{2\pi} \sqrt{1 + \cos 2t}\,dt$

55. $\displaystyle\int \frac{x^2}{x^2 + 4}\,dx$

56. $\displaystyle\int \frac{x^3}{9 + x^2}\,dx$

57. $\displaystyle\int \frac{4x^2 + 3}{2x - 1}\,dx$

58. $\displaystyle\int \frac{2x}{x - 4}\,dx$

59. $\displaystyle\int \frac{2y - 1}{y^2 + 4}\,dy$

60. $\displaystyle\int \frac{y + 4}{y^2 + 1}\,dy$

61. $\displaystyle\int \frac{t + 2}{\sqrt{4 - t^2}}\,dt$

62. $\displaystyle\int \frac{2t^2 + \sqrt{1 - t^2}}{t\sqrt{1 - t^2}}\,dt$

63. $\displaystyle\int \frac{\tan x\,dx}{\tan x + \sec x}$

64. $\displaystyle\int \frac{\cot x}{\cot x + \csc x}\,dx$

65. $\displaystyle\int \sec(5 - 3x)\,dx$

66. $\displaystyle\int x\csc(x^2 + 3)\,dx$

67. $\displaystyle\int \cot\left(\frac{x}{4}\right)\,dx$

68. $\displaystyle\int \tan(2x - 7)\,dx$

69. $\displaystyle\int x\sqrt{1 - x}\,dx$

70. $\displaystyle\int 3x\sqrt{2x + 1}\,dx$

71. $\displaystyle\int \sqrt{z^2 + 1}\,dz$

72. $\displaystyle\int (16 + z^2)^{-3/2}\,dz$

73. $\displaystyle\int \frac{dy}{\sqrt{25 + y^2}}$

74. $\displaystyle\int \frac{dy}{\sqrt{25 + 9y^2}}$

75. $\displaystyle\int \frac{dx}{x^2\sqrt{1 - x^2}}$

76. $\displaystyle\int \frac{x^3\,dx}{\sqrt{1 - x^2}}$

77. $\displaystyle\int \frac{x^2\,dx}{\sqrt{1 - x^2}}$

78. $\displaystyle\int \sqrt{4 - x^2}\,dx$

79. $\displaystyle\int \frac{dx}{\sqrt{x^2 - 9}}$

80. $\displaystyle\int \frac{12\,dx}{(x^2 - 1)^{3/2}}$

81. $\displaystyle\int \frac{\sqrt{w^2 - 1}}{w}\,dw$

82. $\displaystyle\int \frac{\sqrt{z^2 - 16}}{z}\,dz$

Integration by Parts

Evaluate the integrals in Exercises 83–90 using integration by parts.

83. $\displaystyle\int \ln(x + 1)\,dx$

84. $\displaystyle\int x^2 \ln x\,dx$

85. $\displaystyle\int \tan^{-1} 3x\,dx$

86. $\displaystyle\int \cos^{-1}\left(\frac{x}{2}\right)\,dx$

87. $\displaystyle\int (x + 1)^2 e^x\,dx$

88. $\displaystyle\int x^2 \sin(1 - x)\,dx$

89. $\displaystyle\int e^x \cos 2x\,dx$

90. $\displaystyle\int e^{-2x} \sin 3x\,dx$

Partial Fractions

Evaluate the integrals in Exercises 91–110. It may be necessary to use a substitution first.

91. $\displaystyle\int \frac{x\,dx}{x^2 - 3x + 2}$

92. $\displaystyle\int \frac{x\,dx}{x^2 + 4x + 3}$

93. $\displaystyle\int \frac{dx}{x(x + 1)^2}$

94. $\displaystyle\int \frac{x + 1}{x^2(x - 1)}\,dx$

95. $\displaystyle\int \frac{\sin\theta\,d\theta}{\cos^2\theta + \cos\theta - 2}$

96. $\displaystyle\int \frac{\cos\theta\,d\theta}{\sin^2\theta + \sin\theta - 6}$

97. $\displaystyle\int \frac{3x^2 + 4x + 4}{x^3 + x}\,dx$

98. $\displaystyle\int \frac{4x\,dx}{x^3 + 4x}$

99. $\displaystyle\int \frac{v + 3}{2v^3 - 8v}\,dv$

100. $\displaystyle\int \frac{(3v - 7)\,dv}{(v - 1)(v - 2)(v - 3)}$

101. $\displaystyle\int \frac{dt}{t^4 + 4t^2 + 3}$

102. $\displaystyle\int \frac{t\,dt}{t^4 - t^2 - 2}$

103. $\displaystyle\int \frac{x^3 + x^2}{x^2 + x - 2}\,dx$

104. $\displaystyle\int \frac{x^3 + 1}{x^3 - x}\,dx$

105. $\displaystyle\int \frac{x^3 + 4x^2}{x^2 + 4x + 3}\,dx$

106. $\displaystyle\int \frac{2x^3 + x^2 - 21x + 24}{x^2 + 2x - 8}\,dx$

107. $\displaystyle\int \frac{dx}{x\left(3\sqrt{x} + 1\right)}$

108. $\displaystyle\int \frac{dx}{x\left(1 + \sqrt[3]{x}\right)}$

109. $\displaystyle\int \frac{ds}{e^s - 1}$

110. $\displaystyle\int \frac{ds}{\sqrt{e^s + 1}}$

Improper Integrals

Evaluate the improper integrals in Exercises 111–120.

111. $\displaystyle\int_0^3 \frac{dx}{\sqrt{9 - x^2}}$

112. $\displaystyle\int_0^1 \ln x \, dx$

113. $\displaystyle\int_{-1}^1 \frac{dy}{y^{2/3}}$

114. $\displaystyle\int_{-2}^0 \frac{d\theta}{(\theta + 1)^{3/5}}$

115. $\displaystyle\int_3^\infty \frac{2 \, du}{u^2 - 2u}$

116. $\displaystyle\int_1^\infty \frac{3v - 1}{4v^3 - v^2} \, dv$

117. $\displaystyle\int_0^\infty x^2 e^{-x} \, dx$

118. $\displaystyle\int_{-\infty}^0 x e^{3x} \, dx$

119. $\displaystyle\int_{-\infty}^\infty \frac{dx}{4x^2 + 9}$

120. $\displaystyle\int_{-\infty}^\infty \frac{4 \, dx}{x^2 + 16}$

Convergence or Divergence

Which of the improper integrals in Exercises 121–126 converge and which diverge?

121. $\displaystyle\int_6^\infty \frac{d\theta}{\sqrt{\theta^2 + 1}}$

122. $\displaystyle\int_0^\infty e^{-u} \cos u \, du$

123. $\displaystyle\int_1^\infty \frac{\ln z}{z} \, dz$

124. $\displaystyle\int_1^\infty \frac{e^{-t}}{\sqrt{t}} \, dt$

125. $\displaystyle\int_{-\infty}^\infty \frac{dx}{e^x + e^{-x}}$

126. $\displaystyle\int_{-\infty}^\infty \frac{dx}{x^2(1 + e^x)}$

Trigonometric Substitutions

Evaluate the integrals in Exercises 127–130 (a) without using a trigonometric substitution, (b) using a trigonometric substitution.

127. $\displaystyle\int \frac{y \, dy}{\sqrt{16 - y^2}}$

128. $\displaystyle\int \frac{x \, dx}{\sqrt{4 + x^2}}$

129. $\displaystyle\int \frac{x \, dx}{4 - x^2}$

130. $\displaystyle\int \frac{t \, dt}{\sqrt{4t^2 - 1}}$

Quadratic Terms

Evaluate the integrals in Exercises 131–134.

131. $\displaystyle\int \frac{x \, dx}{9 - x^2}$

132. $\displaystyle\int \frac{dx}{x(9 - x^2)}$

133. $\displaystyle\int \frac{dx}{9 - x^2}$

134. $\displaystyle\int \frac{dx}{\sqrt{9 - x^2}}$

Assorted Integrations

Evaluate the integrals in Exercises 135–202. The integrals are listed in random order.

135. $\displaystyle\int \frac{x \, dx}{1 + \sqrt{x}}$

136. $\displaystyle\int \frac{x^3 + 2}{4 - x^2} \, dx$

137. $\displaystyle\int \frac{dx}{x(x^2 + 1)^2}$

138. $\displaystyle\int \frac{\cos \sqrt{x}}{\sqrt{x}} \, dx$

139. $\displaystyle\int \frac{dx}{\sqrt{-2x - x^2}}$

140. $\displaystyle\int \frac{(t - 1) \, dt}{\sqrt{t^2 - 2t}}$

141. $\displaystyle\int \frac{du}{\sqrt{1 + u^2}}$

142. $\displaystyle\int e^t \cos e^t \, dt$

143. $\displaystyle\int \frac{2 - \cos x + \sin x}{\sin^2 x} \, dx$

144. $\displaystyle\int \frac{\sin^2 \theta}{\cos^2 \theta} \, d\theta$

145. $\displaystyle\int \frac{9 \, dv}{81 - v^4}$

146. $\displaystyle\int \frac{\cos x \, dx}{1 + \sin^2 x}$

147. $\displaystyle\int \theta \cos (2\theta + 1) \, d\theta$

148. $\displaystyle\int_2^\infty \frac{dx}{(x - 1)^2}$

149. $\displaystyle\int \frac{x^3 \, dx}{x^2 - 2x + 1}$

150. $\displaystyle\int \frac{d\theta}{\sqrt{1 + \sqrt{\theta}}}$

151. $\displaystyle\int \frac{2 \sin \sqrt{x} \, dx}{\sqrt{x} \sec \sqrt{x}}$

152. $\displaystyle\int \frac{x^5 \, dx}{x^4 - 16}$

153. $\displaystyle\int \frac{dy}{\sin y \cos y}$

154. $\displaystyle\int \frac{d\theta}{\theta^2 - 2\theta + 4}$

155. $\displaystyle\int \frac{\tan x}{\cos^2 x} \, dx$

156. $\displaystyle\int \frac{dr}{(r + 1)\sqrt{r^2 + 2r}}$

157. $\displaystyle\int \frac{(r + 2) \, dr}{\sqrt{-r^2 - 4r}}$

158. $\displaystyle\int \frac{y \, dy}{4 + y^4}$

159. $\displaystyle\int \frac{\sin 2\theta \, d\theta}{(1 + \cos 2\theta)^2}$

160. $\displaystyle\int \frac{dx}{(x^2 - 1)^2}$

161. $\displaystyle\int_{\pi/4}^{\pi/2} \sqrt{1 + \cos 4x} \, dx$

162. $\displaystyle\int (15)^{2x+1} \, dx$

163. $\displaystyle\int \frac{x \, dx}{\sqrt{2 - x}}$

164. $\displaystyle\int \frac{\sqrt{1 - v^2}}{v^2} \, dv$

165. $\displaystyle\int \frac{dy}{y^2 - 2y + 2}$

166. $\displaystyle\int \ln \sqrt{x - 1} \, dx$

167. $\displaystyle\int \theta^2 \tan (\theta^3) \, d\theta$

168. $\displaystyle\int \frac{x \, dx}{\sqrt{8 - 2x^2 - x^4}}$

169. $\displaystyle\int \frac{z + 1}{z^2(z^2 + 4)} \, dz$

170. $\displaystyle\int x^3 e^{(x^2)} \, dx$

171. $\displaystyle\int \frac{t \, dt}{\sqrt{9 - 4t^2}}$

172. $\displaystyle\int_0^{\pi/10} \sqrt{1 + \cos 5\theta} \, d\theta$

173. $\displaystyle\int \frac{\cot \theta \, d\theta}{1 + \sin^2 \theta}$

174. $\displaystyle\int \frac{\tan^{-1} x}{x^2} \, dx$

175. $\displaystyle\int \frac{\tan \sqrt{y}}{2\sqrt{y}} \, dy$

176. $\displaystyle\int \frac{e^t \, dt}{e^{2t} + 3e^t + 2}$

177. $\displaystyle\int \frac{\theta^2 \, d\theta}{4 - \theta^2}$

178. $\displaystyle\int \frac{1 - \cos 2x}{1 + \cos 2x} \, dx$

179. $\displaystyle\int \frac{\cos (\sin^{-1} x)}{\sqrt{1 - x^2}} \, dx$

180. $\displaystyle\int \frac{\cos x \, dx}{\sin^3 x - \sin x}$

181. $\int \sin\dfrac{x}{2}\cos\dfrac{x}{2}\,dx$

182. $\int \dfrac{x^2 - x + 2}{(x^2 + 2)^2}\,dx$

199. $\int \dfrac{8\,dm}{m\sqrt{49m^2 - 4}}$

183. $\int \dfrac{e^t\,dt}{1 + e^t}$

184. $\int \tan^3 t\,dt$

200. $\int \dfrac{dt}{t(1 + \ln t)\sqrt{(\ln t)(2 + \ln t)}}$

185. $\int_1^\infty \dfrac{\ln y}{y^3}\,dy$

186. $\int \dfrac{3 + \sec^2 x + \sin x}{\tan x}\,dx$

201. $\int_0^1 3(x-1)^2\left(\int_0^x \sqrt{1 + (t-1)^4}\,dt\right)dx$

187. $\int \dfrac{\cot v\,dv}{\ln \sin v}$

188. $\int \dfrac{dx}{(2x-1)\sqrt{x^2 - x}}$

202. $\int_2^\infty \dfrac{4v^3 + v - 1}{v^2(v-1)(v^2+1)}\,dv$

189. $\int e^{\ln\sqrt{x}}\,dx$

190. $\int e^\theta\sqrt{3 + 4e^\theta}\,d\theta$

203. Suppose for a certain function f it is known that
$$f'(x) = \frac{\cos x}{x}, \quad f(\pi/2) = a, \quad \text{and} \quad f(3\pi/2) = b.$$
Use integration by parts to evaluate
$$\int_{\pi/2}^{3\pi/2} f(x)\,dx.$$

191. $\int \dfrac{\sin 5t\,dt}{1 + (\cos 5t)^2}$

192. $\int \dfrac{dv}{\sqrt{e^{2v} - 1}}$

193. $\int (27)^{3\theta+1}\,d\theta$

194. $\int x^5 \sin x\,dx$

195. $\int \dfrac{dr}{1 + \sqrt{r}}$

196. $\int \dfrac{4x^3 - 20x}{x^4 - 10x^2 + 9}\,dx$

204. Find a positive number a satisfying
$$\int_0^a \frac{dx}{1 + x^2} = \int_a^\infty \frac{dx}{1 + x^2}.$$

197. $\int \dfrac{8\,dy}{y^3(y+2)}$

198. $\int \dfrac{(t+1)\,dt}{(t^2 + 2t)^{2/3}}$

CHAPTER 7 ADDITIONAL EXERCISES–THEORY, EXAMPLES, APPLICATIONS

Challenging Integrals

Evaluate the integrals in Exercises 1–10.

1. $\int (\sin^{-1} x)^2\,dx$

2. $\int \dfrac{dx}{x(x+1)(x+2)\cdots(x+m)}$

3. $\int x\sin^{-1} x\,dx$

4. $\int \sin^{-1}\sqrt{y}\,dy$

5. $\int \dfrac{d\theta}{1 - \tan^2\theta}$

6. $\int \ln(\sqrt{x} + \sqrt{1+x})\,dx$

7. $\int \dfrac{dt}{t - \sqrt{1 - t^2}}$

8. $\int \dfrac{(2e^{2x} - e^x)\,dx}{\sqrt{3e^{2x} - 6e^x - 1}}$

9. $\int \dfrac{dx}{x^4 + 4}$

10. $\int \dfrac{dx}{x^6 - 1}$

Limits

Evaluate the limits in Exercises 11 and 12.

11. $\lim\limits_{x\to\infty} \int_{-x}^x \sin t\,dt$

12. $\lim\limits_{x\to 0^+} x\int_x^1 \dfrac{\cos t}{t^2}\,dt$

Evaluate the limits in Exercises 13 and 14 by identifying them with definite integrals and evaluating the integrals.

13. $\lim\limits_{n\to\infty} \sum\limits_{k=1}^n \ln\sqrt[n]{1 + \dfrac{k}{n}}$

14. $\lim\limits_{n\to\infty} \sum\limits_{k=0}^{n-1} \dfrac{1}{\sqrt{n^2 - k^2}}$

Theory and Applications

15. Find the length of the curve
$$y = \int_0^x \sqrt{\cos 2t}\,dt, \quad 0 \le x \le \pi/4.$$

16. Find the length of the curve $y = \ln(1 - x^2)$, $0 \le x \le 1/2$.

17. The region in the first quadrant that is enclosed by the x-axis and the curve $y = 3x\sqrt{1-x}$ is revolved about the y-axis to generate a solid. Find the volume of the solid.

18. The region in the first quadrant that is enclosed by the x-axis, the curve $y = 5/(x\sqrt{5-x})$, and the lines $x = 1$ and $x = 4$ is revolved about the x-axis to generate a solid. Find the volume of the solid.

19. The region in the first quadrant enclosed by the coordinate axes, the curve $y = e^x$, and the line $x = 1$ is revolved about the y-axis to generate a solid. Find the volume of the solid.

20. The region in the first quadrant that is bounded above by the curve $y = e^x - 1$, below by the x-axis, and on the right by the line $x = \ln 2$ is revolved about the line $x = \ln 2$ to generate a solid. Find the volume of the solid.

21. Let R be the "triangular" region in the first quadrant that is bounded above by the line $y = 1$, below by the curve $y = \ln x$, and on the left by the line $x = 1$. Find the volume of the solid generated by revolving R about

  a) the x-axis
  b) the line $y = 1$.

22. (*Continuation of Exercise 21.*) Find the volume of the solid generated by revolving the shaded region about (a) the y-axis, (b) the line $x = 1$.

23. The region between the curve

$$y = f(x) = \begin{cases} 0, & x = 0 \\ x \ln x, & 0 < x \le 2 \end{cases}$$

is revolved about the x-axis to generate the solid shown here.

  a) Show that f is continuous at $x = 0$.
  b) Find the volume of the solid.

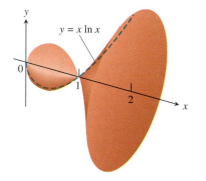

24. The infinite region bounded by the coordinate axes and the curve $y = -\ln x$ in the first quadrant is revolved about the x-axis to generate a solid. Find the volume of the solid.

25. Find the centroid of the region in the first quadrant that is bounded below by the x-axis, above by the curve $y = \ln x$, and on the right by the line $x = e$.

26. Find the centroid of the region in the plane enclosed by the curves $y = \pm(1 - x^2)^{-1/2}$ and the lines $x = 0$ and $x = 1$.

27. Find the length of the curve $y = \ln x$ from $x = 1$ to $x = e$.

28. Find the area of the surface generated by revolving the curve in Exercise 27 about the y-axis.

29. *The length of an astroid.* The graph of the equation $x^{2/3} + y^{2/3} = 1$ is one of a family of curves called *astroids* (not "asteroids") because of their starlike appearance (Fig. 7.21). Find the length of this particular astroid.

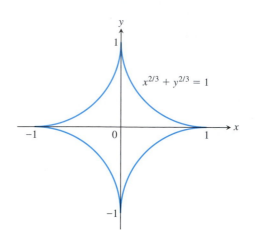

7.21 The astroid in Exercises 29 and 30.

30. *The surface generated by an astroid.* Find the area of the surface generated by revolving the curve in Fig. 7.21 about the x-axis.

31. Find a curve through the origin whose length is

$$\int_0^4 \sqrt{1 + \frac{1}{4x}} \, dx.$$

32. Without evaluating either integral, explain why

$$2 \int_{-1}^1 \sqrt{1 - x^2} \, dx = \int_{-1}^1 \frac{dx}{\sqrt{1 - x^2}}.$$

(Source: Peter A. Lindstrom, *Mathematics Magazine*, Vol. 45, No. 1, January 1972, p. 47.)

33. a) GRAPHER Graph the function $f(x) = e^{(x - e^x)}, -5 \le x \le 3$.

  b) Show that $\displaystyle\int_{-\infty}^{\infty} f(x) \, dx$ converges and find its value.

34. Find $\displaystyle\lim_{n \to \infty} \int_0^1 \frac{n \, y^{n-1}}{1 + y} \, dy$.

35. Derive the integral formula

$$\int x \left(\sqrt{x^2 - a^2} \right)^n dx = \frac{\left(\sqrt{x^2 - a^2} \right)^{n+2}}{n + 2} + C, \quad n \ne -2.$$

36. Prove that

$$\frac{\pi}{6} < \int_0^1 \frac{dx}{\sqrt{4 - x^2 - x^3}} < \frac{\pi \sqrt{2}}{8}.$$

(*Hint:* Observe that for $0 < x < 1$, we have $4 - x^2 > 4 - x^2 - x^3 > 4 - 2x^2$, with the left-hand side becoming an equality for $x = 0$ and the right-hand side becoming an equality for $x = 1$.)

37. For what value or values of a does

$$\int_1^{\infty} \left(\frac{ax}{x^2 + 1} - \frac{1}{2x} \right) dx$$

converge? Evaluate the corresponding integral(s).

38. For each $x > 0$, let $G(x) = \int_0^\infty e^{-xt}\,dt$. Prove that $xG(x) = 1$ for each $x > 0$.

39. *Infinite area and finite volume.* What values of p have the following property: The area of the region between the curve $y = x^{-p}$, $1 \le x < \infty$, and the x-axis is infinite but the volume of the solid generated by revolving the region about the x-axis is finite.

40. *Infinite area and finite volume.* What values of p have the following property: The area of the region in the first quadrant enclosed by the curve $y = x^{-p}$, the y-axis, the line $x = 1$, and the interval $[0, 1]$ on the x-axis is infinite but the volume of the solid generated by revolving the region about one of the coordinate axes is finite.

Tabular Integration

The technique of tabular integration also applies to integrals of the form $\int f(x)\,g(x)\,dx$ when neither function can be differentiated repeatedly to become zero. For example, to evaluate

$$\int e^{2x}\cos x\,dx$$

we begin as before with a table listing successive derivatives of e^{2x} and integrals of $\cos x$:

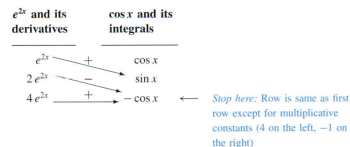

e^{2x} and its derivatives		$\cos x$ and its integrals
e^{2x}	$+$	$\cos x$
$2e^{2x}$	$-$	$\sin x$
$4e^{2x}$	$+$	$-\cos x$

Stop here: Row is same as first row except for multiplicative constants (4 on the left, -1 on the right)

We stop differentiating and integrating as soon as we reach a row that is the same as the first row except for multiplicative constants. We interpret the table as saying

$$\int e^{2x}\cos x\,dx$$

$$= +(e^{2x}\sin x) - (2e^{2x}(-\cos x)) + \int (4e^{2x})(-\cos x)\,dx.$$

We take signed products from the diagonal arrows and a signed integral for the last horizontal arrow. Transposing the integral on the right-hand side over to the left-hand side now gives

$$5\int e^{2x}\cos x\,dx = e^{2x}\sin x + 2e^{2x}\cos x$$

or

$$\int e^{2x}\cos x\,dx = \frac{e^{2x}\sin x + 2e^{2x}\cos x}{5} + C,$$

after dividing by 5 and adding the constant of integration.

Use tabular integration to evaluate the integrals in Exercises 41–48.

41. $\displaystyle\int e^{2x}\cos 3x\,dx$

42. $\displaystyle\int e^{3x}\sin 4x\,dx$

43. $\displaystyle\int \sin 3x \sin x\,dx$

44. $\displaystyle\int \cos 5x \sin 4x\,dx$

45. $\displaystyle\int e^{ax}\sin bx\,dx$

46. $\displaystyle\int e^{ax}\cos bx\,dx$

47. $\displaystyle\int \ln(ax)\,dx$

48. $\displaystyle\int x^2 \ln(ax)\,dx$

The Gamma Function and Stirling's Formula

Euler's gamma function $\Gamma(x)$ ("gamma of x"; Γ is a Greek capital g) uses an integral to extend the factorial function from the nonnegative integers to other real values. The formula is

$$\Gamma(x) = \int_0^\infty t^{x-1}e^{-t}\,dt, \quad x > 0.$$

For each positive x, the number $\Gamma(x)$ is the integral of $t^{x-1}e^{-t}$ with respect to t from 0 to ∞. Figure 7.22 shows the graph of Γ near the origin. You will see how to calculate $\Gamma(1/2)$ if you do Additional Exercise 31 in Chapter 13.

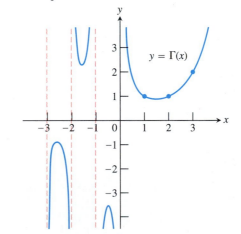

7.22 $\Gamma(x)$ is a continuous function of x whose value at each positive integer $n + 1$ is $n!$. The defining integral formula for Γ is valid only for $x > 0$, but we can extend Γ to negative noninteger values of x with the formula $\Gamma(x) = (\Gamma(x + 1))/x$, which is the subject of Exercise 49.

49. *If n is a nonnegative integer,* $\Gamma(n+1) = n!$

    a)   Show that $\Gamma(1) = 1$.

    b)   Then apply integration by parts to the integral for $\Gamma(x+1)$ to show that $\Gamma(x+1) = x\Gamma(x)$. This gives

$$\Gamma(2) = 1\Gamma(1) = 1$$

$$\Gamma(3) = 2\Gamma(2) = 2$$

$$\Gamma(4) = 3\Gamma(3) = 6$$

$$\vdots$$

$$\Gamma(n+1) = n\Gamma(n) = n! \qquad (1)$$

    c)   Use mathematical induction to verify Eq. (1) for every nonnegative integer n.

50. *Stirling's formula.* Scottish mathematician James Stirling (1692–1770) showed that

$$\lim_{x \to \infty} \left(\frac{e}{x}\right)^x \sqrt{\frac{x}{2\pi}}\, \Gamma(x) = 1,$$

so for large x,

$$\Gamma(x) = \left(\frac{x}{e}\right)^x \sqrt{\frac{2\pi}{x}}(1 + \epsilon(x)), \quad \epsilon(x) \to 0 \text{ as } x \to \infty. \qquad (2)$$

Dropping $\epsilon(x)$ leads to the approximation

$$\Gamma(x) \approx \left(\frac{x}{e}\right)^x \sqrt{\frac{2\pi}{x}}. \qquad \textbf{(Stirling's formula)} \qquad (3)$$

    a)   *Stirling's approximation for n!.* Use Eq. (3) and the fact that $n! = n\Gamma(n)$ to show that

$$n! \approx \left(\frac{n}{e}\right)^n \sqrt{2n\pi}. \qquad \textbf{(Stirling's approximation)} \qquad (4)$$

As you will see if you do Exercise 68 in Section 8.2, Eq. (4) leads to the approximation

$$\sqrt[n]{n!} \approx \frac{n}{e}. \qquad (5)$$

    b)   CALCULATOR Compare your calculator's value for $n!$ with the value given by Stirling's approximation for $n = 10, 20, 30, \ldots$, as far as your calculator can go.

    c)   CALCULATOR A refinement of Eq. (2) gives

$$\Gamma(x) = \left(\frac{x}{e}\right)^x \sqrt{\frac{2\pi}{x}}\, e^{1/(12x)}(1 + \epsilon(x)),$$

or

$$\Gamma(x) \approx \left(\frac{x}{e}\right)^x \sqrt{\frac{2\pi}{x}}\, e^{1/(12x)}$$

which tells us that

$$n! \approx \left(\frac{n}{e}\right)^n \sqrt{2n\pi}\, e^{1/(12n)}. \qquad (6)$$

Compare the values given for 10! by your calculator, Stirling's approximation, and Eq. (6).

CHAPTER 8

Infinite Series

OVERVIEW   In this chapter we develop a remarkable formula that enables us to express many functions as "infinite polynomials" and at the same time tells how much error we will incur if we truncate those polynomials to make them finite. In addition to providing effective polynomial approximations of differentiable functions, these infinite polynomials (called power series) have many other uses. They provide an efficient way to evaluate nonelementary integrals and they solve differential equations that give insight into heat flow, vibration, chemical diffusion, and signal transmission. What you will learn here sets the stage for the roles played by series of functions of all kinds in science and mathematics.

8.1   Limits of Sequences of Numbers

Informally, a sequence is an ordered list of things, but in this chapter the things will usually be numbers. We have seen sequences before, such as the sequence $x_0, x_1, \ldots, x_n, \ldots$ of numbers generated by Newton's method and the sequence $c_1, c_2, \ldots, c_n, \ldots$ of polygons that define Helga von Koch's snowflake. These sequences have limits, but many equally important sequences do not.

Definitions and Notation

We can list the integer multiples of 3 by assigning each multiple a position:

$$
\begin{array}{cccccc}
\text{Domain:} & 1 & 2 & 3\ldots & n\ldots \\
 & \downarrow & \downarrow & \downarrow & \downarrow \\
\text{Range:} & 3 & 6 & 9 & 3n
\end{array}
$$

The first number is 3, the second 6, the third 9, and so on. The assignment is a function that assigns $3n$ to the nth place. And that is the basic idea for constructing sequences. There is a function that tells us where each item is to be placed.

> #### Definition
>
> An **infinite sequence** (or **sequence**) of numbers is a function whose domain is the set of integers greater than or equal to some integer n_0.

Usually n_0 is 1 and the domain of the sequence is the set of positive integers. But sometimes we want to start sequences elsewhere. We take $n_0 = 0$ when we begin Newton's method. We might take $n_0 = 3$ if we were defining a sequence of n-sided polygons.

Sequences are defined the way other functions are, some typical rules being

$$a(n) = \sqrt{n}, \qquad a(n) = (-1)^{n+1}\frac{1}{n}, \qquad a(n) = \frac{n-1}{n}$$

(Example 1 and Fig. 8.1).

To indicate that the domains are sets of integers, we use a letter like n from the middle of the alphabet for the independent variable, instead of the x, y, z, and t used widely in other contexts. The formulas in the defining rules, however, like those above, are often valid for domains larger than the set of positive integers. This can be an advantage, as we will see.

The number $a(n)$ is the **nth term** of the sequence, or the **term with index n.** If $a(n) = (n-1)/n$, we have

First term	Second term	Third term		nth term
$a(1) = 0$	$a(2) = \dfrac{1}{2},$	$a(3) = \dfrac{2}{3},$	$\ldots,$	$a(n) = \dfrac{n-1}{n}.$

When we use the subscript notation a_n for $a(n)$, the sequence is written

$$a_1 = 0, \qquad a_2 = \frac{1}{2}, \qquad a_3 = \frac{2}{3}, \qquad \ldots, \qquad a_n = \frac{n-1}{n}.$$

To describe sequences, we often write the first few terms as well as a formula for the nth term.

EXAMPLE 1

We write	For the sequence whose defining rule is
$1, \sqrt{2}, \sqrt{3}, \sqrt{4}, \ldots, \sqrt{n}, \ldots$	$a_n = \sqrt{n}$
$1, \dfrac{1}{2}, \dfrac{1}{3}, \ldots, \dfrac{1}{n}, \ldots$	$a_n = \dfrac{1}{n}$
$1, -\dfrac{1}{2}, \dfrac{1}{3}, -\dfrac{1}{4}, \ldots, (-1)^{n+1}\dfrac{1}{n}, \ldots$	$a_n = (-1)^{n+1}\dfrac{1}{n}$
$0, \dfrac{1}{2}, \dfrac{2}{3}, \dfrac{3}{4}, \ldots, \dfrac{n-1}{n}, \ldots$	$a_n = \dfrac{n-1}{n}$
$0, -\dfrac{1}{2}, \dfrac{2}{3}, -\dfrac{3}{4}, \ldots, (-1)^{n+1}\left(\dfrac{n-1}{n}\right), \ldots$	$a_n = (-1)^{n+1}\left(\dfrac{n-1}{n}\right)$
$3, 3, 3, \ldots, 3, \ldots$	$a_n = 3$

Notation   We refer to the sequence whose nth term is a_n with the notation $\{a_n\}$ ("the sequence a sub n"). The second sequence in Example 1 is $\{1/n\}$ ("the sequence 1 over n"); the last sequence is $\{3\}$ ("the constant sequence 3").

8.1 The sequences of Example 1 are graphed here in two different ways: by plotting the numbers a_n on a horizontal axis and by plotting the points (n, a_n) in the coordinate plane.

The terms $a_n = \sqrt{n}$ eventually surpass every integer, so the sequence $\{a_n\}$ diverges, . . .

. . . but the terms $a_n = 1/n$ decrease steadily and get arbitrarily close to 0 as n increases, so the sequence $\{a_n\}$ converges to 0.

The terms $a_n = (-1)^{n+1}(1/n)$ alternate in sign but still converge to 0.

The terms $a_n = (n - 1)/n$ approach 1 steadily and get arbitrarily close as n increases, so the sequence $\{a_n\}$ converges to 1.

The terms $a_n = (-1)^{n+1}[(n - 1)/n]$ alternate in sign. The positive terms approach 1. But the negative terms approach -1 as n increases, so the sequence $\{a_n\}$ diverges.

The terms in the sequence of constants $a_n = 3$ have the same value regardless of n, so the sequence $\{a_n\}$ converges to 3.

Convergence and Divergence

As Fig. 8.1 shows, the sequences of Example 1 do not behave the same way. The sequences $\{1/n\}$, $\{(-1)^{n+1}(1/n)\}$, and $\{(n-1)/n\}$ each seem to approach a single limiting value as n increases, and $\{3\}$ is at a limiting value from the very first. On the other hand, terms of $\{(-1)^{n+1}(n-1)/n\}$ seem to accumulate near two different values, -1 and 1, while the terms of $\{\sqrt{n}\}$ become increasingly large and do not accumulate anywhere.

To distinguish sequences that approach a unique limiting value L, as n increases, from those that do not, we say that the former sequences *converge,* according to the following definition.

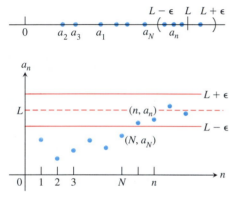

8.2 $a_n \to L$ if $y = L$ is a horizontal asymptote of the sequence of points $\{(n, a_n)\}$. In this figure, all the a_n's after a_N lie within ϵ of L.

Definitions

The sequence $\{a_n\}$ **converges** to the number L if to every positive number ϵ there corresponds an integer N such that for all n,

$$n > N \quad \Rightarrow \quad |a_n - L| < \epsilon.$$

If no such number L exists, we say that $\{a_n\}$ **diverges.**

If $\{a_n\}$ converges to L, we write $\lim_{n\to\infty} a_n = L$, or simply $a_n \to L$, and call L the **limit** of the sequence (Fig. 8.2).

EXAMPLE 2   *Testing the definition*

Show that

a) $\displaystyle\lim_{n\to\infty} \frac{1}{n} = 0$ **b)** $\displaystyle\lim_{n\to\infty} k = k$ (any constant k)

Solution

a) Let $\epsilon > 0$ be given. We must show that there exists an integer N such that for all n,

$$n > N \quad \Rightarrow \quad \left|\frac{1}{n} - 0\right| < \epsilon.$$

This implication will hold if $(1/n) < \epsilon$ or $n > 1/\epsilon$. If N is any integer greater than $1/\epsilon$, the implication will hold for all $n > N$. This proves that $\lim_{n\to\infty}(1/n) = 0$.

b) Let $\epsilon > 0$ be given. We must show that there exists an integer N such that for all n,

$$n > N \quad \Rightarrow \quad |k - k| < \epsilon.$$

Since $k - k = 0$, we can use any positive integer for N and the implication will hold. This proves that $\lim_{n\to\infty} k = k$ for any constant k. ☐

$$a_n = (-1)^{n+1}\left(\frac{n-1}{n}\right)$$

Neither the ϵ-interval about 1 nor the ϵ-interval about -1 contains a complete tail of the sequence.

EXAMPLE 3   Show that $\{(-1)^{n+1}[(n-1)/n]\}$ diverges.

Solution   Take a positive ϵ smaller than 1 so that the bands shown in Fig. 8.3 about the lines $y = 1$ and $y = -1$ do not overlap. Any $\epsilon < 1$ will do. Convergence

8.3 The sequence $\{(-1)^{n+1}[(n-1)/n]\}$ diverges.

to 1 would require every point of the graph beyond a certain index N to lie inside the upper band, but this will never happen. As soon as a point (n, a_n) lies in the upper band, every alternate point starting with $(n + 1, a_{n+1})$ will lie in the lower band. Hence the sequence cannot converge to 1. Likewise, it cannot converge to -1. On the other hand, because the terms of the sequence get alternately closer to 1 and -1, they never accumulate near any other value. Therefore, the sequence diverges. ❑

The behavior of $\{(-1)^{n+1}[(n-1)/n]\}$ is qualitatively different from that of $\{\sqrt{n}\}$, which diverges because it outgrows every real number L. To describe the behavior of $\{\sqrt{n}\}$ we write

$$\lim_{n \to \infty} (\sqrt{n}) = \infty.$$

In speaking of infinity as a limit of a sequence $\{a_n\}$, we do not mean that the difference between a_n and infinity becomes small as n increases. We mean that a_n becomes numerically large as n increases.

Recursive Definitions

Recursion formulas arise regularly in computer programs and numerical routines for solving differential equations.

So far, we have calculated each a_n directly from the value of n. But sequences are often defined **recursively** by giving

1. The value(s) of the initial term or terms, and
2. A rule, called a **recursion formula,** for calculating any later term from terms that precede it.

Factorial notation

The notation $n!$ ("n factorial") means the product $1 \cdot 2 \cdot 3 \cdot \cdots \cdot n$ of the integers from 1 to n. Notice that $(n + 1)! = (n + 1) \cdot n!$. Thus, $4! = 1 \cdot 2 \cdot 3 \cdot 4 = 24$ and $5! = 1 \cdot 2 \cdot 3 \cdot 4 \cdot 5 = 5 \cdot 4! = 120$. We define $0!$ to be 1. Factorials grow even faster than exponentials, as the following table suggests.

n	e^n (rounded)	$n!$
1	3	1
5	148	120
10	22,026	3,628,800
20	4.9×10^8	2.4×10^{18}

EXAMPLE 4   *Sequences constructed recursively*

a) The statements $a_1 = 1$ and $a_n = a_{n-1} + 1$ define the sequence $1, 2, 3, \ldots, n, \ldots$ of positive integers. With $a_1 = 1$, we have $a_2 = a_1 + 1 = 2$, $a_3 = a_2 + 1 = 3$, and so on.

b) The statements $a_1 = 1$ and $a_n = n \cdot a_{n-1}$ define the sequence $1, 2, 6, 24, \ldots, n!, \ldots$ of factorials. With $a_1 = 1$, we have $a_2 = 2 \cdot a_1 = 2$, $a_3 = 3 \cdot a_2 = 6$, $a_4 = 4 \cdot a_3 = 24$, and so on.

c) The statements $a_1 = 1$, $a_2 = 1$, and $a_{n+1} = a_n + a_{n-1}$ define the sequence $1, 1, 2, 3, 5, \ldots$ of **Fibonacci numbers.** With $a_1 = 1$ and $a_2 = 1$, we have $a_3 = 1 + 1 = 2$, $a_4 = 2 + 1 = 3$, $a_5 = 3 + 2 = 5$, and so on.

d) As we can see by applying Newton's method, the statements $x_0 = 1$ and $x_{n+1} = x_n - [(\sin x_n - x_n^2)/(\cos x_n - 2x_n)]$ define a sequence that converges to a solution of the equation $\sin x - x^2 = 0$. ❑

Subsequences

If the terms of one sequence appear in another sequence in their given order, we call the first sequence a **subsequence** of the second.

EXAMPLE 5   *Subsequences of the sequence of positive integers*

a) The subsequence of even integers: $2, 4, 6, \ldots, 2n, \ldots$
b) The subsequence of odd integers: $1, 3, 5, \ldots, 2n - 1, \ldots$
c) The subsequence of primes: $2, 3, 5, 7, 11, \ldots$ ❑

Subsequences are important for two reasons:

1. If a sequence $\{a_n\}$ converges to L, then all of its subsequences converge to L. If we know that a sequence converges, it may be quicker to find or estimate its limit by examining a particular subsequence.

2. If any subsequence of a sequence $\{a_n\}$ diverges, or if two subsequences have different limits, then $\{a_n\}$ diverges. For example, the sequence $\{(-1)^n\}$ diverges because the subsequence $-1, -1, -1, \ldots$ of odd numbered terms converges to -1 while the subsequence $1, 1, 1, \ldots$ of even numbered terms converges to 1, a different limit.

Subsequences also provide a new way to view convergence. A **tail** of a sequence is a subsequence that consists of all terms of the sequence from some index N on. In other words, a tail is one of the sets $\{a_n \mid n \geq N\}$. Another way to say that $a_n \to L$ is to say that every ϵ-interval about L contains a tail of the sequence.

The convergence or divergence of a sequence has nothing to do with how the sequence begins. It depends only on how the tails behave.

Bounded Nondecreasing Sequences

Definition

A sequence $\{a_n\}$ with the property that $a_n \leq a_{n+1}$ for all n is called a **nondecreasing sequence.**

EXAMPLE 6 *Nondecreasing sequences*

a) The sequence $1, 2, 3, \ldots, n, \ldots$ of natural numbers

b) The sequence $\dfrac{1}{2}, \dfrac{2}{3}, \dfrac{3}{4}, \ldots, \dfrac{n}{n+1}, \ldots$

c) The constant sequence $\{3\}$ ❑

There are two kinds of nondecreasing sequences—those whose terms increase beyond any finite bound and those whose terms do not.

Definitions

A sequence $\{a_n\}$ is **bounded from above** if there exists a number M such that $a_n \leq M$ for all n. The number M is an **upper bound** for $\{a_n\}$. If M is an upper bound for $\{a_n\}$ but no number less than M is an upper bound for $\{a_n\}$, then M is the **least upper bound** for $\{a_n\}$.

EXAMPLE 7

a) The sequence $1, 2, 3, \ldots, n, \ldots$ has no upper bound.

b) The sequence $\dfrac{1}{2}, \dfrac{2}{3}, \dfrac{3}{4}, \ldots, \dfrac{n}{n+1}, \ldots$ is bounded above by $M = 1$. No number less than 1 is an upper bound for the sequence, so 1 is the least upper bound (Exercise 47). ❑

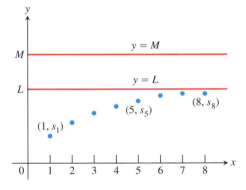

8.4 If the terms of a nondecreasing sequence have an upper bound M, they have a limit $L \leq M$.

A nondecreasing sequence that is bounded from above always has a least upper bound. This fact is a consequence of the completeness property of real numbers but we will not prove it here. Instead, we will prove that if L is the least upper bound, then the sequence converges to L.

Suppose we plot the points $(1, s_1), (2, s_2), \ldots, (n, s_n), \ldots$ in the xy-plane. If M is an upper bound of the sequence, all these points will lie on or below the line $y = M$ (Fig. 8.4). The line $y = L$ is the lowest such line. None of the points (n, s_n) lies above $y = L$, but some do lie above any lower line $y = L - \epsilon$, if ϵ is a positive number. The sequence converges to L because

a) $s_n \leq L$ for *all* values of n and

b) given any $\epsilon > 0$, there exists at least one integer N for which $s_N > L - \epsilon$.

The fact that $\{s_n\}$ is nondecreasing tells us further that

$$s_n \geq s_N > L - \epsilon \qquad \text{for all } n \geq N.$$

Thus, *all* the numbers s_n beyond the Nth number lie within ϵ of L. This is precisely the condition for L to be the limit of the sequence s_n.

The facts for nondecreasing sequences are summarized in the following theorem. A similar result holds for nonincreasing sequences (Exercise 41).

Theorem 1

The Nondecreasing Sequence Theorem

A nondecreasing sequence of real numbers converges if and only if it is bounded from above. If a nondecreasing sequence converges, it converges to its least upper bound.

Exercises 8.1

Finding Terms of a Sequence

Each of Exercises 1–6 gives a formula for the nth term a_n of a sequence $\{a_n\}$. Find the values of a_1, a_2, a_3, and a_4.

1. $a_n = \dfrac{1-n}{n^2}$

2. $a_n = \dfrac{1}{n!}$

3. $a_n = \dfrac{(-1)^{n+1}}{2n-1}$

4. $a_n = 2 + (-1)^n$

5. $a_n = \dfrac{2^n}{2^{n+1}}$

6. $a_n = \dfrac{2^n - 1}{2^n}$

Each of Exercises 7–12 gives the first term or two of a sequence along with a recursion formula for the remaining terms. Write out the first ten terms of the sequence.

7. $a_1 = 1, \quad a_{n+1} = a_n + (1/2^n)$

8. $a_1 = 1, \quad a_{n+1} = a_n/(n+1)$

9. $a_1 = 2, \quad a_{n+1} = (-1)^{n+1} a_n/2$

10. $a_1 = -2, \quad a_{n+1} = n a_n/(n+1)$

11. $a_1 = a_2 = 1, \quad a_{n+2} = a_{n+1} + a_n$

12. $a_1 = 2, \quad a_2 = -1, \quad a_{n+2} = a_{n+1}/a_n$

Finding a Sequence's Formula

In Exercises 13–22, find a formula for the nth term of the sequence.

13. The sequence $1, -1, 1, -1, 1, \ldots$ 1's with alternating signs

14. The sequence $-1, 1, -1, 1, -1, \ldots$ 1's with alternating signs

15. The sequence $1, -4, 9, -16, 25, \ldots$ Squares of the positive integers, with alternating signs

16. The sequence $1, -\dfrac{1}{4}, \dfrac{1}{9}, -\dfrac{1}{16}, \dfrac{1}{25}, \ldots$ Reciprocals of squares of the positive integers, with alternating signs

17. The sequence $0, 3, 8, 15, 24, \ldots$ Squares of the positive integers diminished by 1

18. The sequence $-3, -2, -1, 0, 1, \ldots$ Integers beginning with -3

19. The sequence $1, 5, 9, 13, 17, \ldots$ Every other odd positive integer

20. The sequence $2, 6, 10, 14, 18, \ldots$ Every other even positive integer

21. The sequence $1, 0, 1, 0, 1, \ldots$ Alternating 1's and 0's

22. The sequence $0, 1, 1, 2, 2, 3, 3, 4, \ldots$ Each positive integer repeated

Calculator Explorations of Limits

In Exercises 23–26, experiment with a calculator to find a value of N that will make the inequality hold for all $n > N$. Assuming that the inequality is the one from the formal definition of the limit of a sequence, what sequence is being considered in each case and what is its limit?

23. $|\sqrt[n]{0.5} - 1| < 10^{-3}$ **24.** $|\sqrt[n]{n} - 1| < 10^{-3}$

25. $(0.9)^n < 10^{-3}$ **26.** $2^n/n! < 10^{-7}$

27. *Sequences generated by Newton's method.* Newton's method, applied to a differentiable function $f(x)$, begins with a starting value x_0 and constructs from it a sequence of numbers $\{x_n\}$ that under favorable circumstances converges to a zero of f. The recursion formula for the sequence is

$$x_{n+1} = x_n - \frac{f(x_n)}{f'(x_n)}.$$

a) Show that the recursion formula for $f(x) = x^2 - a$, $a > 0$, can be written as $x_{n+1} = (x_n + a/x_n)/2$.

b) Starting with $x_0 = 1$ and $a = 3$, calculate successive terms of the sequence until the display begins to repeat. What number is being approximated? Explain.

28. (*Continuation of Exercise 27.*) Repeat part (b) of Exercise 27 with $a = 2$ in place of $a = 3$.

29. *A recursive definition of $\pi/2$.* If you start with $x_1 = 1$ and define the subsequent terms of $\{x_n\}$ by the rule $x_n = x_{n-1} + \cos x_{n-1}$, you generate a sequence that converges rapidly to $\pi/2$. (a) Try it. (b) Use the accompanying figure to explain why the convergence is so rapid.

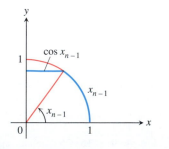

30. According to a front-page article in the December 15, 1992, issue of *The Wall Street Journal,* Ford Motor Company now uses about $7\frac{1}{4}$ hours of labor to produce stampings for the average vehicle, down from an estimated 15 hours in 1980. The Japanese need only about $3\frac{1}{2}$ hours.

Ford's improvement since 1980 represents an average decrease of 6% per year. If that rate continues, then n years from now Ford will use about

$$S_n = 7.25(0.94)^n$$

hours of labor to produce stampings for the average vehicle. Assuming that the Japanese continue to spend $3\frac{1}{2}$ hours per vehicle, how many more years will it take Ford to catch up? Find out two ways:

a) Find the first term of the sequence $\{S_n\}$ that is less than or equal to 3.5.

b) GRAPHER Graph $f(x) = 7.25(0.94)^x$ and use TRACE to find where the graph crosses the line $y = 3.5$.

Theory and Examples

In Exercises 31–34, determine if the sequence is nondecreasing and if it is bounded from above.

31. $a_n = \dfrac{3n+1}{n+1}$ **32.** $a_n = \dfrac{(2n+3)!}{(n+1)!}$

33. $a_n = \dfrac{2^n 3^n}{n!}$ **34.** $a_n = 2 - \dfrac{2}{n} - \dfrac{1}{2^n}$

Which of the sequences in Exercises 35–40 converge, and which diverge? Give reasons for your answers.

35. $a_n = 1 - \dfrac{1}{n}$ **36.** $a_n = n - \dfrac{1}{n}$

37. $a_n = \dfrac{2^n - 1}{2^n}$ **38.** $a_n = \dfrac{2^n - 1}{3^n}$

39. $a_n = ((-1)^n + 1)\left(\dfrac{n+1}{n}\right)$

40. The first term of a sequence is $x_1 = \cos(1)$. The next terms are $x_2 = x_1$ or $\cos(2)$, whichever is larger; and $x_3 = x_2$ or $\cos(3)$, whichever is larger (farther to the right). In general,

$$x_{n+1} = \max\{x_n, \cos(n+1)\}.$$

41. *Nonincreasing sequences.* A sequence of numbers $\{a_n\}$ in which $a_n \geq a_{n+1}$ for every n is called a **nonincreasing sequence**. A sequence $\{a_n\}$ is **bounded from below** if there is a number M with $M \leq a_n$ for every n. Such a number M is called a **lower bound** for the sequence. Deduce from Theorem 1 that a nonincreasing sequence that is bounded from below converges and that a nonincreasing sequence that is not bounded from below diverges.

(*Continuation of Exercise 41.*) Using the conclusion of Exercise 41, determine which of the sequences in Exercises 42–46 converge and which diverge.

42. $a_n = \dfrac{n+1}{n}$

43. $a_n = \dfrac{1 + \sqrt{2n}}{\sqrt{n}}$

44. $a_n = \dfrac{1 - 4^n}{2^n}$

45. $a_n = \dfrac{4^{n+1} + 3^n}{4^n}$

46. $a_1 = 1, \quad a_{n+1} = 2a_n - 3$

47. *The sequence $\{n/(n+1)\}$ has a least upper bound of 1.* Show that if M is a number less than 1, then the terms of $\{n/(n+1)\}$ eventually exceed M. That is, if $M < 1$ there is an integer N such that $n/(n+1) > M$ whenever $n > N$. Since $n/(n+1) < 1$ for every n, this proves that 1 is a least upper bound for $\{n/(n+1)\}$.

48. *Uniqueness of least upper bounds.* Show that if M_1 and M_2 are least upper bounds for the sequence $\{a_n\}$, then $M_1 = M_2$. That is, a sequence cannot have two different least upper bounds.

49. Is it true that a sequence $\{a_n\}$ of positive numbers must converge if it is bounded from above? Give reasons for your answer.

50. Prove that if $\{a_n\}$ is a convergent sequence, then to every positive number ϵ there corresponds an integer N such that for all m and n,

$$m > N \quad \text{and} \quad n > N \quad \Rightarrow \quad |a_m - a_n| < \epsilon.$$

51. *Uniqueness of limits.* Prove that limits of sequences are unique. That is, show that if L_1 and L_2 are numbers such that $a_n \to L_1$ and $a_n \to L_2$, then $L_1 = L_2$.

52. *Limits and subsequences.* Prove that if two subsequences of a sequence $\{a_n\}$ have different limits $L_1 \neq L_2$, then $\{a_n\}$ diverges.

53. For a sequence $\{a_n\}$ the terms of even index are denoted by a_{2k} and the terms of odd index by a_{2k+1}. Prove that if $a_{2k} \to L$ and $a_{2k+1} \to L$, then $a_n \to L$.

54. Prove that a sequence $\{a_n\}$ converges to 0 if and only if the sequence of absolute values $\{|a_n|\}$ converges to 0.

✺ CAS Explorations and Projects

Use a CAS to perform the following steps for the sequences in Exercises 55–66.

a) Calculate and then plot the first 25 terms of the sequence. Does the sequence appear to be bounded from above or below? Does it appear to converge or diverge? If it does converge, what is the limit L?

b) If the sequence converges, find an integer N such that $|a_n - L| \leq 0.01$ for $n \geq N$. How far in the sequence do you have to get for the terms to lie within 0.0001 of L?

55. $a_n = \sqrt[n]{n}$

56. $a_n = \left(1 + \dfrac{0.5}{n}\right)^n$

57. $a_1 = 1, \quad a_{n+1} = a_n + \dfrac{1}{5^n}$

58. $a_1 = 1, \quad a_{n+1} = a_n + (-2)^n$

59. $a_n = \sin n$

60. $a_n = n \sin \dfrac{1}{n}$

61. $a_n = \dfrac{\sin n}{n}$

62. $a_n = \dfrac{\ln n}{n}$

63. $a_n = (0.9999)^n$

64. $a_n = 123456^{1/n}$

65. $a_n = \dfrac{8^n}{n!}$

66. $a_n = \dfrac{n^{41}}{19^n}$

67. *Compound interest, deposits, and withdrawals.* If you invest an amount of money A_0 at a fixed annual interest rate r compounded m times per year, and if the constant amount b is added to the account at the end of each compounding period (or taken from the account if $b < 0$), then the amount you have after $n + 1$ compounding periods is

$$A_{n+1} = \left(1 + \frac{r}{m}\right) A_n + b. \tag{1}$$

a) If $A_0 = 1000$, $r = 0.02015$, $m = 12$, and $b = 50$, calculate and plot the first 100 points (n, A_n). How much money is in your account at the end of 5 years? Does $\{A_n\}$ converge? Is $\{A_n\}$ bounded?

b) Repeat part (a) with $A_0 = 5000$, $r = 0.0589$, $m = 12$, and $b = -50$.

c) If you invest 5000 dollars in a certificate of deposit (CD) that pays 4.5% annually, compounded quarterly, and you make no further investments in the CD, approximately how many years will it take before you have 20,000 dollars? What if the CD earns 6.25%?

d) It can be shown that for any $k \geq 0$, the sequence defined recursively by Eq. (1) satisfies the relation

$$A_k = \left(1 + \frac{r}{m}\right)^k \left(A_0 + \frac{mb}{r}\right) - \frac{mb}{r}. \tag{2}$$

For the values of the constants A_0, r, m, and b given in part (a), validate this assertion by comparing the values of the first 50 terms of both sequences. Then show by direct substitution that the terms in Eq. (2) satisfy the recursion formula (1).

68. *Logistic difference equation.* The recursive relation

$$a_{n+1} = ra_n(1 - a_n)$$

is called the **logistic difference equation,** and when the initial value a_0 is given the equation defines the **logistic sequence** $\{a_n\}$. Throughout this exercise we choose a_0 in the interval $0 < a_0 < 1$, say $a_0 = 0.3$.

a) Choose $r = 3/4$. Calculate and plot the points (n, a_n) for the first 100 terms in the sequence. Does it appear to converge? What do you guess is the limit? Does the limit seem to depend on your choice of a_0?

b) Choose several values of r in the interval $1 < r < 3$ and repeat the procedures in part (a). Be sure to choose some points near the endpoints of the interval. Describe the behavior of the sequences you observe in your plots.

c) Now examine the behavior of the sequence for values of r near the endpoints of the interval $3 < r < 3.45$. The transition value $r = 3$ is called a **bifurcation value** and the new behavior of the sequence in the interval is called an **attracting 2-cycle.** Explain why this reasonably describes the behavior.

d) Next explore the behavior for r values near the endpoints of each of the intervals $3.45 < r < 3.54$ and $3.54 < r < 3.55$. Plot the first 200 terms of the sequences. Describe in your own words the behavior observed in your plots for each interval. Among how many values does the sequence appear to oscillate for each interval? The values $r = 3.45$ and $r = 3.54$ (rounded to 2 decimal places) are also called bifurcation values because the behavior of the sequence changes as r crosses over those values.

e) The situation gets even more interesting. There is actually an increasing sequence of bifurcation values $3 < 3.45 < 3.54 < \cdots < c_n < c_{n+1} \cdots$ such that for $c_n < r < c_{n+1}$ the logistic sequence $\{a_n\}$ eventually oscillates steadily among 2^n values, called an **attracting 2^n-cycle.** Moreover, the bifurcation sequence $\{c_n\}$ is bounded above by 3.57 (so it converges). If you choose a value of $r < 3.57$ you will observe a 2^n-cycle of some sort. Choose $r = 3.5695$ and plot 300 points.

f) Let us see what happens when $r > 3.57$. Choose $r = 3.65$ and calculate and plot the first 300 terms of $\{a_n\}$. Observe how the terms wander around in an unpredictable, chaotic fashion. You cannot predict the value of a_{n+1} from the value of a_n.

g) For $r = 3.65$ choose two starting values of a_0 that are close together, say, $a_0 = 0.3$ and $a_0 = 0.301$. Calculate and plot the first 300 values of the sequences determined by each starting value. Compare the behaviors observed in your plots. How far out do you go before the corresponding terms of your two sequences appear to depart from each other? Repeat the exploration for $r = 3.75$. Can you see how the plots look different depending on your choice of a_0? We say that the logistic sequence is **sensitive to the initial condition** a_0.

8.2 Theorems for Calculating Limits of Sequences

The study of limits would be cumbersome if we had to answer every question about convergence by applying the definition. Fortunately, three theorems make this largely unnecessary. The first is a version of Theorem 1, Section 1.2.

Theorem 2

Let $\{a_n\}$ and $\{b_n\}$ be sequences of real numbers and let A and B be real numbers. The following rules hold if $\lim_{n\to\infty} a_n = A$ and $\lim_{n\to\infty} b_n = B$.

1. *Sum Rule:* $\qquad\qquad\qquad\quad \lim_{n\to\infty} (a_n + b_n) = A + B$
2. *Difference Rule:* $\qquad\qquad\quad \lim_{n\to\infty} (a_n - b_n) = A - B$
3. *Product Rule:* $\qquad\qquad\qquad \lim_{n\to\infty} (a_n \cdot b_n) = A \cdot B$
4. *Constant Multiple Rule:* $\quad \lim_{n\to\infty} (k \cdot b_n) = k \cdot B \quad$ (Any number k)
5. *Quotient Rule:* $\qquad\qquad\qquad \lim_{n\to\infty} \dfrac{a_n}{b_n} = \dfrac{A}{B} \quad$ if $B \neq 0$

EXAMPLE 1 By combining Theorem 2 with the limit results in Example 2 of the preceding section, we have

$$\lim_{n\to\infty} \left(-\frac{1}{n}\right) = -1 \cdot \lim_{n\to\infty} \frac{1}{n} = -1 \cdot 0 = 0$$

$$\lim_{n\to\infty} \left(\frac{n-1}{n}\right) = \lim_{n\to\infty} \left(1 - \frac{1}{n}\right) = \lim_{n\to\infty} 1 - \lim_{n\to\infty} \frac{1}{n} = 1 - 0 = 1$$

$$\lim_{n\to\infty} \frac{5}{n^2} = 5 \cdot \lim_{n\to\infty} \frac{1}{n} \cdot \lim_{n\to\infty} \frac{1}{n} = 5 \cdot 0 \cdot 0 = 0$$

$$\lim_{n\to\infty} \frac{4 - 7n^6}{n^6 + 3} = \lim_{n\to\infty} \frac{(4/n^6) - 7}{1 + (3/n^6)} = \frac{0 - 7}{1 + 0} = -7.$$

One consequence of Theorem 2 is that every nonzero multiple of a divergent sequence $\{a_n\}$ diverges. For suppose, to the contrary, that $\{ca_n\}$ converges for some number $c \neq 0$. Then, by taking $k = 1/c$ in the Constant Multiple Rule in Theorem 2, we see that the sequence

$$\left\{\frac{1}{c} \cdot ca_n\right\} = \{a_n\}$$

converges. Thus, $\{ca_n\}$ cannot converge unless $\{a_n\}$ also converges. If $\{a_n\}$ does not converge, then $\{ca_n\}$ does not converge.

The next theorem is the sequence version of the Sandwich Theorem in Section 1.2.

Theorem 3

The Sandwich Theorem for Sequences

Let $\{a_n\}$, $\{b_n\}$, and $\{c_n\}$ be sequences of real numbers. If $a_n \leq b_n \leq c_n$ holds for all n beyond some index N, and if $\lim_{n\to\infty} a_n = \lim_{n\to\infty} c_n = L$, then $\lim_{n\to\infty} b_n = L$ also.

An immediate consequence of Theorem 3 is that, if $|b_n| \leq c_n$ and $c_n \to 0$, then $b_n \to 0$ because $-c_n \leq b_n \leq c_n$. We use this fact in the next example.

EXAMPLE 2 Since $1/n \to 0$, we know that

a) $\dfrac{\cos n}{n} \to 0$ because $\left|\dfrac{\cos n}{n}\right| = \dfrac{|\cos n|}{n} \leq \dfrac{1}{n}$;

b) $\dfrac{1}{2^n} \to 0$ because $\dfrac{1}{2^n} \leq \dfrac{1}{n}$;

c) $(-1)^n \dfrac{1}{n} \to 0$ because $\left|(-1)^n \dfrac{1}{n}\right| \leq \dfrac{1}{n}$.   ❑

The application of Theorems 2 and 3 is broadened by a theorem stating that applying a continuous function to a convergent sequence produces a convergent sequence. We state the theorem without proof.

Theorem 4

The Continuous Function Theorem for Sequences

Let $\{a_n\}$ be a sequence of real numbers. If $a_n \to L$ and if f is a function that is continuous at L and defined at all a_n, then $f(a_n) \to f(L)$.

EXAMPLE 3 Show that $\sqrt{(n+1)/n} \to 1$.

Solution We know that $(n+1)/n \to 1$. Taking $f(x) = \sqrt{x}$ and $L = 1$ in Theorem 4 gives $\sqrt{(n+1)/n} \to \sqrt{1} = 1$.   ❑

Technology   *The Sequence* $\{2^{1/n}\}$ What happens if you enter 2 in your calculator and take square roots repeatedly? The numbers form a sequence that appears to converge to 1, as suggested in the accompanying table. Try it for yourself.

n	$2^{1/n}$
2	1.4142 13562
4	1.1892 07115
8	1.0905 07733
64	1.0108 89286
256	1.0027 11275
1024	1.0006 77131
16384	1.0000 42307

What is happening in the table above? The sequence $\{1/n\}$ converges to 0. By taking $a_n = 1/n$, $f(x) = 2^x$, and $L = 0$ in Theorem 4, we see that $2^{1/n} = f(1/n) \to f(L) = 2^0 = 1$. Since the successive square roots of 2 form a subsequence $2^{1/2}, 2^{1/4}, 2^{1/8}, \dots$ of $\{2^{1/n}\}$, the square roots must converge to 1 also (Fig. 8.5).

Using l'Hôpital's Rule

The next theorem enables us to use l'Hôpital's rule to find the limits of some sequences.

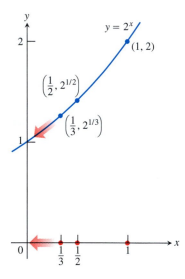

8.5 As $n \to \infty$, $1/n \to 0$ and $2^{1/n} \to 2^0$.

> ### Theorem 5
> Suppose that $f(x)$ is a function defined for all $x \geq n_0$ and that $\{a_n\}$ is a sequence of real numbers such that $a_n = f(n)$ for $n \geq n_0$. Then
> $$\lim_{x \to \infty} f(x) = L \quad \Rightarrow \quad \lim_{n \to \infty} a_n = L.$$

Proof Suppose that $\lim_{x \to \infty} f(x) = L$. Then for each positive number ϵ there is a number M such that for all x,

$$x > M \quad \Rightarrow \quad |f(x) - L| < \epsilon.$$

Let N be an integer greater than M and greater than or equal to n_0. Then

$$n > N \quad \Rightarrow \quad a_n = f(n) \quad \text{and} \quad |a_n - L| = |f(n) - L| < \epsilon. \quad \square$$

EXAMPLE 4   Show that $\lim_{n \to \infty} (\ln n)/n = 0$.

Solution The function $(\ln x)/x$ is defined for all $x \geq 1$ and agrees with the given sequence at positive integers. Therefore, by Theorem 5, $\lim_{n \to \infty} (\ln n)/n$ will equal $\lim_{x \to \infty} (\ln x)/x$ if the latter exists. A single application of l'Hôpital's rule shows that

$$\lim_{x \to \infty} \frac{\ln x}{x} = \lim_{x \to \infty} \frac{1/x}{1} = \frac{0}{1} = 0.$$

We conclude that $\lim_{n\to\infty} (\ln n)/n = 0$.

When we use l'Hôpital's rule to find the limit of a sequence, we often treat n as a continuous real variable and differentiate directly with respect to n. This saves us from having to rewrite the formula for a_n as we did in Example 4.

EXAMPLE 5 Find $\lim_{n\to\infty} (2^n/5n)$.

Solution By l'Hôpital's rule,

$$\lim_{n\to\infty} \frac{2^n}{5n} = \lim_{n\to\infty} \frac{2^n \cdot \ln 2}{5}$$

$$= \infty.$$

Limits That Arise Frequently

Table 8.1

The limits in Table 8.1 arise frequently. The first limit is from Example 4. The next two can be proved by taking logarithms and applying Theorem 4 (Exercises 71 and 72). The remaining proofs can be found in Appendix 6.

Table 8.1

1. $\lim_{n\to\infty} \dfrac{\ln n}{n} = 0$

2. $\lim_{n\to\infty} \sqrt[n]{n} = 1$

3. $\lim_{n\to\infty} x^{1/n} = 1 \qquad (x > 0)$

4. $\lim_{n\to\infty} x^n = 0 \qquad (|x| < 1)$

5. $\lim_{n\to\infty} \left(1 + \dfrac{x}{n}\right)^n = e^x \qquad$ (Any x)

6. $\lim_{n\to\infty} \dfrac{x^n}{n!} = 0 \qquad$ (Any x)

In formulas (3)–(6), x remains fixed as $n \to \infty$.

EXAMPLE 6   *Limits from Table 8.1*

1. $\dfrac{\ln (n^2)}{n} = \dfrac{2 \ln n}{n} \to 2 \cdot 0 = 0$ Formula 1

2. $\sqrt[n]{n^2} = n^{2/n} = (n^{1/n})^2 \to (1)^2 = 1$ Formula 2

3. $\sqrt[n]{3n} = 3^{1/n}(n^{1/n}) \to 1 \cdot 1 = 1$ Formula 3 with $x = 3$, and Formula 2

4. $\left(-\dfrac{1}{2}\right)^n \to 0$ Formula 4 with $x = -\dfrac{1}{2}$

5. $\left(\dfrac{n - 2}{n}\right)^n = \left(1 + \dfrac{-2}{n}\right)^n \to e^{-2}$ Formula 5 with $x = -2$

6. $\dfrac{100^n}{n!} \to 0$ Formula 6 with $x = 100$

EXAMPLE 7   Does the sequence whose nth term is

$$a_n = \left(\frac{n + 1}{n - 1}\right)^n$$

converge? If so, find $\lim_{n\to\infty} a_n$.

Solution The limit leads to the indeterminate form 1^∞. We can apply l'Hôpital's rule if we first change the form to $\infty \cdot 0$ by taking the natural logarithm of a_n:

$$\ln a_n = \ln \left(\frac{n + 1}{n - 1}\right)^n$$

$$= n \ln \left(\frac{n + 1}{n - 1}\right).$$

Then,

$$\lim_{n \to \infty} \ln a_n = \lim_{n \to \infty} n \ln \left(\frac{n+1}{n-1} \right) \qquad \infty \cdot 0$$

$$= \lim_{n \to \infty} \frac{\ln \left(\dfrac{n+1}{n-1} \right)}{1/n} \qquad \frac{0}{0}$$

$$= \lim_{n \to \infty} \frac{-2/(n^2 - 1)}{-1/n^2} \qquad \text{l'Hôpital's rule}$$

$$= \lim_{n \to \infty} \frac{2n^2}{n^2 - 1} = 2.$$

Since $\ln a_n \to 2$, and $f(x) = e^x$ is continuous, Theorem 4 tells us that

$$a_n = e^{\ln a_n} \to e^2.$$

The sequence $\{a_n\}$ converges to e^2. ❑

✳ Picard's Method for Finding Roots

The problem of solving the equation

$$f(x) = 0 \tag{1}$$

is equivalent to that of solving the equation

$$g(x) = f(x) + x = x, \tag{2}$$

obtained by adding x to both sides of Eq. (1). By this simple change, we cast Eq. (1) into a form that may render it solvable on a computer by a powerful method called **Picard's method** (after the French mathematician Charles Émile Picard, 1856–1941).

If the domain of g contains the range of g, we can start with a point x_0 in the domain and apply g repeatedly to get

$$x_1 = g(x_0), \qquad x_2 = g(x_1), \qquad x_3 = g(x_2), \qquad \dots \tag{3}$$

Under simple restrictions that we will describe shortly, the sequence generated by the recursion formula $x_{n+1} = g(x_n)$ will converge to a point x for which $g(x) = x$. This point solves the equation $f(x) = 0$ because

$$f(x) = g(x) - x = x - x = 0. \tag{4}$$

A point x for which $g(x) = x$ is a **fixed point** of g. We see in Eq. (4) that the fixed points of g are precisely the roots of f.

EXAMPLE 8   *Testing the method*

Solve the equation

$$\frac{1}{4}x + 3 = x.$$

Solution   By algebra, we know that the solution is $x = 4$. To apply Picard's method, we take

$$g(x) = \frac{1}{4}x + 3,$$

8.6 The Picard solution of the equation $g(x) = (1/4)x + 3 = x$ (Example 8).

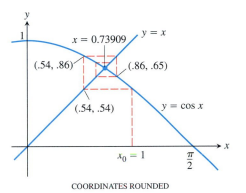

COORDINATES ROUNDED

8.7 The solution of $\cos x = x$ by Picard's method starting at $x_0 = 1$ (Example 9).

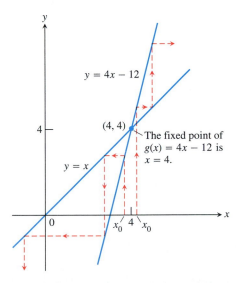

8.8 Applying the Picard method to $g(x) = 4x - 12$ will not find the fixed point unless x_0 is the fixed point 4 itself (Example 10).

choose a starting point, say $x_0 = 1$, and calculate the initial terms of the sequence $x_{n+1} = g(x_n)$. Table 8.2 lists the results. In 10 steps, the solution of the original equation is found with an error of magnitude less than 3×10^{-6}.

Figure 8.6 shows the geometry of the solution. We start with $x_0 = 1$ and calculate the first value $g(x_0)$. This becomes the second x-value x_1. The second y-value $g(x_1)$ becomes the third x-value x_2, and so on. The process is shown as a path (called the *iteration path*) that starts at $x_0 = 1$, moves up to $(x_0, g(x_0)) = (x_0, x_1)$, over to (x_1, x_1), up to $(x_1, g(x_1))$, and so on. The path converges to the point where the graph of g meets the line $y = x$. This is the point where $g(x) = x$. ❑

Table 8.2 Successive iterates of $g(x) = (1/4)x + 3$, starting with $x_0 = 1$

x_n	$x_{n+1} = g(x_n) = (1/4)x_n + 3$
$x_0 = 1$	$x_1 = g(x_0) = (1/4)(1) + 3 = 3.25$
$x_1 = 3.25$	$x_2 = g(x_1) = (1/4)(3.25) + 3 = 3.8125$
$x_2 = 3.8125$	$x_3 = g(x_2) = 3.9531\ 25$
$x_3 = 3.9531\ 25$	$x_4 = 3.9882\ 8125$
$\vdots$	$x_5 = 3.9970\ 70313$
	$x_6 = 3.9992\ 67578$
	$x_7 = 3.9998\ 16895$
	$x_8 = 3.9999\ 54224$
	$x_9 = 3.9999\ 88556$
	$x_{10} = 3.9999\ 97139$
	$\vdots$

EXAMPLE 9   Solve the equation $\cos x = x$.

Solution   We take $g(x) = \cos x$, choose $x_0 = 1$ as a starting value, and use the recursion formula $x_{n+1} = g(x_n)$ to find

$$x_0 = 1, \qquad x_1 = \cos 1, \qquad x_2 = \cos (x_1), \dots .$$

We can approximate the first 50 terms or so on a calculator in radian mode by entering 1 and taking the cosine repeatedly. The display stops changing when $\cos x = x$ to the number of decimal places in the display.

Try it for yourself. As you continue to take the cosine, the successive approximations lie alternately above and below the fixed point $x = 0.739085133\dots .$

Figure 8.7 shows that the values oscillate this way because the path of the procedure spirals around the fixed point. ❑

EXAMPLE 10   Picard's method will not solve the equation

$$g(x) = 4x - 12 = x.$$

As Fig. 8.8 shows, any choice of x_0 except $x_0 = 4$, the solution itself, generates a divergent sequence that moves away from the solution. ❑

The difficulty in Example 10 can be traced to the fact that the slope of the line $y = 4x - 12$ exceeds 1, the slope of the line $y = x$. Conversely, the process worked in Example 8 because the slope of the line $y = (1/4)x + 3$ was numerically less than 1. A theorem from advanced calculus tells us that if $g'(x)$ is continuous on a closed interval I whose interior contains a solution of the equation $g(x) = x$, and if $|g'(x)| < 1$ on I, then any choice of x_0 in the interior of I will lead to the solution. (See the introduction to Exercises 83 and 84 about what to do if $|g'(x)| > 1$.)

Exercises 8.2

Finding Limits

Which of the sequences $\{a_n\}$ in Exercises 1–62 converge, and which diverge? Find the limit of each convergent sequence.

1. $a_n = 2 + (0.1)^n$

2. $a_n = \dfrac{n + (-1)^n}{n}$

3. $a_n = \dfrac{1 - 2n}{1 + 2n}$

4. $a_n = \dfrac{2n + 1}{1 - 3\sqrt{n}}$

5. $a_n = \dfrac{1 - 5n^4}{n^4 + 8n^3}$

6. $a_n = \dfrac{n + 3}{n^2 + 5n + 6}$

7. $a_n = \dfrac{n^2 - 2n + 1}{n - 1}$

8. $a_n = \dfrac{1 - n^3}{70 - 4n^2}$

9. $a_n = 1 + (-1)^n$

10. $a_n = (-1)^n \left(1 - \dfrac{1}{n}\right)$

11. $a_n = \left(\dfrac{n + 1}{2n}\right)\left(1 - \dfrac{1}{n}\right)$

12. $a_n = \left(2 - \dfrac{1}{2^n}\right)\left(3 + \dfrac{1}{2^n}\right)$

13. $a_n = \dfrac{(-1)^{n+1}}{2n - 1}$

14. $a_n = \left(-\dfrac{1}{2}\right)^n$

15. $a_n = \sqrt{\dfrac{2n}{n + 1}}$

16. $a_n = \dfrac{1}{(0.9)^n}$

17. $a_n = \sin\left(\dfrac{\pi}{2} + \dfrac{1}{n}\right)$

18. $a_n = n\pi \cos(n\pi)$

19. $a_n = \dfrac{\sin n}{n}$

20. $a_n = \dfrac{\sin^2 n}{2^n}$

21. $a_n = \dfrac{n}{2^n}$

22. $a_n = \dfrac{3^n}{n^3}$

23. $a_n = \dfrac{\ln(n + 1)}{\sqrt{n}}$

24. $a_n = \dfrac{\ln n}{\ln 2n}$

25. $a_n = 8^{1/n}$

26. $a_n = (0.03)^{1/n}$

27. $a_n = \left(1 + \dfrac{7}{n}\right)^n$

28. $a_n = \left(1 - \dfrac{1}{n}\right)^n$

29. $a_n = \sqrt[n]{10n}$

30. $a_n = \sqrt[n]{n^2}$

31. $a_n = \left(\dfrac{3}{n}\right)^{1/n}$

32. $a_n = (n + 4)^{1/(n+4)}$

33. $a_n = \dfrac{\ln n}{n^{1/n}}$

34. $a_n = \ln n - \ln(n + 1)$

35. $a_n = \sqrt[n]{4^n n}$

36. $a_n = \sqrt[n]{3^{2n+1}}$

37. $a_n = \dfrac{n!}{n^n}$ (*Hint:* Compare with $1/n$.)

38. $a_n = \dfrac{(-4)^n}{n!}$

39. $a_n = \dfrac{n!}{10^{6n}}$

40. $a_n = \dfrac{n!}{2^n \cdot 3^n}$

41. $a_n = \left(\dfrac{1}{n}\right)^{1/(\ln n)}$

42. $a_n = \ln\left(1 + \dfrac{1}{n}\right)^n$

43. $a_n = \left(\dfrac{3n + 1}{3n - 1}\right)^n$

44. $a_n = \left(\dfrac{n}{n + 1}\right)^n$

45. $a_n = \left(\dfrac{x^n}{2n + 1}\right)^{1/n}, \quad x > 0$

46. $a_n = \left(1 - \dfrac{1}{n^2}\right)^n$

47. $a_n = \dfrac{3^n \cdot 6^n}{2^{-n} \cdot n!}$

48. $a_n = \dfrac{(10/11)^n}{(9/10)^n + (11/12)^n}$

49. $a_n = \tanh n$

50. $a_n = \sinh(\ln n)$

51. $a_n = \dfrac{n^2}{2n - 1} \sin \dfrac{1}{n}$

52. $a_n = n\left(1 - \cos \dfrac{1}{n}\right)$

53. $a_n = \tan^{-1} n$

54. $a_n = \dfrac{1}{\sqrt{n}} \tan^{-1} n$

55. $a_n = \left(\dfrac{1}{3}\right)^n + \dfrac{1}{\sqrt{2^n}}$

56. $a_n = \sqrt[n]{n^2 + n}$

57. $a_n = \dfrac{(\ln n)^{200}}{n}$

58. $a_n = \dfrac{(\ln n)^5}{\sqrt{n}}$

59. $a_n = n - \sqrt{n^2 - n}$

60. $a_n = \dfrac{1}{\sqrt{n^2 - 1} - \sqrt{n^2 + n}}$

61. $a_n = \dfrac{1}{n} \int_1^n \dfrac{1}{x} \, dx$

62. $a_n = \int_1^n \dfrac{1}{x^p} \, dx, \quad p > 1$

Theory and Examples

63. The first term of a sequence is $x_1 = 1$. Each succeeding term is the sum of all those that come before it:

$$x_{n+1} = x_1 + x_2 + \cdots + x_n.$$

Write out enough early terms of the sequence to deduce a general formula for x_n that holds for $n \geq 2$.

64. A sequence of rational numbers is described as follows:

$$\frac{1}{1}, \frac{3}{2}, \frac{7}{5}, \frac{17}{12}, \ldots, \frac{a}{b}, \frac{a+2b}{a+b}, \ldots$$

Here the numerators form one sequence, the denominators form a second sequence, and their ratios form a third sequence. Let x_n and y_n be, respectively, the numerator and the denominator of the nth fraction $r_n = x_n/y_n$.

a) Verify that $x_1^2 - 2y_1^2 = -1$, $x_2^2 - 2y_2^2 = +1$ and, more generally, that if $a^2 - 2b^2 = -1$ or $+1$, then

$$(a + 2b)^2 - 2(a + b)^2 = +1 \quad \text{or} \quad -1,$$

respectively.

b) The fractions $r_n = x_n/y_n$ approach a limit as n increases. What is that limit? (*Hint:* Use part (a) to show that $r_n^2 - 2 = \pm(1/y_n)^2$ and that y_n is not less than n.)

65. *Newton's method.* The following sequences come from the recursion formula for Newton's method,

$$x_{n+1} = x_n - \frac{f(x_n)}{f'(x_n)}.$$

Do the sequences converge? If so, to what value? In each case, begin by identifying the function f that generates the sequence.

a) $x_0 = 1$, $\quad x_{n+1} = x_n - \dfrac{x_n^2 - 2}{2x_n} = \dfrac{x_n}{2} + \dfrac{1}{x_n}$

b) $x_0 = 1$, $\quad x_{n+1} = x_n - \dfrac{\tan x_n - 1}{\sec^2 x_n}$

c) $x_0 = 1$, $\quad x_{n+1} = x_n - 1$

66. a) Suppose that $f(x)$ is differentiable for all x in $[0, 1]$ and that $f(0) = 0$. Define the sequence $\{a_n\}$ by the rule $a_n = nf(1/n)$. Show that $\lim_{n\to\infty} a_n = f'(0)$.

Use the result in part (a) to find the limits of the following sequences $\{a_n\}$.

b) $a_n = n \tan^{-1} \dfrac{1}{n}$ c) $a_n = n(e^{1/n} - 1)$

d) $a_n = n \ln\left(1 + \dfrac{2}{n}\right)$

67. *Pythagorean triples.* A triple of positive integers a, b, and c is called a **Pythagorean triple** if $a^2 + b^2 = c^2$. Let a be an odd positive integer and let

$$b = \left\lfloor \frac{a^2}{2} \right\rfloor \quad \text{and} \quad c = \left\lceil \frac{a^2}{2} \right\rceil$$

be, respectively, the integer floor and ceiling for $a^2/2$.

a) Show that $a^2 + b^2 = c^2$. (*Hint:* Let $a = 2n + 1$ and express b and c in terms of n.)

b) By direct calculation, or by appealing to the figure here, find

$$\lim_{a\to\infty} \frac{\left\lfloor \dfrac{a^2}{2} \right\rfloor}{\left\lceil \dfrac{a^2}{2} \right\rceil}.$$

68. *The nth root of n!*

a) Show that $\lim_{n\to\infty} (2n\pi)^{1/(2n)} = 1$ and hence, using Stirling's approximation (Chapter 7, Additional Exercise 50a), that

$$\sqrt[n]{n!} \approx \frac{n}{e} \quad \text{for large values of } n.$$

b) CALCULATOR Test the approximation in (a) for $n = 40$, 50, 60, ... , as far as your calculator will allow.

69. a) Assuming that $\lim_{n\to\infty} (1/n^c) = 0$ if c is any positive constant, show that

$$\lim_{n\to\infty} \frac{\ln n}{n^c} = 0$$

if c is any positive constant.

b) Prove that $\lim_{n\to\infty} (1/n^c) = 0$ if c is any positive constant. (*Hint:* If $\epsilon = 0.001$ and $c = 0.04$, how large should N be to ensure that $|1/n^c - 0| < \epsilon$ if $n > N$?)

70. *The zipper theorem.* Prove the "zipper theorem" for sequences: If $\{a_n\}$ and $\{b_n\}$ both converge to L, then the sequence

$$a_1, b_1, a_2, b_2, \ldots, a_n, b_n, \ldots$$

converges to L.

71. Prove that $\lim_{n\to\infty} \sqrt[n]{n} = 1$.

72. Prove that $\lim_{n\to\infty} x^{1/n} = 1$, $(x > 0)$.

73. Prove Theorem 3.

74. Prove Theorem 4.

✳ Picard's Method

CALCULATOR Use Picard's method to solve the equations in Exercises 75–80.

75. $\sqrt{x} = x$ **76.** $x^2 = x$

77. $\cos x + x = 0$ **78.** $\cos x = x + 1$

79. $x - \sin x = 0.1$

80. $\sqrt{x} = 4 - \sqrt{1 + x}$ (*Hint:* Square both sides first.)

81. Solving the equation $\sqrt{x} = x$ by Picard's method finds the solution $x = 1$ but not the solution $x = 0$. Why? (*Hint:* Graph $y = x$ and $y = \sqrt{x}$ together.)

82. Solving the equation $x^2 = x$ by Picard's method with $|x_0| \neq 1$ can find the solution $x = 0$ but not the solution $x = 1$. Why? (*Hint:* Graph $y = x^2$ and $y = x$ together.)

Slope greater than 1. Example 10 showed that we cannot apply Picard's method to find a fixed point of $g(x) = 4x - 12$. But we can apply the method to find a fixed point of $g^{-1}(x) = (1/4)x + 3$ because the derivative of g^{-1} is 1/4, whose value is less than 1 in magnitude on any interval. In Example 8, we found the fixed point of g^{-1} to be $x = 4$. Now notice that 4 is also a fixed point of g, since

$$g(4) = 4(4) - 12 = 4.$$

In finding the fixed point of g^{-1}, we found the fixed point of g.

A function and its inverse always have the same fixed points. The graphs of the functions are symmetric about the line $y = x$ and therefore intersect the line at the same points.

We now see that the application of Picard's method is quite broad. For suppose g is one-to-one, with a continuous first derivative whose magnitude is greater than 1 on a closed interval I whose interior contains a fixed point of g. Then the derivative of g^{-1}, being the reciprocal of g', has magnitude less than 1 on I. Picard's method applied to g^{-1} on I will find the fixed point of g. As cases in point, find the fixed points of the functions in Exercises 83 and 84.

83. $g(x) = 2x + 3$

84. $g(x) = 1 - 4x$

8.3 Infinite Series

In mathematics and science we often write functions as infinite polynomials, such as

$$\frac{1}{1 - x} = 1 + x + x^2 + x^3 + \cdots + x^n + \cdots, \qquad |x| < 1,$$

(we will see the importance of doing so as the chapter continues). For any allowable value of x, we evaluate the polynomial as an infinite sum of constants, a sum we call an *infinite series*. The goal of this section and the next four is to familiarize ourselves with infinite series.

Series and Partial Sums

We begin by asking how to assign meaning to an expression like

$$1 + \frac{1}{2} + \frac{1}{4} + \frac{1}{8} + \frac{1}{16} + \cdots.$$

The way to do so is not to try to add all the terms at once (we cannot) but rather to add the terms one at a time from the beginning and look for a pattern in how these partial sums grow.

Partial sum		Value
first:	$s_1 = 1$	$2 - 1$
second:	$s_2 = 1 + \dfrac{1}{2}$	$2 - \dfrac{1}{2}$
third:	$s_3 = 1 + \dfrac{1}{2} + \dfrac{1}{4}$	$2 - \dfrac{1}{4}$
$\vdots$	$\vdots$	$\vdots$
nth:	$s_n = 1 + \dfrac{1}{2} + \dfrac{1}{4} + \cdots + \dfrac{1}{2^{n-1}}$	$2 - \dfrac{1}{2^{n-1}}$

Indeed there is a pattern. The partial sums form a sequence whose nth term is

$$s_n = 2 - \frac{1}{2^{n-1}}.$$

This sequence converges to 2 because $\lim_{n\to\infty} (1/2^n) = 0$. We say

"the sum of the infinite series $1 + \dfrac{1}{2} + \dfrac{1}{4} + \cdots + \dfrac{1}{2^{n-1}} + \cdots$ is 2."

Is the sum of any finite number of terms in this series equal to 2? No. Can we actually add an infinite number of terms one by one? No. But we can still define their sum by defining it to be the limit of the sequence of partial sums as $n \to \infty$, in this case 2 (Fig. 8.9). Our knowledge of sequences and limits enables us to break away from the confines of finite sums.

8.9 As the lengths 1, 1/2, 1/4, 1/8, … are added one by one, the sum approaches 2.

Definitions

Given a sequence of numbers $\{a_n\}$, an expression of the form

$$a_1 + a_2 + a_3 + \cdots + a_n + \cdots$$

is an **infinite series**. The number a_n is the **nth term** of the series. The sequence $\{s_n\}$ defined by

$$s_1 = a_1$$

$$s_2 = a_1 + a_2$$

$$\vdots$$

$$s_n = a_1 + a_2 + \cdots + a_n = \sum_{k=1}^{n} a_k$$

$$\vdots$$

is the **sequence of partial sums** of the series, the number s_n being the **nth partial sum.** If the sequence of partial sums converges to a limit L, we say that the series **converges** and that its **sum** is L. In this case, we also write

$$a_1 + a_2 + \cdots + a_n + \cdots = \sum_{n=1}^{\infty} a_n = L.$$

If the sequence of partial sums of the series does not converge, we say that the series **diverges.**

When we begin to study a given series $a_1 + a_2 + \cdots + a_n + \cdots$, we might not know whether it converges or diverges. In either case, it is convenient to use sigma notation to write the series as

$$\sum_{n=1}^{\infty} a_n, \qquad \sum_{k=1}^{\infty} a_k, \qquad \text{or} \qquad \sum a_n \qquad$$

A useful shorthand when summation from 1 to ∞ is understood

Geometric Series

Geometric series are series of the form

$$a + ar + ar^2 + \cdots + ar^{n-1} + \cdots = \sum_{n=1}^{\infty} ar^{n-1} \qquad (1)$$

in which a and r are fixed real numbers and $a \neq 0$. The **ratio** r can be positive, as in

$$1 + \frac{1}{2} + \frac{1}{4} + \cdots + \left(\frac{1}{2}\right)^{n-1} + \cdots,$$

or negative, as in

$$1 - \frac{1}{3} + \frac{1}{9} - \cdots + \left(-\frac{1}{3}\right)^{n-1} + \cdots.$$

If $r = 1$, the nth partial sum of the series in (1) is

$$s_n = a + a(1) + a(1)^2 + \cdots + a(1)^{n-1} = na,$$

and the series diverges because $\lim_{n \to \infty} s_n = \pm\infty$, depending on the sign of a. If $r = -1$, the series diverges because the nth partial sums alternate between a and 0. If $|r| \neq 1$, we can determine the convergence or divergence of the series in the following way:

$$s_n = a + ar + ar^2 + \cdots + ar^{n-1}$$

$$rs_n = ar + ar^2 + \cdots + ar^{n-1} + ar^n \qquad \text{Multiply } s_n \text{ by } r.$$

$$s_n - rs_n = a - ar^n \qquad \begin{array}{l}\text{Subtract } rs_n \text{ from } s_n.\\ \text{Most of the terms on}\\ \text{the right cancel.}\end{array}$$

$$s_n(1 - r) = a(1 - r^n) \qquad \text{Factor.}$$

$$s_n = \frac{a(1 - r^n)}{1 - r}, \qquad (r \neq 1). \qquad \begin{array}{l}\text{We can solve for}\\ s_n \text{ if } r \neq 1.\end{array}$$

If $|r| < 1$, then $r^n \to 0$ as $n \to \infty$ (as in Section 8.2) and $s_n \to a/(1 - r)$. If $|r| > 1$, then $|r^n| \to \infty$ and the series diverges.

Equation (2) holds *only* if the summation begins with $n = 1$.

> If $|r| < 1$, the geometric series $a + ar + ar^2 + \cdots + ar^{n-1} + \cdots$ converges to $a/(1 - r)$:
>
> $$\sum_{n=1}^{\infty} ar^{n-1} = \frac{a}{1 - r}, \qquad |r| < 1. \qquad (2)$$
>
> If $|r| \geq 1$, the series diverges.

EXAMPLE 1   The geometric series with $a = 1/9$ and $r = 1/3$ is

$$\frac{1}{9} + \frac{1}{27} + \frac{1}{81} + \cdots = \sum_{n=1}^{\infty} \frac{1}{9}\left(\frac{1}{3}\right)^{n-1} = \frac{1/9}{1 - (1/3)} = \frac{1}{6}.$$

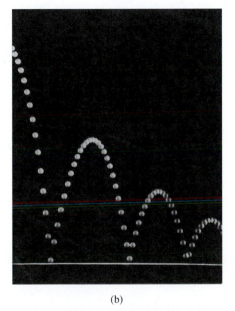

(b)

8.10 (a) Example 3 shows how to use a geometric series to calculate the total vertical distance traveled by a bouncing ball if the height of each rebound is reduced by the factor r. (b) A strobo-scopic photo of a bouncing ball.

EXAMPLE 2 The series

$$\sum_{n=1}^{\infty} \frac{(-1)^n 5}{4^n} = -\frac{5}{4} + \frac{5}{16} - \frac{5}{64} + \cdots$$

is a geometric series with $a = -5/4$ and $r = -1/4$. It converges to

$$\frac{a}{1-r} = \frac{-5/4}{1+(1/4)} = -1.$$

EXAMPLE 3 You drop a ball from a meters above a flat surface. Each time the ball hits the surface after falling a distance h, it rebounds a distance rh, where r is positive but less than 1. Find the total distance the ball travels up and down (Fig. 8.10).

Solution The total distance is

$$s = a + \underbrace{2ar + 2ar^2 + 2ar^3 + \cdots}_{\text{This sum is } 2ar/(1-r).} = a + \frac{2ar}{1-r} = a\frac{1+r}{1-r}.$$

If $a = 6$ m and $r = 2/3$, for instance, the distance is

$$s = 6\frac{1+(2/3)}{1-(2/3)} = 6\left(\frac{5/3}{1/3}\right) = 30 \text{ m.}$$

EXAMPLE 4 *Repeating decimals*

Express the repeating decimal 5.23 23 23 ... as the ratio of two integers.

Solution

$$5.23\ 23\ 23\ldots = 5 + \frac{23}{100} + \frac{23}{(100)^2} + \frac{23}{(100)^3} + \cdots$$

$$= 5 + \frac{23}{100}\underbrace{\left(1 + \frac{1}{100} + \left(\frac{1}{100}\right)^2 + \cdots\right)}_{1/(1-0.01)} \quad a = 1, r = 1/100$$

$$= 5 + \frac{23}{100}\left(\frac{1}{0.99}\right) = 5 + \frac{23}{99} = \frac{518}{99}$$

Telescoping Series

Unfortunately, formulas like the one for the sum of a convergent geometric series are rare and we usually have to settle for an estimate of a series' sum (more about this later). The next example, however, is another case in which we can find the sum exactly.

EXAMPLE 5 Find the sum of the series $\sum_{n=1}^{\infty} \frac{1}{n(n+1)}$.

Solution We look for a pattern in the sequence of partial sums that might lead to

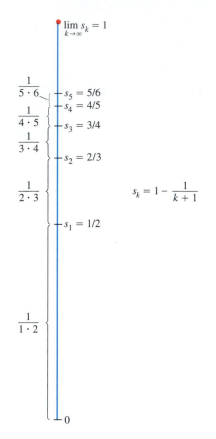

8.11 The partial sums of the series in Example 5.

a formula for s_k. The key, as in the integration

$$\int \frac{dx}{x(x+1)} = \int \frac{dx}{x} - \int \frac{dx}{x+1},$$

is partial fractions. The observation that

$$\frac{1}{k(k+1)} = \frac{1}{k} - \frac{1}{k+1} \tag{3}$$

permits us to write the partial sum

$$\sum_{n=1}^{k} \frac{1}{n(n+1)} = \frac{1}{1 \cdot 2} + \frac{1}{2 \cdot 3} + \cdots + \frac{1}{k \cdot (k+1)}$$

as

$$s_k = \left(\frac{1}{1} - \frac{1}{2}\right) + \left(\frac{1}{2} - \frac{1}{3}\right) + \cdots + \left(\frac{1}{k} - \frac{1}{k+1}\right). \tag{4}$$

Removing parentheses and canceling the terms of opposite sign collapses the sum to

$$s_k = 1 - \frac{1}{k+1}. \tag{5}$$

We now see that $s_k \to 1$ as $k \to \infty$. The series converges, and its sum is 1 (Fig. 8.11).

$$\sum_{n=1}^{\infty} \frac{1}{n(n+1)} = 1.$$

❑

Divergent Series

Geometric series with $|r| \geq 1$ are not the only series to diverge.

EXAMPLE 6   The series

$$\sum_{n=1}^{\infty} n^2 = 1 + 4 + 9 + \cdots + n^2 + \cdots$$

diverges because the partial sums grow beyond every number L. After $n = 1$, the partial sum $s_n = 1 + 4 + 9 + \cdots + n^2$ is greater than n^2.   ❑

EXAMPLE 7   The series

$$\sum_{n=1}^{\infty} \frac{n+1}{n} = \frac{2}{1} + \frac{3}{2} + \frac{4}{3} + \cdots + \frac{n+1}{n} + \cdots$$

diverges because the partial sums eventually outgrow every preassigned number. Each term is greater than 1, so the sum of n terms is greater than n.   ❑

The *n*th-Term Test for Divergence

Observe that $\lim_{n \to \infty} a_n$ must equal zero if the series $\sum_{n=1}^{\infty} a_n$ converges. To see why, let S represent the series' sum and $s_n = a_1 + a_2 + \cdots + a_n$ the nth partial sum. When n is large, both s_n and s_{n-1} are close to S, so their difference, a_n, is close to zero. More formally,

$$a_n = s_n - s_{n-1} \quad \rightarrow \quad S - S = 0. \qquad \text{\color{blue}Difference Rule for sequences}$$

Caution

Theorem 6 *does not say* that $\sum_{n=1}^{\infty} a_n$ converges if $a_n \to 0$. It is possible for a series to diverge when $a_n \to 0$.

Theorem 6

If $\displaystyle\sum_{n=1}^{\infty} a_n$ converges, then $a_n \to 0$.

Theorem 6 leads to a test for detecting the kind of divergence that occurred in Examples 6–8.

The nth-Term Test for Divergence

$\displaystyle\sum_{n=1}^{\infty} a_n$ diverges if $\displaystyle\lim_{n\to\infty} a_n$ fails to exist or is different from zero.

EXAMPLE 8   In applying the nth-Term Test, we can see that

a) $\displaystyle\sum_{n=1}^{\infty} n^2$ diverges because $n^2 \to \infty$

b) $\displaystyle\sum_{n=1}^{\infty} \frac{n+1}{n}$ diverges because $\dfrac{n+1}{n} \to 1$

c) $\displaystyle\sum_{n=1}^{\infty} (-1)^{n+1}$ diverges because $\lim_{n\to\infty} (-1)^{n+1}$ does not exist

d) $\displaystyle\sum_{n=1}^{\infty} \frac{-n}{2n+5}$ diverges because $\lim_{n\to\infty} \dfrac{-n}{2n+5} = -\dfrac{1}{2} \neq 0$. ❑

EXAMPLE 9 $\color{blue}a_n \to 0$ *but the series diverges*

The series

$$1 + \underbrace{\frac{1}{2} + \frac{1}{2}}_{2 \text{ terms}} + \underbrace{\frac{1}{4} + \frac{1}{4} + \frac{1}{4} + \frac{1}{4}}_{4 \text{ terms}} + \cdots + \underbrace{\frac{1}{2^n} + \frac{1}{2^n} + \cdots + \frac{1}{2^n}}_{2^n \text{ terms}} + \cdots$$

diverges even though its terms form a sequence that converges to 0. ❑

Combining Series

Whenever we have two convergent series, we can add them term by term, subtract them term by term, or multiply them by constants to make new convergent series.

> ### Theorem 7
> If $\sum a_n = A$ and $\sum b_n = B$ are convergent series, then
>
> **1.** *Sum Rule:* $\qquad\qquad\qquad \sum (a_n + b_n) = \sum a_n + \sum b_n = A + B$
>
> **2.** *Difference Rule:* $\qquad\qquad \sum (a_n - b_n) = \sum a_n - \sum b_n = A - B$
>
> **3.** *Constant Multiple Rule:* $\quad \sum k a_n = k \sum a_n = kA \qquad$ (Any number k).

Proof The three rules for series follow from the analogous rules for sequences in Theorem 2, Section 8.2. To prove the Sum Rule for series, let

$$A_n = a_1 + a_2 + \cdots + a_n, \quad B_n = b_1 + b_2 + \cdots + b_n.$$

Then the partial sums of $\sum(a_n + b_n)$ are

$$\begin{aligned} S_n &= (a_1 + b_1) + (a_2 + b_2) + \cdots + (a_n + b_n) \\ &= (a_1 + \cdots + a_n) + (b_1 + \cdots + b_n) \\ &= A_n + B_n. \end{aligned}$$

Since $A_n \to A$ and $B_n \to B$, we have $S_n \to A + B$ by the Sum Rule for sequences. The proof of the Difference Rule is similar.

To prove the Constant Multiple Rule for series, observe that the partial sums of $\sum k a_n$ form the sequence

$$S_n = k a_1 + k a_2 + \cdots + k a_n = k(a_1 + a_2 + \cdots + a_n) = k A_n,$$

which converges to kA by the Constant Multiple Rule for sequences. $\qquad \Box$

As corollaries of Theorem 7, we have

1. Every nonzero constant multiple of a divergent series diverges.

2. If $\sum a_n$ converges and $\sum b_n$ diverges, then $\sum(a_n + b_n)$ and $\sum(a_n - b_n)$ both diverge.

We omit the proofs.

EXAMPLE 10   Find the sums of the following series.

a) $\displaystyle\sum_{n=1}^{\infty} \frac{3^{n-1} - 1}{6^{n-1}} = \sum_{n=1}^{\infty} \left(\frac{1}{2^{n-1}} - \frac{1}{6^{n-1}} \right)$

$\qquad\qquad\qquad = \displaystyle\sum_{n=1}^{\infty} \frac{1}{2^{n-1}} - \sum_{n=1}^{\infty} \frac{1}{6^{n-1}}$ Difference Rule

$\qquad\qquad\qquad = \dfrac{1}{1 - (1/2)} - \dfrac{1}{1 - (1/6)}$ Geometric series with $a = 1$ and
$r = 1/2, 1/6$

$\qquad\qquad\qquad = 2 - \dfrac{6}{5}$

$\qquad\qquad\qquad = \dfrac{4}{5}$

b) $\displaystyle\sum_{n=1}^{\infty} \frac{4}{2^{n-1}} = 4 \sum_{n=1}^{\infty} \frac{1}{2^{n-1}}$ Constant Multiple Rule

$\displaystyle = 4 \left(\frac{1}{1 - (1/2)} \right)$ Geometric series with $a = 1$, $r = 1/2$

$= 8$ ❏

Adding or Deleting Terms

We can always add a finite number of terms to a series or delete a finite number of terms without altering the series' convergence or divergence, although in the case of convergence this will usually change the sum. If $\sum_{n=1}^{\infty} a_n$ converges, then $\sum_{n=k}^{\infty} a_n$ converges for any $k > 1$ and

$$\sum_{n=1}^{\infty} a_n = a_1 + a_2 + \cdots + a_{k-1} + \sum_{n=k}^{\infty} a_n. \tag{6}$$

Conversely, if $\sum_{n=k}^{\infty} a_n$ converges for any $k > 1$, then $\sum_{n=1}^{\infty} a_n$ converges. Thus,

$$\sum_{n=1}^{\infty} \frac{1}{5^n} = \frac{1}{5} + \frac{1}{25} + \frac{1}{125} + \sum_{n=4}^{\infty} \frac{1}{5^n} \tag{7}$$

and

$$\sum_{n=4}^{\infty} \frac{1}{5^n} = \left(\sum_{n=1}^{\infty} \frac{1}{5^n} \right) - \frac{1}{5} - \frac{1}{25} - \frac{1}{125}. \tag{8}$$

Reindexing

As long as we preserve the order of its terms, we can reindex any series without altering its convergence. To raise the starting value of the index h units, replace the n in the formula for a_n by $n - h$:

$$\sum_{n=1}^{\infty} a_n = \sum_{n=1+h}^{\infty} a_{n-h} = a_1 + a_2 + a_3 + \cdots.$$

To lower the starting value of the index h units, replace the n in the formula for a_n by $n + h$:

$$\sum_{n=1}^{\infty} a_n = \sum_{n=1-h}^{\infty} a_{n+h} = a_1 + a_2 + a_3 + \cdots.$$

It works like a horizontal shift.

EXAMPLE 11 We can write the geometric series that starts with

$$1 + \frac{1}{2} + \frac{1}{4} + \cdots$$

as

$$\sum_{n=0}^{\infty} \frac{1}{2^n}, \qquad \sum_{n=5}^{\infty} \frac{1}{2^{n-5}}, \qquad \text{or even} \qquad \sum_{n=-4}^{\infty} \frac{1}{2^{n+4}}.$$

The partial sums remain the same no matter what indexing we choose. ❏

We usually give preference to indexings that lead to simple expressions.

Exercises 8.3

Finding *n*th Partial Sums

In Exercises 1–6, find a formula for the *n*th partial sum of each series and use it to find the series' sum if the series converges.

1. $2 + \dfrac{2}{3} + \dfrac{2}{9} + \dfrac{2}{27} + \cdots + \dfrac{2}{3^{n-1}} + \cdots$

2. $\dfrac{9}{100} + \dfrac{9}{100^2} + \dfrac{9}{100^3} + \cdots + \dfrac{9}{100^n} + \cdots$

3. $1 - \dfrac{1}{2} + \dfrac{1}{4} - \dfrac{1}{8} + \cdots + (-1)^{n-1}\dfrac{1}{2^{n-1}} + \cdots$

4. $1 - 2 + 4 - 8 + \cdots + (-1)^{n-1} 2^{n-1} + \cdots$

5. $\dfrac{1}{2 \cdot 3} + \dfrac{1}{3 \cdot 4} + \dfrac{1}{4 \cdot 5} + \cdots + \dfrac{1}{(n+1)(n+2)} + \cdots$

6. $\dfrac{5}{1 \cdot 2} + \dfrac{5}{2 \cdot 3} + \dfrac{5}{3 \cdot 4} + \cdots + \dfrac{5}{n(n+1)} + \cdots$

Series with Geometric Terms

In Exercises 7–14, write out the first few terms of each series to show how the series starts. Then find the sum of the series.

7. $\displaystyle\sum_{n=0}^{\infty} \dfrac{(-1)^n}{4^n}$

8. $\displaystyle\sum_{n=2}^{\infty} \dfrac{1}{4^n}$

9. $\displaystyle\sum_{n=1}^{\infty} \dfrac{7}{4^n}$

10. $\displaystyle\sum_{n=0}^{\infty} (-1)^n \dfrac{5}{4^n}$

11. $\displaystyle\sum_{n=0}^{\infty} \left(\dfrac{5}{2^n} + \dfrac{1}{3^n} \right)$

12. $\displaystyle\sum_{n=0}^{\infty} \left(\dfrac{5}{2^n} - \dfrac{1}{3^n} \right)$

13. $\displaystyle\sum_{n=0}^{\infty} \left(\dfrac{1}{2^n} + \dfrac{(-1)^n}{5^n} \right)$

14. $\displaystyle\sum_{n=0}^{\infty} \left(\dfrac{2^{n+1}}{5^n} \right)$

Telescoping Series

Use partial fractions to find the sum of each series in Exercises 15–22.

15. $\displaystyle\sum_{n=1}^{\infty} \dfrac{4}{(4n-3)(4n+1)}$

16. $\displaystyle\sum_{n=1}^{\infty} \dfrac{6}{(2n-1)(2n+1)}$

17. $\displaystyle\sum_{n=1}^{\infty} \dfrac{40n}{(2n-1)^2(2n+1)^2}$

18. $\displaystyle\sum_{n=1}^{\infty} \dfrac{2n+1}{n^2(n+1)^2}$

19. $\displaystyle\sum_{n=1}^{\infty} \left(\dfrac{1}{\sqrt{n}} - \dfrac{1}{\sqrt{n+1}} \right)$

20. $\displaystyle\sum_{n=1}^{\infty} \left(\dfrac{1}{2^{1/n}} - \dfrac{1}{2^{1/(n+1)}} \right)$

21. $\displaystyle\sum_{n=1}^{\infty} \left(\dfrac{1}{\ln(n+2)} - \dfrac{1}{\ln(n+1)} \right)$

22. $\displaystyle\sum_{n=1}^{\infty} (\tan^{-1}(n) - \tan^{-1}(n+1))$

Convergence or Divergence

Which series in Exercises 23–40 converge, and which diverge? Give reasons for your answers. If a series converges, find its sum.

23. $\displaystyle\sum_{n=0}^{\infty} \left(\dfrac{1}{\sqrt{2}} \right)^n$

24. $\displaystyle\sum_{n=0}^{\infty} (\sqrt{2})^n$

25. $\displaystyle\sum_{n=1}^{\infty} (-1)^{n+1} \dfrac{3}{2^n}$

26. $\displaystyle\sum_{n=1}^{\infty} (-1)^{n+1} n$

27. $\displaystyle\sum_{n=0}^{\infty} \cos n\pi$

28. $\displaystyle\sum_{n=0}^{\infty} \dfrac{\cos n\pi}{5^n}$

29. $\displaystyle\sum_{n=0}^{\infty} e^{-2n}$

30. $\displaystyle\sum_{n=1}^{\infty} \ln \dfrac{1}{n}$

31. $\displaystyle\sum_{n=1}^{\infty} \dfrac{2}{10^n}$

32. $\displaystyle\sum_{n=0}^{\infty} \dfrac{1}{x^n}, \quad |x| > 1$

33. $\displaystyle\sum_{n=0}^{\infty} \dfrac{2^n - 1}{3^n}$

34. $\displaystyle\sum_{n=1}^{\infty} \left(1 - \dfrac{1}{n} \right)^n$

35. $\displaystyle\sum_{n=0}^{\infty} \dfrac{n!}{1000^n}$

36. $\displaystyle\sum_{n=1}^{\infty} \dfrac{n^n}{n!}$

37. $\displaystyle\sum_{n=1}^{\infty} \ln \left(\dfrac{n}{n+1} \right)$

38. $\displaystyle\sum_{n=1}^{\infty} \ln \left(\dfrac{n}{2n+1} \right)$

39. $\displaystyle\sum_{n=0}^{\infty} \left(\dfrac{e}{\pi} \right)^n$

40. $\displaystyle\sum_{n=0}^{\infty} \dfrac{e^{n\pi}}{\pi^{ne}}$

Geometric Series

In each of the geometric series in Exercises 41–44, write out the first few terms of the series to find *a* and *r*, and find the sum of the series. Then express the inequality $|r| < 1$ in terms of *x* and find the values of *x* for which the inequality holds and the series converges.

41. $\displaystyle\sum_{n=0}^{\infty} (-1)^n x^n$

42. $\displaystyle\sum_{n=0}^{\infty} (-1)^n x^{2n}$

43. $\displaystyle\sum_{n=0}^{\infty} 3 \left(\dfrac{x-1}{2} \right)^n$

44. $\displaystyle\sum_{n=0}^{\infty} \dfrac{(-1)^n}{2} \left(\dfrac{1}{3 + \sin x} \right)^n$

In Exercises 45–50, find the values of *x* for which the given geometric series converges. Also, find the sum of the series (as a function of *x*) for those values of *x*.

45. $\displaystyle\sum_{n=0}^{\infty} 2^n x^n$

46. $\displaystyle\sum_{n=0}^{\infty} (-1)^n x^{-2n}$

47. $\displaystyle\sum_{n=0}^{\infty} (-1)^n (x+1)^n$

48. $\displaystyle\sum_{n=0}^{\infty} \left(-\dfrac{1}{2} \right)^n (x-3)^n$

49. $\displaystyle\sum_{n=0}^{\infty} \sin^n x$

50. $\displaystyle\sum_{n=0}^{\infty} (\ln x)^n$

Repeating Decimals

Express each of the numbers in Exercises 51–58 as the ratio of two integers.

51. $0.\overline{23} = 0.23\ 23\ 23 \ldots$

52. $0.\overline{234} = 0.234\ 234\ 234 \ldots$

53. $0.\overline{7} = 0.7777 \ldots$

54. $0.\overline{d} = 0.dddd \ldots ,$ where d is a digit

55. $0.0\overline{6} = 0.06666 \ldots$

56. $1.\overline{414} = 1.414\ 414\ 414 \ldots$

57. $1.24\overline{123} = 1.24\ 123\ 123\ 123 \ldots$

58. $3.\overline{142857} = 3.142857\ 142857 \ldots$

Theory and Examples

59. The series in Exercise 5 can also be written as

$$\sum_{n=1}^{\infty} \frac{1}{(n+1)(n+2)} \quad \text{and} \quad \sum_{n=-1}^{\infty} \frac{1}{(n+3)(n+4)}.$$

Write it as a sum beginning with (a) $n = -2$, (b) $n = 0$, (c) $n = 5$.

60. The series in Exercise 6 can also be written as

$$\sum_{n=1}^{\infty} \frac{5}{n(n+1)} \quad \text{and} \quad \sum_{n=0}^{\infty} \frac{5}{(n+1)(n+2)}.$$

Write it as a sum beginning with (a) $n = -1$, (b) $n = 3$, (c) $n = 20$.

61. Make up an infinite series of nonzero terms whose sum is

a)   1   **b)** -3 **c)**   0.

Can you make an infinite series of nonzero terms that converges to any number you want? Explain.

62. Make up an example of two divergent infinite series whose term-by-term sum converges.

63. Show by example that $\sum (a_n/b_n)$ may diverge even though $\sum a_n$ and $\sum b_n$ converge and no b_n equals 0.

64. Find convergent geometric series $A = \sum a_n$ and $B = \sum b_n$ that illustrate the fact that $\sum a_n b_n$ may converge without being equal to AB.

65. Show by example that $\sum (a_n/b_n)$ may converge to something other than A/B even when $A = \sum a_n$, $B = \sum b_n \neq 0$, and no b_n equals 0.

66. If $\sum a_n$ converges and $a_n > 0$ for all n, can anything be said about $\sum (1/a_n)$? Give reasons for your answer.

67. What happens if you add a finite number of terms to a divergent series or delete a finite number of terms from a divergent series? Give reasons for your answer.

68. If $\sum a_n$ converges and $\sum b_n$ diverges, can anything be said about their term-by-term sum $\sum (a_n + b_n)$? Give reasons for your answer.

69. Make up a geometric series $\sum ar^{n-1}$ that converges to the number 5 if

a) $a = 2$ **b)** $a = 13/2$.

70. Find the value of b for which

$$1 + e^b + e^{2b} + e^{3b} + \cdots = 9.$$

71. For what values of r does the infinite series

$$1 + 2r + r^2 + 2r^3 + r^4 + 2r^5 + r^6 + \cdots$$

converge? Find the sum of the series when it converges.

72. Show that the error $(L - s_n)$ obtained by   placing a convergent geometric series with one of its partial s_n is $ar^n/(1 - r)$.

73. A ball is dropped from a height of 4 m.   ch time it strikes the pavement after falling from a height of   neters it rebounds to a height of $0.75h$ meters. Find the total distance the ball travels up and down.

74. (*Continuation of Exercise 73.*) Find the total number of seconds the ball in Exercise 73 is traveling. (*Hint:* The formula $s = 4.9t^2$ gives $t = \sqrt{s/4.9}$.)

75. The accompanying figure shows the first five of a sequence of squares. The outermost square has an area of 4 m². Each of the other squares is obtained by joining the midpoints of the sides of the squares before it. Find the sum of the areas of all the squares.

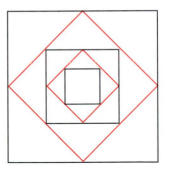

76. The accompanying figure shows the first three rows and part of the fourth row of a sequence of rows of semicircles. There are 2^n semicircles in the nth row, each of radius $1/2^n$. Find the sum of the areas of all the semicircles.

77. *Helga von Koch's snowflake curve.* Helga von Koch's snowflake (p. 167) is a curve of infinite length that encloses a region of finite area. To see why this is so, suppose the curve is generated by starting with an equilateral triangle whose sides have length 1.

a) Find the length L_n of the nth curve C_n and show that $\lim_{n\to\infty} L_n = \infty$.

b) Find the area A_n of the region enclosed by C_n and calculate $\lim_{n\to\infty} A_n$.

78. The accompanying figure provides an informal proof that $\sum_{n=1}^{\infty} (1/n^2)$ is less than 2. Explain what is going on. (Source: "Convergence with Pictures" by P. J. Rippon, *American Mathematical Monthly,* Vol. 93, No. 6, 1986, pp. 476–78.)

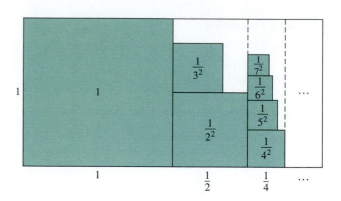

8.4

The Integral Test for Series of Nonnegative Terms

Given a series $\sum a_n$, we have two questions:

1. Does the series converge?

2. If it converges, what is its sum?

Much of the rest of this chapter is devoted to the first question. But as a practical matter, the second question is just as important, and we will return to it later.

In this section and the next two, we study series that do not have negative terms. The reason for this restriction is that the partial sums of these series form nondecreasing sequences, and nondecreasing sequences that are bounded from above always converge (Theorem 1, Section 8.1). To show that a series of nonnegative terms converges, we need only show that its partial sums are bounded from above.

It may at first seem to be a drawback that this approach establishes the fact of convergence without producing the sum of the series in question. Surely it would be better to compute sums of series directly from formulas for their partial sums. But in most cases such formulas are not available, and in their absence we have to turn instead to the two-step procedure of first establishing convergence and then approximating the sum.

Nondecreasing Partial Sums

Suppose that $\sum_{n=1}^{\infty} a_n$ is an infinite series with $a_n \geq 0$ for all n. Then each partial sum is greater than or equal to its predecessor because $s_{n+1} = s_n + a_n$:

$$s_1 \leq s_2 \leq s_3 \leq \cdots \leq s_n \leq s_{n+1} \leq \cdots.$$

Since the partial sums form a nondecreasing sequence, the Nondecreasing Sequence Theorem (Theorem 1, Section 8.1) tells us that the series will converge if and only if the partial sums are bounded from above.

> ### Corollary of Theorem 1
> A series $\sum_{n=1}^{\infty} a_n$ of nonnegative terms converges if and only if its partial sums are bounded from above.

Caution

Notice that the nth-Term Test for divergence does not detect the divergence of the harmonic series. The nth term, $1/n$, goes to zero, but the series still diverges.

Nicole Oresme (1320–1382)

The argument we use to show the divergence of the harmonic series was devised by the French theologian, mathematician, physicist, and bishop Nicole Oresme (pronounced "or-*rem*"). Oresme was a vigorous opponent of astrology, a dynamic preacher, an adviser of princes, a friend of King Charles V, a popularizer of science, and a skillful translator of Latin into French.

Oresme did not believe in Albert of Saxony's generally accepted model of free fall (Chapter 6, Additional Exercise 26) but preferred Aristotle's constant-acceleration model, the model that became popular among Oxford scholars in the 1330s and that Galileo eventually used three hundred years later.

EXAMPLE 1   *The harmonic series*

The series

$$\sum_{n=1}^{\infty} \frac{1}{n} = 1 + \frac{1}{2} + \frac{1}{3} + \cdots + \frac{1}{n} + \cdots$$

is called the **harmonic series.** It diverges because there is no upper bound for its partial sums. To see why, group the terms of the series in the following way:

$$1 + \frac{1}{2} + \underbrace{\left(\frac{1}{3} + \frac{1}{4}\right)}_{> \frac{2}{4} = \frac{1}{2}} + \underbrace{\left(\frac{1}{5} + \frac{1}{6} + \frac{1}{7} + \frac{1}{8}\right)}_{> \frac{4}{8} = \frac{1}{2}} + \underbrace{\left(\frac{1}{9} + \frac{1}{10} + \cdots + \frac{1}{16}\right)}_{> \frac{8}{16} = \frac{1}{2}} + \cdots .$$

The sum of the first two terms is 1.5. The sum of the next two terms is $1/3 + 1/4$, which is greater than $1/4 + 1/4 = 1/2$. The sum of the next four terms is $1/5 + 1/6 + 1/7 + 1/8$, which is greater than $1/8 + 1/8 + 1/8 + 1/8 = 1/2$. The sum of the next eight terms is $1/9 + 1/10 + 1/11 + 1/12 + 1/13 + 1/14 + 1/15 + 1/16$, which is greater than $8/16 = 1/2$. The sum of the next 16 terms is greater than $16/32 = 1/2$, and so on. In general, the sum of 2^n terms ending with $1/2^{n+1}$ is greater than $2^n/2^{n+1} = 1/2$. The sequence of partial sums is not bounded from above: If $n = 2^k$, the partial sum s_n is greater than $k/2$. The harmonic series diverges. ❑

The Integral Test

We introduce the Integral Test with a series that is related to the harmonic series, but whose nth term is $1/n^2$ instead of $1/n$.

EXAMPLE 2   Does the following series converge?

$$\sum_{n=1}^{\infty} \frac{1}{n^2} = 1 + \frac{1}{4} + \frac{1}{9} + \frac{1}{16} + \cdots + \frac{1}{n^2} + \cdots \tag{1}$$

Solution   We determine the convergence of $\sum_{n=1}^{\infty} (1/n^2)$ by comparing it with $\int_1^{\infty} (1/x^2)\, dx$. To carry out the comparison, we think of the terms of the series as values of the function $f(x) = 1/x^2$ and interpret these values as the areas of rectangles under the curve $y = 1/x^2$.

As Fig. 8.12 shows,

$$s_n = \frac{1}{1^2} + \frac{1}{2^2} + \frac{1}{3^2} + \cdots + \frac{1}{n^2}$$

$$= f(1) + f(2) + f(3) + \cdots + f(n)$$

$$< f(1) + \int_1^n \frac{1}{x^2}\, dx$$

$$< 1 + \int_1^{\infty} \frac{1}{x^2}\, dx$$

$$< 1 + 1 = 2. \qquad \text{As in Section 7.6, Example 8,} \quad \int_1^{\infty} (1/x^2)\, dx = 1.$$

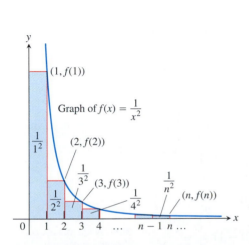

8.12 Figure for the area comparisons in Example 2.

Thus the partial sums of $\sum_{n=1}^{\infty} 1/n^2$ are bounded from above (by 2) and the series converges. The sum of the series is known to be $\pi^2/6 \approx 1.64493$. ❑

The Integral Test

Let $\{a_n\}$ be a sequence of positive terms. Suppose that $a_n = f(n)$, where f is a continuous, positive, decreasing function of x for all $x \geq N$ (N a positive integer). Then the series $\sum_{n=N}^{\infty} a_n$ and the integral $\int_N^{\infty} f(x) \, dx$ both converge or both diverge.

Caution

The series and integral need not have the same value in the convergent case. As we saw in Example 2, $\sum_{n=1}^{\infty} (1/n^2) = \pi^2/6$ while $\int_1^{\infty} (1/x^2) \, dx = 1$.

Proof   We establish the test for the case $N = 1$. The proof for general N is similar.

We start with the assumption that f is a decreasing function with $f(n) = a_n$ for every n. This leads us to observe that the rectangles in Fig. 8.13(a), which have areas $a_1, a_2, \ldots, a_n$, collectively enclose more area than that under the curve $y = f(x)$ from $x = 1$ to $x = n + 1$. That is,

$$\int_1^{n+1} f(x) \, dx \leq a_1 + a_2 + \cdots + a_n.$$

In Fig. 8.13(b) the rectangles have been faced to the left instead of to the right. If we momentarily disregard the first rectangle, of area a_1, we see that

$$a_2 + a_3 + \cdots + a_n \leq \int_1^n f(x) \, dx.$$

If we include a_1, we have

$$a_1 + a_2 + \cdots + a_n \leq a_1 + \int_1^n f(x) \, dx.$$

Combining these results gives

$$\int_1^{n+1} f(x) \, dx \leq a_1 + a_2 + \cdots + a_n \leq a_1 + \int_1^n f(x) \, dx. \tag{2}$$

If $\int_1^{\infty} f(x) \, dx$ is finite, the right-hand inequality shows that $\sum a_n$ is finite. If $\int_1^{\infty} f(x) \, dx$ is infinite, the left-hand inequality shows that $\sum a_n$ is infinite. Hence the series and the integral are both finite or both infinite.   ❑

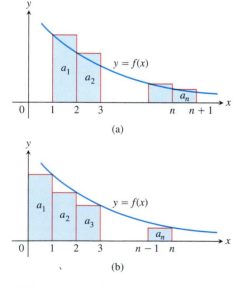

8.13 Subject to the conditions of the Integral Test, the series $\sum_{n=1}^{\infty} a_n$ and the integral $\int_1^{\infty} f(x) \, dx$ both converge or both diverge.

EXAMPLE 3   *The p-series.*   Show that the **p-series**

$$\sum_{n=1}^{\infty} \frac{1}{n^p} = \frac{1}{1^p} + \frac{1}{2^p} + \frac{1}{3^p} + \cdots + \frac{1}{n^p} + \cdots \tag{3}$$

(p a real constant) converges if $p > 1$, and diverges if $p \leq 1$.

Solution   If $p > 1$, then $f(x) = 1/x^p$ is a positive decreasing function of x. Since

$$\int_1^{\infty} \frac{1}{x^p} \, dx = \int_1^{\infty} x^{-p} \, dx = \lim_{b \to \infty} \left[\frac{x^{-p+1}}{-p+1} \right]_1^b$$

$$= \frac{1}{1-p} \lim_{b \to \infty} \left(\frac{1}{b^{p-1}} - 1 \right)$$

$$= \frac{1}{1-p}(0-1) = \frac{1}{p-1}, \qquad \begin{array}{l} b^{p-1} \to \infty \text{ as } b \to \infty \\ \text{because } p - 1 > 0. \end{array}$$

the series converges by the Integral Test.

If $p < 1$, then $1 - p > 0$ and

$$\int_1^\infty \frac{1}{x^p}\, dx = \frac{1}{1-p} \lim_{b\to\infty} (b^{1-p} - 1) = \infty.$$

The series diverges by the Integral Test.

If $p = 1$, we have the (divergent) harmonic series

$$1 + \frac{1}{2} + \frac{1}{3} + \cdots + \frac{1}{n} + \cdots.$$

We have convergence for $p > 1$ but divergence for every other value of p.

❑

Exercises 8.4

Determining Convergence or Divergence

Which of the series in Exercises 1–30 converge, and which diverge? Give reasons for your answers. (When you check an answer, remember that there may be more than one way to determine the series' convergence or divergence.)

1. $\displaystyle\sum_{n=1}^\infty \frac{1}{10^n}$ **2.** $\displaystyle\sum_{n=1}^\infty e^{-n}$ **3.** $\displaystyle\sum_{n=1}^\infty \frac{n}{n+1}$

4. $\displaystyle\sum_{n=1}^\infty \frac{5}{n+1}$ **5.** $\displaystyle\sum_{n=1}^\infty \frac{3}{\sqrt{n}}$ **6.** $\displaystyle\sum_{n=1}^\infty \frac{-2}{n\sqrt{n}}$

7. $\displaystyle\sum_{n=1}^\infty -\frac{1}{8^n}$ **8.** $\displaystyle\sum_{n=1}^\infty \frac{-8}{n}$ **9.** $\displaystyle\sum_{n=2}^\infty \frac{\ln n}{n}$

10. $\displaystyle\sum_{n=2}^\infty \frac{\ln n}{\sqrt{n}}$ **11.** $\displaystyle\sum_{n=1}^\infty \frac{2^n}{3^n}$ **12.** $\displaystyle\sum_{n=1}^\infty \frac{5^n}{4^n + 3}$

13. $\displaystyle\sum_{n=0}^\infty \frac{-2}{n+1}$ **14.** $\displaystyle\sum_{n=1}^\infty \frac{1}{2n-1}$ **15.** $\displaystyle\sum_{n=1}^\infty \frac{2^n}{n+1}$

16. $\displaystyle\sum_{n=1}^\infty \frac{1}{\sqrt{n}(\sqrt{n}+1)}$ **17.** $\displaystyle\sum_{n=2}^\infty \frac{\sqrt{n}}{\ln n}$ **18.** $\displaystyle\sum_{n=1}^\infty \left(1 + \frac{1}{n}\right)^n$

19. $\displaystyle\sum_{n=1}^\infty \frac{1}{(\ln 2)^n}$ **20.** $\displaystyle\sum_{n=1}^\infty \frac{1}{(\ln 3)^n}$

21. $\displaystyle\sum_{n=3}^\infty \frac{(1/n)}{(\ln n)\sqrt{\ln^2 n - 1}}$ **22.** $\displaystyle\sum_{n=1}^\infty \frac{1}{n(1 + \ln^2 n)}$

23. $\displaystyle\sum_{n=1}^\infty n \sin \frac{1}{n}$ **24.** $\displaystyle\sum_{n=1}^\infty n \tan \frac{1}{n}$

25. $\displaystyle\sum_{n=1}^\infty \frac{e^n}{1 + e^{2n}}$ **26.** $\displaystyle\sum_{n=1}^\infty \frac{2}{1 + e^n}$

27. $\displaystyle\sum_{n=1}^\infty \frac{8 \tan^{-1} n}{1 + n^2}$ **28.** $\displaystyle\sum_{n=1}^\infty \frac{n}{n^2 + 1}$

29. $\displaystyle\sum_{n=1}^\infty \operatorname{sech} n$ **30.** $\displaystyle\sum_{n=1}^\infty \operatorname{sech}^2 n$

Theory and Examples

For what values of a, if any, do the series in Exercises 31 and 32 converge?

31. $\displaystyle\sum_{n=1}^\infty \left(\frac{a}{n+2} - \frac{1}{n+4}\right)$ **32.** $\displaystyle\sum_{n=3}^\infty \left(\frac{1}{n-1} - \frac{2a}{n+1}\right)$

33. a) Draw illustrations like those in Figs. 8.12 and 8.13 to show that the partial sums of the harmonic series satisfy the inequalities

$$\ln(n+1) = \int_1^{n+1} \frac{1}{x}\, dx \le 1 + \frac{1}{2} + \cdots + \frac{1}{n}$$

$$\le 1 + \int_1^n \frac{1}{x}\, dx = 1 + \ln n.$$

b) There is absolutely no empirical evidence for the divergence of the harmonic series even though we know it diverges. The partial sums just grow too slowly. To see what we mean, suppose you had started with $s_1 = 1$ the day the universe was formed, 13 billion years ago, and added a new term every *second*. About how large would the partial sum s_n be today, assuming a 365-day year?

34. Are there any values of x for which $\sum_{n=1}^\infty (1/(nx))$ converges? Give reasons for your answer.

35. Is it true that if $\sum_{n=1}^\infty a_n$ is a divergent series of positive numbers then there is also a divergent series $\sum_{n=1}^\infty b_n$ of positive numbers with $b_n < a_n$ for every n? Is there a "smallest" divergent series of positive numbers? Give reasons for your answers.

36. (*Continuation of Exercise 35*) Is there a "largest" convergent series of positive numbers? Explain.

37. *The Cauchy condensation test.* The Cauchy condensation test says: Let $\{a_n\}$ be a nonincreasing sequence ($a_n \ge a_{n+1}$ for all n) of positive terms that converges to 0. Then $\sum a_n$ converges if and only if $\sum 2^n a_{2^n}$ converges. For example, $\sum (1/n)$ diverges because $\sum 2^n \cdot (1/2^n) = \sum 1$ diverges. Show why the test works.

38. Use the Cauchy condensation test from Exercise 37 to show that

a) $\displaystyle\sum_{n=2}^{\infty} \frac{1}{n \ln n}$ diverges;

b) $\displaystyle\sum_{n=1}^{\infty} \frac{1}{n^p}$ converges if $p > 1$ and diverges if $p \le 1$.

39. *Logarithmic p-series*

a) Show that

$$\int_2^{\infty} \frac{dx}{x(\ln x)^p} \qquad (p \text{ a positive constant})$$

converges if and only if $p > 1$.

b) What implications does the fact in (a) have for the convergence of the series

$$\sum_{n=2}^{\infty} \frac{1}{n(\ln n)^p}?$$

Give reasons for your answer.

40. (*Continuation of Exercise 39.*) Use the result in Exercise 39 to determine which of the following series converge and which diverge. Support your answer in each case.

a) $\displaystyle\sum_{n=2}^{\infty} \frac{1}{n(\ln n)}$

b) $\displaystyle\sum_{n=2}^{\infty} \frac{1}{n(\ln n)^{1.01}}$

c) $\displaystyle\sum_{n=2}^{\infty} \frac{1}{n \ln (n^3)}$

d) $\displaystyle\sum_{n=2}^{\infty} \frac{1}{n(\ln n)^3}$

41. *Euler's constant.* Graphs like those in Fig. 8.13 suggest that as n increases there is little change in the difference between the sum

$$1 + \frac{1}{2} + \cdots + \frac{1}{n}$$

and the integral

$$\ln n = \int_1^n \frac{1}{x}\, dx.$$

To explore this idea, carry out the following steps.

a) By taking $f(x) = 1/x$ in inequality (2), show that

$$\ln (n + 1) \le 1 + \frac{1}{2} + \cdots + \frac{1}{n} \le 1 + \ln n$$

or

$$0 < \ln (n + 1) - \ln n \le 1 + \frac{1}{2} + \cdots + \frac{1}{n} - \ln n \le 1.$$

Thus, the sequence

$$a_n = 1 + \frac{1}{2} + \cdots + \frac{1}{n} - \ln n$$

is bounded from below and from above.

b) Show that

$$\frac{1}{n+1} < \int_n^{n+1} \frac{1}{x}\, dx = \ln (n + 1) - \ln n,$$

and use this result to show that the sequence $\{a_n\}$ in part (a) is decreasing.

Since a decreasing sequence that is bounded from below converges (Exercise 41 in Section 8.1), the numbers a_n defined in (a) converge:

$$1 + \frac{1}{2} + \cdots + \frac{1}{n} - \ln n \to \gamma.$$

The number γ, whose value is 0.5772 ..., is called *Euler's constant.* In contrast to other special numbers like π and e, no other expression with a simple law of formulation has ever been found for γ.

42. Use the integral test to show that

$$\sum_{n=0}^{\infty} e^{-n^2}$$

converges.

8.5

Comparison Tests for Series of Nonnegative Terms

The key question in using Corollary 1 in the preceding section is how to determine in any particular instance whether the s_n's are bounded from above. Sometimes we can establish this by showing that each s_n is less than or equal to the corresponding partial sum of a series already known to converge.

EXAMPLE 1   The series

$$\sum_{n=0}^{\infty} \frac{1}{n!} = 1 + \frac{1}{1!} + \frac{1}{2!} + \frac{1}{3!} + \cdots \qquad (1)$$

converges because its terms are all positive and less than or equal to the corresponding terms of

$$1 + \sum_{n=0}^{\infty} \frac{1}{2^n} = 1 + 1 + \frac{1}{2} + \frac{1}{2^2} + \cdots. \tag{2}$$

To see how this relationship leads to an upper bound for the partial sums of $\sum_{n=0}^{\infty}(1/(n!))$, let

$$s_n = 1 + \frac{1}{1!} + \frac{1}{2!} + \cdots + \frac{1}{n!}$$

and observe that, for each n,

$$s_n \leq 1 + 1 + \frac{1}{2} + \frac{1}{2^2} + \cdots + \frac{1}{2^{n-1}} < 1 + \sum_{n=0}^{\infty} \frac{1}{2^n} = 1 + \frac{1}{1 - (1/2)} = 3.$$

Thus the partial sums of $\sum_{n=0}^{\infty}(1/(n!))$ are all less than 3, so $\sum_{n=0}^{\infty}(1/(n!))$ converges.

The fact that 3 is an upper bound for the partial sums of $\sum_{n=0}^{\infty}(1/(n!))$ does not mean that the series converges to 3. As we will see in Section 8.10, the series converges to e. ❑

The Direct Comparison Test

We established the convergence in Example 1 by comparing the terms of the given series with the terms of a series known to converge. This idea can be pursued further to yield a number of tests known as *comparison tests*.

Direct Comparison Test for Series of Nonnegative Terms

Let $\sum a_n$ be a series with no negative terms.

a) $\sum a_n$ converges if there is a convergent series $\sum c_n$ with $a_n \leq c_n$ for all $n > N$, for some integer N.

b) $\sum a_n$ diverges if there is a divergent series of nonnegative terms $\sum d_n$ with $a_n \geq d_n$ for all $n > N$, for some integer N.

Proof In part (a), the partial sums of $\sum a_n$ are bounded above by

$$M = a_1 + a_2 + \cdots + a_n + \sum_{n=N+1}^{\infty} c_n.$$

They therefore form a nondecreasing sequence with a limit $L \leq M$.

In part (b), the partial sums of $\sum a_n$ are not bounded from above. If they were, the partial sums for $\sum d_n$ would be bounded by

$$M' = d_1 + d_2 + \cdots + d_N + \sum_{n=N+1}^{\infty} a_n$$

and $\sum d_n$ would have to converge instead of diverge. ❑

To apply the Direct Comparison Test to a series, we need not include the early terms of the series. We can start the test with any index N provided we include all the terms of the series being tested from there on.

EXAMPLE 2   Does the following series converge?

$$5 + \frac{2}{3} + 1 + \frac{1}{7} + \frac{1}{2} + \frac{1}{3!} + \frac{1}{4!} + \cdots + \frac{1}{k!} + \cdots$$

Solution   We ignore the first four terms and compare the remaining terms with those of the convergent geometric series $\sum_{n=1}^{\infty} 1/2^n$. We see that

$$\frac{1}{2} + \frac{1}{3!} + \frac{1}{4!} + \cdots \leq \frac{1}{2} + \frac{1}{4} + \frac{1}{8} + \cdots.$$

Therefore, the original series converges by the Direct Comparison Test.   □

To apply the Direct Comparison Test, we need to have on hand a list of series whose convergence or divergence we know. Here is what we know so far:

Convergent series	**Divergent series**
Geometric series with $|r| < 1$	Geometric series with $|r| \geq 1$
Telescoping series like $\displaystyle\sum_{n=1}^{\infty} \frac{1}{n(n+1)}$	The harmonic series $\displaystyle\sum_{n=1}^{\infty} \frac{1}{n}$
The series $\displaystyle\sum_{n=0}^{\infty} \frac{1}{n!}$	Any series $\sum a_n$ for which $\lim_{n \to \infty} a_n$ does not exist or $\lim_{n \to \infty} a_n \neq 0$
The p-series $\displaystyle\sum_{n=1}^{\infty} \frac{1}{n^p}$ with $p > 1$	The p-series $\displaystyle\sum_{n=1}^{\infty} \frac{1}{n^p}$ with $p \leq 1$

The Limit Comparison Test

We now introduce a comparison test that is particularly handy for series in which a_n is a rational function of n.

Suppose we wanted to investigate the convergence of the series

a) $\displaystyle\sum_{n=2}^{\infty} \frac{2n}{n^2 - n + 1}$ **b)** $\displaystyle\sum_{n=2}^{\infty} \frac{8n^3 + 100n^2 + 1000}{2n^6 - n + 5}.$

In determining convergence or divergence, only the tails matter. And when n is very large, the highest powers in the numerator and denominator matter the most. So in (a), we might reason this way: For n large,

$$a_n = \frac{2n}{n^2 - n + 1}$$

behaves like $2n/n^2 = 2/n$. Since $\sum 1/n$ diverges, we expect $\sum a_n$ to diverge, too.

In (b) we might reason that for n large

$$a_n = \frac{8n^3 + 100n^2 + 1000}{2n^6 - n + 5}$$

will behave approximately like $(8n^3)/(2n^6) = 4/n^3$. Since $\sum 4/n^3$ converges (it is 4 times a convergent p-series), we expect $\sum a_n$ to converge, too.

Our expectations about $\sum a_n$ in each case are correct, as the following test shows.

Limit Comparison Test

Suppose that $a_n > 0$ and $b_n > 0$ for all $n \geq N$ (N an integer).

1. If $\lim\limits_{n \to \infty} \dfrac{a_n}{b_n} = c > 0$, then $\sum a_n$ and $\sum b_n$ both converge or both diverge.

2. If $\lim\limits_{n \to \infty} \dfrac{a_n}{b_n} = 0$ and $\sum b_n$ converges, then $\sum a_n$ converges.

3. If $\lim\limits_{n \to \infty} \dfrac{a_n}{b_n} = \infty$ and $\sum b_n$ diverges, then $\sum a_n$ diverges.

Proof We will prove part (1). Parts (2) and (3) are left as Exercises 37 (a) and (b). Since $c/2 > 0$, there exists an integer N such that for all n

$$n > N \Rightarrow \left| \frac{a_n}{b_n} - c \right| < \frac{c}{2}.$$

Limit definition with $\epsilon = c/2$, $L = c$, and a_n replaced by a_n/b_n

Thus, for $n > N$,

$$-\frac{c}{2} < \frac{a_n}{b_n} - c < \frac{c}{2},$$

$$\frac{c}{2} < \frac{a_n}{b_n} < \frac{3c}{2},$$

$$\left(\frac{c}{2} \right) b_n < a_n < \left(\frac{3c}{2} \right) b_n.$$

If $\sum b_n$ converges, then $\sum (3c/2) b_n$ converges and $\sum a_n$ converges by the Direct Comparison Test. If $\sum b_n$ diverges, then $\sum (c/2) b_n$ diverges and $\sum a_n$ diverges by the Direct Comparison Test. ∎

EXAMPLE 3 Which of the following series converge, and which diverge?

a) $\dfrac{3}{4} + \dfrac{5}{9} + \dfrac{7}{16} + \dfrac{9}{25} + \cdots = \displaystyle\sum_{n=1}^{\infty} \frac{2n+1}{(n+1)^2} = \sum_{n=1}^{\infty} \frac{2n+1}{n^2 + 2n + 1}$

b) $\dfrac{1}{1} + \dfrac{1}{3} + \dfrac{1}{7} + \dfrac{1}{15} + \cdots = \displaystyle\sum_{n=1}^{\infty} \frac{1}{2^n - 1}$

c) $\dfrac{1 + 2 \ln 2}{9} + \dfrac{1 + 3 \ln 3}{14} + \dfrac{1 + 4 \ln 4}{21} + \cdots = \displaystyle\sum_{n=2}^{\infty} \frac{1 + n \ln n}{n^2 + 5}$

Solution

We could just as well have taken $b_n = 2/n$ but $1/n$ is simpler.

a) Let $a_n = (2n+1)/(n^2 + 2n + 1)$. For n large, we expect a_n to behave like $2n/n^2 = 2/n$, so we let $b_n = 1/n$. Since

$$\sum_{n=1}^{\infty} b_n = \sum_{n=1}^{\infty} \frac{1}{n} \quad \text{diverges}$$

and

$$\lim_{n \to \infty} \frac{a_n}{b_n} = \lim_{n \to \infty} \frac{2n^2 + n}{n^2 + 2n + 1} = 2,$$

$\sum a_n$ diverges by part 1 of the Limit Comparison Test.

b) Let $a_n = 1/(2^n - 1)$. For n large, we expect a_n to behave like $1/2^n$, so we let $b_n = 1/2^n$. Since

$$\sum_{n=1}^{\infty} b_n = \sum_{n=1}^{\infty} \frac{1}{2^n} \text{ converges}$$

and

$$\lim_{n \to \infty} \frac{a_n}{b_n} = \lim_{n \to \infty} \frac{2^n}{2^n - 1}$$

$$= \lim_{n \to \infty} \frac{1}{1 - (1/2^n)}$$

$$= 1,$$

$\sum a_n$ converges by part 1 of the Limit Comparison Test.

c) Let $a_n = (1 + n \ln n)/(n^2 + 5)$. For n large, we expect a_n to behave like $(n \ln n)/n^2 = (\ln n)/n$, which is greater than $1/n$ for $n \geq 3$, so we take $b_n = 1/n$. Since

$$\sum_{n=2}^{\infty} b_n = \sum_{n=2}^{\infty} \frac{1}{n} \text{ diverges}$$

and

$$\lim_{n \to \infty} \frac{a_n}{b_n} = \lim_{n \to \infty} \frac{n + n^2 \ln n}{n^2 + 5}$$

$$= \infty,$$

$\sum a_n$ diverges by part 3 of the Limit Comparison Test. ❑

EXAMPLE 4   Does $\displaystyle\sum_{n=1}^{\infty} \frac{\ln n}{n^{3/2}}$ converge?

Solution   Because $\ln n$ grows more slowly than n^c for any positive constant c (Section 8.2, Exercise 69), we would expect to have

$$\frac{\ln n}{n^{3/2}} < \frac{n^{1/4}}{n^{3/2}} = \frac{1}{n^{5/4}}$$

for n sufficiently large. Indeed, taking $a_n = (\ln n)/n^{3/2}$ and $b_n = 1/n^{5/4}$, we have

$$\lim_{n \to \infty} \frac{a_n}{b_n} = \lim_{n \to \infty} \frac{\ln n}{n^{1/4}}$$

$$= \lim_{n \to \infty} \frac{1/n}{(1/4)\, n^{-3/4}} \qquad \text{l'Hôpital's rule}$$

$$= \lim_{n \to \infty} \frac{4}{n^{1/4}} = 0.$$

Since $\sum b_n = \sum (1/n^{5/4})$ (a p-series with $p > 1$) converges, $\sum a_n$ converges by part 2 of the Limit Comparison Test. ❑

Exercises 8.5

Determining Convergence or Divergence

Which of the series in Exercises 1–36 converge, and which diverge? Give reasons for your answers.

1. $\displaystyle\sum_{n=1}^{\infty} \frac{1}{2\sqrt{n} + \sqrt[3]{n}}$

2. $\displaystyle\sum_{n=1}^{\infty} \frac{3}{n + \sqrt{n}}$

3. $\displaystyle\sum_{n=1}^{\infty} \frac{\sin^2 n}{2^n}$

4. $\displaystyle\sum_{n=1}^{\infty} \frac{1 + \cos n}{n^2}$

5. $\displaystyle\sum_{n=1}^{\infty} \frac{2n}{3n - 1}$

6. $\displaystyle\sum_{n=1}^{\infty} \frac{n + 1}{n^2\sqrt{n}}$

7. $\displaystyle\sum_{n=1}^{\infty} \left(\frac{n}{3n + 1}\right)^n$

8. $\displaystyle\sum_{n=1}^{\infty} \frac{1}{\sqrt{n^3 + 2}}$

9. $\displaystyle\sum_{n=3}^{\infty} \frac{1}{\ln(\ln n)}$

10. $\displaystyle\sum_{n=2}^{\infty} \frac{1}{(\ln n)^2}$

11. $\displaystyle\sum_{n=1}^{\infty} \frac{(\ln n)^2}{n^3}$

12. $\displaystyle\sum_{n=1}^{\infty} \frac{(\ln n)^3}{n^3}$

13. $\displaystyle\sum_{n=2}^{\infty} \frac{1}{\sqrt{n}\ln n}$

14. $\displaystyle\sum_{n=1}^{\infty} \frac{(\ln n)^2}{n^{3/2}}$

15. $\displaystyle\sum_{n=1}^{\infty} \frac{1}{1 + \ln n}$

16. $\displaystyle\sum_{n=1}^{\infty} \frac{1}{(1 + \ln n)^2}$

17. $\displaystyle\sum_{n=2}^{\infty} \frac{\ln(n + 1)}{n + 1}$

18. $\displaystyle\sum_{n=1}^{\infty} \frac{1}{(1 + \ln^2 n)}$

19. $\displaystyle\sum_{n=2}^{\infty} \frac{1}{n\sqrt{n^2 - 1}}$

20. $\displaystyle\sum_{n=1}^{\infty} \frac{\sqrt{n}}{n^2 + 1}$

21. $\displaystyle\sum_{n=1}^{\infty} \frac{1 - n}{n2^n}$

22. $\displaystyle\sum_{n=1}^{\infty} \frac{n + 2^n}{n^2 2^n}$

23. $\displaystyle\sum_{n=1}^{\infty} \frac{1}{3^{n-1} + 1}$

24. $\displaystyle\sum_{n=1}^{\infty} \frac{3^{n-1} + 1}{3^n}$

25. $\displaystyle\sum_{n=1}^{\infty} \sin\frac{1}{n}$

26. $\displaystyle\sum_{n=1}^{\infty} \tan\frac{1}{n}$

27. $\displaystyle\sum_{n=1}^{\infty} \frac{10n + 1}{n(n + 1)(n + 2)}$

28. $\displaystyle\sum_{n=3}^{\infty} \frac{5n^3 - 3n}{n^2(n - 2)(n^2 + 5)}$

29. $\displaystyle\sum_{n=1}^{\infty} \frac{\tan^{-1} n}{n^{1.1}}$

30. $\displaystyle\sum_{n=1}^{\infty} \frac{\sec^{-1} n}{n^{1.3}}$

31. $\displaystyle\sum_{n=1}^{\infty} \frac{\coth n}{n^2}$

32. $\displaystyle\sum_{n=1}^{\infty} \frac{\tanh n}{n^2}$

33. $\displaystyle\sum_{n=1}^{\infty} \frac{1}{n\sqrt[n]{n}}$

34. $\displaystyle\sum_{n=1}^{\infty} \frac{\sqrt[n]{n}}{n^2}$

35. $\displaystyle\sum_{n=1}^{\infty} \frac{1}{1 + 2 + 3 + \cdots + n}$

36. $\displaystyle\sum_{n=1}^{\infty} \frac{1}{1 + 2^2 + 3^2 + \cdots + n^2}$

Theory and Examples

37. Prove (a) Part 2 and (b) Part 3 of the Limit Comparison Test.

38. If $\sum_{n=1}^{\infty} a_n$ is a convergent series of nonnegative numbers, can anything be said about $\sum_{n=1}^{\infty} (a_n/n)$? Explain.

39. Suppose that $a_n > 0$ and $b_n > 0$ for $n \geq N$ (N an integer). If $\lim_{n\to\infty} (a_n/b_n) = \infty$ and $\sum a_n$ converges, can anything be said about $\sum b_n$? Give reasons for your answer.

40. Prove that if $\sum a_n$ is a convergent series of nonnegative terms, then $\sum a_n^2$ converges.

✪ CAS Exploration and Project

41. It is not yet known whether the series

$$\sum_{n=1}^{\infty} \frac{1}{n^3 \sin^2 n}$$

converges or diverges. Use a CAS to explore the behavior of the series by performing the following steps.

a) Define the sequence of partial sums

$$s_k = \sum_{n=1}^{k} \frac{1}{n^3 \sin^2 n}.$$

What happens when you try to find the limit of s_k as $k \to \infty$? Does your CAS find a closed form answer for this limit?

b) Plot the first 100 points (k, s_k) for the sequence of partial sums. Do they appear to converge? What would you estimate the limit to be?

c) Next plot the first 200 points (k, s_k). Discuss the behavior in your own words.

d) Plot the first 400 points (k, s_k). What happens when $k = 355$? Calculate the number 355/113. Explain from your calculation what happened at $k = 355$. For what values of k would you guess this behavior might occur again?

You will find an interesting discussion of this series in Chapter 72 of *Mazes for the Mind* by Clifford A. Pickover, St. Martin's Press, Inc., New York, 1992.

8.6 | The Ratio and Root Tests for Series of Nonnegative Terms

Convergence tests that depend on comparing series with integrals or other series are called *extrinsic* tests. They are useful, but there are reasons to look for tests that do not require comparison. As a practical matter, we may not be able to find the series or functions we need to make a comparison work. And, in principle, all the information about a given series should be contained in its own terms. We therefore turn our attention to *intrinsic* tests—tests that depend only on the series at hand.

The Ratio Test

The first intrinsic test, the Ratio Test, measures the rate of growth (or decline) of a series by examining the ratio a_{n+1}/a_n. For a geometric series $\sum ar^n$, this rate is a constant $((ar^{n+1})/(ar^n) = r)$, and the series converges if and only if its ratio is less than 1 in absolute value. But even if the ratio is not constant, we may be able to find a geometric series for comparison, as in Example 1.

The series in Example 1 converges rapidly, as the following computer data suggest.

n	s_n
5	1.5492 06349
10	1.5702 89085
15	1.5707 83080
20	1.5707 95964
25	1.5707 96317
30	1.5707 96327
35	1.5707 96327

EXAMPLE 1   Let $a_1 = 1$ and let $a_{n+1} = \dfrac{n}{2n+1} a_n$ for all n. Does the series $\sum a_n$ converge?

Solution   We begin by writing a few terms of the series:

$$a_1 = 1, \qquad a_2 = \frac{1}{3}a_1 = \frac{1}{3}, \qquad a_3 = \frac{2}{5}a_2 = \frac{1 \cdot 2}{3 \cdot 5}, \qquad a_4 = \frac{3}{7}a_3 = \frac{1 \cdot 2 \cdot 3}{3 \cdot 5 \cdot 7}.$$

Each term is somewhat less than 1/2 the term before it, because $n/(2n+1)$ is less than 1/2. Therefore the terms of the series are less than or equal to the terms of the geometric series

$$1 + \left(\frac{1}{2}\right) + \left(\frac{1}{2}\right)^2 + \cdots + \left(\frac{1}{2}\right)^{n-1} + \cdots,$$

which converges to 2. So our series also converges, and its sum is less than 2. The table in the margin shows how quickly the series converges to its known limit, $\pi/2$.   ◻

The Ratio Test

Let $\sum a_n$ be a series with positive terms, and suppose that

$$\lim_{n \to \infty} \frac{a_{n+1}}{a_n} = \rho.$$

Then

a)   the series *converges* if $\rho < 1$,
b)   the series *diverges* if $\rho > 1$ or ρ is infinite,
c)   the test is *inconclusive* if $\rho = 1$.

In proving the Ratio Test, we will make a comparison with an appropriate geometric series as in Example 1, but when we *apply* the test there is no need for comparison.

Proof
a) $\rho < 1$. Let r be a number between ρ and 1. Then the number $\epsilon = r - \rho$ is positive. Since

$$\frac{a_{n+1}}{a_n} \to \rho,$$

a_{n+1}/a_n must lie within ϵ of ρ when n is large enough, say for all $n \geq N$. In particular,

$$\frac{a_{n+1}}{a_n} < \rho + \epsilon = r, \qquad \text{when } n \geq N.$$

That is,

$$a_{N+1} < ra_N,$$

$$a_{N+2} < ra_{N+1} < r^2 a_N,$$

$$a_{N+3} < ra_{N+2} < r^3 a_N,$$

$$\vdots$$

$$a_{N+m} < ra_{N+m-1} < r^m a_N.$$

These inequalities show that the terms of our series, after the Nth term, approach zero more rapidly than the terms in a geometric series with ratio $r < 1$. More precisely, consider the series $\sum c_n$, where $c_n = a_n$ for $n = 1, 2, \ldots, N$ and $c_{N+1} = ra_N, c_{N+2} = r^2 a_N, \ldots, c_{N+m} = r^m a_N, \ldots$. Now $a_n \le c_n$ for all n, and

$$\sum_{n=1}^{\infty} c_n = a_1 + a_2 + \cdots + a_{N-1} + a_N + ra_N + r^2 a_N + \cdots$$

$$= a_1 + a_2 + \cdots + a_{N-1} + a_N (1 + r + r^2 + \cdots).$$

The geometric series $1 + r + r^2 + \cdots$ converges because $|r| < 1$, so $\sum c_n$ converges. Since $a_n \le c_n$, $\sum a_n$ also converges.

b) **$1 < \rho \le \infty$.** From some index M on,

$$\frac{a_{n+1}}{a_n} > 1 \qquad \text{and} \qquad a_M < a_{M+1} < a_{M+2} < \cdots.$$

The terms of the series do not approach zero as n becomes infinite, and the series diverges by the nth-Term Test.

c) **$\rho = 1$.** The two series

$$\sum_{n=1}^{\infty} \frac{1}{n} \qquad \text{and} \qquad \sum_{n=1}^{\infty} \frac{1}{n^2}$$

show that some other test for convergence must be used when $\rho = 1$.

$$\text{For } \sum_{n=1}^{\infty} \frac{1}{n}: \qquad \frac{a_{n+1}}{a_n} = \frac{1/(n+1)}{1/n} = \frac{n}{n+1} \to 1.$$

$$\text{For } \sum_{n=1}^{\infty} \frac{1}{n^2}: \qquad \frac{a_{n+1}}{a_n} = \frac{1/(n+1)^2}{1/n^2} = \left(\frac{n}{n+1}\right)^2 \to 1^2 = 1.$$

In both cases $\rho = 1$, yet the first series diverges while the second converges.   ❏

The Ratio Test is often effective when the terms of a series contain factorials of expressions involving n or expressions raised to the nth power.

EXAMPLE 2   Investigate the convergence of the following series.

a) $\displaystyle\sum_{n=0}^{\infty} \frac{2^n + 5}{3^n}$

b) $\displaystyle\sum_{n=1}^{\infty} \frac{(2n)!}{n!n!}$

c) $\displaystyle\sum_{n=1}^{\infty} \frac{4^n n! n!}{(2n)!}$

Solution

a) For the series $\sum_{n=0}^{\infty} (2^n + 5)/3^n$,

$$\frac{a_{n+1}}{a_n} = \frac{(2^{n+1} + 5)/3^{n+1}}{(2^n + 5)/3^n} = \frac{1}{3} \cdot \frac{2^{n+1} + 5}{2^n + 5} = \frac{1}{3} \cdot \left(\frac{2 + 5 \cdot 2^{-n}}{1 + 5 \cdot 2^{-n}}\right) \to \frac{1}{3} \cdot \frac{2}{1} = \frac{2}{3}.$$

The series converges because $\rho = 2/3$ is less than 1.

This does *not* mean that 2/3 is the sum of the series. In fact,

$$\sum_{n=0}^{\infty} \frac{2^n + 5}{3^n} = \sum_{n=0}^{\infty} \left(\frac{2}{3}\right)^n + \sum_{n=0}^{\infty} \frac{5}{3^n} = \frac{1}{1 - (2/3)} + \frac{5}{1 - (1/3)} = \frac{21}{2}.$$

b) If $a_n = \dfrac{(2n)!}{n!n!}$, then $a_{n+1} = \dfrac{(2n+2)!}{(n+1)!(n+1)!}$ and

$$\frac{a_{n+1}}{a_n} = \frac{n!n!(2n+2)(2n+1)(2n)!}{(n+1)!(n+1)!(2n)!}$$

$$= \frac{(2n+2)(2n+1)}{(n+1)(n+1)} = \frac{4n+2}{n+1} \rightarrow 4.$$

The series diverges because $\rho = 4$ is greater than 1.

c) If $a_n = 4^n n!n!/(2n)!$, then

$$\frac{a_{n+1}}{a_n} = \frac{4^{n+1}(n+1)!(n+1)!}{(2n+2)(2n+1)(2n)!} \cdot \frac{(2n)!}{4^n n!n!}$$

$$= \frac{4(n+1)(n+1)}{(2n+2)(2n+1)} = \frac{2(n+1)}{2n+1} \rightarrow 1.$$

Because the limit is $\rho = 1$, we cannot decide from the Ratio Test whether the series converges. However, when we notice that $a_{n+1}/a_n = (2n+2)/(2n+1)$, we conclude that a_{n+1} is always greater than a_n because $(2n+2)/(2n+1)$ is always greater than 1. Therefore, all terms are greater than or equal to $a_1 = 2$, and the nth term does not approach zero as $n \rightarrow \infty$. The series diverges. ❑

The *n*th-Root Test

The convergence tests we have so far for $\sum a_n$ work best when the formula for a_n is relatively simple. But consider the following.

EXAMPLE 3   Let $a_n = \begin{cases} n/2^n, & n \text{ odd} \\ 1/2^n, & n \text{ even.} \end{cases}$ Does $\sum a_n$ converge?

Solution   We write out several terms of the series:

$$\sum_{n=1}^{\infty} a_n = \frac{1}{2^1} + \frac{1}{2^2} + \frac{3}{2^3} + \frac{1}{2^4} + \frac{5}{2^5} + \frac{1}{2^6} + \frac{7}{2^7} + \cdots$$

$$= \frac{1}{2} + \frac{1}{4} + \frac{3}{8} + \frac{1}{16} + \frac{5}{32} + \frac{1}{64} + \frac{7}{128} + \cdots.$$

Clearly, this is not a geometric series. The nth term approaches zero as $n \rightarrow \infty$, so we do not know if the series diverges. The Integral Test does not look promising. The Ratio Test produces

$$\frac{a_{n+1}}{a_n} = \begin{cases} \dfrac{1}{2n}, & n \text{ odd} \\ \dfrac{n+1}{2}, & n \text{ even.} \end{cases}$$

As $n \rightarrow \infty$, the ratio is alternately small and large and has no limit.
    A test that will answer the question (the series converges) is the nth-Root Test.

❑

The nth-Root Test

Let $\sum a_n$ be a series with $a_n \geq 0$ for $n \geq N$, and suppose that

$$\lim_{n\to\infty} \sqrt[n]{a_n} = \rho.$$

Then

a) the series *converges* if $\rho < 1$,

b) the series *diverges* if $\rho > 1$ or ρ is infinite,

c) the test is *inconclusive* if $\rho = 1$.

Proof

a) $\rho < 1.$ Choose an $\epsilon > 0$ so small that $\rho + \epsilon < 1$. Since $\sqrt[n]{a_n} \to \rho$, the terms $\sqrt[n]{a_n}$ eventually get closer than ϵ to ρ. In other words, there exists an index $M \geq N$ such that

$$\sqrt[n]{a_n} < \rho + \epsilon \qquad \text{when } n \geq M.$$

Then it is also true that

$$a_n < (\rho + \epsilon)^n \qquad \text{for } n \geq M.$$

Now, $\sum_{n=M}^{\infty} (\rho + \epsilon)^n$, a geometric series with ratio $(\rho + \epsilon) < 1$, converges. By comparison, $\sum_{n=M}^{\infty} a_n$ converges, from which it follows that

$$\sum_{n=1}^{\infty} a_n = a_1 + \cdots + a_{M-1} + \sum_{n=M}^{\infty} a_n$$

converges.

b) $1 < \rho \leq \infty.$ For all indices beyond some integer M, we have $\sqrt[n]{a_n} > 1$, so that $a_n > 1$ for $n > M$. The terms of the series do not converge to zero. The series diverges by the nth-Term Test.

c) $\rho = 1.$ The series $\sum_{n=1}^{\infty}(1/n)$ and $\sum_{n=1}^{\infty}(1/n^2)$ show that the test is not conclusive when $\rho = 1$. The first series diverges and the second converges, but in both cases $\sqrt[n]{a_n} \to 1$. ❑

EXAMPLE 3 (continued)   Let $a_n = \begin{cases} n/2^n, & n \text{ odd} \\ 1/2^n, & n \text{ even.} \end{cases}$ Does $\sum a_n$ converge?

Solution   We apply the nth-Root Test, finding that

$$\sqrt[n]{a_n} = \begin{cases} \sqrt[n]{n}/2, & n \text{ odd} \\ 1/2, & n \text{ even.} \end{cases}$$

Therefore,

$$\frac{1}{2} \leq \sqrt[n]{a_n} \leq \frac{\sqrt[n]{n}}{2}.$$

Since $\sqrt[n]{n} \to 1$ (Section 8.2, Table 8.1), we have $\lim_{n\to\infty} \sqrt[n]{a_n} = 1/2$ by the Sandwich Theorem. The limit is less than 1, so the series converges by the nth-Root Test. ❑

EXAMPLE 4 Which of the following series converges, and which diverges?

a) $\sum_{n=1}^{\infty} \frac{n^2}{2^n}$ b) $\sum_{n=1}^{\infty} \frac{2^n}{n^2}$

Solution

a) $\sum_{n=1}^{\infty} \frac{n^2}{2^n}$ converges because $\sqrt[n]{\frac{n^2}{2^n}} = \frac{\sqrt[n]{n^2}}{\sqrt[n]{2^n}} = \frac{\left(\sqrt[n]{n}\right)^2}{2} \to \frac{1}{2} < 1.$

b) $\sum_{n=1}^{\infty} \frac{2^n}{n^2}$ diverges because $\sqrt[n]{\frac{2^n}{n^2}} = \frac{2}{\left(\sqrt[n]{n}\right)^2} \to \frac{2}{1} > 1.$ ☐

Exercises 8.6

Determining Convergence or Divergence

Which of the series in Exercises 1–26 converge, and which diverge? Give reasons for your answers. (When checking your answers, remember there may be more than one way to determine a series' convergence or divergence.)

1. $\sum_{n=1}^{\infty} \frac{n^{\sqrt{2}}}{2^n}$

2. $\sum_{n=1}^{\infty} n^2 e^{-n}$

3. $\sum_{n=1}^{\infty} n! \, e^{-n}$

4. $\sum_{n=1}^{\infty} \frac{n!}{10^n}$

5. $\sum_{n=1}^{\infty} \frac{n^{10}}{10^n}$

6. $\sum_{n=1}^{\infty} \left(\frac{n-2}{n}\right)^n$

7. $\sum_{n=1}^{\infty} \frac{2 + (-1)^n}{1.25^n}$

8. $\sum_{n=1}^{\infty} \frac{(-2)^n}{3^n}$

9. $\sum_{n=1}^{\infty} \left(1 - \frac{3}{n}\right)^n$

10. $\sum_{n=1}^{\infty} \left(1 - \frac{1}{3n}\right)^n$

11. $\sum_{n=1}^{\infty} \frac{\ln n}{n^3}$

12. $\sum_{n=1}^{\infty} \frac{(\ln n)^n}{n^n}$

13. $\sum_{n=1}^{\infty} \left(\frac{1}{n} - \frac{1}{n^2}\right)$

14. $\sum_{n=1}^{\infty} \left(\frac{1}{n} - \frac{1}{n^2}\right)^n$

15. $\sum_{n=1}^{\infty} \frac{\ln n}{n}$

16. $\sum_{n=1}^{\infty} \frac{n \ln n}{2^n}$

17. $\sum_{n=1}^{\infty} \frac{(n+1)(n+2)}{n!}$

18. $\sum_{n=1}^{\infty} e^{-n}(n^3)$

19. $\sum_{n=1}^{\infty} \frac{(n+3)!}{3! n! 3^n}$

20. $\sum_{n=1}^{\infty} \frac{n 2^n (n+1)!}{3^n n!}$

21. $\sum_{n=1}^{\infty} \frac{n!}{(2n+1)!}$

22. $\sum_{n=1}^{\infty} \frac{n!}{n^n}$

23. $\sum_{n=2}^{\infty} \frac{n}{(\ln n)^n}$

24. $\sum_{n=2}^{\infty} \frac{n}{(\ln n)^{(n/2)}}$

25. $\sum_{n=1}^{\infty} \frac{n! \ln n}{n(n+2)!}$

26. $\sum_{n=1}^{\infty} \frac{3^n}{n^3 2^n}$

Which of the series $\sum_{n=1}^{\infty} a_n$ defined by the formulas in Exercises 27–38 converge, and which diverge? Give reasons for your answers.

27. $a_1 = 2, \quad a_{n+1} = \frac{1 + \sin n}{n} a_n$

28. $a_1 = 1, \quad a_{n+1} = \frac{1 + \tan^{-1} n}{n} a_n$

29. $a_1 = \frac{1}{3}, \quad a_{n+1} = \frac{3n - 1}{2n + 5} a_n$

30. $a_1 = 3, \quad a_{n+1} = \frac{n}{n + 1} a_n$

31. $a_1 = 2, \quad a_{n+1} = \frac{2}{n} a_n$

32. $a_1 = 5, \quad a_{n+1} = \frac{\sqrt[n]{n}}{2} a_n$

33. $a_1 = 1, \quad a_{n+1} = \frac{1 + \ln n}{n} a_n$

34. $a_1 = \frac{1}{2}, \quad a_{n+1} = \frac{n + \ln n}{n + 10} a_n$

35. $a_1 = \frac{1}{3}, \quad a_{n+1} = \sqrt[n]{a_n}$

36. $a_1 = \frac{1}{2}, \quad a_{n+1} = (a_n)^{n+1}$

37. $a_n = \frac{2^n n! n!}{(2n)!}$

38. $a_n = \frac{(3n)!}{n!(n+1)!(n+2)!}$

Which of the series in Exercises 39–44 converge, and which diverge? Give reasons for your answers.

39. $\displaystyle\sum_{n=1}^{\infty} \frac{(n!)^n}{(n^n)^2}$

40. $\displaystyle\sum_{n=1}^{\infty} \frac{(n!)^n}{n^{(n^2)}}$

41. $\displaystyle\sum_{n=1}^{\infty} \frac{n^n}{2^{(n^2)}}$

42. $\displaystyle\sum_{n=1}^{\infty} \frac{n^n}{(2^n)^2}$

43. $\displaystyle\sum_{n=1}^{\infty} \frac{1 \cdot 3 \cdot \cdots \cdot (2n-1)}{4^n 2^n n!}$

44. $\displaystyle\sum_{n=1}^{\infty} \frac{1 \cdot 3 \cdot \cdots \cdot (2n-1)}{[2 \cdot 4 \cdot \cdots \cdot (2n)](3^n + 1)}$

Theory and Examples

45. Neither the Ratio nor the nth-Root Test helps with p-series. Try them on

$$\sum_{n=1}^{\infty} \frac{1}{n^p}$$

and show that both tests fail to provide information about convergence.

46. Show that neither the Ratio Test nor the nth-Root Test provides information about the convergence of

$$\sum_{n=2}^{\infty} \frac{1}{(\ln n)^p} \qquad (p \text{ constant}).$$

47. Let $a_n = \begin{cases} n/2^n & \text{if } n \text{ is a prime number} \\ 1/2^n & \text{otherwise.} \end{cases}$

Does $\sum a_n$ converge? Give reasons for your answer.

8.7	# Alternating Series, Absolute and Conditional Convergence

A series in which the terms are alternately positive and negative is an **alternating series.**

Here are three examples:

$$1 - \frac{1}{2} + \frac{1}{3} - \frac{1}{4} + \frac{1}{5} - \cdots + \frac{(-1)^{n+1}}{n} + \cdots \tag{1}$$

$$-2 + 1 - \frac{1}{2} + \frac{1}{4} - \frac{1}{8} + \cdots + \frac{(-1)^n 4}{2^n} + \cdots \tag{2}$$

$$1 - 2 + 3 - 4 + 5 - 6 + \cdots + (-1)^{n+1}n + \cdots \tag{3}$$

Series (1), called the **alternating harmonic series,** converges, as we will see in a moment. Series (2), a geometric series with ratio $r = -1/2$, converges to $-2/[1 + (1/2)] = -4/3$. Series (3) diverges because the nth term does not approach zero.

We prove the convergence of the alternating harmonic series by applying the Alternating Series Test.

Theorem 8
The Alternating Series Test (Leibniz's Theorem)
The series

$$\sum_{n=1}^{\infty} (-1)^{n+1} u_n = u_1 - u_2 + u_3 - u_4 + \cdots$$

converges if all three of the following conditions are satisfied:

1. The u_n's are all positive.
2. $u_n \geq u_{n+1}$ for all $n \geq N$, for some integer N,
3. $u_n \to 0$.

Proof If n is an even integer, say $n = 2m$, then the sum of the first n terms is

$$s_{2m} = (u_1 - u_2) + (u_3 - u_4) + \cdots + (u_{2m-1} - u_{2m})$$

$$= u_1 - (u_2 - u_3) - (u_4 - u_5) - \cdots - (u_{2m-2} - u_{2m-1}) - u_{2m}.$$

The first equality shows that s_{2m} is the sum of m nonnegative terms, since each term in parentheses is positive or zero. Hence $s_{2m+2} \geq s_{2m}$, and the sequence $\{s_{2m}\}$ is nondecreasing. The second equality shows that $s_{2m} \leq u_1$. Since $\{s_{2m}\}$ is nondecreasing and bounded from above, it has a limit, say

$$\lim_{m \to \infty} s_{2m} = L. \tag{4}$$

If n is an odd integer, say $n = 2m + 1$, then the sum of the first n terms is $s_{2m+1} = s_{2m} + u_{2m+1}$. Since $u_n \to 0$,

$$\lim_{m \to \infty} u_{2m+1} = 0$$

and, as $m \to \infty$,

$$s_{2m+1} = s_{2m} + u_{2m+1} \to L + 0 = L. \tag{5}$$

Combining the results of (4) and (5) gives $\lim_{n \to \infty} s_n = L$ (Section 8.1, Exercise 53). ❏

EXAMPLE 1   The alternating harmonic series

$$\sum_{n=1}^{\infty} (-1)^{n+1} \frac{1}{n} = 1 - \frac{1}{2} + \frac{1}{3} - \frac{1}{4} + \cdots$$

satisfies the three requirements of Theorem 8 with $N = 1$; it therefore converges. ❏

A graphical interpretation of the partial sums (Fig. 8.14) shows how an alternating series converges to its limit L when the three conditions of Theorem 8 are satisfied with $N = 1$. (Exercise 63 asks you to picture the case $N > 1$.) Starting from the origin of the x-axis, we lay off the positive distance $s_1 = u_1$. To find the point corresponding to $s_2 = u_1 - u_2$, we back up a distance equal to u_2. Since $u_2 \leq u_1$, we do not back up any farther than the origin. We continue in this seesaw fashion, backing up or going forward as the signs in the series demand. But for $n \geq N$, each forward or backward step is shorter than (or at most the same size as) the preceding step, because $u_{n+1} \leq u_n$. And since the nth term approaches zero as n increases, the size of step we take forward or backward gets smaller and smaller. We oscillate across the limit L, and the amplitude of oscillation approaches zero. The limit L lies between any two successive sums s_n and s_{n+1} and hence differs from s_n by an amount less than u_{n+1}.

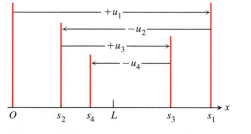

8.14 The partial sums of an alternating series that satisfies the hypotheses of Theorem 8 for $N = 1$ straddle the limit from the beginning.

Because

$$|L - s_n| < u_{n+1} \qquad \text{for } n \geq N,$$

we can make useful estimates of the sums of convergent alternating series.

Theorem 9

The Alternating Series Estimation Theorem

If the alternating series $\sum_{n=1}^{\infty} (-1)^{n+1} u_n$ satisfies the three conditions of Theorem 8, then for $n \geq N$,

$$s_n = u_1 - u_2 + \cdots + (-1)^{n+1} u_n$$

approximates the sum L of the series with an error whose absolute value is less than u_{n+1}, the numerical value of the first unused term. Furthermore, the remainder, $L - s_n$, has the same sign as the first unused term.

We leave the verification of the sign of the remainder for Exercise 53.

EXAMPLE 2   We try Theorem 9 on a series whose sum we know:

$$\sum_{n=0}^{\infty} (-1)^n \frac{1}{2^n} = 1 - \frac{1}{2} + \frac{1}{4} - \frac{1}{8} + \frac{1}{16} - \frac{1}{32} + \frac{1}{64} - \frac{1}{128} + \frac{1}{256} - \cdots.$$

The theorem says that if we truncate the series after the eighth term, we throw away a total that is positive and less than 1/256. The sum of the first eight terms is 0.6640 625. The sum of the series is

$$\frac{1}{1 - (-1/2)} = \frac{1}{3/2} = \frac{2}{3}.$$

The difference, $(2/3) - 0.6640\ 625 = 0.0026\ 04166\ 6\ldots$, is positive and less than $(1/256) = 0.0039\ 0625$.   ☐

Absolute Convergence

Definition

A series $\sum a_n$ **converges absolutely** (is **absolutely convergent**) if the corresponding series of absolute values, $\sum |a_n|$, converges.

The geometric series

$$1 - \frac{1}{2} + \frac{1}{4} - \frac{1}{8} + \cdots$$

converges absolutely because the corresponding series of absolute values

$$1 + \frac{1}{2} + \frac{1}{4} + \frac{1}{8} + \cdots$$

converges. The alternating harmonic series does not converge absolutely. The corresponding series of absolute values is the (divergent) harmonic series.

> ### Definition
> A series that converges but does not converge absolutely **converges conditionally.**

The alternating harmonic series converges conditionally.

Absolute convergence is important for two reasons. First, we have good tests for convergence of series of positive terms. Second, if a series converges absolutely, then it converges. That is the thrust of the next theorem.

Caution

We can rephrase Theorem 10 to say that *every absolutely convergent series converges.* However, the converse statement is false: Many convergent series do not converge absolutely.

> ### Theorem 10
> **The Absolute Convergence Test**
> If $\sum_{n=1}^{\infty} |a_n|$ converges, then $\sum_{n=1}^{\infty} a_n$ converges.

Proof For each n,

$$-|a_n| \leq a_n \leq |a_n|, \qquad \text{so} \qquad 0 \leq a_n + |a_n| \leq 2|a_n|.$$

If $\sum_{n=1}^{\infty} |a_n|$ converges, then $\sum_{n=1}^{\infty} 2|a_n|$ converges and, by the Direct Comparison Test, the nonnegative series $\sum_{n=1}^{\infty}(a_n + |a_n|)$ converges. The equality $a_n = (a_n + |a_n|) - |a_n|$ now lets us express $\sum_{n=1}^{\infty} a_n$ as the difference of two convergent series:

$$\sum_{n=1}^{\infty} a_n = \sum_{n=1}^{\infty}(a_n + |a_n| - |a_n|) = \sum_{n=1}^{\infty}(a_n + |a_n|) - \sum_{n=1}^{\infty} |a_n|.$$

Therefore, $\sum_{n=1}^{\infty} a_n$ converges. ❑

EXAMPLE 3 For $\sum_{n=1}^{\infty}(-1)^{n+1}\dfrac{1}{n^2} = 1 - \dfrac{1}{4} + \dfrac{1}{9} - \dfrac{1}{16} + \cdots$, the corresponding series of absolute values is the convergent series

$$\sum_{n=1}^{\infty} \frac{1}{n^2} = 1 + \frac{1}{4} + \frac{1}{9} + \frac{1}{16} + \cdots.$$

The original series converges because it converges absolutely. ❑

EXAMPLE 4 For $\sum_{n=1}^{\infty} \dfrac{\sin n}{n^2} = \dfrac{\sin 1}{1} + \dfrac{\sin 2}{4} + \dfrac{\sin 3}{9} + \cdots$, the corresponding series of absolute values is

$$\sum_{n=1}^{\infty} \left| \frac{\sin n}{n^2} \right| = \frac{|\sin 1|}{1} + \frac{|\sin 2|}{4} + \cdots,$$

which converges by comparison with $\sum_{n=1}^{\infty} (1/n^2)$ because $|\sin n| \leq 1$ for every n. The original series converges absolutely; therefore it converges. ❑

EXAMPLE 5 *Alternating p-series*

If p is a positive constant, the sequence $\{1/n^p\}$ is a decreasing sequence with limit zero. Therefore the alternating p-series

$$\sum_{n=1}^{\infty} \frac{(-1)^{n-1}}{n^p} = 1 - \frac{1}{2^p} + \frac{1}{3^p} - \frac{1}{4^p} + \cdots, \qquad p > 0$$

converges.

If $p > 1$, the series converges absolutely. If $0 < p \leq 1$, the series converges conditionally.

Conditional convergence: $1 - \dfrac{1}{\sqrt{2}} + \dfrac{1}{\sqrt{3}} - \dfrac{1}{\sqrt{4}} + \cdots$

Absolute convergence: $1 - \dfrac{1}{2^{3/2}} + \dfrac{1}{3^{3/2}} - \dfrac{1}{4^{3/2}} + \cdots$

Rearranging Series

> **Theorem 11**
>
> **The Rearrangement Theorem for Absolutely Convergent Series**
>
> If $\sum_{n=1}^{\infty} a_n$ converges absolutely, and $b_1, b_2, \ldots, b_n, \ldots$ is any arrangement of the sequence $\{a_n\}$, then $\sum b_n$ converges absolutely and
>
> $$\sum_{n=1}^{\infty} b_n = \sum_{n=1}^{\infty} a_n.$$

(For an outline of the proof, see Exercise 60.)

EXAMPLE 6   As we saw in Example 3, the series

$$1 - \frac{1}{4} + \frac{1}{9} - \frac{1}{16} + \cdots + (-1)^{n-1}\frac{1}{n^2} + \cdots$$

converges absolutely. A possible rearrangement of the terms of the series might start with a positive term, then two negative terms, then three positive terms, then four negative terms, and so on: After k terms of one sign, take $k + 1$ terms of the opposite sign. The first ten terms of such a series look like this:

$$1 - \frac{1}{4} - \frac{1}{16} + \frac{1}{9} + \frac{1}{25} + \frac{1}{49} - \frac{1}{36} - \frac{1}{64} - \frac{1}{100} - \frac{1}{144} + \cdots.$$

The Rearrangement Theorem says that both series converge to the same value. In this example, if we had the second series to begin with, we would probably be glad to exchange it for the first, if we knew that we could. We can do even better: The sum of either series is also equal to

$$\sum_{n=1}^{\infty} \frac{1}{(2n-1)^2} - \sum_{n=1}^{\infty} \frac{1}{(2n)^2}.$$

(See Exercise 61.)

Caution

If we rearrange infinitely many terms of a conditionally convergent series, we can get results that are far different from the sum of the original series.

The kind of behavior illustrated by this example is typical of what can happen with any conditionally convergent series. Moral: Add the terms of a conditionally convergent series in the order given.

EXAMPLE 7   *Rearranging the alternating harmonic series*

The alternating harmonic series

$$\frac{1}{1} - \frac{1}{2} + \frac{1}{3} - \frac{1}{4} + \frac{1}{5} - \frac{1}{6} + \frac{1}{7} - \frac{1}{8} + \frac{1}{9} - \frac{1}{10} + \frac{1}{11} - \cdots$$

can be rearranged to diverge or to reach any preassigned sum.

a) *Rearranging $\sum_{n=1}^{\infty}(-1)^{n+1}/n$ to diverge.* The series of terms $\sum[1/(2n-1)]$ diverges to $+\infty$ and the series of terms $\sum(-1/2n)$ diverges to $-\infty$. No matter how far out in the sequence of odd-numbered terms we begin, we can always add enough positive terms to get an arbitrarily large sum. Similarly, with the negative terms, no matter how far out we start, we can add enough consecutive even-numbered terms to get a negative sum of arbitrarily large absolute value. If we wished to do so, we could start adding odd-numbered terms until we had a sum greater than $+3$, say, and then follow that with enough consecutive negative terms to make the new total less than -4. We could then add enough positive terms to make the total greater than $+5$ and follow with consecutive unused negative terms to make a new total less than -6, and so on. In this way, we could make the swings arbitrarily large in either direction.

Flowchart 8.1 Procedure for Determining Convergence

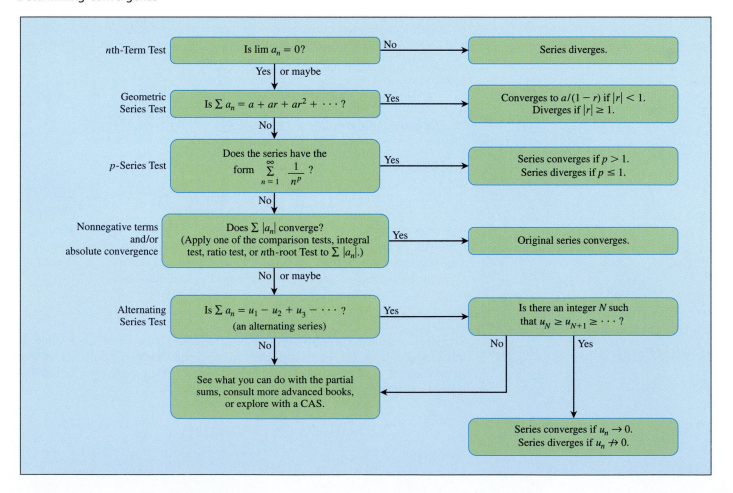

b) *Rearranging* $\sum_{n=1}^{\infty}(-1)^{n+1}/n$ *to converge to* 1. Another possibility is to focus on a particular limit. Suppose we try to get sums that converge to 1. We start with the first term, 1/1, and then subtract 1/2. Next we add 1/3 and 1/5, which brings the total back to 1 or above. Then we add consecutive negative terms until the total is less than 1. We continue in this manner: When the sum is less than 1, add positive terms until the total is 1 or more; then subtract (add negative) terms until the total is again less than 1. This process can be continued indefinitely. Because both the odd-numbered terms and the even-numbered terms of the original series approach zero as $n \to \infty$, the amount by which our partial sums exceed 1 or fall below it approaches zero. So the new series converges to 1. The rearranged series starts like this:

$$\frac{1}{1} - \frac{1}{2} + \frac{1}{3} + \frac{1}{5} - \frac{1}{4} + \frac{1}{7} + \frac{1}{9} - \frac{1}{6} + \frac{1}{11} + \frac{1}{13} - \frac{1}{8} + \frac{1}{15} + \frac{1}{17} - \frac{1}{10}$$

$$+ \frac{1}{19} + \frac{1}{21} - \frac{1}{12} + \frac{1}{23} + \frac{1}{25} - \frac{1}{14} + \frac{1}{27} - \frac{1}{16} + \cdots$$

Exercises 8.7

Determining Convergence or Divergence

Which of the alternating series in Exercises 1–10 converge, and which diverge? Give reasons for your answers.

1. $\sum_{n=1}^{\infty}(-1)^{n+1}\dfrac{1}{n^2}$

2. $\sum_{n=1}^{\infty}(-1)^{n+1}\dfrac{1}{n^{3/2}}$

3. $\sum_{n=1}^{\infty}(-1)^{n+1}\left(\dfrac{n}{10}\right)^n$

4. $\sum_{n=1}^{\infty}(-1)^{n+1}\dfrac{10^n}{n^{10}}$

5. $\sum_{n=2}^{\infty}(-1)^{n+1}\dfrac{1}{\ln n}$

6. $\sum_{n=1}^{\infty}(-1)^{n+1}\dfrac{\ln n}{n}$

7. $\sum_{n=2}^{\infty}(-1)^{n+1}\dfrac{\ln n}{\ln n^2}$

8. $\sum_{n=1}^{\infty}(-1)^{n}\ln\left(1+\dfrac{1}{n}\right)$

9. $\sum_{n=1}^{\infty}(-1)^{n+1}\dfrac{\sqrt{n}+1}{n+1}$

10. $\sum_{n=1}^{\infty}(-1)^{n+1}\dfrac{3\sqrt{n}+1}{\sqrt{n}+1}$

Absolute Convergence

Which of the series in Exercises 11–44 converge absolutely, which converge, and which diverge? Give reasons for your answers.

11. $\sum_{n=1}^{\infty}(-1)^{n+1}(0.1)^n$

12. $\sum_{n=1}^{\infty}(-1)^{n+1}\dfrac{(0.1)^n}{n}$

13. $\sum_{n=1}^{\infty}(-1)^{n}\dfrac{1}{\sqrt{n}}$

14. $\sum_{n=1}^{\infty}\dfrac{(-1)^n}{1+\sqrt{n}}$

15. $\sum_{n=1}^{\infty}(-1)^{n+1}\dfrac{n}{n^3+1}$

16. $\sum_{n=1}^{\infty}(-1)^{n+1}\dfrac{n!}{2^n}$

17. $\sum_{n=1}^{\infty}(-1)^{n}\dfrac{1}{n+3}$

18. $\sum_{n=1}^{\infty}(-1)^{n}\dfrac{\sin n}{n^2}$

19. $\sum_{n=1}^{\infty}(-1)^{n+1}\dfrac{3+n}{5+n}$

20. $\sum_{n=2}^{\infty}(-1)^{n}\dfrac{1}{\ln(n^3)}$

21. $\sum_{n=1}^{\infty}(-1)^{n+1}\dfrac{1+n}{n^2}$

22. $\sum_{n=1}^{\infty}\dfrac{(-2)^{n+1}}{n+5^n}$

23. $\sum_{n=1}^{\infty}(-1)^{n}n^2(2/3)^n$

24. $\sum_{n=1}^{\infty}(-1)^{n+1}(\sqrt[n]{10})$

25. $\sum_{n=1}^{\infty}(-1)^{n}\dfrac{\tan^{-1}n}{n^2+1}$

26. $\sum_{n=2}^{\infty}(-1)^{n+1}\dfrac{1}{n\ln n}$

27. $\sum_{n=1}^{\infty}(-1)^{n}\dfrac{n}{n+1}$

28. $\sum_{n=1}^{\infty}(-1)^{n}\dfrac{\ln n}{n-\ln n}$

29. $\sum_{n=1}^{\infty}\dfrac{(-100)^n}{n!}$

30. $\sum_{n=1}^{\infty}(-5)^{-n}$

31. $\sum_{n=1}^{\infty}\dfrac{(-1)^{n-1}}{n^2+2n+1}$

32. $\sum_{n=2}^{\infty}(-1)^{n}\left(\dfrac{\ln n}{\ln n^2}\right)^n$

33. $\sum_{n=1}^{\infty}\dfrac{\cos n\pi}{n\sqrt{n}}$

34. $\sum_{n=1}^{\infty}\dfrac{\cos n\pi}{n}$

35. $\sum_{n=1}^{\infty}\dfrac{(-1)^{n}(n+1)^n}{(2n)^n}$

36. $\sum_{n=1}^{\infty}\dfrac{(-1)^{n+1}(n!)^2}{(2n)!}$

37. $\sum_{n=1}^{\infty}(-1)^{n}\dfrac{(2n)!}{2^n n! n}$

38. $\sum_{n=1}^{\infty}(-1)^{n}\dfrac{(n!)^2 3^n}{(2n+1)!}$

39. $\sum_{n=1}^{\infty}(-1)^{n}(\sqrt{n+1}-\sqrt{n})$

40. $\sum_{n=1}^{\infty}(-1)^{n}(\sqrt{n^2+n}-n)$

41. $\sum_{n=1}^{\infty}(-1)^{n}\left(\sqrt{n+\sqrt{n}}-\sqrt{n}\right)$

42. $\sum_{n=1}^{\infty}\dfrac{(-1)^{n}}{\sqrt{n}+\sqrt{n+1}}$

43. $\sum_{n=1}^{\infty} (-1)^n \operatorname{sech} n$

44. $\sum_{n=1}^{\infty} (-1)^n \operatorname{csch} n$

Error Estimation

In Exercises 45–48, estimate the magnitude of the error involved in using the sum of the first four terms to approximate the sum of the entire series.

45. $\sum_{n=1}^{\infty} (-1)^{n+1} \dfrac{1}{n}$ It can be shown that the sum is ln 2.

46. $\sum_{n=1}^{\infty} (-1)^{n+1} \dfrac{1}{10^n}$

47. $\sum_{n=1}^{\infty} (-1)^{n+1} \dfrac{(0.01)^n}{n}$ As you will see in Section 8.8, the sum is ln (1.01).

48. $\dfrac{1}{1+t} = \sum_{n=0}^{\infty} (-1)^n t^n, \quad 0 < t < 1$

CALCULATOR Approximate the sums in Exercises 49 and 50 with an error of magnitude less than 5×10^{-6}.

49. $\sum_{n=0}^{\infty} (-1)^n \dfrac{1}{(2n)!}$ As you will see in Section 8.10, the sum is cos 1, the cosine of 1 radian.

50. $\sum_{n=0}^{\infty} (-1)^n \dfrac{1}{n!}$ As you will see in Section 8.10, the sum is e^{-1}.

Theory and Examples

51. a) The series

$$\frac{1}{3} - \frac{1}{2} + \frac{1}{9} - \frac{1}{4} + \frac{1}{27} - \frac{1}{8} + \cdots + \frac{1}{3^n} - \frac{1}{2^n} + \cdots$$

does not meet one of the conditions of Theorem 8. Which one?

b) Find the sum of the series in (a).

52. CALCULATOR The limit L of an alternating series that satisfies the conditions of Theorem 8 lies between the values of any two consecutive partial sums. This suggests using the average

$$\frac{s_n + s_{n+1}}{2} = s_n + \frac{1}{2}(-1)^{n+2} a_{n+1}$$

to estimate L. Compute

$$s_{20} + \frac{1}{2} \cdot \frac{1}{21}$$

as an approximation to the sum of the alternating harmonic series. The exact sum is $\ln 2 = 0.6931\ldots$.

53. *The sign of the remainder of an alternating series that satisfies the conditions of Theorem 8.* Prove the assertion in Theorem 9 that whenever an alternating series satisfying the conditions of Theorem 8 is approximated with one of its partial sums, then the remainder (sum of the unused terms) has the same sign as the first unused term. (*Hint:* Group the remainder's terms in consecutive pairs.)

54. Show that the sum of the first $2n$ terms of the series

$$1 - \frac{1}{2} + \frac{1}{2} - \frac{1}{3} + \frac{1}{3} - \frac{1}{4} + \frac{1}{4} - \frac{1}{5} + \frac{1}{5} - \frac{1}{6} + \cdots$$

is the same as the sum of the first n terms of the series

$$\frac{1}{1 \cdot 2} + \frac{1}{2 \cdot 3} + \frac{1}{3 \cdot 4} + \frac{1}{4 \cdot 5} + \frac{1}{5 \cdot 6} + \cdots.$$

Do these series converge? What is the sum of the first $2n + 1$ terms of the first series? If the series converge, what is their sum?

55. Show that if $\sum_{n=1}^{\infty} a_n$ diverges, then $\sum_{n=1}^{\infty} |a_n|$ diverges.

56. Show that if $\sum_{n=1}^{\infty} a_n$ converges absolutely, then

$$\left| \sum_{n=1}^{\infty} a_n \right| \le \sum_{n=1}^{\infty} |a_n|.$$

57. Show that if $\sum_{n=1}^{\infty} a_n$ and $\sum_{n=1}^{\infty} b_n$ both converge absolutely, then so does

a) $\sum_{n=1}^{\infty} (a_n + b_n)$ **b)** $\sum_{n=1}^{\infty} (a_n - b_n)$

c) $\sum_{n=1}^{\infty} k a_n$ (k any number)

58. Show by example that $\sum_{n=1}^{\infty} a_n b_n$ may diverge even if $\sum_{n=1}^{\infty} a_n$ and $\sum_{n=1}^{\infty} b_n$ both converge.

59. CALCULATOR In Example 7, suppose the goal is to arrange the terms to get a new series that converges to $-1/2$. Start the new arrangement with the first negative term, which is $-1/2$. Whenever you have a sum that is less than or equal to $-1/2$, start introducing positive terms, taken in order, until the new total is greater than $-1/2$. Then add negative terms until the total is less than or equal to $-1/2$ again. Continue this process until your partial sums have been above the target at least three times and finish at or below it. If s_n is the sum of the first n terms of your new series, plot the points (n, s_n) to illustrate how the sums are behaving.

60. *Outline of the proof of the Rearrangement Theorem (Theorem 11).*

a) Let ϵ be a positive real number, let $L = \sum_{n=1}^{\infty} a_n$, and let $s_k = \sum_{n=1}^{k} a_n$. Show that for some index N_1 and for some index $N_2 \ge N_1$,

$$\sum_{n=N_1}^{\infty} |a_n| < \frac{\epsilon}{2} \quad \text{and} \quad |s_{N_2} - L| < \frac{\epsilon}{2}.$$

Since all the terms $a_1, a_2, \ldots, a_{N_2}$ appear somewhere in the sequence $\{b_n\}$, there is an index $N_3 \ge N_2$ such that if $n \ge N_3$, then $\left(\sum_{k=1}^{n} b_k\right) - s_{N_2}$ is at most a sum of terms a_m with $m \ge N_1$. Therefore, if $n \ge N_3$,

$$\left| \sum_{k=1}^{n} b_k - L \right| \le \left| \sum_{k=1}^{n} b_k - s_{N_2} \right| + |s_{N_2} - L|$$

$$\le \sum_{k=N_1}^{\infty} |a_k| + |s_{N_2} - L| < \epsilon.$$

b) The argument in (a) shows that if $\sum_{n=1}^{\infty} a_n$ converges absolutely then $\sum_{n=1}^{\infty} b_n$ converges and $\sum_{n=1}^{\infty} b_n = \sum_{n=1}^{\infty} a_n$. Now show that because $\sum_{n=1}^{\infty} |a_n|$ converges, $\sum_{n=1}^{\infty} |b_n|$ converges to $\sum_{n=1}^{\infty} |a_n|$.

61. *Unzipping absolutely convergent series.*

a) Show that if $\sum_{n=1}^{\infty} |a_n|$ converges and

$$b_n = \begin{cases} a_n & \text{if } a_n \geq 0 \\ 0 & \text{if } a_n < 0, \end{cases}$$

then $\sum_{n=1}^{\infty} b_n$ converges.

b) Use the results in (a) to show likewise that if $\sum_{n=1}^{\infty} |a_n|$ converges and

$$c_n = \begin{cases} 0 & \text{if } a_n \geq 0 \\ a_n & \text{if } a_n < 0, \end{cases}$$

then $\sum_{n=1}^{\infty} c_n$ converges.

In other words, if a series converges absolutely, its positive terms form a convergent series, and so do its negative terms. Furthermore,

$$\sum_{n=1}^{\infty} a_n = \sum_{n=1}^{\infty} b_n + \sum_{n=1}^{\infty} c_n$$

because $b_n = (a_n + |a_n|)/2$ and $c_n = (a_n - |a_n|)/2$.

62. What is wrong here:

Multiply both sides of the alternating harmonic series

$$S = 1 - \frac{1}{2} + \frac{1}{3} - \frac{1}{4} + \frac{1}{5} - \frac{1}{6} +$$

$$\frac{1}{7} - \frac{1}{8} + \frac{1}{9} - \frac{1}{10} + \frac{1}{11} - \frac{1}{12} + \cdots$$

by 2 to get

$$2S = 2 - 1 +$$

$$\frac{2}{3} - \frac{1}{2} + \frac{2}{5} - \frac{1}{3} + \frac{2}{7} - \frac{1}{4} + \frac{2}{9} - \frac{1}{5} + \frac{2}{11} - \frac{1}{6} + \cdots.$$

Collect terms with the same denominator, as the arrows indicate, to arrive at

$$2S = 1 - \frac{1}{2} + \frac{1}{3} - \frac{1}{4} + \frac{1}{5} - \frac{1}{6} + \cdots.$$

The series on the right-hand side of this equation is the series we started with. Therefore, $2S = S$, and dividing by S gives $2 = 1$. (Source: "Riemann's Rearrangement Theorem" by Stewart Galanor, *Mathematics Teacher,* Vol. 80, No. 8, 1987, pp. 675–81.)

63. Draw a figure similar to Fig. 8.14 to illustrate the convergence of the series in Theorem 8 when $N > 1$.

8.8

Power Series

Now that we can test infinite series for convergence we can study the infinite polynomials mentioned at the beginning of Section 8.3. We call these polynomials power series because they are defined as infinite series of powers of some variable, in our case x. Like polynomials, power series can be added, subtracted, multiplied, differentiated, and integrated to give new power series.

Power Series and Convergence

We begin with the formal definition.

Equation (1) is the special case obtained by taking $a = 0$ in Eq. (2).

Definition

A **power series about $x = 0$** is a series of the form

$$\sum_{n=0}^{\infty} c_n x^n = c_0 + c_1 x + c_2 x^2 + \cdots + c_n x^n + \cdots. \tag{1}$$

A **power series about $x = a$** is a series of the form

$$\sum_{n=0}^{\infty} c_n (x - a)^n = c_0 + c_1 (x - a) + c_2 (x - a)^2 + \cdots + c_n (x - a)^n + \cdots \tag{2}$$

in which the **center** a and the **coefficients** $c_0, c_1, c_2, \ldots, c_n, \ldots$ are constants.

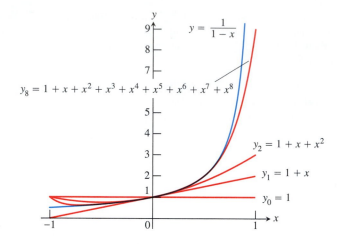

8.15 The graphs of $f(x) = 1/(1-x)$ and four of its polynomial approximations (Example 1).

EXAMPLE 1   Taking all the coefficients to be 1 in Eq. (1) gives the geometric power series

$$\sum_{n=0}^{\infty} x^n = 1 + x + x^2 + \cdots + x^n + \cdots .$$

This is the geometric series with first term 1 and ratio x. It converges to $1/(1-x)$ for $|x| < 1$. We express this fact by writing

$$\frac{1}{1-x} = 1 + x + x^2 + \cdots + x^n + \cdots , \qquad -1 < x < 1. \tag{3}$$

□

Up to now, we have used Eq. (3) as a formula for the sum of the series on the right. We now change the focus: We think of the partial sums of the series on the right as polynomials $P_n(x)$ that approximate the function on the left. For values of x near zero, we need take only a few terms of the series to get a good approximation. As we move toward $x = 1$, or -1, we must take more terms. Figure 8.15 shows the graphs of $f(x) = 1/(1-x)$, and the approximating polynomials $y_n = P_n(x)$ for $n = 0, 1, 2$, and 8.

EXAMPLE 2   The power series

$$1 - \frac{1}{2}(x-2) + \frac{1}{4}(x-2)^2 + \cdots + \left(-\frac{1}{2}\right)^n (x-2)^n + \cdots \tag{4}$$

matches Eq. (2) with $a = 2$, $c_0 = 1$, $c_1 = -1/2$, $c_2 = 1/4, \ldots, c_n = (-1/2)^n$. This is a geometric series with first term 1 and ratio $r = -\dfrac{x-2}{2}$. The series converges for $\left|\dfrac{x-2}{2}\right| < 1$ or $0 < x < 4$. The sum is

$$\frac{1}{1-r} = \frac{1}{1 + \dfrac{x-2}{2}} = \frac{2}{x},$$

so

$$\frac{2}{x} = 1 - \frac{(x-2)}{2} + \frac{(x-2)^2}{4} - \cdots + \left(-\frac{1}{2}\right)^n (x-2)^n + \cdots , \qquad 0 < x < 4.$$

Series (4) generates useful polynomial approximations of $f(x) = 2/x$ for values of x near 2:

$$P_0(x) = 1$$

$$P_1(x) = 1 - \frac{1}{2}(x - 2) = 2 - \frac{x}{2}$$

$$P_2(x) = 1 - \frac{1}{2}(x - 2) + \frac{1}{4}(x - 2)^2 = 3 - \frac{3x}{2} + \frac{x^2}{4},$$

and so on (Fig. 8.16). ❑

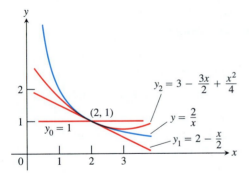

8.16 The graphs of $f(x) = 2/x$ and its first three polynomial approximations (Example 2).

EXAMPLE 3 For what values of x do the following power series converge?

a) $\displaystyle\sum_{n=1}^{\infty} (-1)^{n-1} \frac{x^n}{n} = x - \frac{x^2}{2} + \frac{x^3}{3} - \cdots$

b) $\displaystyle\sum_{n=1}^{\infty} (-1)^{n-1} \frac{x^{2n-1}}{2n-1} = x - \frac{x^3}{3} + \frac{x^5}{5} - \cdots$

c) $\displaystyle\sum_{n=0}^{\infty} \frac{x^n}{n!} = 1 + x + \frac{x^2}{2!} + \frac{x^3}{3!} + \cdots$

d) $\displaystyle\sum_{n=0}^{\infty} n!\, x^n = 1 + x + 2!\, x^2 + 3!\, x^3 + \cdots$

Solution Apply the Ratio Test to the series $\sum |u_n|$, where u_n is the nth term of the series in question.

a) $\left|\dfrac{u_{n+1}}{u_n}\right| = \dfrac{n}{n+1}|x| \to |x|.$

The series converges absolutely for $|x| < 1$. It diverges if $|x| > 1$ because the nth term does not converge to zero. At $x = 1$, we get the alternating harmonic series $1 - 1/2 + 1/3 - 1/4 + \cdots$, which converges. At $x = -1$ we get $-1 - 1/2 - 1/3 - 1/4 - \cdots$, the negative of the harmonic series; it diverges. Series (a) converges for $-1 < x \le 1$ and diverges elsewhere.

b) $\left|\dfrac{u_{n+1}}{u_n}\right| = \dfrac{2n-1}{2n+1}x^2 \to x^2.$

The series converges absolutely for $x^2 < 1$. It diverges for $x^2 > 1$ because the nth term does not converge to zero. At $x = 1$ the series becomes $1 - 1/3 + 1/5 - 1/7 + \cdots$, which converges by the Alternating Series Theorem. It also converges at $x = -1$ because it is again an alternating series that satisfies the conditions for convergence. The value at $x = -1$ is the negative of the value at $x = 1$. Series (b) converges for $-1 \le x \le 1$ and diverges elsewhere.

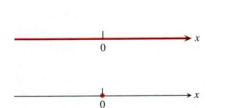

c) $\left|\dfrac{u_{n+1}}{u_n}\right| = \left|\dfrac{x^{n+1}}{(n+1)!} \cdot \dfrac{n!}{x^n}\right| = \dfrac{|x|}{n+1} \to 0$ for every x.

The series converges absolutely for all x.

d) $\left|\dfrac{u_{n+1}}{u_n}\right| = \left|\dfrac{(n+1)!\, x^{n+1}}{n!\, x^n}\right| = (n+1)|x| \to \infty$ unless $x = 0$.

The series diverges for all values of x except $x = 0$. ❑

Example 3 illustrates how we usually test a power series for convergence, and the possible results.

How to Test a Power Series for Convergence

Step 1: Use the Ratio Test (or nth-Root Test) to find the interval where the series converges absolutely. Ordinarily, this is an open interval

$$|x - a| < R \qquad \text{or} \qquad a - R < x < a + R.$$

Step 2: If the interval of absolute convergence is finite, test for convergence or divergence at each endpoint, as in Examples 3(a) and (b). Use a Comparison Test, the Integral Test, or the Alternating Series Test.

Step 3: If the interval of absolute convergence is $a - R < x < a + R$, the series diverges for $|x - a| > R$ (it does not even converge conditionally), because the nth term does not approach zero for those values of x.

To simplify the notation, Theorem 12 deals with the convergence of series of the form $\sum a_n x^n$. For series of the form $\sum a_n(x - a)^n$ we can replace $x - a$ by x' and apply the results to the series $\sum a_n(x')^n$.

Theorem 12

The Convergence Theorem for Power Series

If $\displaystyle\sum_{n=0}^{\infty} a_n x^n = a_0 + a_1 x + a_2 x^2 + \cdots$

converges for $x = c \neq 0$, then it converges absolutely for all $|x| < |c|$. If the series diverges for $x = d$, then it diverges for all $|x| > |d|$.

Proof Suppose the series $\sum_{n=0}^{\infty} a_n c^n$ converges. Then $\lim_{n \to \infty} a_n c^n = 0$. Hence, there is an integer N such that $|a_n c^n| < 1$ for all $n \geq N$. That is,

$$|a_n| < \frac{1}{|c|^n} \qquad \text{for } n \geq N. \tag{5}$$

Now take any x such that $|x| < |c|$ and consider

$$|a_0| + |a_1 x| + \cdots + |a_{N-1} x^{N-1}| + |a_N x^N| + |a_{N+1} x^{N+1}| + \cdots.$$

There are only a finite number of terms prior to $|a_N x^N|$, and their sum is finite. Starting with $|a_N x^N|$ and beyond, the terms are less than

$$\left|\frac{x}{c}\right|^N + \left|\frac{x}{c}\right|^{N+1} + \left|\frac{x}{c}\right|^{N+2} + \cdots \tag{6}$$

because of (5). But the series in (6) is a geometric series with ratio $r = |x/c|$, which is less than 1, since $|x| < |c|$. Hence the series (6) converges, so the original series converges absolutely. This proves the first half of the theorem.

The second half of the theorem follows from the first. If the series diverges at $x = d$ and converges at a value x_0 with $|x_0| > |d|$, we may take $c = x_0$ in the first half of the theorem and conclude that the series converges absolutely at d. But the series cannot converge absolutely and diverge at one and the same time. Hence, if it diverges at d, it diverges for all $|x| > |d|$. ❏

The Radius and Interval of Convergence

The examples we have looked at, and the theorem we just proved, lead to the conclusion that a power series behaves in one of the following three ways.

Possible Behavior of $\sum c_n(x - a)^n$

1. There is a positive number R such that the series diverges for $|x - a| > R$ but converges absolutely for $|x - a| < R$. The series may or may not converge at either of the endpoints $x = a - R$ and $x = a + R$.

2. The series converges absolutely for every x ($R = \infty$).

3. The series converges at $x = a$ and diverges elsewhere ($R = 0$).

In case 1, the set of points at which the series converges is a finite interval, called the **interval of convergence.** We know from the examples that the interval can be open, half-open, or closed, depending on the particular series. But no matter which kind of interval it is, R is called the **radius of convergence** of the series, and $a + R$ is the least upper bound of the set of points at which the series converges. The convergence is absolute at every point in the interior of the interval. If a power series converges absolutely for all values of x, we say that its **radius of convergence is infinite.** If it converges only at $x = a$, the **radius of convergence is zero.**

Term-by-Term Differentiation

A theorem from advanced calculus says that a power series can be differentiated term by term at each interior point of its interval of convergence.

A word of caution

Term-by-term differentiation might not work for other kinds of series. For example, the trigonometric series

$$\sum_{n=1}^{\infty} \frac{\sin (n!\, x)}{n^2}$$

converges for all x. But if we differentiate term by term we get the series

$$\sum_{n=1}^{\infty} \frac{n!\, \cos (n!\, x)}{n^2},$$

which diverges for all x.

Theorem 13

The Term-by-Term Differentiation Theorem

If $\sum c_n(x - a)^n$ converges for $a - R < x < a + R$ for some $R > 0$, it defines a function f:

$$f(x) = \sum_{n=0}^{\infty} c_n(x - a)^n, \qquad a - R < x < a + R.$$

Such a function f has derivatives of all orders inside the interval of convergence. We can obtain the derivatives by differentiating the original series term by term:

$$f'(x) = \sum_{n=1}^{\infty} n c_n(x - a)^{n-1}$$

$$f''(x) = \sum_{n=2}^{\infty} n(n - 1) c_n(x - a)^{n-2},$$

and so on. Each of these derived series converges at every interior point of the interval of convergence of the original series.

EXAMPLE 4   Find series for $f'(x)$ and $f''(x)$ if

$$f(x) = \frac{1}{1-x} = 1 + x + x^2 + x^3 + x^4 + \cdots + x^n + \cdots$$

$$= \sum_{n=0}^{\infty} x^n, \qquad -1 < x < 1$$

Solution

$$f'(x) = \frac{1}{(1-x)^2} = 1 + 2x + 3x^2 + 4x^3 + \cdots + nx^{n-1} + \cdots$$

$$= \sum_{n=1}^{\infty} nx^{n-1}, \qquad -1 < x < 1$$

$$f''(x) = \frac{2}{(1-x)^3} = 2 + 6x + 12x^2 + \cdots + n(n-1)x^{n-2} + \cdots$$

$$= \sum_{n=2}^{\infty} n(n-1)x^{n-2}, \qquad -1 < x < 1$$

Term-by-Term Integration

Another advanced theorem states that a power series can be integrated term by term throughout its interval of convergence.

Theorem 14

The Term-by-Term Integration Theorem

Suppose that

$$f(x) = \sum_{n=0}^{\infty} c_n(x-a)^n$$

converges for $a - R < x < a + R \ (R > 0)$. Then

$$\sum_{n=0}^{\infty} c_n(x-a)^{n+1}/(n+1)$$

converges for $a - R < x < a + R$ and

$$\int f(x)\,dx = \sum_{n=0}^{\infty} c_n \frac{(x-a)^{n+1}}{n+1} + C$$

for $a - R < x < a + R$.

EXAMPLE 5   *A series for* $\tan^{-1} x$, $-1 \le x \le 1$

Identify the function

$$f(x) = x - \frac{x^3}{3} + \frac{x^5}{5} - \cdots, \qquad -1 \le x \le 1.$$

Solution   We differentiate the original series term by term and get

$$f'(x) = 1 - x^2 + x^4 - x^6 + \cdots, \qquad -1 < x < 1.$$

This is a geometric series with first term 1 and ratio $-x^2$, so

$$f'(x) = \frac{1}{1 - (-x^2)} = \frac{1}{1 + x^2}.$$

We can now integrate $f'(x) = 1/(1 + x^2)$ to get

$$\int f'(x)\,dx = \int \frac{dx}{1 + x^2} = \tan^{-1} x + C.$$

The series for $f(x)$ is zero when $x = 0$, so $C = 0$. Hence

$$f(x) = x - \frac{x^3}{3} + \frac{x^5}{5} - \frac{x^7}{7} + \cdots = \tan^{-1} x, \qquad -1 < x < 1. \qquad (7)$$

In Section 8.11, we will see that the series also converges to $\tan^{-1} x$ at $x = \pm 1$.   ❏

Notice that the original series in Example 5 converges at both endpoints of the original interval of convergence, but Theorem 13 can guarantee the convergence of the differentiated series only inside the interval.

EXAMPLE 6 *A series for* ln $(1 + x)$, $-1 < x \le 1$

The series

$$\frac{1}{1 + t} = 1 - t + t^2 - t^3 + \cdots$$

converges on the open interval $-1 < t < 1$. Therefore,

$$\ln (1 + x) = \int_0^x \frac{1}{1 + t}\,dt = t - \frac{t^2}{2} + \frac{t^3}{3} - \frac{t^4}{4} + \cdots \Big]_0^x$$

$$= x - \frac{x^2}{2} + \frac{x^3}{3} - \frac{x^4}{4} + \cdots, \qquad -1 < x < 1.$$

It can also be shown that the series converges at $x = 1$ to the number ln 2, but that was not guaranteed by the theorem.   ❏

Technology *Study of Series* Series are in many ways analogous to integrals. Just as the number of functions with explicit antiderivatives in terms of elementary functions is small compared to the number of integrable functions, the number of power series in x that agree with explicit elementary functions on x-intervals is small compared to the number of power series that converge on some x-interval. Graphing utilities can aid in the study of such series in much the same way that numerical integration aids in the study of definite integrals. The ability to study power series at particular values of x is built into most Computer Algebra Systems.

If a series converges rapidly enough, CAS exploration might give us an idea of the sum. For instance, in calculating the early partial sums of the series $\sum_{n=1}^{\infty}[1/(2^{n-1})]$ (Section 8.5, Example 3b), Maple returns $S_n = 1.6066\ 95152$ for $31 \le n \le 200$. This suggests that the sum of the series is $1.6066\ 95152$ to 10 digits. Indeed,

$$\sum_{n=201}^{\infty} \frac{1}{2^n - 1} = \sum_{n=201}^{\infty} \frac{1}{2^{n-1}(2 - (1/2^{n-1}))} < \sum_{n=201}^{\infty} \frac{1}{2^{n-1}} = \frac{1}{2^{199}} < 1.25 \times 10^{-60}.$$

The remainder after 200 terms is negligible.

However, CAS and calculator exploration cannot do much for us if the series converges or diverges very slowly, and indeed can be downright misleading. For example, try calculating the partial sums of the series $\sum_{n=1}^{\infty}[1/(10^{10}n)]$. The terms are tiny in comparison to the numbers we normally work with and the partial sums, even for hundreds of terms, are miniscule. We might well be fooled into thinking that the series converges. In fact, it diverges, as we can see by writing it as $(1/10^{10}) \sum_{n=1}^{\infty}(1/n)$.

We will know better how to interpret numerical results after studying error estimates in Section 8.10.

Multiplication of Power Series

Still another advanced theorem states that absolutely converging power series can be multiplied the way we multiply polynomials.

Theorem 15

The Series Multiplication Theorem for Power Series

If $A(x) = \sum_{n=0}^{\infty} a_n x^n$ and $B(x) = \sum_{n=0}^{\infty} b_n x^n$ converge absolutely for $|x| < R$, and

$$c_n = a_0 b_n + a_1 b_{n-1} + a_2 b_{n-2} + \cdots + a_{n-1} b_1 + a_n b_0 = \sum_{k=0}^{n} a_k b_{n-k},$$

then $\sum_{n=0}^{\infty} c_n x^n$ converges absolutely to $A(x)B(x)$ for $|x| < R$:

$$\left(\sum_{n=0}^{\infty} a_n x^n \right) \cdot \left(\sum_{n=0}^{\infty} b_n x^n \right) = \sum_{n=0}^{\infty} c_n x^n.$$

EXAMPLE 7   Multiply the geometric series

$$\sum_{n=0}^{\infty} x^n = 1 + x + x^2 + \cdots + x^n + \cdots = \frac{1}{1-x}, \qquad \text{for } |x| < 1,$$

by itself to get a power series for $1/(1-x)^2$, for $|x| < 1$.

Solution   Let

$$A(x) = \sum_{n=0}^{\infty} a_n x^n = 1 + x + x^2 + \cdots + x^n + \cdots = 1/(1-x)$$

$$B(x) = \sum_{n=0}^{\infty} b_n x^n = 1 + x + x^2 + \cdots + x^n + \cdots = 1/(1-x)$$

and

$$c_n = \underbrace{a_0 b_n + a_1 b_{n-1} + \cdots + a_k b_{n-k} + \cdots + a_n b_0}_{n+1 \text{ terms}}$$

$$= \underbrace{1 + 1 + \cdots + 1}_{n+1 \text{ ones}} = n + 1.$$

Then, by the Series Multiplication Theorem,

$$A(x) \cdot B(x) = \sum_{n=0}^{\infty} c_n x^n = \sum_{n=0}^{\infty} (n + 1) x^n$$

$$= 1 + 2x + 3x^2 + 4x^3 + \cdots + (n + 1)x^n + \cdots$$

is the series for $1/(1-x)^2$. The series all converge absolutely for $|x| < 1$.

Notice that Example 4 gives the same answer because

$$\frac{d}{dx} \left(\frac{1}{1-x} \right) = \frac{1}{(1-x)^2}.$$

Exercises 8.8

Intervals of Convergence

In Exercises 1–32, (a) find the series' radius and interval of convergence. For what values of x does the series converge (b) absolutely, (c) conditionally?

1. $\displaystyle\sum_{n=0}^{\infty} x^n$

2. $\displaystyle\sum_{n=0}^{\infty} (x+5)^n$

3. $\displaystyle\sum_{n=0}^{\infty} (-1)^n (4x+1)^n$

4. $\displaystyle\sum_{n=1}^{\infty} \frac{(3x-2)^n}{n}$

5. $\displaystyle\sum_{n=0}^{\infty} \frac{(x-2)^n}{10^n}$

6. $\displaystyle\sum_{n=0}^{\infty} (2x)^n$

7. $\displaystyle\sum_{n=0}^{\infty} \frac{nx^n}{n+2}$

8. $\displaystyle\sum_{n=1}^{\infty} \frac{(-1)^n (x+2)^n}{n}$

9. $\displaystyle\sum_{n=1}^{\infty} \frac{x^n}{n\sqrt{n}\,3^n}$

10. $\displaystyle\sum_{n=1}^{\infty} \frac{(x-1)^n}{\sqrt{n}}$

11. $\displaystyle\sum_{n=0}^{\infty} \frac{(-1)^n x^n}{n!}$

12. $\displaystyle\sum_{n=0}^{\infty} \frac{3^n x^n}{n!}$

13. $\displaystyle\sum_{n=0}^{\infty} \frac{x^{2n+1}}{n!}$

14. $\displaystyle\sum_{n=0}^{\infty} \frac{(2x+3)^{2n+1}}{n!}$

15. $\displaystyle\sum_{n=0}^{\infty} \frac{x^n}{\sqrt{n^2+3}}$

16. $\displaystyle\sum_{n=0}^{\infty} \frac{(-1)^n x^n}{\sqrt{n^2+3}}$

17. $\displaystyle\sum_{n=0}^{\infty} \frac{n(x+3)^n}{5^n}$

18. $\displaystyle\sum_{n=0}^{\infty} \frac{nx^n}{4^n(n^2+1)}$

19. $\displaystyle\sum_{n=0}^{\infty} \frac{\sqrt{n}\,x^n}{3^n}$

20. $\displaystyle\sum_{n=1}^{\infty} \sqrt[n]{n}\,(2x+5)^n$

21. $\displaystyle\sum_{n=1}^{\infty} \left(1+\frac{1}{n}\right)^n x^n$

22. $\displaystyle\sum_{n=1}^{\infty} (\ln n)\, x^n$

23. $\displaystyle\sum_{n=1}^{\infty} n^n x^n$

24. $\displaystyle\sum_{n=0}^{\infty} n!(x-4)^n$

25. $\displaystyle\sum_{n=1}^{\infty} \frac{(-1)^{n+1}(x+2)^n}{n2^n}$

26. $\displaystyle\sum_{n=0}^{\infty} (-2)^n (n+1)(x-1)^n$

27. $\displaystyle\sum_{n=2}^{\infty} \frac{x^n}{n(\ln n)^2}$ $\left(\begin{array}{l}\text{Get the information you need}\\ \text{about } \sum 1/(n(\ln n)^2) \text{ from}\\ \text{Section 8.4, Exercise 39.}\end{array}\right)$

28. $\displaystyle\sum_{n=2}^{\infty} \frac{x^n}{n\ln n}$ $\left(\begin{array}{l}\text{Get the information you need about}\\ \sum 1/(n\ln n) \text{ from}\\ \text{Section 8.4, Exercise 38.}\end{array}\right)$

29. $\displaystyle\sum_{n=1}^{\infty} \frac{(4x-5)^{2n+1}}{n^{3/2}}$

30. $\displaystyle\sum_{n=1}^{\infty} \frac{(3x+1)^{n+1}}{2n+2}$

31. $\displaystyle\sum_{n=1}^{\infty} \frac{(x+\pi)^n}{\sqrt{n}}$

32. $\displaystyle\sum_{n=0}^{\infty} \frac{(x-\sqrt{2})^{2n+1}}{2^n}$

In Exercises 33–38, find the series' interval of convergence and, within this interval, the sum of the series as a function of x.

33. $\displaystyle\sum_{n=0}^{\infty} \frac{(x-1)^{2n}}{4^n}$

34. $\displaystyle\sum_{n=0}^{\infty} \frac{(x+1)^{2n}}{9^n}$

35. $\displaystyle\sum_{n=0}^{\infty} \left(\frac{\sqrt{x}}{2}-1\right)^n$

36. $\displaystyle\sum_{n=0}^{\infty} (\ln x)^n$

37. $\displaystyle\sum_{n=0}^{\infty} \left(\frac{x^2+1}{3}\right)^n$

38. $\displaystyle\sum_{n=0}^{\infty} \left(\frac{x^2-1}{2}\right)^n$

Theory and Examples

39. For what values of x does the series

$$1 - \frac{1}{2}(x-3) + \frac{1}{4}(x-3)^2 + \cdots + \left(-\frac{1}{2}\right)^n (x-3)^n + \cdots$$

converge? What is its sum? What series do you get if you differentiate the given series term by term? For what values of x does the new series converge? What is its sum?

40. If you integrate the series in Exercise 39 term by term, what new series do you get? For what values of x does the new series converge, and what is another name for its sum?

41. The series

$$\sin x = x - \frac{x^3}{3!} + \frac{x^5}{5!} - \frac{x^7}{7!} + \frac{x^9}{9!} - \frac{x^{11}}{11!} + \cdots$$

converges to $\sin x$ for all x.

a) Find the first six terms of a series for $\cos x$. For what values of x should the series converge?

b) By replacing x by $2x$ in the series for $\sin x$, find a series that converges to $\sin 2x$ for all x.

c) Using the result in (a) and series multiplication, calculate the first six terms of a series for $2\sin x \cos x$. Compare your answer with the answer in (b).

42. The series

$$e^x = 1 + x + \frac{x^2}{2!} + \frac{x^3}{3!} + \frac{x^4}{4!} + \frac{x^5}{5!} + \cdots$$

converges to e^x for all x.

a) Find a series for $(d/dx)e^x$. Do you get the series for e^x? Explain your answer.

b) Find a series for $\int e^x dx$. Do you get the series for e^x? Explain your answer.

c) Replace x by $-x$ in the series for e^x to find a series that converges to e^{-x} for all x. Then multiply the series for e^x and e^{-x} to find the first six terms of a series for $e^{-x} \cdot e^x$.

43. The series

$$\tan x = x + \frac{x^3}{3} + \frac{2x^5}{15} + \frac{17x^7}{315} + \frac{62x^9}{2835} + \cdots$$

converges to $\tan x$ for $-\pi/2 < x < \pi/2$.

a) Find the first five terms of the series for $\ln |\sec x|$. For what values of x should the series converge?

b) Find the first five terms of the series for $\sec^2 x$. For what values of x should this series converge?

c) Check your result in (b) by squaring the series given for $\sec x$ in Exercise 44.

44. The series for

$$\sec x = 1 + \frac{x^2}{2} + \frac{5}{24}x^4 + \frac{61}{720}x^6 + \frac{277}{8064}x^8 + \cdots$$

converges to $\sec x$ for $-\pi/2 < x < \pi/2$.

a) Find the first five terms of a power series for the function $\ln|\sec x + \tan x|$. For what values of x should the series converge?

b) Find the first four terms of a series for $\sec x \tan x$. For what values of x should the series converge?

c) Check your result in (b) by multiplying the series for $\sec x$ by the series given for $\tan x$ in Exercise 43.

45. *Uniqueness of convergent power series*

a) Show that if two power series $\sum_{n=0}^{\infty} a_n x^n$ and $\sum_{n=0}^{\infty} b_n x^n$

are convergent and equal for all values of x in an open interval $(-c, c)$, then $a_n = b_n$ for every n. (*Hint:* Let $f(x) = \sum_{n=0}^{\infty} a_n x^n = \sum_{n=0}^{\infty} b_n x^n$. Differentiate term by term to show that a_n and b_n both equal $f^{(n)}(0)/(n!)$.)

b) Show that if $\sum_{n=0}^{\infty} a_n x^n = 0$ for all x in an open interval $(-c, c)$, then $a_n = 0$ for every n.

46. *The sum of the series $\sum_{n=0}^{\infty} (n^2/2^n)$*. To find the sum of this series, express $1/(1 - x)$ as a geometric series, differentiate both sides of the resulting equation with respect to x, multiply both sides of the result by x, differentiate again, multiply by x again, and set x equal to $1/2$. What do you get? (Source: David E. Dobbs' letter to the editor, *Illinois Mathematics Teacher,* Vol. 33, Issue 4, 1982, p. 27.)

47. *Convergence at endpoints.* Show by examples that the convergence of a power series at an endpoint of its interval of convergence may be either conditional or absolute.

48. Make up a power series whose interval of convergence is

a) $(-3, 3)$　　　**b)** $(-2, 0)$　　　**c)** $(1, 5)$.

8.9 Taylor and Maclaurin Series

This section shows how functions that are infinitely differentiable generate power series called Taylor series. In many cases, these series can provide useful polynomial approximations of the generating functions.

Series Representations

We know that within its interval of convergence the sum of a power series is a continuous function with derivatives of all orders. But what about the other way around? If a function $f(x)$ has derivatives of all orders on an interval I, can it be expressed as a power series on I? And if it can, what will its coefficients be?

We can answer the last question readily if we assume that $f(x)$ is the sum of a power series

$$f(x) = \sum_{n=0}^{\infty} a_n(x - a)^n$$

$$= a_0 + a_1(x - a) + a_2(x - a)^2 + \cdots + a_n(x - a)^n + \cdots$$

with a positive radius of convergence. By repeated term-by-term differentiation within the interval of convergence I we obtain

$$f'(x) = a_1 + 2a_2(x - a) + 3a_3(x - a)^2 + \cdots + na_n(x - a)^{n-1} + \cdots$$

$$f''(x) = 1 \cdot 2a_2 + 2 \cdot 3a_3(x - a) + 3 \cdot 4a_4(x - a)^2 + \cdots$$

$$f'''(x) = 1 \cdot 2 \cdot 3a_3 + 2 \cdot 3 \cdot 4a_4(x - a) + 3 \cdot 4 \cdot 5a_5(x - a)^2 + \cdots,$$

with the nth derivative, for all n, being

$$f^{(n)}(x) = n!\,a_n + \text{ a sum of terms with } (x - a) \text{ as a factor.}$$

Since these equations all hold at $x = a$, we have

$$f'(a) = a_1,$$

$$f''(a) = 1 \cdot 2a_2,$$

$$f'''(a) = 1 \cdot 2 \cdot 3a_3,$$

and, in general,

$$f^{(n)}(a) = n!\, a_n.$$

These formulas reveal a marvelous pattern in the coefficients of any power series $\sum_{n=0}^{\infty} a_n(x - a)^n$ that converges to the values of f on I ("represents f on I," we say). If there *is* such a series (still an open question), then there is only one such series and its nth coefficient is

$$a_n = \frac{f^{(n)}(a)}{n!}.$$

If f has a series representation, then the series must be

$$f(x) = f(a) + f'(a)(x - a) + \frac{f''(a)}{2!}(x - a)^2$$

$$+ \cdots + \frac{f^{(n)}(a)}{n!}(x - a)^n + \cdots. \tag{1}$$

But if we start with an arbitrary function f that is infinitely differentiable on an interval I centered at $x = a$ and use it to generate the series in Eq. (1), will the series then converge to $f(x)$ at each x in the interior of I? The answer is maybe—for some functions it will but for other functions it will not, as we will see.

Taylor and Maclaurin Series

Definitions

Let f be a function with derivatives of all orders throughout some interval containing a as an interior point. Then the **Taylor series generated by f at $x = a$** is

$$\sum_{k=0}^{\infty} \frac{f^{(k)}(a)}{k!}(x - a)^k = f(a) + f'(a)(x - a) + \frac{f''(a)}{2!}(x - a)^2$$

$$+ \cdots + \frac{f^{(n)}(a)}{n!}(x - a)^n + \cdots.$$

The **Maclaurin series generated by f** is

$$\sum_{k=0}^{\infty} \frac{f^{(k)}(0)}{k!} x^k = f(0) + f'(0)x + \frac{f''(0)}{2!} x^2 + \cdots + \frac{f^{(n)}(0)}{n!} x^n + \cdots,$$

the Taylor series generated by f at $x = 0$.

EXAMPLE 1   Find the Taylor series generated by $f(x) = 1/x$ at $a = 2$. Where, if anywhere, does the series converge to $1/x$?

Solution We need to find $f(2), f'(2), f''(2), \ldots$. Taking derivatives we get

$$f(x) = x^{-1}, \qquad\qquad f(2) = 2^{-1} = \frac{1}{2},$$

$$f'(x) = -x^{-2}, \qquad\qquad f'(2) = -\frac{1}{2^2},$$

$$f''(x) = 2!\,x^{-3}, \qquad\qquad \frac{f''(2)}{2!} = 2^{-3} = \frac{1}{2^3},$$

$$f'''(x) = -3!\,x^{-4}, \qquad\qquad \frac{f'''(2)}{3!} = -\frac{1}{2^4},$$

$$\vdots \qquad\qquad\qquad \vdots$$

$$f^{(n)}(x) = (-1)^n n!\,x^{-(n+1)}, \qquad \frac{f^{(n)}(2)}{n!} = \frac{(-1)^n}{2^{n+1}}.$$

The Taylor series is

$$f(2) + f'(2)(x-2) + \frac{f''(2)}{2!}(x-2)^2 + \cdots + \frac{f^{(n)}}{n!}(x-2)^n + \cdots$$

$$= \frac{1}{2} - \frac{(x-2)}{2^2} + \frac{(x-2)^2}{2^3} - \cdots + (-1)^n \frac{(x-2)^n}{2^{n+1}} + \cdots.$$

This is a geometric series with first term $1/2$ and ratio $r = -(x-2)/2$. It converges absolutely for $|x-2| < 2$ and its sum is

$$\frac{1/2}{1 + (x-2)/2} = \frac{1}{2 + (x-2)} = \frac{1}{x}.$$

In this example the Taylor series generated by $f(x) = 1/x$ at $a = 2$ converges to $1/x$ for $|x-2| < 2$ or $0 < x < 4$. ◻

Taylor Polynomials

The linearization of a differentiable function f at a point a is the polynomial

$$P_1(x) = f(a) + f'(a)(x-a).$$

If f has derivatives of higher order at a, then it has higher order polynomial approximations as well, one for each available derivative. These polynomials are called the Taylor polynomials of f.

We speak of a Taylor polynomial of *order n* rather than *degree n* because $f^{(n)}(a)$ may be zero. The first two Taylor polynomials of $\cos x$ at $x = 0$, for example, are $P_0(x) = 1$ and $P_1(x) = 1$. The first order polynomial has degree zero, not one.

Definition

Let f be a function with derivatives of order k for $k = 1, 2, \ldots, N$ in some interval containing a as an interior point. Then for any integer n from 0 through N, the **Taylor polynomial of order n** generated by f at $x = a$ is the polynomial

$$P_n(x) = f(a) + f'(a)(x-a) + \frac{f''(a)}{2!}(x-a)^2 + \cdots$$

$$+ \frac{f^{(k)}(a)}{k!}(x-a)^k + \cdots + \frac{f^{(n)}(a)}{n!}(x-a)^n.$$

Just as the linearization of f at $x = a$ provides the best linear approximation of f in the neighborhood of a, the higher order Taylor polynomials provide the best polynomial approximations of their respective degrees. (See Exercise 32.)

EXAMPLE 2   Find the Taylor series and the Taylor polynomials generated by $f(x) = e^x$ at $x = 0$.

Solution Since

$$f(x) = e^x, \qquad f'(x) = e^x, \qquad \dots, \qquad f^{(n)}(x) = e^x, \qquad \dots,$$

we have

$$f(0) = e^0 = 1, \qquad f'(0) = 1, \qquad \dots, \qquad f^{(n)}(0) = 1, \qquad \dots.$$

The Taylor series generated by f at $x = 0$ is

$$f(0) + f'(0)x + \frac{f''(0)}{2!}x^2 + \cdots + \frac{f^{(n)}(0)}{n!}x^n + \cdots$$

$$= 1 + x + \frac{x^2}{2} + \cdots + \frac{x^n}{n!} + \cdots$$

$$= \sum_{k=0}^{\infty} \frac{x^k}{k!}.$$

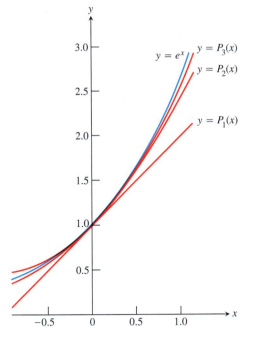

8.17 The graph of $f(x) = e^x$ and its Taylor polynomials

$$P_1(x) = 1 + x,$$

$$P_2(x) = 1 + x + (x^2/2!), \text{ and}$$

$$P_3(x) = 1 + x + (x^2/2!) + (x^3/3!).$$

Notice the very close agreement near the center $x = 0$.

By definition, this is also the Maclaurin series for e^x. In Section 8.10 we will see that the series converges to e^x at every x.

The Taylor polynomial of order n at $x = 0$ is

$$P_n(x) = 1 + x + \frac{x^2}{2} + \cdots + \frac{x^n}{n!}.$$

See Fig. 8.17.   ❑

EXAMPLE 3   Find the Taylor series and Taylor polynomials generated by $f(x) = \cos x$ at $x = 0$.

Solution The cosine and its derivatives are

$$\begin{aligned} f(x) &= & \cos x & \qquad f'(x) &= & -\sin x, \\ f''(x) &= & -\cos x, & \qquad f^{(3)}(x) &= & \sin x, \\ & & \vdots & & & \vdots \\ f^{(2n)}(x) &= & (-1)^n \cos x, & \qquad f^{(2n+1)}(x) &= & (-1)^{n+1} \sin x. \end{aligned}$$

At $x = 0$, the cosines are 1 and the sines are 0, so

$$f^{(2n)}(0) = (-1)^n, \qquad f^{(2n+1)}(0) = 0.$$

The Taylor series generated by f at 0 is

$$f(0) + f'(0)x + \frac{f''(0)}{2!}x^2 + \frac{f'''(0)}{3!}x^3 + \cdots + \frac{f^{(n)}(0)}{n!}x^n + \cdots$$

$$= 1 + 0 \cdot x - \frac{x^2}{2!} + 0 \cdot x^3 + \frac{x^4}{4!} + \cdots + (-1)^n \frac{x^{2n}}{(2n)!} + \cdots$$

$$= \sum_{n=0}^{\infty} \frac{(-1)^n x^{2n}}{(2n)!}.$$

8.18 The polynomials

$$P_{2n}(x) = \sum_{k=0}^{n} [(-1)^k x^{2k}/(2k)!]$$

converge to cos x as $n \to \infty$. We can deduce the behavior of cos x arbitrarily far away solely from knowing the values of the cosine and its derivatives at $x = 0$.

Infinitely differentiable functions that are represented by their Taylor series only at isolated points are, in practice, very rare.

Who invented Taylor series?

Brook Taylor (1685–1731) did not invent Taylor series, and Maclaurin series were not developed by Colin Maclaurin (1698–1746). James Gregory was already working with Taylor series when Taylor was only a few years old, and he published the Maclaurin series for tan x, sec x, $\tan^{-1}x$, and $\sec^{-1}x$ ten years before Maclaurin was born. Nicolaus Mercator discovered the Maclaurin series for ln $(1 + x)$ at about the same time.

Taylor was unaware of Gregory's work when he published his book *Methodus incrementorum directa et inversa* in 1715, containing what we now call Taylor series. Maclaurin quoted Taylor's work in a calculus book he wrote in 1742. The book popularized series representations of functions and although Maclaurin never claimed to have discovered them, Taylor series centered at $x = 0$ became known as Maclaurin series. History evened things up in the end. Maclaurin, a brilliant mathematician, was the original discoverer of the rule for solving systems of equations that we call Cramer's rule.

By definition, this is also the Maclaurin series for cos x. In Section 8.10, we will see that the series converges to cos x at every x.

Because $f^{(2n+1)}(0) = 0$, the Taylor polynomials of orders $2n$ and $2n + 1$ are identical:

$$P_{2n}(x) = P_{2n+1}(x) = 1 - \frac{x^2}{2!} + \frac{x^4}{4!} - \cdots + (-1)^n \frac{x^{2n}}{(2n)!}.$$

Figure 8.18 shows how well these polynomials approximate $f(x) = \cos x$ near $x = 0$. Only the right-hand portions of the graphs are given because the graphs are symmetric about the y-axis. ❑

EXAMPLE 4   *A function f whose Taylor series converges at every x but converges to f(x) only at x = 0*

It can be shown (though not easily) that

$$f(x) = \begin{cases} 0, & x = 0 \\ e^{-1/x^2}, & x \neq 0 \end{cases}$$

(Fig. 8.19) has derivatives of all orders at $x = 0$ and that $f^{(n)}(0) = 0$ for all n. This means that the Taylor series generated by f at $x = 0$ is

$$f(0) + f'(0)x + \frac{f''(0)}{2!}x^2 + \cdots + \frac{f^{(n)}(0)}{n!}x^n + \cdots$$

$$= 0 + 0 \cdot x + 0 \cdot x^2 + \cdots + 0 \cdot x^n + \cdots$$

$$= 0 + 0 + \cdots + 0 + \cdots.$$

The series converges for every x (its sum is 0) but converges to $f(x)$ only at $x = 0$. ❑

Two questions still remain.

1.   For what values of x can we normally expect a Taylor series to converge to its generating function?

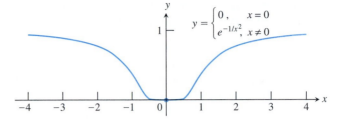

$$y = \begin{cases} 0, & x = 0 \\ e^{-1/x^2}, & x \neq 0 \end{cases}$$

8.19 The graph of the continuous extension of $y = e^{-1/x^2}$ is so flat at the origin that all of its derivatives there are zero (Example 4).

2. How accurately do a function's Taylor polynomials approximate the function on a given interval?

The answers are provided by a theorem of Taylor in the next section.

Exercises 8.9

Finding Taylor Polynomials

In Exercises 1–8, find the Taylor polynomials of orders 0, 1, 2, and 3 generated by f at a.

1. $f(x) = \ln x, \quad a = 1$

2. $f(x) = \ln(1 + x), \quad a = 0$

3. $f(x) = 1/x, \quad a = 2$

4. $f(x) = 1/(x + 2), \quad a = 0$

5. $f(x) = \sin x, \quad a = \pi/4$

6. $f(x) = \cos x, \quad a = \pi/4$

7. $f(x) = \sqrt{x}, \quad a = 4$

8. $f(x) = \sqrt{x + 4}, \quad a = 0$

Finding Maclaurin Series

Find the Maclaurin series for the functions in Exercises 9–20.

9. e^{-x}

10. $e^{x/2}$

11. $\dfrac{1}{1 + x}$

12. $\dfrac{1}{1 - x}$

13. $\sin 3x$

14. $\sin \dfrac{x}{2}$

15. $7 \cos(-x)$

16. $5 \cos \pi x$

17. $\cosh x = \dfrac{e^x + e^{-x}}{2}$

18. $\sinh x = \dfrac{e^x - e^{-x}}{2}$

19. $x^4 - 2x^3 - 5x + 4$

20. $(x + 1)^2$

Finding Taylor Series

In Exercises 21–28, find the Taylor series generated by f at $x = a$.

21. $f(x) = x^3 - 2x + 4, \quad a = 2$

22. $f(x) = 2x^3 + x^2 + 3x - 8, \quad a = 1$

23. $f(x) = x^4 + x^2 + 1, \quad a = -2$

24. $f(x) = 3x^5 - x^4 + 2x^3 + x^2 - 2, \quad a = -1$

25. $f(x) = 1/x^2, \quad a = 1$

26. $f(x) = x/(1 - x), \quad a = 0$

27. $f(x) = e^x, \quad a = 2$

28. $f(x) = 2^x, \quad a = 1$

Theory and Examples

29. Use the Taylor series generated by e^x at $x = a$ to show that

$$e^x = e^a \left[1 + (x - a) + \frac{(x - a)^2}{2!} + \cdots \right].$$

30. *(Continuation of Exercise 29.)* Find the Taylor series generated by e^x at $x = 1$. Compare your answer with the formula in Exercise 29.

31. Let $f(x)$ have derivatives through order n at $x = a$. Show that the Taylor polynomial of order n and its first n derivatives have the same values that f and its first n derivatives have at $x = a$.

32. *Of all polynomials of degree $\leq n$, the Taylor polynomial of order n gives the best approximation.* Suppose that $f(x)$ is differentiable on an interval centered at $x = a$ and that $g(x) = b_0 + b_1(x - a) + \cdots + b_n(x - a)^n$ is a polynomial of degree n with constant coefficients $b_0, \cdots, b_n$. Let $E(x) = f(x) - g(x)$. Show that if we impose on g the conditions

a) $E(a) = 0$ The approximation error is zero at $x = a$.

b) $\lim\limits_{x \to a} \dfrac{E(x)}{(x - a)^n} = 0$,   The error is negligible when compared to $(x - a)^n$.

then

$$g(x) = f(a) + f'(a)(x - a) + \frac{f''(a)}{2!}(x - a)^2 + \cdots$$
$$+ \frac{f^{(n)}(a)}{n!}(x - a)^n.$$

Thus, the Taylor polynomial $P_n(x)$ is the only polynomial of degree less than or equal to n whose error is both zero at $x = a$ and negligible when compared with $(x - a)^n$.

Quadratic Approximations

The Taylor polynomial of order 2 generated by a twice-differentiable function $f(x)$ at $x = a$ is called the **quadratic approximation** of f at $x = a$. In Exercises 33–38, find the (a) linearization (Taylor polynomial of order 1) and (b) quadratic approximation of f at $x = 0$.

33. $f(x) = \ln(\cos x)$ **34.** $f(x) = e^{\sin x}$

35. $f(x) = 1/\sqrt{1 - x^2}$ **36.** $f(x) = \cosh x$

37. $f(x) = \sin x$ **38.** $f(x) = \tan x$

8.10

Convergence of Taylor Series; Error Estimates

This section addresses the two questions left unanswered by Section 8.9:

1.   When does a Taylor series converge to its generating function?
2.   How accurately do a function's Taylor polynomials approximate the function on a given interval?

Taylor's Theorem

We answer these questions with the following theorem.

> **Theorem 16**
> **Taylor's Theorem**
> If f and its first n derivatives $f', f'', \ldots, f^{(n)}$ are continuous on $[a, b]$ or on $[b, a]$, and $f^{(n)}$ is differentiable on (a, b) or on (b, a), then there exists a number c between a and b such that
> $$f(b) = f(a) + f'(a)(b - a) + \frac{f''(a)}{2!}(b - a)^2 + \cdots$$
> $$+ \frac{f^{(n)}(a)}{n!}(b - a)^n + \frac{f^{(n+1)}(c)}{(n + 1)!}(b - a)^{n+1}.$$

Taylor's theorem is a generalization of the Mean Value Theorem (Exercise 39). There is a proof of Taylor's theorem at the end of this section.

When we apply Taylor's theorem, we usually want to hold a fixed and treat b as an independent variable. Taylor's formula is easier to use in circumstances like these if we change b to x. Here is how the theorem reads with this change.

Corollary to Taylor's Theorem
Taylor's Formula

If f has derivatives of all orders in an open interval I containing a, then for each positive integer n and for each x in I,

$$f(x) = f(a) + f'(a)(x - a) + \frac{f''(a)}{2!}(x - a)^2 + \cdots$$

$$+ \frac{f^{(n)}(a)}{n!}(x - a)^n + R_n(x), \tag{1}$$

where

$$R_n(x) = \frac{f^{(n+1)}(c)}{(n+1)!}(x - a)^{n+1} \qquad \text{for some } c \text{ between } a \text{ and } x. \tag{2}$$

When we state Taylor's theorem this way, it says that for each x in I,

$$f(x) = P_n(x) + R_n(x).$$

Pause for a moment to think about how remarkable this equation is. For any value of n we want, the equation gives both a polynomial approximation of f of that order and a formula for the error involved in using that approximation over the interval I.

Equation (1) is called **Taylor's formula.** The function $R_n(x)$ is called the **remainder of order n** or the **error term** for the approximation of f by $P_n(x)$ over I. If $R_n(x) \to 0$ as $n \to \infty$ for all x in I, we say that the Taylor series generated by f at $x = a$ **converges** to f on I, and we write

$$f(x) = \sum_{k=0}^{\infty} \frac{f^{(k)}(a)}{k!}(x - a)^k.$$

EXAMPLE 1   *The Maclaurin series for e^x*

Show that the Taylor series generated by $f(x) = e^x$ at $x = 0$ converges to $f(x)$ for every real value of x.

Solution The function has derivatives of all orders throughout the interval $I = (-\infty, \infty)$. Equations (1) and (2) with $f(x) = e^x$ and $a = 0$ give

$$e^x = 1 + x + \frac{x^2}{2!} + \cdots + \frac{x^n}{n!} + R_n(x) \qquad \text{Polynomial from Section 8.9, Example 2}$$

and

$$R_n(x) = \frac{e^c}{(n+1)!}x^{n+1} \qquad \text{for some } c \text{ between } 0 \text{ and } x.$$

Since e^x is an increasing function of x, e^c lies between $e^0 = 1$ and e^x. When x is negative, so is c, and $e^c < 1$. When x is zero, $e^x = 1$ and $R_n(x) = 0$. When x is positive, so is c, and $e^c < e^x$. Thus,

$$|R_n(x)| \leq \frac{|x|^{n+1}}{(n+1)!} \qquad \text{when } x \leq 0,$$

and

$$|R_n(x)| < e^x \frac{x^{n+1}}{(n+1)!} \qquad \text{when } x > 0.$$

Finally, because

$$\lim_{n \to \infty} \frac{x^{n+1}}{(n+1)!} = 0 \qquad \text{for every } x, \qquad \text{Section 8.2}$$

$\lim_{n \to \infty} R_n(x) = 0$, and the series converges to e^x for every x.

$$e^x = \sum_{k=0}^{\infty} \frac{x^k}{k!} = 1 + x + \frac{x^2}{2!} + \cdots + \frac{x^k}{k!} + \cdots.$$

Estimating the Remainder

It is often possible to estimate $R_n(x)$ as we did in Example 1. This method of estimation is so convenient that we state it as a theorem for future reference.

Theorem 17
The Remainder Estimation Theorem

If there are positive constants M and r such that $|f^{(n+1)}(t)| \leq Mr^{n+1}$ for all t between a and x, inclusive, then the remainder term $R_n(x)$ in Taylor's theorem satisfies the inequality

$$|R_n(x)| \leq M \frac{r^{n+1}|x - a|^{n+1}}{(n+1)!}.$$

If these conditions hold for every n and all the other conditions of Taylor's theorem are satisfied by f, then the series converges to $f(x)$.

In the simplest examples, we can take $r = 1$ provided f and all its derivatives are bounded in magnitude by some constant M. In other cases, we may need to consider r. For example, if $f(x) = 2\cos(3x)$, each time we differentiate we get a factor of 3 and r needs to be greater than 1. In this particular case, we can take $r = 3$ along with $M = 2$.

We are now ready to look at some examples of how the Remainder Estimation Theorem and Taylor's theorem can be used together to settle questions of convergence. As you will see, they can also be used to determine the accuracy with which a function is approximated by one of its Taylor polynomials.

EXAMPLE 2   *The Maclaurin series for* sin *x*

Show that the Maclaurin series for sin x converges to sin x for all x.

Solution   The function and its derivatives are

$$f(x) = \sin x, \qquad f'(x) = \cos x,$$
$$f''(x) = -\sin x, \qquad f'''(x) = -\cos x,$$
$$\vdots \qquad\qquad\qquad \vdots$$
$$f^{(2k)}(x) = (-1)^k \sin x, \qquad f^{(2k+1)}(x) = (-1)^k \cos x,$$

so

$$f^{(2k)}(0) = 0 \qquad \text{and} \qquad f^{(2k+1)}(0) = (-1)^k.$$

The series has only odd-powered terms and, for $n = 2k + 1$, Taylor's theorem gives

$$\sin x = x - \frac{x^3}{3!} + \frac{x^5}{5!} - \cdots + \frac{(-1)^k x^{2k+1}}{(2k+1)!} + R_{2k+1}(x).$$

All the derivatives of $\sin x$ have absolute values less than or equal to 1, so we can apply the Remainder Estimation Theorem with $M = 1$ and $r = 1$ to obtain

$$|R_{2k+1}(x)| \le 1 \cdot \frac{|x|^{2k+2}}{(2k+2)!}.$$

Since $(|x|^{2k+2}/(2k+2)!) \to 0$ as $k \to \infty$, whatever the value of x, $R_{2k+1}(x) \to 0$, and the Maclaurin series for $\sin x$ converges to $\sin x$ for every x.

$$\sin x = \sum_{k=0}^{\infty} \frac{(-1)^k x^{2k+1}}{(2k+1)!} = x - \frac{x^3}{3!} + \frac{x^5}{5!} - \frac{x^7}{7!} + \cdots. \qquad (3)$$

❏

EXAMPLE 3   *The Maclaurin series for cos x*

Show that the Maclaurin series for $\cos x$ converges to $\cos x$ for every value of x.

Solution   We add the remainder term to the Taylor polynomial for $\cos x$ (Section 8.9, Example 3) to obtain Taylor's formula for $\cos x$ with $n = 2k$:

$$\cos x = 1 - \frac{x^2}{2!} + \frac{x^4}{4!} - \cdots + (-1)^k \frac{x^{2k}}{(2k)!} + R_{2k}(x).$$

Because the derivatives of the cosine have absolute value less than or equal to 1, the Remainder Estimation Theorem with $M = 1$ and $r = 1$ gives

$$|R_{2k}(x)| \le 1 \cdot \frac{|x|^{2k+1}}{(2k+1)!}.$$

For every value of x, $R_{2k} \to 0$ as $k \to \infty$. Therefore, the series converges to $\cos x$ for every value of x.

$$\cos x = \sum_{k=0}^{\infty} \frac{(-1)^k x^{2k}}{(2k)!} = 1 - \frac{x^2}{2!} + \frac{x^4}{4!} - \frac{x^6}{6!} + \cdots. \qquad (4)$$

❏

EXAMPLE 4 *Finding a Maclaurin series by substitution*

Find the Maclaurin series for $\cos 2x$.

Solution We can find the Maclaurin series for $\cos 2x$ by substituting $2x$ for x in the Maclaurin series for $\cos x$:

$$\cos 2x = \sum_{k=0}^{\infty} \frac{(-1)^k (2x)^{2k}}{(2k)!} = 1 - \frac{(2x)^2}{2!} + \frac{(2x)^4}{4!} - \frac{(2x)^6}{6!} + \cdots \qquad \text{Eq. (4) with } 2x \text{ for } x$$

$$= 1 - \frac{2^2 x^2}{2!} + \frac{2^4 x^4}{4!} - \frac{2^6 x^6}{6!} + \cdots$$

$$= \sum_{k=0}^{\infty} (-1)^k \frac{2^{2k} x^{2k}}{(2k)!}.$$

Eq. (4) holds for $-\infty < x < \infty$, implying that it holds for $-\infty < 2x < \infty$, so the newly created series converges for all x. Exercise 45 explains why the series is in fact the Maclaurin series for $\cos 2x$. ❏

EXAMPLE 5 *Finding a Maclaurin series by multiplication*

Find the Maclaurin series for $x \sin x$.

Solution We can find the Maclaurin series for $x \sin x$ by multiplying the Maclaurin series for $\sin x$ (Eq. 3) by x:

$$x \sin x = x \left(x - \frac{x^3}{3!} + \frac{x^5}{5!} - \frac{x^7}{7!} + \cdots \right)$$

$$= x^2 - \frac{x^4}{3!} + \frac{x^6}{5!} - \frac{x^8}{7!} + \cdots.$$

The new series converges for all x because the series for $\sin x$ converges for all x. Exercise 45 explains why the series is the Maclaurin series for $x \sin x$. ❏

Truncation Error

The Maclaurin series for e^x converges to e^x for all x. But we still need to decide how many terms to use to approximate e^x to a given degree of accuracy. We get this information from the Remainder Estimation Theorem.

EXAMPLE 6 Calculate e with an error of less than 10^{-6}.

Solution We can use the result of Example 1 with $x = 1$ to write

$$e = 1 + 1 + \frac{1}{2!} + \cdots + \frac{1}{n!} + R_n(1),$$

with

$$R_n(1) = e^c \frac{1}{(n+1)!} \qquad \text{for some } c \text{ between } 0 \text{ and } 1.$$

For the purposes of this example, we assume that we know that $e < 3$. Hence, we are certain that

$$\frac{1}{(n+1)!} < R_n(1) < \frac{3}{(n+1)!}$$

because $1 < e^c < 3$ for $0 < c < 1$.

By experiment we find that $1/9! > 10^{-6}$, while $3/10! < 10^{-6}$. Thus we should take $(n+1)$ to be at least 10, or n to be at least 9. With an error of less than 10^{-6},

$$e = 1 + 1 + \frac{1}{2} + \frac{1}{3!} + \cdots + \frac{1}{9!} \approx 2.7182\ 82.$$ ❑

EXAMPLE 7   For what values of x can we replace $\sin x$ by $x - (x^3/3!)$ with an error of magnitude no greater than 3×10^{-4}?

Solution   Here we can take advantage of the fact that the Maclaurin series for $\sin x$ is an alternating series for every nonzero value of x. According to the Alternating Series Estimation Theorem (Section 8.7), the error in truncating

$$\sin x = x - \frac{x^3}{3!} + \frac{x^5}{5!} - \cdots$$

after $(x^3/3!)$ is no greater than

$$\left| \frac{x^5}{5!} \right| = \frac{|x|^5}{120}.$$

Therefore the error will be less than or equal to 3×10^{-4} if

$$\frac{|x|^5}{120} < 3 \times 10^{-4} \qquad \text{or} \qquad |x| < \sqrt[5]{360 \times 10^{-4}} \approx 0.514. \qquad \text{Rounded down, to be safe}$$

The Alternating Series Estimation Theorem tells us something that the Remainder Estimation Theorem does not: namely, that the estimate $x - (x^3/3!)$ for $\sin x$ is an underestimate when x is positive because then $x^5/120$ is positive.

Figure 8.20 shows the graph of $\sin x$, along with the graphs of a number of its approximating Taylor polynomials. The graph of $P_3(x) = x - (x^3/3!)$ is almost indistinguishable from the sine curve when $-1 \le x \le 1$.

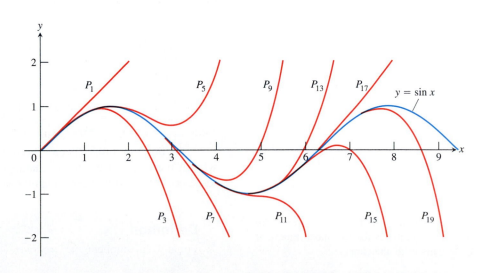

8.20 The polynomials

$$P_{2n+1}(x) = \sum_{k=0}^{n} \frac{(-1)^k x^{2k+1}}{(2k+1)!}$$

converge to $\sin x$ as $n \to \infty$.

You might wonder how the estimate given by the Remainder Estimation Theorem compares with the one just obtained from the Alternating Series Estimation Theorem. If we write

$$\sin x = x - \frac{x^3}{3!} + R_3,$$

then the Remainder Estimation Theorem gives

$$|R_3| \leq 1 \cdot \frac{|x|^4}{4!} = \frac{|x|^4}{24},$$

which is not as good. But if we recognize that $x - (x^3/3!) = 0 + x + 0x^2 - (x^3/3!) + 0x^4$ is the Taylor polynomial of order 4 as well as of order 3, then

$$\sin x = x - \frac{x^3}{3!} + 0 + R_4,$$

and the Remainder Estimation Theorem with $M = r = 1$ gives

$$|R_4| \leq 1 \cdot \frac{|x|^5}{5!} = \frac{|x|^5}{120}.$$

This is what we had from the Alternating Series Estimation Theorem. ❑

Combining Taylor Series

On the intersection of their intervals of convergence, Taylor series can be added, subtracted, and multiplied by constants, and the results are once again Taylor series. The Taylor series for $f(x) + g(x)$ is the sum of the Taylor series for $f(x)$ and $g(x)$ because the nth derivative of $f + g$ is $f^{(n)} + g^{(n)}$, and so on. Thus we obtain the Maclaurin series for $(1 + \cos 2x)/2$ by adding 1 to the Maclaurin series for $\cos 2x$ and dividing the combined results by 2, and the Maclaurin series for $\sin x + \cos x$ is the term-by-term sum of the Maclaurin series for $\sin x$ and $\cos x$.

✳ Euler's Formula

As you may recall, a complex number is a number of the form $a + bi$, where a and b are real numbers and $i = \sqrt{-1}$. If we substitute $x = i\theta$ (θ real) in the Maclaurin series for e^x and use the relations

$$i^2 = -1, \qquad i^3 = i^2 i = -i, \qquad i^4 = i^2 i^2 = 1, \qquad i^5 = i^4 i = i,$$

and so on, to simplify the result, we obtain

$$e^{i\theta} = 1 + \frac{i\theta}{1!} + \frac{i^2\theta^2}{2!} + \frac{i^3\theta^3}{3!} + \frac{i^4\theta^4}{4!} + \frac{i^5\theta^5}{5!} + \frac{i^6\theta^6}{6!} + \cdots$$

$$= \left(1 - \frac{\theta^2}{2!} + \frac{\theta^4}{4!} - \frac{\theta^6}{6!} + \cdots\right) + i\left(\theta - \frac{\theta^3}{3!} + \frac{\theta^5}{5!} - \cdots\right) = \cos\theta + i\sin\theta.$$

This does not *prove* that $e^{i\theta} = \cos\theta + i\sin\theta$ because we have not yet defined what it means to raise e to an imaginary power. But it does say how to define $e^{i\theta}$ to be consistent with other things we know.

One of the amazing consequences of Euler's formula is the equation

$$e^{i\pi} = -1.$$

When written in the form $e^{i\pi} + 1 = 0$, this equation combines the five most important constants in mathematics.

Definition

For any real number θ, $e^{i\theta} = \cos\theta + i\sin\theta$. $\qquad\qquad$ (5)

Equation (5), called **Euler's formula,** enables us to define e^{a+bi} to be $e^a \cdot e^{bi}$ for any complex number $a + bi$.

A Proof of Taylor's Theorem

We prove Taylor's theorem assuming $a < b$. The proof for $a > b$ is nearly the same.

The Taylor polynomial

$$P_n(x) = f(a) + f'(a)(x - a) + \frac{f''(a)}{2!}(x - a)^2 + \cdots + \frac{f^{(n)}(a)}{n!}(x - a)^n$$

and its first n derivatives match the function f and its first n derivatives at $x = a$. We do not disturb that matching if we add another term of the form $K(x - a)^{n+1}$, where K is any constant, because such a term and its first n derivatives are all equal to zero at $x = a$. The new function

$$\phi_n(x) = P_n(x) + K(x - a)^{n+1}$$

and its first n derivatives still agree with f and its first n derivatives at $x = a$.

We now choose the particular value of K that makes the curve $y = \phi_n(x)$ agree with the original curve $y = f(x)$ at $x = b$. In symbols,

$$f(b) = P_n(b) + K(b - a)^{n+1}, \quad \text{or} \quad K = \frac{f(b) - P_n(b)}{(b - a)^{n+1}}. \tag{6}$$

With K defined by Eq. (6), the function

$$F(x) = f(x) - \phi_n(x)$$

measures the difference between the original function f and the approximating function ϕ_n for each x in $[a, b]$.

We now use Rolle's theorem (Section 3.2). First, because $F(a) = F(b) = 0$ and both F and F' are continuous on $[a, b]$, we know that

$$F'(c_1) = 0 \quad \text{for some } c_1 \text{ in } (a, b).$$

Next, because $F'(a) = F'(c_1) = 0$ and both F' and F'' are continuous on $[a, c_1]$, we know that

$$F''(c_2) = 0 \quad \text{for some } c_2 \text{ in } (a, c_1).$$

Rolle's theorem, applied successively to F'', F''', ..., $F^{(n-1)}$ implies the existence of

$$c_3 \quad \text{in } (a, c_2) \quad \text{such that } F'''(c_3) = 0,$$
$$c_4 \quad \text{in } (a, c_3) \quad \text{such that } F^{(4)}(c_4) = 0,$$
$$\vdots$$
$$c_n \quad \text{in } (a, c_{n-1}) \quad \text{such that } F^{(n)}(c_n) = 0.$$

Finally, because $F^{(n)}$ is continuous on $[a, c_n]$ and differentiable on (a, c_n), and $F^{(n)}(a) = F^{(n)}(c_n) = 0$, Rolle's theorem implies that there is a number c_{n+1} in (a, c_n) such that

$$F^{(n+1)}(c_{n+1}) = 0. \tag{7}$$

If we differentiate $F(x) = f(x) - P_n(x) - K(x - a)^{n+1}$ a total of $n + 1$ times, we get

$$F^{(n+1)}(x) = f^{(n+1)}(x) - 0 - (n+1)!K. \qquad (8)$$

Equations (7) and (8) together give

$$K = \frac{f^{(n+1)}(c)}{(n+1)!} \quad \text{for some number } c = c_{n+1} \text{ in } (a, b). \qquad (9)$$

Equations (6) and (9) give

$$f(b) = P_n(b) + \frac{f^{(n+1)}(c)}{(n+1)!}(b - a)^{n+1}.$$

This concludes the proof.   ❑

Exercises 8.10

Maclaurin Series by Substitution

Use substitution (as in Example 4) to find the Maclaurin series of the functions in Exercises 1–6.

1. e^{-5x}

2. $e^{-x/2}$

3. $5 \sin(-x)$

4. $\sin\left(\dfrac{\pi x}{2}\right)$

5. $\cos \sqrt{x}$

6. $\cos(x^{3/2}/\sqrt{2})$

More Maclaurin Series

Find Maclaurin series for the functions in Exercises 7–18.

7. xe^x

8. $x^2 \sin x$

9. $\dfrac{x^2}{2} - 1 + \cos x$

10. $\sin x - x + \dfrac{x^3}{3!}$

11. $x \cos \pi x$

12. $x^2 \cos(x^2)$

13. $\cos^2 x$ (*Hint:* $\cos^2 x = (1 + \cos 2x)/2$.)

14. $\sin^2 x$

15. $\dfrac{x^2}{1 - 2x}$

16. $x \ln(1 + 2x)$

17. $\dfrac{1}{(1 - x)^2}$

18. $\dfrac{2}{(1 - x)^3}$

Error Estimates

19. For approximately what values of x can you replace $\sin x$ by $x - (x^3/6)$ with an error of magnitude no greater than 5×10^{-4}? Give reasons for your answer.

20. If $\cos x$ is replaced by $1 - (x^2/2)$ and $|x| < 0.5$, what estimate can be made of the error? Does $1 - (x^2/2)$ tend to be too large, or too small? Give reasons for your answer.

21. How close is the approximation $\sin x = x$ when $|x| < 10^{-3}$? For which of these values of x is $x < \sin x$?

22. The estimate $\sqrt{1 + x} = 1 + (x/2)$ is used when x is small. Estimate the error when $|x| < 0.01$.

23. The approximation $e^x = 1 + x + (x^2/2)$ is used when x is small. Use the Remainder Estimation Theorem to estimate the error when $|x| < 0.1$.

24. (*Continuation of Exercise 23.*) When $x < 0$, the series for e^x is an alternating series. Use the Alternating Series Estimation Theorem to estimate the error that results from replacing e^x by $1 + x + (x^2/2)$ when $-0.1 < x < 0$. Compare your estimate with the one you obtained in Exercise 23.

25. Estimate the error in the approximation $\sinh x = x + (x^3/3!)$ when $|x| < 0.5$. (*Hint:* Use R_4, not R_3.)

26. When $0 \le h \le 0.01$, show that e^h may be replaced by $1 + h$ with an error of magnitude no greater than 0.6% of h. Use $e^{0.01} = 1.01$.

27. For what positive values of x can you replace $\ln(1 + x)$ by x with an error of magnitude no greater than 1% of the value of x?

28. You plan to estimate $\pi/4$ by evaluating the Maclaurin series for $\tan^{-1} x$ at $x = 1$. Use the Alternating Series Estimation Theorem to determine how many terms of the series you would have to add to be sure the estimate is good to 2 decimal places.

29. a) Use the Maclaurin series for $\sin x$ and the Alternating Series Estimation Theorem to show that

$$1 - \frac{x^2}{6} < \frac{\sin x}{x} < 1, \quad x \ne 0.$$

b) GRAPHER Graph $f(x) = (\sin x)/x$ together with the functions $y = 1 - (x^2/6)$ and $y = 1$ for $-5 \le x \le 5$. Comment on the relationships among the graphs.

30. a) Use the Maclaurin series for $\cos x$ and the Alternating Series Estimation Theorem to show that

$$\frac{1}{2} - \frac{x^2}{24} < \frac{1 - \cos x}{x^2} < \frac{1}{2}, \quad x \ne 0.$$

(This is the inequality in Section 1.2, Exercise 46.)

b)   **GRAPHER** Graph $f(x) = (1 - \cos x)/x^2$ together with $y = (1/2) - (x^2/24)$ and $y = 1/2$ for $-9 \le x \le 9$. Comment on the relationships among the graphs.

Finding and Identifying Maclaurin Series

Each of the series in Exercises 31–34 is the value of the Maclaurin series of a function $f(x)$ at some point. What function and what point? What is the sum of the series?

31. $(0.1) - \dfrac{(0.1)^3}{3!} + \dfrac{(0.1)^5}{5!} - \cdots + \dfrac{(-1)^k (0.1)^{2k+1}}{(2k+1)!} + \cdots$

32. $1 - \dfrac{\pi^2}{4^2 \cdot 2!} + \dfrac{\pi^4}{4^4 \cdot 4!} - \cdots + \dfrac{(-1)^k (\pi)^{2k}}{4^{2k} \cdot (2k!)} + \cdots$

33. $\dfrac{\pi}{3} - \dfrac{\pi^3}{3^3 \cdot 3} + \dfrac{\pi^5}{3^5 \cdot 5} - \cdots + \dfrac{(-1)^k \pi^{2k+1}}{3^{2k+1}(2k+1)} + \cdots$

34. $\pi - \dfrac{\pi^2}{2} + \dfrac{\pi^3}{3} - \cdots + (-1)^{k-1}\dfrac{\pi^k}{k} + \cdots$

35. Multiply the Maclaurin series for e^x and $\sin x$ together to find the first five nonzero terms of the Maclaurin series for $e^x \sin x$.

36. Multiply the Maclaurin series for e^x and $\cos x$ together to find the first five nonzero terms of the Maclaurin series for $e^x \cos x$.

37. Use the identity $\sin^2 x = (1 - \cos 2x)/2$ to obtain the Maclaurin series for $\sin^2 x$. Then differentiate this series to obtain the Maclaurin series for $2 \sin x \cos x$. Check that this is the series for $\sin 2x$.

38. (*Continuation of Exercise 37.*) Use the identity $\cos^2 x = \cos 2x + \sin^2 x$ to obtain a power series for $\cos^2 x$.

Theory and Examples

39. *Taylor's theorem and the Mean Value Theorem.* Explain how the Mean Value Theorem (Section 3.2, Theorem 4) is a special case of Taylor's theorem.

40. *Linearizations at inflection points* (*Continuation of Section 3.7, Exercise 63*). Show that if the graph of a twice-differentiable function $f(x)$ has an inflection point at $x = a$, then the linearization of f at $x = a$ is also the quadratic approximation of f at $x = a$. This explains why tangent lines fit so well at inflection points.

41. *The (second) second derivative test.* Use the equation

$$f(x) = f(a) + f'(a)(x - a) + \frac{f''(c_2)}{2}(x - a)^2$$

to establish the following test.

Let f have continuous first and second derivatives and suppose that $f'(a) = 0$. Then

a) f has a local maximum at a if $f'' \le 0$ throughout an interval whose interior contains a;

b) f has a local minimum at a if $f'' \ge 0$ throughout an interval whose interior contains a.

42. *A cubic approximation.* Use Taylor's formula with $a = 0$ and $n = 3$ to find the standard cubic approximation of $f(x) = 1/(1 - x)$ at $x = 0$. Give an upper bound for the magnitude of the error in the approximation when $|x| \le 0.1$.

43. a)   Use Taylor's formula with $n = 2$ to find the quadratic approximation of $f(x) = (1 + x)^k$ at $x = 0$ (k a constant).

b)   If $k = 3$, for approximately what values of x in the interval $[0, 1]$ will the error in the quadratic approximation be less than $1/100$?

44. *Improving approximations to π.*

a)   Let P be an approximation of π accurate to n decimals. Show that $P + \sin P$ gives an approximation correct to $3n$ decimals. (*Hint:* Let $P = \pi + x$.)

b)   Try it with a calculator.

45. *The Maclaurin series generated by $f(x) = \sum_{n=0}^{\infty} a_n x^n$ is $\sum_{n=0}^{\infty} a_n x^n$.* A function defined by a power series $\sum_{n=0}^{\infty} a_n x^n$ with a radius of convergence $c > 0$ has a Maclaurin series that converges to the function at every point of $(-c, c)$. Show this by showing that the Maclaurin series generated by $f(x) = \sum_{n=0}^{\infty} a_n x^n$ is the series $\sum_{n=0}^{\infty} a_n x^n$ itself.

An immediate consequence of this is that series like

$$x \sin x = x^2 - \frac{x^4}{3!} + \frac{x^6}{5!} - \frac{x^8}{7!} + \cdots$$

and

$$x^2 e^x = x^2 + x^3 + \frac{x^4}{2!} + \frac{x^5}{3!} + \cdots,$$

obtained by multiplying Maclaurin series by powers of x, as well as series obtained by integration and differentiation of convergent power series, are themselves the Maclaurin series generated by the functions they represent.

46. *Maclaurin series for even functions and odd functions* (*Continuation of Section 8.8, Exercise 45*). Suppose that $f(x) = \sum_{n=0}^{\infty} a_n x^n$ converges for all x in an open interval $(-c, c)$. Show that

a)   If f is even, then $a_1 = a_3 = a_5 = \cdots = 0$, i.e., the series for f contains only even powers of x.

b)   If f is odd, then $a_0 = a_2 = a_4 = \cdots = 0$, i.e., the series for f contains only odd powers of x.

47. *Taylor polynomials of periodic functions*

a)   Show that every continuous periodic function $f(x)$, $-\infty < x < \infty$, is bounded in magnitude by showing that there exists a positive constant M such that $|f(x)| \le M$ for all x.

b)   Show that the graph of every Taylor polynomial of positive degree generated by $f(x) = \cos x$ must eventually move away from the graph of $\cos x$ as $|x|$ increases. You can see this in Fig. 8.18. The Taylor polynomials of $\sin x$ behave in a similar way (Fig. 8.20).

48. GRAPHER

a)   Graph the curves $y = (1/3) - (x^2)/5$ and $y = (x - \tan^{-1} x)/x^3$ together with the line $y = 1/3$.

b)   Use a Maclaurin series to explain what you see. What is

$$\lim_{x \to 0} \frac{x - \tan^{-1} x}{x^3} \, ?$$

Euler's Formula

49. Use Eq. (5) to write the following powers of e in the form $a + bi$.

a) $e^{-i\pi}$ **b)** $e^{i\pi/4}$ **c)** $e^{-i\pi/2}$

50. *Euler's identities.* Use Eq. (5) to show that

$$\cos\theta = \frac{e^{i\theta} + e^{-i\theta}}{2} \quad \text{and} \quad \sin\theta = \frac{e^{i\theta} - e^{-i\theta}}{2i}.$$

51. Establish the equations in Exercise 50 by combining the formal Maclaurin series for $e^{i\theta}$ and $e^{-i\theta}$.

52. Show that

a) $\cosh i\theta = \cos\theta$,   **b)** $\sinh i\theta = i\sin\theta$.

53. By multiplying the Maclaurin series for e^x and $\sin x$, find the terms through x^5 of the Maclaurin series for $e^x \sin x$. This series is the imaginary part of the series for

$$e^x \cdot e^{ix} = e^{(1+i)x}.$$

Use this fact to check your answer. For what values of x should the series for $e^x \sin x$ converge?

54. When a and b are real, we define $e^{(a+ib)x}$ with the equation

$$e^{(a+ib)x} = e^{ax} \cdot e^{ibx} = e^{ax}(\cos bx + i\sin bx).$$

Differentiate the right-hand side of this equation to show that

$$\frac{d}{dx} e^{(a+ib)x} = (a + ib)e^{(a+ib)x}.$$

Thus the familiar rule $(d/dx)e^{kx} = ke^{kx}$ holds for k complex as well as real.

55. Use the definition of $e^{i\theta}$ to show that for any real numbers θ, θ_1, and θ_2,

a) $e^{i\theta_1}e^{i\theta_2} = e^{i(\theta_1 + \theta_2)}$,

b) $e^{-i\theta} = 1/e^{i\theta}$.

56. Two complex numbers $a + ib$ and $c + id$ are equal if and only if $a = c$ and $b = d$. Use this fact to evaluate

$$\int e^{ax}\cos bx \, dx \quad \text{and} \quad \int e^{ax}\sin bx \, dx$$

from

$$\int e^{(a+ib)x} \, dx = \frac{a - ib}{a^2 + b^2} e^{(a+ib)x} + C,$$

where $C = C_1 + iC_2$ is a complex constant of integration.

CAS Explorations and Projects—Linear, Quadratic, and Cubic Approximations

Taylor's formula with $n = 1$ and $a = 0$ gives the linearization of a function at $x = 0$. With $n = 2$ and $n = 3$ we obtain the standard quadratic and cubic approximations. In these exercises we explore the errors associated with these approximations. We seek answers to two questions:

a) For what values of x can the function be replaced by each approximation with an error less than 10^{-2}?

b) What is the maximum error we could expect if we replace the function by each approximation over the specified interval?

Using a CAS, perform the following steps to aid in answering questions (a) and (b) for the functions and intervals in Exercises 57–62.

Step 1: Plot the function over the specified interval.

Step 2: Find the Taylor polynomials $P_1(x)$, $P_2(x)$, and $P_3(x)$ at $x = 0$.

Step 3: Calculate the $(n + 1)$st derivative $f^{(n+1)}(c)$ associated with the remainder term for each Taylor polynomial. Plot the derivative as a function of c over the specified interval and estimate its maximum absolute value, M.

Step 4: Calculate the remainder $R_n(x)$ for each polynomial. Using the estimate M from step 3 in place of $f^{(n+1)}(c)$, plot $R_n(x)$ over the specified interval. Then estimate the values of x that answer question (a).

Step 5: Compare your estimated error with the actual error $E_n(x) = |f(x) - P_n(x)|$ by plotting $E_n(x)$ over the specified interval. This will help answer question (b).

Step 6: Graph the function and its three Taylor approximations together. Discuss the graphs in relation to the information discovered in steps 4 and 5.

57. $f(x) = \dfrac{1}{\sqrt{1+x}}, \quad |x| \le \dfrac{3}{4}$

58. $f(x) = (1 + x)^{3/2}, \quad -\dfrac{1}{2} \le x \le 2$

59. $f(x) = \dfrac{x}{x^2 + 1}, \quad |x| \le 2$

60. $f(x) = (\cos x)(\sin 2x), \quad |x| \le 2$

61. $f(x) = e^{-x}\cos 2x, \quad |x| \le 1$

62. $f(x) = e^{x/3}\sin 2x, \quad |x| \le 2$

8.11

Applications of Power Series

This section introduces the binomial series for estimating powers and roots and shows how series are sometimes used to approximate the solution of an initial value problem, to evaluate nonelementary integrals, and to evaluate limits that lead to indeterminate forms. We provide a self-contained derivation of the Maclaurin series for $\tan^{-1} x$ and conclude with a reference table of frequently used series.

The Binomial Series for Powers and Roots

The Maclaurin series generated by $f(x) = (1+x)^m$, when m is constant, is

$$1 + mx + \frac{m(m-1)}{2!} x^2 + \frac{m(m-1)(m-2)}{3!} x^3 + \cdots$$

$$+ \frac{m(m-1)(m-2)\cdots(m-k+1)}{k!} x^k + \cdots. \quad (1)$$

This series, called the **binomial series,** converges absolutely for $|x| < 1$. To derive the series, we first list the function and its derivatives:

$$f(x) = (1+x)^m$$

$$f'(x) = m(1+x)^{m-1}$$

$$f''(x) = m(m-1)(1+x)^{m-2}$$

$$f'''(x) = m(m-1)(m-2)(1+x)^{m-3}$$

$$\vdots$$

$$f^{(k)}(x) = m(m-1)(m-2)\cdots(m-k+1)(1+x)^{m-k}.$$

We then evaluate these at $x = 0$ and substitute into the Maclaurin series formula to obtain the series in (1).

If m is an integer greater than or equal to zero, the series stops after $(m+1)$ terms because the coefficients from $k = m+1$ on are zero.

If m is not a positive integer or zero, the series is infinite and converges for $|x| < 1$. To see why, let u_k be the term involving x^k. Then apply the Ratio Test for absolute convergence to see that

$$\left| \frac{u_{k+1}}{u_k} \right| = \left| \frac{m-k}{k+1} x \right| \rightarrow |x| \qquad \text{as } k \rightarrow \infty.$$

Our derivation of the binomial series shows only that it is generated by $(1+x)^m$ and converges for $|x| < 1$. The derivation does not show that the series converges to $(1+x)^m$. It does, but we assume that part without proof.

For $-1 < x < 1$,

$$(1+x)^m = 1 + \sum_{k=1}^{\infty} \binom{m}{k} x^k, \qquad (2)$$

where we define

$$\binom{m}{1} = m, \qquad \binom{m}{2} = \frac{m(m-1)}{2!},$$

and

$$\binom{m}{k} = \frac{m(m-1)(m-2)\cdots(m-k+1)}{k!} \qquad \text{for } k \geq 3.$$

EXAMPLE 1   If $m = -1$,

$$\binom{-1}{1} = -1, \qquad \binom{-1}{2} = \frac{-1(-2)}{2!} = 1,$$

and

$$\binom{-1}{k} = \frac{-1(-2)(-3)\cdots(-1-k+1)}{k!} = (-1)^k \left(\frac{k!}{k!}\right) = (-1)^k.$$

With these coefficient values, Eq. (2) becomes the geometric series

$$(1+x)^{-1} = 1 + \sum_{k=1}^{\infty}(-1)^k x^k = 1 - x + x^2 - x^3 + \cdots + (-1)^k x^k + \cdots .$$

❑

EXAMPLE 2   We know from Section 3.7, Example 1, that $\sqrt{1-x} \approx 1 + (x/2)$ for $|x|$ small. With $m = 1/2$, the binomial series gives quadratic and higher order approximations as well, along with error estimates that come from the Alternating Series Estimation Theorem:

$$(1+x)^{1/2} = 1 + \frac{x}{2} + \frac{\left(\frac{1}{2}\right)\left(-\frac{1}{2}\right)}{2!}x^2 + \frac{\left(\frac{1}{2}\right)\left(-\frac{1}{2}\right)\left(-\frac{3}{2}\right)}{3!}x^3$$

$$+ \frac{\left(\frac{1}{2}\right)\left(-\frac{1}{2}\right)\left(-\frac{3}{2}\right)\left(-\frac{5}{2}\right)}{4!}x^4 + \cdots$$

$$= 1 + \frac{x}{2} - \frac{x^2}{8} + \frac{x^3}{16} - \frac{5x^4}{128} + \cdots .$$

Substitution for x gives still other approximations. For example,

$$\sqrt{1-x^2} \approx 1 - \frac{x^2}{2} - \frac{x^4}{8} \qquad \text{for } |x^2| \text{ small}$$

$$\sqrt{1 - \frac{1}{x}} \approx 1 - \frac{1}{2x} - \frac{1}{8x^2} \qquad \text{for } \left|\frac{1}{x}\right| \text{ small, i.e., } |x| \text{ large.}$$

❑

Power Series Solutions of Differential Equations and Initial Value Problems

When we cannot find a relatively simple expression for the solution of an initial value problem or differential equation, we try to get information about the solution in other ways. One way is to try to find a power series representation for the solution. If we can do so, we immediately have a source of polynomial approximations of the solution, which may be all that we really need. The first example (Example 3) deals with a first order linear differential equation that could be solved with the methods of Section 6.11. The example shows how, not knowing this, we can solve the equation with power series. The second example (Example 4) deals with an equation that cannot be solved by previous methods.

EXAMPLE 3   Solve the initial value problem

$$y' - y = x, \qquad y(0) = 1.$$

Solution   We assume that there is a solution of the form

$$y = a_0 + a_1 x + a_2 x^2 + \cdots + a_{n-1} x^{n-1} + a_n x^n + \cdots. \tag{3}$$

Our goal is to find values for the coefficients a_k that make the series and its first derivative

$$y' = a_1 + 2a_2 x + 3a_3 x^2 + \cdots + n a_n x^{n-1} + \cdots \tag{4}$$

satisfy the given differential equation and initial condition. The series $y' - y$ is the difference of the series in Eqs. (3) and (4):

$$y' - y = (a_1 - a_0) + (2a_2 - a_1)x + (3a_3 - a_2)x^2 + \cdots$$

$$+ (n a_n - a_{n-1})x^{n-1} + \cdots. \tag{5}$$

If y is to satisfy the equation $y' - y = x$, the series in (5) must equal x. Since power series representations are unique, as you saw if you did Exercise 45 in Section 8.8, the coefficients in Eq. (5) must satisfy the equations

$$a_1 - a_0 = 0 \qquad \text{Constant terms}$$

$$2a_2 - a_1 = 1 \qquad \text{Coefficients of } x$$

$$3a_3 - a_2 = 0 \qquad \text{Coefficients of } x^2$$

$$\vdots \qquad\qquad \vdots$$

$$n a_n - a_{n-1} = 0 \qquad \text{Coefficients of } x^{n-1}$$

$$\vdots \qquad\qquad \vdots$$

We can also see from Eq. (3) that $y = a_0$ when $x = 0$, so that $a_0 = 1$ (this being the initial condition). Putting it all together, we have

$$a_0 = 1, \qquad a_1 = a_0 = 1, \qquad a_2 = \frac{1 + a_1}{2} = \frac{1 + 1}{2} = \frac{2}{2},$$

$$a_3 = \frac{a_2}{3} = \frac{2}{3 \cdot 2} = \frac{2}{3!}, \qquad \cdots, \qquad a_n = \frac{a_{n-1}}{n} = \frac{2}{n!}, \qquad \cdots$$

Substituting these coefficient values into the equation for y (Eq. 3) gives

$$y = 1 + x + 2 \cdot \frac{x^2}{2!} + 2 \cdot \frac{x^3}{3!} + \cdots + 2 \cdot \frac{x^n}{n!} + \cdots$$

$$= 1 + x + 2 \underbrace{\left(\frac{x^2}{2!} + \frac{x^3}{3!} + \cdots + \frac{x^n}{n!} + \cdots \right)}_{\text{the Maclaurin series for } e^x - 1 - x}$$

$$= 1 + x + 2(e^x - 1 - x) = 2e^x - 1 - x.$$

The solution of the initial value problem is $y = 2e^x - 1 - x$.

As a check, we see that

$$y(0) = 2e^0 - 1 - 0 = 2 - 1 = 1$$

and

$$y' - y = (2e^x - 1) - (2e^x - 1 - x) = x.$$

❑

EXAMPLE 4   Find a power series solution for

$$y'' + x^2 y = 0. \tag{6}$$

Solution   We assume that there is a solution of the form

$$y = a_0 + a_1 x + a_2 x^2 + \cdots + a_n x^n + \cdots, \tag{7}$$

and find what the coefficients a_k have to be to make the series and its second derivative

$$y'' = 2a_2 + 3 \cdot 2a_3 x + \cdots + n(n-1)a_n x^{n-2} + \cdots \tag{8}$$

satisfy Eq. (6). The series for $x^2 y$ is x^2 times the right-hand side of Eq. (7):

$$x^2 y = a_0 x^2 + a_1 x^3 + a_2 x^4 + \cdots + a_n x^{n+2} + \cdots. \tag{9}$$

The series for $y'' + x^2 y$ is the sum of the series in Eqs. (8) and (9):

$$y'' + x^2 y = 2a_2 + 6a_3 x + (12a_4 + a_0)x^2 + (20a_5 + a_1)x^3$$
$$+ \cdots + (n(n-1)a_n + a_{n-4})x^{n-2} + \cdots. \tag{10}$$

Notice that the coefficient of x^{n-2} in Eq. (9) is a_{n-4}. If y and its second derivative y'' are to satisfy Eq. (6), the coefficients of the individual powers of x on the right-hand side of Eq. (10) must all be zero:

$$2a_2 = 0, \quad 6a_3 = 0, \quad 12a_4 + a_0 = 0, \quad 20a_5 + a_1 = 0, \tag{11}$$

and for all $n \geq 4$,

$$n(n-1)a_n + a_{n-4} = 0. \tag{12}$$

We can see from Eq. (7) that

$$a_0 = y(0), \quad a_1 = y'(0).$$

In other words, the first two coefficients of the series are the values of y and y' at $x = 0$. The equations in (11) and the recursion formula in (12) enable us to evaluate all the other coefficients in terms of a_0 and a_1.

The first two of Eqs. (11) give

$$a_2 = 0, \quad a_3 = 0.$$

Equation (12) shows that if $a_{n-4} = 0$, then $a_n = 0$; so we conclude that

$$a_6 = 0, \quad a_7 = 0, \quad a_{10} = 0, \quad a_{11} = 0,$$

and whenever $n = 4k + 2$ or $4k + 3$, a_n is zero. For the other coefficients we have

$$a_n = \frac{-a_{n-4}}{n(n-1)}$$

so that

$$a_4 = \frac{-a_0}{4 \cdot 3}, \quad a_8 = \frac{-a_4}{8 \cdot 7} = \frac{a_0}{3 \cdot 4 \cdot 7 \cdot 8}$$

$$a_{12} = \frac{-a_8}{11 \cdot 12} = \frac{-a_0}{3 \cdot 4 \cdot 7 \cdot 8 \cdot 11 \cdot 12}$$

and

$$a_5 = \frac{-a_1}{5 \cdot 4}, \qquad a_9 = \frac{-a_5}{9 \cdot 8} = \frac{a_1}{4 \cdot 5 \cdot 8 \cdot 9}$$

$$a_{13} = \frac{-a_9}{12 \cdot 13} = \frac{-a_1}{4 \cdot 5 \cdot 8 \cdot 9 \cdot 12 \cdot 13}.$$

The answer is best expressed as the sum of two separate series—one multiplied by a_0, the other by a_1:

$$y = a_0 \left(1 - \frac{x^4}{3 \cdot 4} + \frac{x^8}{3 \cdot 4 \cdot 7 \cdot 8} - \frac{x^{12}}{3 \cdot 4 \cdot 7 \cdot 8 \cdot 11 \cdot 12} + \cdots \right)$$

$$+ a_1 \left(x - \frac{x^5}{4 \cdot 5} + \frac{x^9}{4 \cdot 5 \cdot 8 \cdot 9} - \frac{x^{13}}{4 \cdot 5 \cdot 8 \cdot 9 \cdot 12 \cdot 13} + \cdots \right).$$

Both series converge absolutely for all x, as is readily seen by the ratio test.   ❏

Evaluating Nonelementary Integrals

Maclaurin series can be used to express nonelementary integrals in terms of series.

Integrals like $\int \sin x^2 \, dx$ arise in the study of the diffraction of light.

EXAMPLE 5   Express $\int \sin x^2 \, dx$ as a power series.

Solution   From the series for $\sin x$ we obtain

$$\sin x^2 = x^2 - \frac{x^6}{3!} + \frac{x^{10}}{5!} - \frac{x^{14}}{7!} + \frac{x^{18}}{9!} - \cdots.$$

Therefore,

$$\int \sin x^2 dx = C + \frac{x^3}{3} - \frac{x^7}{7 \cdot 3!} + \frac{x^{11}}{11 \cdot 5!} - \frac{x^{15}}{15 \cdot 7!} + \frac{x^{19}}{19 \cdot 9!} - \cdots   ❏$$

EXAMPLE 6   Estimate $\int_0^1 \sin x^2 \, dx$ with an error of less than 0.001.

Solution   From the indefinite integral in Example 5,

$$\int_0^1 \sin x^2 \, dx = \frac{1}{3} - \frac{1}{7 \cdot 3!} + \frac{1}{11 \cdot 5!} - \frac{1}{15 \cdot 7!} + \frac{1}{19 \cdot 9!} - \cdots.$$

The series alternates, and we find by experiment that

$$\frac{1}{11 \cdot 5!} \approx 0.0007 \ 6$$

is the first term to be numerically less than 0.001. The sum of the preceding two terms gives

$$\int_0^1 \sin x^2 \, dx \approx \frac{1}{3} - \frac{1}{42} \approx 0.310.$$

With two more terms we could estimate

$$\int_0^1 \sin x^2 \, dx \approx 0.3102 \ 68$$

with an error of less than 10^{-6}. With only one term beyond that we have

$$\int_0^1 \sin x^2 \, dx \approx \frac{1}{3} - \frac{1}{42} + \frac{1}{1320} - \frac{1}{75600} + \frac{1}{6894720} \approx 0.3102 \ 68303,$$

with an error of about 1.08×10^{-9}. To guarantee this accuracy with the error formula for the trapezoidal rule would require using about 8,000 subintervals. ∎

Arctangents

In Section 8.8, Example 5, we found a series for $\tan^{-1} x$ by differentiating to get

$$\frac{d}{dx} \tan^{-1} x = \frac{1}{1+x^2} = 1 - x^2 + x^4 - x^6 + \cdots$$

and integrating to get

$$\tan^{-1} x = x - \frac{x^3}{3} + \frac{x^5}{5} - \frac{x^7}{7} + \cdots.$$

However, we did not prove the term-by-term integration theorem on which this conclusion depended. We now derive the series again by integrating both sides of the finite formula

$$\frac{1}{1+t^2} = 1 - t^2 + t^4 - t^6 + \cdots + (-1)^n t^{2n} + \frac{(-1)^{n+1} t^{2n+2}}{1+t^2}, \qquad (13)$$

in which the last term comes from adding the remaining terms as a geometric series with first term $a = (-1)^{n+1} t^{2n+2}$ and ratio $r = -t^2$. Integrating both sides of Eq. (13) from $t = 0$ to $t = x$ gives

$$\tan^{-1} x = x - \frac{x^3}{3} + \frac{x^5}{5} - \frac{x^7}{7} + \cdots + (-1)^n \frac{x^{2n+1}}{2n+1} + R(n, x),$$

where

$$R(n, x) = \int_0^x \frac{(-1)^{n+1} t^{2n+2}}{1+t^2} \, dt.$$

The denominator of the integrand is greater than or equal to 1; hence

$$|R(n, x)| \le \int_0^{|x|} t^{2n+2} \, dt = \frac{|x|^{2n+3}}{2n+3}.$$

We take this route instead of finding the Maclaurin series directly because the formulas for the higher order derivatives of $\tan^{-1} x$ are unmanageable.

If $|x| \le 1$, the right side of this inequality approaches zero as $n \to \infty$. Therefore $\lim_{n \to \infty} R(n, x) = 0$ if $|x| \le 1$ and

$$\tan^{-1} x = \sum_{n=0}^{\infty} \frac{(-1)^n x^{2n+1}}{2n+1}, \qquad |x| \le 1.$$

$$\tan^{-1} x = x - \frac{x^3}{3} + \frac{x^5}{5} - \frac{x^7}{7} + \cdots, \qquad |x| \le 1 \qquad (14)$$

When we put $x = 1$ in Eq. (14), we get **Leibniz's formula:**

$$\frac{\pi}{4} = 1 - \frac{1}{3} + \frac{1}{5} - \frac{1}{7} + \frac{1}{9} - \cdots + \frac{(-1)^n}{2n+1} + \cdots.$$

This series converges too slowly to be a useful source of decimal approximations of π. It is better to use a formula like

$$\pi = 48 \tan^{-1} \frac{1}{18} + 32 \tan^{-1} \frac{1}{57} - 20 \tan^{-1} \frac{1}{239},$$

which uses values of x closer to zero.

Evaluating Indeterminate Forms

We can sometimes evaluate indeterminate forms by expressing the functions involved as Taylor series.

EXAMPLE 7 Evaluate $\displaystyle\lim_{x \to 1} \frac{\ln x}{x - 1}$.

Solution We represent $\ln x$ as a Taylor series in powers of $x - 1$. This can be accomplished by calculating the Taylor series generated by $\ln x$ at $x = 1$ directly or by replacing x by $x - 1$ in the series for $\ln x$ in Section 8.8, Example 6. Either way, we obtain

$$\ln x = (x - 1) - \frac{1}{2}(x - 1)^2 + \cdots,$$

from which we find that

$$\lim_{x \to 1} \frac{\ln x}{x - 1} = \lim_{x \to 1} \left(1 - \frac{1}{2}(x - 1) + \cdots \right) = 1. \qquad \square$$

EXAMPLE 8 Evaluate $\displaystyle\lim_{x \to 0} \frac{\sin x - \tan x}{x^3}$.

Solution The Maclaurin series for $\sin x$ and $\tan x$, to terms in x^5, are

$$\sin x = x - \frac{x^3}{3!} + \frac{x^5}{5!} - \cdots, \qquad \tan x = x + \frac{x^3}{3} + \frac{2x^5}{15} + \cdots.$$

Hence,

$$\sin x - \tan x = -\frac{x^3}{2} - \frac{x^5}{8} - \cdots = x^3 \left(-\frac{1}{2} - \frac{x^2}{8} - \cdots \right)$$

and

$$\lim_{x \to 0} \frac{\sin x - \tan x}{x^3} = \lim_{x \to 0} \left(-\frac{1}{2} - \frac{x^2}{8} - \cdots \right)$$

$$= -\frac{1}{2}. \qquad \square$$

If we apply series to calculate $\lim_{x \to 0}((1/\sin x) - (1/x))$, we not only find the limit successfully but also discover an approximation formula for $\csc x$.

EXAMPLE 9 Find $\displaystyle\lim_{x \to 0} \left(\frac{1}{\sin x} - \frac{1}{x} \right)$.

Solution

$$\frac{1}{\sin x} - \frac{1}{x} = \frac{x - \sin x}{x \sin x} = \frac{x - \left(x - \dfrac{x^3}{3!} + \dfrac{x^5}{5!} - \cdots\right)}{x \cdot \left(x - \dfrac{x^3}{3!} + \dfrac{x^5}{5!} - \cdots\right)}$$

$$= \frac{x^3 \left(\dfrac{1}{3!} - \dfrac{x^2}{5!} + \cdots\right)}{x^2 \left(1 - \dfrac{x^2}{3!} + \cdots\right)} = x \, \frac{\dfrac{1}{3!} - \dfrac{x^2}{5!} + \cdots}{1 - \dfrac{x^2}{3!} + \cdots}.$$

Therefore,

$$\lim_{x \to 0} \left(\frac{1}{\sin x} - \frac{1}{x}\right) = \lim_{x \to 0} \left(x \, \frac{\dfrac{1}{3!} - \dfrac{x^2}{5!} + \cdots}{1 - \dfrac{x^2}{3!} + \cdots}\right) = 0.$$

From the quotient on the right, we can see that if $|x|$ is small, then

$$\frac{1}{\sin x} - \frac{1}{x} \approx x \cdot \frac{1}{3!} = \frac{x}{6} \qquad \text{or} \qquad \csc x \approx \frac{1}{x} + \frac{x}{6}. \qquad \square$$

Frequently Used Maclaurin Series

$$\frac{1}{1-x} = 1 + x + x^2 + \cdots + x^n + \cdots = \sum_{n=0}^{\infty} x^n, \qquad |x| < 1$$

$$\frac{1}{1+x} = 1 - x + x^2 - \cdots + (-x)^n + \cdots = \sum_{n=0}^{\infty} (-1)^n x^n, \qquad |x| < 1$$

$$e^x = 1 + x + \frac{x^2}{2!} + \cdots + \frac{x^n}{n!} + \cdots = \sum_{n=0}^{\infty} \frac{x^n}{n!}, \qquad |x| < \infty$$

$$\sin x = x - \frac{x^3}{3!} + \frac{x^5}{5!} - \cdots + (-1)^n \frac{x^{2n+1}}{(2n+1)!} + \cdots = \sum_{n=0}^{\infty} \frac{(-1)^n x^{2n+1}}{(2n+1)!}, \qquad |x| < \infty$$

$$\cos x = 1 - \frac{x^2}{2!} + \frac{x^4}{4!} - \cdots + (-1)^n \frac{x^{2n}}{(2n)!} + \cdots = \sum_{n=0}^{\infty} \frac{(-1)^n x^{2n}}{(2n)!}, \qquad |x| < \infty$$

$$\ln(1+x) = x - \frac{x^2}{2} + \frac{x^3}{3} - \cdots + (-1)^{n-1} \frac{x^n}{n} + \cdots = \sum_{n=1}^{\infty} \frac{(-1)^{n-1} x^n}{n}, \qquad -1 < x \le 1$$

$$\ln \frac{1+x}{1-x} = 2 \tanh^{-1} x = 2 \left(x + \frac{x^3}{3} + \frac{x^5}{5} + \cdots + \frac{x^{2n+1}}{2n+1} + \cdots\right) = 2 \sum_{n=0}^{\infty} \frac{x^{2n+1}}{2n+1}, \qquad |x| < 1$$

$$\tan^{-1} x = x - \frac{x^3}{3} + \frac{x^5}{5} - \cdots + (-1)^n \frac{x^{2n+1}}{2n+1} + \cdots = \sum_{n=0}^{\infty} \frac{(-1)^n x^{2n+1}}{2n+1}, \qquad |x| \le 1$$

(Continued)

Binomial Series

$$(1+x)^m = 1 + mx + \frac{m(m-1)x^2}{2!} + \frac{m(m-1)(m-2)x^3}{3!} + \cdots + \frac{m(m-1)(m-2)\cdots(m-k+1)x^k}{k!} + \cdots$$

$$= 1 + \sum_{k=1}^{\infty} \binom{m}{k} x^k, \qquad |x| < 1,$$

where

$$\binom{m}{1} = m, \qquad \binom{m}{2} = \frac{m(m-1)}{2!}, \qquad \binom{m}{k} = \frac{m(m-1)\cdots(m-k+1)}{k!} \qquad \text{for } k \geq 3.$$

Note: To write the binomial series compactly, it is customary to define $\binom{m}{0}$ to be 1 and to take $x^0 = 1$ (even in the usually excluded case where $x = 0$), yielding $(1+x)^m = \sum_{k=0}^{\infty} \binom{m}{k} x^k$. If m is a *positive integer*, the series terminates at x^m and the result converges for all x.

Exercises 8.11

Binomial Series

Find the first four terms of the binomial series for the functions in Exercises 1–10.

1. $(1+x)^{1/2}$
2. $(1+x)^{1/3}$
3. $(1-x)^{-1/2}$

4. $(1-2x)^{1/2}$
5. $\left(1 + \dfrac{x}{2}\right)^{-2}$
6. $\left(1 - \dfrac{x}{2}\right)^{-2}$

7. $(1+x^3)^{-1/2}$
8. $(1+x^2)^{-1/3}$

9. $\left(1 + \dfrac{1}{x}\right)^{1/2}$
10. $\left(1 - \dfrac{2}{x}\right)^{1/3}$

Find the binomial series for the functions in Exercises 11–14.

11. $(1+x)^4$
12. $(1+x^2)^3$

13. $(1-2x)^3$
14. $\left(1 - \dfrac{x}{2}\right)^4$

Initial Value Problems

Find series solutions for the initial value problems in Exercises 15–32.

15. $y' + y = 0, \quad y(0) = 1$
16. $y' - 2y = 0, \quad y(0) = 1$
17. $y' - y = 1, \quad y(0) = 0$
18. $y' + y = 1, \quad y(0) = 2$
19. $y' - y = x, \quad y(0) = 0$
20. $y' + y = 2x, \quad y(0) = -1$
21. $y' - xy = 0, \quad y(0) = 1$
22. $y' - x^2 y = 0, \quad y(0) = 1$
23. $(1-x)y' - y = 0, \quad y(0) = 2$
24. $(1+x^2)y' + 2xy = 0, \quad y(0) = 3$
25. $y'' - y = 0, \quad y'(0) = 1$ and $y(0) = 0$
26. $y'' + y = 0, \quad y'(0) = 0$ and $y(0) = 1$

27. $y'' + y = x, \quad y'(0) = 1$ and $y(0) = 2$
28. $y'' - y = x, \quad y'(0) = 2$ and $y(0) = -1$
29. $y'' - y = -x, \quad y'(2) = -2$ and $y(2) = 0$
30. $y'' - x^2 y = 0, \quad y'(0) = b$ and $y(0) = a$
31. $y'' + x^2 y = x, \quad y'(0) = b$ and $y(0) = a$
32. $y'' - 2y' + y = 0, \quad y'(0) = 1$ and $y(0) = 0$

Approximations and Nonelementary Integrals

CALCULATOR In Exercises 33–36, use series to estimate the integrals' values with an error of magnitude less than 10^{-3}. (The answer section gives the integrals' values rounded to 5 decimal places.)

33. $\displaystyle\int_0^{0.2} \sin x^2 \, dx$
34. $\displaystyle\int_0^{0.2} \frac{e^{-x} - 1}{x} \, dx$

35. $\displaystyle\int_0^{0.1} \frac{1}{\sqrt{1 + x^4}} \, dx$
36. $\displaystyle\int_0^{0.25} \sqrt[3]{1 + x^2} \, dx$

CALCULATOR Use series to approximate the values of the integrals in Exercises 37–40 with an error of magnitude less than 10^{-8}. (The answer section gives the integrals' values rounded to 10 decimal places.)

37. $\displaystyle\int_0^{0.1} \frac{\sin x}{x} \, dx$
38. $\displaystyle\int_0^{0.1} e^{-x^2} \, dx$

39. $\displaystyle\int_0^{0.1} \sqrt{1 + x^4} \, dx$
40. $\displaystyle\int_0^1 \frac{1 - \cos x}{x^2} \, dx$

41. Estimate the error if $\cos t^2$ is approximated by $1 - \dfrac{t^4}{2} + \dfrac{t^8}{4!}$ in the integral $\int_0^1 \cos t^2 \, dt$.

42. Estimate the error if $\cos \sqrt{t}$ is approximated by $1 - \dfrac{t}{2} + \dfrac{t^2}{4!} - \dfrac{t^3}{6!}$ in the integral $\int_0^1 \cos \sqrt{t}\, dt$.

In Exercises 43–46, find a polynomial that will approximate $F(x)$ throughout the given interval with an error of magnitude less than 10^{-3}.

43. $F(x) = \displaystyle\int_0^x \sin t^2\, dt$, [0, 1]

44. $F(x) = \displaystyle\int_0^x t^2 e^{-t^2}\, dt$, [0, 1]

45. $F(x) = \displaystyle\int_0^x \tan^{-1} t\, dt$, a) [0, 0.5] b) [0, 1]

46. $F(x) = \displaystyle\int_0^x \dfrac{\ln(1+t)}{t}\, dt$, a) [0, 0.5] b) [0, 1]

Indeterminate Forms

Use series to evaluate the limits in Exercises 47–56.

47. $\displaystyle\lim_{x \to 0} \dfrac{e^x - (1+x)}{x^2}$

48. $\displaystyle\lim_{x \to 0} \dfrac{e^x - e^{-x}}{x}$

49. $\displaystyle\lim_{t \to 0} \dfrac{1 - \cos t - (t^2/2)}{t^4}$

50. $\displaystyle\lim_{\theta \to 0} \dfrac{\sin \theta - \theta + (\theta^3/6)}{\theta^5}$

51. $\displaystyle\lim_{y \to 0} \dfrac{y - \tan^{-1} y}{y^3}$

52. $\displaystyle\lim_{y \to 0} \dfrac{\tan^{-1} y - \sin y}{y^3 \cos y}$

53. $\displaystyle\lim_{x \to \infty} x^2 (e^{-1/x^2} - 1)$

54. $\displaystyle\lim_{x \to \infty} (x+1) \sin \dfrac{1}{x+1}$

55. $\displaystyle\lim_{x \to 0} \dfrac{\ln(1+x^2)}{1 - \cos x}$

56. $\displaystyle\lim_{x \to 2} \dfrac{x^2 - 4}{\ln(x-1)}$

Theory and Examples

57. Replace x by $-x$ in the Maclaurin series for $\ln(1+x)$ to obtain a series for $\ln(1-x)$. Then subtract this from the Maclaurin series for $\ln(1+x)$ to show that for $|x| < 1$,

$$\ln \dfrac{1+x}{1-x} = 2\left(x + \dfrac{x^3}{3} + \dfrac{x^5}{5} + \cdots\right).$$

58. How many terms of the Maclaurin series for $\ln(1+x)$ should you add to be sure of calculating $\ln(1.1)$ with an error of magnitude less than 10^{-8}? Give reasons for your answer.

59. According to the Alternating Series Estimation Theorem, how many terms of the Maclaurin series for $\tan^{-1} 1$ would you have to add to be sure of finding $\pi/4$ with an error of magnitude less than 10^{-3}? Give reasons for your answer.

60. Show that the Maclaurin series for $f(x) = \tan^{-1} x$ diverges for $|x| > 1$.

61. CALCULATOR About how many terms of the Maclaurin series for $\tan^{-1} x$ would you have to use to evaluate each term on the right-hand side of the equation

$$\pi = 48 \tan^{-1} \dfrac{1}{18} + 32 \tan^{-1} \dfrac{1}{57} - 20 \tan^{-1} \dfrac{1}{239}$$

with an error of magnitude less than 10^{-6}? In contrast, the convergence of $\sum_{n=1}^{\infty} (1/n^2)$ to $\pi^2/6$ is so slow that even 50 terms will not yield two-place accuracy.

62. Integrate the first three nonzero terms of the Maclaurin series for $\tan t$ from 0 to x to obtain the first three nonzero terms of the Maclaurin series for $\ln \sec x$.

63. a) Use the binomial series and the fact that

$$\dfrac{d}{dx} \sin^{-1} x = (1 - x^2)^{-1/2}$$

to generate the first four nonzero terms of the Maclaurin series for $\sin^{-1} x$. What is the radius of convergence?

b) Use your result in (a) to find the first five nonzero terms of the Maclaurin series for $\cos^{-1} x$.

64. a) Find the first four nonzero terms of the Maclaurin series for

$$\sinh^{-1} x = \int_0^x \dfrac{dt}{\sqrt{1 + t^2}}.$$

b) CALCULATOR Use the first *three* terms of the series in (a) to estimate $\sinh^{-1} 0.25$. Give an upper bound for the magnitude of the estimation error.

65. Obtain the Maclaurin series for $1/(1+x)^2$ from the series for $-1/(1+x)$.

66. Use the Maclaurin series for $1/(1-x^2)$ to obtain a series for $2x/(1-x^2)^2$.

67. CAS The English mathematician Wallis discovered the formula

$$\dfrac{\pi}{4} = \dfrac{2 \cdot 4 \cdot 4 \cdot 6 \cdot 6 \cdot 8 \cdot \cdots}{3 \cdot 3 \cdot 5 \cdot 5 \cdot 7 \cdot 7 \cdot \cdots}.$$

Find π to 2 decimal places with this formula.

68. CALCULATOR Construct a table of natural logarithms $\ln n$ for $n = 1, 2, 3, \ldots, 10$ by using the formula in Exercise 57, but taking advantage of the relationships $\ln 4 = 2 \ln 2$, $\ln 6 = \ln 2 + \ln 3$, $\ln 8 = 3 \ln 2$, $\ln 9 = 2 \ln 3$, and $\ln 10 = \ln 2 + \ln 5$ to reduce the job to the calculation of relatively few logarithms by series. Start by using the following values for x in Exercise 57:

$$\dfrac{1}{3}, \ \dfrac{1}{5}, \ \dfrac{1}{9}, \ \dfrac{1}{13}.$$

69. Integrate the binomial series for $(1 - x^2)^{-1/2}$ to show that for $|x| < 1$,

$$\sin^{-1} x = x + \sum_{n=1}^{\infty} \dfrac{1 \cdot 3 \cdot 5 \cdot \cdots \cdot (2n-1)}{2 \cdot 4 \cdot 6 \cdot \cdots \cdot (2n)} \dfrac{x^{2n+1}}{2n+1}.$$

70. *Series for* $\tan^{-1} x$ *for* $|x| > 1$. Derive the series

$$\tan^{-1} x = \dfrac{\pi}{2} - \dfrac{1}{x} + \dfrac{1}{3x^3} - \dfrac{1}{5x^5} + \cdots, \quad x > 1$$

$$\tan^{-1} x = -\dfrac{\pi}{2} - \dfrac{1}{x} + \dfrac{1}{3x^3} - \dfrac{1}{5x^5} + \cdots, \quad x < -1,$$

by integrating the series

$$\dfrac{1}{1+t^2} = \dfrac{1}{t^2} \cdot \dfrac{1}{1 + (1/t^2)} = \dfrac{1}{t^2} - \dfrac{1}{t^4} + \dfrac{1}{t^6} - \dfrac{1}{t^8} + \cdots$$

in the first case from x to ∞ and in the second case from $-\infty$ to x.

)

71. *The value of $\sum_{n=1}^{\infty} \tan^{-1}(2/n^2)$*

a) Use the formula for the tangent of the difference of two angles to show that

$$\tan(\tan^{-1}(n+1) - \tan^{-1}(n-1)) = \frac{2}{n^2}$$

CHAPTER **8** QUESTIONS TO GUIDE YOUR REVIEW

---

1. What is an infinite sequence? What does it mean for such a sequence to converge? to diverge? Give examples.

2. What uses can be found for subsequences? Give examples.

3. What is a nondecreasing sequence? Under what circumstances does such a sequence have a limit? Give examples.

4. What theorems are available for calculating limits of sequences? Give examples.

5. What theorem sometimes enables us to use l'Hôpital's rule to calculate the limit of a sequence? Give an example.

6. What six sequence limits are likely to arise when you work with sequences and series?

7. What is Picard's method for solving the equation $f(x) = 0$? Give an example.

8. What is an infinite series? What does it mean for such a series to converge? to diverge? Give examples.

9. What is a geometric series? When does such a series converge? diverge? When it does converge, what is its sum? Give examples.

10. Besides geometric series, what other convergent and divergent series do you know?

11. What is the nth-Term Test for Divergence? What is the idea behind the test?

12. What can be said about term-by-term sums and differences of convergent series? about constant multiples of convergent and divergent series?

13. What happens if you add a finite number of terms to a convergent series? a divergent series? What happens if you delete a finite number of terms from a convergent series? a divergent series?

14. How do you reindex a series? Why might you want to do this?

15. Under what circumstances will an infinite series of nonnegative terms converge? diverge? Why study series of nonnegative terms?

16. What is the Integral Test? What is the reasoning behind it? Give an example of its use.

17. When do p-series converge? diverge? How do you know? Give examples of convergent and divergent p-series.

18. What are the Direct Comparison Test and the Limit Comparison Test? What is the reasoning behind these tests? Give examples of their use.

19. What are the Ratio and Root Tests? Do they always give you the information you need to determine convergence or divergence? Give examples.

20. What is an alternating series? What theorem is available for determining the convergence of such a series?

21. How can you estimate the error involved in approximating the sum of an alternating series with one of the series' partial sums? What is the reasoning behind the estimate?

22. What is absolute convergence? conditional convergence? How are the two related?

23. What do you know about rearranging the terms of an absolutely convergent series? of a conditionally convergent series? Give examples.

24. What is a power series? How do you test a power series for convergence? What are the possible outcomes?

25. What are the basic facts about
   a) term-by-term differentiation of power series?
   b) term-by-term integration of power series?
   c) multiplication of power series?
   Give examples.

26. What is the Taylor series generated by a function $f(x)$ at a point $x = a$? What information do you need about f to construct the series? Give an example.

27. What is a Maclaurin series?

28. Does a Taylor series always converge to its generating function? Explain.

29. What are Taylor polynomials? Of what use are they?

30. What is Taylor's formula? What does it say about the errors involved in using Taylor polynomials to approximate functions? In particular, what does Taylor's formula say about the error in a linearization? a quadratic approximation?

31. What is the binomial series? On what interval does it converge? How is it used?

32. How can you sometimes use power series to solve initial value problems?

33. How can you sometimes use power series to estimate the values of nonelementary definite integrals?

34. What are the Maclaurin series for $1/(1-x)$, $1/(1+x)$, e^x, $\sin x$, $\cos x$, $\ln(1+x)$, $\ln[(1+x)/(1-x)]$, and $\tan^{-1} x$? How do you estimate the errors involved in replacing these series with their partial sums?

CHAPTER 8 PRACTICE EXERCISES

Convergent or Divergent Sequences

Which of the sequences whose nth terms appear in Exercises 1–18 converge, and which diverge? Find the limit of each convergent sequence.

1. $a_n = 1 + \dfrac{(-1)^n}{n}$

2. $a_n = \dfrac{1-(-1)^n}{\sqrt{n}}$

3. $a_n = \dfrac{1-2^n}{2^n}$

4. $a_n = 1 + (0.9)^n$

5. $a_n = \sin \dfrac{n\pi}{2}$

6. $a_n = \sin n\pi$

7. $a_n = \dfrac{\ln(n^2)}{n}$

8. $a_n = \dfrac{\ln(2n+1)}{n}$

9. $a_n = \dfrac{n + \ln n}{n}$

10. $a_n = \dfrac{\ln(2n^3+1)}{n}$

11. $a_n = \left(\dfrac{n-5}{n}\right)^n$

12. $a_n = \left(1 + \dfrac{1}{n}\right)^{-n}$

13. $a_n = \sqrt[n]{\dfrac{3^n}{n}}$

14. $a_n = \left(\dfrac{3}{n}\right)^{1/n}$

15. $a_n = n(2^{1/n} - 1)$

16. $a_n = \sqrt[n]{2n+1}$

17. $a_n = \dfrac{(n+1)!}{n!}$

18. $a_n = \dfrac{(-4)^n}{n!}$

Convergent Series

Find the sums of the series in Exercises 19–24.

19. $\displaystyle\sum_{n=3}^{\infty} \dfrac{1}{(2n-3)(2n-1)}$

20. $\displaystyle\sum_{n=2}^{\infty} \dfrac{-2}{n(n+1)}$

21. $\displaystyle\sum_{n=1}^{\infty} \dfrac{9}{(3n-1)(3n+2)}$

22. $\displaystyle\sum_{n=3}^{\infty} \dfrac{-8}{(4n-3)(4n+1)}$

23. $\displaystyle\sum_{n=0}^{\infty} e^{-n}$

24. $\displaystyle\sum_{n=1}^{\infty} (-1)^n \dfrac{3}{4^n}$

Convergent or Divergent Series

Which of the series in Exercises 25–40 converge absolutely, which converge conditionally, and which diverge? Give reasons for your answers.

25. $\displaystyle\sum_{n=1}^{\infty} \dfrac{1}{\sqrt{n}}$

26. $\displaystyle\sum_{n=1}^{\infty} \dfrac{-5}{n}$

27. $\displaystyle\sum_{n=1}^{\infty} \dfrac{(-1)^n}{\sqrt{n}}$

28. $\displaystyle\sum_{n=1}^{\infty} \dfrac{1}{2n^3}$

29. $\displaystyle\sum_{n=1}^{\infty} \dfrac{(-1)^n}{\ln(n+1)}$

30. $\displaystyle\sum_{n=2}^{\infty} \dfrac{1}{n(\ln n)^2}$

31. $\displaystyle\sum_{n=1}^{\infty} \dfrac{\ln n}{n^3}$

32. $\displaystyle\sum_{n=3}^{\infty} \dfrac{\ln n}{\ln(\ln n)}$

33. $\displaystyle\sum_{n=1}^{\infty} \dfrac{(-1)^n}{n\sqrt{n^2+1}}$

34. $\displaystyle\sum_{n=1}^{\infty} \dfrac{(-1)^n\, 3n^2}{n^3+1}$

35. $\displaystyle\sum_{n=1}^{\infty} \dfrac{n+1}{n!}$

36. $\displaystyle\sum_{n=1}^{\infty} \dfrac{(-1)^n(n^2+1)}{2n^2+n-1}$

37. $\displaystyle\sum_{n=1}^{\infty} \dfrac{(-3)^n}{n!}$

38. $\displaystyle\sum_{n=1}^{\infty} \dfrac{2^n\, 3^n}{n^n}$

39. $\displaystyle\sum_{n=1}^{\infty} \dfrac{1}{\sqrt{n(n+1)(n+2)}}$

40. $\displaystyle\sum_{n=2}^{\infty} \dfrac{1}{n\sqrt{n^2-1}}$

Power Series

In Exercises 41–50, (a) find the series' radius and interval of convergence. Then identify the values of x for which the series converges (b) absolutely and (c) conditionally.

41. $\displaystyle\sum_{n=1}^{\infty} \dfrac{(x+4)^n}{n\, 3^n}$

42. $\displaystyle\sum_{n=1}^{\infty} \dfrac{(x-1)^{2n-2}}{(2n-1)!}$

43. $\displaystyle\sum_{n=1}^{\infty} \dfrac{(-1)^{n-1}(3x-1)^n}{n^2}$

44. $\displaystyle\sum_{n=0}^{\infty} \dfrac{(n+1)(2x+1)^n}{(2n+1)2^n}$

45. $\displaystyle\sum_{n=1}^{\infty} \dfrac{x^n}{n^n}$

46. $\displaystyle\sum_{n=1}^{\infty} \dfrac{x^n}{\sqrt{n}}$

47. $\displaystyle\sum_{n=0}^{\infty} \frac{(n+1)x^{2n-1}}{3^n}$

48. $\displaystyle\sum_{n=0}^{\infty} \frac{(-1)^n (x-1)^{2n+1}}{2n+1}$

49. $\displaystyle\sum_{n=1}^{\infty} (\operatorname{csch} n) x^n$

50. $\displaystyle\sum_{n=1}^{\infty} (\coth n) x^n$

Maclaurin Series

Each of the series in Exercises 51–56 is the value of the Maclaurin series of a function $f(x)$ at a particular point. What function and what point? What is the sum of the series?

51. $1 - \dfrac{1}{4} + \dfrac{1}{16} - \cdots + (-1)^n \dfrac{1}{4^n} + \cdots$

52. $\dfrac{2}{3} - \dfrac{4}{18} + \dfrac{8}{81} - \cdots + (-1)^{n-1} \dfrac{2^n}{n3^n} + \cdots$

53. $\pi - \dfrac{\pi^3}{3!} + \dfrac{\pi^5}{5!} - \cdots + (-1)^n \dfrac{\pi^{2n+1}}{(2n+1)!} + \cdots$

54. $1 - \dfrac{\pi^2}{9 \cdot 2!} + \dfrac{\pi^4}{81 \cdot 4!} - \cdots + (-1)^n \dfrac{\pi^{2n}}{3^{2n}(2n)!} + \cdots$

55. $1 + \ln 2 + \dfrac{(\ln 2)^2}{2!} + \cdots + \dfrac{(\ln 2)^n}{n!} + \cdots$

56. $\dfrac{1}{\sqrt{3}} - \dfrac{1}{9\sqrt{3}} + \dfrac{1}{45\sqrt{3}} - \cdots$

$\qquad + (-1)^{n-1} \dfrac{1}{(2n-1)(\sqrt{3})^{2n-1}} + \cdots$

Find Maclaurin series for the functions in Exercises 57–64.

57. $\dfrac{1}{1-2x}$

58. $\dfrac{1}{1+x^3}$

59. $\sin \pi x$

60. $\sin \dfrac{2x}{3}$

61. $\cos (x^{5/2})$

62. $\cos \sqrt{5x}$

63. $e^{(\pi x/2)}$

64. e^{-x^2}

Taylor Series

In Exercises 65–68, find the first four nonzero terms of the Taylor series generated by f at $x = a$.

65. $f(x) = \sqrt{3 + x^2}$ at $x = -1$

66. $f(x) = 1/(1-x)$ at $x = 2$

67. $f(x) = 1/(x+1)$ at $x = 3$

68. $f(x) = 1/x$ at $x = a > 0$

Initial Value Problems

Use power series to solve the initial value problems in Exercises 69–76.

69. $y' + y = 0, \quad y(0) = -1$

70. $y' - y = 0, \quad y(0) = -3$

71. $y' + 2y = 0, \quad y(0) = 3$

72. $y' + y = 1, \quad y(0) = 0$

73. $y' - y = 3x, \quad y(0) = -1$

74. $y' + y = x, \quad y(0) = 0$

75. $y' - y = x, \quad y(0) = 1$

76. $y' - y = -x, \quad y(0) = 2$

Nonelementary Integrals

Use series to approximate the values of the integrals in Exercises 77–80 with an error of magnitude less than 10^{-8}. (The answer section gives the integrals' values rounded to 10 decimal places.)

77. $\displaystyle\int_0^{1/2} e^{-x^3} \, dx$

78. $\displaystyle\int_0^1 x \sin (x^3) \, dx$

79. $\displaystyle\int_0^{1/2} \frac{\tan^{-1} x}{x} \, dx$

80. $\displaystyle\int_0^{1/64} \frac{\tan^{-1} x}{\sqrt{x}} \, dx$

Indeterminate Forms

In Exercises 81–86:

a) Use power series to evaluate the limit.

b) GRAPHER Then use a grapher to support your calculation.

81. $\displaystyle\lim_{x \to 0} \frac{7 \sin x}{e^{2x} - 1}$

82. $\displaystyle\lim_{\theta \to 0} \frac{e^\theta - e^{-\theta} - 2\theta}{\theta - \sin \theta}$

83. $\displaystyle\lim_{t \to 0} \left(\frac{1}{2 - 2\cos t} - \frac{1}{t^2} \right)$

84. $\displaystyle\lim_{h \to 0} \frac{(\sin h)/h - \cos h}{h^2}$

85. $\displaystyle\lim_{z \to 0} \frac{1 - \cos^2 z}{\ln(1-z) + \sin z}$

86. $\displaystyle\lim_{y \to 0} \frac{y^2}{\cos y - \cosh y}$

87. Use a series representation of $\sin 3x$ to find values of r and s for which

$$\lim_{x \to 0} \left(\frac{\sin 3x}{x^3} + \frac{r}{x^2} + s \right) = 0.$$

88. a) Show that the approximation $\csc x \approx 1/x + x/6$ in Section 8.11, Example 9, leads to the approximation $\sin x \approx 6x/(6 + x^2)$.

b) GRAPHER EXPLORATION Compare the accuracies of the approximations $\sin x \approx x$ and $\sin x \approx 6x/(6 + x^2)$ by comparing the graphs of $f(x) = \sin x - x$ and $g(x) = \sin x - (6x/(6 + x^2))$. Describe what you find.

Theory and Examples

89. a) Show that the series

$$\sum_{n=1}^{\infty} \left(\sin \frac{1}{2n} - \sin \frac{1}{2n+1} \right)$$

converges.

b) CALCULATOR Estimate the magnitude of the error involved in using the sum of the sines through $n = 20$ to approximate the sum of the series. Is the approximation too large, or too small? Give reasons for your answer.

90. a) Show that the series $\displaystyle\sum_{n=1}^{\infty} \left(\tan \frac{1}{2n} - \tan \frac{1}{2n+1} \right)$ converges.

b) CALCULATOR Estimate the magnitude of the error in using the sum of the tangents through $-\tan(1/41)$ to approximate the sum of the series. Is the approximation too large, or too small? Give reasons for your answer.

91. Find the radius of convergence of the series

$$\sum_{n=1}^{\infty} \frac{2 \cdot 5 \cdot 8 \cdot \cdots \cdot (3n - 1)}{2 \cdot 4 \cdot 6 \cdot \cdots \cdot (2n)} x^n.$$

92. Find the radius of convergence of the series

$$\sum_{n=1}^{\infty} \frac{3 \cdot 5 \cdot 7 \cdot \cdots \cdot (2n + 1)}{4 \cdot 9 \cdot 14 \cdot \cdots \cdot (5n - 1)} (x - 1)^n.$$

93. Find a closed-form formula for the nth partial sum of the series $\sum_{n=2}^{\infty} \ln (1 - (1/n^2))$ and use it to determine the convergence or divergence of the series.

94. Evaluate $\sum_{k=2}^{\infty} (1/(k^2 - 1))$ by finding the limit as $n \to \infty$ of the series' nth partial sum.

95. a)   Find the interval of convergence of the series

$$y = 1 + \frac{1}{6} x^3 + \frac{1}{180} x^6 + \cdots$$

$$+ \frac{1 \cdot 4 \cdot 7 \cdot \cdots \cdot (3n - 2)}{(3n)!} x^{3n} + \cdots.$$

b)   Show that the function defined by the series satisfies a differential equation of the form

$$\frac{d^2 y}{dx^2} = x^a y + b$$

and find the values of the constants a and b.

96. a)   Find the Maclaurin series for the function $x^2/(1 + x)$.

b)   Does the series converge at $x = 1$? Explain.

97. If $\sum_{n=1}^{\infty} a_n$ and $\sum_{n=1}^{\infty} b_n$ are convergent series of nonnegative numbers, can anything be said about $\sum_{n=1}^{\infty} a_n b_n$? Give reasons for your answer.

98. If $\sum_{n=1}^{\infty} a_n$ and $\sum_{n=1}^{\infty} b_n$ are divergent series of nonnegative numbers, can anything be said about $\sum_{n=1}^{\infty} a_n b_n$? Give reasons for your answer.

99. Prove that the sequence $\{x_n\}$ and the series $\sum_{k=1}^{\infty} (x_{k+1} - x_k)$ both converge or both diverge.

100. Prove that $\sum_{n=1}^{\infty} (a_n/(1 + a_n))$ converges if $a_n > 0$ for all n and $\sum_{n=1}^{\infty} a_n$ converges.

101. (*Continuation of Section 3.8, Exercise 25.*) If you did Exercise 25 in Section 3.8, you saw that in practice Newton's method stopped too far from the root of $f(x) = (x - 1)^{40}$ to give a useful estimate of its value, $x = 1$. Prove that nevertheless, for any starting value $x_0 \neq 1$, the sequence $x_0, x_1, x_2, \ldots, x_n, \ldots$ of approximations generated by Newton's method really does converge to 1.

102. a)   Suppose that $a_1, a_2, a_3, \ldots, a_n$ are positive numbers satisfying the following conditions:

i) $a_1 \geq a_2 \geq a_3 \geq \cdots$;

ii)   the series $a_2 + a_4 + a_8 + a_{16} + \cdots$ diverges.

Show that the series

$$\frac{a_1}{1} + \frac{a_2}{2} + \frac{a_3}{3} + \cdots$$

diverges.

b)   Use the result in (a) to show that

$$1 + \sum_{n=2}^{\infty} \frac{1}{n \ln n}$$

diverges.

103. Suppose you wish to obtain a quick estimate for the value of $\int_0^1 x^2 e^x \, dx$. There are several ways to do this.

a)   Use the trapezoidal rule with $n = 2$ to estimate $\int_0^1 x^2 e^x \, dx$.

b)   Write out the first three nonzero terms of the Maclaurin series for $x^2 e^x$ to obtain the fourth Maclaurin polynomial $P(x)$ for $x^2 e^x$. Use $\int_0^1 P(x) \, dx$ to obtain another estimate for $\int_0^1 x^2 e^x \, dx$.

c)   The second derivative of $f(x) = x^2 e^x$ is positive for all $x > 0$. Explain why this enables you to conclude that the trapezoidal rule estimate obtained in (a) is too large. (*Hint:* What does the second derivative tell you about the graph of a function? How does this relate to the trapezoidal approximation of the area under this graph?)

d)   All the derivatives of $f(x) = x^2 e^x$ are positive for $x > 0$. Explain why this enables you to conclude that all Maclaurin polynomial approximations to $f(x)$ for x in [0, 1] will be too small. (*Hint:* $f(x) = P_n(x) + R_n(x)$.)

e)   Use integration by parts to evaluate $\int_0^1 x^2 e^x \, dx$.

CHAPTER **8**   ADDITIONAL EXERCISES–THEORY, EXAMPLES, APPLICATIONS

Convergence or Divergence

Which of the series $\sum_{n=1}^{\infty} a_n$ defined by the formulas in Exercises 1–4 converge, and which diverge? Give reasons for your answers.

1. $\displaystyle\sum_{n=1}^{\infty} \frac{1}{(3n-2)^{n+(1/2)}}$

2. $\displaystyle\sum_{n=1}^{\infty} \frac{(\tan^{-1} n)^2}{n^2 + 1}$

3. $\displaystyle\sum_{n=1}^{\infty} (-1)^n \tanh n$

4. $\displaystyle\sum_{n=2}^{\infty} \frac{\log_n(n!)}{n^3}$

Which of the series $\sum_{n=1}^{\infty} a_n$ defined by the formulas in Exercises 5–8 converge, and which diverge? Give reasons for your answers.

5. $a_1 = 1, \quad a_{n+1} = \dfrac{n(n+1)}{(n+2)(n+3)} a_n$

(*Hint:* Write out several terms, see which factors cancel, and then generalize.)

6. $a_1 = a_2 = 7, \quad a_{n+1} = \dfrac{n}{(n-1)(n+1)} a_n$ if $n \geq 2$

7. $a_1 = a_2 = 1, \quad a_{n+1} = \dfrac{1}{1 + a_n}$ if $n \geq 2$

8. $a_n = 1/3^n$ if n is odd, $\quad a_n = n/3^n$ if n is even

Choosing Centers for Taylor Series

Taylor's formula

$$f(x) = f(a) + f'(a)(x-a) + \frac{f''(a)}{2!}(x-a)^2 + \cdots$$
$$+ \frac{f^{(n)}(a)}{n!}(x-a)^n + \frac{f^{(n+1)}(c)}{(n+1)!}(x-a)^{n+1}$$

expresses the value of f at x in terms of the values of f and its derivatives at $x = a$. In numerical computations, we therefore need f to be a point where we know the values of f and its derivatives. We also need a to be close enough to the values of f we are interested in to make $(x-a)^{n+1}$ so small we can neglect the remainder.

In Exercises 9–14, what Taylor series would you choose to represent the function near the given value of x? (There may be more than one good answer.) Write out the first four nonzero terms of the series you choose.

9. $\cos x$ near $x = 1$

10. $\sin x$ near $x = 6.3$

11. e^x near $x = 0.4$

12. $\ln x$ near $x = 1.3$

13. $\cos x$ near $x = 69$

14. $\tan^{-1} x$ near $x = 2$

Theory and Examples

15. Let a and b be constants with $0 < a < b$. Does the sequence $\{(a^n + b^n)^{1/n}\}$ converge? If it does converge, what is the limit?

16. Find the sum of the infinite series

$$1 + \frac{2}{10} + \frac{3}{10^2} + \frac{7}{10^3} + \frac{2}{10^4} + \frac{3}{10^5} + \frac{7}{10^6} + \frac{2}{10^7} + \frac{3}{10^8} + \frac{7}{10^9} + \cdots.$$

17. Evaluate

$$\sum_{n=0}^{\infty} \int_n^{n+1} \frac{1}{1+x^2}\, dx.$$

18. Find all values of x for which

$$\sum_{n=1}^{\infty} \frac{n x^n}{(n+1)(2x+1)^n}$$

converges absolutely.

19. *Generalizing Euler's constant.* Figure 8.21 shows the graph of a positive twice-differentiable decreasing function f whose second derivative is positive on $(0, \infty)$. For each n, the number A_n is the area of the lunar region between the curve and the line segment joining the points $(n, f(n))$ and $(n+1, f(n+1))$.

a) Use the figure to show that $\sum_{n=1}^{\infty} A_n < (1/2)(f(1) - f(2))$.

b) Then show the existence of

$$\lim_{n\to\infty} \left[\sum_{k=1}^{n} f(k) - \frac{1}{2}(f(1) + f(n)) - \int_1^n f(x)\, dx \right].$$

c) Then show the existence of

$$\lim_{n\to\infty} \left[\sum_{k=1}^{n} f(k) - \int_1^n f(x)\, dx \right].$$

If $f(x) = 1/x$, the limit in (c) is Euler's constant (Section 8.4, Exercise 41). (Source: "Convergence with Pictures" by P. J. Rippon, *American Mathematical Monthly*, Vol. 93, No. 6, 1986, pp. 476–78.)

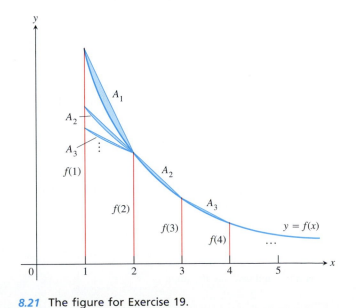

8.21 The figure for Exercise 19.

20. This exercise refers to the "right side up" equilateral triangle with sides of length $2b$ in the accompanying figure. "Upside down" equilateral triangles are removed from the original triangle as the sequence of pictures suggests. The sum of the areas removed from the original triangle forms an infinite series.

a) Find this infinite series.

b) Find the sum of this infinite series and hence find the total area removed from the original triangle.

c) Is every point on the original triangle removed? Explain why or why not.

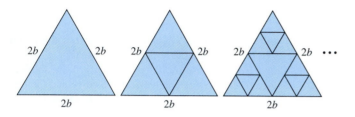

✹ 21. CAS EXPLORATION

a) Does the value of

$$\lim_{n\to\infty}\left(1-\frac{\cos(a/n)}{n}\right)^n, \quad a \text{ constant,}$$

appear to depend on the value of a? If so, how?

b) Does the value of

$$\lim_{n\to\infty}\left(1-\frac{\cos(a/n)}{bn}\right)^n, \quad a \text{ and } b \text{ constant, } b \neq 0,$$

appear to depend on the value of b? If so, how?

c) Use calculus to confirm your findings in (a) and (b).

22. Show that if $\sum_{n=1}^{\infty} a_n$ converges, then

$$\sum_{n=1}^{\infty}\left(\frac{1+\sin(a_n)}{2}\right)^n$$

converges.

23. Find a value for the constant b that will make the radius of convergence of the power series

$$\sum_{n=2}^{\infty}\frac{b^n x^n}{\ln n}$$

equal to 5.

24. How do you know that the functions $\sin x$, $\ln x$, and e^x are not polynomials? Give reasons for your answer.

25. Find the value of a for which the limit

$$\lim_{x\to0}\frac{\sin(ax)-\sin x-x}{x^3}$$

is finite and evaluate the limit.

26. Find values of a and b for which

$$\lim_{x\to0}\frac{\cos(ax)-b}{2x^2}=-1.$$

27. *Raabe's (or Gauss's) test.* The following test, which we state without proof, is an extension of the Ratio Test.

Raabe's test: If $\sum_{n=1}^{\infty} u_n$ is a series of positive constants and there exist constants C, K, and N such that

$$\frac{u_n}{u_{n+1}}=1+\frac{C}{n}+\frac{f(n)}{n^2}, \tag{1}$$

where $|f(n)| < K$ for $n \geq N$, then $\sum_{n=1}^{\infty} u_n$ converges if $C > 1$ and diverges if $C \leq 1$.

Show that the results of Raabe's test agree with what you know about the series $\sum_{n=1}^{\infty}(1/n^2)$ and $\sum_{n=1}^{\infty}(1/n)$.

28. (*Continuation of Exercise 27.*) Suppose that the terms of $\sum_{n=1}^{\infty} u_n$ are defined recursively by the formulas

$$u_1=1, \quad u_{n+1}=\frac{(2n-1)^2}{(2n)(2n+1)}u_n.$$

Apply Raabe's test to determine whether the series converges.

29. If $\sum_{n=1}^{\infty} a_n$ converges, and if $a_n \neq 1$ and $a_n > 0$ for all n,

a) Show that $\sum_{n=1}^{\infty} a_n^2$ converges.

b) Does $\sum_{n=1}^{\infty} a_n/(1-a_n)$ converge? Explain.

30. (*Continuation of Exercise 29.*) If $\sum_{n=1}^{\infty} a_n$ converges, and if $1 > a_n > 0$ for all n, show that $\sum_{n=1}^{\infty} \ln(1-a_n)$ converges. (*Hint:* First show that $|\ln(1-a_n)| \leq a_n/(1-a_n)$.)

31. *Nicole Oresme's theorem.* Prove Nicole Oresme's theorem that

$$1+\frac{1}{2}\cdot2+\frac{1}{4}\cdot3+\cdots+\frac{n}{2^{n-1}}+\cdots=4.$$

(*Hint:* Differentiate both sides of the equation $1/(1-x)=1+\sum_{n=1}^{\infty} x^n$.)

32. a) Show that

$$\sum_{n=1}^{\infty}\frac{n(n+1)}{x^n}=\frac{2x^2}{(x-1)^3}$$

for $|x| > 1$ by differentiating the identity

$$\sum_{n=1}^{\infty} x^{n+1}=\frac{x^2}{1-x}$$

twice, multiplying the result by x, and then replacing x by $1/x$.

b) CALCULATOR Use part (a) to find the real solution greater than 1 of the equation

$$x=\sum_{n=1}^{\infty}\frac{n(n+1)}{x^n}.$$

33. *A fast estimate of $\pi/2$.* As you saw if you did Exercise 29 in Section 8.1, the sequence generated by starting with $x_0 = 1$

and applying the recursion formula $x_{n+1} = x_n + \cos x_n$ converges rapidly to $\pi/2$. To explain the speed of the convergence, let $\epsilon_n = (\pi/2) - x_n$. (See the accompanying figure.) Then

$$\epsilon_{n+1} = \frac{\pi}{2} - x_n - \cos x_n$$

$$= \epsilon_n - \cos\left(\frac{\pi}{2} - \epsilon_n\right)$$

$$= \epsilon_n - \sin \epsilon_n$$

$$= \frac{1}{3!}\left(\epsilon_n\right)^3 - \frac{1}{5!}\left(\epsilon_n\right)^5 + \cdots.$$

Use this equality to show that

$$0 < \epsilon_{n+1} < \frac{1}{6}\left(\epsilon_n\right)^3.$$

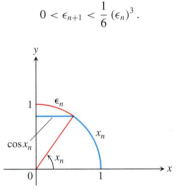

34. If $\sum_{n=1}^{\infty} a_n$ is a convergent series of positive numbers, can anything be said about the convergence of $\sum_{n=1}^{\infty} \ln(1 + a_n)$? Give reasons for your answer.

35. *Quality control*

a) Differentiate the series

$$\frac{1}{1 - x} = 1 + x + x^2 + \cdots + x^n + \cdots$$

to obtain a series for $1/(1 - x)^2$.

b) In one throw of two dice, the probability of getting a roll of 7 is $p = 1/6$. If you throw the dice repeatedly, the probability that a 7 will appear for the first time at the nth throw is $q^{n-1}p$, where $q = 1 - p = 5/6$. The expected number of throws until a 7 first appears is $\sum_{n=1}^{\infty} nq^{n-1}p$. Find the sum of this series.

c) As an engineer applying statistical control to an industrial operation, you inspect items taken at random from the assembly line. You classify each sampled item as either "good" or "bad." If the probability of an item's being good is p and of an item's being bad is $q = 1 - p$, the probability that the first bad item found is the nth one inspected is $p^{n-1}q$. The average number inspected up to and including the first bad item found is $\sum_{n=1}^{\infty} np^{n-1}q$. Evaluate this sum, assuming $0 < p < 1$.

36. *Expected value.* Suppose that a random variable X may assume the values 1, 2, 3, ... , with probabilities $p_1, p_2, p_3, \ldots$, where p_k is the probability that X equals k ($k = 1, 2, 3, \ldots$). Suppose also that $p_k \geq 0$ and that $\sum_{k=1}^{\infty} p_k = 1$. The **expected value** of X,

denoted by $E(X)$, is the number $\sum_{k=1}^{\infty} kp_k$, provided the series converges. In each of the following cases, show that $\sum_{k=1}^{\infty} p_k = 1$ and find $E(X)$ if it exists. (*Hint:* See Exercise 35.)

a) $p_k = 2^{-k}$ **b)** $p_k = \dfrac{5^{k-1}}{6^k}$

c) $p_k = \dfrac{1}{k(k+1)} = \dfrac{1}{k} - \dfrac{1}{k+1}$

37. *Safe and effective dosage.* The concentration in the blood resulting from a single dose of a drug normally decreases with time as the drug is eliminated from the body. Doses may therefore need to be repeated periodically to keep the concentration from dropping below some particular level. One model for the effect of repeated doses gives the residual concentration just before the $(n + 1)$st dose as

$$R_n = C_0 e^{-kt_0} + C_0 e^{-2kt_0} + \cdots + C_0 e^{-nkt_0},$$

where $C_0 = $ the change in concentration achievable by a single dose (mg/ml), $k = $ the *elimination constant* (h^{-1}), and $t_0 = $ time between doses (h). See Fig. 8.22.

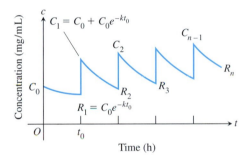

8.22 One possible effect of repeated doses on the concentration of a drug in the bloodstream.

a) Write R_n in closed form as a single fraction, and find $R = \lim_{n \to \infty} R_n$.

b) Calculate R_1 and R_{10} for $C_0 = 1$ mg/ml, $k = 0.1$ h^{-1}, and $t_0 = 10$ h. How good an estimate of R is R_{10}?

c) If $k = 0.01$ h^{-1} and $t_0 = 10$ h, find the smallest n such that $R_n > (1/2)R$.

(Source: *Prescribing Safe and Effective Dosage*, B. Horelick and S. Koont, COMAP, Inc., Lexington, MA.)

38. (*Continuation of Exercise 37.*) If a drug is known to be ineffective below a concentration C_L and harmful above some higher concentration C_H, one needs to find values of C_0 and t_0 that will produce a concentration that is safe (not above C_H) but effective (not below C_L). See Fig. 8.23. We therefore want to find values for C_0 and t_0 for which

$$R = C_L \quad \text{and} \quad C_0 + R = C_H.$$

Thus $C_0 = C_H - C_L$. When these values are substituted in the

8.23 Safe and effective concentrations of a drug. C_0 is the change in concentration produced by one dose; t_0 is the time between doses.

equation for R obtained in part (a) of Exercise 37, the resulting equation simplifies to

$$t_0 = \frac{1}{k} \ln \frac{C_H}{C_L}.$$

To reach an effective level rapidly, one might administer a "loading" dose that would produce a concentration of C_H mg/ml. This could be followed every t_0 hours by a dose that raises the concentration by $C_0 = C_H - C_L$ mg/ml.

a) Verify the preceding equation for t_0.

b) If $k = 0.05\,\text{h}^{-1}$ and the highest safe concentration is e times the lowest effective concentration, find the length of time between doses that will assure safe and effective concentrations.

c) Given $C_H = 2$ mg/ml, $C_L = 0.5$ mg/ml, and $k = 0.02\,\text{h}^{-1}$, determine a scheme for administering the drug.

d) Suppose that $k = 0.2\,\text{h}^{-1}$ and that the smallest effective concentration is 0.03 mg/ml. A single dose that produces a concentration of 0.1 mg/ml is administered. About how long will the drug remain effective?

39. *An infinite product.* The infinite product

$$\prod_{n=1}^{\infty}(1 + a_n) = (1 + a_1)(1 + a_2)(1 + a_3)\cdots$$

is said to converge if the series

$$\sum_{n=1}^{\infty} \ln(1 + a_n),$$

obtained by taking the natural logarithm of the product, converges. Prove that the product converges if $a_n > -1$ for every n and if $\sum_{n=1}^{\infty} |a_n|$ converges. (*Hint:* Show that

$$|\ln(1 + a_n)| \le \frac{|a_n|}{1 - |a_n|} \le 2|a_n|$$

when $|a_n| < 1/2$.)

40. If p is a constant, show that the series

$$1 + \sum_{n=3}^{\infty} \frac{1}{n \cdot \ln n \cdot [\ln(\ln n)]^p}$$

(a) converges if $p > 1$, (b) diverges if $p \le 1$. In general, if $f_1(x) = x$, $f_{n+1}(x) = \ln(f_n(x))$, and n takes on the values 1,

2, 3, ..., we find that $f_2(x) = \ln x$, $f_3(x) = \ln(\ln x)$, and so on. If $f_n(a) > 1$, then

$$\int_a^{\infty} \frac{dx}{f_1(x)f_2(x)\cdots f_n(x)(f_{n+1}(x))^p}$$

converges if $p > 1$ and diverges if $p \le 1$.

41. a) Prove the following theorem: If $\{c_n\}$ is a sequence of numbers such that every sum $t_n = \sum_{k=1}^{n} c_k$ is bounded, then the series $\sum_{n=1}^{\infty} c_n/n$ converges and is equal to $\sum_{n=1}^{\infty} t_n/(n(n+1))$.

　　Outline of proof: Replace c_1 by t_1 and c_n by $t_n - t_{n-1}$ for $n \ge 2$. If $s_{2n+1} = \sum_{k=1}^{2n+1} c_k/k$, show that

$$s_{2n+1} = t_1\left(1 - \frac{1}{2}\right) + t_2\left(\frac{1}{2} - \frac{1}{3}\right)$$

$$+ \cdots + t_{2n}\left(\frac{1}{2n} - \frac{1}{2n+1}\right) + \frac{t_{2n+1}}{2n+1}$$

$$= \sum_{k=1}^{2n} \frac{t_k}{k(k+1)} + \frac{t_{2n+1}}{2n+1}.$$

Because $|t_k| < M$ for some constant M, the series

$$\sum_{k=1}^{\infty} \frac{t_k}{k(k+1)}$$

converges absolutely and s_{2n+1} has a limit as $n \to \infty$. Finally, if $s_{2n} = \sum_{k=1}^{2n} c_k/k$, then $s_{2n+1} - s_{2n} = c_{2n+1}/(2n+1)$ approaches zero as $n \to \infty$ because $|c_{2n+1}| = |t_{2n+1} - t_{2n}| < 2M$. Hence the sequence of partial sums of the series $\sum c_k/k$ converges and the limit is $\sum_{k=1}^{\infty} t_k/(k(k+1))$.

b) Show how the foregoing theorem applies to the alternating harmonic series

$$1 - \frac{1}{2} + \frac{1}{3} - \frac{1}{4} + \frac{1}{5} - \frac{1}{6} + \cdots.$$

c) Show that the series

$$1 - \frac{1}{2} - \frac{1}{3} + \frac{1}{4} + \frac{1}{5} - \frac{1}{6} - \frac{1}{7} + \cdots$$

converges. (After the first term, the signs are two negative, two positive, two negative, two positive, and so on in that pattern.)

42. *The convergence of $\sum_{n=1}^{\infty}[(-1)^{n-1}x^n]/n$ to $\ln(1+x)$ for $-1 < x \le 1$*

a) Show by long division or otherwise that

$$\frac{1}{1+t} = 1 - t + t^2 - t^3 + \cdots + (-1)^n t^n + \frac{(-1)^{n+1}t^{n+1}}{1+t}.$$

b) By integrating the equation of part (a) with respect to t from 0 to x, show that

$$\ln(1+x) = x - \frac{x^2}{2} + \frac{x^3}{3} - \frac{x^4}{4} + \cdots$$

$$+ (-1)^n \frac{x^{n+1}}{n+1} + R_{n+1}.$$

where

$$R_{n+1} = (-1)^{n+1} \int_0^x \frac{t^{n+1}}{1+t} \, dt.$$

c) If $x \geq 0$, show that

$$|R_{n+1}| \leq \int_0^x t^{n+1} \, dt = \frac{x^{n+2}}{n+2}.$$

$\Big($*Hint*: As t varies from 0 to x,

$$1 + t \geq 1 \quad \text{and} \quad t^{n+1}/(1+t) \leq t^{n+1},$$

and

$$\left| \int_0^x f(t) \, dt \right| \leq \int_0^x \left| f(t) \right| dt. \Big)$$

d) If $-1 < x < 0$, show that

$$\left| R_{n+1} \right| \leq \left| \int_0^x \frac{t^{n+1}}{1-|x|} \, dt \right| = \frac{|x|^{n+2}}{(n+2)(1-|x|)}.$$

$\Big($*Hint*: If $x < t \leq 0$, then $|1+t| \geq 1 - |x|$ and

$$\left| \frac{t^{n+1}}{1+t} \right| \leq \frac{|t|^{n+1}}{1-|x|}. \Big)$$

e) Use the foregoing results to prove that the series

$$x - \frac{x^2}{2} + \frac{x^3}{3} - \frac{x^4}{4} + \cdots + \frac{(-1)^n x^{n+1}}{n+1} + \cdots$$

converges to $\ln(1+x)$ for $-1 < x \leq 1$.

Conic Sections, Parametrized Curves, and Polar Coordinates

OVERVIEW The study of motion has been important since ancient times, and calculus provides the mathematics we need to describe it. In this chapter, we extend our ability to analyze motion by showing how to track the position of a moving body as a function of time. We begin with equations for conic sections, since these are the paths traveled by planets, satellites, and other bodies (even electrons) whose motions are driven by inverse square forces. As we will see in Chapter 11, once we know that the path of a moving body is a conic section, we immediately have information about the body's velocity and the force that drives it. Planetary motion is best described with the help of polar coordinates (another of Newton's inventions, although James-Jakob-Jacques Bernoulli (1655–1705) usually gets the credit), so we also investigate curves, derivatives, and integrals in this new coordinate system.

9.1 Conic Sections and Quadratic Equations

This section shows how the conic sections from Greek geometry are described today as the graphs of quadratic equations in the coordinate plane. The Greeks of Plato's time described these curves as the curves formed by cutting a double cone with a plane (Fig. 9.1, on the following page); hence the name *conic section*.

Circles

> **Definitions**
> A **circle** is the set of points in a plane whose distance from a given fixed point in the plane is constant. The fixed point is the **center** of the circle; the constant distance is the **radius.**

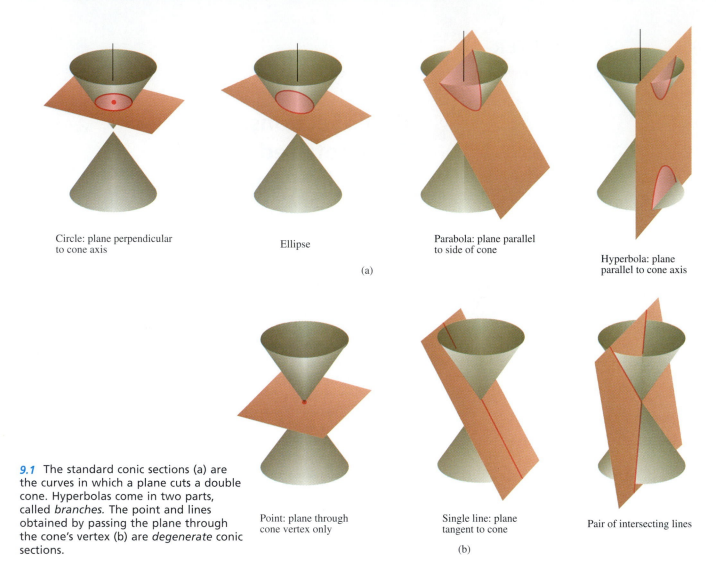

Circle: plane perpendicular
to cone axis

Ellipse

Parabola: plane parallel
to side of cone

(a)

Hyperbola: plane
parallel to cone axis

Point: plane through
cone vertex only

Single line: plane
tangent to cone

Pair of intersecting lines

(b)

9.1 The standard conic sections (a) are the curves in which a plane cuts a double cone. Hyperbolas come in two parts, called *branches.* The point and lines obtained by passing the plane through the cone's vertex (b) are *degenerate* conic sections.

The standard-form equations for circles, derived in Preliminaries, Section 4, from the distance formula $d = \sqrt{(x_2 - x_1)^2 + (y_2 - y_1)^2}$, are these:

Circles

Circle of radius a centered
at the origin:

$$x^2 + y^2 = a^2$$

Circle of radius a centered
at the point (h, k):

$$(x - h)^2 + (y - k)^2 = a^2$$

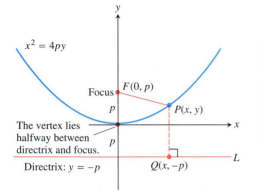

9.2 The parabola $x^2 = 4py$.

Parabolas

> ### Definitions
> A set that consists of all the points in a plane equidistant from a given fixed point and a given fixed line in the plane is a **parabola.** The fixed point is the **focus** of the parabola. The fixed line is the **directrix.**

If the focus F lies on the directrix L, the parabola is the line through F perpendicular to L. We consider this to be a degenerate case and assume henceforth that F does not lie on L.

A parabola has its simplest equation when its focus and directrix straddle one of the coordinate axes. For example, suppose that the focus lies at the point $F(0, p)$ on the positive y-axis and that the directrix is the line $y = -p$ (Fig. 9.2). In the notation of the figure, a point $P(x, y)$ lies on the parabola if and only if $PF = PQ$. From the distance formula,

$$PF = \sqrt{(x - 0)^2 + (y - p)^2} = \sqrt{x^2 + (y - p)^2}$$

$$PQ = \sqrt{(x - x)^2 + (y - (-p))^2} = \sqrt{(y + p)^2}.$$

When we equate these expressions, square, and simplify, we get

$$y = \frac{x^2}{4p} \qquad \text{or} \qquad x^2 = 4py. \qquad \text{\color{blue}Standard form} \qquad (1)$$

These equations reveal the parabola's symmetry about the y-axis. We call the y-axis the **axis** of the parabola (short for "axis of symmetry").

The point where a parabola crosses its axis is the **vertex.** The vertex of the parabola $x^2 = 4py$ lies at the origin (Fig. 9.2). The positive number p is the parabola's **focal length.**

If the parabola opens downward, with its focus at $(0, -p)$ and its directrix the line $y = p$, then Eqs. (1) become

$$y = -\frac{x^2}{4p} \qquad \text{and} \qquad x^2 = -4py$$

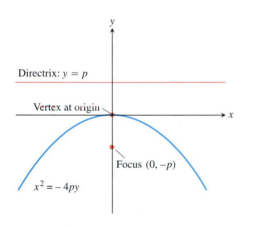

9.3 The parabola $x^2 = -4py$.

(Fig. 9.3). We obtain similar equations for parabolas opening to the right or to the left (Fig. 9.4, on the following page, and Table 9.1).

Table 9.1 Standard-form equations for parabolas with vertices at the origin ($p > 0$)

Equation	Focus	Directrix	Axis	Opens
$x^2 = \ \ 4py$	$(0, p)$	$y = -p$	y-axis	Up
$x^2 = -4py$	$(0, -p)$	$y = \ \ p$	y-axis	Down
$y^2 = \ \ 4px$	$(p, 0)$	$x = -p$	x-axis	To the right
$y^2 = -4px$	$(-p, 0)$	$x = \ \ p$	x-axis	To the left

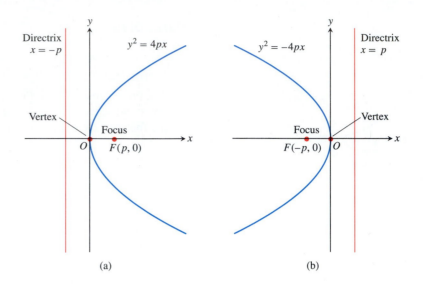

9.4 (a) The parabola $y^2 = 4px$. (b) The parabola $y^2 = -4px$.

(a)   (b)

EXAMPLE 1   Find the focus and directrix of the parabola $y^2 = 10x$.

Solution   We find the value of p in the standard equation $y^2 = 4px$:

$$4p = 10, \quad \text{so} \quad p = \frac{10}{4} = \frac{5}{2}.$$

Then we find the focus and directrix for this value of p:

Focus: $(p, 0) = \left(\dfrac{5}{2}, 0\right)$

Directrix: $x = -p$ or $x = -\dfrac{5}{2}.$

The horizontal and vertical shift formulas in Preliminaries, Section 4, can be applied to the equations in Table 9.1 to give equations for a variety of parabolas in other locations (see Exercises 39, 40, and 45–48).

Ellipses

Definitions
An **ellipse** is the set of points in a plane whose distances from two fixed points in the plane have a constant sum. The two fixed points are the **foci** of the ellipse.

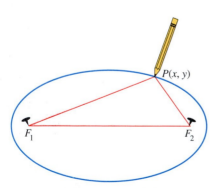

9.5 How to draw an ellipse.

The quickest way to construct an ellipse uses the definition. Put a loop of string around two tacks F_1 and F_2, pull the string taut with a pencil point P, and move the pencil around to trace a closed curve (Fig. 9.5). The curve is an ellipse because the sum $PF_1 + PF_2$, being the length of the loop minus the distance between the tacks, remains constant. The ellipse's foci lie at F_1 and F_2.

9.6 Points on the focal axis of an ellipse.

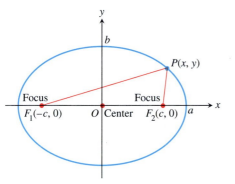

9.7 The ellipse defined by the equation $PF_1 + PF_2 = 2a$ is the graph of the equation $(x^2/a^2) + (y^2/b^2) = 1$.

Definitions

The line through the foci of an ellipse is the ellipse's **focal axis.** The point on the axis halfway between the foci is the **center.** The points where the focal axis and ellipse cross are the ellipse's **vertices** (Fig. 9.6).

If the foci are $F_1(-c, 0)$ and $F_2(c, 0)$ (Fig. 9.7), and $PF_1 + PF_2$ is denoted by $2a$, then the coordinates of a point P on the ellipse satisfy the equation

$$\sqrt{(x + c)^2 + y^2} + \sqrt{(x - c)^2 + y^2} = 2a.$$

To simplify this equation, we move the second radical to the right-hand side, square, isolate the remaining radical, and square again, obtaining

$$\frac{x^2}{a^2} + \frac{y^2}{a^2 - c^2} = 1. \tag{2}$$

Since $PF_1 + PF_2$ is greater than the length $F_1 F_2$ (triangle inequality for triangle $PF_1 F_2$), the number $2a$ is greater than $2c$. Accordingly, $a > c$ and the number $a^2 - c^2$ in Eq. (2) is positive.

The algebraic steps leading to Eq. (2) can be reversed to show that every point P whose coordinates satisfy an equation of this form with $0 < c < a$ also satisfies the equation $PF_1 + PF_2 = 2a$. A point therefore lies on the ellipse if and only if its coordinates satisfy Eq. (2).

If

$$b = \sqrt{a^2 - c^2}, \tag{3}$$

then $a^2 - c^2 = b^2$ and Eq. (2) takes the form

$$\frac{x^2}{a^2} + \frac{y^2}{b^2} = 1. \tag{4}$$

Equation (4) reveals that this ellipse is symmetric with respect to the origin and both coordinate axes. It lies inside the rectangle bounded by the lines $x = \pm a$ and $y = \pm b$. It crosses the axes at the points $(\pm a, 0)$ and $(0, \pm b)$. The tangents at these points are perpendicular to the axes because

$$\frac{dy}{dx} = -\frac{b^2 x}{a^2 y} \qquad \text{Obtained from Eq. (4) by implicit differentiation}$$

is zero if $x = 0$ and infinite if $y = 0$.

The Major and Minor Axes of an Ellipse

The **major axis** of the ellipse in Eq. (4) is the line segment of length $2a$ joining the points $(\pm a, 0)$. The **minor axis** is the line segment of length $2b$ joining the points $(0, \pm b)$. The number a itself is the **semimajor axis,** the number b the **semiminor axis.** The number c, found from Eq. (3) as

$$c = \sqrt{a^2 - b^2},$$

is the **center-to-focus distance** of the ellipse.

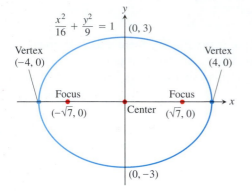

9.8 Major axis horizontal (Example 2).

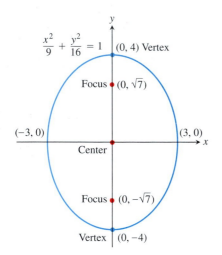

9.9 Major axis vertical (Example 3).

EXAMPLE 2   *Major axis horizontal*

The ellipse

$$\frac{x^2}{16} + \frac{y^2}{9} = 1 \tag{5}$$

(Fig. 9.8) has

Semimajor axis: $a = \sqrt{16} = 4,$ Semiminor axis: $b = \sqrt{9} = 3$

Center-to-focus distance: $c = \sqrt{16 - 9} = \sqrt{7}$

Foci: $(\pm c, 0) = (\pm \sqrt{7}, 0)$

Vertices: $(\pm a, 0) = (\pm 4, 0)$

Center: $(0, 0).$ ❑

EXAMPLE 3   *Major axis vertical*

The ellipse

$$\frac{x^2}{9} + \frac{y^2}{16} = 1, \tag{6}$$

obtained by interchanging x and y in Eq. (5), has its major axis vertical instead of horizontal (Fig. 9.9). With a^2 still equal to 16 and b^2 equal to 9, we have

Semimajor axis: $a = \sqrt{16} = 4,$ Semiminor axis: $b = \sqrt{9} = 3$

Center-to-focus distance: $c = \sqrt{16 - 9} = \sqrt{7}$

Foci: $(0, \pm c) = (0, \pm \sqrt{7})$

Vertices: $(0, \pm a) = (0, \pm 4)$

Center: $(0, 0).$ ❑

There is never any cause for confusion in analyzing equations like (5) and (6). We simply find the intercepts on the coordinate axes; then we know which way the major axis runs because it is the longer of the two axes. The center always lies at the origin and the foci lie on the major axis.

Standard-Form Equations for Ellipses Centered at the Origin

Foci on the x-axis: $\dfrac{x^2}{a^2} + \dfrac{y^2}{b^2} = 1$ $(a > b)$

Center-to-focus distance: $c = \sqrt{a^2 - b^2}$
Foci: $(\pm c, 0)$
Vertices: $(\pm a, 0)$

Foci on the y-axis: $\dfrac{x^2}{b^2} + \dfrac{y^2}{a^2} = 1$ $(a > b)$

Center-to-focus distance: $c = \sqrt{a^2 - b^2}$
Foci: $(0, \pm c)$
Vertices: $(0, \pm a)$

In each case, a is the semimajor axis and b is the semiminor axis.

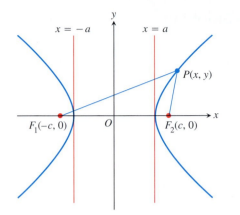

9.10 Hyperbolas have two branches. For points on the right-hand branch of the hyperbola shown here, $PF_1 - PF_2 = 2a$. For points on the left-hand branch, $PF_2 - PF_1 = 2a$.

Hyperbolas

Definitions

A **hyperbola** is the set of points in a plane whose distances from two fixed points in the plane have a constant difference. The two fixed points are the **foci** of the hyperbola.

If the foci are $F_1(-c, 0)$ and $F_2(c, 0)$ (Fig. 9.10) and the constant difference is $2a$, then a point (x, y) lies on the hyperbola if and only if

$$\sqrt{(x + c)^2 + y^2} - \sqrt{(x - c)^2 + y^2} = \pm 2a. \tag{7}$$

To simplify this equation, we move the second radical to the right-hand side, square, isolate the remaining radical, and square again, obtaining

$$\frac{x^2}{a^2} + \frac{y^2}{a^2 - c^2} = 1. \tag{8}$$

So far, this looks just like the equation for an ellipse. But now $a^2 - c^2$ is negative because $2a$, being the difference of two sides of triangle PF_1F_2, is less than $2c$, the third side.

The algebraic steps leading to Eq. (8) can be reversed to show that every point P whose coordinates satisfy an equation of this form with $0 < a < c$ also satisfies Eq. (7). A point therefore lies on the hyperbola if and only if its coordinates satisfy Eq. (8).

If we let b denote the positive square root of $c^2 - a^2$,

$$b = \sqrt{c^2 - a^2}, \tag{9}$$

then $a^2 - c^2 = -b^2$ and Eq. (8) takes the more compact form

$$\frac{x^2}{a^2} - \frac{y^2}{b^2} = 1. \tag{10}$$

The differences between Eq. (10) and the equation for an ellipse (Eq. 4) are the minus sign and the new relation

$$c^2 = a^2 + b^2. \qquad \text{From Eq. (9)}$$

Like the ellipse, the hyperbola is symmetric with respect to the origin and coordinate axes. It crosses the x-axis at the points $(\pm a, 0)$. The tangents at these points are vertical because

$$\frac{dy}{dx} = \frac{b^2 x}{a^2 y} \qquad \text{Obtained from Eq. (10) by implicit differentiation}$$

is infinite when $y = 0$. The hyperbola has no y-intercepts; in fact, no part of the curve lies between the lines $x = -a$ and $x = a$.

Definitions

The line through the foci of a hyperbola is the **focal axis.** The point on the axis halfway between the foci is the hyperbola's **center.** The points where the focal axis and hyperbola cross are the **vertices** (Fig. 9.11).

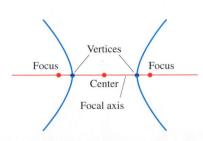

9.11 Points on the focal axis of a hyperbola.

Asymptotes of Hyperbolas—Graphing

The hyperbola

$$\frac{x^2}{a^2} - \frac{y^2}{b^2} = 1 \qquad (11)$$

has two asymptotes, the lines

$$y = \pm \frac{b}{a} x.$$

The asymptotes give us the guidance we need to graph hyperbolas quickly. (See the drawing lesson.) The fastest way to find the equations of the asymptotes is to replace the 1 in Eq. (11) by 0 and solve the new equation for y:

$$\underbrace{\frac{x^2}{a^2} - \frac{y^2}{b^2} = 1}_{\text{hyperbola}} \quad \Rightarrow \quad \underbrace{\frac{x^2}{a^2} - \frac{y^2}{b^2} = 0}_{\text{0 for 1}} \quad \Rightarrow \quad \underbrace{y = \pm \frac{b}{a} x.}_{\text{asymptotes}}$$

Standard-Form Equations for Hyperbolas Centered at the Origin

Foci on the x-axis: $\dfrac{x^2}{a^2} - \dfrac{y^2}{b^2} = 1$ *Foci on the y-axis:* $\dfrac{y^2}{a^2} - \dfrac{x^2}{b^2} = 1$

    Center-to-focus distance: $c = \sqrt{a^2 + b^2}$ Center-to-focus distance: $c = \sqrt{a^2 + b^2}$

    Foci: $(\pm c, 0)$ Foci: $(0, \pm c)$

    Vertices: $(\pm a, 0)$ Vertices: $(0, \pm a)$

    Asymptotes: $\dfrac{x^2}{a^2} - \dfrac{y^2}{b^2} = 0$ or $y = \pm \dfrac{b}{a} x$ Asymptotes: $\dfrac{y^2}{a^2} - \dfrac{x^2}{b^2} = 0$ or $y = \pm \dfrac{a}{b} x$

Notice the difference in the asymptote equations (b/a in the first, a/b in the second).

DRAWING LESSON

How to Graph the Hyperbola $\dfrac{x^2}{a^2} - \dfrac{y^2}{b^2} = 1$

1   Mark the points $(\pm a, 0)$ and $(0, \pm b)$ with line segments and complete the rectangle they determine.

2   Sketch the asymptotes by extending the rectangle's diagonals.

3   Use the rectangle and asymptotes to guide your drawing.

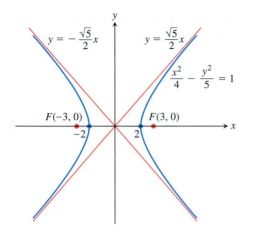

9.12 The hyperbola in Example 4.

EXAMPLE 4   *Foci on the x-axis*

The equation

$$\frac{x^2}{4} - \frac{y^2}{5} = 1 \tag{12}$$

is Eq. (10) with $a^2 = 4$ and $b^2 = 5$ (Fig. 9.12). We have

Center-to-focus distance: $c = \sqrt{a^2 + b^2} = \sqrt{4 + 5} = 3$

Foci: $(\pm c, 0) = (\pm 3, 0)$,   Vertices: $(\pm a, 0) = (\pm 2, 0)$

Center: $(0, 0)$

Asymptotes: $\dfrac{x^2}{4} - \dfrac{y^2}{5} = 0$ or $y = \pm \dfrac{\sqrt{5}}{2} x.$

EXAMPLE 5   *Foci on the y-axis*

The hyperbola

$$\frac{y^2}{4} - \frac{x^2}{5} = 1,$$

obtained by interchanging x and y in Eq. (12), has its vertices on the y-axis instead of the x-axis (Fig. 9.13). With a^2 still equal to 4 and b^2 equal to 5, we have

Center-to-focus distance: $c = \sqrt{a^2 + b^2} = \sqrt{4 + 5} = 3$

Foci: $(0, \pm c) = (0, \pm 3)$,   Vertices: $(0, \pm a) = (0, \pm 2)$

Center: $(0, 0)$

Asymptotes: $\dfrac{y^2}{4} - \dfrac{x^2}{5} = 0$ or $y = \pm \dfrac{2}{\sqrt{5}} x.$

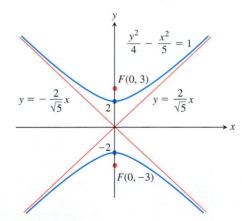

9.13 The hyperbola in Example 5.

Reflective Properties

The chief applications of parabolas involve their use as reflectors of light and radio waves. Rays originating at a parabola's focus are reflected out of the parabola parallel to the parabola's axis (Fig. 9.14, on the following page, and Exercise 90). This property is used by flashlight, headlight, and spotlight reflectors and by microwave broadcast antennas to direct radiation from point sources into narrow beams. Conversely, electromagnetic waves arriving parallel to a parabolic reflector's axis are directed

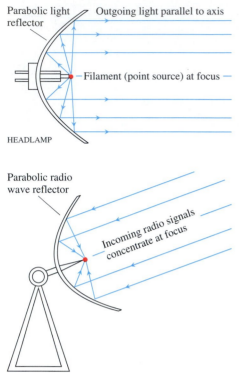

Parabolic light reflector — Outgoing light parallel to axis

Filament (point source) at focus

HEADLAMP

Parabolic radio wave reflector

Incoming radio signals concentrate at focus

RADIO TELESCOPE

9.14 Two of the many uses of parabolic reflectors.

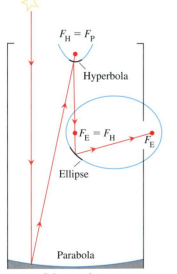

$F_H = F_P$

Hyperbola

$F_E = F_H$

F_E

Ellipse

Parabola

Primary mirror

9.16 Schematic drawing of a reflecting telescope.

toward the reflector's focus. This property is used to intensify signals picked up by radio telescopes and television satellite dishes, to focus arriving light in telescopes, and to concentrate sunlight in solar heaters.

If an ellipse is revolved about its major axis to generate a surface (the surface is called an *ellipsoid*) and the interior is silvered to produce a mirror, light from one focus will be reflected to the other focus (Fig. 9.15). Ellipsoids reflect sound the same way, and this property is used to construct *whispering galleries,* rooms in which a person standing at one focus can hear a whisper from the other focus. Statuary Hall in the U.S. Capitol building is a whispering gallery. Ellipsoids also appear in instruments used to study aircraft noise in wind tunnels (sound at one focus can be received at the other focus with relatively little interference from other sources).

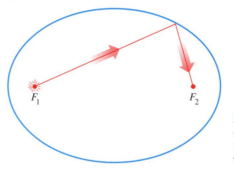

F_1 F_2

9.15 An elliptical mirror (shown here in profile) reflects light from one focus to the other.

Light directed toward one focus of a hyperbolic mirror is reflected toward the other focus. This property of hyperbolas is combined with the reflective properties of parabolas and ellipses in designing modern telescopes. In Fig. 9.16 starlight reflects off a primary parabolic mirror toward the mirror's focus F_P. It is then reflected by a small hyperbolic mirror, whose focus is $F_H = F_P$, toward the second focus of the hyperbola, $F_E = F_H$. Since this focus is shared by an ellipse, the light is reflected by the elliptical mirror to the ellipse's second focus to be seen by an observer.

As recent experience with NASA's Hubble space telescope shows, the mirrors have to be nearly perfect to focus properly. The aberration that caused the malfunction in Hubble's primary mirror (now corrected with additional mirrors) amounted to about half a wavelength of visible light, no more than 1/50 the width of a human hair.

Other Applications

Water pipes are sometimes designed with elliptical cross sections to allow for expansion when the water freezes. The triggering mechanisms in some lasers are elliptical, and stones on a beach become more and more elliptical as they are ground down by waves. There are also applications of ellipses to fossil formation. The ellipsolith, once thought to be a separate species, is now known to be an elliptically deformed nautilus.

Hyperbolic paths arise in Einstein's theory of relativity and form the basis for the (unrelated) LORAN radio navigation system. (LORAN is short for "long range navigation.") Hyperbolas also form the basis for a new system the Burlington Northern Railroad developed for using synchronized electronic signals from satellites to track freight trains. Computers aboard Burlington Northern locomotives in Minnesota have been able to track trains to within one mile per hour of their speed and to within 150 feet of their actual location.

Exercises 9.1

Identifying Graphs

Match the parabolas in Exercises 1–4 with the following equations:

$$x^2 = 2y, \quad x^2 = -6y, \quad y^2 = 8x, \quad y^2 = -4x.$$

Then find the parabola's focus and directrix.

1.

2.

3.

4.

Match each conic section in Exercises 5–8 with one of these equations:

$$\frac{x^2}{4} + \frac{y^2}{9} = 1, \quad \frac{x^2}{2} + y^2 = 1,$$

$$\frac{y^2}{4} - x^2 = 1, \quad \frac{x^2}{4} - \frac{y^2}{9} = 1.$$

Then find the conic section's foci and vertices. If the conic section is a hyperbola, find its asymptotes as well.

5.

6.

7.

8.

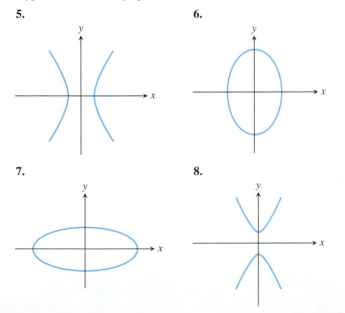

Parabolas

Exercises 9–16 give equations of parabolas. Find each parabola's focus and directrix. Then sketch the parabola. Include the focus and directrix in your sketch.

9. $y^2 = 12x$ **10.** $x^2 = 6y$ **11.** $x^2 = -8y$

12. $y^2 = -2x$ **13.** $y = 4x^2$ **14.** $y = -8x^2$

15. $x = -3y^2$ **16.** $x = 2y^2$

Ellipses

Exercises 17–24 give equations for ellipses. Put each equation in standard form. Then sketch the ellipse. Include the foci in your sketch.

17. $16x^2 + 25y^2 = 400$ **18.** $7x^2 + 16y^2 = 112$

19. $2x^2 + y^2 = 2$ **20.** $2x^2 + y^2 = 4$

21. $3x^2 + 2y^2 = 6$ **22.** $9x^2 + 10y^2 = 90$

23. $6x^2 + 9y^2 = 54$ **24.** $169x^2 + 25y^2 = 4225$

Exercises 25 and 26 give information about the foci and vertices of ellipses centered at the origin of the xy-plane. In each case, find the ellipse's standard-form equation from the given information.

25. Foci: $(\pm\sqrt{2}, 0)$ **26.** Foci: $(0, \pm 4)$
     Vertices: $(\pm 2, 0)$ Vertices: $(0, \pm 5)$

Hyperbolas

Exercises 27–34 give equations for hyperbolas. Put each equation in standard form and find the hyperbola's asymptotes. Then sketch the hyperbola. Include the asymptotes and foci in your sketch.

27. $x^2 - y^2 = 1$ **28.** $9x^2 - 16y^2 = 144$

29. $y^2 - x^2 = 8$ **30.** $y^2 - x^2 = 4$

31. $8x^2 - 2y^2 = 16$ **32.** $y^2 - 3x^2 = 3$

33. $8y^2 - 2x^2 = 16$ **34.** $64x^2 - 36y^2 = 2304$

Exercises 35–38 give information about the foci, vertices, and asymptotes of hyperbolas centered at the origin of the xy-plane. In each case, find the hyperbola's standard-form equation from the information given.

35. Foci: $(0, \pm\sqrt{2})$ **36.** Foci: $(\pm 2, 0)$
     Asymptotes: $y = \pm x$ Asymptotes: $y = \pm\dfrac{1}{\sqrt{3}}x$

37. Vertices: $(\pm 3, 0)$ **38.** Vertices: $(0, \pm 2)$
     Asymptotes: $y = \pm\dfrac{4}{3}x$ Asymptotes: $y = \pm\dfrac{1}{2}x$

Shifting Conic Sections

39. The parabola $y^2 = 8x$ is shifted down 2 units and right 1 unit to generate the parabola $(y + 2)^2 = 8(x - 1)$. (a) Find the new parabola's vertex, focus, and directrix. (b) Plot the new vertex, focus, and directrix, and sketch in the parabola.

40. The parabola $x^2 = -4y$ is shifted left 1 unit and up 3 units to generate the parabola $(x + 1)^2 = -4(y - 3)$. (a) Find the new parabola's vertex, focus, and directrix. (b) Plot the new vertex, focus, and directrix, and sketch in the parabola.

41. The ellipse $(x^2/16) + (y^2/9) = 1$ is shifted 4 units to the right and 3 units up to generate the ellipse

$$\frac{(x - 4)^2}{16} + \frac{(y - 3)^2}{9} = 1.$$

(a) Find the foci, vertices, and center of the new ellipse. (b) Plot the new foci, vertices, and center, and sketch in the new ellipse.

42. The ellipse $(x^2/9) + (y^2/25) = 1$ is shifted 3 units to the left and 2 units down to generate the ellipse

$$\frac{(x + 3)^2}{9} + \frac{(y + 2)^2}{25} = 1.$$

(a) Find the foci, vertices, and center of the new ellipse. (b) Plot the new foci, vertices, and center, and sketch in the new ellipse.

43. The hyperbola $(x^2/16) - (y^2/9) = 1$ is shifted 2 units to the right to generate the hyperbola

$$\frac{(x - 2)^2}{16} - \frac{y^2}{9} = 1.$$

(a) Find the center, foci, vertices, and asymptotes of the new hyperbola. (b) Plot the new center, foci, vertices, and asymptotes, and sketch in the hyperbola.

44. The hyperbola $(y^2/4) - (x^2/5) = 1$ is shifted 2 units down to generate the hyperbola

$$\frac{(y + 2)^2}{4} - \frac{x^2}{5} = 1.$$

(a) Find the center, foci, vertices, and asymptotes of the new hyperbola. (b) Plot the new center, foci, vertices, and asymptotes, and sketch in the hyperbola.

Exercises 45–48 give equations for parabolas and tell how many units up or down and to the right or left each parabola is to be shifted. Find an equation for the new parabola, and find the new vertex, focus, and directrix.

45. $y^2 = 4x$,   left 2, down 3

46. $y^2 = -12x$,   right 4, up 3

47. $x^2 = 8y$,   right 1, down 7

48. $x^2 = 6y$,   left 3, down 2

Exercises 49–52 give equations for ellipses and tell how many units up or down and to the right or left each ellipse is to be shifted. Find an equation for the new ellipse, and find the new foci, vertices, and center.

49. $\dfrac{x^2}{6} + \dfrac{y^2}{9} = 1$,   left 2, down 1

50. $\dfrac{x^2}{2} + y^2 = 1$,   right 3, up 4

51. $\dfrac{x^2}{3} + \dfrac{y^2}{2} = 1$,   right 2, up 3

52. $\dfrac{x^2}{16} + \dfrac{y^2}{25} = 1$,   left 4, down 5

Exercises 53–56 give equations for hyperbolas and tell how many units up or down and to the right or left each hyperbola is to be shifted. Find an equation for the new hyperbola, and find the new center, foci, vertices, and asymptotes.

53. $\dfrac{x^2}{4} - \dfrac{y^2}{5} = 1$,   right 2, up 2

54. $\dfrac{x^2}{16} - \dfrac{y^2}{9} = 1$,   left 5, down 1

55. $y^2 - x^2 = 1$,   left 1, down 1

56. $\dfrac{y^2}{3} - x^2 = 1$,   right 1, up 3

Find the center, foci, vertices, asymptotes, and radius, as appropriate, of the conic sections in Exercises 57–68.

57. $x^2 + 4x + y^2 = 12$

58. $2x^2 + 2y^2 - 28x + 12y + 114 = 0$

59. $x^2 + 2x + 4y - 3 = 0$

60. $y^2 - 4y - 8x - 12 = 0$

61. $x^2 + 5y^2 + 4x = 1$

62. $9x^2 + 6y^2 + 36y = 0$

63. $x^2 + 2y^2 - 2x - 4y = -1$

64. $4x^2 + y^2 + 8x - 2y = -1$

65. $x^2 - y^2 - 2x + 4y = 4$

66. $x^2 - y^2 + 4x - 6y = 6$

67. $2x^2 - y^2 + 6y = 3$

68. $y^2 - 4x^2 + 16x = 24$

Inequalities

Sketch the regions in the xy-plane whose coordinates satisfy the inequalities or pairs of inequalities in Exercises 69–74.

69. $9x^2 + 16y^2 \leq 144$

70. $x^2 + y^2 \geq 1$ and $4x^2 + y^2 \leq 4$

71. $x^2 + 4y^2 \geq 4$ and $4x^2 + 9y^2 \leq 36$

72. $(x^2 + y^2 - 4)(x^2 + 9y^2 - 9) \leq 0$

73. $4y^2 - x^2 \geq 4$

74. $|x^2 - y^2| \leq 1$

Theory and Examples

75. *Archimedes' formula for the volume of a parabolic solid.*
The region enclosed by the parabola $y = (4h/b^2)x^2$ and the line $y = h$ is revolved about the y-axis to generate the solid shown here. Show that the volume of the solid is $3/2$ the volume of the corresponding cone.

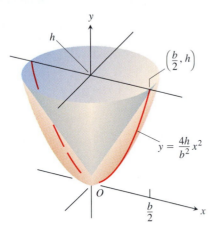

76. *Suspension bridge cables hang in parabolas.* The suspension bridge cable shown here supports a uniform load of w pounds per horizontal foot. It can be shown that if H is the horizontal tension of the cable at the origin, then the curve of the cable satisfies the equation

$$\frac{dy}{dx} = \frac{w}{H}x.$$

Show that the cable hangs in a parabola by solving this differential equation subject to the initial condition that $y = 0$ when $x = 0$.

77. Find an equation for the circle through the points $(1, 0)$, $(0, 1)$, and $(2, 2)$.

78. Find an equation for the circle through the points $(2, 3)$, $(3, 2)$, and $(-4, 3)$.

79. Find an equation for the circle centered at $(-2, 1)$ that passes through the point $(1, 3)$. Is the point $(1.1, 2.8)$ inside, outside, or on the circle?

80. Find equations for the tangents to the circle $(x - 2)^2 + (y - 1)^2 = 5$ at the points where the circle crosses the coordinate axes. (*Hint:* Use implicit differentiation.)

81. If lines are drawn parallel to the coordinate axes through a point P on the parabola $y^2 = kx, k > 0$, the parabola partitions the rectangular region bounded by these lines and the coordinate axes into two smaller regions, A and B.

a) If the two smaller regions are revolved about the y-axis, show that they generate solids whose volumes have the ratio 4:1.

b) What is the ratio of the volumes generated by revolving the regions about the x-axis?

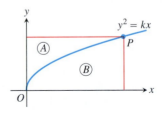

82. Show that the tangents to the curve $y^2 = 4px$ from any point on the line $x = -p$ are perpendicular.

83. Find the dimensions of the rectangle of largest area that can be inscribed in the ellipse $x^2 + 4y^2 = 4$ with its sides parallel to the coordinate axes. What is the area of the rectangle?

84. Find the volume of the solid generated by revolving the region enclosed by the ellipse $9x^2 + 4y^2 = 36$ about the (a) x-axis, (b) y-axis.

85. The "triangular" region in the first quadrant bounded by the x-axis, the line $x = 4$, and the hyperbola $9x^2 - 4y^2 = 36$ is revolved about the x-axis to generate a solid. Find the volume of the solid.

86. The region bounded on the left by the y-axis, on the right by the hyperbola $x^2 - y^2 = 1$, and above and below by the lines $y = \pm 3$ is revolved about the y-axis to generate a solid. Find the volume of the solid.

87. Find the centroid of the region that is bounded below by the x-axis and above by the ellipse $(x^2/9) + (y^2/16) = 1$.

88. The curve $y = \sqrt{x^2 + 1}, 0 \le x \le \sqrt{2}$, which is part of the upper branch of the hyperbola $y^2 - x^2 = 1$, is revolved about the x-axis to generate a surface. Find the area of the surface.

89. The circular waves in the photograph here were made by touching the surface of a ripple tank, first at A and then at B. As the waves expanded, their point of intersection appeared to trace a hyperbola. Did it really do that? To find out, we can model the waves with circles centered at A and B.

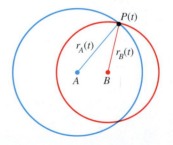

At time t, the point P is $r_A(t)$ units from A and $r_B(t)$ units from B. Since the radii of the circles increase at a constant rate, the rate at which the waves are traveling is

$$\frac{dr_A}{dt} = \frac{dr_B}{dt}.$$

Conclude from this equation that $r_A - r_B$ has a constant value, so that P must lie on a hyperbola with foci at A and B.

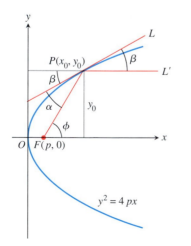

The expanding waves in Exercise 89.

90. *The reflective property of parabolas.* The figure here shows a typical point $P(x_0, y_0)$ on the parabola $y^2 = 4px$. The line L is tangent to the parabola at P. The parabola's focus lies at $F(p, 0)$. The ray L' extending from P to the right is parallel to the x-axis. We show that light from F to P will be reflected out along L' by showing that β equals α. Establish this equality by taking the following steps.

a) Show that $\tan \beta = 2p/y_0$.

b) Show that $\tan \phi = y_0/(x_0 - p)$.

c) Use the identity

$$\tan \alpha = \frac{\tan \phi - \tan \beta}{1 + \tan \phi \tan \beta}$$

to show that $\tan \alpha = 2p/y_0$.

Since α and β are both acute, $\tan \beta = \tan \alpha$ implies $\beta = \alpha$.

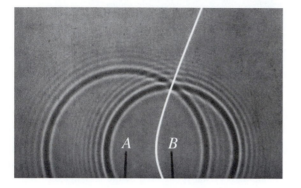

91. *How the astronomer Kepler used string to draw parabolas.* Kepler's method for drawing a parabola (with more modern tools) requires a string the length of a T square and a table whose edge can serve as the parabola's directrix. Pin one end of the string to the point where you want the focus to be and the other end to the upper end of the T square. Then, holding the string taut against the T square with a pencil, slide the T square along the table's edge. As the T square moves, the pencil will trace a parabola. Why?

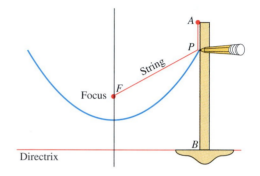

92. *Construction of a hyperbola.* The following diagrams appeared (unlabeled) in Ernest J. Eckert, "Constructions Without Words," *Mathematics Magazine,* Vol. 66, No. 2, April 1993, p. 113. Explain the constructions.

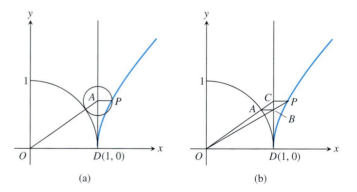

(a)   (b)

93. *The width of a parabola at the focus.* Show that the number $4p$ is the **width** of the parabola $x^2 = 4py$ $(p > 0)$ at the focus by showing that the line $y = p$ cuts the parabola at points that are $4p$ units apart.

94. *The asymptotes of $(x^2/a^2) - (y^2/b^2) = 1$.* Show that the vertical distance between the line $y = (b/a)x$ and the upper half of the right-hand branch $y = (b/a)\sqrt{x^2 - a^2}$ of the hyperbola $(x^2/a^2) - (y^2/b^2) = 1$ approaches 0 by showing that

$$\lim_{x \to \infty} \left(\frac{b}{a}x - \frac{b}{a}\sqrt{x^2 - a^2}\right) = \frac{b}{a} \lim_{x \to \infty} \left(x - \sqrt{x^2 - a^2}\right) = 0.$$

Similar results hold for the remaining portions of the hyperbola and the lines $y = \pm(b/a)x$.

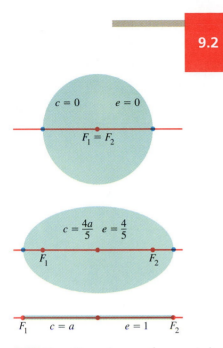

9.17 The ellipse changes from a circle to a line segment as *c* increases from 0 to *a*.

Table 9.2 Eccentricities of planetary orbits

Mercury	0.21	Saturn	0.06
Venus	0.01	Uranus	0.05
Earth	0.02	Neptune	0.01
Mars	0.09	Pluto	0.25
Jupiter	0.05		

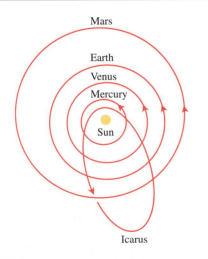

9.18 The orbit of the asteroid Icarus is highly eccentric. Earth's orbit is so nearly circular that its foci lie inside the sun.

Classifying Conic Sections by Eccentricity

We now show how to associate with each conic section a number called the conic section's eccentricity. The eccentricity reveals the conic section's type (circle, ellipse, parabola, or hyperbola) and, in the case of ellipses and hyperbolas, describes the conic section's general proportions.

Eccentricity

Although the center-to-focus distance c does not appear in the equation

$$\frac{x^2}{a^2} + \frac{y^2}{b^2} = 1, \quad (a > b)$$

for an ellipse, we can still determine c from the equation $c = \sqrt{a^2 - b^2}$. If we fix a and vary c over the interval $0 \leq c \leq a$, the resulting ellipses will vary in shape (Fig. 9.17). They are circles if $c = 0$ (so that $a = b$) and flatten as c increases. If $c = a$, the foci and vertices overlap and the ellipse degenerates into a line segment.

We use the ratio of c to a to describe the various shapes the ellipse can take. We call this ratio the ellipse's eccentricity.

Definition

The **eccentricity** of the ellipse $(x^2/a^2) + (y^2/b^2) = 1$ $(a > b)$ is

$$e = \frac{c}{a} = \frac{\sqrt{a^2 - b^2}}{a}.$$

The planets in the solar system revolve around the sun in elliptical orbits with the sun at one focus. Most of the orbits are nearly circular, as can be seen from the eccentricities in Table 9.2. Pluto has a fairly eccentric orbit, with $e = 0.25$, as does Mercury, with $e = 0.21$. Other members of the solar system have orbits that are even more eccentric. Icarus, an asteroid about 1 mile wide that revolves around the sun every 409 Earth days, has an orbital eccentricity of 0.83 (Fig. 9.18).

EXAMPLE 1   The orbit of Halley's comet is an ellipse 36.18 astronomical units long by 9.12 astronomical units wide. (One *astronomical unit* [AU] is 149,597,870 km, the semimajor axis of Earth's orbit.) Its eccentricity is

$$e = \frac{\sqrt{a^2 - b^2}}{a} = \frac{\sqrt{(36.18/2)^2 - (9.12/2)^2}}{(1/2)(36.18)} = \frac{\sqrt{(18.09)^2 - (4.56)^2}}{18.09} \approx 0.97. \quad \square$$

Whereas a parabola has one focus and one directrix, each ellipse has two foci and two directrices. These are the lines perpendicular to the major axis at distances $\pm a/e$ from the center. The parabola has the property that

$$PF = 1 \cdot PD \tag{1}$$

for any point P on it, where F is the focus and D is the point nearest P on the directrix. For an ellipse, it can be shown that the equations that replace (1) are

$$PF_1 = e \cdot PD_1, \qquad PF_2 = e \cdot PD_2. \tag{2}$$

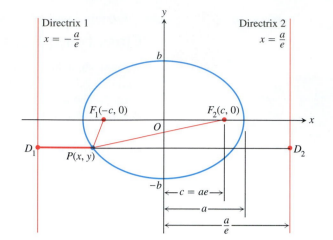

9.19 The foci and directrices of the ellipse $(x^2/a^2) + (y^2/b^2) = 1$. Directrix 1 corresponds to focus F_1, and directrix 2 to focus F_2.

Here, e is the eccentricity, P is any point on the ellipse, F_1 and F_2 are the foci, and D_1 and D_2 are the points on the directrices nearest P (Fig. 9.19).

In each equation in (2) the directrix and focus must correspond; that is, if we use the distance from P to F_1, we must also use the distance from P to the directrix at the same end of the ellipse. The directrix $x = -a/e$ corresponds to $F_1(-c, 0)$, and the directrix $x = a/e$ corresponds to $F_2(c, 0)$.

The eccentricity of a hyperbola is also $e = c/a$, only in this case c equals $\sqrt{a^2 + b^2}$ instead of $\sqrt{a^2 - b^2}$. In contrast to the eccentricity of an ellipse, the eccentricity of a hyperbola is always greater than 1.

Halley's comet

Edmund Halley (1656–1742; pronounced "*haw*-ley"), British biologist, geologist, sea captain, pirate, spy, Antarctic voyager, astronomer, adviser on fortifications, company founder and director, and the author of the first actuarial mortality tables, was also the mathematician who pushed and harried Newton into writing his *Principia*. Despite his accomplishments, Halley is known today chiefly as the man who calculated the orbit of the great comet of 1682: "wherefore if according to what we have already said [the comet] should return again about the year 1758, candid posterity will not refuse to acknowledge that this was first discovered by an Englishman." Indeed, candid posterity did not refuse—ever since the comet's return in 1758, it has been known as Halley's comet.

Last seen rounding the sun during the winter and spring of 1985–86, the comet is due to return in the year 2062. A recent study indicates that the comet has made about 2000 cycles so far with about the same number to go before the sun erodes it away completely.

Definition

The **eccentricity** of the hyperbola $(x^2/a^2) - (y^2/b^2) = 1$ is

$$e = \frac{c}{a} = \frac{\sqrt{a^2 + b^2}}{a}.$$

In both ellipse and hyperbola, the eccentricity is the ratio of the distance between the foci to the distance between the vertices (because $c/a = 2c/2a$).

$$\text{Eccentricity} = \frac{\text{distance between foci}}{\text{distance between vertices}}$$

In an ellipse, the foci are closer together than the vertices and the ratio is less than 1. In a hyperbola, the foci are farther apart than the vertices and the ratio is greater than 1.

EXAMPLE 2   Locate the vertices of an ellipse of eccentricity 0.8 whose foci lie at the points $(0, \pm 7)$.

Solution   Since $e = c/a$, the vertices are the points $(0, \pm a)$ where

$$a = \frac{c}{e} = \frac{7}{0.8} = 8.75,$$

or $(0, \pm 8.75)$.   ❏

EXAMPLE 3   Find the eccentricity of the hyperbola $9x^2 - 16y^2 = 144$.

Solution   We divide both sides of the hyperbola's equation by 144 to put it in standard form, obtaining

$$\frac{9x^2}{144} - \frac{16y^2}{144} = 1 \quad \text{and} \quad \frac{x^2}{16} - \frac{y^2}{9} = 1.$$

With $a^2 = 16$ and $b^2 = 9$, we find that $c = \sqrt{a^2 + b^2} = \sqrt{16 + 9} = 5$, so

$$e = \frac{c}{a} = \frac{5}{4}.$$ ❏

As with the ellipse, it can be shown that the lines $x = \pm a/e$ act as directrices for the hyperbola and that

$$PF_1 = e \cdot PD_1 \quad \text{and} \quad PF_2 = e \cdot PD_2. \tag{3}$$

Here P is any point on the hyperbola, F_1 and F_2 are the foci, and D_1 and D_2 are the points nearest P on the directrices (Fig. 9.20).

To complete the picture, we define the eccentricity of a parabola to be $e = 1$. Equations (1) – (3) then have the common form $PF = e \cdot PD$.

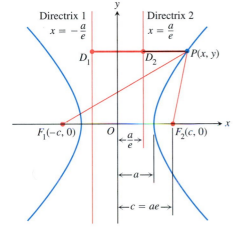

9.20 The foci and directrices of the hyperbola $(x^2/a^2) - (y^2/b^2) = 1$. No matter where P lies on the hyperbola, $PF_1 = e \cdot PD_1$ and $PF_2 = e \cdot PD_2$.

> **Definition**
>
> The **eccentricity** of a parabola is $e = 1$.

The "focus–directrix" equation $PF = e \cdot PD$ unites the parabola, ellipse, and hyperbola in the following way. Suppose that the distance PF of a point P from a fixed point F(the focus) is a constant multiple of its distance from a fixed line (the directrix). That is, suppose

$$PF = e \cdot PD, \tag{4}$$

where e is the constant of proportionality. Then the path traced by P is

a)   a *parabola* if $e = 1$,
b)   an *ellipse* of eccentricity e if $e < 1$, and
c)   a *hyperbola* of eccentricity e if $e > 1$.

Equation (4) may not look like much to get excited about. There are no co-ordinates in it and when we try to translate it into coordinate form it translates in different ways, depending on the size of e. At least, that is what happens in Cartesian coordinates. However, in polar coordinates, as we will see in Section 9.8, the equation $PF = e \cdot PD$ translates into a single equation regardless of the value of e, an equation so simple that it has been the equation of choice of astronomers and space scientists for nearly 300 years.

Given the focus and corresponding directrix of a hyperbola centered at the origin and with foci on the x-axis, we can use the dimensions shown in Fig. 9.20 to find e. Knowing e, we can derive a Cartesian equation for the hyperbola from the equation $PF = e \cdot PD$, as in the next example. We can find equations for ellipses centered at the origin and with foci on the x-axis in a similar way, using dimensions shown in Fig. 9.19.

EXAMPLE 4   Find a Cartesian equation for the hyperbola centered at the origin that has a focus at $(3, 0)$ and the line $x = 1$ as the corresponding directrix.

Solution   We first use the dimensions shown in Fig. 9.20 to find the hyperbola's eccentricity. The focus is

$$(c, 0) = (3, 0), \qquad \text{so} \qquad c = 3.$$

The directrix is the line

$$x = \frac{a}{e} = 1, \qquad \text{so} \qquad a = e.$$

When combined with the equation $e = c/a$ that defines eccentricity, these results give

$$e = \frac{c}{a} = \frac{3}{e}, \qquad \text{so} \qquad e^2 = 3 \quad \text{and} \quad e = \sqrt{3}.$$

Knowing e, we can now derive the equation we want from the equation $PF = e \cdot PD$. In the notation of Fig. 9.21, we have

$$PF = e \cdot PD \qquad\qquad \text{Eq. (4)}$$
$$\sqrt{(x-3)^2 + (y-0)^2} = \sqrt{3}\,|x - 1| \qquad\qquad e = \sqrt{3}$$
$$x^2 - 6x + 9 + y^2 = 3(x^2 - 2x + 1)$$
$$2x^2 - y^2 = 6$$
$$\frac{x^2}{3} - \frac{y^2}{6} = 1.$$

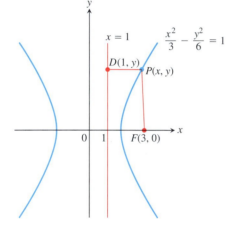

9.21 The hyperbola in Example 4.

Exercises 9.2

Ellipses

In Exercises 1–8, find the eccentricity of the ellipse. Then find and graph the ellipse's foci and directrices.

1. $16x^2 + 25y^2 = 400$

2. $7x^2 + 16y^2 = 112$

3. $2x^2 + y^2 = 2$

4. $2x^2 + y^2 = 4$

5. $3x^2 + 2y^2 = 6$

6. $9x^2 + 10y^2 = 90$

7. $6x^2 + 9y^2 = 54$

8. $169x^2 + 25y^2 = 4225$

Exercises 9–12 give the foci or vertices and the eccentricities of ellipses centered at the origin of the xy-plane. In each case, find the ellipse's standard-form equation.

9. Foci: $(0, \pm 3)$
Eccentricity: 0.5

10. Foci: $(\pm 8, 0)$
Eccentricity: 0.2

11. Vertices: $(0, \pm 70)$
Eccentricity: 0.1

12. Vertices: $(\pm 10, 0)$
Eccentricity: 0.24

Exercises 13–16 give foci and corresponding directrices of ellipses centered at the origin of the xy-plane. In each case, use the dimensions in Fig. 9.19 to find the eccentricity of the ellipse. Then find the ellipse's standard-form equation.

13. Focus: $(\sqrt{5}, 0)$

Directrix: $x = \dfrac{9}{\sqrt{5}}$

14. Focus: $(4, 0)$

Directrix: $x = \dfrac{16}{3}$

15. Focus: $(-4, 0)$
Directrix: $x = -16$

16. Focus: $(-\sqrt{2}, 0)$
Directrix: $x = -2\sqrt{2}$

17. Draw an ellipse of eccentricity 4/5. Explain your procedure.

18. Draw the orbit of Pluto (eccentricity 0.25) to scale. Explain your procedure.

19. The endpoints of the major and minor axes of an ellipse are $(1, 1)$, $(3, 4)$, $(1, 7)$, and $(-1, 4)$. Sketch the ellipse, give its equation in standard form, and find its foci, eccentricity, and directrices.

20. Find an equation for the ellipse of eccentricity 2/3 that has the line $x = 9$ as a directrix and the point $(4, 0)$ as the corresponding focus.

21. What values of the constants a, b, and c make the ellipse

$$4x^2 + y^2 + ax + by + c = 0$$

lie tangent to the x-axis at the origin and pass through the point $(-1, 2)$? What is the eccentricity of the ellipse?

22. *The reflective property of ellipses.* An ellipse is revolved about its major axis to generate an ellipsoid. The inner surface of the ellipsoid is silvered to make a mirror. Show that a ray of light emanating from one focus will be reflected to the other focus. Sound waves also follow such paths, and this property is used in constructing "whispering galleries." (*Hint:* Place the ellipse in standard position in the xy-plane and show that the lines from a point P on the ellipse to the two foci make congruent angles with the tangent to the ellipse at P.)

Hyperbolas

In Exercises 23–30, find the eccentricity of the hyperbola. Then find and graph the hyperbola's foci and directrices.

23. $x^2 - y^2 = 1$

24. $9x^2 - 16y^2 = 144$

25. $y^2 - x^2 = 8$

26. $y^2 - x^2 = 4$

27. $8x^2 - 2y^2 = 16$

28. $y^2 - 3x^2 = 3$

29. $8y^2 - 2x^2 = 16$

30. $64x^2 - 36y^2 = 2304$

Exercises 31–34 give the eccentricities and the vertices or foci of hyperbolas centered at the origin of the xy-plane. In each case, find the hyperbola's standard-form equation.

31. Eccentricity: 3
Vertices: $(0, \pm 1)$

32. Eccentricity: 2
Vertices: $(\pm 2, 0)$

33. Eccentricity: 3
Foci: $(\pm 3, 0)$

34. Eccentricity: 1.25
Foci: $(0, \pm 5)$

Exercises 35–38 give foci and corresponding directrices of hyperbolas centered at the origin of the xy-plane. In each case, find the hyperbola's eccentricity. Then find the hyperbola's standard-form equation.

35. Focus: $(4, 0)$
Directrix: $x = 2$

36. Focus: $(\sqrt{10}, 0)$
Directrix: $x = \sqrt{2}$

37. Focus: $(-2, 0)$

Directrix: $x = -\dfrac{1}{2}$

38. Focus: $(-6, 0)$
Directrix: $x = -2$

39. A hyperbola of eccentricity 3/2 has one focus at $(1, -3)$. The corresponding directrix is the line $y = 2$. Find an equation for the hyperbola.

40. *The effect of eccentricity on a hyperbola's shape.* What happens to the graph of a hyperbola as its eccentricity increases? To find out, rewrite the equation $(x^2/a^2) - (y^2/b^2) = 1$ in terms of a and e instead of a and b. Graph the hyperbola for various values of e and describe what you find.

41. *The reflective property of hyperbolas.* Show that a ray of light directed toward one focus of a hyperbolic mirror, as in the accompanying figure, is reflected toward the other focus. (*Hint:* Show that the tangent to the hyperbola at P bisects the angle made by segments PF_1 and PF_2.)

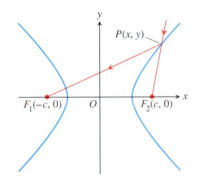

42. *A confocal ellipse and hyperbola.* Show that an ellipse and a hyperbola that have the same foci A and B, as in the accompanying figure, cross at right angles at their point of intersection. (*Hint:* A ray of light from focus A that met the hyperbola at P would be reflected from the hyperbola as if it came directly from B (Exercise 41). The same ray would be reflected off the ellipse to pass through B (Exercise 22).)

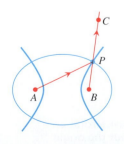

Quadratic Equations and Rotations

In this section, we examine one of the most amazing results in analytic geometry, which is that the Cartesian graph of any equation

$$Ax^2 + Bxy + Cy^2 + Dx + Ey + F = 0, \tag{1}$$

in which A, B, and C are not all zero, is nearly always a conic section. The exceptions are the cases in which there is no graph at all or the graph consists of two parallel lines. It is conventional to call all graphs of Eq. (1), curved or not, **quadratic curves.**

The Cross Product Term

You may have noticed that the term Bxy did not appear in the equations for the conic sections in Section 9.1. This happened because the axes of the conic sections ran parallel to (in fact, coincided with) the coordinate axes.

To see what happens when the parallelism is absent, let us write an equation for a hyperbola with $a = 3$ and foci at $F_1(-3, -3)$ and $F_2(3, 3)$ (Fig. 9.22). The equation $|PF_1 - PF_2| = 2a$ becomes $|PF_1 - PF_2| = 2(3) = 6$ and

$$\sqrt{(x + 3)^2 + (y + 3)^2} - \sqrt{(x - 3)^2 + (y - 3)^2} = \pm 6.$$

When we transpose one radical, square, solve for the radical that still appears, and square again, the equation reduces to

$$2xy = 9, \tag{2}$$

a case of Eq. (1) in which the cross-product term is present. The asymptotes of the hyperbola in Eq. (2) are the x- and y-axes, and the focal axis makes an angle of $\pi/4$ radians with the positive x-axis. As in this example, the cross product term is present in Eq. (1) only when the axes of the conic are tilted.

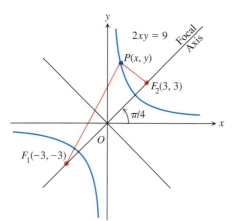

9.22 The focal axis of the hyperbola $2xy = 9$ makes an angle of $\pi/4$ radians with the positive x-axis.

Rotating the Coordinate Axes to Eliminate the Cross Product Term

To eliminate the xy-term from the equation of a conic, we rotate the coordinate axes to eliminate the "tilt" in the axes of the conic. The equations for the rotations we use are derived in the following way. In the notation of Fig. 9.23, which shows a counterclockwise rotation about the origin through an angle α,

$$x = OM = OP \cos(\theta + \alpha) = OP \cos \theta \cos \alpha - OP \sin \theta \sin \alpha$$

$$y = MP = OP \sin(\theta + \alpha) = OP \cos \theta \sin \alpha + OP \sin \theta \cos \alpha. \tag{3}$$

Since

$$OP \cos \theta = OM' = x'$$

and

$$OP \sin \theta = M'P = y',$$

the equations in (3) reduce to the following.

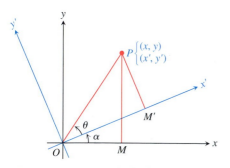

9.23 A counterclockwise rotation through angle α about the origin.

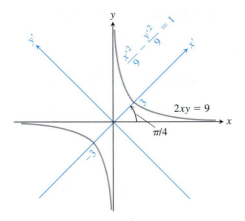

9.24 The hyperbola in Example 1 (x' and y' are the new coordinates).

Equations for Rotating Coordinate Axes

$$x = x' \cos \alpha - y' \sin \alpha$$
$$y = x' \sin \alpha + y' \cos \alpha$$

(4)

EXAMPLE 1   The x- and y-axes are rotated through an angle of $\pi/4$ radians about the origin. Find an equation for the hyperbola $2xy = 9$ in the new coordinates.

Solution   Since $\cos \pi/4 = \sin \pi/4 = 1/\sqrt{2}$, we substitute

$$x = \frac{x' - y'}{\sqrt{2}}, \qquad y = \frac{x' + y'}{\sqrt{2}}$$

from Eqs. (4) into the equation $2xy = 9$ and obtain

$$2 \left(\frac{x' - y'}{\sqrt{2}} \right) \left(\frac{x' + y'}{\sqrt{2}} \right) = 9$$

$$x'^2 - y'^2 = 9$$

$$\frac{x'^2}{9} - \frac{y'^2}{9} = 1.$$

See Fig. 9.24.   ❏

If we apply Eqs. (4) to the quadratic equation (1), we obtain a new quadratic equation

$$A' x'^2 + B' x' y' + C' y'^2 + D' x' + E' y' + F' = 0.$$

(5)

The new and old coefficients are related by the equations

$$A' = A \cos^2 \alpha + B \cos \alpha \sin \alpha + C \sin^2 \alpha$$

$$B' = B \cos 2\alpha + (C - A) \sin 2\alpha$$

$$C' = A \sin^2 \alpha - B \sin \alpha \cos \alpha + C \cos^2 \alpha$$

(6)

$$D' = D \cos \alpha + E \sin \alpha$$

$$E' = -D \sin \alpha + E \cos \alpha$$

$$F' = F.$$

These equations show, among other things, that if we start with an equation for a curve in which the cross product term is present $(B \neq 0)$, we can find a rotation angle α that produces an equation in which no cross product term appears $(B' = 0)$. To find α, we set $B' = 0$ in the second equation in (6) and solve the resulting equation,

$$B \cos 2\alpha + (C - A) \sin 2\alpha = 0,$$

for α. In practice, this means determining α from one of the two equations

---

$$\cot 2\alpha = \frac{A - C}{B} \qquad \text{or} \qquad \tan 2\alpha = \frac{B}{A - C}.$$

(7)

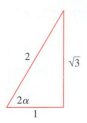

9.25 This triangle identifies
$2\alpha = \cot^{-1}(1/\sqrt{3})$ as $\pi/3$ (Example 2).

EXAMPLE 2   The coordinate axes are to be rotated through an angle α to produce an equation for the curve

$$2x^2 + \sqrt{3}\,xy + y^2 - 10 = 0$$

that has no cross product term. Find α and the new equation. Identify the curve.

Solution   The equation $2x^2 + \sqrt{3}\,xy + y^2 - 10 = 0$ has $A = 2$, $B = \sqrt{3}$, and $C = 1$. We substitute these values into Eq. (7) to find α:

$$\cot 2\alpha = \frac{A - C}{B} = \frac{2 - 1}{\sqrt{3}} = \frac{1}{\sqrt{3}}.$$

From the right triangle in Fig. 9.25, we see that one appropriate choice of angle is $2\alpha = \pi/3$, so we take $\alpha = \pi/6$. Substituting $\alpha = \pi/6$, $A = 2$, $B = \sqrt{3}$, $C = 1$, $D = E = 0$, and $F = -10$ into Eqs. (6) gives

$$A' = \frac{5}{2}, \qquad B' = 0, \qquad C' = \frac{1}{2}, \qquad D' = E' = 0, \qquad F' = -10.$$

Equation (5) then gives

$$\frac{5}{2}x'^2 + \frac{1}{2}y'^2 - 10 = 0, \qquad \text{or} \qquad \frac{x'^2}{4} + \frac{y'^2}{20} = 1.$$

The curve is an ellipse with foci on the new y'-axis (Fig. 9.26).   ◻

Possible Graphs of Quadratic Equations

We now return to the graph of the general quadratic equation.

Since axes can always be rotated to eliminate the cross product term, there is no loss of generality in assuming that this has been done and that our equation has the form

$$Ax^2 + Cy^2 + Dx + Ey + F = 0. \qquad (8)$$

Equation (8) represents

a)   a *circle* if $A = C \neq 0$ (special cases: the graph is a point or there is no graph at all);

b)   a *parabola* if Eq. (8) is quadratic in one variable and linear in the other;

c)   an *ellipse* if A and C are both positive or both negative (special cases: circles, a single point or no graph at all);

d)   a *hyperbola* if A and C have opposite signs (special case: a pair of intersecting lines);

e)   a *straight line* if A and C are zero and at least one of D and E is different from zero;

f)   *one or two straight lines* if the left-hand side of Eq. (8) can be factored into the product of two linear factors.

See Table 9.3 (on page 732) for examples.

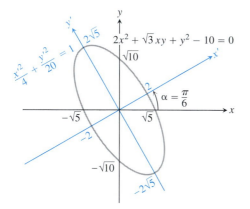

9.26 The conic section in Example 2.

The Discriminant Test

We do not need to eliminate the xy-term from the equation

$$Ax^2 + Bxy + Cy^2 + Dx + Ey + F = 0 \qquad (9)$$

to tell what kind of conic section the equation represents. If this is the only information we want, we can apply the following test instead.

As we have seen, if $B \neq 0$, then rotating the coordinate axes through an angle α that satisfies the equation

$$\cot 2\alpha = \frac{A - C}{B} \tag{10}$$

will change Eq. (9) into an equivalent form

$$A' x'^2 + C' y'^2 + D' x' + E' y' + F' = 0 \tag{11}$$

without a cross product term.

Now, the graph of Eq. (11) is a (real or degenerate)

a) *parabola* if A' or $C' = 0$; that is, if $A' C' = 0$;
b) *ellipse* if A' and C' have the same sign; that is, if $A' C' > 0$;
c) *hyperbola* if A' and C' have opposite signs; that is, if $A' C' < 0$.

It can also be verified from Eqs. (6) that for any rotation of axes,

$$B^2 - 4AC = B'^2 - 4A' C'. \tag{12}$$

This means that the quantity $B^2 - 4AC$ is not changed by a rotation. But when we rotate through the angle α given by Eq. (10), B' becomes zero, so

$$B^2 - 4AC = -4A' C'.$$

Since the curve is a parabola if $A' C' = 0$, an ellipse if $A' C' > 0$, and a hyperbola if $A' C' < 0$, the curve must be a parabola if $B^2 - 4AC = 0$, an ellipse if $B^2 - 4AC < 0$, and a hyperbola if $B^2 - 4AC > 0$. The number $B^2 - 4AC$ is called the **discriminant** of Eq. (9).

The Discriminant Test

With the understanding that occasional degenerate cases may arise, the quadratic curve $Ax^2 + Bxy + Cy^2 + Dx + Ey + F = 0$ is

a) a **parabola** if $B^2 - 4AC = 0$,
b) an **ellipse** if $B^2 - 4AC < 0$,
c) a **hyperbola** if $B^2 - 4AC > 0$.

EXAMPLE 3

a) $3x^2 - 6xy + 3y^2 + 2x - 7 = 0$ represents a parabola because
$$B^2 - 4AC = (-6)^2 - 4 \cdot 3 \cdot 3 = 36 - 36 = 0.$$

b) $x^2 + xy + y^2 - 1 = 0$ represents an ellipse because
$$B^2 - 4AC = (1)^2 - 4 \cdot 1 \cdot 1 = -3 < 0.$$

c) $xy - y^2 - 5y + 1 = 0$ represents a hyperbola because
$$B^2 - 4AC = (1)^2 - 4(0)(-1) = 1 > 0.$$

Table 9.3 Examples of quadratic curves

$Ax^2 + Bxy + Cy^2 + Dx + Ey + F = 0$								
	A	**B**	**C**	**D**	**E**	**F**	**Equation**	**Remarks**
Circle	1		1			−4	$x^2 + y^2 = 4$	$A = C; F < 0$
Parabola			1	−9			$y^2 = 9x$	Quadratic in y, linear in x
Ellipse	4		9			−36	$4x^2 + 9y^2 = 36$	A, C have same sign, $A \neq C; F < 0$
Hyperbola	1		−1			−1	$x^2 - y^2 = 1$	A, C have opposite signs
One line (still a conic section)	1						$x^2 = 0$	y-axis
Intersecting lines (still a conic section)		1		1	−1	−1	$xy + x - y - 1 = 0$	Factors to $(x - 1)(y + 1) = 0$, so $x = 1, y = -1$
Parallel lines (not a conic section)	1			−3		2	$x^2 - 3x + 2 = 0$	Factors to $(x - 1)(x - 2) = 0$, so $x = 1, x = 2$
Point	1		1				$x^2 + y^2 = 0$	The origin
No graph	1					1	$x^2 = -1$	No graph

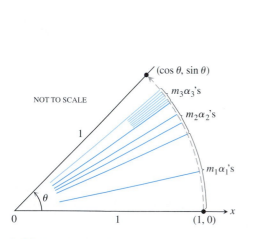

9.27 To calculate the sine and cosine of an angle θ between 0 and 2π, the calculator rotates the point (1, 0) to an appropriate location on the unit circle and displays the resulting coordinates.

Technology   *How Calculators Use Rotations to Evaluate Sines and Cosines*
Some calculators use rotations to calculate sines and cosines of arbitrary angles. The procedure goes something like this: The calculator has, stored,

1. ten angles or so, say

    $$\alpha_1 = \sin^{-1}(10^{-1}), \qquad \alpha_2 = \sin^{-1}(10^{-2}), \qquad \dots, \qquad \alpha_{10} = \sin^{-1}(10^{-10}),$$

    and

2. twenty numbers, the sines and cosines of the angles $\alpha_1, \alpha_2, \dots, \alpha_{10}$.

To calculate the sine and cosine of an arbitrary angle θ, we enter θ (in radians) into the calculator. The calculator subtracts or adds multiples of 2π to θ to replace θ by the angle between 0 and 2π that has the same sine and cosine as θ (we continue to call the angle θ). The calculator then "writes" θ as a sum of multiples of α_1 (as many as possible without overshooting) plus multiples of α_2 (again, as many as possible), and so on, working its way to α_{10}. This gives

$$\theta \approx m_1\alpha_1 + m_2\alpha_2 + \cdots + m_{10}\alpha_{10}.$$

The calculator then rotates the point (1, 0) through m_1 copies of α_1 (through α_1, m_1 times in succession), plus m_2 copies of α_2, and so on, finishing off with m_{10} copies of α_{10} (Fig. 9.27). The coordinates of the final position of (1, 0) on the unit circle are the values the calculator gives for $(\cos\theta, \sin\theta)$.

Exercises 9.3

Using the Discriminant

Use the discriminant $B^2 - 4AC$ to decide whether the equations in Exercises 1–16 represent parabolas, ellipses, or hyperbolas.

1. $x^2 - 3xy + y^2 - x = 0$

2. $3x^2 - 18xy + 27y^2 - 5x + 7y = -4$

3. $3x^2 - 7xy + \sqrt{17}\, y^2 = 1$

4. $2x^2 - \sqrt{15}\, xy + 2y^2 + x + y = 0$

5. $x^2 + 2xy + y^2 + 2x - y + 2 = 0$

6. $2x^2 - y^2 + 4xy - 2x + 3y = 6$

7. $x^2 + 4xy + 4y^2 - 3x = 6$

8. $x^2 + y^2 + 3x - 2y = 10$

9. $xy + y^2 - 3x = 5$

10. $3x^2 + 6xy + 3y^2 - 4x + 5y = 12$

11. $3x^2 - 5xy + 2y^2 - 7x - 14y = -1$

12. $2x^2 - 4.9xy + 3y^2 - 4x = 7$

13. $x^2 - 3xy + 3y^2 + 6y = 7$

14. $25x^2 + 21xy + 4y^2 - 350x = 0$

15. $6x^2 + 3xy + 2y^2 + 17y + 2 = 0$

16. $3x^2 + 12xy + 12y^2 + 435x - 9y + 72 = 0$

Rotating Coordinate Axes

In Exercises 17–26, rotate the coordinate axes to change the given equation into an equation that has no cross product (xy) term. Then identify the graph of the equation. (The new equations will vary with the size and direction of the rotation you use.)

17. $xy = 2$

18. $x^2 + xy + y^2 = 1$

19. $3x^2 + 2\sqrt{3}\, xy + y^2 - 8x + 8\sqrt{3}\, y = 0$

20. $x^2 - \sqrt{3}\, xy + 2y^2 = 1$

21. $x^2 - 2xy + y^2 = 2$

22. $3x^2 - 2\sqrt{3}\, xy + y^2 = 1$

23. $\sqrt{2}\, x^2 + 2\sqrt{2}\, xy + \sqrt{2}\, y^2 - 8x + 8y = 0$

24. $xy - y - x + 1 = 0$

25. $3x^2 + 2xy + 3y^2 = 19$

26. $3x^2 + 4\sqrt{3}\, xy - y^2 = 7$

27. Find the sine and cosine of an angle through which the coordinate axes can be rotated to eliminate the cross product term from the equation

$$14x^2 + 16xy + 2y^2 - 10x + 26{,}370\, y - 17 = 0.$$

Do not carry out the rotation.

28. Find the sine and cosine of an angle through which the coordinate axes can be rotated to eliminate the cross product term from the equation

$$4x^2 - 4xy + y^2 - 8\sqrt{5}\, x - 16\sqrt{5}\, y = 0.$$

Do not carry out the rotation.

Calculator

The conic sections in Exercises 17–26 were chosen to have rotation angles that were "nice" in the sense that once we knew $\cot 2\alpha$ or $\tan 2\alpha$ we could identify 2α and find $\sin \alpha$ and $\cos \alpha$ from familiar triangles. The conic sections encountered in practice may not have such nice rotation angles, and we may have to use a calculator to determine α from the value of $\cot 2\alpha$ or $\tan 2\alpha$.

In Exercises 29–34, use a calculator to find an angle α through which the coordinate axes can be rotated to change the given equation into a quadratic equation that has no cross product term. Then find $\sin \alpha$ and $\cos \alpha$ to 2 decimal places and use Eqs. (6) to find the coefficients of the new equation to the nearest decimal place. In each case, say whether the conic section is an ellipse, a hyperbola, or a parabola.

29. $x^2 - xy + 3y^2 + x - y - 3 = 0$

30. $2x^2 + xy - 3y^2 + 3x - 7 = 0$

31. $x^2 - 4xy + 4y^2 - 5 = 0$

32. $2x^2 - 12xy + 18y^2 - 49 = 0$

33. $3x^2 + 5xy + 2y^2 - 8y - 1 = 0$

34. $2x^2 + 7xy + 9y^2 + 20x - 86 = 0$

Theory and Examples

35. What effect does a 90° rotation about the origin have on the equations of the following conic sections? Give the new equation in each case.

   a) The ellipse $(x^2/a^2) + (y^2/b^2) = 1$ $(a > b)$

   b) The hyperbola $(x^2/a^2) - (y^2/b^2) = 1$

   c) The circle $x^2 + y^2 = a^2$

   d) The line $y = mx$

   e) The line $y = mx + b$

36. What effect does a 180° rotation about the origin have on the equations of the following conic sections? Give the new equation in each case.

   a) The ellipse $(x^2/a^2) + (y^2/b^2) = 1$ $(a > b)$

   b) The hyperbola $(x^2/a^2) - (y^2/b^2) = 1$

   c) The circle $x^2 + y^2 = a^2$

   d) The line $y = mx$

   e) The line $y = mx + b$

37. *The Hyperbola $xy = a$.* The hyperbola $xy = 1$ is one of many hyperbolas of the form $xy = a$ that appear in science and mathematics.

   a) Rotate the coordinate axes through an angle of $45°$ to change the equation $xy = 1$ into an equation with no xy-term. What is the new equation?

   b) Do the same for the equation $xy = a$.

38. Find the eccentricity of the hyperbola $xy = 2$.

39. Can anything be said about the graph of the equation $Ax^2 + Bxy + Cy^2 + Dx + Ey + F = 0$ if $AC < 0$? Give reasons for your answer.

40. Does any nondegenerate conic section $Ax^2 + Bxy + Cy^2 + Dx + Ey + F = 0$ have all of the following properties?

   a) It is symmetric with respect to the origin.

   b) It passes through the point $(1, 0)$.

   c) It is tangent to the line $y = 1$ at the point $(-2, 1)$.

   Give reasons for your answer.

41. Show that the equation $x^2 + y^2 = a^2$ becomes $x'^2 + y'^2 = a^2$ for every choice of the angle α in the rotation equations (4).

42. Show that rotating the axes through an angle of $\pi/4$ radians will eliminate the xy-term from Eq. (1) whenever $A = C$.

43. a) Decide whether the equation

$$x^2 + 4xy + 4y^2 + 6x + 12y + 9 = 0$$

represents an ellipse, a parabola, or a hyperbola.

   b) Show that the graph of the equation in (a) is the line $2y = -x - 3$.

44. a) Decide whether the conic section with equation

$$9x^2 + 6xy + y^2 - 12x - 4y + 4 = 0$$

represents a parabola, an ellipse, or a hyperbola.

   b) Show that the graph of the equation in (a) is the line $y = -3x + 2$.

45. a) What kind of conic section is the curve $xy + 2x - y = 0$?

   b) Solve the equation $xy + 2x - y = 0$ for y and sketch the curve as the graph of a rational function of x.

   c) Find equations for the lines parallel to the line $y = -2x$ that are normal to the curve. Add the lines to your sketch.

46. Prove or find counterexamples to the following statements about the graph of $Ax^2 + Bxy + Cy^2 + Dx + Ey + F = 0$.

   a) If $AC > 0$, the graph is an ellipse.

   b) If $AC > 0$, the graph is a hyperbola.

   c) If $AC < 0$, the graph is a hyperbola.

47. *A nice area formula for ellipses.* When $B^2 - 4AC$ is negative, the equation

$$Ax^2 + Bxy + Cy^2 = 1$$

represents an ellipse. If the ellipse's semi-axes are a and b, its area is πab (a standard formula). Show that the area is also given by the formula $2\pi/\sqrt{4AC - B^2}$. (*Hint:* Rotate the coordinate axes to eliminate the xy-term and apply Eq. (12) to the new equation.)

48. *Other invariants.* We describe the fact that $B'^2 - 4A'C'$ equals $B^2 - 4AC$ after a rotation about the origin by saying that the discriminant of a quadratic equation is an **invariant** of the equation. Use Eqs. (6) to show that the numbers (a) $A + C$ and (b) $D^2 + E^2$ are also invariants, in the sense that

$$A' + C' = A + C \quad \text{and} \quad D'^2 + E'^2 = D^2 + E^2.$$

We can use these equalities to check against numerical errors when we rotate axes. They can also be helpful in shortening the work required to find values for the new coefficients.

49. *A proof that $B'^2 - 4A'C' = B^2 - 4AC$.* Use Eqs. (6) to show that $B'^2 - 4A'C' = B^2 - 4AC$ for any rotation of axes about the origin. The calculation works out nicely but requires patience.

9.4

Parametrizations of Plane Curves

When the path of a particle moving in the plane looks like the curve in Fig. 9.28, we cannot hope to describe it with a Cartesian formula that expresses y directly in terms of x or x directly in terms of y. Instead, we express each of the particle's coordinates as a function of time t and describe the path with a pair of equations, $x = f(t)$ and $y = g(t)$. For studying motion, equations like these are preferable to a Cartesian formula because they tell us the particle's position at any time t.

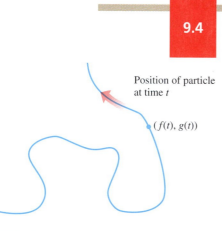

Position of particle at time t

$(f(t), g(t))$

9.28 The path traced by a particle moving in the xy-plane is not always the graph of a function of x or a function of y.

Definitions

If x and y are given as continuous functions

$$x = f(t), \qquad y = g(t)$$

over an interval of t-values, then the set of points $(x, y) = (f(t), g(t))$ defined by these equations is a **curve** in the coordinate plane. The equations are **parametric equations** for the curve. The variable t is a **parameter** for the curve and its domain I is the **parameter interval**. If I is a closed interval, $a \le t \le b$, the point $(f(a), g(a))$ is the **initial point** of the curve and $(f(b), g(b))$ is the **terminal point** of the curve. When we give parametric equations and a parameter interval for a curve in the plane, we say that we have **parametrized** the curve. The equations and interval constitute a **parametrization** of the curve.

In many applications t denotes time, but it might instead denote an angle (as in some of the following examples) or the distance a particle has traveled along its path from its starting point (as it sometimes will when we later study motion).

EXAMPLE 1   *The circle $x^2 + y^2 = 1$*

The equations and parameter interval

$$x = \cos t, \qquad y = \sin t, \qquad 0 \le t \le 2\pi,$$

describe the position $P(x, y)$ of a particle that moves counterclockwise around the circle $x^2 + y^2 = 1$ as t increases (Fig. 9.29).

We know that the point lies on this circle for every value of t because

$$x^2 + y^2 = \cos^2 t + \sin^2 t = 1.$$

But how much of the circle does the point $P(x, y)$ actually traverse?

To find out, we track the motion as t runs from 0 to 2π. The parameter t is the radian measure of the angle that radius OP makes with the positive x-axis. The particle starts at $(1, 0)$, moves up and to the left as t approaches $\pi/2$, and continues around the circle to stop again at $(1, 0)$ when $t = 2\pi$. The particle traces the circle exactly once. ☐

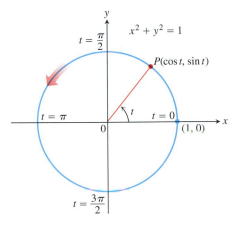

9.29 The equations $x = \cos t$, $y = \sin t$ describe motion on the circle $x^2 + y^2 = 1$. The arrow shows the direction of increasing t (Example 1).

EXAMPLE 2   *A semicircle*

The equations and parameter interval

$$x = \cos t, \qquad y = -\sin t, \qquad 0 \le t \le \pi,$$

describe the position $P(x, y)$ of a particle that moves clockwise around the circle $x^2 + y^2 = 1$ as t increases from 0 to π.

We know that the point P lies on this circle for all t because its coordinates satisfy the circle's equation. How much of the circle does the particle traverse? To find out, we track the motion as t runs from 0 to π. As in Example 1, the particle starts at $(1, 0)$. But now as t increases, y becomes negative, decreasing to -1 when $t = \pi/2$ and then increasing back to 0 as t approaches π. The motion stops at $t = \pi$ with only the lower half of the circle covered (Fig. 9.30). ☐

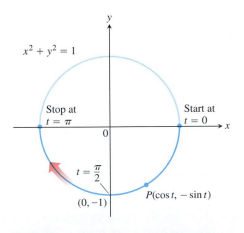

9.30 The point $P(\cos t, -\sin t)$ moves clockwise as t increases from 0 to π (Example 2).

EXAMPLE 3   *Half a parabola*

The position $P(x, y)$ of a particle moving in the xy-plane is given by the equations

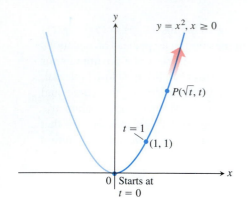

9.31 The equations $x = \sqrt{t}, y = t$ and interval $t \geq 0$ describe the motion of a particle that traces the right-hand half of the parabola $y = x^2$ (Example 3).

and parameter interval

$$x = \sqrt{t}, \qquad y = t, \qquad t \geq 0.$$

Identify the path traced by the particle and describe the motion.

Solution We try to identify the path by eliminating t between the equations $x = \sqrt{t}$ and $y = t$. With any luck, this will produce a recognizable algebraic relation between x and y. We find that

$$y = t = \left(\sqrt{t}\right)^2 = x^2.$$

This means that the particle's position coordinates satisfy the equation $y = x^2$, so the particle moves along the parabola $y = x^2$.

It would be a mistake, however, to conclude that the particle's path is the entire parabola $y = x^2$— it is only half the parabola. The particle's x-coordinate is never negative. The particle starts at $(0, 0)$ when $t = 0$ and rises into the first quadrant as t increases (Fig. 9.31). ❑

EXAMPLE 4   *An entire parabola*

The position $P(x, y)$ of a particle moving in the xy-plane is given by the equations and parameter interval

$$x = t, \qquad y = t^2, \qquad -\infty < t < \infty.$$

Identify the particle's path and describe the motion.

Solution We identify the path by eliminating t between the equations $x = t$ and $y = t^2$, obtaining

$$y = (t)^2 = x^2.$$

The particle's position coordinates satisfy the equation $y = x^2$, so the particle moves along this curve.

In contrast to Example 3, the particle now traverses the entire parabola. As t increases from $-\infty$ to ∞, the particle comes down the left-hand side, passes through the origin, and moves up the right-hand side (Fig. 9.32).

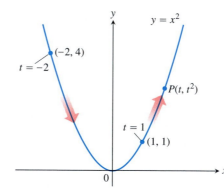

9.32 The path defined by $x = t$, $y = t^2, -\infty < t < \infty$ is the entire parabola $y = x^2$ (Example 4).

As Example 4 illustrates, any curve $y = f(x)$ has the parametrization $x = t$, $y = f(t)$. This is so simple we usually do not use it, but the point of view is occasionally helpful.

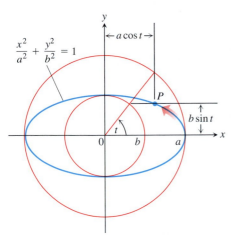

$$\frac{x^2}{a^2} + \frac{y^2}{b^2} = 1$$

9.33 The ellipse in Example 5, drawn for $a > b$. The coordinates of P are $x = a \cos t$, $y = b \sin t$.

EXAMPLE 5   *A parametrization of the ellipse $x^2/a^2 + y^2/b^2 = 1$*

Describe the motion of a particle whose position $P(x, y)$ at time t is given by

$$x = a \cos t, \qquad y = b \sin t, \qquad 0 \le t \le 2\pi.$$

Solution   We find a Cartesian equation for the particle's coordinates by eliminating t between the equations

$$\cos t = \frac{x}{a}, \qquad \sin t = \frac{y}{b}.$$

We accomplish this with the identity $\cos^2 t + \sin^2 t = 1$, which yields

$$\left(\frac{x}{a}\right)^2 + \left(\frac{y}{b}\right)^2 = 1, \qquad \text{or} \qquad \frac{x^2}{a^2} + \frac{y^2}{b^2} = 1.$$

The particle's coordinates (x, y) satisfy the equation $(x^2/a^2) + (y^2/b^2) = 1$, so the particle moves along this ellipse. When $t = 0$, the particle's coordinates are

$$x = a \cos (0) = a, \qquad y = b \sin (0) = 0,$$

so the motion starts at $(a, 0)$. As t increases, the particle rises and moves toward the left, moving counterclockwise. It traverses the ellipse once, returning to its starting position $(a, 0)$ at time $t = 2\pi$ (Fig. 9.33).

EXAMPLE 6   *A parametrization of the circle $x^2 + y^2 = a^2$*

The equations and parameter interval

$$x = a \cos t, \qquad y = a \sin t, \qquad 0 \le t \le 2\pi,$$

obtained by taking $b = a$ in Example 5, describe the circle $x^2 + y^2 = a^2$.

EXAMPLE 7   *A parametrization of the right-hand branch of the hyperbola $x^2 - y^2 = 1$*

Describe the motion of the particle whose position $P(x, y)$ at time t is given by

$$x = \sec t, \qquad y = \tan t, \qquad -\frac{\pi}{2} < t < \frac{\pi}{2}.$$

Solution   We find a Cartesian equation for the coordinates of P by eliminating t between the equations

$$\sec t = x, \qquad \tan t = y.$$

We accomplish this with the identity $\sec^2 t - \tan^2 t = 1$, which yields

$$x^2 - y^2 = 1.$$

Since the particle's coordinates (x, y) satisfy the equation $x^2 - y^2 = 1$, the motion takes place somewhere on this hyperbola. As t runs between $-\pi/2$ and $\pi/2$, $x = \sec t$ remains positive and $y = \tan t$ runs between $-\infty$ and ∞, so P traverses the hyperbola's right-hand branch. It comes in along the branch's lower half as $t \to 0^-$, reaches $(1, 0)$ at $t = 0$, and moves out into the first quadrant as t increases toward $\pi/2$ (Fig. 9.34).

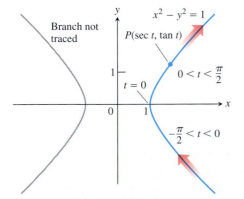

9.34 The equations $x = \sec t$, $y = \tan t$ and interval $-\pi/2 < t < \pi/2$ describe the right-hand branch of the hyperbola $x^2 - y^2 = 1$ (Example 7).

Huygen's clock

The problem with a pendulum clock whose bob swings in a circular arc is that the frequency of the swing depends on the amplitude of the swing. The wider the swing, the longer it takes the bob to return to center.

This does not happen if the bob can be made to swing in a cycloid. In 1673, Christiaan Huygens (1629–1695), the Dutch mathematician, physicist, and astronomer who discovered the rings of Saturn, driven by a need to make accurate determinations of longitude at sea, designed a pendulum clock whose bob would swing in a cycloid. He hung the bob from a fine wire constrained by guards that caused it to draw up as it swung away from center. How were the guards shaped? They were cycloids, too.

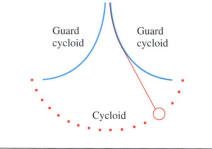

EXAMPLE 8   Cycloids

A wheel of radius a rolls along a horizontal straight line. Find parametric equations for the path traced by a point P on the wheel's circumference. The path is called a **cycloid.**

Solution   We take the line to be the x-axis, mark a point P on the wheel, start the wheel with P at the origin, and roll the wheel to the right. As parameter, we use the angle t through which the wheel turns, measured in radians. Figure 9.35 shows the wheel a short while later, when its base lies at units from the origin. The wheel's center C lies at (at, a) and the coordinates of P are

$$x = at + a \cos \theta, \qquad y = a + a \sin \theta.$$

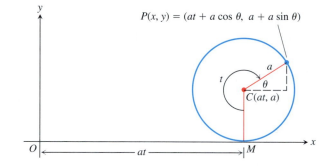

9.35 The position of $P(x, y)$ on the rolling wheel at angle t (Example 8).

To express θ in terms of t, we observe that $t + \theta = 3\pi/2$, so that

$$\theta = \frac{3\pi}{2} - t.$$

This makes

$$\cos \theta = \cos \left(\frac{3\pi}{2} - t \right) = - \sin t, \qquad \sin \theta = \sin \left(\frac{3\pi}{2} - t \right) = - \cos t.$$

The equations we seek are

$$x = at - a \sin t, \qquad y = a - a \cos t.$$

These are usually written with the a factored out:

$$x = a (t - \sin t), \qquad y = a (1 - \cos t). \tag{1}$$

Figure 9.36 shows the first arch of the cycloid and part of the next.   ❏

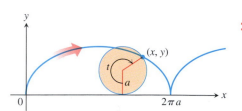

9.36 The cycloid $x = a(t - \sin t)$, $y = a(1 - \cos t)$, for $t \geq 0$.

✳ Brachistochrones and Tautochrones

If we turn Fig. 9.36 upside down, Eqs. (1) still apply and the resulting curve (Fig. 9.37) has two interesting physical properties. The first relates to the origin O and the point B at the bottom of the first arch. Among all smooth curves joining these points, the cycloid is the curve along which a frictionless bead, subject only to the force of gravity, will slide from O to B the fastest. This makes the cycloid a **brachistochrone** ("brah-*kiss*-toe-krone"), or shortest time curve for these points. The second property is that even if you start the bead partway down the curve

The witch of Agnesi

Although l'Hôpital wrote the first text on differential calculus, the first text to include differential and integral calculus along with analytic geometry, infinite series, and differential equations was written in the 1740s by the Italian mathematician Maria Gaetana Agnesi (1718–1799). Agnesi, a gifted scholar and linguist whose Latin essay defending higher education for women was published when she was only nine years old, was a well-published scientist by age 20 and an honorary faculty member of the University of Bologna by age 30.

Today, Agnesi is remembered chiefly for a bell-shaped curve called *the witch of Agnesi*. This name, found only in English texts, is the result of a mistranslation. Agnesi's own name for the curve was *versiera* or "turning curve." John Colson, a noted Cambridge mathematician who felt Agnesi's text so important that he learned Italian to translate it "for the benefit of British youth" (he particularly had in mind young women, for whom he hoped Agnesi would be a role model), probably confused *versiera* with *avversiera*, which means "wife of the devil" and translates into "witch." You can find out more about the witch by doing Exercise 29.

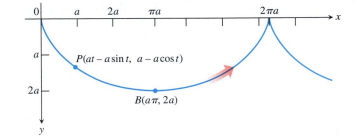

9.37 To study motion along an upside-down cycloid under the influence of gravity, we turn Fig. 9.36 upside down. This points the *y*-axis in the direction of the gravitational force and makes the downward *y*-coordinates positive. The equations and parameter interval for the cycloid are still

$$x = a(t - \sin t),$$
$$y = a(1 - \cos t), \quad t \geq 0.$$

The arrow shows the direction of increasing *t*.

toward B, it will still take the bead the same amount of time to reach B. This makes the cycloid a **tautochrone** ("*taw*-toe-krone"), or same-time curve for O and B.

Are there any other brachistochrones joining O and B, or is the cycloid the only one? We can formulate this as a mathematical question in the following way. At the start, the kinetic energy of the bead is zero, since its velocity is zero. The work done by gravity in moving the bead from $(0, 0)$ to any other point (x, y) in the plane is mgy, and this must equal the change in kinetic energy. That is,

$$mgy = \frac{1}{2}mv^2 - \frac{1}{2}m(0)^2.$$

Thus, the velocity of the bead when it reaches (x, y) has to be

$$v = \sqrt{2gy}.$$

That is,

$$\frac{ds}{dt} = \sqrt{2gy} \qquad \text{\textit{ds} is the arc length differential along the bead's path.}$$

or

$$dt = \frac{ds}{\sqrt{2gy}} = \frac{\sqrt{1 + (dy/dx)^2}\, dx}{\sqrt{2gy}}.$$

The time T_f it takes the bead to slide along a particular path $y = f(x)$ from O to $B(a\pi, 2a)$ is

$$T_f = \int_{x=0}^{x=a\pi} \sqrt{\frac{1 + (dy/dx)^2}{2gy}}\, dx. \tag{2}$$

What curves $y = f(x)$, if any, minimize the value of this integral?

At first sight, we might guess that the straight line joining O and B would give the shortest time, but perhaps not. There might be some advantage in having the bead fall vertically at first to build up its velocity faster. With a higher velocity, the bead could travel a longer path and still reach B first. Indeed, this is the right idea.

The solution, from a branch of mathematics known as the calculus of variations, is that the original cycloid from O to B is the one and only brachistochrone for O and B.

While the solution of the brachistrochrone problem is beyond our present reach, we can still show why the cycloid is a tautochrone. For the cycloid, Eq. (2) takes the form

$$T_{\text{cycloid}} = \int_{x=0}^{x=a\pi} \sqrt{\frac{dx^2 + dy^2}{2gy}}$$

$$= \int_{t=0}^{t=\pi} \sqrt{\frac{a^2(2 - 2\cos t)}{2ga(1 - \cos t)}} \, dt \qquad \begin{array}{l} \text{From Eqs. (1),} \\ dx = a(1 - \cos t)\,dt, \\ dy = a\sin t\,dt, \text{ and} \\ y = a(1 - \cos t) \end{array}$$

$$= \int_0^\pi \sqrt{\frac{a}{g}} \, dt = \pi \sqrt{\frac{a}{g}}.$$

Thus, the amount of time it takes the frictionless bead to slide down the cycloid to B after it is released from rest at O is $\pi\sqrt{a/g}$.

Suppose that instead of starting the bead at O we start it at some lower point on the cycloid, a point (x_0, y_0) corresponding to the parameter value $t_0 > 0$. The bead's velocity at any later point (x, y) on the cycloid is

$$v = \sqrt{2g\,(y - y_0)} = \sqrt{2ga\,(\cos t_0 - \cos t)}. \qquad y = a(1 - \cos t)$$

Accordingly, the time required for the bead to slide from (x_0, y_0) down to B is

$$T = \int_{t_0}^\pi \sqrt{\frac{a^2(2 - 2\cos t)}{2ga\,(\cos t_0 - \cos t)}} \, dt = \sqrt{\frac{a}{g}} \int_{t_0}^\pi \sqrt{\frac{1 - \cos t}{\cos t_0 - \cos t}} \, dt$$

$$= \sqrt{\frac{a}{g}} \int_{t_0}^\pi \sqrt{\frac{2\sin^2(t/2)}{(2\cos^2(t_0/2) - 1) - (2\cos^2(t/2) - 1)}} \, dt$$

$$= \sqrt{\frac{a}{g}} \int_{t_0}^\pi \frac{\sin(t/2)\,dt}{\sqrt{\cos^2(t_0/2) - \cos^2(t/2)}}$$

$$= \sqrt{\frac{a}{g}} \int_{t=t_0}^{t=\pi} \frac{-2\,du}{\sqrt{a^2 - u^2}} \qquad \begin{array}{l} u = \cos(t/2) \\ -2du = \sin(t/2)\,dt \\ c = \cos(t_0/2) \end{array}$$

$$= 2\sqrt{\frac{a}{g}} \left[-\sin^{-1}\frac{u}{c} \right]_{t=t_0}^{t=\pi}$$

$$= 2\sqrt{\frac{a}{g}} \left[-\sin^{-1}\frac{\cos(t/2)}{\cos(t_0/2)} \right]_{t_0}^{\pi}$$

$$= 2\sqrt{\frac{a}{g}} (-\sin^{-1} 0 + \sin^{-1} 1) = \pi\sqrt{\frac{a}{g}}.$$

This is precisely the time it takes the bead to slide to B from O. It takes the bead the same amount of time to reach B no matter where it starts. Beads starting simultaneously from O, A, and C in Fig. 9.38, for instance, will all reach B at the same time.

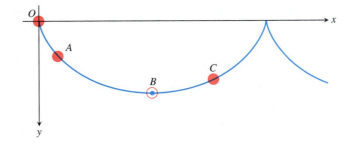

9.38 Beads released simultaneously on the cycloid at O, A, and C will reach B at the same time.

Standard Parametrizations

$Circle\ x^2 + y^2 = a^2$:

$$x = a \cos t$$
$$y = a \sin t$$
$$0 \leq t \leq 2\pi$$

$Ellipse\ \dfrac{x^2}{a^2} + \dfrac{y^2}{b^2} = 1$:

$$x = a \cos t$$
$$y = b \sin t$$
$$0 \leq t \leq 2\pi$$

Cycloid generated by a circle of radius a:

$$x = a(t - \sin t), \qquad y = a(1 - \cos t)$$

Exercises 9.4

Finding Cartesian Equations from Parametric Equations

Exercises 1–24 give parametric equations and parameter intervals for the motion of a particle in the xy-plane. Identify the particle's path by finding a Cartesian equation for it. Graph the Cartesian equation. (The graphs will vary with the equation used.) Indicate the portion of the graph traced by the particle and the direction of motion.

1. $x = \cos t, \quad y = \sin t, \quad 0 \leq t \leq \pi$

2. $x = \cos 2t, \quad y = \sin 2t, \quad 0 \leq t \leq \pi$

3. $x = \sin (2\pi (1 - t)), \quad y = \cos (2\pi (1 - t)), \quad 0 \leq t \leq 1$

4. $x = \cos (\pi - t), \quad y = \sin (\pi - t), \quad 0 \leq t \leq \pi$

5. $x = 4 \cos t, \quad y = 2 \sin t, \quad 0 \leq t \leq 2\pi$

6. $x = 4 \sin t, \quad y = 2 \cos t, \quad 0 \leq t \leq \pi$

7. $x = 4 \cos t, \quad y = 5 \sin t, \quad 0 \leq t \leq \pi$

8. $x = 4 \sin t, \quad y = 5 \cos t, \quad 0 \leq t \leq 2\pi$

9. $x = 3t, \quad y = 9t^2, \quad -\infty < t < \infty$

10. $x = -\sqrt{t}, \quad y = t, \quad t \geq 0$

11. $x = t, \quad y = \sqrt{t}, \quad t \geq 0$

12. $x = \sec^2 t - 1, \quad y = \tan t, \quad -\pi/2 < t < \pi/2$

13. $x = -\sec t, \quad y = \tan t, \quad -\pi/2 < t < \pi/2$

14. $x = \csc t, \quad y = \cot t, \quad 0 < t < \pi$

15. $x = 2t - 5, \quad y = 4t - 7, \quad -\infty < t < \infty$

16. $x = 1 - t, \quad y = 1 + t, \quad -\infty < t < \infty$

17. $x = t, \quad y = 1 - t, \quad 0 \leq t \leq 1$

18. $x = 3 - 3t, \quad y = 2t, \quad 0 \leq t \leq 1$

19. $x = t, \quad y = \sqrt{1 - t^2}, \quad -1 \leq t \leq 0$

20. $x = t, \quad y = \sqrt{4 - t^2}, \quad 0 \leq t \leq 2$

21. $x = t^2, \quad y = \sqrt{t^4 + 1}, \quad t \geq 0$

22. $x = \sqrt{t + 1}, \quad y = \sqrt{t}, \quad t \geq 0$

23. $x = -\cosh t, \quad y = \sinh t, \quad -\infty < t < \infty$

24. $x = 2 \sinh t, \quad y = 2 \cosh t, \quad -\infty < t < \infty$

Determining Parametric Equations

25. Find parametric equations and a parameter interval for the motion of a particle that starts at $(a, 0)$ and traces the circle $x^2 + y^2 = a^2$

a) once clockwise,   **b)** once counterclockwise,

c) twice clockwise,   **d)** twice counterclockwise.

(There are many ways to do these, so your answers may not be the same as the ones in the back of the book.)

26. Find parametric equations and a parameter interval for the motion of a particle that starts at $(a, 0)$ and traces the ellipse $(x^2/a^2) + (y^2/b^2) = 1$

a) once clockwise,   **b)** once counterclockwise,

c) twice clockwise,   **d)** twice counterclockwise.

(As in Exercise 25, there are many correct answers.)

27. Find parametric equations for the semicircle

$$x^2 + y^2 = a^2, \quad y > 0,$$

using as parameter the slope $t = dy/dx$ of the tangent to the curve at (x, y).

28. Find parametric equations for the circle

$$x^2 + y^2 = a^2,$$

using as parameter the arc length s measured counterclockwise from the point $(a, 0)$ to the point (x, y).

29. *The witch of Maria Agnesi.* The bell-shaped witch of Maria Agnesi can be constructed in the following way. Start with a circle of radius 1, centered at the point $(0, 1)$, as shown in the accompanying figure. Choose a point A on the line $y = 2$ and connect it to the origin with a line segment. Call the point where the segment crosses the circle B. Let P be the point where the vertical line through A crosses the horizontal line through B. The witch is the curve traced by P as A moves along the line $y = 2$. Find parametric equations and a parameter interval for the witch by expressing the coordinates of P in terms of t, the radian measure of the angle that segment OA makes with the positive x-axis. The following equalities (which you may assume) will help.

a) $x = AQ$ **b)** $y = 2 - AB \sin t$

c) $AB \cdot OA = (AQ)^2$

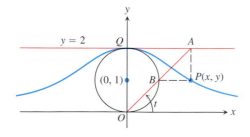

30. *The involute of a circle.* If a string wound around a fixed circle is unwound while held taut in the plane of the circle, its end P

traces an *involute* of the circle. In Fig. 9.39, the circle in question is the circle $x^2 + y^2 = 1$ and the tracing point starts at $(1, 0)$. The unwound portion of the string is tangent to the circle at Q, and t is the radian measure of the angle from the positive x-axis to segment OQ. Derive parametric equations for the involute by expressing the coordinates x and y of P in terms of t for $t \geq 0$.

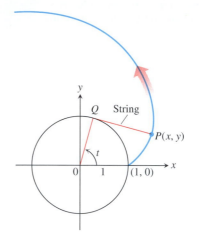

9.39 The involute of a circle of radius 1 (Exercise 30.)

31. *Parametrizations of lines in the plane* (Fig. 9.40).

a) Show that the equations and parameter interval

$$x = x_0 + (x_1 - x_0)t, \quad y = y_0 + (y_1 - y_0)t, \quad -\infty < t < \infty,$$

describe the line through the points (x_0, y_0) and (x_1, y_1).

b) Using the same parameter interval, write parametric equations for the line through a point (x_1, y_1) and the origin.

c) Using the same parameter interval, write parametric equations for the line through $(-1, 0)$ and $(0, 1)$.

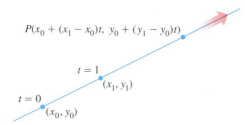

9.40 The line in Exercise 31. The arrow shows the direction of increasing t.

32. *The trammel of Archimedes.* The mechanical system pictured here is called the trammel of Archimedes. It consists of a rigid bar of length L, one end attached to a roller that rolls along the y-axis. At a fixed distance R from this end, the bar is attached to a second roller on the x-axis. Let P be the point at the free end of the bar and let θ be the angle the bar makes with the positive x-axis.

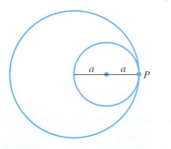

a) Find parametric equations for the path of P in terms of the parameter θ.

b) Find an equation in x and y whose graph is the path of P, and identify this path.

33. Hypocycloids. When a circle rolls on the inside of a fixed circle, any point P on the circumference of the rolling circle describes a *hypocycloid*. Let the fixed circle be $x^2 + y^2 = a^2$, let the radius of the rolling circle be b, and let the initial position of the tracing point P be $A(a, 0)$. Find parametric equations for the hypocycloid, using as the parameter the angle θ from the positive x-axis to the line joining the circles' centers. In particular, if $b = a/4$, as in the accompanying figure, show that the hypocycloid is the astroid

$$x = a \cos^3 \theta, \quad y = a \sin^3 \theta.$$

34. More about hypocycloids. The accompanying figure shows a circle of radius a tangent to the inside of a circle of radius $2a$. The point P, shown as the point of tangency in the figure, is attached to the smaller circle. What path does P trace as the smaller circle rolls around the inside of the larger circle?

35. As the point N moves along the line $y = a$ in the accompanying figure, P moves in such a way that $OP = MN$. Find parametric equations for the coordinates of P as functions of the angle t that the line ON makes with the positive y-axis.

36. Trochoids. A wheel of radius a rolls along a horizontal straight line without slipping. Find parametric equations for the curve traced out by a point P on a spoke of the wheel b units from its center. As parameter, use the angle θ through which the wheel turns. The curve is called a **trochoid,** which is a cycloid when $b = a$.

Distance Using Parametric Equations

37. Find the point on the parabola $x = t$, $y = t^2$, $-\infty < t < \infty$, closest to the point $(2, 1/2)$. (*Hint:* Minimize the square of the distance as a function of t.)

38. Find the point on the ellipse $x = 2 \cos t$, $y = \sin t$, $0 \le t \le 2\pi$ closest to the point $(3/4, 0)$. (*Hint:* Minimize the square of the distance as a function of t.)

Grapher Explorations

If you have a parametric equation grapher, graph the following equations over the given intervals.

39. Ellipse. $x = 4 \cos t$, $y = 2 \sin t$, over

a) $0 \le t \le 2\pi$ **b)** $0 \le t \le \pi$
c) $-\pi/2 \le t \le \pi/2$.

40. Hyperbola branch. $x = \sec t$ (enter as $1/\cos(t)$), $y = \tan t$ (enter as $\sin(t)/\cos(t)$), over

a) $-1.5 \le t \le 1.5$ **b)** $-0.5 \le t \le 0.5$
c) $-0.1 \le t \le 0.1$.

41. Parabola. $x = 2t + 3$, $y = t^2 - 1$, $-2 \le t \le 2$

42. Cycloid. $x = t - \sin t$, $y = 1 - \cos t$, over

a) $0 \le t \le 2\pi$ **b)** $0 \le t \le 4\pi$
c) $\pi \le t \le 3\pi$.

43. Astroid. $x = \cos^3 t$, $y = \sin^3 t$, over

a) $0 \le t \le 2\pi$ **b)** $-\pi/2 \le t \le \pi/2$.

44. *A nice curve (a deltoid)*

$$x = 2\cos t + \cos 2t, \quad y = 2\sin t - \sin 2t, \quad 0 \le t \le 2\pi$$

What happens if you replace 2 with -2 in the equations for x and y? Graph the new equations and find out.

45. *An even nicer curve*

$$x = 3\cos t + \cos 3t, \quad y = 3\sin t - \sin 3t, \quad 0 \le t \le 2\pi$$

What happens if you replace 3 with -3 in the equations for x and y? Graph the new equations and find out.

46. *Projectile motion.* Graph

$$x = (64\cos\alpha)\,t, \quad y = -16t^2 + (64\sin\alpha)\,t, \quad 0 \le t \le 4\sin\alpha$$

for the following firing angles.

a) $\alpha = \pi/4$ **b)** $\alpha = \pi/6$ **c)** $\alpha = \pi/3$
d) $\alpha = \pi/2$ (watch out—here it comes!)

47. *Three beautiful curves*

a) *Epicycloid:*

$$x = 9\cos t - \cos 9t, \quad y = 9\sin t - \sin 9t, \quad 0 \le t \le 2\pi$$

b) *Hypocycloid:*

$$x = 8\cos t + 2\cos 4t, \quad y = 8\sin t - 2\sin 4t, \quad 0 \le t \le 2\pi$$

c) *Hypotrochoid:*

$$x = \cos t + 5\cos 3t, \quad y = 6\cos t - 5\sin 3t, \quad 0 \le t \le 2\pi$$

48. *More beautiful curves*

a) $x = 6\cos t + 5\cos 3t, \quad y = 6\sin t - 5\sin 3t,$
$0 \le t \le 2\pi$
b) $x = 6\cos 2t + 5\cos 6t, \quad y = 6\sin 2t - 5\sin 6t,$
$0 \le t \le \pi$
c) $x = 6\cos t + 5\cos 3t, \quad y = 6\sin 2t - 5\sin 3t,$
$0 \le t \le 2\pi$
d) $x = 6\cos 2t + 5\cos 6t, \quad y = 6\sin 4t - 5\sin 6t,$
$0 \le t \le \pi$

9.5

Calculus with Parametrized Curves

This section shows how to find slopes, lengths, and surface areas associated with parametrized curves.

Slopes of Parametrized Curves

> **Definitions**
>
> A parametrized curve $x = f(t), y = g(t)$ is **differentiable at $t = t_0$** if f and g are differentiable at $t = t_0$. The curve is **differentiable** if it is differentiable at every parameter value. The curve is **smooth** if f' and g' are continuous and not simultaneously zero.

At a point on a differentiable parametrized curve where y is also a differentiable function of x, the derivatives dx/dt, dy/dt, and dy/dx are related by the Chain Rule equation

$$\frac{dy}{dt} = \frac{dy}{dx}\frac{dx}{dt}.$$

If $dx/dt \ne 0$, we may divide both sides of this equation by dx/dt to solve for dy/dx.

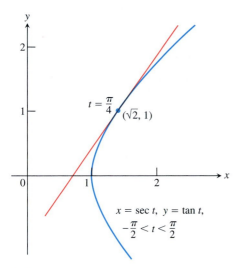

9.41 The hyperbola branch in Example 1.

Formula for Finding *dy/dx* from *dy/dt* and *dx/dt*   (*dx/dt* ≠ 0)

$$\frac{dy}{dx} = \frac{dy/dt}{dx/dt} \qquad (1)$$

EXAMPLE 1   Find the tangent to the right-hand hyperbola branch

$$x = \sec t, \qquad y = \tan t, \qquad -\frac{\pi}{2} < t < \frac{\pi}{2},$$

at the point $(\sqrt{2}, 1)$, where $t = \pi/4$ (Fig. 9.41).

Solution   The slope of the curve at t is

$$\frac{dy}{dx} = \frac{dy/dt}{dx/dt} = \frac{\sec^2 t}{\sec t \tan t} = \frac{\sec t}{\tan t}. \qquad \text{Eq. (1)}$$

Setting t equal to $\pi/4$ gives

$$\left.\frac{dy}{dx}\right|_{t=\pi/4} = \frac{\sec (\pi/4)}{\tan (\pi/4)}$$

$$= \frac{\sqrt{2}}{1} = \sqrt{2}.$$

The point–slope equation of the tangent is

$$y - y_0 = m\,(x - x_0)$$
$$y - 1 = \sqrt{2}\,(x - \sqrt{2})$$
$$y = \sqrt{2}\,x - 2 + 1$$
$$y = \sqrt{2}\,x - 1. \qquad \square$$

Notice the lack of symmetry in Eq. (2). To find d^2y/dx^2, we divide the derivative of y' by the derivative of x, not by the derivative of x'.

The Parametric Formula for *d²y/dx²*

If the parametric equations for a curve define y as a twice-differentiable function of x, we may calculate d^2y/dx^2 as a function of t in the following way:

$$\frac{d^2y}{dx^2} = \frac{d}{dx}\,(y') = \frac{dy'/dt}{dx/dt}. \qquad \text{Eq. (1) with } y \text{ replaced by } y'$$

Formula for Finding *d²y/dx²* from *y'* = *dy/dx* and *dx/dt*   (*dx/dt* ≠ 0)

$$\frac{d^2y}{dx^2} = \frac{dy'/dt}{dx/dt} \qquad (2)$$

How to Express *d²y/dx²* in Terms of *t*

Step 1: Express $y' = dy/dx$ in terms of t.

Step 2: Find dy'/dt.

Step 3: Divide dy'/dt by dx/dt. The quotient is d^2y/dx^2.

EXAMPLE 2   Find d^2y/dx^2 if $x = t - t^2$ and $y = t - t^3$.

Solution

Step 1: *Express y' in terms of t:*

$$y' = \frac{dy}{dx} = \frac{dy/dt}{dx/dt} = \frac{1 - 3t^2}{1 - 2t} \qquad \text{Eq. (1) with } x = t - t^2,$$
$$y = t - t^3$$

Step 2: *Differentiate y' with respect to t:*

$$\frac{dy'}{dt} = \frac{d}{dt}\left(\frac{1 - 3t^2}{1 - 2t}\right)$$
$$= \frac{2 - 6t + 6t^2}{(1 - 2t)^2}$$

Step 3: *Divide dy'/dt by dx/dt.* Since

$$\frac{dx}{dt} = \frac{d}{dt}(t - t^2) = 1 - 2t, \qquad x = t - t^2$$

we have

$$\frac{d^2y}{dx^2} = \frac{dy'/dt}{dx/dt} \qquad \text{Eq. (2)}$$
$$= \frac{2 - 6t + 6t^2}{(1 - 2t)^2} \cdot \frac{1}{1 - 2t}$$
$$= \frac{2 - 6t + 6t^2}{(1 - 2t)^3}.$$

Lengths of Parametrized Curves. Centroids

We find an integral for the length of a smooth curve $x = f(t)$, $y = g(t)$, $a \le t \le b$, by rewriting the integral $L = \int ds$ from Section 5.5 in the following way:

$$L = \int_{t=a}^{t=b} ds = \int_a^b \sqrt{dx^2 + dy^2}$$

$$= \int_a^b \sqrt{\left(\frac{(dx)^2}{(dt)^2} + \frac{(dy)^2}{(dt)^2}\right) dt^2} = \int_a^b \sqrt{\left(\frac{dx}{dt}\right)^2 + \left(\frac{dy}{dt}\right)^2} \, dt.$$

The only requirement besides the continuity of the integrand is that the point $P(x, y) = P(f(t), g(t))$ not trace any portion of the curve more than once as t moves from a to b.

Length

If a smooth curve $x = f(t)$, $y = g(t)$, $a \le t \le b$, is traversed exactly once as t increases from a to b, the curve's length is

$$L = \int_a^b \sqrt{\left(\frac{dx}{dt}\right)^2 + \left(\frac{dy}{dt}\right)^2} \, dt. \qquad (3)$$

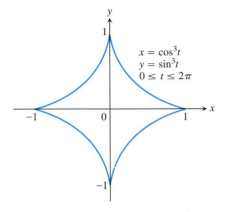

9.42 The astroid in Example 3.

The length formulas in Section 5.5 are special cases of Eq. (3) (Exercises 35 and 36).

What if there are two different parametrizations for a curve whose length we want to find—does it matter which one we use? The answer, from advanced calculus, is no, as long as the parametrization we choose meets the conditions preceding Eq. (3).

EXAMPLE 3   Find the length of the astroid (Fig. 9.42)

$$x = \cos^3 t, \qquad y = \sin^3 t, \qquad 0 \le t \le 2\pi.$$

Solution   Because of the curve's symmetry with respect to the coordinate axes, its length is four times the length of the first-quadrant portion. We have

$$x = \cos^3 t, \qquad y = \sin^3 t$$

$$\left(\frac{dx}{dt}\right)^2 = [3\cos^2 t(-\sin t)]^2 = 9\cos^4 t \,\sin^2 t$$

$$\left(\frac{dy}{dt}\right)^2 = [3\sin^2 t(\cos t)]^2 = 9\sin^4 t \,\cos^2 t$$

$$\sqrt{\left(\frac{dx}{dt}\right)^2 + \left(\frac{dy}{dt}\right)^2} = \sqrt{9\cos^2 t \,\sin^2 t \underbrace{(\cos^2 t + \sin^2 t)}_{1}}$$

$$= \sqrt{9\cos^2 t \,\sin^2 t}$$

$$= 3|\cos t \,\sin t|$$

$$= 3\cos t \,\sin t. \qquad \text{\footnotesize $\cos t \,\sin t \ge 0$ for $0 \le t \le \pi/2$}$$

Therefore,

$$\text{Length of first-quadrant portion} = \int_0^{\pi/2} 3\cos t \,\sin t \,dt$$

$$= \frac{3}{2} \int_0^{\pi/2} \sin 2t \,dt \qquad \text{\footnotesize $\cos t \,\sin t = (1/2)\sin 2t$}$$

$$= -\frac{3}{4} \cos 2t \,\Big]_0^{\pi/2} = \frac{3}{2}.$$

The length of the astroid is four times this: $4(3/2) = 6.$ ☐

EXAMPLE 4   Find the centroid of the first-quadrant arc of the astroid in Example 3.

Solution   We take the curve's density to be $\delta = 1$ and calculate the curve's mass and moments about the coordinate axes as we did at the end of Section 5.7.

The distribution of mass is symmetric about the line $y = x$, so $\bar{x} = \bar{y}$. A typical segment of the curve (Fig. 9.43) has mass

$$dm = 1 \cdot ds = \sqrt{\left(\frac{dx}{dt}\right)^2 + \left(\frac{dy}{dt}\right)^2}\; dt = 3\cos t \,\sin t \,dt. \qquad \text{\footnotesize From Example 3}$$

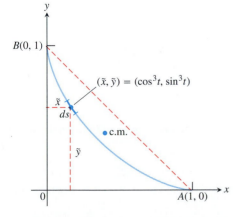

9.43 A typical segment of the arc in Example 4. The centroid (c.m.) of the curve lies about a third of the way toward chord *AB*.

The curve's mass is

$$M = \int_0^{\pi/2} dm = \int_0^{\pi/2} 3 \cos t \sin t \, dt = \frac{3}{2}.$$ Again from Example 3

The curve's moment about the x-axis is

$$M_x = \int \tilde{y} \, dm = \int_0^{\pi/2} \sin^3 t \cdot 3 \cos t \sin t \, dt$$

$$= 3 \int_0^{\pi/2} \sin^4 t \cos t \, dt = 3 \cdot \frac{\sin^5 t}{5} \Bigg]_0^{\pi/2} = \frac{3}{5}.$$

Hence,

$$\overline{y} = \frac{M_x}{M} = \frac{3/5}{3/2} = \frac{2}{5}.$$

The centroid is the point (2/5, 2/5) (Fig. 9.43).

The Area of a Surface of Revolution

For smooth parametrized curves, the length formula in Eq. (3) leads to the following formulas for surfaces of revolution. The derivations are similar to the derivations of the Cartesian formulas in Section 5.6.

Surface Area

If a smooth curve $x = f(t)$, $y = g(t)$, $a \le t \le b$, is traversed exactly once as t increases from a to b, then the areas of the surfaces generated by revolving the curve about the coordinate axes are as follows.

1. Revolution about
    the x-axis ($y \ge 0$): $\qquad S = \int_a^b 2\pi y \sqrt{\left(\frac{dx}{dt}\right)^2 + \left(\frac{dy}{dt}\right)^2} \, dt$ (4)

2. Revolution about
    the y-axis ($x \ge 0$): $\qquad S = \int_a^b 2\pi x \sqrt{\left(\frac{dx}{dt}\right)^2 + \left(\frac{dy}{dt}\right)^2} \, dt$ (5)

As with length, we can calculate surface area from any convenient parametrization that meets the stated criteria.

EXAMPLE 5 The standard parametrization of the circle of radius 1 centered at the point (0, 1) in the xy-plane is

$$x = \cos t, \qquad y = 1 + \sin t, \qquad 0 \le t \le 2\pi.$$

Use this parametrization to find the area of the surface swept out by revolving the circle about the x-axis (Fig. 9.44).

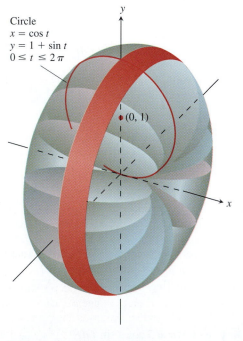

Circle
$x = \cos t$
$y = 1 + \sin t$
$0 \le t \le 2\pi$

9.44 The surface in Example 5.

Solution   We evaluate the formula

$$S = \int_a^b 2\pi \, y \sqrt{\left(\frac{dx}{dt}\right)^2 + \left(\frac{dy}{dt}\right)^2} \, dt \qquad \text{Eq. (4) for revolution about the } x\text{-axis}$$

$$= \int_0^{2\pi} 2\pi \, (1 + \sin t) \sqrt{\underbrace{(-\sin t)^2 + (\cos t)^2}_{1}} \, dt$$

$$= 2\pi \int_0^{2\pi} (1 + \sin t) \, dt$$

$$= 2\pi \Big[t - \cos t \Big]_0^{2\pi} = 4\pi^2.$$

Exercises 9.5

Tangents to Parametrized Curves

In Exercises 1–12, find an equation for the line tangent to the curve at the point defined by the given value of t. Also, find the value of d^2y/dx^2 at this point.

1. $x = 2 \cos t$, $\quad y = 2 \sin t$, $\quad t = \pi/4$

2. $x = \sin 2\pi t$, $\quad y = \cos 2\pi t$, $\quad t = -1/6$

3. $x = 4 \sin t$, $\quad y = 2 \cos t$, $\quad t = \pi/4$

4. $x = \cos t$, $\quad y = \sqrt{3} \cos t$, $\quad t = 2\pi/3$

5. $x = t$, $\quad y = \sqrt{t}$, $\quad t = 1/4$

6. $x = \sec^2 t - 1$, $\quad y = \tan t$, $\quad t = -\pi/4$

7. $x = \sec t$, $\quad y = \tan t$, $\quad t = \pi/6$

8. $x = -\sqrt{t + 1}$, $\quad y = \sqrt{3t}$, $\quad t = 3$

9. $x = 2t^2 + 3$, $\quad y = t^4$, $\quad t = -1$

10. $x = 1/t$, $\quad y = -2 + \ln t$, $\quad t = 1$

11. $x = t - \sin t$, $\quad y = 1 - \cos t$, $\quad t = \pi/3$

12. $x = \cos t$, $\quad y = 1 + \sin t$, $\quad t = \pi/2$

Implicitly Defined Parametrizations

Assuming that the equations in Exercises 13–16 define x and y implicitly as differentiable functions $x = f(t)$, $y = g(t)$, find the slope of the curve $x = f(t)$, $y = g(t)$ at the given value of t.

13. $x^2 - 2tx + 2t^2 = 4$, $\quad 2y^3 - 3t^2 = 4$, $\quad t = 2$

14. $x = \sqrt{5 - \sqrt{t}}$, $\quad y(t - 1) = \ln y$, $\quad t = 1$

15. $x + 2x^{3/2} = t^2 + t$, $\quad y\sqrt{t + 1} + 2t\sqrt{y} = 4$, $\quad t = 0$

16. $x \sin t + 2x = t$, $\quad t \sin t - 2t = y$, $\quad t = \pi$

Lengths of Curves

Find the lengths of the curves in Exercises 17–22.

17. $x = \cos t$, $\quad y = t + \sin t$, $\quad 0 \le t \le \pi$

18. $x = t^3$, $\quad y = 3t^2/2$, $\quad 0 \le t \le \sqrt{3}$

19. $x = t^2/2$, $\quad y = (2t + 1)^{3/2}/3$, $\quad 0 \le t \le 4$

20. $x = (2t + 3)^{3/2}/3$, $\quad y = t + t^2/2$, $\quad 0 \le t \le 3$

21. $x = 8 \cos t + 8t \sin t$ $\qquad$ **22.** $x = \ln (\sec t + \tan t) - \sin t$
   $y = 8 \sin t - 8t \cos t,$ $\qquad\qquad$ $y = \cos t$, $\quad 0 \le t \le \pi/3$
   $0 \le t \le \pi/2$

Surface Area

Find the areas of the surfaces generated by revolving the curves in Exercises 23–26 about the indicated axes.

23. $x = \cos t$, $\quad y = 2 + \sin t$, $\quad 0 \le t \le 2\pi$; $\quad x$-axis

24. $x = (2/3)t^{3/2}$, $\quad y = 2\sqrt{t}$, $\quad 0 \le t \le \sqrt{3}$; $\quad y$-axis

25. $x = t + \sqrt{2}$, $\quad y = (t^2/2) + \sqrt{2}t$, $\quad -\sqrt{2} \le t \le \sqrt{2}$; $\quad y$-axis

26. $x = \ln (\sec t + \tan t) - \sin t$, $\quad y = \cos t$, $\quad 0 \le t \le \pi/3$; x-axis

27. *A cone frustum.* The line segment joining the points $(0, 1)$ and $(2, 2)$ is revolved about the x-axis to generate a frustum of a cone. Find the surface area of the frustum using the parametrization $x = 2t$, $y = t + 1$, $0 \le t \le 1$. Check your result with the geometry formula: Area $= \pi \, (r_1 + r_2)$(slant height).

28. *A cone.* The line segment joining the origin to the point (h, r) is revolved about the x-axis to generate a cone of height h and base radius r. Find the cone's surface area with the parametric equations $x = ht$, $y = rt$, $0 \le t \le 1$. Check your result with the geometry formula: Area $= \pi r$(slant height).

Centroids

29. a) Find the coordinates of the centroid of the curve

$$x = \cos t + t \sin t, \quad y = \sin t - t \cos t, \quad 0 \le t \le \pi/2.$$

b) **CALCULATOR** The curve is a portion of the involute in Fig. 9.39. Sketch the curve. Find the centroid's coordinates to the nearest tenth and add the centroid to your sketch.

30. a) Find the coordinates of the centroid of the curve

$$x = e^t \cos t, \quad y = e^t \sin t, \quad 0 \le t \le \pi.$$

b) **CALCULATOR** Sketch the curve. Find the centroid's coordinates to the nearest tenth and add the centroid to your sketch.

31. a) Find the coordinates of the centroid of the curve

$$x = \cos t, \quad y = t + \sin t, \quad 0 \le t \le \pi.$$

b) Sketch the curve and add the centroid to your sketch.

32. INTEGRAL EVALUATOR Most centroid calculations for curves are done with a calculator or computer that has an integral evaluation program. As a case in point, find, to the nearest hundredth, the coordinates of the centroid of the curve

$$x = t^3, \quad y = 3t^2/2, \quad 0 \le t \le \sqrt{3}.$$

Theory and Examples

33. *Length is independent of parametrization.* To illustrate the fact that the numbers we get for length do not depend on the way we parametrize our curves (except for the mild restrictions mentioned earlier), calculate the length of the semicircle $y = \sqrt{1 - x^2}$ with these two different parametrizations:

a) $x = \cos 2t, \quad y = \sin 2t, \quad 0 \le t \le \pi/2$
b) $x = \sin \pi t, \quad y = \cos \pi t, \quad -1/2 \le t \le 1/2$

34. *Elliptic integrals.* The length of the ellipse

$$x = a \cos t, \quad y = b \sin t, \quad 0 \le t \le 2\pi$$

turns out to be

$$\text{Length} = 4a \int_0^{\pi/2} \sqrt{1 - e^2 \cos^2 t} \, dt,$$

where e is the ellipse's eccentricity. The integral in this formula, called an *elliptic integral*, is nonelementary except when $e = 0$ or 1.

a) **CALCULATOR** Use the trapezoidal rule with $n = 10$ to estimate the length of the ellipse when $a = 1$ and $e = 1/2$.
b) Use the fact that the absolute value of the second derivative of $f(t) = \sqrt{1 - e^2 \cos^2 t}$ is less than 1 to find an upper bound for the error in the estimate you obtained in (a).

35. As mentioned in Section 9.4, the graph of a function $y = f(x)$ over an interval $[a, b]$ automatically has the parametrization

$$x = x, \quad y = f(x), \quad a \le x \le b.$$

The parameter, in this case, is x itself. Show that for this parametrization the parametric length

formula

$$L = \int_a^b \sqrt{\left(\frac{dx}{dt}\right)^2 + \left(\frac{dy}{dt}\right)^2} \, dt$$

reduces to the Cartesian formula

$$L = \int_a^b \sqrt{1 + \left(\frac{dy}{dx}\right)^2} \, dx$$

derived in Section 5.5. This will show that the Cartesian formula is a special case of the parametric formula.

36. (*Continuation of Exercise 35.*) Show that the Cartesian formula

$$L = \int_c^d \sqrt{1 + \left(\frac{dx}{dy}\right)^2} \, dy$$

for the length of the curve $x = g(y), c \le y \le d$ (Section 5.5, Eq. 3), is a special case of the parametric length formula

$$L = \int_a^b \sqrt{\left(\frac{dx}{dt}\right)^2 + \left(\frac{dy}{dt}\right)^2} \, dt.$$

37. Find the area under one arch of the cycloid

$$x = a(\theta - \sin \theta), \quad y = a(1 - \cos \theta).$$

(*Hint:* Use $dx = (dx/d\theta) \, d\theta$.)

38. Find the length of one arch of the cycloid

$$x = a(\theta - \sin \theta), \quad y = a(1 - \cos \theta).$$

39. Find the area of the surface generated by revolving one arch of the cycloid $x = \theta - \sin \theta, y = 1 - \cos \theta$ about the x-axis.

40. Find the volume swept out by revolving the region bounded by the x-axis and one arch of the cycloid $x = \theta - \sin \theta, y = 1 - \cos \theta$ about the x-axis. (*Hint:* $dV = \pi y^2 \, dx = \pi y^2 (dx/d\theta) \, d\theta$.)

Grapher Explorations

The curves in Exercises 41 and 42 are called *Bowditch curves* or *Lissajous figures*. In each case, find the point in the interior of the first quadrant where the tangent to the curve is horizontal, and find the equations of the two tangents at the origin.

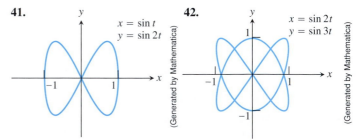

41. $x = \sin t$, $y = \sin 2t$ (Generated by Mathematica)
42. $x = \sin 2t$, $y = \sin 3t$ (Generated by Mathematica)

Graph the parametric curves in Exercises 43–49 over parameter intervals of your choice. The curves are Bowditch curves (Lissajous figures), the general formula being

$$x = a \sin(mt + d), \quad y = b \sin nt,$$

with m and n integers.

43. $x = \sin 2t, \quad y = \sin t$

44. $x = \sin 3t, \quad y = \sin 4t$

45. $x = \sin t, \quad y = \sin 4t$

46. $x = \sin t, \quad y = \sin 5t$

47. $x = \sin 3t, \quad y = \sin 5t$

48. $x = \sin (3t + \pi/2), \quad y = \sin 5t$

49. $x = \sin (3t + \pi/4), \quad y = \sin 5t$

✺ CAS Explorations and Projects

Use a CAS to perform the following steps on the parametrized curves in Exercises 50–55.

a) Plot the curve for the given interval of t values.

b) Find dy/dx and d^2y/dx^2 at the point t_0.

c) Find an equation for the tangent line to the curve at the point defined by the given value t_0. Plot the curve together with the tangent line on a single graph.

d) Find the length of the curve over the interval.

50. $x = \dfrac{1}{3}t^3, \quad y = \dfrac{1}{2}t^2, \quad 0 \le t \le 1, \quad t_0 = 1/2$

51. $x = 2t^3 - 16t^2 + 25t + 5, \quad y = t^2 + t - 3, \quad 0 \le t \le 6,$
$t_0 = 3/2$

52. $x = e^t - t^2, \quad y = t + e^{-t}, \quad -1 \le t \le 2, \quad t_0 = 1$

53. $x = t - \cos t, \quad y = 1 + \sin t, \quad -\pi \le t \le \pi, \quad t_0 = \pi/4$

54. $x = e^t + \sin 2t, \quad y = e^t + \cos (t^2), \quad -\sqrt{2}\pi \le t \le \pi/4,$
$t_0 = -\pi/4$

55. $x = e^t \cos t, \quad y = e^t \sin t, \quad 0 \le t \le \pi, \quad t_0 = \pi/2$

The equations in Exercises 56 and 57 define x and y implicitly as differentiable functions of t. Use a CAS to perform the following steps:

a) Solve the first equation for x and the second equation for y to find $x = f(t)$ and $y = g(t)$.

b) Find the slope of the curve $x = f(t)$ and $y = g(t)$ at t_0.

c) Find an equation for the tangent line to the curve at the point defined by t_0.

d) Plot the curve together with the tangent line over the specified interval of t-values.

56. $x^2 - 2tx + 3t^2 = 4, \quad y^3 - 2t^2 = 7, \quad -1 \le t \le 2, \quad t_0 = 1$

57. $x^2 \cos t + 2x = t, \quad t \sin t + 2\sqrt{y} = y, \quad -2\pi \le t \le 2\pi,$
$t_0 = -\pi/4$

9.6

Polar Coordinates

In this section, we study polar coordinates and their relation to Cartesian coordinates. While a point in the plane has just one pair of Cartesian coordinates, it has infinitely many pairs of polar coordinates. This has interesting consequences for graphing, as we will see in the next section.

Definition of Polar Coordinates

To define polar coordinates, we first fix an **origin** O (called the **pole**) and an **initial ray** from O (Fig. 9.45). Then each point P can be located by assigning to it a **polar coordinate pair** (r, θ) in which r gives the directed distance from O to P and θ gives the directed angle from the initial ray to ray OP.

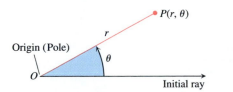

9.45 To define polar coordinates for the plane, we start with an origin, called the pole, and an initial ray.

Polar Coordinates

$$P(r, \theta) \qquad (1)$$

Directed distance from O to P Directed angle from initial ray to OP

As in trigonometry, θ is positive when measured counterclockwise and negative when measured clockwise. The angle associated with a given point is not unique.

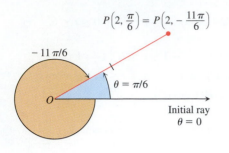

9.46 Polar coordinates are not unique.

For instance, the point 2 units from the origin along the ray $\theta = \pi/6$ has polar coordinates $r = 2, \theta = \pi/6$. It also has coordinates $r = 2, \theta = -11\pi/6$ (Fig. 9.46).

Negative Values of *r*

There are occasions when we wish to allow r to be negative. That is why we use directed distance in (1). The point $P(2, 7\pi/6)$ can be reached by turning $7\pi/6$ rad counterclockwise from the initial ray and going forward 2 units (Fig. 9.47). It can also be reached by turning $\pi/6$ rad counterclockwise from the initial ray and going *backward* 2 units. So the point also has polar coordinates $r = -2, \theta = \pi/6$.

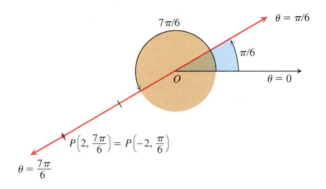

9.47 Polar coordinates can have negative *r*-values.

EXAMPLE 1   Find all the polar coordinates of the point $P(2, \pi/6)$.

Solution   We sketch the initial ray of the coordinate system, draw the ray from the origin that makes an angle of $\pi/6$ rad with the initial ray, and mark the point $(2, \pi/6)$ (Fig. 9.48). We then find the angles for the other coordinate pairs of P in which $r = 2$ and $r = -2$.

For $r = 2$, the complete list of angles is

$$\frac{\pi}{6}, \quad \frac{\pi}{6} \pm 2\pi, \quad \frac{\pi}{6} \pm 4\pi, \quad \frac{\pi}{6} \pm 6\pi, \quad \dots.$$

For $r = -2$, the angles are

$$-\frac{5\pi}{6}, \quad -\frac{5\pi}{6} \pm 2\pi, \quad -\frac{5\pi}{6} \pm 4\pi, \quad -\frac{5\pi}{6} \pm 6\pi, \quad \dots.$$

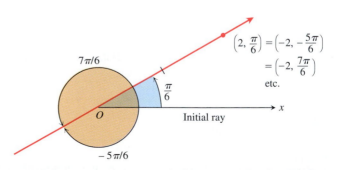

9.48 The point $P(2, \pi/6)$ has infinitely many polar coordinate pairs.

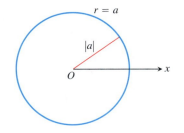

9.49 The polar equation for this circle is $r = a$.

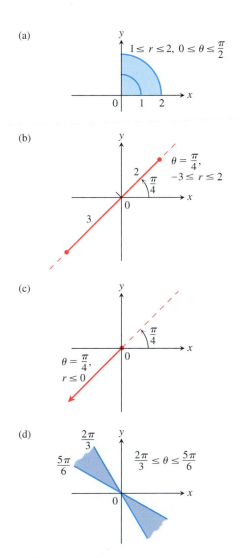

9.50 The graphs of typical inequalities in r and θ (Example 3).

The corresponding coordinate pairs of P are

$$\left(2, \frac{\pi}{6} + 2n\pi\right), \qquad n = 0, \pm 1, \pm 2, \ldots$$

and

$$\left(-2, -\frac{5\pi}{6} + 2n\pi\right), \qquad n = 0, \pm 1, \pm 2, \ldots.$$

When $n = 0$, the formulas give $(2, \pi/6)$ and $(-2, -5\pi/6)$. When $n = 1$, they give $(2, 13\pi/6)$ and $(-2, 7\pi/6)$, and so on. ❏

Elementary Coordinate Equations and Inequalities

If we hold r fixed at a constant value $r = a \neq 0$, the point $P\,(r, \theta)$ will lie $|a|$ units from the origin O. As θ varies over any interval of length 2π, P then traces a circle of radius $|a|$ centered at O (Fig. 9.49).

If we hold θ fixed at a constant value $\theta = \theta_0$ and let r vary between $-\infty$ and ∞, the point $P\,(r, \theta)$ traces the line through O that makes an angle of measure θ_0 with the initial ray.

Equation	Graph
$r = a$	Circle of radius $|a|$ centered at O
$\theta = \theta_0$	Line through O making an angle θ_0 with the initial ray

EXAMPLE 2

a) $r = 1$ and $r = -1$ are equations for the circle of radius 1 centered at O.
b) $\theta = \pi/6$, $\theta = 7\pi/6$, and $\theta = -5\pi/6$ are equations for the line in Fig. 9.48. ❏

Equations of the form $r = a$ and $\theta = \theta_0$ can be combined to define regions, segments, and rays.

EXAMPLE 3   Graph the sets of points whose polar coordinates satisfy the following conditions.

a) $1 \leq r \leq 2$ and $0 \leq \theta \leq \dfrac{\pi}{2}$

b) $-3 \leq r \leq 2$ and $\theta = \dfrac{\pi}{4}$

c) $r \leq 0$ and $\theta = \dfrac{\pi}{4}$

d) $\dfrac{2\pi}{3} \leq \theta \leq \dfrac{5\pi}{6}$ (no restriction on r)

Solution   The graphs are shown in Fig. 9.50. ❏

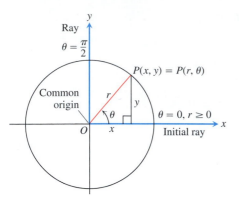

9.51 The usual way to relate polar and Cartesian coordinates.

Cartesian Versus Polar Coordinates

When we use both polar and Cartesian coordinates in a plane, we place the two origins together and take the initial polar ray as the positive x-axis. The ray $\theta = \pi/2$, $r > 0$, becomes the positive y-axis (Fig. 9.51). The two coordinate systems are then related by the following equations.

Equations Relating Polar and Cartesian Coordinates

$$x = r \cos \theta, \qquad y = r \sin \theta, \qquad x^2 + y^2 = r^2, \qquad \frac{y}{x} = \tan \theta \quad (2)$$

We use Eqs. (2) to rewrite polar equations in Cartesian form and vice versa.

EXAMPLE 4

Polar equation	Cartesian equivalent
$r \cos \theta = 2$	$x = 2$
$r^2 \cos \theta \sin \theta = 4$	$xy = 4$
$r^2 \cos^2 \theta - r^2 \sin^2 \theta = 1$	$x^2 - y^2 = 1$
$r = 1 + 2r \cos \theta$	$y^2 - 3x^2 - 4x - 1 = 0$
$r = 1 - \cos \theta$	$x^4 + y^4 + 2x^2y^2 + 2x^3 + 2xy^2 - y^2 = 0$

With some curves, we are better off with polar coordinates; with others, we aren't. ❑

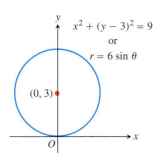

9.52 The circle in Example 5.

EXAMPLE 5   Find a polar equation for the circle $x^2 + (y - 3)^2 = 9$ (Fig. 9.52).

Solution

$$x^2 + y^2 - 6y + 9 = 9 \qquad \text{Expand } (y-3)^2.$$
$$x^2 + y^2 - 6y = 0 \qquad \text{The 9's cancel.}$$
$$r^2 - 6r \sin \theta = 0 \qquad x^2 + y^2 = r^2$$
$$r = 0 \quad \text{or} \quad r - 6 \sin \theta = 0$$
$$r = 6 \sin \theta \qquad \text{Includes both possibilities}$$

We will say more about polar equations of conic sections in Section 9.8.   ❑

EXAMPLE 6   Replace the following polar equations by equivalent Cartesian equations, and identify their graphs.

a) $r \cos \theta = -4$

b) $r^2 = 4r \cos \theta$

c) $r = \dfrac{4}{2 \cos \theta - \sin \theta}$

Solution   We use the substitutions $r \cos \theta = x$, $r \sin \theta = y$, $r^2 = x^2 + y^2$.

a) $r \cos \theta = -4$

   The Cartesian equation: $r \cos \theta = -4$
   $$x = -4$$
   The graph: Vertical line through $x = -4$ on the x-axis

b) $r^2 = 4r \cos \theta$

   The Cartesian equation: $r^2 = 4r \cos \theta$
   $$x^2 + y^2 = 4x$$
   $$x^2 - 4x + y^2 = 0$$
   $$x^2 - 4x + 4 + y^2 = 4 \qquad \text{Completing the square}$$
   $$(x - 2)^2 + y^2 = 4$$
   The graph: Circle, radius 2, center $(h, k) = (2, 0)$

c) $r = \dfrac{4}{2 \cos \theta - \sin \theta}$

   The Cartesian equation: $r(2 \cos \theta - \sin \theta) = 4$
   $$2r \cos \theta - r \sin \theta = 4$$
   $$2x - y = 4$$
   $$y = 2x - 4$$
   The graph: Line, slope $m = 2$, y-intercept $b = -4$

Exercises 9.6

Polar Coordinate Pairs

1. Which polar coordinate pairs label the same point?

   a) $(3, 0)$ **b)** $(-3, 0)$ **c)** $(2, 2\pi/3)$
   d) $(2, 7\pi/3)$ **e)** $(-3, \pi)$ **f)** $(2, \pi/3)$
   g) $(-3, 2\pi)$ **h)** $(-2, -\pi/3)$

2. Which polar coordinate pairs label the same point?

   a) $(-2, \pi/3)$ **b)** $(2, -\pi/3)$ **c)** (r, θ)
   d) $(r, \theta + \pi)$ **e)** $(-r, \theta)$ **f)** $(2, -2\pi/3)$
   g) $(-r, \theta + \pi)$ **h)** $(-2, 2\pi/3)$

3. Plot the following points (given in polar coordinates). Then find all the polar coordinates of each point.

   a) $(2, \pi/2)$ **b)** $(2, 0)$
   c) $(-2, \pi/2)$ **d)** $(-2, 0)$

4. Plot the following points (given in polar coordinates). Then find all the polar coordinates of each point.

   a) $(3, \pi/4)$ **b)** $(-3, \pi/4)$
   c) $(3, -\pi/4)$ **d)** $(-3, -\pi/4)$

Polar to Cartesian Coordinates

5. Find the Cartesian coordinates of the points in Exercise 1.

6. Find the Cartesian coordinates of the following points (given in polar coordinates).

   a) $\left(\sqrt{2}, \pi/4\right)$ **b)** $(1, 0)$

   c) $(0, \pi/2)$ **d)** $\left(-\sqrt{2}, \pi/4\right)$

   e) $(-3, 5\pi/6)$ **f)** $(5, \tan^{-1}(4/3))$

   g) $(-1, 7\pi)$ **h)** $\left(2\sqrt{3}, 2\pi/3\right)$

Graphing Polar Equations and Inequalities

Graph the sets of points whose polar coordinates satisfy the equations and inequalities in Exercises 7–22.

7. $r = 2$ **8.** $0 \le r \le 2$

9. $r \ge 1$ **10.** $1 \le r \le 2$

11. $0 \le \theta \le \pi/6, \quad r \ge 0$ **12.** $\theta = 2\pi/3, \quad r \le -2$

13. $\theta = \pi/3, \quad -1 \le r \le 3$

14. $\theta = 11\pi/4, \quad r \geq -1$

15. $\theta = \pi/2, \quad r \geq 0$

16. $\theta = \pi/2, \quad r \leq 0$

17. $0 \leq \theta \leq \pi, \quad r = 1$

18. $0 \leq \theta \leq \pi, \quad r = -1$

19. $\pi/4 \leq \theta \leq 3\pi/4, \quad 0 \leq r \leq 1$

20. $-\pi/4 \leq \theta \leq \pi/4, \quad -1 \leq r \leq 1$

21. $-\pi/2 \leq \theta \leq \pi/2, \quad 1 \leq r \leq 2$

22. $0 \leq \theta \leq \pi/2, \quad 1 \leq |r| \leq 2$

Polar to Cartesian Equations

Replace the polar equations in Exercises 23–48 by equivalent Cartesian equations. Then describe or identify the graph.

23. $r \cos \theta = 2$

24. $r \sin \theta = -1$

25. $r \sin \theta = 0$

26. $r \cos \theta = 0$

27. $r = 4 \csc \theta$

28. $r = -3 \sec \theta$

29. $r \cos \theta + r \sin \theta = 1$

30. $r \sin \theta = r \cos \theta$

31. $r^2 = 1$

32. $r^2 = 4r \sin \theta$

33. $r = \dfrac{5}{\sin \theta - 2 \cos \theta}$

34. $r^2 \sin 2\theta = 2$

35. $r = \cot \theta \csc \theta$

36. $r = 4 \tan \theta \sec \theta$

37. $r = \csc \theta \, e^{r \cos \theta}$

38. $r \sin \theta = \ln r + \ln \cos \theta$

39. $r^2 + 2r^2 \cos \theta \sin \theta = 1$

40. $\cos^2 \theta = \sin^2 \theta$

41. $r^2 = -4r \cos \theta$

42. $r^2 = -6r \sin \theta$

43. $r = 8 \sin \theta$

44. $r = 3 \cos \theta$

45. $r = 2 \cos \theta + 2 \sin \theta$

46. $r = 2 \cos \theta - \sin \theta$

47. $r \sin \left(\theta + \dfrac{\pi}{6} \right) = 2$

48. $r \sin \left(\dfrac{2\pi}{3} - \theta \right) = 5$

Cartesian to Polar Equations

Replace the Cartesian equations in Exercises 49–62 by equivalent polar equations.

49. $x = 7$

50. $y = 1$

51. $x = y$

52. $x - y = 3$

53. $x^2 + y^2 = 4$

54. $x^2 - y^2 = 1$

55. $\dfrac{x^2}{9} + \dfrac{y^2}{4} = 1$

56. $xy = 2$

57. $y^2 = 4x$

58. $x^2 + xy + y^2 = 1$

59. $x^2 + (y - 2)^2 = 4$

60. $(x - 5)^2 + y^2 = 25$

61. $(x - 3)^2 + (y + 1)^2 = 4$

62. $(x + 2)^2 + (y - 5)^2 = 16$

Theory and Examples

63. Find all polar coordinates of the origin.

64. *Vertical and horizontal lines*

   a) Show that every vertical line in the xy-plane has a polar equation of the form $r = a \sec \theta$.

   b) Find the analogous polar equation for horizontal lines in the xy-plane.

9.7	# Graphing in Polar Coordinates

This section describes techniques for graphing equations in polar coordinates.

Symmetry

Figure 9.53 illustrates the standard polar coordinate tests for symmetry.

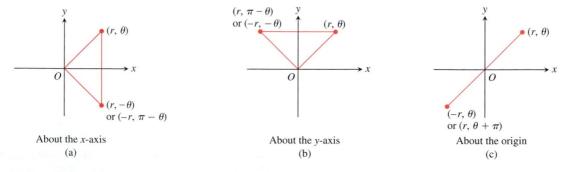

About the x-axis
(a)

About the y-axis
(b)

About the origin
(c)

9.53 Three tests for symmetry.

Symmetry Tests for Polar Graphs

1. *Symmetry about the x-axis:* If the point (r, θ) lies on the graph, the point $(r, -\theta)$ or $(-r, \pi - \theta)$ lies on the graph (Fig. 9.53a).

2. *Symmetry about the y-axis:* If the point (r, θ) lies on the graph, the point $(r, \pi - \theta)$ or $(-r, -\theta)$ lies on the graph (Fig. 9.53b).

3. *Symmetry about the origin:* If the point (r, θ) lies on the graph, the point $(-r, \theta)$ or $(r, \theta + \pi)$ lies on the graph (Fig. 9.53c).

Slope

The slope of a polar curve $r = f(\theta)$ is given by dy/dx, not by $r' = df/d\theta$. To see why, think of the graph of f as the graph of the parametric equations

$$x = r \cos \theta = f(\theta) \cos \theta, \qquad y = r \sin \theta = f(\theta) \sin \theta.$$

If f is a differentiable function of θ, then so are x and y and, when $dx/d\theta \neq 0$, we can calculate dy/dx from the parametric formula

$$\frac{dy}{dx} = \frac{dy/d\theta}{dx/d\theta} \qquad \text{Section 9.5, Eq. (1) with } t = \theta$$

$$= \frac{\dfrac{d}{d\theta}(f(\theta) \cdot \sin \theta)}{\dfrac{d}{d\theta}(f(\theta) \cdot \cos \theta)}$$

$$= \frac{\dfrac{df}{d\theta} \sin \theta + f(\theta) \cos \theta}{\dfrac{df}{d\theta} \cos \theta - f(\theta) \sin \theta} \qquad \text{Product Rule for Derivatives}$$

Slope of the Curve $r = f(\theta)$

$$\frac{dy}{dx}\bigg|_{(r,\theta)} = \frac{f'(\theta) \sin \theta + f(\theta) \cos \theta}{f'(\theta) \cos \theta - f(\theta) \sin \theta}, \qquad (1)$$

provided $dx/d\theta \neq 0$ at (r, θ).

If the curve $r = f(\theta)$ passes through the origin at $\theta = \theta_0$, then $f(\theta_0) = 0$, and Eq. (1) gives

$$\frac{dy}{dx}\bigg|_{(0,\theta_0)} = \frac{f'(\theta_0) \sin \theta_0}{f'(\theta_0) \cos \theta_0} = \tan \theta_0.$$

If the graph of $r = f(\theta)$ passes through the origin at the value $\theta = \theta_0$, the slope of the curve there is $\tan \theta_0$. The reason we say "slope at $(0, \theta_0)$" and not just "slope at the origin" is that a polar curve may pass through the origin more than once, with different slopes at different θ-values. This is not the case in our first example, however.

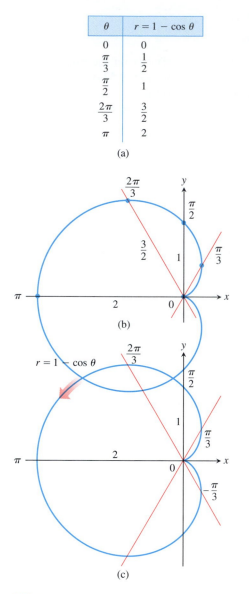

θ	$r = 1 - \cos\theta$
0	0
$\dfrac{\pi}{3}$	$\dfrac{1}{2}$
$\dfrac{\pi}{2}$	1
$\dfrac{2\pi}{3}$	$\dfrac{3}{2}$
π	2

(a)

(b)

(c)

9.54 The steps in graphing the cardioid $r = 1 - \cos\theta$ (Example 1). The arrow shows the direction of increasing θ.

EXAMPLE 1   *A cardioid*

Graph the curve $r = 1 - \cos\theta$.

Solution   The curve is symmetric about the x-axis because

$$(r, \theta) \text{ on the graph} \implies r = 1 - \cos\theta$$

$$\implies r = 1 - \cos(-\theta) \qquad \cos\theta = \cos(-\theta)$$

$$\implies (r, -\theta) \text{ on the graph.}$$

As θ increases from 0 to π, $\cos\theta$ decreases from 1 to -1, and $r = 1 - \cos\theta$ increases from a minimum value of 0 to a maximum value of 2. As θ continues on from π to 2π, $\cos\theta$ increases from -1 back to 1 and r decreases from 2 back to 0. The curve starts to repeat when $\theta = 2\pi$ because the cosine has period 2π.

The curve leaves the origin with slope $\tan(0) = 0$ and returns to the origin with slope $\tan(2\pi) = 0$.

We make a table of values from $\theta = 0$ to $\theta = \pi$, plot the points, draw a smooth curve through them with a horizontal tangent at the origin, and reflect the curve across the x-axis to complete the graph (Fig. 9.54). The curve is called a *cardioid* because of its heart shape. Cardioid shapes appear in the cams that direct the even layering of thread on bobbins and reels, and in the signal-strength pattern of certain radio antennae.   ❑

EXAMPLE 2   Graph the curve $r^2 = 4\cos\theta$.

Solution   The equation $r^2 = 4\cos\theta$ requires $\cos\theta \geq 0$, so we get the entire graph by running θ from $-\pi/2$ to $\pi/2$. The curve is symmetric about the x-axis because

$$(r, \theta) \text{ on the graph} \implies r^2 = 4\cos\theta$$

$$\implies r^2 = 4\cos(-\theta) \qquad \cos\theta = \cos(-\theta)$$

$$\implies (r, -\theta) \text{ on the graph.}$$

The curve is also symmetric about the origin because

$$(r, \theta) \text{ on the graph} \implies r^2 = 4\cos\theta$$

$$\implies (-r)^2 = 4\cos\theta$$

$$\implies (-r, \theta) \text{ on the graph.}$$

Together, these two symmetries imply symmetry about the y-axis.

The curve passes through the origin when $\theta = -\pi/2$ and $\theta = \pi/2$. It has a vertical tangent both times because $\tan\theta$ is infinite.

For each value of θ in the interval between $-\pi/2$ and $\pi/2$, the formula $r^2 = 4\cos\theta$ gives two values of r:

$$r = \pm2\sqrt{\cos\theta}.$$

We make a short table of values, plot the corresponding points, and use information about symmetry and tangents to guide us in connecting the points with a smooth curve (Fig. 9.55).   ❑

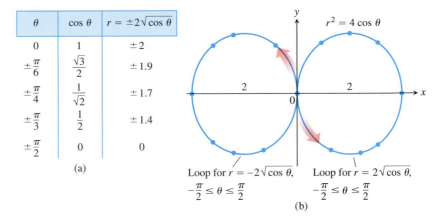

θ	$\cos\theta$	$r = \pm2\sqrt{\cos\theta}$
0	1	±2
$\pm\dfrac{\pi}{6}$	$\dfrac{\sqrt{3}}{2}$	±1.9
$\pm\dfrac{\pi}{4}$	$\dfrac{1}{\sqrt{2}}$	±1.7
$\pm\dfrac{\pi}{3}$	$\dfrac{1}{2}$	±1.4
$\pm\dfrac{\pi}{2}$	0	0

(a)

Loop for $r = -2\sqrt{\cos\theta}$, $-\dfrac{\pi}{2} \le \theta \le \dfrac{\pi}{2}$ Loop for $r = 2\sqrt{\cos\theta}$, $-\dfrac{\pi}{2} \le \theta \le \dfrac{\pi}{2}$

(b)

9.55 The graph of $r^2 = 4\cos\theta$ (Example 2). The arrows show the direction of increasing θ. The values of r in the table are rounded.

Steps for Faster Graphing

Step 1: First graph $r = f(\theta)$ in the *Cartesian $r\theta$-plane* (that is, plot the values of θ on a horizontal axis and the corresponding values of r along a vertical axis).

Step 2: Then use the Cartesian graph as a "table" and guide to sketch the *polar* coordinate graph.

Faster Graphing

One way to graph a polar equation $r = f(\theta)$ is to make a table of (r, θ) values, plot the corresponding points, and connect them in order of increasing θ. This can work well if there are enough points to reveal all the loops and dimples in the graph. Here we describe another method of graphing that is usually quicker and more reliable. The steps are listed at left.

This method is better than simple point plotting because the Cartesian graph, even when hastily drawn, shows at a glance where r is positive, negative, and nonexistent, as well as where r is increasing and decreasing. As examples, we graph $r = 1 + \cos(\theta/2)$ and $r^2 = \sin 2\theta$.

EXAMPLE 3   Graph the curve

$$r = 1 + \cos\frac{\theta}{2}.$$

Solution   We first graph r as a function of θ in the Cartesian $r\theta$-plane. Since the cosine has period 2π, we must let θ run from 0 to 4π to produce the entire graph (Fig. 9.56a, on the following page). The arrows from the θ-axis to the curve give radii for graphing $r = 1 + \cos(\theta/2)$ in the polar plane (Fig. 9.56b, on the following page). ☐

EXAMPLE 4   *A lemniscate*

Graph the curve $r^2 = \sin 2\theta$.

Solution   Here we begin by plotting r^2 (not r) as a function of θ in the Cartesian $r^2\theta$-plane, treating r^2 as a variable that may have negative as well as positive values (Fig. 9.57a, on the following page). We pass from there to the graph of $r = \pm\sqrt{\sin 2\theta}$ in the $r\theta$-plane (Fig. 9.57b, on the following page) and then draw the polar graph (Fig. 9.57c, on the following page). The graph in Fig. 9.57(b) "covers" the final polar graph in Fig. 9.57(c) twice. We could have managed with either loop alone, with the two upper halves, or with the two lower halves. The double covering does no harm, however, and we learn a little more about the behavior of the function this way. ☐

DRAWING LESSON
How to Use Cartesian Graphs to Draw Polar Graphs

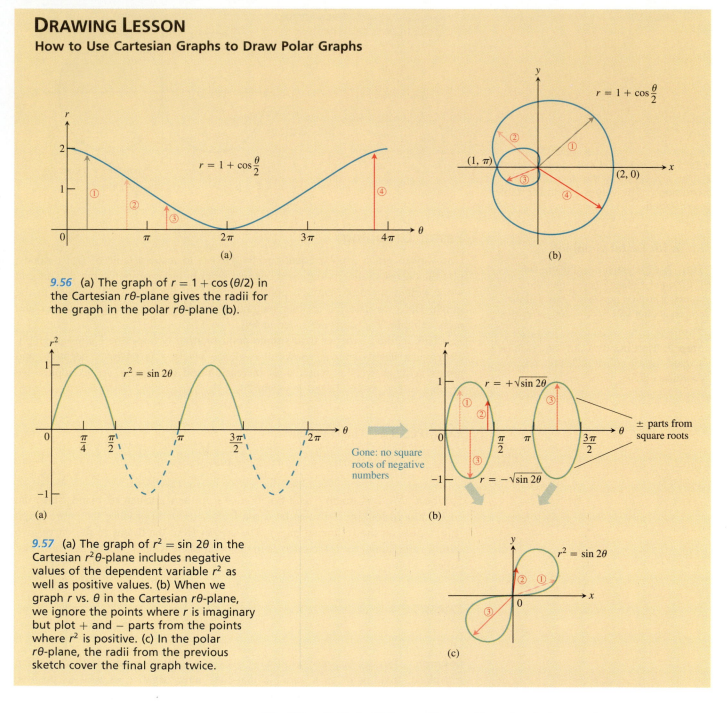

9.56 (a) The graph of $r = 1 + \cos(\theta/2)$ in the Cartesian $r\theta$-plane gives the radii for the graph in the polar $r\theta$-plane (b).

9.57 (a) The graph of $r^2 = \sin 2\theta$ in the Cartesian $r^2\theta$-plane includes negative values of the dependent variable r^2 as well as positive values. (b) When we graph r vs. θ in the Cartesian $r\theta$-plane, we ignore the points where r is imaginary but plot $+$ and $-$ parts from the points where r^2 is positive. (c) In the polar $r\theta$-plane, the radii from the previous sketch cover the final graph twice.

Finding Points Where Polar Graphs Intersect

The fact that we can represent a point in different ways in polar coordinates makes extra care necessary in deciding when a point lies on the graph of a polar equation and in determining the points in which polar graphs intersect. The problem is that a point of intersection may satisfy the equation of one curve with polar coordinates that are different from the ones with which it satisfies the equation of another curve.

Thus, solving the equations of two curves simultaneously may not identify all their points of intersection. The only sure way to identify all the points of intersection is to graph the equations.

EXAMPLE 5   *Deceptive coordinates*

Show that the point $(2, \pi/2)$ lies on the curve $r = 2 \cos 2\theta$.

Solution It may seem at first that the point $(2, \pi/2)$ does not lie on the curve because substituting the given coordinates into the equation gives

$$2 = 2 \cos 2 \left(\frac{\pi}{2} \right) = 2 \cos \pi = -2,$$

which is not a true equality. The magnitude is right, but the sign is wrong. This suggests looking for a pair of coordinates for the given point in which r is negative, for example, $(-2, -(\pi/2))$. If we try these in the equation $r = 2 \cos 2\theta$, we find

$$-2 = 2 \cos 2 \left(-\frac{\pi}{2} \right) = 2(-1) = -2,$$

and the equation is satisfied. The point $(2, \pi/2)$ does lie on the curve.

EXAMPLE 6   *Elusive intersection points*

Find the points of intersection of the curves

$$r^2 = 4 \cos \theta \qquad \text{and} \qquad r = 1 - \cos \theta.$$

Solution In Cartesian coordinates, we can always find the points where two curves cross by solving their equations simultaneously. In polar coordinates, the story is different. Simultaneous solution may reveal some intersection points without revealing others. In this example, simultaneous solution reveals only two of the four intersection points. The others are found by graphing. (Also, see Exercise 49.)

If we substitute $\cos \theta = r^2/4$ in the equation $r = 1 - \cos \theta$, we get

$$r = 1 - \cos \theta = 1 - \frac{r^2}{4}$$

$$4r = 4 - r^2$$

$$r^2 + 4r - 4 = 0$$

$$r = -2 \pm 2\sqrt{2}. \qquad \text{Quadratic formula}$$

The value $r = -2 - 2\sqrt{2}$ has too large an absolute value to belong to either curve. The values of θ corresponding to $r = -2 + 2\sqrt{2}$ are

$$\theta = \cos^{-1}(1 - r) \qquad \text{From } r = 1 - \cos \theta$$

$$= \cos^{-1} \left(1 - \left(2\sqrt{2} - 2 \right) \right) \qquad \text{Set } r = 2\sqrt{2} - 2.$$

$$= \cos^{-1} \left(3 - 2\sqrt{2} \right)$$

$$= \pm 80°. \qquad \text{Rounded to the nearest degree}$$

We have thus identified two intersection points: $(r, \theta) = (2\sqrt{2} - 2, \pm 80°)$.

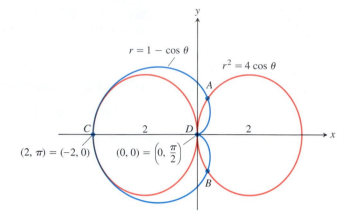

9.58 The four points of intersection of the curves $r = 1 - \cos\theta$ and $r^2 = 4\cos\theta$ (Example 6). Only A and B were found by simultaneous solution. The other two were disclosed by graphing.

If we graph the equations $r^2 = 4\cos\theta$ and $r = 1 - \cos\theta$ together (Fig. 9.58), as we can now do by combining the graphs in Figs. 9.54 and 9.55, we see that the curves also intersect at the point $(2, \pi)$ and the origin. Why weren't the r-values of these points revealed by the simultaneous solution? The answer is that the points $(0, 0)$ and $(2, \pi)$ are not on the curves "simultaneously." They are not reached at the same value of θ. On the curve $r = 1 - \cos\theta$, the point $(2, \pi)$ is reached when $\theta = \pi$. On the curve $r^2 = 4\cos\theta$, it is reached when $\theta = 0$, where it is identified not by the coordinates $(2, \pi)$, which do not satisfy the equation, but by the coordinates $(-2, 0)$, which do. Similarly, the cardioid reaches the origin when $\theta = 0$, but the curve $r^2 = 4\cos\theta$ reaches the origin when $\theta = \pi/2$.   ❏

Technology   *Finding Intersections*   The *simultaneous mode* of a graphing utility gives new meaning to the *simultaneous solution* of a pair of polar coordinate equations. A simultaneous solution occurs only where the two graphs "collide" while they are being drawn simultaneously and not where one graph intersects the other at a point that had been illuminated earlier. The distinction is particularly important in the areas of traffic control or missile defense. For example, in traffic control the only issue is whether two aircraft are in the same place at the same time. The question of whether the curves the craft follow intersect is unimportant.

To illustrate, graph the polar equations

$$r = \cos 2\theta \qquad \text{and} \qquad r = \sin 2\theta$$

in simultaneous mode with $0 \le \theta < 2\pi$, θ Step $= 0.1$, and view dimensions [xmin, xmax] $= [-1, 1]$ by [ymin, ymax] $= [-1, 1]$. *While the graphs are being drawn on the screen*, count the number of times the two graphs illuminate a single pixel simultaneously. Explain why these points of intersection of the two graphs correspond to simultaneous solutions of the equations. (You may find it helpful to slow down the graphing by making θ Step smaller, say 0.05, for example.) In how many points total do the graphs actually intersect?

$r_1 = \sin 2\theta$ and $r_2 = \cos 2\theta$ graphed together.

Exercises 9.7

Symmetries and Polar Graphs

Identify the symmetries of the curves in Exercises 1–12. Then sketch the curves.

1. $r = 1 + \cos \theta$

2. $r = 2 - 2 \cos \theta$

3. $r = 1 - \sin \theta$

4. $r = 1 + \sin \theta$

5. $r = 2 + \sin \theta$

6. $r = 1 + 2 \sin \theta$

7. $r = \sin (\theta/2)$

8. $r = \cos (\theta/2)$

9. $r^2 = \cos \theta$

10. $r^2 = \sin \theta$

11. $r^2 = -\sin \theta$

12. $r^2 = -\cos \theta$

Graph the lemniscates in Exercises 13–16. What symmetries do these curves have?

13. $r^2 = 4 \cos 2\theta$

14. $r^2 = 4 \sin 2\theta$

15. $r^2 = -\sin 2\theta$

16. $r^2 = -\cos 2\theta$

Slopes of Polar Curves

Use Eq. (1) to find the slopes of the curves in Exercises 17–20 at the given points. Sketch the curves along with their tangents at these points.

17. *Cardioid.* $r = -1 + \cos \theta; \quad \theta = \pm\pi/2$

18. *Cardioid.* $r = -1 + \sin \theta; \quad \theta = 0, \pi$

19. *Four-leaved rose.* $r = \sin 2\theta; \quad \theta = \pm\pi/4, \pm 3\pi/4$

20. *Four-leaved rose.* $r = \cos 2\theta; \quad \theta = 0, \pm\pi/2, \pi$

Limaçons

Graph the limaçons in Exercises 21–24. Limaçon ("*lee*-ma-sahn") is Old French for "snail." You will understand the name when you graph the limaçons in Exercise 21. Equations for limaçons have the form $r = a \pm b \cos \theta$ or $r = a \pm b \sin \theta$. There are four basic shapes.

21. *Limaçons with an inner loop*

    a) $\quad r = \dfrac{1}{2} + \cos \theta$
    b) $\quad r = \dfrac{1}{2} + \sin \theta$

22. *Cardioids*

    a) $\quad r = 1 - \cos \theta$
    b) $\quad r = -1 + \sin \theta$

23. *Dimpled limaçons*

    a) $\quad r = \dfrac{3}{2} + \cos \theta$
    b) $\quad r = \dfrac{3}{2} - \sin \theta$

24. *Oval limaçons*

    a) $\quad r = 2 + \cos \theta$
    b) $\quad r = -2 + \sin \theta$

Graphing Polar Inequalities

25. Sketch the region defined by the inequalities $-1 \le r \le 2$ and $-\pi/2 \le \theta \le \pi/2$.

26. Sketch the region defined by the inequalities $0 \le r \le 2 \sec \theta$ and $-\pi/4 \le \theta \le \pi/4$.

In Exercises 27 and 28, sketch the region defined by the inequality.

27. $0 \le r \le 2 - 2 \cos \theta$

28. $0 \le r^2 \le \cos \theta$

Intersections

29. Show that the point $(2, 3\pi/4)$ lies on the curve $r = 2 \sin 2\theta$.

30. Show that $(1/2, 3\pi/2)$ lies on the curve $r = -\sin (\theta/3)$.

Find the points of intersection of the pairs of curves in Exercises 31–38.

31. $r = 1 + \cos \theta, \quad r = 1 - \cos \theta$

32. $r = 1 + \sin \theta, \quad r = 1 - \sin \theta$

33. $r = 2 \sin \theta, \quad r = 2 \sin 2\theta$

34. $r = \cos \theta, \quad r = 1 - \cos \theta$

35. $r = \sqrt{2}, \quad r^2 = 4 \sin \theta$

36. $r^2 = \sqrt{2} \sin \theta, \quad r^2 = \sqrt{2} \cos \theta$

37. $r = 1, \quad r^2 = 2 \sin 2\theta$

38. $r^2 = \sqrt{2} \cos 2\theta, \quad r^2 = \sqrt{2} \sin 2\theta$

▨ **GRAPHER** Find the points of intersection of the pairs of curves in Exercises 39–42.

39. $r^2 = \sin 2\theta, \quad r^2 = \cos 2\theta$

40. $r = 1 + \cos \dfrac{\theta}{2}, \quad r = 1 - \sin \dfrac{\theta}{2}$

41. $r = 1, \quad r = 2 \sin 2\theta$

42. $r = 1, \quad r^2 = 2 \sin 2\theta$

▨ Grapher Explorations

43. Which of the following has the same graph as $r = 1 - \cos \theta$?

    a) $\quad r = -1 - \cos \theta$
    b) $\quad r = 1 + \cos \theta$

Confirm your answer with algebra.

44. Which of the following has the same graph as $r = \cos 2\theta$?

    a) $\quad r = -\sin (2\theta + \pi/2)$
    b) $\quad r = -\cos (\theta/2)$

Confirm your answer with algebra.

45. *A rose within a rose.* Graph the equation $r = 1 - 2 \sin 3\theta$.

46. *The nephroid of Freeth.* Graph the nephroid of Freeth:

$$r = 1 + 2 \sin \frac{\theta}{2}.$$

47. *Roses.* Graph the roses $r = \cos m\theta$ for $m = 1/3, 2, 3,$ and 7.

48. *Spirals.* Polar coordinates are just the thing for defining spirals. Graph the following spirals.

a) $r = \theta$

b) $r = -\theta$

c) *A logarithmic spiral:* $r = e^{\theta/10}$

d) *A hyperbolic spiral:* $r = 8/\theta$

e) *An equilateral hyperbola:* $r = \pm 10/\sqrt{\theta}$

(Use different colors for the two branches.)

Theory and Examples

49. (*Continuation of Example 6.*) The simultaneous solution of the equations

$$r^2 = 4 \cos \theta \tag{2}$$

$$r = 1 - \cos \theta \tag{3}$$

in the text did not reveal the points $(0, 0)$ and $(2, \pi)$ in which their graphs intersected.

a) We could have found the point $(2, \pi)$, however, by replacing the (r, θ) in Eq. (2) by the equivalent $(-r, \theta + \pi)$ to obtain

$$r^2 = 4 \cos \theta$$

$$(-r)^2 = 4 \cos (\theta + \pi) \tag{4}$$

$$r^2 = -4 \cos \theta.$$

Solve Eqs. (3) and (4) simultaneously to show that $(2, \pi)$ is a common solution. (This will still not reveal that the graphs intersect at $(0, 0)$.)

b) The origin is still a special case. (It often is.) Here is one way to handle it: Set $r = 0$ in Eqs. (2) and (3) and solve each equation for a corresponding value of θ. Since $(0, \theta)$ is the origin for *any* θ, this will show that both curves pass through the origin even if they do so for different θ-values.

50. If a curve has any two of the symmetries listed at the beginning of the section, can anything be said about its having or not having the third symmetry? Give reasons for your answer.

*51. Find the maximum width of the petal of the four-leaved rose $r = \cos 2\theta$, which lies along the *x*-axis.

*52. Find the maximum height above the *x*-axis of the cardioid $r = 2(1 + \cos \theta)$.

9.8

Polar Equations for Conic Sections

Polar coordinates are important in astronomy and astronautical engineering because the ellipses, parabolas, and hyperbolas along which satellites, moons, planets, and comets move can all be described with a single relatively simple coordinate equation. We develop that equation here.

Lines

Suppose the perpendicular from the origin to line L meets L at the point $P_0(r_0, \theta_0)$, with $r_0 \geq 0$ (Fig. 9.59). Then, if $P(r, \theta)$ is any other point on L, the points P, P_0, and O are the vertices of a right triangle, from which we can read the relation

$$\frac{r_0}{r} = \cos (\theta - \theta_0)$$

or

$$r \cos (\theta - \theta_0) = r_0.$$

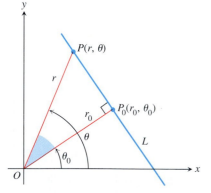

9.59 We can obtain a polar equation for line L by reading the relation $r_0/r = \cos (\theta - \theta_0)$ from triangle OP_0P.

The Standard Polar Equation for Lines

If the point $P_0(r_0, \theta_0)$ is the foot of the perpendicular from the origin to the line L, and $r_0 \geq 0$, then an equation for L is

$$r \cos (\theta - \theta_0) = r_0.$$

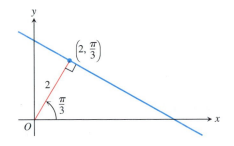

9.60 The standard polar equation of this line is

$$r \cos \left(\theta - \frac{\pi}{3}\right) = 2$$

(Example 1).

EXAMPLE 1 Use the identity $\cos (A - B) = \cos A \cos B + \sin A \sin B$ to find a Cartesian equation for the line in Fig. 9.60.

Solution

$$r \cos \left(\theta - \frac{\pi}{3}\right) = 2$$

$$r \left(\cos \theta \cos \frac{\pi}{3} + \sin \theta \sin \frac{\pi}{3}\right) = 2$$

$$\frac{1}{2} r \cos \theta + \frac{\sqrt{3}}{2} r \sin \theta = 2$$

$$\frac{1}{2} x + \frac{\sqrt{3}}{2} y = 2$$

$$x + \sqrt{3} y = 4 \qquad \square$$

Circles

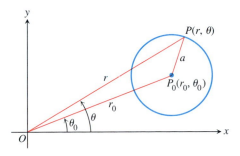

9.61 We can get a polar equation for this circle by applying the Law of Cosines to triangle OP_0P.

To find a polar equation for the circle of radius a centered at $P_0(r_0, \theta_0)$, we let $P(r, \theta)$ be a point on the circle and apply the Law of Cosines to triangle OP_0P (Fig. 9.61). This gives

$$a^2 = r_0^2 + r^2 - 2r_0 r \cos (\theta - \theta_0). \tag{1}$$

If the circle passes through the origin, then $r_0 = a$ and Eq. (1) simplifies to

$$a^2 = a^2 + r^2 - 2ar \cos (\theta - \theta_0) \qquad \text{Eq. (1) with } r_0 = a$$

$$r^2 = 2ar \cos (\theta - \theta_0)$$

$$r = 2a \cos (\theta - \theta_0). \tag{2}$$

If the circle's center lies on the positive x-axis, $\theta_0 = 0$ and Eq. (2) becomes

$$r = 2a \cos \theta. \tag{3}$$

If the center lies on the positive y-axis, $\theta = \pi/2$, $\cos (\theta - \pi/2) = \sin \theta$, and Eq. (2) becomes

$$r = 2a \sin \theta. \tag{4}$$

Equations for circles through the origin centered on the negative x- and y-axes can be obtained from Eqs. (3) and (4) by replacing r with $-r$.

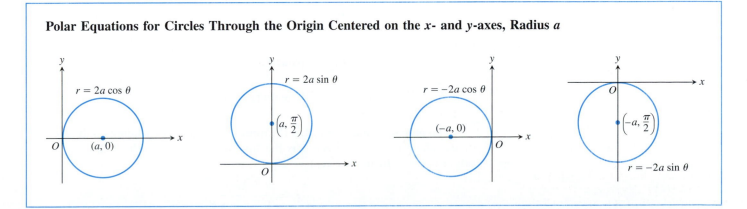

Polar Equations for Circles Through the Origin Centered on the x- and y-axes, Radius a

EXAMPLE 2   *Circles through the origin*

| | **Center** | |
Radius	**(polar coordinates)**	**Equation**
3	$(3, 0)$	$r = 6 \cos \theta$
2	$(2, \pi/2)$	$r = 4 \sin \theta$
1/2	$(-1/2, 0)$	$r = -\cos \theta$
1	$(-1, \pi/2)$	$r = -2 \sin \theta$

Ellipses, Parabolas, and Hyperbolas Unified

To find polar equations for ellipses, parabolas, and hyperbolas, we place one focus at the origin and the corresponding directrix to the right of the origin along the vertical line $x = k$ (Fig. 9.62). This makes

$$PF = r$$

and

$$PD = k - FB = k - r \cos \theta.$$

The conic's focus–directrix equation $PF = e \cdot PD$ then becomes

$$r = e(k - r \cos \theta),$$

which can be solved for r to obtain

$$r = \frac{ke}{1 + e \cos \theta}. \qquad (5)$$

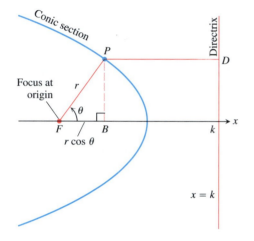

9.62 If a conic section is put in this position, then $PF = r$ and $PD = k - r \cos \theta$.

This equation represents an ellipse if $0 < e < 1$, a parabola if $e = 1$, and a hyperbola if $e > 1$. And there we have it—ellipses, parabolas, and hyperbolas all with the same basic equation.

EXAMPLE 3   *Typical conics from Eq. (5)*

$$e = \frac{1}{2}: \quad \text{ellipse} \quad r = \frac{k}{2 + \cos \theta}$$

$$e = 1: \quad \text{parabola} \quad r = \frac{k}{1 + \cos \theta}$$

$$e = 2: \quad \text{hyperbola} \quad r = \frac{2k}{1 + 2 \cos \theta}$$

You may see variations of Eq. (5) from time to time, depending on the location of the directrix. If the directrix is the line $x = -k$ to the left of the origin (the origin is still a focus), we replace Eq. (5) by

$$r = \frac{ke}{1 - e \cos \theta}.$$

Table 9.4 Equations for conic sections ($e > 0$)

A.
$$r = \frac{ke}{1 + e \cos \theta}$$

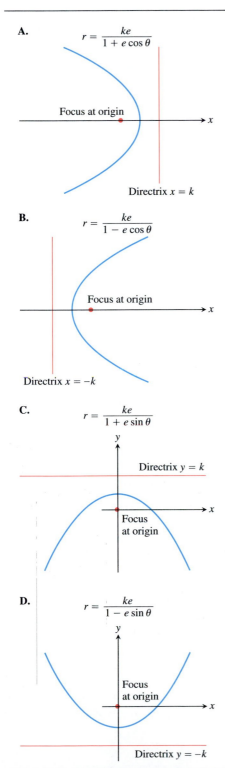

B.
$$r = \frac{ke}{1 - e \cos \theta}$$

C.
$$r = \frac{ke}{1 + e \sin \theta}$$

D.
$$r = \frac{ke}{1 - e \sin \theta}$$

The denominator now has a $(-)$ instead of a $(+)$. If the directrix is either of the lines $y = k$ or $y = -k$, the equations we get have sines in them instead of cosines, as shown in Table 9.4.

EXAMPLE 4   Find an equation for the hyperbola with eccentricity $3/2$ and directrix $x = 2$.

Solution   We use Eq. (A) in Table 9.4 with $k = 2$ and $e = 3/2$ to get

$$r = \frac{2(3/2)}{1 + (3/2)\cos \theta} \quad \text{or} \quad r = \frac{6}{2 + 3 \cos \theta}.$$ ☐

EXAMPLE 5   Find the directrix of the parabola

$$r = \frac{25}{10 + 10 \cos \theta}.$$

Solution   We divide the numerator and denominator by 10 to put the equation in standard form:

$$r = \frac{5/2}{1 + \cos \theta}.$$

This is the equation

$$r = \frac{ke}{1 + e \cos \theta}$$

with $k = 5/2$ and $e = 1$. The equation of the directrix is $x = 5/2$.   ☐

From the ellipse diagram in Fig. 9.63, we see that k is related to the eccentricity e and the semimajor axis a by the equation

$$k = \frac{a}{e} - ea. \tag{6}$$

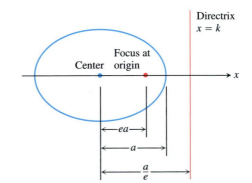

9.63 In an ellipse with semimajor axis a, the focus–directrix distance is $k = (a/e) - ea$, so $ke = a(1 - e^2)$.

From this, we find that $ke = a(1 - e^2)$. Replacing ke in Eq. (5) by $a(1 - e^2)$ gives the standard polar equation for an ellipse.

Ellipse with Eccentricity e and Semimajor Axis a

$$r = \frac{a(1 - e^2)}{1 + e \cos \theta} \tag{7}$$

Notice that when $e = 0$, Eq. (7) becomes $r = a$, which represents a circle.

Equation (7) is the starting point for calculating planetary orbits.

EXAMPLE 6 Find a polar equation for an ellipse with semimajor axis 39.44 AU (astronomical units) and eccentricity 0.25. This is the approximate size of Pluto's orbit around the sun.

Solution We use Eq. (7) with $a = 39.44$ and $e = 0.25$ to find

$$r = \frac{39.44(1 - (0.25)^2)}{1 + 0.25 \cos \theta} = \frac{147.9}{4 + \cos \theta}.$$

At its point of closest approach (perihelion), Pluto is

$$r = \frac{147.9}{4 + 1} = 29.58 \text{ AU}$$

from the sun. At its most distant point (aphelion), Pluto is

$$r = \frac{147.9}{4 - 1} = 49.3 \text{ AU}$$

from the sun (Fig. 9.64).

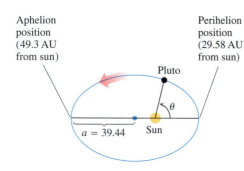

Aphelion position (49.3 AU from sun)

Perihelion position (29.58 AU from sun)

Pluto

θ

$a = 39.44$ Sun

9.64 The orbit of Pluto (Example 6).

EXAMPLE 7 Find the distance from one focus of the ellipse in Example 6 to the associated directrix.

Solution We use Eq. (6) with $a = 39.44$ and $e = 0.25$ to find

$$k = 39.44 \left(\frac{1}{0.25} - 0.25 \right) = 147.9 \text{ AU}.$$

Exercises 9.8

Lines

Find polar and Cartesian equations for the lines in Exercises 1–4.

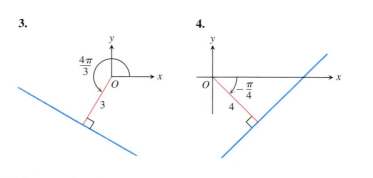

1.

2.

3.

4.

Sketch the lines in Exercises 5–8 and find Cartesian equations for them.

5. $r \cos \left(\theta - \dfrac{\pi}{4} \right) = \sqrt{2}$

6. $r \cos \left(\theta + \dfrac{3\pi}{4} \right) = 1$

7. $r \cos \left(\theta - \dfrac{2\pi}{3} \right) = 3$

8. $r \cos \left(\theta + \dfrac{\pi}{3} \right) = 2$

Find a polar equation in the form $r \cos (\theta - \theta_0) = r_0$ for each of the lines in Exercises 9–12.

9. $\sqrt{2} x + \sqrt{2} y = 6$

10. $\sqrt{3} x - y = 1$

11. $y = -5$

12. $x = -4$

Circles

Find polar equations for the circles in Exercises 13–16.

13.

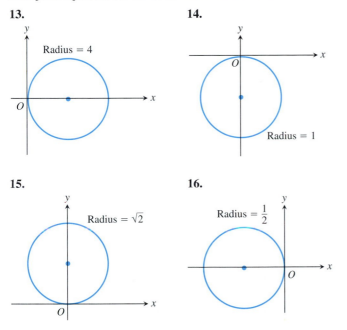

Radius = 4

14.

Radius = 1

15.

Radius = $\sqrt{2}$

16.

Radius = $\dfrac{1}{2}$

Sketch the circles in Exercises 17–20. Give polar coordinates for their centers and identify their radii.

17. $r = 4 \cos \theta$

18. $r = 6 \sin \theta$

19. $r = -2 \cos \theta$

20. $r = -8 \sin \theta$

Find polar equations for the circles in Exercises 21–28. Sketch each circle in the coordinate plane and label it with both its Cartesian and polar equations.

21. $(x - 6)^2 + y^2 = 36$

22. $(x + 2)^2 + y^2 = 4$

23. $x^2 + (y - 5)^2 = 25$

24. $x^2 + (y + 7)^2 = 49$

25. $x^2 + 2x + y^2 = 0$

26. $x^2 - 16x + y^2 = 0$

27. $x^2 + y^2 + y = 0$

28. $x^2 + y^2 - \dfrac{4}{3} y = 0$

Conic Sections from Eccentricities and Directrices

Exercises 29–36 give the eccentricities of conic sections with one focus at the origin, along with the directrix corresponding to that focus. Find a polar equation for each conic section.

29. $e = 1, \quad x = 2$

30. $e = 1, \quad y = 2$

31. $e = 5, \quad y = -6$

32. $e = 2, \quad x = 4$

33. $e = 1/2, \quad x = 1$

34. $e = 1/4, \quad x = -2$

35. $e = 1/5, \quad y = -10$

36. $e = 1/3, \quad y = 6$

Parabolas and Ellipses

Sketch the parabolas and ellipses in Exercises 37–44. Include the directrix that corresponds to the focus at the origin. Label the vertices with appropriate polar coordinates. Label the centers of the ellipses as well.

37. $r = \dfrac{1}{1 + \cos \theta}$

38. $r = \dfrac{6}{2 + \cos \theta}$

39. $r = \dfrac{25}{10 - 5 \cos \theta}$

40. $r = \dfrac{4}{2 - 2 \cos \theta}$

41. $r = \dfrac{400}{16 + 8 \sin \theta}$

42. $r = \dfrac{12}{3 + 3 \sin \theta}$

43. $r = \dfrac{8}{2 - 2 \sin \theta}$

44. $r = \dfrac{4}{2 - \sin \theta}$

Graphing Inequalities

Sketch the regions defined by the inequalities in Exercises 45 and 46.

45. $0 \le r \le 2 \cos \theta$

46. $-3 \cos \theta \le r \le 0$

Grapher Explorations

Graph the lines and conic sections in Exercises 47–56.

47. $r = 3 \sec (\theta - \pi/3)$

48. $r = 4 \sec (\theta + \pi/6)$

49. $r = 4 \sin \theta$

50. $r = -2 \cos \theta$

51. $r = 8/(4 + \cos \theta)$

52. $r = 8/(4 + \sin \theta)$

53. $r = 1/(1 - \sin \theta)$

54. $r = 1/(1 + \cos \theta)$

55. $r = 1/(1 + 2 \sin \theta)$

56. $r = 1/(1 + 2 \cos \theta)$

Theory and Examples

57. *Perihelion and aphelion.* A planet travels about its sun in an ellipse whose semimajor axis has length a.

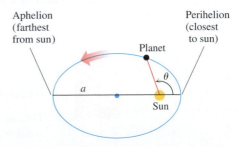

Aphelion (farthest from sun)

Perihelion (closest to sun)

Planet

a

θ

Sun

a) Show that $r = a(1 - e)$ when the planet is closest to the sun and that $r = a(1 + e)$ when the planet is farthest from the sun.

b) Use the data in the table below to find how close each planet in our solar system comes to the sun and how far away each planet gets from the sun.

58. *Planetary orbits.* In Example 6, we found a polar equation for the orbit of Pluto. Use the data in the table below to find polar equations for the orbits of the other planets.

Planet	Semimajor axis (astronomical units)	Eccentricity
Mercury	0.3871	0.2056
Venus	0.7233	0.0068
Earth	1.000	0.0167
Mars	1.524	0.0934
Jupiter	5.203	0.0484
Saturn	9.539	0.0543
Uranus	19.18	0.0460
Neptune	30.06	0.0082
Pluto	39.44	0.2481

59. a) Find Cartesian equations for the curves $r = 4 \sin \theta$ and $r = \sqrt{3} \sec \theta$.

b) Sketch the curves together and label their points of intersection in both Cartesian and polar coordinates.

60. Repeat Exercise 59 for $r = 8 \cos \theta$ and $r = 2 \sec \theta$.

61. Find a polar equation for the parabola with focus $(0, 0)$ and directrix $r \cos \theta = 4$.

62. Find a polar equation for the parabola with focus $(0, 0)$ and directrix $r \cos (\theta - \pi/2) = 2$.

63. a) *The space engineer's formula for eccentricity.* The space engineer's formula for the eccentricity of an elliptical orbit is

$$e = \frac{r_{max} - r_{min}}{r_{max} + r_{min}},$$

where r is the distance from the space vehicle to the attracting focus of the ellipse along which it travels. Why does the formula work?

b) *Drawing ellipses with string.* You have a string with a knot in each end that can be pinned to a drawing board. The string is 10 in. long from the center of one knot to the center of the other. How far apart should the pins be to

use the method illustrated in Fig. 9.5 (Section 9.1) to draw an ellipse of eccentricity 0.2? The resulting ellipse would resemble the orbit of Mercury.

64. *Halley's comet* (See Section 9.2, Example 1.)

a) Write an equation for the orbit of Halley's comet in a coordinate system in which the sun lies at the origin and the other focus lies on the negative x-axis, scaled in astronomical units.

b) How close does the comet come to the sun in astronomical units? in kilometers?

c) What is the farthest the comet gets from the sun in astronomical units? in kilometers?

In Exercises 65–68, find a polar equation for the given curve. In each case, sketch a typical curve.

65. $x^2 + y^2 - 2ay = 0$

66. $y^2 = 4ax + 4a^2$

67. $x \cos \alpha + y \sin \alpha = p$ (α, p constant)

68. $(x^2 + y^2)^2 + 2ax(x^2 + y^2) - a^2 y^2 = 0$

❂ CAS Explorations and Projects

69. Use a CAS to plot the polar equation

$$r = \frac{ke}{1 + e \cos \theta}$$

for various values of k and e, $-\pi \le \theta \le \pi$. Answer the following questions.

a) Take $k = -2$. Describe what happens to the plots as you take e to be 3/4, 1, and 5/4. Repeat for $k = 2$.

b) Take $k = -1$. Describe what happens to the plots as you take e to be 7/6, 5/4, 4/3, 3/2, 2, 3, 5, 10, and 20. Repeat for $e = 1/2, 1/3, 1/4, 1/10,$ and $1/20$.

c) Now keep $e > 0$ fixed and describe what happens as you take k to be $-1, -2, -3, -4,$ and -5. Be sure to look at graphs for parabolas, ellipses, and hyperbolas.

70. Use a CAS to plot the polar ellipse

$$r = \frac{a(1 - e^2)}{1 + e \cos \theta}$$

for various values of $a > 0$ and $0 < e < 1$, $-\pi \le \theta \le \pi$.

a) Take $e = 9/10$. Describe what happens to the plots as you let a equal 1, 3/2, 2, 3, 5, and 10. Repeat with $e = 1/4$.

b) Take $a = 2$. Describe what happens as you take e to be 9/10, 8/10, 7/10, $\ldots$, 1/10, 1/20, and 1/50.

9.9

Integration in Polar Coordinates

This section shows how to calculate areas of plane regions, lengths of curves, and areas of surfaces of revolution in polar coordinates.

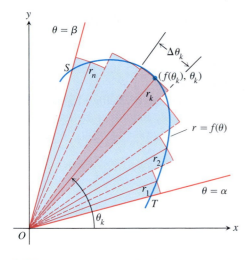

9.65 To derive a formula for the area of region *OTS*, we approximate the region with fan-shaped circular sectors.

Area in the Plane

The region OTS in Fig. 9.65 is bounded by the rays $\theta = \alpha$ and $\theta = \beta$ and the curve $r = f(\theta)$. We approximate the region with n nonoverlapping fan-shaped circular sectors based on a partition P of angle TOS. The typical sector has radius $r_k = f(\theta_k)$ and central angle of radian measure $\Delta\theta_k$. Its area is

$$A_k = \frac{1}{2} r_k^2 \Delta\theta_k = \frac{1}{2} \left(f(\theta_k) \right)^2 \Delta\theta_k.$$

The area of region OTS is approximately

$$\sum_{k=1}^{n} A_k = \sum_{k=1}^{n} \frac{1}{2} \left(f(\theta_k) \right)^2 \Delta\theta_k.$$

If f is continuous, we expect the approximations to improve as $\| P \| \to 0$, and we are led to the following formula for the region's area:

$$A = \lim_{\|P\| \to 0} \sum_{k=1}^{n} \frac{1}{2} \left(f(\theta_k) \right)^2 \Delta\theta_k$$

$$= \int_{\alpha}^{\beta} \frac{1}{2} (f(\theta))^2 \, d\theta.$$

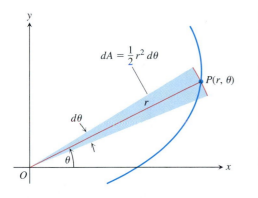

9.66 The area differential *dA*.

Area of the Fan-shaped Region Between the Origin and the Curve $r = f(\theta)$, $\alpha \le \theta \le \beta$

$$A = \int_{\alpha}^{\beta} \frac{1}{2} r^2 \, d\theta.$$

This is the integral of the **area differential** (Fig. 9.66)

$$dA = \frac{1}{2} r^2 \, d\theta.$$

EXAMPLE 1   Find the area of the region in the plane enclosed by the cardioid $r = 2(1 + \cos\theta)$.

Solution   We graph the cardioid (Fig. 9.67) and determine that the radius OP sweeps out the region exactly once as θ runs from 0 to 2π. The area is therefore

$$\int_{\theta=0}^{\theta=2\pi} \frac{1}{2} r^2 \, d\theta = \int_{0}^{2\pi} \frac{1}{2} \cdot 4(1 + \cos\theta)^2 \, d\theta$$

$$= \int_{0}^{2\pi} 2(1 + 2\cos\theta + \cos^2\theta) \, d\theta$$

$$= \int_{0}^{2\pi} \left(2 + 4\cos\theta + 2\,\frac{1 + \cos 2\theta}{2} \right) d\theta$$

$$= \int_{0}^{2\pi} (3 + 4\cos\theta + \cos 2\theta) \, d\theta$$

$$= \left[3\theta + 4\sin\theta + \frac{\sin 2\theta}{2} \right]_{0}^{2\pi} = 6\pi - 0 = 6\pi.$$

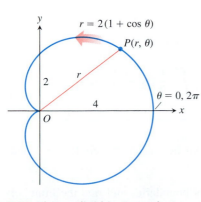

9.67 The cardioid in Example 1.

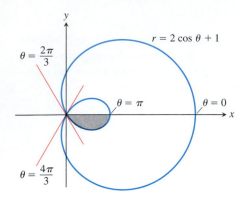

9.68 The limaçon in Example 2.

EXAMPLE 2   Find the area inside the smaller loop of the limaçon

$$r = 2 \cos \theta + 1.$$

Solution   After sketching the curve (Fig. 9.68), we see that the smaller loop is traced out by the point (r, θ) as θ increases from $\theta = 2\pi/3$ to $\theta = 4\pi/3$. Since the curve is symmetric about the x-axis (the equation is unaltered when we replace θ by $-\theta$), we may calculate the area of the shaded half of the inner loop by integrating from $\theta = 2\pi/3$ to $\theta = \pi$. The area we seek will be twice the resulting integral:

$$A = 2 \int_{2\pi/3}^{\pi} \frac{1}{2} r^2 \, d\theta = \int_{2\pi/3}^{\pi} r^2 \, d\theta.$$

Since

$$r^2 = (2 \cos \theta + 1)^2 = 4 \cos^2 \theta + 4 \cos \theta + 1$$

$$= 4 \cdot \frac{1 + \cos 2\theta}{2} + 4 \cos \theta + 1$$

$$= 2 + 2 \cos 2\theta + 4 \cos \theta + 1$$

$$= 3 + 2 \cos 2\theta + 4 \cos \theta,$$

we have

$$A = \int_{2\pi/3}^{\pi} (3 + 2 \cos 2\theta + 4 \cos \theta) \, d\theta$$

$$= \left[3\theta + \sin 2\theta + 4 \sin \theta \right]_{2\pi/3}^{\pi}$$

$$= (3\pi) - \left(2\pi - \frac{\sqrt{3}}{2} + 4 \cdot \frac{\sqrt{3}}{2} \right)$$

$$= \pi - \frac{3\sqrt{3}}{2}. \qquad \square$$

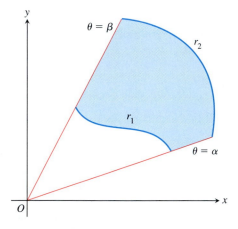

9.69 The area of the shaded region is calculated by subtracting the area of the region between r_1 and the origin from the area of the region between r_2 and the origin.

To find the area of a region like the one in Fig. 9.69, which lies between two polar curves $r_1 = r_1(\theta)$ and $r_2 = r_2(\theta)$ from $\theta = \alpha$ to $\theta = \beta$, we subtract the integral of $(1/2)r_1^2 \, d\theta$ from the integral of $(1/2)r_2^2 \, d\theta$. This leads to the following formula.

Area of the Region $0 \le r_1(\theta) \le r \le r_2(\theta), \quad \alpha \le \theta \le \beta$

$$A = \int_{\alpha}^{\beta} \frac{1}{2} r_2^2 \, d\theta - \int_{\alpha}^{\beta} \frac{1}{2} r_1^2 \, d\theta = \int_{\alpha}^{\beta} \frac{1}{2} (r_2^2 - r_1^2) \, d\theta \qquad (1)$$

EXAMPLE 3   Find the area of the region that lies inside the circle $r = 1$ and outside the cardioid $r = 1 - \cos \theta$.

Solution   We sketch the region to determine its boundaries and find the limits of integration (Fig. 9.70). The outer curve is $r_2 = 1$, the inner curve is $r_1 = 1 - \cos \theta$,

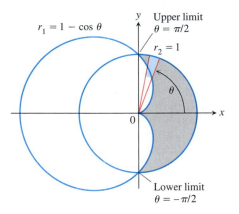

9.70 The region and limits of integration in Example 3.

and θ runs from $-\pi/2$ to $\pi/2$. The area, from Eq. (1), is

$$A = \int_{-\pi/2}^{\pi/2} \frac{1}{2} (r_2^2 - r_1^2) \, d\theta$$

$$= 2 \int_0^{\pi/2} \frac{1}{2} (r_2^2 - r_1^2) \, d\theta \qquad \text{Symmetry}$$

$$= \int_0^{\pi/2} (1 - (1 - 2\cos\theta + \cos^2\theta)) \, d\theta$$

$$= \int_0^{\pi/2} (2\cos\theta - \cos^2\theta) \, d\theta = \int_0^{\pi/2} \left(2\cos\theta - \frac{1 + \cos 2\theta}{2}\right) d\theta$$

$$= \left[2\sin\theta - \frac{\theta}{2} - \frac{\sin 2\theta}{4}\right]_0^{\pi/2} = 2 - \frac{\pi}{4}.$$

The Length of a Curve

We can obtain a polar coordinate formula for the length of a curve $r = f(\theta)$, $\alpha \leq \theta \leq \beta$, by parametrizing the curve as

$$x = r\cos\theta = f(\theta)\cos\theta, \qquad y = r\sin\theta = f(\theta)\sin\theta, \qquad \alpha \leq \theta \leq \beta. \quad \textbf{(2)}$$

The parametric length formula, Eq. (3) from Section 9.5, then gives the length as

$$L = \int_\alpha^\beta \sqrt{\left(\frac{dx}{d\theta}\right)^2 + \left(\frac{dy}{d\theta}\right)^2} \, d\theta.$$

This equation becomes

$$L = \int_\alpha^\beta \sqrt{r^2 + \left(\frac{dr}{d\theta}\right)^2} \, d\theta$$

when Eqs. (2) are substituted for x and y (Exercise 33).

Length of a Curve

If $r = f(\theta)$ has a continuous first derivative for $\alpha \leq \theta \leq \beta$ and if the point $P(r, \theta)$ traces the curve $r = f(\theta)$ exactly once as θ runs from α to β, then the length of the curve is

$$L = \int_\alpha^\beta \sqrt{r^2 + \left(\frac{dr}{d\theta}\right)^2} \, d\theta. \qquad \textbf{(3)}$$

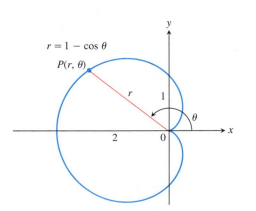

9.71 Example 4 calculates the length of this cardioid.

EXAMPLE 4   Find the length of the cardioid $r = 1 - \cos\theta$.

Solution   We sketch the cardioid to determine the limits of integration (Fig. 9.71). The point $P(r, \theta)$ traces the curve once, counterclockwise as θ runs from 0 to 2π, so these are the values we take for α and β.

With

$$r = 1 - \cos\theta, \qquad \frac{dr}{d\theta} = \sin\theta,$$

we have

$$r^2 + \left(\frac{dr}{d\theta}\right)^2 = (1 - \cos\theta)^2 + (\sin\theta)^2$$

$$= 1 - 2\cos\theta + \underbrace{\cos^2\theta + \sin^2\theta}_{1} = 2 - 2\cos\theta$$

and

$$L = \int_\alpha^\beta \sqrt{r^2 + \left(\frac{dr}{d\theta}\right)^2}\,d\theta = \int_0^{2\pi} \sqrt{2 - 2\cos\theta}\,d\theta$$

$$= \int_0^{2\pi} \sqrt{4\sin^2\frac{\theta}{2}}\,d\theta \qquad 1 - \cos\theta = 2\sin^2\frac{\theta}{2}$$

$$= \int_0^{2\pi} 2\left|\sin\frac{\theta}{2}\right|\,d\theta$$

$$= \int_0^{2\pi} 2\sin\frac{\theta}{2}\,d\theta \qquad \sin\frac{\theta}{2} \geq 0 \quad \text{for} \quad 0 \leq \theta \leq 2\pi$$

$$= \left[-4\cos\frac{\theta}{2}\right]_0^{2\pi} = 4 + 4 = 8.$$

The Area of a Surface of Revolution

To derive polar coordinate formulas for the area of a surface of revolution, we parametrize the curve $r = f(\theta)$, $\alpha \leq \theta \leq \beta$, with Eqs. (2) and apply the surface area equations in Section 9.5.

Area of a Surface of Revolution

If $r = f(\theta)$ has a continuous first derivative for $\alpha \leq \theta \leq \beta$ and if the point $P(r, \theta)$ traces the curve $r = f(\theta)$ exactly once as θ runs from α to β, then the areas of the surfaces generated by revolving the curve about the x- and y-axes are given by the following formulas:

1. Revolution about the x-axis $(y \geq 0)$:
$$S = \int_\alpha^\beta 2\pi r \sin\theta \sqrt{r^2 + \left(\frac{dr}{d\theta}\right)^2}\,d\theta \qquad (4)$$

2. Revolution about the y-axis $(x \geq 0)$:
$$S = \int_\alpha^\beta 2\pi r \cos\theta \sqrt{r^2 + \left(\frac{dr}{d\theta}\right)^2}\,d\theta \qquad (5)$$

EXAMPLE 5 Find the area of the surface generated by revolving the right-hand loop of the lemniscate $r^2 = \cos 2\theta$ about the y-axis.

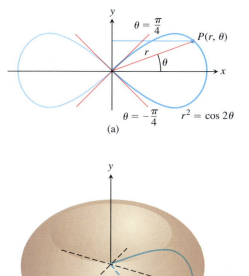

(a)

(b)

9.72 The right-hand half of a lemniscate (a) is revolved about the y-axis to generate a surface (b), whose area is calculated in Example 5.

Solution We sketch the loop to determine the limits of integration (Fig. 9.72). The point $P(r, \theta)$ traces the curve once, counterclockwise as θ runs from $-\pi/4$ to $\pi/4$, so these are the values we take for α and β.

We evaluate the area integrand in Eq. (4) in stages. First,

$$2\pi r \cos \theta \sqrt{r^2 + \left(\frac{dr}{d\theta}\right)^2} = 2\pi \cos \theta \sqrt{r^4 + \left(r\frac{dr}{d\theta}\right)^2}. \tag{6}$$

Next, $r^2 = \cos 2\theta$, so

$$2r\frac{dr}{d\theta} = -2 \sin 2\theta$$

$$r\frac{dr}{d\theta} = -\sin 2\theta$$

$$\left(r\frac{dr}{d\theta}\right)^2 = \sin^2 2\theta.$$

Finally, $r^4 = (r^2)^2 = \cos^2 2\theta$, so the square root on the right-hand side of Eq. (6) simplifies to

$$\sqrt{r^4 + \left(r\frac{dr}{d\theta}\right)^2} = \sqrt{\cos^2 2\theta + \sin^2 2\theta} = 1.$$

All together, we have

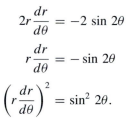

$$S = \int_\alpha^\beta 2\pi r \cos \theta \sqrt{r^2 + \left(\frac{dr}{d\theta}\right)^2} \, d\theta \qquad \text{Eq. (4)}$$

$$= \int_{-\pi/4}^{\pi/4} 2\pi \cos \theta \cdot (1) \, d\theta$$

$$= 2\pi \left[\sin \theta\right]_{-\pi/4}^{\pi/4}$$

$$= 2\pi \left[\frac{\sqrt{2}}{2} + \frac{\sqrt{2}}{2}\right] = 2\pi \sqrt{2}.$$

Exercises 9.9

Areas Inside Polar Curves

Find the areas of the regions in Exercises 1–6.

1. Inside the oval limaçon $r = 4 + 2 \cos \theta$

2. Inside the cardioid $r = a(1 + \cos \theta), \quad a > 0$

3. Inside one leaf of the four-leaved rose $r = \cos 2\theta$

4. Inside the lemniscate $r^2 = 2a^2 \cos 2\theta, \quad a > 0$

5. Inside one loop of the lemniscate $r^2 = 4 \sin 2\theta$

6. Inside the six-leaved rose $r^2 = 2 \sin 3\theta$

Areas Shared by Polar Regions

Find the areas of the regions in Exercises 7–16.

7. Shared by the circles $r = 2 \cos \theta$ and $r = 2 \sin \theta$

8. Shared by the circles $r = 1$ and $r = 2 \sin \theta$

9. Shared by the circle $r = 2$ and the cardioid $r = 2(1 - \cos \theta)$

10. Shared by the cardioids $r = 2(1 + \cos \theta)$ and $r = 2(1 - \cos \theta)$

11. Inside the lemniscate $r^2 = 6 \cos 2\theta$ and outside the circle $r = \sqrt{3}$

12. Inside the circle $r = 3a \cos \theta$ and outside the cardioid $r = a(1 + \cos \theta)$, $a > 0$

13. Inside the circle $r = -2 \cos \theta$ and outside the circle $r = 1$

14. **a)** Inside the outer loop of the limaçon $r = 2 \cos \theta + 1$ (See Fig. 9.68.)

    b) Inside the outer loop and outside the inner loop of the limaçon $r = 2 \cos \theta + 1$

15. Inside the circle $r = 6$ above the line $r = 3 \csc \theta$

16. Inside the lemniscate $r^2 = 6 \cos 2\theta$ to the right of the line $r = (3/2) \sec \theta$

17. **a)** Find the area of the shaded region in Fig. 9.73.

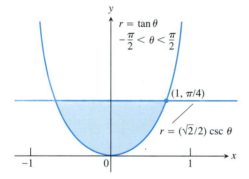

9.73 The region in Exercise 17.

b) It looks as if the graph of $r = \tan \theta$, $-\pi/2 < \theta < \pi/2$, could be asymptotic to the lines $x = 1$ and $x = -1$. Is it? Give reasons for your answer.

18. The area of the region that lies inside the cardioid curve $r = \cos \theta + 1$ and outside the circle $r = \cos \theta$ is not

$$\frac{1}{2} \int_0^{2\pi} [(\cos \theta + 1)^2 - \cos^2 \theta] \, d\theta = \pi.$$

Why not? What *is* the area? Give reasons for your answers.

Lengths of Polar Curves

Find the lengths of the curves in Exercises 19–27.

19. The spiral $r = \theta^2$, $0 \le \theta \le \sqrt{5}$

20. The spiral $r = e^\theta/\sqrt{2}$, $0 \le \theta \le \pi$

21. The cardioid $r = 1 + \cos \theta$

22. The curve $r = a \sin^2 (\theta/2)$, $0 \le \theta \le \pi$, $a > 0$

23. The parabolic segment $r = 6/(1 + \cos \theta)$, $0 \le \theta \le \pi/2$

24. The parabolic segment $r = 2/(1 - \cos \theta)$, $\pi/2 \le \theta \le \pi$

25. The curve $r = \cos^3 (\theta/3)$, $0 \le \theta \le \pi/4$

26. The curve $r = \sqrt{1 + \sin 2\theta}$, $0 \le \theta \le \pi \sqrt{2}$

27. The curve $r = \sqrt{1 + \cos 2\theta}$, $0 \le \theta \le \pi \sqrt{2}$

28. *Circumferences of circles.* As usual, when faced with a new formula, it is a good idea to try it on familiar objects to be sure it gives results consistent with past experience. Use the length

formula in Eq. (3) to calculate the circumferences of the following circles ($a > 0$):

a) $r = a$ **b)** $r = a \cos \theta$ **c)** $r = a \sin \theta$

Surface Area

Find the areas of the surfaces generated by revolving the curves in Exercises 29–32 about the indicated axes.

29. $r = \sqrt{\cos 2\theta}$, $0 \le \theta \le \pi/4$, y-axis

30. $r = \sqrt{2} e^{\theta/2}$, $0 \le \theta \le \pi/2$, x-axis

31. $r^2 = \cos 2\theta$, x-axis

32. $r = 2a \cos \theta, a > 0$, y-axis

Theory and Examples

33. *The length of the curve $r = f(\theta)$, $\alpha \le \theta \le \beta$.* Assuming that the necessary derivatives are continuous, show how the substitutions

$$x = f(\theta) \cos \theta, \quad y = f(\theta) \sin \theta$$

(Eqs. 2 in the text) transform

$$L = \int_\alpha^\beta \sqrt{\left(\frac{dx}{d\theta}\right)^2 + \left(\frac{dy}{d\theta}\right)^2} \, d\theta$$

into

$$L = \int_\alpha^\beta \sqrt{r^2 + \left(\frac{dr}{d\theta}\right)^2} \, d\theta.$$

34. *Average value.* If f is continuous, the average value of the polar coordinate r over the curve $r = f(\theta)$, $\alpha \le \theta \le \beta$, with respect to θ is given by the formula

$$r_{av} = \frac{1}{\beta - \alpha} \int_\alpha^\beta f(\theta) \, d\theta.$$

Use this formula to find the average value of r with respect to θ over the following curves ($a > 0$).

a) The cardioid $r = a(1 - \cos \theta)$

b) The circle $r = a$

c) The circle $r = a \cos \theta$, $-\pi/2 \le \theta \le \pi/2$

35. *$r = f(\theta)$ vs. $r = 2 f(\theta)$.* Can anything be said about the relative lengths of the curves $r = f(\theta)$, $\alpha \le \theta \le \beta$, and $r = 2 f(\theta)$, $\alpha \le \theta \le \beta$? Give reasons for your answer.

36. *$r = f(\theta)$ vs. $r = 2 f(\theta)$.* The curves $r = f(\theta)$, $\alpha \le \theta \le \beta$, and $r = 2 f(\theta)$, $\alpha \le \theta \le \beta$, are revolved about the x-axis to generate surfaces. Can anything be said about the relative areas of these surfaces? Give reasons for your answer.

Centroids of Fan-Shaped Regions

Since the centroid of a triangle is located on each median, two-thirds of the way from the vertex to the opposite base, the lever arm for the moment about the x-axis of the thin triangular region in Fig. 9.74 is about $(2/3)r \sin \theta$. Similarly, the lever arm for the moment of the triangular region about the y-axis is about $(2/3)r \cos \theta$. These approximations improve as $\Delta\theta \to 0$ and lead to the following formulas

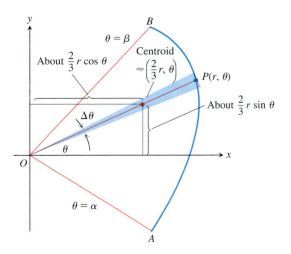

for the coordinates of the centroid of region AOB:

$$\bar{x} = \frac{\int \frac{2}{3}r\cos\theta \cdot \frac{1}{2}r^2\,d\theta}{\int \frac{1}{2}r^2\,d\theta} = \frac{\frac{2}{3}\int r^3\cos\theta\,d\theta}{\int r^2\,d\theta},$$

$$\bar{y} = \frac{\int \frac{2}{3}r\sin\theta \cdot \frac{1}{2}r^2\,d\theta}{\int \frac{1}{2}r^2\,d\theta} = \frac{\frac{2}{3}\int r^3\sin\theta\,d\theta}{\int r^2\,d\theta},$$

with limits $\theta = \alpha$ to $\theta = \beta$ on all integrals.

37. Find the centroid of the region enclosed by the cardioid $r = a(1 + \cos\theta)$.

38. Find the centroid of the semicircular region $0 \le r \le a$, $0 \le \theta \le \pi$.

9.74 The moment of the thin triangular sector about the x-axis is approximately

$$\frac{2}{3}r\sin\theta\,dA = \frac{2}{3}r\sin\theta \cdot \frac{1}{2}r^2\,d\theta = \frac{1}{3}r^3\sin\theta\,d\theta.$$

CHAPTER   9   QUESTIONS TO GUIDE YOUR REVIEW

1. What is a parabola? What are the Cartesian equations for parabolas whose vertices lie at the origin and whose foci lie on the coordinate axes? How can you find the focus and directrix of such a parabola from its equation?

2. What is an ellipse? What are the Cartesian equations for ellipses centered at the origin with foci on one of the coordinate axes? How can you find the foci, vertices, and directrices of such an ellipse from its equation?

3. What is a hyperbola? What are the Cartesian equations for hyperbolas centered at the origin with foci on one of the coordinate axes? How can you find the foci, vertices, and directrices of such an ellipse from its equation?

4. What is the eccentricity of a conic section? How can you classify conic sections by eccentricity? How are an ellipse's shape and eccentricity related?

5. Explain the equation $PF = e \cdot PD$.

6. What is a quadratic curve in the xy-plane? Give examples of degenerate and nondegenerate quadratic curves.

7. How can you find a Cartesian coordinate system in which the new equation for a conic section in the plane has no xy-term? Give an example.

8. How can you tell what kind of graph to expect from a quadratic equation in x and y? Give examples.

9. What is a parametrized curve in the xy-plane? If you find a Cartesian equation for the path of a particle whose motion in the plane is described parametrically, what kind of match can you expect between the Cartesian equation's graph and the path of motion? Give examples.

10. What are some typical parametrizations for conic sections?

11. What is a cycloid? What are typical parametric equations for cycloids? What physical properties account for the importance of cycloids?

12. What is the formula for the slope dy/dx of a parametrized curve $x = f(t)$, $y = g(t)$? When does the formula apply? When can you expect to be able to find d^2y/dx^2 as well? Give examples.

13. How do you find the length of a smooth parametrized curve $x = f(t)$, $y = g(t)$, $a \le t \le b$? What does smoothness have to do with length? What else do you need to know about the parametrization in order to find the curve's length? Give examples.

14. Under what conditions can you find the area of the surface generated by revolving a curve $x = f(t)$, $y = g(t)$, $a \le t \le b$, about the x-axis? the y-axis? Give examples.

15. How do you find the centroid of a smooth parametrized curve $x = f(t)$, $y = g(t)$, $a \le t \le b$? Give an example.

16. What are polar coordinates? What equations relate polar coordinates to Cartesian coordinates? Why might you want to change from one coordinate system to the other?

17. What consequence does the lack of uniqueness of polar coordinates have for graphing? Give an example.

18. How do you graph equations in polar coordinates? Include in your discussion symmetry, slope, behavior at the origin, and the use of Cartesian graphs. Give examples.

19. What are the standard equations for lines and conic sections in polar coordinates? Give examples.

20. How do you find the area of a region $0 \leq r_1(\theta) \leq r \leq r_2(\theta)$, $\alpha \leq \theta \leq \beta$, in the polar coordinate plane? Give examples.

21. Under what conditions can you find the length of a curve $r = f(\theta)$, $\alpha \leq \theta \leq \beta$, in the polar coordinate plane? Give an example of a typical calculation.

22. Under what conditions can you find the area of the surface generated by revolving a curve $r = f(\theta)$, $\alpha \leq \theta \leq \beta$, about the x-axis? the y-axis? Give examples of typical calculations.

CHAPTER 9 PRACTICE EXERCISES

Graphing Conic Sections

Sketch the parabolas in Exercises 1–4. Include the focus and directrix in each sketch.

1. $x^2 = -4y$

2. $x^2 = 2y$

3. $y^2 = 3x$

4. $y^2 = -(8/3)x$

Find the eccentricities of the ellipses and hyperbolas in Exercises 5–8. Sketch each conic section. Include the foci, vertices, and asymptotes (as appropriate) in your sketch.

5. $16x^2 + 7y^2 = 112$

6. $x^2 + 2y^2 = 4$

7. $3x^2 - y^2 = 3$

8. $5y^2 - 4x^2 = 20$

Shifting Conic Sections

Exercises 9–14 give equations for conic sections and tell how many units up or down and to the right or left each curve is to be shifted. Find an equation for the new conic section and find the new foci, vertices, centers, and asymptotes, as appropriate. If the curve is a parabola, find the new directrix as well.

9. $x^2 = -12y$, right 2, up 3

10. $y^2 = 10x$, left 1/2, down 1

11. $\dfrac{x^2}{9} + \dfrac{y^2}{25} = 1$, left 3, down 5

12. $\dfrac{x^2}{169} + \dfrac{y^2}{144} = 1$, right 5, up 12

13. $\dfrac{y^2}{8} - \dfrac{x^2}{2} = 1$, right 2, up $2\sqrt{2}$

14. $\dfrac{x^2}{36} - \dfrac{y^2}{64} = 1$, left 10, down 3

Identifying Conic Sections

Identify the conic sections in Exercises 15–22 and find their foci, vertices, centers, and asymptotes (as appropriate). If the curve is a parabola, find its directrix as well.

15. $x^2 - 4x - 4y^2 = 0$

16. $4x^2 - y^2 + 4y = 8$

17. $y^2 - 2y + 16x = -49$

18. $x^2 - 2x + 8y = -17$

19. $9x^2 + 16y^2 + 54x - 64y = -1$

20. $25x^2 + 9y^2 - 100x + 54y = 44$

21. $x^2 + y^2 - 2x - 2y = 0$

22. $x^2 + y^2 + 4x + 2y = 1$

Using the Discriminant

What conic sections or degenerate cases do the equations in Exercises 23–28 represent? Give a reason for your answer in each case.

23. $x^2 + xy + y^2 + x + y + 1 = 0$

24. $x^2 + 4xy + 4y^2 + x + y + 1 = 0$

25. $x^2 + 3xy + 2y^2 + x + y + 1 = 0$

26. $x^2 + 2xy - 2y^2 + x + y + 1 = 0$

27. $x^2 - 2xy + y^2 = 0$

28. $x^2 - 3xy + 4y^2 = 0$

Rotating Conic Sections

Identify the conic sections in Exercises 29–32. Then rotate the coordinate axes to find a new equation for the conic section that has no cross product term. (The new equations will vary with the size and direction of the rotations used.)

29. $2x^2 + xy + 2y^2 - 15 = 0$

30. $3x^2 + 2xy + 3y^2 = 19$

31. $x^2 + 2\sqrt{3}xy - y^2 + 4 = 0$

32. $x^2 - 3xy + y^2 = 5$

Identifying Parametric Equations in the Plane

Exercises 33–38 give parametric equations and parameter intervals for the motion of a particle in the xy-plane. Identify the particle's path by finding a Cartesian equation for it. Graph the Cartesian equation and indicate the direction of motion and the portion traced by the particle.

33. $x = t/2, \quad y = t + 1; \quad -\infty < t < \infty$

34. $x = \sqrt{t}, \quad y = 1 - \sqrt{t}; \quad t \geq 0$

35. $x = (1/2) \tan t, \quad y = (1/2) \sec t; \quad -\pi/2 < t < \pi/2$

36. $x = -2 \cos t, \quad y = 2 \sin t; \quad 0 \leq t \leq \pi$

37. $x = -\cos t, \quad y = \cos^2 t; \quad 0 \leq t \leq \pi$

38. $x = 4 \cos t, \quad y = 9 \sin t; \quad 0 \leq t \leq 2\pi$

Finding Parametric Equations and Tangent Lines

39. Find parametric equations and a parameter interval for the motion of a particle in the xy-plane that traces the ellipse $16x^2 + 9y^2 = 144$ once counterclockwise. (There are many ways to do this, so your answer may not be the same as the one in the back of the book.)

40. Find parametric equations and a parameter interval for the motion of a particle that starts at the point $(-2, 0)$ in the xy-plane and traces the circle $x^2 + y^2 = 4$ three times clockwise. (There are many ways to do this.)

In Exercises 41 and 42, find an equation for the line in the xy-plane that is tangent to the curve at the point corresponding to the given value of t. Also, find the value of d^2y/dx^2 at this point.

41. $x = (1/2) \tan t, \quad y = (1/2) \sec t; \quad t = \pi/3$

42. $x = 1 + 1/t^2, \quad y = 1 - 3/t; \quad t = 2$

Lengths of Parametrized Curves

Find the lengths of the curves in Exercises 43 and 44.

43. $x = e^{2t} - \dfrac{t}{8}, \quad y = e^t; \quad 0 \leq t \leq \ln 2$

44. The enclosed loop in Fig. 9.75.

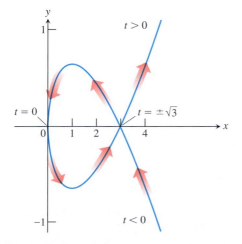

9.75 Exercise 44 refers to the curve $x = t^2$, $y = (t^3/3) - t$ shown here. The loop starts at $t = -\sqrt{3}$ and ends at $t = \sqrt{3}$.

Surface Areas

Find the areas of the surfaces generated by revolving the curves in Exercises 45 and 46 about the indicated axes.

45. $x = t^2/2, \quad y = 2t, \quad 0 \leq t \leq \sqrt{5}; \quad x$-axis

46. $x = t^2 + 1/(2t), \quad y = 4\sqrt{t}, \quad 1/\sqrt{2} \leq t \leq 1; \quad y$-axis

Graphs in the Polar Plane

Sketch the regions defined by the polar coordinate inequalities in Exercises 47 and 48.

47. $0 \leq r \leq 6 \cos \theta$

48. $-4 \sin \theta \leq r \leq 0$

Match each graph in Exercises 49–56 with the appropriate equation (a)–(l). There are more equations than graphs, so some equations will not be matched.

a) $r = \cos 2\theta$ **b)** $r \cos \theta = 1$

c) $r = \dfrac{6}{1 - 2 \cos \theta}$ **d)** $r = \sin 2\theta$

e) $r = \theta$ **f)** $r^2 = \cos 2\theta$

g) $r = 1 + \cos \theta$ **h)** $r = 1 - \sin \theta$

i) $r = \dfrac{2}{1 - \cos \theta}$ **j)** $r^2 = \sin 2\theta$

k) $r = -\sin \theta$ **l)** $r = 2 \cos \theta + 1$

49. Four-leaved rose

50. Spiral

51. Limaçon

52. Lemniscate

53. Circle

54. Cardioid

55. Parabola

56. Lemniscate

Intersections of Graphs in the Polar Plane

Find the points of intersection of the curves given by the polar coordinate equations in Exercises 57–64.

57. $r = \sin \theta, \quad r = 1 + \sin \theta$

58. $r = \cos \theta, \quad r = 1 - \cos \theta$

59. $r = 1 + \cos \theta, \quad r = 1 - \cos \theta$

60. $r = 1 + \sin \theta, \quad r = 1 - \sin \theta$

61. $r = 1 + \sin \theta, \quad r = -1 + \sin \theta$

62. $r = 1 + \cos \theta, \quad r = -1 + \cos \theta$

63. $r = \sec \theta, \quad r = 2 \sin \theta$

64. $r = -2 \csc \theta, \quad r = -4 \cos \theta$

Tangent Lines in the Polar Plane

In Exercises 65 and 66, find equations for the lines that are tangent to the polar coordinate curves at the origin.

65. The lemniscate $r^2 = \cos 2\theta$

66. The limaçon $r = 2 \cos \theta + 1$

67. Find polar coordinate equations for the lines that are tangent to the tips of the petals of the four-leaved rose $r = \sin 2\theta$.

68. Find polar coordinate equations for the lines that are tangent to the cardioid $r = 1 + \sin \theta$ at the points where it crosses the x-axis.

Polar to Cartesian Equations

Sketch the lines in Exercises 69–74. Also, find a Cartesian equation for each line.

69. $r \cos \left(\theta + \dfrac{\pi}{3} \right) = 2 \sqrt{3}$

70. $r \cos \left(\theta - \dfrac{3\pi}{4} \right) = \dfrac{\sqrt{2}}{2}$

71. $r = 2 \sec \theta$

72. $r = -\sqrt{2} \sec \theta$

73. $r = -(3/2) \csc \theta$

74. $r = \left(3\sqrt{3} \right) \csc \theta$

Find Cartesian equations for the circles in Exercises 75–78. Sketch each circle in the coordinate plane and label it with both its Cartesian and polar equations.

75. $r = -4 \sin \theta$

76. $r = 3 \sqrt{3} \sin \theta$

77. $r = 2 \sqrt{2} \cos \theta$

78. $r = -6 \cos \theta$

Cartesian to Polar Equations

Find polar equations for the circles in Exercises 79–82. Sketch each circle in the coordinate plane and label it with both its Cartesian and polar equations.

79. $x^2 + y^2 + 5y = 0$

80. $x^2 + y^2 - 2y = 0$

81. $x^2 + y^2 - 3x = 0$

82. $x^2 + y^2 + 4x = 0$

Conic Sections in Polar Coordinates

Sketch the conic sections whose polar coordinate equations are given in Exercises 83–86. Give polar coordinates for the vertices and, in the case of ellipses, for the centers as well.

83. $r = \dfrac{2}{1 + \cos \theta}$

84. $r = \dfrac{8}{2 + \cos \theta}$

85. $r = \dfrac{6}{1 - 2 \cos \theta}$

86. $r = \dfrac{12}{3 + \sin \theta}$

Exercises 87–90 give the eccentricities of conic sections with one focus at the origin of the polar coordinate plane, along with the directrix for that focus. Find a polar equation for each conic section.

87. $e = 2, \quad r \cos \theta = 2$

88. $e = 1, \quad r \cos \theta = -4$

89. $e = 1/2, \quad r \sin \theta = 2$

90. $e = 1/3, \quad r \sin \theta = -6$

Area, Length, and Surface Area in the Polar Plane

Find the areas of the regions in the polar coordinate plane described in Exercises 91–94.

91. Enclosed by the limaçon $r = 2 - \cos \theta$

92. Enclosed by one leaf of the three-leaved rose $r = \sin 3\theta$

93. Inside the "figure eight" $r = 1 + \cos 2\theta$ and outside the circle $r = 1$

94. Inside the cardioid $r = 2(1 + \sin \theta)$ and outside the circle $r = 2 \sin \theta$

Find the lengths of the curves given by the polar coordinate equations in Exercises 95–98.

95. $r = -1 + \cos \theta$

96. $r = 2 \sin \theta + 2 \cos \theta, \quad 0 \le \theta \le \pi/2$

97. $r = 8 \sin^3 (\theta/3), \quad 0 \le \theta \le \pi/4$

98. $r = \sqrt{1 + \cos 2\theta}, \quad -\pi/2 \le \theta \le \pi/2$

Find the areas of the surfaces generated by revolving the polar coordinate curves in Exercises 99 and 100 about the indicated axes.

99. $r = \sqrt{\cos 2\theta}, \quad 0 \le \theta \le \pi/4, \quad x$-axis

100. $r^2 = \sin 2\theta, \quad y$-axis

Theory and Examples

101. Find the volume of the solid generated by revolving the region enclosed by the ellipse $9x^2 + 4y^2 = 36$ about (a) the x-axis, (b) the y-axis.

102. The "triangular" region in the first quadrant bounded by the x-axis, the line $x = 4$, and the hyperbola $9x^2 - 4y^2 = 36$ is revolved about the x-axis to generate a solid. Find the volume of the solid.

103. A ripple tank is made by bending a strip of tin around the perimeter of an ellipse for the wall of the tank and soldering a flat bottom onto this. An inch or two of water is put in the tank and you drop a marble into it, right at one focus of the ellipse. Ripples radiate outward through the water, reflect from the strip around the edge of the tank, and a few seconds later a drop of water spurts up at the second focus. Why?

104. *LORAN.* A radio signal was sent simultaneously from towers A and B, located several hundred miles apart on the northern California coast. A ship offshore received the signal from A 1400 microseconds before receiving the signal from B. Assuming that the signals traveled at the rate of 980 ft/microsecond, what can be said about the location of the ship relative to the two towers?

105. On a level plane, at the same instant, you hear the sound of a rifle and that of the bullet hitting the target. What can be said about your location relative to the rifle and target?

106. *Archimedes spirals.* The graph of an equation of the form $r = a\theta$, where a is a nonzero constant, is called an **Archimedes spiral.** Is there anything special about the widths between the successive turns of such a spiral?

107. a) Show that the equations $x = r \cos\theta, y = r \sin\theta$ transform the polar equation

$$r = \frac{k}{1 + e \cos\theta}$$

into the Cartesian equation

$$(1 - e^2)x^2 + y^2 + 2kex - k^2 = 0.$$

b) Then apply the criteria of Section 9.3 to show that

$$e = 0 \quad \Rightarrow \quad \text{circle}$$
$$0 < e < 1 \quad \Rightarrow \quad \text{ellipse}$$
$$e = 1 \quad \Rightarrow \quad \text{parabola}$$
$$e > 1 \quad \Rightarrow \quad \text{hyperbola.}$$

108. *A satellite orbit.* A satellite is in an orbit that passes over the North and South Poles of the earth. When it is over the South Pole it is at the highest point of its orbit, 1000 miles above the earth's surface. Above the North Pole it is at the lowest point of its orbit, 300 miles above the earth's surface.

a) Assuming that the orbit is an ellipse with one focus at the center of the earth, find its eccentricity. (Take the diameter of the earth to be 8000 miles.)

b) Using the north–south axis of the earth as the x-axis and the center of the earth as origin, find a polar equation for the orbit.

The Angle Between the Radius Vector and the Tangent Line to a Polar Coordinate Curve

In Cartesian coordinates, when we want to discuss the direction of a curve at a point, we use the angle ϕ measured counterclockwise from the positive x-axis to the tangent line. In polar coordinates, it is more convenient to calculate the angle ψ from the *radius vector* to the tangent line (Fig. 9.76). The angle ϕ can then be calculated from the relation

$$\phi = \theta + \psi, \tag{1}$$

which comes from applying the exterior angle theorem to the triangle in Fig. 9.76.

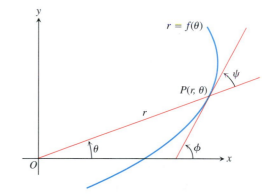

9.76 The angle ψ between the tangent line and the radius vector.

Suppose the equation of the curve is given in the form $r = f(\theta)$, where $f(\theta)$ is a differentiable function of θ. Then

$$x = r \cos\theta \quad \text{and} \quad y = r \sin\theta \tag{2}$$

are differentiable functions of θ with

$$\frac{dx}{d\theta} = -r \sin\theta + \cos\theta \frac{dr}{d\theta},$$
$$\frac{dy}{d\theta} = r \cos\theta + \sin\theta \frac{dr}{d\theta}. \tag{3}$$

Since $\psi = \phi - \theta$ from (1),

$$\tan\psi = \tan(\phi - \theta) = \frac{\tan\phi - \tan\theta}{1 + \tan\phi \tan\theta}.$$

Furthermore,

$$\tan \phi = \frac{dy}{dx} = \frac{dy/d\theta}{dx/d\theta}$$

because $\tan \phi$ is the slope of the curve at P. Also,

$$\tan \theta = \frac{y}{x}.$$

Hence

$$\tan \psi = \frac{\dfrac{dy/d\theta}{dx/d\theta} - \dfrac{y}{x}}{1 + \dfrac{y}{x}\dfrac{dy/d\theta}{dx/d\theta}} \tag{4}$$

$$= \frac{x\dfrac{dy}{d\theta} - y\dfrac{dx}{d\theta}}{x\dfrac{dx}{d\theta} + y\dfrac{dy}{d\theta}}.$$

The numerator in the last expression in Eq. (4) is found from Eqs. (2) and (3) to be

$$x\frac{dy}{d\theta} - y\frac{dx}{d\theta} = r^2.$$

Similarily, the denominator is

$$x\frac{dx}{d\theta} + y\frac{dy}{d\theta} = r\frac{dr}{d\theta}.$$

When we substitute these into Eq. (4), we obtain

$$\tan \psi = \frac{r}{dr/d\theta}. \tag{5}$$

This is the equation we use for finding ψ as a function of θ.

109. Show, by reference to a figure, that the angle β between the tangents to two curves at a point of intersection may be found from the formula

$$\tan \beta = \frac{\tan \psi_2 - \tan \psi_1}{1 + \tan \psi_2 \tan \psi_1}. \tag{6}$$

When will the two curves intersect at right angles?

110. Find the value of $\tan \psi$ for the curve $r = \sin^4(\theta/4)$.

111. Find the angle between the radius vector to the curve $r = 2a \sin 3\theta$ and its tangent when $\theta = \pi/6$.

112. a) GRAPHER Graph the hyperbolic spiral $r\theta = 1$. What appears to happen to ψ as the spiral winds in around the origin?

b) Confirm your finding in (a) analytically.

113. The circles $r = \sqrt{3} \cos \theta$ and $r = \sin \theta$ intersect at the point $(\sqrt{3}/2, \pi/3)$. Show that their tangents are perpendicular there.

114. Sketch the cardioid $r = a(1 + \cos \theta)$ and circle $r = 3a \cos \theta$ in one diagram and find the angle between their tangents at the point of intersection that lies in the first quadrant.

115. Find the points of intersection of the parabolas

$$r = \frac{1}{1 - \cos \theta} \quad \text{and} \quad r = \frac{3}{1 + \cos \theta}$$

and the angles between their tangents at these points.

116. Find points on the cardioid $r = a(1 + \cos \theta)$ where the tangent line is (a) horizontal, (b) vertical.

117. Show that parabolas $r = a/(1 + \cos \theta)$ and $r = b/(1 - \cos \theta)$ are orthogonal at each point of intersection ($ab \neq 0$).

118. Find the angle at which the cardioid $r = a(1 - \cos \theta)$ crosses the ray $\theta = \pi/2$.

119. Find the angle between the line $r = 3 \sec \theta$ and the cardioid $r = 4(1 + \cos \theta)$ at one of their intersections.

120. Find the slope of the tangent line to the curve $r = a \tan(\theta/2)$ at $\theta = \pi/2$.

121. Find the angle at which the parabolas $r = 1/(1 - \cos \theta)$ and $r = 1/(1 - \sin \theta)$ intersect in the first quadrant.

122. The equation $r^2 = 2 \csc 2\theta$ represents a curve in polar coordinates.

a) Sketch the curve.

b) Find an equivalent Cartesian equation for the curve.

c) Find the angle at which the curve intersects the ray $\theta = \pi/4$.

123. Suppose that the angle ψ from the radius vector to the tangent line of the curve $r = f(\theta)$ has the constant value α.

a) Show that the area bounded by the curve and two rays $\theta = \theta_1, \theta = \theta_2$, is proportional to $r_2^2 - r_1^2$, where (r_1, θ_1) and (r_2, θ_2) are polar coordinates of the ends of the arc of the curve between these rays. Find the factor of proportionality.

b) Show that the length of the arc of the curve in part (a) is proportional to $r_2 - r_1$, and find the proportionality constant.

124. Let P be a point on the hyperbola $r^2 \sin 2\theta = 2a^2$. Show that the triangle formed by OP, the tangent at P, and the initial line is isosceles.

Finding Conic Sections

1. Find an equation for the parabola with focus $(4, 0)$ and directrix $x = 3$. Sketch the parabola together with its vertex, focus, and directrix.

2. Find the vertex, focus, and directrix of the parabola

$$x^2 - 6x - 12y + 9 = 0.$$

3. Find an equation for the curve traced by the point $P(x, y)$ if the distance from P to the vertex of the parabola $x^2 = 4y$ is twice the distance from P to the focus. Identify the curve.

4. A line segment of length $a + b$ runs from the x-axis to the y-axis. The point P on the segment lies a units from one end and b units from the other end. Show that P traces an ellipse as the ends of the segment slide along the axes.

5. The vertices of an ellipse of eccentricity 0.5 lie at the points $(0, \pm 2)$. Where do the foci lie?

6. Find an equation for the ellipse of eccentricity 2/3 that has the line $x = 2$ as a directrix and the point $(4, 0)$ as the corresponding focus.

7. One focus of a hyperbola lies at the point $(0, -7)$ and the corresponding directrix is the line $y = -1$. Find an equation for the hyperbola if its eccentricity is (a) 2, (b) 5.

8. Find an equation for the hyperbola with foci $(0, -2)$ and $(0, 2)$ that passes through the point $(12, 7)$.

Orthogonal Curves

Two curves are said to be **orthogonal** if their tangents cross at right angles at every point where the curves intersect. Exercises 9–12 are about orthogonal conic sections.

9. Sketch the curves $xy = 2$ and $x^2 - y^2 = 3$ together and show that they are orthogonal.

10. Sketch the curves $y^2 = 4x + 4$ and $y^2 = 64 - 16x$ together and show that they are orthogonal.

11. Show that the curves $2x^2 + 3y^2 = a^2$ and $ky^2 = x^3$ are orthogonal for all values of the constants a and k ($a \neq 0, k \neq 0$). Sketch the four curves corresponding to $a = 2, a = 4, k = 1/2, k = -2$ in one diagram.

12. Show that the parabolas

$$y^2 = 4a(a - x), \quad a > 0$$

and

$$y^2 = 4b(b + x), \quad b > 0$$

have a common focus, the same for any a and b. Show that the parabolas intersect at the points $(a - b, \pm 2\sqrt{ab})$ and that each a-parabola is orthogonal to every b-parabola. By varying a and b, we obtain two families of confocal parabolas. Each family is said to be a set of **orthogonal trajectories** of the other family (Fig. 9.77).

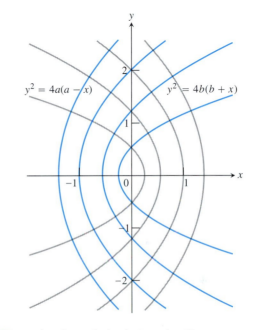

9.77 Confocal parabolas in Exercise 12.

Tangents to Conic Sections

13. *Constructing tangents to parabolas.* Show that the tangent to the parabola $y^2 = 4px$ at the point $P(x_1, y_1) \neq (0, 0)$ on the parabola meets the axis of symmetry x_1 units to the left of the vertex. This provides an accurate way to construct a tangent to the parabola at any point other than the origin (where we already have the y-axis): Mark the point $P(x_1, y_1)$ in question, drop a perpendicular from P to the x-axis, measure $2x_1$ units to the left, mark that point, and draw a line from there through P.

14. Show that no tangent can be drawn from the origin to the hyperbola $x^2 - y^2 = 1$. (*Hint:* If the tangent to a curve at a point $P(x, y)$ on the curve passes through the origin, then the slope of the curve at P is y/x.)

15. Show that any tangent to the hyperbola $xy = a^2$ makes a triangle of area $2a^2$ with the hyperbola's asymptotes.

16. a) Show that the line

$$b^2 xx_1 + a^2 yy_1 - a^2 b^2 = 0$$

is tangent to the ellipse $b^2 x^2 + a^2 y^2 - a^2 b^2 = 0$ at the point (x_1, y_1) on the ellipse.

b) Show that the line

$$b^2 xx_1 - a^2 yy_1 - a^2 b^2 = 0$$

is tangent to the hyperbola $b^2 x^2 - a^2 y^2 - a^2 b^2 = 0$ at the point (x_1, y_1) on the hyperbola.

c) Show that the tangent to the conic section

$$Ax^2 + B xy + C y^2 + Dx + Ey + F = 0$$

at a point (x_1, y_1) on it has an equation that may be written in the form

$$Axx_1 + B\left(\frac{x_1 y + xy_1}{2}\right) + C yy_1 + D\left(\frac{x + x_1}{2}\right)$$

$$+ E\left(\frac{y + y_1}{2}\right) + F = 0.$$

Equations and Inequalities

What points in the xy-plane satisfy the equations and inequalities in Exercises 17–24? Draw a figure for each exercise.

17. $(x^2 - y^2 - 1)(x^2 + y^2 - 25)(x^2 + 4y^2 - 4) = 0$

18. $(x + y)(x^2 + y^2 - 1) = 0$

19. $(x^2/9) + (y^2/16) \leq 1$

20. $(x^2/9) - (y^2/16) \leq 1$

21. $(9x^2 + 4y^2 - 36)(4x^2 + 9y^2 - 16) \leq 0$

22. $(9x^2 + 4y^2 - 36)(4x^2 + 9y^2 - 16) > 0$

23. $x^4 - (y^2 - 9)^2 = 0$

24. $x^2 + xy + y^2 < 3$

Parametric Equations

25. *Epicycloids.* When a circle rolls externally along the circumference of a second, fixed circle, any point P on the circumference of the rolling circle describes an *epicycloid,* as shown here. Let the fixed circle have its center at the origin O and have radius a.

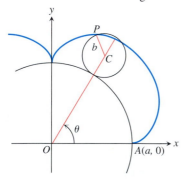

Let the radius of the rolling circle be b and let the initial position of the tracing point P be $A(a, 0)$. Find parametric equations for the epicycloid, using as the parameter the angle θ from the positive x-axis to the line through the circles' centers.

26. Find parametric equations and a Cartesian equation for the curve traced by the point $P(x, y)$ if its coordinates satisfy the differential equations

$$\frac{dx}{dt} = -2y, \quad \frac{dy}{dt} = \cos t,$$

subject to the conditions that $x = 3$ and $y = 0$ when $t = 0$. Identify the curve.

27. *Pythagorean triples.* Suppose that the coordinates of a particle $P(x, y)$ moving in the plane are

$$x = \frac{1 - t^2}{1 + t^2} \quad \text{and} \quad y = \frac{2t}{1 + t^2}$$

for $-\infty < t < \infty$. Show that $x^2 + y^2 = 1$ and hence that the motion takes place on the unit circle. What one point of the circle is not covered by the motion? Sketch the circle and indicate the direction of motion for increasing t. For what values of t does $(x, y) = (0, -1)$? $(1, 0)$? $(0, 1)$?

From $x^2 + y^2 = 1$, we obtain

$$(t^2 - 1)^2 + (2t)^2 = (t^2 + 1)^2,$$

an equation of interest in number theory because it generates *Pythagorean triples* of integers. When t is an integer greater than 1, $a = t^2 - 1$, $b = 2t$, and $c = t^2 + 1$ are positive integers that satisfy the equation $a^2 + b^2 = c^2$.

28. a) Find the centroid of the region enclosed by the x-axis and the cycloid arch

$$x = a(t - \sin t), \quad y = a(1 - \cos t), \quad 0 \leq t \leq 2\pi.$$

b) Find the first moments about the coordinate axes of the curve

$$x = (2/3) t^{3/2}, \quad y = 2\sqrt{t}, \quad 0 \leq t \leq \sqrt{3}.$$

Polar Coordinates

29. a) Find an equation in polar coordinates for the curve

$$x = e^{2t} \cos t, \quad y = e^{2t} \sin t, \quad -\infty < t < \infty.$$

b) Find the length of the curve from $t = 0$ to $t = 2\pi$.

30. Find the length of the curve $r = 2 \sin^3(\theta/3), 0 \leq \theta \leq 3\pi$, in the polar coordinate plane.

31. Find the area of the surface generated by revolving the first-quadrant portion of the cardioid $r = 1 + \cos \theta$ about the x-axis. (*Hint:* Use the identities $1 + \cos \theta = 2 \cos^2(\theta/2)$ and $\sin \theta = 2 \sin(\theta/2) \cos(\theta/2)$ to simplify the integral.)

32. Sketch the regions enclosed by the curves $r = 2a \cos^2(\theta/2)$ and $r = 2a \sin^2(\theta/2)$, $a > 0$, in the polar coordinate plane and find the area of the portion of the plane they have in common.

Exercises 33–36 give the eccentricities of conic sections with one focus at the origin of the polar coordinate plane, along with the directrix for that focus. Find a polar equation for each conic section.

33. $e = 2$, $r \cos \theta = 2$

34. $e = 1$, $r \cos \theta = -4$

35. $e = 1/2$, $r \sin \theta = 2$

36. $e = 1/3$, $r \sin \theta = -6$

Theory and Examples

37. A rope with a ring in one end is looped over two pegs in a horizontal line. The free end, after being passed through the ring, has a weight suspended from it to make the rope hang taut. If the rope slips freely over the pegs and through the ring, the weight will descend as far as possible. Assume that the length of the rope is at least four times as great as the distance between the pegs and that the configuration of the rope is symmetric with respect to the line of the vertical part of the rope.

a) Find the angle formed at the bottom of the loop (Fig. 9.78).

b) Show that for each fixed position of the ring on the rope, the possible locations of the ring in space lie on an ellipse with foci at the pegs.

c) Justify the original symmetry assumption by combining the result in (b) with the assumption that the rope and weight will take a rest position of minimal potential energy.

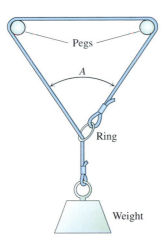

9.78 Exercise 37 asks how large the angle *A* will be when the frictionless rope shown here is pulled tight by the weight.

38. Two radar stations lie 20 km apart along an east–west line. A low-flying plane traveling from west to east is known to have a speed of v_0 km/sec. At $t = 0$ a signal is sent from the station at $(-10, 0)$, bounces off the plane, and is received at $(10, 0)$ $30/c$

seconds later (c is the velocity of the signal). When $t = 10/v_0$, another signal is sent out from the station at $(-10, 0)$, reflects off the plane, and is once again received $30/c$ seconds later by the other station. Find the position of the plane when it reflects the second signal under the assumption that v_0 is much less than c.

39. A comet moves in a parabolic orbit with the sun at the focus. When the comet is 4×10^7 miles from the sun, the line from the comet to the sun makes a $60°$ angle with the orbit's axis, as shown here. How close will the comet come to the sun?

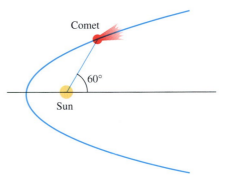

40. Find the points on the parabola $x = 2t$, $y = t^2$, $-\infty < t < \infty$, closest to the point $(0, 3)$.

41. Find the eccentricity of the ellipse $x^2 + xy + y^2 = 1$ to the nearest hundredth.

42. Find the eccentricity of the hyperbola $xy = 1$.

43. Is the curve $\sqrt{x} + \sqrt{y} = 1$ part of a conic section? If so, what kind of conic section? If not, why not?

44. Show that the curve $2xy - \sqrt{2}\,y + 2 = 0$ is a hyperbola. Find the hyperbola's center, vertices, foci, axes, and asymptotes.

45. Find a polar coordinate equation for

a) the parabola with focus at the origin and vertex at $(a, \pi/4)$;

b) the ellipse with foci at the origin and $(2, 0)$ and one vertex at $(4, 0)$;

c) the hyperbola with one focus at the origin, center at $(2, \pi/2)$, and a vertex at $(1, \pi/2)$.

46. Any line through the origin will intersect the ellipse $r = 3/(2 + \cos \theta)$ in two points P_1 and P_2. Let d_1 be the distance between P_1 and the origin and let d_2 be the distance between P_2 and the origin. Compute $(1/d_1) + (1/d_2)$.

47. *Generating a cardioid with circles.* Cardioids are special epicycloids (Exercise 25). Show that if you roll a circle of radius a about another circle of radius a in the polar coordinate plane, as in Fig. 9.79, the original point of contact P will trace a cardioid. (*Hint:* Start by showing that angles *OBC* and *PAD* both have measure θ.)

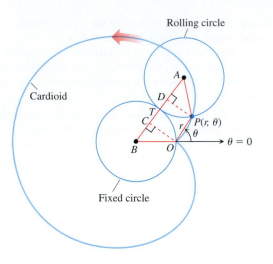

9.79 As the circle centered at A rolls around the circle centered at B, the point P traces a cardioid (Exercise 47).

48. *A bifold closet door.* A bifold closet door consists of two one-foot-wide panels, hinged at point P. The outside bottom corner of one panel rests on a pivot at O (see the accompanying figure). The outside bottom corner of the other panel, denoted by Q, slides along a straight track, shown in the figure as a portion of the x-axis. Assume that as Q moves back and forth, the bottom of the door rubs against a thick carpet. What shape will the door sweep out on the surface of the carpet?

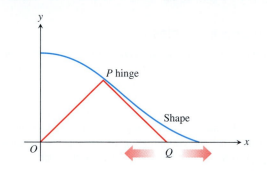

49. GRAPHER EXPLORATION Graph the curve $r = \cos 5\theta + n \cos \theta$, $0 \le \theta \le \pi$ for integers $n = -5$ (heart) to $n = 5$ (bell). (Source: *The College Mathematics Journal*, Vol. 25, No. 1, Jan. 1994.)

CHAPTER 10

Vectors and Analytic Geometry in Space

OVERVIEW   This chapter introduces vectors and three-dimensional coordinate systems. Just as the coordinate plane is the natural place to study functions of a single variable, coordinate space is the place to study functions of two variables (or more). We establish coordinates in space by adding a third axis that measures distance above and below the xy-plane. This builds on what we already know without forcing us to start over again.

10.1 Vectors in the Plane

Some of the things we measure are determined by their magnitudes. To record mass, length, or time, for example, we need only write down a number and name an appropriate unit of measure. But we need more information to describe a force, displacement, or velocity. To describe a force, we need to record the direction in which it acts as well as how large it is. To describe a body's displacement, we have to say in what direction it moves as well as how far. To describe a body's velocity, we have to know where the body is headed as well as how fast it is going.

Quantities that have direction as well as magnitude are usually represented by arrows that point in the direction of the action and whose lengths give the magnitude of the action in terms of a suitably chosen unit.

When we discuss these arrows abstractly, we think of them as directed line segments and we call them *vectors*.

Definitions
A **vector** in the plane is a directed line segment. Two vectors are **equal** or **the same** if they have the same length and direction.

Thus, the arrows we use when we draw vectors are understood to represent the same vector if they have the same length, are parallel, and point in the same direction.

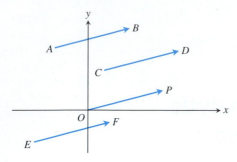

10.1 Arrows with the same length and direction represent the same vector (Example 1).

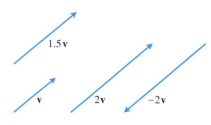

10.2 Scalar multiples of **v**.

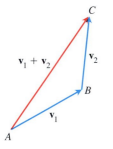

10.3 The sum of $\mathbf{v}_1$ and $\mathbf{v}_2$.

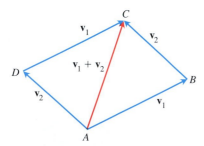

10.4 The Parallelogram Law of Addition. Quadrilateral *ABCD* is a parallelogram because opposite sides have equal lengths. The law was used by Aristotle to describe the combined action of two forces.

In print, vectors are usually described with single boldface roman letters, as in **v** ("vector vee"). The vector defined by the directed line segment from point *A* to point *B* is written as $\overrightarrow{AB}$ ("vector *ab*").

EXAMPLE 1   The four arrows in Fig. 10.1 have the same length and direction. They therefore represent the same vector, and we write

$$\overrightarrow{AB} = \overrightarrow{CD} = \overrightarrow{OP} = \overrightarrow{EF}.$$ ❑

Scalars and Scalar Multiples

We multiply a vector by a positive real number by multiplying its length by the number (Fig. 10.2). To multiply a vector by 2, we double its length. To multiply a vector by 1.5, we increase its length by 50%, and so on. We multiply a vector by a negative number by reversing the vector's direction and multiplying the length by the number's absolute value.

If *c* is a nonzero real number and **v** a vector, the direction of *c***v** agrees with that of **v** if *c* is positive and is opposite to that of **v** if *c* is negative. Since real numbers work like scaling factors in this context, we call them **scalars** and call multiples like *c***v** **scalar multiples** of **v**.

To include multiplication by zero, we adopt the convention that multiplying a vector by zero produces the **zero vector 0,** consisting of points that are degenerate line segments of zero length. Unlike other vectors, the vector **0** has no direction.

Geometric Addition: The Parallelogram Law

Two nonzero vectors $\mathbf{v}_1$ and $\mathbf{v}_2$ can be added geometrically by drawing a representative of $\mathbf{v}_1$, say from *A* to *B* as in Fig. 10.3, and then a representative of $\mathbf{v}_2$ starting from the terminal point *B* of $\mathbf{v}_1$. In Fig. 10.3, $\mathbf{v}_2 = \overrightarrow{BC}$. The sum $\mathbf{v}_1 + \mathbf{v}_2$ is then the vector represented by the arrow from the initial point *A* of $\mathbf{v}_1$ to the terminal point *C* of $\mathbf{v}_2$. That is, if

$$\mathbf{v}_1 = \overrightarrow{AB} \qquad \text{and} \qquad \mathbf{v}_2 = \overrightarrow{BC},$$

then

$$\mathbf{v}_1 + \mathbf{v}_2 = \overrightarrow{AB} + \overrightarrow{BC} = \overrightarrow{AC}.$$

This description of addition is sometimes called the **Parallelogram Law** of addition because $\mathbf{v}_1 + \mathbf{v}_2$ is given by the diagonal of the parallelogram determined by $\mathbf{v}_1$ and $\mathbf{v}_2$ (Fig. 10.4).

Components

Two vectors are said to be **parallel** if they are nonzero scalar multiples of one another or, equivalently, if the line segments representing them are parallel.

Whenever a vector **v** can be written as a sum

$$\mathbf{v} = \mathbf{v}_1 + \mathbf{v}_2$$

of two *nonparallel* vectors, the vectors $\mathbf{v}_1$ and $\mathbf{v}_2$ are said to be **components** of **v**. We also say that we have **represented** or **resolved** **v** in terms of $\mathbf{v}_1$ and $\mathbf{v}_2$.

The most common algebra of vectors is based on representing each vector in terms of components parallel to the Cartesian coordinate axes and writing each

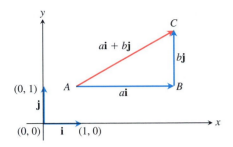

10.5 The basic vectors **i** and **j**. Any vector $\overrightarrow{AC}$ in the plane can be expressed as a scalar multiple of **i** plus a scalar multiple of **j**.

component as an appropriate multiple of a **basic** vector of length 1. The basic vector in the positive x-direction is the vector **i** determined by the directed line segment that runs from $(0, 0)$ to $(1, 0)$. The basic vector in the positive y-direction is the vector **j** determined by the directed line segment from $(0, 0)$ to $(0, 1)$. Then $a\,\mathbf{i}$, a being a scalar, represents a vector of length $|a|$ parallel to the x-axis, pointing to the right if $a > 0$ and to the left if $a < 0$. Similarly, $b\,\mathbf{j}$ is a vector of length $|b|$ parallel to the y-axis, pointing up if $b > 0$ and down if $b < 0$. Figure 10.5 shows a vector $\mathbf{v} = \overrightarrow{AC}$ resolved into its **i**- and **j**-components as the sum

$$\mathbf{v} = a\,\mathbf{i} + b\,\mathbf{j}.$$

Definitions

If $\mathbf{v} = a\,\mathbf{i} + b\,\mathbf{j}$, the vectors $a\,\mathbf{i}$ and $b\,\mathbf{j}$ are the **vector components of v in the directions of i and j.** The numbers a and b are the **scalar components of v in the directions of i and j.**

Components enable us to define the equality of vectors algebraically.

Definition

Equality of Vectors (Algebraic Definition)

$$a\,\mathbf{i} + b\,\mathbf{j} = a'\,\mathbf{i} + b'\,\mathbf{j} \qquad \Leftrightarrow \qquad a = a' \text{ and } b = b' \qquad (1)$$

Two vectors are equal if and only if their scalar components in the directions of **i** and **j** are identical.

Algebraic Addition

Vectors may be added algebraically by adding their corresponding scalar components, as shown in Fig. 10.6.

If $\mathbf{v}_1 = a_1\,\mathbf{i} + b_1\,\mathbf{j}$ and $\mathbf{v}_2 = a_2\,\mathbf{i} + b_2\,\mathbf{j}$, then

$$\mathbf{v}_1 + \mathbf{v}_2 = (a_1 + a_2)\,\mathbf{i} + (b_1 + b_2)\,\mathbf{j}. \qquad (2)$$

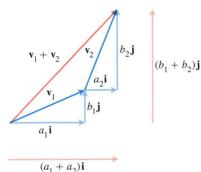

10.6 If $\mathbf{v}_1 = a_1\,\mathbf{i} + b_1\,\mathbf{j}$ and $\mathbf{v}_2 = a_2\,\mathbf{i} + b_2\,\mathbf{j}$, then $\mathbf{v}_1 + \mathbf{v}_2 = (a_1 + a_2)\,\mathbf{i} + (b_1 + b_2)\,\mathbf{j}$.

EXAMPLE 2

$$(2\,\mathbf{i} - 4\,\mathbf{j}) + (5\,\mathbf{i} + 3\,\mathbf{j}) = (2 + 5)\,\mathbf{i} + (-4 + 3)\,\mathbf{j} = 7\,\mathbf{i} - \mathbf{j} \qquad \square$$

Subtraction

The negative of a vector $\mathbf{v}$ is the vector $-\mathbf{v} = (-1)\,\mathbf{v}$. It has the same length as $\mathbf{v}$ but points in the opposite direction. To subtract a vector $\mathbf{v}_2$ from a vector $\mathbf{v}_1$, we add $-\mathbf{v}_2$ to $\mathbf{v}_1$. This can be done geometrically by drawing $-\mathbf{v}_2$ from the tip of $\mathbf{v}_1$

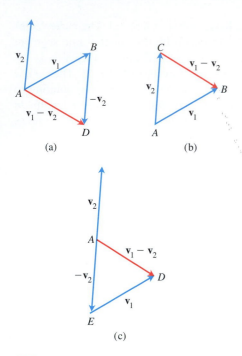

10.7 Three ways to draw $\mathbf{v}_1 - \mathbf{v}_2$ (there are others): (a) as $\mathbf{v}_1 + (-\mathbf{v}_2)$; (b) as the vector from the tip of $\mathbf{v}_2$ to the tip of $\mathbf{v}_1$; and (c) as $-\mathbf{v}_2 + \mathbf{v}_1$.

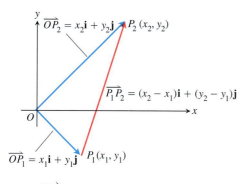

10.8 $\overrightarrow{P_1P_2} = (x_2 - x_1)\,\mathbf{i} + (y_2 - y_1)\,\mathbf{j}$

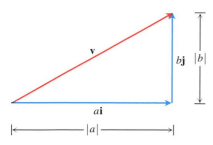

10.9 The length of $\mathbf{v}$ is $\sqrt{|a|^2 + |b|^2} = \sqrt{a^2 + b^2}$.

and then drawing the vector from the initial point of $\mathbf{v}_1$ to the tip of $-\mathbf{v}_2$, as shown in Fig. 10.7(a), where

$$\overrightarrow{AD} = \overrightarrow{AB} + \overrightarrow{BD} = \mathbf{v}_1 + (-\mathbf{v}_2) = \mathbf{v}_1 - \mathbf{v}_2.$$

Another way to draw $\mathbf{v}_1 - \mathbf{v}_2$ is to draw $\mathbf{v}_1$ and $\mathbf{v}_2$ with a common initial point and then draw $\mathbf{v}_1 - \mathbf{v}_2$ as the vector from the tip of $\mathbf{v}_2$ to the tip of $\mathbf{v}_1$. This is illustrated in Fig. 10.7(b), where

$$\overrightarrow{CB} = \overrightarrow{CA} + \overrightarrow{AB} = -\mathbf{v}_2 + \mathbf{v}_1 = \mathbf{v}_1 - \mathbf{v}_2.$$

Still another way is to draw $\mathbf{v}_1$ from the tip of $-\mathbf{v}_2$ (Fig. 10.7c).

In terms of components, vector subtraction follows the algebraic law

$$\mathbf{v}_1 - \mathbf{v}_2 = (a_1 - a_2)\,\mathbf{i} + (b_1 - b_2)\,\mathbf{j}, \tag{3}$$

which says that corresponding scalar components are subtracted.

EXAMPLE 3

$$(6\,\mathbf{i} + 2\,\mathbf{j}) - (3\,\mathbf{i} - 5\,\mathbf{j}) = (6 - 3)\,\mathbf{i} + (2 - (-5))\,\mathbf{j} = 3\,\mathbf{i} + 7\,\mathbf{j} \qquad \square$$

We find the components of the vector from a point $P_1(x_1, y_1)$ to a point $P_2(x_2, y_2)$ by subtracting the components of $\overrightarrow{OP_1} = x_1\,\mathbf{i} + y_1\,\mathbf{j}$ from the components of $\overrightarrow{OP_2} = x_2\,\mathbf{i} + y_2\,\mathbf{j}$ (Fig. 10.8).

The vector from $P_1(x_1, y_1)$ to $P_2(x_2, y_2)$ is

$$\overrightarrow{P_1P_2} = (x_2 - x_1)\,\mathbf{i} + (y_2 - y_1)\,\mathbf{j}. \tag{4}$$

EXAMPLE 4 The vector from $P_1(3, 4)$ to $P_2(5, 1)$ is

$$\overrightarrow{P_1P_2} = (5 - 3)\,\mathbf{i} + (1 - 4)\,\mathbf{j} = 2\,\mathbf{i} - 3\,\mathbf{j}. \qquad \square$$

Magnitude

The **magnitude** or **length** of $\mathbf{v} = a\,\mathbf{i} + b\,\mathbf{j}$ is $|\mathbf{v}| = \sqrt{a^2 + b^2}$. We arrive at this number by applying the Pythagorean theorem to the right triangle determined by $\mathbf{v}$ and its two vector components (Fig. 10.9). The bars in $|\mathbf{v}|$ (read "the magnitude of $\mathbf{v}$" or "the length of $\mathbf{v}$") are the same bars we use for absolute values.

The magnitude or length of $\mathbf{v} = a\,\mathbf{i} + b\,\mathbf{j}$ is $|\mathbf{v}| = \sqrt{a^2 + b^2}$. $\tag{5}$

EXAMPLE 5 You push a loaded supermarket cart by applying a 20-lb force $\mathbf{F}$ that makes a 30° angle with the horizontal. Resolve $\mathbf{F}$ into its horizontal and vertical components. (The horizontal component is the effective force in the direction of motion. The vertical component just adds weight to the cart.)

$|\mathbf{F}| = 20$

$30°$

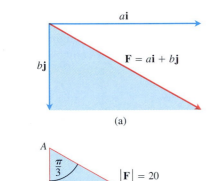

$a\,\mathbf{i}$

$b\,\mathbf{j}$

$\mathbf{F} = a\,\mathbf{i} + b\,\mathbf{j}$

(a)

A

$\dfrac{\pi}{3}$

$|b\,\mathbf{j}| = |b|$

$|\mathbf{F}| = 20$

$\dfrac{\pi}{6}$

B

$|a\,\mathbf{i}| = |a|$

C

(b)

10.10 (a) The horizontal and vertical components of the vector **F** in Example 5. (b) The right triangle determined by **F** and its components.

Solution We draw a vector triangle for $\mathbf{F} = a\,\mathbf{i} + b\,\mathbf{j}$ and its vector components along with the right triangle determined by their magnitudes (Fig. 10.10). The triangle is a 30-60-90 triangle, so $|a| = 10\sqrt{3}$ and $|b| = 10$. The horizontal component of **F** is $10\sqrt{3}\,\mathbf{i}$. The vertical component is $-10\,\mathbf{j}$ (negative because it points down). That is, $\mathbf{F} = 10\sqrt{3}\,\mathbf{i} - 10\,\mathbf{j}$. ❑

Scalar Multiplication

Scalar multiplication can be accomplished component by component.

If c is a scalar and $\mathbf{v} = a\,\mathbf{i} + b\,\mathbf{j}$ is a vector, then

$$c\,\mathbf{v} = c(a\,\mathbf{i} + b\,\mathbf{j}) = (ca)\,\mathbf{i} + (cb)\,\mathbf{j}. \qquad (6)$$

The length of $c\,\mathbf{v}$ is $|c|$ times the length of **v**:

$$|c\,\mathbf{v}| = |(ca)\,\mathbf{i} + (cb)\,\mathbf{j}| \qquad \text{Eq. (6)}$$

$$= \sqrt{(ca)^2 + (cb)^2} \qquad \begin{array}{l}\text{Eq. (5) with } ca \text{ and } cb \text{ in}\\ \text{place of } a \text{ and } b\end{array}$$

$$= \sqrt{c^2(a^2 + b^2)}$$

$$= \sqrt{c^2}\sqrt{a^2 + b^2}$$

$$= |c||\mathbf{v}|.$$

If c is a scalar and **v** is a vector, then $|c\,\mathbf{v}| = |c||\mathbf{v}|$.

EXAMPLE 6 If $c = -2$ and $\mathbf{v} = -3\,\mathbf{i} + 4\,\mathbf{j}$, then

$$|\mathbf{v}| = |-3\,\mathbf{i} + 4\,\mathbf{j}| = \sqrt{(-3)^2 + (4)^2} = \sqrt{9 + 16} = \sqrt{25} = 5$$

$$|-2\,\mathbf{v}| = |(-2)(-3\,\mathbf{i} + 4\,\mathbf{j})| = |6\,\mathbf{i} - 8\,\mathbf{j}| = \sqrt{(6)^2 + (-8)^2} = \sqrt{36 + 64}$$

$$= \sqrt{100} = 10 = |-2|5 = |c||\mathbf{v}|. ❑$$

The Zero Vector

In terms of components, the zero vector is the vector

$$\mathbf{0} = 0\,\mathbf{i} + 0\,\mathbf{j}.$$

It is the only vector whose length is zero, as we can see from the fact that

$$|a\,\mathbf{i} + b\,\mathbf{j}| = \sqrt{a^2 + b^2} = 0 \quad \Leftrightarrow \quad a = b = 0.$$

Unit Vectors

Any vector whose length is 1 is a **unit vector.** The vectors $\mathbf{i}$ and $\mathbf{j}$ are unit vectors.

$$|\mathbf{i}| = |1\,\mathbf{i} + 0\,\mathbf{j}| = \sqrt{1^2 + 0^2} = 1, \qquad |\mathbf{j}| = |0\,\mathbf{i} + 1\,\mathbf{j}| = \sqrt{0^2 + 1^2} = 1$$

If $\mathbf{u}$ is the unit vector obtained by rotating $\mathbf{i}$ through an angle θ in the positive direction, then $\mathbf{u}$ has a horizontal component $\cos\theta$ and vertical component $\sin\theta$ (Fig. 10.11), so that

$$\mathbf{u} = (\cos\theta)\,\mathbf{i} + (\sin\theta)\,\mathbf{j}. \tag{7}$$

As θ varies from 0 to 2π, the point P in Fig. 10.11 traces the circle $x^2 + y^2 = 1$ counterclockwise. This takes in all possible directions, so *Eq. (7) gives every unit vector in the plane.*

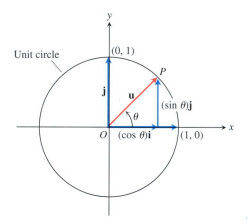

10.11 The unit vector that makes an angle of measure θ with the positive x-axis. Every unit vector in the plane has the form

$$\mathbf{u} = (\cos\theta)\,\mathbf{i} + (\sin\theta)\,\mathbf{j}$$

for some θ.

Length vs. Direction

If $\mathbf{v} \neq \mathbf{0}$, then

$$\left|\frac{\mathbf{v}}{|\mathbf{v}|}\right| = \left|\frac{1}{|\mathbf{v}|}\mathbf{v}\right| = \frac{1}{|\mathbf{v}|}|\mathbf{v}| = 1,$$

so $\mathbf{v}/|\mathbf{v}|$ is a unit vector in the direction of $\mathbf{v}$. We can therefore express $\mathbf{v}$ in terms of its two important features, length and direction, by writing $\mathbf{v} = |\mathbf{v}|(\mathbf{v}/|\mathbf{v}|)$.

> If $\mathbf{v} \neq \mathbf{0}$, then
>
> 1. $\dfrac{\mathbf{v}}{|\mathbf{v}|}$ is a unit vector in the direction of $\mathbf{v}$;
>
> 2. the equation $\mathbf{v} = |\mathbf{v}|\dfrac{\mathbf{v}}{|\mathbf{v}|}$ expresses $\mathbf{v}$ in terms of its length and direction.

EXAMPLE 7 Express $\mathbf{v} = 3\,\mathbf{i} - 4\,\mathbf{j}$ as a product of its length and direction.

Solution Length of $\mathbf{v}$: $\qquad |\mathbf{v}| = \sqrt{(3)^2 + (-4)^2} = \sqrt{9 + 16} = 5$

Direction of $\mathbf{v}$: $\qquad \dfrac{\mathbf{v}}{|\mathbf{v}|} = \dfrac{3\,\mathbf{i} - 4\,\mathbf{j}}{5} = \dfrac{3}{5}\mathbf{i} - \dfrac{4}{5}\mathbf{j}$

$$\mathbf{v} = 3\,\mathbf{i} - 4\,\mathbf{j} = 5\underbrace{\left(\frac{3}{5}\mathbf{i} - \frac{4}{5}\mathbf{j}\right)}_{\text{direction}}$$

length

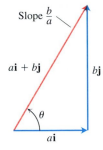

Slope $\dfrac{b}{a}$

$a\mathbf{i} + b\mathbf{j}$ $b\mathbf{j}$

θ

$a\mathbf{i}$

10.12 If $a \neq 0$, the vector $a\mathbf{i} + b\mathbf{j}$ has slope $b/a = \tan\theta$.

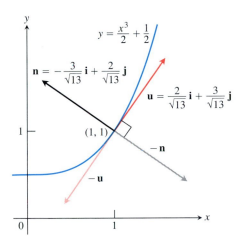

$y = \dfrac{x^3}{2} + \dfrac{1}{2}$

$\mathbf{n} = -\dfrac{3}{\sqrt{13}}\mathbf{i} + \dfrac{2}{\sqrt{13}}\mathbf{j}$

$\mathbf{u} = \dfrac{2}{\sqrt{13}}\mathbf{i} + \dfrac{3}{\sqrt{13}}\mathbf{j}$

$(1, 1)$

$-\mathbf{n}$

$-\mathbf{u}$

10.13 The unit tangent and normal vectors at the point (1, 1) on the curve $y = (x^3/2) + 1/2$.

If $\mathbf{v} = a\mathbf{i} + b\mathbf{j}$, then $\mathbf{p} = -b\mathbf{i} + a\mathbf{j}$ and $\mathbf{q} = b\mathbf{i} - a\mathbf{j}$ are perpendicular to $\mathbf{v}$ because their slopes are both $-a/b$, the negative reciprocal of $\mathbf{v}$'s slope.

Slopes, Tangents, and Normals

A vector is **parallel** to a line if the segments that represent the vector are parallel to the line. The **slope** of a vector that is not vertical is the slope shared by the lines parallel to the vector. Thus, if $a \neq 0$, the vector $\mathbf{v} = a\mathbf{i} + b\mathbf{j}$ has a well-defined slope, which can be calculated from the components of $\mathbf{v}$ as the number b/a (Fig. 10.12).

A vector is **tangent** or **normal** to a curve at a point if it is parallel or normal to the line that is tangent to the curve at that point. The next example shows how to find such vectors.

EXAMPLE 8   Find unit vectors tangent and normal to the curve

$$y = \frac{x^3}{2} + \frac{1}{2}$$

at the point (1, 1).

Solution   We find the unit vectors that are parallel and normal to the curve's tangent line at (1, 1) (Fig. 10.13).

The slope of the line tangent to the curve at (1, 1) is

$$y' = \frac{3x^2}{2}\bigg|_{x=1} = \frac{3}{2}.$$

We look for a unit vector with this slope. The vector $\mathbf{v} = 2\mathbf{i} + 3\mathbf{j}$ has slope 3/2, as does every nonzero multiple of $\mathbf{v}$. To find a multiple of $\mathbf{v}$ that is a unit vector, we divide $\mathbf{v}$ by

$$|\mathbf{v}| = \sqrt{2^2 + 3^2} = \sqrt{13},$$

obtaining

$$\mathbf{u} = \frac{\mathbf{v}}{|\mathbf{v}|} = \frac{2}{\sqrt{13}}\mathbf{i} + \frac{3}{\sqrt{13}}\mathbf{j}.$$

The vector $\mathbf{u}$ is tangent to the curve at (1, 1) because it has the same direction as $\mathbf{v}$. Of course,

$$-\mathbf{u} = -\frac{2}{\sqrt{13}}\mathbf{i} - \frac{3}{\sqrt{13}}\mathbf{j},$$

which points in the opposite direction, is also tangent to the curve at (1, 1). Without some additional requirement, there is no reason to prefer one of these vectors to the other.

To find unit vectors normal to the curve at (1, 1), we look for unit vectors whose slopes are the negative reciprocal of the slope of $\mathbf{u}$. This is quickly done by interchanging the scalar components of $\mathbf{u}$ and changing the sign of one of them. We obtain

$$\mathbf{n} = -\frac{3}{\sqrt{13}}\mathbf{i} + \frac{2}{\sqrt{13}}\mathbf{j} \quad \text{and} \quad -\mathbf{n} = \frac{3}{\sqrt{13}}\mathbf{i} - \frac{2}{\sqrt{13}}\mathbf{j}.$$

Again, either one will do. The vectors have opposite directions but both are normal to the curve at (1, 1). ❑

Exercises 10.1

Geometry and Calculation

1. The vectors **A**, **B**, and **C** in the figure here lie in a plane. Copy them on a sheet of paper. Then, by arranging vectors head to tail, as in Figs. 10.3 and 10.6, sketch

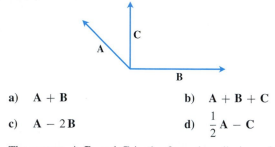

   a)   **A** + **B**

   b)   **A** + **B** + **C**

   c)   **A** − 2 **B**

   d) $\frac{1}{2}$ **A** − **C**

2. The vectors **A**, **B**, and **C** in the figure here lie in a plane. Copy them on a sheet of paper. Then, by arranging vectors head to tail, as in Figs. 10.3 and 10.6, sketch

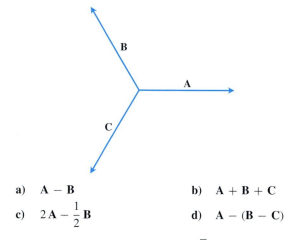

   a)   **A** − **B**

   b)   **A** + **B** + **C**

   c)   2 **A** − $\frac{1}{2}$ **B**

   d)   **A** − (**B** − **C**)

Let **A** = 2 **i** − 7 **j**, **B** = **i** + 6 **j**, and **C** = $\sqrt{3}$ **i** − π **j**. Write each of the vectors in Exercises 3–6 in the form a **i** + b **j**.

3. **A** + 2 **B**

4. **A** + **B** − **C**

5. 3 **A** − $\frac{1}{\pi}$ **C**

6. 2 **A** − 3 **B** + 32 **j**

7. Vectors **u**, **v**, and **w** are determined by the sides of triangle ABC as shown.

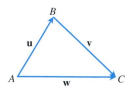

   a)   Express **w** in terms of **u** and **v**.

   b)   Express **v** in terms of **u** and **w**.

8. Vectors **u** and **w** are determined by the sides of triangle ABC as shown, and P is the midpoint of side BC. Express **a** in terms of **u** and **w**.

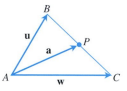

Express the vectors in Exercises 9–16 in the form a **i** + b **j** and sketch them as arrows in the coordinate plane beginning at the origin.

9. $\overrightarrow{P_1 P_2}$ if P_1 is the point (5, 7) and P_2 is the point (2, 9)

10. $\overrightarrow{P_1 P_2}$ if P_1 is the point (1, 2) and P_2 is the point (−3, 5)

11. $\overrightarrow{AB}$ if A is the point (−5, 3) and B is the point (−10, 8)

12. $\overrightarrow{AB}$ if A is the point (−7, −8) and B is the point (6, 11)

13. $\overrightarrow{P_1 P_2}$ if P_1 is the point (1, 3) and P_2 is the point (2, −1)

14. $\overrightarrow{P_3 P_4}$ if P_3 is the point (1, 3) and P_4 is the midpoint of the line segment $P_1 P_2$ joining $P_1(2, −1)$ and $P_2(−4, 3)$

15. The sum of the vectors $\overrightarrow{AB}$ and $\overrightarrow{CD}$, given the four points $A(1, −1)$, $B(2, 0)$, $C(−1, 3)$, and $D(−2, 2)$

16. The vector from the point A to the origin where $\overrightarrow{AB} = 4$ **i** − 2 **j** and B is the point (−2, 5)

17. Given the vector $\overrightarrow{AB} = 3$ **i** − **j** and A is the point (2, 9), find the point B.

18. Given the vector $\overrightarrow{PQ} = −6$ **i** − 4 **j** and Q is the point (3, 3), find the point P.

Unit Vectors

Sketch the vectors in Exercises 19–22 and express each vector in the form a **i** + b **j**.

19. The unit vectors **u** = $(\cos\theta)$ **i** + $(\sin\theta)$ **j** for $\theta = \pi/6$ and $\theta = 2\pi/3$. Include the circle $x^2 + y^2 = 1$ in your sketch.

20. The unit vectors **u** = $(\cos\theta)$ **i** + $(\sin\theta)$ **j** for $\theta = −\pi/4$ and $\theta = −3\pi/4$. Include the circle $x^2 + y^2 = 1$ in your sketch.

21. The unit vector obtained by rotating **j** counterclockwise $3\pi/4$ rad about the origin

22. The unit vector obtained by rotating **j** clockwise $2\pi/3$ rad about the origin

For the vectors in Exercises 23 and 24, find unit vectors **u** = $(\cos\theta)$ **i** + $(\sin\theta)$ **j** in the same direction.

23. 6 **i** − 8 **j**

24. −**i** + 3 **j**

In Exercises 25–28, find the unit vectors that are tangent and normal to the curve at the given point (four vectors in all). Then sketch the vectors and curve together.

25. $y = x^2$,   (2, 4)

26. $x^2 + 2y^2 = 6$,   (2, 1)

27. $y = \tan^{-1}x$, $(1, \pi/4)$

28. $y = \sum_{n=0}^{\infty} \dfrac{x^n}{n!}$,   (0, 1)

In Exercises 29–32, find the unit vectors that are tangent and normal to the curve at the given point (four vectors in all).

29. $3x^2 + 8xy + 2y^2 - 3 = 0$,   (1, 0)

30. $x^2 - 6xy + 8y^2 - 2x - 1 = 0$,   (1, 1)

31. $y = \int_0^x \sqrt{3 + t^4}\, dt$,   (0, 0)

32. $y = \int_e^x \ln(\ln t)\, dt$,   (e, 0)

Length and Direction

In Exercises 33 and 34, express each vector as a product of its length and direction.

33. $5\mathbf{i} + 12\mathbf{j}$

34. $2\mathbf{i} - 3\mathbf{j}$

35. Find the unit vectors that are parallel to the vector $3\mathbf{i} - 4\mathbf{j}$ (two vectors in all).

36. Find a vector of length 2 whose direction is the opposite of the direction of the vector $\mathbf{A} = -\mathbf{i} + 2\mathbf{j}$. How many such vectors are there?

37. Show that $\mathbf{A} = 3\mathbf{i} + 6\mathbf{j}$ and $\mathbf{B} = -\mathbf{i} - 2\mathbf{j}$ have opposite directions. Sketch $\mathbf{A}$ and $\mathbf{B}$ together.

38. Show that $\mathbf{A} = 3\mathbf{i} + 6\mathbf{j}$ and $\mathbf{B} = (1/2)\mathbf{i} + \mathbf{j}$ have the same direction.

Theory and Applications

39. You are pulling on a suitcase with a force $\mathbf{F}$ (pictured here) whose magnitude is $|\mathbf{F}| = 10$ lb. Find the x- and y-components of $\mathbf{F}$.

40. A kite string exerts a 12-lb pull ($|\mathbf{F}| = 12$) on a kite and makes a 45° angle with the horizontal. Find the horizontal and vertical components of $\mathbf{F}$.

41. Let $\mathbf{A} = 2\mathbf{i} + \mathbf{j}$, $\mathbf{B} = \mathbf{i} + \mathbf{j}$, and $\mathbf{C} = \mathbf{i} - \mathbf{j}$. Find scalars α, β, such that $\mathbf{A} = \alpha\mathbf{B} + \beta\mathbf{C}$.

42. Let $\mathbf{A} = \mathbf{i} - 2\mathbf{j}$, $\mathbf{B} = 2\mathbf{i} + 3\mathbf{j}$, and $\mathbf{C} = \mathbf{i} + \mathbf{j}$. Write $\mathbf{A} = \mathbf{A}_1 + \mathbf{A}_2$ where $\mathbf{A}_1$ is parallel to $\mathbf{B}$ and $\mathbf{A}_2$ is parallel to $\mathbf{C}$. (See Exercise 41.)

43. A bird flies from its nest 5 km in the direction 60° north of east, where it stops to rest on a tree. It then flies 10 km in the direction due southeast to land atop a telephone pole. Place an xy-coordinate system so that the origin is the bird's nest, the x-axis points east, and the y-axis points north.

  a) At what point is the tree located?
  b) At what point is the telephone pole?

44. A bird flies from its nest 7 km in the direction northeast, where it stops to rest on a tree. It then flies 8 km in the direction 30° south of west to land atop a telephone pole. Place an xy-coordinate system so that the origin is the bird's nest, the x-axis points east, and the y-axis points north.

  a) At what point is the tree located?
  b) At what point is the telephone pole?

45. Let $\mathbf{v}$ be a vector in the plane not parallel to the y-axis. How is the slope of $-\mathbf{v}$ related to the slope of $\mathbf{v}$? Give reasons for your answer.

10.2

Cartesian (Rectangular) Coordinates and Vectors in Space

Our goal now is to describe the three-dimensional Cartesian coordinate system and learn our way around in space. This means defining distance, practicing with the arithmetic of vectors in space (the rules are the same as in the plane but with an extra term), and making connections between sets of points and equations and inequalities.

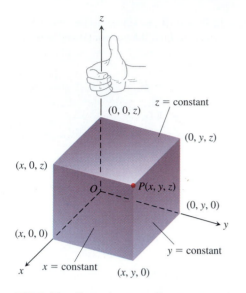

10.14 The Cartesian coordinate system is right-handed.

Cartesian Coordinates

To locate points in space, we use three mutually perpendicular coordinate axes, arranged as in Fig. 10.14. The axes Ox, Oy, and Oz shown there make a *right-handed* coordinate frame. When you hold your right hand so that the fingers curl from the positive x-axis toward the positive y-axis, your thumb points along the positive z-axis.

The Cartesian coordinates (x, y, z) of a point P in space are the numbers at which the planes through P perpendicular to the axes cut the axes.

Points on the x-axis have y- and z-coordinates equal to zero. That is, they have coordinates of the form $(x, 0, 0)$. Similarly, points on the y-axis have coordinates of the form $(0, y, 0)$. Points on the z-axis have coordinates of the form $(0, 0, z)$.

The points in a plane perpendicular to the x-axis all have the same x-coordinate, this being the number at which that plane cuts the x-axis. The y- and z-coordinates can be any numbers. Similarly, the points in a plane perpendicular to the y-axis have a common y-coordinate and the points in a plane perpendicular to the z-axis have a common z-coordinate. To write equations for these planes, we name the common coordinate's value. The plane $x = 2$ is the plane perpendicular to the x-axis at $x = 2$. The plane $y = 3$ is the plane perpendicular to the y-axis at $y = 3$. The plane $z = 5$ is the plane perpendicular to the z-axis at $z = 5$. Figure 10.15 shows the planes $x = 2$, $y = 3$, and $z = 5$, together with their intersection point $(2, 3, 5)$.

The planes $x = 2$ and $y = 3$ in Fig. 10.15 intersect in a line parallel to the z-axis. This line is described by the *pair* of equations $x = 2$, $y = 3$. A point (x, y, z) lies on the line if and only if $x = 2$ and $y = 3$. Similarly, the line of intersection of the planes $y = 3$ and $z = 5$ is described by the equation pair $y = 3$, $z = 5$. This line runs parallel to the x-axis. The line of intersection of the planes $x = 2$ and $z = 5$, parallel to the y-axis, is described by the equation pair $x = 2$, $z = 5$.

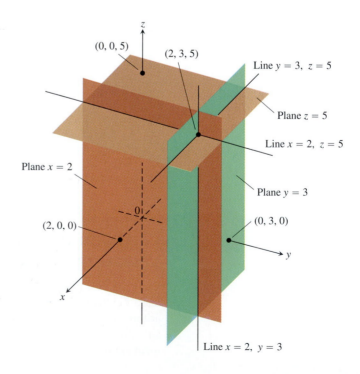

10.15 The planes $x = 2$, $y = 3$, and $z = 5$ determine three lines through the point $(2, 3, 5)$.

The planes determined by the coordinate axes are the **xy-plane,** whose standard equation is $z = 0$; the **yz-plane,** whose standard equation is $x = 0$; and the **xz-plane,** whose standard equation is $y = 0$. They meet at the **origin** $(0, 0, 0)$ (Fig. 10.16).

The three **coordinate planes** $x = 0$, $y = 0$, and $z = 0$ divide space into eight cells called **octants.** The octant in which the point coordinates are all positive is called the **first octant;** there is no conventional numbering for the other seven octants.

Cartesian coordinates for space are also called **rectangular coordinates** because the axes that define them meet at right angles.

In the following examples, we match coordinate equations and inequalities with the sets of points they define in space.

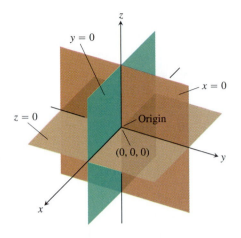

10.16 The planes $x = 0$, $y = 0$, and $z = 0$ divide space into eight octants.

EXAMPLE 1

Defining equations and inequalities	Verbal description
$z \geq 0$	The half-space consisting of the points on and above the xy-plane.
$x = -3$	The plane perpendicular to the x-axis at $x = -3$. This plane lies parallel to the yz-plane and 3 units behind it.
$z = 0, x \leq 0, y \geq 0$	The second quadrant of the xy-plane.
$x \geq 0, y \geq 0, z \geq 0$	The first octant.
$-1 \leq y \leq 1$	The slab between the planes $y = -1$ and $y = 1$ (planes included).
$y = -2, z = 2$	The line in which the planes $y = -2$ and $z = 2$ intersect. Alternatively, the line through the point $(0, -2, 2)$ parallel to the x-axis. ❑

EXAMPLE 2   What points $P(x, y, z)$ satisfy the equations

$$x^2 + y^2 = 4 \qquad \text{and} \qquad z = 3?$$

Solution   The points lie in the horizontal plane $z = 3$ and, in this plane, make up the circle $x^2 + y^2 = 4$. We call this set of points "the circle $x^2 + y^2 = 4$ in the plane $z = 3$" or, more simply, "the circle $x^2 + y^2 = 4, z = 3$" (Fig. 10.17).   ❑

Vectors in Space

The sets of equivalent directed line segments that we use to represent forces, displacements, and velocities in space are called vectors, just as in the plane. The same rules of addition, subtraction, and scalar multiplication apply.

The vectors represented by the directed line segments from the origin to the points $(1, 0, 0)$, $(0, 1, 0)$, and $(0, 0, 1)$ are the **basic vectors** (Fig. 10.18, on the following page). We denote them by **i, j,** and **k.** The **position vector r** from the origin O to the typical point $P(x, y, z)$ is

$$\mathbf{r} = \overrightarrow{OP} = x\,\mathbf{i} + y\,\mathbf{j} + z\,\mathbf{k}. \tag{1}$$

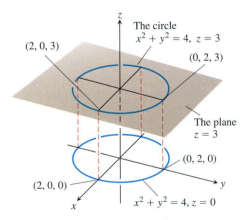

10.17 The circle $x^2 + y^2 = 4$, $z = 3$.

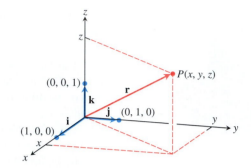

10.18 The position vector of a point in space.

Definition

Addition and Subtraction for Vectors in Space

For any vectors $\mathbf{A} = a_1\,\mathbf{i} + a_2\,\mathbf{j} + a_3\,\mathbf{k}$ and $\mathbf{B} = b_1\,\mathbf{i} + b_2\,\mathbf{j} + b_3\,\mathbf{k}$,

$$\mathbf{A} + \mathbf{B} = (a_1 + b_1)\,\mathbf{i} + (a_2 + b_2)\,\mathbf{j} + (a_3 + b_3)\,\mathbf{k}$$

$$\mathbf{A} - \mathbf{B} = (a_1 - b_1)\,\mathbf{i} + (a_2 - b_2)\,\mathbf{j} + (a_3 - b_3)\,\mathbf{k}.$$

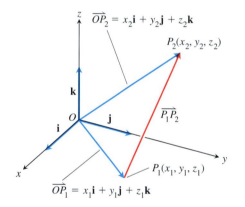

10.19 The vector from P_1 to P_2 is $\overrightarrow{P_1 P_2}$ $= (x_2 - x_1)\,\mathbf{i} + (y_2 - y_1)\,\mathbf{j} + (z_2 - z_1)\,\mathbf{k}.$

The Vector Between Two Points

We can express the vector $\overrightarrow{P_1 P_2}$ from the point $P_1(x_1, y_1, z_1)$ to the point $P_2(x_2, y_2, z_2)$ in terms of the coordinates of P_1 and P_2 because (Fig. 10.19)

$$\overrightarrow{P_1 P_2} = \overrightarrow{OP_2} - \overrightarrow{OP_1}$$

$$= (x_2\,\mathbf{i} + y_2\,\mathbf{j} + z_2\,\mathbf{k}) - (x_1\,\mathbf{i} + y_1\,\mathbf{j} + z_1\,\mathbf{k}) \qquad (2)$$

$$= (x_2 - x_1)\,\mathbf{i} + (y_2 - y_1)\,\mathbf{j} + (z_2 - z_1)\,\mathbf{k}.$$

The vector from $P_1(x_1, y_1, z_1)$ to $P_2(x_2, y_2, z_2)$ is

$$\overrightarrow{P_1 P_2} = (x_2 - x_1)\,\mathbf{i} + (y_2 - y_1)\,\mathbf{j} + (z_2 - z_1)\,\mathbf{k}. \qquad (3)$$

Magnitude

As always, the important features of a vector are its magnitude and direction. We find a formula for the magnitude (length) of $a_1\,\mathbf{i} + a_2\,\mathbf{j} + a_3\,\mathbf{k}$ by applying the Pythagorean theorem to the right triangles in Fig. 10.20. From triangle ABC,

$$|\overrightarrow{AC}| = \sqrt{a_1{}^2 + a_2{}^2}$$

and from triangle ACD,

$$|a_1\,\mathbf{i} + a_2\,\mathbf{j} + a_3\,\mathbf{k}| = |\overrightarrow{AD}| = \sqrt{|\overrightarrow{AC}|^2 + |\overrightarrow{CD}|^2} = \sqrt{a_1{}^2 + a_2{}^2 + a_3{}^2}.$$

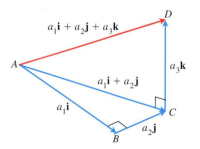

10.20 We find the length of $\overrightarrow{AD}$ by applying the Pythagorean theorem to the right triangles ABC and ACD.

The **magnitude (length)** of $\mathbf{A} = a_1\,\mathbf{i} + a_2\,\mathbf{j} + a_3\,\mathbf{k}$ is

$$|\mathbf{A}| = |a_1\,\mathbf{i} + a_2\,\mathbf{j} + a_3\,\mathbf{k}| = \sqrt{a_1{}^2 + a_2{}^2 + a_3{}^2}. \qquad (4)$$

Scalar Multiplication

> ### Definition
> If c is a scalar and $\mathbf{A} = a_1\,\mathbf{i} + a_2\,\mathbf{j} + a_3\,\mathbf{k}$ is a vector, then
> $$c\,\mathbf{A} = (ca_1)\,\mathbf{i} + (ca_2)\,\mathbf{j} + (ca_3)\,\mathbf{k}.$$

EXAMPLE 3   The length of $\mathbf{A} = \mathbf{i} - 2\,\mathbf{j} + 3\,\mathbf{k}$ is

$$|\mathbf{A}| = \sqrt{(1)^2 + (-2)^2 + (3)^2} = \sqrt{1 + 4 + 9} = \sqrt{14}.$$ ❑

If we multiply $\mathbf{A} = a_1\mathbf{i} + a_2\mathbf{j} + a_3\mathbf{k}$ by a scalar c, the length of $c\mathbf{A}$ is $|c|$ times the length of $\mathbf{A}$, as in the plane. The reason is the same, as well:

$$c\,\mathbf{A} = ca_1\,\mathbf{i} + ca_2\,\mathbf{j} + ca_3\,\mathbf{k},$$

$$|c\,\mathbf{A}| = \sqrt{(ca_1)^2 + (ca_2)^2 + (ca_3)^2} = \sqrt{c^2a_1{}^2 + c^2a_2{}^2 + c^2a_3{}^2} \tag{5}$$

$$= |c|\sqrt{a_1{}^2 + a_2{}^2 + a_3{}^2} = |c|\,|\mathbf{A}|.$$

EXAMPLE 4   If $\mathbf{A}$ is the vector of Example 3, then the length of

$$2\,\mathbf{A} = 2(\mathbf{i} - 2\,\mathbf{j} + 3\,\mathbf{k}) = 2\,\mathbf{i} - 4\,\mathbf{j} + 6\,\mathbf{k}$$

is

$$\sqrt{(2)^2 + (-4)^2 + (6)^2} = \sqrt{4 + 16 + 36} = \sqrt{56}$$

$$= \sqrt{4 \cdot 14} = 2\sqrt{14} = 2|\mathbf{A}|.$$ ❑

The Zero Vector

The **zero vector** in space is the vector $\mathbf{0} = 0\,\mathbf{i} + 0\,\mathbf{j} + 0\,\mathbf{k}$. As in the plane, $\mathbf{0}$ has zero length, and no direction.

Unit Vectors

A **unit vector** in space is a vector of length 1. The basic vectors are unit vectors because

$$|\mathbf{i}| = |1\,\mathbf{i} + 0\,\mathbf{j} + 0\,\mathbf{k}| = \sqrt{1^2 + 0^2 + 0^2} = 1,$$

$$|\mathbf{j}| = |0\,\mathbf{i} + 1\,\mathbf{j} + 0\,\mathbf{k}| = \sqrt{0^2 + 1^2 + 0^2} = 1,$$

$$|\mathbf{k}| = |0\,\mathbf{i} + 0\,\mathbf{j} + 1\,\mathbf{k}| = \sqrt{0^2 + 0^2 + 1^2} = 1.$$

Magnitude and Direction

If $\mathbf{A} \neq \mathbf{0}$, then $\mathbf{A}/|\mathbf{A}|$ is a unit vector in the direction of $\mathbf{A}$ and we can use the equation

$$\mathbf{A} = |\mathbf{A}|\,\frac{\mathbf{A}}{|\mathbf{A}|} \tag{6}$$

to express $\mathbf{A}$ as a product of its magnitude and direction.

EXAMPLE 5   Express $\mathbf{A} = \mathbf{i} - 2\,\mathbf{j} + 3\,\mathbf{k}$ as a product of its magnitude and direction.

Solution

$$\mathbf{A} = |\mathbf{A}| \cdot \frac{\mathbf{A}}{|\mathbf{A}|} \qquad\qquad \text{Eq. (6)}$$

$$= \sqrt{14} \cdot \frac{\mathbf{i} - 2\,\mathbf{j} + 3\,\mathbf{k}}{\sqrt{14}} \qquad \text{From Example 3}$$

$$= \sqrt{14}\left(\frac{1}{\sqrt{14}}\mathbf{i} - \frac{2}{\sqrt{14}}\mathbf{j} + \frac{3}{\sqrt{14}}\mathbf{k}\right) = (\text{length of } \mathbf{A}) \cdot (\text{direction of } \mathbf{A}) \quad \square$$

EXAMPLE 6   Find a unit vector $\mathbf{u}$ in the direction of the vector from $P_1(1, 0, 1)$ to $P_2(3, 2, 0)$.

Solution   We divide $\overrightarrow{P_1 P_2}$ by its length:

$$\overrightarrow{P_1 P_2} = (3 - 1)\mathbf{i} + (2 - 0)\mathbf{j} + (0 - 1)\mathbf{k} = 2\,\mathbf{i} + 2\,\mathbf{j} - \mathbf{k}$$

$$|\overrightarrow{P_1 P_2}| = \sqrt{(2)^2 + (2)^2 + (-1)^2} = \sqrt{4 + 4 + 1} = \sqrt{9} = 3$$

$$\mathbf{u} = \frac{\overrightarrow{P_1 P_2}}{|\overrightarrow{P_1 P_2}|} = \frac{2\,\mathbf{i} + 2\,\mathbf{j} - \mathbf{k}}{3} = \frac{2}{3}\mathbf{i} + \frac{2}{3}\mathbf{j} - \frac{1}{3}\mathbf{k}. \quad \square$$

EXAMPLE 7   Find a vector 6 units long in the direction of $\mathbf{A} = 2\,\mathbf{i} + 2\,\mathbf{j} - \mathbf{k}$.

Solution   The vector we want is

$$6\,\frac{\mathbf{A}}{|\mathbf{A}|} = 6\,\frac{2\,\mathbf{i} + 2\,\mathbf{j} - \mathbf{k}}{\sqrt{2^2 + 2^2 + (-1)^2}} = 6\,\frac{2\,\mathbf{i} + 2\,\mathbf{j} - \mathbf{k}}{3} = 4\,\mathbf{i} + 4\,\mathbf{j} - 2\,\mathbf{k}. \quad \square$$

Distance in Space

The distance between two points P_1 and P_2 in space is the length of $\overrightarrow{P_1 P_2}$.

The Distance Between $P_1(x_1, y_1, z_1)$ and $P_2(x_2, y_2, z_2)$

$$|\overrightarrow{P_1 P_2}| = \sqrt{(x_2 - x_1)^2 + (y_2 - y_1)^2 + (z_2 - z_1)^2} \tag{7}$$

EXAMPLE 8   The distance between $P_1(2, 1, 5)$ and $P_2(-2, 3, 0)$ is

$$|\overrightarrow{P_1 P_2}| = \sqrt{(-2 - 2)^2 + (3 - 1)^2 + (0 - 5)^2}$$

$$= \sqrt{16 + 4 + 25}$$

$$= \sqrt{45} = 3\sqrt{5}. \quad \square$$

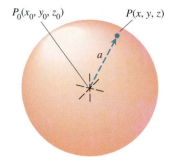

$P_0(x_0, y_0, z_0)$ $P(x, y, z)$

a

10.21 The standard equation of the sphere of radius a centered at (x_0, y_0, z_0) is

$(x - x_0)^2 + (y - y_0)^2 + (z - z_0)^2 = a^2.$

Spheres

We use Eq. (7) to write equations for spheres in space (Fig. 10.21). A point $P(x, y, z)$ lies on the sphere of radius a centered at $P_0(x_0, y_0, z_0)$ precisely when $|\overrightarrow{P_0 P}| = a$ or

$$(x - x_0)^2 + (y - y_0)^2 + (z - z_0)^2 = a^2.$$

The Standard Equation for the Sphere of Radius a and Center (x_0, y_0, z_0)

$$(x - x_0)^2 + (y - y_0)^2 + (z - z_0)^2 = a^2 \qquad (8)$$

EXAMPLE 9   Find the center and radius of the sphere

$$x^2 + y^2 + z^2 + 3x - 4z + 1 = 0.$$

Solution   We find the center and radius of a sphere the way we find the center and radius of a circle: Complete the squares on the x-, y-, and z-terms as necessary and write each quadratic as a squared linear expression. Then, from the equation in standard form, read off the center and radius. For the sphere here we have

$$x^2 + y^2 + z^2 + 3x - 4z + 1 = 0$$

$$(x^2 + 3x \quad) + y^2 + (z^2 - 4z \quad) = -1$$

$$\left(x^2 + 3x + \left(\frac{3}{2}\right)^2\right) + y^2 + \left(z^2 - 4z + \left(\frac{-4}{2}\right)^2\right) = -1 + \left(\frac{3}{2}\right)^2 + \left(\frac{-4}{2}\right)^2$$

$$\left(x + \frac{3}{2}\right)^2 + y^2 + (z - 2)^2 = -1 + \frac{9}{4} + 4 = \frac{21}{4}.$$

This is Eq. (8) with $x_0 = -3/2$, $y_0 = 0$, $z_0 = 2$, and $a = \sqrt{21}/2$. The center is $(-3/2, 0, 2)$. The radius is $\sqrt{21}/2$.   ❏

EXAMPLE 10   *Sets bounded by spheres or portions of spheres*

Defining equations and inequalities	Description
a) $x^2 + y^2 + z^2 < 4$	The interior of the sphere $x^2 + y^2 + z^2 = 4$.
b) $x^2 + y^2 + z^2 \leq 4$	The solid ball bounded by the sphere $x^2 + y^2 + z^2 = 4$. Alternatively, the sphere $x^2 + y^2 + z^2 = 4$ together with its interior.
c) $x^2 + y^2 + z^2 > 4$	The exterior of the sphere $x^2 + y^2 + z^2 = 4$.
d) $x^2 + y^2 + z^2 = 4,$ $z \leq 0$	The lower hemisphere cut from the sphere $x^2 + y^2 + z^2 = 4$ by the xy-plane (the plane $z = 0$).

❏

DRAWING LESSON
How to Draw Three-dimensional Objects to Look Three-dimensional

1　Break lines. When one line passes behind another, break it to show that it doesn't touch and that part of it is hidden.

2　Make the angle between the positive *x*-axis and the positive *y*-axis large enough.

3　Draw planes parallel to the coordinate planes as if they were rectangles with sides parallel to the coordinate axes.

4　Dash or omit hidden portions of lines. Don't let the line touch the boundary of the parallelogram that represents the plane, unless the line lies in the plane.

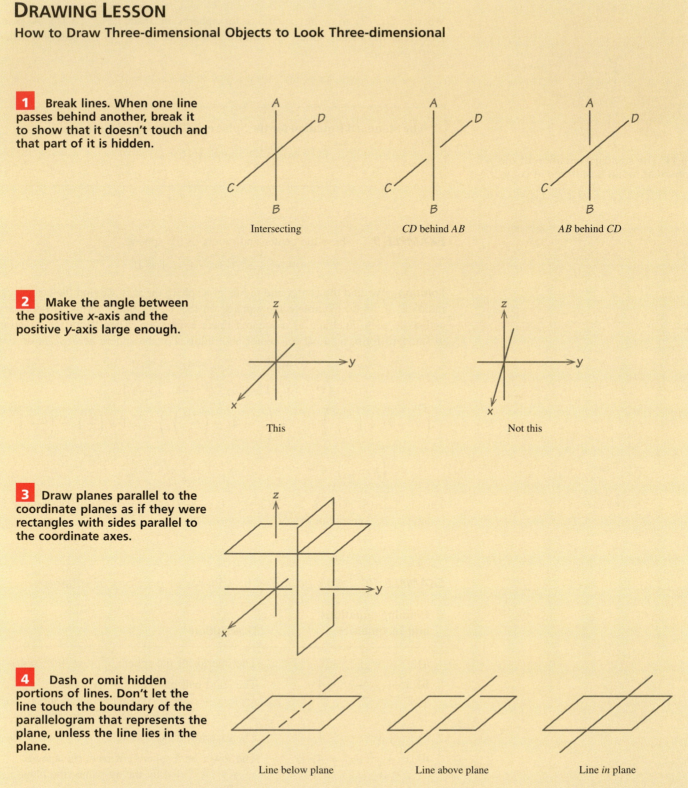

Intersecting　　　　*CD* behind *AB*　　　　*AB* behind *CD*

This　　　　Not this

Line below plane　　　　Line above plane　　　　Line *in* plane

5 Spheres: Draw the sphere first (outline and equator); draw axes, if any, later. Use line breaks and dashed lines.

Sphere first

Axes later

6 A general rule for perspective: Draw the object as if it lies some distance away, below, and to your left.

Advice ignored

Advice followed

7 To draw a plane that crosses all three coordinate axes, follow the steps shown here:
(a) Sketch the axes and mark the intercepts. (b) Connect the intercepts to form two sides of a parallelogram. (c) Complete the parallelogram and enlarge it by drawing lines parallel to its sides. (d) Darken the exposed parts, break hidden lines, and, if desired, dash hidden portions of the axes. You may wish to erase the smaller parallelogram at this point.

(a)

(b)

(c)

(d)

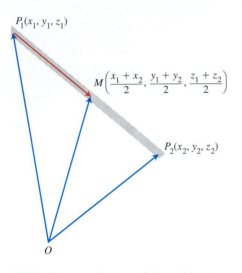

10.22 The coordinates of the midpoint are the averages of the coordinates of P_1 and P_2.

Midpoints

The coordinates of the midpoint of a line segment are found by averaging.

The midpoint M of the line segment joining points $P_1(x_1, y_1, z_1)$ and $P_2(x_2, y_2, z_2)$ is the point

$$\left(\frac{x_1 + x_2}{2}, \frac{y_1 + y_2}{2}, \frac{z_1 + z_2}{2} \right).$$

To see why, observe (Fig. 10.22) that

$$\overrightarrow{OM} = \overrightarrow{OP_1} + \frac{1}{2}(\overrightarrow{P_1P_2}) = \overrightarrow{OP_1} + \frac{1}{2}(\overrightarrow{OP_2} - \overrightarrow{OP_1})$$

$$= \frac{1}{2}(\overrightarrow{OP_1} + \overrightarrow{OP_2})$$

$$= \frac{x_1 + x_2}{2}\mathbf{i} + \frac{y_1 + y_2}{2}\mathbf{j} + \frac{z_1 + z_2}{2}\mathbf{k}.$$

EXAMPLE 11   The midpoint of the segment joining $P_1(3, -2, 0)$ and $P_2(7, 4, 4)$ is

$$\left(\frac{3 + 7}{2}, \frac{-2 + 4}{2}, \frac{0 + 4}{2} \right) = (5, 1, 2).$$ ∎

Exercises 10.2

Sets, Equations, and Inequalities

In Exercises 1–12, give a geometric description of the set of points in space whose coordinates satisfy the given pairs of equations.

1. $x = 2$, $y = 3$

2. $x = -1$, $z = 0$

3. $y = 0$, $z = 0$

4. $x = 1$, $y = 0$

5. $x^2 + y^2 = 4$, $z = 0$

6. $x^2 + y^2 = 4$, $z = -2$

7. $x^2 + z^2 = 4$, $y = 0$

8. $y^2 + z^2 = 1$, $x = 0$

9. $x^2 + y^2 + z^2 = 1$, $x = 0$

10. $x^2 + y^2 + z^2 = 25$, $y = -4$

11. $x^2 + y^2 + (z + 3)^2 = 25$, $z = 0$

12. $x^2 + (y - 1)^2 + z^2 = 4$, $y = 0$

In Exercises 13–18, describe the sets of points in space whose coordinates satisfy the given inequalities or combinations of equations and inequalities.

13. a) $x \geq 0$, $y \geq 0$, $z = 0$

   b) $x \geq 0$, $y \leq 0$, $z = 0$

14. a) $0 \leq x \leq 1$

   b) $0 \leq x \leq 1$, $0 \leq y \leq 1$

   c) $0 \leq x \leq 1$, $0 \leq y \leq 1$, $0 \leq z \leq 1$

15. a) $x^2 + y^2 + z^2 \leq 1$

   b) $x^2 + y^2 + z^2 > 1$

16. a) $x^2 + y^2 \leq 1$, $z = 0$

   b) $x^2 + y^2 \leq 1$, $z = 3$

   c) $x^2 + y^2 \leq 1$,   no restriction on z

17. a) $x^2 + y^2 + z^2 = 1$, $z \geq 0$

   b) $x^2 + y^2 + z^2 \leq 1$, $z \geq 0$

18. a) $x = y$, $z = 0$

   b) $x = y$,   no restriction on z

In Exercises 19–28, describe the given set with a single equation or with a pair of equations.

19. The plane perpendicular to the

   a) x-axis at $(3, 0, 0)$

   b) y-axis at $(0, -1, 0)$

   c) z-axis at $(0, 0, -2)$

20. The plane through the point $(3, -1, 2)$ perpendicular to the

    a) x-axis   **b)** y-axis   **c)** z-axis

21. The plane through the point $(3, -1, 1)$ parallel to the

    a) xy-plane   **b)** yz-plane   **c)** xz-plane

22. The circle of radius 2 centered at $(0, 0, 0)$ and lying in the

    a) xy-plane   **b)** yz-plane   **c)** xz-plane

23. The circle of radius 2 centered at $(0, 2, 0)$ and lying in the

    a) xy-plane   **b)** yz-plane   **c)** plane $y = 2$

24. The circle of radius 1 centered at $(-3, 4, 1)$ and lying in a plane parallel to the

    a) xy-plane   **b)** yz-plane   **c)** xz-plane

25. The line through the point $(1, 3, -1)$ parallel to the

    a) x-axis   **b)** y-axis   **c)** z-axis

26. The set of points in space equidistant from the origin and the point $(0, 2, 0)$

27. The circle in which the plane through the point $(1, 1, 3)$ perpendicular to the z-axis meets the sphere of radius 5 centered at the origin

28. The set of points in space that lie 2 units from the point $(0, 0, 1)$ and, at the same time, 2 units from the point $(0, 0, -1)$

Write inequalities to describe the sets in Exercises 29–34.

29. The slab bounded by the planes $z = 0$ and $z = 1$ (planes included)

30. The solid cube in the first octant bounded by the coordinate planes and the planes $x = 2$, $y = 2$, and $z = 2$

31. The half-space consisting of the points on and below the xy-plane

32. The upper hemisphere of the sphere of radius 1 centered at the origin

33. The (a) interior and (b) exterior of the sphere of radius 1 centered at the point $(1, 1, 1)$

34. The closed region bounded by the spheres of radius 1 and radius 2 centered at the origin. (*Closed* means the spheres are to be included. Had we wanted the spheres left out, we would have asked for the *open* region bounded by the spheres. This is analogous to the way we use *closed* and *open* to describe intervals: *closed* means endpoints included, *open* means endpoints left out. Closed sets include boundaries; open sets leave them out.)

Length and Direction

In Exercises 35–44, express each vector as a product of its length and direction.

35. $2\mathbf{i} + \mathbf{j} - 2\mathbf{k}$ **36.** $3\mathbf{i} - 6\mathbf{j} + 2\mathbf{k}$

37. $\mathbf{i} + 4\mathbf{j} - 8\mathbf{k}$ **38.** $9\mathbf{i} - 2\mathbf{j} + 6\mathbf{k}$

39. $5\mathbf{k}$ **40.** $-4\mathbf{j}$

41. $\dfrac{3}{5}\mathbf{i} + \dfrac{4}{5}\mathbf{k}$ **42.** $\dfrac{1}{\sqrt{2}}\mathbf{i} - \dfrac{1}{\sqrt{2}}\mathbf{k}$

43. $\dfrac{1}{\sqrt{6}}\mathbf{i} - \dfrac{1}{\sqrt{6}}\mathbf{j} - \dfrac{1}{\sqrt{6}}\mathbf{k}$ **44.** $\dfrac{\mathbf{i}}{\sqrt{3}} + \dfrac{\mathbf{j}}{\sqrt{3}} + \dfrac{\mathbf{k}}{\sqrt{3}}$

45. Find the vectors whose lengths and directions are given. Try to do the calculations without writing.

	Length	Direction
a)	2	$\mathbf{i}$
b)	$\sqrt{3}$	$-\mathbf{k}$
c)	$\dfrac{1}{2}$	$\dfrac{3}{5}\mathbf{j} + \dfrac{4}{5}\mathbf{k}$
d)	7	$\dfrac{6}{7}\mathbf{i} - \dfrac{2}{7}\mathbf{j} + \dfrac{3}{7}\mathbf{k}$

46. Find the vectors whose lengths and directions are given. Try to do the calculations without writing.

	Length	Direction
a)	7	$-\mathbf{j}$
b)	$\sqrt{2}$	$-\dfrac{3}{5}\mathbf{i} - \dfrac{4}{5}\mathbf{k}$
c)	$\dfrac{13}{12}$	$\dfrac{3}{13}\mathbf{i} - \dfrac{4}{13}\mathbf{j} - \dfrac{12}{13}\mathbf{k}$
d)	$a > 0$	$\dfrac{1}{\sqrt{2}}\mathbf{i} + \dfrac{1}{\sqrt{3}}\mathbf{j} - \dfrac{1}{\sqrt{6}}\mathbf{k}$

47. Find a vector of magnitude 7 in the direction of $\mathbf{A} = 12\mathbf{i} - 5\mathbf{k}$.

48. Find a vector $\sqrt{5}$ units long in the direction of $\mathbf{A} = \mathbf{i} + \mathbf{j} + \mathbf{k}$.

49. Find a vector 5 units long in the direction opposite to the direction of $\mathbf{A} = 2\mathbf{i} - 3\mathbf{j} + 6\mathbf{k}$.

50. Find a vector of magnitude 3 in the direction opposite to the direction of $\mathbf{A} = (1/2)\mathbf{i} - (1/2)\mathbf{j} - (1/2)\mathbf{k}$.

Vectors Determined by Points; Midpoints and Distance

In Exercises 51–56, find

  a) the distance between points P_1 and P_2,
  b) the direction of $\overrightarrow{P_1 P_2}$,
  c) the midpoint of line segment $P_1 P_2$.

51. $P_1(1, 1, 1)$, $P_2(3, 3, 0)$

52. $P_1(-1, 1, 5)$, $P_2(2, 5, 0)$

53. $P_1(1, 4, 5)$, $P_2(4, -2, 7)$

54. $P_1(3, 4, 5)$, $P_2(2, 3, 4)$

55. $P_1(0, 0, 0)$, $P_2(2, -2, -2)$

56. $P_1(5, 3, -2)$, $P_2(0, 0, 0)$

57. If $\overrightarrow{AB} = \mathbf{i} + 4\mathbf{j} - 2\mathbf{k}$ and B is the point $(5, 1, 3)$, find A.

58. If $\overrightarrow{AB} = -7\mathbf{i} + 3\mathbf{j} + 8\mathbf{k}$ and A is the point $(-2, -3, 6)$, find B.

Spheres

Find the centers and radii of the spheres in Exercises 59–62.

59. $(x + 2)^2 + y^2 + (z - 2)^2 = 8$

60. $\left(x + \dfrac{1}{2}\right)^2 + \left(y + \dfrac{1}{2}\right)^2 + \left(z + \dfrac{1}{2}\right)^2 = \dfrac{21}{4}$

61. $\left(x - \sqrt{2}\right)^2 + \left(y - \sqrt{2}\right)^2 + \left(z + \sqrt{2}\right)^2 = 2$

62. $x^2 + \left(y + \dfrac{1}{3}\right)^2 + \left(z - \dfrac{1}{3}\right)^2 = \dfrac{29}{9}$

Find equations for the spheres whose centers and radii are given in Exercises 63–66.

	Center	Radius
63.	$(1, 2, 3)$	$\sqrt{14}$
64.	$(0, -1, 5)$	2
65.	$(-2, 0, 0)$	$\sqrt{3}$
66.	$(0, -7, 0)$	7

Find the centers and radii of the spheres in Exercises 67–70.

67. $x^2 + y^2 + z^2 + 4x - 4z = 0$

68. $x^2 + y^2 + z^2 - 6y + 8z = 0$

69. $2x^2 + 2y^2 + 2z^2 + x + y + z = 9$

70. $3x^2 + 3y^2 + 3z^2 + 2y - 2z = 9$

71. Find a formula for the distance from the point $P(x, y, z)$ to the

   a) x-axis   **b)** y-axis   **c)** z-axis

72. Find a formula for the distance from the point $P(x, y, z)$ to the

   a) xy-plane   **b)** yz-plane   **c)** xz-plane

Geometry with Vectors

73. Suppose that A, B, and C are the corner points of the thin triangular plate of constant density shown here.

   a) Find the vector from C to the midpoint M of side AB.

   b) Find the vector from C to the point that lies two-thirds of the way from C to M on the median CM.

   c) Find the coordinates of the point in which the medians of $\triangle ABC$ intersect. According to Exercise 31, Section 5.7, this point is the plate's center of mass.

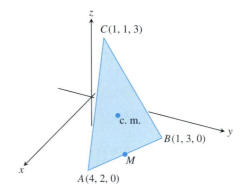

74. Find the vector from the origin to the point of intersection of the medians of the triangle whose vertices are

$$A(1, -1, 2), \quad B(2, 1, 3), \quad \text{and} \quad C(-1, 2, -1).$$

75. Let $ABCD$ be a general, not necessarily planar, quadrilateral in space. Show that the two segments joining the midpoints of opposite sides of $ABCD$ bisect each other. (*Hint:* Show that the segments have the same midpoint.)

76. Vectors are drawn from the center of a regular n-sided polygon in the plane to the vertices of the polygon. Show that the sum of the vectors is zero. (*Hint:* What happens to the sum if you rotate the polygon about its center?)

77. Suppose that A, B, and C are vertices of a triangle and that a, b, and c are, respectively, the midpoints of the opposite sides. Show that $\overrightarrow{Aa} + \overrightarrow{Bb} + \overrightarrow{Cc} = 0$.

10.3

Dot Products

We now introduce the dot product, the first of two methods we will learn for multiplying vectors together. Dot products are also called *scalar* products because the multiplication results in a scalar, not a vector. The products in the next section are vectors.

Scalar Products

10.23 The angle between **A** and **B**.

When two nonzero vectors **A** and **B** are placed so their initial points coincide, they form an angle θ of measure $0 \le \theta \le \pi$ (Fig. 10.23). This angle is called the **angle between A and B.**

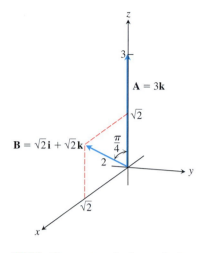

10.24 The vectors in Example 1.

Definition

The **scalar product** (**dot product**) $\mathbf{A} \cdot \mathbf{B}$ ("**A** dot **B**") of vectors **A** and **B** is the number

$$\mathbf{A} \cdot \mathbf{B} = |\mathbf{A}||\mathbf{B}| \cos \theta, \tag{1}$$

where θ is the angle between **A** and **B**.

In words, $\mathbf{A} \cdot \mathbf{B}$ is the length of **A** times the length of **B** times the cosine of the angle between **A** and **B**.

EXAMPLE 1   If $\mathbf{A} = 3\,\mathbf{k}$ and $\mathbf{B} = \sqrt{2}\,\mathbf{i} + \sqrt{2}\,\mathbf{k}$ (Fig. 10.24), then

$$\mathbf{A} \cdot \mathbf{B} = |\mathbf{A}||\mathbf{B}| \cos \theta = (3)(2) \cos \frac{\pi}{4} = 6 \cdot \frac{\sqrt{2}}{2} = 3\sqrt{2}. \qquad \square$$

Since the sign of $\mathbf{A} \cdot \mathbf{B}$ is determined by $\cos \theta$, the scalar product is positive if the angle between the vectors is acute, negative if the angle is obtuse. (We look at right angles in a moment.)

Since the angle a vector **A** makes with itself is zero, and $\cos 0 = 1$,

$$\mathbf{A} \cdot \mathbf{A} = |\mathbf{A}||\mathbf{A}| \cos 0 = |\mathbf{A}||\mathbf{A}|(1) = |\mathbf{A}|^2, \qquad \text{or} \qquad |\mathbf{A}| = \sqrt{\mathbf{A} \cdot \mathbf{A}}. \tag{2}$$

Calculation

To calculate $\mathbf{A} \cdot \mathbf{B}$ from the components of **A** and **B** in a Cartesian coordinate system with unit vectors **i, j,** and **k,** we let

$$\mathbf{A} = a_1\,\mathbf{i} + a_2\,\mathbf{j} + a_3\,\mathbf{k},$$

$$\mathbf{B} = b_1\,\mathbf{i} + b_2\,\mathbf{j} + b_3\,\mathbf{k},$$

and

$$\mathbf{C} = \mathbf{B} - \mathbf{A} = (b_1 - a_1)\,\mathbf{i} + (b_2 - a_2)\,\mathbf{j} + (b_3 - a_3)\,\mathbf{k}.$$

The law of cosines for the triangle whose sides represent **A, B,** and **C** (Fig. 10.25) is

$$|\mathbf{C}|^2 = |\mathbf{A}|^2 + |\mathbf{B}|^2 - 2|\mathbf{A}||\mathbf{B}| \cos \theta,$$

$$|\mathbf{A}||\mathbf{B}| \cos \theta = \frac{|\mathbf{A}|^2 + |\mathbf{B}|^2 - |\mathbf{C}|^2}{2}.$$

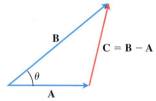

10.25 We obtain Eq. (3) by applying the law of cosines to a triangle whose sides represent **A, B,** and **C = B − A**.

The left side of this equation is $\mathbf{A} \cdot \mathbf{B}$. We evelute the right side by squaring the components of **A, B,** and **C** (Eq. 4, Section 10.2). The resulting cancellations give

$$\mathbf{A} \cdot \mathbf{B} = a_1 b_1 + a_2 b_2 + a_3 b_3. \tag{3}$$

Thus, to find the scalar product of two given vectors we multiply their corresponding **i-, j-,** and **k-** components and add the results.

Solving Eq. (1) for θ gives a formula for finding angles between vectors.

The angle between two nonzero vectors $\mathbf{A}$ and $\mathbf{B}$ is

$$\theta = \cos^{-1}\left(\frac{\mathbf{A} \cdot \mathbf{B}}{|\mathbf{A}||\mathbf{B}|}\right). \qquad (4)$$

Since the values of the arc cosine lie in $[0, \pi]$, Eq. (4) automatically gives the angle made by $\mathbf{A}$ and $\mathbf{B}.$

EXAMPLE 2   Find the angle between $\mathbf{A} = \mathbf{i} - 2\mathbf{j} - 2\mathbf{k}$ and $\mathbf{B} = 6\mathbf{i} + 3\mathbf{j} + 2\mathbf{k}.$

Solution   We use Eq. (4):

$$\mathbf{A} \cdot \mathbf{B} = (1)(6) + (-2)(3) + (-2)(2) = 6 - 6 - 4 = -4$$

$$|\mathbf{A}| = \sqrt{(1)^2 + (-2)^2 + (-2)^2} = \sqrt{9} = 3$$

$$|\mathbf{B}| = \sqrt{(6)^2 + (3)^2 + (2)^2} = \sqrt{49} = 7$$

$$\theta = \cos^{-1}\left(\frac{\mathbf{A} \cdot \mathbf{B}}{|\mathbf{A}||\mathbf{B}|}\right) \qquad \text{Eq. (4)}$$

$$= \cos^{-1}\left(\frac{-4}{(3)(7)}\right) = \cos^{-1}\left(-\frac{4}{21}\right) \approx 1.76 \text{ rad} \qquad \text{About } 101°$$

$\square$

Laws of the Dot Product

From the equation $\mathbf{A} \cdot \mathbf{B} = a_1b_1 + a_2b_2 + a_3b_3$, we can see right away that

---

$$\mathbf{A} \cdot \mathbf{B} = \mathbf{B} \cdot \mathbf{A}. \qquad (5)$$

In other words, the dot product is commutative. We can also see from Eq. (3) that if c is any number (or scalar), then

---

$$(c\,\mathbf{A}) \cdot \mathbf{B} = \mathbf{A} \cdot (c\,\mathbf{B}) = c\,(\mathbf{A} \cdot \mathbf{B}). \qquad (6)$$

If $\mathbf{C} = c_1\mathbf{i} + c_2\mathbf{j} + c_3\mathbf{k}$ is any third vector, then

$$\mathbf{A} \cdot (\mathbf{B} + \mathbf{C}) = a_1(b_1 + c_1) + a_2(b_2 + c_2) + a_3(b_3 + c_3)$$

$$= (a_1b_1 + a_2b_2 + a_3b_3) + (a_1c_1 + a_2c_2 + a_3c_3)$$

$$= \mathbf{A} \cdot \mathbf{B} + \mathbf{A} \cdot \mathbf{C}.$$

Hence dot products obey the distributive law:

$$\mathbf{A} \cdot (\mathbf{B} + \mathbf{C}) = \mathbf{A} \cdot \mathbf{B} + \mathbf{A} \cdot \mathbf{C}. \tag{7}$$

If we combine this with the commutative law, Eq. (5), it is also evident that

$$(\mathbf{A} + \mathbf{B}) \cdot \mathbf{C} = \mathbf{A} \cdot \mathbf{C} + \mathbf{B} \cdot \mathbf{C}. \tag{8}$$

Equations (7) and (8) together permit us to multiply sums of vectors by the familiar laws of algebra. For example,

$$(\mathbf{A} + \mathbf{B}) \cdot (\mathbf{C} + \mathbf{D}) = \mathbf{A} \cdot \mathbf{C} + \mathbf{A} \cdot \mathbf{D} + \mathbf{B} \cdot \mathbf{C} + \mathbf{B} \cdot \mathbf{D}. \tag{9}$$

Perpendicular (Orthogonal) Vectors

Two nonzero vectors $\mathbf{A}$ and $\mathbf{B}$ are perpendicular or **orthogonal** if the angle between them is $\pi/2$. For such vectors, we automatically have $\mathbf{A} \cdot \mathbf{B} = 0$ because $\cos(\pi/2) = 0$. The converse is also true. If $\mathbf{A}$ and $\mathbf{B}$ are nonzero vectors with $\mathbf{A} \cdot \mathbf{B} = |\mathbf{A}||\mathbf{B}| \cos\theta = 0$, then $\cos\theta = 0$ and $\theta = \cos^{-1} 0 = \pi/2$.

> Nonzero vectors $\mathbf{A}$ and $\mathbf{B}$ are perpendicular (orthogonal) if and only if $\mathbf{A} \cdot \mathbf{B} = 0$.

EXAMPLE 3 $\mathbf{A} = 3\mathbf{i} - 2\mathbf{j} + \mathbf{k}$ and $\mathbf{B} = 2\mathbf{j} + 4\mathbf{k}$ are orthogonal because

$$\mathbf{A} \cdot \mathbf{B} = (3)(0) + (-2)(2) + (1)(4) = 0. \qquad \square$$

Vector Projections

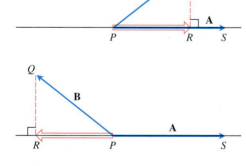

The **vector projection** of $\mathbf{B} = \overrightarrow{PQ}$ onto a nonzero vector $\mathbf{A} = \overrightarrow{PS}$ (Fig. 10.26) is the vector $\overrightarrow{PR}$ determined by dropping a perpendicular from Q to the line PS. The notation for this vector is

$$\text{proj}_{\mathbf{A}} \mathbf{B} \qquad \text{("the vector projection of } \mathbf{B} \text{ onto } \mathbf{A}\text{").}$$

If $\mathbf{B}$ represents a force, then $\text{proj}_{\mathbf{A}} \mathbf{B}$ represents the effective force in the direction of $\mathbf{A}$ (Fig. 10.27).

10.26 The vector projection of $\mathbf{B}$ onto $\mathbf{A}$.

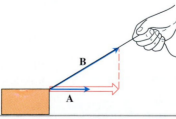

10.27 If we pull on a box with force $\mathbf{B}$, the effective force in the direction of $\mathbf{A}$ is the vector projection of $\mathbf{B}$ onto $\mathbf{A}$.

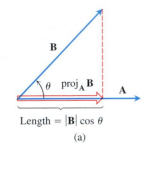

Length $= |\mathbf{B}| \cos \theta$

(a)

Length $= -|\mathbf{B}| \cos \theta$

(b)

10.28 The length of proj$_\mathbf{A}$ **B** is (a) $|\mathbf{B}| \cos \theta$ if $\cos \theta \geq 0$ and (b) $-|\mathbf{B}| \cos \theta$ if $\cos \theta < 0$.

If the angle θ between **A** and **B** is acute, proj$_\mathbf{A}$ **B** has length $|\mathbf{B}| \cos \theta$ and direction $\mathbf{A}/|\mathbf{A}|$(Fig. 10.28). If θ is obtuse, $\cos \theta < 0$ and proj$_\mathbf{A}$ **B** has length $-|\mathbf{B}| \cos \theta$ and direction $-\mathbf{A}/|\mathbf{A}|$. In any case,

$$\text{proj}_\mathbf{A}\ \mathbf{B} = (|\mathbf{B}| \cos \theta) \frac{\mathbf{A}}{|\mathbf{A}|}$$

$$= \left(\frac{\mathbf{A} \cdot \mathbf{B}}{|\mathbf{A}|} \right) \frac{\mathbf{A}}{|\mathbf{A}|} \qquad \begin{aligned} |\mathbf{B}| \cos \theta &= \frac{|\mathbf{A}||\mathbf{B}| \cos \theta}{|\mathbf{A}|} \\ &= \frac{\mathbf{A} \cdot \mathbf{B}}{|\mathbf{A}|} \end{aligned}$$

$$= \left(\mathbf{B} \cdot \frac{\mathbf{A}}{|\mathbf{A}|} \right) \frac{\mathbf{A}}{|\mathbf{A}|}.$$

$$\boxed{\text{proj}_\mathbf{A}\ \mathbf{B} = \left(\mathbf{B} \cdot \frac{\mathbf{A}}{|\mathbf{A}|} \right) \frac{\mathbf{A}}{|\mathbf{A}|} = \left(\frac{\mathbf{B} \cdot \mathbf{A}}{\mathbf{A} \cdot \mathbf{A}} \right) \mathbf{A}} \qquad (10)$$

The number $|\mathbf{B}| \cos \theta$ is called the **scalar component of B in the direction of A.** Since

$$|\mathbf{B}| \cos \theta = \mathbf{B} \cdot \frac{\mathbf{A}}{|\mathbf{A}|}, \qquad (11)$$

we can find the scalar component by "dotting" **B** with the direction of **A.** Equation (10) says that the vector projection of **B** onto **A** is the scalar component of **B** in the direction of **A** times the direction of **A.**

While the first part of Eq. (10) describes the effect of **B** in the direction of **A,** the second part is better for calculation because it avoids square roots.

EXAMPLE 4   Find the vector projection of $\mathbf{B} = 6\,\mathbf{i} + 3\,\mathbf{j} + 2\,\mathbf{k}$ onto $\mathbf{A} = \mathbf{i} - 2\,\mathbf{j} - 2\,\mathbf{k}$ and the scalar component of **B** in the direction of **A.**

Solution   We find proj$_\mathbf{A}$ **B** from Eq. (10):

$$\text{proj}_\mathbf{A}\ \mathbf{B} = \frac{\mathbf{B} \cdot \mathbf{A}}{\mathbf{A} \cdot \mathbf{A}}\mathbf{A} = \frac{6 - 6 - 4}{1 + 4 + 4}\,(\mathbf{i} - 2\,\mathbf{j} - 2\,\mathbf{k})$$

$$= -\frac{4}{9}\,(\mathbf{i} - 2\,\mathbf{j} - 2\,\mathbf{k}) = -\frac{4}{9}\,\mathbf{i} + \frac{8}{9}\,\mathbf{j} + \frac{8}{9}\,\mathbf{k}.$$

We find the scalar component of **B** in the direction of **A** from Eq. (11):

$$|\mathbf{B}| \cos \theta = \mathbf{B} \cdot \frac{\mathbf{A}}{|\mathbf{A}|} = (6\,\mathbf{i} + 3\,\mathbf{j} + 2\,\mathbf{k}) \cdot \left(\frac{1}{3}\,\mathbf{i} - \frac{2}{3}\,\mathbf{j} - \frac{2}{3}\,\mathbf{k} \right)$$

$$= 2 - 2 - \frac{4}{3} = -\frac{4}{3}. \qquad \square$$

Writing a Vector as a Sum of Orthogonal Vectors

In mechanics, we often need to express a vector **B** as a sum of a vector parallel to a vector **A** and a vector orthogonal to **A.** We can accomplish this with the equation

$$\mathbf{B} = \text{proj}_\mathbf{A}\ \mathbf{B} + (\mathbf{B} - \text{proj}_\mathbf{A}\ \mathbf{B}), \qquad (12)$$

shown in Fig. 10.29.

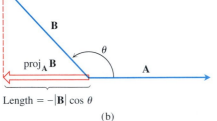

10.29 Writing **B** as the sum of vectors parallel and orthogonal to **A.**

Where vectors came from

Although Aristotle used vectors to describe the effects of forces, the idea of resolving vectors into geometric components parallel to the coordinate axes came from Descartes. The algebra of vectors we use today was developed simultaneously and independently in the 1870s by Josiah Willard Gibbs (1839–1903), a mathematical physicist at Yale University, and by the English mathematical physicist Oliver Heaviside (1850–1925), the Heaviside of Heaviside layer fame. The works of Gibbs and Heaviside grew out of more complicated mathematical theories developed some years earlier by the Irish mathematician William Hamilton (1805–1865) and the German linguist, physicist, and geometer Hermann Grassman (1809–1877).

How to Write B as a Vector Parallel to A Plus a Vector Orthogonal to A

$$\mathbf{B} = \text{proj}_{\mathbf{A}}\, \mathbf{B} + (\mathbf{B} - \text{proj}_{\mathbf{A}}\, \mathbf{B})$$

$$= \underbrace{\left(\frac{\mathbf{B}\cdot\mathbf{A}}{\mathbf{A}\cdot\mathbf{A}}\right)\mathbf{A}}_{\text{parallel to }\mathbf{A}} + \underbrace{\left(\mathbf{B} - \left(\frac{\mathbf{B}\cdot\mathbf{A}}{\mathbf{A}\cdot\mathbf{A}}\right)\mathbf{A}\right)}_{\text{orthogonal to }\mathbf{A}} \qquad (13)$$

EXAMPLE 5   Express $\mathbf{B} = 2\,\mathbf{i} + \mathbf{j} - 3\,\mathbf{k}$ as the sum of a vector parallel to $\mathbf{A} = 3\,\mathbf{i} - \mathbf{j}$ and a vector orthogonal to $\mathbf{A}$.

Solution   We use Eq. (13). With

$$\mathbf{A}\cdot\mathbf{B} = 6 - 1 = 5 \qquad \text{and} \qquad \mathbf{A}\cdot\mathbf{A} = 9 + 1 = 10,$$

Eq. (13) gives

$$\mathbf{B} = \frac{\mathbf{B}\cdot\mathbf{A}}{\mathbf{A}\cdot\mathbf{A}}\mathbf{A} + \left(\mathbf{B} - \frac{\mathbf{B}\cdot\mathbf{A}}{\mathbf{A}\cdot\mathbf{A}}\mathbf{A}\right) = \frac{5}{10}(3\,\mathbf{i} - \mathbf{j}) + \left(2\,\mathbf{i} + \mathbf{j} - 3\,\mathbf{k} - \frac{5}{10}(3\,\mathbf{i} - \mathbf{j})\right)$$

$$= \left(\frac{3}{2}\,\mathbf{i} - \frac{1}{2}\,\mathbf{j}\right) + \left(\frac{1}{2}\,\mathbf{i} + \frac{3}{2}\,\mathbf{j} - 3\,\mathbf{k}\right).$$

Check: The first vector in the sum is parallel to $\mathbf{A}$ because it is $(1/2)\,\mathbf{A}$. The second vector in the sum is orthogonal to $\mathbf{A}$ because

$$\left(\frac{1}{2}\,\mathbf{i} + \frac{3}{2}\,\mathbf{j} - 3\,\mathbf{k}\right)\cdot(3\,\mathbf{i} - \mathbf{j}) = \frac{3}{2} - \frac{3}{2} = 0.$$

10.30 The work done by a constant force **F** during a displacement **D** is $(|\mathbf{F}|\cos\theta)|\mathbf{D}|$.

Work

In Section 5.8, we calculated the work done by a constant force of magnitude F in moving an object through a distance d as $W = Fd$. That formula holds only if the force is directed along the line of motion. If a force $\mathbf{F}$ moving an object through a displacement $\mathbf{D} = \overrightarrow{PQ}$ has some other direction, the work is performed by the component of $\mathbf{F}$ in the direction of $\mathbf{D}$. If θ is the angle between $\mathbf{F}$ and $\mathbf{D}$ (Fig. 10.30), then

$$\text{Work} = \left(\begin{array}{c}\text{scalar component of } \mathbf{F} \\ \text{in the direction of } \mathbf{D}\end{array}\right)\left(\text{length of } \mathbf{D}\right)$$

$$= (|\mathbf{F}|\cos\theta)\,|\mathbf{D}|$$

$$= \mathbf{F}\cdot\mathbf{D}$$

Definition

The **work** done by a constant force $\mathbf{F}$ acting through a displacement $\mathbf{D} = \overrightarrow{PQ}$ is

$$W = \mathbf{F}\cdot\mathbf{D} = |\mathbf{F}||\mathbf{D}|\cos\theta, \qquad (14)$$

where θ is the angle between $\mathbf{F}$ and $\mathbf{D}$.

Work

The standard units of work are the foot-pound and newton-meter, both force-distance units. The newton-meter is usually called a *joule*. For more about the notion of work and the units involved, read the first few paragraphs of Section 5.8.

EXAMPLE 6 If $|\mathbf{F}| = 40$ N (newtons), $|\mathbf{D}| = 3$ m, and $\theta = 60°$, the work done by $\mathbf{F}$ in acting from P to Q is

$$\text{Work} = |\mathbf{F}||\mathbf{D}|\cos\theta \qquad \text{Eq. (14)}$$
$$= (40)(3)\cos 60° \qquad \text{Given values}$$
$$= (120)(1/2)$$
$$= 60 \text{ J (joules)} \qquad \square$$

We will encounter more interesting work problems in Chapter 14 when we can find the work done by a variable force along a path in space.

Exercises 10.3

Calculations

In Exercises 1–10, find

a) $\mathbf{A} \cdot \mathbf{B}$, $|\mathbf{A}|$, $|\mathbf{B}|$;
b) the cosine of the angle between $\mathbf{A}$ and $\mathbf{B}$;
c) the scalar component of $\mathbf{B}$ in the direction of $\mathbf{A}$;
d) the vector $\text{proj}_\mathbf{A}\mathbf{B}$.

1. $\mathbf{A} = 2\mathbf{i} - 4\mathbf{j} + \sqrt{5}\,\mathbf{k}$, $\quad \mathbf{B} = -2\mathbf{i} + 4\mathbf{j} - \sqrt{5}\,\mathbf{k}$

2. $\mathbf{A} = (3/5)\mathbf{i} + (4/5)\mathbf{k}$, $\quad \mathbf{B} = 5\mathbf{i} + 12\mathbf{j}$

3. $\mathbf{A} = 10\mathbf{i} + 11\mathbf{j} - 2\mathbf{k}$, $\quad \mathbf{B} = 3\mathbf{j} + 4\mathbf{k}$

4. $\mathbf{A} = 2\mathbf{i} + 10\mathbf{j} - 11\mathbf{k}$, $\quad \mathbf{B} = 2\mathbf{i} + 2\mathbf{j} + \mathbf{k}$

5. $\mathbf{A} = -2\mathbf{i} + 7\mathbf{j}$, $\quad \mathbf{B} = \mathbf{k}$

6. $\mathbf{A} = \dfrac{1}{\sqrt{2}}\mathbf{i} + \dfrac{1}{\sqrt{3}}\mathbf{j} + \dfrac{1}{\sqrt{6}}\mathbf{k}$, $\quad \mathbf{B} = \dfrac{1}{\sqrt{2}}\mathbf{j} - \mathbf{k}$

7. $\mathbf{A} = 5\mathbf{j} - 3\mathbf{k}$, $\quad \mathbf{B} = \mathbf{i} + \mathbf{j} + \mathbf{k}$

8. $\mathbf{A} = \mathbf{i} + \mathbf{k}$, $\quad \mathbf{B} = \mathbf{i} + \mathbf{j} + \mathbf{k}$

9. $\mathbf{A} = -\mathbf{i} + \mathbf{j}$, $\quad \mathbf{B} = \sqrt{2}\,\mathbf{i} + \sqrt{3}\,\mathbf{j} + 2\mathbf{k}$

10. $\mathbf{A} = -5\mathbf{i} + \mathbf{j}$, $\quad \mathbf{B} = 2\mathbf{i} + \sqrt{17}\,\mathbf{j} + 10\mathbf{k}$

11. Write $\mathbf{B} = 3\mathbf{j} + 4\mathbf{k}$ as the sum of a vector parallel to $\mathbf{A} = \mathbf{i} + \mathbf{j}$ and a vector orthogonal to $\mathbf{A}$.

12. Write $\mathbf{B} = \mathbf{j} + \mathbf{k}$ as the sum of a vector parallel to $\mathbf{A} = \mathbf{i} + \mathbf{j}$ and a vector orthogonal to $\mathbf{A}$.

13. Write $\mathbf{B} = 8\mathbf{i} + 4\mathbf{j} - 12\mathbf{k}$ as the sum of a vector parallel to $\mathbf{A} = \mathbf{i} + 2\mathbf{j} - \mathbf{k}$ and a vector orthogonal to $\mathbf{A}$.

14. $\mathbf{B} = \mathbf{i} + (\mathbf{j} + \mathbf{k})$ is already the sum of a vector parallel to $\mathbf{i}$ and a vector orthogonal to $\mathbf{i}$. If you use Eq. (13) with $\mathbf{A} = \mathbf{i}$, do you get $\mathbf{B}_1 = \mathbf{i}$ and $\mathbf{B}_2 = \mathbf{j} + \mathbf{k}$? Try it and find out.

Geometry

15. *Sums and differences.* In the accompanying figure, it looks as

if $\mathbf{v}_1 + \mathbf{v}_2$ and $\mathbf{v}_1 - \mathbf{v}_2$ are orthogonal. Is this mere coincidence, or are there circumstances under which we may expect the sum of two vectors to be orthogonal to their difference? Give reasons for your answer.

16. Suppose that AB is the diameter of a circle with center O and that C is a point on one of the two arcs joining A and B. Show that $\overrightarrow{CA}$ and $\overrightarrow{CB}$ are orthogonal.

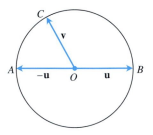

17. Show that the diagonals of a rhombus (parallelogram with sides of equal length) are perpendicular.

18. Show that squares are the only rectangles with perpendicular diagonals.

19. Prove the fact, often exploited by carpenters, that a parallelogram is a rectangle if and only if its diagonals are equal in length.

20. Show that the indicated diagonal of the parallelogram determined by vectors **u** and **v** bisects the angle between **u** and **v** if $|\mathbf{u}| = |\mathbf{v}|$.

21. The accompanying figure shows a pyramid $OABCD$ with a square base whose sides are 1 unit long. The pyramid's height is also 1 unit, and the point D stands directly above the midpoint of the diagonal OB. Find the angle between $\overrightarrow{OB}$ and $\overrightarrow{OD}$.

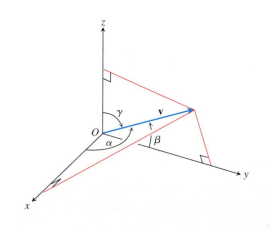

22. *Direction angles and direction cosines.* The direction angles α, β, and γ of a vector $\mathbf{v} = a\,\mathbf{i} + b\,\mathbf{j} + c\,\mathbf{k}$ are defined as follows:

α is the angle between **v** and the positive x-axis ($0 \le \alpha \le \pi$),

β is the angle between **v** and the positive y-axis ($0 \le \beta \le \pi$),

γ is the angle between **v** and the positive z-axis ($0 \le \gamma \le \pi$).

a) Show that

$$\cos\alpha = \frac{a}{|\mathbf{v}|}, \quad \cos\beta = \frac{b}{|\mathbf{v}|}, \quad \cos\gamma = \frac{c}{|\mathbf{v}|},$$

and $\cos^2\alpha + \cos^2\beta + \cos^2\gamma = 1$. These cosines are called the direction cosines of **v.**

b) *Unit vectors are built from direction cosines.* Show that if $\mathbf{v} = a\,\mathbf{i} + b\,\mathbf{j} + c\,\mathbf{k}$ is a unit vector then a, b, and c are the direction cosines of **v.**

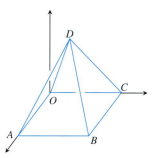

Angles Between Vectors

CALCULATOR Find the angles between the vectors in Exercises 23–26 to the nearest hundredth of a radian.

23. $\mathbf{A} = 2\,\mathbf{i} + \mathbf{j}, \quad \mathbf{B} = \mathbf{i} + 2\,\mathbf{j} - \mathbf{k}$

24. $\mathbf{A} = 2\,\mathbf{i} - 2\,\mathbf{j} + \mathbf{k}, \quad \mathbf{B} = 3\,\mathbf{i} + 4\,\mathbf{k}$

25. $\mathbf{A} = \sqrt{3}\,\mathbf{i} - 7\,\mathbf{j}, \quad \mathbf{B} = \sqrt{3}\,\mathbf{i} + \mathbf{j} - 2\,\mathbf{k}$

26. $\mathbf{A} = \mathbf{i} + \sqrt{2}\,\mathbf{j} - \sqrt{2}\,\mathbf{k}, \quad \mathbf{B} = -\mathbf{i} + \mathbf{j} + \mathbf{k}$

CALCULATOR Find the angles in Exercises 27–29 to the nearest hundredth of a radian.

27. The interior angles of the triangle ABC whose vertices are $A(-1, 0, 2)$, $B(2, 1, -1)$, and $C(1, -2, 2)$

28. The angle between $\mathbf{A} = 2\,\mathbf{i} + 2\,\mathbf{j} + \mathbf{k}$ and $\mathbf{B} = 2\,\mathbf{i} + 10\,\mathbf{j} - 11\,\mathbf{k}$

29. The angle between the diagonal of a cube and the diagonal of one of its faces. (*Hint:* Use a cube whose edges represent **i**, **j**, and **k**.)

30. **CALCULATOR** A water main is to be constructed with a 20% grade in the north direction and a 10% grade in the east direction. Determine the angle θ required in the water main for the turn from north to east. (See Preliminaries, Section 2, Exercise 46.)

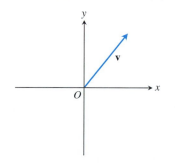

Theory and Examples

31. a) Use the fact that $\mathbf{u} \cdot \mathbf{v} = |\mathbf{u}||\mathbf{v}| \cos\theta$ to show that the inequality $|\mathbf{u} \cdot \mathbf{v}| \le |\mathbf{u}||\mathbf{v}|$ for any vectors **u** and **v**.

b) Under what circumstances, if any, does $|\mathbf{u} \cdot \mathbf{v}|$ equal $|\mathbf{u}||\mathbf{v}|$? Give reasons for your answer.

32. Copy the axes and vector shown here. Then shade in the points (x, y) for which $(x\,\mathbf{i} + y\,\mathbf{j}) \cdot \mathbf{v} \le 0$. Justify your answer.

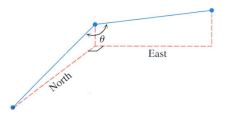

33. If $\mathbf{u}_1$ and $\mathbf{u}_2$ are orthogonal unit vectors and $\mathbf{v} = a\,\mathbf{u}_1 + b\,\mathbf{u}_2$, find $\mathbf{v} \cdot \mathbf{u}_1$.

34. *Cancellation in dot products.* In real-number multiplication, if $ab_1 = ab_2$ and a is not zero, we can cancel the a and conclude that $b_1 = b_2$. Does the same rule hold for vector multiplication: If $\mathbf{A} \cdot \mathbf{B}_1 = \mathbf{A} \cdot \mathbf{B}_2$ and $\mathbf{A} \neq 0$, can you conclude that $\mathbf{B}_1 = \mathbf{B}_2$? Give reasons for your answer.

35. Suppose $\mathbf{A}$, $\mathbf{B}$, and $\mathbf{C}$ are mutually orthogonal vectors. Let $\mathbf{D} = 5\mathbf{A} - 6\mathbf{B} + 3\mathbf{C}$.

   a) If $\mathbf{A}$, $\mathbf{B}$, and $\mathbf{C}$ are unit vectors, find $|\mathbf{D}|$, the magnitude of $\mathbf{D}$.

   b) If $|\mathbf{A}| = 2$, $|\mathbf{B}| = 3$, and $|\mathbf{C}| = 4$, find $|\mathbf{D}|$.

36. Suppose $\mathbf{A}$, $\mathbf{B}$, and $\mathbf{C}$ are mutually orthogonal unit vectors. If $\mathbf{D}$ is a vector such that $\mathbf{D} = \alpha \mathbf{A} + \beta \mathbf{B} + \gamma \mathbf{C}$ where α, β, and γ are scalars, prove that $\alpha = \mathbf{D} \cdot \mathbf{A}$, $\beta = \mathbf{D} \cdot \mathbf{B}$, and $\gamma = \mathbf{D} \cdot \mathbf{C}$.

Work

37. Find the work done by a force $\mathbf{F} = 5\mathbf{k}$ (magnitude 5 N) in moving an object along the line from the origin to the point $(1, 1, 1)$ (distance in meters).

38. The union Pacific's *Big Boy* locomotive could pull 6000-ton trains with a tractive effort (pull) of 602,148 N (135,375 lb). At this level of effort, about how much work did *Big Boy* do on the (approximately straight) 605-km journey from San Francisco to Los Angeles?

39. How much work does it take to slide a crate 20 m along a loading dock by pulling on it with a 200-N force at an angle of 30° from the horizontal?

40. The wind passing over a boat's sail exerted a 1000-lb magnitude force $\mathbf{F}$ as shown here. How much work did the wind perform in moving the boat forward 1 mi? Answer in foot-pounds.

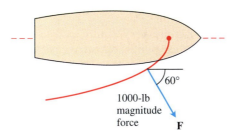

1000-lb magnitude force **F**

Equations for Lines in the Plane

41. Show that the vector $\mathbf{v} = a\mathbf{i} + b\mathbf{j}$ is perpendicular to the line $ax + by = c$ by establishing that the slope of $\mathbf{v}$ is the negative reciprocal of the slope of the given line.

42. Show that the vector $\mathbf{v} = a\mathbf{i} + b\mathbf{j}$ is parallel to the line $bx - ay = c$ by establishing that the slope of the line segment representing $\mathbf{v}$ is the same as the slope of the given line.

In Exercises 43–46, use the result of Exercise 41 to find an equation for the line through P perpendicular to $\mathbf{v}$. Then sketch the line. Include $\mathbf{v}$ in your sketch *as a vector starting at the origin*.

43. $P(2, 1)$, $\quad \mathbf{v} = \mathbf{i} + 2\mathbf{j}$

44. $P(-1, 2)$, $\quad \mathbf{v} = -2\mathbf{i} - \mathbf{j}$

45. $P(-2, -7)$, $\quad \mathbf{v} = -2\mathbf{i} + \mathbf{j}$

46. $P(11, 10)$, $\quad \mathbf{v} = 2\mathbf{i} - 3\mathbf{j}$

In Exercises 47–50, use the result of Exercise 42 to find an equation for the line through P parallel to $\mathbf{v}$. Then sketch the line. Include $\mathbf{v}$ in your sketch *as a vector starting at the origin*.

47. $P(-2, 1)$, $\quad \mathbf{v} = \mathbf{i} - \mathbf{j}$

48. $P(0, -2)$, $\quad \mathbf{v} = 2\mathbf{i} + 3\mathbf{j}$

49. $P(1, 2)$, $\quad \mathbf{v} = -\mathbf{i} - 2\mathbf{j}$

50. $P(1, 3)$, $\quad \mathbf{v} = 3\mathbf{i} - 2\mathbf{j}$

Angles Between Lines in the Plane

The acute angle between intersecting lines that do not cross at right angles is the same as the angle determined by vectors normal to the lines or by the vectors parallel to the lines.

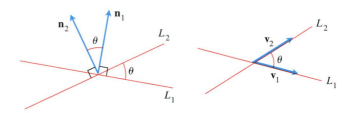

Use this fact and the results of Exercise 41 or 42 to find the acute angles between the lines in Exercises 51–54.

51. $3x + y = 5$, $\quad 2x - y = 4$

52. $y = \sqrt{3}x - 1$, $\quad y = -\sqrt{3}x + 2$

53. $\sqrt{3}x - y = -2$, $\quad x - \sqrt{3}y = 1$

54. $x + \sqrt{3}y = 1$, $\quad (1 - \sqrt{3})x + (1 + \sqrt{3})y = 8$

CALCULATOR In Exercises 55 and 56, find the acute angle between the lines to the nearest hundredth of a radian.

55. $3x - 4y = 3$, $\quad x - y = 7$

56. $12x + 5y = 1$, $\quad 2x - 2y = 3$

Angles Between Differentiable Curves

The angles between two differentiable curves at a point of intersection are the angles between the curves' tangent lines at these points. Find the angles between the curves in Exercises 57–60. (You will not need a calculator.)

57. $y = (3/2) - x^2$, $\quad y = x^2$ (two points of intersection)

58. $x = (3/4) - y^2$, $\quad x = y^2 - (3/4)$ (two points of intersection)

59. $y = x^3$, $\quad x = y^2$ (two points of intersection)

60. $y = -x^2$, $\quad y = \sqrt[3]{x}$ (two points of intersection)

Cross Products

In studying lines in the plane, we needed to describe how a line was tilting, and we did so with the notions of slope and angle of inclination. In space, we need to be able to describe how a plane is tilting. We accomplish this by multiplying two vectors in the plane together to get a third vector perpendicular to the plane. The direction of this third vector tells us the "inclination" of the plane. The product we use to multiply the vectors together is the *vector* or *cross product,* the second of the two vector multiplication methods we study in calculus.

Cross products are widely used to describe the effects of forces in studies of electricity, magnetism, fluid flows, and orbital mechanics. This section presents the mathematical properties that account for the use of cross products in these fields.

The Cross Product of Two Vectors in Space

We start with two nonzero vectors **A** and **B** in space. If **A** and **B** are not parallel, they determine a plane. We select a unit vector **n** perpendicular to the plane by the **right-hand rule.** This means we choose **n** to be the unit (normal) vector that points the way your right thumb points when your fingers curl through the angle θ from **A** to **B** (Fig. 10.31). We then define the **vector product** $\mathbf{A} \times \mathbf{B}$ ("A cross B") to be the *vector*

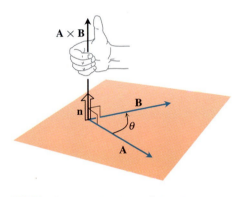

10.31 The construction of $\mathbf{A} \times \mathbf{B}$.

Definition

$$\mathbf{A} \times \mathbf{B} = (|\mathbf{A}||\mathbf{B}| \sin\theta)\, \mathbf{n} \qquad (1)$$

The vector $\mathbf{A} \times \mathbf{B}$ is orthogonal to both **A** and **B** because it is a scalar multiple of **n**. The vector product of **A** and **B** is often called the **cross product** of **A** and **B** because of the cross in the notation $\mathbf{A} \times \mathbf{B}$.

Since the sines of 0 and π are both zero in Eq. (1), it makes sense to define the cross product of two parallel nonzero vectors to be **0.**

If one or both of **A** and **B** are zero, we also define $\mathbf{A} \times \mathbf{B}$ to be zero. This way, the cross product of two vectors **A** and **B** is zero if and only if **A** and **B** are parallel or one or both of them are zero.

> Nonzero vectors **A** and **B** are parallel if and only if $\mathbf{A} \times \mathbf{B} = \mathbf{0}$.

10.32 The construction of $\mathbf{B} \times \mathbf{A}$.

A x B vs. B x A

Reversing the order of the factors in a nonzero cross product reverses the direction of the product. When the fingers of our right hand curl through the angle θ from **B** to **A,** our thumb points the opposite way and the unit vector we choose in forming $\mathbf{B} \times \mathbf{A}$ is the negative of the one we choose in forming $\mathbf{A} \times \mathbf{B}$ (Fig. 10.32). Thus,

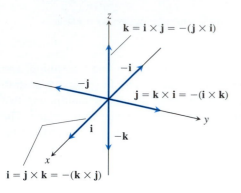

10.33 The pairwise cross products of **i**, **j**, and **k**.

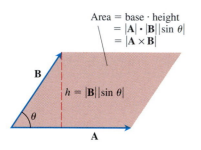

Area = base · height
= |**A**| · |**B**||sin θ|
= |**A** × **B**|

$h = |\mathbf{B}||\sin\theta|$

10.34 The parallelogram determined by **A** and **B**.

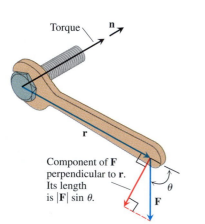

Torque **n**

r

Component of **F** perpendicular to **r**. Its length is |**F**| sin θ.

θ

F

10.35 The torque vector describes the tendency of the force **F** to drive the bolt forward.

for all vectors **A** and **B**,

$$\mathbf{B} \times \mathbf{A} = -(\mathbf{A} \times \mathbf{B}). \tag{2}$$

Unlike the dot product, the cross product is not commutative.

When we apply the definition to calculate the pairwise cross products of **i**, **j**, and **k**, we find (Fig. 10.33)

$$\mathbf{i} \times \mathbf{j} = -(\mathbf{j} \times \mathbf{i}) = \mathbf{k}$$
$$\mathbf{j} \times \mathbf{k} = -(\mathbf{k} \times \mathbf{j}) = \mathbf{i} \tag{3}$$
$$\mathbf{k} \times \mathbf{i} = -(\mathbf{i} \times \mathbf{k}) = \mathbf{j}$$

Diagram for recalling these products

and

$$\mathbf{i} \times \mathbf{i} = \mathbf{j} \times \mathbf{j} = \mathbf{k} \times \mathbf{k} = \mathbf{0}.$$

| A x B | Is the Area of a Parallelogram

Because **n** is a unit vector, the magnitude of **A** × **B** is

$$|\mathbf{A} \times \mathbf{B}| = |\mathbf{A}||\mathbf{B}||\sin\theta||\mathbf{n}| = |\mathbf{A}||\mathbf{B}|\sin\theta. \tag{4}$$

This is the area of the parallelogram determined by **A** and **B** (Fig. 10.34), |**A**| being the base of the parallelogram and |**B**||sin θ| the height.

Torque

When we turn a bolt by applying a force **F** to a wrench (Fig. 10.35), the torque we produce acts along the axis of the bolt to drive the bolt forward. The magnitude of the torque depends on how far out on the wrench the force is applied and on how much of the force is perpendicular to the wrench at the point of application. The number we use to measure the torque's magnitude is the product of the length of the lever arm **r** and the scalar component of **F** perpendicular to **r**. In the notation of Fig. 10.35,

$$\text{Magnitude of torque vector} = |\mathbf{r}||\mathbf{F}|\sin\theta,$$

or |**r** × **F**|. If we let **n** be a unit vector along the axis of the bolt in the direction of the torque, then a complete description of the torque vector is **r** × **F**, or

$$\text{Torque vector} = (|\mathbf{r}||\mathbf{F}|\sin\theta)\,\mathbf{n}.$$

Recall that we defined **A** × **B** to be **0** when **A** and **B** are parallel. This is consistent with the torque interpretation as well. If the force **F** in Fig. 10.35 is parallel to the wrench, meaning that we are trying to turn the bolt by pushing or pulling along the line of the wrench's handle, the torque produced is zero.

EXAMPLE 1   The magnitude of the torque exerted by force **F** about the pivot

10.36 The magnitude of the torque exerted by **F** and P is about 56.4 ft-lb (Example 1).

point P in Fig. 10.36 is

$$|\overrightarrow{PQ} \times \mathbf{F}| = |\overrightarrow{PQ}||\mathbf{F}| \sin 70° \qquad \text{Eq. (4)}$$

$$\approx (3)(20)(0.94)$$

$$\approx 56.4 \text{ ft-lb.} \qquad \square$$

The Associative and Distributive Laws

As a rule, cross-product multiplication is *not associative* because $(\mathbf{A} \times \mathbf{B}) \times \mathbf{C}$ lies in the plane of **A** and **B** whereas $\mathbf{A} \times (\mathbf{B} \times \mathbf{C})$ lies in the plane of **B** and **C**. However, the following laws do hold.

Scalar Distributive Law

$$(r\mathbf{A}) \times (s\mathbf{B}) = (rs)(\mathbf{A} \times \mathbf{B}) \qquad (5)$$

Vector Distributive Laws

$$\mathbf{A} \times (\mathbf{B} + \mathbf{C}) = \mathbf{A} \times \mathbf{B} + \mathbf{A} \times \mathbf{C} \qquad (6)$$

$$(\mathbf{B} + \mathbf{C}) \times \mathbf{A} = \mathbf{B} \times \mathbf{A} + \mathbf{C} \times \mathbf{A} \qquad (7)$$

As a special case of Eq. (5) we also have

$$(-\mathbf{A}) \times \mathbf{B} = \mathbf{A} \times (-\mathbf{B}) = -(\mathbf{A} \times \mathbf{B}). \qquad (8)$$

The Scalar Distributive Law can be verified by applying Eq. (1) to the products on both sides of Eq. (5) and comparing the results. The Vector Distributive Law in Eq. (6) is not easy to prove. We will assume it here and leave the proof to Appendix 7. Equation (7) follows from Eq. (6): Multiply both sides of Eq. (6) by -1 and reverse the orders of the products.

The Determinant Formula for A x B

Our next objective is to calculate $\mathbf{A} \times \mathbf{B}$ from the components of **A** and **B** relative to a Cartesian coordinate system.

Suppose

$$\mathbf{A} = a_1 \mathbf{i} + a_2 \mathbf{j} + a_3 \mathbf{k}, \qquad \mathbf{B} = b_1 \mathbf{i} + b_2 \mathbf{j} + b_3 \mathbf{k}.$$

Then the distributive laws and the rules for multiplying **i**, **j**, and **k** tell us that

$$\mathbf{A} \times \mathbf{B} = (a_1 \mathbf{i} + a_2 \mathbf{j} + a_3 \mathbf{k}) \times (b_1 \mathbf{i} + b_2 \mathbf{j} + b_3 \mathbf{k})$$

$$= a_1 b_1 \mathbf{i} \times \mathbf{i} + a_1 b_2 \mathbf{i} \times \mathbf{j} + a_1 b_3 \mathbf{i} \times \mathbf{k}$$

$$+ a_2 b_1 \mathbf{j} \times \mathbf{i} + a_2 b_2 \mathbf{j} \times \mathbf{j} + a_2 b_3 \mathbf{j} \times \mathbf{k}$$

$$+ a_3 b_1 \mathbf{k} \times \mathbf{i} + a_3 b_2 \mathbf{k} \times \mathbf{j} + a_3 b_3 \mathbf{k} \times \mathbf{k}$$

$$= (a_2 b_3 - a_3 b_2) \mathbf{i} - (a_1 b_3 - a_3 b_1) \mathbf{j} + (a_1 b_2 - a_2 b_1) \mathbf{k}.$$

Determinants

(For more information, see Appendix 8.)

$$\begin{vmatrix} a & b \\ c & d \end{vmatrix} = ad - bc$$

EXAMPLE

$$\begin{vmatrix} 2 & 1 \\ -4 & 3 \end{vmatrix} = (2)(3) - (1)(-4)$$

$$= 6 + 4 = 10 \qquad \square$$

$$\begin{vmatrix} a_1 & a_2 & a_3 \\ b_1 & b_2 & b_3 \\ c_1 & c_2 & c_3 \end{vmatrix}$$

$$= a_1 \begin{vmatrix} b_2 & b_3 \\ c_2 & c_3 \end{vmatrix} - a_2 \begin{vmatrix} b_1 & b_3 \\ c_1 & c_3 \end{vmatrix} + a_3 \begin{vmatrix} b_1 & b_2 \\ c_1 & c_2 \end{vmatrix}$$

EXAMPLE

$$\begin{vmatrix} -5 & 3 & 1 \\ 2 & 1 & 1 \\ -4 & 3 & 1 \end{vmatrix}$$

$$= (-5) \begin{vmatrix} 1 & 1 \\ 3 & 1 \end{vmatrix} - (3) \begin{vmatrix} 2 & 1 \\ -4 & 1 \end{vmatrix}$$

$$+ (1) \begin{vmatrix} 2 & 1 \\ -4 & 3 \end{vmatrix}$$

$$= -5(1 - 3) - 3(2 + 4) + 1(6 + 4)$$

$$= 10 - 18 + 10 = 2 \qquad \square$$

The terms in the last line are the same as the terms in the expansion of the symbolic determinant

$$\begin{vmatrix} \mathbf{i} & \mathbf{j} & \mathbf{k} \\ a_1 & a_2 & a_3 \\ b_1 & b_2 & b_3 \end{vmatrix}.$$

We therefore have the following rule.

> If $\mathbf{A} = a_1\,\mathbf{i} + a_2\,\mathbf{j} + a_3\,\mathbf{k}$ and $\mathbf{B} = b_1\,\mathbf{i} + b_2\,\mathbf{j} + b_3\,\mathbf{k}$, then
>
> $$\mathbf{A} \times \mathbf{B} = \begin{vmatrix} \mathbf{i} & \mathbf{j} & \mathbf{k} \\ a_1 & a_2 & a_3 \\ b_1 & b_2 & b_3 \end{vmatrix}. \tag{9}$$

EXAMPLE 2   Find $\mathbf{A} \times \mathbf{B}$ and $\mathbf{B} \times \mathbf{A}$ if

$$\mathbf{A} = 2\,\mathbf{i} + \mathbf{j} + \mathbf{k}, \qquad \mathbf{B} = -4\,\mathbf{i} + 3\,\mathbf{j} + \mathbf{k}.$$

Solution

$$\mathbf{A} \times \mathbf{B} = \begin{vmatrix} \mathbf{i} & \mathbf{j} & \mathbf{k} \\ 2 & 1 & 1 \\ -4 & 3 & 1 \end{vmatrix} = \begin{vmatrix} 1 & 1 \\ 3 & 1 \end{vmatrix} \mathbf{i} - \begin{vmatrix} 2 & 1 \\ -4 & 1 \end{vmatrix} \mathbf{j} + \begin{vmatrix} 2 & 1 \\ -4 & 3 \end{vmatrix} \mathbf{k} \qquad \text{Eq. (9)}$$

$$= -2\,\mathbf{i} - 6\,\mathbf{j} + 10\,\mathbf{k}$$

$$\mathbf{B} \times \mathbf{A} = -(\mathbf{A} \times \mathbf{B}) = 2\,\mathbf{i} + 6\,\mathbf{j} - 10\,\mathbf{k} \qquad \text{Eq. (2)}$$

$$\square$$

EXAMPLE 3   Find a vector perpendicular to the plane of $P(1, -1, 0)$, $Q(2, 1, -1)$, and $R(-1, 1, 2)$.

Solution   The vector $\overrightarrow{PQ} \times \overrightarrow{PR}$ is perpendicular to the plane because it is perpendicular to both vectors. In terms of components,

$$\overrightarrow{PQ} = (2 - 1)\,\mathbf{i} + (1 + 1)\,\mathbf{j} + (-1 - 0)\,\mathbf{k} = \mathbf{i} + 2\,\mathbf{j} - \mathbf{k}$$

$$\overrightarrow{PR} = (-1 - 1)\,\mathbf{i} + (1 + 1)\,\mathbf{j} + (2 - 0)\,\mathbf{k} = -2\,\mathbf{i} + 2\,\mathbf{j} + 2\,\mathbf{k}$$

$$\overrightarrow{PQ} \times \overrightarrow{PR} = \begin{vmatrix} \mathbf{i} & \mathbf{j} & \mathbf{k} \\ 1 & 2 & -1 \\ -2 & 2 & 2 \end{vmatrix} = \begin{vmatrix} 2 & -1 \\ 2 & 2 \end{vmatrix} \mathbf{i} - \begin{vmatrix} 1 & -1 \\ -2 & 2 \end{vmatrix} \mathbf{j} + \begin{vmatrix} 1 & 2 \\ -2 & 2 \end{vmatrix} \mathbf{k}$$

$$= 6\,\mathbf{i} + 6\,\mathbf{k}. \qquad \square$$

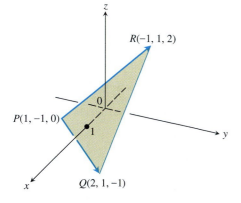

10.37 The area of triangle *PQR* is half of $|\overrightarrow{PQ} \times \overrightarrow{PR}|$ (Example 4).

EXAMPLE 4   Find the area of the triangle with vertices $P(1, -1, 0)$, $Q(2, 1, -1)$, and $R(-1, 1, 2)$ (Fig. 10.37).

Solution The area of the parallelogram determined by P, Q, and R is

$$|\overrightarrow{PQ} \times \overrightarrow{PR}| = |6\mathbf{i} + 6\mathbf{k}| \qquad \text{Values from Example 3}$$

$$= \sqrt{(6)^2 + (6)^2} = \sqrt{2 \cdot 36} = 6\sqrt{2}.$$

The triangle's area is half of this, or $3\sqrt{2}$. ❏

EXAMPLE 5 Find a unit vector perpendicular to the plane of $P(1, -1, 0)$, $Q(2, 1, -1)$, and $R(-1, 1, 2)$.

Solution Since $\overrightarrow{PQ} \times \overrightarrow{PR}$ is perpendicular to the plane, its direction $\mathbf{n}$ is a unit vector perpendicular to the plane. Taking values from Examples 3 and 4, we have

$$\mathbf{n} = \frac{\overrightarrow{PQ} \times \overrightarrow{PR}}{|\overrightarrow{PQ} \times \overrightarrow{PR}|} = \frac{6\mathbf{i} + 6\mathbf{k}}{6\sqrt{2}} = \frac{1}{\sqrt{2}}\mathbf{i} + \frac{1}{\sqrt{2}}\mathbf{k}.$$

❏

The Triple Scalar or Box Product

The product $(\mathbf{A} \times \mathbf{B}) \cdot \mathbf{C}$ is called the **triple scalar product** of $\mathbf{A}$, $\mathbf{B}$, and $\mathbf{C}$ (in that order). As you can see from the formula

$$|(\mathbf{A} \times \mathbf{B}) \cdot \mathbf{C}| = |\mathbf{A} \times \mathbf{B}||\mathbf{C}||\cos\theta|, \qquad (10)$$

the absolute value of the product is the volume of the parallelepiped (parallelogram-sided box) determined by $\mathbf{A}$, $\mathbf{B}$, and $\mathbf{C}$ (Fig. 10.38). The number $|\mathbf{A} \times \mathbf{B}|$ is the area of the base parallelogram. The number $|\mathbf{C}||\cos\theta|$ is the parallelepiped's height. Because of this geometry, $(\mathbf{A} \times \mathbf{B}) \cdot \mathbf{C}$ is also called the **box product** of $\mathbf{A}$, $\mathbf{B}$, and $\mathbf{C}$.

| The dot and cross may be interchanged in a triple scalar product without altering its value. |

By treating the planes of $\mathbf{B}$ and $\mathbf{C}$ and of $\mathbf{C}$ and $\mathbf{A}$ as the base planes of the parallelepiped determined by $\mathbf{A}$, $\mathbf{B}$, and $\mathbf{C}$, we see that

$$(\mathbf{A} \times \mathbf{B}) \cdot \mathbf{C} = (\mathbf{B} \times \mathbf{C}) \cdot \mathbf{A} = (\mathbf{C} \times \mathbf{A}) \cdot \mathbf{B}. \qquad (11)$$

Since the dot product is commutative, Eq. (11) also gives

$$(\mathbf{A} \times \mathbf{B}) \cdot \mathbf{C} = \mathbf{A} \cdot (\mathbf{B} \times \mathbf{C}). \qquad (12)$$

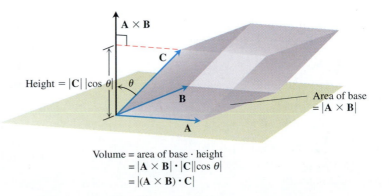

Volume = area of base · height
$$= |\mathbf{A} \times \mathbf{B}| \cdot |\mathbf{C}||\cos\theta|$$
$$= |(\mathbf{A} \times \mathbf{B}) \cdot \mathbf{C}|$$

10.38 The number $|(\mathbf{A} \times \mathbf{B}) \cdot \mathbf{C}|$ is the volume of a parallelepiped.

The triple scalar product can be evaluated as a determinant:

$$\mathbf{A} \cdot (\mathbf{B} \times \mathbf{C}) = \mathbf{A} \cdot \left[\begin{vmatrix} b_2 & b_3 \\ c_2 & c_3 \end{vmatrix} \mathbf{i} - \begin{vmatrix} b_1 & b_3 \\ c_1 & c_3 \end{vmatrix} \mathbf{j} + \begin{vmatrix} b_1 & b_2 \\ c_1 & c_2 \end{vmatrix} \mathbf{k} \right]$$

$$= a_1 \begin{vmatrix} b_2 & b_3 \\ c_2 & c_3 \end{vmatrix} - a_2 \begin{vmatrix} b_1 & b_3 \\ c_1 & c_3 \end{vmatrix} + a_3 \begin{vmatrix} b_1 & b_2 \\ c_1 & c_2 \end{vmatrix}$$

$$= \begin{vmatrix} a_1 & a_2 & a_3 \\ b_1 & b_2 & b_3 \\ c_1 & c_2 & c_3 \end{vmatrix}$$

$$\mathbf{A} \cdot (\mathbf{B} \times \mathbf{C}) = (\mathbf{A} \times \mathbf{B}) \cdot \mathbf{C} = \begin{vmatrix} a_1 & a_2 & a_3 \\ b_1 & b_2 & b_3 \\ c_1 & c_2 & c_3 \end{vmatrix} \tag{13}$$

EXAMPLE 6 Find the volume of the box (parallelepiped) determined by $\mathbf{A} = \mathbf{i} + 2\mathbf{j} - \mathbf{k}$, $\mathbf{B} = -2\mathbf{i} + 3\mathbf{k}$, and $\mathbf{C} = 7\mathbf{j} - 4\mathbf{k}$.

Solution

$$\mathbf{A} \cdot (\mathbf{B} \times \mathbf{C}) = \begin{vmatrix} 1 & 2 & -1 \\ -2 & 0 & 3 \\ 0 & 7 & -4 \end{vmatrix} = \begin{vmatrix} 0 & 3 \\ 7 & -4 \end{vmatrix} - 2 \begin{vmatrix} -2 & 3 \\ 0 & -4 \end{vmatrix} - \begin{vmatrix} -2 & 0 \\ 0 & 7 \end{vmatrix}$$

$$= -21 - 16 + 14 = -23$$

The volume is $|\mathbf{A} \cdot (\mathbf{B} \times \mathbf{C})| = 23$. ❑

Exercises 10.4

Calculations

In Exercises 1–8, find the length and direction (when defined) of $\mathbf{A} \times \mathbf{B}$ and $\mathbf{B} \times \mathbf{A}$.

1. $\mathbf{A} = 2\mathbf{i} - 2\mathbf{j} - \mathbf{k}$, $\mathbf{B} = \mathbf{i} - \mathbf{k}$

2. $\mathbf{A} = 2\mathbf{i} + 3\mathbf{j}$, $\mathbf{B} = -\mathbf{i} + \mathbf{j}$

3. $\mathbf{A} = 2\mathbf{i} - 2\mathbf{j} + 4\mathbf{k}$, $\mathbf{B} = -\mathbf{i} + \mathbf{j} - 2\mathbf{k}$

4. $\mathbf{A} = \mathbf{i} + \mathbf{j} - \mathbf{k}$, $\mathbf{B} = 0$

5. $\mathbf{A} = 2\mathbf{i}$, $\mathbf{B} = -3\mathbf{j}$

6. $\mathbf{A} = \mathbf{i} \times \mathbf{j}$, $\mathbf{B} = \mathbf{j} \times \mathbf{k}$

7. $\mathbf{A} = -8\mathbf{i} - 2\mathbf{j} - 4\mathbf{k}$, $\mathbf{B} = 2\mathbf{i} + 2\mathbf{j} + \mathbf{k}$

8. $\mathbf{A} = \dfrac{3}{2}\mathbf{i} - \dfrac{1}{2}\mathbf{j} + \mathbf{k}$, $\mathbf{B} = \mathbf{i} + \mathbf{j} + 2\mathbf{k}$

In Exercises 9–14, sketch the coordinate axes and then include the vectors $\mathbf{A}$, $\mathbf{B}$, and $\mathbf{A} \times \mathbf{B}$ as vectors starting at the origin.

9. $\mathbf{A} = \mathbf{i}$, $\mathbf{B} = \mathbf{j}$

10. $\mathbf{A} = \mathbf{i} - \mathbf{k}$, $\mathbf{B} = \mathbf{j}$

11. $\mathbf{A} = \mathbf{i} - \mathbf{k}$, $\mathbf{B} = \mathbf{j} + \mathbf{k}$

12. $\mathbf{A} = 2\mathbf{i} - \mathbf{j}$, $\mathbf{B} = \mathbf{i} + 2\mathbf{j}$

13. $\mathbf{A} = \mathbf{i} + \mathbf{j}$, $\mathbf{B} = \mathbf{i} - \mathbf{j}$

14. $\mathbf{A} = \mathbf{j} + 2\mathbf{k}$, $\mathbf{B} = \mathbf{i}$

In Exercises 15–18:

a) Find the area of the triangle determined by the points P, Q, and R.

b) Find a unit vector perpendicular to plane PQR.

15. $P(1, -1, 2)$, $Q(2, 0, -1)$, $R(0, 2, 1)$

16. $P(1, 1, 1)$, $Q(2, 1, 3)$, $R(3, -1, 1)$

17. $P(2, -2, 1)$, $Q(3, -1, 2)$, $R(3, -1, 1)$

18. $P(-2, 2, 0)$, $Q(0, 1, -1)$, $R(-1, 2, -2)$

19. Let $\mathbf{A} = 5\mathbf{i} - \mathbf{j} + \mathbf{k}$, $\mathbf{B} = \mathbf{j} - 5\mathbf{k}$, $\mathbf{C} = -15\mathbf{i} + 3\mathbf{j} - 3\mathbf{k}$. Which vectors, if any, are (a) perpendicular, (b) parallel? Give reasons for your answers.

20. Let $\mathbf{A} = \mathbf{i} + 2\mathbf{j} - \mathbf{k}$, $\mathbf{B} = -\mathbf{i} + \mathbf{j} + \mathbf{k}$, $\mathbf{C} = \mathbf{i} + \mathbf{k}$, $\mathbf{D} = -(\pi/2)\mathbf{i} - \pi\mathbf{j} + (\pi/2)\mathbf{k}$. Which vectors, if any, are (a) perpendicular, (b) parallel? Give reasons for your answers.

In Exercises 21 and 22, find the magnitude of the torque exerted by $\mathbf{F}$ on the bolt at P if $|\overrightarrow{PQ}| = 8$ in. and $|\mathbf{F}| = 30$ lb. Answer in foot-pounds.

21.   **22.**

In Exercises 23–26, verify that $(\mathbf{A} \times \mathbf{B}) \cdot \mathbf{C} = (\mathbf{B} \times \mathbf{C}) \cdot \mathbf{A} = (\mathbf{C} \times \mathbf{A}) \cdot \mathbf{B}$ and find the volume of the parallelepiped (box) determined by $\mathbf{A}$, $\mathbf{B}$, and $\mathbf{C}$.

	A	B	C
23.	$2\mathbf{i}$	$2\mathbf{j}$	$2\mathbf{k}$
24.	$\mathbf{i} - \mathbf{j} + \mathbf{k}$	$2\mathbf{i} + \mathbf{j} - 2\mathbf{k}$	$-\mathbf{i} + 2\mathbf{j} - \mathbf{k}$
25.	$2\mathbf{i} + \mathbf{j}$	$2\mathbf{i} - \mathbf{j} + \mathbf{k}$	$\mathbf{i} + 2\mathbf{k}$
26.	$\mathbf{i} + \mathbf{j} - 2\mathbf{k}$	$-\mathbf{i} - \mathbf{k}$	$2\mathbf{i} + 4\mathbf{j} - 2\mathbf{k}$

Theory and Examples

27. Which of the following are *always true* and which are *not always true*? Give reasons for your answers.

a) $|\mathbf{A}| = \sqrt{\mathbf{A} \cdot \mathbf{A}}$
b) $\mathbf{A} \cdot \mathbf{A} = |\mathbf{A}|$
c) $\mathbf{A} \times \mathbf{0} = \mathbf{0} \times \mathbf{A} = \mathbf{0}$
d) $\mathbf{A} \times (-\mathbf{A}) = \mathbf{0}$
e) $\mathbf{A} \times \mathbf{B} = \mathbf{B} \times \mathbf{A}$
f) $\mathbf{A} \times (\mathbf{B} + \mathbf{C}) = \mathbf{A} \times \mathbf{B} + \mathbf{A} \times \mathbf{C}$
g) $(\mathbf{A} \times \mathbf{B}) \cdot \mathbf{B} = 0$
h) $(\mathbf{A} \times \mathbf{B}) \cdot \mathbf{C} = \mathbf{A} \cdot (\mathbf{B} \times \mathbf{C})$

28. Which of the following are *always true* and which are *not always true*? Give reasons for your answers.

a) $\mathbf{A} \cdot \mathbf{B} = \mathbf{B} \cdot \mathbf{A}$
b) $\mathbf{A} \times \mathbf{B} = -(\mathbf{B} \times \mathbf{A})$
c) $(-\mathbf{A}) \times \mathbf{B} = -(\mathbf{A} \times \mathbf{B})$
d) $(c\mathbf{A}) \cdot \mathbf{B} = \mathbf{A} \cdot (c\mathbf{B}) = c(\mathbf{A} \cdot \mathbf{B})$ (any number c)
e) $c(\mathbf{A} \times \mathbf{B}) = (c\mathbf{A}) \times \mathbf{B} = \mathbf{A} \times (c\mathbf{B})$ (any number c)
f) $\mathbf{A} \cdot \mathbf{A} = |\mathbf{A}|^2$
g) $(\mathbf{A} \times \mathbf{A}) \cdot \mathbf{A} = 0$
h) $(\mathbf{A} \times \mathbf{B}) \cdot \mathbf{A} = \mathbf{B} \cdot (\mathbf{A} \times \mathbf{B})$

29. Given nonzero vectors $\mathbf{A}$, $\mathbf{B}$, and $\mathbf{C}$, use dot-product and cross-product notation, as appropriate, to describe the following.

a) The vector projection of $\mathbf{A}$ onto $\mathbf{B}$
b) A vector orthogonal to $\mathbf{A}$ and $\mathbf{B}$
c) A vector orthogonal to $\mathbf{A} \times \mathbf{B}$ and $\mathbf{C}$
d) The volume of the parallelepiped determined by $\mathbf{A}$, $\mathbf{B}$, and $\mathbf{C}$

30. Given nonzero vectors $\mathbf{A}$, $\mathbf{B}$, and $\mathbf{C}$, use dot-product and cross-product notation to describe the following.

a) A vector orthogonal to $\mathbf{A} \times \mathbf{B}$ and $\mathbf{A} \times \mathbf{C}$
b) A vector orthogonal to $\mathbf{A} + \mathbf{B}$ and $\mathbf{A} - \mathbf{B}$
c) A vector of length $|\mathbf{A}|$ in the direction of $\mathbf{B}$
d) The area of the parallelogram determined by $\mathbf{A}$ and $\mathbf{C}$

31. Let $\mathbf{A}$, $\mathbf{B}$, and $\mathbf{C}$ be vectors. Which of the following make sense, and which do not? Give reasons for your answers.

a) $(\mathbf{A} \times \mathbf{B}) \cdot \mathbf{C}$
b) $\mathbf{A} \times (\mathbf{B} \cdot \mathbf{C})$
c) $\mathbf{A} \times (\mathbf{B} \times \mathbf{C})$
d) $\mathbf{A} \cdot (\mathbf{B} \cdot \mathbf{C})$

32. Show that except in degenerate cases $(\mathbf{A} \times \mathbf{B}) \times \mathbf{C}$ lies in the plane of $\mathbf{A}$ and $\mathbf{B}$ while $\mathbf{A} \times (\mathbf{B} \times \mathbf{C})$ lies in the plane of $\mathbf{B}$ and $\mathbf{C}$. What *are* the degenerate cases?

33. *Cancellation in cross products.* If $\mathbf{A} \times \mathbf{B} = \mathbf{A} \times \mathbf{C}$ and $\mathbf{A} \neq \mathbf{0}$, then does $\mathbf{B} = \mathbf{C}$? Give reasons for your answer.

34. *Double cancellation.* If $\mathbf{A} \neq \mathbf{0}$, and if $\mathbf{A} \times \mathbf{B} = \mathbf{A} \times \mathbf{C}$ and $\mathbf{A} \cdot \mathbf{B} = \mathbf{A} \cdot \mathbf{C}$, then does $\mathbf{B} = \mathbf{C}$? Give reasons for your answers.

Area in the Plane

Find the areas of the parallelograms whose vertices are given in Exercises 35–38.

35. $A(1, 0)$, $B(0, 1)$, $C(-1, 0)$, $D(0, -1)$

36. $A(0, 0)$, $B(7, 3)$, $C(9, 8)$, $D(2, 5)$

37. $A(-1, 2)$, $B(2, 0)$, $C(7, 1)$, $D(4, 3)$

38. $A(-6, 0)$, $B(1, -4)$, $C(3, 1)$, $D(-4, 5)$

Find the areas of the triangles whose vertices are given in Exercises 39–42.

39. $A(0, 0)$, $B(-2, 3)$, $C(3, 1)$

40. $A(-1, -1)$, $B(3, 3)$, $C(2, 1)$

41. $A(-5, 3)$, $B(1, -2)$, $C(6, -2)$

42. $A(-6, 0)$, $B(10, -5)$, $C(-2, 4)$

43. Find a formula for the area of the triangle in the xy-plane with vertices at $(0, 0)$, (a_1, a_2), and (b_1, b_2). Explain your work.

44. Find a concise formula for the area of a triangle with vertices (a_1, a_2), (b_1, b_2), and (c_1, c_2).

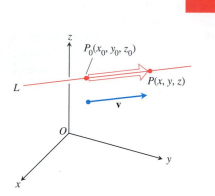

10.39 A point P lies on the line through P_0 parallel to $\mathbf{v}$ if and only if $\overrightarrow{P_0P}$ is a scalar multiple of $\mathbf{v}$.

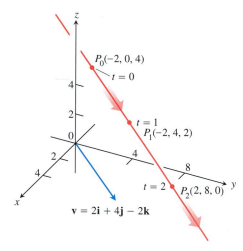

10.40 Selected points and parameter values on the line $x = -2 + 2t$, $y = 4t$, $z = 4 - 2t$. The arrows show the direction of increasing t (Example 1).

10.5 Lines and Planes in Space

This section shows how to use scalar and vector products to write equations for lines, line segments, and planes in space.

Lines and Line Segments in Space

Suppose L is a line in space passing through a point $P_0(x_0, y_0, z_0)$ parallel to a vector $\mathbf{v} = A\mathbf{i} + B\mathbf{j} + C\mathbf{k}$. Then L is the set of all points $P(x, y, z)$ for which $\overrightarrow{P_0P}$ is parallel to $\mathbf{v}$ (Fig. 10.39). That is, P lies on L if and only if $\overrightarrow{P_0P}$ is a scalar multiple of $\mathbf{v}$.

Vector Equation for the Line Through $P_0(x_0, y_0, z_0)$ Parallel to v

$$\overrightarrow{P_0P} = t\mathbf{v}, \qquad -\infty < t < \infty \qquad (1)$$

Equating the corresponding components of the two sides of Eq. (1) gives three scalar equations involving the parameter t:

$$(x - x_0)\mathbf{i} + (y - y_0)\mathbf{j} + (z - z_0)\mathbf{k} = t(A\mathbf{i} + B\mathbf{j} + C\mathbf{k}), \qquad \text{Eq. (1) expanded}$$

$$x - x_0 = tA, \qquad y - y_0 = tB, \qquad z - z_0 = tC. \qquad \text{Components equated}$$

When rearranged, these equations give us the standard parametrization of the line for the parameter interval $-\infty < t < \infty$.

Standard Parametrization of the Line Through $P_0(x_0, y_0, z_0)$ Parallel to $\mathbf{v} = A\mathbf{i} + B\mathbf{j} + C\mathbf{k}$

$$x = x_0 + tA, \qquad y = y_0 + tB, \qquad z = z_0 + tC, \qquad -\infty < t < \infty \quad (2)$$

EXAMPLE 1   Find parametric equations for the line through $(-2, 0, 4)$ parallel to $\mathbf{v} = 2\mathbf{i} + 4\mathbf{j} - 2\mathbf{k}$ (Fig. 10.40).

Solution   With $P_0(x_0, y_0, z_0)$ equal to $(-2, 0, 4)$ and $A\mathbf{i} + B\mathbf{j} + C\mathbf{k}$ equal to $2\mathbf{i} + 4\mathbf{j} - 2\mathbf{k}$, Eqs. (2) become

$$x = -2 + 2t, \qquad y = 4t, \qquad z = 4 - 2t. \qquad \square$$

EXAMPLE 2   Find parametric equations for the line through $P(-3, 2, -3)$ and $Q(1, -1, 4)$.

Solution   The vector

$$\overrightarrow{PQ} = (1 - (-3))\mathbf{i} + (-1 - 2)\mathbf{j} + (4 - (-3))\mathbf{k}$$

$$= 4\mathbf{i} - 3\mathbf{j} + 7\mathbf{k}$$

is parallel to the line, and Eqs. (2) with $(x_0, y_0, z_0) = (-3, 2, -3)$ give

$$x = -3 + 4t, \qquad y = 2 - 3t, \qquad z = -3 + 7t.$$

We could have chosen $Q(1, -1, 4)$ as the "base point" and written

$$x = 1 + 4t, \qquad y = -1 - 3t, \qquad z = 4 + 7t.$$

These equations serve as well as the first; they simply place you at a different point for a given value of t. ❑

To parametrize a line segment joining two points, we first parametrize the line through the points. We then find the t-values for the endpoints and restrict t to lie in the closed interval bounded by these values. The line equations together with this added restriction parametrize the segment.

Notice that parametrizations are not unique.

EXAMPLE 3   Parametrize the line segment joining the points $P(-3, 2, -3)$ and $Q(1, -1, 4)$ (Fig. 10.41).

Solution   We begin with equations for the line through P and Q, taking them, in this case, from Example 2:

$$x = -3 + 4t, \qquad y = 2 - 3t, \qquad z = -3 + 7t. \tag{3}$$

We observe that the point

$$(x, y, z) = (-3 + 4t, 2 - 3t, -3 + 7t)$$

passes through $P(-3, 2, -3)$ at $t = 0$ and $Q(1, -1, 4)$ at $t = 1$. We add the restriction $0 \le t \le 1$ to Eqs. (3) to parametrize the segment:

$$x = -3 + 4t, \qquad y = 2 - 3t, \qquad z = -3 + 7t, \qquad 0 \le t \le 1. \quad ❑$$

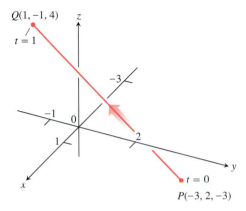

10.41 Example 3 derives a parametrization of line segment PQ. The arrow shows the direction of increasing t.

The Distance from a Point to a Line in Space

To find the distance from a point S to a line that passes through a point P parallel to a vector $\mathbf{v}$, we find the length of the component of $\overrightarrow{PS}$ normal to the line (Fig. 10.42). In the notation of the figure, the length is $|\overrightarrow{PS}| \sin \theta$, which is $|\overrightarrow{PS} \times \mathbf{v}|/|\mathbf{v}|$.

Distance from a Point S to a Line Through P Parallel to $\mathbf{v}$

$$d = \frac{|\overrightarrow{PS} \times \mathbf{v}|}{|\mathbf{v}|} \tag{4}$$

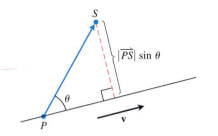

10.42 The distance from S to the line through P parallel to $\mathbf{v}$ is $|\overrightarrow{PS}| \sin \theta$, where θ is the angle between $\overrightarrow{PS}$ and $\mathbf{v}$.

EXAMPLE 4   Find the distance from the point $S(1, 1, 5)$ to the line

$$L: \qquad x = 1 + t, \quad y = 3 - t, \quad z = 2t.$$

Solution   We see from the equations for L that L passes through $P(1, 3, 0)$ parallel to $\mathbf{v} = \mathbf{i} - \mathbf{j} + 2\mathbf{k}$. With

$$\overrightarrow{PS} = (1 - 1)\mathbf{i} + (1 - 3)\mathbf{j} + (5 - 0)\mathbf{k} = -2\mathbf{j} + 5\mathbf{k}$$

and

$$\vec{PS} \times \mathbf{v} = \begin{vmatrix} \mathbf{i} & \mathbf{j} & \mathbf{k} \\ 0 & -2 & 5 \\ 1 & -1 & 2 \end{vmatrix} = \mathbf{i} + 5\mathbf{j} + 2\mathbf{k},$$

Eq. (4) gives

$$d = \frac{|\vec{PS} \times \mathbf{v}|}{|\mathbf{v}|} = \frac{\sqrt{1+25+4}}{\sqrt{1+1+4}} = \frac{\sqrt{30}}{\sqrt{6}} = \sqrt{5}. \qquad ❑$$

Equations for Planes in Space

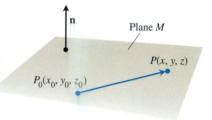

10.43 The standard equation for a plane in space is defined in terms of a vector normal to the plane: A point P lies in the plane through P_0 normal to $\mathbf{n}$ if and only if $\mathbf{n} \cdot \vec{P_0 P} = 0$.

Suppose plane M passes through a point $P_0(x_0, y_0, z_0)$ and is normal (perpendicular) to the nonzero vector $\mathbf{n} = A\mathbf{i} + B\mathbf{j} + C\mathbf{k}$. Then M is the set of all points $P(x, y, z)$ for which $\vec{P_0 P}$ is orthogonal to $\mathbf{n}$ (Fig. 10.43). That is, P lies on M if and only if $\mathbf{n} \cdot \vec{P_0 P} = 0$. This equation is equivalent to

$$(A\mathbf{i} + B\mathbf{j} + C\mathbf{k}) \cdot [(x - x_0)\mathbf{i} + (y - y_0)\mathbf{j} + (z - z_0)\mathbf{k}] = 0$$

or

$$A(x - x_0) + B(y - y_0) + C(z - z_0) = 0.$$

Plane Through $P_0(x_0, y_0, z_0)$ Normal to $\mathbf{n} = A\mathbf{i} + B\mathbf{j} + C\mathbf{k}$

Vector equation: $\qquad \mathbf{n} \cdot \vec{P_0 P} = 0 \qquad\qquad$ (5)

Component equation: $\qquad A(x - x_0) + B(y - y_0) + C(z - z_0) = 0 \qquad$ (6)

EXAMPLE 5   Find an equation for the plane through $P_0(-3, 0, 7)$ perpendicular to $\mathbf{n} = 5\mathbf{i} + 2\mathbf{j} - \mathbf{k}$.

Solution

$$5(x - (-3)) + 2(y - 0) + (-1)(z - 7) = 0 \qquad \text{Eq. (5) with the given } \mathbf{n} \text{ and } P_0$$

$$5x + 15 + 2y - z + 7 = 0$$

$$5x + 2y - z = -22 \qquad ❑$$

Notice in Example 5 how the components of $\mathbf{n} = 5\mathbf{i} + 2\mathbf{j} - \mathbf{k}$ became the coefficients of x, y, and z in the equation $5x + 2y - z = -22$. This was no accident, because Eq. (6), when rearranged, takes the form

$$Ax + By + Cz = D,$$

where $D = Ax_0 + By_0 + Cz_0$.

$A\mathbf{i} + B\mathbf{j} + C\mathbf{k}$ is normal to the plane $Ax + By + Cz = D$.

EXAMPLE 6   *A plane determined by three points*

Find the plane through $A(0, 0, 1)$, $B(2, 0, 0)$, and $C(0, 3, 0)$.

Solution   We find a vector normal to the plane and use it with one of the points (it does not matter which) to write an equation for the plane.

The cross product

$$\overrightarrow{AB} \times \overrightarrow{AC} = \begin{vmatrix} \mathbf{i} & \mathbf{j} & \mathbf{k} \\ 2 & 0 & -1 \\ 0 & 3 & -1 \end{vmatrix} = 3\mathbf{i} + 2\mathbf{j} + 6\mathbf{k}$$

is normal to the plane. We substitute the components of this vector and the coordinates of the point $(0, 0, 1)$ into Eq. (6) to get

$$3(x - 0) + 2(y - 0) + 6(z - 1) = 0,$$

$$3x + 2y + 6z = 6.$$ ❏

EXAMPLE 7   *The intersection of a line and a plane*

Find the point where the line

$$x = \frac{8}{3} + 2t, \qquad y = -2t, \qquad z = 1 + t$$

intersects the plane $3x + 2y + 6z = 6$.

Solution   The point

$$\left(\frac{8}{3} + 2t, -2t, 1 + t \right)$$

lies in the plane if its coordinates satisfy the equation of the plane; that is, if

$$3\left(\frac{8}{3} + 2t \right) + 2(-2t) + 6(1 + t) = 6$$

$$8 + 6t - 4t + 6 + 6t = 6$$

$$8t = -8$$

$$t = -1.$$

The point of intersection is

$$(x, y, z)\big|_{t=-1} = \left(\frac{8}{3} - 2, 2, 1 - 1 \right) = \left(\frac{2}{3}, 2, 0 \right).$$ ❏

EXAMPLE 8   *The distance from a point to a plane*

Find the distance from $S(1, 1, 3)$ to the plane $3x + 2y + 6z = 6$.

Solution   We find a point P in the plane and calculate the length of the vector projection of $\overrightarrow{PS}$ onto a vector $\mathbf{n}$ normal to the plane (Fig. 10.44, on the following page).

The coefficients in the equation $3x + 2y + 6z = 6$ give

$$\mathbf{n} = 3\mathbf{i} + 2\mathbf{j} + 6\mathbf{k}.$$

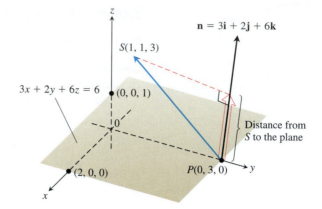

10.44 The distance from S to the plane is the length of the vector projection of $\overrightarrow{PS}$ onto $\mathbf{n}$ (Example 8).

The points on the plane easiest to find from the plane's equation are the intercepts. If we take P to be the y-intercept $(0, 3, 0)$, then

$$\overrightarrow{PS} = (1 - 0)\mathbf{i} + (1 - 3)\mathbf{j} + (3 - 0)\mathbf{k}$$

$$= \mathbf{i} - 2\mathbf{j} + 3\mathbf{k},$$

$$|\mathbf{n}| = \sqrt{(3)^2 + (2)^2 + (6)^2} = \sqrt{49} = 7.$$

The distance from S to the plane is

> **How to Find the Distance from a Point to a Plane**
>
> To find the distance from a point S to a plane $Ax + By + Cz = D$, find
>
> 1. a point P on the plane,
> 2. $\overrightarrow{PS}$,
> 3. the direction of $\mathbf{n} = A\mathbf{i} + B\mathbf{j} + C\mathbf{k}$.
>
> Then calculate the distance as
>
> $$d = \left| \overrightarrow{PS} \cdot \frac{\mathbf{n}}{|\mathbf{n}|} \right|.$$

$$d = \left| \overrightarrow{PS} \cdot \frac{\mathbf{n}}{|\mathbf{n}|} \right| \qquad \text{length of proj}_\mathbf{n} \ \overrightarrow{PS}$$

$$= \left| (\mathbf{i} - 2\mathbf{j} + 3\mathbf{k}) \cdot \left(\frac{3}{7}\mathbf{i} + \frac{2}{7}\mathbf{j} + \frac{6}{7}\mathbf{k} \right) \right|$$

$$= \left| \frac{3}{7} - \frac{4}{7} + \frac{18}{7} \right| = \frac{17}{7}.$$

Angles Between Planes; Lines of Intersection

The angle between two intersecting planes is defined to be the (acute) angle determined by their normal vectors (Fig. 10.45).

EXAMPLE 9   Find the angle between the planes $3x - 6y - 2z = 15$ and $2x + y - 2z = 5$.

Solution   The vectors

$$\mathbf{n}_1 = 3\mathbf{i} - 6\mathbf{j} - 2\mathbf{k}, \qquad \mathbf{n}_2 = 2\mathbf{i} + \mathbf{j} - 2\mathbf{k}$$

are normals to the planes. The angle between them is

$$\theta = \cos^{-1}\left(\frac{\mathbf{n}_1 \cdot \mathbf{n}_2}{|\mathbf{n}_1||\mathbf{n}_2|} \right) \qquad \text{Eq. (4), Section 10.3}$$

$$= \cos^{-1}\left(\frac{4}{21} \right)$$

$$\approx 1.38 \text{ rad} \qquad \text{About } 79°$$

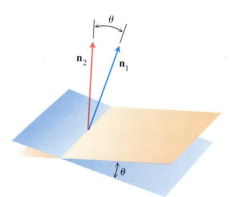

10.45 The angle between two planes is obtained from the angle between their normals.

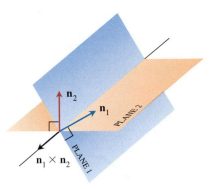

10.46 How the line of intersection of two planes is related to the planes' normal vectors (Example 10).

EXAMPLE 10   Find a vector parallel to the line of intersection of the planes $3x - 6y - 2z = 15$ and $2x + y - 2z = 5$.

Solution   The line of intersection of two planes is perpendicular to the planes' normal vectors $\mathbf{n}_1$ and $\mathbf{n}_2$ (Fig. 10.46), and therefore parallel to $\mathbf{n}_1 \times \mathbf{n}_2$. Turning this around, $\mathbf{n}_1 \times \mathbf{n}_2$ is a vector parallel to the planes' line of intersection. In our case,

$$\mathbf{n}_1 \times \mathbf{n}_2 = \begin{vmatrix} \mathbf{i} & \mathbf{j} & \mathbf{k} \\ 3 & -6 & -2 \\ 2 & 1 & -2 \end{vmatrix} = 14\,\mathbf{i} + 2\,\mathbf{j} + 15\,\mathbf{k}.$$

Any nonzero scalar multiple of $\mathbf{n}_1 \times \mathbf{n}_2$ will do as well.   ❑

EXAMPLE 11   Find parametric equations for the line in which the planes $3x - 6y - 2z = 15$ and $2x + y - 2z = 5$ intersect.

Solution   We find a vector parallel to the line and a point on the line and use Eqs. (2).

Example 10 identifies $\mathbf{v} = 14\,\mathbf{i} + 2\,\mathbf{j} + 15\,\mathbf{k}$ as a vector parallel to the line. To find a point on the line, we can take any point common to the two planes. Substituting $z = 0$ in the plane equations and solving for x and y simultaneously identifies one of these points as $(3, -1, 0)$. The line is

$$x = 3 + 14t, \qquad y = -1 + 2t, \qquad z = 15t.$$ ❑

Exercises 10.5

Lines and Line Segments

Find parametric equations for the lines in Exercises 1–12.

1. The line through the point $P(3, -4, -1)$ parallel to the vector $\mathbf{i} + \mathbf{j} + \mathbf{k}$

2. The line through $P(1, 2, -1)$ and $Q(-1, 0, 1)$

3. The line through $P(-2, 0, 3)$ and $Q(3, 5, -2)$

4. The line through $P(1, 2, 0)$ and $Q(1, 1, -1)$

5. The line through the origin parallel to the vector $2\,\mathbf{j} + \mathbf{k}$

6. The line through the point $(3, -2, 1)$ parallel to the line $x = 1 + 2t, y = 2 - t, z = 3t$

7. The line through $(1, 1, 1)$ parallel to the z-axis

8. The line through $(2, 4, 5)$ perpendicular to the plane $3x + 7y - 5z = 21$

9. The line through $(0, -7, 0)$ perpendicular to the plane $x + 2y + 2z = 13$

10. The line through $(2, 3, 0)$ perpendicular to the vectors $\mathbf{A} = \mathbf{i} + 2\,\mathbf{j} + 3\,\mathbf{k}$ and $\mathbf{B} = 3\,\mathbf{i} + 4\,\mathbf{j} + 5\,\mathbf{k}$

11. The x-axis

12. The z-axis

Find parametrizations for the line segments joining the points in Exercises 13–20. Draw coordinate axes and sketch each segment, indicating the direction of increasing t for your parametrization.

13. $(0, 0, 0)$, $(1, 1, 3/2)$

14. $(0, 0, 0)$, $(1, 0, 0)$

15. $(1, 0, 0)$, $(1, 1, 0)$

16. $(1, 1, 0)$, $(1, 1, 1)$

17. $(0, 1, 1)$, $(0, -1, 1)$

18. $(0, 2, 0)$, $(3, 0, 0)$

19. $(2, 0, 2)$, $(0, 2, 0)$

20. $(1, 0, -1)$, $(0, 3, 0)$

Planes

Find equations for the planes in Exercises 21–26.

21. The plane through $P_0(0, 2, -1)$ normal to $\mathbf{n} = 3\,\mathbf{i} - 2\,\mathbf{j} - \mathbf{k}$

22. The plane through $(1, -1, 3)$ parallel to the plane $3x + y + z = 7$

23. The plane through $(1, 1, -1)$, $(2, 0, 2)$, and $(0, -2, 1)$

24. The plane through $(2, 4, 5)$, $(1, 5, 7)$, and $(-1, 6, 8)$

25. The plane through $P_0(2, 4, 5)$ perpendicular to the line
$$x = 5 + t, \quad y = 1 + 3t, \quad z = 4t$$

26. The plane through $A(1, -2, 1)$ perpendicular to the vector from the origin to A

27. Find the point of intersection of the lines $x = 2t + 1, y = 3t + 2, z = 4t + 3$, and $x = s + 2, y = 2s + 4, z = -4s - 1$, and then find the plane determined by these lines.

28. Find the point of intersection of the lines $x = t, y = -t + 2, z = t + 1$, and $x = 2s + 2, y = s + 3, z = 5s + 6$, and then find the plane determined by these lines.

In Exercises 29 and 30, find the plane determined by the intersecting lines.

29. $L1: \quad x = -1 + t, \ y = 2 + t, \ z = 1 - t, \ -\infty < t < \infty$
    $L2: \quad x = 1 - 4s, \ y = 1 + 2s, \ z = 2 - 2s, \ -\infty < s < \infty$

30. $L1: \quad x = t, \ y = 3 - 3t, \ z = -2 - t, \ -\infty < t < \infty$
    $L2: \quad x = 1 + s, \ y = 4 + s, \ z = -1 + s, \ -\infty < s < \infty$

31. Find a plane through $P_0(2, 1, -1)$ and perpendicular to the line of intersection of the planes $2x + y - z = 3, x + 2y + z = 2$.

32. Find a plane through the points $P_1(1, 2, 3), P_2(3, 2, 1)$ and perpendicular to the plane $4x - y + 2z = 7$.

Distances

In Exercises 33–38, find the distance from the point to the line.

33. $(0, 0, 12); \quad x = 4t, \quad y = -2t, \quad z = 2t$

34. $(0, 0, 0); \quad x = 5 + 3t, \quad y = 5 + 4t, \quad z = -3 - 5t$

35. $(2, 1, 3); \quad x = 2 + 2t, \quad y = 1 + 6t, \quad z = 3$

36. $(2, 1, -1); \quad x = 2t, \quad y = 1 + 2t, \quad z = 2t$

37. $(3, -1, 4); \quad x = 4 - t, \quad y = 3 + 2t, \quad z = -5 + 3t$

38. $(-1, 4, 3); \quad x = 10 + 4t, \quad y = -3, \quad z = 4t$

In Exercises 39–44, find the distance from the point to the plane.

39. $(2, -3, 4), \quad x + 2y + 2z = 13$

40. $(0, 0, 0), \quad 3x + 2y + 6z = 6$

41. $(0, 1, 1), \quad 4y + 3z = -12$

42. $(2, 2, 3), \quad 2x + y + 2z = 4$

43. $(0, -1, 0), \quad 2x + y + 2z = 4$

44. $(1, 0, -1), \quad -4x + y + z = 4$

45. Find the distance from the plane $x + 2y + 6z = 1$ to the plane $x + 2y + 6z = 10$.

46. Find the distance from the line $x = 2 + t, y = 1 + t, z = -(1/2) - (1/2)t$ to the plane $x + 2y + 6z = 10$.

Angles

Find the angles between the planes in Exercises 47 and 48. (You will not need a calculator.)

47. $x + y = 1, \quad 2x + y - 2z = 2$

48. $5x + y - z = 10, \quad x - 2y + 3z = -1$

CALCULATOR Use a calculator to find the acute angles between the planes in Exercises 49–52 to the nearest hundredth of a radian.

49. $2x + 2y + 2z = 3, \quad 2x - 2y - z = 5$

50. $x + y + z = 1, \quad z = 0$ (the xy-plane)

51. $2x + 2y - z = 3, \quad x + 2y + z = 2$

52. $4y + 3z = -12, \quad 3x + 2y + 6z = 6$

Intersecting Lines and Planes

In Exercises 53–56, find the point in which the line meets the plane.

53. $x = 1 - t, \quad y = 3t, \quad z = 1 + t; \quad 2x - y + 3z = 6$

54. $x = 2, \quad y = 3 + 2t, \quad z = -2 - 2t; \quad 6x + 3y - 4z = -12$

55. $x = 1 + 2t, \quad y = 1 + 5t, \quad z = 3t; \quad x + y + z = 2$

56. $x = -1 + 3t, \quad y = -2, \quad z = 5t; \quad 2x - 3z = 7$

Find parametrizations for the lines in which the planes in Exercises 57–60 intersect.

57. $x + y + z = 1, \quad x + y = 2$

58. $3x - 6y - 2z = 3, \quad 2x + y - 2z = 2$

59. $x - 2y + 4z = 2, \quad x + y - 2z = 5$

60. $5x - 2y = 11, \quad 4y - 5z = -17$

Given two lines in space, either they are parallel, or they intersect, or they are skew (imagine, for example, the flight paths of two planes in the sky). Exercises 61 and 62 each give three lines. In each exercise, determine whether the lines, taken two at a time, are parallel, intersect, or are skew. If they intersect, find the point of intersection.

61. $L1: \quad x = 3 + 2t, y = -1 + 4t, z = 2 - t, -\infty < t < \infty$
    $L2: \quad x = 1 + 4s, y = 1 + 2s, z = -3 + 4s, -\infty < s < \infty$
    $L3: \quad x = 3 + 2r, y = 2 + r, z = -2 + 2r, -\infty < r < \infty$

62. $L1: \quad x = 1 + 2t, y = -1 - t, z = 3t, -\infty < t < \infty$
    $L2: \quad x = 2 - s, y = 3s, z = 1 + s, -\infty < s < \infty$
    $L3: \quad x = 5 + 2r, y = 1 - r, z = 8 + 3r, -\infty < r < \infty$

Theory and Examples

63. Use Eqs. (2) to generate a parametrization of the line through $P(2, -4, 7)$ parallel to $\mathbf{v}_1 = 2\mathbf{i} - \mathbf{j} + 3\mathbf{k}$. Then generate another parametrization of the line using the point $P_2(-2, -2, 1)$ and the vector $\mathbf{v}_2 = -\mathbf{i} + (1/2)\mathbf{j} - (3/2)\mathbf{k}$.

64. Use Eq. (6) to generate an equation for the plane through $P_1(4, 1, 5)$ normal to $\mathbf{n}_1 = \mathbf{i} - 2\mathbf{j} + \mathbf{k}$. Then generate another equation for the same plane using the point $P_2(3, -2, 0)$ and the normal vector $\mathbf{n}_2 = -\sqrt{2}\mathbf{i} + 2\sqrt{2}\mathbf{j} - \sqrt{2}\mathbf{k}$.

65. Find the points in which the line $x = 1 + 2t, y = -1 - t, z = 3t$ meets the coordinate planes. Describe the reasoning behind your answer.

66. Find equations for the line in the plane $z = 3$ that makes an angle of $\pi/6$ rad with $\mathbf{i}$ and an angle of $\pi/3$ rad with $\mathbf{j}$. Describe the reasoning behind your answer.

67. Is the line $x = 1 - 2t$, $y = 2 + 5t$, $z = -3t$ parallel to the plane $2x + y - z = 8$? Give reasons for your answer.

68. How can you tell when two planes $A_1x + B_1y + C_1z = D_1$ and $A_2x + B_2y + C_2z = D_2$ are parallel? perpendicular? Give reasons for your answer.

69. Find two different planes whose intersection is the line $x = 1 + t$, $y = 2 - t$, $z = 3 + 2t$. Write equations for each plane in the form $Ax + By + Cz = D$.

70. Find a plane through the origin that meets the plane M: $2x + 3y + z = 12$ in a right angle. How do you know that your plane is perpendicular to M?

71. For any nonzero numbers a, b, and c, the graph of $(x/a) + (y/b) + (z/c) = 1$ is a plane. Which planes have an equation of this form?

72. Suppose L_1 and L_2 are disjoint (nonintersecting) nonparallel lines. Is it possible for a nonzero vector to be perpendicular to both L_1 and L_2? Give reasons for your answer.

Computer Graphics

73. *Perspective in computer graphics.* In computer graphics and perspective drawing we need to represent objects seen by the eye in space as images on a two-dimensional plane. Suppose the eye is at $E(x_0, 0, 0)$ as shown here and that we want to represent a point $P_1(x_1, y_1, z_1)$ as a point on the $yz-$plane. We do this by projecting P_1 onto the plane with a ray from E. The point P_1 will be portrayed as the point $P(0, y, z)$. The problem for us as graphics designers is to find y and z given E and P_1.

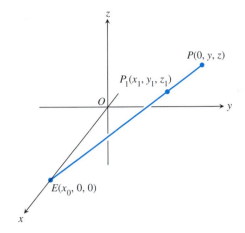

a) Write a vector equation that holds between $\overrightarrow{EP}$ and $\overrightarrow{EP_1}$. Use the equation to express y and z in terms of x_0, x_1, y_1, and z_1.

b) Test the formulas obtained for y and z in part (a) by investigating their behavior at $x_1 = 0$ and $x_1 = x_0$ and by seeing what happens as $x_0 \to \infty$.

74. *Hidden lines.* Here is another typical problem in computer graphics. Your eye is at $(4, 0, 0)$. You are looking at a triangular plate whose vertices are at $(1, 0, 1)$, $(1, 1, 0)$, and $(-2, 2, 2)$. The line segment from $(1, 0, 0)$ to $(0, 2, 2)$ passes through the plate. What portion of the line segment is hidden from your view by the plate? (This is an exercise in finding intersections of lines and planes.)

10.6

Cylinders and Quadric Surfaces

In the calculus of functions of a single variable, we began with lines and used our knowledge of lines to study curves in the plane. We investigated tangents and found that, when highly magnified, differentiable curves were effectively linear. Among the curves of special interest were the conic sections, curves defined by second-degree equations in x and y.

To study the calculus of functions of more than one variable, we begin much the same way. We start with planes and use our knowledge of planes to study surfaces in space. Among the surfaces of special interest are cylinders and quadric surfaces, surfaces defined by second-degree equations in x, y, and z. We described the planes in Section 10.5. In this section, we describe the surfaces.

Cylinders

A **cylinder** is the surface composed of all the lines that (1) lie parallel to a given line in space and (2) pass through a given plane curve. The curve is a **generating curve** for the cylinder (Fig. 10.47, on the following page). In solid geometry, where *cylinder* means *circular cylinder*, the generating curves are circles, but now we allow generating curves of any kind. The cylinder in our first example is generated by a parabola.

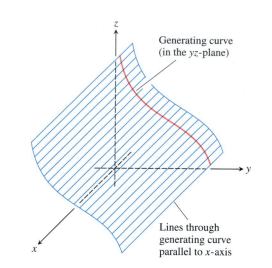

10.47 A cylinder and generating curve.

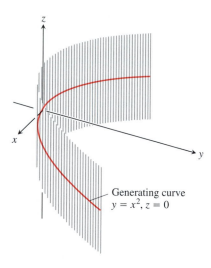

10.48 The cylinder of lines passing through the parabola $y = x^2$ in the xy-plane parallel to the z-axis (Example 1).

When graphing a cylinder or other surface by hand or analyzing one generated by a computer, it helps to look at the curves formed by intersecting the surface with planes parallel to the coordinate planes. These curves are called **cross sections** or **traces.**

EXAMPLE 1   *The parabolic cylinder $y = x^2$*

Find an equation for the cylinder made by the lines parallel to the z-axis that pass through the parabola $y = x^2$, $z = 0$ (Fig. 10.48).

Solution   Suppose that the point $P_0(x_0, x_0^2, 0)$ lies on the parabola $y = x^2$ in the xy-plane. Then, for any value of z, the point $Q(x_0, x_0^2, z)$ will lie on the cylinder because it lies on the line $x = x_0$, $y = x_0^2$ through P_0 parallel to the z-axis. Conversely, any point $Q(x_0, x_0^2, z)$ whose y-coordinate is the square of its x-coordinate lies on the cylinder because it lies on the line $x = x_0$, $y = x_0^2$ through P_0 parallel to the z-axis (Fig. 10.49).

Regardless of the value of z, therefore, the points on the surface are the points whose coordinates satisfy the equation $y = x^2$. This makes $y = x^2$ an equation for the cylinder. Because of this, we call the cylinder "the cylinder $y = x^2$."

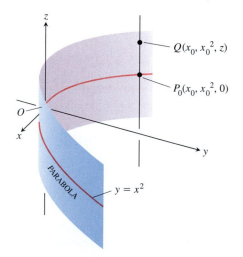

10.49 Every point of the cylinder in Fig. 10.48 has coordinates of the form (x_0, x_0^2, z). We call the cylinder "the cylinder $y = x^2$."

DRAWING LESSON
How to Draw Cylinders Parallel to the Coordinate Axes

$$x^2 + y^2 = 1 \qquad\qquad z = y^2$$

1 Sketch all three coordinate axes *very lightly*.

2 Sketch the trace of the cylinder in the coordinate plane of the two variables that appear in the cylinder's equation. Sketch *very lightly*.

3 Sketch traces in parallel planes on either side (again, *lightly*).

4 Add parallel outer edges to give the shape definition.

5 If more definition is required, darken the parts of the lines that are exposed to view. Leave the hidden parts light. Use line breaks when you can.

Elliptical trace
(cross section)

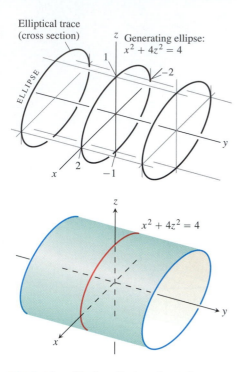

10.50 The elliptic cylinder $x^2 + 4z^2 = 4$ is made of lines parallel to the y-axis and passing through the ellipse $x^2 + 4z^2 = 4$ in the xz-plane. The cross sections or "traces" of the cylinder in planes perpendicular to the y-axis are ellipses congruent to the generating ellipse. The cylinder extends along the entire y-axis.

As Example 1 suggests, any curve $f(x, y) = c$ in the xy-plane defines a cylinder parallel to the z-axis whose equation is also $f(x, y) = c$. The equation $x^2 + y^2 = 1$ defines the circular cylinder made by the lines parallel to the z-axis that pass through the circle $x^2 + y^2 = 1$ in the xy-plane. The equation $x^2 + 4y^2 = 9$ defines the elliptical cylinder made by the lines parallel to the z-axis that pass through the ellipse $x^2 + 4y^2 = 9$ in the xy-plane.

In a similar way, any curve $g(x, z) = c$ in the xz-plane defines a cylinder parallel to the y-axis whose space equation is also $g(x, z) = c$ (Fig. 10.50). Any curve $h(y, z) = c$ defines a cylinder parallel to the x-axis whose space equation is also $h(y, z) = c$ (Fig. 10.51).

10.51 The hyperbolic cylinder $y^2 - z^2 = 1$ is made of lines parallel to the x-axis and passing through the hyperbola $y^2 - z^2 = 1$ in the yz-plane. The cross sections of the cylinder in planes perpendicular to the x-axis are hyperbolas congruent to the generating hyperbola.

> An equation in any two of the three Cartesian coordinates defines a cylinder parallel to the axis of the third coordinate.

Quadric Surfaces

A **quadric surface** is the graph in space of a second-degree equation in x, y, and z. The most general form is

$$Ax^2 + By^2 + Cz^2 + Dxy + Eyz + Fxz + Gx + Hy + Jz + K = 0,$$

where A, B, C, and so on are constants, but the equation can be simplified by translation and rotation, as in the two-dimensional case in Section 9.3. We will study only the simpler equations. Although the definition did not require it, the cylinders in Figs. 10.49–10.51 were also examples of quadric surfaces. We now examine ellipsoids (these include spheres), paraboloids, cones, and hyperboloids.

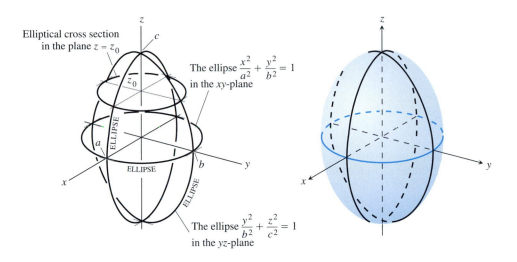

10.52 The ellipsoid

$$\frac{x^2}{a^2} + \frac{y^2}{b^2} + \frac{z^2}{c^2} = 1$$

in Example 2.

EXAMPLE 2   The **ellipsoid**

$$\frac{x^2}{a^2} + \frac{y^2}{b^2} + \frac{z^2}{c^2} = 1 \tag{1}$$

(Fig. 10.52) cuts the coordinate axes at $(\pm a, 0, 0)$, $(0, \pm b, 0)$, and $(0, 0, \pm c)$. It lies within the rectangular box defined by the inequalities $|x| \le a$, $|y| \le b$, and $|z| \le c$. The surface is symmetric with respect to each of the coordinate planes because the variables in the defining equation are squared.

The curves in which the three coordinate planes cut the surface are ellipses. For example,

$$\frac{x^2}{a^2} + \frac{y^2}{b^2} = 1 \qquad \text{when} \qquad z = 0.$$

The section cut from the surface by the plane $z = z_0$, $|z_0| < c$, is the ellipse

$$\frac{x^2}{a^2(1 - (z_0/c)^2)} + \frac{y^2}{b^2(1 - (z_0/c)^2)} = 1. \tag{2}$$

If any two of the semiaxes a, b, and c are equal, the surface is an **ellipsoid of revolution.** If all three are equal, the surface is a sphere. ❑

Technology *Visualizing in Space* A Computer Algebra System (CAS) or other computer graphing utility can help in visualizing surfaces in space. It can draw traces in different planes with far more patience than most people can muster. Many computer graphing systems can rotate the figure so you can see it as if it were a physical model you could turn in your hand. Hidden-line algorithms (see Exercise 74, Section 10.5) are used to block out portions of the surface that you would not see from your current viewing angle. Often a CAS will require surfaces to be entered in parametric form, as discussed in Section 14.6 (see also CAS Exercises 57–60 in Section 12.1). Sometimes you may have to manipulate the grid mesh to see all portions of a surface.

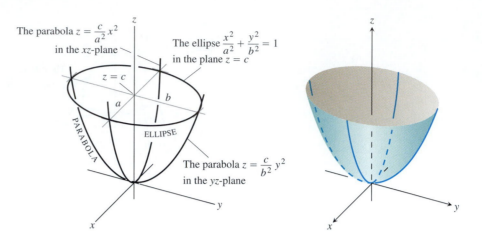

10.53 The elliptic paraboloid $(x^2/a^2) + (y^2/b^2) = z/c$ in Example 3, shown for $c > 0$. The cross sections perpendicular to the z-axis above the xy-plane are ellipses. The cross sections in the planes that contain the z-axis are parabolas.

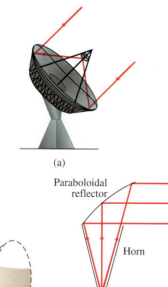

(a)

Paraboloidal reflector

Horn

Focus

(b)   (c)

10.54 Many antennas are shaped like pieces of paraboloids of revolution. (a) Radio telescopes use the same principles as optical telescopes. (b) A "rectangular-cut" radar reflector. (c) The profile of a horn antenna in a microwave radio link.

EXAMPLE 3   The **elliptic paraboloid**

$$\frac{x^2}{a^2} + \frac{y^2}{b^2} = \frac{z}{c} \tag{3}$$

is symmetric with respect to the planes $x = 0$ and $y = 0$ (Fig. 10.53). The only intercept on the axes is the origin. Except for this point, the surface lies above or entirely below the xy-plane, depending on the sign of c. The sections cut by the coordinate planes are

$x = 0$:   the parabola $z = \dfrac{c}{b^2} y^2$

$y = 0$:   the parabola $z = \dfrac{c}{a^2} x^2$ (4)

$z = 0$:   the point $(0, 0, 0)$.

Each plane $z = z_0$ above the xy-plane cuts the surface in the ellipse

$$\frac{x^2}{a^2} + \frac{y^2}{b^2} = \frac{z_0}{c}. \qquad \square$$

EXAMPLE 4   The **circular paraboloid** or **paraboloid of revolution**

$$\frac{x^2}{a^2} + \frac{y^2}{a^2} = \frac{z}{c} \tag{5}$$

is obtained by taking $b = a$ in Eq. (3) for the elliptic paraboloid. The cross sections of the surface by planes perpendicular to the z-axis are circles centered on the z-axis. The cross sections by planes containing the z-axis are congruent parabolas with a common focus at the point $(0, 0, a^2/4c)$.

Shapes cut from circular paraboloids are used for antennas in radio telescopes, satellite trackers, and microwave radio links (Fig. 10.54). $\square$

EXAMPLE 5   The **elliptic cone**

$$\frac{x^2}{a^2} + \frac{y^2}{b^2} = \frac{z^2}{c^2} \tag{6}$$

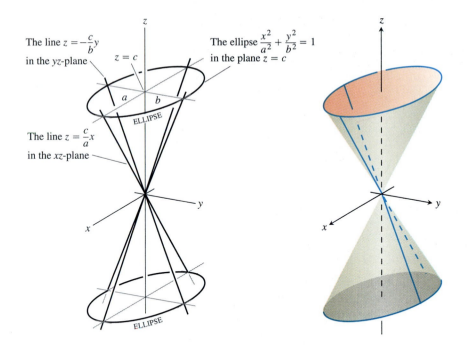

10.55 The elliptic cone $(x^2/a^2) + (y^2/b^2) = (z^2/c^2)$ in Example 5. Planes perpendicular to the z-axis cut the cone in ellipses above and below the xy-plane. Vertical planes that contain the z-axis cut it in pairs of intersecting lines.

is symmetric with respect to the three coordinate planes (Fig. 10.55). The sections cut by the coordinate planes are

$x = 0$:   the lines $z = \pm \dfrac{c}{b} y$ (7)

$y = 0$:   the lines $z = \pm \dfrac{c}{a} x$ (8)

$z = 0$:   the point $(0, 0, 0)$.

The sections cut by planes $z = z_0$ above and below the xy-plane are ellipses whose centers lie on the z-axis and whose vertices lie on the lines in Eqs. (7) and (8).
    If $a = b$, the cone is a right circular cone. ❏

EXAMPLE 6   The **hyperboloid of one sheet**

$$\frac{x^2}{a^2} + \frac{y^2}{b^2} - \frac{z^2}{c^2} = 1 \qquad (9)$$

is symmetric with respect to each of the three coordinate planes (Fig. 10.56, on the following page). The sections cut out by the coordinate planes are

$x = 0$:   the hyperbola $\dfrac{y^2}{b^2} - \dfrac{z^2}{c^2} = 1$

$y = 0$:   the hyperbola $\dfrac{x^2}{a^2} - \dfrac{z^2}{c^2} = 1$ (10)

$z = 0$:   the ellipse $\dfrac{x^2}{a^2} + \dfrac{y^2}{b^2} = 1.$

The plane $z = z_0$ cuts the surface in an ellipse with center on the z-axis and vertices on one of the hyperbolas in (10).

10.56 The hyperboloid $(x^2/a^2) + (y^2/b^2) - (z^2/c^2) = 1$ in Example 6. Planes perpendicular to the z-axis cut it in ellipses. Vertical planes containing the z-axis cut it in hyperbolas.

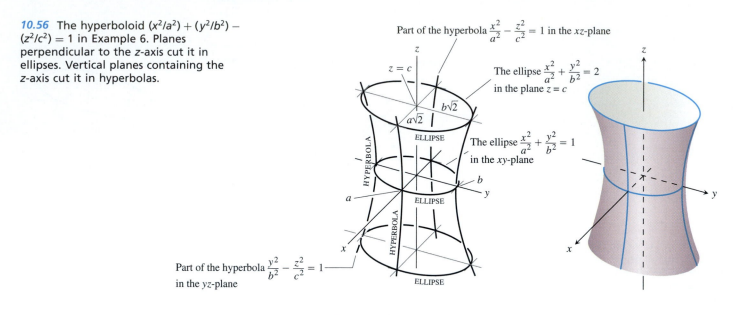

Part of the hyperbola $\dfrac{x^2}{a^2} - \dfrac{z^2}{c^2} = 1$ in the xz-plane

The ellipse $\dfrac{x^2}{a^2} + \dfrac{y^2}{b^2} = 2$ in the plane $z = c$

The ellipse $\dfrac{x^2}{a^2} + \dfrac{y^2}{b^2} = 1$ in the xy-plane

Part of the hyperbola $\dfrac{y^2}{b^2} - \dfrac{z^2}{c^2} = 1$ in the yz-plane

The surface is connected, meaning that it is possible to travel from one point on it to any other without leaving the surface. For this reason, it is said to have *one* sheet, in contrast to the hyperboloid in the next example, which has two sheets.

If $a = b$, the hyperboloid is a surface of revolution.

EXAMPLE 7   The **hyperboloid of two sheets**

$$\frac{z^2}{c^2} - \frac{x^2}{a^2} - \frac{y^2}{b^2} = 1 \tag{11}$$

is symmetric with respect to the three coordinate planes (Fig. 10.57). The plane $z = 0$ does not intersect the surface; in fact, for a horizontal plane to intersect the

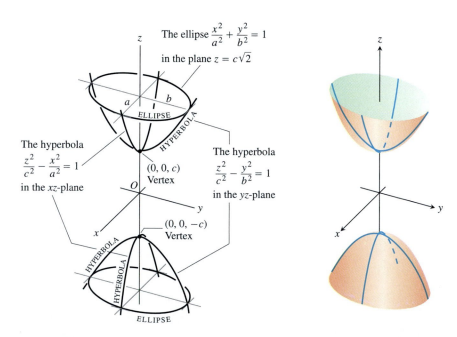

The ellipse $\dfrac{x^2}{a^2} + \dfrac{y^2}{b^2} = 1$ in the plane $z = c\sqrt{2}$

The hyperbola $\dfrac{z^2}{c^2} - \dfrac{x^2}{a^2} = 1$ in the xz-plane

$(0, 0, c)$ Vertex

The hyperbola $\dfrac{z^2}{c^2} - \dfrac{y^2}{b^2} = 1$ in the yz-plane

$(0, 0, -c)$ Vertex

10.57 The hyperboloid $(z^2/c^2) - (x^2/a^2) - (y^2/b^2) = 1$ in Example 7. Planes perpendicular to the z-axis above and below the vertices cut it in ellipses. Vertical planes containing the z-axis cut it in hyperbolas.

surface, we must have $|z| \geq c$. The hyperbolic sections

$$x = 0: \quad \frac{z^2}{c^2} - \frac{y^2}{b^2} = 1, \qquad y = 0: \quad \frac{z^2}{c^2} - \frac{x^2}{a^2} = 1,$$

have their vertices and foci on the z-axis. The surface is separated into two portions, one above the plane $z = c$ and the other below the plane $z = -c$. This accounts for its name.

Equations (9) and (11) have different numbers of negative terms. The number in each case is the same as the number of sheets of the hyperboloid. If we replace the 1 on the right side of either Eq. (9) or Eq. (11) by 0, we obtain the equation

$$\frac{x^2}{a^2} + \frac{y^2}{b^2} = \frac{z^2}{c^2}$$

for an elliptic cone (Eq. 6). The hyperboloids are asymptotic to this cone (Fig. 10.58) in the same way that the hyperbolas

$$\frac{x^2}{a^2} - \frac{y^2}{b^2} = \pm 1$$

are asymptotic to the lines

$$\frac{x^2}{a^2} - \frac{y^2}{b^2} = 0$$

in the xy-plane. ❑

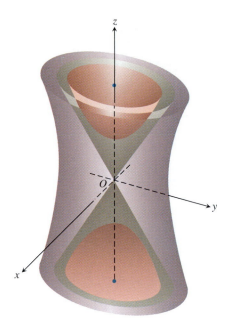

10.58 Both hyperboloids are asymptotic to the cone (Example 7).

EXAMPLE 8 The **hyperbolic paraboloid**

$$\frac{y^2}{b^2} - \frac{x^2}{a^2} = \frac{z}{c}, \qquad c > 0 \tag{12}$$

has symmetry with respect to the planes $x = 0$ and $y = 0$ (Fig. 10.59). The sections in these planes are

$$x = 0: \quad \text{the parabola } z = \frac{c}{b^2} y^2, \tag{13}$$

$$y = 0: \quad \text{the parabola } z = -\frac{c}{a^2} x^2. \tag{14}$$

10.59 The hyperbolic paraboloid $(y^2/b^2) - (x^2/a^2) = z/c$, $c > 0$ in Example 8. The cross sections in planes perpendicular to the z-axis above and below the xy-plane are hyperbolas. The cross sections in planes perpendicular to the other axes are parabolas.

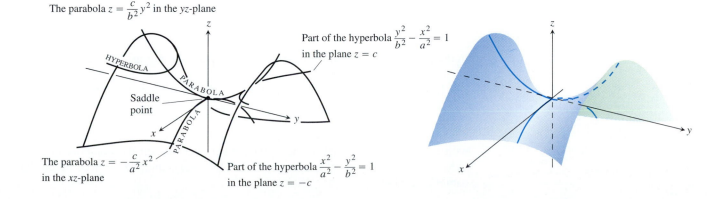

DRAWING LESSON
How to Draw Quadric Surfaces

$$x^2 + \frac{y^2}{4} + z^2 = 1 \qquad\qquad z = 4 - x^2 - y^2$$

1 Lightly sketch the three coordinate axes.

2 Decide on a scale and mark the intercepts on the axes.

3 Sketch cross sections in the coordinate planes and in a few parallel planes, but don't clutter the picture. Use tangent lines as guides.

4 If more is required, darken the parts exposed to view. Leave the rest light. Use line breaks when you can.

In the plane $x = 0$, the parabola opens upward from the origin. The parabola in the plane $y = 0$ opens downward.

If we cut the surface by a plane $z = z_0 > 0$, the section is a hyperbola,

$$\frac{y^2}{b^2} - \frac{x^2}{a^2} = \frac{z_0}{c}, \tag{15}$$

with its focal axis parallel to the y-axis and its vertices on the parabola in (13). If z_0 is negative, the focal axis is parallel to the x-axis and the vertices lie on the parabola in (14).

Near the origin, the surface is shaped like a saddle. To a person traveling along the surface in the yz-plane, the origin looks like a minimum. To a person traveling in the xz-plane, the origin looks like a maximum. Such a point is called a **minimax** or **saddle point** of a surface. ❑

Liquid Mirror Telescopes

When a circular pan of liquid is rotated about its vertical axis, the surface of the liquid does not stay flat. Instead, it assumes the shape of a paraboloid of revolution, exactly what is needed for the primary mirror of a reflecting telescope. At the turn of the century, attempts to make reliable mirrors with revolving mercury failed because of surface ripples and focus losses caused by variations in the speed of rotation. Today, these difficulties can be overcome with synchronous motors driven by oscillator-stabilized power supplies and checked against constant clocks.

Using the same idea, astronomers at the Steward Observatory's Mirror Laboratory in Tucson, Arizona, have used a large heated spinning turntable to cast borosilicate glass blanks for lightweight mirrors. Spincast mirrors cost less than traditionally cast mirrors and can be made larger. Their shorter focal lengths also allow them to be installed in compact telescope frames that are less expensive to house and less likely to flex in strong winds.

Exercises 10.6

Matching Equations with Surfaces

In Exercises 1–12, match the equation with the surface it defines. Also, identify each surface by type (paraboloid, ellipsoid, etc.) The surfaces are labeled (a)–(l).

1. $x^2 + y^2 + 4z^2 = 10$ **2.** $z^2 + 4y^2 - 4x^2 = 4$

3. $9y^2 + z^2 = 16$ **4.** $y^2 + z^2 = x^2$

5. $x = y^2 - z^2$ **6.** $x = -y^2 - z^2$

7. $x^2 + 2z^2 = 8$ **8.** $z^2 + x^2 - y^2 = 1$

9. $x = z^2 - y^2$ **10.** $z = -4x^2 - y^2$

11. $x^2 + 4z^2 = y^2$ **12.** $9x^2 + 4y^2 + 2z^2 = 36$

a)

b)

c)

d)

e)

f)

g)

h)

i)

j)

k)

l)

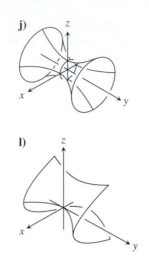

Drawing

Sketch the surfaces in Exercises 13–76.

CYLINDERS

13. $x^2 + y^2 = 4$ **14.** $x^2 + z^2 = 4$

15. $z = y^2 - 1$ **16.** $x = y^2$

17. $x^2 + 4z^2 = 16$ **18.** $4x^2 + y^2 = 36$

19. $z^2 - y^2 = 1$ **20.** $yz = 1$

ELLIPSOIDS

21. $9x^2 + y^2 + z^2 = 9$ **22.** $4x^2 + 4y^2 + z^2 = 16$

23. $4x^2 + 9y^2 + 4z^2 = 36$ **24.** $9x^2 + 4y^2 + 36z^2 = 36$

PARABOLOIDS

25. $z = x^2 + 4y^2$ **26.** $z = x^2 + 9y^2$

27. $z = 8 - x^2 - y^2$ **28.** $z = 18 - x^2 - 9y^2$

29. $x = 4 - 4y^2 - z^2$ **30.** $y = 1 - x^2 - z^2$

CONES

31. $x^2 + y^2 = z^2$ **32.** $y^2 + z^2 = x^2$

33. $4x^2 + 9z^2 = 9y^2$ **34.** $9x^2 + 4y^2 = 36z^2$

HYPERBOLOIDS

35. $x^2 + y^2 - z^2 = 1$ **36.** $y^2 + z^2 - x^2 = 1$

37. $(y^2/4) + (z^2/9) - (x^2/4) = 1$

38. $(x^2/4) + (y^2/4) - (z^2/9) = 1$

39. $z^2 - x^2 - y^2 = 1$ **40.** $(y^2/4) - (x^2/4) - z^2 = 1$

41. $x^2 - y^2 - (z^2/4) = 1$ **42.** $(x^2/4) - y^2 - (z^2/4) = 1$

HYPERBOLIC PARABOLOIDS

43. $y^2 - x^2 = z$ **44.** $x^2 - y^2 = z$

ASSORTED

45. $x^2 + y^2 + z^2 = 4$ **46.** $4x^2 + 4y^2 = z^2$

47. $z = 1 + y^2 - x^2$ **48.** $y^2 - z^2 = 4$

49. $y = -(x^2 + z^2)$ **50.** $z^2 - 4x^2 - 4y^2 = 4$

51. $16x^2 + 4y^2 = 1$ **52.** $z = x^2 + y^2 + 1$

53. $x^2 + y^2 - z^2 = 4$ **54.** $x = 4 - y^2$

55. $x^2 + z^2 = y$ **56.** $z^2 - (x^2/4) - y^2 = 1$

57. $x^2 + z^2 = 1$ **58.** $4x^2 + 4y^2 + z^2 = 4$

59. $16y^2 + 9z^2 = 4x^2$ **60.** $z = x^2 - y^2 - 1$

61. $9x^2 + 4y^2 + z^2 = 36$ **62.** $4x^2 + 9z^2 = y^2$

63. $x^2 + y^2 - 16z^2 = 16$ **64.** $z^2 + 4y^2 = 9$

65. $z = -(x^2 + y^2)$ **66.** $y^2 - x^2 - z^2 = 1$

67. $x^2 - 4y^2 = 1$ **68.** $z = 4x^2 + y^2 - 4$

69. $4y^2 + z^2 - 4x^2 = 4$ **70.** $z = 1 - x^2$

71. $x^2 + y^2 = z$ **72.** $(x^2/4) + y^2 - z^2 = 1$

73. $yz = 1$ **74.** $36x^2 + 9y^2 + 4z^2 = 36$

75. $9x^2 + 16y^2 = 4z^2$ **76.** $4z^2 - x^2 - y^2 = 4$

Theory and Examples

77. a) Express the area A of the cross section cut from the ellipsoid

$$x^2 + \frac{y^2}{4} + \frac{z^2}{9} = 1$$

by the plane $z = c$ as a function of c. (The area of an ellipse with semiaxes a and b is πab.)

b) Use slices perpendicular to the z-axis to find the volume of the ellipsoid in (a).

c) Now find the volume of the ellipsoid

$$\frac{x^2}{a^2} + \frac{y^2}{b^2} + \frac{z^2}{c^2} = 1.$$

Does your formula give the volume of a sphere of radius a if $a = b = c$?

78. The barrel shown here is shaped like an ellipsoid with equal pieces cut from the ends by planes perpendicular to the z-axis. The cross sections perpendicular to the z-axis are circular. The barrel is $2h$ units high, its midsection radius is R, and its end radii are both r. Find a formula for the barrel's volume. Then check two things. First, suppose the sides of the barrel are straightened to turn the barrel into a cylinder of radius R and height $2h$. Does your formula give the cylinder's volume? Second, suppose $r = 0$

and $h = R$ so the barrel is a sphere. Does your formula give the sphere's volume?

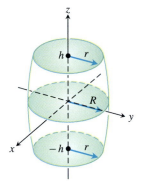

79. Show that the volume of the segment cut from the paraboloid

$$\frac{x^2}{a^2} + \frac{y^2}{b^2} = \frac{z}{c}$$

by the plane $z = h$ equals half the segment's base times its altitude. (Figure 10.53 shows the segment for the special case $h = c$.)

80. a) Find the volume of the solid bounded by the hyperboloid

$$\frac{x^2}{a^2} + \frac{y^2}{b^2} - \frac{z^2}{c^2} = 1$$

and the planes $z = 0$ and $z = h, h > 0$.

b) Express your answer in (a) in terms of h and the areas A_0 and A_h of the regions cut by the hyperboloid from the planes $z = 0$ and $z = h$.

c) Show that the volume in (a) is also given by the formula

$$V = \frac{h}{6}\left(A_0 + 4A_m + A_h\right),$$

where A_m is the area of the region cut by the hyperboloid from the plane $z = h/2$.

81. If the hyperbolic paraboloid $(y^2/b^2) - (x^2/a^2) = z/c$ is cut by the plane $y = y_1$, the resulting curve is a parabola. Find its vertex and focus.

82. Suppose you set $z = 0$ in the equation

$$A x^2 + B y^2 + C z^2 + D xy + E yz + F xz + G x + H y + J z + K = 0$$

to obtain a curve in the xy-plane. What will the curve be like? Give reasons for your answer.

83. Every time we found the trace of a quadric surface in a plane parallel to one of the coordinate planes, it turned out to be a conic section. Was this mere coincidence? Did it have to happen? Give reasons for your answer.

84. Suppose you intersect a quadric surface with a plane that is *not* parallel to one of the coordinate planes. What will the trace in the plane be like? Give reasons for your answer.

▦ Computer Grapher Explorations

Plot the surfaces in Exercises 85–88 over the indicated domains. If you can, rotate the surface into different viewing positions.

85. $z = y^2, \quad -2 \le x \le 2, \quad -0.5 \le y \le 2$

86. $z = 1 - y^2, \quad -2 \le x \le 2, \quad -2 \le y \le 2$

87. $z = x^2 + y^2, \quad -3 \le x \le 3, \quad -3 \le y \le 3$

88. $z = x^2 + 2y^2 \quad$ over

    a) $-3 \le x \le 3, \quad -3 \le y \le 3$
    b) $-1 \le x \le 1, \quad -2 \le y \le 3$
    c) $-2 \le x \le 2, \quad -2 \le y \le 2$
    d) $-2 \le x \le 2, \quad -1 \le y \le 1$

✹ CAS Explorations and Projects

Use a CAS to plot the surfaces in Exercises 89–94. Identify the type of quadric surface from your graph.

89. $\dfrac{x^2}{9} + \dfrac{y^2}{36} = 1 - \dfrac{z^2}{25}$

90. $\dfrac{x^2}{9} - \dfrac{z^2}{9} = 1 - \dfrac{y^2}{16}$

91. $5x^2 = z^2 - 3y^2$

92. $\dfrac{y^2}{16} = 1 - \dfrac{x^2}{9} + z$

93. $\dfrac{x^2}{9} - 1 = \dfrac{y^2}{16} + \dfrac{z^2}{2}$

94. $y - \sqrt{4 - z^2} = 0$

10.7

Cylindrical and Spherical Coordinates

This section introduces two new coordinate systems for space: the cylindrical coordinate system and the spherical coordinate system. Cylindrical coordinates simplify the equations of cylinders. Spherical coordinates simplify the equations of spheres and cones. We will use cylindrical coordinates to study planetary motion in Section 11.5.

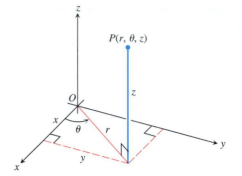

10.60 The cylindrical coordinates of a point in space are r, θ, and z.

Cylindrical Coordinates

We obtain cylindrical coordinates for space by combining polar coordinates in the xy-plane with the usual z-axis. This assigns to every point in space one or more coordinate triples of the form (r, θ, z), as shown in Fig. 10.60.

Definition

Cylindrical coordinates represent a point P in space by ordered triples (r, θ, z) in which

1. r and θ are polar coordinates for the vertical projection of P on the xy-plane,

2. z is the rectangular vertical coordinate.

The values of x, y, r, and θ in rectangular and cylindrical coordinates are related by the usual equations.

Equations Relating Rectangular (x, y, z) and Cylindrical (r, θ, z) Coordinates

$$x = r\cos\theta, \qquad y = r\sin\theta, \qquad z = z$$
$$r^2 = x^2 + y^2, \qquad \tan\theta = y/x$$

(1)

In cylindrical coordinates, the equation $r = a$ describes not just a circle in the xy-plane but an entire cylinder about the z-axis (Fig. 10.61). The z-axis is given by $r = 0$. The equation $\theta = \theta_0$ describes the plane that contains the z-axis and makes an angle θ_0 with the positive x-axis. And, just as in rectangular coordinates, the equation $z = z_0$ describes a plane perpendicular to the z-axis.

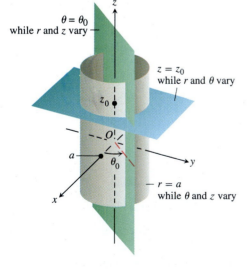

10.61 Constant-coordinate equations in cylindrical coordinates yield cylinders and planes.

EXAMPLE 1 What points satisfy the equations

$$r = 2, \qquad \theta = \frac{\pi}{4}?$$

Along this line, z varies while r and θ have the constant values $r = 2$ and $\theta = \pi/4$.

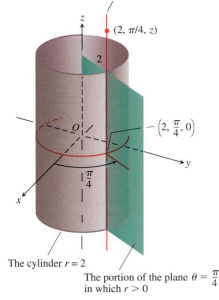

The cylinder $r = 2$

The portion of the plane $\theta = \dfrac{\pi}{4}$ in which $r > 0$

10.62 The points whose first two cylindrical coordinates are $r = 2$ and $\theta = \pi/4$ form a line parallel to the z-axis (Example 1).

Solution These points make up the line in which the cylinder $r = 2$ cuts the portion of the plane $\theta = \pi/4$ where r is positive (Fig. 10.62). This is the line through the point $(2, \pi/4, 0)$ parallel to the z-axis. ❏

EXAMPLE 2 Sketch the surface $r = 1 + \cos\theta$.

Solution The equation involves only r and θ; the coordinate variable z is missing. Therefore, the surface is a cylinder of lines that pass through the cardioid $r = 1 + \cos\theta$ in the $r\theta$-plane and lie parallel to the z-axis. The rules for sketching the cylinder are the same as always: sketch the x-, y-, and z-axes, draw a few perpendicular cross sections, connect the cross sections with parallel lines, and darken the exposed parts (Fig. 10.63). ❏

EXAMPLE 3 Find a Cartesian equation for the surface $z = r^2$ and identify the surface.

Solution From Eqs. (1) we have $z = r^2 = x^2 + y^2$. The surface is the circular paraboloid $x^2 + y^2 = z$. ❏

EXAMPLE 4 Find an equation for the circular cylinder $4x^2 + 4y^2 = 9$ in cylindrical coordinates.

Solution The cylinder consists of the points whose distance from the z-axis is $\sqrt{x^2 + y^2} = 3/2$. The corresponding equation in cylindrical coordinates is $r = 3/2$. ❏

EXAMPLE 5 Find an equation for the cylinder $x^2 + (y - 3)^2 = 9$ in cylindrical coordinates (Fig. 10.64).

Solution The equation for the cylinder in cylindrical coordinates is the same as the polar equation for the cylinder's base in the xy-plane:

$$r = 6\sin\theta.$$ ❏

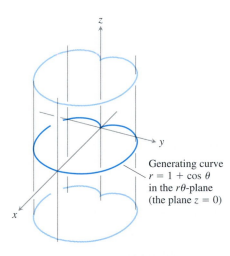

Generating curve $r = 1 + \cos\theta$ in the $r\theta$-plane (the plane $z = 0$)

10.63 The cylindrical coordinate equation $r = 1 + \cos\theta$ defines a cylinder in space whose cross sections perpendicular to the z-axis are cardioids (Example 2).

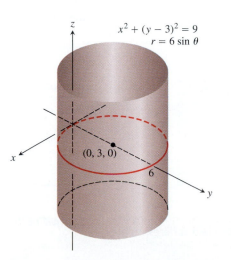

$x^2 + (y - 3)^2 = 9$
$r = 6\sin\theta$

$(0, 3, 0)$

10.64 The cylinder in Example 5.

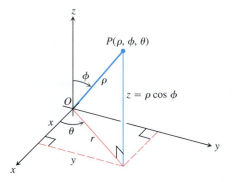

10.65 The spherical coordinates $\rho, \phi,$ and θ and their relation to *x, y, z,* and *r.*

A few books give spherical coordinates in the order (ρ, θ, ϕ), with θ and ϕ reversed. In some cases, you may also find *r* being used for ρ. Watch out for this when you read elsewhere.

Spherical Coordinates

Spherical coordinates locate points in space with angles and a distance, as shown in Fig. 10.65.

The first coordinate, $\rho = |\overrightarrow{OP}|$, is the point's distance from the origin. Unlike *r, the variable ρ is never negative.* The second coordinate, ϕ, is the angle $\overrightarrow{OP}$ makes with the positive *z*-axis. It is required to lie in the interval $[0, \pi]$. The third coordinate is the angle θ as measured in cylindrical coordinates.

> **Definition**
>
> **Spherical coordinates** represent a point *P* in space by ordered triples (ρ, ϕ, θ) in which
>
> **1.** ρ is the distance from *P* to the origin,
> **2.** ϕ is the angle $\overrightarrow{OP}$ makes with the positive *z*-axis $(0 \le \phi \le \pi)$,
> **3.** θ is the angle from cylindrical coordinates.

The equation $\rho = a$ describes the sphere of radius *a* centered at the origin (Fig. 10.66). The equation $\phi = \phi_0$ describes a single cone whose vertex lies at the origin and whose axis lies along the *z*-axis. (We broaden our interpretation to include the *xy*-plane as the cone $\phi = \pi/2$.) If ϕ_0 is greater than $\pi/2$, the cone $\phi = \phi_0$ opens downward.

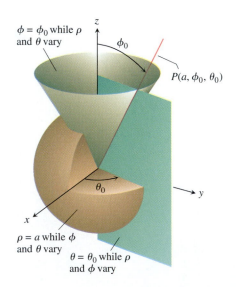

$\phi = \phi_0$ while ρ and θ vary

$P(a, \phi_0, \theta_0)$

$\rho = a$ while ϕ and θ vary

$\theta = \theta_0$ while ρ and ϕ vary

10.66 Constant-coordinate equations in spherical coordinates yield spheres, single cones, and half-planes.

> **Equations Relating Spherical Coordinates to Cartesian and Cylindrical Coordinates**
>
> $$r = \rho \sin\phi, \qquad x = r\cos\theta = \rho\sin\phi\cos\theta,$$
> $$z = \rho\cos\phi, \qquad y = r\sin\theta = \rho\sin\phi\sin\theta, \tag{2}$$
> $$\rho = \sqrt{x^2 + y^2 + z^2} = \sqrt{r^2 + z^2}$$

EXAMPLE 6 Find a spherical coordinate equation for the sphere

$$x^2 + y^2 + (z - 1)^2 = 1.$$

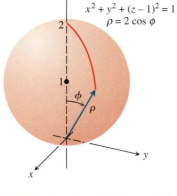

$x^2 + y^2 + (z-1)^2 = 1$
$\rho = 2\cos\phi$

10.67 The sphere in Example 6.

Solution   We use Eqs. (2) to substitute for x, y, and z:

$$x^2 + y^2 + (z-1)^2 = 1$$

$$\rho^2 \sin^2\phi \cos^2\theta + \rho^2 \sin^2\phi \sin^2\theta + (\rho\cos\phi - 1)^2 = 1 \qquad \text{Eqs. (2)}$$

$$\rho^2 \sin^2\phi \underbrace{(\cos^2\theta + \sin^2\theta)}_{1} + \rho^2 \cos^2\phi - 2\rho\cos\phi + 1 = 1$$

$$\rho^2 \underbrace{(\sin^2\phi + \cos^2\phi)}_{1} = 2\rho\cos\phi$$

$$\rho^2 = 2\rho\cos\phi$$

$$\rho = 2\cos\phi$$

See Fig. 10.67.   ❏

EXAMPLE 7   Find a spherical coordinate equation for the cone $z = \sqrt{x^2 + y^2}$ (Fig. 10.68).

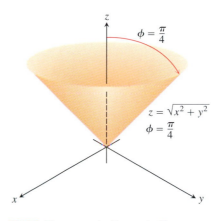

$\phi = \dfrac{\pi}{4}$

$z = \sqrt{x^2 + y^2}$
$\phi = \dfrac{\pi}{4}$

10.68 The cone in Example 7.

Solution 1   Use geometry. The cone is symmetric with respect to the z-axis and cuts the first quadrant of the yz-plane along the line $z = y$. The angle between the cone and the positive z-axis is therefore $\pi/4$ radians. The cone consists of the points whose spherical coordinates have ϕ equal to $\pi/4$, so its equation is $\phi = \pi/4$.

Solution 2   Use algebra. If we use Eqs. (2) to substitute for x, y, and z we obtain the same result:

$$z = \sqrt{x^2 + y^2}$$

$$\rho\cos\phi = \sqrt{\rho^2 \sin^2\phi} \qquad \text{Example 6}$$

$$\rho\cos\phi = \rho\sin\phi \qquad \rho \geq 0,\ \sin\phi \geq 0$$

$$\cos\phi = \sin\phi$$

$$\phi = \frac{\pi}{4} \qquad 0 \leq \phi \leq \pi$$

❏

Exercises 10.7

Converting Point Coordinates

The following table gives the coordinates of specific points in space in one of three coordinate systems. In Exercises 1–10, find coordinates for each point in the other two systems. There may be more than one right answer because points in cylindrical and spherical coordinates may have more than one coordinate triple.

Rectangular (x, y, z)	Cylindrical $(r,\ \theta,\ z)$	Spherical $(\rho, \phi,\ \theta)$
1. $(0, 0, 0)$		
2. $(1, 0, 0)$		
3. $(0, 1, 0)$		
4. $(0, 0, 1)$		
5.	$(1, 0, 0)$	
6.	$(\sqrt{2}, 0, 1)$	
7.	$(1, \pi/2, 1)$	
8.		$(\sqrt{3}, \pi/3, -\pi/2)$
9.		$(2\sqrt{2}, \pi/2, 3\pi/2)$
10.		$(\sqrt{2}, \pi, 2\pi/2)$

Converting Equations and Inequalities; Associated Figures

In Exercises 11–36, translate the equations and inequalities from the given coordinate system (rectangular, cylindrical, spherical) into equations and inequalities in the other two systems. Also, identify the figure being defined.

11. $r = 0$

12. $x^2 + y^2 = 5$

13. $z = 0$

14. $z = -2$

15. $z = \sqrt{x^2 + y^2}, \quad z \le 1$

16. $z = \sqrt{x^2 + y^2}, \quad 1 \le z \le 2$

17. $\rho \sin\phi \cos\theta = 0$

18. $\tan^2 \phi = 1$

19. $x^2 + y^2 + z^2 = 4$

20. $x^2 + y^2 + \left(z - \dfrac{1}{2}\right)^2 = \dfrac{1}{4}$

21. $\rho = 5\cos\phi$

22. $\rho = -6\cos\phi$

23. $r = \csc\theta$

24. $r = -3\sec\theta$

25. $\rho = \sqrt{2}\sec\phi$

26. $\rho = 9\csc\phi$

27. $x^2 + y^2 + (z - 1)^2 = 1, \quad z \le 1$

28. $r^2 + z^2 = 4, \quad z \le -\sqrt{2}$

29. $\rho = 3, \quad \pi/3 \le \phi \le 2\pi/3$

30. $x^2 + y^2 + z^2 = 3, \quad 0 \le z \le \sqrt{3}/2$

31. $z = 4 - 4r^2, \quad 0 \le r \le 1$

32. $z = 4 - r, \quad 0 \le r \le 4$

33. $\phi = 3\pi/4, \quad 0 \le \rho \le \sqrt{2}$

34. $\phi = \pi/2, \quad 0 \le \rho \le \sqrt{7}$

35. $z + r^2 \cos 2\theta = 0$

36. $z^2 - r^2 = 1$

37. Find the rectangular coordinates of the center of the sphere
$$r^2 + z^2 = 4r\cos\theta + 6r\sin\theta + 2z.$$

38. Find the rectangular coordinates of the center of the sphere
$$\rho = 2\sin\phi \ (\cos\theta - 2\sin\theta).$$

Sets Defined by Equations

In Exercises 39–42, describe the set of points whose cylindrical coordinates satisfy the given equation. Sketch each set.

39. $r = -2\sin\theta$

40. $r = 2\cos\theta$

41. $r = 1 - \cos\theta$

42. $r = 1 + \sin\theta$

In Exercises 43 and 44, describe the set of points in space whose spherical coordinates satisfy the given equation. Sketch each set.

43. $\rho = 1 - \cos\phi$

44. $\rho = 1 + \cos\phi$

Theory and Examples

45. *Horizontal planes in cylindrical and spherical coordinates.* (a) Show that the plane whose equation is $z = c \ (c \ne 0)$ in rectangular and cylindrical coordinates has the equation $\rho = c\sec\phi$ in spherical coordinates. (b) Find an equation for the xy-plane in spherical coordinates.

46. *Vertical circular cylinders in spherical coordinates.* Find an equation of the form $\rho = f(\phi)$ for the cylinder $x^2 + y^2 = a^2$.

47. *Vertical planes in cylindrical coordinates.* (a) Show that planes perpendicular to the x-axis have equations of the form $r = a\sec\theta$ in cylindrical coordinates. (b) Show that planes perpendicular to the y-axis have equations of the form $r = b\csc\theta$.

48. *(Continuation of Exercise 47.)* Find an equation of the form $r = f(\theta)$ in cylindrical coordinates for the plane $ax + by = c, c \ne 0$.

49. *Symmetry.* What symmetry will you find in a surface that has an equation of the form $r = f(z)$ in cylindrical coordinates? Give reasons for your answer.

50. *Symmetry.* What symmetry will you find in a surface that has an equation of the form $\rho = f(\phi)$ in spherical coordinates? Give reasons for your answer.

✿ CAS Explorations and Projects

Use a CAS to perform the following steps in Exercises 51 and 52.

a) Solve the given equation for r (cylindrical coordinates) or ρ (spherical coordinates). Choose the expression with the positive square root and simplify it.

b) Plot r as a function of θ and z, or ρ as a function of θ and ϕ, as appropriate. Use the specified ranges of the variables for your plots. (Other ranges can present difficulties for a computer grapher in producing the complete surface. Experiment with other ranges later if you wish.)

c) From your plot in (b), estimate the center and radius of the sphere you graphed to the nearest integer value.

d) Convert the given equation from cylindrical or spherical coordinates to rectangular coordinates. Simplify the results. You may need to complete the final simplification by hand to obtain an equation for the sphere in the factored form $(x - x_0)^2 + (y - y_0)^2 + (z - z_0)^2 = a^2$. Compare this result against your estimates from (c).

e) Plot the implicit equation obtained in (d). How does it compare with the plot produced in (b)? Can you explain any discrepancies from the way in which a computer grapher plots surfaces?

51. $r^2 + z^2 = 2r(\cos \theta + \sin \theta) + 2, \quad \dfrac{\pi}{4} \le \theta \le \dfrac{9\pi}{4}, \quad -2 \le z \le 2$

52. $\rho^2 = 2\rho(\cos \theta \sin \phi - \cos \phi) + 2, \quad 0 \le \theta \le 2\pi, \quad 0 \le \phi \le \pi$

CHAPTER 10 QUESTIONS TO GUIDE YOUR REVIEW

1. When do directed line segments (in the plane or space) represent the same vector?

2. How are vectors added and subtracted geometrically? algebraically?

3. How do you find a vector's magnitude and direction?

4. If a vector is multiplied by a positive scalar, how is the result related to the original vector? What if the scalar is zero? negative?

5. Define the *dot product* (*scalar product*) of two vectors. Which algebraic laws (commutative, associative, distributive, cancellation) are satisfied by dot products, and which, if any, are not? Give examples. When is the dot product of two vectors equal to zero?

6. What geometric or physical interpretations do dot products have? Give examples.

7. What is the vector projection of a vector **B** onto a vector **A**? How do you write **B** as the sum of a vector parallel to **A** and a vector orthogonal to **A**?

8. Define the *cross product* (*vector product*) of two vectors. Which algebraic laws (commutative, associative, distributive, cancellation) are satisfied by cross products, and which are not? Give examples. When is the cross product of two vectors equal to zero?

9. What geometric or physical interpretations do cross products have? Give examples.

10. What is the determinant formula for calculating the cross product of two vectors relative to the Cartesian **i, j, k**-coordinate system? Use it in an example.

11. How do you find equations for lines, line segments, and planes in space? Give examples. Can you express a line in space by a single equation? A plane?

12. How do you find the distance from a point to a line in space? From a point to a plane? Give examples.

13. What are box products? What significance do they have? How are they evaluated? Give an example.

14. How do you find equations for spheres in space? Give examples.

15. How do you find the intersection of two lines in space? a line and a plane? two planes? Give examples.

16. What is a cylinder? Give examples of equations that define cylinders in Cartesian coordinates; in cylindrical coordinates. What advice can you give about drawing cylinders?

17. What are quadric surfaces? Give examples of different kinds of ellipsoids, paraboloids, cones, and hyperboloids (equations and sketches). What advice can you give about sketching quadric surfaces?

18. How are cylindrical and spherical coordinates defined? Draw diagrams that show how cylindrical and spherical coordinates are related to Cartesian coordinates.

19. Are cylindrical and spherical coordinates for a point in space unique? Explain.

20. What sets have constant-coordinate equations (like $x = 1$ or $r = 1$ or $\phi = \pi/3$) in the three coordinate systems for space?

21. What equations are available for changing from one space coordinate system to another?

CHAPTER **10** PRACTICE EXERCISES

Vector Calculations

1. Draw the unit vectors $\mathbf{u} = (\cos\theta)\,\mathbf{i} + (\sin\theta)\,\mathbf{j}$ for $\theta = 0, \pi/2$, $2\pi/3, 5\pi/4$, and $5\pi/3$, together with the coordinate axes and unit circle.

2. Find the unit vector obtained by rotating
   a) $\mathbf{i}$ clockwise through an angle $\pi/4$;
   b) $\mathbf{j}$ counterclockwise through an angle $2\pi/3$.

Express the vectors in Exercises 3–6 in terms of their lengths and directions.

3. $\sqrt{2}\,\mathbf{i} + \sqrt{2}\,\mathbf{j}$

4. $-\mathbf{i} - \mathbf{j}$

5. $2\,\mathbf{i} - 3\,\mathbf{j} + 6\,\mathbf{k}$

6. $\mathbf{i} + 2\,\mathbf{j} - \mathbf{k}$

7. Find a vector 2 units long in the direction of $\mathbf{A} = 4\,\mathbf{i} - \mathbf{j} + 4\,\mathbf{k}$.

8. Find a vector 5 units long in the direction opposite to the direction of $\mathbf{A} = (3/5)\,\mathbf{i} + (4/5)\,\mathbf{k}$.

9. The accompanying figure shows parallelogram $ABCD$ and the midpoint P of diagonal BD.
   a) Express $\overrightarrow{BD}$ in terms of $\overrightarrow{AB}$ and $\overrightarrow{AD}$.
   b) Express $\overrightarrow{AP}$ in terms of $\overrightarrow{AB}$ and $\overrightarrow{AD}$.
   c) Prove that P is also the midpoint of diagonal AC.

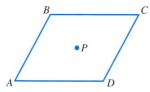

10. The points O, A, B, C, D, and E are vertices of the rectangular box shown here. Express $\overrightarrow{OD}$ and $\overrightarrow{OE}$ in terms of $\overrightarrow{OA}$, $\overrightarrow{OB}$, and $\overrightarrow{OC}$.

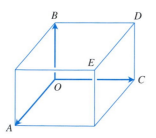

11. Copy the vectors $\mathbf{u}$ and $\mathbf{v}$ and sketch the vector projection of $\mathbf{v}$ onto $\mathbf{u}$.

12. Express vectors $\mathbf{a}$, $\mathbf{b}$, and $\mathbf{c}$ in terms of $\mathbf{u}$ and $\mathbf{v}$.

In Exercises 13 and 14, find $|\mathbf{A}|$, $|\mathbf{B}|$, $\mathbf{A} \cdot \mathbf{B}$, $\mathbf{B} \cdot \mathbf{A}$, $\mathbf{A} \times \mathbf{B}$, $\mathbf{B} \times \mathbf{A}$, $|\mathbf{A} \times \mathbf{B}|$, the angle between $\mathbf{A}$ and $\mathbf{B}$, the scalar component of $\mathbf{B}$ in the direction of $\mathbf{A}$, and the vector projection of $\mathbf{B}$ onto $\mathbf{A}$.

13. $\mathbf{A} = \mathbf{i} + \mathbf{j}$
    $\mathbf{B} = 2\,\mathbf{i} + \mathbf{j} - 2\,\mathbf{k}$

14. $\mathbf{A} = \mathbf{i} + \mathbf{j} + 2\,\mathbf{k}$
    $\mathbf{B} = -\mathbf{i} - \mathbf{k}$

In Exercises 15 and 16, write $\mathbf{B}$ as the sum of a vector parallel to $\mathbf{A}$ and a vector orthogonal to $\mathbf{A}$.

15. $\mathbf{A} = 2\,\mathbf{i} + \mathbf{j} - \mathbf{k}$,
    $\mathbf{B} = \mathbf{i} + \mathbf{j} - 5\,\mathbf{k}$

16. $\mathbf{A} = \mathbf{i} - 2\,\mathbf{j}$,
    $\mathbf{B} = \mathbf{i} + \mathbf{j} + \mathbf{k}$

In Exercises 17 and 18, draw coordinate axes and then sketch $\mathbf{A}$, $\mathbf{B}$, and $\mathbf{A} \times \mathbf{B}$ as vectors at the origin.

17. $\mathbf{A} = \mathbf{i}$, $\mathbf{B} = \mathbf{i} + \mathbf{j}$

18. $\mathbf{A} = \mathbf{i} - \mathbf{j}$, $\mathbf{B} = \mathbf{i} + \mathbf{j}$

In Exercises 19 and 20, find the unit vectors that are tangent and normal to the curve at point P.

19. $y = \tan x$, $P(\pi/4, 1)$

20. $x^2 + y^2 = 25$, $P(3, 4)$

21. For any vectors $\mathbf{A}$ and $\mathbf{B}$, show that $|\mathbf{A} + \mathbf{B}|^2 + |\mathbf{A} - \mathbf{B}|^2 = 2\,|\mathbf{A}|^2 + 2\,|\mathbf{B}|^2$.

22. Let ABC be the triangle determined by vectors $\mathbf{u}$ and $\mathbf{v}$.
    a) Express the area of $\triangle ABC$ in terms of $\mathbf{u}$ and $\mathbf{v}$.
    b) Express the triangle's altitude h in terms of $\mathbf{u}$ and $\mathbf{v}$.
    c) Find the area and altitude if $\mathbf{u} = \mathbf{i} - \mathbf{j} + \mathbf{k}$ and $\mathbf{v} = 2\,\mathbf{i} + \mathbf{k}$.

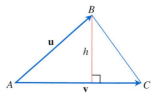

23. If $|\mathbf{v}| = 2$, $|\mathbf{w}| = 3$, and the angle between $\mathbf{v}$ and $\mathbf{w}$ is $\pi/3$, find $|\mathbf{v} - 2\mathbf{w}|$.

24. For what value or values of a will the vectors $\mathbf{u} = 2\mathbf{i} + 4\mathbf{j} - 5\mathbf{k}$ and $\mathbf{v} = -4\mathbf{i} - 8\mathbf{j} + a\mathbf{k}$ be parallel?

In Exercises 25 and 26, find (a) the area of the parallelogram determined by vectors $\mathbf{A}$ and $\mathbf{B}$, (b) the volume of the parallelepiped determined by the vectors $\mathbf{A}$, $\mathbf{B}$, and $\mathbf{C}$.

25. $\mathbf{A} = \mathbf{i} + \mathbf{j} - \mathbf{k}$, $\quad \mathbf{B} = 2\mathbf{i} + \mathbf{j} + \mathbf{k}$, $\quad \mathbf{C} = -\mathbf{i} - 2\mathbf{j} + 3\mathbf{k}$

26. $\mathbf{A} = \mathbf{i} + \mathbf{j}$, $\quad \mathbf{B} = \mathbf{j}$, $\quad \mathbf{C} = \mathbf{i} + \mathbf{j} + \mathbf{k}$

Lines, Planes, and Distances

27. Suppose that $\mathbf{n}$ is normal to a plane and that $\mathbf{v}$ is parallel to the plane. Describe how you would find a vector $\mathbf{u}$ that is both perpendicular to $\mathbf{v}$ and parallel to the plane.

28. Find a vector in the plane parallel to the line $ax + by = c$.

In Exercises 29 and 30, find the distance from the point to the line.

29. $(2, 2, 0)$; $\quad x = -t$, $\quad y = t$, $\quad z = -1 + t$

30. $(0, 4, 1)$; $\quad x = 2 + t$, $\quad y = 2 + t$, $\quad z = t$

31. Parametrize the line that passes through the point $(1, 2, 3)$ parallel to the vector $\mathbf{v} = -3\mathbf{i} + 7\mathbf{k}$.

32. Parametrize the line segment joining the points $P(1, 2, 0)$ and $Q(1, 3, -1)$.

In Exercises 33 and 34, find the distance from the point to the plane.

33. $(6, 0, -6)$, $\quad x - y = 4$

34. $(3, 0, 10)$, $\quad 2x + 3y + z = 2$

35. Find an equation for the plane that passes through the point $(3, -2, 1)$ normal to the vector $\mathbf{n} = 2\mathbf{i} + \mathbf{j} + \mathbf{k}$.

36. Find an equation for the plane that passes through the point $(-1, 6, 0)$ perpendicular to the line $x = -1 + t, y = 6 - 2t$, $z = 3t$.

In Exercises 37 and 38, find an equation for the plane through points P, Q, and R.

37. $P(1, -1, 2)$, $\quad Q(2, 1, 3)$, $\quad R(-1, 2, -1)$

38. $P(1, 0, 0)$, $\quad Q(0, 1, 0)$, $\quad R(0, 0, 1)$

39. Find the points in which the line $x = 1 + 2t, y = -1 - t$, $z = 3t$ meets the three coordinate planes.

40. Find the point in which the line through the origin perpendicular to the plane $2x - y - z = 4$ meets the plane $3x - 5y + 2z = 6$.

41. Find the acute angle between the planes $x = 7$ and $x + y + \sqrt{2}z = -3$.

42. Find the acute angle between the planes $x + y = 1$ and $y + z = 1$.

43. Find parametric equations for the line in which the planes $x + 2y + z = 1$ and $x - y + 2z = -8$ intersect.

44. Show that the line in which the planes

$$x + 2y - 2z = 5 \quad \text{and} \quad 5x - 2y - z = 0$$

intersect is parallel to the line

$$x = -3 + 2t, \quad y = 3t, \quad z = 1 + 4t.$$

45. The planes $3x + 6z = 1$ and $2x + 2y - z = 3$ intersect in a line.

a) Show that the planes are orthogonal.

b) Find equations for the line of intersection.

46. Find an equation for the plane that passes through the point $(1, 2, 3)$ parallel to $\mathbf{u} = 2\mathbf{i} + 3\mathbf{j} + \mathbf{k}$ and $\mathbf{v} = \mathbf{i} - \mathbf{j} + 2\mathbf{k}$.

47. Is $\mathbf{v} = 2\mathbf{i} - 4\mathbf{j} + \mathbf{k}$ related in any special way to the plane $2x + y = 5$? Give reasons for your answer.

48. The equation $\mathbf{n} \cdot \overrightarrow{P_0 P} = 0$ represents the plane through P_0 normal to $\mathbf{n}$. What set does the inequality $\mathbf{n} \cdot \overrightarrow{P_0 P} > 0$ represent?

49. Find the distance from the point $P(1, 4, 0)$ to the plane through $A(0, 0, 0)$, $B(2, 0, -1)$ and $C(2, -1, 0)$.

50. Find the distance from the point $(2, 2, 3)$ to the plane $2x + 3y + 5z = 0$.

51. Find a vector parallel to the plane $2x - y - z = 4$ and orthogonal to $\mathbf{i} + \mathbf{j} + \mathbf{k}$.

52. Find a unit vector orthogonal to $\mathbf{A}$ in the plane of $\mathbf{B}$ and $\mathbf{C}$ if $\mathbf{A} = 2\mathbf{i} - \mathbf{j} + \mathbf{k}$, $\mathbf{B} = \mathbf{i} + 2\mathbf{j} + \mathbf{k}$, and $\mathbf{C} = \mathbf{i} + \mathbf{j} - 2\mathbf{k}$.

53. Find a vector of magnitude 2 parallel to the line of intersection of the planes $x + 2y + z - 1 = 0$ and $x - y + 2z + 7 = 0$.

54. Find the point in which the line through the origin perpendicular to the plane $2x - y - z = 4$ meets the plane $3x - 5y + 2z = 6$.

55. Find the point in which the line through $P(3, 2, 1)$ normal to the plane $2x - y + 2z = -2$ meets the plane.

56. What angle does the line of intersection of the planes $2x + y - z = 0$ and $x + y + 2z = 0$ make with the positive x-axis?

57. The line

$$L: \quad x = 3 + 2t, \quad y = 2t, \quad z = t$$

intersects the plane $x + 3y - z = -4$ in a point P. Find the coordinates of P and find equations for the line through P perpendicular to L.

58. Show that for every real number k the plane

$$x - 2y + z + 3 + k(2x - y - z + 1) = 0$$

contains the line of intersection of the planes

$$x - 2y + z + 3 = 0 \quad \text{and} \quad 2x - y - z + 1 = 0.$$

59. Find an equation for the plane through $A(-2, 0, -3)$ and $B(1, -2, 1)$ that lies parallel to the line through $C(-2, -13/5, 26/5)$ and $D(16/5, -13/5, 0)$.

60. Is the line $x = 1 + 2t, y = -2 + 3t, z = -5t$ related in any way to the plane $-4x - 6y + 10z = 9$? Give reasons for your answer.

61. Which of the following are equations for the plane through the points $P(1, 1, -1)$, $Q(3, 0, 2)$, and $R(-2, 1, 0)$?

a) $(2\mathbf{i} - 3\mathbf{j} + 3\mathbf{k}) \cdot ((x + 2)\mathbf{i} + (y - 1)\mathbf{j} + z\mathbf{k}) = 0$

b) $x = 3 - t, \quad y = -11t, \quad z = 2 - 3t$

c) $(x + 2) + 11(y - 1) = 3z$

d) $(2\mathbf{i} - 3\mathbf{j} + 3\mathbf{k}) \times ((x + 2)\mathbf{i} + (y - 1)\mathbf{j} + z\mathbf{k}) = \mathbf{0}$

e) $(2\mathbf{i} - \mathbf{j} + 3\mathbf{k}) \times (-3\mathbf{i} + \mathbf{k}) \cdot ((x + 2)\mathbf{i} + (y - 1)\mathbf{j} + z\mathbf{k}) = 0$

62. The parallelogram shown here has vertices at $A(2, -1, 4)$, $B(1, 0, -1)$, $C(1, 2, 3)$, and D. Find

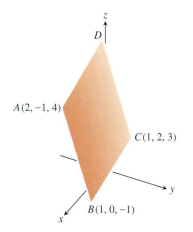

a) the coordinates of D,

b) the cosine of the interior angle at B,

c) the vector projection of $\overrightarrow{BA}$ onto $\overrightarrow{BC}$,

d) the area of the parallelogram,

e) an equation for the plane of the parallelogram,

f) the areas of the orthogonal projections of the parallelogram on the three coordinate planes.

63. *Distance between lines.* Find the distance between the line L_1 through the points $A(1, 0, -1)$ and $B(-1, 1, 0)$ and the line L_2 through the points $C(3, 1, -1)$ and $D(4, 5, -2)$. The distance is to be measured along the line perpendicular to the two lines. First find a vector $\mathbf{n}$ perpendicular to both lines. Then project $\overrightarrow{AC}$ onto $\mathbf{n}$.

64. *(Continuation of Exercise 63.)* Find the distance between the line through $A(4, 0, 2)$ and $B(2, 4, 1)$ and the line through $C(1, 3, 2)$ and $D(2, 2, 4)$.

Quadric Surfaces

Identify and sketch the surfaces in Exercises 65–76.

65. $x^2 + y^2 + z^2 = 4$

66. $x^2 + (y - 1)^2 + z^2 = 1$

67. $4x^2 + 4y^2 + z^2 = 4$

68. $36x^2 + 9y^2 + 4z^2 = 36$

69. $z = -(x^2 + y^2)$

70. $y = -(x^2 + z^2)$

71. $x^2 + y^2 = z^2$

72. $x^2 + z^2 = y^2$

73. $x^2 + y^2 - z^2 = 4$

74. $4y^2 + z^2 - 4x^2 = 4$

75. $y^2 - x^2 - z^2 = 1$

76. $z^2 - x^2 - y^2 = 1$

Coordinate Systems

The equations in Exercises 77–86 define sets both in the plane and in three-dimensional space. Identify both sets for each equation.

RECTANGULAR COORDINATES

77. $x = 0$

78. $x + y = 1$

79. $x^2 + y^2 = 4$

80. $x^2 + 4y^2 = 16$

81. $x = y^2$

82. $y^2 - x^2 = 1$

CYLINDRICAL COORDINATES

83. $r = 1 - \cos\theta$

84. $r = \sin\theta$

85. $r^2 = 2\cos 2\theta$

86. $r = \cos 2\theta$

Describe the sets defined by the spherical coordinate equations and inequalities in Exercises 87–92.

87. $\rho = 2$

88. $\theta = \pi/4$

89. $\phi = \pi/6$

90. $\rho = 1, \quad \phi = \pi/2$

91. $\rho = 1, \quad 0 \le \phi \le \pi/2$

92. $1 \le \rho \le 2$

The following table gives the coordinates of points in space in one of three coordinate systems. In Exercises 93–98, find coordinates for each point in the other two systems. There may be more than one right answer because cylindrical and spherical coordinates are not unique.

	Rectangular (x, y, z)	Cylindrical (r, θ, z)	Spherical (ρ, ϕ, θ)
93.		$(1, 0, 0)$	
94.		$\left(1, \dfrac{\pi}{2}, 0\right)$	
95.			$\left(\sqrt{2}, \dfrac{\pi}{4}, \dfrac{\pi}{2}\right)$
96.			$\left(2, \dfrac{5\pi}{6}, 0\right)$
97.	$(-1, 0, -1)$		
98.	$(0, -1, 1)$		

In Exercises 99–110, translate the equations from the given coordinate system (rectangular, cylindrical, spherical) into the other two systems. Identify the set of points in space defined by the equation.

RECTANGULAR

99. $z = 2$

100. $z = \sqrt{3x^2 + 3y^2}$

101. $x^2 + y^2 + (z + 1)^2 = 1$

102. $x^2 + y^2 + (z - 3)^2 = 9$

CYLINDRICAL

103. $z = r^2$

104. $z = |r|$

105. $r = 7 \sin \theta$

106. $r = 4 \cos \theta$

SPHERICAL

107. $\rho = 4$

108. $\rho = \sqrt{3} \sec \phi$

109. $\phi = 3\pi/4$

110. $\rho \cos \phi + \rho^2 \sin^2 \phi = 1$

CHAPTER 10   ADDITIONAL EXERCISES–THEORY, EXAMPLES, APPLICATIONS

Applications

1. *Submarine hunting.* Two surface ships on maneuvers are trying to determine a submarine's course and speed to prepare for an aircraft intercept. As shown here, ship A is located at $(4, 0, 0)$ while ship B is located at $(0, 5, 0)$. All coordinates are given in thousands of feet. Ship A locates the submarine in the direction of the vector $2\mathbf{i} + 3\mathbf{j} - (1/3)\mathbf{k}$, and ship B locates it in the direction of the vector $18\mathbf{i} - 6\mathbf{j} - \mathbf{k}$. Four minutes ago, the submarine was located at $(2, -1, -1/3)$. The aircraft is due in 20 min. Assuming the submarine moves in a straight line at a constant speed, to what position should the surface ships direct the aircraft?

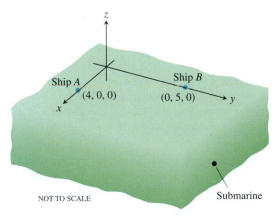

Ship A — $(4, 0, 0)$ Ship B — $(0, 5, 0)$

NOT TO SCALE   Submarine

2. *A helicopter rescue.* Two helicopters, H_1 and H_2, are traveling together. At time $t = 0$ hours, they separate and follow different straight-line paths given by

H_1: $x = 6 + 40t$, $y = -3 + 10t$, $z = -3 + 2t$

H_2: $x = 6 + 110t$, $y = -3 + 4t$, $z = -3 + t$,

all coordinates measured in miles. Due to system malfunctions, H_2 stops its flight at $(446, 13, 1)$ and, in a negligible amount of time, lands at $(446, 13, 0)$. Two hours later, H_1 is advised of this fact and heads toward H_2 at 150 mph. How long will it take H_1 to reach H_2?

3. *Work.* Find the work done in pushing a car 250 m with a force of magnitude 160 N directed at an angle of $\pi/6$ rad downward from the horizontal against the back of the car.

4. *Torque.* The operator's manual for the Toro® 21-in. lawnmower says "tighten the spark plug to 15 ft-lb (20.4 N · m)." If you are installing the plug with a 10.5-in. socket wrench that places the center of your hand 9 in. from the axis of the spark plug, about how hard should you pull? Answer in pounds.

9 in.

Theory and Examples

5. Show that $|\mathbf{A} + \mathbf{B}| \le |\mathbf{A}| + |\mathbf{B}|$ for any vectors $\mathbf{A}$ and $\mathbf{B}$.

6. Suppose that vectors $\mathbf{A}$ and $\mathbf{B}$ are not parallel and that $\mathbf{A} = \mathbf{C} + \mathbf{D}$, where $\mathbf{C}$ is parallel to $\mathbf{B}$ and $\mathbf{D}$ is orthogonal to $\mathbf{B}$. Express $\mathbf{C}$ and $\mathbf{D}$ in terms of $\mathbf{A}$ and $\mathbf{B}$.

7. Show that $\mathbf{C} = |\mathbf{B}|\mathbf{A} + |\mathbf{A}|\mathbf{B}$ bisects the angle between $\mathbf{A}$ and $\mathbf{B}$.

8. Show that $|\mathbf{B}|\mathbf{A} + |\mathbf{A}|\mathbf{B}$ and $|\mathbf{B}|\mathbf{A} - |\mathbf{A}|\mathbf{B}$ are orthogonal.

9. *Dot multiplication is positive definite.* Show that dot multiplication of vectors is *positive definite*; that is, show that $\mathbf{A} \cdot \mathbf{A} \ge 0$ for every vector $\mathbf{A}$ and that $\mathbf{A} \cdot \mathbf{A} = 0$ if and only if $\mathbf{A} = \mathbf{0}$.

10. By forming the cross product of two appropriate vectors, derive the trigonometric identity

$$\sin(A - B) = \sin A \cos B - \cos A \sin B.$$

11. Use vectors to prove that

$$(a^2 + b^2)(c^2 + d^2) \ge (ac + bd)^2$$

for any four numbers a, b, c, and d. (*Hint:* Let $\mathbf{A} = a\mathbf{i} + b\mathbf{j}$ and $\mathbf{B} = c\mathbf{i} + d\mathbf{j}$.)

12. In the figure here, D is the midpoint of side AB of triangle ABC, and E is one-third of the way between C and B. Use vectors to prove that F is the midpoint of line segment CD.

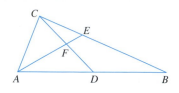

13. a) Show that

$$\begin{vmatrix} x_1 - x & y_1 - y & z_1 - z \\ x_2 - x & y_2 - y & z_2 - z \\ x_3 - x & y_3 - y & z_3 - z \end{vmatrix} = 0$$

   is an equation for the plane through the three noncollinear points $P_1(x_1, y_1, z_1)$, $P_2(x_2, y_2, z_2)$, and $P_3(x_3, y_3, z_3)$.

   b) What set of points in space is described by the equation

$$\begin{vmatrix} x & y & z & 1 \\ x_1 & y_1 & z_1 & 1 \\ x_2 & y_2 & z_2 & 1 \\ x_3 & y_3 & z_3 & 1 \end{vmatrix} = 0?$$

14. Show that the lines

$$x = a_1 s + b_1, y = a_2 s + b_2, z = a_3 s + b_3, -\infty < s < \infty,$$

and

$$x = c_1 t + d_1, y = c_2 t + d_2, z = c_3 t + d_3, -\infty < t < \infty,$$

intersect or are parallel if and only if

$$\begin{vmatrix} a_1 & c_1 & b_1 - d_1 \\ a_2 & c_2 & b_2 - d_2 \\ a_3 & c_3 & b_3 - d_3 \end{vmatrix} = 0.$$

15. Use vectors to show that the distance from $P_1(x_1, y_1)$ to the line $ax + by = c$ is

$$d = \frac{|ax_1 + by_1 - c|}{\sqrt{a^2 + b^2}}.$$

16. a) Use vectors to show that the distance from $P_1(x_1, y_1, z_1)$ to the plane $Ax + By + Cz = D$ is

$$d = \frac{|Ax_1 + By_1 + Cz_1 - D|}{\sqrt{A^2 + B^2 + C^2}}.$$

   b) Find an equation for the sphere that is tangent to the planes $x + y + z = 3$ and $x + y + z = 9$ if the planes $2x - y = 0$ and $3x - z = 0$ pass through the center of the sphere.

17. a) Show that the distance between the parallel planes $Ax + By + Cz = D_1$ and $Ax + By + Cz = D_2$ is

$$d = \frac{|D_1 - D_2|}{|A\mathbf{i} + B\mathbf{j} + C\mathbf{k}|}. \qquad (1)$$

   b) Use Eq. (1) to find the distance between the planes $2x + 3y - z = 6$ and $2x + 3y - z = 12$.

   c) Find an equation for the plane parallel to the plane $2x - y + 2z = -4$ if the point $(3, 2, -1)$ is equidistant from the two planes.

d) Write equations for the planes that lie parallel to and 5 units away from the plane $x - 2y + z = 3$.

18. Prove that four points A, B, C, and D are coplanar (lie in a common plane) if and only if $\overrightarrow{AD} \cdot (\overrightarrow{AB} \times \overrightarrow{BC}) = 0$.

19. *Triple vector products.* The **triple vector products** $(\mathbf{A} \times \mathbf{B}) \times \mathbf{C}$ and $\mathbf{A} \times (\mathbf{B} \times \mathbf{C})$ are usually not equal, although the formulas for evaluating them from components are similar:

$$(\mathbf{A} \times \mathbf{B}) \times \mathbf{C} = (\mathbf{A} \cdot \mathbf{C})\mathbf{B} - (\mathbf{B} \cdot \mathbf{C})\mathbf{A}. \qquad (2)$$

$$\mathbf{A} \times (\mathbf{B} \times \mathbf{C}) = (\mathbf{A} \cdot \mathbf{C})\mathbf{B} - (\mathbf{A} \cdot \mathbf{B})\mathbf{C}. \qquad (3)$$

Verify each formula for the following vectors by evaluating its two sides and comparing the results.

	A	B	C
a)	$2\mathbf{i}$	$2\mathbf{j}$	$2\mathbf{k}$
b)	$\mathbf{i} - \mathbf{j} + \mathbf{k}$	$2\mathbf{i} + \mathbf{j} - 2\mathbf{k}$	$-\mathbf{i} + 2\mathbf{j} - \mathbf{k}$
c)	$2\mathbf{i} + \mathbf{j}$	$2\mathbf{i} - \mathbf{j} + \mathbf{k}$	$\mathbf{i} + 2\mathbf{k}$
d)	$\mathbf{i} + \mathbf{j} - 2\mathbf{k}$	$-\mathbf{i} - \mathbf{k}$	$2\mathbf{i} + 4\mathbf{j} - 2\mathbf{k}$

20. Show that if $\mathbf{A}$, $\mathbf{B}$, $\mathbf{C}$, and $\mathbf{D}$ are any vectors, then

   a) $\mathbf{A} \times (\mathbf{B} \times \mathbf{C}) + \mathbf{B} \times (\mathbf{C} \times \mathbf{A}) + \mathbf{C} \times (\mathbf{A} \times \mathbf{B}) = \mathbf{0}$,

   b) $\mathbf{A} \times \mathbf{B} = (\mathbf{A} \cdot \mathbf{B} \times \mathbf{i})\mathbf{i} + (\mathbf{A} \cdot \mathbf{B} \times \mathbf{j})\mathbf{j} + (\mathbf{A} \cdot \mathbf{B} \times \mathbf{k})\mathbf{k}$,

   c) $(\mathbf{A} \times \mathbf{B}) \cdot (\mathbf{C} \times \mathbf{D}) = \begin{vmatrix} \mathbf{A} \cdot \mathbf{C} & \mathbf{B} \cdot \mathbf{C} \\ \mathbf{A} \cdot \mathbf{D} & \mathbf{B} \cdot \mathbf{D} \end{vmatrix}$.

21. Prove or disprove the formula

$$\mathbf{A} \times (\mathbf{A} \times (\mathbf{A} \times \mathbf{B})) \cdot \mathbf{C} = -|\mathbf{A}|^2 \mathbf{A} \cdot \mathbf{B} \times \mathbf{C}.$$

22. *The projection of a vector on a plane.* Let P be a plane in space and let $\mathbf{v}$ be a vector. The vector projection of $\mathbf{v}$ onto the plane P, $\text{proj}_P \mathbf{v}$, can be defined informally as follows. Suppose the sun is shining so that its rays are normal to the plane P. Then $\text{proj}_P \mathbf{v}$ is the "shadow" of $\mathbf{v}$ onto P. If P is the plane $x + 2y + 6z = 6$ and $\mathbf{v} = \mathbf{i} + \mathbf{j} + \mathbf{k}$, find $\text{proj}_P \mathbf{v}$.

23. The accompanying figure shows nonzero vectors $\mathbf{v}$, $\mathbf{w}$, and $\mathbf{z}$, with $\mathbf{z}$ orthogonal to the line L, and $\mathbf{v}$ and $\mathbf{w}$ making equal angles β with L. Assuming $|\mathbf{v}| = |\mathbf{w}|$, find $\mathbf{w}$ in terms of $\mathbf{v}$ and $\mathbf{z}$.

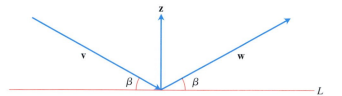

24. *The parabolic coordinate system.* This exercise introduces a new coordinate system for space, the **parabolic coordinate system.** A point P is determined by an ordered triple (α, β, γ) in which

   i) $\alpha\beta$ is the square of the distance from P to the z-axis,

   ii) $|\alpha - \beta|$ is twice the distance from P to the xy-plane, and P lies above the xy-plane if $\alpha - \beta > 0$ and below it if $\alpha - \beta < 0$,

iii) $\gamma = \theta$, where θ has the same meaning as in cylindrical and spherical coordinate systems, except that we restrict γ to lie in $[0, 2\pi]$.

a) What are the equations for changing from parabolic coordinates to Cartesian coordinates?

b) Why is "parabolic coordinate system" an appropriate name?

*25. *Point masses and gravitation.* In physics the law of gravitation says that if P and Q are (point) masses with mass M and m, respectively, then P is attracted to Q by the force

$$\mathbf{F} = \frac{GMm\,\mathbf{r}}{|\mathbf{r}|^3},$$

where $\mathbf{r}$ is the vector from P to Q and G is a constant (the gravitational constant). Moreover, if $Q_1, \ldots, Q_k$ are (point) masses with mass $m_1, \ldots, m_k$, respectively, then the force on P due to all the Q_i's is

$$\mathbf{F} = \sum_{i=1}^{k} \frac{GMm_i}{|\mathbf{r}_i|^3}\,\mathbf{r}_i,$$

where $\mathbf{r}_i$ is the vector from P to Q_i.

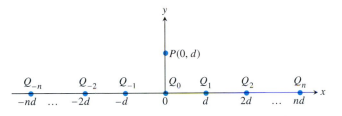

a) Let point P with mass M be located at the point $(0, d)$, $d > 0$, in the coordinate plane. For $i = -n, -n+1, \ldots, -1, 0, 1, \ldots, n$, let Q_i be located at the point $(id, 0)$ and have mass m. Find the magnitude of the gravitational force on P due to all the Q_i's.

b) Is the limit as $n \to \infty$ of the magnitude of the force on P finite? Why, or why not?

*26. *Relativistic sums.* Einstein's special theory of relativity roughly says that with respect to a reference frame (coordinate system) no material object can travel as fast as c, the speed of light. So, if $\overrightarrow{x}$ and $\overrightarrow{y}$ are two velocities such that $|\overrightarrow{x}| < c$ and $|\overrightarrow{y}| < c$, then the **relativistic sum** $\overrightarrow{x} \oplus \overrightarrow{y}$ of $\overrightarrow{x}$ and $\overrightarrow{y}$ must have length less than c. Einstein's special theory of relativity says

$$\overrightarrow{x} \oplus \overrightarrow{y} = \frac{\overrightarrow{x} + \overrightarrow{y}}{1 + \dfrac{\overrightarrow{x} \cdot \overrightarrow{y}}{c^2}} + \frac{1}{c^2} \cdot \frac{\gamma_x}{\gamma_x + 1} \cdot \frac{\overrightarrow{x} \times (\overrightarrow{x} \times \overrightarrow{y})}{1 + \dfrac{\overrightarrow{x} \cdot \overrightarrow{y}}{c^2}},$$

where

$$\gamma_x = \frac{1}{\sqrt{1 - \dfrac{\overrightarrow{x} \cdot \overrightarrow{x}}{c^2}}}.$$

It can be shown that if $|\overrightarrow{x}| < c$ and $|\overrightarrow{y}| < c$, then $|\overrightarrow{x} \oplus \overrightarrow{y}| < c$. This exercise deals with two special cases.

a) Prove that if $\overrightarrow{x}$ and $\overrightarrow{y}$ are orthogonal, $|\overrightarrow{x}| < c$, $|\overrightarrow{y}| < c$, then $|\overrightarrow{x} \oplus \overrightarrow{y}| < c$.

b) Prove that if $\overrightarrow{x}$ and $\overrightarrow{y}$ are parallel, $|\overrightarrow{x}| < c$, $|\overrightarrow{y}| < c$, then $|\overrightarrow{x} \oplus \overrightarrow{y}| < c$.

c) Compute $\lim_{c \to \infty} \overrightarrow{x} \oplus \overrightarrow{y}$.

Vector-Valued Functions and Motion in Space

OVERVIEW   When a body travels through space, the equations $x = f(t)$, $y = g(t)$, and $z = h(t)$ that give the body's coordinates as functions of time serve as parametric equations for the body's motion and path. With vector notation, we can condense these into a single equation $\mathbf{r}(t) = f(t)\,\mathbf{i} + g(t)\,\mathbf{j} + h(t)\,\mathbf{k}$ that gives the body's position as a vector function of time.

In this chapter, we show how to use calculus to study the paths, velocities, and accelerations of moving bodies. As we go along, we will see how our work answers the standard questions about the paths and motions of projectiles, planets, and satellites. In the final section, we use our new vector calculus to derive Kepler's laws of planetary motion from Newton's laws of motion and gravitation.

11.1 Vector-Valued Functions and Space Curves

To track a particle moving in space, we run a vector $\mathbf{r}$ from the origin to the particle (Fig. 11.1) and study the changes in $\mathbf{r}$. If the particle's position coordinates are twice-differentiable functions of time, then so is $\mathbf{r}$, and we can find the particle's velocity and acceleration vectors at any time by differentiating $\mathbf{r}$. Conversely, if we know either the particle's velocity vector or acceleration vector as a continuous function of time, and if we have enough information about the particle's initial velocity and position, we can find $\mathbf{r}$ as a function of time by integration.

Definitions

When a particle moves through space during a time interval I, we think of the particle's coordinates as functions defined on I:

$$x = f(t), \qquad y = g(t), \qquad z = h(t), \qquad t \,\varepsilon\, I. \tag{1}$$

The points $(x, y, z) = (f(t), g(t), h(t)), t \,\varepsilon\, I$, make up the **curve** in space that we call the particle's **path.** The equations and interval in (1) **parametrize** the curve. The vector

$$\mathbf{r}(t) = \overrightarrow{OP} = f(t)\,\mathbf{i} + g(t)\,\mathbf{j} + h(t)\,\mathbf{k}$$

from the origin to the particle's **position** $P(f(t), g(t), h(t))$ at time t is the particle's **position vector.** The functions f, g, and h are the **component functions**

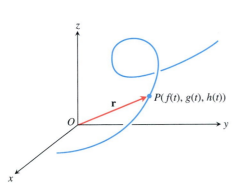

11.1 The position vector $\mathbf{r} = \overrightarrow{OP}$ of a particle moving through space is a function of time.

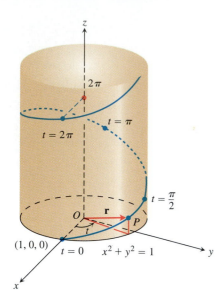

11.2 The upper half of the helix $\mathbf{r}(t) = (\cos t)\mathbf{i} + (\sin t)\mathbf{j} + t\mathbf{k}$.

(**components**) of the position vector. We think of the particle's path as the **curve traced by r** during the time interval *I*.

Equation (1) defines **r** as a vector function of the real variable *t* on the interval *I*. More generally, a **vector function** or **vector-valued function** on a domain set *D* is a rule that assigns a vector in space to each element in *D*. For now, the domains will be intervals of real numbers. Later, in Chapter 14, the domains will be regions in the plane or in space. Vector functions will then be called "vector fields."

We refer to real-valued functions as **scalar functions** to distinguish them from vector functions. The components of **r** are scalar functions of *t*. When we define a vector-valued function by giving its component functions, we assume the vector function's domain to be the common domain of the components.

EXAMPLE 1   A Helix

The vector function

$$\mathbf{r}(t) = (\cos t)\mathbf{i} + (\sin t)\mathbf{j} + t\mathbf{k}$$

is defined for all real values of *t*. The curve traced by **r** is a helix (from an old Greek word for "spiral") that winds around the circular cylinder $x^2 + y^2 = 1$ (Fig. 11.2). The curve lies on the cylinder because the **i**- and **j**-components of **r**, being the *x*- and *y*-coordinates of the tip of **r**, satisfy the cylinder's equation:

$$x^2 + y^2 = (\cos t)^2 + (\sin t)^2 = 1.$$

The curve rises as the **k**-component $z = t$ increases. Each time *t* increases by 2π, the curve completes one turn around the cylinder. The equations

$$x = \cos t, \qquad y = \sin t, \qquad z = t$$

parametrize the helix, the interval $-\infty < t < \infty$ being understood. You will find more helices in Fig. 11.3.   ❑

Limits and Continuity

The way we define limits of vector-valued functions is similar to the way we define limits of real-valued functions.

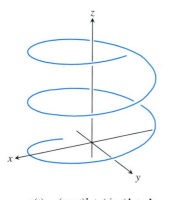

$\mathbf{r}(t) = (\cos t)\mathbf{i} + (\sin t)\mathbf{j} + t\mathbf{k}$

(Generated by Mathematica)

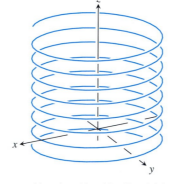

$\mathbf{r}(t) = (\cos t)\mathbf{i} + (\sin t)\mathbf{j} + 0.3t\mathbf{k}$

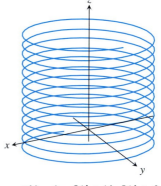

$\mathbf{r}(t) = (\cos 5t)\mathbf{i} + (\sin 5t)\mathbf{j} + t\mathbf{k}$

11.3 Helices drawn by computer.

Definition

Let $\mathbf{r}(t) = f(t)\,\mathbf{i} + g(t)\,\mathbf{j} + h(t)\,\mathbf{k}$ be a vector function and $\mathbf{L}$ a vector. We say that $\mathbf{r}$ has **limit L** as t approaches t_0 and write

$$\lim_{t \to t_0} \mathbf{r}(t) = \mathbf{L}$$

if, for every number $\epsilon > 0$, there exists a corresponding number $\delta > 0$ such that for all t

$$0 < |t - t_0| < \delta \quad \Rightarrow \quad |\mathbf{r}(t) - \mathbf{L}| < \epsilon.$$

If $\mathbf{L} = L_1\,\mathbf{i} + L_2\,\mathbf{j} + L_3\,\mathbf{k}$, then $\lim_{t \to t_0} \mathbf{r}(t) = \mathbf{L}$ precisely when

$$\lim_{t \to t_0} f(t) = L_1, \qquad \lim_{t \to t_0} g(t) = L_2, \qquad \text{and} \qquad \lim_{t \to t_0} h(t) = L_3.$$

The equation

$$\lim_{t \to t_0} \mathbf{r}(t) = \left(\lim_{t \to t_0} f(t) \right)\mathbf{i} + \left(\lim_{t \to t_0} g(t) \right)\mathbf{j} + \left(\lim_{t \to t_0} h(t) \right)\mathbf{k} \tag{2}$$

provides a practical way to calculate limits of vector functions.

EXAMPLE 2 If $\mathbf{r}(t) = (\cos t)\,\mathbf{i} + (\sin t)\,\mathbf{j} + t\,\mathbf{k}$, then

$$\lim_{t \to \pi/4} \mathbf{r}(t) = \left(\lim_{t \to \pi/4} \cos t \right)\mathbf{i} + \left(\lim_{t \to \pi/4} \sin t \right)\mathbf{j} + \left(\lim_{t \to \pi/4} t \right)\mathbf{k}$$

$$= \frac{\sqrt{2}}{2}\,\mathbf{i} + \frac{\sqrt{2}}{2}\,\mathbf{j} + \frac{\pi}{4}\,\mathbf{k}.$$

We define continuity for vector functions the same way we define continuity for scalar functions.

Definition

A vector function $\mathbf{r}(t)$ is **continuous at a point** $t = t_0$ in its domain if $\lim_{t \to t_0} \mathbf{r}(t) = \mathbf{r}(t_0)$. The function is **continuous** if it is continuous at every point in its domain.

Since limits can be expressed in terms of components, we can test vector functions for continuity by examining their components (Exercise 51).

Component Test for Continuity at a Point

The vector function $\mathbf{r}(t) = f(t)\,\mathbf{i} + g(t)\,\mathbf{j} + h(t)\,\mathbf{k}$ is continuous at $t = t_0$ if and only if f, g, and h are continuous at t_0.

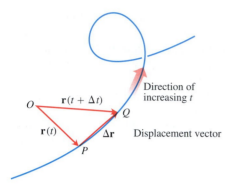

11.4 Between time t and time $t + \Delta t$, the particle moving along the path shown here undergoes the displacement $\overrightarrow{PQ} = \Delta \mathbf{r}$. The vector sum $\mathbf{r}(t) + \Delta \mathbf{r}$ gives the new position, $\mathbf{r}(t + \Delta t)$.

EXAMPLE 3

a) The function

$$\mathbf{r}(t) = (\cos t)\mathbf{i} + (\sin t)\mathbf{j} + t\,\mathbf{k}$$

is continuous because $\cos t$, $\sin t$, and t are continuous.

b) The function

$$\mathbf{g}(t) = (\cos t)\mathbf{i} + (\sin t)\mathbf{j} + \lfloor t \rfloor \mathbf{k}$$

is discontinuous at every integer. ∎

Derivatives and Motion

Suppose that $\mathbf{r}(t) = f(t)\mathbf{i} + g(t)\mathbf{j} + h(t)\mathbf{k}$ is the position vector of a particle moving along a curve in space and that f, g, and h are differentiable functions of t. Then the difference between the particle's positions at time t and time $t + \Delta t$ is

$$\Delta \mathbf{r} = \mathbf{r}(t + \Delta t) - \mathbf{r}(t)$$

(Fig. 11.4). In terms of components,

$$\Delta \mathbf{r} = \mathbf{r}(t + \Delta t) - \mathbf{r}(t)$$
$$= [f(t + \Delta t)\mathbf{i} + g(t + \Delta t)\mathbf{j} + h(t + \Delta t)\mathbf{k}]$$
$$-[f(t)\mathbf{i} + g(t)\mathbf{j} + h(t)\mathbf{k}]$$
$$= [f(t + \Delta t) - f(t)]\mathbf{i} + [g(t + \Delta t) - g(t)]\mathbf{j} + [h(t + \Delta t) - h(t)]\mathbf{k}.$$

As Δt approaches zero, three things seem to happen simultaneously. First, Q approaches P along the curve. Second, the secant line PQ seems to approach a limiting position tangent to the curve at P. Third, the quotient $\Delta \mathbf{r}/\Delta t$ approaches the limit

$$\lim_{\Delta t \to 0} \frac{\Delta \mathbf{r}}{\Delta t} = \left[\lim_{\Delta t \to 0} \frac{f(t + \Delta t) - f(t)}{\Delta t}\right]\mathbf{i} + \left[\lim_{\Delta t \to 0} \frac{g(t + \Delta t) - g(t)}{\Delta t}\right]\mathbf{j}$$
$$+ \left[\lim_{\Delta t \to 0} \frac{h(t + \Delta t) - h(t)}{\Delta t}\right]\mathbf{k}$$
$$= \left[\frac{df}{dt}\right]\mathbf{i} + \left[\frac{dg}{dt}\right]\mathbf{j} + \left[\frac{dh}{dt}\right]\mathbf{k}.$$

We are therefore led by past experience to the following definitions.

Definitions

The vector function $\mathbf{r}(t) = f(t)\mathbf{i} + g(t)\mathbf{j} + h(t)\mathbf{k}$ is **differentiable at $t = t_0$** if f, g, and h are differentiable at t_0. Also, $\mathbf{r}$ is said to be **differentiable** if it is differentiable at every point of its domain. At any point t at which $\mathbf{r}$ is differentiable, its **derivative** is the vector

$$\frac{d\mathbf{r}}{dt} = \lim_{\Delta t \to 0} \frac{\mathbf{r}(t + \Delta t) - \mathbf{r}(t)}{\Delta t} = \frac{df}{dt}\mathbf{i} + \frac{dg}{dt}\mathbf{j} + \frac{dh}{dt}\mathbf{k}.$$

The curve traced by $\mathbf{r}$ is **smooth** if $d\mathbf{r}/dt$ is continuous and never $\mathbf{0}$, i.e., if f, g, and h have continuous first derivatives that are not simultaneously 0.

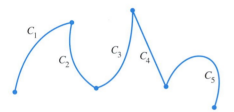

11.5 A piecewise smooth curve made up of five smooth curves connected end to end in continuous fashion.

The vector $d\mathbf{r}/dt$, when different from $\mathbf{0}$, is also a vector *tangent* to the curve. The **tangent line** to the curve at a point $(f(t_0), g(t_0), h(t_0))$ is defined to be the line through the point parallel to $d\mathbf{r}/dt$ at $t = t_0$. We require $d\mathbf{r}/dt \neq \mathbf{0}$ for a smooth curve to make sure the curve has a continuously turning tangent at each point. On a smooth curve there are no sharp corners or cusps.

A curve that is made up of a finite number of smooth curves pieced together in a continuous fashion is called **piecewise smooth** (Fig. 11.5).

Look once again at Fig. 11.4. We drew the figure for Δt positive, so $\Delta\mathbf{r}$ points forward, in the direction of the motion. The vector $\Delta\mathbf{r}/\Delta t$ (not shown), having the same direction as $\Delta\mathbf{r}$, points forward too. Had Δt been negative, $\Delta\mathbf{r}$ would have pointed backward, against the direction of motion. The quotient $\Delta\mathbf{r}/\Delta t$, however, being a negative scalar multiple of $\Delta\mathbf{r}$, would once again have pointed forward. No matter how $\Delta\mathbf{r}$ points, $\Delta\mathbf{r}/\Delta t$ points forward and we expect the vector $d\mathbf{r}/dt = \lim_{\Delta t \to 0} \Delta\mathbf{r}/\Delta t$, when different from $\mathbf{0}$, to do the same. This means that the derivative $d\mathbf{r}/dt$ is just what we want for modeling a particle's velocity. It points in the direction of motion and gives the rate of change of position with respect to time. For a smooth curve, the velocity is never zero; the particle does not stop or reverse direction.

> **Definitions**
>
> If $\mathbf{r}$ is the position vector of a particle moving along a smooth curve in space, then
>
> $$\mathbf{v}(t) = \frac{d\mathbf{r}}{dt}$$
>
> is the particle's **velocity vector**, tangent to the curve. At any time t, the direction of $\mathbf{v}$ is the **direction of motion**, the magnitude of $\mathbf{v}$ is the particle's **speed,** and the derivative $\mathbf{a} = d\mathbf{v}/dt$, when it exists, is the particle's **acceleration vector.** In short,
>
> 1. Velocity is the derivative of position: $\mathbf{v} = \dfrac{d\mathbf{r}}{dt}.$
>
> 2. Speed is the magnitude of velocity: Speed $= |\mathbf{v}|$.
>
> 3. Acceleration is the derivative of velocity: $\mathbf{a} = \dfrac{d\mathbf{v}}{dt} = \dfrac{d^2\mathbf{r}}{dt^2}.$
>
> 4. The vector $\mathbf{v}/|\mathbf{v}|$ is the direction of motion at time t.

We can express the velocity of a moving particle as the product of its speed and direction.

$$\text{Velocity} = |\mathbf{v}|\left(\frac{\mathbf{v}}{|\mathbf{v}|}\right) = (\text{speed})\,(\text{direction})$$

EXAMPLE 4   The vector

$$\mathbf{r}(t) = (3\cos t)\,\mathbf{i} + (3\sin t)\,\mathbf{j} + t^2\,\mathbf{k}$$

gives the position of a moving body at time t. Find the body's speed and direction when $t = 2$. At what times, if any, are the body's velocity and acceleration orthogonal?

Solution

$$\mathbf{r} = (3 \cos t)\,\mathbf{i} + (3 \sin t)\,\mathbf{j} + t^2\,\mathbf{k}$$

$$\mathbf{v} = \frac{d\mathbf{r}}{dt} = -(3 \sin t)\,\mathbf{i} + (3 \cos t)\,\mathbf{j} + 2t\,\mathbf{k}$$

$$\mathbf{a} = \frac{d^2\mathbf{r}}{dt^2} = -(3 \cos t)\,\mathbf{i} - (3 \sin t)\,\mathbf{j} + 2\,\mathbf{k}$$

At $t = 2$, the body's speed and direction are

Speed: $\qquad |\mathbf{v}(2)| = \sqrt{(-3 \sin 2)^2 + (3 \cos 2)^2 + (4)^2} = 5$

Direction: $\qquad \dfrac{\mathbf{v}(2)}{|\mathbf{v}(2)|} = -\left(\dfrac{3}{5} \sin 2\right)\mathbf{i} + \left(\dfrac{3}{5} \cos 2\right)\mathbf{j} + \dfrac{4}{5}\,\mathbf{k}.$

To find the times when $\mathbf{v}$ and $\mathbf{a}$ are orthogonal, we look for values of t for which

$$\mathbf{v} \cdot \mathbf{a} = 9 \sin t \cos t - 9 \cos t \sin t + 4t = 4t = 0.$$

The only value is $t = 0$. $\qquad\qquad\qquad\qquad\qquad\qquad\qquad\qquad\qquad$ ❑

Differentiation Rules

Because the derivatives of vector functions may be computed component by component, the rules for differentiating vector functions have the same form as the rules for differentiating scalar functions.

Differentiation Rules for Vector Functions

Constant Function Rule: $\qquad \dfrac{d}{dt}\mathbf{C} = \mathbf{0} \qquad$ (any constant vector $\mathbf{C}$)

If $\mathbf{u}$ and $\mathbf{v}$ are differentiable vector functions of t, then

Scalar Multiple Rules: $\qquad \dfrac{d}{dt}(c\,\mathbf{u}) = c\dfrac{d\mathbf{u}}{dt} \qquad$ (any number c)

$$\dfrac{d}{dt}(f\,\mathbf{u}) = \dfrac{df}{dt}\mathbf{u} + f\dfrac{d\mathbf{u}}{dt} \qquad \text{(any differentiable scalar function } f(t))$$

Sum Rule: $\qquad \dfrac{d}{dt}(\mathbf{u} + \mathbf{v}) = \dfrac{d\mathbf{u}}{dt} + \dfrac{d\mathbf{v}}{dt}$

Difference Rule: $\qquad \dfrac{d}{dt}(\mathbf{u} - \mathbf{v}) = \dfrac{d\mathbf{u}}{dt} - \dfrac{d\mathbf{v}}{dt}$

Dot Product Rule: $\qquad \dfrac{d}{dt}(\mathbf{u} \cdot \mathbf{v}) = \dfrac{d\mathbf{u}}{dt} \cdot \mathbf{v} + \mathbf{u} \cdot \dfrac{d\mathbf{v}}{dt}$

Cross Product Rule: $\qquad \dfrac{d}{dt}(\mathbf{u} \times \mathbf{v}) = \dfrac{d\mathbf{u}}{dt} \times \mathbf{v} + \mathbf{u} \times \dfrac{d\mathbf{v}}{dt}$

When you use the Cross Product Rule, remember to preserve the order of the factors. If $\mathbf{u}$ comes first on the left side of the equation, it must also come first on the right or the signs will be wrong.

As an algebraic convenience, we sometimes write the product of a scalar c and a vector $\mathbf{v}$ as $\mathbf{v}c$ instead of $c\mathbf{v}$. This permits us, for instance, to write the Chain Rule in a familiar form:

$$\frac{d\mathbf{r}}{ds} = \frac{d\mathbf{r}}{dt}\frac{dt}{ds}.$$

Chain Rule (Short Form): If $\mathbf{r}$ is a differentiable function of t and t is a differentiable function of s, then

$$\frac{d\mathbf{r}}{ds} = \frac{d\mathbf{r}}{dt}\frac{dt}{ds}.$$

We will prove the product rules and Chain Rule but leave the rules for constants, scalar multiples, sums, and differences as exercises.

Proof of the Dot Product Rule Suppose that

$$\mathbf{u} = u_1(t)\,\mathbf{i} + u_2(t)\,\mathbf{j} + u_3(t)\,\mathbf{k}$$

and

$$\mathbf{v} = v_1(t)\,\mathbf{i} + v_2(t)\,\mathbf{j} + v_3(t)\,\mathbf{k}.$$

Then

$$\frac{d}{dt}(\mathbf{u}\cdot\mathbf{v}) = \frac{d}{dt}(u_1v_1 + u_2v_2 + u_3v_3)$$

$$= \underbrace{u_1'v_1 + u_2'v_2 + u_3'v_3}_{\mathbf{u}'\cdot\mathbf{v}} + \underbrace{u_1v_1' + u_2v_2' + u_3v_3'}_{\mathbf{u}\cdot\mathbf{v}'}.\qquad\square$$

Proof of the Cross Product Rule We model the proof after the proof of the product rule for scalar functions. According to the definition of derivative,

$$\frac{d}{dt}(\mathbf{u}\times\mathbf{v}) = \lim_{h\to 0}\frac{\mathbf{u}(t+h)\times\mathbf{v}(t+h) - \mathbf{u}(t)\times\mathbf{v}(t)}{h}.$$

To change this fraction into an equivalent one that contains the difference quotients for the derivatives of $\mathbf{u}$ and $\mathbf{v}$, we subtract and add $\mathbf{u}(t)\times\mathbf{v}(t+h)$ in the numerator. Then

$$\frac{d}{dt}(\mathbf{u}\times\mathbf{v})$$

$$= \lim_{h\to 0}\frac{\mathbf{u}(t+h)\times\mathbf{v}(t+h) - \mathbf{u}(t)\times\mathbf{v}(t+h) + \mathbf{u}(t)\times\mathbf{v}(t+h) - \mathbf{u}(t)\times\mathbf{v}(t)}{h}$$

$$= \lim_{h\to 0}\left[\frac{\mathbf{u}(t+h) - \mathbf{u}(t)}{h}\times\mathbf{v}(t+h) + \mathbf{u}(t)\times\frac{\mathbf{v}(t+h) - \mathbf{v}(t)}{h}\right]$$

$$= \lim_{h\to 0}\frac{\mathbf{u}(t+h) - \mathbf{u}(t)}{h}\times\lim_{h\to 0}\mathbf{v}(t+h) + \lim_{h\to 0}\mathbf{u}(t)\times\lim_{h\to 0}\frac{\mathbf{v}(t+h) - \mathbf{v}(t)}{h}.$$

The last of these equalities holds because the limit of the cross product of two vector functions is the cross product of their limits if the latter exist (Exercise 52). As h approaches zero, $\mathbf{v}(t+h)$ approaches $\mathbf{v}(t)$ because $\mathbf{v}$, being differentiable at t, is continuous at t (Exercise 53). The two fractions approach the values of $d\mathbf{u}/dt$ and $d\mathbf{v}/dt$ at t. In short,

$$\frac{d}{dt}(\mathbf{u}\times\mathbf{v}) = \frac{d\mathbf{u}}{dt}\times\mathbf{v} + \mathbf{u}\times\frac{d\mathbf{v}}{dt}.\qquad\square$$

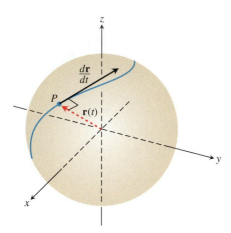

11.6 If a particle moves on a sphere in such a way that its position **r** is a differentiable function of time, then $\mathbf{r} \cdot (d\mathbf{r}/dt) = 0$.

We will use this observation repeatedly in Section 11.4.

Proof of the Chain Rule Suppose that $\mathbf{r}(t) = f(t)\,\mathbf{i} + g(t)\,\mathbf{j} + h(t)\,\mathbf{k}$ is a differentiable vector function of t and that t is a differentiable scalar function of some other variable s. Then f, g, and h are differentiable functions of s, and the Chain Rule for differentiable real-valued functions gives

$$\frac{d\mathbf{r}}{ds} = \frac{df}{ds}\,\mathbf{i} + \frac{dg}{ds}\,\mathbf{j} + \frac{dh}{ds}\,\mathbf{k}$$

$$= \frac{df}{dt}\frac{dt}{ds}\,\mathbf{i} + \frac{dg}{dt}\frac{dt}{ds}\,\mathbf{j} + \frac{dh}{dt}\frac{dt}{ds}\,\mathbf{k}$$

$$= \left(\frac{df}{dt}\,\mathbf{i} + \frac{dg}{dt}\,\mathbf{j} + \frac{dh}{dt}\,\mathbf{k} \right) \frac{dt}{ds}$$

$$= \frac{d\mathbf{r}}{dt}\frac{dt}{ds}. \qquad \square$$

Vector Functions of Constant Length

When we track a particle moving on a sphere centered at the origin (Fig. 11.6), the position vector has a constant length equal to the radius of the sphere. The velocity vector $d\mathbf{r}/dt$, tangent to the path of motion, is tangent to the sphere and hence perpendicular to **r**. This is always the case for a differentiable vector function of constant length: The vector and its first derivative are orthogonal. With the length constant, the change in the function is a change in direction only, and direction changes take place at right angles.

> If **u** is a differentiable vector function of t of constant length, then
>
> $$\mathbf{u} \cdot \frac{d\mathbf{u}}{dt} = 0. \tag{3}$$

To see why Eq. (3) holds, suppose that **u** is a differentiable function of t and that $|\mathbf{u}|$ is constant. Then $\mathbf{u} \cdot \mathbf{u} = |\mathbf{u}|^2$ is constant and we may differentiate both sides of this equation to get

$$\frac{d}{dt}(\mathbf{u} \cdot \mathbf{u}) = \frac{d}{dt}(\text{constant}) = 0$$

$$\frac{d\mathbf{u}}{dt} \cdot \mathbf{u} + \mathbf{u} \cdot \frac{d\mathbf{u}}{dt} = 0 \qquad \text{Dot Product Rule with } \mathbf{v} = \mathbf{u}$$

$$2\mathbf{u} \cdot \frac{d\mathbf{u}}{dt} = 0 \qquad \text{Dot multiplication is commutative.}$$

$$\mathbf{u} \cdot \frac{d\mathbf{u}}{dt} = 0.$$

EXAMPLE 5   Show that

$$\mathbf{u}(t) = (\sin t)\,\mathbf{i} + (\cos t)\,\mathbf{j} + \sqrt{3}\,\mathbf{k}$$

has constant length and is orthogonal to its derivative.

Solution

$$\mathbf{u}(t) = (\sin t)\,\mathbf{i} + (\cos t)\,\mathbf{j} + \sqrt{3}\,\mathbf{k}$$

$$|\mathbf{u}(t)| = \sqrt{(\sin t)^2 + (\cos t)^2 + (\sqrt{3})^2} = \sqrt{1+3} = 2$$

$$\frac{d\mathbf{u}}{dt} = (\cos t)\,\mathbf{i} - (\sin t)\,\mathbf{j}$$

$$\mathbf{u}\cdot\frac{d\mathbf{u}}{dt} = \sin t \cos t - \sin t \cos t = 0$$

Integrals of Vector Functions

A differentiable vector function $\mathbf{R}(t)$ is an **antiderivative** of a vector function $\mathbf{r}(t)$ on an interval I if $d\mathbf{R}/dt = \mathbf{r}$ at each point of I. If $\mathbf{R}$ is an antiderivative of $\mathbf{r}$ on I, it can be shown, working one component at a time, that every antiderivative of $\mathbf{r}$ on I has the form $\mathbf{R} + \mathbf{C}$ for some constant vector $\mathbf{C}$ (Exercise 56). The set of all antiderivatives of $\mathbf{r}$ on I is the **indefinite integral** of $\mathbf{r}$ on I.

Definition

The **indefinite integral** of $\mathbf{r}$ with respect to t is the set of all antiderivatives of $\mathbf{r}$, denoted by $\int \mathbf{r}(t)\,dt$. If $\mathbf{R}$ is any antiderivative of $\mathbf{r}$, then

$$\int \mathbf{r}(t)\,dt = \mathbf{R}(t) + \mathbf{C}.$$

The usual arithmetic rules for indefinite integrals apply.

EXAMPLE 6

$$\int ((\cos t)\,\mathbf{i} + \mathbf{j} - 2t\,\mathbf{k})\,dt$$

$$= \left(\int \cos t\,dt\right)\mathbf{i} + \left(\int dt\right)\mathbf{j} - \left(\int 2t\,dt\right)\mathbf{k} \tag{4}$$

$$= (\sin t + C_1)\,\mathbf{i} + (t + C_2)\,\mathbf{j} - (t^2 + C_3)\,\mathbf{k} \tag{5}$$

$$= (\sin t)\,\mathbf{i} + t\,\mathbf{j} - t^2\,\mathbf{k} + \mathbf{C} \qquad \mathbf{C} = C_1\mathbf{i} + C_2\mathbf{j} - C_3\mathbf{k}$$

As in the integration of scalar functions, we recommend that you skip the steps in (4) and (5) and go directly to the final form. Find an antiderivative for each component and add a constant vector at the end.

Definite integrals of vector functions are defined in terms of components.

Definition

If the components of $\mathbf{r}(t) = f(t)\,\mathbf{i} + g(t)\,\mathbf{j} + h(t)\,\mathbf{k}$ are integrable over $[a, b]$ then so is $\mathbf{r}$, and the **definite integral** of $\mathbf{r}$ from a to b is

$$\int_a^b \mathbf{r}(t)\,dt = \left(\int_a^b f(t)\,dt\right)\mathbf{i} + \left(\int_a^b g(t)\,dt\right)\mathbf{j} + \left(\int_a^b h(t)\,dt\right)\mathbf{k}.$$

The usual arithmetic rules for definite integrals apply (Exercise 54).

EXAMPLE 7

$$\int_0^\pi ((\cos t)\,\mathbf{i} + \mathbf{j} - 2t\,\mathbf{k})\,dt = \left(\int_0^\pi \cos t\,dt\right)\mathbf{i} + \left(\int_0^\pi dt\right)\mathbf{j} - \left(\int_0^\pi 2t\,dt\right)\mathbf{k}$$

$$= \left[\sin t\right]_0^\pi \mathbf{i} + \left[t\right]_0^\pi \mathbf{j} - \left[t^2\right]_0^\pi \mathbf{k}$$

$$= [0 - 0]\mathbf{i} + [\pi - 0]\mathbf{j} - [\pi^2 - 0^2]\mathbf{k}$$

$$= \pi\,\mathbf{j} - \pi^2\,\mathbf{k} \qquad \square$$

EXAMPLE 8 *Finding a particle's position function from its velocity function and initial position*

The velocity of a particle moving in space is

$$\frac{d\mathbf{r}}{dt} = (\cos t)\,\mathbf{i} - (\sin t)\,\mathbf{j} + \mathbf{k}.$$

Find the particle's position as a function of t if $\mathbf{r} = 2\mathbf{i} + \mathbf{k}$ when $t = 0$.

Solution Our goal is to solve the initial value problem that consists of

The differential equation: $\quad \dfrac{d\mathbf{r}}{dt} = (\cos t)\,\mathbf{i} - (\sin t)\,\mathbf{j} + \mathbf{k}$

The initial condition: $\quad \mathbf{r}(0) = 2\mathbf{i} + \mathbf{k}$

Integrating both sides of the differential equation with respect to t gives

$$\mathbf{r}(t) = (\sin t)\,\mathbf{i} + (\cos t)\,\mathbf{j} + t\,\mathbf{k} + \mathbf{C}.$$

We then use the initial condition to find the right value for $\mathbf{C}$:

$$(\sin 0)\,\mathbf{i} + (\cos 0)\,\mathbf{j} + (0)\,\mathbf{k} + \mathbf{C} = 2\mathbf{i} + \mathbf{k} \qquad \mathbf{r}(0) = 2\mathbf{i} + \mathbf{k}$$

$$\mathbf{j} + \mathbf{C} = 2\mathbf{i} + \mathbf{k}$$

$$\mathbf{C} = 2\mathbf{i} - \mathbf{j} + \mathbf{k}.$$

The particle's position as a function of t is

$$\mathbf{r}(t) = (\sin t + 2)\,\mathbf{i} + (\cos t - 1)\,\mathbf{j} + (t + 1)\,\mathbf{k}.$$

To check (always a good idea), we can see from this formula that

$$\frac{d\mathbf{r}}{dt} = (\cos t + 0)\,\mathbf{i} + (-\sin t - 0)\,\mathbf{j} + (1 + 0)\,\mathbf{k}$$

$$= (\cos t)\,\mathbf{i} - (\sin t)\,\mathbf{j} + \mathbf{k}$$

and

$$\mathbf{r}(0) = (\sin 0 + 2)\,\mathbf{i} + (\cos 0 - 1)\,\mathbf{j} + (0 + 1)\,\mathbf{k}$$

$$= 2\mathbf{i} + \mathbf{k}. \qquad \square$$

Exercises 11.1

Motion in the *xy*-plane

In Exercises 1–4, $\mathbf{r}(t)$ is the position of a particle in the *xy*-plane at time *t*. Find an equation in *x* and *y* whose graph is the path of the particle. Then find the particle's velocity and acceleration vectors at the given value of *t*.

1. $\mathbf{r}(t) = (t+1)\mathbf{i} + (t^2-1)\mathbf{j}, \quad t = 1$

2. $\mathbf{r}(t) = (t^2+1)\mathbf{i} + (2t-1)\mathbf{j}, \quad t = 1/2$

3. $\mathbf{r}(t) = e^t\mathbf{i} + \dfrac{2}{9}e^{2t}\mathbf{j}, \quad t = \ln 3$

4. $\mathbf{r}(t) = (\cos 2t)\mathbf{i} + (3\sin 2t)\mathbf{j}, \quad t = 0$

Exercises 5–8 give the position vectors of particles moving along various curves in the *xy*-plane. In each case, find the particle's velocity and acceleration vectors at the stated times and sketch them as vectors on the curve.

5. *Motion on the circle $x^2 + y^2 = 1$*

$$\mathbf{r}(t) = (\sin t)\mathbf{i} + (\cos t)\mathbf{j}; \quad t = \pi/4 \text{ and } \pi/2$$

6. *Motion on the circle $x^2 + y^2 = 16$*

$$\mathbf{r}(t) = \left(4\cos\frac{t}{2}\right)\mathbf{i} + \left(4\sin\frac{t}{2}\right)\mathbf{j}; \quad t = \pi \text{ and } 3\pi/2$$

7. *Motion on the cycloid $x = t - \sin t,\ y = 1 - \cos t$*

$$\mathbf{r}(t) = (t-\sin t)\mathbf{i} + (1-\cos t)\mathbf{j}; \quad t = \pi \text{ and } 3\pi/2$$

8. *Motion on the parabola $y = x^2 + 1$*

$$\mathbf{r}(t) = t\mathbf{i} + (t^2+1)\mathbf{j}; \quad t = -1, 0, \text{ and } 1$$

Velocity and Acceleration in Space

In Exercises 9–14, $\mathbf{r}(t)$ is the position of a particle in space at time *t*. Find the particle's velocity and acceleration vectors. Then find the particle's speed and direction of motion at the given value of *t*. Write the particle's velocity at that time as the product of its speed and direction.

9. $\mathbf{r}(t) = (t+1)\mathbf{i} + (t^2-1)\mathbf{j} + 2t\,\mathbf{k}, \quad t = 1$

10. $\mathbf{r}(t) = (1+t)\mathbf{i} + \dfrac{t^2}{\sqrt{2}}\mathbf{j} + \dfrac{t^3}{3}\mathbf{k}, \quad t = 1$

11. $\mathbf{r}(t) = (2\cos t)\mathbf{i} + (3\sin t)\mathbf{j} + 4t\,\mathbf{k}, \quad t = \pi/2$

12. $\mathbf{r}(t) = (\sec t)\mathbf{i} + (\tan t)\mathbf{j} + \dfrac{4}{3}t\,\mathbf{k}, \quad t = \pi/6$

13. $\mathbf{r}(t) = (2\ln(t+1))\mathbf{i} + t^2\mathbf{j} + \dfrac{t^2}{2}\mathbf{k}, \quad t = 1$

14. $\mathbf{r}(t) = (e^{-t})\mathbf{i} + (2\cos 3t)\mathbf{j} + (2\sin 3t)\mathbf{k}, \quad t = 0$

In Exercises 15–18, $\mathbf{r}(t)$ is the position of a particle in space at time *t*. Find the angle between the velocity and acceleration vectors at time $t = 0$.

15. $\mathbf{r}(t) = (3t+1)\mathbf{i} + \sqrt{3}\,t\mathbf{j} + t^2\mathbf{k}$

16. $\mathbf{r}(t) = \left(\dfrac{\sqrt{2}}{2}t\right)\mathbf{i} + \left(\dfrac{\sqrt{2}}{2}t - 16t^2\right)\mathbf{j}$

17. $\mathbf{r}(t) = (\ln(t^2+1))\mathbf{i} + (\tan^{-1} t)\mathbf{j} + \sqrt{t^2+1}\,\mathbf{k}$

18. $\mathbf{r}(t) = \dfrac{4}{9}(1+t)^{3/2}\mathbf{i} + \dfrac{4}{9}(1-t)^{3/2}\mathbf{j} + \dfrac{1}{3}t\,\mathbf{k}$

In Exercises 19 and 20, $\mathbf{r}(t)$ is the position vector of a particle in space at time *t*. Find the time or times in the given time interval when the velocity and acceleration vectors are orthogonal.

19. $\mathbf{r}(t) = (t - \sin t)\mathbf{i} + (1 - \cos t)\mathbf{j}, \quad 0 \le t \le 2\pi$

20. $\mathbf{r}(t) = (\sin t)\mathbf{i} + t\mathbf{j} + (\cos t)\mathbf{k}, \quad t \ge 0$

Integrating Vector-valued Functions

Evaluate the integrals in Exercises 21–26.

21. $\displaystyle\int_0^1 [t^3\mathbf{i} + 7\mathbf{j} + (t+1)\mathbf{k}]\,dt$

22. $\displaystyle\int_1^2 \left[(6 - 6t)\mathbf{i} + 3\sqrt{t}\,\mathbf{j} + \left(\frac{4}{t^2}\right)\mathbf{k}\right]dt$

23. $\displaystyle\int_{-\pi/4}^{\pi/4} [(\sin t)\mathbf{i} + (1 + \cos t)\mathbf{j} + (\sec^2 t)\mathbf{k}]\,dt$

24. $\displaystyle\int_0^{\pi/3} [(\sec t \tan t)\mathbf{i} + (\tan t)\mathbf{j} + (2\sin t \cos t)\mathbf{k}]\,dt$

25. $\displaystyle\int_1^4 \left[\frac{1}{t}\mathbf{i} + \frac{1}{5-t}\mathbf{j} + \frac{1}{2t}\mathbf{k}\right]dt$

26. $\displaystyle\int_0^1 \left[\frac{2}{\sqrt{1-t^2}}\mathbf{i} + \frac{\sqrt{3}}{1+t^2}\mathbf{k}\right]dt$

Initial Value Problems for Vector–valued Functions

Solve the initial value problems in Exercises 27–32 for $\mathbf{r}$ as a vector function of *t*.

27. Differential equation: $\dfrac{d\mathbf{r}}{dt} = -t\mathbf{i} - t\mathbf{j} - t\mathbf{k}$

    Initial condition: $\mathbf{r}(0) = \mathbf{i} + 2\mathbf{j} + 3\mathbf{k}$

28. Differential equation: $\dfrac{d\mathbf{r}}{dt} = (180t)\mathbf{i} + (180t - 16t^2)\mathbf{j}$

    Initial condition: $\mathbf{r}(0) = 100\mathbf{j}$

29. Differential equation: $\dfrac{d\mathbf{r}}{dt} = \dfrac{3}{2}(t+1)^{1/2}\mathbf{i} + e^{-t}\mathbf{j} + \dfrac{1}{t+1}\mathbf{k}$

    Initial condition: $\mathbf{r}(0) = \mathbf{k}$

30. Differential equation: $\dfrac{d\mathbf{r}}{dt} = (t^3 + 4t)\mathbf{i} + t\mathbf{j} + 2t^2\mathbf{k}$

Initial condition: $\mathbf{r}(0) = \mathbf{i} + \mathbf{j}$

31. Differential equation: $\dfrac{d^2\mathbf{r}}{dt^2} = -32\mathbf{k}$

Initial conditions: $\mathbf{r}(0) = 100\mathbf{k}$ and

$$\dfrac{d\mathbf{r}}{dt}\bigg|_{t=0} = 8\mathbf{i} + 8\mathbf{j}$$

32. Differential equation: $\dfrac{d^2\mathbf{r}}{dt^2} = -(\mathbf{i} + \mathbf{j} + \mathbf{k})$

Initial conditions: $\mathbf{r}(0) = 10\mathbf{i} + 10\mathbf{j} + 10\mathbf{k}$ and

$$\dfrac{d\mathbf{r}}{dt}\bigg|_{t=0} = \mathbf{0}$$

Tangent Lines to Smooth Curves

As mentioned in the text, the tangent line to a smooth curve $\mathbf{r}(t) = f(t)\mathbf{i} + g(t)\mathbf{j} + h(t)\mathbf{k}$ at $t = t_0$ is the line that passes through the point $(f(t_0), g(t_0), h(t_0))$ parallel to $\mathbf{v}(t_0)$, the curve's velocity vector at t_0. In Exercises 33–36, find parametric equations for the line that is tangent to the given curve at the given parameter value $t = t_0$.

33. $\mathbf{r}(t) = (\sin t)\mathbf{i} + (t^2 - \cos t)\mathbf{j} + e^t\mathbf{k}, \quad t_0 = 0$

34. $\mathbf{r}(t) = (2\sin t)\mathbf{i} + (2\cos t)\mathbf{j} + 5t\mathbf{k}, \quad t_0 = 4\pi$

35. $\mathbf{r}(t) = (a\sin t)\mathbf{i} + (a\cos t)\mathbf{j} + bt\mathbf{k}, \quad t_0 = 2\pi$

36. $\mathbf{r}(t) = (\cos t)\mathbf{i} + (\sin t)\mathbf{j} + (\sin 2t)\mathbf{k}, \quad t_0 = \dfrac{\pi}{2}$

Motion on Circular Paths

37. Each of the following equations (a)–(e) describes the motion of a particle having the same path, namely the unit circle $x^2 + y^2 = 1$. Although the path of each particle in (a)–(e) is the same, the behavior, or "dynamics," of each particle is different. For each particle, answer the following questions.

   i) Does the particle have constant speed? If so, what is its constant speed?

   ii) Is the particle's acceleration vector always orthogonal to its velocity vector?

   iii) Does the particle move clockwise or counterclockwise around the circle?

   iv) Does the particle begin at the point $(1, 0)$?

  a) $\mathbf{r}(t) = (\cos t)\mathbf{i} + (\sin t)\mathbf{j}, \quad t \geq 0$

  b) $\mathbf{r}(t) = \cos(2t)\mathbf{i} + \sin(2t)\mathbf{j}, \quad t \geq 0$

  c) $\mathbf{r}(t) = \cos(t - \pi/2)\mathbf{i} + \sin(t - \pi/2)\mathbf{j}, \quad t \geq 0$

  d) $\mathbf{r}(t) = (\cos t)\mathbf{i} - (\sin t)\mathbf{j}, \quad t \geq 0$

  e) $\mathbf{r}(t) = \cos(t^2)\mathbf{i} + \sin(t^2)\mathbf{j}, \quad t \geq 0$

38. Show that the vector-valued function

$$\mathbf{r}(t) = (2\mathbf{i} + 2\mathbf{j} + \mathbf{k})$$
$$+ \cos t\left(\frac{1}{\sqrt{2}}\mathbf{i} - \frac{1}{\sqrt{2}}\mathbf{j}\right) + \sin t\left(\frac{1}{\sqrt{3}}\mathbf{i} + \frac{1}{\sqrt{3}}\mathbf{j} + \frac{1}{\sqrt{3}}\mathbf{k}\right)$$

describes the motion of a particle moving in the circle of radius 1 centered at the point $(2, 2, 1)$ and lying in the plane $x + y - 2z = 2$.

Motion along a Straight Line

39. At time $t = 0$, a particle is located at the point $(1, 2, 3)$. It travels in a straight line to the point $(4, 1, 4)$, has speed 2 at $(1, 2, 3)$ and constant acceleration $3\mathbf{i} - \mathbf{j} + \mathbf{k}$. Find an equation for the position vector $\mathbf{r}(t)$ of the particle at time t.

40. A particle traveling in a straight line is located at the point $(1, -1, 2)$ and has speed 2 at time $t = 0$. The particle moves toward the point $(3, 0, 3)$ with constant acceleration $2\mathbf{i} + \mathbf{j} + \mathbf{k}$. Find its position vector $\mathbf{r}(t)$ at time t.

Theory and Examples

41. A particle moves along the top of the parabola $y^2 = 2x$ from left to right at a constant speed of 5 units per second. Find the velocity of the particle as it moves through the point $(2, 2)$.

42. A particle moves on a cycloid in the xy-plane in such a way that its position at time t is

$$\mathbf{r}(t) = (t - \sin t)\mathbf{i} + (1 - \cos t)\mathbf{j}.$$

Find the maximum and minimum values of $|\mathbf{v}|$ and $|\mathbf{a}|$. (*Hint:* Find the extreme values of $|\mathbf{v}|^2$ and $|\mathbf{a}|^2$ first and take square roots later.)

43. A particle moves around the ellipse $(y/3)^2 + (z/2)^2 = 1$ in the yz-plane in such a way that its position at time t is

$$\mathbf{r}(t) = (3\cos t)\mathbf{j} + (2\sin t)\mathbf{k}.$$

Find the maximum and minimum values of $|\mathbf{v}|$ and $|\mathbf{a}|$. (See the hint in Exercise 42.)

44. *A satellite in circular orbit.* A satellite of mass m is revolving at a constant speed v around a body of mass M (Earth, for example) in a circular orbit of radius r_0 (measured from the body's center of mass). Determine the satellite's orbital period T (the time to complete one full orbit), as follows:

  a) Coordinatize the orbital plane by placing the origin at the body's center of mass, with the satellite on the x-axis at $t = 0$ and moving counterclockwise, as in the accompanying figure.

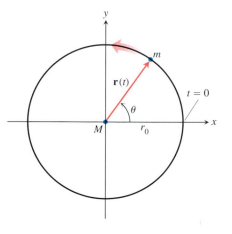

Let $\mathbf{r}(t)$ be the satellite's position vector at time t. Show that $\theta = vt/r_0$ and hence that

$$\mathbf{r}(t) = \left(r_0 \cos \frac{vt}{r_0}\right)\mathbf{i} + \left(r_0 \sin \frac{vt}{r_0}\right)\mathbf{j}.$$

b) Find the acceleration of the satellite.

c) According to Newton's Law of Gravitation, the gravitational force exerted on the satellite is directed toward M and is given by

$$\mathbf{F} = \left(-\frac{GmM}{r_0^2}\right)\frac{\mathbf{r}}{r_0},$$

where G is the universal constant of gravitation. Using Newton's second law, $\mathbf{F} = m\,\mathbf{a}$, show that $v^2 = GM/r_0$.

d) Show that the orbital period T satisfies $vT = 2\pi r_0$.

e) From parts (c) and (d), deduce that

$$T^2 = \frac{4\pi^2}{GM} r_0^3.$$

That is, the square of the period of a satellite in circular orbit is proportional to the cube of the radius from the orbital center.

45. Let $\mathbf{v}$ be a differentiable vector function of t. Show that if $\mathbf{v} \cdot (d\mathbf{v}/dt) = 0$ for all t, then $|\mathbf{v}|$ is constant.

46. *Derivatives of triple scalar products*

a) Show that if $\mathbf{u}$, $\mathbf{v}$, and $\mathbf{w}$ are differentiable vector functions of t, then

$$\frac{d}{dt}(\mathbf{u} \cdot \mathbf{v} \times \mathbf{w}) \tag{6}$$

$$= \frac{d\mathbf{u}}{dt} \cdot \mathbf{v} \times \mathbf{w} + \mathbf{u} \cdot \frac{d\mathbf{v}}{dt} \times \mathbf{w} + \mathbf{u} \cdot \mathbf{v} \times \frac{d\mathbf{w}}{dt}.$$

b) Show that Eq. (6) is equivalent to

$$\frac{d}{dt}\begin{vmatrix} u_1 & u_2 & u_3 \\ v_1 & v_2 & v_3 \\ w_1 & w_2 & w_3 \end{vmatrix} = \begin{vmatrix} \dfrac{du_1}{dt} & \dfrac{du_2}{dt} & \dfrac{du_3}{dt} \\ v_1 & v_2 & v_3 \\ w_1 & w_2 & w_3 \end{vmatrix} + \begin{vmatrix} u_1 & u_2 & u_3 \\ \dfrac{dv_1}{dt} & \dfrac{dv_2}{dt} & \dfrac{dv_3}{dt} \\ w_1 & w_2 & w_3 \end{vmatrix}$$

$$+ \begin{vmatrix} u_1 & u_2 & u_3 \\ v_1 & v_2 & v_3 \\ \dfrac{dw_1}{dt} & \dfrac{dw_2}{dt} & \dfrac{dw_3}{dt} \end{vmatrix}. \tag{7}$$

Equation (7) says that the derivative of a 3 by 3 determinant of differentiable functions is the sum of the three determinants obtained from the original by differentiating one row at a time. The result extends to determinants of any order.

47. *(Continuation of Exercise 46.)* Suppose that $\mathbf{r}(t) = f(t)\mathbf{i} + g(t)\mathbf{j} + h(t)\mathbf{k}$ and that f, g, and h have derivatives through order three. Use Eq. (6) or (7) to show that

$$\frac{d}{dt}\left(\mathbf{r} \cdot \frac{d\mathbf{r}}{dt} \times \frac{d^2\mathbf{r}}{dt^2}\right) = \mathbf{r} \cdot \left(\frac{d\mathbf{r}}{dt} \times \frac{d^3\mathbf{r}}{dt^3}\right). \tag{8}$$

(Hint: Differentiate on the left and look for vectors whose products are zero.)

48. *The Constant Function Rule.* Prove that if $\mathbf{u}$ is the vector function with the constant value $\mathbf{C}$, then $d\mathbf{u}/dt = \mathbf{0}$.

49. *The Scalar Multiple Rules*

a) Prove that if $\mathbf{u}$ is a differentiable function of t and c is any real number, then

$$\frac{d(c\mathbf{u})}{dt} = c\frac{d\mathbf{u}}{dt}.$$

b) Prove that if $\mathbf{u}$ is a differentiable function of t and f is a differentiable scalar function of t, then

$$\frac{d}{dt}(f\,\mathbf{u}) = \frac{df}{dt}\mathbf{u} + f\frac{d\mathbf{u}}{dt}.$$

50. *The Sum and Difference Rules.* Prove that if $\mathbf{u}$ and $\mathbf{v}$ are differentiable functions of t, then

$$\frac{d}{dt}(\mathbf{u} + \mathbf{v}) = \frac{d\mathbf{u}}{dt} + \frac{d\mathbf{v}}{dt}$$

and

$$\frac{d}{dt}(\mathbf{u} - \mathbf{v}) = \frac{d\mathbf{u}}{dt} - \frac{d\mathbf{v}}{dt}.$$

51. *The component test for continuity at a point.* Show that the vector function $\mathbf{r}$ defined by the rule $\mathbf{r}(t) = f(t)\mathbf{i} + g(t)\mathbf{j} + h(t)\mathbf{k}$ is continuous at $t = t_0$ if and only if f, g, and h are continuous at t_0.

52. *Limits of cross products of vector functions.* Suppose that $\mathbf{r}_1(t) = f_1(t)\mathbf{i} + f_2(t)\mathbf{j} + f_3(t)\mathbf{k}$, $\mathbf{r}_2(t) = g_1(t)\mathbf{i} + g_2(t)\mathbf{j} + g_3(t)\mathbf{k}$, $\lim_{t \to t_0} \mathbf{r}_1(t) = \mathbf{A}$, and $\lim_{t \to t_0} \mathbf{r}_2(t) = \mathbf{B}$. Use the determinant formula for cross products and the Limit Product Rule for scalar functions to show that

$$\lim_{t \to t_0} (\mathbf{r}_1(t) \times \mathbf{r}_2(t)) = \mathbf{A} \times \mathbf{B}$$

53. *Differentiable vector functions are continuous.* Show that if $\mathbf{r}(t) = f(t)\mathbf{i} + g(t)\mathbf{j} + h(t)\mathbf{k}$ is differentiable at $t = t_0$, then it is continuous at t_0 as well.

54. Establish the following properties of integrable vector functions.

a) The *Constant Scalar Multiple Rule:*

$$\int_a^b k\,\mathbf{r}(t)\,dt = k\int_a^b \mathbf{r}(t)\,dt \quad \text{(any scalar } k\text{)}$$

The *Rule for Negatives,*

$$\int_a^b (-\mathbf{r}(t))\,dt = -\int_a^b \mathbf{r}(t)\,dt,$$

is obtained by taking $k = -1$.

b) The *Sum and Difference Rules:*

$$\int_a^b (\mathbf{r}_1(t) \pm \mathbf{r}_2(t))\,dt = \int_a^b \mathbf{r}_1(t)\,dt \pm \int_a^b \mathbf{r}_2(t)\,dt$$

c) The *Constant Vector Multiple Rules:*

$$\int_a^b \mathbf{C} \cdot \mathbf{r}(t)\,dt = \mathbf{C} \cdot \int_a^b \mathbf{r}(t)\,dt \quad \text{(any constant vector } \mathbf{C}\text{)}$$

and

$$\int_a^b \mathbf{C} \times \mathbf{r}(t)\, dt = \mathbf{C} \times \int_a^b \mathbf{r}(t)\, dt \quad \text{(any constant vector } \mathbf{C})$$

55. *Products of scalar and vector functions.* Suppose that the scalar function $u(t)$ and the vector function $\mathbf{r}(t)$ are both defined for $a \le t \le b$.

a) Show that $u\,\mathbf{r}$ is continuous on $[a, b]$ if u and $\mathbf{r}$ are continuous on $[a, b]$.

b) If u and $\mathbf{r}$ are both differentiable on $[a, b]$, show that $u\,\mathbf{r}$ is differentiable on $[a, b]$ and that

$$\frac{d}{dt}(u\,\mathbf{r}) = u\,\frac{d\mathbf{r}}{dt} + \mathbf{r}\,\frac{du}{dt}.$$

56. *Antiderivatives of vector functions*

a) Use Corollary 2 of the Mean Value Theorem for scalar functions to show that if two vector functions $\mathbf{R}_1(t)$ and $\mathbf{R}_2(t)$ have identical derivatives on an interval I, then the functions differ by a constant vector value throughout I.

b) Use the result in (a) to show that if $\mathbf{R}(t)$ is any antiderivative of $\mathbf{r}(t)$ on I, then every other antiderivative of $\mathbf{r}$ on I equals $\mathbf{R}(t) + \mathbf{C}$ for some constant vector $\mathbf{C}$.

57. *The Fundamental Theorem of Calculus.* The Fundamental Theorem of Calculus for scalar functions of a real variable holds for vector functions of a real variable as well. Prove this by using the theorem for scalar functions to show first that if a vector function $\mathbf{r}(t)$ is continuous for $a \le t \le b$, then

$$\frac{d}{dt}\int_a^t \mathbf{r}(\tau)\, d\tau = \mathbf{r}(t)$$

at every point t of $[a, b]$. Then use the conclusion in part (b) of Exercise 56 to show that if $\mathbf{R}$ is any antiderivative of $\mathbf{r}$ on $[a, b]$ then

$$\int_a^b \mathbf{r}(t)\, dt = \mathbf{R}(b) - \mathbf{R}(a).$$

CAS Explorations and Projects

Use a CAS to perform the following steps in Exercises 58–61.

a) Plot the space curve traced out by the position vector $\mathbf{r}$.

b) Find the components of the velocity vector $d\mathbf{r}/dt$.

c) Evaluate $d\mathbf{r}/dt$ at the given point t_0 and determine the equation of the tangent line to the curve at $\mathbf{r}(t_0)$.

d) Plot the tangent line together with the curve over the given interval.

58. $\mathbf{r}(t) = (\sin t - t \cos t)\,\mathbf{i} + (\cos t + t \sin t)\,\mathbf{j} + t^2\,\mathbf{k}, \quad 0 \le t \le 6\pi, \quad t_0 = 3\pi/2$

59. $\mathbf{r}(t) = \sqrt{2}\, t\,\mathbf{i} + e^t\,\mathbf{j} + e^{-t}\,\mathbf{k}, \quad -2 \le t \le 3, \quad t_0 = 1$

60. $\mathbf{r}(t) = (\sin 2t)\,\mathbf{i} + (\ln(1+t))\,\mathbf{j} + t\,\mathbf{k}, \quad 0 \le t \le 4\pi, \quad t_0 = \pi/4$

61. $\mathbf{r}(t) = (\ln(t^2 + 2))\,\mathbf{i} + (\tan^{-1} 3t)\,\mathbf{j} + \sqrt{t^2 + 1}\,\mathbf{k}, \quad -3 \le t \le 5, \quad t_0 = 3$

In Exercises 62 and 63, you will explore graphically the behavior of the helix

$$\mathbf{r}(t) = (\cos at)\,\mathbf{i} + (\sin at)\,\mathbf{j} + bt\,\mathbf{k}$$

as you change the values of the constants a and b. Use a CAS to perform the steps in each exercise.

62. Set $b = 1$. Plot the helix $\mathbf{r}(t)$ together with the tangent line to the curve at $t = 3\pi/2$ for $a = 1, 2, 4$, and 6 over the interval $0 \le t \le 4\pi$. Describe in your own words what happens to the graph of the helix and the position of the tangent line as a increases through these positive values.

63. Set $a = 1$. Plot the helix $\mathbf{r}(t)$ together with the tangent line to the curve at $t = 3\pi/2$ for $b = 1/4, 1/2, 2$, and 4 over the interval $0 \le t \le 4\pi$. Describe in your own words what happens to the graph of the helix and the position of the tangent line as b increases through these positive values.

11.2

Modeling Projectile Motion

When we shoot a projectile into the air we usually want to know beforehand how far it will go (will it reach the target?), how high it will rise (will it clear the hill?), and when it will land (when do we get results?). We get this information from the direction and magnitude of the projectile's initial velocity vector, using Newton's Second Law of Motion.

The Vector and Parametric Equations for Ideal Projectile Motion

To derive equations for projectile motion, we assume that the projectile behaves like a particle moving in a vertical coordinate plane and that the only force acting on the projectile during its flight is the constant force of gravity, which always points

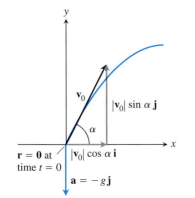

(a) Position, velocity, acceleration, and launch angle at $t = 0$

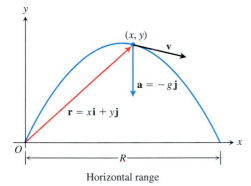

Horizontal range

(b) Position, velocity, and acceleration at a later time t

11.7 The flight of an ideal projectile.

straight down. In practice, none of these assumptions really holds. The ground moves beneath the projectile as the earth turns, the air creates a frictional force that varies with the projectile's speed and altitude, and the force of gravity changes as the projectile moves along. All this must be taken into account by applying corrections to the predictions of the *ideal* equations we are about to derive. The corrections, however, are not the subject of this section.

We assume that our projectile is launched from the origin at time $t = 0$ into the first quadrant with an initial velocity $\mathbf{v}_0$ (Fig. 11.7). If $\mathbf{v}_0$ makes an angle α with the horizontal, then

$$\mathbf{v}_0 = (|\mathbf{v}_0| \cos \alpha)\,\mathbf{i} + (|\mathbf{v}_0| \sin \alpha)\,\mathbf{j}. \tag{1}$$

If we use the simpler notation v_0 for the initial speed $|\mathbf{v}_0|$, then

$$\mathbf{v}_0 = (v_0 \cos \alpha)\,\mathbf{i} + (v_0 \sin \alpha)\,\mathbf{j}. \tag{2}$$

The projectile's initial position is

$$\mathbf{r}_0 = 0\,\mathbf{i} + 0\,\mathbf{j} = \mathbf{0}. \tag{3}$$

Newton's Second Law of Motion says that the force acting on the projectile is its mass m times its acceleration, or $m(d^2\mathbf{r}/dt^2)$ if $\mathbf{r}$ is the projectile's position vector and t is time. If the force is solely the gravitational force $-mg\,\mathbf{j}$, then

$$m\frac{d^2\mathbf{r}}{dt^2} = -mg\,\mathbf{j} \qquad \text{and} \qquad \frac{d^2\mathbf{r}}{dt^2} = -g\,\mathbf{j}. \tag{4}$$

We find $\mathbf{r}$ as a function of t by solving the following initial value problem:

Differential equation: $\qquad \dfrac{d^2\mathbf{r}}{dt^2} = -g\,\mathbf{j}$

Initial conditions: $\qquad \mathbf{r} = \mathbf{r}_0 \quad \text{and} \quad \dfrac{d\mathbf{r}}{dt} = \mathbf{v}_0 \quad \text{when} \quad t = 0$

The first integration gives

$$\frac{d\mathbf{r}}{dt} = -(gt)\,\mathbf{j} + \mathbf{v}_0.$$

A second integration gives

$$\mathbf{r} = -\frac{1}{2}gt^2\,\mathbf{j} + \mathbf{v}_0 t + \mathbf{r}_0.$$

Substituting the values of $\mathbf{v}_0$ and $\mathbf{r}_0$ from Eqs. (2) and (3) gives

$$\mathbf{r} = -\frac{1}{2}gt^2\,\mathbf{j} + \underbrace{(v_0 \cos \alpha)t\,\mathbf{i} + (v_0 \sin \alpha)t\,\mathbf{j}}_{\mathbf{v}_0 t} + \mathbf{0}$$

or

$$\mathbf{r} = (v_0 \cos \alpha)t\,\mathbf{i} + \left((v_0 \sin \alpha)t - \frac{1}{2}gt^2\right)\mathbf{j}. \tag{5}$$

Equation (5) is the *vector equation* for ideal projectile motion. The angle α is the projectile's **launch angle (firing angle, angle of elevation)**, and v_0, as we said before, is the projectile's **initial speed.**

Equation (5) is equivalent to a pair of scalar equations,

$$x = (v_0 \cos \alpha)t, \qquad y = (v_0 \sin \alpha)t - \frac{1}{2}gt^2. \tag{6}$$

These are known as the *parametric equations* for ideal projectile motion. If time is measured in seconds and distance in meters, g is 9.8 m/sec^2 and Eqs. (6) give x and y in meters. With feet in place of meters, g is 32 ft/sec^2 and Eqs. (6) give x and y in feet.

EXAMPLE 1   A projectile is fired from the origin over horizontal ground at an initial speed of 500 m/sec at a launch angle of 60°. Where will the projectile be 10 sec later?

Solution   We use Eqs. (6) with $v_0 = 500$, $\alpha = 60°$, $g = 9.8$, and $t = 10$ to find the projectile's coordinates to the nearest meter 10 sec after firing:

$$x = (v_0 \cos \alpha)t = 500 \cdot \frac{1}{2} \cdot 10 = 2500 \text{ m}$$

$$y = (v_0 \sin \alpha)t - \frac{1}{2}gt^2$$

$$= 500 \cdot \frac{\sqrt{3}}{2} \cdot 10 - \frac{1}{2} \cdot 9.8 \cdot (10)^2$$

$$= 2500\sqrt{3} - 490$$

$$\approx 3840 \text{ m.}$$

Ten seconds after firing, the projectile is 3840 m in the air and 2500 m downrange.   ❑

Height, Flight Time, and Range

Equations (6) enable us to answer most questions about an ideal projectile fired from the origin.

The projectile reaches its highest point when its vertical velocity component is zero, that is, when

$$\frac{dy}{dt} = v_0 \sin \alpha - gt = 0, \qquad \text{or} \qquad t = \frac{v_0 \sin \alpha}{g}.$$

For this value of t, the value of y is

$$y_{max} = (v_0 \sin \alpha)\left(\frac{v_0 \sin \alpha}{g}\right) - \frac{1}{2}g\left(\frac{v_0 \sin \alpha}{g}\right)^2 = \frac{(v_0 \sin \alpha)^2}{2g}. \qquad (7)$$

To find when the projectile lands, when fired over horizontal ground, we set y equal to zero in Eqs. (6) and solve for t:

$$(v_0 \sin \alpha)t - \frac{1}{2}gt^2 = 0$$

$$t\left(v_0 \sin \alpha - \frac{1}{2}gt\right) = 0$$

$$t = 0, \qquad t = \frac{2v_0 \sin \alpha}{g}. \qquad (8)$$

Since 0 is the time the projectile is fired, $(2v_0 \sin \alpha)/g$ must be the time when the projectile strikes the ground.

To find the projectile's range R, the distance from the origin to the point of impact on horizontal ground, we find the value of x when $t = (2v_0 \sin \alpha)/g$:

$$x = (v_0 \cos \alpha)t$$

$$R = (v_0 \cos \alpha)\left(\frac{2v_0 \sin \alpha}{g}\right)$$

$$= \frac{v_0^2}{g}(2 \sin \alpha \cos \alpha) = \frac{v_0^2}{g} \sin 2\alpha. \tag{9}$$

The range is largest when $\sin 2\alpha = 1$ or $\alpha = 45°$.

EXAMPLE 2   Find the maximum height, flight time, and range of a projectile fired from the origin over horizontal ground at an initial speed of 500 m/sec and a launch angle of $60°$ (same projectile as in Example 1).

Solution

Maximum height (Eq. 7): $y_{max} = \dfrac{(v_0 \sin \alpha)^2}{2g}$

$$= \frac{(500 \sin 60°)^2}{2(9.8)} \approx 9566 \text{ m}$$

Flight time (Eq. 8): $t = \dfrac{2v_0 \sin \alpha}{g}$

$$= \frac{2(500) \sin 60°}{9.8} \approx 88 \text{ sec}$$

Range (Eq. 9): $R = \dfrac{v_0^2}{g} \sin 2\alpha$

$$= \frac{(500)^2 \sin 120°}{9.8} \approx 22{,}092 \text{ m} \qquad \square$$

Ideal Trajectories Are Parabolic

It is often claimed that water from a hose traces a parabola in the air, but anyone who looks closely enough will see this is not so. The air slows the water down, and its forward progress is too slow at the end to match the rate at which it falls.

What is really being claimed is that ideal projectiles move along parabolas, and this we can see from Eqs. (6). If we substitute $t = x/(v_0 \cos \alpha)$ from the first equation into the second, we obtain the Cartesian coordinate equation

$$y = -\left(\frac{g}{2v_0^2 \cos^2 \alpha}\right)x^2 + (\tan \alpha)x.$$

This equation has the form $y = ax^2 + bx$, so its graph is a parabola.

Firing from (x_0, y_0)

If we fire our ideal projectile from the point (x_0, y_0) instead of the origin (Fig. 11.8), the equations that replace Eqs. (6) are

$$x = x_0 + (v_0 \cos \alpha)t, \qquad y = y_0 + (v_0 \sin \alpha)t - \frac{1}{2}gt^2, \tag{10}$$

as you will be invited to show in Exercise 19.

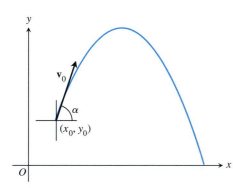

11.8 The path of a projectile fired from (x_0, y_0) with an initial velocity $\mathbf{v}_0$ at an angle of α degrees with the horizontal.

EXAMPLE 3   To open the 1992 Summer Olympics in Barcelona, bronze medalist archer Antonio Rebollo lit the Olympic torch with a flaming arrow (Fig. 11.9).

Suppose that Rebollo wanted the arrow to reach its maximum height exactly 4 ft above the center of the cauldron (Fig. 11.10).

a) If he shot the arrow at a height of 6 ft above ground level 30 yd from the 70-ft-high cauldron, express y_{max} in terms of the initial speed v_0 and firing angle α.

b) If $y_{max} = 74$ ft (Fig. 11.10), use the results of part (a) to find the value of $v_0 \sin \alpha$.

c) When the arrow reaches y_{max}, the horizontal distance traveled to the center of the cauldron is $x = 90$ ft. Use this fact to find the value of $v_0 \cos \alpha$.

d) Find the initial firing angle of the arrow.

Solution

a) We use a coordinate system in which the x-axis lies along the ground toward the left (to match the photograph in Fig. 11.9) and the coordinates of the flaming arrow at $t = 0$ are $x_0 = 0$ and $y_0 = 6$ (Fig. 11.10). We have

$$y = y_0 + (v_0 \sin \alpha)t - \frac{1}{2}gt^2 \qquad \text{Eqs. (10)}$$

$$= 6 + (v_0 \sin \alpha)t - \frac{1}{2}gt^2. \qquad y_0 = 6$$

We find the time when the arrow reaches its highest point by setting $dy/dt = 0$ and solving for t, obtaining

$$t = \frac{v_0 \sin \alpha}{g}.$$

For this value of t, the value of y is

$$y_{max} = 6 + (v_0 \sin \alpha)\left(\frac{v_0 \sin \alpha}{g}\right) - \frac{1}{2}g\left(\frac{v_0 \sin \alpha}{g}\right)^2 = 6 + \frac{(v_0 \sin \alpha)^2}{2g}.$$

b) Using $y_{max} = 74$ and $g = 32$, we see from part (a) that

$$74 = 6 + \frac{(v_0 \sin \alpha)^2}{(2)(32)}$$

11.9 Spanish archer Antonio Rebollo lights the Olympic torch in Barcelona with a flaming arrow.

or

$$v_0 \sin \alpha = \sqrt{(68)(64)}.$$

c) We substitute the time to reach y_{max} from part (a) and the horizontal distance $x = 90$ ft into Eqs. (10) to obtain

$$x = x_0 + (v_0 \cos \alpha)t \qquad \text{Eqs. (10)}$$

$$90 = 0 + (v_0 \cos \alpha)t \qquad x = 90, \quad x_0 = 0$$

$$= (v_0 \cos \alpha)\left(\frac{v_0 \sin \alpha}{g}\right). \qquad t = (v_0 \sin \alpha)/g$$

11.10 Ideal path of the arrow that lit the Olympic torch.

Solving this equation for $v_0 \cos \alpha$ and using the result from part (b), we have

$$v_0 \cos \alpha = \frac{90g}{v_0 \sin \alpha}$$

d) Parts (b) and (c) together tell us that

$$\tan \alpha = \frac{v_0 \sin \alpha}{v_0 \cos \alpha}$$

$$= \frac{(v_0 \sin \alpha)^2}{90g}$$

$$= \frac{(68)(64)}{(90)(32)} = \frac{68}{45}$$

or

$$\alpha \approx \tan^{-1}\left(\frac{68}{45}\right) \approx 57°.$$

This is Rebollo's firing angle.   ❑

Exercises 11.2

The projectiles in the following exercises are to be treated as ideal projectiles whose behavior is faithfully portrayed by the equations derived in the text. Most of the arithmetic, however, is realistic and is best done with a calculator. All launch angles are assumed to be measured from the horizontal. All projectiles are assumed to be fired from the origin over horizontal ground, unless stated otherwise.

1. A projectile is fired at a speed of 840 m/sec at an angle of 60°. How long will it take to get 21 km downrange?

2. Find the muzzle speed of a gun whose maximum range is 24.5 km.

3. A projectile is fired with an initial speed of 500 m/sec at an angle of elevation of 45°.

   a) When and how far away will the projectile strike?
   b) How high overhead will the projectile be when it is 5 km downrange?
   c) What is the highest the projectile will go?

4. A baseball is thrown from the stands 32 ft above the field at an angle of 30° up from horizontal. When and how far away will the ball strike the ground if its initial speed is 32 ft/sec?

5. An athlete throws a 16-lb shot at an angle of 45° to the horizontal from 6.5 ft above the ground at an initial speed of 44 ft/sec. How long after launch and how far from the inner edge of the stopboard does the shot land?

6. Because of its initial evaluation, the shot in Exercise 5 would have gone slightly farther if it had been launched at a 40° angle. How much farther? Answer in inches.

7. A spring gun at ground level fires a golf ball at an angle of 45°. The ball lands 10 m away. What was the ball's initial speed? For the same initial speed, find the two firing angles that make the range 6 m.

8. An electron in a TV tube is beamed horizontally at a speed of 5×10^6 m/sec toward the face of the tube 40 cm away. About how far will the electron drop before it hits?

9. Laboratory tests designed to find how far golf balls of different hardness go when hit with a driver showed that a 100-compression ball hit with a club head speed of 100 mph at a launch angle of 9° carried 248.8 yd. What was the launch speed of the ball? (It was more than 100 mph. At the same time the club head was moving forward, the compressed ball was kicking away from the club face, adding to the ball's forward speed.)

10. A human cannonball is to be fired with an initial speed of $v_0 = 80\sqrt{10}/3$ ft/sec. The circus performer (of the right caliber, naturally) hopes to land on a special cushion located 200 ft downrange. The circus is being held in a large room with a flat ceiling 75 ft high. Can the performer be fired to the cushion without striking the ceiling? If so, what should the cannon's angle of elevation be?

11. A golf ball leaves the ground at a 30° angle at a speed of 90 ft/sec. Will it clear the top of a 30-ft tree 135 ft away?

12. A golf ball is hit from the tee to a green elevated 45 ft above the tee with an initial speed of 116 ft/sec at an angle of elevation of 45°. Assuming that the pin, 369 ft downrange, does not get in the way, where will the ball land in relation to the pin?

13. In Moscow in 1987, Natalya Lisouskaya set a women's world record by putting an 8-lb 13-oz shot 73 ft 10 in. Assuming that she launched the shot at a 40° angle to the horizontal 6.5 ft above the ground, what was the shot's initial speed?

14. A baseball hit by a Boston Red Sox player at a 20° angle from 3 ft above the ground just cleared the left end of the "Green Monster," the left-field wall in Fenway Park (Fig. 11.11). This wall is 37 ft high and 315 ft from home plate. About how fast was the ball going? How long did it take the ball to reach the wall?

15. Show that a projectile fired at an angle of α degrees, $0 < \alpha < 90$, has the same range as a projectile fired at the same speed at an angle of $(90 - \alpha)$ degrees. (In models that take air resistance into account, this symmetry is lost.)

16. What two angles of elevation will enable a projectile to reach a target 16 km downrange on the same level as the gun if the projectile's initial speed is 400 m/sec?

17. Show that doubling a projectile's initial speed at a given launch angle multiplies its range by 4. By about what percentage should you increase the initial speed to double the height and range?

18. Show that a projectile attains three-quarters of its maximum height in half the time it takes to reach the maximum height.

19. Derive the equations

$$x = x_0 + (v_0 \cos \alpha)t$$

$$y = y_0 + (v_0 \sin \alpha)t - \frac{1}{2}gt^2$$

11.11 The "Green Monster," the left-field wall at Fenway Park in Boston (Exercise 14).

(Eqs. 10 in the text) by solving the following initial value problem for a vector $\mathbf{r}$ in the plane.

Differential equation: $\dfrac{d^2\mathbf{r}}{dt^2} = -g\mathbf{j}$

Initial conditions: $\mathbf{r} = x_0\mathbf{i} + y_0\mathbf{j}$

and

$$\frac{d\mathbf{r}}{dt} = (v_0\cos\alpha)\mathbf{i} + (v_0\sin\alpha)\mathbf{j}$$

when $t = 0$

20. Using the firing angle $\alpha = 57°$ in Example 3, find the speed with which the flaming arrow left Rebollo's bow. See Fig. 11.10.

21. The cauldron in Example 3 is 12 ft in diameter. Using Eqs. (10) and Example 3(c), find how long it takes the flaming arrow to cover the horizontal distance to the rim. How high is the arrow at this time?

22. The multiflash photograph shown below shows a model train engine moving at a constant speed on a straight track. As the engine moved along, a marble was fired into the air by a spring in the engine's smokestack. The marble, which continued to move with the same forward speed as the engine, rejoined the engine 1 sec after it was fired. Measure the angle the marble's path made with the horizontal and use the information to find how high the marble went and how fast the engine was moving.

The train in Exercise 22.

23. Figure 11.12 shows an experiment with two marbles. Marble A was launched toward marble B with launch angle α and initial speed v_0. At the same instant, marble B was released to fall from rest $R\tan\alpha$ units directly above a spot R units downrange from A. The marbles were found to collide regardless of the value of v_0. Was this mere coincidence, or must this happen? Give reasons for your answer.

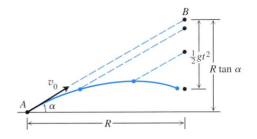

11.12 The marbles in Exercise 23.

24. An ideal projectile is launched straight down an inclined plane, as shown in profile in Fig. 11.13.

a)   Show that the greatest downhill range is achieved when the initial velocity vector bisects angle AOR.

b)   If the projectile were fired uphill instead of down, what launch angle would maximize its range? Give reasons for your answer.

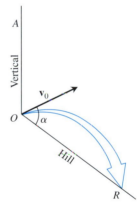

11.13 Maximum downhill range occurs when the velocity vector bisects angle AOR (Exercise 24).

25. An ideal projectile, launched from the origin into the first octant at time $t = 0$ with initial velocity $\mathbf{v}_0$, experiences a constant downward acceleration $\mathbf{a} = -g\mathbf{k}$ from gravity. Find the projectile's velocity and position as functions of t.

26. *Air resistance proportional to velocity.* If a projectile of mass m launched with initial velocity $\mathbf{v}_0$ encounters an air resistance proportional to its velocity, the total force $m(d^2\mathbf{r}/dt^2)$ on the projectile satisfies the equation

$$m\frac{d^2\mathbf{r}}{dt^2} = -mg\mathbf{j} - k\frac{d\mathbf{r}}{dt},$$

where k is the proportionality constant. Show that one integration of this equation gives

$$\frac{d\mathbf{r}}{dt} + \frac{k}{m}\mathbf{r} = \mathbf{v}_0 - gt\mathbf{j}.$$

Solve this equation. To do so, multiply both sides of the equation by $e^{(k/m)t}$. The left-hand side will then be the derivative of a product, with the result that both sides of the equation can now be integrated.

The function $e^{(k/m)t}$ is called an *integrating factor* for the differential equation because multiplying the equation by it makes the equation integrable.

27. For a projectile fired from the ground at launch angle α with initial speed v_0, consider α as a variable and v_0 as a fixed constant. For each α, $0 < \alpha \leq \pi/2$, we obtain a parabolic trajectory (Fig 11.14). Show that the points in the plane that give the maximum heights of these parabolic trajectories all lie on the ellipse

$$x^2 + 4\left(y - \frac{v_0^2}{4g}\right)^2 = \frac{v_0^4}{4g^2},$$

where $x \geq 0$.

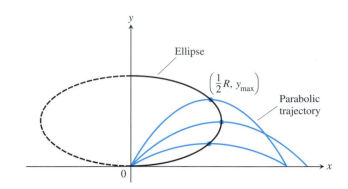

11.14 The parabolas in Exercise 27.

28. GRAPHER If you have access to a parametric equation grapher and have not yet done Exercise 46 in Section 9.4, do it now. It is about ideal projectile motion.

11.3 Arc Length and the Unit Tangent Vector **T**

As you can imagine, differentiable curves, especially those with continuous first and second derivatives, have been subjects of intense study, for their mathematical interest as well as their applications to motion in space. In this section and the next, we study some of the features that account for the importance of these curves.

Arc Length Along a Curve

11.15 Smooth curves can be scaled like number lines, the coordinate of each point being its directed distance from a preselected base point.

One of the special features of smooth space curves is that they have a measurable length. This enables us to locate points along these curves by giving their directed distance s along the curve from some **base point,** the way we locate points on coordinate axes by giving their directed distance from the origin (Fig. 11.15). Time is the natural parameter for describing a moving body's velocity and acceleration, but s is the natural parameter for studying a curve's shape. Both parameters appear in analyses of space flight.

To measure distance along a smooth curve in space, we add a z-term to the formula we use for curves in the plane.

> ### Definition
> The **length** of a smooth curve $\mathbf{r}(t) = f(t)\mathbf{i} + g(t)\mathbf{j} + h(t)\mathbf{k}$, $a \leq t \leq b$, that is traced exactly once as t increases from $t = a$ to $t = b$ is
>
> $$L = \int_a^b \sqrt{\left(\frac{df}{dt}\right)^2 + \left(\frac{dg}{dt}\right)^2 + \left(\frac{dh}{dt}\right)^2}\, dt$$
>
> $$= \int_a^b \sqrt{\left(\frac{dx}{dt}\right)^2 + \left(\frac{dy}{dt}\right)^2 + \left(\frac{dz}{dt}\right)^2}\, dt. \qquad (1)$$

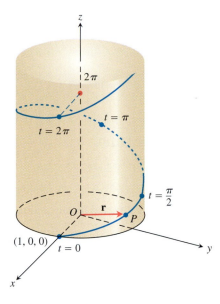

11.16 The helix $\mathbf{r}(t) = (\cos t)\,\mathbf{i} + (\sin t)\,\mathbf{j}$ $+t\,\mathbf{k}$ in Example 1.

We use the Greek letter τ ("tau") as the variable of integration in Eq. (3) because the letter t is already in use as the upper limit.

Just as for plane curves, we can calculate the length of a curve in space from any convenient parametrization that meets the stated conditions. Again, we omit the proof.

The square root in Eq. (1) is $|\mathbf{v}|$, the length of the velocity vector $d\mathbf{r}/dt$. This enables us to write the formula for length a shorter way.

Length Formula (Short Form)

$$L = \int_a^b |\mathbf{v}|\, dt \qquad (2)$$

EXAMPLE 1   Find the length of one turn of the helix

$$\mathbf{r}(t) = (\cos t)\,\mathbf{i} + (\sin t)\,\mathbf{j} + t\,\mathbf{k}.$$

Solution   The helix makes one full turn as t runs from 0 to 2π (Fig. 11.16). The length of this portion of the curve is

$$L = \int_a^b |\mathbf{v}|\, dt = \int_0^{2\pi} \sqrt{(-\sin t)^2 + (\cos t)^2 + (1)^2}\, dt$$

$$= \int_0^{2\pi} \sqrt{2}\, dt = 2\pi\sqrt{2}.$$

This is $\sqrt{2}$ times the length of the circle in the xy-plane over which the helix stands. ❑

If we choose a base point $P(t_0)$ on a smooth curve C parametrized by t, each value of t determines a point $P(t) = (x(t), y(t), z(t))$ on C and a "directed distance"

$$s(t) = \int_{t_0}^t |\mathbf{v}(\tau)|\, d\tau, \qquad (3)$$

measured along C from the base point (Fig. 11.17). If $t > t_0$, $s(t)$ is the distance from $P(t_0)$ to $P(t)$. If $t < t_0$, $s(t)$ is the negative of the distance. Each value of s determines a point on C and this parametrizes C with respect to s. We call s an **arc length parameter** for the curve. The parameter's value increases in the direction of increasing t.

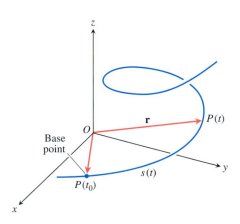

11.17 The directed distance along the curve from $P(t_0)$ to any point $P(t)$ is

$$s(t) = \int_{t_0}^t |\mathbf{v}(\tau)|\, d\tau.$$

Arc Length Parameter with Base Point $P(t_0)$

$$s(t) = \int_{t_0}^t \sqrt{[x'(\tau)]^2 + [y'(\tau)]^2 + [z'(\tau)]^2}\, d\tau = \int_{t_0}^t |\mathbf{v}(\tau)|\, d\tau \qquad (4)$$

EXAMPLE 2   If $t_0 = 0$, the arc length parameter along the helix

$$\mathbf{r}(t) = (\cos t)\,\mathbf{i} + (\sin t)\,\mathbf{j} + t\,\mathbf{k}$$

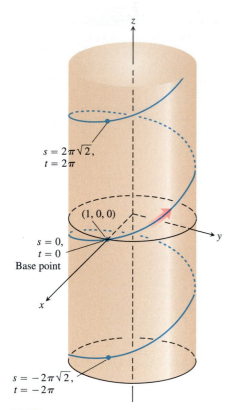

$s = 2\pi\sqrt{2},$
$t = 2\pi$

$(1, 0, 0)$

$s = 0,$
$t = 0$
Base point

$s = -2\pi\sqrt{2},$
$t = -2\pi$

11.18 Arc length parameter values on the helix $\mathbf{r}(t) = (\cos t)\,\mathbf{i} + (\sin t)\,\mathbf{j} + t\,\mathbf{k}$ (Example 2).

from t_0 to t is

$$s(t) = \int_{t_0}^{t} |\mathbf{v}(\tau)|\, d\tau \qquad \text{Eq. (4)}$$

$$= \int_{0}^{t} \sqrt{2}\, d\tau \qquad \text{Value from Example 1}$$

$$= \sqrt{2}\, t.$$

Thus, $s(2\pi) = 2\pi\sqrt{2},\, s(-2\pi) = -2\pi\sqrt{2}$, and so on (Fig. 11.18). ❑

EXAMPLE 3 *Distance Along a Line.*

Show that if $\mathbf{u} = u_1\,\mathbf{i} + u_2\,\mathbf{j} + u_3\,\mathbf{k}$ is a unit vector, then the directed distance along the line

$$\mathbf{r}(t) = (x_0 + tu_1)\,\mathbf{i} + (y_0 + tu_2)\,\mathbf{j} + (z_0 + tu_3)\,\mathbf{k}$$

from the point $P_0(x_0, y_0, z_0)$ where $t = 0$ is t itself.

Solution

$$\mathbf{v} = \frac{d}{dt}(x_0 + tu_1)\,\mathbf{i} + \frac{d}{dt}(y_0 + tu_2)\,\mathbf{j} + \frac{d}{dt}(z_0 + tu_3)\,\mathbf{k} = u_1\,\mathbf{i} + u_2\,\mathbf{j} + u_3\,\mathbf{k} = \mathbf{u},$$

so

$$s(t) = \int_{0}^{t} |\mathbf{v}|\, d\tau = \int_{0}^{t} |\mathbf{u}|\, d\tau = \int_{0}^{t} 1\, d\tau = t.$$ ❑

Speed on a Smooth Curve

Since the derivatives beneath the radical in Eq. (4) are continuous (the curve is smooth), the Fundamental Theorem of Calculus tells us that s is a differentiable function of t with derivative

$$\frac{ds}{dt} = |\mathbf{v}(t)|. \qquad (5)$$

As we expect, the speed with which the particle moves along its path is the magnitude of $\mathbf{v}$.

Notice that while the base point $P(t_0)$ plays a role in defining s in Eq. (4), it plays no role in Eq. (5). The rate at which a moving particle covers distance along its path has nothing to do with how far away the base point is.

Notice also that $ds/dt > 0$ since, by definition, $|\mathbf{v}|$ is never zero for a smooth curve. We see once again that s is an increasing function of t.

The Unit Tangent Vector **T**

Since $ds/dt > 0$ for the curves we are considering, s is one-to-one and has an inverse that gives t as a differentiable function of s (Section 6.1). The derivative of the inverse is

$$\frac{dt}{ds} = \frac{1}{ds/dt} = \frac{1}{|\mathbf{v}|}. \qquad (6)$$

This makes $\mathbf{r}$ a differentiable function of s whose derivative can be calculated with

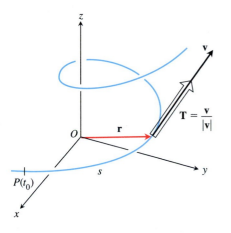

11.19 We find the unit tangent vector **T** by dividing **v** by |**v**|.

the Chain Rule to be

$$\frac{d\mathbf{r}}{ds} = \frac{d\mathbf{r}}{dt}\frac{dt}{ds} = \mathbf{v}\frac{1}{|\mathbf{v}|} = \frac{\mathbf{v}}{|\mathbf{v}|}. \tag{7}$$

Equation (7) says that $d\mathbf{r}/ds$ is a unit vector in the direction of **v**. We call $d\mathbf{r}/ds$ the unit tangent vector of the curve traced by **r** and denote it by **T** (Fig. 11.19).

> **Definition**
>
> The **unit tangent vector** of a differentiable curve $\mathbf{r}(t)$ is
>
> $$\mathbf{T} = \frac{d\mathbf{r}}{ds} = \frac{d\mathbf{r}/dt}{ds/dt} = \frac{\mathbf{v}}{|\mathbf{v}|}. \tag{8}$$

The unit tangent vector **T** is a differentiable function of t whenever **v** is a differentiable function of t. As we will see in the next section, **T** is one of three unit vectors in a traveling reference frame that is used to describe the motion of space vehicles and other bodies moving in three dimensions.

EXAMPLE 4   Find the unit tangent vector of the helix

$$\mathbf{r}(t) = (\cos t)\,\mathbf{i} + (\sin t)\,\mathbf{j} + t\,\mathbf{k}.$$

Solution

$$\mathbf{v} = (-\sin t)\,\mathbf{i} + (\cos t)\,\mathbf{j} + \mathbf{k}$$

$$|\mathbf{v}| = \sqrt{(-\sin t)^2 + (\cos t)^2 + (1)^2} = \sqrt{2}$$

$$\mathbf{T} = \frac{\mathbf{v}}{|\mathbf{v}|} = -\frac{\sin t}{\sqrt{2}}\,\mathbf{i} + \frac{\cos t}{\sqrt{2}}\,\mathbf{j} + \frac{1}{\sqrt{2}}\,\mathbf{k}$$

EXAMPLE 5   *The involute of a circle (Fig. 11.20)*

Find the unit tangent vector of the curve

$$\mathbf{r}(t) = (\cos t + t \sin t)\,\mathbf{i} + (\sin t - t \cos t)\,\mathbf{j}, \qquad t > 0.$$

Solution

$$\mathbf{v} = \frac{d\mathbf{r}}{dt} = (-\sin t + \sin t + t \cos t)\,\mathbf{i} + (\cos t - \cos t + t \sin t)\,\mathbf{j}$$

$$= (t \cos t)\,\mathbf{i} + (t \sin t)\,\mathbf{j}$$

$$|\mathbf{v}| = \sqrt{t^2 \cos^2 t + t^2 \sin^2 t} = \sqrt{t^2} = |t| = t \qquad {\scriptstyle |t| = t \text{ because } t > 0}$$

$$\mathbf{T} = \frac{\mathbf{v}}{|\mathbf{v}|} = \frac{\mathbf{v}}{t} = (\cos t)\,\mathbf{i} + (\sin t)\,\mathbf{j}$$

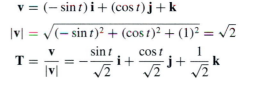

11.20 The *involute* of a circle is the path traced by the endpoint P of a string unwinding from a circle, here the unit circle in the xy-plane.

EXAMPLE 6   For the counterclockwise motion

$$\mathbf{r}(t) = (\cos t)\,\mathbf{i} + (\sin t)\,\mathbf{j}$$

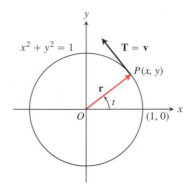

11.21 The motion $\mathbf{r}(t) = (\cos t)\,\mathbf{i} + (\sin t)\,\mathbf{j}$ (Example 6).

around the unit circle,

$$\mathbf{v} = (-\sin t)\,\mathbf{i} + (\cos t)\,\mathbf{j}$$

is already a unit vector, so $\mathbf{T} = \mathbf{v}$ (Fig. 11.21).   ❑

Exercises 11.3

In Exercises 1–8, find the curve's unit tangent vector. Also, find the length of the indicated portion of the curve.

1. $\mathbf{r}(t) = (2\cos t)\,\mathbf{i} + (2\sin t)\,\mathbf{j} + \sqrt{5}\,t\,\mathbf{k}, \quad 0 \le t \le \pi$

2. $\mathbf{r}(t) = (6\sin 2t)\,\mathbf{i} + (6\cos 2t)\,\mathbf{j} + 5t\,\mathbf{k}, \quad 0 \le t \le \pi$

3. $\mathbf{r}(t) = t\,\mathbf{i} + (2/3)t^{3/2}\,\mathbf{k}, \quad 0 \le t \le 8$

4. $\mathbf{r}(t) = (2+t)\,\mathbf{i} - (t+1)\,\mathbf{j} + t\,\mathbf{k}, \quad 0 \le t \le 3$

5. $\mathbf{r}(t) = (\cos^3 t)\,\mathbf{j} + (\sin^3 t)\,\mathbf{k}, \quad 0 \le t \le \pi/2$

6. $\mathbf{r}(t) = 6t^3\,\mathbf{i} - 2t^3\,\mathbf{j} - 3t^3\,\mathbf{k}, \quad 1 \le t \le 2$

7. $\mathbf{r}(t) = (t\cos t)\,\mathbf{i} + (t\sin t)\,\mathbf{j} + (2\sqrt{2}/3)t^{3/2}\,\mathbf{k}, \quad 0 \le t \le \pi$

8. $\mathbf{r}(t) = (t\sin t + \cos t)\,\mathbf{i} + (t\cos t - \sin t)\,\mathbf{j}, \quad \sqrt{2} \le t \le 2$

9. Find the point on the curve

$$\mathbf{r}(t) = (5\sin t)\,\mathbf{i} + (5\cos t)\,\mathbf{j} + 12t\,\mathbf{k}$$

at a distance 26π units along the curve from the origin in the direction of increasing arc length.

10. Find the point on the curve

$$\mathbf{r}(t) = (12\sin t)\,\mathbf{i} - (12\cos t)\,\mathbf{j} + 5t\,\mathbf{k}$$

at a distance 13π units along the curve from the origin in the direction opposite to the direction of increasing arc length.

In Exercises 11–14, find the arc length parameter along the curve from the point where $t = 0$ by evaluating the integral

$$s = \int_0^t |\mathbf{v}(\tau)|\,d\tau$$

from Eq. (3). Then find the length of the indicated portion of the curve.

11. $\mathbf{r}(t) = (4\cos t)\,\mathbf{i} + (4\sin t)\,\mathbf{j} + 3t\,\mathbf{k}, \quad 0 \le t \le \pi/2$

12. $\mathbf{r}(t) = (\cos t + t\sin t)\,\mathbf{i} + (\sin t - t\cos t)\,\mathbf{j}, \quad \pi/2 \le t \le \pi$

13. $\mathbf{r}(t) = (e^t\cos t)\,\mathbf{i} + (e^t\sin t)\,\mathbf{j} + e^t\,\mathbf{k}, \quad -\ln 4 \le t \le 0$

14. $\mathbf{r}(t) = (1+2t)\,\mathbf{i} + (1+3t)\,\mathbf{j} + (6-6t)\,\mathbf{k}, \quad -1 \le t \le 0$

15. Find the length of the curve

$$\mathbf{r}(t) = (\sqrt{2}\,t)\,\mathbf{i} + (\sqrt{2}\,t)\,\mathbf{j} + (1 - t^2)\,\mathbf{k}$$

from $(0, 0, 1)$ to $(\sqrt{2}, \sqrt{2}, 0)$.

16. The length $2\pi\sqrt{2}$ of the turn of the helix in Example 1 is also the length of the diagonal of a square 2π units on a side. Show how to obtain this square by cutting away and flattening a portion of the cylinder around which the helix winds.

17. a) Show that the curve $\mathbf{r}(t) = (\cos t)\,\mathbf{i} + (\sin t)\,\mathbf{j} + (1 - \cos t)\,\mathbf{k}$, $0 \le t \le 2\pi$, is an ellipse by showing that it is the intersection of a right circular cylinder and a plane. Find equations for the cylinder and plane.

   b) Sketch the ellipse on the cylinder. Add to your sketch the unit tangent vectors at $t = 0, \pi/2, \pi$, and $3\pi/2$.

   c) Show that the acceleration vector always lies parallel to the plane (orthogonal to a vector normal to the plane). Thus, if you draw the acceleration as a vector attached to the ellipse, it will lie in the plane of the ellipse. Add the acceleration vectors for $t = 0, \pi/2, \pi$, and $3\pi/2$ to your sketch.

   d) Write an integral for the length of the ellipse. Do not try to evaluate the integral—it is nonelementary.

e) NUMERICAL INTEGRATOR Estimate the length of the ellipse to two decimal places.

18. *Length is independent of parametrization.* To illustrate the fact that the length of a smooth space curve does not depend on the parametrization you use to compute it, calculate the length of one turn of the helix in Example 1 with the following parametrizations.

a) $\mathbf{r}(t) = (\cos 4t)\,\mathbf{i} + (\sin 4t)\,\mathbf{j} + 4t\,\mathbf{k}, \quad 0 \le t \le \pi/2$

b) $\mathbf{r}(t) = [\cos (t/2)]\,\mathbf{i} + [\sin (t/2)]\,\mathbf{j} + (t/2)\,\mathbf{k}, \quad 0 \le t \le 4\pi$

c) $\mathbf{r}(t) = (\cos t)\,\mathbf{i} - (\sin t)\,\mathbf{j} - t\,\mathbf{k}, \quad -2\pi \le t \le 0$

11.4 Curvature, Torsion, and the **TNB** Frame

In this section we define a frame of mutually orthogonal unit vectors that always travels with a body moving along a curve in space (Fig. 11.22). The frame has three vectors. The first is **T**, the unit tangent vector. The second is **N**, the unit vector that gives the direction of $d\mathbf{T}/ds$. The third is $\mathbf{B} = \mathbf{T} \times \mathbf{N}$. These vectors and their derivatives, when available, give useful information about a vehicle's orientation in space and about how the vehicle's path turns and twists.

For example, $|d\mathbf{T}/ds|$ tells how much a vehicle's path turns to the left or right as it moves along; it is called the *curvature* of the vehicle's path. The number $-(d\mathbf{B}/ds) \cdot \mathbf{N}$ tells how much a vehicle's path rotates or twists out of its plane of motion as the vehicle moves along; it is called the *torsion* of the vehicle's path. Look at Fig. 11.22 again. If P is a train climbing up a curved track, the rate at which the headlight turns from side to side per unit distance is the curvature of the track. The rate at which the engine tends to twist out of the plane formed by **T** and **N** is the torsion.

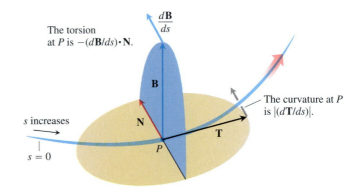

The torsion at P is $-(d\mathbf{B}/ds) \cdot \mathbf{N}$.

The curvature at P is $|(d\mathbf{T}/ds)|$.

11.22 Every moving body travels with a **TNB** frame that characterizes the geometry of its path of motion.

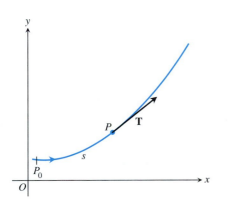

11.23 As P moves along the curve in the direction of increasing arc length, the unit tangent vector turns. The value of $|d\mathbf{T}/ds|$ at P is called the *curvature* of the curve at P.

The Curvature of a Plane Curve

As a particle moves along a smooth curve in the plane, $\mathbf{T} = d\mathbf{r}/ds$ turns as the curve bends. Since **T** is a unit vector, its length remains constant and only its direction changes as the particle moves along the curve. The rate at which **T** turns per unit of length along the curve is called the *curvature* (Fig. 11.23). The traditional symbol for the curvature function is the Greek letter κ ("kappa").

Definition

If **T** is the unit tangent vector of a smooth curve, the **curvature** function of the curve is

$$\kappa = \left|\frac{d\mathbf{T}}{ds}\right|.$$

If $|d\mathbf{T}/ds|$ is large, **T** turns sharply as the particle passes through P, and the curvature at P is large. If $|d\mathbf{T}/ds|$ is close to zero, **T** turns more slowly and the curvature at P is smaller. Testing the definition, we see in Examples 1 and 2 that the curvature is constant for straight lines and circles.

EXAMPLE 1   *The curvature of a straight line is zero*

On a straight line, the unit tangent vector **T** always points in the same direction, so its components are constants. Therefore $|d\mathbf{T}/ds| = |\mathbf{0}| = 0$ (Fig. 11.24).   ❑

EXAMPLE 2   *The curvature of a circle of radius a is 1/a*

To see why (Fig. 11.25), start with the parametrization

$$\mathbf{r}(\theta) = (a\cos\theta)\mathbf{i} + (a\sin\theta)\mathbf{j}$$

and substitute $\theta = s/a$ to parametrize in terms of arc length s.

$$\mathbf{r} = \left(a\cos\frac{s}{a}\right)\mathbf{i} + \left(a\sin\frac{s}{a}\right)\mathbf{j}.$$

Then

$$\mathbf{T} = \frac{d\mathbf{r}}{ds} = \left(-\sin\frac{s}{a}\right)\mathbf{i} + \left(\cos\frac{s}{a}\right)\mathbf{j}$$

and

$$\frac{d\mathbf{T}}{ds} = \left(-\frac{1}{a}\cos\frac{s}{a}\right)\mathbf{i} - \left(\frac{1}{a}\sin\frac{s}{a}\right)\mathbf{j}.$$

Hence, for any value of s,

$$\kappa = \left|\frac{d\mathbf{T}}{ds}\right|$$

$$= \sqrt{\frac{1}{a^2}\cos^2\left(\frac{s}{a}\right) + \frac{1}{a^2}\sin^2\left(\frac{s}{a}\right)}$$

$$= \frac{1}{\sqrt{a^2}} = \frac{1}{|a|} = \frac{1}{a}. \quad \text{Since } a > 0, \ |a| = a.$$ ❑

11.24 Along a straight line, **T** always points in the same direction. The curvature, $|d\mathbf{T}/ds|$, is zero (Example 1).

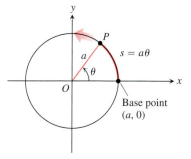

11.25 The point P has coordinates $(a\cos\theta, a\sin\theta) = (a\cos(s/a), \ a\sin(s/a))$ (Example 2).

The Principal Unit Normal Vector for Plane Curves

Since **T** has constant length, the vector $d\mathbf{T}/ds$ is orthogonal to **T** (Section 11.1). Therefore, if we divide $d\mathbf{T}/ds$ by the length κ, we obtain a *unit* vector orthogonal to **T** (Fig. 11.26).

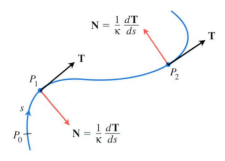

11.26 The vector $d\mathbf{T}/ds$, normal to the curve, always points in the direction in which $\mathbf{T}$ is turning. The vector $\mathbf{N}$ is the direction of $d\mathbf{T}/ds$.

Definition

At a point where $\kappa \neq 0$, the **principal unit normal** vector for a curve in the plane is

$$\mathbf{N} = \frac{1}{\kappa}\frac{d\mathbf{T}}{ds}.$$

The vector $d\mathbf{T}/ds$ points in the direction in which $\mathbf{T}$ turns as the curve bends. Therefore, if we face in the direction of increasing arc length, the vector $d\mathbf{T}/ds$ points toward the right if $\mathbf{T}$ turns clockwise and toward the left if $\mathbf{T}$ turns counterclockwise. In other words, the principal normal vector $\mathbf{N}$ will point toward the concave side of the curve (Fig. 11.26). Exercise 10 illustrates what happens when $\kappa = 0$ at a point.

Because the arc length parameter for a smooth curve $\mathbf{r}(t) = f(t)\mathbf{i} + g(t)\mathbf{j}$ is defined with ds/dt positive, $ds/dt = |ds/dt|$, and the Chain Rule gives

$$\mathbf{N} = \frac{d\mathbf{T}/ds}{|d\mathbf{T}/ds|}$$

$$= \frac{(d\mathbf{T}/dt)(dt/ds)}{|d\mathbf{T}/dt||dt/ds|}$$

$$= \frac{d\mathbf{T}/dt}{|d\mathbf{T}/dt|}. \tag{1}$$

This formula enables us to find $\mathbf{N}$ without having to find κ and s first.

EXAMPLE 3   Find $\mathbf{T}$ and $\mathbf{N}$ for the circular motion

$$\mathbf{r}(t) = (\cos 2t)\mathbf{i} + (\sin 2t)\mathbf{j}.$$

Solution   We first find $\mathbf{T}$:

$$\mathbf{v} = -(2\sin 2t)\mathbf{i} + (2\cos 2t)\mathbf{j},$$

$$|\mathbf{v}| = \sqrt{4\sin^2 2t + 4\cos^2 2t} = 2,$$

$$\mathbf{T} = \frac{\mathbf{v}}{|\mathbf{v}|}$$

$$= -(\sin 2t)\mathbf{i} + (\cos 2t)\mathbf{j}.$$

From this we find

$$\frac{d\mathbf{T}}{dt} = -(2\cos 2t)\mathbf{i} - (2\sin 2t)\mathbf{j},$$

$$\left|\frac{d\mathbf{T}}{dt}\right| = \sqrt{4\cos^2 2t + 4\sin^2 2t} = 2,$$

and

$$\mathbf{N} = \frac{d\mathbf{T}/dt}{|d\mathbf{T}/dt|}$$

$$= -(\cos 2t)\mathbf{i} - (\sin 2t)\mathbf{j}. \qquad \text{Eq. (1)} \qquad \square$$

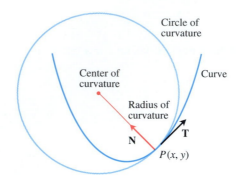

11.27 The osculating circle at $P(x, y)$ lies toward the inner side of the curve.

Circle of Curvature and Radius of Curvature

The **circle of curvature** or **osculating circle** at a point P on a plane curve where $\kappa \neq 0$ is the circle in the plane of the curve that

1. is tangent to the curve at P (has the same tangent line the curve has);
2. has the same curvature the curve has at P; and
3. lies toward the concave or inner side of the curve (as in Fig. 11.27).

The **radius of curvature** of the curve at P is the radius of the circle of curvature, which, according to Example 2, is

$$\text{Radius of curvature} = \rho = \frac{1}{\kappa}. \tag{2}$$

To find ρ, we find κ and take the reciprocal. The **center of curvature** of the curve at P is the center of the circle of curvature.

Curvature and Normal Vectors for Space Curves

Just as it does for a curve in the plane, the arc length parameter s gives the unit tangent vector $\mathbf{T} = d\mathbf{r}/ds$ for a smooth curve in space. We again define the curvature to be

$$\kappa = \left| \frac{d\mathbf{T}}{ds} \right|. \tag{3}$$

The vector $d\mathbf{T}/ds$ is orthogonal to $\mathbf{T}$ and we define the principal unit normal to be

$$\mathbf{N} = \frac{1}{\kappa} \frac{d\mathbf{T}}{ds} = \frac{d\mathbf{T}/dt}{|d\mathbf{T}/dt|}. \tag{4}$$

EXAMPLE 4   Find the curvature for the helix (Fig. 11.28)

$$\mathbf{r}(t) = (a \cos t)\mathbf{i} + (a \sin t)\mathbf{j} + bt\,\mathbf{k}, \qquad a, b \geq 0, \ a^2 + b^2 \neq 0.$$

Solution   We calculate $\mathbf{T}$ from the velocity vector $\mathbf{v}$:

$$\mathbf{v} = -(a \sin t)\mathbf{i} + (a \cos t)\mathbf{j} + b\,\mathbf{k}$$

$$|\mathbf{v}| = \sqrt{a^2 \sin^2 t + a^2 \cos^2 t + b^2} = \sqrt{a^2 + b^2}$$

$$\mathbf{T} = \frac{\mathbf{v}}{|\mathbf{v}|} = \frac{1}{\sqrt{a^2 + b^2}} [-(a \sin t)\mathbf{i} + (a \cos t)\mathbf{j} + b\,\mathbf{k}].$$

Then, using the Chain Rule, we find $d\mathbf{T}/ds$ as

$$\frac{d\mathbf{T}}{ds} = \frac{d\mathbf{T}}{dt} \frac{dt}{ds} \qquad \textcolor{blue}{\text{Chain Rule}}$$

$$= \frac{d\mathbf{T}}{dt} \cdot \frac{1}{|\mathbf{v}|} \qquad \textcolor{blue}{\frac{ds}{dt} = |\mathbf{v}|, \ \text{so} \ \frac{dt}{ds} = \frac{1}{|\mathbf{v}|}}$$

$$= \frac{1}{\sqrt{a^2 + b^2}} [-(a \cos t)\mathbf{i} - (a \sin t)\mathbf{j}] \cdot \left(\frac{1}{\sqrt{a^2 + b^2}} \right)$$

$$= \frac{a}{a^2 + b^2} [-(\cos t)\mathbf{i} - (\sin t)\mathbf{j}].$$

11.28 The helix $\mathbf{r}(t) = (a \cos t)\mathbf{i} + (a \sin t)\mathbf{j} + bt\,\mathbf{k}$, drawn with a and b positive and $t \geq 0$ (Example 4).

Therefore,

$$\kappa = \left| \frac{d\mathbf{T}}{ds} \right|$$

$$= \frac{a}{a^2 + b^2} \left| -(\cos t)\mathbf{i} - (\sin t)\mathbf{j} \right|$$

$$= \frac{a}{a^2 + b^2} \sqrt{(\cos t)^2 + (\sin t)^2} = \frac{a}{a^2 + b^2}. \tag{5}$$

From Eq. (5) we see that increasing b for a fixed a decreases the curvature. Decreasing a for a fixed b eventually decreases the curvature as well. Stretching a spring tends to straighten it.

If $b = 0$, the helix reduces to a circle of radius a and its curvature reduces to $1/a$, as it should. If $a = 0$, the helix becomes the z-axis, and its curvature reduces to 0, again as it should. ☐

EXAMPLE 5   Find **N** for the helix in Example 4.

Solution   We have

$$\frac{d\mathbf{T}}{dt} = -\frac{1}{\sqrt{a^2 + b^2}}[(a \cos t)\mathbf{i} + (a \sin t)\mathbf{j}] \qquad \text{Example 4}$$

$$\left| \frac{d\mathbf{T}}{dt} \right| = \frac{1}{\sqrt{a^2 + b^2}} \sqrt{a^2 \cos^2 t + a^2 \sin^2 t} = \frac{a}{\sqrt{a^2 + b^2}}$$

$$\mathbf{N} = \frac{d\mathbf{T}/dt}{|d\mathbf{T}/dt|} \qquad \text{Eq. (4)}$$

$$= -\frac{\sqrt{a^2 + b^2}}{a} \cdot \frac{1}{\sqrt{a^2 + b^2}}[(a \cos t)\mathbf{i} + (a \sin t)\mathbf{j}]$$

$$= -(\cos t)\mathbf{i} - (\sin t)\mathbf{j}. \qquad \text{☐}$$

Torsion and the Binormal Vector

The **binormal vector** of a curve in space is $\mathbf{B} = \mathbf{T} \times \mathbf{N}$, a unit vector orthogonal to both **T** and **N** (Fig. 11.29). Together **T, N,** and **B** define a moving right-handed vector frame that plays a significant role in calculating the flight paths of space vehicles.

How does $d\mathbf{B}/ds$ behave in relation to **T, N,** and **B**? From the rule for differentiating a cross product, we have

$$\frac{d\mathbf{B}}{ds} = \frac{d\mathbf{T}}{ds} \times \mathbf{N} + \mathbf{T} \times \frac{d\mathbf{N}}{ds}.$$

Since **N** is the direction of $d\mathbf{T}/ds$, $(d\mathbf{T}/ds) \times \mathbf{N} = \mathbf{0}$ and

$$\frac{d\mathbf{B}}{ds} = \mathbf{0} + \mathbf{T} \times \frac{d\mathbf{N}}{ds} = \mathbf{T} \times \frac{d\mathbf{N}}{ds}. \tag{6}$$

From this we see that $d\mathbf{B}/ds$ is orthogonal to **T** since a cross product is orthogonal to its factors.

Since $d\mathbf{B}/ds$ is also orthogonal to **B** (the latter has constant length), it follows that $d\mathbf{B}/ds$ is orthogonal to the plane of **B** and **T**. In other words, $d\mathbf{B}/ds$ is parallel

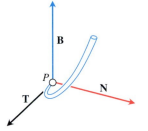

11.29 The vectors **T, N,** and **B** (in that order) make a right-handed frame of mutually orthogonal unit vectors in space. You can call it the **Frenet** ("fre-*nay*") **frame** (after Jean-Frédéric Frenet, 1816–1900), or you can call it the **TNB frame.**

to **N,** so $d\mathbf{B}/ds$ is a scalar multiple of **N.** In symbols,

$$\frac{d\mathbf{B}}{ds} = -\tau\,\mathbf{N}.$$

The minus sign in this equation is traditional. The scalar τ is called the torsion along the curve. Notice that

$$\frac{d\mathbf{B}}{ds} \cdot \mathbf{N} = -\tau\,\mathbf{N} \cdot \mathbf{N} = -\tau\,(1) = -\tau,$$

so that

$$\tau = -\frac{d\mathbf{B}}{ds} \cdot \mathbf{N}.$$

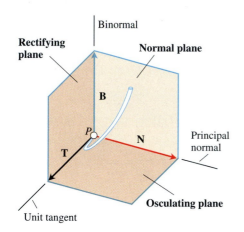

Binormal

Rectifying plane

Normal plane

B

P

N

Principal normal

T

Osculating plane

Unit tangent

11.30 The names of the three planes determined by **T, N,** and **B.**

> **Definition**
>
> Let $\mathbf{B} = \mathbf{T} \times \mathbf{N}$. The **torsion** function of a smooth curve is
>
> $$\tau = -\frac{d\mathbf{B}}{ds} \cdot \mathbf{N}.$$

Unlike the curvature κ, which is never negative, the torsion τ may be positive, negative, or zero.

The three planes determined by **T, N,** and **B** are shown in Fig. 11.30. The curvature $\kappa = |d\mathbf{T}/ds|$ can be thought of as the rate at which the normal plane turns as the point P moves along the curve. Similarly, the torsion $\tau = -(d\mathbf{B}/ds) \cdot \mathbf{N}$ is the rate at which the osculating plane turns about **T** as P moves along the curve. Torsion measures how the curve twists.

The Tangential and Normal Components of Acceleration

When a body is accelerated by gravity, brakes, a combination of rocket motors, or whatever, we usually want to know how much of the acceleration acts to move the body straight ahead in the direction of motion, in the tangential direction **T.** We can find out if we use the Chain Rule to rewrite **v** as

$$\mathbf{v} = \frac{d\mathbf{r}}{dt} = \frac{d\mathbf{r}}{ds}\frac{ds}{dt} = \mathbf{T}\frac{ds}{dt}$$

and differentiate both ends of this string of equalities to get

$$\mathbf{a} = \frac{d\mathbf{v}}{dt} = \frac{d}{dt}\left(\mathbf{T}\frac{ds}{dt}\right) = \frac{d^2s}{dt^2}\mathbf{T} + \frac{ds}{dt}\frac{d\mathbf{T}}{dt}$$

$$= \frac{d^2s}{dt^2}\mathbf{T} + \frac{ds}{dt}\left(\frac{d\mathbf{T}}{ds}\frac{ds}{dt}\right) = \frac{d^2s}{dt^2}\mathbf{T} + \frac{ds}{dt}\left(\kappa\,\mathbf{N}\frac{ds}{dt}\right)$$

$$= \frac{d^2s}{dt^2}\mathbf{T} + \kappa\left(\frac{ds}{dt}\right)^2\mathbf{N}.$$

$$\mathbf{a} = a_{\mathrm{T}}\mathbf{T} + a_{\mathrm{N}}\mathbf{N}, \tag{7}$$

where

$$a_{\mathrm{T}} = \frac{d^2 s}{dt^2} = \frac{d}{dt}|\mathbf{v}| \quad \text{and} \quad a_{\mathrm{N}} = \kappa \left(\frac{ds}{dt}\right)^2 = \kappa |\mathbf{v}|^2 \tag{8}$$

are the **tangential** and **normal** scalar components of acceleration.

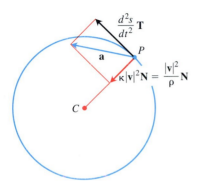

11.31 The tangential and normal components of acceleration. The acceleration **a** always lies in the plane of **T** and **N**, orthogonal to **B**.

11.32 The tangential and normal components of the acceleration of a body that is speeding up as it moves counterclockwise around a circle of radius ρ.

Equation (7) is remarkable in that **B** does not appear. No matter how the path of the moving body we are watching may appear to twist and turn in space, the acceleration **a** *always lies in the plane of* **T** and **N** orthogonal to **B**. The equation also tells us exactly how much of the acceleration takes place tangent to the motion $(d^2 s/dt^2)$ and how much takes place normal to the motion $[\kappa(ds/dt)^2]$ (Fig. 11.31).

What information can we read from Eqs. (8)? By definition, acceleration **a** is the rate of change of velocity **v,** and in general both the length and direction of **v** change as a body moves along its path. The tangential component of acceleration a_{T} measures the rate of change of the *length* of **v** (that is, the change in the speed). The normal component of acceleration a_{N} measures the rate of change of the *direction* of **v**.

Notice that the normal scalar component of the acceleration is the curvature times the *square* of the speed. This explains why you have to hold on when your car makes a sharp (large κ), high-speed (large $|\mathbf{v}|$) turn. If you double the speed of your car, you will experience four times the normal component of acceleration for the same curvature.

If a body moves in a circle at a constant speed, $d^2 s/dt^2$ is zero and all the acceleration points along **N** toward the circle's center. If the body is speeding up or slowing down, **a** has a nonzero tangential component (Fig. 11.32).

To calculate a_{N} we usually use the formula $a_{\mathrm{N}} = \sqrt{|\mathbf{a}|^2 - a_{\mathrm{T}}^2}$, which comes from solving the equation $|\mathbf{a}|^2 = \mathbf{a} \cdot \mathbf{a} = a_{\mathrm{T}}^2 + a_{\mathrm{N}}^2$ for a_{N}. With this formula we can find a_{N} without having to calculate κ first.

$$a_{\mathrm{N}} = \sqrt{|\mathbf{a}|^2 - a_{\mathrm{T}}^2} \tag{9}$$

EXAMPLE 6   Without finding **T** and **N,** write the acceleration of the motion

$$\mathbf{r}(t) = (\cos t + t \sin t)\,\mathbf{i} + (\sin t - t \cos t)\,\mathbf{j}, \quad t > 0$$

in the form $\mathbf{a} = a_{\mathrm{T}}\mathbf{T} + a_{\mathrm{N}}\mathbf{N}$. (The path of the motion is the involute of the circle in Fig. 11.33, on the following page.)

Solution   We use the first of Eqs. (8) to find a_{T}:

$$\mathbf{v} = (t \cos t)\,\mathbf{i} + (t \sin t)\,\mathbf{j}$$

$$|\mathbf{v}| = \sqrt{t^2 \cos^2 t + t^2 \sin^2 t} = \sqrt{t^2} = |t| = t \qquad t > 0$$

$$a_{\mathrm{T}} = \frac{d}{dt}|\mathbf{v}| = \frac{d}{dt}(t) = 1. \qquad \text{Eq. (8)}$$

Value from Section 11.3, Example 5

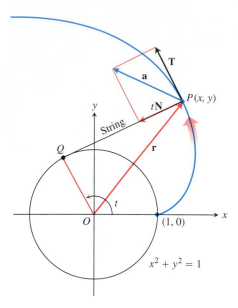

11.33 The tangential and normal components of the acceleration of the motion $\mathbf{r}(t) = (\cos t + t \sin t)\mathbf{i} + (\sin t - t \cos t)\mathbf{j}$, for $t > 0$ (Example 6).

Knowing a_T, we use Eq. (9) to find a_N:

$$\mathbf{a} = (\cos t - t \sin t)\mathbf{i} + (\sin t + t \cos t)\mathbf{j}$$

$$|\mathbf{a}|^2 = t^2 + 1 \qquad \text{After some algebra}$$

$$a_N = \sqrt{|\mathbf{a}|^2 - a_T{}^2}$$

$$= \sqrt{(t^2 + 1) - (1)} = \sqrt{t^2} = t.$$

We then use Eq. (7) to find $\mathbf{a}$:

$$\mathbf{a} = a_T\,\mathbf{T} + a_N\,\mathbf{N} = (1)\,\mathbf{T} + (t)\,\mathbf{N} = \mathbf{T} + t\,\mathbf{N}.$$

See Fig. 11.33.   ❏

Formulas for Computing Curvature and Torsion

We now give some easy-to-use formulas for computing the curvature and torsion of a smooth curve. From Eq. (7), we have

$$\mathbf{v} \times \mathbf{a} = \left(\frac{ds}{dt}\,\mathbf{T}\right) \times \left[\frac{d^2s}{dt^2}\,\mathbf{T} + \kappa\left(\frac{ds}{dt}\right)^2 \mathbf{N}\right] \qquad \begin{array}{l}\text{From Section}\\ \text{11.3, Eq. (8),}\\ \mathbf{v} = d\mathbf{r}/dt = \\ (ds/dt)\,\mathbf{T}\end{array}$$

$$= \left(\frac{ds}{dt}\frac{d^2s}{dt^2}\right)(\mathbf{T} \times \mathbf{T}) + \kappa\left(\frac{ds}{dt}\right)^3 (\mathbf{T} \times \mathbf{N})$$

$$= \kappa\left(\frac{ds}{dt}\right)^3 \mathbf{B}. \qquad \begin{array}{l}\mathbf{T} \times \mathbf{T} = \mathbf{0} \text{ and}\\ \mathbf{T} \times \mathbf{N} = \mathbf{B}\end{array}$$

It follows that

$$|\mathbf{v} \times \mathbf{a}| = \kappa\left|\frac{ds}{dt}\right|^3 |\mathbf{B}| = \kappa\,|\mathbf{v}|^3. \qquad \frac{ds}{dt} = |\mathbf{v}| \text{ and } |\mathbf{B}| = 1$$

Solving for κ gives the following formula.

> **A Vector Formula for Curvature**
>
> $$\kappa = \frac{|\mathbf{v} \times \mathbf{a}|}{|\mathbf{v}|^3} \qquad\qquad (10)$$

Equation (10) calculates the curvature, a geometric property of the curve, from the velocity and acceleration of any vector representation of the curve in which $|\mathbf{v}|$ is different from zero. Take a moment to think about how remarkable this really is: From any formula for motion along a curve, no matter how variable the motion may be (as long as $\mathbf{v}$ is never zero), we can calculate a physical property of the curve that seems to have nothing to do with the way the curve is traversed.

The most widely used formula for torsion, derived in more advanced texts, is

Newton's dot notation for derivatives

The dots in Eq. (11) denote differentiation with respect to t, one derivative for each dot. Thus, $\dot{x}$ ("x dot") means dx/dt, $\ddot{x}$ ("x double dot") means d^2x/dt^2, and $\dddot{x}$ ("x triple dot") means d^3x/dt^3. Similarly, $\dot{y} = dy/dt$ and so on.

$$\tau = \frac{\begin{vmatrix} \dot{x} & \dot{y} & \dot{z} \\ \ddot{x} & \ddot{y} & \ddot{z} \\ \dddot{x} & \dddot{y} & \dddot{z} \end{vmatrix}}{|\mathbf{v} \times \mathbf{a}|^2} \qquad \text{(if } \mathbf{v} \times \mathbf{a} \neq \mathbf{0}\text{).} \qquad (11)$$

This formula calculates the torsion directly from the derivatives of the component functions $x = f(t), y = g(t), z = h(t)$ that make up **r.** The determinant's first row comes from **v,** the second row comes from **a,** and the third row comes from $\dot{\mathbf{a}} = d\mathbf{a}/dt$.

EXAMPLE 7   Use Eqs. (10) and (11) to find κ and τ for the helix

$$\mathbf{r}(t) = (a\cos t)\,\mathbf{i} + (a\sin t)\,\mathbf{j} + bt\,\mathbf{k}, \qquad a, b \geq 0, \ a^2 + b^2 \neq 0.$$

Solution   We calculate the curvature with Eq. (10):

$$\mathbf{v} = -(a\sin t)\,\mathbf{i} + (a\cos t)\,\mathbf{j} + b\,\mathbf{k},$$

$$\mathbf{a} = -(a\cos t)\,\mathbf{i} - (a\sin t)\,\mathbf{j},$$

$$\mathbf{v} \times \mathbf{a} = \begin{vmatrix} \mathbf{i} & \mathbf{j} & \mathbf{k} \\ -a\sin t & a\cos t & b \\ -a\cos t & -a\sin t & 0 \end{vmatrix}$$

$$= (ab\sin t)\,\mathbf{i} - (ab\cos t)\,\mathbf{j} + a^2\,\mathbf{k},$$

$$\kappa = \frac{|\mathbf{v} \times \mathbf{a}|}{|\mathbf{v}|^3} = \frac{\sqrt{a^2 b^2 + a^4}}{(a^2 + b^2)^{3/2}} = \frac{a\sqrt{a^2 + b^2}}{(a^2 + b^2)^{3/2}} = \frac{a}{a^2 + b^2}. \tag{12}$$

Notice that Eq. (12) agrees with Eq. (5) in Example 4, where we calculated the curvature directly from its definition.

To evaluate Eq. (11) for the torsion, we find the entries in the determinant by differentiating **r** with respect to t. We already have **v** and **a,** and

$$\dot{\mathbf{a}} = \frac{d\mathbf{a}}{dt} = (a\sin t)\,\mathbf{i} - (a\cos t)\,\mathbf{j}.$$

Hence,

$$\tau = \frac{\begin{vmatrix} \dot{x} & \dot{y} & \dot{z} \\ \ddot{x} & \ddot{y} & \ddot{z} \\ \dddot{x} & \dddot{y} & \dddot{z} \end{vmatrix}}{|\mathbf{v} \times \mathbf{a}|^2} = \frac{\begin{vmatrix} -a\sin t & a\cos t & b \\ -a\cos t & -a\sin t & 0 \\ a\sin t & -a\cos t & 0 \end{vmatrix}}{(a\sqrt{a^2 + b^2})^2} \quad \begin{array}{l}\text{\color{blue}Value of}\\ \text{\color{blue}$|\mathbf{v} \times \mathbf{a}|$}\\ \text{\color{blue}from Eq. (12)}\end{array}$$

$$= \frac{b(a^2 \cos^2 t + a^2 \sin^2 t)}{a^2(a^2 + b^2)}$$

$$= \frac{b}{a^2 + b^2}. \tag{13}$$

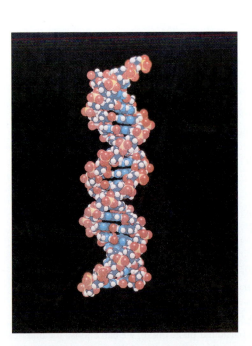

11.34 The helical shape of a DNA molecule is characterized by its constant curvature and torsion.

From Eq. (13), we see that the torsion of a helix about a circular cylinder is constant. In fact, constant curvature and constant torsion characterize the helix among all curves in space.

The DNA molecule, the basic building block of life forms, is designed in the form of two helices winding around each other, a little like the rungs and sides of a twisted rope ladder (Fig. 11.34). Not only is the space occupied by the DNA molecule very much smaller than it would be if it were unraveled, but when the molecule is damaged the imperfect piece can be snipped out by a kind of molecular scissors (because the curvature and torsion functions are constant) and the DNA made right again.

Formulas for Curves in Space

Unit tangent vector: $\mathbf{T} = \dfrac{\mathbf{v}}{|\mathbf{v}|}$

Principal unit normal vector: $\mathbf{N} = \dfrac{d\mathbf{T}/dt}{|d\mathbf{T}/dt|}$

Binormal vector: $\mathbf{B} = \mathbf{T} \times \mathbf{N}$

Curvature: $\kappa = \left| \dfrac{d\mathbf{T}}{ds} \right| = \dfrac{|\mathbf{v} \times \mathbf{a}|}{|\mathbf{v}|^3}$

Torsion: $\tau = -\dfrac{d\mathbf{B}}{ds} \cdot \mathbf{N} = \dfrac{\begin{vmatrix} \dot{x} & \dot{y} & \dot{z} \\ \ddot{x} & \ddot{y} & \ddot{z} \\ \dddot{x} & \dddot{y} & \dddot{z} \end{vmatrix}}{|\mathbf{v} \times \mathbf{a}|^2}$

Tangential and normal scalar components of acceleration:
$\mathbf{a} = a_{\mathrm{T}}\mathbf{T} + a_{\mathrm{N}}\mathbf{N}$

$a_{\mathrm{T}} = \dfrac{d}{dt}|\mathbf{v}|$

$a_{\mathrm{N}} = \kappa\,|\mathbf{v}|^2 = \sqrt{|\mathbf{a}|^2 - a_{\mathrm{T}}^2}$

Exercises 11.4

Plane Curves

Find **T**, **N**, and κ for the plane curves in Exercises 1–4.

1. $\mathbf{r}(t) = t\,\mathbf{i} + (\ln\cos t)\,\mathbf{j}, \quad -\pi/2 < t < \pi/2$

2. $\mathbf{r}(t) = (\ln\sec t)\,\mathbf{i} + t\,\mathbf{j}, \quad -\pi/2 < t < \pi/2$

3. $\mathbf{r}(t) = (2t + 3)\,\mathbf{i} + (5 - t^2)\,\mathbf{j}$

4. $\mathbf{r}(t) = (\cos t + t\sin t)\,\mathbf{i} + (\sin t - t\cos t)\,\mathbf{j}, \quad t > 0$

In Exercises 5 and 6, write **a** in the form $\mathbf{a} = a_{\mathrm{T}}\mathbf{T} + a_{\mathrm{N}}\mathbf{N}$ without finding **T** and **N**.

5. $\mathbf{r}(t) = (2t + 3)\,\mathbf{i} + (t^2 - 1)\,\mathbf{j}$

6. $\mathbf{r}(t) = \ln(t^2 + 1)\,\mathbf{i} + (t - 2\tan^{-1} t)\,\mathbf{j}$

7. *A formula for the curvature of the graph of a function in the xy-plane*

a) The graph $y = f(x)$ in the xy-plane automatically has the parametrization $x = x$, $y = f(x)$, and the vector formula $\mathbf{r}(x) = x\,\mathbf{i} + f(x)\,\mathbf{j}$. Use this formula to show that if f is a twice-differentiable function of x, then

$$\kappa(x) = \frac{|f''(x)|}{\left[1 + (f'(x))^2\right]^{3/2}}.$$

b) Use the formula for κ in (a) to find the curvature of $y = \ln(\cos x)$, $-\pi/2 < x < \pi/2$. Compare your answer with the answer in Exercise 1.

c) Show that the curvature is zero at a point of inflection.

8. *A formula for the curvature of a parametrized plane curve*

a) Show that the curvature of a smooth curve $\mathbf{r}(t) = f(t)\,\mathbf{i} + g(t)\,\mathbf{j}$ defined by twice-differentiable functions $x = f(t)$ and $y = g(t)$ is given by the formula

$$\kappa = \frac{|\dot{x}\ddot{y} - \dot{y}\ddot{x}|}{(\dot{x}^2 + \dot{y}^2)^{3/2}}.$$

Apply the formula to find the curvatures of the following curves.

b) $\mathbf{r}(t) = t\,\mathbf{i} + (\ln\sin t)\,\mathbf{j}, \quad 0 < t < \pi$

c) $\mathbf{r}(t) = [\tan^{-1}(\sinh t)]\,\mathbf{i} + (\ln\cosh t)\,\mathbf{j}.$

9. *Normals to plane curves*

a) Show that $\mathbf{n}(t) = -g'(t)\,\mathbf{i} + f'(t)\,\mathbf{j}$ and $-\mathbf{n}(t) = g'(t)\,\mathbf{i} - f'(t)\,\mathbf{j}$ are both normal to the curve $\mathbf{r}(t) = f(t)\,\mathbf{i} + g(t)\,\mathbf{j}$ at the point $(f(t), g(t))$.

To obtain **N** for a particular plane curve, we can choose the one of **n** or −**n** from part (a) that points toward the concave side of the curve, and make it into a unit vector. (See Fig. 11.26.) Apply this method to find **N** for the following curves.

b) $\mathbf{r}(t) = t\,\mathbf{i} + e^{2t}\,\mathbf{j}$

c) $\mathbf{r}(t) = \sqrt{4 - t^2}\,\mathbf{i} + t\,\mathbf{j}, \quad -2 \le t \le 2$

10. (*Continuation of Exercise 9.*)

a) Use the method of Exercise 9 to find **N** for the curve $\mathbf{r}(t) = t\,\mathbf{i} + (1/3)t^3\,\mathbf{j}$ when $t < 0$; when $t > 0$.

b) Calculate

$$\mathbf{N} = \frac{d\mathbf{T}/dt}{|d\mathbf{T}/dt|}, t \ne 0,$$

for the curve in part (a). Does **N** exist at $t = 0$? Graph the curve and explain what is happening to **N** as t passes from negative to positive values.

Space Curves

Find **T**, **N**, **B**, κ, and τ for the space curves in Exercises 11–18.

11. $\mathbf{r}(t) = (3\sin t)\,\mathbf{i} + (3\cos t)\,\mathbf{j} + 4t\,\mathbf{k}$

12. $\mathbf{r}(t) = (\cos t + t\sin t)\,\mathbf{i} + (\sin t - t\cos t)\,\mathbf{j} + 3\,\mathbf{k}$

13. $\mathbf{r}(t) = (e^t\cos t)\,\mathbf{i} + (e^t\sin t)\,\mathbf{j} + 2\,\mathbf{k}$

14. $\mathbf{r}(t) = (6\sin 2t)\,\mathbf{i} + (6\cos 2t)\,\mathbf{j} + 5t\,\mathbf{k}$

15. $\mathbf{r}(t) = (t^3/3)\,\mathbf{i} + (t^2/2)\,\mathbf{j}, \quad t > 0$

16. $\mathbf{r}(t) = (\cos^3 t)\,\mathbf{i} + (\sin^3 t)\,\mathbf{j}, \quad 0 < t < \pi/2$

17. $\mathbf{r}(t) = t\,\mathbf{i} + (a\cosh(t/a))\,\mathbf{j}, \quad a > 0$

18. $\mathbf{r}(t) = (\cosh t)\,\mathbf{i} - (\sinh t)\,\mathbf{j} + t\,\mathbf{k}$

In Exercises 19 and 20, write **a** in the form $a_\mathrm{T}\,\mathbf{T} + a_\mathrm{N}\,\mathbf{N}$ without finding **T** and **N**.

19. $\mathbf{r}(t) = (a\cos t)\,\mathbf{i} + (a\sin t)\,\mathbf{j} + bt\,\mathbf{k}$

20. $\mathbf{r}(t) = (1 + 3t)\,\mathbf{i} + (t - 2)\,\mathbf{j} - 3t\,\mathbf{k}$

In Exercises 21–24, write **a** in the form $\mathbf{a} = a_\mathrm{T}\,\mathbf{T} + a_\mathrm{N}\,\mathbf{N}$ at the given value of t without finding **T** and **N**.

21. $\mathbf{r}(t) = (t + 1)\,\mathbf{i} + 2t\,\mathbf{j} + t^2\,\mathbf{k}, \quad t = 1$

22. $\mathbf{r}(t) = (t\cos t)\,\mathbf{i} + (t\sin t)\,\mathbf{j} + t^2\,\mathbf{k}, \quad t = 0$

23. $\mathbf{r}(t) = t^2\,\mathbf{i} + (t + (1/3)t^3)\,\mathbf{j} + (t - (1/3)t^3)\,\mathbf{k}, \quad t = 0$

24. $\mathbf{r}(t) = (e^t\cos t)\,\mathbf{i} + (e^t\sin t)\,\mathbf{j} + \sqrt{2}e^t\,\mathbf{k}, \quad t = 0$

In Exercises 25 and 26, find **r**, **T**, **N**, and **B** at the given value of t. Then find equations for the osculating, normal, and rectifying planes at that value of t.

25. $\mathbf{r}(t) = (\cos t)\,\mathbf{i} + (\sin t)\,\mathbf{j} - \mathbf{k}, \quad t = \pi/4$

26. $\mathbf{r}(t) = (\cos t)\,\mathbf{i} + (\sin t)\,\mathbf{j} + t\,\mathbf{k}, \quad t = 0$

Physical Applications

27. The speedometer on your car reads a steady 35 mph. Could you be accelerating? Explain.

28. Can anything be said about the acceleration of a particle that is moving at a constant speed? Give reasons for your answer.

29. Can anything be said about the speed of a particle whose acceleration is always orthogonal to its velocity? Give reasons for your answer.

30. An object of mass m travels along the parabola $y = x^2$ with a constant speed of 10 units/sec. What is the force on the object due to its acceleration at $(0, 0)$? at $(2^{1/2}, 2)$? Write your answers in terms of **i** and **j**. (Remember Newton's law, $\mathbf{F} = m\,\mathbf{a}$.)

31. The following is a quotation from an article in *The American Mathematical Monthly*, titled "Curvature in the Eighties" by Robert Osserman (October 1990, page 731):

> Curvature also plays a key role in physics. The magnitude of a force required to move an object at constant speed along a curved path is, according to Newton's laws, a constant multiple of the curvature of the trajectories.

Explain mathematically why the second sentence of the quotation is true.

32. Show that a moving particle will move in a straight line if the normal component of its acceleration is zero.

More on Curvature

33. Show that the parabola $y = ax^2, a \ne 0$, has its largest curvature at its vertex and has no minimum curvature. (*Note:* Since the curvature of a curve remains the same if the curve is translated or rotated, this result is true for any parabola.)

34. Show that the ellipse $x = a\cos t, y = b\sin t, a > b > 0$, has its largest curvature on its major axis and its smallest curvature on its minor axis. (As in Exercise 33, the same is true for any ellipse.)

35. *Maximizing the curvature of a helix.* In Example 4, we found the curvature of the helix $\mathbf{r}(t) = (a\cos t)\,\mathbf{i} + (a\sin t)\,\mathbf{j} + bt\,\mathbf{k}$ $(a, b \ge 0)$ to be $\kappa = a/(a^2 + b^2)$. What is the largest value κ can have for a given value of b? Give reasons for your answer.

36. *A sometime shortcut to curvature.* If you already know $|a_\mathrm{N}|$ and $|\mathbf{v}|$, then the formula $a_\mathrm{N} = \kappa\,|\mathbf{v}|^2$ gives a convenient way to find the curvature. Use it to find the curvature and radius of curvature of the curve

$$\mathbf{r}(t) = (\cos t + t\sin t)\,\mathbf{i} + (\sin t - t\cos t)\,\mathbf{j}, \quad t > 0.$$

(Take a_N and $|\mathbf{v}|$ from Example 6.)

37. Show that κ and τ are both zero for the line

$$\mathbf{r}(t) = (x_0 + At)\,\mathbf{i} + (y_0 + Bt)\,\mathbf{j} + (z_0 + Ct)\,\mathbf{k}.$$

38. *Total curvature.* We find the **total curvature** of the portion of a smooth curve that runs from $s = s_0$ to $s = s_1 > s_0$ by integrating κ from s_0 to s_1. If the curve has some other parameter, say t, then the total curvature is

$$K = \int_{s_0}^{s_1} \kappa\,ds = \int_{t_0}^{t_1} \kappa\,\frac{ds}{dt}\,dt = \int_{t_0}^{t_1} \kappa\,|\mathbf{v}|\,dt,$$

where t_0 and t_1 correspond to s_0 and s_1. Find the total curvature of the portion of the helix $\mathbf{r}(t) = (3\cos t)\,\mathbf{i} + (3\sin t)\,\mathbf{j} + t\,\mathbf{k}$, $0 \le t \le 4\pi$.

39. (*Continuation of Exercise 38.*) Find the total curvatures of the following curves.

    a) The involute of the unit circle: $\mathbf{r}(t) = (\cos t + t \sin t)\mathbf{i} + (\sin t - t \cos t)\mathbf{j}$, $a \le t \le b$ ($a > 0$). (Exercise 36 gives a convenient way to find κ. Use values from Example 6.)

    b) The parabola $y = x^2$, $-\infty < x < \infty$

40. **a)** Find an equation for the circle of curvature of the curve $\mathbf{r}(t) = t\mathbf{i} + (\sin t)\mathbf{j}$ at the point $(\pi/2, 1)$. (The curve parametrizes the graph of $y = \sin x$ in the xy-plane.)

    b) Find an equation for the circle of curvature of the curve $\mathbf{r}(t) = (2 \ln t)\mathbf{i} - [t + (1/t)]\mathbf{j}$, $e^{-2} \le t \le e^2$, at the point $(0, -2)$, where $t = 1$.

Theory and Examples

41. What can be said about the torsion of a (sufficiently differentiable) plane curve $\mathbf{r}(t) = f(t)\mathbf{i} + g(t)\mathbf{j}$? Give reasons for your answer.

42. *The torsion of a helix.* In Example 7, we found the torsion of the helix

$$\mathbf{r}(t) = (a \cos t)\mathbf{i} + (a \sin t)\mathbf{j} + bt\,\mathbf{k}, \quad a, b \ge 0$$

to be $\tau = b/(a^2 + b^2)$. What is the largest value τ can have for a given value of a? Give reasons for your answer.

43. *Differentiable curves with zero torsion lie in planes.* That a sufficiently differentiable curve with zero torsion lies in a plane is a special case of the fact that a particle whose velocity remains perpendicular to a fixed vector $\mathbf{C}$ moves in a plane perpendicular to $\mathbf{C}$. This, in turn, can be viewed as the solution of the following problem in calculus.

    Suppose $\mathbf{r}(t) = f(t)\mathbf{i} + g(t)\mathbf{j} + h(t)\mathbf{k}$ is twice differentiable for all t in an interval $[a, b]$, that $\mathbf{r} = 0$ when $t = a$, and that $\mathbf{v} \cdot \mathbf{k} = 0$ for all t in $[a, b]$. Then $h(t) = 0$ for all t in $[a, b]$.

    Solve this problem. (*Hint:* Start with $\mathbf{a} = d^2\mathbf{r}/dt^2$ and apply the initial conditions in reverse order.)

44. *A formula that calculates τ from $\mathbf{B}$ and $\mathbf{v}$.* If we start with the definition $\tau = -(d\mathbf{B}/ds) \cdot \mathbf{N}$ and apply the Chain Rule to rewrite $d\mathbf{B}/ds$ as

$$\frac{d\mathbf{B}}{ds} = \frac{d\mathbf{B}}{dt}\frac{dt}{ds} = \frac{d\mathbf{B}}{dt}\frac{1}{|\mathbf{v}|},$$

we arrive at the formula

$$\tau = -\frac{1}{|\mathbf{v}|}\left(\frac{d\mathbf{B}}{dt} \cdot \mathbf{N}\right).$$

The advantage of this formula over Eq. (11) is that it is easier to derive and state. The disadvantage is that it can take a lot of work to evaluate without a computer. Use the new formula to find the torsion of the helix in Example 7.

Grapher Explorations

The formula

$$\kappa(x) = \frac{|f''(x)|}{\left[1 + (f'(x))^2\right]^{3/2}},$$

derived in Exercise 7, expresses the curvature $\kappa(x)$ of a twice-differentiable plane curve $y = f(x)$ as a function of x. Find the curvature function of each of the curves in Exercises 45–48. Then graph $f(x)$ together with $\kappa(x)$ over the given interval. You will find some surprises.

45. $y = x^2$, $-2 \le x \le 2$

46. $y = x^4/4$, $-2 \le x \le 2$

47. $y = \sin x$, $0 \le x \le 2\pi$

48. $y = e^x$, $-1 \le x \le 2$

CAS Explorations and Projects—Circles of Curvature

In Exercises 49–56 you will use a CAS to explore the osculating circle at a point P on a plane curve where $\kappa \ne 0$. Use a CAS to perform the following steps:

    a) Plot the plane curve given in parametric or function form over the specified interval to see what it looks like.

    b) Calculate the curvature κ of the curve at the given value t_0 using the appropriate formula from Exercise 7 or 8. Use the parametrization $x = t$ and $y = f(t)$ if the curve is given as a function $y = f(x)$.

    c) Find the unit normal vector $\mathbf{N}$ at t_0. Notice that the signs of the components of $\mathbf{N}$ depend on whether the unit tangent vector $\mathbf{T}$ is turning clockwise or counterclockwise at $t = t_0$. (See Exercise 9.)

    d) If $\mathbf{C} = a\mathbf{i} + b\mathbf{j}$ is the vector from the origin to the center (a, b) of the osculating circle, find the center $\mathbf{C}$ from the vector equation

$$\mathbf{C} = \mathbf{r}(t_0) + \frac{1}{\kappa(t_0)}\mathbf{N}(t_0).$$

The point $P(x_0, y_0)$ on the curve is given by the position vector $\mathbf{r}(t_0)$.

    e) Plot implicitly the equation $(x - a)^2 + (y - b)^2 = 1/\kappa^2$ of the osculating circle. Then plot the curve and osculating circle together. You may need to experiment with the size of the viewing window, but be sure it is square.

49. $\mathbf{r}(t) = (3 \cos t)\mathbf{i} + (5 \sin t)\mathbf{j}$, $0 \le t \le 2\pi$, $t_0 = \pi/4$

50. $\mathbf{r}(t) = (\cos^3 t)\mathbf{i} + (\sin^3 t)\mathbf{j}$, $0 \le t \le 2\pi$, $t_0 = \pi/4$

51. $\mathbf{r}(t) = t^2\mathbf{i} + (t^3 - 3t)\mathbf{j}$, $-4 \le t \le 4$, $t_0 = 3/5$

52. $\mathbf{r}(t) = (t^3 - 2t^2 - t)\mathbf{i} + \dfrac{3t}{\sqrt{1 + t^2}}\mathbf{j}$, $-2 \le t \le 5$, $t_0 = 1$

53. $\mathbf{r}(t) = (2t - \sin t)\mathbf{i} + (2 - 2 \cos t)\mathbf{j}$, $0 \le t \le 3\pi$, $t_0 = 3\pi/2$

54. $\mathbf{r}(t) = (e^{-t} \cos t)\mathbf{i} + (e^{-t} \sin t)\mathbf{j}$, $0 \le t \le 6\pi$, $t_0 = \pi/4$

55. $y = x^2 - x$, $-2 \le x \le 5$, $x_0 = 1$

56. $y = x(1 - x)^{2/5}$, $-1 \le x \le 2$, $x_0 = 1/2$

Rounding the answers to four decimal places, use a CAS to find **v**, **a**, speed, **T**, **N**, **B**, κ, τ, and the tangential and normal components of acceleration for the curves in Exercises 57–60 at the given values of t.

57. $\mathbf{r}(t) = (t \cos t)\mathbf{i} + (t \sin t)\mathbf{j} + t\mathbf{k}, \quad t = \sqrt{3}$

58. $\mathbf{r}(t) = (e^t \cos t)\mathbf{i} + (e^t \sin t)\mathbf{j} + e^t\mathbf{k}, \quad t = \ln 2$

59. $\mathbf{r}(t) = (t - \sin t)\mathbf{i} + (1 - \cos t)\mathbf{j} + \sqrt{-t}\,\mathbf{k}, \quad t = -3\pi$

60. $\mathbf{r}(t) = (3t - t^2)\mathbf{i} + (3t^2)\mathbf{j} + (3t + t^3)\mathbf{k}, \quad t = 1$

11.5

Planetary Motion and Satellites

In this section, we derive Kepler's laws of planetary motion from Newton's laws of motion and gravitation and discuss the orbits of Earth satellites. The derivation of Kepler's laws from Newton's is one of the triumphs of calculus. It draws on almost everything we have studied so far, including the algebra and geometry of vectors in space, the calculus of vector functions, the solutions of differential equations and initial value problems, and the polar coordinate description of conic sections.

Vector Equations for Motion in Polar and Cylindrical Coordinates

When a particle moves along a curve in the polar coordinate plane, we express its position, velocity, and acceleration in terms of the moving unit vectors

$$\mathbf{u}_r = (\cos\theta)\mathbf{i} + (\sin\theta)\mathbf{j}, \qquad \mathbf{u}_\theta = -(\sin\theta)\mathbf{i} + (\cos\theta)\mathbf{j}, \tag{1}$$

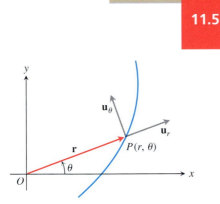

11.35 The length of **r** is the positive polar coordinate r of the point P. Thus, $\mathbf{u}_r$, which is $\mathbf{r}/|\mathbf{r}|$, is also $\mathbf{r}/r$. Equations (1) express $\mathbf{u}_r$ and $\mathbf{u}_\theta$ in terms of **i** and **j**.

shown in Fig. 11.35. The vector $\mathbf{u}_r$ points along the position vector $\overrightarrow{OP}$, so $\mathbf{r} = r\mathbf{u}_r$. The vector $\mathbf{u}_\theta$, orthogonal to $\mathbf{u}_r$, points in the direction of increasing θ.

We find from (1) that

$$\frac{d\mathbf{u}_r}{d\theta} = -(\sin\theta)\mathbf{i} + (\cos\theta)\mathbf{j} = \mathbf{u}_\theta$$

$$\frac{d\mathbf{u}_\theta}{d\theta} = -(\cos\theta)\mathbf{i} - (\sin\theta)\mathbf{j} = -\mathbf{u}_r. \tag{2}$$

When we differentiate $\mathbf{u}_r$ and $\mathbf{u}_\theta$ with respect to t to find how they change with time, the Chain Rule gives

$$\dot{\mathbf{u}}_r = \frac{d\mathbf{u}_r}{d\theta}\dot{\theta} = \dot{\theta}\,\mathbf{u}_\theta, \qquad \dot{\mathbf{u}}_\theta = \frac{d\mathbf{u}_\theta}{d\theta}\dot{\theta} = -\dot{\theta}\,\mathbf{u}_r. \tag{3}$$

Hence,

$$\mathbf{v} = \dot{\mathbf{r}} = \frac{d}{dt}\left(r\,\mathbf{u}_r\right) = \dot{r}\,\mathbf{u}_r + r\dot{\mathbf{u}}_r = \dot{r}\,\mathbf{u}_r + r\dot{\theta}\,\mathbf{u}_\theta. \tag{4}$$

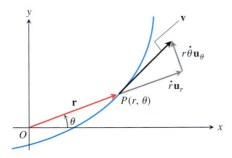

11.36 In polar coordinates, the velocity vector is

$$\mathbf{v} = \dot{r}\,\mathbf{u}_r + r\dot{\theta}\,\mathbf{u}_\theta.$$

See Fig. 11.36.

The acceleration is

$$\mathbf{a} = \dot{\mathbf{v}} = (\ddot{r}\,\mathbf{u}_r + \dot{r}\,\dot{\mathbf{u}}_r) + (\dot{r}\dot{\theta}\,\mathbf{u}_\theta + r\ddot{\theta}\,\mathbf{u}_\theta + r\dot{\theta}\,\dot{\mathbf{u}}_\theta). \tag{5}$$

As in the previous section, we use Newton's dot notation for time derivatives to keep the formulas as simple as we can: $\dot{\mathbf{u}}_r$ means $d\mathbf{u}_r/dt$, $\dot{\theta}$ means $d\theta/dt$, and so on.

When Eqs. (3) are used to evaluate $\dot{\mathbf{u}}_r$ and $\dot{\mathbf{u}}_\theta$ and the components are separated, the equation for acceleration becomes

$$\mathbf{a} = (\ddot{r} - r\dot{\theta}^2)\,\mathbf{u}_r + (r\ddot{\theta} + 2\dot{r}\dot{\theta})\,\mathbf{u}_\theta. \tag{6}$$

Notice that $|\mathbf{r}| \neq r$ if $z \neq 0$.

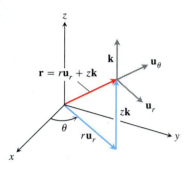

11.37 Position vector and basic unit vectors in cylindrical coordinates.

To extend these equations of motion to space, we add $z\mathbf{k}$ to the right-hand side of the equation $\mathbf{r} = r\mathbf{u}_r$. Then, in cylindrical coordinates,

$$\mathbf{r} = r\,\mathbf{u}_r + z\,\mathbf{k}$$

$$\mathbf{v} = \dot{r}\,\mathbf{u}_r + r\dot{\theta}\,\mathbf{u}_\theta + \dot{z}\,\mathbf{k} \tag{7}$$

$$\mathbf{a} = (\ddot{r} - r\dot{\theta}^2)\,\mathbf{u}_r + (r\ddot{\theta} + 2\dot{r}\dot{\theta})\mathbf{u}_\theta + \ddot{z}\,\mathbf{k}.$$

The vectors $\mathbf{u}_r$, $\mathbf{u}_\theta$, and $\mathbf{k}$ make a right-handed frame (Fig. 11.37) in which

$$\mathbf{u}_r \times \mathbf{u}_\theta = \mathbf{k}, \qquad \mathbf{u}_\theta \times \mathbf{k} = \mathbf{u}_r, \qquad \mathbf{k} \times \mathbf{u}_r = \mathbf{u}_\theta. \tag{8}$$

Planets Move in Planes

Newton's Law of Gravitation says that if $\mathbf{r}$ is the radius vector from the center of a sun of mass M to the center of a planet of mass m, then the force $\mathbf{F}$ of the gravitational attraction between the planet and sun is

$$\mathbf{F} = -\frac{GmM}{|\mathbf{r}|^2}\frac{\mathbf{r}}{|\mathbf{r}|} \tag{9}$$

(Fig. 11.38). The number G is the (universal) **gravitational constant.** If we measure mass in kilograms, force in newtons, and distance in meters, G is about 6.6720×10^{-11} Nm^{2}kg^{-2}.

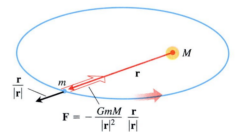

11.38 The force of gravity is directed along the line joining the centers of mass.

Combining Eq. (9) with Newton's second law, $\mathbf{F} = m\ddot{\mathbf{r}}$, for the force acting on the planet gives

$$m\ddot{\mathbf{r}} = -\frac{GmM}{|\mathbf{r}|^2}\frac{\mathbf{r}}{|\mathbf{r}|},$$

$$\ddot{\mathbf{r}} = -\frac{GM}{|\mathbf{r}|^2}\frac{\mathbf{r}}{|\mathbf{r}|}. \tag{10}$$

The planet is accelerated toward the sun's center at all times.

Equation (10) says that $\ddot{\mathbf{r}}$ is a scalar multiple of $\mathbf{r}$, so that

$$\mathbf{r} \times \ddot{\mathbf{r}} = \mathbf{0}. \tag{11}$$

A routine calculation shows $\mathbf{r} \times \ddot{\mathbf{r}}$ to be the derivative of $\mathbf{r} \times \dot{\mathbf{r}}$:

$$\frac{d}{dt}(\mathbf{r} \times \dot{\mathbf{r}}) = \underbrace{\dot{\mathbf{r}} \times \dot{\mathbf{r}}}_{\mathbf{0}} + \mathbf{r} \times \ddot{\mathbf{r}} = \mathbf{r} \times \ddot{\mathbf{r}}. \tag{12}$$

Hence Eq. (11) is equivalent to

$$\frac{d}{dt}(\mathbf{r} \times \dot{\mathbf{r}}) = \mathbf{0}, \tag{13}$$

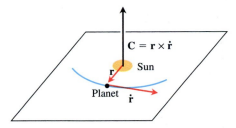

11.39 A planet that obeys Newton's laws of gravitation and motion travels in the plane through the sun's center of mass perpendicular to $\mathbf{C} = \mathbf{r} \times \dot{\mathbf{r}}$.

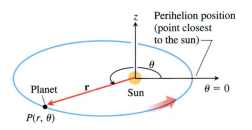

11.40 The coordinate system for planetary motion. The motion is counterclockwise when viewed from above, as it is here, and $\dot{\theta} > 0$.

which integrates to

$$\mathbf{r} \times \dot{\mathbf{r}} = \mathbf{C} \qquad (14)$$

for some constant vector $\mathbf{C}$.

Equation (14) tells us that $\mathbf{r}$ and $\dot{\mathbf{r}}$ always lie in a plane perpendicular to $\mathbf{C}$. Hence, the planet moves in a fixed plane through the center of its sun (Fig. 11.39).

Coordinates and Initial Conditions

We now introduce cylindrical coordinates in a way that places the origin at the sun's center of mass and makes the plane of the planet's motion the polar coordinate plane. This makes $\mathbf{r}$ the planet's polar coordinate position vector and makes $|\mathbf{r}|$ equal to r and $\mathbf{r}/|\mathbf{r}|$ equal to $\mathbf{u}_r$. We also position the z-axis in a way that makes $\mathbf{k}$ the direction of $\mathbf{C}$. Thus, $\mathbf{k}$ has the same right-hand relation to $\mathbf{r} \times \dot{\mathbf{r}}$ that $\mathbf{C}$ does, and the planet's motion is counterclockwise when viewed from the positive z-axis. This makes θ increase with t, so that $\dot{\theta} > 0$ for all t. Finally, we rotate the polar coordinate plane about the z-axis, if necessary, to make the initial ray coincide with the direction $\mathbf{r}$ has when the planet is closest to the sun. This runs the ray through the planet's **perihelion** position (Fig. 11.40).

If we measure time so that $t = 0$ at perihelion, we have the following initial conditions for the planet's motion.

1. $r = r_0$, the minimum radius, when $t = 0$
2. $\dot{r} = 0$ when $t = 0$ (because r has a minimum value then)
3. $\theta = 0$ when $t = 0$
4. $|\mathbf{v}| = v_0$ when $t = 0$

Since

$$
\begin{aligned}
v_0 &= |\mathbf{v}|_{t=0} \\
&= \left| \dot{r}\,\mathbf{u}_r + r\dot{\theta}\,\mathbf{u}_\theta \right|_{t=0} & \text{Eq. (4)} \\
&= \left| r\dot{\theta}\,\mathbf{u}_\theta \right|_{t=0} & \dot{r} = 0 \text{ when } t = 0 \\
&= \left(|r\dot{\theta}| \|\mathbf{u}_\theta| \right)_{t=0} \\
&= \left| r\dot{\theta} \right|_{t=0} & |\mathbf{u}_\theta| = 1 \\
&= (r\dot{\theta})_{t=0}, & r \text{ and } \dot{\theta} \text{ both positive}
\end{aligned}
$$

we also know that

5. $r\dot{\theta} = v_0$ when $t = 0$.

Statement of Kepler's First Law (The Conic Section Law)

Kepler's first law says that a planet's path is a conic section with the sun at one focus. The eccentricity of the conic is

$$e = \frac{r_0 v_0^2}{GM} - 1 \qquad (15)$$

and the polar equation is

$$r = \frac{(1+e)r_0}{1 + e\cos\theta}. \tag{16}$$

The derivation uses Kepler's second law, so we will state and prove the second law before proving the first law.

Kepler's Second Law (The Equal Area Law)

Kepler's second law says that the radius vector from the sun to a planet (the vector **r** in our model) sweeps out equal areas in equal times (Fig. 11.41). To derive the law, we use Eq. (4) to evaluate the cross product $\mathbf{C} = \mathbf{r} \times \dot{\mathbf{r}}$ from Eq. (14):

$$
\begin{aligned}
\mathbf{C} = \mathbf{r} \times \dot{\mathbf{r}} &= \mathbf{r} \times \mathbf{v} \\
&= r\,\mathbf{u}_r \times (\dot{r}\,\mathbf{u}_r + r\dot{\theta}\,\mathbf{u}_\theta) \qquad \text{Eq. (4)} \\
&= r\dot{r}\,\underbrace{(\mathbf{u}_r \times \mathbf{u}_r)}_{\mathbf{0}} + r(r\dot{\theta})\,\underbrace{(\mathbf{u}_r \times \mathbf{u}_\theta)}_{\mathbf{k}} \qquad (17) \\
&= r(r\dot{\theta})\,\mathbf{k}.
\end{aligned}
$$

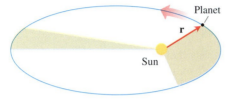

11.41 The line joining a planet to its sun sweeps over equal areas in equal times.

Setting t equal to zero shows that

$$\mathbf{C} = [r(r\dot{\theta})]_{t=0}\,\mathbf{k} = r_0 v_0\,\mathbf{k}. \tag{18}$$

Substituting this value for **C** in Eq. (17) gives

$$r_0 v_0\,\mathbf{k} = r^2\dot{\theta}\,\mathbf{k}, \qquad \text{or} \qquad r^2\dot{\theta} = r_0 v_0. \tag{19}$$

This is where the area comes in. The area differential in polar coordinates is

$$dA = \frac{1}{2}r^2 d\theta$$

(Section 9.9). Accordingly, dA/dt has the constant value

$$\frac{dA}{dt} = \frac{1}{2}r^2\dot{\theta} = \frac{1}{2}r_0 v_0, \tag{20}$$

which is Kepler's second law.

For Earth, r_0 is about 150,000,000 km, v_0 is about 30 km/sec, and dA/dt is about 2,250,000,000 km²/sec. Every time your heart beats, Earth advances 30 km along its orbit, and the radius joining Earth to the sun sweeps out 2,250,000,000 km² of area.

The German astronomer, mathematician, and physicist Johannes Kepler (1571–1630) was the first, and until Descartes the only, scientist to demand physical (as opposed to theological) explanations of celestial phenomena. His three laws of motion, the results of a lifetime of work, changed the course of astronomy forever and played a crucial role in the development of Newton's physics.

Proof of Kepler's First Law

To prove that a planet moves along a conic section with one focus at its sun, we need to express the planet's radius r as a function of θ. This requires a long sequence of calculations and some substitutions that are not altogether obvious.

We begin with the equation that comes from equating the coefficients of $\mathbf{u}_r = \mathbf{r}/|\mathbf{r}|$ in Eqs. (6) and (10):

$$\ddot{r} - r\dot{\theta}^2 = -\frac{GM}{r^2}. \tag{21}$$

We eliminate $\dot{\theta}$ temporarily by replacing it with $r_0 v_0/r^2$ from Eq. (19) and rearrange the resulting equation to get

$$\ddot{r} = \frac{r_0{}^2 v_0{}^2}{r^3} - \frac{GM}{r^2}. \tag{22}$$

We change this into a first order equation by a change of variable. With

$$p = \frac{dr}{dt}, \qquad \frac{d^2 r}{dt^2} = \frac{dp}{dt} = \frac{dp}{dr}\frac{dr}{dt} = p\frac{dp}{dr}, \qquad \text{Chain Rule}$$

Eq. (22) becomes

$$p\frac{dp}{dr} = \frac{r_0{}^2 v_0{}^2}{r^3} - \frac{GM}{r^2}. \tag{23}$$

Multiplying through by 2 and integrating with respect to r gives

$$p^2 = (\dot{r})^2 = -\frac{r_0{}^2 v_0{}^2}{r^2} + \frac{2GM}{r} + C_1. \tag{24}$$

The initial conditions that $r = r_0$ and $\dot{r} = 0$ when $t = 0$ determine the value of C_1 to be

$$C_1 = v_0{}^2 - \frac{2GM}{r_0}.$$

Accordingly, Eq. (24), after a suitable rearrangement, becomes

$$\dot{r}^2 = v_0{}^2 \left(1 - \frac{r_0{}^2}{r^2} \right) + 2GM \left(\frac{1}{r} - \frac{1}{r_0} \right). \tag{25}$$

The effect of going from Eq. (21) to Eq. (25) has been to replace a second order differential equation in r by a first order differential equation in r. Our goal is still to express r in terms of θ, so we now bring θ back into the picture. To accomplish this, we divide both sides of Eq. (25) by the squares of the corresponding sides of the equation $r^2\dot{\theta} = r_0 v_0$ (Eq. 19) and use the fact that $\dot{r}/\dot{\theta} = (dr/dt)/(d\theta/dt) = dr/d\theta$ to get

$$\frac{1}{r^4} \left(\frac{dr}{d\theta} \right)^2 = \frac{1}{r_0{}^2} - \frac{1}{r^2} + \frac{2GM}{r_0{}^2 v_0{}^2} \left(\frac{1}{r} - \frac{1}{r_0} \right)$$

$$= \frac{1}{r_0{}^2} - \frac{1}{r^2} + 2h \left(\frac{1}{r} - \frac{1}{r_0} \right). \qquad h = \frac{GM}{r_0{}^2 v_0{}^2} \tag{26}$$

To simplify further, we substitute

$$u = \frac{1}{r}, \qquad u_0 = \frac{1}{r_0}, \qquad \frac{du}{d\theta} = -\frac{1}{r^2}\frac{dr}{d\theta}, \qquad \left(\frac{du}{d\theta} \right)^2 = \frac{1}{r^4}\left(\frac{dr}{d\theta} \right)^2,$$

obtaining

$$\left(\frac{du}{d\theta} \right)^2 = u_0{}^2 - u^2 + 2hu - 2hu_0 = (u_0 - h)^2 - (u - h)^2, \tag{27}$$

$$\frac{du}{d\theta} = \pm\sqrt{(u_0 - h)^2 - (u - h)^2}. \tag{28}$$

Which sign do we take? We know that $\dot{\theta} = r_0 v_0/r^2$ is positive. Also, r starts from a minimum value at $t = 0$, so it cannot immediately decrease, and $\dot{r} \geq 0$, at least for early positive values of t. Therefore,

$$\frac{dr}{d\theta} = \frac{\dot{r}}{\dot{\theta}} \geq 0 \qquad \text{and} \qquad \frac{du}{d\theta} = -\frac{1}{r^2}\frac{dr}{d\theta} \leq 0.$$

The correct sign for Eq. (28) is the negative sign. With this determined, we rearrange Eq. (28) and integrate both sides with respect to θ:

$$\frac{-1}{\sqrt{(u_0 - h)^2 - (u - h)^2}}\frac{du}{d\theta} = 1$$

$$\cos^{-1}\left(\frac{u - h}{u_0 - h}\right) = \theta + C_2. \tag{29}$$

The constant C_2 is zero because $u = u_0$ when $\theta = 0$ and $\cos^{-1}(1) = 0$. Therefore,

$$\frac{u - h}{u_0 - h} = \cos\theta$$

and

$$\frac{1}{r} = u = h + (u_0 - h)\cos\theta. \tag{30}$$

A few more algebraic maneuvers produce the final equation

$$r = \frac{(1 + e)r_0}{1 + e\cos\theta}, \tag{31}$$

where

$$e = \frac{1}{r_0 h} - 1 = \frac{r_0 v_0^2}{GM} - 1. \tag{32}$$

Together, Eqs. (31) and (32) say that the path of the planet is a conic section with one focus at the sun and with eccentricity $(r_0 v_0^2 / GM) - 1$. This is the modern formulation of Kepler's first law.

Statement of Kepler's Third Law (The Time–Distance Law)

The time T it takes a planet to go around its sun once is the planet's **orbital period.** *Kepler's third law* says that T and the orbit's semimajor axis a are related by the equation

$$\frac{T^2}{a^3} = \frac{4\pi^2}{GM}. \tag{33}$$

Since the right-hand side of this equation is constant within a given solar system, the ratio of T^2 to a^3 is *the same for every planet in the system.*

Kepler's third law is the starting point for working out the size of our solar system. It allows the semimajor axis of each planetary orbit to be expressed in astronomical units, Earth's semimajor axis being one unit. The distance between any two planets at any time can then be predicted in astronomical units and all that remains is to find one of these distances in kilometers. This can be done by bouncing radar waves off Venus, for example. The astronomical unit is now known, after a series of such measurements, to be 149,597,870 km.

We derive Kepler's third law by combining two formulas for the area enclosed by the planet's elliptical orbit:

Formula 1: $\quad \text{Area} = \pi ab \qquad$ The geometry formula in which a is the semimajor axis and b is the semiminor axis

Formula 2: $\quad \displaystyle \text{Area} = \int_0^T dA$

$$= \int_0^T \frac{1}{2} r_0 v_0 \, dt \qquad \text{Eq. (20)}$$

$$= \frac{1}{2} T r_0 v_0.$$

Equating these gives

$$T = \frac{2\pi ab}{r_0 v_0} = \frac{2\pi a^2}{r_0 v_0} \sqrt{1 - e^2}. \qquad \begin{array}{l}\text{For any ellipse,}\\ b = a\sqrt{1 - e^2}\end{array} \qquad (34)$$

It remains only to express a and e in terms of r_0, v_0, G, and M. Equation (32) does this for e. For a, we observe that setting θ equal to π in Eq. (31) gives

$$r_{\max} = r_0 \frac{1 + e}{1 - e}.$$

Hence,

$$2a = r_0 + r_{\max} = \frac{2r_0}{1 - e} = \frac{2r_0 GM}{2GM - r_0 v_0{}^2}. \qquad (35)$$

Squaring both sides of Eq. (34) and substituting the results of Eqs. (32) and (35) now produces Kepler's third law (Exercise 14).

11.42 This multiflash photograph shows a body being deflected by an inverse square law force. It moves along a hyperbola.

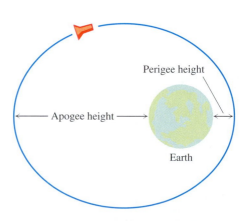

11.43 The orbit of an Earth satellite: $2a$ = diameter of earth + perigee height + apogee height.

Orbit Data

Although Kepler discovered his laws empirically and stated them only for the six planets known at the time, the modern derivations of Kepler's laws show that they apply to any body driven by a force that obeys an inverse square law. They apply to Halley's comet and the asteroid Icarus. They apply to the moon's orbit about Earth, and they applied to the orbit of the spacecraft *Apollo 8* about the moon. They also applied to the air puck shown in Fig. 11.42 being deflected by an inverse square law force—its path is a hyperbola. Charged particles fired at the nuclei of atoms scatter along hyperbolic paths.

Tables 11.1–11.3 give additional data for planetary orbits and for the orbits of seven of Earth's artificial satellites (Fig. 11.43). *Vanguard 1* sent back data that revealed differences between the levels of Earth's oceans and provided the first determination of the precise locations of some of the more isolated Pacific islands. The data also verified that the gravitation of the sun and moon would affect the orbits of Earth's satellites and that solar radiation could exert enough pressure to deform an orbit.

Syncom 3 is one of a series of U.S. Department of Defense telecommunications satellites. *Tiros 11* (for "television infrared observation satellite") is one of a series of weather satellites. *GOES 4* (for "geostationary operational environmental satellite") is one of a series of satellites designed to gather information about Earth's atmosphere. Its orbital period, 1436.2 minutes, is nearly the same as Earth's rotational period of 1436.1 minutes, and its orbit is nearly circular ($e = 0.0003$). *Intelsat 5* is a heavy-capacity commercial telecommunications satellite.

Table 11.1 Values of *a*, *e*, and *T* for the major planets

Planet	Semimajor axis $a^\dagger$	Eccentricity *e*	Period *T*
Mercury	57.95	0.2056	87.967 days
Venus	108.11	0.0068	224.701 days
Earth	149.57	0.0167	365.256 days
Mars	227.84	0.0934	1.8808 years
Jupiter	778.14	0.0484	11.8613 years
Saturn	1427.0	0.0543	29.4568 years
Uranus	2870.3	0.0460	84.0081 years
Neptune	4499.9	0.0082	164.784 years
Pluto	5909	0.2481	248.35 years

† Millions of kilometers

Table 11.2 Data on Earth's satellites

Name	Launch date	Time or expected time aloft	Mass at launch (kg)	Period (min)	Perigee height (km)	Apogee height (km)	Semimajor axis *a* (km)	Eccentricity
Sputnik 1	Oct. 1957	57.6 days	83.6	96.2	215	939	6,955	0.052
Vanguard 1	March 1958	300 years	1.47	138.5	649	4,340	8,872	0.208
Syncom 3	Aug. 1964	$>10^6$ years	39	1436.2	35,718	35,903	42,189	0.002
Skylab 4	Nov. 1973	84.06 days	13,980	93.11	422	437	6,808	0.001
Tiros 11	Oct. 1978	500 years	734	102.12	850	866	7,236	0.001
GOES 4	Sept. 1980	$>10^6$ years	627	1436.2	35,776	35,800	42,166	0.0003
Intelsat 5	Dec. 1980	$>10^6$ years	1,928	1417.67	35,143	35,707	41,803	0.007

Table 11.3 Numerical data

Gravitational constant: $G = 6.6720 \times 10^{-11} \text{ Nm}^2\text{kg}^{-2}$
(When you use this value of *G* in a calculation, remember to express force in newtons, distance in meters, mass in kilograms, and time in seconds.)

Sun's mass: 1.99×10^{30} kg

Earth's mass: 5.975×10^{24} kg

Equatorial radius of Earth: 6378.533 km

Polar radius of Earth: 6356.912 km

Earth's rotational period: 1436.1 min

Earth's orbital period: 1 year $= 365.256$ days

Exercises 11.5

Reminder: When a calculation involves the gravitational constant G, express force in newtons, distance in meters, mass in kilograms, and time in seconds.

Calculator Exercises

1. Since the orbit of *Skylab 4* had a semimajor axis of $a = 6808$ km, Kepler's third law with M equal to the earth's mass should give the period. Calculate it. Compare your result with the value in Table 11.2.

2. Earth's distance from the sun at perihelion is approximately 149,577,000 km, and the eccentricity of the earth's orbit about the sun is 0.0167. Find the velocity v_0 of Earth in its orbit at perihelion. (Use Eq. 15.)

3. In July 1965, the USSR launched *Proton I,* weighing 12,200 kg (at launch), with a perigee height of 183 km, an apogee height of 589 km, and a period of 92.25 min. Using the relevant data for the mass of Earth and the gravitational constant G, find the semimajor axis a of the orbit from Eq. (33). Compare your answer with the number you get by adding the perigee and apogee heights to the diameter of the earth.

4. **a)** The *Viking 1* orbiter, which surveyed Mars from August 1975 to June 1976, had a period of 1639 min. Use this and the fact that the mass of Mars is 6.418×10^{23} kg to find the semimajor axis of the *Viking 1* orbit.
   b) The *Viking 1* orbiter was 1499 km from the surface of Mars at its closest point and 35,800 km from the surface at its farthest point. Use this information together with the value you obtained in part (a) to estimate the average diameter of Mars.

5. The *Viking 2* orbiter, which surveyed Mars from September 1975 to August 1976, moved in an ellipse whose semimajor axis was 22,030 km. What was the orbital period? (Express your answer in minutes.)

6. *Geosynchronous orbits.* Several satellites in the earth's equatorial plane have nearly circular orbits whose periods are the same as the earth's rotational period. Such orbits are called **geosynchronous** or **geostationary** because they hold the satellite over the same spot on Earth's surface.

   a) Approximately what is the semimajor axis of a geosynchronous orbit? Give reasons for your answer.
   b) About how high is a geosynchronous orbit above the Earth's surface?
   c) Which of the satellites in Table 11.2 have (nearly) geosynchronous orbits?

7. The mass of Mars is 6.418×10^{23} kg. If a satellite revolving about Mars is to hold a stationary orbit (have the same period as the period of Mars's rotation, which is 1477.4 min), what must the semimajor axis of its orbit be? Give reasons for your answer.

8. The period of the moon's rotation about Earth is 2.36055×10^6 sec. About how far away is the moon?

9. A satellite moves around Earth in a circular orbit. Express the satellite's speed as a function of the orbit's radius.

10. If T is measured in seconds and a in meters, what is the value of T^2/a^3 for planets in our solar system? for satellites orbiting Earth? for satellites orbiting the moon? (The moon's mass is 7.354×10^{22} kg.)

Noncalculator Exercises

11. For what values of v_0 in Eq. (15) is the orbit in Eq. (16) a circle? an ellipse? a parabola? a hyperbola?

12. Show that a planet in a circular orbit moves with a constant speed. (*Hint:* This is a consequence of one of Kepler's laws.)

13. Suppose that $\mathbf{r}$ is the position vector of a particle moving along a plane curve and dA/dt is the rate at which the vector sweeps out area. Without introducing coordinates, and assuming the necessary derivatives exist, give a geometric argument based on increments and limits for the validity of the equation

    $$\frac{dA}{dt} = \frac{1}{2}|\mathbf{r} \times \dot{\mathbf{r}}|.$$

14. Complete the derivation of Kepler's third law (the part following Eq. 34).

15. Two planets, planet A and planet B, are orbiting their sun in circular orbits with A being the inner planet and B being farther away from the sun. Suppose the positions of A and B at time t are

    $$\mathbf{r}_A(t) = 2\cos(2\pi t)\mathbf{i} + 2\sin(2\pi t)\mathbf{j}$$

    and

    $$\mathbf{r}_B(t) = 3\cos(\pi t)\mathbf{i} + 3\sin(\pi t)\mathbf{j},$$

    respectively, where the sun is assumed to be located at the origin and distance is measured in astronomical units. (Notice that planet A moves faster than planet B.)

    The people on planet A regard their planet, not the sun, as the center of their planetary system (their solar system).

    a) Using planet A as the origin of a new coordinate system, give parametric equations for the location of planet B at time t. Write your answer in terms of $\cos(\pi t)$ and $\sin(\pi t)$.
    b) GRAPHER Using planet A as the origin, sketch a graph of the path of planet B.

    This exercise illustrates the difficulty that people before Kepler's time, with an earth-centered (planet A) view of our solar system, had in understanding the motions of the planets (i.e., planet B = Mars). See D. G. Saari's article in *The American Mathematical Monthly,* Vol. 97, Feb. 1990, pp. 105–119.

16. Kepler discovered that the path of the earth around the sun is an ellipse with the sun at one of the foci. Let $\mathbf{r}(t)$ be the position vector from the center of the sun to the center of the earth at time t. Let $\mathbf{w}$ be the vector from the earth's South Pole to North Pole. It is known that $\mathbf{w}$ is constant and not orthogonal to the plane of the ellipse (the earth's axis is tilted). In terms of $\mathbf{r}(t)$ and $\mathbf{w}$, give the mathematical meaning of (i) perihelion, (ii) aphelion, (iii) equinox, (iv) summer solstice, (v) winter solstice.

CHAPTER **11** QUESTIONS TO GUIDE YOUR REVIEW

1. State the rules for differentiating and integrating vector functions. Give examples.

2. How do you define and calculate the velocity, speed, direction of motion, and acceleration of a body moving along a sufficiently differentiable space curve? Give an example.

3. What is special about the derivatives of vector functions of constant length? Give an example.

4. What are the vector and parametric equations for ideal projectile motion? How do you find a projectile's maximum height, flight time, and range? Give examples.

5. How do you define and calculate the length of a segment of a smooth space curve? Give an example. What mathematical assumptions are involved in the definition?

6. How do you measure distance along a smooth curve in space from a preselected base point? Give an example.

7. What is a differentiable curve's unit tangent vector? Give an example.

8. Define curvature, circle of curvature (osculating circle), center of curvature, and radius of curvature for twice-differentiable curves in the plane. Give examples. What curves have zero curvature? constant curvature?

9. What is a plane curve's principal normal vector? When is it defined? Which way does it point? Give an example.

10. How do you define $\mathbf{N}$ and κ for curves in space? How are these quantities related? Give examples.

11. What is a curve's binormal vector? Give an example. How is this vector related to the curve's torsion? Give an example.

12. What formulas are available for writing a moving body's acceleration as a sum of its tangential and normal components? Give an example. Why might one want to write the acceleration this way? What if the body moves at a constant speed? at a constant speed around a circle?

13. State Kepler's laws. To what do they apply?

CHAPTER **11** PRACTICE EXERCISES

Motion in a Cartesian Plane

In Exercises 1 and 2, graph the curves and sketch their velocity and acceleration vectors at the given values of t. Then write $\mathbf{a}$ in the form $\mathbf{a} = a_T \mathbf{T} + a_N \mathbf{N}$ without finding $\mathbf{T}$ and $\mathbf{N}$, and find the value of κ at the given values of t.

1. $\mathbf{r}(t) = (4 \cos t) \mathbf{i} + (\sqrt{2} \sin t) \mathbf{j}, \quad t = 0$ and $\pi/4$

2. $\mathbf{r}(t) = (\sqrt{3} \sec t) \mathbf{i} + (\sqrt{3} \tan t) \mathbf{j}, \quad t = 0$

3. The position of a particle in the plane at time t is

$$\mathbf{r} = \frac{1}{\sqrt{1+t^2}} \mathbf{i} + \frac{t}{\sqrt{1+t^2}} \mathbf{j}.$$

Find the particle's highest speed.

4. Suppose $\mathbf{r}(t) = (e^t \cos t) \mathbf{i} + (e^t \sin t) \mathbf{j}$. Show that the angle between $\mathbf{r}$ and $\mathbf{a}$ never changes. What *is* the angle?

5. At point P, the velocity and acceleration of a particle moving in the plane are $\mathbf{v} = 3 \mathbf{i} + 4 \mathbf{j}$ and $\mathbf{a} = 5 \mathbf{i} + 15 \mathbf{j}$. Find the curvature of the particle's path at P.

6. Find the point on the curve $y = e^x$ where the curvature is greatest.

7. A particle moves around the unit circle in the xy-plane. Its position at time t is $\mathbf{r} = x \mathbf{i} + y \mathbf{j}$, where x and y are differentiable functions of t. Find dy/dt if $\mathbf{v} \cdot \mathbf{i} = y$. Is the motion clockwise, or counterclockwise?

8. You send a message through a pneumatic tube that follows the curve $9y = x^3$ (distance in meters). At the point $(3, 3)$, $\mathbf{v} \cdot \mathbf{i} = 4$ and $\mathbf{a} \cdot \mathbf{i} = -2$. Find the values of $\mathbf{v} \cdot \mathbf{j}$ and $\mathbf{a} \cdot \mathbf{j}$ at $(3, 3)$.

9. A particle moves in the plane so that its velocity and position vectors are always orthogonal. Show that the particle moves in a circle centered at the origin.

10. A circular wheel with radius 1 ft and center C rolls to the right along the x-axis at a half-turn per second. (See accompanying figure.) At time t seconds, the position vector of the point P on the wheel's circumference is

$$\mathbf{r} = (\pi t - \sin \pi t)\,\mathbf{i} + (1 - \cos \pi t)\,\mathbf{j}.$$

a)   Sketch the curve traced by P during the interval $0 \le t \le 3$.
b)   Find $\mathbf{v}$ and $\mathbf{a}$ at $t = 0, 1, 2,$ and 3 and add these vectors to your sketch.
c)   At any given time, what is the forward speed of the topmost point of the wheel? of C?

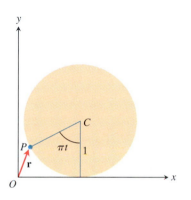

Projectile Motion and Motion in a Plane

11. *Shot put.* A shot leaves the thrower's hand 6.5 ft above the ground at a $45°$ angle at 44 ft/sec. Where is it 3 sec later?

12. *Javelin.* A javelin leaves the thrower's hand 7 ft above the ground at a $45°$ angle at 80 ft/sec. How high does it go?

13. A golf ball is hit with an initial speed v_0 at an angle α to the horizontal from a point that lies at the foot of a straight-sided hill that is inclined at an angle ϕ to the horizontal, where

$$0 < \phi < \alpha < \frac{\pi}{2}.$$

Show that the ball lands at a distance

$$\frac{2v_0{}^2 \cos \alpha}{g \cos^2 \phi} \sin(\alpha - \phi),$$

measured up the face of the hill. Hence, show that the greatest range that can be achieved for a given v_0 occurs when $\alpha = (\phi/2) + (\pi/4)$, i.e., when the initial velocity vector bisects the angle between the vertical and the hill.

14. *The Dictator.* The Civil War mortar Dictator weighed so much (17,120 lb) that it had to be mounted on a railroad car. It had a 13-in. bore and used a 20-lb powder charge to fire a 200-lb shell. The mortar was made by Mr. Charles Knapp in his ironworks

in Pittsburgh, Pennsylvania, and was used by the Union army in 1864 in the siege of Petersburg, Virginia. How far did it shoot? Here we have a difference of opinion. The ordnance manual claimed 4325 yd, while field officers claimed 4752 yd. Assuming a $45°$ firing angle, what muzzle speeds are involved here?

15. *The world's record for popping a champagne cork*

a)   Until 1988, the world's record for popping a champagne cork was 109 ft. 6 in., once held by Captain Michael Hill of the British Royal Artillery (of course). Assuming Cpt. Hill held the bottle neck at ground level at a $45°$ angle, and the cork behaved like an ideal projectile, how fast was the cork going as it left the bottle?
b)   A new world record of 177 ft. 9 in. was set on June 5, 1988, by Prof. Emeritus Heinrich of Rensselaer Polytechnic Institute, firing from 4 ft. above ground level at the Woodbury Vineyards Winery, New York. Assuming an ideal trajectory, what was the cork's initial speed?

16. *Javelin.* In Potsdam in 1988, Petra Felke of (then) East Germany set a women's world record by throwing a javelin 262 ft 5 in.

a)   Assuming that Felke launched the javelin at a $40°$ angle to the horizontal 6.5 ft above the ground, what was the javelin's initial speed?
b)   How high did the javelin go?

17. *Synchronous curves.* By eliminating α from the ideal projectile equations

$$x = (v_0 \cos \alpha)t, \qquad y = (v_0 \sin \alpha)t - \frac{1}{2}gt^2,$$

show that $x^2 + (y + gt^2/2)^2 = v_0{}^2 t^2$. This shows that projectiles launched simultaneously from the origin at the same initial speed will, at any given instant, all lie on the circle of radius $v_0 t$ centered at $(0, -gt^2/2)$, regardless of their launch angle. These circles are the *synchronous curves* of the launching.

18. Show that the radius of curvature of a twice-differentiable plane curve $\mathbf{r}(t) = f(t)\,\mathbf{i} + g(t)\,\mathbf{j}$ is given by the formula

$$\rho = \frac{\dot{x}^2 + \dot{y}^2}{\sqrt{\ddot{x}^2 + \ddot{y}^2 - \ddot{s}^2}}, \quad \text{where } \ddot{s} = \frac{d}{dt}\sqrt{\dot{x}^2 + \dot{y}^2}.$$

19. Express the curvature of the curve

$$\mathbf{r}(t) = \left(\int_0^t \cos\left(\frac{1}{2}\pi\theta^2\right)d\theta\right)\mathbf{i} + \left(\int_0^t \sin\left(\frac{1}{2}\pi\theta^2\right)d\theta\right)\mathbf{j}$$

as a function of the directed distance s measured along the curve from the origin (Fig. 11.44).

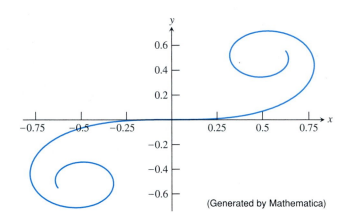

11.44 The curve in Exercise 19.

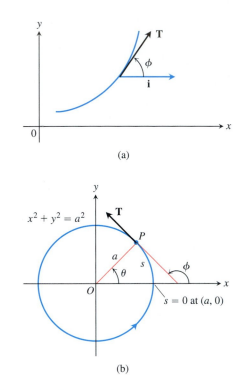

(a)

(b)

11.45 Figures for Exercise 20.

20. *An alternative definition of curvature in the plane.* An alternative definition gives the curvature of a sufficiently differentiable plane curve to be $|d\phi/ds|$, where ϕ is the angle between $\mathbf{T}$ and $\mathbf{i}$ (Fig. 11.45(a)). Figure 11.45(b) shows the distance s measured counterclockwise around the circle $x^2 + y^2 = a^2$ from the point $(a, 0)$ to a point P, along with the angle ϕ at P. Calculate the circle's curvature using the alternative definition. (*Hint :* $\phi = \theta + \pi/2$.)

Motion in Space

Find the lengths of the curves in Exercises 21 and 22.

21. $\mathbf{r}(t) = (2\cos t)\,\mathbf{i} + (2\sin t)\,\mathbf{j} + t^2\,\mathbf{k}, \quad 0 \le t \le \pi/4$

22. $\mathbf{r}(t) = (3\cos t)\,\mathbf{i} + (3\sin t)\,\mathbf{j} + 2t^{3/2}\,\mathbf{k}, \quad 0 \le t \le 3$

In Exercises 23–26, find $\mathbf{T}, \mathbf{N}, \mathbf{B}, \kappa,$ and τ at the given value of t.

23. $\mathbf{r}(t) = \frac{4}{9}(1+t)^{3/2}\,\mathbf{i} + \frac{4}{9}(1-t)^{3/2}\,\mathbf{j} + \frac{1}{3}t\,\mathbf{k}, \quad t = 0$

24. $\mathbf{r}(t) = (e^t \sin 2t)\,\mathbf{i} + (e^t \cos 2t)\,\mathbf{j} + 2e^t\,\mathbf{k}, \quad t = 0$

25. $\mathbf{r}(t) = t\,\mathbf{i} + \frac{1}{2}e^{2t}\,\mathbf{j}, \quad t = \ln 2$

26. $\mathbf{r}(t) = (3\cosh 2t)\,\mathbf{i} + (3\sinh 2t)\,\mathbf{j} + 6t\,\mathbf{k}, \quad t = \ln 2$

In Exercises 27 and 28, write $\mathbf{a}$ in the form $\mathbf{a} = a_{\mathrm{T}}\,\mathbf{T} + a_{\mathrm{N}}\,\mathbf{N}$ at $t = 0$ without finding $\mathbf{T}$ and $\mathbf{N}$.

27. $\mathbf{r}(t) = (2 + 3t + 3t^2)\,\mathbf{i} + (4t + 4t^2)\,\mathbf{j} - (6\cos t)\,\mathbf{k}$

28. $\mathbf{r}(t) = (2 + t)\,\mathbf{i} + (t + 2t^2)\,\mathbf{j} + (1 + t^2)\,\mathbf{k}$

29. Find $\mathbf{T}, \mathbf{N}, \mathbf{B}, \kappa,$ and τ as functions of t if $\mathbf{r}(t) = (\sin t)\,\mathbf{i} + (\sqrt{2}\cos t)\,\mathbf{j} + (\sin t)\,\mathbf{k}$.

30. At what times in the interval $0 \le t \le \pi$ are the velocity and acceleration vectors of the motion $\mathbf{r}(t) = \mathbf{i} + (5\cos t)\,\mathbf{j} + (3\sin t)\,\mathbf{k}$ orthogonal?

31. The position of a particle moving in space at time $t \ge 0$ is

$$\mathbf{r}(t) = 2\,\mathbf{i} + \left(4\sin\frac{t}{2}\right)\mathbf{j} + \left(3 - \frac{t}{\pi}\right)\mathbf{k}.$$

Find the first time $\mathbf{r}$ is orthogonal to the vector $\mathbf{i} - \mathbf{j}$.

32. Find equations for the osculating, normal, and rectifying planes of the curve $\mathbf{r}(t) = t\,\mathbf{i} + t^2\,\mathbf{j} + t^3\,\mathbf{k}$ at the point $(1, 1, 1)$.

33. Find parametric equations for the line that is tangent to the curve $\mathbf{r}(t) = e^t\,\mathbf{i} + (\sin t)\,\mathbf{j} + \ln(1 - t)\,\mathbf{k}$ at $t = 0$.

34. Find parametric equations for the line tangent to the helix $\mathbf{r}(t) = (\sqrt{2}\cos t)\,\mathbf{i} + (\sqrt{2}\sin t)\,\mathbf{j} + t\,\mathbf{k}$ at the point where $t = \pi/4$.

35. *The view from Skylab 4.* What percentage of Earth's surface area could the astronauts see when *Skylab 4* was at its apogee height, 437 km above the surface? To find out, model the visible surface as the surface generated by revolving the circular arc *GT*, shown here, about the y-axis. Then carry out these steps:

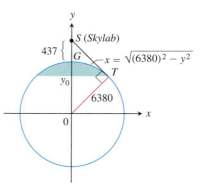

1. Use similar triangles in the figure to show that $y_0/6380 = 6380/(6380 + 437)$. Solve for y_0.

2. To four significant digits, calculate the visible area as

$$VA = \int_{y_0}^{6380} 2\pi x \sqrt{1 + \left(\frac{dx}{dy}\right)^2}\, dy.$$

3. Express the result as a percentage of Earth's surface area.

CHAPTER **11**   ADDITIONAL EXERCISES–THEORY, EXAMPLES, APPLICATIONS

Applications

1. A straight river is 100 m wide. A rowboat leaves the far shore at time $t = 0$. The person in the boat rows at a rate of 20 m/min, always toward the near shore. The velocity of the river at (x, y) is $\mathbf{v} = \left(-\dfrac{1}{250}(y - 50)^2 + 10\right)\mathbf{i}$ m/min, $0 < y < 100$.

    a) Given that $\mathbf{r}(0) = 0\mathbf{i} + 100\mathbf{j}$, what is the position of the boat at time t?

    b) How far downstream will the boat land on the near shore?

2. A straight river is 20 m wide. The velocity of the river at (x, y) is

$$\mathbf{v} = -\frac{3x(20 - x)}{100}\mathbf{j} \text{ m/min}, \quad 0 \le x \le 20.$$

A boat leaves the shore at $(0, 0)$ and travels through the water with a constant velocity. It arrives at the opposite shore at $(20, 0)$. The speed of the boat is always $\sqrt{20}$ m/min.

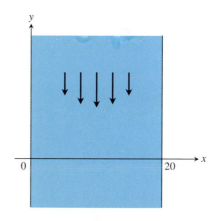

a) Find the velocity of the boat.
b) Find the location of the boat at time t.
c) Sketch the path of the boat.

3. The line through the origin and the point $A(1, 1, 1)$ is the axis of rotation of a rigid body rotating with a constant angular speed of 3/2 rad/sec. The rotation appears to be clockwise when we look toward the origin from A. Find the velocity $\mathbf{v}$ of the point of the body that is at the position $B(1, 3, 2)$.

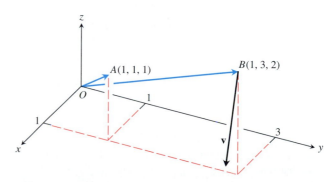

4. The curve

$$\mathbf{r}(t) = (2\sqrt{t}\cos t)\,\mathbf{i} + (3\sqrt{t}\sin t)\,\mathbf{j} + \sqrt{1-t}\,\mathbf{k}, \quad 0 \le t \le 1,$$

lies on a quadric surface. Describe the surface and find an equation for it.

5. A frictionless particle P, starting from rest at time $t = 0$ at the point $(a, 0, 0)$, slides down the helix

$$\mathbf{r}(\theta) = (a\cos\theta)\,\mathbf{i} + (a\sin\theta)\,\mathbf{j} + b\theta\,\mathbf{k} \quad (a, b > 0)$$

under the influence of gravity, as in Fig. 11.46. The θ in this equation is the cylindrical coordinate θ and the helix is the curve $r = a, z = b\theta, \theta \ge 0$, in cylindrical coordinates. We assume θ to be a differentiable function of t for the motion. The law of conservation of energy tells us that the particle's speed after it has fallen straight down a distance z is $\sqrt{2gz}$, where g is the constant acceleration of gravity.

a) Find the angular velocity $d\theta/dt$ when $\theta = 2\pi$.

b) Express the particle's θ- and z-coordinates as functions of t.

c) Express the tangential and normal components of the velocity $d\mathbf{r}/dt$ and acceleration $d^2\mathbf{r}/dt^2$ as functions of t. Does the acceleration have any nonzero component in the direction of the binormal vector $\mathbf{B}$?

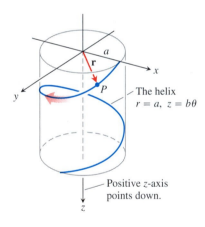

11.46 The circular helix in Exercise 5.

6. Suppose the curve in Exercise 5 is replaced by the conical helix $r = a\theta, z = b\theta$ shown in Fig. 11.47.

a) Express the angular velocity $d\theta/dt$ as a function of θ.

b) Express the distance the particle travels along the helix as a function of θ.

Polar Coordinate Systems and Motion in Space

7. Deduce from the orbit equation

$$r = \frac{(1+e)r_0}{1 + e\cos\theta}$$

that a planet is closest to its sun when $\theta = 0$ and show that $r = r_0$ at that time.

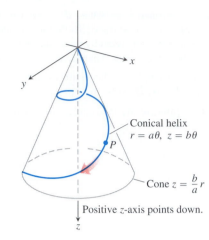

Conical helix
$r = a\theta,\ z = b\theta$

Cone $z = \dfrac{b}{a}r$

Positive z-axis points down.

11.47 The conical helix in Exercise 6.

8. *A Kepler equation.* The problem of locating a planet in its orbit at a given time and date eventually leads to solving "Kepler" equations of the form

$$f(x) = x - 1 - \frac{1}{2}\sin x = 0.$$

a) Show that this particular equation has a solution between $x = 0$ and $x = 2$.

b) With your computer or calculator in radian mode, use Newton's method to find the solution to as many places as you can.

9. In Section 11.5, we found the velocity of a particle moving in the plane to be

$$\mathbf{v} = \dot{x}\,\mathbf{i} + \dot{y}\,\mathbf{j} = \dot{r}\,\mathbf{u}_r + r\dot{\theta}\,\mathbf{u}_\theta.$$

a) Express $\dot{x}$ and $\dot{y}$ in terms of $\dot{r}$ and $r\dot{\theta}$ by evaluating the dot products $\mathbf{v} \cdot \mathbf{i}$ and $\mathbf{v} \cdot \mathbf{j}$.

b) Express $\dot{r}$ and $r\dot{\theta}$ in terms of $\dot{x}$ and $\dot{y}$ by evaluating the dot products $\mathbf{v} \cdot \mathbf{u}_r$ and $\mathbf{v} \cdot \mathbf{u}_\theta$.

10. Express the curvature of a twice-differentiable curve $r = f(\theta)$ in the polar coordinate plane in terms of f and its derivatives.

11. A slender rod through the origin of the polar coordinate plane rotates (in the plane) about the origin at the rate of 3 rad/min. A beetle starting from the point $(2, 0)$ crawls along the rod toward the origin at the rate of 1 in./min.

a) Find the beetle's acceleration and velocity in polar form when it is halfway to (1 in. from) the origin.

b) CALCULATOR To the nearest tenth of an inch, what will be the length of the path the beetle has traveled by the time it reaches the origin?

12. *Conservation of angular momentum.* Let $\mathbf{r}(t)$ denote the position in space of a moving object at time t. Suppose the force acting on the object at time t is

$$\mathbf{F}(t) = -\frac{c}{|\mathbf{r}(t)|^3}\mathbf{r}(t),$$

where c is a constant. In physics the **angular momentum** of an object at time t is defined to be $\mathbf{L}(t) = \mathbf{r}(t) \times m\mathbf{v}(t)$, where m is the mass of the object and $\mathbf{v}(t)$ is the velocity. Prove that angular momentum is a conserved quantity; i.e., prove that $\mathbf{L}(t)$ is a constant vector, independent of time. Remember Newton's law $\mathbf{F} = m\mathbf{a}$. (This is a calculus problem, not a physics problem.)

Cylindrical Coordinate Systems

13. *Unit vectors for position and motion in cylindrical coordinates.* When the position of a particle moving in space is given in cylindrical coordinates, the unit vectors we use to describe its position and motion are

$$\mathbf{u}_r = (\cos\theta)\,\mathbf{i} + (\sin\theta)\,\mathbf{j}, \quad \mathbf{u}_\theta = -(\sin\theta)\,\mathbf{i} + (\cos\theta)\,\mathbf{j},$$

and $\mathbf{k}$ (Fig. 11.48). The particle's position vector is then $\mathbf{r} = r\,\mathbf{u}_r + z\,\mathbf{k}$, where r is the positive polar distance coordinate of the particle's position.

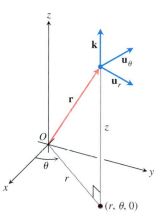

11.48 The unit vectors for describing motion in cylindrical coordinates (Exercise 13).

a) Show that $\mathbf{u}_r$, $\mathbf{u}_\theta$, and $\mathbf{k}$, in this order, form a right-handed frame of unit vectors.

b) Show that

$$\frac{d\mathbf{u}_r}{d\theta} = \mathbf{u}_\theta \quad \text{and} \quad \frac{d\mathbf{u}_\theta}{d\theta} = -\mathbf{u}_r.$$

c) Assuming that the necessary derivatives with respect to t exist, express $\mathbf{v} = \dot{\mathbf{r}}$ and $\mathbf{a} = \ddot{\mathbf{r}}$ in terms of $\mathbf{u}_r$, $\mathbf{u}_\theta$, $\mathbf{k}$, $\dot{r}$, and $\dot{\theta}$. (The dots indicate derivatives with respect to t: $\dot{\mathbf{r}}$ means $d\mathbf{r}/dt$, $\ddot{\mathbf{r}}$ means $d^2\mathbf{r}/dt^2$, and so on.) Section 11.5 derives these formulas and shows how the vectors mentioned here are used in describing planetary motion.

14. *Arc length in cylindrical coordinates*

a) Show that when you express $ds^2 = dx^2 + dy^2 + dz^2$ in terms of cylindrical coordinates, you get $ds^2 = dr^2 + r^2 d\theta^2 + dz^2$.

b) Interpret this result geometrically in terms of the edges and a diagonal of a box. Sketch the box.

c) Use the result in (a) to find the length of the curve $r = e^\theta, z = e^\theta, 0 \le \theta \le \ln 8$.

Spherical Coordinate Systems

15. *Unit vectors for position and motion in spherical coordinates.* Hold two of the three spherical coordinates ρ, ϕ, θ of a point P in space constant while letting the remaining coordinate increase. Let $\mathbf{u}$, with a subscript corresponding to the increasing coordinate, be the unit vector that points in the direction in which P then starts to move. The three resulting unit vectors, $\mathbf{u}_\rho, \mathbf{u}_\phi$, and $\mathbf{u}_\theta$ at P are shown in Fig. 11.49.

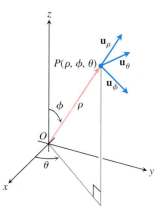

11.49 The unit vectors for describing motion in spherical coordinates (Exercise 15).

a) Express $\mathbf{u}_\rho$, $\mathbf{u}_\phi$, and $\mathbf{u}_\theta$ in terms of $\mathbf{i}$, $\mathbf{j}$, and $\mathbf{k}$.

b) Show that $\mathbf{u}_\rho \cdot \mathbf{u}_\phi = 0$.

c) Show that $\mathbf{u}_\theta = \mathbf{u}_\rho \times \mathbf{u}_\phi$.

d) Show that $\mathbf{u}_\rho$, $\mathbf{u}_\phi$, and $\mathbf{u}_\theta$, in that order, make a right-handed frame of mutually orthogonal vectors.

16. *Arc length in spherical coordinates*

a) Show that when you express $ds^2 = dx^2 + dy^2 + dz^2$ in terms of spherical coordinates, you get $ds^2 = d\rho^2 + \rho^2 d\phi^2 + \rho^2 \sin^2\phi\, d\theta^2$.

b) Interpret this result geometrically in terms of the edges and a diagonal of a "box" cut from a solid sphere. Sketch the box.

c) Use the result in (a) to find the length of the curve $\rho = 2e^\theta, \phi = \pi/6, 0 \le \theta \le \ln 8$.

Multivariable Functions and Partial Derivatives

OVERVIEW Functions with two or more independent variables appear more often in science than functions of a single variable, and their calculus is even richer. Their derivatives are more varied and more interesting because of the different ways in which the variables can interact. Their integrals lead to a greater variety of applications. The studies of probability, statistics, fluid dynamics, and electricity, to mention only a few, all lead in natural ways to functions of more than one variable. The mathematics of these functions is one of the finest achievements in science.

12.1 Functions of Several Variables

Many functions depend on more than one independent variable. The function $V = \pi r^2 h$ calculates the volume of a right circular cylinder from its radius and height. The function $f(x, y) = x^2 + y^2$ calculates the height of the paraboloid $z = x^2 + y^2$ above the point $P(x, y)$ from the two coordinates of P. In this section, we define functions of more than one independent variable and discuss ways to graph them.

Functions and Variables

Real-valued functions of several independent real variables are defined much the way you would imagine from the single-variable case. The domains are sets of ordered pairs (triples, quadruples, whatever) of real numbers, and the ranges are sets of real numbers of the kind we have worked with all along.

> **Definitions**
>
> Suppose D is a set of n-tuples of real numbers $(x_1, x_2, \ldots, x_n)$. A **real-valued function** f on D is a rule that assigns a real number
>
> $$w = f(x_1, x_2, \ldots, x_n)$$
>
> to each element in D. The set D is the function's **domain.** The set of w-values taken on by f is the function's **range.** The symbol w is the **dependent variable** of f, and f is said to be a function of the n **independent variables** x_1 to x_n. We also call the x's the function's **input variables** and call w the function's **output variable.**

If f is a function of two independent variables, we usually call the independent variables x and y and picture the domain of f as a region in the xy-plane. If f is a function of three independent variables, we call the variables x, y, and z and picture the domain as a region in space.

In applications, we tend to use letters that remind us of what the variables stand for. To say that the volume of a right circular cylinder is a function of its radius and height, we might write $V = f(r, h)$. To be more specific, we might replace the notation $f(r, h)$ by the formula that calculates the value of V from the values of r and h, and write $V = \pi r^2 h$. In either case, r and h would be the independent variables and V the dependent variable of the function.

As usual, we evaluate functions defined by formulas by substituting the values of the independent variables in the formula and calculating the corresponding value of the dependent variable.

EXAMPLE 1   The value of $f(x, y, z) = \sqrt{x^2 + y^2 + z^2}$ at the point $(3, 0, 4)$ is

$$f(3, 0, 4) = \sqrt{(3)^2 + (0)^2 + (4)^2} = \sqrt{25} = 5.$$ ❑

Domains

In defining functions of more than one variable, we follow the usual practice of excluding inputs that lead to complex numbers or division by zero. If $f(x, y) = \sqrt{y - x^2}$, y cannot be less than x^2. If $f(x, y) = 1/(xy)$, xy cannot be zero. The domains of functions are otherwise assumed to be the largest sets for which the defining rules generate real numbers.

EXAMPLE 2   *Functions of two variables*

Function	Domain	Range
$w = \sqrt{y - x^2}$	$y \geq x^2$	$[0, \infty)$
$w = \dfrac{1}{xy}$	$xy \neq 0$	$(-\infty, 0) \cup (0, \infty)$
$w = \sin xy$	Entire plane	$[-1, 1]$

❑

EXAMPLE 3   *Functions of three variables*

Function	Domain	Range
$w = \sqrt{x^2 + y^2 + z^2}$	Entire space	$[0, \infty)$
$w = \dfrac{1}{x^2 + y^2 + z^2}$	$(x, y, z) \neq (0, 0, 0)$	$(0, \infty)$
$w = xy \ln z$	Half-space $z > 0$	$(-\infty, \infty)$

❑

The domains of functions defined on portions of the plane can have interior points and boundary points just the way the domains of functions defined on intervals of the real line can.

(a) Interior point

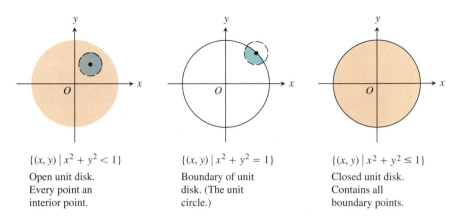

$\{(x, y) \mid x^2 + y^2 < 1\}$
Open unit disk.
Every point an
interior point.

$\{(x, y) \mid x^2 + y^2 = 1\}$
Boundary of unit
disk. (The unit
circle.)

$\{(x, y) \mid x^2 + y^2 \le 1\}$
Closed unit disk.
Contains all
boundary points.

12.2 Interior points and boundary points of the unit disk in the plane.

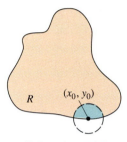

(b) Boundary point

12.1 Interior points and boundary points of a plane region R. An interior point is necessarily a point of R. A boundary point of R need not belong to R.

Definitions

A point (x_0, y_0) in a region (set) R in the xy-plane is an **interior point** of R if it is the center of a disk that lies entirely in R (Fig. 12.1). A point (x_0, y_0) is a **boundary point** of R if every disk centered at (x_0, y_0) contains points that lie outside of R as well as points that lie in R. (The boundary point itself need not belong to R.)

The interior points of a region, as a set, make up the **interior** of the region. The region's boundary points make up its **boundary**. A region is **open** if it consists entirely of interior points. A region is **closed** if it contains all of its boundary points (Fig. 12.2).

As with intervals of real numbers, some regions in the plane are neither open nor closed. If you start with the open disk in Fig. 12.2 and add to it some but not all of its boundary points, the resulting set is neither open nor closed. The boundary points that *are* there keep the set from being open. The absence of the remaining boundary points keeps the set from being closed.

Definitions

A region in the plane is **bounded** if it lies inside a disk of fixed radius. A region is **unbounded** if it is not bounded.

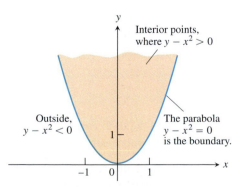

12.3 The domain of $f(x, y) = \sqrt{y - x^2}$ consists of the shaded region and its bounding parabola $y = x^2$.

EXAMPLE 4

Bounded sets in the plane: Line segments, triangles, interiors of triangles, rectangles, disks

Unbounded sets in the plane: Lines, coordinate axes, the graphs of functions defined on infinite intervals, quadrants, half-planes, the plane itself   ❑

EXAMPLE 5   The domain of the function $f(x, y) = \sqrt{y - x^2}$ is closed and unbounded (Fig. 12.3). The parabola $y = x^2$ is the boundary of the domain. The points above the parabola make up the domain's interior.   ❑

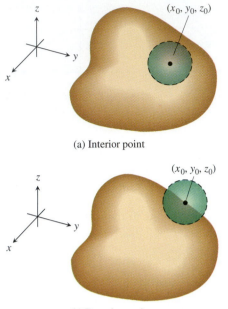

(a) Interior point

(b) Boundary point

12.4 Interior points and boundary points of a region in space.

The definitions of interior, boundary, open, closed, bounded, and unbounded for regions in space are similar to those for regions in the plane. To accommodate the extra dimension, we use balls instead of disks. A **closed ball** consists of the region of points inside a sphere together with the sphere. An **open ball** is the region of points inside a sphere without the bounding sphere.

> **Definitions**
> A point (x_0, y_0, z_0) in a region D in space is an **interior point** of D if it is the center of a ball that lies entirely in D (Fig. 12.4). A point (x_0, y_0, z_0) is a **boundary point** of D if every sphere centered at (x_0, y_0, z_0) encloses points that lie outside D as well as points that lie inside D. The **interior** of D is the set of interior points of D. The **boundary** of D is the set of boundary points of D.
>
> A region D is **open** if it consists entirely of interior points. A region is **closed** if it contains its entire boundary.

EXAMPLE 6

Open sets in space:	Open balls; the open half-space $z > 0$; the first octant (bounding planes absent); space itself
Closed sets in space:	Lines; planes; closed balls; the closed half-space $z \geq 0$; the first octant together with its bounding planes; space itself
Neither open nor closed:	A closed ball with part of its bounding sphere removed; solid cube with a missing face, edge, or corner point

Graphs and Level Curves of Functions of Two Variables

There are two standard ways to picture the values of a function $f(x, y)$. One is to draw and label curves in the domain on which f has a constant value. The other is to sketch the surface $z = f(x, y)$ in space.

> **Definitions**
> The set of points in the plane where a function $f(x, y)$ has a constant value $f(x, y) = c$ is called a **level curve** of f. The set of all points $(x, y, f(x, y))$ in space, for (x, y) in the domain of f, is called the **graph** of f. The graph of f is also called the **surface $z = f(x, y)$.**

EXAMPLE 7 Graph $f(x, y) = 100 - x^2 - y^2$ and plot the level curves $f(x, y) = 0$, $f(x, y) = 51$, and $f(x, y) = 75$ in the domain of f in the plane.

Solution The domain of f is the entire xy-plane, and the range of f is the set of real numbers less than or equal to 100. The graph is the paraboloid $z = 100 - x^2 - y^2$, a portion of which is shown in Fig. 12.5.

The level curve $f(x, y) = 0$ is the set of points in the xy-plane at which

$$f(x, y) = 100 - x^2 - y^2 = 0, \qquad \text{or} \qquad x^2 + y^2 = 100,$$

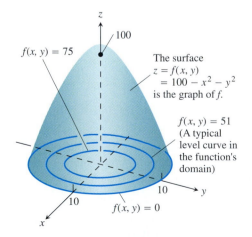

12.5 The graph and selected level curves of the function $f(x, y) = 100 - x^2 - y^2$.

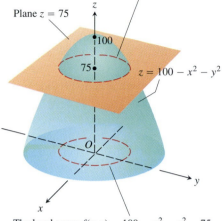

The contour line $f(x, y) = 100 - x^2 - y^2 = 75$ is the circle $x^2 + y^2 = 25$ in the plane $z = 75$.

Plane $z = 75$

$z = 100 - x^2 - y^2$

The level curve $f(x, y) = 100 - x^2 - y^2 = 75$ is the circle $x^2 + y^2 = 25$ in the xy-plane.

12.6 The graph of $f(x, y) = 100 - x^2 - y^2$ and its intersection with the plane $z = 75$.

which is the circle of radius 10 centered at the origin. Similarly, the level curves $f(x, y) = 51$ and $f(x, y) = 75$ (Fig. 12.5) are the circles

$$f(x, y) = 100 - x^2 - y^2 = 51, \quad \text{or} \quad x^2 + y^2 = 49,$$
$$f(x, y) = 100 - x^2 - y^2 = 75, \quad \text{or} \quad x^2 + y^2 = 25.$$

The level curve $f(x, y) = 100$ consists of the origin alone. (It is still a level curve.)

Contour Lines

The curve in space in which the plane $z = c$ cuts a surface $z = f(x, y)$ is made up of the points that represent the function value $f(x, y) = c$. It is called the **contour line** $f(x, y) = c$ to distinguish it from the level curve $f(x, y) = c$ in the domain of f. Figure 12.6 shows the contour line $f(x, y) = 75$ on the surface $z = 100 - x^2 - y^2$ defined by the function $f(x, y) = 100 - x^2 - y^2$. The contour line lies directly above the circle $x^2 + y^2 = 25$, which is the level curve $f(x, y) = 75$ in the function's domain.

Not everyone makes this distinction, however, and you may wish to call both kinds of curves by a single name and rely on context to convey which one you have in mind. On most maps, for example, the curves that represent constant elevation (height above sea level) are called contours, not level curves (Fig. 12.7).

Level Surfaces of Functions of Three Variables

In the plane, the points where a function of two independent variables has a constant value $f(x, y) = c$ make a curve in the function's domain. In space, the points where a function of three independent variables has a constant value $f(x, y, z) = c$ make a surface in the function's domain.

Definition
The set of points (x, y, z) in space where a function of three independent variables has a constant value $f(x, y, z) = c$ is called a **level surface** of f.

12.7 Contours on Mt. Washington in north central New Hampshire. The streams, which follow paths of steepest descent, run perpendicular to the contours. So does the Cog Railway.

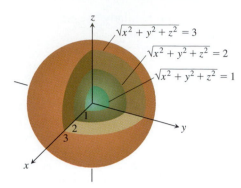

12.8 The level surfaces of $f(x, y, z) = \sqrt{x^2 + y^2 + z^2}$ are concentric spheres.

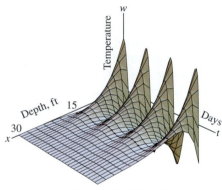

(Generated by Mathematica)

12.9 This computer-generated graph of

$$w = \cos(1.7 \times 10^{-2}t - 0.2x)e^{-0.2x}$$

shows the seasonal variation of the temperature below ground as a fraction of surface temperature. At $x = 15$ ft the variation is only 5% of the variation at the surface. At $x = 30$ ft the variation is less than 0.25% of the surface variation. (Adapted from art provided by Norton Starr for G. C. Berresford's "Differential Equations and Root Cellars," *The UMAP Journal*, Vol. 2, No. 3 [1981], pp. 53–75.)

EXAMPLE 8   Describe the level surfaces of the function

$$f(x, y, z) = \sqrt{x^2 + y^2 + z^2}.$$

Solution   The value of f is the distance from the origin to the point (x, y, z). Each level surface $\sqrt{x^2 + y^2 + z^2} = c$, $c > 0$, is a sphere of radius c centered at the origin. Figure 12.8 shows a cutaway view of three of these spheres. The level surface $\sqrt{x^2 + y^2 + z^2} = 0$ consists of the origin alone.

We are not graphing the function here. The graph of the function, made up of the points $(x, y, z, \sqrt{x^2 + y^2 + z^2})$, lies in a four-variable space. Instead, we are looking at level surfaces in the function's domain.

The function's level surfaces show how the function's values change as we move through its domain. If we remain on a sphere of radius c centered at the origin, the function maintains a constant value, namely c. If we move from one sphere to another, the function's value changes. It increases if we move away from the origin and decreases if we move toward the origin. The way the function's values change depends on the direction we take. The dependence of change on direction is important. We will return to it in Section 12.7.   ❑

Computer Graphing

The three-dimensional graphing programs for computers make it possible to graph functions of two variables with only a few keystrokes. We can often get information more quickly from a graph than from a formula.

EXAMPLE 9   Figure 12.9 shows a computer-generated graph of the function $w = \cos(1.7 \times 10^{-2}t - 0.2x)\,e^{-0.2x}$, where t is in days and x is in feet. The graph shows how the temperature beneath the earth's surface varies with time. The variation is given as a fraction of the variation at the surface. At a depth of 15 ft, the variation (change in vertical amplitude in the figure) is about 5 percent of the surface variation. At 30 ft, there is almost no variation during the year.

The graph also shows that the temperature 15 ft below the surface is about half a year out of phase with the surface temperature. When the temperature is lowest on the surface (late January, say) it is at its highest 15 ft below. Fifteen feet below the ground, the seasons are reversed.   ❑

Exercises 12.1

Domain, Range, and Level Curves

In Exercises 1–12, (a) find the function's domain, (b) find the function's range, (c) describe the function's level curves, (d) find the boundary of the function's domain, (e) determine if the domain is an open region, a closed region, or neither, and (f) decide if the domain is bounded or unbounded.

1. $f(x, y) = y - x$

2. $f(x, y) = \sqrt{y - x}$

3. $f(x, y) = 4x^2 + 9y^2$

4. $f(x, y) = x^2 - y^2$

5. $f(x, y) = xy$

6. $f(x, y) = y/x^2$

7. $f(x, y) = \dfrac{1}{\sqrt{16 - x^2 - y^2}}$

8. $f(x, y) = \sqrt{9 - x^2 - y^2}$

9. $f(x, y) = \ln(x^2 + y^2)$

10. $f(x, y) = e^{-(x^2 + y^2)}$

11. $f(x, y) = \sin^{-1}(y - x)$

12. $f(x, y) = \tan^{-1}\left(\dfrac{y}{x}\right)$

Identifying Surfaces and Level Curves

Exercises 13–18 show level curves for the functions graphed in (a)–(f). Match each set of curves with the appropriate function.

13.

14.

15.

16.

17.

18.

(a)

$$z = (\cos x)(\cos y)\, e^{-\sqrt{x^2+y^2}/4}$$

(b)

$$z = -\frac{xy^2}{x^2 + y^2}$$

(c)

$$z = \frac{1}{(4x^2 + y^2)}$$

(d)

$$z = e^{-y}\cos x$$

(e)

$$z = \frac{xy(x^2 - y^2)}{(x^2 + y^2)}$$

(f)

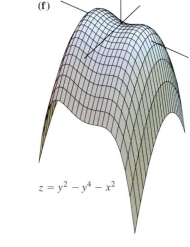

$$z = y^2 - y^4 - x^2$$

Identifying Functions of Two Variables

Display the values of the functions in Exercises 19–28 in two ways: (a) by sketching the surface $z = f(x, y)$ and (b) by drawing an assortment of level curves in the function's domain. Label each level curve with its function value.

19. $f(x, y) = y^2$ **20.** $f(x, y) = 4 - y^2$

21. $f(x, y) = x^2 + y^2$ **22.** $f(x, y) = \sqrt{x^2 + y^2}$

23. $f(x, y) = -(x^2 + y^2)$ **24.** $f(x, y) = 4 - x^2 - y^2$

25. $f(x, y) = 4x^2 + y^2$ **26.** $f(x, y) = 4x^2 + y^2 + 1$

27. $f(x, y) = 1 - |y|$ **28.** $f(x, y) = 1 - |x| - |y|$

Level Surfaces

In Exercises 29–36, sketch a typical level surface for the function.

29. $f(x, y, z) = x^2 + y^2 + z^2$

30. $f(x, y, z) = \ln(x^2 + y^2 + z^2)$

31. $f(x, y, z) = x + z$ **32.** $f(x, y, z) = z$

33. $f(x, y, z) = x^2 + y^2$ **34.** $f(x, y, z) = y^2 + z^2$

35. $f(x, y, z) = z - x^2 - y^2$

36. $f(x, y, z) = (x^2/25) + (y^2/16) + (z^2/9)$

Finding a Level Curve

In Exercises 37–40, find an equation for the level curve of the function $f(x, y)$ that passes through the given point.

37. $f(x, y) = 16 - x^2 - y^2$, $\left(2\sqrt{2}, \sqrt{2}\right)$

38. $f(x, y) = \sqrt{x^2 - 1}$, $(1, 0)$

39. $f(x, y) = \int_x^y \dfrac{dt}{1 + t^2}$, $\left(-\sqrt{2}, \sqrt{2}\right)$

40. $f(x, y) = \sum_{n=0}^{\infty} \left(\dfrac{x}{y}\right)^n$, $(1, 2)$

Finding a Level Surface

In Exercises 41–44, find an equation for the level surface of the function through the given point.

41. $f(x, y, z) = \sqrt{x - y} - \ln z$, $(3, -1, 1)$

42. $f(x, y, z) = \ln(x^2 + y + z^2)$, $(-1, 2, 1)$

43. $g(x, y, z) = \sum_{n=0}^{\infty} \dfrac{(x + y)^n}{n! \, z^n}$, $(\ln 2, \ln 4, 3)$

44. $g(x, y, z) = \int_x^y \dfrac{d\theta}{\sqrt{1 - \theta^2}} + \int_{\sqrt{2}}^z \dfrac{dt}{t\sqrt{t^2 - 1}}$, $(0, 1/2, 2)$

Theory and Examples

45. *The maximum value of a function on a line in space.* Does the function $f(x, y, z) = xyz$ have a maximum value on the line $x = 20 - t$, $y = t$, $z = 20$? If so, what is it? Give reasons for your answer. (*Hint:* Along the line, $w = f(x, y, z)$ is a differentiable function of t.)

46. *The minimum value of a function on a line in space.* Does the function $f(x, y, z) = xy - z$ have a minimum value on the line $x = t - 1$, $y = t - 2$, $z = t + 7$? If so, what is it? Give reasons for your answer. (*Hint:* Along the line, $w = f(x, y, z)$ is a differentiable function of t.)

47. *The Concorde's sonic booms.* The width w of the region in which people on the ground hear the *Concorde's* sonic boom directly, not reflected from a layer in the atmosphere, is a function of

$T = $ air temperature at ground level (in degrees Kelvin),

$h = $ the *Concorde's* altitude (in km),

$d = $ the vertical temperature gradient (temperature drop in degrees Kelvin per km).

The formula for w is

$$w = 4(Th/d)^{1/2}.$$

See Fig. 12.10.

The Washington-bound *Concorde* approaches the United States from Europe on a course that takes it south of Nantucket Island at an altitude of 16.8 km. If the surface temperature is 290 K and the vertical temperature gradient is 5 K/km, how many

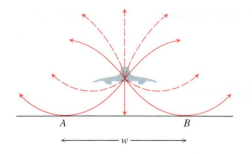

Sonic boom carpet

12.10 Sound waves from the *Concorde* bend as the temperature changes above and below the altitude at which the plane flies. The sonic boom carpet is the region on the ground that receives shock waves directly from the plane, not reflected from the atmosphere or diffracted along the ground. The carpet is determined by the grazing rays striking the ground from the point directly under the plane (Exercise 47).

kilometers south of Nantucket must the plane be flown to keep its sonic boom carpet away from the island? (From "Concorde Sonic Booms as an Atmospheric Probe" by N. K. Balachandra, W. L. Donn, and D. H. Rind, *Science*, July 1, 1977, Vol. 197, pp. 47–49).

48. As you know, the graph of a real-valued function of a single real variable is a set in a two-coordinate space. The graph of a real-valued function of two independent real variables is a set in a three-coordinate space. The graph of a real-valued function of three independent real variables is a set in a four-coordinate space. How would you define the graph of a real-valued function $f(x_1, x_2, x_3, x_4)$ of four independent real variables? How would you define the graph of a real-valued function $f(x_1, x_2, x_3, \ldots, x_n)$ of n independent real variables?

⊛ CAS Explorations and Projects—Explicit Surfaces

Use a CAS to perform the following steps for each of the functions in Exercises 49–52.

a) Plot the surface over the given rectangle.

b) Plot several level curves in the rectangle.

c) Plot the level curve of f through the given point.

49. $f(x, y) = x \sin \dfrac{y}{2} + y \sin 2x$, $0 \le x \le 5\pi$, $0 \le y \le 5\pi$, $P(3\pi, 3\pi)$

50. $f(x, y) = (\sin x)(\cos y) \, e^{\sqrt{x^2 + y^2}/8}$, $0 \le x \le 5\pi$, $0 \le y \le 5\pi$, $P(4\pi, 4\pi)$

51. $f(x, y) = \sin(x + 2 \cos y)$, $-2\pi \le x \le 2\pi$, $-2\pi \le y \le 2\pi$, $P(\pi, \pi)$

52. $f(x, y) = e^{(x^{0.1} - y)} \sin(x^2 + y^2)$, $0 \le x \le 2\pi$, $-2\pi \le y \le \pi$, $P(\pi, -\pi)$

✪ CAS Explorations and Projects—Implicit Surfaces

Use a CAS to plot the level surfaces in Exercises 53–56.

53. $4 \ln (x^2 + y^2 + z^2) = 1$ **54.** $x^2 + z^2 = 1$

55. $x + y^2 - 3z^2 = 1$

56. $\sin \left(\dfrac{x}{2} \right) - (\cos y)\sqrt{x^2 + z^2} = 2$

✪ CAS Explorations and Projects— Parametrized Surfaces

Just as you describe curves in the plane parametrically with a pair of equations $x = f(t)$, $y = g(t)$ defined on some parameter interval I, you can sometimes describe surfaces in space with a triple of equations $x = f(u, v)$, $y = g(u, v)$, $z = h(u, v)$ defined on some parameter rectangle $a \leq u \leq b, c \leq v \leq d$. Many computer algebra systems permit you to plot such surfaces in *parametric mode*. (Parametrized surfaces are discussed in detail in Section 14.6.) Use a CAS to plot the surfaces in Exercises 57–60. Also plot several level curves in the xy-plane.

57. $x = u \cos v,$ $y = u \sin v,$ $z = u,$ $0 \leq u \leq 2,$ $0 \leq v \leq 2\pi$

58. $x = u \cos v,$ $y = u \sin v,$ $z = v,$ $0 \leq u \leq 2,$ $0 \leq v \leq 2\pi$

59. $x = (2 + \cos u) \cos v,$ $y = (2 + \cos u) \sin v,$ $z = \sin u,$ $0 \leq u \leq 2\pi,$ $0 \leq v \leq 2\pi$

60. $x = 2 \cos u \cos v,$ $y = 2 \cos u \sin v,$ $z = 2 \sin u,$ $0 \leq u \leq 2\pi,$ $0 \leq v \leq \pi$

12.2

Limits and Continuity

This section treats limits and continuity for multivariable functions.

Limits

If the values of $f(x, y)$ lie arbitrarily close to a fixed real number L for all points (x, y) sufficiently close to a point (x_0, y_0), we say that f approaches the limit L as (x, y) approaches (x_0, y_0). This is similar to the informal definition for the limit of a function of a single variable. Notice, however, that if (x_0, y_0) lies in the interior of f's domain, (x, y) can approach (x_0, y_0) from any direction. The direction of approach can be an issue, as in some of the examples that follow.

Definition

We say that a function $f(x, y)$ approaches the **limit** L as (x, y) approaches (x_0, y_0), and write

$$\lim_{(x,y) \to (x_0, y_0)} f(x, y) = L$$

if, for every number $\epsilon > 0$, there exists a corresponding number $\delta > 0$ such that for all (x, y) in the domain of f,

$$0 < \sqrt{(x - x_0)^2 + (y - y_0)^2} < \delta \quad \Rightarrow \quad |f(x, y) - L| < \epsilon. \quad (1)$$

The δ-ϵ requirement in the definition of limit is equivalent to the requirement that, given $\epsilon > 0$, there exists a corresponding $\delta > 0$ such that for all x,

$$0 < |x - x_0| < \delta \quad \text{and} \quad 0 < |y - y_0| < \delta \quad \Rightarrow \quad |f(x, y) - L| < \epsilon \quad (2)$$

(Exercise 59). Thus, in calculating limits we can think either in terms of distance in the plane or in terms of differences in coordinates.

The definition of limit applies to boundary points (x_0, y_0) as well as interior points of the domain of f. The only requirement is that the point (x, y) remain in the domain at all times.

It can be shown, as for functions of a single variable, that

$$\lim_{(x,y)\to(x_0,y_0)} x = x_0$$

$$\lim_{(x,y)\to(x_0,y_0)} y = y_0 \qquad\qquad (3)$$

$$\lim_{(x,y)\to(x_0,y_0)} k = k. \qquad \text{(Any number } k\text{)}$$

It can also be shown that the limit of the sum of two functions is the sum of their limits (when they both exist), with similar results for the limits of the differences, products, constant multiples, quotients, and powers.

Theorem 1

Properties of Limits of Functions of Two Variables

The following rules hold if

$$\lim_{(x,y)\to(x_0,y_0)} f(x,y) = L \qquad \text{and} \qquad \lim_{(x,y)\to(x_0,y_0)} g(x,y) = M.$$

1. *Sum Rule:* $\qquad\qquad\qquad \lim [f(x,y) + g(x,y)] = L + M$

2. *Difference Rule:* $\qquad\quad\; \lim [f(x,y) - g(x,y)] = L - M$

3. *Product Rule:* $\qquad\qquad\; \lim f(x,y) \cdot g(x,y) = L \cdot M$

4. *Constant Multiple Rule:* $\quad \lim kf(x,y) = kL \quad$ (Any number k)

5. *Quotient Rule:* $\qquad\qquad \lim \dfrac{f(x,y)}{g(x,y)} = \dfrac{L}{M} \quad \text{if} \quad M \neq 0.$

6. *Power Rule:* $\qquad\qquad\;$ If m and n are integers, then
$$\lim [f(x,y)]^{m/n} = L^{m/n},$$
provided $L^{m/n}$ is a real number.

All limits are to be taken as $(x,y) \to (x_0, y_0)$, and L and M are to be real numbers.

When we apply Theorem 1 to the limits in Eqs. (3), we obtain the useful result that the limits of polynomials and rational functions as $(x,y) \to (x_0, y_0)$ can be calculated by evaluating the functions at (x_0, y_0). The only requirement is that the functions be defined at (x_0, y_0).

EXAMPLE 1

a) $\displaystyle \lim_{(x,y)\to(0,1)} \frac{x - xy + 3}{x^2 y + 5xy - y^3} = \frac{0 - (0)(1) + 3}{(0)^2(1) + 5(0)(1) - (1)^3} = -3$

b) $\displaystyle \lim_{(x,y)\to(3,-4)} \sqrt{x^2 + y^2} = \sqrt{(3)^2 + (-4)^2} = \sqrt{25} = 5$ $\square$

EXAMPLE 2   Find

$$\lim_{(x,y)\to(0,0)} \frac{x^2 - xy}{\sqrt{x} - \sqrt{y}}.$$

Solution   Since the denominator $\sqrt{x} - \sqrt{y}$ approaches 0 as $(x,y) \to (0,0)$, we cannot use the Quotient Rule from Theorem 1. However, if we multiply numerator and denominator by $\sqrt{x} + \sqrt{y}$, we produce an equivalent fraction whose limit we *can* find:

$$\lim_{(x,y)\to(0,0)} \frac{x^2 - xy}{\sqrt{x} - \sqrt{y}} = \lim_{(x,y)\to(0,0)} \frac{(x^2 - xy)(\sqrt{x} + \sqrt{y})}{(\sqrt{x} - \sqrt{y})(\sqrt{x} + \sqrt{y})}$$

$$= \lim_{(x,y)\to(0,0)} \frac{x(x - y)(\sqrt{x} + \sqrt{y})}{x - y} \qquad \text{Algebra}$$

$$= \lim_{(x,y)\to(0,0)} x(\sqrt{x} + \sqrt{y}) \qquad \begin{array}{l}\text{Cancel the} \\ \text{factor} \\ (x - y).\end{array}$$

$$= 0(\sqrt{0} + \sqrt{0}) = 0 \qquad \square$$

Continuity

As with functions of a single variable, continuity is defined in terms of limits.

> **Definitions**
>
> A function $f(x, y)$ is **continuous at the point (x_0, y_0)** if
>
> 1. f is defined at (x_0, y_0),
> 2. $\displaystyle\lim_{(x,y)\to(x_0,y_0)} f(x, y)$ exists,
> 3. $\displaystyle\lim_{(x,y)\to(x_0,y_0)} f(x, y) = f(x_0, y_0)$.
>
> A function is **continuous** if it is continuous at every point of its domain.

As with the definition of limit, the definition of continuity applies at boundary points as well as interior points of the domain of f. The only requirement is that the point (x, y) remain in the domain at all times.

As you may have guessed, one of the consequences of Theorem 1 is that algebraic combinations of continuous functions are continuous at every point at which all the functions involved are defined. This means that sums, differences, products, constant multiples, quotients, and powers of continuous functions are continuous where defined. In particular, polynomials and rational functions of two variables are continuous at every point at which they are defined.

If $z = f(x, y)$ is a continuous function of x and y, and $w = g(z)$ is a continuous function of z, then the composite $w = g(f(x, y))$ is continuous. Thus,

$$e^{x-y}, \qquad \cos\frac{xy}{x^2 + 1}, \qquad \ln(1 + x^2y^2)$$

are continuous at every point (x, y).

As with functions of a single variable, the general rule is that composites of continuous functions are continuous. The only requirement is that each function be continuous where it is applied.

12.11 (a) The graph of

$$f(x,y) = \begin{cases} \dfrac{2xy}{x^2 + y^2}, & (x, y) \neq (0, 0) \\ 0, & (x, y) = (0, 0) \end{cases}$$

The function is continuous at every point except the origin. (b) The level curves of f.

EXAMPLE 3   Show that

$$f(x, y) = \begin{cases} \dfrac{2xy}{x^2 + y^2}, & (x, y) \neq (0, 0) \\ 0, & (x, y) = (0, 0) \end{cases}$$

is continuous at every point except the origin (Fig. 12.11).

(Generated by Mathematica)

(a)

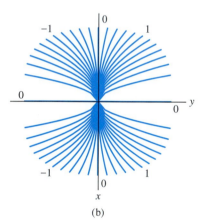

(b)

12.12 (a) The graph of $f(x, y) = 2x^2y/(x^4 + y^2)$. As the graph suggests and the level-curve values in (b) confirm, $\lim_{(x,y)\to(0,0)} f(x, y)$ does not exist.

Solution The function f is continuous at any point $(x, y) \neq (0, 0)$ because its values are then given by a rational function of x and y.

At (0, 0) the value of f is defined, but f, we claim, has no limit as $(x, y) \to (0, 0)$. The reason is that different paths of approach to the origin can lead to different results, as we will now see.

For every value of m, the function f has a constant value on the "punctured" line $y = mx$, $x \neq 0$, because

$$f(x, y)\bigg|_{y=mx} = \frac{2xy}{x^2 + y^2}\bigg|_{y=mx} = \frac{2x(mx)}{x^2 + (mx)^2} = \frac{2mx^2}{x^2 + m^2x^2} = \frac{2m}{1 + m^2}.$$

Therefore, f has this number as its limit as (x, y) approaches $(0, 0)$ along the line:

$$\lim_{\substack{(x,y)\to(0,0) \\ \text{along } y=mx}} f(x, y) = \lim_{(x,y)\to(0,0)} \left[f(x, y)\bigg|_{y=mx} \right] = \frac{2m}{1 + m^2}.$$

This limit changes with m. There is therefore no single number we may call the limit of f as (x, y) approaches the origin. The limit fails to exist, and the function is not continuous. ❑

Example 3 illustrates an important point about limits of functions of two variables (or even more variables, for that matter). For a limit to exist at a point, the limit must be the same along every approach path. Therefore, if we ever find paths with different limits, we know the function has no limit at the point they approach.

The Two-Path Test for the Nonexistence of a Limit

If a function $f(x, y)$ has different limits along two different paths as (x, y) approaches (x_0, y_0), then $\lim_{(x,y)\to(x_0,y_0)} f(x, y)$ does not exist.

EXAMPLE 4 Show that the function

$$f(x, y) = \frac{2x^2y}{x^4 + y^2}$$

(Fig. 12.12) has no limit as (x, y) approaches $(0, 0)$.

Solution Along the curve $y = kx^2$, $x \neq 0$, the function has the constant value

$$f(x, y)\bigg|_{y=kx^2} = \frac{2x^2y}{x^4 + y^2}\bigg|_{y=kx^2} = \frac{2x^2(kx^2)}{x^4 + (kx^2)^2} = \frac{2kx^4}{x^4 + k^2x^4} = \frac{2k}{1 + k^2}.$$

Therefore,

$$\lim_{\substack{(x,y)\to(0,0) \\ \text{along } y=kx^2}} f(x, y) = \lim_{(x,y)\to(0,0)} \left[f(x, y)\bigg|_{y=kx^2} \right] = \frac{2k}{1 + k^2}.$$

This limit varies with the path of approach. If (x, y) approaches $(0, 0)$ along the parabola $y = x^2$, for instance, $k = 1$ and the limit is 1. If (x, y) approaches $(0, 0)$ along the x-axis, $k = 0$ and the limit is 0. By the two-path test, f has no limit as (x, y) approaches $(0, 0)$.

The language here may seem contradictory. You might well ask, "What do you mean f has no limit as (x, y) approaches the origin—it has lots of limits." But that is the point. There is no single path-independent limit, and therefore, by the definition, $\lim_{(x,y)\to(0,0)} f(x, y)$ does not exist. It is our translating this formal statement into the more colloquial "has no limit" that creates the apparent contradiction. The mathematics is fine. The problem arises in how we talk about it. We need the formality to keep things straight. ∎

Functions of More Than Two Variables

The definitions of limit and continuity for functions of two variables and the conclusions about limits and continuity for sums, products, quotients, powers, and composites all extend to functions of three or more variables. Functions like

$$\ln(x + y + z) \qquad \text{and} \qquad \frac{y \sin z}{x - 1}$$

are continuous throughout their domains, and limits like

$$\lim_{P\to(1,0,-1)} \frac{e^{x+z}}{z^2 + \cos\sqrt{xy}} = \frac{e^{1-1}}{(-1)^2 + \cos 0} = \frac{1}{2},$$

where P denotes the point (x, y, z), may be found by direct substitution.

Exercises 12.2

Evaluating Limits

Find the limits in Exercises 1–12.

1. $\displaystyle\lim_{(x,y)\to(0,0)} \frac{3x^2 - y^2 + 5}{x^2 + y^2 + 2}$

2. $\displaystyle\lim_{(x,y)\to(0,4)} \frac{x}{\sqrt{y}}$

3. $\displaystyle\lim_{(x,y)\to(3,4)} \sqrt{x^2 + y^2 - 1}$

4. $\displaystyle\lim_{(x,y)\to(2,-3)} \left(\frac{1}{x} + \frac{1}{y}\right)^2$

5. $\displaystyle\lim_{(x,y)\to(0,\pi/4)} \sec x \tan y$

6. $\displaystyle\lim_{(x,y)\to(0,0)} \cos\frac{x^2 + y^3}{x + y + 1}$

7. $\displaystyle\lim_{(x,y)\to(0,\ln 2)} e^{x-y}$

8. $\displaystyle\lim_{(x,y)\to(1,1)} \ln|1 + x^2 y^2|$

9. $\displaystyle\lim_{(x,y)\to(0,0)} \frac{e^y \sin x}{x}$

10. $\displaystyle\lim_{(x,y)\to(1,1)} \cos\sqrt[3]{|xy| - 1}$

11. $\displaystyle\lim_{(x,y)\to(1,0)} \frac{x \sin y}{x^2 + 1}$

12. $\displaystyle\lim_{(x,y)\to(\pi/2,0)} \frac{\cos y + 1}{y - \sin x}$

Limits of Quotients

Find the limits in Exercises 13–20 by rewriting the fractions first.

13. $\displaystyle\lim_{\substack{(x,y)\to(1,1) \\ x\neq y}} \frac{x^2 - 2xy + y^2}{x - y}$

14. $\displaystyle\lim_{\substack{(x,y)\to(1,1) \\ x\neq y}} \frac{x^2 - y^2}{x - y}$

15. $\displaystyle\lim_{\substack{(x,y)\to(1,1) \\ x\neq1}} \frac{xy - y - 2x + 2}{x - 1}$

16. $\displaystyle\lim_{\substack{(x,y)\to(2,-4) \\ y\neq-4,\ x\neq x^2}} \frac{y + 4}{x^2 y - xy + 4x^2 - 4x}$

17. $\displaystyle\lim_{\substack{(x,y)\to(0,0) \\ x\neq y}} \frac{x - y + 2\sqrt{x} - 2\sqrt{y}}{\sqrt{x} - \sqrt{y}}$

18. $\displaystyle\lim_{\substack{(x,y)\to(2,2) \\ x+y\neq4}} \frac{x + y - 4}{\sqrt{x + y} - 2}$

19. $\displaystyle\lim_{\substack{(x,y)\to(2,0) \\ 2x-y\neq4}} \frac{\sqrt{2x - y} - 2}{2x - y - 4}$

20. $\displaystyle\lim_{\substack{(x,y)\to(4,3) \\ x\neq y+1}} \frac{\sqrt{x} - \sqrt{y + 1}}{x - y - 1}$

Limits with Three Variables

Find the limits in Exercises 21–26.

21. $\displaystyle\lim_{P\to(1,3,4)} \left(\frac{1}{x} + \frac{1}{y} + \frac{1}{z}\right)$

22. $\displaystyle\lim_{P\to(1,-1,-1)} \frac{2xy + yz}{x^2 + z^2}$

23. $\displaystyle\lim_{P\to(3,3,0)} (\sin^2 x + \cos^2 y + \sec^2 z)$

24. $\displaystyle\lim_{P\to(-1/4,\pi/2,2)} \tan^{-1} xyz$

25. $\displaystyle\lim_{P\to(\pi,0,3)} ze^{-2y}\cos 2x$

26. $\displaystyle\lim_{P\to(0,-2,0)} \ln\sqrt{x^2 + y^2 + z^2}$

Continuity in the Plane

At what points (x, y) in the plane are the functions in Exercises 27–30 continuous?

27. a) $f(x, y) = \sin(x + y)$ **b)** $f(x, y) = \ln(x^2 + y^2)$

28. a) $f(x, y) = \dfrac{x + y}{x - y}$ **b)** $f(x, y) = \dfrac{y}{x^2 + 1}$

29. a) $g(x, y) = \sin \dfrac{1}{xy}$ **b)** $g(x, y) = \dfrac{x + y}{2 + \cos x}$

30. a) $g(x, y) = \dfrac{x^2 + y^2}{x^2 - 3x + 2}$ **b)** $g(x, y) = \dfrac{1}{x^2 - y}$

Continuity in Space

At what points (x, y, z) in space are the functions in Exercises 31–34 continuous?

31. a) $f(x, y, z) = x^2 + y^2 - 2z^2$

     b) $f(x, y, z) = \sqrt{x^2 + y^2 - 1}$

32. a) $f(x, y, z) = \ln xyz$ **b)** $f(x, y, z) = e^{x+y} \cos z$

33. a) $h(x, y, z) = xy \sin \dfrac{1}{z}$ **b)** $h(x, y, z) = \dfrac{1}{x^2 + z^2 - 1}$

34. a) $h(x, y, z) = \dfrac{1}{|y| + |z|}$ **b)** $h(x, y, z) = \dfrac{1}{|xy| + |z|}$

No Limit at a Point

By considering different paths of approach, show that the functions in Exercises 35–42 have no limit as $(x, y) \to (0, 0)$.

35. $f(x, y) = -\dfrac{x}{\sqrt{x^2 + y^2}}$ **36.** $f(x, y) = \dfrac{x^4}{x^4 + y^2}$

(Generated by Mathematica)   (Generated by Mathematica)

37. $f(x, y) = \dfrac{x^4 - y^2}{x^4 + y^2}$ **38.** $f(x, y) = \dfrac{xy}{|xy|}$

39. $g(x, y) = \dfrac{x - y}{x + y}$ **40.** $g(x, y) = \dfrac{x + y}{x - y}$

41. $h(x, y) = \dfrac{x^2 + y}{y}$ **42.** $h(x, y) = \dfrac{x^2}{x^2 - y}$

Theory and Examples

43. If $\lim_{(x,y) \to (x_0, y_0)} f(x, y) = L$, must f be defined at (x_0, y_0)? Give reasons for your answer.

44. If $f(x_0, y_0) = 3$, what can you say about

$$\lim_{(x,y) \to (x_0, y_0)} f(x, y)$$

if f is continuous at (x_0, y_0)? if f is not continuous at (x_0, y_0)? Give reasons for your answer.

The Sandwich Theorem for functions of two variables states that if $g(x, y) \le f(x, y) \le h(x, y)$ for all $(x, y) \ne (x_0, y_0)$ in a disk centered at (x_0, y_0) and if g and h have the same finite limit L as $(x, y) \to (x_0, y_0)$, then

$$\lim_{(x,y) \to (x_0, y_0)} f(x, y) = L.$$

Use this result to support your answers to the questions in Exercises 45–48.

45. Does knowing that

$$1 - \frac{x^2 y^2}{3} < \frac{\tan^{-1} xy}{xy} < 1$$

tell you anything about

$$\lim_{(x,y) \to (0,0)} \frac{\tan^{-1} xy}{xy}?$$

Give reasons for your answer.

46. Does knowing that

$$2|xy| - \frac{x^2 y^2}{6} < 4 - 4 \cos \sqrt{|xy|} < 2|xy|$$

tell you anything about

$$\lim_{(x,y) \to (0,0)} \frac{4 - 4 \cos \sqrt{|xy|}}{|xy|}?$$

Give reasons for your answer.

47. Does knowing that $|\sin(1/x)| \le 1$ tell you anything about

$$\lim_{(x,y) \to (0,0)} y \sin \frac{1}{x}?$$

Give reasons for your answer.

48. Does knowing that $|\cos(1/y)| \le 1$ tell you anything about

$$\lim_{(x,y) \to (0,0)} x \cos \frac{1}{y}?$$

Give reasons for your answer.

49. (*Continuation of Example 3.*)

     a) Reread Example 3. Then substitute $m = \tan \theta$ into the formula

$$f(x, y) \Big|_{y=mx} = \frac{2m}{1 + m^2}$$

and simplify the result to show how the value of f varies with the line's angle of inclination.

     b) Use the formula you obtained in (a) to show that the limit of f as $(x, y) \to (0, 0)$ along the line $y = mx$ varies from -1 to 1 depending on the angle of approach.

50. Define $f(0, 0)$ in a way that extends

$$f(x, y) = xy \frac{x^2 - y^2}{x^2 + y^2}$$

to be continuous at the origin.

Changing to Polar Coordinates

If you cannot make any headway with $\lim_{(x,y)\to(0,0)} f(x,y)$ in rectangular coordinates, try changing to polar coordinates. Substitute $x = r\cos\theta$, $y = r\sin\theta$, and investigate the limit of the resulting expression as $r \to 0$. In other words, try to decide whether there exists a number L satisfying the following criterion:

Given $\epsilon > 0$, there exists a $\delta > 0$ such that for all r and θ,

$$|r| < \delta \quad \Rightarrow \quad |f(r,\theta) - L| < \epsilon. \qquad (4)$$

If such an L exists, then

$$\lim_{(x,y)\to(0,0)} f(x,y) = \lim_{r\to 0} f(r,\theta) = L.$$

For instance,

$$\lim_{(x,y)\to(0,0)} \frac{x^3}{x^2+y^2} = \lim_{r\to 0} \frac{r^3\cos^3\theta}{r^2} = \lim_{r\to 0} r\cos^3\theta = 0.$$

To verify the last of these equalities, we need to show that (4) is satisfied with $f(r,\theta) = r\cos^3\theta$ and $L = 0$. That is, we need to show that given any $\epsilon > 0$ there exists a $\delta > 0$ such that for all r and θ,

$$|r| < \delta \quad \Rightarrow \quad |r\cos^3\theta - 0| < \epsilon.$$

Since

$$|r\cos^3\theta| = |r||\cos^3\theta| \le |r| \cdot 1 = |r|,$$

the implication holds for all r and θ if we take $\delta = \epsilon$.

In contrast,

$$\frac{x^2}{x^2+y^2} = \frac{r^2\cos^2\theta}{r^2} = \cos^2\theta$$

takes on all values from 0 to 1 regardless of how small $|r|$ is, so that $\lim_{(x,y)\to(0,0)} x^2/(x^2+y^2)$ does not exist.

In each of these instances, the existence or nonexistence of the limit as $r \to 0$ is fairly clear. Shifting to polar coordinates does not always help, however, and may even tempt us to false conclusions. For example, the limit may exist along every straight line (or ray) $\theta = $ constant and yet fail to exist in the broader sense. Example 4 illustrates this point. In polar coordinates, $f(x,y) = (2x^2y)/(x^4+y^2)$ becomes

$$f(r\cos\theta, r\sin\theta) = \frac{r\cos\theta\sin 2\theta}{r^2\cos^4\theta + \sin^2\theta}$$

for $r \neq 0$. If we hold θ constant and let $r \to 0$, the limit is 0. On the path $y = x^2$, however, we have $r\sin\theta = r^2\cos^2\theta$ and

$$f(r\cos\theta, r\sin\theta) = \frac{r\cos\theta\sin 2\theta}{r^2\cos^4\theta + (r\cos^2\theta)^2}$$

$$= \frac{2r\cos^2\theta\sin\theta}{2r^2\cos^4\theta} = \frac{r\sin\theta}{r^2\cos^2\theta} = 1.$$

In Exercises 51–56, find the limit of f as $(x,y) \to (0,0)$ or show that the limit does not exist.

51. $f(x,y) = \dfrac{x^3 - xy^2}{x^2 + y^2}$

52. $f(x,y) = \cos\left(\dfrac{x^3 - y^3}{x^2 + y^2}\right)$

53. $f(x,y) = \dfrac{y^2}{x^2 + y^2}$

54. $f(x,y) = \dfrac{2x}{x^2 + x + y^2}$

55. $f(x,y) = \tan^{-1}\left(\dfrac{|x| + |y|}{x^2 + y^2}\right)$

56. $f(x,y) = \dfrac{x^2 - y^2}{x^2 + y^2}$

In Exercises 57 and 58, define $f(0,0)$ in a way that extends f to be continuous at the origin.

57. $f(x,y) = \ln\left(\dfrac{3x^2 - x^2y^2 + 3y^2}{x^2 + y^2}\right)$ **58.** $f(x,y) = \dfrac{2xy^2}{x^2 + y^2}$

Using the δ-ϵ Definitions

59. Show that the δ-ϵ requirement in the definition of limit expressed in Eq. (1) is equivalent to the requirement expressed in Eq. (2).

60. Using the formal δ-ϵ definition of limit of a function $f(x,y)$ as $(x,y) \to (x_0, y_0)$ as a guide, state a formal definition for the limit of a function $g(x,y,z)$ as $(x,y,z) \to (x_0, y_0, z_0)$. What would be the analogous definition for a function $h(x,y,z,t)$ of four independent variables?

Each of Exercises 61–64 gives a function $f(x,y)$ and a positive number ϵ. In each exercise, either show that there exists a $\delta > 0$ such that for all (x,y),

$$\sqrt{x^2 + y^2} < \delta \quad \Rightarrow \quad |f(x,y) - f(0,0)| < \epsilon$$

or show that there exists a $\delta > 0$ such that for all (x,y),

$$|x| < \delta \text{ and } |y| < \delta \quad \Rightarrow \quad |f(x,y) - f(0,0)| < \epsilon.$$

Do either one or the other, whichever seems more convenient. There is no need to do both.

61. $f(x,y) = x^2 + y^2$, $\epsilon = 0.01$

62. $f(x,y) = y/(x^2 + 1)$, $\epsilon = 0.05$

63. $f(x,y) = (x + y)/(x^2 + 1)$, $\epsilon = 0.01$

64. $f(x,y) = (x + y)/(2 + \cos x)$, $\epsilon = 0.02$

Each of Exercises 65–68 gives a function $f(x,y,z)$ and a positive number ϵ. In each exercise, either show that there exists a $\delta > 0$ such that for all (x,y,z),

$$\sqrt{x^2 + y^2 + z^2} < \delta \quad \Rightarrow \quad |f(x,y,z) - f(0,0,0)| < \epsilon$$

or show that there exists a $\delta > 0$ such that for all (x,y,z),

$$|x| < \delta, \quad |y| < \delta, \quad \text{and}$$

$$|z| < \delta \quad \Rightarrow \quad |f(x,y,z) - f(0,0,0)| < \epsilon.$$

Do either one or the other, whichever seems more convenient. There is no need to do both.

65. $f(x,y,z) = x^2 + y^2 + z^2$, $\epsilon = 0.015$

66. $f(x,y,z) = xyz$, $\epsilon = 0.008$

67. $f(x,y,z) = \dfrac{x + y + z}{x^2 + y^2 + z^2 + 1}$, $\epsilon = 0.015$

68. $f(x,y,z) = \tan^2 x + \tan^2 y + \tan^2 z$, $\epsilon = 0.03$

69. Show that $f(x,y,z) = x + y - z$ is continuous at every point (x_0, y_0, z_0).

70. Show that $f(x,y,z) = x^2 + y^2 + z^2$ is continuous at the origin.

12.3 Partial Derivatives

When we hold all but one of the independent variables of a function constant and differentiate with respect to that one variable, we get a "partial" derivative. This section shows how partial derivatives arise and how to calculate partial derivatives by applying the rules for differentiating functions of a single variable.

Definitions and Notation

If (x_0, y_0) is a point in the domain of a function $f(x, y)$, the vertical plane $y = y_0$ will cut the surface $z = f(x, y)$ in the curve $z = f(x, y_0)$ (Fig. 12.13). This curve is the graph of the function $z = f(x, y_0)$ in the plane $y = y_0$. The horizontal coordinate in this plane is x; the vertical coordinate is z.

We define the partial derivative of f with respect to x at the point (x_0, y_0) as the ordinary derivative of $f(x, y_0)$ with respect to x at the point $x = x_0$.

Definition

The **partial derivative of $f(x, y)$ with respect to x** at the point (x_0, y_0) is

$$\left.\frac{\partial f}{\partial x}\right|_{(x_0, y_0)} = \left.\frac{d}{dx} f(x, y_0)\right|_{x=x_0} = \lim_{h \to 0} \frac{f(x_0 + h, y_0) - f(x_0, y_0)}{h}, \quad (1)$$

provided the limit exists. (Think of ∂ as a kind of d.)

The slope of the curve $z = f(x, y_0)$ at the point $P(x_0, y_0, f(x_0, y_0))$ in the plane $y = y_0$ is the value of the partial derivative of f with respect to x at (x_0, y_0). The tangent line to the curve at P is the line in the plane $y = y_0$ that passes through P with this slope. The partial derivative $\partial f / \partial x$ at (x_0, y_0) gives the rate of change of f with respect to x when y is held fixed at the value y_0. This is the rate of change of f in the direction of $\mathbf{i}$ at (x_0, y_0).

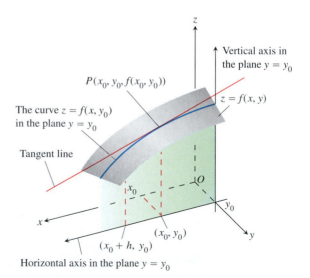

12.13 The intersection of the plane $y = y_0$ with the surface $z = f(x, y)$, viewed from a point above the first quadrant of the xy-plane.

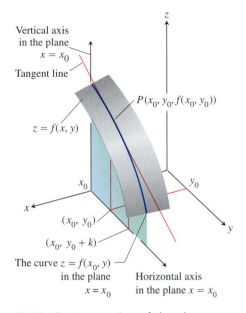

Vertical axis in the plane $x = x_0$

Tangent line

$P(x_0, y_0, f(x_0, y_0))$

$z = f(x, y)$

x_0

y_0

x

(x_0, y_0)

$(x_0, y_0 + k)$

The curve $z = f(x_0, y)$
in the plane
$x = x_0$

Horizontal axis
in the plane $x = x_0$

y

12.14 The intersection of the plane $x = x_0$ with the surface $z = f(x, y)$, viewed from above the first quadrant of the xy-plane.

The notation for a partial derivative depends on what we want to emphasize:

$\dfrac{\partial f}{\partial x}(x_0, y_0)$ or $f_x(x_0, y_0)$

"Partial derivative of f with respect to x at (x_0, y_0)" or "f sub x at (x_0, y_0)." Convenient for stressing the point (x_0, y_0).

$\dfrac{\partial z}{\partial x}\Big|_{(x_0, y_0)}$

"Partial derivative of z with respect to x at (x_0, y_0)." Common in science and engineering when you are dealing with variables and do not mention the function explicitly.

$f_x,\quad \dfrac{\partial f}{\partial x},\quad z_x,\quad$ or $\dfrac{\partial z}{\partial x}$

"Partial derivative of f (or z) with respect to x." Convenient when you regard the partial derivative as a function in its own right.

The definition of the partial derivative of $f(x, y)$ with respect to y at a point (x_0, y_0) is similar to the definition of the partial derivative of f with respect to x. We hold x fixed at the value x_0 and take the ordinary derivative of $f(x_0, y)$ with respect to y at y_0.

> **Definition**
>
> The **partial derivative of $f(x, y)$ with respect to y** at the point (x_0, y_0) is
>
> $$\frac{\partial f}{\partial y}\Big|_{(x_0, y_0)} = \frac{d}{dy} f(x_0, y)\Big|_{y=y_0}$$
>
> $$= \lim_{h \to 0} \frac{f(x_0, y_0 + h) - f(x_0, y_0)}{h}, \qquad (2)$$
>
> provided the limit exists.

The slope of the curve $z = f(x_0, y)$ at the point $P(x_0, y_0, f(x_0, y_0))$ in the vertical plane $x = x_0$ (Fig. 12.14) is the partial derivative of f with respect to y at (x_0, y_0). The tangent line to the curve at P is the line in the plane $x = x_0$ that passes through P with this slope. The partial derivative gives the rate of change of f with respect to y at (x_0, y_0) when x is held fixed at the value x_0. This is the rate of change of f in the direction of $\mathbf{j}$ at (x_0, y_0).

The partial derivative with respect to y is denoted the same way as the partial derivative with respect to x:

$$\frac{\partial f}{\partial y}(x_0, y_0), \qquad f_y(x_0, y_0), \qquad \frac{\partial f}{\partial y}, \qquad f_y.$$

Notice that we now have two tangent lines associated with the surface $z = f(x, y)$ at the point $P(x_0, y_0, f(x_0, y_0))$ (Fig. 12.15, on the following page). Is the plane they determine tangent to the surface at P? It would be nice if it were, but we have to learn more about partial derivatives before we can find out.

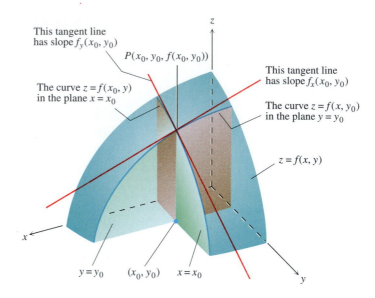

12.15 Figures 12.13 and 12.14 combined. The tangent lines at the point $(x_0, y_0, f(x_0, y_0))$ determine a plane that, in this picture at least, appears to be tangent to the surface.

Calculations

As Eq. (1) shows, we calculate $\partial f/\partial x$ by differentiating f with respect to x in the usual way while treating y as a constant. As Eq. (2) shows, we can calculate $\partial f/\partial y$ by differentiating f with respect to y in the usual way while holding x constant.

EXAMPLE 1   Find the values of $\partial f/\partial x$ and $\partial f/\partial y$ at the point $(4, -5)$ if

$$f(x, y) = x^2 + 3xy + y - 1.$$

Solution   To find $\partial f/\partial x$, we regard y as a constant and differentiate with respect to x:

$$\frac{\partial f}{\partial x} = \frac{\partial}{\partial x}(x^2 + 3xy + y - 1) = 2x + 3 \cdot 1 \cdot y + 0 - 0 = 2x + 3y.$$

The value of $\partial f/\partial x$ at $(4, -5)$ is $2(4) + 3(-5) = -7$.

To find $\partial f/\partial y$, we regard x as a constant and differentiate with respect to y:

$$\frac{\partial f}{\partial y} = \frac{\partial}{\partial y}(x^2 + 3xy + y - 1) = 0 + 3 \cdot x \cdot 1 + 1 - 0 = 3x + 1.$$

The value of $\partial f/\partial y$ at $(4, -5)$ is $3(4) + 1 = 13$.   ❑

EXAMPLE 2   Find $\partial f/\partial y$ if $f(x, y) = y \sin xy$.

Solution   We treat x as a constant and f as a product of y and $\sin xy$:

$$\frac{\partial f}{\partial y} = \frac{\partial}{\partial y}(y \sin xy) = y \frac{\partial}{\partial y} \sin xy + (\sin xy) \frac{\partial}{\partial y}(y)$$

$$= (y \cos xy) \frac{\partial}{\partial y}(xy) + \sin xy = xy \cos xy + \sin xy.$$ ❑

Technology   *Partial Differentiation*   A simple grapher can support your calculations even in multiple dimensions. If you specify the values of all but one independent variable, the grapher can calculate partial derivatives and can plot traces with respect to that remaining variable. Typically a Computer Algebra System can compute partial derivatives symbolically and numerically as easily as it can compute simple derivatives. Most systems use the same command to differentiate a function, regardless of the number of variables. (Simply specify the variable with which differentiation is to take place.)

EXAMPLE 3   Find f_x if $f(x, y) = \dfrac{2y}{y + \cos x}$.

Solution   We treat f as a quotient. With y held constant, we get

$$f_x = \frac{\partial}{\partial x}\left(\frac{2y}{y + \cos x}\right) = \frac{(y + \cos x)\dfrac{\partial}{\partial x}(2y) - 2y\dfrac{\partial}{\partial x}(y + \cos x)}{(y + \cos x)^2}$$

$$= \frac{(y + \cos x)(0) - 2y(-\sin x)}{(y + \cos x)^2} = \frac{2y \sin x}{(y + \cos x)^2}.$$

EXAMPLE 4   The plane $x = 1$ intersects the paraboloid $z = x^2 + y^2$ in a parabola. Find the slope of the tangent to the parabola at $(1, 2, 5)$ (Fig. 12.16).

Solution   The slope is the value of the partial derivative $\partial z / \partial y$ at $(1, 2)$:

$$\frac{\partial z}{\partial y}\bigg|_{(1, 2)} = \frac{\partial}{\partial y}(x^2 + y^2)\bigg|_{(1, 2)} = 2y\bigg|_{(1, 2)} = 2(2) = 4.$$

As a check, we can treat the parabola as the graph of the single-variable function $z = (1)^2 + y^2 = 1 + y^2$ in the plane $x = 1$ and ask for the slope at $y = 2$. The slope, calculated now as an ordinary derivative, is

$$\frac{dz}{dy}\bigg|_{y=2} = \frac{d}{dy}(1 + y^2)\bigg|_{y=2} = 2y\bigg|_{y=2} = 4.$$

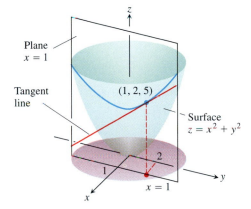

12.16 The tangent to the curve of intersection of the plane $x = 1$ and surface $z = x^2 + y^2$ at the point $(1, 2, 5)$ (Example 4).

Implicit differentiation works for partial derivatives the way it works for ordinary derivatives.

EXAMPLE 5   Find $\partial z / \partial x$ if the equation

$$yz - \ln z = x + y$$

defines z as a function of the two independent variables x and y and the partial derivative exists.

Solution   We differentiate both sides of the equation with respect to x, holding y constant and treating z as a differentiable function of x:

$$\frac{\partial}{\partial x}(yz) - \frac{\partial}{\partial x}\ln z = \frac{\partial x}{\partial x} + \frac{\partial y}{\partial x}$$

$$y\frac{\partial z}{\partial x} - \frac{1}{z}\frac{\partial z}{\partial x} = 1 + 0 \qquad \text{With } y \text{ constant,}$$
$$\frac{\partial}{\partial x}(yz) = y\frac{\partial z}{\partial x}.$$

$$\left(y - \frac{1}{z}\right)\frac{\partial z}{\partial x} = 1$$

$$\frac{\partial z}{\partial x} = \frac{z}{yz - 1}.$$

Functions of More Than Two Variables

The definitions of the partial derivatives of functions of more than two independent variables are like the definitions for functions of two variables. They are ordinary derivatives with respect to one variable, taken while the other independent variables are held constant.

EXAMPLE 6   If x, y, and z are independent variables and

$$f(x, y, z) = x \sin (y + 3z),$$

then

$$\frac{\partial f}{\partial z} = \frac{\partial}{\partial z} [x \sin (y + 3z)] = x \frac{\partial}{\partial z} \sin (y + 3z)$$

$$= x \cos (y + 3z) \frac{\partial}{\partial z} (y + 3z) = 3x \cos (y + 3z).$$ ❏

EXAMPLE 7   *Electrical resistors in parallel*

If resistors of R_1, R_2, and R_3 ohms are connected in parallel to make an R-ohm resistor, the value of R can be found from the equation

$$\frac{1}{R} = \frac{1}{R_1} + \frac{1}{R_2} + \frac{1}{R_3} \tag{3}$$

(Fig. 12.17). Find the value of $\partial R / \partial R_2$ when $R_1 = 30$, $R_2 = 45$, and $R_3 = 90$ ohms.

Solution   To find $\partial R / \partial R_2$, we treat R_1 and R_3 as constants and differentiate both sides of Eq. (3) with respect to R_2:

$$\frac{\partial}{\partial R_2} \left(\frac{1}{R} \right) = \frac{\partial}{\partial R_2} \left(\frac{1}{R_1} + \frac{1}{R_2} + \frac{1}{R_3} \right)$$

$$-\frac{1}{R^2} \frac{\partial R}{\partial R_2} = 0 - \frac{1}{R_2^2} + 0$$

$$\frac{\partial R}{\partial R_2} = \frac{R^2}{R_2^2} = \left(\frac{R}{R_2} \right)^2.$$

When $R_1 = 30$, $R_2 = 45$, and $R_3 = 90$,

$$\frac{1}{R} = \frac{1}{30} + \frac{1}{45} + \frac{1}{90} = \frac{3 + 2 + 1}{90} = \frac{6}{90} = \frac{1}{15},$$

so $R = 15$ and

$$\frac{\partial R}{\partial R_2} = \left(\frac{15}{45} \right)^2 = \left(\frac{1}{3} \right)^2 = \frac{1}{9}.$$ ❏

12.17 Resistors arranged this way are said to be connected in parallel (Example 7). Each resistor lets a portion of the current through. Their combined resistance R is calculated with the formula $\frac{1}{R} = \frac{1}{R_1} + \frac{1}{R_2} + \frac{1}{R_3}$.

The Relationship Between Continuity and the Existence of Partial Derivatives

A function $f(x, y)$ can have partial derivatives with respect to both x and y at a point without being continuous there. This is different from functions of a single variable, where the existence of a derivative implies continuity. However, if the partial derivatives of $f(x, y)$ exist and are continuous throughout a disk centered at (x_0, y_0), then f is continuous at (x_0, y_0), as we will see in the next section.

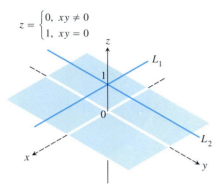

$$z = \begin{cases} 0, & xy \neq 0 \\ 1, & xy = 0 \end{cases}$$

12.18 The graph of

$$f(x, y) = \begin{cases} 0, & xy \neq 0 \\ 1, & xy = 0 \end{cases}$$

consists of the lines L_1 and L_2 and the four open quadrants of the *xy*-plane. The function has partial derivatives at the origin but is not continuous there.

EXAMPLE 8   The function

$$f(x, y) = \begin{cases} 0, & xy \neq 0 \\ 1, & xy = 0 \end{cases}$$

(Fig. 12.18) is not continuous at (0, 0). The limit of f as (x, y) approaches (0, 0) along the line $y = x$ is 0, but $f(0, 0) = 1$. The partial derivatives f_x and f_y, being the slopes of the horizontal lines L_1 and L_2 in Fig. 12.18, both exist at (0, 0).   ❑

Second Order Partial Derivatives

When we differentiate a function $f(x, y)$ twice, we produce its second order derivatives. These derivatives are usually denoted by

$\dfrac{\partial^2 f}{\partial x^2}$ "d squared f d x squared"   or f_{xx} "f sub x x"

$\dfrac{\partial^2 f}{\partial y^2}$ "d squared f d y squared" f_{yy} "f sub y y"

$\dfrac{\partial^2 f}{\partial x \partial y}$ "d squared f d x d y" f_{yx} "f sub y x"

$\dfrac{\partial^2 f}{\partial y \partial x}$ "d squared f d y d x" f_{xy} "f sub x y"

The defining equations are

$$\frac{\partial^2 f}{\partial x^2} = \frac{\partial}{\partial x}\left(\frac{\partial f}{\partial x}\right), \qquad \frac{\partial^2 f}{\partial x\, \partial y} = \frac{\partial}{\partial x}\left(\frac{\partial f}{\partial y}\right),$$

and so on. Notice the order in which the derivatives are taken:

$\dfrac{\partial^2 f}{\partial x \partial y}$ Differentiate first with respect to y, then with respect to x.

$f_{yx} = (f_y)_x$ Means the same thing.

EXAMPLE 9   If $f(x, y) = x \cos y + y e^x$, then

$$\frac{\partial f}{\partial x} = \cos y + y e^x$$

$$\frac{\partial^2 f}{\partial y\, \partial x} = \frac{\partial}{\partial y}\left(\frac{\partial f}{\partial x}\right) = -\sin y + e^x$$

$$\frac{\partial^2 f}{\partial x^2} = \frac{\partial}{\partial x}\left(\frac{\partial f}{\partial x}\right) = y e^x.$$

Also,

$$\frac{\partial f}{\partial y} = -x \sin y + e^x$$

$$\frac{\partial^2 f}{\partial x\, \partial y} = \frac{\partial}{\partial x}\left(\frac{\partial f}{\partial y}\right) = -\sin y + e^x$$

$$\frac{\partial^2 f}{\partial y^2} = \frac{\partial}{\partial y}\left(\frac{\partial f}{\partial y}\right) = -x \cos y.$$

❑

Euler's Theorem

You may have noticed that the "mixed" second order partial derivatives

$$\frac{\partial^2 f}{\partial y \, \partial x} \quad \text{and} \quad \frac{\partial^2 f}{\partial x \, \partial y}$$

in Example 9 were equal. This was not a coincidence. They must be equal whenever f, f_x, f_y, f_{xy}, and f_{yx} are continuous.

Theorem 2

Euler's Theorem (The Mixed Derivative Theorem)

If $f(x, y)$ and its partial derivatives f_x, f_y, f_{xy}, and f_{yx} are defined throughout an open region containing a point (a, b) and are all continuous at (a, b), then

$$f_{xy}(a, b) = f_{yx}(a, b). \tag{4}$$

You can find a proof of Theorem 2 in Appendix 9.

Theorem 2 says that to calculate a mixed second order derivative we may differentiate in either order. This can work to our advantage.

EXAMPLE 10   Find $\partial^2 w / \partial x \, \partial y$ if

$$w = xy + \frac{e^y}{y^2 + 1}.$$

Solution   The symbol $\partial^2 w / \partial x \, \partial y$ tells us to differentiate first with respect to y and then with respect to x. However, if we postpone the differentiation with respect to y and differentiate first with respect to x, we get the answer more quickly. In two steps,

$$\frac{\partial w}{\partial x} = y \quad \text{and} \quad \frac{\partial^2 w}{\partial y \, \partial x} = 1.$$

We are in for more work if we differentiate first with respect to y. (Just try it.)   ❑

Partial Derivatives of Still Higher Order

Although we will deal mostly with first and second order partial derivatives, because these appear the most frequently in applications, there is no theoretical limit to how many times we can differentiate a function as long as the derivatives involved exist. Thus we get third and fourth order derivatives denoted by symbols like

$$\frac{\partial^3 f}{\partial x \, \partial y^2} = f_{yyx},$$

$$\frac{\partial^4 f}{\partial x^2 \, \partial y^2} = f_{yyxx},$$

and so on. As with second order derivatives, the order of differentiation is immaterial as long as the derivatives through the order in question are continuous.

Exercises 12.3

Calculating First Order Partial Derivatives

In Exercises 1–22, find $\partial f/\partial x$ and $\partial f/\partial y$.

1. $f(x, y) = 2x^2 - 3y - 4$ **2.** $f(x, y) = x^2 - xy + y^2$

3. $f(x, y) = (x^2 - 1)(y + 2)$

4. $f(x, y) = 5xy - 7x^2 - y^2 + 3x - 6y + 2$

5. $f(x, y) = (xy - 1)^2$ **6.** $f(x, y) = (2x - 3y)^3$

7. $f(x, y) = \sqrt{x^2 + y^2}$ **8.** $f(x, y) = (x^3 + (y/2))^{2/3}$

9. $f(x, y) = 1/(x + y)$ **10.** $f(x, y) = x/(x^2 + y^2)$

11. $f(x, y) = (x + y)/(xy - 1)$

12. $f(x, y) = \tan^{-1}(y/x)$ **13.** $f(x, y) = e^{(x+y+1)}$

14. $f(x, y) = e^{-x} \sin(x + y)$ **15.** $f(x, y) = \ln(x + y)$

16. $f(x, y) = e^{xy} \ln y$ **17.** $f(x, y) = \sin^2(x - 3y)$

18. $f(x, y) = \cos^2(3x - y^2)$ **19.** $f(x, y) = x^y$

20. $f(x, y) = \log_y x$

21. $f(x, y) = \displaystyle\int_x^y g(t)\, dt$ (g continuous for all t)

22. $f(x, y) = \displaystyle\sum_{n=0}^{\infty} (xy)^n$ ($|xy| < 1$)

In Exercises 23–34, find f_x, f_y, and f_z.

23. $f(x, y, z) = 1 + xy^2 - 2z^2$ **24.** $f(x, y, z) = xy + yz + xz$

25. $f(x, y, z) = x - \sqrt{y^2 + z^2}$

26. $f(x, y, z) = (x^2 + y^2 + z^2)^{-1/2}$

27. $f(x, y, z) = \sin^{-1}(xyz)$ **28.** $f(x, y, z) = \sec^{-1}(x + yz)$

29. $f(x, y, z) = \ln(x + 2y + 3z)$

30. $f(x, y, z) = yz \ln(xy)$ **31.** $f(x, y, z) = e^{-(x^2+y^2+z^2)}$

32. $f(x, y, z) = e^{-xyz}$

33. $f(x, y, z) = \tanh(x + 2y + 3z)$

34. $f(x, y, z) = \sinh(xy - z^2)$

In Exercises 35–40, find the partial derivative of the function with respect to each variable.

35. $f(t, \alpha) = \cos(2\pi t - \alpha)$ **36.** $g(u, v) = v^2 e^{(2u/v)}$

37. $h(\rho, \phi, \theta) = \rho \sin \phi \cos \theta$

38. $g(r, \theta, z) = r(1 - \cos \theta) - z$

39. *Work done by the heart.* (Section 3.7, Exercise 56)

$$W(P, V, \delta, v, g) = PV + \frac{V\delta v^2}{2g}$$

40. *Wilson lot size formula.* (Section 3.6, Exercise 57)

$$A(c, h, k, m, q) = \frac{km}{q} + cm + \frac{hq}{2}$$

Calculating Second Order Partial Derivatives

Find all the second order partial derivatives of the functions in Exercises 41–46.

41. $f(x, y) = x + y + xy$ **42.** $f(x, y) = \sin xy$

43. $g(x, y) = x^2 y + \cos y + y \sin x$

44. $h(x, y) = xe^y + y + 1$ **45.** $r(x, y) = \ln(x + y)$

46. $s(x, y) = \tan^{-1}(y/x)$

Mixed Partial Derivatives

In Exercises 47–50, verify that $w_{xy} = w_{yx}$.

47. $w = \ln(2x + 3y)$ **48.** $w = e^x + x \ln y + y \ln x$

49. $w = xy^2 + x^2 y^3 + x^3 y^4$ **50.** $w = x \sin y + y \sin x + xy$

51. Which order of differentiation will calculate f_{xy} faster: x first, or y first? Try to answer without writing anything down.

  a) $f(x, y) = x \sin y + e^y$
  b) $f(x, y) = 1/x$
  c) $f(x, y) = y + (x/y)$
  d) $f(x, y) = y + x^2 y + 4y^3 - \ln(y^2 + 1)$
  e) $f(x, y) = x^2 + 5xy + \sin x + 7e^x$
  f) $f(x, y) = x \ln xy$

52. The fifth order partial derivative $\partial^5 f/\partial x^2 \partial y^3$ is zero for each of the following functions. To show this as quickly as possible, which variable would you differentiate with respect to first: x, or y? Try to answer without writing anything down.

  a) $f(x, y) = y^2 x^4 e^x + 2$
  b) $f(x, y) = y^2 + y(\sin x - x^4)$
  c) $f(x, y) = x^2 + 5xy + \sin x + 7e^x$
  d) $f(x, y) = xe^{y^2/2}$

Using the Partial Derivative Definition

In Exercises 53 and 54, use the limit definition of partial derivative to compute the partial derivatives of the functions at the specified points.

53. $f(x, y) = 1 - x + y - 3x^2 y$, $\dfrac{\partial f}{\partial x}$ and $\dfrac{\partial f}{\partial y}$ at $(1, 2)$

54. $f(x, y) = 4 + 2x - 3y - xy^2$, $\dfrac{\partial f}{\partial x}$ and $\dfrac{\partial f}{\partial y}$ at $(-2, 1)$

55. Let $w = f(x, y, z)$ be a function of three independent variables, and write the formal definition of the partial derivative $\partial f/\partial z$ at (x_0, y_0, z_0). Use this definition to find $\partial f/\partial z$ at $(1, 2, 3)$ for $f(x, y, z) = x^2 yz^2$.

56. Let $w = f(x, y, z)$ be a function of three independent variables and write the formal definition of the partial derivative $\partial f/\partial y$ at (x_0, y_0, z_0). Use this definition to find $\partial f/\partial y$ at $(-1, 0, 3)$ for $f(x, y, z) = -2xy^2 + yz^2$.

Differentiating Implicitly

57. Find the value of $\partial z/\partial x$ at the point $(1, 1, 1)$ if the equation

$$xy + z^3x - 2yz = 0$$

defines z as a function of the two independent variables x and y and the partial derivative exists.

58. Find the value of $\partial x/\partial z$ at the point $(1, -1, -3)$ if the equation

$$xz + y\ln x - x^2 + 4 = 0$$

defines x as a function of the two independent variables y and z and the partial derivative exists.

Exercises 59 and 60 are about the triangle shown here.

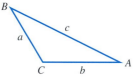

59. Express A implicitly as a function of $a, b,$ and c and calculate $\partial A/\partial a$ and $\partial A/\partial b$.

60. Express a implicitly as a function of $A, b,$ and B and calculate $\partial a/\partial A$ and $\partial a/\partial B$.

61. Express v_x in terms of u and v if the equations $x = v\ln u$ and $y = u\ln v$ define u and v as functions of the independent variables x and y, and if v_x exists. (*Hint:* Differentiate both equations with respect to x and solve for v_x with Cramer's rule.)

62. Find $\partial x/\partial u$ and $\partial y/\partial u$ if the equations $u = x^2 - y^2$ and $v = x^2 - y$ define x and y as functions of the independent variables u and v, and the partial derivatives exist. (See the hint in Exercise 61.) Then let $s = x^2 + y^2$ and find $\partial s/\partial u$.

Laplace Equations

The three-dimensional Laplace equation

$$\frac{\partial^2 f}{\partial x^2} + \frac{\partial^2 f}{\partial y^2} + \frac{\partial^2 f}{\partial z^2} = 0 \tag{5}$$

is satisfied by steady-state temperature distributions $T = f(x, y, z)$ in space, by gravitational potentials, and by electrostatic potentials. The *two-dimensional Laplace equation*

$$\frac{\partial^2 f}{\partial x^2} + \frac{\partial^2 f}{\partial y^2} = 0, \tag{6}$$

obtained by dropping the $\partial^2 f/\partial z^2$ term from Eq. (5), describes potentials and steady-state temperature distributions in a plane (Fig. 12.19).

Show that each function in Exercises 63–68 satisfies a Laplace equation.

63. $f(x, y, z) = x^2 + y^2 - 2z^2$

64. $f(x, y, z) = 2z^3 - 3(x^2 + y^2)z$

65. $f(x, y) = e^{-2y}\cos 2x$

66. $f(x, y) = \ln\sqrt{x^2 + y^2}$

67. $f(x, y, z) = (x^2 + y^2 + z^2)^{-1/2}$

68. $f(x, y, z) = e^{3x+4y}\cos 5z$

(a)

(b)

Boundary temperatures controlled

12.19 Steady-state temperature distributions in planes and solids satisfy Laplace equations. The plane (a) may be treated as a thin slice of the solid (b) perpendicular to the z-axis.

The Wave Equation

If we stand on an ocean shore and take a snapshot of the waves, the picture shows a regular pattern of peaks and valleys in an instant of time. We see periodic vertical motion in space, with respect to distance. If we stand in the water, we can feel the rise and fall of the water as the waves go by. We see periodic vertical motion in time. In physics, this beautiful symmetry is expressed by the *one-dimensional wave equation*

$$\frac{\partial^2 w}{\partial t^2} = c^2\frac{\partial^2 w}{\partial x^2}, \tag{7}$$

where w is the wave height, x is the distance variable, t is the time variable, and c is the velocity with which the waves are propagated.

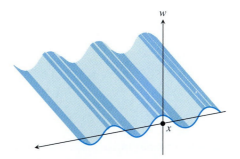

In our example, x is the distance across the ocean's surface, but in other applications x might be the distance along a vibrating string, distance through air (sound waves), or distance through space (light waves). The number c varies with the medium and type of wave.

Show that the functions in Exercises 69–75 are all solutions of the wave equation.

69. $w = \sin(x + ct)$

70. $w = \cos(2x + 2ct)$

71. $w = \sin(x + ct) + \cos(2x + 2ct)$

72. $w = \ln(2x + 2ct)$

73. $w = \tan(2x - 2ct)$

74. $w = 5\cos(3x + 3ct) + e^{x+ct}$

75. $w = f(u)$, where f is a differentiable function of u and $u = a(x + ct)$, where a is a constant.

| 12.4 | # Differentiability, Linearization, and Differentials |

In this section, we define differentiability and proceed from there to linearizations and differentials. The mathematical results of the section stem from the Increment Theorem. As we will see in the next section, this theorem also underlies the Chain Rule for multivariable functions.

Differentiability

Surprising as it may seem, the starting point for differentiability is not Fermat's difference quotient but rather the idea of increment. You may recall from our work with functions of a single variable that if $y = f(x)$ is differentiable at $x = x_0$, then the change in the value of f that results from changing x from x_0 to $x_0 + \Delta x$ is given by an equation of the form

$$\Delta y = f'(x_0)\Delta x + \epsilon\,\Delta x \tag{1}$$

in which $\epsilon \to 0$ as $\Delta x \to 0$. For functions of two variables, the analogous property becomes the definition of differentiability. The Increment Theorem (from advanced calculus) tells us when to expect the property to hold.

Theorem 3
The Increment Theorem for Functions of Two Variables

Suppose that the first partial derivatives of $f(x, y)$ are defined throughout an open region R containing the point (x_0, y_0) and that f_x and f_y are continuous at (x_0, y_0). Then the change

$$\Delta z = f(x_0 + \Delta x, y_0 + \Delta y) - f(x_0, y_0)$$

in the value of f that results from moving from (x_0, y_0) to another point $(x_0 + \Delta x, y_0 + \Delta y)$ in R satisfies an equation of the form

$$\Delta z = f_x(x_0, y_0)\Delta x + f_y(x_0, y_0)\Delta y + \epsilon_1\Delta x + \epsilon_2\Delta y, \tag{2}$$

in which $\epsilon_1, \epsilon_2 \to 0$ as $\Delta x, \Delta y \to 0$.

You will see where the epsilons come from if you read the proof in Appendix 10. You will also see that similar results hold for functions of more than two independent variables.

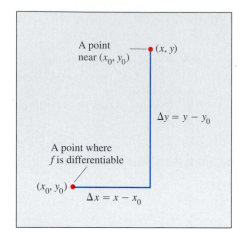

12.20 If f is differentiable at (x_0, y_0), then the value of f at any point (x, y) nearby is approximately $f(x_0, y_0) + f_x(x_0, y_0)\Delta x + f_y(x_0, y_0)\Delta y$.

As we can see from Theorems 3 and 4, a function $f(x, y)$ must be continuous at a point (x_0, y_0) if f_x and f_y are continuous throughout an open region containing (x_0, y_0). But remember that it is still possible for a function of two variables to be discontinuous at a point where its first partial derivatives exist, as we saw in Section 12.3, Example 8. Existence alone is not enough.

Definition

A function $f(x, y)$ is **differentiable at (x_0, y_0)** if $f_x(x_0, y_0)$ and $f_y(x_0, y_0)$ exist and Eq. (2) holds for f at (x_0, y_0). We call f **differentiable** if it is differentiable at every point in its domain.

In light of this definition, we have the immediate corollary of Theorem 3 that a function is differentiable if its first partial derivatives are *continuous*.

Corollary of Theorem 3

If the partial derivatives f_x and f_y of a function $f(x, y)$ are continuous throughout an open region R, then f is differentiable at every point of R.

If we replace the Δz in Eq. (2) by the expression $f(x, y) - f(x_0, y_0)$ and rewrite the equation as

$$f(x, y) = f(x_0, y_0) + f_x(x_0, y_0)\Delta x + f_y(x_0, y_0)\Delta y + \epsilon_1\Delta x + \epsilon_2\Delta y, \qquad (3)$$

we see that the right-hand side of the new equation approaches $f(x_0, y_0)$ as Δx and Δy approach 0. This tells us that a function $f(x, y)$ is continuous at every point where it is differentiable.

Theorem 4

If a function $f(x, y)$ is differentiable at (x_0, y_0), then f is continuous at (x_0, y_0).

How to Linearize a Function of Two Variables

Functions of two variables can be complicated, and we sometimes need to replace them with simpler ones that give the accuracy required for specific applications without being so hard to work with. We do this in a way that is similar to the way we find linear replacements for functions of a single variable (Section 3.7).

Suppose the function we wish to replace is $z = f(x, y)$ and that we want the replacement to be effective near a point (x_0, y_0) at which we know the values of f, f_x, and f_y and at which f is differentiable. Since f is differentiable, Eq. (3) holds for f at (x_0, y_0). Therefore, if we move from (x_0, y_0) to any point (x, y) by increments $\Delta x = x - x_0$ and $\Delta y = y - y_0$ (Fig. 12.20), the new value of f will be

$$f(x, y) = f(x_0, y_0) + f_x(x_0, y_0)(x - x_0)$$
$$+ f_y(x_0, y_0)(y - y_0) + \epsilon_1\Delta x + \epsilon_2\Delta y, \qquad \begin{array}{l}\text{Eq. (3), with} \\ \Delta x = x - x_0 \\ \text{and } \Delta y = y - y_0\end{array}$$

where $\epsilon_1, \epsilon_2 \to 0$ as $\Delta x, \Delta y \to 0$. If the increments Δx and Δy are small, the products $\epsilon_1\Delta x$ and $\epsilon_2\Delta y$ will eventually be smaller still and we will have

$$f(x, y) \approx \underbrace{f(x_0, y_0) + f_x(x_0, y_0)(x - x_0) + f_y(x_0, y_0)(y - y_0)}_{L(x, y)}.$$

In other words, as long as Δx and Δy are small, f will have approximately the same value as the linear function L. If f is hard to use, and our work can tolerate the error involved, we may safely replace f by L.

Definitions

The **linearization** of a function $f(x, y)$ at a point (x_0, y_0) where f is differentiable is the function

$$L(x, y) = f(x_0, y_0) + f_x(x_0, y_0)(x - x_0) + f_y(x_0, y_0)(y - y_0). \quad \textbf{(4)}$$

The approximation

$$f(x, y) \approx L(x, y)$$

is the **standard linear approximation** of f at (x_0, y_0).

In Section 12.8 we will see that the plane $z = L(x, y)$ is tangent to the surface $z = f(x, y)$ at the point (x_0, y_0). Thus, the linearization of a function of two variables is a tangent-*plane* approximation in the same way that the linearization of a function of a single variable is a tangent-*line* approximation.

EXAMPLE 1   Find the linearization of

$$f(x, y) = x^2 - xy + \frac{1}{2}y^2 + 3$$

at the point $(3, 2)$.

Solution   We evaluate Eq. (4) with

$$f(x_0, y_0) = \left(x^2 - xy + \frac{1}{2}y^2 + 3 \right)_{(3,2)} = 8,$$

$$f_x(x_0, y_0) = \frac{\partial}{\partial x} \left(x^2 - xy + \frac{1}{2}y^2 + 3 \right)_{(3,2)} = (2x - y)_{(3,2)} = 4,$$

$$f_y(x_0, y_0) = \frac{\partial}{\partial y} \left(x^2 - xy + \frac{1}{2}y^2 + 3 \right)_{(3,2)} = (-x + y)_{(3,2)} = -1,$$

getting

$$L(x, y) = f(x_0, y_0) + f_x(x_0, y_0)(x - x_0) + f_y(x_0, y_0)(y - y_0) \qquad \text{Eq. (4)}$$

$$= 8 + (4)(x - 3) + (-1)(y - 2) = 4x - y - 2.$$

The linearization of f at $(3, 2)$ is $L(x, y) = 4x - y - 2$.   ❏

How Accurate Is the Standard Linear Approximation?

To find the error in the approximation $f(x, y) \approx L(x, y)$, we use the second order partial derivatives of f. Suppose that the first and second order partial derivatives of f are continuous throughout an open set containing a closed rectangular region R centered at (x_0, y_0) and given by the inequalities

$$|x - x_0| \le h, \qquad |y - y_0| \le k$$

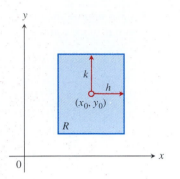

12.21 The rectangular region R: $|x - x_0| \le h, |y - y_0| \le k$ in the xy-plane. On this kind of region, we can find useful error bounds for our approximations.

(Fig. 12.21). Since R is closed and bounded, the second partial derivatives all take on absolute maximum values on R. If B is the largest of these values, then, as explained in Section 12.10, the error $E(x, y) = f(x, y) - L(x, y)$ in the standard linear approximation satisfies the inequality

$$|E(x, y)| \le \frac{1}{2} B \left(|x - x_0| + |y - y_0|\right)^2$$

throughout R.

When we use this inequality to estimate E, we usually cannot find the values of f_{xx}, f_{yy}, and f_{xy} that determine B and we have to settle for an upper bound or "worst-case" value instead. If M is any common upper bound for $|f_{xx}|, |f_{yy}|$, and $|f_{xy}|$ on R, then B will be less than or equal to M and we will know that

$$|E(x, y)| \le \frac{1}{2} M \left(|x - x_0| + |y - y_0|\right)^2 .$$

This is the inequality normally used in estimating E. When we need to make $|E(x, y)|$ small for a given M, we just make $|x - x_0|$ and $|y - y_0|$ small.

The Error in the Standard Linear Approximation

If f has continuous first and second partial derivatives throughout an open set containing a rectangle R centered at (x_0, y_0) and if M is any upper bound for the values of $|f_{xx}|, |f_{yy}|$, and $|f_{xy}|$ on R, then the error $E(x, y)$ incurred in replacing $f(x, y)$ on R by its linearization

$$L(x, y) = f(x_0, y_0) + f_x(x_0, y_0)(x - x_0) + f_y(x_0, y_0)(y - y_0)$$

satisfies the inequality

$$|E(x, y)| \le \frac{1}{2} M \left(|x - x_0| + |y - y_0|\right)^2 . \tag{5}$$

EXAMPLE 2   In Example 1, we found the linearization of

$$f(x, y) = x^2 - xy + \frac{1}{2} y^2 + 3$$

at $(3, 2)$ to be

$$L(x, y) = 4x - y - 2.$$

Find an upper bound for the error in the approximation $f(x, y) \approx L(x, y)$ over the rectangle

$$R: \quad |x - 3| \le 0.1, \quad |y - 2| \le 0.1.$$

Express the upper bound as a percentage of $f(3, 2)$, the value of f at the center of the rectangle.

Solution   We use the inequality

$$|E(x, y)| \le \frac{1}{2} M \left(|x - x_0| + |y - y_0|\right)^2 . \qquad \text{Eq. (5)}$$

To find a suitable value for M, we calculate f_{xx}, f_{xy}, and f_{yy}, finding, after a routine differentiation, that all three derivatives are constant, with values

$$|f_{xx}| = |2| = 2, \qquad |f_{xy}| = |-1| = 1, \qquad |f_{yy}| = |1| = 1.$$

The largest of these is 2, so we may safely take M to be 2. With $(x_0, y_0) = (3, 2)$, we then know that, throughout R,

$$|E(x, y)| \leq \frac{1}{2}(2)\,(|x - 3| + |y - 2|)^2 = (|x - 3| + |y - 2|)^2.$$

Finally, since $|x - 3| \leq 0.1$ and $|y - 2| \leq 0.1$ on R, we have

$$|E(x, y)| \leq (0.1 + 0.1)^2 = 0.04.$$

As a percentage of $f(3, 2) = 8$, the error is no greater than

$$\frac{0.04}{8} \times 100 = 0.5\%.$$

As long as (x, y) stays in R, the approximation $f(x, y) \approx L(x, y)$ will be in error by no more than 0.04, which is 1/2% of the value of f at the center of R. ❑

Predicting Change with Differentials

Suppose we know the values of a differentiable function $f(x, y)$ and its first partial derivatives at a point (x_0, y_0) and we want to predict how much the value of f will change if we move to a point $(x_0 + \Delta x, y_0 + \Delta y)$ nearby. If Δx and Δy are small, f and its linearization at (x_0, y_0) will change by nearly the same amount, so the change in L will give a practical estimate of the change in f.

The change in f is

$$\Delta f = f(x_0 + \Delta x, y_0 + \Delta y) - f(x_0, y_0).$$

A straightforward calculation with Eq. (4), using the notation $x - x_0 = \Delta x$ and $y - y_0 = \Delta y$, shows that the corresponding change in L is

$$\Delta L = L(x_0 + \Delta x, y_0 + \Delta y) - L(x_0, y_0)$$

$$= f_x(x_0, y_0)\Delta x + f_y(x_0, y_0)\Delta y.$$

The formula for Δf is usually as hard to work with as the formula for f. The change in L, however, is just a known constant times Δx plus a known constant times Δy.

The change ΔL is usually described in the more suggestive notation

$$df = f_x(x_0, y_0)\,dx + f_y(x_0, y_0)\,dy,$$

in which df denotes the change in the linearization that results from the changes dx and dy in x and y. As usual, we call dx and dy differentials of x and y, and call df the corresponding differential of f.

Definition

If we move from (x_0, y_0) to a point $(x_0 + dx, y_0 + dy)$ nearby, the resulting differential in f is

$$df = f_x(x_0, y_0)\,dx + f_y(x_0, y_0)\,dy. \tag{6}$$

This change in the linearization of f is called the **total differential of f.**

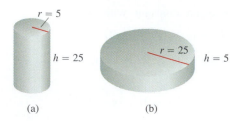

$r = 5$

$h = 25$

$r = 25$ $h = 5$

(a)   (b)

12.22 The volume of cylinder (a) is more sensitive to a small change in r than it is to an equally small change in h. The volume of cylinder (b) is more sensitive to small changes in h than it is to small changes in r.

Absolute change vs. relative change

If you measure a 20-volt potential with an error of 10 volts, your reading is probably too crude to be useful. You are off by 50%. But if you measure a 200,000-volt potential with an error of 10 volts, your reading is within 0.005% of the true value. An absolute error of 10 volts is significant in the first case but of no consequence in the second because the relative error is so small.

In other cases, a small relative error—say, traveling a few meters too far in a journey of hundreds of thousands of meters—can have spectacular consequences.

EXAMPLE 3 *Sensitivity to change*

Your company manufactures right circular cylindrical molasses storage tanks that are 25 ft high with a radius of 5 ft. How sensitive are the tanks' volumes to small variations in height and radius?

Solution As a function of radius r and height h, the typical tank's volume is

$$V = \pi r^2 h.$$

The change in volume caused by small changes dr and dh in radius and height is approximately

$$dV = V_r(5, 25)\, dr + V_h(5, 25)\, dh \qquad \text{Eq. (6) with } f = V \text{ and } (x_0, y_0) = (5, 25)$$

$$= (2\pi rh)_{(5,25)}\, dr + (\pi r^2)_{(5,25)}\, dh$$

$$= 250\pi\, dr + 25\pi\, dh.$$

Thus, a 1-unit change in r will change V by about 250π units. A 1-unit change in h will change V by about 25π units. The tank's volume is 10 times more sensitive to a small change in r than it is to a small change of equal size in h. As a quality control engineer concerned with being sure the tanks have the correct volume, you would want to pay special attention to their radii.

In contrast, if the values of r and h are reversed to make $r = 25$ and $h = 5$, then the total differential in V becomes

$$dV = (2\pi rh)_{(25,5)}\, dr + (\pi r^2)_{(25,5)}\, dh = 250\pi\, dr + 625\pi\, dh.$$

Now the volume is more sensitive to changes in h than to changes in r (Fig. 12.22).

The general rule to be learned from this example is that functions are most sensitive to small changes in the variables that generate the largest partial derivatives. ❑

Absolute, Relative, and Percentage Change

When we move from (x_0, y_0) to a point nearby, we can describe the corresponding change in the value of a function $f(x, y)$ in three different ways.

	True	**Estimate**
Absolute change:	Δf	df
Relative change:	$\dfrac{\Delta f}{f(x_0, y_0)}$	$\dfrac{df}{f(x_0, y_0)}$
Percentage change:	$\dfrac{\Delta f}{f(x_0, y_0)} \times 100$	$\dfrac{df}{f(x_0, y_0)} \times 100$

EXAMPLE 4 Suppose that the variables r and h change from the initial values of $(r_0, h_0) = (1, 5)$ by the amounts $dr = 0.03$ and $dh = -0.1$. Estimate the resulting absolute, relative, and percentage changes in the values of the function $V = \pi r^2 h$.

Solution To estimate the absolute change in V, we evaluate

$$dV = V_r(r_0, h_0)\, dr + V_h(r_0, h_0)\, dh$$

to get $dV = 2\pi r_0 h_0\, dr + \pi r_0^2\, dh = 2\pi(1)(5)(0.03) + \pi(1)^2(-0.1)$

$$= 0.3\pi - 0.1\pi = 0.2\pi.$$

We divide this by $V(r_0, h_0)$ to estimate the relative change:

$$\frac{dV}{V(r_0, h_0)} = \frac{0.2\,\pi}{\pi r_0{}^2 h_0} = \frac{0.2\,\pi}{\pi (1)^2 (5)} = 0.04.$$

We multiply this by 100 to estimate the percentage change:

$$\frac{dV}{V(r_0, h_0)} \times 100 = 0.04 \times 100 = 4\%.$$

EXAMPLE 5   The volume $V = \pi r^2 h$ of a right circular cylinder is to be calculated from measured values of r and h. Suppose that r is measured with an error of no more than 2% and h with an error of no more than 0.5%. Estimate the resulting possible percentage error in the calculation of V.

Solution   We are told that

$$\left| \frac{dr}{r} \times 100 \right| \le 2 \qquad \text{and} \qquad \left| \frac{dh}{h} \times 100 \right| \le 0.5.$$

Since

$$\frac{dV}{V} = \frac{2\pi r h\, dr + \pi r^2 dh}{\pi r^2 h} = \frac{2\, dr}{r} + \frac{dh}{h},$$

we have

$$\left| \frac{dV}{V} \times 100 \right| = \left| 2\frac{dr}{r} \times 100 + \frac{dh}{h} \times 100 \right|$$

$$\le 2 \left| \frac{dr}{r} \times 100 \right| + \left| \frac{dh}{h} \times 100 \right| \le 2(2) + 0.5 = 4.5.$$

We estimate the error in the volume calculation to be at most 4.5%.

How accurately do we have to measure r and h to have a reasonable chance of calculating $V = \pi r^2 h$ with an error, say, of less than 2%? Questions like this are hard to answer because there is usually no single right answer. Since

$$\frac{dV}{V} = 2\frac{dr}{r} + \frac{dh}{h},$$

we see that dV/V is controlled by a combination of dr/r and dh/h. If we can measure h with great accuracy, we might come out all right even if we are sloppy about measuring r. On the other hand, our measurement of h might have so large a dh that the resulting dV/V would be too crude an estimate of $\Delta V/V$ to be useful even if dr were zero.

What we do in such cases is look for a reasonable square about the measured values (r_0, h_0) in which V will not vary by more than the allowed amount from $V_0 = \pi r_0{}^2 h_0$.

EXAMPLE 6   Find a reasonable square about the point $(r_0, h_0) = (5, 12)$ in which the value of $V = \pi r^2 h$ will not vary by more than ± 0.1.

Solution   We approximate the variation ΔV by the differential

$$dV = 2\pi r_0 h_0\, dr + \pi r_0{}^2 dh = 2\pi(5)(12)\, dr + \pi(5)^2 dh = 120\pi\, dr + 25\pi\, dh.$$

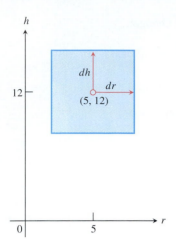

12.23 A small square about the point (5, 12) in the rh-plane (Example 6).

Since the region to which we are restricting our attention is a square (Fig. 12.23), we may set $dh = dr$ to get

$$dV = 120\pi \, dr + 25\pi \, dr = 145\pi \, dr.$$

We then ask, How small must we take dr to be sure that $|dV|$ is no larger than 0.1? To answer, we start with the inequality

$$|dV| \leq 0.1,$$

express dV in terms of dr,

$$|145\pi \, dr| \leq 0.1,$$

and find a corresponding upper bound for dr:

$$|dr| \leq \frac{0.1}{145\pi} \approx 2.1 \times 10^{-4}.$$

Rounding down to make sure dr won't accidentally be too big

With $dh = dr$, then, the square we want is described by the inequalities

$$|r - 5| \leq 2.1 \times 10^{-4}, \qquad |h - 12| \leq 2.1 \times 10^{-4}.$$

As long as (r, h) stays in this square, we may expect $|dV|$ to be less than or equal to 0.1 and we may expect $|\Delta V|$ to be approximately the same size. ∎

Functions of More Than Two Variables

Analogous results hold for differentiable functions of more than two variables.

1. The **linearization** of $f(x, y, z)$ at a point $P_0(x_0, y_0, z_0)$ is

$$L(x, y, z) = f(P_0) + f_x(P_0)(x - x_0) + f_y(P_0)(y - y_0) + f_z(P_0)(z - z_0). \quad (7)$$

2. Suppose that R is a closed rectangular solid centered at P_0 and lying in an open region on which the second partial derivatives of f are continuous. Suppose also that $|f_{xx}|, |f_{yy}|, |f_{zz}|, |f_{xy}|, |f_{xz}|$, and $|f_{yz}|$ are all less than or equal to M throughout R. Then the **error** $E(x, y, z) = f(x, y, z) - L(x, y, z)$ in the approximation of f by L is bounded throughout R by the inequality

$$|E| \leq \frac{1}{2} M \left(|x - x_0| + |y - y_0| + |z - z_0|\right)^2. \quad (8)$$

3. If the second partial derivatives of f are continuous and if x, y, and z change from x_0, y_0, and z_0 by small amounts dx, dy, and dz, the **total differential**

$$df = f_x(P_0) \, dx + f_y(P_0) \, dy + f_z(P_0) \, dz$$

gives a good approximation of the resulting change in f.

EXAMPLE 7   Find the linearization $L(x, y, z)$ of

$$f(x, y, z) = x^2 - xy + 3 \sin z$$

at the point $(x_0, y_0, z_0) = (2, 1, 0)$. Find an upper bound for the error incurred in replacing f by L on the rectangle

$$R: \quad |x - 2| \leq 0.01, \quad |y - 1| \leq 0.02, \quad |z| \leq 0.01.$$

Solution   A routine evaluation gives

$$f(2, 1, 0) = 2, \qquad f_x(2, 1, 0) = 3, \qquad f_y(2, 1, 0) = -2, \qquad f_z(2, 1, 0) = 3.$$

With these values, Eq. (7) becomes

$$L(x, y, z) = 2 + 3(x - 2) + (-2)(y - 1) + 3(z - 0) = 3x - 2y + 3z - 2.$$

Equation (8) gives an upper bound for the error incurred by replacing f by L on R. Since

$$f_{xx} = 2, \qquad f_{yy} = 0, \qquad f_{zz} = -3 \sin z,$$

$$f_{xy} = -1, \qquad f_{xz} = 0, \qquad f_{yz} = 0,$$

we may safely take M to be $\max |-3 \sin z| = 3$. Hence

$$|E| \leq \frac{1}{2}(3)(0.01 + 0.02 + 0.01)^2 = 0.0024.$$

The error will be no greater than 0.0024.   ❏

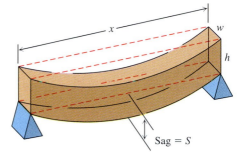

12.24 A beam supported at its two ends before and after loading. Example 8 shows how the sag S is related to the weight of the load and the dimensions of the beam.

EXAMPLE 8 *Controlling sag in uniformly loaded beams*

A horizontal rectangular beam, supported at both ends, will sag when subjected to a uniform load (constant weight per linear foot). The amount S of sag (Fig. 12.24) is calculated with the formula

$$S = C \frac{px^4}{wh^3}.$$

In this equation,

> $p = $ the load (newtons per meter of beam length),
>
> $x = $ the length between supports (m),
>
> $w = $ the width of the beam (m),
>
> $h = $ the height of the beam (m),
>
> $C = $ a constant that depends on the units of measurement and on the material from which the beam is made.

$x_0 = 4$ m $w_0 = 10$ cm

$h_0 = 20$ cm

12.25 The dimensions of the beam in Example 8.

Find dS for a beam 4 m long, 10 cm wide, and 20 cm high that is subjected to a load of 100 N/m (Fig. 12.25). What conclusions can be drawn about the beam from the expression for dS?

Solution Since S is a function of the four independent variables p, x, w, and h, its total differential dS is given by the equation

$$dS = S_p \, dp + S_x \, dx + S_w \, dw + S_h \, dh.$$

When we write this out for a particular set of values p_0, x_0, w_0, and h_0 and simplify the result, we find that

$$dS = S_0 \left(\frac{dp}{p_0} + \frac{4dx}{x_0} - \frac{dw}{w_0} - \frac{3dh}{h_0} \right),$$

where $S_0 = S(p_0, x_0, w_0, h_0) = Cp_0x_0^4/(w_0h_0^3)$.

If $p_0 = 100$ N/m, $x_0 = 4$ m, $w_0 = 0.1$ m, and $h_0 = 0.2$ m, then

$$dS = S_0 \left(\frac{dp}{100} + dx - 10dw - 15dh \right). \tag{9}$$

Here is what we can learn from Eq. (9). Since dp and dx appear with positive coefficients, increases in p and x will increase the sag. But dw and dh appear with negative coefficients, so increases in w and h will decrease the sag (make the beam stiffer). The sag is not very sensitive to changes in load because the coefficient of dp is 1/100. The magnitude of the coefficient of dh is greater than the magnitude of the coefficient of dw. Making the beam 1 cm higher will therefore decrease the sag more than making the beam 1 cm wider. ◻

Exercises 12.4

Finding Linearizations

In Exercises 1–6, find the linearization $L(x, y)$ of the function at each point.

1. $f(x, y) = x^2 + y^2 + 1$ at (a) (0, 0), (b) (1, 1)

2. $f(x, y) = (x + y + 2)^2$ at (a) (0, 0), (b) (1, 2)

3. $f(x, y) = 3x - 4y + 5$ at (a) (0, 0), (b) (1, 1)

4. $f(x, y) = x^3 y^4$ at (a) (1, 1), (b) (0, 0)

5. $f(x, y) = e^x \cos y$ at (a) (0, 0), (b) $(0, \pi/2)$

6. $f(x, y) = e^{2y-x}$ at (a) (0, 0), (b) (1, 2)

Upper Bounds for Errors in Linear Approximations

In Exercises 7–12, find the linearization $L(x, y)$ of the function $f(x, y)$ at P_0. Then use inequality (5) to find an upper bound for the magnitude $|E|$ of the error in the approximation $f(x, y) \approx L(x, y)$ over the rectangle R.

7. $f(x, y) = x^2 - 3xy + 5$ at $P_0(2, 1)$,
   R: $|x - 2| \le 0.1$, $|y - 1| \le 0.1$

8. $f(x, y) = (1/2)x^2 + xy + (1/4)y^2 + 3x - 3y + 4$ at $P_0(2, 2)$,
   R: $|x - 2| \le 0.1$, $|y - 2| \le 0.1$

9. $f(x, y) = 1 + y + x \cos y$ at $P_0(0, 0)$,
   R: $|x| \le 0.2$, $|y| \le 0.2$
   (Use $|\cos y| \le 1$ and $|\sin y| \le 1$ in estimating E.)

10. $f(x, y) = xy^2 + y \cos(x - 1)$ at $P_0(1, 2)$,
    R: $|x - 1| \le 0.1$, $|y - 2| \le 0.1$

11. $f(x, y) = e^x \cos y$ at $P_0(0, 0)$,
    R: $|x| \le 0.1$, $|y| \le 0.1$
    (Use $e^x \le 1.11$ and $|\cos y| \le 1$ in estimating E.)

12. $f(x, y) = \ln x + \ln y$ at $P_0(1, 1)$,
    R: $|x - 1| \le 0.2$, $|y - 1| \le 0.2$

Sensitivity to Change. Estimates

13. You plan to calculate the area of a long, thin rectangle from measurements of its length and width. Which dimension should you measure more carefully? Give reasons for your answer.

14. a) Around the point (1, 0), is $f(x, y) = x^2(y + 1)$ more sensitive to changes in x, or to changes in y? Give reasons for your answer.

   b) What ratio of dx to dy will make df equal zero at (1, 0)?

15. Suppose T is to be found from the formula $T = x(e^y + e^{-y})$ where x and y are found to be 2 and ln 2 with maximum possible errors of $|dx| = 0.1$ and $|dy| = 0.02$. Estimate the maximum possible error in the computed value of T.

16. About how accurately may $V = \pi r^2 h$ be calculated from measurements of r and h that are in error by 1%?

17. If $r = 5.0$ cm and $h = 12.0$ cm to the nearest millimeter, what should we expect the maximum percentage error in calculating $V = \pi r^2 h$ to be?

18. To estimate the volume of a cylinder of radius about 2 m and height about 3 m, about how accurately should the radius and height be measured so that the error in the volume estimate will not exceed 0.1 m³? Assume that the possible error dr in measuring r is equal to the possible error dh in measuring h.

19. Give a reasonable square centered at (1, 1) over which the value of $f(x, y) = x^3 y^4$ will not vary by more than ±0.1.

20. *Variation in electrical resistance.* The resistance R produced by wiring resistors of R_1 and R_2 ohms in parallel (Fig. 12.26) can be calculated from the formula
$$\frac{1}{R} = \frac{1}{R_1} + \frac{1}{R_2}.$$

12.26 The circuit in Exercises 20 and 21.

a) Show that
$$dR = \left(\frac{R}{R_1}\right)^2 dR_1 + \left(\frac{R}{R_2}\right)^2 dR_2.$$

b) You have designed a two-resistor circuit like the one in Fig. 12.26 to have resistances of $R_1 = 100$ ohms and $R_2 = 400$

ohms, but there is always some variation in manufacturing and the resistors received by your firm will probably not have these exact values. Will the value of R be more sensitive to variation in R_1, or to variation in R_2? Give reasons for your answer.

21. (*Continuation of Exercise 20.*) In another circuit like the one in Fig. 12.26, you plan to change R_1 from 20 to 20.1 ohms and R_2 from 25 to 24.9 ohms. By about what percentage will this change R?

22. *Error carry-over in coordinate changes*

a) If $x = 3 \pm 0.01$ and $y = 4 \pm 0.01$, as shown here, with approximately what accuracy can you calculate the polar coordinates r and θ of the point $P(x, y)$ from the formulas $r^2 = x^2 + y^2$ and $\theta = \tan^{-1}(y/x)$? Express your estimates as percentage changes of the values that r and θ have at the point $(x_0, y_0) = (3, 4)$.

b) At the point $(x_0, y_0) = (3, 4)$, are the values of r and θ more sensitive to changes in x, or to changes in y? Give reasons for your answer.

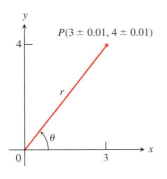

$P(3 \pm 0.01, 4 \pm 0.01)$

Functions of Three Variables

Find the linearizations $L(x, y, z)$ of the functions in Exercises 23–28 at the given points.

23. $f(x, y, z) = xy + yz + xz$ at

    a) $(1, 1, 1)$ **b)** $(1, 0, 0)$ **c)** $(0, 0, 0)$

24. $f(x, y, z) = x^2 + y^2 + z^2$ at

    a) $(1, 1, 1)$ **b)** $(0, 1, 0)$ **c)** $(1, 0, 0)$

25. $f(x, y, z) = \sqrt{x^2 + y^2 + z^2}$ at

    a) $(1, 0, 0)$ **b)** $(1, 1, 0)$ **c)** $(1, 2, 2)$

26. $f(x, y, z) = (\sin xy)/z$ at

    a) $(\pi/2, 1, 1)$ **b)** $(2, 0, 1)$

27. $f(x, y, z) = e^x + \cos(y + z)$ at

    a) $(0, 0, 0)$ **b)** $\left(0, \dfrac{\pi}{2}, 0\right)$ **c)** $\left(0, \dfrac{\pi}{4}, \dfrac{\pi}{4}\right)$

28. $f(x, y, z) = \tan^{-1}(xyz)$ at

    a) $(1, 0, 0)$ **b)** $(1, 1, 0)$ **c)** $(1, 1, 1)$

In Exercises 29–32, find the linearization $L(x, y, z)$ of the function $f(x, y, z)$ at P_0. Then use inequality (8) to find an upper bound for the

magnitude of the error E in the approximation $f(x, y, z) \approx L(x, y, z)$ over the region R.

29. $f(x, y, z) = xz - 3yz + 2$ at $P_0(1, 1, 2)$
    $R:$ $|x - 1| \le 0.01$, $|y - 1| \le 0.01$, $|z - 2| \le 0.02$

30. $f(x, y, z) = x^2 + xy + yz + (1/4)z^2$ at $P_0(1, 1, 2)$
    $R:$ $|x - 1| \le 0.01$, $|y - 1| \le 0.01$, $|z - 2| \le 0.08$

31. $f(x, y, z) = xy + 2yz - 3xz$ at $P_0(1, 1, 0)$
    $R:$ $|x - 1| \le 0.01$, $|y - 1| \le 0.01$, $|z| \le 0.01$

32. $f(x, y, z) = \sqrt{2} \cos x \sin(y + z)$ at $P_0(0, 0, \pi/4)$
    $R:$ $|x| \le 0.01$, $|y| \le 0.01$, $|z - \pi/4| \le 0.01$

Theory and Examples

33. The beam of Example 8 is tipped on its side so that $h = 0.1$ m and $w = 0.2$ m.

    a) What is the value of dS now?

    b) Compare the sensitivity of the newly positioned beam to a small change in height with its sensitivity to an equally small change in width.

34. A standard 12-fl oz can of soda is essentially a cylinder of radius $r = 1$ in. and height $h = 5$ in.

    a) At these dimensions, how sensitive is the can's volume to a small change in the radius versus a small change in the height?

    b) Could you design a soda can that *appears* to hold more soda but in fact holds the same 12 fl oz? What might its dimensions be? (There is more than one correct answer.)

35. If $|a|$ is much greater than $|b|$, $|c|$, and $|d|$, to which of a, b, c, and d is the value of the determinant

$$f(a, b, c, d) = \begin{vmatrix} a & b \\ c & d \end{vmatrix}$$

most sensitive? Give reasons for your answer.

36. Estimate how strongly simultaneous errors of 2% in a, b, and c might affect the calculation of the product

$$p(a, b, c) = abc.$$

37. Estimate how much wood it takes to make a hollow rectangular box whose inside measurements are 5 ft long by 3 ft wide by 2 ft deep if the box is made of lumber 1/2-in. thick and the box has no top.

38. The area of a triangle is $(1/2)ab \sin C$, where a and b are the lengths of two sides of the triangle and C is the measure of the included angle. In surveying a triangular plot, you have measured

$a = 150 \pm \dfrac{1}{2}$ ft

$C = 60° \pm 2°$

$b = 200 \pm \dfrac{1}{2}$ ft

$a, b,$ and C to be 150 ft, 200 ft, and $60°$, respectively. By about how much could your area calculation be in error if your values of a and b are off by half a foot each and your measurement of C is off by $2°$? See the figure. Remember to use radians.

39. Suppose that $u = xe^y + y \sin z$ and that $x, y,$ and z can be measured with maximum possible errors of $\pm 0.2, \pm 0.6,$ and $\pm \pi/180,$ respectively. Estimate the maximum possible error in calculating u from the measured values $x = 2, y = \ln 3, z = \pi/2.$

40. *The Wilson lot size formula.* The Wilson lot size formula in economics says that the most economical quantity Q of goods (radios, shoes, brooms, whatever) for a store to order is given by the formula $Q = \sqrt{2KM/h},$ where K is the cost of placing the order, M is the number of items sold per week, and h is the weekly holding cost for each item (cost of space, utilities, security, and so on). To which of the variables $K, M,$ and h is Q most sensitive near the point $(K_0, M_0, h_0) = (2, 20, 0.05)$? Give reasons for your answer.

41. Does a function $f(x, y)$ with continuous first partial derivatives throughout an open region R have to be continuous on R? Give reasons for your answer.

42. If a function $f(x, y)$ has continuous second partial derivatives throughout an open region R, must the first order partial derivatives of f be continuous on R? Give reasons for your answer.

12.5 The Chain Rule

When we are interested in the temperature $w = f(x, y, z)$ at points along a curve $x = g(t), y = h(t), z = k(t)$ in space, or in the pressure or density along a path through a gas or fluid, we may think of f as a function of the single variable t. For each value of t, the temperature at the point $(g(t), h(t), k(t))$ is the value of the composite function $f(g(t), h(t), k(t))$. If we then wish to know the rate at which f changes with respect to t along the path, we have only to differentiate this composite with respect to t, provided, of course, the derivative exists.

Sometimes we can find the derivative by substituting the formulas for $g, h,$ and k into the formula for f and differentiating directly with respect to t. But we often have to work with functions whose formulas are too complicated for convenient substitution or for which formulas are not readily available. To find a function's derivatives under circumstances like these, we use the Chain Rule. The form the Chain Rule takes depends on how many variables are involved but, except for the presence of additional variables, it works just like the Chain Rule in Section 2.5.

The Chain Rule for Functions of Two Variables

In Section 2.5, we used the Chain Rule when $w = f(x)$ was a differentiable function of x and $x = g(t)$ was a differentiable function of t. This made w a differentiable function of t and the Chain Rule said that dw/dt could be calculated with the formula

$$\frac{dw}{dt} = \frac{dw}{dx}\frac{dx}{dt}.$$

The analogous formula for a function $w = f(x, y)$ is given in Theorem 5.

Theorem 5

Chain Rule for Functions of Two Independent Variables

If $w = f(x, y)$ is differentiable and x and y are differentiable functions of t, then w is a differentiable function of t and

$$\frac{dw}{dt} = \frac{\partial f}{\partial x}\frac{dx}{dt} + \frac{\partial f}{\partial y}\frac{dy}{dt}. \tag{1}$$

Proof The proof consists of showing that if x and y are differentiable at $t = t_0$, then w is differentiable at t_0 and

$$\left(\frac{dw}{dt}\right)_{t_0} = \left(\frac{\partial w}{\partial x}\right)_{P_0} \left(\frac{dx}{dt}\right)_{t_0} + \left(\frac{\partial w}{\partial y}\right)_{P_0} \left(\frac{dy}{dt}\right)_{t_0}, \tag{2}$$

where $P_0 = (x(t_0), y(t_0))$.

Let Δx, Δy, and Δw be the increments that result from changing t from t_0 to $t_0 + \Delta t$. Since f is differentiable (remember the definition in Section 12.4),

$$\Delta w = \left(\frac{\partial w}{\partial x}\right)_{P_0} \Delta x + \left(\frac{\partial w}{\partial y}\right)_{P_0} \Delta y + \epsilon_1 \Delta x + \epsilon_2 \Delta y, \tag{3}$$

where $\epsilon_1, \epsilon_2 \to 0$ as $\Delta x, \Delta y \to 0$. To find dw/dt, we divide Eq. (3) through by Δt and let Δt approach zero. The division gives

$$\frac{\Delta w}{\Delta t} = \left(\frac{\partial w}{\partial x}\right)_{P_0} \frac{\Delta x}{\Delta t} + \left(\frac{\partial w}{\partial y}\right)_{P_0} \frac{\Delta y}{\Delta t} + \epsilon_1 \frac{\Delta x}{\Delta t} + \epsilon_2 \frac{\Delta y}{\Delta t},$$

and letting Δt approach zero gives

$$\left(\frac{dw}{dt}\right)_{t_0} = \lim_{\Delta t \to 0} \frac{\Delta w}{\Delta t}$$

$$= \left(\frac{\partial w}{\partial x}\right)_{P_0} \left(\frac{dx}{dt}\right)_{t_0} + \left(\frac{\partial w}{\partial y}\right)_{P_0} \left(\frac{dy}{dt}\right)_{t_0} + 0 \cdot \left(\frac{dx}{dt}\right)_{t_0} + 0 \cdot \left(\frac{dy}{dt}\right)_{t_0}.$$

This establishes Eq. (2) and completes the proof. ❑

The **tree diagram** in the margin provides a convenient way to remember the Chain Rule. From the diagram you see that when $t = t_0$ the derivatives dx/dt and dy/dt are evaluated at t_0. The value of t_0 then determines the value x_0 for the differentiable function x and the value y_0 for the differentiable function y. The partial derivatives $\partial w/\partial x$ and $\partial w/\partial y$ (which are themselves functions of x and y) are evaluated at the point $P_0(x_0, y_0)$ corresponding to t_0. The "true" independent variable is t, while x and y are *intermediate variables* (controlled by t) and w is the dependent variable.

A more precise notation for the Chain Rule shows how the various derivatives in Eq. (1) are evaluated:

$$\frac{dw}{dt}(t_0) = \frac{\partial f}{\partial x}(x_0, y_0) \cdot \frac{dx}{dt}(t_0) + \frac{\partial f}{\partial y}(x_0, y_0) \cdot \frac{dy}{dt}(t_0).$$

EXAMPLE 1 Use the Chain Rule to find the derivative of

$$w = xy$$

with respect to t along the path $x = \cos t$, $y = \sin t$. What is the derivative's value at $t = \pi/2$?

Solution We evaluate the right-hand side of Eq. (1) with $w = xy$, $x = \cos t$, and $y = \sin t$:

$$\frac{\partial w}{\partial x} = y = \sin t, \qquad \frac{\partial w}{\partial y} = x = \cos t, \qquad \frac{dx}{dt} = -\sin t, \qquad \frac{dy}{dt} = \cos t$$

$$\frac{dw}{dt} = \frac{\partial w}{\partial x}\frac{dx}{dt} + \frac{\partial w}{\partial y}\frac{dy}{dt} = (\sin t)(-\sin t) + (\cos t)(\cos t) \qquad \text{Eq. (1) with values from above}$$

$$= -\sin^2 t + \cos^2 t = \cos 2t.$$

The way to remember the Chain Rule is to picture the diagram below. To find dw/dt, start at w and read down each route to t, multiplying derivatives along the way. Then add the products.

Chain Rule

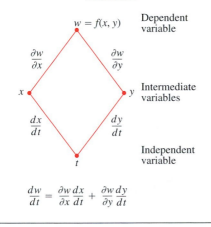

$w = f(x, y)$ — Dependent variable

$\dfrac{\partial w}{\partial x}$ $\dfrac{\partial w}{\partial y}$

x y — Intermediate variables

$\dfrac{dx}{dt}$ $\dfrac{dy}{dt}$

t — Independent variable

$$\frac{dw}{dt} = \frac{\partial w}{\partial x}\frac{dx}{dt} + \frac{\partial w}{\partial y}\frac{dy}{dt}$$

Here we have three routes from w to t instead of two. But finding dw/dt is still the same. Read down each route, multiplying derivatives along the way; then add.

Chain Rule

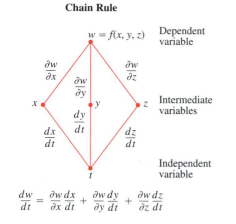

$$\frac{dw}{dt} = \frac{\partial w}{\partial x}\frac{dx}{dt} + \frac{\partial w}{\partial y}\frac{dy}{dt} + \frac{\partial w}{\partial z}\frac{dz}{dt}$$

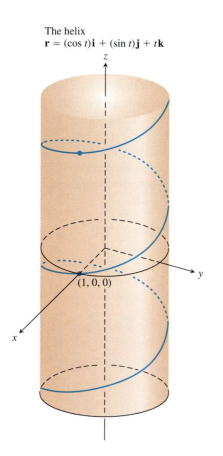

The helix
$$\mathbf{r} = (\cos t)\mathbf{i} + (\sin t)\mathbf{j} + t\mathbf{k}$$

12.27 Example 2 shows how the values of $w = xy + z$ vary with t along this helix.

Notice in our calculation that we have substituted the functional expressions $x = \cos t$ and $y = \sin t$ in the partial derivatives $\partial w/\partial x$ and $\partial w/\partial y$. The resulting derivative dw/dt is then expressed in terms of the independent variable t (so the intermediate variables x and y do not appear).

In this example we can check the result with a more direct calculation. As a function of t,

$$w = xy = \cos t \sin t = \frac{1}{2}\sin 2t,$$

so

$$\frac{dw}{dt} = \frac{d}{dt}\left(\frac{1}{2}\sin 2t\right) = \frac{1}{2}\cdot 2\cos 2t = \cos 2t.$$

In either case,

$$\left(\frac{dw}{dt}\right)_{t=\pi/2} = \cos\left(2\cdot\frac{\pi}{2}\right) = \cos\pi = -1. \qquad \square$$

The Chain Rule for Functions of Three Variables

To get the Chain Rule for functions of three variables, we add a term to Eq. (1).

> **Chain Rule for Functions of Three Independent Variables**
>
> If $w = f(x, y, z)$ is differentiable and x, y, and z are differentiable functions of t, then w is a differentiable function of t and
>
> $$\frac{dw}{dt} = \frac{\partial f}{\partial x}\frac{dx}{dt} + \frac{\partial f}{\partial y}\frac{dy}{dt} + \frac{\partial f}{\partial z}\frac{dz}{dt}. \qquad (4)$$

The derivation is identical with the derivation of Eq. (1) except that there are now three intermediate variables instead of two. The diagram we use for remembering the new equation is similar as well.

EXAMPLE 2   *Changes in a function's values along a helix*

Find dw/dt if

$$w = xy + z, \qquad x = \cos t, \qquad y = \sin t, \qquad z = t$$

(Fig. 12.27). What is the derivative's value at $t = 0$?

Solution

$$\frac{dw}{dt} = \frac{\partial w}{\partial x}\frac{dx}{dt} + \frac{\partial w}{\partial y}\frac{dy}{dt} + \frac{\partial w}{\partial z}\frac{dz}{dt} \qquad \text{Eq. (4)}$$

$$= (y)(-\sin t) + (x)(\cos t) + (1)(1)$$

$$= (\sin t)(-\sin t) + (\cos t)(\cos t) + 1 \qquad \text{Substitute for the intermediate variables.}$$

$$= -\sin^2 t + \cos^2 t + 1 = 1 + \cos 2t$$

$$\left(\frac{dw}{dt}\right)_{t=0} = 1 + \cos(0) = 2. \qquad \square$$

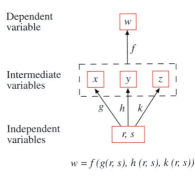

Dependent variable

Intermediate variables

Independent variables

$$w = f(g(r, s), h(r, s), k(r, s))$$

(a)

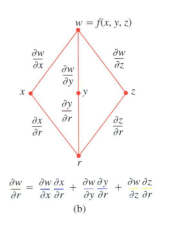

$$\frac{\partial w}{\partial r} = \frac{\partial w}{\partial x}\frac{\partial x}{\partial r} + \frac{\partial w}{\partial y}\frac{\partial y}{\partial r} + \frac{\partial w}{\partial z}\frac{\partial z}{\partial r}$$

(b)

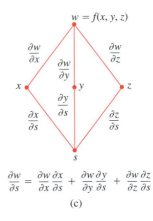

$$\frac{\partial w}{\partial s} = \frac{\partial w}{\partial x}\frac{\partial x}{\partial s} + \frac{\partial w}{\partial y}\frac{\partial y}{\partial s} + \frac{\partial w}{\partial z}\frac{\partial z}{\partial s}$$

(c)

12.28 Composite function and tree diagrams for Eqs. (5) and (6).

The Chain Rule for Functions Defined on Surfaces

If we are interested in the temperature $w = f(x, y, z)$ at points (x, y, z) on a globe in space, we might prefer to think of x, y, and z as functions of the variables r and s that give the points' longitudes and latitudes. If $x = g(r, s)$, $y = h(r, s)$, and $z = k(r, s)$, we could then express the temperature as a function of r and s with the composite function

$$w = f(g(r, s), h(r, s), k(r, s)).$$

Under the right conditions, w would have partial derivatives with respect to both r and s that could be calculated in the following way.

Chain Rule for Two Independent Variables and Three Intermediate Variables

Suppose that $w = f(x, y, z)$, $x = g(r, s)$, $y = h(r, s)$, and $z = k(r, s)$. If all four functions are differentiable, then w has partial derivatives with respect to r and s, given by the formulas

$$\frac{\partial w}{\partial r} = \frac{\partial w}{\partial x}\frac{\partial x}{\partial r} + \frac{\partial w}{\partial y}\frac{\partial y}{\partial r} + \frac{\partial w}{\partial z}\frac{\partial z}{\partial r}, \qquad (5)$$

$$\frac{\partial w}{\partial s} = \frac{\partial w}{\partial x}\frac{\partial x}{\partial s} + \frac{\partial w}{\partial y}\frac{\partial y}{\partial s} + \frac{\partial w}{\partial z}\frac{\partial z}{\partial s}. \qquad (6)$$

Equation (5) can be derived from Eq. (4) by holding s fixed and setting r equal to t. Similarly, Eq. (6) can be derived by holding r fixed and setting s equal to t. The tree diagrams for Eqs. (5) and (6) are shown in Fig. 12.28.

EXAMPLE 3   Express $\partial w/\partial r$ and $\partial w/\partial s$ in terms of r and s if

$$w = x + 2y + z^2, \qquad x = \frac{r}{s}, \qquad y = r^2 + \ln s, \qquad z = 2r.$$

Solution

$$\frac{\partial w}{\partial r} = \frac{\partial w}{\partial x}\frac{\partial x}{\partial r} + \frac{\partial w}{\partial y}\frac{\partial y}{\partial r} + \frac{\partial w}{\partial z}\frac{\partial z}{\partial r} \qquad \text{Eq. (5)}$$

$$= (1)\left(\frac{1}{s}\right) + (2)(2r) + (2z)(2)$$

$$= \frac{1}{s} + 4r + (4r)(2) = \frac{1}{s} + 12r \qquad \text{Substitute for intermediate variable } z.$$

$$\frac{\partial w}{\partial s} = \frac{\partial w}{\partial x}\frac{\partial x}{\partial s} + \frac{\partial w}{\partial y}\frac{\partial y}{\partial s} + \frac{\partial w}{\partial z}\frac{\partial z}{\partial s} \qquad \text{Eq. (6)}$$

$$= (1)\left(-\frac{r}{s^2}\right) + (2)\left(\frac{1}{s}\right) + (2z)(0) = \frac{2}{s} - \frac{r}{s^2} \qquad \square$$

If f is a function of two variables instead of three, Eqs. (5) and (6) become one term shorter, because the intermediate variable z doesn't appear.

Chain Rule

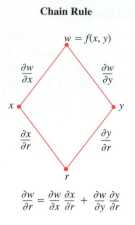

12.29 Tree diagram for the first of Eqs. (7).

Chain Rule

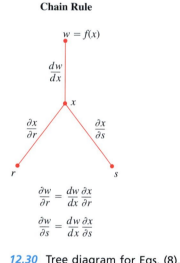

12.30 Tree diagram for Eqs. (8).

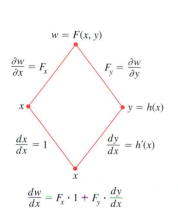

12.31 Tree diagram for Eq. (9).

If $w = f(x, y)$, $x = g(r, s)$, and $y = h(r, s)$, then

$$\frac{\partial w}{\partial r} = \frac{\partial w}{\partial x}\frac{\partial x}{\partial r} + \frac{\partial w}{\partial y}\frac{\partial y}{\partial r} \quad \text{and} \quad \frac{\partial w}{\partial s} = \frac{\partial w}{\partial x}\frac{\partial x}{\partial s} + \frac{\partial w}{\partial y}\frac{\partial y}{\partial s}. \quad (7)$$

Figure 12.29 shows the tree diagram for the first of Eqs. (7). The diagram for the second equation is similar—just replace r with s.

EXAMPLE 4 Express $\partial w/\partial r$ and $\partial w/\partial s$ in terms of r and s if

$$w = x^2 + y^2, \qquad x = r - s, \qquad y = r + s.$$

Solution We use Eqs. (7):

$$\frac{\partial w}{\partial r} = \frac{\partial w}{\partial x}\frac{\partial x}{\partial r} + \frac{\partial w}{\partial y}\frac{\partial y}{\partial r} \qquad \frac{\partial w}{\partial s} = \frac{\partial w}{\partial x}\frac{\partial x}{\partial s} + \frac{\partial w}{\partial y}\frac{\partial y}{\partial s}$$

$$= (2x)(1) + (2y)(1) \qquad = (2x)(-1) + (2y)(1)$$

$$= 2(r - s) + 2(r + s) \qquad = -2(r - s) + 2(r + s)$$

$$= 4r \qquad\qquad = 4s$$

Substitute for the intermediate variables.

If f is a function of x alone, Eqs. (5) and (6) simplify still further.

If $w = f(x)$ and $x = g(r, s)$, then

$$\frac{\partial w}{\partial r} = \frac{dw}{dx}\frac{\partial x}{\partial r} \quad \text{and} \quad \frac{\partial w}{\partial s} = \frac{dw}{dx}\frac{\partial x}{\partial s}. \quad (8)$$

Here dw/dx is the ordinary (single-variable) derivative (Fig. 12.30).

Implicit Differentiation (Continued from Chapter 2)

Believe it or not, the two-variable Chain Rule in Eq. (1) leads to a formula that takes most of the work out of implicit differentiation. Suppose:

1. The function $F(x, y)$ is differentiable and
2. The equation $F(x, y) = 0$ defines y implicitly as a differentiable function of x, say $y = h(x)$.

Since $w = F(x, y) = 0$, the derivative dw/dx must be zero. Computing the derivative from the Chain Rule (tree diagram in Fig. 12.31), we find

$$0 = \frac{dw}{dx} = F_x\frac{dx}{dx} + F_y\frac{dy}{dx} \qquad \text{Eq. (1) with } t = x \text{ and } f = F$$

$$= F_x \cdot 1 + F_y \cdot \frac{dy}{dx}. \quad (9)$$

If $F_y = \partial w / \partial y \neq 0$, we can solve Eq. (9) for dy/dx to get

$$\frac{dy}{dx} = -\frac{F_x}{F_y}.$$

> Suppose that $F(x, y)$ is differentiable and that the equation $F(x, y) = 0$ defines y as a differentiable function of x. Then, at any point where $F_y \neq 0$,
>
> $$\frac{dy}{dx} = -\frac{F_x}{F_y}. \qquad (10)$$

EXAMPLE 5   Find dy/dx if $x^2 + \sin y - 2y = 0$.

Solution   Take $F(x, y) = x^2 + \sin y - 2y$. Then

$$\frac{dy}{dx} = -\frac{F_x}{F_y} = -\frac{2x}{\cos y - 2}. \qquad \text{Eq. (10)}$$

This calculation is significantly shorter than the single-variable calculation with which we found dy/dx in Section 2.6, Example 3. ◻

Remembering the Different Forms of the Chain Rule

How are we to remember all the different forms of the Chain Rule? The answer is that there is no need to remember them all. The best thing to do is to draw the appropriate tree diagram by placing the dependent variable on top, the intermediate variables in the middle, and the selected independent variable at the bottom. To find the derivative of the dependent variable with respect to the selected independent variable, start at the dependent variable and read down each branch of the tree to the independent variable, calculating and multiplying the derivatives along the branch. Then add the products you found for the different branches. Let us summarize.

The Chain Rule for Functions of Many Variables

Suppose $w = f(x, y, \ldots, v)$ is a differentiable function of the variables $x, y, \ldots, v$ (a finite set) and the $x, y, \ldots, v$ are differentiable functions of $p, q, \ldots, t$ (another finite set). Then w is a differentiable function of the variables p through t and the partial derivatives of w with respect to these variables are given by equations of the form

$$\frac{\partial w}{\partial p} = \frac{\partial w}{\partial x}\frac{\partial x}{\partial p} + \frac{\partial w}{\partial y}\frac{\partial y}{\partial p} + \cdots + \frac{\partial w}{\partial v}\frac{\partial v}{\partial p}. \qquad (11)$$

The other equations are obtained by replacing p by $q, \ldots, t$, one at a time.

One way to remember Eq. (11) is to think of the right-hand side as the dot product of two vectors with components

$$\underbrace{\left(\frac{\partial w}{\partial x}, \frac{\partial w}{\partial y}, \ldots, \frac{\partial w}{\partial v}\right)}_{\substack{\text{Derivatives of } w \text{ with}\\ \text{respect to the}\\ \text{intermediate variables}}} \quad \text{and} \quad \underbrace{\left(\frac{\partial x}{\partial p}, \frac{\partial y}{\partial p}, \ldots, \frac{\partial v}{\partial p}\right)}_{\substack{\text{Derivatives of the intermediate}\\ \text{variables with respect to the}\\ \text{selected independent variable}}}.$$

Exercises 12.5

Chain Rule: One Independent Variable

In Exercises 1–6, (a) express dw/dt as a function of t, both by using the Chain Rule and by expressing w in terms of t and differentiating directly with respect to t. Then (b) evaluate dw/dt at the given value of t.

1. $w = x^2 + y^2$, $x = \cos t$, $y = \sin t$; $t = \pi$

2. $w = x^2 + y^2$, $x = \cos t + \sin t$, $y = \cos t - \sin t$; $t = 0$

3. $w = \dfrac{x}{z} + \dfrac{y}{z}$, $x = \cos^2 t$, $y = \sin^2 t$, $z = 1/t$; $t = 3$

4. $w = \ln(x^2 + y^2 + z^2)$, $x = \cos t$, $y = \sin t$, $z = 4\sqrt{t}$; $t = 3$

5. $w = 2ye^x - \ln z$, $x = \ln(t^2 + 1)$, $y = \tan^{-1} t$, $z = e^t$; $t = 1$

6. $w = z - \sin xy$, $x = t$, $y = \ln t$, $z = e^{t-1}$; $t = 1$

Chain Rule: Two and Three Independent Variables

In Exercises 7 and 8, (a) express $\partial z/\partial r$ and $\partial z/\partial \theta$ as functions of r and θ both by using the Chain Rule and by expressing z directly in terms of r and θ before differentiating. Then (b) evaluate $\partial z/\partial r$ and $\partial z/\partial \theta$ at the given point (r, θ).

7. $z = 4e^x \ln y$, $x = \ln(r \cos \theta)$, $y = r \sin \theta$; $(r, \theta) = (2, \pi/4)$

8. $z = \tan^{-1}(x/y)$, $x = r \cos \theta$, $y = r \sin \theta$; $(r, \theta) = (1.3, \pi/6)$

In Exercises 9 and 10, (a) express $\partial w/\partial u$ and $\partial w/\partial v$ as functions of u and v both by using the Chain Rule and by expressing w directly in terms of u and v before differentiating. Then (b) evaluate $\partial w/\partial u$ and $\partial w/\partial v$ at the given point (u, v).

9. $w = xy + yz + xz$, $x = u + v$, $y = u - v$, $z = uv$; $(u, v) = (1/2, 1)$

10. $w = \ln(x^2 + y^2 + z^2)$, $x = ue^v \sin u$, $y = ue^v \cos u$, $z = ue^v$; $(u, v) = (-2, 0)$

In Exercises 11 and 12, (a) express $\partial u/\partial x$, $\partial u/\partial y$, and $\partial u/\partial z$ as functions of x, y, and z both by using the Chain Rule and by expressing u directly in terms of x, y, and z before differentiating. Then (b) evaluate $\partial u/\partial x$, $\partial u/\partial y$, and $\partial u/\partial z$ at the given point (x, y, z).

11. $u = \dfrac{p - q}{q - r}$, $p = x + y + z$, $q = x - y + z$, $r = x + y - z$; $(x, y, z) = (\sqrt{3}, 2, 1)$

12. $u = e^{qr} \sin^{-1} p$, $p = \sin x$, $q = z^2 \ln y$, $r = 1/z$; $(x, y, z) = (\pi/4, 1/2, -1/2)$

Using a Tree Diagram

In Exercises 13–24, draw a tree diagram and write a Chain Rule formula for each derivative.

13. $\dfrac{dz}{dt}$ for $z = f(x, y)$, $x = g(t)$, $y = h(t)$

14. $\dfrac{dz}{dt}$ for $z = f(u, v, w)$, $u = g(t)$, $v = h(t)$, $w = k(t)$

15. $\dfrac{\partial w}{\partial u}$ and $\dfrac{\partial w}{\partial v}$ for $w = h(x, y, z)$, $x = f(u, v)$, $y = g(u, v)$, $z = k(u, v)$

16. $\dfrac{\partial w}{\partial x}$ and $\dfrac{\partial w}{\partial y}$ for $w = f(r, s, t)$, $r = g(x, y)$, $s = h(x, y)$, $t = k(x, y)$

17. $\dfrac{\partial w}{\partial u}$ and $\dfrac{\partial w}{\partial v}$ for $w = g(x, y)$, $x = h(u, v)$, $y = k(u, v)$

18. $\dfrac{\partial w}{\partial x}$ and $\dfrac{\partial w}{\partial y}$ for $w = g(u, v)$, $u = h(x, y)$, $v = k(x, y)$

19. $\dfrac{\partial z}{\partial t}$ and $\dfrac{\partial z}{\partial s}$ for $z = f(x, y)$, $x = g(t, s)$, $y = h(t, s)$

20. $\dfrac{\partial y}{\partial r}$ for $y = f(u)$, $u = g(r, s)$

21. $\dfrac{\partial w}{\partial s}$ and $\dfrac{\partial w}{\partial t}$ for $w = g(u)$, $u = h(s, t)$

22. $\dfrac{\partial w}{\partial p}$ for $w = f(x, y, z, v)$, $x = g(p, q)$, $y = h(p, q)$, $z = j(p, q)$, $v = k(p, q)$

23. $\dfrac{\partial w}{\partial r}$ and $\dfrac{\partial w}{\partial s}$ for $w = f(x, y)$, $x = g(r)$, $y = h(s)$

24. $\dfrac{\partial w}{\partial s}$ for $w = g(x, y)$, $x = h(r, s, t)$, $y = k(r, s, t)$

Implicit Differentiation

Assuming that the equations in Exercises 25–28 define y as a differentiable function of x, use Eq. (10) to find the value of dy/dx at the given point.

25. $x^3 - 2y^2 + xy = 0$, $(1, 1)$

26. $xy + y^2 - 3x - 3 = 0$, $(-1, 1)$

27. $x^2 + xy + y^2 - 7 = 0$, $(1, 2)$

28. $xe^y + \sin xy + y - \ln 2 = 0$, $(0, \ln 2)$

Equation (10) can be generalized to functions of three variables and even more. The three-variable version goes like this:

If the equation $F(x, y, z) = 0$ determines z as a differentiable function of x and y, then, at points where $F_z \neq 0$,

$$\frac{\partial z}{\partial x} = -\frac{F_x}{F_z} \quad \text{and} \quad \frac{\partial z}{\partial y} = -\frac{F_y}{F_z}. \tag{12}$$

Use these equations to find the values of $\partial z/\partial x$ and $\partial z/\partial y$ at the points in Exercises 29–32.

29. $z^3 - xy + yz + y^3 - 2 = 0$, $(1, 1, 1)$

30. $\dfrac{1}{x} + \dfrac{1}{y} + \dfrac{1}{z} - 1 = 0, \quad (2, 3, 6)$

31. $\sin(x+y) + \sin(y+z) + \sin(x+z) = 0, \quad (\pi, \pi, \pi)$

32. $xe^y + ye^z + 2\ln x - 2 - 3\ln 2 = 0, \quad (1, \ln 2, \ln 3)$

Finding Specified Partial Derivatives

33. Find $\partial w/\partial r$ when $r = 1, \quad s = -1$ if $w = (x+y+z)^2$, $x = r - s, \quad y = \cos(r+s), \quad z = \sin(r+s)$.

34. Find $\partial w/\partial v$ when $u = -1, \quad v = 2$ if $w = xy + \ln z$, $x = v^2/u, \quad y = u + v, \quad z = \cos u$.

35. Find $\partial w/\partial v$ when $u = 0, \quad v = 0$ if $w = x^2 + (y/x)$, $x = u - 2v + 1, \quad y = 2u + v - 2$.

36. Find $\partial z/\partial u$ when $u = 0, \quad v = 1$ if $z = \sin xy + x \sin y$, $x = u^2 + v^2, \quad y = uv$.

37. Find $\partial z/\partial u$ and $\partial z/\partial v$ when $u = \ln 2, \quad v = 1$ if $z = 5\tan^{-1}x$ and $x = e^u + \ln v$.

38. Find $\partial z/\partial u$ and $\partial z/\partial v$ when $u = 1$ and $v = -2$ if $z = \ln q$ and $q = \sqrt{v+3}\tan^{-1}u$.

Theory and Examples

39. *Changing voltage in a circuit.* The voltage V in a circuit that satisfies the law $V = IR$ is slowly dropping as the battery wears out. At the same time, the resistance R is increasing as the resistor heats up. Use the equation

$$\frac{dV}{dt} = \frac{\partial V}{\partial I}\frac{dI}{dt} + \frac{\partial V}{\partial R}\frac{dR}{dt}$$

to find how the current is changing at the instant when $R = 600$ ohms, $I = 0.04$ amp, $dR/dt = 0.5$ ohm/sec, and $dV/dt = -0.01$ volt/sec.

40. *Changing dimensions in a box.* The lengths a, b, and c of the edges of a rectangular box are changing with time. At the instant in question, $a = 1$ m, $b = 2$ m, $c = 3$ m, $da/dt = db/dt = 1$ m/sec, and $dc/dt = -3$ m/sec. At what rates are the box's volume V and surface area S changing at that instant? Are the box's interior diagonals increasing in length, or decreasing?

41. If $f(u, v, w)$ is differentiable and $u = x - y, v = y - z$, and $w = z - x$, show that

$$\frac{\partial f}{\partial x} + \frac{\partial f}{\partial y} + \frac{\partial f}{\partial z} = 0.$$

42. a) Show that if we substitute polar coordinates $x = r\cos\theta$ and $y = r\sin\theta$ in a differentiable function $w = f(x, y)$, then

$$\frac{\partial w}{\partial r} = f_x \cos\theta + f_y \sin\theta$$

and

$$\frac{1}{r}\frac{\partial w}{\partial\theta} = -f_x \sin\theta + f_y \cos\theta.$$

b) Solve the equations in (a) to express f_x and f_y in terms of $\partial w/\partial r$ and $\partial w/\partial\theta$.

c) Show that

$$(f_x)^2 + (f_y)^2 = \left(\frac{\partial w}{\partial r}\right)^2 + \frac{1}{r^2}\left(\frac{\partial w}{\partial\theta}\right)^2.$$

43. Show that if $w = f(u, v)$ satisfies the Laplace equation $f_{uu} + f_{vv} = 0$, and if $u = (x^2 - y^2)/2$ and $v = xy$, then w satisfies the Laplace equation $w_{xx} + w_{yy} = 0$.

44. Let $w = f(u) + g(v)$, where $u = x + iy$ and $v = x - iy$ and $i = \sqrt{-1}$. Show that w satisfies the Laplace equation $w_{xx} + w_{yy} = 0$ if all the necessary functions are differentiable.

Changes in Functions along Curves

45. Suppose that the partial derivatives of a function $f(x, y, z)$ at points on the helix $x = \cos t, y = \sin t, z = t$ are

$$f_x = \cos t, \quad f_y = \sin t, \quad f_z = t^2 + t - 2.$$

At what points on the curve, if any, can f take on extreme values?

46. Let $w = x^2e^{2y}\cos 3z$. Find the value of dw/dt at the point $(1, \ln 2, 0)$ on the curve $x = \cos t, y = \ln(t+2), z = t$.

47. Let $T = f(x, y)$ be the temperature at the point (x, y) on the circle $x = \cos t, y = \sin t, 0 \le t \le 2\pi$, and suppose that

$$\frac{\partial T}{\partial x} = 8x - 4y, \quad \frac{\partial T}{\partial y} = 8y - 4x.$$

a) Find where the maximum and minimum temperatures on the circle occur by examining the derivatives dT/dt and d^2T/dt^2.

b) Suppose $T = 4x^2 - 4xy + 4y^2$. Find the maximum and minimum values of T on the circle.

48. Let $T = g(x, y)$ be the temperature at the point (x, y) on the ellipse

$$x = 2\sqrt{2}\cos t, \quad y = \sqrt{2}\sin t, \quad 0 \le t \le 2\pi,$$

and suppose that

$$\frac{\partial T}{\partial x} = y, \quad \frac{\partial T}{\partial y} = x.$$

a) Locate the maximum and minimum temperatures on the ellipse by examining dT/dt and d^2T/dt^2.

b) Suppose that $T = xy - 2$. Find the maximum and minimum values of T on the ellipse.

Differentiating Integrals

Under mild continuity restrictions, it is true that if

$$F(x) = \int_a^b g(t, x)\, dt,$$

then $F'(x) = \int_a^b g_x(t, x)\, dt$. Using this fact and the Chain Rule, we can find the derivative of

$$F(x) = \int_a^{f(x)} g(t, x)\, dt$$

by letting

$$G(u, x) = \int_a^u g(t, x)\, dt,$$

where $u = f(x)$. Find the derivatives of the functions in Exercises 49 and 50.

49. $F(x) = \int_0^{x^2} \sqrt{t^4 + x^3}\, dt$

50. $F(x) = \int_{x^2}^1 \sqrt{t^3 + x^2}\, dt$

12.6

*Partial Derivatives with Constrained Variables

In finding partial derivatives of functions like $w = f(x, y)$, we have assumed x and y to be independent. But in many applications this is not the case. For example, the internal energy U of a gas may be expressed as a function $U = f(P, V, T)$ of pressure P, volume V, and temperature T. If the individual molecules of the gas do not interact, however, P, V, and T obey the ideal gas law

$$PV = nRT \qquad (n \text{ and } R \text{ constant})$$

and so fail to be independent. Finding partial derivatives in situations like these can be complicated. But it is better to face the complication now than to meet it for the first time while you are also trying to learn economics, engineering, or physics.

Decide Which Variables Are Dependent and Which Are Independent

If the variables in a function $w = f(x, y, z)$ are constrained by a relation like the one imposed on x, y, and z by the equation $z = x^2 + y^2$, the geometric meanings and the numerical values of the partial derivatives of f will depend on which variables are chosen to be dependent and which are chosen to be independent. To see how this choice can affect the outcome, we consider the calculation of $\partial w/\partial x$ when $w = x^2 + y^2 + z^2$ and $z = x^2 + y^2$.

EXAMPLE 1   Find $\partial w/\partial x$ if $w = x^2 + y^2 + z^2$ and $z = x^2 + y^2$.

Solution   We are given two equations in the four unknowns x, y, z, and w. Like many such systems, this one can be solved for two of the unknowns (the dependent variables) in terms of the others (the independent variables). In being asked for $\partial w/\partial x$, we are told that w is to be a dependent variable and x an independent variable. The possible choices for the other variables come down to

Dependent	Independent
w, z	x, y
w, y	x, z

In either case, we can express w explicitly in terms of the selected independent

---

*This section is based on notes written for MIT by Arthur P. Mattuck.

variables. We do this by using the second equation to eliminate the remaining dependent variable in the first equation.

In the first case, the remaining dependent variable is z. We eliminate it from the first equation by replacing it by $x^2 + y^2$. The resulting expression for w is

$$w = x^2 + y^2 + z^2 = x^2 + y^2 + (x^2 + y^2)^2$$
$$= x^2 + y^2 + x^4 + 2x^2y^2 + y^4$$

and

$$\frac{\partial w}{\partial x} = 2x + 4x^3 + 4xy^2. \tag{1}$$

This is the formula for $\partial w / \partial x$ when x and y are the independent variables.

In the second case, where the independent variables are x and z and the remaining dependent variable is y, we eliminate the dependent variable y in the expression for w by replacing y^2 by $z - x^2$. This gives

$$w = x^2 + y^2 + z^2 = x^2 + (z - x^2) + z^2 = z + z^2$$

and

$$\frac{\partial w}{\partial x} = 0. \tag{2}$$

This is the formula for $\partial w / \partial x$ when x and z are the independent variables.

The formulas for $\partial w / \partial x$ in Eqs. (1) and (2) are genuinely different. We cannot change either formula into the other by using the relation $z = x^2 + y^2$. There is not just one $\partial w / \partial x$, there are two, and we see that the original instruction to find $\partial w / \partial x$ was incomplete. *Which* $\partial w / \partial x$? we ask.

The geometric interpretations of Eqs. (1) and (2) help to explain why the equations differ. The function $w = x^2 + y^2 + z^2$ measures the square of the distance from the point (x, y, z) to the origin. The condition $z = x^2 + y^2$ says that the point (x, y, z) lies on the paraboloid of revolution shown in Fig. 12.32. What does it mean to calculate $\partial w / \partial x$ at a point $P(x, y, z)$ that can move only on this surface? What is the value of $\partial w / \partial x$ when the coordinates of P are, say, $(1, 0, 1)$?

If we take x and y to be independent, then we find $\partial w / \partial x$ by holding y fixed (at $y = 0$ in this case) and letting x vary. This means that P moves along the parabola $z = x^2$ in the xz-plane. As P moves on this parabola, w, which is the square of the distance from P to the origin, changes. We calculate $\partial w / \partial x$ in this case (our first solution above) to be

$$\frac{\partial w}{\partial x} = 2x + 4x^3 + 4xy^2.$$

At the point $P(1, 0, 1)$, the value of this derivative is

$$\frac{\partial w}{\partial x} = 2 + 4 + 0 = 6.$$

If we take x and z to be independent, then we find $\partial w / \partial x$ by holding z fixed while x varies. Since the z-coordinate of P is 1, varying x moves P along a circle in the plane $z = 1$. As P moves along this circle, its distance from the origin remains constant, and w, being the square of this distance, does not change. That is,

$$\frac{\partial w}{\partial x} = 0,$$

as we found in our second solution.

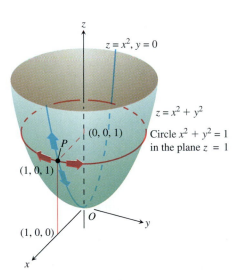

12.32 If P is constrained to lie on the paraboloid $z = x^2 + y^2$, the value of the partial derivative of $w = x^2 + y^2 + z^2$ with respect to x at P depends on the direction of motion (Example 1). (a) As x changes, with $y = 0$, P moves up or down the surface on the parabola $z = x^2$ in the xz-plane with $\partial w/\partial x = 2x + 4x^3$. (b) As x changes, with $z = 1$, P moves on the circle $x^2 + y^2 = 1, z = 1$, and $\partial w/\partial x = 0$.

> **How to Find $\partial w/\partial x$ When the Variables in $w = f(x, y, z)$ Are Constrained by Another Equation**
>
> As we saw in Example 1, a typical routine for finding $\partial w/\partial x$ when the variables in the function $w = f(x, y, z)$ are related by another equation has three steps. These steps apply to finding $\partial w/\partial y$ and $\partial w/\partial z$ as well.
>
> **Step 1** *Decide* which variables are to be dependent and which are to be independent. (In practice, the decision is based on the physical or theoretical context of our work. In the exercises at the end of this section, we say which variables are which.)
>
> **Step 2** *Eliminate* the other dependent variable(s) in the expression for w.
>
> **Step 3** *Differentiate* as usual.

If we cannot carry out step 2 after deciding which variables are dependent, we differentiate the equations as they are and try to solve for $\partial w/\partial x$ afterward. The next example shows how this is done.

EXAMPLE 2   Find $\partial w/\partial x$ at the point $(x, y, z) = (2, -1, 1)$ if

$$w = x^2 + y^2 + z^2, \qquad z^3 - xy + yz + y^3 = 1,$$

and x and y are the independent variables.

Solution   It is not convenient to eliminate z in the expression for w. We therefore differentiate both equations implicitly with respect to x, treating x and y as independent variables and w and z as dependent variables. This gives

$$\frac{\partial w}{\partial x} = 2x + 2z\frac{\partial z}{\partial x} \tag{3}$$

and

$$3z^2\frac{\partial z}{\partial x} - y + y\frac{\partial z}{\partial x} + 0 = 0. \tag{4}$$

These equations may now be combined to express $\partial w/\partial x$ in terms of x, y, and z. We solve Eq. (4) for $\partial z/\partial x$ to get

$$\frac{\partial z}{\partial x} = \frac{y}{y + 3z^2}$$

and substitute into Eq. (3) to get

$$\frac{\partial w}{\partial x} = 2x + \frac{2yz}{y + 3z^2}.$$

The value of this derivative at $(x, y, z) = (2, -1, 1)$ is

$$\left(\frac{\partial w}{\partial x}\right)_{(2, -1, 1)} = 2(2) + \frac{2(-1)(1)}{-1 + 3(1)^2} = 4 + \frac{-2}{2} = 3. \qquad \square$$

To show what variables are assumed to be independent in calculating a derivative, we can use the following notation:

$$\left(\frac{\partial w}{\partial x}\right)_y \qquad \partial w/\partial x \text{ with } x \text{ and } y \text{ independent}$$

$$\left(\frac{\partial f}{\partial y}\right)_{x,t} \qquad \partial f/\partial y \text{ with } y, x, \text{ and } t \text{ independent.}$$

EXAMPLE 3 Find $\left(\dfrac{\partial w}{\partial x}\right)_{y,z}$ if $w = x^2 + y - z + \sin t$ and $x + y = t$.

Solution

With x, y, z independent, we have

$$t = x + y, \qquad w = x^2 + y - z + \sin(x + y)$$

$$\left(\frac{\partial w}{\partial x}\right)_{y,z} = 2x + 0 - 0 + \cos(x + y)\frac{\partial}{\partial x}(x + y)$$

$$= 2x + \cos(x + y). \qquad \blacksquare$$

Arrow Diagrams

In solving problems like the one in Example 3, it often helps to start with an arrow diagram that shows how the variables and functions are related. If

$$w = x^2 + y - z + \sin t \qquad \text{and} \qquad x + y = t$$

and we are asked to find $\partial w/\partial x$ when $x, y,$ and z are independent, the appropriate diagram is one like this:

$$\begin{pmatrix} x \\ y \\ z \end{pmatrix} \quad \rightarrow \quad \begin{pmatrix} x \\ y \\ z \\ t \end{pmatrix} \quad \rightarrow \quad w \qquad\qquad (5)$$

$$\begin{array}{ccc} \text{independent} & \text{intermediate} & \text{dependent} \\ \text{variables} & \text{variables and} & \text{variable} \\ & \text{relations} & \\ & x = x & \\ & y = y & \\ & z = z & \\ & t = x + y & \end{array}$$

The diagram shows the independent variables on the left, the intermediate variables and their relation to the independent variables in the middle, and the dependent variable on the right.

To find $\partial w/\partial x$, we first apply the four-variable form of the Chain Rule to w, getting

$$\frac{\partial w}{\partial x} = \frac{\partial w}{\partial x}\frac{\partial x}{\partial x} + \frac{\partial w}{\partial y}\frac{\partial y}{\partial x} + \frac{\partial w}{\partial z}\frac{\partial z}{\partial x} + \frac{\partial w}{\partial t}\frac{\partial t}{\partial x}. \qquad (6)$$

We then use the formula for $w = x^2 + y - z + \sin t$ to evaluate the partial derivatives of w that appear on the right-hand side of Eq. (6). This gives

$$\frac{\partial w}{\partial x} = 2x\frac{\partial x}{\partial x} + (1)\frac{\partial y}{\partial x} + (-1)\frac{\partial z}{\partial x} + \cos t\frac{\partial t}{\partial x}$$

$$= 2x\frac{\partial x}{\partial x} + \frac{\partial y}{\partial x} - \frac{\partial z}{\partial x} + \cos t\frac{\partial t}{\partial x}. \qquad (7)$$

To calculate the remaining partial derivatives, we apply what we know about the dependence and independence of the variables involved. As shown in the diagram (5), the variables x, y, and z are independent and $t = x + y$. Hence,

$$\frac{\partial x}{\partial x} = 1, \quad \frac{\partial y}{\partial x} = 0, \quad \frac{\partial z}{\partial x} = 0, \quad \frac{\partial t}{\partial x} = \frac{\partial}{\partial x}(x + y) = (1 + 0) = 1.$$

We substitute these values into Eq. (7) to find $\partial w / \partial x$:

$$\left(\frac{\partial w}{\partial x}\right)_{y, z} = 2x(1) + 0 - 0 + (\cos t)(1)$$

$$= 2x + \cos t$$

$$= 2x + \cos (x + y). \qquad \text{In terms of the independent variables}$$

Exercises 12.6

Finding Partial Derivatives with Constrained Variables

In Exercises 1–3, begin by drawing a diagram that shows the relations among the variables.

1. If $w = x^2 + y^2 + z^2$ and $z = x^2 + y^2$, find

a) $\left(\dfrac{\partial w}{\partial y}\right)_z$ b) $\left(\dfrac{\partial w}{\partial z}\right)_x$ c) $\left(\dfrac{\partial w}{\partial z}\right)_y$

2. If $w = x^2 + y - z + \sin t$ and $x + y = t$, find

a) $\left(\dfrac{\partial w}{\partial y}\right)_{x, z}$ b) $\left(\dfrac{\partial w}{\partial y}\right)_{z, t}$ c) $\left(\dfrac{\partial w}{\partial z}\right)_{x, y}$

d) $\left(\dfrac{\partial w}{\partial z}\right)_{y, t}$ e) $\left(\dfrac{\partial w}{\partial t}\right)_{x, z}$ f) $\left(\dfrac{\partial w}{\partial t}\right)_{y, z}$

3. Let $U = f(P, V, T)$ be the internal energy of a gas that obeys the ideal gas law $PV = nRT$ (n and R constant). Find

a) $\left(\dfrac{\partial U}{\partial P}\right)_V$ b) $\left(\dfrac{\partial U}{\partial T}\right)_V$

4. Find

a) $\left(\dfrac{\partial w}{\partial x}\right)_y$ b) $\left(\dfrac{\partial w}{\partial z}\right)_y$

at the point $(x, y, z) = (0, 1, \pi)$ if

$$w = x^2 + y^2 + z^2 \quad \text{and} \quad y \sin z + z \sin x = 0.$$

5. Find

a) $\left(\dfrac{\partial w}{\partial y}\right)_x$ b) $\left(\dfrac{\partial w}{\partial y}\right)_z$

at the point $(w, x, y, z) = (4, 2, 1, -1)$ if

$$w = x^2 y^2 + yz - z^3 \quad \text{and} \quad x^2 + y^2 + z^2 = 6.$$

6. Find $\left(\dfrac{\partial u}{\partial y}\right)_x$ at the point $(u, v) = (\sqrt{2}, 1)$ if $x = u^2 + v^2$ and $y = uv$.

7. Suppose that $x^2 + y^2 = r^2$ and $x = r \cos \theta$, as in polar coordinates. Find

$$\left(\frac{\partial x}{\partial r}\right)_\theta \quad \text{and} \quad \left(\frac{\partial r}{\partial x}\right)_y.$$

8. Suppose that

$$w = x^2 - y^2 + 4z + t \quad \text{and} \quad x + 2z + t = 25.$$

Show that the equations

$$\frac{\partial w}{\partial x} = 2x - 1 \quad \text{and} \quad \frac{\partial w}{\partial x} = 2x - 2$$

each give $\partial w / \partial x$, depending on which variables are chosen to be dependent and which variables are chosen to be independent. Identify the independent variables in each case.

Partial Derivatives without Specific Formulas

9. Establish the fact, widely used in hydrodynamics, that if $f(x, y, z) = 0$, then

$$\left(\frac{\partial x}{\partial y}\right)_z \left(\frac{\partial y}{\partial z}\right)_x \left(\frac{\partial z}{\partial x}\right)_y = -1.$$

(*Hint:* Express all the derivatives in terms of the formal partial derivatives $\partial f / \partial x$, $\partial f / \partial y$, and $\partial f / \partial z$.)

10. If $z = x + f(u)$, where $u = xy$, show that

$$x \frac{\partial z}{\partial x} - y \frac{\partial z}{\partial y} = x.$$

11. Suppose that the equation $g(x, y, z) = 0$ determines z as a differentiable function of the independent variables x and y and that $g_z \neq 0$. Show that

$$\left(\frac{\partial z}{\partial y}\right)_x = -\frac{\partial g / \partial y}{\partial g / \partial z}.$$

12. Suppose that $f(x, y, z, w) = 0$ and $g(x, y, z, w) = 0$ determine z and w as differentiable functions of the independent variables

x and y, and suppose that

$$\frac{\partial f}{\partial z}\frac{\partial g}{\partial w} - \frac{\partial f}{\partial w}\frac{\partial g}{\partial z} \neq 0.$$

Show that

$$\left(\frac{\partial z}{\partial x}\right)_y = -\frac{\dfrac{\partial f}{\partial x}\dfrac{\partial g}{\partial w} - \dfrac{\partial f}{\partial w}\dfrac{\partial g}{\partial x}}{\dfrac{\partial f}{\partial z}\dfrac{\partial g}{\partial w} - \dfrac{\partial f}{\partial w}\dfrac{\partial g}{\partial z}}$$

and

$$\left(\frac{\partial w}{\partial y}\right)_x = -\frac{\dfrac{\partial f}{\partial z}\dfrac{\partial g}{\partial y} - \dfrac{\partial f}{\partial y}\dfrac{\partial g}{\partial z}}{\dfrac{\partial f}{\partial z}\dfrac{\partial g}{\partial w} - \dfrac{\partial f}{\partial w}\dfrac{\partial g}{\partial z}}.$$

12.7 Directional Derivatives, Gradient Vectors, and Tangent Planes

We know from Section 12.5 that if $f(x, y)$ is differentiable, then the rate at which f changes with respect to t along a differentiable curve $x = g(t), y = h(t)$ is

$$\frac{df}{dt} = \frac{\partial f}{\partial x}\frac{dx}{dt} + \frac{\partial f}{\partial y}\frac{dy}{dt}.$$

At any point $P_0(x_0, y_0) = P_0(g(t_0), h(t_0))$, this equation gives the rate of change of f with respect to increasing t and therefore depends, among other things, on the direction of motion along the curve. This observation is particularly important when the curve is a straight line and t is the arc length parameter along the line measured from P_0 in the direction of a given unit vector $\mathbf{u}$. For then df/dt is the rate of change of f with respect to distance in its domain in the direction of $\mathbf{u}$. By varying $\mathbf{u}$, we find the rates at which f changes with respect to distance as we move through P_0 in different directions. These "directional derivatives" have useful interpretations in science and engineering as well as in mathematics. This section develops a formula for calculating them and proceeds from there to find equations for tangent planes and normal lines on surfaces in space.

Directional Derivatives in the Plane

Suppose that the function $f(x, y)$ is defined throughout a region R in the xy-plane, that $P_0(x_0, y_0)$ is a point in R, and that $\mathbf{u} = u_1\mathbf{i} + u_2\mathbf{j}$ is a unit vector. Then the equations

$$x = x_0 + su_1, \qquad y = y_0 + su_2$$

parametrize the line through P_0 parallel to $\mathbf{u}$. The parameter s measures arc length from P_0 in the direction of $\mathbf{u}$. We find the rate of change of f at P_0 in the direction of $\mathbf{u}$ by calculating df/ds at P_0 (Fig. 12.33):

Definition

The **derivative of f at $P_0(x_0, y_0)$ in the direction of the unit vector** $\mathbf{u} = u_1\mathbf{i} + u_2\mathbf{j}$ is the number

$$\left(\frac{df}{ds}\right)_{\mathbf{u}, P_0} = \lim_{s \to 0} \frac{f(x_0 + su_1, y_0 + su_2) - f(x_0, y_0)}{s}, \tag{1}$$

provided the limit exists.

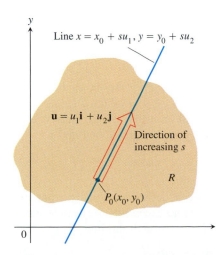

12.33 The rate of change of f in the direction of $\mathbf{u}$ at a point P_0 is the rate at which f changes along this line at P_0.

The directional derivative is also denoted by

$$(D_{\mathbf{u}}f)_{P_0}.$$ "The derivative of f at P_0 in the direction of $\mathbf{u}$"

EXAMPLE 1   Find the derivative of

$$f(x, y) = x^2 + xy$$

at $P_0(1, 2)$ in the direction of the unit vector $\mathbf{u} = \left(1/\sqrt{2}\right)\mathbf{i} + \left(1/\sqrt{2}\right)\mathbf{j}.$

Solution

$$\left(\frac{df}{ds}\right)_{\mathbf{u}, P_0} = \lim_{s \to 0} \frac{f(x_0 + su_1, y_0 + su_2) - f(x_0, y_0)}{s} \qquad \text{Eq. (1)}$$

$$= \lim_{s \to 0} \frac{f\left(1 + s \cdot \dfrac{1}{\sqrt{2}}, \, 2 + s \cdot \dfrac{1}{\sqrt{2}}\right) - f(1, 2)}{s}$$

$$= \lim_{s \to 0} \frac{\left(1 + \dfrac{s}{\sqrt{2}}\right)^2 + \left(1 + \dfrac{s}{\sqrt{2}}\right)\left(2 + \dfrac{s}{\sqrt{2}}\right) - (1^2 + 1 \cdot 2)}{s}$$

$$= \lim_{s \to 0} \frac{\left(1 + \dfrac{2s}{\sqrt{2}} + \dfrac{s^2}{2}\right) + \left(2 + \dfrac{3s}{\sqrt{2}} + \dfrac{s^2}{2}\right) - 3}{s}$$

$$= \lim_{s \to 0} \frac{\dfrac{5s}{\sqrt{2}} + s^2}{s} = \lim_{s \to 0} \left(\frac{5}{\sqrt{2}} + s\right) = \left(\frac{5}{\sqrt{2}} + 0\right) = \frac{5}{\sqrt{2}}.$$

The rate of change of $f(x, y) = x^2 + xy$ at $P_0(1, 2)$ in the direction $\mathbf{u} = \left(1/\sqrt{2}\right)\mathbf{i} + \left(1/\sqrt{2}\right)\mathbf{j}$ is $5/\sqrt{2}$.   ❏

Geometric Interpretation of the Directional Derivative

The equation $z = f(x, y)$ represents a surface S in space. If $z_0 = f(x_0, y_0)$, then the point $P(x_0, y_0, z_0)$ lies on S. The vertical plane that passes through P and $P_0(x_0, y_0)$ parallel to $\mathbf{u}$ intersects S in a curve C (Fig. 12.34). The rate of change of f in the direction of $\mathbf{u}$ is the slope of the tangent to C at P.

Notice that when $\mathbf{u} = \mathbf{i}$ the directional derivative at P_0 is $\partial f/\partial x$ evaluated at (x_0, y_0). When $\mathbf{u} = \mathbf{j}$ the directional derivative at P_0 is $\partial f/\partial y$ evaluated at (x_0, y_0). The directional derivative generalizes the two partial derivatives. We can now ask for the rate of change of f in any direction $\mathbf{u}$, not just the directions $\mathbf{i}$ and $\mathbf{j}$.

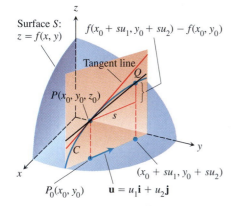

12.34 The slope of curve C at P_0 is

$$\lim_{Q \to P} \text{slope}(PQ)$$

$$= \lim_{s \to 0} \frac{f(x_0 + su_1, y_0 + su_2) - f(x_0, y_0)}{s}$$

$$= \left(\frac{df}{ds}\right)_{\mathbf{u}, P_0}.$$

Calculation

As you know, it is rarely convenient to calculate a derivative directly from its definition as a limit, and the directional derivative is no exception. We can develop a more efficient formula in the following way. We begin with the line

$$x = x_0 + su_1, \qquad y = y_0 + su_2, \qquad (2)$$

through $P_0(x_0, y_0)$, parametrized with the arc length parameter s increasing in the direction of the unit vector $\mathbf{u} = u_1\,\mathbf{i} + u_2\,\mathbf{j}$. Then

$$
\begin{aligned}
\left(\frac{df}{ds}\right)_{\mathbf{u},\,P_0} &= \left(\frac{\partial f}{\partial x}\right)_{P_0}\frac{dx}{ds} + \left(\frac{\partial f}{\partial y}\right)_{P_0}\frac{dy}{ds} && \text{Chain Rule} \\[2mm]
&= \left(\frac{\partial f}{\partial x}\right)_{P_0}\cdot u_1 + \left(\frac{\partial f}{\partial y}\right)_{P_0}\cdot u_2 && \begin{array}{l}\text{From Eqs. (2),}\\ dx/ds = u_1 \text{ and}\\ dy/ds = u_2\end{array} \\[2mm]
&= \underbrace{\left[\left(\frac{\partial f}{\partial x}\right)_{P_0}\mathbf{i} + \left(\frac{\partial f}{\partial y}\right)_{P_0}\mathbf{j}\right]}_{\text{gradient of } f \text{ at } P_0} \cdot \underbrace{\left[u_1\,\mathbf{i} + u_2\,\mathbf{j}\right]}_{\text{direction } \mathbf{u}}. && (3)
\end{aligned}
$$

The notation ∇f is read "grad f" as well as "gradient of f" and "del f." The symbol ∇ by itself is read "del." Another notation for the gradient is grad f, read the way it is written.

Definition

The **gradient vector (gradient)** of $f(x, y)$ at a point $P_0(x_0, y_0)$ is the vector

$$
\nabla f = \frac{\partial f}{\partial x}\mathbf{i} + \frac{\partial f}{\partial y}\mathbf{j}
$$

obtained by evaluating the partial derivatives of f at P_0.

Equation (3) says that the derivative of f in the direction of $\mathbf{u}$ at P_0 is the dot product of $\mathbf{u}$ with the gradient of f at P_0.

Theorem 6

If the partial derivatives of $f(x, y)$ are defined at $P_0(x_0, y_0)$, then

$$
\left(\frac{df}{ds}\right)_{\mathbf{u},\,P_0} = (\nabla f)_{P_0}\cdot\mathbf{u}, \tag{4}
$$

the scalar product of the gradient f at P_0 and $\mathbf{u}$.

EXAMPLE 2 Find the derivative of $f(x, y) = xe^y + \cos(xy)$ at the point $(2, 0)$ in the direction of $\mathbf{A} = 3\,\mathbf{i} - 4\,\mathbf{j}$.

Solution The direction of $\mathbf{A}$ is obtained by dividing $\mathbf{A}$ by its length:

$$
\mathbf{u} = \frac{\mathbf{A}}{|\mathbf{A}|} = \frac{\mathbf{A}}{5} = \frac{3}{5}\mathbf{i} - \frac{4}{5}\mathbf{j}.
$$

The partial derivatives of f at $(2, 0)$ are

$$
f_x(2, 0) = (e^y - y\sin(xy))_{(2,0)} = e^0 - 0 = 1
$$

$$
f_y(2, 0) = (xe^y - x\sin(xy))_{(2,0)} = 2e^0 - 2\cdot 0 = 2.
$$

The gradient of f at $(2, 0)$ is

$$
\nabla f\big|_{(2,0)} = f_x(2, 0)\,\mathbf{i} + f_y(2, 0)\,\mathbf{j} = \mathbf{i} + 2\,\mathbf{j}
$$

(Generated by Mathematica)

12.35 It is customary to picture ∇f as a vector in the domain of f. In the case of $f(x, y) = xe^y + \cos(xy)$, the domain is the entire plane. The rate at which f changes in the direction $\mathbf{u} = (3/5)\mathbf{i} - (4/5)\mathbf{j}$ is $\nabla f \cdot \mathbf{u} = -1$ (Example 2).

(Fig. 12.35). The derivative of f at (2, 0) in the direction of $\mathbf{A}$ is therefore

$$(D_{\mathbf{u}}f)\big|_{(2,0)} = \nabla f\big|_{(2,0)} \cdot \mathbf{u} \qquad \text{Eq. (4)}$$

$$= (\mathbf{i} + 2\mathbf{j}) \cdot \left(\frac{3}{5}\mathbf{i} - \frac{4}{5}\mathbf{j}\right) = \frac{3}{5} - \frac{8}{5} = -1. \qquad \square$$

Properties of Directional Derivatives

Evaluating the dot product in the formula

$$D_{\mathbf{u}}f = \nabla f \cdot \mathbf{u} = |\nabla f||\mathbf{u}| \cos\theta = |\nabla f| \cos\theta$$

reveals the following properties.

Properties of the Directional Derivative $D_{\mathbf{u}}\, f = \nabla f \cdot \mathbf{u} = |\nabla f| \cos\theta$

1. The function f increases most rapidly when $\cos\theta = 1$, or when $\mathbf{u}$ is the direction of ∇f. That is, at each point P in its domain, f increases most rapidly in the direction of the gradient vector ∇f at P. The derivative in this direction is

   $$D_{\mathbf{u}}f = |\nabla f| \cos(0) = |\nabla f|.$$

2. Similarly, f decreases most rapidly in the direction of $-\nabla f$. The derivative in this direction is $D_{\mathbf{u}}f = |\nabla f| \cos(\pi) = -|\nabla f|$.

3. Any direction $\mathbf{u}$ orthogonal to the gradient is a direction of zero change in f because θ then equals $\pi/2$ and

   $$D_{\mathbf{u}}f = |\nabla f| \cos(\pi/2) = |\nabla f| \cdot 0 = 0.$$

As we will discuss later, these properties hold in three dimensions as well as two.

EXAMPLE 3 Find the directions in which $f(x, y) = (x^2/2) + (y^2/2)$ (a) increases most rapidly and (b) decreases most rapidly at the point (1, 1). (c) What are the directions of zero change in f at (1, 1)?

Solution

a) The function increases most rapidly in the direction of ∇f at (1, 1). The gradient is

$$(\nabla f)_{(1,1)} = (x\mathbf{i} + y\mathbf{j})_{(1,1)} = \mathbf{i} + \mathbf{j}.$$

Its direction is

$$\mathbf{u} = \frac{\mathbf{i} + \mathbf{j}}{|\mathbf{i} + \mathbf{j}|} = \frac{\mathbf{i} + \mathbf{j}}{\sqrt{(1)^2 + (1)^2}} = \frac{1}{\sqrt{2}}\mathbf{i} + \frac{1}{\sqrt{2}}\mathbf{j}.$$

b) The function decreases most rapidly in the direction of $-\nabla f$ at (1, 1), which is

$$-\mathbf{u} = -\frac{1}{\sqrt{2}}\mathbf{i} - \frac{1}{\sqrt{2}}\mathbf{j}.$$

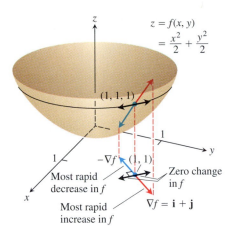

12.36 The direction in which $f(x, y) = (x^2/2) + (y^2/2)$ increases most rapidly at (1, 1) is the direction of $\nabla f|_{(1,1)} = \mathbf{i} + \mathbf{j}$. It corresponds to the direction of steepest ascent on the surface at (1, 1, 1).

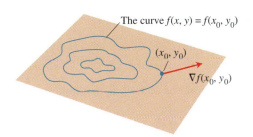

12.37 The gradient of a differentiable function of two variables at a point is always normal to the function's level curve through that point.

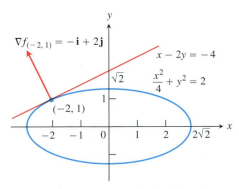

12.38 We can find the tangent to the ellipse $(x^2/4) + y^2 = 2$ by treating the ellipse as a level curve of the function $f(x, y) = (x^2/4) + y^2$ (Example 4).

c) The directions of zero change at (1, 1) are the directions orthogonal to ∇f:

$$\mathbf{n} = -\frac{1}{\sqrt{2}}\mathbf{i} + \frac{1}{\sqrt{2}}\mathbf{j} \quad \text{and} \quad -\mathbf{n} = \frac{1}{\sqrt{2}}\mathbf{i} - \frac{1}{\sqrt{2}}\mathbf{j}.$$

See Fig. 12.36.

Gradients and Tangents to Level Curves

If a differentiable function $f(x, y)$ has a constant value c along a smooth curve $\mathbf{r} = g(t)\mathbf{i} + h(t)\mathbf{j}$ (making the curve a level curve of f), then $f(g(t), h(t)) = c$. Differentiating both sides of this equation with respect to t leads to the equations

$$\frac{d}{dt}f(g(t), h(t)) = \frac{d}{dt}(c)$$

$$\frac{\partial f}{\partial x}\frac{dg}{dt} + \frac{\partial f}{\partial y}\frac{dh}{dt} = 0 \qquad \text{Chain Rule}$$

$$\underbrace{\left(\frac{\partial f}{\partial x}\mathbf{i} + \frac{\partial f}{\partial y}\mathbf{j}\right)}_{\nabla f} \cdot \underbrace{\left(\frac{dg}{dt}\mathbf{i} + \frac{dh}{dt}\mathbf{j}\right)}_{\frac{d\mathbf{r}}{dt}} = 0. \qquad (5)$$

Equation (5) says that ∇f is normal to the tangent vector $d\mathbf{r}/dt$, so it is normal to the curve.

> At every point (x_0, y_0) in the domain of $f(x, y)$, the gradient of f is normal to the level curve through (x_0, y_0) (Fig. 12.37).

This observation enables us to find equations for tangent lines to level curves. They are the lines normal to the gradients. The line through a point $P_0(x_0, y_0)$ normal to a vector $\mathbf{N} = A\mathbf{i} + B\mathbf{j}$ has the equation

$$A(x - x_0) + B(y - y_0) = 0$$

(Exercise 59). If $\mathbf{N}$ is the gradient $(\nabla f)_{(x_0, y_0)} = f_x(x_0, y_0)\mathbf{i} + f_y(x_0, y_0)\mathbf{j}$, the equation becomes

$$f_x(x_0, y_0)(x - x_0) + f_y(x_0, y_0)(y - y_0) = 0. \qquad (6)$$

EXAMPLE 4   Find an equation for the tangent to the ellipse

$$\frac{x^2}{4} + y^2 = 2$$

(Fig. 12.38) at the point $(-2, 1)$.

Solution   The ellipse is a level curve of the function

$$f(x, y) = \frac{x^2}{4} + y^2.$$

The gradient of f at $(-2, 1)$ is

$$\nabla f\big|_{(-2,1)} = \left(\frac{x}{2}\mathbf{i} + 2y\mathbf{j}\right)_{(-2,1)} = -\mathbf{i} + 2\mathbf{j}.$$

The tangent is the line

$$(-1)(x+2) + (2)(y-1) = 0 \qquad \text{Eq. (6)}$$

$$x - 2y = -4.$$

Functions of Three Variables

We obtain three-variable formulas by adding the z-terms to the two-variable formulas. For a differentiable function $f(x, y, z)$ and a unit vector $\mathbf{u} = u_1\mathbf{i} + u_2\mathbf{j} + u_3\mathbf{k}$ in space, we have

$$\nabla f = \frac{\partial f}{\partial x}\mathbf{i} + \frac{\partial f}{\partial y}\mathbf{j} + \frac{\partial f}{\partial z}\mathbf{k}$$

and

$$D_{\mathbf{u}}f = \nabla f \cdot \mathbf{u} = \frac{\partial f}{\partial x}u_1 + \frac{\partial f}{\partial y}u_2 + \frac{\partial f}{\partial z}u_3.$$

The directional derivative can once again be written in the form

$$D_{\mathbf{u}}f = \nabla f \cdot \mathbf{u} = |\nabla f||u|\cos\theta = |\nabla f|\cos\theta,$$

so the properties listed earlier for functions of two variables continue to hold. At any given point, f increases most rapidly in the direction of ∇f and decreases most rapidly in the direction of $-\nabla f$. In any direction orthogonal to ∇f, the derivative is zero.

EXAMPLE 5

a) Find the derivative of $f(x, y, z) = x^3 - xy^2 - z$ at $P_0(1, 1, 0)$ in the direction of $\mathbf{A} = 2\mathbf{i} - 3\mathbf{j} + 6\mathbf{k}.$

b) In what directions does f change most rapidly at P_0, and what are the rates of change in these directions?

Solution

a) The direction of $\mathbf{A}$ is obtained by dividing $\mathbf{A}$ by its length:

$$|\mathbf{A}| = \sqrt{(2)^2 + (-3)^2 + (6)^2} = \sqrt{49} = 7$$

$$\mathbf{u} = \frac{\mathbf{A}}{|\mathbf{A}|} = \frac{2}{7}\mathbf{i} - \frac{3}{7}\mathbf{j} + \frac{6}{7}\mathbf{k}.$$

The partial derivatives of f at P_0 are

$$f_x = 3x^2 - y^2\big|_{(1,1,0)} = 2,$$

$$f_y = -2xy\big|_{(1,1,0)} = -2, \qquad f_z = -1\big|_{(1,1,0)} = -1.$$

The gradient of f at P_0 is

$$\nabla f\big|_{(1,1,0)} = 2\mathbf{i} - 2\mathbf{j} - \mathbf{k}.$$

The derivative of f at P_0 in the direction of $\mathbf{A}$ is therefore

$$(D_{\mathbf{u}}f)\big|_{(1,1,0)} = \nabla f\big|_{(1,1,0)} \cdot \mathbf{u} = (2\mathbf{i} - 2\mathbf{j} - \mathbf{k}) \cdot \left(\frac{2}{7}\mathbf{i} - \frac{3}{7}\mathbf{j} + \frac{6}{7}\mathbf{k}\right)$$

$$= \frac{4}{7} + \frac{6}{7} - \frac{6}{7} = \frac{4}{7}.$$

b) The function increases most rapidly in the direction of $\nabla f = 2\mathbf{i} - 2\mathbf{j} - \mathbf{k}$, and decreases most rapidly in the direction of $-\nabla f$. The rates of change in the directions are, respectively,

$$|\nabla f| = \sqrt{(2)^2 + (-2)^2 + (-1)^2} = \sqrt{9} = 3 \qquad \text{and} \qquad -|\nabla f| = -3. \quad \square$$

Equations for Tangent Planes and Normal Lines

If $\mathbf{r} = g(t)\mathbf{i} + h(t)\mathbf{j} + k(t)\mathbf{k}$ is a smooth curve on the level surface $f(x, y, z) = c$ of a differentiable function f, then $f(g(t), h(t), k(t)) = c$. Differentiating both sides of this equation with respect to t leads to

$$\frac{d}{dt} f(g(t), h(t), k(t)) = \frac{d}{dt}(c)$$

$$\frac{\partial f}{\partial x}\frac{dg}{dt} + \frac{\partial f}{\partial y}\frac{dh}{dt} + \frac{\partial f}{\partial z}\frac{dk}{dt} = 0 \qquad \text{Chain Rule}$$

$$\underbrace{\left(\frac{\partial f}{\partial x}\mathbf{i} + \frac{\partial f}{\partial y}\mathbf{j} + \frac{\partial f}{\partial z}\mathbf{k}\right)}_{\nabla f} \cdot \underbrace{\left(\frac{dg}{dt}\mathbf{i} + \frac{dh}{dt}\mathbf{j} + \frac{dk}{dt}\mathbf{k}\right)}_{d\mathbf{r}/dt} = 0. \tag{7}$$

At every point along the curve, ∇f is orthogonal to the curve's velocity vector.

Now let us restrict our attention to the curves that pass through P_0 (Fig. 12.39). All the velocity vectors at P_0 are orthogonal to ∇f at P_0, so the curves' tangent lines all lie in the plane through P_0 normal to ∇f. We call this plane the tangent plane of the surface at P_0. The line through P_0 perpendicular to the plane is the surface's normal line at P_0.

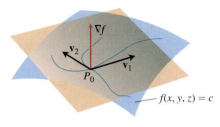

12.39 ∇f is orthogonal to the velocity vector of every smooth curve in the surface through P_0. The velocity vectors at P_0 therefore lie in a common plane, which we call the tangent plane at P_0.

Definitions

The **tangent plane** at the point $P_0(x_0, y_0, z_0)$ on the level surface $f(x, y, z) = c$ is the plane through P_0 normal to $\nabla f\big|_{P_0}$.

The **normal line** of the surface at P_0 is the line through P_0 parallel to $\nabla f\big|_{P_0}$.

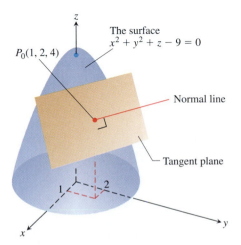

12.40 The tangent plane and normal line to the surface $x^2 + y^2 + z - 9 = 0$ at $P_0(1, 2, 4)$ (Example 6).

Thus, from Section 10.5, the tangent plane and normal line, respectively, have the following equations:

$$f_x(P_0)(x - x_0) + f_y(P_0)(y - y_0) + f_z(P_0)(z - z_0) = 0 \tag{8}$$

$$x = x_0 + f_x(P_0)t, \qquad y = y_0 + f_y(P_0)t, \qquad z = z_0 + f_z(P_0)t. \tag{9}$$

EXAMPLE 6 Find the tangent plane and normal line of the surface

$$f(x, y, z) = x^2 + y^2 + z - 9 = 0 \qquad \text{A circular paraboloid}$$

at the point $P_0(1, 2, 4)$.

Solution The surface is shown in Fig. 12.40.

The tangent plane is the plane through P_0 perpendicular to the gradient of f at P_0. The gradient is

$$\nabla f\big|_{P_0} = (2x\mathbf{i} + 2y\mathbf{j} + \mathbf{k})_{(1,2,4)} = 2\mathbf{i} + 4\mathbf{j} + \mathbf{k}.$$

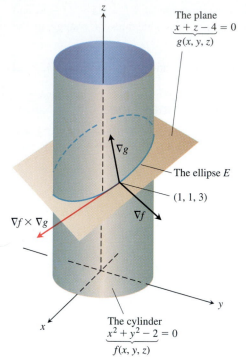

12.41 The cylinder $f(x, y, z) = x^2 + y^2 - 2 = 0$ and the plane $g(x, y, z) = x + z - 4 = 0$ intersect in an ellipse E (Example 7).

The plane is therefore the plane

$$2(x - 1) + 4(y - 2) + (z - 4) = 0, \qquad \text{or} \qquad 2x + 4y + z = 14.$$

The line normal to the surface at P_0 is

$$x = 1 + 2t, \qquad y = 2 + 4t, \qquad z = 4 + t.$$

EXAMPLE 7   The surfaces

$$f(x, y, z) = x^2 + y^2 - 2 = 0 \qquad \text{A cylinder}$$

and

$$g(x, y, z) = x + z - 4 = 0 \qquad \text{A plane}$$

meet in an ellipse E (Fig. 12.41). Find parametric equations for the line tangent to E at the point $P_0(1, 1, 3)$.

Solution   The tangent line is orthogonal to both ∇f and ∇g at P_0, and therefore parallel to $\mathbf{v} = \nabla f \times \nabla g$. The components of $\mathbf{v}$ and the coordinates of P_0 give us equations for the line. We have

$$\nabla f_{(1,1,3)} = (2x\,\mathbf{i} + 2y\,\mathbf{j})_{(1,1,3)} = 2\mathbf{i} + 2\mathbf{j}$$

$$\nabla g_{(1,1,3)} = (\mathbf{i} + \mathbf{k})_{(1,1,3)} = \mathbf{i} + \mathbf{k}$$

$$\mathbf{v} = (2\mathbf{i} + 2\mathbf{j}) \times (\mathbf{i} + \mathbf{k}) = \begin{vmatrix} \mathbf{i} & \mathbf{j} & \mathbf{k} \\ 2 & 2 & 0 \\ 1 & 0 & 1 \end{vmatrix} = 2\mathbf{i} - 2\mathbf{j} - 2\mathbf{k}.$$

The line is

$$x = 1 + 2t, \qquad y = 1 - 2t, \qquad z = 3 - 2t.$$

Planes Tangent to a Surface $z = f(x, y)$

To find an equation for the plane tangent to a surface $z = f(x, y)$ at a point $P_0(x_0, y_0, z_0)$ where $z_0 = f(x_0, y_0)$, we first observe that the equation $z = f(x, y)$ is equivalent to $f(x, y) - z = 0$. The surface $z = f(x, y)$ is therefore the zero level surface of the function $F(x, y, z) = f(x, y) - z$. The partial derivatives of F are

$$F_x = \frac{\partial}{\partial x}(f(x, y) - z) = f_x - 0 = f_x$$

$$F_y = \frac{\partial}{\partial y}(f(x, y) - z) = f_y - 0 = f_y$$

$$F_z = \frac{\partial}{\partial z}(f(x, y) - z) = 0 - 1 = -1.$$

The formula

$$F_x(P_0)(x - x_0) + F_y(P_0)(y - y_0) + F_z(P_0)(z - z_0) = 0 \qquad \begin{array}{l}\text{Eq. (8) restated} \\ \text{for } F(x, y, z)\end{array}$$

for the plane tangent to the level surface at P_0 therefore reduces to

$$f_x(x_0, y_0)(x - x_0) + f_y(x_0, y_0)(y - y_0) - (z - z_0) = 0.$$

The plane tangent to the surface $z = f(x, y)$ at the point $P_0(x_0, y_0, z_0) = (x_0, y_0, f(x_0, y_0))$ is

$$f_x(x_0, y_0)(x - x_0) + f_y(x_0, y_0)(y - y_0) - (z - z_0) = 0. \qquad (10)$$

EXAMPLE 8 Find the plane tangent to the surface $z = x \cos y - ye^x$ at $(0, 0, 0)$.

Solution We calculate the partial derivatives of $f(x, y) = x \cos y - ye^x$ and use Eq. (10):

$$f_x(0, 0) = (\cos y - ye^x)_{(0,0)} = 1 - 0 \cdot 1 = 1$$

$$f_y(0, 0) = (-x \sin y - e^x)_{(0,0)} = 0 - 1 = -1.$$

The tangent plane is therefore

$$1 \cdot (x - 0) - 1 \cdot (y - 0) - (z - 0) = 0, \qquad \text{Eq. (10)}$$

or

$$x - y - z = 0. \qquad \square$$

Increments and Distance

The directional derivative plays the role of an ordinary derivative when we want to estimate how much a function f changes if we move a small distance ds from a point P_0 to another point nearby. If f were a function of a single variable, we would have

$$df = f'(P_0) \, ds. \qquad \text{Ordinary derivative} \times \text{increment}$$

For a function of two or more variables, we use the formula

$$df = (\nabla f|_{P_0} \cdot \mathbf{u}) \, ds, \qquad \text{Directional derivative} \times \text{increment}$$

where $\mathbf{u}$ is the direction of the motion away from P_0.

Estimating the Change in f in a Direction u

To estimate the change in the value of a function f when we move a small distance ds from a point P_0 in a particular direction $\mathbf{u}$, use the formula

$$df = \underbrace{(\nabla f|_{P_0} \cdot \mathbf{u})}_{\substack{\text{directional} \\ \text{derivative}}} \cdot \underbrace{ds}_{\substack{\text{distance} \\ \text{increment}}}$$

EXAMPLE 9 Estimate how much the value of

$$f(x, y, z) = xe^y + yz$$

will change if the point $P(x, y, z)$ moves 0.1 unit from $P_0(2, 0, 0)$ straight toward $P_1(4, 1, -2)$.

Solution We first find the derivative of f at P_0 in the direction of the vector

$$\overrightarrow{P_0 P_1} = 2\mathbf{i} + \mathbf{j} - 2\mathbf{k}.$$

The direction of this vector is

$$\mathbf{u} = \frac{\overrightarrow{P_0 P_1}}{|\overrightarrow{P_0 P_1}|} = \frac{\overrightarrow{P_0 P_1}}{3} = \frac{2}{3}\mathbf{i} + \frac{1}{3}\mathbf{j} - \frac{2}{3}\mathbf{k}.$$

The gradient of f at P_0 is

$$\nabla f\big|_{(2,0,0)} = (e^y \mathbf{i} + (xe^y + z)\mathbf{j} + y\mathbf{k})\big|_{(2,0,0)} = \mathbf{i} + 2\mathbf{j}.$$

Therefore,

$$\nabla f\big|_{P_0} \cdot \mathbf{u} = (\mathbf{i} + 2\mathbf{j}) \cdot \left(\frac{2}{3}\mathbf{i} + \frac{1}{3}\mathbf{j} - \frac{2}{3}\mathbf{k}\right) = \frac{2}{3} + \frac{2}{3} = \frac{4}{3}.$$

The change df in f that results from moving $ds = 0.1$ unit away from P_0 in the direction of $\mathbf{u}$ is approximately

$$df = (\nabla f\big|_{P_0} \cdot \mathbf{u})(ds) = \left(\frac{4}{3}\right)(0.1) \approx 0.13.$$

Algebra Rules for Gradients

If we know the gradients of two functions f and g, we automatically know the gradients of their constant multiples, sum, difference, product, and quotient.

These rules have the same form as the corresponding rules for derivatives, as they should (Exercise 65).

Algebra Rules for Gradients

1. *Constant Multiple Rule:* $\nabla(kf) = k\nabla f$ (any number k)

2. *Sum Rule:* $\nabla(f + g) = \nabla f + \nabla g$

3. *Difference Rule:* $\nabla(f - g) = \nabla f - \nabla g$

4. *Product Rule:* $\nabla(fg) = f\nabla g + g\nabla f$

5. *Quotient Rule:* $\nabla\left(\dfrac{f}{g}\right) = \dfrac{g\nabla f - f\nabla g}{g^2}$

EXAMPLE 10   We illustrate the rules with

$$f(x, y, z) = x - y \qquad g(x, y, z) = z$$
$$\nabla f = \mathbf{i} - \mathbf{j} \qquad\qquad \nabla g = \mathbf{k}.$$

We have:

1. $\nabla(2f) = \nabla(2x - 2y) = 2\mathbf{i} - 2\mathbf{j} = 2\nabla f$

2. $\nabla(f + g) = \nabla(x - y + z) = \mathbf{i} - \mathbf{j} + \mathbf{k} = \nabla f + \nabla g$

3. $\nabla(f - g) = \nabla(x - y - z) = \mathbf{i} - \mathbf{j} - \mathbf{k} = \nabla f - \nabla g$

4. $\nabla(fg) = \nabla(xz - yz) = z\mathbf{i} - z\mathbf{j} + (x - y)\mathbf{k} = g\nabla f + f\nabla g$

5. $\displaystyle \nabla\left(\frac{f}{g}\right) = \nabla\left(\frac{x-y}{z}\right) = \frac{\partial}{\partial x}\left(\frac{x-y}{z}\right)\mathbf{i} + \frac{\partial}{\partial y}\left(\frac{x-y}{z}\right)\mathbf{j} + \frac{\partial}{\partial z}\left(\frac{x-y}{z}\right)\mathbf{k}$

$\displaystyle = \frac{1}{z}\mathbf{i} - \frac{1}{z}\mathbf{j} + \frac{z\cdot 0 - (x-y)\cdot 1}{z^2}\mathbf{k}$

$\displaystyle = \frac{z\mathbf{i} - z\mathbf{j} - (x-y)\mathbf{k}}{z^2} = \frac{g\nabla f - f\nabla g}{g^2}$

Exercises 12.7

Calculating Gradients at Points

In Exercises 1–4, find the gradient of the function at the given point. Then sketch the gradient together with the level curve that passes through the point.

1. $f(x, y) = y - x$, (2, 1)

2. $f(x, y) = \ln(x^2 + y^2)$, (1, 1)

3. $g(x, y) = y - x^2$, (−1, 0)

4. $g(x, y) = \dfrac{x^2}{2} - \dfrac{y^2}{2}$, $(\sqrt{2}, 1)$

In Exercises 5–8, find ∇f at the given point.

5. $f(x, y, z) = x^2 + y^2 - 2z^2 + z\ln x$, (1, 1, 1)

6. $f(x, y, z) = 2z^3 - 3(x^2 + y^2)z + \tan^{-1} xz$, (1, 1, 1)

7. $f(x, y, z) = (x^2 + y^2 + z^2)^{-1/2} + \ln(xyz)$, (−1, 2, −2)

8. $f(x, y, z) = e^{x+y}\cos z + (y+1)\sin^{-1} x$, $(0, 0, \pi/6)$

Finding Directional Derivatives in the *xy*-Plane

In Exercises 9–16, find the derivative of the function at P_0 in the direction of **A**.

9. $f(x, y) = 2xy - 3y^2$, $P_0(5, 5)$, $\mathbf{A} = 4\mathbf{i} + 3\mathbf{j}$

10. $f(x, y) = 2x^2 + y^2$, $P_0(-1, 1)$, $\mathbf{A} = 3\mathbf{i} - 4\mathbf{j}$

11. $g(x, y) = x - (y^2/x) + \sqrt{3}\sec^{-1}(2xy)$, $P_0(1, 1)$, $\mathbf{A} = 12\mathbf{i} + 5\mathbf{j}$

12. $h(x, y) = \tan^{-1}(y/x) + \sqrt{3}\sin^{-1}(xy/2)$, $P_0(1, 1)$, $\mathbf{A} = 3\mathbf{i} - 2\mathbf{j}$

13. $f(x, y, z) = xy + yz + zx$, $P_0(1, -1, 2)$, $\mathbf{A} = 3\mathbf{i} + 6\mathbf{j} - 2\mathbf{k}$

14. $f(x, y, z) = x^2 + 2y^2 - 3z^2$, $P_0(1, 1, 1)$, $\mathbf{A} = \mathbf{i} + \mathbf{j} + \mathbf{k}$

15. $g(x, y, z) = 3e^x\cos yz$, $P_0(0, 0, 0)$, $\mathbf{A} = 2\mathbf{i} + \mathbf{j} - 2\mathbf{k}$

16. $h(x, y, z) = \cos xy + e^{yz} + \ln zx$, $P_0(1, 0, 1/2)$, $\mathbf{A} = \mathbf{i} + 2\mathbf{j} + 2\mathbf{k}$

Directions of Most Rapid Increase and Decrease

In Exercises 17–22, find the directions in which the functions increase and decrease most rapidly at P_0. Then find the derivatives of the functions in these directions.

17. $f(x, y) = x^2 + xy + y^2$, $P_0(-1, 1)$

18. $f(x, y) = x^2 y + e^{xy}\sin y$, $P_0(1, 0)$

19. $f(x, y, z) = (x/y) - yz$, $P_0(4, 1, 1)$

20. $g(x, y, z) = xe^y + z^2$, $P_0(1, \ln 2, 1/2)$

21. $f(x, y, z) = \ln xy + \ln yz + \ln xz$, $P_0(1, 1, 1)$

22. $h(x, y, z) = \ln(x^2 + y^2 - 1) + y + 6z$, $P_0(1, 1, 0)$

Estimating Change

23. By about how much will
$$f(x, y, z) = \ln\sqrt{x^2 + y^2 + z^2}$$
change if the point $P(x, y, z)$ moves from $P_0(3, 4, 12)$ a distance of $ds = 0.1$ units in the direction of $3\mathbf{i} + 6\mathbf{j} - 2\mathbf{k}$?

24. By about how much will
$$f(x, y, z) = e^x\cos yz$$
change as the point $P(x, y, z)$ moves from the origin a distance of $ds = 0.1$ units in the direction of $2\mathbf{i} + 2\mathbf{j} - 2\mathbf{k}$?

25. By about how much will
$$g(x, y, z) = x + x\cos z - y\sin z + y$$
change if the point $P(x, y, z)$ moves from $P_0(2, -1, 0)$ a distance of $ds = 0.2$ units toward the point $P_1(0, 1, 2)$?

26. By about how much will
$$h(x, y, z) = \cos(\pi xy) + xz^2$$
change if the point $P(x, y, z)$ moves from $P_0(-1, -1, -1)$ a distance of $ds = 0.1$ units toward the origin?

Tangent Planes and Normal Lines to Surfaces

In Exercises 27–34, find equations for the (a) tangent plane and (b) normal line at the point P_0 on the given surface.

27. $x^2 + y^2 + z^2 = 3$, $P_0(1, 1, 1)$

28. $x^2 + y^2 - z^2 = 18$, $P_0(3, 5, -4)$

29. $2z - x^2 = 0$, $P_0(2, 0, 2)$

30. $x^2 + 2xy - y^2 + z^2 = 7$, $P_0(1, -1, 3)$

31. $\cos \pi x - x^2 y + e^{xz} + yz = 4$, $P_0(0, 1, 2)$

32. $x^2 - xy - y^2 - z = 0$, $P_0(1, 1, -1)$

33. $x + y + z = 1$, $P_0(0, 1, 0)$

34. $x^2 + y^2 - 2xy - x + 3y - z = -4$, $P_0(2, -3, 18)$

In Exercises 35–38, find an equation for the plane that is tangent to the given surface at the given point.

35. $z = \ln(x^2 + y^2)$, $(1, 0, 0)$ **36.** $z = e^{-(x^2 + y^2)}$, $(0, 0, 1)$

37. $z = \sqrt{y - x}$, $(1, 2, 1)$ **38.** $z = 4x^2 + y^2$, $(1, 1, 5)$

Tangent Lines to Curves

In Exercises 39–42, sketch the curve $f(x, y) = c$ together with ∇f and the tangent line at the given point. Then write an equation for the tangent line.

39. $x^2 + y^2 = 4$, $(\sqrt{2}, \sqrt{2})$

40. $x^2 - y = 1$, $(\sqrt{2}, 1)$

41. $xy = -4$, $(2, -2)$

42. $x^2 - xy + y^2 = 7$, $(-1, 2)$ (This is the curve in Section 2.6, Example 4.)

In Exercises 43–48, find parametric equations for the line tangent to the curve of intersection of the surfaces at the given point.

43. Surfaces: $x + y^2 + 2z = 4$, $x = 1$
Point: $(1, 1, 1)$

44. Surfaces: $xyz = 1$, $x^2 + 2y^2 + 3z^2 = 6$
Point: $(1, 1, 1)$

45. Surfaces: $x^2 + 2y + 2z = 4$, $y = 1$
Point: $(1, 1, 1/2)$

46. Surfaces: $x + y^2 + z = 2$, $y = 1$
Point: $(1/2, 1, 1/2)$

47. Surfaces: $x^3 + 3x^2 y^2 + y^3 + 4xy - z^2 = 0$,
$x^2 + y^2 + z^2 = 11$
Point: $(1, 1, 3)$

48. Surfaces: $x^2 + y^2 = 4$, $x^2 + y^2 - z = 0$
Point: $\left(\sqrt{2}, \sqrt{2}, 4\right)$

Theory and Examples

49. In what directions is the derivative of $f(x, y) = xy + y^2$ at $P(3, 2)$ equal to zero?

50. In what two directions is the derivative of $f(x, y) = (x^2 - y^2)/(x^2 + y^2)$ at $P(1, 1)$ equal to zero?

51. Is there a direction $\mathbf{A}$ in which the rate of change of $f(x, y) = x^2 - 3xy + 4y^2$ at $P(1, 2)$ equals 14? Give reasons for your answer.

52. Is there a direction $\mathbf{A}$ in which the rate of change of the temperature function $T(x, y, z) = 2xy - yz$ (temperature in degrees Celsius, distance in feet) at $P(1, -1, 1)$ is $-3°$C/ft? Give reasons for your answer.

53. The derivative of $f(x, y)$ at $P_0(1, 2)$ in the direction of $\mathbf{i} + \mathbf{j}$ is $2\sqrt{2}$ and in the direction of $-2\mathbf{j}$ is -3. What is the derivative of f in the direction of $-\mathbf{i} - 2\mathbf{j}$? Give reasons for your answer.

54. The derivative of $f(x, y, z)$ at a point P is greatest in the direction of $\mathbf{A} = \mathbf{i} + \mathbf{j} - \mathbf{k}$. In this direction the value of the derivative is $2\sqrt{3}$.

   k) What is ∇f at P? Give reasons for your answer.
   l) What is the derivative of f at P in the direction of $\mathbf{i} + \mathbf{j}$?

55. *Temperature change along a circle.* Suppose that the Celsius temperature at the point (x, y) in the xy-plane is $T(x, y) = x \sin 2y$ and that distance in the xy-plane is measured in meters. A particle is moving *clockwise* around the circle of radius 1 m centered at the origin at the constant rate of 2 m/sec.

   a) How fast is the temperature experienced by the particle changing in °C/m at the point $P(1/2, \sqrt{3}/2)$?
   b) How fast is the temperature experienced by the particle changing in °C/sec at P?

56. *Change along the involute of a circle.* Find the derivative of $f(x, y) = x^2 + y^2$ in the direction of the unit tangent vector of the curve

$$\mathbf{r}(t) = (\cos t + t \sin t)\mathbf{i} + (\sin t - t \cos t)\mathbf{j}, \quad t > 0$$

(Fig. 12.42).

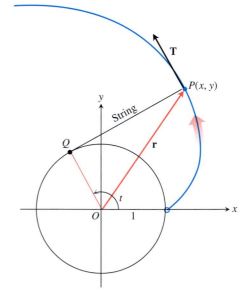

12.42 The involute of the unit circle from Section 11.3, Example 5. If you move out along the involute, covering distance along the curve at a constant rate, your distance from the origin will increase at a constant rate as well. (This is how to interpret the result of your calculation in Exercise 56.)

57. *Change along a helix.* Find the derivative of $f(x, y, z) = x^2 + y^2 + z^2$ in the direction of the unit tangent vector of the helix

$$\mathbf{r}(t) = (\cos t)\,\mathbf{i} + (\sin t)\,\mathbf{j} + t\,\mathbf{k}$$

at the points where $t = -\pi/4, 0$, and $\pi/4$. The function f gives the square of the distance from a point $P(x, y, z)$ on the helix to the origin. The derivatives calculated here give the rates at which the square of the distance is changing with respect to t as P moves through the points where $t = -\pi/4, 0$, and $\pi/4$.

58. The Celsius temperature in a region in space is given by $T(x, y, z) = 2x^2 - xyz$. A particle is moving in this region and its position at time t is given by $x = 2t^2, y = 3t, z = -t^2$, where time is measured in seconds and distance in meters.

a) How fast is the temperature experienced by the particle changing in °C/m when the particle is at the point $P(8, 6, -4)$?

b) How fast is the temperature experienced by the particle changing in °C/sec at P?

59. Show that $A(x - x_0) + B(y - y_0) = 0$ is an equation for the line in the xy-plane through the point (x_0, y_0) normal to the vector $\mathbf{N} = A\,\mathbf{i} + B\,\mathbf{j}$.

60. *Normal curves and tangent curves.* A curve is *normal* to a surface $f(x, y, z) = c$ at a point of intersection if the curve's velocity vector is a scalar multiple of ∇f at the point. The curve is *tangent* to the surface at a point of intersection if its velocity vector is orthogonal to ∇f there.

a) Show that the curve

$$\mathbf{r}(t) = \sqrt{t}\,\mathbf{i} + \sqrt{t}\,\mathbf{j} - \frac{1}{4}(t + 3)\,\mathbf{k}$$

is normal to the surface $x^2 + y^2 - z = 3$ when $t = 1$.

b) Show that the curve

$$\mathbf{r}(t) = \sqrt{t}\,\mathbf{i} + \sqrt{t}\,\mathbf{j} + (2t - 1)\,\mathbf{k}$$

is tangent to the surface $x^2 + y^2 - z = 1$ when $t = 1$.

61. *Another way to see why gradients are normal to level curves.* Suppose that a differentiable function $f(x, y)$ has a constant value c along the differentiable curve $x = g(t), y = h(t)$ for all values of t. Differentiate both sides of the equation $f(g(t), h(t)) = c$ with respect to t to show that ∇f is normal to the curve's tangent vector at every point.

62. *The linearization of f(x, y) is a tangent-plane approxima-tion.* Show that the tangent plane at the point $P_0(x_0, y_0, f(x_0, y_0))$ on the surface $z = f(x, y)$ defined by a differentiable function f is the plane

$$f_x(x_0, y_0)(x - x_0) + f_y(x_0, y_0)(y - y_0) - (z - f(x_0, y_0)) = 0$$

or

$$z = f(x_0, y_0) + f_x(x_0, y_0)(x - x_0) + f_y(x_0, y_0)(y - y_0).$$

Thus the tangent plane at P_0 is the graph of the linearization of f at P_0 (Fig. 12.43).

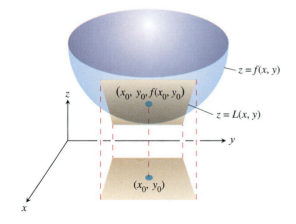

12.43 The graph of a function $z = f(x, y)$ and its linearization at a point (x_0, y_0). The plane defined by L is tangent to the surface at the point above the point (x_0, y_0). This furnishes a geometric explanation of why the values of L lie close to those of f in the immediate neighborhood of (x_0, y_0) (Exercise 62).

63. *Directional derivatives and scalar components.* How is the derivative of a differentiable function $f(x, y, z)$ at a point P_0 in the direction of a unit vector $\mathbf{u}$ related to the scalar component of $(\nabla f)_{P_0}$ in the direction of $\mathbf{u}$? Give reasons for your answer.

64. *Directional derivatives and partial derivatives.* Assuming that the necessary derivatives of $f(x, y, z)$ are defined, how are $D_{\mathbf{i}}f$, $D_{\mathbf{j}}f$, and $D_{\mathbf{k}}f$ related to f_x, f_y, and f_z? Give reasons for your answer.

65. *The algebra rules for gradients.* Given a constant k and the gradients

$$\nabla f = \frac{\partial f}{\partial x}\,\mathbf{i} + \frac{\partial f}{\partial y}\,\mathbf{j} + \frac{\partial f}{\partial z}\,\mathbf{k}$$

and

$$\nabla g = \frac{\partial g}{\partial x}\,\mathbf{i} + \frac{\partial g}{\partial y}\,\mathbf{j} + \frac{\partial g}{\partial z}\,\mathbf{k},$$

use the scalar equations

$$\frac{\partial}{\partial x}(kf) = k\frac{\partial f}{\partial x}, \quad \frac{\partial}{\partial x}(f \pm g) = \frac{\partial f}{\partial x} \pm \frac{\partial g}{\partial x},$$

$$\frac{\partial}{\partial x}(fg) = f\frac{\partial g}{\partial x} + g\frac{\partial f}{\partial x}, \quad \frac{\partial}{\partial x}\left(\frac{f}{g}\right) = \frac{g\dfrac{\partial f}{\partial x} - f\dfrac{\partial g}{\partial x}}{g^2},$$

and so on, to establish the following rules:

a) $\nabla(kf) = k\nabla f$

b) $\nabla(f + g) = \nabla f + \nabla g$

c) $\nabla(f - g) = \nabla f - \nabla g$

d) $\nabla(fg) = f\nabla g + g\nabla f$

e) $\nabla\left(\dfrac{f}{g}\right) = \dfrac{g\nabla f - f\nabla g}{g^2}$

12.44 The function

$$z = (\cos x)(\cos y)e^{-\sqrt{x^2+y^2}}$$

has a maximum value of 1 and a minimum value of about −0.067 on the square region $|x| \le 3\pi/2$, $|y| \le 3\pi/2$.

(Generated by Mathematica)

12.45 The "roof surface"

$$z = \frac{1}{2}(|\,|x| - |y|\,| - |x| - |y|)$$

viewed from the point (10, 15, 20). The defining function has a maximum value of 0 and a minimum value of −a on the square region $|x| \le a$, $|y| \le a$.

12.46 A local maximum is a mountain peak and a local minimum is a valley low.

<table>
<tr><td>12.8</td></tr>
</table>

Extreme Values and Saddle Points

Continuous functions defined on closed bounded regions in the xy-plane take on absolute maximum and minimum values on these domains (Figs. 12.44 and 12.45). It is important to be able to find these values and to know where they occur. We can often accomplish this by examining partial derivatives.

The Derivative Tests

To find the local extreme values of a function of a single variable, we look for points where the graph has a horizontal tangent line. At such points we then look for local maxima, local minima, and points of inflection. For a function $f(x, y)$ of two variables, we look for points where the surface $z = f(x, y)$ has a horizontal tangent *plane*. At such points we then look for local maxima, local minima, and saddle points (more about saddle points in a moment).

> **Definitions**
>
> Let $f(x, y)$ be defined on a region R containing the point (a, b). Then
>
> 1. $f(a, b)$ is a **local maximum** value of f if $f(a, b) \ge f(x, y)$ for all domain points (x, y) in an open disk centered at (a, b).
> 2. $f(a, b)$ is a **local minimum** value of f if $f(a, b) \le f(x, y)$ for all domain points (x, y) in an open disk centered at (a, b).

Local maxima correspond to mountain peaks on the surface $z = f(x, y)$ and local minima correspond to valley bottoms (Fig. 12.46). At such points the tangent planes, when they exist, are horizontal. Local extrema are also called **relative extrema.**

As with functions of a single variable, the key to identifying the local extrema is a first derivative test.

> **Theorem 7**
>
> **First Derivative Test for Local Extreme Values**
>
> If $f(x, y)$ has a local maximum or minimum value at an interior point (a, b) of its domain, and if the first partial derivatives exist there, then $f_x(a, b) = 0$ and $f_y(a, b) = 0$.

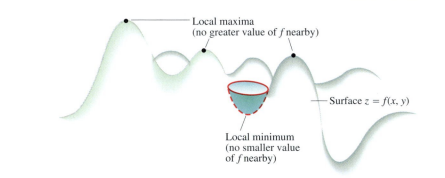

Local maxima
(no greater value of f nearby)

Surface $z = f(x, y)$

Local minimum
(no smaller value
of f nearby)

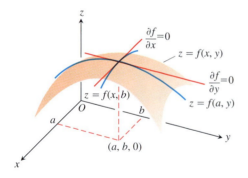

12.47 The maximum of f occurs at $x = a, y = b$.

Proof   Suppose that f has a local maximum value at an interior point (a, b) of its domain. Then

1. $x = a$ is an interior point of the domain of the curve $z = f(x, b)$ in which the plane $y = b$ cuts the surface $z = f(x, y)$ (Fig. 12.47).
2. The function $z = f(x, b)$ is a differentiable function of x at $x = a$ (the derivative is $f_x(a, b)$).
3. The function $z = f(x, b)$ has a local maximum value at $x = a$.
4. The value of the derivative of $z = f(x, b)$ at $x = a$ is therefore zero (Theorem 2, Section 3.1). Since this derivative is $f_x(a, b)$, we conclude that $f_x(a, b) = 0$.

A similar argument with the function $z = f(a, y)$ shows that $f_y(a, b) = 0$.

This proves the theorem for local maximum values. The proof for local minimum values is left as Exercise 48.   ◻

If we substitute the values $f_x(a, b) = 0$ and $f_y(a, b) = 0$ into the equation

$$f_x(a, b)(x - a) + f_y(a, b)(y - b) - (z - f(a, b)) = 0$$

for the tangent plane to the surface $z = f(x, y)$ at (a, b), the equation reduces to

$$0 \cdot (x - a) + 0 \cdot (y - b) - z + f(a, b) = 0$$

or $$z = f(a, b).$$

Thus, Theorem 7 says that the surface does indeed have a horizontal tangent plane at a local extremum, provided there is a tangent plane there.

As in the single-variable case, Theorem 7 says that the only places a function $f(x, y)$ can ever have an extreme value are

1. Interior points where $f_x = f_y = 0$,
2. Interior points where one or both of f_x and f_y do not exist,
3. Boundary points of the function's domain.

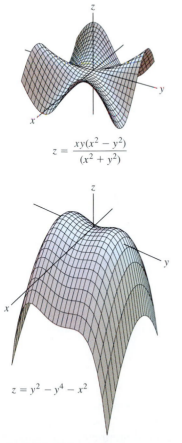

$$z = \frac{xy(x^2 - y^2)}{(x^2 + y^2)}$$

$$z = y^2 - y^4 - x^2$$

(Generated by Mathematica)

12.48 Saddle points at the origin.

Definition

An interior point of the domain of a function $f(x, y)$ where both f_x and f_y are zero or where one or both of f_x and f_y do not exist is a **critical point** of f.

Thus, the only points where a function $f(x, y)$ can assume extreme values are critical points and boundary points. As with differentiable functions of a single variable, not every critical point gives rise to a local extremum. A differentiable function of a single variable might have a point of inflection. A differentiable function of two variables might have a saddle point.

Definition

A differentiable function $f(x, y)$ has a **saddle point** at a critical point (a, b) if in every open disk centered at (a, b) there are domain points (x, y) where $f(x, y) > f(a, b)$ and domain points (x, y) where $f(x, y) < f(a, b)$. The corresponding point $(a, b, f(a, b))$ on the surface $z = f(x, y)$ is called a saddle point of the surface (Fig. 12.48).

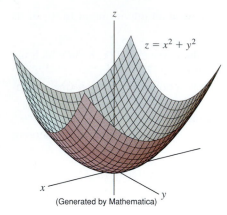

12.49 The graph of the function $f(x, y) = x^2 + y^2$ is the paraboloid $z = x^2 + y^2$. The function has only one critical point, the origin, which gives rise to a local minimum value of 0 (Example 1).

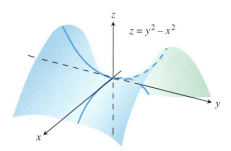

12.50 The origin is a saddle point of the function $f(x, y) = y^2 - x^2$. There are no local extreme values (Example 2).

EXAMPLE 1   Find the local extreme values of $f(x, y) = x^2 + y^2$.

Solution   The domain of f is the entire plane (so there are no boundary points) and the partial derivatives $f_x = 2x$ and $f_y = 2y$ exist everywhere. Therefore, local extreme values can occur only where

$$f_x = 2x = 0 \qquad \text{and} \qquad f_y = 2y = 0.$$

The only possibility is the origin, where the value of f is zero. Since f is never negative, we see that the origin gives a local minimum (Fig. 12.49). ❏

EXAMPLE 2   Find the local extreme values (if any) of $f(x, y) = y^2 - x^2$.

Solution   The domain of f is the entire plane (so there are no boundary points) and the partial derivatives $f_x = -2x$ and $f_y = 2y$ exist everywhere. Therefore, local extrema can occur only at the origin $(0, 0)$. However, along the positive x-axis f has the value $f(x, 0) = -x^2 < 0$; along the positive y-axis f has the value $f(0, y) = y^2 > 0$. Therefore every open disk in the xy-plane centered at $(0, 0)$ contains points where the function is positive and points where it is negative. The function has a saddle point at the origin (Fig. 12.50) instead of a local extreme value. We conclude that the function has no local extreme values. ❏

The fact that $f_x = f_y = 0$ at an interior point (a, b) of R does not tell us enough to be sure f has a local extreme value there. However, if f and its first and second partial derivatives are continuous on R, we may be able to learn the rest from the following theorem, proved in Section 12.10.

Theorem 8

Second Derivative Test for Local Extreme Values

Suppose $f(x, y)$ and its first and second partial derivatives are continuous throughout a disk centered at (a, b) and that $f_x(a, b) = f_y(a, b) = 0$. Then

i) f has a **local maximum** at (a, b) if $f_{xx} < 0$ and $f_{xx} f_{yy} - f_{xy}^2 > 0$ at (a, b);

ii) f has a **local minimum** at (a, b) if $f_{xx} > 0$ and $f_{xx} f_{yy} - f_{xy}^2 > 0$ at (a, b);

iii) f has a **saddle point** at (a, b) if $f_{xx} f_{yy} - f_{xy}^2 < 0$ at (a, b).

iv) **The test is inconclusive** at (a, b) if $f_{xx} f_{yy} - f_{xy}^2 = 0$ at (a, b). In this case, we must find some other way to determine the behavior of f at (a, b).

The expression $f_{xx} f_{yy} - f_{xy}^2$ is called the **discriminant** of f. It is sometimes easier to remember the determinant form,

$$f_{xx} f_{yy} - f_{xy}^2 = \begin{vmatrix} f_{xx} & f_{xy} \\ f_{xy} & f_{yy} \end{vmatrix}.$$

Theorem 8 says that if the discriminant is positive at the point (a, b), then the surface curves the same way in all directions: downwards if $f_{xx} < 0$, giving rise to a local maximum, and upwards if $f_{xx} > 0$, giving a local minimum. On the other hand, if the discriminant is negative at (a, b), then the surface curves up in some directions and down in others, so we have a saddle point.

EXAMPLE 3   Find the local extreme values of the function

$$f(x, y) = xy - x^2 - y^2 - 2x - 2y + 4.$$

Solution   The function is defined and differentiable for all x and y and its domain has no boundary points. The function therefore has extreme values only at the points where f_x and f_y are simultaneously zero. This leads to

$$f_x = y - 2x - 2 = 0, \qquad f_y = x - 2y - 2 = 0,$$

or

$$x = y = -2.$$

Therefore, the point $(-2, -2)$ is the only point where f may take on an extreme value. To see if it does so, we calculate

$$f_{xx} = -2, \qquad f_{yy} = -2, \qquad f_{xy} = 1.$$

The discriminant of f at $(a, b) = (-2, -2)$ is

$$f_{xx} f_{yy} - f_{xy}{}^2 = (-2)(-2) - (1)^2 = 4 - 1 = 3.$$

The combination

$$f_{xx} < 0 \qquad \text{and} \qquad f_{xx} f_{yy} - f_{xy}{}^2 > 0$$

tells us that f has a local maximum at $(-2, -2)$. The value of f at this point is $f(-2, -2) = 8$.   ❏

EXAMPLE 4   Find the local extreme values of $f(x, y) = xy$.

Solution   Since f is differentiable everywhere (Fig. 12.51), it can assume extreme values only where

$$f_x = y = 0 \qquad \text{and} \qquad f_y = x = 0.$$

Thus, the origin is the only point where f might have an extreme value. To see what happens there, we calculate

$$f_{xx} = 0, \qquad f_{yy} = 0, \qquad f_{xy} = 1.$$

The discriminant,

$$f_{xx} f_{yy} - f_{xy}{}^2 = -1,$$

is negative. Therefore the function has a saddle point at $(0, 0)$. We conclude that $f(x, y) = xy$ has no local extreme values.   ❏

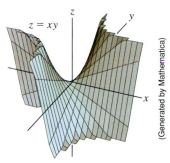

(Generated by Mathematica)

12.51 The surface $z = xy$ has a saddle point at the origin (Example 4).

Absolute Maxima and Minima on Closed Bounded Regions

We organize the search for the absolute extrema of a continuous function $f(x, y)$ on a closed and bounded region R into three steps.

Step 1: *List the interior points* of R where f may have local maxima and minima and evaluate f at these points. These are the points where $f_x = f_y = 0$ or where one or both of f_x and f_y fail to exist (the critical points of f).

Step 2: *List the boundary points* of R where f has local maxima and minima and evaluate f at these points. We will show how to do this shortly.

Step 3: Look through the lists for the maximum and minimum values of f. These will be the absolute maximum and minimum values of f on R. Since absolute maxima and minima are also local maxima and minima, the absolute maximum and minimum values of f already appear somewhere in the lists made in steps 1 and 2. We have only to glance at the lists to see what they are.

EXAMPLE 5 Find the absolute maximum and minimum values of

$$f(x, y) = 2 + 2x + 2y - x^2 - y^2$$

on the triangular plate in the first quadrant bounded by the lines $x = 0$, $y = 0$, $y = 9 - x$.

Solution Since f is differentiable, the only places where f can assume these values are points inside the triangle (Fig. 12.52) where $f_x = f_y = 0$ and points on the boundary.

Interior points. For these we have

$$f_x = 2 - 2x = 0, \qquad f_y = 2 - 2y = 0,$$

yielding the single point $(x, y) = (1, 1)$. The value of f there is

$$f(1, 1) = 4.$$

Boundary points. We take the triangle one side at a time:

1. On the segment OA, $y = 0$. The function

$$f(x, y) = f(x, 0) = 2 + 2x - x^2$$

may now be regarded as a function of x defined on the closed interval $0 \le x \le 9$. Its extreme values (we know from Chapter 3) may occur at the endpoints

$$x = 0 \quad \text{where} \quad f(0, 0) = 2$$
$$x = 9 \quad \text{where} \quad f(9, 0) = 2 + 18 - 81 = -61$$

and at the interior points where $f'(x, 0) = 2 - 2x = 0$. The only interior point where $f'(x, 0) = 0$ is $x = 1$, where

$$f(x, 0) = f(1, 0) = 3.$$

2. On the segment OB, $x = 0$ and

$$f(x, y) = f(0, y) = 2 + 2y - y^2.$$

We know from the symmetry of f in x and y and from the analysis we just carried out that the candidates on this segment are

$$f(0, 0) = 2, \qquad f(0, 9) = -61, \qquad f(0, 1) = 3.$$

3. We have already accounted for the values of f at the endpoints of AB, so we need only look at the interior points of AB. With $y = 9 - x$, we have

$$f(x, y) = 2 + 2x + 2(9 - x) - x^2 - (9 - x)^2 = -61 + 18x - 2x^2.$$

Setting $f'(x, 9 - x) = 18 - 4x = 0$ gives

$$x = \frac{18}{4} = \frac{9}{2}.$$

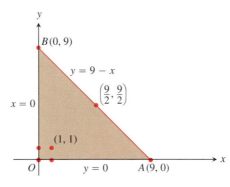

12.52 This triangular plate is the domain of the function in Example 5.

At this value of x,

$$y = 9 - \frac{9}{2} = \frac{9}{2} \quad \text{and} \quad f(x, y) = f\left(\frac{9}{2}, \frac{9}{2}\right) = -\frac{41}{2}.$$

Summary. We list all the candidates: 4, 2, −61, 3, −(41/2). The maximum is 4, which f assumes at (1, 1). The minimum is −61, which f assumes at (0, 9) and (9, 0).

Conclusion

Despite the power of Theorem 7, we urge you to remember its limitations. It does not apply to boundary points of a function's domain, where it is possible for a function to have extreme values along with nonzero derivatives. And it does not apply to points where either f_x or f_y fails to exist.

Summary of Max-Min Tests

The extreme values of $f(x, y)$ can occur only at

(i) **boundary points** of the domain of f,

(ii) **critical points** (interior points where $f_x = f_y = 0$ or points where f_x or f_y fail to exist).

If the first and second order partial derivatives of f are continuous throughout a disk centered at a point (a, b), and $f_x(a, b) = f_y(a, b) = 0$, you may be able to classify $f(a, b)$ with the **second derivative test:**

(i) $f_{xx} < 0$ and $f_{xx} f_{yy} - f_{xy}{}^2 > 0$ at (a, b) $\Rightarrow$ **local maximum,**

(ii) $f_{xx} > 0$ and $f_{xx} f_{yy} - f_{xy}{}^2 > 0$ at (a, b) $\Rightarrow$ **local minimum,**

(iii) $f_{xx} f_{yy} - f_{xy}{}^2 < 0$ at (a, b) $\Rightarrow$ **saddle point,**

(iv) $f_{xx} f_{yy} - f_{xy}{}^2 = 0$ at (a, b) $\Rightarrow$ **test is inconclusive.**

Exercises 12.8

Finding Local Extrema

Find all the local maxima, local minima, and saddle points of the functions in Exercises 1–30.

1. $f(x, y) = x^2 + xy + y^2 + 3x - 3y + 4$

2. $f(x, y) = x^2 + 3xy + 3y^2 - 6x + 3y - 6$

3. $f(x, y) = 2xy - 5x^2 - 2y^2 + 4x + 4y - 4$

4. $f(x, y) = 2xy - 5x^2 - 2y^2 + 4x - 4$

5. $f(x, y) = x^2 + xy + 3x + 2y + 5$

6. $f(x, y) = y^2 + xy - 2x - 2y + 2$

7. $f(x, y) = 5xy - 7x^2 + 3x - 6y + 2$

8. $f(x, y) = 2xy - x^2 - 2y^2 + 3x + 4$

9. $f(x, y) = x^2 - 4xy + y^2 + 6y + 2$

10. $f(x, y) = 3x^2 + 6xy + 7y^2 - 2x + 4y$

11. $f(x, y) = 2x^2 + 3xy + 4y^2 - 5x + 2y$

12. $f(x, y) = 4x^2 - 6xy + 5y^2 - 20x + 26y$

13. $f(x, y) = x^2 - y^2 - 2x + 4y + 6$

14. $f(x, y) = x^2 - 2xy + 2y^2 - 2x + 2y + 1$

15. $f(x, y) = x^2 + 2xy$

16. $f(x, y) = 3 + 2x + 2y - 2x^2 - 2xy - y^2$

17. $f(x, y) = x^3 - y^3 - 2xy + 6$

18. $f(x, y) = x^3 + 3xy + y^3$

19. $f(x, y) = 6x^2 - 2x^3 + 3y^2 + 6xy$

20. $f(x, y) = 3y^2 - 2y^3 - 3x^2 + 6xy$

21. $f(x, y) = 9x^3 + y^3/3 - 4xy$

22. $f(x, y) = 8x^3 + y^3 + 6xy$

23. $f(x, y) = x^3 + y^3 + 3x^2 - 3y^2 - 8$

24. $f(x, y) = 2x^3 + 2y^3 - 9x^2 + 3y^2 - 12y$

25. $f(x, y) = 4xy - x^4 - y^4$

26. $f(x, y) = x^4 + y^4 + 4xy$

27. $f(x, y) = \dfrac{1}{x^2 + y^2 - 1}$

28. $f(x, y) = \dfrac{1}{x} + xy + \dfrac{1}{y}$

29. $f(x, y) = y \sin x$

30. $f(x, y) = e^{2x} \cos y$

Finding Absolute Extrema

In Exercises 31–38, find the absolute maxima and minima of the functions on the given domains.

31. $f(x, y) = 2x^2 - 4x + y^2 - 4y + 1$ on the closed triangular plate bounded by the lines $x = 0$, $y = 2$, $y = 2x$ in the first quadrant

32. $D(x, y) = x^2 - xy + y^2 + 1$ on the closed triangular plate in the first quadrant bounded by the lines $x = 0$, $y = 4$, $y = x$

33. $f(x, y) = x^2 + y^2$ on the closed triangular plate bounded by the lines $x = 0$, $y = 0$, $y + 2x = 2$ in the first quadrant

34. $T(x, y) = x^2 + xy + y^2 - 6x$ on the rectangular plate $0 \le x \le 5$, $-3 \le y \le 3$

35. $T(x, y) = x^2 + xy + y^2 - 6x + 2$ on the rectangular plate $0 \le x \le 5$, $-3 \le y \le 0$

36. $f(x, y) = 48xy - 32x^3 - 24y^2$ on the rectangular plate $0 \le x \le 1$, $0 \le y \le 1$

37. $f(x, y) = (4x - x^2) \cos y$ on the rectangular plate $1 \le x \le 3$, $-\pi/4 \le y \le \pi/4$ (Fig. 12.53)

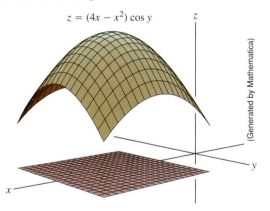

$z = (4x - x^2) \cos y$

12.53 The function and domain in Exercise 37.

38. $f(x, y) = 4x - 8xy + 2y + 1$ on the triangular plate bounded by the lines $x = 0$, $y = 0$, $x + y = 1$ in the first quadrant

39. Find two numbers a and b with $a \le b$ such that

$$\int_a^b (6 - x - x^2) \, dx$$

has its largest value.

40. Find two numbers a and b with $a \le b$ such that

$$\int_a^b (24 - 2x - x^2)^{1/3} dx$$

has its largest value.

41. *Temperatures.* The flat circular plate in Fig. 12.54 has the shape of the region $x^2 + y^2 \le 1$. The plate, including the boundary where $x^2 + y^2 = 1$, is heated so that the temperature at the point (x, y) is

$$T(x, y) = x^2 + 2y^2 - x.$$

Find the temperatures at the hottest and coldest points on the plate.

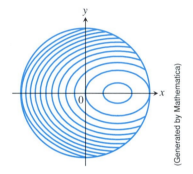

12.54 Curves of constant temperature are called isotherms. The figure shows isotherms of the temperature function $T(x, y) = x^2 + 2y^2 - x$ on the disk $x^2 + y^2 \le 1$ in the xy-plane. Exercise 41 asks you to locate the extreme temperatures.

42. Find the critical point of

$$f(x, y) = xy + 2x - \ln x^2 y$$

in the open first quadrant ($x > 0$, $y > 0$) and show that f takes on a minimum there (Fig. 12.55).

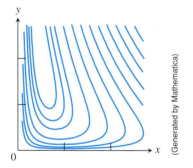

12.55 The function $f(x, y) = xy + 2x - \ln x^2 y$ (selected level curves shown here) takes on a minimum value somewhere in the open first quadrant $x > 0$, $y > 0$ (Exercise 42).

Theory and Examples

43. Find the maxima, minima, and saddle points of $f(x, y)$, if any, given that

a) $f_x = 2x - 4y$ and $f_y = 2y - 4x$
b) $f_x = 2x - 2$ and $f_y = 2y - 4$
c) $f_x = 9x^2 - 9$ and $f_y = 2y + 4$

Describe your reasoning in each case.

44. The discriminant $f_{xx}f_{yy} - f_{xy}^2$ is zero at the origin for each of the following functions, so the second derivative test fails there. Determine whether the function has a maximum, a minimum, or neither at the origin by imagining what the surface $z = f(x, y)$ looks like. Describe your reasoning in each case.

a) $f(x, y) = x^2 y^2$ b) $f(x, y) = 1 - x^2 y^2$
c) $f(x, y) = xy^2$ d) $f(x, y) = x^3 y^2$
e) $f(x, y) = x^3 y^3$ f) $f(x, y) = x^4 y^4$

45. Show that $(0, 0)$ is a critical point of $f(x, y) = x^2 + kxy + y^2$ no matter what value the constant k has. (*Hint:* Consider two cases: $k = 0$ and $k \neq 0$.)

46. For what values of the constant k does the second derivative test guarantee that $f(x, y) = x^2 + kxy + y^2$ will have a saddle point at $(0, 0)$? a local minimum at $(0, 0)$? For what values of k is the second derivative test inconclusive? Give reasons for your answers.

47. a) If $f_x(a, b) = f_y(a, b) = 0$, must f have a local maximum or minimum value at (a, b)? Give reasons for your answer.
b) Can you conclude anything about $f(a, b)$ if f and its first and second partial derivatives are continuous throughout a disk centered at (a, b) and $f_{xx}(a, b)$ and $f_{yy}(a, b)$ differ in sign? Give reasons for your answer.

48. Using the proof of Theorem 7 given in the text for the case in which f has a local maximum at (a, b), prove the theorem for the case in which f has a local minimum at (a, b).

49. Among all the points on the graph of $z = 10 - x^2 - y^2$ that lie above the plane $x + 2y + 3z = 0$, find the point farthest from the plane.

50. Find the point on the graph of $z = x^2 + y^2 + 10$ nearest the plane $x + 2y - z = 0$.

51. The function $f(x, y) = x + y$ fails to have an absolute maximum value in the closed first quadrant $x \geq 0$ and $y \geq 0$. Does this contradict the discussion on finding absolute extrema given in the text? Give reasons for your answer.

52. Consider the function $f(x, y) = x^2 + y^2 + 2xy - x - y + 1$ over the square $0 \leq x \leq 1$ and $0 \leq y \leq 1$.

a) Show that f has an absolute minimum along the line segment $2x + 2y = 1$ in this square. What *is* the absolute minimum value?
b) Find the absolute maximum value of f over the square.

Extreme Values on Parametrized Curves

To find the extreme values of a function $f(x, y)$ on a curve $x =$

$x(t), y = y(t)$, we treat f as a function of the single variable t and use the Chain Rule to find where df/dt is zero. As in any other single-variable case, the extreme values of f are then found among the values at the

a) critical points (points where df/dt is zero or fails to exist), and
b) endpoints of the parameter domain.

Find the absolute maximum and minimum values of the following functions on the given curves.

53. Functions:

a) $f(x, y) = x + y$ b) $g(x, y) = xy$
c) $h(x, y) = 2x^2 + y^2$

Curves:

i)   The semicircle $x^2 + y^2 = 4$, $y \geq 0$
ii) The quarter circle $x^2 + y^2 = 4$, $x \geq 0$, $y \geq 0$

Use the parametric equations $x = 2 \cos t, y = 2 \sin t$.

54. Functions:

a) $f(x, y) = 2x + 3y$ b) $g(x, y) = xy$
c) $h(x, y) = x^2 + 3y^2$

Curves:

i)   The semi-ellipse $(x^2/9) + (y^2/4) = 1$, $y \geq 0$
ii) The quarter ellipse $(x^2/9) + (y^2/4) = 1$, $x \geq 0$, $y \geq 0$

Use the parametric equations $x = 3 \cos t, y = 2 \sin t$.

55. Function: $f(x, y) = xy$
Curves:

i)   The line $x = 2t$, $y = t + 1$
ii)   The line segment $x = 2t$, $y = t + 1$, $-1 \leq t \leq 0$
iii) The line segment $x = 2t$, $y = t + 1$, $0 \leq t \leq 1$

56. Functions:

a) $f(x, y) = x^2 + y^2$ b) $g(x, y) = 1/(x^2 + y^2)$

Curves:

i)   The line $x = t$, $y = 2 - 2t$
ii) The line segment $x = t$, $y = 2 - 2t$, $0 \leq t \leq 1$

Least Squares and Regression Lines

When we try to fit a line $y = mx + b$ to a set of numerical data points $(x_1, y_1), (x_2, y_2), \ldots, (x_n, y_n)$ (Fig. 12.56, on the following page), we usually choose the line that minimizes the sum of the squares of the vertical distances from the points to the line. In theory, this means finding the values of m and b that minimize the value of the function

$$w = (mx_1 + b - y_1)^2 + \cdots + (mx_n + b - y_n)^2. \qquad (1)$$

The values of m and b that do this are found with the first and second derivative tests to be

$$m = \frac{\left(\sum x_k\right)\left(\sum y_k\right) - n \sum x_k y_k}{\left(\sum x_k\right)^2 - n \sum x_k^2}, \qquad (2)$$

$$b = \frac{1}{n}\left(\sum y_k - m \sum x_k\right), \qquad (3)$$

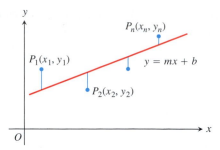

12.56 To fit a line to noncollinear points, we choose the line that minimizes the sum of the squares of the deviations.

with all sums running from $k = 1$ to $k = n$. Many scientific calculators have these formulas built in, enabling you to find m and b with only a few key presses after you have entered the data.

The line $y = mx + b$ determined by these values of m and b is called the **least squares line, regression line,** or **trend line** for the data under study. Finding a least squares line lets you

1. summarize data with a simple expression,
2. predict values of y for other, experimentally untried values of x,
3. handle data analytically.

EXAMPLE   Find the least squares line for the points $(0, 1)$, $(1, 3)$, $(2, 2)$, $(3, 4)$, $(4, 5)$.

Solution   We organize the calculations in a table:

k	x_k	y_k	x_k^2	$x_k y_k$
1	0	1	0	0
2	1	3	1	3
3	2	2	4	4
4	3	4	9	12
5	4	5	16	20
Σ	10	15	30	39

Then we find

$$m = \frac{(10)(15) - 5(39)}{(10)^2 - 5(30)} = 0.9 \qquad \text{Eq. (2) with } n = 5 \text{ and data from the table}$$

and use the value of m to find

$$b = \frac{1}{5}(15 - (0.9)(10)) = 1.2. \qquad \text{Eq. (3) with } n = 5, m = 0.9$$

The least squares line is $y = 0.9x + 1.2$ (Fig. 12.57). ☐

In Exercises 57–60, use Eqs. (2) and (3) to find the least squares line for each set of data points. Then use the linear equation you obtain to predict the value of y that would correspond to $x = 4$.

57. $(-1, 2)$, $(0, 1)$, $(3, -4)$ **58.** $(-2, 0)$, $(0, 2)$, $(2, 3)$

59. $(0, 0)$, $(1, 2)$, $(2, 3)$ **60.** $(0, 1)$, $(2, 2)$, $(3, 2)$

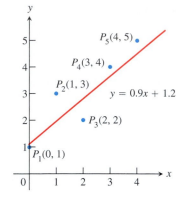

12.57 The least squares line for the data in the example.

61. Write a linear equation for the effect of irrigation on the yield of alfalfa by fitting a least squares line to the data in Table 12.1 (from the University of California Experimental Station, *Bulletin* No. 450, p. 8). Plot the data and draw the line.

Table 12.1 Growth of alfalfa

x (total seasonal depth of water applied, in.)	y (average alfalfa yield, tons/acre)
12	5.27
18	5.68
24	6.25
30	7.21
36	8.20
42	8.71

62. *Craters of Mars.* One theory of crater formation suggests that the frequency of large craters should fall off as the square of the diameter (Marcus, *Science*, June 21, 1968, p. 1334). Pictures from *Mariner IV* show the frequencies listed in Table 12.2. Fit a line of the form $F = m(1/D^2) + b$ to the data. Plot the data and draw the line.

Table 12.2 Crater sizes on Mars

Diameter in km, D	$1/D^2$ (for left value of class interval)	Frequency, F
32–45	0.001	51
45–64	0.0005	22
64–90	0.00024	14
90–128	0.000123	4

Table 12.3 Compositions by Mozart

Köchel number, K	Year composed, y
1	1761
75	1771
155	1772
219	1775
271	1777
351	1780
425	1783
503	1786
575	1789
626	1791

Table 12.4 Sinkings of German submarines by U.S. during 16 consecutive months of WWII

Month	Guesses by U.S. (reported sinkings) x	Actual number y
1	3	3
2	2	2
3	4	6
4	2	3
5	5	4
6	5	3
7	9	11
8	12	9
9	8	10
10	13	16
11	14	13
12	3	5
13	4	6
14	13	19
15	10	15
16	16	15
	123	140

63. *Köchel numbers.* In 1862, the German musicologist Ludwig von Köchel made a chronological list of the musical works of Wolf-gang Amadeus Mozart. This list is the source of the Köchel numbers, or "K numbers," that now accompany the titles of Mozart's pieces (Sinfonia Concertante in E-flat major, K.364, for example). Table 12.3 gives the Köchel numbers and composition dates (y) of ten of Mozart's works.

a) Plot y vs. K to show that y is close to being a linear function of K.

b) Find a least squares line $y = mK + b$ for the data and add the line to your plot in (a).

c) K.364 was composed in 1779. What date is predicted by the least squares line?

64. *Submarine sinkings.* The data in Table 12.4 show the results of a historical study of German submarines sunk by the U.S. Navy during 16 consecutive months of World War II. The data given for each month are the number of reported sinkings and the number of actual sinkings. The number of submarines sunk was slightly greater than the Navy's reports implied. Find a least squares line for estimating the number of actual sinkings from the number of reported sinkings.

CAS Explorations and Projects

In Exercises 65–70, you will explore functions to identify their local extrema. Use a CAS to perform the following steps:

a) Plot the function over the given rectangle.

b) Plot some level curves in the rectangle.

c) Calculate the function's first partial derivatives and use the CAS equation solver to find the critical points. How do the critical points relate to the level curves plotted in (b)? Which critical points, if any, appear to give a saddle point? Give reasons for your answer.

d) Calculate the function's second partial derivatives and find the discriminant $f_{xx}f_{yy} - f_{xy}{}^2$.

e) Using the max-min tests, classify the critical points found in (c). Are your findings consistent with your discussion in (c)?

65. $f(x, y) = x^2 + y^3 - 3xy, \quad -5 \le x \le 5, \quad -5 \le y \le 5$

66. $f(x, y) = x^3 - 3xy^2 + y^2, \quad -2 \le x \le 2, \quad -2 \le y \le 2$

67. $f(x, y) = x^4 + y^2 - 8x^2 - 6y + 16, \quad -3 \le x \le 3, \\ -6 \le y \le 6$

68. $f(x, y) = 2x^4 + y^4 - 2x^2 - 2y^2 + 3, \quad -3/2 \le x \le 3/2, \\ -3/2 \le y \le 3/2$

69. $f(x, y) = 5x^6 + 18x^5 - 30x^4 + 30xy^2 - 120x^3, \quad -4 \le x \le 3, \\ -2 \le y \le 2$

70. $f(x, y) = \begin{cases} x^5 \ln(x^2 + y^2), & (x, y) \neq (0, 0) \\ 0, & (x, y) = (0, 0) \end{cases},$

$-2 \le x \le 2, \quad -2 \le y \le 2$

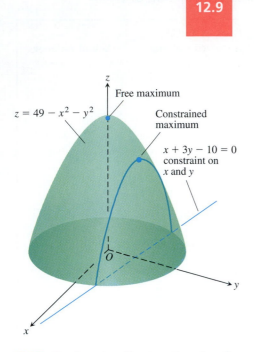

12.58 The function $f(x, y) = 49 - x^2 - y^2$, subject to the constraint $g(x, y) = x + 3y - 10 = 0$.

(labels on figure:)
Free maximum

$z = 49 - x^2 - y^2$

Constrained maximum

$x + 3y - 10 = 0$
constraint on
x and y

12.9

Lagrange Multipliers

As we saw in Section 12.8, we sometimes need to find the extreme values of a function whose domain is constrained to lie within some particular subset of the plane—a disk, for example, or a closed triangular region. But, as Fig. 12.58 suggests, a function may be subject to other kinds of constraints as well.

In this section, we explore a powerful method for finding extreme values of constrained functions: the method of *Lagrange multipliers*. Lagrange developed the method in 1755 to solve max-min problems in geometry. Today the method is important in economics, in engineering (where it is used in designing multistage rockets, for example), and in mathematics.

Constrained Maxima and Minima

EXAMPLE 1   Find the point $P(x, y, z)$ closest to the origin on the plane $2x + y - z - 5 = 0$.

Solution   The problem asks us to find the minimum value of the function

$$|\overrightarrow{OP}| = \sqrt{(x - 0)^2 + (y - 0)^2 + (z - 0)^2}$$
$$= \sqrt{x^2 + y^2 + z^2}$$

subject to the constraint that

$$2x + y - z - 5 = 0.$$

Since $|\overrightarrow{OP}|$ has a minimum value wherever the function

$$f(x, y, z) = x^2 + y^2 + z^2$$

has a minimum value, we may solve the problem by finding the minimum value of $f(x, y, z)$ subject to the constraint $2x + y - z - 5 = 0$. If we regard x and y as the independent variables in this equation and write z as

$$z = 2x + y - 5,$$

our problem reduces to one of finding the points (x, y) at which the function

$$h(x, y) = f(x, y, 2x + y - 5) = x^2 + y^2 + (2x + y - 5)^2$$

has its minimum value or values. Since the domain of h is the entire xy-plane, the first derivative test of Section 12.8 tells us that any minima that h might have must occur at points where

$$h_x = 2x + 2(2x + y - 5)(2) = 0, \qquad h_y = 2y + 2(2x + y - 5) = 0.$$

This leads to

$$10x + 4y = 20, \qquad 4x + 4y = 10,$$

and the solution

$$x = \frac{5}{3}, \qquad y = \frac{5}{6}.$$

We may apply a geometric argument together with the second derivative test to show that these values minimize h. The z-coordinate of the corresponding point on

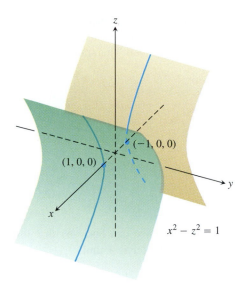

12.59 The hyperbolic cylinder $x^2 - z^2 - 1 = 0$ in Example 2.

The hyperbolic cylinder $x^2 - z^2 = 1$

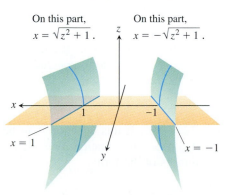

12.60 The region in the xy-plane from which the first two coordinates of the points (x, y, z) on the hyperbolic cylinder $x^2 - z^2 = 1$ are selected excludes the band $-1 < x < 1$ in the xy-plane.

the plane $z = 2x + y - 5$ is

$$z = 2\left(\frac{5}{3}\right) + \frac{5}{6} - 5 = -\frac{5}{6}.$$

Therefore, the point we seek is

$$\text{Closest point:} \qquad P\left(\frac{5}{3}, \frac{5}{6}, -\frac{5}{6}\right).$$

The distance from P to the origin is $5/\sqrt{6} \approx 2.04$.   ❑

Attempts to solve a constrained maximum or minimum problem by substitution, as we might call the method of Example 1, do not always go smoothly. This is one of the reasons for learning the new method of this section.

EXAMPLE 2   Find the points closest to the origin on the hyperbolic cylinder $x^2 - z^2 - 1 = 0$.

Solution 1   The cylinder is shown in Fig. 12.59. We seek the points on the cylinder closest to the origin. These are the points whose coordinates minimize the value of the function

$$f(x, y, z) = x^2 + y^2 + z^2 \qquad \text{Square of the distance}$$

subject to the constraint that $x^2 - z^2 - 1 = 0$. If we regard x and y as independent variables in the constraint equation, then

$$z^2 = x^2 - 1$$

and the values of $f(x, y, z) = x^2 + y^2 + z^2$ on the cylinder are given by the function

$$h(x, y) = x^2 + y^2 + (x^2 - 1) = 2x^2 + y^2 - 1.$$

To find the points on the cylinder whose coordinates minimize f, we look for the points in the xy-plane whose coordinates minimize h. The only extreme value of h occurs where

$$h_x = 4x = 0 \qquad \text{and} \qquad h_y = 2y = 0,$$

that is, at the point $(0, 0)$. But now we're in trouble—there are no points on the cylinder where both x and y are zero. What went wrong?

What happened was that the first derivative test found (as it should have) the point *in the domain of h* where h has a minimum value. We, on the other hand, want the points *on the cylinder* where h has a minimum value. While the domain of h is the entire xy-plane, the domain from which we can select the first two coordinates of the points (x, y, z) on the cylinder is restricted to the "shadow" of the cylinder on the xy-plane; it does not include the band between the lines $x = -1$ and $x = 1$ (Fig. 12.60).

We can avoid this problem if we treat y and z as independent variables (instead of x and y) and express x in terms of y and z as

$$x^2 = z^2 + 1.$$

With this substitution, $f(x, y, z) = x^2 + y^2 + z^2$ becomes

$$k(y, z) = (z^2 + 1) + y^2 + z^2 = 1 + y^2 + 2z^2$$

and we look for the points where k takes on its smallest value. The domain of

k in the yz-plane now matches the domain from which we select the y- and z-coordinates of the points (x, y, z) on the cylinder. Hence, the points that minimize k in the plane will have corresponding points on the cylinder. The smallest values of k occur where

$$k_y = 2y = 0 \qquad \text{and} \qquad k_z = 4z = 0,$$

or where $y = z = 0$. This leads to

$$x^2 = z^2 + 1 = 1, \qquad x = \pm 1.$$

The corresponding points on the cylinder are $(\pm 1, 0, 0)$. We can see from the inequality

$$k(y, z) = 1 + y^2 + 2z^2 \geq 1$$

that the points $(\pm 1, 0, 0)$ give a minimum value for k. We can also see that the minimum distance from the origin to a point on the cylinder is 1 unit.

Solution 2 Another way to find the points on the cylinder closest to the origin is to imagine a small sphere centered at the origin expanding like a soap bubble until it just touches the cylinder (Fig. 12.61). At each point of contact, the cylinder and sphere have the same tangent plane and normal line. Therefore, if the sphere and cylinder are represented as the level surfaces obtained by setting

$$f(x, y, z) = x^2 + y^2 + z^2 - a^2 \qquad \text{and} \qquad g(x, y, z) = x^2 - z^2 - 1$$

equal to 0, then the gradients ∇f and ∇g will be parallel where the surfaces touch. At any point of contact we should therefore be able to find a scalar λ ("lambda") such that

$$\nabla f = \lambda \nabla g,$$

or

$$2x\,\mathbf{i} + 2y\,\mathbf{j} + 2z\,\mathbf{k} = \lambda(2x\,\mathbf{i} - 2z\,\mathbf{k}).$$

Thus, the coordinates x, y, and z of any point of tangency will have to satisfy the three scalar equations

$$2x = 2\lambda x, \qquad 2y = 0, \qquad 2z = -2\lambda z. \tag{1}$$

For what values of λ will a point (x, y, z) whose coordinates satisfy the equations in (1) also lie on the surface $x^2 - z^2 - 1 = 0$? To answer this question, we use the fact that no point on the surface has a zero x-coordinate to conclude that $x \neq 0$ in the first equation in (1). This means that $2x = 2\lambda x$ only if

$$2 = 2\lambda, \qquad \text{or} \qquad \lambda = 1.$$

For $\lambda = 1$, the equation $2z = -2\lambda z$ becomes $2z = -2z$. If this equation is to be satisfied as well, z must be zero. Since $y = 0$ also (from the equation $2y = 0$), we conclude that the points we seek all have coordinates of the form

$$(x, 0, 0).$$

What points on the surface $x^2 - z^2 = 1$ have coordinates of this form? The points $(x, 0, 0)$ for which

$$x^2 - (0)^2 = 1, \qquad x^2 = 1, \qquad \text{or} \qquad x = \pm 1.$$

The points on the cylinder closest to the origin are the points $(\pm 1, 0, 0)$. ❑

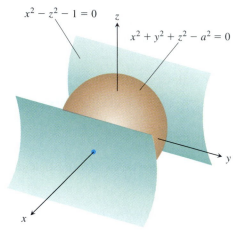

12.61 A sphere expanding like a soap bubble centered at the origin until it just touches the hyperbolic cylinder

$$x^2 - z^2 - 1 = 0.$$

See Solution 2 of Example 2.

The Method of Lagrange Multipliers

In Solution 2 of Example 2, we solved the problem by the **method of Lagrange multipliers.** In general terms, the method says that the extreme values of a function $f(x, y, z)$ whose variables are subject to a constraint $g(x, y, z) = 0$ are to be found on the surface $g = 0$ at the points where

$$\nabla f = \lambda \nabla g$$

for some scalar λ (called a **Lagrange multiplier**).

To explore the method further and see why it works, we first make the following observation, which we state as a theorem.

Theorem 9

The Orthogonal Gradient Theorem

Suppose that $f(x, y, z)$ is differentiable in a region whose interior contains a smooth curve

$$C: \quad \mathbf{r} = g(t)\,\mathbf{i} + h(t)\,\mathbf{j} + k(t)\,\mathbf{k}.$$

If P_0 is a point on C where f has a local maximum or minimum relative to its values on C, then ∇f is orthogonal to C at P_0.

Proof We show that ∇f is orthogonal to the curve's velocity vector at P_0. The values of f on C are given by the composite $f(g(t), h(t), k(t))$, whose derivative with respect to t is

$$\frac{df}{dt} = \frac{\partial f}{\partial x}\frac{dg}{dt} + \frac{\partial f}{\partial y}\frac{dh}{dt} + \frac{\partial f}{\partial z}\frac{dk}{dt} = \nabla f \cdot \mathbf{v}.$$

At any point P_0 where f has a local maximum or minimum relative to its values on the curve, $df/dt = 0$, so

$$\nabla f \cdot \mathbf{v} = 0. \qquad \square$$

By dropping the z-terms in Theorem 9, we obtain a similar result for functions of two variables.

Corollary of Theorem 9

At the points on a smooth curve $\mathbf{r} = g(t)\,\mathbf{i} + h(t)\,\mathbf{j}$ where a differentiable function $f(x, y)$ takes on its local maxima and minima relative to its values on the curve, $\nabla f \cdot \mathbf{v} = 0$.

Theorem 9 is the key to the method of Lagrange multipliers. Suppose that $f(x, y, z)$ and $g(x, y, z)$ are differentiable and that P_0 is a point on the surface $g(x, y, z) = 0$ where f has a local maximum or minimum value relative to its other values on the surface. Then f takes on a local maximum or minimum at P_0 relative to its values on every differentiable curve through P_0 on the surface $g(x, y, z) = 0$. Therefore, ∇f is orthogonal to the velocity vector of every such differentiable curve through P_0. But so is ∇g (because ∇g is orthogonal to the level surface $g = 0$, as we saw in Section 12.7). Therefore, at P_0, ∇f is some scalar multiple λ of ∇g.

The Method of Lagrange Multipliers

Suppose that $f(x, y, z)$ and $g(x, y, z)$ are differentiable. To find the local maximum and minimum values of f subject to the constraint $g(x, y, z) = 0$, find the values of x, y, z, and λ that simultaneously satisfy the equations

$$\nabla f = \lambda \nabla g \quad \text{and} \quad g(x, y, z) = 0.$$

For functions of two independent variables, the appropriate equations are

$$\nabla f = \lambda \nabla g \quad \text{and} \quad g(x, y) = 0.$$

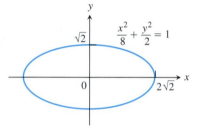

12.62 Example 3 shows how to find the largest and smallest values of the product xy on this ellipse.

EXAMPLE 3   Find the greatest and smallest values that the function

$$f(x, y) = xy$$

takes on the ellipse (Fig. 12.62)

$$\frac{x^2}{8} + \frac{y^2}{2} = 1.$$

Solution   We want the extreme values of $f(x, y) = xy$ subject to the constraint

$$g(x, y) = \frac{x^2}{8} + \frac{y^2}{2} - 1 = 0.$$

To do so, we first find the values of x, y, and λ for which

$$\nabla f = \lambda \nabla g \quad \text{and} \quad g(x, y) = 0.$$

The gradient equation gives

$$y\,\mathbf{i} + x\,\mathbf{j} = \frac{\lambda}{4}x\,\mathbf{i} + \lambda y\,\mathbf{j},$$

from which we find

$$y = \frac{\lambda}{4}x, \qquad x = \lambda y, \qquad \text{and} \qquad y = \frac{\lambda}{4}(\lambda y) = \frac{\lambda^2}{4}y,$$

so that $y = 0$ or $\lambda = \pm 2$. We now consider these two cases.

Case 1: If $y = 0$, then $x = y = 0$. But $(0, 0)$ is not on the ellipse. Hence, $y \neq 0$.

Case 2: If $y \neq 0$, then $\lambda = \pm 2$ and $x = \pm 2y$. Substituting this in the equation $g(x, y) = 0$ gives

$$\frac{(\pm 2y)^2}{8} + \frac{y^2}{2} = 1, \qquad 4y^2 + 4y^2 = 8, \qquad \text{and} \qquad y = \pm 1.$$

The function $f(x, y) = xy$ therefore takes on its extreme values on the ellipse at the four points $(\pm 2, 1), (\pm 2, -1)$. The extreme values are $xy = 2$ and $xy = -2$.

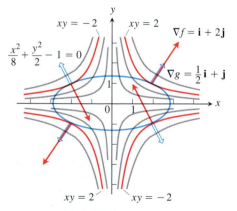

12.63 When subjected to the constraint $g(x, y) = x^2/8 + y^2/2 - 1 = 0$, the function $f(x, y) = xy$ takes on extreme values at the four points $(\pm 2, \pm 1)$. These are the points on the ellipse when ∇f (red) is a scalar multiple of ∇g (blue) (Example 3).

The Geometry of the Solution   The level curves of the function $f(x, y) = xy$ are the hyperbolas $xy = c$ (Fig. 12.63). The farther the hyperbolas lie from the origin, the larger the absolute value of f. We want to find the extreme values of $f(x, y)$, given that the point (x, y) also lies on the ellipse $x^2 + 4y^2 = 8$. Which hyperbolas intersecting the ellipse lie farthest from the origin? The hyperbolas that just graze the ellipse, the ones that are tangent to it. At these points, any vector normal to the hyperbola is normal to the ellipse, so $\nabla f = y\mathbf{i} + x\mathbf{j}$ is a multiple ($\lambda = \pm 2$) of $\nabla g = (x/4)\mathbf{i} + y\mathbf{j}$. At the point $(2, 1)$, for example,

$$\nabla f = \mathbf{i} + 2\mathbf{j}, \qquad \nabla g = \frac{1}{2}\mathbf{i} + \mathbf{j}, \qquad \text{and} \qquad \nabla f = 2\nabla g.$$

At the point $(-2, 1)$,

$$\nabla f = \mathbf{i} - 2\mathbf{j}, \qquad \nabla g = -\frac{1}{2}\mathbf{i} + \mathbf{j}, \qquad \text{and} \qquad \nabla f = -2\nabla g. \qquad \square$$

EXAMPLE 4 Find the maximum and minimum values of the function $f(x, y) = 3x + 4y$ on the circle $x^2 + y^2 = 1$.

Solution We model this as a Lagrange multiplier problem with

$$f(x, y) = 3x + 4y, \qquad g(x, y) = x^2 + y^2 - 1$$

and look for the values of x, y, and λ that satisfy the equations

$$\nabla f = \lambda \nabla g : \qquad 3\mathbf{i} + 4\mathbf{j} = 2x\lambda\mathbf{i} + 2y\lambda\mathbf{j},$$
$$g(x, y) = 0 : \qquad x^2 + y^2 - 1 = 0.$$

The gradient equation implies that $\lambda \neq 0$ and gives

$$x = \frac{3}{2\lambda}, \qquad y = \frac{2}{\lambda}.$$

These equations tell us, among other things, that x and y have the same sign. With these values for x and y, the equation $g(x, y) = 0$ gives

$$\left(\frac{3}{2\lambda}\right)^2 + \left(\frac{2}{\lambda}\right)^2 - 1 = 0,$$

so $\dfrac{9}{4\lambda^2} + \dfrac{4}{\lambda^2} = 1$, $\qquad 9 + 16 = 4\lambda^2$, $\qquad 4\lambda^2 = 25$, $\qquad$ and $\qquad \lambda = \pm\dfrac{5}{2}$.

Thus,

$$x = \frac{3}{2\lambda} = \pm\frac{3}{5}, \qquad y = \frac{2}{\lambda} = \pm\frac{4}{5},$$

and $f(x, y) = 3x + 4y$ has extreme values at $(x, y) = \pm(3/5, 4/5)$.

By calculating the value of $3x + 4y$ at the points $\pm(3/5, 4/5)$, we see that its maximum and minimum values on the circle $x^2 + y^2 = 1$ are

$$3\left(\frac{3}{5}\right) + 4\left(\frac{4}{5}\right) = \frac{25}{5} = 5 \qquad \text{and} \qquad 3\left(-\frac{3}{5}\right) + 4\left(-\frac{4}{5}\right) = -\frac{25}{5} = -5.$$

The Geometry of the Solution (Fig. 12.64) The level curves of $f(x, y) = 3x + 4y$ are the lines $3x + 4y = c$. The farther the lines lie from the origin, the larger the absolute value of f. We want to find the extreme values of $f(x, y)$ given that the point (x, y) also lies on the circle $x^2 + y^2 = 1$. Which lines intersecting the circle lie farthest from the origin? The lines tangent to the circle. At the points of tangency, any vector normal to the line is normal to the circle, so the gradient $\nabla f = 3\mathbf{i} + 4\mathbf{j}$ is a multiple ($\lambda = \pm 5/2$) of the gradient $\nabla g = 2x\mathbf{i} + 2y\mathbf{j}$. At the point $(3/5, 4/5)$, for example,

$$\nabla f = 3\mathbf{i} + 4\mathbf{j}, \qquad \nabla g = \frac{6}{5}\mathbf{i} + \frac{8}{5}\mathbf{j}, \qquad \text{and} \qquad \nabla f = \frac{5}{2}\nabla g. \qquad \square$$

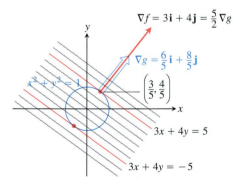

12.64 The function $f(x, y) = 3x + 4y$ takes on its largest value on the unit circle $g(x, y) = x^2 + y^2 - 1 = 0$ at the point $(3/5, 4/5)$ and its smallest value at the point $(-3/5, -4/5)$ (Example 4). At each of these points, ∇f is a scalar multiple of ∇g. The figure shows the gradients at the first point but not the second.

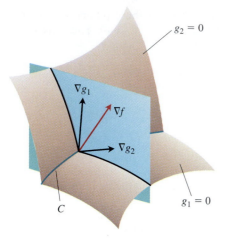

12.65 The vectors ∇g_1 and ∇g_2 lie in a plane perpendicular to the curve C because ∇g_1 is normal to the surface $g_1 = 0$ and ∇g_2 is normal to the surface $g_2 = 0$.

Lagrange Multipliers with Two Constraints

Many problems require us to find the extreme values of a differentiable function $f(x, y, z)$ whose variables are subject to two constraints. If the constraints are

$$g_1(x, y, z) = 0 \qquad \text{and} \qquad g_2(x, y, z) = 0$$

and g_1 and g_2 are differentiable, with ∇g_1 not parallel to ∇g_2, we find the constrained local maxima and minima of f by introducing two Lagrange multipliers λ and μ (mu, pronounced "mew"). That is, we locate the points $P(x, y, z)$ where f takes on its constrained extreme values by finding the values of $x, y, z, \lambda,$ and μ that simultaneously satisfy the equations

$$\nabla f = \lambda \nabla g_1 + \mu \nabla g_2, \qquad g_1(x, y, z) = 0, \qquad g_2(x, y, z) = 0. \qquad (2)$$

The equations in (2) have a nice geometric interpretation. The surfaces $g_1 = 0$ and $g_2 = 0$ (usually) intersect in a smooth curve, say C (Fig. 12.65), and along this curve we seek the points where f has local maximum and minimum values relative to its other values on the curve. These are the points where ∇f is normal to C, as we saw in Theorem 9. But ∇g_1 and ∇g_2 are also normal to C at these points because C lies in the surfaces $g_1 = 0$ and $g_2 = 0$. Therefore ∇f lies in the plane determined by ∇g_1 and ∇g_2, which means that $\nabla f = \lambda \nabla g_1 + \mu \nabla g_2$ for some λ and μ. Since the points we seek also lie in both surfaces, their coordinates must satisfy the equations $g_1(x, y, z) = 0$ and $g_2(x, y, z) = 0$, which are the remaining requirements in Eqs. (2).

EXAMPLE 5   The plane $x + y + z = 1$ cuts the cylinder $x^2 + y^2 = 1$ in an ellipse (Fig. 12.66). Find the points on the ellipse that lie closest to and farthest from the origin.

Solution   We find the extreme values of

$$f(x, y, z) = x^2 + y^2 + z^2$$

(the square of the distance from (x, y, z) to the origin) subject to the constraints

$$g_1(x, y, z) = x^2 + y^2 - 1 = 0 \qquad (3)$$

$$g_2(x, y, z) = x + y + z - 1 = 0. \qquad (4)$$

The gradient equation in (2) then gives

$$\nabla f = \lambda \nabla g_1 + \mu \nabla g_2 \qquad \text{Eq. (2)}$$

$$2x\,\mathbf{i} + 2y\,\mathbf{j} + 2z\,\mathbf{k} = \lambda(2x\,\mathbf{i} + 2y\,\mathbf{j}) + \mu(\mathbf{i} + \mathbf{j} + \mathbf{k})$$

$$2x\,\mathbf{i} + 2y\,\mathbf{j} + 2z\,\mathbf{k} = (2\lambda x + \mu)\,\mathbf{i} + (2\lambda y + \mu)\,\mathbf{j} + \mu\,\mathbf{k}$$

or

$$2x = 2\lambda x + \mu, \qquad 2y = 2\lambda y + \mu, \qquad 2z = \mu. \qquad (5)$$

The scalar equations in (5) yield

$$2x = 2\lambda x + 2z \qquad \Rightarrow \qquad (1 - \lambda)x = z,$$
$$2y = 2\lambda y + 2z \qquad \Rightarrow \qquad (1 - \lambda)y = z. \qquad (6)$$

Equations (6) are satisfied simultaneously if either $\lambda = 1$ and $z = 0$ or $\lambda \neq 1$ and $x = y = z/(1 - \lambda)$.

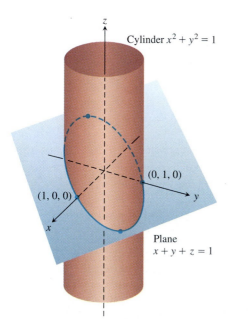

12.66 On the ellipse where the plane and cylinder meet, what are the points closest to and farthest from the origin (Example 5)?

If $z = 0$, then solving Eqs. (3) and (4) simultaneously to find the corresponding points on the ellipse gives the two points $(1, 0, 0)$ and $(0, 1, 0)$. This makes sense when you look at Fig. 12.66.

If $x = y$, then Eqs. (3) and (4) give

$$x^2 + x^2 - 1 = 0 \qquad\qquad x + x + z - 1 = 0$$
$$2x^2 = 1 \qquad\qquad z = 1 - 2x$$
$$x = \pm\frac{\sqrt{2}}{2} \qquad\qquad z = 1 \mp \sqrt{2}.$$

The corresponding points on the ellipse are

$$P_1 = \left(\frac{\sqrt{2}}{2}, \frac{\sqrt{2}}{2}, 1 - \sqrt{2}\right) \qquad \text{and} \qquad P_2 = \left(-\frac{\sqrt{2}}{2}, -\frac{\sqrt{2}}{2}, 1 + \sqrt{2}\right).$$

But here we need to be careful. While P_1 and P_2 both give local maxima of f on the ellipse, P_2 is farther from the origin than P_1.

The points on the ellipse closest to the origin are $(1, 0, 0)$ and $(0, 1, 0)$. The point on the ellipse farthest from the origin is P_2. ▫

Exercises 12.9

Two Independent Variables with One Constraint

1. Find the points on the ellipse $x^2 + 2y^2 = 1$ where $f(x, y) = xy$ has its extreme values.

2. Find the extreme values of $f(x, y) = xy$ subject to the constraint $g(x, y) = x^2 + y^2 - 10 = 0$.

3. Find the maximum value of $f(x, y) = 49 - x^2 - y^2$ on the line $x + 3y = 10$ (Fig. 12.58).

4. Find the local extreme values of $f(x, y) = x^2y$ on the line $x + y = 3$.

5. Find the points on the curve $xy^2 = 54$ nearest the origin.

6. Find the points on the curve $x^2y = 2$ nearest the origin.

7. Use the method of Lagrange multipliers to find

   a) the minimum value of $x + y$, subject to the constraints $xy = 16, x > 0, y > 0$;

   b) the maximum value of xy, subject to the constraint $x + y = 16$.

   Comment on the geometry of each solution.

8. Find the points on the curve $x^2 + xy + y^2 = 1$ in the xy-plane that are nearest to and farthest from the origin.

9. Find the dimensions of the closed right circular cylindrical can of smallest surface area whose volume is 16π cm^3.

10. Find the radius and height of the open right circular cylinder of largest surface area that can be inscribed in a sphere of radius a. What *is* the largest surface area?

11. Use the method of Lagrange multipliers to find the dimensions of the rectangle of greatest area that can be inscribed in the ellipse $x^2/16 + y^2/9 = 1$ with sides parallel to the coordinate axes.

12. Find the dimensions of the rectangle of largest perimeter that can be inscribed in the ellipse $x^2/a^2 + y^2/b^2 = 1$ with sides parallel to the coordinate axes. What *is* the largest perimeter?

13. Find the maximum and minimum values of $x^2 + y^2$ subject to the constraint $x^2 - 2x + y^2 - 4y = 0$.

14. Find the maximum and minimum values of $3x - y + 6$ subject to the constraint $x^2 + y^2 = 4$.

15. The temperature at a point (x, y) on a metal plate is $T(x, y) = 4x^2 - 4xy + y^2$. An ant on the plate walks around the circle of radius 5 centered at the origin. What are the highest and lowest temperatures encountered by the ant?

16. Your firm has been asked to design a storage tank for liquid petroleum gas. The customer's specifications call for a cylindrical tank with hemispherical ends, and the tank is to hold 8000 m^3 of gas. The customer also wants to use the smallest amount of material possible in building the tank. What radius and height do you recommend for the cylindrical portion of the tank?

Three Independent Variables with One Constraint

17. Find the point on the plane $x + 2y + 3z = 13$ closest to the point $(1, 1, 1)$.

18. Find the point on the sphere $x^2 + y^2 + z^2 = 4$ which is farthest from the point $(1, -1, 1)$.

19. Find the minimum distance from the surface $x^2 + y^2 - z^2 = 1$ to the origin.

20. Find the point on the surface $z = xy + 1$ nearest the origin.

21. Find the points on the surface $z^2 = xy + 4$ closest to the origin.

22. Find the point(s) on the surface $xyz = 1$ closest to the origin.

23. Find the maximum and minimum values of

$$f(x, y, z) = x - 2y + 5z$$

on the sphere $x^2 + y^2 + z^2 = 30$.

24. Find the points on the sphere $x^2 + y^2 + z^2 = 25$ where $f(x, y, z) = x + 2y + 3z$ has its maximum and minimum values.

25. Find three real numbers whose sum is 9 and the sum of whose squares is as small as possible.

26. Find the largest product the positive numbers x, y, and z can have if $x + y + z^2 = 16$.

27. Find the dimensions of the closed rectangular box with maximum volume that can be inscribed in the unit sphere.

28. Find the volume of the largest closed rectangular box in the first octant having three faces in the coordinate planes and a vertex on the plane $x/a + y/b + z/c = 1$, where $a > 0, b > 0$, and $c > 0$.

29. A space probe in the shape of the ellipsoid

$$4x^2 + y^2 + 4z^2 = 16$$

enters the earth's atmosphere and its surface begins to heat. After one hour, the temperature at the point (x, y, z) on the probe's surface is

$$T(x, y, z) = 8x^2 + 4yz - 16z + 600.$$

Find the hottest point on the probe's surface.

30. Suppose that the Celsius temperature at the point (x, y, z) on the sphere $x^2 + y^2 + z^2 = 1$ is $T = 400xyz^2$. Locate the highest and lowest temperatures on the sphere.

31. *An example from economics.* In economics, the usefulness or *utility* of amounts x and y of two capital goods G_1 and G_2 is sometimes measured by a function $U(x, y)$. For example, G_1 and G_2 might be two chemicals a pharmaceutical company needs to have on hand and $U(x, y)$ the gain from manufacturing a product whose synthesis requires different amounts of the chemicals depending on the process used. If G_1 costs a dollars per kilogram, G_2 costs b dollars per kilogram, and the total amount allocated for the purchase of G_1 and G_2 together is c dollars, then the company's managers want to maximize $U(x, y)$ given that $ax + by = c$. Thus, they need to solve a typical Lagrange multiplier problem.

Suppose that

$$U(x, y) = xy + 2x$$

and that the equation $ax + by = c$ simplifies to

$$2x + y = 30.$$

Find the maximum value of U and the corresponding values of x and y subject to this latter constraint.

32. You are in charge of erecting a radio telescope on a newly discovered planet. To minimize interference, you want to place it where the magnetic field of the planet is weakest. The planet is spherical, with a radius of 6 units. Based on a coordinate system whose origin is at the center of the planet, the strength of the magnetic field is given by $M(x, y, z) = 6x - y^2 + xz + 60$. Where should you locate the radio telescope?

Lagrange Multipliers with Two Constraints

33. Maximize the function $f(x, y, z) = x^2 + 2y - z^2$ subject to the constraints $2x - y = 0$ and $y + z = 0$.

34. Minimize the function $f(x, y, z) = x^2 + y^2 + z^2$ subject to the constraints $x + 2y + 3z = 6$ and $x + 3y + 9z = 9$.

35. Find the point closest to the origin on the line of intersection of the planes $y + 2z = 12$ and $x + y = 6$.

36. Find the maximum value that $f(x, y, z) = x^2 + 2y - z^2$ can have on the line of intersection of the planes $2x - y = 0$ and $y + z = 0$.

37. Find the extreme values of $f(x, y, z) = x^2yz + 1$ on the intersection of the plane $z = 1$ with the sphere $x^2 + y^2 + z^2 = 10$.

38. a) Find the maximum value of $w = xyz$ on the line of intersection of the two planes $x + y + z = 40$ and $x + y - z = 0$.

b) Give a geometric argument to support your claim that you have found a maximum, and not a minimum, value of w.

39. Find the extreme values of the function $f(x, y, z) = xy + z^2$ on the circle in which the plane $y - x = 0$ intersects the sphere $x^2 + y^2 + z^2 = 4$.

40. Find the point closest to the origin on the curve of intersection of the plane $2y + 4z = 5$ and the cone $z^2 = 4x^2 + 4y^2$.

Theory and Examples

41. *The condition $\nabla f = \lambda \nabla g$ is not sufficient.* While $\nabla f = \lambda \nabla g$ is a necessary condition for the occurrence of an extreme value of $f(x, y)$ subject to the condition $g(x, y) = 0$, it does not in itself guarantee that one exists. As a case in point, try using the method of Lagrange multipliers to find a maximum value of $f(x, y) = x + y$ subject to the constraint that $xy = 16$. The method will identify the two points $(4, 4)$ and $(-4, -4)$ as candidates for the location of extreme values. Yet the sum $(x + y)$ has no maximum value on the hyperbola $xy = 16$. The farther you go from the origin on this hyperbola in the first quadrant, the larger the sum $f(x, y) = x + y$ becomes.

42. *A least squares plane.* The plane $z = Ax + By + C$ is to be "fitted" to the following points (x_k, y_k, z_k):

$$(0, 0, 0), \quad (0, 1, 1), \quad (1, 1, 1), \quad (1, 0, -1).$$

Find the values of A, B, and C that minimize the sum

$$\sum_{k=1}^{4} (Ax_k + By_k + C - z_k)^2,$$

the sum of the squares of the deviations.

43. a) Show that the maximum value of $a^2b^2c^2$ on a sphere of radius r centered at the origin of a Cartesian abc-coordinate system is $(r^2/3)^3$.

b) Using part (a), show that for nonnegative numbers $a, b,$ and c,

$$(abc)^{1/3} \le \frac{a+b+c}{3}.$$

That is, the *geometric mean* of three numbers is less than or equal to the *arithmetic mean*.

44. Let $a_1, a_2, \ldots, a_n$ be n positive numbers. Find the maximum of $\sum_{i=1}^{n} a_i x_i$ subject to the constraint $\sum_{i=1}^{n} x_i^2 = 1$.

✪ CAS Explorations and Projects

In Exercises 45–50, use a CAS to perform the following steps implementing the method of Lagrange multipliers for finding constrained extrema:

a) Form the function $h = f - \lambda_1 g_1 - \lambda_2 g_2$, where f is the function to optimize subject to the constraints $g_1 = 0$ and $g_2 = 0$.

b) Determine all the first partial derivatives of h, including the partials with respect to λ_1 and λ_2, and set them equal to 0.

c) Solve the system of equations found in (b) for all the unknowns, including λ_1 and λ_2.

d) Evaluate f at each of the solution points found in (c) and select the extreme value subject to the constraints asked for in the exercise.

45. Minimize $f(x, y, z) = xy + yz$ subject to the constraints $x^2 + y^2 - 2 = 0$ and $x^2 + z^2 - 2 = 0$.

46. Minimize $f(x, y, z) = xyz$ subject to the constraints $x^2 + y^2 - 1 = 0$ and $x - z = 0$.

47. Maximize $f(x, y, z) = x^2 + y^2 + z^2$ subject to the constraints $2y + 4z - 5 = 0$ and $4x^2 + 4y^2 - z^2 = 0$.

48. Minimize $f(x, y, z) = x^2 + y^2 + z^2$ subject to the constraints $x^2 - xy + y^2 - z^2 - 1 = 0$ and $x^2 + y^2 - 1 = 0$.

49. Minimize $f(x, y, z, w) = x^2 + y^2 + z^2 + w^2$ subject to the constraints $2x - y + z - w - 1 = 0$ and $x + y - z + w - 1 = 0$.

50. Determine the distance from the line $y = x + 1$ to the parabola $y^2 = x$. (*Hint:* Let (x, y) be a point on the line and (w, z) a point on the parabola. You want to minimize $(x - w)^2 + (y - z)^2$.)

12.10 Taylor's Formula

This section uses Taylor's formula (Section 8.10) to derive the second derivative test for local extreme values (Section 12.8) and the error formula for linearizations of functions of two independent variables (Section 12.4, Eq. 5). The use of Taylor's formula in these derivations leads to an extension of the formula that provides polynomial approximations of all orders for functions of two independent variables.

The Derivation of the Second Derivative Test

Let $f(x, y)$ have continuous partial derivatives in an open region R containing a point $P(a, b)$ where $f_x = f_y = 0$ (Fig. 12.67). Let h and k be increments small enough to put the point $S(a + h, b + k)$ and the line segment joining it to P inside R. We parametrize the segment PS as

$$x = a + th, \qquad y = b + tk, \qquad 0 \le t \le 1.$$

If $F(t) = f(a + th, b + tk)$, the Chain Rule gives

$$F'(t) = f_x \frac{dx}{dt} + f_y \frac{dy}{dt} = h f_x + k f_y.$$

Since f_x and f_y are differentiable (they have continuous partial derivatives), F' is a differentiable function of t and

$$F'' = \frac{\partial F'}{\partial x}\frac{dx}{dt} + \frac{\partial F'}{\partial y}\frac{dy}{dt} = \frac{\partial}{\partial x}\left(h f_x + k f_y\right) \cdot h + \frac{\partial}{\partial y}\left(h f_x + k f_y\right) \cdot k$$

$$= h^2 f_{xx} + 2hk f_{xy} + k^2 f_{yy}. \qquad f_{xy} = f_{yx}$$

Since F and F' are continuous on $[0, 1]$ and F' is differentiable on $(0, 1)$, we can

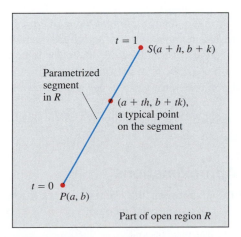

12.67 We begin the derivation of the second derivative test at $P(a, b)$ by parametrizing a typical line segment from P to a point S nearby.

apply Taylor's formula with $n = 2$ and $a = 0$ to obtain

$$F(1) = F(0) + F'(0)(1 - 0) + F''(c)\frac{(1 - 0)^2}{2}$$

$$F(1) = F(0) + F'(0) + \frac{1}{2}F''(c) \tag{1}$$

for some c between 0 and 1. Writing Eq. (1) in terms of f gives

$$f(a + h, b + k) = f(a, b) + hf_x(a, b) + kf_y(a, b)$$

$$+ \frac{1}{2}\left(h^2 f_{xx} + 2hk f_{xy} + k^2 f_{yy}\right)\bigg|_{(a+ch, b+ck)}. \tag{2}$$

Since $f_x(a, b) = f_y(a, b) = 0$, this reduces to

$$f(a + h, b + k) - f(a, b) = \frac{1}{2}\left(h^2 f_{xx} + 2hk f_{xy} + k^2 f_{yy}\right)\bigg|_{(a+ch, b+ck)}. \tag{3}$$

The presence of an extremum of f at (a, b) is determined by the sign of $f(a + h, b + k) - f(a, b)$. By Eq. (3), this is the same as the sign of

$$Q(c) = (h^2 f_{xx} + 2hk f_{xy} + k^2 f_{yy})|_{(a+ch, b+ck)}.$$

Now, if $Q(0) \neq 0$, the sign of $Q(c)$ will be the same as the sign of $Q(0)$ for sufficiently small values of h and k. We can predict the sign of

$$Q(0) = h^2 f_{xx}(a, b) + 2hk f_{xy}(a, b) + k^2 f_{yy}(a, b) \tag{4}$$

from the signs of f_{xx} and $f_{xx}f_{yy} - f_{xy}^2$ at (a, b). Multiply both sides of Eq. (3) by f_{xx} and rearrange the right-hand side to get

$$f_{xx}Q(0) = (hf_{xx} + kf_{xy})^2 + (f_{xx}f_{yy} - f_{xy}^2)k^2. \tag{5}$$

From Eq. (5) we see that

1. If $f_{xx} < 0$ and $f_{xx}f_{yy} - f_{xy}^2 > 0$ at (a, b), then $Q(0) < 0$ for all sufficiently small nonzero values of h and k, and f has a *local maximum* value at (a, b).
2. If $f_{xx} > 0$ and $f_{xx}f_{yy} - f_{xy}^2 > 0$ at (a, b), then $Q(0) > 0$ for all sufficiently small nonzero values of h and k, and f has a *local minimum* value at (a, b).
3. If $f_{xx}f_{yy} - f_{xy}^2 < 0$ at (a, b), there are combinations of arbitrarily small nonzero values of h and k for which $Q(0) > 0$, and other values for which $Q(0) < 0$. Arbitrarily close to the point $P_0(a, b, f(a, b))$ on the surface $z = f(x, y)$ there are points above P_0 and points below P_0, so f has a *saddle point* at (a, b).
4. If $f_{xx}f_{yy} - f_{xy}^2 = 0$, another test is needed. The possibility that $Q(0)$ equals zero prevents us from drawing conclusions about the sign of $Q(c)$.

The Error Formula for Linear Approximations

We want to show that the difference $E(x, y)$ between the values of a function $f(x, y)$ and its linearization $L(x, y)$ at (x_0, y_0) satisfies the inequality

$$|E(x, y)| \leq \frac{1}{2}B(|x - x_0| + |y - y_0|)^2.$$

The function f is assumed to have continuous second partial derivatives throughout an open set containing a closed rectangular region R centered at (x_0, y_0).

The number B is the largest value that any of $|f_{xx}|$, $|f_{yy}|$, and $|f_{xy}|$ take on R.

The inequality we want comes from Eq. (2). We substitute x_0 and y_0 for a and b, and $x - x_0$ and $y - y_0$ for h and k, respectively, and rearrange the result as

$$f(x, y) = \underbrace{f(x_0, y_0) + f_x(x_0, y_0)(x - x_0) + f_y(x_0, y_0)(y - y_0)}_{\text{linearization } L(x, y)}$$

$$+ \underbrace{\frac{1}{2} \left((x - x_0)^2 f_{xx} + 2(x - x_0)(y - y_0) f_{xy} + (y - y_0)^2 f_{yy} \right) \Big|_{(x_0 + c(x - x_0), y_0 + c(y - y_0))}}_{\text{error } E(x, y)}.$$

This remarkable equation reveals that

$$|E| \leq \frac{1}{2} \left(|x - x_0|^2 |f_{xx}| + 2|x - x_0||y - y_0||f_{xy}| + |y - y_0|^2 |f_{yy}| \right).$$

Hence, if B is an upper bound for the values of $|f_{xx}|$, $|f_{xy}|$, and $|f_{yy}|$ on R,

$$|E| \leq \frac{1}{2} \left(|x - x_0|^2 B + 2|x - x_0||y - y_0| B + |y - y_0|^2 B \right)$$

$$\leq \frac{1}{2} B \left(|x - x_0| + |y - y_0| \right)^2.$$

Taylor's Formula for Functions of Two Variables

The formulas derived earlier for F' and F'' can be obtained by applying to $f(x, y)$ the operators

$$\left(h \frac{\partial}{\partial x} + k \frac{\partial}{\partial y} \right) \quad \text{and} \quad \left(h \frac{\partial}{\partial x} + k \frac{\partial}{\partial y} \right)^2 = h^2 \frac{\partial^2}{\partial x^2} + 2hk \frac{\partial^2}{\partial x \, \partial y} + k^2 \frac{\partial^2}{\partial y^2}.$$

These are the first two instances of a more general formula,

$$F^{(n)}(t) = \frac{d^n}{dt^n} F(t) = \left(h \frac{\partial}{\partial x} + k \frac{\partial}{\partial y} \right)^n f(x, y), \tag{6}$$

which says that applying d^n/dt^n to $F(t)$ gives the same result as applying the operator

$$\left(h \frac{\partial}{\partial x} + k \frac{\partial}{\partial y} \right)^n$$

to $f(x, y)$ after expanding it by the binomial theorem.

If partial derivatives of f through order $n + 1$ are continuous throughout a rectangular region centered at (a, b), we may extend the Taylor formula for $F(t)$ to

$$F(t) = F(0) + F'(0) t + \frac{F''(0)}{2!} t^2 + \cdots + \frac{F^{(n)}(0)}{n!} t^n + \text{remainder},$$

and take $t = 1$ to obtain

$$F(1) = F(0) + F'(0) + \frac{F''(0)}{2!} + \cdots + \frac{F^{(n)}(0)}{n!} + \text{remainder}.$$

When we replace the first n derivatives on the right of this last series by their equivalent expressions from Eq. (6) evaluated at $t = 0$ and add the appropriate remainder term, we arrive at the following formula.

Taylor's Formula for $f(x, y)$ at the Point (a, b)

Suppose $f(x, y)$ and its partial derivatives through order $n + 1$ are continuous throughout an open rectangular region R centered at a point (a, b). Then, throughout R,

$$f(a + h, b + k) = f(a, b) + (hf_x + kf_y)|_{(a,b)} + \frac{1}{2!}(h^2 f_{xx} + 2hkf_{xy} + k^2 f_{yy})|_{(a,b)}$$

$$+ \frac{1}{3!}(h^3 f_{xxx} + 3h^2 k f_{xxy} + 3hk^2 f_{xyy} + k^3 f_{yyy})|_{(a,b)} + \cdots + \frac{1}{n!}\left(h\frac{\partial}{\partial x} + k\frac{\partial}{\partial y}\right)^n f\bigg|_{(a,b)}$$

$$+ \frac{1}{(n+1)!}\left(h\frac{\partial}{\partial x} + k\frac{\partial}{\partial y}\right)^{n+1} f\bigg|_{(a+ch,b+ck)}. \tag{7}$$

The first n derivative terms are evaluated at (a, b). The last term is evaluated at some point $(a + ch, b + ck)$ on the line segment joining (a, b) and $(a + h, b + k)$.

If $(a, b) = (0, 0)$ and we treat h and k as independent variables (denoting them now by x and y), then Eq. (7) assumes the following simpler form.

Taylor's Formula for $f(x, y)$ at the Origin

$$f(x, y) = f(0, 0) + xf_x + yf_y + \frac{1}{2!}(x^2 f_{xx} + 2xyf_{xy} + y^2 f_{yy})$$

$$+ \frac{1}{3!}(x^3 f_{xxx} + 3x^2 y f_{xxy} + 3xy^2 f_{xyy} + y^3 f_{yyy}) + \cdots + \frac{1}{n!}\left(x\frac{\partial}{\partial x} + y\frac{\partial}{\partial y}\right)^n f$$

$$+ \frac{1}{(n+1)!}\left(x\frac{\partial}{\partial x} + y\frac{\partial}{\partial y}\right)^{n+1} f\bigg|_{(cx,cy)}. \tag{8}$$

The first n derivative terms are evaluated at $(0, 0)$. The last term is evaluated at a point on the line segment joining the origin and (x, y).

Taylor's formula provides polynomial approximations of two-variable functions. The first n derivative terms give the polynomial; the last term gives the approximation error. The first three terms of Taylor's formula give the function's linearization. To improve on the linearization, we add higher power terms.

EXAMPLE 1   Find a quadratic $f(x, y) = \sin x \sin y$ near the origin. How accurate is the approximation if $|x| \leq 0.1$ and $|y| \leq 0.1$?

Solution   We take $n = 2$ in Eq. (8):

$$f(x, y) = f(0, 0) + (xf_x + yf_y) + \frac{1}{2}(x^2 f_{xx} + 2xyf_{xy} + y^2 f_{yy})$$

$$+ \frac{1}{6}(x^3 f_{xxx} + 3x^2 y f_{xxy} + 3xy^2 f_{xyy} + y^3 f_{yyy})_{(cx,cy)}$$

with

$$\begin{aligned} f(0, 0) &= \sin x \sin y|_{(0,0)} = 0, & f_{xx}(0, 0) &= -\sin x \sin y|_{(0,0)} = 0, \\ f_x(0, 0) &= \cos x \sin y|_{(0,0)} = 0, & f_{xy}(0, 0) &= \cos x \cos y|_{(0,0)} = 1, \\ f_y(0, 0) &= \sin x \cos y|_{(0,0)} = 0, & f_{yy}(0, 0) &= -\sin x \sin y|_{(0,0)} = 0, \end{aligned}$$

we have

$$\sin x \sin y \approx 0 + 0 + 0 + \frac{1}{2}(x^2(0) + 2xy(1) + y^2(0)),$$

$$\sin x \sin y \approx xy.$$

The error in the approximation is

$$E(x, y) = \frac{1}{6}(x^3 f_{xxx} + 3x^2 y f_{xxy} + 3xy^2 f_{xyy} + y^3 f_{yyy})|_{(cx,cy)}.$$

The third derivatives never exceed 1 in absolute value because they are products of sines and cosines. Also, $|x| \le 0.1$ and $|y| \le 0.1$. Hence

$$|E(x, y)| \le \frac{1}{6}((0.1)^3 + 3(0.1)^3 + 3(0.1)^3 + (0.1)^3) \le \frac{8}{6}(0.1)^3 \le 0.00134$$

(rounded up). The error will not exceed 0.00134 if $|x| \le 0.1$ and $|y| \le 0.1$.   ❏

Exercises 12.10

Finding Quadratic and Cubic Approximations

In Exercises 1–10, use Taylor's formula for $f(x, y)$ at the origin to find quadratic and cubic approximations of f near the origin.

1. $f(x, y) = x e^y$

2. $f(x, y) = e^x \cos y$

3. $f(x, y) = y \sin x$

4. $f(x, y) = \sin x \cos y$

5. $f(x, y) = e^x \ln (1 + y)$

6. $f(x, y) = \ln (2x + y + 1)$

7. $f(x, y) = \sin (x^2 + y^2)$

8. $f(x, y) = \cos (x^2 + y^2)$

9. $f(x, y) = \dfrac{1}{1 - x - y}$

10. $f(x, y) = \dfrac{1}{1 - x - y + xy}$

11. Use Taylor's formula to find a quadratic approximation of $f(x, y) = \cos x \cos y$ at the origin. Estimate the error in the approximation if $|x| \le 0.1$ and $|y| \le 0.1$.

12. Use Taylor's formula to find a quadratic approximation of $e^x \sin y$ at the origin. Estimate the error in the approximation if $|x| \le 0.1$ and $|y| \le 0.1$.

CHAPTER **12** QUESTIONS TO GUIDE YOUR REVIEW

1. What is a real-valued function of two independent variables? three independent variables? Give examples.

2. What does it mean for sets in the plane or in space to be open? closed? Give examples. Give examples of sets that are neither open nor closed.

3. How can you display the values of a function $f(x, y)$ of two independent variables graphically? How do you do the same for a function $f(x, y, z)$ of three independent variables?

4. What does it mean for a function $f(x, y)$ to have limit L as $(x, y) \to (x_0, y_0)$? What are the basic properties of limits of functions of two independent variables?

5. When is a function of two (three) independent variables continuous at a point in its domain? Give examples of functions that are continuous at some points but not others.

6. What can be said about algebraic combinations and composites of continuous functions?

7. Explain the two-path test for nonexistence of limits.

8. How are the partial derivatives $\partial f/\partial x$ and $\partial f/\partial y$ of a function $f(x, y)$ defined? How are they interpreted and calculated?

9. How does the relation between first partial derivatives and continuity of functions of two independent variables differ from the relation between first derivatives and continuity for real-valued functions of a single independent variable? Give an example.

10. What is Euler's theorem for mixed second order partial derivatives? How can it help in calculating partial derivatives of second and higher orders? Give examples.

11. What does it mean for a function $f(x, y)$ to be differentiable? What does the Increment Theorem say about differentiability?

12. How can you sometimes decide from examining f_x and f_y that a function $f(x, y)$ is differentiable? What is the relation between the differentiability of f and the continuity of f at a point?

13. How do you linearize a function $f(x, y)$ of two independent variables at a point (x_0, y_0)? Why might you want to do this? How do you linearize a function of three independent variables?

14. What can you say about the accuracy of linear approximations of functions of two (three) independent variables?

15. If (x, y) moves from (x_0, y_0) to a point $(x_0 + dx, y_0 + dy)$ nearby, how can you estimate the resulting change in the value of a differentiable function $f(x, y)$? Give an example.

16. What is the Chain Rule? What form does it take for functions of two independent variables? three independent variables? functions defined on surfaces? How do you diagram these different forms? Give examples. What pattern enables one to remember all the different forms?

17. What is the derivative of a function $f(x, y)$ at a point P_0 in the direction of a unit vector $\mathbf{u}$? What rate does it describe? What geometric interpretation does it have? Give examples.

18. What is the gradient vector of a function $f(x, y)$? How is it related to the function's directional derivatives? State the analogous results for functions of three independent variables.

19. How do you find the tangent line at a point on a level curve of a differentiable function $f(x, y)$? How do you find the tangent plane and normal line at a point on a level surface of a differentiable function $f(x, y, z)$? Give examples.

20. How can you use directional derivatives to estimate change?

21. How do you define local maxima, local minima, and saddle points for a differentiable function $f(x, y)$? Give examples.

22. What derivative tests are available for determining the local extreme values of a function $f(x, y)$? How do they enable you to narrow your search for these values? Give examples.

23. How do you find the extrema of a continuous function $f(x, y)$ on a closed bounded region of the xy-plane? Give an example.

24. Describe the method of Lagrange multipliers and give examples.

25. How does Taylor's formula for a function $f(x, y)$ generate polynomial approximations and error estimates?

CHAPTER 12 PRACTICE EXERCISES

Domain, Range, and Level Curves

In Exercises 1–4, find the domain and range of the given function and identify its level curves. Sketch a typical level curve.

1. $f(x, y) = 9x^2 + y^2$
2. $f(x, y) = e^{x+y}$
3. $g(x, y) = 1/xy$
4. $g(x, y) = \sqrt{x^2 - y}$

In Exercise 5–8, find the domain and range of the given function and identify its level surfaces. Sketch a typical level surface.

5. $f(x, y, z) = x^2 + y^2 - z$

6. $g(x, y, z) = x^2 + 4y^2 + 9z^2$

7. $h(x, y, z) = \dfrac{1}{x^2 + y^2 + z^2}$

8. $k(x, y, z) = \dfrac{1}{x^2 + y^2 + z^2 + 1}$

Evaluating Limits

Find the limits in Exercises 9–14.

9. $\displaystyle\lim_{(x,y)\to(\pi,\,\ln 2)} e^y \cos x$

10. $\displaystyle\lim_{(x,y)\to(0,0)} \dfrac{2+y}{x + \cos y}$

11. $\displaystyle\lim_{\substack{(x,y)\to(1,1)\\x\neq y}} \dfrac{x-y}{x^2 - y^2}$

12. $\displaystyle\lim_{(x,y)\to(1,1)} \dfrac{x^3 y^3 - 1}{xy - 1}$

13. $\displaystyle\lim_{P\to(1,-1,e)} \ln|x + y + z|$

14. $\displaystyle\lim_{P\to(1,-1,-1)} \tan^{-1}(x + y + z)$

By considering different paths of approach, show that the limits in Exercises 15 and 16 do not exist.

15. $\displaystyle\lim_{\substack{(x,y)\to(0,0)\\y\neq x^2}} \dfrac{y}{x^2 - y}$

16. $\displaystyle\lim_{\substack{(x,y)\to(0,0)\\xy\neq 0}} \dfrac{x^2 + y^2}{xy}$

17. **a)** Let $f(x, y) = (x^2 - y^2)/(x^2 + y^2)$ for $(x, y) \neq (0, 0)$. Is it possible to define $f(0, 0)$ in a way that makes f continuous at the origin? Why?

b) Let
$$f(x, y) = \begin{cases} \dfrac{\sin(x-y)}{|x| + |y|}, & |x| + |y| \neq 0, \\ 0, & (x, y) = (0, 0). \end{cases}$$
Is f continuous at the origin? Why?

18. Let
$$f(r, \theta) = \begin{cases} \dfrac{\sin 6r}{6r}, & r \neq 0, \\ 1, & r = 0, \end{cases}$$
where r and θ are polar coordinates. Find

a) $\displaystyle\lim_{r\to 0} f(r, \theta)$ **b)** $f_r(0, 0)$ **c)** $f_\theta(r, \theta)$, $r \neq 0$

$z = f(r, \theta)$

(Generated by Mathematica)

Partial Derivatives

In Exercises 19–24, find the partial derivative of the function with respect to each variable.

19. $g(r, \theta) = r \cos \theta + r \sin \theta$

20. $f(x, y) = \dfrac{1}{2} \ln (x^2 + y^2) + \tan^{-1} \dfrac{y}{x}$

21. $f(R_1, R_2, R_3) = \dfrac{1}{R_1} + \dfrac{1}{R_2} + \dfrac{1}{R_3}$

22. $h(x, y, z) = \sin (2\pi x + y - 3z)$

23. $P(n, R, T, V) = \dfrac{nRT}{V}$ (the Ideal Gas Law)

24. $f(r, l, T, w) = \dfrac{1}{2rl} \sqrt{\dfrac{T}{\pi w}}$

Second Order Partials

Find the second order partial derivatives of the functions in Exercises 25–28.

25. $g(x, y) = y + \dfrac{x}{y}$

26. $g(x, y) = e^x + y \sin x$

27. $f(x, y) = x + xy - 5x^3 + \ln (x^2 + 1)$

28. $f(x, y) = y^2 - 3xy + \cos y + 7e^y$

Linearizations

In Exercises 29 and 30, find the linearization $L(x, y)$ of the function $f(x, y)$ at the point P_0. Then find an upper bound for the magnitude of the error E in the approximation $f(x, y) \approx L(x, y)$ over the rectangle R.

29. $f(x, y) = \sin x \cos y, \quad P_0(\pi/4, \pi/4)$

$R: \quad \left| x - \dfrac{\pi}{4} \right| \le 0.1, \quad \left| y - \dfrac{\pi}{4} \right| \le 0.1$

30. $f(x, y) = xy - 3y^2 + 2, \quad P_0(1, 1)$

$R: \quad |x - 1| \le 0.1, \quad |y - 1| \le 0.2$

Find the linearizations of the functions in Exercises 31 and 32 at the given points.

31. $f(x, y, z) = xy + 2yz - 3xz$ at $(1, 0, 0)$ and $(1, 1, 0)$

32. $f(x, y, z) = \sqrt{2} \cos x \sin (y + z)$ at $(0, 0, \pi/4)$ and $(\pi/4, \pi/4, 0)$

Estimates and Sensitivity to Change

33. You plan to calculate the volume inside a stretch of pipeline that is about 36 in. in diameter and 1 mi long. With which measurement should you be more careful—the length, or the diameter? Why?

34. Near the point $(1, 2)$, is $f(x, y) = x^2 - xy + y^2 - 3$ more sensitive to changes in x, or to changes in y? How do you know?

35. Suppose that the current I (amperes) in an electrical circuit is related to the voltage V (volts) and the resistance R (ohms) by the equation $I = V/R$. If the voltage drops from 24 to 23 volts and the resistance drops from 100 to 80 ohms, will I increase, or decrease? By about how much? Express the changes in V and R and the estimated change in I as percentages of their original values.

36. If $a = 10$ cm and $b = 16$ cm to the nearest millimeter, what should you expect the maximum percentage error to be in the calculated area $A = \pi ab$ of the ellipse $x^2/a^2 + y^2/b^2 = 1$?

37. Let $y = uv$ and $z = u + v$, where u and v are positive independent variables.

a) If u is measured with an error of 2% and v with an error of 3%, about what is the percentage error in the calculated value of y?

b) Show that the percentage error in the calculated value of z is less than the percentage error in the value of y.

38. *Cardiac index.* To make different people comparable in studies of cardiac output (Section 2.7, Exercise 25), researchers divide the measured cardiac output by the body surface area to find the *cardiac index C:*

$$C = \frac{\text{cardiac output}}{\text{body surface area}}.$$

The body surface area B is calculated with the formula

$$B = 71.84 w^{0.425} h^{0.725},$$

which gives B in square centimeters when w is measured in kilograms and h in centimeters. You are about to calculate the cardiac index of a person with the following measurements:

Cardiac output:	7 L/min
Weight:	70 kg
Height:	180 cm

Which will have a greater effect on the calculation, a 1-kg error in measuring the weight, or a 1-cm error in measuring the height?

Chain Rule Calculations

39. Find dw/dt at $t = 0$ if $w = \sin (xy + \pi), x = e^t$, and $y = \ln (t + 1)$.

40. Find dw/dt at $t = 1$ if $w = xe^y + y \sin z - \cos z, x = 2\sqrt{t}$, $y = t - 1 + \ln t, z = \pi t$.

41. Find $\partial w/\partial r$ and $\partial w/\partial s$ when $r = \pi$ and $s = 0$ if $w = \sin(2x - y)$, $x = r + \sin s$, $y = rs$.

42. Find $\partial w/\partial u$ and $\partial w/\partial v$ when $u = v = 0$ if $w = \ln\sqrt{1 + x^2} - \tan^{-1} x$ and $x = 2e^u \cos v$.

43. Find the value of the derivative of $f(x, y, z) = xy + yz + xz$ with respect to t on the curve $x = \cos t$, $y = \sin t$, $z = \cos 2t$ at $t = 1$.

44. Show that if $w = f(s)$ is any differentiable function of s and if $s = y + 5x$, then

$$\frac{\partial w}{\partial x} - 5\frac{\partial w}{\partial y} = 0.$$

Implicit Differentiation

Assuming that the equations in Exercises 45 and 46 define y as a differentiable function of x, find the value of dy/dx at point P.

45. $1 - x - y^2 - \sin xy = 0$, $P(0, 1)$

46. $2xy + e^{x+y} - 2 = 0$, $P(0, \ln 2)$

Partial Derivatives with Constrained Variables

In Exercises 47 and 48, begin by drawing a diagram that shows the relations among the variables.

47. If $w = x^2 e^{yz}$ and $z = x^2 - y^2$, find

a) $\left(\dfrac{\partial w}{\partial y}\right)_z$ **b)** $\left(\dfrac{\partial w}{\partial z}\right)_x$ **c)** $\left(\dfrac{\partial w}{\partial z}\right)_y$

48. Let $U = f(P, V, T)$ be the internal energy of a gas that obeys the ideal gas law $PV = nRT$ (n and R constant). Find

a) $\left(\dfrac{\partial U}{\partial T}\right)_P$ **b)** $\left(\dfrac{\partial U}{\partial V}\right)_T$

Directional Derivatives

In Exercises 49–52, find the directions in which f increases and decreases most rapidly at P_0 and find the derivative of f in each direction. Also, find the derivative of f at P_0 in the direction of the vector $\mathbf{A}$.

49. $f(x, y) = \cos x \cos y$, $P_0(\pi/4, \pi/4)$, $\mathbf{A} = 3\mathbf{i} + 4\mathbf{j}$

50. $f(x, y) = x^2 e^{-2y}$, $P_0(1, 0)$, $\mathbf{A} = \mathbf{i} + \mathbf{j}$

51. $f(x, y, z) = \ln(2x + 3y + 6z)$, $P_0(-1, -1, 1)$, $\mathbf{A} = 2\mathbf{i} + 3\mathbf{j} + 6\mathbf{k}$

52. $f(x, y, z) = x^2 + 3xy - z^2 + 2y + z + 4$, $P_0(0, 0, 0)$, $\mathbf{A} = \mathbf{i} + \mathbf{j} + \mathbf{k}$

53. Find the derivative of $f(x, y, z) = xyz$ in the direction of the velocity vector of the helix

$$\mathbf{r}(t) = (\cos 3t)\mathbf{i} + (\sin 3t)\mathbf{j} + 3t\,\mathbf{k}$$

at $t = \pi/3$.

54. What is the largest value that the directional derivative of $f(x, y, z) = xyz$ can have at the point $(1, 1, 1)$?

55. At the point $(1, 2)$ the function $f(x, y)$ has a derivative of 2 in the direction toward $(2, 2)$ and derivative of -2 in the direction toward $(1, 1)$.

a) Find $f_x(1, 2)$ and $f_y(1, 2)$.
b) Find the derivative of f at $(1, 2)$ in the direction toward the point $(4, 6)$.

56. Which of the following statements are true if $f(x, y)$ is differentiable at (x_0, y_0)?

a) If $\mathbf{u}$ is a unit vector, the derivative of f at (x_0, y_0) in the direction of $\mathbf{u}$ is $(f_x(x_0, y_0)\mathbf{i} + f_y(x_0, y_0)\mathbf{j}) \cdot \mathbf{u}$.
b) The derivative of f at (x_0, y_0) in the direction of $\mathbf{u}$ is a vector.
c) The directional derivative of f at (x_0, y_0) has its greatest value in the direction of ∇f.
d) At (x_0, y_0), vector ∇f is normal to the curve $f(x, y) = f(x_0, y_0)$.

Gradients, Tangent Planes, and Normal Lines

In Exercises 57 and 58, sketch the surface $f(x, y, z) = c$ together with ∇f at the given points.

57. $x^2 + y + z^2 = 0$; $(0, -1, \pm 1)$, $(0, 0, 0)$

58. $y^2 + z^2 = 4$; $(2, \pm 2, 0)$, $(2, 0, \pm 2)$

In Exercises 59 and 60, find an equation for the plane tangent to the level surface $f(x, y, z) = c$ at the point P_0. Also, find parametric equations for the line that is normal to the surface at P_0.

59. $x^2 - y - 5z = 0$, $P_0(2, -1, 1)$

60. $x^2 + y^2 + z = 4$, $P_0(1, 1, 2)$

In Exercises 61 and 62, find an equation for the plane tangent to the surface $z = f(x, y)$ at the given point

61. $z = \ln(x^2 + y^2)$, $(0, 1, 0)$

62. $z = 1/(x^2 + y^2)$, $(1, 1, 1/2)$

In Exercises 63 and 64, find equations for the lines that are tangent and normal to the level curve $f(x, y) = c$ at the point P_0. Then sketch the lines and level curve together with ∇f at P_0.

63. $y - \sin x = 1$, $P_0(\pi, 1)$

64. $\dfrac{y^2}{2} - \dfrac{x^2}{2} = \dfrac{3}{2}$, $P_0(1, 2)$

Tangent Lines to Curves

In Exercises 65 and 66, find parametric equations for the line that is tangent to the curve of intersection of the surfaces at the given point.

65. Surfaces: $x^2 + 2y + 2z = 4$, $y = 1$
   Point: $(1, 1, 1/2)$

66. Surfaces: $x + y^2 + z = 2$, $y = 1$
   Point: $(1/2, 1, 1/2)$

Local Extrema

Test the functions in Exercises 67–72 for local maxima and minima and saddle points. Find each function's values at these points.

67. $f(x, y) = x^2 - xy + y^2 + 2x + 2y - 4$

68. $f(x, y) = 5x^2 + 4xy - 2y^2 + 4x - 4y$

69. $f(x, y) = 2x^3 + 3xy + 2y^3$

70. $f(x, y) = x^3 + y^3 - 3xy + 15$

71. $f(x, y) = x^3 + y^3 + 3x^2 - 3y^2$

72. $f(x, y) = x^4 - 8x^2 + 3y^2 - 6y$

Absolute Extrema

In Exercises 73–80, find the absolute maximum and minimum values of f on the region R.

73. $f(x, y) = x^2 + xy + y^2 - 3x + 3y$
R: The triangular region cut from the first quadrant by the line $x + y = 4$

74. $f(x, y) = x^2 - y^2 - 2x + 4y + 1$
R: The rectangular region in the first quadrant bounded by the coordinate axes and the lines $x = 4$ and $y = 2$

75. $f(x, y) = y^2 - xy - 3y + 2x$
R: The square region enclosed by the lines $x = \pm 2$ and $y = \pm 2$

76. $f(x, y) = 2x + 2y - x^2 - y^2$
R: The square bounded by the coordinate axes and the lines $x = 2$, $y = 2$ in the first quadrant

77. $f(x, y) = x^2 - y^2 - 2x + 4y$
R: The triangular region bounded below by the x-axis, above by the line $y = x + 2$, and on the right by the line $x = 2$

78. $f(x, y) = 4xy - x^4 - y^4 + 16$
R: The triangular region bounded below by the line $y = -2$, above by the line $y = x$, and on the right by the line $x = 2$

79. $f(x, y) = x^3 + y^3 + 3x^2 - 3y^2$
R: The square region enclosed by the lines $x = \pm 1$ and $y = \pm 1$

80. $f(x, y) = x^3 + 3xy + y^3 + 1$
R: The square region enclosed by the lines $x = \pm 1$ and $y = \pm 1$

Lagrange Multipliers

81. Find the extreme values of $f(x, y) = x^3 + y^2$ on the circle $x^2 + y^2 = 1$.

82. Find the extreme values of $f(x, y) = xy$ on the circle $x^2 + y^2 = 1$.

83. Find the extreme values of $f(x, y) = x^2 + 3y^2 + 2y$ on the unit disk $x^2 + y^2 \leq 1$.

84. Find the extreme values of $f(x, y) = x^2 + y^2 - 3x - xy$ on the disk $x^2 + y^2 \leq 9$.

85. Find the extreme values of $f(x, y, z) = x - y + z$ on the unit sphere $x^2 + y^2 + z^2 = 1$.

86. Find the points on the surface $z^2 - xy = 4$ closest to the origin.

87. A closed rectangular box is to have volume V cm^3. The cost of the material used in the box is a cents/cm^2 for top and bottom, b cents/cm^2 for front and back, and c cents/cm^2 for the remaining sides. What dimensions minimize the total cost of materials?

88. Find the plane $x/a + y/b + z/c = 1$ that passes through the point $(2, 1, 2)$ and cuts off the least volume from the first octant.

89. Find the extreme values of $f(x, y, z) = x(y + z)$ on the curve of intersection of the right circular cylinder $x^2 + y^2 = 1$ and the hyperbolic cylinder $xz = 1$.

90. Find the point closest to the origin on the curve of intersection of the plane $x + y + z = 1$ and the cone $z^2 = 2x^2 + 2y^2$.

Theory and Examples

91. Let $w = f(r, \theta)$, $r = \sqrt{x^2 + y^2}$, and $\theta = \tan^{-1}(y/x)$. Find $\partial w/\partial x$ and $\partial w/\partial y$ and express your answers in terms of r and θ.

92. Let $z = f(u, v)$, $u = ax + by$, and $v = ax - by$. Express z_x and z_y in terms of f_u, f_v, and the constants a and b.

93. If a and b are constants, $w = u^3 + \tanh u + \cos u$, and $u = ax + by$, show that

$$a\frac{\partial w}{\partial y} = b\frac{\partial w}{\partial x}.$$

94. If $w = \ln(x^2 + y^2 + 2z)$, $x = r + s$, $y = r - s$, and $z = 2rs$, find w_r and w_s by the Chain Rule. Then check your answer another way.

95. The equations $e^u \cos v - x = 0$ and $e^u \sin v - y = 0$ define u and v as differentiable functions of x and y. Show that the angle between the vectors

$$\frac{\partial u}{\partial x}\mathbf{i} + \frac{\partial u}{\partial y}\mathbf{j} \quad \text{and} \quad \frac{\partial v}{\partial x}\mathbf{i} + \frac{\partial v}{\partial y}\mathbf{j}$$

is constant.

96. Introducing polar coordinates $x = r\cos\theta$ and $y = r\sin\theta$ changes $f(x, y)$ to $g(r, \theta)$. Find the value of $\partial^2 g/\partial\theta^2$ at the point $(r, \theta) = (2, \pi/2)$, given that

$$\frac{\partial f}{\partial x} = \frac{\partial f}{\partial y} = \frac{\partial^2 f}{\partial x^2} = \frac{\partial^2 f}{\partial y^2} = 1$$

at that point.

97. Find the points on the surface

$$(y + z)^2 + (z - x)^2 = 16$$

where the normal line is parallel to the yz-plane.

98. Find the points on the surface

$$xy + yz + zx - x - z^2 = 0$$

where the tangent plane is parallel to the xy-plane.

99. Suppose that $\nabla f(x, y, z)$ is always parallel to the position vector $x\mathbf{i} + y\mathbf{j} + z\mathbf{k}$. Show that $f(0, 0, a) = f(0, 0, -a)$ for any a.

100. Show that the directional derivative of

$$f(x, y, z) = \sqrt{x^2 + y^2 + z^2}$$

at the origin equals 1 in any direction but that f has no gradient vector at the origin.

101. Show that the line normal to the surface $xy + z = 2$ at the point $(1, 1, 1)$ passes through the origin.

102. a) Sketch the surface $x^2 - y^2 + z^2 = 4$.
   b) Find a vector normal to the surface at $(2, -3, 3)$. Add the vector to your sketch.
   c) Find the equations for the tangent plane and normal line at $(2, -3, 3)$.

CHAPTER 12 ADDITIONAL EXERCISES–THEORY, EXAMPLES, APPLICATIONS

Partial Derivatives

1. If you did Exercise 50 in Section 12.2, you know that the function

$$f(x, y) = \begin{cases} xy\dfrac{x^2 - y^2}{x^2 + y^2}, & (x, y) \neq (0, 0) \\ 0, & (x, y) = (0, 0) \end{cases}$$

(see the accompanying figure) is continuous at $(0, 0)$. Find $f_{xy}(0, 0)$ and $f_{yx}(0, 0)$.

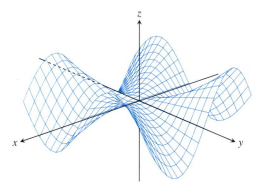

(Generated by Mathematica)

2. Find a function $w = f(x, y)$ whose first partial derivatives are $\partial w/\partial x = 1 + e^x \cos y$ and $\partial w/\partial y = 2y - e^x \sin y$, and whose value at the point $(\ln 2, 0)$ is $\ln 2$.

3. *A proof of Leibniz's rule.* Leibniz's rule says that if f is continuous on $[a, b]$ and if $u(x)$ and $v(x)$ are differentiable functions of x whose values lie in $[a, b]$, then

$$\frac{d}{dx} \int_{u(x)}^{v(x)} f(t)\, dt = f(v(x))\frac{dv}{dx} - f(u(x))\frac{du}{dx}.$$

Prove the rule by setting

$$g(u, v) = \int_u^v f(t)\, dt, \quad u = u(x), \quad v = v(x)$$

and calculating dg/dx with the Chain Rule.

4. Suppose that f is a twice-differentiable function of r, that $r = \sqrt{x^2 + y^2 + z^2}$, and that

$$f_{xx} + f_{yy} + f_{zz} = 0.$$

Show that for some constants a and b,

$$f(r) = \frac{a}{r} + b.$$

5. *Homogeneous functions.* A function $f(x, y)$ is *homogeneous of degree n* (n a nonnegative integer) if $f(tx, ty) = t^n f(x, y)$ for all $t, x,$ and y. For such a function (sufficiently differentiable), prove that

a) $x\dfrac{\partial f}{\partial x} + y\dfrac{\partial f}{\partial y} = nf(x, y)$

b) $x^2\left(\dfrac{\partial^2 f}{\partial x^2}\right) + 2xy\left(\dfrac{\partial^2 f}{\partial x \partial y}\right) + y^2\left(\dfrac{\partial^2 f}{\partial y^2}\right) = n(n-1)f.$

6. *Spherical coordinates.* Let $\mathbf{r} = x\mathbf{i} + y\mathbf{j} + z\mathbf{k}$. Express $x, y,$ and z as functions of the spherical coordinates $\rho, \phi,$ and θ and calculate $\partial\mathbf{r}/\partial\rho$, $\partial\mathbf{r}/\partial\phi$, and $\partial\mathbf{r}/\partial\theta$. Then express these derivatives in terms of the unit vectors

$$\mathbf{u}_\rho = (\sin\phi\cos\theta)\mathbf{i} + (\sin\phi\sin\theta)\mathbf{j} + (\cos\phi)\mathbf{k}$$

$$\mathbf{u}_\phi = (\cos\phi\cos\theta)\mathbf{i} + (\cos\phi\sin\theta)\mathbf{j} - (\sin\phi)\mathbf{k}$$

$$\mathbf{u}_\theta = -(\sin\theta)\mathbf{i} + (\cos\theta)\mathbf{j}.$$

Gradients and Tangents

7. Let $\mathbf{r} = x\mathbf{i} + y\mathbf{j} + z\mathbf{k}$ and let $r = |\mathbf{r}|$.

a) Show that $\nabla r = \mathbf{r}/r$.
b) Show that $\nabla(r^n) = nr^{n-2}\mathbf{r}$.
c) Find a function whose gradient equals $\mathbf{r}$.
d) Show that $\mathbf{r} \cdot d\mathbf{r} = r\, dr$.
e) Show that $\nabla(\mathbf{A} \cdot \mathbf{r}) = \mathbf{A}$ for any constant vector $\mathbf{A}$.

8. Suppose that a differentiable function $f(x, y)$ has the constant value c along the differentiable curve $x = g(t), y = h(t)$; that is,

$$f(g(t), h(t)) = c$$

for all values of t. Differentiate both sides of this equation with respect to t to show that ∇f is orthogonal to the curve's tangent vector at every point on the curve.

9. Show that the curve

$$\mathbf{r}(t) = (\ln t)\mathbf{i} + (t\ln t)\mathbf{j} + t\mathbf{k}$$

is tangent to the surface

$$xz^2 - yz + \cos xy = 1$$

at $(0, 0, 1)$.

10. Show that the curve

$$\mathbf{r}(t) = \left(\frac{t^3}{4} - 2\right)\mathbf{i} + \left(\frac{4}{t} - 3\right)\mathbf{j} + \cos(t - 2)\mathbf{k}$$

is tangent to the surface

$$x^3 + y^3 + z^3 - xyz = 0$$

at $(0, -1, 1)$.

11. *The gradient in cylindrical coordinates.* Suppose cylindrical coordinates r, θ, z are introduced into a function $w = f(x, y, z)$ to yield $w = F(r, \theta, z)$. Show that

$$\nabla w = \frac{\partial w}{\partial r}\mathbf{u}_r + \frac{1}{r}\frac{\partial w}{\partial \theta}\mathbf{u}_\theta + \frac{\partial w}{\partial z}\mathbf{k}, \qquad (1)$$

where

$$\mathbf{u}_r = (\cos\theta)\mathbf{i} + (\sin\theta)\mathbf{j}$$

$$\mathbf{u}_\theta = -(\sin\theta)\mathbf{i} + (\cos\theta)\mathbf{j}.$$

(*Hint:* Express the right-hand side of Eq. (1) in terms of $\mathbf{i}, \mathbf{j}$, and $\mathbf{k}$ and use the Chain Rule to express the components of $\mathbf{i}, \mathbf{j}$, and $\mathbf{k}$ in rectangular coordinates.)

12. *The gradient in spherical coordinates.* Suppose spherical coordinates ρ, ϕ, θ are introduced into a function $w = f(x, y, z)$ to yield a function $w = F(\rho, \phi, \theta)$. Show that

$$\nabla w = \frac{\partial w}{\partial \rho}\mathbf{u}_\rho + \frac{1}{\rho}\frac{\partial w}{\partial \phi}\mathbf{u}_\phi + \frac{1}{\rho\sin\phi}\frac{\partial w}{\partial \theta}\mathbf{u}_\theta, \qquad (2)$$

where

$$\mathbf{u}_\rho = (\sin\phi\cos\theta)\mathbf{i} + (\sin\phi\sin\theta)\mathbf{j} + (\cos\phi)\mathbf{k}$$

$$\mathbf{u}_\phi = (\cos\phi\cos\theta)\mathbf{i} + (\cos\phi\sin\theta)\mathbf{j} - (\sin\phi)\mathbf{k}$$

$$\mathbf{u}_\theta = -(\sin\theta)\mathbf{i} + (\cos\theta)\mathbf{j}.$$

(*Hint:* Express the right-hand side of Eq. (2) in terms of $\mathbf{i}, \mathbf{j}$, and $\mathbf{k}$ and use the Chain Rule to express the components of $\mathbf{i}, \mathbf{j}$, and $\mathbf{k}$ in rectangular coordinates.)

Extreme Values

13. Show that the only possible maxima and minima of z on the surface $z = x^3 + y^3 - 9xy + 27$ occur at $(0, 0)$ and $(3, 3)$. Show that neither a maximum nor a minimum occurs at $(0, 0)$. Determine whether z has a maximum or a minimum at $(3, 3)$.

14. Find the maximum value of $f(x, y) = 6xye^{-(2x+3y)}$ in the closed first quadrant (includes the nonnegative axes).

15. Find the minimum volume for a region bounded by the planes $x = 0, y = 0, z = 0$ and a plane tangent to the ellipsoid

$$\frac{x^2}{a^2} + \frac{y^2}{b^2} + \frac{z^2}{c^2} = 1$$

at a point in the first octant.

16. By minimizing the function $f(x, y, u, v) = (x - u)^2 + (y - v)^2$ subject to the constraints $y = x + 1$ and $u = v^2$, find the minimum distance in the xy-plane from the line $y = x + 1$ to the parabola $y^2 = x$.

Theory and Examples

17. Prove the following theorem: If $f(x, y)$ is defined in an open region R of the xy-plane, and if f_x and f_y are bounded on R, then $f(x, y)$ is continuous on R. (The assumption of boundedness is essential.)

18. Suppose $\mathbf{r}(t) = g(t)\mathbf{i} + h(t)\mathbf{j} + k(t)\mathbf{k}$ is a smooth curve in the domain of a differentiable function $f(x, y, z)$. Describe the relation between df/dt, ∇f, and $\mathbf{v} = d\mathbf{r}/dt$. What can be said about ∇f and $\mathbf{v}$ at interior points of the curve where f has extreme values relative to its other values on the curve? Give reasons for your answer.

19. Suppose that f and g are functions of x and y such that

$$\frac{\partial f}{\partial y} = \frac{\partial g}{\partial x} \quad \text{and} \quad \frac{\partial f}{\partial x} = \frac{\partial g}{\partial y},$$

and suppose that

$$\frac{\partial f}{\partial x} = 0, \quad f(1, 2) = g(1, 2) = 5, \quad \text{and} \quad f(0, 0) = 4.$$

Find $f(x, y)$ and $g(x, y)$.

20. We know that if $f(x, y)$ is a function of two variables and if $\mathbf{u} = a\mathbf{i} + b\mathbf{j}$ is a unit vector, then $D_\mathbf{u}f(x, y) = f_x(x, y)a + f_y(x, y)b$ is the rate of change of $f(x, y)$ at (x, y) in the direction of $\mathbf{u}$. Give a similar formula for the rate of change *of the rate of change* of $f(x, y)$ at (x, y) in the direction $\mathbf{u}$.

21. *Path of a heat-seeking particle.* A heat-seeking particle has the property that at any point (x, y) in the plane it moves in the direction of maximum temperature increase. If the temperature at (x, y) is $T(x, y) = -e^{-2y}\cos x$, find an equation $y = f(x)$ for the path of a heat-seeking particle at the point $(\pi/4, 0)$.

22. A particle traveling in a straight line with constant velocity $\mathbf{i} + \mathbf{j} - 5\mathbf{k}$ passes through the point $(0, 0, 30)$ and hits the surface $z = 2x^2 + 3y^2$. The particle ricochets off the surface, the angle of reflection being equal to the angle of incidence. Assuming no loss of speed, what is the velocity of the particle after the ricochet? Simplify your answer.

23. Let S be the surface that is the graph of $f(x, y) = 10 - x^2 - y^2$. Suppose the temperature in space at each point (x, y, z) is $T(x, y, z) = x^2y + y^2z + 4x + 14y + z$.

  a) Among all of the possible directions tangential to the surface S at the point $(0, 0, 10)$, which direction will make the rate of change of temperature at $(0, 0, 10)$ a maximum?

  b) Which direction tangential to S at the point $(1, 1, 8)$ will make the rate of change of temperature a maximum?

24. On a flat surface of land, geologists drilled a borehole straight down and hit a mineral deposit at 1000 ft. They drilled a second borehole 100 ft to the north of the first and hit the mineral deposit at 950 ft. A third borehole 100 ft east of the first borehole struck the mineral deposit at 1025 ft. The geologists have reasons to believe that the mineral deposit is in the shape of a dome and for the sake of economy they would like to find where the deposit is closest to the surface. Assuming the surface to be the xy-plane, in what direction from the first borehole would you suggest the geologists drill their fourth borehole?

The One-Dimensional Heat Equation

If $w(x, t)$ represents the temperature at position x at time t in a uniform conducting rod with perfectly insulated sides (see the accompanying figure), then the partial derivatives w_{xx} and w_t satisfy a differential equation of the form

$$w_{xx} = \frac{1}{c^2} w_t. \tag{3}$$

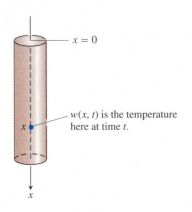

$x = 0$

$w(x, t)$ is the temperature here at time t.

x

x

This equation is called the **one-dimensional heat equation.** The value of the positive constant c^2 is determined by the material from which

the rod is made. It has been determined experimentally for a broad range of materials. For a given application one finds the appropriate value in a table. For dry soil, for example, $c^2 = 0.19$ ft²/day.

In chemistry and biochemistry, the heat equation is known as the **diffusion equation.** In this context, $w(x, t)$ represents the concentration of a dissolved substance, a salt for instance, diffusing along a tube filled with liquid. The value of $w(x, t)$ is the concentration at point x at time t. In other applications, $w(x, t)$ represents the diffusion of a gas down a long, thin pipe.

In electrical engineering, the heat equation appears in the forms

$$v_{xx} = RCv_t \tag{4}$$

and

$$i_{xx} = RCi_t, \tag{5}$$

which are known as the **telegraph equations.** These equations describe the voltage v and the flow of current i in a coaxial cable or in any other cable in which leakage and inductance are negligible. The functions and constants in these equations are

$$v(x, t) = \text{voltage at point } x \text{ at time } t$$

$$R = \text{resistance per unit length}$$

$$C = \text{capacitance to ground per unit of cable length}$$

$$i(x, t) = \text{current at point } x \text{ at time } t.$$

25. Find all solutions of the one-dimensional heat equation of the form $w = e^{rt} \sin \pi x$, where r is a constant.

26. Find all solutions of the one-dimensional heat equation that have the form $w = e^{rt} \sin kx$ and satisfy the conditions that $w(0, t) = 0$ and $w(L, t) = 0$. What happens to these solutions as $t \to \infty$?

CHAPTER 13

Multiple Integrals

OVERVIEW The problems we can solve by integrating functions of two and three variables are similar to the problems solved by single-variable integration, but more general. As in the previous chapter, we can perform the necessary calculations by drawing on our experience with functions of a single variable.

13.1 Double Integrals

We now show how to integrate a continuous function $f(x, y)$ over a bounded region in the xy-plane. There are many similarities between the "double" integrals we define here and the "single" integrals we defined in Chapter 4 for functions of a single variable. Every double integral can be evaluated in stages, using the single-integration methods already at our command.

Double Integrals over Rectangles

Suppose that $f(x, y)$ is defined on a rectangular region R given by

$$R: \quad a \le x \le b, \quad c \le y \le d.$$

We imagine R to be covered by a network of lines parallel to the x- and y-axes (Fig. 13.1). These lines divide R into small pieces of area $\Delta A = \Delta x \Delta y$. We number these in some order $\Delta A_1, \Delta A_2, \ldots, \Delta A_n$, choose a point (x_k, y_k) in each piece ΔA_k, and form the sum

$$S_n = \sum_{k=1}^{n} f(x_k, y_k) \, \Delta A_k. \tag{1}$$

If f is continuous throughout R, then, as we refine the mesh width to make both Δx and Δy go to zero, the sums in (1) approach a limit called the **double integral** of f over R. The notation for it is

$$\iint_R f(x, y) \, dA \quad \text{or} \quad \iint_R f(x, y) \, dx \, dy.$$

Thus,

$$\iint_R f(x, y) \, dA = \lim_{\Delta A \to 0} \sum_{k=1}^{n} f(x_k, y_k) \Delta A_k. \tag{2}$$

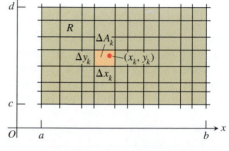

13.1 Rectangular grid partitioning the region R into small rectangles of area $\Delta A_k = \Delta x_k \Delta y_k$.

As with functions of a single variable, the sums approach this limit no matter how the intervals $[a, b]$ and $[c, d]$ that determine R are partitioned, as long as the norms of the partitions both go to zero. The limit in (2) is also independent of the order in which the areas ΔA_k are numbered and independent of the choice of the point (x_k, y_k) within each ΔA_k. The values of the individual approximating sums S_n depend on these choices, but the sums approach the same limit in the end. The proof of the existence and uniqueness of this limit for a continuous function f is given in more advanced texts. The continuity of f is a sufficient condition for the existence of the double integral, but not a necessary one. The limit in question exists for many discontinuous functions as well.

Properties of Double Integrals

Like single integrals, double integrals of continuous functions have algebraic properties that are useful in computations and applications.

1. $$\iint_R kf(x, y)\, dA = k \iint_R f(x, y)\, dA \qquad \text{(any number } k\text{)}$$

2. $$\iint_R (f(x, y) \pm g(x, y))\, dA = \iint_R f(x, y)\, dA \pm \iint_R g(x, y)\, dA$$

3. $$\iint_R f(x, y)\, dA \geq 0 \quad \text{if} \quad f(x, y) \geq 0 \text{ on } R$$

4. $$\iint_R f(x, y)\, dA \geq \iint_R g(x, y)\, dA \quad \text{if} \quad f(x, y) \geq g(x, y) \text{ on } R$$

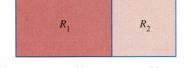

$$\iint_{R_1 \cup R_2} f(x, y)\, dA = \iint_{R_1} f(x, y)\, dA + \iint_{R_2} f(x, y)\, dA$$

13.2 Double integrals have the same kind of domain additivity property that single integrals have.

These are like the single-integral properties in Section 4.5. There is also an additivity property:

5. $$\iint_R f(x, y)\, dA = \iint_{R_1} f(x, y)\, dA + \iint_{R_2} f(x, y)\, dA.$$

It holds when R is the union of two nonoverlapping rectangles R_1 and R_2 (Fig. 13.2). Again, we omit the proof.

Double Integrals as Volumes

When $f(x, y)$ is positive, we may interpret the double integral of f over a rectangular region R as the volume of the solid prism bounded below by R and above by the surface $z = f(x, y)$ (Fig. 13.3). Each term $f(x_k, y_k)\, \Delta A_k$ in the sum $S_n = \sum f(x_k, y_k)\, \Delta A_k$ is the volume of a vertical rectangular prism that approximates the volume of the portion of the solid that stands directly above the base ΔA_k. The sum S_n thus approximates what we want to call the total volume of the solid. We *define* this volume to be

$$\text{Volume} = \lim S_n = \iint_R f(x, y)\, dA. \qquad (3)$$

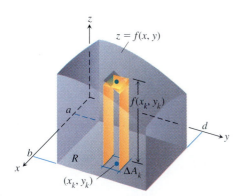

13.3 Approximating solids with rectangular prisms leads us to define the volumes of more general prisms as double integrals. The volume of the prism shown here is the double integral of $f(x, y)$ over the base region R.

As you might expect, this more general method of calculating volume agrees with the methods in Chapter 5, but we will not prove this here.

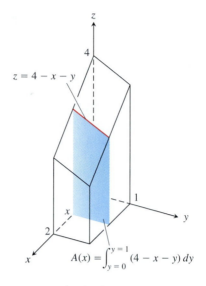

13.4 To obtain the cross-section area $A(x)$, we hold x fixed and integrate with respect to y.

Fubini's Theorem for Calculating Double Integrals

Suppose we wish to calculate the volume under the plane $z = 4 - x - y$ over the rectangular region $R: 0 \le x \le 2, \ 0 \le y \le 1$ in the xy-plane. If we apply the method of slicing from Section 5.2, with slices perpendicular to the x-axis (Fig. 13.4), then the volume is

$$\int_{x=0}^{x=2} A(x) \ dx, \qquad (4)$$

where $A(x)$ is the cross-section area at x. For each value of x we may calculate $A(x)$ as the integral

$$A(x) = \int_{y=0}^{y=1} (4 - x - y) \, dy, \qquad (5)$$

which is the area under the curve $z = 4 - x - y$ in the plane of the cross section at x. In calculating $A(x)$, x is held fixed and the integration takes place with respect to y. Combining (4) and (5), we see that the volume of the entire solid is

$$\text{Volume} = \int_{x=0}^{x=2} A(x) \, dx = \int_{x=0}^{x=2} \left(\int_{y=0}^{y=1} (4 - x - y) \, dy \right) dx$$

$$= \int_{x=0}^{x=2} \left[4y - xy - \frac{y^2}{2} \right]_{y=0}^{y=1} dx = \int_{x=0}^{x=2} \left(\frac{7}{2} - x \right) dx = \left[\frac{7}{2}x - \frac{x^2}{2} \right]_0^2 = 5. \qquad (6)$$

If we had just wanted to write instructions for calculating the volume, without carrying out any of the integrations, we could write

$$\text{Volume} = \int_0^2 \int_0^1 (4 - x - y) \, dy \, dx.$$

The expression on the right, called an **iterated** or **repeated integral,** says that the volume is obtained by integrating $4 - x - y$ with respect to y from $y = 0$ to $y = 1$, holding x fixed, and then integrating the resulting expression in x with respect to x from $x = 0$ to $x = 2$.

What would have happened if we had calculated the volume by slicing with planes perpendicular to the y-axis (Fig. 13.5)? As a function of y, the typical cross-section area is

$$A(y) = \int_{x=0}^{x=2} (4 - x - y) \, dx = \left[4x - \frac{x^2}{2} - xy \right]_{x=0}^{x=2} = 6 - 2y. \qquad (7)$$

The volume of the entire solid is therefore

$$\text{Volume} = \int_{y=0}^{y=1} A(y) \, dy = \int_{y=0}^{y=1} (6 - 2y) \, dy = \left[6y - y^2 \right]_0^1 = 5,$$

in agreement with our earlier calculation.

Again, we may give instructions for calculating the volume as an iterated integral by writing

$$\text{Volume} = \int_0^1 \int_0^2 (4 - x - y) \, dx \, dy.$$

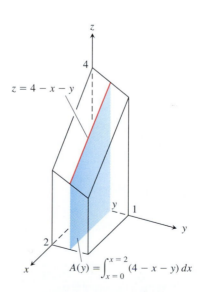

13.5 To obtain the cross-section area $A(y)$, we hold y fixed and integrate with respect to x.

The expression on the right says we can find the volume by integrating $4 - x - y$ with respect to x from $x = 0$ to $x = 2$ (as in Eq. 7) and integrating the result with respect to y from $y = 0$ to $y = 1$. In this iterated integral the order of integration is first x and then y, the reverse of the order in Eq. (6).

What do these two volume calculations with iterated integrals have to do with the double integral

$$\iint\limits_{R} (4 - x - y) \, dA$$

over the rectangle $R : 0 \le x \le 2, \ 0 \le y \le 1$? The answer is that they both give the value of the double integral. A theorem published in 1907 by Guido Fubini (1879–1943) says that the double integral of any continuous function over a rectangle can be calculated as an iterated integral in either order of integration. (Fubini proved his theorem in greater generality, but this is how it translates into what we're doing now.)

Theorem 1

Fubini's Theorem (First Form)

If $f(x, y)$ is continuous on the rectangular region $R : a \le x \le b, \ c \le y \le d$, then

$$\iint\limits_{R} f(x, y) \, dA = \int_{c}^{d} \int_{a}^{b} f(x, y) \, dx \, dy = \int_{a}^{b} \int_{c}^{d} f(x, y) \, dy \, dx.$$

Fubini's theorem says that double integrals over rectangles can be calculated as iterated integrals. This means we can evaluate a double integral by integrating with respect to one variable at a time.

Fubini's theorem also says that we may calculate the double integral by integrating in *either* order, a genuine convenience, as we will see in Example 3. In particular, when we calculate a volume by slicing, we may use either planes perpendicular to the x-axis or planes perpendicular to the y-axis.

EXAMPLE 1   Calculate $\iint_{R} f(x, y) \, dA$ for

$$f(x, y) = 1 - 6x^2 y \qquad \text{and} \qquad R : \quad 0 \le x \le 2, \quad -1 \le y \le 1.$$

Solution   By Fubini's theorem,

$$\iint\limits_{R} f(x, y) \, dA = \int_{-1}^{1} \int_{0}^{2} (1 - 6x^2 y) \, dx \, dy = \int_{-1}^{1} \left[x - 2x^3 y \right]_{x=0}^{x=2} dy$$

$$= \int_{-1}^{1} (2 - 16y) \, dy = \left[2y - 8y^2 \right]_{-1}^{1} = 4.$$

Reversing the order of integration gives the same answer:

$$\int_{0}^{2} \int_{-1}^{1} (1 - 6x^2 y) \, dy \, dx = \int_{0}^{2} \left[y - 3x^2 y^2 \right]_{y=-1}^{y=1} dx$$

$$= \int_{0}^{2} \left[(1 - 3x^2) - (-1 - 3x^2) \right] dx = \int_{0}^{2} 2 \, dx = 4.$$

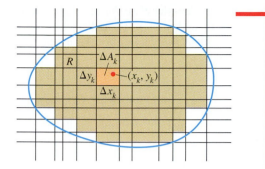

13.6 A rectangular grid partitioning a bounded nonrectangular region into cells.

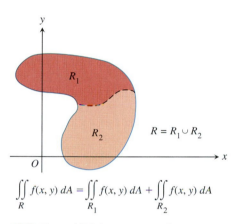

$$\iint\limits_R f(x, y)\, dA = \iint\limits_{R_1} f(x, y)\, dA + \iint\limits_{R_2} f(x, y)\, dA$$

13.7 The additivity property for rectangular regions holds for regions bounded by continuous curves.

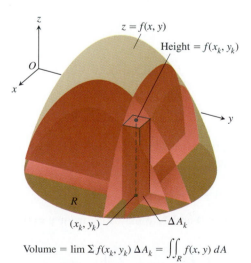

Volume $= \lim \Sigma f(x_k, y_k)\, \Delta A_k = \iint_R f(x, y)\, dA$

13.8 We define the volumes of solids with curved bases the same way we define the volumes of solids with rectangular bases.

Technology *Multiple Integration* Most Computer Algebra Systems can calculate both multiple and iterated integrals. The typical procedure is to apply the CAS integrate command in nested iterations according to the order of integration you specify:

Integral	Typical CAS Formulation
$\displaystyle \iint x^2 y\, dx\, dy$	int(int(x ^ 2 * y, x), y);
$\displaystyle \int_{-\pi/3}^{\pi/4} \int_0^1 x \cos y\, dx\, dy$	int(int(x* cos(y), x = 0 . . 1), y = −Pi/3 . . Pi/4);

If a CAS cannot produce an exact value for a definite integral, it can usually find an approximate value numerically.

Double Integrals over Bounded Nonrectangular Regions

To define the double integral of a function $f(x, y)$ over a bounded nonrectangular region, like the one shown in Fig. 13.6, we again imagine R to be covered by a rectangular grid, but we include in the partial sum only the small pieces of area $\Delta A = \Delta x\, \Delta y$ that lie entirely within the region (shaded in the figure). We number the pieces in some order, choose an arbitrary point (x_k, y_k) in each ΔA_k, and form the sum

$$S_n = \sum_{k=1}^n f(x_k, y_k)\, \Delta A_k.$$

The only difference between this sum and the one in Eq. (1) for rectangular regions is that now the areas ΔA_k may not cover all of R. But as the mesh becomes increasingly fine and the number of terms in S_n increases, more and more of R is included. If f is continuous and the boundary of R is made from the graphs of a finite number of continuous functions of x and/or continuous functions of y joined end to end, then the sums S_n will have a limit as the norms of the partitions that define the rectangular grid independently approach zero. We call the limit the **double integral** of f over R:

$$\iint\limits_R f(x, y)\, dA = \lim_{\Delta A \to 0} \sum f(x_k, y_k)\, \Delta A_k.$$

This limit may also exist under less restrictive circumstances.

Double integrals of continuous functions over nonrectangular regions have the same algebraic properties as integrals over rectangular regions. The domain additivity property corresponding to property 5 says that if R is decomposed into nonoverlapping regions R_1 and R_2 with boundaries that are again made of a finite number of line segments or smooth curves (see Fig. 13.7 for an example), then

$$\iint\limits_R f(x, y)\, dA = \iint\limits_{R_1} f(x, y)\, dA + \iint\limits_{R_2} f(x, y)\, dA.$$

If $f(x, y)$ is positive and continuous over R (Fig. 13.8), we define the volume of the solid region between R and the surface $z = f(x, y)$ to be $\iint_R f(x, y)\, dA$, as before.

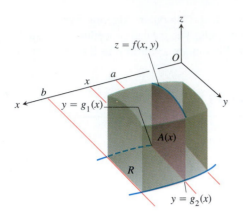

13.9 The area of the vertical slice shown here is

$$A(x) = \int_{g_1(x)}^{g_2(x)} f(x, y)\, dy.$$

To calculate the volume of the solid, we integrate this area from $x = a$ to $x = b$.

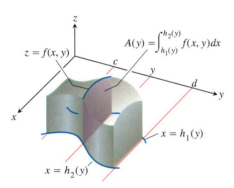

13.10 The volume of the solid shown here is

$$\int_{c}^{d} A(y)\, dy = \int_{c}^{d} \int_{h_1(y)}^{h_2(y)} f(x, y)\, dx\, dy.$$

If R is a region like the one shown in the xy-plane in Fig. 13.9, bounded "above" and "below" by the curves $y = g_2(x)$ and $y = g_1(x)$ and on the sides by the lines $x = a$, $x = b$, we may again calculate the volume by the method of slicing. We first calculate the cross-section area

$$A(x) = \int_{y=g_1(x)}^{y=g_2(x)} f(x, y)\, dy$$

and then integrate $A(x)$ from $x = a$ to $x = b$ to get the volume as an iterated integral:

$$V = \int_{a}^{b} A(x)\, dx = \int_{a}^{b} \int_{g_1(x)}^{g_2(x)} f(x, y)\, dy\, dx. \tag{8}$$

Similarly, if R is a region like the one shown in Fig. 13.10, bounded by the curves $x = h_2(y)$ and $x = h_1(y)$ and the lines $y = c$ and $y = d$, then the volume calculated by slicing is given by the iterated integral

$$\text{Volume} = \int_{c}^{d} \int_{h_1(y)}^{h_2(y)} f(x, y)\, dx\, dy. \tag{9}$$

The fact that the iterated integrals in Eqs. (8) and (9) both give the volume that we defined to be the double integral of f over R is a consequence of the following stronger form of Fubini's theorem.

> ### Theorem 2
> ### Fubini's Theorem (Stronger Form)
>
> Let $f(x, y)$ be continuous on a region R.
>
> **1.** If R is defined by $a \le x \le b$, $g_1(x) \le y \le g_2(x)$, with g_1 and g_2 continuous on $[a, b]$, then
>
> $$\iint_R f(x, y)\, dA = \int_{a}^{b} \int_{g_1(x)}^{g_2(x)} f(x, y)\, dy\, dx.$$
>
> **2.** If R is defined by $c \le y \le d$, $h_1(y) \le x \le h_2(y)$, with h_1 and h_2 continuous on $[c, d]$, then
>
> $$\iint_R f(x, y)\, dA = \int_{c}^{d} \int_{h_1(y)}^{h_2(y)} f(x, y)\, dx\, dy.$$

EXAMPLE 2   Find the volume of the prism whose base is the triangle in the xy-plane bounded by the x-axis and the lines $y = x$ and $x = 1$ and whose top lies in the plane

$$z = f(x, y) = 3 - x - y.$$

Solution   See Fig. 13.11. For any x between 0 and 1, y may vary from $y = 0$ to $y = x$ (Fig. 13.11b). Hence,

$$V = \int_{0}^{1} \int_{0}^{x} (3 - x - y)\, dy\, dx = \int_{0}^{1} \left[3y - xy - \frac{y^2}{2} \right]_{y=0}^{y=x} dx$$

$$= \int_{0}^{1} \left(3x - \frac{3x^2}{2} \right) dx = \left[\frac{3x^2}{2} - \frac{x^3}{2} \right]_{x=0}^{x=1} = 1.$$

(a)

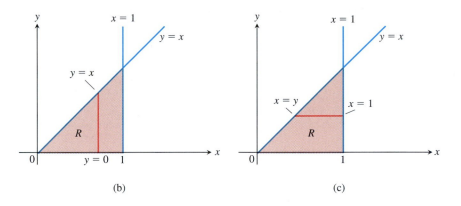

(b)

(c)

13.11 (a) Prism with a triangular base in the *xy*-plane. The volume of this prism is defined as a double integral over *R*. To evaluate it as an iterated integral, we may integrate first with respect to *y* and then with respect to *x*, or the other way around (Example 2). (b) Integration limits of

$$\int_{x=0}^{x=1} \int_{y=0}^{y=x} f(x, y)\, dy\, dx.$$

If we integrate first with respect to *y*, we integrate along a vertical line through *R* and then integrate from left to right to include all the vertical lines in *R*. (c) Integration limits of

$$\int_{y=0}^{y=1} \int_{x=y}^{x=1} f(x, y)\, dx\, dy.$$

If we integrate first with respect to *x*, we integrate along a horizontal line through *R* and then integrate from bottom to top to include all the horizontal lines in *R*.

When the order of integration is reversed (Fig. 13.11c), the integral for the volume is

$$V = \int_0^1 \int_y^1 (3 - x - y)\, dx\, dy = \int_0^1 \left[3x - \frac{x^2}{2} - xy \right]_{x=y}^{x=1} dy$$

$$= \int_0^1 \left(3 - \frac{1}{2} - y - 3y + \frac{y^2}{2} + y^2 \right) dy$$

$$= \int_0^1 \left(\frac{5}{2} - 4y + \frac{3}{2}y^2 \right) dy = \left[\frac{5}{2}y - 2y^2 + \frac{y^3}{2} \right]_{y=0}^{y=1} = 1.$$

The two integrals are equal, as they should be. ❑

While Fubini's theorem assures us that a double integral may be calculated as an iterated integral in either order of integration, the value of one integral may be easier to find than the value of the other. The next example shows how this can happen.

EXAMPLE 3 Calculate

$$\iint_R \frac{\sin x}{x}\, dA,$$

where *R* is the triangle in the *xy*-plane bounded by the *x*-axis, the line *y = x*, and the line *x = 1*.

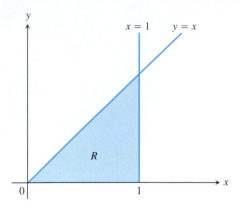

13.12 The region of integration in Example 3.

Solution The region of integration is shown in Fig. 13.12. If we integrate first with respect to y and then with respect to x, we find

$$\int_0^1 \left(\int_0^x \frac{\sin x}{x} \, dy \right) dx = \int_0^1 \left(y \, \frac{\sin x}{x} \right]_{y=0}^{y=x} \right) dx = \int_0^1 \sin x \, dx$$

$$= -\cos(1) + 1 \approx 0.46.$$

If we reverse the order of integration and attempt to calculate

$$\int_0^1 \int_y^1 \frac{\sin x}{x} \, dx \, dy,$$

we are stopped by the fact that $\int ((\sin x)/x) \, dx$ cannot be expressed in terms of elementary functions.

There is no general rule for predicting which order of integration will be the good one in circumstances like these, so don't worry about how to start your integrations. Just forge ahead and if the order you first choose doesn't work, try the other. □

Finding the Limits of Integration

The hardest part of evaluating a double integral can be finding the limits of integration. Fortunately, there is a good procedure to follow.

Procedure for Finding Limits of Integration

A. To evaluate $\iint_R f(x, y) \, dA$ over a region R, integrating first with respect to y and then with respect to x, take the following steps:

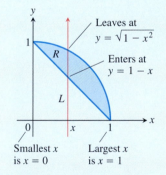

1. *A sketch.* Sketch the region of integration and label the bounding curves.

2. *The y-limits of integration.* Imagine a vertical line L cutting through R in the direction of increasing y. Mark the y-values where L enters and leaves. These are the y-limits of integration.

3. *The x-limits of integration.* Choose x-limits that include all the vertical lines through R. The integral is

$$\iint_R f(x, y) \, dA =$$

$$\int_{x=0}^{x=1} \int_{y=1-x}^{y=\sqrt{1-x^2}} f(x, y) \, dy \, dx.$$

B. To evaluate the same double integral as an iterated integral with the order of integration reversed, use horizontal lines instead of vertical lines. The integral is

$$\iint\limits_{R} f(x, y)\, dA = \int_{0}^{1} \int_{1-y}^{\sqrt{1-y^2}} f(x, y)\, dx\, dy.$$

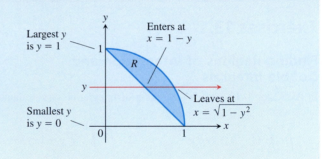

EXAMPLE 4   Sketch the region of integration for the integral

$$\int_{0}^{2} \int_{x^2}^{2x} (4x + 2)\, dy\, dx$$

and write an equivalent integral with the order of integration reversed.

Solution   The region of integration is given by the inequalities $x^2 \leq y \leq 2x$ and $0 \leq x \leq 2$. It is therefore the region bounded by the curves $y = x^2$ and $y = 2x$ between $x = 0$ and $x = 2$ (Fig. 13.13a).

To find limits for integrating in the reverse order, we imagine a horizontal line passing from left to right through the region. It enters at $x = y/2$ and leaves at $x = \sqrt{y}$. To include all such lines, we let y run from $y = 0$ to $y = 4$ (Fig. 13.13b). The integral is

$$\int_{0}^{4} \int_{y/2}^{\sqrt{y}} (4x + 2)\, dx\, dy.$$

The common value of these integrals is 8.

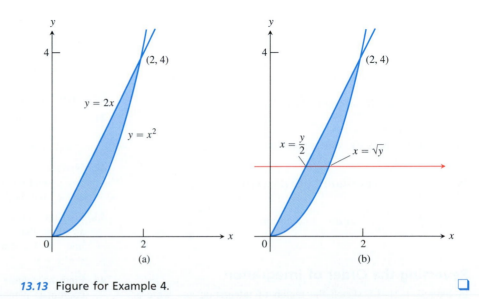

13.13 Figure for Example 4.

Exercises 13.1

Finding Regions of Integration and Double Integrals

In Exercises 1–10, sketch the region of integration and evaluate the integral.

1. $\displaystyle\int_0^3 \int_0^2 (4 - y^2)\, dy\, dx$

2. $\displaystyle\int_0^3 \int_{-2}^0 (x^2 y - 2xy)\, dy\, dx$

3. $\displaystyle\int_{-1}^0 \int_{-1}^1 (x + y + 1)\, dx\, dy$

4. $\displaystyle\int_\pi^{2\pi} \int_0^\pi (\sin x + \cos y)\, dx\, dy$

5. $\displaystyle\int_0^\pi \int_0^x x \sin y\, dy\, dx$

6. $\displaystyle\int_0^\pi \int_0^{\sin x} y\, dy\, dx$

7. $\displaystyle\int_1^{\ln 8} \int_0^{\ln y} e^{x+y}\, dx\, dy$

8. $\displaystyle\int_1^2 \int_y^{y^2} dx\, dy$

9. $\displaystyle\int_0^1 \int_0^{y^2} 3y^3 e^{xy}\, dx\, dy$

10. $\displaystyle\int_1^4 \int_0^{\sqrt{x}} \frac{3}{2} e^{y/\sqrt{x}}\, dy\, dx$

In Exercises 11–16, integrate f over the given region.

11. $f(x, y) = x/y$ over the region in the first quadrant bounded by the lines $y = x$, $y = 2x$, $x = 1$, $x = 2$

12. $f(x, y) = 1/(xy)$ over the square $1 \le x \le 2, 1 \le y \le 2$

13. $f(x, y) = x^2 + y^2$ over the triangular region with vertices $(0, 0)$, $(1, 0)$, and $(0, 1)$

14. $f(x, y) = y \cos xy$ over the rectangle $0 \le x \le \pi, 0 \le y \le 1$

15. $f(u, v) = v - \sqrt{u}$ over the triangular region cut from the first quadrant of the uv-plane by the line $u + v = 1$

16. $f(s, t) = e^s \ln t$ over the region in the first quadrant of the st-plane that lies above the curve $s = \ln t$ from $t = 1$ to $t = 2$

Each of Exercises 17–20 gives an integral over a region in a Cartesian coordinate plane. Sketch the region and evaluate the integral.

17. $\displaystyle\int_{-2}^0 \int_v^{-v} 2\, dp\, dv$ (the pv-plane)

18. $\displaystyle\int_0^1 \int_0^{\sqrt{1-s^2}} 8t\, dt\, ds$ (the st-plane)

19. $\displaystyle\int_{-\pi/3}^{\pi/3} \int_0^{\sec t} 3 \cos t\, du\, dt$ (the tu-plane)

20. $\displaystyle\int_0^3 \int_{-2}^{4-2u} \frac{4 - 2u}{v^2}\, dv\, du$ (the uv-plane)

Reversing the Order of Integration

In Exercises 21–30, sketch the region of integration and write an equivalent double integral with the order of integration reversed.

21. $\displaystyle\int_0^1 \int_2^{4-2x} dy\, dx$

22. $\displaystyle\int_0^2 \int_{y-2}^0 dx\, dy$

23. $\displaystyle\int_0^1 \int_y^{\sqrt{y}} dx\, dy$

24. $\displaystyle\int_0^1 \int_{1-x}^{1-x^2} dy\, dx$

25. $\displaystyle\int_0^1 \int_1^{e^x} dy\, dx$

26. $\displaystyle\int_0^{\ln 2} \int_{e^y}^2 dx\, dy$

27. $\displaystyle\int_0^{3/2} \int_0^{9-4x^2} 16x\, dy\, dx$

28. $\displaystyle\int_0^2 \int_0^{4-y^2} y\, dx\, dy$

29. $\displaystyle\int_0^1 \int_{-\sqrt{1-y^2}}^{\sqrt{1-y^2}} 3y\, dx\, dy$

30. $\displaystyle\int_0^2 \int_{-\sqrt{4-x^2}}^{\sqrt{4-x^2}} 6x\, dy\, dx$

Evaluating Double Integrals

In Exercises 31–40, sketch the region of integration, determine the order of integration, and evaluate the integral.

31. $\displaystyle\int_0^\pi \int_x^\pi \frac{\sin y}{y}\, dy\, dx$

32. $\displaystyle\int_0^2 \int_x^2 2y^2 \sin xy\, dy\, dx$

33. $\displaystyle\int_0^1 \int_y^1 x^2 e^{xy}\, dx\, dy$

34. $\displaystyle\int_0^2 \int_0^{4-x^2} \frac{xe^{2y}}{4 - y}\, dy\, dx$

35. $\displaystyle\int_0^{2\sqrt{\ln 3}} \int_{y/2}^{\sqrt{\ln 3}} e^{x^2}\, dx\, dy$

36. $\displaystyle\int_0^3 \int_{\sqrt{x/3}}^1 e^{y^3}\, dy\, dx$

37. $\displaystyle\int_0^{1/16} \int_{y^{1/4}}^{1/2} \cos(16\pi x^5)\, dx\, dy$

38. $\displaystyle\int_0^8 \int_{\sqrt[3]{x}}^2 \frac{dy\, dx}{y^4 + 1}$

39. $\displaystyle\iint_R (y - 2x^2)\, dA$ where R is the region inside the square $|x| + |y| = 1$

40. $\displaystyle\iint_R xy\, dA$ where R is the region bounded by the lines $y = x$, $y = 2x$, and $x + y = 2$

Volume Beneath a Surface $z = f(x, y)$

41. Find the volume of the region that lies under the paraboloid $z = x^2 + y^2$ and above the triangle enclosed by the lines $y = x$, $x = 0$, and $x + y = 2$ in the xy-plane.

42. Find the volume of the solid that is bounded above by the cylinder $z = x^2$ and below by the region enclosed by the parabola $y = 2 - x^2$ and the line $y = x$ in the xy-plane.

43. Find the volume of the solid whose base is the region in the xy-plane that is bounded by the parabola $y = 4 - x^2$ and the line $y = 3x$, while the top of the solid is bounded by the plane $z = x + 4$.

44. Find the volume of the solid in the first octant bounded by the coordinate planes, the cylinder $x^2 + y^2 = 4$, and the plane $z + y = 3$.

45. Find the volume of the solid in the first octant bounded by the coordinate planes, the plane $x = 3$, and the parabolic cylinder $z = 4 - y^2$.

46. Find the volume of the solid cut from the first octant by the surface $z = 4 - x^2 - y$.

47. Find the volume of the wedge cut from the first octant by the cylinder $z = 12 - 3y^2$ and the plane $x + y = 2$.

48. Find the volume of the solid cut from the square column $|x| + |y| \le 1$ by the planes $z = 0$ and $3x + z = 3$.

49. Find the volume of the solid that is bounded on the front and back by the planes $x = 2$ and $x = 1$, on the sides by the cylinders $y = \pm 1/x$, and above and below by the planes $z = x + 1$ and $z = 0$.

50. Find the volume of the solid that is bounded on the front and back by the planes $x = \pm \pi/3$, on the sides by the cylinders $y = \pm \sec x$, above by the cylinder $z = 1 + y^2$, and below by the xy-plane.

Integrals over Unbounded Regions

Evaluate the improper integrals in Exercises 51–54 as iterated integrals.

51. $\displaystyle \int_1^{\infty} \int_{e^{-x}}^1 \frac{1}{x^3 y} \, dy \, dx$

52. $\displaystyle \int_{-1}^1 \int_{-1/\sqrt{1-x^2}}^{1/\sqrt{1-x^2}} (2y + 1) \, dy \, dx$

53. $\displaystyle \int_{-\infty}^{\infty} \int_{-\infty}^{\infty} \frac{1}{(x^2 + 1)(y^2 + 1)} \, dx \, dy$

54. $\displaystyle \int_0^{\infty} \int_0^{\infty} x e^{-(x+2y)} \, dx \, dy$

Approximating Double Integrals

In Exercises 55 and 56, approximate the double integral of $f(x, y)$ over the region R partitioned by the given vertical lines $x = a$ and horizontal lines $y = c$. In each subrectangle use (x_k, y_k) as indicated for your approximation.

$$\iint_R f(x, y) \, dA \approx \sum_{k=1}^n f(x_k, y_k) \, \Delta A_k$$

55. $f(x, y) = x + y$ over the region R bounded above by the semicircle $y = \sqrt{1 - x^2}$ and below by the x-axis, using the partition $x = -1, -1/2, 0, 1/4, 1/2, 1$ and $y = 0, 1/2, 1$ with (x_k, y_k) the lower left corner in the kth subrectangle (provided the subrectangle lies within R)

56. $f(x, y) = x + 2y$ over the region R inside the circle $(x - 2)^2 + (y - 3)^2 = 1$ using the partition $x = 1, 3/2, 2, 5/2, 3$ and $y = 2, 5/2, 3, 7/2, 4$ with (x_k, y_k) the center (centroid) in the kth subrectangle (provided it lies within R)

Theory and Examples

57. Integrate $f(x, y) = \sqrt{4 - x^2}$ over the smaller sector cut from the disk $x^2 + y^2 \le 4$ by the rays $\theta = \pi/6$ and $\theta = \pi/2$.

58. Integrate $f(x, y) = 1/[(x^2 - x)(y - 1)^{2/3}]$ over the infinite rectangle $2 \le x < \infty$, $0 \le y \le 2$.

59. A solid right (noncircular) cylinder has its base R in the xy-plane and is bounded above by the paraboloid $z = x^2 + y^2$. The cylinder's volume is

$$V = \int_0^1 \int_0^y (x^2 + y^2) \, dx \, dy + \int_1^2 \int_0^{2-y} (x^2 + y^2) \, dx \, dy.$$

Sketch the base region R and express the cylinder's volume as a single iterated integral with the order of integration reversed. Then evaluate the integral to find the volume.

60. Evaluate the integral

$$\int_0^2 (\tan^{-1} \pi x - \tan^{-1} x) \, dx.$$

(*Hint:* Write the integrand as an integral.)

61. What region R in the xy-plane maximizes the value of

$$\iint_R (4 - x^2 - 2y^2) \, dA?$$

Give reasons for your answer.

62. What region R in the xy-plane minimizes the value of

$$\iint_R (x^2 + y^2 - 9) \, dA?$$

Give reasons for your answer.

63. Is it all right to evaluate the integral of a continuous function $f(x, y)$ over a rectangular region in the xy-plane and get different answers depending on the order of integration? Give reasons for your answer.

64. How would you evaluate the double integral of a continuous function $f(x, y)$ over the region R in the xy-plane enclosed by the triangle with vertices $(0, 1)$, $(2, 0)$, and $(1, 2)$? Give reasons for your answer.

65. Prove that

$$\int_{-\infty}^{\infty} \int_{-\infty}^{\infty} e^{-x^2 - y^2} dx \, dy = \lim_{b \to \infty} \int_{-b}^b \int_{-b}^b e^{-x^2 - y^2} dx \, dy$$

$$= 4 \left(\int_0^{\infty} e^{-x^2} dx \right)^2.$$

66. Evaluate the improper integral $\displaystyle \int_0^1 \int_0^3 \frac{x^2}{(y - 1)^{2/3}} \, dy \, dx$.

Numerical Evaluation

Use a double-integral evaluator to estimate the values of the integrals in Exercises 67–70.

67. $\displaystyle \int_1^3 \int_1^x \frac{1}{xy} \, dy \, dx$

68. $\displaystyle \int_0^1 \int_0^1 e^{-(x^2 + y^2)} \, dy \, dx$

69. $\displaystyle \int_0^1 \int_0^1 \tan^{-1} xy \, dy \, dx$

70. $\displaystyle \int_{-1}^1 \int_0^{\sqrt{1-x^2}} 3\sqrt{1 - x^2 - y^2} \, dy \, dx$

13.2

Areas, Moments, and Centers of Mass

In this section we show how to use double integrals to define and calculate the areas of bounded regions in the plane and the masses, moments, centers of mass, and radii of gyration of thin plates covering these regions. The calculations are similar to the ones in Chapter 5, but now we can handle a greater variety of shapes.

Areas of Bounded Regions in the Plane

If we take $f(x, y) = 1$ in the definition of the double integral over a region R in the preceding section, the partial sums reduce to

$$S_n = \sum_{k=1}^{n} f(x_k, y_k)\Delta A_k = \sum_{k=1}^{n} \Delta A_k. \tag{1}$$

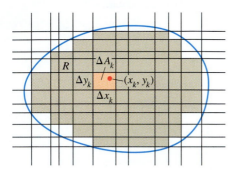

This approximates what we would like to call the area of R. As Δx and Δy approach zero, the coverage of R by the ΔA_k's (Fig. 13.14) becomes increasingly complete, and we define the area of R to be the limit

$$\text{Area} = \lim_{n \to \infty} \sum_{k=1}^{n} \Delta A_k = \iint_R dA. \tag{2}$$

13.14 The first step in defining the area of a region is to partition the interior of the region into cells.

Definition
The **area** of a closed, bounded plane region R is

$$A = \iint_R dA. \tag{3}$$

As with the other definitions in this chapter, the definition here applies to a greater variety of regions than does the earlier single-variable definition of area, but it agrees with the earlier definition on regions to which they both apply.

To evaluate the integral in (3), we integrate the constant function $f(x, y) = 1$ over R.

EXAMPLE 1   Find the area of the region R bounded by $y = x$ and $y = x^2$ in the first quadrant.

Solution   We sketch the region (Fig. 13.15) and calculate the area as

$$A = \int_0^1 \int_{x^2}^{x} dy\, dx = \int_0^1 \Big[y\Big]_{x^2}^{x} dx = \int_0^1 (x - x^2)\, dx = \left[\frac{x^2}{2} - \frac{x^3}{3}\right]_0^1 = \frac{1}{6}. \quad \square$$

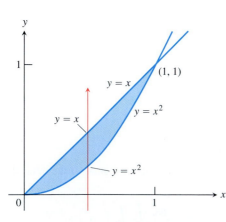

13.15 The area of the region between the parabola and the line in Example 1 is

$$\int_0^1 \int_{x^2}^{x} dy\, dx.$$

EXAMPLE 2   Find the area of the region R enclosed by the parabola $y = x^2$ and the line $y = x + 2$.

Solution   If we divide R into the regions R_1 and R_2 shown in Fig. 13.16(a), we may

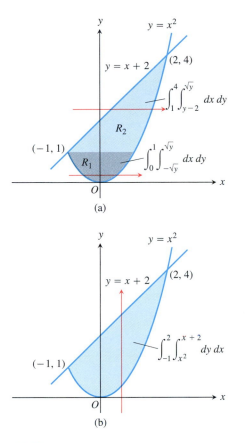

13.16 Calculating this area takes (a) two double integrals if the first integration is with respect to *x*, but (b) only one if the first integration is with respect to *y* (Example 2).

Global warming

The "global warming" controversy deals with whether the average air temperature over the surface of the earth is increasing.

---

calculate the area as

$$A = \iint_{R_1} dA + \iint_{R_2} dA = \int_0^1 \int_{-\sqrt{y}}^{\sqrt{y}} dx\, dy + \int_1^4 \int_{y-2}^{\sqrt{y}} dx\, dy.$$

On the other hand, reversing the order of integration (Fig. 13.16b) gives

$$A = \int_{-1}^2 \int_{x^2}^{x+2} dy\, dx.$$

This result is simpler and is the only one we would bother to write down in practice. The area is

$$A = \int_{-1}^2 \left[y \right]_{x^2}^{x+2} dx = \int_{-1}^2 (x + 2 - x^2)\, dx = \left[\frac{x^2}{2} + 2x - \frac{x^3}{3} \right]_{-1}^2 = \frac{9}{2}.$$

Average Value

The average value of an integrable function of a single variable on a closed interval is the integral of the function over the interval divided by the length of the interval. For an integrable function of two variables defined on a closed and bounded region that has a measurable area, the average value is the integral over the region divided by the area of the region. If f is the function and R the region, then

$$\textbf{Average value of } f \text{ over } R = \frac{1}{\text{area of } R} \iint_R f\, dA. \tag{4}$$

If f is the area density of a thin plate covering R, then the double integral of f over R divided by the area of R is the plate's average density in units of mass per unit area. If $f(x, y)$ is the distance from the point (x, y) to a fixed point P, then the average value of f over R is the average distance of points in R from P.

EXAMPLE 3   Find the average value of $f(x, y) = x \cos xy$ over the rectangle $R: 0 \le x \le \pi$, $0 \le y \le 1$.

Solution   The value of the integral of f over R is

$$\int_0^\pi \int_0^1 x \cos xy\, dy\, dx = \int_0^\pi \left[\sin xy \right]_{y=0}^{y=1} dx$$

$$= \int_0^\pi (\sin x - 0)\, dx = -\cos x \Big]_0^\pi = 1 + 1 = 2.$$

The area of R is π. The average value of f over R is $2/\pi$.

First and Second Moments and Centers of Mass

To find the moments and centers of mass of thin sheets and plates, we use formulas similar to those in Chapter 5. The main difference is that now, with double integrals, we can accommodate a greater variety of shapes and density functions. The formulas are given in Table 13.1, on the following page. The examples that follow show how the formulas are used.

The mathematical difference between the **first moments** M_x and M_y and the **moments of inertia,** or **second moments,** I_x and I_y is that the second moments use the *squares* of the "lever-arm" distances x and y.

Table 13.1 Mass and moment formulas for thin plates covering regions in the *xy*-plane

---

Density: $\delta(x, y)$

Mass: $M = \iint \delta(x, y)\, dA$ **First moments:** $M_x = \iint y\delta(x, y)\, dA, \quad M_y = \iint x\delta(x, y)\, dA$

Center of mass: $\bar{x} = \dfrac{M_y}{M}, \quad \bar{y} = \dfrac{M_x}{M}$

Moments of inertia (second moments):

   About the *x*-axis: $I_x = \iint y^2\,\delta(x, y)\, dA$ About the origin $I_0 = \iint (x^2 + y^2)\,\delta(x, y)\, dA = I_x + I_y$
   (polar moment):

   About the *y*-axis: $I_y = \iint x^2\,\delta(x, y)\, dA$

   About a line *L*: $I_L = \iint r^2(x, y)\delta(x, y)\, dA$, where $r(x, y)$ = distance from (x, y) to L

Radii of gyration: About the *x*-axis: $R_x = \sqrt{I_x/M}$

   About the *y*-axis: $R_y = \sqrt{I_y/M}$

   About the origin: $R_0 = \sqrt{I_0/M}$

The moment I_0 is also called the **polar moment** of inertia about the origin. It is calculated by integrating the density $\delta(x, y)$ (mass per unit area) times $r^2 = x^2 + y^2$, the square of the distance from a representative point (x, y) to the origin. Notice that $I_0 = I_x + I_y$; once we find two, we get the third automatically. (The moment I_0 is sometimes called I_z, for moment of inertia about the z-axis. The identity $I_z = I_x + I_y$ is then called the **Perpendicular Axis Theorem.**)

The **radius of gyration** R_x is defined by the equation

$$I_x = MR_x{}^2.$$

It tells how far from the x-axis the entire mass of the plate might be concentrated to give the same I_x. The radius of gyration gives a convenient way to express the moment of inertia in terms of a mass and a length. The radii R_y and R_0 are defined in a similar way, with

$$I_y = MR_y{}^2 \qquad \text{and} \qquad I_0 = MR_0{}^2.$$

We take square roots to get the formulas in Table 13.1.

Why the interest in moments of inertia? A body's first moments tell us about balance and about the torque the body exerts about different axes in a gravitational field. But if the body is a rotating shaft, we are more likely to be interested in how much energy is stored in the shaft or about how much energy it will take to accelerate the shaft to a particular angular velocity. This is where the second moment or moment of inertia comes in.

Think of partitioning the shaft into small blocks of mass Δm_k and let r_k denote the distance from the kth block's center of mass to the axis of rotation (Fig. 13.17). If the shaft rotates at an angular velocity of $\omega = d\theta/dt$ radians per second, the block's center of mass will trace its orbit at a linear speed of

$$v_k = \frac{d}{dt}\left(r_k\theta\right) = r_k\frac{d\theta}{dt} = r_k\,\omega. \tag{5}$$

13.17 To find an integral for the amount of energy stored in a rotating shaft, we first imagine the shaft to be partitioned into small blocks. Each block has its own kinetic energy. We add the contributions of the individual blocks to find the kinetic energy of the shaft.

The block's kinetic energy will be approximately

$$\frac{1}{2}\Delta m_k v_k{}^2 = \frac{1}{2}\Delta m_k \left(r_k\,\omega\right)^2 = \frac{1}{2}\,\omega^2 r_k{}^2\,\Delta m_k. \tag{6}$$

The kinetic energy of the shaft will be approximately

$$\sum \frac{1}{2}\,\omega^2 r_k{}^2 \Delta m_k. \tag{7}$$

The integral approached by these sums as the shaft is partitioned into smaller and smaller blocks gives the shaft's kinetic energy:

$$\mathrm{KE}_{\mathrm{shaft}} = \int \frac{1}{2}\omega^2 r^2\,dm = \frac{1}{2}\omega^2 \int r^2\,dm. \tag{8}$$

The factor

$$I = \int r^2\,dm \tag{9}$$

is the moment of inertia of the shaft about its axis of rotation, and we see from Eq. (8) that the shaft's kinetic energy is

$$\mathrm{KE}_{\mathrm{shaft}} = \frac{1}{2}I\omega^2. \tag{10}$$

To start a shaft of inertial moment I rotating at an angular velocity ω, we need to provide a kinetic energy of $\mathrm{KE} = (1/2)I\omega^2$. To stop the shaft, we have to take this amount of energy back out. To start a locomotive with mass m moving at a linear velocity v, we need to provide a kinetic energy of $\mathrm{KE} = (1/2)\,mv^2$. To stop the locomotive, we have to remove this amount of energy. The shaft's moment of inertia is analogous to the locomotive's mass. What makes the locomotive hard to start or stop is its mass. What makes the shaft hard to start or stop is its moment of inertia. The moment of inertia takes into account not only the mass but also its distribution.

The moment of inertia also plays a role in determining how much a horizontal metal beam will bend under a load. The stiffness of the beam is a constant times I, the polar moment of inertia of a typical cross section of the beam perpendicular to the beam's longitudinal axis. The greater the value of I, the stiffer the beam and the less it will bend under a given load. That is why we use I beams instead of beams whose cross sections are square. The flanges at the top and bottom of the beam hold most of the beam's mass away from the longitudinal axis to maximize the value of I (Fig. 13.18).

If you want to see the moment of inertia at work, try the following experiment. Tape two coins to the ends of a pencil and twiddle the pencil about the center of mass. The moment of inertia accounts for the resistance you feel each time you change the direction of motion. Now move the coins an equal distance toward the

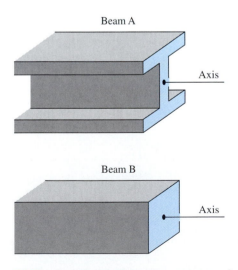

13.18 The greater the polar moment of inertia of the cross section of a beam about the beam's longitudinal axis, the stiffer the beam. Beams A and B have the same cross-section area, but A is stiffer.

First moments are "balancing" moments.
Second moments are "turning" moments.

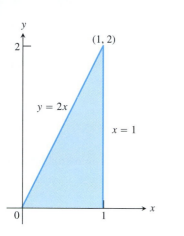

13.19 The triangular region covered by the plate in Example 4.

center of mass and twiddle the pencil again. The system has the same mass and the same center of mass but now offers less resistance to the changes in motion. The moment of inertia has been reduced. The moment of inertia is what gives a baseball bat, golf club, or tennis racket its "feel." Tennis rackets that weigh the same, look the same, and have identical centers of mass will feel different and behave differently if their masses are not distributed the same way.

EXAMPLE 4   A thin plate covers the triangular region bounded by the x-axis and the lines $x = 1$ and $y = 2x$ in the first quadrant. The plate's density at the point (x, y) is $\delta(x, y) = 6x + 6y + 6$. Find the plate's mass, first moments, center of mass, moments of inertia, and radii of gyration about the coordinate axes.

Solution   We sketch the plate and put in enough detail to determine the limits of integration for the integrals we have to evaluate (Fig. 13.19).

The plate's mass is

$$M = \int_0^1 \int_0^{2x} \delta(x, y) \, dy \, dx = \int_0^1 \int_0^{2x} (6x + 6y + 6) \, dy \, dx$$

$$= \int_0^1 \left[6xy + 3y^2 + 6y \right]_{y=0}^{y=2x} dx$$

$$= \int_0^1 (24x^2 + 12x) \, dx = \left[8x^3 + 6x^2 \right]_0^1 = 14.$$

The first moment about the x-axis is

$$M_x = \int_0^1 \int_0^{2x} y \delta(x, y) \, dy \, dx = \int_0^1 \int_0^{2x} (6xy + 6y^2 + 6y) \, dy \, dx$$

$$= \int_0^1 \left[3xy^2 + 2y^3 + 3y^2 \right]_{y=0}^{y=2x} dx = \int_0^1 (28x^3 + 12x^2) \, dx$$

$$= \left[7x^4 + 4x^3 \right]_0^1 = 11.$$

A similar calculation gives

$$M_y = \int_0^1 \int_0^{2x} x \delta(x, y) \, dy \, dx = 10.$$

The coordinates of the center of mass are therefore

$$\overline{x} = \frac{M_y}{M} = \frac{10}{14} = \frac{5}{7}, \qquad \overline{y} = \frac{M_x}{M} = \frac{11}{14}.$$

The moment of inertia about the x-axis is

$$I_x = \int_0^1 \int_0^{2x} y^2 \delta(x, y) \, dy \, dx = \int_0^1 \int_0^{2x} (6xy^2 + 6y^3 + 6y^2) \, dy \, dx$$

$$= \int_0^1 \left[2xy^3 + \frac{3}{2}y^4 + 2y^3 \right]_{y=0}^{y=2x} dx = \int_0^1 (40x^4 + 16x^3) \, dx$$

$$= \left[8x^5 + 4x^4 \right]_0^1 = 12.$$

Similarly, the moment of inertia about the y-axis is

$$I_y = \int_0^1 \int_0^{2x} x^2 \delta(x, y) \, dy \, dx = \frac{39}{5}.$$

Since we know I_x and I_y, we do not need to evaluate an integral to find I_0; we can use the equation $I_0 = I_x + I_y$ instead:

$$I_0 = 12 + \frac{39}{5} = \frac{60 + 39}{5} = \frac{99}{5}.$$

The three radii of gyration are

$$R_x = \sqrt{I_x/M} = \sqrt{12/14} = \sqrt{6/7}$$

$$R_y = \sqrt{I_y/M} = \sqrt{\left(\frac{39}{5}\right)/14} = \sqrt{39/70}$$

$$R_0 = \sqrt{I_0/M} = \sqrt{\left(\frac{99}{5}\right)/14} = \sqrt{99/70}.$$

Centroids of Geometric Figures

When the density of an object is constant, it cancels out of the numerator and denominator of the formulas for $\bar{x}$ and $\bar{y}$. As far as $\bar{x}$ and $\bar{y}$ are concerned, δ might as well be 1. Thus, when δ is constant, the location of the center of mass becomes a feature of the object's shape and not of the material of which it is made. In such cases, engineers may call the center of mass the **centroid** of the shape. To find a centroid, we set δ equal to 1 and proceed to find $\bar{x}$ and $\bar{y}$ as before, by dividing first moments by masses.

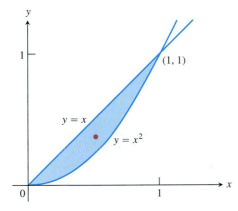

13.20 Example 5 finds the centroid of the region shown here.

EXAMPLE 5   Find the centroid of the region in the first quadrant that is bounded above by the line $y = x$ and below by the parabola $y = x^2$.

Solution   We sketch the region and include enough detail to determine the limits of integration (Fig. 13.20). We then set δ equal to 1 and evaluate the appropriate formulas from Table 13.1:

$$M = \int_0^1 \int_{x^2}^x 1 \, dy \, dx = \int_0^1 \Big[y \Big]_{y=x^2}^{y=x} dx = \int_0^1 (x - x^2) \, dx = \left[\frac{x^2}{2} - \frac{x^3}{3} \right]_0^1 = \frac{1}{6}$$

$$M_x = \int_0^1 \int_{x^2}^x y \, dy \, dx = \int_0^1 \left[\frac{y^2}{2} \right]_{y=x^2}^{y=x} dx$$

$$= \int_0^1 \left(\frac{x^2}{2} - \frac{x^4}{2} \right) dx = \left[\frac{x^3}{6} - \frac{x^5}{10} \right]_0^1 = \frac{1}{15}$$

$$M_y = \int_0^1 \int_{x^2}^x x \, dy \, dx = \int_0^1 \Big[xy \Big]_{y=x^2}^{y=x} dx = \int_0^1 (x^2 - x^3) \, dx = \left[\frac{x^3}{3} - \frac{x^4}{4} \right]_0^1 = \frac{1}{12}.$$

From these values of M, M_x, and M_y, we find

$$\bar{x} = \frac{M_y}{M} = \frac{1/12}{1/6} = \frac{1}{2} \quad \text{and} \quad \bar{y} = \frac{M_x}{M} = \frac{1/15}{1/6} = \frac{2}{5}.$$

The centroid is the point $\left(\frac{1}{2}, \frac{2}{5} \right)$.

Exercises 13.2

Area by Double Integration

In Exercises 1–8, sketch the region bounded by the given lines and curves. Then express the region's area as an iterated double integral and evaluate the integral.

1. The coordinate axes and the line $x + y = 2$

2. The lines $x = 0$, $y = 2x$, and $y = 4$

3. The parabola $x = -y^2$ and the line $y = x + 2$

4. The parabola $x = y - y^2$ and the line $y = -x$

5. The curve $y = e^x$ and the lines $y = 0$, $x = 0$, and $x = \ln 2$

6. The curves $y = \ln x$ and $y = 2 \ln x$ and the line $x = e$, in the first quadrant

7. The parabolas $x = y^2$ and $x = 2y - y^2$

8. The parabolas $x = y^2 - 1$ and $x = 2y^2 - 2$

The integrals and sums of integrals in Exercises 9–14 give the areas of regions in the xy-plane. Sketch each region, label each bounding curve with its equation, and give the coordinates of the points where the curves intersect. Then find the area of the region.

9. $\int_0^6 \int_{y^2/3}^{2y} dx \, dy$

10. $\int_0^3 \int_{-x}^{x(2-x)} dy \, dx$

11. $\int_0^{\pi/4} \int_{\sin x}^{\cos x} dy \, dx$

12. $\int_{-1}^2 \int_{y^2}^{y+2} dx \, dy$

13. $\int_{-1}^0 \int_{-2x}^{1-x} dy \, dx + \int_0^2 \int_{-x/2}^{1-x} dy \, dx$

14. $\int_0^2 \int_{x^2-4}^0 dy \, dx + \int_0^4 \int_0^{\sqrt{x}} dy \, dx$

Average Values

15. Find the average value of $f(x, y) = \sin(x + y)$ over
    a) the rectangle $0 \le x \le \pi$, $0 \le y \le \pi$,
    b) the rectangle $0 \le x \le \pi$, $0 \le y \le \pi/2$.

16. Which do you think will be larger, the average value of $f(x, y) = xy$ over the square $0 \le x \le 1$, $0 \le y \le 1$, or the average value of f over the quarter circle $x^2 + y^2 \le 1$ in the first quadrant? Calculate them to find out.

17. Find the average height of the paraboloid $z = x^2 + y^2$ over the square $0 \le x \le 2$, $0 \le y \le 2$.

18. Find the average value of $f(x, y) = 1/(xy)$ over the square $\ln 2 \le x \le 2 \ln 2$, $\ln 2 \le y \le 2 \ln 2$.

Constant Density

19. Find the center of mass of a thin plate of density $\delta = 3$ bounded by the lines $x = 0$, $y = x$, and the parabola $y = 2 - x^2$ in the first quadrant.

20. Find the moments of inertia and radii of gyration about the co-ordinate axes of a thin rectangular plate of constant density δ bounded by the lines $x = 3$ and $y = 3$ in the first quadrant.

21. Find the centroid of the region in the first quadrant bounded by the x-axis, the parabola $y^2 = 2x$, and the line $x + y = 4$.

22. Find the centroid of the triangular region cut from the first quadrant by the line $x + y = 3$.

23. Find the centroid of the semicircular region bounded by the x-axis and the curve $y = \sqrt{1 - x^2}$.

24. The area of the region in the first quadrant bounded by the parabola $y = 6x - x^2$ and the line $y = x$ is 125/6 square units. Find the centroid.

25. Find the centroid of the region cut from the first quadrant by the circle $x^2 + y^2 = a^2$.

26. Find the moment of inertia about the x-axis of a thin plate of density $\delta = 1$ bounded by the circle $x^2 + y^2 = 4$. Then use your result to find I_y and I_0 for the plate.

27. Find the centroid of the region between the x-axis and the arch $y = \sin x$, $0 \le x \le \pi$.

28. Find the moment of inertia with respect to the y-axis of a thin sheet of constant density $\delta = 1$ bounded by the curve $y = (\sin^2 x)/x^2$ and the interval $\pi \le x \le 2\pi$ of the x-axis.

29. *The centroid of an infinite region.* Find the centroid of the infinite region in the second quadrant enclosed by the coordinate axes and the curve $y = e^x$. (Use improper integrals in the mass-moment formulas.)

30. *The first moment of an infinite plate.* Find the first moment about the y-axis of a thin plate of density $\delta(x, y) = 1$ covering the infinite region under the curve $y = e^{-x^2/2}$ in the first quadrant.

Variable Density

31. Find the moment of inertia and radius of gyration about the x-axis of a thin plate bounded by the parabola $x = y - y^2$ and the line $x + y = 0$ if $\delta(x, y) = x + y$.

32. Find the mass of a thin plate occupying the smaller region cut from the ellipse $x^2 + 4y^2 = 12$ by the parabola $x = 4y^2$ if $\delta(x, y) = 5x$.

33. Find the center of mass of a thin triangular plate bounded by the y-axis and the lines $y = x$ and $y = 2 - x$ if $\delta(x, y) = 6x + 3y + 3$.

34. Find the center of mass and moment of inertia about the x-axis of a thin plate bounded by the curves $x = y^2$ and $x = 2y - y^2$ if the density at the point (x, y) is $\delta(x, y) = y + 1$.

35. Find the center of mass and the moment of inertia and radius of gyration about the y-axis of a thin rectangular plate cut from the first quadrant by the lines $x = 6$ and $y = 1$ if $\delta(x, y) = x + y + 1$.

36. Find the center of mass and the moment of inertia and radius of gyration about the y-axis of a thin plate bounded by the line $y = 1$ and the parabola $y = x^2$ if the density is $\delta(x, y) = y + 1$.

37. Find the center of mass and the moment of inertia and radius of gyration about the y-axis of a thin plate bounded by the x-axis, the lines $x = \pm 1$, and the parabola $y = x^2$ if $\delta(x, y) = 7y + 1$.

38. Find the center of mass and the moment of inertia and radius of gyration about the x-axis of a thin rectangular plate bounded by the lines $x = 0$, $x = 20$, $y = -1$, and $y = 1$ if $\delta(x, y) = 1 + (x/20)$.

39. Find the center of mass, the moments of inertia and radii of gyration about the coordinate axes, and the polar moment of inertia and radius of gyration of a thin triangular plate bounded by the lines $y = x$, $y = -x$, and $y = 1$ if $\delta(x, y) = y + 1$.

40. Repeat Exercise 39 for $\delta(x, y) = 3x^2 + 1$.

Theory and Examples

41. If $f(x, y) = (10{,}000\,e^y)/(1 + |x|/2)$ represents the "population density" of a certain bacteria on the xy-plane, where x and y are measured in centimeters, find the total population of bacteria within the rectangle $-5 \le x \le 5$ and $-2 \le y \le 0$.

42. If $f(x, y) = 100\,(y + 1)$ represents the population density of a planar region on Earth, where x and y are measured in miles, find the number of people in the region bounded by the curves $x = y^2$ and $x = 2y - y^2$.

43. *Appliance design.* When we design an appliance, one of the concerns is how hard the appliance will be to tip over. When tipped, it will right itself as long as its center of mass lies on the correct side of the *fulcrum,* the point on which the appliance is riding as it tips. Suppose the profile of an appliance of approximately constant density is parabolic, like an old-fashioned radio. It fills the region $0 \le y \le a(1 - x^2)$, $-1 \le x \le 1$, in the xy-plane (Fig. 13.21). What values of a will guarantee that the appliance will have to be tipped more than $45°$ to fall over?

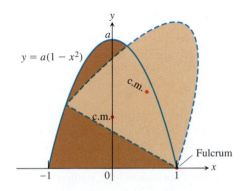

13.21 The profile of the appliance in Exercise 43.

44. *Minimizing a moment of inertia.* A rectangular plate of constant density $\delta(x, y) = 1$ occupies the region bounded by the lines $x = 4$ and $y = 2$ in the first quadrant. The moment of inertia I_a of the rectangle about the line $y = a$ is given by the integral

$$I_a = \int_0^4 \int_0^2 (y - a)^2 \, dy \, dx.$$

Find the value of a that minimizes I_a.

45. Find the centroid of the infinite region in the xy-plane bounded by the curves $y = 1/\sqrt{1 - x^2}$, $y = -1/\sqrt{1 - x^2}$, and the lines $x = 0$, $x = 1$.

46. Find the radius of gyration of a slender rod of constant linear density δ gm/cm and length L cm with respect to an axis

a) through the rod's center of mass perpendicular to the rod's axis;

b) perpendicular to the rod's axis at one end of the rod.

47. A thin plate of constant density δ occupies the region R in the xy-plane bounded by the curves $x = y^2$ and $x = 2y - y^2$ (see Exercise 34).

a) Find δ such that the plate has the same mass as the plate in Exercise 34.

b) Compare the value of δ found in part (a) with the average value of $\delta(x, y) = y + 1$ over R.

48. According to the *Texas Almanac,* Texas has 254 counties and a National Weather Service station in each county. Assume that at time t_0 each of the 254 weather stations recorded the local temperature. Find a formula that would give a reasonable approximation to the average temperature in Texas at time t_0. Your answer should involve information that is readily available in the *Texas Almanac.*

The Parallel Axis Theorem

Let $L_{\text{c.m.}}$ be a line in the xy-plane that runs through the center of mass of a thin plate of mass m covering a region in the plane. Let L be a line in the plane parallel to and h units away from $L_{\text{c.m.}}$. The **Parallel Axis Theorem** says that under these conditions the moments of inertia I_L and $I_{\text{c.m.}}$ of the plate about L and $L_{\text{c.m.}}$ satisfy the equation

$$I_L = I_{\text{c.m.}} + mh^2. \qquad (1)$$

This equation gives a quick way to calculate one moment when the other moment and the mass are known.

49. *Proof of the Parallel Axis Theorem*

a) Show that the first moment of a thin flat plate about any line in the plane of the plate through the plate's center of mass is zero. (*Hint:* Place the center of mass at the origin with the line along the y-axis. What does the formula $\bar{x} = M_y/M$ then tell you?)

b) Use the result in (a) to derive the Parallel Axis Theorem. Assume that the plane is coordinatized in a way that makes $L_{\text{c.m.}}$ the y-axis and L the line $x = h$. Then expand the integrand of the integral for I_L to rewrite the integral as the sum of integrals whose values you recognize.

50. a) Use the Parallel Axis Theorem and the results of Example 4 to find the moments of inertia of the plate in Example 4 about the vertical and horizontal lines through the plate's center of mass.

b) Use the results in (a) to find the plate's moments of inertia about the lines $x = 1$ and $y = 2$.

Pappus's Formula

In addition to stating the centroid theorems in Section 5.10, Pappus knew that the centroid of the union of two nonoverlapping plane regions lies on the line segment joining their individual centroids. More specifically, suppose that m_1 and m_2 are the masses of thin plates P_1 and P_2 that cover nonoverlapping regions in the xy-plane. Let $\mathbf{c}_1$ and $\mathbf{c}_2$ be the vectors from the origin to the respective centers of mass of P_1 and P_2. Then the center of mass of the union $P_1 \cup P_2$ of the two plates is determined by the vector

$$\mathbf{c} = \frac{m_1\mathbf{c}_1 + m_2\mathbf{c}_2}{m_1 + m_2}. \tag{2}$$

Equation (2) is known as **Pappus's formula.** For more than two nonoverlapping plates, as long as their number is finite, the formula generalizes to

$$\mathbf{c} = \frac{m_1\mathbf{c}_1 + m_2\mathbf{c}_2 + \cdots + m_n\mathbf{c}_n}{m_1 + m_2 + \cdots + m_n}. \tag{3}$$

This formula is especially useful for finding the centroid of a plate of irregular shape that is made up of pieces of constant density whose centroids we know from geometry. We find the centroid of each piece and apply Eq. (3) to find the centroid of the plate.

51. Derive Pappus's formula (Eq. 2). (*Hint:* Sketch the plates as regions in the first quadrant and label their centers of mass as $(\overline{x}_1, \overline{y}_1)$ and $(\overline{x}_2, \overline{y}_2)$. What are the moments of $P_1 \cup P_2$ about the coordinate axes?)

52. Use Eq. (2) and mathematical induction to show that Eq. (3) holds for any positive integer $n > 2$.

53. Let A, B, and C be the shapes indicated in Fig. 13.22(a). Use Pappus's formula to find the centroid of

a) $A \cup B$ **b)** $A \cup C$ **c)** $B \cup C$

d) $A \cup B \cup C$

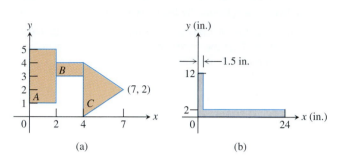

13.22 The figures for Exercises 53 and 54.

54. Locate the center of mass of the carpenter's square in Fig. 13.22(b).

55. An isosceles triangle T has base $2a$ and altitude h. The base lies along the diameter of a semicircular disk D of radius a so that the two together make a shape resembling an ice cream cone. What relation must hold between a and h to place the centroid of $T \cup D$ on the common boundary of T and D? inside T?

56. An isosceles triangle T of altitude h has as its base one side of a square Q whose edges have length s. (The square and triangle do not overlap.) What relation must hold between h and s to place the centroid of $T \cup Q$ on the base of the triangle? Compare your answer with the answer to Exercise 55.

13.3

Double Integrals in Polar Form

Integrals are sometimes easier to evaluate if we change to polar coordinates. This section shows how to accomplish the change and how to evaluate integrals over regions whose boundaries are given by polar equations.

Integrals in Polar Coordinates

When we defined the double integral of a function over a region R in the xy-plane, we began by cutting R into rectangles whose sides were parallel to the coordinate axes. These were the natural shapes to use because their sides have either constant x-values or constant y-values. In polar coordinates, the natural shape is a "polar rectangle" whose sides have constant r- and θ-values.

Suppose that a function $f(r, \theta)$ is defined over a region R that is bounded by the rays $\theta = \alpha$ and $\theta = \beta$ and by the continuous curves $r = g_1(\theta)$ and $r = g_2(\theta)$. Suppose also that $0 \le g_1(\theta) \le g_2(\theta) \le a$ for every value of θ between α and β. Then R lies in a fan-shaped region Q defined by the inequalities $0 \le r \le a$ and $\alpha \le \theta \le \beta$. See Fig. 13.23.

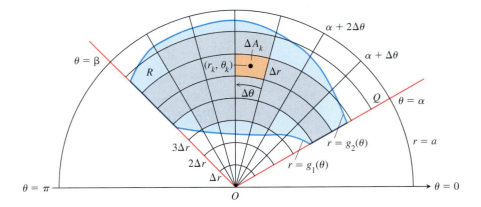

13.23 The region R: $g_1(\theta) \le r \le g_2(\theta)$, $\alpha \le \theta \le \beta$ is contained in the fan-shaped region Q: $0 \le r \le a$, $\alpha \le \theta \le \beta$. The partition of Q by circular arcs and rays induces a partition of R.

We cover Q by a grid of circular arcs and rays. The arcs are cut from circles centered at the origin, with radii Δr, $2\Delta r$, ..., $m\Delta r$, where $\Delta r = a/m$. The rays are given by

$$\theta = \alpha, \quad \theta = \alpha + \Delta\theta, \quad \theta = \alpha + 2\Delta\theta, \quad \ldots, \quad \theta = \alpha + m'\Delta\theta = \beta,$$

where $\Delta\theta = (\beta - \alpha)/m'$. The arcs and rays partition Q into small patches called "polar rectangles."

We number the polar rectangles that lie inside R (the order does not matter), calling their areas ΔA_1, ΔA_2, ..., ΔA_n.

We let (r_k, θ_k) be the center of the polar rectangle whose area is ΔA_k. By "center" we mean the point that lies halfway between the circular arcs on the ray that bisects the arcs. We then form the sum

$$S_n = \sum_{k=1}^{n} f(r_k, \theta_k) \Delta A_k. \tag{1}$$

If f is continuous throughout R, this sum will approach a limit as we refine the grid to make Δr and $\Delta\theta$ go to zero. The limit is called the double integral of f over R. In symbols,

$$\lim_{n \to \infty} S_n = \iint_R f(r, \theta) \, dA.$$

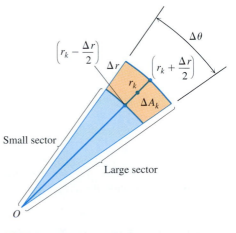

13.24 The observation that

$$\Delta A_k = \begin{pmatrix} \text{area of} \\ \text{large sector} \end{pmatrix} - \begin{pmatrix} \text{area of} \\ \text{small sector} \end{pmatrix}$$

leads to the formula $\Delta A_k = r_k \Delta r \Delta\theta$. The text explains why.

To evaluate this limit, we first have to write the sum S_n in a way that expresses ΔA_k in terms of Δr and $\Delta\theta$. The radius of the inner arc bounding ΔA_k is $r_k - (\Delta r/2)$ (Fig. 13.24). The radius of the outer arc is $r_k + (\Delta r/2)$. The areas of the circular sectors subtended by these arcs at the origin are

$$\text{Inner radius:} \quad \frac{1}{2}\left(r_k - \frac{\Delta r}{2}\right)^2 \Delta\theta \qquad \text{Outer radius:} \quad \frac{1}{2}\left(r_k + \frac{\Delta r}{2}\right)^2 \Delta\theta. \tag{2}$$

Therefore,

$$\Delta A_k = \text{Area of large sector} - \text{Area of small sector}$$

$$= \frac{\Delta\theta}{2}\left[\left(r_k + \frac{\Delta r}{2}\right)^2 - \left(r_k - \frac{\Delta r}{2}\right)^2\right] = \frac{\Delta\theta}{2}(2r_k \Delta r) = r_k \Delta r \Delta\theta.$$

Combining this result with Eq. (1) gives

$$S_n = \sum_{k=1}^{n} f(r_k, \theta_k) r_k \Delta r \Delta\theta. \tag{3}$$

A version of Fubini's theorem now says that the limit approached by these sums can be evaluated by repeated single integrations with respect to r and θ as

$$\iint_R f(r,\theta)\,dA = \int_{\theta=\alpha}^{\theta=\beta} \int_{r=g_1(\theta)}^{r=g_2(\theta)} f(r,\theta)\,r\,dr\,d\theta. \tag{4}$$

Limits of Integration

The procedure for finding limits of integration in rectangular coordinates also works for polar coordinates.

How to Integrate in Polar Coordinates

To evaluate $\iint_R f(r,\theta)\,dA$ over a region R in polar coordinates, integrating first with respect to r and then with respect to θ, take the following steps.

1. *A sketch.* Sketch the region and label the bounding curves.

2. *The r-limits of integration.* Imagine a ray L from the origin cutting through R in the direction of increasing r. Mark the r-values where L enters and leaves R. These are the r-limits of integration. They usually depend on the angle θ that L makes with the positive x-axis.

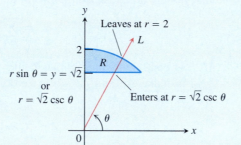

3. *The θ-limits of integration.* Find the smallest and largest θ-values that bound R. These are the θ-limits of integration.

The integral is

$$\iint_R f(r,\theta)\,dA = \int_{\theta=\pi/4}^{\theta=\pi/2} \int_{r=\sqrt{2}\csc\theta}^{r=2} f(r,\theta)\,r\,dr\,d\theta.$$

EXAMPLE 1 Find the limits of integration for integrating $f(r,\theta)$ over the region R that lies inside the cardioid $r = 1 + \cos\theta$ and outside the circle $r = 1$.

Solution

Step 1: *A sketch.* We sketch the region and label the bounding curves (Fig. 13.25).

Step 2: *The r-limits of integration.* A typical ray from the origin enters R where $r = 1$ and leaves where $r = 1 + \cos\theta$.

Step 3: *The θ-limits of integration.* The rays from the origin that intersect R run from

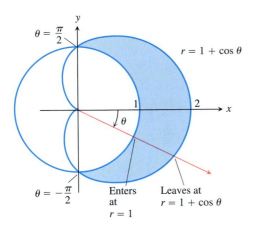

13.25 The sketch for Example 1.

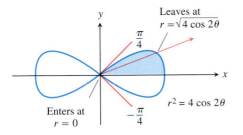

13.26 To integrate over the shaded region, we run r from 0 to $\sqrt{4\cos 2\theta}$ and θ from 0 to $\pi/4$ (Example 2).

$\theta = -\pi/2$ to $\theta = \pi/2$. The integral is

$$\int_{-\pi/2}^{\pi/2} \int_{1}^{1+\cos\theta} f(r, \theta)\, r\, dr\, d\theta. \qquad \square$$

If $f(r, \theta)$ is the constant function whose value is 1, then the integral of f over R is the area of R.

Area in Polar Coordinates

The area of a closed and bounded region R in the polar coordinate plane is

$$A = \iint_R r\, dr\, d\theta. \qquad (5)$$

As you might expect, this formula for area is consistent with all earlier formulas, although we will not prove the fact.

EXAMPLE 2   Find the area enclosed by the lemniscate $r^2 = 4\cos 2\theta$.

Solution   We graph the lemniscate to determine the limits of integration (Fig. 13.26) and see that the total area is 4 times the first-quadrant portion.

$$A = 4\int_0^{\pi/4} \int_0^{\sqrt{4\cos 2\theta}} r\, dr\, d\theta = 4\int_0^{\pi/4} \left[\frac{r^2}{2}\right]_{r=0}^{r=\sqrt{4\cos 2\theta}} d\theta$$

$$= 4\int_0^{\pi/4} 2\cos 2\theta\, d\theta = 4\sin 2\theta\bigg]_0^{\pi/4} = 4. \qquad \square$$

Changing Cartesian Integrals into Polar Integrals

The procedure for changing a Cartesian integral $\iint_R f(x, y)\, dx\, dy$ into a polar integral has two steps.

Step 1: Substitute $x = r\cos\theta$ and $y = r\sin\theta$, and replace $dx\, dy$ by $r\, dr\, d\theta$ in the Cartesian integral.

Step 2: Supply polar limits of integration for the boundary of R.

The Cartesian integral then becomes

$$\iint_R f(x, y)\, dx\, dy = \iint_G f(r\cos\theta, r\sin\theta)\, r\, dr\, d\theta, \qquad (6)$$

where G denotes the region of integration in polar coordinates. This is like the substitution method in Chapter 4 except that there are now two variables to substitute for instead of one. Notice that $dx\, dy$ is not replaced by $dr\, d\theta$ but by $r\, dr\, d\theta$. We will see why in Section 13.7.

EXAMPLE 3   Find the polar moment of inertia about the origin of a thin plate of density $\delta(x, y) = 1$ bounded by the quarter circle $x^2 + y^2 = 1$ in the first quadrant.

13.27 In polar coordinates, this region is described by simple inequalities:

$$0 \le r \le 1 \quad \text{and} \quad 0 \le \theta \le \pi/2$$

(Example 3).

Solution We sketch the plate to determine the limits of integration (Fig. 13.27). In Cartesian coordinates, the polar moment is the value of the integral

$$\int_0^1 \int_0^{\sqrt{1-x^2}} (x^2 + y^2)\, dy\, dx.$$

Integration with respect to y gives

$$\int_0^1 \left(x^2\sqrt{1-x^2} + \frac{(1-x^2)^{3/2}}{3} \right) dx,$$

an integral difficult to evaluate without tables.

Things go better if we change the original integral to polar coordinates. Substituting $x = r\cos\theta$, $y = r\sin\theta$, and replacing $dx\,dy$ by $r\,dr\,d\theta$, we get

$$\int_0^1 \int_0^{\sqrt{1-x^2}} (x^2 + y^2)\, dy\, dx = \int_0^{\pi/2} \int_0^1 (r^2)\, r\, dr\, d\theta$$

$$= \int_0^{\pi/2} \left[\frac{r^4}{4} \right]_{r=0}^{r=1} d\theta = \int_0^{\pi/2} \frac{1}{4}\, d\theta = \frac{\pi}{8}.$$

Why is the polar coordinate transformation so effective? One reason is that $x^2 + y^2$ simplifies to r^2. Another is that the limits of integration become constants. ❏

EXAMPLE 4 Evaluate

$$\iint_R e^{x^2 + y^2}\, dy\, dx,$$

where R is the semicircular region bounded by the x-axis and the curve $y = \sqrt{1-x^2}$ (Fig. 13.28).

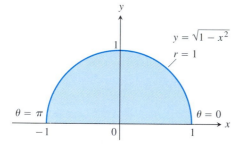

13.28 The semicircular region in Example 4 is the region

$$0 \le r \le 1, \quad 0 \le \theta \le \pi.$$

Solution In Cartesian coordinates, the integral in question is a nonelementary integral and there is no direct way to integrate $e^{x^2+y^2}$ with respect to either x or y. Yet this integral and others like it are important in mathematics—in statistics, for example—and we must find a way to evaluate it. Polar coordinates save the day. Substituting $x = r\cos\theta$, $y = r\sin\theta$, and replacing $dy\,dx$ by $r\,dr\,d\theta$ enables us to evaluate the integral as

$$\iint_R e^{x^2 + y^2}\, dy\, dx = \int_0^\pi \int_0^1 e^{r^2}\, r\, dr\, d\theta = \int_0^\pi \left[\frac{1}{2} e^{r^2} \right]_0^1 d\theta$$

$$= \int_0^\pi \frac{1}{2}(e-1)\, d\theta = \frac{\pi}{2}(e-1).$$

The r in the $r\,dr\,d\theta$ was just what we needed to integrate e^{r^2}. Without it we would have been stuck, as we were at the beginning. ❏

Exercises 13.3

Evaluating Polar Integrals

In Exercises 1–16, change the Cartesian integral into an equivalent polar integral. Then evaluate the polar integral.

1. $\displaystyle\int_{-1}^1 \int_0^{\sqrt{1-x^2}} dy\, dx$

2. $\displaystyle\int_{-1}^1 \int_{-\sqrt{1-x^2}}^{\sqrt{1-x^2}} dy\, dx$

3. $\int_0^1 \int_0^{\sqrt{1-y^2}} (x^2 + y^2)\, dx\, dy$

4. $\int_{-1}^1 \int_{-\sqrt{1-y^2}}^{\sqrt{1-y^2}} (x^2 + y^2)\, dx\, dy$

5. $\int_{-a}^a \int_{-\sqrt{a^2-x^2}}^{\sqrt{a^2-x^2}} dy\, dx$

6. $\int_0^2 \int_0^{\sqrt{4-y^2}} (x^2 + y^2)\, dx\, dy$

7. $\int_0^6 \int_0^y x\, dx\, dy$ **8.** $\int_0^2 \int_0^x y\, dy\, dx$

9. $\int_{-1}^0 \int_{-\sqrt{1-x^2}}^0 \dfrac{2}{1+\sqrt{x^2+y^2}}\, dy\, dx$

10. $\int_{-1}^1 \int_{-\sqrt{1-y^2}}^0 \dfrac{4\sqrt{x^2+y^2}}{1+x^2+y^2}\, dx\, dy$

11. $\int_0^{\ln 2} \int_0^{\sqrt{(\ln 2)^2-y^2}} e^{\sqrt{x^2+y^2}}\, dx\, dy$

12. $\int_0^1 \int_0^{\sqrt{1-x^2}} e^{-(x^2+y^2)}\, dy\, dx$

13. $\int_0^2 \int_0^{\sqrt{1-(x-1)^2}} \dfrac{x+y}{x^2+y^2}\, dy\, dx$

14. $\int_0^2 \int_{-\sqrt{1-(y-1)^2}}^0 xy^2\, dx\, dy$

15. $\int_{-1}^1 \int_{-\sqrt{1-y^2}}^{\sqrt{1-y^2}} \ln(x^2+y^2+1)\, dx\, dy$

16. $\int_{-1}^1 \int_{-\sqrt{1-x^2}}^{\sqrt{1-x^2}} \dfrac{2}{(1+x^2+y^2)^2}\, dy\, dx$

Finding Area in Polar Coordinates

17. Find the area of the region cut from the first quadrant by the curve $r = 2(2 - \sin 2\theta)^{1/2}$.

18. Find the area of the region that lies inside the cardioid $r = 1 + \cos\theta$ and outside the circle $r = 1$.

19. Find the area enclosed by one leaf of the rose $r = 12\cos 3\theta$.

20. Find the area of the region enclosed by the positive x-axis and spiral $r = 4\theta/3, 0 \le \theta \le 2\pi$. The region looks like a snail shell.

21. Find the area of the region cut from the first quadrant by the cardioid $r = 1 + \sin\theta$.

22. Find the area of the region common to the interiors of the cardioids $r = 1 + \cos\theta$ and $r = 1 - \cos\theta$.

Masses and Moments

23. Find the first moment about the x-axis of a thin plate of constant

density $\delta(x, y) = 3$, bounded below by the x-axis and above by the cardioid $r = 1 - \cos\theta$.

24. Find the moment of inertia about the x-axis and the polar moment of inertia about the origin of a thin disk bounded by the circle $x^2 + y^2 = a^2$ if the disk's density at the point (x, y) is $\delta(x, y) = k(x^2 + y^2)$, k a constant.

25. Find the mass of a thin plate covering the region outside the circle $r = 3$ and inside the circle $r = 6\sin\theta$ if the plate's density function is $\delta(x, y) = 1/r$.

26. Find the polar moment of inertia about the origin of a thin plate covering the region that lies inside the cardioid $r = 1 - \cos\theta$ and outside the circle $r = 1$ if the plate's density function is $\delta(x, y) = 1/r^2$.

27. Find the centroid of the region enclosed by the cardioid $r = 1 + \cos\theta$.

28. Find the polar moment of inertia about the origin of a thin plate enclosed by the cardioid $r = 1 + \cos\theta$ if the plate's density function is $\delta(x, y) = 1$.

Average Values

29. Find the average height of the hemisphere $z = \sqrt{a^2 - x^2 - y^2}$ above the disk $x^2 + y^2 \le a^2$ in the xy-plane.

30. Find the average height of the (single) cone $z = \sqrt{x^2 + y^2}$ above the disk $x^2 + y^2 \le a^2$ in the xy-plane.

31. Find the average distance from a point $P(x, y)$ in the disk $x^2 + y^2 \le a^2$ to the origin.

32. Find the average value of the *square* of the distance from the point $P(x, y)$ in the disk $x^2 + y^2 \le 1$ to the boundary point $A(1, 0)$.

Theory and Examples

33. Integrate $f(x, y) = [\ln(x^2 + y^2)]/\sqrt{x^2 + y^2}$ over the region $1 \le x^2 + y^2 \le e$.

34. Integrate $f(x, y) = [\ln(x^2 + y^2)]/(x^2 + y^2)$ over the region $1 \le x^2 + y^2 \le e^2$.

35. The region that lies inside the cardioid $r = 1 + \cos\theta$ and outside the circle $r = 1$ is the base of a solid right cylinder. The top of the cylinder lies in the plane $z = x$. Find the cylinder's volume.

36. The region enclosed by the lemniscate $r^2 = 2\cos 2\theta$ is the base of a solid right cylinder whose top is bounded by the sphere $z = \sqrt{2 - r^2}$. Find the cylinder's volume.

37. a) The usual way to evaluate the improper integral $I = \int_0^\infty e^{-x^2}\, dx$ is first to calculate its square:

$$I^2 = \left(\int_0^\infty e^{-x^2}\, dx\right)\left(\int_0^\infty e^{-y^2}\, dy\right) = \int_0^\infty \int_0^\infty e^{-(x^2+y^2)}\, dx\, dy.$$

Evaluate the last integral using polar coordinates and solve the resulting equation for I.

b) (*Continuation of Section 7.6, Exercise 92.*) Evaluate

$$\lim_{x\to\infty} \operatorname{erf}(x) = \lim_{x\to\infty} \int_0^x \frac{2e^{-t^2}}{\sqrt{\pi}}\, dt.$$

38. Evaluate the integral

$$\int_0^\infty \int_0^\infty \frac{1}{(1+x^2+y^2)^2} \, dx \, dy.$$

39. Integrate the function $f(x, y) = 1/(1 - x^2 - y^2)$ over the disk $x^2 + y^2 \le 3/4$. Does the integral of $f(x, y)$ over the disk $x^2 + y^2 \le 1$ exist? Give reasons for your answer.

40. Use the double integral in polar coordinates to derive the formula

$$A = \int_\alpha^\beta \frac{1}{2} r^2 \, d\theta$$

for the area of the fan-shaped region between the origin and polar curve $r = f(\theta), \alpha \le \theta \le \beta$.

41. Let P_0 be a point inside a circle of radius a and let h denote the distance from P_0 to the center of the circle. Let d denote the distance from an arbitrary point P to P_0. Find the average value of d^2 over the region enclosed by the circle. (*Hint:* Simplify your work by placing the center of the circle at the origin and P_0 on the x-axis.)

42. Suppose that the area of a region in the polar coordinate plane is

$$A = \int_{\pi/4}^{3\pi/4} \int_{\csc\theta}^{2\sin\theta} r \, dr \, d\theta.$$

a) Sketch the region and find its area.

b) Use one of Pappus's theorems together with the centroid information in Exercise 26 of Section 5.10 to find the volume of the solid generated by revolving the region about the x-axis.

✷ CAS Explorations and Projects

In Exercises 43–46, use a CAS to change the Cartesian integrals into an equivalent polar integral and evaluate the polar integral. Perform the following steps in each exercise.

a) Plot the Cartesian region of integration in the xy-plane.

b) Change each boundary curve of the Cartesian region in (a) to its polar representation by solving its Cartesian equation for r and θ.

c) Using the results in (b), plot the polar region of integration in the $r\theta$-plane.

d) Change the integrand from Cartesian to polar coordinates. Determine the limits of integration from your plot in (c) and evaluate the polar integral using the CAS integration utility.

43. $\displaystyle\int_0^1 \int_x^1 \frac{y}{x^2 + y^2} \, dy \, dx$

44. $\displaystyle\int_0^1 \int_0^{x/2} \frac{x}{x^2 + y^2} \, dy \, dx$

45. $\displaystyle\int_0^1 \int_{-y/3}^{y/3} \frac{y}{\sqrt{x^2 + y^2}} \, dx \, dy$

46. $\displaystyle\int_0^1 \int_y^{2-y} \sqrt{x + y} \, dx \, dy$

13.4

Triple Integrals in Rectangular Coordinates

We use triple integrals to find the volumes of three-dimensional shapes, the masses and moments of solids, and the average values of functions of three variables. In Chapter 14, we will also see how these integrals arise in the studies of vector fields and fluid flow.

Triple Integrals

If $F(x, y, z)$ is a function defined on a closed bounded region D in space—the region occupied by a solid ball, for example, or a lump of clay—then the integral of F over D may be defined in the following way. We partition a rectangular region containing D into rectangular cells by planes parallel to the coordinate planes (Fig. 13.29). We number the cells that lie inside D from 1 to n in some order, a typical cell having dimensions Δx_k by Δy_k by Δz_k and volume ΔV_k. We choose a point (x_k, y_k, z_k) in each cell and form the sum

$$S_n = \sum_{k=1}^{n} F(x_k, y_k, z_k) \Delta V_k. \tag{1}$$

If F is continuous and the bounding surface of D is made of smooth surfaces joined along continuous curves, then as Δx_k, Δy_k, and Δz_k approach zero independently the sums S_n approach a limit

$$\lim_{n\to\infty} S_n = \iiint_D F(x, y, z) \, dV. \tag{2}$$

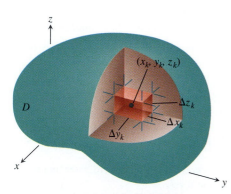

13.29 Partitioning a solid with rectangular cells of volume ΔV_k.

We call this limit the **triple integral of F over D.** The limit also exists for some discontinuous functions.

Properties of Triple Integrals

Triple integrals have the same algebraic properties as double and single integrals. If $F = F(x, y, z)$ and $G = G(x, y, z)$ are continuous, then

1. $\displaystyle\iiint_D kF \, dV = k \iiint_D F \, dV$ (any number k)

2. $\displaystyle\iiint_D (F \pm G) \, dV = \iiint_D F \, dV \pm \iiint_D G \, dV$

3. $\displaystyle\iiint_D F \, dV \geq 0$ if $F \geq 0$ on D

4. $\displaystyle\iiint_D F \, dV \geq \iiint_D G \, dV$ if $F \geq G$ on D.

Triple integrals also have an additivity property, used in physics and engineering as well as in mathematics. If the domain D of a continuous function F is partitioned by smooth surfaces into a finite number of nonoverlapping cells $D_1, D_2, \ldots, D_n$, then

5. $\displaystyle\iiint_D F \, dV = \iiint_{D_1} F \, dV + \iiint_{D_2} F \, dV + \cdots + \iiint_{D_n} F \, dV.$

Volume of a Region in Space

If F is the constant function whose value is 1, then the sums in Eq. (1) reduce to

$$S_n = \sum F(x_k, y_k, z_k) \Delta V_k = \sum 1 \cdot \Delta V_k = \sum \Delta V_k. \tag{3}$$

As Δx, Δy, and Δz approach zero, the cells ΔV_k become smaller and more numerous and fill up more and more of D. We therefore define the volume of D to be the triple integral

$$\lim_{n \to \infty} \sum_{k=1}^{n} \Delta V_k = \iiint_D dV.$$

Definition

The **volume** of a closed, bounded region D in space is

$$V = \iiint_D dV. \tag{4}$$

As we will see in a moment, this integral enables us to calculate the volumes of solids enclosed by curved surfaces.

Evaluation

We seldom evaluate a triple integral from its definition as a limit. Instead, we apply a three-dimensional version of Fubini's theorem to evaluate it by repeated single integrations. As with double integrals, there is a geometric procedure for finding the limits of integration.

How to Find Limits of Integration in Triple Integrals

To evaluate

$$\iiint\limits_{D} F(x, y, z)\, dV$$

over a region D, integrating first with respect to z, then with respect to y, finally with x, take the following steps.

1. *A sketch.* Sketch the region D along with its "shadow" R (vertical projection) in the xy-plane. Label the upper and lower bounding surfaces of D and the upper and lower bounding curves of R.

2. *The z-limits of integration.* Draw a line M passing through a typical point (x, y) in R parallel to the z-axis. As z increases, M enters D at $z = f_1(x, y)$ and leaves at $z = f_2(x, y)$. These are the z-limits of integration.

3. *The y-limits of integration.* Draw a line L through (x, y) parallel to the y-axis. As y increases, L enters R at $y = g_1(x)$ and leaves at $y = g_2(x)$. These are the y-limits of integration.

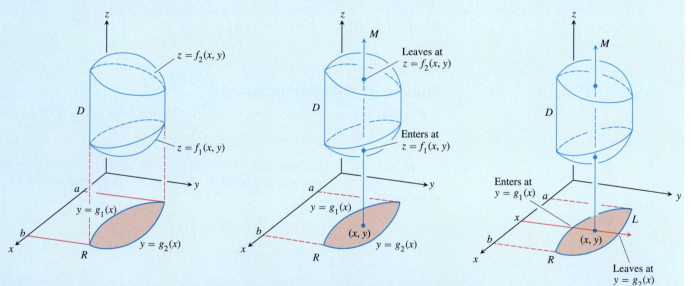

4. *The x-limits of integration.* Choose x-limits that include all lines through R parallel to the y-axis ($x = a$ and $x = b$ in the preceding figure). These are the x-limits of integration. The integral is

$$\int_{x=a}^{x=b} \int_{y=g_1(x)}^{y=g_2(x)} \int_{z=f_1(x,y)}^{z=f_2(x,y)} F(x, y, z)\, dz\, dy\, dx.$$

Follow similar procedures if you change the order of integration. The "shadow" of region D lies in the plane of the last two variables with respect to which the iterated integration takes place.

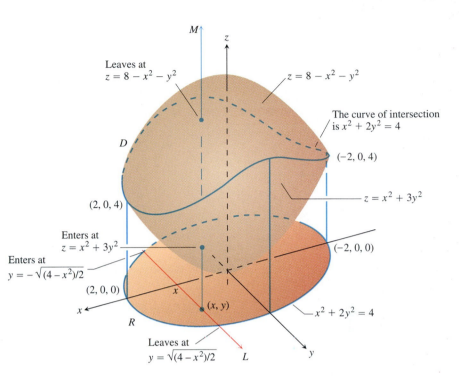

13.30 The volume of the region enclosed by these two paraboloids is calculated in Example 1.

EXAMPLE 1 Find the volume of the region D enclosed by the surfaces $z = x^2 + 3y^2$ and $z = 8 - x^2 - y^2$.

Solution The volume is

$$V = \iiint_D dz\, dy\, dx,$$

the integral of $F(x, y, z) = 1$ over D. To find the limits of integration for evaluating the integral, we take these steps.

Step 1: *A sketch.* The surfaces (Fig. 13.30) intersect on the elliptical cylinder $x^2 + 3y^2 = 8 - x^2 - y^2$ or $x^2 + 2y^2 = 4$. The boundary of the region R, the projection of D onto the xy-plane, is an ellipse with the same equation: $x^2 + 2y^2 = 4$. The "upper" boundary of R is the curve $y = \sqrt{(4 - x^2)/2}$. The lower boundary is the curve $y = -\sqrt{(4 - x^2)/2}$.

Step 2: *The z-limits of integration.* The line M passing through a typical point (x, y) in R parallel to the z-axis enters D at $z = x^2 + 3y^2$ and leaves at $z = 8 - x^2 - y^2$.

Step 3: *The y-limits of integration.* The line L through (x, y) parallel to the y-axis enters R at $y = -\sqrt{(4 - x^2)/2}$ and leaves at $y = \sqrt{(4 - x^2)/2}$.

Step 4: *The x-limits of integration.* As L sweeps across R, the value of x varies from $x = -2$ at $(-2, 0, 0)$ to $x = 2$ at $(2, 0, 0)$. The volume of D is

$$V = \iiint_D dz\, dy\, dx$$

$$= \int_{-2}^{2} \int_{-\sqrt{(4-x^2)/2}}^{\sqrt{(4-x^2)/2}} \int_{x^2+3y^2}^{8-x^2-y^2} dz\, dy\, dx$$

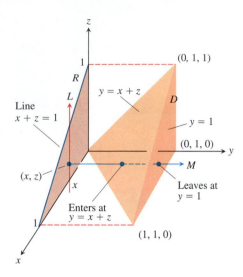

13.31 The tetrahedron in Example 2.

$$= \int_{-2}^{2} \int_{-\sqrt{(4-x^2)/2}}^{\sqrt{(4-x^2)/2}} (8 - 2x^2 - 4y^2)\, dy\, dx$$

$$= \int_{-2}^{2} \left[(8 - 2x^2)y - \frac{4}{3}y^3 \right]_{y=-\sqrt{(4-x^2)/2}}^{y=\sqrt{(4-x^2)/2}} dx$$

$$= \int_{-2}^{2} \left(2(8 - 2x^2)\sqrt{\frac{4 - x^2}{2}} - \frac{8}{3}\left(\frac{4 - x^2}{2}\right)^{3/2} \right) dx$$

$$= \int_{-2}^{2} \left[8\left(\frac{4 - x^2}{2}\right)^{3/2} - \frac{8}{3}\left(\frac{4 - x^2}{2}\right)^{3/2} \right] dx = \frac{4\sqrt{2}}{3}\int_{-2}^{2}(4 - x^2)^{3/2}\, dx$$

$$= 8\pi\sqrt{2}. \qquad \text{After integration with the substitution } x = 2\sin u \qquad \Box$$

In the next example, we project D onto the xz-plane instead of the xy-plane.

EXAMPLE 2   Set up the limits of integration for evaluating the triple integral of a function $F(x, y, z)$ over the tetrahedron D with vertices $(0, 0, 0)$, $(1, 1, 0)$, $(0, 1, 0)$, and $(0, 1, 1)$.

Solution

Step 1: *A sketch.* We sketch D along with its "shadow" R in the xz-plane (Fig. 13.31). The upper (right-hand) bounding surface of D lies in the plane $y = 1$. The lower (left-hand) bounding surface lies in the plane $y = x + z$. The upper boundary of R is the line $z = 1 - x$. The lower boundary is the line $z = 0$.

Step 2: *The y-limits of integration.* The line through a typical point (x, z) in R parallel to the y-axis enters D at $y = x + z$ and leaves at $y = 1$.

Step 3: *The z-limits of integration.* The line L through (x, z) parallel to the z-axis enters R at $z = 0$ and leaves at $z = 1 - x$.

Step 4: *The x-limits of integration.* As L sweeps across R, the value of x varies from $x = 0$ to $x = 1$. The integral is

$$\int_0^1 \int_0^{1-x} \int_{x+z}^1 F(x, y, z)\, dy\, dz\, dx. \qquad \Box$$

As we know, there are sometimes (but not always) two different orders in which the single integrations for evaluating a double integral may be worked. For triple integrals, there could be as many as *six.*

EXAMPLE 3   Each of the following integrals gives the volume of the solid shown in Fig. 13.32.

a) $\displaystyle\int_0^1 \int_0^{1-z} \int_0^2 dx\, dy\, dz$ b) $\displaystyle\int_0^1 \int_0^{1-y} \int_0^2 dx\, dz\, dy$

c) $\displaystyle\int_0^1 \int_0^2 \int_0^{1-z} dy\, dx\, dz$ d) $\displaystyle\int_0^2 \int_0^1 \int_0^{1-z} dy\, dz\, dx$

e) $\displaystyle\int_0^1 \int_0^2 \int_0^{1-y} dz\, dx\, dy$ f) $\displaystyle\int_0^2 \int_0^1 \int_0^{1-y} dz\, dy\, dx$ $\Box$

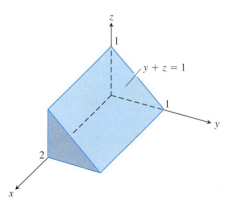

13.32 Example 3 gives six different iterated triple integrals for the volume of this prism.

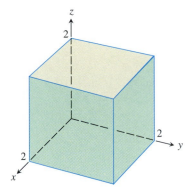

13.33 The region of integration in Example 4.

Average Value of a Function in Space

The average value of a function F over a region D in space is defined by the formula

$$\textbf{Average value of } F \text{ over } D = \frac{1}{\text{volume of } D} \iiint_D F\,dV. \qquad (5)$$

For example, if $F(x, y, z) = \sqrt{x^2 + y^2 + z^2}$, then the average value of F over D is the average distance of points in D from the origin. If $F(x, y, z)$ is the density of a solid that occupies a region D in space, then the average value of F over D is the average density of the solid in units of mass per unit volume.

EXAMPLE 4   Find the average value of $F(x, y, z) = xyz$ over the cube bounded by the coordinate planes and the planes $x = 2$, $y = 2$, and $z = 2$ in the first octant.

Solution   We sketch the cube with enough detail to show the limits of integration (Fig. 13.33). We then use Eq. (5) to calculate the average value of F over the cube. The volume of the cube is $(2)(2)(2) = 8$. The value of the integral of F over the cube is

$$\int_0^2 \int_0^2 \int_0^2 xyz\,dx\,dy\,dz = \int_0^2 \int_0^2 \left[\frac{x^2}{2} yz \right]_{x=0}^{x=2} dy\,dz = \int_0^2 \int_0^2 2yz\,dy\,dz$$

$$= \int_0^2 \left[y^2 z \right]_{y=0}^{y=2} dz = \int_0^2 4z\,dz = \left[2z^2 \right]_0^2 = 8.$$

With these values, Eq. (5) gives

$$\begin{array}{c} \text{Average value of} \\ xyz \text{ over the cube} \end{array} = \frac{1}{\text{volume}} \iiint_{\text{cube}} xyz\,dV = \left(\frac{1}{8} \right)(8) = 1.$$

In evaluating the integral, we chose the order dx, dy, dz, but any of the other five possible orders would have done as well.   ❑

Exercises 13.4

Evaluating Triple Integrals in Different Iterations

1. Find the common value of the integrals in Example 3.

2. Write six different iterated triple integrals for the volume of the rectangular solid in the first octant bounded by the coordinate planes and the planes $x = 1$, $y = 2$, and $z = 3$. Evaluate one of the integrals.

3. Write six different iterated triple integrals for the volume of the tetrahedron cut from the first octant by the plane $6x + 3y + 2z = 6$. Evaluate one of the integrals.

4. Write six different iterated triple integrals for the volume of the region in the first octant enclosed by the cylinder $x^2 + z^2 = 4$ and the plane $y = 3$. Evaluate one of the integrals.

5. Let D be the region bounded by the paraboloids $z = 8 - x^2 - y^2$ and $z = x^2 + y^2$. Write six different triple iterated integrals for the volume of D. Evaluate one of the integrals.

6. Let D be the region bounded by the paraboloid $z = x^2 + y^2$ and the plane $z = 2y$. Write triple iterated integrals in the order $dz\,dx\,dy$ and $dz\,dy\,dx$ that give the volume of D. Do not evaluate either integral.

Evaluating Triple Iterated Integrals

Evaluate the integrals in Exercises 7–20.

7. $\int_0^1 \int_0^1 \int_0^1 (x^2 + y^2 + z^2)\,dz\,dy\,dx$

8. $\int_0^{\sqrt{2}} \int_0^{3y} \int_{x^2+3y^2}^{8-x^2-y^2} dz\,dx\,dy$ **9.** $\int_1^e \int_1^e \int_1^e \frac{1}{xyz}\,dx\,dy\,dz$

10. $\int_0^1 \int_0^{3-3x} \int_0^{3-3x-y} dz\,dy\,dx$

11. $\int_0^1 \int_0^\pi \int_0^\pi y \sin z\,dx\,dy\,dz$

12. $\int_{-1}^1 \int_{-1}^1 \int_{-1}^1 (x+y+z)\,dy\,dx\,dz$

13. $\int_0^3 \int_0^{\sqrt{9-x^2}} \int_0^{\sqrt{9-x^2}} dz\,dy\,dx$

14. $\int_0^2 \int_{-\sqrt{4-y^2}}^{\sqrt{4-y^2}} \int_0^{2x+y} dz\,dx\,dy$

15. $\int_0^1 \int_0^{2-x} \int_0^{2-x-y} dz\,dy\,dx$

16. $\int_0^1 \int_0^{1-x^2} \int_3^{4-x^2-y} x\,dz\,dy\,dx$

17. $\int_0^\pi \int_0^\pi \int_0^\pi \cos(u+v+w)\,du\,dv\,dw$ $(uvw\text{-space})$

18. $\int_1^e \int_1^e \int_1^e \ln r \ln s \ln t\,dt\,dr\,ds$ $(rst\text{-space})$

19. $\int_0^{\pi/4} \int_0^{\ln \sec v} \int_{-\infty}^{2t} e^x\,dx\,dt\,dv$ $(tvx\text{-space})$

20. $\int_0^7 \int_0^2 \int_0^{\sqrt{4-q^2}} \frac{q}{r+1}\,dp\,dq\,dr$ $(pqr\text{-space})$

Volumes Using Triple Integrals

21. Here is the region of integration of the integral

$$\int_{-1}^1 \int_{x^2}^1 \int_0^{1-y} dz\,dy\,dx.$$

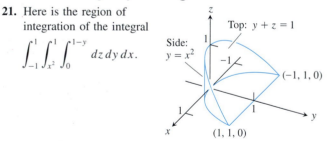

Side:
y = x²

Top: y + z = 1

(−1, 1, 0)

(1, 1, 0)

Rewrite the integral as an equivalent iterated integral in the order

a) $dy\,dz\,dx$ b) $dy\,dx\,dz$
c) $dx\,dy\,dz$ d) $dx\,dz\,dy$
e) $dz\,dx\,dy$

22. Here is the region of integration of the integral

$$\int_0^1 \int_{-1}^0 \int_0^{y^2} dz\,dy\,dx.$$

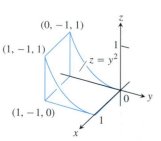

(0, −1, 1)

(1, −1, 1)

z = y²

(1, −1, 0)

Rewrite the integral as an equivalent iterated integral in the order

a) $dy\,dz\,dx$ b) $dy\,dx\,dz$
c) $dx\,dy\,dz$ d) $dx\,dz\,dy$
e) $dz\,dx\,dy$

Find the volumes of the regions in Exercises 23–36.

23. The region between the cylinder $z = y^2$ and the xy-plane that is bounded by the planes $x = 0, x = 1,$ $y = -1, y = 1$

24. The region in the first octant bounded by the coordinate planes and the planes $x + z = 1, y + 2z = 2$

25. The region in the first octant bounded by the coordinate planes, the plane $y + z = 2,$ and the cylinder $x = 4 - y^2$

26. The wedge cut from the cylinder $x^2 + y^2 = 1$ by the planes $z = -y$ and $z = 0$

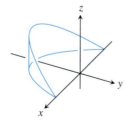

27. The tetrahedron in the first octant bounded by the coordinate planes and the plane $x + y/2 + z/3 = 1$

28. The region in the first octant bounded by the coordinate planes, the plane $y = 1 - x$, and the surface $z = \cos(\pi x/2)$, $0 \le x \le 1$

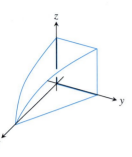

29. The region common to the interiors of the cylinders $x^2 + y^2 = 1$ and $x^2 + z^2 = 1$ (Fig. 13.34)

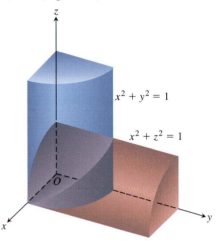

$x^2 + y^2 = 1$

$x^2 + z^2 = 1$

13.34 One-eighth of the region common to the cylinders $x^2 + y^2 = 1$ and $x^2 + z^2 = 1$ in Exercise 29.

30. The region in the first octant bounded by the coordinate planes and the surface $z = 4 - x^2 - y$

31. The region in the first octant bounded by the coordinate planes, the plane $x + y = 4$, and the cylinder $y^2 + 4z^2 = 16$

32. The region cut from the cylinder $x^2 + y^2 = 4$ by the plane $z = 0$ and the plane $x + z = 3$

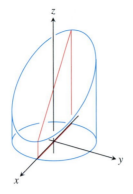

33. The region between the planes $x + y + 2z = 2$ and $2x + 2y + z = 4$ in the first octant

34. The finite region bounded by the planes $z = x$, $x + z = 8$, $z = y$, $y = 8$, and $z = 0$.

35. The region cut from the solid elliptical cylinder $x^2 + 4y^2 \le 4$ by the xy-plane and the plane $z = x + 2$

36. The region bounded in back by the plane $x = 0$, on the front and sides by the parabolic cylinder $x = 1 - y^2$, on the top by the paraboloid $z = x^2 + y^2$, and on the bottom by the xy-plane

Average Values

In Exercises 37–40, find the average value of $F(x, y, z)$ over the given region.

37. $F(x, y, z) = x^2 + 9$ over the cube in the first octant bounded by the coordinate planes and the planes $x = 2$, $y = 2$, and $z = 2$

38. $F(x, y, z) = x + y - z$ over the rectangular solid in the first octant bounded by the coordinate planes and the planes $x = 1$, $y = 1$, and $z = 2$

39. $F(x, y, z) = x^2 + y^2 + z^2$ over the cube in the first octant bounded by the coordinate planes and the planes $x = 1$, $y = 1$, and $z = 1$

40. $F(x, y, z) = xyz$ over the cube in the first octant bounded by the coordinate planes and the planes $x = 2$, $y = 2$, and $z = 2$

Changing the Order of Integration

Evaluate the integrals in Exercises 41–44 by changing the order of integration in an appropriate way.

41. $\displaystyle\int_0^4 \int_0^1 \int_{2y}^2 \frac{4\cos(x^2)}{2\sqrt{z}}\, dx\, dy\, dz$

42. $\displaystyle\int_0^1 \int_0^1 \int_{x^2}^1 12xz\, e^{zy^2}\, dy\, dx\, dz$

43. $\displaystyle\int_0^1 \int_{\sqrt[3]{z}}^1 \int_0^{\ln 3} \frac{\pi e^{2x} \sin \pi y^2}{y^2}\, dx\, dy\, dz$

44. $\displaystyle\int_0^2 \int_0^{4-x^2} \int_0^x \frac{\sin 2z}{4 - z}\, dy\, dz\, dx$

Theory and Examples

45. Solve for a:

$$\int_0^1 \int_0^{4-a-x^2} \int_a^{4-x^2-y} dz\, dy\, dx = \frac{4}{15}.$$

46. For what value of c is the volume of the ellipsoid $x^2 + (y/2)^2 + (z/c)^2 = 1$ equal to 8π?

47. What domain D in space minimizes the value of the integral

$$\iiint_D (4x^2 + 4y^2 + z^2 - 4)\, dV?$$

Give reasons for your answer.

48. What domain D in space maximizes the value of the integral

$$\iiint_D (1 - x^2 - y^2 - z^2)\, dV?$$

Give reasons for your answer.

CAS Explorations and Projects

In Exercises 49–52, use a CAS integration utility to evaluate the triple integral of the given function over the specified solid region.

49. $F(x, y, z) = x^2y^2z$ over the solid cylinder bounded by $x^2 + y^2 = 1$ and the planes $z = 0$ and $z = 1$.

50. $F(x, y, z) = |xyz|$ over the solid bounded below by the paraboloid $z = x^2 + y^2$ and above by the plane $z = 1$.

51. $F(x, y, z) = \dfrac{z}{(x^2 + y^2 + z^2)^{3/2}}$ over the solid bounded below by the cone $z = \sqrt{x^2 + y^2}$ and above by the plane $z = 1$.

52. $F(x, y, z) = x^4 + y^2 + z^2$ over the solid sphere $x^2 + y^2 + z^2 \le 1$.

13.5

Masses and Moments in Three Dimensions

This section shows how to calculate the masses and moments of three-dimensional objects in Cartesian coordinates. The formulas are similar to those for two-dimensional objects. For calculations in spherical and cylindrical coordinates, see Section 13.6.

Masses and Moments

If $\delta(x, y, z)$ is the density of an object occupying a region D in space (mass per unit volume), the integral of δ over D gives the mass of the object. To see why, imagine partitioning the object into n mass elements like the one in Fig. 13.35. The object's mass is the limit

$$M = \lim_{n\to\infty} \sum_{k=1}^n \Delta m_k = \lim_{n\to\infty} \sum_{k=1}^n \delta(x_k, y_k, z_k)\Delta V_k = \iiint_D \delta(x, y, z)\, dV. \quad (1)$$

If $r(x, y, z)$ is the distance from the point (x, y, z) in D to a line L, then the moment of inertia of the mass $\Delta m_k = \delta(x_k, y_k, z_k)\Delta V_k$ about the line L (shown in Fig. 13.35) is approximately $\Delta I_k = r^2(x_k, y_k, z_k)\Delta m_k$. The moment of inertia of the entire object about L is

$$I_L = \lim_{n\to\infty} \sum_{k=1}^n \Delta I_k = \lim_{n\to\infty} \sum_{k=1}^n r^2(x_k, y_k, z_k)\,\delta(x_k, y_k, z_k)\Delta V_k = \iiint_D r^2\delta\, dV.$$

If L is the x-axis, then $r^2 = y^2 + z^2$ (Fig. 13.36) and

$$I_x = \iiint_D (y^2 + z^2)\,\delta\, dV.$$

Similarly,

$$I_y = \iiint_D (x^2 + z^2)\,\delta\, dV \qquad \text{and} \qquad I_z = \iiint_D (x^2 + y^2)\,\delta\, dV.$$

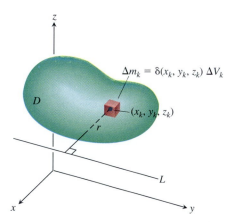

$\Delta m_k = \delta(x_k, y_k, z_k)\, \Delta V_k$

(x_k, y_k, z_k)

13.35 To define an object's mass and moment of inertia about a line, we first imagine it to be partitioned into a finite number of mass elements Δm_k.

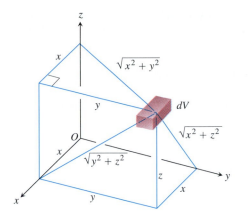

13.36 Distances from dV to the coordinate planes and axes.

These and other useful formulas are summarized in Table 13.2.

Table 13.2 Mass and moment formulas for objects in space

Mass: $\qquad M = \iiint_D \delta \, dV \qquad (\delta = \text{density})$

First moments about the coordinate planes:

$$M_{yz} = \iiint_D x \, \delta \, dV, \qquad M_{xz} = \iiint_D y \, \delta \, dV, \qquad M_{xy} = \iiint_D z \, \delta \, dV$$

Center of mass:

$$\overline{x} = \frac{M_{yz}}{M}, \qquad \overline{y} = \frac{M_{xz}}{M}, \qquad \overline{z} = \frac{M_{xy}}{M}$$

Moments of inertia (second moments):

$$I_x = \iiint (y^2 + z^2) \, \delta \, dV$$
$$I_y = \iiint (x^2 + z^2) \, \delta \, dV$$
$$I_z = \iiint (x^2 + y^2) \, \delta \, dV$$

Moment of inertia about a line L:

$I_L = \iiint r^2 \, \delta \, dV \qquad (r(x, y, z) = \text{distance from points } (x, y, z) \text{ to line } L)$

Radius of gyration about a line L:

$$R_L = \sqrt{I_L / M}$$

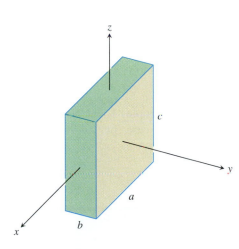

13.37 Example 1 calculates I_x, I_y, and I_z for the block shown here. The origin lies at the center of the block.

EXAMPLE 1 Find I_x, I_y, I_z for the rectangular solid of constant density δ shown in Fig. 13.37.

Solution The preceding formula for I_x gives

$$I_x = \int_{-c/2}^{c/2} \int_{-b/2}^{b/2} \int_{-a/2}^{a/2} (y^2 + z^2) \, \delta \, dx \, dy \, dz. \qquad (2)$$

We can avoid some of the work of integration by observing that $(y^2 + z^2) \delta$ is an even function of x, y, and z and therefore

$$I_x = 8 \int_0^{c/2} \int_0^{b/2} \int_0^{a/2} (y^2 + z^2) \, \delta \, dx \, dy \, dz = 4a\delta \int_0^{c/2} \int_0^{b/2} (y^2 + z^2) \, dy \, dz$$

$$= 4a\delta \int_0^{c/2} \left[\frac{y^3}{3} + z^2 y \right]_{y=0}^{y=b/2} dz$$

$$= 4a\delta \int_0^{c/2} \left(\frac{b^3}{24} + \frac{z^2 b}{2} \right) dz$$

$$= 4a\delta \left(\frac{b^3 c}{48} + \frac{c^3 b}{48} \right) = \frac{abc\delta}{12} (b^2 + c^2) = \frac{M}{12} (b^2 + c^2).$$

Similarly,

$$I_y = \frac{M}{12} (a^2 + c^2) \qquad \text{and} \qquad I_z = \frac{M}{12} (a^2 + b^2). \qquad \square$$

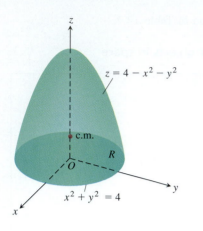

13.38 Example 2 finds the center of mass of this solid.

EXAMPLE 2   Find the center of mass of a solid of constant density δ bounded below by the disk R: $x^2 + y^2 \leq 4$ in the plane $z = 0$ and above by the paraboloid $z = 4 - x^2 - y^2$ (Fig. 13.38).

Solution   By symmetry, $\bar{x} = \bar{y} = 0$. To find $\bar{z}$, we first calculate

$$M_{xy} = \iiint_R \int_{z=0}^{z=4-x^2-y^2} z \, \delta \, dz \, dy \, dx = \iint_R \left[\frac{z^2}{2}\right]_{z=0}^{z=4-x^2-y^2} \delta \, dy \, dx$$

$$= \frac{\delta}{2} \iint_R (4 - x^2 - y^2)^2 \, dy \, dx$$

$$= \frac{\delta}{2} \int_0^{2\pi} \int_0^2 (4 - r^2)^2 \, r \, dr \, d\theta \qquad \text{\color{blue}{Polar coordinates}}$$

$$= \frac{\delta}{2} \int_0^{2\pi} \left[-\frac{1}{6}(4 - r^2)^3\right]_{r=0}^{r=2} \, d\theta = \frac{16\delta}{3} \int_0^{2\pi} d\theta = \frac{32\pi\delta}{3}.$$

A similar calculation gives

$$M = \iiint_R \int_0^{4-x^2-y^2} \delta \, dz \, dy \, dx = 8\pi \, \delta.$$

Therefore $\bar{z} = (M_{xy}/M) = 4/3$, and the center of mass is $(\bar{x}, \bar{y}, \bar{z}) = (0, 0, 4/3)$. ☐

When the density of a solid object is constant (as in Examples 1 and 2), the center of mass is called the **centroid** of the object (as was the case for two-dimensional shapes in Section 13.2).

Exercises 13.5

Constant Density

The solids in Exercises 1–12 all have constant density $\delta = 1$.

1. Evaluate the integral for I_x in Eq. (2) directly to show that the shortcut in Example 1 gives the same answer. Use the results in Example 1 to find the radius of gyration of the rectangular solid about each coordinate axis.

2. The coordinate axes in the figure to the right run through the centroid of a solid wedge parallel to the labeled edges. Find I_x, I_y, and I_z if $a = b = 6$ and $c = 4$.

3. Find the moments of inertia of the rectangular solid shown here with respect to its edges by calculating I_x, I_y, and I_z.

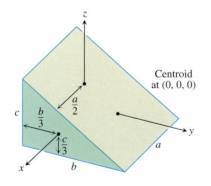

Figure for Exercise 2

4. **a)**   Find the centroid and the moments of inertia I_x, I_y, and I_z of the tetrahedron whose vertices are the points $(0, 0, 0)$, $(1, 0, 0)$, $(0, 1, 0)$, and $(0, 0, 1)$.

   b)   Find the radius of gyration of the tetrahedron about the x-axis. Compare it with the distance from the centroid to the x-axis.

5. A solid "trough" of constant density is bounded below by the surface $z = 4y^2$, above by the plane $z = 4$, and on the ends by the planes $x = 1$ and $x = -1$. Find the center of mass and the moments of inertia with respect to the three axes.

6. A solid of constant density is bounded below by the plane $z = 0$, on the sides by the elliptic cylinder $x^2 + 4y^2 = 4$, and above by the plane $z = 2 - x$ (see the figure).

a) Find $\bar{x}$ and $\bar{y}$.

b) Evaluate the integral

$$M_{xy} = \int_{-2}^{2} \int_{-(1/2)\sqrt{4-x^2}}^{(1/2)\sqrt{4-x^2}} \int_{0}^{2-x} z \, dz \, dy \, dx,$$

using integral tables to carry out the final integration with respect to x. Then divide M_{xy} by M to verify that $\bar{z} = 5/4$.

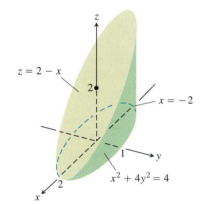

7. a) Find the center of mass of a solid of constant density bounded below by the paraboloid $z = x^2 + y^2$ and above by the plane $z = 4$.

b) Find the plane $z = c$ that divides the solid into two parts of equal volume. This plane does not pass through the center of mass.

8. A solid cube, 2 units on a side, is bounded by the planes $x = \pm 1$, $z = \pm 1$, $y = 3$, and $y = 5$. Find the center of mass and the moments of inertia and radii of gyration about the coordinate axes.

9. A wedge like the one in Exercise 2 has $a = 4$, $b = 6$, and $c = 3$. Make a quick sketch to check for yourself that the square of the distance from a typical point (x, y, z) of the wedge to the line $L: z = 0$, $y = 6$ is $r^2 = (y - 6)^2 + z^2$. Then calculate the moment of inertia and radius of gyration of the wedge about L.

10. A wedge like the one in Exercise 2 has $a = 4$, $b = 6$, and $c = 3$. Make a quick sketch to check for yourself that the square of the distance from a typical point (x, y, z) of the wedge to the line $L: x = 4$, $y = 0$ is $r^2 = (x - 4)^2 + y^2$. Then calculate the moment of inertia and radius of gyration of the wedge about L.

11. A solid like the one in Exercise 3 has $a = 4$, $b = 2$, and $c = 1$. Make a quick sketch to check for yourself that the square of the distance between a typical point (x, y, z) of the solid and the line $L: y = 2$, $z = 0$ is $r^2 = (y - 2)^2 + z^2$. Then find the moment of inertia and radius of gyration of the solid about L.

12. A solid like the one in Exercise 3 has $a = 4$, $b = 2$, and $c = 1$. Make a quick sketch to check for yourself that the square of the distance between a typical point (x, y, z) of the solid and the line $L: x = 4$, $y = 0$ is $r^2 = (x - 4)^2 + y^2$. Then find the moment of inertia and radius of gyration of the solid about L.

Variable Density

In Exercises 13 and 14, find (a) the mass of the solid and (b) the center of mass.

13. A solid region in the first octant is bounded by the coordinate planes and the plane $x + y + z = 2$. The density of the solid is $\delta(x, y, z) = 2x$.

14. A solid in the first octant is bounded by the planes $y = 0$ and $z = 0$ and by the surfaces $z = 4 - x^2$ and $x = y^2$ (see the figure). Its density function is $\delta(x, y, z) = kxy$.

In Exercises 15 and 16, find

a) the mass of the solid

b) the center of mass

c) the moments of inertia about the coordinate axes

d) the radii of gyration about the coordinate axes.

15. A solid cube in the first octant is bounded by the coordinate planes and by the planes $x = 1$, $y = 1$, and $z = 1$. The density of the cube is $\delta(x, y, z) = x + y + z + 1$.

16. A wedge like the one in Exercise 2 has dimensions $a = 2$, $b = 6$, and $c = 3$. The density is $\delta(x, y, z) = x + 1$. Notice that if the density is constant, the center of mass will be $(0, 0, 0)$.

17. Find the mass of the solid bounded by the planes $x + z = 1$, $x - z = -1$, $y = 0$ and the surface $y = \sqrt{z}$. The density of the solid is $\delta(x, y, z) = 2y + 5$.

18. Find the mass of the solid region bounded by the parabolic surfaces $z = 16 - 2x^2 - 2y^2$ and $z = 2x^2 + 2y^2$ if the density of the solid is $\delta(x, y, z) = \sqrt{x^2 + y^2}$.

Work

In Exercises 19 and 20, calculate the following.

a) The amount of work done by (constant) gravity g in moving the liquid filled in the container to the xy-plane (*Hint:* Partition the liquid in the container into small volume elements ΔV_i and find the work done (approximately) by gravity on each element.

Summation and passage to the limit gives a triple integral to evaluate.)

b) The work done by gravity in moving the center of mass down to the xy-plane

19. The container is a cubical box in the first octant bounded by the coordinate planes and the planes $x = 1$, $y = 1$, and $z = 1$. The density of the liquid filling the box is $\delta(x, y, z) = x + y + z + 1$ (refer to Exercise 15).

20. The container is in the shape of the region bounded by $y = 0$, $z = 0$, $z = 4 - x^2$, and $x = y^2$. The density of the liquid filling the region is $\delta(x, y, z) = kxy$ (see Exercise 14).

The Parallel Axis Theorem

The Parallel Axis Theorem (Exercises 13.2) holds in three dimensions as well as in two. Let $L_{c.m.}$ be a line through the center of mass of a body of mass m and let L be a parallel line h units away from $L_{c.m.}$. The **Parallel Axis Theorem** says that the moments of inertia $I_{c.m.}$ and I_L of the body about $L_{c.m.}$ and L satisfy the equation

$$I_L = I_{c.m.} + mh^2. \tag{1}$$

As in the two-dimensional case, the theorem gives a quick way to calculate one moment when the other moment and the mass are known.

21. *Proof of the Parallel Axis Theorem*

a) Show that the first moment of a body in space about any plane through the body's center of mass is zero. (*Hint:* Place the body's center of mass at the origin and let the plane be the yz-plane. What does the formula $\bar{x} = M_{yz}/M$ then tell you?)

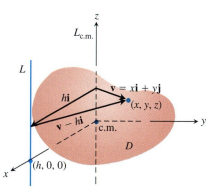

b) To prove the Parallel Axis Theorem, place the body with its center of mass at the origin, with the line $L_{c.m.}$ along the z-axis and the line L perpendicular to the xy-plane at the point $(h, 0, 0)$. Let D be the region of space occupied by the body. Then, in the notation of the figure,

$$I_L = \iiint\limits_{D} |\mathbf{v} - h\mathbf{i}|^2 \, dm. \tag{2}$$

Expand the integrand in this integral and complete the proof.

22. The moment of inertia about a diameter of a solid sphere of constant density and radius a is $(2/5)ma^2$, where m is the mass

of the sphere. Find the moment of inertia about a line tangent to the sphere.

23. The moment of inertia of the solid in Exercise 3 about the z-axis is $I_z = abc(a^2 + b^2)/3$.

a) Use Eq. (1) to find the moment of inertia and radius of gyration of the solid about the line parallel to the z-axis through the solid's center of mass.

b) Use Eq. (1) and the result in (a) to find the moment of inertia and radius of gyration of the solid about the line $x = 0$, $y = 2b$.

24. If $a = b = 6$ and $c = 4$, the moment of inertia of the solid wedge in Exercise 2 about the x-axis is $I_x = 208$. Find the moment of inertia of the wedge about the line $y = 4$, $z = -4/3$ (the edge of the wedge's narrow end).

Pappus's Formula

Pappus's formula (Exercises 13.2) holds in three dimensions as well as in two. Suppose that bodies B_1 and B_2 of mass m_1 and m_2, respectively, occupy nonoverlapping regions in space and that $\mathbf{c}_1$ and $\mathbf{c}_2$ are the vectors from the origin to the bodies' respective centers of mass. Then the center of mass of the union $B_1 \cup B_2$ of the two bodies is determined by the vector

$$\mathbf{c} = \frac{m_1 \mathbf{c}_1 + m_2 \mathbf{c}_2}{m_1 + m_2}. \tag{3}$$

As before, this formula is called **Pappus's formula.** As in the two-dimensional case, the formula generalizes to

$$\mathbf{c} = \frac{m_1 \mathbf{c}_1 + m_2 \mathbf{c}_2 + \cdots + m_n \mathbf{c}_n}{m_1 + m_2 + \cdots + m_n} \tag{4}$$

for n bodies.

25. Derive Pappus's formula (Eq. 3). (*Hint:* Sketch B_1 and B_2 as nonoverlapping regions in the first octant and label their centers of mass $(\bar{x}_1, \bar{y}_1, \bar{z}_1)$ and $(\bar{x}_2, \bar{y}_2, \bar{z}_2)$. Express the moments of $B_1 \cup B_2$ about the coordinate planes in terms of the masses m_1 and m_2 and the coordinates of these centers.)

26. The figure below shows a solid made from three rectangular solids of constant density $\delta = 1$. Use Pappus's formula to find the center of mass of

a) $A \cup B$ **b)** $A \cup C$

c) $B \cup C$ **d)** $A \cup B \cup C$.

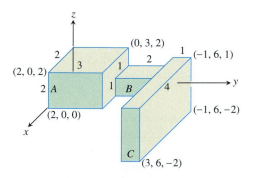

27. a) Suppose that a solid right circular cone C of base radius a and altitude h is constructed on the circular base of a solid hemisphere S of radius a so that the union of the two solids resembles an ice cream cone. The centroid of a solid cone lies one-fourth of the way from the base toward the vertex. The centroid of a solid hemisphere lies three-eighths of the way from the base to the top. What relation must hold between h and a to place the centroid of $C \cup S$ in the common base of the two solids?

b) If you have not already done so, answer the analogous ques-

tion about a triangle and a semicircle (Section 13.2, Exercise 55). The answers are not the same.

28. A solid pyramid P with height h and four congruent sides is built with its base as one face of a solid cube C whose edges have length s. The centroid of a solid pyramid lies one-fourth of the way from the base toward the vertex. What relation must hold between h and s to place the centroid of $P \cup C$ in the base of the pyramid? Compare your answer with the answer to Exercise 27. Also compare it to the answer to Exercise 56 in Section 13.2.

13.6 Triple Integrals in Cylindrical and Spherical Coordinates

When a calculation in physics, engineering, or geometry involves a cylinder, cone, or sphere, we can often simplify our work by using cylindrical or spherical coordinates.

Cylindrical Coordinates

Cylindrical coordinates (Fig. 13.39) are good for describing cylinders whose axes run along the z-axis and planes that either contain the z-axis or lie perpendicular to the z-axis. As we saw in Section 10.7, surfaces like these have equations of constant coordinate value:

$$r = 4 \qquad \text{Cylinder, radius 4, axis the } z\text{-axis}$$

$$\theta = \frac{\pi}{3} \qquad \text{Plane containing the } z\text{-axis}$$

$$z = 2 \qquad \text{Plane perpendicular to the } z\text{-axis}$$

The volume element for subdividing a region in space with cylindrical coordinates is

$$dV = dz\, r\, dr\, d\theta \qquad (1)$$

(Fig. 13.40). Triple integrals in cylindrical coordinates are then evaluated as iterated integrals, as in the following example.

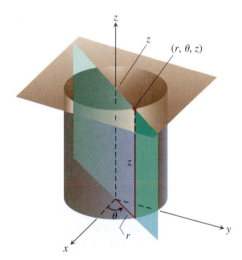

13.39 Cylindrical coordinates and typical surfaces of constant coordinate value.

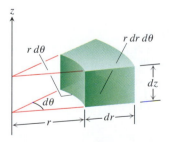

13.40 The volume element in cylindrical coordinates is $dV = dz\, r\, dr\, d\theta$.

EXAMPLE 1 Find the limits of integration in cylindrical coordinates for integrating a function $f(r, \theta, z)$ over the region D bounded below by the plane $z = 0$, laterally by the circular cylinder $x^2 + (y - 1)^2 = 1$, and above by the paraboloid $z = x^2 + y^2$.

Solution

Step 1: *A sketch* (Fig. 13.41). The base of D is also the region's projection R on the xy-plane. The boundary of R is the circle $x^2 + (y - 1)^2 = 1$. Its polar coordinate equation is

$$x^2 + (y - 1)^2 = 1$$

$$x^2 + y^2 - 2y + 1 = 1$$

$$r^2 - 2r \sin \theta = 0$$

$$r = 2 \sin \theta.$$

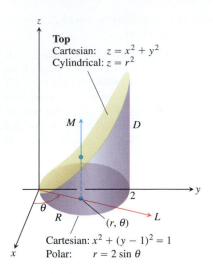

Top
Cartesian: $z = x^2 + y^2$
Cylindrical: $z = r^2$

M D

(r, θ) L

R

Cartesian: $x^2 + (y - 1)^2 = 1$
Polar: $r = 2 \sin \theta$

13.41 The figure for Example 1.

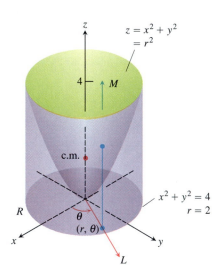

$z = x^2 + y^2$
$= r^2$

M

c.m.

$x^2 + y^2 = 4$
$r = 2$

R

(r, θ)

L

13.42 Example 2 shows how to find the centroid of this solid.

Step 2: *The z-limits of integration.* A line M through a typical point (r, θ) in R parallel to the z-axis enters D at $z = 0$ and leaves at $z = x^2 + y^2 = r^2$.

Step 3: *The r-limits of integration.* A ray L through (r, θ) from the origin enters R at $r = 0$ and leaves at $r = 2 \sin \theta$.

Step 4: *The θ-limits of integration.* As L sweeps across R, the angle θ it makes with the positive x-axis runs from $\theta = 0$ to $\theta = \pi$. The integral is

$$\iiint_D f(r, \theta, z) \, dV = \int_0^\pi \int_0^{2 \sin \theta} \int_0^{r^2} f(r, \theta, z) \, dz \, r \, dr \, d\theta.$$

Example 1 illustrates a good procedure for finding limits of integration in cylindrical coordinates. The procedure is summarized in the box on the following page.

EXAMPLE 2   Find the centroid ($\delta = 1$) of the solid enclosed by the cylinder $x^2 + y^2 = 4$, bounded above by the paraboloid $z = x^2 + y^2$ and below by the xy-plane.

Solution   We sketch the solid, bounded above by the paraboloid $z = r^2$ and below by the plane $z = 0$ (Fig. 13.42). Its base R is the disk $|r| \le 2$ in the xy-plane.

The solid's centroid $(\overline{x}, \overline{y}, \overline{z})$ lies on its axis of symmetry, here the z-axis. This makes $\overline{x} = \overline{y} = 0$. To find $\overline{z}$, we divide the first moment M_{xy} by the mass M.

To find the limits of integration for the mass and moment integrals, we continue with the four basic steps. We completed step 1 with our initial sketch. The remaining steps give the limits of integration.

Step 2: *The z-limits.* A line M through a typical point (r, θ) in the base parallel to the z-axis enters the solid at $z = 0$ and leaves at $z = r^2$.

Step 3: *The r-limits.* A ray L through (r, θ) from the origin enters R at $r = 0$ and leaves at $r = 2$.

Step 4: *The θ-limits.* As L sweeps over the base like a clock hand, the angle θ it makes with the positive x-axis runs from $\theta = 0$ to $\theta = 2\pi$. The value of M_{xy} is

$$M_{xy} = \int_0^{2\pi} \int_0^2 \int_0^{r^2} z \, dz \, r \, dr \, d\theta = \int_0^{2\pi} \int_0^2 \left[\frac{z^2}{2} \right]_0^{r^2} r \, dr \, d\theta$$

$$= \int_0^{2\pi} \int_0^2 \frac{r^5}{2} \, dr \, d\theta = \int_0^{2\pi} \left[\frac{r^6}{12} \right]_0^2 d\theta = \int_0^{2\pi} \frac{16}{3} \, d\theta = \frac{32\pi}{3}.$$

The value of M is

$$M = \int_0^{2\pi} \int_0^2 \int_0^{r^2} dz \, r \, dr \, d\theta = \int_0^{2\pi} \int_0^2 \left[z \right]_0^{r^2} r \, dr \, d\theta$$

$$= \int_0^{2\pi} \int_0^2 r^3 \, dr \, d\theta = \int_0^{2\pi} \left[\frac{r^4}{4} \right]_0^2 d\theta = \int_0^{2\pi} 4 \, d\theta = 8\pi.$$

Therefore,

$$\overline{z} = \frac{M_{xy}}{M} = \frac{32\pi}{3} \frac{1}{8\pi} = \frac{4}{3},$$

and the centroid is $(0, 0, 4/3)$. Notice that the centroid lies outside the solid.

How to Integrate in Cylindrical Coordinates

To evaluate

$$\iiint_D f(r,\theta,z)\ dV$$

over a region D in space in cylindrical coordinates, integrating first with respect to z, then with respect to r, and finally with respect to θ, take the following steps.

1. *A sketch.* Sketch the region D along with its projection R on the xy-plane. Label the surfaces and curves that bound D and R.

2. *The z-limits of integration.* Draw a line M through a typical point (r,θ) of R parallel to the z-axis. As z increases, M enters D at $z = g_1(r,\theta)$ and leaves at $z = g_2(r,\theta)$. These are the z-limits of integration.

3. *The r-limits of integration.* Draw a ray L through (r,θ) from the origin. The ray enters R at $r = h_1(\theta)$ and leaves at $r = h_2(\theta)$. These are the r-limits of integration.

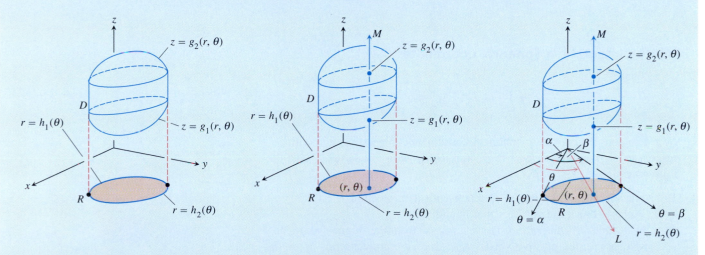

4. *The θ-limits of integration.* As L sweeps across R, the angle θ it makes with the positive x-axis runs from $\theta = \alpha$ to $\theta = \beta$. These are the θ-limits of integration. The integral is

$$\iiint_D f(r,\theta,z)\,dV = \int_{\theta=\alpha}^{\theta=\beta} \int_{r=h_1(\theta)}^{r=h_2(\theta)} \int_{z=g_1(r,\theta)}^{z=g_2(r,\theta)} f(r,\theta,z)\,dz\,r\,dr\,d\theta. \tag{2}$$

Spherical Coordinates

Spherical coordinates (Fig. 13.43, on the following page) are good for describing spheres centered at the origin, half-planes hinged along the z-axis, and single-napped cones whose vertices lie at the origin and whose axes lie along the z-axis. Surfaces like these have equations of constant coordinate value:

$\rho = 4$ Sphere, radius 4, center at origin

$\phi = \dfrac{\pi}{3}$ Cone opening up from the origin, making an angle of $\pi/3$ radians with the positive z-axis

$\theta = \dfrac{\pi}{3}$ Half-plane, hinged along the z-axis, making an angle of $\pi/3$ radians with the positive x-axis

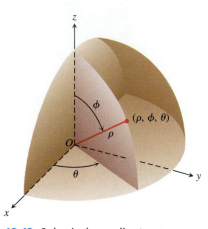

13.43 Spherical coordinates are measured with a distance and two angles.

The volume element in spherical coordinates is the volume of a **spherical wedge** defined by the differentials $d\rho$, $d\phi$, and $d\theta$ (Fig. 13.44). The wedge is approximately a rectangular box with one side a circular arc of length $\rho\,d\phi$, another side a circular arc of length $\rho\sin\phi\,d\theta$, and thickness $d\rho$. Therefore the volume element in spherical coordinates is

$$dV = \rho^2 \sin\phi\,d\rho\,d\phi\,d\theta, \tag{3}$$

and triple integrals take the form

$$\iiint F(\rho, \phi, \theta)\,dV = \iiint F(\rho, \phi, \theta)\,\rho^2 \sin\phi\,d\rho\,d\phi\,d\theta. \tag{4}$$

To evaluate these integrals, we usually integrate first with respect to ρ. The procedure for finding the limits of integration is shown in the following box. We restrict our attention to integrating over domains that are solids of revolution about the z-axis (or portions thereof) and for which the limits for θ and ϕ are constant.

How to Integrate in Spherical Coordinates

To evaluate

$$\iiint_D f(\rho, \phi, \theta)\ dV$$

over a region D in space in spherical coordinates, integrating first with respect to ρ, then with respect to ϕ, and finally with respect to θ, take the following steps.

1. *A sketch.* Sketch the region D along with its projection R on the xy-plane. Label the surfaces that bound D.
2. *The ρ-limits of integration.* Draw a ray M from the origin making an angle ϕ with the positive z-axis. Also draw the projection of M on the xy-plane (call the projection L). The ray L makes an angle θ with the positive x-axis. As ρ increases, M enters D at $\rho = g_1(\phi, \theta)$ and leaves at $\rho = g_2(\phi, \theta)$. These are the ρ-limits of integration.
3. *The ϕ-limits of integration.* For any given θ, the angle ϕ that M makes with the z-axis runs from $\phi = \phi_{\min}$ to $\phi = \phi_{\max}$. These are the ϕ-limits of integration.
4. *The θ-limits of integration.* The ray L sweeps over R as θ runs from α to β. These are the θ-limits of integration. The integral is

$$\iiint_D f(\rho, \phi, \theta)\,dV = \int_{\theta=\alpha}^{\theta=\beta} \int_{\phi=\phi_{\min}}^{\phi=\phi_{\max}} \int_{\rho=g_1(\phi,\theta)}^{\rho=g_2(\phi,\theta)} f(\rho, \phi, \theta)\,\rho^2 \sin\phi\,d\rho\,d\phi\,d\theta. \tag{5}$$

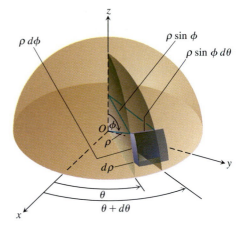

13.44 The volume element in spherical coordinates is

$$dV = d\rho \cdot \rho\, d\phi \cdot \rho \sin \phi\, d\theta$$
$$= \rho^2 \sin \phi\, d\rho\, d\phi\, d\theta.$$

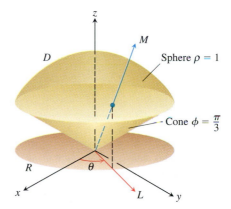

13.45 The solid in Example 3.

EXAMPLE 3   Find the volume of the upper region D cut from the solid sphere $\rho \leq 1$ by the cone $\phi = \pi/3$.

Solution   The volume is $V = \iiint_D \rho^2 \sin \phi\, d\rho\, d\phi\, d\theta$, the integral of $f(\rho, \phi, \theta)$ $= 1$ over D.

To find the limits of integration for evaluating the integral, we take the following steps.

Step 1: *A sketch.* We sketch D and its projection R on the xy-plane (Fig. 13.45).

Step 2: *The ρ-limits of integration.* We draw a ray M from the origin making an angle ϕ with the positive z-axis. We also draw L, the projection of M on the xy-plane, along with the angle θ that L makes with the positive x-axis. Ray M enters D at $\rho = 0$ and leaves at $\rho = 1$.

Step 3: *The ϕ-limits of integration.* The cone $\phi = \pi/3$ makes an angle of $\pi/3$ with the positive z-axis. For any given θ, the angle ϕ can run from $\phi = 0$ to $\phi = \pi/3$.

Step 4: *The θ-limits of integration.* The ray L sweeps over R as θ runs from 0 to 2π. The volume is

$$V = \iiint_D \rho^2 \sin \phi\, d\rho\, d\phi\, d\theta = \int_0^{2\pi} \int_0^{\pi/3} \int_0^1 \rho^2 \sin \phi\, d\rho\, d\phi\, d\theta$$

$$= \int_0^{2\pi} \int_0^{\pi/3} \left[\frac{\rho^3}{3} \right]_0^1 \sin \phi\, d\phi\, d\theta = \int_0^{2\pi} \int_0^{\pi/3} \frac{1}{3} \sin \phi\, d\phi\, d\theta$$

$$= \int_0^{2\pi} \left[-\frac{1}{3} \cos \phi \right]_0^{\pi/3} d\theta = \int_0^{2\pi} \left(-\frac{1}{6} + \frac{1}{3} \right) d\theta = \frac{1}{6}(2\pi) = \frac{\pi}{3}. \quad \square$$

EXAMPLE 4   A solid of constant density $\delta = 1$ occupies the region D in Example 3. Find the solid's moment of inertia about the z-axis.

Solution   In rectangular coordinates, the moment is

$$I_z = \iiint (x^2 + y^2)\, dV.$$

In spherical coordinates, $x^2 + y^2 = (\rho \sin \phi \cos \theta)^2 + (\rho \sin \phi \sin \theta)^2 = \rho^2 \sin^2 \phi$. Hence,

$$I_z = \iiint (\rho^2 \sin^2 \phi)\, \rho^2 \sin \phi\, d\rho\, d\phi\, d\theta = \iiint \rho^4 \sin^3 \phi\, d\rho\, d\phi\, d\theta.$$

For the region in Example 3, this becomes

$$I_z = \int_0^{2\pi} \int_0^{\pi/3} \int_0^1 \rho^4 \sin^3 \phi\, d\rho\, d\phi\, d\theta = \int_0^{2\pi} \int_0^{\pi/3} \left[\frac{\rho^5}{5} \right]_0^1 \sin^3 \phi\, d\phi\, d\theta$$

$$= \frac{1}{5} \int_0^{2\pi} \int_0^{\pi/3} (1 - \cos^2 \phi) \sin \phi\, d\phi\, d\theta = \frac{1}{5} \int_0^{2\pi} \left[-\cos \phi + \frac{\cos^3 \phi}{3} \right]_0^{\pi/3} d\theta$$

$$= \frac{1}{5} \int_0^{2\pi} \left(-\frac{1}{2} + 1 + \frac{1}{24} - \frac{1}{3} \right) d\theta = \frac{1}{5} \int_0^{2\pi} \frac{5}{24} d\theta = \frac{1}{24}(2\pi) = \frac{\pi}{12}. \quad \square$$

Coordinate Conversion Formulas (from Section 10.8)

Cylindrical to Rectangular	Spherical to Rectangular	Spherical to Cylindrical
$x = r \cos \theta$	$x = \rho \sin \phi \cos \theta$	$r = \rho \sin \phi$
$y = r \sin \theta$	$y = \rho \sin \phi \sin \theta$	$z = \rho \cos \phi$
$z = z$	$z = \rho \cos \phi$	$\theta = \theta$

Corresponding volume elements

$$dV = dx\,dy\,dz$$
$$= dz\,r\,dr\,d\theta$$
$$= \rho^2 \sin \phi\,d\rho\,d\phi\,d\theta$$

Exercises 13.6

Cylindrical Coordinates

Evaluate the cylindrical coordinate integrals in Exercises 1–6.

1. $\displaystyle\int_0^{2\pi}\int_0^1\int_r^{\sqrt{2-r^2}} dz\,r\,dr\,d\theta$

2. $\displaystyle\int_0^{2\pi}\int_0^3\int_{r^2/3}^{\sqrt{18-r^2}} dz\,r\,dr\,d\theta$

3. $\displaystyle\int_0^{2\pi}\int_0^{\theta/2\pi}\int_0^{3+24r^2} dz\,r\,dr\,d\theta$

4. $\displaystyle\int_0^{\pi}\int_0^{\theta/\pi}\int_{-\sqrt{4-r^2}}^{3\sqrt{4-r^2}} z\,dz\,r\,dr\,d\theta$

5. $\displaystyle\int_0^{2\pi}\int_0^1\int_r^{1/\sqrt{2-r^2}} 3\,dz\,r\,dr\,d\theta$

6. $\displaystyle\int_0^{2\pi}\int_0^1\int_{-1/2}^{1/2} (r^2 \sin^2 \theta + z^2)\,dz\,r\,dr\,d\theta$

The integrals we have seen so far suggest that there are preferred orders of integration for cylindrical coordinates, but other orders usually work well and are occasionally easier to evaluate. Evaluate the integrals in Exercises 7–10.

7. $\displaystyle\int_0^{2\pi}\int_0^3\int_0^{z/3} r^3\,dr\,dz\,d\theta$

8. $\displaystyle\int_{-1}^1\int_0^{2\pi}\int_0^{1+\cos\theta} 4r\,dr\,d\theta\,dz$

9. $\displaystyle\int_0^1\int_0^{\sqrt{z}}\int_0^{2\pi} (r^2 \cos^2 \theta + z^2)\,r\,d\theta\,dr\,dz$

10. $\displaystyle\int_0^2\int_{r-2}^{\sqrt{4-r^2}}\int_0^{2\pi} (r \sin \theta + 1)\,r\,d\theta\,dz\,dr$

11. Let D be the region bounded below by the plane $z = 0$, above by the sphere $x^2 + y^2 + z^2 = 4$, and on the sides by the cylinder $x^2 + y^2 = 1$. Set up the triple integrals in cylindrical coordinates that give the volume of D using the following orders of integration.

a) $dz\,dr\,d\theta$
b) $dr\,dz\,d\theta$
c) $d\theta\,dz\,dr$

12. Let D be the region bounded below by the cone $z = \sqrt{x^2 + y^2}$ and above by the paraboloid $z = 2 - x^2 - y^2$. Set up the triple integrals in cylindrical coordinates that give the volume of D using the following orders of integration.

a) $dz\,dr\,d\theta$
b) $dr\,dz\,d\theta$
c) $d\theta\,dz\,dr$

13. Give the limits of integration for evaluating the integral

$$\iiint f(r,\theta,z)\;dz\,r\,dr\,d\theta$$

as an iterated integral over the region that is bounded below by the plane $z = 0$, on the side by the cylinder $r = \cos \theta$, and on top by the paraboloid $z = 3r^2$.

14. Convert the integral

$$\int_{-1}^1\int_0^{\sqrt{1-y^2}}\int_0^x (x^2 + y^2)\,dz\,dx\,dy$$

to an equivalent integral in cylindrical coordinates and evaluate the result.

In Exercises 15–20, set up the iterated integral for evaluating $\iiint_D f(r,\theta,z)\,dz\,r\,dr\,d\theta$ over the given region D.

15. D is the right circular cylinder whose base is the circle $r = 2 \sin \theta$ in the xy-plane and whose top lies in the plane $z = 4 - y$.

$z = 4 - y$
$r = 2 \sin \theta$

16. D is the right circular cylinder whose base is the circle $r = 3 \cos \theta$ and whose top lies in the plane $z = 5 - x$.

$z = 5 - x$
$r = 3 \cos \theta$

17. D is the solid right cylinder whose base is the region in the xy-plane that lies inside the cardioid $r = 1 + \cos \theta$ and outside the circle $r = 1$ and whose top lies in the plane $z = 4$.

$r = 1$
$r = 1 + \cos \theta$

18. D is the solid right cylinder whose base is the region between the circles $r = \cos \theta$ and $r = 2 \cos \theta$, and whose top lies in the plane $z = 3 - y$.

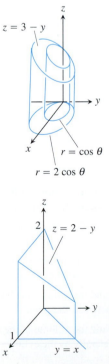

$z = 3 - y$
$r = \cos \theta$
$r = 2 \cos \theta$

19. D is the prism whose base is the triangle in the xy-plane bounded by the x-axis and the lines $y = x$ and $x = 1$ and whose top lies in the plane $z = 2 - y$.

$z = 2 - y$
$y = x$

20. D is the prism whose base is the triangle in the xy-plane bounded by the y-axis and the lines $y = x$ and $y = 1$ and whose top lies in the plane $z = 2 - x$.

$z = 2 - x$
$y = x$

Spherical Coordinates

Evaluate the spherical coordinate integrals in Exercises 21–26.

21. $\displaystyle\int_0^\pi \int_0^\pi \int_0^{2 \sin \phi} \rho^2 \sin \phi \, d\rho \, d\phi \, d\theta$

22. $\displaystyle\int_0^{2\pi} \int_0^{\pi/4} \int_0^2 (\rho \cos \phi) \rho^2 \sin \phi \, d\rho \, d\phi \, d\theta$

23. $\displaystyle\int_0^{2\pi} \int_0^\pi \int_0^{(1-\cos \phi)/2} \rho^2 \sin \phi \, d\rho \, d\phi \, d\theta$

24. $\displaystyle\int_0^{3\pi/2} \int_0^\pi \int_0^1 5\rho^3 \sin^3 \phi \, d\rho \, d\phi \, d\theta$

25. $\displaystyle\int_0^{2\pi} \int_0^{\pi/3} \int_{\sec \phi}^2 3\rho^2 \sin \phi \, d\rho \, d\phi \, d\theta$

26. $\displaystyle\int_0^{2\pi} \int_0^{\pi/4} \int_0^{\sec \phi} (\rho \cos \phi) \rho^2 \sin \phi \, d\rho \, d\phi \, d\theta$

The previous integrals suggest there are preferred orders of integration for spherical coordinates, but other orders are possible and occasionally easier to evaluate. Evaluate the integrals in Exercises 27–30.

27. $\displaystyle\int_0^2 \int_{-\pi}^0 \int_{\pi/4}^{\pi/2} \rho^3 \sin 2\phi \, d\phi \, d\theta \, d\rho$

28. $\displaystyle\int_{\pi/6}^{\pi/3} \int_{\csc \phi}^{2 \csc \phi} \int_0^{2\pi} \rho^2 \sin \phi \, d\theta \, d\rho \, d\phi$

29. $\displaystyle\int_0^1 \int_0^\pi \int_0^{\pi/4} 12\rho \sin^3 \phi \, d\phi \, d\theta \, d\rho$

30. $\displaystyle\int_{\pi/6}^{\pi/2} \int_{-\pi/2}^{\pi/2} \int_{\csc \phi}^2 5 \rho^4 \sin^3 \phi \, d\rho \, d\theta \, d\phi$

31. Let D be the region in Exercise 11. Set up the triple integrals in spherical coordinates that give the volume of D using the following orders of integration.

a) $d\rho \, d\phi \, d\theta$ b) $d\phi \, d\rho \, d\theta$

32. Let D be the region bounded below by the cone $z = \sqrt{x^2 + y^2}$ and above by the plane $z = 1$. Set up the triple integrals in spherical coordinates that give the volume of D using the following orders of integration.

a) $d\rho \, d\phi \, d\theta$ b) $d\phi \, d\rho \, d\theta$

In Exercises 33–38, (a) find the spherical coordinate limits for the integral that calculates the volume of the given solid, and (b) then evaluate the integral.

33. The solid between the sphere $\rho = \cos \phi$ and the hemisphere $\rho = 2, z \geq 0$

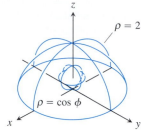

34. The solid bounded below by the hemisphere $\rho = 1, z \geq 0$, and above by the cardioid of revolution $\rho = 1 + \cos \phi$

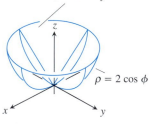

35. The solid enclosed by the cardioid of revolution $\rho = 1 - \cos \phi$

36. The upper portion cut from the solid in Exercise 35 by the xy-plane

37. The solid bounded below by the sphere $\rho = 2 \cos \phi$ and above by the cone $z = \sqrt{x^2 + y^2}$

38. The solid bounded below by the xy-plane, on the sides by the sphere $\rho = 2$, and above by the cone $\phi = \pi/3$

Rectangular, Cylindrical, and Spherical Coordinates

39. Set up triple integrals for the volume of the sphere $\rho = 2$ in (a) spherical, (b) cylindrical, and (c) rectangular coordinates.

40. Let D be the region in the first octant that is bounded below by the cone $\phi = \pi/4$ and above by the sphere $\rho = 3$. Express the volume of D as an iterated triple integral in (a) cylindrical and (b) spherical coordinates. Then (c) find V.

41. Let D be the smaller cap cut from a solid ball of radius 2 units by a plane 1 unit from the center of the sphere. Express the volume of D as an iterated triple integral in (a) spherical, (b) cylindrical, and (c) rectangular coordinates. Then (d) find the volume by evaluating one of the three triple integrals.

42. Express the moment of inertia I_z of the solid hemisphere $x^2 + y^2 + z^2 \leq 1, z \geq 0$, as an iterated integral in (a) cylindrical and (b) spherical coordinates. Then (c) find I_z.

Volumes

Find the volumes of the solids in Exercises 43–48.

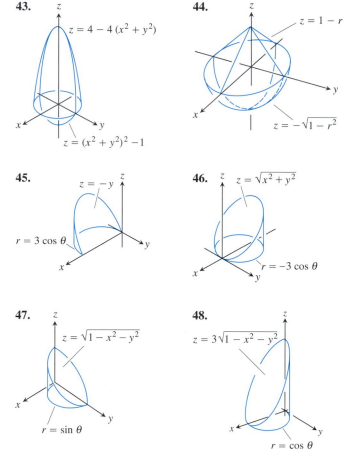

49. Find the volume of the portion of the solid sphere $\rho \leq a$ that lies between the cones $\phi = \pi/3$ and $\phi = 2\pi/3$.

50. Find the volume of the region cut from the solid sphere $\rho \leq a$ by the half-planes $\theta = 0$ and $\theta = \pi/6$ in the first octant.

51. Find the volume of the smaller region cut from the solid sphere $\rho \leq 2$ by the plane $z = 1$.

52. Find the volume of the solid enclosed by the cone $z = \sqrt{x^2 + y^2}$ between the planes $z = 1$ and $z = 2$.

53. Find the volume of the region bounded below by the plane $z = 0$, laterally by the cylinder $x^2 + y^2 = 1$, and above by the paraboloid $z = x^2 + y^2$.

54. Find the volume of the region bounded below by the paraboloid $z = x^2 + y^2$, laterally by the cylinder $x^2 + y^2 = 1$, and above by the paraboloid $z = x^2 + y^2 + 1$.

55. Find the volume of the solid cut from the thick-walled cylinder $1 \le x^2 + y^2 \le 2$ by the cones $z = \pm\sqrt{x^2 + y^2}$.

56. Find the volume of the region that lies inside the sphere $x^2 + y^2 + z^2 = 2$ and outside the cylinder $x^2 + y^2 = 1$.

57. Find the volume of the region enclosed by the cylinder $x^2 + y^2 = 4$ and the planes $z = 0$ and $y + z = 4$.

58. Find the volume of the region enclosed by the cylinder $x^2 + y^2 = 4$ and the planes $z = 0$ and $x + y + z = 4$.

59. Find the volume of the region bounded above by the paraboloid $z = 5 - x^2 - y^2$ and below by the paraboloid $z = 4x^2 + 4y^2$.

60. Find the volume of the region bounded above by the paraboloid $z = 9 - x^2 - y^2$, below by the xy-plane, and lying *outside* the cylinder $x^2 + y^2 = 1$.

61. Find the volume of the region cut from the solid cylinder $x^2 + y^2 \le 1$ by the sphere $x^2 + y^2 + z^2 = 4$.

62. Find the volume of the region bounded above by the sphere $x^2 + y^2 + z^2 = 2$ and below by the paraboloid $z = x^2 + y^2$.

Average Values

63. Find the average value of the function $f(r, \theta, z) = r$ over the region bounded by the cylinder $r = 1$ between the planes $z = -1$ and $z = 1$.

64. Find the average value of the function $f(r, \theta, z) = r$ over the solid ball bounded by the sphere $r^2 + z^2 = 1$. (This is the sphere $x^2 + y^2 + z^2 = 1$.)

65. Find the average value of the function $f(\rho, \phi, \theta) = \rho$ over the solid ball $\rho \le 1$.

66. Find the average value of the function $f(\rho, \phi, \theta) = \rho \cos \phi$ over the solid upper ball $\rho \le 1$, $0 \le \phi \le \pi/2$.

Masses, Moments, and Centroids

67. A solid of constant density is bounded below by the plane $z = 0$, above by the cone $z = r$, $r \ge 0$, and on the sides by the cylinder $r = 1$. Find the center of mass.

68. Find the centroid of the region in the first octant that is bounded above by the cone $z = \sqrt{x^2 + y^2}$, below by the plane $z = 0$, and on the sides by the cylinder $x^2 + y^2 = 4$ and the planes $x = 0$ and $y = 0$.

69. Find the centroid of the solid in Exercise 38.

70. Find the centroid of the solid bounded above by the sphere $\rho = a$ and below by the cone $\phi = \pi/4$.

71. Find the centroid of the region that is bounded above by the surface $z = \sqrt{r}$, on the sides by the cylinder $r = 4$, and below by the xy-plane.

72. Find the centroid of the region cut from the solid ball $r^2 + z^2 \le 1$ by the half-planes $\theta = -\pi/3, r \ge 0$, and $\theta = \pi/3, r \ge 0$.

73. Find the moment of inertia and radius of gyration about the z-axis of a thick-walled right circular cylinder bounded on the inside by the cylinder $r = 1$, on the outside by the cylinder $r = 2$, and on the top and bottom by the planes $z = 4$ and $z = 0$. (Take $\delta = 1$.)

74. Find the moment of inertia of a solid circular cylinder of radius 1 and height 2 (a) about the axis of the cylinder, (b) about a line through the centroid perpendicular to the axis of the cylinder. (Take $\delta = 1$.)

75. Find the moment of inertia of a right circular cone of base radius 1 and height 1 about an axis through the vertex parallel to the base. (Take $\delta = 1$.)

76. Find the moment of inertia of a solid sphere of radius a about a diameter. (Take $\delta = 1$.)

77. Find the moment of inertia of a right circular cone of base radius a and height h about its axis. (*Hint:* Place the cone with its vertex at the origin and its axis along the z-axis.)

78. A solid is bounded on the top by the paraboloid $z = r^2$, on the bottom by the plane $z = 0$, and on the sides by the cylinder $r = 1$. Find the center of mass and the moment of inertia and radius of gyration about the z-axis if the density is (a) $\delta(r, \theta, z) = z$; (b) $\delta(r, \theta, z) = r$.

79. A solid is bounded below by the cone $z = \sqrt{x^2 + y^2}$ and above by the plane $z = 1$. Find the center of mass and the moment of inertia and radius of gyration about the z-axis if the density is (a) $\delta(r, \theta, z) = z$; (b) $\delta(r, \theta, z) = z^2$.

80. A solid ball is bounded by the sphere $\rho = a$. Find the moment of inertia and radius of gyration about the z-axis if the density is

a) $\delta(\rho, \phi, \theta) = \rho^2$,   **b)** $\delta(\rho, \phi, \theta) = r = \rho \sin \phi$.

81. Show that the centroid of the solid semi-ellipsoid of revolution $(r^2/a^2) + (z^2/h^2) \le 1, z \ge 0$, lies on the z-axis three-eighths of the way from the base to the top. The special case $h = a$ gives a solid hemisphere. Thus the centroid of a solid hemisphere lies on the axis of symmetry three-eighths of the way from the base to the top.

82. Show that the centroid of a solid right circular cone is one-fourth of the way from the base to the vertex. (In general, the centroid of a solid cone or pyramid is one-fourth of the way from the centroid of the base to the vertex.)

83. A solid right circular cylinder is bounded by the cylinder $r = a$ and the planes $z = 0$ and $z = h, h > 0$. Find the center of mass and the moment of inertia and radius of gyration about the z-axis if the density is $\delta(r, \theta, z) = z + 1$.

84. A spherical planet of radius R has an atmosphere whose density is $\mu = \mu_0 e^{-ch}$, where h is the altitude above the surface of the planet, μ_0 is the density at sea level, and c is a positive constant. Find the mass of the planet's atmosphere.

85. A planet is in the shape of a sphere of radius R and total mass M with spherically symmetric density distribution that increases linearly as one approaches its center. What is the density at the center of this planet if the density at its edge (surface) is taken to be zero?

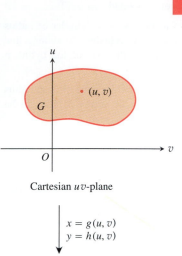

Cartesian uv-plane

$x = g(u, v)$
$y = h(u, v)$

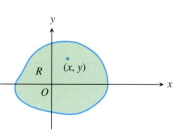

Cartesian xy-plane

13.46 The equations $x = g(u, v)$ and $y = h(u, v)$ allow us to change an integral over a region R in the xy-plane into an integral over a region G in the uv-plane.

13.7 Substitutions in Multiple Integrals

This section shows how to evaluate multiple integrals by substitution. As in single integration, the goal of substitution is to replace complicated integrals by ones that are easier to evaluate. Substitutions accomplish this by simplifying the integrand, the limits of integration, or both.

Substitutions in Double Integrals

The polar coordinate substitution of Section 13.3 is a special case of a more general substitution method for double integrals, a method that pictures changes in variables as transformations of regions.

Suppose that a region G in the uv-plane is transformed one-to-one into the region R in the xy-plane by equations of the form

$$x = g(u, v), \qquad y = h(u, v),$$

as suggested in Fig. 13.46. We call R the **image** of G under the transformation, and G the **preimage** of R. Any function $f(x, y)$ defined on R can be thought of as a function $f(g(u, v), h(u, v))$ defined on G as well. How is the integral of $f(x, y)$ over R related to the integral of $f(g(u, v), h(u, v))$ over G?

The answer is: If g, h, and f have continuous partial derivatives and $J(u, v)$ (to be discussed in a moment) is zero only at isolated points, if at all, then

$$\iint\limits_{R} f(x, y)\, dx\, dy = \iint\limits_{G} f(g(u, v), h(u, v))\, |J(u, v)|\, du\, dv. \tag{1}$$

The factor $J(u, v)$, whose absolute value appears in Eq. (1), is the *Jacobian* of the coordinate transformation, named after the mathematician Carl Jacobi.

> **Definition**
> The **Jacobian determinant** or **Jacobian** of the coordinate transformation $x = g(u, v), y = h(u, v)$ is
> $$J(u, v) = \begin{vmatrix} \dfrac{\partial x}{\partial u} & \dfrac{\partial x}{\partial v} \\[2ex] \dfrac{\partial y}{\partial u} & \dfrac{\partial y}{\partial v} \end{vmatrix} = \frac{\partial x}{\partial u}\frac{\partial y}{\partial v} - \frac{\partial y}{\partial u}\frac{\partial x}{\partial v}. \tag{2}$$

Notice the "Reversed" Order

The transforming equations $x = g(u, v)$ and $y = h(u, v)$ go from G to R, but we use them to change an integral over R into an integral over G.

The Jacobian is also denoted by

$$J(u, v) = \frac{\partial(x, y)}{\partial(u, v)}$$

to help remember how the determinant in Eq. (2) is constructed from the partial derivatives of x and y. The derivation of Eq. (1) is intricate and properly belongs to a course in advanced calculus. We will not give the derivation here.

For polar coordinates, we have r and θ in place of u and v. With $x = r \cos \theta$

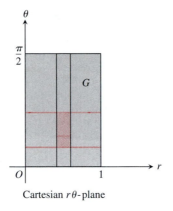

Cartesian $r\theta$-plane

$$x = r\cos\theta$$
$$y = r\sin\theta$$

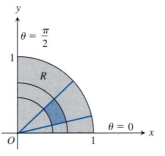

Cartesian xy-plane

13.47 The equations $x = r\cos\theta, y = r\sin\theta$ transform G into R.

and $y = r\sin\theta$, the Jacobian is

$$J(r, \theta) = \begin{vmatrix} \dfrac{\partial x}{\partial r} & \dfrac{\partial x}{\partial \theta} \\[2mm] \dfrac{\partial y}{\partial r} & \dfrac{\partial y}{\partial \theta} \end{vmatrix} = \begin{vmatrix} \cos\theta & -r\sin\theta \\ \sin\theta & r\cos\theta \end{vmatrix} = r(\cos^2\theta + \sin^2\theta) = r.$$

Hence, Eq. (1) becomes

$$\iint\limits_{R} f(x, y)\, dx\, dy = \iint\limits_{G} f(r\cos\theta, r\sin\theta)\, |r|\, dr\, d\theta$$

$$= \iint\limits_{G} f(r\cos\theta, r\sin\theta)\, r\, dr\, d\theta, \qquad \text{If } r \geq 0 \quad (3)$$

which is Eq. (6) in Section 13.3.

Figure 13.47 shows how the equations $x = r\cos\theta$, $y = r\sin\theta$ transform the rectangle $G: 0 \leq r \leq 1$, $0 \leq \theta \leq \pi/2$ into the quarter circle R bounded by $x^2 + y^2 = 1$ in the first quadrant of the xy-plane.

Notice that the integral on the right-hand side of Eq. (3) is not the integral of $f(r\cos\theta, r\sin\theta)$ over a region in the polar coordinate plane. It is the integral of the product of $f(r\cos\theta, r\sin\theta)$ and r over a region G in the *Cartesian $r\theta$-plane*.

Here is an example of another substitution.

EXAMPLE 1 Evaluate

$$\int_0^4 \int_{x=y/2}^{x=(y/2)+1} \frac{2x-y}{2}\, dx\, dy$$

by applying the transformation

$$u = \frac{2x-y}{2}, \qquad v = \frac{y}{2} \qquad (4)$$

and integrating over an appropriate region in the uv-plane.

Solution We sketch the region R of integration in the xy-plane and identify its boundaries (Fig. 13.48).

To apply Eq. (1), we need to find the corresponding uv-region G and the Jacobian of the transformation. To find them, we first solve Eqs. (4) for x and y in terms of u and v. Routine algebra gives

$$x = u + v, \qquad y = 2v. \qquad (5)$$

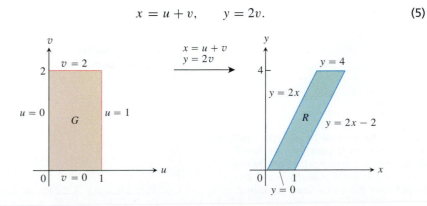

13.48 The equations $x = u + v$ and $y = 2v$ transform G into R. Reversing the transformation by the equations $u = (2x - y)/2$ and $v = y/2$ transforms R into G. See Example 1.

We then find the boundaries of G by substituting these expressions into the equations for the boundaries of R (Fig. 13.48).

xy-equations for the boundary of R	Corresponding uv-equations for the boundary of G	Simplified uv-equations
$x = y/2$	$u + v = 2v/2 = v$	$u = 0$
$x = (y/2) + 1$	$u + v = (2v/2) + 1 = v + 1$	$u = 1$
$y = 0$	$2v = 0$	$v = 0$
$y = 4$	$2v = 4$	$v = 2$

The Jacobian of the transformation (again from Eqs. 5) is

$$J(u, v) = \begin{vmatrix} \dfrac{\partial x}{\partial u} & \dfrac{\partial x}{\partial v} \\ \dfrac{\partial y}{\partial u} & \dfrac{\partial y}{\partial v} \end{vmatrix} = \begin{vmatrix} \dfrac{\partial}{\partial u}(u + v) & \dfrac{\partial}{\partial v}(u + v) \\ \dfrac{\partial}{\partial u}(2v) & \dfrac{\partial}{\partial v}(2v) \end{vmatrix} = \begin{vmatrix} 1 & 1 \\ 0 & 2 \end{vmatrix} = 2.$$

We now have everything we need to apply Eq. (1):

$$\int_0^4 \int_{x=y/2}^{x=(y/2)+1} \frac{2x - y}{2} \, dx \, dy = \int_{v=0}^{v=2} \int_{u=0}^{u=1} u |J(u, v)| \, du \, dv$$

$$= \int_0^2 \int_0^1 (u)(2) \, du \, dv = \int_0^2 \left[u^2 \right]_0^1 dv = \int_0^2 dv = 2. \qquad \square$$

EXAMPLE 2 Evaluate

$$\int_0^1 \int_0^{1-x} \sqrt{x + y} \, (y - 2x)^2 \, dy \, dx.$$

Solution We sketch the region R of integration in the xy-plane and identify its boundaries (Fig. 13.49). The integrand suggests the transformation $u = x + y$ and $v = y - 2x$. Routine algebra produces x and y as functions of u and v:

$$x = \frac{u}{3} - \frac{v}{3}, \qquad y = \frac{2u}{3} + \frac{v}{3}. \tag{6}$$

From Eqs. (6) we can find the boundaries of the uv-region G (Fig. 13.49).

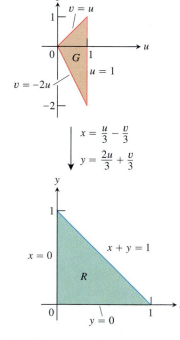

$x = \dfrac{u}{3} - \dfrac{v}{3}$

$y = \dfrac{2u}{3} + \dfrac{v}{3}$

13.49 The equations $x = (u/3) - (v/3)$ and $y = (2u/3) + (v/3)$ transform G into R. Reversing the transformation by the equations $u = x + y$ and $v = y - 2x$ transforms R into G. See Example 2.

xy-equations for the boundary of R	Corresponding uv-equations for the boundary of G	Simplified uv-equations
$x + y = 1$	$\left(\dfrac{u}{3} - \dfrac{v}{3}\right) + \left(\dfrac{2u}{3} + \dfrac{v}{3}\right) = 1$	$u = 1$
$x = 0$	$\dfrac{u}{3} - \dfrac{v}{3} = 0$	$v = u$
$y = 0$	$\dfrac{2u}{3} + \dfrac{v}{3} = 0$	$v = -2u$

The Jacobian of the transformation in Eq. (6) is

$$J(u, v) = \begin{vmatrix} \dfrac{\partial x}{\partial u} & \dfrac{\partial x}{\partial v} \\[2mm] \dfrac{\partial y}{\partial u} & \dfrac{\partial y}{\partial v} \end{vmatrix} = \begin{vmatrix} \dfrac{1}{3} & -\dfrac{1}{3} \\[2mm] \dfrac{2}{3} & \dfrac{1}{3} \end{vmatrix} = \dfrac{1}{3}.$$

Applying Eq. (1), we evaluate the integral:

$$\int_0^1 \int_0^{1-x} \sqrt{x+y}\,(y-2x)^2\,dy\,dx = \int_{u=0}^{u=1} \int_{v=-2u}^{v=u} u^{1/2} v^2\, |J(u, v)|\, dv\, du$$

$$= \int_0^1 \int_{-2u}^{u} u^{1/2} v^2 \left(\frac{1}{3}\right) dv\, du = \frac{1}{3} \int_0^1 u^{1/2} \left(\frac{1}{3} v^3\right)\Big|_{v=-2u}^{v=u} du$$

$$= \frac{1}{9} \int_0^1 u^{1/2}(u^3 + 8u^3)\, du = \int_0^1 u^{7/2}\, du = \frac{2}{9} u^{9/2} \Big|_0^1 = \frac{2}{9}.$$ ❑

Substitutions in Triple Integrals

The cylindrical and spherical coordinate substitutions are special cases of a substitution method that pictures changes of variables in triple integrals as transformations of three-dimensional regions. The method is like the method for double integrals except that now we work in three dimensions instead of two.

Suppose that a region G in uvw-space is transformed one-to-one into the region D in xyz-space by differentiable equations of the form

$$x = g\,(u, v, w), \qquad y = h\,(u, v, w), \qquad z = k\,(u, v, w),$$

as suggested in Fig. 13.50. Then any function $F\,(x, y, z)$ defined on D can be thought of as a function

$$F\,(g(u, v, w), h\,(u, v, w), k\,(u, v, w)) = H\,(u, v, w)$$

defined on G. If g, h, and k have continuous first partial derivatives, then the integral of $F\,(x, y, z)$ over D is related to the integral of $H\,(u, v, w)$ over G by the equation

$$\iiint_D F\,(x, y, z)\, dx\, dy\, dz = \iiint_G H\,(u, v, w)\, |J\,(u, v, w)|\, du\, dv\, dw. \quad (7)$$

The factor $J\,(u, v, w)$, whose absolute value appears in this equation, is the **Jacobian determinant**

$$J(u, v, w) = \begin{vmatrix} \dfrac{\partial x}{\partial u} & \dfrac{\partial x}{\partial v} & \dfrac{\partial x}{\partial w} \\[2mm] \dfrac{\partial y}{\partial u} & \dfrac{\partial y}{\partial v} & \dfrac{\partial y}{\partial w} \\[2mm] \dfrac{\partial z}{\partial u} & \dfrac{\partial z}{\partial v} & \dfrac{\partial z}{\partial w} \end{vmatrix} = \dfrac{\partial(x, y, z)}{\partial(u, v, w)}. \quad (8)$$

As in the two-dimensional case, the derivation of the change-of-variable formula in Eq. (7) is complicated and we will not go into it here.

For cylindrical coordinates, r, θ, and z take the place of u, v, and w. The transformation from Cartesian $r\theta z$-space to Cartesian xyz-space is given by the equations

$$x = r\cos\theta, \qquad y = r\sin\theta, \qquad z = z$$

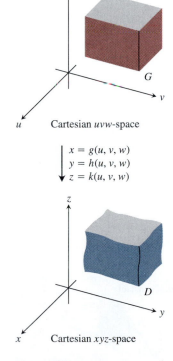

Cartesian uvw-space

$$\begin{array}{l} x = g(u, v, w) \\ y = h(u, v, w) \\ z = k(u, v, w) \end{array}$$

Cartesian xyz-space

13.50 The equations $x = g(u, v, w)$, $y = h(u, v, w)$, and $z = k(u, v, w)$ allow us to change an integral over a region D in Cartesian xyz-space into an integral over a region G in Cartesian uvw-space.

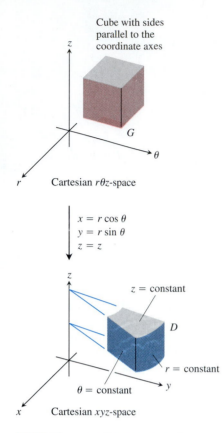

Cube with sides parallel to the coordinate axes

Cartesian $r\theta z$-space

$x = r \cos \theta$
$y = r \sin \theta$
$z = z$

$z = $ constant

D

$r = $ constant

$\theta = $ constant

Cartesian xyz-space

13.51 The equations $x = r\cos\theta$, $y = r\sin\theta$, and $z = z$ transform G into D.

(Fig. 13.51). The Jacobian of the transformation is

$$J(r, \theta, z) = \begin{vmatrix} \dfrac{\partial x}{\partial r} & \dfrac{\partial x}{\partial \theta} & \dfrac{\partial x}{\partial z} \\ \dfrac{\partial y}{\partial r} & \dfrac{\partial y}{\partial \theta} & \dfrac{\partial y}{\partial z} \\ \dfrac{\partial z}{\partial r} & \dfrac{\partial z}{\partial \theta} & \dfrac{\partial z}{\partial z} \end{vmatrix} = \begin{vmatrix} \cos\theta & -r\sin\theta & 0 \\ \sin\theta & r\cos\theta & 0 \\ 0 & 0 & 1 \end{vmatrix}$$

$$= r\cos^2\theta + r\sin^2\theta = r.$$

The corresponding version of Eq. (7) is

$$\iiint_D F(x, y, z)\, dx\, dy\, dz = \iiint_G H(r, \theta, z)\, |r|\, dr\, d\theta\, dz. \qquad (9)$$

We can drop the absolute value signs whenever $r \geq 0$.

For spherical coordinates, ρ, ϕ, and θ take the place of u, v, and w. The transformation from Cartesian $\rho\phi\theta$-space to Cartesian xyz-space is given by

$$x = \rho\sin\phi\cos\theta, \qquad y = \rho\sin\phi\sin\theta, \qquad z = \rho\cos\phi$$

(Fig. 13.52). The Jacobian of the transformation is

$$J(\rho, \phi, \theta) = \begin{vmatrix} \dfrac{\partial x}{\partial \rho} & \dfrac{\partial x}{\partial \phi} & \dfrac{\partial x}{\partial \theta} \\ \dfrac{\partial y}{\partial \rho} & \dfrac{\partial y}{\partial \phi} & \dfrac{\partial y}{\partial \theta} \\ \dfrac{\partial z}{\partial \rho} & \dfrac{\partial z}{\partial \phi} & \dfrac{\partial z}{\partial \theta} \end{vmatrix} = \rho^2 \sin\phi \qquad (10)$$

(Exercise 17). The corresponding version of Eq. (7) is

$$\iiint_D F(x, y, z)\, dx\, dy\, dz = \iiint_G H(\rho, \phi, \theta)\, |\rho^2 \sin\phi|\, d\rho\, d\phi\, d\theta. \qquad (11)$$

We can drop the absolute value signs because $\sin\phi$ is never negative.

Here is an example of another substitution.

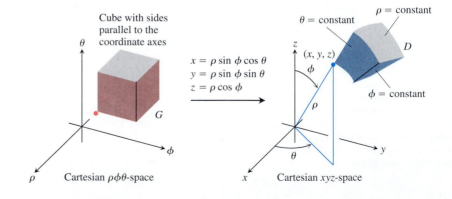

13.52 The equations $x = \rho\sin\phi\cos\theta$, $y = \rho\sin\phi\sin\theta$, and $z = \rho\cos\phi$ transform G into D.

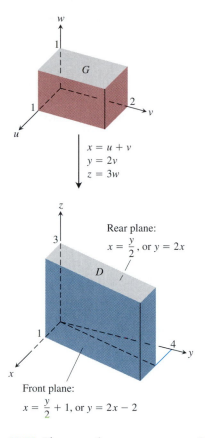

$x = u + v$
$y = 2v$
$z = 3w$

Rear plane:
$x = \dfrac{y}{2}$, or $y = 2x$

D

Front plane:
$x = \dfrac{y}{2} + 1$, or $y = 2x - 2$

13.53 The equations $x = u + v$, $y = 2v$, and $z = 3w$ transform G into D. Reversing the transformation by the equations $u = (2x - y)/2$, $v = y/2$, and $w = z/3$ transforms D into G. See Example 3.

Carl Gustav Jacob Jacobi

Jacobi (1804–1851), one of nineteenth-century Germany's most accomplished scientists, developed the theory of determinants and transformations into a powerful tool for evaluating multiple integrals and solving differential equations. He also applied transformation methods to study nonelementary integrals like the ones that arise in the calculation of arc length. Like Euler, Jacobi was a prolific writer and an even more prolific calculator and worked in a variety of mathematical and applied fields.

EXAMPLE 3   Evaluate

$$\int_0^3 \int_0^4 \int_{x=y/2}^{x=(y/2)+1} \left(\frac{2x - y}{2} + \frac{z}{3} \right) dx \, dy \, dz$$

by applying the transformation

$$u = (2x - y)/2, \qquad v = y/2, \qquad w = z/3 \tag{12}$$

and integrating over an appropriate region in uvw-space.

Solution   We sketch the region D of integration in xyz-space and identify its boundaries (Fig. 13.53). In this case, the bounding surfaces are planes.

To apply Eq. (7), we need to find the corresponding uvw-region G and the Jacobian of the transformation. To find them, we first solve Eqs. (12) for x, y, and z in terms of u, v, and w. Routine algebra gives

$$x = u + v, \qquad y = 2v, \qquad z = 3w. \tag{13}$$

We then find the boundaries of G by substituting these expressions into the equations for the boundaries of D:

xyz-equations for the boundary of D	Corresponding uvw-equations for the boundary of G	Simplified uvw-equations
$x = y/2$	$u + v = 2v/2 = v$	$u = 0$
$x = (y/2) + 1$	$u + v = (2v/2) + 1 = v + 1$	$u = 1$
$y = 0$	$2v = 0$	$v = 0$
$y = 4$	$2v = 4$	$v = 2$
$z = 0$	$3w = 0$	$w = 0$
$z = 3$	$3w = 3$	$w = 1$

The Jacobian of the transformation, again from Eqs. (13), is

$$J(u, v, w) = \begin{vmatrix} \dfrac{\partial x}{\partial u} & \dfrac{\partial x}{\partial v} & \dfrac{\partial x}{\partial w} \\[2mm] \dfrac{\partial y}{\partial u} & \dfrac{\partial y}{\partial v} & \dfrac{\partial y}{\partial w} \\[2mm] \dfrac{\partial z}{\partial u} & \dfrac{\partial z}{\partial v} & \dfrac{\partial z}{\partial w} \end{vmatrix} = \begin{vmatrix} 1 & 1 & 0 \\ 0 & 2 & 0 \\ 0 & 0 & 3 \end{vmatrix} = 6.$$

We now have everything we need to apply Eq. (7):

$$\int_0^3 \int_0^4 \int_{x=y/2}^{x=(y/2)+1} \left(\frac{2x - y}{2} + \frac{z}{3} \right) dx \, dy \, dz$$

$$= \int_0^1 \int_0^2 \int_0^1 (u + w) \, |J(u, v, w)| \, du \, dv \, dw$$

$$= \int_0^1 \int_0^2 \int_0^1 (u + w)(6) \, du \, dv \, dw = 6 \int_0^1 \int_0^2 \left[\frac{u^2}{2} + uw \right]_0^1 dv \, dw$$

$$= 6 \int_0^1 \int_0^2 \left(\frac{1}{2} + w \right) dv \, dw = 6 \int_0^1 \left[\frac{v}{2} + vw \right]_0^2 dw = 6 \int_0^1 (1 + 2w) \, dw$$

$$= 6 \left[w + w^2 \right]_0^1 = 6(2) = 12. \qquad \square$$

Exercises 13.7

Transformations of Coordinates

1. a) Solve the system

$$u = x - y, \qquad v = 2x + y$$

for x and y in terms of u and v. Then find the value of the Jacobian $\partial(x, y)/\partial(u, v)$.

b) Find the image under the transformation $u = x - y, v = 2x + y$ of the triangular region with vertices $(0, 0)$, $(1, 1)$, and $(1, -2)$ in the xy-plane. Sketch the transformed region in the uv-plane.

2. a) Solve the system

$$u = x + 2y, \qquad v = x - y$$

for x and y in terms of u and v. Then find the value of the Jacobian $\partial(x, y)/\partial(u, v)$.

b) Find the image under the transformation $u = x + 2y, v = x - y$ of the triangular region in the xy-plane bounded by the lines $y = 0$, $y = x$, and $x + 2y = 2$. Sketch the transformed region in the uv-plane.

3. a) Solve the system

$$u = 3x + 2y, \qquad v = x + 4y$$

for x and y in terms of u and v. Then find the value of the Jacobian $\partial(x, y)/\partial(u, v)$.

b) Find the image under the transformation $u = 3x + 2y, v = x + 4y$ of the triangular region in the xy-plane bounded by the x-axis, the y-axis, and the line $x + y = 1$. Sketch the transformed region in the uv-plane.

4. a) Solve the system

$$u = 2x - 3y, \qquad v = -x + y$$

for x and y in terms of u and v. Then find the value of the Jacobian $\partial(x, y)/\partial(u, v)$.

b) Find the image under the transformation $u = 2x - 3y, v = -x + y$ of the parallelogram R in the xy-plane with boundaries $x = -3$, $x = 0$, $y = x$, and $y = x + 1$. Sketch the transformed region in the uv-plane.

5. Find the Jacobian $\partial(x, y)/\partial(u, v)$ for the transformation

a) $x = u \cos v, \quad y = u \sin v$.

b) $x = u \sin v, \quad y = u \cos v$.

6. Find the Jacobian $\partial(x, y, z)/\partial(u, v, w)$ of the transformation

a) $x = u \cos v, \quad y = u \sin v, \quad z = w$

b) $x = 2u - 1, \quad y = 3v - 4, \quad z = \dfrac{1}{2}(w - 4)$.

Double Integrals

7. Evaluate the integral

$$\int_0^4 \int_{x=y/2}^{x=(y/2)+1} \frac{2x - y}{2} dx\, dy$$

from Example 1 directly by integration with respect to x and y to confirm that its value is 2.

8. Use the transformation in Exercise 1 to evaluate the integral

$$\iint_R (2x^2 - xy - y^2)\, dx\, dy$$

for the region R in the first quadrant bounded by the lines $y = -2x + 4$, $y = -2x + 7$, $y = x - 2$, and $y = x + 1$.

9. Use the transformation in Exercise 3 to evaluate the integral

$$\iint_R (3x^2 + 14xy + 8y^2)\, dx\, dy$$

for the region R in the first quadrant bounded by the lines $y = -\dfrac{3}{2}x + 1$, $y = -\dfrac{3}{2}x + 3$, $y = -\dfrac{1}{4}x$, and $y = -\dfrac{1}{4}x + 1$.

10. Use the transformation and parallelogram R in Exercise 4 to evaluate the integral

$$\iint_R 2(x - y)\, dx\, dy.$$

11. Let R be the region in the first quadrant of the xy-plane bounded by the hyperbolas $xy = 1, xy = 9$ and the lines $y = x, y = 4x$. Use the transformation $x = u/v, y = uv$ with $u > 0$ and $v > 0$ to rewrite

$$\iint_R \left(\sqrt{\frac{y}{x}} + \sqrt{xy} \right) dx\, dy$$

as an integral over an appropriate region G in the uv-plane. Then evaluate the uv-integral over G.

12. a) Find the Jacobian of the transformation $x = u, y = uv$, and sketch the region $G: 1 \le u \le 2, 1 \le uv \le 2$ in the uv-plane.

b) Then use Eq. (1) to transform the integral

$$\int_1^2 \int_1^2 \frac{y}{x} dy\, dx$$

into an integral over G, and evaluate both integrals.

13. A thin plate of constant density covers the region bounded by the ellipse $x^2/a^2 + y^2/b^2 = 1, a > 0, b > 0$, in the xy-plane. Find the first moment of the plate about the origin. (*Hint:* Use the transformation $x = ar \cos \theta, y = br \sin \theta$.)

14. The area πab of the ellipse $x^2/a^2 + y^2/b^2 = 1$ can be found by integrating the function $f(x, y) = 1$ over the region bounded by the ellipse in the xy-plane. Evaluating the integral directly requires a trigonometric substitution. An easier way to evaluate the integral is to use the transformation $x = au, y = bv$ and evaluate the transformed integral over the disk $G: u^2 + v^2 \le 1$ in the uv-plane. Find the area this way.

15. Use the transformation in Exercise 2 to evaluate the integral

$$\int_0^{2/3} \int_y^{2-2y} (x + 2y) e^{(y-x)} \, dx \, dy$$

by first writing it as an integral over a region G in the uv-plane.

16. Use the transformation $x = u + (1/2)v$, $y = v$ to evaluate the integral

$$\int_0^2 \int_{y/2}^{(y+4)/2} y^3 (2x - y) e^{(2x-y)^2} \, dx \, dy$$

by first writing it as an integral over a region G in the uv-plane.

Triple Integrals

17. Evaluate the determinant in Eq. (10) to show that the Jacobian of the transformation from Cartesian $\rho\phi\theta$-space to Cartesian xyz-space is $\rho^2 \sin \phi$.

18. Evaluate the integral in Example 3 by integrating with respect to x, y, and z.

19. Find the volume of the ellipsoid

$$\frac{x^2}{a^2} + \frac{y^2}{b^2} + \frac{z^2}{c^2} = 1.$$

(*Hint:* Let $x = au$, $y = bv$, and $z = cw$. Then find the volume of an appropriate region in uvw-space.)

20. Evaluate

$$\int\int\int |xyz| \, dx \, dy \, dz$$

over the solid ellipsoid

$$\frac{x^2}{a^2} + \frac{y^2}{b^2} + \frac{z^2}{c^2} \le 1.$$

(*Hint:* Let $x = au$, $y = bv$, and $z = cw$. Then integrate over an appropriate region in uvw-space.)

21. Let D be the region in xyz-space defined by the inequalities

$$1 \le x \le 2, \qquad 0 \le xy \le 2, \qquad 0 \le z \le 1.$$

Evaluate

$$\int\int\int_D (x^2 y + 3xyz) \, dx \, dy \, dz$$

by applying the transformation

$$u = x, \qquad v = xy, \qquad w = 3z$$

and integrating over an appropriate region G in uvw-space.

22. Assuming the result that the center of mass of a solid hemisphere lies on the axis of symmetry three-eighths of the way from the base toward the top, show, by transforming the appropriate integrals, that the center of mass of a solid semi-ellipsoid $(x^2/a^2) + (y^2/b^2) + (z^2/c^2) \le 1$, $z \ge 0$, lies on the z-axis three-eighths of the way from the base toward the top. (You can do this without evaluating any of the integrals.)

Single Integrals

23. *Substitutions in single integrals.* How can substitutions in single definite integrals be viewed as transformations of regions? What is the Jacobian in such a case? Illustrate with an example.

CHAPTER 13 QUESTIONS TO GUIDE YOUR REVIEW

1. Define the double integral of a function of two variables over a bounded region in the coordinate plane.

2. How are double integrals evaluated as iterated integrals? Does the order of integration matter? How are the limits of integration determined? Give examples.

3. How are double integrals used to calculate areas, average values, masses, moments, centers of mass, and radii of gyration? Give examples.

4. How can you change a double integral in rectangular coordinates into a double integral in polar coordinates? Why might it be worthwhile to do so? Give an example.

5. Define the triple integral of a function $f(x, y, z)$ over a bounded region in space.

6. How are triple integrals in rectangular coordinates evaluated? How are the limits of integration determined? Give an example.

7. How are triple integrals in rectangular coordinates used to calculate volumes, average values, masses, moments, centers of mass, and radii of gyration? Give examples.

8. How are triple integrals defined in cylindrical and spherical coordinates? Why might one prefer working in one of these coordinate systems to working in rectangular coordinates?

9. How are triple integrals in cylindrical and spherical coordinates evaluated? How are the limits of integration found? Give examples.

10. How are substitutions in double integrals pictured as transformations of two-dimensional regions? Give a sample calculation.

11. How are substitutions in triple integrals pictured as transformations of three-dimensional regions? Give a sample calculation.

CHAPTER 13 PRACTICE EXERCISES

Planar Regions of Integration

In Exercises 1–4, sketch the region of integration and evaluate the double integral.

1. $\displaystyle\int_1^{10}\int_0^{1/y} ye^{xy}\,dx\,dy$

2. $\displaystyle\int_0^1\int_0^{x^3} e^{y/x}\,dy\,dx$

3. $\displaystyle\int_0^{3/2}\int_{-\sqrt{9-4t^2}}^{\sqrt{9-4t^2}} t\,ds\,dt$

4. $\displaystyle\int_0^1\int_{\sqrt{y}}^{2-\sqrt{y}} x\,y\,dx\,dy$

Reversing the Order of Integration

In Exercises 5–8, sketch the region of integration and write an equivalent integral with the order of integration reversed. Then evaluate both integrals.

5. $\displaystyle\int_0^4\int_{-\sqrt{4-y}}^{(y-4)/2} dx\,dy$

6. $\displaystyle\int_0^1\int_{x^2}^{x} \sqrt{x}\,dy\,dx$

7. $\displaystyle\int_0^{3/2}\int_{-\sqrt{9-4y^2}}^{\sqrt{9-4y^2}} y\,dx\,dy$

8. $\displaystyle\int_0^2\int_0^{4-x^2} 2x\,dy\,dx$

Evaluating Double Integrals

Evaluate the integrals in Exercises 9–12.

9. $\displaystyle\int_0^1\int_{2y}^2 4\cos(x^2)\,dx\,dy$

10. $\displaystyle\int_0^2\int_{y/2}^1 e^{x^2}\,dx\,dy$

11. $\displaystyle\int_0^8\int_{\sqrt[3]{x}}^2 \frac{dy\,dx}{y^4+1}$

12. $\displaystyle\int_0^1\int_{\sqrt[3]{y}}^1 \frac{2\pi\sin\pi x^2}{x^2}\,dx\,dy$

Areas and Volumes

13. Find the area of the region enclosed by the line $y = 2x + 4$ and the parabola $y = 4 - x^2$ in the xy-plane.

14. Find the area of the "triangular" region in the xy-plane that is bounded on the right by the parabola $y = x^2$, on the left by the line $x + y = 2$, and above by the line $y = 4$.

15. Find the volume under the paraboloid $z = x^2 + y^2$ above the triangle enclosed by the lines $y = x$, $x = 0$, and $x + y = 2$ in the xy-plane.

16. Find the volume under the parabolic cylinder $z = x^2$ above the region enclosed by the parabola $y = 6 - x^2$ and the line $y = x$ in the xy-plane.

Average Values

Find the average value of $f(x, y) = xy$ over the regions in Exercises 17 and 18.

17. The square bounded by the lines $x = 1$, $y = 1$ in the first quadrant

18. The quarter circle $x^2 + y^2 \le 1$ in the first quadrant

Masses and Moments

19. Find the centroid of the "triangular" region bounded by the lines $x = 2$, $y = 2$ and the hyperbola $xy = 2$ in the xy-plane.

20. Find the centroid of the region between the parabola $x + y^2 - 2y = 0$ and the line $x + 2y = 0$ in the xy-plane.

21. Find the polar moment of inertia about the origin of a thin triangular plate of constant density $\delta = 3$, bounded by the y-axis and the lines $y = 2x$ and $y = 4$ in the xy-plane.

22. Find the polar moment of inertia about the center of a thin rectangular sheet of constant density $\delta = 1$ bounded by the lines

   a) $x = \pm 2$, $y = \pm 1$ in the xy-plane

   b) $x = \pm a$, $y = \pm b$ in the xy-plane.

   (*Hint:* Find I_x. Then use the formula for I_x to find I_y and add the two to find I_0.)

23. Find the moment of inertia and radius of gyration about the x-axis of a thin plate of constant density δ covering the triangle with vertices $(0, 0)$, $(3, 0)$, and $(3, 2)$ in the xy-plane.

24. Find the center of mass and the moments of inertia and radii of gyration about the coordinate axes of a thin plate bounded by the line $y = x$ and the parabola $y = x^2$ in the xy-plane if the density is $\delta(x, y) = x + 1$.

25. Find the mass and first moments about the coordinate axes of a thin square plate bounded by the lines $x = \pm 1$, $y = \pm 1$ in the xy-plane if the density is $\delta(x, y) = x^2 + y^2 + 1/3$.

26. Find the moment of inertia and radius of gyration about the x-axis of a thin triangular plate of constant density δ whose base lies along the interval $[0, b]$ on the x-axis and whose vertex lies on the line $y = h$ above the x-axis. As you will see, it does not matter where on the line this vertex lies. All such triangles have the same moment of inertia and radius of gyration.

Polar Coordinates

Evaluate the integrals in Exercises 27 and 28 by changing to polar coordinates.

27. $\displaystyle\int_{-1}^1\int_{-\sqrt{1-x^2}}^{\sqrt{1-x^2}} \frac{2\,dy\,dx}{(1+x^2+y^2)^2}$

28. $\displaystyle\int_{-1}^1\int_{-\sqrt{1-y^2}}^{\sqrt{1-y^2}} \ln(x^2+y^2+1)\,dx\,dy$

29. Find the centroid of the region in the polar coordinate plane defined by the inequalities $0 \le r \le 3$ and $-\pi/3 \le \theta \le \pi/3$.

30. Find the centroid of the region in the first quadrant bounded by the rays $\theta = 0$ and $\theta = \pi/2$ and the circles $r = 1$ and $r = 3$.

31. a) Find the centroid of the region in the polar coordinate plane that lies inside the cardioid $r = 1 + \cos\theta$ and outside the circle $r = 1$.

b) CALCULATOR Sketch the region and show the centroid in your sketch.

32. a) Find the centroid of the plane region defined by the polar coordinate inequalities $0 \le r \le a$, $-\alpha \le \theta \le \alpha$ $(0 < \alpha \le \pi)$. How does the centroid move as $\alpha \to \pi^-$?

b) CALCULATOR Sketch the region for $\alpha = 5\pi/6$ and show the centroid in your sketch.

33. Integrate the function $f(x, y) = 1/(1 + x^2 + y^2)^2$ over the region enclosed by one loop of the lemniscate $(x^2 + y^2)^2 - (x^2 - y^2) = 0$.

34. Integrate $f(x, y) = 1/(1 + x^2 + y^2)^2$ over

a) the triangle with vertices $(0, 0)$, $(1, 0)$, $(1, \sqrt{3})$;

b) the first quadrant of the xy-plane.

Triple Integrals in Cartesian Coordinates

Evaluate the integrals in Exercises 35–38.

35. $\displaystyle\int_0^\pi \int_0^\pi \int_0^\pi \cos(x + y + z)\, dx\, dy\, dz$

36. $\displaystyle\int_{\ln 6}^{\ln 7} \int_0^{\ln 2} \int_{\ln 4}^{\ln 5} e^{(x+y+z)}\, dz\, dy\, dx$

37. $\displaystyle\int_0^1 \int_0^{x^2} \int_0^{x+y} (2x - y - z)\, dz\, dy\, dx$

38. $\displaystyle\int_1^e \int_1^x \int_0^z \frac{2y}{z^3}\, dy\, dz\, dx$

39. Find the volume of the wedge-shaped region enclosed on the side by the cylinder $x = -\cos y$, $-\pi/2 \le y \le \pi/2$, on the top by the plane $z = -2x$, and below by the xy-plane.

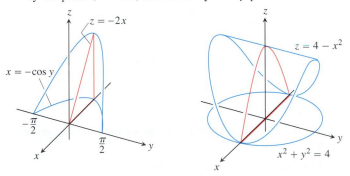

40. Find the volume of the solid that is bounded above by the cylinder $z = 4 - x^2$, on the sides by the cylinder $x^2 + y^2 = 4$, and below by the xy-plane.

41. Find the average value of $f(x, y, z) = 30xz\sqrt{x^2 + y}$ over the rectangular solid in the first octant bounded by the coordinate planes and the planes $x = 1$, $y = 3$, $z = 1$.

42. Find the average value of ρ over the solid sphere $\rho \le a$ (spherical coordinates).

Cylindrical and Spherical Coordinates

43. Convert

$$\int_0^{2\pi} \int_0^{\sqrt{2}} \int_r^{\sqrt{4-r^2}} 3\, dz\, r\, dr\, d\theta, \quad r \ge 0$$

to (a) rectangular coordinates with the order of integration $dz\, dx\, dy$, and (b) spherical coordinates. Then (c) evaluate one of the integrals.

44. (a) Convert to cylindrical coordinates. Then **(b)** evaluate the new integral.

$$\int_0^1 \int_{-\sqrt{1-x^2}}^{\sqrt{1-x^2}} \int_{-(x^2+y^2)}^{(x^2+y^2)} 21\, xy^2\, dz\, dy\, dx$$

45. (a) Convert to spherical coordinates. Then **(b)** evaluate the new integral.

$$\int_{-1}^1 \int_{-\sqrt{1-x^2}}^{\sqrt{1-x^2}} \int_{\sqrt{x^2+y^2}}^1 dz\, dy\, dx$$

46. Write an iterated triple integral for the integral of $f(x, y, z) = 6 + 4y$ over the region in the first octant bounded by the cone $z = \sqrt{x^2 + y^2}$, the cylinder $x^2 + y^2 = 1$, and the coordinate planes in **(a)** rectangular coordinates, **(b)** cylindrical coordinates, **(c)** spherical coordinates. Then **(d)** find the integral of f by evaluating one of the triple integrals.

47. Set up an integral in rectangular coordinates equivalent to the integral

$$\int_0^{\pi/2} \int_1^{\sqrt{3}} \int_1^{\sqrt{4-r^2}} r^3 \sin\theta \cos\theta\, z^2\, dz\, dr\, d\theta.$$

Arrange the order of integration to be z first, then y, then x.

48. The volume of a solid is

$$\int_0^2 \int_0^{\sqrt{2x-x^2}} \int_{-\sqrt{4-x^2-y^2}}^{\sqrt{4-x^2-y^2}} dz\, dy\, dx.$$

a) Describe the solid by giving equations for the surfaces that form its boundary.

b) Convert the integral to cylindrical coordinates but do not evaluate the integral.

49. Let D be the smaller spherical cap cut from a solid ball of radius 2 by a plane 1 unit from the center of the sphere. Express the volume of D as an iterated triple integral in **(a)** rectangular, **(b)** cylindrical, and **(c)** spherical coordinates. *Do not evaluate the integrals.*

50. Express the moment of inertia I_z of the solid hemisphere bounded below by the plane $z = 0$ and above by the sphere $x^2 + y^2 + z^2 = 1$ as an iterated integral in **(a)** rectangular, **(b)** cylindrical, and **(c)** spherical coordinates. *Do not evaluate the integrals.*

51. *Spherical vs. cylindrical coordinates.* Triple integrals involving spherical shapes do not always require spherical coordinates for convenient evaluation. Some calculations may be accomplished more easily with cylindrical coordinates. As a case in point, find the volume of the region bounded above by the sphere $x^2 + y^2 + z^2 = 8$ and below by the plane $z = 2$ by using (a) cylindrical coordinates, (b) spherical coordinates.

52. Find the moment of inertia about the z-axis of a solid of constant density $\delta = 1$ that is bounded above by the sphere $\rho = 2$ and below by the cone $\phi = \pi/3$ (spherical coordinates).

53. Find the moment of inertia of a solid of constant density δ bounded by two concentric spheres of radii a and b ($a < b$) about a diameter.

54. Find the moment of inertia about the z-axis of a solid of density $\delta = 1$ enclosed by the spherical coordinate surface $\rho = 1 - \cos \phi$.

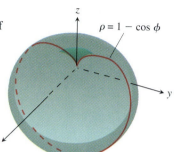

Substitutions

55. Show that if $u = x - y$ and $v = y$, then

$$\int_0^\infty \int_0^x e^{-sx} f(x - y, y) \, dy \, dx = \int_0^\infty \int_0^\infty e^{-s(u+v)} f(u, v) \, du \, dv.$$

56. What relationship must hold between the constants a, b, and c to make

$$\int_{-\infty}^\infty \int_{-\infty}^\infty e^{-(ax^2 + 2bxy + cy^2)} \, dx \, dy = 1?$$

(*Hint:* Let $s = \alpha x + \beta y$ and $t = \gamma x + \delta y$, where $(\alpha \delta - \beta \gamma)^2 = ac - b^2$. Then $ax^2 + 2bxy + cy^2 = s^2 + t^2$.)

CHAPTER 13 ADDITIONAL EXERCISES–THEORY, EXAMPLES, APPLICATIONS

Volumes

1. The base of a sand pile covers the region in the xy-plane that is bounded by the parabola $x^2 + y = 6$ and the line $y = x$. The height of the sand above the point (x, y) is x^2. Express the volume of sand as (a) a double integral, (b) a triple integral. Then (c) find the volume.

2. A hemispherical bowl of radius 5 cm is filled with water to within 3 cm of the top. Find the volume of water in the bowl.

3. Find the volume of the portion of the solid cylinder $x^2 + y^2 \leq 1$ that lies between the planes $z = 0$ and $x + y + z = 2$.

4. Find the volume of the region bounded above by the sphere $x^2 + y^2 + z^2 = 2$ and below by the paraboloid $z = x^2 + y^2$.

5. Find the volume of the region bounded above by the paraboloid $z = 3 - x^2 - y^2$ and below by the paraboloid $z = 2x^2 + 2y^2$.

6. Find the volume of the region enclosed by the spherical coordinate surface $\rho = 2 \sin \phi$ (Fig. 13.54).

7. A circular cylindrical hole is bored through a solid sphere, the axis of the hole being a diameter of the sphere. The volume of the remaining solid is

$$V = 2 \int_0^{2\pi} \int_0^{\sqrt{3}} \int_1^{\sqrt{4-z^2}} r \, dr \, dz \, d\theta.$$

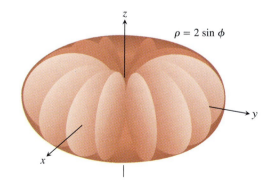

13.54 The surface in Exercise 6.

a) Find the radius of the hole and the radius of the sphere.

b) Evaluate the integral.

8. Find the volume of material cut from the solid sphere $r^2 + z^2 \leq 9$ by the cylinder $r = 3 \sin \theta$.

9. Find the volume of the region enclosed by the surfaces $z = x^2 + y^2$ and $z = (x^2 + y^2 + 1)/2$.

10. Find the volume of the region in the first octant that lies between the cylinders $r = 1$ and $r = 2$ and that is bounded below by the xy-plane and above by the surface $z = xy$.

Changing the Order of Integration

In Exercises 11 and 12, sketch the region of integration and write an equivalent iterated integral with the order of integration reversed.

11. $\displaystyle\int_0^1 \int_{x^2}^x f(x, y) \, dy \, dx$ **12.** $\displaystyle\int_0^4 \int_y^{2\sqrt{y}} f(x, y) \, dx \, dy$

13. Evaluate the integral

$$\int_0^\infty \frac{e^{-ax} - e^{-bx}}{x} \, dx.$$

(*Hint:* Use the relation

$$\frac{e^{-ax} - e^{-bx}}{x} = \int_a^b e^{-xy} \, dy.$$

to form a double integral and evaluate the integral by changing the order of integration.)

14. a)   Show, by changing to polar coordinates, that

$$\int_0^{a\sin\beta} \int_{y\cot\beta}^{\sqrt{a^2-y^2}} \ln(x^2 + y^2) \, dx \, dy = a^2 \beta \left(\ln a - \frac{1}{2} \right),$$

where $a > 0$ and $0 < \beta < \pi/2$.

b)   Rewrite the Cartesian integral with the order of integration reversed.

15. By changing the order of integration, show that the following double integral can be reduced to a single integral:

$$\int_0^x \int_0^u e^{m(x-t)} f(t) \, dt \, du = \int_0^x (x - t) e^{m(x-t)} f(t) \, dt.$$

Similarly, it can be shown that

$$\int_0^x \int_0^v \int_0^u e^{m(x-t)} f(t) \, dt \, du \, dv = \int_0^x \frac{(x - t)^2}{2} e^{m(x-t)} f(t) \, dt.$$

16. Sometimes a multiple integral with variable limits can be changed into one with constant limits. By changing the order of integration, show that

$$\int_0^1 f(x) \left(\int_0^x g(x - y) f(y) \, dy \right) dx$$

$$= \int_0^1 f(y) \left(\int_y^1 g(x - y) f(x) \, dx \right) dy$$

$$= \frac{1}{2} \int_0^1 \int_0^1 g(|x - y|) f(x) \, f(y) \, dx \, dy.$$

Masses and Moments

17. A thin plate of constant density is to occupy the triangular region in the first quadrant of the xy-plane having vertices $(0, 0)$, $(a, 0)$, and $(a, 1/a)$. What value of a will minimize the plate's polar moment of inertia about the origin?

18. Find the polar moment of inertia about the origin of a thin triangular plate of constant density $\delta = 3$ bounded by the y-axis and the lines $y = 2x$ and $y = 4$ in the xy-plane.

19. Find the centroid of the region in the polar coordinate plane

that lies inside the cardioid $r = 1 + \cos\theta$ and outside the circle $r = 1$.

20. Find the centroid of the boomerang-shaped region between the parabolas $y^2 = -4(x - 1)$ and $y^2 = -2(x - 2)$ in the xy-plane.

21. The counterweight of a flywheel of constant density 1 has the form of the smaller segment cut from a circle of radius a by a chord at a distance b from the center $(b < a)$. Find the mass of the counterweight and its polar moment of inertia about the center of the wheel.

22. Find the radii of gyration about the x- and y-axes of a thin plate of density $\delta = 1$ enclosed by one loop of the lemniscate $r^2 = 2a^2 \cos 2\theta$.

23. A solid is bounded on the top by the paraboloid $z = r^2$, on the bottom by the plane $z = 0$, and on the sides by the cylinder $r = 1$. Find the center of mass and the moment of inertia and radius of gyration about the z-axis if the density is (a) $\delta(r, \theta, z) = z$; (b) $\delta(r, \theta, z) = r$.

24. A solid is bounded below by the cone $z = \sqrt{x^2 + y^2}$ and above by the plane $z = 1$. Find the center of mass and the moment of inertia and radius of gyration about the z-axis if the density is (a) $\delta(r, \theta, z) = z$; (b) $\delta(r, \theta, z) = z^2$.

25. Use spherical coordinates to find the centroid of a solid hemisphere of radius a.

26. Find the moment of inertia and radius of gyration of a solid sphere of radius a and density $\delta = 1$ about a diameter of the sphere.

Theory and Applications

27. Evaluate

$$\int_0^a \int_0^b e^{\max(b^2 x^2, a^2 y^2)} \, dy \, dx,$$

where a and b are positive numbers and

$$\max(b^2 x^2, a^2 y^2) = \begin{cases} b^2 x^2 & \text{if } b^2 x^2 \geq a^2 y^2 \\ a^2 y^2 & \text{if } b^2 x^2 < a^2 y^2. \end{cases}$$

28. Show that

$$\iint \frac{\partial^2 F(x, y)}{\partial x \, \partial y} \, dx \, dy$$

over the rectangle $x_0 \leq x \leq x_1$, $y_0 \leq y \leq y_1$, is

$$F(x_1, y_1) - F(x_0, y_1) - F(x_1, y_0) + F(x_0, y_0).$$

29. Suppose that $f(x, y)$ can be written as a product $f(x, y) = F(x)G(y)$ of a function of x and a function of y. Then the integral of f over the rectangle $R: a \leq x \leq b, c \leq y \leq d$ can be evaluated as a product as well, by the formula

$$\iint_R f(x, y) \, dA = \left(\int_a^b F(x) \, dx \right) \left(\int_c^d G(y) \, dy \right). \qquad (1)$$

The argument is that

$$\iint_R f(x, y)\, dA = \int_c^d \left(\int_a^b F(x) G(y)\, dx \right) dy \qquad \text{(i)}$$

$$= \int_c^d \left(G(y) \int_a^b F(x)\, dx \right) dy \qquad \text{(ii)}$$

$$= \int_c^d \left(\int_a^b F(x)\, dx \right) G(y)\, dy \qquad \text{(iii)}$$

$$= \left(\int_a^b F(x)\, dx \right) \int_c^d G(y)\, dy. \qquad \text{(iv)}$$

a) Give reasons for steps (i)–(iv).

When it applies, Eq. (1) can be a time saver. Use it to evaluate the following integrals.

b) $\displaystyle \int_0^{\ln 2} \int_0^{\pi/2} e^x \cos y \, dy \, dx$

c) $\displaystyle \int_1^2 \int_{-1}^1 \frac{x}{y^2} \, dx \, dy$

30. Let $D_\mathbf{u} f$ denote the derivative of $f(x, y) = (x^2 + y^2)/2$ in the direction of the unit vector $\mathbf{u} = u_1 \mathbf{i} + u_2 \mathbf{j}$.

a) Find the average value of $D_\mathbf{u} f$ over the triangular region cut from the first quadrant by the line $x + y = 1$.

b) Show in general that the average value of $D_\mathbf{u} f$ over a region in the xy-plane is the value of $D_\mathbf{u} f$ at the centroid of the region.

31. *The value of $\Gamma(1/2)$.* As we saw in Additional Exercises 49 and 50 in Chapter 7, the gamma function,

$$\Gamma(x) = \int_0^\infty t^{x-1} e^{-t} \, dt,$$

extends the factorial function from the nonnegative integers to other real values. Of particular interest in the theory of differential equations is the number

$$\Gamma\left(\frac{1}{2}\right) = \int_0^\infty t^{(1/2)-1} e^{-t} \, dt = \int_0^\infty \frac{e^{-t}}{\sqrt{t}} \, dt. \qquad \text{(2)}$$

a) If you have not yet done Exercise 37 in Section 13.3, do it now to show that

$$I = \int_0^\infty e^{-y^2} \, dy = \frac{\sqrt{\pi}}{2}.$$

b) Substitute $y = \sqrt{t}$ in Eq. (2) to show that $\Gamma(1/2) = 2I = \sqrt{\pi}$.

32. The electrical charge distribution on a circular plate of radius R meters is $\sigma(r, \theta) = kr(1 - \sin \theta)$ coulomb/m^2 (k a constant). Integrate σ over the plate to find the total charge Q.

33. *A parabolic rain gauge.* A bowl is in the shape of the graph of $z = x^2 + y^2$ from $z = 0$ to $z = 10$ in. You plan to calibrate the bowl to make it into a rain gauge. What height in the bowl would correspond to 1 in. of rain? 3 in. of rain?

34. *Water in a satellite dish.* A parabolic satellite dish is 2 m wide and 1/2 m deep. Its axis of symmetry is tilted 30 degrees from the vertical.

a) Set up, but do not evaluate, a triple integral in rectangular coordinates that gives the amount of water the satellite dish will hold. (*Hint:* Put your coordinate system so that the satellite dish is in "standard position" and the plane of the water level is slanted.) (*Caution:* The limits of integration are not "nice.")

b) What would be the smallest tilt of the satellite dish so that it holds no water?

35. *Cylindrical shells.* In Section 5.4, we learned how to find the volume of a solid of revolution using the shell method, namely if the region between the curve $y = f(x)$ and the x-axis from a to b ($0 < a < b$) is revolved about the y-axis the volume of the resulting solid is $\int_a^b 2\pi x \, f(x) \, dx$. Prove that finding volumes by using triple integrals gives the same result. (*Hint:* Use cylindrical coordinates with the roles of y and z changed.)

36. *An infinite half-cylinder.* Let D be the interior of the infinite right circular half-cylinder of radius 1 with its single-end face suspended 1 unit above the origin and its axis the ray from $(0, 0, 1)$ to ∞. Use cylindrical coordinates to evaluate

$$\iiint_D z(r^2 + z^2)^{-5/2} \, dV.$$

37. *Hypervolume.* We have learned that $\int_a^b 1 \, dx$ is the length of the interval $[a, b]$ on the number line (one-dimensional space), $\iint_R 1 \, dA$ is the area of region R in the xy-plane (two-dimensional space), and $\iiint_D 1 \, dV$ is the volume of the region D in three-dimensional space (xyz-space). We could continue: If Q is a region in 4-space ($xyzw$-space), then $\iiiint_Q 1 \, dV$ is the "hypervolume" of Q. Use your generalizing abilities and a Cartesian coordinate system of 4-space to find the hypervolume inside the unit 4-sphere $x^2 + y^2 + z^2 + w^2 = 1$.

Integration in Vector Fields

OVERVIEW   This chapter treats integration in vector fields. The mathematics in this chapter is the mathematics that is used to describe the properties of electromagnetism, explain the flow of heat in stars, and calculate the work it takes to put a satellite in orbit.

14.1 Line Integrals

When a curve $\mathbf{r}(t) = g(t)\,\mathbf{i} + h(t)\,\mathbf{j} + k(t)\,\mathbf{k}$, $a \leq t \leq b$, passes through the domain of a function $f(x, y, z)$ in space, the values of f along the curve are given by the composite function $f(g(t), h(t), k(t))$. If we integrate this composite with respect to arc length from $t = a$ to $t = b$, we calculate the so-called line integral of f along the curve. Despite the three-dimensional geometry, the line integral is an ordinary integral of a real-valued function over an interval of real numbers.

The importance of line integrals lies in their application. These are the integrals with which we calculate the work done by variable forces along paths in space and the rates at which fluids flow along curves and across boundaries.

Definitions and Notation

Suppose that $f(x, y, z)$ is a function whose domain contains the curve $\mathbf{r}(t) = g(t)\,\mathbf{i} + h(t)\,\mathbf{j} + k(t)\,\mathbf{k}$, $a \leq t \leq b$. We partition the curve into a finite number of subarcs (Fig. 14.1). The typical subarc has length Δs_k. In each subarc we choose a point (x_k, y_k, z_k) and form the sum

$$S_n = \sum_{k=1}^{n} f(x_k, y_k, z_k)\,\Delta s_k. \tag{1}$$

If f is continuous and the functions g, h, and k have continuous first derivatives, then the sums in (1) approach a limit as n increases, and the lengths Δs_k approach zero. We call this limit the **integral of f over the curve from a to b.** If the curve is denoted by a single letter, C for example, the notation for the integral is

$$\int_C f(x, y, z)\,ds \qquad \text{"The integral of } f \text{ over } C\text{"} \tag{2}$$

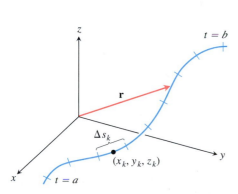

14.1 The curve $\mathbf{r} = g(t)\,\mathbf{i} + h(t)\,\mathbf{j} + k(t)\,\mathbf{k}$, partitioned into small arcs from $t = a$ to $t = b$. The length of a typical subarc is Δs_k.

Evaluation for Smooth Curves

If $\mathbf{r}(t)$ is smooth for $a \leq t \leq b$ ($\mathbf{v} = d\mathbf{r}/dt$ is continuous and never $\mathbf{0}$), we can use the equation

$$s(t) = \int_a^t |\mathbf{v}(\tau)| \, d\tau \qquad \text{Eq. (4) of Section 11.3, with } t_0 = a$$

to express ds in Eq. (2) as $ds = |\mathbf{v}(t)| \, dt$. A theorem from advanced calculus says that we can then evaluate the integral of f over C as

$$\int_C f(x, y, z) \, ds = \int_a^b f(g(t), h(t), k(t)) |\mathbf{v}(t)| \, dt.$$

This formula will evaluate the integral correctly no matter what parametrization we use, as long as the parametrization is smooth.

How to Evaluate a Line Integral

To integrate a continuous function $f(x, y, z)$ over a curve C:

1. Find a smooth parametrization of C,

$$\mathbf{r}(t) = g(t) \mathbf{i} + h(t) \mathbf{j} + k(t) \mathbf{k}, \qquad a \leq t \leq b.$$

2. Evaluate the integral as

$$\int_C f(x, y, z) \, ds = \int_a^b f(g(t), h(t), k(t)) \, |\mathbf{v}(t)| \, dt. \qquad (3)$$

Notice that if f has the constant value 1, then the integral of f over C gives the length of C.

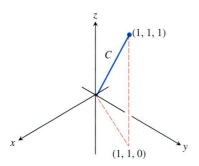

14.2 The integration path in Example 1.

EXAMPLE 1 Integrate $f(x, y, z) = x - 3y^2 + z$ over the line segment C joining the origin and the point $(1, 1, 1)$ (Fig. 14.2).

Solution We choose the simplest parametrization we can think of:

$$\mathbf{r}(t) = t\mathbf{i} + t\mathbf{j} + t\mathbf{k}, \qquad 0 \leq t \leq 1.$$

The components have continuous first derivatives and $|\mathbf{v}(t)| = \sqrt{1^2 + 1^2 + 1^2} = \sqrt{3}$ is never 0, so the parametrization is smooth. The integral of f over C is

$$\int_C f(x, y, z) \, ds = \int_0^1 f(t, t, t) \left(\sqrt{3}\right) dt \qquad \text{Eq. (3)}$$

$$= \int_0^1 (t - 3t^2 + t) \sqrt{3} \, dt$$

$$= \sqrt{3} \int_0^1 (2t - 3t^2) \, dt = \sqrt{3} \left[t^2 - t^3 \right]_0^1 = 0. \qquad \square$$

Additivity

Line integrals have the useful property that if a curve C is made by joining a finite number of curves $C_1, C_2, \ldots, C_n$ end to end, then the integral of a function over

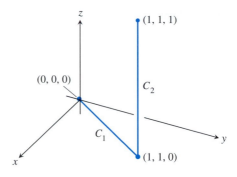

14.3 The path of integration in Example 2.

C is the sum of the integrals over the curves that make it up:

$$\int_C f \, ds = \int_{C_1} f \, ds + \int_{C_2} f \, ds + \cdots + \int_{C_n} f \, ds. \tag{4}$$

EXAMPLE 2 Figure 14.3 shows another path from the origin to $(1, 1, 1)$, the union of line segments C_1 and C_2. Integrate $f(x, y, z) = x - 3y^2 + z$ over $C_1 \cup C_2$.

Solution We choose the simplest parametrizations for C_1 and C_2 we can think of, checking the lengths of the velocity vectors as we go along:

$$C_1: \quad \mathbf{r}(t) = t\,\mathbf{i} + t\,\mathbf{j}, \quad 0 \le t \le 1; \quad |\mathbf{v}| = \sqrt{1^2 + 1^2} = \sqrt{2}$$

$$C_2: \quad \mathbf{r}(t) = \mathbf{i} + \mathbf{j} + t\,\mathbf{k}, \quad 0 \le t \le 1; \quad |\mathbf{v}| = \sqrt{0^2 + 0^2 + 1^2} = 1.$$

With these parametrizations we find that

$$\int_{C_1 \cup C_2} f(x, y, z) \, ds = \int_{C_1} f(x, y, z) \, ds + \int_{C_2} f(x, y, z) \, ds \qquad \text{Eq. (4)}$$

$$= \int_0^1 f(t, t, 0) \sqrt{2} \, dt + \int_0^1 f(1, 1, t)(1) \, dt \qquad \text{Eq. (3)}$$

$$= \int_0^1 (t - 3t^2 + 0) \sqrt{2} \, dt + \int_0^1 (1 - 3 + t)(1) \, dt$$

$$= \sqrt{2}\left[\frac{t^2}{2} - t^3\right]_0^1 + \left[\frac{t^2}{2} - 2t\right]_0^1 = -\frac{\sqrt{2}}{2} - \frac{3}{2}. \qquad \square$$

Notice three things about the integrations in Examples 1 and 2. First, as soon as the components of the appropriate curve were substituted into the formula for f, the integration became a standard integration with respect to t. Second, the integral of f over $C_1 \cup C_2$ was obtained by integrating f over each section of the path and adding the results. Third, the integrals of f over C and $C_1 \cup C_2$ had different values. For most functions, the value of the integral along a path joining two points changes if you change the path between them. For some functions, however, the value remains the same, as we will see in Section 14.3.

Mass and Moment Calculations

We treat coil springs and wires like masses distributed along smooth curves in space. The distribution is described by a continuous density function $\delta(x, y, z)$ (mass per unit length). The spring's or wire's mass, center-of-mass, and moments are then calculated with the formulas in Table 14.1, on the following page. The formulas also apply to thin rods.

EXAMPLE 3 A coil spring lies along the helix

$$\mathbf{r}(t) = (\cos 4t)\,\mathbf{i} + (\sin 4t)\,\mathbf{j} + t\,\mathbf{k}, \quad 0 \le t \le 2\pi.$$

The spring's density is a constant, $\delta = 1$. Find the spring's mass and center of mass, and its moment of inertia and radius of gyration about the z-axis.

Solution We sketch the spring (Fig. 14.4). Because of the symmetries involved, the center of mass lies at the point $(0, 0, \pi)$ on the z-axis.

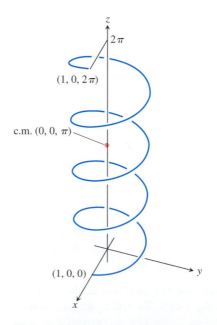

14.4 The helical spring in Example 3.

Table 14.1 Mass and moment formulas for coil springs, thin rods, and wires lying along a smooth curve C in space

---

Mass: $\quad M = \displaystyle\int_C \delta(x, y, z)\, ds$

First moments about the coordinate planes:

$$M_{yz} = \int_C x\, \delta\, ds, \qquad M_{xz} = \int_C y\, \delta\, ds, \qquad M_{xy} = \int_C z\, \delta\, ds$$

Coordinates of the center of mass:

$$\bar{x} = M_{yz}/M, \qquad \bar{y} = M_{xz}/M, \qquad \bar{z} = M_{xy}/M$$

Moments of inertia:

$$I_x = \int_C (y^2 + z^2)\, \delta\, ds, \qquad I_y = \int_C (x^2 + z^2)\, \delta\, ds$$

$$I_z = \int_C (x^2 + y^2)\, \delta\, ds, \qquad I_L = \int_C r^2\, \delta\, ds$$

$$r(x, y, z) = \text{distance from point } (x, y, z) \text{ to line } L$$

Radius of gyration about a line L: $\quad R_L = \sqrt{I_L/M}$

For the remaining calculations, we first find $|\mathbf{v}(t)|$:

$$|\mathbf{v}(t)| = \sqrt{\left(\frac{dx}{dt}\right)^2 + \left(\frac{dy}{dt}\right)^2 + \left(\frac{dz}{dt}\right)^2}$$

$$= \sqrt{(-4\sin 4t)^2 + (4\cos 4t)^2 + 1} = \sqrt{17}.$$

We then evaluate the formulas from Table 14.1 using Eq. (3):

$$M = \int_{\text{Helix}} \delta\, ds = \int_0^{2\pi} (1)\sqrt{17}\, dt = 2\pi\sqrt{17}$$

$$I_z = \int_{\text{Helix}} (x^2 + y^2)\delta\, ds = \int_0^{2\pi} (\cos^2 4t + \sin^2 4t)(1)\sqrt{17}\, dt$$

$$= \int_0^{2\pi} \sqrt{17}\, dt = 2\pi\sqrt{17}$$

$$R_z = \sqrt{I_z/M} = \sqrt{2\pi\sqrt{17}/(2\pi\sqrt{17})} = 1.$$

Notice that the radius of gyration about the z-axis is the radius of the cylinder around which the helix winds. ❑

EXAMPLE 4 A slender metal arch, denser at the bottom than top, lies along the semicircle $y^2 + z^2 = 1$, $z \geq 0$, in the yz-plane (Fig. 14.5). Find the center of the arch's mass if the density at the point (x, y, z) on the arch is $\delta(x, y, z) = 2 - z$.

Solution We know that $\bar{x} = 0$ and $\bar{y} = 0$ because the arch lies in the yz-plane with its mass distributed symmetrically about the z-axis. To find $\bar{z}$, we parametrize the circle as

$$\mathbf{r}(t) = (\cos t)\mathbf{j} + (\sin t)\mathbf{k}, \qquad 0 \leq t \leq \pi.$$

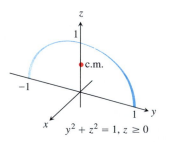

14.5 Example 4 shows how to find the center of mass of a circular arch of variable density.

For this parametrization,

$$|\mathbf{v}(t)| = \sqrt{\left(\frac{dx}{dt}\right)^2 + \left(\frac{dy}{dt}\right)^2 + \left(\frac{dz}{dt}\right)^2} = \sqrt{(0)^2 + (-\sin t)^2 + (\cos t)^2} = 1.$$

The formulas in Table 14.1 then give

$$M = \int_C \delta \, ds = \int_C (2 - z) \, ds = \int_0^\pi (2 - \sin t) \, dt = 2\pi - 2$$

$$M_{xy} = \int_C z \delta \, ds = \int_C z(2 - z) \, ds = \int_0^\pi (\sin t)(2 - \sin t) \, dt$$

$$= \int_0^\pi (2 \sin t - \sin^2 t) \, dt = \frac{8 - \pi}{2}$$

$$\bar{z} = \frac{M_{xy}}{M} = \frac{8 - \pi}{2} \cdot \frac{1}{2\pi - 2} = \frac{8 - \pi}{4\pi - 4} \approx 0.57.$$

With $\bar{z}$ to the nearest hundredth, the center of mass is $(0, 0, 0.57)$.   ❏

Exercises 14.1

Graphs of Vector Equations

Match the vector equations in Exercises 1–8 with the graphs in Fig. 14.6.

14.6 The graphs for Exercises 1–8.

1. $\mathbf{r}(t) = t \mathbf{i} + (1 - t) \mathbf{j}, \quad 0 \le t \le 1$

2. $\mathbf{r}(t) = \mathbf{i} + \mathbf{j} + t \mathbf{k}, \quad -1 \le t \le 1$

3. $\mathbf{r}(t) = (2 \cos t) \mathbf{i} + (2 \sin t) \mathbf{j}, \quad 0 \le t \le 2\pi$

4. $\mathbf{r}(t) = t \mathbf{i}, \quad -1 \le t \le 1$

5. $\mathbf{r}(t) = t \mathbf{i} + t \mathbf{j} + t \mathbf{k}, \quad 0 \le t \le 2$

6. $\mathbf{r}(t) = t \mathbf{j} + (2 - 2t) \mathbf{k}, \quad 0 \le t \le 1$

7. $\mathbf{r}(t) = (t^2 - 1) \mathbf{j} + 2t \mathbf{k}, \quad -1 \le t \le 1$

8. $\mathbf{r}(t) = (2 \cos t) \mathbf{i} + (2 \sin t) \mathbf{k}, \quad 0 \le t \le \pi$

Evaluating Line Integrals over Space Curves

9. Evaluate $\int_C (x + y)\,ds$ where C is the straight-line segment $x = t$, $y = (1 - t)$, $z = 0$, from $(0, 1, 0)$ to $(1, 0, 0)$.

10. Evaluate $\int_C (x - y + z - 2)\,ds$ where C is the straight-line segment $x = t$, $y = (1 - t)$, $z = 1$, from $(0, 1, 1)$ to $(1, 0, 1)$.

11. Evaluate $\int_C (xy + y + z)\,ds$ along the curve $\mathbf{r}(t) = 2t\,\mathbf{i} + t\,\mathbf{j} + (2 - 2t)\,\mathbf{k}$, $0 \le t \le 1$.

12. Evaluate $\int_C \sqrt{x^2 + y^2}\,ds$ along the curve $\mathbf{r}(t) = (4 \cos t)\,\mathbf{i} + (4 \sin t)\,\mathbf{j} + 3t\,\mathbf{k}$, $-2\pi \le t \le 2\pi$.

13. Find the line integral of $f(x, y, z) = x + y + z$ over the straight-line segment from $(1, 2, 3)$ to $(0, -1, 1)$.

14. Find the line integral of $f(x, y, z) = \sqrt{3}/(x^2 + y^2 + z^2)$ over the curve $\mathbf{r}(t) = t\,\mathbf{i} + t\,\mathbf{j} + t\,\mathbf{k}$, $1 \le t \le \infty$.

15. Integrate $f(x, y, z) = x + \sqrt{y} - z^2$ over the path from $(0, 0, 0)$ to $(1, 1, 1)$ (Fig. 14.7a) given by
    C_1: $\mathbf{r}(t) = t\,\mathbf{i} + t^2\,\mathbf{j}$, $0 \le t \le 1$
    C_2: $\mathbf{r}(t) = \mathbf{i} + \mathbf{j} + t\,\mathbf{k}$, $0 \le t \le 1$

16. Integrate $f(x, y, z) = x + \sqrt{y} - z^2$ over the path from $(0, 0, 0)$ to $(1, 1, 1)$ (Fig. 14.7b) given by
    C_1: $\mathbf{r}(t) = t\,\mathbf{k}$, $0 \le t \le 1$
    C_2: $\mathbf{r}(t) = t\,\mathbf{j} + \mathbf{k}$, $0 \le t \le 1$
    C_3: $\mathbf{r}(t) = t\,\mathbf{i} + \mathbf{j} + \mathbf{k}$, $0 \le t \le 1$

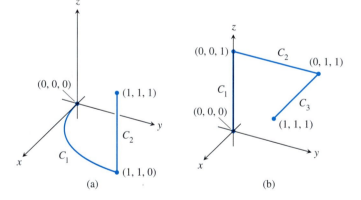

14.7 The paths of integration for Exercises 15 and 16.

17. Integrate $f(x, y, z) = (x + y + z)/(x^2 + y^2 + z^2)$ over the path $\mathbf{r}(t) = t\,\mathbf{i} + t\,\mathbf{j} + t\,\mathbf{k}$, $0 < a \le t \le b$.

18. Integrate $f(x, y, z) = -\sqrt{x^2 + z^2}$ over the circle
    $$\mathbf{r}(t) = (a \cos t)\,\mathbf{j} + (a \sin t)\,\mathbf{k}, \quad 0 \le t \le 2\pi.$$

Line Integrals over Plane Curves

In Exercises 19–22, integrate f over the given curve.

19. $f(x, y) = x^3/y$, C: $y = x^2/2$, $0 \le x \le 2$

20. $f(x, y) = (x + y^2)/\sqrt{1 + x^2}$, C: $y = x^2/2$ from $(1, 1/2)$ to $(0, 0)$

21. $f(x, y) = x + y$, C: $x^2 + y^2 = 4$ in the first quadrant from $(2, 0)$ to $(0, 2)$

22. $f(x, y) = x^2 - y$, C: $x^2 + y^2 = 4$ in the first quadrant from $(0, 2)$ to $(\sqrt{2}, \sqrt{2})$

Mass and Moments

23. Find the mass of a wire that lies along the curve $\mathbf{r}(t) = (t^2 - 1)\,\mathbf{j} + 2t\,\mathbf{k}$, $0 \le t \le 1$, if the density is $\delta = (3/2)\,t$.

24. A wire of density $\delta(x, y, z) = 15\sqrt{y + 2}$ lies along the curve $\mathbf{r}(t) = (t^2 - 1)\,\mathbf{j} + 2t\,\mathbf{k}$, $-1 \le t \le 1$. Find its center of mass. Then sketch the curve and center of mass together.

25. Find the mass of a thin wire lying along the curve $\mathbf{r}(t) = \sqrt{2}\,t\,\mathbf{i} + \sqrt{2}\,t\,\mathbf{j} + (4 - t^2)\,\mathbf{k}$, $0 \le t \le 1$, if the density is (a) $\delta = 3t$, (b) $\delta = 1$.

26. Find the center of mass of a thin wire lying along the curve $\mathbf{r}(t) = t\,\mathbf{i} + 2t\,\mathbf{j} + (2/3)t^{3/2}\,\mathbf{k}$, $0 \le t \le 2$, if the density is $\delta = 3\sqrt{5 + t}$.

27. A circular wire hoop of constant density δ lies along the circle $x^2 + y^2 = a^2$ in the xy-plane. Find the hoop's moment of inertia and radius of gyration about the z-axis.

28. A slender rod of constant density lies along the line segment $\mathbf{r}(t) = t\,\mathbf{j} + (2 - 2t)\,\mathbf{k}$, $0 \le t \le 1$, in the yz-plane. Find the moments of inertia and radii of gyration of the rod about the three coordinate axes.

29. A spring of constant density δ lies along the helix
    $$\mathbf{r}(t) = (\cos t)\,\mathbf{i} + (\sin t)\,\mathbf{j} + t\,\mathbf{k}, \quad 0 \le t \le 2\pi.$$
    a) Find I_z and R_z.
    b) Suppose you have another spring of constant density δ that is twice as long as the spring in (a) and lies along the helix for $0 \le t \le 4\pi$. Do you expect I_z and R_z for the longer spring to be the same as those for the shorter one, or should they be different? Check your predictions by calculating I_z and R_z for the longer spring.

30. A wire of constant density $\delta = 1$ lies along the curve
    $$\mathbf{r}(t) = (t \cos t)\,\mathbf{i} + (t \sin t)\,\mathbf{j} + (2\sqrt{2}/3)\,t^{3/2}\,\mathbf{k}, \quad 0 \le t \le 1.$$
    Find $\bar{z}$, I_z, and R_z.

31. Find I_x and R_x for the arch in Example 4.

32. Find the center of mass, and the moments of inertia and radii of gyration about the coordinate axes of a thin wire lying along the curve
    $$\mathbf{r}(t) = t\,\mathbf{i} + \frac{2\sqrt{2}}{3}\,t^{3/2}\,\mathbf{j} + \frac{t^2}{2}\,\mathbf{k}, \quad 0 \le t \le 2,$$
    if the density is $\delta = 1/(t + 1)$.

✪ CAS Explorations and Projects

In Exercises 33–36, use a CAS to perform the following steps to evaluate the line integrals:

a) Find $ds = |\mathbf{v}(t)|\,dt$ for the path $\mathbf{r}(t) = g(t)\,\mathbf{i} + h(t)\,\mathbf{j} + k(t)\,\mathbf{k}$.
b) Express the integrand $f(g(t), h(t), k(t))\,|\mathbf{v}(t)|$ as a function of the parameter t.
c) Evaluate $\int_C f\,ds$ using Eq. (3) in the text.

33. $f(x, y, z) = \sqrt{1 + 30x^2 + 10y}$; $\quad \mathbf{r}(t) = t\,\mathbf{i} + t^2\,\mathbf{j} + 3t^2\,\mathbf{k}$,
$0 \le t \le 2$

34. $f(x, y, z) = \sqrt{1 + x^3 + 5y^3}$; $\quad \mathbf{r}(t) = t\,\mathbf{i} + \frac{1}{3}t^2\,\mathbf{j} + \sqrt{t}\,\mathbf{k}$,
$0 \le t \le 2$

35. $f(x, y, z) = x\sqrt{y} - 3z^2$; $\quad \mathbf{r}(t) = \cos 2t\,\mathbf{i} + \sin 2t\,\mathbf{j} + 5t\,\mathbf{k}$,
$0 \le t \le 2\pi$

36. $f(x, y, z) = \left(1 + \frac{9}{4}z^{1/3}\right)^{1/4}$; $\quad \mathbf{r}(t) = \cos 2t\,\mathbf{i} + \sin 2t\,\mathbf{j} + t^{5/2}\,\mathbf{k}$,
$0 \le t \le 2\pi$

14.2 Vector Fields, Work, Circulation, and Flux

When we study physical phenomena that are represented by vectors, we replace integrals over closed intervals by integrals over paths through vector fields. We use such integrals to find the work done in moving an object along a path against a variable force (a vehicle sent into space against Earth's gravitational field) or to find the work done by a vector field in moving an object along a path through the field (the work done by an accelerator in raising the energy of a particle). We also use line integrals to find the rates at which fluids flow along and across curves.

Vector Fields

A **vector field** on a domain in the plane or in space is a function that assigns a vector to each point in the domain. A field of three-dimensional vectors might have a formula like

$$\mathbf{F}(x, y, z) = M(x, y, z)\,\mathbf{i} + N(x, y, z)\,\mathbf{j} + P(x, y, z)\,\mathbf{k}.$$

The field is **continuous** if the **component functions** M, N, and P are continuous, **differentiable** if M, N, and P are differentiable, and so on. A field of two-dimensional vectors might have a formula like

$$\mathbf{F}(x, y) = M(x, y)\,\mathbf{i} + N(x, y)\,\mathbf{j}.$$

If we attach a projectile's velocity vector to each point of the projectile's trajectory in the plane of motion, we have a two-dimensional field defined along the trajectory. If we attach the gradient vector of a scalar function to each point of a level surface of the function, we have a three-dimensional field on the surface. If we attach the velocity vector to each point of a flowing fluid, we have a three-dimensional field defined on a region in space. These and other fields are illustrated in Figs. 14.8–14.16. Some of the illustrations give formulas for the fields as well.

To sketch the fields that had formulas, we picked a representative selection of domain points and sketched the vectors attached to them. Notice the convention that the arrows representing the vectors are drawn with their tails, not their heads, at the points where the vector functions are evaluated. This is different from the way we drew the position vectors of the planets and projectiles in Chapter 11, with their tails at the origin and their heads at the planet's and projectile's locations.

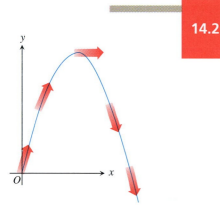

14.8 The velocity vectors $\mathbf{v}(t)$ of a projectile's motion make a vector field along the trajectory.

$f(x, y, z) = c$

14.9 The field of gradient vectors ∇f on a surface $f(x, y, z) = c$.

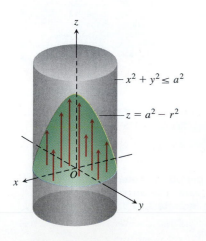

14.10 The flow of fluid in a long cylindrical pipe. The vectors $\mathbf{v} = (a^2 - r^2)\,\mathbf{k}$ inside the cylinder that have their bases in the xy-plane have their tips on the paraboloid $z = a^2 - r^2$.

14.11 Velocity vectors of a flow around an airfoil in a wind tunnel. The streamlines were made visible by kerosene smoke. (Adapted from *NCFMF Book of Film Notes,* 1974, MIT Press with Education Development Center, Inc., Newton, Massachusetts.)

14.12 Streamlines in a contracting channel. The water speeds up as the channel narrows and the velocity vectors increase in length. (Adapted from *NCFMF Book of Film Notes,* 1974, MIT Press with Education Development Center, Inc., Newton, Massachusetts.)

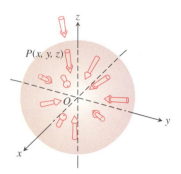

14.13 Vectors in the gravitational field
$$\mathbf{F} = -\frac{GM(x\,\mathbf{i} + y\,\mathbf{j} + z\,\mathbf{k})}{(x^2 + y^2 + z^2)^{3/2}}.$$

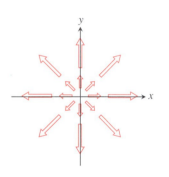

14.14 The radial field $\mathbf{F} = x\,\mathbf{i} + y\,\mathbf{j}$ of position vectors of points in the plane. Notice the convention that an arrow is drawn with its tail, not its head, at the point where $\mathbf{F}$ is evaluated.

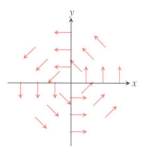

14.15 The circumferential or "spin" field of unit vectors

$$\mathbf{F} = (-y\,\mathbf{i} + x\,\mathbf{j})/(x^2 + y^2)^{1/2}$$

in the plane. The field is not defined at the origin.

WIND SPEED, M/S

0 2 4 6 8 10 12 14 16+

14.16 NASA's *Seasat* used radar during a 3-day period in September 1978 to take 350,000 wind measurements over the world's oceans. The arrows show wind direction; their length and the color contouring indicate speed. Notice the heavy storm south of Greenland.

Gradient Fields

Definition

The **gradient field** of a differentiable function $f(x, y, z)$ is the field of gradient vectors

$$\nabla f = \frac{\partial f}{\partial x}\mathbf{i} + \frac{\partial f}{\partial y}\mathbf{j} + \frac{\partial f}{\partial z}\mathbf{k}.$$

EXAMPLE 1   Find the gradient field of $f(x, y, z) = xyz$.

Solution   The gradient field of f is the field $\mathbf{F} = \nabla f = yz\,\mathbf{i} + xz\,\mathbf{j} + xy\,\mathbf{k}$.

As we will see in Section 14.3, gradient fields are of special importance in engineering, mathematics, and physics.

The Work Done by a Force over a Curve in Space

Suppose that the vector field $\mathbf{F} = M(x, y, z)\mathbf{i} + N(x, y, z)\mathbf{j} + P(x, y, z)\mathbf{k}$ represents a force throughout a region in space (it might be the force of gravity or an electromagnetic force of some kind) and that

$$\mathbf{r}(t) = g(t)\mathbf{i} + h(t)\mathbf{j} + k(t)\mathbf{k}, \quad a \leq t \leq b,$$

is a smooth curve in the region. Then the integral of $\mathbf{F} \cdot \mathbf{T}$, the scalar component of $\mathbf{F}$ in the direction of the curve's unit tangent vector, over the curve is called the work done by $\mathbf{F}$ over the curve from a to b (Fig. 14.17).

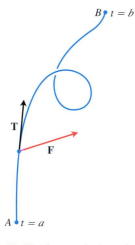

14.17 The work done by a continuous field $\mathbf{F}$ over a smooth path $\mathbf{r} = g(t)\,\mathbf{i} + h(t)\,\mathbf{j} + k(t)\,\mathbf{k}$ from A to B is the integral of $\mathbf{F} \cdot \mathbf{T}$ over the path from $t = a$ to $t = b$.

Definition

The **work** done by a force $\mathbf{F} = M(x, y, z)\mathbf{i} + N(x, y, z)\mathbf{j} + P(x, y, z)\mathbf{k}$ over a smooth curve $\mathbf{r}(t) = g(t)\mathbf{i} + h(t)\mathbf{j} + k(t)\mathbf{k}$ from $t = a$ to $t = b$ is

$$W = \int_{t=a}^{t=b} \mathbf{F} \cdot \mathbf{T}\,ds. \tag{1}$$

We motivate Eq. (1) with the same kind of reasoning we used in Section 5.8 to derive the formula $W = \int_a^b F(x)\,dx$ for the work done by a continuous force of magnitude $F(x)$ directed along an interval of the x-axis. We divide the curve into short segments, apply the constant-force-times-distance formula for work to approximate the work over each curved segment, add the results to approximate the work over the entire curve, and calculate the work as the limit of the approximating sums as the segments become shorter and more numerous. To find exactly what the limiting integral should be, we partition the parameter interval $I = [a, b]$ in the usual way and choose a point c_k in each subinterval $[t_k, t_{k+1}]$. The partition of I determines ("induces," we say) a partition of the curve, with the point P_k being the tip of the position vector $\mathbf{r}$ at $t = t_k$ and Δs_k being the length of the curve segment $P_k P_{k+1}$ (Fig. 14.18). If $\mathbf{F}_k$ denotes the value of $\mathbf{F}$ at the point on the curve corresponding to $t = c_k$, and $\mathbf{T}_k$ denotes the curve's tangent vector at this point, then $\mathbf{F}_k \cdot \mathbf{T}_k$ is the scalar component of $\mathbf{F}$ in the direction of $\mathbf{T}$ at $t = c_k$

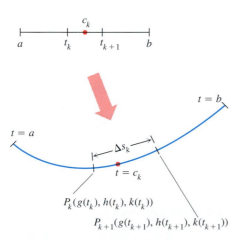

14.18 Each partition of a parameter interval $a \leq t \leq b$ induces a partition of the curve $\mathbf{r} = g(t)\,\mathbf{i} + h(t)\,\mathbf{j} + k(t)\,\mathbf{k}$.

14.19 An enlarged view of the curve segment $P_k P_{k+1}$ in Fig. 14.18, showing the force vector and unit tangent vector at the point on the curve where $t = c_k$.

(Fig. 14.19). The work done by **F** along the curve segment $P_k P_{k+1}$ will be approximately

$$\left(\begin{array}{c} \text{force component in} \\ \text{direction of motion} \end{array} \right) \times \left(\begin{array}{c} \text{distance} \\ \text{applied} \end{array} \right) = \mathbf{F}_k \cdot \mathbf{T}_k \Delta s_k.$$

The work done by **F** along the curve from $t = a$ to $t = b$ will be approximately

$$\sum_{k=1}^{n} \mathbf{F}_k \cdot \mathbf{T}_k \Delta s_k.$$

As the norm of the partition of $[a, b]$ approaches zero, the norm of the induced partition of the curve approaches zero and these sums approach the line integral

$$\int_{t=a}^{t=b} \mathbf{F} \cdot \mathbf{T}\, ds.$$

The sign of the number we calculate with this integral depends on the direction in which the curve is traversed as t increases. If we reverse the direction of motion, we reverse the direction of **T** and change the sign of $\mathbf{F} \cdot \mathbf{T}$ and its integral.

Notation and Evaluation

Table 14.2 shows six ways to write the work integral in Eq. (1).

Table 14.2 Different ways to write the work integral

$\mathbf{W} = \displaystyle\int_{t=a}^{t=b} \mathbf{F} \cdot \mathbf{T}\, ds$	The definition
$= \displaystyle\int_{t=a}^{t=b} \mathbf{F} \cdot d\mathbf{r}$	Compact differential form
$= \displaystyle\int_{a}^{b} \mathbf{F} \cdot \dfrac{d\mathbf{r}}{dt}\, dt$	Expanded to include dt; emphasizes the parameter t and velocity vector $d\mathbf{r}/dt$
$= \displaystyle\int_{a}^{b} \left(M\dfrac{dg}{dt} + N\dfrac{dh}{dt} + P\dfrac{dk}{dt} \right) dt$	Emphasizes the component functions
$= \displaystyle\int_{a}^{b} \left(M\dfrac{dx}{dt} + N\dfrac{dy}{dt} + P\dfrac{dz}{dt} \right) dt$	Abbreviates the components of **r**
$= \displaystyle\int_{a}^{b} M\,dx + N\,dy + P\,dz$	dt's canceled; the most common form

Despite their variety, the formulas in Table 14.2 are all evaluated the same way.

How to Evaluate a Work Integral

To evaluate the work integral, take these steps:

1. Evaluate **F** on the curve as a function of the parameter t.
2. Find $d\mathbf{r}/dt$.
3. Dot **F** with $d\mathbf{r}/dt$.
4. Integrate from $t = a$ to $t = b$.

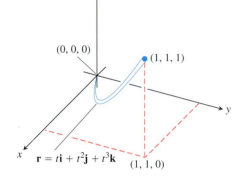

14.20 The curve in Example 2.

EXAMPLE 2 Find the work done by $\mathbf{F} = (y - x^2)\mathbf{i} + (z - y^2)\mathbf{j} + (x - z^2)\mathbf{k}$ over the curve $\mathbf{r}(t) = t\mathbf{i} + t^2\mathbf{j} + t^3\mathbf{k}$, $0 \le t \le 1$, from $(0, 0, 0)$ to $(1, 1, 1)$ (Fig. 14.20).

Solution

Step 1: *Evaluate* $\mathbf{F}$ *on the curve.*

$$\mathbf{F} = (y - x^2)\mathbf{i} + (z - y^2)\mathbf{j} + (x - z^2)\mathbf{k}$$
$$= \underbrace{(t^2 - t^2)}_{0}\mathbf{i} + (t^3 - t^4)\mathbf{j} + (t - t^6)\mathbf{k}$$

Step 2: *Find* $d\mathbf{r}/dt$.

$$\frac{d\mathbf{r}}{dt} = \frac{d}{dt}(t\mathbf{i} + t^2\mathbf{j} + t^3\mathbf{k}) = \mathbf{i} + 2t\mathbf{j} + 3t^2\mathbf{k}$$

Step 3: *Dot* $\mathbf{F}$ *with* $d\mathbf{r}/dt$.

$$\mathbf{F} \cdot \frac{d\mathbf{r}}{dt} = [(t^3 - t^4)\mathbf{j} + (t - t^6)\mathbf{k}] \cdot (\mathbf{i} + 2t\mathbf{j} + 3t^2\mathbf{k})$$
$$= (t^3 - t^4)(2t) + (t - t^6)(3t^2) = 2t^4 - 2t^5 + 3t^3 - 3t^8$$

Step 4: *Integrate from* $t = 0$ *to* $t = 1$.

$$\text{Work} = \int_0^1 (2t^4 - 2t^5 + 3t^3 - 3t^8)\,dt$$
$$= \left[\frac{2}{5}t^5 - \frac{2}{6}t^6 + \frac{3}{4}t^4 - \frac{3}{9}t^9\right]_0^1 = \frac{29}{60} \qquad \square$$

Flow Integrals and Circulation

Instead of being a force field, suppose that $\mathbf{F} = M\mathbf{i} + N\mathbf{j} + P\mathbf{k}$ represents the velocity field of a fluid flowing through a region in space (a tidal basin or the turbine chamber of a hydroelectric generator, for example). Under these circumstances, the integral of $\mathbf{F} \cdot \mathbf{T}$ along a curve in the region gives the fluid's flow along the curve.

Definitions

If $\mathbf{r}(t) = g(t)\mathbf{i} + h(t)\mathbf{j} + k(t)\mathbf{k}$, $a \le t \le b$, is a smooth curve in the domain of a continuous velocity field $\mathbf{F} = M(x, y, z)\mathbf{i} + N(x, y, z)\mathbf{j} + P(x, y, z)\mathbf{k}$, the **flow** along the curve from $t = a$ to $t = b$ is the integral of $\mathbf{F} \cdot \mathbf{T}$ over the curve from a to b:

$$\text{Flow} = \int_a^b \mathbf{F} \cdot \mathbf{T}\,ds. \qquad (2)$$

The integral in this case is called a **flow integral**. If the curve is a closed loop, the flow is called the **circulation** around the curve.

We evaluate flow integrals the same way we evaluate work integrals.

EXAMPLE 3 A fluid's velocity field is $\mathbf{F} = x\mathbf{i} + z\mathbf{j} + y\mathbf{k}$. Find the flow along the helix $\mathbf{r}(t) = (\cos t)\mathbf{i} + (\sin t)\mathbf{j} + t\mathbf{k}$, $0 \le t \le \pi/2$.

Solution

Step 1: *Evaluate* **F** *on the curve.*

$$\mathbf{F} = x\,\mathbf{i} + z\,\mathbf{j} + y\,\mathbf{k} = (\cos t)\,\mathbf{i} + t\,\mathbf{j} + (\sin t)\,\mathbf{k}$$

Step 2: *Find* $d\mathbf{r}/dt.$
$$\frac{d\mathbf{r}}{dt} = (-\sin t)\,\mathbf{i} + (\cos t)\,\mathbf{j} + \mathbf{k}$$

Step 3: *Find* $\mathbf{F} \cdot (d\mathbf{r}/dt).$
$$\mathbf{F} \cdot \frac{d\mathbf{r}}{dt} = (\cos t)(-\sin t) + (t)(\cos t) + (\sin t)(1)$$
$$= -\sin t \cos t + t \cos t + \sin t$$

Step 4: *Integrate from* $t = a$ *to* $t = b.$
$$\text{Flow} = \int_{t=a}^{t=b} \mathbf{F} \cdot \frac{d\mathbf{r}}{dt}\,dt = \int_0^{\pi/2} (-\sin t \cos t + t \cos t + \sin t)\,dt$$
$$= \left[\frac{\cos^2 t}{2} + t \sin t\right]_0^{\pi/2} = \left(0 + \frac{\pi}{2}\right) - \left(\frac{1}{2} + 0\right) = \frac{\pi}{2} - \frac{1}{2} \qquad \square$$

EXAMPLE 4   Find the circulation of the field $\mathbf{F} = (x - y)\,\mathbf{i} + x\,\mathbf{j}$ around the circle $\mathbf{r}(t) = (\cos t)\,\mathbf{i} + (\sin t)\,\mathbf{j}, 0 \le t \le 2\pi.$

Solution

1. On the circle, $\mathbf{F} = (x - y)\,\mathbf{i} + x\,\mathbf{j} = (\cos t - \sin t)\,\mathbf{i} + (\cos t)\,\mathbf{j}.$

2. $\dfrac{d\mathbf{r}}{dt} = (-\sin t)\,\mathbf{i} + (\cos t)\,\mathbf{j}$

3. $\mathbf{F} \cdot \dfrac{d\mathbf{r}}{dt} = -\sin t \cos t + \underbrace{\sin^2 t + \cos^2 t}_{1}$

4. $\text{Circulation} = \displaystyle\int_0^{2\pi} \mathbf{F} \cdot \frac{d\mathbf{r}}{dt}\,dt = \int_0^{2\pi} (1 - \sin t \cos t)\,dt$
$$= \left[t - \frac{\sin^2 t}{2}\right]_0^{2\pi} = 2\pi. \qquad \square$$

Flux Across a Plane Curve

To find the rate at which a fluid is entering or leaving a region enclosed by a smooth curve C in the xy-plane, we calculate the line integral over C of $\mathbf{F} \cdot \mathbf{n},$ the scalar component of the fluid's velocity field in the direction of the curve's outward-pointing normal vector. The value of this integral is called the flux of $\mathbf{F}$ across $C.$ *Flux* is Latin for *flow*, but many flux calculations involve no motion at all. If $\mathbf{F}$ were an electric field or a magnetic field, for instance, the integral of $\mathbf{F} \cdot \mathbf{n}$ would still be called the flux of the field across $C.$

Definition

If C is a smooth closed curve in the domain of a continuous vector field $\mathbf{F} = M(x, y)\,\mathbf{i} + N(x, y)\,\mathbf{j}$ in the plane, and if $\mathbf{n}$ is the outward-pointing unit

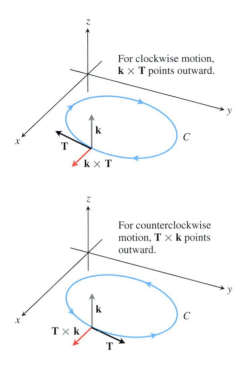

14.21 To find an outward unit normal vector for a smooth curve C in the xy-plane that is traversed counter-clockwise as t increases, we take $\mathbf{n} = \mathbf{T} \times \mathbf{k}$.

normal vector on C, the **flux** of $\mathbf{F}$ across C is given by the following line integral:

$$\text{Flux of } \mathbf{F} \text{ across } C = \int_C \mathbf{F} \cdot \mathbf{n} \, ds. \qquad (3)$$

Notice the difference between flux and circulation. The flux of $\mathbf{F}$ across C is the line integral with respect to arc length of $\mathbf{F} \cdot \mathbf{n}$, the scalar component of $\mathbf{F}$ in the direction of the outward normal. The circulation of $\mathbf{F}$ around C is the line integral with respect to arc length of $\mathbf{F} \cdot \mathbf{T}$, the scalar component of $\mathbf{F}$ in the direction of the unit tangent vector. Flux is the integral of the normal component of $\mathbf{F}$; circulation is the integral of the tangential component of $\mathbf{F}$.

To evaluate the integral in (3), we begin with a parametrization

$$x = g(t), \qquad y = h(t), \qquad a \le t \le b,$$

that traces the curve C exactly once as t increases from a to b. We can find the outward unit normal vector $\mathbf{n}$ by crossing the curve's unit tangent vector $\mathbf{T}$ with the vector $\mathbf{k}$. But which order do we choose, $\mathbf{T} \times \mathbf{k}$ or $\mathbf{k} \times \mathbf{T}$? Which one points outward? It depends on which way C is traversed as the parameter t increases. If the motion is clockwise, then $\mathbf{k} \times \mathbf{T}$ points outward; if the motion is counterclockwise, then $\mathbf{T} \times \mathbf{k}$ points outward (Fig. 14.21). The usual choice is $\mathbf{n} = \mathbf{T} \times \mathbf{k}$, the choice that assumes counterclockwise motion. Thus, while the value of the arc length integral in the definition of flux in Eq. (3) does not depend on which way C is traversed, the formulas we are about to derive for evaluating the integral in Eq. (3) will assume counterclockwise motion.

In terms of components,

$$\mathbf{n} = \mathbf{T} \times \mathbf{k} = \left(\frac{dx}{ds} \mathbf{i} + \frac{dy}{ds} \mathbf{j} \right) \times \mathbf{k} = \frac{dy}{ds} \mathbf{i} - \frac{dx}{ds} \mathbf{j}.$$

If $\mathbf{F} = M(x, y) \mathbf{i} + N(x, y) \mathbf{j}$, then

$$\mathbf{F} \cdot \mathbf{n} = M(x, y) \frac{dy}{ds} - N(x, y) \frac{dx}{ds}.$$

Hence,

$$\int_C \mathbf{F} \cdot \mathbf{n} \, ds = \int_C \left(M \frac{dy}{ds} - N \frac{dx}{ds} \right) ds = \oint_C M \, dy - N \, dx.$$

We put a directed circle $\circlearrowleft$ on the last integral as a reminder that the integration around the closed curve C is to be in the counterclockwise direction. To evaluate this integral, we express M, dy, N, and dx in terms of t and integrate from $t = a$ to $t = b$. We do not need to know either $\mathbf{n}$ or ds to find the flux.

The Formula for Calculating Flux Across a Smooth Closed Plane Curve

$$(\text{Flux of } \mathbf{F} = M \mathbf{i} + N \mathbf{j} \text{ across } C) = \oint_C M \, dy - N \, dx \qquad (4)$$

The integral can be evaluated from any smooth parametrization $x = g(t)$, $y = h(t), a \le t \le b$, that traces C counterclockwise exactly once.

EXAMPLE 5   Find the flux of $\mathbf{F} = (x - y)\mathbf{i} + x\mathbf{j}$ across the circle $x^2 + y^2 = 1$ in the xy-plane.

Solution   The parametrization $\mathbf{r}(t) = (\cos t)\mathbf{i} + (\sin t)\mathbf{j}$, $0 \le t \le 2\pi$, traces the circle counterclockwise exactly once. We can therefore use this parametrization in Eq. (4). With

$$M = x - y = \cos t - \sin t, \qquad dy = d(\sin t) = \cos t \, dt$$

$$N = x = \cos t, \qquad dx = d(\cos t) = -\sin t \, dt,$$

we find

$$\text{Flux} = \int_C M \, dy - N \, dx = \int_0^{2\pi} (\cos^2 t - \sin t \cos t + \cos t \sin t) \, dt \qquad \text{Eq. (4)}$$

$$= \int_0^{2\pi} \cos^2 t \, dt = \int_0^{2\pi} \frac{1 + \cos 2t}{2} \, dt = \left[\frac{t}{2} + \frac{\sin 2t}{4} \right]_0^{2\pi} = \pi.$$

The flux of $\mathbf{F}$ across the circle is π. Since the answer is positive, the net flow across the curve is outward. A net inward flow would have given a negative flux.   ❑

Exercises 14.2

Vector and Gradient Fields

Find the gradient fields of the functions in Exercises 1–4.

1. $f(x, y, z) = (x^2 + y^2 + z^2)^{-1/2}$

2. $f(x, y, z) = \ln \sqrt{x^2 + y^2 + z^2}$

3. $g(x, y, z) = e^z - \ln(x^2 + y^2)$

4. $g(x, y, z) = xy + yz + xz$

5. Give a formula $\mathbf{F} = M(x, y)\mathbf{i} + N(x, y)\mathbf{j}$ for the vector field in the plane that has the property that $\mathbf{F}$ points toward the origin with magnitude inversely proportional to the square of the distance from (x, y) to the origin. (The field is not defined at $(0, 0)$.)

6. Give a formula $\mathbf{F} = M(x, y)\mathbf{i} + N(x, y)\mathbf{j}$ for the vector field in the plane that has the properties that $\mathbf{F} = \mathbf{0}$ at $(0, 0)$ and that at any other point (a, b), $\mathbf{F}$ is tangent to the circle $x^2 + y^2 = a^2 + b^2$ and points in the clockwise direction with magnitude $|\mathbf{F}| = \sqrt{a^2 + b^2}$.

Work

In Exercises 7–12, find the work done by force $\mathbf{F}$ from $(0, 0, 0)$ to $(1, 1, 1)$ over each of the following paths (Fig. 14.22):

a) The straight-line path C_1: $\mathbf{r}(t) = t\mathbf{i} + t\mathbf{j} + t\mathbf{k}$, $0 \le t \le 1$

b) The curved path C_2: $\mathbf{r}(t) = t\mathbf{i} + t^2\mathbf{j} + t^4\mathbf{k}$, $0 \le t \le 1$

c) The path $C_3 \cup C_4$ consisting of the line segment from $(0, 0, 0)$ to $(1, 1, 0)$ followed by the segment from $(1, 1, 0)$ to $(1, 1, 1)$

7. $\mathbf{F} = 3y\mathbf{i} + 2x\mathbf{j} + 4z\mathbf{k}$ **8.** $\mathbf{F} = [1/(x^2 + 1)]\mathbf{j}$

9. $\mathbf{F} = \sqrt{z}\,\mathbf{i} - 2x\mathbf{j} + \sqrt{y}\,\mathbf{k}$ **10.** $\mathbf{F} = xy\mathbf{i} + yz\mathbf{j} + xz\mathbf{k}$

11. $\mathbf{F} = (3x^2 - 3x)\mathbf{i} + 3z\mathbf{j} + \mathbf{k}$

12. $\mathbf{F} = (y + z)\mathbf{i} + (z + x)\mathbf{j} + (x + y)\mathbf{k}$

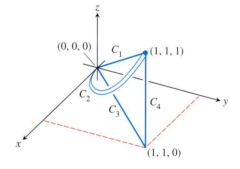

14.22 The paths from $(0, 0, 0)$ to $(1, 1, 1)$.

In Exercises 13–16, find the work done by $\mathbf{F}$ over the curve in the direction of increasing t.

13. $\mathbf{F} = xy\mathbf{i} + y\mathbf{j} - yz\mathbf{k}$
$\mathbf{r}(t) = t\mathbf{i} + t^2\mathbf{j} + t\mathbf{k}$, $0 \le t \le 1$

14. $\mathbf{F} = 2y\mathbf{i} + 3x\mathbf{j} + (x + y)\mathbf{k}$
$\mathbf{r}(t) = (\cos t)\mathbf{i} + (\sin t)\mathbf{j} + (t/6)\mathbf{k}$, $0 \le t \le 2\pi$

15. $\mathbf{F} = z\mathbf{i} + x\mathbf{j} + y\mathbf{k}$
$\mathbf{r}(t) = (\sin t)\mathbf{i} + (\cos t)\mathbf{j} + t\mathbf{k}$, $0 \le t \le 2\pi$

16. $\mathbf{F} = 6z\mathbf{i} + y^2\mathbf{j} + 12x\mathbf{k}$
$\mathbf{r}(t) = (\sin t)\mathbf{i} + (\cos t)\mathbf{j} + (t/6)\mathbf{k}$, $0 \le t \le 2\pi$

Line Integrals and Vector Fields in the Plane

17. Evaluate $\int_C xy \, dx + (x + y) \, dy$ along the curve $y = x^2$ from $(-1, 1)$ to $(2, 4)$.

18. Evaluate $\int_C (x - y)\,dx + (x + y)\,dy$ counterclockwise around the triangle with vertices $(0, 0)$, $(1, 0)$, and $(0, 1)$.

19. Evaluate $\int_C \mathbf{F} \cdot \mathbf{T}\,ds$ for the vector field $\mathbf{F} = x^2\mathbf{i} - y\mathbf{j}$ along the curve $x = y^2$ from $(4, 2)$ to $(1, -1)$.

20. Evaluate $\int_C \mathbf{F} \cdot d\mathbf{r}$ for the vector field $\mathbf{F} = y\mathbf{i} - x\mathbf{j}$ counterclockwise along the unit circle $x^2 + y^2 = 1$ from $(1, 0)$ to $(0, 1)$.

21. Find the work done by the force $\mathbf{F} = xy\mathbf{i} + (y - x)\mathbf{j}$ over the straight line from $(1, 1)$ to $(2, 3)$.

22. Find the work done by the gradient of $f(x, y) = (x + y)^2$ counterclockwise around the circle $x^2 + y^2 = 4$ from $(2, 0)$ to itself.

23. Find the circulation and flux of the fields

$$\mathbf{F}_1 = x\mathbf{i} + y\mathbf{j} \quad \text{and} \quad \mathbf{F}_2 = -y\mathbf{i} + x\mathbf{j}$$

around and across each of the following curves.

a) The circle $\mathbf{r}(t) = (\cos t)\mathbf{i} + (\sin t)\mathbf{j}, \quad 0 \le t \le 2\pi$

b) The ellipse $\mathbf{r}(t) = (\cos t)\mathbf{i} + (4 \sin t)\mathbf{j}, \quad 0 \le t \le 2\pi$

24. Find the flux of the fields

$$\mathbf{F}_1 = 2x\mathbf{i} - 3y\mathbf{j} \quad \text{and} \quad \mathbf{F}_2 = 2x\mathbf{i} + (x - y)\mathbf{j}$$

across the circle

$$\mathbf{r}(t) = (a \cos t)\mathbf{i} + (a \sin t)\mathbf{j}, \quad 0 \le t \le 2\pi.$$

In Exercises 25–28, find the circulation and flux of the field $\mathbf{F}$ around and across the closed semicircular path that consists of the semicircular arch $\mathbf{r}_1(t) = (a \cos t)\mathbf{i} + (a \sin t)\mathbf{j}, 0 \le t \le \pi$, followed by the line segment $\mathbf{r}_2(t) = t\mathbf{i}, -a \le t \le a$.

25. $\mathbf{F} = x\mathbf{i} + y\mathbf{j}$

26. $\mathbf{F} = x^2\mathbf{i} + y^2\mathbf{j}$

27. $\mathbf{F} = -y\mathbf{i} + x\mathbf{j}$

28. $\mathbf{F} = -y^2\mathbf{i} + x^2\mathbf{j}$

29. Evaluate the flow integral of the velocity field $\mathbf{F} = (x + y)\mathbf{i} - (x^2 + y^2)\mathbf{j}$ along each of the following paths from $(1, 0)$ to $(-1, 0)$ in the xy-plane.

a) The upper half of the circle $x^2 + y^2 = 1$

b) The line segment from $(1, 0)$ to $(-1, 0)$

c) The line segment from $(1, 0)$ to $(0, -1)$ followed by the line segment from $(0, -1)$ to $(-1, 0)$

30. Find the flux of the field $\mathbf{F}$ in Exercise 29 outward across the triangle with vertices $(1, 0)$, $(0, 1)$, $(-1, 0)$.

Sketching and Finding Fields in the Plane

31. Draw the spin field

$$\mathbf{F} = -\frac{y}{\sqrt{x^2 + y^2}}\mathbf{i} + \frac{x}{\sqrt{x^2 + y^2}}\mathbf{j}$$

(see Fig. 14.15) along with its horizontal and vertical components at a representative assortment of points on the circle $x^2 + y^2 = 4$.

32. Draw the radial field

$$\mathbf{F} = x\mathbf{i} + y\mathbf{j}$$

(see Fig. 14.14) along with its horizontal and vertical components at a representative assortment of points on the circle $x^2 + y^2 = 1$.

33. a) Find a field $\mathbf{G} = P(x, y)\mathbf{i} + Q(x, y)\mathbf{j}$ in the xy-plane with the property that at any point $(a, b) \ne (0, 0)$, $\mathbf{G}$ is a vector of magnitude $\sqrt{a^2 + b^2}$ tangent to the circle $x^2 + y^2 = a^2 + b^2$ and pointing in the counterclockwise direction. (The field is undefined at $(0, 0)$.)

b) How is $\mathbf{G}$ related to the spin field $\mathbf{F}$ in Fig. 14.15?

34. a) Find a field $\mathbf{G} = P(x, y)\mathbf{i} + Q(x, y)\mathbf{j}$ in the xy-plane with the property that at any point $(a, b) \ne (0, 0)$, $\mathbf{G}$ is a unit vector tangent to the circle $x^2 + y^2 = a^2 + b^2$ and pointing in the clockwise direction.

b) How is $\mathbf{G}$ related to the spin field $\mathbf{F}$ in Fig. 14.15?

35. Find a field $\mathbf{F} = M(x, y)\mathbf{i} + N(x, y)\mathbf{j}$ in the xy-plane with the property that at each point $(x, y) \ne (0, 0)$, $\mathbf{F}$ is a unit vector pointing toward the origin. (The field is undefined at $(0, 0)$.)

36. Find a field $\mathbf{F} = M(x, y)\mathbf{i} + N(x, y)\mathbf{j}$ in the xy-plane with the property that at each point $(x, y) \ne (0, 0)$, $\mathbf{F}$ points toward the origin and $|\mathbf{F}|$ is (a) the distance from (x, y) to the origin, (b) inversely proportional to the distance from (x, y) to the origin. (The field is undefined at $(0, 0)$.)

Flow Integrals in Space

In Exercises 37–40, $\mathbf{F}$ is the velocity field of a fluid flowing through a region in space. Find the flow along the given curve in the direction of increasing t.

37. $\mathbf{F} = -4xy\mathbf{i} + 8y\mathbf{j} + 2\mathbf{k}$
$\mathbf{r}(t) = t\mathbf{i} + t^2\mathbf{j} + \mathbf{k}, \quad 0 \le t \le 2$

38. $\mathbf{F} = x^2\mathbf{i} + yz\mathbf{j} + y^2\mathbf{k}$
$\mathbf{r}(t) = 3t\mathbf{j} + 4t\mathbf{k}, \quad 0 \le t \le 1$

39. $\mathbf{F} = (x - z)\mathbf{i} + x\mathbf{k}$
$\mathbf{r}(t) = (\cos t)\mathbf{i} + (\sin t)\mathbf{k}, \quad 0 \le t \le \pi$

40. $\mathbf{F} = -y\mathbf{i} + x\mathbf{j} + 2\mathbf{k}$
$\mathbf{r}(t) = (-2 \cos t)\mathbf{i} + (2 \sin t)\mathbf{j} + 2t\mathbf{k}, \quad 0 \le t \le 2\pi$

41. Find the circulation of $\mathbf{F} = 2x\mathbf{i} + 2z\mathbf{j} + 2y\mathbf{k}$ around the closed path consisting of the following three curves traversed in the direction of increasing t:

$C_1: \quad \mathbf{r}(t) = (\cos t)\mathbf{i} + (\sin t)\mathbf{j} + t\mathbf{k}, \quad 0 \le t \le \pi/2$

$C_2: \quad \mathbf{r}(t) = \mathbf{j} + (\pi/2)(1 - t)\mathbf{k}, \quad 0 \le t \le 1$

$C_3: \quad \mathbf{r}(t) = t\mathbf{i} + (1 - t)\mathbf{j}, \quad 0 \le t \le 1$

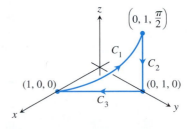

42. Let C be the ellipse in which the plane $2x + 3y - z = 0$ meets the cylinder $x^2 + y^2 = 12$. Show, without evaluating either line integral directly, that the circulation of the field $\mathbf{F} = x\mathbf{i} + y\mathbf{j} + z\mathbf{k}$ around C in either direction is zero.

43. The field $\mathbf{F} = xy\mathbf{i} + y\mathbf{j} - yz\mathbf{k}$ is the velocity field of a flow in space. Find the flow from $(0, 0, 0)$ to $(1, 1, 1)$ along the curve of intersection of the cylinder $y = x^2$ and the plane $z = x$. (*Hint:* Use $t = x$ as the parameter.)

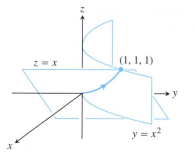

44. Find the flow of the field $\mathbf{F} = \nabla(xy^2z^3)$

a) once around the curve C in Exercise 42, clockwise as viewed from above.

b) along the line segment from $(1, 1, 1)$ to $(2, 1, -1)$.

Theory and Examples

45. Suppose $f(t)$ is differentiable and positive for $a \leq t \leq b$. Let C be the path $\mathbf{r}(t) = t\mathbf{i} + f(t)\mathbf{j}$, $a \leq t \leq b$, and $\mathbf{F} = y\mathbf{i}$. Is there any relation between the value of the work integral

$$\int_C \mathbf{F} \cdot d\mathbf{r}$$

and the area of the region bounded by the t-axis, the graph of f, and the lines $t = a$ and $t = b$? Give reasons for your answer.

46. A particle moves along the smooth curve $y = f(x)$ from $(a, f(a))$ to $(b, f(b))$. The force moving the particle has constant magnitude k and always points away from the origin. Show that the work done by the force is

$$\int_C \mathbf{F} \cdot \mathbf{T}\, ds = k\left[(b^2 + (f(b))^2)^{1/2} - (a^2 + (f(a))^2)^{1/2} \right].$$

✿ CAS Explorations and Projects

In Exercises 47–52, use a CAS to perform the following steps for finding the work done by force $\mathbf{F}$ over the given path:

a) Find $d\mathbf{r}$ for the path $\mathbf{r}(t) = g(t)\mathbf{i} + h(t)\mathbf{j} + k(t)\mathbf{k}$.

b) Evaluate the force $\mathbf{F}$ along the path.

c) Evaluate $\displaystyle\int_C \mathbf{F} \cdot d\mathbf{r}$.

47. $\mathbf{F} = xy^6\mathbf{i} + 3x(xy^5 + 2)\mathbf{j}$; $\mathbf{r}(t) = 2\cos t\,\mathbf{i} + \sin t\,\mathbf{j},\, 0 \leq t \leq 2\pi$

48. $\mathbf{F} = \dfrac{3}{1+x^2}\mathbf{i} + \dfrac{2}{1+y^2}\mathbf{j}$; $\mathbf{r}(t) = \cos t\,\mathbf{i} + \sin t\,\mathbf{j}$, $0 \leq t \leq \pi$

49. $\mathbf{F} = (y + yz\cos xyz)\mathbf{i} + (x^2 + xz\cos xyz)\mathbf{j} + (z + xy\cos xyz)\mathbf{k}$; $\mathbf{r}(t) = 2\cos t\,\mathbf{i} + 3\sin t\,\mathbf{j} + \mathbf{k}$, $0 \leq t \leq 2\pi$

50. $\mathbf{F} = 2xy\mathbf{i} - y^2\mathbf{j} + ze^x\mathbf{k}$; $\mathbf{r}(t) = -t\mathbf{i} + \sqrt{t}\,\mathbf{j} + 3t\,\mathbf{k}$, $1 \leq t \leq 4$

51. $\mathbf{F} = (2y + \sin x)\mathbf{i} + (z^2 + (1/3)\cos y)\mathbf{j} + x^4\mathbf{k}$; $\mathbf{r}(t) = \sin t\,\mathbf{i} + \cos t\,\mathbf{j} + \sin 2t\,\mathbf{k}$, $-\pi/2 \leq t \leq \pi/2$

52. $\mathbf{F} = (x^2y)\mathbf{i} + \dfrac{1}{3}x^3\mathbf{j} + xy\mathbf{k}$; $\mathbf{r}(t) = \cos t\,\mathbf{i} + \sin t\,\mathbf{j} + (2\sin^2(t) - 1)\mathbf{k}$, $0 \leq t \leq 2\pi$

14.3

Path Independence, Potential Functions, and Conservative Fields

In gravitational and electric fields, the amount of work it takes to move a mass or a charge from one point to another depends only on the object's initial and final positions and not on the path taken in between. This section discusses the notion of path independence of work integrals and describes the remarkable properties of fields in which work integrals are path independent.

Path Independence

If A and B are two points in an open region D in space, the work $\int \mathbf{F} \cdot d\mathbf{r}$ done in moving a particle from A to B by a field $\mathbf{F}$ defined on D usually depends on the path taken. For some special fields, however, the integral's value is the same for all paths from A to B. If this is true for all points A and B in D, we say that the integral $\int \mathbf{F} \cdot d\mathbf{r}$ is path independent in D and that $\mathbf{F}$ is conservative on D.

Definitions

Let **F** be a field defined on an open region D in space, and suppose that for any two points A and B in D the work $\int_A^B \mathbf{F} \cdot d\mathbf{r}$ done in moving from A to B is the same over all paths from A to B. Then the integral $\int \mathbf{F} \cdot d\mathbf{r}$ is **path independent in D** and the field **F** is **conservative on D.**

The word *conservative* comes from physics, where it refers to fields in which the principle of conservation of energy holds (it does, in conservative fields).

Under conditions normally met in practice, a field **F** is conservative if and only if it is the gradient field of a scalar function f; that is, if and only if $\mathbf{F} = \nabla f$ for some f. The function f is then called a potential function for **F**.

Definition

If **F** is a field defined on D and $\mathbf{F} = \nabla f$ for some scalar function f on D, then f is called a **potential function** for **F**.

An electric potential is a scalar function whose gradient field is an electric field. A gravitational potential is a scalar function whose gradient field is a gravitational field, and so on. As we will see, once we have found a potential function f for a field **F,** we can evaluate all the work integrals in the domain of **F** by

$$\int_A^B \mathbf{F} \cdot d\mathbf{r} = \int_A^B \nabla f \cdot d\mathbf{r} = f(B) - f(A). \tag{1}$$

If you think of ∇f for functions of several variables as being something like the derivative f' for functions of a single variable, then you see that Eq. (1) is the vector calculus analogue of the Fundamental Theorem of Calculus formula

$$\int_a^b f'(x)\,dx = f(b) - f(a).$$

Conservative fields have other remarkable properties we will study as we go along. For example, saying that **F** is conservative on D is equivalent to saying that the integral of **F** around every closed path in D is zero. Naturally, we need to impose conditions on the curves, fields, and domains to make Eq. (1) and its implications hold.

We assume that all curves are **piecewise smooth,** i.e., made up of finitely many smooth pieces connected end to end, as discussed in Section 11.1. We also assume that the components of **F** have continuous first partial derivatives. When $\mathbf{F} = \nabla f$, this continuity requirement guarantees that the mixed second derivatives of the potential function f are equal, a result we will find revealing in studying conservative fields **F.**

We assume D to be an *open* region in space. This means that every point in D is the center of a ball that lies entirely in D. We also assume D to be **connected,** which in an open region means that every point can be connected to every other point by a smooth curve that lies in the region.

Line Integrals in Conservative Fields

The following result provides a convenient way to evaluate a line integral in a conservative field. The result establishes that the value of the integral depends only on the endpoints and not on the specific path joining them.

> ### Theorem 1
> ### The Fundamental Theorem of Line Integrals
> 1. Let $\mathbf{F} = M\,\mathbf{i} + N\,\mathbf{j} + P\,\mathbf{k}$ be a vector field whose components are continuous throughout an open connected region D in space. Then there exists a differentiable function f such that
>
> $$\mathbf{F} = \nabla f = \frac{\partial f}{\partial x}\,\mathbf{i} + \frac{\partial f}{\partial y}\,\mathbf{j} + \frac{\partial f}{\partial z}\,\mathbf{k}$$
>
> if and only if for all points A and B in D the value of $\int_A^B \mathbf{F} \cdot d\mathbf{r}$ is independent of the path joining A to B in D.
> 2. If the integral is independent of the path from A to B, its value is
>
> $$\int_A^B \mathbf{F} \cdot d\mathbf{r} = f(B) - f(A).$$

Proof That $\mathbf{F} = \nabla f$ ***Implies Path Independence of the Integral*** Suppose that A and B are two points in D and that $C: \quad \mathbf{r}(t) = g(t)\,\mathbf{i} + h(t)\,\mathbf{j} + k(t)\,\mathbf{k}, \quad a \leq t \leq b,$ is a smooth curve in D joining A and B. Along the curve, f is a differentiable function of t and

$$\frac{df}{dt} = \frac{\partial f}{\partial x}\frac{dx}{dt} + \frac{\partial f}{\partial y}\frac{dy}{dt} + \frac{\partial f}{\partial z}\frac{dz}{dt} \qquad \text{Chain Rule}$$

$$= \nabla f \cdot \left(\frac{dx}{dt}\,\mathbf{i} + \frac{dy}{dt}\,\mathbf{j} + \frac{dz}{dt}\,\mathbf{k} \right) = \nabla f \cdot \frac{d\mathbf{r}}{dt} = \mathbf{F} \cdot \frac{d\mathbf{r}}{dt}. \qquad \begin{array}{l}\text{Because}\\ \mathbf{F} = \nabla f\end{array} \quad (2)$$

Therefore, $\displaystyle\int_C \mathbf{F} \cdot d\mathbf{r} = \int_{t=a}^{t=b} \mathbf{F} \cdot \frac{d\mathbf{r}}{dt}\,dt = \int_a^b \frac{df}{dt}\,dt \qquad \text{Eq. (2)}$

$$= f(g(t), h(t), k(t)) \Big]_a^b = f(B) - f(A).$$

Thus, the value of the work integral depends only on the values of f at A and B and not on the path in between. This proves Part 2 as well as the forward implication in Part 1. We omit the more technical proof of the reverse implication. ❏

EXAMPLE 1 Find the work done by the conservative field

$$\mathbf{F} = yz\,\mathbf{i} + xz\,\mathbf{j} + xy\,\mathbf{k} = \nabla(xyz)$$

along any smooth curve C joining the point $(-1, 3, 9)$ to $(1, 6, -4)$.

Solution With $f(x, y, z) = xyz$, we have

$$\int_A^B \mathbf{F} \cdot d\mathbf{r} = \int_A^B \nabla f \cdot d\mathbf{r} \qquad \mathbf{F} = \nabla f$$

$$= f(B) - f(A) \qquad \text{Fundamental Theorem, Part 2}$$

$$= xyz\big|_{(1,6,-4)} - xyz\big|_{(-1,3,9)}$$

$$= (1)(6)(-4) - (-1)(3)(9)$$

$$= -24 + 27 = 3. \qquad ❏$$

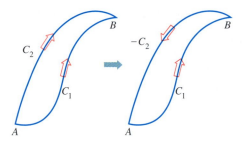

14.23 If we have two paths from A to B, one of them can be reversed to make a loop.

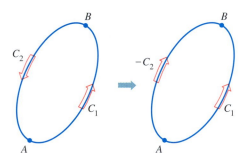

14.24 If A and B lie on a loop, we can reverse part of the loop to make two paths from A to B.

Theorem 2

The following statements are equivalent:

1. $\int \mathbf{F} \cdot d\mathbf{r} = 0$ around every closed loop in D.
2. The field $\mathbf{F}$ is conservative on D.

Proof That **(1)** $\Rightarrow$ **(2)** We want to show that for any two points A and B in D the integral of $\mathbf{F} \cdot d\mathbf{r}$ has the same value over any two paths C_1 and C_2 from A to B. We reverse the direction on C_2 to make a path $-C_2$ from B to A (Fig. 14.23). Together, C_1 and $-C_2$ make a closed loop C, and

$$\int_{C_1} \mathbf{F} \cdot d\mathbf{r} - \int_{C_2} \mathbf{F} \cdot d\mathbf{r} = \int_{C_1} \mathbf{F} \cdot d\mathbf{r} + \int_{-C_2} \mathbf{F} \cdot d\mathbf{r} = \int_{C} \mathbf{F} \cdot d\mathbf{r} = 0.$$

Thus the integrals over C_1 and C_2 give the same value.

Proof That **(2)** $\Rightarrow$ **(1)** We want to show that the integral of $\mathbf{F} \cdot d\mathbf{r}$ is zero over any closed loop C. We pick two points A and B on C and use them to break C into two pieces: C_1 from A and B followed by C_2 from B back to A (Fig. 14.24). Then

$$\oint_{C} \mathbf{F} \cdot d\mathbf{r} = \int_{C_1} \mathbf{F} \cdot d\mathbf{r} + \int_{C_2} \mathbf{F} \cdot d\mathbf{r} = \int_{A}^{B} \mathbf{F} \cdot d\mathbf{r} - \int_{A}^{B} \mathbf{F} \cdot d\mathbf{r} = 0. \qquad \square$$

The following diagram summarizes the results of Theorems 1 and 2.

Theorem 1		Theorem 2	
$\mathbf{F} = \nabla f$ on D	$\Leftrightarrow$	$\mathbf{F}$ conservative on D	$\Leftrightarrow$ $\oint_{C} \mathbf{F} \cdot d\mathbf{r} = 0$ over any closed path in D

Now that we see how convenient it is to evaluate line integrals in conservative fields, two questions remain:

1. How do we know when a given field $\mathbf{F}$ is conservative?
2. If $\mathbf{F}$ is in fact conservative, how do we find a potential function f (so that $\mathbf{F} = \nabla f$)?

Finding Potentials for Conservative Fields

The test for being conservative is this:

The Component Test for Conservative Fields

Let $\mathbf{F} = M(x, y, z)\,\mathbf{i} + N(x, y, z)\,\mathbf{j} + P(x, y, z)\,\mathbf{k}$ be a field whose component functions have continuous first partial derivatives. Then, $\mathbf{F}$ is conservative if and only if

$$\frac{\partial P}{\partial y} = \frac{\partial N}{\partial z}, \qquad \frac{\partial M}{\partial z} = \frac{\partial P}{\partial x}, \qquad \text{and} \qquad \frac{\partial N}{\partial x} = \frac{\partial M}{\partial y}. \qquad (3)$$

Proof We show that Eqs. (3) must hold if $\mathbf{F}$ is conservative. There is a potential function f such that

$$\mathbf{F} = M\,\mathbf{i} + N\,\mathbf{j} + P\,\mathbf{k} = \frac{\partial f}{\partial x}\mathbf{i} + \frac{\partial f}{\partial y}\mathbf{j} + \frac{\partial f}{\partial z}\mathbf{k}.$$

Hence

$$\frac{\partial P}{\partial y} = \frac{\partial}{\partial y}\left(\frac{\partial f}{\partial z}\right) = \frac{\partial^2 f}{\partial y\,\partial z}$$

$$= \frac{\partial^2 f}{\partial z\,\partial y}$$

Continuity implies that the mixed partial derivatives are equal.

$$= \frac{\partial}{\partial z}\left(\frac{\partial f}{\partial y}\right) = \frac{\partial N}{\partial z}.$$

The other two equations in (3) are proved similarly.

The second half of the proof, that Eqs. (3) imply that $\mathbf{F}$ is conservative, is a consequence of Stokes's theorem, taken up in Section 14.7. ❑

When we know that $\mathbf{F}$ is conservative, we usually want to find a potential function for $\mathbf{F}$. This requires solving the equation $\nabla f = \mathbf{F}$ or

$$\frac{\partial f}{\partial x}\mathbf{i} + \frac{\partial f}{\partial y}\mathbf{j} + \frac{\partial f}{\partial z}\mathbf{k} = M\,\mathbf{i} + N\,\mathbf{j} + P\,\mathbf{k}$$

for f. We accomplish this by integrating the three equations

$$\frac{\partial f}{\partial x} = M, \qquad \frac{\partial f}{\partial y} = N, \qquad \frac{\partial f}{\partial z} = P.$$

EXAMPLE 2   Show that $\mathbf{F} = (e^x \cos y + yz)\,\mathbf{i} + (xz - e^x \sin y)\,\mathbf{j} + (xy + z)\,\mathbf{k}$ is conservative and find a potential function for it.

Solution   We apply the test in Eqs. (3) to

$$M = e^x \cos y + yz, \qquad N = xz - e^x \sin y, \qquad P = xy + z$$

and calculate

$$\frac{\partial P}{\partial y} = x = \frac{\partial N}{\partial z}, \qquad \frac{\partial M}{\partial z} = y = \frac{\partial P}{\partial x}, \qquad \frac{\partial N}{\partial x} = -e^x \sin y + z = \frac{\partial M}{\partial y}.$$

Together, these equalities tell us that there is a function f with $\nabla f = \mathbf{F}$.

We find f by integrating the equations

$$\frac{\partial f}{\partial x} = e^x \cos y + yz, \qquad \frac{\partial f}{\partial y} = xz - e^x \sin y, \qquad \frac{\partial f}{\partial z} = xy + z. \qquad (4)$$

We integrate the first equation with respect to x, holding y and z fixed, to get

$$f(x, y, z) = e^x \cos y + xyz + g(y, z).$$

We write the constant of integration as a function of y and z because its value may change if y and z change. We then calculate $\partial f/\partial y$ from this equation and match it with the expression for $\partial f/\partial y$ in Eq. (4). This gives

$$-e^x \sin y + xz + \frac{\partial g}{\partial y} = xz - e^x \sin y,$$

so $\partial g/\partial y = 0$. Therefore, g is a function of z alone, and

$$f(x, y, z) = e^x \cos y + xyz + h(z).$$

We now calculate $\partial f/\partial z$ from this equation and match it to the formula for $\partial f/\partial z$ in Eq. (4). This gives

$$xy + \frac{dh}{dz} = xy + z, \qquad \text{or} \qquad \frac{dh}{dz} = z,$$

so

$$h(z) = \frac{z^2}{2} + C.$$

Hence,

$$f(x, y, z) = e^x \cos y + xyz + \frac{z^2}{2} + C.$$

We have found infinitely many potential functions for **F**, one for each value of C.

❑

EXAMPLE 3   Show that $\mathbf{F} = (2x - 3)\,\mathbf{i} - z\,\mathbf{j} + (\cos z)\,\mathbf{k}$ is not conservative.

Solution   We apply the component test in Eqs. (3) and find right away that

$$\frac{\partial P}{\partial y} = \frac{\partial}{\partial y}(\cos z) = 0, \qquad \frac{\partial N}{\partial z} = \frac{\partial}{\partial z}(-z) = -1.$$

The two are unequal, so **F** is not conservative. No further testing is required.   ❑

Exact Differential Forms

As we will see in the next section and again later on, it is often convenient to express work and circulation integrals in the "differential" form

$$\int_A^B M\,dx + N\,dy + P\,dz$$

mentioned in Section 14.2. Such integrals are relatively easy to evaluate if $M\,dx + N\,dy + P\,dz$ is the differential of a function f. For then

$$\int_A^B M\,dx + N\,dy + P\,dz = \int_A^B \frac{\partial f}{\partial x}\,dx + \frac{\partial f}{\partial y}\,dy + \frac{\partial f}{\partial z}\,dz$$

$$= \int_A^B \nabla f \cdot d\mathbf{r}$$

$$= f(B) - f(A). \qquad \text{Theorem 1}$$

Thus

$$\int_A^B df = f(B) - f(A),$$

just as with differentiable functions of a single variable.

Definitions
The form $M(x, y, z)\,dx + N(x, y, z)\,dy + P(x, y, z)\,dz$ is called a **differential form**. A differential form is **exact** on a domain D in space if

$$M\,dx + N\,dy + P\,dz = \frac{\partial f}{\partial x}\,dx + \frac{\partial f}{\partial y}\,dy + \frac{\partial f}{\partial z}\,dz = df$$

for some (scalar) function f throughout D.

Notice that if $M\,dx + N\,dy + P\,dz = df$ on D, then $\mathbf{F} = M\,\mathbf{i} + N\,\mathbf{j} + P\,\mathbf{k}$ is the gradient field of f on D. Conversely, if $\mathbf{F} = \nabla f$, then the form $M\,dx + N\,dy + P\,dz$ is exact. The test for the form's being exact is therefore the same as the test for $\mathbf{F}$'s being conservative.

The Test for Exactness of $M\,dx + N\,dy + P\,dz$

The differential form $M\,dx + N\,dy + P\,dz$ is exact if and only if

$$\frac{\partial P}{\partial y} = \frac{\partial N}{\partial z}, \qquad \frac{\partial M}{\partial z} = \frac{\partial P}{\partial x}, \qquad \text{and} \qquad \frac{\partial N}{\partial x} = \frac{\partial M}{\partial y}. \tag{5}$$

This is equivalent to saying that the field $\mathbf{F} = M\,\mathbf{i} + N\,\mathbf{j} + P\,\mathbf{k}$ is conservative.

EXAMPLE 4 Show that $y\,dx + x\,dy + 4\,dz$ is exact, and evaluate the integral

$$\int_{(1,1,1)}^{(2,3,-1)} y\,dx + x\,dy + 4\,dz$$

over the line segment from $(1, 1, 1)$ to $(2, 3, -1)$.

Solution We let $M = y$, $N = x$, $P = 4$ and apply the test of Eq. (5):

$$\frac{\partial P}{\partial y} = 0 = \frac{\partial N}{\partial z}, \qquad \frac{\partial M}{\partial z} = 0 = \frac{\partial P}{\partial x}, \qquad \frac{\partial N}{\partial x} = 1 = \frac{\partial M}{\partial y}.$$

These equalities tell us that $y\,dx + x\,dy + 4\,dz$ is exact, so

$$y\,dx + x\,dy + 4\,dz = df$$

for some function f, and the integral's value is $f(2, 3, -1) - f(1, 1, 1)$.

We find f up to a constant by integrating the equations

$$\frac{\partial f}{\partial x} = y, \qquad \frac{\partial f}{\partial y} = x, \qquad \frac{\partial f}{\partial z} = 4. \tag{6}$$

From the first equation we get

$$f(x, y, z) = xy + g(y, z).$$

The second equation tells us that

$$\frac{\partial f}{\partial y} = x + \frac{\partial g}{\partial y} = x, \qquad \text{or} \qquad \frac{\partial g}{\partial y} = 0.$$

Hence, g is a function of z alone, and

$$f(x, y, z) = xy + h(z).$$

The third of Eqs. (6) tells us that

$$\frac{\partial f}{\partial z} = 0 + \frac{dh}{dz} = 4, \qquad \text{or} \qquad h(z) = 4z + C.$$

Therefore, $f(x, y, z) = xy + 4z + C.$

The value of the integral is

$$f(2, 3, -1) - f(1, 1, 1) = 2 + C - (5 + C) = -3. \qquad \square$$

Exercises 14.3

Testing for Conservative Fields

Which fields in Exercises 1–6 are conservative, and which are not?

1. $\mathbf{F} = yz\,\mathbf{i} + xz\,\mathbf{j} + xy\,\mathbf{k}$

2. $\mathbf{F} = (y \sin z)\,\mathbf{i} + (x \sin z)\,\mathbf{j} + (xy \cos z)\,\mathbf{k}$

3. $\mathbf{F} = y\,\mathbf{i} + (x + z)\,\mathbf{j} - y\,\mathbf{k}$

4. $\mathbf{F} = -y\,\mathbf{i} + x\,\mathbf{j}$

5. $\mathbf{F} = (z + y)\,\mathbf{i} + z\,\mathbf{j} + (y + x)\,\mathbf{k}$

6. $\mathbf{F} = (e^x \cos y)\,\mathbf{i} - (e^x \sin y)\,\mathbf{j} + z\,\mathbf{k}$

Finding Potential Functions

In Exercises 7–12, find a potential function f for the field $\mathbf{F}$.

7. $\mathbf{F} = 2x\,\mathbf{i} + 3y\,\mathbf{j} + 4z\,\mathbf{k}$

8. $\mathbf{F} = (y + z)\,\mathbf{i} + (x + z)\,\mathbf{j} + (x + y)\,\mathbf{k}$

9. $\mathbf{F} = e^{y+2z}(\mathbf{i} + x\,\mathbf{j} + 2x\,\mathbf{k})$

10. $\mathbf{F} = (y \sin z)\,\mathbf{i} + (x \sin z)\,\mathbf{j} + (xy \cos z)\,\mathbf{k}$

11. $\mathbf{F} = (\ln x + \sec^2(x + y))\,\mathbf{i} +$
$$\left(\sec^2(x + y) + \frac{y}{y^2 + z^2}\right)\mathbf{j} + \frac{z}{y^2 + z^2}\,\mathbf{k}$$

12. $\mathbf{F} = \dfrac{y}{1 + x^2 y^2}\,\mathbf{i} + \left(\dfrac{x}{1 + x^2 y^2} + \dfrac{z}{\sqrt{1 - y^2 z^2}}\right)\mathbf{j} +$
$$\left(\frac{y}{\sqrt{1 - y^2 z^2}} + \frac{1}{z}\right)\mathbf{k}$$

Evaluating Line Integrals

In Exercises 13–22, show that the differential forms in the integrals are exact. Then evaluate the integrals.

13. $\displaystyle\int_{(0,0,0)}^{(2,3,-6)} 2x\,dx + 2y\,dy + 2z\,dz$

14. $\displaystyle\int_{(1,1,2)}^{(3,5,0)} yz\,dx + xz\,dy + xy\,dz$

15. $\displaystyle\int_{(0,0,0)}^{(1,2,3)} 2xy\,dx + (x^2 - z^2)\,dy - 2yz\,dz$

16. $\displaystyle\int_{(0,0,0)}^{(3,3,1)} 2x\,dx - y^2\,dy - \frac{4}{1 + z^2}\,dz$

17. $\displaystyle\int_{(1,0,0)}^{(0,1,1)} \sin y \cos x\,dx + \cos y \sin x\,dy + dz$

18. $\displaystyle\int_{(0,2,1)}^{(1,\pi/2,2)} 2 \cos y\,dx + \left(\frac{1}{y} - 2x \sin y\right)dy + \frac{1}{z}\,dz$

19. $\displaystyle\int_{(1,1,1)}^{(1,2,3)} 3x^2\,dx + \frac{z^2}{y}\,dy + 2z \ln y\,dz$

20. $\displaystyle\int_{(1,2,1)}^{(2,1,1)} (2x \ln y - yz)\,dx + \left(\frac{x^2}{y} - xz\right)dy - xy\,dz$

21. $\displaystyle\int_{(1,1,1)}^{(2,2,2)} \frac{1}{y}\,dx + \left(\frac{1}{z} - \frac{x}{y^2}\right)dy - \frac{y}{z^2}\,dz$

22. $\displaystyle\int_{(-1,-1,-1)}^{(2,2,2)} \frac{2x\,dx + 2y\,dy + 2z\,dz}{x^2 + y^2 + z^2}$

23. Evaluate the integral
$$\int_{(1,1,1)}^{(2,3,-1)} y\,dx + x\,dy + 4\,dz$$
from Example 4 by finding parametric equations for the line segment from $(1, 1, 1)$ to $(2, 3, -1)$ and evaluating the line integral of $\mathbf{F} = y\,\mathbf{i} + x\,\mathbf{j} + 4\,\mathbf{k}$ along the segment. Since $\mathbf{F}$ is conservative, the integral is independent of the path.

24. Evaluate $\displaystyle\int_C x^2\,dx + yz\,dy + (y^2/2)\,dz$

along the line segment C joining $(0, 0, 0)$ to $(0, 3, 4)$.

Theory, Applications, and Examples

Show that the values of the integrals in Exercises 25 and 26 do not depend on the path taken from A to B.

25. $\displaystyle\int_A^B z^2\,dx + 2y\,dy + 2xz\,dz$

26. $\displaystyle\int_A^B \frac{x\,dx + y\,dy + z\,dz}{\sqrt{x^2 + y^2 + z^2}}$

In Exercises 27 and 28, express $\mathbf{F}$ in the form ∇f.

27. $\mathbf{F} = \dfrac{2x}{y}\,\mathbf{i} + \left(\dfrac{1 - x^2}{y^2}\right)\mathbf{j}$

28. $\mathbf{F} = (e^x \ln y)\,\mathbf{i} + \left(\dfrac{e^x}{y} + \sin z\right)\mathbf{j} + (y \cos z)\,\mathbf{k}$

29. Find the work done by $\mathbf{F} = (x^2 + y)\,\mathbf{i} + (y^2 + x)\,\mathbf{j} + ze^z\,\mathbf{k}$ over the following paths from $(1, 0, 0)$ to $(1, 0, 1)$.

   a)   The line segment $x = 1, y = 0, 0 \le z \le 1$

   b)   The helix $\mathbf{r}(t) = (\cos t)\,\mathbf{i} + (\sin t)\,\mathbf{j} + (t/2\pi)\,\mathbf{k}, 0 \le t \le 2\pi$

   c)   The x-axis from $(1, 0, 0)$ to $(0, 0, 0)$ followed by the parabola $z = x^2, y = 0$ from $(0, 0, 0)$ to $(1, 0, 1)$

30. Find the work done by $\mathbf{F} = e^{yz}\,\mathbf{i} + (xz\,e^{yz} + z \cos y)\,\mathbf{j} + (xy\,e^{yz} + \sin y)\,\mathbf{k}$ over the following paths from $(1, 0, 1)$ to $(1, \pi/2, 0)$.

   a)   The line segment $x = 1, y = \pi t/2, z = 1 - t, 0 \le t \le 1$

   b)   The line segment from $(1, 0, 1)$ to the origin followed by the line segment from the origin to $(1, \pi/2, 0)$

   c)   The line segment from $(1, 0, 1)$ to $(1, 0, 0)$, followed by the x-axis from $(1, 0, 0)$ to the origin, followed by the parabola $y = \pi x^2/2, z = 0$

31. Let $\mathbf{F} = \nabla(x^3 y^2)$ and let C be the path in the xy-plane from $(-1, 1)$ to $(1, 1)$ that consists of the line segment from $(-1, 1)$ to $(0, 0)$ followed by the line segment from $(0, 0)$ to $(1, 1)$. Evaluate $\int_C \mathbf{F} \cdot d\mathbf{r}$ in two ways:

a) Find parametrizations for the segments that make up C and evaluate the integral;

b) Use the fact that $f(x, y) = x^3 y^2$ is a potential function for $\mathbf{F}$.

32. Evaluate $\int_C 2x \cos y \, dx - x^2 \sin y \, dy$ along the following paths C in the xy-plane.

a) The parabola $y = (x - 1)^2$ from $(1, 0)$ to $(0, 1)$

b) The line segment from $(-1, \pi)$ to $(1, 0)$

c) The x-axis from $(-1, 0)$ to $(1, 0)$

d) The astroid $\mathbf{r}(t) = (\cos^3 t)\mathbf{i} + (\sin^3 t)\mathbf{j}$, $0 \leq t \leq 2\pi$, counterclockwise from $(1, 0)$ back to $(1, 0)$

33. Find a potential function for the gravitational field

$$\mathbf{F} = -GmM \frac{x\mathbf{i} + y\mathbf{j} + z\mathbf{k}}{(x^2 + y^2 + z^2)^{3/2}} \qquad (G, m, \text{ and } M \text{ are constants}).$$

34. (*Continuation of Exercise 33.*) Let P_1 and P_2 be points at distances s_1 and s_2 from the origin. Show that the work done by the gravitational field in Exercise 33 in moving a particle from P_1 to P_2 is the quantity

$$GmM \left(\frac{1}{s_2} - \frac{1}{s_1} \right).$$

35. a) How are the constants a, b, and c related if the following differential form is exact?

$$(ay^2 + 2czx) \, dx + y(bx + cz) \, dy + (ay^2 + cx^2) \, dz$$

b) For what values of b and c will

$$\mathbf{F} = (y^2 + 2czx)\mathbf{i} + y(bx + cz)\mathbf{j} + (y^2 + cx^2)\mathbf{k}$$

be a gradient field?

36. Suppose that $\mathbf{F} = \nabla f$ is a conservative vector field and

$$g(x, y, z) = \int_{(0,0,0)}^{(x,y,z)} \mathbf{F} \cdot d\mathbf{r}.$$

Show that $\nabla g = \mathbf{F}$.

37. You have been asked to find the path along which a force field $\mathbf{F}$ will perform the least work in moving a particle between two locations. A quick calculation on your part shows $\mathbf{F}$ to be conservative. How should you respond? Give reasons for your answer.

38. By experiment, you find that a force field $\mathbf{F}$ performs only half as much work in moving an object along path C_1 from A to B as it does in moving the object along path C_2 from A to B. What can you conclude about $\mathbf{F}$? Give reasons for your answer.

14.4

Green's Theorem in the Plane

We now come to a theorem that can be used to describe the relationship between the way an incompressible fluid flows along or across the boundary of a plane region and the way it moves inside the region. The connection between the fluid's boundary behavior and its internal behavior is made possible by the notions of divergence and curl. The divergence of a fluid's velocity field measures the rate at which fluid is being piped into or out of the region at any given point. The curl measures the fluid's rate of rotation at each point.

Green's theorem states that, under conditions usually met in practice, the outward flux of a vector field across the boundary of a plane region equals the double integral of the divergence of the field over the interior of the region. In another form, it states that the counterclockwise circulation of a field around the boundary of a region equals the double integral of the curl of the field over the region.

Green's theorem is one of the great theorems of calculus. It is deep and surprising and has far-reaching consequences. In pure mathematics, it ranks in importance with the Fundamental Theorem of Calculus. In applied mathematics, the generalizations of Green's theorem to three dimensions provide the foundation for theorems about electricity, magnetism, and fluid flow.

We talk in terms of velocity fields of fluid flows because fluid flows are easy to picture. We would like you to be aware, however, that Green's theorem applies to any vector field satisfying certain mathematical conditions. It does not depend for its validity on the field's having a particular physical interpretation.

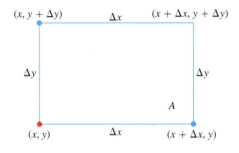

$(x, y + \Delta y)$ Δx $(x + \Delta x, y + \Delta y)$

Δy Δy

A

(x, y) Δx $(x + \Delta x, y)$

14.25 The rectangle for defining the flux density (divergence) of a vector field at a point (x, y).

Flux Density at a Point: Divergence

We need two new ideas for Green's theorem. The first is the idea of the flux density of a vector field at a point, which in mathematics is called the *divergence* of the vector field. We obtain it in the following way.

Suppose that $\mathbf{F}(x, y) = M(x, y)\,\mathbf{i} + N(x, y)\,\mathbf{j}$ is the velocity field of a fluid flow in the plane and that the first partial derivatives of M and N are continuous at each point of a region R. Let (x, y) be a point in R and let A be a small rectangle with one corner at (x, y) that, along with its interior, lies entirely in R (Fig. 14.25). The sides of the rectangle, parallel to the coordinate axes, have lengths of Δx and Δy. The rate at which fluid leaves the rectangle across the bottom edge is approximately

$$\mathbf{F}(x, y) \cdot (-\mathbf{j})\Delta x = -N(x, y)\Delta x. \tag{1}$$

This is the scalar component of the velocity at (x, y) in the direction of the outward normal times the length of the segment. If the velocity is in meters per second, for example, the exit rate will be in meters per second times meters or square meters per second. The rates at which the fluid crosses the other three sides in the directions of their outward normals can be estimated in a similar way. All told, we have

Top: $\mathbf{F}(x, y + \Delta y) \cdot \mathbf{j}\,\Delta x = N(x, y + \Delta y)\Delta x$

Bottom: $\mathbf{F}(x, y) \cdot (-\mathbf{j})\,\Delta x = -N(x, y)\Delta x$

Right: $\mathbf{F}(x + \Delta x, y) \cdot \mathbf{i}\,\Delta y = M(x + \Delta x, y)\,\Delta y$ (2)

Left: $\mathbf{F}(x, y) \cdot (-\mathbf{i})\,\Delta y = -M(x, y)\,\Delta y.$

Combining opposite pairs gives

Top and bottom: $(N(x, y + \Delta y) - N(x, y))\,\Delta x \approx \left(\dfrac{\partial N}{\partial y}\Delta y\right)\Delta x$ (3)

Right and left: $(M(x + \Delta x, y) - M(x, y))\,\Delta y \approx \left(\dfrac{\partial M}{\partial x}\Delta x\right)\Delta y.$ (4)

Adding (3) and (4) gives

$$\text{Flux across rectangle boundary} \approx \left(\frac{\partial M}{\partial x} + \frac{\partial N}{\partial y}\right)\Delta x\,\Delta y.$$

We now divide by $\Delta x\,\Delta y$ to estimate the total flux per unit area or flux density for the rectangle:

$$\frac{\text{Flux across rectangle boundary}}{\text{Rectangle area}} \approx \left(\frac{\partial M}{\partial x} + \frac{\partial N}{\partial y}\right).$$

Finally, we let Δx and Δy approach zero to define what we call the *flux density* of $\mathbf{F}$ at the point (x, y).

In mathematics, we call the flux density the *divergence* of $\mathbf{F}$. The symbol for it is div $\mathbf{F}$, pronounced "divergence of $\mathbf{F}$" or "div $\mathbf{F}$."

Definition

The **flux density** or **divergence** of a vector field $\mathbf{F} = M\,\mathbf{i} + N\,\mathbf{j}$ at the point (x, y) is

$$\text{div } \mathbf{F} = \frac{\partial M}{\partial x} + \frac{\partial N}{\partial y}. \tag{5}$$

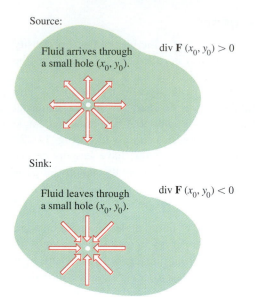

Source:

Fluid arrives through a small hole (x_0, y_0).

div $\mathbf{F}(x_0, y_0) > 0$

Sink:

Fluid leaves through a small hole (x_0, y_0).

div $\mathbf{F}(x_0, y_0) < 0$

14.26 In the flow of an incompressible fluid across a plane region, the divergence is positive at a "source," a point where fluid enters the system, and negative at a "sink," a point where the fluid leaves the system.

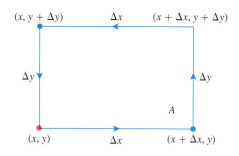

$(x, y + \Delta y)$ Δx $(x + \Delta x, y + \Delta y)$

Δy Δy

A

(x, y) Δx $(x + \Delta x, y)$

14.27 The rectangle for defining the circulation density (curl) of a vector field at a point (x, y).

Intuitively, if water were flowing into a region through a small hole at the point (x_0, y_0), the lines of flow would diverge there (hence the name) and, since water would be flowing out of a small rectangle about (x_0, y_0), the divergence of $\mathbf{F}$ at (x_0, y_0) would be positive. If the water were draining out instead of flowing in, the divergence would be negative. See Fig. 14.26.

EXAMPLE 1   Find the divergence of $\mathbf{F}(x, y) = (x^2 - y)\mathbf{i} + (xy - y^2)\mathbf{j}$.

Solution   We use the formula in Eq. (5):

$$\text{div } \mathbf{F} = \frac{\partial M}{\partial x} + \frac{\partial N}{\partial y} = \frac{\partial}{\partial x}(x^2 - y) + \frac{\partial}{\partial y}(xy - y^2)$$

$$= 2x + x - 2y = 3x - 2y. \qquad \square$$

Circulation Density at a Point: The Curl

The second of the two new ideas we need for Green's theorem is the idea of circulation density of a vector field $\mathbf{F}$ at a point, which in mathematics is called the *curl* of $\mathbf{F}$. To obtain it, we return to the velocity field

$$\mathbf{F}(x, y) = M(x, y)\mathbf{i} + N(x, y)\mathbf{j}$$

and the rectangle A. The rectangle is redrawn here as Fig. 14.27.

The counterclockwise circulation of $\mathbf{F}$ around the boundary of A is the sum of flow rates along the sides. For the bottom edge, the flow rate is approximately

$$\mathbf{F}(x, y) \cdot \mathbf{i}\, \Delta x = M(x, y)\, \Delta x. \qquad (6)$$

This is the scalar component of the velocity $\mathbf{F}(x, y)$ in the direction of the tangent vector $\mathbf{i}$ times the length of the segment. The rates of flow along the other sides in the counterclockwise direction are expressed in a similar way. In all, we have

Top: $\mathbf{F}(x, y + \Delta y) \cdot (-\mathbf{i})\, \Delta x = -M(x, y + \Delta y)\, \Delta x$

Bottom: $\mathbf{F}(x, y) \cdot \mathbf{i}\, \Delta x = M(x, y)\, \Delta x$

Right: $\mathbf{F}(x + \Delta x, y) \cdot \mathbf{j}\, \Delta y = N(x + \Delta x, y)\, \Delta y$ (7)

Left: $\mathbf{F}(x, y) \cdot (-\mathbf{j})\, \Delta y = -N(x, y)\, \Delta y.$

We add opposite pairs to get

Top and bottom:

$$-(M(x, y + \Delta y) - M(x, y))\, \Delta x \approx -\left(\frac{\partial M}{\partial y}\Delta y\right)\Delta x \qquad (8)$$

Right and left:

$$(N(x + \Delta x, y) - N(x, y))\, \Delta y \approx \left(\frac{\partial N}{\partial x}\Delta x\right)\Delta y. \qquad (9)$$

Adding (8) and (9) and dividing by $\Delta x \Delta y$ gives an estimate of the circulation density for the rectangle:

$$\frac{\text{Circulation around rectangle}}{\text{Rectangle area}} \approx \frac{\partial N}{\partial x} - \frac{\partial M}{\partial y}.$$

Finally, we let Δx and Δy approach zero to define what we call the *circulation density* of $\mathbf{F}$ at the point (x, y).

Curl $\mathbf{F}\,(x_0, y_0) > 0$
Counterclockwise circulation

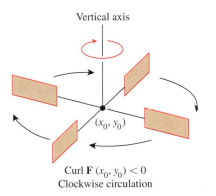

Curl $\mathbf{F}\,(x_0, y_0) < 0$
Clockwise circulation

14.28 In the flow of an incompressible fluid over a plane region, the curl measures the rate of the fluid's rotation at a point. The curl is positive at points where the rotation is counterclockwise and negative where the rotation is clockwise.

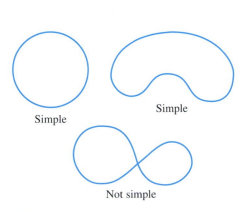

14.29 In proving Green's theorem, we distinguish between two kinds of closed curves, simple and not simple. Simple curves do not cross themselves. A circle is simple but a figure 8 is not.

Definition
The **circulation density** or **curl** of a vector field $\mathbf{F} = M\,\mathbf{i} + N\,\mathbf{j}$ at the point (x, y) is

$$\text{curl } \mathbf{F} = \frac{\partial N}{\partial x} - \frac{\partial M}{\partial y}. \qquad (10)$$

If water is moving about a region in the xy-plane in a thin layer, then the circulation, or curl, at a point (x_0, y_0) gives a way to measure how fast and in what direction a small paddle wheel will spin if it is put into the water at (x_0, y_0) with its axis perpendicular to the plane (Fig. 14.28).

EXAMPLE 2 Find the curl of the vector field

$$\mathbf{F}(x, y) = (x^2 - y)\,\mathbf{i} + (xy - y^2)\,\mathbf{j}.$$

Solution We use the formula in Eq. (10):

$$\text{curl } \mathbf{F} = \frac{\partial N}{\partial x} - \frac{\partial M}{\partial y} = \frac{\partial}{\partial x}(xy - y^2) - \frac{\partial}{\partial y}(x^2 - y) = y + 1.$$

Green's Theorem in the Plane

In one form, Green's theorem says that under suitable conditions the outward flux of a vector field across a simple closed curve in the plane (Fig. 14.29) equals the double integral of the divergence of the field over the region enclosed by the curve. Recall the formulas for flux in Eqs. (3) and (4) in Section 14.2.

Theorem 3
Green's Theorem (Flux-Divergence or Normal Form)
The outward flux of a field $\mathbf{F} = M\,\mathbf{i} + N\,\mathbf{j}$ across a simple closed curve C equals the double integral of div $\mathbf{F}$ over the region R enclosed by C.

$$\underbrace{\oint_C \mathbf{F} \cdot \mathbf{n}\,ds = \oint_C M\,dy - N\,dx}_{\text{outward flux}} = \underbrace{\iint_R \left(\frac{\partial M}{\partial x} + \frac{\partial N}{\partial y} \right) dx\,dy}_{\text{divergence integral}} \qquad (11)$$

In another form, Green's theorem says that the counterclockwise circulation of a vector field around a simple closed curve is the double integral of the curl of the field over the region enclosed by the curve.

Theorem 4
Green's Theorem (Circulation-Curl or Tangential Form)
The counterclockwise circulation of a field $\mathbf{F} = M\,\mathbf{i} + N\,\mathbf{j}$ around a simple closed curve C in the plane equals the double integral of curl $\mathbf{F}$ over the

(continued)

For a two-dimensional field $\mathbf{F} = M\mathbf{i} + N\mathbf{j}$, the integral in Eq. (2), Section 14.2, for circulation takes the equivalent form

$$\oint_C \mathbf{F} \cdot \mathbf{T}\,ds = \oint_C M\,dx + N\,dy.$$

region R enclosed by C.

$$\underbrace{\oint_C \mathbf{F} \cdot \mathbf{T}\,ds = \oint_C M\,dx + N\,dy}_{\substack{\text{counterclockwise} \\ \text{circulation}}} = \underbrace{\iint_R \left(\frac{\partial N}{\partial x} - \frac{\partial M}{\partial y}\right) dx\,dy}_{\text{curl integral}} \qquad (12)$$

The two forms of Green's theorem are equivalent. Applying Eq. (11) to the field $\mathbf{G}_1 = N\mathbf{i} - M\mathbf{j}$ gives Eq. (12), and applying Eq. (12) to $\mathbf{G}_2 = -N\mathbf{i} + M\mathbf{j}$ gives Eq. (11).

We need two kinds of assumptions for Green's theorem to hold. First, we need conditions on M and N to ensure the existence of the integrals. The usual assumptions are that M, N, and their first partial derivatives are continuous at every point of some open region containing C and R. Second, we need geometric conditions on the curve C. It must be simple, closed, and made up of pieces along which we can integrate M and N. The usual assumptions are that C is piecewise smooth. The proof we give for Green's theorem, however, assumes things about the shape of R as well. You can find proofs that are less restrictive in more advanced texts. First let's look at some examples.

EXAMPLE 3   Verify both forms of Green's theorem for the field

$$\mathbf{F}(x, y) = (x - y)\mathbf{i} + x\mathbf{j}$$

and the region R bounded by the unit circle

$$C: \qquad \mathbf{r}(t) = (\cos t)\mathbf{i} + (\sin t)\mathbf{j}, \quad 0 \le t \le 2\pi.$$

Solution   We first express all functions, derivatives, and differentials in terms of t:

$$M = \cos t - \sin t, \qquad dx = d(\cos t) = -\sin t\,dt,$$

$$N = \cos t, \qquad dy = d(\sin t) = \cos t\,dt,$$

$$\frac{\partial M}{\partial x} = 1, \qquad \frac{\partial M}{\partial y} = -1, \qquad \frac{\partial N}{\partial x} = 1, \qquad \frac{\partial N}{\partial y} = 0.$$

The two sides of Eq. (11):

$$\oint_C M\,dy - N\,dx = \int_{t=0}^{t=2\pi} (\cos t - \sin t)(\cos t\,dt) - (\cos t)(-\sin t\,dt)$$

$$= \int_0^{2\pi} \cos^2 t\,dt = \pi$$

$$\iint_R \left(\frac{\partial M}{\partial x} + \frac{\partial N}{\partial y}\right) dx\,dy = \iint_R (1 + 0)\,dx\,dy$$

$$= \iint_R dx\,dy = \text{area of unit circle} = \pi.$$

The two sides of Eq. (12):

$$\oint_C M\, dx + N\, dy = \int_{t=0}^{t=2\pi} (\cos t - \sin t)(-\sin t\, dt) + (\cos t)(\cos t\, dt)$$

$$= \int_0^{2\pi} (-\sin t\, \cos t + 1)\, dt = 2\pi$$

$$\iint_R \left(\frac{\partial N}{\partial x} - \frac{\partial M}{\partial y} \right) dx\, dy = \iint_R (1 - (-1))\, dx\, dy = 2 \iint_R dx\, dy = 2\pi. \qquad \square$$

Using Green's Theorem to Evaluate Line Integrals

If we construct a closed curve C by piecing a number of different curves end to end, the process of evaluating a line integral over C can be lengthy because there are so many different integrals to evaluate. However, if C bounds a region R to which Green's theorem applies, we can use Green's theorem to change the line integral around C into one double integral over R.

EXAMPLE 4   Evaluate the integral

$$\oint_C xy\, dy - y^2\, dx,$$

where C is the square cut from the first quadrant by the lines $x = 1$ and $y = 1$.

Solution   We can use either form of Green's theorem to change the line integral into a double integral over the square.

1. *With Eq. (11):* Taking $M = xy$, $N = y^2$, and C and R as the square's boundary and interior gives

$$\oint_C xy\, dy - y^2\, dx = \iint_R (y + 2y)\, dx\, dy = \int_0^1 \int_0^1 3y\, dx\, dy$$

$$= \int_0^1 \left[3xy \right]_{x=0}^{x=1} dy = \int_0^1 3y\, dy = \frac{3}{2} y^2 \bigg]_0^1 = \frac{3}{2}.$$

2. *With Eq. (12):* Taking $M = -y^2$ and $N = xy$ gives the same result:

$$\oint_C -y^2\, dx + xy\, dy = \iint_R (y - (-2y))\, dx\, dy = \frac{3}{2}. \qquad \square$$

The Green of Green's Theorem

The Green of Green's theorem was George Green (1793–1841), a self-taught scientist in Nottingham, England. Green's work on the mathematical foundations of gravitation, electricity, and magnetism was published privately in 1828 in a short book entitled *An Essay on the Application of Mathematical Analysis to Electricity and Magnetism.* The book sold all of fifty-two copies (fewer than one hundred were printed), the copies going mostly to Green's patrons and personal friends. A few weeks before Green's death in 1841, William Thomson noticed a reference to Green's book and in 1845 was finally able to locate a copy. Excited by what he read, Thomson shared Green's ideas with other scientists and had the book republished in a series of journal articles. Green's mathematics provided the foundation on which Thomson, Stokes, Rayleigh, and Maxwell built the present-day theory of electromagnetism.

EXAMPLE 5   Calculate the outward flux of the field $\mathbf{F}(x, y) = x\,\mathbf{i} + y^2\,\mathbf{j}$ across the square bounded by the lines $x = \pm 1$ and $y = \pm 1$.

Solution   Calculating the flux with a line integral would take four integrations, one for each side of the square. With Green's theorem, we can change the line integral to one double integral. With $M = x$, $N = y^2$, C the square, and R the square's interior, we have

$$\text{Flux} = \oint_C \mathbf{F} \cdot \mathbf{n}\, ds = \oint_C M\, dy - N\, dx$$

$$= \iint_R \left(\frac{\partial M}{\partial x} + \frac{\partial N}{\partial y} \right) dx\, dy \qquad \text{Green's theorem}$$

$$= \int_{-1}^{1} \int_{-1}^{1} (1 + 2y)\, dx\, dy = \int_{-1}^{1} \left[x + 2xy \right]_{x=-1}^{x=1} dy$$

$$= \int_{-1}^{1} (2 + 4y)\, dy = \left[2y + 2y^2 \right]_{-1}^{1} = 4. \qquad \blacksquare$$

A Proof of Green's Theorem (Special Regions)

Let C be a smooth simple closed curve in the xy-plane with the property that lines parallel to the axes cut it in no more than two points. Let R be the region enclosed by C and suppose that M, N, and their first partial derivatives are continuous at every point of some open region containing C and R. We want to prove the circulation-curl form of Green's theorem,

$$\oint_C M\, dx + N\, dy = \iint_R \left(\frac{\partial N}{\partial x} - \frac{\partial M}{\partial y} \right) dx\, dy. \qquad (13)$$

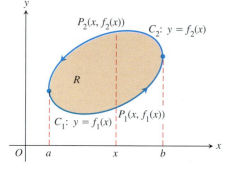

14.30 The boundary curve C is made up of C_1, the graph of $y = f_1(x)$, and C_2, the graph of $y = f_2(x)$.

Figure 14.30 shows C made up of two directed parts:

$$C_1: \quad y = f_1(x), \quad a \le x \le b, \qquad C_2: \quad y = f_2(x), \quad b \ge x \ge a.$$

For any x between a and b, we can integrate $\partial M / \partial y$ with respect to y from $y = f_1(x)$ to $y = f_2(x)$ and obtain

$$\int_{f_1(x)}^{f_2(x)} \frac{\partial M}{\partial y}\, dy = M(x, y) \Big]_{y=f_1(x)}^{y=f_2(x)} = M(x, f_2(x)) - M(x, f_1(x)). \qquad (14)$$

We can then integrate this with respect to x from a to b:

$$\int_a^b \int_{f_1(x)}^{f_2(x)} \frac{\partial M}{\partial y}\, dy\, dx = \int_a^b [M(x, f_2(x)) - M(x, f_1(x))]\, dx$$

$$= -\int_b^a M(x, f_2(x))\, dx - \int_a^b M(x, f_1(x))\, dx$$

$$= -\int_{C_2} M\, dx - \int_{C_1} M\, dx$$

$$= -\oint_C M\, dx.$$

Therefore

$$\oint_C M\, dx = \iint_R \left(-\frac{\partial M}{\partial y} \right) dx\, dy. \qquad (15)$$

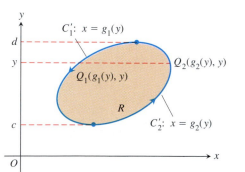

14.31 The boundary curve C is made up of C_1', the graph of $x = g_1(y)$, and C_2', the graph of $x = g_2(y)$.

Equation (15) is half the result we need for Eq. (13). We derive the other half by integrating $\partial N / \partial x$ first with respect to x and then with respect to y, as suggested by Fig. 14.31. This shows the curve C of Fig. 14.30 decomposed into the two

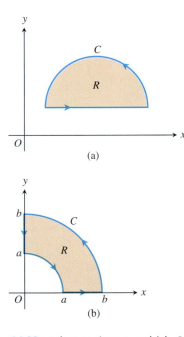

14.32 To prove Green's theorem for a rectangle, we divide the boundary into four directed line segments.

directed parts C_1': $x = g_1(y)$, $d \geq y \geq c$ and C_2': $x = g_2(y)$, $c \leq y \leq d$. The result of this double integration is

$$\oint_C N \, dy = \iint_R \frac{\partial N}{\partial x} \, dx \, dy. \tag{16}$$

Combining Eqs. (15) and (16) gives Eq. (13). This concludes the proof. $\qquad \square$

Extending the Proof to Other Regions

The argument we just gave does not apply directly to the rectangular region in Fig. 14.32 because the lines $x = a$, $x = b$, $y = c$, and $y = d$ meet the region's boundary in more than two points. However, if we divide the boundary C into four directed line segments,

$$C_1: \quad y = c, \quad a \leq x \leq b, \qquad C_2: \quad x = b, \quad c \leq y \leq d,$$

$$C_3: \quad y = d, \quad b \geq x \geq a, \qquad C_4: \quad x = a, \quad d \geq y \geq c,$$

we can modify the argument in the following way.

Proceeding as in the proof of Eq. (16), we have

$$\int_c^d \int_a^b \frac{\partial N}{\partial x} \, dx \, dy = \int_c^d (N(b, y) - N(a, y)) \, dy$$

$$= \int_c^d N(b, y) \, dy + \int_d^c N(a, y) \, dy \tag{17}$$

$$= \int_{C_2} N \, dy + \int_{C_4} N \, dy.$$

Because y is constant along C_1 and C_3, $\int_{C_1} N \, dy = \int_{C_3} N \, dy = 0$, so we can add $\int_{C_1} N \, dy + \int_{C_3} N \, dy$ to the right-hand side of Eq. (17) without changing the equality. Doing so, we have

$$\int_c^d \int_a^b \frac{\partial N}{\partial x} \, dx \, dy = \oint_C N \, dy. \tag{18}$$

Similarly, we can show that

$$\int_a^b \int_c^d \frac{\partial M}{\partial y} \, dy \, dx = -\oint_C M \, dx. \tag{19}$$

Subtracting Eq. (19) from Eq. (18), we again arrive at

$$\oint_C M \, dx + N \, dy = \iint_R \left(\frac{\partial N}{\partial x} - \frac{\partial M}{\partial y} \right) \, dx \, dy.$$

Regions like those in Fig. 14.33 can be handled with no greater difficulty. Equation (13) still applies. It also applies to the horseshoe-shaped region R shown in Fig. 14.34, as we see by putting together the regions R_1 and R_2 and their boundaries. Green's theorem applies to C_1, R_1 and to C_2, R_2, yielding

$$\int_{C_1} M \, dx + N \, dy = \iint_{R_1} \left(\frac{\partial N}{\partial x} - \frac{\partial M}{\partial y} \right) \, dx \, dy$$

$$\int_{C_2} M \, dx + N \, dy = \iint_{R_2} \left(\frac{\partial N}{\partial x} - \frac{\partial M}{\partial y} \right) \, dx \, dy.$$

14.33 Other regions to which Green's theorem applies.

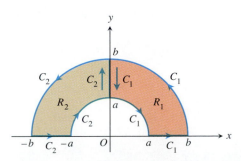

14.34 A region R that combines regions R_1 and R_2.

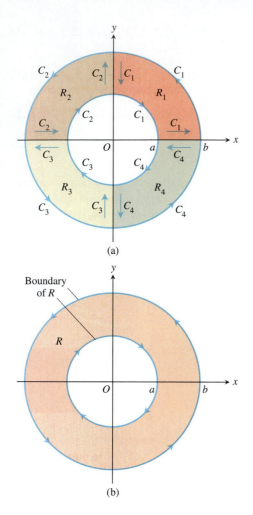

(a)

(b)

14.35 The annular region R combines four smaller regions. In polar coordinates, $r = a$ for the inner circle, $r = b$ for the outer circle, and $a \le r \le b$ for the region itself.

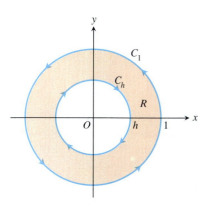

14.36 Green's theorem may be applied to the annular region R by integrating along the boundaries as shown (Example 6).

When we add these two equations, the line integral along the y-axis from b to a for C_1 cancels the integral over the same segment but in the opposite direction for C_2. Hence

$$\oint_C M\,dx + N\,dy = \iint_R \left(\frac{\partial N}{\partial x} - \frac{\partial M}{\partial y}\right) dx\,dy,$$

where C consists of the two segments of the x-axis from $-b$ to $-a$ and from a to b and of the two semicircles, and where R is the region inside C.

The device of adding line integrals over separate boundaries to build up an integral over a single boundary can be extended to any finite number of subregions. In Fig. 14.35(a), let C_1 be the boundary, oriented counterclockwise, of the region R_1 in the first quadrant. Similarly for the other three quadrants: C_i is the boundary of the region R_i, $i = 1, 2, 3, 4$. By Green's theorem,

$$\oint_{C_i} M\,dx + N\,dy = \iint_{R_i} \left(\frac{\partial N}{\partial x} - \frac{\partial M}{\partial y}\right) dx\,dy. \tag{20}$$

We add Eqs. (20) for $i = 1, 2, 3, 4$, and get (Fig. 14.35b):

$$\oint_{r=b} (M\,dx + N\,dy) + \oint_{r=a} (M\,dx + N\,dy) = \iint_{a \le r \le b} \left(\frac{\partial N}{\partial x} - \frac{\partial M}{\partial y}\right) dx\,dy. \tag{21}$$

Equation (21) says that the double integral of $(\partial N/\partial x) - (\partial M/\partial y)$ over the annular ring R equals the line integral of $M\,dx + N\,dy$ over the complete boundary of R in the direction that keeps R on our left as we progress (Fig. 14.35b).

EXAMPLE 6   Verify the circulation form of Green's theorem (Eq. 12) on the annular ring R: $h^2 \le x^2 + y^2 \le 1, 0 < h < 1$ (Fig. 14.36), if

$$M = \frac{-y}{x^2 + y^2}, \qquad N = \frac{x}{x^2 + y^2}.$$

Solution   The boundary of R consists of the circle

$$C_1: \qquad x = \cos t, \qquad y = \sin t, \qquad 0 \le t \le 2\pi,$$

traversed counterclockwise as t increases, and the circle

$$C_h: \qquad x = h\cos\theta, \qquad y = -h\sin\theta, \qquad 0 \le \theta \le 2\pi,$$

traversed clockwise as θ increases. The functions M and N and their partial derivatives are continuous throughout R. Moreover,

$$\frac{\partial M}{\partial y} = \frac{(x^2 + y^2)(-1) + y(2y)}{(x^2 + y^2)^2}$$

$$= \frac{y^2 - x^2}{(x^2 + y^2)^2} = \frac{\partial N}{\partial x},$$

so

$$\iint_R \left(\frac{\partial N}{\partial x} - \frac{\partial M}{\partial y}\right) dx\,dy = \iint_R 0\,dx\,dy = 0.$$

The integral of $M\,dx + N\,dy$ over the boundary of R is

$$\int_C M\,dx + N\,dy = \oint_{C_1} \frac{x\,dy - y\,dx}{x^2 + y^2} + \oint_{C_h} \frac{x\,dy - y\,dx}{x^2 + y^2}$$

$$= \int_0^{2\pi} (\cos^2 t + \sin^2 t)\,dt - \int_0^{2\pi} \frac{h^2(\cos^2 \theta + \sin^2 \theta)}{h^2}\,d\theta$$

$$= 2\pi - 2\pi = 0. \qquad \square$$

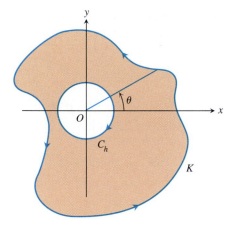

14.37 The region bounded by the circle C_h and the curve K.

The functions M and N in Example 6 are discontinuous at $(0, 0)$, so we cannot apply Green's theorem to the circle C_1 and the region inside it. We must exclude the origin. We do so by excluding the points inside C_h.

We could replace the circle C_1 in Example 6 by an ellipse or any other simple closed curve K surrounding C_h (Fig. 14.37). The result would still be

$$\oint_K (M\,dx + N\,dy) + \oint_{C_h} (M\,dx + N\,dy) = \iint_R \left(\frac{\partial N}{\partial x} - \frac{\partial M}{\partial y} \right) dy\,dx = 0,$$

which leads to the surprising conclusion that

$$\oint_K (M\,dx + N\,dy) = 2\pi$$

for any such curve K. We can explain this result by changing to polar coordinates. With

$$x = r \cos \theta \qquad\qquad y = r \sin \theta$$

$$dx = -r \sin \theta\,d\theta + \cos \theta\,dr, \qquad dy = r \cos \theta\,d\theta + \sin \theta\,dr,$$

we have

$$\frac{x\,dy - y\,dx}{x^2 + y^2} = \frac{r^2(\cos^2 \theta + \sin^2 \theta)\,d\theta}{r^2} = d\theta,$$

and θ increases by 2π as we traverse K once counterclockwise.

Exercises 14.4

Verifying Green's Theorem

In Exercises 1–4, verify Green's theorem by evaluating both sides of Eqs. (11) and (12) for the field $\mathbf{F} = M\,\mathbf{i} + N\,\mathbf{j}$. Take the domains of integration in each case to be the disk $R: x^2 + y^2 \le a^2$ and its bounding circle $C: \mathbf{r} = (a \cos t)\,\mathbf{i} + (a \sin t)\,\mathbf{j}, 0 \le t \le 2\pi$.

1. $\mathbf{F} = -y\,\mathbf{i} + x\,\mathbf{j}$ **2.** $\mathbf{F} = y\,\mathbf{i}$

3. $\mathbf{F} = 2x\,\mathbf{i} - 3y\,\mathbf{j}$ **4.** $\mathbf{F} = -x^2 y\,\mathbf{i} + xy^2\,\mathbf{j}$

Counterclockwise Circulation and Outward Flux

In Exercises 5–10, use Green's theorem to find the counterclockwise circulation and outward flux for the field $\mathbf{F}$ and curve C.

5. $\mathbf{F} = (x - y)\,\mathbf{i} + (y - x)\,\mathbf{j}$
   C: The square bounded by $x = 0, x = 1, y = 0, y = 1$

6. $\mathbf{F} = (x^2 + 4y)\,\mathbf{i} + (x + y^2)\,\mathbf{j}$
   C: The square bounded by $x = 0, x = 1, y = 0, y = 1$

7. $\mathbf{F} = (y^2 - x^2)\,\mathbf{i} + (x^2 + y^2)\,\mathbf{j}$
   C: The triangle bounded by $y = 0, x = 3$, and $y = x$

8. $\mathbf{F} = (x + y)\,\mathbf{i} - (x^2 + y^2)\,\mathbf{j}$
   C: The triangle bounded by $y = 0, x = 1$, and $y = x$

9. $\mathbf{F} = (x + e^x \sin y)\,\mathbf{i} + (x + e^x \cos y)\,\mathbf{j}$
   C: The right-hand loop of the lemniscate $r^2 = \cos 2\theta$

10. $\mathbf{F} = \left(\tan^{-1} \dfrac{y}{x} \right) \mathbf{i} + \ln (x^2 + y^2)\,\mathbf{j}$

   C: The boundary of the region defined by the polar coordinate inequalities $1 \le r \le 2, 0 \le \theta \le \pi$

11. Find the counterclockwise circulation and outward flux of the field $\mathbf{F} = xy\,\mathbf{i} + y^2\,\mathbf{j}$ around and over the boundary of the region enclosed by the curves $y = x^2$ and $y = x$ in the first quadrant.

12. Find the counterclockwise circulation and the outward flux of the field $\mathbf{F} = (-\sin y)\,\mathbf{i} + (x\cos y)\,\mathbf{j}$ around and over the square cut from the first quadrant by the lines $x = \pi/2$ and $y = \pi/2$.

13. Find the outward flux of the field

$$\mathbf{F} = \left(3xy - \frac{x}{1+y^2}\right)\mathbf{i} + (e^x + \tan^{-1} y)\,\mathbf{j}$$

across the cardioid $r = a(1 + \cos\theta),\ a > 0$.

14. Find the counterclockwise circulation of $\mathbf{F} = (y + e^x \ln y)\,\mathbf{i} + (e^x/y)\,\mathbf{j}$ around the boundary of the region that is bounded above by the curve $y = 3 - x^2$ and below by the curve $y = x^4 + 1$.

Work

In Exercises 15 and 16, find the work done by $\mathbf{F}$ in moving a particle once counterclockwise around the given curve.

15. $\mathbf{F} = 2xy^3\,\mathbf{i} + 4x^2y^2\,\mathbf{j}$

    C: The boundary of the "triangular" region in the first quadrant enclosed by the x-axis, the line $x = 1$, and the curve $y = x^3$

16. $\mathbf{F} = (4x - 2y)\,\mathbf{i} + (2x - 4y)\,\mathbf{j}$

    C: The circle $(x - 2)^2 + (y - 2)^2 = 4$

Evaluating Line Integrals in the Plane

Apply Green's theorem to evaluate the integrals in Exercises 17–20.

17. $\oint_C (y^2\,dx + x^2\,dy)$

    C: The triangle bounded by $x = 0,\ x + y = 1,\ y = 0$

18. $\oint_C (3y\,dx + 2x\,dy)$

    C: The boundary of $0 \le x \le \pi,\ 0 \le y \le \sin x$

19. $\oint_C (6y + x)\,dx + (y + 2x)\,dy$

    C: The circle $(x - 2)^2 + (y - 3)^2 = 4$

20. $\oint_C (2x + y^2)\,dx + (2xy + 3y)\,dy$

    C: Any simple closed curve in the plane for which Green's theorem holds

Calculating Area with Green's Theorem

If a simple closed curve C in the plane and the region R it encloses satisfy the hypotheses of Green's theorem, the area of R is given by:

Green's Theorem Area Formula

$$\text{Area of } R = \frac{1}{2}\oint_C x\,dy - y\,dx \qquad (22)$$

The reason is that by Eq. (11), run backward,

$$\text{Area of } R = \iint_R dy\,dx = \iint_R \left(\frac{1}{2} + \frac{1}{2}\right) dy\,dx$$

$$= \oint_C \frac{1}{2}x\,dy - \frac{1}{2}y\,dx.$$

Use the Green's theorem area formula (Eq. 22) to find the areas of the regions enclosed by the curves in Exercises 21–24.

21. The circle $\mathbf{r}(t) = (a\cos t)\,\mathbf{i} + (a\sin t)\,\mathbf{j},\quad 0 \le t \le 2\pi$

22. The ellipse $\mathbf{r}(t) = (a\cos t)\,\mathbf{i} + (b\sin t)\,\mathbf{j},\quad 0 \le t \le 2\pi$

23. The astroid (Fig. 9.42) $\mathbf{r}(t) = (\cos^3 t)\,\mathbf{i} + (\sin^3 t)\,\mathbf{j},\quad 0 \le t \le 2\pi$

24. The curve (Fig. 9.75) $\mathbf{r}(t) = t^2\,\mathbf{i} + ((t^3/3) - t)\,\mathbf{j},$
    $-\sqrt{3} \le t \le \sqrt{3}$

Theory and Examples

25. Let C be the boundary of a region on which Green's theorem holds. Use Green's theorem to calculate

    a) $\oint_C f(x)\,dx + g(y)\,dy,$

    b) $\oint_C ky\,dx + hx\,dy \quad (k$ and h constants).

26. Show that the value of

    $$\oint_C xy^2\,dx + (x^2y + 2x)\,dy$$

    around any square depends only on the area of the square and not on its location in the plane.

27. What is special about the integral

    $$\oint_C 4x^3 y\,dx + x^4\,dy?$$

    Give reasons for your answer.

28. What is special about the integral

    $$\oint_C -y^3\,dx + x^3\,dy?$$

    Give reasons for your answer.

29. Show that if R is a region in the plane bounded by a piecewise smooth simple closed curve C, then

    $$\text{Area of } R = \oint_C x\,dy = -\oint_C y\,dx.$$

30. Suppose that a nonnegative function $y = f(x)$ has a continuous first derivative on $[a, b]$. Let C be the boundary of the region in the xy-plane that is bounded below by the x-axis, above by the graph of f, and on the sides by the lines $x = a$ and $x = b$. Show

that

$$\int_a^b f(x)\,dx = -\oint_C y\,dx.$$

31. Let A be the area and $\bar{x}$ the x-coordinate of the centroid of a region R that is bounded by a piecewise smooth simple closed curve C in the xy-plane. Show that

$$\frac{1}{2}\oint_C x^2\,dy = -\oint_C xy\,dx = \frac{1}{3}\oint_C x^2\,dy - xy\,dx = A\bar{x}.$$

32. Let I_y be the moment of inertia about the y-axis of the region in Exercise 31. Show that

$$\frac{1}{3}\oint_C x^3\,dy = -\oint_C x^2y\,dx = \frac{1}{4}\oint_C x^3\,dy - x^2y\,dx = I_y.$$

33. *Green's theorem and Laplace's equation.* Assuming that all the necessary derivatives exist and are continuous, show that if $f(x, y)$ satisfies the Laplace equation

$$\frac{\partial^2 f}{\partial x^2} + \frac{\partial^2 f}{\partial y^2} = 0,$$

then

$$\oint_C \frac{\partial f}{\partial y}\,dx - \frac{\partial f}{\partial x}\,dy = 0$$

for all closed curves C to which Green's theorem applies. (The converse is also true: If the line integral is always zero, then f satisfies the Laplace equation.)

34. Among all smooth simple closed curves in the plane, oriented counterclockwise, find the one along which the work done by

$$\mathbf{F} = \left(\frac{1}{4}x^2y + \frac{1}{3}y^3\right)\mathbf{i} + x\,\mathbf{j}$$

is greatest. (*Hint:* Where is curl $\mathbf{F}$ positive?)

35. Green's theorem holds for a region R with any finite number of holes as long as the bounding curves are smooth, simple, and closed and we integrate over each component of the boundary in the direction that keeps R on our immediate left as we go along (Fig. 14.38).

a) Let $f(x, y) = \ln(x^2 + y^2)$ and let C be the circle $x^2 + y^2 = a^2$. Evaluate the flux integral

$$\oint_C \nabla f \cdot \mathbf{n}\,ds.$$

b) Let K be an arbitrary smooth simple closed curve in the plane that does not pass through $(0, 0)$. Use Green's theorem to show that

$$\oint_K \nabla f \cdot \mathbf{n}\,ds$$

has two possible values, depending on whether $(0, 0)$ lies inside K or outside K.

14.38 Green's theorem holds for regions with more than one hole (Exercise 35).

36. *Bendixson's criterion.* The **streamlines** of a planar fluid flow are the smooth curves traced by the fluid's individual particles. The vectors $\mathbf{F} = M(x, y)\mathbf{i} + N(x, y)\mathbf{j}$ of the flow's velocity field are the tangent vectors of the streamlines. Show that if the flow takes place over a *simply connected* region R (no holes or missing points) and that if $M_x + N_y \neq 0$ throughout R, then none of the streamlines in R is closed. In other words, no particle of fluid ever has a closed trajectory in R. The criterion $M_x + N_y \neq 0$ is called **Bendixson's criterion** for the nonexistence of closed trajectories.

37. Establish Eq. (16) to finish the proof of the special case of Green's theorem.

38. Establish Eq. (19) to complete the argument for the extension of Green's theorem.

39. Can anything be said about the curl of a conservative two-dimensional vector field? Give reasons for your answer.

40. Does Green's theorem give any information about the circulation of a conservative field? Does this agree with anything else you know? Give reasons for your answer.

⊗ CAS Explorations and Projects

In Exercises 41–44, use a CAS and Green's theorem to find the counterclockwise circulation of the field $\mathbf{F}$ around the simple closed curve C. Perform the following CAS steps:

a) Plot C in the xy-plane.

b) Determine the integrand $\dfrac{\partial N}{\partial x} - \dfrac{\partial M}{\partial y}$ for the curl form of Green's theorem.

c) Determine the (double integral) limits of integration from your plot in (a) and evaluate the curl integral for the circulation.

41. $\mathbf{F} = (2x - y)\mathbf{i} + (x + 3y)\mathbf{j}$, C: The ellipse $x^2 + 4y^2 = 4$

42. $\mathbf{F} = (2x^3 - y^3)\mathbf{i} + (x^3 + y^3)\mathbf{j}$, C: The ellipse $\dfrac{x^2}{4} + \dfrac{y^2}{9} = 1$

43. $\mathbf{F} = x^{-1}e^y\,\mathbf{i} + (e^y \ln x + 2x)\,\mathbf{j}$,
C: The boundary of the region defined by $y = 1 + x^4$ (below) and $y = 2$ (above)

44. $\mathbf{F} = x\,e^y\,\mathbf{i} + 4x^2 \ln y\,\mathbf{j}$, C: The triangle with vertices $(0, 0)$, $(2, 0)$, and $(0, 4)$

Surface $f(x, y, z) = c$

S

R

The vertical projection or "shadow" of S on a coordinate plane

14.39 As we will soon see, the integral of a function $g(x, y, z)$ over a surface S in space can be calculated by evaluating a related double integral over the vertical projection or "shadow" of S on a coordinate plane.

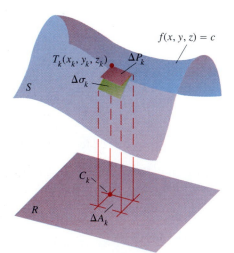

$f(x, y, z) = c$

$T_k(x_k, y_k, z_k)$ ΔP_k

$\Delta \sigma_k$

S

C_k

R

ΔA_k

14.40 A surface S and its vertical projection onto a plane beneath it. You can think of R as the shadow of S on the plane. The tangent plate ΔP_k approximates the surface patch $\Delta \sigma_k$ above ΔA_k.

14.5

Surface Area and Surface Integrals

We know how to integrate a function over a flat region in a plane, but what if the function is defined over a curved surface? How do we calculate its integral then? The trick to evaluating one of these so-called surface integrals is to rewrite it as a double integral over a region in a coordinate plane beneath the surface (Fig. 14.39). In Sections 14.7 and 14.8 we will see how surface integrals provide just what we need to generalize the two forms of Green's theorem to three dimensions.

The Definition of Surface Area

Figure 14.40 shows a surface S lying above its "shadow" region R in a plane beneath it. The surface is defined by the equation $f(x, y, z) = c$. If the surface is **smooth** (∇f is continuous and never vanishes on S), we can define and calculate its area as a double integral over R.

The first step in defining the area of S is to partition the region R into small rectangles ΔA_k of the kind we would use if we were defining an integral over R. Directly above each ΔA_k lies a patch of surface $\Delta \sigma_k$ that we may approximate with a portion ΔP_k of the tangent plane. To be specific, we suppose that ΔP_k is a portion of the plane that is tangent to the surface at the point $T_k(x_k, y_k, z_k)$ directly above the back corner C_k of ΔA_k. If the tangent plane is parallel to R, then ΔP_k will be congruent to ΔA_k. Otherwise, it will be a parallelogram whose area is somewhat larger than the area of ΔA_k.

Figure 14.41 gives a magnified view of $\Delta \sigma_k$ and ΔP_k, showing the gradient vector $\nabla f(x_k, y_k, z_k)$ at T_k and a unit vector $\mathbf{p}$ that is normal to R. The figure also shows the angle γ_k between ∇f and $\mathbf{p}$. The other vectors in the picture, $\mathbf{u}_k$ and $\mathbf{v}_k$, lie along the edges of the patch ΔP_k in the tangent plane. Thus, both $\mathbf{u}_k \times \mathbf{v}_k$ and ∇f are normal to the tangent plane.

We now need the fact from advanced vector geometry that $|(\mathbf{u}_k \times \mathbf{v}_k) \cdot \mathbf{p}|$ is the area of the projection of the parallelogram determined by $\mathbf{u}_k$ and $\mathbf{v}_k$ onto any plane whose normal is $\mathbf{p}$. In our case, this translates into the statement

$$|(\mathbf{u}_k \times \mathbf{v}_k) \cdot \mathbf{p}| = \Delta A_k. \tag{1}$$

Now, $|\mathbf{u}_k \times \mathbf{v}_k|$ itself is the area ΔP_k (standard fact about cross products) so Eq. (1) becomes

$$\underbrace{|\mathbf{u}_k \times \mathbf{v}_k|}_{\Delta P_k} \underbrace{|\mathbf{p}|}_{1} \underbrace{|\cos (\text{angle between } \mathbf{u}_k \times \mathbf{v}_k \text{ and } \mathbf{p})|}_{\substack{\text{same as } |\cos \gamma_k| \text{ because} \\ \nabla f \text{ and } \mathbf{u}_k \times \mathbf{v}_k \text{ are both} \\ \text{normal to the tangent plane}}} = \Delta A_k \tag{2}$$

or

$$\Delta P_k |\cos \gamma_k| = \Delta A_k$$

or

$$\Delta P_k = \frac{\Delta A_k}{|\cos \gamma_k|},$$

provided $\cos \gamma_k \neq 0$. We will have $\cos \gamma_k \neq 0$ as long as ∇f is not parallel to the ground plane and $\nabla f \cdot \mathbf{p} \neq 0$.

Since the patches ΔP_k approximate the surface patches $\Delta \sigma_k$ that fit together to make S, the sum

$$\sum \Delta P_k = \sum \frac{\Delta A_k}{|\cos \gamma_k|} \tag{3}$$

looks like an approximation of what we might like to call the surface area of S. It also looks as if the approximation would improve if we refined the partition of R. In fact, the sums on the right-hand side of Eq. (3) are approximating sums for the double integral

$$\iint\limits_{R} \frac{1}{|\cos \gamma|} \, dA. \tag{4}$$

We therefore define the **area** of S to be the value of this integral whenever it exists.

A Practical Formula

For any surface $f(x, y, z) = c$, we have $|\nabla f \cdot \mathbf{p}| = |\nabla f||\mathbf{p}||\cos \gamma|$, so

$$\frac{1}{|\cos \gamma|} = \frac{|\nabla f|}{|\nabla f \cdot \mathbf{p}|}.$$

This combines with Eq. (4) to give a practical formula for area.

The Formula for Surface Area

The area of the surface $f(x, y, z) = c$ over a closed and bounded plane region R is

$$\text{Surface area} = \iint\limits_{R} \frac{|\nabla f|}{|\nabla f \cdot \mathbf{p}|} \, dA, \tag{5}$$

where $\mathbf{p}$ is a unit vector normal to R and $\nabla f \cdot \mathbf{p} \neq 0$.

Thus, the area is the double integral over R of the magnitude of ∇f divided by the magnitude of the scalar component of ∇f normal to R.

We reached Eq. (5) under the assumption that $\nabla f \cdot \mathbf{p} \neq 0$ throughout R and that ∇f is continuous. Whenever the integral exists, however, we define its value to be the area of the portion of the surface $f(x, y, z) = c$ that lies over R.

EXAMPLE 1   Find the area of the surface cut from the bottom of the paraboloid $x^2 + y^2 - z = 0$ by the plane $z = 4$.

Solution   We sketch the surface S and the region R below it in the xy-plane (Fig. 14.42). The surface S is part of the level surface $f(x, y, z) = x^2 + y^2 - z = 0$, and R is the disk $x^2 + y^2 \leq 4$ in the xy-plane. To get a unit vector normal to the plane of R, we can take $\mathbf{p} = \mathbf{k}$.

At any point (x, y, z) on the surface, we have

$$f(x, y, z) = x^2 + y^2 - z$$
$$\nabla f = 2x\,\mathbf{i} + 2y\,\mathbf{j} - \mathbf{k}$$
$$|\nabla f| = \sqrt{(2x)^2 + (2y)^2 + (-1)^2}$$
$$= \sqrt{4x^2 + 4y^2 + 1}$$
$$|\nabla f \cdot \mathbf{p}| = |\nabla f \cdot \mathbf{k}| = |-1| = 1.$$

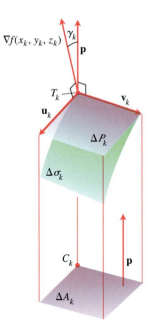

14.41 Magnified view from the preceding figure. The vector $\mathbf{u}_k \times \mathbf{v}_k$ (not shown) is parallel to the vector ∇f because both vectors are normal to the plane of ΔP_k.

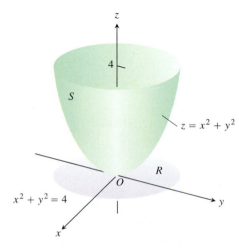

14.42 The area of this parabolic surface is calculated in Example 1.

In the region R, $dA = dx\,dy$. Therefore,

$$\text{Surface area} = \iint_R \frac{|\nabla f|}{|\nabla f \cdot \mathbf{p}|}\,dA \qquad \text{Eq. (5)}$$

$$= \iint_{x^2+y^2\le 4} \sqrt{4x^2 + 4y^2 + 1}\,dx\,dy$$

$$= \int_0^{2\pi}\int_0^2 \sqrt{4r^2+1}\,r\,dr\,d\theta \qquad \text{Polar coordinates}$$

$$= \int_0^{2\pi}\left[\frac{1}{12}(4r^2+1)^{3/2}\right]_0^2 d\theta$$

$$= \int_0^{2\pi}\frac{1}{12}(17^{3/2}-1)\,d\theta = \frac{\pi}{6}(17\sqrt{17}-1). \qquad \square$$

EXAMPLE 2 Find the area of the cap cut from the hemisphere $x^2 + y^2 + z^2 = 2$, $z \ge 0$, by the cylinder $x^2 + y^2 = 1$ (Fig. 14.43).

Solution The cap S is part of the level surface $f(x, y, z) = x^2 + y^2 + z^2 = 2$. It projects one-to-one onto the disk $R: x^2 + y^2 \le 1$ in the xy-plane. The vector $\mathbf{p} = \mathbf{k}$ is normal to the plane of R.

At any point on the surface,

$$f(x, y, z) = x^2 + y^2 + z^2$$

$$\nabla f = 2x\,\mathbf{i} + 2y\,\mathbf{j} + 2z\,\mathbf{k}$$

$$|\nabla f| = 2\sqrt{x^2 + y^2 + z^2} = 2\sqrt{2} \qquad \begin{array}{l}\text{Because } x^2+y^2+z^2=2 \\ \text{at points of } S\end{array}$$

$$|\nabla f \cdot \mathbf{p}| = |\nabla f \cdot \mathbf{k}| = |2z| = 2z.$$

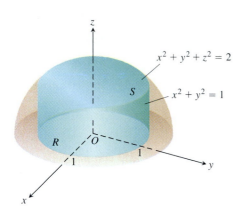

14.43 The cap cut from the hemisphere by the cylinder projects vertically onto the disk $R: x^2 + y^2 \le 1$ (Example 2).

Therefore,

$$\text{Surface area} = \iint_R \frac{|\nabla f|}{|\nabla f \cdot \mathbf{p}|}\,dA = \iint_R \frac{2\sqrt{2}}{2z}\,dA = \sqrt{2}\iint_R \frac{dA}{z}. \qquad (6)$$

What do we do about the z?

Since z is the z-coordinate of a point on the sphere, we can express it in terms of x and y as

$$z = \sqrt{2 - x^2 - y^2}.$$

We continue the work of Eq. (6) with this substitution:

$$\text{Surface area} = \sqrt{2}\iint_R \frac{dA}{z} = \sqrt{2}\iint_{x^2+y^2\le 1} \frac{dA}{\sqrt{2 - x^2 - y^2}}$$

$$= \sqrt{2}\int_0^{2\pi}\int_0^1 \frac{r\,dr\,d\theta}{\sqrt{2 - r^2}} \qquad \text{Polar coordinates}$$

$$= \sqrt{2}\int_0^{2\pi}\left[-(2 - r^2)^{1/2}\right]_{r=0}^{r=1} d\theta$$

$$= \sqrt{2}\int_0^{2\pi}(\sqrt{2} - 1)\,d\theta = 2\pi\,(2 - \sqrt{2}). \qquad \square$$

$f(x, y, z) = c$

(x_k, y_k, z_k) ΔP_k

$\Delta \sigma_k$

S

R ΔA_k

14.44 If we know how an electrical charge is distributed over a surface, we can find the total charge with a suitably modified surface integral.

Surface Integrals

We now show how to integrate a function over a surface, using the ideas just developed for calculating surface area.

Suppose, for example, that we have an electrical charge distributed over a surface $f(x, y, z) = c$ like the one shown in Fig. 14.44 and that the function $g(x, y, z)$ gives the charge per unit area (charge density) at each point on S. Then we may calculate the total charge on S as an integral in the following way.

We partition the shadow region R on the ground plane beneath the surface into small rectangles of the kind we would use if we were defining the surface area of S. Then directly above each ΔA_k lies a patch of surface $\Delta \sigma_k$ that we approximate with a parallelogram-shaped portion of tangent plane, ΔP_k.

Up to this point the construction proceeds as in the definition of surface area, but now we take one additional step: We evaluate g at (x_k, y_k, z_k) and then approximate the total charge on the surface patch $\Delta \sigma_k$ by the product $g(x_k, y_k, z_k) \Delta P_k$. The rationale is that when the partition of R is sufficiently fine, the value of g throughout $\Delta \sigma_k$ is nearly constant and ΔP_k is nearly the same as $\Delta \sigma_k$. The total charge over S is then approximated by the sum

$$\text{Total charge} \approx \sum g(x_k, y_k, z_k) \Delta P_k = \sum g(x_k, y_k, z_k) \frac{\Delta A_k}{|\cos \gamma_k|}. \qquad (7)$$

If f, the function defining the surface S, and its first partial derivatives are continuous, and if g is continuous over S, then the sums on the right-hand side of Eq. (7) approach the limit

$$\iint_R g(x, y, z) \frac{dA}{|\cos \gamma|} = \iint_R g(x, y, z) \frac{|\nabla f|}{|\nabla f \cdot \mathbf{p}|} dA \qquad (8)$$

as the partition of R is refined in the usual way. This limit is called the integral of g over the surface S and is calculated as a double integral over R. The value of the integral is the total charge on the surface S.

As you might expect, the formula in Eq. (8) defines the integral of *any* function g over the surface S as long as the integral exists.

Definitions

If R is the shadow region of a surface S defined by the equation $f(x, y, z) = c$, and g is a continuous function defined at the points of S, then the **integral of g over S** is the integral

$$\iint_R g(x, y, z) \frac{|\nabla f|}{|\nabla f \cdot \mathbf{p}|} dA, \qquad (9)$$

where $\mathbf{p}$ is a unit vector normal to R and $\nabla f \cdot \mathbf{p} \neq 0$. The integral itself is called a **surface integral.**

The integral in (9) takes on different meanings in different applications. If g has the constant value 1, the integral gives the area of S. If g gives the mass density of a thin shell of material modeled by S, the integral gives the mass of the shell.

Algebraic Properties: The Surface Area Differential

We can abbreviate the integral in (9) by writing $d\sigma$ for $(|\nabla f|/|\nabla f \cdot \mathbf{p}|) dA$.

The Surface Area Differential and the Differential Form
for Surface Integrals

$$d\sigma = \frac{|\nabla f|}{|\nabla f \cdot \mathbf{p}|}\, dA \qquad \iint_S g\, d\sigma \qquad (10)$$

surface area   differential formula
differential   for surface integrals

Surface integrals behave like other double integrals, the integral of the sum of two functions being the sum of their integrals and so on. The domain additivity property takes the form

$$\iint_S g\, d\sigma = \iint_{S_1} g\, d\sigma + \iint_{S_2} g\, d\sigma + \cdots + \iint_{S_n} g\, d\sigma.$$

The idea is that if S is partitioned by smooth curves into a finite number of nonoverlapping smooth patches (i.e., if S is **piecewise smooth**), then the integral over S is the sum of the integrals over the patches. Thus, the integral of a function over the surface of a cube is the sum of the integrals over the faces of the cube. We integrate over a turtle shell of welded plates by integrating one plate at a time and adding the results.

EXAMPLE 3   Integrate $g(x, y, z) = xyz$ over the surface of the cube cut from the first octant by the planes $x = 1$, $y = 1$, and $z = 1$ (Fig. 14.45).

Solution   We integrate xyz over each of the six sides and add the results. Since $xyz = 0$ on the sides that lie in the coordinate planes, the integral over the surface of the cube reduces to

$$\iint_{\substack{\text{cube}\\\text{surface}}} xyz\, d\sigma = \iint_{\text{side } A} xyz\, d\sigma + \iint_{\text{side } B} xyz\, d\sigma + \iint_{\text{side } C} xyz\, d\sigma.$$

Side A is the surface $f(x, y, z) = z = 1$ over the square region $R_{xy}: 0 \le x \le 1$, $0 \le y \le 1$, in the xy-plane. For this surface and region,

$$\mathbf{p} = \mathbf{k}, \qquad \nabla f = \mathbf{k}, \qquad |\nabla f| = 1, \qquad |\nabla f \cdot \mathbf{p}| = |\mathbf{k} \cdot \mathbf{k}| = 1,$$

$$d\sigma = \frac{|\nabla f|}{|\nabla f \cdot \mathbf{p}|}\, dA = \frac{1}{1}\, dx\, dy = dx\, dy,$$

$$xyz = xy(1) = xy,$$

and

$$\iint_{\text{side } A} xyz\, d\sigma = \iint_{R_{xy}} xy\, dx\, dy = \int_0^1 \int_0^1 xy\, dx\, dy = \int_0^1 \frac{y}{2}\, dy = \frac{1}{4}.$$

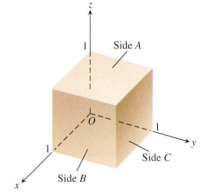

14.45 To integrate a function over the surface of a cube, we integrate over each face and add the results (Example 3).

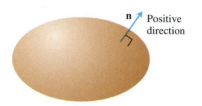

14.46 Smooth closed surfaces in space are orientable. The outward unit normal vector defines the positive direction at each point.

14.47 To make a Möbius band, take a rectangular strip of paper *abcd*, give the end *bc* a single twist, and paste the ends of the strip together to match *a* with *c* and *b* with *d*. The Möbius band is a nonorientable or one-sided surface.

Symmetry tells us that the integrals of xyz over sides B and C are also 1/4. Hence,

$$\iint_{\substack{\text{cube}\\\text{surface}}} xyz \, d\sigma = \frac{1}{4} + \frac{1}{4} + \frac{1}{4} = \frac{3}{4}.$$

Orientation

We call a smooth surface S **orientable** or **two-sided** if it is possible to define a field $\mathbf{n}$ of unit normal vectors on S that varies continuously with position. Any patch or subportion of an orientable surface is orientable. Spheres and other smooth closed surfaces in space (smooth surfaces that enclose solids) are orientable. By convention, we choose $\mathbf{n}$ on a closed surface to point outward.

Once $\mathbf{n}$ has been chosen, we say that we have **oriented** the surface, and we call the surface together with its normal field an **oriented surface.** The vector $\mathbf{n}$ at any point is called the **positive direction** at that point (Fig. 14.46).

The Möbius band in Fig. 14.47 is not orientable. No matter where you start to construct a continuous unit normal field (shown as the shaft of a thumbtack in the figure), moving the vector continuously around the surface in the manner shown will return it to the starting point with a direction opposite to the one it had when it started out. The vector at that point cannot point both ways and yet it must if the field is to be continuous. We conclude that no such field exists.

The Surface Integral for Flux

Suppose that $\mathbf{F}$ is a continuous vector field defined over an oriented surface S and that $\mathbf{n}$ is the chosen unit normal field on the surface. We call the integral of $\mathbf{F} \cdot \mathbf{n}$ over S the flux across S in the positive direction. Thus, the flux is the integral over S of the scalar component of $\mathbf{F}$ in the direction of $\mathbf{n}$.

> **Definition**
>
> The **flux** of a three-dimensional vector field $\mathbf{F}$ across an oriented surface S in the direction of $\mathbf{n}$ is given by the formula
>
> $$\text{Flux} = \iint_S \mathbf{F} \cdot \mathbf{n} \, d\sigma. \qquad (11)$$

The definition is analogous to the flux of a two-dimensional field $\mathbf{F}$ across a plane curve C. In the plane (Section 14.2), the flux is

$$\int_C \mathbf{F} \cdot \mathbf{n} \, ds,$$

the integral of the scalar component of $\mathbf{F}$ normal to the curve.

If $\mathbf{F}$ is the velocity field of a three-dimensional fluid flow, the flux of $\mathbf{F}$ across S is the net rate at which fluid is crossing S in the chosen positive direction. We will discuss such flows in more detail in Section 14.7.

If $\mathbf{S}$ is part of a level surface $g(x, y, z) = c$, then $\mathbf{n}$ may be taken to be one of the two fields

$$\mathbf{n} = \pm \frac{\nabla g}{|\nabla g|}, \qquad (12)$$

depending on which one gives the preferred direction. The corresponding flux is

$$\text{Flux} = \iint_S \mathbf{F} \cdot \mathbf{n} \, d\sigma \qquad \text{Eq. (11)}$$

$$= \iint_R \left(\mathbf{F} \cdot \frac{\pm \nabla g}{|\nabla g|} \right) \frac{|\nabla g|}{|\nabla g \cdot \mathbf{p}|} \, dA \qquad \text{Eqs. (12) and (10)}$$

$$= \iint_R \mathbf{F} \cdot \frac{\pm \nabla g}{|\nabla g \cdot \mathbf{p}|} \, dA. \qquad (13)$$

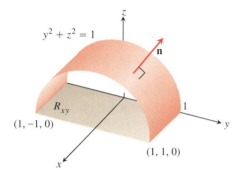

14.48 Example 4 calculates the flux of a vector field outward through this surface. The area of the shadow region R_{xy} is 2.

EXAMPLE 4 Find the flux of $\mathbf{F} = yz\,\mathbf{j} + z^2\,\mathbf{k}$ outward through the surface S cut from the cylinder $y^2 + z^2 = 1$, $z \geq 0$, by the planes $x = 0$ and $x = 1$.

Solution The outward normal field on S (Fig. 14.48) may be calculated from the gradient of $g(x, y, z) = y^2 + z^2$ to be

$$\mathbf{n} = +\frac{\nabla g}{|\nabla g|} = \frac{2y\,\mathbf{j} + 2z\,\mathbf{k}}{\sqrt{4y^2 + 4z^2}} = \frac{2y\,\mathbf{j} + 2z\,\mathbf{k}}{2\sqrt{1}} = y\,\mathbf{j} + z\,\mathbf{k}.$$

With $\mathbf{p} = \mathbf{k}$, we also have

$$d\sigma = \frac{|\nabla g|}{|\nabla g \cdot \mathbf{k}|} \, dA = \frac{2}{|2z|} \, dA = \frac{1}{z} \, dA.$$

We can drop the absolute value bars because $z \geq 0$ on S.

The value of $\mathbf{F} \cdot \mathbf{n}$ on the surface is given by the formula

$$\mathbf{F} \cdot \mathbf{n} = (yz\,\mathbf{j} + z^2\,\mathbf{k}) \cdot (y\,\mathbf{j} + z\,\mathbf{k})$$

$$= y^2 z + z^3 = z(y^2 + z^2)$$

$$= z. \qquad y^2 + z^2 = 1 \text{ on } S$$

Therefore, the flux of $\mathbf{F}$ outward through S is

$$\iint_S \mathbf{F} \cdot \mathbf{n} \, d\sigma = \iint_S (z) \left(\frac{1}{z} \, dA \right) = \iint_{R_{xy}} dA = \text{area}(R_{xy}) = 2.$$

Moments and Masses of Thin Shells

Thin shells of material like bowls, metal drums, and domes are modeled with surfaces. Their moments and masses are calculated with the formulas in Table 14.3.

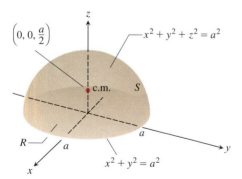

14.49 The center of mass of a thin hemispherical shell of constant density lies on the axis of symmetry halfway from the base to the top (Example 5).

EXAMPLE 5 Find the center of mass of a thin hemispherical shell of radius a and constant density δ.

Solution We model the shell with the hemisphere

$$f(x, y, z) = x^2 + y^2 + z^2 = a^2, \qquad z \geq 0$$

(Fig. 14.49). The symmetry of the surface about the z-axis tells us that $\bar{x} = \bar{y} = 0$. It remains only to find $\bar{z}$ from the formula $\bar{z} = M_{xy}/M$.

The mass of the shell is

$$M = \iint_S \delta \, d\sigma = \delta \iint_S d\sigma = (\delta)(\text{area of } S) = 2\pi a^2 \delta.$$

Table 14.3 Mass and moment formulas for very thin shells

Mass: $M = \displaystyle\iint_S \delta(x, y, z)\, d\sigma$ ($\delta(x, y, z) =$ density at (x, y, z),
mass per unit area)

First moments about the coordinate planes:

$$M_{yz} = \iint_S x\delta\, d\sigma, \qquad M_{xz} = \iint_S y\delta\, d\sigma, \qquad M_{xy} = \iint_S z\delta\, d\sigma$$

Coordinates of center of mass:

$$\bar{x} = M_{yz}/M, \qquad \bar{y} = M_{xz}/M, \qquad \bar{z} = M_{xy}/M$$

Moments of inertia:

$$I_x = \iint_S (y^2 + z^2)\, \delta\, d\sigma, \qquad I_y = \iint_S (x^2 + z^2)\, \delta\, d\sigma,$$

$$I_z = \iint_S (x^2 + y^2)\, \delta\, d\sigma, \qquad I_L = \iint_S r^2\, \delta\, d\sigma,$$

$$r(x, y, z) = \text{distance from point } (x, y, z) \text{ to line } L$$

Radius of gyration about a line L: $R_L = \sqrt{I_L/M}$

To evaluate the integral for M_{xy}, we take $\mathbf{p} = \mathbf{k}$ and calculate

$$|\nabla f| = |2x\,\mathbf{i} + 2y\,\mathbf{j} + 2z\,\mathbf{k}| = 2\sqrt{x^2 + y^2 + z^2} = 2a$$

$$|\nabla f \cdot \mathbf{p}| = |\nabla f \cdot \mathbf{k}| = |2z| = 2z$$

$$d\sigma = \frac{|\nabla f|}{|\nabla f \cdot \mathbf{p}|}\, dA = \frac{a}{z}\, dA.$$

Then

$$M_{xy} = \iint_S z\delta\, d\sigma = \delta \iint_R z \frac{a}{z}\, dA = \delta a \iint_R dA = \delta a(\pi a^2) = \delta\pi a^3$$

$$\bar{z} = \frac{M_{xy}}{M} = \frac{\pi a^3 \delta}{2\pi a^2 \delta} = \frac{a}{2}.$$

The shell's center of mass is the point $(0, 0, a/2)$.   ❏

Exercises 14.5

Surface Area

1. Find the area of the surface cut from the paraboloid $x^2 + y^2 - z = 0$ by the plane $z = 2$.

2. Find the area of the band cut from the paraboloid $x^2 + y^2 - z = 0$ by the planes $z = 2$ and $z = 6$.

3. Find the area of the region cut from the plane $x + 2y + 2z = 5$ by the cylinder whose walls are $x = y^2$ and $x = 2 - y^2$.

4. Find the area of the portion of the surface $x^2 - 2z = 0$ that lies above the triangle bounded by the lines $x = \sqrt{3}$, $y = 0$, and $y = x$ in the xy-plane.

5. Find the area of the surface $x^2 - 2y - 2z = 0$ that lies above the triangle bounded by the lines $x = 2$, $y = 0$, and $y = 3x$ in the xy-plane.

6. Find the area of the cap cut from the sphere $x^2 + y^2 + z^2 = 2$ by the cone $z = \sqrt{x^2 + y^2}$.

7. Find the area of the ellipse cut from the plane $z = cx$ by the cylinder $x^2 + y^2 = 1$.

8. Find the area of the upper portion of the cylinder $x^2 + z^2 = 1$ that lies between the planes $x = \pm 1/2$ and $y = \pm 1/2$.

9. Find the area of the portion of the paraboloid $x = 4 - y^2 - z^2$ that lies above the ring $1 \le y^2 + z^2 \le 4$ in the yz-plane.

10. Find the area of the surface cut from the paraboloid $x^2 + y + z^2 = 2$ by the plane $y = 0$.

11. Find the area of the surface $x^2 - 2\ln x + \sqrt{15}y - z = 0$ above the square R: $1 \le x \le 2, 0 \le y \le 1$, in the xy-plane.

12. Find the area of the surface $2x^{3/2} + 2y^{3/2} - 3z = 0$ above the square R: $0 \le x \le 1, 0 \le y \le 1$, in the xy-plane.

Surface Integrals

13. Integrate $g(x, y, z) = x + y + z$ over the surface of the cube cut from the first octant by the planes $x = a$, $y = a$, $z = a$.

14. Integrate $g(x, y, z) = y + z$ over the surface of the wedge in the first octant bounded by the coordinate planes and the planes $x = 2$ and $y + z = 1$.

15. Integrate $g(x, y, z) = xyz$ over the surface of the rectangular solid cut from the first octant by the planes $x = a$, $y = b$, and $z = c$.

16. Integrate $g(x, y, z) = xyz$ over the surface of the rectangular solid bounded by the planes $x = \pm a$, $y = \pm b$, and $z = \pm c$.

17. Integrate $g(x, y, z) = x + y + z$ over the portion of the plane $2x + 2y + z = 2$ that lies in the first octant.

18. Integrate $g(x, y, z) = x\sqrt{y^2 + 4}$ over the surface cut from the parabolic cylinder $y^2 + 4z = 16$ by the planes $x = 0$, $x = 1$, and $z = 0$.

Flux Across a Surface

In Exercises 19 and 20, find the flux of the field $\mathbf{F}$ across the portion of the given surface in the specified direction.

19. $\mathbf{F}(x, y, z) = -\mathbf{i} + 2\mathbf{j} + 3\mathbf{k}$
    S: rectangular surface $z = 0$, $0 \le x \le 2$, $0 \le y \le 3$, direction $\mathbf{k}$

20. $\mathbf{F}(x, y, z) = yx^2\mathbf{i} - 2\mathbf{j} + xz\mathbf{k}$
    S: rectangular surface $y = 0$, $-1 \le x \le 2$, $2 \le z \le 7$, direction $-\mathbf{j}$

In Exercises 21–26, find the flux of the field $\mathbf{F}$ across the portion of the sphere $x^2 + y^2 + z^2 = a^2$ in the first octant in the direction away from the origin.

21. $\mathbf{F}(x, y, z) = z\mathbf{k}$

22. $\mathbf{F}(x, y, z) = -y\mathbf{i} + x\mathbf{j}$

23. $\mathbf{F}(x, y, z) = y\mathbf{i} - x\mathbf{j} + \mathbf{k}$

24. $\mathbf{F}(x, y, z) = zx\mathbf{i} + zy\mathbf{j} + z^2\mathbf{k}$

25. $\mathbf{F}(x, y, z) = x\mathbf{i} + y\mathbf{j} + z\mathbf{k}$

26. $\mathbf{F}(x, y, z) = \dfrac{x\mathbf{i} + y\mathbf{j} + z\mathbf{k}}{\sqrt{x^2 + y^2 + z^2}}$

27. Find the flux of the field $\mathbf{F}(x, y, z) = z^2\mathbf{i} + x\mathbf{j} - 3z\mathbf{k}$ upward through the surface cut from the parabolic cylinder $z = 4 - y^2$ by the planes $x = 0, x = 1$, and $z = 0$.

28. Find the flux of the field $\mathbf{F}(x, y, z) = 4x\mathbf{i} + 4y\mathbf{j} + 2\mathbf{k}$ outward (away from the z-axis) through the surface cut from the bottom of the paraboloid $z = x^2 + y^2$ by the plane $z = 1$.

29. Let S be the portion of the cylinder $y = e^x$ in the first octant that projects parallel to the x-axis onto the rectangle R_{yz} : $1 \le y \le 2$, $0 \le z \le 1$ in the yz-plane (Fig. 14.50). Let $\mathbf{n}$ be the unit vector normal to S that points away from the yz-plane. Find the flux of the field $\mathbf{F}(x, y, z) = -2\mathbf{i} + 2y\mathbf{j} + z\mathbf{k}$ across S in the direction of $\mathbf{n}$.

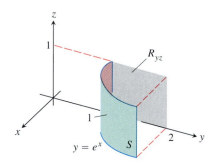

14.50 The surface and region in Exercise 29.

30. Let S be the portion of the cylinder $y = \ln x$ in the first octant whose projection parallel to the y-axis onto the xz-plane is the rectangle R_{xz}: $1 \le x \le e, 0 \le z \le 1$. Let $\mathbf{n}$ be the unit vector normal to S that points away from the xz-plane. Find the flux of $\mathbf{F} = 2y\mathbf{j} + z\mathbf{k}$ through S in the direction of $\mathbf{n}$.

31. Find the outward flux of the field $\mathbf{F} = 2xy\mathbf{i} + 2yz\mathbf{j} + 2xz\mathbf{k}$ across the surface of the cube cut from the first octant by the planes $x = a$, $y = a$, $z = a$.

32. Find the outward flux of the field $\mathbf{F} = xz\mathbf{i} + yz\mathbf{j} + \mathbf{k}$ across the surface of the upper cap cut from the solid sphere $x^2 + y^2 + z^2 \le 25$ by the plane $z = 3$.

Moments and Masses

33. Find the centroid of the portion of the sphere $x^2 + y^2 + z^2 = a^2$ that lies in the first octant.

34. Find the centroid of the surface cut from the cylinder $y^2 + z^2 = 9$, $z \ge 0$, by the planes $x = 0$ and $x = 3$ (resembles the surface in Example 4).

35. Find the center of mass and the moment of inertia and radius of gyration about the z-axis of a thin shell of constant density δ cut from the cone $x^2 + y^2 - z^2 = 0$ by the planes $z = 1$ and $z = 2$.

36. Find the moment of inertia about the z-axis of a thin shell of constant density δ cut from the cone $4x^2 + 4y^2 - z^2 = 0$, $z \geq 0$, by the circular cylinder $x^2 + y^2 = 2x$ (Fig. 14.51).

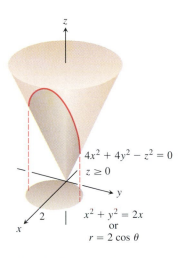

4x² + 4y² − z² = 0
z ≥ 0

x² + y² = 2x
or
r = 2 cos θ

14.51 The surface in Exercise 36.

37. **a)** Find the moment of inertia about a diameter of a thin spherical shell of radius a and constant density δ. (Work with a hemispherical shell and double the result.)

b) Use the Parallel Axis Theorem (Exercises 13.5) and the result in (a) to find the moment of inertia about a line tangent to the shell.

38. **a)** Find the centroid of the lateral surface of a solid cone of base radius a and height h (cone surface minus the base).

b) Use Pappus's formula (Exercises 13.5) and the result in (a) to find the centroid of the complete surface of a solid cone (side plus base).

c) A cone of radius a and height h is joined to a hemisphere of radius a to make a surface S that resembles an ice cream cone. Use Pappus's formula and the results in (a) and Example 5 to find the centroid of S. How high does the cone have to be to place the centroid in the plane shared by the bases of the hemisphere and cone?

Special Formulas for Surface Area

If S is the surface defined by a function $z = f(x, y)$ that has continuous first partial derivatives throughout a region R_{xy} in the xy-plane (Fig. 14.52), then S is also the level surface $F(x, y, z) = 0$ of the function $F(x, y, z) = f(x, y) - z$. Taking the unit normal to R_{xy} to be $\mathbf{p} = \mathbf{k}$ then gives

$$|\nabla F| = |f_x \mathbf{i} + f_y \mathbf{j} - \mathbf{k}| = \sqrt{f_x^2 + f_y^2 + 1},$$
$$|\nabla F \cdot \mathbf{p}| = |(f_x \mathbf{i} + f_y \mathbf{j} - \mathbf{k}) \cdot \mathbf{k}| = |-1| = 1,$$

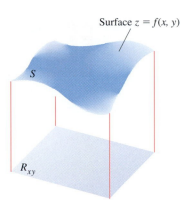

Surface z = f(x, y)

S

R_{xy}

14.52 For a surface $z = f(x, y)$, the surface area formula in Eq. (5) takes the form

$$A = \iint_{R_{xy}} \sqrt{f_x^2 + f_y^2 + 1}\, dx\, dy.$$

and

$$A = \iint_{R_{xy}} \frac{|\nabla F|}{|\nabla F \cdot \mathbf{p}|}\, dA = \iint_{R_{xy}} \sqrt{f_x^2 + f_y^2 + 1}\, dx\, dy. \quad (14)$$

Similarly, the area of a smooth surface $x = f(y, z)$ over a region R_{yz} in the yz-plane is

$$A = \iint_{R_{yz}} \sqrt{f_y^2 + f_z^2 + 1}\, dy\, dz, \quad (15)$$

and the area of a smooth $y = f(x, z)$ over a region R_{xz} in the xz-plane is

$$A = \iint_{R_{xz}} \sqrt{f_x^2 + f_z^2 + 1}\, dx\, dz. \quad (16)$$

Use Eqs. (14)–(16) to find the areas of the surfaces in Exercises 39–44.

39. The surface cut from the bottom of the paraboloid $z = x^2 + y^2$ by the plane $z = 3$

40. The surface cut from the "nose" of the paraboloid $x = 1 - y^2 - z^2$ by the yz-plane

41. The portion of the cone $z = \sqrt{x^2 + y^2}$ that lies over the region between the circle $x^2 + y^2 = 1$ and the ellipse $9x^2 + 4y^2 = 36$ in the xy-plane. (*Hint:* Use formulas from geometry to find the area of the region.)

42. The triangle cut from the plane $2x + 6y + 3z = 6$ by the bounding planes of the first octant. Calculate the area three ways, once with each area formula

43. The surface in the first octant cut from the cylinder $y = (2/3)z^{3/2}$ by the planes $x = 1$ and $y = 16/3$

44. The portion of the plane $y + z = 4$ that lies above the region cut from the first quadrant of the xz-plane by the parabola $x = 4 - z^2$

14.53 A parametrized surface.

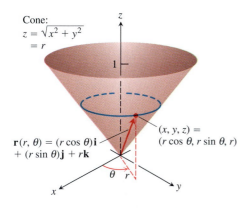

$\mathbf{r}(r, \theta) = (r \cos \theta)\mathbf{i}$
$+ (r \sin \theta)\mathbf{j} + r\mathbf{k}$

14.54 The cone in Example 1.

14.6

Parametrized Surfaces

We have defined curves in the plane in three different ways:

Explicit form:	$y = f(x)$
Implicit form:	$F(x, y) = 0$
Parametric vector form:	$\mathbf{r}(t) = f(t)\mathbf{i} + g(t)\mathbf{j}, \quad a \leq t \leq b.$

We have analogous definitions of surfaces in space:

| Explicit form: | $z = f(x, y)$ |
| Implicit form: | $F(x, y, z) = 0.$ |

There is also a parametric form that gives the position of a point on the surface as a vector function of two variables. The present section extends the investigation of surface area and surface integrals to surfaces described parametrically.

Parametrizations of Surfaces

Let

$$\mathbf{r}(u, v) = f(u, v)\mathbf{i} + g(u, v)\mathbf{j} + h(u, v)\mathbf{k} \tag{1}$$

be a continuous vector function that is defined on a region R in the uv-plane and one-to-one on the interior of R (Fig. 14.53). We call the range of $\mathbf{r}$ the **surface** S defined or traced by $\mathbf{r}$, and Eq. (1) together with the domain R constitute a **parametrization** of the surface. The variables u and v are the **parameters**, and R is the **parameter domain.** To simplify our discussion, we will take R to be a rectangle defined by inequalities of the form $a \leq u \leq b, c \leq v \leq d$. The requirement that $\mathbf{r}$ be one-to-one on the interior of R ensures that S does not cross itself. Notice that Eq. (1) is the vector equivalent of *three* parametric equations:

$$x = f(u, v), \qquad y = g(u, v), \qquad z = h(u, v).$$

EXAMPLE 1   Find a parametrization of the cone

$$z = \sqrt{x^2 + y^2}, \qquad 0 \leq z \leq 1.$$

Solution   Here, cylindrical coordinates provide everything we need. A typical point (x, y, z) on the cone (Fig. 14.54) has $x = r \cos \theta$, $y = r \sin \theta$, and $z = \sqrt{x^2 + y^2} = r$, with $0 \leq r \leq 1$ and $0 \leq \theta \leq 2\pi$. Taking $u = r$ and $v = \theta$ in Eq. (1) gives the parametrization

$$\mathbf{r}(r, \theta) = (r \cos \theta)\mathbf{i} + (r \sin \theta)\mathbf{j} + r\mathbf{k}, \qquad 0 \leq r \leq 1, \quad 0 \leq \theta \leq 2\pi. \quad \square$$

EXAMPLE 2   Find a parametrization of the sphere $x^2 + y^2 + z^2 = a^2$.

Solution   Spherical coordinates provide what we need. A typical point (x, y, z) on the sphere (Fig. 14.55) has $x = a \sin \phi \cos \theta$, $y = a \sin \phi \sin \theta$, and $z = a \cos \phi$, $0 \leq \phi \leq \pi$, $0 \leq \theta \leq 2\pi$. Taking $u = \phi$ and $v = \theta$ in Eq. (1) gives the parametrization

$$\mathbf{r}(\phi, \theta) = (a \sin \phi \cos \theta)\mathbf{i} + (a \sin \phi \sin \theta)\mathbf{j} + (a \cos \phi)\mathbf{k},$$

$$0 \leq \phi \leq \pi, \quad 0 \leq \theta \leq 2\pi. \quad \square$$

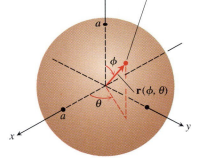

$(x, y, z) = (a \sin \phi \cos \theta, a \sin \phi \sin \theta, a \cos \phi)$

14.55 The sphere in Example 2.

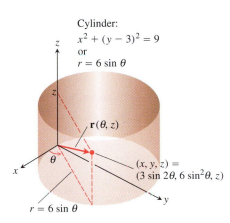

Cylinder:
$x^2 + (y - 3)^2 = 9$
or
$r = 6 \sin \theta$

$\mathbf{r}(\theta, z)$

$(x, y, z) =$
$(3 \sin 2\theta, 6 \sin^2\theta, z)$

$r = 6 \sin \theta$

14.56 The cylinder in Example 3.

EXAMPLE 3 Find a parametrization of the cylinder

$$x^2 + (y - 3)^2 = 9, \qquad 0 \le z \le 5.$$

Solution In cylindrical coordinates, a point (x, y, z) has $x = r \cos \theta$, $y = r \sin \theta$, and $z = z$. For points on the cylinder $x^2 + (y - 3)^2 = 9$ (Fig. 14.56), $r = 6 \sin \theta$, $0 \le \theta \le \pi$ (Section 10.7, Example 5). A typical point on the cylinder therefore has

$$x = r \cos \theta = 6 \sin \theta \cos \theta = 3 \sin 2\theta$$

$$y = r \sin \theta = 6 \sin^2 \theta$$

$$z = z.$$

Taking $u = \theta$ and $v = z$ in Eq. (1) gives the parametrization

$$\mathbf{r}(\theta, z) = (3 \sin 2\theta)\,\mathbf{i} + (6 \sin^2 \theta)\,\mathbf{j} + z\,\mathbf{k}, \qquad 0 \le \theta \le \pi, \quad 0 \le z \le 5. \quad \square$$

Surface Area

Our goal is to find a double integral for calculating the area of a curved surface S based on the parametrization

$$\mathbf{r}(u, v) = f(u, v)\,\mathbf{i} + g(u, v)\,\mathbf{j} + h(u, v)\,\mathbf{k}, \qquad a \le u \le b, \quad c \le v \le d.$$

We need to assume that S is smooth enough for the construction we are about to carry out. The definition of smoothness involves the partial derivatives of $\mathbf{r}$ with respect to u and v:

$$\mathbf{r}_u = \frac{\partial \mathbf{r}}{\partial u} = \frac{\partial f}{\partial u}\,\mathbf{i} + \frac{\partial g}{\partial u}\,\mathbf{j} + \frac{\partial h}{\partial u}\,\mathbf{k}$$

$$\mathbf{r}_v = \frac{\partial \mathbf{r}}{\partial v} = \frac{\partial f}{\partial v}\,\mathbf{i} + \frac{\partial g}{\partial v}\,\mathbf{j} + \frac{\partial h}{\partial v}\,\mathbf{k}.$$

> **Definition**
>
> A parametrized surface $\mathbf{r}(u, v) = f(u, v)\,\mathbf{i} + g(u, v)\,\mathbf{j} + h(u, v)\,\mathbf{k}$ is **smooth** if $\mathbf{r}_u$ and $\mathbf{r}_v$ are continuous and $\mathbf{r}_u \times \mathbf{r}_v$ is never zero on the parameter domain.

Now consider a small rectangle ΔA_{uv} in R with sides on the lines $u = u_0$, $u = u_0 + \Delta u$, $v = v_0$, and $v = v_0 + \Delta v$ (Fig. 14.57, on the following page). Each side of ΔA_{uv} maps to a curve on the surface S, and together these four curves bound a "curved area element" $\Delta \sigma_{uv}$. In the notation of the figure, the side $v = v_0$ maps to curve C_1, the side $u = u_0$ maps to C_2, and their common vertex (u_0, v_0) maps to P_0. Figure 14.58 (on the following page) shows an enlarged view of $\Delta \sigma_{uv}$. The vector $\mathbf{r}_u(u_0, v_0)$ is tangent to C_1 at P_0. Likewise, $\mathbf{r}_v(u_0, v_0)$ is tangent to C_2 at P_0. The cross product $\mathbf{r}_u \times \mathbf{r}_v$ is normal to the surface at P_0. (Here is where we begin to use the assumption that S is smooth. We want to be sure that $\mathbf{r}_u \times \mathbf{r}_v \ne \mathbf{0}$.)

We next approximate the surface element $\Delta \sigma_{uv}$ by the parallelogram on the tangent plane whose sides are determined by the vectors $\Delta u\,\mathbf{r}_u$ and $\Delta v\,\mathbf{r}_v$ (Fig. 14.59, on the following page). The area of this parallelogram is

$$|\Delta u\,\mathbf{r}_u \times \Delta v\,\mathbf{r}_v| = |\mathbf{r}_u \times \mathbf{r}_v|\Delta u\,\Delta v. \tag{2}$$

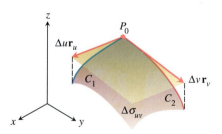

14.57 A rectangular area element ΔA_{uv} in the uv-plane maps onto a curved area element $\Delta\sigma_{uv}$ on S.

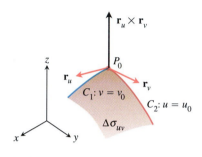

14.58 A magnified view of a surface area element $\Delta\sigma_{uv}$.

14.59 The parallelogram determined by the vectors $\Delta u\,\mathbf{r}_u$ and $\Delta v\,\mathbf{r}_v$ approximates the surface area element $\Delta\sigma_{uv}$.

A partition of the region R in the uv-plane by rectangular regions ΔA_{uv} generates a partition of the surface S into surface area elements $\Delta\sigma_{uv}$. We approximate the area of each surface element $\Delta\sigma_{uv}$ by the parallelogram area in Eq. (2) and sum these areas together to obtain an approximation of the area of S:

$$\sum_u \sum_v |\mathbf{r}_u \times \mathbf{r}_v|\,\Delta u\,\Delta v. \tag{3}$$

As Δu and Δv approach zero independently, the continuity of $\mathbf{r}_u$ and $\mathbf{r}_v$ guarantees that the sum in Eq. (3) approaches the double integral $\int_c^d \int_a^b |\mathbf{r}_u \times \mathbf{r}_v|\,du\,dv$. This double integral gives the area of the surface S.

Parametric Formula for the Area of a Smooth Surface

The area of the smooth surface

$$\mathbf{r}(u, v) = f(u, v)\,\mathbf{i} + g(u, v)\,\mathbf{j} + h(u, v)\,\mathbf{k}, \qquad a \le u \le b, \quad c \le v \le d$$

is

$$A = \int_c^d \int_a^b |\mathbf{r}_u \times \mathbf{r}_v|\,du\,dv. \tag{4}$$

As in Section 14.5, we can abbreviate the integral in (4) by writing $d\sigma$ for $|\mathbf{r}_u \times \mathbf{r}_v|\,du\,dv$.

Surface Area Differential and the Differential Formula for Surface Area

$$d\sigma = |\mathbf{r}_u \times \mathbf{r}_v|\,du\,dv \qquad \iint_S d\sigma \tag{5}$$

surface area differential

differential formula for surface area

EXAMPLE 4 Find the surface area of the cone in Example 1 (Fig. 14.54).

Solution In Example 1 we found the parametrization

$$\mathbf{r}(r, \theta) = (r \cos \theta)\,\mathbf{i} + (r \sin \theta)\,\mathbf{j} + r\,\mathbf{k}, \qquad 0 \le r \le 1, \qquad 0 \le \theta \le 2\pi.$$

To apply Eq. (4) we first find $\mathbf{r}_r \times \mathbf{r}_\theta$:

$$\mathbf{r}_r \times \mathbf{r}_\theta = \begin{vmatrix} \mathbf{i} & \mathbf{j} & \mathbf{k} \\ \cos \theta & \sin \theta & 1 \\ -r \sin \theta & r \cos \theta & 0 \end{vmatrix}$$

$$= -(r \cos \theta)\,\mathbf{i} - (r \sin \theta)\,\mathbf{j} + \underbrace{(r \cos^2 \theta + r \sin^2 \theta)}_{r}\,\mathbf{k}.$$

Thus, $|\mathbf{r}_r \times \mathbf{r}_\theta| = \sqrt{r^2 \cos^2 \theta + r^2 \sin^2 \theta + r^2} = \sqrt{2r^2} = \sqrt{2}\,r$. The area of the cone is

$$A = \int_0^{2\pi} \int_0^1 |\mathbf{r}_r \times \mathbf{r}_\theta|\,dr\,d\theta \qquad \text{Eq. (4) with } u = r, v = \theta$$

$$= \int_0^{2\pi} \int_0^1 \sqrt{2}\,r\,dr\,d\theta = \int_0^{2\pi} \frac{\sqrt{2}}{2}\,d\theta = \frac{\sqrt{2}}{2}\,(2\pi) = \pi\sqrt{2}. \qquad \square$$

EXAMPLE 5 Find the surface area of a sphere of radius a.

Solution We use the parametrization from Example 2:

$$\mathbf{r}(\phi, \theta) = (a \sin \phi \cos \theta)\,\mathbf{i} + (a \sin \phi \sin \theta)\,\mathbf{j} + (a \cos \phi)\,\mathbf{k},$$

$$0 \le \phi \le \pi, \quad 0 \le \theta \le 2\pi.$$

For $\mathbf{r}_\phi \times \mathbf{r}_\theta$ we get

$$\mathbf{r}_\phi \times \mathbf{r}_\theta = \begin{vmatrix} \mathbf{i} & \mathbf{j} & \mathbf{k} \\ a \cos \phi \cos \theta & a \cos \phi \sin \theta & -a \sin \phi \\ -a \sin \phi \sin \theta & a \sin \phi \cos \theta & 0 \end{vmatrix}$$

$$= (a^2 \sin^2 \phi \cos \theta)\,\mathbf{i} + (a^2 \sin^2 \phi \sin \theta)\,\mathbf{j} + (a^2 \sin \phi \cos \phi)\,\mathbf{k}.$$

Thus,

$$|\mathbf{r}_\phi \times \mathbf{r}_\theta| = \sqrt{a^4 \sin^4 \phi \cos^2 \theta + a^4 \sin^4 \phi \sin^2 \theta + a^4 \sin^2 \phi \cos^2 \phi}$$

$$= \sqrt{a^4 \sin^4 \phi + a^4 \sin^2 \phi \cos^2 \phi} = \sqrt{a^4 \sin^2 \phi\,(\sin^2 \phi + \cos^2 \phi)}$$

$$= a^2 \sqrt{\sin^2 \phi} = a^2 \sin \phi,$$

since $\sin \phi \ge 0$ for $0 \le \phi \le \pi$. Therefore the area of the sphere is

$$A = \int_0^{2\pi} \int_0^\pi a^2 \sin \phi\,d\phi\,d\theta$$

$$= \int_0^{2\pi} \Big[-a^2 \cos \phi \Big]_0^\pi\,d\theta = \int_0^{2\pi} 2a^2\,d\theta = 4\pi a^2. \qquad \square$$

Surface Integrals

Having found the formula for calculating the area of a parametrized surface, we can now integrate a function over the surface using the parametrized form.

Definition

If S is a smooth surface defined parametrically as $\mathbf{r}(u, v) = f(u, v)\mathbf{i} + g(u, v)\mathbf{j} + h(u, v)\mathbf{k}$, $a \le u \le b$, $c \le v \le d$, and $G(x, y, z)$ is a continuous function defined on S, then the **integral of G over S** is

$$\iint\limits_{S} G(x, y, z)\, d\sigma = \int_{c}^{d}\int_{a}^{b} G(f(u, v), g(u, v), h(u, v))|\mathbf{r}_u \times \mathbf{r}_v|\, du\, dv.$$

EXAMPLE 6　Integrate $G(x, y, z) = x^2$ over the cone $z = \sqrt{x^2 + y^2}$, $0 \le z \le 1$.

Solution　Continuing the work in Examples 1 and 4, we have $|\mathbf{r}_r \times \mathbf{r}_\theta| = \sqrt{2}\, r$ and

$$\iint\limits_{S} x^2\, d\sigma = \int_{0}^{2\pi}\int_{0}^{1} (r^2 \cos^2 \theta)(\sqrt{2}\, r)\, dr\, d\theta \qquad x = r\cos\theta$$

$$= \sqrt{2}\int_{0}^{2\pi}\int_{0}^{1} r^3 \cos^2 \theta\, dr\, d\theta$$

$$= \frac{\sqrt{2}}{4}\int_{0}^{2\pi} \cos^2 \theta\, d\theta = \frac{\sqrt{2}}{4}\left[\frac{\theta}{2} + \frac{1}{4}\sin 2\theta\right]_{0}^{2\pi} = \frac{\pi\sqrt{2}}{4}.$$

EXAMPLE 7　Find the flux of $\mathbf{F} = yz\,\mathbf{i} + x\,\mathbf{j} - z^2\,\mathbf{k}$ outward through the parabolic cylinder $y = x^2$, $0 \le x \le 1$, $0 \le z \le 4$ (Fig. 14.60).

Solution　On the surface we have $x = x$, $y = x^2$, and $z = z$, so we automatically have the parametrization $\mathbf{r}(x, z) = x\,\mathbf{i} + x^2\,\mathbf{j} + z\,\mathbf{k}$, $0 \le x \le 1$, $0 \le z \le 4$. The cross product of tangent vectors is

$$\mathbf{r}_x \times \mathbf{r}_z = \begin{vmatrix} \mathbf{i} & \mathbf{j} & \mathbf{k} \\ 1 & 2x & 0 \\ 0 & 0 & 1 \end{vmatrix} = 2x\,\mathbf{i} - \mathbf{j}.$$

The unit normal pointing outward from the surface is

$$\mathbf{n} = \frac{\mathbf{r}_x \times \mathbf{r}_z}{|\mathbf{r}_x \times \mathbf{r}_z|} = \frac{2x\,\mathbf{i} - \mathbf{j}}{\sqrt{4x^2 + 1}}.$$

On the surface, $y = x^2$, so the vector field is

$$\mathbf{F} = yz\,\mathbf{i} + x\,\mathbf{j} - z^2\,\mathbf{k} = x^2 z\,\mathbf{i} + x\,\mathbf{j} - z^2\,\mathbf{k}.$$

Thus,

$$\mathbf{F} \cdot \mathbf{n} = \frac{1}{\sqrt{4x^2 + 1}}\left((x^2 z)(2x) + (x)(-1) + (-z^2)(0)\right)$$

$$= \frac{2x^3 z - x}{\sqrt{4x^2 + 1}}.$$

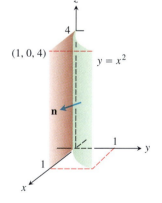

14.60 The parabolic surface in Example 7.

The flux of $\mathbf{F}$ outward through the surface is

$$\iint_S \mathbf{F} \cdot \mathbf{n} \, d\sigma = \int_0^4 \int_0^1 \frac{2x^3 z - x}{\sqrt{4x^2 + 1}} \, |\mathbf{r}_x \times \mathbf{r}_z| \, dx \, dz$$

$$= \int_0^4 \int_0^1 \frac{2x^3 z - x}{\sqrt{4x^2 + 1}} \sqrt{4x^2 + 1} \, dx \, dz$$

$$= \int_0^4 \int_0^1 (2x^3 z - x) \, dx \, dz = \int_0^4 \left[\frac{1}{2} x^4 z - \frac{1}{2} x^2 \right]_{x=0}^{x=1} dz$$

$$= \int_0^4 \frac{1}{2} (z - 1) \, dz = \frac{1}{4} (z - 1)^2 \bigg]_0^4$$

$$= \frac{1}{4} (9) - \frac{1}{4} (1) = 2.$$

EXAMPLE 8   Find the center of mass of a thin shell of constant density δ cut from the cone $z = \sqrt{x^2 + y^2}$ by the planes $z = 1$ and $z = 2$ (Fig. 14.61).

Solution   The symmetry of the surface about the z-axis tells us that $\bar{x} = \bar{y} = 0$. We find $\bar{z} = M_{xy}/M$. Working as in Examples 1 and 4 we have

$$\mathbf{r}(r, \theta) = r \cos \theta \, \mathbf{i} + r \sin \theta \, \mathbf{j} + r \, \mathbf{k}, \qquad 1 \le r \le 2, \quad 0 \le \theta \le 2\pi,$$

and

$$|\mathbf{r}_r \times \mathbf{r}_\theta| = \sqrt{2}\, r.$$

Therefore,

$$M = \iint_S \delta \, d\sigma = \int_0^{2\pi} \int_1^2 \delta \sqrt{2}\, r \, dr \, d\theta$$

$$= \delta \sqrt{2} \int_0^{2\pi} \left[\frac{r^2}{2} \right]_1^2 d\theta = \delta \sqrt{2} \int_0^{2\pi} \left(2 - \frac{1}{2} \right) d\theta$$

$$= \delta \sqrt{2} \left[\frac{3\theta}{2} \right]_0^{2\pi} = 3\pi \, \delta \sqrt{2}$$

$$M_{xy} = \iint_S \delta z \, d\sigma = \int_0^{2\pi} \int_1^2 \delta r \sqrt{2}\, r \, dr \, d\theta$$

$$= \delta \sqrt{2} \int_0^{2\pi} \int_1^2 r^2 \, dr \, d\theta = \delta \sqrt{2} \int_0^{2\pi} \left[\frac{r^3}{3} \right]_1^2 d\theta$$

$$= \delta \sqrt{2} \int_0^{2\pi} \frac{7}{3} \, d\theta = \frac{14}{3} \pi \, \delta \sqrt{2}$$

$$\bar{z} = \frac{M_{xy}}{M} = \frac{14 \pi \, \delta \sqrt{2}}{3(3 \pi \, \delta \sqrt{2})} = \frac{14}{9}.$$

The shell's center of mass is the point $(0, 0, 14/9)$.

$z = \sqrt{x^2 + y^2}$

14.61 The cone frustum in Example 8.

Exercises 14.6

Finding Parametrizations for Surfaces

In Exercises 1–16, find a parametrization of the surface. (There are many correct ways to do these, so your answers may not be the same as those in the back of the book.)

1. The paraboloid $z = x^2 + y^2$, $z \le 4$

2. The paraboloid $z = 9 - x^2 - y^2$, $z \ge 0$

3. The first-octant portion of the cone $z = \sqrt{x^2 + y^2}/2$ between the planes $z = 0$ and $z = 3$

4. The portion of the cone $z = 2\sqrt{x^2 + y^2}$ between the planes $z = 2$ and $z = 4$

5. The cap cut from the sphere $x^2 + y^2 + z^2 = 9$ by the cone $z = \sqrt{x^2 + y^2}$

6. The portion of the sphere $x^2 + y^2 + z^2 = 4$ in the first octant between the xy-plane and the cone $z = \sqrt{x^2 + y^2}$

7. The portion of the sphere $x^2 + y^2 + z^2 = 3$ between the planes $z = \sqrt{3}/2$ and $z = -\sqrt{3}/2$

8. The upper portion cut from the sphere $x^2 + y^2 + z^2 = 8$ by the plane $z = -2$

9. The surface cut from the parabolic cylinder $z = 4 - y^2$ by the planes $x = 0$, $x = 2$, and $z = 0$

10. The surface cut from the parabolic cylinder $y = x^2$ by the planes $z = 0$, $z = 3$, and $y = 2$

11. The portion of the cylinder $y^2 + z^2 = 9$ between the planes $x = 0$ and $x = 3$

12. The portion of the cylinder $x^2 + z^2 = 4$ above the xy-plane between the planes $y = -2$ and $y = 2$

13. The portion of the plane $x + y + z = 1$
    a) inside the cylinder $x^2 + y^2 = 9$
    b) inside the cylinder $y^2 + z^2 = 9$

14. The portion of the plane $x - y + 2z = 2$
    a) inside the cylinder $x^2 + z^2 = 3$
    b) inside the cylinder $y^2 + z^2 = 2$

15. The portion of the cylinder $(x - 2)^2 + z^2 = 4$ between the planes $y = 0$ and $y = 3$

16. The portion of the cylinder $y^2 + (z - 5)^2 = 25$ between the planes $x = 0$ and $x = 10$

17. The portion of the plane $y + 2z = 2$ inside the cylinder $x^2 + y^2 = 1$

18. The portion of the plane $z = -x$ inside the cylinder $x^2 + y^2 = 4$

19. The portion of the cone $z = 2\sqrt{x^2 + y^2}$ between the planes $z = 2$ and $z = 6$

20. The portion of the cone $z = \sqrt{x^2 + y^2}/3$ between the planes $z = 1$ and $z = 4/3$

21. The portion of the cylinder $x^2 + y^2 = 1$ between the planes $z = 1$ and $z = 4$

22. The portion of the cylinder $x^2 + z^2 = 10$ between the planes $y = -1$ and $y = 1$

23. The cap cut from the paraboloid $z = 2 - x^2 - y^2$ by the cone $z = \sqrt{x^2 + y^2}$

24. The portion of the paraboloid $z = x^2 + y^2$ between the planes $z = 1$ and $z = 4$

25. The lower portion cut from the sphere $x^2 + y^2 + z^2 = 2$ by the cone $z = \sqrt{x^2 + y^2}$

26. The portion of the sphere $x^2 + y^2 + z^2 = 4$ between the planes $z = -1$ and $z = \sqrt{3}$

Parametrized Surface Integrals

In Exercises 27–34, integrate the given function over the given surface.

27. $G(x, y, z) = x$, over the parabolic cylinder $y = x^2$, $0 \le x \le 2$, $0 \le z \le 3$

28. $G(x, y, z) = z$, over the cylindrical surface $y^2 + z^2 = 4$, $z \ge 0$, $1 \le x \le 4$

29. $G(x, y, z) = x^2$, over the unit sphere $x^2 + y^2 + z^2 = 1$

30. $G(x, y, z) = z^2$, over the hemisphere $x^2 + y^2 + z^2 = a^2$, $z \ge 0$

31. $F(x, y, z) = z$, over the portion of the plane $x + y + z = 4$ that lies above the square $0 \le x \le 1$, $0 \le y \le 1$, in the xy-plane

32. $F(x, y, z) = z - x$, over the cone $z = \sqrt{x^2 + y^2}$, $0 \le z \le 1$

33. $H(x, y, z) = x^2\sqrt{5 - 4z}$, over the parabolic dome $z = 1 - x^2 - y^2$, $z \ge 0$

34. $H(x, y, z) = yz$, over the part of the sphere $x^2 + y^2 + z^2 = 4$ that lies above the cone $z = \sqrt{x^2 + y^2}$

Areas of Parametrized Surfaces

In Exercises 17–26, use a parametrization to express the area of the surface as a double integral. Then evaluate the integral. (There are many correct ways to set up the integrals, so your integrals may not be the same as those in the back of the book. They should have the same values, however.)

Flux Across Parametrized Surfaces

In Exercises 35–44, use a parametrization to find the flux $\iint_S \mathbf{F} \cdot \mathbf{n}\, d\sigma$ across the surface in the given direction.

35. $\mathbf{F} = z^2\mathbf{i} + x\mathbf{j} - 3z\mathbf{k}$ outward (normal away from the x-axis) through the surface cut from the parabolic cylinder $z = 4 - y^2$ by the planes $x = 0$, $x = 1$, and $z = 0$

36. $\mathbf{F} = x^2\mathbf{j} - xz\mathbf{k}$ outward (normal away from the yz-plane) through the surface cut from the parabolic cylinder $y = x^2$, $-1 \le x \le 1$, by the planes $z = 0$ and $z = 2$

37. $\mathbf{F} = z\mathbf{k}$ across the portion of the sphere $x^2 + y^2 + z^2 = a^2$ in the first octant in the direction away from the origin

38. $\mathbf{F} = x\mathbf{i} + y\mathbf{j} + z\mathbf{k}$ across the sphere $x^2 + y^2 + z^2 = a^2$ in the direction away from the origin

39. $\mathbf{F} = 2xy\mathbf{i} + 2yz\mathbf{j} + 2xz\mathbf{k}$ upward across the portion of the plane $x + y + z = 2a$ that lies above the square $0 \le x \le a, 0 \le y \le a$, in the xy-plane

40. $\mathbf{F} = x\mathbf{i} + y\mathbf{j} + z\mathbf{k}$ outward through the portion of the cylinder $x^2 + y^2 = 1$ cut by the planes $z = 0$ and $z = a$

41. $\mathbf{F} = xy\mathbf{i} - z\mathbf{k}$ outward (normal away from the z-axis) through the cone $z = \sqrt{x^2 + y^2}, 0 \le z \le 1$

42. $\mathbf{F} = y^2\mathbf{i} + xz\mathbf{j} - \mathbf{k}$ outward (normal away from the z-axis) through the cone $z = 2\sqrt{x^2 + y^2}, 0 \le z \le 2$

43. $\mathbf{F} = -x\mathbf{i} - y\mathbf{j} + z^2\mathbf{k}$ outward (normal away from the z-axis) through the portion of the cone $z = \sqrt{x^2 + y^2}$ between the planes $z = 1$ and $z = 2$

44. $\mathbf{F} = 4x\mathbf{i} + 4y\mathbf{j} + 2\mathbf{k}$ outward (normal away from the z-axis) through the surface cut from the bottom of the paraboloid $z = x^2 + y^2$ by the plane $z = 1$

Moments and Masses

45. Find the centroid of the portion of the sphere $x^2 + y^2 + z^2 = a^2$ that lies in the first octant.

46. Find the center of mass and the moment of inertia and radius of gyration about the z-axis of a thin shell of constant density δ cut from the cone $x^2 + y^2 - z^2 = 0$ by the planes $z = 1$ and $z = 2$.

47. Find the moment of inertia about the z-axis of a thin spherical shell $x^2 + y^2 + z^2 = a^2$ of constant density δ.

48. Find the moment of inertia about the z-axis of a thin conical shell $z = \sqrt{x^2 + y^2}, 0 \le z \le 1$, of constant density δ.

Tangent Planes to Parametrized Surfaces

The tangent plane at a point $P_0(f(u_0, v_0), g(u_0, v_0), h(u_0, v_0))$ on a parametrized surface $\mathbf{r}(u, v) = f(u, v)\mathbf{i} + g(u, v)\mathbf{j} + h(u, v)\mathbf{k}$ is the plane through P_0 normal to the vector $\mathbf{r}_u(u_0, v_0) \times \mathbf{r}_v(u_0, v_0)$, which is the cross product of the tangent vectors $\mathbf{r}_u(u_0, v_0)$ and $\mathbf{r}_v(u_0, v_0)$ at P_0. In Exercises 49–52, find an equation for the plane that is tangent to the surface at the given point P_0. Then find a Cartesian equation for the surface and sketch the surface and tangent plane together.

49. The cone $\mathbf{r}(r, \theta) = (r \cos \theta)\mathbf{i} + (r \sin \theta)\mathbf{j} + r\mathbf{k}$, $r \ge 0, 0 \le \theta \le 2\pi$ at the point $P_0\left(\sqrt{2}, \sqrt{2}, 2\right)$ corresponding to $(r, \theta) = (2, \pi/4)$

50. The hemisphere surface
$$\mathbf{r}(\phi, \theta) = (4 \sin \phi \cos \theta)\mathbf{i} + (4 \sin \phi \sin \theta)\mathbf{j} + (4 \cos \phi)\mathbf{k},$$
$0 \le \phi \le \pi/2, 0 \le \theta \le 2\pi$, at the point $P_0\left(\sqrt{2}, \sqrt{2}, 2\sqrt{3}\right)$ corresponding to $(\phi, \theta) = (\pi/6, \pi/4)$

51. The circular cylinder $\mathbf{r}(\theta, z) = (3 \sin 2\theta)\mathbf{i} + (6 \sin^2 \theta)\mathbf{j} + z\mathbf{k}$, $0 \le \theta \le \pi$, at the point $P_0\left(\dfrac{3\sqrt{3}}{2}, \dfrac{9}{2}, 0\right)$ corresponding to $(\theta, z) = (\pi/3, 0)$ (See Example 3.)

52. The parabolic cylinder surface $\mathbf{r}(x, y) = x\mathbf{i} + y\mathbf{j} - x^2\mathbf{k}, -\infty < x < \infty, -\infty < y < \infty$, at the point $P_0(1, 2, -1)$ corresponding to $(x, y) = (1, 2)$

Further Examples of Parametrizations

53. a) A *torus of revolution* (doughnut) is the surface obtained by rotating a circle C in the xz-plane about the z-axis in space. If the radius of C is $r > 0$ and the center is $(R, 0, 0)$, show that a parametrization of the torus is
$$\mathbf{r}(u, v) = ((R + r \cos u) \cos v)\mathbf{i}$$
$$+ ((R + r \cos u) \sin v)\mathbf{j} + (r \sin u)\mathbf{k},$$
where $0 \le u \le 2\pi$ and $0 \le v \le 2\pi$ are the angles in Fig. 14.62.

b) Show that the surface area of the torus is $A = 4\pi^2 R r$.

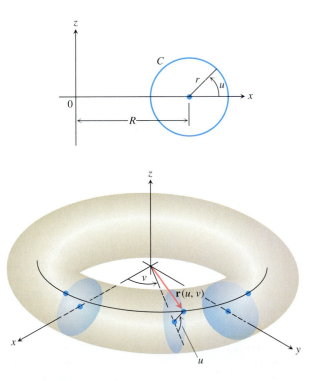

14.62 The torus surface in Exercise 53.

54. *Parametrization of a surface of revolution.* Suppose the parametrized curve $C: (f(u), g(u))$ is revolved about the x-axis, where $g(u) > 0$ for $a \le u \le b$.

a) Show that
$$\mathbf{r}(u, v) = f(u)\mathbf{i} + (g(u) \cos v)\mathbf{j} + (g(u) \sin v)\mathbf{k}$$

is a parametrization of the resulting surface of revolution, where $0 \le v \le 2\pi$ is the angle from the xy-plane to the point $\mathbf{r}(u, v)$ on the surface. (See the accompanying figure.) Notice that $f(u)$ measures distance *along* the axis of revolution and $g(u)$ measures distance *from* the axis of revolution.

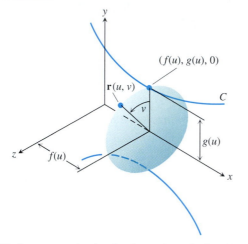

b) Find a parametrization for the surface obtained by revolving the curve $x = y^2$, $y \ge 0$, about the x-axis.

55. a) Recall the parametrization $x = a \cos\theta$, $y = b \sin\theta$, $0 \le \theta \le 2\pi$ for the ellipse $(x^2/a^2) + (y^2/b^2) = 1$ (Section 9.4, Example 5). Using the angles θ and ϕ as defined in spherical coordinates, show that

$$\mathbf{r}(\theta, \phi) = (a \cos\theta \cos\phi)\mathbf{i}$$
$$+ (b \sin\theta \cos\phi)\mathbf{j} + (c \sin\phi)\mathbf{k}$$

is a parametrization of the ellipsoid $(x^2/a^2) + (y^2/b^2) + (z^2/c^2) = 1$.

b) Write an integral for the surface area of the ellipsoid, but do not evaluate the integral.

56. a) Find a parametrization for the hyperboloid of one sheet $x^2 + y^2 - z^2 = 1$ in terms of the angle θ associated with the circle $x^2 + y^2 = r^2$ and the hyperbolic parameter u associated with the hyperbolic function $r^2 - z^2 = 1$. (See Section 6.10, Exercise 86.)

b) Generalize the result in (a) to the hyperboloid $(x^2/a^2) + (y^2/b^2) - (z^2/c^2) = 1$.

57. (*Continuation of Exercise 56.*) Find a Cartesian equation for the plane tangent to the hyperboloid $x^2 + y^2 - z^2 = 25$ at the point $(x_0, y_0, 0)$, where $x_0^2 + y_0^2 = 25$.

58. Find a parametrization of the hyperboloid of two sheets $(z^2/c^2) - (x^2/a^2) - (y^2/b^2) = 1$.

<table>
<tr><td>**14.7**</td></tr>
</table>

Stokes's Theorem

As we saw in Section 14.4, the circulation density or curl of a two-dimensional field $\mathbf{F} = M\mathbf{i} + N\mathbf{j}$ at a point (x, y) is described by the scalar quantity $(\partial N/\partial x - \partial M/\partial y)$. In three dimensions, the circulation around a point P in a plane is described with a vector. This vector is normal to the plane of the circulation (Fig. 14.63) and points in the direction that gives it a right-hand relation to the circulation line. The length of the vector gives the rate of the fluid's rotation, which usually varies as the circulation plane is tilted about P. It turns out that the vector of greatest circulation in a flow with velocity field $\mathbf{F} = M\mathbf{i} + N\mathbf{j} + P\mathbf{k}$ is

$$\text{curl } \mathbf{F} = \left(\frac{\partial P}{\partial y} - \frac{\partial N}{\partial z}\right)\mathbf{i} + \left(\frac{\partial M}{\partial z} - \frac{\partial P}{\partial x}\right)\mathbf{j} + \left(\frac{\partial N}{\partial x} - \frac{\partial M}{\partial y}\right)\mathbf{k}. \qquad (1)$$

We get this information from Stokes's theorem, the generalization of the circulation-curl form of Green's theorem to space.

Del Notation

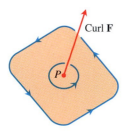

14.63 The circulation vector at a point P in a plane in a three-dimensional fluid flow. Notice its right-hand relation to the circulation line.

The formula for curl $\mathbf{F}$ in Eq. (1) is usually written using the symbolic operator

$$\nabla = \mathbf{i}\frac{\partial}{\partial x} + \mathbf{j}\frac{\partial}{\partial y} + \mathbf{k}\frac{\partial}{\partial z}. \qquad (2)$$

George Gabriel Stokes

Sir George Gabriel Stokes (1819–1903), one of the most influential scientific figures of his century, was Lucasian Professor of Mathematics at Cambridge University from 1849 until his death in 1903. His theoretical and experimental investigations covered hydrodynamics, elasticity, light, gravity, sound, heat, meteorology, and solar physics. He left electricity and magnetism to his friend William Thomson, Baron Kelvin of Largs. It is another one of those delightful quirks of history that the theorem we call Stokes's theorem isn't his theorem at all. He learned of it from Thomson in 1850 and a few years later included it among the questions on an examination he wrote for the Smith Prize. It has been known as Stokes's theorem ever since. As usual, things have balanced out. Stokes was the original discoverer of the principles of spectrum analysis that we now credit to Bunsen and Kirchhoff.

(The symbol ∇ is pronounced "del.") The curl of $\mathbf{F}$ is $\nabla \times \mathbf{F}$:

$$\nabla \times \mathbf{F} = \begin{vmatrix} \mathbf{i} & \mathbf{j} & \mathbf{k} \\ \dfrac{\partial}{\partial x} & \dfrac{\partial}{\partial y} & \dfrac{\partial}{\partial z} \\ M & N & P \end{vmatrix}$$

$$= \left(\frac{\partial P}{\partial y} - \frac{\partial N}{\partial z} \right) \mathbf{i} + \left(\frac{\partial M}{\partial z} - \frac{\partial P}{\partial x} \right) \mathbf{j} + \left(\frac{\partial N}{\partial x} - \frac{\partial M}{\partial y} \right) \mathbf{k} \qquad (3)$$

$$= \text{curl } \mathbf{F}.$$

$$\text{curl } \mathbf{F} = \nabla \times \mathbf{F} \qquad (4)$$

EXAMPLE 1   Find the curl of $\mathbf{F} = (x^2 - y)\mathbf{i} + 4z\mathbf{j} + x^2\mathbf{k}$.

Solution

$$\text{curl } \mathbf{F} = \nabla \times \mathbf{F} \qquad \text{Eq. (4)}$$

$$= \begin{vmatrix} \mathbf{i} & \mathbf{j} & \mathbf{k} \\ \dfrac{\partial}{\partial x} & \dfrac{\partial}{\partial y} & \dfrac{\partial}{\partial z} \\ x^2 - y & 4z & x^2 \end{vmatrix}$$

$$= \left(\frac{\partial}{\partial y}(x^2) - \frac{\partial}{\partial z}(4z) \right) \mathbf{i} - \left(\frac{\partial}{\partial x}(x^2) - \frac{\partial}{\partial z}(x^2 - y) \right) \mathbf{j}$$

$$+ \left(\frac{\partial}{\partial x}(4z) - \frac{\partial}{\partial y}(x^2 - y) \right) \mathbf{k}$$

$$= (0 - 4)\mathbf{i} - (2x - 0)\mathbf{j} + (0 + 1)\mathbf{k}$$

$$= -4\mathbf{i} - 2x\mathbf{j} + \mathbf{k}$$

As we will see, the operator ∇ has a number of other applications. For instance, when applied to a scalar function $f(x, y, z)$, it gives the gradient of f:

$$\nabla f = \frac{\partial f}{\partial x}\mathbf{i} + \frac{\partial f}{\partial y}\mathbf{j} + \frac{\partial f}{\partial z}\mathbf{k}.$$

This may now be read as "del f" as well as "grad f."

Stokes's Theorem

Stokes's theorem says that, under conditions normally met in practice, the circulation of a vector field around the boundary of an oriented surface in space in the direction counterclockwise with respect to the surface's unit normal vector field $\mathbf{n}$ (Fig. 14.64) equals the integral of the normal component of the curl of the field over the surface.

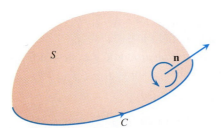

14.64 The orientation of the bounding curve C gives it a right-handed relation to the normal field $\mathbf{n}$.

> **Theorem 5**
> **Stokes's Theorem**
>
> The circulation of $\mathbf{F} = M\mathbf{i} + N\mathbf{j} + P\mathbf{k}$ around the boundary C of an oriented surface S in the direction counterclockwise with respect to the surface's unit normal vector $\mathbf{n}$ equals the integral of $\nabla \times \mathbf{F} \cdot \mathbf{n}$ over S.
>
> $$\underbrace{\oint_C \mathbf{F} \cdot d\mathbf{r}}_{\substack{\text{counterclockwise} \\ \text{circulation}}} = \underbrace{\iint_S \nabla \times \mathbf{F} \cdot \mathbf{n}\, d\sigma}_{\text{curl integral}} \qquad (5)$$

Notice from Eq. (5) that if two different oriented surfaces S_1 and S_2 have the same boundary C, then their curl integrals are equal:

$$\iint_{S_1} \nabla \times \mathbf{F} \cdot \mathbf{n}_1\, d\sigma = \iint_{S_2} \nabla \times \mathbf{F} \cdot \mathbf{n}_2\, d\sigma.$$

Both curl integrals equal the counterclockwise circulation integral on the left side of Eq. (5) as long as the unit normal vectors $\mathbf{n}_1$ and $\mathbf{n}_2$ correctly orient the surfaces.

Naturally, we need some mathematical restrictions on $\mathbf{F}$, C, and S to ensure the existence of the integrals in Stokes's equation. The usual restrictions are that all the functions and derivatives involved be continuous.

If C is a curve in the xy-plane, oriented counterclockwise, and R is the region in the xy-plane bounded by C, then $d\sigma = dx\,dy$ and

$$(\nabla \times \mathbf{F}) \cdot \mathbf{n} = (\nabla \times \mathbf{F}) \cdot \mathbf{k} = \left(\frac{\partial N}{\partial x} - \frac{\partial M}{\partial y} \right). \qquad (6)$$

Under these conditions, Stokes's equation becomes

$$\oint_C \mathbf{F} \cdot d\mathbf{r} = \iint_R \left(\frac{\partial N}{\partial x} - \frac{\partial M}{\partial y} \right) dx\,dy,$$

which is the circulation-curl form of the equation in Green's theorem. Conversely, by reversing these steps we can rewrite the circulation-curl form of Green's theorem for two-dimensional fields in del notation as

$$\oint_C \mathbf{F} \cdot d\mathbf{r} = \iint_R \nabla \times \mathbf{F} \cdot \mathbf{k}\, dA. \qquad (7)$$

See Fig. 14.65.

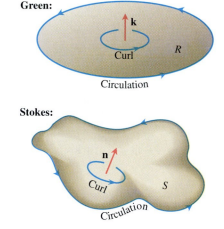

14.65 Green's theorem vs. Stokes's theorem.

EXAMPLE 2 Evaluate Eq. (5) for the hemisphere S: $x^2 + y^2 + z^2 = 9$, $z \geq 0$, its bounding circle C: $x^2 + y^2 = 9$, $z = 0$, and the field $\mathbf{F} = y\mathbf{i} - x\mathbf{j}$.

Solution We calculate the counterclockwise circulation around C (as viewed from above) using the parametrization $\mathbf{r}(\theta) = (3\cos\theta)\mathbf{i} + (3\sin\theta)\mathbf{j}$, $0 \leq \theta \leq 2\pi$:

$$d\mathbf{r} = (-3\sin\theta\, d\theta)\mathbf{i} + (3\cos\theta\, d\theta)\mathbf{j}$$

$$\mathbf{F} = y\mathbf{i} - x\mathbf{j} = (3\sin\theta)\mathbf{i} - (3\cos\theta)\mathbf{j}$$

$$\mathbf{F} \cdot d\mathbf{r} = -9\sin^2\theta\, d\theta - 9\cos^2\theta\, d\theta = -9\, d\theta$$

$$\oint_C \mathbf{F} \cdot d\mathbf{r} = \int_0^{2\pi} -9\, d\theta = -18\pi.$$

For the curl integral of **F,** we have

$$\nabla \times \mathbf{F} = \left(\frac{\partial P}{\partial y} - \frac{\partial N}{\partial z} \right) \mathbf{i} + \left(\frac{\partial M}{\partial z} - \frac{\partial P}{\partial x} \right) \mathbf{j} + \left(\frac{\partial N}{\partial x} - \frac{\partial M}{\partial y} \right) \mathbf{k}$$

$$= (0 - 0)\mathbf{i} + (0 - 0)\mathbf{j} + (-1 - 1)\mathbf{k} = -2\mathbf{k}$$

$$\mathbf{n} = \frac{x\mathbf{i} + y\mathbf{j} + z\mathbf{k}}{\sqrt{x^2 + y^2 + z^2}} = \frac{x\mathbf{i} + y\mathbf{j} + z\mathbf{k}}{3} \qquad \text{Outer unit normal}$$

$$d\sigma = \frac{3}{z} \, dA \qquad \text{Section 14.5, Example 5, with } a = 3$$

$$\nabla \times \mathbf{F} \cdot \mathbf{n} \, d\sigma = -\frac{2z}{3}\frac{3}{z} \, dA = -2 \, dA$$

and

$$\iint_S \nabla \times \mathbf{F} \cdot \mathbf{n} \, d\sigma = \iint_{x^2+y^2 \le 9} -2 \, dA = -18\pi.$$

The circulation around the circle equals the integral of the curl over the hemisphere, as it should.   ◻

EXAMPLE 3   Find the circulation of the field $\mathbf{F} = (x^2 - y)\mathbf{i} + 4z\mathbf{j} + x^2\mathbf{k}$ around the curve C in which the plane $z = 2$ meets the cone $z = \sqrt{x^2 + y^2}$, counterclockwise as viewed from above (Fig. 14.66).

Solution   Stokes's theorem enables us to find the circulation by integrating over the surface of the cone. Traversing C in the counterclockwise direction viewed from above corresponds to taking the *inner* normal **n** to the cone (which has a positive z-component).

We parametrize the cone as

$$\mathbf{r}(r, \theta) = (r \cos \theta)\mathbf{i} + (r \sin \theta)\mathbf{j} + r\mathbf{k}, \qquad 0 \le r \le 2, \quad 0 \le \theta \le 2\pi.$$

We then have

$$\mathbf{n} = \frac{\mathbf{r}_r \times \mathbf{r}_\theta}{|\mathbf{r}_r \times \mathbf{r}_\theta|} = \frac{-(r \cos \theta)\mathbf{i} - (r \sin \theta)\mathbf{j} + r\mathbf{k}}{r\sqrt{2}} \qquad \text{Section 14.6, Example 4}$$

$$= \frac{1}{\sqrt{2}} (-(\cos \theta)\mathbf{i} - (\sin \theta)\mathbf{j} + \mathbf{k})$$

$$d\sigma = r\sqrt{2} \, dr \, d\theta \qquad \text{Section 14.6, Example 4}$$

$$\nabla \times \mathbf{F} = -4\mathbf{i} - 2x\mathbf{j} + \mathbf{k} \qquad \text{Example 1}$$

$$= -4\mathbf{i} - 2r \cos \theta \, \mathbf{j} + \mathbf{k}. \qquad x = r \cos \theta$$

Accordingly,

$$\nabla \times \mathbf{F} \cdot \mathbf{n} = \frac{1}{\sqrt{2}} (4 \cos \theta + 2r \cos \theta \sin \theta + 1)$$

$$= \frac{1}{\sqrt{2}} (4 \cos \theta + r \sin 2\theta + 1)$$

14.66 The curve C and cone S in Example 3.

and the circulation is

$$\oint_C \mathbf{F} \cdot d\mathbf{r} = \iint_S \nabla \times \mathbf{F} \cdot \mathbf{n}\, d\sigma \qquad \text{Stokes's theorem}$$

$$= \int_0^{2\pi} \int_0^2 \frac{1}{\sqrt{2}}(4\cos\theta + r\sin 2\theta + 1)(r\sqrt{2}\, dr\, d\theta) = 4\pi.$$

An Interpretation of $\nabla \times \mathbf{F}$

Suppose that $\mathbf{v}(x, y, z)$ is the velocity of a moving fluid whose density at (x, y, z) is $\delta(x, y, z)$, and let $\mathbf{F} = \delta\mathbf{v}$. Then

$$\oint_C \mathbf{F} \cdot d\mathbf{r}$$

is the circulation of the fluid around the closed curve C. By Stokes's theorem, the circulation is equal to the flux of $\nabla \times \mathbf{F}$ through a surface S spanning C:

$$\oint_C \mathbf{F} \cdot d\mathbf{r} = \iint_S \nabla \times \mathbf{F} \cdot \mathbf{n}\, d\sigma.$$

Suppose we fix a point Q in the domain of $\mathbf{F}$ and a direction $\mathbf{u}$ at Q. Let C be a circle of radius ρ, with center at Q, whose plane is normal to $\mathbf{u}$. If $\nabla \times \mathbf{F}$ is continuous at Q, then the average value of the $\mathbf{u}$-component of $\nabla \times \mathbf{F}$ over the circular disk S bounded by C approaches the $\mathbf{u}$-component of $\nabla \times \mathbf{F}$ at Q as $\rho \to 0$:

$$(\nabla \times \mathbf{F} \cdot \mathbf{u})_Q = \lim_{\rho \to 0} \frac{1}{\pi\rho^2} \iint_S \nabla \times \mathbf{F} \cdot \mathbf{u}\, d\sigma. \qquad (8)$$

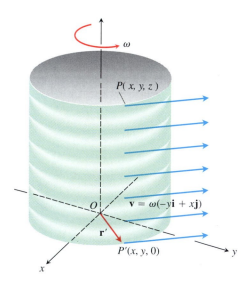

14.67 The paddle wheel interpretation of curl **F**.

If we replace the double integral in Eq. (8) by the circulation, we get

$$(\nabla \times \mathbf{F} \cdot \mathbf{u})_Q = \lim_{\rho \to 0} \frac{1}{\pi\rho^2} \oint_C \mathbf{F} \cdot d\mathbf{r}. \qquad (9)$$

The left-hand side of Eq. (9) has its maximum value when $\mathbf{u}$ is the direction of $\nabla \times \mathbf{F}$. When ρ is small, the limit on the right-hand side of Eq. (9) is approximately

$$\frac{1}{\pi\rho^2} \oint_C \mathbf{F} \cdot d\mathbf{r},$$

which is the circulation around C divided by the area of the disk (circulation density). Suppose that a small paddle wheel of radius ρ is introduced into the fluid at Q, with its axle directed along $\mathbf{u}$. The circulation of the fluid around C will affect the rate of spin of the paddle wheel. The wheel will spin fastest when the circulation integral is maximized; therefore it will spin fastest when the axle of the paddle wheel points in the direction of $\nabla \times \mathbf{F}$ (Fig. 14.67).

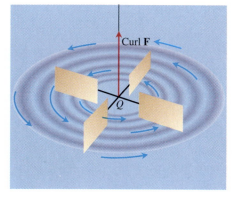

14.68 A steady rotational flow parallel to the *xy*-plane, with constant angular velocity ω in the positive (counterclockwise) direction.

EXAMPLE 4 A fluid of constant density δ rotates around the z-axis with velocity $\mathbf{v} = \omega(-y\,\mathbf{i} + x\,\mathbf{j})$, where ω is a positive constant called the *angular velocity* of the rotation (Fig. 14.68). If $\mathbf{F} = \delta\mathbf{v}$, find $\nabla \times \mathbf{F}$ and relate it to the circulation density.

Solution With $\mathbf{F} = \delta\mathbf{v} = -\delta\omega y\,\mathbf{i} + \delta\omega x\,\mathbf{j}$,

$$\nabla \times \mathbf{F} = \left(\frac{\partial P}{\partial y} - \frac{\partial N}{\partial z}\right)\mathbf{i} + \left(\frac{\partial M}{\partial z} - \frac{\partial P}{\partial x}\right)\mathbf{j} + \left(\frac{\partial N}{\partial x} - \frac{\partial M}{\partial y}\right)\mathbf{k}$$

$$= (0 - 0)\mathbf{i} + (0 - 0)\mathbf{j} + (\delta\omega - (-\delta\omega))\mathbf{k} = 2\,\delta\omega\,\mathbf{k}.$$

By Stokes's theorem, the circulation of $\mathbf{F}$ around a circle C of radius ρ bounding a disk S in a plane normal to $\nabla \times \mathbf{F}$, say the xy-plane, is

$$\oint_C \mathbf{F} \cdot d\mathbf{r} = \iint_S \nabla \times \mathbf{F} \cdot \mathbf{n}\, d\sigma = \iint_S 2\,\delta\omega\, \mathbf{k} \cdot \mathbf{k}\, dx\, dy = (2\,\delta\omega)(\pi\rho^2).$$

Thus,
$$(\nabla \times \mathbf{F}) \cdot \mathbf{k} = 2\,\delta\omega = \frac{1}{\pi\rho^2} \oint_C \mathbf{F} \cdot d\mathbf{r},$$

in agreement with Eq. (9) with $\mathbf{u} = \mathbf{k}$.

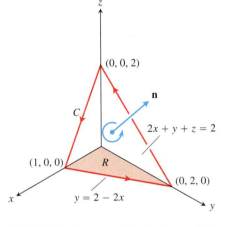

14.69 The planar surface in Example 5.

EXAMPLE 5   Use Stokes's theorem to evaluate $\int_C \mathbf{F} \cdot d\mathbf{r}$, if $\mathbf{F} = xz\,\mathbf{i} + xy\,\mathbf{j} + 3xz\,\mathbf{k}$ and C is the boundary of the portion of the plane $2x + y + z = 2$ in the first octant, traversed counterclockwise as viewed from above (Fig. 14.69).

Solution   The plane is the level surface $f(x, y, z) = 2$ of the function $f(x, y, z) = 2x + y + z$. The unit normal vector

$$\mathbf{n} = \frac{\nabla f}{|\nabla f|} = \frac{(2\mathbf{i} + \mathbf{j} + \mathbf{k})}{|2\mathbf{i} + \mathbf{j} + \mathbf{k}|} = \frac{1}{\sqrt{6}}(2\mathbf{i} + \mathbf{j} + \mathbf{k})$$

is consistent with the counterclockwise motion around C. To apply Stokes's theorem, we find

$$\text{curl } \mathbf{F} = \nabla \times \mathbf{F} = \begin{vmatrix} \mathbf{i} & \mathbf{j} & \mathbf{k} \\ \dfrac{\partial}{\partial x} & \dfrac{\partial}{\partial y} & \dfrac{\partial}{\partial z} \\ xz & xy & 3xz \end{vmatrix} = (x - 3z)\,\mathbf{j} + y\,\mathbf{k}.$$

On the plane, z equals $2 - 2x - y$, so

$$\nabla \times \mathbf{F} = (x - 3(2 - 2x - y))\,\mathbf{j} + y\,\mathbf{k} = (7x + 3y - 6)\,\mathbf{j} + y\,\mathbf{k}$$

and

$$\nabla \times \mathbf{F} \cdot \mathbf{n} = \frac{1}{\sqrt{6}}(7x + 3y - 6 + y) = \frac{1}{\sqrt{6}}(7x + 4y - 6).$$

The surface area element is

$$d\sigma = \frac{|\nabla f|}{|\nabla f \cdot \mathbf{k}|}\, dA = \frac{\sqrt{6}}{1}\, dx\, dy.$$

The circulation is $\displaystyle \oint_C \mathbf{F} \cdot d\mathbf{r} = \iint_S \nabla \times \mathbf{F} \cdot \mathbf{n}\, d\sigma$ Stokes's theorem

$$= \int_0^1 \int_0^{2-2x} \frac{1}{\sqrt{6}}(7x + 4y - 6)\sqrt{6}\, dy\, dx$$

$$= \int_0^1 \int_0^{2-2x} (7x + 4y - 6)\, dy\, dx = -1.$$

Proof of Stokes's Theorem for Polyhedral Surfaces

Let S be a polyhedral surface consisting of a finite number of plane regions. (Think of one of Buckminster Fuller's geodesic domes.) We apply Green's theorem to each

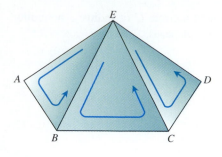

14.70 Part of a polyhedral surface.

separate panel of S. There are two types of panels:

1. those that are surrounded on all sides by other panels and
2. those that have one or more edges that are not adjacent to other panels.

The boundary Δ of S consists of those edges of the type 2 panels that are not adjacent to other panels. In Fig. 14.70, the triangles EAB, BCE, and CDE represent a part of S, with $ABCD$ part of the boundary Δ. Applying Green's theorem to the three triangles in turn and adding the results, we get

$$\left(\oint_{EAB} + \oint_{BCE} + \oint_{CDE} \right) \mathbf{F} \cdot d\mathbf{r} = \left(\iint_{EAB} + \iint_{BCE} + \iint_{CDE} \right) \nabla \times \mathbf{F} \cdot \mathbf{n}\, d\sigma. \quad (10)$$

The three line integrals on the left-hand side of Eq. (10) combine into a single line integral taken around the periphery $ABCDE$ because the integrals along interior segments cancel in pairs. For example, the integral along segment BE in triangle ABE is opposite in sign to the integral along the same segment in triangle EBC. Similarly for segment CE. Hence (10) reduces to

$$\oint_{ABCDE} \mathbf{F} \cdot d\mathbf{r} = \iint_{ABCDE} \nabla \times \mathbf{F} \cdot \mathbf{n}\, d\sigma.$$

When we apply Green's theorem to all the panels and add the results, we get

$$\oint_{\Delta} \mathbf{F} \cdot d\mathbf{r} = \iint_{S} \nabla \times \mathbf{F} \cdot \mathbf{n}\, d\sigma. \quad (11)$$

This is Stokes's theorem for a polyhedral surface S. You can find proofs for more general surfaces in advanced calculus texts. ❏

Stokes's Theorem for Surfaces with Holes

Stokes's theorem can be extended to an oriented surface S that has one or more holes (Fig. 14.71), in a way analogous to the extension of Green's theorem: The surface integral over S of the normal component of $\nabla \times \mathbf{F}$ equals the sum of the line integrals around all the boundary curves of the tangential component of $\mathbf{F}$, where the curves are to be traced in the direction induced by the orientation of S.

An Important Identity

The following identity arises frequently in mathematics and the physical sciences.

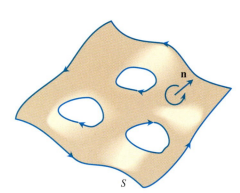

14.71 Stokes's theorem also holds for oriented surfaces with holes.

$$\text{curl grad } f = \mathbf{0} \qquad \text{or} \qquad \nabla \times \nabla f = \mathbf{0} \quad (12)$$

This identity holds for any function $f(x, y, z)$ whose second partial derivatives are continuous. The proof goes like this:

$$\nabla \times \nabla f = \begin{vmatrix} \mathbf{i} & \mathbf{j} & \mathbf{k} \\ \dfrac{\partial}{\partial x} & \dfrac{\partial}{\partial y} & \dfrac{\partial}{\partial z} \\ \dfrac{\partial f}{\partial x} & \dfrac{\partial f}{\partial y} & \dfrac{\partial f}{\partial z} \end{vmatrix} = (f_{zy} - f_{yz})\mathbf{i} - (f_{zx} - f_{xz})\mathbf{j} + (f_{yx} - f_{xy})\mathbf{k}.$$

If the second partial derivatives are continuous, the mixed second derivatives in parentheses are equal (Euler's theorem, Section 12.3) and the vector is zero.

Conservative Fields and Stokes's Theorem

In Section 14.3, we found that saying that a field $\mathbf{F}$ is conservative in an open region D in space is equivalent to saying that the integral of $\mathbf{F}$ around every closed loop in D is zero. This, in turn, is equivalent in *simply connected* open regions to saying that $\nabla \times \mathbf{F} = \mathbf{0}$. A region D is **simply connected** if every closed path in D can be contracted to a point in D without ever leaving D. If D consisted of space with a line removed, for example, D would not be simply connected. There would be no way to contract a loop around the line to a point without leaving D. On the other hand, space itself *is* simply connected (Fig. 14.72).

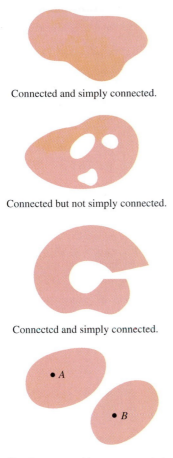

Connected and simply connected.

Connected but not simply connected.

Connected and simply connected.

• A

• B

Simply connected but not connected.
No path from A to B lies entirely in the region.

14.72 Connectivity and simple connectivity are not the same. Neither implies the other, as these pictures of plane regions illustrate. To make three-dimensional regions with these properties, thicken the plane regions into cylinders.

> ### Theorem 6
> If $\nabla \times \mathbf{F} = \mathbf{0}$ at every point of a simply connected open region D in space, then on any piecewise smooth closed path C in D,
> $$\oint_C \mathbf{F} \cdot d\mathbf{r} = 0.$$

Sketch of a Proof Theorem 6 is usually proved in two steps. The first step is for simple closed curves. A theorem from topology, a branch of advanced mathematics, states that every differentiable simple closed curve C in a simply connected open region D is the boundary of a smooth two-sided surface S that also lies in D. Hence, by Stokes's theorem,

$$\oint_C \mathbf{F} \cdot d\mathbf{r} = \iint_S \nabla \times \mathbf{F} \cdot \mathbf{n}\, d\sigma = 0.$$

The second step is for curves that cross themselves, like the one in Fig. 14.73. The idea is to break these into simple loops spanned by orientable surfaces, apply Stokes's theorem one loop at a time, and add the results. ❑

The following diagram summarizes the results for conservative fields defined on connected, simply connected open regions.

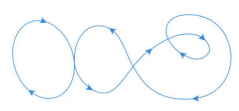

14.73 In a simply connected open region in space, differentiable curves that cross themselves can be divided into loops to which Stokes's theorem applies.

Exercises 14.7

Using Stokes's Theorem to Calculate Circulation

In Exercises 1–6, use the surface integral in Stokes's theorem to calculate the circulation of the field $\mathbf{F}$ around the curve C in the indicated direction.

1. $\mathbf{F} = x^2\mathbf{i} + 2x\mathbf{j} + z^2\mathbf{k}$
   C: The ellipse $4x^2 + y^2 = 4$ in the xy-plane, counterclockwise when viewed from above

2. $\mathbf{F} = 2y\mathbf{i} + 3x\mathbf{j} - z^2\mathbf{k}$
   C: The circle $x^2 + y^2 = 9$ in the xy-plane, counterclockwise when viewed from above

3. $\mathbf{F} = y\mathbf{i} + xz\mathbf{j} + x^2\mathbf{k}$
   C: The boundary of the triangle cut from the plane $x + y + z = 1$ by the first octant, counterclockwise when viewed from above

4. $\mathbf{F} = (y^2 + z^2)\mathbf{i} + (x^2 + z^2)\mathbf{j} + (x^2 + y^2)\mathbf{k}$
   C: The boundary of the triangle cut from the plane $x + y + z = 1$ by the first octant, counterclockwise when viewed from above

5. $\mathbf{F} = (y^2 + z^2)\mathbf{i} + (x^2 + y^2)\mathbf{j} + (x^2 + y^2)\mathbf{k}$
   C: The square bounded by the lines $x = \pm 1$ and $y = \pm 1$ in the xy-plane, counterclockwise when viewed from above

6. $\mathbf{F} = x^2y^3\mathbf{i} + \mathbf{j} + z\mathbf{k}$
   C: The intersection of the cylinder $x^2 + y^2 = 4$ and the hemisphere $x^2 + y^2 + z^2 = 16$, $z \geq 0$

Flux of the Curl

7. Let $\mathbf{n}$ be the outer unit normal of the elliptical shell
   $$S: \quad 4x^2 + 9y^2 + 36z^2 = 36, \quad z \geq 0,$$
   and let
   $$\mathbf{F} = y\mathbf{i} + x^2\mathbf{j} + (x^2 + y^4)^{3/2}\sin e^{\sqrt{xyz}}\mathbf{k}.$$
   Find the value of
   $$\iint_S \nabla \times \mathbf{F} \cdot \mathbf{n}\,d\sigma.$$

   (Hint: One parametrization of the ellipse at the base of the shell is $x = 3\cos t$, $y = 2\sin t$, $0 \leq t \leq 2\pi$.*)*

8. Let $\mathbf{n}$ be the outer unit normal (normal away from the origin) of the parabolic shell
   $$S: \quad 4x^2 + y + z^2 = 4, \quad y \geq 0$$
   and let
   $$\mathbf{F} = \left(-z + \frac{1}{2 + x}\right)\mathbf{i} + (\tan^{-1} y)\mathbf{j} + \left(x + \frac{1}{4 + z}\right)\mathbf{k}.$$
   Find the value of
   $$\iint_S \nabla \times \mathbf{F} \cdot \mathbf{n}\,d\sigma.$$

9. Let S be the cylinder $x^2 + y^2 = a^2$, $0 \leq z \leq h$, together with its top, $x^2 + y^2 \leq a^2$, $z = h$. Let $\mathbf{F} = -y\mathbf{i} + x\mathbf{j} + x^2\mathbf{k}$. Use Stokes's theorem to calculate the flux of $\nabla \times \mathbf{F}$ outward through S.

10. Evaluate
    $$\iint_S \nabla \times (y\mathbf{i}) \cdot \mathbf{n}\,d\sigma,$$
    where S is the hemisphere $x^2 + y^2 + z^2 = 1$, $z \geq 0$.

11. Show that
    $$\iint_S \nabla \times \mathbf{F} \cdot \mathbf{n}\,d\sigma$$
    has the same value for all oriented surfaces S that span C and that induce the same positive direction on C.

12. Let $\mathbf{F}$ be a differentiable vector field defined on a region containing a smooth closed oriented surface S and its interior. Let $\mathbf{n}$ be the unit normal vector field on S. Suppose that S is the union of two surfaces S_1 and S_2 joined along a smooth simple closed curve C. Can anything be said about
    $$\iint_S \nabla \times \mathbf{F} \cdot \mathbf{n}\,d\sigma?$$
    Give reasons for your answer.

Stokes's Theorem for Parametrized Surfaces

In Exercises 13–18, use the surface integral in Stokes's theorem to calculate the flux of the curl of the field $\mathbf{F}$ across the surface S in the direction of the outward unit normal $\mathbf{n}$.

13. $\mathbf{F} = 2z\mathbf{i} + 3x\mathbf{j} + 5y\mathbf{k}$
    S: $\mathbf{r}(r, \theta) = (r\cos\theta)\mathbf{i} + (r\sin\theta)\mathbf{j} + (4 - r^2)\mathbf{k}$, $0 \leq r \leq 2$, $0 \leq \theta \leq 2\pi$

14. $\mathbf{F} = (y - z)\mathbf{i} + (z - x)\mathbf{j} + (x + z)\mathbf{k}$
    S: $\mathbf{r}(r, \theta) = (r\cos\theta)\mathbf{i} + (r\sin\theta)\mathbf{j} + (9 - r^2)\mathbf{k}$, $0 \leq r \leq 3$, $0 \leq \theta \leq 2\pi$

15. $\mathbf{F} = x^2y\mathbf{i} + 2y^3z\mathbf{j} + 3z\mathbf{k}$
    S: $\mathbf{r}(r, \theta) = (r\cos\theta)\mathbf{i} + (r\sin\theta)\mathbf{j} + r\mathbf{k}$, $0 \leq r \leq 1$, $0 \leq \theta \leq 2\pi$

16. $\mathbf{F} = (x - y)\mathbf{i} + (y - z)\mathbf{j} + (z - x)\mathbf{k}$
    S: $\mathbf{r}(r, \theta) = (r\cos\theta)\mathbf{i} + (r\sin\theta)\mathbf{j} + (5 - r)\mathbf{k}$, $0 \leq r \leq 5$, $0 \leq \theta \leq 2\pi$

17. $\mathbf{F} = 3y\mathbf{i} + (5 - 2x)\mathbf{j} + (z^2 - 2)\mathbf{k}$
    S: $\mathbf{r}(\phi, \theta) = (\sqrt{3}\sin\phi\cos\theta)\mathbf{i} + (\sqrt{3}\sin\phi\sin\theta)\mathbf{j} + (\sqrt{3}\cos\phi)\mathbf{k}$, $0 \leq \phi \leq \pi/2$, $0 \leq \theta \leq 2\pi$

18. $\mathbf{F} = y^2\mathbf{i} + z^2\mathbf{j} + x\mathbf{k}$
    S: $\mathbf{r}(\phi, \theta) = (2\sin\phi\cos\theta)\mathbf{i} + (2\sin\phi\sin\theta)\mathbf{j} + (2\cos\phi)\mathbf{k}$, $0 \leq \phi \leq \pi/2$, $0 \leq \theta \leq 2\pi$

Theory and Examples

19. Use the identity $\nabla \times \nabla f = \mathbf{0}$ (Eq. 12 in the text) and Stokes's theorem to show that the circulations of the following fields around the boundary of any smooth orientable surface in space are zero.

   a) $\mathbf{F} = 2x\,\mathbf{i} + 2y\,\mathbf{j} + 2z\,\mathbf{k}$
   b) $\mathbf{F} = \nabla(xy^2z^3)$
   c) $\mathbf{F} = \nabla \times (x\,\mathbf{i} + y\,\mathbf{j} + z\,\mathbf{k})$
   d) $\mathbf{F} = \nabla f$

20. Let $f(x, y, z) = (x^2 + y^2 + z^2)^{-1/2}$. Show that the clockwise circulation of the field $\mathbf{F} = \nabla f$ around the circle $x^2 + y^2 = a^2$ in the xy-plane is zero

   a) by taking $\mathbf{r} = (a \cos t)\,\mathbf{i} + (a \sin t)\,\mathbf{j}, 0 \le t \le 2\pi$, and integrating $\mathbf{F} \cdot d\mathbf{r}$ over the circle, and
   b) by applying Stokes's theorem.

21. Let C be a simple closed smooth curve in the plane $2x + 2y + z = 2$, oriented as shown here. Show that

$$\oint_C 2y\,dx + 3z\,dy - x\,dz$$

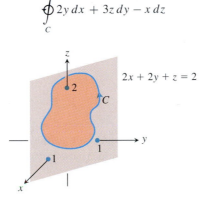

$2x + 2y + z = 2$

depends only on the area of the region enclosed by C and not on the position or shape of C.

22. Show that if $\mathbf{F} = x\,\mathbf{i} + y\,\mathbf{j} + z\,\mathbf{k}$, then $\nabla \times \mathbf{F} = \mathbf{0}$.

23. Find a vector field with twice-differentiable components whose curl is $x\,\mathbf{i} + y\,\mathbf{j} + z\,\mathbf{k}$ or prove that no such field exists.

24. Does Stokes's theorem say anything special about circulation in a field whose curl is zero? Give reasons for your answer.

25. Let R be a region in the xy-plane that is bounded by a piecewise smooth simple closed curve C, and suppose that the moments of inertia of R about the x- and y-axes are known to be I_x and I_y. Evaluate the integral

$$\oint_C \nabla(r^4) \cdot \mathbf{n}\,ds,$$

where $r = \sqrt{x^2 + y^2}$, in terms of I_x and I_y.

26. Show that the curl of

$$\mathbf{F} = \frac{-y}{x^2 + y^2}\,\mathbf{i} + \frac{x}{x^2 + y^2}\,\mathbf{j} + z\,\mathbf{k}$$

is zero but that

$$\oint_C \mathbf{F} \cdot d\mathbf{r}$$

is not zero if C is the circle $x^2 + y^2 = 1$ in the xy-plane. (Theorem 6 does not apply here because the domain of $\mathbf{F}$ is not simply connected. The field $\mathbf{F}$ is not defined along the z-axis so there is no way to contract C to a point without leaving the domain of $\mathbf{F}$.)

14.8

The Divergence Theorem and a Unified Theory

The divergence form of Green's theorem in the plane states that the net outward flux of a vector field across a simple closed curve can be calculated by integrating the divergence of the field over the region enclosed by the curve. The corresponding theorem in three dimensions, called the Divergence Theorem, states that the net outward flux of a vector field across a closed surface in space can be calculated by integrating the divergence of the field over the region enclosed by the surface. In this section, we prove the Divergence Theorem and show how it simplifies the calculation of flux. We also derive Gauss's law for flux in an electric field and the continuity equation of hydrodynamics. Finally, we unify the chapter's vector integral theorems into a single fundamental theorem.

Divergence in Three Dimensions

The **divergence** of a vector field $\mathbf{F} = M(x, y, z)\,\mathbf{i} + N(x, y, z)\,\mathbf{j} + P(x, y, z)\,\mathbf{k}$ is the scaler function

$$\operatorname{div} \mathbf{F} = \nabla \cdot \mathbf{F} = \frac{\partial M}{\partial x} + \frac{\partial N}{\partial y} + \frac{\partial P}{\partial z}. \tag{1}$$

The Divergence Theorem

Mikhail Vassilievich Ostrogradsky (1801–1862) was the first mathematician to publish a proof of the Divergence Theorem. Upon being denied his degree at Kharkhov University by the minister for religious affairs and national education (for atheism), Ostrogradsky left Russia for Paris in 1822, attracted by the presence of Laplace, Legendre, Fourier, Poisson, and Cauchy. While working on the theory of heat in the mid-1820s, he formulated the Divergence Theorem as a tool for converting volume integrals to surface integrals.

Carl Friedrich Gauss (1777–1855) had already proved the theorem while working on the theory of gravitation, but his notebooks were not to be published until many years later. (The theorem is sometimes called Gauss's theorem.) The list of Gauss's accomplishments in science and mathematics is truly astonishing, ranging from the invention of the electric telegraph (with Wilhelm Weber in 1833) to the development of a wonderfully accurate theory of planetary orbits and to work in non-Euclidean geometry that later became fundamental to Einstein's general theory of relativity.

The symbol "div **F**" is read as "divergence of **F**" or "div **F**." The notation $\nabla \cdot \mathbf{F}$ is read "del dot **F**."

Div **F** has the same physical interpretation in three dimensions that it does in two. If **F** is the velocity field of a fluid flow, the value of div **F** at a point (x, y, z) is the rate at which fluid is being piped in or drained away at (x, y, z). The divergence is the flux per unit volume or flux density at the point.

EXAMPLE 1 Find the divergence of $\mathbf{F} = 2xz\,\mathbf{i} - xy\,\mathbf{j} - z\,\mathbf{k}$.

Solution The divergence of **F** is

$$\nabla \cdot \mathbf{F} = \frac{\partial}{\partial x}(2xz) + \frac{\partial}{\partial y}(-xy) + \frac{\partial}{\partial z}(-z) = 2z - x - 1.$$

The Divergence Theorem

The Divergence Theorem says that under suitable conditions the outward flux of a vector field across a closed surface (oriented outward) equals the triple integral of the divergence of the field over the region enclosed by the surface.

Theorem 7

The Divergence Theorem

The flux of a vector field $\mathbf{F} = M\,\mathbf{i} + N\,\mathbf{j} + P\,\mathbf{k}$ across a closed oriented surface S in the direction of the surface's outward unit normal field **n** equals the integral of $\nabla \cdot \mathbf{F}$ over the region D enclosed by the surface:

$$\underset{S}{\iint} \mathbf{F} \cdot \mathbf{n}\,d\sigma = \underset{D}{\iiint} \nabla \cdot \mathbf{F}\,dV. \qquad (2)$$

$$\underset{\text{outward flux}}{\underbrace{}} \qquad \underset{\text{divergence integral}}{\underbrace{}}$$

EXAMPLE 2 Evaluate both sides of Eq. (2) for the field $\mathbf{F} = x\,\mathbf{i} + y\,\mathbf{j} + z\,\mathbf{k}$ over the sphere $x^2 + y^2 + z^2 = a^2$.

Solution The outer unit normal to S, calculated from the gradient of $f(x, y, z) = x^2 + y^2 + z^2 - a^2$, is

$$\mathbf{n} = \frac{2(x\,\mathbf{i} + y\,\mathbf{j} + z\,\mathbf{k})}{\sqrt{4(x^2 + y^2 + z^2)}} = \frac{x\,\mathbf{i} + y\,\mathbf{j} + z\,\mathbf{k}}{a}.$$

Hence

$$\mathbf{F} \cdot \mathbf{n}\,d\sigma = \frac{x^2 + y^2 + z^2}{a}\,d\sigma = \frac{a^2}{a}\,d\sigma = a\,d\sigma$$

because $x^2 + y^2 + z^2 = a^2$ on the surface. Therefore

$$\underset{S}{\iint} \mathbf{F} \cdot \mathbf{n}\,d\sigma = \underset{S}{\iint} a\,d\sigma = a \underset{S}{\iint} d\sigma = a\,(4\pi a^2) = 4\pi a^3.$$

The divergence of $\mathbf{F}$ is

$$\nabla \cdot \mathbf{F} = \frac{\partial}{\partial x}(x) + \frac{\partial}{\partial y}(y) + \frac{\partial}{\partial z}(z) = 3,$$

so

$$\iiint_D \nabla \cdot \mathbf{F}\, dV = \iiint_D 3\, dV = 3\left(\frac{4}{3}\pi a^3\right) = 4\pi a^3.$$

❑

EXAMPLE 3 Find the flux of $\mathbf{F} = xy\,\mathbf{i} + yz\,\mathbf{j} + xz\,\mathbf{k}$ outward through the surface of the cube cut from the first octant by the planes $x = 1$, $y = 1$, and $z = 1$.

Solution Instead of calculating the flux as a sum of six separate integrals, one for each face of the cube, we can calculate the flux by integrating the divergence

$$\nabla \cdot \mathbf{F} = \frac{\partial}{\partial x}(xy) + \frac{\partial}{\partial y}(yz) + \frac{\partial}{\partial z}(xz) = y + z + x$$

over the cube's interior:

$$\text{Flux} = \underset{\substack{\text{cube}\\\text{surface}}}{\iint} \mathbf{F} \cdot \mathbf{n}\, d\sigma = \underset{\substack{\text{cube}\\\text{interior}}}{\iiint} \nabla \cdot \mathbf{F}\, dV \qquad \text{The Divergence Theorem}$$

$$= \int_0^1 \int_0^1 \int_0^1 (x + y + z)\, dx\, dy\, dz = \frac{3}{2}. \qquad \text{Routine integration}$$

❑

Proof of the Divergence Theorem (Special Regions)

To prove the Divergence Theorem, we assume that the components of $\mathbf{F}$ have continuous first partial derivatives. We also assume that D is a convex region with no holes or bubbles, such as a solid sphere, cube, or ellipsoid, and that S is a piecewise smooth surface. In addition, we assume that any line perpendicular to the xy-plane at an interior point of the region R_{xy} that is the projection of D on the xy-plane intersects the surface S in exactly two points, producing surfaces

$$S_1: \quad z = f_1(x, y), \quad (x, y) \text{ in } R_{xy}$$
$$S_2: \quad z = f_2(x, y), \quad (x, y) \text{ in } R_{xy},$$

with $f_1 \le f_2$. We make similar assumptions about the projection of D onto the other coordinate planes. See Fig. 14.74.

The components of the unit normal vector $\mathbf{n} = n_1\,\mathbf{i} + n_2\,\mathbf{j} + n_3\,\mathbf{k}$ are the cosines of the angles α, β, and γ that $\mathbf{n}$ makes with $\mathbf{i}$, $\mathbf{j}$, and $\mathbf{k}$ (Fig. 14.75). This is true because all the vectors involved are unit vectors. We have

$$n_1 = \mathbf{n} \cdot \mathbf{i} = |\mathbf{n}|\,|\mathbf{i}| \cos \alpha = \cos \alpha$$
$$n_2 = \mathbf{n} \cdot \mathbf{j} = |\mathbf{n}|\,|\mathbf{j}| \cos \beta = \cos \beta$$
$$n_3 = \mathbf{n} \cdot \mathbf{k} = |\mathbf{n}|\,|\mathbf{k}| \cos \gamma = \cos \gamma.$$

Thus,

$$\mathbf{n} = (\cos \alpha)\,\mathbf{i} + (\cos \beta)\,\mathbf{j} + (\cos \gamma)\,\mathbf{k}$$

and

$$\mathbf{F} \cdot \mathbf{n} = M \cos \alpha + N \cos \beta + P \cos \gamma.$$

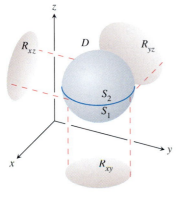

14.74 We first prove the Divergence Theorem for the kind of three-dimensional region shown here. We then extend the theorem to other regions.

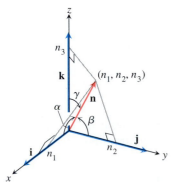

14.75 The scalar components of a unit normal vector $\mathbf{n}$ are the cosines of the angles α, β, and γ that it makes with $\mathbf{i}$, $\mathbf{j}$, and $\mathbf{k}$.

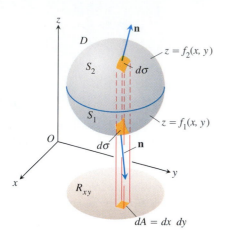

14.76 The three-dimensional region D enclosed by the surfaces S_1 and S_2 shown here projects vertically onto a two-dimensional region R_{xy} in the xy-plane.

In component form, the Divergence Theorem states that

$$\iint_S (M \cos \alpha + N \cos \beta + P \cos \gamma)\, d\sigma = \iiint_D \left(\frac{\partial M}{\partial x} + \frac{\partial N}{\partial y} + \frac{\partial P}{\partial z} \right) dx\, dy\, dz.$$

We prove the theorem by proving the three following equalities:

$$\iint_S M \cos \alpha\, d\sigma = \iiint_D \frac{\partial M}{\partial x}\, dx\, dy\, dz \qquad (3)$$

$$\iint_S N \cos \beta\, d\sigma = \iiint_D \frac{\partial N}{\partial y}\, dx\, dy\, dz \qquad (4)$$

$$\iint_S P \cos \gamma\, d\sigma = \iiint_D \frac{\partial P}{\partial z}\, dx\, dy\, dz \qquad (5)$$

We prove Eq. (5) by converting the surface integral on the left to a double integral over the projection R_{xy} of D on the xy-plane (Fig. 14.76). The surface S consists of an upper part S_2 whose equation is $z = f_2(x, y)$ and a lower part S_1 whose equation is $z = f_1(x, y)$. On S_2, the outer normal $\mathbf{n}$ has a positive $\mathbf{k}$-component and

$$\cos \gamma\, d\sigma = dx\, dy \quad \text{because} \quad d\sigma = \frac{dA}{|\cos \gamma|} = \frac{dx\, dy}{\cos \gamma}.$$

See Fig. 14.77. On S_1, the outer normal $\mathbf{n}$ has a negative $\mathbf{k}$-component and

$$\cos \gamma\, d\sigma = -dx\, dy.$$

Therefore,

$$\iint_S P \cos \gamma\, d\sigma = \iint_{S_2} P \cos \gamma\, d\sigma + \iint_{S_1} P \cos \gamma\, d\sigma$$

$$= \iint_{R_{xy}} P(x, y, f_2(x, y))\, dx\, dy - \iint_{R_{xy}} P(x, y, f_1(x, y))\, dx\, dy$$

$$= \iint_{R_{xy}} [P(x, y, f_2(x, y)) - P(x, y, f_1(x, y))]\, dx\, dy$$

$$= \iint_{R_{xy}} \left[\int_{f_1(x,y)}^{f_2(x,y)} \frac{\partial P}{\partial z}\, dz \right] dx\, dy = \iiint_D \frac{\partial P}{\partial z}\, dz\, dx\, dy.$$

This proves Eq. (5).

The proofs for Eqs. (3) and (4) follow the same pattern; or just permute $x, y, z;\ M, N, P;\ \alpha, \beta, \gamma$, in order, and get those results from Eq. (5).

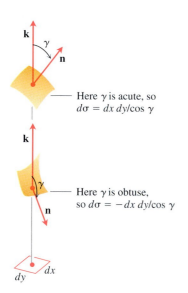

14.77 An enlarged view of the area patches in Fig. 14.76. The relations $d\sigma = \pm\, dx\, dy/\cos \gamma$ are derived in Section 14.5.

The Divergence Theorem for Other Regions

The Divergence Theorem can be extended to regions that can be partitioned into a finite number of simple regions of the type just discussed and to regions that can be defined as limits of simpler regions in certain ways. For example, suppose that D is the region between two concentric spheres and that $\mathbf{F}$ has continuously differentiable components throughout D and on the bounding surfaces. Split D by an equatorial plane and apply the Divergence Theorem to each half separately. The

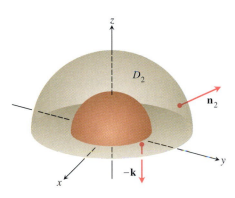

14.78 The lower half of the solid region between two concentric spheres.

14.79 The upper half of the solid region between two concentric spheres.

bottom half, D_1, is shown in Fig. 14.78. The surface that bounds D_1 consists of an outer hemisphere, a plane washer-shaped base, and an inner hemisphere. The Divergence Theorem says that

$$\iint_{S_1} \mathbf{F} \cdot \mathbf{n}_1 \, d\sigma_1 = \iiint_{D_1} \nabla \cdot \mathbf{F} \, dV_1. \tag{6}$$

The unit normal $\mathbf{n}_1$ that points outward from D_1 points away from the origin along the outer surface, equals $\mathbf{k}$ along the flat base, and points toward the origin along the inner surface. Next apply the Divergence Theorem to D_2, as shown in Fig. 14.79:

$$\iint_{S_2} \mathbf{F} \cdot \mathbf{n}_2 \, d\sigma_2 = \iiint_{D_2} \nabla \cdot \mathbf{F} \, dV_2. \tag{7}$$

As we follow $\mathbf{n}_2$ over S_2, pointing outward from D_2, we see that $\mathbf{n}_2$ equals $-\mathbf{k}$ along the washer-shaped base in the xy-plane, points away from the origin on the outer sphere, and points toward the origin on the inner sphere. When we add Eqs. (6) and (7), the integrals over the flat base cancel because of the opposite signs of $\mathbf{n}_1$ and $\mathbf{n}_2$. We thus arrive at the result

$$\iint_{S} \mathbf{F} \cdot \mathbf{n} \, d\sigma = \iiint_{D} \nabla \cdot \mathbf{F} \, dV,$$

with D the region between the spheres, S the boundary of D consisting of two spheres, and $\mathbf{n}$ the unit normal to S directed outward from D.

EXAMPLE 4   Find the net outward flux of the field

$$\mathbf{F} = \frac{x\,\mathbf{i} + y\,\mathbf{j} + z\,\mathbf{k}}{\rho^3}, \qquad \rho = \sqrt{x^2 + y^2 + z^2}$$

across the boundary of the region D: $0 < a^2 \le x^2 + y^2 + z^2 \le b^2$.

Solution   The flux can be calculated by integrating $\nabla \cdot \mathbf{F}$ over D. We have

$$\frac{\partial \rho}{\partial x} = \frac{1}{2}(x^2 + y^2 + z^2)^{-1/2}(2x) = \frac{x}{\rho}$$

and

$$\frac{\partial M}{\partial x} = \frac{\partial}{\partial x}(x\rho^{-3}) = \rho^{-3} - 3x\rho^{-4}\frac{\partial \rho}{\partial x} = \frac{1}{\rho^3} - \frac{3x^2}{\rho^5}.$$

Similarly,

$$\frac{\partial N}{\partial y} = \frac{1}{\rho^3} - \frac{3y^2}{\rho^5} \quad \text{and} \quad \frac{\partial P}{\partial z} = \frac{1}{\rho^3} - \frac{3z^2}{\rho^5}.$$

Hence,

$$\text{div } \mathbf{F} = \frac{3}{\rho^3} - \frac{3}{\rho^5}(x^2 + y^2 + z^2) = \frac{3}{\rho^3} - \frac{3\rho^2}{\rho^5} = 0$$

and

$$\iiint_{D} \nabla \cdot \mathbf{F} \, dV = 0.$$

So the integral of $\nabla \cdot \mathbf{F}$ over D is zero and the net outward flux across the boundary of D is zero. But there is more to learn from this example. The flux leaving D across the inner sphere S_a is the negative of the flux leaving D across the outer sphere S_b (because the sum of these fluxes is zero). This means that the flux of $\mathbf{F}$ across S_a in the direction away from the origin equals the flux of $\mathbf{F}$

across S_b in the direction away from the origin. Thus, the flux of $\mathbf{F}$ across a sphere centered at the origin is independent of the radius of the sphere. What is this flux?

To find it, we evaluate the flux integral directly. The outward unit normal on the sphere of radius a is

$$\mathbf{n} = \frac{x\,\mathbf{i} + y\,\mathbf{j} + z\,\mathbf{k}}{\sqrt{x^2 + y^2 + z^2}} = \frac{x\,\mathbf{i} + y\,\mathbf{j} + z\,\mathbf{k}}{a}.$$

Hence, on the sphere,

$$\mathbf{F} \cdot \mathbf{n} = \frac{x\,\mathbf{i} + y\,\mathbf{j} + z\,\mathbf{k}}{a^3} \cdot \frac{x\,\mathbf{i} + y\,\mathbf{j} + z\,\mathbf{k}}{a} = \frac{x^2 + y^2 + z^2}{a^4} = \frac{a^2}{a^4} = \frac{1}{a^2}$$

and

$$\iint_{S_a} \mathbf{F} \cdot \mathbf{n}\, d\sigma = \frac{1}{a^2} \iint_{S_a} d\sigma = \frac{1}{a^2}\,(4\pi a^2) = 4\pi.$$

The outward flux of $\mathbf{F}$ across any sphere centered at the origin is 4π. ❑

Gauss's Law—One of the Four Great Laws of Electromagnetic Theory

There is more to be learned from Example 4. In electromagnetic theory, the electric field created by a point charge q located at the origin is the inverse square field

$$\mathbf{E}(x, y, z) = \frac{1}{4\pi\,\epsilon_0} \frac{q}{|\mathbf{r}|^2}\left(\frac{\mathbf{r}}{|\mathbf{r}|}\right) = \frac{q}{4\pi\,\epsilon_0}\frac{\mathbf{r}}{|\mathbf{r}|^3} = \frac{q}{4\pi\,\epsilon_0}\frac{x\,\mathbf{i} + y\,\mathbf{j} + z\,\mathbf{k}}{\rho^3}$$

where ϵ_0 is a physical constant, $\mathbf{r}$ is the position vector of the point (x, y, z), and $\rho = |\mathbf{r}| = \sqrt{x^2 + y^2 + z^2}$. In the notation of Example 4,

$$\mathbf{E} = \frac{q}{4\pi\,\epsilon_0}\,\mathbf{F}.$$

The calculations in Example 4 show that the outward flux of $\mathbf{E}$ across any sphere centered at the origin is q/ϵ_0. But this result is not confined to spheres. The outward flux of $\mathbf{E}$ across any closed surface S that encloses the origin (and to which the Divergence Theorem applies) is also q/ϵ_0. To see why, we have only to imagine a large sphere S_a centered at the origin and enclosing the surface S. Since

$$\nabla \cdot \mathbf{E} = \nabla \cdot \frac{q}{4\pi\,\epsilon_0}\,\mathbf{F} = \frac{q}{4\pi\,\epsilon_0}\,\nabla \cdot \mathbf{F} = 0$$

when $\rho > 0$, the integral of $\nabla \cdot \mathbf{E}$ over the region D between S and S_a is zero. Hence, by the Divergence Theorem,

$$\iint_{\substack{\text{boundary}\\ \text{of } D}} \mathbf{E} \cdot \mathbf{n}\, d\sigma = 0,$$

and the flux of $\mathbf{E}$ across S in the direction away from the origin must be the same as the flux of $\mathbf{E}$ across S_a in the direction away from the origin, which is q/ϵ_0. This statement, called *Gauss's law*, also applies to charge distributions that are more general than the one assumed here, as you will see in nearly any physics text.

$$\text{Gauss's Law:}\qquad \iint_S \mathbf{E} \cdot \mathbf{n}\, d\sigma = \frac{q}{\epsilon_0}$$

The Continuity Equation of Hydrodynamics

Let D be a region in space bounded by a closed oriented surface S. If $\mathbf{v}(x, y, z)$ is the velocity field of a fluid flowing smoothly through D, $\delta = \delta(t, x, y, z)$ is the fluid's density at (x, y, z) at time t, and $\mathbf{F} = \delta \mathbf{v}$, then the **continuity equation** of hydrodynamics states that

$$\nabla \cdot \mathbf{F} + \frac{\partial \delta}{\partial t} = 0.$$

If the functions involved have continuous first partial derivatives, the equation evolves naturally from the Divergence Theorem, as we will now see.

First, the integral

$$\iint\limits_S \mathbf{F} \cdot \mathbf{n} \, d\sigma$$

14.80 The fluid that flows upward through the patch $\Delta\sigma$ in a short time Δt fills a "cylinder" whose volume is approximately base × height $= \mathbf{v} \cdot \mathbf{n} \, \Delta\sigma \, \Delta t$.

is the rate at which mass leaves D across S (leaves because $\mathbf{n}$ is the outer normal). To see why, consider a patch of area $\Delta\sigma$ on the surface (Fig. 14.80). In a short time interval Δt, the volume ΔV of fluid that flows across the patch is approximately equal to the volume of a cylinder with base area $\Delta\sigma$ and height $(\mathbf{v}\,\Delta t) \cdot \mathbf{n}$, where $\mathbf{v}$ is a velocity vector rooted at a point of the patch:

$$\Delta V \approx \mathbf{v} \cdot \mathbf{n}\,\Delta\sigma\,\Delta t.$$

The mass of this volume of fluid is about

$$\Delta m \approx \delta \mathbf{v} \cdot \mathbf{n}\,\Delta\sigma\,\Delta t,$$

so the rate at which mass is flowing out of D across the patch is about

$$\frac{\Delta m}{\Delta t} \approx \delta \mathbf{v} \cdot \mathbf{n}\,\Delta\sigma.$$

This leads to the approximation

$$\frac{\Sigma\,\Delta m}{\Delta t} \approx \sum \delta \mathbf{v} \cdot \mathbf{n}\,\Delta\sigma$$

as an estimate of the average rate at which mass flows across S. Finally, letting $\Delta\sigma \to 0$ and $\Delta t \to 0$ gives the instantaneous rate at which mass leaves D across S as

$$\frac{dm}{dt} = \iint\limits_S \delta \mathbf{v} \cdot \mathbf{n}\,d\sigma,$$

which for our particular flow is

$$\frac{dm}{dt} = \iint\limits_S \mathbf{F} \cdot \mathbf{n}\,d\sigma,$$

Now let B be a solid sphere centered at a point Q in the flow. The average value of $\nabla \cdot \mathbf{F}$ over B is

$$\frac{1}{\text{volume of } B} \iiint\limits_B \nabla \cdot \mathbf{F}\,dV.$$

It is a consequence of the continuity of the divergence that $\nabla \cdot \mathbf{F}$ actually takes on this value at some point P in B. Thus,

$$(\nabla \cdot \mathbf{F})_P = \frac{1}{\text{volume of } B} \iiint_B \nabla \cdot \mathbf{F} \, dV = \frac{\iint_S \mathbf{F} \cdot \mathbf{n} \, d\sigma}{\text{volume of } B}$$

$$= \frac{\text{rate at which mass leaves } B \text{ across its surface } S}{\text{volume of } B}. \qquad (8)$$

The fraction on the right describes decrease in mass per unit volume.

Now let the radius of B approach zero while the center Q stays fixed. The left-hand side of Eq. (8) converges to $(\nabla \cdot \mathbf{F})_Q$, the right side to $(-\partial \delta / \partial t)_Q$. The equality of these two limits is the continuity equation

$$\nabla \cdot \mathbf{F} = -\frac{\partial \delta}{\partial t}.$$

The continuity equation "explains" $\nabla \cdot \mathbf{F}$: The divergence of $\mathbf{F}$ at a point is the rate at which the density of the fluid is decreasing there.

The Divergence Theorem

$$\iint_S \mathbf{F} \cdot \mathbf{n} \, d\sigma = \iiint_D \nabla \cdot \mathbf{F} \, dV$$

now says that the net decrease in density of the fluid in region D is accounted for by the mass transported across the surface S. In a way, the theorem is a statement about conservation of mass.

Unifying the Integral Theorems

If we think of a two-dimensional field $\mathbf{F} = M(x, y)\mathbf{i} + N(x, y)\mathbf{j}$ as a three-dimensional field whose $\mathbf{k}$-component is zero, then $\nabla \cdot \mathbf{F} = (\partial M / \partial x) + (\partial N / \partial y)$ and the normal form of Green's theorem can be written as

$$\oint_C \mathbf{F} \cdot \mathbf{n} \, ds = \iint_R \left(\frac{\partial M}{\partial x} + \frac{\partial N}{\partial y} \right) dx \, dy = \iint_R \nabla \cdot \mathbf{F} \, dA.$$

Similarly, $\nabla \times \mathbf{F} \cdot \mathbf{k} = (\partial N / \partial x) - (\partial M / \partial y)$, so the tangential form of Green's theorem can be written as

$$\oint_C \mathbf{F} \cdot d\mathbf{r} = \iint_R \left(\frac{\partial N}{\partial x} - \frac{\partial M}{\partial y} \right) dx \, dy = \iint_R \nabla \times \mathbf{F} \cdot \mathbf{k} \, dA.$$

With the equations of Green's theorem now in del notation, we can see their relationships to the equations in Stokes's theorem and the Divergence Theorem.

Green's Theorem and Its Generalization to Three Dimensions

Normal form of
Green's theorem:
$$\oint_C \mathbf{F} \cdot \mathbf{n} \, ds = \iint_R \nabla \cdot \mathbf{F} \, dA$$

Divergence Theorem:
$$\iint_S \mathbf{F} \cdot \mathbf{n} \, d\sigma = \iiint_D \nabla \cdot \mathbf{F} \, dV$$

> Tangential form of Green's theorem:
> $$\oint_C \mathbf{F} \cdot d\mathbf{r} = \iint_R \nabla \times \mathbf{F} \cdot \mathbf{k} \, dA$$
>
> Stokes's theorem:
> $$\oint_C \mathbf{F} \cdot d\mathbf{r} = \iint_S \nabla \times \mathbf{F} \cdot \mathbf{n} \, d\sigma$$

Notice how Stokes's theorem generalizes the tangential (curl) form of Green's theorem from a flat surface in the plane to a surface in three-dimensional space. In each case, the integral of the normal component of curl $\mathbf{F}$ over the interior of the surface equals the circulation of $\mathbf{F}$ around the boundary.

Likewise, the Divergence Theorem generalizes the normal (flux) form of Green's theorem from a two-dimensional region in the plane to a three-dimensional region in space. In each case, the integral of $\nabla \cdot \mathbf{F}$ over the interior of the region equals the total flux of the field across the boundary.

There is still more to be learned here. All of these results can be thought of as forms of a *single fundamental theorem*. Think back to the Fundamental Theorem of Calculus in Section 4.7. It says that if $f(x)$ is differentiable on $[a, b]$ then

$$\int_a^b \frac{df}{dx}\, dx = f(b) - f(a).$$

$$\mathbf{n} = -\mathbf{i} \qquad \mathbf{n} = \mathbf{i}$$

14.81 The outward unit normals at the boundary of $[a, b]$ in one-dimensional space.

If we let $\mathbf{F} = f(x)\mathbf{i}$ throughout $[a, b]$, then $(df/dx) = \nabla \cdot \mathbf{F}$. If we define the unit vector $\mathbf{n}$ normal to the boundary of $[a, b]$ to be $\mathbf{i}$ at b and $-\mathbf{i}$ at a (Fig. 14.81) then

$$f(b) - f(a) = f(b)\mathbf{i} \cdot (\mathbf{i}) + f(a)\mathbf{i} \cdot (-\mathbf{i})$$

$$= \mathbf{F}(b) \cdot \mathbf{n} + \mathbf{F}(a) \cdot \mathbf{n}$$

$$= \text{total outward flux of } \mathbf{F} \text{ across the boundary of } [a, b].$$

The Fundamental Theorem now says that

$$\int_{[a,\, b]} \nabla \cdot \mathbf{F} \, dx = \text{total outward flux of } \mathbf{F} \text{ across the boundary.}$$

The Fundamental Theorem of Calculus, the flux form of Green's theorem, and the Divergence Theorem all say that the integral of the differential operator $\nabla \cdot$ operating on a field $\mathbf{F}$ over a region equals the sum of the normal field components over the boundary of the region.

Stokes's theorem and the circulation form of Green's theorem say that, when things are properly oriented, the integral of the normal component of the curl operating on a field equals the sum of the tangential field components on the boundary of the surface.

The beauty of these interpretations is the observance of a marvelous underlying principle, which we might state as follows.

> The integral of a differential operator acting on a field over a region equals the sum of the field components appropriate to the operator over the boundary of the region.

Exercises 14.8

Calculating Divergence

In Exercises 1–4, find the divergence of the field.

1. The spin field in Fig. 14.15.

2. The radial field in Fig. 14.14.

3. The gravitational field in Fig. 14.13.

4. The velocity field in Fig. 14.10.

Using the Divergence Theorem to Calculate Outward Flux

In Exercises 5–16, use the Divergence Theorem to find the outward flux of **F** across the boundary of the region D.

5. $\mathbf{F} = (y - x)\mathbf{i} + (z - y)\mathbf{j} + (y - x)\mathbf{k}$
   D: The cube bounded by the planes $x = \pm 1$, $y = \pm 1$, and $z = \pm 1$

6. $\mathbf{F} = x^2\mathbf{i} + y^2\mathbf{j} + z^2\mathbf{k}$
   a) D: The cube cut from the first octant by the planes $x = 1$, $y = 1$, and $z = 1$
   b) D: The cube bounded by the planes $x = \pm 1$, $y = \pm 1$, and $z = \pm 1$
   c) D: The region cut from the solid cylinder $x^2 + y^2 \le 4$ by the planes $z = 0$ and $z = 1$

7. $\mathbf{F} = y\mathbf{i} + xy\mathbf{j} - z\mathbf{k}$
   D: The region inside the solid cylinder $x^2 + y^2 \le 4$ between the plane $z = 0$ and the paraboloid $z = x^2 + y^2$

8. $\mathbf{F} = x^2\mathbf{i} + xz\mathbf{j} + 3z\mathbf{k}$
   D: the solid sphere $x^2 + y^2 + z^2 \le 4$

9. $\mathbf{F} = x^2\mathbf{i} - 2xy\mathbf{j} + 3xz\mathbf{k}$
   D: The region cut from the first octant by the sphere $x^2 + y^2 + z^2 = 4$

10. $\mathbf{F} = (6x^2 + 2xy)\mathbf{i} + (2y + x^2z)\mathbf{j} + 4x^2y^3\mathbf{k}$
    D: The region cut from the first octant by the cylinder $x^2 + y^2 = 4$ and the plane $z = 3$

11. $\mathbf{F} = 2xz\mathbf{i} - xy\mathbf{j} - z^2\mathbf{k}$
    D: The wedge cut from the first octant by the plane $y + z = 4$ and the elliptical cylinder $4x^2 + y^2 = 16$

12. $\mathbf{F} = x^3\mathbf{i} + y^3\mathbf{j} + z^3\mathbf{k}$
    D: The solid sphere $x^2 + y^2 + z^2 \le a^2$

13. $\mathbf{F} = \sqrt{x^2 + y^2 + z^2}\,(x\mathbf{i} + y\mathbf{j} + z\mathbf{k})$
    D: The region $1 \le x^2 + y^2 + z^2 \le 2$

14. $\mathbf{F} = (x\mathbf{i} + y\mathbf{j} + z\mathbf{k})/\sqrt{x^2 + y^2 + z^2}$
    D: The region $1 \le x^2 + y^2 + z^2 \le 4$

15. $\mathbf{F} = (5x^3 + 12xy^2)\mathbf{i} + (y^3 + e^y \sin z)\mathbf{j} + (5z^3 + e^y \cos z)\mathbf{k}$
    D: The solid region between the spheres $x^2 + y^2 + z^2 = 1$ and $x^2 + y^2 + z^2 = 2$

16. $\mathbf{F} = \ln(x^2 + y^2)\mathbf{i} - \left(\dfrac{2z}{x}\tan^{-1}\dfrac{y}{x}\right)\mathbf{j} + z\sqrt{x^2 + y^2}\,\mathbf{k}$
    D: The thick-walled cylinder $1 \le x^2 + y^2 \le 2$, $-1 \le z \le 2$

Properties of Curl and Divergence

17. div (curl **G**) = 0
    a) Show that if the necessary partial derivatives of the components of the field $\mathbf{G} = M\mathbf{i} + N\mathbf{j} + P\mathbf{k}$ are continuous, then $\nabla \cdot \nabla \times \mathbf{G} = 0$.
    b) What, if anything, can you conclude about the flux of the field $\nabla \times \mathbf{G}$ across a closed surface? Give reasons for your answer.

18. Let $\mathbf{F}_1$ and $\mathbf{F}_2$ be differentiable vector fields, and let a and b be arbitrary real constants. Verify the following identities.
    a) $\nabla \cdot (a\mathbf{F}_1 + b\mathbf{F}_2) = a\nabla \cdot \mathbf{F}_1 + b\nabla \cdot \mathbf{F}_2$
    b) $\nabla \times (a\mathbf{F}_1 + b\mathbf{F}_2) = a\nabla \times \mathbf{F}_1 + b\nabla \times \mathbf{F}_2$
    c) $\nabla \cdot (\mathbf{F}_1 \times \mathbf{F}_2) = \mathbf{F}_2 \cdot \nabla \times \mathbf{F}_1 - \mathbf{F}_1 \cdot \nabla \times \mathbf{F}_2$

19. Let **F** be a differentiable vector field and let $g(x, y, z)$ be a differentiable scalar function. Verify the following identities.
    a) $\nabla \cdot (g\mathbf{F}) = g\nabla \cdot \mathbf{F} + \nabla g \cdot \mathbf{F}$
    b) $\nabla \times (g\mathbf{F}) = g\nabla \times \mathbf{F} + \nabla g \times \mathbf{F}$

20. If $\mathbf{F} = M\mathbf{i} + N\mathbf{j} + P\mathbf{k}$ is a differentiable vector field, we define the notation $\mathbf{F} \cdot \nabla$ to mean

    $$M\frac{\partial}{\partial x} + N\frac{\partial}{\partial y} + P\frac{\partial}{\partial z}.$$

    For differentiable vector fields $\mathbf{F}_1$ and $\mathbf{F}_2$ verify the following identities.
    a) $\nabla \times (\mathbf{F}_1 \times \mathbf{F}_2) = (\mathbf{F}_2 \cdot \nabla)\mathbf{F}_1 - (\mathbf{F}_1 \cdot \nabla)\mathbf{F}_2 + (\nabla \cdot \mathbf{F}_2)\mathbf{F}_1 - (\nabla \cdot \mathbf{F}_1)\mathbf{F}_2$
    b) $\nabla(\mathbf{F}_1 \cdot \mathbf{F}_2) = (\mathbf{F}_1 \cdot \nabla)\mathbf{F}_2 + (\mathbf{F}_2 \cdot \nabla)\mathbf{F}_1 + \mathbf{F}_1 \times (\nabla \times \mathbf{F}_2) + \mathbf{F}_2 \times (\nabla \times \mathbf{F}_1)$

Theory and Examples

21. Let **F** be a field whose components have continuous first partial derivatives throughout a portion of space containing a region D bounded by a smooth closed surface S. If $|\mathbf{F}| \le 1$, can any bound be placed on the size of

    $$\iiint_D \nabla \cdot \mathbf{F}\, dV?$$

    Give reasons for your answer.

22. The base of the closed cubelike surface shown here is the unit square in the xy-plane. The four sides lie in the planes $x = 0$, $x = 1$, $y = 0$, and $y = 1$. The top is an arbitrary smooth surface whose identity is unknown. Let $\mathbf{F} = x\mathbf{i} - 2y\mathbf{j} + (z + 3)\mathbf{k}$, and suppose the outward flux of **F** through side A is 1 and through side B is -3. Can you conclude anything about the outward flux through the top? Give reasons for your answer.

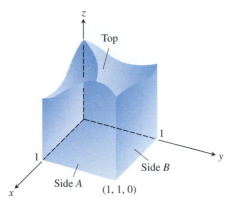

23. a) Show that the flux of the position vector field $\mathbf{F} = x\,\mathbf{i} + y\,\mathbf{j} + z\,\mathbf{k}$ outward through a smooth closed surface S is three times the volume of the region enclosed by the surface.

b) Let $\mathbf{n}$ be the outward unit normal vector field on S. Show that it is not possible for $\mathbf{F}$ to be orthogonal to $\mathbf{n}$ at every point of S.

24. Among all rectangular solids defined by the inequalities $0 \le x \le a, 0 \le y \le b, 0 \le z \le 1$, find the one for which the total flux of $\mathbf{F} = (-x^2 - 4xy)\,\mathbf{i} - 6yz\,\mathbf{j} + 12z\,\mathbf{k}$ outward through the six sides is greatest. What *is* the greatest flux?

25. Let $\mathbf{F} = x\,\mathbf{i} + y\,\mathbf{j} + z\,\mathbf{k}$ and suppose that the surface S and region D satisfy the hypotheses of the Divergence Theorem. Show that the volume of D is given by the formula

$$\text{Volume of } D = \frac{1}{3} \iint_S \mathbf{F} \cdot \mathbf{n}\,d\sigma.$$

26. Show that the outward flux of a constant vector field $\mathbf{F} = \mathbf{C}$ across any closed surface to which the Divergence Theorem applies is zero.

27. *Harmonic functions.* A function $f(x, y, z)$ is said to be **harmonic** in a region D in space if it satisfies the Laplace equation

$$\nabla^2 f = \nabla \cdot \nabla f = \frac{\partial^2 f}{\partial x^2} + \frac{\partial^2 f}{\partial y^2} + \frac{\partial^2 f}{\partial z^2} = 0$$

throughout D.

a) Suppose that f is harmonic throughout a bounded region D enclosed by a smooth surface S and that $\mathbf{n}$ is the chosen unit normal vector on S. Show that the integral over S of $\nabla f \cdot \mathbf{n}$, the derivative of f in the direction of $\mathbf{n}$, is zero.

b) Show that if f is harmonic on D, then

$$\iint_S f\,\nabla f \cdot \mathbf{n}\,d\sigma = \iiint_D |\nabla f|^2\,dV.$$

28. Let S be the surface of the portion of the solid sphere $x^2 + y^2 + z^2 \le a^2$ that lies in the first octant and let $f(x, y, z) = \ln\sqrt{x^2 + y^2 + z^2}$. Calculate

$$\iint_S \nabla f \cdot \mathbf{n}\,d\sigma.$$

($\nabla f \cdot \mathbf{n}$ is the derivative of f in the direction of $\mathbf{n}$.)

29. *Green's first formula.* Suppose that f and g are scalar functions with continuous first- and second-order partial derivatives throughout a closed region D that is bounded by a piecewise smooth surface S. Show that

$$\iint_S f\,\nabla g \cdot \mathbf{n}\,d\sigma = \iiint_D (f\,\nabla^2 g + \nabla f \cdot \nabla g)\,dV. \qquad (9)$$

Equation (9) is **Green's first formula.** (*Hint:* Apply the Divergence Theorem to the field $\mathbf{F} = f\,\nabla g$.)

30. *Green's second formula.* (*Continuation of Exercise 29*). Interchange f and g in Eq. (9) to obtain a similar formula. Then subtract this formula from Eq. (9) to show that

$$\iint_S (f\,\nabla g - g\,\nabla f) \cdot \mathbf{n}\,d\sigma$$

$$= \iiint_D (f\,\nabla^2 g - g\,\nabla^2 f)\,dV. \qquad (10)$$

This equation is **Green's second formula.**

31. *Conservation of mass.* Let $\mathbf{v}(t, x, y, z)$ be a continuously differentiable vector field over the region D in space and let $p(t, x, y, z)$ be a continuously differentiable scalar function. The variable t represents the time domain. The Law of Conservation of Mass asserts that

$$\frac{d}{dt} \iiint_D p(t, x, y, z)\,dV = -\iint_S p\,\mathbf{v} \cdot \mathbf{n}\,d\sigma,$$

where S is the surface enclosing D.

a) Give a physical interpretation of the conservation of mass law if $\mathbf{v}$ is a velocity flow field and p represents the density of the fluid at point (x, y, z) at time t.

b) Use the Divergence Theorem and Leibniz's rule,

$$\frac{d}{dt} \iiint_D p(t, x, y, z)\,dV = \iiint_D \frac{\partial p}{\partial t}\,dV,$$

to show that the Law of Conservation of Mass is equivalent to the continuity equation,

$$\nabla \cdot p\,\mathbf{v} + \frac{\partial p}{\partial t} = 0.$$

(In the first term $\nabla \cdot p\,\mathbf{v}$ the variable t is held fixed and in the second term $\partial p / \partial t$ it is assumed that the point (x, y, z) in D is held fixed.)

32. *General diffusion equation.* Let $T(t, x, y, z)$ be a function with continuous second derivatives giving the temperature at time t at the point (x, y, z) of a solid occupying a region D in space. If the solid's specific heat and mass density are denoted by the constants c and ρ respectively, the quantity $c\rho T$ is called the solid's **heat energy per unit volume.**

a) Explain why $-\nabla T$ points in the direction of heat flow.

b) Let $-k\nabla T$ denote the **energy flux vector.** (Here the constant k is called the **conductivity.**) Assuming the Law of

Conservation of Mass with $-k\nabla T = \mathbf{v}$ and $c\rho T = p$ in Exercise 31, derive the diffusion (heat) equation

$$\frac{\partial T}{\partial t} = K\nabla^2 T,$$

where $K = k/(c\rho) > 0$ is the *diffusivity* constant. (Notice

that if $T(t, x)$ represents the temperature at time t at position x in a uniform conducting rod with perfectly insulated sides, then $\nabla^2 T = \partial^2 T/\partial x^2$ and the diffusion equation reduces to the one-dimensional heat equation in the Chapter 12 Additional Exercises.)

CHAPTER **14**   QUESTIONS TO GUIDE YOUR REVIEW

1. What are line integrals? How are they evaluated? Give examples.

2. How can you use line integrals to find the centers of mass of springs? Explain.

3. What is a vector field? a gradient field? Give examples.

4. How do you calculate the work done by a force in moving a particle along a curve? Give an example.

5. What are flow, circulation, and flux?

6. What is special about path independent fields?

7. How can you tell when a field is conservative?

8. What is a potential function? Show by example how to find a potential function for a conservative field.

9. What is a differential form? What does it mean for such a form to be exact? How do you test for exactness? Give examples.

10. What is the divergence of a vector field? How can you interpret it?

11. What is the curl of a vector field? How can you interpret it?

12. What is Green's theorem? How can you interpret it?

13. How do you calculate the area of a curved surface in space? Give an example.

14. What is an oriented surface? How do you calculate the flux of a three-dimensional vector field across an oriented surface? Give an example.

15. What are surface integrals? What can you calculate with them? Give an example.

16. What is a parametrized surface? How do you find the area of such a surface? Give examples.

17. How do you integrate a function over a parametrized surface? Give an example.

18. What is Stokes's theorem? How can you interpret it?

19. Summarize the chapter's results on conservative fields.

20. What is the Divergence Theorem? How can you interpret it?

21. How does the Divergence Theorem generalize Green's theorem?

22. How does Stokes's theorem generalize Green's theorem?

23. How can Green's theorem, Stokes's theorem, and the Divergence Theorem be thought of as forms of a single fundamental theorem?

CHAPTER **14**   PRACTICE EXERCISES

Evaluating Line Integrals

1. Figure 14.82 shows two polygonal paths in space joining the origin to the point $(1, 1, 1)$. Integrate $f(x, y, z) = 2x - 3y^2 - 2z + 3$ over each path.

2. Figure 14.83 shows three polygonal paths joining the origin to the point $(1, 1, 1)$. Integrate $f(x, y, z) = x^2 + y - z$ over each path.

3. Integrate $f(x, y, z) = \sqrt{x^2 + z^2}$ over the circle

$$\mathbf{r}(t) = (a\cos t)\mathbf{j} + (a\sin t)\mathbf{k}, \quad 0 \le t \le 2\pi.$$

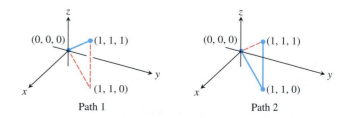

14.82 The paths in Exercise 1.

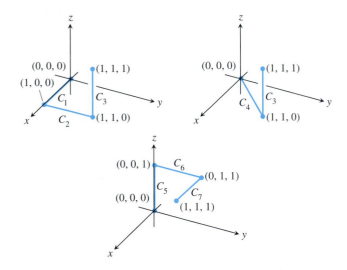

14.83 The paths in Exercise 2.

4. Integrate $f(x, y, z) = \sqrt{x^2 + y^2}$ over the involute curve
$\mathbf{r}(t) = (\cos t + t \sin t)\mathbf{i} + (\sin t - t \cos t)\mathbf{j}, \ 0 \le t \le \sqrt{3}$.
(See Fig. 11.20.)

Evaluate the integrals in Exercises 5 and 6.

5. $\displaystyle\int_{(-1,1,1)}^{(4,-3,0)} \frac{dx + dy + dz}{\sqrt{x + y + z}}$

6. $\displaystyle\int_{(1,1,1)}^{(10,3,3)} dx - \sqrt{\frac{z}{y}}\, dy - \sqrt{\frac{y}{z}}\, dz$

7. Integrate $\mathbf{F} = -(y \sin z)\mathbf{i} + (x \sin z)\mathbf{j} + (xy \cos z)\mathbf{k}$ around the circle cut from the sphere $x^2 + y^2 + z^2 = 5$ by the plane $z = -1$, clockwise as viewed from above.

8. Integrate $\mathbf{F} = 3x^2 y\mathbf{i} + (x^3 + 1)\mathbf{j} + 9z^2\mathbf{k}$ around the circle cut from the sphere $x^2 + y^2 + z^2 = 9$ by the plane $x = 2$.

Evaluate the line integrals in Exercises 9 and 10.

9. $\displaystyle\int_C 8x \sin y \, dx - 8y \cos x \, dy$

C is the square cut from the first quadrant by the lines $x = \pi/2$ and $y = \pi/2$.

10. $\displaystyle\int_C y^2 \, dx + x^2 \, dy$

C is the circle $x^2 + y^2 = 4$.

Evaluating Surface Integrals

11. Find the area of the elliptical region cut from the plane $x + y + z = 1$ by the cylinder $x^2 + y^2 = 1$.

12. Find the area of the cap cut from the paraboloid $y^2 + z^2 = 3x$ by the plane $x = 1$.

13. Find the area of the cap cut from the top of the sphere $x^2 + y^2 + z^2 = 1$ by the plane $z = \sqrt{2}/2$.

14. a) Find the area of the surface cut from the hemisphere $x^2 + y^2 + z^2 = 4, z \ge 0$, by the cylinder $x^2 + y^2 = 2x$.
b) Find the area of the portion of the cylinder that lies inside the hemisphere. (*Hint:* Project onto the xz-plane. Or evaluate the integral $\int h \, ds$, where h is the altitude of the cylinder and ds is the element of arc length on the circle $x^2 + y^2 = 2x$ in the xy-plane.)

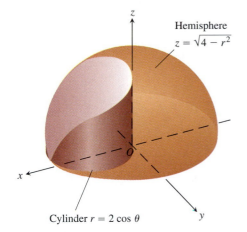

15. Find the area of the triangle in which the plane $(x/a) + (y/b) + (z/c) = 1 \ (a, b, c > 0)$ intersects the first octant. Check your answer with an appropriate vector calculation.

16. Integrate

a) $g(x, y, z) = \dfrac{yz}{\sqrt{4y^2 + 1}}$ **b)** $g(x, y, z) = \dfrac{z}{\sqrt{4y^2 + 1}}$

over the surface cut from the parabolic cylinder $y^2 - z = 1$ by the planes $x = 0, x = 3$, and $z = 0$.

17. Integrate $g(x, y, z) = x^4 y(y^2 + z^2)$ over the portion of the cylinder $y^2 + z^2 = 25$ that lies in the first octant between the planes $x = 0$ and $x = 1$ and above the plane $z = 3$.

18. CALCULATOR The state of Wyoming is bounded by the meridians $111°\ 3'$ and $104°\ 3'$ west longitude and by the circles $41°$ and $45°$ north latitude. Assuming that the earth is a sphere of radius $R = 3959$ mi, find the area of Wyoming.

Parametrized Surfaces

Find the parametrizations for the surfaces in Exercises 19–24. (There are many ways to do these, so your answers may not be the same as those in the back of the book.)

19. The portion of the sphere $x^2 + y^2 + z^2 = 36$ between the planes $z = -3$ and $z = 3\sqrt{3}$

20. The portion of the paraboloid $z = -(x^2 + y^2)/2$ above the plane $z = -2$

21. The cone $z = 1 + \sqrt{x^2 + y^2}, z \le 3$

22. The portion of the plane $4x + 2y + 4z = 12$ that lies above the square $0 \le x \le 2, 0 \le y \le 2$ in the first quadrant

23. The portion of the paraboloid $y = 2(x^2 + z^2), y \le 2$, that lies above the xy-plane

24. The portion of the hemisphere $x^2 + y^2 + z^2 = 10$, $y \geq 0$, in the first octant

25. Find the area of the surface

$$\mathbf{r}(u, v) = (u + v)\mathbf{i} + (u - v)\mathbf{j} + v\mathbf{k}, \quad 0 \leq u \leq 1, \quad 0 \leq v \leq 1.$$

26. Integrate $f(x, y, z) = xy - z^2$ over the surface in Exercise 25.

27. Find the surface area of the helicoid

$$\mathbf{r}(r, \theta) = r \cos \theta \, \mathbf{i} + r \sin \theta \, \mathbf{j} + \theta \mathbf{k}, \quad 0 \leq \theta \leq 2\pi \text{ and } 0 \leq r \leq 1,$$

in the accompanying figure.

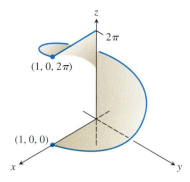

28. Evaluate the integral $\iint_S \sqrt{x^2 + y^2 + 1} \, d\sigma$, where S is the helicoid in Exercise 27.

Conservative Fields

Which of the fields in Exercises 29–32 are conservative, and which are not?

29. $\mathbf{F} = x\mathbf{i} + y\mathbf{j} + z\mathbf{k}$

30. $\mathbf{F} = (x\mathbf{i} + y\mathbf{j} + z\mathbf{k})/(x^2 + y^2 + z^2)^{3/2}$

31. $\mathbf{F} = x e^y \mathbf{i} + y e^z \mathbf{j} + z e^x \mathbf{k}$

32. $\mathbf{F} = (\mathbf{i} + z\mathbf{j} + y\mathbf{k})/(x + yz)$

Find potential functions for the fields in Exercises 33 and 34.

33. $\mathbf{F} = 2\mathbf{i} + (2y + z)\mathbf{j} + (y + 1)\mathbf{k}$

34. $\mathbf{F} = (z \cos xz)\mathbf{i} + e^y \mathbf{j} + (x \cos xz)\mathbf{k}$

Work and Circulation

In Exercises 35 and 36, find the work done by each field along the paths from $(0, 0, 0)$ to $(1, 1, 1)$ in Fig. 14.82.

35. $\mathbf{F} = 2xy\mathbf{i} + \mathbf{j} + x^2\mathbf{k}$ **36.** $\mathbf{F} = 2xy\mathbf{i} + x^2\mathbf{j} + \mathbf{k}$

37. Find the work done by

$$\mathbf{F} = \frac{x\mathbf{i} + y\mathbf{j}}{(x^2 + y^2)^{3/2}}$$

over the plane curve $\mathbf{r}(t) = (e^t \cos t)\mathbf{i} + (e^t \sin t)\mathbf{j}$ from the point $(1, 0)$ to the point $(e^{2\pi}, 0)$ in two ways:

a) by using the parametrization of the curve to evaluate the work integral.

b) by evaluating a potential function for $\mathbf{F}$.

38. Find the flow of the field $\mathbf{F} = \nabla(x^2 z e^y)$

a) once around the ellipse C in which the plane $x + y + z = 1$ intersects the cylinder $x^2 + z^2 = 25$, clockwise as viewed from the positive y-axis.

b) along the curved boundary of the helicoid in Exercise 27 from $(1, 0, 0)$ to $(1, 0, 2\pi)$.

39. Suppose $\mathbf{F}(x, y) = (x + y)\mathbf{i} - (x^2 + y^2)\mathbf{j}$ is the velocity field of a fluid flowing across the xy-plane. Find the flow along each of the following paths from $(1, 0)$ to $(-1, 0)$.

a) The upper half of the circle $x^2 + y^2 = 1$

b) The line segment from $(1, 0)$ to $(-1, 0)$

c) The line segment from $(1, 0)$ to $(0, -1)$ followed by the line segment from $(0, -1)$ to $(-1, 0)$

40. Find the circulation of $\mathbf{F} = 2x\mathbf{i} + 2z\mathbf{j} + 2y\mathbf{k}$ along the closed path consisting of the helix $\mathbf{r}_1(t) = (\cos t)\mathbf{i} + (\sin t)\mathbf{j} + t\mathbf{k}$, $0 \leq t \leq \pi/2$, followed by the line segments $\mathbf{r}_2(t) = \mathbf{j} + (\pi/2)(1 - t)\mathbf{k}$, $0 \leq t \leq 1$, and $\mathbf{r}_3(t) = t\mathbf{i} + (1 - t)\mathbf{j}$, $0 \leq t \leq 1$.

In Exercises 41 and 42, use the surface integral in Stokes's theorem to find the circulation of the field $\mathbf{F}$ around the curve C in the indicated direction.

41. $\mathbf{F} = y^2\mathbf{i} - y\mathbf{j} + 3z^2\mathbf{k}$

C: The ellipse in which the plane $2x + 6y - 3z = 6$ meets the cylinder $x^2 + y^2 = 1$, counterclockwise as viewed from above

42. $\mathbf{F} = (x^2 + y)\mathbf{i} + (x + y)\mathbf{j} + (4y^2 - z)\mathbf{k}$

C: The circle in which the plane $z = -y$ meets the sphere $x^2 + y^2 + z^2 = 4$, counterclockwise as viewed from above

Mass and Moments

43. Find the mass of a thin wire lying along the curve $\mathbf{r}(t) = \sqrt{2}\,t\mathbf{i} + \sqrt{2}\,t\mathbf{j} + (4 - t^2)\mathbf{k}$, $0 \leq t \leq 1$, if the density at t is (a) $\delta = 3t$, (b) $\delta = 1$.

44. Find the center of mass of a thin wire lying along the curve $\mathbf{r}(t) = t\mathbf{i} + 2t\mathbf{j} + (2/3)t^{3/2}\mathbf{k}$, $0 \leq t \leq 2$, if the density at t is $\delta = 3\sqrt{5 + t}$.

45. Find the center of mass and the moments of inertia and radii of gyration about the coordinate axes of a thin wire lying along the curve

$$\mathbf{r}(t) = t\mathbf{i} + \frac{2\sqrt{2}}{3}t^{3/2}\mathbf{j} + \frac{t^2}{2}\mathbf{k}, \quad 0 \leq t \leq 2,$$

if the density at t is $\delta = 1/(t + 1)$.

46. A slender metal arch lies along the semicircle $y = \sqrt{a^2 - x^2}$ in the xy-plane. The density at the point (x, y) on the arch is $\delta(x, y) = 2a - y$. Find the center of mass.

47. A wire of constant density $\delta = 1$ lies along the curve $\mathbf{r}(t) = (e^t \cos t)\mathbf{i} + (e^t \sin t)\mathbf{j} + e^t\mathbf{k}$, $0 \leq t \leq \ln 2$. Find $\bar{z}$, I_z, and R_z.

48. Find the mass and center of mass of a wire of constant density δ that lies along the helix $\mathbf{r}(t) = (2 \sin t)\mathbf{i} + (2 \cos t)\mathbf{j} + 3t\mathbf{k}$, $0 \leq t \leq 2\pi$.

49. Find $I_z, R_z,$ and the center of mass of a thin shell of density $\delta(x, y, z) = z$ cut from the upper portion of the sphere $x^2 + y^2 + z^2 = 25$ by the plane $z = 3$.

50. Find the moment of inertia about the z-axis of the surface of the cube cut from the first octant by the planes $x = 1, y = 1,$ and $z = 1$ if the density is $\delta = 1$.

Flux Across a Plane Curve or Surface

Use Green's theorem to find the counterclockwise circulation and outward flux for the fields and curves in Exercises 51 and 52.

51. $\mathbf{F} = (2xy + x)\mathbf{i} + (xy - y)\mathbf{j}$
   C: The square bounded by $x = 0, x = 1, y = 0, y = 1$

52. $\mathbf{F} = (y - 6x^2)\mathbf{i} + (x + y^2)\mathbf{j}$
   C: The triangle made by the lines $y = 0, y = x,$ and $x = 1$

53. Show that

$$\oint_C \ln x \, \sin y \, dy - \frac{\cos y}{x} \, dx = 0$$

for any closed curve C to which Green's theorem applies.

54. a)   Show that the outward flux of the position vector field $\mathbf{F} = x\mathbf{i} + y\mathbf{j}$ across any closed curve to which Green's theorem applies is twice the area of the region enclosed by the curve.

b)   Let $\mathbf{n}$ be the outward unit normal vector to a closed curve to which Green's theorem applies. Show that it is not possible for $\mathbf{F} = x\mathbf{i} + y\mathbf{j}$ to be orthogonal to $\mathbf{n}$ at every point of C.

In Exercises 55–58, find the outward flux of $\mathbf{F}$ across the boundary of D.

55. $\mathbf{F} = 2xy\mathbf{i} + 2yz\mathbf{j} + 2xz\mathbf{k}$
   D: The cube cut from the first octant by the planes $x = 1,$ $y = 1, z = 1$

56. $\mathbf{F} = xz\mathbf{i} + yz\mathbf{j} + \mathbf{k}$
   D: The entire surface of the upper cap cut from the solid sphere $x^2 + y^2 + z^2 \le 25$ by the plane $z = 3$

57. $\mathbf{F} = -2x\mathbf{i} - 3y\mathbf{j} + z\mathbf{k}$
   D: The upper region cut from the solid sphere $x^2 + y^2 + z^2 \le 2$ by the paraboloid $z = x^2 + y^2$

58. $\mathbf{F} = (6x + y)\mathbf{i} - (x + z)\mathbf{j} + 4yz\mathbf{k}$
   D: The region in the first octant bounded by the cone $z = \sqrt{x^2 + y^2}$, the cylinder $x^2 + y^2 = 1$, and the coordinate planes

59. Let S be the surface that is bounded on the left by the hemisphere $x^2 + y^2 + z^2 = a^2, y \le 0$, in the middle by the cylinder $x^2 + z^2 = a^2, 0 \le y \le a$, and on the right by the plane $y = a$. Find the flux of the field $\mathbf{F} = y\mathbf{i} + z\mathbf{j} + x\mathbf{k}$ outward across S.

60. Find the outward flux of the field $\mathbf{F} = 3xz^2\mathbf{i} + y\mathbf{j} - z^3\mathbf{k}$ across the surface of the solid in the first octant that is bounded by the cylinder $x^2 + 4y^2 = 16$ and the planes $y = 2z, x = 0,$ and $z = 0$.

61. Use the Divergence Theorem to find the flux of $\mathbf{F} = xy^2\mathbf{i} + x^2y\mathbf{j} + y\mathbf{k}$ outward through the region enclosed by the cylinder $x^2 + y^2 = 1$ and the planes $z = 1$ and $z = -1$.

62. Find the flux of $\mathbf{F} = (3z + 1)\mathbf{k}$ upward across the hemisphere $x^2 + y^2 + z^2 = a^2, z \ge 0$ (a) with the Divergence Theorem, (b) by evaluating the flux integral directly.

CHAPTER | **14** | ADDITIONAL EXERCISES–THEORY, EXAMPLES, APPLICATIONS

Finding Areas with Green's Theorem

Use the Green's theorem area formula, Eq. (22) in Exercises 14.4, to find the areas of the regions enclosed by the curves in Exercises 1–4.

1. The limaçon $x = 2 \cos t - \cos 2t, y = 2 \sin t - \sin 2t,$ $0 \le t \le 2\pi$

2. The deltoid $x = 2 \cos t + \cos 2t, y = 2 \sin t - \sin 2t,$ $0 \le t \le 2\pi$

3. The eight curve $x = (1/2) \sin 2t, y = \sin t, 0 \le t \le \pi$ (one loop)

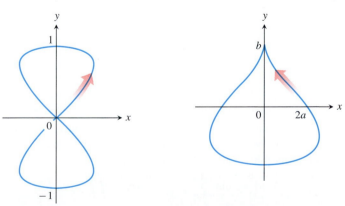

4. The teardrop $x = 2a \cos t - a \sin 2t, y = b \sin t, 0 \le t \le 2\pi$

Theory and Applications

5. a) Give an example of a vector field $\mathbf{F}(x, y, z)$ that has value $\mathbf{0}$ at only one point and such that curl $\mathbf{F}$ is nonzero everywhere. Be sure to identify the point and compute the curl.

b) Give an example of a vector field $\mathbf{F}(x, y, z)$ that has value $\mathbf{0}$ on precisely one line and such that curl $\mathbf{F}$ is nonzero everywhere. Be sure to identify the line and compute the curl.

c) Give an example of a vector field $\mathbf{F}(x, y, z)$ that has value $\mathbf{0}$ on a surface and such that curl $\mathbf{F}$ is nonzero everywhere. Be sure to identify the surface and compute the curl.

6. Find all points (a, b, c) on the sphere $x^2 + y^2 + z^2 = R^2$ where the vector field $\mathbf{F} = yz^2\,\mathbf{i} + xz^2\,\mathbf{j} + 2xyz\,\mathbf{k}$ is normal to the surface and $\mathbf{F}(a, b, c) \neq \mathbf{0}$.

7. Find the mass of a spherical shell of radius R such that at each point (x, y, z) on the surface the mass density $\delta(x, y, z)$ is its distance to some fixed point (a, b, c) on the surface.

8. Find the mass of a helicoid

$$\mathbf{r}(r, \theta) = (r \cos \theta)\,\mathbf{i} + (r \sin \theta)\,\mathbf{j} + \theta\,\mathbf{k},$$

$0 \le r \le 1$, $0 \le \theta \le 2\pi$, if the density function is $\delta(x, y, z) = 2\sqrt{x^2 + y^2}$. See Practice Exercise 27 for a figure.

9. Among all rectangular regions $0 \le x \le a$, $0 \le y \le b$, find the one for which the total outward flux of $\mathbf{F} = (x^2 + 4xy)\,\mathbf{i} - 6y\,\mathbf{j}$ across the four sides is least. What *is* the least flux?

10. Find an equation for the plane through the origin such that the circulation of the flow field $\mathbf{F} = z\,\mathbf{i} + x\,\mathbf{j} + y\,\mathbf{k}$ around the circle of intersection of the plane with the sphere $x^2 + y^2 + z^2 = 4$ is a maximum.

11. A string lies along the circle $x^2 + y^2 = 4$ from $(2, 0)$ to $(0, 2)$ in the first quadrant. The density of the string is $\rho(x, y) = xy$.

a) Partition the string into a finite number of subarcs to show that the work done by gravity to move the string straight down to the x-axis is given by

$$\text{Work} = \lim_{n \to \infty} \sum_{k=1}^{n} g\,x_k\,y_k^{\,2}\Delta s_k = \int_C g\,xy^2\,ds,$$

where g is the gravitational constant.

b) Find the total work done by evaluating the line integral in part (a).

c) Show that the total work done equals the work required to move the string's center of mass $(\bar{x}, \bar{y})$ straight down to the x-axis.

12. A thin sheet lies along the portion of the plane $x + y + z = 1$ in the first octant. The density of the sheet is $\delta(x, y, z) = xy$.

a) Partition the sheet into a finite number of subpieces to show that the work done by gravity to move the sheet straight down to the xy-plane is given by

$$\text{Work} = \lim_{n \to \infty} \sum_{k=1}^{n} g\,x_k\,y_k\,z_k\Delta\sigma_k = \iint_S g\,xyz\,d\sigma,$$

where g is the gravitational constant.

b) Find the total work done by evaluating the surface integral in part (a).

c) Show that the total work done equals the work required to move the sheet's center of mass $(\bar{x}, \bar{y}, \bar{z})$ straight down to the xy-plane.

13. *Archimedes' principle.* If an object such as a ball is placed in a liquid, it will either sink to the bottom, float, or sink a certain distance and remain suspended in the liquid. Suppose a fluid has constant weight density w and that the fluid's surface coincides with the plane $z = 4$. A spherical ball remains suspended in the fluid and occupies the region $x^2 + y^2 + (z - 2)^2 \le 1$.

a) Show that the surface integral giving the magnitude of the total force on the ball due to the fluid's pressure is

$$\text{Force} = \lim_{n \to \infty} \sum_{k=1}^{n} w\,(4 - z_k)\,\Delta\sigma_k = \iint_S w\,(4 - z)\,d\sigma.$$

b) Since the ball is not moving, it is being held up by the buoyant force of the liquid. Show that the magnitude of the buoyant force on the sphere is

$$\text{Buoyant force} = \iint_S w\,(z - 4)\,\mathbf{k} \cdot \mathbf{n}\,d\sigma,$$

where $\mathbf{n}$ is the outer unit normal at (x, y, z). This illustrates Archimedes' principle that the magnitude of the buoyant force on a submerged solid equals the weight of the displaced fluid.

c) Use the Divergence Theorem to find the magnitude of the buoyant force in part (b).

14. *Fluid force on a curved surface.* A cone in the shape of the surface $z = \sqrt{x^2 + y^2}$, $0 \le z \le 2$, is filled with a liquid of constant weight density w. Assuming the xy-plane is "ground level," show that the total force on the portion of the cone from $z = 1$ to $z = 2$ due to liquid pressure is the surface integral

$$F = \iint_S w\,(2 - z)\,d\sigma.$$

Evaluate the integral.

15. *Faraday's law.* If $\mathbf{E}(t, x, y, z)$ and $\mathbf{B}(t, x, y, z)$ represent the electric and magnetic fields at point (x, y, z) at time t, a basic principle of electromagnetic theory says that $\nabla \times \mathbf{E} = -\partial\mathbf{B}/\partial t$. In this expression $\nabla \times \mathbf{E}$ is computed with t held fixed and $\partial\mathbf{B}/\partial t$ is calculated with (x, y, z) fixed. Use Stokes's theorem to derive Faraday's law

$$\oint_C \mathbf{E} \cdot d\mathbf{r} = -\frac{\partial}{\partial t} \iint_S \mathbf{B} \cdot \mathbf{n}\,d\sigma,$$

where C represents a wire loop through which current flows counterclockwise with respect to the surface's unit normal $\mathbf{n}$, giving rise to the voltage

$$\oint_C \mathbf{E} \cdot d\mathbf{r}$$

around C. The surface integral on the right side of the equation is called the **magnetic flux,** and S is any oriented surface with boundary C.

16. Let

$$\mathbf{F} = -\frac{GmM}{|\mathbf{r}|^3}\,\mathbf{r}$$

be the gravitational force field defined for $\mathbf{r} \neq \mathbf{0}$. Use Gauss's law in Section 14.8 to show that there is no continuously differentiable vector field $\mathbf{H}$ satisfying $\mathbf{F} = \nabla \times \mathbf{H}$.

17. If $f(x, y, z)$ and $g(x, y, z)$ are continuously differentiable scalar functions defined over the oriented surface S with boundary curve C, prove that

$$\iint_S (\nabla f \times \nabla g) \cdot \mathbf{n}\,d\sigma = \oint_C f\,\nabla g \cdot d\mathbf{r}.$$

18. Suppose that $\nabla \cdot \mathbf{F}_1 = \nabla \cdot \mathbf{F}_2$ and $\nabla \times \mathbf{F}_1 = \nabla \times \mathbf{F}_2$ over a region D enclosed by the oriented surface S with outward unit normal $\mathbf{n}$ and that $\mathbf{F}_1 \cdot \mathbf{n} = \mathbf{F}_2 \cdot \mathbf{n}$ on S. Prove that $\mathbf{F}_1 = \mathbf{F}_2$ throughout D.

19. Prove or disprove that if $\nabla \cdot \mathbf{F} = 0$ and $\nabla \times \mathbf{F} = \mathbf{0}$, then $\mathbf{F} = \mathbf{0}$.

20. Let S be an oriented surface parametrized by $\mathbf{r}(u, v)$. Define the notation $d\boldsymbol{\sigma} = \mathbf{r}_u\,du \times \mathbf{r}_v\,dv$ so that $d\boldsymbol{\sigma}$ is a vector normal to the surface. Also, the magnitude $d\sigma = |d\boldsymbol{\sigma}|$ is the element of surface area (by Eq. 5 in Section 14.6). Derive the identity

$$d\sigma = (EG - F^2)^{1/2}\,du\,dv$$

where

$$E = |\mathbf{r}_u|^2, \quad F = \mathbf{r}_u \cdot \mathbf{r}_v, \quad \text{and} \quad G = |\mathbf{r}_v|^2.$$

21. Show that the volume V of a region D in space enclosed by the oriented surface S with outward normal $\mathbf{n}$ satisfies the identity

$$V = \frac{1}{3}\iint_S \mathbf{r} \cdot \mathbf{n}\,d\sigma,$$

where $\mathbf{r}$ is the position vector of the point (x, y, z) in D.

Appendices

Mathematical Induction

Many formulas, like

$$1 + 2 + \cdots + n = \frac{n(n+1)}{2},$$

can be shown to hold for every positive integer n by applying an axiom called the *mathematical induction principle*. A proof that uses this axiom is called a *proof by mathematical induction* or a *proof by induction*.

The steps in proving a formula by induction are the following.

Step 1: Check that the formula holds for $n = 1$.

Step 2: Prove that if the formula holds for any positive integer $n = k$, then it also holds for the next integer, $n = k + 1$.

Once these steps are completed (the axiom says), we know that the formula holds for all positive integers n. By step 1 it holds for $n = 1$. By step 2 it holds for $n = 2$, and therefore by step 2 also for $n = 3$, and by step 2 again for $n = 4$, and so on. If the first domino falls, and the kth domino always knocks over the $(k + 1)$st when it falls, all the dominoes fall.

From another point of view, suppose we have a sequence of statements S_1, $S_2, \ldots, S_n, \ldots$, one for each positive integer. Suppose we can show that assuming any one of the statements to be true implies that the next statement in line is true. Suppose that we can also show that S_1 is true. Then we may conclude that the statements are true from S_1 on.

EXAMPLE 1   Show that for every positive integer n,

$$1 + 2 + \cdots + n = \frac{n(n+1)}{2}.$$

Solution   We accomplish the proof by carrying out the two steps above.

Step 1: The formula holds for $n = 1$ because

$$1 = \frac{1(1+1)}{2}.$$

Step 2: If the formula holds for $n = k$, does it also hold for $n = k + 1$? The answer is yes, and here's why: If

$$1 + 2 + \cdots + k = \frac{k(k+1)}{2},$$

then

$$1 + 2 + \cdots + k + (k+1) = \frac{k(k+1)}{2} + (k+1) = \frac{k^2 + k + 2k + 2}{2}$$

$$= \frac{(k+1)(k+2)}{2} = \frac{(k+1)((k+1)+1)}{2}.$$

The last expression in this string of equalities is the expression $n(n+1)/2$ for $n = (k+1)$.

The mathematical induction principle now guarantees the original formula for all positive integers n. Notice that all *we* have to do is carry out steps 1 and 2. The mathematical induction principle does the rest.   ❏

EXAMPLE 2   Show that for all positive integers n,

$$\frac{1}{2^1} + \frac{1}{2^2} + \cdots + \frac{1}{2^n} = 1 - \frac{1}{2^n}.$$

Solution   We accomplish the proof by carrying out the two steps of mathematical induction.

Step 1: The formula holds for $n = 1$ because

$$\frac{1}{2^1} = 1 - \frac{1}{2^1}.$$

Step 2: If

$$\frac{1}{2^1} + \frac{1}{2^2} + \cdots + \frac{1}{2^k} = 1 - \frac{1}{2^k},$$

then

$$\frac{1}{2^1} + \frac{1}{2^2} + \cdots + \frac{1}{2^k} + \frac{1}{2^{k+1}} = 1 - \frac{1}{2^k} + \frac{1}{2^{k+1}} = 1 - \frac{1 \cdot 2}{2^k \cdot 2} + \frac{1}{2^{k+1}}$$

$$= 1 - \frac{2}{2^{k+1}} + \frac{1}{2^{k+1}} = 1 - \frac{1}{2^{k+1}}.$$

Thus, the original formula holds for $n = (k+1)$ whenever it holds for $n = k$.

With these steps verified, the mathematical induction principle now guarantees the formula for every positive integer n.   ❏

Other Starting Integers

Instead of starting at $n = 1$, some induction arguments start at another integer. The steps for such an argument are as follows.

Step 1: Check that the formula holds for $n = n_1$ (the first appropriate integer).

Step 2: Prove that if the formula holds for any integer $n = k \geq n_1$, then it also holds for $n = (k+1)$.

Once these steps are completed, the mathematical induction principle guarantees the formula for all $n \geq n_1$.

EXAMPLE 3   Show that $n! > 3^n$ if n is large enough.

Solution   How large is large enough? We experiment:

n	1	2	3	4	5	6	7
$n!$	1	2	6	24	120	720	5040
3^n	3	9	27	81	243	729	2187

It looks as if $n! > 3^n$ for $n \geq 7$. To be sure, we apply mathematical induction. We take $n_1 = 7$ in step 1 and try for step 2.

Suppose $k! > 3^k$ for some $k \geq 7$. Then

$$(k+1)! = (k+1)(k!) > (k+1)3^k > 7 \cdot 3^k > 3^{k+1}.$$

Thus, for $k \geq 7$,

$$k! > 3^k \quad \Rightarrow \quad (k+1)! > 3^{k+1}.$$

The mathematical induction principle now guarantees $n! \geq 3^n$ for all $n \geq 7$.   ❏

Exercises A.1

1. Assuming that the triangle inequality $|a + b| \leq |a| + |b|$ holds for any two numbers a and b, show that

$$|x_1 + x_2 + \cdots + x_n| \leq |x_1| + |x_2| + \cdots + |x_n|$$

for any n numbers.

2. Show that if $r \neq 1$, then

$$1 + r + r^2 + \cdots + r^n = \frac{1 - r^{n+1}}{1 - r}$$

for every positive integer n.

3. Use the Product Rule, $\dfrac{d}{dx}(uv) = u\dfrac{dv}{dx} + v\dfrac{du}{dx}$,

and the fact that $\dfrac{d}{dx}(x) = 1$

to show that $\dfrac{d}{dx}(x^n) = nx^{n-1}$

for every positive integer n.

4. Suppose that a function $f(x)$ has the property that $f(x_1 x_2) = f(x_1) + f(x_2)$ for any two positive numbers x_1 and x_2. Show that

$$f(x_1 x_2 \cdots x_n) = f(x_1) + f(x_2) + \cdots + f(x_n)$$

for the product of any n positive numbers $x_1, x_2 \ldots, x_n$.

5. Show that

$$\frac{2}{3^1} + \frac{2}{3^2} + \cdots + \frac{2}{3^n} = 1 - \frac{1}{3^n}$$

for all positive integers n.

6. Show that $n! > n^3$ if n is large enough.

7. Show that $2^n > n^2$ if n is large enough.

8. Show that $2^n \geq 1/8$ for $n \geq -3$.

9. *Sums of squares.* Show that the sum of the squares of the first n positive integers is

$$\frac{n\left(n + \dfrac{1}{2}\right)(n + 1)}{3}.$$

10. *Sums of cubes.* Show that the sum of the cubes of the first n positive integers is $(n(n + 1)/2)^2$.

11. *Rules for finite sums.* Show that the following finite sum rules hold for every positive integer n.

a) $\displaystyle\sum_{k=1}^{n}(a_k + b_k) = \sum_{k=1}^{n}a_k + \sum_{k=1}^{n}b_k$

b) $\displaystyle\sum_{k=1}^{n}(a_k - b_k) = \sum_{k=1}^{n}a_k - \sum_{k=1}^{n}b_k$

c) $\displaystyle\sum_{k=1}^{n}ca_k = c \cdot \sum_{k=1}^{n}a_k$ \qquad (Any number c)

d) $\displaystyle\sum_{k=1}^{n}a_k = n \cdot c$ \qquad (if a_k has the constant value c)

12. Show that $|x^n| = |x|^n$ for every positive integer n and every real number x.

A.2 Proofs of Limit Theorems in Section 1.2

This appendix proves Theorem 1, Parts 2–5, and Theorem 4 from Section 1.2.

Theorem 1

Properties of Limits

The following rules hold if $\lim_{x \to c} f(x) = L$ and $\lim_{x \to c} g(x) = M$ (L and M real numbers).

1. *Sum Rule:* $\displaystyle\lim_{x \to c} [f(x) + g(x)] = L + M$

2. *Difference Rule:* $\displaystyle\lim_{x \to c} [f(x) - g(x)] = L - M$

3. *Product Rule:* $\displaystyle\lim_{x \to c} f(x) \cdot g(x) = L \cdot M$

4. *Constant Multiple Rule:* $\displaystyle\lim_{x \to c} kf(x) = kL$ (any number k)

5. *Quotient Rule:* $\displaystyle\lim_{x \to c} \frac{f(x)}{g(x)} = \frac{L}{M},$ if $M \neq 0$

6. *Power Rule:* If m and n are integers, then

$$\lim_{x \to c} [f(x)]^{m/n} = L^{m/n}$$

provided $L^{m/n}$ is a real number.

We proved the Sum Rule in Section 1.3 and the Power Rule is proved in more advanced texts. We obtain the Difference Rule by replacing $g(x)$ by $-g(x)$ and M by $-M$ in the Sum Rule. The Constant Multiple Rule is the special case $g(x) = k$ of the Product Rule. This leaves only the Product and Quotient Rules.

Proof of the Limit Product Rule We show that for any $\epsilon > 0$ there exists a $\delta > 0$ such that for all x in the intersection D of the domains of f and g,

$$0 < |x - c| < \delta \quad \Rightarrow \quad |f(x)\, g(x) - LM| < \epsilon.$$

Suppose then that ϵ is a positive number, and write $f(x)$ and $g(x)$ as

$$f(x) = L + (f(x) - L), \qquad g(x) = M + (g(x) - M).$$

Multiply these expressions together and subtract LM:

$$\begin{aligned}
f(x) \cdot g(x) - LM &= (L + (f(x) - L))(M + (g(x) - M)) - LM \\
&= LM + L(g(x) - M) + M(f(x) - L) \\
&\quad + (f(x) - L)(g(x) - M) - LM \\
&= L(g(x) - M) + M(f(x) - L) + (f(x) - L)(g(x) - M).
\end{aligned} \tag{1}$$

Since f and g have limits L and M as $x \to c$, there exist positive numbers $\delta_1, \delta_2, \delta_3$, and δ_4 such that for all x in D

$$0 < |x - c| < \delta_1 \quad \Rightarrow \quad |f(x) - L| < \sqrt{\epsilon/3}$$

$$0 < |x - c| < \delta_2 \quad \Rightarrow \quad |g(x) - M| < \sqrt{\epsilon/3}$$

$$0 < |x - c| < \delta_3 \quad \Rightarrow \quad |f(x) - L| < \epsilon/(3(1 + |M|))$$

$$0 < |x - c| < \delta_4 \quad \Rightarrow \quad |g(x) - M| < \epsilon/(3(1 + |L|))$$

(2)

If we take δ to be the smallest numbers δ_1 through δ_4, the inequalities on the right-hand side of (2) will hold simultaneously for $0 < |x - c| < \delta$. Therefore, for all x in D, $0 < |x - c| < \delta$ implies

$$|f(x) \cdot g(x) - LM|$$

$$\leq |L||g(x) - M| + |M||f(x) - L| + |f(x) - L||g(x) - M|$$

$$\leq (1 + |L|)|g(x) - M| + (1 + |M|)|f(x) - L| + |f(x) - L||g(x) - M|$$

$$\leq \frac{\epsilon}{3} + \frac{\epsilon}{3} + \sqrt{\frac{\epsilon}{3}}\sqrt{\frac{\epsilon}{3}} = \epsilon.$$

> Triangle inequality applied to Eq. (1)

> Values from (2)

This completes the proof of the Limit Product Rule. ❏

Proof of the Limit Quotient Rule   We show that $\lim_{x \to c}(1/g(x)) = 1/M$. We can then conclude that

$$\lim_{x \to c} \frac{f(x)}{g(x)} = \lim_{x \to c} \left(f(x) \cdot \frac{1}{g(x)} \right) = \lim_{x \to c} f(x) \cdot \lim_{x \to c} g(x) = L \cdot \frac{1}{M} = \frac{L}{M}$$

by the Limit Product Rule.

Let $\epsilon > 0$ be given. To show that $\lim_{x \to c}(1/g(x)) = 1/M$, we need to show that there exists a $\delta > 0$ such that for all x

$$0 < |x - c| < \delta \quad \Rightarrow \quad \left| \frac{1}{g(x)} - \frac{1}{M} \right| < \epsilon.$$

Since $|M| > 0$, there exists a positive number δ_1 such that for all x

$$0 < |x - c| < \delta_1 \quad \Rightarrow \quad |g(x) - M| < \frac{M}{2}.$$

(3)

For any numbers A and B it can be shown that $|A| - |B| \leq |A - B|$ and $|B| - |A| \leq |A - B|$, from which it follows that $||A| - |B|| \leq |A - B|$. With $A = g(x)$ and $B = M$, this becomes

$$||g(x)| - |M|| \leq |g(x) - M|,$$

which can be combined with the inequality on the right in (3) to get, in turn,

$$||g(x)| - |M|| < \frac{|M|}{2}$$

$$-\frac{|M|}{2} < |g(x)| - |M| < \frac{|M|}{2}$$

$$\frac{|M|}{2} < |g(x)| < \frac{3|M|}{2}$$

$$|M| < 2|g(x)| < 3|M|$$

$$\frac{1}{|g(x)|} < \frac{2}{|M|} < \frac{3}{|g(x)|}$$

(4)

Therefore, $0 < |x - c| < \delta_1$ implies that

$$\left| \frac{1}{g(x)} - \frac{1}{M} \right| = \left| \frac{M - g(x)}{Mg(x)} \right| \leq \frac{1}{|M|} \cdot \frac{1}{|g(x)|} \cdot |M - g(x)|$$

$$< \frac{1}{|M|} \cdot \frac{2}{|M|} \cdot |M - g(x)|. \qquad \text{Inequality (4)} \qquad (5)$$

Since $(1/2)|M|^2 \epsilon > 0$, there exists a number $\delta_2 > 0$ such that for all x

$$0 < |x - c| < \delta_2 \quad \Rightarrow \quad |M - g(x)| < \frac{\epsilon}{2}|M|^2. \qquad (6)$$

If we take δ to be the smaller of δ_1 and δ_2, the conclusions in (5) and (6) both hold for all x such that $0 < |x - c| < \delta$. Combining these conclusions gives

$$0 < |x - c| < \delta \quad \Rightarrow \quad \left| \frac{1}{g(x)} - \frac{1}{M} \right| < \epsilon.$$

This concludes the proof of the Limit Quotient Rule.   ❏

Theorem 4
The Sandwich Theorem

Suppose that $g(x) \leq f(x) \leq h(x)$ for all x in some open interval containing c, except possibly at $x = c$ itself. Suppose also that $\lim_{x \to c} g(x) = \lim_{x \to c} h(x) = L$. Then $\lim_{x \to c} f(x) = L$.

Proof for Right-hand Limits   Suppose $\lim_{x \to c^+} g(x) = \lim_{x \to c^+} h(x) = L$. Then for any $\epsilon > 0$ there exists a $\delta > 0$ such that for all x the inequality $c < x < c + \delta$ implies

$$L - \epsilon < g(x) < L + \epsilon \quad \text{and} \quad L - \epsilon < h(x) < L + \epsilon. \qquad (7)$$

These inequalities combine with the inequality $g(x) \leq f(x) \leq h(x)$ to give

$$L - \epsilon < g(x) \leq f(x) \leq h(x) < L + \epsilon,$$

$$L - \epsilon < f(x) < L + \epsilon, \qquad (8)$$

$$-\epsilon < f(x) - L < \epsilon.$$

Therefore, for all x, the inequality $c < x < c + \delta$ implies $|f(x) - L| < \epsilon$.   ❏

Proof for Left-hand Limits   Suppose $\lim_{x \to c^-} g(x) = \lim_{x \to c^-} h(x) = L$. Then for any $\epsilon > 0$ there exists a $\delta > 0$ such that for all x the inequality $c - \delta < x < c$ implies

$$L - \epsilon < g(x) < L + \epsilon \quad \text{and} \quad L - \epsilon < h(x) < L + \epsilon. \qquad (9)$$

We conclude as before that for all x, $c - \delta < x < c$ implies $|f(x) - L| < \epsilon$.   ❏

Proof for Two-sided Limits   If $\lim_{x \to c} g(x) = \lim_{x \to c} h(x) = L$, then $g(x)$ and $h(x)$ both approach L as $x \to c^+$ and as $x \to c^-$; so $\lim_{x \to c^+} f(x) = L$ and $\lim_{x \to c^-} f(x) = L$. Hence $\lim_{x \to c} f(x)$ exists and equals L.   ❏

Exercises A.2

1. Suppose that functions $f_1(x)$, $f_2(x)$, and $f_3(x)$ have limits L_1, L_2, and L_3, respectively, as $x \to c$. Show that their sum has limit $L_1 + L_2 + L_3$. Use mathematical induction (Appendix 1) to generalize this result to the sum of any finite number of functions.

2. Use mathematical induction and the Limit Product Rule in Theorem 1 to show that if functions $f_1(x)$, $f_2(x)$, ..., $f_n(x)$ have limits L_1, L_2, ..., L_n as $x \to c$, then

   $$\lim_{x \to c} f_1(x) f_2(x) \cdot \cdots \cdot f_n(x) = L_1 \cdot L_2 \cdot \cdots \cdot L_n.$$

3. Use the fact that $\lim_{x \to c} x = c$ and the result of Exercise 2 to show that $\lim_{x \to c} x^n = c^n$ for any integer $n > 1$.

4. *Limits of polynomials.* Use the fact that $\lim_{x \to c}(k) = k$ for any number k together with the results of Exercises 1 and 3 to show that $\lim_{x \to c} f(x) = f(c)$ for any polynomial function

   $$f(x) = a_0 x^n + a_1 x^{n-1} + \cdots + a_{n-1}x + a_n.$$

5. *Limits of rational functions.* Use Theorem 1 and the result of Exercise 4 to show that if $f(x)$ and $g(x)$ are polynomial functions and $g(c) \neq 0$, then

   $$\lim_{x \to c} \frac{f(x)}{g(x)} = \frac{f(c)}{g(c)}.$$

6. *Composites of continuous functions.* Figure A.1 gives the diagram for a proof that the composite of two continuous functions is continuous. Reconstruct the proof from the diagram. The statement to be proved is this: If f is continuous at $x = c$ and g is continuous at $f(c)$, then $g \circ f$ is continuous at c.

   Assume that c is an interior point of the domain of f and that $f(c)$ is an interior point of the domain of g. This will make the limits involved two-sided. (The arguments for the cases that involve one-sided limits are similar.)

A.1 The diagram for a proof that the composite of two continuous functions is continuous. The continuity of composites holds for any finite number of functions. The only requirement is that each function be continuous where it is applied. In the figure, f is to be continuous at c and g at $f(c)$.

A.3

Complex Numbers

Complex numbers are expressions of the form $a + ib$, where a and b are real numbers and i is a symbol for $\sqrt{-1}$. Unfortunately, the words "real" and "imaginary" have connotations that somehow place $\sqrt{-1}$ in a less favorable position in our minds than $\sqrt{2}$. As a matter of fact, a good deal of imagination, in the sense of *inventiveness,* has been required to construct the *real* number system, which forms the basis of the calculus. In this appendix we review the various stages of this invention. The further invention of a complex number system will then not seem so strange.

The Development of the Real Numbers

The earliest stage of number development was the recognition of the **counting numbers** 1, 2, 3, . . . , which we now call the **natural numbers** or the **positive integers.** Certain simple arithmetical operations can be performed with these numbers without getting outside the system. That is, the system of positive integers is **closed** under the operations of addition and multiplication. By this we mean that if m and n are any positive integers, then

$$m + n = p \qquad \text{and} \qquad mn = q \tag{1}$$

are also positive integers. Given the two positive integers on the left-hand side of either equation in (1), we can find the corresponding positive integer on the right. More than this, we can sometimes specify the positive integers m and p and find a positive integer n such that $m + n = p$. For instance, $3 + n = 7$ can be solved when the only numbers we know are the positive integers. But the equation $7 + n = 3$ cannot be solved unless the number system is enlarged.

The number zero and the negative integers were invented to solve equations like $7 + n = 3$. In a civilization that recognizes all the **integers**

$$\ldots, -3, -2, -1, 0, 1, 2, 3, \ldots, \tag{2}$$

an educated person can always find the missing integer that solves the equation $m + n = p$ when given the other two integers in the equation.

Suppose our educated people also know how to multiply any two of the integers in (2). If, in Eqs. (1), they are given m and q, they discover that sometimes they can find n and sometimes they cannot. If their imagination is still in good working order, they may be inspired to invent still more numbers and introduce fractions, which are just ordered pairs m/n of integers m and n. The number zero has special properties that may bother them for a while, but they ultimately discover that it is handy to have all ratios of integers m/n, excluding only those having zero in the denominator. This system, called the set of **rational numbers,** is now rich enough for them to perform the so-called **rational operations** of arithmetic:

1. a) addition
   b) subtraction
2. a) multiplication
   b) division

on any two numbers in the system, *except that they cannot divide by zero.*

The geometry of the unit square (Fig. A.2) and the Pythagorean theorem showed that they could construct a geometric line segment that, in terms of some basic unit of length, has length equal to $\sqrt{2}$. Thus they could solve the equation

$$x^2 = 2$$

by a geometric construction. But then they discovered that the line segment representing $\sqrt{2}$ and the line segment representing the unit of length 1 were incommensurable quantities. This means that the ratio $\sqrt{2}/1$ cannot be expressed as the ratio of two *integer* multiples of some other, presumably more fundamental, unit of length. That is, our educated people could not find a rational number solution of the equation $x^2 = 2$.

There *is* no rational number whose square is 2. To see why, suppose that there were such a rational number. Then we could find integers p and q with no common factor other than 1, and such that

$$p^2 = 2q^2. \tag{3}$$

Since p and q are integers, p must be even; otherwise its product with itself would be odd. In symbols, $p = 2p_1$, where p_1 is an integer. This leads to $2p_1^2 = q^2$, which says q must be even, say $q = 2q_1$, where q_1 is an integer. This makes 2 a factor of both p and q, contrary to our choice of p and q as integers with no common factor other than 1. Hence there is no rational number whose square is 2.

Although our educated people could not find a rational solution of the equation $x^2 = 2$, they could get a sequence of rational numbers

$$\frac{1}{1}, \frac{7}{5}, \frac{41}{29}, \frac{239}{169}, \ldots \tag{4}$$

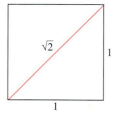

A.2 With a straightedge and compass, it is possible to construct a segment of irrational length.

whose squares form a sequence

$$\frac{1}{1}, \frac{49}{25}, \frac{1681}{841}, \frac{57,121}{28,561}, \quad \ldots, \tag{5}$$

that converges to 2 as its limit. This time their imagination suggested that they needed the concept of a limit of a sequence of rational numbers. If we accept the fact that an increasing sequence that is bounded from above always approaches a limit and observe that the sequence in (4) has these properties, then we want it to have a limit L. This would also mean, from (5), that $L^2 = 2$, and hence L is *not* one of our rational numbers. If to the rational numbers we further add the limits of all bounded increasing sequences of rational numbers, we arrive at the system of all "real" numbers. The word *real* is placed in quotes because there is nothing that is either "more real" or "less real" about this system than there is about any other mathematical system.

The Complex Numbers

Imagination was called upon at many stages during the development of the real number system. In fact, the art of invention was needed at least three times in constructing the systems we have discussed so far:

1. The *first invented* system: the set of *all integers* as constructed from the counting numbers.
2. The *second invented* system: the set of *rational numbers m/n* as constructed from the integers.
3. The *third invented* system: the set of all *real numbers x* as constructed from the rational numbers.

These invented systems form a hierarchy in which each system contains the previous system. Each system is also richer than its predecessor in that it permits additional operations to be performed without going outside the system:

1. In the system of all integers, we can solve all equations of the form

$$x + a = 0, \tag{6}$$

   where a can be any integer.
2. In the system of all rational numbers, we can solve all equations of the form

$$ax + b = 0, \tag{7}$$

   provided a and b are rational numbers and $a \neq 0$.
3. In the system of all real numbers, we can solve all of the equations in (6) and (7) and, in addition, all quadratic equations

$$ax^2 + bx + c = 0 \quad \text{having} \quad a \neq 0 \quad \text{and} \quad b^2 - 4ac \geq 0. \tag{8}$$

You are probably familiar with the formula that gives the solutions of (8), namely,

$$x = \frac{-b \pm \sqrt{b^2 - 4ac}}{2a}, \tag{9}$$

and are familiar with the further fact that when the discriminant, $d = b^2 - 4ac$, is negative, the solutions in (9) do *not* belong to any of the systems discussed above. In fact, the very simple quadratic equation

$$x^2 + 1 = 0$$

is impossible to solve if the only number systems that can be used are the three invented systems mentioned so far.

Thus we come to the *fourth invented* system, the set of all complex numbers $a + ib$. We could dispense entirely with the symbol i and use a notation like (a, b). We would then speak simply of a pair of real numbers a and b. Since, under algebraic operations, the numbers a and b are treated somewhat differently, it is essential to keep the *order* straight. We therefore might say that the **complex number system** consists of the set of all ordered pairs of real numbers (a, b), together with the rules by which they are to be equated, added, multiplied, and so on, listed below. We will use both the (a, b) notation and the notation $a + ib$ in the discussion that follows. We call a the **real part** and b the **imaginary part** of the complex number (a, b).

We make the following definitions.

Equality
$$a + ib = c + id$$
if and only if
$$a = c \quad \text{and} \quad b = d.$$

Two complex numbers (a, b) and (c, d) are *equal* if and only if $a = c$ and $b = d$.

Addition
$$(a + ib) + (c + id)$$
$$= (a + c) + i(b + d)$$

The sum of the two complex numbers (a, b) and (c, d) is the complex number $(a + c, b + d)$.

Multiplication
$$(a + ib)(c + id)$$
$$= (ac - bd) + i(ad + bc)$$

The product of two complex numbers (a, b) and (c, d) is the complex number $(ac - bd, ad + bc)$.

$$c(a + ib) = ac + i(bc)$$

The product of a real number c and the complex number (a, b) is the complex number (ac, bc).

The set of all complex numbers (a, b) in which the second number b is zero has all the properties of the set of real numbers a. For example, addition and multiplication of $(a, 0)$ and $(c, 0)$ give

$$(a, 0) + (c, 0) = (a + c, 0),$$

$$(a, 0) \cdot (c, 0) = (ac, 0),$$

which are numbers of the same type with imaginary part equal to zero. Also, if we multiply a "real number" $(a, 0)$ and the complex number (c, d), we get

$$(a, 0) \cdot (c, d) = (ac, ad) = a(c, d).$$

In particular, the complex number $(0, 0)$ plays the role of zero in the complex number system, and the complex number $(1, 0)$ plays the role of unity.

The number pair $(0, 1)$, which has real part equal to zero and imaginary part equal to one, has the property that its square,

$$(0, 1)(0, 1) = (-1, 0),$$

has real part equal to minus one and imaginary part equal to zero. Therefore, in the system of complex numbers (a, b), there is a number $x = (0, 1)$ whose square can be added to unity $= (1, 0)$ to produce zero $= (0, 0)$; that is,

$$(0, 1)^2 + (1, 0) = (0, 0).$$

The equation

$$x^2 + 1 = 0$$

therefore has a solution $x = (0, 1)$ in this new number system.

You are probably more familiar with the $a + ib$ notation than you are with the notation (a, b). And since the laws of algebra for the ordered pairs enable us to write

$$(a, b) = (a, 0) + (0, b) = a(1, 0) + b(0, 1),$$

while $(1, 0)$ behaves like unity and $(0, 1)$ behaves like a square root of minus one, we need not hesitate to write $a + ib$ in place of (a, b). The i associated with b is like a tracer element that tags the imaginary part of $a + ib$. We can pass at will from the realm of ordered pairs (a, b) to the realm of expressions $a + ib$, and conversely. But there is nothing less "real" about the symbol $(0, 1) = i$ than there is about the symbol $(1, 0) = 1$, once we have learned the laws of algebra in the complex number system (a, b).

To reduce any rational combination of complex numbers to a single complex number, we apply the laws of elementary algebra, replacing i^2 wherever it appears by -1. Of course, we cannot divide by the complex number $(0, 0) = 0 + i\,0$. But if $a + ib \neq 0$, then we may carry out a division as follows:

$$\frac{c + id}{a + ib} = \frac{(c + id)(a - ib)}{(a + ib)(a - ib)} = \frac{(ac + bd) + i(ad - bc)}{a^2 + b^2}.$$

The result is a complex number $x + iy$ with

$$x = \frac{ac + bd}{a^2 + b^2}, \qquad y = \frac{ad - bc}{a^2 + b^2},$$

and $a^2 + b^2 \neq 0$, since $a + ib = (a, b) \neq (0, 0)$.

The number $a - ib$ that is used as multiplier to clear the i from the denominator is called the **complex conjugate** of $a + ib$. It is customary to use $\bar{z}$ (read "z bar") to denote the complex conjugate of z; thus

$$z = a + ib, \qquad \bar{z} = a - ib.$$

Multiplying the numerator and denominator of the fraction $(c + id)/(a + ib)$ by the complex conjugate of the denominator will always replace the denominator by a real number.

EXAMPLE 1

a) $(2 + 3i) + (6 - 2i) = (2 + 6) + (3 - 2)i = 8 + i$

b) $(2 + 3i) - (6 - 2i) = (2 - 6) + (3 - (-2))i = -4 + 5i$

c) $(2 + 3i)(6 - 2i) = (2)(6) + (2)(-2i) + (3i)(6) + (3i)(-2i)$

$$= 12 - 4i + 18i - 6i^2 = 12 + 14i + 6 = 18 + 14i$$

d) $\dfrac{2 + 3i}{6 - 2i} = \dfrac{2 + 3i}{6 - 2i} \dfrac{6 + 2i}{6 + 2i}$

$$= \frac{12 + 4i + 18i + 6i^2}{36 + 12i - 12i - 4i^2}$$

$$= \frac{6 + 22i}{40} = \frac{3}{20} + \frac{11}{20} i$$

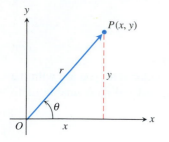

A.3 This Argand diagram represents $z = x + iy$ both as a point $P(x, y)$ and as a vector $\overrightarrow{OP}$.

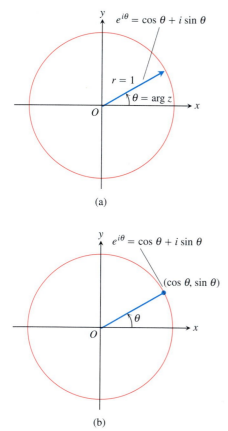

(a)

(b)

A.4 Argand diagrams for $e^{i\theta} = \cos\theta + i\sin\theta$ (a) as a vector, (b) as a point.

Argand Diagrams

There are two geometric representations of the complex number $z = x + iy$:

a) as the point $P(x, y)$ in the xy-plane and

b) as the vector $\overrightarrow{OP}$ from the origin to P.

In each representation, the x-axis is called the **real axis** and the y-axis is the **imaginary axis.** Both representations are **Argand diagrams** for $x + iy$ (Fig. A.3).

In terms of the polar coordinates of x and y, we have

$$x = r\cos\theta, \qquad y = r\sin\theta,$$

and

$$z = x + iy = r(\cos\theta + i\sin\theta). \tag{10}$$

We define the **absolute value** of a complex number $x + iy$ to be the length r of a vector $\overrightarrow{OP}$ from the origin to $P(x, y)$. We denote the absolute value by vertical bars, thus:

$$|x + iy| = \sqrt{x^2 + y^2}.$$

If we always choose the polar coordinates r and θ so that r is nonnegative, then

$$r = |x + iy|.$$

The polar angle θ is called the **argument** of z and is written $\theta = \arg z$. Of course, any integer multiple of 2π may be added to θ to produce another appropriate angle.

The following equation gives a useful formula connecting a complex number z, its conjugate $\bar{z}$, and its absolute value $|z|$, namely,

$$z \cdot \bar{z} = |z|^2.$$

Euler's Formula, Products, and Quotients

The identity

$$e^{i\theta} = \cos\theta + i\sin\theta,$$

called **Euler's formula,** enables us to rewrite Eq. (10) as

$$z = re^{i\theta}.$$

This, in turn, leads to the following rules for calculating products, quotients, powers, and roots of complex numbers. It also leads to Argand diagrams for $e^{i\theta}$. Since $\cos\theta + i\sin\theta$ is what we get from Eq. (10) by taking $r = 1$, we can say that $e^{i\theta}$ is represented by a unit vector that makes an angle θ with the positive x-axis, as shown in Fig. A.4.

Products   To multiply two complex numbers, we multiply their absolute values and add their angles. Let

$$z_1 = r_1 e^{i\theta_1}, \qquad z_2 = r_2 e^{i\theta_2}, \tag{11}$$

so that

$$|z_1| = r_1, \quad \arg z_1 = \theta_1; \qquad |z_2| = r_2, \quad \arg z_2 = \theta_2.$$

Then

$$z_1 z_2 = r_1 e^{i\theta_1} \cdot r_2 e^{i\theta_2} = r_1 r_2 e^{i(\theta_1 + \theta_2)}$$

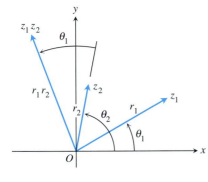

A.5 When z_1 and z_2 are multiplied, $|z_1 z_2| = r_1 \cdot r_2$ and $\arg(z_1 z_2) = \theta_1 + \theta_2$.

exp (A) stands for e^A.

and hence

$$|z_1 z_2| = r_1 r_2 = |z_1| \cdot |z_2|,$$

$$\arg(z_1 z_2) = \theta_1 + \theta_2 = \arg z_1 + \arg z_2. \tag{12}$$

Thus the product of two complex numbers is represented by a vector whose length is the product of the lengths of the two factors and whose argument is the sum of their arguments (Fig. A.5). In particular, a vector may be rotated counterclockwise through an angle θ by multiplying it by $e^{i\theta}$. Multiplication by i rotates $90°$, by -1 rotates $180°$, by $-i$ rotates $270°$, etc.

EXAMPLE 2   Let $z_1 = 1 + i$, $z_2 = \sqrt{3} - i$. We plot these complex numbers in an Argand diagram (Fig. A.6) from which we read off the polar representations

$$z_1 = \sqrt{2} e^{i\pi/4}, \qquad z_2 = 2 e^{-i\pi/6}.$$

Then

$$z_1 z_2 = 2\sqrt{2} \exp\left(\frac{i\pi}{4} - \frac{i\pi}{6}\right) = 2\sqrt{2} \exp\left(\frac{i\pi}{12}\right)$$

$$= 2\sqrt{2} \left(\cos\frac{\pi}{12} + i \sin\frac{\pi}{12}\right) \approx 2.73 + 0.73i.$$

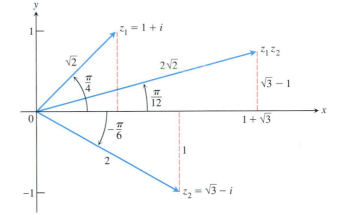

A.6 To multiply two complex numbers, multiply their absolute values and add their arguments.

Quotients

Suppose $r_2 \neq 0$ in Eq. (11). Then

$$\frac{z_1}{z_2} = \frac{r_1 e^{i\theta_1}}{r_2 e^{i\theta_2}} = \frac{r_1}{r_2} e^{i(\theta_1 - \theta_2)}.$$

Hence

$$\left|\frac{z_1}{z_2}\right| = \frac{r_1}{r_2} = \frac{|z_1|}{|z_2|} \quad \text{and} \quad \arg\left(\frac{z_1}{z_2}\right) = \theta_1 - \theta_2 = \arg z_1 - \arg z_2.$$

That is, we divide lengths and subtract angles.

EXAMPLE 3   Let $z_1 = 1 + i$ and $z_2 = \sqrt{3} - i$, as in Example 2. Then

$$\frac{1+i}{\sqrt{3}-i} = \frac{\sqrt{2}e^{i\pi/4}}{2e^{-i\pi/6}} = \frac{\sqrt{2}}{2}e^{5\pi i/12} \approx 0.707\left(\cos\frac{5\pi}{12} + i\,\sin\frac{5\pi}{12}\right)$$

$$\approx 0.183 + 0.\overline{683}i.$$ ❑

Powers

If n is a positive integer, we may apply the product formulas in (12) to find

$$z^n = z \cdot z \cdot \cdots \cdot z.$$ n factors

With $z = re^{i\theta}$, we obtain

$$z^n = (re^{i\theta})^n = r^n e^{i(\theta+\theta+\cdots+\theta)}$$ n summands

$$= r^n e^{in\theta}.$$ (13)

The length $r = |z|$ is raised to the nth power and the angle $\theta = \arg z$ is multiplied by n.

If we take $r = 1$ in Eq. (13), we obtain De Moivre's theorem.

> **De Moivre's Theorem**
>
> $$(\cos\theta + i\,\sin\theta)^n = \cos n\theta + i\,\sin n\theta.$$ (14)

If we expand the left-hand side of De Moivre's equation (Eq. 14) by the binomial theorem and reduce it to the form $a + ib$, we obtain formulas for $\cos n\theta$ and $\sin n\theta$ as polynomials of degree n in $\cos\theta$ and $\sin\theta$.

EXAMPLE 4   If $n = 3$ in Eq. (14), we have

$$(\cos\theta + i\,\sin\theta)^3 = \cos 3\theta + i\,\sin 3\theta.$$

The left-hand side of this equation is

$$\cos^3\theta + 3i\,\cos^2\theta\,\sin\theta - 3\cos\theta\,\sin^2\theta - i\,\sin^3\theta.$$

The real part of this must equal $\cos 3\theta$ and the imaginary part must equal $\sin 3\theta$. Therefore,

$$\cos 3\theta = \cos^3\theta - 3\cos\theta\,\sin^2\theta,$$

$$\sin 3\theta = 3\cos^2\theta\,\sin\theta - \sin^3\theta.$$ ❑

Roots   If $z = re^{i\theta}$ is a complex number different from zero and n is a positive integer, then there are precisely n different complex numbers $w_0, w_1, \ldots, w_{n-1}$, that are nth roots of z. To see why, let $w = \rho e^{i\alpha}$ be an nth root of $z = re^{i\theta}$, so that

$$w^n = z$$

or

$$\rho^n e^{in\alpha} = re^{i\theta}.$$

Then

$$\rho = \sqrt[n]{r}$$

is the real, positive nth root of r. As regards the angle, although we cannot say that $n\alpha$ and θ must be equal, we can say that they may differ only by an integer multiple of 2π. That is,

$$n\alpha = \theta + 2k\pi, \qquad k = 0, \pm 1, \pm 2, \dots .$$

Therefore,

$$\alpha = \frac{\theta}{n} + k\frac{2\pi}{n}.$$

Hence all nth roots of $z = re^{i\theta}$ are given by

$$\sqrt[n]{re^{i\theta}} = \sqrt[n]{r} \exp i\left(\frac{\theta}{n} + k\frac{2\pi}{n}\right), \qquad k = 0, \pm 1, \pm 2, \dots . \qquad (15)$$

There might appear to be infinitely many different answers corresponding to the infinitely many possible values of k. But $k = n + m$ gives the same answer as $k = m$ in Eq. (15). Thus we need only take n consecutive values for k to obtain all the different nth roots of z. For convenience, we take

$$k = 0, 1, 2, \dots, n - 1.$$

All the nth roots of $re^{i\theta}$ lie on a circle centered at the origin O and having radius equal to the real, positive nth root of r. One of them has argument $\alpha = \theta/n$. The others are uniformly spaced around the circle, each being separated from its neighbors by an angle equal to $2\pi/n$. Figure A.7 illustrates the placement of the three cube roots, w_0, w_1, w_2, of the complex number $z = re^{i\theta}$.

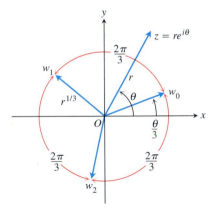

A.7 The three cube roots of $z = re^{i\theta}$.

EXAMPLE 5   Find the four fourth roots of -16.

Solution   As our first step, we plot the number -16 in an Argand diagram (Fig. A.8) and determine its polar representation $re^{i\theta}$. Here, $z = -16, r = +16$, and $\theta = \pi$. One of the fourth roots of $16e^{i\pi}$ is $2e^{i\pi/4}$. We obtain others by successive additions of $2\pi/4 = \pi/2$ to the argument of this first one. Hence

$$\sqrt[4]{16 \exp i\pi} = 2 \exp i\left(\frac{\pi}{4}, \frac{3\pi}{4}, \frac{5\pi}{4}, \frac{7\pi}{4}\right),$$

and the four roots are

$$w_0 = 2\left[\cos\frac{\pi}{4} + i\sin\frac{\pi}{4}\right] = \sqrt{2}(1 + i),$$

$$w_1 = 2\left[\cos\frac{3\pi}{4} + i\sin\frac{3\pi}{4}\right] = \sqrt{2}(-1 + i),$$

$$w_2 = 2\left[\cos\frac{5\pi}{4} + i\sin\frac{5\pi}{4}\right] = \sqrt{2}(-1 - i),$$

$$w_3 = 2\left[\cos\frac{7\pi}{4} + i\sin\frac{7\pi}{4}\right] = \sqrt{2}(1 - i). \qquad \square$$

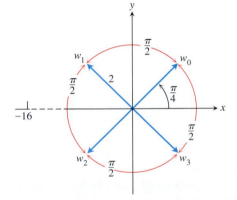

A.8 The four fourth roots of -16.

The Fundamental Theorem of Algebra   One may well say that the invention of $\sqrt{-1}$ is all well and good and leads to a number system that is richer than the real number system alone; but where will this process end? Are we also going to invent still more systems so as to obtain $\sqrt[4]{-1}$, $\sqrt[6]{-1}$, and so on? By now it should be clear that this is not necessary. These numbers are already expressible in terms of the complex number system $a + ib$. In fact, the Fundamental Theorem of Algebra says that with the introduction of the complex numbers we now have enough numbers to factor every polynomial into a product of linear factors and hence enough numbers to solve every possible polynomial equation.

The Fundamental Theorem of Algebra

Every polynomial equation of the form

$$a_0 z^n + a_1 z^{n-1} + a_2 z^{n-2} + \cdots + a_{n-1} z + a_n = 0,$$

in which the coefficients $a_0, a_1, \ldots, a_n$ are any complex numbers, whose degree n is greater than or equal to one, and whose leading coefficient a_0 is not zero, has exactly n roots in the complex number system, provided each multiple root of multiplicity m is counted as m roots.

A proof of this theorem can be found in almost any text on the theory of functions of a complex variable.

Exercises A.3

Operations with Complex Numbers

1. *How computers multiply complex numbers*
   Find $(a, b) \cdot (c, d) = (ac - bd, ad + bc)$.

   a) $(2, 3) \cdot (4, -2)$ b) $(2, -1) \cdot (-2, 3)$
   c) $(-1, -2) \cdot (2, 1)$

   (This is how complex numbers are multiplied by computers.)

2. Solve the following equations for the real numbers, x and y.

   a) $(3 + 4i)^2 - 2(x - iy) = x + iy$

   b) $\left(\dfrac{1+i}{1-i}\right)^2 + \dfrac{1}{x + iy} = 1 + i$

   c) $(3 - 2i)(x + iy) = 2(x - 2iy) + 2i - 1$

Graphing and Geometry

3. How may the following complex numbers be obtained from $z = x + iy$ geometrically? Sketch.

   a) $\bar{z}$ b) $\overline{(-z)}$
   c) $-z$ d) $1/z$

4. Show that the distance between the two points z_1 and z_2 in an Argand diagram is equal to $|z_1 - z_2|$.

In Exercises 5–10, graph the points $z = x + iy$ that satisfy the given conditions.

5. a) $|z| = 2$ b) $|z| < 2$ c) $|z| > 2$

6. $|z - 1| = 2$ 7. $|z + 1| = 1$

8. $|z + 1| = |z - 1|$ 9. $|z + i| = |z - 1|$

10. $|z + 1| \geq |z|$

Express the complex numbers in Exercises 11–14 in the form $re^{i\theta}$, with $r \geq 0$ and $-\pi < \theta \leq \pi$. Draw an Argand diagram for each calculation.

11. $(1 + \sqrt{-3})^2$ 12. $\dfrac{1+i}{1-i}$

13. $\dfrac{1 + i\sqrt{3}}{1 - i\sqrt{3}}$ 14. $(2 + 3i)(1 - 2i)$

Theory and Examples

15. Show with an Argand diagram that the law for adding complex numbers is the same as the parallelogram law for adding vectors.

16. Show that the conjugate of the sum (product, or quotient) of two complex numbers z_1 and z_2 is the same as the sum (product, or quotient) of their conjugates.

17. *Complex roots of polynomials with real coefficients come in complex-conjugate pairs.*

    a) Extend the results of Exercise 16 to show that $f(\bar{z}) = \overline{f(z)}$

if

$$f(z) = a_0 z^n + a_1 z^{n-1} + \cdots + a_{n-1} z + a_n$$

is a polynomial with real coefficients $a_0, \ldots, a_n$.

b) If z is a root of the equation $f(z) = 0$, where $f(z)$ is a polynomial with real coefficients as in part (a), show that the conjugate $\bar{z}$ is also a root of the equation. (*Hint:* Let $f(z) = u + iv = 0$; then both u and v are zero. Now use the fact that $f(\bar{z}) = \overline{f(z)} = u - iv$.)

18. Show that $|\bar{z}| = |z|$.

19. If z and $\bar{z}$ are equal, what can you say about the location of the point z in the complex plane?

20. Let $Re(z)$ denote the real part of z and $Im(z)$ the imaginary part. Show that the following relations hold for any complex numbers z, z_1, and z_2.

a) $z + \bar{z} = 2Re(z)$
b) $z - \bar{z} = 2i \, Im(z)$
c) $|Re(z)| \leq |z|$

d) $|z_1 + z_2|^2 = |z_1|^2 + |z_2|^2 + 2Re(z_1\bar{z}_2)$
e) $|z_1 + z_2| \leq |z_1| + |z_2|$

Use De Moivre's theorem to express the trigonometric functions in Exercises 21 and 22 in terms of $\cos \theta$ and $\sin \theta$.

21. $\cos 4\theta$ 22. $\sin 4\theta$

Roots

23. Find the three cube roots of 1.

24. Find the two square roots of i.

25. Find the three cube roots of $-8i$.

26. Find the six sixth roots of 64.

27. Find the four solutions of the equation $z^4 - 2z^2 + 4 = 0$.

28. Find the six solutions of the equation $z^6 + 2z^3 + 2 = 0$.

29. Find all solutions of the equation $x^4 + 4x^2 + 16 = 0$.

30. Solve the equation $x^4 + 1 = 0$.

| A.4 | # Simpson's One-Third Rule |

Simpson's rule for approximating $\int_a^b f(x)\,dx$ is based on approximating the graph of f with parabolic arcs.

The area of the shaded region under the parabola in Fig. A.9 is

$$\text{Area} = \frac{h}{3}(y_0 + 4y_1 + y_2).$$

This formula is known as Simpson's one-third rule.

We can derive the formula as follows. To simplify the algebra, we use the coordinate system in Fig. A.10. The area under the parabola is the same no matter where the y-axis is, as long as we preserve the vertical scale. The parabola has an equation of the form $y = Ax^2 + Bx + C$, so the area under it from $x = -h$ to

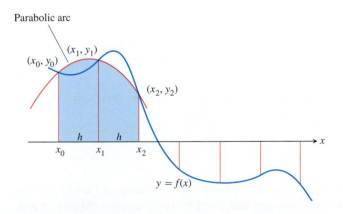

A.9 Simpson's rule approximates short stretches of curve with parabolic arcs.

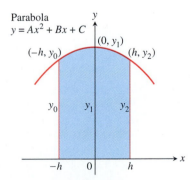

A.10 By integrating from $-h$ to h, the shaded area is found to be

$$\frac{h}{3}(y_0 + 4y_1 + y_2).$$

$x = h$ is

$$\text{Area} = \int_{-h}^{h} (Ax^2 + Bx + C)\, dx = \left[\frac{Ax^3}{3} + \frac{Bx^2}{2} + Cx \right]_{-h}^{h}$$

$$= \frac{2Ah^3}{3} + 2Ch$$

$$= \frac{h}{3}(2Ah^2 + 6C).$$

Since the curve passes through $(-h, y_0)$, $(0, y_1)$, and (h, y_2), we also have

$$y_0 = Ah^2 - Bh + C, \qquad y_1 = C, \qquad y_2 = Ah^2 + Bh + C.$$

From these equations we obtain

$$C = y_1,$$

$$Ah^2 - Bh = y_0 - y_1,$$

$$Ah^2 + Bh = y_2 - y_1,$$

$$2Ah^2 = y_0 + y_2 - 2y_1.$$

These substitutions for C and $2Ah^2$ give

$$\text{Area} = \frac{h}{3}(2Ah^2 + 6C) = \frac{h}{3}((y_0 + y_2 - 2y_1) + 6y_1) = \frac{h}{3}(y_0 + 4y_1 + y_2).$$

A.5 Cauchy's Mean Value Theorem and the Stronger Form of l'Hôpital's Rule

This appendix proves the finite-limit case of the stronger form of l'Hôpital's Rule (Section 6.6, Theorem 3).

L'Hôpital's Rule (Stronger Form)

Suppose that

$$f(x_0) = g(x_0) = 0$$

and that the functions f and g are both differentiable on an open interval (a, b) that contains the point x_0. Suppose also that $g' \neq 0$ at every point in (a, b) except possibly x_0. Then

$$\lim_{x \to x_0} \frac{f(x)}{g(x)} = \lim_{x \to x_0} \frac{f'(x)}{g'(x)}, \tag{1}$$

provided the limit on the right exists.

The proof of the stronger form of l'Hôpital's rule is based on Cauchy's Mean Value Theorem, a mean value theorem that involves two functions instead of one. We prove Cauchy's theorem first and then show how it leads to l'Hôpital's rule.

Cauchy's Mean Value Theorem

Suppose that functions f and g are continuous on $[a, b]$ and differentiable throughout (a, b) and suppose also that $g' \neq 0$ throughout (a, b). Then there exists a number c in (a, b) at which

$$\frac{f'(c)}{g'(c)} = \frac{f(b) - f(a)}{g(b) - g(a)}.$$ (2)

The ordinary Mean Value Theorem (Section 3.2, Theorem 4) is the case $g(x) = x$.

Proof of Cauchy's Mean Value Theorem   We apply the Mean Value Theorem of Section 3.2 twice. First we use it to show that $g(a) \neq g(b)$. For if $g(b)$ did equal $g(a)$, then the Mean Value Theorem would give

$$g'(c) = \frac{g(b) - g(a)}{b - a} = 0$$

for some c between a and b. This cannot happen because $g'(x) \neq 0$ in (a, b).

We next apply the Mean Value Theorem to the function

$$F(x) = f(x) - f(a) - \frac{f(b) - f(a)}{g(b) - g(a)} [g(x) - g(a)].$$

This function is continuous and differentiable where f and g are, and $F(b) = F(a) = 0$. Therefore there is a number c between a and b for which $F'(c) = 0$. In terms of f and g this says

$$F'(c) = f'(c) - \frac{f(b) - f(a)}{g(b) - g(a)} [g'(c)] = 0,$$

or

$$\frac{f'(c)}{g'(c)} = \frac{f(b) - f(a)}{g(b) - g(a)},$$

which is Eq. (2).   ❏

Proof of the Stronger Form of l'Hôpital's Rule   We first establish Eq. (1) for the case $x \to x_0^+$. The method needs almost no change to apply to $x \to x_0^-$, and the combination of these two cases establishes the result.

Suppose that x lies to the right of x_0. Then $g'(x) \neq 0$ and we can apply Cauchy's Mean Value Theorem to the closed interval from x_0 to x. This produces a number c between x_0 and x such that

$$\frac{f'(c)}{g'(c)} = \frac{f(x) - f(x_0)}{g(x) - g(x_0)}.$$

But $f(x_0) = g(x_0) = 0$, so

$$\frac{f'(c)}{g'(c)} = \frac{f(x)}{g(x)}.$$

As x approaches x_0, c approaches x_0 because it lies between x and x_0. Therefore,

$$\lim_{x \to x_0^+} \frac{f(x)}{g(x)} = \lim_{c \to x_0^+} \frac{f'(c)}{g'(c)} = \lim_{x \to x_0^+} \frac{f'(x)}{g'(x)}.$$

This establishes l'Hôpital's rule for the case where x approaches x_0 from above. The case where x approaches x_0 from below is proved by applying Cauchy's Mean Value Theorem to the closed interval $[x, x_0], x < x_0$. ❏

A.6 Limits That Arise Frequently

This appendix verifies limits (4)–(6) in Section 8.2, Table 1.

Limit 4: If $|x| < 1$, $\lim\limits_{n \to \infty} x^n = 0$ We need to show that to each $\epsilon > 0$ there corresponds an integer N so large that $|x^n| < \epsilon$ for all n greater than N. Since $\epsilon^{1/n} \to 1$, while $|x| < 1$, there exists an integer N for which $\epsilon^{1/N} > |x|$. In other words,

$$|x^N| = |x|^N < \epsilon. \tag{1}$$

This is the integer we seek because, if $|x| < 1$, then

$$|x^n| < |x^N| \text{ for all } n > N. \tag{2}$$

Combining (1) and (2) produces $|x^n| < \epsilon$ for all $n > N$, concluding the proof.

Limit 5: For any number x, $\lim\limits_{n \to \infty} \left(1 + \dfrac{x}{n}\right)^n = e^x$ Let

$$a_n = \left(1 + \frac{x}{n}\right)^n.$$

Then

$$\ln a_n = \ln \left(1 + \frac{x}{n}\right)^n = n \ln \left(1 + \frac{x}{n}\right) \to x,$$

as we can see by the following application of l'Hôpital's rule, in which we differentiate with respect to n:

$$\lim_{n \to \infty} n \ln \left(1 + \frac{x}{n}\right) = \lim_{n \to \infty} \frac{\ln (1 + x/n)}{1/n}$$

$$= \lim_{n \to \infty} \frac{\left(\dfrac{1}{1 + x/n}\right) \cdot \left(-\dfrac{x}{n^2}\right)}{-1/n^2} = \lim_{n \to \infty} \frac{x}{1 + x/n} = x.$$

Apply Theorem 4, Section 8.2, with $f(x) = e^x$ to conclude that

$$\left(1 + \frac{x}{n}\right)^n = a_n = e^{\ln a_n} \to e^x.$$

Limit 6: For any number x, $\lim\limits_{n \to \infty} \dfrac{x^n}{n!} = 0$ Since

$$-\frac{|x|^n}{n!} \le \frac{x^n}{n!} \le \frac{|x|^n}{n!},$$

all we need to show is that $|x|^n/n! \to 0$. We can then apply the Sandwich Theorem for Sequences (Section 8.2, Theorem 3) to conclude that $x^n/n! \to 0$.

The first step in showing that $|x|^n/n! \to 0$ is to choose an integer $M > |x|$, so that $(|x|/M) < 1$. By Limit 4, just proved, we then have $(|x|/M)^n \to 0$. We then restrict our attention to values of $n > M$. For these values of n, we can write

$$\frac{|x|^n}{n!} = \frac{|x|^n}{1 \cdot 2 \cdot \cdots \cdot M \cdot \underbrace{(M+1)(M+2) \cdot \cdots \cdot n}_{(n-M) \text{ factors}}}$$

$$\le \frac{|x|^n}{M! M^{n-M}} = \frac{|x|^n M^M}{M! M^n} = \frac{M^M}{M!} \left(\frac{|x|}{M}\right)^n.$$

Thus,

$$0 \le \frac{|x|^n}{n!} \le \frac{M^M}{M!} \left(\frac{|x|}{M}\right)^n.$$

Now, the constant $M^M/M!$ does not change as n increases. Thus the Sandwich Theorem tell us that $|x|^n/n! \to 0$ because $(|x|/M)^n \to 0$.

A.7 The Distributive Law for Vector Cross Products

In this appendix we prove the distributive law

$$\mathbf{A} \times (\mathbf{B} + \mathbf{C}) = \mathbf{A} \times \mathbf{B} + \mathbf{A} \times \mathbf{C} \tag{1}$$

from Eq. (6) in Section 10.4.

Proof To derive Eq. (1), we construct $\mathbf{A} \times \mathbf{B}$ a new way. We draw $\mathbf{A}$ and $\mathbf{B}$ from the common point O and construct a plane M perpendicular to $\mathbf{A}$ at O (Fig. A.11). We then project $\mathbf{B}$ orthogonally onto M, yielding a vector $\mathbf{B}'$ with length $|\mathbf{B}| \sin \theta$. We rotate $\mathbf{B}'$ 90° about $\mathbf{A}$ in the positive sense to produce a vector $\mathbf{B}''$. Finally, we multiply $\mathbf{B}''$ by the length of $\mathbf{A}$. The resulting vector $|\mathbf{A}|\mathbf{B}''$ is equal to $\mathbf{A} \times \mathbf{B}$ since $\mathbf{B}''$ has the same direction as $\mathbf{A} \times \mathbf{B}$ by its construction (Fig. A.11) and

$$|\mathbf{A}||\mathbf{B}''| = |\mathbf{A}||\mathbf{B}'| = |\mathbf{A}||\mathbf{B}| \sin \theta = |\mathbf{A} \times \mathbf{B}|.$$

Now each of these three operations, namely,

1. projection onto M,
2. rotation about $\mathbf{A}$ through 90°,
3. multiplication by the scalar $|\mathbf{A}|$,

A.11 As explained in the text, $\mathbf{A} \times \mathbf{B} = |\mathbf{A}|\mathbf{B}''$.

A.12 The vectors, **B, C, B + C,** and their projections onto a plane perpendicular to **A.**

when applied to a triangle whose plane is not parallel to **A,** will produce another triangle. If we start with the triangle whose sides are **B, C,** and **B + C** (Fig. A.12) and apply these three steps, we successively obtain

1. a triangle whose sides are **B′, C′,** and **(B + C)′** satisfying the vector equation

    $$\mathbf{B'} + \mathbf{C'} = (\mathbf{B} + \mathbf{C})';$$

2. a triangle whose sides are **B″, C″,** and **(B + C)″** satisfying the vector equation

    $$\mathbf{B''} + \mathbf{C''} = (\mathbf{B} + \mathbf{C})''$$

    (the double prime on each vector has the same meaning as in Fig. A.11); and, finally,

3. a triangle whose sides are **|A|B″, |A|C″,** and **|A|(B + C)″** satisfying the vector equation

    $$|\mathbf{A}|\mathbf{B''} + |\mathbf{A}|\mathbf{C''} = |\mathbf{A}|(\mathbf{B} + \mathbf{C})''. \tag{2}$$

    Substituting $|\mathbf{A}|\mathbf{B''} = \mathbf{A} \times \mathbf{B}$, $|\mathbf{A}|\mathbf{C''} = \mathbf{A} \times \mathbf{C}$, and $|\mathbf{A}|(\mathbf{B} + \mathbf{C})'' = \mathbf{A} \times (\mathbf{B} + \mathbf{C})$ from our discussion above into Eq. (2) gives

    $$\mathbf{A} \times \mathbf{B} + \mathbf{A} \times \mathbf{C} = \mathbf{A} \times (\mathbf{B} + \mathbf{C}),$$

which is the law we wanted to establish.   ❏

A.8   Determinants and Cramer's Rule

A rectangular array of numbers like

$$A = \begin{bmatrix} 2 & 1 & 3 \\ 1 & 0 & -2 \end{bmatrix}$$

is called a **matrix.** We call A a 2 by 3 matrix because it has two rows and three columns. An m by n matrix has m rows and n columns, and the **entry** or **element** (number) in the ith row and jth column is denoted by a_{ij}. The matrix

$$A = \begin{bmatrix} 2 & 1 & 3 \\ 1 & 0 & -2 \end{bmatrix}$$

has

$$a_{11} = 2, \qquad a_{12} = 1, \qquad a_{13} = 3,$$
$$a_{21} = 1, \qquad a_{22} = 0, \qquad a_{23} = -2.$$

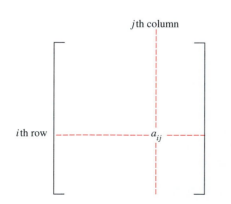

jth column

ith row — a_{ij}

A matrix with the same number of rows as columns is a **square matrix.** It is a **matrix of order n** if the number of rows and columns is n.

With each square matrix A we associate a number det A or $|a_{ij}|$, called the **determinant** of A, calculated from the entries of A in the following way. For $n = 1$ and $n = 2$, we define

The vertical bars in the notation $|a_{ij}|$ do not mean absolute value.

$$\det [a] = a, \tag{1}$$

$$\det \begin{bmatrix} a_{11} & a_{12} \\ a_{21} & a_{22} \end{bmatrix} = a_{11}a_{22} - a_{21}a_{12}. \tag{2}$$

For a matrix of order 3, we define

$$\det A = \det \begin{bmatrix} a_{11} & a_{12} & a_{13} \\ a_{21} & a_{22} & a_{23} \\ a_{31} & a_{32} & a_{33} \end{bmatrix} = \begin{array}{l} \text{Sum of all signed products} \\ \text{of the form } \pm a_{1i}a_{2j}a_{3k}, \end{array} \tag{3}$$

where i, j, k is a permutation of 1, 2, 3 in some order. There are $3! = 6$ such permutations, so there are six terms in the sum. The sign is positive when the index of the permutation is even and negative when the index is odd.

Definition
Index of a Permutation

Given any permutation of the numbers $1, 2, 3, \ldots, n$, denote the permutation by $i_1, i_2, i_3, \ldots, i_n$. In this arrangement, some of the numbers following i_1 may be less than i_1, and the number of these is called the **number of inversions** in the arrangement pertaining to i_1. Likewise, there are a number of inversions pertaining to each of the other i's; it is the number of indices that come after that particular i in the arrangement and are less than it. The **index** of the permutation is the sum of all of the numbers of inversions pertaining to the separate indices.

EXAMPLE 1   For $n = 5$, the permutation

$$5 \quad 3 \quad 1 \quad 2 \quad 4$$

has 4 inversions pertaining to the first element, 5, 2 inversions pertaining to the second element, 3, and no further inversions, so the index is $4 + 2 = 6$. ❏

The following table shows the permutations of 1, 2, 3, the index of each permutation, and the signed product in the determinant of Eq. (3).

Permutation	Index	Signed product
1 2 3	0	$+a_{11}a_{22}a_{33}$
1 3 2	1	$-a_{11}a_{23}a_{32}$
2 1 3	1	$-a_{12}a_{21}a_{33}$
2 3 1	2	$+a_{12}a_{23}a_{31}$
3 1 2	2	$+a_{13}a_{21}a_{32}$
3 2 1	3	$-a_{13}a_{22}a_{31}$

$$\tag{4}$$

The sum of the six signed products is

$$a_{11}(a_{22}a_{33} - a_{23}a_{32}) - a_{12}(a_{21}a_{33} - a_{23}a_{31}) + a_{13}(a_{21}a_{32} - a_{22}a_{31})$$

$$= a_{11} \begin{vmatrix} a_{22} & a_{23} \\ a_{32} & a_{33} \end{vmatrix} - a_{12} \begin{vmatrix} a_{21} & a_{23} \\ a_{31} & a_{33} \end{vmatrix} + a_{13} \begin{vmatrix} a_{21} & a_{22} \\ a_{31} & a_{32} \end{vmatrix} = \begin{vmatrix} a_{11} & a_{12} & a_{13} \\ a_{21} & a_{22} & a_{23} \\ a_{31} & a_{32} & a_{33} \end{vmatrix}. \quad (5)$$

The formula

$$\begin{vmatrix} a_{11} & a_{12} & a_{13} \\ a_{21} & a_{22} & a_{23} \\ a_{31} & a_{32} & a_{33} \end{vmatrix} = a_{11} \begin{vmatrix} a_{22} & a_{23} \\ a_{32} & a_{33} \end{vmatrix} - a_{12} \begin{vmatrix} a_{21} & a_{23} \\ a_{31} & a_{33} \end{vmatrix} + a_{13} \begin{vmatrix} a_{21} & a_{22} \\ a_{31} & a_{32} \end{vmatrix} \quad (6)$$

reduces the calculation of a 3 by 3 determinant to the calculation of three 2 by 2 determinants.

Many people prefer to remember the following scheme for calculating the six signed products in the determinant of a 3 by 3 matrix:

$$(7)$$

Minors and Cofactors

The second order determinants on the right-hand side of Eq. (6) are called the **minors** (short for "minor determinants") of the entries they multiply. Thus,

$$\begin{vmatrix} a_{22} & a_{23} \\ a_{32} & a_{33} \end{vmatrix} \text{ is the minor of } a_{11}, \qquad \begin{vmatrix} a_{21} & a_{23} \\ a_{31} & a_{33} \end{vmatrix} \text{ is the minor of } a_{12},$$

and so on. The minor of the element a_{ij} in a matrix A is the determinant of the matrix that remains after we delete the row and column containing a_{ij}:

$$\begin{vmatrix} a_{11} & a_{12} & a_{13} \\ a_{21} & a_{22} & a_{23} \\ a_{31} & a_{32} & a_{33} \end{vmatrix}. \qquad \text{The minor of } a_{22} \text{ is } \begin{vmatrix} a_{11} & a_{13} \\ a_{31} & a_{33} \end{vmatrix}.$$

$$\begin{vmatrix} a_{11} & a_{12} & a_{13} \\ a_{21} & a_{22} & a_{23} \\ a_{31} & a_{32} & a_{33} \end{vmatrix}. \qquad \text{The minor of } a_{23} \text{ is } \begin{vmatrix} a_{11} & a_{12} \\ a_{31} & a_{32} \end{vmatrix}.$$

The **cofactor** A_{ij} of a_{ij} is $(-1)^{i+j}$ times the minor of a_{ij}. Thus,

$$A_{22} = (-1)^{2+2} \begin{vmatrix} a_{11} & a_{13} \\ a_{31} & a_{33} \end{vmatrix} = \begin{vmatrix} a_{11} & a_{13} \\ a_{31} & a_{33} \end{vmatrix},$$

$$A_{23} = (-1)^{2+3} \begin{vmatrix} a_{11} & a_{12} \\ a_{31} & a_{32} \end{vmatrix} = - \begin{vmatrix} a_{11} & a_{12} \\ a_{31} & a_{32} \end{vmatrix}.$$

The factor $(-1)^{i+j}$ changes the sign of the minor when $i + j$ is odd. There is a checkerboard pattern for remembering these changes:

$$\begin{matrix} + & - & + \\ - & + & - \\ + & - & + \end{matrix}$$

In the upper left corner, $i = 1$, $j = 1$ and $(-1)^{1+1} = +1$. In going from any cell to an adjacent cell in the same row or column, we change i by 1 or j by 1, but not both, so we change the exponent from even to odd or from odd to even, which changes the sign from + to − or from − to +.

When we rewrite Eq. (6) in terms of cofactors we get

$$\det A = a_{11}A_{11} + a_{12}A_{12} + a_{13}A_{13}. \tag{8}$$

EXAMPLE 2   Find the determinant of

$$A = \begin{bmatrix} 2 & 1 & 3 \\ 3 & -1 & -2 \\ 2 & 3 & 1 \end{bmatrix}.$$

Solution 1 The cofactors are

$$A_{11} = (-1)^{1+1} \begin{vmatrix} -1 & -2 \\ 3 & 1 \end{vmatrix}, \qquad A_{12} = (-1)^{1+2} \begin{vmatrix} 3 & -2 \\ 2 & 1 \end{vmatrix},$$

$$A_{13} = (-1)^{1+3} \begin{vmatrix} 3 & -1 \\ 2 & 3 \end{vmatrix}.$$

To find $\det A$, we multiply each element of the first row of A by its cofactor and add:

$$\det A = 2 \begin{vmatrix} -1 & -2 \\ 3 & 1 \end{vmatrix} + (-1) \begin{vmatrix} 3 & -2 \\ 2 & 1 \end{vmatrix} + 3 \begin{vmatrix} 3 & -1 \\ 2 & 3 \end{vmatrix}$$

$$= 2(-1 + 6) - 1(3 + 4) + 3(9 + 2) = 10 - 7 + 33 = 36.$$

Solution 2 From (7) we find

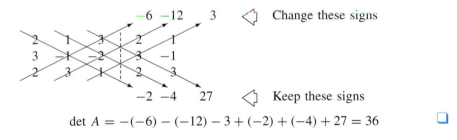

$$\det A = -(-6) - (-12) - 3 + (-2) + (-4) + 27 = 36 \qquad \square$$

Expanding by Columns or by Other Rows

The determinant of a square matrix can be calculated from the cofactors of any row or any column.

If we were to expand the determinant in Example 2 by cofactors according to elements of its third column, say, we would get

$$+3 \begin{vmatrix} 3 & -1 \\ 2 & 3 \end{vmatrix} - (-2) \begin{vmatrix} 2 & 1 \\ 2 & 3 \end{vmatrix} + 1 \begin{vmatrix} 2 & 1 \\ 3 & -1 \end{vmatrix}$$

$$= 3(9 + 2) + 2(6 - 2) + 1(-2 - 3) = 33 + 8 - 5 = 36.$$

Useful Facts About Determinants

Fact 1: If two rows (or columns) are identical, the determinant is zero.

Fact 2: Interchanging two rows (or columns) changes the sign of the determinant.

Fact 3: The determinant is the sum of the products of the elements of the ith row (or column) by their cofactors, for any i.

Fact 4: The determinant of the transpose of a matrix is the same as the determinant of the original matrix. (The **transpose** of a matrix is obtained by writing the rows as columns.)

Fact 5: Multiplying each element of some row (or column) by a constant c multiplies the determinant by c.

Fact 6: If all elements above the main diagonal (or all below it) are zero, the determinant is the product of the elements on the main diagonal. (The **main diagonal** is the diagonal from upper left to lower right.)

EXAMPLE 3

$$\begin{vmatrix} 3 & 4 & 7 \\ 0 & -2 & 5 \\ 0 & 0 & 5 \end{vmatrix} = (3)(-2)(5) = -30$$

Fact 7: If the elements of any row are multiplied by the cofactors of the corresponding elements of a different row and these products are summed, the sum is zero.

EXAMPLE 4

If A_{11}, A_{12}, A_{13} are the cofactors of the elements of the first row of $A = (a_{ij})$, then the sums

$$a_{21}A_{11} + a_{22}A_{12} + a_{23}A_{13}$$

(elements of second row times cofactors of elements of first row) and

$$a_{31}A_{11} + a_{32}A_{12} + a_{33}A_{13}$$

are both zero.

Fact 8: If the elements of any column are multiplied by the cofactors of the corresponding elements of a different column and these products are summed, the sum is zero.

Fact 9: If each element of a row is multiplied by a constant c and the results added to a different row, the determinant is not changed. A similar result holds for columns.

EXAMPLE 5

If we start with

$$A = \begin{bmatrix} 2 & 1 & 3 \\ 3 & -1 & -2 \\ 2 & 3 & 1 \end{bmatrix}$$

and add -2 times row 1 to row 2 (subtract 2 times row 1 from row 2), we get

$$B = \begin{bmatrix} 2 & 1 & 3 \\ -1 & -3 & -8 \\ 2 & 3 & 1 \end{bmatrix}.$$

Since det $A = 36$ (Example 2), we should find that det $B = 36$ as well. Indeed we do, as the following calculation shows:

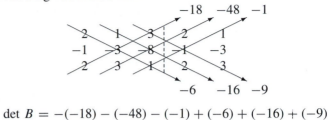

$$\det B = -(-18) - (-48) - (-1) + (-6) + (-16) + (-9)$$
$$= 18 + 48 + 1 - 6 - 16 - 9 = 67 - 31 = 36.$$ ❑

EXAMPLE 6 Evaluate the fourth order determinant

$$D = \begin{vmatrix} 1 & -2 & 3 & 1 \\ 2 & 1 & 0 & 2 \\ -1 & 2 & 1 & -2 \\ 0 & 1 & 2 & 1 \end{vmatrix}.$$

Solution We subtract 2 times row 1 from row 2 and add row 1 to row 3 to get

$$D = \begin{vmatrix} 1 & -2 & 3 & 1 \\ 0 & 5 & -6 & 0 \\ 0 & 0 & 4 & -1 \\ 0 & 1 & 2 & 1 \end{vmatrix}.$$

We then multiply the elements of the first column by their cofactors to get

$$D = \begin{vmatrix} 5 & -6 & 0 \\ 0 & 4 & -1 \\ 1 & 2 & 1 \end{vmatrix} = 5(4 + 2) - (-6)(0 + 1) + 0 = 36.$$ ❑

Cramer's Rule

If the determinant $D = \det A = \begin{vmatrix} a_{11} & a_{12} \\ a_{21} & a_{22} \end{vmatrix} = 0$, the system

$$a_{11}x + a_{12}y = b_1,$$
$$a_{21}x + a_{22}y = b_2$$

(9)

has either infinitely many solutions or no solution at all. The system

$$x + y = 0,$$
$$2x + 2y = 0$$

whose determinant is

$$D = \begin{vmatrix} 1 & 1 \\ 2 & 2 \end{vmatrix} = 2 - 2 = 0$$

has infinitely many solutions. We can find an x to match any given y. The system

$$x + y = 0,$$
$$2x + 2y = 2$$

has no solution. If $x + y = 0$, then $2x + 2y = 2(x + y)$ cannot be 2.

If $D \neq 0$, the system (9) has a unique solution, and Cramer's rule states that it may be found from the formulas

$$x = \frac{\begin{vmatrix} b_1 & a_{12} \\ b_2 & a_{22} \end{vmatrix}}{D}, \qquad y = \frac{\begin{vmatrix} a_{11} & b_1 \\ a_{21} & b_2 \end{vmatrix}}{D}. \tag{10}$$

The numerator in the formula for x comes from replacing the first column in A (the x-column) by the column of constants b_1 and b_2 (the b-column). Replacing the y-column by the b-column gives the numerator of the y-solution.

EXAMPLE 7   Solve the system

$$3x - y = 9,$$
$$x + 2y = -4.$$

Solution   We use Eqs. (10). The determinant of the coefficient matrix is

$$D = \begin{vmatrix} 3 & -1 \\ 1 & 2 \end{vmatrix} = 6 + 1 = 7.$$

Hence,

$$x = \frac{\begin{vmatrix} 9 & -1 \\ -4 & 2 \end{vmatrix}}{D} = \frac{18 - 4}{7} = \frac{14}{7} = 2,$$

$$y = \frac{\begin{vmatrix} 3 & 9 \\ 1 & -4 \end{vmatrix}}{D} = \frac{-12 - 9}{7} = \frac{-21}{7} = -3.$$

Systems of three equations in three unknowns work the same way. If

$$D = \det A = \begin{vmatrix} a_{11} & a_{12} & a_{13} \\ a_{21} & a_{22} & a_{23} \\ a_{31} & a_{32} & a_{33} \end{vmatrix} = 0,$$

the system

$$a_{11}x + a_{12}y + a_{13}z = b_1,$$
$$a_{21}x + a_{22}y + a_{23}z = b_2, \tag{11}$$
$$a_{31}x + a_{32}y + a_{33}z = b_3$$

has either infinitely many solutions or no solution at all. If $D \neq 0$, the system has a unique solution, given by Cramer's rule:

$$x = \frac{1}{D} \begin{vmatrix} b_1 & a_{12} & a_{13} \\ b_2 & a_{22} & a_{23} \\ b_3 & a_{32} & a_{33} \end{vmatrix}, \qquad y = \frac{1}{D} \begin{vmatrix} a_{11} & b_1 & a_{13} \\ a_{21} & b_2 & a_{23} \\ a_{31} & b_3 & a_{33} \end{vmatrix}.$$

$$z = \frac{1}{D} \begin{vmatrix} a_{11} & a_{12} & b_1 \\ a_{21} & a_{22} & b_2 \\ a_{31} & a_{32} & b_3 \end{vmatrix}.$$

The pattern continues in higher dimensions.

Exercises A.8

Evaluating Determinants

Evaluate the following determinants.

1. $\begin{vmatrix} 2 & 3 & 1 \\ 4 & 5 & 2 \\ 1 & 2 & 3 \end{vmatrix}$

2. $\begin{vmatrix} 2 & -1 & -2 \\ -1 & 2 & 1 \\ 3 & 0 & -3 \end{vmatrix}$

3. $\begin{vmatrix} 1 & 2 & 3 & 4 \\ 0 & 1 & 2 & 3 \\ 0 & 0 & 2 & 1 \\ 0 & 0 & 3 & 2 \end{vmatrix}$

4. $\begin{vmatrix} 1 & -1 & 2 & 3 \\ 2 & 1 & 2 & 6 \\ 1 & 0 & 2 & 3 \\ -2 & 2 & 0 & -5 \end{vmatrix}$

Evaluate the following determinants by expanding according to the cofactors of (a) the third row and (b) the second column.

5. $\begin{vmatrix} 2 & -1 & 2 \\ 1 & 0 & 3 \\ 0 & 2 & 1 \end{vmatrix}$

6. $\begin{vmatrix} 1 & 0 & -1 \\ 0 & 2 & -2 \\ 2 & 0 & 1 \end{vmatrix}$

7. $\begin{vmatrix} 1 & 1 & 0 & 0 \\ 0 & 0 & -2 & 1 \\ 0 & -1 & 0 & 7 \\ 3 & 0 & 2 & 1 \end{vmatrix}$

8. $\begin{vmatrix} 0 & 1 & 0 & 0 \\ 0 & 1 & 1 & 0 \\ 1 & 1 & 1 & 1 \\ 1 & 1 & 0 & 0 \end{vmatrix}$

Systems of Equations

Solve the following systems of equations by Cramer's rule.

9.
$$x + 8y = 4$$
$$3x - y = -13$$

10.
$$2x + 3y = 5$$
$$3x - y = 2$$

11.
$$4x - 3y = 6$$
$$3x - 2y = 5$$

12.
$$x + y + z = 2$$
$$2x - y + z = 0$$
$$x + 2y - z = 4$$

13.
$$2x + y - z = 2$$
$$x - y + z = 7$$
$$2x + 2y + z = 4$$

14.
$$2x - 4y = 6$$
$$x + y + z = 1$$
$$5y + 7z = 10$$

15.
$$x \quad\quad - z = 3$$
$$2y - 2z = 2$$
$$2x \quad\quad + z = 3$$

16.
$$x_1 + x_2 - x_3 + x_4 = 2$$
$$x_1 - x_2 + x_3 + x_4 = -1$$
$$x_1 + x_2 + x_3 - x_4 = 2$$
$$x_1 \quad\quad + x_3 + x_4 = -1$$

Theory and Examples

17. Find values of h and k for which the system

$$2x + hy = 8,$$
$$x + 3y = k$$

has (a) infinitely many solutions, (b) no solution at all.

18. For what value of x will

$$\begin{vmatrix} x & x & 1 \\ 2 & 0 & 5 \\ 6 & 7 & 1 \end{vmatrix} = 0?$$

19. Suppose u, v, and w are twice-differentiable functions of x that satisfy the relation $au + bv + cw = 0$, where a, b, and c are constants, not all zero. Show that

$$\begin{vmatrix} u & v & w \\ u' & v' & w' \\ u'' & v'' & w'' \end{vmatrix} = 0.$$

20. *Partial fractions.* Expanding the quotient

$$\frac{ax + b}{(x - r_1)(x - r_2)}$$

by partial fractions calls for finding the values of C and D that make the equation

$$\frac{ax + b}{(x - r_1)(x - r_2)} = \frac{C}{x - r_1} + \frac{D}{x - r_2}$$

hold for all x.

a) Find a system of linear equations that determines C and D.

b) Under what circumstances does the system of equations in part (a) have a unique solution? That is, when is the determinant of the coefficient matrix of the system different from zero?

A.9	

Euler's Theorem and the Increment Theorem

This appendix derives Euler's Theorem (Theorem 2, Section 12.3) and the Increment Theorem for Functions of Two Variables (Theorem 3, Section 12.4). Euler first published his theorem in 1734, in a series of papers he wrote on hydrodynamics.

Euler's Theorem

If $f(x, y)$ and its partial derivatives f_x, f_y, f_{xy}, and f_{yx} are defined throughout an open region containing a point (a, b) and are all continuous at (a, b), then $f_{xy}(a, b) = f_{yx}(a, b)$.

Proof   The equality of $f_{xy}(a, b)$ and $f_{yx}(a, b)$ can be established by four applications of the Mean Value Theorem (Theorem 4, Section 3.2). By hypothesis, the point (a, b) lies in the interior of a rectangle R in the xy-plane on which f, f_x, f_y, f_{xy}, and f_{yx} are all defined. We let h and k be numbers such that the point $(a + h, b + k)$ also lies in the rectangle R, and we consider the difference

$$\Delta = F(a + h) - F(a), \tag{1}$$

where

$$F(x) = f(x, b + k) - f(x, b). \tag{2}$$

We apply the Mean Value Theorem to F (which is continuous because it is differentiable), and Eq. (1) becomes

$$\Delta = hF'(c_1), \tag{3}$$

where c_1 lies between a and $a + h$. From Eq. (2),

$$F'(x) = f_x(x, b + k) - f_x(x, b),$$

so Eq. (3) becomes

$$\Delta = h[f_x(c_1, b + k) - f_x(c_1, b)]. \tag{4}$$

Now we apply the Mean Value Theorem to the function $g(y) = f_x(c_1, y)$ and have

$$g(b + k) - g(b) = kg'(d_1),$$

or

$$f_x(c_1, b + k) - f_x(c_1, b) = kf_{xy}(c_1, d_1),$$

for some d_1 between b and $b + k$. By substituting this into Eq. (4), we get

$$\Delta = hkf_{xy}(c_1, d_1), \tag{5}$$

for some point (c_1, d_1) in the rectangle R' whose vertices are the four points (a, b), $(a + h, b)$, $(a + h, b + k)$, and $(a, b + k)$. (See Fig. A.13.)

By substituting from Eq. (2) into Eq. (1), we may also write

$$\Delta = f(a + h, b + k) - f(a + h, b) - f(a, b + k) + f(a, b)$$
$$= [f(a + h, b + k) - f(a, b + k)] - [f(a + h, b) - f(a, b)] \tag{6}$$
$$= \phi(b + k) - \phi(b),$$

where $$\phi(y) = f(a + h, y) - f(a, y). \tag{7}$$

The Mean Value Theorem applied to Eq. (6) now gives

$$\Delta = k\phi'(d_2), \tag{8}$$

for some d_2 between b and $b + k$. By Eq. (7),

$$\phi'(y) = f_y(a + h, y) - f_y(a, y). \tag{9}$$

Substituting from Eq. (9) into Eq. (8) gives

$$\Delta = k[f_y(a + h, d_2) - f_y(a, d_2)].$$

Finally, we apply the Mean Value Theorem to the expression in brackets and get

$$\Delta = khf_{yx}(c_2, d_2), \tag{10}$$

for some c_2 between a and $a + h$.

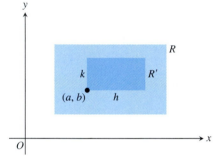

A.13 The key to proving $f_{xy}(a, b) = f_{yx}(a, b)$ is the fact that no matter how small R' is, f_{xy} and f_{yx} take on equal values somewhere inside R' (although not necessarily at the same point).

Together, Eqs. (5) and (10) show that

$$f_{xy}(c_1, d_1) = f_{yx}(c_2, d_2), \tag{11}$$

where (c_1, d_1) and (c_2, d_2) both lie in the rectangle R' (Fig. A.13). Equation (11) is not quite the result we want, since it says only that f_{xy} has the same value at (c_1, d_1) that f_{yx} has at (c_2, d_2). But the numbers h and k in our discussion may be made as small as we wish. The hypothesis that f_{xy} and f_{yx} are both continuous at (a, b) means that $f_{xy}(c_1, d_1) = f_{xy}(a, b) + \epsilon_1$ and $f_{yx}(c_2, d_2) = f_{yx}(a, b) + \epsilon_2$, where $\epsilon_1, \epsilon_2 \to 0$ as $h, k \to 0$. Hence, if we let h and $k \to 0$, we have $f_{xy}(a, b) = f_{yx}(a, b)$. $\qquad \blacksquare$

The equality of $f_{xy}(a, b)$ and $f_{yx}(a, b)$ can be proved with hypotheses weaker than the ones we assumed. For example, it is enough for f, f_x, and f_y to exist in R and for f_{xy} to be continuous at (a, b). Then f_{yx} will exist at (a, b) and will equal f_{xy} at that point.

A.14 The rectangular region T in the proof of the Increment Theorem. The figure is drawn for Δx and Δy positive, but either increment might be zero or negative.

> ### The Increment Theorem for Functions of Two Variables
>
> Suppose that the first partial derivatives of $z = f(x, y)$ are defined throughout an open region R containing the point (x_0, y_0) and that f_x and f_y are continuous at (x_0, y_0). Then the change $\Delta z = f(x_0 + \Delta x, y_0 + \Delta y) - f(x_0, y_0)$ in the value of f that results from moving from (x_0, y_0) to another point $(x_0 + \Delta x, y_0 + \Delta y)$ in R satisfies an equation of the form
>
> $$\Delta z = f_x(x_0, y_0)\Delta x + f_y(x_0, y_0)\Delta y + \epsilon_1 \Delta x + \epsilon_2 \Delta y,$$
>
> in which $\epsilon_1, \epsilon_2 \to 0$ as $\Delta x, \Delta y \to 0$.

Proof We work within a rectangle T centered at $A(x_0, y_0)$ and lying within R, and we assume that Δx and Δy are already so small that the line segment joining A to $B(x_0 + \Delta x, y_0)$ and the line segment joining B to $C(x_0 + \Delta x, y_0 + \Delta y)$ lie in the interior of T (Fig. A.14).

We may think of Δz as the sum $\Delta z = \Delta z_1 + \Delta z_2$ of two increments, where

$$\Delta z_1 = f(x_0 + \Delta x, y_0) - f(x_0, y_0)$$

is the change in the value of f from A to B and

$$\Delta z_2 = f(x_0 + \Delta x, y_0 + \Delta y) - f(x_0 + \Delta x, y_0)$$

is the change in the value of f from B to C (Fig. A.15, on the following page).

On the closed interval of x-values joining x_0 to $x_0 + \Delta x$, the function $F(x) = f(x, x_0)$ is a differentiable (and hence continuous) function of x, with derivative

$$F'(x) = f_x(x, y_0).$$

By the Mean Value Theorem (Theorem 4, Section 3.2), there is an x-value c between x_0 and $x_0 + \Delta x$ at which

$$F(x_0 + \Delta x) - F(x_0) = F'(c)\Delta x$$

or

$$f(x_0 + \Delta x, y_0) - f(x_0, y_0) = f_x(c, y_0)\Delta x$$

A.15 Part of the surface $z = f(x, y)$ near $P_0(x_0, y_0, f(x_0, y_0))$. The points P_0, P', and P'' have the same height $z_0 = f(x_0, y_0)$ above the xy-plane. The change in z is $\Delta z = P'S$. The change

$$\Delta z_1 = f(x_0 + \Delta x, y_0) - f(x_0, y_0),$$

shown as $P''Q = P'Q'$, is caused by changing x from x_0 to $x_0 + \Delta x$ while holding y equal to y_0. Then, with x held equal to $x_0 + \Delta x$,

$$\Delta z_2 = f(x_0 + \Delta x, y_0 + \Delta y)$$
$$-f(x_0 + \Delta x, y_0)$$

is the change in z caused by changing y from y_0 to $y_0 + \Delta y$. This is represented by $Q'S$. The total change in z is the sum of Δz_1 and Δz_2.

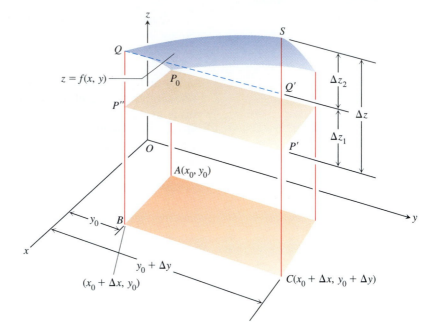

or

$$\Delta z_1 = f_x(c, y_0)\Delta x. \tag{12}$$

Similarly, $G(y) = f(x_0 + \Delta x, y)$ is a differentiable (and hence continuous) function of y on the closed y-interval joining y_0 and $y_0 + \Delta y$, with derivative

$$G'(y) = f_y(x_0 + \Delta x, y).$$

Hence there is a y-value d between y_0 and $y_0 + \Delta y$ at which

$$G(y_0 + \Delta y) - G(y_0) = G'(d)\Delta y$$

or

$$f(x_0 + \Delta x, y_0 + \Delta y) - f(x_0 + \Delta x, y) = f_y(x_0 + \Delta x, d)\Delta y$$

or

$$\Delta z_2 = f_y(x_0 + \Delta x, d)\Delta y. \tag{13}$$

Now, as Δx and $\Delta y \to 0$, we know $c \to x_0$ and $d \to y_0$. Therefore, since f_x and f_y are continuous at (x_0, y_0), the quantities

$$\epsilon_1 = f_x(c, y_0) - f_x(x_0, y_0),$$
$$\epsilon_2 = f_y(x_0 + \Delta x, d) - f_y(x_0, y_0) \tag{14}$$

both approach zero as Δx and $\Delta y \to 0$.

Finally,

$$\begin{aligned} \Delta z &= \Delta z_1 + \Delta z_2 \\ &= f_x(c, y_0)\Delta x + f_y(x_0 + \Delta x, d)\,\Delta y &&\text{From (12) and (13)} \\ &= [f_x(x_0, y_0) + \epsilon_1]\,\Delta x + [f_y(x_0, y_0) + \epsilon_2]\,\Delta y &&\text{From (14)} \\ &= f_x(x_0, y_0)\,\Delta x + f_y(x_0, y_0)\,\Delta y + \epsilon_1\,\Delta x + \epsilon_2\,\Delta y, \end{aligned}$$

where ϵ_1 and $\epsilon_2 \to 0$ as Δx and $\Delta y \to 0$. This is what we set out to prove. ❑

Analogous results hold for functions of any finite number of independent variables. Suppose that the first partial derivatives of

$$w = f(x, y, z)$$

are defined throughout an open region containing the point (x_0, y_0, z_0) and that f_x, f_y, and f_z are continuous at (x_0, y_0). Then

$$\Delta w = f(x_0 + \Delta x, y_0 + \Delta y, z_0 + \Delta z) - f(x_0, y_0, z_0)$$

$$= f_x \Delta x + f_y \Delta y + f_z \Delta z + \epsilon_1 \Delta x + \epsilon_2 \Delta y + \epsilon_3 \Delta z, \tag{15}$$

where

$$\epsilon_1, \epsilon_2, \epsilon_3 \to 0 \qquad \text{when} \qquad \Delta x, \Delta y, \text{ and } \Delta z \to 0.$$

The partial derivatives f_x, f_y, f_z in this formula are to be evaluated at the point (x_0, y_0, z_0).

The result (15) can be proved by treating Δw as the sum of three increments,

$$\Delta w_1 = f(x_0 + \Delta x, y_0, z_0) - f(x_0, y_0, z_0), \tag{16}$$

$$\Delta w_2 = f(x_0 + \Delta x, y_0 + \Delta y, z_0) - f(x_0 + \Delta x, y_0, z_0), \tag{17}$$

$$\Delta w_3 = f(x_0 + \Delta x, y_0 + \Delta y, z_0 + \Delta z) - f(x_0 + \Delta x, y_0 + \Delta y, z_0), \tag{18}$$

and applying the Mean Value Theorem to each of these separately. Two coordinates remain constant and only one varies in each of these partial increments Δw_1, Δw_2, Δw_3. In (17), for example, only y varies, since x is held equal to $x_0 + \Delta x$ and z is held equal to z_0. Since $f(x_0 + \Delta x, y, z_0)$ is a continuous function of y with a derivative f_y, it is subject to the Mean Value Theorem, and we have

$$\Delta w_2 = f_y(x_0 + \Delta x, y_1, z_0) \Delta y$$

for some y_1 between y_0 and $y_0 + \Delta y$.

Answers

PRELIMINARY CHAPTER

Section 1, pp. 7–8

1. $0.\overline{1}, 0.\overline{2}, 0.\overline{3}, 0.\overline{8}$ **3.** a) Not necessarily true   b) True   c) True
d) True   e) True   f) True   g) True   h) True
5. $x < -2$

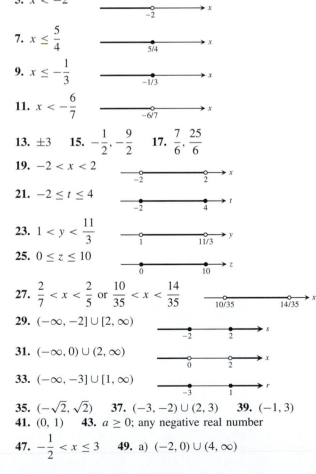

7. $x \le \dfrac{5}{4}$

9. $x \le -\dfrac{1}{3}$

11. $x < -\dfrac{6}{7}$

13. ± 3 **15.** $-\dfrac{1}{2}, -\dfrac{9}{2}$ **17.** $\dfrac{7}{6}, \dfrac{25}{6}$

19. $-2 < x < 2$

21. $-2 \le t \le 4$

23. $1 < y < \dfrac{11}{3}$

25. $0 \le z \le 10$

27. $\dfrac{2}{7} < x < \dfrac{2}{5}$ or $\dfrac{10}{35} < x < \dfrac{14}{35}$

29. $(-\infty, -2] \cup [2, \infty)$

31. $(-\infty, 0) \cup (2, \infty)$

33. $(-\infty, -3] \cup [1, \infty)$

35. $(-\sqrt{2}, \sqrt{2})$ **37.** $(-3, -2) \cup (2, 3)$ **39.** $(-1, 3)$
41. $(0, 1)$ **43.** $a \ge 0$; any negative real number
47. $-\dfrac{1}{2} < x \le 3$ **49.** a) $(-2, 0) \cup (4, \infty)$

Section 2, pp. 15–17

1. $2, -4; 2\sqrt{5}$ **3.** $-4.9, 0; 4.9$ **5.** Unit circle
7. The circle centered at the origin with points less than a radius of $\sqrt{3}$ and its interior

9. $m_\perp = -\dfrac{1}{3}$ 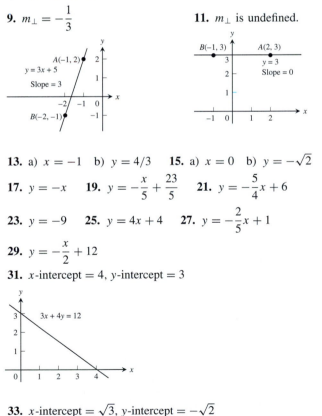   **11.** $m_\perp$ is undefined.

13. a) $x = -1$ b) $y = 4/3$ **15.** a) $x = 0$ b) $y = -\sqrt{2}$

17. $y = -x$ **19.** $y = -\dfrac{x}{5} + \dfrac{23}{5}$ **21.** $y = -\dfrac{5}{4}x + 6$

23. $y = -9$ **25.** $y = 4x + 4$ **27.** $y = -\dfrac{2}{5}x + 1$

29. $y = -\dfrac{x}{2} + 12$

31. x-intercept $= 4$, y-intercept $= 3$

33. x-intercept $= \sqrt{3}$, y-intercept $= -\sqrt{2}$

35. Yes. The lines are perpendicular because their slopes, $-A/B$ and B/A, are negative reciprocals of one another.
37. $(3, -3)$ **39.** $(-2, -9)$ **41.** a) ≈ -2.5 degrees/inch
b) ≈ -16.1 degrees/inch   c) ≈ -8.3 degrees/inch
43. 5.97 atm
45. Yes: $C = F = -40°$ **51.**

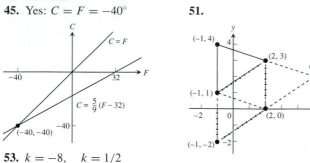

53. $k = -8, \quad k = 1/2$

Section 3, pp. 25–27

1. $D : (-\infty, \infty), \quad R : [1, \infty)$ **3.** $D : (0, \infty), \quad R : (0, \infty)$
5. $D : [-2, 2], \quad R : [0, 2]$
7. a) Not a function of x because some values of x have two values of y
b) A function of x because for every x there is only one possible y
9. $A = \dfrac{\sqrt{3}}{4} x^2, \quad p = 3x$ **11.** $x = \dfrac{d}{\sqrt{3}}, \quad A = 2d^2, \quad V = \dfrac{d^3}{3\sqrt{3}}$
13. Symmetric about the origin   **15.** Symmetric about the origin

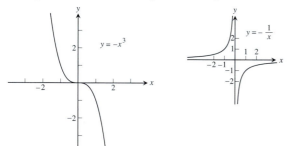

17. Symmetric about the y-axis

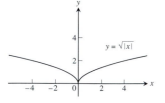

19. Symmetric about the origin

21. No symmetry

23. Symmetric about the y-axis

25. a) For each positive value of x, there are two values of y.   b) For each value of $x \neq 0$, there are two values of y.

27. Even   **29.** Even   **31.** Odd   **33.** Even   **35.** Neither
37. Neither
39. $D_f : -\infty < x < \infty, \quad D_g : x \geq 1, \quad R_f : -\infty < y < \infty,$
$R_g : y \geq 0, \quad D_{f+g} = D_{f \cdot g} = D_g, \quad R_{f+g} : y \geq 1, \quad R_{f \cdot g} : y \geq 0$
41. $D_f : -\infty < x < \infty, \quad D_g : -\infty < x < \infty, \quad R_f : y = 2,$
$R_g : y \geq 1, \quad D_{f/g} : -\infty < x < \infty, \quad R_{f/g} : 0 < y \leq 2,$
$D_{g/f} : -\infty < x < \infty, \quad R_{g/f} : y \geq 1/2$
43. a) 2   b) 22   c) $x^2 + 2$ d) $x^2 + 10x + 22$ e) 5   f) -2
g) $x + 10$ h) $x^4 - 6x^2 + 6$
45. a) $\dfrac{4}{x^2} - 5$ b) $\dfrac{4}{x^2} - 5$ c) $\left(\dfrac{4}{x} - 5\right)^2$ d) $\left(\dfrac{1}{4x - 5}\right)^2$

e) $\dfrac{1}{4x^2 - 5}$ f) $\dfrac{1}{(4x - 5)^2}$
47. a) $f(g(x))$ b) $j(g(x))$ c) $g(g(x))$ d) $j(j(x))$
e) $g(h(f(x)))$ f) $h(j(f(x)))$
49.

	$g(x)$	$f(x)$	$f \circ g(x)$
a)	$x - 7$	$\sqrt{x}$	$\sqrt{x - 7}$
b)	$x + 2$	$3x$	$3x + 6$
c)	x^2	$\sqrt{x - 5}$	$\sqrt{x^2 - 5}$
d)	$\dfrac{x}{x - 1}$	$\dfrac{x}{x - 1}$	x
e)	$\dfrac{1}{x - 1}$	$1 + \dfrac{1}{x}$	x
f)	$\dfrac{1}{x}$	$\dfrac{1}{x}$	x

51.

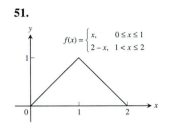

$$f(x) = \begin{cases} x, & 0 \le x \le 1 \\ 2-x, & 1 < x \le 2 \end{cases}$$

53.

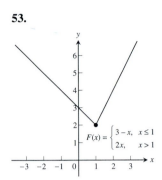

$$F(x) = \begin{cases} 3-x, & x \le 1 \\ 2x, & x > 1 \end{cases}$$

55. a) $y = \begin{cases} x, & 0 \le x \le 1 \\ 2-x, & 1 < x \le 2 \end{cases}$

b) $y = \begin{cases} 2, & 0 \le x < 1 \text{ or } 2 \le x < 3 \\ 0, & 1 \le x < 2 \text{ or } 3 \le x \le 4 \end{cases}$

57. a) $0 \le x < 1$ b) $-1 < x \le 0$ **59.** Yes
61. a) Odd   b) Odd   c) Odd   d) Even   e) Even   f) Even
g) Even   h) Even   i) Odd

Section 4, pp. 32–35

1. a) $y = -(x+7)^2$ b) $y = -(x-4)^2$
3. a) Position 4   b) Position 1   c) Position 2   d) Position 3
5. $(x+2)^2 + (y+3)^2 = 49$ **7.** $y+1 = (x+1)^3$

9. $y = \sqrt{x+0.81}$

11. $y = 2x$

13. $x+1 = y^2$

15. $y-1 = \dfrac{1}{x-1}$

17.

$y = \sqrt{x+4}$

19.

$y = |x-2|$

21.

$y = 1 + \sqrt{x-1}$

23.

$y = (x+1)^{2/3}$

25.

$y = 1 - x^{2/3}$

27.

$y = \sqrt[3]{x-1} - 1$

29.

$y = \dfrac{1}{x-2}$

31.

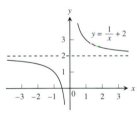

$y = \dfrac{1}{x} + 2$

33.

$y = \dfrac{1}{(x-1)^2}$

35.

$y = \dfrac{1}{x^2} + 1$

37. a) $D : [0, 2]$, $R : [2, 3]$

$y = f(x) + 2$

b) $D : [0, 2]$, $R : [-1, 0]$

$y = f(x) - 1$

c) $D : [0, 2], \quad R : [0, 2]$

d) $D : [0, 2], \quad R : [-1, 0]$

47. $x^2 + (y - 3/2)^2 = 25/4$

e) $D : [-2, 0], \quad R : [0, 1]$

f) $D : [1, 3], \quad R : [0, 1]$

49. $(x - 2)^2 + (y + 2)^2 = 8$

g) $D : [-2, 0], \quad R : [0, 1]$

h) $D : [-1, 1], \quad R : [0, 1]$

51.

53.

39. $x^2 + (y - 2)^2 = 4$

41. $(x + 1)^2 + (y - 5)^2 = 10$

55.

57.

43. $(x + \sqrt{3})^2 + (y + 2)^2 = 4$

45. $(x + 2)^2 + (y - 2)^2 = 4$

59. $D : 0 \leq x \leq 1, \quad R : 0 \leq y \leq 1/2$

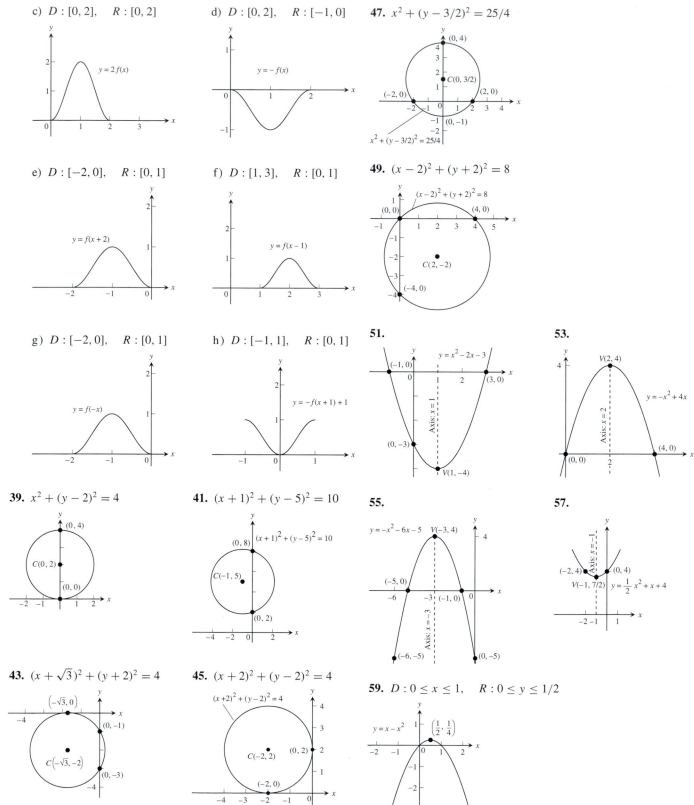

61. Exterior points of a circle of radius $\sqrt{7}$, centered at the origin
63. A circle of radius 2, centered at $(1, 0)$, together with its interior
65. The washer between the circles $x^2 + y^2 = 1$ and $x^2 + y^2 = 4$ (points with distance from the origin between 1 and 2)
67. The interior points of a circle centered at $(0, -3)$ with a radius of 3 that lie above the line $y = -3$
69. $(x + 2)^2 + (y - 1)^2 < 6$ **71.** $x^2 + y^2 \le 2, \quad x \ge 1$

73. $y = y_0 + m(x - x_0)$ **75.** $\left(\dfrac{1}{\sqrt{5}}, \dfrac{2}{\sqrt{5}} \right), \quad \left(-\dfrac{1}{\sqrt{5}}, -\dfrac{2}{\sqrt{5}} \right)$

77. $\left(\dfrac{1 - \sqrt{5}}{2}, \dfrac{3 - \sqrt{5}}{2} \right), \quad \left(\dfrac{1 + \sqrt{5}}{2}, \dfrac{3 + \sqrt{5}}{2} \right)$

79. $\left(-\dfrac{1}{\sqrt{3}}, -\dfrac{1}{3} \right), \quad \left(\dfrac{1}{\sqrt{3}}, -\dfrac{1}{3} \right)$

81. $\left(\dfrac{1}{2}, -\dfrac{\sqrt{3}}{2} \right), \quad \left(\dfrac{1}{2}, \dfrac{\sqrt{3}}{2} \right)$

Section 5, pp. 43–47

1. a) 8π m   b) $\dfrac{55\pi}{9}$ m   **3.** 8.4 in.

5.

θ	$-\pi$	$-2\pi/3$	0	$\pi/2$	$3\pi/4$
$\sin\theta$	0	$-\dfrac{\sqrt{3}}{2}$	0	1	$\dfrac{1}{\sqrt{2}}$
$\cos\theta$	-1	$-\dfrac{1}{2}$	1	0	$-\dfrac{1}{\sqrt{2}}$
$\tan\theta$	0	$\sqrt{3}$	0	UND	-1
$\cot\theta$	UND	$\dfrac{1}{\sqrt{3}}$	UND	0	-1
$\sec\theta$	-1	-2	1	UND	$-\sqrt{2}$
$\csc\theta$	UND	$-\dfrac{2}{\sqrt{3}}$	UND	1	$\sqrt{2}$

7. $\cos x = -4/5, \tan x = -3/4$ **9.** $\sin x = -\dfrac{\sqrt{8}}{3}, \tan x = -\sqrt{8}$

11. $\sin x = -\dfrac{1}{\sqrt{5}}, \cos x = -\dfrac{2}{\sqrt{5}}$

13. Period π

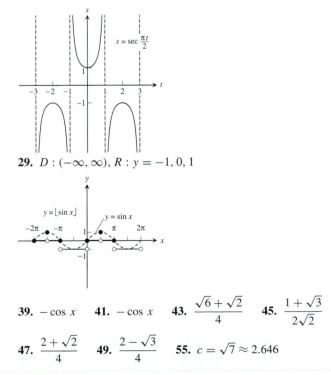

15. Period 2

17. Period 6

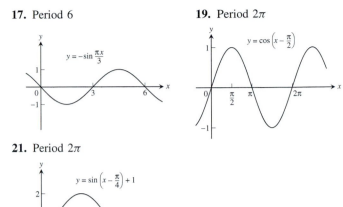

19. Period 2π

21. Period 2π

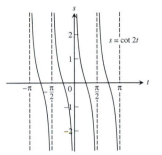

23. Period $\pi/2$, symmetric about the origin

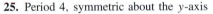

25. Period 4, symmetric about the y-axis

29. $D : (-\infty, \infty), R : y = -1, 0, 1$

39. $-\cos x$ **41.** $-\cos x$ **43.** $\dfrac{\sqrt{6} + \sqrt{2}}{4}$ **45.** $\dfrac{1 + \sqrt{3}}{2\sqrt{2}}$

47. $\dfrac{2 + \sqrt{2}}{4}$ **49.** $\dfrac{2 - \sqrt{3}}{4}$ **55.** $c = \sqrt{7} \approx 2.646$

59. $a = 1.464$

61. $A = 2, B = 2\pi, C = -\pi, D = -1$

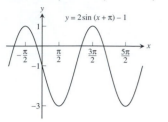

$y = 2\sin(x + \pi) - 1$

63. $A = -\dfrac{2}{\pi}, B = 4, C = 0, D = \dfrac{1}{\pi}$

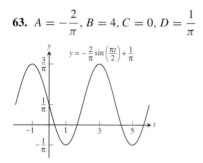

$y = -\dfrac{2}{\pi}\sin\left(\dfrac{\pi t}{2}\right) + \dfrac{1}{\pi}$

65. a) 37   b) 365   c) Right 101   d) Up 25

Practice Exercises, pp. 48–49

1. $(0, 11)$

3. No, no two sides have the same length; no, no two sides are perpendicular

5. $A = \pi r^2, C = 2\pi r, A = \dfrac{C^2}{4\pi}$ **7.** $x = \tan\theta, y = \tan^2\theta$

9. Replaces the portion for $x < 0$ with mirror image of portion for $x > 0$, to make the new graph symmetric with respect to the y-axis.

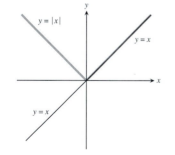

$y = |x|$

$y = x$

$y = x$

11. It does not change it.

13. Adds the mirror image of the portion for $x > 0$ to make the new graph symmetric with respect to the y-axis.

15. Reflects the portion for $y < 0$ across the x-axis.

17. Reflects the portion for $y < 0$ across the x-axis.

19. Reflects the portion for $y < 0$ across the x-axis.

$y = |x^2 + x|$

$y = x^2 + x$

21. Period π

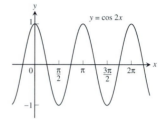

$y = \cos 2x$

23. Period 2

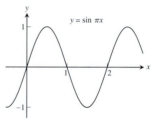

$y = \sin \pi x$

25.

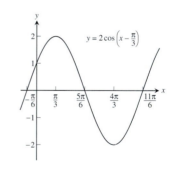

$y = 2\cos\left(x - \dfrac{\pi}{3}\right)$

27. a) $a = 1, \quad b = \sqrt{3}$ b) $a = 2\sqrt{3}/3, \quad c = 4\sqrt{3}/3$

29. a) $a = \dfrac{b}{\tan B}$ b) $c = \dfrac{a}{\sin A}$ **31.** 16.98 m

33. $3\sin x \cos^2 x - \sin^3 x$ **35.** b) 4π

Additional Exercises, pp. 49–50

3. Yes. For instance: $f(x) = 1/x$ and $g(x) = 1/x$, or $f(x) = 2x$ and $g(x) = x/2$, or $f(x) = e^x$ and $g(x) = \ln x$.

5. If $f(x)$ is odd, then $g(x) = f(x) - 2$ is not odd. Nor is $g(x)$ even, unless $f(x) = 0$ for all x. If f is even, then $g(x) = f(x) - 2$ is also even.

7.

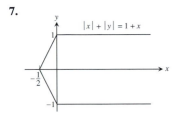

$|x| + |y| = 1 + x$

9. $\sqrt{2}$ **11.** 3/4   **13.** $3\sqrt{15}/16$ **27.** $-4 < m < 0$

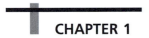

CHAPTER 1

Section 1.1, pp. 57–60

1. a) Does not exist. As x approaches 1 from the right, $g(x)$ approaches 0. As x approaches 1 from the left, $g(x)$ approaches 1. There is no single number L that all the values $g(x)$ get arbitrarily close to as $x \to 1$.   b) 1   c) 0
3. a) True   b) True   c) False   d) False   e) False   f) True
5. As x approaches 0 from the left, $x/|x|$ approaches -1. As x approaches 0 from the right, $x/|x|$ approaches 1. There is no single number L that the function values all get arbitrarily close to as $x \to 0$.
11. a) $f(x) = (x^2 - 9)/(x + 3)$

x	-3.1	-3.01	-3.001	-3.0001	-3.00001	-3.000001
$f(x)$	-6.1	-6.01	-6.001	-6.0001	-6.00001	-6.000001

x	-2.9	-2.99	-2.999	-2.9999	-2.99999	-2.999999
$f(x)$	-5.9	-5.99	-5.999	-5.9999	-5.99999	-5.999999

c) $\lim\limits_{x \to -3} f(x) = -6$

13. a) $G(x) = (x + 6)/(x^2 + 4x - 12)$

x	-5.9	-5.99	-5.999	-5.9999	-5.99999	-5.999999
$G(x)$	$-.126582$	$-.1251564$	$-.1250156$	$-.1250015$	$-.1250001$	$-.1250000$

x	-6.1	-6.01	-6.001	-6.0001	-6.00001	-6.000001
$G(x)$	$-.123456$	$-.124843$	$-.124984$	$-.124998$	$-.124999$	$-.124999$

c) $\lim\limits_{x \to -6} G(x) = -1/8 = -0.125$

15. a) $f(x) = (x^2 - 1)/(|x| - 1)$

x	-1.1	-1.01	-1.001	-1.0001	-1.00001	-1.000001
$f(x)$	2.1	2.01	2.001	2.0001	2.00001	2.000001

x	$-.9$	$-.99$	$-.999$	$-.9999$	$-.99999$	$-.999999$
$f(x)$	1.9	1.99	1.999	1.9999	1.99999	1.999999

c) $\lim\limits_{x \to -1} f(x) = 2$

17. a) $g(\theta) = (\sin \theta)/\theta$

θ	.1	.01	.001	.0001	.00001	.000001
$g(\theta)$	.998334	.999983	.999999	.999999	.999999	.999999

θ	$-.1$	$-.01$	$-.001$	$-.0001$	$-.00001$	$-.000001$
$g(\theta)$	.998334	.999983	.999999	.999999	.999999	.999999

$\lim\limits_{\theta \to 0} g(\theta) = 1$

19. a) $f(x) = x^{1/(1-x)}$

x	.9	.99	.999	.9999	.99999	.999999
$f(x)$	.348678	.366032	.367695	.367861	.367877	.367879

x	1.1	1.01	1.001	1.0001	1.00001	1.000001
$f(x)$	.385543	.369711	.368063	.367897	.367881	.367878

$\lim\limits_{x \to 1} f(x) \approx 0.36788$

21. 4   **23.** 0   **25.** 9   **27.** $\pi/2$ **29.** a) 19   b) 1

31. a) $-\dfrac{4}{\pi}$ b) $-\dfrac{3\sqrt{3}}{\pi}$ **33.** 1

35. Graphs can shift during a press run, so your estimates may not completely agree with these.

a)

PQ_1	PQ_2	PQ_3	PQ_4
43	46	49	50

The appropriate units are m/sec.

b) ≈ 50 m/sec or 180 km/h

37. a)

b) $\approx \$56,000$/year
c) $\approx \$42,000$/year

39. a) $0.414213, 0.449489, (\sqrt{1+h}-1)/h$ b) $g(x) = \sqrt{x}$

$1+h$	1.1	1.01	1.001	1.0001	1.00001	1.000001
$\sqrt{1+h}$	1.04880	1.004987	1.0004998	1.0000499	1.000005	1.0000005
$(\sqrt{1+h}-1)/h$	0.4880	0.4987	0.4998	0.499	0.5	0.5

c) 0.5 d) 0.5

Section 1.2, pp. 65–66

1. -9 **3.** 4 **5.** -8 **7.** 5/8 **9.** 5/2 **11.** 27 **13.** 16
15. 3/2 **17.** 1/10 **19.** -7 **21.** 3/2 **23.** $-1/2$ **25.** 4/3
27. 1/6 **29.** 4
31. a) Quotient rule b) Difference and Power rules
c) Sum and Constant Multiple rules
33. a) -10 b) -20 c) -1 d) 5/7
35. a) 4 b) -21 c) -12 d) $-7/3$
37. 2 **39.** 3 **41.** $1/(2\sqrt{7})$ **43.** $\sqrt{5}$ **45.** a) The limit is 1.
49. 7 **51.** a) 5 b) 5

Section 1.3, pp. 74–77

1. $\delta = 2$

3. $\delta = 1/2$

5. $\delta = 1/18$

7. $\delta = 0.1$ **9.** $\delta = 7/16$ **11.** $\delta = \sqrt{5} - 2$ **13.** $\delta = 0.36$
15. $(3.99, 4.01), \ \delta = 0.01$ **17.** $(-0.19, 0.21), \delta = 0.19$
19. $(3, 15), \delta = 5$ **21.** $(10/3, 5), \delta = 2/3$
23. $(-\sqrt{4.5}, -\sqrt{3.5}), \delta = \sqrt{4.5} - 2 \approx 0.12$
25. $(\sqrt{15}, \sqrt{17}), \delta = \sqrt{17} - 4 \approx 0.12$

27. $\left(2 - \dfrac{0.03}{m}, 2 + \dfrac{0.03}{m}\right), \delta = \dfrac{0.03}{m}$

29. $\left(\dfrac{1}{2} - \dfrac{c}{m}, \dfrac{c}{m} + \dfrac{1}{2}\right), \delta = \dfrac{c}{m}$

31. $L = -3, \delta = 0.01$ **33.** $L = 4, \delta = 0.05$
35. $L = 4, \delta = 0.75$
55. [3.384, 3.387]. To be safe, the left endpoint was rounded up and the right endpoint rounded down.

Section 1.4, pp. 83–86

1. a) True b) True c) False d) True e) True f) True
g) False h) False i) False j) False k) True l) False
3. a) 2, 1 b) No, $\lim\limits_{x \to 2^+} f(x) \neq \lim\limits_{x \to 2^-} f(x)$ c) 3, 3 d) Yes, 3
5. a) No b) Yes, 0 c) No
7. a)

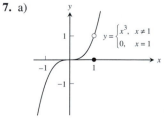

b) 1, 1 c) Yes, 1
9. a) $D : 0 \le x \le 2, R : 0 < y \le 1$ and $y = 2$ b) $(0, 1) \cup (1, 2)$
c) $x = 2$ d) $x = 0$

11. $\sqrt{3}$ **13.** 1 **15.** $2/\sqrt{5}$ **17.** a) 1 b) -1 **19.** a) 1
b) 2/3 **21.** ∞ **23.** $-\infty$ **25.** $-\infty$ **27.** ∞ **29.** a) ∞
b) $-\infty$ **31.** ∞ **33.** ∞ **35.** $-\infty$ **37.** a) ∞ b) $-\infty$
c) $-\infty$ d) ∞ **39.** a) $-\infty$ b) ∞ c) 0 d) 3/2
41. a) $-\infty$ b) 1/4 c) 1/4 d) 1/4 e) It will be $-\infty$.
43. a) $-\infty$ b) ∞ **45.** a) ∞ b) ∞ c) ∞ d) ∞
51. $\delta = \epsilon^2, \ \lim\limits_{x \to 5^+} \sqrt{x-5} = 0$ **55.** a) 400 b) 399
c) The limit does not exist.
61. a) For every positive real number B there exists a corresponding number $\delta > 0$ such that for all x

$$x_0 - \delta < x < x_0 \Rightarrow f(x) > B.$$

b) For every negative real number $-B$ there exists a corresponding number $\delta > 0$ such that for all x

$$x_0 < x < x_0 + \delta \Rightarrow f(x) < -B.$$

c) For every negative real number $-B$ there exists a corresponding number $\delta > 0$ such that for all x

$$x_0 - \delta < x < x_0 \Rightarrow f(x) < -B.$$

Section 1.5, pp. 95–97

1. No; discontinuous at $x = 2$; not defined at $x = 2$
3. Continuous **5.** a) Yes b) Yes c) Yes d) Yes
7. a) No b) No **9.** 0 **11.** 1, nonremovable; 0, removable
13. All x except $x = 2$ **15.** All x except $x = 3, x = 1$
17. All x **19.** All x except $x = 0$
21. All x except $x = n\pi/2, n$ any integer
23. All x except $n\pi/2, n$ an odd integer
25. All $x > -3/2$ **27.** All x **29.** 0 **31.** 1 **33.** $\sqrt{2}/2$
35. $g(3) = 6$ **37.** $f(1) = 3/2$ **39.** $a = 4/3$
63. $x \approx 1.8794, -1.5321, -0.3473$ **65.** $x \approx 1.7549$
67. $x \approx 3.5156$ **69.** $x \approx 0.7391$

Section 1.6, pp. 101–103

1. $P_1 : m_1 = 1, P_2 : m_2 = 5$ **3.** $P_1 : m_1 = 5/2, P_2 : m_2 = -1/2$
5. $y = 2x + 5$ **7.** $y = x + 1$

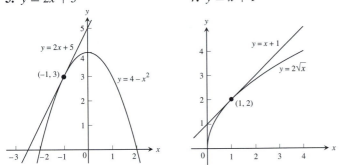

9. $y = 12x + 16$

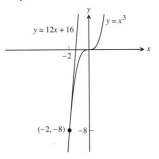

11. $m = 4, y - 5 = 4(x - 2)$ **13.** $m = -2, y - 3 = -2(x - 3)$

15. $m = 12, y - 8 = 12(t - 2)$ **17.** $m = \dfrac{1}{4}, y - 2 = \dfrac{1}{4}(x - 4)$

19. $m = -10$ **21.** $m = -1/4$ **23.** $(-2, -5)$
25. $y = -(x + 1), y = -(x - 3)$ **27.** 19.6 m/sec
29. 6π **31.** Yes   **33.** Yes   **35.** a) Nowhere
37. a) At $x = 0$ **39.** a) Nowhere   **41.** a) At $x = 1$
43. a) At $x = 0$

Chapter 1 Practice Exercises, pp. 104–105

1.

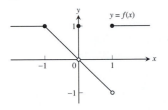

At $x = -1$: $\lim\limits_{x \to -1^-} f(x) = \lim\limits_{x \to -1^+} f(x) = 1$, so
          $\lim\limits_{x \to -1} f(x) = 1 = f(-1)$; continuous at $x = -1$
At $x = 0$: $\lim\limits_{x \to 0^-} f(x) = \lim\limits_{x \to 0^+} f(x) = 0$, so $\lim\limits_{x \to 0} f(x) = 0$.
          However, $f(0) \neq 0$, so f is discontinuous at $x = 0$.
          The discontinuity can be removed by redefining $f(0)$
          to be 0.

At $x = 1$: $\lim\limits_{x \to 1^-} f(x) = -1$ and $\lim\limits_{x \to 1^+} f(x) = 1$, so $\lim\limits_{x \to 1} f(x)$
          does not exist. The function is discontinuous at
          $x = 1$ and the discontinuity is not removable.

3. a) -21 b) 49   c) 0   d) 1   e) 1   f) 7   **5.** 4   **7.** 2
9. 0   **11.** a) Does not exist   b) 0   **13.** 1/2   **15.** $2x$
17. $-1/4$ **19.** a) $(-\infty, +\infty)$ b) $[0, \infty)$
c) $(-\infty, 0)$ and $(0, \infty)$ d) $(0, \infty)$ **25.** b) 1.3247 17957 24

Chapter 1 Additional Exercises, pp. 105–107

1. a) No   b) No
5. 0; the left-hand limit was needed because the function is undefined
for $v > c$.
7. a) B b) A c) A d) A **11.** a) $2.56 < x < 5.76$
b) $3.24 < x < 4.84$ **21.** b) $\min(a, b) = \dfrac{a + b}{2} - \dfrac{|a - b|}{2}$

CHAPTER 2

Section 2.1, pp. 117–120

1. $-2x, \; 6, \; 0, \; -2$ **3.** $-\dfrac{2}{t^3}, \; 2, \; -\dfrac{1}{4}, \; -\dfrac{2}{3\sqrt{3}}$

5. $\dfrac{3}{2\sqrt{3\theta}}, \; \dfrac{3}{2\sqrt{3}}, \; \dfrac{1}{2}, \; \dfrac{3}{2\sqrt{2}}$ **7.** $6x^2$ **9.** $\dfrac{1}{(2t + 1)^2}$

11. $\dfrac{-1}{2(q + 1)\sqrt{q + 1}}$ **13.** $1 - \dfrac{9}{x^2}, \; 0$ **15.** $3t^2 - 2t, \; 5$

17. $\dfrac{-4}{(x - 2)\sqrt{x - 2}}, \; y - 4 = -\dfrac{1}{2}(x - 6)$ **19.** 6   **21.** 1/8

23. -1 **25.** $-1/4$ **27.** b   **29.** d
31. a) $x = 0, 1, 4$ **33.**
b)

35. Since $\lim\limits_{x \to 0^+} f'(x) = 1$ while $\lim\limits_{x \to 0^-} f'(x) = 0$, $f(x)$ is not
differentiable at $x = 0$.
37. Since $\lim\limits_{x \to 1^+} f'(x) = 2$ while $\lim\limits_{x \to 1^-} f'(x) = 1/2$, $f(x)$ is not
differentiable at $x = 1$.
39. a) $-3 \leq x \leq 2$ b) None   c) None
41. a) $-3 \leq x < 0, 0 < x \leq 3$ b) None   c) $x = 0$
43. a) $-1 \leq x < 0, 0 < x \leq 2$ b) $x = 0$ c) None
45. a) $y' = -2x$
c) $x < 0, x = 0, x > 0$ d) $-\infty < x < 0, 0 < x < \infty$

47. a) $y' = x^2$
b)

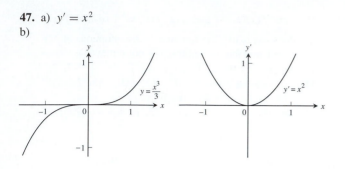

c) $x \neq 0, x = 0$, none   d) $-\infty < x < \infty$, none
49. $y' = 3x^2$ is never negative
51. Yes, $y + 16 = -(x - 3)$ is tangent at $(3, -16)$
53. No, the function $y = \lfloor x \rfloor$ does not satisfy the intermediate value property of derivatives.
55. Yes, $(-f)'(x) = -(f'(x))$
57. For $g(t) = mt$ and $h(t) = t$, $\lim\limits_{t \to 0} \dfrac{g(t)}{h(t)} = m$, which need not be zero.

Section 2.2, pp. 129–131

1. $\dfrac{dy}{dx} = -2x, \quad \dfrac{d^2y}{dx^2} = -2$

3. $\dfrac{ds}{dt} = 15t^2 - 15t^4, \quad \dfrac{d^2s}{dt^2} = 30t - 60t^3$

5. $\dfrac{dy}{dx} = 4x^2 - 1, \quad \dfrac{d^2y}{dx^2} = 8x$

7. $\dfrac{dw}{dz} = -6z^{-3} + \dfrac{1}{z^2}, \quad \dfrac{d^2w}{dz^2} = 18z^{-4} - \dfrac{2}{z^3}$

9. $\dfrac{dy}{dx} = 12x - 10 + 10x^{-3}, \quad \dfrac{d^2y}{dx^2} = 12 - 30x^{-4}$

11. $\dfrac{dr}{ds} = \dfrac{-2}{3s^3} + \dfrac{5}{2s^2}, \quad \dfrac{d^2r}{ds^2} = \dfrac{2}{s^4} - \dfrac{5}{s^3}$

13. $y' = -5x^4 + 12x^2 - 2x - 3$ **15.** $y' = 3x^2 + 10x + 2 - \dfrac{1}{x^2}$

17. $y' = \dfrac{-19}{(3x - 2)^2}$ **19.** $g'(x) = \dfrac{x^2 + x + 4}{(x + 0.5)^2}$

21. $\dfrac{dv}{dt} = \dfrac{t^2 - 2t - 1}{(1 + t^2)^2}$ **23.** $f'(s) = \dfrac{1}{\sqrt{s}(\sqrt{s} + 1)^2}$

25. $v' = -\dfrac{1}{x^2} + 2x^{-3/2}$ **27.** $y' = \dfrac{-4x^3 - 3x^2 + 1}{(x^2 - 1)^2(x^2 + x + 1)^2}$

29. $y' = 2x^3 - 3x - 1, \ y'' = 6x^2 - 3, \ y''' = 12x, \ y^{(4)} = 12,$
$y^{(n)} = 0$ for $n \geq 5$
31. $y' = 2x - 7x^{-2}, \ y'' = 2 + 14x^{-3}$

33. $\dfrac{dr}{d\theta} = 3\theta^{-4}, \quad \dfrac{d^2r}{d\theta^2} = -12\theta^{-5}$

35. $\dfrac{dw}{dz} = -z^{-2} - 1, \quad \dfrac{d^2w}{dz^2} = 2z^{-3}$

37. $\dfrac{dp}{dq} = \dfrac{1}{6}q + \dfrac{1}{6}q^{-3} + q^{-5}, \quad \dfrac{d^2p}{dq^2} = \dfrac{1}{6} - \dfrac{1}{2}q^{-4} - 5q^{-6}$

39. a) 13 b) -7 c) 7/25 d) 20   **41.** a) $y = -\dfrac{x}{8} + \dfrac{5}{4}$
b) $m = -4$ at $(0, 1)$ c) $y = 8x - 15, \ y = 8x + 17$
43. $y = 4x, \ y = 2$ **45.** $a = 1, b = 1, c = 0$ **47.** a) $y = 2x + 2,$

c) $(2, 6)$ **49.** $\dfrac{dP}{dV} = -\dfrac{nRT}{(V - nb)^2} + \dfrac{2an^2}{V^3}$

51. The Product Rule is then the Constant Multiple Rule, so the latter is a special case of the Product Rule.

55. a) $\dfrac{3}{2}x^{1/2},$ b) $\dfrac{5}{2}x^{3/2},$ c) $\dfrac{7}{2}x^{5/2},$ d) $\dfrac{d}{dx}\left(x^{n/2}\right) = \dfrac{n}{2}x^{(n/2)-1}$

Section 2.3, pp. 139–143

1. a) 80 m, 8 m/sec b) 0 m/sec, 16 m/sec; 1.6 m/sec², 1.6 m/sec²
c) no change in direction
3. a) -9 m, -3 m/sec b) 3 m/sec, 12 m/sec; 6 m/sec², -12 m/sec²
c) no change in direction
5. a) -20 m, -5 m/sec b) 45 m/sec, (1/5) m/sec; 140 m/sec²,
(4/25) m/sec² c) no change in direction
7. a) $a(1) = -6$ m/sec², $a(3) = 6$ m/sec² b) $v(2) = 3$ m/sec
c) 6 m
9. Mars: ≈ 7.5 sec, Jupiter: ≈ 1.2 sec
11. a) $24 - 9.8t$ m/sec, -9.8 m/sec² b) 2.4 sec c) 29.4 m
d) 0.7 sec going up, 4.2 sec going down e) 4.9 sec
13. 320 sec on the moon, 52 sec on Earth; $\approx 66{,}560$ ft on the moon,
$\approx 10{,}816$ ft on Earth
15. a) $9.8t$ m/sec, b) 9.8 m/sec²
17. a) $t = 2, t = 7$ b) $3 \leq t \leq 6$
c)

d)

19. a) 190 ft/sec b) 2 sec c) 8 sec, 0 ft/sec, d) 10.8 sec,
90 ft/sec e) 2.8 sec f) greatest acceleration happens 2 sec after
launch g) constant acceleration between 2 and 10.8 sec, -32 ft/sec²
21. a) Answers will vary.

b)

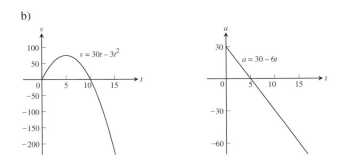

23. C = position, A = velocity, B = acceleration
25. a) \$110/machine   b) \$80   c) \$79.90
27. a) 10^4 bacteria/h   b) 0 bacteria/h   c) -10^4 bacteria/h
29. a) $\dfrac{t}{12} - 1$ b) Fastest ($dy/dt = -1$ m/h) when $t = 0$, Slowest
($dy/dt = 0$ m/h) when $t = 12$
31. $t = 25$ sec, $D = 6250/9$ m
33. a) $t = 6.25$ sec   b) Up on [0, 6.25), down on (6.25, 12.5]
c) $t = 6.25$ sec   d) Speeds up on (6.25, 12.5], slows down on
[0, 6.25)   e) Fastest at $t = 0, 12.5$, slowest at $t = 6.25$
f) $t = 6.25$ sec
35. a) $t = (6 \pm \sqrt{15})/3$ b) left on $((6 - \sqrt{15})/3, (6 + \sqrt{15})/3))$;
right on $[0, (6 - \sqrt{15})/3) \cup ((6 + \sqrt{15})/3, 4]$ c) $t = (6 \pm \sqrt{15})/3$
d) Speeds up on $((6 - \sqrt{15})/3, 2) \cup ((6 + \sqrt{15})/3, 4]$,
slows down on $[0, (6 - \sqrt{15})/3) \cup (2, (6 + \sqrt{15})/3)$
e) Fastest at $t = 0, 4$; slowest at $t = (6 \pm \sqrt{15})/3)$
f) $t = (6 + \sqrt{15})/3$

Section 2.4, pp. 152–154

1. $-10 - 3 \sin x$ **3.** $-\csc x \cot x - \dfrac{2}{\sqrt{x}}$ **5.** 0

7. $\dfrac{-\csc^2 x}{(1 + \cot x)^2}$ **9.** $4 \tan x \sec x - \csc^2 x$ **11.** $x^2 \cos x$

13. $\sec^2 t - 1$ **15.** $\dfrac{-2 \csc t \cot t}{(1 - \csc t)^2}$ **17.** $-\theta (\theta \cos \theta + 2 \sin \theta)$

19. $\sec \theta \csc \theta (\tan \theta - \cot \theta) = \sec^2 \theta - \csc^2 \theta$ **21.** $\sec^2 q$
23. $\sec^2 q$ **25.** a) $2 \csc^3 x - \csc x$ b) $2 \sec^3 x - \sec x$
27. 0   **29.** -1 **31.** 0   **33.** 1   **35.** 3/4   **37.** 2   **39.** 1/2
41. 2   **43.** 1   **45.** 1/2   **47.** 3/8
49.   **51.**

53. Yes, at $x = \pi$ **55.** No

57. $\left(-\dfrac{\pi}{4}, -1\right); \left(\dfrac{\pi}{4}, 1\right)$

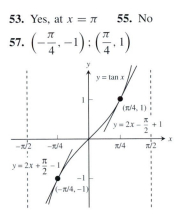

59. a) $y = -x + \pi/2 + 2$ b) $y = 4 - \sqrt{3}$
61. $-\sqrt{2}$ m/sec, $\sqrt{2}$ m/sec, $\sqrt{2}$ m/sec^2, $\sqrt{2}$ m/sec^3 **63.** $c = 9$
65. $\sin x$

Section 2.5, pp. 160–163

1. $12x^3$ **3.** $3 \cos (3x + 1)$ **5.** $-\sin (\sin x) \cos x$
7. $10 \sec^2 (10x - 5)$

9. With $u = (2x + 1)$, $y = u^5$: $\dfrac{dy}{dx} = \dfrac{dy}{du} \dfrac{du}{dx} = 5u^4 \cdot 2 =$
$10(2x + 1)^4$

11. With $u = (1 - (x/7))$, $y = u^{-7}$: $\dfrac{dy}{dx} = \dfrac{dy}{du} \dfrac{du}{dx} = -7u^{-8} \cdot \left(-\dfrac{1}{7}\right)$
$= \left(1 - \dfrac{x}{7}\right)^{-8}$

13. With $u = ((x^2/8) + x - (1/x))$, $y = u^4$: $\dfrac{dy}{dx} = \dfrac{dy}{du} \dfrac{du}{dx} =$
$4u^3 \cdot \left(\dfrac{x}{4} + 1 + \dfrac{1}{x^2}\right) = 4\left(\dfrac{x^2}{8} + x - \dfrac{1}{x}\right)^3 \left(\dfrac{x}{4} + 1 + \dfrac{1}{x^2}\right)$

15. With $u = \tan x$, $y = \sec u$: $\dfrac{dy}{dx} = \dfrac{dy}{du} \dfrac{du}{dx} =$
$(\sec u \tan u)(\sec^2 x) = \sec (\tan x) \tan (\tan x) \sec^2 x$

17. With $u = \sin x$, $y = u^3$: $\dfrac{dy}{dx} = \dfrac{dy}{du} \dfrac{du}{dx} = 3u^2 \cos x =$
$3 \sin^2 x(\cos x)$

19. $-\dfrac{1}{2\sqrt{3 - t}}$ **21.** $\dfrac{4}{\pi}(\cos 3t - \sin 5t)$ **23.** $\dfrac{\csc \theta}{\cot \theta + \csc \theta}$

25. $2x \sin^4 x + 4x^2 \sin^3 x \cos x + \cos^{-2} x + 2x \cos^{-3} x \sin x$

27. $(3x - 2)^6 - \dfrac{1}{x^3 \left(4 - \dfrac{1}{2x^2}\right)^2}$ **29.** $\dfrac{(4x + 3)^3 (4x + 7)}{(x + 1)^4}$

31. $\sqrt{x} \sec^2 (2\sqrt{x}) + \tan (2\sqrt{x})$ **33.** $\dfrac{2 \sin \theta}{(1 + \cos \theta)^2}$

35. $\dfrac{dr}{d\theta} = -2 \sin (\theta^2) \sin 2\theta + 2\theta \cos (2\theta) \cos (\theta^2)$

37. $\dfrac{dq}{dt} = \left(\dfrac{t + 2}{2(t + 1)^{3/2}}\right) \cos \left(\dfrac{t}{\sqrt{t + 1}}\right)$

39. $2\pi \sin(\pi t - 2) \cos(\pi t - 2)$ **41.** $\dfrac{8 \sin(2t)}{(1 + \cos 2t)^5}$

43. $-2 \cos(\cos(2t - 5))(\sin(2t - 5))$

45. $\left(1 + \tan^4\left(\dfrac{t}{12}\right)\right)^2 \left(\tan^3\left(\dfrac{t}{12}\right)\sec^2\left(\dfrac{t}{12}\right)\right)$

47. $-\dfrac{t \sin(t^2)}{\sqrt{1 + \cos(t^2)}}$ **49.** $\dfrac{6}{x^3}\left(1 + \dfrac{1}{x}\right)\left(1 + \dfrac{2}{x}\right)$

51. $2 \csc^2(3x - 1) \cot(3x - 1)$ **53.** $5/2$ **55.** $-\pi/4$ **57.** 0

59. a) $2/3$ b) $2\pi + 5$ c) $15 - 8\pi$ d) $37/6$ e) -1
f) $\sqrt{2}/24$ g) $5/32$ h) $-5/(3\sqrt{17})$

61. 5 **63.** a) 1 b) 1 **65.** a) $y = \pi x + 2 - \pi$ b) $\pi/2$

67. It multiplies the velocity, acceleration, and jerk by 2, 4, and 8 respectively.

69. $v = \dfrac{2}{5}$ m/sec, $a = -\dfrac{4}{125}$ m/sec^2

53. a) $y = -\dfrac{\pi}{2}x + \pi$, b) $y = \dfrac{2}{\pi}x - \dfrac{2}{\pi} + \dfrac{\pi}{2}$

55. a) $y = 2\pi x - 2\pi$, b) $y = -\dfrac{x}{2\pi} + \dfrac{1}{2\pi}$

57. Points: $(-\sqrt{7}, 0)$ and $(\sqrt{7}, 0)$, Slope: -2

59. $m = -1$ at $\left(\dfrac{\sqrt{3}}{4}, \dfrac{\sqrt{3}}{2}\right)$, $m = \sqrt{3}$ at $\left(\dfrac{\sqrt{3}}{4}, \dfrac{1}{2}\right)$

61. $(-3, 2): m = -\dfrac{27}{8}$; $(-3, -2): m = \dfrac{27}{8}$; $(3, 2): m = \dfrac{27}{8}$;
$(3, -2): m = -\dfrac{27}{8}$

63. a) False b) True c) True d) True **65.** $(3, -1)$

69. $\dfrac{dy}{dx} = -\dfrac{y^3 + 2xy}{x^2 + 3xy^2}$, $\dfrac{dx}{dy} = -\dfrac{x^2 + 3xy^2}{y^3 + 2xy}$, $\dfrac{dx}{dy} = \dfrac{1}{dy/dx}$

Section 2.6, pp. 170–172

1. $\dfrac{9}{4}x^{5/4}$ **3.** $\dfrac{2^{1/3}}{3x^{2/3}}$ **5.** $\dfrac{7}{2(x + 6)^{1/2}}$ **7.** $-(2x + 5)^{-3/2}$

9. $\dfrac{2x^2 + 1}{(x^2 + 1)^{1/2}}$ **11.** $\dfrac{ds}{dt} = \dfrac{2}{7}t^{-5/7}$

13. $\dfrac{dy}{dt} = -\dfrac{4}{3}(2t + 5)^{-5/3} \cos\left[(2t + 5)^{-2/3}\right]$

15. $f'(x) = \dfrac{-1}{4\sqrt{x}(1 - \sqrt{x})}$

17. $h'(\theta) = -\dfrac{2}{3}(\sin 2\theta)(1 + \cos 2\theta)^{-2/3}$ **19.** $\dfrac{-2xy - y^2}{x^2 + 2xy}$

21. $\dfrac{1 - 2y}{2x + 2y - 1}$ **23.** $\dfrac{-2x^3 + 3x^2y - xy^2 + x}{x^2y - x^3 + y}$ **25.** $\dfrac{1}{y(x + 1)^2}$

27. $\cos^2 y$ **29.** $\dfrac{-\cos^2(xy) - y}{x}$

31. $\dfrac{-y^2}{y \sin\left(\dfrac{1}{y}\right) - \cos\left(\dfrac{1}{y}\right) + xy}$ **33.** $-\dfrac{\sqrt{r}}{\sqrt{\theta}}$ **35.** $\dfrac{-r}{\theta}$

37. $y' = -\dfrac{x}{y}$, $y'' = \dfrac{-y^2 - x^2}{y^3}$

39. $y' = \dfrac{x + 1}{y}$, $y'' = \dfrac{y^2 - (x + 1)^2}{y^3}$

41. $y' = \dfrac{\sqrt{y}}{\sqrt{y} + 1}$, $y'' = \dfrac{1}{2(\sqrt{y} + 1)^3}$ **43.** -2

45. $(-2, 1): m = -1$, $(-2, -1): m = 1$ **47.** a) $y = \dfrac{7}{4}x - \dfrac{1}{2}$,
b) $y = -\dfrac{4}{7}x + \dfrac{29}{7}$ **49.** a) $y = 3x + 6$, b) $y = -\dfrac{1}{3}x + \dfrac{8}{3}$

51. a) $y = \dfrac{6}{7}x + \dfrac{6}{7}$, b) $y = -\dfrac{7}{6}x - \dfrac{7}{6}$

Section 2.7, pp. 176–180

1. $\dfrac{dA}{dt} = 2\pi r \dfrac{dr}{dt}$ **3.** a) $\dfrac{dV}{dt} = \pi r^2 \dfrac{dh}{dt}$ b) $\dfrac{dV}{dt} = 2\pi hr \dfrac{dr}{dt}$
c) $\dfrac{dV}{dt} = \pi r^2 \dfrac{dh}{dt} + 2\pi hr \dfrac{dr}{dt}$

5. a) 1 volt/sec b) $-\dfrac{1}{3}$ amp/sec c) $\dfrac{dR}{dt} = \dfrac{1}{I}\left(\dfrac{dV}{dt} - \dfrac{V}{I}\dfrac{dI}{dt}\right)$
d) 3/2 ohms/sec, R is increasing.

7. a) $\dfrac{dS}{dt} = \dfrac{x}{\sqrt{x^2 + y^2}}\dfrac{dx}{dt}$
b) $\dfrac{dS}{dt} = \dfrac{x}{\sqrt{x^2 + y^2}}\dfrac{dx}{dt} + \dfrac{y}{\sqrt{x^2 + y^2}}\dfrac{dy}{dt}$ c) $\dfrac{dx}{dt} = -\dfrac{y}{x}\dfrac{dy}{dt}$

9. a) $\dfrac{dA}{dt} = \dfrac{1}{2}ab \cos\theta \dfrac{d\theta}{dt}$
b) $\dfrac{dA}{dt} = \dfrac{1}{2}ab \cos\theta \dfrac{d\theta}{dt} + \dfrac{1}{2}b \sin\theta \dfrac{da}{dt}$
c) $\dfrac{dA}{dt} = \dfrac{1}{2}ab \cos\theta \dfrac{d\theta}{dt} + \dfrac{1}{2}b \sin\theta \dfrac{da}{dt} + \dfrac{1}{2}a \sin\theta \dfrac{db}{dt}$

11. a) 14 cm^2/sec, increasing b) 0 cm/sec, constant
c) $-14/13$ cm/sec, decreasing

13. a) -12 ft/sec b) -59.5 ft^2/sec c) -1 rad/sec

15. 20 ft/sec **17.** a) $\dfrac{dh}{dt} = 11.19$ cm/min
b) $\dfrac{dr}{dt} = 14.92$ cm/min **19.** a) $\dfrac{-1}{24\pi}$ m/min
b) $r = \sqrt{26y - y^2}$ m, c) $\dfrac{dr}{dt} = -\dfrac{5}{288\pi}$ m/min

21. 1 ft/min, 40π ft^2/min **23.** 11 ft/sec
25. Increasing at 466/1681 L/min **27.** 1 rad/sec **29.** -5 m/sec
31. -1500 ft/sec **33.** $\dfrac{5}{72\pi}$ in/min, $\dfrac{10}{3}$ in^2/min **35.** 7.1 in/min
37. a) $-32\sqrt{13} \approx -8.875$ ft/sec,
b) $d\theta_1/dt = -8/65$ rad/sec, $d\theta_2/dt = 8/65$ rad/sec

c) $d\theta_1/dt = -1/6$ rad/sec, $d\theta_2/dt = 1/6$ rad/sec

39. 29.5 knots

Chapter 2 Practice Exercises, pp. 181–185

1. $5x^4 - .25x + .25$ **3.** $3x(x-2)$ **5.** $2(x+1)(2x^2+4x+1)$

7. $3(\theta^2 + \sec\theta + 1)^2(2\theta + \sec\theta\tan\theta)$ **9.** $\dfrac{1}{2\sqrt{t}(1+\sqrt{t})^2}$

11. $2\sec^2 x\tan x$ **13.** $8\cos^3(1-2t)\sin(1-2t)$

15. $5(\sec t)(\sec t + \tan t)^5$ **17.** $\dfrac{\theta\cos\theta + \sin\theta}{\sqrt{2\theta}\,\sin\theta}$ **19.** $\dfrac{\cos\sqrt{2\theta}}{\sqrt{2\theta}}$

21. $x\csc\left(\dfrac{2}{x}\right) + \csc\left(\dfrac{2}{x}\right)\cot\left(\dfrac{2}{x}\right)$

23. $\dfrac{1}{2}x^{1/2}\sec(2x)^2\left[16\tan(2x)^2 - x^{-2}\right]$ **25.** $-10x\csc^2(x^2)$

27. $8x^3\sin(2x^2)\cos(2x^2) + 2x\sin^2(2x^2)$ **29.** $\dfrac{-(t+1)}{8t^3}$

31. $\dfrac{1-x}{(x+1)^3}$ **33.** $-\dfrac{1}{2x^2\left(1+\dfrac{1}{x}\right)^{1/2}}$ **35.** $\dfrac{-2\sin\theta}{(\cos\theta - 1)^2}$

37. $3\sqrt{2x+1}$ **39.** $-9\left(\dfrac{5x+\cos 2x}{(5x^2+\sin 2x)^{5/2}}\right)$ **41.** $-\dfrac{y+2}{x+3}$

43. $\dfrac{-3x^2-4y+2}{4x-4y^{1/3}}$ **45.** $-y/x$ **47.** $\dfrac{1}{2y(x+1)^2}$

49. $\dfrac{dp}{dq} = \dfrac{6q-4p}{3p^2+4q}$ **51.** $\dfrac{dr}{ds} = (2r-1)(\tan 2s)$

53. a) $\dfrac{d^2y}{dx^2} = \dfrac{-2xy^3 - 2x^4}{y^5}$ b) $\dfrac{d^2y}{dx^2} = \dfrac{-2xy^2 - 1}{x^4y^3}$

55. a) 1   b) 6   c) 1   d) $-1/9$ e) $-40/3$ f) 2   g) $-4/9$

57. 0   **59.** $\sqrt{3}$ **61.** $-1/2$ **63.** $\dfrac{-2}{(2t+1)^2}$

65. a)   b) Yes   c) Yes

$f(x) = \begin{cases} x^2, & -1 \le x < 0 \\ -x^2, & 0 \le x \le 1 \end{cases}$

67. a)   b) Yes   c) No

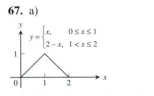

$y = \begin{cases} x, & 0 \le x \le 1 \\ 2-x, & 1 < x \le 2 \end{cases}$

69. $\left(\dfrac{5}{2}, \dfrac{9}{4}\right)$ and $\left(\dfrac{3}{2}, -\dfrac{1}{4}\right)$ **71.** $(-1, 27)$ and $(2, 0)$

73. a) $(-2, 16), (3, 11)$ b) $(0, 20), (1, 7)$

75.

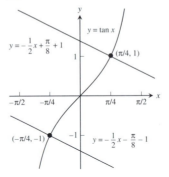

77. 1/4   **79.** 4

81. Tangent: $y = -\dfrac{1}{4}x + \dfrac{9}{4}$; Normal: $y = 4x - 2$

83. Tangent: $y = 2x - 4$; Normal: $y = -\dfrac{1}{2}x + \dfrac{7}{2}$

85. Tangent: $y = -\dfrac{5}{4}x + 6$; Normal: $y = \dfrac{4}{5}x - \dfrac{11}{5}$

87. $(1, 1) : m = -1/2$, $(1, -1)$; m not defined

89. $B = $ graph of f, $A = $ graph of f'

91.

93. a) 0, 0   b) 1700 rabbits, ≈ 1400 rabbits   **95.** 3/2   **97.** -1

99. 1/2   **101.** 4   **103.** 1   **107.** Yes, $k = 1/2$

109. a) $\dfrac{dS}{dt} = (4\pi r + 2\pi h)\dfrac{dr}{dt}$ b) $\dfrac{dS}{dt} = 2\pi r\dfrac{dh}{dt}$

c) $\dfrac{dS}{dt} = (4\pi r + 2\pi h)\dfrac{dr}{dt} + 2\pi r\dfrac{dh}{dt}$ d) $\dfrac{dr}{dt} = -\dfrac{r}{2r+h}\dfrac{dh}{dt}$

111. -40 m^2/sec   **113.** 0.02 ohm/sec   **115.** 5 m/sec^2

117. a) $r = \dfrac{2}{5}$ h   b) $-\dfrac{125}{144\pi}$ ft/min

119. a) $\dfrac{3}{5}$ km/sec or 600 m/sec   b) $\dfrac{18}{\pi}$ RPM

Chapter 2 Additional Exercises, pp. 185–187

1. a) $\sin 2\theta = 2\sin\theta\cos\theta$; $2\cos 2\theta$
$= 2\sin\theta(-\sin\theta) + \cos\theta(2\cos\theta)$; $2\cos 2\theta$
$= -2\sin^2\theta + 2\cos^2\theta$; $\cos 2\theta = \cos^2\theta - \sin^2\theta$
b) $\cos 2\theta = \cos^2\theta - \sin^2\theta$; $-2\sin 2\theta$
$= 2\cos\theta(-\sin\theta) - 2\sin\theta(\cos\theta)$; $\sin 2\theta$
$= \cos\theta\sin\theta + \sin\theta\cos\theta$; $\sin 2\theta = 2\sin\theta\cos\theta$
3. a) $a = 1, b = 0, c = -1/2$ b) $b = \cos a, c = \sin a$

5. $h = -4, k = 9/2, a = \dfrac{5\sqrt{5}}{2}$

7. a) $0.09y$ b) increasing at 1% per year
9. Answers will vary. Here is one possibility.

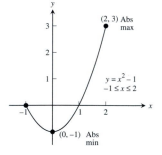

11. a) 2 sec, 64 ft/sec b) 12.31 sec, 393.85 ft.

15. a) $m = -\dfrac{b}{\pi}$, b) $m = -1, b = \pi$ **17.** a) $a = 3/4, b = 9/4$

23. h' is defined but not continuous at $x = 0$; k' is defined *and* continuous at $x = 0$.

CHAPTER 3

Section 3.1, pp. 195–196

1. Absolute minimum at $x = c_2$, absolute maximum at $x = b$
3. Absolute maximum at $x = c$, no absolute minimum
5. Absolute minimum at $x = a$, absolute maximum at $x = c$
7. Absolute maximum: -3, absolute minimum: $-19/3$

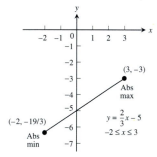

9. Absolute maximum: 3, absolute minimum: -1

11. Absolute maximum: -0.25, absolute minimum: -4

13. Absolute maximum: 2, absolute minimum: -1

15. Absolute maximum: 2, absolute minimum: 0

17. Absolute maximum: 1, absolute minimum: -1

19. Absolute maximum: $2/\sqrt{3}$, absolute minimum: 1

21. Absolute maximum: 2, absolute minimum: -1

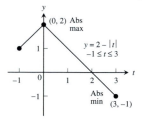

23. Increasing on (0, 8), decreasing on (−1, 0), absolute maximum: 16 at $x = 8$, absolute minimum: 0 at $x = 0$
25. Increasing on (−32, 1), absolute maximum: 1 at $\theta = 1$, absolute minimum: −8 at $\theta = -32$
27. a) Local maximum: 0 at $x = \pm 2$, local minimum: −4 at $x = 0$, absolute maximum: 0, absolute minimum: −4
b) Local maximum: 0 at $x = -2$, local minimum: −4 at $x = 0$, absolute maximum: 0, absolute minimum: −4
c) No local maximum, local minimum: −4 at $x = 0$, absolute minimum: −4
d) Local maximum: 0 at $x = -2$, local minimum: −4 at $x = 0$, absolute minimum: −4
e) No local extrema, no absolute extrema   **29.** Yes

Section 3.2, pp. 203–205

1. 1/2   **3.** 1
5. Does not; f is not differentiable at the interior domain point $x = 0$.
7. Does
11. a)

i.

ii.

iii.

iv.

27. $1.09999 \le f(0.1) \le 1.1$ **31.** Yes   **33.** a) 4   b) 3   c) 3
35. a) $\dfrac{x^2}{2} + C$ b) $\dfrac{x^3}{3} + C$ c) $\dfrac{x^4}{4} + C$
37. a) $\dfrac{1}{x} + C$ b) $x + \dfrac{1}{x} + C$ c) $5x - \dfrac{1}{x} + C$
39. a) $-\dfrac{1}{2}\cos 2t + C$ b) $2 \sin \dfrac{t}{2} + C$
c) $-\dfrac{1}{2}\cos 2t + 2 \sin \dfrac{t}{2} + C$
41. $f(x) = x^2 - x$ **43.** $r(\theta) = 8\theta + \cot \theta - 2\pi - 1$

Section 3.3, pp. 208–209

1. a) 0, 1   b) increasing on (−∞, 0) and (1, ∞), decreasing on (0, 1)   c) local maximum at $x = 0$, local minimum at $x = 1$
3. a) −2, 1   b) increasing on (−2, 1) and (1, ∞), decreasing on (−∞, −2)   c) No local maximum, local minimum at $x = -2$
5. a) −2, 1, 3   b) increasing on (−2, 1) and (3, ∞), decreasing on (−∞, −2) and (1, 3)   c) local maximum at $x = 1$, local minima at $x = -2, 3$
7. a) −2, 0   b) increasing on (−∞, −2) and (0, ∞), decreasing on (−2, 0)   c) Local maximum at $x = -2$, local minimum at $x = 0$
9. a) increasing on (−∞, −1.5), decreasing on (−1.5, ∞)

b) local maximum: 5.25 at $t = -1.5$
c) Absolute maximum: 5.25 at $t = -1.5$
11. a) Decreasing on (−∞, 0), increasing on (0, 4/3), decreasing on (4/3, ∞)   b) local minimum at $x = 0$ (0, 0), local maximum at $x = 4/3$ (4/3, 32/27)   c) no absolute extrema
13. a) Decreasing on (−∞, 0), increasing on (0, 1/2), decreasing on (1/2, ∞)   b) local minimum at $\theta = 0$ (0, 0), local maximum at $\theta = 1/2$ (1/2, 1/4)   c) no absolute extrema
15. a) Increasing on (−∞, ∞), never decreasing   b) no local extrema   c) no absolute extrema
17. a) Increasing on (−2, 0) and (2, ∞), decreasing on (−∞, −2) and (0, 2)   b) local maximum: 16 at $x = 0$, local minimum: 0 at $x = \pm 2$ c) no absolute maximum, absolute minimum: 0 at $x = \pm 2$
19. a) Increasing on (−∞, −1), decreasing on (−1, 0), increasing on (0, 1), decreasing on (1, ∞)
b) local maximum at $x = \pm 1$ (1, 0.5), (−1, 0.5), local minimum at $x = 0$ (0, 0)   c) absolute maximum: 1/2 at $x = \pm 1$, no absolute minimum
21. a) Decreasing on $(-2\sqrt{2}, -2)$, increasing on (−2, 2), decreasing on $(2, 2\sqrt{2})$ b) local minima: $g(-2) = -4$, $g(2\sqrt{2}) = 0$; local maxima: $g(-2\sqrt{2}) = 0$, $g(2) = 4$ c) absolute maximum: 4 at $x = 2$, absolute minimum: −4 at $x = -2$
23. a) Increasing on (−∞, 1), decreasing when $1 < x < 2$, decreasing when $2 < x < 3$, discontinuous at $x = 2$, increasing on (3, ∞), b) local minimum at $x = 3$ (3, 6), local maximum at $x = 1$ (1, 2)
c) no absolute extrema
25. a) Increasing on (−2, 0) and (0, ∞), decreasing on (−∞, −2)
b) local minimum: $-6\sqrt[3]{2}$ at $x = -2$ c) no absolute maximum, absolute minimum: $-6\sqrt[3]{2}$ at $x = -2$
27. a) Increasing on $(-\infty, -2/\sqrt{7})$ and $(2/\sqrt{7}, \infty)$, decreasing on $(-2/\sqrt{7}, 2/\sqrt{7})$ b) local maximum: $24\sqrt[3]{2}/7^{7/6} \approx 3.12$ at $x = -2/\sqrt{7}$, local minimum: $-24\sqrt[3]{2}/7^{7/6} \approx -3.12$ at $x = 2/\sqrt{7}$
c) no absolute extrema
29. a) Local maximum: 1 at $x = 1$, local minimum: 0 at $x = 2$
b) absolute maximum: 1 at $x = 1$; no absolute minimum
31. a) Local maximum: 1 at $x = 1$, local minimum: 0 at $x = 2$
b) no absolute maximum, absolute minimum: 0 at $x = 2$
33. a) Local maxima: −9 at $t = -3$ and 16 at $t = 2$, local minimum: −16 at $t = -2$ b) absolute maximum: 16 at $t = 2$, no absolute minimum
35. a) Local minimum: 0 at $x = 0$ b) no absolute maximum, absolute minimum: 0 at $x = 0$
37. a) Local minimum: $(\pi/3) - \sqrt{3}$ at $x = 2\pi/3$, local maximum: 0 at $x = 0$, local maximum: π at $x = 2\pi$
39. a) Local minimum: 0 at $x = \pi/4$
41. Local maximum: 3 at $\theta = 0$, Local minimum: −3 at $\theta = 2\pi$
43.

45. a)

b)

47. Rising

Section 3.4, pp. 217–220

1. Local maximum: 3/2 at $x = -1$, local minimum: -3 at $x = 2$, point of inflection at $(1/2, -3/4)$, rising on $(-\infty, -1)$ and $(2, \infty)$, falling on $(-1, 2)$, concave up on $(1/2, \infty)$, concave down on $(-\infty, 1/2)$

3. Local maximum: 3/4 at $x = 0$, local minimum: 0 at $x = \pm 1$, points of inflection at $\left(-\sqrt{3}, \dfrac{3\sqrt[3]{4}}{4}\right)$ and $\left(\sqrt{3}, \dfrac{3\sqrt[3]{4}}{4}\right)$, rising on $(-1, 0)$ and $(1, \infty)$, falling on $(-\infty, -1)$ and $(0, 1)$, concave up on $(-\infty, -\sqrt{3})$ and $(\sqrt{3}, \infty)$, concave down on $(-\sqrt{3}, \sqrt{3})$

5. Local maxima: $-2\pi/3 + \sqrt{3}/2$ at $x = -2\pi/3$; $\dfrac{\pi}{3} + \dfrac{\sqrt{3}}{2}$ at $x = \dfrac{\pi}{3}$, local minima: $-\dfrac{\pi}{3} - \dfrac{\sqrt{3}}{2}$ at $x = -\dfrac{\pi}{3}$; $2\pi/3 - \sqrt{3}/2$ at $x = \dfrac{2\pi}{3}$, points of inflection at $(-\pi/2, -\pi/2)$, $(0, 0)$, and $(\pi/2, \pi/2)$, rising on $(-\pi/3, \pi/3)$, falling on $(-2\pi/3, -\pi/3)$ and $(\pi/3, 2\pi/3)$, concave up on $(-\pi/2, 0)$ and $(\pi/2, 2\pi/3)$, concave down on $(-2\pi/3, -\pi/2)$ and $(0, \pi/2)$

7. Local maxima: 1 at $x = -\dfrac{\pi}{2}$ and $x = \dfrac{\pi}{2}$; 0 at $x = -2\pi$ and $x = 2\pi$; local minima: -1 at $x = -\dfrac{3\pi}{2}$ and $x = \dfrac{3\pi}{2}$, 0 at $x = 0$, points of inflection at $(-\pi, 0)$ and $(\pi, 0)$, rising on $(-3\pi/2, -\pi/2)$, $(0, \pi/2)$ and $(3\pi/2, 2\pi)$, falling on $(-2\pi, -3\pi/2)$, $(-\pi/2, 0)$ and $(\pi/2, 3\pi/2)$, concave up on $(-2\pi, -\pi)$ and $(\pi, 2\pi)$, concave down on $(-\pi, \pi)$

9.

11.

13.

15.

17.

19.

21.

23.

25.

27.

29.

31.

33.

35.

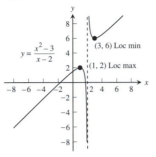

53. $y'' = 2 \tan \theta \sec^2 \theta, -\dfrac{\pi}{2} < \theta < \dfrac{\pi}{2}$

37.

39.

55. $y'' = -\sin t, 0 \le t \le 2\pi$

57. $y'' = -\dfrac{2}{3}(x+1)^{-5/3}$

41. $y'' = 1 - 2x$

43. $y'' = 3(x-3)(x-1)$

59. $y'' = \dfrac{1}{3}x^{-2/3} + \dfrac{2}{3}x^{-5/3}$

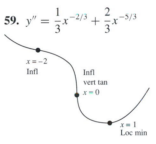

61. $y'' = \begin{cases} -2, & x < 0 \\ 2, & x > 0 \end{cases}$

45. $y'' = 3(x-2)(x+2)$

47. $y'' = 4(4-x)(5x^2 - 16x + 8)$

63.

65.

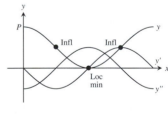

49. $y'' = 2 \sec^2 x \tan x$

51. $y'' = -\dfrac{1}{2}\csc^2 \dfrac{\theta}{2}, 0 < \theta < 2\pi$

67.

Point	y'	y''
P	$-$	$+$
Q	$+$	0
R	$+$	$-$
S	0	$-$
T	$-$	$-$

69.

73. ≈ 60 thousand units
75. Local minimum at $x = 2$, inflection points at $x = 1$ and $x = 5/3$
79. $b = -3$

81. a) $\left(-\dfrac{b}{2a}, \dfrac{4ac - b^2}{4a}\right)$ b) concave up if $a > 0$, concave down if $a < 0$

85. The zeros of $y' = 0$ and $y'' = 0$ are extrema and points of inflection, respectively.

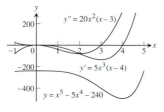

87. The zeros of $y' = 0$ and $y'' = 0$ are extrema and points of inflection, respectively. Inflection at $x = -\sqrt[3]{2}$, local maximum at $x = -2$, local minimum at $x = 0$.

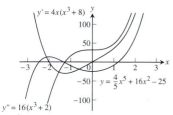

91. b) $f'(x) = 3x^2 + k$; $-12k$; positive if $k < 0$, negative if $k > 0$, 0 if $k = 0$; f' has two zeros if $k < 0$, one zero if $k = 0$, no zeros if $k > 0$
93. b) A cusp since $\lim\limits_{x \to 0^-} y' = \infty$ and $\lim\limits_{x \to 0^+} y' = -\infty$
95. Yes, the graph of y' crosses through zero near -3, so y has a horizontal tangent near -3.

Section 3.5, pp. 230–233

1. a) -3 b) -3 **3.** a) $1/2$ b) $1/2$ **5.** a) $-5/3$ b) $-5/3$
7. 0   **9.** -1 **11.** a) $2/5$ b) $2/5$ **13.** a) 0   b) 0
15. a) $-\infty$ b) ∞ **17.** a) 7   b) 7   **19.** a) $-\infty$ b) ∞
21. a) ∞ b) $-\infty$ **23.** a) $-2/3$ b) $-2/3$ **25.** 0   **27.** 1
29. ∞

31. Here is one possibility.

33. Here is one possibility.

35. Here is one possibility.

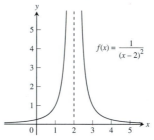

37. Here is one possibility.

39.

41.

43.

45.

47.

49.

63.

65.

51.

53.

67.

69.

71.

73.

55.

57.

75.

59.

61.

79. Increasing   **83.** 2
85. b) One possibility is $f(x) = 2 + (1/x) \sin(x^2)$.
89. $x = -1$, $y = 1 - x$ **91.** $x = 1$, $x = -1$, $y = x - 1$
99. a) $y \to \infty$, b) $y \to \infty$, c) cusp at $x = \pm 1$
101. The distance in part (c) is so great that small movements are not visible.
103. 1   **105.** 3/2   **107.** 3

Section 3.6, pp. 242–247

1. $r = 25$ m, $s = 50$ m **3.** 16 in. **5.** a) $(x, 1 - x)$

b) $A(x) = 2x(1 - x)$ c) 1/2 square units **7.** $\dfrac{14}{3} \times \dfrac{35}{3} \times \dfrac{5}{3}$ in.

9. 80,000 m² **11.** base: 10 ft, height: 5 ft **13.** 9×18 in.

15. $\pi/2$ **17.** $r = h = \dfrac{10}{\sqrt[3]{\pi}}$ cm **19.** a) $18 \times 18 \times 36$ in.

21. a) 12 cm, 6 cm b) 12 cm, 6 cm
23. a) The circumference of the circle is 4 m.
25. If r is the radius of the semicircle, $2r$ is the base of the rectangle, and h is the height of the rectangle, then $(2r)/h = 8/(4 + \pi)$.

27. $\pi/6$ **29.** $\dfrac{v_0^2}{2g} + s_0$ **31.** a) $4\sqrt{3} \times 4\sqrt{6}$ in. **33.** $2\sqrt{2}$ amps

35. a) When t is an integer multiple of π

b) $t = \dfrac{2\pi}{3}, t = \dfrac{4\pi}{3}, 3\dfrac{\sqrt{3}}{2}$ **37.** a) $t = \dfrac{8}{5}, t = 4$

b) $\dfrac{8}{5} < t < 4$ c) $\dfrac{2187}{125}$ units/time
41. No. The function has an absolute minimum of 3/4.

43. a) $\left(c - \dfrac{1}{2}, \sqrt{c - \dfrac{1}{2}}\right)$ b) $(0, 0)$ **45.** a) $a = -3, b = -9$

b) $a = -3, b = -24$ **47.** a) $y = -1$ **49.** $(7/2)\sqrt{17}$

53. $M = c/2$ **55.** $\dfrac{c}{2} + 50$ **57.** $\sqrt{\dfrac{2km}{h}}$

Section 3.7, pp. 257–260

1. $4x - 3$ **3.** $2x - 2$ **5.** $\dfrac{1}{4}x + 1$ **7.** $2x$ **9.** -5

11. $\dfrac{1}{12}x + \dfrac{4}{3}$ **13.** a) $L(x) = x$ b) $L(x) = \pi - x$

15. a) $L(x) = 1$ b) $L(x) = 2 - 2\sqrt{3}\left(x + \dfrac{\pi}{3}\right)$ **17.** a) $1 + 2x$

b) $1 - 5x$ c) $2 + 2x$ d) $1 - 6x$ e) $3 + x$ f) $1 - \dfrac{x}{2}$

19. $\dfrac{3}{2}x + 1$. It is equal to their sum. **21.** $\left(3x^2 - \dfrac{3}{2\sqrt{x}}\right) dx$

23. $\dfrac{2 - 2x^2}{(1 + x^2)^2} dx$ **25.** $\dfrac{1 - y}{3\sqrt{y} + x} dx$ **27.** $\dfrac{5}{2\sqrt{x}} \cos(5\sqrt{x}) dx$

29. $(4x^2) \sec^2\left(\dfrac{x^3}{3}\right) dx$

31. $\dfrac{3}{\sqrt{x}}(\csc(1 - 2\sqrt{x}) \cot(1 - 2\sqrt{x})) dx$ **33.** a) .21 b) .2

c) .01 **35.** a) .231 b) .2 c) .031 **37.** a) $-1/3$ b) $-2/5$
c) 1/15 **39.** $dV = 4\pi r_0^2 dr$ **41.** $dS = 12x_0 dx$
43. $dV = 2\pi r_0 h dr$ **45.** a) $.08\pi$ m² b) 2% **47.** 3%
49. 3% **51.** 1/3% **53.** .05%

57. Volume $= (x + \Delta x)^3 = x^3 + 3x^2(\Delta x) + 3x(\Delta x)^2 + (\Delta x)^3$

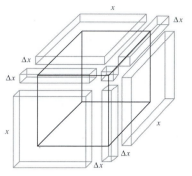

59. $\displaystyle\lim_{x \to 0} \dfrac{\sqrt{1 + x}}{1 + \left(\dfrac{x}{2}\right)} = \dfrac{\sqrt{1 + 0}}{1 + \left(\dfrac{0}{2}\right)} = \dfrac{1}{1} = 1$

Section 3.8, pp. 266–268

1. $x_2 = 13/21, -5/3$ **3.** $x_2 = 5763/4945, -51/31$
5. $x_2 = 2387/2000$ **7.** $x \approx 0.45$ **9.** The root is 1.17951
13.

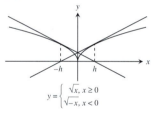

$$y = \begin{cases} \sqrt{x}, & x \geq 0 \\ \sqrt{-x}, & x < 0 \end{cases}$$

15. a) The points of intersection of $y = x^3$ and $y = 3x + 1$ or $y = x^3 - 3x$ and $y = 1$ have the same x-values as the roots of part (i) or the solutions of part (iv). b) $-1.53209, -0.34730$
17. 2.45, 0.000245 **19.** 1.1655 61185 **21.** a) Two
b) 0.3500 35015 05249 and -1.0261 73161 5301
23. ± 1.3065 62964 8764, ± 0.5411 96100 14619
25. Answers will vary with machine used.
27.

x_0	Approximation of corresponding root
-1.0	-0.976823589
0.1	0.100363332
0.6	0.642746671
2.0	1.98371387

Chapter 3 Practice Exercises, pp. 269–272

1. No
3. No minimum, absolute maximum: $f(1) = 16$, critical points: $x = 1$ and 11/3
7. No **11.** b) one **13.** b) 0.8555 99677 2

19.

21.

b)

23.

25.

39.

41.

27.

43.

45.

29. a) Local maximum at $x = 4$, local minimum at $x = -4$, inflection point at $x = 0$

b)

47.

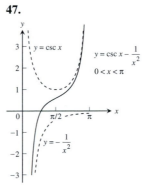

31. a) Local maximum at $x = 0$, local minima at $x = -1$ and $x = 2$, inflection points at $x = (1 \pm \sqrt{7})/3$

b)

33. a) Local maximum at $x = -\sqrt{2}$, local minimum at $x = \sqrt{2}$, inflection points at $x = \pm 1$ and 0

49. a) $t = 0, 6, 12$ b) $t = 3, 9$ c) $6 < t < 12$ d) $0 < t < 6$, $12 < t < 14$

51. 2/5 **53.** 0 **55.** $-\infty$ **57.** 0 **59.** 1 **61.** a) 0, 36
b) 18, 18 **63.** 54 square units **65.** height = 2, radius = $\sqrt{2}$

67. $x = 15$ mi, $y = 9$ mi **69.** $x = 5 - \sqrt{5}$ hundred ≈ 276 tires, $y = 2(5 - \sqrt{5})$ hundred ≈ 553 tires
71. a) $L(x) = 2x + (\pi - 2)/2$

b) $L(x) = -\sqrt{2}\,x + \sqrt{2}(4 - \pi)/4$

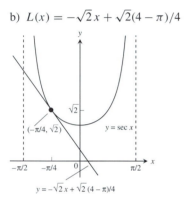

73. $L(x) = 1.5x + 0.5$ **75.** $dV = \dfrac{2}{3}\pi r_0 h \, dr$

77. a) error $< 1\%$ b) 3% **79.** $dh \approx \pm 2.3271$ ft
81. $x_5 = 2.1958\ 23345$

Chapter 3 Additional Exercises, pp. 272–274

3. The extreme points will not be at the end of an open interval.
5. a) A local minimum at $x = -1$, points of inflection at $x = 0$ and $x = 2$ b) A local maximum at $x = 0$ and local minima at $x = -1$ and $x = 2$, points of inflection at $x = \dfrac{1 \pm \sqrt{7}}{3}$
11. $a = 1, b = 0, c = 1$ **13.** Yes **15.** Drill the hole at $y = h/2$.
17. $r = \dfrac{RH}{2(H - R)}$ for $H < 2R$, $r = R$ if $H \le 2R$
21. a) 0.8156 ft b) 0.00613 sec c) It will lose about 8.83 min/day.

CHAPTER 4

Section 4.1, pp. 280–282

1. a) x^2 b) $\dfrac{x^3}{3}$ c) $\dfrac{x^3}{3} - x^2 + x$ **3.** a) x^{-3} b) $-\dfrac{1}{3}x^{-3}$
c) $-\dfrac{1}{3}x^{-3} + x^2 + 3x$ **5.** a) $-\dfrac{1}{x}$ b) $-\dfrac{5}{x}$ c) $2x + \dfrac{5}{x}$

7. a) $\sqrt{x^3}$ b) $\sqrt{x}$ c) $\dfrac{2\sqrt{x^3}}{3} + 2\sqrt{x}$ **9.** a) $x^{2/3}$ b) $x^{1/3}$
c) $x^{-1/3}$ **11.** a) $\cos(\pi x)$ b) $-3\cos x$
c) $-\dfrac{1}{\pi}\cos(\pi x) + \cos(3x)$ **13.** a) $\tan x$ b) $2\tan\left(\dfrac{x}{3}\right)$
c) $-\dfrac{2}{3}\tan\left(\dfrac{3x}{2}\right)$ **15.** a) $-\csc x$ b) $\dfrac{1}{5}\csc(5x)$ c) $2\csc\left(\dfrac{\pi x}{2}\right)$
17. $x + \dfrac{\cos(2x)}{2}$ **19.** $\dfrac{x^2}{2} + x + C$ **21.** $t^3 + \dfrac{t^2}{4} + C$
23. $\dfrac{x^4}{2} - \dfrac{5x^2}{2} + 7x + C$ **25.** $-\dfrac{1}{x} - \dfrac{x^3}{3} - \dfrac{x}{3} + C$
27. $\dfrac{3}{2}x^{2/3} + C$ **29.** $\dfrac{2}{3}x^{3/2} + \dfrac{3}{4}x^{4/3} + C$ **31.** $4y^2 - \dfrac{8}{3}y^{3/4} + C$
33. $x^2 + \dfrac{2}{x} + C$ **35.** $2\sqrt{t} - \dfrac{2}{\sqrt{t}} + C$ **37.** $-2\sin t + C$
39. $-21\cos\dfrac{\theta}{3} + C$ **41.** $3\cot x + C$ **43.** $-\dfrac{1}{2}\csc\theta + C$
45. $4\sec x - 2\tan x + C$ **47.** $-\dfrac{1}{2}\cos 2x + \cot x + C$
49. $2y - \sin 2y + C$ **51.** $\dfrac{t}{2} + \dfrac{\sin 4t}{8} + C$ **53.** $\tan\theta + C$
55. $-\cot x - x + C$ **57.** $-\cos\theta + \theta + C$
65. a) Wrong: $\dfrac{d}{dx}\left(\dfrac{x^2}{2}\sin x + C\right) = \dfrac{2x}{2}\sin x + \dfrac{x^2}{2}\cos x = $
$x\sin x + \dfrac{x^2}{2}\cos x$
b) Wrong: $\dfrac{d}{dx}(-x\cos x + C) = -\cos x + x\sin x$
c) Right: $\dfrac{d}{dx}(-x\cos x + \sin x + C) = -\cos x + x\sin x + \cos x$
$= x\sin x$
67. a) Wrong: $\dfrac{d}{dx}\left(\dfrac{(2x+1)^3}{3} + C\right) = \dfrac{3(2x+1)^2(2)}{3} = 2(2x+1)^2$
b) Wrong: $\dfrac{d}{dx}((2x+1)^3 + C) = 3(2x+1)^2(2) = 6(2x+1)^2$
c) Right: $\dfrac{d}{dx}((2x+1)^3 + C) = 6(2x+1)^2$
69. a) $-\sqrt{x} + C$ b) $x + C$ c) $\sqrt{x} + C$ d) $-x + C$
e) $x - \sqrt{x} + C$ f) $-x - \sqrt{x} + C$ g) $\dfrac{x^2}{2} - \sqrt{x} + C$
h) $-3x + C$

Section 4.2, pp. 288–290

1. b **3.** $y = x^2 - 7x + 10$ **5.** $y = -\dfrac{1}{x} + \dfrac{x^2}{2} - \dfrac{1}{2}$
7. $y = 9x^{1/3} + 4$ **9.** $s = t + \sin t + 4$ **11.** $r = \cos(\pi\theta) - 1$

13. $v = \dfrac{1}{2}\sec t + \dfrac{1}{2}$ **15.** $y = x^2 - x^3 + 4x + 1$

17. $r = \dfrac{1}{t} + 2t - 2$ **19.** $y = x^3 - 4x^2 + 5$

21. $y = -\sin t + \cos t + t^3 - 1$ **23.** $s = 4.9t^2 + 5t + 10$

25. $s = \dfrac{1 - \cos(\pi t)}{\pi}$ **27.** $s = 16t^2 + 20t + 5$

29. $s = \sin(2t) - 3$ **31.** $y = 2x^{3/2} - 50$ **33.** $y = x - x^{4/3} + \dfrac{1}{2}$

35. $y = -\sin x - \cos x - 2$

37.   **39.**

41.

43.

45. 48 m/sec   **47.** 14 m/sec   **49.** $t = 88/k$, $k = 16$

51. a) $v = 10t^{3/2} - 6t^{1/2}$ b) $s = 4t^{5/2} - 4t^{3/2}$

55. a) 1: 33.2 units, 2: 33.2 units, 3: 33.2 units   b) True

Section 4.3, p. 296

1. $-\dfrac{1}{3}\cos 3x + C$ **3.** $\dfrac{1}{2}\sec 2t + C$ **5.** $-(7x - 2)^{-4} + C$

7. $-6(1 - r^3)^{1/2} + C$ **9.** $\dfrac{1}{3}(x^{3/2} - 1) - \dfrac{1}{6}\sin(2x^{3/2} - 2) + C$

11. a) $-\dfrac{1}{4}(\cot^2 2\theta) + C$ b) $-\dfrac{1}{4}(\csc^2 2\theta) + C$

13. $-\dfrac{1}{3}(3 - 2s)^{3/2} + C$ **15.** $\dfrac{2}{5}(5s + 4)^{1/2} + C$

17. $-\dfrac{2}{5}(1 - \theta^2)^{5/4} + C$ **19.** $-\dfrac{1}{3}(7 - 3y^2)^{3/2} + C$

21. $(-2/(1 + \sqrt{x})) + C$ **23.** $\dfrac{1}{3}\sin(3z + 4) + C$

25. $\dfrac{1}{3}\tan(3x + 2) + C$ **27.** $\dfrac{1}{2}\sin^6\left(\dfrac{x}{3}\right) + C$

29. $\left(\dfrac{r^3}{18} - 1\right)^6 + C$ **31.** $-\dfrac{2}{3}\cos(x^{3/2} + 1) + C$

33. $\sec\left(v + \dfrac{\pi}{2}\right) + C$ **35.** $\dfrac{1}{2\cos(2t + 1)} + C$

37. $-\dfrac{2}{3}(\cot^3 y)^{1/2} + C$ **39.** $-\sin\left(\dfrac{1}{t} - 1\right) + C$

41. $-\dfrac{\sin^2(1/\theta)}{2} + C$ **43.** $\dfrac{(s^3 + 2s^2 - 5s + 5)^2}{2} + C$

45. $\dfrac{1}{16}(1 + t^4)^4 + C$ **47.** a) $-\dfrac{6}{2 + \tan^3 x} + C$

b) $-\dfrac{6}{2 + \tan^3 x} + C$ c) $-\dfrac{6}{2 + \tan^3 x} + C$

49. $\dfrac{1}{6}\sin\sqrt{3(2r - 1)^2 + 6} + C$ **51.** $s = \dfrac{1}{2}(3t^2 - 1)^4 - 5$

53. $s = 4t - 2\sin\left(2t + \dfrac{\pi}{6}\right) + 9$

55. $s = \sin\left(2t - \dfrac{\pi}{2}\right) + 100t + 1$ **57.** 6 m

Section 4.4, pp. 305–309

1. ≈ 44.8, 6.7 L/min   **3.** a) 87 in.   b) 87 in.   **5.** a) 3,490 ft
b) 3,840 ft   **7.** a) 112   b) 9%   **9.** a) 80π b) 6%
11. a) $93\pi/2$, overestimate   b) 9%   **13.** a) 40   b) 25%
c) 36, 12.5%   **15.** a) 118.5π or ≈ 372.28 m^3 b) error $\approx 11\%$
17. a) 10π, underestimate   b) 20%   **19.** 31/16   **21.** 1
23. a) 74.65 ft/sec   b) 45.28 ft/sec   c) 146.59 ft
25. a) upper $= 758$ gal, lower $= 543$ gal   b) upper $= 2363$ gal,
lower $= 1693$ gal   c) ≈ 31.4 hours, ≈ 32.4 hours

Section 4.5, pp. 320–323

1. $\dfrac{6(1)}{1 + 1} + \dfrac{6(2)}{2 + 1} = 7$

3. $\cos(1)\pi + \cos(2)\pi + \cos(3)\pi + \cos(4)\pi = 0$

5. $\sin\pi - \sin\dfrac{\pi}{2} + \sin\dfrac{\pi}{3} = \dfrac{\sqrt{3} - 2}{2}$

7. All of them   **9.** b   **11.** $\displaystyle\sum_{k=1}^{6} k$ **13.** $\displaystyle\sum_{k=1}^{4} \dfrac{1}{2^k}$

15. $\displaystyle\sum_{k=1}^{5} (-1)^{k+1}\dfrac{1}{k}$ **17.** a) -15 b) 1   c) 1   d) -11 e) 16

19. a) 55   b) 385   c) 3025   **21.** -56 **23.** -73 **25.** 240
27. 3376

29. a)

b)

c) 31. a)

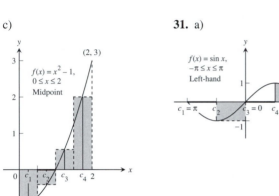

b) c)

33. 1.2 **35.** $\int_0^2 x^2\,dx$ **37.** $\int_{-7}^5 (x^2 - 3x)\,dx$

39. $\int_2^3 \frac{1}{1-x}\,dx$ **41.** $\int_{-\pi/4}^0 \sec x\,dx$ **43.** 15 **45.** −480

47. 2.75 **49.** Area = 21 square units

51. Area = $9\pi/2$ square units **53.** Area = 2.5 square units

55. Area = 3 square units **57.** $b^2/2$ **59.** $b^2 - a^2$ **61.** 1/2

63. $3\pi^2/2$ **65.** 7/3 **67.** 1/24 **69.** $3a^2/2$ **71.** $b/3$

73. Using n subintervals of length $\Delta x = b/n$ and right-endpoint values:

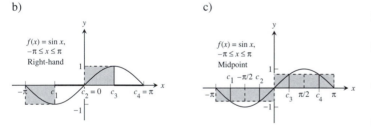

$$\text{Area} = \int_0^b 3x^2\,dx = b^3$$

75. Using n subintervals of length $\Delta x = b/n$ and right-endpoint values:

$$\text{Area} = \int_0^b 2x\,dx = b^2$$

77. $a = 0$ and $b = 1$ maximize the integral. **81.** $b^3/3$

Section 4.6, pp. 330–332

1. a) 0 b) −8 c) −12 d) 10 e) −2 f) 16 **3.** a) 5
b) $5\sqrt{3}$ c) −5 d) −5 **5.** a) 4 b) −4 **7.** −14 **9.** 10

11. −2 **13.** −7/4 **15.** 7 **17.** 0 **19.** 5 1/3 **21.** 19/3

23. a) 6 b) $7\frac{1}{3}$ **25.** a) 0 b) 8/3

27. av(f) = 0, assumed at $x = 1$

29. av(f) = −2, assumed at $x = \sqrt{3}/3$

31. av(f) = 1, assumed at $t = 0$ and $t = 2$

33. a) av(g) = −1/2, assumed at $x = \pm1/2$ b) av(g) = 1, assumed at $x = 2$ c) av(g) = 1/4, assumed at $x = 5/4$

35. 3/2 **37.** 0 **39.** Upper bound = 1, lower bound = 1/2

47. Upper bound = 1/2 **51.** 37.5 mi/hr

Section 4.7, pp. 338–342

1. 6 **3.** 8 **5.** 1 **7.** 5/2 **9.** 2 **11.** $2\sqrt{3}$ **13.** 0

15. $-\pi/4$ **17.** $\frac{2\pi^3}{3}$ **19.** −8/3 **21.** −3/4

23. $\sqrt{2} - \sqrt[4]{8} + 1$ **25.** 16 **27.** 0 **29.** $\frac{1}{3}(2\sqrt{2} - 1)$

31. $\frac{\pi}{2} + \sin 2$ **33.** $\sqrt{2}/3$ **35.** 28/3 **37.** 1/2 **39.** 51/4

41. π **43.** $\frac{\sqrt{2}\,\pi}{2}$ **45.** $(\cos\sqrt{x})\left(\frac{1}{2\sqrt{x}}\right)$ **47.** $4t^5$

49. $\sqrt{1 + x^2}$ **51.** $\frac{1}{2}x^{-1/2}\sin x$ **53.** 1

55. d, since $y' = \frac{1}{x}$ and $y(\pi) = \int_\pi^\pi \frac{1}{t}\,dt - 3 = -3$

57. b, since $y' = \sec x$ and $y(0) = \int_0^0 \sec t\,dt + 4 = 4$

59. $y = \int_2^x \sec t\,dt + 3$ **61.** $s = \int_{t_0}^t f(x)\,dx + s_0$

63. a) 125/6 b) $h = 25/4$ d) $\frac{2}{3}bh$ **65.** a) $9.00 b) $10.00

67. a) $v = \frac{ds}{dt} = \frac{d}{dt}\int_0^1 f(x)\,dx = f(t) \Rightarrow v(5) = f(5) = 2$ m/sec

b) $a = df/dt$ is negative since the slope of the tangent line at $t = 5$ is negative.

c) $s = \int_0^3 f(x)\,dx = \frac{1}{2}(3)(3) = \frac{9}{2}$ m since the integral is the area
of the triangle formed by $y = f(x)$, the x-axis, and $x = 3$.
d) $t = 6$ since after $t = 6$ to $t = 9$, the region lies below the x-axis.
e) At $t = 4$ and $t = 7$, since there are horizontal tangents there.
f) Toward the origin between $t = 6$ and $t = 9$ since the velocity is
negative on this interval. Away from the origin between $t = 0$ and
$t = 6$ since the velocity is positive there.
g) Right or positive side, because the integral of f from 0 to 9 is
positive, there being more area above the x-axis than below.

69. $\int_{-2}^{2} 4(9 - x^2)\,dx = 368/3$ **71.** $\int_{4}^{8} \pi(64 - x^2)\,dx = 320\pi/3$

75. $2x - 2$ **77.** $-3x + 5$
79. a) True. Since f is continuous, g is differentiable by Part 1 of
the Fundamental Theorem of Calculus.
b) True: g is continuous because it is differentiable.
c) True, since $g'(1) = f(1) = 0$.
d) False, since $g''(1) = f'(1) > 0$.
e) True, since $g'(1) = 0$ and $g''(1) = f'(1) > 0$.
f) False: $g''(x) = f'(x) > 0$, so g'' never changes sign.
g) True, since $g'(1) = f(1) = 0$ and $g'(x) = f(x)$ is an increasing
function of x (because $f'(x) > 0$).

Section 4.8, pp. 344–345

1. a) $14/3$ b) $2/3$ **3.** a) $1/2$ b) $-1/2$ **5.** a) $15/16$ b) 0
7. a) 0 b) $1/8$ **9.** a) 4 b) 0 **11.** a) $1/6$ b) $1/2$
13. a) 0 b) 0 **15.** $2\sqrt{3}$ **17.** $3/4$ **19.** $9^{5/4} - 1$ **21.** 3
23. $\pi/3$ **25.** $16/3$ **27.** $2^{5/2}$ **29.** $F(6) - F(2)$ **31.** a) -3
b) 3 **33.** $I = a/2$

Section 4.9, pp. 353–356

1. I: a) $1.5, 0$ b) $1.5, 0$ c) 0%
II: a) $1.5, 0$ b) $1.5, 0$ c) 0%
3. I: a) $2.75, 0.08$ b) $2.67, 0.08$ c) $0.0312 \approx 3\%$
II: a) $2.67, 0$ b) $2.67, 0$ c) 0%
5. I: a) $6.25, 0.5$ b) $6, 0.25$ c) $0.0417 \approx 4\%$
II: a) $6, 0$ b) $6, 0$ c) 0%
7. I: a) $0.509, 0.03125$ b) $0.5, 0.009$ c) $0.018 \approx 2\%$
II: a) $0.5, 0.002604$ b) $0.5, 0.0004$ c) 0%
9. I: a) $1.8961, 0.161$ b) $2, 0.1039$ c) $0.052 \approx 5\%$
II: a) $2.0045, 0.0066$ b) $2, 0.00454$ c) 0%
11. a) 0.31929 b) 0.32812 c) $1/3, 0.01404, 0.00521$
13. a) 1.95643 b) 2.00421 c) $2, 0.04357, -0.00421$ **15.** a) 1
b) 2 **17.** a) 116 b) 2 **19.** a) 283 b) 2 **21.** a) 71
b) 10 **23.** a) 76 b) 12 **25.** a) 82 b) 8 **27.** 1013
29. ≈ 466.7 in^2 **31.** $4, 4$ **33.** a) 3.11571 b) 0.02588
c) With $M = 3.11$, we get $|E_T| \le (\pi^3/1200)(3.11) < 0.081$
37. 1.08943 **39.** 0.82812

Chapter 4 Practice Exercises, pp. 357–360

1. a) about 680 ft

b)

h (feet)

t (sec)

9. a) 4 b) 2 c) -2 d) -2π e) $8/5$

11. $8/3$ **13.** 62 **15.** $y = x - \frac{1}{x} - 1$

17. $r = 4t^{5/2} + 4t^{3/2} - 8t$ **21.** $y = \int_{5}^{x}\left(\frac{\sin t}{t}\right)dt - 3$

23. $\frac{x^4}{4} + \frac{5}{2}x^2 - 7x + C$ **25.** $2t^{3/2} - \frac{4}{t} + C$

27. $-\frac{1}{2(r^2 + 5)} + C$ **29.** $-(2 - \theta^2)^{3/2} + C$

31. $\frac{1}{3}(1 + x^4)^{3/4} + C$ **33.** $10\tan\frac{s}{10} + C$

35. $-\frac{1}{\sqrt{2}}\csc\sqrt{2}\theta + C$ **37.** $\frac{1}{2}x - \sin\frac{x}{2} + C$

39. $-4(\cos x)^{1/2} + C$ **41.** $\theta^2 + \theta + \sin(2\theta + 1) + C$

43. $\frac{t^3}{3} + \frac{4}{t} + C$ **45.** 16 **47.** 2 **49.** 1 **51.** 8
53. $27\sqrt{3}/160$ **55.** $\pi/2$ **57.** $\sqrt{3}$ **59.** $6\sqrt{3} - 2\pi$ **61.** -1
63. 2 **65.** -2 **67.** 1 **69.** $\sqrt{2} - 1$ **71.** a) b b) b
75. At least 16 **77.** $T = \pi, S = \pi$ **79.** $25°$F **81.** Yes
83. $-\sqrt{1 + x^2}$ **85.** cost $\approx \$12,518.10$ (trapezoidal rule), no
87. $600, \$18.00$ **89.** $300, \$6.00$

Chapter 4 Additional Exercises, pp. 360–364

1. a) Yes b) No **5.** a) $1/4$ b) $\sqrt[3]{12}$ **7.** $f(x) = \dfrac{x}{\sqrt{x^2 + 1}}$

9. $y = x^3 + 2x - 4$
11. $36/5$

3. a) $-1/2$ b) 31 c) 13 d) 0

5. $\int_{1}^{5}(2x - 1)^{-1/2}\,dx = 2$

7. $\int_{-\pi}^{0}\cos\frac{x}{2}\,dx = 2$

13. $\dfrac{1}{2} - \dfrac{2}{\pi}$

15. $13/3$

17. 1/2   **19.** 2/x   **21.** $\dfrac{\sin 4y}{\sqrt{y}} - \dfrac{\sin y}{2\sqrt{y}}$ **23.** 1/6

25. $\displaystyle\int_0^1 f(x)\,dx$

CHAPTER 5

Section 5.1, pp. 371–373

1. $\pi/2$ **3.** 1/12   **5.** 128/15   **7.** 5/6   **9.** 38/3   **11.** 49/6
13. 32/3   **15.** 48/5   **17.** 8/3   **19.** 8
21. 5/3 (There are three intersection points.)   **23.** 18   **25.** 243/8
27. 8/3   **29.** 2   **31.** 104/15   **33.** 56/15   **35.** 4
37. $\dfrac{4}{3} - \dfrac{4}{\pi}$ **39.** $\pi/2$ **41.** 2   **43.** 1/2   **45.** 1
47. a) $(\pm\sqrt{c}, c)$ b) $c = 4^{2/3}$ c) $c = 4^{2/3}$ **49.** 11/3   **51.** 3/4
53. Neither

Section 5.2, pp. 377–378

1. a) $A(x) = \pi(1 - x^2)$ b) $A(x) = 4(1 - x^2)$
c) $A(x) = 2(1 - x^2)$ d) $A(x) = \sqrt{3}(1 - x^2)$ **3.** 16   **5.** 16/3
7. a) $2\sqrt{3}$ b) 8   **9.** 8π **11.** a) $s^2 h$ b) $s^2 h$

Section 5.3, pp. 385–387

1. $2\pi/3$ **3.** $4 - \pi$ **5.** $32\pi/5$ **7.** 36π **9.** π
11. $\pi\left(\dfrac{\pi}{2} + 2\sqrt{2} - \dfrac{11}{3}\right)$ **13.** 2π **15.** 2π **17.** 3π
19. $\pi^2 - 2\pi$ **21.** $\dfrac{2\pi}{3}$ **23.** 2π **25.** $117\pi/5$ **27.** $\pi(\pi - 2)$
29. $4\pi/3$ **31.** 8π **33.** $\pi(\sqrt{3})$ **35.** $7\pi/6$ **37.** a) 8π
b) $32\pi/5$ c) $8\pi/3$ d) $224\pi/15$ **39.** a) $16\pi/15$ b) $56\pi/15$
c) $64\pi/15$ **41.** $V = 1053\pi$ cm³   **43.** a) $c = 2/\pi$ b) $c = 0$
45. $V = 2a^2b\pi^2$ **47.** b) $V = \dfrac{\pi r^2 h}{3}$

Section 5.4, pp. 392–393

1. 6π **3.** 2π **5.** $14\pi/3$ **7.** 8π **9.** $5\pi/6$ **11.** $128\pi/5$
13. 3π **15.** $\dfrac{16\pi}{15}(3\sqrt{2} + 5)$ **17.** $8\pi/3$ **19.** $4\pi/3$
21. $16\pi/3$ **23.** a) $6\pi/5$ b) $4\pi/5$ c) 2π d) 2π
25. a) $5\pi/3$ b) $4\pi/3$ c) 2π d) $2\pi/3$ **27.** a) $11\pi/15$
b) $97\pi/105$ c) $121\pi/210$ d) $23\pi/30$ **29.** a) $512\pi/21$
b) $832\pi/21$ **31.** a) $\pi/6$ b) $\pi/6$ **33.** $9\pi/16$ **35.** b) 4π
37. Disk: 2 integrals; washer: 2 integrals; shell: 1 integral   **39.** $3x$

Section 5.5, pp. 398–400

1. a) $\displaystyle\int_{-1}^2 \sqrt{1 + 4x^2}\,dx$ c) ≈ 6.13 **3.** a) $\displaystyle\int_0^\pi \sqrt{1 + \cos^2 y}\,dy$
c) ≈ 3.82 **5.** a) $\displaystyle\int_{-1}^3 \sqrt{1 + (y+1)^2}\,dy$ c) ≈ 9.29
7. a) $\displaystyle\int_0^{\pi/6} \sec x\,dx$ c) ≈ 0.55 **9.** 12   **11.** 53/6
13. 123/32   **15.** 99/8   **17.** 2
19. a) $y = \sqrt{x}$ or $y = -\sqrt{x} + 2$ b) Two   **21.** 1   **23.** 21.07 in.

Section 5.6, pp. 405–407

1. a) $2\pi \displaystyle\int_0^{\pi/4} \tan x\sqrt{1 + \sec^4 x}\,dx$ c) ≈ 3.84
3. a) $2\pi \displaystyle\int_1^2 \dfrac{1}{y}\sqrt{1 + y^{-4}}\,dy$ c) ≈ 5.02
5. a) $2\pi \displaystyle\int_1^4 (3 - \sqrt{x})^2\sqrt{1 + (1 - 3x^{-1/2})^2}\,dx$ c) ≈ 63.37
7. a) $2\pi \displaystyle\int_0^{\pi/3} \left(\displaystyle\int_0^y \tan t\,dt\right)\sec y\,dy$ c) ≈ 2.08 **9.** $4\pi\sqrt{5}$
11. $3\pi\sqrt{5}$ **13.** $98\pi/81$ **15.** 2π **17.** $\pi(\sqrt{8} - 1)/9$
19. $35\pi\sqrt{5}/3$ **21.** $253\pi/20$
25. a) $2\pi \displaystyle\int_{-\pi/2}^{\pi/2} (\cos x)\sqrt{1 + \sin^2 x}\,dx$ b) ≈ 14.4236
27. Order 226.2 liters of each color.   **31.** $5\sqrt{2}\,\pi$ **33.** 14.4
35. 54.9

Section 5.7, pp. 416–418

1. 4 ft   **3.** $(L/4, L/4)$ **5.** $M_0 = 8, M = 8, \bar{x} = 1$
7. $M_0 = 15/2, M = 9/2, \bar{x} = 5/3$
9. $M_0 = 73/6, M = 5, \bar{x} = 73/30$ **11.** $M_0 = 3, M = 3, \bar{x} = 1$
13. $\bar{x} = 0, \bar{y} = 12/5$ **15.** $\bar{x} = 1, \bar{y} = -3/5$
17. $\bar{x} = 16/105, \bar{y} = 8/15$ **19.** $\bar{x} = 0, \bar{y} = \pi/8$
21. $\bar{x} = 1, \bar{y} = -2/5$ **23.** $\bar{x} = \bar{y} = \dfrac{2}{4 - \pi}$
25. $\bar{x} = 3/2, \bar{y} = 1/2$ **27.** a) $\dfrac{224\pi}{3}$ b) $\bar{x} = 2, \bar{y} = 0$
c)

31. $\bar{x} = \bar{y} = 1/3$ **33.** $\bar{x} = a/3$, $\bar{y} = b/3$ **35.** $13\delta/6$

37. $\bar{x} = 0$, $\bar{y} = \dfrac{a\pi}{4}$

Section 5.8, pp. 424–427

1. 400 ft·lb   **3.** 780 J   **5.** 72,900 ft·lb   **9.** 400 N/m
11. 4 cm, 0.08 J   **13.** a) 7238 lb/in.
b) 905 in·lb, 2714 in·lb
15. a) 1,497,600 ft·lb   b) 1 hr, 40 min
d) At 62.26 lb/ft³: a) 1,494,240 ft·lb   b) 1 hr, 40 min
At 62.59 lb/ft³: a) 1,502,160 ft·lb   b) 1 hr, 40 min
17. 38,484,510 J   **19.** 7,238,229.48 ft·lb   **21.** 91.32 in·oz
23. 21,446,605.9 J   **25.** 967,611 ft·lb, at a cost of \$4838.05
27. 5.144×10^{10} J   **31.** ≈ 85.1 ft·lb   **33.** ≈ 64.6 ft·lb
35. ≈ 110.6 ft·lb

Section 5.9, pp. 432–434

1. 114,511,052 lb, 28,627,763 lb   **5.** 2808 lb   **7.** a) 1164.8 lb
b) 1194.7 lb   **9.** a) 374.4 lb   b) 7.5 in.   c) No   **11.** 1309 lb
13. 4.2 lb   **15.** 41.6 lb   **17.** a) 93.33 lb   b) 3 ft   **19.** 1035 ft³
21. $wb/2$
23. No. The tank will overflow because the movable end will have
moved only $3\frac{1}{3}$ ft by the time the tank is full.

Section 5.10, pp. 441–443

1. b) 20 m   c) 0 m   **3.** b) 6 m   c) 2 m   **5.** b) 245 m
c) 0 m   **7.** b) 6 m   c) 4 m   **9.** b) $2 < t < 4$ c) 6 m

d) $\dfrac{22}{3}$ m   **11.** a) Total distance = 7, displacement = 3

b) Total distance = 19.5, displacement = -4.5 **13.** About 65%
15. $\sqrt{3}\pi$ **17.** a) 210 ft³   b) 13,440 lb

19. $V = 32\pi$, $S = 32\sqrt{2}\pi$ **21.** $4\pi^2$ **23.** $\bar{x} = 0$, $\bar{y} = \dfrac{2a}{\pi}$

25. $\bar{x} = 0$, $\bar{y} = \dfrac{4b}{3\pi}$ **27.** $\sqrt{2}\pi a^3(4 + 3\pi)/6$ **29.** $\dfrac{2a^3}{3}$

Chapter 5 Practice Exercises, pp. 444–447

1. 1   **3.** 1/6   **5.** 18   **7.** 9/8   **9.** $\dfrac{\pi^2}{32} + \dfrac{\sqrt{2}}{2} - 1$ **11.** 4

13. $\dfrac{8\sqrt{2} - 7}{6}$ **15.** Min: -4, max: 0, area: 27/4   **17.** 6/5

19. $9\pi/280$ **21.** π^2 **23.** $72\pi/35$ **25.** a) 2π b) π
c) $12\pi/5$ d) $26\pi/5$ **27.** a) 8π b) $1088\pi/15$ c) $512\pi/15$
29. $\pi(3\sqrt{3} - \pi)/3$ **31.** a) $16\pi/15$ b) $8\pi/5$ c) $8\pi/3$
d) $32\pi/5$ **33.** $28\pi/3$ **35.** 10/3   **37.** 285/8
39. $28\pi\sqrt{2}/3$ **41.** 4π **43.** $\bar{x} = 0$, $\bar{y} = 8/5$
45. $\bar{x} = 3/2$, $\bar{y} = 12/5$ **47.** $\bar{x} = 9/5$, $\bar{y} = 11/10$ **49.** 4640 J
51. 10 ft·lb, 30 ft·lb   **53.** 418,208.81 ft·lb
55. 22,500π ft·lb, 257 sec   **57.** 332.8 lb   **59.** 2196.48 lb
61. $216w_1 + 360w_2$ **63.** a) 64/3 m   b) 0 m
65. a) 15 m   b) -5 m

Chapter 5 Additional Exercises, pp. 447–448

1. $f(x) = \sqrt{\dfrac{2x - a}{\pi}}$ **3.** $f(x) = \sqrt{C^2 - 1}\, x + a$, where $C \geq 1$

5. $\bar{x} = 0$, $\bar{y} = \dfrac{n}{2n + 1}$, $(0, 1/2)$

9. a) $\bar{x} = \bar{y} = 4(a^2 + ab + b^2)/(3\pi(a + b))$ b) $(2a/\pi, 2a/\pi)$

11. 28/3   **13.** $\dfrac{4h\sqrt{3mh}}{3}$ **15.** $\approx 2{,}329.6$ lb

17. a) $2h/3$ b) $(6a^2 + 8ah + 3h^2)/(6a + 4h)$

CHAPTER 6

Section 6.1, pp. 454–457

1. One-to-one   **3.** Not one-to-one   **5.** One-to-one
7. D: $(0, 1]$ R: $[0, \infty)$

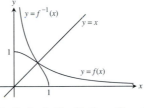

9. D: $[-1, 1]$ R: $[-\pi/2, \pi/2]$

11. a) Symmetric about the line $y = x$

13. $f^{-1}(x) = \sqrt{x - 1}$ **15.** $f^{-1}(x) = \sqrt[3]{x + 1}$
17. $f^{-1}(x) = \sqrt{x} - 1$
19. $f^{-1}(x) = \sqrt[5]{x}$; domain: $-\infty < x < \infty$, range: $-\infty < y < \infty$
21. $f^{-1}(x) = \sqrt[3]{x - 1}$; domain: $-\infty < x < \infty$, range: $-\infty < y < \infty$

23. $f^{-1}(x) = \dfrac{1}{\sqrt{x}}$; domain: $x > 0$, range: $y > 0$

25. a) $f^{-1}(x) = \dfrac{x}{2} - \dfrac{3}{2}$

b)

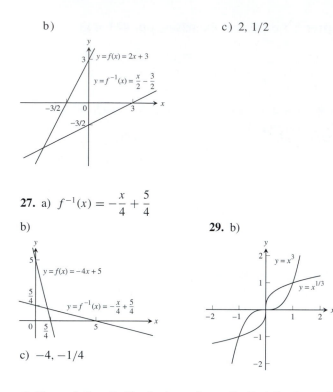

c) 2, 1/2

27. a) $f^{-1}(x) = -\dfrac{x}{4} + \dfrac{5}{4}$

b)   29. b)

c) $-4, -1/4$

c) Slope of f at $(1, 1)$: 3, slope of g at $(1, 1)$: $1/3$, slope of f at $(-1, -1)$: 3, slope of g at $(-1, -1)$: $1/3$
d) $y = 0$ is tangent to $y = x^3$ at $x = 0$; $x = 0$ is tangent to $y = \sqrt[3]{x}$ at $x = 0$

31. $1/9$ 33. 3   35. a) $f^{-1}(x) = \dfrac{1}{m}x$

b) The graph of f^{-1} is the line through the origin with slope $1/m$.
37. a) $f^{-1}(x) = x - 1$

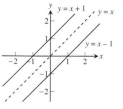

b) $f^{-1}(x) = x - b$. The graph of f^{-1} is a line parallel to the graph of f. The graphs of f and f^{-1} lie on opposite sides of the line $y = x$ and are equidistant from that line.
c) Their graphs will be parallel to one another and lie on opposite sides of the line $y = x$ equidistant from that line.

41. Increasing, therefore one-to-one; $df^{-1}/dx = \dfrac{1}{9}x^{-2/3}$

43. Decreasing, therefore one-to-one; $df^{-1}/dx = -\dfrac{1}{3}x^{-2/3}$

Section 6.2, pp. 465–467

1. a) $\ln 3 - 2\ln 2$ b) $2(\ln 2 - \ln 3)$ c) $-\ln 2$,   d) $\dfrac{2}{3}\ln 3$

e) $\ln 3 + \dfrac{1}{2}\ln 2$ f) $\dfrac{1}{2}(3\ln 3 - \ln 2)$ 3. a) $\ln 5$ b) $\ln(x - 3)$

c) $\ln(t^2)$ 5. $1/x$ 7. $2/t$ 9. $-1/x$ 11. $\dfrac{1}{\theta + 1}$

13. $3/x$ 15. $2(\ln t) + (\ln t)^2$ 17. $x^3 \ln x$ 19. $\dfrac{1 - \ln t}{t^2}$

21. $\dfrac{1}{x(1 + \ln x)^2}$ 23. $\dfrac{1}{x \ln x}$ 25. $2\cos(\ln \theta)$

27. $-\dfrac{3x + 2}{2x(x + 1)}$ 29. $\dfrac{2}{t(1 - \ln t)^2}$ 31. $\dfrac{\tan(\ln \theta)}{\theta}$

33. $\dfrac{10x}{x^2 + 1} + \dfrac{1}{2(1 - x)}$ 35. $2x \ln |x| - x \ln \dfrac{|x|}{\sqrt{2}}$

37. $\left(\dfrac{1}{2}\right)\sqrt{x(x + 1)}\left(\dfrac{1}{x} + \dfrac{1}{x + 1}\right) = \dfrac{2x + 1}{2\sqrt{x(x + 1)}}$

39. $\left(\dfrac{1}{2}\right)\sqrt{\dfrac{t}{t + 1}}\left(\dfrac{1}{t} - \dfrac{1}{t + 1}\right) = \dfrac{1}{2\sqrt{t}(t + 1)^{3/2}}$

41. $\sqrt{\theta + 3}\,(\sin \theta)\left(\dfrac{1}{2(\theta + 3)} + \cot \theta\right)$

43. $t(t + 1)(t + 2)\left[\dfrac{1}{t} + \dfrac{1}{t + 1} + \dfrac{1}{t + 2}\right] = 3t^2 + 6t + 2$

45. $\dfrac{\theta + 5}{\theta \cos \theta}\left[\dfrac{1}{\theta + 5} - \dfrac{1}{\theta} + \tan \theta\right]$

47. $\dfrac{x\sqrt{x^2 + 1}}{(x + 1)^{2/3}}\left[\dfrac{1}{x} + \dfrac{x}{x^2 + 1} - \dfrac{2}{3(x + 1)}\right]$

49. $\dfrac{1}{3}\sqrt[3]{\dfrac{x(x - 2)}{x^2 + 1}}\left(\dfrac{1}{x} + \dfrac{1}{x - 2} - \dfrac{2x}{x^2 + 1}\right)$ 51. $\ln\left(\dfrac{2}{3}\right)$

53. $\ln |y^2 - 25| + C$ 55. $\ln 3$ 57. $(\ln 2)^2$ 59. $\dfrac{1}{\ln 4}$

61. $\ln |6 + 3\tan t| + C$ 63. $\ln 2$ 65. $\ln 27$
67. $\ln(1 + \sqrt{x}) + C$
69. a) Max $= 0$ at $x = 0$, min $= -\ln 2$ at $x = \pi/3$
b) Max $= 1$ at $x = 1$, min $= \cos(\ln 2)$ at $x = 1/2$ and $x = 2$
71. $\ln 16$ 73. $4\pi \ln 4$ 75. $\pi \ln 16$
77. a) $6 + \ln 2$ b) $8 + \ln 9$
79. a) $\bar{x} \approx 1.44, \bar{y} \approx 0.36$
b)

81. $y = x + \ln |x| + 2$ 83. b) 0.00469   85. 2

Section 6.3, pp. 472–474

1. a) 7.2   b) $\dfrac{1}{x^2}$ c) $\dfrac{x}{y}$ 3. a) 1   b) 1   c) $-x^2 - y^2$

5. e^{2t+4} **7.** $e^{5t}+40$ **9.** $y=2xe^x+1$ **11.** a) $k=\ln 2$
b) $k=(1/10)\ln 2$ c) $k=1000\ln a$ **13.** a) $t=-10\ln 3$
b) $t=-\dfrac{\ln 2}{k}$ c) $t=\dfrac{\ln .4}{\ln .2}$ **15.** $4(\ln x)^2$ **17.** $-5e^{-5x}$
19. $-7e^{(5-7x)}$ **21.** xe^x **23.** x^2e^x **25.** $2e^\theta\cos\theta$
27. $2\theta\, e^{-\theta^2}\sin(e^{-\theta^2})$ **29.** $\dfrac{1-t}{t}$ **31.** $1/(1+e^\theta)$

33. $e^{\cos t}(1-t\sin t)$ **35.** $(\sin x)/x$ **37.** $\dfrac{ye^y\cos x}{1-ye^y\sin x}$

39. $\dfrac{2e^{2x}-\cos(x+3y)}{3\cos(x+3y)}$ **41.** $\dfrac{1}{3}e^{3x}-5e^{-x}+C$ **43.** 1

45. $8e^{(x+1)}+C$ **47.** 2   **49.** $2e^{\sqrt r}+C$ **51.** $-e^{-t^2}+C$

53. $-e^{1/x}+C$ **55.** e **57.** $\dfrac{1}{\pi}e^{\sec\pi t}+C$ **59.** 1

61. $\ln(1+e^r)+C$ **63.** $y=1-\cos(e^t-2)$
65. $y=2(e^{-x}+x)-1$
67. Maximum: 1 at $x=0$, minimum: $2-2\ln 2$ at $x=\ln 2$
69. Abs max of $1/(2e)$ assumed at $x=1/\sqrt e$ **71.** 2
73. $y=e^{x/2}-1$

75. a) $\dfrac{d}{dx}(x\ln x-x+C)=x\cdot\dfrac{1}{x}+\ln x-1+0=\ln x$

b) $\dfrac{1}{e-1}$ **77.** b) $|\text{error}|\approx 0.02140$ **79.** 2.71828183

Section 6.4, pp. 480–482

1. a) 7 b) $\sqrt 2$ c) 75 d) 2 e) 0.5 f) -1 **3.** a) $\sqrt x$
b) x^2 c) $\sin x$ **5.** a) $\dfrac{\ln 3}{\ln 2}$ b) 3 c) 2   **7.** $x=12$

9. $x=3$ or $x=2$ **11.** $2^x\ln x$ **13.** $\left(\dfrac{\ln 5}{2\sqrt s}\right)5^{\sqrt s}$ **15.** $\pi x^{(\pi-1)}$
17. $-\sqrt 2\cos\theta^{(\sqrt 2-1)}\sin\theta$ **19.** $7^{\sec\theta}(\ln 7)^2(\sec\theta\tan\theta)$
21. $(3\cos 3t)(2^{\sin 3t})\ln 2$ **23.** $\dfrac{1}{\theta\ln 2}$ **25.** $\dfrac{3}{x\ln 4}$
27. $\dfrac{2(\ln r)}{r(\ln 2)(\ln 4)}$ **29.** $\dfrac{-2}{(x+1)(x-1)}$
31. $\sin(\log_7\theta)+\dfrac{1}{\ln 7}\cos(\log_7\theta)$ **33.** $\dfrac{1}{\ln 5}$ **35.** $\dfrac{1}{t}(\log_2 3)3^{\log_2 t}$
37. $\dfrac{1}{t}$ **39.** $(x+1)^x\left(\dfrac{x}{x+1}+\ln(x+1)\right)$

41. $(\sqrt t)^t\left(\dfrac{\ln t}{2}+\dfrac{1}{2}\right)$ **43.** $(\sin x)^x(\ln\sin x+x\cot x)$

45. $\left(x^{\ln x}\right)\left(\dfrac{\ln x^2}{x}\right)$ **47.** $\dfrac{5^x}{\ln 5}+C$ **49.** $\dfrac{1}{2\ln 2}$ **51.** $\dfrac{1}{\ln 2}$

53. $\dfrac{6}{\ln 7}$ **55.** 32760   **57.** $\dfrac{3x^{(\sqrt 3+1)}}{\sqrt 3+1}+C$ **59.** $3^{\sqrt 2+1}$

61. $\dfrac{1}{\ln 10}\left(\dfrac{(\ln x)^2}{2}\right)+C$ **63.** $2(\ln 2)^2$ **65.** $\dfrac{3\ln 2}{2}$
67. $\ln 10$ **69.** $(\ln 10)\ln|\ln x|+C$ **71.** $\ln(\ln x),\ x>1$

73. $-\ln x$ **75.** $2\ln 5$ **77.** $\left[10^{-7.44},10^{-7.37}\right]$ **79.** $k=10$
81. a) 10^{-7} b) 7 c) $1:1$ **83.** $x\approx-0.76666$
85. a) $L(x)=1+(\ln 2)x\approx 0.69x+1$
87. a) 1.89279 b) -0.35621 c) 0.94575 d) -2.80735
e) 5.29595 f) 0.97041 g) -1.03972 h) -1.61181

Section 6.5, pp. 488–491

1. a) -0.00001 b) 10,536 years c) 82%   **3.** 54.88 g
5. 59.8 ft   **7.** 2.8147497×10^{14} **9.** a) 8 years b) 32.02 years
11. 15.28 years   **13.** a) $A_0e^{0.2}$ b) 17.33 years c) 27.47 years
15. 4.50%   **17.** 0.585 days   **21.** a) 17.5 min. b) 13.26 min.
23. $-3°C$ **25.** About 6658 years   **27.** 41 years old

Section 6.6, pp. 496–498

1. $1/4$ **3.** $-23/7$ **5.** $5/7$ **7.** 0   **9.** -16 **11.** -2
13. $1/4$ **15.** 2   **17.** 3   **19.** -1 **21.** $\ln 3$ **23.** $\dfrac{1}{\ln 2}$
25. $\ln 2$ **27.** 1   **29.** $1/2$ **31.** $\ln 2$ **33.** 0   **35.** $-1/2$
37. $\ln 2$ **39.** -1 **41.** 1   **43.** $1/e$ **45.** 1   **47.** $1/e$
49. $e^{1/2}$ **51.** 1   **53.** 3   **55.** 1   **57.** (b) is correct.

59. (d) is correct.   **61.** $c=\dfrac{27}{10}$

Section 6.7, pp. 503–504

1. a) Slower b) slower c) slower d) faster e) slower
f) slower g) same h) slower
3. a) Same b) faster c) same d) same e) slower f) faster
g) slower h) same
5. a) Same b) same c) same d) faster e) faster f) same
g) slower h) faster   **7.** d, a, c, b
9. a) False b) false c) true d) true e) true f) true
g) false h) true
13. When the degree of f is less than or equal to the degree of g.
15. Polynomials of a greater degree grow at a greater rate than polynomials of a lesser degree. Polynomials of the same degree grow at the same rate.
21. b) $\ln\left(e^{17000000}\right)=17,000,000<\left(e^{17\times 10^6}\right)^{1/10^6}$
$=e^{17}\approx 24,154,952.75$
c) $x\approx 3.4306311\times 10^{15}$ d) They cross at $x\approx 3.4306311\times 10^{15}$
23. a) The algorithm that takes O $(n\log_2 n)$ steps
25. It could take one million for a sequential search; at most 20 steps for a binary search.

Section 6.8, pp. 510–513

1. a) $\pi/4$ b) $-\pi/3$ c) $\pi/6$ **3.** a) $-\pi/6$ b) $\pi/4$ c) $-\pi/3$
5. a) $\pi/3$ b) $3\pi/4$ c) $\pi/6$ **7.** a) $3\pi/4$ b) $\pi/6$ c) $2\pi/3$
9. a) $\pi/4$ b) $-\pi/3$ c) $\pi/6$ **11.** a) $3\pi/4$ b) $\pi/6$ c) $2\pi/3$
13. $\cos\alpha=\dfrac{12}{13}$, $\tan\alpha=\dfrac{5}{12}$, $\sec\alpha=\dfrac{13}{12}$, $\csc\alpha=\dfrac{13}{5}$,

$\cot\alpha=\dfrac{12}{5}$

15. $\sin \alpha = \dfrac{2}{\sqrt{5}}$, $\cos \alpha = -\dfrac{1}{\sqrt{5}}$, $\tan \alpha = -2$, $\csc \alpha = \dfrac{\sqrt{5}}{2}$, $\cot \alpha = -\dfrac{1}{2}$

17. $1/\sqrt{2}$ **19.** $-1/\sqrt{3}$ **21.** $\dfrac{4+\sqrt{3}}{2\sqrt{3}}$ **23.** 1   **25.** $-\sqrt{2}$

27. $\pi/6$ **29.** $\dfrac{\sqrt{x^2+4}}{2}$ **31.** $\sqrt{9y^2-1}$ **33.** $\sqrt{1-x^2}$

35. $\dfrac{\sqrt{x^2-2x}}{x-1}$ **37.** $\dfrac{\sqrt{9-4y^2}}{3}$ **39.** $\dfrac{\sqrt{x^2-16}}{x}$ **41.** $\pi/2$

43. $\pi/2$ **45.** $\pi/2$ **47.** 0   **51.** $\theta = \cos^{-1}\left(\dfrac{1}{\sqrt{3}}\right) \approx 54.7°$

57. a) Defined; there is an angle whose tangent is 2.
b) Not defined; there is no angle whose cosine is 2.
59. a) Not defined; no angle has secant 0.
b) Not defined; no angle has sine $\sqrt{2}$.
61. a) 0.84107   **b)** -0.72973 **c)** 0.46365
63. a) Domain; all real numbers except those having the form $\dfrac{\pi}{2} + k\pi$ where k is an integer; range: $-\pi/2 < y < \pi/2$.
b) Domain: $-\infty < x < \infty$; range: $-\infty < y < \infty$
65. a) Domain; $-\infty < x < \infty$; range: $0 \le y \le \pi$
b) Domain: $-1 \le x \le 1$; range: $-1 \le y \le 1$
67. The graphs are identical.

Section 6.9, pp. 518–520

1. $\dfrac{-2x}{\sqrt{1-x^4}}$ **3.** $\dfrac{\sqrt{2}}{\sqrt{1-2t^2}}$ **5.** $\dfrac{1}{|2s+1|\sqrt{s^2+s}}$

7. $\dfrac{-2x}{(x^2+1)\sqrt{x^4+2x^2}}$ **9.** $\dfrac{-1}{\sqrt{1-t^2}}$ **11.** $\dfrac{-1}{2\sqrt{t}(1+t)}$

13. $\dfrac{1}{\tan^{-1}(x(1+x^2))}$ **15.** $\dfrac{-e^t}{|e^t|\sqrt{(e^t)^2-1}} = \dfrac{-1}{\sqrt{e^{2t}-1}}$

17. $\dfrac{-2s^2}{\sqrt{1-s^2}}$ **19.** 0   **21.** $\sin^{-1}x$ **23.** $\sin^{-1}\dfrac{x}{7} + C$

25. $\dfrac{1}{\sqrt{17}}\tan^{-1}\dfrac{x}{\sqrt{17}} + C$ **27.** $\dfrac{1}{\sqrt{2}}\sec^{-1}\left|\dfrac{5x}{\sqrt{2}}\right| + C$ **29.** $2\pi/3$

31. $\pi/16$ **33.** $-\pi/12$ **35.** $\dfrac{3}{2}\sin^{-1}2(r-1) + C$

37. $\dfrac{\sqrt{2}}{2}\tan^{-1}\left(\dfrac{x-1}{\sqrt{2}}\right) + C$ **39.** $\dfrac{1}{4}\sec^{-1}\left|\dfrac{2x-1}{2}\right| + C$

41. π **43.** $\pi/12$ **45.** $\dfrac{1}{2}\sin^{-1}y^2 + C$ **47.** $\sin^{-1}(x-2) + C$

49. π **51.** $\dfrac{1}{2}\tan^{-1}\left(\dfrac{y-1}{2}\right) + C$ **53.** 2π

55. $\sec^{-1}|x+1| + C$ **57.** $e^{\sin^{-1}x} + C$ **59.** $\dfrac{1}{3}(\sin^{-1}x)^3 + C$

61. $\ln|\tan^{-1}y| + C$ **63.** $\sqrt{3}-1$ **65.** 5   **67.** 2

73. $y = \sin^{-1}(x)$ **75.** $y = \sec^{-1}(x) + \dfrac{2\pi}{3}$, $x > 1$ **77.** $3\sqrt{5}$ ft.

79. Yes, $\sin^{-1}(x)$ and $\cos^{-1}(x)$ differ by the constant $\pi/2$.
89. $\pi^2/2$ **91. a)** $\pi^2/2$ **b)** 2π

Section 6.10, pp. 525–529

1. $\cosh x = 5/4$, $\tanh x = -3/5$, $\coth x = -5/3$, $\operatorname{sech} x = 4/5$, $\operatorname{csch} x = -4/3$
3. $\sinh x = 8/15$, $\tanh x = 8/17$, $\coth x = 17/8$, $\operatorname{sech} x = 15/17$, $\operatorname{csch} x = 15/8$

5. $x + \dfrac{1}{x}$ **7.** e^{5x} **9.** e^{4x} **13.** $2\cosh\dfrac{x}{3}$

15. $\operatorname{sech}^2\sqrt{t} + \dfrac{\tanh\sqrt{t}}{\sqrt{t}}$ **17.** $\coth z$

19. $(\ln\operatorname{sech}\theta)(\operatorname{sech}\theta\tanh\theta)$ **21.** $\tanh^3 v$ **23.** 2

25. $\dfrac{1}{2\sqrt{x(1+x)}}$ **27.** $\dfrac{1}{1+\theta} - \tanh^{-1}\theta$ **29.** $\dfrac{1}{2\sqrt{t}} - \coth^{-1}\sqrt{t}$

31. $-\operatorname{sech}^{-1}x$ **33.** $\dfrac{\ln 2}{\sqrt{1+\left(\dfrac{1}{2}\right)^{2\theta}}}$ **35.** $|\sec x|$

41. $\dfrac{\cosh 2x}{2} + C$ **43.** $12\sinh\left(\dfrac{x}{2} - \ln 3\right) + C$

45. $7\ln\left|e^{x/7} + e^{-x/7}\right| + C$ **47.** $\tanh\left(x - \dfrac{1}{2}\right) + C$

49. $-2\operatorname{sech}\sqrt{t} + C$ **51.** $\ln\dfrac{5}{2}$ **53.** $\dfrac{3}{32} + \ln 2$ **55.** $e - e^{-1}$

57. 3/4   **59.** $\dfrac{3}{8} + \ln\sqrt{2}$ **61.** $\ln(2/3)$ **63.** $\dfrac{-\ln 3}{2}$ **65.** $\ln 3$

67. a) $\sinh^{-1}(\sqrt{3})$ **b)** $\ln(\sqrt{3}+2)$

69. a) $\coth^{-1}(2) - \coth^{-1}(5/4)$ **b)** $\left(\dfrac{1}{2}\right)\ln\left(\dfrac{1}{3}\right)$

71. a) $-\operatorname{sech}^{-1}\left(\dfrac{12}{13}\right) + \operatorname{sech}^{-1}\left(\dfrac{4}{5}\right)$

b) $-\ln\left(\dfrac{1+\sqrt{1-(12/13)^2}}{(12/13)}\right) + \ln\left(\dfrac{1+\sqrt{1-(4/5)^2}}{(4/5)}\right)$

$= -\ln\left(\dfrac{3}{2}\right) + \ln(2) = \ln(4/3)$

73. a) 0   **b)** 0

75. b) i) $f(x) = \dfrac{2f(x)}{2} + 0 = f(x)$, **ii)** $f(x) = 0 + \dfrac{2f(x)}{2} = f(x)$

77. b) $\sqrt{\dfrac{mg}{k}}$ **c)** $80\sqrt{5} \approx 178.89$ ft/sec

79. $y = \operatorname{sech}^{-1}(x) - \sqrt{1-x^2}$ **81.** 2π **83. a)** $\dfrac{6}{5}$ **b)** $\dfrac{\sinh ab}{a}$

85. a) $\bar{x} = 0$, $\bar{y} = \dfrac{5}{8} + \dfrac{\ln 4}{3} \approx 1.09$

b)

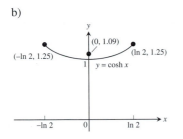

89. c) $a \approx 0.0417525$ d) ≈ 47.90 lb

Section 6.11, pp. 537–540

9. $y = \tan(x^2 + C)$ **11.** $\frac{2}{3}y^{3/2} - x^{1/2} = C$ **13.** $e^y - e^x = C$

15. $y = \dfrac{e^x + C}{x}$ **17.** $y = \dfrac{C - \cos x}{x^3}, x > 0$

19. $y = \dfrac{1}{2} - \dfrac{1}{x} + \dfrac{C}{x^2}, x > 0$ **21.** $y = \dfrac{1}{2}xe^{x/2} + Ce^{x/2}$

23. $y = x^2 e^{-2x} + Ce^{-2x}$ **25.** $-e^{-y} - e^{\sin x} = C$

27. $s = \dfrac{t^3}{3(t-1)^4} - \dfrac{t}{(t-1)^4} + \dfrac{C}{(t-1)^4}$ **29.** $2\tan\sqrt{x} = t + C$

31. $r = \csc\theta\,(\ln|\sec\theta| + C)$ **33.** $y = -e^{-x}\operatorname{sech} x + C\operatorname{sech} x$

35. $y = \dfrac{3}{2} - \dfrac{1}{2}e^{-2t}$ **37.** $y = -\dfrac{1}{\theta}\cos\theta + \dfrac{\pi}{2\theta}$

39. $y = 6e^{x^2} - \dfrac{e^{x^2}}{x+1}$ **41.** $y = y_0 e^{kt}$

43. (b) is correct, but (a) is not.

45. a) $c = \dfrac{G}{100Vk} + \left(c_0 - \dfrac{G}{100Vk}\right)e^{-kt}$ b) $\dfrac{G}{100Vk}$

47. 1 hour **49.** a) 550 ft b) $25\ln 22 \approx 77$ sec

51. $t = \dfrac{L}{R}\ln 2$ seconds

53. a) $i = \dfrac{V}{R} - \dfrac{V}{R}e^{-3} = \dfrac{V}{R}(1 - e^{-3}) \approx 0.95\dfrac{V}{R}$ amp b) 86%

55. a) 10 lb/min b) $100 + t$ gal c) $4\left(\dfrac{y}{100+t}\right)$ lb/min

d) $\dfrac{dy}{dt} = 10 - \dfrac{4y}{100+t}$, $y(0) = 50$, $y = 2(100+t) - \dfrac{150}{\left(1+\dfrac{t}{100}\right)^4}$

e) concentration $= \dfrac{y(25)}{\text{amt. brine in tank}} = \dfrac{188.6}{125} \approx 1.5$ lb/gal

57. $y(27.8) \approx 14.8$ lb, $t \approx 27.8$ min

Section 6.12, pp. 545–546

1. y (exact) $= \dfrac{x}{2} - \dfrac{4}{x}$, $y_1 = -0.25$, $y_2 = 0.3$, $y_3 = 0.75$

3. y (exact) $= 3e^{x(x+2)}$, $y_1 = 4.2$, $y_2 = 6.216$, $y_3 = 9.697$

5. y (exact) $= e^{x^2} + 1$, $y_1 = 2.0$, $y_2 = 2.0202$, $y_3 = 2.0618$

7. $y \approx 2.48832$, exact value is e

9. $y \approx -0.2272$, exact value is $1/(1 - 2\sqrt{5}) \approx -0.2880$

11. b **13.** a

15.

17.

19.

Chapter 6 Practice Exercises, pp. 548–551

1. $-2e^{-x/5}$ **3.** xe^{4x} **5.** $\dfrac{2\sin\theta\cos\theta}{\sin^2\theta} = 2\cot\theta$ **7.** $\dfrac{2}{(\ln 2)x}$

9. $-8^{-t}(\ln 8)$ **11.** $18x^{2.6}$ **13.** $(x+2)^{x+2}(\ln(x+2) + 1)$

15. $-\dfrac{1}{\sqrt{1-u^2}}$ **17.** $\dfrac{-1}{\sqrt{1-x^2}\,\cos^{-1}x}$

19. $\tan^{-1}(t) + \dfrac{t}{1+t^2} - \dfrac{1}{2t}$ **21.** $\dfrac{1-z}{\sqrt{z^2-1}} + \sec^{-1}z$ **23.** -1

25. $\dfrac{2(x^2+1)}{\sqrt{\cos 2x}}\left[\dfrac{2x}{x^2+1} + \tan 2x\right]$

27. $5\left[\dfrac{(t+1)(t-1)}{(t-2)(t+3)}\right]^5\left[\dfrac{1}{t+1} + \dfrac{1}{t-1} - \dfrac{1}{t-2} - \dfrac{1}{t-3}\right]$

29. $\dfrac{1}{\sqrt{\theta}}(\sin\theta)^{\sqrt{\theta}}(\ln\sqrt{\sin\theta} + \theta\cot\theta)$ **31.** $-\cos e^x + C$

33. $\tan(e^x - 7) + C$ **35.** $e^{\tan x} + C$ **37.** $\dfrac{-\ln 7}{3}$ **39.** $\ln 8$

41. $\ln(9/25)$ **43.** $-[\ln|\cos(\ln v)|] + C$ **45.** $-\dfrac{1}{2}(\ln x)^{-2} + C$

47. $-\cot(1 + \ln r) + C$ **49.** $\dfrac{1}{2\ln 3}\left(3^{x^2}\right) + C$ **51.** $3\ln 7$

53. $15/16 + \ln 2$ **55.** $e - 1$ **57.** $1/6$ **59.** $9/14$

61. $\dfrac{1}{3}[(\ln 4)^3 - (\ln 2)^3]$ or $\dfrac{7}{3}(\ln 2)^3$ **63.** $\dfrac{9\ln 2}{4}$ **65.** π

67. $\pi/\sqrt{3}$ **69.** $\sec^{-1}2y + C$ **71.** $\pi/12$

73. $\sin^{-1}(x+1) + C$ **75.** $\pi/2$ **77.** $\dfrac{1}{3}\sec^{-1}\left(\dfrac{t+1}{3}\right) + C$

79. $y = \dfrac{\ln 2}{\ln(3/2)}$ **81.** $y = \ln x - \ln 3$ **83.** $y = \dfrac{1}{1-e^x}$

85. $\ln 10$ **87.** $\ln 2$ **89.** 5 **91.** $-\infty$ **93.** 1 **95.** e^3

97. a) Same rate   b) same rate   c) faster   d) faster   e) same rate
f) same rate   **99.** a) True   b) false   c) false   d) true   e) true
f) true   **101.** 1/3
103. Absolute maximum $= 0$ at $x = e/2$, absolute minimum $= -0.5$
at $x = 0.5$
105. 1   **107.** $1/e$ m/sec   **109.** $1/\sqrt{2}$ units long by $1/\sqrt{e}$ units
high, $A = 1/\sqrt{2e} \approx 0.43$ units²
111. $\ln 5x - \ln 3x = \ln (5/3)$ **113.** 1/2
115. a) Absolute maximum of $2/e$ at $x = e^2$, inflection point
$(e^{8/3}, (8/3)e^{-4/3})$, concave up on $(e^{8/3}, \infty)$, concave down on $(0, e^{8/3})$
b) Absolute maximum of 1 at $x = 0$, inflection points
$(\pm 1/\sqrt{2}, 1/\sqrt{e})$, concave up on $(-\infty, -1/\sqrt{2}) \cup (1/\sqrt{2}, \infty)$, concave down on $(-1/\sqrt{2}, 1/\sqrt{2})$
c) Absolute maximum of 1 at $x = 0$, inflection point $(1, 2/e)$, concave up on $(1, \infty)$, concave down on $(-\infty, 1)$
117. 18,935 years   **119.** $20(5 - \sqrt{17})$ m
121. $y = \ln (-e^{-x-2} + 2e^{-2})$

123. $y = \dfrac{1}{(x + 1)^2} \cdot \left(\dfrac{x^3}{3} + \dfrac{x^2}{2} + 1 \right)$

125. y (exact) $= \dfrac{1}{2}x^2 - \dfrac{3}{2}$; $y \approx 0.4$; exact value is 1/2

127. y (exact) $= -e^{(x^2-1)/2}$; $y \approx -3.4192$; exact value is
$-e^{3/2} \approx -4.4817$

Chapter 6 Additional Exercises, pp. 551–553

1. $\pi/2$ **3.** $1/\sqrt{e}$ **5.** $\ln 2$ **7.** a) 1   b) $\pi/2$ c) π
9. $a = 2$, $b = -2$ **11.** $\dfrac{1}{\ln 2}$, $\dfrac{1}{2 \ln 2}$, $2:1$ **13.** $x = 2$

15. 2/17   **23.** $\bar{x} = \dfrac{\ln 4}{\pi}$, $\bar{y} = 0$ **27.** b) 61°
29. a) $c - (c - y_0)e^{-(kA/V)t}$ b) c

CHAPTER 7

Section 7.1, pp. 560–561

1. $2\sqrt{8x^2 + 1} + C$ **3.** $2(\sin v)^{3/2} + C$ **5.** $\ln 5$
7. $2 \ln (\sqrt{x} + 1) + C$ **9.** $-\dfrac{1}{7} \ln |\sin (3 - 7x)| + C$
11. $-\ln |\csc (e^\theta + 1) + \cot (e^\theta + 1)| + C$

13. $3 \ln \left| \sec \dfrac{t}{3} + \tan \dfrac{t}{3} \right| + C$

15. $-\ln |\csc (s - \pi) + \cot (s - \pi)| + C$ **17.** 1   **19.** $e^{\tan v} + C$

21. $\dfrac{3^{(x+1)}}{\ln 3} + C$ **23.** $\dfrac{2^{\sqrt{w}}}{\ln 2} + C$ **25.** $3 \tan^{-1} 3u + C$

27. $\pi/18$ **29.** $\sin^{-1} s^2 + C$ **31.** $6 \sec^{-1} |5x| + C$
33. $\tan^{-1} e^x + C$ **35.** $\ln (2 + \sqrt{3})$ **37.** 2π
39. $\sin^{-1}(t - 2) + C$
41. $\sec^{-1} |x + 1| + C$, when $|x + 1| > 1$

43. $\tan x - 2 \ln |\csc x + \cot x| - \cot x - x + C$
45. $x + \sin 2x + C$ **47.** $x - \ln |x + 1| + C$ **49.** $7 + \ln 8$
51. $2t^2 - t + 2 \tan^{-1} \left(\dfrac{t}{2} \right) + C$ **53.** $\sin^{-1} x + \sqrt{1 - x^2} + C$
55. $\sqrt{2}$ **57.** $\tan x - \sec x + C$ **59.** $\ln |1 + \sin \theta| + C$
61. $\cot x + x + \csc x + C$ **63.** 4   **65.** $\sqrt{2}$ **67.** 2
69. $\ln |\sqrt{2} + 1| - \ln |\sqrt{2} - 1|$ **71.** $4 - \dfrac{\pi}{2}$

73. $-\ln |\csc (\sin \theta) + \cot (\sin \theta)| + C$
75. $\ln |\sin x| + \ln |\cos x| + C$ **77.** $12 \tan^{-1}(\sqrt{y}) + C$

79. $\sec^{-1} \left| \dfrac{x - 1}{7} \right| + C$ **81.** $\ln |\sec (\tan t)| + C$

83. a) $\sin \theta - \dfrac{1}{3} \sin^3 \theta + C$ b) $\sin \theta - \dfrac{2}{3} \sin^3 \theta + \dfrac{1}{5} \sin^5 \theta + C$

c) $\displaystyle \int \cos^9 \theta \, d\theta = \int \cos^8 \theta (\cos \theta) \, d\theta = \int (1 - \sin^2 \theta)^4 (\cos \theta) \, d\theta$

85. a) $\displaystyle \int \tan^3 \theta \, d\theta = \dfrac{1}{2} \tan^2 \theta - \int \tan \theta \, d\theta = \dfrac{1}{2} \tan^2 \theta +$

$\ln |\cos \theta| + C$

b) $\displaystyle \int \tan^5 \theta \, d\theta = \dfrac{1}{4} \tan^4 \theta - \int \tan^3 \theta \, d\theta$

c) $\displaystyle \int \tan^7 \theta \, d\theta = \dfrac{1}{6} \tan^6 \theta - \int \tan^5 \theta \, d\theta$

d) $\displaystyle \int \tan^{2k+1} \theta \, d\theta = \dfrac{1}{2k} \tan^{2k} \theta - \int \tan^{2k-1} \theta \, d\theta$

87. $2\sqrt{2} - \ln (3 + 2\sqrt{2})$ **89.** π^2 **91.** $\ln (2 + \sqrt{3})$

93. $\bar{x} = 0$, $\bar{y} = \dfrac{1}{\ln (2\sqrt{2} + 3)}$

Section 7.2, pp. 567–569

1. $-2x \cos (x/2) + 4 \sin (x/2) + C$

3. $t^2 \sin t + 2t \cos t - 2 \sin t + C$ **5.** $\ln 4 - \dfrac{3}{4}$

7. $y \tan^{-1}(y) - \ln \sqrt{1 + y^2} + C$ **9.** $x \tan x + \ln |\cos x| + C$
11. $(x^3 - 3x^2 + 6x - 6)e^x + C$ **13.** $(x^2 - 7x + 7)e^x + C$

15. $(x^5 - 5x^4 + 20x^3 - 60x^2 + 120x - 120)e^x + C$ **17.** $\dfrac{\pi^2 - 4}{8}$

19. $\dfrac{5\pi - 3\sqrt{3}}{9}$ **21.** $\dfrac{1}{2}(-e^\theta \cos \theta + e^\theta \sin \theta) + C$

23. $\dfrac{e^{2x}}{13}(3 \sin 3x + 2 \cos 3x) + C$

25. $\dfrac{2}{3} \left(\sqrt{3s + 9} \, e^{\sqrt{3s+9}} - e^{\sqrt{3s+9}} \right) + C$ **27.** $\dfrac{\pi\sqrt{3}}{3} - \ln(2) - \dfrac{\pi^2}{18}$

29. $\dfrac{1}{2}[-x \cos (\ln x) + x \sin (\ln x)] + C$ **31.** a) π b) 3π c) 5π

d) $(2n + 1)\pi$ **33.** $2\pi(1 - \ln 2)$ **35.** a) $\pi(\pi - 2)$ b) 2π

37. a) $\bar{x} = \dfrac{6-2e}{e-2} \approx 0.78, \bar{y} = \dfrac{e^2-3}{8(e-2)} \approx 0.76$

b)

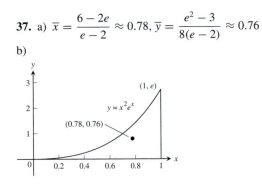

39. $\pi^2 + \pi - 4$ **41.** a) $\dfrac{1}{2\pi}\left(1 - e^{-2\pi}\right)$

43. $x \sin^{-1} x + \cos(\sin^{-1} x) + C$

45. $x \sec^{-1} x - \ln\left|x + \sqrt{x^2-1}\right| + C$ **47.** Yes

49. a) $x \sinh^{-1} x - \cosh(\sinh^{-1} x) + C$
b) $x \sinh^{-1} x + (1+x^2)^{1/2} + C$

Section 7.3, pp. 576–578

1. $\dfrac{2}{x-3} + \dfrac{3}{x-2}$ **3.** $\dfrac{1}{x+1} + \dfrac{3}{(x+1)^2}$

5. $\dfrac{-2}{z} + \dfrac{-1}{z^2} + \dfrac{2}{z-1}$ **7.** $1 + \dfrac{17}{t-3} + \dfrac{-12}{t-2}$

9. $\dfrac{1}{2}[\ln|1+x| - \ln|1-x|] + C$

11. $\dfrac{1}{7}\ln|(x+6)^2(x-1)^5| + C$ **13.** $(\ln 15)/2$

15. $-\dfrac{1}{2}\ln|t| + \dfrac{1}{6}\ln|t+2| + \dfrac{1}{3}\ln|t-1| + C$ **17.** $3\ln 2 - 2$

19. $\dfrac{1}{4}\ln\left|\dfrac{x+1}{x-1}\right| - \dfrac{x}{2(x^2-1)} + C$ **21.** $(\pi + 2\ln 2)/8$

23. $\tan^{-1} y - \dfrac{1}{y^2+1} + C$

25. $-(s-1)^{-2} + (s-1)^{-1} + \tan^{-1} s + C$

27. $\dfrac{-1}{\theta^2 + 2\theta + 2} + \ln|\theta^2 + 2\theta + 2| - \tan^{-1}(\theta+1) + C$

29. $x^2 + \ln\left|\dfrac{x-1}{x}\right| + C$

31. $9x + 2\ln|x| + \dfrac{1}{x} + 7\ln|x-1| + C$

33. $\dfrac{y^2}{2} - \ln|y| + \dfrac{1}{2}\ln(1+y^2) + C$ **35.** $\ln\left|\dfrac{e^t+1}{e^t+2}\right| + C$

37. $\dfrac{1}{5}\ln\left|\dfrac{\sin y - 2}{\sin y + 3}\right| + C$

39. $\dfrac{(\tan^{-1} 2x)^2}{4} - 3\ln|x-2| + \dfrac{6}{x-2} + C$

41. $x = \ln|t-2| - \ln|t-1| + \ln 2$ **43.** $x = \dfrac{6t}{t+2} - 1$

45. $3\pi \ln 25$ **47.** 1.10   **49.** a) $x = \dfrac{1000e^{4t}}{499 + e^{4t}}$ b) 1.55 days

51. a) $\dfrac{22}{7} - \pi$ b) 0.04%   c) The area is less than 0.003.

Section 7.4, pp. 582–583

1. $\ln\left|\sqrt{9+y^2} + y\right| + C$ **3.** $\pi/4$ **5.** $\pi/6$

7. $\dfrac{25}{2}\sin^{-1}\left(\dfrac{t}{5}\right) + \dfrac{t\sqrt{25-t^2}}{2} + C$

9. $\dfrac{1}{2}\ln\left|\dfrac{2x}{7} + \dfrac{\sqrt{4x^2-49}}{7}\right| + C$

11. $7\left[\dfrac{\sqrt{y^2-49}}{7} - \sec^{-1}\left(\dfrac{y}{7}\right)\right] + C$ **13.** $\dfrac{\sqrt{x^2-1}}{x} + C$

15. $\dfrac{1}{3}(x^2+4)^{3/2} - 4\sqrt{x^2+4} + C$ **17.** $\dfrac{-2\sqrt{4-w^2}}{w} + C$

19. $4\sqrt{3} - 4\pi/3$ **21.** $-\dfrac{x}{\sqrt{x^2-1}} + C$

23. $-\dfrac{1}{5}\left(\dfrac{\sqrt{1-x^2}}{x}\right)^5 + C$ **25.** $2\tan^{-1} 2x + \dfrac{4x}{(4x^2+1)} + C$

27. $\dfrac{1}{3}\left(\dfrac{v}{\sqrt{1-v^2}}\right)^3 + C$ **29.** $\ln 9 - \ln(1+\sqrt{10})$ **31.** $\pi/6$

33. $\sec^{-1}|x| + C$ **35.** $\sqrt{x^2-1} + C$

37. $y = 2\left[\dfrac{\sqrt{x^2-4}}{2} - \sec^{-1}\left(\dfrac{x}{2}\right)\right]$ **39.** $y = \dfrac{3}{2}\tan^{-1}\left(\dfrac{x}{2}\right) - \dfrac{3\pi}{8}$

41. $3\pi/4$ **43.** $\dfrac{2}{1-\tan(x/2)} + C$ **45.** 1   **47.** $\dfrac{\sqrt{3}\pi}{9}$

49. $\dfrac{1}{\sqrt{2}}\ln\left|\dfrac{\tan(t/2) + 1 - \sqrt{2}}{\tan(t/2) + 1 + \sqrt{2}}\right| + C$ **51.** $\ln\left|\dfrac{1 + \tan(\theta/2)}{1 - \tan(\theta/2)}\right| + C$

Section 7.5, pp. 591–594

1. $\dfrac{2}{\sqrt{3}}\left(\tan^{-1}\sqrt{\dfrac{x-3}{3}}\right) + C$ **3.** $\sqrt{x-2}\left(\dfrac{2(x-2)}{3} + 4\right) + C$

5. $\dfrac{(2x-3)^{3/2}(x+1)}{5} + C$

7. $\dfrac{-\sqrt{9-4x}}{x} - \dfrac{2}{3}\ln\left|\dfrac{\sqrt{9-4x} - 3}{\sqrt{9-4x} + 3}\right| + C$

9. $\dfrac{(x+2)(2x-6)\sqrt{4x-x^2}}{6} + 4\sin^{-1}\left(\dfrac{x-2}{2}\right) + C$

11. $-\dfrac{1}{\sqrt{7}} \ln \left| \dfrac{\sqrt{7}+\sqrt{7+x^2}}{x} \right| + C$

13. $\sqrt{4-x^2} - 2 \ln \left| \dfrac{2+\sqrt{4-x^2}}{x} \right| + C$

15. $\dfrac{p}{2}\sqrt{25-p^2} + \dfrac{25}{2} \sin^{-1} \dfrac{p}{5} + C$

17. $2 \sin^{-1} \dfrac{r}{2} - \dfrac{1}{2} r \sqrt{4-r^2} + C$

19. $-\dfrac{1}{3} \tan^{-1} \left[\dfrac{1}{3} \tan \left(\dfrac{\pi}{4} - \theta \right) \right] + C$

21. $\dfrac{e^{2t}}{13}(2 \cos 3t + 3 \sin 3t) + C$

23. $\dfrac{x^2}{2} \cos^{-1}(x) + \dfrac{1}{4} \sin^{-1}(x) - \dfrac{1}{4} x \sqrt{1-x^2} + C$

25. $\dfrac{s}{18(9-s^2)} + \dfrac{1}{108} \ln \left| \dfrac{s+3}{s-3} \right| + C$

27. $-\dfrac{\sqrt{4x+9}}{x} + \dfrac{2}{3} \ln \left| \dfrac{\sqrt{4x+9}-3}{\sqrt{4x+9}+3} \right| + C$

29. $2\sqrt{3t-4} - 4 \tan^{-1} \sqrt{\dfrac{3t-4}{4}} + C$

31. $\dfrac{x^3}{3} \tan^{-1} x - \dfrac{x^2}{6} + \dfrac{1}{6} \ln (1+x^2) + C$

33. $-\dfrac{\cos 5x}{10} - \dfrac{\cos x}{2} + C$ **35.** $8 \left[\dfrac{\sin(7t/2)}{7} - \dfrac{\sin(9t/2)}{9} \right] + C$

37. $6 \sin(\theta/12) + \dfrac{6}{7} \sin(7\theta/12) + C$

39. $\dfrac{1}{2} \ln \left| x^2+1 \right| + \dfrac{x}{2(1+x^2)} + \dfrac{1}{2} \tan^{-1} x + C$

41. $\left(x - \dfrac{1}{2} \right) \sin^{-1} \sqrt{x} + \dfrac{1}{2} \sqrt{x-x^2} + C$

43. $\sin^{-1} \sqrt{x} - \sqrt{x-x^2} + C$

45. $\sqrt{1-\sin^2 t} - \ln \left| \dfrac{1+\sqrt{1-\sin^2 t}}{\sin t} \right| + C$

47. $\ln \left| \ln y + \sqrt{3 + (\ln y)^2} \right| + C$ **49.** $\ln \left| 3r + \sqrt{9r^2 - 1} \right| + C$

51. $x \cos^{-1} \sqrt{x} + \dfrac{1}{2} \sin^{-1} \sqrt{x} - \dfrac{1}{2} \sqrt{x-x^2} + C$

53. $-\dfrac{\sin^4 2x \cos 2x}{10} - \dfrac{2 \sin^2 2x \cos 2x}{15} - \dfrac{4 \cos 2x}{15} + C$

55. $\dfrac{\cos^3 2\pi t \sin 2\pi t}{\pi} + \dfrac{3}{2} \dfrac{\cos 2\pi t \sin 2\pi t}{\pi} + 3t + C$

57. $\dfrac{\sin^3 2\theta \cos^2 2\theta}{10} + \dfrac{\sin^3 2\theta}{15} + C$ **59.** $\dfrac{2}{3} \tan^3 t + C$

61. $\tan^2 2x - 2 \ln |\sec 2x| + C$

63. $8 \left[-\dfrac{1}{3} \cot^3 t + \cot t + t \right] + C$

65. $\dfrac{(\sec \pi x)(\tan \pi x)}{\pi} + \dfrac{1}{\pi} \ln |\sec \pi x + \tan \pi x| + C$

67. $\dfrac{\sec^2 3x \tan 3x}{3} + \dfrac{2}{3} \tan 3x + C$

69. $\dfrac{-\csc^3 x \cot x}{4} - \dfrac{3 \csc x \cot x}{8} - \dfrac{3}{8} \ln |\csc x + \cot x| + C$

71. $4x^4 (\ln x)^2 - 2x^4 (\ln x) + \dfrac{x^2}{2} + C$ **73.** $\dfrac{e^{3x}}{9}(3x-1) + C$

75. $2x^3 e^{x/2} - 12x^2 e^{x/2} + 96 e^{x/2} \left(\dfrac{x}{2} - 1 \right) + C$

77. $\dfrac{x^2 2^x}{\ln 2} - \dfrac{2}{\ln 2} \left[\dfrac{x 2^x}{\ln 2} - \dfrac{2^x}{(\ln 2)^2} \right] + C$ **79.** $\dfrac{x \pi^x}{\ln \pi} - \dfrac{\pi^x}{(\ln \pi)^2} + C$

81. $\dfrac{1}{2} \left[\sec(e^t - 1) \tan(e^t - 1) + \ln |\sec(e^t - 1) + \tan(e^t - 1)| \right] + C$

83. $\sqrt{2} + \ln(\sqrt{2}+1)$ **85.** $\pi/3$

87. $\dfrac{1}{120} \sinh^4 3x \cosh 3x - \dfrac{1}{90} \sinh^2 3x \cosh 3x + \dfrac{2}{90} \cosh 3x + C$

89. $\dfrac{x^2}{3} \sinh 3x - \dfrac{2x}{9} \cosh 3x + \dfrac{2}{27} \sinh 3x + C$

91. $-\dfrac{\operatorname{sech}^7 x}{7} + C$ **101.** $\pi(2\sqrt{3}+\sqrt{2}) \ln(\sqrt{2}+\sqrt{3})$

103. $\bar{x} = 4/3,\ \bar{y} = \ln \sqrt{2}$ **105.** 7.62 **107.** $\pi/8$ **111.** $\pi/4$

Section 7.6, pp. 603–605

1. $\pi/2$ **3.** 2 **5.** 6 **7.** $\pi/2$ **9.** $\ln 3$ **11.** $\ln 4$ **13.** 0

15. $\sqrt{3}$ **17.** π **19.** $\ln \left(1 + \dfrac{\pi}{2} \right)$ **21.** -1 **23.** 1

25. $-1/4$ **27.** $\pi/2$ **29.** $\pi/3$ **31.** 6 **33.** $\ln 2$
35. Diverges **37.** Converges **39.** Converges **41.** Converges
43. Diverges **45.** Converges **47.** Converges **49.** Diverges
51. Converges **53.** Converges **55.** Diverges **57.** Converges
59. Diverges **61.** Converges **63.** Converges
65. b) ≈ 0.88621 **69.** 1 **71.** 2π **73.** $\ln 2$ **79.** Diverges
81. Converges **83.** Converges **85.** Diverges **91.** b) $\pi/2$

Chapter 7 Practice Exercises, pp. 606–609

1. $\dfrac{1}{12}(4x^2 - 9)^{3/2} + C$ **3.** $\dfrac{(2x+1)^{5/2}}{10} - \dfrac{(2x+1)^{3/2}}{6} + C$

5. $\dfrac{\sqrt{8x^2+1}}{8} + C$ **7.** $\dfrac{1}{2} \ln(25+y^2) + C$

9. $\dfrac{-\sqrt{9-4t^4}}{8} + C$ **11.** $\dfrac{9}{25}(z^{5/3}+1)^{5/3} + C$

13. $-\dfrac{1}{2(1-\cos 2\theta)} + C$ **15.** $-\dfrac{1}{4}\ln|3 + 4\cos t| + C$

17. $-\dfrac{1}{2}e^{\cos 2x} + C$ **19.** $-\dfrac{1}{3}\cos^3(e^\theta) + C$ **21.** $\dfrac{2^{x-1}}{\ln 2} + C$

23. $\ln|\ln v| + C$ **25.** $\ln|2 + \tan^{-1}x| + C$ **27.** $\sin^{-1}(2x) + C$

29. $\dfrac{1}{3}\sin^{-1}\left(\dfrac{3t}{4}\right) + C$ **31.** $\dfrac{1}{3}\tan^{-1}\left(\dfrac{t}{3}\right) + C$

33. $\dfrac{1}{5}\sec^{-1}\left|\dfrac{5x}{4}\right| + C$ **35.** $\sin^{-1}\left(\dfrac{x-2}{2}\right) + C$

37. $\dfrac{1}{2}\tan^{-1}\left(\dfrac{y-2}{2}\right) + C$ **39.** $\sec^{-1}|x-1| + C$

41. $\dfrac{x}{2} - \dfrac{\sin 2x}{4} + C$

43. $\dfrac{2}{3}\cos^3\left(\dfrac{\theta}{2}\right) - 2\cos\left(\dfrac{\theta}{2}\right) + C$

45. $\dfrac{\tan^2(2t)}{4} - \dfrac{1}{2}\ln|\sec 2t| + C$

47. $-\dfrac{1}{2}\ln|\csc(2x) + \cot(2x)| + C$ **49.** $\ln\sqrt{2}$ **51.** 2

53. $2\sqrt{2}$ **55.** $x - 2\tan^{-1}\left(\dfrac{x}{2}\right) + C$

57. $x + x^2 + 2\ln|2x-1| + C$

59. $\ln(y^2+4) - \dfrac{1}{2}\tan^{-1}\left(\dfrac{y}{2}\right) + C$

61. $-\sqrt{4-t^2} + 2\sin^{-1}\left(\dfrac{t}{2}\right) + C$ **63.** $x - \tan x + \sec x + C$

65. $-\dfrac{1}{3}\ln|\sec(5-3x) + \tan(5-3x)| + C$

67. $4\ln\left|\sin\left(\dfrac{x}{4}\right)\right| + C$ **69.** $-2\left(\dfrac{(\sqrt{1-x})^3}{3} - \dfrac{(\sqrt{1-x})^5}{5}\right) + C$

71. $\dfrac{1}{2}\left(z\sqrt{z^2+1} + \ln\left|z + \sqrt{z^2+1}\right|\right) + C$

73. $\ln\left|y + \sqrt{25+y^2}\right| + C$ **75.** $\dfrac{-\sqrt{1-x^2}}{x} + C$

77. $\dfrac{\sin^{-1}x}{2} - \dfrac{x\sqrt{1-x^2}}{2} + C$ **79.** $\ln\left|\dfrac{x}{3} + \dfrac{\sqrt{x^2-9}}{3}\right| + C$

81. $\sqrt{w^2-1} - \sec^{-1}(w) + C$

83. $[(x+1)(\ln(x+1)) - (x+1)] + C$

85. $x\tan^{-1}(3x) - \dfrac{1}{6}\ln(1+9x^2) + C$

87. $(x+1)^2 e^x - 2(x+1)e^x + 2e^x + C$

89. $\dfrac{2e^x \sin 2x}{5} + \dfrac{e^x \cos 2x}{5} + C$

91. $2\ln|x-2| - \ln|x-1| + C$

93. $\ln|x| - \ln|x+1| + \dfrac{1}{x+1} + C$ **95.** $-\dfrac{1}{3}\ln\left|\dfrac{\cos\theta - 1}{\cos\theta + 2}\right| + C$

97. $4\ln|x| - \dfrac{1}{2}\ln(x^2+1) + 4\tan^{-1}x + C$

99. $\dfrac{1}{16}\ln\left|\dfrac{(v-2)^5(v+2)}{v^6}\right| + C$

101. $\dfrac{1}{2}\tan^{-1}t - \dfrac{\sqrt{3}}{6}\tan^{-1}\dfrac{t}{\sqrt{3}} + C$

103. $\dfrac{x^2}{2} + \dfrac{4}{3}\ln|x+2| + \dfrac{2}{3}\ln|x-1| + C$

105. $\dfrac{x^2}{2} - \dfrac{9}{2}\ln|x+3| + \dfrac{3}{2}\ln|x+1| + C$

107. $\dfrac{1}{3}\ln\left|\dfrac{\sqrt{x+1}-1}{\sqrt{x+1}+1}\right| + C$ **109.** $\ln|1-e^{-x}| + C$ **111.** $\pi/2$

113. 6 **115.** $\ln 3$ **117.** 2 **119.** $\pi/6$ **121.** Diverges

123. Diverges   **125.** Converges   **127.** $-\sqrt{16-y^2} + C$

129. $-\dfrac{1}{2}\ln|4-x^2| + C$ **131.** $\ln\dfrac{1}{\sqrt{9-x^2}} + C$

133. $\dfrac{1}{6}\ln\left|\dfrac{x+3}{x-3}\right| + C$

135. $\dfrac{2x^{3/2}}{3} - x + 2\sqrt{x} - 2\ln(\sqrt{x}+1) + C$

137. $\ln\left|\dfrac{x}{\sqrt{x^2+1}}\right| - \dfrac{1}{2}\left(\dfrac{x}{\sqrt{x^2+1}}\right)^2 + C$ **139.** $\sin^{-1}(x+1) + C$

141. $\ln|u + \sqrt{1+u^2}| + C$

143. $-2\cot x - \ln|\csc x + \cot x| + \csc x + C$

145. $\dfrac{1}{12}\ln\left|\dfrac{3+v}{3-v}\right| + \dfrac{1}{6}\tan^{-1}\dfrac{v}{3} + C$

147. $\dfrac{\theta\sin(2\theta+1)}{2} + \dfrac{\cos(2\theta+1)}{4} + C$

149. $\dfrac{x^2}{2} + 2x + 3\ln|x-1| - \dfrac{1}{x-1} + C$ **151.** $-\cos(2\sqrt{x}) + C$

153. $-\ln|\csc(2y) + \cot(2y)| + C$ **155.** $\dfrac{1}{2}\tan^2 x + C$

157. $-\sqrt{4-(r+2)^2} + C$ **159.** $\dfrac{1}{4}\sec^2\theta + C$ **161.** $\dfrac{\sqrt{2}}{2}$

163. $2\left(\dfrac{(\sqrt{2-x})^3}{3} - 2\sqrt{2-x}\right) + C$ **165.** $\tan^{-1}(y-1) + C$

167. $\dfrac{1}{3}\ln|\sec\theta^3| + C$

169. $\dfrac{1}{4}\ln|z| - \dfrac{1}{4z} - \dfrac{1}{4}\left[\dfrac{1}{2}\ln(z^2+4) + \dfrac{1}{2}\tan^{-1}\left(\dfrac{z}{2}\right)\right] + C$

171. $-\dfrac{1}{4}\sqrt{9-4t^2} + C$ **173.** $\ln|\sin\theta| - \dfrac{1}{2}\ln(1+\sin^2\theta) + C$

175. $\ln|\sec\sqrt{y}| + C$ **177.** $-\theta + \ln\left|\dfrac{\theta+2}{\theta-2}\right| + C$ **179.** $x + C$

181. $-\dfrac{\cos x}{2} + C$ **183.** $\ln(1 + e^t) + C$ **185.** $1/4$

187. $\ln|\ln\sin v| + C$ **189.** $\dfrac{2}{3}x^{3/2} + C$

191. $-\dfrac{1}{5}\tan^{-1}\cos(5t) + C$ **193.** $\dfrac{1}{3}\left(\dfrac{27^{3\theta+1}}{\ln 27}\right) + C$

195. $2\sqrt{r} - 2\ln(1 + \sqrt{r}) + C$ **197.** $\ln\left|\dfrac{y}{y+2}\right| + \dfrac{2}{y} - \dfrac{2}{y^2} + C$

199. $4\sec^{-1}\left(\dfrac{7m}{2}\right) + C$ **201.** $\dfrac{\sqrt{8}-1}{6}$ **203.** $\dfrac{\pi}{2}(3b - a) + 2$

Chapter 7 Additional Exercises, pp. 609–612

1. $x(\sin^{-1}x)^2 + 2(\sin^{-1}x)\sqrt{1-x^2} - 2x + C$

3. $\dfrac{x^2\sin^{-1}x}{2} + \dfrac{x\sqrt{1-x^2} - \sin^{-1}x}{4} + C$

5. $\dfrac{\ln|\sec 2\theta + \tan 2\theta| + 2\theta}{4} + C$

7. $\dfrac{1}{2}\left(\ln\left(t - \sqrt{1-t^2}\right) - \sin^{-1}t\right) + C$

9. $\dfrac{1}{16}\ln\left|\dfrac{x^2+2x+2}{x^2-2x+2}\right| + \dfrac{1}{8}\left(\tan^{-1}(x+1) + \tan^{-1}(x-1)\right) + C$

11. 0 **13.** $\ln(4) - 1$ **15.** 1 **17.** $32\pi/35$ **19.** 2π

21. a) π b) $\pi(2e - 5)$ **23.** b) $\pi\left(\dfrac{8(\ln 2)^2}{3} - \dfrac{16(\ln 2)}{9} + \dfrac{16}{27}\right)$

25. $\left(\dfrac{e^2+1}{4}, \dfrac{e-2}{2}\right)$

27. $\sqrt{1+e^2} - \ln\left(\dfrac{\sqrt{1+e^2}}{e} + \dfrac{1}{e}\right) - \sqrt{2} + \ln(1 + \sqrt{2})$ **29.** 6

31. $y = \sqrt{x}, \quad 0 \le x \le 4$ **33.** b) 1 **37.** $a = \dfrac{1}{2}, -\dfrac{\ln 2}{4}$

39. $\dfrac{1}{2} < p \le 1$

41. $\dfrac{e^{2x}}{13}(3\sin 3x + 2\cos 3x) + C$

43. $\dfrac{\cos x \sin 3x - 3\sin x \cos 3x}{8} + C$

45. $\dfrac{e^{ax}}{a^2+b^2}(a\sin bx - b\cos bx) + C$

47. $x\ln(ax) - x + C$

CHAPTER 8

Section 8.1, pp. 619–622

1. $a_1 = 0, \ a_2 = -1/4, \ a_3 = -2/9, \ a_4 = -3/16$
3. $a_1 = 1, \ a_2 = -1/3, \ a_3 = 1/5, \ a_4 = -1/7$
5. $a_1 = 1/2, \ a_2 = 1/2, \ a_3 = 1/2, \ a_4 = 1/2$

7. $1, \dfrac{3}{2}, \dfrac{7}{4}, \dfrac{15}{8}, \dfrac{31}{16}, \dfrac{63}{32}, \dfrac{127}{64}, \dfrac{255}{128}, \dfrac{511}{256}, \dfrac{1023}{512}$

9. $2, \ 1, \ -\dfrac{1}{2}, \ -\dfrac{1}{4}, \dfrac{1}{8}, \dfrac{1}{16}, -\dfrac{1}{32}, -\dfrac{1}{64}, \dfrac{1}{128}, \dfrac{1}{256}$

11. $1, 1, 2, 3, 5, 8, 13, 21, 34, 55$ **13.** $a_n = (-1)^{n+1}, \ n \ge 1$
15. $a_n = (-1)^{n+1}(n)^2, \ n \ge 1$ **17.** $a_n = n^2 - 1, \ n \ge 1$

19. $a_n = 4n - 3, \ n \ge 1$ **21.** $a_n = \dfrac{1 + (-1)^{n+1}}{2}, \ n \ge 1$

23. $N = 692, \ a_n = \sqrt[n]{0.5}, \ L = 1$
25. $N = 65, \ a_n = (0.9)^n, \ L = 0$ **27.** b) $\sqrt{3}$
31. Nondecreasing, bounded   **33.** Not nondecreasing, bounded
35. Converges, nondecreasing sequence theorem
37. Converges, nondecreasing sequence theorem
39. Diverges, definition of divergence   **43.** Converges
45. Converges

Section 8.2, pp. 628–630

1. Converges, 2 **3.** Converges, -1 **5.** Converges, -5
7. Diverges   **9.** Diverges   **11.** Converges, $1/2$
13. Converges, 0 **15.** Converges, $\sqrt{2}$ **17.** Converges, 1
19. Converges, 0 **21.** Converges, 0 **23.** Converges, 0
25. Converges, 1 **27.** Converges, e^7 **29.** Converges, 1
31. Converges, 1 **33.** Diverges   **35.** Converges, 4
37. Converges, 0 **39.** Diverges   **41.** Converges, e^{-1}
43. Converges, $e^{2/3}$ **45.** Converges, $x \ (x > 0)$ **47.** Converges, 0
49. Converges, 1 **51.** Converges, $1/2$ **53.** Converges, $\pi/2$
55. Converges, 0 **57.** Converges, 0 **59.** Converges, $1/2$
61. Converges, 0 **63.** $x_n = 2^{n-2}$
65. a) $f(x) = x^2 - 2, \ 1.414213562 \approx \sqrt{2}$
b) $f(x) = \tan(x) - 1, \ 0.7853981635 \approx \pi/4$
c) $f(x) = e^x$, diverges   **67.** b) 1 **75.** 1 **77.** -0.73908456
79. 0.853748068 **83.** -3

Section 8.3, pp. 638–640

1. $s_n = \dfrac{2(1 - (1/3)^n)}{1 - (1/3)}, \ 3$ **3.** $s_n = \dfrac{1 - (-1/2)^n}{1 - (-1/2)}, \ 2/3$

5. $s_n = \dfrac{1}{2} - \dfrac{1}{n+2}, \ \dfrac{1}{2}$ **7.** $1 - \dfrac{1}{4} + \dfrac{1}{16} - \dfrac{1}{64} + \cdots, \ \dfrac{4}{5}$

9. $\dfrac{7}{4} + \dfrac{7}{16} + \dfrac{7}{64} + \cdots, \ \dfrac{7}{3}$

11. $(5 + 1) + \left(\dfrac{5}{2} + \dfrac{1}{3}\right) + \left(\dfrac{5}{4} + \dfrac{1}{9}\right) + \left(\dfrac{5}{8} + \dfrac{1}{27}\right) + \cdots, \ \dfrac{23}{2}$

13. $(1 + 1) + \left(\dfrac{1}{2} - \dfrac{1}{5}\right) + \left(\dfrac{1}{4} + \dfrac{1}{25}\right) + \left(\dfrac{1}{8} - \dfrac{1}{125}\right) + \cdots, \ \dfrac{17}{6}$

15. 1 **17.** 5 **19.** Converges, 1 **21.** Converges, $-\dfrac{1}{\ln 2}$

23. Converges, $2 + \sqrt{2}$ **25.** Converges, 1 **27.** Diverges

29. Converges, $\dfrac{e^2}{e^2 - 1}$ **31.** Converges, $2/9$ **33.** Converges, $3/2$

35. Diverges   **37.** Diverges   **39.** Converges, $\dfrac{\pi}{\pi - e}$

41. $a = 1, r = -x$; converges to $1/(1 + x)$ for $|x| < 1$

43. $a = 3, r = (x - 1)/2$; converges to $6/(3 - x)$ for x in $(-1, 3)$

45. $|x| < \dfrac{1}{2}, \dfrac{1}{1 - 2x}$ **47.** $-2 < x < 0, \ \dfrac{1}{2 + x}$

49. $x \neq (2k + 1)\dfrac{\pi}{2}$, k an integer; $\dfrac{1}{1 - \sin x}$ **51.** 23/99

53. 7/9   **55.** 1/15   **57.** 41251/33300

59. a) $\displaystyle\sum_{n=-2}^{\infty} \dfrac{1}{(n + 4)(n + 5)}$ b) $\displaystyle\sum_{n=0}^{\infty} \dfrac{1}{(n + 2)(n + 3)}$

c) $\displaystyle\sum_{n=5}^{\infty} \dfrac{1}{(n - 3)(n - 2)}$ **61.** a) Answers may vary.

b) Answers may vary.   c) Answers may vary.   **69.** a) $r = 3/5$

b) $r = -3/10$ **71.** $|r| < 1, \dfrac{1 + 2r}{1 - r^2}$ **73.** 28 m   **75.** 8 m²

77. a) $3\left(\dfrac{4}{3}\right)^{n-1}$

b) $A_n = A + \dfrac{1}{3}A + \dfrac{1}{3}\left(\dfrac{4}{9}\right)A + \cdots + \dfrac{1}{3}\left(\dfrac{4}{9}\right)^{n-2}A$,

$\displaystyle\lim_{n \to \infty} A_n = 2\sqrt{3}/5$

Section 8.4, pp. 643–644

1. Converges; geometric series, $r = \dfrac{1}{10} < 1$

3. Diverges; $\displaystyle\lim_{n \to \infty} \dfrac{n}{n + 1} = 1 \neq 0$

5. Diverges; p-series, $p < 1$

7. Converges; geometric series, $r = \dfrac{1}{8} < 1$

9. Diverges; Integral Test

11. Converges; geometric series, $r = 2/3 < 1$

13. Diverges; Integral Test

15. Diverges; $\displaystyle\lim_{n \to \infty} \dfrac{2^n}{n + 1} \neq 0$

17. Diverges; $\lim_{n\to\infty} (\sqrt{n}/\ln n) \neq 0$

19. Diverges; geometric series, $r = \dfrac{1}{\ln 2} > 1$

21. Converges; Integral Test
23. Diverges; nth-Term Test
25. Converges; Integral Test
27. Converges; Integral Test
29. Converges; Integral Test   **31.** $a = 1$ **33.** b) About 41.55
35. True

Section 8.5, p. 649

1. Diverges; limit comparison with $\sum(1/\sqrt{n})$
3. Converges; compare with $\sum(1/2^n)$ **5.** Diverges; nth-Term Test

7. Converges; $\left(\dfrac{n}{3n + 1}\right)^n < \left(\dfrac{n}{3n}\right)^n = \left(\dfrac{1}{3}\right)^n$

9. Diverges; direct comparison with $\sum(1/n)$
11. Converges; limit comparison with $\sum(1/n^2)$
13. Diverges; limit comparison with $\sum(1/n)$
15. Diverges; limit comparison with $\sum(1/n)$
17. Diverges; Integral Test
19. Converges; compare with $\sum(1/n^{3/2})$

21. Converges; $\dfrac{1}{n2^n} \leq \dfrac{1}{2^n}$

23. Converges; $\dfrac{1}{3^{n-1} + 1} < \dfrac{1}{3^{n-1}}$

25. Diverges; limit comparison with $\sum(1/n)$
27. Converges; compare with $\sum(1/n^2)$

29. Converges; $\dfrac{\tan^{-1} n}{n^{1.1}} < \dfrac{\pi/2}{n^{1.1}}$

31. Converges; compare with $\sum(1/n^2)$

33. Diverges; $3n > n\sqrt[n]{n} \Rightarrow \dfrac{1}{3n} < \dfrac{1}{n\sqrt[n]{n}} \Rightarrow \displaystyle\sum_{n=1}^{\infty} \dfrac{1}{n\sqrt[n]{n}}$ diverges

35. Converges; limit comparison with $\sum(1/n^2)$

Section 8.6, pp. 654–655

1. Converges; Ratio Test   **3.** Diverges; Ratio Test
5. Converges; Ratio Test   **7.** Converges; compare with $\sum(3/(1.25)^n)$

9. Diverges; $\displaystyle\lim_{n \to \infty} \left(1 - \dfrac{3}{n}\right)^n = e^{-3} \neq 0$

11. Converges; compare with $\sum(1/n^2)$
13. Diverges; compare with $\sum(1/(2n))$
15. Diverges; compare with $\sum(1/n)$ **17.** Converges; Ratio Test
19. Converges; Ratio Test   **21.** Converges; Ratio Test
23. Converges; Root Test   **25.** Converges; compare with $\sum(1/n^2)$
27. Converges; Ratio Test   **29.** Diverges; Ratio Test
31. Converges; Ratio Test   **33.** Converges; Ratio Test

35. Diverges; $a_n = \left(\dfrac{1}{3}\right)^{(1/n!)} \to 1$ **37.** Converges; Ratio Test

39. Diverges; Root Test   **41.** Converges; Root Test
43. Converges; Ratio Test   **47.** Yes

Section 8.7, pp. 661–663

1. Converges by Theorem 8   **3.** Diverges; $a_n \not\to 0$
5. Converges by Theorem 8   **7.** Diverges; $a_n \to 1/2$
9. Converges by Theorem 8

11. Converges absolutely. Series of absolute values is a convergent geometric series.

13. Converges conditionally. $1/\sqrt{n} \to 0$ but $\sum_{n=1}^{\infty} \dfrac{1}{\sqrt{n}}$ diverges.

15. Converges absolutely. Compare with $\sum_{n=1}^{\infty}(1/n^2)$.

17. Converges conditionally. $1/(n+3) \to 0$ but $\sum_{n=1}^{\infty} \dfrac{1}{n+3}$ diverges (compare with $\sum_{n=1}^{\infty}(1/n)$).

19. Diverges; $\dfrac{3+n}{5+n} \to 1$

21. Converges conditionally; $\left(\dfrac{1}{n^2} + \dfrac{1}{n}\right) \to 0$ but $(1+n)/n^2 > 1/n$

23. Converges absolutely; Root Test

25. Converges absolutely by Integral Test   **27.** Diverges; $a_n \not\to 0$

29. Converges absolutely by the Ratio Test

31. Converges absolutely; $\dfrac{1}{n^2 + 2n + 1} < \dfrac{1}{n^2}$

33. Converges absolutely since $\left|\dfrac{\cos n\pi}{n\sqrt{n}}\right| = \left|\dfrac{(-1)^{n+1}}{n^{3/2}}\right| = \dfrac{1}{n^{3/2}}$ (convergent p-series)

35. Converges absolutely by Root Test   **37.** Diverges; $a_n \to \infty$

39. Converges conditionally; $\sqrt{n+1} - \sqrt{n} = 1/(\sqrt{n} + \sqrt{n+1}) \to 0$, but series of absolute values diverges (compare with $\sum(1/\sqrt{n})$)

41. Diverges, $a_n \to 1/2 \neq 0$

43. Converges absolutely; $\operatorname{sech} n = \dfrac{2}{e^n + e^{-n}} = \dfrac{2e^n}{e^{2n} + 1} < \dfrac{2e^n}{e^{2n}}$
$= \dfrac{2}{e^n}$, a term from a convergent geometric series.

45. $|\text{Error}| < 0.2$ **47.** $|\text{Error}| < 2 \times 10^{-11}$ **49.** 0.54030

51. a) $a_n \geq a_{n+1}$ b) $-1/2$

Section 8.8, pp. 671–672

1. a) 1, $-1 < x < 1$ b) $-1 < x < 1$ c) none

3. a) 1/4, $-1/2 < x < 0$ b) $-1/2 < x < 0$ c) none

5. a) 10, $-8 < x < 12$ b) $-8 < x < 12$ c) none

7. a) 1, $-1 < x < 1$ b) $-1 < x < 1$ c) none

9. a) 3, $[-3, 3]$ b) $[-3, 3]$ c) none   **11.** a) ∞, for all x
b) for all x c) none   **13.** a) ∞, for all x b) for all x c) none

15. a) 1, $-1 \leq x < 1$ b) $-1 < x < 1$ c) $x = -1$

17. a) 5, $-8 < x < 2$ b) $-8 < x < 2$ c) none

19. a) 3, $-3 < x < 3$ b) $-3 < x < 3$ c) none

21. a) 1, $-1 < x < 1$ b) $-1 < x < 1$ c) none

23. a) 0, $x = 0$ b) $x = 0$ c) none   **25.** a) 2, $-4 < x \leq 0$
b) $-4 < x < 0$ c) $x = 0$ **27.** a) 1, $-1 \leq x \leq 1$
b) $-1 \leq x \leq 1$ c) none   **29.** a) 1/4, $1 \leq x \leq 3/2$
b) $1 \leq x \leq 3/2$ c) none   **31.** a) 1, $(-1 - \pi) \leq x < (1 - \pi)$
b) $(-1 - \pi) < x < (1 - \pi)$ c) $x = -1 - \pi$

33. $-1 < x < 3$, $4/(3 + 2x - x^2)$ **35.** $0 < x < 16$, $2/(4 - \sqrt{x})$

37. $-\sqrt{2} < x < \sqrt{2}$, $3/(2 - x^2)$

39. $1 < x < 5$, $2/(x - 1)$, $1 < x < 5$, $-2/(x - 1)^2$

41. a) $\cos x = 1 - \dfrac{x^2}{2!} + \dfrac{x^4}{4!} - \dfrac{x^6}{6!} + \dfrac{x^8}{8!} - \dfrac{x^{10}}{10!} + \cdots$; converges for all x

b) and c) $2x - \dfrac{2^3 x^3}{3!} + \dfrac{2^5 x^5}{5!} - \dfrac{2^7 x^7}{7!} + \dfrac{2^9 x^9}{9!} - \dfrac{2^{11} x^{11}}{11!} + \cdots$

43. a) $\dfrac{x^2}{2} + \dfrac{x^4}{12} + \dfrac{x^6}{45} + \dfrac{17 x^8}{2520} + \dfrac{31 x^{10}}{14175}$, $-\dfrac{\pi}{2} < x < \dfrac{\pi}{2}$

b) $1 + x^2 + \dfrac{2x^4}{3} + \dfrac{17 x^6}{45} + \dfrac{62 x^8}{315} + \cdots$, $-\dfrac{\pi}{2} < x < \dfrac{\pi}{2}$

Section 8.9, pp. 677–678

1. $P_0(x) = 0$, $P_1(x) = x - 1$, $P_2(x) = (x - 1) - \dfrac{1}{2}(x - 1)^2$,

$P_3(x) = (x - 1) - \dfrac{1}{2}(x - 1)^2 + \dfrac{1}{3}(x - 1)^3$

3. $P_0(x) = \dfrac{1}{2}$, $P_1(x) = \dfrac{1}{2} - \dfrac{1}{4}(x - 2)$,

$P_2(x) = \dfrac{1}{2} - \dfrac{1}{4}(x - 2) + \dfrac{1}{8}(x - 2)^2$,

$P_3(x) = \dfrac{1}{2} - \dfrac{1}{4}(x - 2) + \dfrac{1}{8}(x - 2)^2 - \dfrac{1}{16}(x - 2)^3$

5. $P_0(x) = \dfrac{\sqrt{2}}{2}$, $P_1(x) = \dfrac{\sqrt{2}}{2} + \dfrac{\sqrt{2}}{2}\left(x - \dfrac{\pi}{4}\right)$,

$P_2(x) = \dfrac{\sqrt{2}}{2} + \dfrac{\sqrt{2}}{2}\left(x - \dfrac{\pi}{4}\right) - \dfrac{\sqrt{2}}{4}\left(x - \dfrac{\pi}{4}\right)^2$,

$P_3(x) = \dfrac{\sqrt{2}}{2} + \dfrac{\sqrt{2}}{2}\left(x - \dfrac{\pi}{4}\right) - \dfrac{\sqrt{2}}{4}\left(x - \dfrac{\pi}{4}\right)^2 - \dfrac{\sqrt{2}}{12}\left(x - \dfrac{\pi}{4}\right)^3$

7. $P_0(x) = 2$, $P_1(x) = 2 + \dfrac{1}{4}(x - 4)$,

$P_2(x) = 2 + \dfrac{1}{4}(x - 4) - \dfrac{1}{64}(x - 4)^2$,

$P_3(x) = 2 + \dfrac{1}{4}(x - 4) - \dfrac{1}{64}(x - 4)^2 + \dfrac{1}{512}(x - 4)^3$

9. $\displaystyle\sum_{n=0}^{\infty} \dfrac{(-x)^n}{n!} = 1 - x + \dfrac{x^2}{2!} - \dfrac{x^3}{3!} + \dfrac{x^4}{4!} - \cdots$

11. $\displaystyle\sum_{n=0}^{\infty}(-1)^n x^n = 1 - x + x^2 - x^3 + \cdots$

13. $\displaystyle\sum_{n=0}^{\infty} \dfrac{(-1)^n 3^{2n+1} x^{2n+1}}{(2n+1)!}$ **15.** $7\displaystyle\sum_{n=0}^{\infty} \dfrac{(-1)^n x^{2n}}{(2n)!}$ **17.** $\displaystyle\sum_{n=0}^{\infty} \dfrac{x^{2n}}{(2n)!}$

19. $x^4 - 2x^3 - 5x + 4$ **21.** $8 + 10(x - 2) + 6(x - 2)^2 + (x - 2)^3$

23. $21 - 36(x + 2) + 25(x + 2)^2 - 8(x + 2)^3 + (x + 2)^4$

25. $\displaystyle\sum_{n=0}^{\infty}(-1)^n(n + 1)(x - 1)^n$ **27.** $\displaystyle\sum_{n=0}^{\infty} \dfrac{e^2}{n!}(x - 2)^n$

33. $L(x) = 0$, $Q(x) = -x^2/2$ **35.** $L(x) = 1$, $Q(x) = 1 + x^2/2$

37. $L(x) = x$, $Q(x) = x$

Section 8.10, pp. 686–688

1. $\displaystyle\sum_{n=0}^{\infty} \dfrac{(-5x)^n}{n!} = 1 - 5x + \dfrac{5^2 x^2}{2!} - \dfrac{5^3 x^3}{3!} + \cdots$

3. $\displaystyle\sum_{n=0}^{\infty} \frac{5(-1)^n(-x)^{2n+1}}{(2n+1)!} = \sum_{n=0}^{\infty} \frac{5(-1)^{n+1}x^{2n+1}}{(2n+1)!}$

$\displaystyle = -5x + \frac{5x^3}{3!} - \frac{5x^5}{5!} + \frac{5x^7}{7!} + \cdots$

5. $\displaystyle\sum_{n=0}^{\infty} \frac{(-1)^n x^n}{(2n)!}$ **7.** $\displaystyle\sum_{n=0}^{\infty} \frac{x^{n+1}}{n!} = x + x^2 + \frac{x^3}{2!} + \frac{x^4}{3!} + \frac{x^5}{4!} + \cdots$

9. $\displaystyle\sum_{n=2}^{\infty} \frac{(-1)^n x^{2n}}{(2n)!} = \frac{x^4}{4!} - \frac{x^6}{6!} + \frac{x^8}{8!} - \frac{x^{10}}{10!} + \cdots$

11. $\displaystyle x - \frac{\pi^2 x^3}{2!} + \frac{\pi^4 x^5}{4!} - \frac{\pi^6 x^7}{6!} + \cdots = \sum_{n=0}^{\infty} \frac{(-1)^n \pi^{2n} x^{2n+1}}{(2n)!}$

13. $\displaystyle 1 + \sum_{n=1}^{\infty} \frac{(-1)^n (2x)^{2n}}{2 \cdot (2n)!} = 1 - \frac{(2x)^2}{2 \cdot 2!} +$

$\displaystyle \frac{(2x)^4}{2 \cdot 4!} - \frac{(2x)^6}{2 \cdot 6!} + \frac{(2x)8}{2 \cdot 8!} - \cdots$

15. $\displaystyle\sum_{n=0}^{\infty} (2x)^{n+2} = 2^2 x^2 + 2^3 x^3 + 2^4 x^4 + \cdots$

17. $\displaystyle\sum_{n=1}^{\infty} n x^{n-1} = 1 + 2x + 3x^2 + 4x^3 + \cdots$

19. $|x| < (0.06)^{1/5} < 0.56968$

21. $|\text{Error}| < (10^{-3})^3/6 < 1.67 \times 10^{-10}, \quad -10^{-3} < x < 0$

23. $|\text{Error}| < (3^{0.1})(0.1)^3/6 < 1.87 \times 10^{-5}$ **25.** 0.000293653

27. $|x| < 0.02$ **31.** $\sin x, \ x = 0.1; \sin(0.1)$

33. $\tan^{-1} x, \ x = \pi/3; \sqrt{3}$

35. $\displaystyle e^x \sin x = x + x^2 + \frac{x^3}{3} - \frac{x^5}{30} - \frac{x^6}{90} \cdots$

43. a) $\displaystyle Q(x) = 1 + kx + \frac{k(k-1)}{2} x^2$ b) for $0 \le x < 100^{-1/3}$

49. a) -1 b) $(1/\sqrt{2})(1+i)$ c) $-i$

53. $\displaystyle x + x^2 + \frac{1}{3}x^3 - \frac{1}{30}x^5 \cdots$; will converge for all x

Section 8.11, pp. 697–699

1. $\displaystyle 1 + \frac{x}{2} - \frac{x^2}{8} + \frac{x^3}{16}$ **3.** $\displaystyle 1 + \frac{1}{2}x + \frac{3}{8}x^2 + \frac{5}{16}x^3 + \cdots$

5. $\displaystyle 1 - x + \frac{3x^2}{4} - \frac{x^3}{2}$ **7.** $\displaystyle 1 - \frac{x^3}{2} + \frac{3x^6}{8} - \frac{5x^9}{16}$

9. $\displaystyle 1 + \frac{1}{2x} - \frac{1}{8x^2} + \frac{1}{16x^3}$

11. $(1+x)^4 = 1 + 4x + 6x^2 + 4x^3 + x^4$

13. $(1-2x)^3 = 1 - 6x + 12x^2 - 8x^3$

15. $\displaystyle y = \sum_{n=0}^{\infty} \frac{(-1)^n}{n!} x^n = e^{-x}$ **17.** $\displaystyle y = \sum_{n=1}^{\infty} (x^n/n!) = e^x - 1$

19. $\displaystyle y = \sum_{n=2}^{\infty} (x^n/n!) = e^x - x - 1$ **21.** $\displaystyle y = \sum_{n=0}^{\infty} \frac{x^{2n}}{2^n n!} = e^{x^2/2}$

23. $\displaystyle y = \sum_{n=0}^{\infty} 2x^n = \frac{2}{1-x}$ **25.** $\displaystyle y = \sum_{n=0}^{\infty} \frac{x^{2n+1}}{(2n+1)!} = \sinh x$

27. $\displaystyle y = 2 + x - 2 \sum_{n=1}^{\infty} \frac{(-1)^{n+1} x^{2n}}{(2n)!}$

29. $\displaystyle y = \sum_{n=0}^{\infty} \frac{-2(x-2)^{2n+1}}{(2n+1)!}$

31. $\displaystyle y = a + bx + \frac{1}{6}x^3 - \frac{ax^4}{3 \cdot 4} - \frac{bx^5}{4 \cdot 5} - \frac{x^7}{6 \cdot 6 \cdot 7} + \frac{ax^8}{3 \cdot 4 \cdot 7 \cdot 8}$

$\displaystyle + \frac{bx^9}{4 \cdot 5 \cdot 8 \cdot 9} \cdots$. For $n \ge 6, \ a_n = (n-2)(n-3) a_{n-4}$.

33. 0.00267 **35.** 0.1 **37.** $0.0999 \ 44461 \ 1$ **39.** $0.1000 \ 01$

41. $1/(13 \cdot 6!) \approx 0.00011$ **43.** $\displaystyle \frac{t^3}{3} - \frac{t^7}{7 \cdot 3!} + \frac{t^{11}}{11 \cdot 5!}$

45. a) $\displaystyle \frac{x^2}{2} - \frac{x^4}{12}$

b) $\displaystyle \frac{x^2}{2} - \frac{x^4}{3 \cdot 4} + \frac{x^6}{5 \cdot 6} - \frac{x^8}{7 \cdot 8} + \cdots + (-1)^{15} \frac{x^{32}}{31 \cdot 32}$

47. $1/2$ **49.** $-1/24$ **51.** $1/3$ **53.** -1 **55.** 2

59. 500 terms

61. 3 terms

63. a) $\displaystyle x + \frac{x^3}{6} + \frac{3x^5}{40} + \frac{5x^7}{112}$, radius of convergence $= 1$

b) $\displaystyle \frac{\pi}{2} - x - \frac{x^3}{6} - \frac{3x^5}{40} - \frac{5x^7}{112}$ **65.** $1 - 2x + 3x^2 - 4x^3 + \cdots$

71. c) $3\pi/4$

Chapter 8 Practice Exercises, pp. 700–702

1. Converges to 1 **3.** Converges to -1 **5.** Diverges

7. Converges to 0 **9.** Converges to 1 **11.** Converges to e^{-5}

13. Converges to 3 **15.** Converges to $\ln 2$ **17.** Diverges

19. $1/6$ **21.** $3/2$ **23.** $e/(e-1)$ **25.** Diverges

27. Converges conditionally   **29.** Converges conditionally

31. Converges absolutely   **33.** Converges absolutely

35. Converges absolutely   **37.** Converges absolutely

39. Converges absolutely   **41.** a) $3, \ -7 \le x < -1$

b) $-7 < x < -1$ c) $x = -7$ **43.** a) $1/3, \ 0 \le x \le 2/3$

b) $0 \le x \le 2/3$ c) none   **45.** a) ∞, for all x b) for all x

c) none   **47.** a) $\sqrt{3}, \ -\sqrt{3} < x < \sqrt{3}$ b) $-\sqrt{3} < x < \sqrt{3}$

c) none   **49.** a) $e, \ (-e, e)$ b) $(-e, \ e)$ c) $\{ \ \}$

51. $\displaystyle \frac{1}{1+x}, \ \frac{1}{4}, \ \frac{4}{5}$ **53.** $\sin x, \ \pi, \ 0$ **55.** $e^x, \ \ln 2, \ 2$

57. $\displaystyle\sum_{n=0}^{\infty} 2^n x^n$ **59.** $\displaystyle\sum_{n=0}^{\infty} \frac{(-1)^n \pi^{2n+1} x^{2n+1}}{(2n+1)!}$ **61.** $\displaystyle\sum_{n=0}^{\infty} \frac{(-1)^n x^{5n}}{(2n)!}$

63. $\displaystyle\sum_{n=0}^{\infty} \frac{((\pi x)/2)^n}{n!}$

65. $\displaystyle 2 - \frac{(x+1)}{2 \cdot 1!} + \frac{3(x+1)^2}{2^3 \cdot 2!} - \frac{9(x+1)^3}{2^5 \cdot 3!} + \cdots$

67. $\displaystyle \frac{1}{4} - \frac{1}{4^2}(x-3) + \frac{1}{4^3}(x-3)^2 - \frac{1}{4^4}(x-3)^3$

69. $y = \sum_{n=0}^{\infty} \frac{(-1)^{n+1}}{n!} x^n = -e^{-x}$

71. $y = 3\sum_{n=0}^{\infty} \frac{(-1)^n 2^n}{n!} x^n = 3e^{-2x}$

73. $y = -1 - x + 2\sum_{n=2}^{\infty} (x^n/n!) = 2e^x - 3x - 3$

75. $y = 1 + x + 2\sum_{n=0}^{\infty} (x^n/n!) = 2e^x - 1 - x$ **77.** 0.4849 17143 1

79. $\approx 0.4872\ 22358\ 3$ **81.** 7/2   **83.** 1/12   **85.** -2
87. $r = -3,\ s = 9/2$
89. b) $|\text{error}| < |\sin(1/42)| < 0.02381$; an underestimate because the remainder is positive

91. 2/3   **93.** $\ln\left(\frac{n+1}{2n}\right)$; the series converges to $\ln\left(\frac{1}{2}\right)$.

95. a) ∞ b) $a = 1,\quad b = 0$ **97.** It converges.

Chapter 8 Additional Exercises, pp. 703–707

1. Converges; Comparison Test   **3.** Diverges; nth Term Test
5. Converges; Comparison Test   **7.** Diverges; nth Term Test

9. With $a = \pi/3,\ \cos x = \frac{1}{2} - \frac{\sqrt{3}}{2}(x - \pi/3) - \frac{1}{4}(x - \pi/3)^2 +$
$\frac{\sqrt{3}}{12}(x - \pi/3)^3 + \cdots$

11. With $a = 0,\ e^x = 1 + x + \frac{x^2}{2!} + \frac{x^3}{3!} + \cdots$

13. With $a = 22\pi,\ \cos x = 1 - \frac{1}{2}(x - 22\pi)^2 + \frac{1}{4!}(x - 22\pi)^4 - \frac{1}{6!}(x - 22\pi)^6 + \cdots$

15. Converges, limit $= b$ **17.** $\pi/2$ **23.** $b = \pm\frac{1}{5}$

25. $a = 2,\ L = -7/6$ **29.** b) Yes

35. a) $\sum_{n=1}^{\infty} nx^{n-1}$ b) 6   c) $1/q$

37. a) $R_n = C_0 e^{-kt_0}\left(1 - e^{-nkt_0}\right)/\left(1 - e^{-kt_0}\right)$,
$R = C_0\left(e^{-kt_0}\right)/\left(1 - e^{-kt_0}\right) = C_0/\left(e^{kt_0} - 1\right)$
b) $R_1 = 1/e \approx 0.368$,
$R_{10} = R(1 - e^{-10}) \approx R(0.9999546) \approx 0.58195$;
$R \approx 0.58198;\ 0 < (R - R_{10})/R < 0.0001$ c) 7

CHAPTER 9

Section 9.1, pp. 719–722

1. $y^2 = 8x,\ F(2, 0)$, directrix: $x = -2$
3. $x^2 = -6y,\ F(0, -3/2)$, directrix: $y = 3/2$

5. $\frac{x^2}{4} - \frac{y^2}{9} = 1,\ F(\pm\sqrt{13}, 0),\ V(\pm 2, 0)$,
asymptotes: $y = \pm\frac{3}{2}x$

7. $\frac{x^2}{2} + y^2 = 1,\ F(\pm 1, 0),\ V(\pm\sqrt{2}, 0)$

9.

11.

13.

15.

17.

19.

21.

23.

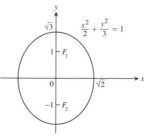

25. $\frac{x^2}{4} + \frac{y^2}{2} = 1$

27. Asymptotes: $y = \pm x$

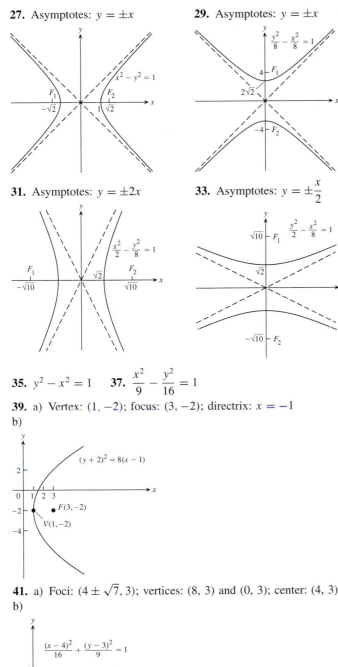

29. Asymptotes: $y = \pm x$

b)

31. Asymptotes: $y = \pm 2x$

33. Asymptotes: $y = \pm \dfrac{x}{2}$

35. $y^2 - x^2 = 1$ **37.** $\dfrac{x^2}{9} - \dfrac{y^2}{16} = 1$

39. a) Vertex: $(1, -2)$; focus: $(3, -2)$; directrix: $x = -1$

b)

41. a) Foci: $(4 \pm \sqrt{7}, 3)$; vertices: $(8, 3)$ and $(0, 3)$; center: $(4, 3)$

b)

43. a) Center: $(2, 0)$; foci: $(7, 0)$ and $(-3, 0)$; vertices: $(6, 0)$ and $(-2, 0)$; asymptotes: $y = \pm \dfrac{3}{4}(x - 2)$

45. $(y + 3)^2 = 4(x + 2)$, $V(-2, -3)$, $F(-1, -3)$, directrix: $x = -3$

47. $(x - 1)^2 = 8(y + 7)$, $V(1, -7)$, $F(1, -5)$, directrix: $y = -9$

49. $\dfrac{(x + 2)^2}{6} + \dfrac{(y + 1)^2}{9} = 1$, $F(-2, \pm\sqrt{3} - 1)$, $V(-2, \pm 3 - 1)$, $C(-2, -1)$

51. $\dfrac{(x - 2)^2}{3} + \dfrac{(y - 3)^2}{2} = 1$, $F(3, 3)$ and $F(1, 3)$, $V(\pm\sqrt{3} + 2, 3)$, $C(2, 3)$

53. $\dfrac{(x - 2)^2}{4} - \dfrac{(y - 2)^2}{5} = 1$, $C(2, 2)$, $F(5, 2)$ and $F(-1, 2)$, $V(4, 2)$ and $V(0, 2)$; asymptotes: $(y - 2) = \pm\dfrac{\sqrt{5}}{2}(x - 2)$

55. $(y + 1)^2 - (x + 1)^2 = 1$, $C(-1, -1)$, $F(-1, \sqrt{2} - 1)$ and $F(-1, -\sqrt{2} - 1)$, $V(-1, 0)$ and $V(-1, -2)$; asymptotes: $(y + 1) = \pm(x + 1)$

57. $C(-2, 0)$, $a = 4$ **59.** $V(-1, 1)$, $F(-1, 0)$

61. Ellipse: $\dfrac{(x + 2)^2}{5} + y^2 = 1$, $C(-2, 0)$, $F(0, 0)$ and $F(-4, 0)$, $V(\sqrt{5} - 2, 0)$ and $V(-\sqrt{5} - 2, 0)$

63. Ellipse: $\dfrac{(x - 1)^2}{2} + (y - 1)^2 = 1$, $C(1, 1)$, $F(2, 1)$ and $F(0, 1)$, $V(\sqrt{2} + 1, 1)$ and $V(-\sqrt{2} + 1, 1)$

65. Hyperbola: $(x - 1)^2 - (y - 2)^2 = 1$, $C(1, 2)$, $F(1 + \sqrt{2}, 2)$ and $F(1 - \sqrt{2}, 2)$, $V(2, 2)$ and $V(0, 2)$; asymptotes: $(y - 2) = \pm(x - 1)$

67. Hyperbola: $\dfrac{(y - 3)^2}{6} - \dfrac{x^2}{3} = 1$, $C(0, 3)$, $F(0, 6)$ and $F(0, 0)$, $V(0, \sqrt{6} + 3)$ and $V(0, -\sqrt{6} + 3)$; asymptotes: $y = \sqrt{2}x + 3$ or $y = -\sqrt{2}x + 3$

69.

71.

$x^2 + 4y^2 \geq 4$ and $4x^2 + 9y^2 \leq 36$

73.

$4y^2 - x^2 \geq 4$

77. $3x^2 + 3y^2 - 7x - 7y + 4 = 0$
79. $(x + 2)^2 + (y - 1)^2 = 13$. The point is inside the circle.
81. b) $1 : 1$ **83.** Length $= 2\sqrt{2}$, width $= \sqrt{2}$, area $= 4$
85. 24π **87.** $(0, 16/(3\pi))$

Section 9.2, pp. 726–727

1. $e = 3/5$, $F(\pm 3, 0)$, $x = \pm 25/3$
3. $e = 1/\sqrt{2}$, $F(0, \pm 1)$, $y = \pm 2$
5. $e = 1/\sqrt{3}$, $F(0, \pm 1)$, $y = \pm 3$
7. $e = \sqrt{3}/3$, $F(\pm\sqrt{3}, 0)$, $x = \pm 3\sqrt{3}$
9. $\dfrac{x^2}{27} + \dfrac{y^2}{36} = 1$ **11.** $\dfrac{x^2}{4851} + \dfrac{y^2}{4900} = 1$
13. $e = \dfrac{\sqrt{5}}{3}$, $\dfrac{x^2}{9} + \dfrac{y^2}{4} = 1$ **15.** $e = 1/2$, $\dfrac{x^2}{64} + \dfrac{y^2}{48} = 1$
19. $\dfrac{(x-1)^2}{4} + \dfrac{(y-4)^2}{9} = 1$, $F(1, 4 \pm \sqrt{5})$, $e = \sqrt{5}/3$,
$y = 4 \pm (9\sqrt{5}/5)$
21. $a = 0$, $b = -4$, $c = 0$, $e = \sqrt{3}/2$
23. $e = \sqrt{2}$, $F(\pm\sqrt{2}, 0)$, $x = \pm 1/\sqrt{2}$
25. $e = \sqrt{2}$, $F(0, \pm 4)$, $y = \pm 2$
27. $e = \sqrt{5}$, $F(\pm\sqrt{10}, 0)$, $x = \pm 2/\sqrt{10}$
29. $e = \sqrt{5}$, $F(0, \pm\sqrt{10})$, $y = \pm 2/\sqrt{10}$ **31.** $y^2 - \dfrac{x^2}{8} = 1$
33. $x^2 - \dfrac{y^2}{8} = 1$ **35.** $e = \sqrt{2}$, $\dfrac{x^2}{8} - \dfrac{y^2}{8} = 1$
37. $e = 2$, $x^2 - \dfrac{y^2}{3} = 1$ **39.** $\dfrac{(y-6)^2}{36} - \dfrac{(x-1)^2}{45} = 1$

Section 9.3, pp. 733–734

1. Hyperbola   **3.** Ellipse   **5.** Parabola   **7.** Parabola
9. Hyperbola   **11.** Hyperbola   **13.** Ellipse   **15.** Ellipse
17. $x'^2 - y'^2 = 4$, hyperbola   **19.** $4x'^2 + 16y' = 0$, parabola
21. $y'^2 = 1$, parallel lines   **23.** $2\sqrt{2}x'^2 + 8\sqrt{2}y' = 0$, parabola
25. $4x'^2 + 2y'^2 = 19$, ellipse
27. $\sin \alpha = 1/\sqrt{5}$, $\cos \alpha = 2/\sqrt{5}$; or $\sin \alpha = -2/\sqrt{5}$,
$\cos \alpha = 1/\sqrt{5}$
29. $A' = 0.88$, $B' = 0.00$, $C' = 3.10$, $D' = 0.74$, $E' = -1.20$,
$F' = -3$, $0.88x'^2 + 3.10y'^2 + 0.74x' - 1.20y' - 3 = 0$, ellipse
31. $A' = 0.00$, $B' = 0.00$, $C' = 5.00$, $D' = 0$, $E' = 0$, $F' = -5$,
$5.00y'^2 - 5 = 0$ or $y' = \pm 1.00$, parallel lines

33. $A' = 5.05$, $B' = 0.00$, $C' = -0.05$, $D' = -5.07$, $E' = -6.18$,
$F' = -1$, $5.05x'^2 - 0.05y'^2 - 5.07x' - 6.18y' - 1 = 0$, hyperbola
35. a) $\dfrac{x'^2}{b^2} + \dfrac{y'^2}{a^2} = 1$ b) $\dfrac{y'^2}{a^2} - \dfrac{x'^2}{b^2} = 1$ c) $x'^2 + y'^2 = a^2$
d) $y' = -\dfrac{1}{m}x'$ e) $y' = -\dfrac{1}{m}x' + \dfrac{b}{m}$
37. a) $x'^2 - y'^2 = 2$ b) $x'^2 - y'^2 = 2a$ **43.** a) Parabola
45. a) Hyperbola
b)

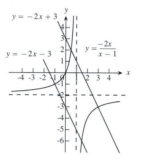

$y = -2x + 3$

$y = -2x - 3$

$y = \dfrac{-2x}{x - 1}$

c) $y = -2x - 3$, $y = -2x + 3$

Section 9.4, pp. 741–744

1.

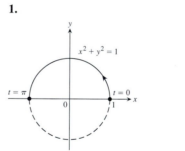

$x^2 + y^2 = 1$

$t = \pi$ $t = 0$

3.

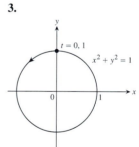

$t = 0, 1$

$x^2 + y^2 = 1$

5.

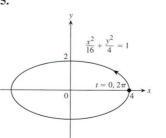

$\dfrac{x^2}{16} + \dfrac{y^2}{4} = 1$

$t = 0, 2\pi$

7.

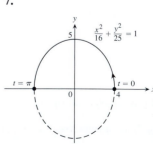

$\dfrac{x^2}{16} + \dfrac{y^2}{25} = 1$

$t = \pi$ $t = 0$

9.

11.

13.

15.

17.

19.

21.

23.

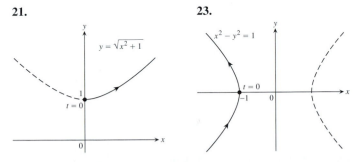

25. a) $x = a \cos t$, $y = -a \sin t$, $0 \le t \le 2\pi$ b) $x = a \cos t$, $y = a \sin t$, $0 \le t \le 2\pi$ c) $x = a \cos t$, $y = -a \sin t$, $0 \le t \le 4\pi$ d) $x = a \cos t$, $y = a \sin t$, $0 \le t \le 4\pi$

27. $x = \dfrac{-at}{\sqrt{1+t^2}}$, $y = \dfrac{a}{\sqrt{1+t^2}}$, $-\infty < t < \infty$

29. $x = 2 \cot t$, $y = 2 \sin^2 t$, $0 < t < \pi$

31. b) $x = x_1 t$, $y = y_1 t$ (answer not unique)   c) $x = -1 + t$, $y = t$ (answer not unique)

33. $x = (a - b) \cos \theta + b \cos \left(\dfrac{a - b}{b} \theta \right)$,

$y = (a - b) \sin \theta - b \sin \left(\dfrac{a - b}{b} \theta \right)$

35. $x = a \sin^2 t \tan t$, $y = a \sin^2 t$ **37.** $(1, 1)$

Section 9.5, pp. 749–751

1. $y = -x + 2\sqrt{2}$, $\dfrac{d^2y}{dx^2} = -\sqrt{2}$

3. $y = -\dfrac{1}{2}x + 2\sqrt{2}$, $\dfrac{d^2y}{dx^2} = -\dfrac{\sqrt{2}}{4}$ **5.** $y = x + \dfrac{1}{4}$, $\dfrac{d^2y}{dx^2} = -2$

7. $y = 2x - \sqrt{3}$, $\dfrac{d^2y}{dx^2} = -3\sqrt{3}$ **9.** $y = x - 4$, $\dfrac{d^2y}{dx^2} = \dfrac{1}{2}$

11. $y = \sqrt{3}x - \dfrac{\pi\sqrt{3}}{3} + 2$, $\dfrac{d^2y}{dx^2} = -4$ **13.** 0 **15.** -6 **17.** 4

19. 12 **21.** π^2 **23.** $8\pi^2$ **25.** $52\pi/3$ **27.** $3\pi\sqrt{5}$

29. a) $(\bar{x}, \bar{y}) = \left(\dfrac{12}{\pi} - \dfrac{24}{\pi^2}, \dfrac{24}{\pi^2} - 2 \right)$ b) Centroid: $(1.4, 0.4)$

31. a) $(\bar{x}, \bar{y}) = \left(\dfrac{1}{3}, \pi - \dfrac{4}{3} \right)$ **33.** a) π b) π **37.** $3\pi a^2$

39. $64\pi/3$ **41.** $\left(\dfrac{\sqrt{2}}{2}, 1 \right)$, $y = 2x$ at $t = 0$, $y = -2x$ at $t = \pi$

Section 9.6, pp. 755–756

1. a) e; b) g; c) h; d) f

3.

a) $\left(2, \dfrac{\pi}{2} + 2n\pi \right)$ and $\left(-2, \dfrac{\pi}{2} + (2n + 1)\pi \right)$, n an integer

b) $(2, 2n\pi)$ and $(-2, (2n + 1)\pi)$, n an integer

c) $\left(2, \dfrac{3\pi}{2} + 2n\pi \right)$ and $\left(-2, \dfrac{3\pi}{2} + (2n + 1)\pi \right)$, n an integer

d) $(2, (2n + 1)\pi)$ and $(-2, 2n\pi)$, n an integer

5. a) $(3, 0)$ b) $(-3, 0)$ c) $(-1, \sqrt{3})$ d) $(1, \sqrt{3})$,   e) $(3, 0)$
f) $(1, \sqrt{3})$ g) $(-3, 0)$ h) $(-1, \sqrt{3})$

7.   **9.**

11.

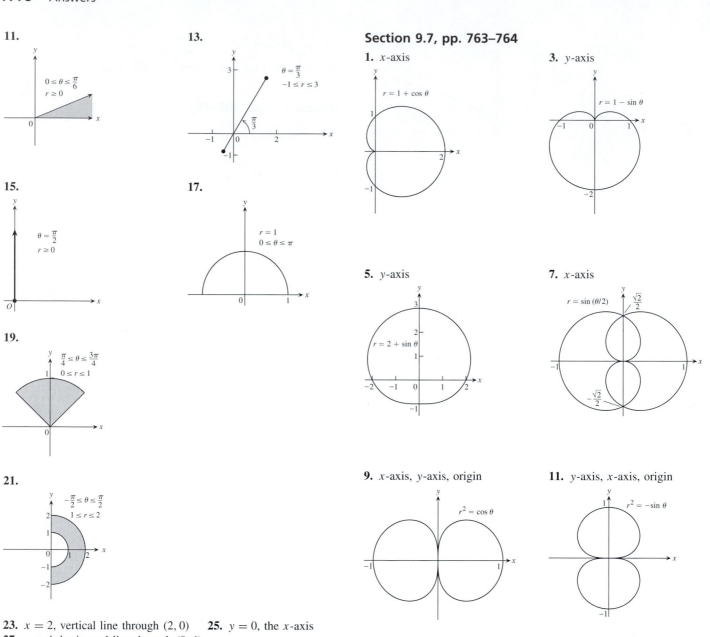

$0 \leq \theta \leq \frac{\pi}{6}$
$r \geq 0$

13.

$\theta = \frac{\pi}{3}$
$-1 \leq r \leq 3$

15.

$\theta = \frac{\pi}{2}$
$r \geq 0$

17.

$r = 1$
$0 \leq \theta \leq \pi$

19.

$\frac{\pi}{4} \leq \theta \leq \frac{3\pi}{4}$
$0 \leq r \leq 1$

21.

$-\frac{\pi}{2} \leq \theta \leq \frac{\pi}{2}$
$1 \leq r \leq 2$

23. $x = 2$, vertical line through $(2, 0)$ **25.** $y = 0$, the x-axis
27. $y = 4$, horizontal line through $(0, 4)$
29. $x + y = 1$, line, $m = -1$, $b = 1$
31. $x^2 + y^2 = 1$, circle, $C(0, 0)$, radius 1
33. $y - 2x = 5$, line, $m = 2$, $b = 5$
35. $y^2 = x$, parabola, vertex $(0, 0)$, opens right
37. $y = e^x$, graph of natural exponential function
39. $x + y = \pm 1$, two straight lines of slope -1, y-intercepts $b = \pm 1$
41. $(x + 2)^2 + y^2 = 4$, circle, $C(-2, 0)$, radius 2
43. $x^2 + (y - 4)^2 = 16$, circle, $C(0, 4)$, radius 4
45. $(x - 1)^2 + (y - 1)^2 = 2$, circle, $C(1, 1)$, radius $\sqrt{2}$
47. $\sqrt{3}y + x = 4$ **49.** $r \cos \theta = 7$ **51.** $\theta = \pi/4$
53. $r = 2$ or $r = -2$ **55.** $4r^2 \cos^2 \theta + 9r^2 \sin^2 \theta = 36$
57. $r \sin^2 \theta = 4 \cos \theta$ **59.** $r = 4 \sin \theta$
61. $r^2 = 6r \cos \theta - 2r \sin \theta - 6$ **63.** $(0, \theta)$, where θ is any angle

Section 9.7, pp. 763–764

1. x-axis

$r = 1 + \cos \theta$

3. y-axis

$r = 1 - \sin \theta$

5. y-axis

$r = 2 + \sin \theta$

7. x-axis

$r = \sin (\theta/2)$ $\frac{\sqrt{2}}{2}$
$-\frac{\sqrt{2}}{2}$

9. x-axis, y-axis, origin

$r^2 = \cos \theta$

11. y-axis, x-axis, origin

$r^2 = -\sin \theta$

13. x-axis, y-axis, origin   **15.** Origin
17. The slope at $(-1, \pi/2)$ is -1, at $(-1, -\pi/2)$ is 1.

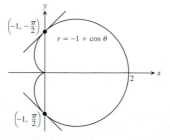

$\left(-1, -\frac{\pi}{2}\right)$
$r = -1 + \cos \theta$
$\left(-1, \frac{\pi}{2}\right)$

19. The slope at $(1, \pi/4)$ is -1, at $(-1, -\pi/4)$ is 1, at $(-1, 3\pi/4)$ is 1, at $(1, -3\pi/4)$ is -1.

21. a)

b)

23. a)

b)

25.

27.

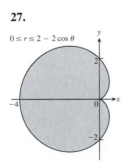

31. $(0, 0)$, $(1, \pi/2)$, $(1, 3\pi/2)$
33. $(0, 0)$, $(\sqrt{3}, \pi/3)$, $(-\sqrt{3}, -\pi/3)$
35. $(\sqrt{2}, \pm\pi/6)$, $(\sqrt{2}, \pm 5\pi/6)$
37. $(1, \pi/12)$, $(1, 5\pi/12)$, $(1, 13\pi/12)$, $(1, 17\pi/12)$ **43.** a

51. $2y = \dfrac{2\sqrt{6}}{9}$

Section 9.8, pp. 768-770

1. $r\cos(\theta - \pi/6) = 5$, $y = -\sqrt{3}x + 10$
3. $r\cos(\theta - 4\pi/3) = 3$, $y = -(\sqrt{3}/3)x - 2\sqrt{3}$ **5.** $y = 2 - x$

7. $y = (\sqrt{3}/3)x + 2\sqrt{3}$ **9.** $r\cos\left(\theta - \dfrac{\pi}{4}\right) = 3$

11. $r\cos\left(\theta + \dfrac{\pi}{2}\right) = 5$ **13.** $r = 8\cos\theta$ **15.** $r = 2\sqrt{2}\sin\theta$

17. $C(2, 0)$, radius $= 2$ **19.** $C(1, \pi)$, radius $= 1$
21. $(x - 6)^2 + y^2 = 36$, $r = 12\cos\theta$
23. $x^2 + (y - 5)^2 = 25$, $r = 10\sin\theta$
25. $(x + 1)^2 + y^2 = 1$, $r = -2\cos\theta$
27. $x^2 + (y + 1/2)^2 = 1/4$, $r = -\sin\theta$ **29.** $r = 2/(1 + \cos\theta)$
31. $r = 30/(1 - 5\sin\theta)$ **33.** $r = 1/(2 + \cos\theta)$
35. $r = 10/(5 - \sin\theta)$

37.

39.

41.

43.

45.

57. b)

Planet	Perihelion	Aphelion
Mercury	0.3075 AU	0.4667 AU
Venus	0.7184 AU	0.7282 AU
Earth	0.9833 AU	1.0167 AU
Mars	1.3817 AU	1.6663 AU
Jupiter	4.9512 AU	5.4548 AU
Saturn	9.0210 AU	10.0570 AU
Uranus	18.2977 AU	20.0623 AU
Neptune	29.8135 AU	30.3065 AU
Pluto	29.6549 AU	49.2251 AU

59. a) $x^2 + (y - 2)^2 = 4$, $x = \sqrt{3}$

b)

61. $r = 4/(1 + \cos \theta)$ **63.** b) The pins should be 2 in. apart.
65. $r = 2a \sin \theta$ (a circle)   **67.** $r \cos (\theta - \alpha) = p$ (a line)

Section 9.9, pp. 775–777

1. 18π **3.** $\pi/8$ **5.** 2   **7.** $\dfrac{\pi}{2} - 1$ **9.** $5\pi - 8$

11. $3\sqrt{3} - \pi$ **13.** $\dfrac{\pi}{3} + \dfrac{\sqrt{3}}{2}$ **15.** $12\pi - 9\sqrt{3}$ **17.** a) $\dfrac{3}{2} - \dfrac{\pi}{4}$

19. $19/3$ **21.** 8   **23.** $3(\sqrt{2} + \ln (1 + \sqrt{2}))$ **25.** $\dfrac{\pi}{8} + \dfrac{3}{8}$

27. 2π **29.** $\pi\sqrt{2}$ **31.** $2\pi (2 - \sqrt{2})$ **37.** $\left(\dfrac{5}{6}a, 0\right)$

Chapter 9 Practice Exercises, pp. 778–782

1.

3.

5. $e = 3/4$

7. $e = 2$

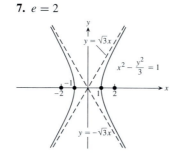

9. $(x - 2)^2 = -12(y - 3)$, $V(2, 3)$, $F(2, 0)$; directrix: $y = 6$
11. $\dfrac{(x + 3)^2}{9} + \dfrac{(y + 5)^2}{25} = 1$, $C(-3, -5)$, $V(-3, 0)$ and
$V(-3, -10)$, $F(-3, -1)$ and $F(-3, -9)$

13. $\dfrac{(y - 2\sqrt{2})^2}{8} - \dfrac{(x - 2)^2}{2} = 1$, $C(2, 2\sqrt{2})$, $V(2, 4\sqrt{2})$ and
$V(2, 0)$, $F(2, \sqrt{10} + 2\sqrt{2})$ and $F(2, -\sqrt{10} + 2\sqrt{2})$; asymptotes:
$y = 2x - 4 + 2\sqrt{2}$ and $y = -2x + 4 + 2\sqrt{2}$

15. Hyperbola: $\dfrac{(x - 2)^2}{4} - y^2 = 1$, $F(2 \pm \sqrt{5}, 0)$, $V(2 \pm 2, 0)$,
$C(2, 0)$; asymptotes: $y = \pm\dfrac{1}{2}(x - 2)$

17. Parabola: $(y - 1)^2 = -16(x + 3)$, $V(-3, 1)$, $F(-7, 1)$;
directrix: $x = 1$

19. Ellipse: $\dfrac{(x + 3)^2}{16} + \dfrac{(y - 2)^2}{9} = 1$, $F(\pm\sqrt{7} - 3, 2)$,
$V(\pm 4 - 3, 2)$, $C(-3, 2)$
21. Circle: $(x - 1)^2 + (y - 1)^2 = 2$, $C(1, 1)$, radius $= \sqrt{2}$
23. Ellipse   **25.** Hyperbola   **27.** Line
29. Ellipse, $5x'^2 + 3y'^2 = 30$ **31.** Hyperbola, $x'^2 - y'^2 = 2$
33.   **35.**

37.

39. $x = 3 \cos t$, $y = 4 \sin t$, $0 \le t \le 2\pi$ **41.** $y = \dfrac{\sqrt{3}}{2}x + \dfrac{1}{4}$, $\dfrac{1}{4}$

43. $3 + \dfrac{\ln 2}{8}$ **45.** $76\pi/3$

47.

49. d   **51.** l   **53.** k   **55.** i   **57.** $(0, 0)$
59. $(0, 0)$, $(1, \pm\pi/2)$ **61.** The graphs coincide.   **63.** $(\sqrt{2}, \pi/4)$
65. $y = x$, $y = -x$

67. At $(1, \pi/4) : r \cos(\theta - \pi/4) = 1$, At $(1, 3\pi/4) :$
$r \cos(\theta - 3\pi/4) = 1$, At $(1, 5\pi/4) : r \cos(\theta - 5\pi/4) = 1$,
At $(1, 7\pi/4) : r \cos(\theta - 7\pi/4) = 1$
69. $y = (\sqrt{3}/3)x - 4$ **71.** $x = 2$ **73.** $y = -3/2$
75. $x^2 + (y + 2)^2 = 4$ **77.** $(x - \sqrt{2})^2 + y^2 = 2$
79. $r = -5 \sin \theta$ **81.** $r = 3 \cos \theta$
83.   **85.**

87. $r = \dfrac{4}{1 + 2 \cos \theta}$ **89.** $r = \dfrac{2}{2 + \sin \theta}$ **91.** $9\pi/2$
93. $2 + \pi/4$ **95.** 8 **97.** $\pi - 3$ **99.** $(2 - \sqrt{2})\pi$
101. a) 24π b) 16π **111.** $\pi/2$ **115.** $\left(2, \pm\dfrac{\pi}{3}\right), \dfrac{\pi}{2}$
119. $\pi/2$ **121.** $\pi/4$

Chapter 9 Additional Exercises, pp. 783–786

1.

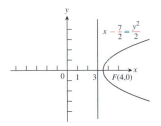

3. $3x^2 + 3y^2 - 8y + 4 = 0$ **5.** $(0, \pm 1)$

7. a) $\dfrac{(y - 1)^2}{16} - \dfrac{x^2}{48} = 1$ b) $\dfrac{16\left(y + \dfrac{3}{4}\right)^2}{25} - \dfrac{2x^2}{75} = 1$

17.   **19.**

21.   **23.**

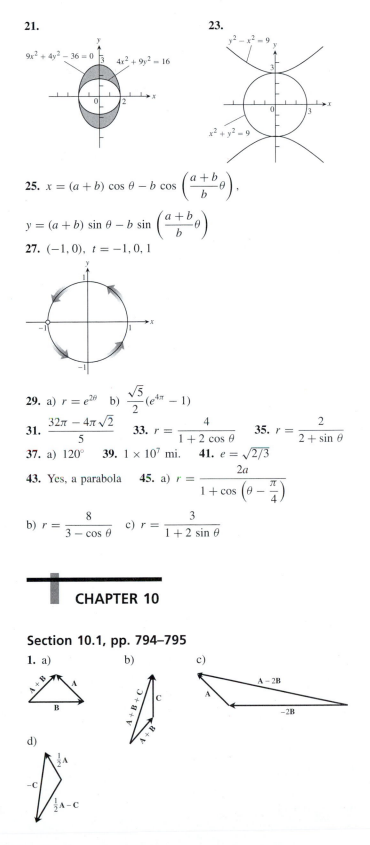

25. $x = (a + b) \cos \theta - b \cos\left(\dfrac{a + b}{b}\theta\right),$

$y = (a + b) \sin \theta - b \sin\left(\dfrac{a + b}{b}\theta\right)$

27. $(-1, 0),\ t = -1, 0, 1$

29. a) $r = e^{2\theta}$ b) $\dfrac{\sqrt{5}}{2}(e^{4\pi} - 1)$

31. $\dfrac{32\pi - 4\pi\sqrt{2}}{5}$ **33.** $r = \dfrac{4}{1 + 2 \cos \theta}$ **35.** $r = \dfrac{2}{2 + \sin \theta}$

37. a) $120°$ **39.** 1×10^7 mi. **41.** $e = \sqrt{2/3}$

43. Yes, a parabola **45.** a) $r = \dfrac{2a}{1 + \cos\left(\theta - \dfrac{\pi}{4}\right)}$

b) $r = \dfrac{8}{3 - \cos \theta}$ c) $r = \dfrac{3}{1 + 2 \sin \theta}$

CHAPTER 10

Section 10.1, pp. 794–795

1. a)   b)   c)

d)

3. $4\mathbf{i} + 5\mathbf{j}$ **5.** $(6 - (\sqrt{3}/\pi))\mathbf{i} - 20\mathbf{j}$

7. a) $\mathbf{w} = \mathbf{v} + \mathbf{u}$ b) $\mathbf{v} = \mathbf{w} - \mathbf{u}$

9.   **11.**

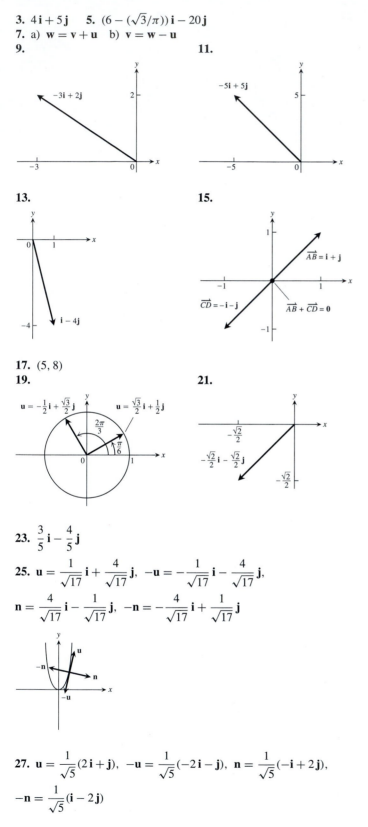

13.   **15.**

17. $(5, 8)$

19.   **21.**

23. $\dfrac{3}{5}\mathbf{i} - \dfrac{4}{5}\mathbf{j}$

25. $\mathbf{u} = \dfrac{1}{\sqrt{17}}\mathbf{i} + \dfrac{4}{\sqrt{17}}\mathbf{j}, \ -\mathbf{u} = -\dfrac{1}{\sqrt{17}}\mathbf{i} - \dfrac{4}{\sqrt{17}}\mathbf{j},$

$\mathbf{n} = \dfrac{4}{\sqrt{17}}\mathbf{i} - \dfrac{1}{\sqrt{17}}\mathbf{j}, \ -\mathbf{n} = -\dfrac{4}{\sqrt{17}}\mathbf{i} + \dfrac{1}{\sqrt{17}}\mathbf{j}$

27. $\mathbf{u} = \dfrac{1}{\sqrt{5}}(2\mathbf{i} + \mathbf{j}), \ -\mathbf{u} = \dfrac{1}{\sqrt{5}}(-2\mathbf{i} - \mathbf{j}), \ \mathbf{n} = \dfrac{1}{\sqrt{5}}(-\mathbf{i} + 2\mathbf{j}),$

$-\mathbf{n} = \dfrac{1}{\sqrt{5}}(\mathbf{i} - 2\mathbf{j})$

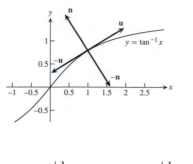

29. $\mathbf{u} = \dfrac{\pm 1}{5}(-4\mathbf{i} + 3\mathbf{j}), \ \mathbf{v} = \dfrac{\pm 1}{5}(3\mathbf{i} + 4\mathbf{j})$

31. $\mathbf{u} = \dfrac{\pm 1}{2}(\mathbf{i} + \sqrt{3}\mathbf{j}), \ \mathbf{v} = \dfrac{\pm 1}{2}(-\sqrt{3}\mathbf{i} + \mathbf{j})$

33. $13\left(\dfrac{5}{13}\mathbf{i} + \dfrac{12}{13}\mathbf{j}\right)$

35. $\dfrac{3}{5}\mathbf{i} - \dfrac{4}{5}\mathbf{j}$ and $-\dfrac{3}{5}\mathbf{i} + \dfrac{4}{5}\mathbf{j}$ **39.** $5\sqrt{3}\mathbf{i}, \ 5\mathbf{j}$

41. $\alpha = 3/2, \ \beta = 1/2$

43. a) $(5\cos 60°, 5\sin 60°) = \left(\dfrac{5}{2}, \dfrac{5\sqrt{3}}{2}\right)$

b) $(5\cos 60° + 10\cos 315°, \ 5\sin 60° + 10\sin 315°)$

$= \left(\dfrac{5 + \sqrt{2}}{2}, \dfrac{5\sqrt{3} - 10\sqrt{2}}{2}\right)$

45. The slope of $-\mathbf{v} = -a\mathbf{i} - b\mathbf{j}$ is $(-b)/(-a) = b/a$, the same as the slope of $\mathbf{v}$.

Section 10.2, pp. 804–806

1. The line through the point $(2, 3, 0)$ parallel to the z-axis

3. The x-axis   **5.** The circle $x^2 + y^2 = 4$ in the xy-plane

7. The circle $x^2 + z^2 = 4$ in the xz-plane

9. The circle $y^2 + z^2 = 1$ in the yz-plane

11. The circle $x^2 + y^2 = 16$ in the xy-plane

13. a) The first quadrant of the xy-plane

b) The fourth quadrant of the xy-plane

15. a) The ball of radius 1 centered at the origin

b) All points greater than 1 unit from the origin

17. a) The upper hemisphere of radius 1 centered at the origin

b) The solid upper hemisphere of radius 1 centered at the origin

19. a) $x = 3$ b) $y = -1$ c) $z = -2$ **21.** a) $z = 1$ b) $x = 3$

c) $y = -1$ **23.** a) $x^2 + (y - 2)^2 = 4, \ z = 0$

b) $(y - 2)^2 + z^2 = 4, \ x = 0$ c) $x^2 + z^2 = 4, \ y = 2$

25. a) $y = 3, \ z = -1$ b) $x = 1, \ z = -1$ c) $x = 1, \ y = 3$

27. $x^2 + y^2 + z^2 = 25, \ z = 3$ **29.** $0 \le z \le 1$ **31.** $z \le 0$

33. a) $(x - 1)^2 + (y - 1)^2 + (z - 1)^2 < 1$

b) $(x - 1)^2 + (y - 1)^2 + (z - 1)^2 > 1$ **35.** $3\left(\dfrac{2}{3}\mathbf{i} + \dfrac{1}{3}\mathbf{j} - \dfrac{2}{3}\mathbf{k}\right)$

37. $9\left(\dfrac{1}{9}\mathbf{i} + \dfrac{4}{9}\mathbf{j} - \dfrac{8}{9}\mathbf{k}\right)$ **39.** $5(\mathbf{k})$ **41.** $1\left(\dfrac{3}{5}\mathbf{i} + \dfrac{4}{5}\mathbf{k}\right)$

43. $\sqrt{\dfrac{1}{2}}\left(\dfrac{1}{\sqrt{3}}\mathbf{i}-\dfrac{1}{\sqrt{3}}\mathbf{j}-\dfrac{1}{\sqrt{3}}\mathbf{k}\right)$ **45.** a) $2\mathbf{i}$ b) $-\sqrt{3}\mathbf{k}$

c) $\dfrac{3}{10}\mathbf{j}+\dfrac{2}{5}\mathbf{k}$ d) $6\mathbf{i}-2\mathbf{j}+3\mathbf{k}$ **47.** $\dfrac{7}{13}(12\mathbf{i}-5\mathbf{k})$

49. $-\dfrac{10}{7}\mathbf{i}+\dfrac{15}{7}\mathbf{j}-\dfrac{30}{7}\mathbf{k}$ **51.** a) 3   b) $\dfrac{2}{3}\mathbf{i}+\dfrac{2}{3}\mathbf{j}-\dfrac{1}{3}\mathbf{k}$

c) $(2, 2, 1/2)$ **53.** a) 7   b) $\dfrac{3}{7}\mathbf{i}-\dfrac{6}{7}\mathbf{j}+\dfrac{2}{7}\mathbf{k}$ c) $(5/2, 1, 6)$

55. a) $2\sqrt{3}$ b) $\dfrac{1}{\sqrt{3}}\mathbf{i}-\dfrac{1}{\sqrt{3}}\mathbf{j}-\dfrac{1}{\sqrt{3}}\mathbf{k}$ c) $(1, -1, -1)$

57. $A(4, -3, 5)$ **59.** $C(-2, 0, 2)$, $a = 2\sqrt{2}$
61. $C(\sqrt{2}, \sqrt{2}, -\sqrt{2})$, $a = \sqrt{2}$
63. $(x-1)^2 + (y-2)^2 + (z-3)^2 = 14$
65. $(x+2)^2 + y^2 + z^2 = 3$ **67.** $C(-2, 0, 2)$, $a = \sqrt{8}$
69. $C\left(-\dfrac{1}{4}, -\dfrac{1}{4}, -\dfrac{1}{4}\right)$, $a = \dfrac{5\sqrt{3}}{4}$
71. a) $\sqrt{y^2+z^2}$ b) $\sqrt{x^2+z^2}$ c) $\sqrt{x^2+y^2}$
73. a) $\dfrac{3}{2}\mathbf{i}+\dfrac{3}{2}\mathbf{j}-3\mathbf{k}$ b) $\mathbf{i}+\mathbf{j}-2\mathbf{k}$ c) $(2, 2, 1)$

Section 10.3, pp. 812–814

1. a) $-25, 5, 5$ b) -1 c) -5 d) $-2\mathbf{i}+4\mathbf{j}-\sqrt{5}\mathbf{k}$

3. a) $25, 15, 5$ b) $1/3$ c) $5/3$ d) $\dfrac{1}{9}(10\mathbf{i}+11\mathbf{j}-2\mathbf{k})$

5. a) $0, \sqrt{53}, 1$ b) 0   c) 0   d) 0   **7.** a) $2, \sqrt{34}, \sqrt{3}$
b) $\dfrac{2}{\sqrt{3}\sqrt{34}}$ c) $\dfrac{2}{\sqrt{34}}$ d) $\dfrac{1}{17}(5\mathbf{j}-3\mathbf{k})$ **9.** a) $\sqrt{3}-\sqrt{2}, \sqrt{2}, 3$
b) $\dfrac{\sqrt{3}-\sqrt{2}}{3\sqrt{2}}$ c) $\dfrac{\sqrt{3}-\sqrt{2}}{\sqrt{2}}$ d) $\dfrac{\sqrt{3}-\sqrt{2}}{2}(-\mathbf{i}+\mathbf{j})$

11. $\left(\dfrac{3}{2}\mathbf{i}+\dfrac{3}{2}\mathbf{j}\right)+\left(-\dfrac{3}{2}\mathbf{i}+\dfrac{3}{2}\mathbf{j}+4\mathbf{k}\right)$

13. $\left(\dfrac{14}{3}\mathbf{i}+\dfrac{28}{3}\mathbf{j}-\dfrac{14}{3}\mathbf{k}\right)+\left(\dfrac{10}{3}\mathbf{i}-\dfrac{16}{3}\mathbf{j}-\dfrac{22}{3}\mathbf{k}\right)$

15. The sum of two vectors of equal length is *always* orthogonal to their difference, as we can see from the equation

$$(\mathbf{v}_1+\mathbf{v}_2)\cdot(\mathbf{v}_1-\mathbf{v}_2)=\mathbf{v}_1\cdot\mathbf{v}_1+\mathbf{v}_2\cdot\mathbf{v}_1-\mathbf{v}_1\cdot\mathbf{v}_2-\mathbf{v}_2\cdot\mathbf{v}_2$$

$$=|\mathbf{v}_1|^2-|\mathbf{v}_2|^2.$$

21. $\tan^{-1}\sqrt{2}$ **23.** 0.75 rad   **25.** 1.77 rad
27. $\angle A \approx 1.24$ rad, $\angle B \approx 0.66$ rad, $\angle C \approx 1.24$ rad   **29.** 0.62 rad
31. a) Since $|\cos\theta| \le 1$, we have

$$|\mathbf{u}\cdot\mathbf{v}|=|\mathbf{u}||\mathbf{v}||\cos\theta|\le|\mathbf{u}||\mathbf{v}|(1)=|\mathbf{u}||\mathbf{v}|.$$

b) We have equality precisely when $|\cos\theta|=1$ or when one or both of $\mathbf{u}$ and $\mathbf{v}$ are $\mathbf{0}$. In the case of nonzero vectors, we have equality when $\theta=0$ or π, i.e., when the vectors are parallel.
33. a   **35.** a) $\sqrt{70}$ b) $\sqrt{568}$ **37.** 5 J   **39.** 3464.10 J

43. $x + 2y = 4$

45. $-2x + y = -3$

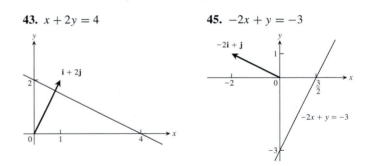

47. $x + y = -1$

49. $2x - y = 0$

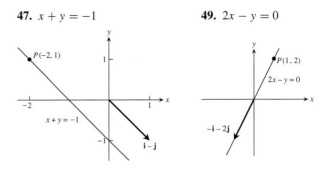

51. $\pi/4$ **53.** $\pi/6$ **55.** 0.14   **57.** $\pi/3$ and $2\pi/3$ at each point
59. At $(0, 0)$: $\pi/2$, at $(1, 1)$: $\pi/4$ and $3\pi/4$

Section 10.4, pp. 820–821

1. $|\mathbf{A}\times\mathbf{B}|=3$, direction is $\dfrac{2}{3}\mathbf{i}+\dfrac{1}{3}\mathbf{j}+\dfrac{2}{3}\mathbf{k}$; $|\mathbf{B}\times\mathbf{A}|=3$, direction is $-\dfrac{2}{3}\mathbf{i}-\dfrac{1}{3}\mathbf{j}-\dfrac{2}{3}\mathbf{k}$
3. $|\mathbf{A}\times\mathbf{B}|=0$, no direction; $|\mathbf{B}\times\mathbf{A}|=0$, no direction
5. $|\mathbf{A}\times\mathbf{B}|=6$, direction is $-\mathbf{k}$; $|\mathbf{B}\times\mathbf{A}|=6$, direction is $\mathbf{k}$
7. $|\mathbf{A}\times\mathbf{B}|=6\sqrt{5}$, direction is $\dfrac{1}{\sqrt{5}}\mathbf{i}-\dfrac{2}{\sqrt{5}}\mathbf{k}$; $|\mathbf{B}\times\mathbf{A}|=6\sqrt{5}$, direction is $-\dfrac{1}{\sqrt{5}}\mathbf{i}+\dfrac{2}{\sqrt{5}}\mathbf{k}$

9.

11.

13.

15. a) $2\sqrt{6}$ b) $\pm\dfrac{1}{\sqrt{6}}(2\mathbf{i}+\mathbf{j}+\mathbf{k})$ **17.** a) $\dfrac{\sqrt{2}}{2}$ b) $\pm\dfrac{1}{\sqrt{2}}(\mathbf{i}-\mathbf{j})$

19. a) None b) **A** and **C** **21.** $10\sqrt{3}$ ft·lb **23.** 8 **25.** 7

27. a) True b) Not always true c) True d) True
e) Not always true f) True g) True h) True

29. a) $\text{proj}_\mathbf{B}\,\mathbf{A} = \dfrac{\mathbf{A}\cdot\mathbf{B}}{\mathbf{B}\cdot\mathbf{B}}\,\mathbf{B}$ b) $\pm\mathbf{A}\times\mathbf{B}$ c) $\pm(\mathbf{A}\times\mathbf{B})\times\mathbf{C}$

d) $|(\mathbf{A}\times\mathbf{B})\cdot\mathbf{C}|$ **31.** a) Yes b) No c) Yes d) No

33. No, **B** need not equal **C**. For example, $\mathbf{i}+\mathbf{j}\neq-\mathbf{i}+\mathbf{j}$, but

$$\mathbf{i}\times(\mathbf{i}+\mathbf{j}) = \mathbf{i}\times\mathbf{i}+\mathbf{i}\times\mathbf{j} = \mathbf{0}+\mathbf{k}=\mathbf{k}$$

$$\mathbf{i}\times(-\mathbf{i}+\mathbf{j}) = -\mathbf{i}\times\mathbf{i}+\mathbf{i}\times\mathbf{j} = \mathbf{0}+\mathbf{k}=\mathbf{k}$$

35. 2 **37.** 13 **39.** 11/2 **41.** 25/2

43. If $\mathbf{A} = a_1\mathbf{i}+a_2\mathbf{j}$ and $\mathbf{B} = b_1\mathbf{i}+b_2\mathbf{j}$, then

$$\mathbf{A}\times\mathbf{B} = \begin{vmatrix} \mathbf{i} & \mathbf{j} & \mathbf{k} \\ a_1 & a_2 & 0 \\ b_1 & b_2 & 0 \end{vmatrix} = \begin{vmatrix} a_1 & a_2 \\ b_1 & b_2 \end{vmatrix}\mathbf{k}$$

and the triangle's area is

$$\frac{1}{2}|\mathbf{A}\times\mathbf{B}| = \pm\frac{1}{2}\begin{vmatrix} a_1 & a_2 \\ b_1 & b_2 \end{vmatrix}.$$

The applicable sign is $(+)$ if the acute angle from **A** to **B** runs counterclockwise in the xy-plane, and $(-)$ if it runs clockwise.

Section 10.5, pp. 827–829

1. $x = 3+t,\ y = -4+t,\ z = -1+t$

3. $x = -2+5t,\ y = 5t,\ z = 3-5t$ **5.** $x = 0,\ y = 2t,\ z = t$

7. $x = 1,\ y = 1,\ z = 1+t$ **9.** $x = t,\ y = -7+2t,\ z = 2t$

11. $x = t,\ y = 0,\ z = 0$

13. $x = t,\ y = t,\ z = \dfrac{3}{2}t,\ 0 \le t \le 1$

15. $x = 1,\ y = 1+t,\ z = 0,\ -1 \le t \le 0$

17. $x = 0,\ y = 1-2t,\ z = 1,\ 0 \le t \le 1$

19. $x = 2-2t,\ y = 2t,\ z = 2-2t,\ 0 \le t \le 1$

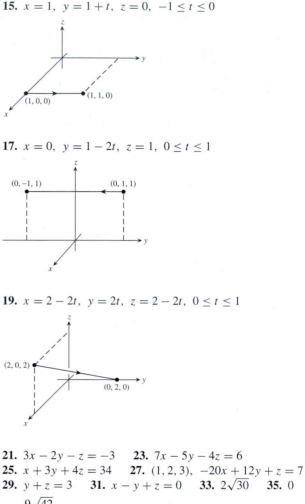

21. $3x-2y-z = -3$ **23.** $7x-5y-4z = 6$

25. $x+3y+4z = 34$ **27.** $(1,2,3),\ -20x+12y+z = 7$

29. $y+z = 3$ **31.** $x-y+z = 0$ **33.** $2\sqrt{30}$ **35.** 0

37. $\dfrac{9\sqrt{42}}{7}$ **39.** 3 **41.** 19/5 **43.** 5/3 **45.** $9/\sqrt{41}$

47. $\pi/4$ **49.** 1.76 rad **51.** 0.82 rad **53.** $\left(\dfrac{3}{2}, -\dfrac{3}{2}, \dfrac{1}{2}\right)$

55. $(1,1,0)$ **57.** $x = 1-t,\ y = 1+t,\ z = -1$

59. $x = 4,\ y = 3+6t,\ z = 1+3t$

61. $L1$ intersects $L2$; $L2$ is parallel to $L3$; $L1$ and $L3$ are skew.

63. $x = 2+2t,\ y = -4-t,\ z = 7+3t;\ x = -2-t,$
$y = -2+(1/2)t,\ z = 1-(3/2)t$

65. $\left(0, -\dfrac{1}{2}, -\dfrac{3}{2}\right),\ (-1,0,-3),\ (1,-1,0)$

69. Many possible answers. One possibility: $x+y = 3$ and $2y+z = 7$

71. $(x/a) + (y/b) + (z/c) = 1$ describes all planes *except* those through the origin or parallel to a coordinate axis.

Section 10.6, pp. 839–841

1. d, ellipsoid **3.** a, cylinder **5.** l, hyperbolic paraboloid
7. b, cylinder **9.** k, hyperbolic paraboloid **11.** h, cone

13.

$x^2 + y^2 = 4$

15.

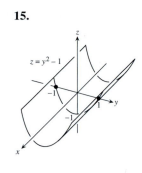

$z = y^2 - 1$

29.

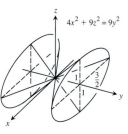

$x = 4 - 4y^2 - z^2$

31.

$x^2 + y^2 = z^2$

33.

$4x^2 + 9z^2 = 9y^2$

35.

$x^2 + y^2 - z^2 = 1$

17.

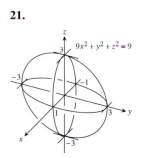

$x^2 + 4z^2 = 16$

19.

$z^2 - y^2 = 1$

37.

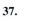

$\dfrac{y^2}{4} + \dfrac{z^2}{9} - \dfrac{x^2}{4} = 1$

39.

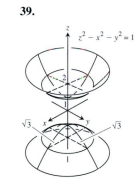

$z^2 - x^2 - y^2 = 1$

21.

$9x^2 + y^2 + z^2 = 9$

23.

$4x^2 + 9y^2 + 4z^2 = 36$

25.

$z = x^2 + 4y^2$

27.

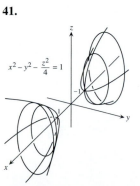

$z = 8 - x^2 - y^2$

41.

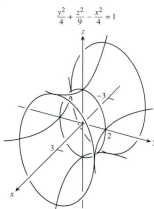

$x^2 - y^2 - \dfrac{z^2}{4} = 1$

43.

$y^2 - x^2 = z$

45.

$x^2 + y^2 + z^2 = 4$

47.

$z = 1 + y^2 - x^2$

49.

$y = -(x^2 + z^2)$

51.

$16x^2 + 4y^2 = 1$

53.

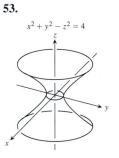

$x^2 + y^2 - z^2 = 4$

55.

$x^2 + z^2 = y$

57.

$x^2 + z^2 = 1$

59.

$16y^2 + 9z^2 = 4x^2$

61.

$9x^2 + 4y^2 + z^2 = 36$

63.

$x^2 + y^2 - 16z^2 = 16$

65.

$z = -(x^2 + y^2)$

67.

$x^2 - 4y^2 = 1$

69.

$4y^2 + z^2 - 4x^2 = 4$

71.

$x^2 + y^2 = z$

73.

$yz = 1$

75.

$9x^2 + 16y^2 = 4z^2$

77. a) $\dfrac{2\pi(9 - c^2)}{9}$ b) 8π c) $\dfrac{4\pi abc}{3}$

81. Vertex $(0, y_1, cy_1^2/b^2)$, focus $(0, y_1, c(y_1^2/b^2) - a^2/(4c))$

Section 10.7, pp. 846–847

	Rectangular	Cylindrical	Spherical
1.	$(0, 0, 0)$	$(0, 0, 0)$	$(0, 0, 0)$
3.	$(0, 1, 0)$	$(1, \pi/2, 0)$	$(1, \pi/2, \pi/2)$
5.	$(1, 0, 0)$	$(1, 0, 0)$	$(1, \pi/2, 0)$
7.	$(0, 1, 1)$	$(1, \pi/2, 1)$	$(\sqrt{2}, \pi/4, \pi/2)$
9.	$(0, -2\sqrt{2}, 0)$	$(2\sqrt{2}, 3\pi/2, 0)$	$(2\sqrt{2}, \pi/2, 3\pi/2)$

11. $x^2 + y^2 = 0$, $\theta = 0$ or $\theta = \pi$, the z-axis

13. $z = 0$, $\phi = \pi/2$, the xy-plane

15. $z = r$, $0 \le r \le 1$; $\phi = \pi/4$, $0 \le \rho \le \sqrt{2}$; a (finite) cone

17. $x = 0$, $\theta = \pi/2$, the yz-plane

19. $r^2 + z^2 = 4$, $\rho = 2$, sphere of radius 2 centered at the origin

21. $x^2 + y^2 + \left(z - \dfrac{5}{2}\right)^2 = \dfrac{25}{4}$, $r^2 + z^2 = 5z$, sphere of radius $5/2$ centered at $(0, 0, 5/2)$ (rectangular)

23. $y = 1$, $\rho \sin \phi \sin \theta = 1$, the plane $y = 1$
25. $z = \sqrt{2}$, the plane $z = \sqrt{2}$
27. $r^2 + z^2 = 2z$, $z \leq 1$; $\rho = 2 \cos \phi$, $\pi/4 \leq \phi \leq \pi/2$; lower half (hemisphere) of the sphere of radius 1 centered at $(0, 0, 1)$ (rectangular)
29. $x^2 + y^2 + z^2 = 9$, $-3/2 \leq z \leq 3/2$; $r^2 + z^2 = 9$, $-3/2 \leq z \leq 3/2$; the portion of the sphere of radius 3 centered at the origin between the planes $z = -3/2$ and $z = 3/2$
31. $z = 4 - 4(x^2 + y^2)$, $0 \leq z \leq 4$; $\rho \cos \phi = 4 - 4\rho^2 \sin^2 \phi$, $0 \leq \phi \leq \pi/2$; the upper portion cut from the paraboloid $z = 4 - 4(x^2 + y^2)$ by the xy-plane
33. $z = -\sqrt{x^2 + y^2}$, $-1 \leq z \leq 0$; $z = -r$, $0 \leq r \leq 1$; cone, vertex at origin, base the circle $x^2 + y^2 = 1$ in the plane $z = -1$
35. $z + x^2 - y^2 = 0$ or $z = y^2 - x^2$, $\cos \phi + \rho \sin^2 \phi \cos 2\theta = 0$, hyperbolic paraboloid
37. $(2, 3, 1)$
39. Right circular cylinder parallel to the z-axis generated by the circle $r = -2 \sin \theta$ in the $r\theta$-plane

41. Cylinder of lines parallel to the z-axis generated by the cardioid $r = 1 - \cos \theta$ in the $r\theta$-plane

43. Cardioid of revolution symmetric about the y-axis, cusp at the origin pointing down

45. b) $\phi = \pi/2$
49. The surface's equation $r = f(z)$ tells us that the point $(r, \theta, z) = (f(z), \theta, z)$ will lie on the surface for all θ. In particular $(f(z), \theta +$
$\pi, z)$ lies on the surface whenever $(f(z), \theta, z)$ lies on the surface, so the surface is symmetric with respect to the z-axis.

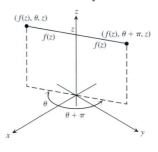

Chapter 10 Practice Exercises, pp. 848–851

1.

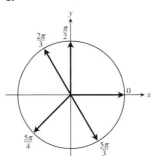

3. $2 \cdot \left(\dfrac{1}{\sqrt{2}} \mathbf{i} + \dfrac{1}{\sqrt{2}} \mathbf{j} \right)$ **5.** $7 \cdot \left(\dfrac{2}{7} \mathbf{i} - \dfrac{3}{7} \mathbf{j} + \dfrac{6}{7} \mathbf{k} \right)$

7. $\dfrac{8}{\sqrt{33}} \mathbf{i} - \dfrac{2}{\sqrt{33}} \mathbf{j} + \dfrac{8z}{\sqrt{33}} \mathbf{k}$

9. a) $\overrightarrow{BD} = \overrightarrow{AD} - \overrightarrow{AB}$

  b) $\overrightarrow{AP} = \dfrac{1}{2} (\overrightarrow{AB} + \overrightarrow{AD})$

11.

13. $|A| = \sqrt{2}$, $|B| = 3$, $\mathbf{A} \cdot \mathbf{B} = \mathbf{B} \cdot \mathbf{A} = 3$, $\mathbf{A} \times \mathbf{B} = -2\mathbf{i} + 2\mathbf{j} - \mathbf{k}$, $\mathbf{B} \times \mathbf{A} = 2\mathbf{i} - 2\mathbf{j} + \mathbf{k}$, $|\mathbf{A} \times \mathbf{B}| = 3$, $\theta = \pi/4$, $|\mathbf{B}| \cos \theta = 3/\sqrt{2}$, $\text{proj}_A \mathbf{B} = (3/2)(\mathbf{i} + \mathbf{j})$
15. $\dfrac{4}{3}(2\mathbf{i} + \mathbf{j} - \mathbf{k}) - \dfrac{1}{3}(5\mathbf{i} + \mathbf{j} + 11\mathbf{k})$

17.

19. unit tangents $\pm\left(\dfrac{1}{\sqrt{5}}\mathbf{i}+\dfrac{2}{\sqrt{5}}\mathbf{j}\right)$, unit normals

$\pm\left(-\dfrac{2}{\sqrt{5}}\mathbf{i}+\dfrac{1}{\sqrt{5}}\mathbf{j}\right)$

23. $2\sqrt{7}$ **25.** a) $A=\sqrt{14}$ b) $V=1$ **29.** $\sqrt{78}/3$
31. $x=1-3t,\ y=2,\ z=3+7t$ **33.** $\sqrt{2}$
35. $2x+y-z=3$ **37.** $-9x+y+7z=4$

39. $\left(0,-\dfrac{1}{2},-\dfrac{3}{2}\right)$, $(-1,0,-3)$, $(1,-1,0)$

41. $\pi/3$ **43.** $x=-5+5t,\ y=3-t,\ z=-3t$
45. b) $x=-12t,\ y=19/12+15t,\ z=1/6+6t$
47. Yes; $\mathbf{v}$ is parallel to the plane.   **49.** 3   **51.** $-3\mathbf{j}+3\mathbf{k}$

53. $\dfrac{2}{\sqrt{35}}(5\mathbf{i}-\mathbf{j}-3\mathbf{k})$ **55.** $\left(\dfrac{11}{9},\dfrac{26}{9},\dfrac{7}{9}\right)$

57. $(1,-2,-1)$; $x=1-5t,\ y=-2+3t,\ z=-1+4t$
59. $2x+7y+2z+10=0$ **61.** a) No b) no c) no d) no
e) yes   **63.** $11/\sqrt{107}$
65.

67.

69.

71.

73.

75.

77. The y-axis in the xy-plane; the yz-plane in three dimensional space
79. The circle centered at $(0,0)$ with radius 2 in the xy-plane; the cylinder parallel to the z-axis in three dimensional space with the circle as a generating curve
81. The parabola $x=y^2$ in the xy-plane; the cylinder parallel to the z-axis in three dimensional space with the parabola as a generating curve
83. A cardioid in the $r\theta$-plane; a cylinder parallel with the z-axis in three dimensional space with the cardioid as a generating curve
85. A horizontal lemniscate of length $2\sqrt{2}$ in the $r\theta$-plane; the cylinder parallel to the z-axis in three dimensional space with the lemniscate as a generating curve
87. The sphere of radius 2 centered at the origin
89. The upper nappe of the cone having its vertex at the origin and making a $\pi/6$ angle with the z-axis
91. The upper hemisphere of the sphere of radius 1 centered at the origin

	Rectangular	Cylindrical	Spherical
93.	$(1,0,0)$	$(1,0,0)$	$(1,\pi/2,0)$
95.	$(0,1,1)$	$(1,\pi/2,1)$	$(\sqrt{2},\pi/4,\pi/2)$
97.	$(-1,0,-1)$	$(1,\pi,-1)$	$(\sqrt{2},3\pi/4,\pi)$

99. Cylindrical: $z=2$, spherical: $\rho\cos\phi=2$, a plane parallel with the xy-plane
101. Cylindrical: $r^2+z^2=-2z$, spherical: $\rho=-2\cos\phi$, sphere of radius 1 centered at $(0,0,-1)$ (rectangular)

103. Rectangular: $z=x^2+y^2$, spherical: $\rho=0$ or $\rho=\dfrac{\cos\phi}{\sin^2\phi}$ when
$0<\phi<\pi/2$, a paraboloid symmetric to the z-axis, opening upward, vertex at the origin
105. Rectangular: $x^2+(y-7/2)^2=49/4$, spherical: $\rho\sin\phi=7\sin\theta$, cylinder parallel to the z-axis generated by the circle
107. Rectangular: $x^2+y^2+z^2=16$, cylindrical: $r^2+z^2=16$, sphere of radius 4 centered at the origin
109. Rectangular: $-\sqrt{x^2+y^2}=z$, cylindrical: $z=-r,\ r\geq0$, single cone making an angle of $3\pi/4$ with the positive z-axis, vertex at the origin

Chapter 10 Additional Exercises, pp. 851–853

1. $(26,23,-1/3)$ **3.** $\approx34{,}641$ J

17. b) $6/\sqrt{14}$ c) $2x - y + 2z = 8$
d) $x - 2y + z = 3 + 5\sqrt{6}$ and $x - 2y + z = 3 - 5\sqrt{6}$

23. $\mathbf{v} - 2\dfrac{\mathbf{v \cdot z}}{|\mathbf{z}|^2}\,\mathbf{z}$ **25.** a) $|\mathbf{F}| = \dfrac{GMm}{d^2}\left(1 + \displaystyle\sum_{i=1}^{n} \dfrac{2}{(i^2+1)^{3/2}}\right)$
b) Yes

CHAPTER 11

Section 11.1, pp. 865–868

1. $y = x^2 - 2x$, $\mathbf{v} = \mathbf{i} + 2\mathbf{j}$, $\mathbf{a} = 2\mathbf{j}$

3. $y = \dfrac{2}{9}x^2$, $\mathbf{v} = 3\mathbf{i} + 4\mathbf{j}$, $\mathbf{a} = 3\mathbf{i} + 8\mathbf{j}$

5. $t = \dfrac{\pi}{4} : \mathbf{v} = \dfrac{\sqrt{2}}{2}\mathbf{i} - \dfrac{\sqrt{2}}{2}\mathbf{j}$, $\mathbf{a} = \dfrac{-\sqrt{2}}{2}\mathbf{i} - \dfrac{\sqrt{2}}{2}\mathbf{j}$;
$t = \pi/2 : \mathbf{v} = -\mathbf{j}$, $\mathbf{a} = -\mathbf{i}$

7. $t = \pi : \mathbf{v} = 2\mathbf{i}$, $\mathbf{a} = -\mathbf{j}$; $t = \dfrac{3\pi}{2} : \mathbf{v} = \mathbf{i} - \mathbf{j}$, $\mathbf{a} = -\mathbf{i}$

9. $\mathbf{v} = \mathbf{i} + 2t\,\mathbf{j} + 2\mathbf{k}$; $\mathbf{a} = 2\mathbf{j}$; speed: 3; direction: $\dfrac{1}{3}\mathbf{i} + \dfrac{2}{3}\mathbf{j} + \dfrac{2}{3}\mathbf{k}$;

$\mathbf{v}(1) = 3\left(\dfrac{1}{3}\mathbf{i} + \dfrac{2}{3}\mathbf{j} + \dfrac{2}{3}\mathbf{k}\right)$

11. $\mathbf{v} = (-2\sin t)\mathbf{i} + (3\cos t)\mathbf{j} + 4\mathbf{k}$;
$\mathbf{a} = (-2\cos t)\mathbf{i} - (3\sin t)\mathbf{j}$; speed: $2\sqrt{5}$;
direction: $(-1/\sqrt{5})\mathbf{i} + (2/\sqrt{5})\mathbf{k}$;
$\mathbf{v}(\pi/2) = 2\sqrt{5}[(-1/\sqrt{5})\mathbf{i} + (2/\sqrt{5})\mathbf{k}]$

13. $\mathbf{v} = \left(\dfrac{2}{t+1}\right)\mathbf{i} + 2t\,\mathbf{j} + t\,\mathbf{k}$; $\mathbf{a} = \left(\dfrac{-2}{(t+1)^2}\right)\mathbf{i} + 2\mathbf{j} + \mathbf{k}$;

speed: $\sqrt{6}$; direction: $\dfrac{1}{\sqrt{6}}\mathbf{i} + \dfrac{2}{\sqrt{6}}\mathbf{j} + \dfrac{1}{\sqrt{6}}\mathbf{k}$;

$\mathbf{v}(1) = \sqrt{6}\left(\dfrac{1}{\sqrt{6}}\mathbf{i} + \dfrac{2}{\sqrt{6}}\mathbf{j} + \dfrac{1}{\sqrt{6}}\mathbf{k}\right)$

15. $\pi/2$ **17.** $\pi/2$ **19.** $t = 0$, π, 2π

21. $(1/4)\mathbf{i} + 7\mathbf{j} + (3/2)\mathbf{k}$ **23.** $\left(\dfrac{\pi + 2\sqrt{2}}{2}\right)\mathbf{j} + 2\mathbf{k}$

25. $(\ln 4)\mathbf{i} + (\ln 4)\mathbf{j} + (\ln 2)\mathbf{k}$

27. $\mathbf{r}(t) = \left(\dfrac{-t^2}{2} + 1\right)\mathbf{i} + \left(\dfrac{-t^2}{2} + 2\right)\mathbf{j} + \left(\dfrac{-t^2}{2} + 3\right)\mathbf{k}$

29. $\mathbf{r}(t) = ((t+1)^{3/2} - 1)\mathbf{i} + (-e^{-t} + 1)\mathbf{j} + (\ln(t+1) + 1)\mathbf{k}$

31. $\mathbf{r}(t) = 8t\,\mathbf{i} + 8t\,\mathbf{j} + (-16t^2 + 100)\mathbf{k}$

33. $x = t$, $y = -1$, $z = 1 + t$

35. $x = at$, $y = a$, $z = 2\pi b + bt$

37. a) (i): It has constant speed 1   (ii): Yes   (iii): Counterclockwise
(iv): Yes   b) (i): It has constant speed 2   (ii): Yes   (iii): Counterclockwise   (iv): Yes   c) (i): It has constant speed 1   (ii): Yes
(iii): Counterclockwise   (iv): It starts at $(0, -1)$ instead of $(1, 0)$
d) (i): It has constant speed 1   (ii): Yes   (iii): Clockwise   (iv): Yes
e) (i): It has variable speed   (ii): No   (iii): Counterclockwise
(iv): Yes

39. $\mathbf{r}(t) = \left(\dfrac{3}{2}t^2 + \dfrac{6}{\sqrt{11}}t + 1\right)\mathbf{i} - \left(\dfrac{1}{2}t^2 + \dfrac{2}{\sqrt{11}}t - 2\right)\mathbf{j} +$

$\left(\dfrac{1}{2}t^2 + \dfrac{2}{\sqrt{11}}t + 3\right)\mathbf{k} = \left(\dfrac{1}{2}t^2 + \dfrac{2t}{\sqrt{11}}\right)(3\mathbf{i} - \mathbf{j} + \mathbf{k}) +$

$(\mathbf{i} + 2\mathbf{j} + 3\mathbf{k})$

41. $\mathbf{v} = 2\sqrt{5}\,\mathbf{i} + \sqrt{5}\,\mathbf{j}$

43. max $|\mathbf{v}| = 3$, min $|\mathbf{v}| = 2$, max $|\mathbf{a}| = 3$, min $|\mathbf{a}| = 2$

Section 11.2, pp. 873–876

1. 50 sec   **3.** a) 72.2 sec, 25,510 m   b) 4020 m   c) 6378 m
5. $t \approx 2.135$ sec, $x \approx 66.42$ ft
7. $v_0 = 9.9$ m/sec, $\alpha = 18.4°$ or $71.6°$ **9.** 190 mph
11. The golf ball will clip the leaves at the top.   **13.** 46.6 ft/sec
17. 141%   **21.** 1.92 sec, 73.7 ft (approx.)

25. $\mathbf{v}(t) = -gt\,\mathbf{k} + \mathbf{v}_0$, $\mathbf{r}(t) = -\dfrac{1}{2}gt^2\,\mathbf{k} + \mathbf{v}_0 t$

Section 11.3, pp. 880–881

1. $\mathbf{T} = \left(-\dfrac{2}{3}\sin t\right)\mathbf{i} + \left(\dfrac{2}{3}\cos t\right)\mathbf{j} + \dfrac{\sqrt{5}}{3}\mathbf{k}$, 3π

3. $\mathbf{T} = \dfrac{1}{\sqrt{1+t}}\mathbf{i} + \dfrac{\sqrt{t}}{\sqrt{1+t}}\mathbf{k}$, $\dfrac{52}{3}$ **5.** $\mathbf{T} = -\cos t\,\mathbf{j} + \sin t\,\mathbf{k}$, $\dfrac{3}{2}$

7. $\mathbf{T} = \left(\dfrac{\cos t - t \sin t}{t + 1}\right)\mathbf{i} + \left(\dfrac{\sin t + t \cos t}{t + 1}\right)\mathbf{j} + \left(\dfrac{\sqrt{2}\,t^{1/2}}{t + 1}\right)\mathbf{k}$,

$\dfrac{\pi^2}{2} + \pi$

9. $(0, 5, 24\pi)$ **11.** $s(t) = 5t$, $L = \dfrac{5\pi}{2}$

13. $s(t) = \sqrt{3}e^t - \sqrt{3}$, $L = \dfrac{3\sqrt{3}}{4}$ **15.** $\sqrt{2} + \ln(1 + \sqrt{2})$

17. a) Cylinder is $x^2 + y^2 = 1$, plane is $x + z = 1$

b) and c)

(0, −1, 1) (−1, 0, 2) (1, 0, 0) (0, 1, 1)

d) $L = \int_0^{2\pi} \sqrt{1 + \sin^2 t}\, dt$ e) $L \approx 7.64$

Section 11.4, pp. 890–893

1. $\mathbf{T} = (\cos t)\mathbf{i} - (\sin t)\mathbf{j}$, $\mathbf{N} = (-\sin t)\mathbf{i} - (\cos t)\mathbf{j}$, $\kappa = \cos t$

3. $\mathbf{T} = \dfrac{1}{\sqrt{1+t^2}}\mathbf{i} - \dfrac{t}{\sqrt{1+t^2}}\mathbf{j}$, $\mathbf{N} = \dfrac{-t}{\sqrt{1+t^2}}\mathbf{i} - \dfrac{1}{\sqrt{1+t^2}}\mathbf{j}$,

$\kappa = \dfrac{1}{2\left(\sqrt{1+t^2}\right)^3}$

5. $\mathbf{a} = \dfrac{2t}{\sqrt{1+t^2}}\mathbf{T} + \dfrac{2}{\sqrt{1+t^2}}\mathbf{N}$ **7.** b) $\cos x$

9. b) $\mathbf{N} = \dfrac{-2e^{2t}}{\sqrt{1+4e^{4t}}}\mathbf{i} + \dfrac{1}{\sqrt{1+4e^{4t}}}\mathbf{j}$

c) $\mathbf{N} = -\dfrac{1}{2}\left(\sqrt{4-t^2}\,\mathbf{i} + t\,\mathbf{j}\right)$

11. $\mathbf{T} = \dfrac{3\cos t}{5}\mathbf{i} - \dfrac{3\sin t}{5}\mathbf{j} + \dfrac{4}{5}\mathbf{k}$, $\mathbf{N} = (-\sin t)\mathbf{i} - (\cos t)\mathbf{j}$,

$\mathbf{B} = \left(\dfrac{4}{5}\cos t\right)\mathbf{i} - \left(\dfrac{4}{5}\sin t\right)\mathbf{j} - \dfrac{3}{5}\mathbf{k}$, $\kappa = \dfrac{3}{25}$, $\tau = -\dfrac{4}{25}$

13. $\mathbf{T} = \left(\dfrac{\cos t - \sin t}{\sqrt{2}}\right)\mathbf{i} + \left(\dfrac{\cos t + \sin t}{\sqrt{2}}\right)\mathbf{j}$,

$\mathbf{N} = \left(\dfrac{-\cos t - \sin t}{\sqrt{2}}\right)\mathbf{i} + \left(\dfrac{-\sin t + \cos t}{\sqrt{2}}\right)\mathbf{j}$,

$\mathbf{B} = \mathbf{k}$, $\kappa = \dfrac{1}{e^t \sqrt{2}}$, $\tau = 0$

15. $\mathbf{T} = \dfrac{t}{\sqrt{t^2+1}}\mathbf{i} + \dfrac{1}{\sqrt{t^2+1}}\mathbf{j}$, $\mathbf{N} = \dfrac{\mathbf{i}}{\sqrt{t^2+1}} - \dfrac{t\,\mathbf{j}}{\sqrt{t^2+1}}$,

$\mathbf{B} = -\mathbf{k}$, $\kappa = \dfrac{1}{t(t^2+1)^{3/2}}$, $\tau = 0$

17. $\mathbf{T} = \left(\operatorname{sech}\dfrac{t}{a}\right)\mathbf{i} + \left(\tanh\dfrac{t}{a}\right)\mathbf{j}$,

$\mathbf{N} = \left(-\tanh\dfrac{t}{a}\right)\mathbf{i} + \left(\operatorname{sech}\dfrac{t}{a}\right)\mathbf{j}$,

$\mathbf{B} = \mathbf{k}$, $\kappa = \dfrac{1}{a}\operatorname{sech}^2\dfrac{t}{a}$, $\tau = 0$

19. $\mathbf{a} = |a|\,\mathbf{N}$ **21.** $\mathbf{a}(1) = \dfrac{4}{3}\mathbf{T} + \dfrac{2\sqrt{5}}{3}\mathbf{N}$ **23.** $\mathbf{a}(0) = 2\mathbf{N}$

25. $\mathbf{r}\left(\dfrac{\pi}{4}\right) = \dfrac{\sqrt{2}}{2}\mathbf{i} + \dfrac{\sqrt{2}}{2}\mathbf{j} - \mathbf{k}$, $\mathbf{T}\left(\dfrac{\pi}{4}\right) = -\dfrac{\sqrt{2}}{2}\mathbf{i} + \dfrac{\sqrt{2}}{2}\mathbf{j}$,

$\mathbf{N}\left(\dfrac{\pi}{4}\right) = -\dfrac{\sqrt{2}}{2}\mathbf{i} - \dfrac{\sqrt{2}}{2}\mathbf{j}$, $\mathbf{B}\left(\dfrac{\pi}{4}\right) = \mathbf{k}$; osculating plane: $z = -1$;

normal plane: $-x + y = 0$; rectifying plane: $x + y = \sqrt{2}$

27. Yes. If the car is moving on a curved path ($\kappa \neq 0$), then $a_N = \kappa |\mathbf{v}|^2 \neq 0$ and $\mathbf{a} \neq \mathbf{0}$.

31. $|\mathbf{F}| = \kappa\left(m\left(\dfrac{ds}{dt}\right)^2\right)$ **35.** $1/(2b)$ **39.** a) $b - a$ b) π

45. $\kappa(x) = 2/(1+4x^2)^{3/2}$ **47.** $\kappa(x) = |\sin x|/(1+\cos^2 x)^{3/2}$

57. Components of $\mathbf{v}$: $-1.8701, 0.7089, 1.0000$

Components of $\mathbf{a}$: $-1.6960, -2.0307, 0$

Speed: 2.2361; Components of $\mathbf{T}$: $-0.8364, 0.3170, 0.4472$

Components of $\mathbf{N}$: $-0.4143, -0.8998, -0.1369$

Components of $\mathbf{B}$: $0.3590, -0.2998, 0.8839$; Curvature: 0.5060

Torsion: 0.2813; Tangential component of acceleration: 0.7746

Normal component of acceleration: 2.5298

59. Components of $\mathbf{v}$: $2.0000, 0, 0.1629$

Components of $\mathbf{a}$: $0, -1.0000, 0.0086$; Speed: 2.0066

Components of $\mathbf{T}$: $0.9967, 0, 0.0812$

Components of $\mathbf{N}$: $-0.0007, -1.0000, 0.0086$

Components of $\mathbf{B}$: $0.0812, -0.0086, -0.9967$; Curvature: 0.2484

Torsion: -0.0411; Tangential component of acceleration: 0.0007

Normal component of acceleration: 1.0000

Section 11.5, pp. 901–902

1. $T = 93.2$ min   **3.** $a = 6763$ km   **5.** $T = 1655$ min

7. $a = 20{,}430$ km   **9.** $|v| = 1.9966 \times 10^7\, r^{-1/2}$ m/sec

11. Circle: $v_0 = \sqrt{\dfrac{GM}{r_0}}$; ellipse: $\sqrt{\dfrac{GM}{r_0}} < v_0 < \sqrt{\dfrac{2GM}{r_0}}$;

parabola: $v_0 = \sqrt{\dfrac{2GM}{r_0}}$; hyperbola: $v_0 > \sqrt{\dfrac{2GM}{r_0}}$

15. a) $x(t) = 2 + (3 - 4\cos(\pi t))\cos(\pi t)$, $y(t) = (3 - 4\cos(\pi t))\sin(\pi t)$

Chapter 11 Practice Exercises, pp. 902–905

1. $\dfrac{x^2}{16} + \dfrac{y^2}{2} = 1$

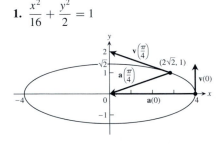

At $t = 0$: $a_\mathbf{T} = 0$, $a_\mathbf{N} = 4$, $\kappa = 2$;

At $t = \dfrac{\pi}{4}$: $a_T = \dfrac{7}{3}$, $a_N = \dfrac{4\sqrt{2}}{3}$, $\kappa = \dfrac{4\sqrt{2}}{27}$

3. $|\mathbf{v}|_{\max} = 1$ **5.** $\kappa = 1/5$ **7.** $dy/dt = -x$; clockwise

11. Shot put is on the ground, about 66 ft, 5 in. from the stopboard.

15. a) 59.19 ft/sec   b) 74.58 ft/sec   **19.** $\kappa = \pi s$

21. Length $= \dfrac{\pi}{4}\sqrt{1 + \dfrac{\pi^2}{16}} + \ln\left(\dfrac{\pi}{4} + \sqrt{1 + \dfrac{\pi^2}{16}}\right)$

23. $\mathbf{T}(0) = \dfrac{2}{3}\mathbf{i} - \dfrac{2}{3}\mathbf{j} + \dfrac{1}{3}\mathbf{k}$; $\mathbf{N}(0) = \dfrac{1}{\sqrt{2}}\mathbf{i} + \dfrac{1}{\sqrt{2}}\mathbf{j}$;

$\mathbf{B}(0) = -\dfrac{1}{3\sqrt{2}}\mathbf{i} + \dfrac{1}{3\sqrt{2}}\mathbf{j} + \dfrac{4}{3\sqrt{2}}\mathbf{k}$; $\kappa = \dfrac{\sqrt{2}}{3}$; $\tau = \dfrac{1}{6}$

25. $\mathbf{T}(\ln 2) = \dfrac{1}{\sqrt{17}}\mathbf{i} + \dfrac{4}{\sqrt{17}}\mathbf{j}$; $\mathbf{N}(\ln 2) = -\dfrac{4}{\sqrt{17}}\mathbf{i} + \dfrac{1}{\sqrt{17}}\mathbf{j}$;

$\mathbf{B}(\ln 2) = \mathbf{k}$; $\kappa = \dfrac{8}{17\sqrt{17}}$; $\tau = 0$

27. $\mathbf{a}(0) = 10\,\mathbf{T} + 6\,\mathbf{N}$

29. $\mathbf{T} = \left(\dfrac{1}{\sqrt{2}}\cos t\right)\mathbf{i} - (\sin t)\mathbf{j} + \left(\dfrac{1}{\sqrt{2}}\cos t\right)\mathbf{k}$;

$\mathbf{N} = \left(-\dfrac{1}{\sqrt{2}}\sin t\right)\mathbf{i} - (\cos t)\mathbf{j} - \left(\dfrac{1}{\sqrt{2}}\sin t\right)\mathbf{k}$;

$\mathbf{B} = \dfrac{1}{\sqrt{2}}\mathbf{i} - \dfrac{1}{\sqrt{2}}\mathbf{k}$; $\kappa = \dfrac{1}{\sqrt{2}}$; $\tau = 0$

31. $\pi/3$ **33.** $x = 1 + t$, $y = t$, $z = -t$

35. 5971 km, 1.639×10^7 km^2, 3.21% visible

Chapter 11 Additional Exercises, pp. 905–907

1. a) $\mathbf{r}(t) = \left(-\dfrac{8}{15}t^3 + 4t^2\right)\mathbf{i} + (-20t + 100)\mathbf{j}$;   b) $\dfrac{100}{3}$ m

3. $\mathbf{v} = \dfrac{\sqrt{3}}{2}\mathbf{i} + \dfrac{\sqrt{3}}{2}\mathbf{j} - \sqrt{3}\mathbf{k}$ **5.** a) $\dfrac{d\theta}{dt}\Big|_{\theta=2\pi} = 2\sqrt{\dfrac{\pi g b}{a^2 + b^2}}$

b) $\theta = \dfrac{gbt^2}{2(a^2 + b^2)}$, $z = \dfrac{gb^2 t^2}{2(a^2 + b^2)}$

c) $\mathbf{v}(t) = \dfrac{gbt}{\sqrt{a^2 + b^2}}\,\mathbf{T}$; $\dfrac{d^2\mathbf{r}}{dt^2} = \dfrac{bg}{\sqrt{a^2 + b^2}}\,\mathbf{T} + a\left(\dfrac{bgt}{a^2 + b^2}\right)^2\mathbf{N}$

There is no component in the direction of $\mathbf{B}$.

9. a) $\dfrac{dx}{dt} = \dot{r}\cos\theta - r\dot{\theta}\sin\theta$, $\dfrac{dy}{dt} = \dot{r}\sin\theta + r\dot{\theta}\cos\theta$

b) $\dfrac{dr}{dt} = \dot{x}\cos\theta + \dot{y}\sin\theta$, $r\dfrac{d\theta}{dt} = -\dot{x}\sin\theta + \dot{y}\cos\theta$

11. a) $\mathbf{a}(1) = -9\mathbf{u}_r - 6\mathbf{u}_\theta$, $\mathbf{v}(1) = -\mathbf{u}_r + 3\mathbf{u}_\theta$ b) 6.5 in.

13. c) $\mathbf{v} = \dot{r}\,\mathbf{u}_r + r\dot{\theta}\,\mathbf{u}_\theta + \dot{z}\,\mathbf{k}$, $\mathbf{a} = (\ddot{r} - r\dot{\theta}^2)\mathbf{u}_r + (r\ddot{\theta} + 2\dot{r}\dot{\theta})\mathbf{u}_\theta + \ddot{z}\,\mathbf{k}$

15. a) $\mathbf{u}_\rho = \sin\phi\cos\theta\,\mathbf{i} + \sin\phi\sin\theta\,\mathbf{j} + \cos\phi\,\mathbf{k}$,

$\mathbf{u}_\phi = \cos\phi\cos\theta\,\mathbf{i} + \cos\phi\sin\theta\,\mathbf{j} - \sin\phi\,\mathbf{k}$,

$\mathbf{u}_\theta = -\sin\theta\,\mathbf{i} + \cos\theta\,\mathbf{j}$

CHAPTER 12

Section 12.1, pp. 914–917

1. a) All points in the xy-plane   b) All reals   c) The lines $y - x = c$ d) No boundary points   e) Both open and closed   f) Unbounded

3. a) All points in the xy-plane   b) $z \geq 0$ c) For $f(x, y) = 0$, the origin; for $f(x, y) \neq 0$, ellipses with the center $(0, 0)$, and major and minor axes, along the x- and y-axes, respectively   d) No boundary points   e) Both open and closed   f) Unbounded

5. a) All points in the xy-plane   b) All reals   c) For $f(x, y) = 0$, the x- and y-axes; for $f(x, y) \neq 0$, hyperbolas with the x- and y-axes as asymptotes   d) No boundary points   e) Both open and closed   f) Unbounded

7. a) All (x, y) satisfying $x^2 + y^2 < 16$ b) $z \geq 1/4$ c) Circles centered at the origin with radii $r < 4$ d) Boundary is the circle $x^2 + y^2 = 16$ e) Open   f) Bounded

9. a) $(x, y) \neq (0, 0)$ b) All reals   c) The circles with center $(0, 0)$ and radii $r > 0$ d) Boundary is the single point $(0, 0)$ e) Open   f) Unbounded

11. a) All (x, y) satisfying $-1 \leq y - x \leq 1$ b) $-\pi/2 \leq z \leq \pi/2$ c) Straight lines of the form $y - x = c$ where $-1 \leq c \leq 1$ d) Boundary is two straight lines $y = 1 + x$ and $y = -1 + x$ e) Closed   f) Unbounded

13. f   **15.** a   **17.** d

19. a)

b)

21. a)

b)

23. a) b)

25. a) b)

27. a) b)

29. **31.**

33. **35.**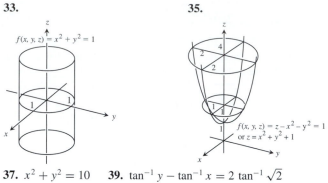

37. $x^2 + y^2 = 10$ **39.** $\tan^{-1} y - \tan^{-1} x = 2 \tan^{-1} \sqrt{2}$

41. $\sqrt{x - y} - \ln z = 2$ **43.** $\dfrac{x + y}{z} = \ln 2$ **45.** Yes, 2000

47. 63 km

Section 12.2, pp. 921–923

1. 5/2   **3.** $2\sqrt{6}$ **5.** 1   **7.** 1/2   **9.** 1   **11.** 0   **13.** 0
15. −1   **17.** 2   **19.** 1/4   **21.** 19/12   **23.** 2   **25.** 3
27. a) All (x, y) b) All (x, y) except $(0, 0)$
29. a) All (x, y) except where $x = 0$ or $y = 0$ b) All (x, y)
31. a) All (x, y, z) b) All (x, y, z) except the interior of the cylinder $x^2 + y^2 = 1$
33. a) All (x, y, z) with $z \neq 0$ b) All (x, y, z) with $x^2 + z^2 \neq 1$
35. Consider paths along $y = x, x > 0$, and along $y = x, x < 0$
37. Consider the paths $y = kx^2$, k a constant
39. Consider the paths $y = mx$, m a constant, $m \neq -1$
41. Consider the paths $y = kx^2$, k a constant, $k \neq 0$ **43.** No
45. The limit is 1   **47.** The limit is 0
49. a) $f(x, y)|_{y=mx} = \sin 2\theta$ where $\tan \theta = m$ **51.** 0
53. Does not exist   **55.** $\pi/2$ **57.** $f(0, 0) = \ln 3$ **61.** $\delta = 0.1$
63. $\delta = 0.005$ **65.** $\delta = \sqrt{0.015}$ **67.** $\delta = 0.005$

Section 12.3, pp. 931–933

1. $\dfrac{\partial f}{\partial x} = 4x, \quad \dfrac{\partial f}{\partial y} = -3$ **3.** $\dfrac{\partial f}{\partial x} = 2x(y + 2), \quad \dfrac{\partial f}{\partial y} = x^2 - 1$

5. $\dfrac{\partial f}{\partial x} = 2y(xy - 1), \quad \dfrac{\partial f}{\partial y} = 2x(xy - 1)$

7. $\dfrac{\partial f}{\partial x} = \dfrac{x}{\sqrt{x^2 + y^2}}, \quad \dfrac{\partial f}{\partial y} = \dfrac{y}{\sqrt{x^2 + y^2}}$

9. $\dfrac{\partial f}{\partial x} = \dfrac{-1}{(x + y)^2}, \quad \dfrac{\partial f}{\partial y} = \dfrac{-1}{(x + y)^2}$

11. $\dfrac{\partial f}{\partial x} = \dfrac{-y^2 - 1}{(xy - 1)^2}, \quad \dfrac{\partial f}{\partial y} = \dfrac{-x^2 - 1}{(xy - 1)^2}$

13. $\dfrac{\partial f}{\partial x} = e^{x+y+1}, \quad \dfrac{\partial f}{\partial y} = e^{x+y+1}$ **15.** $\dfrac{\partial f}{\partial x} = \dfrac{1}{x + y}, \quad \dfrac{\partial f}{\partial y} = \dfrac{1}{x + y}$

17. $\dfrac{\partial f}{\partial x} = 2 \sin (x - 3y) \cos (x - 3y),$

$\dfrac{\partial f}{\partial y} = -6 \sin (x - 3y) \cos (x - 3y)$

19. $\dfrac{\partial f}{\partial x} = yx^{y-1}$, $\dfrac{\partial f}{\partial y} = x^y \ln x$ **21.** $\dfrac{\partial f}{\partial x} = -g(x)$, $\dfrac{\partial f}{\partial y} = g(y)$

23. $f_x = y^2$, $f_y = 2xy$, $f_z = -4z$

25. $f_x = 1$, $f_y = -y(y^2 + z^2)^{-1/2}$, $f_z = -z(y^2 + z^2)^{-1/2}$

27. $f_x = \dfrac{yz}{\sqrt{1 - x^2 y^2 z^2}}$, $f_y = \dfrac{xz}{\sqrt{1 - x^2 y^2 z^2}}$, $f_z = \dfrac{xy}{\sqrt{1 - x^2 y^2 z^2}}$

29. $f_x = \dfrac{1}{x + 2y + 3z}$, $f_y = \dfrac{2}{x + 2y + 3z}$, $f_z = \dfrac{3}{x + 2y + 3z}$

31. $f_x = -2xe^{-(x^2+y^2+z^2)}$, $f_y = -2ye^{-(x^2+y^2+z^2)}$, $f_z = -2ze^{-(x^2+y^2+z^2)}$

33. $f_x = \text{sech}^2(x + 2y + 3z)$, $f_y = 2\text{sech}^2(x + 2y + 3z)$, $f_z = 3\text{sech}^2(x + 2y + 3z)$

35. $\dfrac{\partial f}{\partial t} = -2\pi \sin(2\pi t - \alpha)$, $\dfrac{\partial f}{\partial \alpha} = \sin(2\pi t - \alpha)$

37. $\dfrac{\partial h}{\partial \rho} = \sin \phi \cos \theta$, $\dfrac{\partial h}{\partial \phi} = \rho \cos \phi \cos \theta$, $\dfrac{\partial h}{\partial \theta} = -\rho \sin \phi \sin \theta$

39. $W_P(P, V, \delta, v, g) = V$, $W_V(P, V, \delta, v, g) = P + \dfrac{\delta v^2}{2g}$,

$W_P(P, V, \delta, v, g) = \dfrac{Vv^2}{2g}$, $W_v(P, V, \delta, v, g) = \dfrac{V\delta v}{g}$,

$W_g(P, V, \delta, v, g) = -\dfrac{V\delta v^2}{2g^2}$

41. $\dfrac{\partial f}{\partial x} = 1 + y$, $\dfrac{\partial f}{\partial y} = 1 + x$, $\dfrac{\partial^2 f}{\partial x^2} = 0$, $\dfrac{\partial^2 f}{\partial y^2} = 0$,

$\dfrac{\partial^2 f}{\partial y\, \partial x} = \dfrac{\partial^2 f}{\partial x\, \partial y} = 1$

43. $\dfrac{\partial g}{\partial x} = 2xy + y \cos x$, $\dfrac{\partial g}{\partial y} = x^2 - \sin y + \sin x$,

$\dfrac{\partial^2 g}{\partial x^2} = 2y - y \sin x$, $\dfrac{\partial^2 g}{\partial y^2} = -\cos y$, $\dfrac{\partial^2 g}{\partial y\, \partial x} = \dfrac{\partial^2 g}{\partial x\, \partial y} = 2x + \cos x$

45. $\dfrac{\partial r}{\partial x} = \dfrac{1}{x + y}$, $\dfrac{\partial r}{\partial y} = \dfrac{1}{x + y}$, $\dfrac{\partial^2 r}{\partial x^2} = \dfrac{-1}{(x + y)^2}$,

$\dfrac{\partial^2 r}{\partial y^2} = \dfrac{-1}{(x + y)^2}$, $\dfrac{\partial^2 r}{\partial y\, \partial x} = \dfrac{\partial^2 r}{\partial x\, \partial y} = \dfrac{-1}{(x + y)^2}$

47. $\dfrac{\partial w}{\partial x} = \dfrac{2}{2x + 3y}$, $\dfrac{\partial w}{\partial y} = \dfrac{3}{2x + 3y}$, $\dfrac{\partial^2 w}{\partial y\, \partial x} = \dfrac{-6}{(2x + 3y)^2}$,

$\dfrac{\partial^2 w}{\partial x\, \partial y} = \dfrac{-6}{(2x + 3y)^2}$

49. $\dfrac{\partial w}{\partial x} = y^2 + 2xy^3 + 3x^2y^4$, $\dfrac{\partial w}{\partial y} = 2xy + 3x^2y^2 + 4x^3y^3$,

$\dfrac{\partial^2 w}{\partial y\, \partial x} = 2y + 6xy^2 + 12x^2y^3$, $\dfrac{\partial^2 w}{\partial x\, \partial y} = 2y + 6xy^2 + 12x^2y^3$

51. a) x first **b)** y first **c)** x first **d)** x first **e)** y first
f) y first   **53.** $f_x(1, 2) = -13$, $f_y(1, 2) = -2$ **55.** 12

57. -2 **59.** $\dfrac{\partial A}{\partial a} = \dfrac{a}{bc \sin A}$, $\dfrac{\partial A}{\partial b} = \dfrac{c \cos A - b}{bc \sin A}$

61. $v_x = \dfrac{\ln v}{(\ln u)(\ln v) - 1}$

Section 12.4, pp. 942–944

1. a) $L(x, y) = 1$ **b)** $L(x, y) = 2x + 2y - 1$
3. a) $L(x, y) = 3x - 4y + 5$ **b)** $L(x, y) = 3x - 4y + 5$
5. a) $L(x, y) = 1 + x$ **b)** $L(x, y) = -y + (\pi/2)$
7. $L(x, y) = 7 + x - 6y$; 0.06 **9.** $L(x, y) = x + y + 1$; 0.08
11. $L(x, y) = 1 + x$; 0.0222
13. Pay more attention to the smaller of the two dimensions. It will generate the larger partial derivative.
15. Maximum error (estimate) ≤ 0.31 in magnitude
17. Maximum percentage error $= \pm 4.83\%$
19. Let $|x - 1| \leq 0.014$, $|y - 1| \leq 0.014$ **21.** $\approx 0.1\%$
23. a) $L(x, y, z) = 2x + 2y + 2z - 3$ **b)** $L(x, y, z) = y + z$
c) $L(x, y, z) = 0$ **25. a)** $L(x, y, z) = x$

b) $L(x, y, z) = \dfrac{1}{\sqrt 2}x + \dfrac{1}{\sqrt 2}y$ **c)** $L(x, y, z) = \dfrac{1}{3}x + \dfrac{2}{3}y + \dfrac{2}{3}z$

27. a) $L(x, y, z) = 2 + x$ **b)** $L(x, y, z) = x - y - z + \dfrac{\pi}{2} + 1$

c) $L(x, y, z) = x - y - z + \dfrac{\pi}{2} + 1$

29. $L(x, y, z) = 2x - 6y - 2z + 6$, 0.0024
31. $L(x, y, z) = x + y - z - 1$, 0.00135

33. a) $S_0\left(\dfrac{1}{100} dp + dx - 5dw - 30 dh\right)$

b) More sensitive to a change in height
35. f is most sensitive to a change in d.   **37.** $(47/24)$ ft^3
39. Magnitude of possible error ≤ 4.8 **41.** Yes

Section 12.5, pp. 950–952

1. $\dfrac{dw}{dt} = 0$, $\dfrac{dw}{dt}(\pi) = 0$ **3.** $\dfrac{dw}{dt} = 1$, $\dfrac{dw}{dt}(3) = 1$

5. $\dfrac{dw}{dt} = 4t \tan^{-1} t + 1$, $\dfrac{dw}{dt}(1) = \pi + 1$

7. a) $\dfrac{\partial z}{\partial r} = 4 \cos \theta \ln(r \sin \theta) + 4 \cos \theta$,

$\dfrac{\partial z}{\partial \theta} = -4r \sin \theta \ln(r \sin \theta) + \dfrac{4r \cos^2 \theta}{\sin \theta}$

b) $\dfrac{\partial z}{\partial r} = \sqrt 2 (\ln 2 + 2)$, $\dfrac{\partial z}{\partial \theta} = -2\sqrt 2 \ln 2 + 4\sqrt 2$

9. a) $\dfrac{\partial w}{\partial u} = 2u + 4uv$, $\dfrac{\partial w}{\partial v} = -2v + 2u^2$ **b)** $\dfrac{\partial w}{\partial u} = 3$, $\dfrac{\partial w}{\partial v} = -\dfrac{3}{2}$

11. a) $\dfrac{\partial u}{\partial x} = 0$, $\dfrac{\partial u}{\partial y} = \dfrac{z}{(z - y)^2}$, $\dfrac{\partial u}{\partial z} = \dfrac{-y}{(z - y)^2}$

b) $\dfrac{\partial u}{\partial x} = 0$, $\dfrac{\partial u}{\partial y} = 1$, $\dfrac{\partial u}{\partial z} = -2$

13. $\dfrac{dz}{dt} = \dfrac{\partial z}{\partial x}\dfrac{dx}{dt} + \dfrac{\partial z}{\partial y}\dfrac{dy}{dt}$

15. $\dfrac{\partial w}{\partial u} = \dfrac{\partial w}{\partial x}\dfrac{\partial x}{\partial u} + \dfrac{\partial w}{\partial y}\dfrac{\partial y}{\partial u} + \dfrac{\partial w}{\partial z}\dfrac{\partial z}{\partial u}$,

$\dfrac{\partial w}{\partial v} = \dfrac{\partial w}{\partial x}\dfrac{\partial x}{\partial v} + \dfrac{\partial w}{\partial y}\dfrac{\partial y}{\partial v} + \dfrac{\partial w}{\partial z}\dfrac{\partial z}{\partial v}$

17. $\dfrac{\partial w}{\partial u} = \dfrac{\partial w}{\partial x}\dfrac{\partial x}{\partial u} + \dfrac{\partial w}{\partial y}\dfrac{\partial y}{\partial u}$, $\quad \dfrac{\partial w}{\partial v} = \dfrac{\partial w}{\partial x}\dfrac{\partial x}{\partial v} + \dfrac{\partial w}{\partial y}\dfrac{\partial y}{\partial v}$

19. $\dfrac{\partial z}{\partial t} = \dfrac{\partial z}{\partial x}\dfrac{\partial x}{\partial t} + \dfrac{\partial z}{\partial y}\dfrac{\partial y}{\partial t}$, $\quad \dfrac{\partial z}{\partial s} = \dfrac{\partial z}{\partial x}\dfrac{\partial x}{\partial s} + \dfrac{\partial z}{\partial y}\dfrac{\partial y}{\partial s}$

21. $\dfrac{\partial w}{\partial s} = \dfrac{dw}{du}\dfrac{\partial u}{\partial s}$, $\quad \dfrac{\partial w}{\partial t} = \dfrac{dw}{du}\dfrac{\partial u}{\partial t}$

23. $\dfrac{\partial w}{\partial r} = \dfrac{\partial w}{\partial x}\dfrac{dx}{dr} + \dfrac{\partial w}{\partial y}\dfrac{dy}{dr} = \dfrac{\partial w}{\partial x}\dfrac{dx}{dr}$ since $\dfrac{dy}{dr} = 0$,

$\dfrac{\partial w}{\partial s} = \dfrac{\partial w}{\partial x}\dfrac{dx}{ds} + \dfrac{\partial w}{\partial y}\dfrac{dy}{ds} = \dfrac{\partial w}{\partial y}\dfrac{dy}{ds}$ since $\dfrac{dx}{ds} = 0$

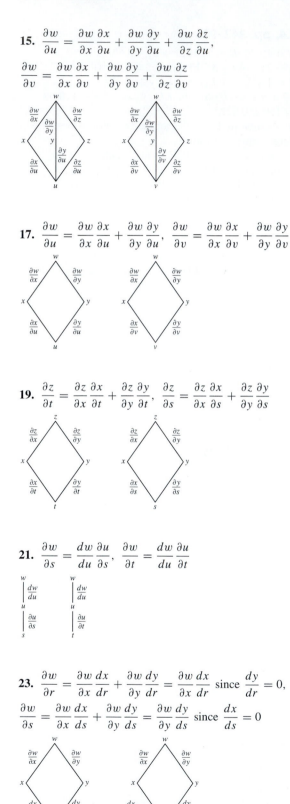

25. 4/3   **27.** $-4/5$ **29.** $\dfrac{\partial z}{\partial x} = \dfrac{1}{4}$, $\dfrac{\partial z}{\partial y} = -\dfrac{3}{4}$

31. $\dfrac{\partial z}{\partial x} = -1$, $\dfrac{\partial z}{\partial y} = -1$ **33.** 12   **35.** -7

37. $\dfrac{\partial z}{\partial u} = 2$, $\dfrac{\partial z}{\partial v} = 1$ **39.** -0.00005 amps/sec

45. $(\cos 1, \sin 1, 1)$ and $(\cos(-2), \sin(-2), -2)$

47. a) Maximum at $\left(-\dfrac{\sqrt{2}}{2}, \dfrac{\sqrt{2}}{2}\right)$ and $\left(\dfrac{\sqrt{2}}{2}, -\dfrac{\sqrt{2}}{2}\right)$, minimum at

$\left(\dfrac{\sqrt{2}}{2}, \dfrac{\sqrt{2}}{2}\right)$ and $\left(-\dfrac{\sqrt{2}}{2}, -\dfrac{\sqrt{2}}{2}\right)$

b) max $= 6$, min $= 2$ **49.** $2x\sqrt{x^8 + x^3} + \displaystyle\int_0^{x^2} \dfrac{3x^2}{2\sqrt{t^4 + x^3}}\, dt$

Section 12.6, pp. 956–957

1. a) 0   b) $1 + 2z$ c) $1 + 2z$ **3.** a) $\dfrac{\partial U}{\partial P} + \dfrac{\partial U}{\partial T}\left(\dfrac{V}{nR}\right)$

b) $\dfrac{\partial U}{\partial P}\left(\dfrac{nR}{V}\right) + \dfrac{\partial U}{\partial T}$ **5.** a) 5   b) 5   **7.** $\dfrac{x}{\sqrt{x^2 + y^2}}$

Section 12.7, pp. 967–969

1.   **3.**

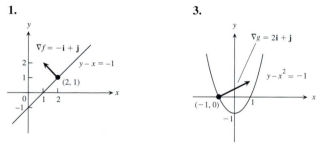

5. $\nabla f = 3\mathbf{i} + 2\mathbf{j} - 4\mathbf{k}$ **7.** $\nabla f = -\dfrac{26}{27}\mathbf{i} + \dfrac{23}{54}\mathbf{j} - \dfrac{23}{54}\mathbf{k}$ **9.** -4

11. $31/13$ **13.** 3   **15.** 2

17. $\mathbf{u} = -\dfrac{1}{\sqrt{2}}\mathbf{i} + \dfrac{1}{\sqrt{2}}\mathbf{j}$, $(D_{\mathbf{u}}f)_{P_0} = \sqrt{2}$; $-\mathbf{u} = \dfrac{1}{\sqrt{2}}\mathbf{i} - \dfrac{1}{\sqrt{2}}\mathbf{j}$,

$(D_{-\mathbf{u}}f)_{P_0} = -\sqrt{2}$

19. $\mathbf{u} = \dfrac{1}{3\sqrt{3}}\mathbf{i} - \dfrac{5}{3\sqrt{3}}\mathbf{j} - \dfrac{1}{3\sqrt{3}}\mathbf{k}$, $(D_{\mathbf{u}}f)_{P_0} = 3\sqrt{3}$;

$-\mathbf{u} = -\dfrac{1}{3\sqrt{3}}\mathbf{i} + \dfrac{5}{3\sqrt{3}}\mathbf{j} + \dfrac{1}{3\sqrt{3}}\mathbf{k}$, $(D_{-\mathbf{u}}f)_{P_0} = -3\sqrt{3}$

21. $\mathbf{u} = \dfrac{1}{\sqrt{3}}(\mathbf{i} + \mathbf{j} + \mathbf{k})$, $(D_{\mathbf{u}}f)_{P_0} = 2\sqrt{3}$; $-\mathbf{u} = -\dfrac{1}{\sqrt{3}}(\mathbf{i} + \mathbf{j} + \mathbf{k})$,

$(D_{-\mathbf{u}}f)_{P_0} = -2\sqrt{3}$

23. $df = \dfrac{9}{910} \approx 0.01$ **25.** $dg = 0$

27. Tangent: $x + y + z = 3$, normal line: $x = 1 + 2t$, $y = 1 + 2t$, $z = 1 + 2t$

29. Tangent: $2x - z - 2 = 0$, normal line: $x = 2 - 4t$, $y = 0$, $z = 2 + 2t$

31. Tangent: $2x + 2y + z - 4 = 0$, normal line: $x = 2t$, $y = 1 + 2t$, $z = 2 + t$

33. Tangent: $x + y + z - 1 = 0$, normal line: $x = t$, $y = 1 + t$, $z = t$

35. $2x - z - 2 = 0$ **37.** $x - y + 2z - 1 = 0$

39.   **41.**

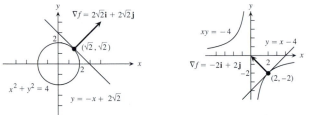

43. $x = 1$, $y = 1 + 2t$, $z = 1 - 2t$

45. $x = 1 - 2t$, $y = 1$, $z = \dfrac{1}{2} + 2t$

47. $x = 1 + 90t$, $y = 1 - 90t$, $z = 3$

49. $\mathbf{u} = \dfrac{7}{\sqrt{53}}\mathbf{i} - \dfrac{2}{\sqrt{53}}\mathbf{j}$, $-\mathbf{u} = -\dfrac{7}{\sqrt{53}}\mathbf{i} + \dfrac{2}{\sqrt{53}}\mathbf{j}$

51. No, the maximum rate of change is $\sqrt{185} < 14$.   **53.** $-\dfrac{7}{\sqrt{5}}$

55. a) $\dfrac{\sqrt{3}}{2}\sin\sqrt{3} - \dfrac{1}{2}\cos\sqrt{3} \approx 0.935°\text{C/ft}$

b) $\sqrt{3}\sin\sqrt{3} - \cos\sqrt{3} \approx 1.87°\text{C/sec}$

57. At $-\dfrac{\pi}{4}$, $-\dfrac{\pi}{2\sqrt{2}}$; at $0, 0$; at $\dfrac{\pi}{4}$, $\dfrac{\pi}{2\sqrt{2}}$

Section 12.8, pp. 975–979

1. $f(-3, 3) = -5$, local minimum

3. $f\left(\dfrac{2}{3}, \dfrac{4}{3}\right) = 0$, local maximum   **5.** $f(-2, 1)$, saddle point

7. $f\left(\dfrac{6}{5}, \dfrac{69}{25}\right)$, saddle point   **9.** $f(2, 1)$, saddle point

11. $f(2, -1) = -6$, local minimum   **13.** $f(1, 2)$, saddle point

15. $f(0, 0)$, saddle point

17. $f(0, 0)$, saddle point; $f\left(-\dfrac{2}{3}, \dfrac{2}{3}\right) = \dfrac{170}{27}$, local maximum

19. $f(0, 0) = 0$, local minimum; $f(1, -1)$, saddle point

21. $f(0, 0)$, saddle point; $f\left(\dfrac{4}{9}, \dfrac{4}{3}\right) = -\dfrac{64}{81}$, local minimum

23. $f(0, 0)$, saddle point; $f(0, 2) = -12$, local minimum; $f(-2, 0) = -4$, local maximum; $f(-2, 2)$, saddle point

25. $f(0, 0)$, saddle point; $f(1, 1) = 2$, $f(-1, -1) = 2$, local maxima

27. $f(0, 0) = -1$, local maximum

29. $f(n\pi, 0)$, saddle point; $f(n\pi, 0) = 0$ for every n

31. Absolute maximum: 1 at $(0, 0)$; absolute minimum: -5 at $(1, 2)$

33. Absolute maximum: 4 at $(0, 2)$; absolute minimum: 0 at $(0, 0)$

35. Absolute maximum: 11 at $(0, -3)$; absolute minimum: -10 at $(4, -2)$

37. Absolute maximum: 4 at $(2, 0)$; absolute minimum: $\dfrac{3\sqrt{2}}{2}$ at $\left(3, -\dfrac{\pi}{4}\right)$, $\left(3, \dfrac{\pi}{4}\right)$, $\left(1, -\dfrac{\pi}{4}\right)$, and $\left(1, \dfrac{\pi}{4}\right)$ **39.** $a = -3$, $b = 2$

41. Hottest: $2\dfrac{1}{4}°$ at $\left(-\dfrac{1}{2}, \dfrac{\sqrt{3}}{2}\right)$ and $\left(-\dfrac{1}{2}, -\dfrac{\sqrt{3}}{2}\right)$; coldest: $-\dfrac{1}{4}°$ at $\left(\dfrac{1}{2}, 0\right)$

43. a) $f(0, 0)$, saddle point   b) $f(1, 2)$, local minimum
c) $f(1, -2)$, local minimum; $f(-1, -2)$, saddle point

49. $\left(\dfrac{1}{6}, \dfrac{1}{3}, \dfrac{355}{36}\right)$

53. a) On the semicircle, max $f = 2\sqrt{2}$ at $t = \pi/4$, min $f = -2$ at $t = \pi$. On the quarter circle, max $f = 2\sqrt{2}$ at $t = \pi/4$, min $f = 2$ at $t = 0, \pi/2$.
b) On the semicircle, max $g = 2$ at $t = \pi/4$, min $g = -2$ at $t = 3\pi/4$. On the quarter circle, max $g = 2$ at $t = \pi/4$, min $g = 0$ at $t = 0, \pi/2$.
c) On the semicircle, max $h = 8$ at $t = 0, \pi$; min $h = 4$ at $t = \pi/2$. On the quarter circle, max $h = 8$ at $t = 0$, min $h = 4$ at $t = \pi/2$.

55. i) min $f = -1/2$ at $t = -1/2$; no max   ii) max $f = 0$ at $t = -1, 0$; min $f = -1/2$ at $t = -1/2$ iii) max $f = 4$ at $t = 1$; min $f = 0$ at $t = 0$

57. $y = -\dfrac{20}{13}x + \dfrac{9}{13}$, $y|_{x=4} = -\dfrac{71}{13}$

59. $y = \dfrac{3}{2}x + \dfrac{1}{6}$, $y|_{x=4} = \dfrac{37}{6}$

61. $y = 0.122x + 3.58$ **63.** a)

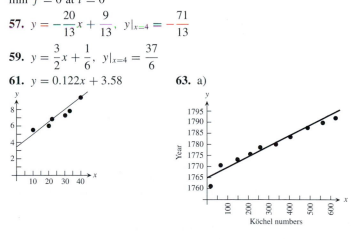

b) $y = 0.0427\text{K} + 1764.8$ c) 1780

Section 12.9, pp. 987–989

1. $\left(\pm\dfrac{1}{\sqrt{2}}, \dfrac{1}{2}\right)$, $\left(\pm\dfrac{1}{\sqrt{2}}, -\dfrac{1}{2}\right)$ **3.** 39   **5.** $(3, \pm 3\sqrt{2})$ **7.** a) 8
b) 64   **9.** $r = 2$ cm, $h = 4$ cm   **11.** $l = 4\sqrt{2}$, $w = 3\sqrt{2}$
13. $f(0, 0) = 0$ is minimum, $f(2, 4) = 20$ is maximum

15. Minimum $= 0°$, maximum $= 125°$ **17.** $\left(\dfrac{3}{2}, 2, \dfrac{5}{2}\right)$ **19.** 1

21. $(0, 0, 2)$, $(0, 0, -2)$

23. $f(1, -2, 5) = 30$ is maximum, $f(-1, 2, -5) = -30$ is minimum

25. 3, 3, 3   **27.** $\dfrac{2}{\sqrt{3}}$ by $\dfrac{2}{\sqrt{3}}$ by $\dfrac{2}{\sqrt{3}}$ units   **29.** $\left(\pm\dfrac{4}{3}, -\dfrac{4}{3}, -\dfrac{4}{3}\right)$

31. $U(8, 14) = \$128$ **33.** $f\left(\dfrac{2}{3}, \dfrac{4}{3}, -\dfrac{4}{3}\right) = \dfrac{4}{3}$ **35.** $(2, 4, 4)$

37. Maximum is $1 + 6\sqrt{3}$ at $(\pm\sqrt{6}, \sqrt{3}, 1)$, minimum is $1 - 6\sqrt{3}$ at $(\pm\sqrt{6}, -\sqrt{3}, 1)$

39. Maximum is 4 at $(0, 0, \pm2)$, minimum is 2 at $(\pm\sqrt{2}, \pm\sqrt{2}, 0)$

Section 12.10, p. 993

1. Quadratic: $x + xy$; cubic: $x + xy + \dfrac{1}{2}xy^2$

3. Quadratic: xy; cubic: xy

5. Quadratic: $y + \dfrac{1}{2}(2xy - y^2)$;

cubic: $y + \dfrac{1}{2}(2xy - y^2) + \dfrac{1}{6}(3x^2y - 3xy^2 + 2y^3)$

7. Quadratic: $\dfrac{1}{2}(2x^2 + 2y^2) = x^2 + y^2$; cubic: $x^2 + y^2$

9. Quadratic: $1 + (x + y) + (x + y)^2$;
cubic: $1 + (x + y) + (x + y)^2 + (x + y)^3$

11. Quadratic: $1 - \dfrac{1}{2}x^2 - \dfrac{1}{2}y^2$, $E(x, y) \leq 0.00134$

Chapter 12 Practice Exercises, pp. 994–998

1.

Domain: all points in the xy-plane; range: $z \geq 0$. Level curves are ellipses with major axis along the y-axis and minor axis along the x-axis.

3.

Domain: all (x, y) such that $x \neq 0$ and $y \neq 0$; range: $z \neq 0$. Level curves are hyperbolas with the x- and y-axes as asymptotes.

5.

Domain: all (x, y, z) such that $(x, y, z) \neq (0, 0, 0)$; range: all real numbers. Level surfaces are paraboloids of revolution with the z-axis as axis.

7.

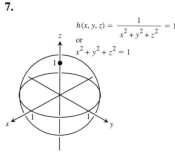

Domain: all (x, y, z) such that $(x, y, z) \neq (0, 0, 0)$; range: positive real numbers. Level surfaces are spheres with center $(0, 0, 0)$ and radius $r > 0$.

9. -2 **11.** $1/2$ **13.** 1   **15.** Let $y = kx^2$, $k \neq 1$

17. a) Does not exist   b) Not continuous at $(0, 0)$

19. $\dfrac{\partial g}{\partial r} = \cos\theta + \sin\theta$, $\dfrac{\partial g}{\partial\theta} = -r\sin\theta + r\cos\theta$

21. $\dfrac{\partial f}{\partial R_1} = -\dfrac{1}{R_1^2}$, $\dfrac{\partial f}{\partial R_2} = -\dfrac{1}{R_2^2}$, $\dfrac{\partial f}{\partial R_3} = -\dfrac{1}{R_3^2}$

23. $\dfrac{\partial P}{\partial n} = \dfrac{RT}{V}$, $\dfrac{\partial P}{\partial R} = \dfrac{nT}{V}$, $\dfrac{\partial P}{\partial T} = \dfrac{nR}{V}$, $\dfrac{\partial P}{\partial V} = -\dfrac{nRT}{V^2}$

25. $\dfrac{\partial^2 g}{\partial x^2} = 0$, $\dfrac{\partial^2 g}{\partial y^2} = \dfrac{2x}{y^3}$, $\dfrac{\partial^2 g}{\partial y\,\partial x} = \dfrac{\partial^2 g}{\partial x\,\partial y} = -\dfrac{1}{y^2}$

27. $\dfrac{\partial^2 f}{\partial x^2} = -30x + \dfrac{2 - 2x^2}{(x^2 + 1)^2}$, $\dfrac{\partial^2 f}{\partial y^2} = 0$, $\dfrac{\partial^2 f}{\partial y\,\partial x} = \dfrac{\partial^2 f}{\partial x\,\partial y} = 1$

29. Answers will depend on the upper bound used for $|f_{xx}|$, $|f_{xy}|$, $|f_{yy}|$. With $M = \sqrt{2}/2$, $|E| \leq 0.0142$. With $M = 1$, $|E| \leq 0.02$.

31. $L(x, y, z) = y - 3z$, $L(x, y, z) = x + y - z - 1$

33. Be more careful with the diameter.

35. $dl = 0.038$, % change in $V = -4.17\%$, % change in $R = -20\%$, % change in $l = 15.83\%$

37. a) 5%   **39.** $\left.\dfrac{dw}{dt}\right|_{t=0} = -1$

41. $\left.\dfrac{\partial w}{\partial r}\right|_{(r,s)=(\pi,0)} = 2$, $\left.\dfrac{\partial w}{\partial s}\right|_{(r,s)=(\pi,0)} = 2 - \pi$

43. $\left.\dfrac{df}{dt}\right|_{t=1} = -(\sin 1 + \cos 2)\sin 1 + (\cos 1 + \cos 2)\cos 1 - 2(\sin 1 + \cos 1)\sin 2$

45. $\dfrac{dy}{dx}\Big|_{(x,y)=(0,1)} = -1$

47. a) $(2y + x^2 z)e^{yz}$ b) $x^2 e^{yz}\left(y - \dfrac{z}{2y}\right)$ c) $(1 + x^2 y)e^{yz}$

49. Increases most rapidly in the direction $\mathbf{u} = -\dfrac{\sqrt{2}}{2}\mathbf{i} - \dfrac{\sqrt{2}}{2}\mathbf{j}$;

decreases most rapidly in the direction $-\mathbf{u} = \dfrac{\sqrt{2}}{2}\mathbf{i} + \dfrac{\sqrt{2}}{2}\mathbf{j}$;

$D_{\mathbf{u}}f = \dfrac{\sqrt{2}}{2},\ D_{-\mathbf{u}}f = -\dfrac{\sqrt{2}}{2},\ D_{\mathbf{u}_1}f = -\dfrac{7}{10}$

51. Increases most rapidly in the direction $\mathbf{u} = \dfrac{2}{7}\mathbf{i} + \dfrac{3}{7}\mathbf{j} + \dfrac{6}{7}\mathbf{k}$;

decreases most rapidly in the direction $-\mathbf{u} = -\dfrac{2}{7}\mathbf{i} - \dfrac{3}{7}\mathbf{j} - \dfrac{6}{7}\mathbf{k}$;
$D_{\mathbf{u}}f = 7,\ D_{-\mathbf{u}}f = -7,\ D_{\mathbf{u}_1}f = 7$
53. $\pi/\sqrt{2}$ **55.** a) $f_x(1, 2) = f_y(1, 2) = 2$ b) $14/5$
57.

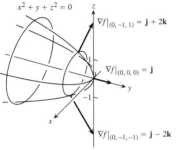

59. Tangent: $4x - y - 5z = 4$, normal line: $x = 2 + 4t$,
$y = -1 - t,\ z = 1 - 5t$
61. $2y - z - 2 = 0$
63. Tangent: $x + y = \pi + 1$, normal line: $y = x - \pi + 1$

65. $x = 1 - 2t,\ y = 1,\ z = \dfrac{1}{2} + 2t$

67. Local minimum of -8 at $(-2, -2)$
69. Saddle point at $(0, 0)$, $f(0, 0) = 0$; local maximum of $1/4$ at
$\left(-\dfrac{1}{2}, -\dfrac{1}{2}\right)$

71. Saddle point at $(0, 0)$, $f(0, 0) = 0$; local minimum of -4 at
$(0, 2)$; local maximum of 4 at $(-2, 0)$; saddle point at $(-2, 2)$,
$f(-2, 2) = 0$
73. Absolute maximum: 28 at $(0, 4)$, absolute minimum: $-9/4$ at
$(3/2, 0)$
75. Absolute maximum: 18 at $(2, -2)$, absolute minimum: $-17/4$ at
$(-2, 1/2)$
77. Absolute maximum: 8 at $(-2, 0)$, absolute minimum: -1 at $(1, 0)$

79. Absolute maximum: 4 at $(1, 0)$, absolute minimum: -4 at $(0, -1)$
81. Absolute maximum: 1 at $(0, \pm 1)$ and $(1, 0)$, absolute minimum:
-1 at $(-1, 0)$
83. Maximum: 5 at $(0, 1)$, minimum: $-1/3$ at $(0, -1/3)$
85. Maximum: $\sqrt{3}$ at $(1/\sqrt{3}, -1/\sqrt{3}, 1/\sqrt{3})$, minimum: $-\sqrt{3}$ at
$(-1/\sqrt{3}, 1/\sqrt{3}, -1/\sqrt{3})$
87. Width $= \left(\dfrac{c^2 V}{ab}\right)^{1/3}$, depth $= \left(\dfrac{b^2 V}{ac}\right)^{1/3}$, height $= \left(\dfrac{a^2 V}{bc}\right)^{1/3}$
89. Maximum $= 3/2$ at $(1/\sqrt{2}, 1/\sqrt{2}, \sqrt{2})$ and
$(-1/\sqrt{2}, -1/\sqrt{2}, -\sqrt{2})$, minimum $= 1/2$ at $(-1/\sqrt{2}, 1/\sqrt{2}, -\sqrt{2})$
and $(1/\sqrt{2}, -1/\sqrt{2}, \sqrt{2})$
91. $\dfrac{\partial w}{\partial x} = \cos\theta\,\dfrac{\partial w}{\partial r} - \dfrac{\sin\theta}{r}\dfrac{\partial w}{\partial\theta},\ \dfrac{\partial w}{\partial y} = \sin\theta\,\dfrac{\partial w}{\partial r} + \dfrac{\cos\theta}{r}\dfrac{\partial w}{\partial\theta}$
97. $(t, -t \pm 4, t)$, t a real number

Chapter 12 Additional Exercises, pp. 998–1000

1. $f_{xy}(0, 0) = -1,\ f_{yx}(0, 0) = 1$ **7.** c) $r^2 = \dfrac{1}{2}(x^2 + y^2 + z^2)$

15. $V = \dfrac{\sqrt{3}abc}{2}$ **19.** $f(x, y) = \dfrac{y}{2} + 4,\ g(x, y) = \dfrac{x}{2} + \dfrac{9}{2}$

21. $y = 2\ln|\sin x| + \ln 2$ **23.** a) $\dfrac{1}{\sqrt{53}}(2\mathbf{i} + 7\mathbf{j})$

b) $\dfrac{-1}{\sqrt{29097}}(98\mathbf{i} - 127\mathbf{j} + 58\mathbf{k})$ **25.** $w = e^{-c^2 \pi^2 t}\sin\pi x$

27. 0.213%

CHAPTER 13

Section 13.1, pp. 1010–1011

1. 16 **3.** 1

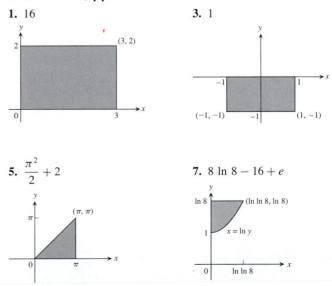

5. $\dfrac{\pi^2}{2} + 2$ **7.** $8\ln 8 - 16 + e$

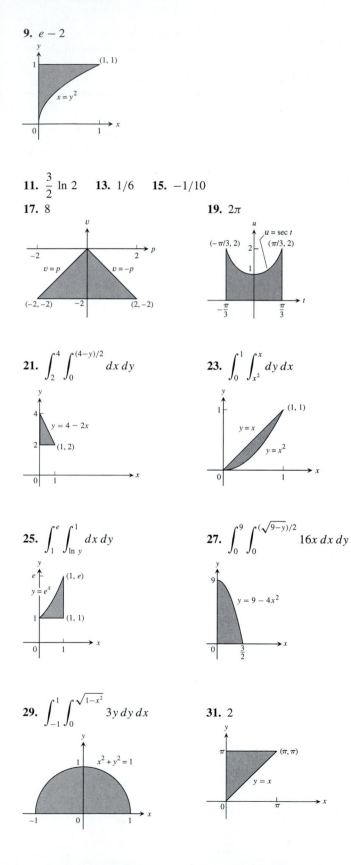

9. $e - 2$

11. $\dfrac{3}{2} \ln 2$ **13.** $1/6$ **15.** $-1/10$

17. 8

19. 2π

21. $\displaystyle\int_{2}^{4} \int_{0}^{(4-y)/2} dx\, dy$

23. $\displaystyle\int_{0}^{1} \int_{x^2}^{x} dy\, dx$

25. $\displaystyle\int_{1}^{e} \int_{\ln y}^{1} dx\, dy$

27. $\displaystyle\int_{0}^{9} \int_{0}^{(\sqrt{9-y})/2} 16x\, dx\, dy$

29. $\displaystyle\int_{-1}^{1} \int_{0}^{\sqrt{1-x^2}} 3y\, dy\, dx$

31. 2

33. $\dfrac{e-2}{2}$

35. 2

37. $\dfrac{1}{80\pi}$

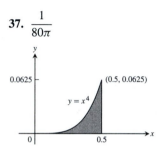

39. $-2/3$ **41.** $4/3$ **43.** $625/12$ **45.** 16   **47.** 20
49. $2(1 + \ln 2)$ **51.** 1   **53.** π^2 **55.** $-1/4$ **57.** $20\sqrt{3}/9$
59. $\displaystyle\int_{0}^{1} \int_{x}^{2-x} (x^2 + y^2)\, dy\, dx = \dfrac{4}{3}$

67. 0.603   **69.** 0.233

Section 13.2, pp. 1018–1020

1. $\displaystyle\int_{0}^{2} \int_{0}^{2-x} dy\, dx = 2$ or $\displaystyle\int_{0}^{2} \int_{0}^{2-y} dx\, dy = 2$

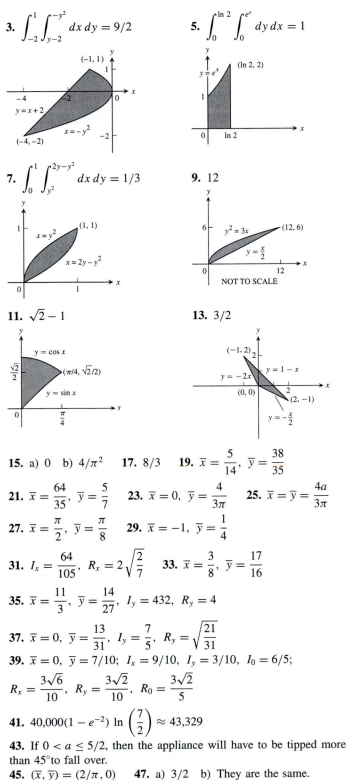

3. $\displaystyle\int_{-2}^{1}\int_{y-2}^{-y^2} dx\,dy = 9/2$

5. $\displaystyle\int_{0}^{\ln 2}\int_{0}^{e^x} dy\,dx = 1$

7. $\displaystyle\int_{0}^{1}\int_{y^2}^{2y-y^2} dx\,dy = 1/3$

9. 12

11. $\sqrt{2}-1$

13. 3/2

15. a) 0 b) $4/\pi^2$ **17.** 8/3   **19.** $\bar{x} = \dfrac{5}{14},\ \bar{y} = \dfrac{38}{35}$

21. $\bar{x} = \dfrac{64}{35},\ \bar{y} = \dfrac{5}{7}$ **23.** $\bar{x} = 0,\ \bar{y} = \dfrac{4}{3\pi}$ **25.** $\bar{x} = \bar{y} = \dfrac{4a}{3\pi}$

27. $\bar{x} = \dfrac{\pi}{2},\ \bar{y} = \dfrac{\pi}{8}$ **29.** $\bar{x} = -1,\ \bar{y} = \dfrac{1}{4}$

31. $I_x = \dfrac{64}{105},\ R_x = 2\sqrt{\dfrac{2}{7}}$ **33.** $\bar{x} = \dfrac{3}{8},\ \bar{y} = \dfrac{17}{16}$

35. $\bar{x} = \dfrac{11}{3},\ \bar{y} = \dfrac{14}{27},\ I_y = 432,\ R_y = 4$

37. $\bar{x} = 0,\ \bar{y} = \dfrac{13}{31},\ I_y = \dfrac{7}{5},\ R_y = \sqrt{\dfrac{21}{31}}$

39. $\bar{x} = 0,\ \bar{y} = 7/10;\ I_x = 9/10,\ I_y = 3/10,\ I_0 = 6/5;$
$R_x = \dfrac{3\sqrt{6}}{10},\ R_y = \dfrac{3\sqrt{2}}{10},\ R_0 = \dfrac{3\sqrt{2}}{5}$

41. $40{,}000(1 - e^{-2})\ln\left(\dfrac{7}{2}\right) \approx 43{,}329$

43. If $0 < a \le 5/2$, then the appliance will have to be tipped more than 45° to fall over.

45. $(\bar{x}, \bar{y}) = (2/\pi, 0)$ **47.** a) 3/2 b) They are the same.

53. a) $\left(\dfrac{7}{5}, \dfrac{31}{10}\right)$ b) $\left(\dfrac{19}{7}, \dfrac{18}{7}\right)$ c) $\left(\dfrac{9}{2}, \dfrac{19}{8}\right)$ d) $\left(\dfrac{11}{4}, \dfrac{43}{16}\right)$

55. In order for c.m. to be on the common boundary, $h = a\sqrt{2}$. In order for c.m. to be inside T, $h > a\sqrt{2}$.

Section 13.3, pp. 1024–1026

1. $\pi/2$ **3.** $\pi/8$ **5.** πa^2 **7.** 36   **9.** $(1 - \ln 2)\pi$

11. $(2\ln 2 - 1)(\pi/2)$ **13.** $\dfrac{\pi}{2} + 1$ **15.** $\pi(\ln(4) - 1)$

17. $2(\pi - 1)$ **19.** 12π **21.** $\dfrac{3\pi}{8} + 1$ **23.** 4   **25.** $6\sqrt{3} - 2\pi$

27. $\bar{x} = 5/6,\ \bar{y} = 0$ **29.** $2a/3$ **31.** $2a/3$ **33.** 2π

35. $\dfrac{4}{3} + \dfrac{5\pi}{8}$ **37.** a) $\sqrt{\pi}/2$ b) 1   **39.** $\pi\ln 4$, no

41. $\dfrac{1}{2}(a^2 + 2h^2)$

Section 13.4, pp. 1031–1034

1. 1

3. $\displaystyle\int_{0}^{1}\int_{0}^{2-2x}\int_{0}^{3-3x-3y/2} dz\,dy\,dx,\quad \int_{0}^{2}\int_{0}^{1-y/2}\int_{0}^{3-3x-3y/2} dz\,dx\,dy,$
$\displaystyle\int_{0}^{1}\int_{0}^{3-3x}\int_{0}^{2-2x-2z/3} dy\,dz\,dx,\quad \int_{0}^{3}\int_{0}^{1-z/3}\int_{0}^{2-2x-2z/3} dy\,dx\,dz,$
$\displaystyle\int_{0}^{2}\int_{0}^{3-3y/2}\int_{0}^{1-y/2-z/3} dx\,dz\,dy,\quad \int_{0}^{3}\int_{0}^{2-2z/3}\int_{0}^{1-y/2-z/3} dx\,dy\,dz.$
The value of all six integrals is 1.

5. $\displaystyle\int_{-2}^{2}\int_{-\sqrt{4-x^2}}^{\sqrt{4-x^2}}\int_{x^2+y^2}^{8-x^2-y^2} 1\,dz\,dy\,dx,$
$\displaystyle\int_{-2}^{2}\int_{-\sqrt{4-y^2}}^{\sqrt{4-y^2}}\int_{x^2+y^2}^{8-x^2-y^2} 1\,dz\,dx\,dy,$
$\displaystyle\int_{-2}^{2}\int_{4}^{8-y^2}\int_{-\sqrt{8-z-y^2}}^{\sqrt{8-z-y^2}} 1\,dx\,dz\,dy + \int_{-2}^{2}\int_{y^2}^{4}\int_{-\sqrt{z-y^2}}^{\sqrt{z-y^2}} 1\,dx\,dz\,dy,$
$\displaystyle\int_{4}^{8}\int_{-\sqrt{8-z}}^{\sqrt{8-z}}\int_{-\sqrt{8-z-y^2}}^{\sqrt{8-z-y^2}} 1\,dx\,dy\,dz + \int_{0}^{4}\int_{-\sqrt{z}}^{\sqrt{z}}\int_{-\sqrt{z-y^2}}^{\sqrt{z-y^2}} 1\,dx\,dy\,dz,$
$\displaystyle\int_{-2}^{2}\int_{4}^{8-x^2}\int_{-\sqrt{8-z-x^2}}^{\sqrt{8-z-x^2}} 1\,dy\,dz\,dx + \int_{-2}^{2}\int_{x^2}^{4}\int_{-\sqrt{z-x^2}}^{\sqrt{z-x^2}} 1\,dy\,dz\,dx,$
$\displaystyle\int_{4}^{8}\int_{-\sqrt{8-z}}^{\sqrt{8-z}}\int_{-\sqrt{8-z-x^2}}^{\sqrt{8-z-x^2}} 1\,dy\,dx\,dz + \int_{0}^{4}\int_{-\sqrt{z}}^{\sqrt{z}}\int_{-\sqrt{z-x^2}}^{\sqrt{z-x^2}} 1\,dy\,dx\,dz.$
The value of all six integrals is 16π.

7. 1   **9.** 1   **11.** $\dfrac{\pi^3}{2}(1 - \cos 1)$ **13.** 18   **15.** 7/6   **17.** 0

19. $\dfrac{1}{2} - \dfrac{\pi}{8}$ **21.** a) $\displaystyle\int_{-1}^{1}\int_{0}^{1-x^2}\int_{x^2}^{1-z} dy\,dz\,dx$

b) $\displaystyle\int_{0}^{1}\int_{-\sqrt{1-z}}^{\sqrt{1-z}}\int_{x^2}^{1-z} dy\,dx\,dz$ c) $\displaystyle\int_{0}^{1}\int_{0}^{1-z}\int_{-\sqrt{y}}^{\sqrt{y}} dy\,dz\,dx$

d) $\int_0^1 \int_0^{1-y} \int_{-\sqrt{y}}^{\sqrt{y}} dx\, dz\, dy$ e) $\int_0^1 \int_{-\sqrt{y}}^{\sqrt{y}} \int_0^{1-y} dz\, dx\, dy$ **23.** 2/3

25. 20/3 **27.** 1 **29.** 16/3 **31.** $8\pi - \dfrac{32}{3}$ **33.** 2 **35.** 4π

37. 31/3 **39.** 1 **41.** $2\sin 4$ **43.** 4 **45.** $a = 3$ or $a = \dfrac{13}{3}$

Section 13.5, pp. 1036–1039

1. $R_x = \sqrt{\dfrac{b^2 + c^2}{12}}$, $R_y = \sqrt{\dfrac{a^2 + c^2}{12}}$, $R_z = \sqrt{\dfrac{a^2 + b^2}{12}}$

3. $I_x = \dfrac{M}{3}(b^2 + c^2)$, $I_y = \dfrac{M}{3}(a^2 + c^2)$, $I_z = \dfrac{M}{3}(a^2 + b^2)$

5. $\bar{x} = \bar{y} = 0$, $\bar{z} = \dfrac{12}{5}$, $I_x = \dfrac{7904}{105} \approx 75.28$, $I_y = \dfrac{4832}{63} \approx 76.70$,

$I_z = \dfrac{256}{45} \approx 5.69$

7. a) $\bar{x} = \bar{y} = 0$, $\bar{z} = \dfrac{8}{3}$ b) $c = 2\sqrt{2}$

9. $I_L = 1386$, $R_L = \sqrt{\dfrac{77}{2}}$ **11.** $I_L = \dfrac{40}{3}$, $R_L = \sqrt{\dfrac{5}{3}}$ **13.** a) $\dfrac{4}{3}$

b) $\bar{x} = \dfrac{4}{5}$, $\bar{y} = \bar{z} = \dfrac{2}{5}$ **15.** a) $\dfrac{5}{2}$ b) $\bar{x} = \bar{y} = \bar{z} = \dfrac{8}{15}$

c) $I_x = I_y = I_z = \dfrac{11}{6}$ d) $R_x = R_y = R_z = \sqrt{\dfrac{11}{15}}$ **17.** 3

19. a) $\dfrac{4}{3}g$ b) $\dfrac{4}{3}g$

23. a) $I_{\text{c.m.}} = \dfrac{abc(a^2 + b^2)}{12}$, $R_{\text{c.m.}} = \sqrt{\dfrac{a^2 + b^2}{12}}$

b) $I_L = \dfrac{abc(a^2 + 7b^2)}{3}$, $R_L = \sqrt{\dfrac{a^2 + 7b^2}{3}}$

27. a) $h = a\sqrt{3}$ b) $h = a\sqrt{2}$

Section 13.6, pp. 1044–1047

1. $4\pi(\sqrt{2} - 1)/3$ **3.** $17\pi/5$ **5.** $\pi(6\sqrt{2} - 8)$ **7.** $3\pi/10$

9. $\pi/3$ **11.** a) $\int_0^{2\pi} \int_0^1 \int_0^{\sqrt{4-r^2}} r\, dz\, dr\, d\theta$

b) $\int_0^{2\pi} \int_0^{\sqrt{3}} \int_0^1 r\, dr\, dz\, d\theta + \int_0^{2\pi} \int_{\sqrt{3}}^2 \int_0^{\sqrt{4-z^2}} r\, dr\, dz\, d\theta$

c) $\int_0^1 \int_0^{\sqrt{4-r^2}} \int_0^{2\pi} r\, d\theta\, dz\, dr$

13. $\int_{-\pi/2}^{\pi/2} \int_0^{\cos\theta} \int_0^{3r^2} f(r, \theta, z)\, r\, dz\, dr\, d\theta$

15. $\int_0^\pi \int_0^{2\sin\theta} \int_0^{4-r\sin\theta} f(r, \theta, z)\, dz\, r\, dr\, d\theta$

17. $\int_{-\pi/2}^{\pi/2} \int_1^{1+\cos\theta} \int_0^4 f(r, \theta, z)\, dz\, r\, dr\, d\theta$

19. $\int_0^{\pi/4} \int_0^{\sec\theta} \int_0^{2-r\sin\theta} f(r, \theta, z)\, dz\, r\, dr\, d\theta$ **21.** π^2 **23.** $\pi/3$

25. 5π **27.** 2π **29.** $\left(\dfrac{8 - 5\sqrt{2}}{2}\right)\pi$

31. a) $\int_0^{2\pi} \int_0^{\pi/6} \int_0^2 \rho^2 \sin\phi\, d\rho\, d\phi\, d\theta\ +$

$\int_0^{2\pi} \int_{\pi/6}^{\pi/2} \int_0^{\csc\phi} \rho^2 \sin\phi\, d\rho\, d\phi\, d\theta$

b) $\int_0^{2\pi} \int_1^2 \int_{\pi/6}^{\sin^{-1}(1/\rho)} \rho^2 \sin\phi\, d\phi\, d\rho\, d\theta\ +$

$\int_0^{2\pi} \int_0^2 \int_0^{\pi/6} \rho^2 \sin\phi\, d\phi\, d\rho\, d\theta$

33. $\int_0^{2\pi} \int_0^{\pi/2} \int_{\cos\phi}^2 \rho^2 \sin\phi\, d\rho\, d\phi\, d\theta = \dfrac{31\pi}{6}$

35. $\int_0^{2\pi} \int_0^\pi \int_0^{1-\cos\phi} \rho^2 \sin\phi\, d\rho\, d\phi\, d\theta = \dfrac{8\pi}{3}$

37. $\int_0^{2\pi} \int_{\pi/4}^{\pi/2} \int_0^{2\cos\phi} \rho^2 \sin\phi\, d\rho\, d\phi\, d\theta = \dfrac{\pi}{3}$

39. a) $8\int_0^{\pi/2} \int_0^{\pi/2} \int_0^2 \rho^2 \sin\phi\, d\rho\, d\phi\, d\theta$

b) $8\int_0^{\pi/2} \int_0^2 \int_0^{\sqrt{4-r^2}} r\, dz\, dr\, d\theta$

c) $8\int_0^2 \int_0^{\sqrt{4-x^2}} \int_0^{\sqrt{4-x^2-y^2}} dz\, dy\, dx$

41. a) $\int_0^{2\pi} \int_0^{\pi/3} \int_{\sec\phi}^2 \rho^2 \sin\phi\, d\rho\, d\phi\, d\theta$

b) $\int_0^{2\pi} \int_0^{\sqrt{3}} \int_1^{\sqrt{4-r^2}} r\, dz\, dr\, d\theta$

c) $\int_{-\sqrt{3}}^{\sqrt{3}} \int_{-\sqrt{3-x^2}}^{\sqrt{3-x^2}} \int_1^{\sqrt{4-x^2-y^2}} dz\, dy\, dx$ d) $\dfrac{5\pi}{3}$ **43.** $8\pi/3$

45. 9/4 **47.** $(3\pi - 4)/18$ **49.** $\dfrac{2\pi a^3}{3}$ **51.** $5\pi/3$ **53.** $\pi/2$

55. $\dfrac{4(2\sqrt{2} - 1)\pi}{3}$ **57.** 16π **59.** $5\pi/2$ **61.** $\dfrac{4\pi(8 - 3\sqrt{3})}{3}$

63. 2/3 **65.** 3/4 **67.** $\bar{x} = \bar{y} = 0$, $\bar{z} = 3/8$

69. $(\bar{x}, \bar{y}, \bar{z}) = (0, 0, 3/8)$ **71.** $\bar{x} = \bar{y} = 0$, $\bar{z} = 5/6$

73. $I_z = 30\pi$, $R_z = \sqrt{\dfrac{5}{2}}$ **75.** $I_x = \pi/4$ **77.** $\dfrac{a^4 h\pi}{10}$

79. a) $(\bar{x}, \bar{y}, \bar{z}) = \left(0, 0, \dfrac{4}{5}\right)$, $I_z = \dfrac{\pi}{12}$, $R_z = \sqrt{\dfrac{1}{3}}$

b) $(\bar{x}, \bar{y}, \bar{z}) = \left(0, 0, \dfrac{5}{6}\right)$, $I_z = \dfrac{\pi}{14}$, $R_z = \sqrt{\dfrac{5}{14}}$

83. $(\bar{x}, \bar{y}, \bar{z}) = \left(0, 0, \dfrac{2h^2 + 3h}{3h + 6}\right)$, $I_z = \dfrac{\pi a^4 (h^2 + 2h)}{4}$, $R_z = \dfrac{a}{\sqrt{2}}$

85. $\dfrac{3M}{\pi R^3}$

Section 13.7, pp. 1054–1055

1. a) $x = \dfrac{u + v}{3}$, $y = \dfrac{v - 2u}{3}$; $\dfrac{1}{3}$

b) Triangular region with boundaries $u = 0$, $v = 0$, and $u + v = 3$

3. a) $x = \dfrac{1}{5}(2u - v)$, $y = \dfrac{1}{10}(3v - u)$; $\dfrac{1}{10}$

b) Triangular region with boundaries $3v = u$, $v = 2u$, and $3u + v = 10$

5. a) $\begin{vmatrix} \cos v & -u \sin v \\ \sin v & u \cos v \end{vmatrix} = u \cos^2 v + u \sin^2 v = u$

b) $\begin{vmatrix} \sin v & u \cos v \\ \cos v & -u \sin v \end{vmatrix} = -u \sin^2 v - u \cos^2 v = -u$

9. $64/5$ **11.** $\displaystyle\int_1^2 \int_1^3 (u + v)\dfrac{2u}{v}\, du\, dv = 8 + \dfrac{52}{3}\ln 2$

13. $\dfrac{\pi ab(a^2 + b^2)}{4}$ **15.** $\dfrac{1}{3}\left(1 + \dfrac{3}{e^2}\right) \approx 0.4687$ **19.** $\dfrac{4\pi\, abc}{3}$

21. $\displaystyle\int_0^3 \int_0^2 \int_1^2 \left(\dfrac{v}{3} + \dfrac{vw}{3u}\right) du\, dv\, dw = 2 + \ln 8$

Chapter 13 Practice Exercises, pp. 1056–1058

1. $9e - 9$ **3.** $9/2$

5. $\displaystyle\int_{-2}^0 \int_{2x+4}^{4-x^2} dy\, dx = \dfrac{4}{3}$ **7.** $\displaystyle\int_{-3}^3 \int_0^{(1/2)\sqrt{9-x^2}} y\, dy\, dx = \dfrac{9}{2}$

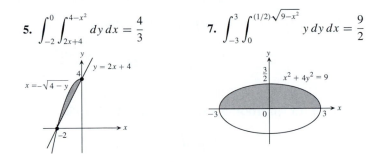

9. $\sin 4$ **11.** $\dfrac{\ln 17}{4}$ **13.** $4/3$ **15.** $4/3$ **17.** $1/4$

19. $\bar{x} = \bar{y} = \dfrac{1}{2 - \ln 4}$ **21.** $I_0 = 104$ **23.** $I_x = 2\delta$, $R_x = \sqrt{\dfrac{2}{3}}$

25. $M = 4$, $M_x = 0$, $M_y = 0$ **27.** π **29.** $\bar{x} = \dfrac{3\sqrt{3}}{\pi}$, $\bar{y} = 0$

31. a) $\bar{x} = \dfrac{15\pi + 32}{6\pi + 48}$, $\bar{y} = 0$

b)

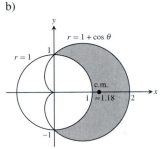

33. $\dfrac{\pi - 2}{4}$ **35.** 0 **37.** $8/35$ **39.** $\pi/2$ **41.** $\dfrac{2(31 - 3^{5/2})}{3}$

43. a) $\displaystyle\int_{-\sqrt{2}}^{\sqrt{2}} \int_{-\sqrt{2-y^2}}^{\sqrt{2-y^2}} \int_{\sqrt{x^2+y^2}}^{\sqrt{4-x^2-y^2}} 3\, dz\, dx\, dy$

b) $\displaystyle\int_0^{2\pi} \int_0^{\pi/4} \int_0^2 3\rho^2 \sin\phi\, d\rho\, d\phi\, d\theta$ c) $2\pi(8 - 4\sqrt{2})$

45. $\displaystyle\int_0^{2\pi} \int_0^{\pi/4} \int_0^{\sec\phi} \rho^2 \sin\phi\, d\rho\, d\phi\, d\theta = \dfrac{\pi}{3}$

47. $\displaystyle\int_0^1 \int_{\sqrt{1-x^2}}^{\sqrt{3-x^2}} \int_1^{\sqrt{4-x^2-y^2}} z^2 xy\, dz\, dy\, dx +$

$\displaystyle\int_1^{\sqrt{3}} \int_0^{\sqrt{3-x^2}} \int_1^{\sqrt{4-x^2-y^2}} z^2 xy\, dz\, dy\, dx$

49. a) $\displaystyle\int_{-\sqrt{3}}^{\sqrt{3}} \int_{-\sqrt{3-x^2}}^{\sqrt{3-x^2}} \int_1^{\sqrt{4-x^2-y^2}} dz\, dy\, dx$

b) $\displaystyle\int_0^{2\pi} \int_0^{\sqrt{3}} \int_1^{\sqrt{4-r^2}} r\, dz\, dr\, d\theta$

c) $\int_0^{2\pi} \int_0^{\pi/3} \int_{\sec\phi}^2 \rho^2 \sin\phi \, d\rho \, d\phi \, d\theta$

51. a) $\dfrac{8\pi(4\sqrt{2}-5)}{3}$ b) $\dfrac{8\pi(4\sqrt{2}-5)}{3}$ **53.** $I_z = \dfrac{8\pi\,\delta(b^5-a^5)}{15}$

Chapter 13 Additional Exercises, pp. 1058–1060

1. a) $\int_{-3}^2 \int_x^{6-x^2} x^2 \, dy \, dx$ b) $\int_{-3}^2 \int_x^{6-x^2} \int_0^{x^2} dz \, dy \, dx$ c) $\dfrac{125}{4}$

3. 2π **5.** $3\pi/2$ **7.** a) Hole radius $= 1$, sphere radius $= 2$ b) $4\sqrt{3}\pi$ **9.** $\pi/4$

11. $\int_0^1 \int_y^{\sqrt{y}} f(x,y) \, dx \, dy$

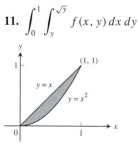

13. $\ln\left(\dfrac{b}{a}\right)$ **17.** $1/\sqrt[4]{3}$ **19.** $\bar{x} = \dfrac{15\pi + 32}{6\pi + 48}$, $\bar{y} = 0$

21. Mass $= a^2 \cos^{-1}\left(\dfrac{b}{a}\right) - b\sqrt{a^2 - b^2}$,

$I_0 = \dfrac{a^4}{2} \cos^{-1}\left(\dfrac{b}{a}\right) - \dfrac{b^3}{2}\sqrt{a^2-b^2} - \dfrac{b^3}{6}(a^2-b^2)^{3/2}$

23. a) $\bar{x} = \bar{y} = 0$, $\bar{z} = 1/2$; $I_z = \dfrac{\pi}{8}$, $R_z = \dfrac{\sqrt{3}}{2}$

b) $\bar{x} = \bar{y} = 0$, $\bar{z} = 5/14$; $I_z = \dfrac{2\pi}{7}$, $R_z = \sqrt{\dfrac{5}{7}}$

25. $\bar{x} = \bar{y} = 0$, $\bar{z} = \dfrac{3a}{8}$ **27.** $\dfrac{1}{ab}\left(e^{a^2b^2} - 1\right)$ **29.** b) 1   c) 0

33. $h = \sqrt{20}$ in., $h = \sqrt{60}$ in.   **37.** $\dfrac{1}{2}\pi^2$

CHAPTER 14

Section 14.1, pp. 1065–1067

1. c   **3.** g   **5.** d   **7.** f   **9.** $\sqrt{2}$ **11.** $13/2$ **13.** $3\sqrt{14}$

15. $(1/6)(5\sqrt{5}+9)$ **17.** $\sqrt{3}\ln(b/a)$ **19.** $\dfrac{10\sqrt{5}-2}{3}$ **21.** 8

23. $2\sqrt{2}-1$ **25.** a) $4\sqrt{2}-2$ b) $\sqrt{2}+\ln(1+\sqrt{2})$

27. $I_z = 2\pi\,\delta a^3$, $R_z = a$ **29.** a) $I_z = 2\pi\sqrt{2}\,\delta$, $R_z = 1$

b) $I_z = 4\pi\sqrt{2}\,\delta$, $R_z = 1$ **31.** $I_x = 2\pi - 2$, $R_x = 1$

Section 14.2, pp. 1074–1076

1. $\nabla f = -(x\mathbf{i} + y\mathbf{j} + z\mathbf{k})(x^2+y^2+z^2)^{-3/2}$

3. $\nabla g = -(2x/(x^2+y^2))\mathbf{i} - (2y/(x^2+y^2))\mathbf{j} + e^z\mathbf{k}$

5. $\mathbf{F} = -\dfrac{kx}{(x^2+y^2)^{3/2}}\mathbf{i} - \dfrac{ky}{(x^2+y^2)^{3/2}}\mathbf{j}$, any $k > 0$

7. a) $\dfrac{9}{2}$ b) $\dfrac{13}{3}$ c) $\dfrac{9}{2}$ **9.** a) $\dfrac{1}{3}$ b) $-\dfrac{1}{5}$ c) 0   **11.** a) 2

b) $\dfrac{3}{2}$ c) $\dfrac{1}{2}$ **13.** $1/2$ **15.** $-\pi$ **17.** $207/12$ **19.** $-39/2$

21. $25/6$ **23.** a) $\text{Circ}_1 = 0$, $\text{circ}_2 = 2\pi$, $\text{flux}_1 = 2\pi$, $\text{flux}_2 = 0$ b) $\text{Circ}_1 = 0$, $\text{circ}_2 = 8\pi$, $\text{flux}_1 = 8\pi$, $\text{flux}_2 = 0$

25. $\text{Circ} = 0$, $\text{flux} = a^2\pi$ **27.** $\text{Circ} = a^2\pi$, $\text{flux} = 0$

29. a) $-\pi/2$ b) 0   c) 1

31.

33. a) $\mathbf{G} = -y\mathbf{i} + x\mathbf{j}$ b) $\mathbf{G} = \sqrt{x^2+y^2}\,\mathbf{F}$

35. $\mathbf{F} = -(x\mathbf{i}+y\mathbf{j})/\sqrt{x^2+y^2}$ **37.** 48   **39.** π **41.** 0

43. $1/2$

Section 14.3, pp. 1083–1084

1. Conservative   **3.** Not conservative   **5.** Not conservative

7. $f(x,y,z) = x^2 + \dfrac{3y^2}{2} + 2z^2 + C$ **9.** $f(x,y,z) = xe^{y+2z} + C$

11. $f(x,y,z) = x\ln x - x + \tan(x+y) + \dfrac{1}{2}\ln(y^2+z^2) + C$

13. 49   **15.** -16 **17.** 1   **19.** $9\ln 2$ **21.** 0   **23.** -3

27. $\mathbf{F} = \nabla\left(\dfrac{x^2-1}{y}\right)$ **29.** a) 1   b) 1   c) 1   **31.** a) 2   b) 2

33. $f(x,y,z) = \dfrac{Gm\,M}{(x^2+y^2+z^2)^{1/2}}$ **35.** a) $c = b = 2a$

b) $c = b = 2$

37. It does not matter what path you use. The work will be the same on any path because the field is conservative.

Section 14.4, pp. 1093–1095

1. $\text{Flux} = 0$, $\text{circ} = 2\pi a^2$ **3.** $\text{Flux} = -\pi a^2$, $\text{circ} = 0$

5. $\text{Flux} = 2$, $\text{circ} = 0$ **7.** $\text{Flux} = -9$, $\text{circ} = 9$

9. $\text{Flux} = 1/2$, $\text{circ} = 1/2$ **11.** $\text{Flux} = 1/5$, $\text{circ} = -1/12$

13. 0   **15.** $2/33$ **17.** 0   **19.** -16π **21.** πa^2 **23.** $\dfrac{3}{8}\pi$

25. a) 0   b) $(h-k)(\text{area of the region})$ **35.** a) 0

Section 14.5, pp. 1103–1105

1. $\dfrac{13}{3}\pi$ **3.** 4   **5.** $6\sqrt{6}-2\sqrt{2}$ **7.** $\pi\sqrt{c^2+1}$

9. $\dfrac{\pi}{6}(17\sqrt{17}-5\sqrt{5})$ **11.** $3+2\ln 2$ **13.** $9a^3$

15. $\dfrac{abc}{4}(ab+ac+bc)$ **17.** 2   **19.** 18   **21.** $\pi a^3/6$

23. $\pi a^2/4$ **25.** $\pi a^3/2$ **27.** -32 **29.** -4 **31.** $3a^4$

33. $\left(\dfrac{a}{2},\dfrac{a}{2},\dfrac{a}{2}\right)$

35. $(\bar{x},\bar{y},\bar{z})=\left(0,0,\dfrac{14}{9}\right)$, $I_z=\dfrac{15\pi\sqrt{2}}{2}\delta$, $R_z=\dfrac{\sqrt{10}}{2}$

37. a) $\dfrac{8\pi}{3}a^4\delta$ b) $\dfrac{20\pi}{3}a^4\delta$ **39.** $\dfrac{\pi}{6}(13\sqrt{13}-1)$ **41.** $5\pi\sqrt{2}$

43. $\dfrac{2}{3}(5\sqrt{5}-1)$

Section 14.6, pp. 1112–1114

1. $\mathbf{r}(r,\theta)=(r\cos\theta)\mathbf{i}+(r\sin\theta)\mathbf{j}+r^2\mathbf{k}$, $0\le r\le 2$, $0\le\theta\le 2\pi$
3. $\mathbf{r}(r,\theta)=(r\cos\theta)\mathbf{i}+(r\sin\theta)\mathbf{j}+(r/2)\mathbf{k}$, $0\le r\le 6$,
$0\le\theta\le\pi/2$
5. $\mathbf{r}(r,\theta)=(r\cos\theta)\mathbf{i}+(r\sin\theta)\mathbf{j}+\sqrt{9-r^2}\,\mathbf{k}$, $0\le r\le 3\sqrt{2}/2$,
$0\le\theta\le 2\pi$; Also: $\mathbf{r}(\phi,\theta)=(3\sin\phi\cos\theta)\mathbf{i}+(3\sin\phi\sin\theta)\mathbf{j}+$
$(3\cos\phi)\mathbf{k}$, $0\le\phi\le\pi/4$, $0\le\theta\le 2\pi$
7. $\mathbf{r}(\phi,\theta)=(\sqrt{3}\sin\phi\cos\theta)\mathbf{i}+(\sqrt{3}\sin\phi\sin\theta)\mathbf{j}+(\sqrt{3}\cos\phi)\mathbf{k}$,
$\pi/3\le\phi\le 2\pi/3$, $0\le\theta\le 2\pi$
9. $\mathbf{r}(x,y)=x\mathbf{i}+y\mathbf{j}+(4-y^2)\mathbf{k}$, $0\le x\le 2$, $-2\le y\le 2$
11. $\mathbf{r}(u,v)=u\mathbf{i}+(3\cos v)\mathbf{j}+(3\sin v)\mathbf{k}$, $0\le u\le 3$, $0\le v\le 2\pi$
13. a) $\mathbf{r}(r,\theta)=(r\cos\theta)\mathbf{i}+(r\sin\theta)\mathbf{j}+(1-r\cos\theta-r\sin\theta)\mathbf{k}$,
$0\le r\le 3$, $0\le\theta\le 2\pi$
b) $\mathbf{r}(u,v)=(1-u\cos v-u\sin v)\mathbf{i}+(u\cos v)\mathbf{j}+(u\sin v)\mathbf{k}$,
$0\le u\le 3, 0\le v\le 2\pi$
15. $\mathbf{r}(u,v)=(4\cos^2 v)\mathbf{i}+u\mathbf{j}+(4\cos v\sin v)\mathbf{k}$, $0\le u\le 3$,
$-(\pi/2)\le v\le(\pi/2)$; Another way: $\mathbf{r}(u,v)=(2+2\cos v)\mathbf{i}+u\mathbf{j}+$
$(2\sin v)\mathbf{k}$, $0\le u\le 3, 0\le v\le 2\pi$

17. $\displaystyle\int_0^{2\pi}\int_0^1\dfrac{\sqrt{5}}{2}r\,dr\,d\theta=\dfrac{\pi\sqrt{5}}{2}$

19. $\displaystyle\int_0^{2\pi}\int_1^3 r\sqrt{5}\,dr\,d\theta=8\pi\sqrt{5}$

21. $\displaystyle\int_0^{2\pi}\int_1^4 1\,du\,dv=6\pi$

23. $\displaystyle\int_0^{2\pi}\int_0^1 u\sqrt{4u^2+1}\,du\,dv=\dfrac{(5\sqrt{5}-1)}{6}\pi$

25. $\displaystyle\int_0^{2\pi}\int_{\pi/4}^\pi 2\sin\phi\,d\phi\,d\theta=(4+2\sqrt{2})\pi$

27. $\displaystyle\iint_S x\,d\sigma=\int_0^3\int_0^2 u\sqrt{4u^2+1}\,du\,dv=\dfrac{17\sqrt{17}-1}{4}$

29. $\displaystyle\iint_S x^2\,d\sigma=\int_0^{2\pi}\int_0^\pi\sin^3\phi\cos^2\theta\,d\phi\,d\theta=\dfrac{4\pi}{3}$

31. $\displaystyle\iint_S z\,d\sigma=\int_0^1\int_0^1(4-u-v)\sqrt{3}\,dv\,du=3\sqrt{3}$

(for $x=u$, $y=v$)

33. $\displaystyle\iint_S x^2\sqrt{5-4z}\,d\sigma=\int_0^1\int_0^{2\pi}u^2\cos^2 v\cdot\sqrt{4u^2+1}\,\cdot$

$u\sqrt{4u^2+1}\,dv\,du=\int_0^1\int_0^{2\pi}u^3(4u^2+1)\cos^2 v\,dv\,du=\dfrac{11\pi}{12}$

35. -32 **37.** $\pi a^3/6$ **39.** $13a^4/6$ **41.** $2\pi/3$ **43.** $-73\pi/6$
45. $(a/2,a/2,a/2)$ **47.** $8\delta\pi a^4/3$
49.   **51.**

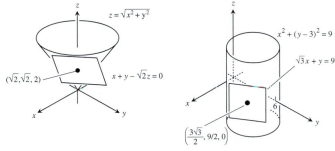

55. b) $A=\displaystyle\int_0^{2\pi}\int_0^\pi[a^2b^2\sin^2\phi\cos^2\phi+b^2c^2\cos^4\phi\cos^2\theta+$
$a^2c^2\cos^4\phi\sin^2\theta]^{1/2}\,d\phi\,d\theta$

57. $x_0 x+y_0 y=25$

Section 14.7, pp. 1122–1123

1. 4π **3.** $-5/6$ **5.** 0   **7.** -6π **9.** $2\pi a^2$ **13.** 12π

15. $-\dfrac{\pi}{4}$ **17.** -15π **25.** $16\,I_y+16\,I_x$

Section 14.8, pp. 1132–1134

1. 0   **3.** 0   **5.** -16 **7.** -8π **9.** 3π **11.** $-40/3$
13. 12π **15.** $12\pi(4\sqrt{2}-1)$
21. The integral's value never exceeds the surface area of S.

Chapter 14 Practice Exercises, pp. 1134–1137

1. Path 1: $2\sqrt{3}$, Path 2: $1+3\sqrt{2}$ **3.** $4a^2$ **5.** 0   **7.** 0

9. 0   **11.** $\pi\sqrt{3}$ **13.** $2\pi\left(1-\dfrac{1}{\sqrt{2}}\right)$ **15.** $\dfrac{abc}{2}\sqrt{\dfrac{1}{a^2}+\dfrac{1}{b^2}+\dfrac{1}{c^2}}$

17. 50
19. $\mathbf{r}(\phi,\theta)=(6\sin\phi\cos\theta)\mathbf{i}+(6\sin\phi\sin\theta)\mathbf{j}+(6\cos\phi)\mathbf{k}$,
$(\pi/6)\le\phi\le 2\pi/3$, $0\le\theta\le 2\pi$
21. $\mathbf{r}(r,\theta)=(r\cos\theta)\mathbf{i}+(r\sin\theta)\mathbf{j}+(1+r)\mathbf{k}$, $0\le r\le 2$,
$0\le\theta\le 2\pi$

23. $\mathbf{r}(u, v) = (u \cos v)\mathbf{i} + 2u^2\mathbf{j} + (u \sin v)\mathbf{k}, 0 \le u \le 1,$
$0 \le v \le \pi$
25. $\sqrt{6}$ **27.** $\pi[\sqrt{2} + \ln(1 + \sqrt{2})]$ **29.** Conservative
31. Not conservative   **33.** $f(x, y, z) = y^2 + yz + 2x + z$
35. Path 1: 2, Path 2: 8/3   **37.** a) $1 - e^{-2\pi}$ b) $1 - e^{-2\pi}$
39. a) $-\pi/2$ b) 0   c) 1   **41.** 0   **43.** a) $4\sqrt{2} - 2$
b) $\sqrt{2} + \ln(1 + \sqrt{2})$

45. $(\bar{x}, \bar{y}, \bar{z}) = \left(1, \dfrac{16}{15}, \dfrac{2}{3}\right); I_x = \dfrac{232}{45}, I_y = \dfrac{64}{15}, I_z = \dfrac{56}{9};$

$R_x = \sqrt{\dfrac{116}{45}}, R_y = \sqrt{\dfrac{32}{15}}, R_z = \sqrt{\dfrac{28}{9}}$

47. $\bar{z} = \dfrac{3}{2}, I_z = \dfrac{7\sqrt{3}}{3}, R_z = \sqrt{\dfrac{7}{3}}$

49. $(\bar{x}, \bar{y}, \bar{z}) = (0, 0, 49/12), I_z = 640\pi, R_z = 2\sqrt{2}$

51. Flux: 3/2, Circ: $-1/2$ **55.** 3   **57.** $\dfrac{2\pi}{3}(7 - 8\sqrt{2})$ **59.** 0

61. π

Chapter 14 Additional Exercises, pp. 1137–1139

1. 6π **3.** 2/3   **5.** a) $\mathbf{F}(x, y, z) = z\mathbf{i} + x\mathbf{j} + y\mathbf{k}$

b) $\mathbf{F}(x, y, z) = z\mathbf{i} + y\mathbf{k}$ c) $\mathbf{F}(x, y, z) = z\mathbf{i}$ **7.** $\dfrac{16\pi R^3}{3}$

9. $a = 2, b = 1$. The minimum flux is -4.   **11.** b) $\dfrac{16}{3}g$

c) Work $= \left(\displaystyle\int_C g \, xy \, ds\right) \bar{y} = g \displaystyle\int_C xy^2 \, ds$ **13.** c) $\dfrac{4}{3}\pi w$

19. False if $\mathbf{F} = y\mathbf{i} + x\mathbf{j}$

APPENDICES

Appendix A.3, pp. A-16–A-17

1. a) $(14, 8)$ b) $(-1, 8)$ c) $(0, -5)$
3. a) By reflecting z across the real axis   b) By reflecting z across the imaginary axis   c) By reflecting z in the origin   d) By reflecting z in the real axis and then multiplying the length of the vector by $1/|z|^2$
5. a) Points on the circle $x^2 + y^2 = 4$ b) points inside the circle $x^2 + y^2 = 4$ c) points outside the circle $x^2 + y^2 = 4$
7. Points on a circle of radius 1, center $(-1, 0)$
9. Points on the line $y = -x$ **11.** $4\,e^{2\pi i/3}$ **13.** $1\,e^{2\pi i/3}$

21. $\cos^4 \theta - 6 \cos^2 \theta \sin^2 \theta + \sin^4 \theta$ **23.** $1, -\dfrac{1}{2} \pm \dfrac{\sqrt{3}}{2}i$

25. $2i, -\sqrt{3} - i, \sqrt{3} - i$ **27.** $\dfrac{\sqrt{6}}{2} \pm \dfrac{\sqrt{2}}{2}i, -\dfrac{\sqrt{6}}{2} \pm \dfrac{\sqrt{2}}{2}i$

29. $1 \pm \sqrt{3}i, -1 \pm \sqrt{3}i$

Appendix A.8, p. A-29

1. -5 **3.** 1   **5.** -7 **7.** 38   **9.** $x = -4, y = 1$
11. $x = 3, y = 2$ **13.** $x = 3, y = -2, z = 2$
15. $x = 2, y = 0, z = -1$ **17.** a) $h = 6, k = 4$
b) $h = 6, k \neq 4$

Index

Note: Numbers in parentheses refer to exercises on the pages indicated.

A Brief Table of Integrals

1. $\displaystyle\int u\, dv = uv - \int v\, du$

2. $\displaystyle\int a^u\, du = \frac{a^u}{\ln a} + C, \quad a \neq 1, \quad a > 0$

3. $\displaystyle\int \cos u\, du = \sin u + C$

4. $\displaystyle\int \sin u\, du = -\cos u + C$

5. $\displaystyle\int (ax+b)^n\, dx = \frac{(ax+b)^{n+1}}{a(n+1)} + C, \quad n \neq -1$

6. $\displaystyle\int (ax+b)^{-1}\, dx = \frac{1}{a} \ln |ax+b| + C$

7. $\displaystyle\int x(ax+b)^n\, dx = \frac{(ax+b)^{n+1}}{a^2}\left[\frac{ax+b}{n+2} - \frac{b}{n+1}\right] + C, \quad n \neq -1, -2$

8. $\displaystyle\int x(ax+b)^{-1}\, dx = \frac{x}{a} - \frac{b}{a^2} \ln |ax+b| + C$

9. $\displaystyle\int x(ax+b)^{-2}\, dx = \frac{1}{a^2}\left[\ln |ax+b| + \frac{b}{ax+b}\right] + C$

10. $\displaystyle\int \frac{dx}{x(ax+b)} = \frac{1}{b} \ln \left|\frac{x}{ax+b}\right| + C$

11. $\displaystyle\int (\sqrt{ax+b})^n\, dx = \frac{2}{a}\frac{(\sqrt{ax+b})^{n+2}}{n+2} + C, \quad n \neq -2$

12. $\displaystyle\int \frac{\sqrt{ax+b}}{x}\, dx = 2\sqrt{ax+b} + b\int \frac{dx}{x\sqrt{ax+b}}$

13. a) $\displaystyle\int \frac{dx}{x\sqrt{ax-b}} = \frac{2}{\sqrt{b}} \tan^{-1}\sqrt{\frac{ax-b}{b}} + C$

   b) $\displaystyle\int \frac{dx}{x\sqrt{ax+b}} = \frac{1}{\sqrt{b}} \ln \left|\frac{\sqrt{ax+b} - \sqrt{b}}{\sqrt{ax+b} + \sqrt{b}}\right| + C$

14. $\displaystyle\int \frac{\sqrt{ax+b}}{x^2}\, dx = -\frac{\sqrt{ax+b}}{x} + \frac{a}{2}\int \frac{dx}{x\sqrt{ax+b}} + C$

15. $\displaystyle\int \frac{dx}{x^2\sqrt{ax+b}} = -\frac{\sqrt{ax+b}}{bx} - \frac{a}{2b}\int \frac{dx}{x\sqrt{ax+b}} + C$

16. $\displaystyle\int \frac{dx}{a^2+x^2} = \frac{1}{a} \tan^{-1}\frac{x}{a} + C$

17. $\displaystyle\int \frac{dx}{(a^2+x^2)^2} = \frac{x}{2a^2(a^2+x^2)} + \frac{1}{2a^3} \tan^{-1}\frac{x}{a} + C$

18. $\displaystyle\int \frac{dx}{a^2-x^2} = \frac{1}{2a} \ln \left|\frac{x+a}{x-a}\right| + C$

19. $\displaystyle\int \frac{dx}{(a^2-x^2)^2} = \frac{x}{2a^2(a^2-x^2)} + \frac{1}{4a^3} \ln \left|\frac{x+a}{x-a}\right| + C$

20. $\displaystyle\int \frac{dx}{\sqrt{a^2+x^2}} = \sinh^{-1}\frac{x}{a} + C = \ln(x + \sqrt{a^2+x^2}) + C$

21. $\displaystyle\int \sqrt{a^2+x^2}\, dx = \frac{x}{2}\sqrt{a^2+x^2} +$

22. $\displaystyle\int x^2\sqrt{a^2+x^2}\, dx = \frac{x}{8}(a^2+2x^2)\sqrt{a^2+x^2} - \frac{a^4}{8} \ln(x + \sqrt{a^2+x^2}) + C$
$\displaystyle\frac{a^2}{2} \ln(x + \sqrt{a^2+x^2}) + C$

23. $\displaystyle\int \frac{\sqrt{a^2+x^2}}{x}\, dx = \sqrt{a^2+x^2} - a \ln\left|\frac{a + \sqrt{a^2+x^2}}{x}\right| + C$

24. $\displaystyle\int \frac{\sqrt{a^2+x^2}}{x^2}\, dx = \ln(x + \sqrt{a^2+x^2}) - \frac{\sqrt{a^2+x^2}}{x} + C$

25. $\displaystyle\int \frac{x^2}{\sqrt{a^2+x^2}}\, dx = -\frac{a^2}{2} \ln(x + \sqrt{a^2+x^2}) + \frac{x\sqrt{a^2+x^2}}{2} + C$

26. $\displaystyle\int \frac{dx}{x\sqrt{a^2+x^2}} = -\frac{1}{a} \ln\left|\frac{a + \sqrt{a^2+x^2}}{x}\right| + C$

27. $\displaystyle\int \frac{dx}{x^2\sqrt{a^2+x^2}} = -\frac{\sqrt{a^2+x^2}}{a^2x} + C$

28. $\displaystyle\int \frac{dx}{\sqrt{a^2-x^2}} = \sin^{-1}\frac{x}{a} + C$

29. $\displaystyle\int \sqrt{a^2-x^2}\, dx = \frac{x}{2}\sqrt{a^2-x^2} + \frac{a^2}{2} \sin^{-1}\frac{x}{a} + C$

30. $\int x^2\sqrt{a^2 - x^2}\, dx = \dfrac{a^4}{8} \sin^{-1} \dfrac{x}{a} - \dfrac{1}{8}x\sqrt{a^2 - x^2}(a^2 - 2x^2) + C$

31. $\int \dfrac{\sqrt{a^2 - x^2}}{x}\, dx = \sqrt{a^2 - x^2} - a \ln \left| \dfrac{a + \sqrt{a^2 - x^2}}{x} \right| + C$ **32.** $\int \dfrac{\sqrt{a^2 - x^2}}{x^2}\, dx = -\sin^{-1} \dfrac{x}{a} - \dfrac{\sqrt{a^2 - x^2}}{x} + C$

33. $\int \dfrac{x^2}{\sqrt{a^2 - x^2}}\, dx = \dfrac{a^2}{2} \sin^{-1} \dfrac{x}{a} - \dfrac{1}{2}x\sqrt{a^2 - x^2} + C$ **34.** $\int \dfrac{dx}{x\sqrt{a^2 - x^2}} = -\dfrac{1}{a} \ln \left| \dfrac{a + \sqrt{a^2 - x^2}}{x} \right| + C$

35. $\int \dfrac{dx}{x^2\sqrt{a^2 - x^2}} = -\dfrac{\sqrt{a^2 - x^2}}{a^2 x} + C$ **36.** $\int \dfrac{dx}{\sqrt{x^2 - a^2}} = \cosh^{-1} \dfrac{x}{a} + C = \ln \left| x + \sqrt{x^2 - a^2} \right| + C$

37. $\int \sqrt{x^2 - a^2}\, dx = \dfrac{x}{2}\sqrt{x^2 - a^2} - \dfrac{a^2}{2} \ln \left| x + \sqrt{x^2 - a^2} \right| + C$

38. $\int (\sqrt{x^2 - a^2})^n\, dx = \dfrac{x(\sqrt{x^2 - a^2})^n}{n + 1} - \dfrac{na^2}{n + 1} \int (\sqrt{x^2 - a^2})^{n-2} dx, \quad n \neq -1$

39. $\int \dfrac{dx}{(\sqrt{x^2 - a^2})^n} = \dfrac{x(\sqrt{x^2 - a^2})^{2-n}}{(2 - n)a^2} - \dfrac{n - 3}{(n - 2)a^2} \int \dfrac{dx}{(\sqrt{x^2 - a^2})^{n-2}}, \quad n \neq 2$

40. $\int x(\sqrt{x^2 - a^2})^n\, dx = \dfrac{(\sqrt{x^2 - a^2})^{n+2}}{n + 2} + C, \quad n \neq -2$

41. $\int x^2\sqrt{x^2 - a^2}\, dx = \dfrac{x}{8}(2x^2 - a^2)\sqrt{x^2 - a^2} - \dfrac{a^4}{8} \ln \left| x + \sqrt{x^2 - a^2} \right| + C$

42. $\int \dfrac{\sqrt{x^2 - a^2}}{x}\, dx = \sqrt{x^2 - a^2} - a \sec^{-1} \left| \dfrac{x}{a} \right| + C$ **43.** $\int \dfrac{\sqrt{x^2 - a^2}}{x^2}\, dx = \ln \left| x + \sqrt{x^2 - a^2} \right| - \dfrac{\sqrt{x^2 - a^2}}{x} + C$

44. $\int \dfrac{x^2}{\sqrt{x^2 - a^2}}\, dx = \dfrac{a^2}{2} \ln \left| x + \sqrt{x^2 - a^2} \right| + \dfrac{x}{2}\sqrt{x^2 - a^2} + C$

45. $\int \dfrac{dx}{x\sqrt{x^2 - a^2}} = \dfrac{1}{a} \sec^{-1} \left| \dfrac{x}{a} \right| + C = \dfrac{1}{a} \cos^{-1} \left| \dfrac{a}{x} \right| + C$ **46.** $\int \dfrac{dx}{x^2\sqrt{x^2 - a^2}} = \dfrac{\sqrt{x^2 - a^2}}{a^2 x} + C$

47. $\int \dfrac{dx}{\sqrt{2ax - x^2}} = \sin^{-1} \left(\dfrac{x - a}{a} \right) + C$

48. $\int \sqrt{2ax - x^2}\, dx = \dfrac{x - a}{2}\sqrt{2ax - x^2} + \dfrac{a^2}{2} \sin^{-1} \left(\dfrac{x - a}{a} \right) + C$

49. $\int (\sqrt{2ax - x^2})^n dx = \dfrac{(x - a)(\sqrt{2ax - x^2})^n}{n + 1} + \dfrac{na^2}{n + 1} \int (\sqrt{2ax - x^2})^{n-2} dx$

50. $\int \dfrac{dx}{(\sqrt{2ax - x^2})^n} = \dfrac{(x - a)(\sqrt{2ax - x^2})^{2-n}}{(n - 2)a^2} + \dfrac{n - 3}{(n - 2)a^2} \int \dfrac{dx}{(\sqrt{2ax - x^2})^{n-2}}$

51. $\int x\sqrt{2ax - x^2}\, dx = \dfrac{(x + a)(2x - 3a)\sqrt{2ax - x^2}}{6} + \dfrac{a^3}{2} \sin^{-1} \left(\dfrac{x - a}{a} \right) + C$

52. $\int \dfrac{\sqrt{2ax - x^2}}{x}\, dx = \sqrt{2ax - x^2} + a \sin^{-1} \left(\dfrac{x - a}{a} \right) + C$ **53.** $\int \dfrac{\sqrt{2ax - x^2}}{x^2}\, dx = -2\sqrt{\dfrac{2a - x}{x}} - \sin^{-1} \left(\dfrac{x - a}{a} \right) + C$

54. $\int \dfrac{x\, dx}{\sqrt{2ax - x^2}} = a \sin^{-1} \left(\dfrac{x - a}{a} \right) - \sqrt{2ax - x^2} + C$ **55.** $\int \dfrac{dx}{x\sqrt{2ax - x^2}} = -\dfrac{1}{a}\sqrt{\dfrac{2a - x}{x}} + C$

56. $\displaystyle\int \sin ax\, dx = -\frac{1}{a}\cos ax + C$

57. $\displaystyle\int \cos ax\, dx = \frac{1}{a}\sin ax + C$

58. $\displaystyle\int \sin^2 ax\, dx = \frac{x}{2} - \frac{\sin 2ax}{4a} + C$

59. $\displaystyle\int \cos^2 ax\, dx = \frac{x}{2} + \frac{\sin 2ax}{4a} + C$

60. $\displaystyle\int \sin^n ax\, dx = -\frac{\sin^{n-1} ax \cos ax}{na} + \frac{n-1}{n}\int \sin^{n-2} ax\, dx$

61. $\displaystyle\int \cos^n ax\, dx = \frac{\cos^{n-1} ax \sin ax}{na}$
$$+ \frac{n-1}{n}\int \cos^{n-2} ax\, dx$$

62. a) $\displaystyle\int \sin ax \cos bx\, dx = -\frac{\cos (a+b)x}{2(a+b)} - \frac{\cos (a-b)x}{2(a-b)} + C, \quad a^2 \neq b^2$

b) $\displaystyle\int \sin ax \sin bx\, dx = \frac{\sin (a-b)x}{2(a-b)} - \frac{\sin (a+b)x}{2(a+b)} + C, \quad a^2 \neq b^2$

c) $\displaystyle\int \cos ax \cos bx\, dx = \frac{\sin (a-b)x}{2(a-b)} + \frac{\sin (a+b)x}{2(a+b)} + C, \quad a^2 \neq b^2$

63. $\displaystyle\int \sin ax \cos ax\, dx = -\frac{\cos 2ax}{4a} + C$

64. $\displaystyle\int \sin^n ax \cos ax\, dx = \frac{\sin^{n+1} ax}{(n+1)a} + C, \quad n \neq -1$

65. $\displaystyle\int \frac{\cos ax}{\sin ax}\, dx = \frac{1}{a}\ln |\sin ax| + C$

66. $\displaystyle\int \cos^n ax \sin ax\, dx = -\frac{\cos^{n+1} ax}{(n+1)a} + C, \quad n \neq -1$

67. $\displaystyle\int \frac{\sin ax}{\cos ax}\, dx = -\frac{1}{a}\ln |\cos ax| + C$

68. $\displaystyle\int \sin^n ax \cos^m ax\, dx = -\frac{\sin^{n-1} ax \cos^{m+1} ax}{a(m+n)} + \frac{n-1}{m+n}\int \sin^{n-2} ax \cos^m ax\, dx, \quad n \neq -m \quad \text{(reduces } \sin^n ax\text{)}$

69. $\displaystyle\int \sin^n ax \cos^m ax\, dx = \frac{\sin^{n+1} ax \cos^{m-1} ax}{a(m+n)} + \frac{m-1}{m+n}\int \sin^n ax \cos^{m-2} ax\, dx, \quad m \neq -n \quad \text{(reduces } \cos^m ax\text{)}$

70. $\displaystyle\int \frac{dx}{b + c \sin ax} = \frac{-2}{a\sqrt{b^2 - c^2}}\tan^{-1}\left[\sqrt{\frac{b-c}{b+c}}\tan\left(\frac{\pi}{4} - \frac{ax}{2}\right)\right] + C, \quad b^2 > c^2$

71. $\displaystyle\int \frac{dx}{b + c \sin ax} = \frac{-1}{a\sqrt{c^2 - b^2}}\ln\left|\frac{c + b \sin ax + \sqrt{c^2 - b^2}\cos ax}{b + c \sin ax}\right| + C, \quad b^2 < c^2$

72. $\displaystyle\int \frac{dx}{1 + \sin ax} = -\frac{1}{a}\tan\left(\frac{\pi}{4} - \frac{ax}{2}\right) + C$

73. $\displaystyle\int \frac{dx}{1 - \sin ax} = \frac{1}{a}\tan\left(\frac{\pi}{4} + \frac{ax}{2}\right) + C$

74. $\displaystyle\int \frac{dx}{b + c \cos ax} = \frac{2}{a\sqrt{b^2 - c^2}}\tan^{-1}\left[\sqrt{\frac{b-c}{b+c}}\tan\frac{ax}{2}\right] + C, \quad b^2 > c^2$

75. $\displaystyle\int \frac{dx}{b + c \cos ax} = \frac{1}{a\sqrt{c^2 - b^2}}\ln\left|\frac{c + b \cos ax + \sqrt{c^2 - b^2}\sin ax}{b + c \cos ax}\right| + C, \quad b^2 < c^2$

76. $\displaystyle\int \frac{dx}{1 + \cos ax} = \frac{1}{a}\tan\frac{ax}{2} + C$

77. $\displaystyle\int \frac{dx}{1 - \cos ax} = -\frac{1}{a}\cot\frac{ax}{2} + C$

78. $\displaystyle\int x \sin ax\, dx = \frac{1}{a^2}\sin ax - \frac{x}{a}\cos ax + C$

79. $\displaystyle\int x \cos ax\, dx = \frac{1}{a^2}\cos ax + \frac{x}{a}\sin ax + C$

80. $\displaystyle\int x^n \sin ax \, dx = -\frac{x^n}{a} \cos ax + \frac{n}{a} \int x^{n-1} \cos ax \, dx$

81. $\displaystyle\int x^n \cos ax \, dx = \frac{x^n}{a} \sin ax - \frac{n}{a} \int x^{n-1} \sin ax \, dx$

82. $\displaystyle\int \tan ax \, dx = \frac{1}{a} \ln |\sec ax| + C$

83. $\displaystyle\int \cot ax \, dx = \frac{1}{a} \ln |\sin ax| + C$

84. $\displaystyle\int \tan^2 ax \, dx = \frac{1}{a} \tan ax - x + C$

85. $\displaystyle\int \cot^2 ax \, dx = -\frac{1}{a} \cot ax - x + C$

86. $\displaystyle\int \tan^n ax \, dx = \frac{\tan^{n-1} ax}{a(n-1)} - \int \tan^{n-2} ax \, dx, \quad n \neq 1$

87. $\displaystyle\int \cot^n ax \, dx = -\frac{\cot^{n-1} ax}{a(n-1)} - \int \cot^{n-2} ax \, dx, \quad n \neq 1$

88. $\displaystyle\int \sec ax \, dx = \frac{1}{a} \ln |\sec ax + \tan ax| + C$

89. $\displaystyle\int \csc ax \, dx = -\frac{1}{a} \ln |\csc ax + \cot ax| + C$

90. $\displaystyle\int \sec^2 ax \, dx = \frac{1}{a} \tan ax + C$

91. $\displaystyle\int \csc^2 ax \, dx = -\frac{1}{a} \cot ax + C$

92. $\displaystyle\int \sec^n ax \, dx = \frac{\sec^{n-2} ax \tan ax}{a(n-1)} + \frac{n-2}{n-1} \int \sec^{n-2} ax \, dx, \quad n \neq 1$

93. $\displaystyle\int \csc^n ax \, dx = -\frac{\csc^{n-2} ax \cot ax}{a(n-1)} + \frac{n-2}{n-1} \int \csc^{n-2} ax \, dx, \quad n \neq 1$

94. $\displaystyle\int \sec^n ax \tan ax \, dx = \frac{\sec^n ax}{na} + C, \quad n \neq 0$

95. $\displaystyle\int \csc^n ax \cot ax \, dx = -\frac{\csc^n ax}{na} + C, \quad n \neq 0$

96. $\displaystyle\int \sin^{-1} ax \, dx = x \sin^{-1} ax + \frac{1}{a}\sqrt{1 - a^2x^2} + C$

97. $\displaystyle\int \cos^{-1} ax \, dx = x \cos^{-1} ax - \frac{1}{a}\sqrt{1 - a^2x^2} + C$

98. $\displaystyle\int \tan^{-1} ax \, dx = x \tan^{-1} ax - \frac{1}{2a} \ln(1 + a^2x^2) + C$

99. $\displaystyle\int x^n \sin^{-1} ax \, dx = \frac{x^{n+1}}{n+1} \sin^{-1} ax - \frac{a}{n+1} \int \frac{x^{n+1} \, dx}{\sqrt{1 - a^2x^2}}, \quad n \neq -1$

100. $\displaystyle\int x^n \cos^{-1} ax \, dx = \frac{x^{n+1}}{n+1} \cos^{-1} ax + \frac{a}{n+1} \int \frac{x^{n+1} \, dx}{\sqrt{1 - a^2x^2}}, \quad n \neq -1$

101. $\displaystyle\int x^n \tan^{-1} ax \, dx = \frac{x^{n+1}}{n+1} \tan^{-1} ax - \frac{a}{n+1} \int \frac{x^{n+1} \, dx}{1 + a^2x^2}, \quad n \neq -1$

102. $\displaystyle\int e^{ax} \, dx = \frac{1}{a} e^{ax} + C$

103. $\displaystyle\int b^{ax} \, dx = \frac{1}{a} \frac{b^{ax}}{\ln b} + C, \quad b > 0, \ b \neq 1$

104. $\displaystyle\int xe^{ax} \, dx = \frac{e^{ax}}{a^2}(ax - 1) + C$

105. $\displaystyle\int x^n e^{ax} \, dx = \frac{1}{a} x^n e^{ax} - \frac{n}{a} \int x^{n-1} e^{ax} \, dx$

106. $\displaystyle\int x^n b^{ax} \, dx = \frac{x^n b^{ax}}{a \ln b} - \frac{n}{a \ln b} \int x^{n-1} b^{ax} \, dx, \quad b > 0, b \neq 1$

107. $\displaystyle\int e^{ax} \sin bx \, dx = \frac{e^{ax}}{a^2 + b^2}(a \sin bx - b \cos bx) + C$

108. $\displaystyle\int e^{ax} \cos bx \, dx = \frac{e^{ax}}{a^2 + b^2}(a \cos bx + b \sin bx) + C$

109. $\displaystyle\int \ln ax \, dx = x \ln ax - x + C$

110. $\displaystyle\int x^n (\ln ax)^m \, dx = \frac{x^{n+1}(\ln ax)^m}{n+1} - \frac{m}{n+1} \int x^n (\ln ax)^{m-1} dx, \quad n \neq -1$

111. $\displaystyle\int x^{-1}(\ln ax)^m \, dx = \frac{(\ln ax)^{m+1}}{m+1} + C, \quad m \neq -1$

112. $\displaystyle\int \frac{dx}{x \ln ax} = \ln |\ln ax| + C$

113. $\displaystyle\int \sinh ax\, dx = \frac{1}{a}\cosh ax + C$

114. $\displaystyle\int \cosh ax\, dx = \frac{1}{a}\sinh ax + C$

115. $\displaystyle\int \sinh^2 ax\, dx = \frac{\sinh 2ax}{4a} - \frac{x}{2} + C$

116. $\displaystyle\int \cosh^2 ax\, dx = \frac{\sinh 2ax}{4a} + \frac{x}{2} + C$

117. $\displaystyle\int \sinh^n ax\, dx = \frac{\sinh^{n-1} ax \cosh ax}{na} - \frac{n-1}{n}\int \sinh^{n-2} ax\, dx, \quad n \neq 0$

118. $\displaystyle\int \cosh^n ax\, dx = \frac{\cosh^{n-1} ax \sinh ax}{na} + \frac{n-1}{n}\int \cosh^{n-2} ax\, dx, \quad n \neq 0$

119. $\displaystyle\int x\sinh ax\, dx = \frac{x}{a}\cosh ax - \frac{1}{a^2}\sinh ax + C$

120. $\displaystyle\int x\cosh ax\, dx = \frac{x}{a}\sinh ax - \frac{1}{a^2}\cosh ax + C$

121. $\displaystyle\int x^n \sinh ax\, dx = \frac{x^n}{a}\cosh ax - \frac{n}{a}\int x^{n-1}\cosh ax\, dx$

122. $\displaystyle\int x^n \cosh ax\, dx = \frac{x^n}{a}\sinh ax - \frac{n}{a}\int x^{n-1}\sinh ax\, dx$

123. $\displaystyle\int \tanh ax\, dx = \frac{1}{a}\ln(\cosh ax) + C$

124. $\displaystyle\int \coth ax\, dx = \frac{1}{a}\ln|\sinh ax| + C$

125. $\displaystyle\int \tanh^2 ax\, dx = x - \frac{1}{a}\tanh ax + C$

126. $\displaystyle\int \coth^2 ax\, dx = x - \frac{1}{a}\coth ax + C$

127. $\displaystyle\int \tanh^n ax\, dx = -\frac{\tanh^{n-1} ax}{(n-1)a} + \int \tanh^{n-2} ax\, dx, \quad n \neq 1$

128. $\displaystyle\int \coth^n ax\, dx = -\frac{\coth^{n-1} ax}{(n-1)a} + \int \coth^{n-2} ax\, dx, \quad n \neq 1$

129. $\displaystyle\int \operatorname{sech} ax\, dx = \frac{1}{a}\sin^{-1}(\tanh ax) + C$

130. $\displaystyle\int \operatorname{csch} ax\, dx = \frac{1}{a}\ln\left|\tanh \frac{ax}{2}\right| + C$

131. $\displaystyle\int \operatorname{sech}^2 ax\, dx = \frac{1}{a}\tanh ax + C$

132. $\displaystyle\int \operatorname{csch}^2 ax\, dx = -\frac{1}{a}\coth ax + C$

133. $\displaystyle\int \operatorname{sech}^n ax\, dx = \frac{\operatorname{sech}^{n-2} ax \tanh ax}{(n-1)a} + \frac{n-2}{n-1}\int \operatorname{sech}^{n-2} ax\, dx, \quad n \neq 1$

134. $\displaystyle\int \operatorname{csch}^n ax\, dx = -\frac{\operatorname{csch}^{n-2} ax \coth ax}{(n-1)a} - \frac{n-2}{n-1}\int \operatorname{csch}^{n-2} ax\, dx, \quad n \neq 1$

135. $\displaystyle\int \operatorname{sech}^n ax \tanh ax\, dx = -\frac{\operatorname{sech}^n ax}{na} + C, \quad n \neq 0$

136. $\displaystyle\int \operatorname{csch}^n ax \coth ax\, dx = -\frac{\operatorname{csch}^n ax}{na} + C, \quad n \neq 0$

137. $\displaystyle\int e^{ax}\sinh bx\, dx = \frac{e^{ax}}{2}\left[\frac{e^{bx}}{a+b} - \frac{e^{-bx}}{a-b}\right] + C, \quad a^2 \neq b^2$

138. $\displaystyle\int e^{ax}\cosh bx\, dx = \frac{e^{ax}}{2}\left[\frac{e^{bx}}{a+b} + \frac{e^{-bx}}{a-b}\right] + C, \quad a^2 \neq b^2$

139. $\displaystyle\int_0^\infty x^{n-1}e^{-x}\, dx = \Gamma(n) = (n-1)!, \quad n > 0$

140. $\displaystyle\int_0^\infty e^{-ax^2}\, dx = \frac{1}{2}\sqrt{\frac{\pi}{a}}, \quad a > 0$

141. $\displaystyle\int_0^{\pi/2} \sin^n x\, dx = \int_0^{\pi/2} \cos^n x\, dx = \begin{cases} \dfrac{1\cdot 3\cdot 5\cdots(n-1)}{2\cdot 4\cdot 6\cdots n}\cdot \dfrac{\pi}{2}, & \text{if } n \text{ is an even integer} \geq 2 \\[2ex] \dfrac{2\cdot 4\cdot 6\cdots(n-1)}{3\cdot 5\cdot 7\cdots n}, & \text{if } n \text{ is an odd integer} \geq 3 \end{cases}$

Trigonometry Formulas

1. Definitions and Fundamental Identities

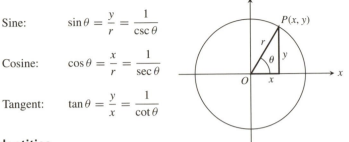

Sine: $\quad \sin\theta = \dfrac{y}{r} = \dfrac{1}{\csc\theta}$

Cosine: $\quad \cos\theta = \dfrac{x}{r} = \dfrac{1}{\sec\theta}$

Tangent: $\quad \tan\theta = \dfrac{y}{x} = \dfrac{1}{\cot\theta}$

$$\tan(A+B) = \frac{\tan A + \tan B}{1 - \tan A \tan B}$$

$$\tan(A-B) = \frac{\tan A - \tan B}{1 + \tan A \tan B}$$

$$\sin\left(A - \frac{\pi}{2}\right) = -\cos A, \qquad \cos\left(A - \frac{\pi}{2}\right) = \sin A$$

$$\sin\left(A + \frac{\pi}{2}\right) = \cos A, \qquad \cos\left(A + \frac{\pi}{2}\right) = -\sin A$$

2. Identities

$$\sin(-\theta) = -\sin\theta, \qquad \cos(-\theta) = \cos\theta$$

$$\sin^2\theta + \cos^2\theta = 1, \qquad \sec^2\theta = 1 + \tan^2\theta, \qquad \csc^2\theta = 1 + \cot^2\theta$$

$$\sin 2\theta = 2\sin\theta\cos\theta, \qquad \cos 2\theta = \cos^2\theta - \sin^2\theta$$

$$\cos^2\theta = \frac{1 + \cos 2\theta}{2}, \qquad \sin^2\theta = \frac{1 - \cos 2\theta}{2}$$

$$\sin(A+B) = \sin A \cos B + \cos A \sin B$$

$$\sin(A-B) = \sin A \cos B - \cos A \sin B$$

$$\cos(A+B) = \cos A \cos B - \sin A \sin B$$

$$\cos(A-B) = \cos A \cos B + \sin A \sin B$$

$$\sin A \sin B = \tfrac{1}{2}\cos(A-B) - \tfrac{1}{2}\cos(A+B)$$

$$\cos A \cos B = \tfrac{1}{2}\cos(A-B) + \tfrac{1}{2}\cos(A+B)$$

$$\sin A \cos B = \tfrac{1}{2}\sin(A-B) + \tfrac{1}{2}\sin(A+B)$$

$$\sin A + \sin B = 2\sin\tfrac{1}{2}(A+B)\cos\tfrac{1}{2}(A-B)$$

$$\sin A - \sin B = 2\cos\tfrac{1}{2}(A+B)\sin\tfrac{1}{2}(A-B)$$

$$\cos A + \cos B = 2\cos\tfrac{1}{2}(A+B)\cos\tfrac{1}{2}(A-B)$$

$$\cos A - \cos B = -2\sin\tfrac{1}{2}(A+B)\sin\tfrac{1}{2}(A-B)$$

Trigonometric Functions

Radian Measure

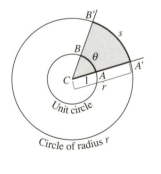

Unit circle

Circle of radius r

$$\frac{s}{r} = \frac{\theta}{1} = \theta \qquad \text{or} \qquad \theta = \frac{s}{r}.$$

$180° = \pi$ radians.

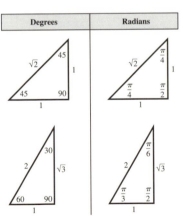

Degrees	Radians

The angles of two common triangles, in degrees and radians.

$y = \sin x$

Domain: $(-\infty, \infty)$
Range: $[-1, 1]$

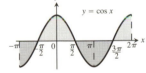

$y = \cos x$

Domain: $(-\infty, \infty)$
Range: $[-1, 1]$

$y = \tan x$

Domain: All real numbers except odd integer multiples of $\pi/2$
Range: $(-\infty, \infty)$

$y = \sec x$

Domain: $x \neq \pm\dfrac{\pi}{2}, \pm\dfrac{3\pi}{2}, \dots$
Range: $(-\infty, -1] \cup [1, \infty)$

$y = \csc x$

Domain: $x \neq 0, \pm\pi, \pm 2\pi, \dots$
Range: $(-\infty, -1] \cup [1, \infty)$

$y = \cot x$

Domain: $x \neq 0, \pm\pi, \pm 2\pi, \dots$
Range: $(-\infty, \infty)$

Values of sin θ, cos θ, and tan θ for selected values of θ

Degrees	−180	−135	−90	−45	0	30	45	60	90	135	180
θ (radians)	$-\pi$	$-3\pi/4$	$-\pi/2$	$-\pi/4$	0	$\pi/6$	$\pi/4$	$\pi/3$	$\pi/2$	$3\pi/4$	π
sin θ	0	$-\sqrt{2}/2$	−1	$-\sqrt{2}/2$	0	1/2	$\sqrt{2}/2$	$\sqrt{3}/2$	1	$\sqrt{2}/2$	0
cos θ	−1	$-\sqrt{2}/2$	0	$\sqrt{2}/2$	1	$\sqrt{3}/2$	$\sqrt{2}/2$	1/2	0	$-\sqrt{2}/2$	−1
tan θ	0	1		−1	0	$\sqrt{3}/3$	1	$\sqrt{3}$		−1	0

Symmetry in the *xy*-Plane

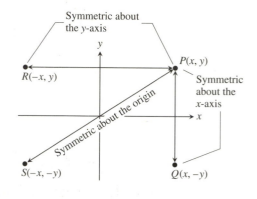

The coordinate formulas for symmetry with respect to the origin and axes in the coordinate plane.

Symmetry Tests for Graphs

1. *Symmetry about the x-axis:*
   If the point (x, y) lies on the graph, then the point $(x, -y)$ lies on the graph.
2. *Symmetry about the y-axis:*
   If the point (x, y) lies on the graph, the point $(-x, y)$ lies on the graph.
3. *Symmetry about the origin:*
   If the point (x, y) lies on the graph, the point $(-x, -y)$ lies on the graph.

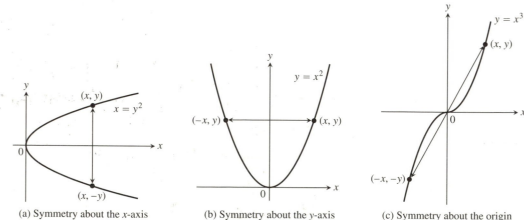

(a) Symmetry about the x-axis   (b) Symmetry about the y-axis   (c) Symmetry about the origin

Symmetry tests for graphs in the *xy*-plane.